Minerals of Britain and Ireland

TITLES OF RELATED INTEREST

CLASSIC GEOLOGY IN EUROPE SERIES

1. *Italian volcanoes* Chris Kilburn & Bill McGuire
2. *Auvergne* Peter Cattermole
3. *Iceland* Thor Thordarson & Armann Hoskuldsson
4. *Canary Islands* Juan Carlos Carracedo & Simon Day
5. *The north of Ireland* Paul Lyle
6. *Leinster* Chris Stillman & George Sevastopulo
7. *Cyprus* Stephen Edwards, Karen Hudson-Edwards, Joe Cann, John Malpas, Costas Xenophontos
8. *The Northwest Highlands of Scotland* Con Gillen
9. *The Inner Hebrides of Scotland* Con Gillen
11. *The Gulf of Corinth* Mike Leeder, Julian Andrews, Richard Collier, Rob Gawthorpe, Lisa McNeill, Clive Portman, Peter Rowe

OTHER TITLES OF RELATED INTEREST

Confronting catastrophe: new perspectives on natural disasters David Alexander

Geology and landscapes of Scotland Con Gillen

La catastrophe: Mount Pelée and the destruction of Saint-Pierre, Martinique Alwyn Scarth

Monitoring the Earth: physical geology in action Claudio Vita-Finzi

Planetary geology Claudio Vita-Finzi

Principles of emergency planning and management David Alexander

Vesuvius: a biography Alwyn Scarth

Volcanoes of Europe Alwyn Scarth & Jean-Claude Tanguy

Minerals of Britain and Ireland

A. G. Tindle
The Open University

TERRA

First published in 2008 by Terra Publishing

Terra Publishing
PO Box 315, Harpenden, Hertfordshire AL5 2ZD, England
Telephone: +44 (0)1582 762413
Fax: +44 (0)870 055 8105
Website: http://www.terrapublishing.net

ISBN-13: 978-1-903544-22-8

British Library Cataloguing-in-Publication Data
A CIP record for this book is available from the British Library

Library of Congress Cataloging-in-Publication Data are available

Typeset in Times New Roman and Helvetica
Printed and bound by L.E.G.O SpA, Vicenza, Italy

CONTENTS

Preface ... vi
Acknowledgements ... vii
Abbreviations ... viii

Introduction ... 1

THE MINERALS

A: abhurite to azurite ... 18
B: babel quartz to bytownite ... 66
C: cacholong to cyrilovite ... 104
D: dahllite to dyscrasite ... 177
E: earthy cobalt to ewaldite ... 191
F: fahlerz to funkite ... 204
G: gadolinite group to gyrolite ... 226
H: haarkies to hystatite ... 262
I: ice to izoklakeite ... 286
J: jacobsite to jurinite ... 292
K: kaersutite to kyanite ... 296
L: labradorite to lyellite ... 305
M: macaulayite to muscovite ... 321
N: nacrite to nováčekite ... 364
O: oakstone to ourayite ... 371
P: palagonite to pyrrhotite ... 377
Q: qandilite to quenselite ... 421
R: radiolite to rutile ... 428
S: sadanagaite to szomolnokite ... 442
T: tacharanite to tyuyamunite ... 494
U: ugrandite to uytenbogaardtite ... 517
V: vaesite to vreckite ... 521
W: wad to wurtzite ... 529
X: xanthiosite to xonotlite ... 542
Y: yarrowite to yugawaralite ... 544
Z: zálesíite to zwieselite ... 545

Appendix 1: Tables ... 551
Appendix 2: Museum collections ... 559
Glossary ... 564
Bibliography ... 567

PREFACE

Minerals of Britain and Ireland (*MBI*) is a summary of the endeavours of many mineralogists, both amateur and professional, who have worked over the past 150 years to establish these islands as a major centre of academic and economic mineralogy. At one time Britain was a world leader in the production of copper, the mining areas of Cornwall and Devon became important producers of both tin and copper, and, more recently, Ireland became a major player in world zinc production. The development of mining (in particular during the 18th and 19th centuries) became the springboard for the discoveries that were to follow. What many readers may not realize is that these discoveries continue unabated despite there now being very little mining activity in Britain or Ireland.

My original intention was to compile, for my personal satisfaction, a list of all the known mineral species from Britain and Ireland. Work began in a relaxed manner with the recording of information from easily accessible sources. Authors such as Peter Golley & Richard Williams (*Cornish mineral reference manual*, 1995) had previously published smaller-scale compilations that made my list grow rapidly, but I became frustrated that these and other authors did not always reference the sources of their information. I made it a stipulation that each entry in my list would be fully referenced. After eight years the list had reached 35 000 entries.

With the purchase of a facsimile copy of *Manual of the mineralogy of Great Britain and Ireland* by Greg & Lettsom (1858), I realized that what I was doing was not only long overdue, but that there were others in the mineralogical community who would appreciate this information being shared. So began *Minerals of Britain and Ireland.*

The original intention of *MBI* was to record where each mineral had been found and to note its mode of occurrence and associated species. It wasn't long before a few photographs were collected and a decision was made to make *MBI* an illustrated volume. The search was then on to find a wide range of images to reflect the diversity of species. It was very pleasing to find that everyone I approached offered to help and, in many cases, images were donated to the cause free of charge. I was given tremendous support, not only from my colleagues at the Open University, but also from elsewhere in the mineral world. David Green and Roy Starkey were truly outstanding in providing help, and through this project they have become close friends.

It didn't take long before it was suggested that *MBI* should cover other aspects of mineralogy in Britain and Ireland. Brief summaries of why minerals occur where they do (see pp. 4–14) and descriptions of museum collections were duly written. There is no substitute for actually seeing minerals to appreciate their beauty, colour and diverse range of forms. Britain (less so Ireland) has many wonderful collections and it has been saddening to find that these are increasingly going into storage. It is rare to find systematic collections, and even rarer to find them well lit and well displayed. No longer does the museum world see the display of objects as its highest priority, as modern trends in edutainment and visitor interactivity consume space once taken up by specimens. Fortunately, there is some positive news to report too, with the rediscovery and investigative cataloguing of an historic mineral collection at Chatsworth House in Derbyshire (Cooper 2005) and the promise of new mineral galleries in Edinburgh in 2011. I hope that these and all other organizations appreciate the importance of having many specimens on display.

The next issue to surface during the writing of *MBI* was completely unexpected and has caused considerable distress in the museum and collector community. It is the issue of provenance (where a mineral was found). For many reasons, it is important that provenance is correctly reported. In 1998 the work of George Ryback and his colleagues at the Natural History Museum (NHM), London, provided evidence that systematic problems had occurred with the reporting and labelling of specimens now in the Arthur Kingsbury collection (at the NHM). This has become the Kingsbury problem (some have said "fraud") – a problem so serious that it deserves a book on its own. All I have been able to do in *MBI* is flag and comment on affected species in the hope that others will confirm or disprove the provenance of samples in this major collection (of 10 000–15 000 specimens) sometime in the future. Sadly, it means it is no longer possible to believe all you read in publications describing British minerals.

Another surprise has been the number of British and Irish type localities, and the number of minerals (species and non-species) that are named after people with British or Irish links. It is a reflection of the long history of the study of mineralogy in these countries. I have tried to provide brief biographical details of all these people, but am conscious that there are some gaps remaining to be filled. There are even some individuals (such as Christopher James and Mathew Rogers) for whom nationality could not be confirmed and they are therefore excluded in this edition of the book. I would be very pleased to hear from readers who are able to contribute further historical details.

With a contract signed, the pressure was on to turn my list into a book. Little did I know what I was letting myself in for. It is no exaggeration to say the book is 100% longer and 50% later than originally planned (although a motocross racing accident is much to blame for the latter). It was suggested at the proposal stage of *MBI* that focusing only on Britain and Ireland might be too restrictive, and that what was really needed was a book on the minerals of Europe. I declined the offer (with no regrets), but hope that *MBI* will encourage other authors to take up the challenge, even it if requires more than one author and results in many volumes.

In the final stages of this work I eventually had to break my golden rule of formal referencing, when I personally discovered new species to Britain and Ireland, and colleagues permitted me to record their unpublished work. By including these discoveries, I have been able make *MBI* as up to date and comprehensive as possible.

The book is now finished. It's a relief and a delight to see it. I hope you will enjoy it as much as I have enjoyed writing it.

Andy Tindle
Newport Pagnell
May 2008

ACKNOWLEDGEMENTS

MBI may have only one author, but it could not have been written without the help of many other individuals. The list is a long one, given the complexity of the book. Help has been sought for mineralogical advice and specimens, for unpublished information, for photographs, for proofreading, for obscure references, and for support from my head of department and secretary. In all cases, help has been abundant and enthusiastic. To those persons listed below and others who I may have missed, very many thanks.

Mineralogical advice and specimens: Dave Alderton, David Anderson, Julie Ballard, Andy Banthorpe, Alan Barnes, Richard Bell, John Betterton, Richard Bevins, Richard Braithwaite, Fred Breaks, Trevor Bridges, Peter Briscoe, Steve Burchmore, Simon Camm, Nick Carruth, Tim Colman, Mick Cooper, Tom Cotterell, Peter Davidson, Ian Dossett, Giles Droop, Alan Dyer, John Ebner, John Faithfull, Jeffrey deFourestier, Harry Foy, Ralph Gertisser, Peter Golley, David Green, Alan Hart, Jack Hartley, John Holbrook, Nigel Hoppe, Jana Horák, Mike Howe, Neil Hubbard, Frank Ince, Rob Ixer, Brian Jackson, Chris Jewson, Diane Johnson, Ian Jones, Jim Knight, Steve Laurie, John Lawson, Robert Lawson, Mike Leppington, Alec Livingstone, Ed Loye, David McCallum, Ray Macdonald, Michael McMullen, David Middleton, Aberra Mogessie, Norman Moles, Stephen Moreton, David Moulding, Alastair Napier, Tim Neall, John Pearce, Graham Pearson, Nick Peters, Matt Power, Monica Price, Hazel Prichard, Mike Rumsey, Steve Rust, Richard Scrivener, Julie Selway, Tim Sherburn, Jean Spence, Chris Stanley, Roy Starkey, David Stephenson, Martin Stolworthy, Peter Tandy, Richard Tayler, Norman Thomson, Gordon Todd, Rick Turner, John Watson, Peter Webb, Helen Wilkinson, Chris Wheatley, Dave Williams, Max Wirth, Mike Wood, Brian Young.

Photographs: Dave Alderton, David Anderson, Sotaro Baba, Richard Bevins, Richard Braithwaite, Sara Chambers, Eric Condliffe, Mick Cooper, Tom Cotterell, Nick Elton, John Faithfull, Jesse Fisher, Ken Gardner, David Green, Jeremy Hooper, Jana Horák, Rob Ixer, Brian Jackson, Steve Laurie, John Mason, Tony Milodowski, Bruce Paterson, Matt Power, Hazel Prichard, Mick Rothwell, Chris Stanley, Roy Starkey, John Still, Martin Stolworthy, Dave Whipp.

Institutional assistance: Mineralogical advice, specimens and in some cases photographs have been provided by: Amgueddfa Cymru (National Museum of Wales, Cardiff); British Geological Survey, Edinburgh and Keyworth; Camborne School of Mines, University of Exeter, Penryn; Department of Earth Sciences, University of Leeds; Hunterian Museum, University of Glasgow; Manchester Museum; Mineralogical Society, London; Natural History Museum, London; National Museums of Scotland (Royal Museum and Museum of Scotland), Edinburgh; Oxford University Museum of Natural History; Royal Cornwall Museum, Truro; Sedgwick Museum, University of Cambridge.

Assistance with proofreading: Mick Cooper, Roy Starkey, Peter Webb.

Secretarial support: Liz Lomas.

Sample preparation: Kay Green, Michelle Higgins, John Watson.

Assistance with Gaelic placenames: Con Gillen, Simon Taylor.

Library staff: Wendy Cawthorne, Nicola Dowson, Steve Gray, James Hatton, Danielle Hughes, Sylvia Humphrey, Jennifer Kelly, Ann Lum, Vikki Richardson, Rachel Scott, Debbie Snook, Judy Thomas, Polly Tucker, Kati Vamos. Also, all curators who assisted with descriptions of mineral collections in museums.

Heads of the Department of Earth Sciences at the Open University: Simon Kelley, Phil Potts, Bob Spicer, Chris Wilson.

Also special thanks to curators of all the museums listed in the Introduction, to Jolyon Ralph, the creator of MinDat – the online mineral and locality database (http://www.mindat.org/index.php), to Michel Outrequin of Cameca, France, for advice on the SX100 electron microprobe, and to Chris Kennedy of Leica Microsystems for microphotographic advice.

MBI owes a great debt to one other individual, without whom the entire project would never have reached completion. Roger Jones of Terra Publishing has been unquestionably the most helpful, understanding and enthusiastic publisher I have ever had the pleasure of working with. I know that he joins with me in expressing heartfelt thanks for the support we have received from all the many friends of this project.

Finally, *MBI* was only possible because of the sacrifices made by Margaret, Lucy, Ian and Ailsa. For a considerable time the project felt like a never ending black hole. In many different ways you all helped to pull me free. Now the job is done, I look forward to spending more time with you all. Time for a walk, Macs?

Andy Tindle
May 2008

In spite of the care taken in preparing *MBI*, it is inevitable that some errors may be found in a work of its size and complexity. Readers are requested to report any apparent errors to the author, as well as any verified new discoveries. Corrections will be posted on the Terra Publishing website – www.terrapublishing.net.

ABBREVIATIONS

Initials and acronyms

□ Used to denote a structural site vacancy in a mineral. Some members of the amphibole and tourmaline groups have structural vacancies.

apfu Atoms per formula unit.

BGS British Geological Survey.

BIF Banded iron formation – a sedimentary rock type consisting of repeated thin layers of hematite or magnetite, alternating with chert or shale. Many BIFs are over 1.8 billion years old.

BM(NH) British Museum (Natural History), renamed the Natural History Museum in 1992.

BMS British Micromount Society.

BSE Backscattered electron (imaging) – a type of image generated by a scanning electron microscope or electron microprobe when high-energy electrons rebound from a sample surface. The quantity of electrons backscattered is proportional to the mean atomic number of the mineral.

CNMMN Commission on New Minerals and Mineral Names (replaced by CNMNC in 2006).

CNMNC Commission on New Minerals, Nomenclature and Classification.

Co. County.

EMP Electron microprobe – an instrument used to determine the chemical composition of a mineral to a resolution of a few microns in diameter, and also capable of taking BSE photographs. Works with flat, polished specimens.

FRS Fellow of the Royal Society of London.

IMA International Mineralogical Association.

MBI Minerals of Britain and Ireland.

MinSoc Mineralogical Society of Great Britain and Ireland.

MVT Mississippi Valley type (see Glossary).

NHM Natural History Museum, London – known until 1992 as the British Museum (Natural History).

NMS National Museum of Scotland (Royal Museum and Museum of Scotland), Edinburgh.

NMW National Museum Wales, Cardiff.

PGE Platinum group element.

PGM Platinum group mineral.

PPL Plane-polarized light using a petrological microscope.

P–T Pressure–temperature.

REE Rare earth element – a group of elements with similar behaviour from La to Lu (atomic numbers 57–71).

SEM Scanning electron microscope – an instrument used to obtain three dimensional (secondary electron) photographs of very small specimens. An SEM is also capable of taking BSE images.

syn. Synonym.

TL Type locality – the place or source where a holotype or type specimen was found.

var. Variety.

XPL Light viewed between crossed polars using a petrological microscope.

XRD X-ray diffraction.

XRF X-ray fluorescence.

Chemical symbols

Symbol	Element	Atomic no.	Symbol	Element	Atomic no.
Ag	Silver	47	Na	Sodium	11
Al	Aluminium	13	Nb	Niobium	41
Ar	Argon	18	Nd	Neodymium	60
As	Arsenic	33	Ne	Neon	10
Au	Gold	79	Ni	Nickel	28
B	Boron	5	O	Oxygen	8
Ba	Barium	56	Os	Osmium	76
Be	Beryllium	4	P	Phosphorus	15
Bi	Bismuth	83	Pb	Lead	82
Br	Bromine	35	Pd	Palladium	46
C	Carbon	6	Pm	Promethium	61
Ca	Calcium	20	Pr	Praseodymium	59
Cd	Cadmium	48	Pt	Platinum	78
Ce	Cerium	58	Ra	Radium	88
Cl	Chlorine	17	Rb	Rubidium	37
Co	Cobalt	27	Re	Rhenium	75
Cr	Chromium	24	Rh	Rhodium	45
Cs	Cesium	55	Rn	Radon	86
Cu	Copper	29	Ru	Ruthenium	44
Dy	Dysprosium	66	S	Sulphur	16
Er	Erbium	68	Sb	Antimony	51
Eu	Europium	63	Sc	Scandium	21
F	Fluorine	9	Se	Selenium	34
Fe	Iron	26	Si	Silicon	14
Ga	Gallium	31	Sm	Samarium	62
Gd	Gadolinium	64	Sn	Tin	50
Ge	Germanium	32	Sr	Strontium	38
H	Hydrogen	1	Ta	Tantalum	73
He	Helium	2	Tb	Terbium	65
Hf	Hafnium	72	Tc	Technetium	43
Hg	Mercury	80	Te	Tellurium	52
Ho	Holmium	67	Th	Thorium	90
I	Iodine	53	Ti	Titanium	22
In	Indium	49	Tl	Thallium	81
Ir	Iridium	77	Tm	Thulium	69
K	Potassium	19	U	Uranium	92
Kr	Krypton	36	V	Vanadium	23
La	Lanthanum	57	W	Tungsten	74
Li	Lithium	3	Xe	Xenon	54
Lu	Lutetium	71	Y	Yttrium	39
Mg	Magnesium	12	Yb	Ytterbium	70
Mn	Manganese	25	Zn	Zinc	30
Mo	Molybdenum	42	Zr	Zirconium	40
N	Nitrogen	7			

INTRODUCTION

Preamble

Minerals of Britain and Ireland (*MBI*) is the culmination of many years' work, sparked by the purchase of a copy of Greg & Lettsom's famous book, *Manual of the mineralogy of Great Britain and Ireland*, published in 1858. Such was the renown of this book that a facsimile copy was produced in 1977 and that too is now a collector's item. "Greg & Lettsom", as the book is commonly called, documents all 241 mineral species that had been discovered in Britain and Ireland at the time of its publication. Remarkably, since 1858 no single-volume work has sought to update Greg & Lettsom, despite its being a dream of many mineralogists, not least Sir Arthur Russell, Leonard Spencer, Arthur Kingsbury and George Ryback. In the intervening years, the Mineralogical Society of Great Britain and Ireland produced a series of supplementary lists of British minerals to fill in the gaps. The first list in 1898 took the total number of known British species to 293, the third in 1958 to 490, and the final (eighth) list, produced in 1992, increased the number to more than 850. Today, the number of confirmed mineral species described from Britain and Ireland stands at 1046; a further 154 species are unconfirmed. If all the group, series, variety and discredited species names are added together, the total number of mineral names in *MBI* is 2268.

Considering the limited surface area of Britain and Ireland, this is a remarkable number of species (more that 25% of the worldwide total) and it is a reflection of not only the varied geology of the region but also a once-thriving minerals industry and the enthusiasm of many dedicated mineralogists and mineral collectors. Although the mines (and some museums) may have closed, and the teaching of mineralogy is becoming increasingly marginalized, the rate of discovery of new species continues unabated. Neither has the role of amateur collectors changed: unlike in many modern sciences, it is through their keen eyes that many new discoveries have been made.

The other major factor in the discovery of new mineral species is the advancement in analytical instrumentation and in particular the techniques of X-ray diffraction (XRD) and electron-microprobe (EMP) analysis, the former providing information on crystal structure, and the latter chemical data on the smallest of samples. In many cases, crystals only a few microns across can easily be analyzed.

MBI seeks to provide information on all known species from the area and to acknowledge the contributions of individuals to the science of mineralogy. In some cases their contributions resulted in new minerals being named in their honour, but there are many others whose work is acknowledged with citations throughout the book. *MBI* also documents many British and Irish type localities, including 37 confirmed species named after some of those locations.

Mineralogy is not an exact science and over the course of time many minerals have been defined. Some are true species, some are varieties, and some eventually become discredited because further examination has shown the original analysis or definition to be faulty. Changes to mineral classifications have also led to many mineral names being decommissioned. *MBI* addresses all these changes and it divides the names into well defined groupings that are discussed further below.

Colour descriptions are generally based on the original bibliographical sources. In some instances, these are less than helpful (e.g. "wine yellow", "aurora red" and "flesh red"), but it seems unwise to substitute another term that may not be accurate, the vagueness of the original notwithstanding. Hence, such terms should be treated with caution, and perhaps they may serve as a salutary reminder of why colour is not used as a definitive component in the formal definitions of minerals.

A desire for the work to be confined to a single volume has resulted in a few compromises. Perhaps the greatest of these was the decision not to include (apart from a few exceptions) descriptions of work earlier than 1858. The rationale behind this is partly because *MBI* was planned as a follow-up volume to Greg & Lettsom, but also because, the older the description, the more scope there is for errors, as mineral names, methods of identification and the quality of chemical data have changed over time. This is not to say that all old work is valueless: the observation of Hey (1973) is still valid that there are 150-year-old analyses as accurate as any made today. Other exclusions are meteorites and hydrocarbons (again with a few exceptions). The former are composed of minerals (but do not have a British origin), whereas the latter are not minerals (as they have a predominantly biogenic origin and a non-crystalline structure). Anthropogenic substances (e.g. substances found in smelter slag material) are also not considered to be minerals, although they can have identical compositions to true minerals.

MBI does not simply report on published works, but also takes one further step in reporting many unpublished results, and it includes species that have never before been described from Britain or Ireland (e.g. rossmanite, smrkovecite and wodginite). There would have been even more new information to report but for the 13-year gestation period that *MBI* has undergone; so, although the confirmation of species such as arsenogoyazite and brewsterite-Ba were made as part of the *MBI* project, descriptions have also been published elsewhere.

Other vital components of *MBI* are a series of maps to give the mineral occurrences both a spatial and geological context, and some wonderful photography (illustrating over 400 species) contributed by some of the world's best mineral photographers, by museums and by individuals. A glossary and more than 3000 bibliographical references complete the book.

Topographical mineralogy: sources of uncertainty

Such is the value of some mineral specimens or the kudos attached to their discovery that there is a long history of being economical with the truth regarding locality information. The most common reason for this is to protect the source of specimens. A mineral dealer may for financial reasons be vague with locality information or, more seriously, change the locality details of a specimen to a more famous one in the hope of increasing revenue. There are also cases where scientists are deliberately vague because they are worried that unscrupulous collectors will destroy small and vulnerable localities. Negligence has also led to errors, with older fluorite specimens from Weardale in County Durham being a case in point. Many of these specimens are labelled "Cumberland", "Alston" or "Alston Moor" (Cumberland as a county no longer exists, as it was merged with Westmorland in the mid-1970s to form the new county of Cumbria). The confusion appears to have occurred during the late 19th and early 20th centuries, when specimens were acquired through dealers operating in Alston, and, for whatever reason, they opted to change the county of origin (Fisher 2004). The Cornish dealer Richard Talling (1820–83) was notorious for protecting his sources by holding back precise locality information. He took the secret of many of his finds to the grave (Cooper 2007). Less easy to explain are the provenancing problems perpetrated by an academic mineralogist, Arthur Kingsbury.

The Kingsbury problem

Arthur W. G. Kingsbury was born in 1906 at East Meon, Hampshire. He received a classical education and for ten years was a lawyer. During this time he developed a great interest in mineralogy and, after inheriting an extensive family mineral collection from his grandfather, and publishing his first work on the minerals of the Mendip Hills (Kingsbury 1935), he aspired to become a professional mineralogist. His work in the Mendips brought him into contact with Leonard Spencer (keeper of minerals at the British Museum [Natural History]) and (later Sir) Arthur Russell, who at the time was Britain's greatest topographical mineralogist. It was through these contacts that, in 1947, Kingsbury obtained a position as research assistant to the reader in mineralogy at the University Museum, Oxford, a position he retained until his death in 1968. Kingsbury published some 30 papers, mainly in the *Mineralogical Magazine*, on minerals from localities in Cornwall, Devon, Cumbria and Yorkshire. He was renowned for his compendious knowledge, his observational skills and his ability to find new species. The mineral arthurite was named in his honour jointly with Sir Arthur Russell (Davis & Hey 1964). Unquestionably, he became the leading authority on British minerals and on his death he left behind an extensive mineral collection, together with manuscript notes and annotated maps that are now held in the Natural History Museum (NHM), London.

Table 1 A. W. G. Kingsbury's contributions to the British mineral list: first British records of the species listed. Asterisks indicate first and only British records. Species in parentheses were identified later on specimens in the Kingsbury collection at the NHM (based on Ryback et al. 1998 and published with permission of *The Geological Curator*).

Species	Claimed locality
adamite	Several localities in the Lake District and Cornwall
aerugite*	South Terras mine, St Stephen-in-Brannel, Cornwall
aikinite	Carrock mine, Grainsgill, Caldbeck Fells, Cumbria
alunite	Embleton quarry, nr Cockermouth, Cumbria; Grainsgill, Caldbeck Fells, Cumbria
amblygonite	Meldon aplite quarry, nr Okehampton, Devon
antlerite	Potts Gill mine, Caldbeck Fells, Cumbria; Old Gunnislake mine, Calstock, Cornwall
argentojarosite	West Turf Pits mine, Grassington, Yorkshire; Treore mine, St Endellion, Cornwall
(argentopyrite)*	Tynebottom mine, Alston, Cumbria
arseniosiderite	Several localities in the Lake District and Cornwall
arthurite	Hingston Downs Consols, Calstock, Cornwall
beaverite	Several localities in the Lake District and Cornwall
beryllonite	Meldon aplite quarry, nr Okehampton, Devon
beudantite	Several localities in the Lake District and Cornwall
bismutoferrite	South Terras mine, St Stephen-in-Brannel, Cornwall
boulangerite	Carrock mine, Grainsgill, Caldbeck Fells, Cumbria; St Endellion, Cornwall
brandtite*	Meldon aplite quarry, nr Okehampton, Devon
breithauptite	Potts Gill mine, Caldbeck Fells, Cumbria
calomel*	Chatsworth mine, Grassington, Yorkshire
carpholite*	Several localities in the Lake District and Cornwall
(černýite)*	Treore mine, St Endellion, Cornwall
chrysoberyl	Meldon aplite quarry, nr Okehampton, Devon
conichalcite	Bedford United mine, Tavistock, Devon; and a locality in Cornwall
cornetite*	Wheal Phoenix, Linkinhorne, Cornwall
cosalite	Carrock mine, Grainsgill, Caldbeck Fells, Cumbria
cuprotungstite	Wheal Gorland, Gwennap, Cornwall
danburite	Meldon aplite quarry, nr Okehampton, Devon; Cheesewring quarry, Linkinhorne, Cornwall
descloizite	Brandy Gill, Caldbeck Fells, Cumbria; and other localities
dioptase	Potts Gill mine, Caldbeck Fells, Cumbria
durangite*	Cheesewring quarry, Linkinhorne, Cornwall
euclase*	Cligga Head, Perranzabuloe, Cornwall; Meldon aplite quarry, nr Okehampton, Devon
eucryptite*	Meldon aplite quarry, nr Okehampton, Devon
eudidymite*	Meldon aplite quarry, nr Okehampton, Devon
(fairfieldite)*	Meldon railway quarry, nr Okehampton, Devon
genthelvite	Treburland mine, Altarnun, Cornwall
(haidingerite)	Wanthwaite mine, St John's in the Vale, Cumbria
helvine*	Red-a-ven mine, Meldon, Okehampton, Devon; and three localities in Cornwall
herderite	Wheal Cock, St Just, Cornwall; Colcerrow quarry, Luxulyan, Cornwall
heterogenite*	Engine Vein mine, Alderley Edge ridge, Cheshire
hübnerite	Carrock mine, Grainsgill, Caldbeck Fells, Cumbria
ilvaite	Wheal Messer, Lanivet, Cornwall; Ramsley mine, Okehampton, and Smallcombe mine, Ilsington, both Devon
(izoklakeite)*	Wheal Cock, St Just, Cornwall
jarosite	Several localities in the Lake District
(jaskólskiite)*	Wheal Cock, St Just, Cornwall
(kentrolite)	Higher Pitts mine, nr Priddy, Somerset
(leucophanite)*	Meldon aplite quarry, nr Okehampton, Devon
lindgrenite*	Brandy Gill, Caldbeck Fells, Cumbria
meneghinite	Pengenna mine, St Kew, Cornwall; Shallowfield Bridge, South Molton, Devon
mercury	Chatsworth mine, Grassington, Yorkshire
metacinnabar	Turf Pits mine, Grassington and Cockhill mine, Bewerley, Yorkshire
milarite*	Meldon aplite quarry, nr Okehampton, Devon; Cheesewring quarry, Linkinhorne, Cornwall
montebrasite	Meldon aplite quarry, nr Okehampton, Devon
natrojarosite	Wanthwaite mine, St John's in the Vale, Cumbria
(paracostibite)*	Wheal Cock, St Just, Cornwall
parahopeite*	Turf Pits mine, Grassington & Cockhill mine, Bewerley, Yorkshire; Roughton Gill, Caldbeck Fells, Cumbria
pharmacolite	Wanthwaite mine, St John's in the Vale, Cumbria
picropharmacolite	Wanthwaite mine, St John's in the Vale, Cumbria
planchéite	Driggith mine, Caldbeck Fells, Cumbria; Alderley Edge ridge, Cheshire; Gunheath pit, St Austell, Cornwall
plumbojarosite	Several localities in Cornwall, the Lake District and Yorkshire
pollucite	Meldon aplite quarry, nr Okehampton, Devon
priceite*	Meldon aplite quarry, nr Okehampton, Devon
(pyrostilpnite)*	St Teath, Cornwall
rhodizite*	Meldon aplite quarry, nr Okehampton, Devon
rockbridgeite	Wheal Phoenix, Linkinhorne, Cornwall; East Wheal Russell, Tavistock, Devon
rosasite	Several localities in Cornwall and Cumbria
spencerite*	Turf Pit mine, Grassington, Yorkshire
(sternbergite)*	Tynebottom mine, NW of Garrigill, nr Alston, Cumbria
strengite	Wheal Phoenix, Linkinhorne, Cornwall; Burdell Gill, Caldbeck Fells, Cumbria
stromeyerite	Pengenna mine, St Kew, Cornwall
svanbergite*	Wheal Coates, St Agnes, Cornwall
tellurobismuthite	Clogau and Vigra mines, nr Dolgellau, Gwynedd, Wales
thaumasite	Embleton quarry, nr Cockermouth, Cumbria
variscite	High Down quarry, Filleigh, Devon; Pitts Cleave quarry, Tavistock, Devon; Treore mine, St Endellion, Cornwall
weilite	Wanthwaite mine, St John's in the Vale, Cumbria
whewellite	Mealbank quarry, Ingleton, Yorkshire
xanthiosite*	South Terras mine, St Stephen-in-Brannel, Cornwall
zinkenite	Carrock mine, Grainsgill, Caldbeck Fells, Cumbria
zunyite*	Embleton quarry, nr Cockermouth, Cumbria

Peter Embrey's obituary of Kingsbury (Embrey 1973) gives an insight into what was to come, by pointing out potential problems inherent in the labelling of specimens:

> I never tire of repeating the lesson that I learned from Arthur Kingsbury almost from the first time we met in 1947: there is no way of being absolutely certain of the locality from which a specimen came if this is missing from the label – nearly everything else can be determined by instrumental examination. The locality problem crops up again and again in specimen mineralogy, and is the curator's bane; not only are there cases where locality information is missing, frequently in poorly preserved collections, but all too often the labelling is wrong. The reasons for the latter range from deceit, where a collector wishes to conceal the true locality from others or to enhance the value of the specimen, through to carelessness and ignorance, and to my lasting regret I shall never be in the same league as Arthur Kingsbury in the ability to spot discrepancies.

For decades Kingsbury's reputation remained intact, although collectors became more and more suspicious as they searched in vain for species that Kingsbury had described in his publications. A major work published during this period was the *Minerals of the English Lake District: Caldbeck Fells* (Cooper & Stanley 1990). Reading between the lines, it is clear the authors were very concerned about the Kingsbury problem, but it was too early to lay too much emphasis on it, although they later voiced their doubts in an article on Caldbeck in the German magazine *Lapis* (Cooper & Stanley 1991a). Michael Cooper recalls that his moment of truth came when he discovered in the British Geological Survey collections vanadinite specimens from Hillsboro, New Mexico, exactly matching those in Kingsbury's collection and labelled "Carrock mine, Caldbeck". At about this time Alan Hart (collections leader and a member of the curatorial staff at the NHM was making a computerized catalogue of the collections and, while working through the gold specimens, noted a remarkable similarity between specimens of crystallized gold from Kanowna, Western Australia, and Kingsbury's specimens labelled "Porthcurnick Beach, Portscatho, Cornwall". As a result of these and other finds, a small suite of specimens comparing some of Kingsbury's "British" minerals with authentic pieces from foreign locations was placed in a table case in the corridor of the Department of Mineralogy and it caused a stir among those that saw it. These compelling circumstances finally forced doubts about Kingsbury's credibility into the open and something had to be done to lay this longstanding problem to rest. As a result, the late George Ryback voluntarily spent long hours in the NHM Department of Mineralogy comparing Kingsbury's specimens with those from other localities. In 1998 the truth was out and Ryback et al. (1998) were able to report:

> Detailed study of eight specimens supposedly collected by A. W. G. Kingsbury from localities in Great Britain has shown beyond reasonable doubt that they came from classic foreign localities. Many species reported by Kingsbury from Britain must now be regarded as of dubious provenance unless validated by further studies.

Further investigations (Ryback et al. 2001) revealed more problems, including 26 occurrences that were discredited or very doubtful. But for the untimely death of George Ryback in 2003, yet more problems would have been reported by now, and it is hoped that his co-workers will be able to clarify the situation further some time in the future. In the meantime, publications on mineralization in the Caldbeck Fells, Cumbria, at Brae Fell mine (Bridges et al. 2006), and at Driggith and Sandbed mines (Green et al. 2006a), provide valuable clarification of the Kingsbury problem at these sites.

Such was Kingsbury's impact on British topographical mineralogy that he described the first and often the only occurrence of 65 "British" species. A further 11 British first occurrences have since been recorded from specimens held in the Kingsbury collection (Table 1). Perhaps the most serious of these is the case of aerugite and xanthiosite, which were first described by Bergemann in 1858 from Johanngeorgenstadt, Saxony, Germany, and named by G. J. Adams in 1869. They were considered to be doubtful species until their validity was established by Davis et al. (1965). In this study, the authors (with Kingsbury as co-author) examined two German specimens in the collection of the BM(NH) and another provided by Kingsbury labelled "South Terras mine, St Stephen-in-Brannel, Cornwall". As a result of this work, South Terras mine became the type locality for both species, and Kingsbury was able to add two more species to his increasing list of British first occurrences. What went unnoticed at the time was that the three specimens actually fitted together and at one time had been part of the same sample; this did not come to light until 2001. To date there are 28 species described by Kingsbury, or discovered on Kingsbury-labelled samples, that have not otherwise been found in Britain or Ireland.

There are many other examples where Kingsbury collaborated with other mineralogists, and their joint work became part of the deception. Unwittingly, they assisted Kingsbury in increasing the number of dubious new British species. Perhaps Jack Hartley is the person whose work (as chief collaborator on Kingsbury's Lake District "discoveries") is most affected by the problem. In a letter to the present author, he states:

> I worked with AK from 1948 to our final field season, which must have been 1952 . . . Soon after the '52 field season he ceased to communicate in any way with me and I now assume that this was when he started to "cook the books" as it were. We were close friends and I liked him very much, and am thankful that he did not implicate me in any way with the deception. At the best of times, minerals were an obsession for him; he had no other topic of conversation. Obviously in the end the obsession overcame his better judgement. I had no suspicions whatsoever that he was acting in this way and it came as a great surprise when the truth came to light.

It is perhaps significant that Kingsbury's own numbered catalogue of his collection ended in 1950 with specimen number 1858. This may be a further line of evidence to suggest that the early 1950s was when Kingsbury's behaviour changed.

Other works affected by Kingsbury's mislabelled samples include: age determinations on pitchblende from South Terras mine (Pockley 1964, Darnley et al. 1965); a study of Ag–Co–Ni minerals from Tynebottom mine, Cumbria (Ixer & Stanley 1987); a glossary of Lake District minerals (Young 1987a), which referenced many of Kingsbury's publications and listed samples in the BM(NH) (but did not mention that they were from the Kingsbury collection); a study of durangite from Cheesewring quarry, Cornwall (Foord et al. 1985); a study of lithium minerals from Meldon aplite quarry, Devon (Drysdale 1991); and a major five-volume *Handbook of geochemistry* (Anthony et al. 1990–2003), which includes now dubious Kingsbury localities in every volume. Kingsbury also donated specimens to other collectors, including Sir Arthur Russell, and these specimens are now in national collections – so the problem has grown. Because the NHM is custodian of the Kingsbury collection, it is not unusual for its name to be mentioned rather than that of Kingsbury.

Alan Hart has uncovered documentary evidence of the mineral species that Kingsbury was targeting next to add to the British species list. The only reason they appear not to have reached publication was that Kingsbury had not been able to acquire the specimens. There are also in the collection incorrectly labelled specimens that had not been written about at all, almost as if they were landmines left for the unwary feet of passing mineralogists (Cooper & Stanley [1990] reported several of these, including breithauptite and prehnite).

It is the number of Kingsbury's publications and the size of his collection that makes separating the dubious labels from the truthful labels so difficult, because his descriptions are very plausible and most of his collection (10 000–15 000 specimens) is probably labelled correctly. However, the value of the collection is now enormously diminished, since genuine and suspect material are so intermixed. We shall never know the motive for Kingsbury's behaviour, but the status he acquired through his unique discoveries clearly played a part. Being a layman among academics at the University of Oxford may also be relevant. *MBI* documents 197 different species with a Kingsbury connection.

Classification

The classification of species used in *MBI* is based on Joseph Mandarino & Malcolm Back's *Fleischer's glossary of mineral species*, a readily available and up-to-date listing of all known species. The book is so popular and the science moving so fast that nine editions have been released in the past 35 years. *MBI* uses the 2004 edition for species names, but as a promised revision to mineral-group nomenclature has only recently been published (Jan. 2008), most group names in *MBI* are based on the 1999 edition. In a few instances the 2008 IMA-CNMNC listing of approved species has been used to update chemical formulae. Minerals in *MBI* are divided into four categories:

- **Confirmed species** Minerals with this status are IMA-approved species for which there is reasonable evidence, mainly from XRD or chemical analysis (or both) to support their occurrence. *MBI* documents 1046 species with confirmed status, but also includes an additional 19 "named species". These are mainly amphiboles whose names are published without previous approval by the IMA Commission on New Minerals and Mineral Names (IMA-CNMMN) (later changed to Commission on New Minerals, Nomenclature and Classification [CNMNC]). However, the names are, recognized by the commission (Burke & Leake 2005). The concept of named species has also been extended to include two tourmaline compositions for which chemical data are available, but for which formal descriptions have not yet been published. This category also includes 17 mineral series, mostly belonging to the zeolite and mica groups. Subheadings for confirmed species appear in **bold type**.
- **Unconfirmed species** These are IMA-approved species for which some doubt exists about their occurrence in Britain or Ireland. Many do not have supporting chemical data, some are identified only on their appearance, whereas others are mentioned in the literature but have later been shown to be misidentifications. *MBI* documents 126 unconfirmed species. There are an additional 28 species in this category that are potentially incorrectly provenanced, as they have been described only by A. W. G. Kingsbury or found later on Kingsbury specimens. Subheadings for unconfirmed species appear in Roman type followed by a question mark.
- **Other mineral names** These are mainly synonyms, variety names and discredited species, which are included for the benefit of those collectors with historically labelled specimens. The names are not meant to be exhaustive, but are merely those encountered during the preparation of the species descriptions. There are 939 names in this category. Some non-crystalline hydrocarbons are also listed under this heading, although they are not recognized as minerals (see asphaltite, bernalite, cloustonite, hattchetite and hydrocarbons). Subheadings for other mineral names appear in grey Roman type.
- **Mineral groups** Species with similar crystallographic properties and related chemical compositions are collectively placed into groups. *MBI* documents 93 mineral groups, listing the members of each group. Subheadings for mineral groups appear in CAPITALS.

British and Irish contributions to mineral nomenclature

Minerals were once classified in the same way as organisms, using taxonomic schemes based on that of Carl von Linné (Linnaeus 1735). At the time, the identities of their component elements were poorly known and it was possible to distinguish mineral species only by their appearance or physical properties. It was not long before such classifications were shown to be unsatisfactory, and chemical classification systems were introduced, albeit with limited knowledge of elemental and analytical chemistry (Wallerius 1747, Werner 1774, 1817). The basic chemical classification that we know today stems from Berzelius (1819), who recognized that silicates were a distinct group of minerals and was able to devise a method for deriving a chemical formula from the results of a chemical analysis.

Until the late 18th century most minerals were named according to their appearance or physical properties, with a few named after a source location (e.g. magnesite after Magnesia, in Thessaly, Greece). Prehnite was the first species named after a person (Colonel Hendrik von Prehn) and, despite objections, this practice is still very popular. William Withering (1741–1799) was the first Briton to be honoured in this way, with the naming of witherite. Since 1970, mineral names have been subject to approval by the IMA-CNMNC, and procedural guidelines have been compiled by Nickel & Mandarino (1987). There are no strict rules and the describer of a new species is free to name it in honour of a person or group, to identify the locality where the mineral was discovered, to describe the chemical or physical aspects of the substance, or to recognize an event. The downside to this arrangement is inconsistency, particularly in the use of hyphens and spaces in mineral names. The resulting nomenclature is both rich and complex, as the tables of British and Irish species show (see Appendix 1: Tables A1–A3). Within these lists there are minerals named after a mythical character, a museum and a soil research institute. With the introduction of the new zeolite classification by Coombs et al. (1997), some Britons have been fortunate enough to have more than one species named after them (e.g. William Philips who has three and John Henry Heuland who has five): Robert Ferguson has four additional species named after him – fergusonite-(Ce), fergusonite-beta-(Ce), fergusonite-beta-(Nd), fergusonite-beta-(Y) – but none is found in Britain. Intriguingly, Arthur Kingsbury is one of the few individuals to have half a mineral named after him, as he shares the name of arthurite with Sir Arthur Russell.

Table A1 shows that most mineral species first described from Britain or Ireland are named after people, and that university mineralogists and museum staff are the most common recipients of this honour. It is pleasing to find that collectors, both amateur or others, are also acknowledged. The tables also include a few non-natives who spent much of their working lives in Britain or Ireland and who made significant contributions to British mineralogy. What came as a surprise when compiling the book was the paucity of published information on the nationality of people who have been honoured with a mineral name. This is a particular problem for minerals discovered in areas once part of the British Empire and almost certainly some British and Irish subjects have not been included here for this reason.

Many other mineral species first described from Britain or Ireland are named after the location (the type locality – TL) where they were first found (Table A2, p. 552). There are a similar number of species with a British or Irish TL, but with a name that does not reflect this (e.g. bazirite), where the name is derived from the chemical composition of the mineral (Table A3). Considering all the type localities in Britain and Ireland, it is Cornwall, England, where most discoveries have been made (35 species, excluding Kingsbury claims), followed by Strathclyde, Scotland (13 species). Scawt Hill, near Larne, Co. Antrim, Ireland, is the individual locality where the largest number of new species (6) have been found, followed by Benallt mine, Lleyn Peninsula, Gwynedd, Wales (4). Three localities (one in England and two in Scotland) are the TL for three species: Merehead (Torr Works) quarry, Cranmore; Kilchoan, Ardnamurchan, Highland Region; and Susanna mine, Leadhills, Strathclyde.

As the science of mineralogy has evolved, there has been a need to expand, rationalize and correct existing nomenclature. This occurs when classifications change (most notably the amphibole group) or when new analytical data confirm that a species (e.g. harringtonite) is actually a mixture of two or more species. In cases where a mineral is known by two or more names, one of these names is defined as the species name and the other names (no longer to be used) are regarded as synonyms (e.g. black jack and blende, which are synonyms of sphalerite). Some mineral names never attained species status, instead being either varietal names (e.g. Blue John), gemological terms (e.g. sapphire), or local names used in certain orefields (e.g. brown linnets, a Derbyshire name for mimetite). Classification changes have left a trail of obsolete names, including many with British or Irish connections as summarized in Tables A4–A6. All these minerals are discussed further in the main text of *MBI*.

British and Irish citizens have also given their names to species and varieties not found in Britain or Ireland. The names are documented in Tables A7 and A8, but are not discussed further. There are also a few mineral names (non-species name) whose type locality might be expected to be British or Irish, but whose origin is elsewhere in the world (many in Canada and USA). They include: ashburtonite, boltonite, bostonite, chelmsfordite, chesterite, clinotobermorite, derbylite, douglasite, doverite, dudleyite, englishite, ferropickeringite, hampshireite, harrisite, hastingsite, lancasterite, lewisite, lincolnite, londonite, magnesiohastingsite, magnesiosussexite, maldonite, mansfieldite, montroseite, newportite, paramontroseite, peckhamite, pickeringite, richmondite, riponite, shannonite, stirlingite, sussexite, wakefieldite-(Ce), wakefieldite-(Y), warwickite, welshite and wiltshireite. Of these, hastingsite, magnesiohastingsite, pickeringite and wakefieldite-(Ce) have been positively identified in Britain or Ireland. Tables A1–A8 collectively contain 464 mineral names that in one way or another have a British or Irish connection.

Spatial and geological context of localities

The two main political areas covered by *MBI* are the United Kingdom of Great Britain and Northern Ireland (governed from London), and the Republic of Ireland (governed from Dublin). Although the Republic of Ireland and Northern Ireland are governed by different authorities, from a geological and mineralogical point of view there are no boundaries between the two and *MBI* therefore treats Ireland as a single entity (prior to 1921 the Republic of Ireland was a part of the United Kingdom). The remaining parts of the United Kingdom, including England, Scotland and Wales, are here referred to as Britain. The offshore territories of the Isle of Man, and the Channel Islands off the coast of France (Alderney, Guernsey, Herm, Jersey, Jethou and Sark), are Crown dependencies of Britain. Figure 1 shows the major political boundaries and the offshore-territory locations. What is not shown is one of Britain's most remote locations, the island of Rockall, which is located in the North Atlantic, some 440 km W of northern Skye, off the northwest coast of Scotland.

Because there have been many changes in the designation of administrative areas in recent years, it has not been easy to decide on the best nomenclature to use in *MBI*. Prior to 1974, Scotland had a fine series of county names (Clackmannanshire and Peebles-shire, to name but two), but these, along with many others in Wales, were swept away with the introduction of administrative regions. England suffered too, but on a much smaller scale. Change occurred again in 1997 with the introduction of unitary authorities. The decision on how to divide up the area was eventually made on the grounds of simplicity and common usage. Generally, the divisions used are those of administrative regions, with minor changes applied to parts of England (e.g. Yorkshire and Sussex are drawn as single regions: North, West and South Yorkshire are combined, as are East and West Sussex). Figures 2–7 illustrate the broad regional divisions used in *MBI*. Table A9 traces changes in regional nomenclature over time, to assist those collectors with old labelled specimens.

MBI does not discuss the fine details of mineralization in Britain and Ireland (this can be found in Pattrick & Polya 1993), but a brief summary is given here of the major (and some minor) orefields, so that the context of many of the mineral occurrences can be better understood. The accompanying figures also show a selection of the most important mineral localities and all the type localities.

The place-names of mineral localities used in *MBI* are derived from many sources and cover a timespan of almost 150 years. There are unavoidable mismatches between *MBI* and other sources, not least because of the constantly evolving policies of central and local government (and its agencies such as the Ordnance Survey) in respect of choice of alternative place-names. Considerable effort has been made to check names in *MBI* with those on current Ordnance Survey topographical maps. However, it has not been possible to do this for all locations. Problems were encountered with old mine names that no longer feature on modern maps, and in areas such as northwest Scotland some names were not found because Gaelic equivalents are now being used. The settlement of Northton on the Isle of Harris (Western Isles) is one such example, where the Ordnance Survey map shows Taobh Tuath instead.

Cornwall and Devon

Lizard complex

The Lizard complex sequence of mainly serpentinites, with gabbro and metamorphosed basic igneous rocks, represents a dismembered obducted ophiolite, originally a fragment of oceanic crust. Complex hydrothermal processes operating on the serpentinites produced copper mineralization at Ghostcroft mine, Mullion, and were responsible for the well known secondary analcime–natrolite–prehnite mineralization associated with gabbro at Dean quarry, St Keverne. Kennack Cove on the Lizard Peninsula is the type locality for lizardite, and Lizard Point is the type locality of saponite. Both occur as alteration products of basic igneous rocks.

Pre-granitic vein deposits

Polymetallic (antimony–lead mainly) veins have been mined from the Middle Devonian to Lower Carboniferous volcano-sedimentary basinal succession, especially in the parish of St Endellion (e.g. at Wheal Boys). The type localities of bournonite (the lead–copper sulphantimonite) and jamesonite (the lead–iron sulphantimonite) are Wheal Boys and St Endellion respectively. In addition to bournonite, stibnite and jamesonite, some mines produced subordinate quantities of galena and gold–silver ores. Bounds Cliff, east of Port Isaac, is the type locality for the lead chloride carbonate, barstowite.

Mineralization associated with the southwest England granite batholith

Much of the mineralization in Cornwall and Devon is related in one way or another to the intrusion of a series of granite plutons that extend westwards from the Dartmoor granite in the east to the major granitic plutons of Bodmin Moor, St Austell, Carnmenellis (and its satellites Carn Brea and Carn Marth), Tregonning–Godolphin, Land's End and the Isles of Scilly, a distance of over 150 km. The region is famous for its tin deposits, the mining of which reached a peak in 1870, when it produced 10 000 tonnes, half the world's supply. What is not quite so well known is that the area was the world's main source of copper in the first half of the 19th century and

Figure 1 Britain and Ireland, showing major topographical features and islands.

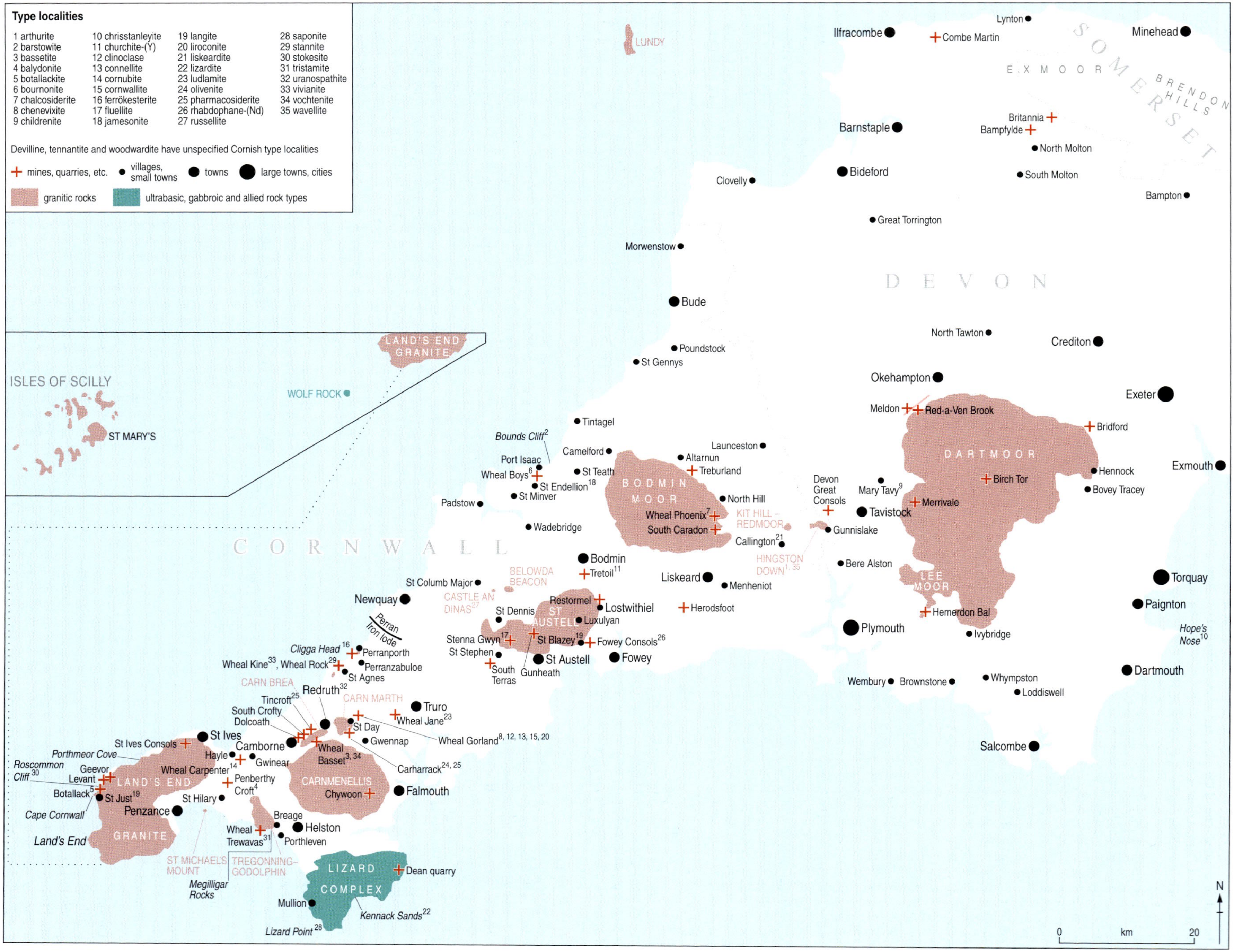

Figure 2 Southwest England.

that ores of arsenic, iron, lead, manganese, silver, sulphur, tungsten, uranium and zinc have all been mined there at one time or another. The importance of the region cannot be underestimated: one of the more common definitions of a mine is "a hole in the ground with a Cornishman at the bottom", for there is no inhabited continent on Earth, and few metalliferous mining regions of the world, where Cornishmen have not worked.

Over the years there have been over 2000 working mines in the region, and no area is more important than that surrounding the towns and villages of Camborne, Redruth, St Day and Gwennap. These include the most recently working mine of South Crofty (closed 1998), which had absorbed the earlier mining operations of Dolcoath (Cornwall's most productive tin mine), Dudnance, East Pool, Longclose, New Cook's Kitchen, Tincroft, Wheal Crofty and Wheal Seton. Geevor mine and Wheal Jane are others that operated between 1970 and 1990. In eastern Cornwall, important tin producers in the late 19th century included the Phoenix United mines, where many exceptionally well crystallized specimens of cuprite were found. The oxidized zones of these mines produced a variety of phosphate species, including andrewsite, chalcosiderite and very fine libethenite. In many cases these mines were initially copper mines and only at depth did tin become a mineable ore. In recent times the main targets were tin–tungsten stockworks or sheeted-vein systems associated with granite cusps such as those at Mulberry pit, Hemerdon and Redmoor. Near to Redmoor is Hingston Down Consols mine, located 2 km west of Gunnislake, Calstock, the type locality of arthurite. Unquestionably the most productive tungsten mine in Cornwall was the Castle-an-Dinas mine, to the north of St Austell (also the type locality for russellite). The largest single producer of both arsenic and copper was Devon Great Consols mine, near Gunnislake, whereas much of the uranium mined in Cornwall came from South Terras mine, in the metamorphic aureole of the St Austell granite. The greatest abundance of lead (as galena) came from East Wheal Rose, near Newlyn East, where two large cross-course veins were worked. Other cross-course veins were worked for iron or manganese oxides, notably the Perran Iron lode and two lodes worked at Restormel (Royal) mine, near Lostwithiel.

The primary ore minerals (e.g. cassiterite and chalcopyrite) exploited by miners are not the only species of interest to mineralogists. Three prime examples with Cornish type localities are the supergene species botallackite (from Botallack mine, St Just), liroconite (from Wheal Gorland, St Day, 2 km E of Redruth, where the world's largest crystal was found), and from the aureole of the Land's End granite at Roscommon Cliff, St Just, the rare hydrated calcium tin silicate, stokesite. Tables A1–A6 reveal many other species from the area, and the main section of *MBI* describes even more.

A discussion of Cornish granite would not be complete without a mention of pegmatites – coarse-grain granites that are a source of large and beautiful crystals. Chywoon quarry, near Longdowns, has produced remarkable crystals of feldspar, ferrocolumbite, fluorapatite and fluorite, and a stroll along the beach at Megiliggar Rocks, near Porthleven, will reveal more attractive pegmatites, albeit not with such spectacular crystals. Although its has relatively fine grain, the unique Meldon aplite, near Okehampton, a 20 m-wide granitic dyke intruded into the metamorphic aureole of the Dartmoor granite, deserves a mention here, as its origin is similar to that of granite pegmatites. It is one of the most chemically evolved intrusive rocks in Britain and it hosts many rare species (e.g. elbaite, herderite, manganotantalite, microlite and petalite), some found nowhere else in Britain.

Metal mining is all but finished in southwest England, but economic activity continues with the exploitation of china clay (mainly for kaolinite), particularly in the St Austell granite. Hydrothermal alteration of the granite has produced one of the finest kaolinite deposits in the world and, in the process of extracting it, a wide range of species have been found in related pegmatites, stockworks and greisen zones, and in oxidation zones, where secondary minerals such as cyrilovite, dufrénite, leucophosphite, libethenite, torbernite and turquoise are but a few of the species that form attractive collector pieces. Gunheath–Hensbarrow pit, north of St Austell, is perhaps the most notable of these localities. However, the future of china-clay production is not assured: in July 2006, Imerys, the major china-clay employer in the region, announced 785 redundancies in Cornwall and Devon (out of a workforce of about 2000), with production being moved to Brazil.

North Devon

The mineralized area between Morte Bay, near Ilfracombe, to Bridgwater in Somerset (including a large area of Exmoor) contains low-temperature mineral deposits, which are difficult to relate to the southwest England granite batholith. The country rocks are mainly Devonian shales, sandstones and grits, which host barite, galena, siderite and sphalerite mineralization. Iron ore was by far the most important resource, with primary siderite and its weathering products, goethite and limonite, being worked from Roman times. Argentiferous galena production was recorded from the Combe Martin area from as early as 1293, and copper–iron-rich veins were worked at nearby North Molton (e.g. Bampfylde mine) at a similar time. Barite appears to have been worked only at Cannington Park, near Bridgwater (Fig. 3).

The remainder of southern and central England

Most of southern England, particularly the southeastern part, is notable for its lack of mineralization. However, there are a few exceptions.

Mendips orefield

The base-metal mineralization consists of joint-infill deposits in Carboniferous Limestone and Triassic Dolomitic Conglomerate, belonging to a class of mineralization referred to as Mississippi Valley type (MVT) deposits. Exploitation of the orefield is ancient, with galena known to have been worked by the Romans. Smithsonite, a supergene replacement product of sphalerite, was worked much later (16th century onwards). Small supergene iron and manganese deposits within Carboniferous Limestone are host to many rare lead oxychloride species, with Merehead (Torr Works) quarry, Cranmore, the type locality for mereheadite, parkinsonite and symesite. However, it is mendipite (another lead oxychloride), with a type locality at Churchill, for which the region is best known.

Derbyshire (Southern Pennine) orefield

Mineralization in Derbyshire is hosted by Carboniferous Limestone and is also MVT, although subtly different from that of the Mendips. The limestone formed at a time when periodic volcanic activity occurred in the region, and vesicular basalt flows, ashes and tuffs, together with a few vents and basaltic sills, now interfinger the limestone sequence. Alteration of these volcanic rocks has produced a rock locally known as toadstone. Galena was the main ore species and this was mined intermittently from Roman times until relatively recently. Barite and fluorite are associated species whose value did not become significant until the mid-19th century. The mineralization occurs in major E–W veins (rakes), stratabound replacement deposits (flats) and, less commonly, cave-infill deposits (pipes). Millclose mine, to the northwest of Matlock, is the most famous mine in the orefield. It began production in 1744 and continued operations intermittently until 1940. Blue John and Treak Cliff caverns, both near Castleton, are well known for the deep purplish-blue and white banded Blue John variety of fluorite. The type localities of the lead fluorchloride, matlockite, and two zinc hydroxides (ashoverite and sweetite) are within the orefield.

West Shropshire orefield

The West Shropshire orefield is centred on the village of Shelve and it embraces the Habberley–Shelve mining district and the Breidden Hills. Mineralization occurs in veins cutting Precambrian and Ordovician sediments, volcanics and intrusive rocks. The area once supported some of the largest lead producers in Britain (e.g. Roman Gravels mine, Shelve and Snailbeach mine, between Shelve and Habberley). Ore from this small and ancient orefield was extracted during Roman times, and evidence exists of 12th- and 13th-century operations before the main period of activity began between 1845 and 1948. Originally, the region was principally a producer of lead ore (galena), with zinc ore (sphalerite) and spar minerals (barite and calcite) becoming important later.

Cheshire Basin

The Cheshire Basin contains several small sediment-hosted Triassic redbed copper–(lead) deposits associated with post-Triassic faulting. The largest deposit, on Alderley Edge ridge, was first worked by the Romans (or possibly by Bronze Age miners) and last worked in 1919. Another deposit at Mottram St Andrew to the E of Alderley Edge is the type locality for the lead–copper vanadate, mottramite. Elsewhere in the Cheshire Basin, the area around Nantwich is famous for its salt deposits, worked from Roman times to the present day.

East Midlands sedimentary iron ore

Jurassic bedded ironstones occur from Cleveland in the north (Fig. 4) to near Banbury in Oxfordshire. They consist of laterally discontinuous deposits of berthierine, chamosite, limonite and siderite, with minor pyrite and varying amounts of calcite, collophane (carbonate-fluorapatite), kaolinite and quartz. Major opencast mines were developed at Frodingham in Lincolnshire (Fig. 4) and Corby in Northamptonshire. A peak in production of 20 million tonnes was reached in 1942, with production at Banbury ending in 1992. The Northampton Ironstone at Lodge pit, Irchester, near Wellingborough, is the type locality of basaluminite and hydrobasaluminite,

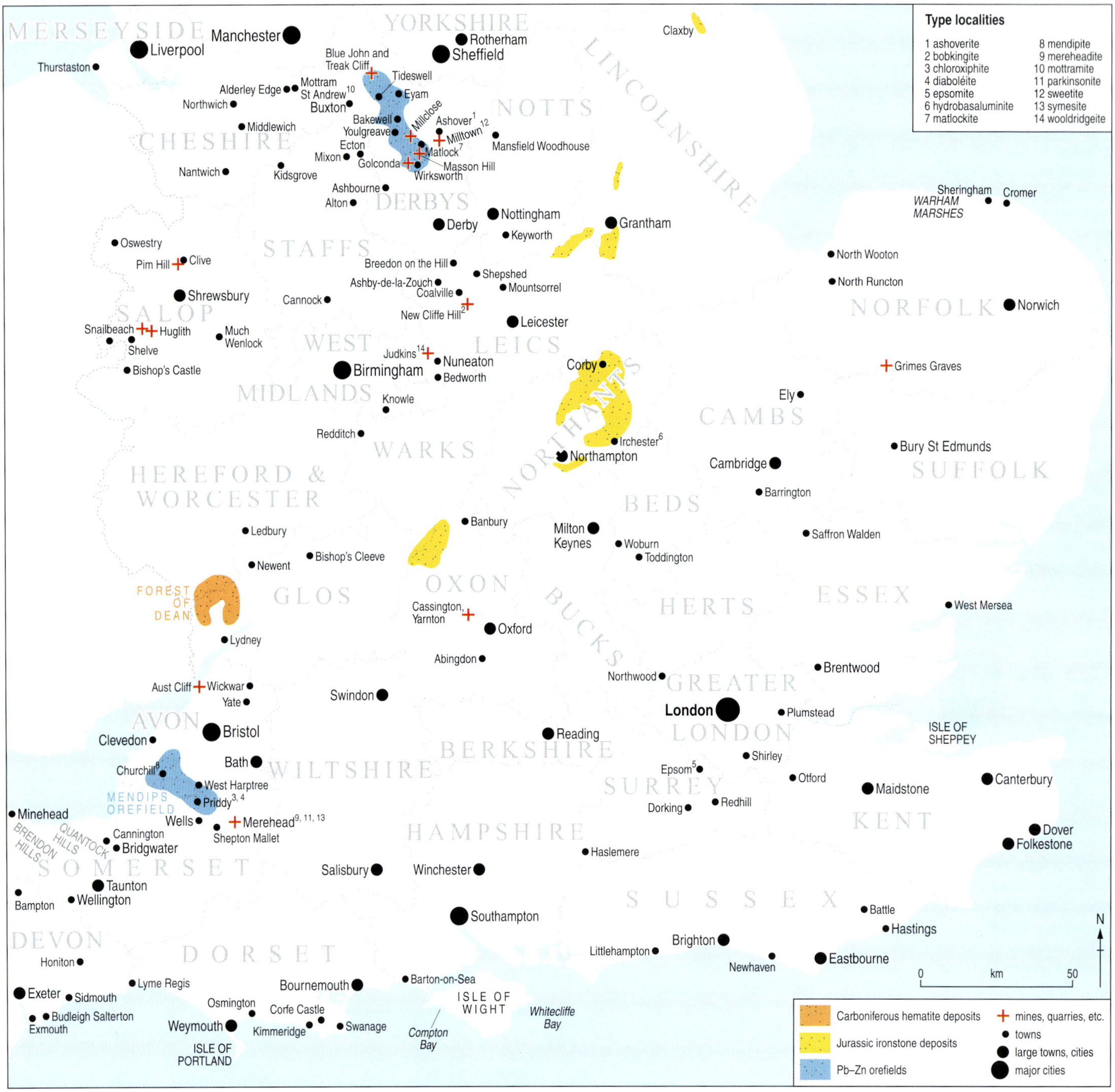

Figure 3 The Midlands and the South of England.

two hydrated aluminium sulphates found as white coatings on joint surfaces. Ironstones have also been mined from Carboniferous rocks in the coalfields of Derbyshire, Nottinghamshire, Shropshire, Staffordshire and Warwickshire, and from Cretaceous rocks in Sussex. Further deposits are found in northern England, and Wales and Scotland. Iron was extracted from Carboniferous rocks in the Forest of Dean from Roman times to 1925, but this was from non-sedimentary hematite deposits.

Other deposits

Between 1875 and 1968 the Northavon district of Avon around Yate produced 50–70% of the world's celestine from evaporite occurrences in the Triassic Mercia Mudstone Group (Keuper Marl) and Lower Coal Measures. In Leicestershire, relatively large quantities of native copper and cuprite have been exposed at the unconformity between the overlying Triassic sediments and the Precambrian basement. One of these locations, New Cliffe Hill quarry at Stanton under Bardon, is the type locality for bobkingite, a mineral found nowhere else in the world.

The North of England

Northern Pennine orefield

The Northern Pennine orefield is divided into two sections: the Alston Block to the north and the Askrigg Block to the south. Both sections contain MVT lead–zinc–(barite–fluorite) mineralization within limestones and associated sediments of Carboniferous age. The orefields date back at least to the Roman occupation of Britain (*c.* 50 BC), with the mining of galena (for lead and its associated silver) being the most important activity until the 19th century. The importance of fluorite (in the chemical and steelmaking industries) and barite (in the glass and paint industries, and as a lubricating agent in drilling muds) kept mines open for much of the 20th century. Both species (particularly fluorite) have produced world-class collector's specimens

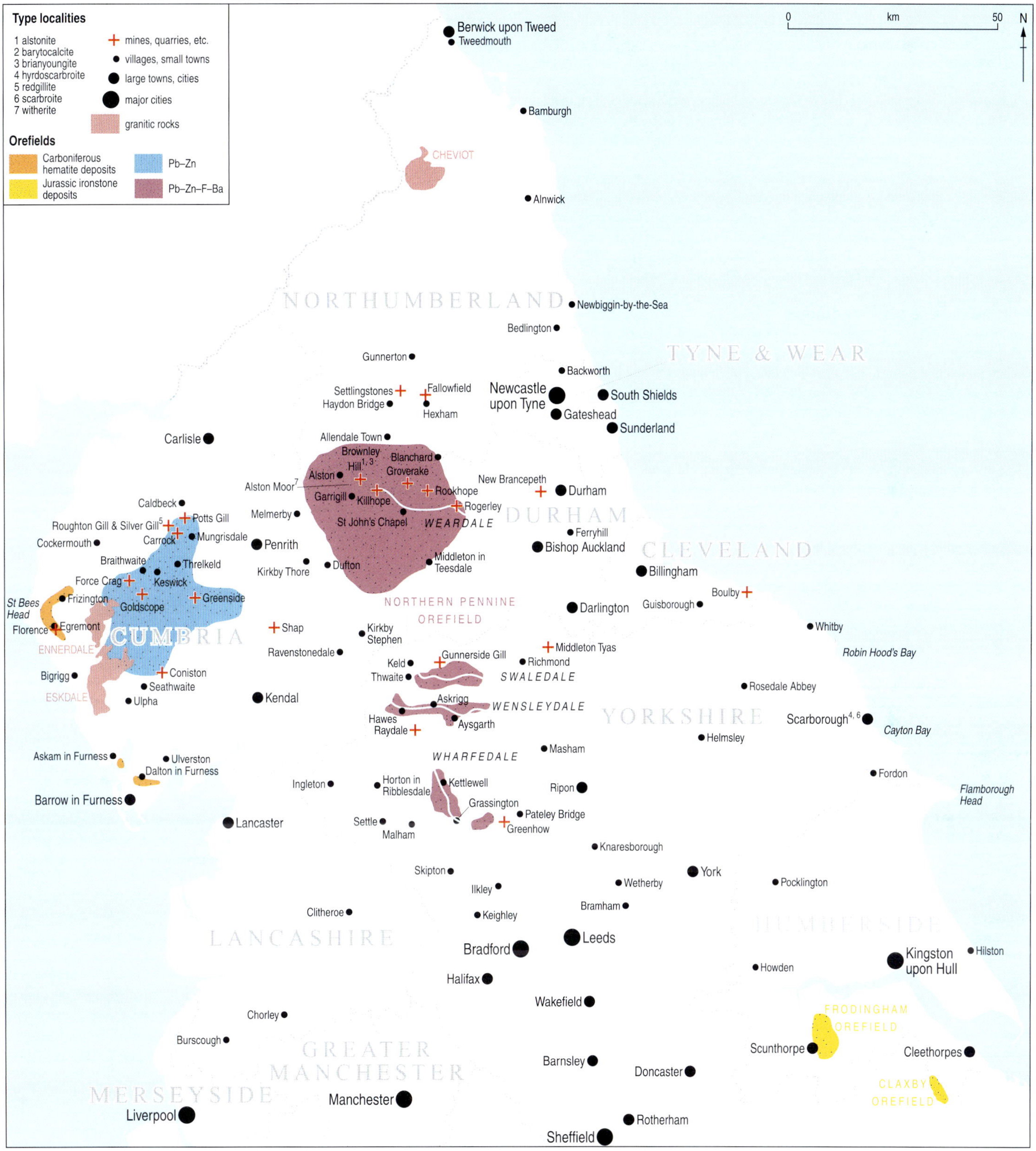

Figure 4 The North of England.

over the years, with the mines of Weardale in the Alston Block perhaps most famous. The high quality of fluorite at Rogerley mine, near Stanhope in Weardale, enabled it to become Britain's first (and only) commercial mine operated solely for specimens. Killhope (or Park Level) mine was one of many in Weardale. The mine was started in 1853, closed in about 1910 (reopening briefly during the First World War) and for a few short years in the 1870s was one of the richest in the whole of Britain. To help separate out the lead ore from waste, Park Level Mill, with its giant waterwheel, was built at Killhope in the late 1870s. Restoration of Killhope started in 1980, and today Killhope is open to the general public as the most complete lead-mining site in Britain. Large deposits of iron oxides, created by the weathering of iron-rich carbonates, are also a feature of Weardale, and it attracted mining activity from Medieval times until the 19th century.

Moving farther westwards, the character of the veins changes, with fluorite replaced by ankerite as the dominant gangue mineral. Nenthead is the centre of the famous lead-mining district of Alston Moor (the type locality of witherite). The rare species brianyoungite is a post-mining oxidation product of a zinc-rich orebody found at Brownley Hill mine (Wellgill Cross vein), Nenthead. Fine witherite specimens have also been found around the periphery of the Alston Block, and for more than a hundred years (1846–1969) this was the source of most of the world's supply of witherite. The main occurrences of witherite are in Northumberland and Durham, with Settlingstone mine being the largest producer for many years. Here, as in many other locations, witherite formed in Carboniferous sediments beneath the Whin Sill, a regionally extensive dolerite intrusion.

The Askrigg and Alston Blocks are similar in many ways, including having basement rocks intruded by Devonian granite intrusions (the Wensleydale and Weardale granites respectively) and having the same type of mineralization. However, whereas the Alston Block has a well characterized concentric zonal sequence of mineralization, the pattern in the Askrigg Block is more irregular. Greenhow Hill, near Pateley Bridge, is one of the more notable locations for collectors' specimens.

Northeast England evaporite deposits

Boulby mine, Loftus, in Cleveland, is one of Britain's deepest mines, and it works Permian evaporite deposits (halite + sylvite) at a depth of 1200 m below surface. The primary products are potash (KCl), sold to the fertilizer industry, and salt for winter road use. Borate nodules found within the base of the Boulby Potash contain beautiful pale-blue boracite crystals and other evaporite minerals.

Caldbeck Fells

Some of the world's best specimens of linarite, mimetite, plumbogummite and pyromorphite have come from the Caldbeck Fells part of the Lake District. Prior to the recently introduced permit system, the region was one of the most popular collecting areas in Britain. The history of the orefield dates back to the 16th century, when Elizabeth I contracted German miners to work copper and silver mineralization there. Although the mines producing many fine specimens closed over a century ago, new discoveries are still being made, including Britain's newest species, redgillite, whose type locality is Silver Gill.

The wide variety of mineralization is attributed to a variety of chemically different mineralizing fluids, to the number of episodes of mineralization (from Devonian to Jurassic times), to the many different host rocks, and to the intense alteration of primary mineral assemblages and host rocks. Tungsten-vein mineralization (last worked at Carrock mine in the 1980s) is related to the intrusion of the Skiddaw granite, but also cuts Ordovician sediments and the Carrock Fell gabbro, and represents one of the earliest phases of mineralization. Copper mineralization (mainly chalcopyrite) of a similar age was worked at Potts Gill, and, at both this location and nearby Sandbed, later Carboniferous to Permian vein mineralization was an important source of commercial barite. Economically and historically, early Carboniferous lead–zinc vein mineralization (as exemplified by Roughton Gill mine) has been the most important, as well as being the primary source for many fine specimens.

Coniston orefield

Coniston orefield is hosted by Ordovician Borrowdale Volcanic rocks (andesites, rhyolites and tuffs) and it consists of disseminated sulphides (with minor chalcopyrite, arsenopyrite and pyrite) in quartz-filled veins that formed in Lower Devonian times. A genetic link with the underlying Eskdale granite has been proposed, but studies of fluid inclusions indicate a class of hydrothermal deposit in which both magmatic and meteoric fluids were involved.

Mining appears to have started in about 1599 with the introduction of German miners by Elizabeth I. Coniston mine was one of the most important mines in the Lake District and in its heyday (in the 1850s) it produced 3000–4000 tonnes of dressed copper ore per year. It alone produced three-quarters of all Lake District copper, more than half of it from one chalcopyrite-rich vein (the Bonser vein). Bonser mine closed in 1908 and mining in the area ceased completely in 1942. Azurite and chrysocolla were once worked in small amounts from secondary deposits in the Levers Water region to the NW of Coniston.

Cumbrian hematite orefield

The Cumbrian hematite orefield is divided into two main sections (West Cumbria and South Cumbria) and is one of the oldest and most productive mining districts in Britain. Its origins can be traced back to the 13th century, and past production is thought to have exceeded 183 million tonnes. The majority of the deposits are fault-controlled metasomatic hematite replacements of Carboniferous Limestone around the northwest and south flanks of the Lake District Massif. There are three types of deposits: flats – tabular flat-lying replacements of limestone; sops – large inverted cone-shape dissolution hollows ranging in size up to 350 m wide and 180 m deep and found only in South Cumbria; and veins – narrow subvertical orebodies formed by replacement of limestone adjacent to faults. Massive compact ore is dominant, but a botryoidal variety of hematite (kidney ore) partially filling cavities is much prized by collectors. The last hematite mine, the linked Florence–Beckermet mine near Egremont, closed in 1980. Florence mine was then reopened as a tourist attraction, and modest production of hematite continued, not for ironmaking but as a pigment and for heat-treating certain types of cast iron. Although the mine is still operational and continues to conduct underground tours, the associated heritage centre closed in March 2007.

Wales

Parys Mountain

The history of mining at Parys Mountain extends from the Bronze Age to the present day. Evidence of copper mining during the Roman occupation has also been found, but it was not until 1768 that the Great Lode was discovered. The deposit proved to be of vast economic importance and, within a few years, Parys Mountain became Europe's premier copper mine. The discovery of other lodes enabled the mine to continue operations until 1890, by which time most large-scale working had ceased. The most recent exploratory mining activity began in 1948 and, since that time, a succession of international mining companies have explored the property, in the process demonstrating probable reserves of six million tonnes of polymetallic sulphide ore. Chalcopyrite, galena and sphalerite are the main ore species thought to have formed from the exhalation of hot mineralizing fluids into the sea during Ordovician times, probably in the form of black smokers. The deposit is a classic volcanogenic massive sulphide (VMS) orebody. It was in the gossan above the Great Lode that anglesite was first found.

Snowdon orefield

Many mines, such as Britannia (the type locality for lanthanite-(Ce)), Hafod-y-Llan and Sygun, have worked variable copper-rich sulphide deposits in the area around Snowdon. The mineralization has a hydrothermal origin and is intimately associated with Ordovician volcanic rocks, with metals probably derived from the underlying volcanic succession. A wide range of secondary species developed later, such as brochantite, chalcoalumite, georgeite and posnjakite. The orefield is surrounded by several old mines, including Drws-y-Coed, that worked related Cu–Pb–Zn-rich veins with different secondary assemblages.

Northeast Wales orefield

The lead–zinc–(barite–fluorite)-enriched section of the Northeast Wales orefield was the most important lead producer in Wales and is divided into two areas: Halkyn–Maeshafn in the north and Minera in the south. Both are examples of MVT deposits found within Carboniferous limestones and associated sediments. Initially mined for lead by the Romans (ingots found in Clwyd date the earliest smelting of galena to AD 74–96), the orefield was later exploited for both lead and zinc. Final closure of the mines came in 1914 in the south and 1959 in the north. Fluorite is an important gangue mineral, particularly around Halkyn. The secondary species cerussite, hemimorphite and smithsonite have all attracted collector interest.

Carboniferous Limestone is also host to copper–lead (chalcopyrite–galena) mineralization on the Great Orme, a headland to the west of Llandudno. Mining here has an extended history dating from the early–middle Bronze Age, and intermittent activity continued until 1880. Mineralization is concentrated in veins and in void spaces within coarse-grain dolomitized limestones close to small faults. Extensive supergene alteration has resulted in a wide range of secondary species being formed, including azurite and malachite.

Central Wales orefield

Over 130 mines were once active in the Central Wales orefield, extracting mainly lead and zinc from fault-related sulphide-bearing veins cutting Ordovician and Silurian sedimentary rocks. Silver was an important byproduct of lead mining. Exploitation began as early as the Bronze Age at Cwmystwyth mine, which was also one of the last mines to close (1943). Most of the production was between the mid-17th and late 19th centuries. Mineralization divides into two main assemblages: "early complex" (A1), of early Devonian age and "late simple" (A2), of Upper Devonian to Carboniferous age. Lead–(silver) and copper (in the form of galena and chalcopyrite respectively) were the main metals mined from A1 assemblages, whereas in A2 assemblages the galena is silver poor, and sphalerite is more common. In the Llanidloes area, barite was also exploited commercially. Repetitive fracturing

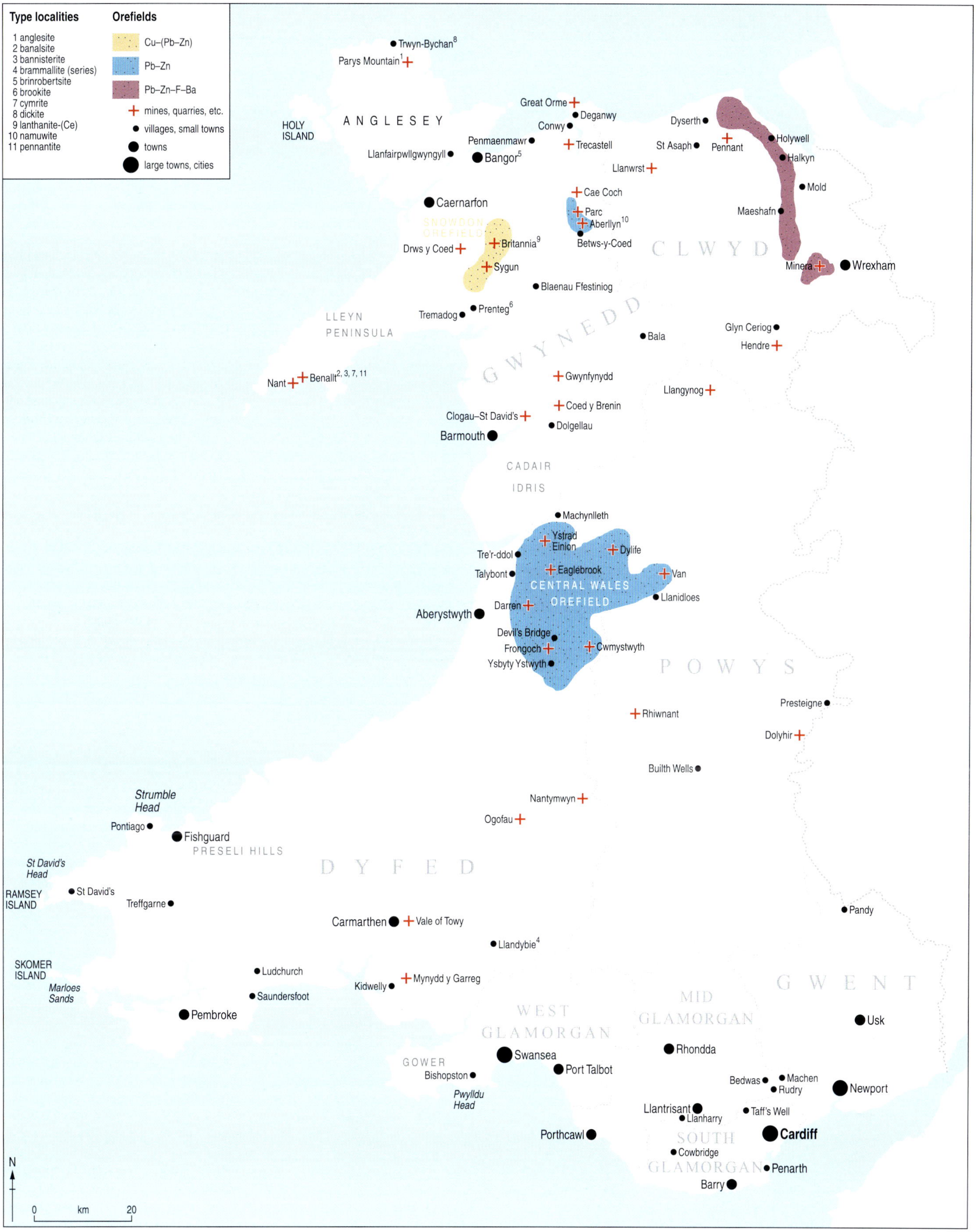

Figure 5 Wales.

and mineralizing events are a feature of the orefield. In addition to the main ore species, over 75 rarer species are recorded from the area, including cobalt pentlandite, siegenite and tučekite.

Benallt orefield

The Benallt orefield is a small region near the tip of the Lleyn Peninsula and is remarkable for the number of new species discovered there. Nant mine operated from 1886 to 1925, whereas Benallt mine operated at various times between 1886 and 1928, and for a short period during the Second World War. Benallt mine is the type locality for banalsite, bannisterite, cymrite and pennantite. The manganese deposit with which these species are associated occurs in lenses lying within Ordovician mudstones close to basic lavas and sills, and is probably related to submarine exhalation of manganese-rich fluids into the Ordovician sea. The area is host to many other rare species, such as alleghanyite, bixbyite, feitknechtite, ganophyllite, neotocite and roxbyite, but these are not usually of collector quality.

Dolgellau and Ogofau gold mines

The Dolgellau and Ogofau gold mines on the south and east sides of the Harlech Dome, such as Clogau–St David and Gwynfynedd, have dominated gold production in Britain since reliable records have been kept. The gold occurs in fault-controlled polymetallic quartz–sulphide veins cutting Cambro-Ordovician clastic rocks. Usually, gold is notable only where the veins intersect graphitic horizons, and this is where extremely localized "bonanza gold" assemblages have been found. High-grade gold–silver mineralization is accompanied by a suite of telluride species, such as aleksite, hedleyite and pilsenite. Mining peaked in the late 19th and early 20th centuries, with diminishing activity ever since. Exploration for gold is still active in the area.

About 75 km to the south, near Pumsaint, are the Dolaucothi gold mines (including the Ogofau deposit), where gold occurs in pyritic shales, quartz veins and reefs within tightly folded and sheared Lower Palaeozoic clastic rocks. The complex history of the area ranges from open pits and trenches worked in Roman times (AD 75–150) to 20th-century workings (1935–1939). In both Dolgellau and Dolaucothi the mineralization is typical of turbidite-hosted mesothermal deposits.

Scotland

Dalradian stratabound mineralization

Bedded barite and lead–zinc deposits in Dalradian rocks are among the most important mineral discoveries in Britain in the 20th century, with Foss mine (50 000 tonnes of barite annually since 1984) and Ben Eagach quarry (10 000 tonnes of barite annually since 1990) being the important producers. The area around Aberfeldy in Tayside Region where these deposits occur represents one of the world's largest barium concentrations, with the element present in barite, celsian and barium-rich muscovite. Mineralization is of the sedimentary-exhalative (sedex) style and is thought to have formed as hydrothermal solutions from submarine springs on the sea floor percolated and precipitated their metal-rich loads into host sediments. Another fossil hydrothermal system occurs in the Dalradian Ben Lawers Schist Formation, which hosts stratiform and irregular-vein chalcopyrite–pyrite mineralization in Abhainn Srathain to the south of Meall Mòr, South Knapdale, Strathclyde. The main copper mineralization is focused on an epidotized metabasite sheet outcropping on the banks of the Abhainn Srathain. A further stratabound deposit occurs at Lecht, Grampian, where manganese-rich ironstone containing principally goethite and cryptomelane was controlled primarily by an underlying porous breccia pipe. This mineralization was caused by deep weathering of the local Dalradian metasediments.

Mineralization associated with Caledonian igneous activity

Several granite complexes within Dalradian metasediments including Kilmelford and Lagalochan, both in Strathclyde and Comrie, Tayside Region, contain gold associated with sulphide-bearing fault and breccia zones. Another gold-bearing complex is at Hare Hill in Strathclyde, where gold is found in zones of disseminated arsenopyrite hosted by hydrothermally altered and quartz-veined granodioritic intrusives. At Fore Burn, Strathclyde, gold occurs in chalcopyrite-rich veins and in a disseminated arsenopyrite–chalcopyrite stockwork within diorite, whereas at Glenhead Burn, Dumfries & Galloway, gold occurs in arsenopyrite-bearing quartz veins. Both are close to the margins of the Loch Doon granite complex. Gold is also found in chalcopyrite-dominated veins close to the margin of the Cairnsmore of Carsphairn granite complex at Moorbrock Hill, Dumfries & Galloway. In Highland Region, quartz–fluorite–calcite veins associated with the Ratagain granitic intrusion contain sporadically distributed gold (electrum) and a suite of silver- and tellurium-rich species. The most recent commercial interest in gold has been centred on Cononish, 4 km SW of Tyndrum, Central Region, where pyrite-rich quartz veins cutting Dalradian metasediments and subsidiary metavolcanics are associated with a late Caledonian granite.

In the far north of Scotland, Kildonan is a well known gold location close to the Helmsdale granite, where gold is found mainly in alluvial deposits. At Rhynie, Grampian, gold is found in a small outlier of Old Red Sandstone sediments, lavas and volcaniclastic rocks. The host is the world-famous fossiliferous sinter (Rhynie Chert) of Devonian age, best known for the early plants it contains. It is a volcanic-hosted gold-rich hot-spring deposit.

Black Stockarton Moor, to the west of the Criffell granite complex, Dumfries & Galloway, is the location of copper–molybdenum mineralization associated with sill-like minor granodiorite intrusives and swarms of porphyrite dykes intruded into Silurian mudstones, siltstones and greywackes. It is a subvolcanic porphyry system. Other notable copper–molybdenum deposits associated with late Caledonian intrusions are found in Strathclyde at Garbh Achadh (near Inveraray) and Kilmelford. At Coire Buidhe, Glen Creran, Strathclyde, molybdenite occurs in veins, disseminated in porphyry rocks, and in some skarn rocks associated with the Cruachan granite. Molybdenite and pyrite are widespread in quartz veins associated with the Grudie and Loch Shin granites in Highland Region.

Tungsten mineralization is rare in Scotland, but at Gairnshiel Bridge, Glen Gairn, near Ballater, Grampian, a unique zinnwaldite-bearing granite is host to a wolframite–(cassiterite)-bearing quartz vein and associated lithium–fluorine-enriched greisen. Acanthite, cosalite, koechlinite, manganocolumbite, molybdenite, russellite, stolzite and a later sulphide assemblage are also reported from this site.

At the southwest margin of the Cairnsmore of Fleet granite complex, Dumfries & Galloway, many mineral veins cut Silurian greywackes. They were worked for their lead (or copper) content and later for zinc. The Blackcraig mines worked galena-rich ore intermittently from 1763 until 1882. At Talnotry, a small magmatic segregation in a sill-like intrusion of diorite was worked for its nickel (nickeline) content around 1885. Later investigations revealed the deposit to be "rich" in platinum-group minerals (PGMs), dominated by the species irarsite and sperrylite. Other magmatic copper–nickel–PGE sulphide deposits have been found in the Knock and Huntly basic–ultrabasic complexes in Grampian. The Shetland ophiolite complex of Unst, Shetland (Fig. 1), contains mineralization with a similar magmatic origin, but with relatively high concentrations of PGMs reported in chromite-rich dunite units.

Uranium-rich mineralization is a feature of host rocks near the southern margin of the Criffell granite complex, Dumfries & Galloway. Here, over 60 veins cutting Silurian shales contain a range of species, including native bismuth, chalcopyrite, löllingite, nickeline, pyrite and uraninite (var. pitchblende).

Strontian orefield

Lead–zinc deposits were discovered in veins cutting Moinian-age granitic gneiss or alongside hydrothermally altered basic dykes near to the Strontian granite intrusion in 1722 and were worked for seven years. Three main periods of activity followed: 1817–1836, 1850–1872 and 1901–1904. The mines were visited regularly by collectors and dealers from 1790 onwards and, as a result, many 18th- and 19th-century mineral collections contain specimens from Strontian. The mines acquired a new lease of life when they were reworked for barite from 1983 to 1991. Thus, many of the older workings, such as Bellsgrove and Middleshop mines, were disturbed by the later larger-scale opencast operations. The region is the type locality for strontianite (and the element strontium, discovered within it by Charles Hope in 1790) and for the rare zeolite, brewsterite (now a series name). Strontian is also famous for abundant and remarkable harmotome, including some of the world's finest examples of the species.

Leadhills–Wanlockhead orefield

The villages of Leadhills and Wanlockhead owe their origins to the exploitation of the mineral wealth of the Lowther Hills, the most important lead–zinc deposit in Scotland. The resources, which also include copper, gold and silver, were the basis for an industry that has left a rich mining heritage in the area. Mining was from Palaeozoic rocks cut by over 70 fissure-fill veins that formed during two periods of mineralization. Between 1700 and 1958 the veins produced about 0.4 million tonnes of lead, 10 000 tonnes of zinc and 25 tonnes of silver. Documentary evidence indicates that mining took place as early as 1239 at Leadhills, although the finding of bronze and stone implements in some surface workings at Leadhills and Wanlockhead suggests an even earlier period of activity. The region is also well known for alluvial gold. Over 70 mineral species have been described from the Leadhills–Wanlockhead orefield, and the area is the type locality for caledonite, chenite,

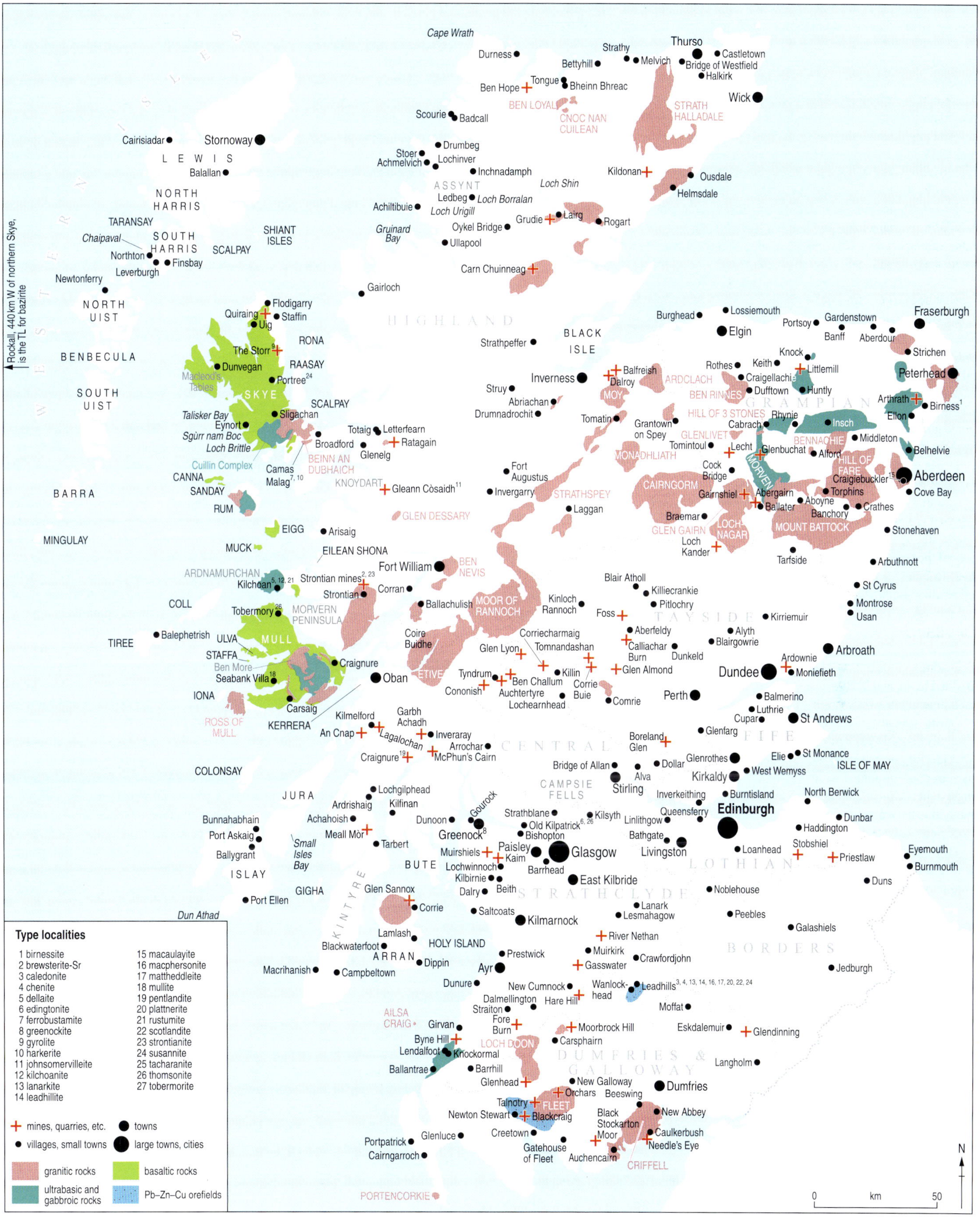

Figure 6 Scotland.

lanarkite, leadhillite, macphersonite, mattheddleite, plattnerite, susannite and scotlandite. These discoveries have established the worldwide reputation of the orefield.

Mineralization associated with Tertiary igneous activity

The economic potential of Tertiary rocks in Scotland is relatively low, although the island of Ailsa Craig in the Firth of Clyde, Strathclyde, has for centuries been a premier source of granite for the production of curling stones. Another important granite (the Beinn an Dubhaich granite) on the Isle of Skye, Highland Region, provided the type locality for both ferrobustamite and harkerite, which occur in a metasomatic skarn zone around the margin of the granite. This contact zone was investigated for its skarn-related magnetite during the 1939–1945 war, but, because of copper contamination, exploitation did not take place. At Kilchoan, on the Ardnamurchan Peninsula, Highland Region, thermal metamorphism of limestone by doleritic and gabbroic intrusions has produced another alteration zone in which the new species dellaite, kilchoanite and rustumite were found.

Tertiary volcanic rocks have also hosted new mineral species, with gyrolite being found at the Storr, 10 km N of Portree, Isle of Skye, Highland Region, and mullite at Seabank Villa, Isle of Mull, Strathclyde. Gyrolite infills vesicles in basaltic rocks, whereas mullite occurs in fused argillaceous sediment (buchite) inclusions (xenoliths) in dolerite intrusions. Scotland has two other type localities where volcanic rocks are the host. Both edingtonite (Old Kilpatrick, near Glasgow) and greenockite (Bishopton railway tunnel, near Renfrew, Strathclyde) occur in pockets and fractures within hydrothermally altered Clyde Plateau Lavas of Carboniferous age.

Ireland

Mineralization in the Irish Midlands Basin

In recent times (1960 onwards), Ireland has seen a major expansion of mining and mineral-exploration activity. The activity is focused on base metals in Lower Carboniferous carbonate rocks and in 1993 the region was producing 3.5% of the world's zinc, with significant lead (1.4%), barite and silver. In 1970, the fifth-largest zinc deposit in the world (and the largest in Europe) was discovered at Navan, Co. Meath, and mining began in 1976. Other important mines were developed at Tynagh, Co. Galway (1965–1981), Silvermines, Co. Tipperary (1968–1982), and Gortdrum, Co. Tipperary (1968–1975). Gortdrum mine is the type locality for the rare copper–mercury sulphide, gortdrumite. More recently, similar deposits have been discovered at Galmoy, Co. Kilkenny (1986), and Lisheen, Co. Tipperary (1990). The most significant deposit of stratiform barite is at Ballynoe, Co. Tipperary, where 6 million tonnes of 85% barite represent a remarkable concentration.

These deposits are enigmatic, as they have features considered diagnostic of both MVT and sedimentary-exhalative deposits. However, it is clear that the occurrence of vent fauna and isotopic evidence for near-surface fluid interactions suggests a near-seafloor origin for them. The deposits formed by replacement of lithified host rocks and have enough shared characteristics, as well as differences from other carbonate-hosted lead–zinc deposits worldwide, to have been given the status of Irish type lead–zinc mineralization.

One of the characteristics of Irish-type mineralization is a metal suite that contains more copper, silver and iron than most MVTs (probably the result of higher hydrothermal fluid temperatures). At Silvermines the discovery of a huge vug in 1978 resulted in many fine specimens of bournonite, galena and sphalerite being recovered. What is disappointing is that, although there is a history of mining in the area dating back to 1289, considerable mining activity in the 17th to 19th centuries, and exploitation of calamine in the mid-20th century, precious little specimen material survives. However, 49 species, mainly primary sulphides, sulphosalts and secondary metalliferous minerals, have been reported from the area. The specimen situation at Tynagh is little different, although over 100 species are recorded, including azurite, cerussite, dundasite, malachite, mimetite and many other unusual (but usually minute) supergene minerals.

Lower Carboniferous limestones in the southwest of Ireland are penecontemporaneous with the lead–zinc ore-bearing deposits of the Irish Midlands, but host polymetallic (copper-dominated) mineralization instead. This is probably because the underlying lithologies are different – Lower Palaeozoic greywackes in the north and Devonian redbeds in the southwest. Some copper deposits are also found in the redbeds.

A few of these deposits have been mined intermittently since the Bronze Age, although the main period of extraction was the 19th century. One of the earliest deposits to be worked was at Ross Island, Co. Kerry, where a stratiform chalcopyrite–pyrite ore underlies sphalerite-rich banded ore containing minor amounts of chalcopyrite, galena and tennantite. Epigenetic vein mineralization (chalcopyrite–tennantite with minor copper and nickel-bearing cobaltite) accompanies the stratiform ore deposits.

Deposits hosted by Devonian redbeds occur at Allihies, in the Bearhaven (Berehaven) Peninsula (1 million tonnes of copper ore mined at 3% copper grade), and in the West Carbery mining district, both in Co. Cork. Ore types include disseminated and vein copper ores dominated by bornite, chalcopyrite and djurleite, and major polymetallic quartz veins in which arsenopyrite, bournonite, chalcopyrite, galena, meneghinite, molybdenite, pyrite and tetrahedrite have all been reported. Glandore iron mine, near Skibbereen, Co. Cork, worked copper, iron (hematite) and manganese ores and is the type locality for corkite.

Mineralization associated with Caledonian igneous activity

A variety of mineral deposits occur along the eastern flank of the Leinster granite batholith. Most distinctive are a suite of lithium-enriched granite pegmatites found between Borris in Co. Carlow and Shillelagh in Co. Wicklow. They contain spodumene, accessory amounts of manganese-rich apatite, beryl, cassiterite, lithiophilite and spessartine, and various niobium–tantalum-rich species. *MBI* reports ferrocolumbite, ferrotantalite, ferrotapiolite, manganocolumbite, manganotantalite and wodginite from drill-core and alluvial deposits related to these pegmatites. The pegmatites are spatially associated with tourmalinites (quartz–tourmaline rocks). Lepidolite-rich greisen at Aclare, Co. Carlow, and Stranakelly, Co. Wicklow, have a similar origin, forming under volatile-rich magmatic conditions. The Leinster granite also plays host to galena–sphalerite-bearing veins (two veins have been worked at Ballycorus, near Carrickgollogan, Co. Dublin) and scheelite mineralization associated with narrow granitic sheets (in the Aughrim–Tinahley area, Co. Wicklow).

A major Caledonian stratabound sulphide orebody, also along the eastern flank of the Leinster granite, has been worked for its copper at Avoca mine, Co. Wicklow. The mineralization is hosted by volcanic rocks and has a long history of production: the total to 1982 was approximately 16 million tonnes, mainly as chalcopyrite concentrates. The tabular or lens-shape orebodies comprise banded, vein and disseminated sulphide types, and are thought to have a volcanogenic exhalative origin.

At Murvey and Mace Head in Co. Galway, molybdenite and chalcopyrite–molybdenite mineralization has been found at the western end of the Galway granite batholith. The deposits formed from hydrothermal or late-magmatic fluids in fault zones cutting granite.

Mineralization associated with Tertiary igneous activity

Tertiary igneous rocks are not major sources of metals, but the basaltic lavas of Co. Antrim are an important attraction for specimen collectors. The basaltic lava columns of the Giant's Causeway are famous for their structure, but less well known are the secondary minerals found in gas cavities (vesicles) within the surrounding lavas. These include a wide range of zeolite species, including garronite, gmelinite and the exceedingly rare gobbinsite, all with type localities in the Larne area. Intrusion of basaltic magmas into adjacent chalk near Larne is also responsible for a range of metasomatically altered rock types and the crystallization of some rare and unusual species. Scawt Hill is the most famous locality, being the type locality for five such species – hydrocalumite, larnite, portlandite, rankinite and scawtite. Ballycraigy, near Larne, is another chalk/dolerite contact region and is the type locality for vaterite. Reaction of a Tertiary dolerite plug with sandstone produced the new species osumilite-(Mg) at Tieveragh, near Cushedall, Co. Antrim, and thermal metamorphism of limestone by Tertiary dolerite dykes produced the calcium silicate, killalaite, whose type locality is at Killala Bay, near Inishcrone, Co. Sligo.

Meteorites

Meteorites are natural objects from outer space that have survived a fall to Earth. They divide into three basic groups (irons, stony irons, and stones) on the basis of their metal and silicate contents. Irons are recognized as being representative of material from the cores of evolved asteroids; stony irons are from the core and mantle of a large evolved body; and stones are derived from asteroids, comets, the Moon or Mars. Although all three groups have been found in Britain and Ireland, they cannot be said to have a British or Irish origin, and their minerals are therefore not listed in the main section of the book.

Meteorites have been found in the following places:

- England – Aldsworth, Cirencester, Gloucestershire (700 g); Appley Bridge, Wigan, Lancashire (15 kg); Ashdon, Saffron Walden, Essex (1.3 kg); Barwell, Leicestershire (44 kg); Danebury, near Andover, Hampshire (30 g); Glatton, Cambridgeshire (767 g); Hambleton and Wold Cottage (near Scarborough),

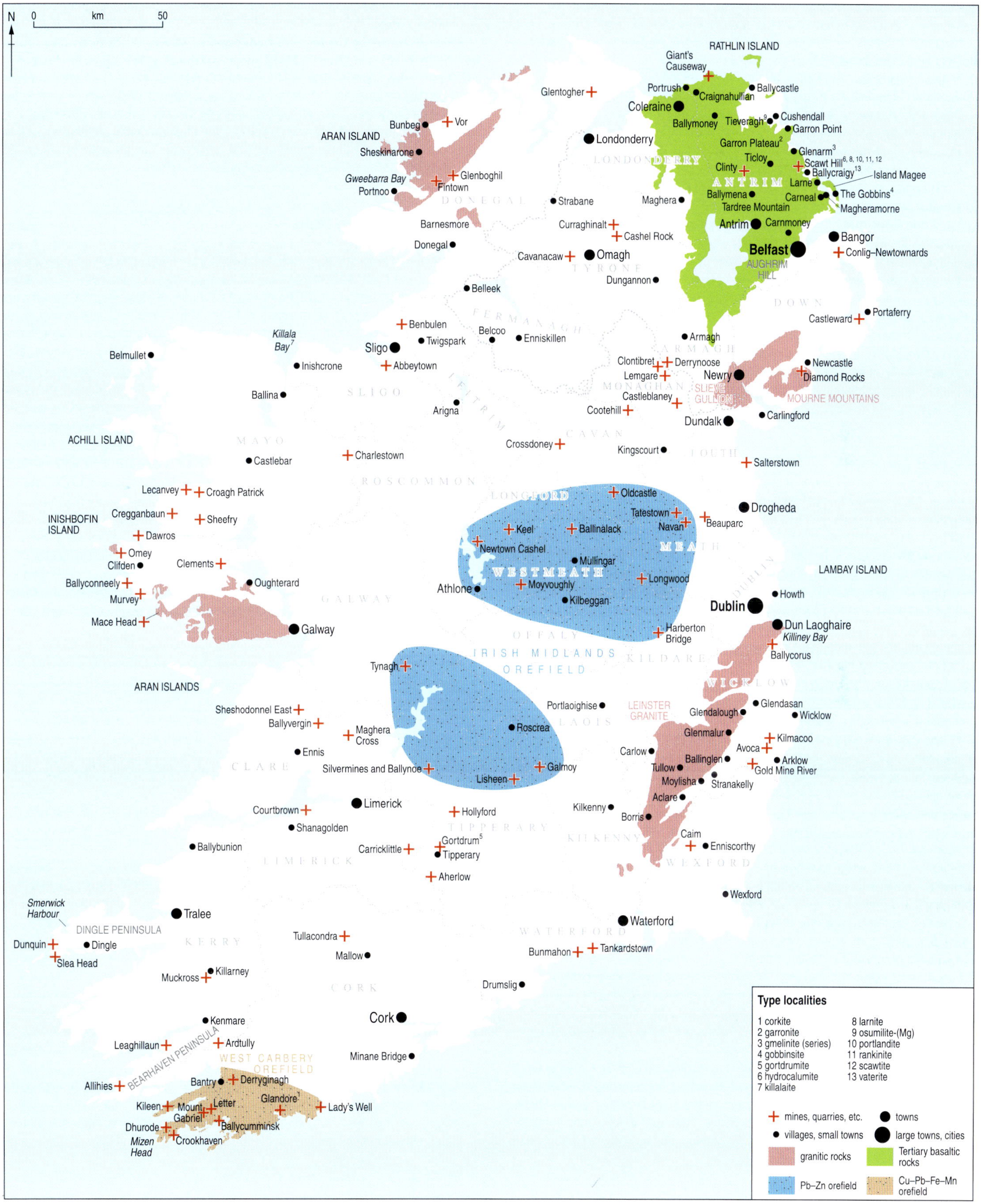

Figure 7 Ireland.

both in Yorkshire (17.6 kg and 25 kg); Hatford, Berkshire (nothing preserved); Launton, Bicester, Oxfordshire (1.05 kg); Middlesborough, Cleveland (1.59 kg); Rowton, Wellington, Shropshire (3.5 kg); Stretchleigh, Ermington, Devon (nothing preserved).

- Wales – Beddgelert and Pontlyfni, both in Gwynedd (794 g and 142 g).
- Scotland – Glenrothes, Fife (13 g); High Possil, Glasgow, Strathclyde (4.5 kg); Perth and Strathmore, both Tayside Region (2 g and 13.3 kg).
- Ireland – Bovedy, Co. Londonderry (5.4 kg); Crumlin, Co. Antrim (4.3 kg); Dundrum and Mooresfort, both in Co. Tipperary (2.2 kg and 3.53 kg); Killeter, Co. Tyrone (140 g); Leighlinbridge, Co. Carlow (271 g); Limerick, Co. Limerick; Pettiswood, Co. Westmeath (nothing preserved).

Although most are stony meteorites (except Hambleton and Rowton, which are stony iron and iron meteorites respectively), none appear to be from the Moon or Mars.

Species identified in these meteorites include: albite (and also oligoclase), augite, chlorapatite, chromite, cohenite $[(Fe,Ni,Co)_3C]^\dagger$, native copper, corundum, daubréelite $[FeCr_2S_4]^\dagger$, diopside, enstatite, ferrosilite, forsterite, graphite, haxonite $[(Fe,Ni)_{23}C_6]^\dagger$, hedenbergite, ilmenite, kamacite $[\alpha\text{-}(Fe,Ni)]^\dagger$ (name discredited and replaced with nickel-rich iron; Burke 2006), magnetite, pigeonite, schreibersite $[(Fe,Ni)_3P]^\dagger$, spinel, taenite $[\gamma\text{-}(Fe,Ni)]^\dagger$, tridymite, troilite and whitlockite (merrillite) $[(Ca_9Mg(PO_3OH)(PO_4)_6]$. Additionally, weathering and alteration has resulted in the formation of vivianite and wüstite on some these meteorites. Species marked "†" are not usually found in terrestrial samples (D. Johnson, Planetary and Space Sciences Research Institute, Open University, pers. comm. 2006).

Mineral societies

Britain is fortunate in having three well established national mineral societies. The Mineralogical Society of Great Britain and Ireland (also informally known as the MinSoc) is an international society for all those working in the mineral sciences. The society, instituted in 1876, has the general objective of advancing the science of mineralogy and its application to other subjects, including crystallography, geochemistry, petrology, environmental science and economic geology. More specifically, the society's activities include: publishing scientific journals (including *Mineralogical Magazine*), books and monographs; supporting and holding scientific meetings; liaising with other societies with similar objectives, such as the IMA, the European Mineralogical Union, the Mineralogical Society of America, the Mineralogical Society of Canada and the Geological Society of London; supporting special-interest groups; providing awards and bursaries to members and non-members; in conjunction with other societies, publishing *Elements*, a bimonthly international magazine of mineralogy, geochemistry and petrology; maintaining MINABS-online, an electronic data service giving users access to over 120 000 abstracts on the mineral sciences dating back to 1982 (over 500 abstracts of new material are added each month); maintaining an active website allowing access to information on mineralogy and online journals (www.minersoc.org).

The Russell Society is Britain's leading society specializing in topographical mineralogy. It is named after the eminent amateur mineralogist, Sir Arthur Russell (1878–1964). The wide spectrum of membership ranges from beginners to professional mineralogists and geologists. Currently over 500 members are spread throughout the United Kingdom as well as overseas. The principal aims of the society are the study, understanding, recording and conservation of minerals and mineralogical sites in Britain. The society works closely with museums and other like-minded organizations in the pursuit of these aims. It encourages responsible collecting and exploration, and works to maintain access to sites that might otherwise be unavailable to collectors and researchers. The society arranges an extensive programme of field trips and behind-the-scenes museum visits, and publishes the *Journal of the Russell Society*. Interested parties can obtain information about the society via its website (www.russellsoc.org).

The British Micromount Society (BMS) was formed in 1981 to provide a forum for the study of minerals using the stereo microscope. Membership numbers more than 200, with a nationwide distribution, and a network of local groups that arrange meetings on an occasional basis. Micromounting – the study of small mineral specimens requiring the use of magnification and illumination, and generally mounted in a plastic box approximately 25 mm in each dimension – has grown in popularity as the availability of larger specimens has diminished. Many rare species simply do not occur as larger specimens, and the crystallographic perfection obtainable in a microspecimen is typically far superior to that of a hand- or cabinet-size specimen. The BMS publishes a regular newsletter, organizes an annual symposium weekend, and maintains a comprehensive National Reference Collection of British Microminerals donated by members. This well established collection (in excess of 2000 specimens) is a superb reference source, and, together with a full catalogue, is building into an important record of contemporary mineral collecting. Further information can be found on the BMS website (www.micromounters.org.uk).

Collecting in the field

Many of the sites described in *MBI* are no longer accessible (particularly mine sites) and others are potentially dangerous (mines, quarries and some coastal locations). The conscientious collector will be aware of safety issues (wear goggles, not hammer in close proximity to others, not enter mine workings without proper arrangements being in place, etc.) and environmental issues (so will ask permission where necessary, not collect more than is reasonable, follow the Country Code) – in other words, will be responsible. Even with all these attributes in place, a collector will still encounter access problems. Health and safety legislation has now made collecting in working quarries increasingly difficult, and other sites are out of bounds for collecting because they are Sites of Special Scientific Interest (SSSI), Geological Conservation Review (GCR) sites or Regionally Important Geological–Geomorphological sites (RIGS). A review of these sites in Cornwall is provided by Selwood et al. (1998). The Caldbeck Fells are a particularly sensitive area, where a permit system operates for collectors.

The Russell Society publishes a Code of Conduct for Collectors:

- Always obtain permission before collecting on private property, and have a letter of authorization with you during the visit. Follow any instructions given by the landowner.
- If you wish to remove specimens, ensure that the owner has given you permission to do this. Removal of specimens without such consent is theft and may lead to prosecution.
- Be aware of the effect that your collecting may have on the environment, wildlife and plants. Limit your collecting to the boundaries of existing quarries and dumps. If collecting in a quarry or pit, only collect from those areas designated by the manager.
- If it is necessary to dig holes in a dump in your search for material, ensure that these are properly filled in before you leave. Ensure that any holes you do make will not affect the stability of trackways, walls or other structures. Do not leave rocks or other excavated material on fields or trackways, where they may cause injury to people or animals.
- Do not dispose of material collected from one site on another different site. This may cause confusion to future mineralogists in their study of specimens reputedly collected from a particular site.
- Do not remove excessive quantities of material from a site or cause damage to interesting geological exposures, fossil locations or mineral veins. The preservation of such sites for the future is vitally important.
- Specimens should never be collected purely for commercial gain unless you have specifically received permission to do so. Avoid buying specimens that you suspect have been obtained without permission, particularly from sensitive areas.
- Ensure that there is some scientific or educational purpose in your visit to a site and that, wherever possible, specimens are added to your own or your society's properly labelled and curated collection. Consideration should be given to offering any particular interesting or rare specimens to national or local museums. Let the site owner know this beforehand.
- Ensure that all specimens found are properly recorded. If your finds are of sufficient interest to warrant being written up for inclusion in a scientific journal, obtain the landowner's agreement to this, and ensure the necessary acknowledgements for access are given.

IMA-CNMNC nomenclature changes first reported in *Fleischer's glossary of mineral species* 2008

The tenth edition of Fleischer's glossary of mineral species by Back & Mandarino (published in January 2008) and Burke (2008) introduced changes to some species names listed in *MBI*. To allow for this, the names are shown here (old/new); the new names are also listed (as /"new name") in the A–Z species section of the book, alongside the more established names.

akaganéite/akaganeite
barium-pharmacosiderite/bariopharmacosiderite bóleite/boleite
calcium-catapleiite/calciocatapleiite chlorapatite/apatite-(CaCl)
carbonate-cyanotrichite/carbonatecyanotrichite cobalt pentlandite/cobaltpentlandite
cumengéite/cumengeite
diaboléite/diaboleite
felsöbányaite/felsőbányaite ferric-nyböite/nybøite ferro-axinite/axinite-(Fe)
ferrocolumbite/columbite-(Fe) ferrotantalite/tantalite-(Fe) ferrotapiolite/tapiolite-(Fe)
fluorapatite/apatite-(CaF) fluorapophyllite/apophyllite-(KF)
hydronium jarosite/hydroniumjarosite hydroxylapatite/apatite-(CaOH)
hydroxylapophyllite/apophyllite-(KOH)
kankite/kaňkite
levyne-Ca/lévyne-Ca levyne-Na/lévyne-Na
mahlmoodite/malhmoodite manganaxinite/axinite-(Mn)
manganocolumbite/columbite-(Mn) manganpyrosmalite/pyrosmalite-(Mn)
manganotantalite/tantalite-(Mn) meta-uranocircite/metauranocircite
meta-uranopilite/metauranopilite moganite/mogánite
nickel-skutterudite/nickelskutterudite
offrétite/offretite
potassium alum/alum-(K) pseudoboléite/pseudoboleite planchéite/plancheite
sal ammoniac/salammoniac senarmontite/sénarmontite
sodium-uranospinite/natrouranospinite sodium-zippeite/natrozippeite
szaibelyite/szaibélyite

The 2008 glossary also introduces a revised group nomenclature. New groups (names in **bold**) that contain confirmed British or Irish species are listed below.

ADELITE–DESCLOIZITE: ADELITE SUBGROUP – arsendescloizite, conichalcite and duftite; DESCLOIZITE SUBGROUP – descloizite, mottramite and pyrobelonite **AESCHYNITE:** aeschynite-(Y) **AIKINITE:** aikinite, gladite, hammarite, lindströmite and meneghinite **ALLUAUDITE:** alluaudite **ALUM:** alum-(K) **ALUNITE–JAROSITE:** ALUNITE SUBGROUP – alunite and osarizawaite SVANBERGITE SUBGROUP: hidalgoite and hinsdalite; PLUMBOGUMMITE SUBGROUP – arsenogoyazite, crandallite, gorceixite, goyazite, philipsbornite and plumbogummite – FLORENCITE SUBGROUP: arsenoflorencite-(Ce), florencite-(Ce) and waylandite; JAROSITE SUBGROUP – beaverite, hydroniumjarosite, jarosite, natrojarosite and plumbojarosite; BEUDANTITE SUBGROUP – beudantite and corkite; DUSSERTITE SUBGROUP: kintoreite and segnitite **ANCYLITE**: ancylite-(Ce) **ANDALUSITE:** andalusite and sillimanite **APTHITALITE:** apthitalite **APOPHYLLITE:** apophyllite-(KF) and apophyllite-(KOH) **ARGYRODITE:** argyrodite and canfieldite **ARSENIOSIDERITE:** arseniosiderite and mitridatite **ATACAMITE:** atacamite **ATELESTITE**: smrkovecite **AXINITE:** axinite-(Fe) and axinite-(Mn)

BASTNÄSITE: bastnäsite-(Ce) **BENITOITE:** bazirite **BERTHIERITE:** berthierite **BERYL:** beryl **BORACITE:** boracite **BOURNONITE:** bournonite and seligmannite **BRAUNITE:** braunite

CANCRINITE–SODALITE: CANCRINITE SUBGROUP – cancrinite and vishnevite; SODALITE SUBGROUP – bicchulite, danalite, genthelvite, helvine nosean and sodalite **CELESTINE** [replaces barite group]: anglesite, barite and celestine **CERVANTITE:** cervantite **CHALCOALUMITE:** chalcoalumite **CHALCOPHANITE:** chalcophanite **CHEVKINITE:** chevkinite-(Ce) and perrierite-(Ce) **COLUMBITE–EUXENITE:** COLUMBITE SUBGROUP – columbite-(Fe), columbite-(Mn), fersmite, qitianlingite, tantalite-(Fe) and tantalite-(Mn); EUXENITE SUBGROUP – euxenite-(Y) **COOPERITE:** braggite and cooperite **COPPER:** copper, gold and silver **CORONADITE** [replaces cryptomelane group]: akaganeite, coronadite, cryptomelane, hollandite, romanèchite and todorokite **CYLINDRITE:** cylindrite and franckeite

DANBURITE: danburite **DEVILLINE:** devilline, ktenasite, lautenthalite and serpierite **DIASPORE:** böhmite, diaspore, goethite, groutite and montroseite **DICKINSONITE:** arrojadite **DJERFISHERITE:** djerfisherite **DUFRÉNITE:** dufrénite **DUNDASITE:** dundasite

EPSOMITE: epsomite, goslarite and morenosite **EUCAIRITE:** eucairite and naumannite

FERGUSONITE: fergusonite-(Y) **FILLOWITE:** johnsomervilleite

GADOLINITE–DATOLITE: GADOLINITE SUBGROUP – gadolinite-(Y); DATOLITE SUBGROUP – datolite **GALENA:** alabandite, altaite, clausthalite and galena **GANOPHYLLITE:** eggletonite, ganophyllite and tamaite **GILLESPITE:** cuprorivaite

HALITE: chlorargyrite, halite and sylvite **HAUSMANNITE:** hausmannite, hetaerolite and hydrohetaerolite **HERDERITE:** herderite **HÖGBOMITE:** högbomite **HUMBOLDTINE:** glushinskite and lindbergite **HUMITE:** HUMITE SUBGROUP – chondrodite, clinohumite, humite and norbergite; MANGANHUMITE SUBGROUP – alleghanyite, manganhumite and sonalite **HYDROMAGNESITE:** dypingite and hydromagnesite

ILESITE [replaces rozenite group]: rozenite **IRON:** iron

JAHNSITE [replaces whiteite group]: jahnsite-(CaMnFe)

KAOLINITE–SERPENTINE: KAOLINITE SUBGROUP – antigorite, chrysotile, dickite, halloysite, lizardite and nacrite; SERPENTINE SUBGROUP – allophane, amesite, berthierine, cronstedtite, fraiponite and greenalite **KOBELLITE:** kobellite **KOECHLINITE:** koechlinite and russellite

LANGITE: langite, posnjakite and wroewolfeite **LANTHANITE:** lanthanite-(Ce) **LAWSONITE:** lawsonite **LEUCOPHOSPHITE:** leucophosphite **LUDWIGITE:** ludwigite

MANESSEITE: brugnatellite **MARCASITE–LÖLLINGITE:** MARCASITE SUBGROUP – kullerudite and marcasite; LÖLLINGITE SUBGROUP – löllingite, rammelsbergite and safflorite **MATILDITE:** bohdanowiczite, matildite and volynskite **MATLOCKITE:** bismoclite and matlockite **MILARITE** [replaces osumilite group]: armenite, milarite and osumilite-(Mg) **MOLYBDENITE:** molybdenite

NEPHELINE: kalsilite and nepheline **NEPTUNITE:** neptunite **NICKELINE:** breithauptite, kotulskite and nickeline

OLIVENITE: adamite, libethenite, olivenite and zincolivenite **OLIVINE:** fayalite, forsterite, monticellite and tephroite **OSMIUM:** osmium, rutheniridosmine and ruthenium **OTTRÉLITE:** chloritoid

PALYGORSKITE: palygorskite **PARISITE:** parisite-(Ce) **PEROVSKITE:** perovskite **PHARMACOLITE:** brushite and churchite-(Y) **PHARMACOSIDERITE:** bariopharmacosiderite and pharmacosiderite **PHOSPHOSIDERITE:** metavariscite and phosphosiderite **PHOSPHURANYLITE:** dewindtite and phurcalite **PICROMERITE:** picromerite **PLAGIONITE:** fülöppite, plagionite and semseyite **PLATINUM:** iridium, palladium and platinum **POLYBASITE:** pearceite and polybasite **PREISINGERITE:** petitjeanite and preisingerite **PYROCHLORE:** PYROCHLORE SUBGROUP – pyrochlore; MICROLITE SUBGROUP – microlite **PYROSMALITE:** pyrosmalite-(Mn)

REYERITE: reyerite **RHODONITE:** rhodonite **ROSASITE:** malachite and rosasite **ROSCHERITE:** griefensteinite

SAMARSKITE: ixiolite **SCHEELITE**: powellite, scheelite, stolzite and wulfenite **SCHOEPITE:** schoepite **SCORODITE** [replaces variscite group]: mansfieldite, scorodite, strengite and variscite **SEPIOLITE:** sepiolite **SKUTTERUDITE:** nickelskutterudite and skutterudite **STAUROLITE:** staurolite **STIBNITE:** bismuthinite and stibnite **SYNCHYSITE:** synchyite-(Ce) and synchysite-(Y) **SPANGOLITE:** jamborite, schulenbergite and spangolite

TALC: minnesotaite, pyrophyllite and talc **TAPIOLITE:** tapiolite-(Fe) **TENNANTITE** [replaces tetrahedrite group]: argentotennantite, freibergite, tennantite and tetrahedrite **TITANITE:** malayaite and titanite **TRIPHYLITE:** lithiophilite, sicklerite and triphylite **TRIPLITE:** triplite and zwieselite **TRITOMITE:** britholite-(Y) and mattheddleite **TULAMEENITE:** tulameenite

URANINITE: thorianite and uraninite

VALLERIITE: valleriite **VESUVIANITE:** vesuvianite

WAGNERITE: triploidite **WARDITE:** cyrilovite **WICKMANITE** [replaces schoenfleisite group]: natanite and wickmanite **WILLEMITE:** phenakite **WODGINITE:** wodginite **WÖHLERITE** [replaces cuspidine group]: baghdadite and cuspidine **WOLFRAMITE:** ferberite and hübnerite **WOLLASTONITE:** bustamite, ferrobustamite, pectolite and wollastonite **WOODWARDITE:** honessite, hydrohonessite, hydrowoodwardite and woodwardite

XENOTIME: wakefieldite-(Ce) and xenotime-(Y)

ZIRCON: coffinite, thorite, thorogummite and zircon

The smectite group now contains bannisterite, brinrobertsite, corrensite, rectorite, stilpnomelane, tosudite and vermiculite; the tetradymite group now contains hedleyite, ikunolite, ingodite, joséite-A, joséite-B, laitakarite, pilsenite, rucklidgeite and tsumoite; and members of the meta-autunite group have been moved to the autunite group – bassetite, meta-autunite, metakahlerite, metanováčekite, metatorbernite, meta-uranocircite and metazeunerite. Uranospathite and vochtenite also become members of the autunite group.

Additionally, the species carbonate-fluorapatite, carbonate-hydroxylapatite, hyalophane, ilmenorutile and melilite are discredited, and it is noted that osumilite-(Mg) is not an IMA-approved species.

A

abhurite?

trigonal $Sn_3O(OH)_{12}Cl_2$

A rare tin oxychloride found within concretions on ingots of refined Cornish tin and discovered on the wreck of the *SS Cheerful*, which sank in 1885, 29 km NNW of St Ives, Cornwall (Wolfe 1996a). In this case, because abhurite is growing on manmade material, it is considered to be an anthropogenic substance rather than a true mineral species (Nickel 1995).

abichit

A synonym of clinoclase.

abriachanite

TL Abriachan, N of Drumnadrochit, Grampian, Scotland

A synonym of magnesioriebeckite, amphibole group.

acadialite

A synonym of chabazite, zeolite group.

acanthite

monoclinic Ag_2S

A rare silver sulphide, the lower-temperature dimorph of argentite. Acanthite occurs in hydrothermal veins, usually in association with other silver-bearing minerals and also in zones of secondary enrichment. Commonly associated minerals include covellite, galena, polybasite, proustite, pyrargyrite, pyrite, sphalerite and tetrahedrite. In early works, acanthite is sometimes described as argentite.

England

Cheshire Mineralized fractures in silicified barite–calcite-rich fault rock, associated with the West mine boundary fault, Alderley Edge ridge, contain relict traces of acanthite (argentite), chalcopyrite and cinnabar, with secondary dioptase, chrysocolla, goethite, malachite and Mn–(Fe) oxyhydroxides (Plant et al. 1999).

Cornwall Acanthite (argentite) and ruby-silver (either proustite or pyrargyrite) were economically mined from the Silver lode of Dolcoath mine, Camborne and a "rich bunch of silver ores associated with siderite", consisting of acanthite (mainly), proustite, pyrargyrite, capillary native silver and stephanite, were found at Wheal Wrey and Ludcott United mines, near Liskeard (Dines 1956). Some of the best acanthite in Britain or Ireland came from Wheal Ludcott, with cubo-octahedral crystals up to 12 mm across being found (Rudler 1905, Rust 1983a). A similar assemblage of acanthite, pyrargyrite, native silver and stephanite was worked along another Silver lode at Wheal Newton, Harrowbarrow, SW of Gunnislake (Dines 1956). At Wheal Herland, Gwinear, copper–tin ores were intersected by a cross course containing arsenopyrite and native silver (Collins 1904a, Hosking 1951), where "small and well defined crystals" of acanthite were found (Greg & Lettsom 1858). A specimen of native silver from Greystone quarry, Lezant, near Launceston appears, to have been gradually coated and replaced by black acicular crystals of acanthite over a period of ten years. Careful storage in a dry atmosphere is clearly necessary to slow down this alteration process (Elton et al. 2004). Other occurrences include: Dolcoath and Mexico mines, Wheal Ann, Wheal Basset, Wheal Brothers, Wheal Duchy, Wheal Vincent (Greg & Lettsom 1858); Wheal Jewel, St Day, Gwennap (Collins 1904a, Rust 1982a); North Trelawny mine, St Ive; Wheal Langford, Calstock; Mount mine, Perranuthnoe and Greystone quarry, Lezant (Golley & Williams 1995); and Perran Silver mine (Wheal Sedgman lode), Perranuthnoe (Butler 1907).

Cumbria Acanthite occurs on cerussite at Force Crag mine, probably as a result of post-mining oxidation of silver (Green et al. 1997). At Piers Gill, 1 km N of Scafell summit, Wasdale, a 60 cm-thick polyphase quartz vein cemented by ankerite, within the Scafell dacite, contains microscopic inclusions of acanthite (argentite) in an Ag–As–Au–Cu–Pb–Zn assemblage similar in many respects to that found in a porphyry-copper system. Associated species include native gold (electrum), pyrargyrite, tetrahedrite, uytenbogaardtite and an unknown selenium–silver mineral (Aldridge 2005). A further description of acanthite, from Tynebottom mine, NW of Garrigill, near Alston, is dubious, as it is based on samples of doubtful provenance from the Kingsbury collection at the NHM (Ixer & Stanley 1987, Ryback et al. 1998, 2001).

Devon Antimony–gold mineralization in quartz–carbonate veins cutting Devonian slates at Loddiswell mine, 3 km N of Loddiswell, contains minor acanthite as a late-forming species (Stanley et al. 1990b).

Wales

Acanthite is noted from Gwynfynydd mine, near Dolgellau, Gwynedd (Pryor 1988, Leake et al. 1993).

Scotland

Central Acanthite and native silver were worked from a bonanza-type silver deposit hosted by Lower Devonian andesitic lavas at the Silver Glen mine, near Alva, Ochil Hills (Parnell 1988a, Moreton et al. 1998, 2007). Associated species are chalcopyrite, erythrite and skutterudite (Heddle 1901). Other occurrences: Airthrie Hill mine, Bridge of Allan, Ochil Hills (Dickie & Forster 1974); and Tyndrum (Pattrick 1985).

Grampian A late-stage sulphide assemblage associated with a hübnerite-bearing quartz vein at Gairnshiel Bridge, Glen Gairn, contains acanthite, cosalite, koechlinite and molybdenite in a greisenized zinnwaldite granite host (Webb et al. 1992).

Lothian Acanthite was one of the principal silver minerals found in association with barytocalcite and witherite mineralization at Hilderston silver mine, Bathgate Hills (Meikle 1994, Moreton et al. 1998, Green & Todd 2001).

Strathclyde Boyleston quarry, SW of Barrhead, and Hartfield Moss, 9 km WSW of Barrhead (Meikle 1992, Livingstone 1993); and Lochy Trib (Eas Daimh), W of Tyndrum (Chapman et al. 2000c).

Tayside Urlar Burn, near Aberfeldy (Ixer et al. 1997).

Large octahedra of acanthite with edges up to 18 mm. These pseudo-crystals are paramorphs after argentite. Wheal Newton, Harrowbarrow, SW of Gunnislake, Cornwall. (© The Natural History Museum, London)

Ireland

Tipperary Acanthite is reported from a high-grade pod of polymetallic silver (assaying up to 6.5% Ag) discovered at Silvermines. The assemblage also includes argyrodite, barite, galena, proustite, smithite, sphalerite and xanthoconite (Taylor & Andrew 1979, Moreton 1999).

Offshore territories

Sark Sark's Hope mine, Little Sark – minor amounts of acanthite in a galena–pyrite-dominated vein deposit (Greg & Lettsom 1858, Mourant & Warren 1934, Ixer & Stanley 1983, Laffoley 1998).

acerdèse

A synonym of manganite.

achroite

A colourless variety of elbaite, tourmaline group; see Collins (1876c).

acmite

A synonym of aegirine. The term was reserved for a variety of aegirine, but is now used for the $NaFe^{3+}Si_2O_6$ molecule in pyroxenes. "Acmite" in the aegirine–riebeckite granite of Rockall island, in the North Atlantic, 440 km W of northern Skye, NW of Scotland (Sabine 1960, Hawkes et al. 1975), is actually a calcium-poor aegirine containing significant amounts of niobium and zirconium (BGS specimen, Tindle unpub. EMP data). Rare acmite is reported from an arfvedsonite-bearing minette at St Hélier, Jersey (Wagner & Velde 1985).

actinolite

AMPHIBOLE GROUP

monoclinic $\square Ca_2(Mg,Fe^{2+})_5Si_8O_{22}(OH)_2$

A common calcium–magnesium–iron silicate hydroxide that forms a series with tremolite and ferro-actinolite. Actinolite, in association with epidote and chlorite, typically develops in metamorphosed basic igneous rocks that have been subjected to the effects of low- to medium-grade (greenschist-facies) metamorphism. It is also found in contact-metamorphosed magnesium carbonate, mafic or ultramafic rocks and glaucophane-bearing blueschists, where it is associated with albite, chlorite, epidote, lawsonite, pumpellyite and stilpnomelane. In pyroxene-bearing igneous rocks, actinolite may form as part of the late- to post-magmatic process of crystallization. Where possible, identification is based on the 1997 classification of the amphibole group (Leake et al. 1997a, Mogessie et al. 2004). Unconfirmed occurrences listed below usually refer to petrographic observations with no supporting analytical data.

England

Cornwall All unconfirmed: Botallack, St Just (Robson 1949b); Wheal Cock Zawn, Roscommon Cliffs, St Just (Couper & Barstow 1977); Carrick Du mine, near Porthmeor beach, St Ives and Tremore (quarry), near Wenford Bridge (both Collins 1911a); Great Perhaver Point, Roseland (Barnes & Andrews 1981); Great Retallack mine, near St Agnes (Dewey 1925); Land's End granite aureole (Mitropoulos 1984); and Porthallow, Lizard Peninsula (Barnes & Andrews 1981). Golley & Williams (1995) report a further six locations: Cadgwith, Coverack and Mullion, all three on the Lizard Peninsula; Wheal Maudlin, 3 km NW of Lostwithiel; Clicker Tor, Menheniot; and Delabole slate quarry, S of Tintagel.

Cumbria Actinolite, edenite and ferropargasite, with chlorite and epidote, are metamorphic replacement products of pyroxene in dolerite dykes cutting, or cut by, the Eskdale granite in the southwest Lake District. Macdonald et al. (1988a) interpret this as a contact-metamorphic effect. An appinitic microdiorite at Scawgill Bridge, 4 km W of Braithwaite, contains parvo-actinolite associated with augite, chromite, magnesiohornblende, olivine, pargasite, potassium feldspar and quartz (Fortey et al. 1994). Unconfirmed: Carrock Fell mine, Grainsgill, Caldbeck Fells (Fortey 1978); Cawdale Moor, Patterdale and Walla Crag, Derwentwater, near Keswick (Dakyns et al. 1897); Gate Crag associated with the Eskdale granite (Young et al. 1986); Honister Slate quarry, Honister Pass (Young 1987a); borehole material from the Sellafield area (Milodowski et al. 1998); and Shap Blue quarry (Firman 1953b).

Devon An albite–epidote–hornblende schist from the Start district contains actinolite (Tilley 1938a). All other occurrences are from the Meldon area, near Okehampton, and are unconfirmed: Ivy Tor, Belstone; Wheal Forest, Meldon Gorge; Sticklepath (all Hall 1868b); Meldon–British Railways quarry complex (Dearman 1966b).

Derbyshire Unconfirmed: Great Rocks Dale dyke (Ford & Sarjeant 1964a).

Wales

Gwynedd As part of a study of low-grade metamorphism of metabasites from the paratectonic Caledonides, Bevins & Rowbotham (1983) describe actinolite from Mynydd Preseli, Cadair Idris and the Aran Mountains. It occurs as a groundmass mineral, commonly as small needles overprinting chlorite, in association with albite, epidote, titanite, calcite, and, on Cadair Idris, also stilpnomelane. Similar metamorphism of the Tal y Fan intrusion, some 7 km SW of Conwy, produced secondary actinolite (after olivine) with variable compositions towards ferro-actinolite, along with prehnite and epidote (Bevins & Merriman 1988). Fibrous and acicular prisms of actinolite also crystallized in albite–quartz veins at the contact of this intrusion. Actinolite is also reported from the Eglwys Rhobell Formation, Rhobell Volcanic Complex, Rhobell Fawr area, near Dolgellau, where it occurs as inclusions and rims on pargasite in cumulo-phyric basalt (Allen & Jackson 1985, Kokelaar 1986). Gyopari (1984) confirms actinolite, in greenstones from the Penmynydd/Gwna boundary exposed in a roadcut near Llanfairpwllgwyngyll, southeast Anglesey. A related occurrence of actinolite grading into glaucophane compositions in greenschists from the Mona Complex at Penrhyn Nefyn on the Lleyn Peninsula, is reported by Gibbons (1981). Unconfirmed: Pwllpillo, Holy Island, Anglesey (Greenly 1919, Maltman 1977); Cadair Idris area, Arthog–Dolgellau district (Cox & Wells 1920); Capel Curig district, Snowdonia (Williams 1922); Gwynfynydd mine, near Dolgellau (Pryor 1988); and Ogof Owain Glyndwr, north of Moel Hebog, near Beddgelert (Bevins 1994).

Powys Tourmaline-bearing amphibolitized gabbro from the Hanter Hill–Stanner igneous complex contains interstitial patches of actinolite and magnesiohornblende (Holgate 1977, Tindle unpub. EMP data).

Scotland

Central A Dalradian actinolite schist/amphibolite is associated with stratabound sulphide mineralization near Tyndrum (Fortey & Smith 1986), but actinolite is unconfirmed.

Dumfries & Galloway Diorites from the Loch Doon granite complex (partly in Strathclyde) contain late-magmatic actinolite and parvo-actinolite replacing pyroxene (Tindle 1982, Tindle & Pearce 1983). At Glenhead Burn, Loch Dee, metasomatically altered biotite-rich hornfels at the margin of the Loch Doon granite complex contains a mineral assemblage consisting of actinolite, diopside, magnetite, pyrrhotite, pyrite and titanite. Lenses of green actinolite-rich rock with a quartz-rich core are also noted (Leake et al. 1981). Actinolite with compositions comparable to those from porphyry-style copper mineralization is reported from the Black Stockarton Moor subvolcanic complex, adjacent to the Criffell plutonic complex (Leake & Brown 1979, Leake et al. 1981).

Grampian At Corrie Loch Kander, near Braemar, actinolite occurs with tremolite, diopside–hedenbergite, hyalophane, pyrrhotite, scapolite and zoisite in schistose bands, within a rock interpreted as a Dalradian exhalative sulphidic dolomitic chert (Fortey et al. 1993). Amphibolite-facies regional metamorphism of metabasalts and ultrabasic rocks from the Dalradian Blackwater Formation, near Cabrach, has produced actinolite, magnesiohornblende and tremolite as alteration products of primary clinopyroxene (Macdonald et al. 2005).

Highland Actinolite makes a most unusual appearance in chlorite–talc-rich ultramafic rocks around An Fharaid Mhòr, near Achmelvich, where it is found in the cores of complex grains intergrown with cummingtonite and overgrown by clinojimthompsonite and jimthompsonite (Droop 1994). Ultramafic granulites at Drumbeg and Gorm Loch, NE of Scourie, Assynt, contain granulite-facies pargasite that exsolved opaque oxides and recrystallized as actinolite (described as tremolite) during retrograde metamorphism (Sills 1982). In the Ballachulish granite complex, late-magmatic, pyroxene-consuming reactions have produced diorites containing actinolite (Weiss & Troll 1989). Amphibole-rich clots in microgranular enclaves within the Strontian granite complex have actinolite cores and magnesiohornblende rims (Castro & Stephens 1992). In the Lewisian of the Moine Nappe at Camas Baravaig on the south coast of Skye, a serpentinite body contains zoned pods in which large green acicular crystals of actinolite have grown. The pods are infilled with dolomite and talc (Matthews 1967). Fenitized granodiorite from the Foyers area, Loch Ness, contains blue fibrous actinolite in vein fillings (Garson et al. 1984). Unconfirmed: actinolite (and tremolite) is reported as rhythmic overgrowths over hornblende in appinites associated with pipes of explosion breccia at Ardsheal Hill, Back Settlement, Gleann Chàrnan and Kiloran Bay, all near Kentallen. The zonation represents growth under alternating physical conditions during the episodic release of trapped volatiles (Bowes et al. 1964); Cnoc nan Cuilean syenite, W of Loch Loyal (King 1942).

Prisms of actinolite, up to 1 mm across, from a roadcut on the A890 near Auchtertyre, Kyleakin, Highland Region, Scotland. (© D. I. Green)

Strathclyde An amphibole in an albite–chlorite–epidote amphibolite from SE of Loch Gair hotel, referred to as "hornblende" by Wiseman (1934) is actually actinolite. It has a low-grade metamorphic origin and was derived from a basic igneous rock (epidiorite). Post-obduction dyke-like doleritic–gabbroic (metadolerite) intrusions associated with the Girvan–Ballantrae ophiolite complex contain actinolite, ferrohornblende and magnesiohornblende that grew during low-grade metamorphism of these rocks (Holub et al. 1984). Within the Kilmelford igneous complex, magnesiohornblende is reported from four separate diorite intrusions at An Cnap, Ceann Mòr, Meall Mòr and Tom Soilleir. Elsewhere in the complex, actinolite and pargasite are also found (Zhou 1987). Unconfirmed: Coire Buidhe, Glen Creran, Appin (Ingram et al. 1993).
Tayside An investigation of amphibole compositions from Dalradian Argyll and Southern Highland Group metabasites around and to the NE of Aberfeldy (centred on the classic area of Barrovian metamorphic zones that developed during the Grampian orogeny) shows that actinolite occurs in the cores of crystals within the chlorite, biotite, garnet and andalusite zones (Zenk & Schulz 2004).
Western Isles Scara Ruadh, Loch Meurach, South Harris – matted aggregates of dark-green crystals up to 5 cm long associated with a serpentinite body (Robinson & Dossett 2002, R. E. Starkey specimen, Tindle unpub. EMP data). Unconfirmed: Mullach Bi and Mullach Sgar, St Kilda (Hirta) (Harrison et al. 1984).
Shetland Gabbros from the Shetland ophiolite at Sandwick Bay, Unst, contain actinolite and magnesiohornblende, formed as a late magmatic replacement of pyroxene (Maynard 1993). At Balta, also on Unst, very pale pea-green actinolite occurs in nephrite, and large, dark-green and dull, somewhat foliated actinolite (described as hornblende) occurs in an augite-rich rock (diallage). A soft white (slight tint of green) schistose actinolite is also noted from a talc schist at Leegarth, on the eastern side of Trista Voe, Fetlar (Heddle 1878c).

Ireland

Cork Unconfirmed: Black Ball Head, near Cahermore (Pracht & Kinnaird 1995).
Donegal A fine-grain schist found in boulders and *in situ* on the shoreline to the N of Maghery Strand, contains small veins and lenses of dark-green actinolite forming blades up to several cm in length. Actinolite is accompanied by minor calcite, pyrrhotite, rose quartz and feldspar (S. Moreton specimen, Tindle unpub. EMP data). Unconfirmed: S of Bunbeg quay, E of Loughlin Island (Whitten 1951).
Galway At Barnahallia Lough, Claddaghduff, west Connemara, the metamorphic aureole of the Omey granite hosts an amphibolite skarn containing manganese-rich actinolite. It forms dark-green acicular masses and radiating tufts of fibrous crystals up to 12 mm long (Moreton & Green 2006a).
Louth Unconfirmed: Carlingford area, especially at Barnavave quarry (Osborne 1932b).
Mayo Some Dalradian Supergroup blueschist-facies garnetiferous metabasites from South Achill, N of Clew Bay, contain glaucophane or ferroglaucophane with actinolite rims (Chew et al. 2003).
Wicklow Distinctly bluish fibrous actinolite, associated with calcite, epidote and quartz, has been found in thin veins cutting a pyroxene-diorite intrusion at Parnell quarry, S of Arklow. An unusual manganese-rich actinolite is present in the Carrigmore diorite at Ballinclare quarry, near Glenealy (S. Moreton specimen, Tindle unpub. EMP data).

Offshore territories

Guernsey Injection of granitic-pipe intrusions into dioritic host magma within the Northern igneous complex of Guernsey, at Beaucette (Bishop & French 1982) and Port Soif (D'Lemos 1992) has resulted in the formation of actinolite. Around the margins of the pipes, well formed zoned crystals of green to pale-green actinolite, magnesiohornblende and pargasite grew in response to sudden and marked changes in magma conditions during this magma-mingling event.
Jersey Unconfirmed: pale-green fibrous actinolite in a syenite at Côtil Point and in veins with andradite garnet at St Clement's. An asbestiform variety was also found in veins within a weathered dolerite dyke, N of Mont Rossignol (Mourant 1961); actinolite is one of many amphibole species described from an igneous complex (olivine-gabbro to aplogranite compositions) at Sorel Point in north Jersey (Salmon 1991). Ferri-actinolite (described as actinolite) is reported from the Precambrian layered diorite–gabbro complex at Le Nez, southeast Jersey (Bishop & Key 1983).

adamite

orthorhombic $Zn_2(AsO_4)(OH)$

A zinc arsenate hydroxide, dimorphous with paradamite, that forms a series with olivenite and is found in the oxidized zone of arsenic- and zinc-bearing hydrothermal deposits.

Adamite occurs in the shallow opencast trials of Higher Longrigg mine, S of Cote Garth, Hartley Birkett, near Kirkby Stephen, Cumbria. At this locality, adamite is found in small barite–calcite–fluorite veins and is the copper-bearing variety, cuproadamite (possibly zincolivenite). It occurs as translucent pale-green crystals to about 0.1 mm long growing on azurite in cavities in barite and is commonly associated with a grey massive sulphide of the tennantite–tetrahedrite group (Braithwaite 1988). Adamite has also been found as white to pale-yellow subhedral crystals up to 0.5 mm long on a limonite-encrusted fluorite–quartz specimen containing minor galena–gersdorffite–pyrite from the nickel deposit in the Dow Scar level flats of Hilton mine, Scordale, Cumbria (Bridges & Green 2005b). Very pale bluish-green acicular microcrystals of adamite have been found with azurite, duftite and mimetite in barite–tennantite ore from Stennerskeugh Clouds mine, Kirkby Stephen, Cumbria (R. E. Starkey pers. comm. 2005).

In the Caldbeck Fells, Cumbria, small quantities of a globular emerald-green mineral, probably copper-rich adamite, has been found associated with agardite–mixite and aurichalcite at the Deer Hills mine (Cooper & Stanley 1990). Well crystallized adamite, accompanied by caledonite, linarite, leadhillite, philipsburgite, ramsbeckite and schulenbergite were collected from the large dumps of the 30-fathom level of Driggith mine (Green & Briscoe 2002). A later investigation showed that drusy crusts of pale-mint to apple-green crystals from Driggith and Sandbed mines actually contain more copper than zinc, whereas some colourless, pale-green and pale-blue crusts are zinc dominant (Green et al. 2006a). Depending on the Cu/Zn ratio, it is possible that both varieties could be the new species zincolivenite. Small dumps up stream from Short Grain mine contain adamite, agardite and philipsburgite (Green & Briscoe 2002). At the Old Potts Gill mine, adamite, hemimorphite, philipsburgite, olivenite and agardite form a supergene assemblage associated with cinnabar-bearing quartz in dump material (Neall & Leppington 1994). Cooper & Stanley (1990) report globular green adamite with serpierite or malachite from this same locality.

Copper-rich adamite (probably zincolivenite) is reported from Tynagh mine, near Killimor, Co. Galway, Ireland. It is found in an unconsolidated, partially oxidized cap, enriched in other supergene minerals above a primary sulphide orebody consisting predominantly of galena, sphalerite and pyrite. In total, 104 species have been described from this locality, including some new to Ireland (bayldonite, beudantite, chalcophyllite, hedyphane, olivenite, parnauite and strashimirite) and some new to

Blocky copper-rich adamite (probably zincolivenite) with a porous orientated adamite overgrowth. From a cavity in mineralized hornfels at Penberthy Croft mine, St Hilary, Cornwall; width 240 μm; SEM image. (© E. Condliffe, University of Leeds)

both Britain and Ireland (clinotyrolite, orthoserpierite and theisite) (Ryback 1988a, Ryback & Moreton 1992).

A discreditation of adamite has been given by Ryback et al. (2001) on specimens in the Kingsbury collection at the NHM, supposedly collected from the Sandbed, Potts Gill (both Caldbeck Fells) and Wanthwaite mines, all in Cumbria (some discussed in Hartley 1984), and Wheal Carpenter, Wheal Trelawny, Penberthy Croft and Pengenna mines in Cornwall. However, a copper-rich variety of adamite (entirely different from the Kingsbury material) forming drusy green crusts and pale-green crystallites up to 1 mm in length has been found at Sandbed mine (Cooper & Stanley 1990, S. A. Rust pers. comm. 2004).

A common theme showing through these occurrences is an elevated copper content. An investigation of "adamite" has shown that samples from the following locations all have significantly more copper than zinc: Penberthy Croft, St Hilary, Cornwall; Dolyhir quarry, Old Radnor, near Presteigne, Powys, Wales; and Tynagh mine, near Killimor, Co. Galway, Ireland (D. I. Green and D. Moulding samples, D. I. Green XRD, Tindle unpub. EMP data). This investigation found only undoubted adamite (a sodium-rich variety) near Howthwaite trial, in the same area as Potts Gill mine. It would therefore appear that many of the copper-bearing adamite specimens are actually the new species zincolivenite (Chukanov et al. 2007), and a full review of British and Irish adamite is now desirable.

Micro-crystals of pale-green copper-rich adamite (probably zincolivenite) coating blue azurite and colourless barite; Dolyhir quarry, Old Radnor, Powys, Wales; width 7 mm. (© D. I. Green)

ADELITE GROUP

Orthorhombic arsenates and vanadates with the general formula $AB^{2+}(XO_4)(OH)$, where A = Ca, Pb; B^{2+} = Co, Cu, Fe, Mg, Ni, Zn; and X = As^{5+}, V^{5+}. Conichalcite and duftite are members of the group found in Britain or Ireland. Tangeite may also be present. A further member of the group, calciovolborthite, was formally discredited in 2006.

adularia

A more completely ordered low-temperature variety of orthoclase, feldspar group.

aegirine

PYROXENE GROUP
monoclinic
$NaFe^{3+}Si_2O_6$

A relatively uncommon sodium–iron silicate found in alkalic igneous rocks (mainly syenites and alkali granites), in regionally metamorphosed schists and in sodium-metasomatized granulites. In most examples described here, aegirine is identified optically, but, as aegirine-augite is very similar, analytical confirmation is desirable.

Aegirine forms a reaction rim between analcime and titanaugite in borehole material from a teschenite sill at Landywood, near Cannock, Staffordshire (Barton 1963). A microscopic examination of riebeckite microgranite from Mynydd Mawr, 5 km W of Snowdon, Gwynedd, Wales, revealed small acicular crystals (<0.3 mm long) of colourless to pale-yellow acmite – a variety of aegirine (Nockolds 1938b).

In Scotland, aegirine is reported from several alkaline intrusive rocks centred on the Assynt area, Highland Region: in a nepheline-syenite pegmatite dyke cutting the Cnoc-na-Sròine mass at Cnoc Bad na h-Achlaise, S of Ledmore (Shand 1910b, Sabine 1950, Guppy 1956); in aegirine–quartz aplites, close to the margin of the Borralan syenite complex, particularly at Allt a' Mhuilinn and along the Ledmore River, W of Loch Borralan (Shand 1910b, Tilley 1958); in melanite syenite from the summit of Beinn Gharbh, NW of Inchnadamph (Sabine 1953); in ledmorite dykes on An Fharaid Mhòr, between Loch Roe and Achmelvich, near Lochinver, and at Camas Eilean Ghlais, Rubha Mòr, Coigach (Sabine 1952, 1953, Tilley 1958); in felsite dykes from Poll an Droighinn, near Inchnadamph, and from Sgonnan More (Teall 1900, Sabine 1953, Thompson & Fowler 1986); and abundant aegirine occurs in "felsite" from the Sail an Ruathair ridge, Benmore Forest, N of Loch Ailsh (Sabine 1953, Macgregor & Phemister 1972). Elsewhere in Highland Region, aegirine occurs with plagioclase and titanite in xenoliths and at the metasomatized contact between gabbro and feldspathic sandstone at Rubha Carrach, N of Ardnamurchan Point, Ardnamurchan (Wells 1951). Middle Devonian finely laminated siltstones and mudstones, near the village of Halkirk, contain aegirine (possibly as an authigenic species) forming green euhedral to subhedral prismatic crystals, up to 2 cm in length (Fortey & Michie 1978). Garson et al. (1984) report veins of aegirine and blue amphiboles in fenitized amphibolites from the Black Isle–Rosemarkie coastal section N of Inverness, and in Moine metasediments near breccia dykes at Dochfour, S of Inverness.

In Strathclyde, very unusual zirconium-rich alkali pyroxenes are reported from minor occurrences of nepheline syenite in the dominantly alkali olivine-dolerite intrusion at Cnoc Rhaonastil, Isle of Islay. Crystals are strongly zoned from diopside–hedenbergite through aegirine-augite to aegirine compositions. Aenigmatite, zircon-rich arfvedsonite, calcium catapleiite, fluorapatite and monazite are associated species (Preston et al. 1998a, 2000a,b). Bright-green subhedral or interstitial aegirine and aegirine-augite in microgranite from Ailsa Craig in the Firth of Clyde is similarly zirconium rich and strongly zoned (Harding 1983, Harrison et al. 1987). Thermal metamorphism of vesicular basalt lavas in the aureole of the volcanic plug at 'S Airde Beinn, 5 km W of Tobermory, Isle of Mull, has replaced original gyrolite in amygdales with reyerite and then wollastonite. Marginal to the wollastonite is a zone of aegirine-augite and, where this abuts feldspathic regions, pure aegirine is formed (Cann 1965). A Lower Carboniferous sill of trachyte at Biglees quarry, 500 m SSE of Biglees farm, 4 km N of West Kilbride, Glasgow, contains deep-green to pale yellow-green aegirine associated with arfvedsonite, calcite, orthoclase and quartz (Farrow et al. 1982). Perfectly fresh subhedral crystals of aegirine with frayed-out needle-like terminations have been found at Howford Bridge, SE of Mauchline, near Prestwick. They occur with apatite and a sodic amphibole in the analcime syenite that forms part of an alkali dolerite (crinanite) sill (Tyrrell 1928b).

Aegirine is a common mineral in the aegirine–riebeckite granite on Rockall island, 440 km W of northern Skye, NW of Scotland. In thin section, it shows patchy zoned, bladed prismatic crystals with green cores and yellowish-brown (acmite) rims. Isolated crystals of both varieties also occur (Sabine 1960, Hawkes et al. 1975). Both varieties are zirconium rich (Sedgwick Museum specimen, Tindle unpub. EMP data).

Other reports of aegirine in Scotland include: in riebeckite-bearing felsite dykes

Zoned aegirine in granite. The beige outer rim has previously been referred to as acmite. Rockall, 440 km W of northern Skye, off the NW coast of Scotland; width 1.3 mm; viewed in PPL and XPL. (© K. Gardner)

N of Colla Firth and Ronas Voe, Mainland, Shetland, where aegirine (acmite) forms stout subhedral prisms or is intergrown with riebeckite in spherules (Phemister 1950).

In Co. Antrim, Ireland, aegirine has a restricted occurrence in hybrid dolerites metasomatically altered during intrusion into limestone at Carneal (Sabine & Young 1975) and Scawt Hill (Tilley & Harwood 1931, Tilley & Alderman 1934), both near Larne. At the former it is found in an apatite–nepheline assemblage, whereas at the latter it forms needles in melilite–wollastonite-rich assemblages and is associated with wollastonite and xonotlite in metasomatized flint nodules. Ryback & Moreton (1992) report aegirine with arfvedsonite, quartz and riebeckite in cavities within a gabbro dyke at Ross Strand, Killala Bay, Co. Mayo.

Dark prisms of aegirine are reported from the pegmatitic portions of veins cutting the mica lamprophyre on South Hill, Jersey (Smith 1936a, Mourant 1961). Pale-green aegirine has also been found on Jersey in lamprophyre dykes at Grève de Lecq and Val Rouget, and in a granitic xenolith within another lamprophyre at Le Doüet de la Mer, St Ouën (Smith 1936b).

aegirine-augite

PYROXENE GROUP

monoclinic $(Ca,Na)(Mg,Fe^{2+},Fe^{3+})Si_2O_6$

A relatively uncommon calcium–sodium–magnesium–iron silicate found in alkalic igneous rocks, in regionally metamorphosed schists including blueschists and in sodium metasomatized granulites. Aegirine-augite is really an intermediate composition and can be considered to be either an Mg–Fe–Ca-rich aegirine or an Na–Al–Fe-rich augite. It appears to have gained species status in the latest classification of the pyroxene group (Morimoto 1988), but this status is not universally accepted (Gaines et al. 1997) and the mineral is not recognized by Anthony et al. (1995).

Aegirine-augite is an essential component of the microporphyritic phonolite on Wolf Rock, 15 km SW of Land's End, Cornwall, where it forms acicular and tabular subhedral crystals dispersed through the groundmass. In thin section it is pleochroic in shades of green and yellowish-brown (Harrison et al. 1977). It had previously been referred to as aegirine (Adye 1907). Rare aegirine-augite forms prisms interstitial to feldspar in borehole material from a teschenite sill at Landywood, near Cannock, Staffordshire (Barton 1963).

In the Highland Region of Scotland, phenocrysts of aegirine-augite occur with albite–oligoclase, biotite, diopside and titanite in the Canisp porphyry, exposed on Canisp, Meall na Braclaich and Suilven (Sabine 1953). Deep-green zoned aegirine-augite is the most abundant coloured mineral in the Cnoc nan Cuilean syenite intrusion, near Ben Loyal. In the main syenite it forms stout subhedral prisms up to 1.5 mm in length and is associated with albite and anorthoclase. Some quartz–feldspar granulites in adjacent metasomatized aureole rocks also contain aegirine-augite (King 1942). Fenitized quartzites, metasomatized by the Borralan syenite complex, contain stubby greenish pyroxene prisms identified as aegirine-augite (Woolley et al. 1972, Rock 1977). From the same complex, Tyler & Howie (1967) present aegirine-augite data from a xenolith in borolanite, and Woolley (1973) describes it in a nepheline-rich borolanite, to the E of Aultivullin. Fenitization of the Abriachan granite, near Loch Ness has produced veins of aegirine-augite and radiating crystals at the edges of quartz grains (Garson et al. 1984). Aegirine-augite also occurs in an alkali pyroxene-bearing facies of the Beinn an Dubhaich granite a few cm from the contact with a skarn formation, 2 km ESE of Camas Malag, Broadford area, Isle of Skye (Tilley 1949). Pyroxene with the optical properties of aegirine-augite is described from the Glen Dessary syenite, W of Loch Arkaig, although chemically the mineral is diopside (Thompson & Fowler 1986, Fowler 1992).

South of Derry Lodge in Glen Lui, Cairngorm Mountains, Grampian, aegirine-augite occurs in banded granulites close to the margin of the Cairngorm granite complex. It forms interstitial masses in pyroxene-rich bands, skeletal crystals to 3 mm in aggregates with biotite, chlorite and feldspar, or minute, often radiating, acicular crystals in aggregates or bands (McLachan 1951). Nepheline syenite within the alkali olivine dolerite intrusion at Cnoc Rhaonastil, Isle of Islay, Strathclyde, contains zirconium-rich pyroxene crystals that are strongly zoned from diopside–hedenbergite through aegirine augite to aegirine compositions (Preston et al. 1998a). Zoned crystals of aegirine and aegirine-augite occur in microgranite from Ailsa Craig in the Firth of Clyde, Strathclyde (Harding 1983, Harrison et al. 1987). Thin rims of green aegirine-augite mantle hedenbergite in alkali dolerite (crinanite) in the Shiant Isles Main Sill, Western Isles. It is also found with aenigmatite in late-stage pegmatites associated with this intrusion (Gibb & Henderson 1996).

At Barnavave, Carlingford, Co. Louth, Ireland, aegirine-augite forms microporphyritic, strongly pleochroic, euhedral crystals in syenite. Albite, microperthitic alkali feldspar, quartz with accessory eudialyte, neptunite and titanite are the associated species (Nockolds 1937, 1950). In Co. Antrim, irregular patches of pyroxenite pegmatite associated with a composite lava flow at McIlroy's Port, Island Magee, are reported to contain aegirine-augite, augite (titanaugite), plagioclase feldspar and sodalite (Walker 1959b). See aegirine for other examples.

aegirite

A synonym of aegirine, pyroxene group.

AENIGMATITE GROUP

Triclinic silicates with the general formula $A_2B_6X_6O_{20}$, where A=Ca, Na; B=Al, Cr^{3+}, Fe^{2+}, Fe^{3+}, Mg, Sb^{5+}, Ti; and X=Al, B, Be, Si. Only aenigmatite is reported from Britain or Ireland.

aenigmatite

AENIGMATITE GROUP

triclinic $Na_2Fe^{2+}_5TiSi_6O_{20}$

An uncommon sodium–iron–titanium silicate found as an accessory constituent in the mainly peralkaline Southern porphyritic epigranite, Marsco area, Isle of Skye, Highland Region, Scotland. The epigranite is unusual in containing no sodic feldspar, but microperthitic potassium feldspar and local concentrations of almandine, biotite, fayalite, hematite, ilmenite and magnetite are reported by Thompson (1969); arfvedsonite and sodic ferrohedenbergite may also be present. In the Ailsa Craig microgranite, Firth of Clyde, Strathclyde, brown-to-black grains of aenigmatite occur as spongy patches associated with an alkali amphibole termed "riebeckitic arfvedsonite" by Howie & Walsh (1981). In thin section, aenigmatite forms variably pleochroic dark reddish-brown or opaque grains that have anhedral form with fringes enclosing euhedral alkali feldspar (Harding 1983, Harrison et al. 1987).

Aenigmatite is an accessory component in the Shiant Isles Main Sill, Western Isles (24 km N of the Isle of Skye). This composite differentiated alkaline intrusion is a classic example of *in situ* fractional crystallization in which aenigmatite and aegirine crystallized in a cross-cutting gabbroic pegmatite at the end of the magmatic event (Gibb & Henderson 1996). A further occurrence of aenigmatite is from the Cnoc Rhaonastil dolerite intrusion on Islay, Strathclyde, which represents a single body of alkali olivine-basalt magma that differentiated *in situ*, much like on the

Shiant Isles. Late-stage alkali differentiates of this magma produced syenites with a minor phase assemblage of aenigmatite, zirconium-bearing aegirine, calcium catapleiite, fluorapatite, monazite and zirconolite (Preston et al. 1998a, 2000a,b).

aerugite?

trigonal $Ni_{8.5}As_3O_{16}$

A rare green nickel arsenate reported with arsenides (arsenopyrite, löllingite, gersdorffite, rammelsbergite, skutterudite), quartz, uraninite (pitchblende) and yellow xanthiosite in a low-temperature uranium lode from near the South shaft at South Terras mine, St Stephen-in-Brannel, Cornwall (Davis et al. 1965, Clark et al. 1995). The type locality was considered to be South Terras mine, although the mineral had been found earlier in Johanngeorgenstadt, Saxony, Germany (Adam 1869). The samples are from the Kingsbury collection, and Ryback et al. (2001) consider the occurrence to be very doubtful. The material most probably came from Germany.

aeschynite-(Y)

orthorhombic $(Y,Ca,Fe,Th)(Ti,Nb)_2(O,OH)_6$

A rare hydrated yttrium-dominant titanium–niobium oxyhydroxide that forms a series with aeschynite-(Ce) and with tantalaeschynite-(Y). Aeschynite is found in altered Ordovician rhyolites in central Snowdonia, Wales, where it forms small anhedral crystals in the groundmass of the rhyolites and in thin veins, for instance at Gallt y Wenallt, Carnedd y Cribau and Beddgelert Forest. Other accessory minerals in the rhyolites include allanite, apatite, lanthanite, larsenite–fayalite, monazite, parisite, synchysite, thorite, titanite, xenotime, zircon, and possibly gadolinite or spencite. Aeschynite is positively identified, but, although the heavy REEs are dominant, the exact species is still to be determined. It is likely that the species is aeschynite-(Y) (Howells et al. 1991, Bevins 1994).

afwillite

monoclinic $Ca_3(SiO_3OH)_2 \cdot H_2O$

Tabular afwillite in a vein cutting a calcite–spurrite rock. Scawt Hill, near Larne, Co. Antrim, Ireland; width 1.3 mm; viewed in PPL (upper) and XPL (lower). (© K. Gardner)

A rare hydrated calcium silicate hydroxide found in the chalk/dolerite contact zone at Scawt Hill, Larne, Co. Antrim, Ireland. Afwillite lines a cavity in a thin calcitic crust covering a spurrite contact rock. Crystals are usually under 0.5 mm in length and they form tabular clusters of subparallel individuals (Tilley 1930a). Within melilite-bearing rocks of the same contact zone, afwillite has been observed in infillings accompanied by calcite, ettringite and portlandite (Tilley 1933). Further investigations of vug minerals from Scawt Hill revealed afwillite intimately associated with hydrocalumite in a larnite-bearing rock carrying melilite–wollastonite aggregates (Tilley 1934).

At Ballycraigy, Larne, Co. Antrim, larnite-bearing rocks from an environment similar to those at Scawt Hill contain hydration cavities infilled with afwillite, portlandite and a calcium silicate hydrogel (Long & McConnell 1959). At Killala Bay, near Inishcrone, Co. Sligo, thermal metamorphism of limestone by basalt–dolerite dykes has produced spurrite-rich rocks in which afwillite occurs with calcite, cuspidine, killalaite, scawtite, xonotlite and other hydrated calcium silicates. The hydrothermal formation of afwillite at 350–550°C in a CO_2-deficient environment is proposed (Nawaz 1974).

A secondary assemblage of afwillite, calcite, killalaite, plombièrite and portlandite forming veinlets and cavities in larnite-rich rocks occurs at Carneal, Larne, Co. Antrim, again from a chalk/dolerite contact zone. Afwillite with a minimum grainsize of 0.5 mm forms anhedral-bladed crystal aggregates, with portlandite occupying cavities, or as radiated-bladed aggregates set in a calcite matrix (Nawaz 1977).

agalmatolite

A synonym of pinite, pyrophyllite or talc (see Macadam 1886b).

agardite-(La)

MIXITE GROUP

hexagonal $(La,Ca)Cu_6(AsO_4)_3(OH)_6 \cdot 3H_2O$

A rare hydrated lanthanum–calcium–copper arsenate hydroxide found at Penberthy Croft mine, St Hilary, Cornwall, where it forms small sprays of radiating acicular crystals to 0.3 mm, varying from a bright yellowish green through vivid green to a paler whitish green. Most commonly, it is found in an altered clinochlore–goethite–manganese-oxide matrix associated with arsenopyrite and quartz, but it also occurs with vein quartz and beudantite as small tufts on bayldonite and manganese oxides. The REE and yttrium content of agardite is variable, and agardite-(Ce), agardite-(La) and agardite-(Y) now have species status. Agardite is visually indistinguishable from mixite at Penberthy Croft (Betterton et al. 2000). Variable REE contents are apparently common, as cerium- and neodymium-rich agardite is noted from Wheal Alfred, Phillack, near Hayle, Cornwall (Embrey 1978, Embrey & Symes 1987, Scheding 1991) and calcium-rich agardite (see zálesíite) is noted from Penberthy Croft and Croft Gothal mines, St Hilary, Cornwall (Golley & Williams 1995). However, calcium-rich agardite appears to be in error (J. Betterton pers. comm. 2005). An undifferentiated lustrous orange agardite mineral from Marke Valley mine, near Liskeard, Cornwall, is illustrated by Moore & Pearce (1997).

Undifferentiated agardite minerals with a pale-green colouration have also been collected from the Caldbeck Fells, Cumbria, at Brandy Gill lead mine and Deer Hills barite mine (Cooper & Stanley 1990). At nearby Potts Gill barite mine, Neall & Leppington (1994) found locally common agardite and olivenite in a supergene assemblage that includes adamite, cinnabar, hemimorphite and philipsburgite. Other undifferentiated agardite minerals are reported from Driggith and Sandbed mines in the Caldbeck Fells (Green et al. 2006a).

agardite-(Nd)

An incompletely characterized member of the mixite group reported from Wheal Alfred, Phillack, near Hayle, Cornwall (Embrey & Symes 1987).

agardite-(Y)

MIXITE GROUP

hexagonal $(Y,Ca)Cu_6(AsO_4)_3(OH)_6 \cdot 3H_2O$

A rare hydrated yttrium–calcium–copper arsenate hydroxide found at Gwaithyrafon mine, SE of Tal-y-bont, Central Wales orefield, Dyfed. Agardite-(Y) is a supergene species that was produced by oxidation of a chalcopyrite–quartz vein and forms pale- to mid-green lustrous acicular crystal clusters up to 0.5 mm across on a black iron-oxide matrix. Associated species include minor chrysocolla and malachite, and rare mimetite (Bevins 1994, Rust & Mason 1994). Anthony et al. (2000) note agardite-(Y) from Wheal Alfred, Phillack, near Hayle, Cornwall.

agaric mineral

A soft variety of calcite (rock milk).

agate

A fine-grain banded variety of quartz (chalcedony) ±moganite.

agnesite

TL Wheal Coates, St Agnes, Cornwall, England

A synonym of bismutite.

aheylite?

triclinic $(Fe^{2+},Zn)Al_6(PO_4)_4(OH)_8 \cdot 4H_2O$

A rare hydrated iron–zinc–aluminium phosphate hydroxide briefly described from Goonbarrow china-clay pit, St Austell, Cornwall; requires confirmation (Golley & Williams 1995).

aikinite

orthorhombic $CuPbBiS_3$

Named after Arthur Aikin (1773–1854), English chemist and mineralogist, and one of the founders of the Geological Society of London; born in Warrington, Cheshire. Aikinite is a copper–lead sulphobismuthite typical of moderate-temperature hydrothermal mineralization, and was first described as a British species from cosalite-bearing quartz stringers within the greisen at Carrock mine, Grainsgill, Caldbeck Fells, Cumbria (Kingsbury & Hartley 1956b). This provenance is dubious (Ryback et al. 1998).

Aikinite has been found in fluorite–quartz veinstone on the dumps of North Devon United (East Wheal Friendship) mine, Peter Tavy, Devon, within the metamorphic aureole of the Dartmoor granite. Small grains of aikinite are associated with arsenopyrite, cassiterite, cobaltite, cosalite, erythrite, gersdorffite, parkerite, sphalerite, stannite, ullmannite and a variety of bismuth-rich minerals, including bismite, native bismuth, bismuthinite, ikunolite and an unnamed bismuth telluride (Rumsey & Savage 2005). Aikinite (aikenite) is mentioned by Greg & Lettsom (1858) from Wheal Maudlin, 3 km NW of Lostwithiel, Cornwall, but this was actually wolframite in the form of pyramidal scheelite.

An aikinite–bismuthinite phase associated with molybdenite mineralization was identified from the Shin granite and associated quartz veins near Lairg, Highland Region (Gallagher et al. 1974, Lowry et al. 1994). Bladed crystals of aikinite commonly occur as inclusions in galena from calcite–fluorite–quartz veins associated with a quartz monzonite intrusion at Ratagain, Highland Region. Chalcopyrite, galena, pyrite, and sphalerite dominate the ore assemblage, and electrum, gersdorffite, hessite, marcasite and tennantite are found as minor phases along with the aikinite (Alderton 1988b). At Tomnadashan, on the southern shore of Loch Tay, Tayside Region, aikinite occurs as inclusions in the major ore mineral pyrite, which is found with native bismuth, chalcopyrite, molybdenite and tennantite–tetrahedrite in granitic lenses intruded into a small diorite body (Pattrick 1984).

Ixer et al. (1990) report aikinite as a minor component of primary gold mineralization in metamorphosed BIF at Ballycoog, Goldmine River valley, near Avoca, Co. Wicklow, Ireland. It is associated with an ore assemblage of chalcopyrite, galena, hematite, magnetite, pyrite and sphalerite.

ajoite?

triclinic $(K,Na)Cu_7^{2+}AlSi_9O_{24}(OH)_6 \cdot 3H_2O$

An hydrated potassium–sodium–copper–aluminium silicate hydroxide. A specimen labelled "ajoite" from East Wheal Russell, Tavistock, Devon, is in the British Micromount Society reference collection, but its composition could not be confirmed by XRD (D. I. Green pers. comm. 2005).

akaganéite/akaganeite

tetragonal $(Fe^{3+},Ni^{2+})_8(OH,O)_{16}Cl_{1.25} \cdot nH_2O$

An uncommon brown hydrated iron–nickel oxyhydroxide chloride that is a significant component of actively forming calcite–iron-sulphide–siderite concretions in the reduced intertidal marsh and sandflat sediments at Wareham Marshes, Norfolk. Concretions grow to 0.4 m in diameter and some have military shell fragments as their nucleus – evidence that the area was once a bombing range (Pye 1988, Al-Agha et al. 1995). Akaganéite also occurs as a secondary mineral at Gravel Hill mine, Perranzabuloe, Cornwall (Ryback & Tandy 1992).

An XRD trace of an iron pan from soil collected at Killiechronan, SW of Salen, Isle of Mull, Strathclyde, Scotland is described as having intensities consistent with akaganéite, hematite and lepidocrocite (Goodman et al. 1988).

åkermanite

MELILITE GROUP

tetragonal $Ca_2MgSi_2O_7$

A rare calcium–magnesium silicate that forms a series with gehlenite and is found in nodular limestones that have been thermally metamorphosed by doleritic or gabbroic intrusions at Kilchoan, Ardnamurchan, Highland Region, Scotland. Åkermanite forms faintly brown crystals in various assemblages that reflect differences in the original composition of the sediments. Associated species include cuspidine, grossular, kilchoanite, larnite, merwinite, monticellite, rustumite, spurrite, vesuvianite and wollastonite (Agrell 1965). Åkermanite is reported without detail from Scawt Hill, near Larne, Co. Antrim, Ireland (Anthony et al. 1995), presumably from the well known chalk/dolerite contact zone at the locality. A sample labelled åkermanite from a nearby contact zone at Carneal proved to be melilite (H. J. Foy specimen, Tindle unpub. EMP data).

alabandite

cubic $Mn^{2+}S$

Alabandite is noted with rhodochrosite in a mineral veinlet from Tintagel, Cornwall (Spencer 1958). Alabandite is also reported from Wheal Mary, Lanivet, near Bodmin, Cornwall (Kingsbury 1961), but the occurrence is suspect.

alabaster

A colourless massive variety of gypsum. Alabaster marble and alabaster onyx are names applied to dendritic or fine-grain banded calcite.

albite

FELDSPAR GROUP

triclinic $NaAlSi_3O_8$

A common sodium aluminosilicate and the sodium-rich end member of the plagioclase solid-solution series. With increasing calcium content, intermediate compositions (and hence not species) have been given the names oligoclase, andesine, labradorite and bytownite. The species anorthite is the calcium-rich end member of the series. Albite is found in acid igneous rocks of both intrusive (granite, granite pegmatite) and extrusive (dacite, rhyolite) origin. It also occurs in alpine-type quartz veins and is widely developed in low-grade metamorphic rocks of basic composition, where it replaces high-temperature calcic plagioclase.

England

Cornwall Albite is an essential component of late-stage volatile-rich topaz granites from the central part of the St Austell granite complex. Topaz granites also contain amblygonite, columbite, fluorapatite, fluorite, ilmenorutile, lepidolite or zinnwaldite, orthoclase and quartz. Hydrothermal alteration of these and other granites in the complex has produced one of the world's largest kaolinite deposits, primarily because of the decomposition of albite and other feldspars. Albite is described from Goonbarrow china-clay pit, near Bugle; Goonvean and Rostowrack china-clay pits, W of Nanpean; Gunheath china-clay pit (Hensbarrow section); and Trethosa china-clay pit, both near Carthew (Bray & Spooner 1983, Manning & Hill 1990, Manning et al. 1996, Psyrillos et al. 1998, Scott et al. 1998, Williamson et al. 2002). Albite–oligoclase compositions are noted in topaz-bearing granite from St Michael's Mount, Marazion (Hosking 1954c). At Carn Clodgy, near Rinsey and at Megiliggar Rocks, near Breage, albite–topaz-bearing granite (representing the roof zone of the Tregonning granite) contains 40% euhedral to subhedral albite (Stone & George 1978, Stone 1992). Albite is also present in related aplites and pegmatites at Megiliggar Rocks (Stone 1969), and at Wheal Metal, Breage, it is associated with cassiterite (including the variety wood tin), chalcedony, chlorite, dolomite, pyrite and sphalerite (Hosking 1951). Complex pegmatites in the St Austell granite at Luxulyan quarry, near Lostwithiel (Hosking 1954b), and in the Carnmenellis granite at Kessel Downs (Halvasso) quarry, near Longdowns, contain appreciable albite associated with fluorapatite, fluorite, muscovite (or lithium mica), orthoclase and quartz (Jackson 1978). At Chywoon quarry, also near Longdowns, albite forms masses of individual crystals up to 2 cm long in pegmatitic cavities associated with potassium feldspar, fluorapatite and quartz (Moulding & Hooper 2005). Other occurrences of albite include: in cavities in an alpine-type quartz vein, associated with chlorite, monazite-(Ce), muscovite, pyrite and rare apatite and anatase at Croft Gothal mine, St Hilary (Betterton 1996d); in a metasomatic albite-dolerite (adinole) on Dinas Head, near Padstow (Agrell 1939, 1941); in a quartz vein associated with monazite and rutile at Lanterdan quarry, near Tintagel (Bowman 1900, Rudler 1905); in a Devonian spilitic suite sill termed minverite at St Minver (Kempe 1968); and in metadolerites (greenstones) metamorphosed to prehnite–pumpellyite facies at Trevone Bay, W of Padstow (Floyd & Rowbotham 1982).

Cumbria High-temperature alteration of diabase adjacent to the graphite mine at

Seathwaite has produced an albite–biotite–calcite–chlorite–hematite–orthoclase–quartz assemblage (Strens 1965). Pervasive hydrothermal alteration of microdiorite at Scawgill Bridge, 4 km W of Braithwaite, converted original plagioclase into an intimate mix of albite and clinozoisite (Fortey et al. 1994). At the Shap granite quarry, albite forms colourless transparent acute diamond-shape crystals in miarolitic cavities within the granite (Wirth & Green 2002). Other occurrences of albite are documented by Young (1987a), but in many cases it is uncertain whether it is albite–oligoclase or end member albite that is being referred to.

Derbyshire Freely growing crystals of albite have been found in vesicles in the Carboniferous dolerite at Calton Hill quarry, Taddington (Sargent 1918, Tomkeieff 1926, Wilcockson 1932, Ford et al. 1993). Xenoliths in the Great Rocks Dale dyke, near Buxton, contain the species albite, microcline, orthoclase, and intermediate compositions with the names labradorite and oligoclase (Cope 1933, Ford et al. 1993).

Devon Albite is common in lithium-mica–topaz-bearing granite at Lee Moor china-clay pit, near Wotter, southwest Dartmoor (Knox & Jackson 1990). Within the metamorphic aureole of the Dartmoor granite at Meldon, near Okehampton, albite is one of the main constituents of a sodic aplite, where it is associated with lithium mica, orthoclase and quartz with accessory elbaite, fluorapatite, fluorite, topaz and occasionally petalite. There are also smaller amounts of amblygonite–montebrasite, a suite of beryllium minerals, manganocolumbite and microlite. In the earliest-formed part of the intrusion (the blue aplite), albite reaches over 80% by volume and approaches an albitite composition. Pegmatitic veins cutting the aplite also contain albite (McLintock 1923, Chaudhry 1971, Chaudhry & Howie 1973a, Chaudhry & Mahmood 1979, von Knorring & Condliffe 1984). Porphyroblasts of albite containing carbonaceous inclusions and associated with rutile occur in a remarkable mica schist in the Bolt area, near Starehole Bay and at Bolt Head, both on the south coast near Start Point (Tilley 1923). Albite is the main sodic feldspar in the Lundy granite (Thorpe et al. 1990, Tindle unpub. EMP data).

Fine, lamellar-twinned albite from the granite on Lundy island, Bristol Channel, off the north coast of Devon; width 2.8mm; viewed in XPL. (© A. G. Tindle)

Leicestershire Albite forming colourless translucent, occasionally greyish-white, crystals has been found in alpine-type veins cutting an andesite dyke at Bardon Hill quarry, Coalville (King 1967). Similar veins cutting volcaniclastic rocks at the same location have produced pinkish tabular crystals of albite (Ince 2007).

Merseyside Gabbroic glacial erratics on Thurstaston beach, Wirral, contain small cavities in which albite has formed. Colourless-to-white, single or twinned crystals occur with magnetite, oligoclase and titanite (Rothwell & Dickinson 1989).

Wales

Clwyd Albite forms colourless-to-white twinned crystals up to 2 mm across in calcite-infilled joints within dolerite at Hendre quarry, Pandy, near Glyn Ceiriog (Starkey et al. 1991).

Dyfed Albite is an important component of both the St David's granophyre (Bloxam & Dirk 1988) and the granophyric gabbros and aplite veins of the St David's Head intrusion, Carn Llidi (Roach 1969, Bevins et al. 1994). Phenocrysts of albite and sanidine are found in nearby rhyolitic lavas at Pen Pedol and Pen-y-Foel to the W of St David's (Davies & Bloxam 1990). Bevins & Rowbotham (1983) report the formation of albite attributable to low-grade metamorphism in altered Ordovician metabasic rocks from Garn Fawr, near Strumble Head; Garn Turn Rocks, E of Wolf's Castle; Mynydd Preseli; and Y Garn, near Pontiago. An alpine-type quartz vein at Brynyrafr mine in the Central Wales orefield also contains albite (Bevins 1994).

Gwynedd Albite is common in Lower Palaeozoic anatase-bearing alpine-type quartz veins at Fron-oleu, Prenteg, Tremadog, where it forms lustrous, colourless-to-white, tabular crystals to 10 mm (Starkey & Robinson 1992) and at Tanygrisiau, Blaenau Ffestiniog, where it forms lustrous, white, equant to thick-tabular crystals to 3 mm (Green & Middleton 1996). The dolerite mass at Hendre Henydd, Penrhyndeudraeth, near Porthmadog, is cut by alpine-type veins containing colourless tabular albite to 3 mm with anatase and brookite (R. E. Starkey pers. comm. 2005). Other alpine-type veins occur at: Arenig Station quarry and Manod quarry, near Blaenau Ffestiniog; Llyn-y-dywarchen, Rhyd-ddu; and Minffordd quarry, Minffordd, near Porthmadog (Bevins 1994). Albite is found in altered Ordovician metabasic rocks at: Aberdaron, Lleyn Peninsula; Cadair Idris; Cwm Eigiau, 6 km W of Trefriw; Dolgellau; Hafodty, 3 km SW of Conwy; and Llanfachreth on Anglesey (Bevins & Rowbotham 1983). Alteration of regionally metamorphosed felsic and mafic volcanic rocks associated with the Parys Mountain polymetallic sulphide deposit on Anglesey has produced an assemblage consisting of albite, chlorite, epidote and sericite (Tennant & Steed 1997). The Tal y Fan olivine dolerite intrusion SW of Conwy contains albite as a replacement product of primary labradorite feldspar (Merriman et al. 1986).

Powys An alteration assemblage of actinolite, albite, clinochlore, epidote and tourmaline has developed in amphibolitized gabbro from 300 m SW of the summit of Hanter Hill (Holgate 1977) and, at Builth Wells, albite, calcite, chlorite, prehnite, pumpellyite and titanite are all alteration products recorded in Ordovician metabasic rocks (Bevins & Rowbotham 1983).

Scotland

Central Albite has been recognized in copper–zinc stratabound mineralization within the Dalradian Ben Challum Quartzite at Creag Bhocan, SSW of Tyndrum (Scott et al. 1988).

Dumfries & Galloway A pegmatitic and aplitic vein associated with the Cairnsmore of Fleet granite in a small quarry near Clatteringshaws Loch contains an albite, microcline, quartz and spessartine assemblage with minor biotite, muscovite and oligoclase (Macleod 1992).

Grampian Plagioclase feldspar is zoned from oligoclase cores to albite rims in wollastonite marble from Kinnaird Head, Fraserburgh (Johnson et al. 2000). Granite containing albite in miarolitic cavities at Upper Craigton quarry, Hill of Fare, and Murdoch's Cairn quarry, near Peterhead, are reported by Heddle (1883). Albite is the dominant constituent of the Peterhead granite (Heddle 1901). Albite is the only sodic feldspar found in zinnwaldite granite and associated greisen at Gairnshiel, Glen Gairn, W of Ballater, and in lepidolite-bearing pegmatite (var. cleavelandite) at Peatfold, Glenbuchat (Tindle unpub. EMP data).

Highland Albite occurs in granitic veins at Sgòr Chaonasaid, on the north slope of the Ben Loyal (Beinn Laoghal) syenite complex, near Lettermore, and as cleavelandite (a bladed variety of albite) in the giant erratic syenite boulder at Beinn Bhreac, 2 km E of Tongue, which is thought to be derived from Ben Loyal, 7 km to the S (Heddle 1883, 1901). The main syenite of the Cnoc nan Cuilean intrusion, adjacent to the Ben Loyal complex, contains clear albite and cloudy anorthoclase intergrown into beautiful perthitic textures (King 1942). Albite is one of the rock-forming minerals in the Glen Dessary syenite and it occurs with biotite, diopside, edenite, ferro-edenite, orthoclase and hypersolvus alkali feldspar (Fowler & Williams 1986, Fowler 1992). Pure albite and orthoclase with complex sutured margins occur in metasomatized quartzites at Allt Bealach Choinnich, near the Loch Borralan syenite complex, Assynt (Woolley et al. 1972), and in the same general area, at Meall a' Bhraghaid to the N of Loch Ailsh, an almost pure albite rock (hornstone porphyry or albitite) is described by Heddle (1883). Moine mylonites in the Dundonnell and Knockan parts of the Assynt–Ullapool section of the Moine thrust zone contain deformed albite crystals with so-called tails of oligoclase (Freeman et al. 1998). Unusual highly brecciated and radioactive red feldspar-rich rocks, termed albitites, cut Moine Series gneiss at Moniack Burn, WSW of Inverness (Garson et al. 1984).

Strathclyde Vesicles in Carboniferous Clyde Plateau Lavas at Boyleston quarry, SW of Barrhead contain transparent, terminated crystals of albite up 10 mm in length (Meikle 1989a) and at Reilly quarry, Houston, sub-mm brownish-orange albite forms intergrowths with quartz (Meikle 2003). Albite is also fairly common in vesicles within Tertiary basaltic lavas on Ben More, Isle of Mull. It forms a massive layer underlying prehnite or scolecite on An Gearna and is very common on Beinn Fhada and Maol nan Damh (McLintock 1910b). Dense aggregates of albite, calcite, pumpellyite, quartz and an indeterminate material replace altered plagioclase in Ordovician spilitic lavas (part of the Girvan–Ballantrae ophiolite complex) near Knocknormal farm, 2 km E of Lendalfoot (Bloxam 1958, Bloxam & Allen 1959). Miarolitic cavities in the Northern granite on the Isle of Arran contain albite and orthoclase on which allanite-(Ce), fergusonite, gadolinite and zircon have grown (Hyslop et al. 1999). Albite co-exists with apatite, calcium catapleiite, fayalite, kaersutite, monazite, zirconium-rich pyroxenes and a variety of zeolites in the Cnoc Rhaonastil nepheline syenite on the Isle of Islay (Preston et al. 1998a). Albite porphyroblasts with complex inclusion-trail morphology are present in Dalradian schists from the Loch Lomond area (Mathavan & Bowes 2005).

Tayside A wide range of plagioclase compositions, including albite, are reported from the Dalradian stratiform barite–base-metal deposit at Foss, near Aberfeldy, although the locality is better known for its celsian and hyalophane content (Moles 1985). Albite in stratiform barite deposits of the Ben Eagach area, near Aberfeldy, is discussed more generally by Coats et al. (1984).

Western Isles Albite is a major constituent of the Lewisian Chaipaval (Ceapabhal) granitic pegmatite found 3 km NW of Northton (Taobh Tuath), South Harris (von Knorring & Dearnley 1959b). Hydrothermal alteration of brecciated dolerites and gabbros on St Kilda (Hirta), has produced an albite–carbonate–chlorite–epidote–titanite– assemblage (Harding et al. 1984). Heddle (1901) reports huge twin crystals of albite from near Northton on the Isle of Stromey (Sròmaigh), off the coast of South Harris.

Shetland On Mainland, pure white albite occurs with quartz in a vein at Colla Firth Voe; pink albite occurs at Hillswick, at the Banks of Niddister; and a snow-white variety occurs at the southwest end of Bixter Voe in a pegmatitic vein (Heddle 1901). At Ve Skerries, a small group of rocks 5 km WSW of Papa Stour, albite forms the main constituent (90% by volume) of a remarkable white albitite (Walker 1932).

Ireland

Carlow Albite is an essential component of spodumene-bearing pegmatites (related to the Leinster granite) that occur in a belt extending from Killedmond for 30 km into Co. Wicklow. The largest pegmatite is at Aclare, near Myshall, where examination of boulders and drill core has revealed characteristic zones including a fine-grain (saccharoidal) albite unit containing muscovite, porphyroblastic quartz and many tiny spessartine garnets. Cavities in this rock are filled with albite, bertrandite, kaolinite and quartz. Other notable pegmatites occur at Borris and Seskinnamadra (Steiger & von Knorring 1975, Scoon 1978, Luecke 1981, McArdle 1984, McArdle & Kennan 1988, Whitworth & Rankin 1989, Whitworth 1992).

Cork Albite occurs in quartz veins that were once mined for chalcopyrite, minor bornite and tennantite–tetrahedrite at the Berehaven (Bearhaven) mines, Allihies (Moreton & Aspen 1993).

Down Blocky white albite commonly lines drusy cavities in the granite at Lindsay's Leap quarry, S of Newcastle, northeast Mourne Mountains. Beryl, chabazite-Ca, danalite, fluorite, heulandite-Ca, monazite-(Ce), orthoclase, smoky quartz, stilbite-Ca and topaz are associated cavity species (Green & Moreton 2001, Green et al. 2005a). Very perfect, translucent, twinned white crystals of albite are described with topaz and smoky quartz from the granite of Slieve Corra, also in the Mourne Mountains (Greg & Lettsom 1858). White masses and subhedral crystals of albite, up to several mm across, commonly occur with epidote and quartz in lenses within hornfelsed greywacke in the aureole of the Mourne Mountains granite at Aughrim quarry, Kilkeel (S. Moreton pers. comm. 2005).

Galway Small amounts of albite occur with andradite–grossular garnet, diopside, vesuvianite and wollastonite in the Fountain Hill skarn – a thermally metamorphosed impure limestone in the Omey granite aureole (Ahmed-Said & Leake 1996). Moreton & Green (2006a) examined an amphibolite skarn near Barnahallia Lough, Claddaghduff, in the same metamorphic aureole, and found cavities containing small blocky white crystals of albite, up to 1 mm long, associated with chabazite-Ca, epidote, grossular, heulandite-Ca and stilbite-Ca.

Louth Albite occurs in thermally metamorphosed limestone close to a syenite intrusion in the vicinity of Barnavave, Carlingford (Osborne 1932b), and as beautiful spindles in microperthite in the syenite itself (Nockolds 1937).

Wicklow Albite is an essential component of spodumene-bearing pegmatites at Moylisha, near Shillelagh and Stranakelly. Part of the Stranakelly pegmatite is greisenized and it contains albite (cleavelandite), lithium mica and quartz. Cassiterite, columbite and microlite are common accessory minerals (Steiger & von Knorring 1975, Scoon 1978, Luecke 1981, McArdle 1984, McArdle & Kennan 1988). Narrow veins cutting amphibolite at the top of Camaderry Mountain, between Glendasan and Glendalough, contain albite, clinozoisite, prehnite, tremolite and titanite (Johnston 1949, Ryback & Moreton 1993).

Offshore territories

Rockall Albite is common in the aegirine–riebeckite granite on Rockall, 440 km W of northern Skye, NW of Scotland, and includes stumpy twinned prisms and lath-shape crystals up to 20 mm in length (Hawkes et al. 1975, Young et al. 1978).

aleksite

pseudo-trigonal $PbBi_2Te_2S_2$

A very rare lead–bismuth sulphotelluride found at Clogau–St David's mine, Bontddu, near Dolgellau, Gwynedd, Wales, where it is intergrown with galena and tellurobismuthite in gold-bearing quartz veins (Bevins & Stanley 1990). Mason et al. (2002) add that the aleksite forms 200 μm laths and that it is part of a bonanza gold assemblage that also includes trace amounts of native bismuth, bismuthinite, boulangerite, bournonite, hedleyite, nagyágite and pilsenite. At this location, Cook et al. (2007) note that aleksite is intergrown with saddlebackite, or with tellurobismuthite and an unnamed phase (corresponding to $PbBi_4Te_4S_3$).

algodonite

orthorhombic (pseudo-hexagonal) Cu_6As

A rare copper arsenide found with coffinite, native copper, lavendulan, malachite, maucherite, nickeline, rammelsbergite and native silver in Permian nodular concretions from the Littleham Cove area, Budleigh Salterton, Devon (Harrison et al. 1975b). Unusual sheet-like "plates" of native copper occurring with the concretions show a complex sequence of alteration and subsequent mineral growth on their surfaces that includes the species algodonite, maucherite and nickeline. De-watering of Permo-Triassic strata during burial compaction is favoured as a mechanism for producing metalliferous pore fluids capable of transporting the copper and associated Ag–As–Co–Cu–Ni–U–V mineralization (Milodowski et al. 2000).

Algodonite has been found in copper-rich void-infilling mineralization near the Precambrian/Triassic unconformity at New Cliffe Hill quarry, Stanton under Bardon, Leicestershire. It occurs in trace amounts infilling voids (< 40 μm across) in cuprite, as rims to large masses of native copper, and with domeykite in 10 μm-wide veinlets cutting native copper (Ixer et al. 2005a).

allagite

A synonym of rhodonite ± rhodochrosite.

allanite-(Ce)

EPIDOTE GROUP

monoclinic $(Ca,Ce,La)_2(Al,Fe^{2+},Fe^{3+})_3Si_3O_{12}(OH)$

Named after Thomas Allan FRS (1777–1833), a collector who discovered the mineral in the cargo of an impounded Danish ship at Leith; born in Edinburgh. The cargo was part of a collection of Greenland minerals belonging to Sir Charles Lewis Geisecke (1761–1833). Allanite-(Ce) is an uncommon calcium–cerium–lanthanum–aluminium–iron silicate hydroxide found as an accessory constituent in some granites, granite pegmatites and syenites. In many cases identification is optical and the dominant ion is therefore not measured. Examples of allanite *sensu lato* are listed at the end of the section, as most are probably allanite-(Ce), but caution must be applied to its identification, as allanite is considered to be a poorly defined collection of species (Ercit 2002).

In the Highland Region of Scotland, quartzose amphibolite masses within the Fort Augustus granite gneiss contain abundant subhedral crystals of allanite-(Ce) up to 1 mm in length, which exhibit striking optical zoning and twinning (Highton 1994). Small masses of hornblendic rocks of Moinian age at White Bridge, near Invergarry, contain altered allanite-(Ce) (Rock 1984). Accessory allanite-(Ce) is reported from miarolitic cavities in the following Tertiary granites on the Isle of Skye: Coire Uaigneich epigranite, Camasunary; Glamaig epigranite, stream bed of Allt Daraich; and Marsco epigranite, Coire nam Laogh. These occurrences have an igneous origin, but data from other intrusions with appreciably lower cerium contents are considered to have a secondary origin, being formed as the granites were altered during a later meteoric–hydrothermal event (Exley 1980).

Reddish-brown allanite-(Ce) has been found on St Kilda (Hirta), Western Isles, with zirconolite (zirkelite) in a granitic vein on Abhainn Mhòr and in a gabbroic pegmatite cutting the Glen Bay gabbro (Harding et al. 1982, 1984). A single 5 mm-diameter yellowish-brown detrital grain of allanite-(Ce) is reported from a sulphide-banded quartzite in the vicinity of the Foss stratiform barite deposit, near Aberfeldy, Tayside Region (Moles 1985).

In Dumfries & Galloway, rare, euhedral, crystals of allanite-(Ce), up to 1 mm in length and yellow in thin section, occur with apatite, titanite and zircon in granodiorite and granite from the Loch Doon granite complex (Tindle & Pearce 1981, Tindle 1982) and from the Cairnsmore of Carsphairn granitic intrusion (Tindle et al.

1988). Barritt (1983) reports accessory allanite-(Ce), apatite, titanite and zircon from the northeast unit of the Cairngorm granite, Grampian and from the Outer Starav granite, a part of the Etive granite complex, Strathclyde. Allanite-(Ce) has also been found in miarolitic cavities in granite on the Isle of Arran, Strathclyde, where it forms black tabular and prismatic crystals up to 10 mm in length. Fergusonite, gadolinite and zircon are associated species (Hyslop et al. 1999).

In Ireland, accessory allanite-(Ce) forms phenocrysts in the perlitic obsidian of Sandy Braes, Co. Antrim (Sabine 1970, Brooks et al. 1981).

England

Cornwall Granite at Holman's quarry, near Camborne (Alderton et al. 1980); phonolite at Wolf Rock, SW of Land's End (Harrison et al. 1977); allanite-(Ce) is tentatively identified at Geevor mine, near Pendeen, St Just (Golley & Williams 1995).
Cumbria Coarse bladed crystals in a pegmatitic patch or vein at Rannerdale Knots, Crummock Water (Young 1987a); in granite and rarely in cavities as free-standing crystals to 2 mm in length, in the Shap granite quarry (Morrison 1918, Basham et al. 1982, Wirth & Green 2002).
Devon Biotite–muscovite granite between Birch Tor and Vitifer iron–tin mines, Dartmoor (Alderton 1988a); Dartmoor granite (Ward et al. 1992).

Wales

Gwynedd Zoned crystals of allanite associated with zoisite in the Crogenen (Cregennan) granophyre, near Dolgellau (Cox & Wells 1920); in the Tan-y-Grisiau microgranite, and up to 3 cm long in associated veins at Ffestiniog granite quarry, Cefn Bychan, near Ffestiniog, Snowdonia (Thomas 1909b, Fearnsides 1910, Bromley 1964, Bevins 1994); in Ordovician acidic lavas at Carneddau y Cribau, Castell Yr Arddu and Cerrig Coehion, all in Snowdonia (Howells et al. 1991, Bevins 1994); forming golden-yellow inclusions in amphibole within hornblende gneiss of the Mona Complex of Anglesey at Craig yr Allor and Y Werthyr (Horák 1993).

Scotland

Dumfries & Galloway Single brown crystals, up to 4 mm in length, in granite from Kinharvie and Glen Burns, 6 km W of New Abbey, near Criffell (McMullen & Todd 1990); allanite, thorite and uraninite are the less common accessory species in the Criffell granodiorite at Needle's Eye, SW of Caulkerbush (Basham et al. 1989).
Highland Within erratic boulders of syenite embedded in till on Beinn Bhreac, 2 km E of Tongue and 7 km N of Ben Loyal (Beinn Laoghal) (Heddle 1883); rare in post-Cambrian minor intrusions in Assynt (e.g. the Canisp porphyry), Breabag porphyrite and in nordmarkite (Sabine 1953); in syenite at Glen Dessary, W of Loch Arkaig (Fowler 1981, Fowler & Williams 1986, Thompson & Fowler 1986); as an accessory species in granodiorite from Kentallen northern quarry, Ballachulish (Muir 1953); single grain in a pegmatitic patch within a pyroxene-rich xenolith in diorite from the Glenmore River, NNE of Suardalan, Ratagain complex (Glenelg–Ratagain), Loch Duich, near Glenelg (Nicholls 1951a); mantled by epidote in a hastingsite gneissic granite belonging to the Carn Chuinneag–Inchbae complex in the Black Water, 3 km up stream of Inchbae Lodge, Easter Ross (Harker & Leahy 2001); and as a ubiquitous radioactive accessory species in syenite of the Cnoc nan Cuilean complex, near Ben Loyal (Gallagher et al. 1971); as a primary magmatic phase and in endoskarn associated with the Beinn an Dubhaich granite, Isle of Skye (Smith et al. 2002).
Strathclyde Small euhedral crystals up to 0.13 mm across (rarely reaching 0.3 mm), sometimes intergrown with or surrounded by epidote, in syenite and appinite at Port Easdale, Kiloran Bay, Colonsay (Reynolds 1936).
Western Isles In the Lewisian granitic pegmatites at Chaipaval (Ceapabhal), 3 km NW of Northton (Taobh Tuath); Loch a' Sgùrr, near Finsbay (Fionnsabhagh); Loch Vallarip (Balla-riop) and Sletteval, all South Harris (von Knorring & Dearnley 1959b, 1960a).

Allanite-(Ce) crystals up to 1 mm long on fluorite from Ardragh quarry, Maam Cross, Co. Galway, Ireland. (© D. I. Green)

Ireland

Donegal Large (1 mm) euhedral, zoned allanite in xenoliths of limestone (now a zoned grossular–hornblende–pyroxene–vesuvianite skarn) within granite at Ballynacarrick (Gindy 1953).
Down In granodiorite at Croreagh quarry (Meighan et al. 1979); as inclusions in apophyllite in greisen-like fissure veins within granite in Silent Valley quarry, near Newcastle, Mourne Mountains (Nockolds & Richey 1939).
Galway Sprays of translucent mm-size brown crystals on fluorite at Ardragh quarry, Maam Cross (Wolfe 2006); as an accessory species with apatite, fluorite, thorite, titanite and zircon in the Murvey granite, Inveran area (Coats & Wilson 1971, O'Connor 1986); in quartz–diorite gneiss from the Roundstone district (Wager 1932).
Kerry In rhyolite from Lough Guitane (Avison 1984).
Wicklow In quartz–mica diorite from Glendalough (Thomson 1908).

Offshore territories

Jersey Best developed (in small quantities) in granite pegmatites intruding diorite, notably at Ronez quarry, Samarès and Grève d'Azette. Allanite is also found at Côtil Point, Gigoulande, Grouville Bay, La Moye, La Pulente, Le Lecq, Mont Mado, Pontac and Portelet (Mourant 1961).

allanite-(Y)

EPIDOTE GROUP

monoclinic $(Y,Ce,Ca)_2(Al,Fe^{3+})_3(SiO_4)_3(OH)$

Allanite-(Y) is a rare yttrium–cerium–calcium–aluminium–iron silicate hydroxide found in hydrothermally altered Tertiary epigranite on the Isle of Skye, Highland Region, Scotland. Late meteoric–hydrothermal processes released heavy REEs from Torridonian arkose at Sconser quarry, and hydrothermal allanite-(Y) was able to form in 1 cm-wide andradite–calcite–chlorite–epidote–quartz veins, possibly attributable to the breakdown of primary zircon (Exley 1980).

alleghanyite

HUMITE GROUP

monoclinic $Mn^{2+}_5(SiO_4)_2(OH)_2$

A rare manganese silicate hydroxide, dimorphous with ribbeite and found in the No. 1 orebody at Benallt mine, Rhiw, Lleyn Peninsula, Gwynedd, Wales. Alleghanyite forms clove-brown bladed crystals up to 2 cm in length and varying in thickness down from 2 mm. It forms in narrow bands cutting dark-purple manganese ore and is associated with banalsite, barite, biotite, calcite, harmotome, jacobsite and natrolite, but most particularly with tephroite, with which it forms composite crystals (Campbell Smith et al. 1944, White & Hyde 1982).

allemontite

A mixture of arsenic–antimony alloys.

allevardite

Shown to be rectorite with approximately 12.5% dickite as an impurity.

alloclasite

orthorhombic (Co,Fe)AsS

A cobalt–iron–arsenide sulphide, dimorphous with glaucodot. In Cornwall, alloclasite and cobaltite are found in material described as smaltite from Wherry mine, Penzance (Embrey & Symes 1987). In Cumbria, alloclasite occurs in the Scar Crag apatite–chlorite–quartz vein on Causey Pike as overgrowths on arsenopyrite. It is associated with glaucodot, cobalt-rich arsenopyrite, cobaltite and skutterudite (Ixer et al. 1979a). Most of the specimens described in this work were collected *in situ* or from the dumps. A few additional samples, which did not differ in any respect from the other samples, came from the Kingsbury collection at the NHM (C. J. Stanley pers. comm. 2005).

Alloclasite has been found at Ireland's most significant cobalt occurrence, Muckross mine, Killarney National Park, Co. Kerry, where it is disseminated in calcite in a primary ore consisting of arsenopyrite, chalcopyrite, galena, pyrite, sphalerite and tennantite (Moreton et al. 1999, Green & Moreton 2001).

allophane

amorphous $Al_2SiO_5 \cdot H_2O$

A relatively common hydrated aluminosilicate found in hydrothermally altered igneous rocks and derived from the breakdown of feldspar, often related to copper deposits, and in sedimentary rocks, particularly chalk and coal beds.

Botryoidal brownish iron-stained allophane, cerussite and dundasite, together with the antimony-rich species jamesonite and senarmontite, occurred on the dumps of Port Quin antimony mine, near Port Quin, St Minver Highlands, Cornwall (Russell 1944, 1949b) and it is also noted from Garras, SE of Helston, Lizard Peninsula, Cornwall (Robson 1949b). A delicate blue, pearly, botryoidal variety of allophane, probably coloured by admixture of chrysocolla was found on black calcareous shale at Wheal Hamblyn, near Bridestone, Devon (Rudler 1905). East Wheal Russell, Tavistock Hamlets, Devon (Pearce 1994b), and Hawkswood mine, North Hill, Cornwall (Gaines et al. 1997) are additional locations in the region.

Allophane is one of the most abundant secondary minerals at East Harptree, N of the Mendip Hills, and Hampstead Farm quarry, Chipping Sodbury, both in Avon, and is always associated with smithsonite- or pyrite-bearing oxidized mineral veins. It is glassy, massive or waxy looking, translucent to transparent, colourless to brown or black, and fills small cavities or takes botryoidal, microstalactitic or coralline form, lining larger cavities. An opaque chalky banded variety intimately mixed with scarbroite and quartz also occurs at Hampstead Farm quarry (Alabaster 1990b). At Woodleaze quarry, Tytherington, Bristol, Avon, glassy allophane and alumino-hydrocalcite occur with chalky-looking material, which appears to be a mix of allophane, gibbsite, scarbroite and quartz (Alabaster 1989a).

Alteration of igneous rocks at Odin lead mine, Castleton, Derbyshire, has produced a creamy deposit that could be allophane. It has a stalactitic form and it coats the roof and walls of a small branch-vein adit (Ford & Sarjeant 1964a, Ford et al. 1993). A soft white material found in collapsed Tertiary sediments filling solution pipes in the Upper Chalk near the coastguard station at Newhaven, Sussex, may be a mix of allophane and imogolite, but infrared analysis has failed to confirm this (Wilmot & Young 1985). Uncertainty also surrounds the allophane described from Driggith and Potts Gill mines, Caldbeck Fells, Cumbria, as this is based on A. W. G. Kingsbury specimens (Hartley 1984, Young 1987a, Cooper & Stanley 1990, Green et al. 2006a). Green & Briscoe (2002) express concern that allophane does not occur in the Caldbeck Fells and it is possible that the specimens are incorrectly provenanced (see Ryback et al. 1998, 2001). The allophane from Brandelhow mine, Derwent Water, Cumbria, noted by Young (1987a) is thought to be a Kingsbury specimen.

Other English occurrences of allophane include Charlton Hill chalk pits, near Woolwich, Kent (Greg & Lettsom 1858, Rudler 1905), Pitsford ironstone quarry, Northamptonshire (Hollingworth & Bannister 1950), and Old Gang mine, E of Gunnerside Gill, Swaledale, Yorkshire (Green 2003).

In Wales, allophane forms beautiful small glassy spherules associated with cerussite and dundasite at New Pandora (Welsh Foxdale) mine, between Betws-y-coed and Trefriw, Gwynedd (Prior 1906), and at the nearby Gorlan mine, gum-like brownish-yellow allophane occurs with the same associated species. Both mines worked a galena–marcasite–sphalerite lode in slate (Russell 1944). Bevins (1994) notes a specimen of allophane from a marcasite-bearing vein at Cwm Dwyfnant, near Llangammarch Wells, Powys. Allophane is found as a post-mining oxidation product in the iron mines at Betws Garmon, Snowdonia, Gwynedd (Jenkins & Johnson 1993), and in the Parys Mountain copper deposit on Anglesey (Jenkins et al. 2000).

Allophane, intermixed with hydrohonessite and copper-analogues of hydrohonessite, is present as a white coating on dump material at an old nickel mine, 12 km SW of Inveraray, Strathclyde, Scotland (Livingstone 2002). The species is also reported as a dull earthy encrustation coating quartz in a lead vein at Kings Laggan mine, Lauchentyre, W of Gatehouse of Fleet, Dumfries & Galloway (Heddle 1901).

Allophane associated with cerussite, dundasite, limonite, pyrite and quartz is noted from Clements lead mine, Carrowgarriff, near Maam, Co. Galway, Ireland (Russell 1912b). At Carneal, Larne, Co. Antrim, allophane occurs with tacharanite in amygdales up to 5 mm across, within hybrid pyroxenites (Sabine & Young 1975).

In a gully at La Cotte de St Brelade, Jersey, Channel Isles, allophane forms a light-brown stalactitic deposit on a cliff face exposed by archaeological excavations (Mourant 1961).

ALLUAUDITE GROUP

Monoclinic phosphates and arsenates with the general formula $NaACD_2(XO_4)_3$, where A = Ca, Mg, Pb; $C = Fe^{2+}, Mn^{2+}$, Mg; $D = Mn^{2+}, Fe^{2+}, Fe^{3+}$, Mg; and X = P, As. In Britain and Ireland, only alluaudite has been found.

alluaudite

ALLUAUDITE GROUP

monoclinic $(Na,Ca)_2(Mn,Mg,Fe^{2+})(Fe^{3+},Mn^{2+})_2(PO_4)_3$

A rare sodium–calcium–manganese–magnesium–iron phosphate that forms a series with ferro-alluaudite and is found in complex granitic pegmatites in the layered roof zone of the Tregonning granite at Megiliggar Rocks, near Breage, Cornwall. It is green and it replaces triphylite and possibly triplite (Stone & George 1983).

almandine

GARNET GROUP

cubic $Fe^{2+}_3Al_2(SiO_4)_3$

A very common iron aluminosilicate that forms series with pyrope and with spessartine, and is found in mica schists and gneisses derived from argillaceous sediments. It also occurs in contact-metamorphic aureoles, in eclogites, granites and in some sedimentary rocks as a detrital species. Like all other species of garnet, end-member compositions are rare and most examples show some degree of solid solution towards other members of the group. Where possible the proportions of the two most abundant end members are listed below (al = almandine, gr = grossular, py = pyrope, sp = spessartine).

England

Cornwall Many fine deep-red icositetrahedral crystals of almandine (al_{90}, gr_5), up to 2 cm across, have been recovered from metabasaltic hornfels (greenstones) in the metamorphic aureole of the Land's End granite. They were found in cliffs close to Zawn a Bal, between the Crowns Rock and Wheal Edward, Botallack, St Just. Biotite, grunerite and quartz are associated species (Alderman 1935, Tilley 1935, Wolloxall 1989). Almandine is also noted from Troon, near Camborne (Robson 1949b).

Cumbria Almandine-rich garnets (1–2 mm across, rarely up to 1 cm) occur as euhedral phenocrysts in andesitic and dacitic lavas, tuffs and intrusive rocks belonging to the Ordovician Borrowdale Volcanic Group. Almandine from Allen Crags, Armboth Fell, Borrowdale, Cam Spout Crag, Cockley Beck Bridge, Crinkle Crags, Great Gable, Green Gable, Haweswater, Langdale, Lower Saddle near Blea Tarn, Mickledore, Ore Gap, northeast Scafell, Stonethwaite in Borrowdale, Sty Head, Threlkeld, Ullswater and Wasdale Head ranges in composition from (al_{64-77}, py_{9-28}) to (al_{72}, gr_{11}) and, at Langdale, zoned crystals have iron-rich cores and magnesium-rich rims (Oliver 1956, Fitton 1972). Green (1915) reports claret-red almandine (al_{74}, py_{13}) from Kidsty Pike, Riggindale.

Devon Almandine (al_{72}, sp_{17}) is reported as an accessory component of biotite granite at Chinkwell Tor, Bonehill Down, Dartmoor (Brammall & Harwood 1932), whereas cordierite granite from the same intrusion at Sweltor quarry, Dartmoor, contains almandine (al_{81}, sp_{10}) (Brammall & Bracewell 1936) and (al_{78}, py_{10}) (Stone 1988). Granite from North West Point, and the VC quarry on Lundy island, 18 km NNW of Hartland Point, contains accessory amounts of small (up to 2 mm or 6 mm across respectively) red anhedral almandine (al_{80-81}, sp_{10-11}) crystals associated with biotite (actually siderophyllite), potassium feldspar, muscovite, plagioclase, quartz and topaz (Stone 1988). Another granite quarry to the N of the VC quarry produced almandine (al_{84-88}, sp_{7-14}) associated with tungsten-rich columbite and thorium-rich uraninite (Thorpe et al. 1990, Tindle unpub. EMP data).

Wales

There appear to be no published analyses of almandine from Wales, although Claxton (1963) reports almandine garnet in the Dutch Gin Schists, a suite of Precambrian regionally metamorphosed mudstones and sandstones close to the Johnston plutonic complex, near Talbenny, Dyfed, and almandine is also noted in rhyolitic lavas from the Capel Curig district, Snowdonia, Gwynedd (Williams 1922).

Scotland

Central Garnet in a mica schist associated with stratiform sulphide mineralization in the Dalradian Ben Challum Quartzite to the E of Tyndrum is mostly almandine, with very variable compositions: (al_{48-68}, gr_{22-26}), (al_{71-76}, py_{11-16}) and (al_{44}, sp_{27}) (Fortey & Smith 1986). A pelitic schist from the base of the Dalradian Ben Lui Schist at Ben Challum, Tyndrum, contains almandine (al_{69}, gr_{15}) and (al_{69-70}, py_{15-16}) crystals in which exsolved magnetite and inclusion trails of chloritoid, quartz, staurolite and white mica have been found (Brearley & Champness 1986).

Fife Almandine (al_{49}, py_{38}) and pyrope occur in volcanic agglomerate at Elie (Heddle 1901).

Grampian Aplites cutting Dalradian pelitic schists at Kinnaird Head, Fraserburgh, contain euhedral pink almandine (al_{63-65}, sp_{31-33}) crystals up to 0.5 mm across (Johnson et al. 2000). Magmatic almandine (al_{53-65}, sp_{31-46}) occurs in accessory amounts as small pale-pink euhedral crystals in the western part of the Cairngorm granite complex (Harrison 1988). Magmatic almandine has also been found in the following locations: (al_{72-78}, sp_{14-21}) in granite at Quarry Hill quarry, Aberhirder; (al_{54}, sp_{38}) in tourmaline pegmatite float from Perkie Burn, Cabrach; (al_{81}, sp_6) in gneissose granite from Hill of Three Stones, N of Glenbuchat; (al_{90-91}, sp_{3-4}) in micrigranite and (al_{54}, sp_{45}) in muscovite pegmatite at Peatfold, Glenbuchat; (al_{58-66}, sp_{31-39}) in tourmaline pegmatite from Pitscurrie quarry, Pitcaple, Insch; (al_{57-62}, sp_{33-40}) in tourmaline-bearing aplite associated with pegmatite at East Head, 1 km NE of Portsoy; (al_{54}, sp_{44}) in the Moy granite (all Tindle unpub. EMP data). Almandine is one of many heavy minerals identified in panned concentrates from the southeast margin of the Cairngorm granite in Luibeg Burn, Derry Cairngorm (Beer & Bennett 1988, Haynes 1993). Localized metasomatism of schists has produced aegirine-augite and epidote granulites containing yellow andradite moulded on colourless cores of almandine (al_{69}, gr_{29}) 2 km to the SSE of Derry Lodge in Glen Lui, Cairngorms (McLachan 1951). Large, well formed wine-red to purplish-red dodecahedral crystals of almandine (al_{74}, py_{21}), ranking as some of the finest in Scotland, have been found near Cairnie church, 7 km from Huntly. They occur in banded biotite–sillimanite gneisses, particularly in those bands rich in quartz and feldspar (Stewart 1950). Garnet in hornfels within the contact aureole at Black Hill, Loch Muick, is almandine (Livingstone 2002). Dalradian migmatites adjoining Caledonian intrusions at Barry Hill, Castle Bridge and Cumrie, in the Huntly–Portsoy area, contain coexisting almandine (al_{65-75}, py_{19-29}) and cordierite, which indicates formation under low-pressure regional conditions. Similar rocks from Black Hill, Lochnagar contain almandine (al_{75-80}, py_{13-16}) (Ashworth & Chinner 1978). Almandine has also been reported in regionally metamorphosed Dalradian metasedimentary rocks: (al_{66-67}, gr_{13-16}) and (al_{67-69}, sp_{12-14}) at Creag nam Ban, near Ballater, Deeside (Smith & Phillips 2002); (al_{54-63}, gr_{22-30}) and (al_{68-78}, py_{12-18}) to the SE of Lochnagar (Goodman et al. 1990, Goodman & Lappin 1996); (al_{53-76}, gr_{5-30}), (al_{66-81}, py_{10-24}) and (al_{58-76}, sp_{11-27}) along the Portsoy–Duchray Hill lineament and Banff coast (Beddoe-Stephens 1990); (al_{75}, py_{16}) from Sparcraigs, near Whitecairns, N of Aberdeen (Stewart 1942); (al_{76}, py_{10}) and (al_{71-83}, sp_{9-18}) along the Stonehaven coast (Baltatzis 1979). Heddle (1901) reports port-wine-colour almandine (al_{79}, py_{12}) from a diorite on Knock Hill, W of Glenbarry.

Highland Descriptions of well formed reddish-brown almandine crystals associated with omphacite from Totaig Ferry, Loch Duich, represented the first discovery of eclogite in Britain (Teall 1891a). Since then, a well exposed mass of eclogite containing round red almandine (al_{58}, py_{23}) grains has also been found on the southwest shore of Loch Duich, near Fern Villa (Mercy & O'Hara 1968). At Flowerdale, near Gairloch, almandine (al_{65}, sp_{21}) is a constituent of grunerite-bearing BIF rocks belonging to the Lewisian Loch Maree Group. The almandine crystals are often poikiloblastic, skeletal or develop unusual prismatic or tabular forms (Williams 1986). Moine Series pelitic hornfels and schists surrounding the Carn Chuinneag–Inchbae igneous complex have been subjected to both contact and regional metamorphism, and now contain small (< 8 mm) anhedral crystals of almandine (al_{79}, py_{12}) associated with albite, andalusite, biotite, kyanite, muscovite, quartz and staurolite (Harker 1954). Moinian metasediments to the N of Loch Sunart and in the northwest section of the contact aureole around the Strontian granodiorite contain zoned almandine (al_{75-79}, py_{11-12}) and (al_{71-73}, sp_{16-17}). For individual crystals, it has been possible to distinguish between growth under thermal and regional metamorphic conditions (Tyler & Ashworth 1981). Spongy almandine (al_{65}, gr_{18}) and (al_{49}, sp_{22}), full of quartz inclusions, occurs in ironstone in the first inlet W of the pier on the N side of Scourie Bay (Barnicoat & O'Hara 1979) and, within the nearby Scourie quartz-dolerite dyke at Creag a' Mhàil, euhedral almandine (al_{64}, gr_{23}) crystals forming clusters up to 3 cm across have been found (O'Hara 1961). Other localities include: Coill an Achaidh Mhòr, Loch Garve, near Dingwall (King 1999); (al_{77}, py_{12}) in a Moinian mica schist from Glenfinnan (Rieder 1995); (al_{53-64}, gr_{28-38}) in Moinian metapelitic gneisses from the Glenfinnan Group in Glen Doe (Zeh & Millar 2001); (al_{70-75}, py_{20-24}) and (al_{73-75}, sp_{11-14}) in Moinian migmatites from Glen Scaddle, near Fort William, and Sròn na Saobhaidh, near Strontian, respectively (Ashworth & Chinner 1978); deep-red almandine (al_{51}, py_{33}) in an eclogite facies amphibole–pyroxene rock from Glen More, Glenelg (O'Hara 1960); (al_{57-61}, gr_{19-25}) in Moinian and Lewisian gneisses close to the Moine thrust plane at Camas Baravaig and Cruard, southeast Isle of Skye (Matthews & Cheeney 1968); accessory amounts of (al_{90}, gr_4) in epigranites from the Marsco area, Western Redhills, Isle of Skye (Thompson 1969); (al_{70-71}, sp_{24-25}) forms perfectly dodecahedral gem-like reddish-brown porphyroblasts, up to 2 cm across (averaging 4 mm) in mica-rich bands of the North Quarry pegmatite (part of the Loch Nevis mica prospect), N of Tarbert, Knoydart (R. E. Starkey specimen, Tindle unpub. EMP data). Heddle (1901) reports: (al_{70}, sp_{18}) forming brown crystals in granitic veins near Black Water, Loch Garve; (al_{61}, sp_{33}) forming red crystals near the summit of Beinn Resipol, NW of Strontian; (al_{61}, gr_{28}) associated with biotite (haughtonite), chlorite, ilmenite and rutile from Clach an Eòin, 2 km W of Bettyhill; (al_{59}, gr_{19}) in a mica slate from Leitir Mhuiseil, 3 km S of Ben Hope; (al_{59-61}, sp_{32-36}) forming red and brown crystals up to 13 cm across in Glensgaich railway cutting, near Strathpeffer; (al_{70}, sp_{27}) forming red crystals in a granite belt in gneiss at Struy Bridge, Strathglass.

Strathclyde Almandine (al_{48}, py_{33}) occurs with pyrope in ultramafic (garnet pyroxenite) xenoliths within a Permian basanite diatreme at Holmbyre, 2 km SW of Dalry (Upton et al. 2003). To the E and W of Loch Lomond, almandine (al_{43-59}, gr_{28-37}) and (al_{51-67}, py_{24-31}) crystals, some with pyrope cores, have been found with amphibole and sodic feldspar, in xenoliths within a lamprophyre dyke (Dempster & Bluck 1991). Other occurrences include: (al_{57}, sp_{16}) and (al_{79-80}, gr_{12-13}) in Dalradian pelites surrounding the southern sector of the Etive granite complex (Droop & Moazzen 2007); (al_{38-42}, py_{20-35}) in garnet metapyroxenite from an aureole assemblage associated with the Girvan–Ballantrae ophiolite complex (Treloar et al. 1980); (al_{53-62}, gr_{19-29}) in epidiorites from Loch-na-Craige, near Achahoish, South Knapdale (Wiseman 1934); (al_{38-55}, sp_{25-38}) and (al_{54-75}, gr_{20-29}) in Dalradian albite schist from around and to the N of Loch Lomond, some porphyroblasts with spessartine cores (Mathavan & Bowes 2005); (al_{73}, gr_{22}) in Dalradian mica schist from Clachan–Shira hydroelectric tunnel, Brannie Burn, near Loch Fyne (Guppy 1956); (al_{70-77}, py_{11-14}) and (al_{43-76}, sp_{9-31}) in Moinian pelitic gneiss; and hornfels in and close to the aureole of the Ross of Mull granite, Isle of Mull (Brearley & Champness 1986, Wheeler et al. 2004).

Tayside Regional metamorphism of Dalradian pelitic and semi-pelitic rocks has produced zoned almandine crystals whose composition changes with increasing metamorphic grade. Investigations of individual crystals show they vary from manganese-rich cores to iron-rich rims in the range (al_{58-75}, gr_{14-24}) and (al_{72-81}, py_{12-13}). Localities include: Alt Girnaig and Sran na Innerach, near Blair Atholl; Amulree, Kindallachan and Pitleoch Burn, all near Dunkeld; Ballinluig and Logierait Wood, between Pitlochry and Dunkeld; Dunfallandy Hill, near Pitlochry; Enochdhu, NE of Pitlochry (Atherton 1968). Staurolite–mica schist from the classic Barrovian metamorphic zones at Craig of Finnoch, Glen Lethnot, Angus, contain large (4 mm) porphyroblasts of almandine (al_{80}, py_9) marginally altered to chlorite (Dempster & Jackson 1996). These are similar to almandine (al_{69}, gr_{17}) and (al_{69-75}, py_{10-13}) in staurolite–mica schists from Cadham Crags and Elf Hillock, Inchdowrie, Glen Clova, Angus (Chinner 1965). Anhedral or spongy grains of almandine (al_{79}, gr_{11} core to al_{85}, py_{11} rim), up to 1.2 mm across and dusted with minute inclusions of graphite, have also been found in kyanite-bearing spotted hornfels within the aureole of the Ben Vuirich granite, 10 km NE of Pitlochry (Ahmed-Said & Tanner 2000). Almandine (al_{75-78}, gr_{11-13}) is also present with zinc-rich staurolite in Dalradian gneisses from the southern contact aureole of the Glen Doll diorite (Schumacher 1985) and at the southern end of Lochan na Lairige, near the dam wall, almandine, epidote and ilmenite have all been noted in Dalradian schists (Green & Todd 2001). Heddle (1901) reports almandine (al_{65}, py_{14}) in a paste of mica from Killicrankie, N of Pitlochry, and (al_{71}, gr_{13}) at Meall Luaidhe, 5 km NW of Ben Lawers. Dalradian schists termed "Green Beds" from Dalnacarn Craig, NE of Pitlochry, contain almandine (al_{62}, gr_{19}) (Phillips 1930).

Western Isles Reddish-brown almandine (al_{53}, gr_{31}) forms rounded crystals, 0.5 mm to 8 cm (or more) across, in eclogite from Rodilpark, South Harris. Almandine (al_{62}, gr_{21}) is also found in an associated metagabbro (Davidson 1943). To the S and SW of Roineabhal, South Harris, sapphirine-bearing and sapphirine-free kyanite (or sillimanite)–orthopyroxene granulites contain zoned almandine (al_{47-54}, py_{39-47})-pyrope (py_{46-58}, al_{31-46}) porphyroblasts (Baba 1999, 2003, Hollis et al. 2006). Panned concentrates from Abhainn Mhòr on St Kilda (Hirta) contain almandine (Harding et al. 1984).

Shetland A well formed, although pitted, 6 cm-diameter almandine crystal from a mica schist at Colvister, north Yell, is illustrated by Livingstone (2002). Heddle (1878b, 1901) reports almandine (al_{68}, py_{14}) from a gneiss at Burra Voe on Yell. On Fetlar, pink almandine (al_{48-52}, gr_{33-37}) crystals, up to 1 cm across, have been found in amphibolites beneath the basal thrust of the Shetland ophiolite complex (Gass et al. 1982).

Ireland

Carlow Schists intruded by lithium pegmatites and aplites at Aclare, near Myshall contain almandine (al_{75-78}, sp_{11-13}) (Whitworth 1992).

Donegal Porphyroblasts of almandine (al_{74}, py_{16}) are reported from graphite–hercynite–staurolite-bearing pelites in the aureole of the Ardara granite at Narin, Portnoo harbour (Atkin 1978). Other metapelites from the Ardara aureole, near Clooney, were examined by Kerrick (1987), who reports almandine compositions spanning the range (al_{59-66}, gr_{18-21}), (al_{80-81}, py_{8-14}) and (al_{65-80}, sp_{9-20}), and notes that some garnet is retrograded to biotite and sillimanite (fibrolite). Garnet is a common

accessory mineral in muscovite granites, aplites and pegmatites of the Donegal granite batholith. Most are spessartine, but, near Dooey Point, garnet from a muscovite granite in the Trawenagh Bay pluton is almandine dominant (al_{50}, sp_{47}) (Hall 1965b).
Galway Almandine (al_{71-77}, py_{18-23}) is found with biotite, cordierite, magnetite, plagioclase, quartz and sillimanite in pelitic hornfels surrounding the Cashel–Lough Wheelaun ultrabasic intrusion of Connemara (Trelour 1981). Elsewhere in Connemara, almandine is noted in Dalradian garnet amphibolites (al_{46-64}, gr_{16-36}) and (al_{43-64}, py_{15-22}) (Leake 1972) and migmatites (al_{75-81}, py_{15-16}) (Barber & Yardley 1985).
Mayo Blueschist-facies metabasites belonging to the Dalradian Supergroup from central and south Achill Island, N of Clew Bay, contain zoned garnets with almandine (al_{41-78}, gr_{16-37}) compositions (Chew et al. 2003). On south Achill Island, low-grade metasediments of the Clew Bay Complex include garnet-bearing schists containing homogeneous almandine–pyrope-rich cores (al_{53-56}, py_{24-43}) with heterogeneous rims of two distinct but intimately intergrown phases with a total range in composition of almandine–grossular (al_{55-75}, gr_{19-28}) (Yardley et al. 1996).
Sligo Regionally metamorphosed Dalradian pelitic rocks containing almandine (al_{70-83}, gr_{8-17}) have later suffered thermal metamorphism around the Easky adamellite, near Easky Lake in the Ox Mountains. This has caused the almandine to break down and form magnetite (Yardley & Long 1981).
Wicklow A schist xenolith in granite from the head of Glendalough contains almandine (al_{63-65}, sp_{22-27}) (S. Moreton specimen, Tindle unpub. EMP data).

alstonite

triclinic (pseudo-orthorhombic) $BaCa(CO_3)_2$

TL Brownley Hill mine, Nenthead, near Alston, Cumbria, and Fallowfield mine, near Hexham, Northumberland, England

A rare barium–calcium carbonate, trimorphous with barytocalcite and paralstonite, and first described as a variety of barytocalcite (Johnston 1835). Museum specimens of alstonite from Brownley Hill, collected in the 19th century, typically form distinctive white pseudo-hexagonal doubly terminated pyramids associated with thin calcite plates on a matrix of calcite and pink barite. Unfortunately, not all specimens from this locality are correctly labelled and some are actually from the co-type locality at Fallowfield mine. However, alstonite from Fallowfield is often associated with pseudo-hexagonal witherite prisms, making discrimination of the two localities relatively easy (Green et al. 2000). Some old specimens from Fallowfield are pale pink and light sensitive. Recent exploration has shown that the exact site of the type locality at Brownley Hill is on High Cross vein between Holme's rise and Richardson's rise (Young et al. 1990b). Alstonite occurs here as a late-stage primary mineral in calcite-lined cavities up to 1.5 m in length within the central portion of calcite–barite fissure veins. Translucent pink bipyramidal crystals, exceptionally up to 2.5 cm, long have been collected from here and almost certainly represent the finest known examples of the species (Green et al. 2000, Green & Briscoe 2002). Alstonite from the co-type localities is further described by Greg & Lettsom (1858), Goodchild (1884), Rudler (1905), Smythe (1922), Davidson & Thomson (1951), Young (1985c), Briscoe (1986b), Green & McCallum (1993) and Green & Young (2006).

Bipyramidal crystals of alstonite up to 25 mm across; High Cross vein, Brownley Hill mine, Nenthead, near Alston, Cumbria. (© D. I. Green)

Bipyramidal alstonite crystal; 2.5 mm long, from Dolyhir quarry, Old Radnor, Powys, Wales. (© D. I. Green)

A few specimens of alstonite associated with ullmannite were recovered from a barite–witherite vein at New Brancepeth colliery, near Durham (Spencer 1910), and rare alstonite is noted from Nentsberry mine, Nenthead, near Alston, Cumbria (Young 1985c). A brief mention of alstonite in a witherite vein at Settlingstones mine, near Hexham, Northumberland, by Dunham (1948) appears to be doubtful (Young 1985c, Dunham 1990).

Alstonite is reported from an old mine site and a bypass roadcut near Llantrisant, 10 km NW of Cardiff, Mid Glamorgan, Wales. It is associated with barite, calcite, dolomite, quartz and witherite in low-temperature hydrothermal Cu–Fe–Pb–Zn sulphide-bearing veins and cavities in dolomitized Triassic conglomerate. Colourless to pale-pink alstonite forms bipyramidal striated crystals up to 6 mm long (Alabaster 1990). Alstonite and other barium-rich minerals (barite, barytocalcite, harmotome and witherite) have been found in veins cutting Precambrian sandstones and siltstones at Dolyhir quarry, Old Radnor, near Presteigne, Powys, Wales. Most of the species occur as microcrystalline crusts (Wolfe 2004, Green et al. 2005d). As quarrying continued, alstonite began to be found as individual euhedral bipyramidal crystals (exceptionally up to 5 mm long) in association with ewaldite, harmotome and witherite (T. F. Cotterell pers. comm. 2007).

altaite

cubic PbTe

A rare lead telluride found in gold-bearing veins at Clogau mine, Bontddu, near Dolgellau, Gwynedd, Wales. Altaite occurs as granular to sub-rounded inclusions in tellurobismuthite in association with hessite (Gilbey 1968, Naden 1988, Bevins 1994).

At Urlar Burn, 5 km SW of Aberfeldy, Tayside Region, Scotland, quartz–sulphide–gold veins contain altaite as abundant (but volumetrically minor) 10–80 μm inclusions in galena. Associated inclusions are hessite ±coloradoite ±volynskite (Livingstone 1993, Ixer et al. 1997).

Micron-scale crystals of altaite with hessite and a mercury telluride are also reported as inclusions in pyrite from Cavanacaw, near Omagh, and Curraghinalt, near Gortin, in the Sperrin Mountains, Co. Tyrone, Ireland. These locations are vein-hosted gold prospects within Dalradian rocks (Parnell et al. 2000).

alum

A general term for hydrous sulphates of aluminium and an alkali. The alum of Heddle (1901) is referred to as potassium alum by Macpherson & Livingstone (1982).

aluminite

monoclinic (pseudo-orthorhombic) $Al_2(SO_4)(OH)_4 \cdot 7H_2O$

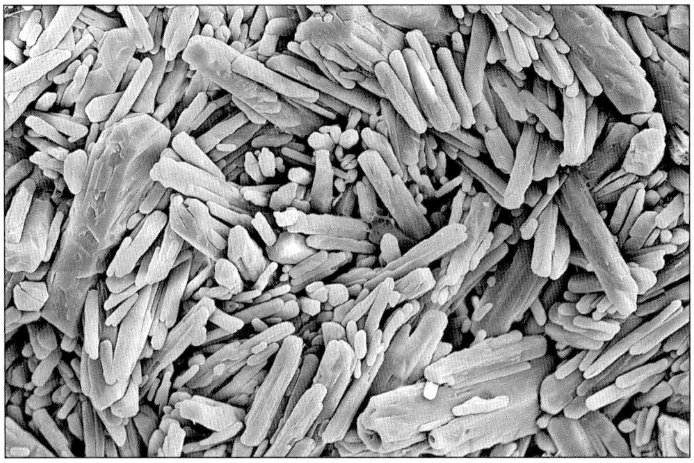

Elongate rods of aluminite from Newhaven, Sussex; width 160 µm; SEM image. (© E. Condliffe, University of Leeds)

An hydrated aluminium sulphate hydroxide found in collapsed Tertiary sediments filling solution pipes in the Upper Chalk, near the coastguard station at Newhaven, Sussex. Aluminite occurs as pure white nodular masses with a crudely botryoidal surface. Masses of pure aluminite up to 15 cm across are embedded into clayey sand and gypsum (selenite) is present locally, where it forms irregular pockets within the pipe fillings. Other aluminium-rich species found in the pipes include basaluminite, bayerite, gibbsite and nordstrandite. All appear to have formed by the reaction of acid groundwater with Tertiary clays and the Chalk (Howell 1878, Wilmot & Young 1985).

A brief mention of aluminite (websterite) on calcite, from near Portrush, Co. Antrim, Ireland, is given by Greg & Lettsom (1858).

aluminocopiapite

COPIAPITE GROUP

triclinic $(Al_{2/3}\square_{1/3})Fe_4^{3+}(SO_4)_6(OH)_2 \cdot 20H_2O$

An hydrated aluminium–iron sulphate hydroxide noted by Golley & Williams (1995) from Dolcoath mine, Camborne, Cornwall. At Hawthorn Hive, a wide bay on the Durham coast (at and around Hive Point and Shippersea Bay), aluminocopiapite has been found in mine waste derived mainly from Dawdon colliery. Aluminocopiapite forms ephemeral, pale-yellow to yellowish-white "cauliflower-like" efflorescent crusts, up to 4 cm across. Apjohnite, calcite, ferricopiapite, gypsum, natrojarosite and sideronatrite are associated species (Young et al. 2008).

alumino-ferro-edenite*

AMPHIBOLE GROUP

monoclinic no balanced formula

A rare sodium–calcium–iron aluminosilicate hydroxide that forms series with both edenite and ferro-edenite, but for which a simple balancing formula cannot be written, although logically the formula approaches $NaCa_2(Fe_4^{2+}Al)(Si_7Al)O_{22}(OH)_2$. It represents a "named amphibole" rather than a true species (Burke & Leake 2005).

Alumino-ferro-edenite is identified (on the based of a single analysis) in a fine-grain granular gabbro in the Insch basic layered intrusion, to the E of Insch, Grampian, Scotland. Edenite, ferro-edenite, ferropargasite and pargasite are other amphibole group members found in these rocks (Wadsworth 1988).

At Beaucette in north Guernsey, Channel Isles, alumino-ferro-edenite occurs in a pegmatitic pocket in gabbro within veined diorite belonging to the late Precambrian Cadomian igneous complex (Bishop & French 1982).

alumino-ferrohornblende

AMPHIBOLE GROUP

monoclinic $Ca_2(Fe_4^{2+})Al(Si_7Al)O_{22}(OH)_2$

A rare calcium–iron aluminosilicate hydroxide that forms series with ferrohornblende and with magnesiohornblende.

Samples from a Dalradian psammite-semipelite sequence at Creag nam Ban, near Ballater, Deeside, Grampian (Smith & Phillips 2002), that had originally been described as cummingtonite, are actually grunerite, apart from one analysis which is alumino-ferrohornblende, when reprocessed using the amphibole classification of Leake et al. (1997a) and Mogessie et al. (2004).

alumino-ferropargasite*

AMPHIBOLE GROUP

monoclinic no balanced formula

A rare sodium–calcium–iron aluminosilicate hydroxide that forms a series with ferropargasite, but for which a simple balancing formula cannot be written, although logically the formula approaches $NaCa_2(Fe_3^{2+}Al_2)(Si_6Al_2)O_{22}(OH)_2$. It represents a named amphibole rather than a true species (Burke & Leake 2005).

Alumino-ferropargasite occurs in skarn deposits near the western margin of the Land's End granite at the Crowns and Gryll's Bunny, Botallack, St Just, Cornwall. Amphibole-rich veins cutting metapelite and metabasalt host rocks contain amphibole with compositions transitional between pargasite and sadanagaite (van Marcke de Lummen 1985). On the basis of the reclassification of the amphibole group (Leake et al. 1997a, Mogessie et al. 2004), some of the pargasite is renamed alumino-ferropargasite.

The new classification also confirms alumino-ferropargasite in the Highland Region of Scotland: forming rims around grunerite aggregates (originally described as bluish-green hornblende and ferro-pargasite) in iron-rich silicate–magnetite–quartz rocks (BIF) of Lewisian age from Flowerdale, near Gairloch (Williams 1986); with clinozoisite, chlorite, epidote and margarite in ellipsoidal bodies of plagioclase (anorthosite xenoliths) within metabasalts of Lewisian age at Meall Mheinnidh, to the NE of Loch Maree (Cattell & Williams 1988).

alumino-ferrowinchite

AMPHIBOLE GROUP

monoclinic $\square NaCa(Fe^{2+}Al)_5Si_8O_{22}(OH)_2$

A rare sodium–calcium–iron aluminosilicate hydroxide that forms a series with ferropargasite. It is not listed as a species by Mandarino & Back (2004). The IMA lists the species, noting that its description was not made with its approval.

Alumino-ferrowinchite is found in blueschists belonging to the Mona Complex from the Penmynydd/Gwna boundary exposed in a roadcut near Llanfairpwllgwyngyll, southeast Anglesey, Gwynedd, Wales. Actinolite, aluminowinchite, ferrobarroisite, ferroglaucophane, ferrowinchite, glaucophane, katophorite and winchite are associated species (Gyopari 1984). Alumino-ferrowinchite is also reported from Ordovician rocks with ophiolite affinities from the Falahill Formation, near Heriot, Galashiels area, Borders (Kassi & Weir 1993, A. Livingstone unpub. EMP data).

aluminokatophorite

AMPHIBOLE GROUP

monoclinic $NaNaCa(Fe_4^{2+}Al)(Si_7Al)O_{22}(OH)_2$

A rare sodium–calcium–iron aluminosilicate hydroxide that forms series with magnesiokatophorite and with katophorite. It is mentioned in the IMA-approved classifications of the amphibole group (Leake et al. 1978, 1997a), but is not listed as a species by Mandarino & Back (2004). A single analysis of aluminokatophorite from the Ordovician Falahill Formation, near Heriot, N of Galashiels, Borders, Scotland, is given by Livingstone (1993).

aluminosadanagaite*

AMPHIBOLE GROUP

monoclinic $(NaK)Ca_2(Fe_3^{2+}Al_2)(Si_5Al_3)O_{22}(OH)_2$

A rare sodium–potassium–calcium–iron aluminosilicate hydroxide that forms series with magnesiosadanagaite and with sadanagaite. It is not listed as a species by Mandarino & Back (2004), suggesting it is a "named amphibole" (Burke & Leake 2005).

Aluminosadanagaite occurs in skarn deposits near the western margin of the Land's End granite at the Crowns and Gryll's Bunny, Botallack, St Just, Cornwall. Amphibole-rich veins cutting metapelite and metabasalt host rocks contain amphibole with compositions transitional between pargasite and sadanagaite (van Marcke de Lummen 1985). Sadanagaite is considered to be a potassium-rich species by Anthony et al. (1995), but a sodium-rich species by Leake et al. (1997a). The Cornish example is intermediate between the two, but according to the IMA-approved reclassification of the amphiboles (Leake et al. 1997a, Mogessie et al. 2004) the composition should now be referred to as potassian aluminosadanagaite.

aluminotschermakite

AMPHIBOLE GROUP

monoclinic $\square Ca_2(Mg_3Al_2)(Si_6Al_2)O_{22}(OH)_2$

A rare calcium–magnesium aluminosilicate hydroxide that forms series with tschermakite and with ferrotschermakite. It is not listed as a species by Mandarino & Back (2004), but is IMA approved.

An investigation of amphibole compositions from Dalradian Argyll and Southern Highland Group metabasites around and to the NE of Aberfeldy, Tayside Region, Scotland (centred on the classic area of Barrovian metamorphic zones that developed during the Grampian orogeny), shows that aluminotschermakite is restricted to the garnet and kyanite zones, with tschermakite found in the sillimanite and andalusite zones (Zenk & Schulz 2004). Data from a Lewisian epidote–garnet–hornblende gneiss from Cruard, Isle of Skye, Highland Region, suggest that aluminotschermakite is present (Matthews & Cheeney 1968), but re-analysis of the same material using a different method (Leake 1968) produced a tschermakite composition. Data was reprocessed using the amphibole classification of Leake et al. (1997a) and the software described in Mogessie et al. (2004).

aluminowinchite*

AMPHIBOLE GROUP

monoclinic $NaCa(Mg_4Al)Si_8O_{22}(OH)_2$

A rare sodium–calcium–magnesium aluminosilicate hydroxide that forms series with both winchite and ferrowinchite. It is not listed as a species by Mandarino & Back (2004), suggesting that it is a "named amphibole" (Burke & Leake 2005).

Strongly foliated blueschists exposed near the Marquis of Anglesey's monument, near Llanfairpwllgwyngyll, Anglesey, Gwynedd, Wales, contain zoned crystals of amphibole, which Horák & Gibbons (1986) have described as barroisite (green cores) and crossite–ferroglaucophane–glaucophane (all blue rims). The current amphibole classification (Leake et al. 1997a, Mogessie et al. 2004) suggests a katophorite core and a rim consisting of aluminowinchite and glaucophane. Blueschists belonging to the Mona Complex from the Penmynydd/Gwna boundary exposed in a roadcut near Llanfairpwllgwyngyll is a related occurrence. A study of amphibole compositions from these rocks has revealed the following species: actinolite, aluminoferrowinchite, aluminowinchite, ferrobarroisite, ferroglaucophane, ferrowinchite, glaucophane, katophorite and winchite (Gyopari 1984). This author had described aluminowinchite as crossite. The Tal y Fan intrusion, a subconcordant sheet of altered olivine dolerite located 7 km SW of Conwy, Gwynedd, also contains aluminowinchite, although this had previously been referred to as actinolite (Bevins & Merriman 1988).

Crossite (reclassified as barroisite), glaucophane and hornblende (reclassified as ferro-edenite) have previously been reported from amphibolites and schists associated with the Girvan–Ballantrae ophiolite complex to the N of Knockormal, Strathclyde, Scotland (Bloxam & Allen 1959). New data indicate the range of amphibole species in these rocks also includes aluminowinchite, edenite, pargasite and winchite (A. Livingstone unpub. EMP data). This work also confirmed aluminowinchite in Ordovician rocks with ophiolite affinities from the Falahill Formation, near Heriot, Galashiels area, Borders.

alumo-berezovite

An aluminium–magnesium-rich variety of chromite.

alumohydrocalcite

triclinic $CaAl_2(CO_3)_2(OH)_4 \cdot 3H_2O$

A rare hydrated calcium–aluminium carbonate hydroxide identified on museum specimens of supposed scarbroite collected 65–110 years ago from the type locality for scarbroite at White Nab, South Bay, Scarborough, Yorkshire. Some specimens proved to be alumohydrocalcite; others were basaluminite or scarbroite. Alumohydrocalcite forms irregular white powdery, slightly lustrous, patches up to several mm across and is mixed with allophane in veins cutting ferruginous sandstone. Loose powdery alumohydrocalcite on scarbroite specimens from the infill of solution pipes in sandstone is also reported from Weston Favell, Northamptonshire, where it forms small (<1 mm) pure soft white masses with a silky lustre (Ryback 1988b).

At Woodlease quarry, a part of the Tytherington quarries, near Grovesend, Bristol, alumohydrocalcite is found in cavities within a sandstone, filling a sinkhole in limestone. It is associated with allophane, gibbsite and an unidentified carbonate, and it appears to have formed through the action of $CaHCO_3$-rich groundwater on allophane and the carbonate, possibly under periglacial conditions (Alabaster 1989a).

Alumohydrocalcite has also been found as white hairs and minute light-blue balls, associated with dundasite, within cavities in manganese oxides from Merehead (Torr Works) quarry, Cranmore, Mendip Hills, Somerset (M. S. Rumsey & R. W. Turner pers. comms 2006).

alumotungstite

cubic or pseudo-cubic? $(W,Al)(O,OH)_3$

Bipyramidal crystals of alumotungstite; Burdell Gill, Caldbeck Fells, Cumbria; width 190 µm; SEM image. (© D. I. Green)

A rare tungsten–aluminium oxyhydroxide that requires further characterization and is produced by the weathering and oxidation of tungsten-bearing minerals such as ferberite and scheelite. Crystalline curtains made up of intergrown octahedral crystals of alumotungstite up to 0.1 mm on edge occur in cavities within wolframite-rich quartz at Hemerdon Bal, Plympton, Devon (Weiss 2000). Alumotungstite also occurs as minute octahedral crystals on partly oxidized ferberite in quartz veins exposed at the head of Burdell Gill, Caldbeck Fells, Cumbria. Unpublished data suggest that the chemical formula for the British examples of this species may be $W_3(Al,Fe)(O_9OH_3) \cdot 4H_2O$ (D. I. Green XRD, Tindle unpub. EMP data).

alumyte

TL Glenarm and Ballintoy, Co. Antrim, Ireland

A synonym of halloysite-7Å (metahalloysite).

ALUNITE GROUP

Trigonal sulphates with the general formula $AB_6(SO_4)_4(OH)_{12}$, where $A = Ag_2$, $(H_3O)_2$, K_2, Na_2, $(NH)_4$, Tl, Pb; and $B = Al$, Cu^{2+}, Fe^{3+}. Alunite, beaverite, hydronium jarosite, jarosite and osarizawaite are members of the group found in Britain or Ireland. Doubtful occurrences include argentojarosite and natrojarosite.

Spray of alumohydrocalcite crystals in a cavity within a scarbroite-bearing vein; South Bay, Scarborough, Yorkshire; width 70 µm; SEM image. (© E. Condliffe, University of Leeds)

alunite

ALUNITE GROUP

trigonal $KAl_3(SO_2)_2(OH)_6$

A potassium–aluminium sulphate hydroxide. In Wales, altered rhyolitic rocks of the Roch Rhyolite Group recovered from a borehole near Treffgarne, Dyfed, contain alunite associated with barite, corundum, kaolinite, pyrite, pyrophyllite, quartz, titanium oxides and possibly illite or sericite (Brown et al. 1987b).

Alunite is also reported from Embleton quarry, near Cockermouth, Cumbria, as finely granular masses and flattened rhombs in a quartz–carbonate vein cutting a quartz-dolerite sill, where it is associated with dickite, gypsum, thaumasite and zunyite (Embrey footnote to Knill 1960). Earthy ferruginous alunite associated with jarosite is also reported from Carrock mine, Grainsgill, Caldbeck Fells, Cumbria (Kingsbury & Hartley 1958). Both Cumbrian occurrences have Kingsbury links and are considered to be dubious British occurrences.

alunogen

triclinic $Al_2(SO_4)_3 \cdot 17H_2O$

An hydrated aluminium sulphate chiefly found as an efflorescence or cavity filling in slates, shales or coal formations containing iron sulphides. White-to-yellowish efflorescent substances on specimens of erythrite from the Wherry mine, Penzance, Cornwall, were presumed by Russell to be alunogen (1949a). Alunogen is also briefly reported by Golley & Williams (1995) from Dolcoath mine, Camborne.

Minute clear platy crystals and yellowish-brown aggregrates of alunogen occur with glassy potassium alum and fluffy tamarugite at Hurlet, near Paisley, Strathclyde, Scotland (Macpherson & Livingstone 1982).

In the NMW collection, samples labelled alunogen – from Parys Mountain, Anglesey, Gwynedd; Cymmer colliery, Porth, Mid Glamorgan; and Deep Navigation colliery, Treharris, Mid Glamorgan – have all been shown to be halotrichite (Bevins 1994).

alurgite

A variety of muscovite, mica group.

amalgam

A silver–mercury alloy (see native silver).

amazonite (amazonstone)

A green variety of microcline, feldspar group.

AMBLYGONITE GROUP

Triclinic phosphates with the general formula $AB(PO_4)X$, where A = Li, Na; B = Al, Fe^{3+}; and X = (OH), F. Amblygonite and montebrasite are members of the group found in Britain or Ireland. Natromontebrasite is also reported, but the mineral has lost its species status and is now discredited, having been shown to be a mixture.

amblygonite

AMBLYGONITE GROUP

triclinic (Li,Na)$Al(PO_4)$(F,OH)

A relatively uncommon lithium–sodium–aluminium phosphate fluorhydroxide that forms a series with montebrasite and is found in lithium-rich granites and granitic pegmatites. In Cornwall, it occurs at Carn Clodgy, near Rinsey, in the roof of the Tregonning granite associated with albite, lithium mica, topaz and tourmaline (Exley & Stone 1964, Stone & George 1978) and at Megiliggar Rocks in aplite–pegmatite sheets (Stone & George 1978, 1983). In the kaolinized lepidolite–topaz–zinnwaldite granite of the St Austell granite complex, Cornwall, amblygonite is a minor constituent along with manganese-rich apatite, columbite, elbaite, ilmenorutile, uraninite and zircon (Manning & Hill 1990, Manning et al. 1996).

Amblygonite is reported with albite, apatite, elbaite, fluorite, lithium mica, montebrasite, topaz and occasionally petalite from a pegmatite vein in the old aplite quarry at Meldon, Okehampton, Devon (Kingsbury 1964b, Moss et al. 1969). Ryback et al. (1998, 2001) express doubt about the provenance of many Kingsbury samples; however, von Knorring & Condliffe (1984) also found amblygonite from the same locality. EMP data confirm the occurrence of amblygonite at Meldon, in association with montebrasite and a secondary mica from a single sample of fine-grain aplite (Tindle unpub. EMP data).

At Stranakelly, near Shillelagh, Co. Wicklow, amblygonite is found within massive quartz partly enveloped by albite (cleavelandite) in the core zone of a spodumene pegmatite (Steiger & von Knorring 1975, Scoon 1978).

amblystegite

A variety of pyroxene (hypersthene).

amesite

KAOLINITE–SERPENTINE GROUP

triclinic (pseudo-hexagonal) $Mg_2Al(Si,Al)O_5(OH)_4$

A magnesium aluminosilicate hydroxide found as one of the matrix minerals in an altered composite Ag–Cu–Pb–Zn mineralized feldspar–quartz porphyry from near Charlestown, Co. Mayo, Ireland (Poustie 1982). Amesite (or possibly chrysotile) is also reported from the Dulais Seam, Cefn Coed colliery, West Glamorgan, Wales (Bevins 1994). Amesite requires confirmation at Tideswell Dale, Derbyshire (Sarjeant 1967).

amethyst

A purple variety of quartz.

amethystoline

A term used by Greg & Lettsom (1858) to name liquid inclusions in amethyst.

amianthus

A synonym of asbestos; may be actinolite, anthophyllite, crocidolite, chrysotile or tremolite.

ammonium muriate

A synonym of sal-ammoniac.

AMPHIBOLE GROUP

Silicates with the general formula $A_{0-1}B_2C_5T_8O_{22}(OH,F,Cl)_2$, where A, B, C and T indicate elements having similar ionic radii and being capable of replacing each other in the structure. This large group falls in the orthorhombic and monoclinic systems, and has similar crystallography and other physical properties, as well as chemical composition. It is common in many igneous and metamorphic rock types.

Amphiboles have been the subject of two major reclassification schemes in recent times (Leake et al. 1978, 1997a, 2004) and to place a species name on an amphibole now requires a substantial chemical analysis. Although important for understanding amphibole paragenesis, the impact of the classification is very significant and has resulted in the redefinition and extinction of many species names. The scientific literature is now littered with amphibole names that are incorrect. To address this problem in part, all available British and Irish analyses have been run through the Hypercard program AMPH-IMA04 (Mogessie et al. 2004) to arrive at new names. The names followed by an asterisk are "named amphiboles" rather than true species (Burke & Leake 2005), since they have not been submitted for IMA-CNMNC approval. Names followed by a question mark require further confirmation. British and Irish species include:

- Mg–Fe–Mn–Li (orthorhombic) subgroup: anthophyllite, gedrite.
- Mg–Fe–Mn–Li (monoclinic) subgroup: cummingtonite, grunerite.
- Calcic subgroup: actinolite, ferri-actinolite*, ferro-actinolite, cannilloite?, aluminocannilloite?* (see cannilloite?), edenite, alumino-ferro-edenite, ferro-edenite, ferrohornblende, alumino-ferrohornblende, magnesiohornblende, ferri-magnesiohornblende*, hastingsite, ferri-magnesiohastingsite*, magnesiohastingsite, kaersutite, pargasite, alumino-ferropargasite*, ferropargasite, potassicpargasite, sadanagaite?, aluminosadanagaite*, magnesiosadanagaite, tschermakite, aluminotschermakite, ferrotschermakite, tremolite.
- Sodic-calcic subgroup: barroisite, ferribarroisite?, katophorite, aluminokatophorite, magnesiokatophorite, magnesiotaramite, richterite, ferrorichterite, winchite, aluminowinchite*, ferriwinchite, alumino-ferrowinchite*, ferri-ferrowinchite*, ferrowinchite.
- Sodic subgroup: arfvedsonite, magnesio-arfvedsonite, potassic magnesio-arfvedsonite, potassic-ferro-eckermannite*, glaucophane, ferroglaucophane, ferric-nyböite*, riebeckite, magnesioriebeckite.
- Na–Ca–Mg–Fe–Mn–Li subgroup: parvo-actinolite*, parvo-cannilloite*, parvo-edenite*, parvo-ferro-edenite*, parvo-magnesiohornblende*, parvo-ferrorichterite*, parvo-tschermakite*. The minerals in this subgroup are all described under their root names (e.g. for parvo-actinolite see actinolite).

Amphibole names found in the British and Irish literature and recommended for extinction include abriachanite, anophorite, asbestos, barkevikite, breadalbanite, byssolite, crocidolite, crossite, kirwanite, magnesio-anthophyllite, magnesio-cummingtonite, mountain silk, nephrite, smaragdite and uralite.

Hornblende is retained as a general or colloquial term for coloured calcic amphiboles, but it too has lost it species status. Most recently, two more classification schemes for the amphibole group have been proposed and each involves major changes to nomenclature (Hawthorne & Oberti 2006). At the moment the schemes have only reached the discussion stage and are not IMA approved.

analcime

ZEOLITE GROUP

cubic, tetragonal, trigonal, orthorhombic, triclinic $NaAlSi_2O_6 \cdot H_2O$

A common hydrated sodium aluminosilicate that forms series with pollucite and possibly with wairakite, and is found in veins, in vesicles or in the groundmass of mafic igneous rocks, typically basalt, dolerite or gabbro. It may originate in deuteric alteration or the passage of late-stage hydrothermal solutions.

England

Cornwall Colourless or china-white subtransparent or translucent crystals of analcime, ranging in size from a few mm to 5 cm are found in hydrothermal veins cutting gabbro at Dean quarry, 2 km SE of St Keverne, Lizard Peninsula. They grow on white calcite and prehnite, and are overgrown by brown calcite, heulandite and natrolite. Bluish-grey pseudomorphs of montmorillonite after analcime are also found at this location (Seager 1969, 1971, 1978a, Hall & Jackson 1975, Wolters & Wolters 1986). At Dean quarry and Porthkerris Point, some 3 km N of Dean quarry, pseudomorphs of adularia after analcime are also reported (Halliday & Mitchell 1976, Seager 1978a). Golley & Williams (1995) note analcime from other locations on the Lizard Peninsula: Carn Barrow; Holseer Cove; Kildown; Parc Bean Cove, Mullion; Porthoustock quarry, St Keverne; Porthkerris quarry; Predannack Head; and Pol Cornick. Analcime was said to have been found in the cliffs near Botallack, St Just (Collins 1892), but this is considered doubtful (Russell 1910b). Analcime is a constituent of the phonolite on Wolf Rock, 15 km from Gwennap Head, Land's End. Analcime (and cancrinite) were probably formed after prolonged heating of the phonolite (perhaps isothermally) and are not weathering products. Associated species are aegirine, nepheline and sanidine, with minor apatite, fluorite, pyrite and tourmaline also present (Harrison et al. 1977).

Devon Permian basaltic and potassium-rich lavas (lamproites) around Tiverton and Crediton contain analcime as an interstitial or euhedral microphenocrystal component (possibly pseudomorphing leucite). The rocks suffered intense hydrothermal alteration or weathering in a desert environment. Two lamproite localities are Holmead farm, Loxbeare, and Knowle quarry, S of Brandirons Corner, near Knowle village (Velde 1971b, Thorpe et al. 1986). Some Tertiary alkali dolerite (crinanite) dykes cutting the granite on Lundy island, off the Devon coast, contain analcime, calcite, chlorite, natrolite and thomsonite infilling vesicles (Thorpe & Tindle 1992).

Derbyshire At Calton Hill, Taddington, analcime occurs sporadically as a milky-white crystalline vesicle lining in Carboniferous basalt-lava flows (Tomkeieff 1928).

Durham The late Carboniferous quartz dolerite of the Whin Sill is a source of analcime. At Cambokeels mine, Weardale, it has produced analcime, apophyllite and chabazite specimens from many joint surfaces. Crusts of analcime up to 30 cm across are composed of clear colourless trapezohedra, with individual crystals reaching 6 mm in length. Similar crystals were also collected from the Copt Hill quarry, Cowshill, Weardale, and High Force quarry, Upper Teesdale (Young et al. 1991, Green & Briscoe 2002).

Merseyside Gabbroic glacial erratics on Thurstaston beach, Wirral, contain small cavities in which fibrous amphibole and attractive analcime trapezohedra have formed (Rothwell & Dickinson 1989).

Trapezohedral crystals of analcime with an orange coating of iron oxide on decomposed granodiorite; Croft Quarry, Leicestershire; width 20 mm. (© R. E. Starkey)

Somerset Analcime, occurring as small pale-brown trapezohedral crystals or pale-brown masses, is reported growing into and replacing aragonitic shell debris in a heavily faulted 15 km-long coastal exposure of Lower Jurassic marine sediments from Hinkley Point to Warren Bay, 1 km W of Watchet harbour. Associated minerals include authigenic aragonite, barite, calcite, celestine, dolomite, pyrite, quartz, sphalerite and solid hydrocarbon (Alabaster 1980).

Staffordshire Analcime is a minor constituent of an olivine dolerite sill discovered in boreholes at Landywood, near Cannock (Barton 1963).

Leicestershire Colourless trapezohedra of analcime (1–5 mm across) and a later generation of translucent, white or orange-tipped analcime (sometimes coated in hematite) line cavity walls in calcite–zeolite veins cutting diorite at Croft quarry, SW of Leicester (Pearson & Jeffrey 1997). Analcime is noted from two other locations: as small colourless crystals with calcite and quartz at Croft quarry and as white crystals associated with laumontite at Enderby (Tschernich 1992).

Wales

Gwent Analcime occurs in monchiquite intrusions at Golden Hill, near Usk (Boulton 1911), and nearby Glen Court (Haslett 1992).

Gwynedd Perfectly formed faceted crystals of analcime with trapezoidal form have been identified in the contact aureole of the Plas Newydd dyke, Anglesey (Henslow 1822, Greenly 1919). At Benallt mine on the Lleyn Peninsula, analcime occurs with paragonite in veins cutting massive manganese ore (Campbell Smith et al. 1946) and at Gimlet quarry, near Pwllheli, it forms colourless trapezohedra up to 2 cm across (Tschernich 1992, Bevins 1994). Analcime fills vesicles in olivine dolerite dykes exposed between Marchlyn Mawr reservoir and Blaen-y-nant to the W of Nant Ffrancon, Snowdonia (Williams 1924, 1930a).

Powys Veins cutting altered Ordovician lavas at Llanelwedd quarry, near Builth Wells, contain analcime associated with laumontite and quartz (Metcalfe 1990).

Scotland

Central Vesicular basalt at Earlsburn (Earlston) reservoir, Touch Hills, SW of Stirling, contains highly distorted white-to-pink analcime associated with harmotome, laumontite, quartz, stilbite and thomsonite (Heddle 1901, Green & Todd 2001). At Campsie Fells, N of Strathblane, analcime was once found as opaque-white to pink trapezohedra up to 7 cm in diameter (Heddle 1901, Tschernich 1992).

Fife Analcime is noted: as flesh-red crystals in basalt at Craighall, SE of Ceres, and as large transparent grey crystals in agglomerate at Ruddons Point, 4 km W of Elie (Heddle 1901, Tschernich 1992).

Grampian Analcime forms: transparent red trapezohedra, up to 1 cm across, at Bowdens, Gapol, and the Slainges, S of Todhead Point (Tschernich 1992).

Highland Varied origins are indicated by the range of rock types containing analcime: nepheline syenite of the Loch Borralan alkaline pyroxenite–syenite complex, in a quarry at Allt a' Mhuilinn (Stewart 1941) and along the Ledmore River (Tilley 1958); Carboniferous–Permian camptonitic dolerite, camptonite and monchiquite dykes at An Torra Ban, E of Strontian, and Allt Tarsuinn, Corrantee mine, NW of Strontian (Gallagher 1963); contact-metamorphosed limestones (hornfels) adjacent to diorite gneiss of the Moine Series, Ardgour district (Drever 1939); vesicular basalt at Ardtornish Bay, Morvern Peninsula (Curry 1905); contaminated

Analcime crystals to 18 mm from a large cavity in porphyritic basalt; Moonen Bay, Isle of Skye, Highland Region, Scotland. (© D. I. Green)

alkali gabbro dyke cutting Jurassic limestones at Camas Mòr, Isle of Muck (Tilley 1947); contaminated gabbro and pyroxenite at Camphouse, Ardnamurchan (Agrell 1951); and a xenolithic monchiquite dyke at Streap and Streap Comlaidh, 7 km NW of Glennfinnan railway station (Walker & Ross 1954). Elsewhere, analcime is a common species in vesicles within Tertiary basaltic lavas on the Isle of Skye: Allt Coir' a' Ghobhainn, near Drynoch (Livingstone 1988, 1989b); as a rare interstitial species in picritic basalt on the northeast side of An Cruachan and E of Sgùrr Beag (Bell & Williamson 1994); Beinn na Greine, near Portree; Dun Diarmaid, shore of Loch Beag, near Bracadale; 1 km N of Dunvegan; Ros à Mheallain, NE of Bracadale (King 1976); near the Quiraing, including Echoing Rock, near Fir Bhreugach (Heddle 1901, Sweet 1960b); large white trapezohedra up to 25 mm across at Moonen Bay, N of Dunvegan (Green & Todd 1996, 2001); Rubha nan Clach, Loch Bracadale (Heddle 1901); crystals up to 4 cm across at Sgùrr nam Fiadh, 3 km S of Talisker Bay (Heddle 1901); Talisker Bay (Heddle 1901, Green & Wood 1996); the Storr, N of Portree (Heddle 1901, King 1976); and Uig, 26 km N of Portree (Sweet 1959). Heddle (1901) also reports analcime from the islands of Canna, Eigg, Sanday and Muck.

Lothian The sodalite-bearing phonolitic trachyte at Traprain Law, near Haddington contains 20% analcime (MacGregor & Ennos 1922, MacGregor 1960, Battey & Moss 1962, Hatch et al. 1972). Analcime also forms crystals to 12 cm across at Ratho quarry, near Edinburgh, and crystals to 15 cm across at Salisbury Crags, Edinburgh (Greg & Lettsom 1858).

Strathclyde Tertiary igneous activity in this region produced analcime in doleritic intrusions and amygdaloidal basaltic lava flows on the Scottish mainland and on some nearby islands: in a dolerite dyke at Craignish Point, 30 km S of Oban (Allison 1936, Livingstone & Baird 1991); Glecknahavill, 8 km N of Campbeltown; Port an Dunain, near Sunadale and West Loch Tarbert, all on the Kintyre Peninsula (McCallein 1932); in a dolerite sill at Howford Bridge, SE of Mauchline, near Prestwick (Tyrrell 1928b); at four localities on the Isle of Mull: Beinn Chreagach, between Carsaig Arches and Carsaig Bay; Sgùrr Dearg, 9 km W of Duart Point (both Heddle 1901); 'S Airde Beinn, 5 km W of Tobermory (Cann 1965a,b); A484 roadcut to the S of Tobermory (Green & Todd 2001); Bunnahabhain (Dyer et al. 1996) and Cnoc Rhaonastil (Walker & Patterson 1959, Preston et al. 2000b), both on the Isle of Islay; several localities on the Isle of Arran – Bennan Head, Breadalbane, Torrylin shore and Whiting Bay (Dyer et al. 1996); as interstitial crystals in an alkali dolerite sill at Clauchlands Point and Kingscross Point, Lamlash Bay, east Arran (Tindle unpub. EMP data); in the Dippin Sill, southeast Arran (Henderson & Gibb 1977, Gibb & Henderson 1978a,b).

The Carboniferous Clyde Plateau basalts around Glasgow contain many examples of analcime: Boyleston quarry, SW of Barrhead (Heddle 1887a, Hall et al. 1989); Bowling quarry, Kilpatrick Hills, near Glasgow (Greg & Lettsom 1858, Heddle 1901); Hartfield Moss, 9 km WSW of Barrhead (Meikle 1990); King's quarry, the Totherick, near Dunlop (Dyer et al. 2003); unusually with epidote, garnet and prehnite at Loanhead quarry, near Beith (Livingstone 1989a, Meikle 1989b, Meikle & Todd 1995, Green & Todd 2001); Reilly quarry, Houston (Wolfe 1997, Meikle 2003). Analcime is an important constituent of the Carboniferous Lugar sill exposed at Glenmuir Water; and Barshaw, near Paisley (Scott 1914b, Howie 1963, Phillips 1968), and interstitial analcime is also present in a trachybasalt intrusion on Loudoun Hill, near Darvel (Upton et al. 2007). Heddle (1901) reports analcime from the Treshnish Isles and from the islands of Staffa and Ulva.

Tayside Analcime is noted: as isolated flesh-colour crystals, up to 5 cm across, at Glen Farg (Greg & Lettsom 1858, Heddle 1901); and in patches up to 3 mm across, in segregation veins at Friarton quarry, Moncrieffe Hill, S and SE of Perth (Davidson 1932).

Western Isles Analcime occurs interstitially and also forms thin veins cutting picrite within the Tertiary alkali dolerite sill on Garbh Eilean and Eilean an Tighe, Shiant Isles, 24 km N of the Isle of Skye (Walker 1930, Gibb & Henderson 1996). On the north shore of Garbh Eilean it is associated with mesolite and stilbite (Heddle 1901).

Fair Isle Interstitial analcime is found associated with euhedral scapolite in veins near the South Lighthouse and on the Tail of Uran (Mykura & Young 1969).

Orkney Devonian alkaline basic rocks (mainly camptonite and teschenite) contain analcime, usually as an interstitial component, although euhedral crystals have been found in pools of carbonate at Point of Ayre, Deerness, Mainland. Other localities include: Black Holm, Copinsay; Hoxa, South Ronaldsay; Muckle Castle, Stonehall, Deerness; and the Burg, Shapinsay (Flett 1900, Kellock 1969). Heddle (1901) reports analcime infilling amygdales in lavas at Sands Geo, Hoy, Orkney.

Well formed irridescent transparent crystal of analcime, 1.25 mm across, from Little Deer Park, Glenarm, Co. Antrim, Ireland. (© D. I. Green)

Ireland

Antrim Analcime is often found in the Tertiary basalts of Co. Antrim. It occurs in the regional analcime–natrolite zone around the margins of the lava pile and was noted at 31% of the localities examined by Walker (1960a). Analcime forms dull-white trapezohedra, sometimes of appreciable size (2 cm at Cushendall, 2.5 cm on the Garron coast, 3 cm at Collinward) and is associated with apophyllite, calcite, chabazite, gyrolite and levyne (Walker 1960a). Along the east coast, abundant small transparent glassy crystals of analcime are found in the gmelinite zone at Little Deer Park, Glenarm, Island Magee, and at Magheramorne quarry, near Larne. This zone is thought to have developed as sea water penetrated faults during normal regional zeolite crystallization (Walker 1959a, Green 2005a). Basalts at Brown's Bay, Island Magee, and at Galboly, near Garron Point, have also produced analcime (Adair 1987). Analcime is associated with chabazite (phacolite) at Killyflugh quarry, N of Ballymena, Lower Toome, and it occurs with gmelinite and natrolite in a quarry to the S of Whitehead (Herbert Smith 1916). On the north coast, analcime occurs at: the Giant's Causeway; Dunluce Castle; both O'Hara Rocks, near Port Stewart, and Doon Point on Rathlin Island (Greg & Lettsom 1858); Cregganboy, near Kinbane (Ashcroft 1916); and Craigahulliar quarry, near Portrush (Robert 2001). Analcime is an interstitial phase, replacing plagioclase feldspar and it also occurs in vesicles in a pyroxenite within the contaminated dolerite intrusion on Scawt Hill, near Larne (Tilley & Harwood 1931, Tilley 1937d). A magnesium-rich analcime (doranite) is described from the Knock escarpment near Carrickfergus (Dyer et al. 1987). Analcime is confirmed from Maxwell's quarry, near Ballymena (H. J. Foy specimen, Tindle unpub. EMP data).

Londonderry Vesicular basalts containing analcime occur at: Aghanoo; Castlerocks, Magilligan; Craignashoke, Ballinascreen; Donald's Hill; and Slieve Gallion (all Greg & Lettsom 1858); and Sounding Hill quarry, NNW of Magherafelt (Old 1972).

Donegal Traces of analcime and leucite are present in an alkali basalt dyke containing peridotitic xenoliths near Inver, 10 km W of Donegal (Shaw & Edgar 1997).

Down A Tertiary dolerite sill encountered in a borehole at Ballyalton, Newtownards, contains analcime and natrolite in amygdales (Knill 1960).

Offshore territories

Guernsey Brilliant colourless transparent icositetrahedra of analcime encrust prisms of natrolite at the Bouet (Mourant 1939).

anatase

tetragonal TiO_2

A relatively common titanium oxide, trimorphous with brookite and rutile, and found in acidic igneous rocks, metamorphic rocks and alpine-type veins formed as a result of low-grade regional metamorphism. Anatase is also a common detrital mineral and as such is reported in conglomerates, grits, sandstones and shales belonging to the Carboniferous Millstone Grit in Yorkshire. It usually forms colourless tabular outgrowths on leucoxene (after ilmenite) and is reported in heavy-mineral concentrates from: Bramley Fall quarry, Newlay; Grimbald's Crag and Plumpton Rocks, both near Knaresborough; Ingleborough; Penyghent; Rumbles Moor (above Ilkley), Tannery borehole (Meanwood, Leeds); and from five quarries in the Horsforth district (Gilligan 1919). A remarkable heavy-mineral concentrate with a similar authigenic origin and consisting almost entirely of dark-blue anatase is reported from a chamositic oolite of Jurassic age at Sheriff's pit, Rosedale, Yorkshire (Rastall & Hemingway 1949). These authors also note anatase in a pebbly oolitic sandstone from Rosedale Abbey Bank, Rosedale. Usually colourless and sometimes well formed crystals (up to 0.17 mm across) of detrital anatase are a feature of Triassic sandstone in the Midlands: Storeton and Weston, Cheshire; Dale Abbey near Derby; Kingswinford, near Stourbridge, and Sutton Coalfield, West Midlands (Scrivenor 1903a). Anatase has also been noted in fireclays from various locations in Yorkshire: Amber Thorn, Denholme, Elland, Gildersome, Ingleton, Queensbury and Tong; and also Knowlesbury, Shropshire; Castlecary and Glenboig (both Strathclyde) and Leven Seat, S of Fauldhouse, Lothian, Scotland (Brindley & Robinson 1947).

England

Cornwall At Penberthy Croft mine, St Hilary, anatase forms rare, small, euhedral, black tabular crystals up to 0.3 mm across, in groups or as isolated individuals. It is found with chlorite, monazite-(La), siderite and quartz in an association typical of alpine-type veins. A similar assemblage has also been found at Croft Gothal mine, St Hilary (Betterton 1996a,d, Betterton et al. 2000). An association of anatase (forming steel-blue tetragonal bipyramids up to 1 mm across), albite and monazite is reported from joint surfaces in slate at Lambshouse quarry, Tintagel (Greg & Lettsom 1858, Bowman 1900). Other locations include: Cheesewring quarry, SW of Henwood, Linkinhorne (Bowman 1911, Green et al. 1996a); Delabole slate quarry (Collins 1892); Dinas Head, near Padstow (Agrell 1939); Hingston Down quarry, 2 km W of Gunnislake, Calstock (Wolfe 2000); Kessel Downs (Halvasso) quarry, near Longdowns (R. E. Starkey pers. comm. 2005); Looe Mills Hill quarry, near Liskeard (Greg & Lettsom 1858); New Cook's Kitchen mine (North Tincroft lode), Redruth (LeBoutillier et al. 2000); Prince of Wales mine and Wheal Zion, both Calstock, and Prince of Wales quarry, Tintagel (Golley & Williams 1995); Wheal Kitty, St Agnes (Collins 1911a); and Wheal Remfry china-clay pit, near Indian Queens, St Austell (Rust 1998).

Cheshire Detrital grains in host rocks close to copper–lead mineralization at Alderley Edge ridge (Ixer & Vaughan 1982).

Cumbria Detrital grains in host rocks close to polymetallic mineralization in the Great Sulphur vein, Alston Moor (Vaughan & Ixer 1980); detrital and authigenic crystals in the Triassic St Bees Sandstone at St Bees Head, SSW of Whitehaven (Ixer et al. 1979b); small lustrous black octahedra up to about 1 mm scattered in a drusy feldspar rock from the Shap granite quarry (Young 1987a); blue, deep-yellow or colour-zoned striated bipyramids in Shap granite (Grantham 1928); dark-brown to black or light- to dark-green striated blocky bipyramids, tabular crystals or paper-thin blades in miarolitic cavities within the Shap granite (Wirth & Green 2002); dark-blue striated bipyramidal crystals to 0.5 mm in chlorite–quartz veins and golden-brown blocky crystals in calcite–quartz veins at Spout Crag quarry, Langdale; also similar in size and colour from Hodson's Field, Grasmere (R. E. Starkey pers. comm. 2005).

In view of the discreditation of many British minerals supposedly collected by A. W. G. Kingsbury (Ryback et al. 1998, 2001), the report of anatase from Embleton and Fell End quarries, near Cockermouth, by Young (1987a) requires further confirmation.

Derbyshire Found near Matlock as an accessory component in altered olivine dolerite of the Ible Sill and at Millclose mine, Wensley (Garnett 1923a,b, Ford & Sarjeant 1964a, Ford et al. 1993).

Devon "Very fine and perfect" crystals of anatase up to 3 mm across and embedded in chlorite veinstone have been found at Virtuous Lady mine, Buckland Monachorum. Crystals are black and perfectly opaque, with tabular or bipyramidal form (Greg & Lettsom 1858, Russell 1913). Boulders of tourmalinized Dartmoor granite on Hamel Down contain anatase, brookite, cassiterite, topaz and zircon (Brammall & Harwood 1925). Anatase has also been found with brookite in pneumatolyzed granite at Lock Gate Cross, near Ponsworthy, and other locations over the Dartmoor granite outcrop (Brammall & Harwood 1923). At New East Wheal Russell, SSW of Tavistock, thin tabular steel-blue crystals of anatase, up to 0.5 mm on edge, have been found in cavities in quartz veins, associated with calcite and chlorite. In the Meldon aplite quarry, near Okehampton, black to steel-blue bipyramidal anatase (< 1 mm) is associated with axinite, brookite, chlorite, quartz and schorl (R. E. Starkey pers. comm. 2005).

Leicestershire Deep-blue crystals up to 0.12 mm in length, in diorite, gabbro and granodiorite from nine localities in the Mountsorrel district (Taylor 1934).

Northumberland Alteration zones in the Whin Sill (quartz dolerite) adjacent to barite–witherite vein mineralization at Settlingstones mine, near Hexham, contain anatase together with ankerite, calcite, illite, kaolinite, muscovite, pyrite and quartz (Ineson 1972).

Wales

Clwyd Anatase occurs with apatite, brookite, monazite-(Ce), rutile and xenotime-(Y) in calcite-filled joints within a large dolerite sill at Hendre quarry, Pandy, near Glyn Ceiriog (Starkey et al. 1991).

Gwent Anatase is noted to occur at Christchurch quarry, near Newport (Bevins 1994).

Gwynedd Anatase occurs with albite, brookite and fluorapatite in an alpine-type vein assemblage of anastomosing quartz veins and fissures in altered dolerite intruded into Ordovician slaty mudstones at Fron-oleu, Prenteg, Tremadog. Some very fine lustrous bipyramidal crystals up to 4 mm long have been recovered, with colours ranging from steel grey to rich golden brown (Starkey & Robinson 1992). Euhedral transparent-to-translucent inky-blue anatase crystals forming flat pyramids up to 3 mm across and associated with albite, chlorite and quartz have been found in a second Welsh alpine-type vein assemblage at Tany-grisiau (Green & Middleton 1996). Similar assemblages are reported from Arenig Station quarry; Hendrehenydd, near Penrhyndeudraeth, and Manod quarry, 1 km E of Blaenau Ffestiniog (Bevins 1994). Anatase is a secondary species forming 10 μm boxworks or laths after original iron–titanium oxides in altered volcanic rocks at Parys Mountain, Anglesey (Pointon & Ixer 1980). Elsewhere, anatase is confirmed on Crib-goch ridge on Snowdon, and at Cwm Meillionen, near Beddgelert (Bevins 1994).

Tabular anatase, 1.7 mm across, from Tanygrisiau, near Blaenau Ffestiniog, Gwynedd, Wales. (© D. I. Green)

Powys Alpine-type veins cutting Yat Wood Formation sediments at Dolyhir quarry, Old Radnor, near Presteigne, contain small orange-to-blue euhedral crystals of anatase up to 0.5mm on edge (T. F. Cotterell pers. comm. 2007).

Scotland

A study of heavy-mineral assemblages from Caledonian granites of Scotland shows that anatase occurs as a secondary species at many locations. It ranges in colour from steely grey, through golden yellow to dark indigo blue, the last particularly characteristic of the Ben Rinnes granite near Dufftown, Grampian, and granite near Loch Ken, Dumfries & Galloway (Mackie 1928). A similar method adopted to investigate mineralization associated with the Cairngorm granite complex, Grampian, discovered anatase together with cassiterite, chrysoberyl, columbite, fergusonite, ilmenite, ilmenorutile, monazite, topaz, xenotime and zircon in stream-sediment panned concentrates collected from Luibeg Burn, near Ben Macdui (Beer & Bennett 1988).

Dumfries & Galloway Lozenge-shape pseudomorphs of anatase after titanite, up to 300 μm across, occur in the Criffell granodiorite at Needle's Eye, SW of Caulkerbush (Basham et al. 1989).

Grampian Anatase occurs with andalusite, apatite, brookite, dumortierite, monazite and tourmaline in pegmatitic veins cutting schist at Colehill farm, near Ellon, Grampian (Mackie 1925).

Highland Anatase, apatite and pyrite are accessory species in calcite patches within feldspathic gneiss at Learnie quarry, Black Isle. Aegirine, barite, hematite and plagioclase are also present (Garson et al. 1984). Anatase replaces ilmenite in altered olivine dolerite close to a barite–galena vein on Garbh Bheinn to the E of Strontian (Gallagher 1964).

Western Isles The Conachair granite on St Kilda (Hirta), contains anatase, chevkinite, fluorite, magnetite, rutile, titanite and zircon in and adjacent to miarolitic cavities (Harding et al. 1984).

Shetland In a hydrothermally altered syenite exposed in one of the Bakkasetter quarries, on the southeastern flank of Noss Hill, South Mainland, anatase forms granular aggregates with hematite; epidote, nontronite and palygorskite are associated alteration species (Stephen 1954).

Ireland

Anatase is noted from the following locations: in Co. Antrim, in bauxites (Eyles 1952) and in Dalradian schists at Ballycastle (Wilson & Robbie 1966); in rhyolite at Lough Guitane, Co. Kerry (Avison 1984); in a beach pebble from Bunmahon cliffs, Co. Waterford, forming inky-blue crystals included in quartz and in quartz-lined cavities associated with clinochlore (Taylor 1996); in most heavy-mineral samples from the northern portion of the Leinster granite, Co. Wicklow, where it forms blue, yellow or whitish, tabular or bipyramidal crystals up to 0.2 mm long (Smithson 1932).

Offshore territories

Jersey Anatase has been found as minute bluish-black horizontally striated tetragonal bipyramids, with garnet and molybdenite in a geode from the granite quarry at L'Etacq. The species is also described from granitic rocks at Gigoulande and Sorel Point (Mourant 1961). Stanley & Ixer (1982) note 2–10 µm laths of anatase commonly occurring in Brioverian sediments adjacent to lead–zinc veins at Le Pulec on the northwest coast.

ancylite-(Ce)

orthorhombic $SrCe(CO_3)_2(OH)\cdot H_2O$

A rare hydrated strontium–cerium carbonate hydroxide found on the Bellsgrove Lodge dump, near Strontian, Highland Region, Scotland, where it forms pale-violet octahedral microcrystals associated with platy barite, calcite and drusy quartz (Livingstone 1993). Ancylite-(Ce) has also been found on the dumps of the Strontian barite mine (Middleshop section), where it occurs as pale rose-pink crystals up to 2 mm across on scalenohedral calcite (Green & Todd 2001, Wolfe 2004). Blocky, euhedral prismatic crystals of ancylite-(Ce), up to 1 mm long and sometimes encrusted with acicular kainosite-(Y) are described from the nearby Whitesmith mine (Green & McCallum 2005).

Pink, bipyramidal ancylite-(Ce), 0.8 mm across, associated with radiating spherules of kainosite-(Y) on chamosite; Whitesmith mine, near Strontian, Highland Region, Scotland. (© D. I. Green)

andalusite

orthorhombic Al_2SiO_5

A common aluminosilicate, trimorphous with kyanite and sillimanite, often found as a result of contact metamorphism of argillaceous sediments, but also found in regionally metamorphosed schists and some granites and granite pegmatites. Andalusite occurs as a detrital mineral and as such is reported from: Cretaceous and Eocene sands at Holden's Wood, Brentwood, Essex; Hayes Common and Plumstead, Kent (now Greater London); Kennel Wood, Shirley, Surrey (now Greater London); and Northwood, Middlesex (now Greater London). It is associated with other detrital species: apatite, epidote, garnet, hornblende, ilmenite, kyanite, magnetite, rutile, sillimanite, spinel, staurolite, tourmaline and zircon (Davies 1915). Thomas (1909a) indicates that it is an abundant species in the Pliocene sands of St Erth and St Agnes in Cornwall. In Wales, andalusite has been found: in glacial deposits at Castell-toch, near Llansadurnen, Dyfed (Strahan et al. 1909); at Ludchurch, Dyfed (Griffiths & Stuart 1940); and in the area between the rivers Neath (West Glamorgan) and Tywi (Powys) (Griffiths 1939). Thomas (1909a) reports many other Welsh locations. Andalusite is a constituent of alluvial panned concentrates collected on St Kilda (Hirta), Western Isles, Scotland, where it is accompanied by many of the species listed above, together with anatase, clear-blue corundum, chevkinite, diopside, fergusonite, goyazite, titanite and xenotime (Harding et al. 1984). Mackie (1928) found andalusite at four locations in Grampian, Scotland, during an examination of the heavy-mineral contents of Caledonian granites: Auchterellon; Colehill, Ellon; Dorbach, and in a granite pebble in Old Red Sandstone rocks at Auchterless. Andalusite is a detrital component in the Ordovician Upper Derrylea Formation to the S of Clew Bay, Co. Mayo, Ireland (Mange et al. 2003).

In Cornwall and the Isles of Scilly, andalusite occurs in fine- and coarse-grain siderophyllite (biotite) granite, where it is a rare magmatic component, or more abundantly as a constituent of entrained non-igneous xenoliths. The granite occurrences include: Bodmin Moor granite at Cheesewring quarry, SW of Henwood, Linkinhorne (Teall 1887), and De Lank quarry on Bodmin Moor (Stimac et al. 1995); Carnmenellis granite at Carnsew quarry, near Penrhyn, and Chywoon quarry, near Longdowns (Stimac et al. 1995); Land's End granite at Castle-an-Dinas (Stimac et al. 1995) and in a quarry near Knill's monument, Carbis Bay, St Ives (Robson 1949a); St Austell granite at Gready and Golden Point quarry (Richardson 1923), and Luxulyan quarry, near Lostwithiel (Stimac et al. 1995); and Scilly Isles granite at Penninis Cove, St Mary's (Stimac et al. 1995). The cores of a few pelitic xenoliths in the Carnmenellis granite contain andalusite mantling zinc-rich spinel (Jefferies 1985b).

On Trencrom Hill, near the roof of the Land's End granite, pale-pink to colourless (in thin section) andalusite is intergrown with corundum (sapphire) and quartz over a small area and probably represents very aluminous digested inclusions, derived from the surrounding pelites (Wilson 1975). Among other xenolith occurrences are: in the Carnmenellis granite at Carnsew, near Penrhyn, and both Chywoon and Kessel Downs (Halvasso) quarries, near Longdowns (Stimac et al. 1995); and in the Bodmin Moor granite at Cheesewring quarry (Shambrook 1982) and De Lank quarry (Stimac et al. 1995). Metamorphism of host rocks (killas) around the granites has also resulted in the formation of andalusite and is reported from: Bosworgey mine, near Fraddam, St Erth (Le Neve Foster 1877); Cligga Head, Perranzabuloe, and Megiliggar Rocks, near Breage (Robson 1949b); Priest's Cove, near Land's End (Hall & Jackson 1975); and Wheal Coates, St Agnes (Collins 1911a, Robson 1949b).

Chiastolite, a variety of andalusite containing carbonaceous inclusions that give a cruciform pattern in cross section, is found in Devon in Devonian slate at Ivybridge and at Holestock, near Okehampton (Collins 1892). Digestion of host rocks by the Dartmoor granite has left behind xenoliths rich in andalusite at Pithill quarry, Ivybridge (Brammall & Harwood 1932). Andalusite is also noted to be a scant or rare component of the granite itself (Brammall & Harwood 1925). Sector-zoned porphyroblasts of andalusite (chiastolite) have also been found in graphitic rocks and semi-pelitic hornfels in the aureole of the Dartmoor granite at Okehampton, Ramsley mine, Aish Tor and Ivybridge station (Rice 1993).

In the Caldbeck Fells of Cumbria, intrusion of the Skiddaw granite has produced andalusite (chiastolite) in slate within the surrounding thermal aureole at: Bannerdale mine, Mungrisdale (Davidson & Thomson 1951); Blackhazel Beck, Mungrisdale Common, and Gibraltar Crag, Skiddaw (Rastall 1910); and Grainsgill Beck (Roberts 1983). Unusually large crystals of andalusite are noted from the east side

Basal section of andalusite showing a characteristic chiastolite cross; from a cordierite-biotite slate in the aureole of the Skiddaw granite, Caldbeck Fells, Cumbria; width 2.75 mm; viewed in PPL. (© K. Gardner)

of Glenderaterra Valley, near Keswick (Rastall 1910). Other Cumbrian occurrences include: Water Crag, near Devoke Water, Eskdale (Simpson 1934, Young 1985d); and Sharp Knott, Loweswater Fell (Eastwood et al. 1931).

Andalusite is described from a marginal facies of the Mullach nan Coirean granite on the north spur of Meall a' Chaoruinn, near Ben Nevis, Highland Region (Haslam 1971), and from the Cairngorm granite, Grampian (Harrison 1988), but the majority of Scottish (and Irish) occurrences are from thermally metamorphosed meta-pelitic rocks.

Scotland

Central Creag Mhòr, Glen Lochay, 5 km W of Killin (Heddle 1901).
Dumfries & Galloway Northeastern slopes of Knocknairling Hill, near New Galloway (Heddle 1901); Penkiln Burn, N of Newton Stewart (Stone & Leake 1984).
Grampian Ballater area (Goodman et al. 1990); Botriphnie, 9.75 km SW of Keith and Clashnarae Hill, 6 km W of Lumsden (Heddle 1901); Creag nam Ban, near Ballater (Smith & Phillips 2002); S of Auchindour Castle, Glen Fiddich (Beddoe-Stephens 1990); Portsoy and along the lineament to Tomintoul (Beddoe-Stephens 1990); between Rothienorman and Methlick, Ythan Valley (Read 1952).
Highland Aureole of the Ballachulish granite (Pattison 1987); along the coast 2 km S of Cape Wrath (Heddle 1901); SW of Clachaig in the banks of Gleann Leac na Muidhe, Glen Coe (Heddle 1901); Sgòr a' Chaoruinn summit, Carn Chuinneag–Inchbae area (Harker 1954b).

Ireland

Dublin Aureole of the Leinster granite at Ballycorus, near Rathmichael and Killiney (Brindley 1957). At Killiney Bay the chiastolite variety forms confused crystals in mica slate (Greg & Lettsom 1858).
Carlow Aclare, near Myshall (Whitworth 1992); Tomduff Hill, E of Seskinnamadra (Steiger & von Knorring 1975, Ryback & Moreton 1993); generally in the aureole of the Leinster granite in Co. Carlow and Co. Wicklow (Kennan 1997).
Donegal Remarkable crystals of pink chiastolite up to 5 cm long between Clooney and Cashelgolan Hill in the aureole of the Ardara granite complex (Akaad 1956, Atherton 1980), also at Lough Doon (Smellie 1974b); aureole of the Barnsmore granite (Smart 1962); aureole of the Omey granite, near Streamstown Bay (Ferguson & Harvey 1979, Ferguson & Al-Ameen 1985); Loch Salt and Scalp Mountain (Greg & Lettsom 1858).
Galway Roadcuts at Claggan (Ryback & Moreton 1992). The locality was named "Claggau quarry" (Anthony et al. 1995), but the occurrence is actually in two very small adjacent roadside quarries on the road from Maam to Cong, about 4 km SW of Cornamona. The rock is a fine-grain schist with thin lenses or strings of quartz carrying flesh-pink subhedral andalusite crystals associated with cordierite (S. Moreton pers. comm. 2006).
Mayo Near Easky Lake, Ox Mountains, in Co. Mayo and Co. Sligo (Yardley & Long 1981).
Wexford Aureole of the Carnsore granite (Trelour & Max 1984).
Wicklow Glendalough, Glenmalur and Luganure (Greg & Lettsom 1858).

Offshore territories

Grey four-sided prisms of andalusite occur in metamorphosed rhyolite near Côtil Point, Jersey, Channel Isles (Mourant 1961).

andersonite

trigonal $Na_2Ca(UO_2)(CO_3)_3{\cdot}6H_2O$

A rare hydrated sodium–calcium uranyl carbonate found along 4 m of the Peeth lode, Geevor mine, near Pendeen, St Just, Cornwall. Andersonite occurs as small masses of intergrown pseudo-cubic crystals to 3 mm on edge. It is bright yellowish-green and forms transparent crystals that in long-wavelength UV light exhibit very bright yellowish-green fluorescence. Associated species are schröckingerite and gypsum. Elsewhere along the Peeth lode other secondary uranium minerals include cuprosklodowskite, johannite and possibly becquerelite (Elton & Hooper 1992, Elton et al. 1992).

andesine

A very common triclinic sodium–calcium aluminosilicate with an intermediate (An_{30-50} mole %) composition between sodium-rich albite (An_{0-10}) and calcium-rich anorthite (An_{90-100}) end members. It is part of the plagioclase feldspar continuous solid-solution series that also includes oligoclase (An_{10-30}), labradorite (An_{50-70}) and bytownite (An_{70-90}).

andradite

GARNET GROUP

cubic $Ca_3Fe^{3+}_2(SiO_4)_3$

A common calcium–iron silicate that forms series with grossular and with schorlomite, and is found in contact-metamorphic aureoles where calcium-rich rocks such as limestones have been metasomatized. It is also found in chlorite schists, serpentinites and alkalic igneous rocks. Like all other species of garnet, end-member compositions are rare, and most examples show some degree of solid solution towards other members of the group. Where possible the proportions of the two most abundant end members are listed below (an = andradite, gr = grossular).

England

Cinnamon-colour garnets in a contact-altered calc-silicate rock at Treburland manganese mine, 2 km SE of Fivelanes, Altarnun, Bodmin Moor, Cornwall, are considered to be andradite (Russell 1946d). The species is noted from South Terras mine, St Stephen-in-Brannel, and Tretoil mine, Lanivet (Golley & Williams 1995), but no data are presented to confirm the identification. Quartz-rich nodules in a metasomatized basalt within the aureole of the Land's End granite at the Crowns, near Botallack, St Just, contain fluor-bearing hydro-andradite (van Marcke de Lummen 1986a). Around the north margin of the Dartmoor granite in Devon, metasomatically altered calc-silicate hornfels containing andradite are reported from: Red-a-ven mine, Meldon, near Okehampton; Meldon railway quarry; Belstone Consols and Ramsley mines, near Sticklepath (El Sharkawi & Dearman 1966). Secondary hydrothermal alteration of the Dartmoor granite itself has produced an andradite-bearing (an_{67}, gr_{32}) calc-silicate rock at a location between the Birch Tor and Vitifer iron–tin mines on Dartmoor (Alderton 1988a). In Cumbria, andradite is noted from Carrock Fell (Young 1987a) and from the aureole of the Shap granite (Firman 1978).

Wales

Honey-yellow andradite crystals have been found on joint faces in lodes adjacent to two Tertiary dykes at Parc mine (No. 3 level), near Llanrwst, Gwynedd (Archer & Elliott 1965).

Scotland

Dumfries & Galloway Orange-to-brown andradite was once found with chabazite, epidote and hematite in cavities within a dyke at Coatsgate quarry, near Moffat (Green & Todd 2001).
Grampian 2 km SSE of Derry Lodge in Glen Lui, Cairngorms, localized metasomatism of schists has produced aegirine-augite and epidote granulites containing yellow andradite (an_{63}, gr_{25}) moulded on colourless cores of almandine. Andradite also forms large ragged porphyroblasts and in the aegirine-augite granulite develops well formed crystals in trails through the rock. Sector-twinned andradite crystals, up to 2 mm across, are present in associated feldspathic veins (McLachan 1951).
Highland The alkali syenite complex centred on Loch Borralan, Assynt, contains a rock type termed borolanite, essentially composed of orthoclase and melanite (a black, brown in thin section, titanium-rich variety of andradite). Localities include Allt a' Mhuilinn (Stewart 1941, Woolley 1973); Cnoc Bad na h'Achlaise; Ledmore River, W of Loch Borralan; Sgonnan Mòr (all Shand 1910b); Cnoc na Sròine, E of

Cluster of lustrous orange andradite crystals, up to 1.1 mm across, associated with white, acicular byssolite; Coatsgate quarry, near Moffat, Dumfries & Galloway, Scotland. (© D. I. Green)

Complex zonation in an andradite (light grey) to andradite-grossular (darker grey) crystal from an albite-bearing hornfels in the aureole of the Mourne Mountains granite. Aughrim quarry, Kilkeel, Co. Down, Ireland; width 600 μm; BSE image. (© A. G. Tindle)

Ledmore (Horne & Teall 1892, Teall 1900, Shand 1910b, Sabine 1953, Harker 1954a). Teall (1900) also notes that borolanite is extensively developed to the E of Aultnacealgach on the northeast side of the Oykel Bridge to Inchnadamph road. A nepheline syenite from the Loch Borralan complex contains andradite (an_{75}, gr_{13}) (Howie & Woolley 1968), whereas borolanite contains andradite (an_{73-87}, gr_{5-21}) (A. Dyer specimen, Tindle unpub. EMP data). A garnet pyroxenite from the SW of the complex contains coarse-grain anhedral vanadium–zirconium-rich andradite (an_{79}, gr_{13}) (Notholt et al. 1985). In the same general area, andradite (melanite) is reported from ledmorite dykes at Camas Eilean Ghlais, Rubha Mòr, Coigach (Teall 1900, Sabine 1953, Tilley 1958), and Loch Roe, Achmelvich, An Fharaid Mhòr, near Lochinver (Tilley 1958). Melanite is noted from the north and east margins of the Cnoc nan Cuilean syenite intrusion, adjacent to the Ben Loyal complex (Gallagher et al. 1971) and is also found as strongly zoned crystals with variable compositions (an_{85-98}, gr_{2-15}) in skarns within the Beinn an Dubhaich aureole, Broadford area, Isle of Skye (Smith et al. 2004). At Camasunary Bay, Strathaird Peninsula (also on the Isle of Skye), andradite (an_{47-67}, gr_{14-33}) occurs with end-member grossular and schorlomite in calcic skarns found at the margin of the Cuillin gabbro complex. Associated species include calcite, melilite, spurrite, tilleyite, vesuvianite and wollastonite, together with minor amounts of strontium-rich apatite, magnetite and pyrite. Rare baghdadite, djerfisherite, kalsilite, kimzeyite, perovskite and spinel complete the assemblage (Beard & Drake 2007).

Strathclyde Euhedral crystals of andradite up to 10 mm across have been found in what is thought to be metamorphosed basalt associated with the skarn deposit at Coire Buidhe, Glen Creran (Ingram et al. 1993, Green & Todd 2001). Brown and yellow andradite (an_{81-100}, gr_{0-19}) and grossular garnet crystals up to 5 mm across occur as sparkling crystalline crusts lining vesicles in Clyde Plateau Lavas at Loanhead quarry, near Beith, where they form part of a low-temperature hydrothermal assemblage that also includes fibrous amphibole, analcime, calcite, epidote, hematite, prehnite, pumpellyite and thomsonite (Livingstone 1989a, Meikle 1989b, Green & Todd 2001). Similar lavas at Reilly quarry, Houston, contain well formed, pale honey- to yellowish-brown andradite crystals, up to 0.2 mm across, associated with calcite and larger crystals (1.5–2 mm) associated with analcime and epidote (Meikle 2003). At Creag an Fheidh on the Isle of Arran, andradite (an_{61-100}, gr_{0-41}) and grossular occur in a skarn assemblage at the junction between vent pyroclastics and a xenolithic Cretaceous chalk block that subsided into the collapsed caldera of the Central ring complex. Andradite is massive here and it varies from dark yellow through yellowish-green to brown (Cressey 1987).

Ireland

Andradite (melilite) and grossular have both been found in thermally metamorphosed impure limestones close to a syenite intrusion at Grange Cross, near Barnavave, Carlingford, Co. Louth. Specimens of pure andradite are not uncommon (Nockolds 1947). An unusual example with an end member andradite core composition (an_{100}) has been found in an albite-bearing hornfels from the metamorphic aureole of the Mourne Mountain granites at Aughrim quarry, Kilkeel, Co. Down. The euhedral crystal has a complex repeated zonation pattern, between an_{100} and (an_{65-68}, gr_{32-35}) compositions (S. Moreton specimen, Tindle unpub. EMP data). Some vesuvianite described by Osborne (1932b) is actually andradite (melilite). Widespread secondary andradite–grossular–hydrogrossular (an_{62}, gr_{44}) garnet,

Equant crystals of andradite enclosed by prehnite, infilling a vesicle in basalt. Loanhead quarry, near Beith, Strathclyde, Scotland; width 2.4 mm; viewed in PPL (left) and XPL (right). (© A. G. Tindle)

epidote, prehnite and titanite occur throughout the Galway granite and its satellite plutons, and in some surrounding country rocks. The assemblage formed as secondary fluids passed through and altered the rocks (Leake 1998).

andrewsite

TL West Phoenix mine, Liskeard, Cornwall

Named after Thomas Andrews FRS (1813–1885), a medical doctor, fellow of the Royal Societies of Edinburgh and London, and professor of chemistry at Queen's College, Belfast; born in Belfast. Andrewsite is a discredited species, shown by Dunn (1990) to be a mixture of barbosalite, chalcosiderite, hentschelite and rockbridgeite.

anemolite

A variety of stalactitic calcite found in Derbyshire caves.

anglesite

BARITE GROUP
orthorhombic $PbSO_4$

TL Parys Mountain, Anglesey, Gwynedd, Wales

A relatively common lead sulphate found in the oxidized zone of lead deposits, where it replaces galena in acidic environments. Anglesite is sometimes an important ore mineral.

England

Avon At Stancombe quarry, near Flax Bourton, Bristol, anglesite occurs with cerussite, calamine, native sulphur and rare greenockite in fractured galena pods within barite–calcite veins cutting a Carboniferous Limestone cave system (Alabaster 1983).
Cheshire Triassic redbed copper–lead mineralization at Alderley Edge ridge, has been subjected to extensive alteration, resulting in the formation of many secondary carbonates, sulphates and sulphides. At Engine Vein mine this has involved the replacement of galena by anglesite and cerussite (Carlon 1979, Ixer & Vaughan 1982).
Cornwall Anglesite and galena are noted from copper lodes at Angarrack (also known as Mellanoweth mine or Wheal Maggot), E of Hayle, Gwinear (Greg & Lettsom 1858, Dines 1956); anglesite has been found as small greenish crystals in cavities within jamesonite at Bodannon antimony mine, Port Gaverne. Nadorite and senarmontite are associated species (Russell 1927a, 1949b); slag-like specimens collected from the wreck of the fireship *Firebrand* (burned in Falmouth harbour *c.* 1780) contain fused lead on which exceptionally well defined and brilliant crystals of anglesite and cotunnite have grown. The anglesite is colourless or black and reaches 5.5 mm in length (Russell 1920); colourless-to-grey, equant to prismatic, anglesite occurs in oxidized galena–quartz veinstone of the Greystone lead lode at Greystone quarry, Lezant, near Launceston. Opaque white-to-grey and yellowish blocky crystals to 10 mm are also noted (Lloyd & Barstow 1983, Wolfe 1997, Elton et al. 2004); a galena–quartz veinlet at Gunver Head in the area of Padstow Consols (Cuddrabridge) mine was found to contain anglesite associated with cerussite, cumengéite, diaboléite, phosgenite and native sulphur (Braithwaite & Merry 1996); anglesite forms rare colourless striated crystals at Hingston Down Consols mine, 2 km W of Gunnislake, Calstock (Russell 1910a); one fractured crystal of anglesite was found in a trial pit on an antimony lode at Linnick, South Petherwin (Russell 1949b); alteration of a primary Cu–Pb–Sb sulphide vein assemblage at Newporth beach (2 km SW of Falmouth) has produced secondary anglesite, atacamite, bindheimite, cerussite, cumengéite, goethite and malachite (Dean 1982); euhedral, tabular to prismatic and poorly crystallized aggregates of anglesite on galena are reported from Penberthy Croft mine, St Hilary (Betterton et al. 2000); Pendogget mine, near Port Isaac, and Wheal Penrose, Sithney, near Helston (Robson 1949b); small crystals of anglesite associated with senarmontite have been found in cavities within jamesonite at Port Quin mine (Russell 1944, 1949b); at Trevinnick mine, St Kew, anglesite forms bright pale-yellow to colourless tabular chisel-shape crystals in cavities within corroded galena (Fisk et al. 1982); Wheal Boys, Trewetha, St Endellion (Russell 1949b); a large tabular crystal of anglesite associated with laurionite and phosgenite is described from a cavity within limonite on a specimen probably from the lead–silver mine of Wheal Rose, Sithney (Russell & Hutchinson 1922, 1927, Russell 1944, Dines 1956); and white-to-colourless euhedral crystals of anglesite are closely intergrown with metatorbernite, pyrite and uranium-rich hydrocarbons at Wheal Speed, St Ives (Harrison & Taylor 1966).
Cumbria *Caldbeck Fells* The South lode exposed in Balliway Rigg, near Thief Gills, higher Roughton Gill produced abundant small crystals and crusts of anglesite, caledonite, cerussite, leadhillite and susannite (Green & Briscoe 2002); remarkable intergrown masses of cream to black bladed anglesite to 25 mm are associated with bindheimite, cerussite, galena and pyromorphite at Brae Fell mine (Greg & Lettsom 1858, Hartley 1984, Cooper & Stanley 1990, Bridges et al. 2006); anglesite forms small brilliant crystals in decomposing galena at Driggith mine (Davidson & Thomson 1951, Stanley & Vaughan 1981, Hartley 1984); Mexico mine, Todd Gill (Greg & Lettsom 1858); Red Gill mine has produced the best anglesite in the Caldbeck Fells, which occurs as small bladed water-clear lustrous crystals associated with caledonite or linarite and as yellow bipyramidal crystals up to 10 mm long (Davidson & Thomson 1951, Wolloxall 1981, Briscoe 1986a, Cooper & Stanley 1990); rare, but dull, opaque crystals to 25 mm were once found at Roughton Gill (Davidson & Thomson 1951, Cooper & Stanley 1990); small crystals of anglesite were found in altered galena from Short Grain mine (Young 1987a, Cooper & Stanley 1990, Green & Briscoe 2002); anglesite forms transparent tablets with a rhombic outline, occasionally altering to leadhillite and associated with linarite at Silver Gill mine, near Roughton Gill (Cooper & Stanley 1990, Green et al. 2000b). Reports of anglesite from Brandy Gill, Old Potts Gill and Sandbed mines (Hartley 1984, Cooper & Stanley 1990) require further investigation, as they are thought to be based on material that may have been incorrectly provenanced by A. W. G. Kingsbury (see Ryback et al. 1998, 2001). Green et al. (2006a) agree with earlier workers in suspecting the provenance of the Kingsbury specimens, adding that anglesite, forming small, colourless, euhedral, blocky to prismatic crystals associated with oxidized galena, are common at Driggith and Sandbed mines. *Elsewhere in the Lake District:* Brandlehow mine, Derwent Water (Young 1987a); very rare minute lustrous transparent equant crystals associated with copper, lead and zinc sulphates and as powdery encrustations on pyromorphite at Force Crag mine, Braithwaite, near Keswick (Green et al. 1997); Goldscope mine, Newlands Valley, near Keswick (Davidson & Thomson 1951); Greenside mine, Glenridding (Greg & Lettsom 1858). *Northern Pennine orefield:* Small crystals at Brownley Hill mine, Nenthead, and at Tynebottom mine, NW of Garrigill (Bridges & Young 1998); small tabular transparent crystals on decomposing galena at Hilton mine, Scordale, near Appleby (Winrow 1986a, Bridges & Green 2005b).
Derbyshire Well developed crystals of anglesite (up to 10 mm long), native sulphur and wulfenite were found in a small flat (stratiform replacement orebody) of galena in an abandoned fluorite working above Ball Eye quarry in the Via Gellia, near Cromford (Bridges 1983); anglesite, matlockite and phosgenite from the Bage mine area, Bole Hill, Wirksworth, have strongly yellow fluorescence under short-wavelength UV light (Burr 1992); anglesite forms colourless blades in galena at Dene quarry, Cromford (Braithwaite 1983); fluorite–galena veinstone at Milltown quarry, near Ashover contains microcrystals of anglesite, cerussite and native sulphur (Rust 1983b, Clark et al. 1984); the main locality in Derbyshire for anglesite is Ranter mine (Rent Tor), in the Gulph, NW of Wirksworth, where blades up to 10 cm long were once found (Greg & Lettsom 1858, Bancroft 1973); well formed anglesite was found in short-lived exposures at Sallet Hole mine, Longstone Edge, and Mill Dam mine, Hucklow Edge. Anglesite crystals to 50 mm were also found in cavities where galena was encased entirely by barite or fluorite (Jeffrey 2001); small colourless bladed crystals of anglesite to 2 mm are associated with galena and linarite in sandstone from Snelston mine, near Ashbourne (Jackman 1996); anglesite is also found on Brassington Moor; at Crich; and at Ladywash mine (Old Edge vein), Eyam (Ford & Sarjeant 1964a, Ford et al. 1993); and Gaines et al. (1997) and Anthony et al. (2003) list Matlock as an additional anglesite locality.
Devon Antimony–gold mineralization in quartz–carbonate veins cutting Devonian slates at Loddiswell mine, 3 km N of Loddiswell contains minor amounts of anglesite, cerussite and malachite as supergene and oxidative alteration products of primary sulphides (Stanley et al. 1990b). Fine colourless crystals of anglesite are reported from cavities in galena at East Tamar mine, Bere Ferrers, and also at Bere Alston (Hall 1868b). Rare anglesite is noted from the antimony mine of Wheal Emily, Wembury (Russell 1949b).
Durham Crusts of bladed crystals up to 5 mm long and cm-size anglesite crystals have been found at Closehouse opencast barite mine, SW of Middleton-in-Teesdale, Lunedale (Vaughan & Ixer 1980, Bridges & Young 1998, Dunham et al. 2001, Green & Briscoe 2002); lustrous, rhombic and bladed crystals of anglesite to 4 mm occur on the dumps of trial workings on Drypry vein, N of Wolsingham (Bridges & Young 1998); well developed crystals of anglesite have been recovered from old workings at Pike Law, Newbiggin (Dunham 1990, Bridges & Young 1998); at Rogerley mine, near Frosterley, Weardale, galena initially alters to fine-grain banded anglesite and then to cerussite (Ixer 2003); and galena alters to anglesite and cerussite in vein and metasomatic ores from the Teesdale Valley. Localities include Coldberry, Eggleshope mine, Flushiemere mine, Middleton Common and Newbiggin Common (Vaughan & Ixer 1980).
Shropshire Rare anglesite is noted in oxidized chalcopyrite–galena–sphalerite ore at Ladywell mine, Shelve (Braithwaite 1982b); veins and vein breccias at Snailbeach

and Tankerville mines in the West Shropshire orefield contain secondary anglesite, cerussite, covellite, goethite and smithsonite replacing a primary galena–sphalerite ore (Pattrick & Bowell 1991).

Yorkshire Very attractive specimens of anglesite and cerussite have been found in heavily oxidized galena-bearing veins at Coldstones quarry, 5 km W of Pateley Bridge (Bridges 1997a). Anglesite and cerussite replace minor galena at Pikedaw mine, between Malham and Settle (Green et al. 2006c).

Wales

Dyfed *Central Wales orefield:* Bwlchglas mine, near Tal-y-bont – microscopic crystals of anglesite with a long prismatic habit, in cavities within galena (Braithwaite 1982a, Rust 1996c); Bwlchrhennaid mine, near Goginan; Cwmystwyth mine; Darren mine, Pen-bont Rhydybeddau and Glogfawr mine, Devil's Bridge (all Bevins 1994); Eaglebrook (Nantycagl) mine, 1 km NW of Nat-y-moch reservoir – small sharp crystals and blocky white crystals of anglesite up to 7 mm long (Braithwaite 1982b, Jones 1983); Esgair Hir mine, 1 km NW of Nat-y-moch reservoir – selenium-bearing anglesite (Rust 1990c, Bevins 1994); Frongoch mine, between Devil's Bridge and Pont-rhyd-y-groes – attractive colourless microcrystals (< 1 mm) and rarer bladed or prismatic crystals of anglesite up to 3 mm long are associated with covellite; they line cavities in galena or develop in veinlets on drusy quartz (Bevins 1994, Green et al. 1996b); Llechweddhelyg mine, near Clawddmelyn, between Bont-goch and Cwmsymlog – minute gem-like yellow-to-clear crystals on corroded galena (Jones 1987); Ystrad Einion mine, near Tal-y-bont – rare anglesite in oxidized galena (Mason & Rust 1997).

Gwynedd Historical records describing the giant copper deposit at Parys Mountain, Anglesey, indicate that anglesite occurred in considerable quantity, but was mainly restricted to the supergene portion of the Great Lode. It occurred in a range of forms, including pinacoids, bipyramids and prisms, as well as in complex combinations. A fine collection of specimens is in the NMW (Southwood & Bevins 1995). Discrete euhedral milky-white to rich-yellow crystals, up to 10 mm in length, on ferruginous gossan are reported from this locality by Greenly (1919) and Bevins (1994). Anglesite is also noted from Bwlch mine, near Deganwy (Russell 1944, Russell 1949b, Ryback & Francis 2001), and Gorlan mine, between Betws-y-coed and Trefriw (Russell 1944); and Llanengan mine, Lleyn (Bevins 1994).

Mid Glamorgan Impressive euhedral colourless crystals of anglesite were collected from a cavity in a galena-bearing vein cutting Carboniferous Limestone at Machen quarry, 7 km ENE of Caerphilly. Anglesite is also noted from Rhyd-y-Gwern mine, near Machen, and Hendy quarry, Miskin, near Llantrisant (Bevins 1994).

Powys A barite–chalcopyrite–galena–tennantite vein temporarily exposed at Dolyhir quarry, Old Radnor, near Presteigne, has produced anglesite, azurite, cerussite, native copper, cuprite, harmotome, malachite, olivenite and tyrolite (Wolfe 2000). Anglesite is also noted from three mines in the Powys section of the Central Wales orefield: Dylife mine, N of Hafren Forest; Geufron mine, near Llanidloes; and South Nantycar, near Rhayader (Bevins 1994).

Scotland

Dumfries & Galloway and Strathclyde Anglesite is found in many of the lead veins of the Leadhills–Wanlockhead orefield, but it is most impressive from the Susanna and Belton Grain veins, where it forms long bladed crystals. Commonly it is colourless or white-to-grey, but may be yellowish, brown or black, the last attributable

Anglesite from Parys Mountain, Anglesey, Gwynedd, Wales; crystals up to 10 mm in length. (© National Museum of Wales, Cardiff; photographer M. P. Cooper)

Lustrous group of striated anglesite crystals from Susanna mine, Leadhills, Strathclyde, Scotland; specimen is 7 cm tall. (© M. P. Cooper)

to disseminated galena. Cerussite, lanarkite and linarite are associated species. Some fine specimens are in the collection of the NMS (Heddle 1901, Wilson & Flett 1921, Gillanders 1981, Livingstone 2002). Anglesite is also noted from Margaret's (Strait-step) vein (Brown 1919) and High Pirn vein (Heddle 1901), both Wanlockhead.

Ireland

Dublin A galena-bearing barite–quartz vein exposed on the coast at Killiney contains a suite of supergene species including anglesite, cerussite, phosgenite and malachite (Green 1998c). Anglesite is also noted from Ballycorus mine, near Carrickgollogan (Greg & Lettsom 1858).

Galway Limestone-hosted galena–marcasite–pyrite–sphalerite ore at Tynagh mine, near Killimor, contains a supergene assemblage that includes well formed white anglesite crystals up to 25 mm long and, not particularly well formed, translucent crystals up to 3 cm long. Azurite, cerussite, malachite and smithsonite are associated secondary species (Schultz 1971, Wolfe 1997, Green & Moreton 2001).

Tipperary A near-surface gossan at Ballygown South mine, near Silvermines, contains cerussite with minor intermixed anglesite above an epigenetic sulphide zone (Boland et al. 1992, Moreton 1999). A depth sequence of secondary species, from this mine and the nearby Knockanroe mine, puts siderite at the greatest depth, with smithsonite, hemimorphite, anglesite, cerussite and then goethite forming towards the surface (Rhoden 1959).

Wicklow Anglesite is noted with cerussite from the Glenmalure lead mines (Greg & Lettsom 1858).

Offshore territories

Jersey At Le Pulec, on the northwest coast, a polymetallic dolomite–quartz vein system dominated by sphalerite contains thin veinlets of galena partially altered to anglesite, bournonite, cerussite and covellite (Stanley & Ixer 1982).

Sark Anglesite, cerussite and leadhillite have all been found near the surface of Sark's Hope mine, Little Sark, where they have replaced primary silver-rich galena. Examination of a single lead-rich tennantite specimen shows that the secondary assemblage forms a complex network of veinlets entirely filling and cementing fractures in earlier-formed species, with anglesite being one of the last species to crystallize (Mourant & Warren 1934, Bishop et al. 1977).

anhydrite

orthorhombic $CaSO_4$

A very common calcium sulphate that occurs as a major component of sedimentary evaporite deposits. Permian evaporites underlie much of North Yorkshire and are extensive beneath the North Sea. Because of the economic importance of anhydrite in these deposits, a significant volume of research exists on its occurrence. However, the species does not feature highly in the collector community, not least because of its tendency to hydrate at room temperature.

Anhydrite and halite are the major constituents of a Permian evaporite sequence discovered in the Aislaby (Eskdale No. 2) borehole, Eskdale, near Whitby, Yorkshire. The lowest 76 m of this 1540 m borehole is formed almost entirely of anhydrite. Much of the anhydrite is white or grey, but some types have a slightly blue tint. Grainsize varies from 0.002 mm to 1 cm, grains taking the form of irregular interlocking grains, elongated grains in radiating groups, fan-like plates with curved cleavages (up to 1 cm in length), rectangular euhedral crystals and tabular euhedral crystals occurring as porphyroblasts in polyhalite. Recrystallization of anhydrite is widespread, particularly when it is associated with halite. Anhydrite replaces dolomite and is itself replaced by halite, magnesite, polyhalite and native sulphur. Anhydrite can also replace gypsum or be replaced by gypsum (Lee & Taitt 1946, Stewart 1949, 1951a,b). Later borings proved abundant anhydrite at other locations in Eskdale (No. 3 Sleights, No. 4 Sneaton and No. 6 Upgang) and confirmed the extent of the evaporite sequence in the Whitby area (Armstrong et al. 1951). Unusually coarse tabular anhydrite to 3 cm has since been found in veins and vugs in the RHB1 borehole to the S of Whitby at Robin Hood's Bay, Yorkshire (Raymond 1962). Another borehole (F.1) at Fordon in Humberside contained a 430 m-thick evaporite sequence in which anhydrite is a major component, with one subzone consisting of 95% anhydrite. Dolomite, halite, kieserite and polyhalite are the other major constituents, with minor amounts also noted of aphthitalite, celestine, glauberite, magnesite, pyrite, sulphoborite, sulphur and talc (Stewart 1963, 1965). Anhydrite nodules with barite, galena, marcasite and sphalerite mineralization are present in Permian dolomitic limestone at Wormald Green, S of Ripon (Harwood & Smith 1986, Leake et al. 1997b).

The Permian evaporite sequence extends into Cleveland, where a 6 m-thick seam (the Billingham Main Anhydrite) carrying 90.5% anhydrite (plus minor gypsum) was mined until 1971 under the ICI factory site at Billingham-on-Tees (Dunham 1960, 1978). Nodular anhydrite from this mine was studied by Holliday (1973), who concluded it formed pseudomorphically after gypsum to produce a wide range of textures, including ragged laths and a wheatsheaf fabric. Perhaps describing similar material from Billingham and Wilton, near Guisborough, Raymond (1962) notes inclusions of anhydrite, up to 30 cm across, consisting of equant anhedral crystals of anhydrite, sometimes with a fibro-radiate structure. The Billingham Main Anhydrite is also encountered at Boulby potash mine, near Loftus, famous for spectacular blue boracite samples, although these were found higher in the sequence in the Boulby Potash, a unit relatively poor in anhydrite (Milne et al. 1977, Green & Freier 1994, 1996). Woods (1973, 1979) describes the Upper Anhydrite (a 7.5 m-thick unit above the Boulby Potash) at Boulby potash mine as a well bedded, very fine-grain pale-grey anhydrite bed containing lenses of sylvite, with halite pseudomorphs after gypsum occurring locally.

Anhydrite occurs in Upper Permian evaporite rocks in the deeper levels of the gypsum mines and quarries around Kirkby Thore, Cumbria: Acorn Bank mine, Temple Sowerby (Sherlock & Hollingworth 1938); Birks Head mine, Long Marton, Appleby (Dunham & Hollingworth 1947, Davidson & Thomson 1951); quarry at Kirkby Thore (Trechmann 1901); borehole to the E of Kirkby Thore (Worley 2005); New Stamps mine (Sherlock & Hollingworth 1938, Dunham & Hollingworth 1947); and Thistle Plaster mine (Sherlock & Hollingworth 1938). Other Cumbrian occurrences include: Barrowmouth mine, Whitehaven (Eastwood et al. 1931); Cockslake mine, Cumwhinton (Meyer 1965); Knot Hill mine, Cotehill (Davidson & Thomson 1951); Langwathby borehole (Holliday 1973); Long Meg mine, Lazonby (Sherlock & Hollingworth 1938); Saltom Bay, St Bees Head (Arthurton & Hemingway 1972); and Sandwith mine, Whitehaven (Dunham et al. 1978, Harrison 1975). Anhydrite is also found in fractured Borrowdale Volcanic Group rocks, Carboniferous Limestone, Brockram, St Bees Evaporite and St Bees Shale examined in borehole material from the Sellafield area of Cumbria, where it forms fibrous veins or bladed crystals pseudomorphed by calcite (Milodowski et al. 1998).

Elsewhere in England, anhydrite is reported from: Cropwell Bishop, Nottinghamshire (Richardson 1921); Mountfield mine, Battle, Sussex (Young et al. 1985c); Brightling mine, Sussex (Holliday 1973) and from boreholes including: Fairlight bore, Sussex (Holliday 1973); Hathern No. 1 bore, NW of Loughborough, Leicestershire (Llewellyn & Stabbins 1970).

Anhydrite is a common gangue mineral in massive sulphide ore from the Morfa Du lode, Parys Mountain, Anglesey, Gwynedd, where it forms as a late infilling within mineralized siliceous sinter (Pointon & Ixer 1980, Southwood & Bevins 1995). Anhydrite also occurred in certain Triassic evaporite horizons at Sully, South Glamorgan, Wales, but is no longer present because of later replacement (Tucker 1977, Waters & Lawrence 1987).

In Scotland, anhydrite is reported from a phonolite at Traprain Law, near Haddington, Lothian (MacGregor 1960), and from Lower Carboniferous nodular carbonate beds of calcite and dolomite in the Tweed area, Burnmouth, Borders (Livingstone 2002).

Pale-blue fibro-lamellar translucent anhydrite is noted from a trap rock on Cave Hill, near Belfast, Co. Antrim, Ireland (Greg & Lettsom 1858). The Permian Belfast Harbour Evaporite Formation, found in a borehole at Belfast harbour, contains finely crystalline pink or white anhydrite in lenses within red silty mudstone. Gypsum is an associated species (Smith 1986). Anhydrite, gypsum and calcite occur in Triassic sediments adjacent to a Tertiary dolerite sill encountered in a borehole to the SW of Flush Hall, Ballyalton, Newtownards, Co. Down (Knill 1960).

anilite

cubic Cu_7S_4

A rare copper sulphide found in a shattered quartz vein at Castletown mine, Ballimore, SSE of Lochgilphead, Strathclyde, Scotland. Anilite occurs as silvery-to-black metallic masses up to 1 cm across, intergrown with chalcopyrite and quartz, and is extensively altered to green supergene minerals. Associated species include botallackite, clinoatacamite, connellite, goethite, hematite and vésigniéite. Grinding for XRD analysis can cause anilite to convert to digenite or djurleite (Starkey & Hubbard 1998, Green & Todd 2001, Hubbard & Green 2001).

ankerite

DOLOMITE GROUP

trigonal $Ca(Fe^{2+},Mg,Mn)(CO_3)_2$

A common calcium–iron–magnesium–manganese carbonate that forms two series, with dolomite and with kutnohorite, and is easily confused with dolomite. It is found in hydrothermal sulphide veins, in low-grade metamorphosed ironstones, in sedimentary BIFs, and in authigenic and diagenetic carbonate sediments. Ankerite commonly occurs in joints within coal seams and has been reported from the Lancashire,

Leicestershire, Nottinghamshire, Yorkshire and South Wales coalfields (Crook 1912).

England

Cornwall Ankerite forms thin veins cutting tourmaline-bearing adinoles (albite-rich rocks) at Dinas Head, near Padstow (Agrell 1939, 1941); small brown thin saddle-shape crystals of ankerite are associated with galena–stibnite vein mineralization at Dizzard (or Duchy and Stuckey United) mine, Poundstock (Russell 1949b); ankerite occurs with calcite, siderite and quartz in the gangue of galena–sphalerite veins at Old Treburgett mine, near St Teath village, E of Port Isaac (Kingsbury & Hartley 1960); lenticular crystals of ankerite, up to 200 µm across have grown on siderite in association with bismuthinite and clinochlore in cavities within quartz veinstone at Penberthy Croft mine, St Hilary (Betterton et al. 2000). Ankerite is also noted from: Port Quin mine, near Port Quin; Tredinnick mine, Landrake with St Erney; Trenute mine, near Lezant and Wheal Boys, Trewetha, St Endellion (Russell 1949b); Herodsfoot mine, Lanreath, near Liskeard, and Tolpetherwin, NE of Two Bridges, South Petherwin (Dines 1956); Eastern Cliff, Kennack; Greystone quarry, Lezant; Poltreworgey mine, St Endellion; Trebullet mine, Lezant; and Tresungers mine, St Endellion (Golley & Williams 1995).

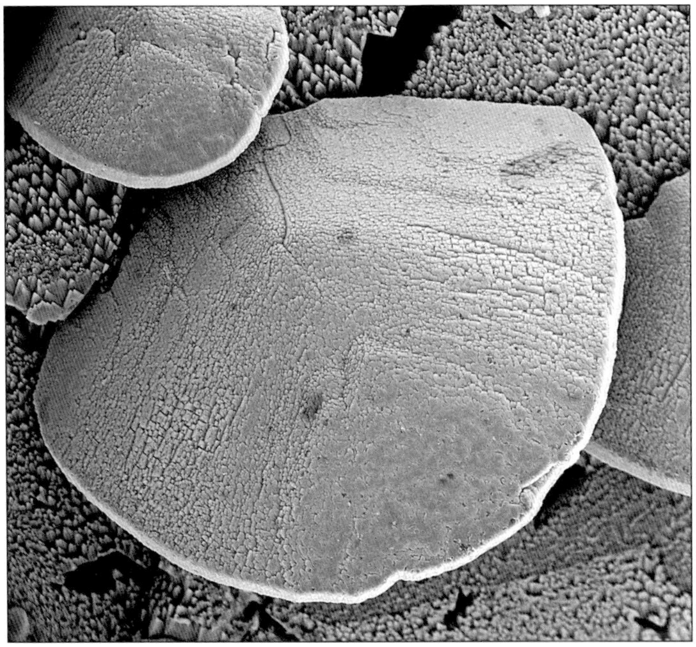

Lenticular crystals of ankerite on siderite; Penberthy Croft mine, St Hilary, Cornwall; width 220 µm; SEM image. (© J. J. Hooper)

Cumbria *Lake District:* Crystallized ankerite at Brandy Gill mine, Caldbeck Fells (Davidson & Thomson 1951); ankerite has reacted with wolframite to form scheelite at Carrock mine, Grainsgill, Caldbeck Fells, and it also occurs with quartz in fault zones and in cross-cutting lead–zinc veins (Smythe & Dunham 1947, Davidson & Thomson 1951, Eastwood 1959, Ewart 1962, Shepherd et al. 1976, Fortey 1980, Beddoe-Stephens & Fortey 1981); Driggith and Sandbed mine, Caldbeck Fells (Green et al. 2006a); Dry Smale Gill, Caldbeck Fells (Shaw 1970); Force Crag mine, Braithwaite, near Keswick, worked a lead–zinc vein in which barite was the major gangue species and where ankerite, fluorite, siderite and quartz were also found (Rastall 1942, Green et al. 1997); Greenside mine, near Glenridding, Patterdale (Firman 1978); Potts Gill mine, Caldbeck Fells (Hartley 1984); Sellafield area boreholes (Milodowski et al. 1998); and ankerite is possibly present as an alteration product of microgranite at Threlkeld quarry, Threlkeld (Rastall 1941). *Northern Pennine orefield:* Most of the workings of Brownley Hill lead mine, Nenthead, near Alston, are dominated by ankerite-bearing veins or replacement-flat deposits. Ankerite commonly lines cavities as white, brown, cream, red or yellow rhombic crystals up to 1 cm across (Young et al. 1990b, Wood 1991, Green & McCallum 1993, Green et al. 2000, Green & Briscoe 2002); well crystallized ankerite (and galena) occurs at Rampgill mine (Proud's Sump flat), Nenthead and Smallcleugh mines, Nenthead, near Alston (Smythe & Dunham 1947, Green & Briscoe 2002); ankerite, calcite, chalcedony, fluorite and quartz are the gangue minerals to galena–sphalerite mineralization at Tynebottom mine, NW of Garrigill, near Alston (Braithwaite 1982b).

Devon Ankerite, siderite (chalybite) and quartz form the gangue to massive jamesonite–semseyite ore at Wheal Emily, Wembury (Russell 1949b).

Dorset Iron-bearing cement infilling pores of uncompacted mud produced during diagenetic alteration of organic material in Kimmeridge Clay at Kimmeridge Bay has a range of compositions from iron-rich dolomite to ankerite (Irwin 1981).

Durham Metasomatic flats in the Carboniferous Great Limestone, once worked for galena, contain ankerite, calcite, fluorite, quartz and siderite gangue at Boltsburn mine, Rookhope, and display interlocking rhombs of ankerite and siderite at Carricks ironstone mine, Wearhead (Smythe & Dunham 1947); ankerite is a minor component of barite mineralization associated with the Whin Sill (quartz dolerite) at Closehouse mine, SW of Middleton-in-Teesdale, Lunedale (Hill & Dunham 1968), and at Cowgreen mine, Harwood-in-Teesdale (Smythe & Dunham 1947). Ankerite is also noted from the following lead veins or mines: Flushiemere mine, Newbiggin Common; Greenhurth mine, Upper Teesdale; Middlehope mine, Weardale; Sedling mine, Weardale; St Peter's mine, N of Allenheads; Westernhope mine, Snowhope Moor; Wynch Bridge, Teesdale (Dunham 1990); West Rigg quarries, N of Westgate, near St John's Chapel (Young 1995); and Frazer's Hush mine, Weardale (Green & Briscoe 2002). In the Durham coalfield, ankerite veinlets have been found in borehole cores from Easington colliery (Smythe & Dunham 1947) and at New Brancepeth colliery, near Durham, ankerite has been replaced by pyrite and marcasite in an assemblage dominated by barite, galena, sphalerite and witherite (Ixer 1986).

Lancashire Ankerite is reported without detail from the Oldham area (Anthony et al. 2003).

Leicestershire Spheroidal carbonate-rich nodules in Triassic sandstone at Newhurst quarry, Shepshed appear to be cemented by ankerite (Ince 2005).

Northumberland Ankerite occurs in partings within coal seams at Scremerston colliery near Tweedmouth, at Newbiggin colliery, and on the foreshore at Newbiggin-by-the-Sea (Smythe & Dunham 1947). A related occurrence is reported from along the coast, to the N of St Mary's Island, where massive brown or white ankerite occurs with calcite in vertical veins cutting Coal Measures sandstones (Hawkes & Smythe 1935). At West Sleekburn colliery, Bedlington, ankerite has been found in a large gas cavity within a dolerite dyke, where it is associated with marcasite and pyrite (Randall 1953). In the Northern Pennine orefield near Hexham, ankerite occurs in close proximity to the Whin Sill at Settlingstones mine (Ineson 1972) and at Stonecroft mine (Dunham 1990).

Yorkshire Brownish dolomite is the commonest mineral in limestones adjacent to lead veins in the Askrigg Block. However, ankerite is reported from: Hartley Birkett opencut; Stony Grooves mine (near Greenhow), Pateley Bridge; and the Hurst mines on Fremington Edge (Dunham & Wilson 1985).

Wales

Dyfed *Central Wales orefield:* Ankerite (iron-rich dolomite) forms light- to dark-brown masses or small rhombohedral crystals. In some orebodies it forms much of the gangue, especially at Castell mine, Ponterwyd, near Aberytstwyth; Cwmbrwyno mine, near Devil's Bridge; Esgair Fraith mine, near Tal-y-bont; and Henfwlch mine, between Cerrig yr Hafan and Nant-y-moch reservoir, E of Tal-y-bont (Jones & Moreton 1977). At Eaglebrook (Nantycagl) mine, Ceulanymaesmawr ankerite is abundant on the dumps as light-brown crystalline cleavage masses (Jones 1983).

Gwynedd Ankerite is one of the cementing minerals in a pyritiferous shale breccia on which hydrozincite and namuwite have later grown at Aberllyn mine, 2 km N of Betws-y-coed, Gwydyr Forest (Bevins et al. 1982b); small rhombs of ankerite occur in altered ashflow tuffs of Ordovician age in association with antimony-rich mineralization (semseyite and stibnite) at Bwlch mine, near Deganwy (Russell 1944, 1949b); and ankerite and chlorite are alteration products of felsic and mafic volcanic rocks at Parys Mountain, Anglesey (Southwood & Bevins 1995, Tennant & Steed 1997).

Powys *Central Wales orefield:* Ankerite is described from Dyfngwm mine, Machynlleth (Rust & Rust 1987) (actually the Pen Dylife section of Dylife mine) and Snowbrook (Nantyreira) mine (Jones & Moreton 1977). Ankerite is also reported from the South Wales coalfield (North & Howarth 1928, Bevins 1994).

Scotland

Central Ankerite is a gangue mineral associated with copper–zinc stratabound mineralization within the Dalradian Ben Challum Quartzite at Creag Bhocan, SSW of Tyndrum (Scott et al. 1988).

Dumfries & Galloway White and brownish-white ankerite crystallized early in galena–nickeline–rammelsbergite-bearing ore found on the dumps of Glencrieff mine, Wanlockhead (Temple 1954b); specimens of antimony–lead ore on the dumps of Glendinning mine (Glenshanna Burn) contain ankerite, calcite, pyrite, semseyite, stibnite and valentinite. Later investigation of unmineralized greywacke and mudstone from the site revealed ankerite to be the major carbonate species present (Herbert Smith 1919, Duller et al. 1997).

Fife Late-stage void-filling ankerite in dolomitized Carboniferous Limestone from Boarhills, Kinghorn, Randerston and St Monans exhibit well developed sector zonation, sometimes combined with concentric zonation (Searl 1990).

Highland Fluid infiltration in magnetite-rich skarn at the margins of the Beinn an Dubhaich granite, Broadford area, Isle of Skye, has produced ankerite in retrogressive veins indicating a hydrothermal system with a significant magmatic component (Holness & Fallick 1994); ankerite occurs in a fault-bounded muscovite (fuchsite)–pyrite–quartz rock at An Leth Allt, north slopes of Sgùrr an Airgid, Loch Duich. Later work suggests that the muscovite is in fact a chromium illite and that formation occurred during low-temperature fault-related CO_2 metasomatism that postdated the regional Caledonian deformation and metamorphism (Clifford 1959, Clifford et al. 1999).

Strathclyde Ankerite is found with barite, calcite, chalcedony, dolomite, fluorite, goethite, gypsum, hematite, prehnite, quartz, romanèchite, tourmaline and wad in geodes within a trachyte sill at Craigmuschat quarry, Gourock (Nicholson & Durant 1986); pink and brownish-white ankerite is noted from the Laverock Hall and Meadowhead lead-bearing veins (respectively) at Leadhills (Temple 1955).

Shetland At the Tang of Norwick, Unst, ankerite forms readily cleavable bluish-grey crystals, up to 25 mm across, associated with talc in a vein at the junction of serpentine with mica slate (Heddle 1878a, 1901).

Ireland

Clare Disseminated framboidal marcasite–pyrite mineralization at Ballyvergin contains a gangue dominated by ankerite, barite, iron-rich calcite, chlorite–pennine, iron-rich dolomite and quartz (Andrew 1986).

Cork Redbed-hosted copper mineralization at Ballycummisk is noted to contain ankerite (Kinnaird et al. 2002).

Galway A range of carbonate compositions from ankerite to iron-free dolomite are found associated with Cu–Pb–Zn–(Ag) mineralization at Tynagh mine, near Killimor (Clifford et al. 1986).

Mayo Ankerite is commonly associated with copper mineralization within an altered acidic porphyry intrusion 8 km to the S of Charlestown (Poustie 1982).

Monaghan Ankerite–quartz veins containing arsenopyrite and pyrite are the principal locus of gold mineralization at Clontibret mine, Tullybuck (Morris et al. 1986).

Tipperary Syn-sedimentary barite–pyrite mineralization at Ballynoe mine, near Silvermines, contains massive patches of chalcopyrite with minor ankerite, galena, pyrrhotite and siderite (Mullane & Kinnaird 1998).

Waterford Ankerite has been found at one of the Knockmahon mines, 22 km SW of Waterford town (Moreton 2005a).

Offshore territories

Jersey An ankerite vein cutting andesite at Petit Port, Trinity, is host to chalcopyrite–pyrite mineralization (Ixer 1980).

annabergite

VIVIANITE GROUP

monoclinic $Ni_3(AsO_4)_2{\cdot}8H_2O$

An uncommon hydrated nickel arsenate that forms a series with erythrite and is formed by the alteration of cobalt–nickel-bearing arsenides and sulphides in the oxidation zone of hydrothermal mineral deposits.

At Penberthy Croft mine, St Hilary, Cornwall, annabergite forms greenish coatings on a dolomitic matrix and forms globular aggregates, to about 1 mm, after millerite in cavities within an arsenopyrite–chalcopyrite–sphalerite-rich matrix (Rust 1996a,d, Betterton et al. 2000). It occurs on nickeline at Wheal Jane, Kea, near Truro, and is also noted from: Pengelly mine, near St Ewe; St Austell Consols; Wheal Chance (South Roskear mine), St Austell (Garby 1848, Greg & Lettsom 1858, Collins 1892); Dolcoath mine, Camborne; East Pool mine, Illogan, near Redruth (Collins 1911a); South Terras mine, St Stephen-in-Brannel (Robson 1949b); and Fowey Consols, Tywardreath (Golley & Williams 1995). All are in Cornwall.

At Hilton lead mine, Scordale, near Appleby, Cumbria, annabergite was found in a small nickel deposit on Dow Scar level. It forms green-to-brown non-crystalline masses in cavities and occasionally clear green crystals. It is an alteration product of the primary species millerite and nickeline (Bridges 1982, Winrow 1986a, Bridges & Young 1998). A large cabinet specimen of annabergite, almost certainly the finest ever collected in Britain, was recovered from Hilton mine in the 1980s (Green & Briscoe 2002). Thin encrustations of annabergite occur at Smallcleugh mine, Nenthead, near Alston, Cumbria, together with other post-mining supergene minerals, notably, brianyoungite, devilline, namuwite, serpierite, schulenbergite and an unnamed zinc analogue of ktenasite (Bridges 1982, Young et al. 1985b, Green & Briscoe 2002). It is also noted in Cumbria from Driggith and Sandbed mines, Caldbeck Fells (Green et al. 2006a); Paddy End mine, Coniston, Eskdale (Young 1987a); Sandbeds Gill mine, near Bassenthwaite (Thomson 1997b, Green & Briscoe 2002); and Tynebottom mine, NW of Garrigill, near Alston (Green & Briscoe 2002). Annabergite forms thin pale-green coatings on calcite–magnetite–nickeline ore on the dumps of Lady's Rake mine, along Harwood Beck, Upper Teesdale, Co. Durham (Young et al. 1985b, Bridges & Young 1998).

Coatings of annabergite are associated with gersdorffite on fractures within hydrocarbons at Great Ormes Head, Llandudno, Gwynedd (Bevins 1994).

In Scotland, efflorescences of annabergite, erythrite, malachite and tyrolite (plus rare conichalcite and picropharmacolite) occur as oxidation products on the dumps of Alva mine, Silver Glen, Ochil Hills, Central Region. The mine is famous for the richest bonanza of native silver ever found in Britain, although the main ore minerals were mostly clinosafflorite and rammelsbergite (Heddle 1901, Moreton 1996, 2007). Another long-abandoned silver mine where annabergite has been reported is at Hilderston, Bathgate Hills, Lothian. Annabergite occurs as apple-green and pale-green to yellow and whitish spherules and coatings on barite with calcite, in a carbonate breccia. It also occurs as pale-yellow earthy balls up to 8 cm across. The primary ore consisted of galena and nickeline in a 1.8 m-wide quartz vein, with annabergite, erythyrite and native silver forming a secondary deposit (Heddle 1901, Wilson & Flett 1921, Ridgway 1980, Meikle 1994).

Around Needle's Eye, SW of Caulkerbush, Dumfries & Galloway, annabergite and erythrite (in stellate intergrowths) with associated zeunerite replace much of the dolomite in the neighbourhood of uraninite (pitchblende) veins. A wide range of supergene species are found associated with uranium-bearing veins on this stretch of the Solway coast and include atacamite, bismutite, boltwoodite, connellite, eulytite, kahlerite, lavendulan, metakahlerite, metazeunerite, mixite, schoepite, uranophane, vandendriesscheite and walpurgite (Williams 1965, Miller & Taylor 1966, Braithwaite & Knight 1990, Green & Todd 2001). Elsewhere in the region, annabergite has also been reported from: Talnotry (Williams 1965); Glencrieff mine, Wanlockhead (Temple 1955); and Menimuir Burn, Cassencarrie, near Creetown (Macpherson & Livingstone 1982, Walker 1988). Dudgeonite (the nickel-poor variety of annabergite), associated with galena, linarite and pyromorphite, is noted in cavities within nickeline from Pibble mine, near Creetown (Heddle 1889a, 1901).

At Sheefry Pb–(Ag) mine, 11 km S of Croagh Patrick, Co. Mayo, Ireland, annabergite and wroewolfeite are two supergene species derived from the oxidation of a primary ore consisting of chalcopyrite, galena and millerite, which are found in a mineralized fault infilled with calcite and minor quartz. Annabergite forms abundant pale-green encrustations on millerite-bearing veinstone (Moreton & Jackson 1991).

Analysis of pale-pink minerals from Sandbed mine, Caldbeck Fells, Cumbria, previously labelled erythrite, has shown them to be annabergite or köttigite (D. I. Green pers. comm. 2006). It is unlikely that these are the only misidentifications.

annite

MICA GROUP

monoclinic $KFe^{2+}_3AlSi_3O_{10}(OH)_2$

An uncommon potassium–iron aluminosilicate hydroxide found in magnesium-poor igneous and metamorphic rocks.

Biotite with an 85% annite component is reported from a body of BIF within amphibolite facies rocks of the supracrustal Lewisian Loch Maree Group at Flowerdale, 1 km W of Gairloch pier, Highland Region, Scotland (Williams 1986). This is sufficient for Livingstone (1993) to describe the mineral as annite, although, like most natural micas, it does not have a pure end-member composition. Annite is most abundant in quartz-rich layers containing garnet and chlorite. In thin section it is strongly pleochroic from brown to pale yellow. The silicate-bearing rocks in Flowerdale are intimately associated with sediments rich in iron oxide, which indicate a very shallow-water environment in which chemically precipitated materials were reworked by current and wave action (Williams 1986). The inadequacies of mica nomenclature are such that, for the purposes of this book, annite is defined to have $Fe/(Fe+Mg)=0.75–1.0$ and $Al^{vi}=0–0.5$, where Fe is total Fe expressed as FeO. All data have been processed using the method described in Tindle & Webb (1990) for trioctahedral micas, and as the Flowerdale mica has $Al^{vi}=0.6$ it is considered to be siderophyllite.

With the annite/biotite boundary set at $Fe/(Fe+Mg)\geq 0.75$, mica from the following locations can be named annite: later fractionates of the Insch gabbroic intrusion, Grampian, Scotland (Wadsworth 1986); foliated granite sheets within and to the W of the Morven–Cabrach gabbro in Glenbuchat, Grampian (Tindle unpub. EMP data); granites from the Ballater (minor annite), Hill of Three Stones, Lochnagar (minor annite) and Peterhead intrusions, Grampian (Tindle unpub. EMP data); coarse syenite from the Cnoc Rhaonastil differentiated dolerite sill, Isle of Islay, Strathclyde (Preston et al. 2000b); blueschist-facies metabasites belonging to the Dalradian Supergroup from central Achill Island, N of Clew Bay, Co. Mayo (Chew et al. 2003); granite from the quarries at Newcastle, northeastern end of the Mourne Mountains, Co. Down (Brown 1956).

Biotite (now a series name), varying in composition from annite to siderophyllite, is described from the St Austell granite, Cornwall (Manning et al. 1996). This probably refers to a solid solution between the two species rather than end-member compositions, although no data are presented that would allow this to be confirmed. It is unlikely that annite occurs in Cornish granites, as trioctahedral mica data from such granites are far removed from annite compositions. According to Stone et al. (1988), all analyzed "biotite" from Cornish granites is actually siderophyllite.

anophorite

A name formally recommended for extinction as part of the reclassification of the amphibole group. The new name titanian calcian magnesio-arfvedsonite was proposed by Leake et al. (1987). A calcio-anophorite composition is described from early Cambrian blueschists at the Marquis of Anglesey's monument (just E of Llanfairpwllgwyngyll, Anglesey, Gwynedd, Wales) and possibly represents a contaminated analysis (Horák & Gibbons 1986). The same analysis had previously been referred to as crossite by Holgate (1951). The 1997 classification of the amphibole group (Leake et al. 1997a, Mogessie et al. 2004) names this mineral katophorite.

anorthite

FELDSPAR GROUP
triclinic
$CaAl_2Si_3O_8$

A common calcium aluminosilicate, trimorphous with dmisteinbergite and svyatoslavite, and found in lavas and intrusive rocks with a basic or ultrabasic composition. It is the high-temperature calcium-rich end member of the plagioclase series, which with increasing sodium content varies through a range of intermediate compositions towards albite.

Anorthite has been optically identified in plagioclase cumulates from the Rhiw layered intrusion, Lleyn Peninsula, Gwynedd, Wales (Cattermole 1976). Anorthite is a minor species in dolerite at Dolyhir quarry, Old Radnor, near Presteigne, Powys (T. F. Cotterell pers. comm. 2007)

Early descriptions (and analyses) of anorthite from Scotland are provided by Heddle (1877, 1901): associated with hornblende in diorite and gabbro on the east side of Wick of Tresta, Isle of Fetlar, Shetland; with pyroxene–amphibole (diallage) at Durness, Highland Region; with diopside (coccolite) in a massive granular vein at Dalnabo limestone quarry, Glen Gairn, 10 km NW of Ballater, Grampian; in a "diallage" rock associated with native copper and pectolite, 2 km N of Lendalfoot, Strathclyde.

At Corrie Loch Kander, near Braemar, Grampian, a calcic plagioclase feldspar (thought to be anorthite), finely intergrown with barium-rich potassium feldspar, occurs in the laminated quartzite bands of a calc-silicate schist. Diopside–hedenbergite, hyalophane, pyrrhotite, scapolite, tremolite–actinolite and zoisite are associated species found in the other schistose bands. The rock is interpreted as a metamorphosed exhalative, sulphidic, dolomitic chert (Fortey et al. 1993). Symplectite-bearing nodules in the Ardgour marble at Coire Dubh, near Corran ferry, Highland Region, are noted to contain anorthite and 23 other minerals, including augite, clinozoisite, meionite, prehnite, vesuvianite and zoisite (Drever 1936).

At Grange Irish in the Carlingford Mountains, Co. Louth, Ireland, an unusual syenite consisting of anorthite and actinolite (described as hornblende) is found where granite has assimilated Carboniferous Limestone (Haughton 1856). Anorthite occurs throughout a sequence of ultrabasic cumulates at the base of the Currywongaun and Doughruagh layered intrusion, Connemara, Co. Galway, Ireland. Augite and enstatite, often with a secondary amphibole, are associated species in these rocks (Angus & Kanaris-Sotiriou 1982).

anorthoclase

FELDSPAR GROUP
triclinic
$(Na,K)AlSi_3O_8$

A relatively common sodium–potassium aluminosilicate found in high-temperature (often alkaline or peralkaline) plutonic and volcanic igneous rocks, such as syenites, trachytes and rhyolites. On Lundy island, off the north coast of Devon, anorthoclase occurs with arfvedsonite and plagioclase feldspar (oligoclase) in peralkaline rhyolite and trachyte dykes, thought to be related to a substantial Tertiary composite volcano (Thorpe & Tindle 1990).

On Anglesey, Gwynedd, Wales, anorthoclase has been found with albite in folded pegmatitic augen in the Precambrian Gwna beds exposed close to Craig-y-don house, near Menai. The two feldspars are also said to occur in vein quartz at Graig Dhu, Amlwch, also on Anglesey (Greenly 1919). Identification is most likely to be optical, and confirmatory chemical analysis is desirable.

In Scotland, anorthoclase compositions are reported from the Sgùrr of Eigg pitchstone on the Isle of Eigg, Highland Region. The rock is strikingly porphyritic and contains phenocrysts of anorthoclase, augite and enstatite phenocrysts, the former up to 5 mm across and strongly resorbed. Some anorthoclase shows Carlsbad or minor tartan twinning, but most crystals are untwinned (Preston et al. 1998b). At Bloodstone Hill, Isle of Rum, Highland Region, mugearite lava contains zoned feldspar phenocrysts with sodic anorthoclase rims and plagioclase (oligoclase) cores. Associated phenocryst species are augite and enstatite. A small gabbro dyke on the nearby island of Muck contains similar coarse-grain zoned feldspar with sodic anorthoclase rims (Ridley 1971). The Shiant Isles Main Sill, Western Isles, is an example of a composite differentiated alkaline basic sill that hosts zoned feldspar crystals, some with plagioclase cores, anorthoclase intermediate zones and sanidine rims. Anorthoclase is most notable in crinanitic pegmatites near the top of the sill where syenitic patches occur (Gibb & Henderson 1996).

Large Carlsbad-twinned phenocrysts of anorthoclase are noted in a bostonite (microsyenite) dyke cutting Old Red Sandstone sediments at Onston Ness, a small promontory on the Loch of Stennes, E of the Bridge of Waithe, near Stromness, Mainland, Orkney (Flett 1900). At Lettermore, Highland Region, syenite from the Cnoc nan Cuilean unit of the Ben Loyal igneous complex contains a cloudy sodic–potassic feldspar termed "anorthoclase" by King (1942). Albite and aegirine-augite are the other major constituents of the rock. A monchiquite dyke on the west side of Camas Nathais, 1 km NE of Rubha Fionn-àird, 4 km W of Benderloch, Strathclyde, contains anorthoclase as a phenocryst phase (Livingstone 2002).

Anorthoclase is also described with apatite, ferrosilite, ilmenite and magnetite from lower crustal xenoliths trapped in a Permian olivine nephelinite dyke at Gribun, west coast of the Isle of Mull, Strathclyde (Upton et al. 1998). Large crystals (xenocrysts) of pink feldspar, up to 4 cm in length are found in the Black Rock, a Late Carboniferous basanitic vent on the west coast of Scotland, 2 km SSW of Fairlie, on the Hunterston Sands, Strathclyde. In thin section, the feldspar exhibits a macroperthitic texture of intergrown anorthoclase and albite (Alexander et al. 1986).

In Fife, generally reddish-brown, but occasionally transparent and usually rounded, fragments of anorthoclase occur as megacrysts in the tuff of the Ardross, Coalyard Hill, Elie Ness, Kellie Law and Ruddons Point volcanic vents along the coast near Elie. Additionally, a perfect doubly terminated crystal 11 cm long is recorded from an associated monchiquitic sheet at Coalyard Hill (Chapman 1976, Chapman & Powell 1976, Macintyre et al. 1981).

Dolerite close to the contact with chalk at Scawt Hill, Larne, Co. Antrim, Ireland, has been hybridized and assemblages including an anorthoclase–fayalite dolerite have been produced. This assemblage occurs in 25 mm-wide segregation veins in the dolerite near the edge of the hybrid zone. Brown augite and plagioclase feldspar (labradorite) are the chief constituents, but, in addition to anorthoclase and fayalite, apatite, magnetite, thomsonite and accessory amounts of an alkali amphibole are also present (Tilley & Harwood 1931).

antarcticite?

trigonal
$CaCl_2 \cdot 6H_2O$

An hydrated calcium chloride reported from Guy's Cliff, about 2 km from Warwick Castle (Warwickshire), where it is described as black slime with the appearance of tar on several parts of the cliff face. It was thought to have formed when rainwater washed through hidden salt beds and is referred to as native calcium chloride by Spiller (1876). Antarcticite requires further confirmation.

anthophyllite

AMPHIBOLE GROUP
orthorhombic
$\square Mg_7Si_8O_{22}(OH)_2$

A relatively common magnesium silicate hydroxide, dimorphous with cummingtonite, that forms a series with ferro-anthophyllite. Anthophyllite is found in amphibolites, gneisses, metaquartzites, iron formations, granulites and schists derived from argillaceous sediments. It also occurs in mafic and ultramafic rocks as a retrograde reaction product. Where possible, identification is based on the 1997 classification of the amphibole group (Leake et al. 1997a, Mogessie et al. 2004).

Anthophyllite–cordierite hornfels rocks developed in Cornwall in sheared pillow lavas and other metabasic rocks (greenstones) at the margin of the Land's End granite (near Kenidjack Castle and from Zawn a Bal, St Just) have been described as the products of both thermal metamorphism and metasomatic processes. Anthophyllite is greyish to greyish-brown in colour, but colourless in thin section. Biotite, cummingtonite, diaspore, ilmenite, magnetite, plagioclase, spinel (pleonaste) and zircon are associated species (Tilley & Flett 1930, Tilley 1935, Chinner & Fox 1974, Mitropoulos 1984). Anthophyllite is also found in altered dolerite from Botallack, St Just (Robson 1949b). Elsewhere in Cornwall, lath-shape crystals of colourless anthophyllite have been described from the Meneage area of the Lizard Peninsula, particularly at the junction of gabbro and serpentinite at Pen Voose, near Landewednack. Elliptical aggregates of crystals up to 5 cm across enclose masses of green hornblende (Teall 1888a). At Kennack Sands, anthophyllite forms honey-colour fibrous vein fill (fibres to 10 cm) within the Lizard peridotite (Power et al. 1997). Golley & Williams (1995) note anthophyllite from Pistil Ogo and Polkernogo, both on the Lizard Peninsula.

In Scotland, anthophyllite appears to have formed as an alteration product of orthopyroxene in dunite–harzburgite peridotite exposed between Rubha Sgeir nan Sgarbh and Loch Meurach in the Langavat Valley, South Harris, Western Isles. This ultramafic complex was subjected to granulite-facies metamorphism, granitization and metasomatism, which produced highly acicular anthophyllite, sometimes as rosettes (Livingstone 1976a). Anthophyllite in Lewisian serpentinite from 0.5 km SE of Borve Lodge, Sound of Taransay, South Harris, appears to be a related occurrence (Guppy 1956). In Highland Region, anthophyllite occurs as a secondary species replacing orthopyroxene in a garnet–hornblende–pyroxene eclogite from Glen More, Glenelg (O'Hara 1960), and in Assynt (around Drumbeg and Gorm Loch, both NE of Scourie) it formed as Scourian ultramafic granulites were retrogressed to amphibolite facies (Sills 1982). Lewisian amphibolite-facies ultramafic rocks from SW of Achmelvich, An Fharaid Mhòr, near Lochinver, contain anthophyllite in prismatic porphyroblasts up to 3 cm long, intimately intergrown with actinolite, chesterite, clinojimthompsonite and jimthompsonite (Droop 1994).

Aureole rocks around the Dorback granite, Braes of Abernethy, 9 km W of Tomintoul, Grampian, include beautiful, perfectly fresh anthophyllite–hornfels containing biotite, cordierite, orthoclase and quartz (Walker & Davidson 1935), and on the Ness of Hillswick, north Mainland, Shetland, anthophyllite occurs in a 60 cm-thick bed, associated with chloritic schists and serpentinite (Heddle 1879b, Livingstone 2002).

In Ireland, anthophyllite occurs in amphibole-rich reaction borders of orthopyroxenite metasomatized by acid pegmatite veins in the Currywongaun–Doughruagh ultrabasic intrusion, Connemara, Co. Galway (Kanaris-Sotiriou & Angus 1979). The Currywongaun intrusion is also host to a pelitic xenolith (cordierite hornfels) in which anthophyllite, biotite, högbomite, ilmenite, magnetite and spinel occur in minor amounts (Angus & Middleton 1985). Haughton (1862) reports anthophyllite in slate resting on granite from Castlecaldwell, Co. Donegal.

anthraconite

A black carbon-rich variety of calcite.

antigorite

KAOLINITE–SERPENTINE GROUP
monoclinic $Mg_3Si_2O_5(OH)_4$

A relatively common magnesium silicate hydroxide, polymorphous with chrysotile and lizardite, and found replacing ultramafic rocks either pervasively or in cross-cutting veinlets.

A white soapy mineral collected from a vein in a serpentine rock at Kennack Sands, Lizard Peninsula, Cornwall, is described by Midgley (1951), who suggests it is probably a variety of antigorite. Golley & Williams (1995) note antigorite at Clicker Tor quarry at Menheniot, and Countybridge quarry on the Lizard Peninsula.

On Holy Island, off the west coast of Anglesey, Gwynedd, Wales, antigorite occurs at a few locations as a product of serpentinization of a suite of basic and ultrabasic intrusions belonging to the Precambrian Mona Complex. At Myndachdy, the serpentinite contains many tiny garnets together with stars of antigorite and chrysotile asbestos (Greenly 1919). Maltman (1977) describes four main textures in the serpentinites of Holy Island; mesh, ribbon, bladed mat and disturbed, and all composed of lizardite or antigorite (or both). All these identifications appear to be based on optical data. Unusually, antigorite is found as extremely rare white to bluish-green platy crystal rosettes and groups up to 1 mm across associated with copper–lead mineralization at Esgair Hir mine, Ceulanymaesmawr, Central Wales orefield, Dyfed. It is aggregated over a thin crust of brochantite with minor linarite and cerussite (Rust & Mason 1988).

In Scotland, perhaps the greatest concentration of antigorite is at Corrycharmaig, Glen Lochay, NW of Killin, Tayside Region, where serpentinization of an ultrabasic intrusion within a regional terrain of Dalradian schists has produced a rock with over 90% antigorite, and minor amounts of chlorite, dolomite, magnesite, tremolite and

Fibrous antigorite (variety picroite) from Corrycharmaig, Glen Lochay, NW of Killin, Tayside Region, Scotland; width 11 cm. (© D. G. Anderson)

opaque oxides (chromite, magnetite and chromium-rich magnetite). The antigorite occurs as fibro-lamellar aggregates and (rarely) in fibrous form. The intrusion also contains a talc–carbonate rock in which antigorite is a minor component (Harrison 1985). Hawson & Hill (1987) suggest that, because chrysotile and lizardite are absent, the serpentinization took place at over 450°C. Chromite-rich layers of the intrusion are host to a rare sulphide and PGM-rich assemblage that includes the species galena, gersdorffite, heazlewoodite, irarsite, laurite, nickeline, native metals (gold, iridium, osmium, ruthenium, silver) and alloys all noted (Power & Pirrie 2000). Similarly altered serpentinites at Arrochymore Point on the shore of Loch Lomond, Central Region (close to the Highland Boundary Fault), contain antigorite, chrysotile and lizardite associated with chromite and magnesiochromite (Henderson et al. 1983).

The williamsite variety of antigorite is reported from Hagdale, Haroldswick, Unst, Shetland (Heddle 1878a). Elsewhere on Unst, antigorite serpentine from ultramafic rocks is spatially associated with the basal thrust of the Shetland ophiolite and is also noted in cumulate pyroxenites. In a talc-carbonate rock, antigorite is intergrown with dolomite, magnesite, talc and quartz or forms cm-thick cross-cutting veins. Antigorite is also found with a massive bladed matt texture with minor relict olivine and disseminated chrome spinel in dunite pods from Cliff quarry, where it is intergrown with carbonate and chlorite (Maynard 1993). Read (1934b) reports solid antigorite with a few layers of tremolite associated with talc schists between Norwick and Clibberswick on the east coast of Unst. At Lunda Wick, along the south coast of Houllnan Ness, northwest Unst, antigorite forms large zoned nodules up to 2 m across in hornblendic and pelitic gneisses. Actinolite, biotite, chlorite and talc are associated species (Read 1934a). Other occurrences of antigorite in serpentinite bodies include Hoo Field, Cunningsburgh, Mainland, Shetland (Read & Dixon 1932); Langavat Valley, South Harris, Western Isles; and Glen Urquhart, Highland Region (Livingstone 2002).

antimonblüthe

A synonym of valentinite.

antimonial lead

A synonym of bournonite.

antimonial ochre

A synonym of cervantite, or stibiconite, or valentinite.

antimonite

A synonym of stibnite.

antimonpearceite

monoclinic $(Ag,Cu)_{16}(Sb,As)_2S_{11}$

A rare silver–copper–antimony–arsenic sulphide possibly present at Pentire (Glaze) mine, near New Polzeath, St Minver, Cornwall. At this site, unorientated rounded or irregular blebs of silver-bearing minerals occur as intergrowths in galena and comprise two distinct types in which tetrahedrite is commonly associated with polybasite or antimonpearceite (Clayton et al. 1990). Further confirmation is required. New nomenclature rules introduced by the IMA in 2006 discredit antimonpearceite and it should now be referred to as polybasite – $Cu(Ag,Cu)_6Ag_9Sb_2S_{11}$.

antimonsaures bleioxyd

A synonym of bindheimite.

antimony

ARSENIC GROUP

trigonal — Sb

A rare metal, found as granules (≤20 µm diameter) in an antimony-rich quartz vein within the metamorphic aureole of the Skiddaw granite at Coombe Height, headwaters of Wet Swine Gill, Caldbeck Fells, Cumbria. The massive white quartz vein contains native antimony in an early-formed suite of As–Fe–Sb minerals (arsenopyrite, berthierite and stibnite mainly). A suite of antimony–lead species including fülöppite, semseyite and zinkenite followed (Fortey et al. 1984, Green & Briscoe 2002). At Driggith mine and the extension of the Driggith vein towards Sandbed, Caldbeck Fells, Cumbria, native antimony and bournonite have been found as inclusions in galena in Upper Devonian to Permian lead–zinc veins. The inclusions are very small (<1 µm rounded grains and rarely to 5 µm) and a near-perfect polish on prepared specimens is required before the fine intergrowths can be seen. Native antimony does occur at other locations in the Lake District, but is considerably less abundant than in galena from the Caldbeck Fells and particularly from the Driggith–Roughton Gill vein system: Barrow mine, WSW of Keswick; Eagle Crag mine, Patterdale; Force Crag mine, Braithwaite, near Keswick; Goldscope mine, Newlands Valley; Greenside mine, near Glenridding, Patterdale; Hartsop Hall mine, Hartsop, Patterdale; Myers Head mine (Low Hartsop mine), Hartsop; and Tilberthwaite Ghyll, near Coniston (Stanley & Vaughan 1981).

An arsenic-rich variety of native antimony, forming discrete hexagonal crystals, has been found in a matrix of native arsenic, arsenolite, elbaite, löllingite and lithium mica in cavities within the Meldon aplite, near Okehampton, Devon (von Knorring & Condliffe 1984). It is also noted by Hosking (1951) that antimony (presumably native antimony) was one of the primary ore minerals found in the Crowns lode of Botallack mine, St Just.

A report of native antimony from the Dolgellau goldbelt, Gwynedd, by Readwin (1888) requires verification (Bevins 1994).

In Ireland, late Hercynian, copper-mineralized quartz–siderite veins at Dhurode, West Carbery orefield, Co. Cork, contain trace amounts of native antimony associated with bournonite, galena and meneghinite, in which complex symplectite textures have developed (Ni Wen et al. 1991). Quartz–carbonate veins and breccias at Bryanlitter, near Clontibret, Co. Monaghan, contain native antimony inclusions in stibnite, associated with sheaves and clusters of pyrrhotite (Steed & Morris 1997). At Tynagh mine, near Killimor, Co. Galway, epigenetic Cu–Pb–Zn–(Ag) mineralization includes a suite of sulphide species that occur in trace amounts: native antimony, arsenopyrite, bournonite, enargite, geocronite, gratonite, jordanite, marcasite, millerite, pyrite, siegenite and tennantite – many occurring as inclusions in galena (Morrissey et al. 1971, Gibson 1979, Clifford et al. 1986).

Native antimony is also reported from veins cutting the Northwest granite on Jersey. At Gigoulande quarry, St Mary, it is found in quartz veins containing chalcopyrite and pyrite with minor amounts of bournonite, marcasite, molybdenite and sphalerite (Ixer 1980, 1992). At Le Pulec (northwest coast) it is found in a polymetallic dolomite–quartz vein system dominated by sphalerite, where it forms intergrowths with bournonite, chalcopyrite, cubanite, galena and tetrahedrite or small discrete anhedral or tabular crystals (2–10 µm diameter) included within coarse-grain galena (Ixer & Stanley 1980).

antimony-glance

A synonym of stibnite.

antimony oxide

A synonym of senarmontite or valentinite.

antimony sulphuret

A synonym of stibnite.

antlerite

orthorhombic — $Cu^{2+}_3(SO_4)(OH)_4$

A very rare copper sulphate hydroxide that occurs as a supergene alteration product of copper ore and is found with atacamite, brochantite, chalcanthite, gypsum, metazeunerite, scorodite and sodium-zippeite from a small area of partially altered and collapsed wallrock traversed by primary sulphide veins on the 17 level, Levant north lode east (No. 2 branch) at Geevor mine, near Pendeen, St Just, Cornwall (Elton & Hooper 1993). Elsewhere in Cornwall, antlerite forms light-green powdery coatings at Perran St George, Perranzabuloe (Bridges et al. 2005). Braithwaite (1994) notes antlerite as a green crust associated with langite and posnjakite, in a barite gossan from Engine Vein mine, Alderley Edge ridge, Cheshire. The collections of the NHM contain a specimen encrusted with an earthy pale-green coating of antlerite on barite–quartz veinstone from Potts Gill mine, Caldbeck Fells, Cumbria (Bridges et al. 2005).

Antlerite is confirmed at Parys Mountain (Mynydd Parys), Anglesey, Gwynedd, Wales, where it is the commonest post-mining copper mineral to grow on chalcopyrite exposed in underground workings. It forms blocky crystals up to 5 µm across, overgrowing brochantite (Jenkins et al. 2000). In Ireland, antlerite is noted from East Avoca mine, Co. Wicklow, as soft amorphous-looking crusts and patches on the walls and floors of mine levels (Bridges et al. 2005).

The remaining descriptions of antlerite are all problematical: Old Gunnislake mine, Calstock, near Gunnislake, Cornwall (Kingsbury 1954) – given that Kingsbury's collection (now at the NHM) contains many incorrectly provenanced specimens (Ryback et al. 1998, 2001), the specimen was examined in detail by Bridges et al. (2005), who concluded that it was probably genuine; Potts Gill barite mine, Caldbeck Fells, Cumbria (Hartley 1984, Young 1987a, Cooper & Stanley 1990) – another Kingsbury specimen, "almost certainly fraudulent" (Bridges et al. 2005); Sandbed mine, Caldbeck Fells (Young 1987a) – later shown to be malachite (Cooper & Stanley 1990, Green et al. 2006a); and Tresavean mine, Gwennap, Cornwall (Golley & Williams 1995) – requires confirmation.

antrimolite

TL Ballintoy, W of Ballycastle, Co. Antrim, Ireland

A variety of thomsonite or mesolite (both zeolite group) on calcite.

APATITE GROUP

Hexagonal or monoclinic pseudo-hexagonal arsenates, phosphates and vanadates with the general formula $A_5(XO_4)_3$(F,Cl,OH), where A = Ba, Ca, Ce, K, Na, Pb, Sr, Y; and X = As^{5+}, P^{5+}, Si^{4+}, V^{5+}. (CO_3) may be partially replaced by (PO_4). Carbonate-fluorapatite, carbonate-hydroxylapatite, chlorapatite, fluorapatite, hedyphane, mimetite, phosphohedyphane, pyromorphite and vanadinite are members of the group found in Britain or Ireland. Apatite is a general term used to describe some members of the group when chemical data required to confirm the species are missing. In many cases, apatite is synonymous with fluorapatite.

aphanèse, aphanesite

Synonyms of clinoclase.

aphrosiderite

A magnesium-rich variety of chamosite, chlorite group.

aphthitalite

trigonal — $(K,Na)_3Na(SO_4)_2$

A rare potassium–sodium sulphate found in the halite–kieserite–anhydrite subzone of the Permian Lower Evaporites in the Fordon F.3 borehole, Humberside (13 km S of Scarborough, Yorkshire). The Permian evaporite sequence underlies much of North Yorkshire and Humberside, and it extends considerably beneath the North Sea. At Fordon, the sequence is about 430 m thick. Aphthitalite is one of the rarest evaporite species reported from Britain, where it replaces kieserite and halite, and forms large euhedral poikiloblasts to 2 cm. It is accompanied by picromerite (Stewart 1954, 1963, 1965). Boulby potash mine, Loftus, Cleveland, is a further locality where aphthitalite has been found (Anthony et al. 2003).

apjohnite

HALOTRICHITE GROUP

monoclinic — $Mn^{2+}Al_2(SO_4)_4{\cdot}22H_2O$

Named after James Apjohn FRS (1796–1886), an Irish chemist and mineralogist at Trinity College, Dublin; born at Grean, Limerick. Apjohnite is an hydrated manganese–aluminium sulphate recorded from Ireland in the supplement to Greg & Lettsom (1858). However, the mineral being described is a brown metallic ore mixed with pyrite, and is clearly not apjohnite; it requires further confirmation.

At Hawthorn Hive, a wide bay on the Co. Durham coast (at and around Hive Point and Shippersea Bay), apjohnite has been found in mine waste derived mainly from Dawdon colliery. Apjohnite forms a thin coating of abundant colourless-to-white acicular crystals, up to 5 mm long, on specimens of pyritized plant fossils that had been stored in sealed containers for months (Young et al. 2008).

aplome

A synonym of andradite, garnet group.

apophyllite

See fluorapophyllite and hydroxyapophyllite. Quite often the species name is unknown and the term apophyllite is used instead.

England

Cornwall Levant mine, St Just (Symes 1977); Terrace (Terras) Hill quarry, Lostwithiel (Russell 1910b); Treburland mine, 1.5 km SE of Fivelanes, Altarnun, Bodmin Moor (Russell 1946d).
Cumbria Shap Blue quarry (Young et al. 1988, 1998).
Devon Meldon aplite quarry, near Okehampton (Symes 1977).
Durham Cambokeels (Cammock Eals) mine, Eastgate and Copt Hill quarry, Cowshill, both Weardale (Young et al. 1991a).
Somerset Merehead (Torr Works) quarry, Cranmore, Mendip Hills (Symes 1977, Symes & Embrey 1977).

Scotland

Fife Orrock quarry, near Burntisland (Green & Todd 2001, Wirth 2003, Moreton 2006).
Highland Apophyllite is reported from many localities: in the Tertiary basalts on the Isle of Skye including: Beinn na Greine, near Portree and Talisker Bay (King 1976); Biod a' Buide; Leacan Fhionn; north shore of Loch Bharcasaig (Varkasaig) and Rubha nan Clach, both Loch Bracadale; Rubha na h-Airde Glaise; Sgùrr an Duine, S of Loch Eynort (Heddle 1901); Biod a' Mhurain, southern Duirinish and Dunvegan quarry (Green & Todd 2001); Creag na Cuall, Duirinish Peninsula (Wolfe 2006); Glen Ollisdal (Wolfe 1996b); Moonen Bay, N of Dunvegan (Green & Todd 1996); slopes of Quiraing (Heddle 1879f, 1901); Sgùrr nam Boc, between Loch Brittle and Loch Eynort (Heddle 1901, Green & Wood 1996, Green & Todd 2001); and the Storr, near Portree (Anderson 1851, Heddle 1901). Other occurrences include: in Moine Series limestones from the Ardgour district (Drever 1939); at the Strontian mines (Scott 1871); and at Beinn an Dubhaich, Broadford area, Isle of Skye (Tilley 1951).
Lothian Ratho quarry, near Edinburgh (Heddle 1901); Traprain Law, near Haddington (Anderson & Elder 1935, Battey & Moss 1962).
Strathclyde Carsaig Arches and at the southern end of the Wilderness, Loch Scridain, both Isle of Mull; Fladda, Treshnish Isles; Auchinstarry quarry, Kilsyth, near Glasgow (all Heddle 1901); Howford Bridge (Tyrell 1928, Anderson & Elder 1935).
Western Isles Mullach Sgar, St Kilda (Hirta) (Harding et al. 1984).

Ireland

Antrim Berk Hill; Carnmoney Hill, N of Belfast; quarry NW of summit of Collin Mountain; quarry, 1.8 km NE of Wolf Hill, Legoniel Roughfort (all Walker 1960a); Catcairn basalt quarry, Legoniel, Belfast (Walker 1960a, Nawaz 1978a).
Londonderry Sounding Hill quarry, NNW of Magherafelt (Old 1972).

apyre

A synonym of andalusite.

aquamarine

A synonym of gem quality, bluish-green iron-rich beryl.

ARAGONITE GROUP

Orthorhombic carbonates with the general formula ACO_3, where A = Ba, Ca, Pb, Sr. Aragonite, cerussite, strontianite and witherite are the members of the group found in Britain or Ireland.

aragonite

ARAGONITE GROUP
orthorhombic $CaCO_3$

A common calcium carbonate, trimorphous with calcite and vaterite, and found in high-temperature blueschist-facies metamorphic rocks, in evaporite deposits, in sinters and hot-spring deposits, in oolitic limestones and muds, and often found in limestone caves and fractures in limestone rocks. It is an important detrital sedimentary component forming the hard parts of many marine shells and skeletons. Over geological time, aragonite alters to calcite.

England

Avon Aragonite forms white rosettes of acicular crystals up to 4 mm long in boulders of banded barite and dolomitic conglomerate on Clevedon beach, near the Pier hotel, 12 km W of Bristol, Avon (Starkey 1984).
Cornwall Aragonite is a fairly common species in the Botallack sett, a group of mines located in the aureole of the Land's End granite near St Just, Cornwall, and which includes Botallack mine, Wheal Cock, Wheal Edward and Wheal Owles. It forms white, grey, red or bluish-green globular concretions, and acicular, fibrous or fine doubly terminated hexagonal prismatic crystals (Bancroft & Weller 1993). Hosking (1951) considers that it may have been one of the primary species in cassiterite-bearing veins of Crowns lode at Botallack. A localized patch of aragonite in the roof of one of the stopes at Wheal Owles is associated with tyuyamunite and uranophane (Elton & Hooper 1994). Aragonite is associated with native silver at the nearby Levant mine (Rust 1982a). At Penberthy Croft mine, St Hilary, aragonite forms colourless-to-white prismatic crystals up to 0.4 mm long as a minor constituent of a basic copper sulphate assemblage. The crystals form crusts on the surface of weathered clinochlore–dolomite–quartz veinstone and are associated with birnessite, brochantite, goethite, erythrite and siderite. Here, aragonite has a post-mining origin (Camm & Merry 1991b, Betterton et al. 2000). Fine sprays of colourless-to-white aragonite crystals up to 80 mm across have been found at Greystone quarry, Lezant, near Launceston (Elton et al. 2004). The species is also noted from: Burraton Coombe quarry, St Stephens by Saltash (Russell 1925); Countybridge quarry, Goonhilly Downs, Lizard Peninsula (Holyer 1996); Parson's beach, Mawnan (Camm 1995); Mawnan Cliffs, Falmouth; Strangles beach, Crackington Haven; Port Isaac; and Tintagel cliffs and Trematon quarry, Saltash (all Golley & Williams 1995).

Aragonite crystals to 15 mm with brown todorokite on dolomite from Port Quin, Cornwall. (© D. I. Green)

Cumbria *Caldbeck Fells:* Beautiful crystals of aragonite have been found at Brandy Gill and Roughton Gill mines (Goodchild 1884, Young 1987a), and well crystallized aragonite is noted from cavities within an ankerite vein at Carrock wolframite mine, Grainsgill (Davidson & Thomson 1951). Radiating sprays of acicular white crystals, up to 1 mm across, and pale blue-green sheaves of aragonite associated with calcium-rich pyromorphite, occur on the dumps of Driggith and Sandbed mines (Green et al. 2006a). *Elsewhere in the Lake District and West Cumbria:* Fine crystals, up to 10 cm long, in limestone at Baron Cross quarry, Stainton, Penrith (Davidson & Thomson 1951); Bigrigg mine, near Egremont (Young 1987a); rare at Brandlehow mine, Derwentwater (Kendall 1884); the very finest British aragonite, forming long slender reddish-brown crystals with acute pyramidal terminations, is said to come from the hematite mines of the Cleator Moor area (Greg & Lettsom 1858, Rudler 1905); good specimens including stellate sprays of white acicular crystals are common in vugs within dolomitized limestone at Eskett quarry, Rowrah, Frizington, Whitehaven. Aragonite is also common in the mines of this area (Young 1987a, Green & Briscoe 2002). *Northern Pennine orefield:* Aragonite is a relatively common primary mineral and it also forms as a supergene species, developing hemispherical aggregates (up to 3 mm across) of compact radiating needles at Blagill and Brownley Hill mines, Nenthead and Tynebottom mine, NW of Garrigill (Bridges 1987, Bridges & Young 1998, Green et al. 2000); massive aragonite with a banded fibrous texture occurs at the Crag Green mines, Alston Moor (Dunham 1990); brittle fibrous bundles, arborescent, columnar and stalactitic (flos ferri) forms of aragonite are reported from Silverband mine, Milburn, near Appleby, and other mines in the Dufton area, and branching sprays up to 45 cm across and small caverns lined with aragonite at Silverband mine are particularly notable (Wallace 1865, Goodchild 1884, Davidson & Thomson 1951). Massive pieces of coralloidal aragonite from Dufton have been described as the satin spar variety of gypsum because of its satiny lustre (Spencer 1897). Aragonite is also recorded (Young 1987a, Green & Briscoe 2002) from: Kelton quarry, Kirkland; Parkhouse quarry, Johnby; and Stockhow quarry, Kirkland.

Derbyshire Aragonite is uncommon in Derbyshire, although it has been recorded as white encrustations (flos ferri) at the Royal and Cumberland Caverns, Masson Hill, Matlock Bath (Greg & Lettsom 1858, Ford et al. 1993) and in the adjoining Wapping mine (Braithwaite 1983, Ince 1991). Aragonite is also noted from Ashover, Bonsall, Chellaston, Hopton Moor, Matlock, Fox Hole Cave in Earl Sterndale, and Treak Cliff Cavern, Castleton (Ford & Sarjeant 1964a).

Devon Beautiful acicular crystals of aragonite have been found along the shores of Torbay and at Buckfastleigh. White compact and fibrous varieties occur at Ilfracombe and a coralloid variety (flos ferri) is known from cavities in limestone at Combe Martin (Greg & Lettsom 1858, Hall 1868b, Collins 1892). Bunches and crusts of small colourless needles of aragonite coat some fracture surfaces and veinlets in blackish Devonian slate outcropping on the beach at St Mary's Bay, Brixham (Braithwaite & Wilson 2001b), and at Hope's Nose, Torquay, the species occurs in gold-bearing carbonate veins cutting Devonian limestones (Stanley et al. 1990a, Camm 1995).

Dorset Aragonite is noted on the shore at Osmington Mills, E of Weymouth (Greg & Lettsom 1858).

Durham Massive aragonite with a banded fibrous texture occurs at the Yorkshire Silverband mine, Cronkley Fell, and a pale bluish-green copper-rich variety is noted from Lunehead mine, near Grains o' th' Beck, Lunedale (Friend & Allchin 1940, Dunham 1990). Aragonite is a primary species associated with calcite and a little quartz in a barite vein and replacement deposit within Carboniferous limestones and sandstones at Closehouse mine, SW of Middleton-in-Teesdale, Lunedale (Young et al. 1994a). It also occurs associated with calcite and bottle-green twinned fluorite at Heights quarry, Weardale (Green & Briscoe 2002).

Leicestershire Radiating groups of colourless-to-white or pale-green acicular aragonite crystals, up to 5 mm long, have been found at Newhurst quarry, Shepshed (Ince 2005). At New Cliffe Hill quarry, Stanton under Bardon, aragonite forms sprays and crusts of colourless-to-white acicular crystals up to 5 mm long in the wallrock surrounding a native copper–cuprite deposit (Hubbard et al. 2005).

Oxfordshire Aragonite is a major constituent of ammonite shells and shell fragments found in calcareous concretions and associated sediments in temporary exposures of Oxford Clay at Cassington and Yarnton (Atkins 1995).

Somerset Horny white aragonite shells of Jurassic fossils (notably ammonites) are replaced by analcime along a 15 km stretch of coastline from Blue Anchor Point to Stolford Rocks and including Hinkley Point (Alabaster 1980). Rare nodules of aragonite have been found in manganese ore at Higher Pitts mine, Priddy, Mendip Hills. It forms compact aggregates of large white prisms altering to calcite and also small needle-like crystals associated with hydrocerussite (Burr 1996). Aragonite is also noted both in the Mendip Hills from Lamb Lair Cavern, Gibbet's Brow, near Compton Martin (Kingsbury 1941), and Merehead (Torr Works) quarry, Cranmore (Alabaster 1975). Greg & Lettsom (1858) record aragonite from a cave at Merridge in the Quantock Hills and a coralloid variety from Halwell Cavern at Froomfield, near Taunton.

Yorkshire Aragonite appears to be much more common in the southern part of the Northern Pennine orefield than it is farther north. It produces massive veinstone, particularly in the Old Gang (NE of Gunnerside, Swaledale) and Arkengarthdale mines, and at Wassen Fell, above Walden, fine acicular masses are reported. Massive banded carbonate at Stool mine, N of Grassington, contains radially fibrous aragonite on fluorite. In Arkengarthdale, specific locations for aragonite include: Faggergill, Sloat Hole and Windegg mines (Dunham & Wilson 1985).

Wales

Clwyd Aragonite is found at Hendre quarry, Pandy, near Glyn Ceiriog, where joints in a dolerite sill cutting Ordovician slates, mudstones and sandstones are predominantly filled with calcite. It forms colourless prismatic crystals up to 3 mm in length, associated with ankerite and quartz. Some are doubly terminated (Starkey et al. 1991, Critchley 1995).

Dyfed *Central Wales orefield:* Aragonite is thought to be more common than the records suggest. It is reported as white to whitish-grey radiating crystals at Esgair Fraith mine, near Tal-y-bont (Jones & Moreton 1977); and at Castell mine, Ponterwyd, near Aberystwyth; Dolaucothi mine, near Pumsaint; and Henfwlch mine, between Cerrig yr Hafan; and Nant-y-moch reservoir, E of Tal-y-bont (all Bevins 1994).

Gwent Speleothems of aragonite occur in Ogof y Daren Cilau cave, Mynydd Llangattwg (Kendall 1988).

Gwynedd Pale-green aragonite is a rare oxidation product found in copper-mineralized Carboniferous Limestone at Great Orme mine, Llandudno. In places, late generations of aragonite, brochantite, calcite and malachite cement Bronze Age spoil material (Dutton et al. 1994, Ixer 2001). Bevins (1994) notes pale-blue botryoidal masses and radiating colourless needles of aragonite associated with malachite at Ty Gwyn mine on Great Ormes Head. In the S of Gwynedd, a trial level on a copper-mineralized quartz vein, 2 km NW of Bontddu, near Dolgellau, has produced rare colourless-to-snow-white botryoidal masses and divergent needles of aragonite associated with erythrite, gypsum and monohydrocalcite (Saich & Rust 1987).

Mid Glamorgan Aragonite replaces alstonite in cavities and veinstone adjacent to hydrothermal Cu–Fe–Pb–Zn mineralization in dolomitized Triassic conglomerates, limestones and marls immediately S of Llantrisant (Alabaster 1990a). Bevins (1994) reports aragonite from collieries in the South Wales coalfield: Bedwas at Trethomas, Cwm at Llantrisant, Deep Navigation at Treharris, Ferndale and Gelli at Rhondda, Marine at Ebbw Vale, and Windsor at Abertridwr, near Bedwas.

Powys Aragonite occurs in altered iron-rich dolomite at South Nantycar mine, S of Rhayader in the Central Wales orefield (Bevins 1994). At Dolyhir quarry, Old Radnor, near Presteigne, aragonite forms small (< 1 mm) sprays on wallrock adjacent to a galena–tennantite vein (T. F. Cotterell pers. comm. 2007).

Scotland

Dumfries & Galloway Aragonite is well represented in the Leadhills–Wanlockhead orefield, with excellent specimens including spicular, radiating crystals up to 1 cm long on calcite from the Bay mine, Wanlockhead, and 4 cm-long fingers of coralline aragonite on a 14 cm-long Wanlockhead specimen now in the Heddle collection at the NMS (Livingstone 2002). Aragonite is noted in vugs within New Glencrieff vein (Temple 1955) and anthropogenic white, pale greenish-blue or pale-yellow botryoidal crusts of aragonite coat slag material on the dumps of the Meadowfoot smelter at Wanlockhead (Green 1987a).

Fife Pink bifurcating veins of aragonite are noted in volcanic rocks along the shore at Kincraig W of Elie (Livingstone 2002).

Lothian Cavities within the limestone quarry at Pertershill, Bathgate Hills, contain aragonite (Livingstone 2002).

Strathclyde Temple (1955) reports late-forming aragonite in vugs within Brown's, Georges Roust's and Laverock Hall veins around Leadhills. A complex growth of aragonite with multiple terminations (crystals up to 7 cm long) on a limonitic boxwork matrix are illustrated by Livingstone (2002), who also describes a bluish-green radiating variety from the dumps of Laverock Hall vein. At Rubha a' Mhill, northeast Isle of Islay, aragonite occurs in cavities within sulphide veins, where it forms sprays of transparent acicular crystals on paratacamite (Green 1988). Vesicles in Tertiary basaltic lavas around Tobermory on the Isle of Mull and the adjacent mainland (Morvern Peninsula) contain a carbonate assemblage in which calcite, aragonite (much paramorphed by calcite), dolomite, chalcedony and quartz are found (Walker 1971).

Orkney Aragonite is noted from narrow veins at Rackwick, northwest Isle of Hoy, and on the island of Fara (Heddle 1901).

Shetland Aragonite has been found as: stellate and radiating groups at Hagdale chromite quarry, as large crystals in veins on a hill N of Uya Sound, and small crystals associated with brucite at Swinna Ness, all on Unst (all Heddle 1901); crystals to 50 mm in length in irregular veins and nodular masses at Sobul Hill, Unst (Thomson 1892). Livingstone (2002) adds Nikka Vord to the Unst localities. At Hill of Clibberswick, Unst, and East Neap, Fetlar, aragonite occurs with brucite and hydromagnesite cementing scree on high serpentinite cliffs. It has formed since the most recent glacial period by the action of solutions derived from rainwater and sea spray dissolving the scree (Flinn & Pentecost 1995).

Ireland

Antrim Aragonite is an occasional amygdale mineral in the Antrim basalts, particularly on Island Magee, where it forms compact pale-pink or lilac radiating aggregates (Walker 1959a, 1960a). Greg & Lettsom (1858) report aragonite from Portrush and the Giant's Causeway, and a fine oily-green radiating variety from Ballintoy.

Kerry Aragonite is one of the secondary species produced by post-mining oxidation of primary Ag–As–Co–Cu–Fe–Pb–Zn mineralization at Muckross mine, Killarney National Park. It forms white to pale-blue mammillated crusts (Moreton et al. 1999). Fine acicular green aragonite is noted from locations referred to by Greg & Lettsom (1858) as "New Copper mines" at Kenmare, S of Killarney.

Terminated sprays of aragonite crystals on a limonitic matrix; Leadhills, Strathclyde, Scotland; crystals to 7 cm. (© M. P. Cooper)

Londonderry The flos ferri variety of aragonite has been found at the Black Path, Down Hill (Greg & Lettsom 1858).

Tipperary Colourless aragonite needles up to 2 mm long overlying azurite are described from cavities in bornite–tennantite ore from Gortdrum mine, near Tipperary town (Braithwaite & Wilson 2003).

arcadialite, arcadiolite

Synonyms of chabazite, zeolite group.

ardennite-(As)

orthorhombic $(Mn^{2+},Ca)_4(Al,Mg,Mn^{3+})_6(AsO_4)(SiO_4)_2(Si_3O_{10})(OH,O)_6$

A complex manganese–calcium–aluminium–magnesium arsenate silicate hydroxide found at Merehead (Torr Works) quarry, Cranmore, Somerset, where it has developed orange-brown cleavages on vein quartz (Embrey 1978). Ardennite-(As) is associated with apophyllite, datolite and nasonite in a late-stage hydrothermal assemblage, and not with the manganese-rich pods at this locality (Turner 2007). The original name of the species (ardennite) was changed after the discovery of the new species ardennite-(V) (Barressi et al. 2007).

arfvedsonite

AMPHIBOLE GROUP

monoclinic $NaNa_2(Fe^{2+}_4Fe^{3+})Si_8O_{22}(OH)_2$

A relatively common sodium–iron silicate hydroxide that forms a series with magnesio-arfvedsonite and is found in alkalic granites and other plutonic rocks and pegmatites. Where possible, identification is based on the 1997 classification of the amphibole group (Leake et al. 1997a, Mogessie et al. 2004).

Accessory amounts of granular to acicular, or rarely prismatic, blue and green arfvedsonite up to 0.14 mm across occur in trachyte and rhyolite (lundyite) dykes cutting granite on Lundy, off the north coast of Devon (Edmonds et al. 1979, Thorpe & Tindle 1992, Tindle unpub. EMP data).

Arfvedsonite is the most common amphibole in a microgranite from Mynydd Mawr, 5 km W of Snowdon, Gwynedd, Wales. In thin section, brown colouration usually indicates arfvedsonite or ferrorichterite compositions, whereas the blue amphibole is riebeckite (NMW specimen, Tindle unpub. EMP data).

Zoned crystals of arfvedsonite and ferrorichterite occur in the Maol na Gainmhich epigranite at Allt Darach, between Maol na Gainmhich and Meall Buide in the Western Redhills, Isle of Skye, Highland Region, Scotland. They form strongly pleochroic (shades of blue, yellow and green) anhedral crystals, up to 2 mm across, enclosing traces of iron–titanium oxides, allanite and zircon (Thompson 1976). Minor arfvedsonite in a pyroxene-bearing alkali granite at the contact of the Beinn an Dubhaich granite with Durness Limestone, at Camas Malag, near Broadford, Isle of Skye, reported by Tilley (1949) requires confirmation. Arfvedsonite (described as riebeckite) is confirmed in a gneissic granite belonging to the Carn Chuinneag–Inchbae complex, Easter Ross, Highland Region (Harker 1962, Harker & Leahy 2001).

In Strathclyde, Scotland, arfvedsonite forms strongly pleochroic (deep bluish-green, greenish-grey or indigo blue) crystals associated with aegirine in a Lower Carboniferous trachyte sill intruding Old Red Sandstone at Biglees quarry, 500 m SSE of Biglees farm, 4 km N of West Kilbride, Glasgow (Farrow et al. 1982). Unusual zirconium-rich arfvedsonite crystals occur within coarse and fine syenites in the differentiated alkali olivine-dolerite intrusion at Cnoc Rhaonastil, Isle of Islay, where ferro-edenite, ferrokaersutite, ferropargasite, hastingsite, kaersutite and magnesio-katophorite are associated amphibole species (Preston et al. 2000b). The irregular dark-blue patches of amphibole in the Ailsa Craig microgranite, Firth of Clyde, reported many times as riebeckite or riebeckitic arfvedsonite (Teall 1891b, Heddle 1901, Howie & Walsh 1981, Harding 1983, Harrison et al. 1987), should now be described as arfvedsonite, although rare fluoro-arfvedsonite is also present (BGS specimen, Tindle unpub. EMP data).

The Rockall alkali granite, about 440 km W of northern Skye, contains crystals of a near-black amphibole up to 1 cm long and which can appear zoned in thin section from greenish blue to deep blue. These crystals have previously been described as riebeckite (Sabine 1960, Hawkes et al. 1975), but actually have arfvedsonite composition (Sedgwick Museum specimen, Tindle unpub. EMP data).

Arfvedsonite is said to occur at Ross Strand, Killala Bay, Co. Mayo, Ireland, in an unusual drusy, alkali-rich rock within gabbro. It forms crystals in cavities associated with aegirine, quartz, riebeckite and accessory minerals (Ryback & Moreton 1992), but further confirmation is desirable. Optical identifications of arfvedsonite in Ireland are reported from a hornblende kersanite dyke at South Bay, SE of Portaferry, Co. Down (Seymour 1900), and from rhyolite intrusions at Cloghaneely, Co. Donegal (Whitten 1954).

argent natif

A synonym of native silver.

argent rouge

A synonym of proustite or pyrargyrite.

argent sulfuré

A synonym of acanthite.

argentite?

See acanthite. Below 177°C, acanthite (monoclinic) forms paramorphs after argentite (cubic). The name "argentite" has confusingly been used to describe pseudo-cubic crystals of silver sulphide (acanthite), in which the original crystal form of the high-temperature argentite has remained.

argentojarosite?

ALUNITE GROUP
trigonal $AgFe_3(SO_4)_2(OH)_6$

A rare silver–iron sulphate hydroxide supposedly found at Treore mine, Port Gaverne, near Port Isaac, St Teath, Cornwall (Kingsbury 1964b), and West Turf Pits mine, Grassington Moor, Yorkshire (Kingsbury 1965a, Dunham & Wilson 1985). However, Ryback et al. (1998, 2001) have expressed serious doubts about the provenance of many Kingsbury samples (now at the NHM), and further confirmation is therefore desirable. A further occurrence of argentojarosite at Wheal Ludcott, St Ive, near Liskeard, Cornwall, does not have confirmatory analytical details (Golley & Williams 1995).

argentopentlandite

PENTLANDITE GROUP
cubic $Ag(Fe,Ni)_8S_8$

A rare silver–iron–nickel sulphide found in bismuthinite-bearing quartz veins at Groverake mine, near Rookhope, Weardale, Co. Durham. The veins outcrop above the buried Weardale granite and are earlier than the main barite–fluorite–galena–sphalerite mineralization of the Northern Pennine orefield. At this locality, argentopentlandite (5–40 µm in diameter) occurs with bismuthinite, cosalite, cubanite, mackinawite, pyrrhotite and synchysite, and all form inclusions in tin-bearing chalcopyrite. Ten other bismuthinite-bearing localities are listed by Ixer et al. (1996), but it is unclear if they too contain argentopentlandite.

Argentopentlandite is one of many sulphide species described from Glen of the Bar, Talnotry, near Newton Stewart, Dumfries & Galloway, Scotland, where copper–nickel mineralization occurs near the base of a diorite intrusion, close to its contact with hornfelsed Ordovician and Silurian shales and greywackes. Argentopentlandite occurs as inclusions (to 70 µm) in chalcopyrite, where it forms euhedral crystals or intergrowths with pyrrhotite (Stanley et al. 1987).

argentopyrite?

orthorhombic $AgFe_2S_3$

A silver–iron sulphide described by Ixer (1986), Ixer & Stanley (1987), Ryback & Tandy (1992) and Criddle & Stanley (1993) from samples in the Kingsbury collection at the NHM. Originally described as having been collected from Tynebottom mine, NW of Garrigill, near Alston, Cumbria, this provenance now requires further investigation, as Ryback et al. (1998, 2001) have shown many samples in this collection to be incorrectly provenanced. Unless further discoveries are made, this species cannot be considered native to Britain or Ireland.

argentotennantite

TETRAHEDRITE GROUP
cubic $Ag_6Cu_4(Fe,Zn)_2As_4S_{13}$

A rare silver–copper–iron–zinc sulpharsenite that forms a series with freibergite and is listed along with 23 other sulphides and sulphosalts from the large sedimentary-exhalative base-metal and barite deposit in the Silvermines district, Co. Tipperary, Ireland (Moreton 1999). Analytical work by Zakrzewski (1989) on material from this location has proved the existence of arsenic–silver-rich tetrahedrite that could be classified as a member of the argentotennantite–freibergite series, and one analysis is antimony-rich argentotennantite. See tennantite for the origin of the name.

argyrodite

orthorhombic (pseudo-cubic) Ag_8GeS_6

A rare silver–germanium sulphide that forms a series with canfieldite. In the B-zone of the Mogul mine, Silvermines, Co. Tipperary, Ireland, an unusual discovery was made of a lenticular pod of massive barite containing high-grade polymetallic silver mineralization, which consisted of patches, grains and intergrowths of argyrodite associated with acanthite, gersdorffite, jordanite, miargyrite, proustite, smithite and xanthoconite, together with galena and sphalerite (Taylor & Andrew 1979, Taylor 1984, Moreton 1999).

A single grain of argyrodite is reported from the Navan lead–zinc orebody, Co. Meath, Ireland, where minor quantities of boulangerite, bournonite, cylindrite, franckeite, freibergite, jordanite, pyrargyrite and semseyite are found in a predominantly galena–sphalerite ore (Andrew & Ashton 1985, Ashton et al. 1986).

arhbarite?

triclinic $Cu_2Mg(AsO_4)(OH)_3$

A rare copper–magnesium arsenate hydroxide, described from an altered native copper sheet enclosed in relict mudstone from 10 cm below the Triassic Budleigh Salterton Pebble Beds at Littleham Cove, near Budleigh Salterton, Devon. Cuprite is the main alteration species within which hydrated copper arsenates have been found. EMP analysis indicates that these are either arhbarite or euchroite (Milodowski et al. 2002).

armenite

OSUMILITE GROUP?
orthorhombic (pseudo-hexagonal) $BaCa_2Al_6Si_9O_{30}·2H_2O$

A rare hydrated barium–calcium aluminosilicate found as porphyroblasts in a sulphidic calc-silicate-rich quartz rock at Corrie Loch Kander, near Braemar, Grampian, Scotland. The main sulphides in the area are galena, pyrite, pyrrhotite and sphalerite, and associated with the armenite are barite, diopside, hyalophane and tremolite–actinolite. The mineralization is interpreted as the result of seafloor exhalative activity related to subsidence and lateral basin extension during the deposition in the Middle Dalradian (Gallagher et al. 1989a, Fortey et al. 1991, 1993).

arragon spar

A synonym of aragonite.

arrojadite

monoclinic $KNa_4CaMn_4^{2+}Fe_{10}^{2+}Al(PO_4)_{12}(OH,F)_2$

A rare complex phosphate that forms a series with dickinsonite and is found in a lepidolite–tourmaline-bearing pegmatite at Glenbuchat, Grampian, Scotland, where it forms honey-brown glassy anhedral grains in bluish-black wad-like areas up to 1 cm across (Macpherson & Livingstone 1982).

A re-examination of the chemistry and structure of all the known occurrences of arrojadite and related minerals has resulted in the establishment of an arrojadite group and the definition of seven individual species – arrojadite-(BaFe), arrojadite-(KFe), arrojadite-(KNa), arrojadite-(PbFe), arrojadite-(SrFe), dickinsonite-(KMnNa) and fluorarrojadite-(BaFe). Other potential members of the group await further characterization (Cámara et al. 2006, Chopin et al. 2006). It is not known which member of the group occurs in Scotland.

arsendescloizite

DESCLOIZITE GROUP
monoclinic $PbZn(AsO_4)(OH)$

A rare lead–zinc arsenate hydroxide found on hydrothermally altered wallrocks on the dumps of Sandbed mine, Caldbeck Fells, Cumbria. Arsendescloizite forms pale apple-green drusy crusts of rounded boat-shape individual crystals, up to 0.05 mm across covering areas of several square mm, and pale-green crusts associated with mimetite (Green et al. 2006a, Neall et al. 2006b).

arseniate of copper

A synonym of liroconite or scorodite.

arseniate of iron

A synonym of pharmacosiderite.

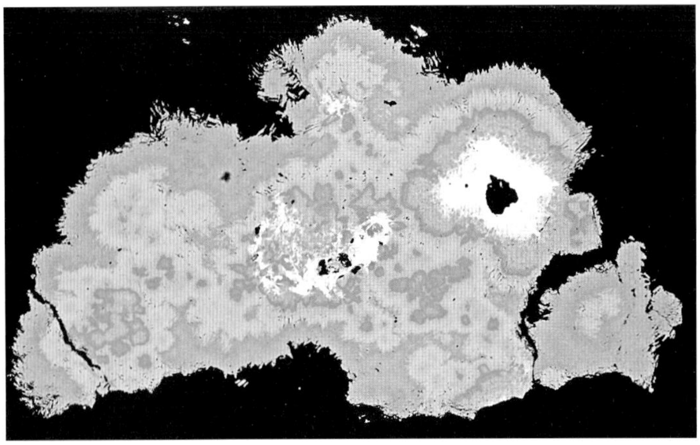

Botryoidal growths of arsendescloizite mantling mimetite (white). Sandbed mine, Caldbeck Fells, Cumbria; width 600 μm; BSE image. (© A. G. Tindle)

arseniate of lead
A synonym of mimetite.

arseniate of lime
A synonym of pharmacolite.

arseniate of nickel
A synonym of annabergite.

ARSENIC GROUP
Trigonal semi-metals that in Britain and Ireland are represented by the native species antimony, arsenic and bismuth.

arsenic

ARSENIC GROUP
trigonal As

Native arsenic has been described from Burraton Coombe quarry, St Stephens by Saltash, Cornwall, where it occurred as large dense masses, externally greyish black and showing a rough columnar jointing. Broken pieces (up to 25 cm across) are bright white with a crystalline fibrous structure and, on some specimens, traces of arsenolite are found. Both species occur in calcite-infilled joints cutting dolerite accompanied by granular arsenopyrite, aragonite and dolomite (Russell 1925, 1944). Smaller lenticular fragments of native arsenic were also found when sinking a new engine shaft at Tolgus mine, Redruth (Russell 1925).

Rare brown encrustations of arsenic on quartz, closely associated with arsenolite, pink and green elbaite, and orthoclase, have been found within cavities in aplite at Meldon quarry, near Okehampton, Devon (von Knorring & Condliffe 1984).

The report (Greg & Lettsom 1858) of arsenic from Dolcoath mine, Camborne, Cornwall, requires further confirmation. Other occurrences in Cornwall, for which no further details are available, include the three in St Just (Geevor, Botallack and Wheal Cock) and Chiverton mine, Perranzabuloe (Golley & Williams 1995).

arsenical cobalt
A synonym of safflorite, löllingite group.

arsenical copper pyrites
A synonym of domeykite.

arsenical iron
A synonym of arsenopyrite or löllingite.

arsenical nickel
A synonym of gersdorffite or nickeline.

arsenical silver
A synonym of dyscrasite, or a mixture of arsenopyrite and dyscrasite.

arsenic oxide
A synonym of arsenolite.

arsenicblüthe
A synonym of arsenolite or pharmacolite.

arseniosiderite

monoclinic $Ca_2Fe^{3+}_3(AsO_4)_3O_2{\cdot}3H_2O$

Scaly, interlocking crystals of arseniosiderite associated with barium-pharmacosiderite (cubes) and scorodite; Howthwaite trial, near Sandbed mine, Caldbeck Fells, Cumbria; width 75 μm; SEM image. (© D. I. Green)

A rare hydrated calcium–iron arsenate found on a single loose block of quartz collected from an abandoned mine dump in Todd Gill, Caldbeck Fells, Cumbria. Arseniosiderite occurs as glistening brown scaly interlocking crystals, individually 2–3 mm long, but which form patches up to 15 mm across. Barium-pharmacosiderite is an associated species (Green & Briscoe 2002; C. M. Leppington pers. comm. 2004). Elsewhere in the Caldbeck Fells, scaly brown crystal groups and dark-brown masses of arseniosiderite occur in the opencut above Sandbed mine, and, at Howthwaite trial, arseniosiderite forms microscopic feathery crystal aggregates associated with scorodite and barium-pharmacosiderite (Green et al. 2006a). The species is also confirmed by XRD from two other locations in the Caldbeck Fells: Sandbeds Gill level, Skiddaw, where it occurs as a thin brown crust in cavities in quartz veinstone, and Dry Gill, where it is found in quartz veinstone and as pale-brown micaceous-like masses on joint planes (D. I. Green pers. comm. 2004).

Arseniosiderite is also reported from: four locations in the Caldbeck Fells, Cumbria – Brandy Gill, Carrock Fell mine, Ingray Gill mine and Nether Row Brow (lower level); Huckworthy Bridge mine, Sampford Spiney, Devon; and Penberthy Croft mine, St Hilary, Cornwall (Kingsbury & Hartley 1957b, 1960). These occurrences are further reported by Hartley (1984), Young (1987a), Camm & Merry (1991b) and Betterton et al. (2000). In all cases, these occurrences deserve further investigation, as Ryback et al. (1998, 2001) have shown many Kingsbury samples, now in the NHM, are incorrectly provenanced.

arsenite
The original form of arsenolite.

arsenkies
A synonym of arsenopyrite.

arsenoflorencite-(Ce)

trigonal $CeAl_3(AsO_4)_2(OH,H_2O)_6$

A rare hydrated cerium–aluminium arsenate hydroxide found as a supergene species in wolframite-bearing quartz veinstone at Hemerdon Bal, Plympton, Devon. Arsenoflorencite-(Ce) forms light-brown rhombohedral crystals, up to 0.5 mm in length, associated with pharmacosiderite and scorodite (Weiss & Merry 2007).

arsenogoyazite

trigonal $SrAl_3(AsO_4)_2(OH,H_2O)_6$

A rare hydrated strontium–aluminium arsenate hydroxide found on the dumps of Wheal Cock, St Just, Cornwall, as white internally zoned pseudo-cubic crystals up

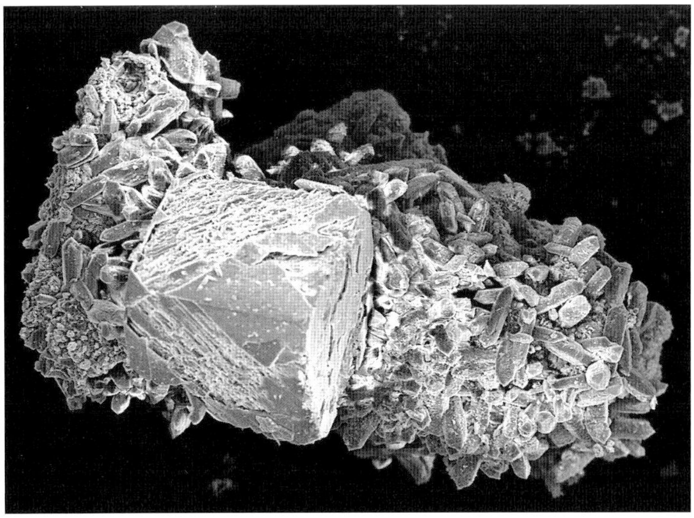

Pseudocubic crystal of arsenogoyazite on matrix; Wheal Cock, St Just, Cornwall; width 600 µm; SEM image. (© D. I. Green)

to about 0.5 mm on edge, in cavities in quartz veinstone (Belson et al. 2005). An arsenogoyazite-like mineral, forming green and yellow-to-white rhombohedral crystals up to 0.25 mm across and clusters to 1 mm, has also been found on a single block from Relistian mine, Gwinear, Cornwall, but XRD analysis proved inconclusive (D. I. Green pers. comm. 2005).

arsenolite

cubic As_2O_3

A rare arsenic oxide, dimorphous with claudetite and described from Burraton Coombe quarry, St Stephens by Saltash, Cornwall, where it occurs as traces on native arsenic, in the form of either very thin white vitreous crusts or minute acicular crystals. Both species were found in calcite-infilled joints cutting dolerite accompanied by granular arsenopyrite, aragonite and dolomite (Russell 1925). Later, realgar and thick creamy-white to yellowish crusts of arsenolite on native arsenic were discovered at this site (Russell 1944). Elsewhere in Cornwall, arsenolite forms colourless octahedral crystals (with white cores) to 0.7 mm associated with annabergite and millerite at Penberthy Croft mine, St Hilary (Rust 1996d, Betterton et al. 2000), and XRD analysis has confirmed its occurrence at Roseworthy, near Camborne, where it is found with gypsum, haidingerite, pharmacolite and weilite as coatings on brickwork and plasterwork on an arsenic calciner (M. R. Power pers. comm. 2004). Other reports of arsenolite in Cornwall are from: Wheal Sparnon, near Redruth (Greg & Lettsom 1858), from Phoenix United mines, Linkinhorne, and from Dolcoath mine, Camborne (Golley & Williams 1995).

At Devon Great Consols mine, near Gunnislake, Devon, arsenolite occurs with gypsum, pharmacolite and scorodite, coating brickwork on an arsenic calciner. Together they develop colloform growths that have clearly formed as weathering products of the arsenic-impregnated masonry (M. R. Power pers. comm. 2004). Accessory arsenolite closely associated with native arsenic is also noted from Meldon aplite quarry, near Okehampton, Devon (von Knorring & Condliffe 1984), and from the Dolgellau goldbelt, Gwynedd, Wales (Readwin 1888).

The occurrences of arsenolite on arsenic calciners are anthropogenic and cannot be considered to be true minerals of Britain.

ARSENOPYRITE GROUP

Monoclinic or orthorhombic sulphides with the general formula *ABS*, where *A* = Co, Fe, Os; and *B* = As, Sb. Arsenopyrite and glaucodot are members of the group found in Britain or Ireland.

arsenopyrite

ARSENOPYRITE GROUP

monoclinic FeAsS

An extremely common and widespread iron–arsenic sulphide of hydrothermal origin, which develops in medium- to high-temperature mineral veins and is often associated with other sulphides such as chalcopyrite, galena and sphalerite. It is also found in tin- and tungsten-bearing veins, gold-bearing quartz veins, granitic pegmatites, gneisses, schists and other metamorphic rocks. Arsenopyrite is the major ore species of arsenic and is described in early literature as mispickel, silver mundic or white mundic. Important sources of arsenopyrite in Britain (mainly southwest England) are documented by Dewey et al. (1920) and Dines (1956).

England

Cornwall Arsenopyrite is generally found in E–W lodes or occasionally in crosscourse veins, as at Treore mine, near Wadebridge. In the early part of the 19th century it was regarded as a contaminant, and large masses were left standing in the stopes on copper lodes. Later it was worked as a byproduct in both copper and tin mines, although it was the predominant mineral at the Calstock and Danescombe mines, NW of Calstock. Arsenic was separated from the other species by roasting the ore in calciners and collecting the soot produced (As_2O_3) from purpose-built flues (Dines 1956, Earls 1996). Dewey (1925) has shown that at Tincroft mine, Illogan, arsenopyrite has a greater affinity with copper and tungsten veins than tin veins and that it is more plentiful in the killas (country rocks) than in the granite. Arsenopyrite is not often found with fluorite. *St Just mining district:* Botallack mine (Robson 1949b, Hosking 1951, Bancroft & Weller 1993); Geevor mine, near Pendeen – early arsenopyrite crystallization was followed by uraninite (pitchblende) intimately associated with bornite and chalcopyrite, then by chalcocite and covellite (Ball et al. 1982, Elton & Hooper 1993); Levant mine and Wheal Castle (Collins 1892); Wheal Cock, near Botallack – metamorphosed greenstone contains arsenopyrite, axinite, calcite, garnet, magnetite and skarn-type assemblages (Kingsbury 1961). *St Ives–Helston mining district:* Carn Barges, near Lamorna Point (Robson 1949a); Carnmeal mine, E of Godolphin; Wheal Providence, near Carbis Bay, St Ives; Wheal Speed, St Ives; Wheal Trewavas, Trewavas Head, Breage (all Ball et al. 1982); Carrick Du mine, near Porthmeor beach, St Ives – excellent arsenopyrite associated with chalcocite, dolomite, erythrite and quartz (Green 1984); Great Wheal Alfred, Phillack, near Hayle, and Relistian mine, Gwinear – arsenopyrite in intrusion breccias (Goode & Taylor 1980); New Rosewarne mine, Gwinear – brecciated lode containing arsenopyrite and cassiterite (Rudler 1905); Penberthy Croft mine, St Hilary – massive steel-grey anhedral aggregates to several cm across and striated tarnished crystals up to 1.2 mm across in quartz-lined cavities (Kingsbury 1954, Camm & Merry 1991b, Rust 1996a,d, Betterton et al. 1998, Betterton et al. 2000); Perran mine, Perranuthnoe, near Marazion – bright arsenopyrite with chalcopyrite, pyrite and sphalerite (Butler 1907); St Michael's Mount, Marazion – minor arsenopyrite with mica, stannite and topaz in cassiterite–quartz–wolframite veins hosted by granite (Hosking 1954c, Jackson & Rankin 1976, Moore & Moore 1979, Alderton 1991, Dominy et al. 1995); Tregurtha Downs, Goldsithney – arsenopyrite, chalcopyrite, covellite and galena in killas traversed by elvan dyke (Ball et al. 1982); Tremearne, near Breage – arsenopyrite (or löllingite) associated with fluorite, lithium mica and topaz in granite pegmatite (Hosking 1951, 1954b); Trevascus mine, near Hayle, Gwinear – arsenopyrite in intrusion breccia (Hosking 1951, Goode & Taylor 1980); Wheal Ann, Phillack – arsenopyrite associated with acanthite (argentite), chlorargyrite, siderite, native silver and cobalt ore (Rust 1982a); Wheal Emma, near Breage – arsenopyrite–quartz lode (Turner & Bannon 1992); Wheal Herland, Gwinear – crosscourse lode containing small amounts of arsenopyrite, native bismuth, galena, stibnite, uraninite (pitchblende), uranyl phosphates and silver, and cobalt minerals (Robson 1949b, Hosking 1951, Ball et al. 1982); Wheal Rose, Sithney, near Porthleven – arsenopyrite was the first species to crystallize in a chalcopyrite–dolomite–galena–quartz–siderite–sphalerite lode (Hosking 1951); Wherry mine, Penzance – well defined tin-white crystals up to 6 mm across in chlorite–quartz-bearing killas (Russell 1949a, Ball et al. 1982). *Camborne–Illogan–Redruth mining district (and southwards into the Carnmenellis granite and its environs):* Bosworgey borehole (in the Carnmenellis granite), near Fraddam – altered granite containing dominantly euhedral crystals of arsenopyrite, often enclosing inclusions of native bismuth, bismuthinite and galena (Ball & Basham 1984); Chywoon quarry, near Longdowns – a few platy crystals of arsenopyrite associated with molybdenite in granite pegmatite (Moulding & Hooper 2005); Creegbrawse and Penkevil mines, Chasewater, near Redruth; West Poldice mine, St Day (all Le Neve Foster 1877); Dolcoath mine, Camborne, and Wheal Unity, St Day, Gwennap – arsenopyrite occurred with many other sulphide species at both locations (Hosking 1951); East Pool and Agar mine, Illogan, near Redruth – arsenopyrite- and chalcopyrite-rich ores at depth within metamorphosed killas overlying granite (Dewey 1925, Ball et al. 1982); New Cook's Kitchen mine (North Tincroft lode), Redruth – arsenopyrite, cassiterite, chalcopyrite and sphalerite occur in a dense non-brecciated chlorite–fluorite–quartz matrix (LeBoutillier et al. 2000); New Tolgus shaft, Redruth – metamorphosed greenstone containing arsenopyrite, axinite, chlorite, garnet, pyrrhotite and siderite (Kingsbury 1961); South Crofty mine, Camborne – early, flat-lying pegmatitic lodes (quartz floors) contain arsenopyrite, löllingite, wolframite and rare stannite (Hosking 1951, 1954b, Ball et al. 1982, Scrivener et al. 1986, Wolters & Wolters 1987, Dominy et al. 1994, LeBoutillier 1996); Ting Tang mine (Morcom's shaft), Carharrack – abundant silvery masses of arsenopyrite on the dumps, in places altered to pharmacosiderite or scorodite (Bruce & Aubrey-Jones 1998); Wheal Basset, Illogan, near Redruth – accessory arsenopyrite is associated with bassetite, cassiterite, chalcocite, chalcopyrite, vochtenite and wolframite at the contact of the Carnmenellis granite with Devonian sediments (Zwaan et al. 1989); Wheal Gorland, St Day, 2 km E of Redruth (Fisher 1988); Wheal Jane, Kea, near Truro – arsenopyrite is associated with cassiterite, muscovite and quartz in an early greisen assemblage and is also found on specimens of pyrite–quartz–siderite–vivianite (Rudler 1905, Rayment et al. 1971, Dominy & Camm 1997); Wheal Pendarves, SW of Camborne – arsenopyrite occurs in a chlorite–quartz assemblage that postdates cassiterite mineralization (Scrivener et al. 1986). *St Agnes mining district:* Cligga Head, near St Agnes – greisen bands in granite contain central quartz veins that host arsenopyrite, cassiterite, chalcopyrite, tourmaline and wolframite (Rudler 1905, Hall 1971, Hall & Jackson 1975, Jackson et al. 1977, Charoy 1979, Chick 1987); Wheal Kitty – bold crystals with a brilliant reddish tarnish associated with chalcopyrite and siderite (Rudler 1905). *Wadebridge–St Endellion mining district:* Barret's Zawn, 5 km E of Port Gaverne – nests of arsenopyrite crystals in slate cliffs (Shrimpton 1993); Bounds Cliff, E of Port Isaac and Pentire (Glaze) mine, near New Polzeath, St Minver – arsenopyrite (marcasite–pyrite) dominated veins formed early in vein systems that went on to develop silver-rich galena–stibnite mineralization (Clayton et al. 1990); Duchy Great Consols mine, Gunnislake; Emmens United mine (Holmbush and Kelly Bray), Callington; Florence and Tonkin mine, Callington (all Le Neve Foster 1877); Old Treburgett mine, St Teath, E of Port Isaac – minor arsenopyrite in galena–sphalerite-bearing quartz–carbonate veins (Kingsbury & Hartley 1960); Treore mine, Port Gaverne, near Port Isaac, St Teath – arsenopyrite-rich vein containing jamesonite, siderite and traces of chalcopyrite, galena and native gold (Dewey et al. 1920, Dewey 1925, Hosking 1951, Kingsbury & Hartley 1960, Shrimpton 1993, Camm 1995). *St Austell mining district:* Bunny (St Austell Hills) mine, Bugle – small amounts of arsenopyrite and pyrite in cassiterite–wolframite lodes (Dines 1956); Castle-an-Dinas mine, St Columb Major – arsenopyrite occurs in quantity in granite and is one of the rarer species found in a lode containing lithium-mica, topaz and wolframite (Hey et al. 1938); Fowey Consols, Tywardreath, St Blazey – newly found material containing rhabdophane on a matrix of arsenopyrite and chalcopyrite is documented by Kingsbury (1964b), but arsenopyrite is not reported by Lewis (1997); Goonvean and Rostowrack china-clay pits, W of Nanpean – arsenopyrite is one of many accessory species found in the waste from the pits (Scott et al. 1998); Gunheath china-clay pit, Carthew – arsenopyrite is a constituent of the greisen, stockwork and quartz-vein mineralization and is also noted in pegmatitic vugs in the roof zone of the granite (Hodkinson & Clark 1977, Manning 1983, Alderton 1991, Weiss 1994); Mulberry mine, Lanivet – a tin stockwork containing arsenopyrite–cassiterite–pyrite–stannite–wolframite-bearing quartz veins (Hosking 1951, Starkey 1986a, Dominy et al. 1995); South Terras mine, St Stephen-in-Brannel – good specimens of massive grey arsenopyrite on the dumps (Robertson & Dines 1929, James 1947, Ball et al. 1982, Shrimpton 1994); St Austell granite – xenoliths in the granite are often ringed with arsenopyrite and covellite (Manning et al. 1996); Treburland mine, 2 km SE of Fivelanes, Altarnun, Bodmin Moor – tin-white short prismatic crystals up to 3 mm in length in a manganese-rich skarn deposit (Russell 1946d, Kingsbury 1961); Wheal Maudlin, 3 km NW of Lostwithiel – worked a complex assemblage that included arsenopyrite, cassiterite, chalcopyrite, fluorite, galena, garnet, pyrite, pyrrhotite, scheelite and sphalerite (Kingsbury 1961). *East Cornwall mining district:* The district extends from the eastern margin of the Bodmin Moor granite, past the granite cupolas of Kit Hill and Hingston Down towards the western margin of the Dartmoor granite in Devon and includes: Cannaframe mine, Bodmin Moor – arsenopyrite–wolframite lode that appears to merge with a pegmatite vein also carrying cassiterite, chalcopyrite and fluorite (Brooks 2001); Cargibbitt mine (Caradon Vale), Caradon town and Penhargett mine (Caradon Great Consols), Penhargett, 7 km NE of Liskeard (Shambrook 1982); Hawkswood mine, North Hill – arsenopyrite is reported from localized areas in wolframite-bearing pegmatite veins close to the Bodmin Moor granite contact (Hosking 1954b, Brooks 2001); Hingston Down Consols mine, 2 km W of Gunnislake, Calstock – massive arsenopyrite associated with anglesite, chalcopyrite, covellite, fluorite, mimetite, pharmacosiderite, scorodite and sphalerite (Russell 1910a); Hingston Down quarry – the rare species jeanbandyite and natanite have been found on an arsenopyrite matrix (Wolfe 1997, Betterton et al. 1998); Okel Tor mine, Calstock, near Gunnislake, and Prince of Wales mine, Calstock – quartz veins containing arsenopyrite (±chalcopyrite) (Turner & Bannon 1992); Redmoor, Holmbush, and Kelly Bray mines, all near Callington – arsenopyrite occurs in greisen-bordered veins, sulphide veins and mineralized cross-course veins (Newall & Newall 1989, Newall 1994); Wheal Langford, Callington – arsenopyrite, chalcopyrite, pyrite and sphalerite in a silver-rich lode (Rust 1982a); and Wheal Mary, Lelant – silver-rich arsenopyrite associated with cobalt ores (Hosking 1951); Wheal Newton, Harrowbarrow, SW of Gunnislake (Le Neve Foster 1877). *Other occurrences:* Burraton Coombe quarry, St Stephens by Saltash – granular arsenopyrite associated with native arsenic and traces of arsenolite in a calcite-filled joint within dolerite (Russell 1925); Camel and Gannel estuaries – arsenopyrite in intertidal sediments (Pirrie et al. 2000b); and Lill Cove mine, Trebarwith Strand, near Tintagel – arsenopyrite, chalcopyrite and pyrite in a quartz lode (Sharpe 1990).

Cumbria In a review of cobalt, copper, lead and zinc mineralization in the Lake District, Stanley & Vaughan (1982a) define two major types of vein deposits:

- arsenopyrite–chalcopyrite–pyrite of Lower Devonian age
- galena–sphalerite of early Carboniferous age.

When the former is closely associated with granite intrusions, more complex mineral assemblages are found involving molybdenum- or tungsten-bearing phases. *Caldbeck Fells:* West bank of Arm o' Grain beck (Green et al. 2006b); Brandy Gill mine (older references probably relate to the present site of Carrock mine) (Greg & Lettsom 1858, Goodchild 1882, Davidson & Thomson 1951, Hartley 1984); Burdell Gill – arsenopyrite was once an important constituent of the quartz veins here, but apart from one wolframite-bearing vein it is now completely altered to pharmacosiderite (Neall & Leppington 2003); Carrock mine, Grainsgill – abundant and well crystallized arsenopyrite associated with scheelite and wolframite close to the Skiddaw granite contact (Finlayson 1910, Hitchin 1934, Dunham & Hollingworth 1947, Davidson & Thomson 1951, Eastwood et al. 1968, Shepherd et al. 1976, Fortey 1980, Beddoe-Stephens & Fortey 1981, Young 1987a, Watson 1988, Shrimpton 1993a, Green & Briscoe 2002); Coombe Height, headwaters of Wet Swine Gill – arsenopyrite occurs with abundant stibnite and a suite of Ag–Fe–Pb–Sb species in a sulphide-bearing quartz vein (Green & Briscoe 2002); Deer Hills vein – massive arsenopyrite much altered to scorodite and pharmacosiderite (Cooper & Stanley 1990, Symes et al. 1990); Driggith mine – arsenopyrite forms inclusions in chalcopyrite and galena (Stanley & Vaughan 1981), and in the opencut between Driggith and Sandbed mines it forms abundant sub-mm size euhedral crystals in volcaniclastic quartz rock (Green et al. 2006a); Grainsgill Beck cutting the northern end of the

Loose striated crystal of arsenopyrite, 20 mm tall, from Smith vein, Carrock mine, Caldbeck Fells, Cumbria. (© M. P. Cooper)

ARSENOPYRITE

Skiddaw granite – flecks of arsenopyrite occur with muscovite and quartz in greisen (Roberts 1983); higher part of Brandy Gill (Kingsbury & Hartley 1960); Ingray Gill mine (Hartley 1984); lower level and top level of Nether Row Brow (near Dumpy Stone), and Potts Gill barite mine (Kingsbury & Hartley 1960, Hartley 1984); Wet Swine Gill – antimony-rich arsenopyrite forms clusters of euhedral tabular crystals in a berthierite–(jamesonite)–stibnite–zinckenite-bearing quartz vein (Fortey et al. 1984, Leppington & Green 1998, Neall & Green 2001b). *Elsewhere in the Lake District and West Cumbria:* Bonser vein, Coniston, one of the most productive sources of copper in the Lake District, also contains arsenopyrite as large fractured euhedral or subhedral grains and early zoned rhombs, some forming inclusions in pyrrhotite (Stanley & Vaughan 1982b); Castle Nook mine, Newlands Valley – small zone of cobalt-rich arsenopyrite; Greenburn mine, Greenburn Valley – arsenopyrite–chalcopyrite–pyrite–sphalerite in Sump Vein; Hellens mine, Tilberthwaite – abundant arsenopyrite on the dumps; Tilberthwaite Gill – vein containing early formed arsenopyrite–chalcopyrite–pyrite and later galena–sphalerite–tetrahedrite (all Stanley 1979); Coniston area – arsenopyrite is an important constituent of many of the copper veins in the area (the primary mineralization is vertically zoned, with chalcopyrite–pyrite near surface, arsenopyrite at greater depth and magnetite in the deepest workings) (Dagger 1977, Firman 1978); Copper Plate (Borrowdale) mine, Borrowdale – particularly rich in arsenopyrite, commonly forming subhedral single crystals or clusters in quartz; Manesty vein, Grange, Borrowdale – arsenopyrite and pyrite have well developed crystal faces, but both are brecciated and overgrown with abundant chalcopyrite (both Strens 1962, Millward et al. 1999); Dale Head mine, Newlands Valley – arsenopyrite occurs as overgrowths on pyrite and as large subhedra (≤5 mm) which have been fractured and infilled with bismuthinite, chalcopyrite, galena, pyrrhotite and sphalerite (Stanley & Vaughan 1980); Embleton quarry – arsenopyrite is associated with dickite, native gold, pyrite, scorodite, stibnite and possibly tetrahedrite in a quartz vein; Fleetwith Pike, Buttermere Fell – arsenopyrite in several small veins; Force Crag mine, Braithwaite; Graphite mine, Seathwaite – uncommon arsenopyrite; Little Mine Crag, Newlands Valley, near Keswick (all Strens 1962); Goldscope mine, Newlands Valley – a well known locality for arsenopyrite, which sometimes occurs in well developed crystals (the gold content of arsenopyrite from this location is doubted by later workers) (Greg & Lettsom 1858, Rastall 1942, Davidson & Thomson 1951, Stanley & Vaughan 1980); Piers Gill, 1 km N of Scafell summit, Wasdale – arsenopyrite occurs with chalcopyrite, galena, pyrite and sphalerite in a 60 cm-thick polyphase quartz vein cemented by ankerite, within the Scafell dacite (rare associated species include acanthite (argentite), native gold (the silver–gold alloy, electrum), tetrahedrite, uytenbogaardtite and an unknown selenium–silver mineral) (Aldridge 2005); Scar Crag cobalt mine, Causey Pike – arsenopyrite forming large euhedral and subhedral grains and aggregates up to 3 mm across is associated with alloclasite, glaucodot and minor amounts of cobaltite and skutterudite in a chlorite-rich quartz vein (some samples examined were from the Kingsbury collection at the NHM) (Davidson & Thomson 1951, Ixer et al. 1979a, Stanley 1979); Seathwaite mine, Seathwaite Tarn, near Coniston – small amounts of arsenopyrite occur in a sulphide assemblage dominated by bornite, digenite and djurleite, and notable wittichenite (Stanley & Criddle 1979); borehole material from the Sellafield area contains evidence of many mineralizing events including a polymetallic As–Cu–Fe–Pb–Zn sulphide assemblage containing arsenopyrite, chalcopyrite, galena, marcasite, pyrite, pyrrhotite, sphalerite and Ag–Bi–Se sulphosalts, in Borrowdale Volcanic Group rocks (Milodowski et al. 1998); Wanthwaite mine, St John's in the Vale, near Threlkeld – massive granular intergrowths of arsenopyrite, galena, pyrite and sphalerite in a thick rib of quartz (Kingsbury & Hartley 1960); Ulpha copper mine, Ulpha – finely crystalline masses of arsenopyrite with chalcopyrite and chlorite in quartz; Whitecombe Beck, Black Combe, Whicham – locally abundant arsenopyrite associated with chalcopyrite in quartz veinstone from several old mines in the valley (both Young 1987a); Wasdale Head – arsenopyrite coating joints in Borrowdale Volcanic Group rocks (Buckley 1984); Water Crag, near Devoke Water – arsenopyrite–bismuthinite–fluorite–molybdenite mineralization occurs in topaz greisen at the margin of the Eskdale granite (Young 1985d, Young et al. 1986); Yewthwaite mine, Newlands Valley – arsenopyrite in a copper-rich vein (Strens 1962).

Derbyshire Arsenopyrite is a rare species in the Southern Pennine orefield, but is listed from Eyam (Hall 1868b); High Rake, Hucklow (Mostagel 1984); and Millclose mine, Wensley, near Matlock (Robinson & Ineson 1979).

Devon Bedford Consols mine, near Calstock; Franco Consols mine, Horrabridge, near Tavistock; Gawton mine, near Calstock; Lady Bertha mine, Buckland Monachorum; Old Bottle Hill mine, Plympton; Tavy Consols mine, Tavistock Hamlets; Wheal Crebor, near Tavistock; Wheal Mary Hutchings mine, Plympton (all Le Neve Foster 1877); Bedford United mine, near Tavistock – zoned deposit with arsenopyrite and chalcopyrite in the centre of the lode and tin–tungsten ores next to the walls (Le Neve Foster 1877, Dewey 1925); Belstone mine, near Okehampton – arsenopyrite associated with axinite, chalcopyrite and garnet in a skarn assemblage (Le Neve Foster 1877, Rudler 1905); Brookwood mine, near Ashburton – banded arsenopyrite–chalcopyrite ore (Durrance & Laming 1982); Devon Great Consols (includes Wheals Anna Maria, Emma, Fanny, Josia, Maria, South Fanny and Watson's mines) – the largest sulphide lode in southwest England produced 540 tonnes of arsenopyrite and 70 800 tonnes of crude arsenic as well as pyrite, copper and tin ores (Dines 1956, Earl 1996); Hemerdon Bal, Plympton – arsenopyrite–cassiterite–wolframite-bearing stockwork (Keats 1981, Ball et al. 1985b); Ivy Tor mine, Belstone, Meldon area, near Okehampton – arsenopyrite and bismuth ores occurred in this mainly copper mine (Hall 1868b, Dines 1956, Braithwaite et al. 1996); Loddiswell Pb–(Ag) mine – arsenopyrite and pyrite were followed by the crystallization of a suite of Ag–Au–Bi–Cu–Pb species (Stanley et al. 1990b); Lundy island – arsenopyrite–cassiterite–chalcopyrite–wolframite-bearing veins cutting granite (Bromley & Holl 1986, Thorpe et al. 1990); North Devon United mine, Peter Tavy – an ore dump near a partially flooded adit was found to be composed dominantly of arsenopyrite, chalcopyrite and scheelite, with minor native bismuth in a quartz gangue (Rumsey & Savage 2005); Meldon (British Railways quarry complex), near Okehampton – small amounts of arsenopyrite, chalcopyrite and molybdenite in veins cutting metasomatized calc-flinta rocks in the aureole of the Dartmoor granite (Dearman 1966b); Red-a-ven mine, Meldon, near Okehampton – traces of arsenopyrite associated with much löllingite and pyrrhotite in a skarn deposit at the margin of the Dartmoor granite (Kingsbury 1961); Sortridge Consols mine, Whitchurch – abundant arsenopyrite in killas on which both beaverite and childrenite have later grown (Braithwaite & Cooper 1982); Tamar mines, near Bere Ferrers and Virtuous Lady mine, Buckland Monachorum – fine crystals at both mines (Hall 1868b, Collins 1892); West Wheal Crebor, Hurlditch Horn, near Tavistock – arsenopyrite,

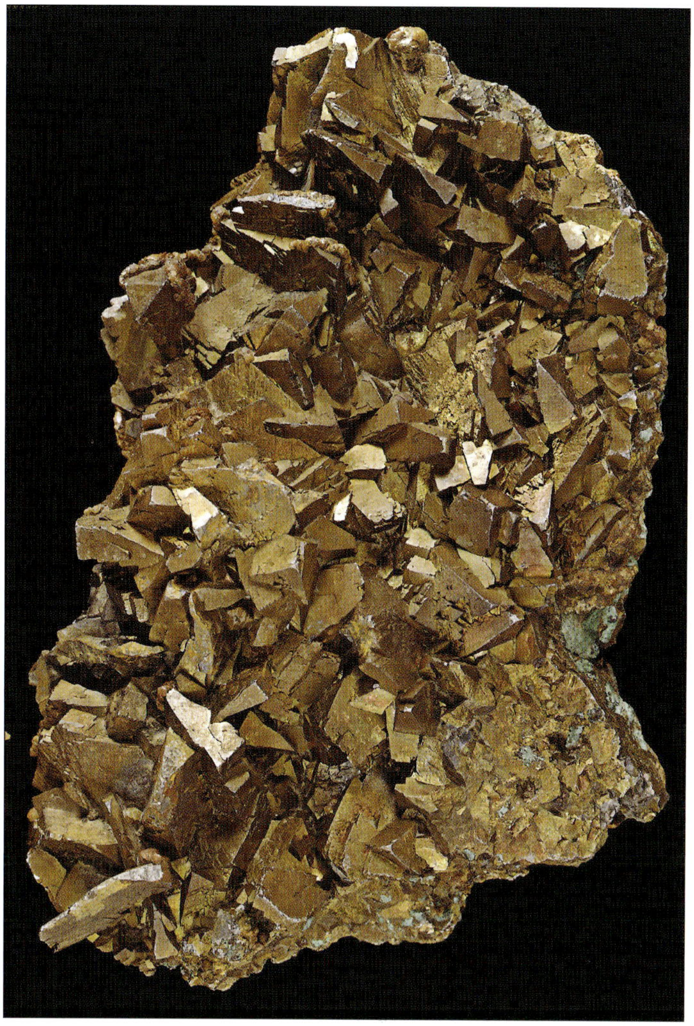

Tapering, intergrown crystals of arsenopyrite up to 32 mm on edge; green encrustation is scorodite; Virtuous Lady mine, Buckland Monachorum, Devon. (© The Natural History Museum, London)

chalcopyrite and pyrite occur in veinstone on dumps around the shafts (Dines 1956); Wheal Forest, Meldon Gorge, near Okehampton – arsenopyrite associated with apatite, heulandite, orthoclase and quartz (Russell 1910b); and Wheal Friendship (Devon Friendship), Mary Tavy – arsenopyrite is one of a wide range of species found at this mine in the metamorphic aureole of the Dartmoor granite (Dewey 1925, Brooke 1986, Braithwaite 1982b).

Durham At Closehouse mine, SW of Middleton-in-Teesdale, minor amounts of arsenopyrite occur in a pyrite–pyrrhotite–sphalerite assemblage. Samples from various locations in Teesdale Valley contain arsenopyrite in a complex Ag–Cu–Fe–Ni–Pb–Sb sulphide assemblage (Vaughan & Ixer 1980). Minor arsenopyrite is also noted in a borehole core that sampled the Weardale granite beneath Rookhope, Weardale (Brown et al. 1987a).

Staffordshire Arsenopyrite is noted from the mines on Ecton Hill (Sarjeant 1957, Robey & Porter 1972).

Wales

Dyfed Arsenopyrite is an important constituent of gold-bearing pyritic shales and quartz veins at the Dolocauthi mines, Pumsaint. The main target for mining was the Roman lode directly underlying the Ogofau pit, where gold was found as inclusions in both arsenopyrite and pyrite (Brammall et al. 1937, Steed et al. 1976, Annels & Roberts 1989). A report of arsenopyrite inclusions in galena, or occasionally in chalcopyrite, from Hyddgen mine, near Nant-y-moch reservoir, Central Wales orefield (Raybould 1974, Jones & Moreton 1977), is considered dubious by Bevins (1994).

Gwynedd Minor arsenopyrite is present in the large copper deposit at Parys Mountain, Anglesey, where it forms zoned, rhombic or lath-shape crystals associated with tetrahedrite group and bismuth-bearing minerals (Pointon & Ixer 1980). It is also noted from Rhôs-mynach mine, Anglesey (Bevins 1994). At Tan-y-Garth mine, Ogwen Valley, near Bethesda, arsenopyrite was mined from a complex sulphide ore found in a series of small veins and within Cambrian sandstones. The ore also contained chalcopyrite, pyrite and pyrrhotite (Dewey et al. 1920). Arsenopyrite is also found at a few locations in Snowdonia, including Drws-y-Coed mine, near Nantlle (Braithwaite 1982b), and Simdde Dyllhuan, Drwys-y-coed, Nantlle (Bevins 1994).

The remaining occurrences of arsenopyrite are from the Dolgellau goldbelt, where it is a relatively minor but widespread component of the gold-bearing vein assemblages: Cefn Coch and Foel Ispri mines (Mason et al. 2002); Clogau mine, Bontddu, near Dolgellau – arsenopyrite in a vein assemblage that also includes cobaltite, pyrite and pyrrhotite (Andrew 1910, Dewey & Eastwood 1925, Gilbey 1968, Allen & Jackson 1985, Naden 1988); Cwm Prysor mine (Gilbey 1968); Dolfrwynog mine, Llanfachreth (Dewey & Eastwood 1925); Glasdir mine, Llanfachreth, near Dolgellau (Allen & Easterbrook 1978, Allen & Jackson 1985); Gwynfynydd mine, near Dolgellau (Allen & Jackson 1985, Dominy et al. 1996c); Prince Edward mine (Gilbey 1968, Mason & Williamson 2001, Mason et al. 2002); St David's mine (Andrew 1910); Turf Copper mine, Coed-y-Brenin, near Dolgellau (Rice & Sharp 1976); Tyddyn Gwladys mine (Forbes 1868); Vigra mine (Andrew 1910, Gilbey 1968); trial level, 2 km NW of Bontddu, near Dolgellau (Saich & Rust 1987).

Striated arsenopyrite crystal (7×15 mm) from St David's mine, near Dolgellau, Gwynedd, Wales. (© National Museum of Wales, Cardiff; photographer M. P. Cooper)

Scotland

Borders Glengaber Burn, SW of Peebles – arsenopyrite inclusions in alluvial gold (Chapman et al. 2000a,c).

Central Arsenopyrite is noted from Airthrie Hill mine, near Bridge of Allan, and Alva mine in the Ochil Hills (Dickie & Forster 1974). However, despite extensive excavations in the mid-1990s, arsenopyrite could not be confirmed at these locations (S. Moreton pers. comm. 2007). At Cononish, near Tyndrum, both alluvial gold and *in situ* vein gold contain arsenopyrite inclusions (Chapman et al. 2000c). Other arsenopyrite inclusions in alluvial gold are recorded from the Ochil Hills (Chapman et al. 2005).

Dumfries & Galloway Small amounts of arsenopyrite have been found with calcite and chalcopyrite in quartz veins cutting diorite 500 m NNW of Barncorkrie, Drummore, Portencorkie area (Cameron & Lawson 1979). Porphyry copper-type mineralization containing some arsenopyrite in fissure veins is described from Black Stockarton Moor to the W of the Criffell granite complex (Fortey 1980). At Talnotry, near Newton Stewart, arsenopyrite occurs in an arsenic-rich magmatic sulphide assemblage that developed near the base of an altered diorite sheet. Associated species include annabergite, bismuthinite, chalcopyrite, froodite, galena, gersdorffite, gold (electrum), hollingworthite, irarsite, nickeline, pentlandite, pyrite, pyrrhotite and sperrylite. Nearby quartz veins at the margin of the Cairnsmore of Fleet granite also carry arsenopyrite (Heddle 1901, Lowry 1993, Lowry et al. 1997, Power et al. 2004). In Palnure Burn, arsenopyrite occurs in a vein close to the contact of the Cairnsmore of Fleet granite with Silurian metasediments (Dudgeon 1895). Disseminated arsenopyrite–pyrite mineralization (most abundant in the wallrocks to stibnite-bearing veins) occurs at Glendinning (Louisa) mine, Glenshanna Burn, 13 km NNW of Langholm (Duller et al. 1997), and in boreholes around the mine (Gallagher et al. 1983). Arsenopyrite, associated with gold mineralization is further noted from Glenhead Burn, Loch Dee (Leake et al. 1981), and from Cairngarroch mine, SE of Portpatrick. At Knipe (Fountainhead) mine, Hare Hill, 5 km SE of New Cumnock, arsenopyrite occurs in a zoned gold-bearing As–Cu–Pb–Sb–Zn assemblage hosted by a late-Caledonian granodioritic intrusion and contains bournonite in late cross-cutting quartz veinlets. Bournonite, chalcopyrite, sphalerite, stibnite and tetrahedrite are associated species (Boast et al. 1990, Duller et al. 1997).

Highland Kildonan Burn, near Helmsdale – arsenopyrite inclusions in alluvial gold (Chapman et al. 2000c); Isle of Rona, off Cape Wrath – arsenopyrite in granite (Greg & Lettsom 1858).

Strathclyde Arsenopyrite is the principal sulphide in a suite of quartz–carbonate veins found in the Fore Burn igneous complex, Tairlow plantation, near Straiton, and together with pyrite it hosts gold inclusions (Allen et al. 1982, Charley et al. 1989). Alluvial gold containing inclusions of arsenopyrite is noted from Fore Burn, from the Leadhills–Wanlockhead area and from Lochy River, W of Tyndrum (Leake et al. 1998, Chapman et al. 2000c). Finely disseminated a porphyry copper or molybdenum-style of mineralization within a subvolcanic complex (comprising a diatreme intruded by breccia pipes, stocks, sheets and dykes) at Lagalochan, near Kilmelford (Fortey 1980, Harris et al. 1988).

Tayside Minor arsenopyrite, chalcopyrite and traces of gold (electrum) occur in polyphase galena–pyrite–sphalerite-bearing quartz veins at Calliachar Burn, Aberfeldy. Similar veins are also present in nearby Urlar Burn (Ixer et al. 1997). Chapman et al. (2000c) describe arsenopyrite inclusions in native gold from Calliachar.

Ireland

Carlow Rare arsenopyrite occurs in spodumene-bearing pegmatite at Aclare, near Myshall (Scoon 1978), and in tourmalinite close to other spodumene pegmatites at Coolasnaghta (McArdle et al. 1989).

Clare Disseminated sulphide deposits in argillaceous limestones and shales at Ballyvergin, Ballyvergin Townland and at the Maghera Cross copper prospect, Maghera Townland, are dominated by chalcopyrite; arsenopyrite forms a minor component of the mineralization (Schultz 1971). Andrew (1986) reports arsenopyrite, chalcopyrite, galena and pyrite as the dominant sulphides at Ballyvergin.

Cork Laterally extensive quartz veins in Devonian redbeds at Dhurode mine, Mizen Head, host arsenopyrite–bournonite–chalcopyrite–meneghinite–tetrahedrite mineralization (Bartlett 1980, Kinnaird et al. 2002). Minor quartz veins at Kilkilleen in the

West Carbery mining district host arsenopyrite–chalcopyrite–galena–tetrahedrite mineralization (Ni Wen et al. 1996).

Galway The large Cu–Pb–Zn–(Ag) cavity-filling and limestone replacement orebody at Tynagh mine, near Killimor, contains minor arsenopyrite replacing chalcopyrite (Morrissey et al. 1971, Schultz 1971, Clifford et al. 1986).

Kerry Old records refer to disseminations and veins at Ballybunnion containing arsenopyrite, chalcopyrite and pyrite, although only the last species was found by Moreton et al. (1995). The primary stratabound Cu–Pb–Zn ore at Muckross mine, Killarney National Park, mostly consists of fine-grain arsenopyrite, chalcopyrite, galena, pyrite, sphalerite and tennantite (Ryback & Moreton 1991, Moreton et al. 1999, Kinnaird et al. 2002). On nearby Crow Island, near Muckross, arsenopyrite forms discrete 10–100 µm-long, subhedral to euhedral rhombic zoned crystals and in places occurs as inclusions in chalcopyrite, galena or sphalerite (Kinnaird et al. 2002, Ixer 2004).

Kilkenny At the southwestern end of the Leinster batholith, near Inistioge, metasomatic tourmaline-rich rocks contain disseminated arsenopyrite (well formed and up to 5 mm long) associated with chalcopyrite, chlorite, fluorite, manganese-rich ilmenite, molybdenite, muscovite, pyrite, pyrrhotite and rutile. The mineralization was produced during alteration of aluminous metasediments and is spatially related to pegmatitic quartz veins showing greisen alteration (Appleby & Williams 1988).

Limerick The Courtbrown epigenetic lead–zinc deposit on the southern shore of the River Shannon near Ballysteen is hosted by Carboniferous reef limestones and contains abundant galena and sphalerite, with arsenopyrite in basal and transitional reefs (Grennan 1986, Reed & Wallace 2001).

Mayo Dalradian quartzite hosting gold-bearing quartz veins at the Lecanvey prospect, Clew Bay, contains abundant disseminated arsenopyrite (Aherne et al. 1989). Arsenopyrite also occurs as inclusions in alluvial gold from: Bunowen River, Glenkeen Bridge on the Carrownisky River, Croagh Patrick, Lugbeg River and Owannadornaun River (all Chapman et al. 2000c).

Monaghan At Clontibret mine, quartz veins are mineralized with stibnite, arsenopyrite and pyrite, the latter two species containing native gold as a lattice constituent (Morris 1984, Morris et al. 1986, McArdle 1989, Ryback & Moreton 1993, Steed & Morris 1997). At Bryanlitter, near Clontibret, the host rocks to gold–stibnite-bearing veins contain disseminated, but locally abundant, arsenopyrite and pyrite (Steed & Morris 1997).

Tipperary Partially dolomitized Carboniferous limestones and shales at Gortdrum mine, near Tipperary town, contain minor arsenopyrite in a major bornite–chalcocite–chalcopyrite–tennantite deposit (Morrissey et al. 1971, Schultz 1971). The large sedimentary-exhalative Ba–Pb–Zn deposits in the Silvermines district have been worked from several mines. Arsenopyrite is described from the following: Knockanroe and Ballygown South mines – abundant needles of arsenopyrite in a thin bed of dolomitic and sandy shale (Rhoden 1959); Mogul mine – subordinate arsenopyrite in the "Lower ore zone" (Schultz 1971, Moreton 1999); Shallee mine – minor arsenopyrite is associated with major amounts of barite, galena and pyrite, and minor sphalerite, tetrahedrite and other silver-rich sulphosalts (Taylor & Andrew 1978); Silvermines (G ore zone) – arsenopyrite is one of the minor and trace species (Morrissey et al. 1971). Minor arsenopyrite is also noted in three stratabound sulphidic lenses dominated by galena, marcasite, pyrite and sphalerite, with minor bornite, chalcopyrite, gersdorffite, gratonite, jordanite, seligmannite and tennantite in the carbonate-hosted Lisheen deposit, between Thurles and Rathdowney (Hitzman et al. 2002, Wilkinson et al. 2005).

Tyrone Arsenopyrite occurs in and adjacent to gold-bearing quartz veins within Dalradian metasediments in the Sperrin Mountains at: Attagh Burn; Cavanacaw; Curraghinalt; Glencurry Burn; and Golan Burn, near Strabane (Clifford 1986, Earls et al. 1989, Woodham et al. 1989, McCaffrey & Johnston 1996, Chapman et al. 2000c, Parnell et al. 2000).

Wicklow At Ballinglen, between Tinahely and Aughrim, greisenization of the Main Leinster granite has produced fluoritized microadamellite containing some arsenopyrite (Steiger & Bowden 1982, Kennan et al. 1986, McArdle & Kennan 1988, Gallagher 1989). Rare arsenopyrite is noted in a low-grade limestone-replacement deposit at St Patrick's mine, Avoca (King & Evans 1964).

Offshore territories

Isle of Man Alderton et al. (1995) note arsenopyrite and other sulphides in gratonite-bearing quartz–carbonate veins from North Laxey mine, by Ballaglass Glen. At the South Bradda mines, N of Port Erin Bay, arsenopyrite occurs with chalcopyrite and pyrite in large blocks of quartz veinstone (Starkey 1998b).

Jersey Arsenopyrite forms zoned, rhombic or tabular crystals up to 600 µm in diameter in carbonate veins cutting granite at Le Doüet de la Mer, St Ouën, whereas in the Southeast quarry, L'Etacq, northwest Jersey, lesser amounts of arsenopyrite are present in chalcopyrite–galena–molybdenite–sphalerite-dominated veins (Ixer 1980, 1992). The Jersey Shale Formation, close to the Northwest granite at Le Pulec, carries arsenide–sulphide mineralization dominated by arsenopyrite, pyrrhotite and pyrite (Mourant 1961, Ixer & Stanley 1980, 1983, Stanley & Ixer 1982, Ixer 1992). A single crystal of arsenopyrite has been found in a geode from Mont Mado (Mourant 1961).

Sark Arsenopyrite is one of many primary ore species in a polymetallic assemblage found in the Pb–(Ag) lode of Sark's Hope mine, Little Sark (Mourant & Warren 1934, Ixer & Stanley 1983).

arsenpolybasite

monoclinic $(Ag,Cu)_{16}(As,Sb)_2S_{11}$

A rare silver–copper arsenic antimony sulphide found as inclusions in galena from the Carboniferous carbonate-hosted syn-sedimentary Cu–Pb–Zn–(Ag) deposit at Tynagh mine, near Killimor, Co. Galway, Ireland. Other species found as inclusions include bournonite, dufrénoysite, gersdorffite, pearceite and tennantite (Boast et al. 1981a). New nomenclature rules introduced by the IMA in 2006 discredit arsenpolybasite and it should now be referred to as pearceite $Cu(Ag,Cu)_6Ag_9As_2S_{11}$.

arsentsumebite

BRACKEBUSCHITE GROUP
monoclinic $Pb_2Cu^{2+}(AsO_4)(SO_4)(OH)$

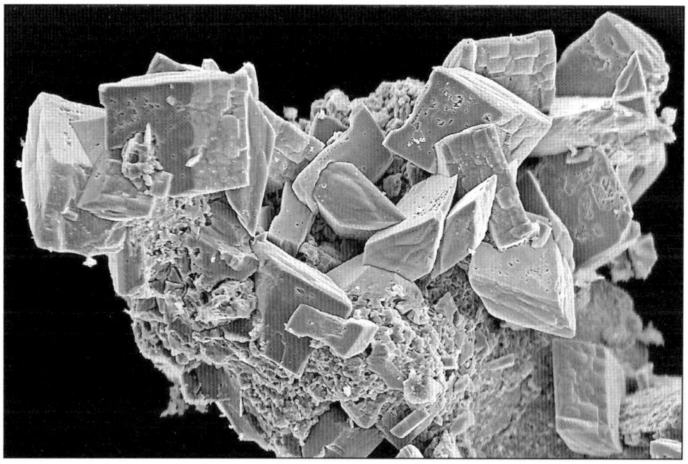

Complex curved monoclinic crystals of arsentsumebite from Dolyhir quarry, Old Radnor, Powys, Wales; width 160 µm; SEM image. (© D. I. Green)

A very rare lead–copper arsenate sulphate hydroxide found between the 60- and 90-fathom levels of Roughton Gill mine, Caldbeck Fells, Cumbria. Arsentsumebite occurs as pale- to apple-green rims (30–50 µm thick) mantling tsumebite spherules (up to 1.5 mm across) on vein quartz. Brochantite and cerussite are associated species (Tindle et al. 2006).

Arsentsumebite, with near-ideal end-member composition, has been found as minute pale-green curved platy to blocky crystals at Dolyhir quarry, Old Radnor, Powys, Wales (Hubbard et al. 2007).

ARTHURITE GROUP

The group consists of monoclinic arsenates and phosphates with the general formula $A^{2+}Fe_2^{3+}(XO_4)_2(O,OH)_2{\cdot}4H_2O$, where A = Ba, Ca, Ce, K, Na, Pb, Sr, Y; and $X = As^{5+}$, P^{5+}, Si^{4+}, V^{5+}. Arthurite is the only species of the group found in Britain or Ireland.

arthurite

ARTHURITE GROUP
monoclinic $Cu^{2+}Fe_2^{3+}(AsO_4)_2(OH)_2{\cdot}4H_2O$

TL Hingston Down Consols mine, 2 km W of Gunnislake, Calstock, Cornwall, England

Named after Sir Arthur Edward Ian Montague Russell (1878–1964), mineral collector of Swallowfield Park, Reading, Berkshire, and Arthur William Gerald Kingsbury (1906–1968), research assistant to the reader in mineralogy at the University of Oxford, who was born at East Meone, near Petersfield, Hampshire (Kingsbury 1966, Embrey 1973, Starkey 1988c). Arthurite is a rare microcrystalline hydrated copper–iron arsenate hydroxide found at the type locality as thin apple-green crusts intimately mixed with pharmacosiderite, quartz and an hidalgoite-like mineral (Davis & Hey

Sprays of arthurite crystals from Hingston Down quarry, Calstock, Cornwall; width 290 µm; SEM–BSE image. (© E. Condliffe, University of Leeds)

1964). Bright-green botryoidal crusts coating narrow fissure veins in kaolinized lithium-rich granite in the Half Stope area of the Gunheath china-clay pit, St Austell, Cornwall, are confirmed as arthurite, and yellowish-green coatings and thin crusts on lithium granite from the Bunny lodes in the same pit are tentatively identified as arthurite (Weiss 1994). Additionally, Golley & Williams (1995) list arthurite from Hingston Down quarry and Cligga Head mine, Perranzabuloe, Cornwall. At Cligga Head, arthurite forms apple-green crusts and spherules to 0.2 mm associated with pharmacosiderite and scorodite, all lining cavities in quartz veinstone (N. Hubbard pers. comm. 2006).

Delicate strands of fibrous arthurite forming an apple green crust on quartz; Gunheath china-clay pit, St Austell, Cornwall; width 28 µm. (© G. Imlach)

asbestos

Fibrous amphibole or chrysotile.

asbolane

hexagonal $(Ni^{2+},Co^{3+})_xMn^{4+}(O,OH)_4 \cdot nH_2O$

A relatively common and locally abundant hydrated nickel–cobalt–manganese oxy-hydroxide found on Alderley Edge ridge, Cheshire: Engine Vein mine, New Venture mine, Opencast mine, Stormy Point area, the Wizard of the Edge, West mine, Wood mine, and also at Mottram St Andrew. Asbolane develops as a result of oxidation of cobalt- and nickel-bearing manganese deposits and is found as black sooty-looking botryoidal patches throughout the mineralized zones of Alderley Edge (Greg & Lettsom 1858, Warrington 1965, Braithwaite 1994). In the same area, the underground workings of "Cobalt mine" (the original name of the mine is unknown) have been recently surveyed and *in situ* X-ray fluorescence analysis of asbolane undertaken to establish the variation in ore grade throughout the mine (Timberlake & Mills 2003).

In Cumbria, black mammilliary masses associated with erythrite and zálesíite from the 60-fathom-level dump at Sandbed mine, Caldbeck Fells, are thought to be asbolane, but their identity could not be confirmed (Green et al. 2006a).

Asbolane, occurring as black masses with an earthy texture and a blue submetallic lustre, is noted in Cornwall from Wheal Gorland, St Day, 2 km E of Redruth (Rust 1981b), and other locations: South Terras mine, St Stephen-in-Brannel; Roscommon Cliff, near Botallack, St Just; St Austell Consols; and Wheal Unity, Gwennap (Greg & Lettsom 1858, Golley & Williams 1995).

In Wales, soft black reniform or botryoidal masses of asbolane with fragments of hematite are described from mineralized veins cutting Carboniferous Limestone at Moel Hiraddug mine, Cwm, near Rhyl, Clwyd (Le Neve Foster 1882). At Aberllyn mine, 2 km N of Betws-y-coed, Gwynedd, post-mining manganese-rich concretions up to 2 cm in diameter have developed over the past century and comprise an unusual zinc analogue of asbolane (Roberts 1986, Jenkins & Johnson 1993).

Asbolane is listed in Scotland from Breeton, near Alva, Ochil Hills, Central Region (Greg & Lettsom 1858), and from the Leadhills–Wanlockhead area near the Strathclyde/Dumfries & Galloway border (Greg & Lettsom 1858, Brown 1927), but these occurrences are not confirmed by later works (Macpherson & Livingstone 1982, Livingstone 2002).

ashoverite

tetragonal $Zn(OH)_2$

TL Milltown quarry, Ashover, Derbyshire, England

A rare zinc hydroxide, trimorphous with sweetite and wülfingite, and found in an oxidized vein exposure cutting limestone at the type locality, where it forms colourless platy crystals associated with fluorite. Ashoverite is one of three polymorphs of $Zn(OH)_2$ occurring at Milltown quarry, the others being sweetite and wülfingite. Other associated species include hemimorphite, hydrocerussite, hydrozincite, smithsonite and highly altered galena. Ashoverite forms thin square plates to 0.7 mm across, as either isolated crystals or radiating aggregates, and is colourless to milky white (Clark et al. 1988, Rust 1991c).

Square translucent colourless-to-grey plates of ashoverite to 0.25 mm encrusting matrix; Milltown quarry, Ashover, Derbyshire. (© M. P. Cooper)

asparagus stone

A synonym of fluorapatite; apatite group.

asphaltite

A group name for solid bituminous hydrocarbons (includes albertite, impsonite, gilsonite, grahamite, nigrite and uintahite). They are not considered to be minerals.

astrapia

TL Isle of Sheppey, Kent, England

A synonym of barite.

atacamite

orthorhombic $Cu_2^{2+}Cl(OH)_3$

An uncommon copper chloride hydroxide, polymorphous with paratacamite, clinoatacamite and botallackite. Atacamite forms as an oxidation product of other copper minerals, especially under arid or saline conditions. Botallack mine, St Just, Cornwall, is situated on a cliff edge overlooking the sea, and interaction of copper minerals with sea water is the likely reason that atacamite forming minute crystals, encrustations and tubular stalactites has been noted from there (Collins 1892, Rudler 1905). Later work has shown the species to be paratacamite (Bannister et al. 1950, Kingsbury 1954), although atacamite is still claimed to be present (Golley & Williams 1995, Anthony et al. 1997, Dossett & Green 1998). Exceptionally fine and well crystallized specimens of atacamite are noted from Botallack, Wheal Hazard and Wheal Cock (1 km N of Botallack) by Bancroft & Weller (1993). At Loe Warren, about 1 km S of Botallack mine, a rich assemblage of supergene Cu–Pb–U minerals occur in a small coastal exposure where atacamite is found with malachite, amorphous copper silicate and rarely botallackite and cerussite. Atacamite forms tiny pale-green to dark-green anhedral crystalline masses (Elton & Hooper 1995a). The habit is identical to that found at Geevor mine, near Pendeen, St Just, where percolating sea water seeped into levels worked beneath the sea bed and allowed both atacamite and paratacamite to form (Elton & Hooper 1993). Elsewhere in Cornwall, atacamite has been described from Penberthy Croft mine, St Hilary (Kingsbury 1954, Kingsbury & Hartley 1956c), but, in view of the doubts about the provenance of many Kingsbury samples (Ryback et al. 1998, 2001), this occurrence is unlikely (Betterton et al. 2000). The reports of atacamite by Camm & Merry (1991b) and Golley & Williams (1995) are based on the work of Kingsbury. Occurrences of atacamite from three other Cornish locations – Perran St George mine, near Perranporth; Wheal Owles, Penwith, St Just and Wheal Prudence, by Hanover Cove, St Agnes (Kingsbury & Hartley 1956c, Kingsbury 1964) – are also in doubt. There does not appear to be any dispute about atacamite from Newporth beach, 2 km SW of Falmouth, where the species is associated with other secondary copper and lead minerals (anglesite, bindheimite, cerussite, cumengéite, goethite, malachite and phosgenite) within a primary Cu–Pb–Sb sulphide vein (Dean 1982, 1983).

A well crystallized specimen of atacamite from Roughton Gill mine, Caldbeck Fells, Cumbria (Davidson & Thomson 1951), is confirmed by Bannister et al. (1950). The provenance of specimens of atacamite claimed to have been found at Hay Gill and Old Potts Gill mines, both in the Caldbeck Fells (Kingsbury & Hartley 1956c, Hartley 1984), is less certain. The second locality in particular is questioned by Green & Briscoe (2002). Young (1987a) and Cooper & Stanley (1990) both quote the work of Kingsbury. To add further confusion, Kingsbury & Hartley (1956c) did not find atacamite at Roughton Gill and they state that the atacamite they found at Potts Gill is so similar to the Roughton Gill material at the BM(NH) that they consider that the latter may have come from Potts Gill.

On the beach between Abersoch and Penrhyn Du, Gwynedd, Wales, dark-green encrustations containing minute drusy crystals of atacamite have been found in the fractures of pebbles from hydrothermally brecciated Ordovician siltstone containing bornite–chalcopyrite–quartz veins. It is associated with botallackite and cumengéite in a supergene assemblage formed from the oxidation of primary sulphides in the presence of sea water (Dossett & Green 1998).

Along the Solway coast, Dumfries & Galloway, Scotland, particularly in the vicinity of Needle's Eye, SW of Caulkerbush, supergene alteration of uranium-rich chalcopyrite-bearing quartz veins has produced green staining at most mineralized localities, and on two specimens is confirmed to be atacamite (Miller & Taylor 1966). At Crammag Head, W of Drummore, atacamite has been identified in narrow discontinuous quartz veins developed along joint planes and crush zones in diorite (Cameron & Lawson 1979).

In Ireland, bright-green, powdery encrustations of atacamite occur with chalcanthite on oxidized chalcopyrite ore at Tankardstown mine, Knockmahon, 22 km SW of Waterford town, Co. Waterford. The locality is close to the sea, and sea spray is the likely source of chlorine required for the formation of atacamite (Moreton 1991, Ryback & Moreton 1991). Moreton (2005a) also found atacamite with botallackite and clinoatacamite at nearby Cuan Cove. On the coast near Dooneen copper mine, Allihies, Co. Cork, atacamite intimately mixed with botallackite forms green coatings at various points along the cliffs (Moreton & Aspen 1993, Green & Moreton 2001). Pale-green distinctly microcrystalline atacamite also occurs with late crusts of chalcanthite a few metres inside the mine (Braithwaite & Wilson 2001b).

Atacamite has been reported from the Cu–Pb–(Ag) mines at Bradda, N of Port Erin Bay, Isle of Man (Smythe 1879–1884, Dewey & Eastwood 1925), but, in view of the misidentifications outlined above, further confirmation is desirable for this occurrence and those at Côtil Point and Bouley Bay, Trinity, northeast Jersey coast (Robinson 1933, Mourant 1961, Ixer 1980), and La Moye, southwest Jersey (Mourant & Warren 1934).

attapulgite

A synonym of palygorskite.

augite

PYROXENE GROUP

monoclinic $(Ca,Na)(Mg,Fe,Al,Ti)(Si,Al)_2O_6$

A very common calcium–sodium–magnesium–iron–titanium aluminosilicate found in mafic igneous rocks, including andesite, basalt, gabbro, pitchstone and some ultramafic rocks. It can also be present in some high-grade metamorphic rocks. In many cases below, identification is based on a chemical analysis.

England

Cornwall Well preserved augite is reported with magnetite and plagioclase feldspar in mafic (greenstone) dykes cutting serpentinite at Lankideen Cove, E of Kennack Sands, Lizard Peninsula (Bonney 1877). It also occurs on the peninsula with diopside, forsterite and chrome spinel in peridotite at Gwenter and Kernewas on Goonhilly Downs and at Coverack Cove (Davies 1983). At St Minver, augite (titanaugite) forms purplish crystals associated with kaersutite in a rock described as minverite (Kempe 1968). Other localities include: dark greenish-black crystals in the cliffs near Botallack and on the dumps of Wheal Cock, St Just (Bancroft & Weller 1993); Clicker Tor quarry close to Menheniot, and Chyandour (Golley & Williams 1995).

Cumbria Augite occurs in andesites at Podhouse Bank, Whicam (Green 1913), and in the metamorphic aureole of the Shap granite, but the colourless lime-augite described from metamorphosed Coniston Limestone at Wasdale Head, Shap (Harker & Marr 1891), is probably diopside. The Carrock gabbro contains augite crystals, locally up to 4 cm long (Eastwood et al. 1968), whereas gabbro at Mecklin Wood in Wasdale contains altered augite crystals up to 13 mm across (Rastall 1906, Clark 1963). Augite is common in doleritic rocks, for instance: the Armathwaite tholeiite dyke, Brampton district (Dixon et al. 1926, Trotter & Hollingworth 1932, Arthurton & Wadge 1981); the Carrock and Skiddaw area (Eastwood et al. 1968); Castle Head, Keswick (Ward 1876); and Graphite mine, Seathwaite (Strens 1962). Augite is less common in granophyre from Carrock Fell (Harker 1895) and from Ennerdale (Rastall 1906). Augite and diopside phenocrysts, up to 10 mm across, occur in microdiorite at Scawgill Bridge, 4 km W of Braithwaite (Fortey et al. 1994).

Derbyshire The Great Rocks Dale dyke to the NE of Buxton contains augite as small anhedral grains up to 0.5 mm across associated with chlorite, ilmenite and plagioclase (Cope 1933) and as granular crystals in the doleritic Waterswallows sill (Ineson et al. 1983). Carboniferous basalts from Calton Hill, Taddington, contain phenocrysts of augite and diopside, together with olivine, plagioclase and iron–titanium oxides (Macdonald et al. 1984). Ultramafic nodules in the Calton Hill basalts contain three pyroxenes: augite, diopside and enstatite (Donaldson 1978).

Leicestershire Zoned augite is a constituent of gabbro at Swithland reservoir, Mountsorrel district, although, confusingly, it is referred to as a colourless diopside (Taylor 1934).

Devon Titanium-rich augite occurs with diopside in Tertiary dolerite dykes cutting the granite on Lundy island, off the north coast of Devon (Durrance & Laming 1982, Thorpe & Tindle 1992, Tindle unpub. EMP data).

Durham The Whin Sill, near High Force, NW of Middleton-in-Teesdale, is a large Carboniferous doleritic intrusion in which augite occurs as zoned crystals associated with olivine, ilmenite and titanomagnetite (Rodgers 1975, Ixer et al. 2005b, Tindle unpub. EMP data).

Shropshire The andesine dolerite on Squilver Hill, 5 km N of Bishop's Castle, contains augite crystals with irregular shape and up to 1 mm in length (Blyth 1948).

Somerset Phenocrysts of augite, enstatite and plagioclase feldspar (andesine) occur in Silurian andesite–latite rocks at Moons Hill quarry, NE of Shepton Mallet, Mendip Hills (van de Kamp 1969).

Euhedral, sector-zoned augite (variety titanaugite) in essexite – a type of alkali gabbro; Crawfordjohn, near Douglas, Dumfries & Galloway, Scotland; width 4.5 mm; viewed in PPL and XPL. (© K. Gardner)

Staffordshire Titanaugite, sometimes fringed with aegirine, is reported from a Carboniferous teschenite sill found in boreholes at Landywood, near Cannock (Barton 1963).

Wales

Dyfed Pillow lavas in the Ordovician Fishguard Volcanic Complex contain spectacular dendritic to fan-shape spherulitic growths of titanaugite formed during quenching of the lavas (Bevins 1982). Augite is the dominant pyroxene in most units (apart from the granophyric gabbro, which has diopside) of the St David's Head intrusion, Carn Llidi, a tholeiitic sill composed principally of gabbros. Pegmatitic patches in this intrusion have augite crystals up to 12 cm in length (Roach 1969, Bevins 1994, Bevins et al. 1994).

Gwynedd Augite and diopside form large ophitic plates in the Tal y Fan intrusion, SW of Conwy (Merriman et al. 1986), and occasionally form phenocrysts or microphenocrysts in basic lavas from the Rhobell Volcanic Complex, Rhobell Fawr area, near Dolgellau (Allen & Jackson 1985, Kokelaar 1986). Other reports of augite are from: dolerite sheets at Bodorgan on the Isle of Anglesey (Maltman 1977); pyroclastic bombs on the summit of Curig Hill, Capel Curig district, Snowdonia (Williams 1922); granitoid rocks from Inner Garnfor, Penrhyn Bodelias and Careg-y-Llam, on the Lleyn Peninsula (Croudace 1982); olivine dolerite dykes at Parc mine (No. 3 level), near Llanrwst (Archer & Elliott 1965).

Scotland

Central Composite groups of pale-brown augite crystals are associated with pigeonite and plagioclase feldspar (labradorite) in a tholeiite dyke at Kinkell, 11 km N of Glasgow (Walker et al. 1952).

Dumfries & Galloway Augite, diopside and enstatite are common species in pyroxene diorites found at the margins of the Loch Doon granite complex (Tindle 1982), and both augite and enstatite (hypersthene) are recorded from diorites in the nearby Cairnsmore of Carsphairn granite complex (Deer 1937b, Tindle et al. 1988). An auto-brecciated basalt at Bail Hill, Sanquhar, contains black augite phenocrysts up to 1.5 cm across (Scott 1914a, Williams 1965). Augite is also reported from: a late-Caledonian minette at Glen Affric, E of Morvich (Canning et al. 1996); a Tertiary tholeiite dyke on the summit of Castle Hill, 16 km NW of Langholm, Eskdalemuir, Snar Valley (Elliot 1956); and a lamprophyre dyke at Newmains, 7 km W of Dumfries (Macdonald et al. 1986).

Fife Augite is present in wehrlite inclusions within a Carboniferous volcanic vent at Coalyard Hill, near Elie. Another vent at Elie Ness contains augite and diopside megacrysts in tuff (Chapman 1976). This report describes augite (salite) from other locations, but changes in nomenclature of the pyroxene group (Morimoto 1988) means they are now classified as diopside. Augite from the Elie Ness vent is also discussed by Upton et al. (2003).

Grampian Augite and diopside are common species in the Insch intrusion, the largest of the Caledonian Newer Gabbro bodies. Similarly, the Boganclogh intrusion to the W of Insch also contains augite and diopside (Wadsworth 1986, 1988). Associated with the Insch intrusion is the Kennethmont diorite–granite complex, which contains both augite and hypersthene in the dioritic rocks (Busrewel et al. 1975). Augite is the only clinopyroxene found in the layered ultramafic, troctolitic and gabbroic intrusion at nearby Belhelvie (Wadsworth 1991). Other examples of augite are reported from: diorite at Craigbuirach (Heddle 1878c); high-grade amphibolites within the metamorphic aureole of the Lochnagar granite (Goodman & Lappin 1996); gabbroic rocks of the Morven–Cabrach intrusion, around Strathdon (Allen 1970); a titanite-rich xenolith within norite at Schivas, NW of Ellon (Guppy 1956).

Highland Augite is an important rock-forming mineral in a variety of rock types from the region, including: gabbro and tonalite from the Tertiary Centre III igneous complex, Ardnamurchan (Walsh 1975); pitchstone from Ardnamurchan and the Isles of Arran and Eigg (Preston et al. 1998b); monzodiorites and quartz diorites from the Ballachulish granite complex (Weiss & Troll 1989); quartzite xenoliths within granodiorite at Kentallen northern quarry, Ballachulish (Muir 1953); quartz diorite from the Ben Nevis granite complex (Haslam 1968); Caledonian minettes from Creag Rhamber and from the Glen Affric–Kintail area (Canning et al. 1996); basaltic lavas from the Isle of Eigg and Bloodstone Hill on the Isle of Rum (Ridley 1971); in picrite dykes from Loch Feith an Leothaid, Lochinver area (Tarney 1969); pyroxene–quartz-bearing ironstone in the first inlet W of the pier on the north shore of Scourie Bay (Barnicoat & O'Hara 1979); a quartz-dolerite dyke at Scourie (O'Hara 1961); and titanaugite in lherzolite xenoliths within a monchiquite dyke at Streap, 7 km NW of Glennfinnan station (Walker & Ross 1954). Augite is reported many times from the mafic–ultramafic layered intrusion, associated minor intrusions and lavas on the Isle of Rum (Heddle 1878c, Tomkeieff 1942, Williams 1978, Wadsworth 1994, Tepley & Davidson 2003). Augite is also common on the Isle of Skye, where it occurs: in olivine tholeiite cone sheets, which invade the Cuillin igneous complex (Bell et al. 1994); in a picritic dyke on Suisnish Point (Parsons 1976); in a ferrodiorite dyke at Harker's Gully, Marsco (Wager & Vincent 1962); in an amygdaloidal olivine dolerite to the N of Portree (Sweet 1961); in the Glamaig epigranite at Eas a' Bradain, near Loch Ainort, between Broadford and the Sligachan hotel (S. A. Rust specimen, Tindle unpub. EMP data); and widely in Tertiary basaltic lavas (Thompson 1974). Heddle (1901) reports other augite occurrences, but without details of rock types.

Lothian Augite is present in a spinel-wehrlite inclusion within a volcanic vent on Fidra island (Chapman 1976).

Strathclyde Augite occurrences include: dolerite dykes cutting microgranite on Ailsa Craig in the Firth of Clyde (Harrison et al. 1987); lherzolite xenoliths in a basanite vent at Black Rock, 2 km SSW of Fairlie, Hunterston Sands (Alexander et al. 1986); dolerite dykes at the southern end of the Craignish Peninsula, SW of Kilmelford, at Craignish Point (Allison 1936) and associated with the Girvan–Ballantrae ophiolite complex (Holub et al. 1984); diorite from the Quarry Diorite unit at the southern end of the Etive granite complex (Moazzen & Droop 2005) and from the Garabal Hill–Glen Fyne igneous complex, W of the head of Loch Lomond (Nockolds 1941, Nockolds & Mitchell 1946); trachyte from Holy Island, Lamlash Bay, Isle of Arran (Tindle unpub. EMP data); teschenite from the Lugar Sill, Glenmuir Water (Phillips 1968); pitchstones at Glen Shurig and in the Judd dykes near Blackwaterfoot, both on the Isle of Arran (Preston et al. 1998b); garnet pyroxenite xenoliths within a Carboniferous basaltic diatreme at Heads of Ayr (Upton et al. 2003); Tertiary tholeiitic dyke at McInroy's Point, SW of Gourock, Firth of Clyde (Herriot 1971); and Devonian basalt, basaltic andesite, andesite and appinite from the Lorne Plateau (Groome & Hall 1974). On the Isle of Mull, augite occurs widely in basaltic lavas belonging to the Tertiary lava succession (Cann 1965a, Fawcett 1965, Kerr 1998). A dark gabbroic rock (termed essexite), which occurs at Crawfordjohn, contains particularly well formed crystals of augite (titanaugite) associated with plagioclase feldspar (labradorite) and olivine (Hatch et al. 1972).

Western Isles Augite (ferro-augite) is present in the Glen Bay granite and a granitic sheet at Na h-Eagan on St Kilda (Hirta) (Harding et al. 1982, 1984). Bowes & Hopgood (1969) note augite (titanaugite) in dolerite dykes on the island of Mingulay.

Orkney A group of alkaline basic rocks from Shapinsay, Rousay and from the Deerness area of Mainland contain pyroxene unsatisfactorily described as diopsidic augite (Flett 1900, Kellock 1969).

Ireland

Antrim Complex zoned crystals of augite with small pigeonite cores occur in unaltered basaltic lava, and also partially melted basalt, close to a dolerite intrusion at Scawt Hill, Larne (Kitchen 1985). Where the dolerite is in contact with chalk, a suite of augite (titanaugite) ±plagioclase ±nepheline-bearing contaminated rocks has developed (Tilley & Harwood 1931). However, confirmation of augite is required, particularly as augite (titanaugite) in contaminated melilite–pyroxene rocks described by these authors from Scawt Hill and by Sabine & Young (1975) from a similar chalk/dolerite contact zone at Carneal, Larne, are classified as diopside and hedenbergite respectively. The tholeiitic basalts of the Giant's Causeway area contain augite and pigeonite (Lyle & Preston 1993). Augite is also present in amygdaloidal basalt from the Gobbins, Island Magee (H. F. Foy specimen, Tindle unpub. EMP data).

Donegal Augite occurs in peridotitic xenoliths within a 30 m-wide alkali basalt (dolerite) dyke that intrudes Lower Carboniferous limestones near Inver, 10 km W of Donegal town. Xenoliths mainly consist of clinopyroxene, alkali feldspar, glass, olivine and plagioclase (Gallagher & Elsden 1990).

Down Dolerite and gabbro from the Tertiary central complex of Slieve Gullion contain augite, enstatite (hypersthene) and pigeonite, together creating complex graphic exsolution textures (Gamble 1982).

Limerick Augite (titanaugite) is present in analcime-bearing basalt from Rathjordan (Prior 1910).

Wicklow Augite, partially resorbed and replaced by magnetite, is a major constituent of a peridotite from Glendalough (Thomson 1908).

Offshore territories

Jersey Mourant (1961) describes augite from dolerites and gabbros including the Wolf's Cave dyke; however, later data indicate that the dyke contains diopside instead (Rowbotham & Lees 1990).

aurichalcite

monoclinic $(Zn,Cu^{2+})_5(CO_3)_2(OH)_6$

A relatively uncommon zinc–copper carbonate hydroxide found in the oxidation zone of many copper and zinc deposits.

England

Avon Barite pebbles on the beach near the Pier hotel at Clevedon, Avon, contain aurichalcite and are derived from fault-filling brecciated veins within Triassic Dolomitic Conglomerate. The pebbles also contain aragonite, azurite, brochantite, cerussite, chalcopyrite, native copper, covellite, cuprite, galena, malachite, phosgenite and native sulphur (Starkey 1984).

Cheshire Rare aurichalcite forms bluish-green radiating blades associated with hemimorphite at Wood mine, Alderley Edge ridge (Braithwaite 1994).

Cornwall Aurichalcite from Greystone quarry, Lezant, near Launceston, is perhaps the finest from anywhere in Britain or Ireland. It occurs in cavities in vein quartz close to the Greystone lead lode associated with chalcopyrite and sphalerite, or as a coating on manganese oxides. Rich blue or bluish-green acicular crystals (up to 10 mm long) produce rosettiform aggregates. Some aurichalcite overlies hemimorphite (Lloyd & Barstow 1983, Elton et al. 2004). At Penberthy Croft mine, St Hilary, occasional minute pale greenish-white spherules of aurichalcite have been found within cavities in cuprite. Associated species include brochantite, connellite and tenorite (Betterton 1996e, Betterton et al. 2000).

Cumbria *Caldbeck Fells:* Aurichalcite forms pale turquoise-blue scales and masses associated with adamite and agardite–mixite at Deer Hills barite mine (Cooper & Stanley 1990); beautiful pale-blue to white aurichalcite crystals forming froth-like aggregates or delicate acicular crystals are described from Roughton Gill mine (Goodchild 1884, Rudler 1905, Davidson & Thomson 1951); radiating sheaves of pale-green aurichalcite are associated with rosasite, and feathery crystals of aurichalcite up to 1 mm in length occur on hemimorphite in partially oxidized sphalerite-rich veinstone from Silver Gill (Hartley 1984, Green et al. 2005b). Aurichalcite from Dry Gill, Potts Gill and Sandbed mines, documented by Kingsbury & Hartley (1957e) and Hartley (1984), requires further confirmation, as many samples in the Kingsbury collection (NHM) are incorrectly provenanced (Ryback et al. 1998, 2001). There are three groups of aurichalcite specimens in the Kingsbury collection labelled "Sandbed mine"; one group appears to be genuine, but the other two are almost certainly incorrectly labelled, as they are similar in appearance to examples from Kelly mine, New Mexico, and Grandview mine, Arizona. Aurichalcite has since been positively identified at Driggith and Sandbed mines, where it forms frothy masses and radiating sprays of pale-blue to bluish-green lath-like crystals up to 2 mm across (Green et al. 2006a). *Elsewhere in the Lake District:* Aurichalcite forms pale-blue crystalline crusts and feathery crystals at Eagle Crag mine, Patterdale (Young 1987a, Green & Briscoe 2002). *Northern Pennine orefield:* Rare pale sky-blue to sea-green sprays of acicular aurichalcite on fluorite occur at Hilton mine, Scordale, near Appleby (Young et al. 1985a, Winrow 1986a, Dunham 1990, Bridges & Young 1998); Rudler (1905) reports aurichalcite from Nenthead, near Alston, but gives no precise location; at Silver Top quarry, Hallbankgate, E of Brampton, altered limestone consisting of calcite, chalcopyrite, limonite and smithsonite contains cavities in which bright turquoise-blue aurichalcite blades occur encrusted with cinnabar (Young et al. 1989, Bridges & Young 1998); bright turquoise-blue aurichalcite crystals, up to 1.5 mm long, encrust fluorite and limestone at Stennerskeugh Clouds, 1 km SW of High Stennerskeugh level, Ravenstonedale. Azurite, duftite and mimetite are associated species in the partially oxidized tennantite-rich ore (Starkey 1987b, Young 1987b); limonitic veinstone on the dumps of an old level driven on Windy Brow vein, Tynehead, near Garrigill, contains small pockets and joint surfaces on which bright turquoise-blue and pale-blue aurichalcite has been found (Young 1985a,c, Dunham 1990, Bridges & Young 1998).

Derbyshire Aurichalcite forms pale-green felted or pearly crusts, or tufts of blades up to 5 mm long, associated with hemimorphite and rosasite, and growing on calcite and fluorite at Rutland Cavern, Heights of Abraham, Matlock Bath (Greg & Lettsom 1858, Braithwaite & Ryback 1963a, Braithwaite 1983); Golconda mine, Brassington (Dunham & Dines 1945, Braithwaite & Ryback 1963a); Millclose mine, Winster, near Matlock (Traill 1939); silky crystals (<2 mm) of aurichalcite occur in vugs in quartz at Snelston mine, near Ashbourne (Jackman 1996); Wapping mine, Cromford (Ince 1991); Black Ox and Long Tor mines, Matlock; Hillocks mine, Monyash; Mouldridge mine, Pikehall; Red Rake, Calver; Watersaw Rake, Longstone Edge (all Ford & Sarjeant 1964a, Ford et al. 1993).

Durham *Northern Pennine orefield:* Closehouse mine, SW of Middleton-in-Teesdale, Lunedale (Dunham 1990, Young et al. 1994a, Bridges & Young 1998); aurichalcite forms small crusts of pale turquoise-blue pearly crystals on barite at Foster's Hush, Lunedale, and has a similar colour in pockets and patches up to 3 mm across in goethite and smithsonite at Greenhurth mine, Upper Teesdale (Young et al. 1985a, Dunham 1990, Bridges & Young 1998).

Leicestershire Pale bluish-green powdery coatings in a fault-breccia zone at Newhurst quarry, Shepshed, are confirmed to be aurichalcite (Ince 2005).

Nottinghamshire Aurichalcite forms pale-green botryoidal aggregates composed of minute lath-like crystals, associated with dolomite and sphalerite on a single specimen from Vale Road quarry, Mansfield Woodhouse (Green & Spence 1988).

Staffordshire Excellent specimens of aurichalcite form feathery bluish-green crusts in pockets within friable barite gangue, associated with chalcopyrite, gypsum, fluorite, hemimorphite, serpierite and smithsonite at Clayton mine, Ecton Hill (Braithwaite et al. 1964a, Braithwaite & Knight 1968, Ford et al. 1993); Dutchman, Good Hope and Waterbank mines on Ecton Hill (Braithwaite & Ryback 1963a); relatively common pale-blue to rich bluish-green aurichalcite forms compact drusy to botryoidal crusts at Waterbank mine (Rust 1995c); and aurichalcite is noted from Mixon (Ford et al. 1993).

Yorkshire *Northern Pennine orefield:* Beautiful turquoise-blue aurichalcite associated with hemimorphite encrusts barite gangue at Brandy Bottle incline, Friarfold vein, NE of Gunnerside, Swaledale. A post-mining origin is suspected (Dunham & Wilson 1985, Young 1987b, Ineson & Young 1994). Traces of supergene aurichalcite and rosasite occur in barite–fluorite–galena–sphalerite veins at Coldstones quarry, 5 km W of Pateley Bridge (Young et al. 1989, 1997). Quartz gangue contains patches of very pale-blue finely crystalline aurichalcite, up to 40 mm across, associated with beaverite, barite, malachite and smithsonite at the head of Stockdale, 0.7 km W of Pikedaw mine, between Malham and Settle (Green et al. 2006c). At Wet Grooves mine, aurichalcite forms bright-blue to bluish-green radial and lath-like crystals on calcite veinstone (Rust 1987b). Other localities include: Hard level, Swaledale; Castaway–Ripley vein, Grassington Moor; Bunting level, Gunnerside Gill; Turf Pits mine, Grassington Moor; Wet Grooves mine, Aysgarth, Wensleydale (all Young 1987b).

Acicular aurichalcite from Machen quarry, Mid Glamorgan, Wales; width 10 mm. (© National Museum of Wales, Cardiff; photographer M. P. Cooper)

Wales

Clwyd Aurichalcite is noted from Nant-y-Plwm mine, Llansannan, and Pen-y-Bryn shaft, Holywell, near Halkyn (Bevins 1994).

Dyfed Small feathery tufts of pale-green aurichalcite occur at Castell mine, Ponterwyd, near Aberystwyth (Jones & Moreton 1977, Rust 1990); Copper level at Cwmystwyth mine (Bevins 1994); light-blue to bluish-green laths of aurichalcite forming semi-transparent to translucent highly striated crystals at Eaglebrook (Nantycagl) mine, Ceulanymaesmawr (Jones 1983); aurichalcite forms small bluish-white masses of lath-like crystals often intergrown with acicular malachite at Frongoch mine, between Devil's Bridge and Pont-rhyd-y-groes (Green et al. 1996); rare pale-blue lath-like crystals of aurichalcite occur at Henfwlch mine, between Cerrig yr Hafan and Nant-y-moch reservoir, E of Tal-y-bont (Bevins 1994); aurichalcite is very rare at Llechweddhelyg mine, near Clawddmelyn (Jones 1987).

Mid Glamorgan Radiating acicular needles of aurichalcite, up to 2 mm long, occur in calcite-lined voids cutting Carboniferous Limestone at Bute quarry, Llantrisant, and in the Locks Common borehole, Porthcawl. Chalcopyrite, galena and sphalerite are associated species found together with secondary zinc-rich malachite and rosasite (Fletcher & Young 1988). Brecciated fracture and infill-type copper–lead-rich barite–calcite veins in dolomitized limestone contain attractive sprays of sky-blue to sea-green lath-like crystals of aurichalcite, associated with hemimorphite, or in some instances malachite or rosasite occur at Machen quarry, 7 km ENE of Caerphilly (Bevins 1994, Plant & Jones 1995). Aurichalcite is noted from old mineworkings in the Cwmleyshon area, near Draethen, and from Hendy quarry, Miskin, near Llantrisant (Bevins 1994).

Powys Aurichalcite forms pale-green feathery crusts at Dylife mine (N of Hafren Forest) and small feathery tufts of pale-green crystals at Snowbrook (also known as Nantyreira) mine (Jones & Moreton 1977, Moreton 1982). Fine specimens of aurichalcite associated with chalcopyrite and smithsonite have been found at Craig Rhiwarth and Ochr Graig mines, Llangynog. Aurichalcite is also noted from the Llanymynech area and at South Nantycar mine, S of Rhayader (Bevins 1994).

Scotland

Dumfries & Galloway Aurichalcite has been found with brochantite, hemimorphite, malachite and veszelyite on the dumps of Straitstep vein (Livingstone 2002). The dumps of West Blackcraig lead–zinc mine, near Newton Stewart, have produced aurichalcite in calcite gangue, as an alteration product of chalcopyrite and sphalerite (Livingstone et al. 1976, Livingstone 2002). Aurichalcite is also noted from Balycary (Heddle 1901).

Strathclyde Aurichalcite associated with lanarkite or smithsonite is noted from unspecified locations around Leadhills (Greg & Lettsom 1858, Rudler 1905). Heddle (1901) notes small incrusting masses of aurichalcite from Susanna vein, but this is doubted by Brown (1927).

Ireland

Kerry Unattractive specimens of aurichalcite, anglesite, hydrozincite, linarite and serpierite can be found at two abandoned copper mines at the southern end of Ross Island (Ryback & Moreton 1991).

Leitrim A specimen of aurichalcite from Twigspark or Lurganboy mine is in the Russell collection at the NHM (Ryback & Moreton 1992).

auriferous pyrites

Pyrite, probably containing inclusions of native gold.

aurostibite

PYRITE GROUP

cubic — $AuSb_2$

A very rare gold antimonide found as inclusions in alluvial gold grains along the Owannadornaun River, Cregganbaun shear zone west, Co. Mayo. Ireland. Aurostibite is associated with arsenopyrite inclusions and is probably derived from sulphide-bearing quartz veins (Leake et al. 1993, Chapman et al. 2000c).

Out of a total of 1794 alluvial gold grains examined by Leake et al. (1998) from the Leadhills area along the Strathclyde/Dumfries & Galloway border, Scotland, one grain is reported to contain aurostibite inclusions.

AUTUNITE GROUP

Tetragonal uranyl arsenate, phosphates and vanadates, all with the general formula $A(UO_2)_2(XO_4)_2 \cdot 8\text{–}12H_2O$, where A = Ba, Ca, Cu, Fe^{2+}, 1/2(H,Al), Mg, Mn^{2+}, $Na_3(UO_2)$; and X = As, P, V. Autunite, fritzscheite, kahlerite, nováčekite, saléeite, sodium-uranospinite, torbernite, uranospinite and zeunerite are members of the group found in Britain or Ireland. Trögerite has also been described, but the provenance of the material is suspect.

autunite

AUTUNITE GROUP

tetragonal — $Ca(UO_2)_2(PO_4)_2 \cdot 10\text{–}12H_2O$

An uncommon hydrated calcium uranyl phosphate formed during alteration of primary uranium-bearing minerals under oxidizing conditions.

Autunite is described from South Terras mine, St Stephen-in-Brannel, Cornwall, as a yellow oxidation product in a uranium lode known locally as Green Jim. In the oxidized zone, autunite was found with torbernite and zippeite in beautiful green, yellow and brown flakes, scales and crystals, and beneath this uraninite (pitchblende) with accessory arsenopyrite, galena and pyrite occurred (Collins 1912, Robertson & Dines 1929, James 1947, Dines 1956, Ball et al. 1982). At the nearby Stenna Gwyn mine, N of St Stephen-in-Brannel, autunite and torbernite associated with apatite, fluellite, fluorite, stannite and wavellite are reported from a cassiterite–wolframite lode cutting the St Austell granite (Greg & Lettsom 1858, James 1947, Dines 1956, Kingsbury 1959b, Ball et al. 1982). Small bright-yellow nearly transparent crystals of autunite are described from Wheal Basset, Illogan, near Redruth, Cornwall, by Collins (1892), who also notes autunite with torbernite from Wheal Edward, St Just. Other Cornish occurrences include: Cheesewring quarry, SW of Henwood, Linkinhorne (Bowman 1911, Shambrook 1982); Chywoon quarry, near Longdowns (Moulding & Hooper 2005); Gunnislake mine, Calstock and Withiel mine, Withiel (James 1947); Ponsanooth, near Penryn (Ghosh 1934); and Tolcarne and South Tolcarne mines, Camborne (Greg & Lettsom 1858).

Merrivale granite quarry on Dartmoor, Devon, is a well known locality for

Interpenetrant, tabular crystals of autunite amongst thin plates of muscovite, from a granite pegmatite at Chywoon quarry, Longdowns, Cornwall; width 560 µm; SEM image. (© J. J. Hooper)

autunite, which occurs as thin crusts of yellow rectangular plates in scattered patches on joint surfaces in the granite. It is closely associated with a deeper-yellow mineral identified by Braithwaite et al. (1989b) as phurcalite. Greenish-yellow platy crystals of autunite are also described from small cavities in the lithium–fluorine-rich Meldon aplite dyke, which is emplaced into the metamorphic aureole of the Dartmoor granite (von Knorring & Condliffe 1984).

Disseminated autunite has been found in a boulder of porphyritic granite at Caen Burn and in wallrock adjacent to a polymetallic hydrocarbon vein cutting granite at Helmsdale, Highland Region, Scotland (Gallagher et al. 1971, Tweedie 1979).

An aplogranite facies of the Barnesmore granite pluton, one of the Donegal granites, located between Donegal and Ballybofey, Co. Donegal, Ireland, hosts 21 separate localities for autunite, all associated with joint surfaces or margins of cross-cutting Tertiary basalt dykes (O'Connor 1986). Mineralized outcrops and boulder trains over the Tullow pluton, at the southern end of the Leinster granite complex, Co. Carlow, contain autunite in quartz breccias, quartz veins and greisen, where it occurs disseminated along fractures, in vugs and in porous desilicated patches. Supergene reworking seems to have produced the autunite concentrations in weathered rock profiles, because borehole information shows the uranium mineralization to die out at shallow depth (O'Connor 1986, Ryback & Moreton 1993).

aventurine

A micaceous variety of quartz.

aventurine feldspar

Calcium-rich albite, feldspar group containing hematite inclusions, or orthoclase, feldspar group with hematite inclusions (sunstone).

awaruite

cubic — Ni_2Fe to Ni_3Fe

A rare nickel–iron alloy found in dunite, west of Hagdale in the Baltasound–Haroldswick area of Unst, Shetland, Scotland. The less common associated species include breithauptite, chalcocite, chalcopyrite, heazlewoodite, magnetite, pentlandite and native copper. Awaruite is one of the common alteration products of pentlandite (Maynard 1993, Prichard et al. 1994, Maynard et al. 1997).

A tentative identification of awaruite has been made in nickel-mineralized chromite-rich lithologies from a serpentinized ultramafic body at Corrycharmaig, Glen Lochay, NW of Killin, Tayside Region. Awaruite occurs with galena, heazlewoodite, nickeline and native silver (Power & Pirrie 2000).

Awaruite has been found at Poundland Burn, within the Ballantrae ophiolite complex, Strathclyde. It occurs rarely with heazlewoodite and pentlandite as small blebs (<10 μm diameter) in serpentinite associated with nodular chromitites. In the nearby Pinbain Bridge area, chromitite contains abundant chrome spinel with inclusions of erlichmanite, irarsite, laurite, Ir–Os–Ru alloys and possibly kashinite (Power & Pirrie 2004).

AXINITE GROUP

Triclinic borosilicates with the general formula $A_3Al_2BSi_4O_{15}(OH)$, where A = Ca, Fe^{2+}, Mg, Mn^{2+}. Ferro-axinite and manganaxinite are the two species of the group found in Britain or Ireland. Axinite-group minerals are often found in contact-metamorphosed rocks around granite plutons. It is highly likely that the axinite occurrences listed below are in fact ferro-axinite, although chemical analysis is required to confirm this.

England

Cornwall Bodmin Wheal Betsy, Lanivet (Golley & Williams 1995); Boscawen Cliffs, St Burian (Collins 1892); Botallack, St Just (Collins 1892, Rudler 1905, Robson 1949b, Hall & Jackson 1975, Alderton & Jackson 1978, Wolloxall 1989, Bancroft & Weller 1993, Merry & Weiss 2005, Wolfe 2006); Carrick Du mine, near Porthmeor beach, St Ives (Robson 1945b, 1949a,b, Howie 1975); Dolcoath mine and East Wheal Crofty, Camborne; Gravel Hill mine, Perranzabuloe; Great Retallack mine, near St Agnes (all Collins 1911a); Lamorna Cove, near Penzance (Collins 1892); Roscommon Cliff, St Just (Hutchinson 1900, Robson 1949b, Gay & Rickson 1960, Embrey & Symes 1987); Rosevanion quarry, St Columb (Hosking 1951); South Terras mine, St Stephen-in-Brannel (Robertson & Dines 1929, Dines 1956); South Tolgus shaft, Redruth (Davey 1948); Terrace (Terras) Hill quarry, Lostwithiel (Collins 1892, Rudler 1905, Russell 1910b); Treburland mine, 2 km SE of Fivelanes, Altarnun (Russell 1946d, Robson 1949b); Tregullan quarry, Bodmin Moor (Barrow & Thomas 1908); Tremore quarries, Lanivet, SW of Bodmin (Barrow & Thomas 1908, Collins 1911a); Trevascus mine, near Hayle, Gwinear (Golley & Williams 1995); Trewellard, near St Just (Rudler 1905); Wheal Cock, Botallack, St Just (Rudler 1905, Kingsbury 1961, Couper & Barstow 1977, Couper & Clark 1977, Alderton & Jackson 1978, Hubbard 1989, Bancroft & Weller 1993); Wheal Mary (Wheal Ding section), Lanivet, near Bodmin (Kingsbury 1961); Wheal Owles, St Just (Jackson et al. 1982); and Wherry mine, Penzance (Russell 1949a).

Devon Okehampton area, around the margins of the Dartmoor granite intrusion: Belstone mine (Solly 1886, Collins 1911a, Rudler 1905); Brent Tor, near Tavistock, and Sticklepath, near Okehampton (Hall 1868b, Collins 1892); Copper Hill mine (Hall 1868b); Fursdon Manor mine (Hall 1868b, Russell 1910b, Collins 1911a); Ivy Tor mine, Belstone, Meldon area (Hall 1868b, Collins 1911a); Meldon (Hall 1868b, Collins 1911a, McLintock 1923, von Knorring 1951, Dearman & Claringbull 1960, Dearman 1966a,b, von Knorring & Condliffe 1984); South Tawton (Collins 1911a); Wheal Forest, Meldon (Hall 1868b, Russell 1910b, Collins 1911a, El Sharkawi & Dearman 1966). Axinite is also noted from Wheal Friendship (Devon Friendship), Mary Tavy (Hall 1868b, Collins 1892, Braithwaite 1982b).

Wales

Dyfed Garn Turn Rocks, S of Fishguard (Bevins 1994).

Gwynedd Carreg Ddu, Porth Dinllaen, Lleyn (Bevins 1994); Graiglwyd quarry, Penmaenmawr (Sargent 1916, 1925); Llanddwyn Bay, Anglesey (Greenly 1919).

Scotland

Highland NE of Loch Maree (Peach et al. 1907).

azurite

monoclinic — $Cu_3^{2+}(CO_3)_2(OH)_2$

A relatively common copper carbonate hydroxide found in the oxidized zone of copper deposits. Chessylite is a synonym of azurite.

England

Avon Azurite has been found with malachite and wulfenite in calcite–chalcopyrite–galena veins at Coombe Farm quarry, Westbury-on-Trym (Alabaster 1978). Pebbles of galena-bearing barite on the beach near the Pier hotel, Clevedon, contain lustrous crystals of azurite, up to 4 mm across, associated with aragonite, aurichalcite, beudantite, brochantite, cerussite, covellite, cuprite, phosgenite, native copper and native sulphur (Starkey 1984).

Cheshire Supergene deep-blue azurite forming impregnations or occasionally small crystals, up to 1 mm long, are common in the mines on Alderley Edge ridge, where redbed style mineralization occurs in coarse-grain clastic sediments of Triassic age. Primary sulphides include chalcopyrite, galena, pyrite (bravoite), sphalerite and tetrahedrite (Ixer & Vaughan 1982). Locations include Doc mine, Engine Vein mine, Mottram mine, New Venture mine, Opencast mine, Stormy Point area and West mine (Carlon 1979, Braithwaite 1994, Plant et al. 1999).

Cornwall Ting Tang mine at Carharrack is one of the more notable localities in Cornwall for azurite and is described by Greg & Lettsom (1858) as being "very prettily crystallized". More recently, spheroidal aggregates of elongated crystals on gossanous quartz veins have been found on the dumps near Morcom's shaft (Bruce & Aubrey-Jones 1998). Massive anhedral aggregates or rare subhedral translucent tabular crystals of dark-blue azurite, up to 1 mm across, occur at Penberthy Croft mine, St Hilary (Camm & Merry 1991b, Betterton et al. 2000). Azurite is also noted from: Bampfylde mine, NNE of South Molton, and Treveddoe mine, NW of Warleggan (Dines 1956); Carharrack mine, Wheal Unity and Wheal Virgin, all Gwennap, and Wheal Mill Pool (Greg & Lettsom 1858); Old Gunnislake mine, Calstock, and Gunheath china-clay pit, St Austell (Golley & Williams 1995); Marke Valley mine, near Liskeard (Collins 1911a); Pentire (Glaze) mine, near New Polzeath, St Minver (Clayton et al. 1990); South Caradon and West Caradon mines, St Cleer, Liskeard (Shambrook 1982, Lawrence 1990); Trevinnick mine, St Kew, and Tolpetherwin quarry, Two Bridges, South Petherwin (Russell 1949b); Wheal Basset, near Redruth, Illogan (Wolfe 1996a); Wheal Buller, near Redruth (Collins 1892); and Wheal Gorland, St Day, 2 km E of Redruth (Rust 1981b, Fisher 1988).

Cumbria *Caldbeck Fells:* Azurite is very rare in this area, but a small amount is recorded from the lead–zinc Driggith vein (Stanley & Vaughan 1981); on the small dumps of a copper trial in Potts Gill (Green & Briscoe 2002); on limonitic barite–quartz matrix from the 60-fathom level of Sandbed mine (Green et al. 2006a), and possibly also at Thief Gills (Cooper & Stanley 1990). Other reports from the Carrock, Dry Gill, Roughton Gill and Sandbed mines (Kingsbury & Hartley 1960, Hartley 1984, Young 1987a, Cooper & Stanley 1990) require verification, because some may actually be linarite and because the descriptions appear to be based on samples with doubtful provenance from the Kingsbury collection (see Ryback et al.

1998, 2001). *Elsewhere in the Lake District and West Cumbria:* Conspicuous azurite and malachite occur in small barite–chalcopyrite–galena veins at Higher Longrigg mine, S of Cote Garth, Hartley Birkett, near Kirkby Stephen (Braithwaite 1988). Azurite is also noted from: Black Scar workings at Levers Water, Coniston, and Dale Head mine, Newlands Valley, near Keswick (Stanley 1979); Eagle Crag mine, Patterdale (Green & Briscoe 2002); Glenderaterra Valley and Goldscope mine in Newlands Valley (Shackleton 1966); Hartley mine, Kirkby Stephen (Davidson & Thomson 1951); Parkhouse quarry, Johnby (Green & Briscoe 2002); Yewthwaite mine, Newlands Valley (Dewey & Eastwood 1925); Sea Wood mine, Bardsea, Ulverston (Young 1987a). *Northern Pennine orefield:* Azurite and malachite are widespread, although usually only in trace amounts. Extensive crusts and good crystals to 2 mm long occur on quartz veinstone on the dumps of the Cornriggs vein, Alston Moor. Crystals of azurite have also been found on the dumps of Copper Slitts vein, Windy Brow (Bridges & Young 1998), and it occurs with aurichalcite encrusting duftite on partially oxidized galena–tennantite ore at Stennerskeugh Clouds, Kirkby Stephen (Starkey 1987b).

Derbyshire Ford et al. (1993) list without detail old azurite occurrences from Goodluck and Middleton Flats mines, Via Gellia; Matlock and Wensley; Rutland mine, Matlock Bath; and Snelston mine (Isolation mine) near Ashbourne. At Snelston mine, beautiful specimens were recovered by A. Russell. They formed crusts of small deep-blue prismatic crystals associated with cerussite and malachite in cavities within sandstone (Jackman 1996).

Devon Excellent specimens of azurite and malachite associated with covellite and native silver have been collected on gossanous material from the Combe Martin mines (Dines 1956, Rust 1982a). Good crystals were once found at Devon Great Consols, near Gunnislake (Collins 1892), and East Tamar mine, Bere Ferrers (Greg & Lettsom 1858). Secondary azurite, malachite, copper arsenates and complex copper chlorides replace sheets of native copper found as concretions in the Littleham Mudstone Formation at Littleham Cove, near Budleigh Salterton (Milodowski et al. 2002).

Durham Epigenetic copper-enriched breccias, hosted within Upper Permian limestones and associated with the Butterknowle fault at Raisby Hill quarry (1 km E of Coxhoe) contain azurite, chalcocite, chalcopyrite, native copper and malachite (Hirst & Smith 1974, Young 2001). Bridges & Young (1998) note spherules of azurite up to 3 mm across in brecciated limestones in a railway cutting S of Raisby Hill quarry, and other occurrences in the Northern Pennine orefield: at Allen's Cleugh, Tynehead; on the spoilheaps of Copper Slitts vein, Tynehead; at Westernhope Old mine, S of Westgate, SW of Stanhope, Weardale; and at Yew Tree mine, Stanhope.

Humberside Beach pebbles derived from the Chalk of Speeton Cliff, near Flamborough Head, contain diffuse spots of irregular shape, up to 1 cm across, consisting of crudely octahedral cuprite intergrown with native copper, mantled by calcite intergrown with malachite and an unidentified copper chloride. Occasionally, minute blue crystals of azurite are found in cavities in this chalk (Roydhouse & Green 2003).

Leicestershire Azurite forms vivid-blue euhedral prismatic crystals or powdery crusts and coatings at Bardon Hill quarry, Coalville. It is associated with chrysocolla, native copper, cuprite and malachite in a basal breccia close to the Precambrian/Triassic unconformity (Ince 2007). Malachite-lined cavities in a native copper–cuprite deposit near the unconformity at New Cliffe Hill quarry, Stanton under Bardon, are sometimes filled with massive to crustiform dark-blue azurite (Hawthorne et al. 2002, Hubbard et al. 2005). At Newhurst quarry, Shepshed, azurite occurs as dark-blue hemispherical aggregates up to 3 mm across on joints in diorite and as lustrous crystals in chalcopyrite-rich veins (Ince 2005).

Shropshire Strings and stains of azurite and malachite were found in the near-surface parts of veins at Wilderley copper mine, Cothercott Hill, N of Wildersley Hill, SSW of Castle Pulverbatch (Dewey & Eastwood 1925). Small blue spherical nodules of azurite up to 3 mm across occur in a clay band at Llynclys quarry, S of Oswestry (R. E. Starkey pers. comm. 2005).

Somerset Crystalline azurite and malachite were once worked for their copper content between Dodington and Nether Stowey in the Quantock Hills (Rudler 1905). Leaching of primary copper sulphides in massive manganese oxide bodies within Carboniferous Limestone at Merehead (Torr Works) quarry, Cranmore, Mendip Hills, has produced a secondary enrichment zone in which azurite, native copper, cuprite and malachite have formed (Symes & Embrey 1977). Rare azurite is noted from Higher Pitts mine, Priddy (Kingsbury 1941, Burr 1996), and small but distinct crystals are recorded from Buckingham mine, near Bridgwater (Greg & Lettsom 1858).

Staffordshire Azurite is noted from Ecton (Sarjeant 1956b, Ford et al. 1993).

Warwickshire Bright-blue crystals of azurite up to 0.1 mm across on calcite and malachite occur at Judkins quarry, Nuneaton, where barium–copper mineralization occurs close to the Precambrian/Cambrian unconformity (Ince et al. 1991).

Yorkshire Very attractive platy crystals of azurite associated with malachite coat joint planes in copper-mineralized veins at Great Sled Dale, 10 km SE of Kirkby Stephen. Rare, lustrous euhedral crystals (up to 3 mm long) and acicular crystals are also reported from this relatively inaccessible location (Briscoe & Green 1989, Wolfe 2006). Rare azurite coats joint planes within limestone and is associated with bornite, chalcocite, chalcopyrite, native copper, cuprite, digenite, djurleite and malachite on dump material from the Middleton Tyas mines, near Scotch Corner (Green 1987b). At Merrybent mine, a few km to the NW of Middleton Tyas, azurite has been found with limonite and malachite, all replacing an earlier assemblage of bornite, chalcocite, chalcopyrite, covellite and digenite, within galena-rich calcite veins (Small 1977, Dunham & Wilson 1985). Spoilheaps near the Kirkby Fell viewpoint, 5 km E of Settle, have yielded specimens of azurite, barite, galena, malachite and sphalerite (Mundy & Arthurton 1994). Raistrick (1983) notes abundant tiny fragments of azurite and malachite in Pikedaw Gill, NW of Malham. Around Pikedaw mine, azurite occurs in quartz gangue associated with locally abundant malachite and lesser amounts of anglesite, barite, beaverite, cerussite and smithsonite (Green et al. 2006c).

Wales

Dyfed Azurite has a restricted distribution in the Central Wales orefield, but is noted as blue coatings and stains associated with malachite at: Copa Hill mine, Cwmystwyth; Darren mine, Pen-bont Rhydybeddau; Eaglebrook (Nantycagl), Ceulanymaesmawr; Esgair Fraith mine, near Tal-y-bont; Llety-Ifan-Hên (Vaughan) mine, SE of Tal-y-bont; Mynyddgorddu mine, W of Bont-goch, Ceulanymaesmawr; and Penrhiw mine, Ystumtuen (Jones & Moreton 1977). Bevins (1994) points out that many of the blue crusts in this area comprise other copper species, in particular, linarite. However, azurite is confirmed at Bwlchrhennaid mine, near Goginan.

Gwynedd Fine azurite has been found associated with malachite on dolomite at the Ty Gwyn mine, Great Ormes Head, Llandudno (Bevins 1994). In reviewing the mineralization at the Great Orme mines, Ixer & Davies (1996) describe an azurite-bearing shale horizon containing azurite spheres and nodules, up to 2 cm in diameter, associated with sulphide grains now oxidized to limonite. Azurite also occurs with chalcopyrite and sphalerite (minor covellite, galena, pyrite) in a small lode at Gamallt mine, Careg y Fran, near Ffestiniog (Lynas 1973); with malachite at Pant-y-Gaseg mine, near Amlwch, Anglesey (Bevins 1994); and with bornite, chalcocite, chalcopyrite, native copper, covellite and tetrahedrite in the Coed-y-Brenin porphyry copper deposit, near Dolgellau (Rice & Sharp 1976).

Mid Glamorgan Fine azurite is noted from Hendy quarry, Miskin, near Llantrisant (Bevins 1994) and in calcite veins encountered in the Locks Common borehole, Porthcawl (Fletcher & Young 1988).

Powys Azurite occurs with harmotome and malachite associated with a galena–tennantite vein at Dolyhir quarry, Old Radnor, near Presteigne (Wolfe 2000). Azurite is (cautiously) identified in the Powys section of the Central Wales orefield at: Dylife mine (Jones & Moreton 1977, Moreton 1982); Snowbrook (Nantyreira) mine (Jones & Moreton 1977); and South Nantycar mine, near Rhayader, and outside the orefield in old copper workings at Llanymynech (Bevins 1994).

Scotland

Central Azurite is noted from three locations in the Ochil Hills: Airthrie Hill mine near Bridge of Allan, and Burn of Sorrow mine near Dollar (Dickie & Forster 1974); Alva silver mine (Moreton 2007).

Dumfries & Galloway Wilson (1921) reported azurite from a fault-bounded chalcopyrite-bearing vein in porphyritic felsite at Colvend mine, near Rockcliffe. However, later work by Moreton & Aspen (1992) suggests that connellite may have been mistaken for azurite. Azurite is recorded from the Leadhills–Wanlockhead orefield (Greg & Lettsom 1858, Brown 1927), although Heddle (1901) thought that this may have been linarite. Azurite and malachite were once found in a barite vein near the house of Orroland, Barlocco, SW of Auchencairn (Heddle 1901).

Highland Wartime investigations into the hematite and manganese ore resources of Britain led to the discovery of a magnetite-rich skarn heavily stained with azurite and malachite at Kilchrist, in the metamorphic aureole of the Beinn an Dubhaich, Broadford area, Isle of Skye (Groves 1952, see also Tilley 1948b, 1951).

Strathclyde Rare azurite is noted from a stratiform chalcopyrite-rich deposit in Middle Dalradian rocks at the Abhainn Srathain mines, S of Meall Mòr, South Knapdale (Smith et al. 1978a). Rare blue crystals of azurite up to 0.5 mm long occur with malachite on chalcopyrite within molybdenite-rich veins at Coire Buidhe, Glen Creran (Ingram et al. 1993). Barite veins at Mannoch Hill mine, SW of Lesmahagow, contain cavities in which rare bright-blue bladed crystals of azurite up to 2 mm long with chalcopyrite, limonite and malachite have been found (Todd & McMullen 1991). Azurite is also noted from: Kilfinan, near Lochgilphead (Berridge 1969); Swinlees mine, 3 km N of Dalry, Renfrewshire Hills (Stephenson & Coats 1983); and a quarry above Torduff reservoir, near Currie (Heddle 1901, Livingstone 2002).

Tayside Azurite, chalcopyrite and malachite were once found in an old quarry at Taymouth Park and at Grandtully, S of Pitlochry (Heddle 1901), although the precise locations are not known.

Orkney Small crystals of azurite were once found in a barite vein at Viera on the Isle of Rousay (Heddle 1901).

Shetland Azurite is noted with kyanite at Hillswick, Grariesum on Mainland (Heddle 1901).

Ireland

Cork Azurite is noted at some copper mines: Brow Head and Streek Head, Mizen Head peninsula (Bartlett 1980); Cappagh and Coosheen mines, near Ballydehob (Ryback & Moreton 1991); Classadaugh, near Four Mile Water (Greg & Lettsom 1858); and Mountain mine, Allihies, Bearhaven (Berehaven) Peninsula (Reilly 1986).

Down Dump-material from copper–lead mines working barite–chalcopyrite–galena mineralization in dolomitized Silurian greywacke and shale breccia between Conlig and Newtownards contains a supergene assemblage that includes azurite, cerussite, gypsum, limonite, malachite and pyromorphite (Moles & Nawaz 1996).

Fermanagh Azurite, chalcopyrite and malachite are recorded from near Belleek (Fowler 1959).

Galway Azurite and malachite are perhaps the best-known collector species from Tynagh mine, near Killimor. They were important constituents of the ore in the upper workings of the mine, which worked a large disseminated Cu–Pb–Zn–(Ag) deposit associated with a faulted Lower Carboniferous carbonate sequence (Schultz 1971, Clifford et al. 1986, Ryback & Moreton 1992, Green & Moreton 2001).

Kerry Russell (1927b) reports azurite (chessylite), bornite, chalcopyrite, native copper, cuprite, malachite, serpierite and tetrahedrite from an old copper mine at Ross Island, Killarney. At Leaghillaun, Bearhaven (Berehaven) Peninsula, a chalcopyrite–tennantite–quartz vein hosted by Devonian sandstones has produced an oxidized suite of supergene minerals, including blue azurite stains, coatings and microcrystalline crusts associated with chrysocolla, clinoclase, cornwallite, malachite, tyrolite and zálesíite (Moreton & Green 2006b).

Limerick The Glen of Aherlow Cu–(Ag) deposit, near Galbally, is a Lower Carboniferous limestone–sandstone-hosted orebody in which azurite and malachite are found within the zone of weathering. Bornite, chalcocite and chalcopyrite are the main ore species (Cameron & Romer 1970).

Tipperary Fractures in sandstones and limestones from the upper surficial oxidation zone of the Cu–(Ag–Hg) orebody at Gortdrum mine, near Tipperary town, carry azurite crystals (up to 2 mm long), malachite and rare cinnabar (Ryback & Moreton 1991). Brochantite, covellite, devilline, digenite, native copper, erythrite, gypsum, lavendulan, posnjakite and tyrolite are other species that occur in this zone (Steed 1986, Braithwaite & Wilson 2003). Azurite is one of many secondary species found in the giant base-metal and barite deposits (e.g. Gortnadyne mine) in the Silvermines district of Tipperary (Russell 1907b, Moreton 1999).

Bright-blue azurite crystals to 0.5 mm on iron-stained matrix; Tynagh mine, near Killimor, Co. Galway, Ireland. (© M. P. Cooper)

Wicklow Greg & Lettsom (1858) report azurite (chessylite) from Audley mine.

Offshore territories

Jersey Small amounts of deep-blue azurite and malachite have been found at La Moye and Westmount (Teilhard de Chardin & Pelletier 1911, Mourant 1961).

Sark Azurite and malachite are oxidation products of chalcopyrite in the Pb–(Ag) lode cutting granite at Sark's Hope mine, Little Sark (Mourant & Warren 1934, Ixer & Stanley 1983).

B

babel quartz
TL Bere Alston, N of Plymouth, Devon, England
A variety of quartz, bearing impressions of partially dissolved fluorite cubes.

babingtonite?
triclinic $Ca_2(Fe^{2+},Mn)Fe^{3+}Si_5O_{14}(OH)$
Named after William Babington FRS (1756–1833), physicist, mineralogist and a founder member of the Geological Society of London, who was born at Port Glenone, near Coleraine, Co. Antrim, Ireland. Babingtonite is a rare calcium–iron–manganese silicate hydroxide that forms a series with manganobabingtonite and is described from the great cliff of Sgòr Chaonasaid, on the north slope of the Ben Loyal (Beinn Laoghal) syenite complex, near Lettermore, Highland Region, Scotland. It is found in a pegmatitic vein associated with galena, lepidocrocite (rubinglimmer), magnetite, microcline (amazonite), titanite (sphene) and topaz (Heddle 1901); the topaz was later shown to be phenakite (Macpherson & Livingstone 1982). At Beinn Bhreac, 2 km E of Tongue and 7 km N of Ben Loyal, babingtonite forms irregular crystals, up to 50 mm across, in a 60 cm-wide granitic vein within a 100-tonne erratic boulder of syenite embedded in till. Heddle (1883) describes it as "mineralogically the most wonderful and interesting stone in Scotland". Unfortunately, modern analysis has failed to confirm the species (B. Jackson pers. comm. 2004). Black to dark-green babingtonite is vaguely described from the Shetland Isles (Greg & Lettsom 1858). Ferrobabingtonite is a synonym of babingtonite.

baddeleyite
monoclinic ZrO_2
Named after Joseph Baddeley, who brought the original specimens back from Ceylon. Baddeleyite is a rare zirconium oxide found as an accessory mineral in basic and ultrabasic rocks; because of its lack of lead inheritance, it is ideal for uranium–lead dating.

Abundant honey-brown striated grains of baddeleyite were recovered from a pegmatitic dolerite at High Force quarry, near Middleton-in-Teesdale, Co. Durham, and were used to show that the intrusion crystallized 297.4 million years ago (D. G. Pearson pers. comm. 2004).

Baddeleyite is found in the layered ultrabasic intrusion on the Isle of Rum, Highland Region, Scotland, where it forms small (20 µm) subhedral-to-anhedral grains in various cumulate rocks. Optically, baddeleyite is colourless and is associated with apatite, chrome spinel, ilmenite, zircon and zirconolite (Williams 1978). In Unit 1 of the Eastern Layered Series of the Rum intrusion, at Allt nam Ba, baddeleyite is locally abundant, occurring within sulphide-rich horizons containing chalcopyrite, galena, pentlandite and pyrite (Power et al. 2000a).

Baddeleyite occurs in trace amounts in the bronzite-picrite Beannach dyke and the olivine-gabbro Strathan dyke (both part of the Scourie Dyke Suite), near Lochinver, Highland Region (Heaman & Tarney 1989, Frick et al. 1994). Baddeleyite also occurs as small (<60 µm-long) light-tan tabular grains, associated with accessory zircon and monazite, in siliceous dolomite within the aureole of the Ballachulish granite, Allt Guibhsachain area, near Ballachulish, S of Fort William, Highland Region (Ferry 1996, Fraser et al. 2004). Striated tabular crystals of baddeleyite were used by Rogers & Dunning (1991) to date the Ratagain pyroxene–mica-diorite pluton, Loch Duich, near Glenelg, Highland Region.

baghdadite
CUSPIDINE GROUP
monoclinic $Ca_6Zr_2(Si_2O_7)_2O_4$
A rare calcium–zirconium silicate found in a high-temperature calcic skarn near the contact of the Cuillin gabbro complex at Camasunary Bay, Strathaird Peninsula, Isle of Skye, Highland Region, Scotland. Associated species include garnet (andradite–grossular), melilite, vesuvianite and wollastonite, together with minor amounts of calcite, magnetite, perovskite, pyrite and tilleyite; djerfisherite and kimzeyite (both rare) complete the assemblage. The occurrence of zirconium-rich minerals suggests that the skarn underwent a degree of metasomatism with the introduction of fluids rich in aluminium, potassium, titanium and zirconium, possibly from the Coire Uaigneich granophyre (Beard & Drake 2007).

baltimorite
A variety of antigorite, kaolinite–serpentine group.

balvraidite
TL Balvraid, Glen Beg, Highland Region, Scotland
A mixture of calcite, altered saponite, mica and much ill defined fibrous mineral matter; see Heddle (1880b).

banalsite
FELDSPAR GROUP
orthorhombic $BaNa_2Al_4Si_4O_{16}$
TL Benallt mine, Rhiw, Lleyn Peninsula, Gwynedd, Wales
A rare barium–sodium aluminosilicate found at the type locality on the footwall side of the no. 1 orebody and later in the no. 2 and no. 5 orebodies. Banalsite is rather coarsely crystalline, sometimes forming veinlets 1–2 cm wide, within dark-purple manganese ore. It is pure white, translucent to transparent in thin flakes, and has a good pearly lustre. Associated minerals include alleghanyite, barite, biotite, calcite, harmotome, jacobsite, natrolite and tephroite (Campbell et al. 1944, Campbell Smith 1945, Liferovich et al. 2006a,b). Dunham et al. (1978) suggest that relict oolitic structures in the manganese ore indicate that it developed through metasomatic alteration of an original chamositic mudstone.

bannisterite
monoclinic $(K,Na,Ca)(Mn,Fe,Mg,Zn)_{10}(Si,Al)_{16}O_{38}(OH)_8 \cdot 2\text{–}6H_2O$
TL Benallt mine, Rhiw, Lleyn Peninsula, Gwynedd, Wales
Named after Frederick Allan Bannister (1901–1970), keeper of minerals at the BM(NH); born in Peckham, southeast London. Bannisterite is a rare hydrated potassium–sodium–iron–calcium–manganese aluminosilicate hydroxide found on the fractured surface of manganese-rich ore from the Ty Canol incline at the type locality. It forms cinnamon-brown crystals associated with calcite and ganophyllite. The formal description of bannisterite was made by Smith & Frondel (1968), based on material originally described as ganophyllite by Campbell Smith & Bannister (1948).

baotite
tetragonal $Ba_4Ti_4(Ti,Nb,Fe)_4(Si_4O_{12})O_{16}Cl$
A very rare barium–titanium–niobium–iron silicate chloride found in sulphidic barite rock at Foss, near Aberfeldy, Tayside Region, Scotland. The Foss stratiform barite–base-metal deposit consists of barite, Ba-muscovite and Ba-feldspar (celsian) in cherts, schists and carbonate rocks with disseminated-to-massive pyrite, sphalerite and galena. Baotite was discovered as two tiny (10–15 µm) inclusions in coarse pyrite and it represents the titanium end member of the niobium–titanium substitution series (Moles 1985).

barbertonite?
MANASSEITE GROUP
hexagonal $Mg_6Cr_2(CO_3)(OH)_{16} \cdot 4H_2O$
A rare hydrated magnesium–chromium carbonate hydroxide, dimorphous with stichtite and reported from an antigorite–serpentinite mass close to talc schists near Hoo Field, Cunningsburgh, Mainland, Shetland, Scotland. Read & Dixon (1932) describe fibrous stichtite forming conspicuous rose-pink patches in aggregates to 0.5 cm associated with antigorite, chromite, dolomite and iron oxide. Stichtite replaces original chromite grains. However, Macpherson & Livingstone (1982) suggest the mineral may be barbertonite instead. The mineral requires confirmation. Barbertonite from Cunningsburgh is also noted by Gaines et al. (1997).

barbosalite

LAZULITE GROUP

monoclinic $Fe^{2+}Fe^{3+}_2(PO_4)_2(OH)_2$

A rare iron phosphate hydroxide found on the type specimen of andrewsite (now discredited) from Stowe's section of the Phoenix United mines, Linkinhorne, Cornwall. Barbosalite and goethite form the core of fibrous radial spherules up to 4 mm across. The outer shell of these spherules was described as pale-green andrewsite, later shown to be a mixture of rockbridgeite and hentschelite (Dunn 1990). In iron-stained fracture veins within partially kaolinized granite at Gunheath china-clay pit, Carthew, St Austell, black barbosalite forms the bulk of a complex matrix of mixed phosphates that includes intermediate members of the frondelite–rockbridgeite and chalcosiderite–turquoise series, and chlorapatite and leucophosphite. The matrix is veined with cassiterite, chalcocite, fluorapatite and quartz (Elton 1997). Elton (1998) describes barbosalite (also from Gunheath), associated with intermediate chalcosiderite–turquoise compositions, fibrous dufrénite, fluorapatite, hentschelite, leucophosphite, stannoidite and varlamoffite. Brochantite, libethenite and torbernite fill the cavities. Barbosalite and the other iron and copper phosphates probably crystallized from phosphatic solutions as they interacted with early-formed iron-rich veins during the later stages of kaolinization.

baricalcite

A barium-rich variety of calcite.

bariopyrochlore?

PYROCHLORE GROUP

cubic $Ba_2Nb_2O_7$

A rare barium–niobium oxide possibly present in lepidolite greisen from Stranakelly, 3 km N of Shillelagh, Co. Wicklow, Ireland. Bariopyrochlore (described as pandaite) has a yellowish appearance and is intergrown with microlite. A partial analysis suggests that its composition is very different from true bariopyrochlore (Scoon 1978) and confirmation of the species is therefore desirable.

BARITE GROUP

Orthorhombic sulphates and chromates with the general formula AXO_4, where A = Ba, Pb, Sr; and $X = Cr^{6+}, S^{6+}$. Anglesite, barite and celestine are members of the group found in Britain or Ireland.

barite

BARITE GROUP

orthorhombic $BaSO_4$

A very common barium sulphate that forms a series with celestine and is found in low-temperature hydrothermal veins, in hematite replacement deposits, and as a primary component of epigenetic carbonate-hosted sulphide deposits. Its use as a heavy mud in drilling wells (exploiting gas and oil) contributed to the development of an important industry around the species (Collins 1972).

England

Avon Pale-cream to pink banded (cockscomb in cavities) barite occurs in a mineralized fault zone within Upper Triassic limestone breccias and as beach pebbles near the Pier hotel at Clevedon beach, 16 km west of Bristol. It is associated with galena (mainly), chalcopyrite, sphalerite and minor hematite, marcasite, pyrite and tennantite (Starkey 1984, 1986c, Ixer et al. 1993, Bridges 2003). Barite lines cavities in iron–manganese-rich pods, replacing Carboniferous Limestone at Coombe Farm quarry, Henbury, near Bristol (Alabaster 1978), and at Wesley Theological College (Wesley mine), Westbury-on-Trym (Alabaster 1989b). It forms aggregates of paper-thin white-to-pink cockscomb crystals, lavender-pink spherules and colourless tabular and prismatic crystals to 6 mm at the last locality. Barite is also recorded from: East Harptree, N of the Mendip Hills – in cavities within silicified Upper Triassic limestone breccias (Alabaster 1990b); Hampstead Farm quarry, Chipping Sodbury – pink lenticular plates (<2 mm) and white to whitish-pink tapering bladed crystals up to 5 cm long (Alabaster 1990b, Brooke 1994b); Southfields quarry, 2 km N of Chipping Sodbury – in Fe–Pb–Zn mineralized veins within faulted Carboniferous Limestone (Curtis 1981); Stancombe quarry, near Flax Bourton, Bristol – some cockscomb barite in cavities within subvertical cinnabar-bearing barite–calcite veins in Carboniferous Limestone (Zachariades 1993); Woodleaze quarry, Tytherington – rare sub-mm-size rosettes and encrustations of colourless tabular barite in sandstone-filled sinkholes within Carboniferous Limestone (Alabaster 1989a).

Cheshire Barite is reported from locations on Alderley Edge ridge, where it occurs in mineralized Triassic New Red Sandstone as pink, grey or creamy-white vein fillings and disseminations associated with chalcopyrite, galena, pyrite and sphalerite. These include: Cobalt mine, Finlow (Timberlake & Mills 2003); Castle Rock; Church quarry; Doc mine; Red Moulding sand quarry (all Braithwaite 1994); Engine Vein, Mottram, New Venture, Opencast, West and Wood mines, and also in the Stormy Point area (Carlon 1979, Braithwaite 1994, Plant et al. 1999). Similar sediments at Bickerton, contain widespread barite associated with bornite, chalcopyrite, pyrite (bravoite) and tennantite mineralization near marginal-basin faults (Naylor et al. 1989). Rudler (1905) reports barite in jointed Triassic breccias and sandstones from Beeston Castle in the Peckforton Hills.

Cleveland Massive, fibrous, white and pink barite associated with calcite and dolomite has been discovered in Magnesian Limestone in a borehole at Middleton Road, West Hartlepool (Fowler 1956).

Cornwall Barite is a common veinstone most often reported from the lead lodes of central and east Cornwall or associated with late uranium mineralization: Ale & Cakes and United mines, St Day, Gwennap – semi-transparent greenish-grey, yellowish or brownish crystals associated with chalcopyrite (Collins 1892, Rudler 1905, Hosking 1951); Gravel Hill mine, Perranzabuloe; Mount Wellington mine, Twelveheads, 7 km W of Truro – translucent greyish-green crystals up to 1.5 cm long and parallel-growth crystal masses up to 3 cm across aggregated over chalcopyrite (Rust 1982e); Herodsfoot mine, Lanreath, near Liskeard – beautifully crystallized specimens (Greg & Lettsom 1858, Dines 1956); Loe Warren Zawn, Botallack, St Just – rare tiny whitish square plates associated with boltwoodite and zeunerite (Elton & Hooper 1995a); Restormel (Royal) iron mine, N of Lostwithiel (Slater & Highley 1976, Ball et al. 1982); St Austell Consols mine – probably from late cross courses and associated with Co–Ni–Pb–U mineralization (Robertson & Dines 1929, James 1947); Wheal Basset, Illogan; Wheal Drea, St Just; Wheal Trelawny, Menheniot (all Golley & Williams 1995); Wheal Buller, near Redruth – associated with chalcopyrite and pitchblende (Ball et al. 1982); Wheal Ludcott, St Ive, near Liskeard (Shrimpton 1993); Wheal Mary Ann, Menheniot – fine transparent crystals, tinted yellow, or tipped with yellow and associated with fluorite (Greg & Lettsom 1858, Rudler 1905, Dewey 1925, Embrey & Symes 1987, Merry & Braithwaite 1995); Wheal Pendarves (Tryphena lode), Camborne – small clear crystals (Barstow & Cooper 1982); Wheal Wrey, St Ive, near Liskeard – large, perfect crystals and quartz epimorphs after tabular barite (Collins 1904b, Dines 1956, Rust 1983a). Barite is also noted in intertidal sediments from the Camel, Fowey and Gannel estuaries (Pirrie et al. 2000b, Pirrie et al. 2002).

Cumbria Barite is a common and widespread mineral in Carboniferous–Permian galena–(sphalerite)-bearing veins. Some of these veins in the Caldbeck Fells produced economic abundances of barite (Stanley & Vaughan 1982a). Particularly fine specimens of barite have also been found in the West Cumbrian (hematite) orefield. Ironstone nodules in the Coal Measures are another localized source of barite.

Barite occurs in all three of the major mineralized areas. *Caldbeck Fells:* Arm o' Grain – white blades in vein quartz (Green et al. 2006b); Bannerdale mine, Mungrisdale – massive barite associated with chalcopyrite, galena and sphalerite on the mine dumps (Ward 1876b, Eastwood 1921, Davidson & Thomson 1951, Eastwood et al. 1968); Brae Fell mine, Dale Beck valley – white bladed crystals up to 20 mm long and crystalline masses (Hartley 1984, Bridges et al. 2006); Brandy Gill and Ingray Gill mines, also sub-economic amounts at Nether Row Brow, near the Dumpy Stone level (Hartley 1984); Burdell Gill – white, radiating aggregates in cellular quartz, associated with goethite and manganese oxides, also supergene barite forming <0.3 mm rhombic tablets (Young 1987a, Neall & Leppington 2003); Carrock mine, Grainsgill (Dewey & Dines 1923, Eastwood 1959, Shepherd et al. 1976, Cooper & Stanley 1990); Deer Hills barite mine (Shaw 1970, Symes et al. 1990); Driggith mine – economic amounts of massive barite (Davidson & Thomson 1951, Eastwood 1959, Eastwood et al. 1968, Shaw 1970, Ineson & Mitchell 1974, Stanley & Vaughan 1981, Cooper & Stanley 1991b); Dry Gill mine – blocky crystals of pale-green to white barite, associated with campylite, also a few large doubly terminated crystals (Greg & Lettsom 1858, Davidson & Thomson 1951, Cooper & Stanley 1990, Green & Briscoe 2002); east slope of High Pike – part of the Driggith–Sandbed vein (Eastwood et al. 1968); Low Pike, near Potts Gill mine – both primary and water-clear supergene barite associated with cornwallite and pseudomalachite (Cooper & Stanley 1990, Young et al. 1990a); Potts Gill barite mine – only small amounts of specimen material remain on the dumps (pale-blue curved tabular crystals to 12 mm, associated with psilomelane), but thousands of tonnes were mined, including fair-grade white barite from South vein (Davidson & Thomson 1951, Eastwood 1959, Shaw 1959, Eastwood et al. 1968, Hartley 1984, Young 1987a, Braithwaite & Ryback 1988, Crowley et al. 1997); Red Gill mine – massive white or pinkish barite (Eastwood 1921, Davidson & Thomson 1951, Cooper & Stanley 1991b); Roughton Gill mine – formerly worked for barite, now uncommon (Greg & Lettsom 1858, Davidson & Thomson 1951, Eastwood et al. 1968); Sandbed mine – barite was mined from the Driggith lode, but mainly from E–W veins (Eastwood 1959, Shaw 1970, 1959, Hartley 1984), plus white rhombic crystals of lead-rich

Barite crystal, 105 mm long, on dolomite. Frizington, Cumbria. (© National Museum of Wales; photographer M. P. Cooper)

barite up to 1.5 mm long (Green et al. 2006a); Short Grain – solid masses and small indistinct tabular crystals with pyromorphite (Cooper et al. 1988, Cooper & Stanley 1990); Silver Gill – thin transparent blades of barite associated with malachite on goethite-encrusted quartz (Green et al. 2005b); Wet Smale Gill (Shaw 1970). *West Cumbria:* Two distinct varieties of barite are described from the hematite deposits (Shepherd 1973):

- a pale bluish-green variety found in vugs in the orebodies
- bladed masses of white or pink barite intimately associated with the ore.

Localities include: Beckermet hematite mine, near Egremont – barite is a minor gangue mineral (Goldring & Greenwood 1990); Crowgarth mine, Cleator Moor – large blue crystals and smaller brilliant crystals of barite associated with hematite, chalcopyrite and pyrite (Sweet 1930, Trotter 1944, 1945, Davidson & Thomson 1951); Dalmellington and Parkside mines, near Frizington – clear, pale-yellow to rich chocolate-brown barite with tabular or prismatic habit, some magnificent specimens (Sweet 1930, Young 1987a); Distington quarries – white barite on joints in Carboniferous Limestone (Eastwood et al. 1931); Ennerdale Hall, Rowrah – vein of white barite (Eastwood et al. 1931); Eskett mine, Rowrah, Frizington – pink barite (Trotter 1944); Florence mine (Ullcoats section), near Egremont – creamy barite associated with fluorite (Green & Briscoe 2002); Frizington – a striking group of large blue barite crystals (Watson 1988, Starkey 1989); Gilcrux mine, Rose Gill, SW of Gilcrux, near Maryport – worked for barite (Eastwood 1959, Crowley et al. 1997); Haile Moor mine, Egremont – semi-crystalline barite (Davidson & Thomson 1951); Kinniside lead mine, Cleator Moor – fine detached translucent-blue barite crystals, also colourless, and white sugary and platy barite (Greg & Lettsom 1858, Goodchild 1885, Eastwood et al. 1931, Young 1987a, Green & Briscoe 2002); Langhorn quarry, Bigrigg (Eastwood et al. 1931, Lewis 1931); Mowbray mine, Frizington – colourless, pale-yellow and yellowish-brown, through many shades of greenish-blue to sky-blue, usually tabular barite, sometimes forming curved crystals (Sweet 1930); Pallaflat mine, Egremont – fine examples of pale-green, brown and pink barite (Young 1987a); Waterblean mine, Millom (Kendall 1893); Sellafield-area boreholes (Milodowski et al. 1998); Winscales mine and Wyndham mine, Bigrigg, near Egremont – barite occurs in lenses within hematite orebodies and in limestone (Trotter 1945). *Elsewhere in the Lake District:* Beacon quarry, 2 km SE of Gilcrux (Crowley et al. 1997); Birk Fell mine, Tilberthwaite (Holland 1981); Blackbeck Tarn, Buttermere (Moorbath 1962); Bleawick, Ullswater (Ward 1876, Wilson et al. 1922); Brandlehow mine, Derwent Water (Davidson & Thomson 1951); Brundleholme mine, Glenderaterra Valley (Eastwood 1921); Carlside, Skiddaw (Ward 1876, Eastwood 1921, Firman 1978); Clints quarry, Bothel – white platy masses and cockscomb aggregates (Young 1987a); Eagle Crag mine, Patterdale – rare cockscomb aggregates (Young 1987a); Force Crag mine, Braithwaite, near Keswick – mined for barite, galena and sphalerite, fine specimens of barite forming groups of large tabular crystals up to 30 cm across (Greg & Lettsom 1858, Davidson & Thomson 1951, Eastwood 1959, Green et al. 1997, Green & Briscoe 2002, Tyler 2005); Garburn road, Troutbeck – veins of white platy barite (Lawrence et al. 1986); Greengill, 6 km N of Cockermouth – economic amounts of barite (Firman 1978); Greenside mine, near Glenridding, Patterdale – abundant barite in the upper stopes (Greg & Lettsom 1858, Eastwood 1959, Firman 1978, Naylor 1981); Hartsop Hall mine, Hartsop, Patterdale – platy aggregates of barite (Russell 1944, Rothwell & Mason 1992); Hause Crag, Ullswater – barite in a hematite vein (Moseley 1960); Hazel Gill, Howgill Fells – barite gangue (Bateson & Johnson 1984); Helvellyn (Wythburn) mine, Thirlmere (Shaw 1970); Kaipe Tarn, Staveley – vein of pink barite with galena (Aveline & Hughes 1872, Eastwood 1921); Paddy End mine, Coniston, Eskdale – cockscomb barite associated with calcite, sphalerite and tennantite (Young 1987a); Ruthwaite mine, Uldale, near Bassenthwaite – mined for barite (Trotter 1944, Eastwood 1959, Eastwood et al. 1968, Crowley et al. 1997); in the aureole of the Shap granite at Shap Blue quarry, at Sherry Gill and in the Birkbeck tunnel (Davidson & Thomson 1951, Firman 1953a, 1978); Shap granite quarry – pink, massive and cockscomb varieties in veins cutting granite (Firman 1953a, 1978, Wirth & Green 2002); Threlkeld mine (Davidson & Thomson 1951, Young 1987a); Tilberthwaite mine (Young 1987a); Yewthwaite mine, Newlands Valley – quartz pseudomorphing barite (Strens 1962). *Northern Pennine orefield:* Dunham (1934, 1948) mapped a series of five concentric zones over the orefield (centred over the buried Weardale granite) and defined a barite–witherite zone, with barytocalcite as the outermost (lowest-temperature) zone. A vertical zonation, with barite at the top, was described later by Ineson (1976). Barite locations include: Amber Hill veins, E of Hilton – worked for barite (Dunham 1990); Blagill mine, near Alston – barite associated with galena and witherite, also pseudomorphs of compact chalky barite after barytocalcite (Davidson & Thomson 1951, Bridges & Young 1998); Brownley Hill mine, Nenthead, near Alston – massive white-to-pink vein-filling barite forming thin transparent colourless plates in vugs, and barite pseudomorphs after harmotome (Young et al. 1990b, Wood 1991, Green & McCallum 1993, Green & Nudds 1998, Green et al. 2000, Green & Briscoe 2002); Dufton Fell mine, near Alston – considerable barite in mineralized flats, some of the finest ever found in Britain, including an exceptionally large crystal (50 kg) displayed in the Great Exhibition in 1851 (Wallace 1865, Rudler 1905, Bramall 1921, Davidson & Thomson 1951, Solomon et al. 1971, Dunham 1990); Hartley mine, Kirkby Stephen (Davidson & Thomson 1951); Higher Longrigg mine, S of Cote Garth, Hartley Birkett, near Kirkby Stephen – pinkish barite of the cawk variety (Braithwaite 1988); Hilton mine, Scordale, near Appleby – worked for galena, then barite and witherite. The deposit contains many vugs with finely crystallized barite and fluorite, also large creamy white tabular crystals of barite up to 12 cm long and dark-grey epimorphs of barite after calcite (Davidson & Thomson 1951, Solomon et al. 1971, Ixer 1986, Winrow 1986a, Green & Briscoe 2002, Bouch et al. 2006); Long Fell barite mine, 4 km ESE of Hilton – veins of nearly pure barite with a few nice crystals (Davidson & Thomson 1951, Dunham 1990); Nentsberry Haggs mine, Nenthead – large transparent, well terminated but etched, barite crystals (to more

BARITE

Tabular barite from Silverband mine, Milburn, near Appleby, Cumbria; the specimen is 125mm tall. (© National Museum of Wales, Cardiff; photographer M. P. Cooper)

than 20cm long) and barite pseudomorphs after alstonite (Davidson & Thomson 1951, Solomon et al. 1971, Green 1993, Green & Briscoe 2002, Bridges & Green 2006b); Rampgill mine, Nenthead – supergene crusts of lustrous white barite crystals up to 2mm across lining vugs in witherite, and barite overgrown by barytocalcite (Bunting 1994, Bridges & Young 1998, Green & Briscoe 2002, Bridges & Green 2006b); Silverband mine, Great Dun Fell, Milburn, near Appleby – former major barite vein deposit that produced fine (large) yellowish-white and clear colourless crystals, some doubly terminated, showing many changes in habit during growth (Wallace 1865, Bramall 1921, Sweet 1930, Dunham & Hollingworth 1947, Davidson & Thomson 1951, Seager & Davidson 1952, Seager 1953, Dunham 1959a, 1990, Patel & Koshy 1969, Solomon et al. 1971, Dunham et al. 1978, Green & Briscoe 2002); Silver Top quarry, Hallbankgate, Brampton – white to pale-pink banded compact fine-grain crystalline nodular masses (Young et al. 1989, Crowley et al. 1997); Stennerskeugh Clouds, Kirkby Stephen – barite–calcite gangue associated with galena and tennantite (Starkey 1987b); Threlkeld Side level, Dufton – well formed transparent crystals with internal phantoms (Bramall 1921, Wolfe 2004); Tynebottom mine, NW of Garrigill, near Alston – sparse secondary barite forming colourless tabular crystals (Rust 1987a).

Derbyshire The most common variety of barite is the creamy massive granular form known as cawk, which is found in almost every mineral vein along with calcite, fluorite and galena. In most cases these veins are found in Carboniferous Limestone. The important occurrences are reviewed here and by Ford & Sarjeant (1993), but the wide distribution of the species makes the reporting of all occurrences impractical. *Castleton–Bradwell area:* Dirtlow Rake, Bradwell Moor, near Castleton – a barite–fluorite replacement deposit close to Dirtlow Rake vein (Jeffrey 1997); Hazard and Hollandtwine mines on Dirtlow Rake – pink-red tabular crystals containing inclusions of hematite in growth zones (Sweet 1930, Ford & Sarjeant 1964a,b); Longstone Edge, Pindale – finely banded pink and white barite (Robinson & Ineson 1979); Odin mine – stalactitic white barite, also barite interbanded with calcite and fluorite (Ford 1955, Ford & Sarjeant 1964b); Pindale area – occurs in a great variety of forms (massive or stalactitic most common) and colours (colourless, greenish, pink, yellow and white), also a pink and white banded variety (Sarjeant 1956b); Old Tor mine – vugs lined with nodular barite coating fluorite (Ford 1955); Treak Cliff – massive, colourless and pink banded barite (Ford 1955, Sarjeant 1956b). *Hucklow–Eyam–Stoney Middleton area:* Calton Hill, Taddington – pink barite usually massive, rarely dendritic (Sarjeant 1956b); High Rake, Longstone Edge, Hucklow, N of Bakewell – pink and buff barite with calcite and galena in a replacement flat deposit (Robinson & Ineson 1979, Jeffrey 2001); Ladywash mine, Eyam – attractive cockscomb barite (Green 1986); Tideslow Rake, Tideswell – pink barite (Robinson & Ineson 1979). *Ashford–Bakewell–Sheldon area:* Near the stone circle at Arbor Low, near Middleton-by-Youlgreave – stalactitic barite (oakstone) mined for ornamental purposes (Adam 1846, Rudler 1905, Ford & Sarjeant 1964a,b, Braithwaite 1983, Sarjeant & Ford 1993), this is probably the locality described as Middleton-in-Youlgreave and Newhaven in Greg & Lettsom (1858); Long Rake, near Middleton-by-Youlgreave – barite–fluorite orebody (Robinson & Ineson 1979, Ball et al. 1985b); Magpie mine on Hard Rake, Sheldon (Robinson & Ineson 1979); Mogshaw Rake, Bakewell – bluish "cat's tooth" barite (Ford et al. 1993); Newhaven, Youlgreave – deep chocolate-brown barite (Sweet 1930); Raper mine, Long Rake, near Middleton-by-Youlgreave – three generations of barite including radiating, acicular and lath-like varieties (Ineson & Al-Kufaishi 1970, Robinson & Ineson 1979). *Bonsall–Matlock Bath–Cromford–Wirksworth area:* Alport Hill, Ashley Hay, near Wirksworth – good white tabular barite crystals in a thin vein within Millstone Grit (Ford & Sarjeant 1964a,b, Braithwaite 1983); Bage mine, Bole Hill, Wirksworth – barite associated with fluorite, galena and matlockite (Burr 1992); Bonsall Moor, Winster – well formed bladed white crystals, massive colourless or white barite, and a white arborescent variety known as brainstone (Rudler 1905, Sarjeant 1956b); Bradhouse mine, between Hopton and Wirksworth, and Cawdor quarry, Matlock (Robinson & Ineson 1979); Bradwell Moor, Middleton Cross, Wirksworth – red and pink varieties with hematite inclusions known as tush (Ford & Sarjeant 1964a); Brownedge, near Winster – brown radiating bladed barite (Braithwaite 1983); Donkey Pastures, above Winster – pink-to-red barite and a stalactitic variety with cockscomb exterior, colour zoned from a white core, through colourless and cream zones, to steel-grey or flesh-pink exterior (Sarjeant & Ford 1993); Gallows Knoll, Hopton, near Brassington – irregular spongy nodules of yellow barite enclosed in silica sand (Ford 1963, Ford & Sarjeant 1964b); Golconda mine, Brassington – a bedded barite–galena deposit, once Derbyshire's largest barite producer (Varvill 1959, Ford & King 1965, Ford & Ineson 1971, Robinson & Ineson 1979); Jug Holes mine and Masson opencast, Masson Hill, Matlock – white-to-orange encrustations on fluorite, cockscomb crystals up to 1.5cm long, bladed white crystals and a grey pseudo-stalactitic or nodular variety of barite from a fluorspar replacement flat deposit (Sarjeant 1956b, Ford & Sarjeant 1964a,b, King 1966, Ixer 1974b, Sarjeant & Ford 1993); Millclose mine, Wensley, Darley Dale, near Matlock – white and buff barite associated with sphalerite (Sweet

Polished section of barite (variety oakstone), 5cm across, with rim of crystalline barite; Arbor Low, Youlgreave, Derbyshire. (© M. P. Cooper)

1930, Robinson & Ineson 1979); Oxclose mine, Snitterton, W of Matlock (Ixer 1974b); Pike Lane opencut, near Winster – white to iron-stained brown cockscomb crystals to 20 mm with calcite in veins associated with a major solution-collapse structure (Uttley 1989); Rutland Cavern, Matlock Bath (Ford & Sarjeant 1964b); Wapping mine, Cromford – hard cream iron-stained nodular cockscomb barite associated with calcite and fluorite (Ince 1991). *Ashover inlier:* Ashover borehole – barite crystals in phosphatic nodules within Namurian shales (Ramsbottom et al. 1962); Butts quarry – dendritic coral-like barite, occasionally used for ornamental purposes (Sweeting 1946); Fall Hill quarry – white cockscomb rosettes associated with pyrite in the Ashover Old (fluorite–galena) vein (Smith 1982); Milltown quarry, near Ashover – uncommon barite forming creamy cockscomb groups to 10 mm on fluorite (Rust 1991c); Starrs Wood veins, Ashover – pink-to-red barite (Ford et al. 1993). *Crich inlier:* Crich quarry – barite–fluorite gangue in galena-bearing veins (Wilson et al. 1995); Glory mine – massive pink-to-red barite in bands associated with fluorite (Sarjeant 1957a, Ford & Sarjeant 1964b). *Ashbourne–Snelston area:* Limestone Hill, between Ashbourne and Stanton – barite in a replacement deposit (Cornwell et al. 2001); Snelston mine, near Ashbourne – barite veins containing pinkish, nodular and cockscomb crystals and thin colourless plates (Jackman 1996). *Other locations:* Bincliffe mine, Manifold Valley (Robinson & Ineson 1979); Kenslow Silica Sand pit, Friden (Ford et al. 1993); Waterswallows, 2 km NE of Buxton (Sarjeant 1956b); Whitwell, NE of Bolsover (Ineson et al. 1972, Robinson & Ineson 1979).

Devon Bridford barite mine is a lode deposit situated in the Teign Valley, near the eastern margin of the Dartmoor granite. Originally a lead mine, it later produced over 40 000 tonnes of colourless, white, yellow or pink, massive, granular, fibrous and reniform barite (Polkinghorne 1951, Dines 1956, Vipan 1959, Beer & Ball 1977). Barite is also reported from: Babbacombe, near Torquay – pink tabular crystals in limestone (Greg & Lettsom 1858, Hall 1868b, Rudler 1905, Embrey & Symes 1987); Birch Ellers mine, near Bridford (Dines 1956, Vipan 1959); Frankmills, S of Christow – beautiful crystals of barite associated with galena, pyrite and sphalerite (Collins 1904b); Hennock, near Chudleigh (Hall 1868b); around Peak Hill, Sidmouth – remarkable honey-brown to colourless lustrous tabular barite in cavities in calcite nodules (Corrie & Green 2003, Green & Betz 2006).

Doubly terminated, colour-zoned barite, 40 mm across, from a cavity in pink Devonian limestone; Petit Tor Point, Babbacombe, near Torquay, Devon. (© M. R. Stolworthy)

Durham *Northern Pennine orefield:* Closehouse mine, SW of Middleton-in-Teesdale, Lunedale worked a fault-bounded vein and replacement deposit in a metasomatized dolerite dyke cutting Carboniferous Limestone and was once Britain's largest barite producer (Dunham 1959a,b, Hill & Dunham 1968, Solomon et al. 1971, Vaughan & Ixer 1980, Young 1985a, Young et al. 1985a, 1994a, Dunham 1990). Other occurrences include: Cow Green mine, Harwood-in-Teesdale – groups of thin, colourless and transparent, rectangular crystals of barite (Seager & Davidson 1952, Dunham 1959a); Foster's Hush, S of Middleton to Brough road – secondary barite after barytocalcite and witherite (Young et al. 1985a); Killhopehead, near Nenthead (Dunham 1992); Lunehead mine, near Grains o' th' Beck, Lunedale (Solomon et al. 1971, Dunham 1990); Yorkshire Silverband mine, Cronkley Fell, near Middleton-in-Teesdale – barite encrusted with pyromorphite (Young 1985a). *Durham coalfield:* Over 200 000 tonnes of barite were mined from a barite–witherite fissure vein discovered in New Brancepeth colliery, near Durham. Barite is usually pure white, opaque and massive, but in small cavities it forms perfectly colourless and transparent thin platy crystals to 10 cm in diameter (Spencer 1910, Smythe 1922, Dunham 1947, 1959a, 1990, Solomon et al. 1971, Young 1985c). Barite is also noted from: Croxdale colliery, S of Durham (Dunham 1990); Healeyfield mine, Castleside (Solomon et al. 1971, Dunham 1990); Lumley colliery, N of Durham (Smythe 1922, Solomon et al. 1971); South Moor colliery – supergene barite as a near-surface capping on a witherite deposit (Dunham 1990, Bridges & Young 1998); Ushaw Moor colliery, Durham – barite–witherite vein (Smythe 1933, Bridges & Young 1998).

Permian Magnesian Limestone occurrences of barite are often associated with calcite and fluorite, and include: boreholes near Fishburn and Rushyford – pinkish-white to pale-pink slender rods and aggregates of curved platy crystals of barite in cavities in dolomitic limestone (Fowler 1943); Chilton quarry, Ferryhill – white and slightly pink massive barite in lenses, nodules and veinlets (Fowler 1956, Solomon et al. 1971, Dunham 1990); Ferryhill area – white or water-clear barite in dolomitic breccias or lining voids in limestone (Hirst & Smith 1974); borehole near Great Stainton – white subhedral crystals to 1 cm in vugs (Raymond 1962); Hesleden Dene, near Castle Eden – white barite on joint planes (Dunham et al. 1948, Fowler 1956); Ten O'Clock Barn No. 2 borehole, between Ferryhill and Hartlepool – whitish or slightly pink massive barite (Fowler 1956).

Kent At Warden Point, Isle of Sheppey, barite forms white or greenish-grey radiating acicular sprays of terminated crystals in the calcite-lined cracks of septarian nodules within London Clay (Greg & Lettsom 1858, Monk 1982).

Colourless elongate crystals of barite on honey-yellow calcite in a septarian nodule from Warden Point, Isle of Sheppey, Kent; width 30 mm. (© R. E. Starkey)

Leicestershire Barite infills cracks in septarian nodules within Middle Liassic rocks at Braunston and Whissendine, both Rutland. It develops a bladed or radiating habit and sometimes forms euhedral crystals (Taylor 1950). At Breedon quarry, Breedon on the Hill, thin films and spheroids of white barite were once found in an ancient breccia-infill cave deposit within Carboniferous Limestone (King 1991). Barite is also noted from: Bardon Hill quarry, Coalville – thin pink tabular crystals or aggregates with a cockscomb habit, and associated with quartz, calcite and pyrolusite, or with chlorite (Ince 2007); New Cliffe Hill quarry, Stanton under Bardon – small colourless tabular crystals (Hubbard et al. 2005); Newhurst quarry, Shepshed – white to pale-pink tabular crystals and hemispherical aggregates of pink crystals associated with hematite and dolomite (Ince 2005); Springhill quarry, Peldar Tor, Whitwick – barite in quartz–chlorite veins (Faithfull & Ince 1992).

Northumberland Barite is an important gangue species in galena–witherite-bearing veins that were once worked in the Haydon Bridge mining district, near Hexham; these include Fallowfield, Langley Barony and Settlingstones mines. Supergene barite is also noted from cavities in witherite at Settlingstones (Spencer 1910, Trestrail 1931, Solomon et al. 1971, Ineson 1972, Dunham 1990, Crowley et al. 1997, Bridges & Young 1998). Other reports of barite are from: Barrasford quarry, 9 km N of Hexham – along mineralized joints in the Whin Sill dolerite (Money et al. 1995);

BARITE

Greenleycleugh mine and Keirsleywell mine, near Mohope Burn (Dunham 1990); Isabella mine, Cowpen, SE of Bedlington (Solomon et al. 1971).

Nottinghamshire Rare traces of barite are associated with galena, wulfenite and uranium-rich asphaltite in Magnesian Limestone at Bulwell, near Nottingham; also, white barite is noted with asphaltite at Mansfield White Stone quarry (Deans 1961). Small pale-blue tabular barite crystals to 7 mm are reported from Vale Road quarry, Mansfield Woodhouse (Green & Spence 1988).

Oxfordshire Snow-white powdery masses and elongate radiating clusters, and brown microcrystalline sheets, of barite have been found in pyritic non-septarian concretions within Oxford Clay at Yarnton (Atkins 1995).

Shropshire The West Shropshire orefield worked Ba–Pb–Zn veins occupying fractures in Ordovician and Precambrian rocks of the Shelve inlier. Barite and galena are dominant at higher levels in the veins, giving way to sphalerite, calcite and quartz at depth (Pattrick & Bowell 1991). Occurrences include: Burgam mine, SSW of Pontesbury – white to pale-cream masses in veins (Starkey 1982); East Wotherton mine, near Misterley – beautifully pellucid crystals (Rudler 1905); Huglith mine, near Habberley – over 500 000 tonnes of barite produced between 1860 to 1944 principally from this mine (Dunham et al. 1978); Rorrington (West Snailbeach) mine, S of Meadowtown – once a small barite producer, includes coarse white masses of tabular crystals and sharply pointed tombstone-shape crystals up to 5 mm long associated with barytocalcite, fluorite and witherite (Starkey et al. 1994); Snailbeach mine, S of Minsterley (Rudler 1905, Smith 1973b). Dewey & Eastwood (1925) note barite at two copper mines in the region: Westcott mine, and Wilderley mine, SSW of Castle Pulverbatch; and Wotherton mine, Chirbury (Russell 1936c). Many other mines that produced barite are listed in Pearce (1994a).

Somerset Water-clear authigenic barite is described from cavities in ammonites within Lower Liassic marine sediments between Blue Anchor Point and Stolford Rocks. Analcime, aragonite, calcite, celestine, dolomite, pyrite and sphalerite are associated species (Alabaster 1980). Stalactitic barite is associated with hematite in a vein at Wick quarry, near Kingswood, between Bath and Bristol (Cowdry 2002). In the Mendip Hills, barite has been found at: Coleman's quarry, Holwell, near Frome (Alabaster 1990a); Higher Pitts mine, Priddy (Burr 1996); Merehead (Torr Works) quarry, Cranmore – barite occurs in sulphide-bearing calcite veins and in manganese-rich deposits, usually as colourless transparent crystalline masses, sometimes with good crystal form (Alabaster 1975, Symes 1977, Symes & Embrey 1977); Whatley quarry, near Frome – bright-orange barite on goethite needles, also white cockscomb barite and isolated crystals of barite in the internal cavities of fossils (Alabaster 1976, Goulding 1997a, House 2002). Barite was once worked from veins in Carboniferous Limestone at Cannington Park, near Bridgwater (Dines 1956).

Staffordshire Rudler (1905) notes barite as a gangue mineral in chalcopyrite-rich ore from Ecton mine and in Keuper sandstone from Peakstones Rock, near Alton.

Surrey Large yellow crystals of barite were once found in fuller's earth pits (Lower Greensand) at Nutfield, near Redhill (Rudler 1905).

Tyne & Wear Barite has been found at: Rising Sun colliery, Backworth – barite forms a rust-colour coating on sulphides in mineralized cavities within the Coal Measures (Randall & Jones 1966); Felling colliery, near Gateshead (Clapham & Dalglish 1864, Dunham 1990); Ryhope, S of Sunderland (Smythe 1922); and Houghton colliery, W of Seaham (Dunham 1990).

Warwickshire Barite occurs as small (< 1 mm) white-to-pink crystals and as extensive crusts on rock surfaces at Judkins quarry, Nuneaton (Ince et al. 1991). Light-pink desert-rose aggregates of barite from this quarry are occasionally sprinkled with wooldridgeite (Hawthorne et al. 1999).

Yorkshire Many occurrences of barite are from galena-bearing veins and flats in Carboniferous rocks within the Askrigg Block (the southern part of the Northern Pennine orefield). *North Swaledale mineral belt:* The dumps of Arkengarthdale and Old Gang mines have been reprocessed for barite and fluorite (Dunham & Dines 1945, Dunham et al. 1959a, 1978, Dunham & Wilson 1985, Green 2003). Excellent specimens can be collected from the dumps of the Brandy Bottle incline adits (workings of the Friarfold–Surrender veins), NE of Gunnerside, where cockscomb barite and coral-like white masses of chisel-shape crystals of barite after witherite occur (Young 1987b, Ineson & Young 1994). Barite is noted elsewhere along the Friarfold–Surrender vein system (Dunham & Wilson 1985, Young et al. 1989, Wood 1993) and from other veins in the Gunnerside Gill area, such as Hard level, Victoria level and Level House Bridge in Flincher Gill, all NE of Gunnerside (Ineson & Young 1994). Barras End mine, near Great Pinseat in the southeastern corner of the Old Gang mining area, has produced unusual banded barite (Briscoe 2006). Other locations include: Dam Rigg level, Arkengarthdale Moor – large well formed clusters of white barite crystals collected in 1996 and said to be the best in Britain (Wolfe 1996a); Great Sled Dale, 10 km SE of Kirkby Stephen – small lustrous transparent crystals of barite, associated with azurite and malachite, are probably supergene in origin (Bateson & Johnson 1984, Briscoe & Green 1989); and Whaw, Arkengarthdale (Rodgers 1975). *Wensleydale:* East Bolton Moor, near Askrigg (Vaughan & Ixer 1980, Young et al. 1992c); Hurst Moor, near Leyburn (Vaughan & Ixer 1980); and Wet Grooves mine, Aysgarth – water-clear crystals and thick chisel-shape crystals up to 3 cm long (Dunham & Wilson 1985, Green 1998b). *Wharfedale & Nidderdale:* Coldstones quarry, 5 km W of Pateley Bridge – white-to-pink cockscomb groups and rosettes on fluorite (Ryde 1987, Young et al. 1989, Young et al. 1997); Duck Street quarry, SE of the summit of Greenhow Hill, near Pateley Bridge (Rankin et al. 1987); Grassington Moor (Young 1987b); Great Almescliffe Crags, near Weeton; Grimbald's Crag and Plumpton Rocks, both near Knaresborough – in the matrix of Millstone Grit (Gilligan 1919); Nussey Knot, Greenhow, near Pateley Bridge – white to dirty-white crystals (Dunham & Wilson 1985); and Yarnbury, near Grassington (Rodgers 1975). *Yorkshire Coalfield:* Exceedingly beautiful crystals of colourless, transparent barite associated with dolomite were found in a large cavity within sandstone at Manvers Main colliery, Wath-upon-Dearne, near Rotherham (Russell 1936b). Similar barite–dolomite–pyrite mineralization was also discovered in a fault zone within siltstone at Hickleton colliery, near Goldthorpe, WNW of Doncaster, and produced slender transparent crystals of barite up to 15 mm long, often coated with dolomite or kaolinite (Uttley 1988). A further occurrence at Silverwood colliery, SW of Doncaster, produced barite crystals (some doubly terminated) reaching 50 mm in length, again associated with dolomite and pyrite (Uttley 1987a, Prowse 1988).

Translucent slice of barite, 1 mm thick and 15 mm across, from Barras End mine, Swaledale, Yorkshire. (© D. I. Green)

Other descriptions of barite include: Hilston Beaches, Holderness – occasional large transparent smoky barite in septarian nodules, and from a roadcut in Magnesian Limestone, near Bramham (Green 1993); Pikedaw Gill, NW of Malham – barite associated with galena and sphalerite on spoilheaps (Raistrick 1983, Mundy & Arthurton 1994); and Whitehall quarry, Butcher Hill, Horsforth – in the matrix of Millstone Grit (Gilligan 1919).

Wales

Clwyd Barite occurs with galena, sphalerite and witherite in a lode deposit cutting Wenlock Shale at Pennant mine, near St Asaph (Carruthers et al. 1915, Archer 1959, Dunham et al. 1978).

Dyfed Barite was once mined commercially from sulphide-rich (chalcopyrite, galena, sphalerite) veins cutting Arenig sandstones at the Vale of Towy (or Nant) mine, near Carmarthen, and from the nearby Cystanog mine (Carruthers 1915, Bevins 1994). Barite is also particularly abundant in veins encountered at mines in the Central Wales orefield, including Bwlchygarreg, near Tal-y-bont, and Esgair Hir mine, Ceulanymaesmawr, where it is often associated with witherite (Bevins 1994).

Gwent Some of the best barite (crystals to 25 mm) ever found in the South Wales coalfield came from the tips of Oakdale colliery and are associated with millerite, siegenite and quartz (Wolfe 1997).

Gwynedd At Pompren mine, near Aberdaron, Lleyn Peninsula, massive barite was mined from a mineralized fault zone between Ordovician and Precambrian rocks (Archer 1959, Dunham et al. 1978). Other brief descriptions of barite include: Benallt mine, Rhiw, Lleyn Peninsula (Spencer 1942, Campbell Smith et al. 1944); Bwlch mine, near Deganwy (Russell 1949b); Parys Mountain, Anglesey (Southwood & Bevins 1995, Jenkins et al. 2000); and Penrhyn Du (or Penrhyn) mine, Lleyn (Hubbard 1991).

Mid Glamorgan Barite occurs in clay–ironstone nodules from the South Wales coalfield at Bedwas colliery, Trethomas; Deep Navigation colliery, Treharris; and Lewis Merthyr colliery, Trehafod (Bevins 1994, Wolfe 1996a). The species is also abundant in many forms from mines and quarries between Llantrisant and Machen including: excellent radiating aggregates of colourless euhedral stubby tabular crystals to 3 cm at Cwmleyshon quarry, near Draethen, and at Mwyndy mine; yellow-stained crystals to 2 cm at Llwyn-saer mine (all Bevins 1994); white or iron-stained cockscomb aggregates forming rosettes up to 25 mm across in galena-bearing calcite veins at Machen quarry, 7 km ENE of Caerphilly (Bridges 1990, Goulding & Price 1995, Plant & Jones 1995); and pseudomorphs after bipyramidal alstonite in sulphide veins cutting Triassic rocks at an old mine described as BK-3, Llantrisant (Alabaster 1990a). Elsewhere barite is reported from: Llangan mine, near Bridgend (Bevins 1994); Tŷ Coch, near Porthcawl (Criddle & Symes 1977); and Vaynor quarry, Cefn Coed, Merthyr Tydfil (Goulding 1995).

Barite in a clay ironstone nodule from the South Wales coalfield at Bedwas colliery, Trethomas, Mid Glamorgan, Wales; the crystal is 12 mm long. (© National Museum of Wales, Cardiff; photographer M. P. Cooper)

Powys *Central Wales orefield:* Barite is found in galena-bearing veins at only a few localities in the Llanidloes area, particularly at Aberdaunant, Bryntail, Gorn and Pen-y-Clun mines – at the last locality forming sparkling colourless euhedral crystals to 2 mm in cavities within witherite (Carruthers et al. 1915, Jones & Moreton 1977, Morgan & Starkey 1991, Mason 1997). Additionally, a solid rib of barite was worked at Van mine in the 1930s (Hughes 1991). Further occurrences are from mines in the Llangynog district at: Craig Rhiwarth (Dewey & Eastwood 1925); Craig-y-Mwyn (Dunham et al. 1978); Cwm Orog and Llangynog (Archer 1959, Dunham et al. 1978); and Galltymain, near Meifod (Bick 1978). Barite is also noted from: Dolyhir quarry, Old Radnor – fine platy crystals up to 1 cm across (Wolfe 2000, 2004, Garfield 2002) and composite crystals up to 4.5 cm long (T. F. Cotterell pers. comm. 2007); and from Weston mine, near Middletown – clear euhedral tabular crystals up to 1.5 cm long (Bevins 1994).

South Glamorgan Excellent pink cockscomb barite has been found associated with calcite at Taff's Well quarry, near Cardiff (Watson 1988, Bevins 1994). Barite is also present in galena-bearing veins at Llangan mine, near Cowbridge (Bevins 1994) and at Ty-pica and Twmpath, between Cowbridge and Bridgend (North 1916). Other localities includes: the coast between Penarth and Sully (Bevins 1994); quarries at Castell Coch and Rhiwbina, on the outskirts of Cardiff, and Nell's Point, Barry Island (North 1916); and Ton Mawr quarry, Pentyrch, Cardiff (Wolfe 2000).

Scotland

Borders Heddle (1901) reports barite lining cavities in what appear to be coprolites in Carboniferous shales at Cockburnspath.

Central *Ochil Hills:* Barite is the most common mineral found in veins cutting Old Red Sandstone sediments, although it has never been worked. It is normally white but often stained pink or red, and has been found at: Airthrie Hill mine, Bridge of Allan – pinkish crystals associated with chalcopyrite and malachite (Wolfe 2004); Alva mine (Heddle 1901, Ineson & Mitchell 1974b, Jassim et al. 1983, Moreton 1996, Moreton et al. 1998); Balquarn Burn, near Menstrie; Burn of Sorrow mine, near Dollar; Tillicoultry (all Dickie & Forster 1974); Blairlogie mines (Smith et al. 1992); and Daiglen Beck, near Tillycoutry (Ineson & Mitchell 1974b). *Tyndrum mining district:* Barite is a gangue species in both lead–zinc vein and vein-breccia mineralization, and in later uranium-rich veins (Pattrick et al. 1983, Pattrick 1985). Heddle (1901) identifies Clifton mine as one of these locations.

Elsewhere, Heddle (1901) reports crystallized barite in a chalcopyrite-bearing vein on the northwestern side of Arndrum Hill, Aberfoyle, and Green & Todd (2001) report it from Earlsburn reservoir and above Baston Burn, Touch Hills, SW of Stirling.

Dumfries & Galloway Small bright orange-pink crystals of barite have been found lining cavities at Auchencairn hematite mine; a bright-red variety was recovered from cliffs at Burrow Head; and veins of white and pinkish massive barite are noted from cliffs between Balcary and Rascarrel (Williams 1965). Barite is widespread in the Wanlockhead mining district, and includes beautiful cockscomb groups and pseudomorphs after witherite from New Glencrieff mine (Wilson & Flett 1921, Brown 1927, Temple 1955, Gillanders 1981). Livingstone (2002) describes many magnificent examples of barite from Wanlockhead, including a 12 cm-long specimen of lustrous, well terminated, translucent-to-white tabular barite with yellow terminations (now in the NMW). Other reports of barite include: in lead–zinc dolomite veins at Callisterhall, near Langholm (Smith et al. 1978b); uncommon radiating pink platy crystals in uranium-rich veins at Marbruie Cove, Southwick (Miller & Taylor 1966); as inclusions in alluvial gold from Glengap Burn, SW of Moffat (Chapman et al. 2000c); associated with chalcopyrite at Balcary copper mine, Heston Island, Auchencairn Bay; and in a barite vein, near the House of Orroland, Barlocco, SW of Auchencairn; associated with copper and lead ores at Pibble mine, Creetown (all Heddle 1901).

Fife Small cockscomb groups of barite crystals have been found in calcite veins cutting Carboniferous dolerite sills at Goat quarry, 2 km NW of Aberdour (Heddle 1901, Ingram et al. 1992). Cavities in minor veins cutting limestone beds at St Monans contain colourless gem-like platy crystals of barite up to 20 mm long, and

Yellow-tipped barite, 50 mm across, from New Glencrieff mine, Wanlockhead, Dumfries & Galloway, Scotland. (© D. I. Green)

at Balmullo quarry, SW of Luthrie, pink cockscomb barite is noted (Moreton 2006). Other locations include: Auchtertool; near Balmerino; Kinkell; Ruddons Point (all Heddle 1901); and Orrock quarry, near Burntisland (Wirth 2003).

Grampian Stratabound barite (rare) and lead–zinc mineralization are reported from the Middle Dalradian Glas Maol Graphitic Schist at Corrie Loch Kander, near Braemar (Gallagher et al. 1989a, Fortey et al. 1991, 1993). Barite is also noted from: Allt Chernie in the Ballater area – as coarse-grain pink tabular crystals of barite up to 2 cm long in a Dalradian fault breccia (Goodman et al. 1990); Arndilly House, Arndilly – as occasional crystals lining voids in a Dalradian fault breccia (Nicholson 1986); Bridge of Don, Aberdeen – associated with manganite (Green & Todd 2001); and Gollachy Burn, Engie (Wallace 1885).

Highland Demand for barite in the 1980s resulted in the reworking of galena-bearing barite veins to the N of Strontian and the development of opencast operations between Bellsgrove and Whitesmith mines. Excellent specimens of galena on cream cockscomb barite were recovered from the dumps (Green & Todd 2001, Wolfe 2004). Barite also occurs in similar veins at the nearby Corrantee and Fee Donald mines (Winrow 1986b), and from Lurga mine, SW of Strontian (Smith et al. 1992). Stratiform and vein barite mineralization occurs in an Old Red Sandstone conglomerate at Balfreish, near Dalroy (Gallagher 1984, Smith et al. 1992). The origin of the deposit is thought to be hydrothermal and possibly related to a manganese-rich hot-spring deposit at Dalroy Burn, Dalroy – where minor barite is also found (Nicholson 1987b, 1989c, 1990). Other locations include: Achanarras farm, near Halkirk – barite associated with galena (Berridge 1969); Allt Tigh Cumhaig Burn, 3 km S of Struy (Russell 1946b); Blar Cnoc na Gaoith, S of Dounreay, northwest Caithness – barite–sphalerite-bearing veins associated with a fault in Old Red Sandstone conglomerate and associated with celestine in faulted syenite at the northeast corner of Loch Loyal (Gallagher et al. 1971); Flodigarry, Isle of Skye – colourless to pale-brown barite laths to 6 cm in calcite–siderite-bearing septarian concretions within Jurassic mudstones (Hendry et al. 2006); Garbh Bheinn, Ardgour, E of Strontian – barite–galena veins (Gallagher 1964); eastern margin of the Grudie granite, 5 km SW of Lairg – barite–hematite–quartz veins (Gallagher 1970, Gallagher et al. 1971, 1974); Helmsdale – barite, calcite, fluorite, pyrite and quartz in veins and shear zones within granite and Old Red Sandstone arkose (Bowie & Gallagher 1970, Gallagher et al. 1971, Tweedie 1979); Loch an Arbhair, Lochinver–Ullapool road, SW of River Kirkaig – red cockscomb barite associated with calcite, harmotome and hematite in veins cutting fault breccia (Waterston 1953); Loch Bruicheach, 12 km SW of Beauly – a compact banded pink barite vein containing layers of fluorite (Wallace 1885, Russell 1946c); the Ord, S of Lairg – barite in veins within red granite (Heddle 1901); Roy Geo, SW of Occumster – white barite in a galena-bearing vein within a vertical crush zone (Berridge 1969); and Struy mines, Loch na Mèine, midway between Strath Glass and Strath Farrar, SW of Struy, near Inverness (Russell 1946b, Berridge 1969, Green 1986).

Lothian Barite is the main gangue species in veins containing galena, native silver and nickeline that cut limestone near a quartz dolerite dyke at Hilderston mine, Bathgate Hills (Heddle 1901, Jassim et al. 1983, Meikle 1994). Heddle (1901) also notes: narrow barite veins in granite on Whiteadder Water between Millknowe and Priestlaw, NW of Cranshaws; an association of barite with calcite, chalcopyrite and ankerite or dolomite (pearlspar) in a coal pit at Addiewell, West Calder; and barite infilling vesicles in basaltic and andesitic lavas and sills at Calton Hill, Crow Hill and Salisbury Crags. The Addiewell occurrence is discussed further by Trechmann (1886), who describes small brilliant tabular crystals of barite up to 18 mm across.

Strathclyde The most important barite mines were at Gasswater, SW of Muirkirk and Muirshiel, near Lochwinnoch, Renfrewshire Hills, W of Paisley (Hobson 1959, Collins 1972, Smith et al. 1992). Other vein deposits were worked at: Cumberhead, Mannoch Hill and Priesthill in the River Nethan district; Myres Burn, near Eaglesham, S of Glasgow (Hobson 1959, Gallagher 1984, Todd & McMullen 1991, Smith et al. 1992); and Sannox Water, Glen Sannox, northeast Isle of Arran (Greg & Lettsom 1858, Heddle 1901, Tyrell 1928a, Collins 1972, MacGregor et al. 1972, Seymour-Smith 1977, Jassim et al. 1983, McKerrow & Atkins 1989). At Gasswater, over 500 000 tonnes of barite was mined from a fissure-infilling deposit found along a series of faults within sediments of Old Red Sandstone to Carboniferous age. The barite is remarkably pure and it forms both a coarsely crystalline white to pale-pink porcellanous variety and a compact non-platy pink variety (Scott 1967, Dunham et al. 1978). The Muirshiel deposit produced massive pink /white banded barite, cockscomb barite and fine lustrous bladed crystals with water-clear terminations from faulted Carboniferous volcanics cut by a Tertiary dyke; strontianite and calcite are associated species (Moore 1979, Todd & Laurence 1989, Green & Todd 2001). Reviewing barite and copper mineralization of the Renfrewshire Hills, Stephenson & Coats (1983) describe the Muirshiel deposit and note other barite veins in the area, including those at Berryglen mine, East and West Kaim mines, and Swinlees mine, Dalry. Other occurrences of barite associated with Carboniferous lavas are from: Boyleston quarry, SW of Barrhead (Meikle 1989a); Hartfield Moss, 9 km WSW of Barrhead (Meikle 1990); Loanhead quarry, near Beith (Meikle 1989b, Meikle & Todd 1995); and Reilly quarry, Houston (Wolfe 1997, Meikle 2003). Barite is also present at: Craigmuschat quarry, Gourock – in geodes within a trachyte sill (Nicholson & Durant 1986, Nicholson 1988c); Green Hill quarry, Kilbirnie – as pale-blue to colourless crystals up to 25 mm long in veins cutting a Carboniferous trachyte plug (Wolfe 1997, Green & Todd 2001); Leadhills – in many veins in the orefield, fine crystals from Wembley shaft (Brown 1927, Temple 1955, Paar et al. 1984, Livingstone 2002); Pockmuir Burn, Auchinstilloch, near Lesmahagow – pink platy crystals in a barite–hematite vein (Evans & El-Nikhely 1982); Rubha a' Mhill, northeast Isle of Islay – tabular crystals in cavities in limestone (Green 1988); Trearne quarry, near Beith – infilling fossil brachiopod shells or in minor veins (Todd 1989); and Wallace's Cave, Cartland Crags, near Lanark (Heddle 1901).

Tayside Barite and barium silicates (celsian, cymrite, hyalophane and barium-rich muscovite) are major constituents of a sediment-hosted submarine exhalative deposit in Middle Dalradian lead–zinc sulphide-bearing calcareous graphitic schists in the Ben Eagach area, near Aberfeldy (Coats et al. 1980, 1984, Fortey & Beddoe-Stephens 1982, Willan & Coleman 1983, Smith et al. 1984). Together with a similar deposit at Foss mine (to the SW of Ben Eagach), they represent one of the world's largest barium concentrations. Foss mine alone is considered to contain over 2 million tonnes of granoblastic barite, which commonly exceeds 90% by volume of the rock (Moles 1983, 1985, Hall 1993b, Green & Todd 2001, Livingstone 2002). The occurrence of barite at Ardownie quarry, Monifieth, is on an entirely different scale and includes cockscomb clusters up to 25 mm across in quartz-lined geodes within Devonian andesite (Ingram 1994). Heddle (1901) notes bright sulphur-yellow barite from Ballintuim, near Bridge of Cally, Strathardle.

Orkney Minor barite and celestine occur with galena and sphalerite in a Middle Devonian algal breccia (part of an evaporite lake system) at Hesta Geo, near Birsay, Mainland (Parnell & Janaway 1990). Barite–galena–(sphalerite) replacement veins and breccia zones have been found in Basal Breccia and Sandstone members of the Old Red Sandstone, near the Moine unconformity at Blackhall and Mill of Cairston, Stromness district, Mainland (Gallagher et al. 1971, Michie 1972). Barite was the main gangue species at Stromness lead mine, in this same area (Wilson & Flett 1921). Elsewhere, Heddle (1901) notes barite at: Wardhill and near Scabra Head on Rousay, and Selwick on Hoy.

Shetland Heddle (1901) reports lamellar and cockscomb barite in druses within a porphyritic eruptive rock at Kirkness Sands on Papa Stour.

Ireland

Cavan Morris (1984) notes a 4 cm-wide sphalerite-bearing barite vein at Muff.

Clare Barite is a minor gangue species in disseminated copper deposits hosted by folded Carboniferous limestones, sandstones and shales at Ballyvergin and Maghera Cross (Schultz 1971, Andrew 1986).

Cork Many barite veins cutting Devonian sandstones and slates have been worked in southwest Cork, including: Audley copper mines, near Clonakilty (Greg & Lettsom 1858); Ballycummisk (Ni Wen et al. 1999, Kinnaird et al. 2002); Derreennalomane mine at Mount Corin, and Duneen Bay (Muckross Head) mines near Clonakilty (Cole 1922, Halliday & Mitchell 1983); Letter and Skeag mines, near Mount Gabriel (Ni Wen et al. 1999, Cole 1922); Derryginagh, Lady's Well and Mount Gabriel West mines (Reilly 1986).

Donegal Barite occurs in lead–zinc veins within Dalradian rocks around the margins of the Donegal granites (McArdle et al. 1986a).

Down Moles & Nawaz (1996) note barite in Carboniferous lead–zinc mineralized veins and dolomite-matrix breccia at Conlig–Newtownards.

Galway Barite and calcite are the major gangue species in the Carboniferous carbonate-hosted syn-sedimentary Cu–Pb–Zn–(Ag) deposit found at Tynagh mine, near Killimor (Schultz 1966b, 1971, Morrissey et al. 1971, Russell 1975, Boast et al. 1981a, Clifford et al. 1986, Banks et al. 2002). Barite is also noted as a gangue species at two lead mines near Oughterard: Cloosh mine – snow-white and transparent crystalline masses (Rudler 1905, Cole 1922); and Glengowla mine (Cole 1922, Halliday & Mitchell 1983).

Limerick Schultz (1971) notes barite in the disseminated Cu–(Ag) deposit at Aherlow, near Galbally.

Londonderry Greg & Lettsom (1858) note barite in thick veins within Old Red Sandstone sediments at Ballynascreen.

Longford Colourless thick bevelled platelets of barite up to 0.2 mm long have been found on the dumps of the breccia zone and fracture-fill cadmium-rich sphalerite–

(galena) deposit at Keel, Ardagh (Patterson 1970, Morrissey et al. 1971, Schultz 1971, Crowe 1986, Slowey 1986, Ryback & Moreton 1993). White barite, sometimes forming rosettes, has been found in late fracture veins associated with a galena–sphalerite deposit in dolomitized limestones at Newtown Cashel (Crowe 1986).

Mayo Barite–galena–sphalerite mineralization occurs in the basal zone of a brecciated and altered acidic porphyry 8 km to the S of Charlestown. It is overlain by chalcopyrite- and pyrite-rich zones (Poustie 1982, O'Connor & Poustie 1986).

Meath The ore at Navan mine comprises fine layers of sphalerite and coarser-grain crystalline galena with varying amounts of barite, marcasite and pyrite. It is a world-class carbonate-hosted epigenetic syn-sedimentary replacement deposit, and in the 1990s it was the largest zinc producer in Europe. Cavities in the ore hosted white tabular barite crystals to 2 cm (Schultz 1971, Boyce et al. 1983, Andrew & Ashton 1985, Ashton et al. 1986, Ryback & Moreton 1993, Anderson et al. 1998, Peace et al. 2003, Wilkinson 2003, Walshaw et al. 2006). Beds stratigraphically equivalent to those at Navan host a sub-economic barite–galena–sphalerite vein deposit at Drumlerry, near Oldcastle (Brand & Emo 1986).

Monaghan Galena-bearing veins contain gangue barite at Cornalough–Carrickagarvan (Hope mine), Lemgare and Lisdrumgormly (Morris 1984). Cole (1922) notes that barite was raised from Carrickartagh mine in about 1898 and in 1919.

Sligo Rare tabular honey-brown barite has been found at the limestone quarry now occupying the site of Abbeytown mine, near Ballysadare (Ryback & Moreton 1992). Barite has been worked from a vein in Carboniferous Limestone at Benbulen Mountain (O'Brien 1959). This was probably from Gleniff (or Glencarbury) mine, although the vein was also worked from an opencut at Tormore (Cole 1922).

Tipperary The large sedimentary-exhalative Ba–Pb–Zn deposits in the Silvermines district have been worked from several mines. Barite is described from the following: Ballygown South mine – radiating and stalactitic masses and aggregations of wedge-shape crystals (Russell 1907b, Rhoden 1959, Boland et al. 1992); Ballynoe (Magcobar) barite open pit (over 5 million tonnes mined) – barite occurs as fine-grain grey or pale-brown microcrystalline or white acicular crystals and tabular crystals up to 3 cm long associated with pyrite (Russell 1907b, Schultz 1971, Larter et al. 1981, Boyce et al. 1983, Halliday & Mitchell 1983, Boland et al. 1992, Wolfe 1996a, Mullane & Kinnaird 1998, Reed & Wallace 2004); Gortnadyne mine – large masses of white barite (Russell 1907b, Rhoden 1959); Mogul mine – white-to-translucent tabular crystals up to 2 cm long (Schultz 1971, Moreton 1999); Shallee East and West mines – massive white barite (Russell 1907b, Rhoden 1959, Halliday & Mitchell 1983); and more generally by other authors – Gages (1860), O'Brien (1959), Morrissey et al. (1971), Taylor & Andrew (1978, 1979), Samson & Russell (1983), Taylor (1984), Russell (1990), Banks et al. (2002). Elsewhere, barite occurs in carbonate-hosted deposits at Gortdrum mine, near Tipperary town (Morrissey et al. 1971), and at Lisheen, between Thurles and Rathdowney (Hitzman 1992, Hitzman et al. 2002, Ixer 2002, Fusciardi et al. 2004, Wilkinson et al. 2005).

Tyrone Fluid inclusions in barite were examined as part of an investigation into gold-bearing vein mineralization at Curraghinalt, near Gortin, Sperrin Mountains (Parnell et al. 2000).

Westmeath Barite occurs in carbonate-hosted lead–zinc deposits at Ballinalack, 15 km NW of Mullingar (Schultz 1971, Jones & Brand 1986), and at Moyvoughly, near Moate, 12 km N of Clara (Poustie & Kucha 1986, Kucha 1988).

Wexford Cole (1922) reports barite veins containing galena at Killiane mine, near Wexford.

Wicklow Sulphide veins containing thin transparent yellow crystals of barite occur at the Luganure mines, Glendalough (Greg & Lettsom 1858, Joly 1886, Rudler 1905). Significant amounts of barite are present at Baravore lead mine, near Clonkeen (Cole 1922).

Offshore territories

Isle of Man Barite is noted from the Foxdale mines (Greg & Lettsom 1858) and from Laxey mine (Lamplugh 1903).

Jersey Mourant (1961) reports white to pale-pink rhombic crystals of barite from small veins at a few localities: Grève-de-Lecq, Grosse Moie, Jeffrey's Leap, north side of Petit Portelet, granite headland N of quarry at Ronez in St Catherine's Bay, and Westmount.

Rockall Accessory amounts of barite occur in aegirine–riebeckite granite (Young et al. 1978).

Sark Mourant (1939) reports barite in a thick pink vein from a cave near the middle of Derrible Bay.

Barium-pharmacosiderite from Driggith mine, Caldbeck Fells, Cumbria; width 140 µm. SEM image. (© D. I. Green)

barium-pharmacosiderite/bariopharmacosiderite

tetragonal $Ba_{0.5}Fe^{3+}_4(AsO_4)_4(OH)_4 \cdot 6H_2O$

A rare hydrated barium–iron arsenate hydroxide found in manganese-rich quartz veins cutting Ordovician greywackes and mudstones that have been thermally metamorphosed by the Skiddaw granite at Burdell Gill, Caldbeck Fells, Cumbria. Within the veins, there are cavities and clefts partially filled with manganese oxides (cryptomelane, pyrolusite and romanèchite) and a wide range of supergene species, such as beudantite, carminite, lepidocrocite, mimetite, parasymplesite, pharmacosiderite, plumbogummite, rhodochrosite, scorodite, switzerite and vivianite. Barium-pharmacosiderite forms yellow to orange-yellow cubes up to 1 mm across and is much less common than the accompanying green pharmacosiderite (Neall & Leppington 2003). Elsewhere in the Caldbeck Fells, barium-pharmacosiderite is noted: as pale-green to greenish-white radiating aggregates to 5 mm across on joint surfaces in a barite–galena vein from an isolated exposure in the opencut above Sandbed mine and as pale greenish-brown cellular earthy to finely crystalline crusts in porous cellular quartz at Dumpy Stone level on the Low Pike vein (Young et al. 1992b); as drusy crusts composed of minute pale-green to yellowish-brown cubic crystals in arsenate-rich gossans between Driggith and Sandbed mines (Green et al. 2006a); as sparse and minute yellow crystals with arseniosiderite on a single loose block of quartz collected from an abandoned mine dump in Todd Gill (C. M. Leppington pers. comm. 2004); from Dry Gill mine (D. I. Green pers. comm. 2006). Barium-pharmacosiderite has also been found above Bannerdale mine, Mungrisdale, Cumbria (D. I. Green pers. comm. 2006).

Small yellow cubes of barium-pharmacosiderite are noted from Tynagh mine, Co. Galway, Ireland (Ryback & Tandy 1992).

barkevikite

A synonym of ferrohornblende, amphibole group. The name has also been applied to kaersutite and other amphibole compositions, but has never been chemically defined.

baroselenite

A synonym of barite.

barroisite

AMPHIBOLE GROUP

monoclinic $\square(Ca,Na)(Mg_3AlFe^{3+})(Si_7Al)O_{22}(OH)_2$

A very uncommon calcium–sodium–magnesium–aluminium–iron aluminosilicate hydroxide that forms a series with ferrobarroisite.

Green cores to blue amphiboles from the Mona Complex at the Marquis of Anglesey's monument, near Llanfairpwllgwyngyll, Anglesey, Gwynedd, were described as hornblende by Greenly (1919). However, analysis of similar material from there and from temporary roadcuts made during construction of the Llanfairpwllgwyngyll bypass indicated that the green cores consist of barroisite, ferrobarroisite and winchite, whereas the rims are pale-blue glaucophane (Gibbons & Gyopari 1986, Bevins 1994). A single analysis of barroisite from blueschists at the Marquis of Anglesey's monument is presented by Horák & Gibbons (1986), but the mineral is magnesio-katophorite if the 1997 classification of the amphibole group is used (Leake et al. 1997a, Mogessie et al. 2004).

Barroisite (originally referred to as crossite) also occurs in dark-green amphibolites associated with the Girvan–Ballantrae ophiolite complex to the N of Knockormal, Strathclyde, Scotland. The barroisite seldom shows good crystal outlines, being mainly confined to mantles around ferro-edenite (Bloxam & Allen 1959).

barstowite

monoclinic $3PbCl_2 \cdot PbCO_3 \cdot H_2O$

TL Bounds Cliff, E of Port Isaac, St Endellion, Cornwall, England

Named after Richard William Barstow (1947–1982), a Cornish mineral dealer born in Bangor, Gwynedd, Wales. Barstowite is an hydrated lead chloride carbonate found in a lead–antimony-rich vein close to sea level at the type locality. It forms small (0.2–0.5 mm) subparallel elongate intergrown crystals (with phosgenite) in aggregates (up to 3 mm) containing minor inclusions of galena and jamesonite. Other associated species are cerussite, chalcopyrite, dolomite, pyrite, quartz and sphalerite. Barstowite is colourless-to-white and transparent and, like phosgenite, probably formed as a result of the action of sea water on the primary assemblage (Stanley et al. 1991, Steele 1999).

baryte

An alternative spelling of barite.

baryt-flusspath

A mixture of barite and fluorite.

barytocalcite

monoclinic $BaCa(CO_3)_2$

TL Blagill mine, near Alston, Cumbria, England?

Barytocalcite is a rare barium–calcium carbonate, trimorphous with alstonite and paralstonite, which typically postdates the primary constituents of barite–calcite–galena–sphalerite–witherite veins. Its first description by Brookes (1824) gave no precise details of the location, although Young (1985a) suggests that the type locality is Blagill mine. Since then barytocalcite has been found at other mines in Cumbria – Ayle Burn, Brownley Hill, Clargill, Heartycleugh, Keirsleywell, Longcleugh, Mohopehead, Nentsberry, Scraithole, Taylor's Groove, Wellhopehead, and additionally at Blackpool Bridge level and Hesleywell Hush. Elsewhere in the northern Pennines, abundant barytocalcite associated with witherite is reported from: Stake Beck mine, Dufton Fell, Cumbria; Foster's Hush, south of the Middleton–Brough road, and Lunehead mine, near Grains o' th' Beck, Lunedale, both Co. Durham; and in Northumberland, rarely at Fallowfield, Greyside and Settlingstones mines, near Hexham (Davidson & Thomson 1951, Ineson 1972, Dunham & Wilson 1985, Young 1985c, Young et al. 1985c, Briscoe 1986b, Dunham 1990, Robinson et al. 1992), and at Ouston mine, near Ninebanks, Northumberland (Young 1993). Green & Briscoe (2002) suggest that the world's best specimens of barytocalcite come from Brownley Hill mine, Nenthead, Cumbria. Excellent specimens of relatively large, complex, almost wave-shape crystals of barytocalcite coated with a crust of minute white calcite and barite are described from the Admiralty Concession flat, 2nd Sun vein, Nentsberry Haggs mine, Nenthead (Bridges & Green 2006b). Additionally, Wolfe (2004) and Bridges & Green (2006b) note an occurrence of barytocalcite with fluorite, galena, gypsum and witherite in well crystallized and sometimes spectacular specimens from Rampgill mine, near Nenthead. The replacement of barite by barytocalcite is discussed by Bridges & Green (2006b), who cite examples from Blagill, Nentsberry Haggs, Rampgill and Scraithole mines. Barytocalcite is briefly noted from the Sir Francis level, Gunnerside Gill, and Old Gang mine, E of Gunnerside Gill, Swaledale, Yorkshire (Rodgers 1975, Ineson & Young 1994).

Honey-brown barytocalcite, 12 mm tall, on witherite; Nentsberry Haggs mine, Nenthead, near Alston, Cumbria. (© D. I. Green)

At Rorrington mine (Engine shaft), S of Meadowtown, Shropshire, barytocalcite associated with chalcopyrite forms compact creamish-white crystalline masses within barite–witherite–calcite veinstone (Starkey et al. 1994). Barytocalcite also forms part of a late-stage mineral assemblage within calcite-lined cavities in a manganese oxide-rich vein at Coleman's quarry, Holwell, near Frome, Somerset (Alabaster 1990a).

Barytocalcite and several other barium-rich minerals (alstonite, barite, harmotome and witherite) have been found in veins cutting Precambrian sandstones and siltstones at Dolyhir quarry, Old Radnor, near Presteigne, Powys, Wales. Most of the species occur as microcrystalline crusts (Wolfe 2004). Later discoveries of barytocalcite at Dolyhir include opaque cream disc-like crystals in rosette-shape aggregates and curved toffee-colour crystals associated with harmotome and witherite (T. F. Cotterell pers. comm. 2007). At Llantrisant in Mid Glamorgan, barytocalcite has been found associated with pyrite, lining a cavity in a siliceous iron deposit (Alabaster 1990a).

In Scotland, barytocalcite has been found associated with barite and manganese-rich calcite from the stratabound barium–zinc mineralization near Aberfeldy, Tayside Region (Moles 1985), and at Hilderston silver mine, Bathgate Hills, Lothian, where it forms well crystallized specimens associated with harmotome and witherite (Meikle 1994, Moreton et al. 1998, Green & Todd 2001).

barytocelestine

A variety of celestine, barite group. See Collie (1879).

baryto-fluate of lime

A mixture of barite and fluorite from Derbyshire.

barytstrontianite

A mixture of barite and strontianite.

basaluminite

monoclinic $Al_4(SO_4)(OH)_{10} \cdot 5H_2O$

TL Lodge Pit, Irchester, near Wellingborough, Northamptonshire, England

An uncommon hydrated aluminium sulphate hydroxide whose species status has been lost. Burke (2006) reporting on an IMA-CNMNC-approved mass discreditation of species states that basaluminite is formally discredited and compositions are now to be referred to as felsöbányaite.

At the type locality, basaluminite coats joint surfaces and fills fissures up to 30 cm wide in Northampton Ironstone (part of the Inferior Oolite). It is a white, powdery to compact, sometimes nodular mineral that with hydrobasaluminite forms the matrix of the infill. Basaluminite is associated with allophane, aragonite, gypsum, goethite and halloysite in discrete layers. The fissures also contain a brecciated chamosite–siderite oolite, above which is a siderite mudstone. It seems probable that the fissures acted as traps for the soluble products of weathering from which the alumina-rich minerals precipitated. Basaluminite is formed when hydrobasaluminite is allowed to dry in air (Bannister & Hollingworth 1948, Hollingworth & Bannister 1950). These authors also note that basaluminite with a similar habit (and formerly labelled aluminite) occurs at Clifton Hill, Brighton, Sussex.

Basaluminite is found in collapsed Tertiary sediments filling solution pipes in the Upper Chalk, near the coastguard station on Castle Hill, Newhaven, Sussex. It occurs as hard compact white-to-greyish masses up to 8 cm across embedded in impure gypsum (var. selenite), or within pipe fillings associated with gibbsite; most commonly it forms crusts developed on the chalk surface. Other aluminium-rich species discovered in the pipes include aluminite, bayerite and nordstrandite. At Falmer, between Newhaven and Brighton, Tertiary sands containing allophane, basaluminite and gibbsite were found in a temporary excavation during the construction of a bypass. In both cases, basaluminite appears to have formed by the reaction of acid groundwater with Tertiary sediments and the Chalk (Wilmot & Young 1985).

Basaluminite and hydrobasaluminite have also been found in the weathering zone of the Oxford Clay at Crook Hill brickyard, Chickerell, near Weymouth, Dorset. The two species occur in reaction rims surrounding carbonate concretions, although basaluminite is found only on those concretions that have fallen to the floor of the pit, suggesting it formed as a dehydration product of hydrobasaluminite (Clayton 1980).

At Hampstead Farm quarry, near Chipping Sodbury, Avon, a suite of secondary aluminium minerals including allophane, basaluminite, gibbsite, para-alumohydrocalcite and scarbroite have developed around heavily oxidized pyritiferous veins in the Carboniferous Limestone close to the Rhaetic/Lower Carboniferous unconformity. Basaluminite forms compact, earthy to porcellanous, monomineralic masses up to 3.5 cm in diameter, closely associated with gypsum (var. selenite). SEM images show that basaluminite forms tabular rhombic crystals very similar in shape to scarbroite from this locality (Alabaster 1990b). A survey of museum specimens by Ryback (1988b) has shown that some specimens labelled scarbroite are actually allophane, alumohydrocalcite or basaluminite. One white opaque earthy specimen of basaluminite came from Scarborough, Yorkshire, and is now in Maidstone Museum.

Middle Cambrian black shales believed to be of the Clogau Group are exposed in a roadside pit on the slopes of Moel Goedog, near Harlech, Gwynedd, Wales. An iron-stained mudstone specimen from here was found to contain a single vesicle (2×1 cm) containing white earthy basaluminite (Ball 1969). Basaluminite is one of 25 post-mining secondary minerals recorded from the copper deposit at Parys Mountain, Anglesey (Jenkins et al. 2000).

Basaluminite intimately associated with gypsum is found coating joints in a highly contorted shale bed at the base of coastal cliffs in a small bay NW of Doon East, 1 km N of Ballybunnion, Co. Kerry, Ireland. The shales contain pyritiferous and carbonaceous bands, and in the local area are known to contain arsenopyrite, chalcanthite, chalcopyrite, melanterite and pyrite. The secondary phosphates – crandallite, turquoise and wavellite – have also been found here (Ryback & Moreton 1993, Moreton et al. 1995).

basanite

A velvety-black flinty jasper used as a touchstone for estimating the purity of precious metal alloys. Basanite is also the name of a sodium- and potassium-rich basaltic extrusive rock.

bassanite

monoclinic (pseudo-hexagonal) $2CaSO_4 \cdot H_2O$

An hydrated calcium sulphate found in ground-up shale samples from the Blue Lias and Shales-with-Beef, and in part of a concretion from the Birchi Nodule bed in the Shales-with-Beef between Lyme Regis and Charmouth, Dorset. Bassanite probably formed as a result of the reaction of calcite with sulphuric acid produced by weathering of sulphides in the shale (Worku & Parker 1992).

bassetite

META-AUTUNITE GROUP

monoclinic $Fe^{2+}(UO_2)_2(PO_4)_2 \cdot 8H_2O$

TL Wheal Basset, near Redruth, Illogan, Cornwall, England

A rare hydrated iron uranyl phosphate found at the type locality where the Carnmenellis granite is in contact with Devonian sandstones. Mineralization consists of three main species: cassiterite, chalcocite and chalcopyrite, with accessory amounts of arsenopyrite, bassetite, pyrite, uranospathite, vochtenite and wolframite. Most commonly yellow, brownish-yellow or bronze-yellow bassetite forms tabular crystals with a pseudo-tetragonal appearance, either in single crystals or as aggregates of criss-crossing laths. A second type of bassetite, found on two specimens in the Harvard mineral collection, forms distinct yellow monoclinic crystals lining cavities and veinlets on oxidized pyritic vein material (Hallimond 1915, Frondel 1954, Zwaan et al. 1989, Penhallurick 1997).

Bassetite supposedly collected by A. W. G. Kingsbury from South Terras mine, St Stephen-in-Brannel, Cornwall (Kingsbury 1954, Golley & Williams 1995), requires confirmation, as it may have a dubious provenance (Ryback et al. 1998, 2001).

bastite

Serpentine resulting from alteration of pyroxene; a synonym of lizardite and scyelite.

Acicular yellow crystals to 4 mm of bassetite from Wheal Basset, Illogan, Cornwall. (© Royal Cornwall Museum, Royal Institution of Cornwall)

bastnäsite-(Ce)

hexagonal $(Ce,La)(CO_3)F$

A rare cerium–lanthanum carbonate fluoride that forms a series with hydroxylbastnäsite-(Ce) and is found in miarolitic cavities within granite at the Shap granite quarry, Cumbria. Bastnäsite-(Ce) produces opaque white hexagonal (but slightly rounded) prisms that are thought to have formed by the weathering of parisite or synchysite. Other crystals are red to white and are mottled in appearance. Associated species found in miarolitic cavities and along joint planes include albite, allanite, anatase, barite, bavenite, bertrandite, bismuthinite, brookite, calcite, cosalite, epidote, fluorapatite, fluorite, gladite, krupkaite, milarite, molybdenite, pyromorphite, rutile, scheelite, sphalerite, synchysite-(Ce) and wulfenite. Much of this assemblage is typical of Bi–Mo–W mineralization found in high-temperature veins that form as hydrothermal fluids pass through granitic intrusions (Wirth & Green 2002).

Bastnäsite reported from bismuthinite-bearing quartz veins in the Alston Block of the Northern Pennine orefield (probably Groverake mine, near Rookhope, Weardale) cannot be given a species name as REE compositions are not noted. Adularia,

Bastnäsite crystals to 0.7 mm long from the Shap granite quarry, Cumbria. (© D. I. Green)

monazite, synchysite and xenotime are also found, the latter closely associated with bastnäsite within quartz or fluorite crystals. The assemblage probably arose during or some time after intrusion of the buried Weardale granite (Ixer et al. 1996).

bauerite

A pseudomorph of silica (opal?) after biotite.

bauxite

A mixture of gibbsite ± böhmite ± diaspore ± goethite.

bavalite

A variety of chamosite, chlorite group.

bavenite

orthorhombic $Ca_4Be_2Al_2Si_9O_{26}(OH)_2$

A rare calcium–beryllium–aluminium silicate hydroxide found in the Meldon aplite quarries, near Okehampton, Devon. The Meldon aplite is a lithium–fluorine-rich dyke emplaced into the metamorphic aureole of the Dartmoor granite. Bavenite occurs in narrow fissures crossing aplite and calc-silicate hornfels adjacent to the aplite, and is associated with small crystals of albite, fluorite and quartz. It forms colourless incomplete rosettes up to 5 cm in diameter, but has a pale lilac-blue hue when massive (Dearman & Claringbull 1960).

Bavenite has also been found at the Shap granite quarry, Cumbria, in a loose block of drusy feldspar-rich granite. Drusy cavities reach 8 mm across and can contain apatite, calcite, epidote, scheelite, titanite and a dark mica, with bavenite forming clear colourless platy crystals up to 0.8 mm across and typically as aggregates up to 5 mm (Cooper 1991, Wirth 1995, Wirth & Green 2002).

Colourless platy crystals of bavenite up to 0.5 mm across; Shap granite quarry, Cumbria. (© D. I. Green)

bayerite

monoclinic $Al(OH)_3$

An aluminium hydroxide, polymorphous with doyleite, gibbsite and nordstrandite, and found in collapsed Tertiary sediments filling solution pipes in the Upper Chalk, near the coastguard station at Newhaven, Sussex. Bayerite occurs in one nodule associated with gibbsite and quartz. Other aluminium-rich species discovered in the pipes include aluminite, felsőbányaite (syn. basaluminite), gibbsite and nordstrandite; all appear to have formed by the reaction of acid groundwater with Tertiary clays and the Chalk (Wilmot & Young 1985).

bayldonite

monoclinic $PbCu_3(AsO_4)_2(OH)_2$

TL Penberthy Croft mine, St Hilary, Cornwall, England

Named after Dr John Bayldon, a physician and friend of A. H. Church, who first described the species. The specimens that Church worked on were found in an old collection by Richard Talling in 1864, who subsequently found the mineral *in situ* at an undivulged location. The type locality is almost certainly Penberthy Croft mine, although this is not noted in the original description (Church 1865b). Bayldonite is an uncommon lead–copper arsenate hydroxide that at Penberthy Croft forms minute globular clusters to 5 mm in diameter, with a grass-green to blackish-green colour and a strong resinous lustre. Rhombohedral crystals up to 0.9 mm long occur individually or in groups within cavities in vein quartz. The cavities have been caused by the leaching out of galena crystals, some of which reached 6 cm across. Caledonite, cerussite, linarite, malachite and mimetite–pyromorphite are also found in the cavities. Pseudomorphs of bayldonite after mimetite are very common. In manganese oxide pods adjacent to the main lode, bayldonite occurs with beudantite, minor beaverite and carminite, and rare pharmacosiderite and scorodite. Duftite–mottramite coats some of this bayldonite. Samples collected underground in the early 1990s are among the best ever found in Europe, and many fine specimens of a size suitable for display in a museum were collected (Camm & Merry 1991b, Betterton et al. 2000). Anthony et al. (2000) note bayldonite in Cornwall from Wheal Gorland, St Day, 2 km E of Redruth, and Wheal Alfred, Phillack, near Hayle.

Bayldonite is also noted in Cornwall from St Day United mine, Gwennap, near Redruth, and Wheal Carpenter, near Hayle, Gwinear (Claringbull 1951, Kingsbury & Hartley 1960, Kingsbury 1964), and on many occasions from the Caldbeck Fells, Cumbria. The Cornish and Caldbeck Fells descriptions by Kingsbury & Hartley

Globular aggregates of bayldonite to 0.5 mm; Penberthy Croft mine, St Hilary, Cornwall. (© M. P. Cooper)

(1956a, 1960), Hartley (1984), Young (1987a) and Cooper & Stanley (1991b), and those in unpublished manuscripts summarized by Cooper & Stanley (1990), are potentially unreliable, as these reports are all based on samples supposedly collected by A. W. G. Kingsbury; Ryback et al. (1998, 2001) have shown that the provenance of many Kingsbury samples is in doubt. The Caldbeck Fells occurrences affected are Brandy Gill mine, Carrock mine, Old Potts Gill mine, Poddy Gill, Sandbed mine and opencast workings between Driggith and Sandbed. Fortunately, descriptions by other authors confirm bayldonite at some of these locations. Green botryoidal bayldonite is reported from Sandbed lead mine (Davidson & Thomson 1951), and at a trial level on the Low Pike vein (near Potts Gill mine) bayldonite is found in a rare supergene assemblage with beudantite, cornwallite, duftite, malachite, mimetite, philipsburgite and pseudomalachite (Cooper & Stanley 1990, Young et al. 1990a, Green & Briscoe 2002, Green et al. 2003). Brandy Gill and the area around Driggith and Sandbed mines are reliable locations for bayldonite, at the latter forming widespread and common dark-green resinous encrustations in cavities in vein quartz (D. I. Green pers. comm. 2004, Green et al. 2006a).

Bayldonite from Gwaithyrafon mine, SE of Tal-y-bont, Central Wales orefield, Dyfed, Wales forms thin emerald-green platy crystals on chalcopyrite and is thought to have a post-mining origin (Rust & Mason 1994). Green glassy balydonite encrusts gartrellite on a specimen from Dolyhir quarry, Old Radnor, near Presteigne, Powys (T. F. Cotterell pers. comm. 2007).

Crusts of bayldonite up to 1 mm thick are associated with azurite in brecciated barite at Tynagh mine, near Killimor, Co. Galway, Ireland (Ryback 1988a, Ryback & Moreton 1992, Green & Moreton 2001, D. I. Green pers. comm. 2004).

bazhenovite?

monoclinic — $Ca_8S_5(S_2O_3)(OH)_{12}{\cdot}20H_2O$

A rare hydrated calcium oxysulphide hydroxide found along the beach, near a slipway, at West Runton, between Sheringham and Cromer, Norfolk. When broken open, the material released a pungent smell like rotten eggs, and vugs within it were found to contain pale-yellow or orange-yellow prismatic blades of bazhenovite up to 8 mm long. It is thought that the material came from a large water and sewage treatment plant that was broken up in 1995 and tipped onto the beach. Bazhenovite probably formed in an old holding tank prior to the break-up of the plant and therefore it not be accepted as a naturally occurring mineral, despite its composition being positively confirmed. Ettringite and possibly portlandite are associated species (P. Smith 2002, British Micromount Society website).

bazirite

hexagonal — $BaZrSi_3O_9$

TL Rockall island, North Atlantic Ocean, 440 km W of northern Skye, NW of Scotland

A very rare barium–zirconium silicate found in aegirine–riebeckite granite from the type locality. At the TL, bazirite is a relatively common late-stage interstitial mineral in the granite, filling voids up to 5 mm across. It forms small, well formed or anhedral, colourless crystals (up to 0.65 mm in length), often with prismatic habit. Elpidite and quartz are close associates, and crushes of the whole rock have yielded additional accessory minerals: apatite, barite, eudialyte, magnetite, pyrochlore, titanite and xenotime. Experimental work indicates that bazirite probably formed under low-temperature hydrous conditions (Hawkes et al. 1975, Young et al. 1978).

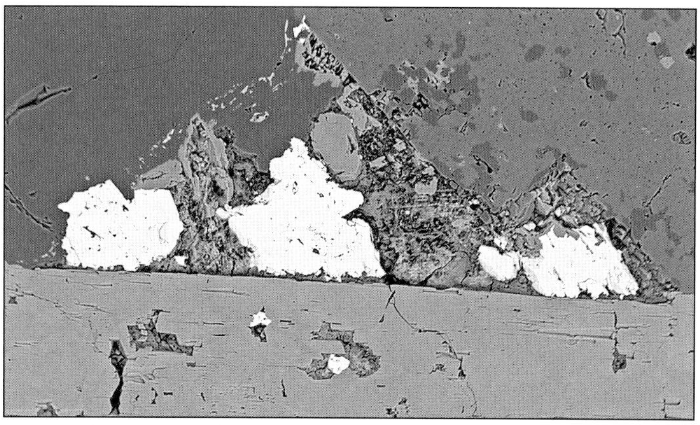

Three irregular crystals of bazirite on riebeckite within a miarolitic cavity in aegirine–riebeckite granite; Rockall, 440 km W of northern Skye, Scotland; width 700 µm; BSE image. (© A. G. Tindle)

beaverite

ALUNITE GROUP

trigonal — $Pb(Fe,Cu)_3(SO_4)_2(OH,H_2O)_6$

An hydrated rare lead–iron–copper sulphate hydroxide formed by the supergene oxidation of lead- and copper-rich minerals, and found at Alderley Edge ridge, Cheshire (Braithwaite 1994). Beaverite was found *in situ* at Wood mine, Alderley Edge, disseminated in a sandstone bed of the Mercia Mudstone Group. Here, nodular galena is coated with green malachite, on which an irregular dissemination of powdery yellow beaverite has grown (Lawton 1994). In an investigation of the zinc analogue of ktenasite from the lead–zinc mines of Brownley Hill and Smallcleugh, Nenthead, near Alston, Cumbria, beaverite was found with ankerite, chalcopyrite, dolomite, gypsum, malachite, namuwite, schulenbergite and serpierite (Livingstone 1991). At Rigg Head quarry, Borrowdale, Cumbria, rich yellow specimens of powdery beaverite with caledonite, cerussite, leadhillite and linarite are noted from an unnamed quartz vein (Green & Briscoe 2002, Green et al. 2006c).

Yellow and green beaverite has been reported from Penberthy Croft mine, St Hilary, Cornwall, by Kingsbury (1954), Robson (1954), Camm & Merry (1991b) and Golley & Williams (1995), but later investigations suggest that these may be misidentifications. Loosely compacted yellow material is segnitite, and similar material with an olive-green colour is hidalgoite. The error is ascribed to the similarities between XRD patterns of minerals in the alunite, beudantite and crandallite groups (Betterton et al. 2000).

Until further confirmation is available, the following claimed occurrences of beaverite from Cornwall and Cumbria must by considered dubious, as they are based on descriptions (or samples collected) by A. W. G. Kingsbury. The provenance of beaverite in samples from the Kingsbury collection (NHM) is outlined by various authors (Ryback et al. 1998, 2001, Green & Briscoe 2002, Green et al. 2006a):

CORNWALL Wheal Carpenter, Gwinear, near Hayle (Kingsbury & Hartley 1960).

CUMBRIA In the Caldbeck Fells: Brandy Gill mine (Kingsbury & Hartley 1957f, 1960, Hartley 1984); Carrock Fell cross cut, Grainsgill, and Thief Gills (Hartley 1984); Driggith mine and Potts Gill barite mine (Kingsbury & Hartley 1957f, Hartley 1984); Ingray Gill mine (Kingsbury & Hartley 1957f, Hartley 1984); Nether Row Brow (Upper and Lower levels) (Kingsbury & Hartley 1957f, 1960, Hartley 1984, Young et al. 1992); Roughton Gill (Kingsbury & Hartley 1957f, 1958, Kingsbury & Hartley 1960, Hartley 1984). These occurrences are also noted by Young (1987a) and Cooper & Stanley (1990).

A further complication is that powdery yellow aggregates of material associated with cerussite, galena and scorodite from the Driggith–Sandbed opencut, once thought to be beaverite, are actually either segnitite or a beudantite–segnitite intermediate composition (D. I. Green pers. comm. 2006). Green et al. (2006a) confirm beaverite as yellowish-brown drusy crusts or powdery encrustations at Driggith and Sandbed mines, but urge caution with the Kingsbury specimens.

Beaverite forms bright-yellow to slightly greenish-yellow compact sub-rounded fragments to 10 mm across, and vivid-yellow compact earthy masses to 15 mm embedded within quartz on the large spoilheaps at Pikedaw mine, between Malham and Settle, Yorkshire. Near by, at the head of Stockdale, irregular earthy patches of yellow beaverite occur with aurichalcite, barite, malachite and smithsonite on quartz gangue (Young et al. 2006).

In Wales, beaverite occurs at Ystrad Einion mine, near Tal-y-bont, Central Wales orefield, Dyfed, where it is relatively common as yellowish-brown botryoidal crusts and scattered globular concretions, and rare as spherical masses of thin platy crystals to 1 mm. It is often associated with goethite (Bevins 1994, Mason & Rust 1997).

bechererite

trigonal — $Zn_7Cu(OH)_{13}[SiO(OH)_3SO_4]$

A very rare zinc–copper sulphate silicate hydroxide that forms as a result of post-mining oxidation of copper and zinc sulphides and is found on the 30-fathom level dumps of Driggith mine, Caldbeck Fells, Cumbria, where it forms pale-blue pyramids on oxidized galena, in association with cerussite and hydrocerussite (Rust et al. 2003). This is corroborated by Green et al. (2006a), who also report it from the 12-fathom level dumps. At Silver Gill, Caldbeck Fells, bechererite occurs as pale-green inverted cone-shape crystals up to 0.1 mm long (Green et al. 2005b). White to very pale-blue crystals of bechererite form a drusy crust on chalcopyrite–pyrite–sphalerite-veined killas at Trevelyn's shaft, Charlotte United Mines, Perranthunoe, Cornwall (Rust et al. 2007).

Bechererite has also been found at five mines in the Central Wales orefield: Bwlchrhennaid mine, near Goginan; Eaglebrook (Nantycagl) mine, Ceulanymaesmawr; Esgair Hir mine, Ceulanymaesmawr; Frongoch mine, between Devil's Bridge and Pont-rhyd-y-groes (all in Dyfed); and Dylife mine in Powys. At Frongoch, bechererite occurs as characteristic colourless to pale-blue or ice-green inverted

Porous cones of bechererite. Frongoch mine, near Pont-rhyd-y-groes, Dyfed, Wales; width 60 µm; SEM image. (© M. Rothwell)

prisms up to 0.8 mm long. Caledonite, cerussite, hemimorphite, hydrocerussite, namuwite, schulenbergite and susannite are associated species (Green et al. 1996b, Rust et al. 2003). Wolfe (2004) considers Frongoch to be the world's premier locality for bechererite.

Bechererite has been found as minute, pale-blue to colourless, striated trigonal crystals up to 0.1 mm long on the dump of Raik's vein, Leadhills, Strathclyde, Scotland. Anglesite and leadhillite are associated species (Rust et al. 2007).

becquerelite?

orthorhombic $Ca(UO_2)_6O_4(OH)_6\cdot 8H_2O$

A rare hydrated calcium uranyl oxide hydroxide tentatively identified from the Peeth lode, Geevor mine, near Pendeen, St Just, Cornwall. Elsewhere along the Peeth lode, other secondary uranium minerals include andersonite, cuprosklodowskite and johannite (Elton & Hooper 1992, Elton et al. 1992). Golley & Williams (1995) report an unconfirmed occurrence of becquerelite from West Wheal Owles, St Just.

A pitchblende-bearing carbonate vein close to the margin of the Criffell granodiorite at Needle's Eye, SW of Caulkerbush, Dumfries & Galloway, Scotland, contains hydrocarbons in which peculiar uranium-rich filaments of apparently organic origin have been found. The walls and cells of the filaments are lined with a calcium–uranium-enriched mineral that may be either becquerelite or liebigite, associated with possible native selenium. Further confirmation is required (Basham et al. 1989).

beekite

TL South Devon, England

Named after Henry Beeke (1751–1837), dean of Bristol cathedral, who first drew attention to the mineral; he was born at Kingsteignton, Devon. Beekite is a variety of quartz (chalcedony ± moganite). See Hughes (1889).

beidellite

SMECTITE GROUP

monoclinic $(Na,Ca_{0.5})_{0.3}Al_2(Si,Al)_4O_{10}(OH)_2\cdot nH_2O$

An uncommon hydrated sodium–calcium–aluminium aluminosilicate hydroxide found in hydrothermally altered vesicular basalts of the Antrim Lava Group at Craigahulliar and White Rocks, both near Portrush, Co. Antrim, Ireland. Beidellite is a clay mineral formed during chemical exchange between basalt and fluid as the volcanic flows cooled. It is associated with saponite and chlorite, often in concentric rings within vesicles (Robert 2001).

bell-metal ore

A synonym of stannite.

bementite

monoclinic $Mn^{2+}_7Si_6O_{15}(OH)_8$

A rare manganese silicate hydroxide found in the manganese ores of Treburland mine, 2 km SE of Fivelanes, Altarnun, Cornwall. The ore lies within, or in contact with, a highly calcareous calc-silicate rock (originally an impure limestone) that has been thermally metamorphosed by the adjacent Bodmin Moor granite pluton. Brownish-yellow bementite occurs in sharply defined parallel veinlets up to 6 mm across, traversing masses of coarsely crystalline rhodonite and as hemispheres up to 4 mm in diameter on joints in rhodonite (Russell 1946d, Tilley 1946).

Bevins (1994) briefly reports bementite from Nant mine, Llanfaelrhys, Rhiw, Lleyn Peninsula, Gwynedd, Wales.

Radiating sheaves of orange-yellow bementite enclosing colourless euhedral crystals of rhodonite. Treburland manganese mine, Altarnun, Cornwall; width 2.25 mm; viewed in PPL (upper) and XPL (lower). (© K. Gardner)

benleonardite

tetragonal $Ag_8(Sb,As)Te_2S_3$

A rare silver–antimony–arsenic–tellurium sulphide found in quartz–sulphide ± gold veins cutting Upper Dalradian gritty quartzites, quartz–mica schists and amphibolites at Calliachar Burn, 4 km to the SW of Aberfeldy, Tayside Region, Scotland.

Benleonardite, gold (electrum) and hessite form small intergrowths in zoned tetrahedrite crystals that are included along the grain boundaries of galena (Ixer et al. 1997). Similar quartz–sulphide veins at the nearby Urlar Burn contain benleonardite as thin rims on hessite, also included in galena (Ixer et al. 1997).

benstonite?

trigonal $(Ba,Sr)_6(Ca,Mn)_6Mg(CO_3)_{13}$

A rare complex barium–strontium–calcium–manganese–magnesium carbonate that may be present in the Foss barite deposit, near Aberfeldy, Tayside Region, Scotland. Analyses of lemon-yellow luminescent barytocalcite grains included in pyrite show a wide range of barium, calcium, magnesium and strontium contents, and some may be benstonite. However, their scarcity and small size (< 100 µm) make confirmation difficult. A sedimentary-exhalative origin has been proposed for the mineralization (Moles 1985).

beraunite

monoclinic $Fe^{2+}Fe^{3+}_5(PO_4)_4(OH)_5{\cdot}4H_2O$

An uncommon hydrated iron phosphate hydroxide found on dumps at the small iron and manganese mine of Roury Glen, between Glandore and Ross Carbery, Co. Cork, Ireland. Vein material consists of shattered quartz and botryoidal iridescent limonite, together with minor psilomelane and pyrolusite. Reddish-brown beraunite forms divergent fibrous crystals up to 1 cm in length, which extend from walls of limonite (Russell 1927a).

In Cornwall, beraunite has been noted from Gravel Hill mine, Perranzabuloe, and Phoenix United mines, Linkinhorne (Golley & Williams 1995). Other descriptions of beraunite are more contentious: Penberthy Croft mine, St Hilary, Cornwall (Golley & Williams 1995) – thought by Betterton et al. (2000) to be an identification error; Wheal Jane, Kea, Cornwall (Greg 1860, Robson 1949b) – discredited by Russell (1927a); Burdell Gill, Caldbeck Fells, Cumbria (Hartley 1984) – could not be confirmed by later investigations (Neall & Leppington 2003), although it is unclear if this was because manganese-rich minerals are notoriously difficult to identify or if the description was based on Kingsbury material of dubious provenance.

berg-crystal

A synonym of quartz (var. rock crystal).

bergmannite

A synonym of natrolite, zeolite group.

bernalite

Named after John Desmond Bernal FRS (1901–1971), a pioneer of X-ray crystallography and professor of physics at Birkbeck College, London; born at Nenagh, Co. Tipperary, Ireland. Bernalite is not a true mineral but a fusible, partially soluble, olefinic bitumen.

berryite

monoclinic $Cu_3Ag_2Pb_3Bi_7S_{16}$

A rare copper–silver–lead sulphobismuthite found in quartz veins cutting monzogranite at Loch Shin, 2 km W of Lairg, Highland Region, Scotland. Elsewhere in the monzogranite, mineralization is disseminated and mainly restricted to fluorite, molybdenite and pyrite. However, in the veins, molybdenite does not occur, and the other species are accompanied by native bismuth, chalcopyrite, covellite, galena, sphalerite and a suite of bismuth sulphosalts: aikinite, berryite, gladite, hammarite, lindströmite, krupkaite, pekoite and tetradymite (Lowry et al. 1994).

berthierine

KAOLINITE–SERPENTINE GROUP

monoclinic or hexagonal $(Fe,Al)_3(Si,Al)_2O_5(OH)_4$

An uncommon iron–aluminium aluminosilicate hydroxide that forms two polymorphs (-1*H* and -1*M*). Berthierine–chamosite is found in many of the vein workings of the Weardale (Co. Durham) fluorspar mines, where it occurs as dark olive-green coatings on vug and vein minerals, and vein breccias. At Redburn mine, near Rookhope, the mineral coats breccias produced by post-mineralization faulting, and at the North Wear aqueduct tunnel it coats nailhead calcite in vuggy ground as well as earlier quartz and siderite veins. At Cambokeels (Cammock Eals) mine, Eastgate, berthierine lines open water-bearing fissures alongside fluorite–pyrite–pyrrhotite–quartz veins. Berthierine was probably deposited from groundwaters that had percolated through the veins and remobilized iron and silica from the vein minerals (Smith & Hardy 1981).

Berthierine, pyrite and siderite are the major iron-rich phases in the Jurassic Northampton Ironstone at Cowthick quarry, near Corby. All three species were formed in a diagenetic marine environment that changed from post-oxic to sulphidic conditions over time. Berthierine forms ooids and also occurs as a fine-grain component of the matrix. It formed early and was partially replaced by pyrite or siderite (Gehring 1990). Berthierine ooids are also present in shallow-marine sandstones and sandy limestones of the Jurassic Hackness Rock Member at Castle Hill (Scarborough), Cayton Bay and Cunstone Nab (Williams 2002). Chamosite from Stanion Lane pit, Corby, and N of Gateside farm, Ayrshire coalfield, 3 km E of Ayr, Strathclyde, Scotland (Brindley 1951, Brindley & Youell 1953), was later redefined as berthierine (Brindley 1982, Clark 1993). Worth farm and Philpotts quarry, Sussex, are further locations for berthierine (Anthony et al. 1995).

berthierite

orthorhombic $FeSb_2S_4$

A rare iron sulphantimonite found in an antimony-rich quartz vein at Coombe Height, headwaters of Wet Swine Gill, Caldbeck Fells, Cumbria. The massive white quartz vein contains an early suite of As–Fe–Sb species (arsenopyrite, berthierite and stibnite) and a later suite of antimony–lead species (fülöppite, semseyite and zinkenite). Berthierite occurs as robust prisms up to 2 mm long (Fortey et al. 1984). Another Cumbrian occurrence is at Hogget Gill, Hartsop Hall, Patterdale, Cumbria, where irridescent blades of berthierite are found within masses of quartz in the walls of an old smelter (Young 1987a).

In Cornwall, berthierite is found in antimony-rich ore at Wheal Prosper, St Ewe, near Mevagissey (Ryback & Tandy 1992); Wheal Boys, Trewetha, St Endellion; and Tresungers mine, St Kew (Golley & Williams 1995). Earlier records from Cornwall are uncertain (Ryback & Tandy 1992).

There is a vague reference to berthierite by Readwin (1888) from the Dolgellau goldbelt, Gwynedd, Wales.

bertrandite

orthorhombic $Be_4Si_2O_7(OH)_2$

A rare beryllium silicate hydroxide found as a late-stage hydrothermal phase in miarolitic cavities and fissures in granite and associated pegmatites.

At Cheesewring quarry, SW of Henwood, Linkinhorne, Cornwall, bertrandite forms small (commonly 2–3 mm in length and 0.3 mm thick, but up to 1 mm in length and 0.8 mm thick) colourless rhombic or oblong tabular crystals on joint surfaces in granite, where it is found in clusters of crystals with fluorite, orthoclase and quartz. Often, bertrandite is penetrated by small needles of greenish tourmaline. The joints also contain anatase, pyrite, sphalerite (blende), torbernite and wolframite. Additionally, bertrandite has been found in a vug in the granite at Cheesewring quarry, associated with fluorite and orthoclase (Bowman 1911). Green et al. (1996a) report other vugs in the granite at Cheesewring quarry containing anatase, bertrandite, brookite, orthoclase, phenakite, schorl and stolzite. At Gold Diggings quarry, 1 km to the W of Cheesewring, clusters of colourless bertrandite, occasionally up to 5 mm in length, have been found on vertical joint faces in granite associated with deep-violet fluorite, orthoclase, pyrite, quartz and sphalerite. In habit, most crystals are identical to those found at Cheesewring quarry, but some from an orthoclase-lined vug are repeatedly twinned (Russell 1913b). Three crystals of bertrandite were found in 1905 on a spoilheap at Kit Hill Great Consols mine, NW of Callington, Cornwall. Associated species are fluorite, pyrite, quartz, sphalerite, black tourmaline, and a white scaly talcose mineral (Russell 1913b). Equant tabular crystals of bertrandite, up to 7 mm across and forming intergrown groups, are described from Hingston Down quarry, 2 km W of Gunnislake, Calstock, Cornwall. They occur in greisenized granite containing chalcopyrite, chlorite, fluorite, pyrite and schorl. On one specimen from an open joint fissure, dozens of equant crystals about 1 mm across were found with fluorite and quartz (Ward 1983b).

Further to the southwest and related to the Tregonning–Godolphin granite, remarkable clusters of colourless platy bertrandite were also discovered on museum specimens of sphalerite from Wheal Metal, Breage, Cornwall. They have a diameter of 5 mm and a thickness of 0.7 mm, and are perched on sphalerite that has grown within cavities in orthoclase (Russell 1913b). A single crystal of bertrandite was also collected from Maen quarry, to the W of Maen Pearne, N of Constantine, Cornwall (Russell 1913b). Hosking (1954a) reports bertrandite in pegmatites cutting the Carnmenellis granite at Trolvis quarry, near Longdowns, where it forms colourless untwinned crystals clustering on apatite, feldspar, fluorite, white mica and quartz. Bertrandite has also been noted from the nearby Kessel Downs (Halvasso) quarry (Golley & Williams 1995) and from Chywoon quarry (Moulding & Hooper 2005), both near Longdowns. A single specimen containing bertrandite and beryl has been collected from the 290-fathom level of South Crofty mine, Camborne, Cornwall.

Colourless plates of bertrandite, 4 mm across, from Chywoon granite quarry, near Longdowns, Cornwall. (© D. I. Green)

Bertrandite occurs in narrow vein-like aggregates of chlorite associated with tourmaline in cassiterite–quartz veinstone and it is also noted to pseudomorph beryl (Phemister 1940). Here, bertrandite has since been discovered in early-formed schorl-bearing quartz veins below the 310-fathom level at South Crofty (Clark 1970b). A description of bertrandite replacing phenakite at Wheal Cock, near Botallack, St Just, Cornwall (Kingsbury 1961), requires further confirmation in the light of evidence that A. W. G. Kingsbury misrepresented locality information (Ryback et al. 1998, 2001).

Bertrandite has been identified in miarolitic cavities at the Shap granite quarry, Cumbria. It occurs as euhedral colourless platy crystals up to 1.5 mm in length and is the third beryllium-bearing species found at this locality, the others being bavenite and milarite (Wirth et al. 2001, Wirth & Green 2002).

Many miarolitic cavities exist in granite at Coire an Lochain in the Cairngorm Mountains, Grampian, Scotland, and in some of them crystals of feldspar, genthelvite and quartz (cairngorm or rock crystal) occur. Pale-brown transparent bertrandite crystals, up to 3 mm in length, were found in one of these cavities attached to genthelvite (Morgan 1967). Later investigations suggest that the species is more widespread throughout the Cairngorm granite (Green & Todd 2001). Colourless-to-white bertrandite, forming euhedral crystals to 1 cm across and associated with fluorite and kaolinite, has been found in the Pass of Ballater, Grampian (Green & Todd 2001). In scree below Creag an t-Seabhaig on the north side of the pass, bertrandite has been found as lustrous colourless crystals up to 1 mm across, encrusting the walls of a 10 mm cavity in coarse pink granite (Starkey 1990).

In Ireland, bertrandite occurs in miarolitic cavities within pegmatites that have formed along the southern margin of the Leinster batholith. At Aclare House, near Myshall, Co. Carlow, boulders of albite–spodumene pegmatite contain cavities hosting platy bertrandite with albite, kaolinite and quartz (Steiger & von Knorring 1975).

beryl

hexagonal $Be_3Al_2Si_6O_{18}$

An uncommon beryllium–aluminium silicate found in granites and granite pegmatites, less commonly in low- to high-temperature hydrothermal veins and in vugs in rhyolite.

In Cornwall, corroded dull-white translucent crystals of beryl, up to 2.5 cm long, have been found at Beam mine, near Hensbarrow Beacon, St Austell. About 25 crystals coat the surface of a specimen collected from an open fissure in granite. Cassiterite and felted black tourmaline are associated species (Hallimond 1939). A single sample containing bertrandite and beryl has been described from the back of the 290-fathom level, E of Robinson's shaft, South Crofty mine, Camborne. The specimen consists of massive quartz containing cassiterite, chlorite and tourmaline, and minor amounts of apatite, fluorite and orthoclase. The beryl forms euhedral prisms reaching 0.5 mm in length or, more frequently, slender needle-like prisms with hollow cores, and in places it is pseudomorphed by chlorite (Phemister 1940). Dirty-green beryl has been found in a pegmatitic vein associated with the Kit Hill granite boss at New Consols mine, Stoke Climsland, Cornwall. Crystals are well defined hexagonal prisms, up to 6 mm long and 4 mm in diameter. Alteration along an irregular network of fractures is to kaolinite. Arsenopyrite, fluorite and tourmaline are associated species (Embrey 1953). Small bluish crystals of beryl associated with cassiterite and topaz are noted from St Michael's Mount, Marazion, Cornwall (Greg & Lettsom 1858, Davison 1920b), but the species was not found by Hosking (1954c). Other Cornish occurrences include: Goonbarrow and Gunheath china-clay pits, St Austell (Golley & Williams 1995); Hawkswood mine, North Hill (Golley & Williams 1995); New Consols mine, Luckett (Golley & Williams 1995); Swit quarry, Craddock Moor, Bodmin Moor (Embrey 1953); and Wheal Castle, St Just (Garby 1848, Greg & Lettsom 1858). Reports of beryl from Longdowns, between Helston and Falmouth (Robson 1949b), and less precisely from the parishes of Mabe and Constantine, near Falmouth (Garby 1848), are disputed by Russell (1924), who suggests there is confusion with apatite. However, since that time the beryllium-rich species bertrandite has been found at Kessel Downs (Halvasso) and Trolvis quarries, both near Longdowns (Hosking 1954c, Golley & Williams 1995), so there is a possibility that beryl could have been found.

In Devon, beryl has been found in veinlets within the lithium–fluorine-rich Meldon aplite, near Okehampton. The beryl is bluish and glassy in appearance, has irregular form and develops crystals up to 2 cm across associated with montebrasite (von Knorring & Condliffe 1984). Other exotic species, such as chrysoberyl, eudidymite and rhodizite reported from this location by Kingsbury (1964) have been called into question by Ryback et al. (1998, 2001) and Wirth et al. (2001), as the samples may have a foreign origin. Beryl has been found in pegmatite-lined drusy cavities within granite at Battery Point and at VC quarry on Lundy island, off the north coast of Devon, where it is said to be comparatively abundant. It occurs as pale-yellow, or more rarely pale-blue or colourless, prisms, the largest of which are 7 mm long. Albite, fluorite, kaolinite, orthoclase, topaz and tourmaline are associated species (McLintock & Hall 1912, Dollar 1941, Edmonds et al. 1979, Thorpe et al. 1990). The granite from Lustleigh, near Bovey, reported to contain rough crystals of beryl associated with garnet (Hall 1868b, Collins 1892), deserves further investigation.

Good beryl crystals are said to have been found at Walla Crag, Derwentwater, near Keswick, Cumbria (Davidson & Thomson 1951), and from Carrock mine, Grainsgill, Caldbeck Fells (Hartley 1984). The latter description is based on Kingsbury material, which in view of the doubts about the accuracy of his provenancing (see above) must be dubious.

Beryl crystal, 7 mm long, from a miarolitic cavity in the Lundy granite, Bristol Channel, off the north coast of Devon. (© D. I. Green)

BERYL

The Caledonian granites and associated pegmatites of Grampian, Scotland, contain beryl at a few locations. Bluish-green crystals of beryl have been found in the cliffs of Coire an Lochain, 2 km WSW of the summit of Cairngorm (Morgan 1967), and are noted more generally from the Cairngorm granite by Harrison (1988), who also reports the presence of abundant pegmatites and aplites throughout the pluton, containing accessory amounts of almandine–spessartine garnet, andalusite, apatite, iron–titanium oxides, monazite, topaz, xenotime, zircon and some other rare beryllium, niobium and yttrium minerals. Cabinet-size specimens of beryl, feldspar, quartz and topaz have all been collected in recent years from Ben Macdui in pegmatites from this intrusion (Green & Todd 2001). Non-pegmatitic beryl has been noted from the top of Beinn a' Bhuird, a high and precipitous summit in the Cairngorm pluton (Burke et al. 1964); however, the Beinn a' Bhuird beryl is now considered to be from a weathered pegmatite. The beryl is pale green, of gem quality (but deeply etched) and forms crystals up to 4 cm in length associated with large smoky quartz crystals (R. E. Starkey pers. comm. 2005). Dunlop (2005) documents further occurrences of beryl from the Cairngorm Mountains. Miarolitic cavities in the nearby Ballater granite, exposed in the Pass of Ballater, contain pale-yellow beryl, bertrandite, fluorite and kaolinite (Heddle 1901, Starkey 1990, Green & Todd 2001). A single specimen of vein quartz, containing a crystal of iron-stained flesh-colour beryl and associated with the Middleton granite, has been described from the northern edge of Knockinglews plantation, near Middleton, between Chapel of Garioch and Inverurie (Colman et al. 1989). Livingstone (2002) notes beryl, topaz and hübnerite from Glen Gairn, near Ballater and a 9 cm-long beryl crystal from Pitcaple, 4 km NW of Inverurie. The last locality is perhaps Pitscurrie (Pitscurry) quarry, Pitcaple, where occasional beryl has been found in tourmaline–garnet-bearing granitic pegmatite veins cutting gabbroic rocks (Leslie 1987). Coarse yellow opaque crystals of beryl (davidsonite), sometimes over 30 cm long, were once found at Rubislaw granite quarry, Aberdeen (Greg & Lettsom 1858).

In the Highland Region of Scotland pale yellowish-green crystals of beryl, up to 75 mm long, occur at Struy Bridge quarry, 25 km WSW of Inverness (Heddle 1901, Berridge 1969), and the species is also noted from a garnet–muscovite–tourmaline pegmatite at Carn Gorm, near Garve and in a hydroelectric tunnel below Loch Fannich, both to the W of Dingwall (Kennedy & Lawrie 1943, Berridge 1969). The Loch Nevis mica prospect to the S and SE of Sgùrr Coire nan Gobhar, and to the S and SE of Beinn Bhuide (on the relatively inaccessible Knoydart Peninsula), consists of lenticular bodies of mica-rich pegmatite containing pale yellowish-green opaque-to-translucent roughly hexagonal crystals of beryl (up to 3×10 cm) associated with gem-like transparent red crystals of almandine–spessartine garnet (Starkey 1987c, Green & Todd 2001, Livingstone 2002).

Beryl, columbite, garnet, topaz and tourmaline are minor minerals in the Lewisian Chaipaval (Ceapabhal) granitic pegmatite that forms a dyke-like intrusion up to 25 m wide and is found 3 km NW of Northton (Taobh Tuath), South Harris, Western Isles (Heddle 1901, von Knorring & Dearnley 1959b).

The Tertiary granites of Scotland are also sources of beryl. A blue variety forming crystals up to 25 mm long in fine-grain granite from near Brodick is reported by Heddle (1901) and, in the Northern Arran granite, miarolitic cavities contain beryl, feldspar, mica, quartz and topaz (Tyrrell 1928a, Emeleus & Gyopari 1992). More recently, Hyslop et al. (1999) have found allanite, fergusonite, gadolinite and zircon in other cavities within the Arran granite, lined by coarsely crystalline albite, orthoclase and smoky quartz. Excellent specimens of aquamarine-blue beryl crystals, at least up to 12 mm long, are described from the central granites of Arran (Green & Todd 2001).

Beryl is briefly mentioned by Heddle (1901) from the following additional locations in Grampian, Scotland: near Blackdog Rock, close to Belhelvie; from a railway cutting between Ellon and Old Deer (these locations are about 16 km apart, although the cuttings are mainly near Ellon); Keig Bridge, NE of Alford; and forming rough yellow crystals at Torry, Aberdeen.

In Ireland, beryl is found in a lithium pegmatite belt on the southeastern margin of the Caledonian Leinster granite batholith. The pegmatites are poorly exposed and often found only as boulder trains or in drill core. At Aclare, near Myshall, Co. Carlow, spodumene–tourmaline pegmatite contains colourless alkali beryl together with accessory amounts of manganese-rich apatite, arsenopyrite, bertrandite, cassiterite, chalcopyrite, columbite–tantalite, fluorite, spessartine–almandine garnet, and zircon (Steiger & von Knorring 1975, Scoon 1978). McArdle & Kennan (1988, 1989) note six significant lithium pegmatites with similar assemblages (including Aclare) between Borris, Co. Carlow and Shillelagh, Co. Wicklow, but do not give details of the occurrences. At the northern end of the batholith, opaque greenish beryl has been described with apatite, tourmaline and garnet from the granite quarries of Glencullen, Co. Dublin (Rudler 1905, Ryback & Moreton 1993).

Prismatic beryl to 14 mm, with orthoclase and smoky quartz; Diamond Rocks, Mourne Mountains, Co. Down, Ireland. (© D. I. Green)

The Rosses granite complex of Co. Donegal is another area in Ireland where beryl has been described. At Sheskinarone, near Dungloe it forms green (occasionally with a shade of blue) crystals in granite and quartz veins (Haughton 1862) and large opaque white crystals in pegmatite (Rudler 1905). Later investigations found beryl in metasomatically altered granite (greisen), in many pegmatites and in quartz veins in the area of Sheskinarone, Meenbannad and Meenmore townships. The main occurrence of beryl is in greisen at Sheskinarone, where it occurs associated with torbernite and uraninite (Burke et al. 1964, Hall & Walsh 1971, Legg et al. 1985). Some 10 cm-long crystals of bluish-green beryl in a quartz matrix were recovered from Sheskinarone in the 1990s (Green & Moreton 2001).

Beryl forming euhedral prismatic crystals in miarolitic cavities is notable from the Tertiary granites of the Mourne Mountains, Co. Down. At Diamond Rocks, W of Slieve Corragh, blue translucent crystals of beryl, up to 20 mm in length, have been found in recent times, but these do not compare with the fine aquamarine-blue beryl crystals in 19th-century collections (among the best ever collected in Britain and Ireland), whose precise origin has never been established (Wolfe 1996b, Green & Moreton 2001). Greg & Lettsom (1858) list beryl from the west side of Rocky Mountain, the northwest side of Binion Hill, Slieve Corra and Slieve Hevila, and fine crystals from Chimney Rock Mountain. The eastern sides of Slieve Bearnagh and Slieve Donard also contain beryl-bearing vugs (Richey 1927). At Lindsay's Leap quarry in the Mourne Mountains, drusy cavities and veins on joint planes in granite contain colourless to pale-blue beryl forming elongated prismatic crystals to 5 mm (Green et al. 2005a). Allanite, chrysoberyl, fluorite, gadolinite and topaz have also been found in vugs within the Mourne granites (Burke et al. 1964, Green & Moreton 2001).

On the southern slopes of Croughaun Hill in Co. Waterford, 10 km S of Carrick on Suir, Co. Tipperary, deep azure-blue beryl has been found in milky-white quartz veins within Upper Devonian conglomerate. Prismatic crystals reach 50 mm in length, but are commonly cracked and bent. Dull-green and colourless varieties are also reported (Sanders & Doff 1991).

beryllonite

monoclinic $NaBePO_4$

A sodium–beryllium phosphate found in complex pegmatite lenses within the lithium–fluorine-rich aplite dyke at Meldon aplite quarry, near Okehampton, Devon. Beryllonite occurs as whitish grains up to 2 mm across and is mostly enclosed in elbaite with some lepidolite and quartz. It is fibrous and silky in appearance, and is closely associated with lepidolite, orthoclase, petalite and pollucite. Beryllonite has also been found intergrown with herderite in the aplitic matrix, close to a pegmatite veinlet (von Knorring & Condliffe 1984). Noted by Kingsbury (1964b, 1965b).

berzelianite

cubic Cu_2Se

A very rare copper selenide, dimorphous with bellidoite and briefly described from the Smid Hope Burn tributary of the River Tweed, near Tweedhope, Borders, Scotland, where it has been found as 2 µm-diameter inclusions associated with native palladium and an unnamed copper telluride in two alluvial gold grains (Leake et al. 1993).

berzelite

A synonym of berzeliite, or mendipite, or petalite.

betafite?

PYROCHLORE GROUP

cubic $(Ca,U,\square)_2(Ti,Nb,Ta)_2(O,OH)_7$

A rare calcium–uranium–titanium–niobium–tantalum oxyhydroxide described from the Lewisian granitic pegmatite at Chaipaval (Ceapabhal), 3 km NW of Northton (Taobh Tuath), South Harris, Western Isles, Scotland. Betafite is an accessory constituent of the pegmatite, along with allanite, beryl, columbite, gahnite, kasolite, magnetite, monazite, pyrite, spessartine, thorite, topaz, tourmaline, uraninite and zircon (von Knorring & Dearnley 1959b). Referring to the same occurrence, von Knorring & Dearnley (1960) describe the species as a metamict betafite-like mineral and, although XRD data were provided, further work is required before the species can be confirmed. Macpherson & Livingstone (1982) suggest that the mineral may instead be pyrochlore.

beta-quartz

A high-temperature form of SiO_2.

betekhtinite

orthorhombic $Cu_{10}(Fe,Pb)S_6$

A rare copper–iron–lead sulphide found in barite–fluorite–galena–hematite–quartz veins cutting the Grudie granite and adjacent Moinian rocks near Lairg, Highland Region, Scotland (Gallagher et al. 1974).

BEUDANTITE GROUP

Trigonal sulphate arsenates and sulphate phosphates, with the general formula $AB_3(XO_4)(SO_4)(OH)_6$, where A = Ba, Ca, Ce, Pb, Sr, (H_3); B = Al, Fe^{3+}, Ga; and X = As^{5+}, P^{5+}. Beudantite, corkite, hidalgoite and hinsdalite are members of the group found in Britain or Ireland. Svanbergite has also been reported, but this is a dubious occurrence.

beudantite

BEUDANTITE GROUP

trigonal $PbFe^{3+}_3(AsO_4)(SO_4)(OH)_6$

A relatively uncommon, but locally abundant, lead–iron–arsenate sulphate hydroxide found in the oxidized zone of polymetallic deposits.

Beudantite occurs in various forms at Penberthy Croft mine, St Hilary, Cornwall; earthy compact masses with drusy areas, radial-globular aggregates and isolated translucent rhombohedral crystals to 0.6 mm. Colour varies from yellow and yellowish-brown, to greenish-brown and dark brown. Associated species include arsenopyrite, bayldonite, carminite, corkite, goethite, hidalgoite, malachite, mimetite, pharmacosiderite and scorodite (Betterton et al. 2000). The species had earlier been noted from this locality by Kingsbury (1954), Kingsbury & Hartley (1960), Camm & Merry (1991b) and Golley & Williams (1995). Beudantite forms small yellow rhombohedral crystals and yellow to greenish-yellow crusts associated with carminite, mimetite and zeunerite at Wheal Gorland, St Day, 2 km E of Redruth, Cornwall (Kingsbury & Hartley 1960, Rust 1981b). Additionally, Golley & Williams (1995) note beudantite from Trevascus mine, near Hayle, Gwinear, Cornwall. Because the provenance of many Kingsbury specimens has been called into question (Ryback et al. 1998, 2001), independent verification is desirable for beudantite described in Cornwall, from: Hingston Down Consols mine, 2 km W of Gunnislake, Calstock; Old Treburgett mine, St Teath, E of Port Isaac; Pengenna (Trewethen) mine, St Kew; Treore mine, Port Gaverne, near Port Isaac, StTeath; Wheal Carpenter, Gwinear, near Hayle (Kingsbury & Hartley 1960).

Beudantite forming two generations of blocky crystals; Howthwaite trial, Caldbeck Fells, Cumbria; width 175 µm; SEM image. (© D. I. Green)

Doubts also exist about the provenance of beudantite specimens described from Cumbria by Kingsbury & Hartley (1957f, 1958, 1960) and referenced by Hartley (1984), Young (1987a), Cooper & Stanley (1990), Young et al. (1992a) and Anthony et al. (2000). These include: Brandy Gill mine; Carrock mine, Grainsgill; Driggith mine; Dry Gill mine; higher part of Brandy Gill; Ingray Gill mine; Lower level, Nether Row Brow; Potts Gill barite mine; higher part of Roughton Gill (all Caldbeck Fells); and Wanthwaite mine, St John's in the Vale, near Threlkeld. Fortunately, independent verification exists for the occurrences at Brandy Gill and Nether Row Brow (Cooper & Stanley 1990) and from Sandbed mine (Young et al. 1992b). The opencut above Sandbed mine has produced yellowish-brown drusy crusts of beudantite on mimetite, and its presence is also noted on the dumps of both Driggith and Sandbed mines. At the nearby Howthwaite trial, a blocky pseudo-cubic mineral with intermediate beudantite–segnitite composition is recorded (Green et al. 2006a).

Cooper & Stanley (1990) have confirmed beudantite at Burdell Gill, Deer Hills and Low Pike, all in the Caldbeck Fells. At Arm o' Grain it occurs as colour-zoned (<0.2 mm) pseudo-cubic crystals that were initially thought to be pharmacosiderite and which are considered to be the best found in the Caldbeck Fells (Green & Briscoe 2002, Green et al. 2006b). At Burdell Gill, beudantite is described as sheafy aggregates of acute rhombohedral crystals to 2 mm, partially embedded in massive beudantite and quartz, and also yellow to dark-brown pseudo-octahedral crystals up to 1 mm across (Neall & Leppington 2003) – but see kintoreite. Cavities produced by the removal of sulphides during supergene alteration of a mineralized quartz vein at Deer Hills are now partially infilled with pale yellowish-green beudantite, some replacing earlier-formed carminite (Symes et al. 1990, Green & Briscoe 2002). Other

Beudantite crystals to 0.4 mm on iron-stained quartz; Arm o' Grain, Caldbeck Fells, Cumbria. (© D. I. Green)

occurrences of beudantite in Cumbria are: Coombe Height, headwaters of Wet Swine Gill, Caldbeck Fells (Neall & Green 2001b, Green & Briscoe 2002); Iron Crag, near Roughton Gill mine, Caldbeck Fells (Wolfe 1996a); Silver Gill, Caldbeck Fells (Green et al. 2005b); Smallcleugh mine, Nenthead, near Alston (Young et al. 2005); Tynebottom mine, NW of Garrigill, near Alston (Bridges & Young 1998).

Beudantite has also been found at Sandford Hill quarry, Somerset, where it is intimately mixed with goethite in light yellowish-brown to dark-brown cellular stalactitic material on which cerussite and mimetite have later crystallized. Beudantite appears to have formed from the alteration of galena as arsenic- and iron-rich solutions percolated through the Carboniferous Limestone host rocks (Livingstone & Cogger 1966). On the beach near the Pier hotel, Clevedon, Avon, powdery yellow beudantite occurs in a mineralized fault containing primary barite, galena, marcasite, pyrite and tennantite–tetrahedrite. Direct alteration of tennantite in and around the other sulphides provided a ready source of arsenic for beudantite to form (Starkey 1986c). Very rare pale greenish-yellow to brownish-yellow beudantite in iron-stained barite-bearing sandstone is noted from Alderley Edge ridge, Cheshire (Braithwaite 1994).

In Wales, beudantite has been found at Ystrad Einion mine, near Tal-y-bont, where it forms minute (<0.15 mm) transparent yellowish-green rhombohedral crystals associated with goethite (Bevins 1994, Mason & Rust 1997), and at Darren mine, Pen-bont Rhydybeddau, where it is associated with gersdorffite (Mason 1998). Both locations are in the Central Wales orefield of Dyfed. Beudantite is one of the supergene alteration species found in a massive sulphide vein at Dolyhir quarry, Old Radnor, near Presteigne, Powys (Bevins & Mason 1997). Beudantite is recorded from two locations in Dumfries & Galloway, Scotland. At High Pirn mine, Wanlockhead, it occurs as a red-brown encrustation on quartz (Temple 1955) and at Penkiln Burn, N of Newton Stewart, it forms complex intergrowths with goethite, limonite, malachite and plumbogummite in a gossan zone close to a swarm of mineralized dioritic minor intrusions (Stone & Leake 1984).

At Tynagh mine, near Killimor, Co. Galway, Ireland, beudantite is one of many supergene species replacing primary galena–sphalerite ore (Ryback 1988a, Ryback & Moreton 1992).

beyerite

tetragonal $(Ca,Pb)Bi_2(CO_3)_2O_2$

A calcium–lead–bismuth carbonate oxide found in a vein on the west side of Buckbarrow Beck, Corney Fell, Waberthwaite, Cumbria. Beyerite forms small radiating clusters of pale yellowish-green needles similar in appearance to russellite (Braithwaite & Ryback 1995). A tentative identification from the Shap granite quarry, Cumbria, is provided by Wirth & Green (2002), who describe a yellow powdery crust composed of minute (<50 µm) tetragonal plates on bismuthinite as possibly being beyerite.

bhreckite

TL Beinn Bhreac (Ben Bhreck), E of Tongue, Highland Region, Scotland

A compound with uncertain composition, perhaps calcium-rich clinochlore.

bialite

A synonym of wavellite and a magnesium-rich variety of tavistockite.

biaxial mica

A synonym of muscovite, mica group.

bicalcareo-carbonate of barytes

TL Brownley Hill, Nenthead, near Alston, Cumbria, England

A synonym of alstonite.

bicchulite

cubic $Ca_2Al_2SiO_6(OH)_2$

A rare calcium–aluminium silicate hydroxide, dimorphous with kamaishilite and found in a chalk/dolerite contact zone associated with a Tertiary volcanic plug at Carneal, Larne, Co. Antrim, Ireland. The chalk is converted by intense contact metamorphism to larnite, merwinite, scawtite, spinel, spurrite, titanite and wollastonite, whereas the dolerite is desilicated to melilite- and wollastonite-bearing pyroxene-rich alkaline hybrids. Hydrated and secondary minerals include allophane, calcite, gyrolite, hibschite (hydrogrossular), portlandite, tacharanite, thomsonite, tobermorite and xonotlite. Bicchulite occurs mixed with hibschite and melilite in a buff coating to a wollastonite rock (Henmi et al. 1973, Sabine & Young 1975).

bieberite?

MELANTERITE GROUP

monoclinic $CoSO_4{\cdot}7H_2O$

A rare hydrated cobalt sulphate briefly mentioned by Robson (1949b) from Wheal Alfred, Phillack, near Hayle, and Penberthy Croft mine, St Hilary, both in Cornwall. At Penberthy Croft its occurrence has not been confirmed subsequently and the species should probably be deleted from further listings (Betterton et al. 2000).

binarkies

A synonym of marcasite.

bindheimite

STIBICONITE GROUP

cubic $Pb_2Sb_2O_6(O,OH)$

A relatively common lead–antimony oxide hydroxide often found as an alteration product of jamesonite. At Trevinnick mine, St Kew, near St Endellion, Cornwall, it formed large detached white, through various shades of yellow, to brown masses with a lamellar structure. Jamesonite and stibiconite are associated species (Greg & Lettsom 1858, Fisk et al. 1982). In the same general area, bindheimite is reported from the Bounds Cliff antimony–lead vein system to the NE of Port Isaac (Clayton et al. 1990), from limonitic gossan at Bodannon mine, Port Gaverne, associated with anglesite, nadorite, senarmontite, native silver and valentinite, and from Wheal Boys, Trewetha, St Endellion, where it is associated with jamesonite, stibnite and stibiconite. At Treore mine, Port Gaverne, near Port Isaac, bindheimite has been found in a quartz vein enriched in arsenopyrite, jamesonite and pyrite. Interestingly, the vein also contains sprigs of native gold, together with boulangerite, bournonite, cerussite, galena, scheelite, sphalerite and tetrahedrite (Russell 1927a, 1949b). Bindheimite is noted by Russell from six other Cornish locations: Linnick, South Petherwin; Pengenna mine, St Kew; Poltreworgy mine, near Port Isaac; Port Quin mine, near Port Quin; Tolpetherwin quarry, Two Bridges, South Petherwin; and Tresungers mine, near Port Isaac. Bindheimite is described as the most abundant secondary species in a bournonite, galena and tetrahedrite-rich quartz vein found at Newporth beach, 2 km SW of Falmouth, Cornwall, where it forms thick, sulphur-yellow to yellow-orange, poorly crystalline crusts and masses with a resinous lustre closely associated with bournonite. Cumengéite frequently overgrows the bindheimite (Dean 1982, 1983, Dean et al. 1983). At Greystone quarry, Lezant, Cornwall, bindheimite specimens found in 1989 form bright yellowish-green crusts, up to 30 mm thick, associated with cerussite, clinochlore, malachite and minor linarite. They are said to be the richest known specimens from Britain (Weiss 2001, Elton et al. 2004).

Antimony–gold mineralization in quartz–carbonate veins cutting Devonian slates at Loddiswell mine, 3 km N of Loddiswell, Devon, contains bismuth-rich bindheimite pseudomorphs after jamesonite (Stanley et al. 1990b).

In Cumbria, yellow bindheimite is abundant in a sulphide-bearing quartz vein in the headwaters of Wet Swine Gill, on Coombe Height, Caldbeck Fells. It occurs with beudantite, pharmacosiderite, scorodite and stibiconite in fractures in the vein, and is considered to be a weathering product. Native antimony, arsenopyrite, berthierite, claudetite, füllöpite, jamesonite, parasymplesite, semseyite, senarmontite, stibnite, native sulphur, valentinite and zinkenite are also present and make up two distinct assemblages dominated by As–Fe–Sb and Sb–Pb species (Fortey et al. 1984, Neall & Green 2001b, Green & Briscoe 2002). Bindheimite forms powdery yellow coatings associated with anglesite or cerussite at Brae Fell mine, Dale Beck Valley, Caldbeck Fells (Bridges et al. 2006). At Mexico mine, Caldbeck Fells, bindheimite occurs as a yellow powder with cerussite, pyromorphite and wulfenite in cavities in quartz from the Lower level dump in Todd Gill (Cooper & Stanley 1990). Yellow powdery crusts of bindheimite also occur in cavities within quartz veinstone at Arm o' Grain (Green et al. 2006b) and Silver Gill, Caldbeck Fells (Green et al. 2005b). Reports of bindheimite from two other localities in the Caldbeck Fells – Driggith mine and on the flanks of Balliway Rigg in higher Roughton Gill (Hartley 1984, Young 1987a, Cooper & Stanley 1990) – are less certain, as they are based on material provenanced by A. W. G. Kingsbury. Ryback et al. (1998, 2001) have shown that in some cases Kingsbury appears to have deliberately falsified the localities of some of his specimens. However, bindheimite is confirmed as yellow powdery crusts associated with partially oxidized galena in quartz veinstone from both Driggith and Sandbed mines (Green et al. 2006a). These authors examined Kingsbury material from Driggith mine and were not confident of its provenance, although it seemed possible that it could be from the Caldbeck Fells.

In the Northern Pennine orefield, pale-yellow powdery bindheimite occurs as a supergene species associated with highly oxidized galena and cerussite at Pike Law, Newbiggin Common, Co. Durham. It may be more common that the confirmed occurrence suggests, as it closely resembles jarosite and native sulphur (Bridges & Young 1998).

Bindheimite is probably widespread in the Central Wales orefield, Dyfed, and has been reported from cavities in quartz at Eaglebrook (Nantycagl) mine, Ceulanymaesmawr (Jones 1983) and Frongoch mine, between Devil's Bridge and Pont-rhyd-y-groes (Jones & Moreton 1977). At Frongoch mine, it is abundant as pale- to canary-yellow crusts covering large areas of veinstone. It appears to have formed by the oxidation of antimony-rich galena and as cubic pseudomorphs after ullmannite (Green et al. 1996b, Mason 2004). Bindheimite is also confirmed at three other locations in the district: Bwlchrhennaid mine and the Goginan mines, near Goginan (Bevins 1994), and Hendrefelen mine, Ysbyty Ystwyth (Clark & Rust 1993).

Elsewhere in Wales, bindheimite is noted as a brownish-yellow crust replacing semseyite at Bwlch antimony mine, near Deganwy, Gwynedd, where it is found in patches and veinlets cutting an Ordovician nodular ignimbrite. Associated species are grey drusy quartz (and chert), stibnite and zinkenite. Anglesite, barite, galena, jamesonite, marcasite, plagionite, pyrite and sphalerite, and possibly boulangerite, robinsonite and stibiconite, are also present in small quantities (Russell 1944, 1949b, Bevins et al. 1988, Ryback & Francis 2001). Plant & Jones (1995) report canary-yellow bindheimite in cavities within galena at Machen quarry, 7 km ENE of Caerphilly, Mid Glamorgan, where it occurs in fault- and joint-bounded veins cutting Lower Carboniferous dolomitized limestones. Powdery yellow bindheimite has been found with cerussite in oxidized galena from a galena–tennantite vein at Dolyhir quarry, Old Radnor, near Presteigne, Powys (T. F. Cotterell pers. comm. 2007).

Yellow earthy encrustations of bindheimite on stibnite are noted at Hare Hill, 5 km SE of New Cumnock, Strathclyde, Scotland (Macpherson & Livingstone 1982).

Specimens of bindheimite from L'Etacq, Jersey, Channel Isles, are said to form yellowish-white acicular crystals about 1 mm long (Mourant 1961), and at Le Pulec it is found in sulphide-rich quartz veins, together with chalcopyrite, galena, jamesonite, molybdenite, pyrite, pyrrhotite and wulfenite (Mourant & Warren 1934).

binnite

A synonym of sartorite and tennantite, and also used as a group name for lead sulpharsenites.

biotite (series)

MICA GROUP

monoclinic $K(Mg,Fe^{2+})_3(Si_3Al)O_{10}(O,F)_2$

A range of very common, dark, lithium-free trioctahedral micas, with compositions between, or close to, the annite–phlogopite and siderophyllite–eastonite. Biotite does not define an end-member composition and is therefore not a true species, but it was given series status in the latest classification of the mica group (Rieder et al. 1998). For the purposes of this work it is desirable to consider separately the compositions intermediate between the end members annite and phlogopite, and it is these data that are described below as biotite. They have compositional boundaries defined as $Fe/(Fe+Mg)=0.25$–0.75 and $Al^{vi}=0.0$–0.5. This is a much tighter definition than the series status and that presented by Thorpe et al. (1990) permits, but in practical terms most analyses described as biotite do fall between these bounds.

England

Cornwall Biotite occurs in metasomatized basaltic rocks in the aureole of the Land's End granite at Gryll's Bunny and the Crowns, Botallack, St Just (van Marcke de Lummen 1985). At Pendennis Point, Falmouth, it forms brown elongated flakes, zoned from phlogopite cores to biotite rims in a strongly peralkaline minette (Hall 1982). Primary titanium-rich biotite in metadolerites (greenstones) from Trevone Bay, W of Padstow, has been altered to a mixture of chlorite, epidote and titanite during prehnite–pumpellyite facies metamorphism (Floyd & Rowbotham 1982). Despite biotite being reported many times from the granites of this region, it is siderophyllite that is usually being described (Tindle unpub. EMP data).

Devon Biotite is reported from microgranular enclaves in granite from Birch Tor on Dartmoor (Stimac et al. 1995) and in a Permian lava from Killerton Park, near Exeter (Tidmarsh 1932).

Northumberland Unusual titanium-rich biotite occurs with diopside and enstatite in the Cheviot granite, SW of Wooler (Haslam 1986).

Wales

Gwent Biotite is confirmed in peridotite nodules from the Great House diatreme, near Usk (R. E. Starkey specimen, Tindle unpub. EMP data).

Scotland

Central A wide range of mica species (many barium-enriched), including eastonite, biotite, muscovite and phlogopite, occur in Middle Dalradian metasediments at Foss, near Aberfeldy. They are associated with syngenetic barium–zinc mineralization (Moles 1985). Biotite and siderophyllite are both present in Dalradian metapelites around Loch Katrine in the Aberfoyle district (Mather 1970).

Dumfries & Galloway Zoned mica crystals (phlogopite cores to biotite rims) commonly occur as phenocrysts in minettes associated with the Caledonian Criffell–Dalbeattie granitic intrusion (Canning et al. 1996). Biotite (dominant), eastonite and siderophyllite are all found in the Criffell main granite, whereas in the smaller western pluton, only biotite occurs (Tindle unpub. EMP data). Biotite is the only mica found in quartz diorite, granodiorite and granite of the Cairnsmore of Carsphairn granite complex (Deer 1937c, Tindle et al. 1988). Other Caledonian granites in the region contain the following dark micas: Cairnsmore of Fleet – biotite and siderophyllite (dominant); Loch Doon – biotite (with siderophyllite in cordierite microgranite and aplites); and Mull of Galloway – biotite (Tindle unpub. EMP data).

Fife Certain pyroxenite nodules found in the Carboniferous–Permian Elie Ness vent contain up to 50% modal titanium-rich biotite associated with augite. Other varieties contain biotite with albite ± pargasite (Chapman 1976).

Grampian Interstitial and poikilitic biotite ± annite in Middle and Upper Zone cumulates and associated gabbroic rocks within the Insch intrusion (the largest of the Caledonian Newer Gabbro bodies) are noted for their elevated barium contents (Wadsworth 1986, 1988). Extensive chemical variation in dark micas from the Cairngorm granite complex has produced a range of micas from biotite through siderophyllite (very common) to zinnwaldite compositions (Harrison 1990). The Glen Avon (or North East) granite is an early-formed part of the Cairngorm complex and it contains abundant biotite (Barritt 1983, Tindle unpub. EMP data). Other Caledonian intrusions in the region contain the following dark micas: Clinterty, Corrennie, Coull, Crathes, Glen Livet, Hill of Fare and Torphins granites – biotite; Glen Gairn quartz diorite, granodiorite and granite – biotite (zinnwaldite in a mineralized unit); Bennachie, Ben Rinnes and Strichen granites – biotite and siderophyllite; Mount Battock – biotite and siderophyllite with minor eastonite; Ballater granite – biotite and siderophyllite with minor annite and zinnwaldite; Lochnagar granite – biotite with very minor annite and siderophyllite; and Peterhead granite – annite and siderophyllite with minor biotite (Tindle unpub. EMP data). Biotite has also been reported in: a grandidierite-bearing pelitic xenolith within norite from the Haddo House complex to the NE of the Wood of Schivas, 32 km N of Aberdeen (Herd et al. 1984); and migmatites from Cumrie and Castle Bridge, both in the Huntly–Portsoy district, and Black Hill, near Lochnagar (Ashworth & Chinner 1978).

Highland Biotite is a constituent of a pyroxenite associated with the Loch Borralan alkali syenite complex (Notholt 1985). It also occurs in the same general area with phlogopite in a melanite-bearing syenite, termed borolanite (Tindle unpub. EMP data) and in metasomatized fenites (Woolley et al. 1972). At Aultivullin quarry, in the southeastern part of the Borralan complex, pegmatite veins contain books of biotite up to 10 cm across (Woolley 1973). Biotite is noted in all varieties of syenite: from minor amounts in some ultramafic blocks and pegmatites, to common magmatic intergrowths with pyroxene in melasyenites and leucosyenites at Glen Dessary, W of Loch Arkaig (Fowler 1992). Other igneous occurrences of biotite include: in gabbro, tonalite and quartz monzonite from the Tertiary Centre III ring complex, Ardnamurchan (Walsh 1975); in diorite, granite and an associated pyroxenite xenolith from the Caledonian Ben Nevis granite complex (Haslam 1968, 1970); in monzodiorite, quartz diorite and granite from the Caledonian Ballachulish complex (Weiss & Troll 1989); in granodiorite, tonalite, granite and associated enclaves from the Caledonian Morvern–Strontian complex (Nockolds & Mitchell 1946, Castro & Stephens 1992); in the Caledonian granite intrusions of Grantown, Monadhliath (also subsidiary eastonite and siderophyllite) and Moy (Tindle unpub. EMP data); in a Permo-Carboniferous lamprophyre (minette) dyke from Coire nan Capull, Loch Sunart, near Strontian (Gallagher 1963); and with phlogopite in Caledonian minettes from the Glen Affric–Kintail area (Canning et al. 1996). Biotite also occurs with ferropargasite in quartzose amphibolite bodies of the Neoproterozoic Moines at White Bridge, Glen Garry SW of Fort Augustus (Rock 1984, Highton 1994).

Strathclyde A megacryst of biotite (described as phlogopite), measuring 5 cm across (actually a cluster of smaller grains), has been found with anorthoclase, kaersutite and pargasite in a Carboniferous basanite vent at Black Rock, 2 km SSW of Fairlie, Hunterston Sands (Alexander et al. 1986). Small quantities of brown biotite occur in an ultramafic (garnet pyroxenite) xenolith within a Permian basanite diatreme at Holmbyre, 2 km SW of Dalry (Upton et al. 2003), and in a pyroxene granulite gneiss xenolith within an olivine-nephelinite dyke at Gribun on the Isle of Mull (Upton et al. 1998). Both are probably derived from the lower crust or mantle. Biotite, mullite, phlogopite, sanidine and spinel have pseudomorphed phengite in Dalradian greenschists thermally metamorphosed by the Sidhean Sluaigh dolerite intrusion, 3 km SW of Strachur, Loch Fyne (Worden et al. 1987). The differentiated alkali olivine-dolerite intrusion at Cnoc Rhaonastil, Isle of Islay, contains deep reddish-brown titanium-rich biotite in gabbroic pegmatite, teschenite and syenite. The intrusion is particularly notable for its range of zirconium-rich amphiboles (Preston et al. 2000b). The Caledonian diorite–granodiorite complexes of Garabal Hill–Glen Fyne (Nockolds 1941), Distinkhorn, 8 km ESE of Galston, and Ross of Mull (Tindle unpub. EMP data), and the Etive diorite–granodiorite–granite complex (Moazzen & Droop 2005) all contain biotite.

Western Isles Metasomatic alteration of a Lewisian basic dyke close to a nearby granitic pegmatite has produced a biotite–magnetite–plagioclase–quartz rock at Garry-a-siar headland, Benbecula (Williams 1987). Biotite and corundum (sapphire) are found in a monchiquite at Cairisiadar (Carishader), Loch Roag, Isle of Lewis (R. E. Starkey specimen, Tindle unpub. EMP data).

Ireland

Down Biotite is a ubiquitous phase in the dolerites and gabbros of the Tertiary central complex of Slieve Gullion. It was late to crystallize and it occurs interstitially or as rims on ilmenite and magnetite (Gamble 1982).

Galway An extensive dataset of mica analyses from the western Galway granite complex indicates biotite is the major dark mica in all the major plutons, except for the Murvey granite, where siderophyllite dominates (Leake 1974, 1998).

Mayo Ophiolite-related Dalradian blueschist-facies garnetiferous metabasites have been investigated by Chew et al. (2003), who also present biotite data from an associated psammite and semi-pelite.

Offshore territories

Guernsey At Beaucette, biotite occurs with eastonite and a suite of calcic amphiboles in meladiorite layers within veined diorite belonging to the late Precambrian Cadomian igneous complex (Bishop & French 1982).

Jersey Titanium-rich biotite is a minor phase in the distinctive coarse-grain olivine-dolerite dyke at Wolf Rock, 0.5 km W of Frémont Point (Rowbotham & Lees 1990). An arfvedsonite-bearing minette at St Hélier contains both biotite and phlogopite, which form large crystals up to 2 mm across with dark reddish-brown cores (Wagner & Velde 1985).

birnessite

monoclinic $(Na,Ca)_{0.5}(Mn^{4+},Mn^{3+})_2O_4\cdot1.5H_2O$

TL Birness, 32 km N of Aberdeen, Grampian, Scotland

A rare hydrated sodium–calcium–manganese oxide found in a 12–38 mm-thick manganese pan from a cutting through fluvioglacial deposits of gravel beside the road at Birness. Birnessite is a soft black mineral forming grains up to 4×8×3 mm, occasionally showing well defined facets. Clay overlying the pan was the probable source of the manganese, and air oxidation of manganese oxides under alkaline conditions is proposed as a mechanism of formation (Jones & Milne 1956). Elsewhere in Scotland, birnessite is reported from exhalative sediments at Noblehouse, near West Linton, Borders Region (associated with pyrolusite and todorokite), and in a manganese-cemented Quaternary raised-beach deposit extending from Auchenmaig to Craignarget Burn around Luce Bay, Dumfries & Galloway. At Luce Bay, birnessite is dominantly soft, powdery and dark-brown to black, and is mixed with lithiophorite and quenselite in an X-ray amorphous wad (Nicholson 1983, 1988b, 1989b).

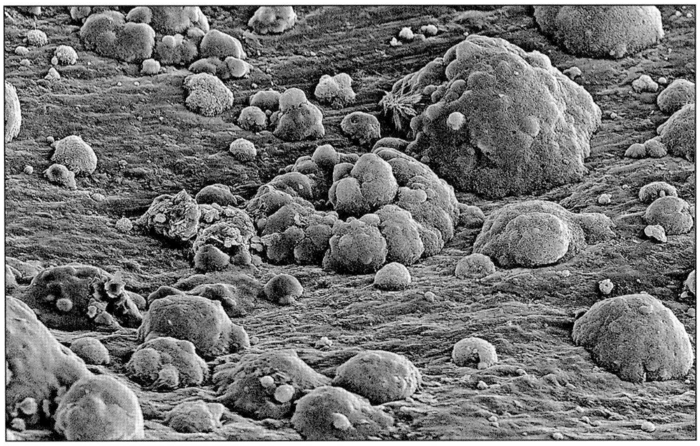

Botryoids of birnessite on a surface of the same species from Penberthy Croft mine, St Hilary, Cornwall; width 320 µm; SEM image. (© E. Condliffe, University of Leeds)

Birnessite is also described from Craigmuschat quarry, Gourock, Strathclyde, where it occurs as (< 1 mm thick) manganese-oxide coatings and as a minor component of (< 10 mm thick) iron–manganese oxide coatings (mainly hematite and goethite) along joint planes and fractures within a trachyte sill. The bluish-black to black colour of birnessite is quite distinctive against the red–blue hues of the iron oxides (Nicholson 1988c). At Dalroy Burn, near Culloden, Highland Region, birnessite is a weathering product developed on a complex stratiform manganese ore (up to 4 m thick). Braunite and lesser bustamite are the main constituents of the ore, which is thought to have a geothermal origin best explained by deposition from hot springs in a sub-lacustrine environment during Devonian times (Nicholson 1990, 1992).

In Cornwall, a hard black glossy massive crust of birnessite has formed as an oxidation product on medium to coarsely crystalline rhodonite and tephroite found on the dumps of Treburland manganese mine, 2 km SE of Fivelanes, Altarnun (Young 1988). It is also reported on weathered blocks of partially cemented slate-rich veinstone from an old dump at Penberthy Croft mine, St Hilary. Here it forms black glossy-to-dull earthy coatings on a matrix of brochantite, connellite, dolomite, goethite, quartz, woodwardite and wroewolfeite (Betterton 1989, Betterton et al. 2000). A report of birnessite from Greystone quarry, Lezant, near Launceston (Golley & Williams 1995), is doubted by Elton et al. (2004), who suggest instead that it is a Ca–Mn–Pb-bearing mineral whose XRD signature cannot be matched with any known species.

A museum specimen of birnessite from Benallt manganese mine, Rhiw, Lleyn Peninsula, Gwynedd, Wales, is briefly mentioned by Bevins (1994).

bisilicate of manganese

A synonym of rhodonite.

bismite

monoclinic Bi_2O_3

An uncommon bismuth oxide found at Carrock mine, Grainsgill, Caldbeck Fells, Cumbria, as a yellowish-white powder replacing bismuthinite (Davidson & Thomson 1951). At the Cross Gill (Great Sulphur vein) outcrop, near Garrigill, Alston Moor, Cumbria, bismite and bismuthinite occur on quartz veinstone. A supergene origin is proposed (Bridges & Young 1998).

In southwest England, bismite has been found in quartz–fluorite veinstone on the dumps of North Devon United (East Wheal Friendship) mine, Peter Tavy, Devon, within the metamorphic aureole of the Dartmoor granite. It surrounds oxidized native bismuth and is associated with arsenopyrite, cassiterite, cobaltite, cosalite, erythrite, gersdorffite, parkerite, sphalerite, stannite and ullmannite, and a variety of bismuth-rich minerals including aikinite, native bismuth, bismuthinite, ikunolite, and an unnamed bismuth telluride (Rumsey & Savage 2005). Bismite is also noted from: Great Dowglas mine, Sticker, near St Austell (Green 1984); Lostwithiel and Roche, both near St Austell (Robson 1949b).

bismoclite

tetragonal BiOCl

A rare bismuth oxychloride found along the shoreline, in the uranium-rich Steps calcite–quartz vein near Marbruie Cove, Southwick, near Dalbeattie, Dumfries & Galloway, Scotland. This and other similar veins are in the aureole of the Criffell granodiorite. Bismoclite and two other chlorine-rich species (atacamite and connellite) are thought to have formed by the action of sea water on primary ore minerals, e.g. native bismuth, chalcopyrite, hematite, pyrite and uraninite (Miller & Taylor 1966). Bismoclite is also found at the Croft Gothal mine, 2 km E of Goldsithney, St Hilary, Cornwall, where it occurs as whitish masses up to about 2 mm across, replacing both native bismuth and bismuthinite (Rust 1996b).

bismuth

ARSENIC GROUP

trigonal Bi

Bismuth (or native bismuth) is a metal found in hydrothermal veins enriched in cobalt, nickel, silver and tin, in granitic pegmatites, and in topaz-bearing tin–tungsten-rich quartz veins.

At Botallack mine, St Just, Cornwall, small rhombohedral crystals of native bismuth, chalcocite, connellite and native copper have been found on drusy quartz within a block of jasper (Greg & Lettsom 1858, Rudler 1905, Wolloxall 1989). The Williams lode at Fowey Consols mine, Tywardreath, St Blazey, contained druses in which native bismuth and bismuthinite were found associated with chalcopyrite, quartz and siderite (Hosking 1951). Crystallized and feathery masses of native bismuth occurred with bismuthinite, erythrite and skutterudite (smaltite) at Wheal Sparnon, near Redruth (Greg & Lettsom 1858, Collins 1892, Hosking 1951). At

Rosewarne and Herland mine (Wheal Herland), Gwinear, cross-course quartz veins intersecting tin lodes contained native bismuth with acanthite, arsenopyrite, bismuthinite, silver-rich galena, pyrargyrite, stibnite and skutterudite (Hosking 1951, Ball et al. 1982, Rust 1982a). Native bismuth is part of a greisen assemblage at Wheal Jane, Kea, near Truro, where it is found with cassiterite, löllingite, muscovite, quartz and wolframite (Holl & Bromley 1988). Another greisen assemblage at St Michael's Mount, Marazion, contains minor native bismuth with a variety of other primary minerals including arsenopyrite, cassiterite, chalcopyrite, kësterite, löllingite, pyrite, stannoidite and wolframite (Alderton 1991). Native bismuth has also been found in fluorite–quartz veinstone on the dumps of North Devon United (East Wheal Friendship) mine, Peter Tavy, within the metamorphic aureole of the Dartmoor granite. It occurs as inclusions in cobaltite and is associated with arsenopyrite, cassiterite, cosalite, erythrite, gersdorffite, parkerite, sphalerite, stannite and ullmannite, and a suite of bismuth-rich minerals, including aikinite, bismite, bismuthinite, ikunolite and an unnamed bismuth telluride (Rumsey & Savage 2005).

Elsewhere in Cornwall, native bismuth is noted: with bismuthinite, cassiterite, fluorite, tourmaline and wolframite at Balleswidden mine, near St Just (Hosking 1951); with bismuthinite at Castle-an-Dinas, St Columb Major (Hey et al. 1938); with bismuthinite, fluorite and safflorite (arsenical cobalt) at Dolcoath mine, Camborne (Greg & Lettsom 1858, Rudler 1905, Hosking 1951); with fluorite at Great Dowgas mine, near Sticker, St Austell (Collins 1892, Hosking 1951); very fine and pure lamellae at Levant mine, St Just (Collins 1892); in an arsenopyrite–chalcopyrite–pyrite-dominated lode assemblage at Mount Wellington mine, Twelveheads, 7 km W of Truro (Kettaneh & Badham 1978); in an anatase–cassiterite–chlorite–ilmenite–monazite–sulphide assemblage at New Cook's Kitchen mine (North Tincroft lode), Redruth (LeBoutillier et al. 2000); at Penlee quarry, Newlyn (Robson 1949b); in a greisen assemblage associated with a sheeted vein system at Redmoor–Kit Hill, N of Callington (Newall & Newall 1989); as a minor species in two iron–manganese lodes at Restormel (Royal) iron mine, N of Lostwithiel (Slater & Highley 1976); in cassiterite–pitchblende-torbernite ore at St Ives Consols, St Ives, where it is said to be lamellar, very fine and pure (Collins 1892, Robson 1945b); at Trugo mine, near St Columb (Greg & Lettsom 1858); with bismuthinite and cobalt ores at Wheal Beauchamp (Pengreep), near Redruth (Hosking 1951); and with pitchblende, galena, molybdenite and sphalerite at Wheal Trenwith, St Ives (Dines 1956, Ball et al. 1982).

In Devon, native bismuth occurs with bismuthinite and molybdenite in quartz veins cutting a basic dyke in the British Railways quarry at Meldon, near Okehampton (Dearman 1965). At Loddiswell mine, 3 km N of Loddiswell, quartz–carbonate veins cutting Devonian slates contain minor native bismuth as inclusions in bournonite (Stanley et al. 1990b). Native bismuth is also noted from: Ivy Tor mine, Belstone, near Okehampton (Braithwaite et al. 1996); Kingswood mine, near Buckfastleigh (Russell 1922); Wheal Friendship (Devon Friendship), Mary Tavy (Braithwaite 1982b); and in a large vein at Atlas mine, W of Ilsington, near Bovey Tracey (Robson 1949b).

Native bismuth, bismuthinite and tetradymite appear to be contemporaneous with wolframite in polymetallic quartz veins associated with the Skiddaw granite at

Massive native bismuth from St Ives Consols, St Ives, Cornwall; width 45 mm. (© Royal Cornwall Museum, Royal Institution of Cornwall; photographer D. I. Green)

Carrock mine, Grainsgill, Caldbeck Fells, Cumbria (Finlayson 1910, Hitchen 1934, Eastwood 1959, Ball et al. 1985a, Cooper & Stanley 1990). Native bismuth is one of the interesting species found in chalcopyrite-bearing quartz veins mined on Coniston Fells, Cumbria; other species being arsenopyrite, bismuthinite, cobaltite, galena, magnetite, marcasite, pyrite, pyrrhotite, safflorite, skutterudite (smaltite), sphalerite and wittichenite (Ineson & Mitchell 1974a, Dagger 1977, Green & Briscoe 2002). The Bonser vein, to the N of Coniston village, was the most productive copper vein in the Lake District and here native bismuth formed inclusions in pyrite or interstitial patches associated with bismuthinite and laitakarite. It is particularly abundant in material derived from deep levels in magnetite-rich portions of the vein (Stanley & Criddle 1979, Stanley & Vaughan 1982b). At Dale Head mine, Newlands Valley, Cumbria, anhedral grains of native bismuth and bismuthinite are found in polymetallic, but pyrrhotite-rich, quartz veins (Stanley & Vaughan 1980). Interstitial native bismuth, bismuthinite and chalcopyrite occur in magnetite-rich veinstone at Long Crag mine, Greenburn Valley, Cumbria (Stanley 1979). Native bismuth occurs as inclusions in arsenopyrite and quartz at Scar Crag cobalt mine on Causey Pike, Cumbria (Ixer et al. 1979a), and is also noted as grains with a well developed cleavage from the Shap granite quarry, Cumbria (Wirth & Green 2002). Native bismuth occurs in small pockets with arsenopyrite, bismuthinite and molybdenite in quartz–topaz greisen associated with the Eskdale granite at Water Crag, near Devoke Water (Young 1985d,e).

In the Northern Pennine orefield, native bismuth is found in high-temperature bismuthinite-bearing chalcopyrite–quartz veins. Ixer et al. (1996) were able to show that this mineralization is restricted to the fluorite zone above or close to the Rookhope cupola of the Weardale granite.

Native bismuth is listed with many other species from the Dolgellau goldbelt, Gwynedd, Wales, by Andrew (1910). It has since been noted as microscopic inclusions in galena in Ag–Au–Bi–Sb–Te bonanza gold assemblages at Clogau mine, Bontddu, near Dolgellau, and Panorama mine, Llanaber, E of Barmouth (Naden 1988, Mason et al. 2002). Minor amounts of native bismuth forming laths 5–10 µm across are enclosed in aggregates of bismuthinite and bismuth sulphosalts within chalcopyrite–galena–tetrahedrite ore from the copper deposit at Parys Mountain, Anglesey (Pointon & Ixer 1980).

At Loch Shin, 2 km W of Lairg, Highland Region, Scotland, quartz veins cutting a monzogranite pluton contain native bismuth and covellite at the margins of chalcopyrite crystals. Other species present include: fluorite, galena, pyrite, sphalerite and a suite of bismuth sulphosalts – aikinite, berryite, gladite, hammarite, krupkaite, lindströmite, pekoite and tetradymite (Lowry et al. 1994). Uranium-rich quartz veins in the vicinity of Needle's Eye, SW of Caulkerbush, Dumfries & Galloway (within the metamorphic aureole of the Criffell granite complex), contain disseminated native bismuth, except in one fissure vein, where a 60 cm-wide lens of granular crystalline native bismuth has been found. Alteration of this lens has produced bismutite, eulytite, mixite and walpurgite (Miller & Taylor 1966, Basham et al. 1989, Braithwaite & Knight 1990, Green & Todd 2001). Native bismuth, bismuthinite, chalcopyrite and native gold inclusions occur within arsenopyrite and pyrite in quartz–carbonate veins associated with the Fore Burn dioritic complex, Tairlow plantation, near Straiton, Strathclyde (Charley et al. 1989).

Other Scottish occurrences of native bismuth include: a reticulated variety associated with erythrite at Alva, Ochil Hills, Central Region (Greg & Lettsom 1858) – requires confirmation; as inclusions in galena–matildite intergrowths within a quartz vein at the contact between hornfels and the Cairnsmore of Fleet granite in Palnure Burn, Talnotry, 5 km NE of Newton Stewart, Dumfries & Galloway (Lowry 1993, Lowry et al. 1997); irregular masses up to 2 mm across intergrown with bismuthinite and chalcopyrite in greywacke-hosted quartz veins at Cairngarroch Bay, Dumfries & Galloway (Lowry et al. 1997); as clusters of irregular grains, up to 20 µm across, associated with froodite and native gold in a borehole cutting gabbro at Littlemill farm, near Huntly, Grampian Region (McKervey et al. 2007); as euhedral crystals included in chrome spinel from chromitite belonging to the Girvan–Ballantrae ophiolite complex, at Pinbain Bridge, Strathclyde (Power & Pirrie 2004); in fractures within galena at Corrie Buie, Meall nan Oighcreag, Loch Tay, Tayside Region (Pattrick 1984, Plant et al. 1989); and in fractures within pyrite found in disseminated and irregular vein mineralization in the Tomnadashan diorite intrusion, Loch Tay, Tayside Region (Pattrick 1984, Plant et al. 1989).

The few reports of native bismuth in Ireland are mostly from localities in Co. Wicklow. Traces of native bismuth have been found with a suite of sulphide species in fluoritized microadamellite, in the Ballingen district, between Tinahely and Aughrim. This greisen-like mineralization is dominated by scheelite and it occurs in the Tullow unit of the Leinster batholith (Steiger & Bowden 1982, 1986, McArdle & Kennan 1988). Ordovician volcanogenic chalcopyrite–pyrite deposits at Kilmacoo (northeastern end of the Avoco district) contain mineralization in which native gold occurs along fractures in pyrite and is typically intergrown with native bismuth, chalcopyrite, galena and kobellite (Milner & McArdle 1992). Another native bismuth–gold association in the Avoca district is at Ballycoog, Goldmine River valley, near Avoca, where mineralization is associated with a metamorphosed BIF and where other bismuth-rich species, aikinite, hammarite and krupkaite are also found (Ixer et al. 1990). Outside Co. Wicklow, Kinnaird et al. (2002) report native bismuth from copper–polymetallic quartz veins hosted by Devonian redbeds at Dhurode, Co. Cork.

bismuth carbonate

A synonym of bismutite.

bismuth glance, bismuthine

Synonyms of bismuthinite.

bismuthinite

orthorhombic Bi_2S_3

A relatively uncommon bismuth sulphide found in low- to high-temperature hydrothermal vein deposits, in tourmaline-bearing copper deposits in granite, and in some high-temperature gold veins.

The Williams lode at Fowey Consols mine, Tywardreath, contained druses in which bismuthinite and native bismuth were found associated with chalcopyrite, siderite and quartz. Museum-quality specimens of bismuthinite, forming long prismatic grey metallic crystals, were recovered from this mine (Hosking 1951, Lewis 1997). In February 1985, probably the most beautiful bismuthinite ever to come out of Cornwall was found on the 260-fathom level of Robinson's section in South Crofty mine, Camborne. Crystals from 2 cm to 6 cm long were recovered from cavities in quartz, occasionally accompanied by green fluorite. Superb specimens were also recovered in the 1970s (Wolters & Wolters 1987, Wolfe 1998). More recently, rare bismuthinite, forming needle-like crystals or blades coating joint surfaces, has been reported from a few other locations in South Crofty mine (LeBoutillier 1996). Dull lath-like crystals and acicular sprays of bismuthinite have been found at Botallack mine and Wheal Cock, St Just, Cornwall, but these are considered to be very inferior to the classic Fowey Consols or South Crofty specimens (Wolloxall 1989, Bancroft & Weller 1993). At Rosewarne and Herland mine (Wheal Herland), Gwinear, cross-course quartz veins intersecting tin lodes were found to contain bismuthinite associated with acanthite, arsenopyrite, native bismuth, silver-rich galena, pyrargyrite, skutterudite and stibnite (Hosking 1951, Ball et al. 1982, Rust 1982a). Grey acicular crystals of bismuthinite, up to 2 mm long, with ankerite, chalcopyrite, clinochlore and siderite have been found in quartz veinstone from Penberthy Croft mine, St Hilary (Betterton 1998, Betterton et al. 2000). Native bismuth has also been found in fluorite–quartz veinstone on the dumps of North Devon United (East Wheal Friendship) mine, Peter Tavy, within the metamorphic aureole of the Dartmoor granite. It occurs as inclusions in cobaltite and is associated with arsenopyrite, cassiterite, cosalite, erythrite, gersdorffite, parkerite, sphalerite, stannite, ullmannite and a suite of bismuth minerals including aikinite, bismite, native bismuth, ikunolite and an unnamed bismuth telluride (Rumsey & Savage 2005).

Elsewhere in Cornwall, bismuthinite occurs: with native bismuth, cassiterite, fluorite, tourmaline and wolframite at Balleswidden mine, near St Just (Hosking 1951); with native bismuth at Castle-an-Dinas, St Columb Major (Hey et al. 1938); as fine 1 mm-long acicular crystals associated with bismoclite and native bismuth at Croft Gothal mine, St Hilary (Rust 1996b); with native bismuth, fluorite and safflorite (arsenical cobalt) at Dolcoath mine, Camborne (Greg & Lettsom 1858, Rudler 1905, Hosking 1951); as an uncommon alluvial species in sediments cored from the Fowey Estuary (Pirrie et al. 2002); with childrenite at George and Charlotte mine, near Callington (Greg & Lettsom 1858); forming shiny grey needles with arsenopyrite, chalcopyrite, digenite and tennantite at Gunheath china-clay pit, Carthew, St Austell (Weiss 1994); at Lanescot mine, St Austell (Greg & Lettsom 1858); at Levant mine, St Just (Golley & Williams 1995); as small needles in quartz vugs within a tin-rich lode at Penhalls mine, St Agnes (Le Neve Foster 1877); at Penlee quarry, Newlyn (Golley & Williams 1995); in a greisen assemblage associated with a sheeted vein system at Redmoor–Kit Hill, N of Callington (Newall & Newall 1989); as a minor species in two iron–manganese lodes at Restormel (Royal) iron mine, N of Lostwithiel (Slater & Highley 1976); at Trugo mine, near St Columb (Golley & Williams 1995); with native bismuth and cobalt ores at Pengreep, near Gwennap (Hosking 1951, Embrey & Symes 1987), and Wheel Beauchamp, near Redruth (Hosking 1951), with arsenopyrite and native bismuth in a greisen assemblage at Wheal Jane, Kea, near Truro (Holl & Bromley 1988); with native bismuth, skutterudite (smaltite) and erythrite at Wheal Sparnon, near Redruth (Greg & Lettsom

Fibrous masses of bismuthinite in vein quartz; Carrock mine, Caldbeck Fells, Cumbria; width 14 mm. (© D. I. Green)

1858, Collins 1892, Hosking 1951); dull metallic elongated crystals up to 6 mm long in vugs of soft altered greenstone in the Wheal Owles sett, St Just (Wolloxall 1988).

In Devon, minute acicular crystals of bismuthinite are noted from Ivy Tor mine, Belstone, near Okehampton (Hall 1868b, Braithwaite et al. 1996), and fine crystals have been found at Devon and Cornwall United mines, near Tavistock (Hall 1868b, Collins 1892). Dearman (1965) notes thin fibres of bismuthinite with native bismuth and molybdenite in quartz veins cutting a basic dyke in the British Railways quarry at Meldon, near Okehampton (Dearman 1965).

In Cumbria, bismuthinite, native bismuth and tetradymite appear to be contemporaneous with wolframite in polymetallic quartz veins associated with the Skiddaw granite at Carrock mine, Grainsgill, Caldbeck Fells (Finlayson 1910, Hitchen 1934, Davidson & Thomson 1951, Ball et al. 1985a, Cooper & Stanley 1990). Bismuthinite is one of the primary ore species found in chalcopyrite-bearing quartz veins mined on Coniston Fells, other species being arsenopyrite, native bismuth, galena, magnetite, marcasite, pyrite, pyrrhotite and sphalerite (Ineson & Mitchell 1974a, Dagger 1977, Green & Briscoe 2002). The Bonser vein (to the N of Coniston village) was the most productive copper vein in the Lake District, and here bismuthinite forms inclusions in pyrite and it occurs along fractures in quartz. It is associated with native bismuth, joséite and laitakarite (Stanley & Vaughan 1982b). Stanley & Criddle (1979) report bismuthinite and native bismuth inclusions in chalcopyrite from both the Bonser and Paddy End veins. At Dale Head mine, Newlands Valley, anhedral grains of bismuthinite are found in polymetallic, but pyrrhotite-rich, quartz veins, where they form aggregates up to 200 μm across, but individual grains are rarely > 50 μm (Stanley & Vaughan 1980).

Elsewhere in Cumbria, bismuthinite occurs: as interstitial grains with native bismuth and chalcopyrite in magnetite-rich veinstone at Long Crag mine, Greenburn Valley (Stanley 1979); as blades up to 4 mm in length with chalcopyrite in quartz from Long Garth, Ulpha (Young 1987a); with arsenopyrite and native bismuth in an apatite–chlorite–quartz vein at Scar Crag cobalt mine on Causey Pike (Ixer et al. 1979a); as long, fibrous or foliated, blade-like aggregates, up to 4 cm in length, associated with quartz and other sulphides on joint faces in granite at Shap granite quarry (Russell 1925, Wirth & Green 2002); with arsenopyrite, fluorite and molybdenite mineralization associated with the Eskdale granite at Water Crag, near Devoke Water (Young 1985d,e, Young et al. 1986).

In the Cumbrian section of the Northern Pennine orefield, bismuthinite is associated with bismite from the Great Sulphur vein at Crossgill (Bridges & Young 1998) and from Sir John's mine, both about 4 km S of Garrigill (Dunham 1990). These and other occurrences of bismuthinite to the W in Co. Durham are mostly associated with high-temperature chalcopyrite-bearing quartz veins within the fluorite zone above or close to the Rookhope and Tynehead cupolas of the Weardale granite. In total, 11 bismuth mineral localities are listed by Ixer et al. (1996), although bismuthinite is not specifically mentioned at all of them. Bismuthinite is recorded from: Cambokeels (Cammock Eals) mine, Eastgate; Groverake mine, near Rookhope; Redburn mine, near Rookhope; Stotfield Burn mine, near Rookhope; and Whiteheaps mine, near Blanchland. Bismuthinite has since been found as sprays of metallic capillary crystals up to 1 cm in length in chalcopyrite-rich veinstone from Frazer's Hush mine close to Groverake mine (Wolfe 1998, Green & Briscoe 2002).

Aggregates of bismuthinite enclosing native bismuth and bismuth sulphosalts occur within chalcopyrite–galena–tetrahedrite ore from the copper deposit at Parys Mountain, Anglesey, Wales (Pointon & Ixer 1980). A specimen of bismuthinite associated with chalcopyrite, pyrite and quartz from Rhôs-mynach-fawr on Anglesey is in the collection of the NMW (Bevins 1994). Bismuthinite is listed with many other species from the Dolgellau goldbelt by Andrew (1910). It has since been noted as microscopic inclusions (50 µm) in galena in Ag–Au–Bi–Sb–Te bonanza gold assemblages at Clogau mine, Bontddu, near Dolgellau, and Ffridd-Goch mine, Llanfachreth (Naden 1988, Mason et al. 2002).

At Loch Shin, 2 km W of Lairg, Highland Region, Scotland, quartz veins cutting a monzogranite pluton contain members of the aikinite–bismuthinite series (aikinite, gladite, hammarite, krupkaite, lindströmite, pekoite), other bismuth-bearing species (berryite, native bismuth and tetradymite), and chalcopyrite, covellite, fluorite, galena, pyrite and sphalerite (Lowry et al. 1994). Prior to this, nearby bismuth mineralization had been noted around the Grudie granite at Allt Sron nan Iarnachan and in Moly Burn, but identification of the species was not attempted (Gallagher et al. 1974, Gallagher & Smith 1975).

Bismuthinite, native bismuth, chalcopyrite and native gold inclusions occur within arsenopyrite and pyrite in quartz–carbonate veins associated with the Fore Burn dioritic complex, Tairlow plantation, near Straiton, Strathclyde (Charley et al. 1989). Other Scottish occurrences of bismuthinite include: minor bismuthinite in an arsenic-rich magmatic sulphide assemblage hosted by a diorite intrusion at, Talnotry, 5 km NE of Newton Stewart, Dumfries & Galloway (Power et al. 2004); with native bismuth and chalcopyrite in greywacke-hosted quartz veins at Cairngarroch Bay, Dumfries & Galloway (Lowry et al. 1997); as blebs in pyrite associated with copper mineralization in granitic lenses intruded into a diorite pluton at Tomnadashan, Loch Tay, Tayside Region (Pattrick 1984); and with molybdenite and pyrite in quartz veins cutting granite on the northeast slopes of Lotus Hill, Beeswing, Dumfries & Galloway (Williams 1973).

In Ireland, bismuthinite appears to have been found only in quartz veins hosted by Old Red Sandstone rocks at Dhurode, West Carbery mining district, Co. Cork, where it occurs as fine intergrowths with chalcopyrite, replacing emplectite and wittichenite. It may be supergene in origin (Ni Wen et al. 1999).

bismuth ochre

The group name for undetermined oxides and carbonates of bismuth, and especially applied to bismite.

bismuth sulfuré

A synonym of bismuthinite.

bismutite

tetragonal $(BiO)_2CO_3$

A rare bismuth carbonate, found in somewhat impure masses in considerable quantity at Wheal Owles, St Just, Cornwall, where it was called agnesite (Greg & Lettsom 1858). Collins (1892, 1911a) also notes its occurrence at Wheal Coates, St Agnes, and this was confirmed by Clark et al. (1986) on Collins's specimens now in the Royal Cornwall Museum, Truro. "Bismuth ochre" from Restormel iron mine, N of Lostwithiel (Greg & Lettsom 1858), was also shown to be bismutite by Clark et al. (1986). More recently, bismutite is recorded from a single specimen from the dumps of Penberthy Croft mine, St Hilary, where it forms tan rounded tabular crystal groups up to 0.75 mm long. It is a supergene species associated with brochantite, native copper, cuprite and goethite (Rust 1995). Bismuthinite has since been found in quartz veinstone at Penberthy Croft (Betterton 1998) and this may have been the primary phase from which bismutite developed. At Wheal Remfry china-clay pit, Indian Queens, near St Austell, bismuth minerals were found in a triangular topaz-rich pod. Bismutite occurs in four habits: as powdery to resinous, creamy to yellowish-brown, masses up to 10 mm across, as pale-yellow botryoidal crusts, as cream-to-brown groups of stacked cone-shape crystals up to 0.5 mm long, and as yellow-to-tan pseudomorphs up to 2 mm long after an unknown tabular mineral. Waylandite is an associated species (Rust 1998). Bismutite is also reported from South Crofty mine, Camborne (Anthony et al. 2003).

Weathering of primary bismuth minerals in molybdenum–tungsten-rich veins at Carrock mine, Grainsgill, Caldbeck Fells, Cumbria, also produced bismutite (Hitchen 1934). It is further noted from the main dump of Carrock mine by Hartley (1984). Bismutite occurs as a supergene species in a narrow chalcopyrite–ferberite–quartz–scheelite vein cutting the Eskdale granodiorite at Buckbarrow Beck, Corney Fell. Other associated supergene species are bismutoferrite, brochantite, chrysocolla, covellite, cuprotungstite, eulytite, goethite, malachite, mixite, namibite and russellite (Young et al. 1986, 1991b, Neall et al. 1993). Pearly scales on corroded bismuthinite from joints within granite at the Shap granite quarry are tentatively identified as bismutite (Wirth & Green 2002).

In Scotland, small (0.2 mm) brown spherules of bismutite occur in a supergene assemblage associated with uranium-bearing veins on the Solway coast in the Needle's Eye area, SW of Caulkerbush, Dumfries & Galloway. Other species include annabergite, boltwoodite, connellite, cornubite, erythrite, eulytite, kahlerite, lavendulan, millerite, mixite, nováčekite, pharmacosiderite, schoepite, tyrolite, uranophane, vandendriesscheite, walpurgite, wulfenite and zeunerite. Most form tiny crystals or encrustations that are of particular interest to micromounters (Braithwaite & Knight 1990, Green & Todd 2001). Bismutite is also reported in alluvial heavy-mineral concentrates from Longwell Water draining the Helmsdale granite, Highland Region (Gallagher et al. 1971), and from the streams in the Coire Dubh area that drain the Etive granite complex to the E and NE of Ben Starav, on the Strathclyde/Highland Region boundary (Haslam et al. 1985).

bismutoferrite

monoclinic $BiFe^{3+}_2Si_2O_8(OH)$

A locally abundant bismuth–iron silicate hydroxide found in a narrow chalcopyrite–ferberite–quartz–scheelite vein cutting the Eskdale granodiorite at Buckbarrow Beck, Corney Fell, Cumbria. Bismutoferrite is one of several supergene minerals developed at this site, others being brochantite, chrysocolla, covellite, cuprotungstite, goethite, malachite and russellite (Young et al. 1986, 1991b). Later investigations added bismutite, eulytite, namibite and mixite to the list of species found in this supergene assemblage (Neall et al. 1993).

Bismutoferrite reported from South Terras mine, St Stephen-in-Brannel, Cornwall (Kingsbury 1964a), is considered doubtful by Ryback et al. (2001), who suggest that it came instead from Schneeberg, Saxony. Another Cornish report of bismutoferrite is from Hingston Down quarry, 2 km W of Gunnislake, Calstock (Golley & Williams 1995).

bitterkalk, bitterspar

Synonyms of dolomite.

bixbyite

cubic $(Mn^{3+},Fe^{3+})_2O_3$

An uncommon manganese–iron oxide identified on hausmannite from the manganese ore deposit at Benallt mine, Rhiw, Lleyn Peninsula, Gwynedd, Wales (Bevins 1994).

black band

A synonym of siderite, calcite group.

black copper

A synonym of tenorite (melaconite).

black hematite

A synonym of psilomelane (now considered to be either birnessite, cryptomelane, hollandite or romanèchite).

black jack

A synonym of sphalerite.

black lead

A synonym of graphite.

black oxide of copper

A synonym of tenorite.

black tin

A synonym of cassiterite.

blaubleibender covellite

A mixture of spionkopite and yarrowite.

bleicarbonat

A synonym of cerussite.

bleiglanz

A synonym of galena.

bleihornerz
A synonym of phosgenite.

bleilazur
A synonym of linarite.

bleinierite
A synonym of bindheimite.

bleisulfat, bleivitriol
A synonym of anglesite.

blende
A synonym of sphalerite.

blister copper
A synonym of chalcopyrite.

blixite
orthorhombic $Pb_2ClO_2(OH)$

A rare lead oxychloride whose crystal structure has been redefined by Krivovichev & Burns (2006). Blixite occurs in iron–manganese-mineralized calcite veins cutting Lower Carboniferous dolomitic limestone at Merehead (Torr Works) quarry, Cranmore, Somerset. It forms sugary to coarsely crystalline pale-yellow to reddish-orange masses in hydrocerussite-lined cavities within iron-rich manganese pods. Brucite, chloroxiphite, crednerite, diaboléite or mendipite occur in association with blixite (Alabaster 1975, 1977). Embrey (1978) casts doubt about the identification, although later investigations seem to confirm blixite and reveal other species in the veins at this locality: cerussite, hematite, paralaurionite, pyrolusite, and three oxychlorides – mereheadite, parkinsonite and symesite (Welsh et al. 1998, 2000, Turner 2007). The lead oxychlorides are thought to have formed by interaction of primary galena with low-temperature metalliferous brines along fractures cutting the original Cu–Fe–Mn–Pb orebody (Symes & Embrey 1977, Humphreys et al. 1980), although it is possible that they precipitated directly from a cooling brine (Welch et al. 2000). See also mereheadite and Turner (2007).

bloodstone
A synonym of heliotrope and of hematite; a green cryptocrystalline variety of quartz containing blood-red spots of jasper (heliotrope).

blue calamine
A synonym of aurichalcite.

blue carbonate of copper
A synonym of azurite.

blue iron earth
An earthy variety of vivianite.

Blue John
TL Castleton, Derbyshire, England

A purple and white, banded variety of fluorite.

blue lead
A synonym of galena.

blue malachite
A synonym of azurite.

blue stone
A synonym of kilmacooite (galena + sphalerite) and chalcanthite. "Blue stone" was also used by Welsh miners to describe a copper–zinc sulphide.

blue vitriol
A synonym of chalcanthite.

Platy aggregates of bobkingite, 0.5 mm across, on cuprite and malachite; New Cliffe Hill quarry, Stanton under Bardon, Leicestershire. (© D. I. Green)

bobkingite
monoclinic $Cu_2^{5+}Pb_2Cl_2(OH)_8(H_2O)_2$

TL New Cliffe Hill quarry, Stanton under Bardon, Leicestershire, England

Named after Robert King (b. 1923), mineralogist (now retired) at the Department of Geology, University of Leicester, and a founding member of the Russell Society; born in Leicester. Bobkingite is a pale-blue hydrated copper oxychloride that forms very thin (≤5 µm) transparent plates up to 0.2 mm across, perched on a compact fibrous crust of malachite and crystallized azurite, attached to massive cuprite. Bobkingite is very rare and only four specimens are known to exist. It was found in a native copper–cuprite deposit close to the Precambrian/Triassic unconformity (Hawthorne et al. 2002, Hubbard et al. 2005).

boehmite
A synonym of böhmite.

bogoslovskite
A synonym of kupferblau (azurite + chrysocolla).

bohdanowiczite
hexagonal $AgBiSe_2$

A very rare silver–bismuth selenide found at Glengap Burn, SW of Moffat, Dumfries & Galloway, Scotland. Bohdanowiczite forms inclusions in two alluvial gold grains thought to be derived from the Annan Permian redbed basin (Chapman et al. 2000c).

böhmite
orthorhombic $AlO(OH)$

An uncommon aluminium hydroxide, dimorphous with diaspore and recognized in microscopic sections of pisolitic bauxitic clays from Saltcoats beach and Dubbs quarry, Kilwinning, Strathclyde, Scotland. Böhmite crystallizes in the drusy centres of pisoliths and forms very small (up to 0.06 mm long) crystals, closely associated with kaolinite and occasionally with diaspore. The clays containing böhmite are derived from the weathering (under a tropical climate) of basalt (de Lapparent 1934).

In Co. Antrim, Ireland, böhmite has been described in bauxitic clays from Ballylig, near Broughshane; Elbinny mine, near Broughshane, Lyles Hill mine, near Templepatrick and Skerry mine, near Newtown Crommelin. All the clays were collected close to basaltic dykes, and the development of böhmite is attributed to the dehydration of gibbsite-rich bauxite by the baking action resulting from the intrusion of these dykes (Eyles 1952).

bole
Impure halloysite.

Interpenetrant cubes of boléite from Gunver Head, near Padstow, Cornwall; width 130 μm; SEM image. (© N. J. Elton)

boléite/boleite

cubic $KPb_{26}Ag_9Cu_{24}Cl_{62}(OH)_{48}$

A potassium–lead–silver–copper chloride hydroxide briefly described from Cornwall with cerussite, cumengéite, malachite, pseudoboléite and phosgenite from Tolcarne beach, Newquay, and Newporth beach, near Falmouth (Ryback & Tandy 1992). Golley & Williams (1995) add two other Cornish occurrences: Padstow Consols and Porthilley mine, St Minver.

boltwoodite

monoclinic $KUO_2(SiO_3OH)\cdot 1.5H_2O$

A rare hydrated potassium uranyl silicate hydroxide found in a mineralized vein within the aureole of the Land's End granite at Loe Warren Zawn, W of Botallack, St Just, Cornwall. Boltwoodite forms radiating and spherulitic crusts of bright greenish-yellow sharply terminated acicular crystals, which may reach 0.05 mm in length. It is associated with chalcocite and uraninite, but is most common around a uraninite nodule. Rutherfordine and vandendriesscheite are two other supergene species with which boltwoodite is often found (Elton & Hooper 1995). In Scotland, boltwoodite has been found in uranium-rich dolomite–quartz veins within the aureole of the Criffell granodiorite in the Needle's Eye area, SW of Caulkerbush, Dumfries & Galloway. Boltwoodite occurs as a supergene alteration product of pitchblende and is found with annabergite, atacamite, bismoclite, bornite, connellite, covellite, erythrite, mixite, uranophane and zeunerite. The primary mineralization consists mainly of native bismuth, chalcopyrite, hematite, pyrite and uraninite (Miller & Taylor 1966).

Acicular sprays of boltwoodite crystals from Loe Warren Zawn, St Just, Cornwall; width 300 μm; SEM image. (© N. J. Elton)

bonamite

A trade name for green smithsonite, calcite group.

boracite

orthorhombic (pseudo-cubic) $Mg_3B_7O_{13}Cl$

A rare magnesium borate with chloride, dimorphous with trembathite that forms a series with ericaite and is found at Boulby potash mine, Loftus, Cleveland, where it occurs in nodules within anhydritic shale overlying halite–sylvite rock of the Permian third Zechstein evaporite cycle. Boracite forms beautiful groups of aquamarine-blue euhedral crystals or spherulitic clusters, some with white compact centres; it may also be pale blue or green. Crystals commonly appear cubic, but they are actually orthorhombic polysynthetic twins. Hilgardite and magnesite are associated species. Matrix specimens of boracite to 20 cm were discovered in 1991 and are probably the finest in the world (Green & Freier 1994, 1996, Green 2005b). The iron-boracite described from this locality (Milne et al. 1977) is actually congolite.

Boracite has also been described from the Aislaby (Eskdale No. 2) borehole, Eskdale, near Whitby, Yorkshire, where it is found in Permian evaporites in association with halite. It forms colourless to greyish-white translucent corroded crystals with a vitreous lustre and pseudo-cubic form. Groups of crystals range up to 10 mm and individual crystals showing complex sector and polysynthetic twinning reach 1 mm (Guppy 1944, 1956, Stewart 1951a, 1965).

Aquamarine-blue boracite with white magnesite platlets; Boulby potash mine, Cleveland; width of specimen 18 mm. (© D. I. Green)

bornite

orthorhombic (pseudo-tetragonal) Cu_5FeS_4

A common copper–iron sulphide found in mafic igneous rocks, contact-metamorphic skarn deposits and intermediate- to high-temperature hydrothermal deposits.

England

Avon Close to the beach near the Pier hotel at Clevedon, near Bristol, barite–sulphide mineralization occurs in metre-wide fault-filling brecciated veins within the Triassic Dolomitic Conglomerate. The earliest mineralization consists of discrete disseminated crystals, up to 80 μm across, of bornite, chalcopyrite, sphalerite and tennantite, with bornite sometimes containing spindle-shape inclusions of chalcopyrite (Ixer et al. 1993).

Cheshire Common bornite occurs as an alteration product of chalcopyrite in Triassic redbed-type Cu–Pb–Zn mineralization at Engine Vein mine, Alderley Edge ridge (Ixer & Vaughan 1982). It also occurs in the nearby Stormy Point area (Carlon 1979, Braithwaite 1994). At Bickerton, also in the Cheshire Basin, bornite is reported from barite-dominated redbed-type mineralization closely associated with marginal basin faults. The paragenetic sequence is: early pyrite (bravoite and nickeloan pyrite) occasionally with Cu–Fe–Ni sulpharsenide rims, followed by bornite and tennantite (the most abundant sulphides), in which chalcopyrite is found as lamellae. Covellite is a secondary replacement species (Naylor et al. 1989).

Cornwall At Gryll's Bunny and the Crowns, Botallack, St Just, skarns formed by replacement of various rocks in the aureole of the Land's End granite contain accessory bornite and chalcopyrite associated with amphibole, cassiterite, epidote, garnet, malayaite and titanite (van Marcke de Lummen 1985). At Geevor mine, near Pendeen, St Just, bornite is intimately associated with chalcopyrite and uraninite (pitchblende) in a phase of mineralization that follows arsenopyrite crystallization (Dines 1956, Ball et al. 1982). Exceptional bornite, often associated with chalcocite, is noted from the nearby Levant mine (Robson 1949b, Wolloxall 1987, Bancroft & Weller 1993). Fine crystals of bornite have also been reported from Carn Brea, Cook's Kitchen and Tincroft mines, all around Redruth (Greg & Lettsom 1858, Rudler 1905). At Gunheath china-clay pit, Carthew, St Austell, bornite occurs in pegmatitic vugs in the roof zone of the Hensbarrow stock, an evolved pluton in the St Austell granite complex. It occurs with apatite, arsenopyrite, cassiterite, chalcopyrite, columbite, mawsonite, muscovite, potassium feldspar, quartz, stannite, stannoidite, tourmaline, wolframite and zinnwaldite (Manning 1983, Alderton 1991). Other reports include: Cligga Head, Perranzabuloe; Polmear mine, St Austell; Craddock Moor, WSW of Caradon Hill; Treveddoe mine, NW of Warleggon (all Dines 1956); Dolcoath mine, Camborne; Wheal Jewel, St Day; Wheal Falmouth (all Greg & Lettsom 1858); Tresavean mine and Pennance Consols mine, both Gwennap (Golley & Williams 1995); East Wheal Lovell (Fatwork section), Wendron and St Ives Consolidated mines, St Ives (Dominy et al. 1996a); New Cook's Kitchen mine (North Tincroft lode), Redruth (LeBoutillier et al. 2000); South Caradon mine, St Cleer, near Liskeard (Dines 1956, Shambrook 1982); South Crofty mine, Camborne (LeBoutillier 1996); South Terras mine, St Stephen-in-Brannel; Wheal Bray, Altarnun; Wheal Trenwith, St Ives (all Ball et al. 1982); Wheal Fanny, Camborne (Penhallurick 1997); and Wheal Gorland, St Day, 2 km E of Redruth (Rust 1981b). The report of bornite from Wheal Mary, Lanivet, near Bodmin (Kingsbury 1961), requires further confirmation.

Iridescent bornite crystal group, 6 mm across, from Cook's Kitchen mine, near Redruth, Cornwall. (© D. I. Green)

Cumbria A classic hydrothermal brecciated sulphide-rich quartz–chlorite vein at Seathwaite mine, Seathwaite Tarn, Coniston, contains early-formed arsenopyrite and cobaltite in an assemblage with a paragenetic sequence of pyrite, digenite, djurleite, wittichenite, gold, bornite, chalcopyrite, spionkopite and yarrowite (these two jointly described as blaubleibender covellite), and covellite. One 20 μm-diameter inclusion of gold in bornite was noted (Stanley & Criddle 1979). Other reports include: bornite cementing fractured quartz at Birk Fell mine, Tilberthwaite, and at the Black Scar workings, Levers Water, Coniston; bornite replaced by goethite and malachite at Dale Head mine, Newlands Valley; bornite–chalcopyrite intergrowths replacing sphalerite at Pave York mine, Greenburn Valley (all Stanley 1979); bornite replaces chalcopyrite at Driggith mine (Thimmaiah 1956) and Sandbed mine (Green et al. 2006a) in the Caldbeck Fells, also noted by Eastwood (1921) at Driggith–Sandbed; bornite is associated with arsenopyrite, chalcocite, chalcopyrite and pyrite at Carrock mine, Caldbeck Fells (Davidson & Thomson 1951); rare bornite, chalcopyrite and cubanite in hematite–pyrite-bearing quartz veins in Gatesgarthdale Beck, W of the summit of Honister Pass, Wasdale (Millward et al. 1999); bornite, chalcopyrite and malachite in old workings on High Pike, near Driggith mine, Caldbeck Fells (Eastwood et al. 1968); massive bornite intergrown with chalcopyrite at Paddy End mine, Coniston, Eskdale; and massive at Roughton Gill mine, Caldbeck Fells and Tilberthwaite Gill (Young 1987a). The report of bornite from Potts Gill barite mine, Caldbeck Fells (Hartley 1984), requires further confirmation.

Derbyshire Bornite is associated with nickel-rich pyrite (bravoite) at Odin mine, Castleton (Ixer 1978), and is further noted with native copper as inclusions in calcite at Riber mine, Matlock (Varvill 1959).

Devon Fine massive specimens of bornite were once found at Britannia and Prince Regent mines, near North Molton (Hall 1868b). Lens-like hematite–quartz veins cutting Devonian shales at Bampfylde mine, North Molton, carry copper ores, chiefly chalcopyrite below the water table and, above it, azurite, bornite, chalcocite and malachite (Dines 1956, Dunham et al. 1978). Concretions in Permian redbeds at Littleham Cove, near Budleigh Salterton, contain minor bornite among a suite of arsenic, cobalt, copper, nickel, silver and uranium minerals (Harrison 1975b, Milodowski et al. 2000). Disseminated bornite occurs with arsenopyrite, chalcopyrite, löllingite and sphalerite in tin-bearing skarns (metasomatized chert and limestone) in the aureole of the Dartmoor granite, near Meldon (El Sharkawi & Dearman 1966).

Durham Small amounts of bornite have been found at Redburn fluorite mine, near Rookhope, Weardale (Day 1979).

Leicestershire Spatially zoned copper-rich quartz–carbonate veins cutting granophyric diorite and volcanic tuffs at Newhurst quarry, near Shepshed, contain bornite–chalcocite intergrowths near the top, chalcopyrite at intermediate levels, and galena at the base. Cavities in the oxidized part of the bornite–chalcocite veins contain rosettes of vésigniéite (King & Wilson 1976). Other bornite occurrences at this quarry are summarized by King (1973) and by Ince (2005), who also reports irridescent purple bornite surrounding chalcopyrite in small calcite veins. Bornite also occurs in a native copper and cuprite deposit near the Precambrian/Triassic unconformity at New Cliffe Hill quarry, Stanton under Bardon, where it typically occurs as an alteration rim within and surrounding chalcopyrite (Hubbard et al. 2005). Bornite is associated with chalcopyrite at Bluebell Wood quarry, near Grosby (King 1998b).

Staffordshire Bornite was one of the principal copper ores worked at Ecton Hill (others being chalcocite and chalcopyrite), where mineralization was concentrated in fracture-filled veins within Carboniferous Limestone (Robey & Porter 1972, Critchley 1979).

Warwickshire A zone of weathering within 10 m below the Precambrian/Cambrian unconformity at Judkins quarry, Nuneaton, contains bornite as irregular masses commonly associated with calcite. Bluish-grey pseudomorphs of bornite after chalcocite also occur with barite, chalcopyrite and wooldridgeite (Ince et al. 1991, Hawthorne et al. 1999a, Ixer et al. 2005a).

Yorkshire Nodules of copper sulphides (like small eggs) are reported from bedding planes in Carboniferous Limestone at Black Scar quarry, S of Middleton Tyas, and at Forcett quarry, East Layton, 12 km N of Richmond. They comprise bornite, chalcocite, chalcopyrite, covellite and digenite (Dunham & Wilson 1985). Bornite is also noted from the Middleton Tyas mines (Green 1987b), Merrybent mine, to the NW of Middleton Tyas (Small 1977, Dunham & Wilson 1985), and around the old copper mines in the Forest of Bowland (Rodgers 1975).

Wales

Dyfed A specimen of bornite from near Devil's Bridge in the Central Wales orefield is in the NMW (Bevins 1994).

Gwynedd Bornite occurs with azurite, chalcocite, chalcopyrite (the economic mineral of interest), native copper, covellite, malachite and tetrahedrite in the porphyry-copper deposit at Coed-y-Brenin, near Dolgellau (Rice & Sharp 1976, Armstrong et al. 2003). At Glasdir mine, Llanfachreth, near Dolgellau, arsenopyrite, bornite, chalcopyrite, covellite, native gold, malachite, marcasite and pyrite mineralization is associated with the intrusion of a breccia pipe that formed during a phase of Ordovician volcanism and which also gave rise to the Rhobell Volcanic Group (Andrew

1910, Dewey & Eastwood 1925, Allen & Easterbrook 1978). Elsewhere bornite is reported from: Dinorwic quarry, near Llanberis; Clogau mine, Bontddu; Tal-y-Sarn quarry, near Llanllyfni (all Bevins 1994); as peacock ore at Cwm Cynfal mine, near Ffestiniog and Drws-y-Coed mine, Pen-y-groes, Beddgelert (Dewey & Eastwood 1925); and Vigra lode, Dolgellau goldbelt (Andrew 1910). Some of these identifications of bornite require confirmation, as they may be tarnished chalcopyrite. However, this is not the case at Llechweddhelyg mine, near Clawddmelyn, between Bontgoch and Cwmsymlog, where bornite, chalcocite and covellite are replacement products after chalcopyrite (Bevins 1994). Bornite is also confirmed from the copper deposit at Parys Mountain, Anglesey (Pointon & Ixer 1980, Southwood & Bevins 1995).

Powys Bornite is noted from Dylife mine, N of Hafren Forest in the Central Wales orefield (Bevins 1994) and from Dolyhir quarry, Old Radnor, near Presteigne (Bevins & Mason 1997).

Scotland

Borders A bornite inclusion has been identified in an alluvial gold grain from Glengaber Burn, SW of Peebles (Chapman et al. 2000c).

Central Bornite has been recovered from the dumps of Alva mine, Silver Glen, Ochil Hills, together with chalcocite, covellite, djurleite, malachite, maucherite and rammelsbergite (Jassim et al. 1983, Moreton et al. 1998, Green & Todd 2001). A late phase of uranium mineralization at Tyndrum Main mine contains minor bornite (Pattrick 1985).

Dumfries & Galloway Low-grade bornite–chalcopyrite occurs in quartz veinlets within an altered granodiorite associated with the porphyry-style copper-mineralized Caledonian subvolcanic complex at Black Stockarton Moor, to the NE of Kirkcudbright (Leake & Brown 1979, Fortey 1980, Lowry et al. 1997). Bornite and covellite replace chalcopyrite in copper–uranium-mineralized quartz veins within the aureole of the Criffell granite complex at Needle's Eye, SW of Caulkerbush (Miller & Taylor 1966, Braithwaite & Knight 1990).

Grampian Disseminated sulphide mineralization, possibly derived from hydrothermal brines, in the Water of Tanar Limestone Formation, Ballater area, consists of pyrite and pyrrhotite with minor bornite and chalcopyrite (Goodman et al. 1990).

Highland A very rich vein of bornite, brochantite and malachite in a crush plane within Durness Limestone was once worked at Rassal (Rapal), N of the head of Loch Kishorn (Heddle 1901, Berridge 1969). On the Isle of Rum, bornite is one of the alteration products of chalcopyrite and pentlandite in the Eastern Layered Series of the Rum layered intrusion. The sulphides occur interstitially to chromite (Butcher et al. 1999, Power et al. 2000a, 2003) or form droplets in the upper part of chromitite units (Dunham & Wilkinson 1985). Bornite is also identified as an accessory mineral with chalcocite, diopside, harkerite and magnetite in a boron–fluorine ore skarn in the aureole of the Beinn an Dubhaich granite, near Broadford, Isle of Skye (Tilley 1948b, 1951, Slater & Highley 1976). Other reports include: small amounts of bornite, covellite and possible linnaeite with chalcopyrite and pyrite in a melanite-pyroxenite dyke (ledmorite) near Cnoc Bad na h'Achlaise, Ledmore, Assynt (Stumpfl 1961); minor disseminated bornite, chalcocite, chalcopyrite, malachite and pyrite mineralization in a sulphide horizon between massive amphibolite and marble bands (part of the Precambrian Loch Maree Group) from the Druim na Fearna area at the western end of Dubh Loch, near Gairloch (Jones et al. 1987); and bornite both disseminated and in quartz veins, sampled by a borehole cutting the Ballachulish granite (Haslam & Kimbell 1981).

Lothian A loose boulder in glacial drift from Loganlee in the Pentland Hills contains masses of bornite up to 1 cm across (Livingstone 2002).

Strathclyde A breccia possibly formed by the explosive release of gases from a magma similar to that which produced the Kilmelford igneous complex at Arduaine, near the southern shore of Loch Melfort, consists of fragmented Dalradian material and vein quartz containing bornite, galena, pyrite and sphalerite (Fortey 1980, Zhou 1987). When work on a trachytic plug at Greenhill quarry (near Kilbirnie) broke into the old workings of Swinlees mine in 1997, large masses of bornite and djurleite were found (Wolfe 1997, Green & Todd 2001). Bornite, greenockite, mottramite and vanadinite form inclusions in prehnite within vesicles in Carboniferous basaltic lavas at Loanhead quarry, near Beith (Meikle 1989b). Bornite is also noted from similar lavas in a temporary borrow pit at Bishopton (Meikle 1992b). Other reports include: minor bornite and covellite associated with chalcopyrite in veinlets cutting epidiorite from the Abhainn Srathain mines, S of Meall Mòr, South Knapdale (Smith et al. 1978a); malachite with bornite and chalcopyrite in barite–calcite–quartz veins associated with a dolerite dyke at the Kaim mines, Renfrewshire Hills, near Lochwinnoch (Stephenson & Coats 1983); disseminated bornite with malachite in sandstones at Drumshantie mine, SE of Gourock (Heddle 1901, Stephenson & Coats 1983); rare bornite in mainly molybdenite mineralization within the Etive granite complex to the E and NE of Ben Starav (Haslam et al. 1985); and fissure veins around Kilfinan, near Lochgilphead, contain azurite, bornite, chalcocite, chalcopyrite and malachite (Berridge 1969).

Tayside Rare bornite replaces chalcopyrite in the Foss and Ben Eagach barite–base-metal deposits, near Aberfeldy (Moles 1985). Bornite is associated with chlorite at the Birnam Hill clay–slate quarry, S of Dunkeld (Heddle 1901).

Shetland A chromite-rich dunite at Cliff on the Isle of Unst contains bornite, chalcocite and covellite, together with a range of nickel and PGM species (Prichard et al. 1994, Maynard et al. 1997). A cumulate pyroxenite containing 5–30 μm-diameter bornite–chalcocite grains around a bornite–chalcopyrite–pentlandite aggregate is also documented from Unst (Maynard 1993), although the precise location is not given. Sulphide nodules in the Devonian Upper Melby Fish Bed at Matta Taing, Walls, and Sandness on West Mainland, contain bornite, chalcopyrite, covellite, galena, pyrite, sphalerite and tennantite (Hall & Donovan 1978).

Ireland

Clare A disseminated sulphide deposit in argillaceous limestones and shales at Ballyvergin, Ballyvergin Townland contains minor bornite in a predominantly chalcopyrite–pyrite ore (Schultz 1971, Andrew 1986).

Cork Disseminated and vein-hosted bornite is reported many times in polymetallic ores from the West Carbery mining district: Ballycummisk mine, near Skull; Brow Head and Streek Head, Mizen Head peninsula; Cappagh (Cappaghglass); Castle Point; Coosheen; Crookhaven; Derrycarhoon; Dhurode; Kilkilleen; Letter; Mount Gabriel (Rudler 1905, Bartlett 1980, Ni Wen et al. 1991, 1996, 1999, Kinnaird et al. 2002). Brilliant-violet bornite in a matrix of malachite and chloritized phyllite from Cappagh is particularly notable. Quartz veins containing chalcopyrite with minor bornite and tennantite–tetrahedrite were once mined at the Berehaven (Bearhaven) mines, Allihies (Moreton & Aspen 1993). At Tullacondra, 10 km NNW of Mallow, disseminated and vein-hosted bornite–chalcopyrite mineralization occurs in a copper–silver deposit close to the Lower Carboniferous/Old Red Sandstone transition (Wilbur & Carter 1986).

Galway At Tynagh mine, near Killimor, epigenetic Pb–Zn–(Ag–Cu) mineralization includes a wide range of vertically zoned sulphide species, with bornite–chalcopyrite occurring beneath chalcopyrite–tennantite (Whitehead 1971, Clifford et al. 1986).

Kerry Epigenetic vein mineralization in limestone at Ardtully (Cloontoo) mine, near Kenmare, consists of bornite, chalcopyrite and tennantite with traces of cobaltite (Cole 1922, Halliday & Mitchell 1983, Kinnaird et al. 2002). At Western mine on Ross Island, a similar deposit contains trace amounts of bornite in a dominantly chalcopyrite–tennantite ore (Kinnaird et al. 2002). Bornite, chalcopyrite and tetrahedrite are also reported from a stratiform deposit near Blue Hole on Ross Island, where pockets, irregular patches and strings of ore are concentrated in thin beds of Carboniferous Limestone (Russell 1927b). A bornite-bearing stratabound arsenic–copper ore has been reported from nearby Crow Island (Kinnaird et al. 2002); however, this is disputed by Ixer (2004), who suggests it is the Ardtully mineralization being described. Disseminated bornite, chalcopyrite and djurleite infills pore spaces in Devonian redbeds at Derrynane (Kinnaird et al. 2002).

Limerick Disseminated sulphides (mainly bornite, chalcocite and chalcopyrite) in fractured Carboniferous Limestone are associated with minor galena, native silver, sphalerite and stromeyerite at Aherlow, near Galbally (Cameron & Romer 1970, Schultz 1971, Halliday & Mitchell 1983). Bornite and tetrahedrite are reported in fragments of quartz gangue at Pallaskentry mine (Cole 1922).

Mayo An Ordovician altered acidic porphyry 8 km to the S of Charlestown hosts disseminated copper mineralization dominated by chalcopyrite, but with bornite in restricted high-grade areas (Poustie 1982).

Meath Bornite does not appear to have been found in the stratiform sphalerite–galena–pyrite orebody at Navan; however, it is found with barite, chalcopyrite and hematite in a nearby Silurian syenite intrusion forming veinlets and disseminations (Andrew & Ashton 1985).

Tipperary Uranium-enriched dolomitized intrusive breccias at Gortdrum mine, near Tipperary town, represent a low-temperature mineral assemblage thought to be derived from Upper Devonian and Lower Carboniferous detritus of granitic provenance. Associated species include bornite, chalcocite, chalcopyrite, pyrite, stromeyerite and tennantite–tetrahedrite. This mineralization is thought to be a precursor to the main epigenetic Cu–(Ag–Hg) sulphide mineralization for which the mine is famous (Duane et al. 1986). Ryback & Moreton (1991) describe microscopic inclusions of gortdrumite in bornite from here. Bornite is also noted from this mine by Morrissey et al. (1971), Halliday & Mitchell (1983), Steed (1983) and Braithwaite & Wilson (2003). The carbonate-hosted Lisheen Pb–Zn–(Ag) deposit, between

Thurles and Rathdowney, consists of three stratabound sulphidic lenses dominated by galena, marcasite, pyrite and sphalerite, in which bornite occurs with arsenopyrite, chalcopyrite, gersdorffite, gratonite, jordanite, seligmannite and tennantite (Hitzman et al. 2002).
Tyrone Gold-bearing quartz veins at Curraghinalt, near Gortin in the Sperrin Mountains, contain massive sulphides dominated by pyrite with subsidiary arsenopyrite, bornite, chalcopyrite, native copper, galena, sphalerite and tennantite–tetrahedrite (Earls et al. 1989).

Offshore territories
Jersey Bornite occurs intergrown with chalcopyrite in rhyolite at Bouley Bay, Trinity, and in arsenopyrite–galena–pyrite–sphalerite-dominated carbonate veins within granite at Le Doüet de la Mer, St Ouën (Ixer 1992). Irridescent grains of bornite have been found in granite at La Moye (Mourant 1961, Ixer 1988), in andesite E of Côtil Point, and in dolerite at Westmount (Mourant 1961). At Le Pulec, on the northwest coast, bornite replaces chalcopyrite in a polymetallic dolomite–quartz vein system dominated by sphalerite (Stanley & Ixer 1982).
Sark Museum samples collected from the copper–silver veins at Le Pot, Little Sark, contain chalcocite, djurleite and a little bornite (Ixer 1992).

botallackite

monoclinic $Cu^{2+}_2Cl(OH)_3$

TL Botallack mine, St Just, Cornwall, England?

A rare copper chloride hydroxide, polymorphous with atacamite, clinoatacamite and paratacamite, and found as a secondary mineral in copper-bearing deposits, often where alteration close to sea water has occurred.

At the type locality, botallackite was found in an area of the mine infiltrated by sea water and it forms thin crusts of minute interlocking crystals with a pale mountain-green (malachite-like) colouration (Church 1865a). Evidence presented by Kingsbury (1964b) suggests that the original specimens were not from Botallack mine, but from Wheal Cock, a neighbouring mine that became part of the Botallack sett around 1840. It was here that miners on the 20-fathom level, working out under the sea, had inadvertently drilled up through the roof of the level and broken through to the sea floor. The offending hole was sealed by the insertion of a wooden plug and, over a period of 30 years, seepage from this plug reacted with the copper ore to form botallackite. Botallack mine is discussed in some detail by Bancroft & Weller (1993), Wolloxall (1989) and Noall (1999).

Light bluish-green thin rectangular plates of botallackite occur sporadically near the base of the seacliff at Loe Warren Zawn, W of Botallack, where surface water draining from rubble at the base of the cliff has trickled over veinlets carrying chalcocite, chalcopyrite and pyrite, and precipitated an assemblage of supergene carbonate and phosphate minerals on exposed surfaces and joint planes near a patch of uraninite mineralization. Associated species include atacamite, cerussite, connellite, cumengéite, dewindtite, kasolite, malachite, widenmannite, wölsendorfite and intermediate members of the torbernite–zeunerite series (Elton & Hooper 1995a,c). At Levant mine, St Just, botallackite forms emerald-green needle-like crystals, up to 4 mm in length (Wolloxall 1987). Elsewhere in Cornwall, botallackite is noted from: Geevor mine, near Pendeen, St Just (Elton & Hooper 1993); St Ives (Robson 1949b); Hanover Cove, Perranzabuloe; Padstow Consols mine (both Golley & Williams 1995); West Wheal Owles, St Just; and Cligga Head, Perranzabuloe (both Kingsbury 1954). Many samples from the Kingsbury collection (now at the NHM) have doubtful provenance (Ryback et al. 1998, 2001), but a spectacular find of botallackite in salt-water pools at Cligga Head in 2007 produced world class quality specimens (D. I. Green pers. comm. 2008).

Botallackite crystals up to 1.5 mm long from Cligga Head, Perranzabuloe, Cornwall. (© D. I. Green)

In Wales, on the beach between Abersoch and Penrhyn Dhu, Gwynedd, acicular to lath-like crystals of botallackite, up to 4 mm long, have been found with atacamite and cumengéite within fractures in pebbles of hydrothermally brecciated Ordovician siltstone containing bornite–chalcopyrite–quartz veins (Dossett & Green 1998). A second occurrence of botallackite in Wales is anthropogenic, as it is from a smelter slag at Halkyn, Clwyd (Bevins 1994).

In Scotland, at Castletown mine, on the eastern side of Loch Gilp, near Lochgilphead, Strathclyde, botallackite has been found in a massive quartz vein exposed along the shoreline. It forms felted masses of microcrystals (up to 0.5 mm long) and larger bladed crystals (up to 3 mm long) associated with anilite, clinoatacamite, connellite, goethite, hematite and vésigniéite (Starkey & Hubbard 1998, Green & Todd 2001). Patches of bluish-green platy botallackite also occur on a radioactive rock containing atacamite and connellite from Southwick Cliffs, Solway Firth, S of Dalbeattie, Dumfries & Galloway (Macpherson & Livingstone 1982).

In Ireland, supergene alteration of a chalcopyrite-bearing quartz lode exposed along a coastal promontory at Dooneen mine, Allihies, Co. Cork, produced emerald-green coatings up to 1 mm thick, essentially of botallackite (with minor atacamite). The coatings are minutely mammillated with a rippled appearance (Moreton & Aspen 1993). However, Braithwaite & Wilson (2001b) have shown that many samples are actually an intimate mix of atacamite with variable amounts of botallackite, and some contain little or no botallackite. These authors go on to describe green crusts containing botallackite with atacamite and clinoatacamite, and small patches of bright-green needle-shape crystals from the Stage lode of Tankardstown copper mine, E of Cuan Cove, Knockmahon, 22 km SW of Waterford town, Co. Waterford.

botryolite

A variety of datolite, gadolinite group.

bottinoite

trigonal $Ni^{2+}(Sb^{5+})_2(OH)_{12}\cdot6H_2O$

A rare hydrated nickel antimonate that has formed as an oxidation product of ullmannite at Brownley Hill mine, Nenthead, near Alston, Cumbria. Bottinoite was found in a single cylindrical cavity over 1 m long within the Brownley Hill North vein (not Tatters String, as originally reported). The cavity was lined with occasional calcite, dolomite, galena and quartz. Bottinoite forms light-blue spherulitic aggregates up to 3 mm in diameter on the galena in association with cerussite and sulphur (Clark & Rust 1993, Green et al. 2000).

Bottinoite is also reported from two localities in Wales, both in the Central Wales orefield, Dyfed: Hendrefelen mine, Ysbyty Ystwyth, and Mynyddgorddu mine, W of Bont-goch, Ceulanymaesmawr. At Hedrefelen, bottinoite occurs in veinstone of brecciated mudstone–sandstone cemented by vuggy quartz, within cavities up to 1 cm in diameter. Calcite, chalcopyrite, galena and ullmannite form euhedral crystals in the cavities, and crystal aggregates of bottinoite up to 3 mm across have formed on the ullmannite. At Mynyddgorddu, bottinoite encrusts cleavage planes on ullmannite enclosed in quartz. Other supergene species from this site include bindheimite, cerussite, chrysocolla, hemimorphite, malachite, pyromorphite and wulfenite (Clark & Rust 1993).

boulangerite

monoclinic $Pb_5Sb_4S_{11}$

A relatively common lead sulphantimonite found in low- to moderate-temperature hydrothermal veins.

In Cornwall, small quantities of boulangerite, either massive or forming unorientated fibres up to 1 mm long, have been found in a quartz vein cutting tuffaceous slate at Port Gaverne, near Port Isaac (Hobson 1972). Near by, at Poltreworgy and Treore mines, both to the SE of Port Gaverne, boulangerite (var. plumosite) occurs with three other antimony-bearing species – bindheimite, jamesonite and stibnite (Russell 1949b). At Bounds Cliff, to the E of Port Gaverne, boulangerite, bournonite, jamesonite, plagionite, stibnite and tetrahedrite occur in antimony–lead-sulphosalt enriched veins, mainly along intergrain fractures and vein intersections, suggesting a late paragenesis for the assemblage. Bindheimite and stibiconite occur as secondary oxidation products of jamesonite and stibnite respectively (Clayton et al. 1990). Other locations for boulangerite in Cornwall include: Old Treburgett, St Teath; Pengenna mine, St Kew; Trevinnick mine, St Kew; and Wheal Boys, Trewetha, St Endellion (all Golley & Williams 1995). Reports of boulangerite from St Endellion, Cornwall (Kingsbury & Hartley 1956b), Shallowford Bridge, 4 km NW of South Molton, Devon (Kingsbury 1965c) and Carrock mine, Grainsgill, Caldbeck Fells, Cumbria (Kingsbury & Hartley 1956b, Hartley 1984), deserve further confirmation, as many samples in the Kingsbury collection have doubtful provenance (Ryback et al. 1998, 2001). Boulangerite is confirmed from other locations in Cumbria: as small fibres in quartz from Nether Row Brow, Caldbeck Fells (Cooper & Stanley 1990), and as intergrowths in galena with native antimony, bournonite and ullmannite from Barrow mine, Braithwaite (Stanley & Vaughan 1981).

Boulangerite is reported as a trace component of polymetallic post-tectonic fracture-hosted vein mineralization at the following mines in the Central Wales orefield, Dyfed: Alltycrib mine, Tal-y-bont; Darren, Pen-bont Rhydybeddau; Eaglebrook (Nantycagl), Ceulanymaesmawr; Erglodd, N of Tal-y-bont; Esgair Hir, Ceulanymaesmawr; and Esgair Fraith, near Tal-y-bont. The most important sulphides in the assemblage are chalcopyrite and galena, which are accompanied by minor bournonite, millerite, siegenite, tetrahedrite and ullmannite, and rare boulangerite, electrum, gersdorffite, pentlandite, pyrite, pyrrhotite and tučekite. Boulangerite often occurs as inclusions in galena (Mason 1997, 1998).

Elsewhere in Wales, boulangerite has been tentatively identified at Bwlch antimony mine, near Deganwy, Gwynedd, where it occurs in patches and veinlets cutting an Ordovician nodular ignimbrite. Associates are chert, grey drusy quartz, stibnite and zinkenite. Anglesite, barite, bindheimite, galena, jamesonite, marcasite, plagionite, pyrite and sphalerite, and possibly robinsonite and stibiconite, are also present in small quantities (Russell 1944, 1949b, Bevins et al. 1988, Ryback & Francis 2001). At Cefn Coch mine, within the Dolgellau goldbelt, Gwynedd, quartz–sulphide veins contain boulangerite as microscopic inclusions in galena (Mason et al. 2002). Boulangerite has also been noted from Clogau mine, Bontddu, near Dolgellau, Gwynedd (Naden 1988, Bevins 1994, Mason 1998, Mason et al. 2002), but the identification is based on samples from the Kingsbury collection at the NHM – a collection now known to contain many samples with incorrectly recorded provenance (Ryback et al. 1998, 2001).

A single inclusion of boulangerite in an alluvial gold grain from the Leadhills–Wanlockhead area, along the Strathclyde/Dumfries & Galloway border, Scotland, is noted by Leake et al. (1998).

In Ireland, boulangerite is one of many primary sulphosalt species found in the large sedimentary-exhalative Ba–Pb–Zn deposit in the Silvermines district, Co. Tipperary. It is a minor species that is concentrated around some of the feeders of the B and Upper G zones of Mogul mine, sometimes forming fine hairs in small vugs (Morrissey et al. 1971, Taylor & Andrew 1978, Taylor 1984, Moreton 1999). Boulangerite also occurs as a "rash" of tiny untwinned laths, up to 40 µm long within galena at the nearby Shallee mine (Rhoden 1959). Wilkinson et al. (2005) note minor boulangerite in the carbonate-hosted Lisheen Pb–Zn–(Ag) deposit, between Thurles and Rathdowney, Co. Tipperary. In Co. Meath, boulangerite is described from the Tara orebody of Navan mine, which worked the largest sedimentary-exhalative lead–zinc orebody in Europe. It occurs as needles in calcite (Ryback & Moreton 1993) and has been identified in polished sections of the ore, together with other minor species – argyrodite, bournonite, cylindrite, franckeite, freibergite, jordanite, pyrargyrite and semseyite (Andrew & Ashton 1985, Ashton et al. 2004). Euhedral laths of boulangerite, up to 150 µm in length, are closely associated with antimony-rich galena that has exsolved very fine needles of boulangerite to form a structurally controlled trellis-work pattern (S. Moreton specimen, Tindle unpub. EMP data).

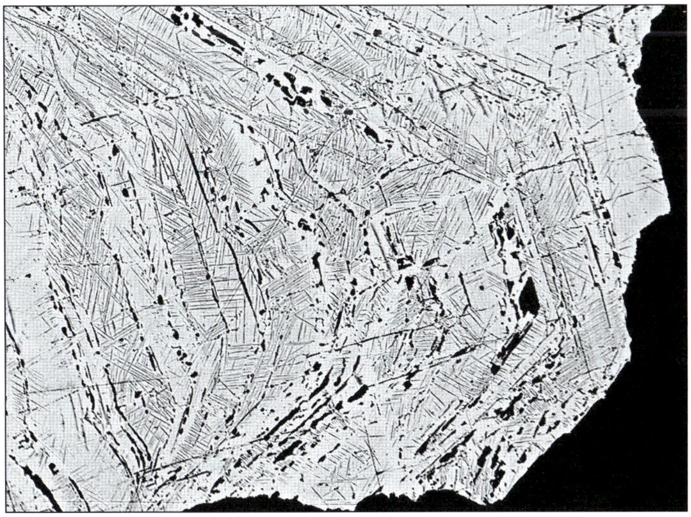

Oriented needles of boulangerite within galena from the Tara mine orebody at Navan, Co. Meath, Ireland; width 580 µm; BSE image. (© A. G. Tindle)

Other occurrences of boulangerite in Ireland include: Abbeytown mine, near Ballysadare, Co. Sligo (Hitzman 1986); Ballinalack, 15 km NW of Mullingar, Co. Westmeath (Jones & Brand 1986); Clontibret mine, Tullybuck, Co. Monaghan (Morris et al. 1986); and Moyvoughly area, Co. Westmeath (Poustie & Kucha 1986).

Joints in granite at L'Etacq quarry, Jersey, Channel Isles, contain molybdenite-rich calcite–quartz veins in which boulangerite (var. plumosite) occurs with arsenopyrite, bournonite, cosalite, chalcopyrite, galena, marcasite, pyrite, pyrrhotite, sphalerite, stannite and tetrahedrite (Mourant 1985, Ixer 1992). Boulangerite was possibly present at Sark's Hope mine on Little Sark (Mourant & Warren 1934). At Foxdale mine, Isle of Man, boulangerite (var. plumosite) forms grey felted aggregates of very minute crystals and wisps and coils of flexible hairs in vugs within quartz, and is closely associated with fine-grain galena, siderite and sphalerite (Smyth 1880, Russell 1949b, Starkey 1998a).

boulder

A common white form of barite.

bournonite

orthorhombic $PbCuSbS_3$

TL Wheal Boys (later Old Trewetha Consols), Trewetha, St Endellion, Cornwall, England

A common lead–copper sulphantimonite that forms a series with seligmannite and is found in hydrothermal veins formed at moderate temperatures. Endellion (later changed to endellionite) is an earlier name that was given to the species by Count de Bournon, a French refugee living in Britain in the 18th century.

At the type locality (once the most important producer of antimony in Britain), finely crystallized bournonite was associated with jamesonite, siderite (chalybite) and sphalerite (Greg & Lettsom 1858, Rudler 1905, Russell 1949, Embrey & Symes 1987). However, the most remarkable occurrence of bournonite is from Herodsfoot mine, Lanreath, near Liskeard, Cornwall, where groups of bright-grey multiple twinned crystals form patterns like cog wheels up to 5 cm across. Specimens of cog-wheel bournonite now at the NHM were collected from cavities within quartz veins and are associated with tetrahedrite. Unquestionably, they are of "noble" quality,

Bournonite (cog-wheel) crystals surrounded by quartz; Herodsfoot mine, Lanreath, near Liskeard, Cornwall; width 110 mm. (© The Natural History Museum, London)

unsurpassed elsewhere in the world (Miers 1884, Rudler 1905, Embrey & Symes 1987, Turner 1990, Bancroft & Weller 1993). In a study of stratabound antimony–lead mineralization in the Endellion–Port Quin district of north Cornwall, Clayton et al. (1990) describe paragenetically late-formed bournonite from all the localities they examined: Bounds Cliff, E of Port Isaac; Pentire (Glaze) mine, near New Polzeath, St Minver; Port Quin; and Reedy Cliff, near Port Quin. In addition, the sulphosalt-dominated assemblages consist of boulangerite, freibergite, jamesonite, plagionite and tetrahedrite. Bournonite also occurs with chalcopyrite, galena, native gold, pyrite, sphalerite and tetrahedrite in shear-zones within brecciated and carbonatized (metasomatized) basaltic volcanics near Port Quin (Clayton 1993). Other Cornish occurrences of bournonite include: crystallized and compact at Budock Vean mine, Constantine, near Falmouth (Greg & Lettsom 1858); black metallic crystals up to 6 mm across, some with a poorly developed cog-wheel habit at Greystone quarry, Lezant, near Launceston (Elton et al. 2004); Nanslow, near Redruth and St Merryn, near Padstow (both Greg & Lettsom 1858); poorly formed crystals up to 2 mm across associated with galena and tetrahedrite (with minor chalcopyrite, pyrite and pyrrhotite) from quartz veinstone on Newporth beach, 2 km SW of Falmouth (Dean 1982, Dean et al. 1983); associated with tetrahedrite on one specimen collected between Swanpool and Maenporth beaches (Dean 1983); associated with jamesonite at Rose mine, St Endellion (Hosking 1951); minor bournonite associated with native gold and tetrahedrite at Tregilden quarry, near Pendoggett, St Teath (Camm 1995); with arsenopyrite, bindheimite, boulangerite, cerussite, galena, native gold, jamesonite, pyrite, scheelite, sphalerite and tetrahedrite at Treore mine, Port Gaverne, near Port Isaac, St Teath (Russell 1949b); and with tetrahedrite at Wheal Anna Maria, Mawnan, near Falmouth (Dean 1982, Camm 1995). Golley & Williams (1995) also note bournonite from: Crednis mine, Padstow; Pengenna mine, St Kew; Tresungers mine, St Endellion; and Swanpool mine, Falmouth.

Occurrences of bourninite in Devon include: bournonite with galena at the Bere Alston mines, near Bere Ferrers (Hall 1868b); bournonite with galena, pyrite and sphalerite in veins at Bridford barite mine, Teign Valley, Christow, 10 km SW of Exeter (Polkinghorne 1951); bournonite containing antimony-rich native bismuth inclusions in an antimony–gold vein deposit at Loddiswell mine, 3 km N of Loddiswell (Stanley et al. 1990b); and rare bournonite in a semseyite-dominated ore associated with anglesite, chalcocite, galena, jamesonite, malachite and pyrite at Wheal Emily, Wembury (Russell 1949b).

The Northern Pennine orefield has produced bournonite at a few locations: with galena, nickeline and ullmannite in a pipe-like deposit close to the Whin Sill at Settlingstones mine, near Hexham, Northumberland (Russell 1927b); Nenthead, near Alston, Cumbria; Sir John's mine, 4 km S of Garrigill, Cumbria; Teesdale Valley, Co. Durham (all Vaughan & Ixer 1980); and with bottinoite, cerussite, galena, hydrozincite, pyrite, sphalerite, native sulphur and ullmannite at Brownley Hill mine, Nenthead, Cumbria (Clark & Rust 1993, Green et al. 2000).

Bournonite is found at Driggith mine, Caldbeck Fells, Cumbria, in a primary vein assemblage that also contains native antimony, arsenopyrite, barite, calcite, chalcopyrite, galena, pyrite, quartz (var. chalcedony) and tetrahedrite. The vein assemblage of the vein is representative of the lead–zinc mineralization of the Caldbeck Fells area. Three forms of bournonite occur: as relatively large grains (< 100 μm) commonly enclosing silver-rich tetrahedrite; along cleavages in galena; as intergrowths in galena, as either platelets or rods usually no more than a few microns in width. Irregular grains of bournonite also occur either associated with native antimony or at chalcopyrite/galena grain boundaries. The former association is further noted from Eagle Crag mine, Patterdale; Greenside mine, near Glenridding, Patterdale; Hartsop Hall and Myers Head mines, Hartsop, Patterdale; Force Crag mine, Braithwaite, near Keswick; and Goldscope and Yewthwaite mines, Newlands Valley (Stanley 1979, Stanley & Vaughan 1981). Elsewhere in Cumbria, bournonite is described as small grains in sugary quartz from Nether Row Brow and from Roughton Gill, both Caldbeck Fells (Cooper & Stanley 1990). These authors note Roughton Gill on the basis of an unpublished A. W. G. Kingsbury manuscript and the location deserves further investigation.

Bournonite is found with tetrahedrite as part of a polymetallic assemblage in the Central Wales orefield of Dyfed, and its occurrence is recorded from the following mines by Mason & Hughes (1990), Bevins (1994) and other workers: Alltycrib; Bronfloyd; Cwmerfin; Cwmsymlog, Darren (Mason 1997, 1998); Eaglebrook (Nantycagl); Erglodd; and Esgair Fraith (all Mason 1997); Esgair Hir (Rust & Mason 1988, Mason 1997); Goginan; Leri Valley (Mason 1998); Llechweddhelyg; Nantyrarian; South Darren; and Pengraigddu. At some of these mines, the sequence chalcopyrite–tetrahedrite–bournonite–galena suggests that bournonite was formed by replacement of tetrahedrite in a lead-rich environment.

In Gwynedd, occurrences of bournonite include: Bwlch mine, near Deganwy (Mason & Hughes 1990, Mason 1998); Ceunant copper mine, Llandegai (Mason & Hughes 1990); 40–80 μm-diameter inclusions in quartz, commonly associated with native gold and pyrite at Clogau mine, Bontddu, near Dolgellau (Naden 1988, Mason et al. 2002); and 5–30 μm grains associated with native bismuth, bismuthinite and other bismuth sulphosalts in the exhalative volcanic–sedimentary copper deposit at Parys Mountain, Anglesey (Wheatley 1971b, Sivaprakash 1977, Pointon & Ixer 1980, Southwood & Bevins 1995, Mason 1998).

Bournonite has been found at only a few locations in Scotland. It forms irregular grains to 1 mm in association with arsenopyrite, galena, pyrite, semseyite, sphalerite, tennantite and tetrahedrite in an ankerite–quartz–sphalerite–stibnite vein at Glendinning antimony mine, Glenshanna Burn, 13 km NNW of Langholm, Dumfries & Galloway (Macpherson & Livingstone 1982, Duller et al. 1997). Examination of borehole cores from the area by Gallagher et al. (1983) revealed bournonite to have a steel-grey colour, a brilliant metallic lustre and a distinctive prismatic habit. At Knipe (Fountainhead) mine, Hare Hill, 5 km SE of New Cumnock, Dumfries & Galloway, gold mineralization associated with a zoned As–Cu–Pb–Sb–Zn assemblage is hosted by a late-Caledonian granodioritic intrusion and contains bournonite in late cross-cutting quartz veinlets. Arsenopyrite, chalcopyrite, sphalerite, stibnite and tetrahedrite are also found at this location (Boast et al. 1990). At Urlar Burn, Aberfeldy, Tayside Region, bournonite occurs in quartz–sulphide ± gold veins that cross-cut gently dipping Upper Dalradian metasediments and metabasic rocks. The assemblage also contains a suite of Co–Fe–Ni arsenides and sulpharsenides. Bournonite replaces galena – the main sulphide phase at this locality (Ixer et al. 1997). Reports of bournonite by Heddle (1901) are unsubstantiated (Macpherson & Livingstone 1982).

Ireland

Clare At Ballyvergin, traces of silver-rich bournonite, geocronite and native silver occur in a sulphide assemblage dominated by arsenopyrite, chalcopyrite, galena and pyrite. Bornite, sphalerite and tennantite are minor species. Mineralization is hosted by a Lower Carboniferous carbonate sequence and it occurs as intergranular disseminations, as infill of diagenetic void spaces, as replacements of bioclasts and in tectonic veinlets (Andrew 1986). Bournonite occurs with geocronite (kilbrickenite) at Kilbricken lead mine, between Ennis and Quin (Russell 1949b).

Cork Bournonite and galena replace meneghinite at Dhurode copper mine, West Carbery mining district, where late-Hercynian mineralized quartz veins cut folded Devonian and Carboniferous rocks of the Munster Basin. Bournonite occurs in an early sulphide assemblage that also contains arsenopyrite, chalcopyrite and tetrahedrite, together with rare molybdenite, pyrite and sphalerite (Ni Wen et al. 1991, Kinnaird et al. 2002).

Galway At Tynagh mine, near Killimor, epigenetic Pb–Zn–(Ag–Cu) mineralization includes a suite of sulphide species that occur in trace amounts and include native antimony, arsenopyrite, bornite, bournonite, chalcopyrite, enargite, geocronite, gratonite, jordanite, marcasite, millerite, polybasite, pyrargyrite, pyrite, siegenite, stephanite and tennantite, many occurring as inclusions in galena. Bournonite is the most common inclusion species after tennantite (Schultz 1966b, Morrissey et al. 1971, Gibson 1979, Boast et al. 1981a, Clifford et al. 1986). Bournonite is also noted from the Caherglassaun lead mine between Gort and Kinvarra (not Kenmare) in south Galway (Greg & Lettsom 1858, Cole 1922).

Longford Bournonite is a minor species forming inclusions up to 20 μm across in galena within epigenetic sphalerite-dominated mineralization found in breccia zones and fracture fillings at Keel, Ardagh (Patterson 1970, Morrissey et al. 1971).

Mayo Bournonite forms mm-size flecks with chalcopyrite, galena, pyrite, sphalerite and tetrahedrite in gold-bearing quartz veins at Cushinyen, within the Cregganbaun shear zone (Thompson et al. 1992).

Meath Bournonite is one of the minor Ag–Cu–Pb sulphosalt species found in the stratiform galena–pyrite–sphalerite ore at Navan. It is a variant of the sedimentary-exhalative class of ore deposit (Andrew & Ashton 1985, Ashton et al. 1986, 2004).

Monaghan A single occurrence of particulate gold in a complex grain of boulangerite–bournonite–galena is reported from the Clontibret mine, Tullybuck. Arsenopyrite, pyrite and stibnite are the main ore species found in ankerite–quartz veins and adjacent host rocks (Morris 1984).

Tipperary Bournonite was one of the many species found in an enormous vug in the Upper G zone of Mogul mine, Silvermines. Sharp blocky crystals up to 2 cm across were recovered, together with fabulous specimens of galena, pyrite and sphalerite. The deposit formed in a classic sedimentary-exhalative environment (Moreton 1999). Bournonite is one of many primary sulphides reported from Shallee and Gortnadyne mines, Silvermines. It forms inclusions in galena at Shallee mine (Rhoden 1959). Bournonite is a constituent of an As–Co–Cu–Ni assemblage found near major faults within the carbonate-hosted Lisheen Pb–Zn–(Ag) deposit, between Thurles and Rathdowney. Associated species are cobaltite–gersdorffite, nickeline, pararammelsbergite, pyrite (bravoite) and tennantite, together with traces of arsenopyrite, barite, chalcopyrite, hematite and ullmannite (Ixer 2002, Fusciardi et al. 2004).

Westmeath Carbonate-hosted barite–galena–marcasite–pyrite–sphalerite mineralization at Ballinalack, 15 km NW of Mullingar, contains minor quantities of bournonite, millerite and possibly boulangerite (Jones & Brand 1986).

Wicklow Bournonite is noted from the Caherglassaun lead mine (Greg & Lettsom 1858, Cole 1922).

Offshore territories

Jersey Bournonite forms lath-like inclusions (up to 20 μm long) in galena within a quartz–sulphide vein found at the German underground hospital, St Peter's Valley, St Lawrence (Ixer 1990b). In another quartz–sulphide vein at Gigoulande quarry, St Mary, bournonite occurs with native antimony, marcasite and molybdenite as trace components in an assemblage dominated by chalcopyrite and pyrite (Ixer 1980, 1992). Vein mineralization at Le Pulec on the northwest coast of Jersey contains 20–75 μm inclusions of bournonite in galena. Bournonite also encloses native antimony, chalcopyrite, galena, pyrite, sphalerite and tetrahedrite (Ixer & Stanley 1980, Stanley & Ixer 1982, Ixer 1992).

Sark Possible bournonite is reported from Sark's Hope mine, Little Sark (Mourant & Warren 1934), but this is not confirmed by Ixer & Stanley (1983).

bowlingite

TL Bowling quarry, Kilpatrick Hills, near Glasgow, Strathclyde, Scotland

A synonym of saponite, smectite group. See Hannay (1877).

BRACKEBUSCHITE GROUP

Monoclinic arsenates, phosphates and vanadates with the general formula $A_2B(XO_4)_2(OH,H_2O)$, where A = Ba, Ca, Pb, Sr; B = Al, Cu^{2+}, Fe^{2+}, Fe^{3+}, Mn^{2+}, Mn^{3+}, Zn; and X = As, P, S, V. Arsentsumebite and tsumebite are members of the group found in Britain or Ireland.

braggite

tetragonal (Pt,Pd,Ni)S

Named after Sir William Henry Bragg FRS (1862–1942), a pioneer in the research on crystal structure by means of X-rays, and his son William Lawrence Bragg (1890–1971), both of the University of Cambridge; born respectively at Westward, near Wigton, Cumbria, and Adelaide, Australia. Braggite is a rare platinum–palladium–nickel sulphide that forms a series with vysotskite and has been identified in the layered mafic–ultramafic intrusion on the Isle of Rum, Highland Region, Scotland. Within certain chromitite layers of the Eastern Layered Series, braggite is found in association with species positively or tentatively identified as cooperite, hongshiite, irarsite, isoferroplatinum, keithconnite, kotulskite, laurite, moncheite, palladoarsenide, platarsite, sperrylite, telargpalite and telluropalladinite, and the native species palladium and platinum. Most of these grains are 0.2–2.0 μm in diameter, although, for some species, grains >15 μm are also present. The majority are in contact with sulphide minerals such as chalcopyrite, bornite or pentlandite. The chromitite layers were formed during magmatic fractionation and crystal settling of an open-system magma chamber, modified by magma mixing and liquid immiscibility of sulphide and silicate melts (Power et al. 2000a, 2003).

Braggite has been tentatively identified as inclusions in garnet and clinopyroxene within an ultramafic (biotite–magnetite pyroxenite) intrusion at Loch Borralan, Assynt, Highland Region. It occurs as ≤5 μm-diameter rounded grains clustered with cooperite. Associated veins contain sperrylite, isomertieite and stibiopalladinite, which may be intergrown with chalcopyrite, galena, hessite and bismuth tellurides (Gunn & Styles 2002, Styles et al. 2004).

brainstone

A white arborescent variety of barite.

brammallite (series)

MICA GROUP

monoclinic $Na_{0.65}Al_{2.0}\square Al_{0.65}Si_{3.35}O_{10}(OH)_2$

TL Llandybie, Dyfed, Wales

Named after Alfred Brammall (1879–1954), English mineralogist, formerly of the Department of Geology, Imperial College, London – a specialist in clay minerals who first found the mineral; born at Boothstown, near Leigh, Lancashire. During a reclassification of the mica group, it was decided that, as brammallite does not define an end-member composition (and is therefore not a true species), it should be given series status to reflect more accurately its current usage in describing a relatively large volume of compositional space (Rieder et al. 1998).

Brammallite is a sodium aluminosilicate hydroxide found along slickensided surfaces or in fissures within shales overlying the Coal Measures. It is a soft white fibrous mineral forming compact tufts of elongated plates up to 0.5 mm long and is reminiscent of the satin-spar variety of gypsum (Bannister 1943). At Pandora mine, near Llyn Geirionydd, Gwynedd, Wales, radiating colourless-to-white fibres have an XRD pattern close to brammallite (Bevins 1994).

brandtite?

monoclinic $Ca_2(Mn^{2+},Mg)(AsO_4)_2{\cdot}2H_2O$

A rare hydrated calcium–manganese–magnesium arsenate, dimorphous with parabrandtite and claimed by A. W. G. Kingsbury to have been found at Meldon aplite quarry, near Okehampton, Devon. This provenance is now considered dubious (Ryback et al. 1998, 2001).

brannerite

monoclinic $(U,Ca,Y,Ce)(Ti,Fe)_2O_6$

A rare uranium–calcium–yttrium–cerium–titanium–iron oxide that forms a series with thorutite and is found in hydrothermal quartz and calcite veins, in silicified pebble conglomerates, and as a primary mineral in granite pegmatite and granite gneiss.

Brannerite is tentatively identified as an accessory component of iron–manganese-rich calcite-filled veins within fractured Borrowdale Volcanic Group rocks and Carboniferous Limestone examined in borehole material from the Sellafield area, Cumbria (Milodowski et al. 1998).

Rare needle-like inclusions of brannerite and uranium-rich leucoxene occur in titanium–uranium-rich bitumen nodules found in Lower and Middle Devonian sandstones and conglomerates at Ness, Quoyelsh and Yesnaby, all on the west coast of Mainland, Orkney, Scotland, and also on the nearby island of Graemsay. Subordinate minerals found in the nodules include barite, calcite, dolomite, goyazite, pyrite, quartz and uraninite, as well as unnamed titanium oxides. Brannerite is a diagenetic species formed by the adsorption of uranium onto titanium, but there is still debate as to whether the primary uranium source (uraninite) was detrital or an authigenic species that co-precipitated with carbon (McReady & Parnell 1998). Additionally, uranium-rich hydrocarbon nodules in Middle Devonian sandstones at Dingwall, Highland Region, are thought to contain brannerite and xenotime as inclusions (Parnell & Eakin 1987). At Little Craig, Glen Gairn, Highland Region, a 6×2 mm metamict metallic grain of brannerite is associated with bismutite, cassiterite, fluorite and molybdenite in massive quartz (Livingstone 2002).

Dolomitized and mineralized intrusive breccias at Gortdrum mine, near Tipperary town, Co. Tipperary, Ireland, contain brannerite with coffinite and more common uraninite (pitchblende) in a low-temperature mineral assemblage thought to be derived from Upper Devonian to Lower Carboniferous detritus of granitic provenance. Associated species include bornite, chalcocite, chalcopyrite, pyrite, stromeyerite and tennantite–tetrahedrite. This mineralization is thought to be a precursor to the main epigenetic Cu–(Ag–Hg) sulphide mineralization for which the mine is famous (Duane et al. 1986, Duane 1988).

brass

An old mining term for pyrite.

braunite

tetragonal $Mn^{2+}Mn_6^{3+}SiO_{12}$

An uncommon manganese silicate, formed by metamorphism or weathering of manganese silicates and oxides.

Braunite occurs with hausmannite, pyrolusite and rhodochrosite at Wyndham mine, Bigrigg, near Egremont, Cumbria (Davidson & Thomson 1951), and is said to occur in the manganese mines, near Launceston, Cornwall (Collins 1892).

The dumps at Tŷ Coch (an abandoned 19th-century manganese mine), near Porthcawl, Mid Glamorgan, Wales, contain massive fine-grain compact very-hard dark-grey manganese ore in which euhedral to subhedral braunite occurs with hausmannite in equal amounts, associated with interstitial and vug-filling barite, calcite, iron-rich calcite, dolomite and manganocalcite. Minor pyrobelonite and vanadinite are also present (Criddle & Symes 1977). Braunite is also confirmed from Nant mine, Llanfaelrhys, Rhiw, Lleyn Peninsula, Gwynedd (Bevins 1994).

Braunite dominates the microbrecciated groundmass of a complex stratiform manganese-rich ore (up to 4 m thick) at Dalroy Burn, near Culloden, Highland Region, Scotland. Bustamite and hematite are minor components of the ore, and birnessite, manganese-rich calcite, manganese-rich ilmenite, rhodochrosite, pyrophanite and possibly jacobsite are also reported. The ore has a geothermal origin best explained by deposition from hot springs in a sub-lacustrine environment during Devonian times (Nicholson 1987b, 1989c, 1990, 1992). Braunite is also present as a trace constituent of pink psammite boulders found in glacial drift, near Kinloch, S of Tongue, Highland Region. The boulders are almost certainly derived from the local Moine rocks. Manganese-rich hematite, manganese-rich muscovite, piemontite and spessartine are associated species (Faithfull 2007).

braunspath

A synonym of ankerite and dolomite (both dolomite group), and siderite (calcite group).

bravoite

A nickel-rich variety of pyrite. See Ixer & Vaughan (1982) for an occurrence in red-bed copper–lead mineralization at Alderley Edge ridge, Cheshire.

brazil

An old mining term for pyrite.

breadalbanite

TL Breadalbane, near Killin, Central Region, Scotland

Named after John Campbell (1796–1862), the second Marquis of Breadalbane, who was a mineral collector; born in Dundee, Tayside Region, Scotland. Breadalbanite is a synonym of hornblende, amphibole group.

bredbergite

A variety of andradite, garnet group.

bredigite

orthorhombic (pseudo-hexagonal) $Ca_7Mg(SiO_4)_4$

TL Scawt Hill, Larne, Co. Antrim, Ireland

A rare high-temperature calcium–magnesium silicate that formed in a chalk/dolerite contact zone at the type locality. Bredigite occurs with gehlenite, larnite and spurrite, and forms euhedral crystals with hexagonal cross sections (up to 0.7 mm across) and barrel- or boat-shape vertical sections. Twinning is commonly seen in thin section (Tilley & Vincent 1948). Anthony et al. (1995) also note bredigite from Ballycraigy, near Larne.

Bredigite has also been found under similar conditions in the Highland Region of Scotland, in the limestone contact zones of the Camas Mòr gabbroic intrusion on the Isle of Muck (Tilley & Vincent 1948), and the gabbroic intrusion at Camphouse, Ardnamurchan (Agrell 1953).

breithauptite

NICKELINE GROUP

hexagonal NiSb

A rare nickel antimonide found in the altered silicate matrix of chromite-rich dunite lenses that formed near the base of the Shetland ophiolite complex at Harold's Grave, Baltasound–Haroldswick area, Unst, Shetland, Scotland. Breithauptite, irarsite, laurite and native osmium form composite grains up to 70 μm across (Prichard & Tarkian 1988). Breithauptite is also reported near by, in a sulphide pod in dunite at Cliff and in a cumulate dunite to the west of Hagdale (Maynard 1993). Here, breithauptite occurs as anhedral, or rounded, 5–75 μm grains that form a lamellar intergrowth with an unidentified lavendar (in reflected light) antimonide. These composite grains are enclosed within awaruite, heazlewoodite or pentlandite. To the N of Baltasound and E of Caldback, sulphide-bearing dunite contains breithauptite, bornite, chalcocite, chalcopyrite, native copper, pentlandite (commonly altered to awaruite, heazlewoodite and magnetite), cobalt-rich pentlandite and a nickel–lead alloy (Prichard et al. 1994).

Breithauptite reported from Potts Gill mine, Caldbeck Fells, Cumbria (Cooper & Stanley 1990), is based on Kingsbury material and may be of dubious provenance (Green & Briscoe 2002).

breunerite or breunnerite

An iron-rich variety of magnesite, calcite group.

brevicite

A synonym of natrolite, zeolite group.

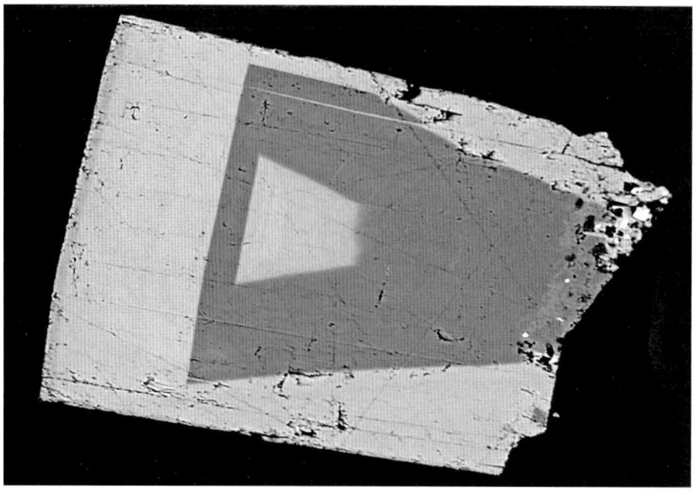

Prismatic crystal of brewsterite-Ba (light grey) with an intermediate zone of brewsterite-Sr (dark grey) from Clashgorm opencut, Strontian, Highland Region, Scotland; width 700 μm; BSE image. (© A. G. Tindle)

brewsterite-Ba

ZEOLITE GROUP

monoclinic $Ba_2(Al_2Si_6)O_{16}{\cdot}5H_2O$

See brewsterite-Sr. At Strontian, Highland Region, Scotland, brewsterite-Ba was found at the margins of compositionally zoned crystals where a maximum atomic ratio Ba/(Ba + Sr) of 0.62 was measured. Other zones in the crystal, particularly the core, were brewsterite-Sr. Crystals are sector zoned, suggesting crystallization at low temperature (Akizuki et al. 1966). Brewsterite-Ba is also confirmed from near Strontian at the Clashgorm opencut and Whitesmith mine, again in zoned crystals (D. I. Green specimen and XRD, Tindle unpub. EMP data).

Brewsterite-Ba has also been identified on a single specimen collected from the dumps near the Deep adit at Old Luganure mine, Glendasan, Co. Wicklow, Ireland. It forms crusts of colourless to pale-yellow, blocky to prismatic crystals, up to 3 mm long, on drusy quartz and is found with harmotome as a minor late-stage primary phase in low-temperature lead–zinc veins (Green et al. 2005g).

brewsterite-Sr

ZEOLITE GROUP

monoclinic $Sr(Si_6Al_2)O_{16}{\cdot}5H_2O$

TL Strontian, Highland Region, Scotland

Named after Sir David Brewster FRS (1781–1868), principal and vice-chancellor of the University of Edinburgh – a natural philosopher who studied the optical properties of minerals and discovered the laws of polarization of light in biaxial crystals; born in Jedburgh, Borders, Scotland. It is likely that type specimens came from either Bellsgrove or Whitesmith mines to the north of the village of Strontian. Brewsterite-Sr is a rare, but locally abundant, hydrated strontium aluminosilicate that forms in hydrothermal lead veins and in the Coombs et al. (1997) classification of the zeolite group was divided into brewsterite-Ba and brewsterite-Sr. The literature does not reflect this new nomenclature, although Coombs et al. (1997) confirm that material from the type locality is brewsterite-Sr. At the type locality, brewsterite-Sr coats fault breccias, lines cavities, and was very late to crystallize, even forming after mining had ceased. It produces flat, prismatic, glassy, white or yellowish-brown crystals up to 4–5 mm long and is associated with barite, harmotome (including the variety morvenite), strontianite and rarely ancylite on barite–calcite quartz gangue (Greg & Lettsom 1858, Mallet 1859, Scott 1871, Heddle 1901, Rudler 1905, Winrow 1986b, Dyer & Wilson 1988, Tschernich 1992, Livingstone 2002). A series of underground and opencast workings between Bellsgrove and Whitesmith mines in the 1980s produced some of the finest specimens of brewsterite (unusually large to 8 mm), calcite and harmotome (Green & Todd 2001). Colourless, prismatic crystals of brewsterite, associated with calcite and kainosite-(Y) from Whitesmith mine are illustrated by Green & McCallum (2005). Brewsterite-Sr is confirmed from Clashgorm opencut and both Bellsgrove and Whitesmith mines, near Strontian (D. I. Green specimen and XRD, Tindle unpub. EMP data).

Greg & Lettsom (1858) report brewsterite from the Giant's Causeway, Co. Antrim, Ireland, but later investigations by Nawaz (1988b) on specimens in the Ulster Museum failed to prove its existence. Other specimens at the NHM have been shown to be stilbite (Tschernich 1992).

Prismatic brewsterite, 3 mm long, from Strontian barite mine, north of Strontian, Highland Region, Scotland. (© D. I. Green)

brewstoline

Liquid inclusions of CO_2 in quartz.

brianyoungite

orthorhombic or monoclinic $Zn_3(CO_3,SO_4)(OH)_4$

TL Brownley Hill mine (Wellgill Cross vein), Nenthead, Cumbria, England

Named after Brian Young (b. 1947), field geologist and mineralogist with the British Geological Survey, author of many papers on the minerals of the Northern Pennine orefield and discoverer of the species; born in Sunderland, Tyne & Wear, England. Brianyoungite is chemically and structurally related to hydrozincite, and it occurs as white rosettes (< 100 μm) of thin bladed crystals (1–2 μm thick) with gypsum on rubbly limestone within the oxidized zone of the disused lead–zinc mine, which is also the type locality for alstonite. Other associated species include goethite, pyrite, smithsonite and a zinc-rich analogue of ktenasite. It is also reported from Smallcleugh mine, 3 km from the type locality. Brianyoungite most probably formed by post-mining oxidation of the overlying zinc-rich orebody (Livingstone & Champness 1993, Bridges & Young 1998, Green et al. 2000, Green & Briscoe 2002). Brianyoungite has also been identified at Driggith and Sandbed mines, Caldbeck Fells, Cumbria, where it forms white crusts of minute pearly lath-like crystals to 0.5 mm in oxidized sphalerite-bearing quartz veinstone (Green et al. 2006a).

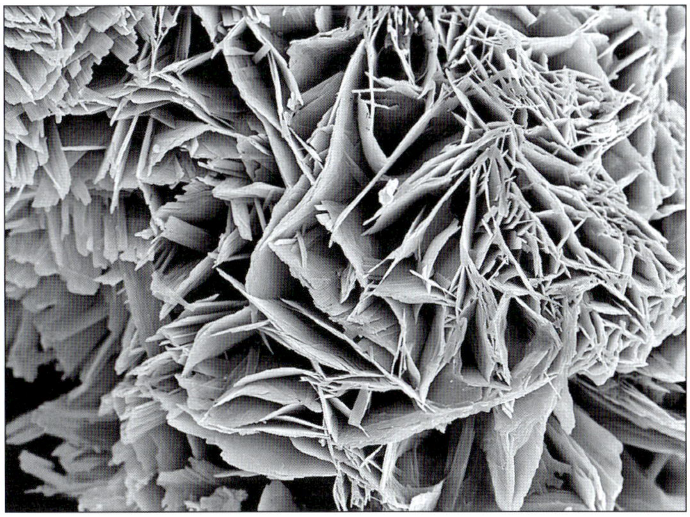

Brianyoungite from Sandbed mine, Caldbeck Fells, Cumbria; width 100 μm; SEM image. (© D. I. Green)

Brianyoungite forms powdery white encrustations on dump material from West Wheal Fortune, 3 km NNE of St Michael's Mount, Cornwall, where it is associated with other secondary species including connellite, erythrite and gypsum (D. I. Green pers. comm. 2003).

Further occurrences of brianyoungite are from Elgar mine, Bow Street (rosettes of pearly white crystals up to 0.3 mm across associated with cerussite), and Frongoch mine, between Devil's Bridge and Pont-rhyd-y-groes (spherulitic aggregates of pearly white lath-like crystals up to 0.5 mm across in a galena matrix), both in the Central Wales orefield, Dyfed, Wales. The primary mineralization in which brianyoungite crystallized belongs to the A2 group of quartz- and galena-bearing veinstones (Cotterell et al. 2003).

bright white cobalt

A synonym of cobaltite.

brinrobertsite

monoclinic $(Na,K,Ca)_{0.3}(Al,Fe,Mg_4)(Si,Al)_8O_{20}(OH)_4{\cdot}3.5H_2O$

TL 750 m NE of Penrhyn Castle, near Bangor, Gwynedd, Wales

Named after Brinley Roberts (Birkbeck College, University of London) in recognition of his extensive research on the mineralogy of pelitic rocks of the United Kingdom; born in Newport, Gwent, Wales. Brinrobertsite is a complex ordered mixed-layered dioctahedral pyrophyllite–smectite clay found in a metabentonite, in association with clinochlore and quartz. Original glass shards and fine-grain glass matrix were leached by hydrothermal fluids, and alteration proceeded in the sequence glass to pyrophyllite to brinrobertsite. The metabentonite is geochemically related to the Caradocian Llewelyn Volcanic Group, which outcrops some 10 km to the SE in Snowdonia. Brinrobertsite is one of a few ordered interstratified clay minerals, others being rectorite (illite–smectite) and corrensite (chlorite–smectite). These minerals are generally the metastable intermediate products of a sequence of reactions involving end-member clay minerals (Dong et al. 2002).

britholite-(Y)

hexagonal $(Y,Ca)_5(SiO_4,PO_4)_3(OH,F)$

A rare yttrium–calcium silicate phosphate fluorhydroxide found in an anorthite–diopside–hastingsite–hedenbergite–magnetite endoskarn associated with the Beinn an Dubhaich granite, Isle of Skye, Highland, Scotland. Britholite-(Y) occurs as discrete crystals (50–100 μm across) or as a metasomatic alteration product of allanite-(Ce) (Smith et al. 2002).

brochantite

monoclinic $Cu^{2+}_4(SO_4)(OH)_6$

A common copper sulphate hydroxide found in the oxidized zone of copper deposits, along with other copper sulphates and carbonates.

England

Avon Pebbles of galena-bearing barite on the beach near the Pier hotel, Clevedon, 16 km west of Bristol, contain brochantite associated with aragonite, aurichalcite, azurite, beudantite, cerussite, native copper, covellite, cuprite, phosgenite and native sulphur (Starkey 1984).

Cheshire Brochantite forms green crusts and crystals up to 0.2 mm long at Engine Vein mine, minute crystals with wulfenite at Wood mine, and bright-green crystals at Doc mine, all Alderley Edge ridge (Braithwaite 1994).

Cornwall At Penberthy Croft mine, St Hilary, brochantite is a common late-stage component of copper-sulphate and arsenate assemblages developed in an extensive zone of supergene enrichment. In the copper-sulphate suite it forms small translucent tabular prismatic to acicular crystals, ranging in colour from light-green through emerald green to dark green. Associated species are birnessite, chrysocolla, clinochlore, connellite, cuprite, dolomite, galena, goethite, langite, quartz, woodwardite and wroewolfeite. A more elongate morphology is noted in the arsenate suite, where crystals up to 1.1 mm long form fan-like groups associated with linarite and mimetite (Kingsbury 1954, Betterton 1989, Betterton et al. 2000).

Other Cornish occurrences include: Botallack, St Just (Bancroft & Weller 1993); Drakewalls mine, Calstock; Gerrans Bay; Levant mine, St Just; Sharptor mine, Linkinhorne, and United mines, Gwennap (all Golley & Williams 1995); Fowey Consols, Tywardreath, St Blazey – rippled green masses of brochantite associated with langite

(Embrey & Symes 1987, Lewis 1997); Geevor mine, near Pendeen, St Just – brochantite associated with atacamite, chalcocite, gypsum and sodium-zippeite (Elton & Hooper 1993); Greystone quarry, Lezant, near Launceston – minute lath-like crystals up to 0.05 mm in length associated with linarite and vauquelinite (Elton et al. 2004); Gunheath china-clay pit, Carthew, St Austell – dirty bluish-green crusts of brochantite on hydrothermally decomposed sulphidic tin ore, intimately intergrown with varlamoffite and as brilliant green powdery fills in cavities in a vein containing cassiterite, stannite and tennantite (Hawkes et al. 1987, Weiss 1994); Marke Valley mine, near Liskeard – brochantite associated with azurite (chessylite), chalcocite, chalcophyllite, chalcopyrite, connellite, cuprite and malachite (Collins 1911a); Mulberry openwork, Lanivet, St Austell – brochantite associated with chalcophyllite, cyanotrichite and scorodite in a tin-rich stockwork (Starkey 1986a); vugs within an anatase–cassiterite–chlorite–ilmenite–monazite–sulphide assemblage in North Tincroft lode, New Cook's Kitchen mine, Redruth, contain a wide range of primary and secondary species, including brochantite (LeBoutillier et al. 2000); Prince of Wales mine, Calstock, near Gunnislake – dump-formed brochantite occurs with linarite (Braithwaite & Cooper 1982); Dolcoath South lode of South Crofty mine, Camborne – brochantite forming crusts and druses on mine walls (LeBoutillier 1996); Ting Tang mine, Carharrack – rare dark emerald-green crusts of brochantite on copper-rich dump material appear to be of post-mining origin (Bruce & Aubrey-Jones 1998); Wheal Edward, St Just – uranium-bearing fissure veins contain brochantite associated with calcite, compreignacite, connellite, cuprosklodowskite and uranophane (Elton et al. 1994b). Brochantite is also noted from: Wheal Cock and Wheal Owles, near Botallack, St Just; Wheal Phoenix (Stowe's section), Linkinhorne; Wheal Providence, near Leedstown, Gwinear; and Wheal Tremayne, Gwinear, near Hayle (Kingsbury 1954). However, but these occurrences deserve further confirmation given the problems with Kingsbury-provenanced samples (Ryback et al. 1998, 2001).

Cumbria *Caldbeck Fells:* Rare, bright emerald-green crystalline crusts of brochantite occur on the dumps of Brae Fell mine, Dale Beck Valley (Bridges et al. 2006); coatings of brochantite have been found at Driggith mine and, similarly, with chrysocolla and malachite in a small copper vein at Potts Gill mine; also at Thief Gills (Davidson & Thomson 1951); uncommon sprays of small dark-green acicular crystals up to 1 mm long, have been found on quartz veinstone from a trial working on the north slopes of Low Pike, near Potts Gill barite mine (Cooper & Stanley 1990, Young et al. 1990a, Green et al. 2003a); beautiful emerald-green crystals of brochantite, sometimes forming pseudomorphs after bladed crystals of linarite or associated with caledonite, occur at Red Gill mine (Davidson & Thomson 1951, Wolloxall 1981, Briscoe 1986a, Cooper & Stanley 1990); small but very perfect and brilliant crystals of brochantite associated with malachite were once found on quartzose rock at Roughton Gill mine (Greg & Lettsom 1858, Rudler 1905, Davidson & Thomson 1951); and brochantite forms minute rounded scales (< 1 mm) and fibrous sprays, apparently altering from langite, and it occurs as a dark emerald-green drusy crust overgrowing cuprite at Silver Gill, near Roughton Gill (Cooper & Stanley 1990, Green et al. 2005b). Reports of brochantite from Brandy Gill, Driggith, Hay Gill, Sandbed (lead) and Mexico mines (Hartley 1984, Cooper & Stanley 1990) may require further confirmation, as they are based on samples with a Kingsbury connection. Ryback et al. (1998, 2001) have shown that many samples in the Kingsbury collection (now at the NHM) have an incorrect provenance. However, brochantite has since been independently confirmed from Driggith and Sandbed mines (Green et al. 2006a) and Potts Gill mine (Young pers. comm. 2005). Green et al. (2006a) note that most of Kingsbury's brochantite specimens are genuine, but at least one probably came from Llaillay, Valparaiso, Chile. *Elsewhere in the Lake District:* Buckbarrow Beck, Corney Fell, Waberthwaite – supergene brochantite, chrysocolla, covellite, goethite and malachite, as well as the rare species bismutoferrite, cuprotungstite and russellite, in chalcopyrite–ferberite–scheelite mineralization in quartz veins within granodiorite (Young et al. 1986, 1991b); Dale Head mine, Newlands Valley – complex translucent mm-size crystals of brochantite (Green & Briscoe 2002); Force Crag mine, Braithwaite, near Keswick – a small area of brecciated vein subjected to supergene processes produced drusy crusts and small blocky crystals of emerald-green brochantite, together with cerussite, langite, namuwite, pyromorphite, serpierite and native silver (Green et al. 1997, Green & Briscoe 2002); Kinniside lead mine, Cleator Moor (Young 1987a); brochantite forms encrustations associated with gypsum, langite and malachite on the walls of old stopes in the Middle level of Paddy End mine, Coniston, Eskdale (Young & Johnson 1985). *Northern Pennine orefield:* Brownley Hill mine, Nenthead, near Alston – rounded green brochantite crystals up to 0.5 mm across occur rarely with harmotome and serpierite in the quartz lining of a large cavity. Brochantite appears to have formed as a result of post-mining oxidation of chalcopyrite (Green et al. 2000); brochantite forms crusts of minute crystals associated with cuprite on the Cornriggs vein and at Cashburn, both on Alston Moor (Bridges & Young 1998); Tynebottom mine, NW of Garrigill, near Alston – small emerald-green crystals of brochantite locally aggregated into dark-green crusts and with a post-mining origin are associated with annabergite, aragonite, chrysocolla, devilline, erythrite, gypsum, hydrozincite, linarite, malachite, serpierite and wroewolfeite (Davidson & Thomson 1951, Braithwaite 1982b, Bridges 1987, Rust 1987a, Bridges & Young 1998, Green & Briscoe 2002).

Derbyshire Very rare emerald-green brochantite occurs at Ball Eye quarry, Bonsall, and at Snelston (also known as Isolation) mine, near Ashbourne (Braithwaite 1983, Ford et al. 1993, Jackman 1996).

Devon Brochantite is associated with connellite and exceptional langite at Bedford United mine, near Tavistock (Watson 1987). It forms vivid- to dark-green crystalline and botryoidal crusts associated with chalcopyrite and devilline at Wheal Friendship (Devon Friendship), Mary Tavy (Brooke 1986).

Durham Brochantite forms crusts in the Firestone level of Groverake mine, near Rookhope, Weardale (Bridges & Young 1998).

Leicestershire Minute prismatic to lath-like crystals and radiating sprays of brochantite are associated with cuprite or malachite in a native copper–cuprite deposit near the Precambrian/Triassic unconformity at New Cliffe Hill quarry, Stanton under Bardon (Hubbard et al. 2005).

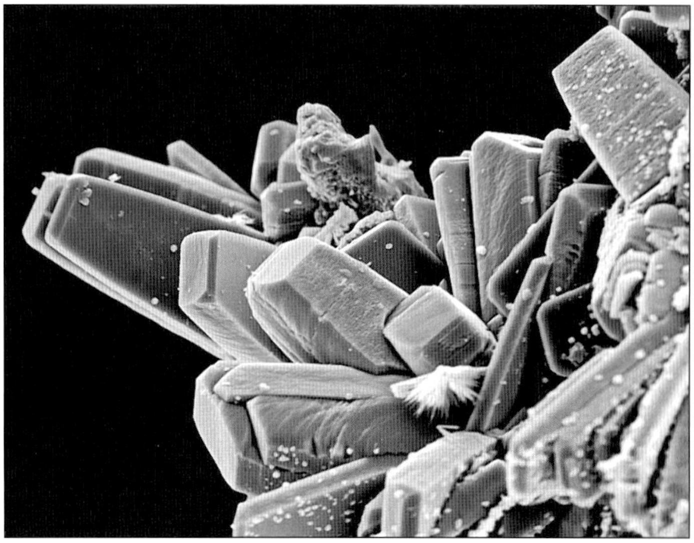

Flat-topped crystals of brochantite from New Cliffe Hill quarry, Stanton under Bardon, Leicestershire; width 110 μm; SEM image. (© D. I. Green)

Staffordshire Supergene oxidation processes acting on a Cu–Pb–Zn assemblage at Waterbank mine, Ecton Hill, have produced rare drusy emerald-green crusts and minute rounded crystals of brochantite up to 0.2 mm across associated with anglesite, hydrozincite, ramsbeckite, serpierite and wulfenite (Rust 1992, 1995c).

Shropshire Beautiful brochantite, langite and linarite were recovered from the dumps of Pool vein, New Engine shaft at Ladywell mine, Shelve, prior to their removal (Braithwaite 1982b).

Wales

Dyfed *Central Wales orefield:* Brochantite is mostly of post-mining origin: Brynarian mine, near Tal-y-bont – brochantite associated with anglesite, cerussite, linarite, ramsbeckite, schulenbergite and langite-group minerals (Rust 1992); Copa Hill mine, Cwmystwyth; Hendrefelen mine, Ysbyty Ystwyth; Mynyddgorddu mine, W of Bont-goch, Ceulanymaesmawr (all Bevins 1994); Eaglebrook (Nantycagl) mine, Ceulanymaesmawr – brochantite forms bright-green platy microcrystals and excellent divergent groups of emerald-green to light-green transparent crystals associated with anglesite, cerussite, native copper, cuprite, devilline, hemimorphite, leadhillite, linarite, malachite, serpierite and wroewolfeite (Jones & Moreton 1977, Braithwaite 1982b, Jones 1983); Esgair Hir mine, Ceulanymaesmawr – very common drusy microcrystalline crusts of brochantite associated with altered chalcopyrite (Rust & Mason 1988); Esgair Fraith mine, near Tal-y-bont (Jones & Moreton 1977); Frongoch mine, between Devil's Bridge and Pont-rhyd-y-groes – brochantite as emerald-green crusts of minute rounded plates in mixed copper–lead sulphide veins and occasionally pseudomorphing langite or wroewolfeite (Green et al. 1996b); Gwaithyrafon mine, SE of Tal-y-bont – emerald-green platy crystals of brochantite

Interlocking sprays of brochantite to 1 mm from Eaglebrook (Nantycagl) mine, Ceulanymaesmawr, Dyfed, Wales. (© D. I. Green)

on chalcopyrite (Rust & Mason 1994); Llechweddhelyg mine, near Clawddmelyn, between Bont-goch and Cwmsymlog – rounded green tabular microcrystals of brochantite in cavities within linarite (Jones 1987); Penrhiw mine, Ystumtuen – very common brochantite forming dark-green microcrystalline coatings on massive sulphide and rarely as euhedral curved tabular crystals up to 0.5 mm in length (Mason & Green 1995); Ystrad Einion mine, near Tal-y-bont – common emerald-green crusts and scattered groups of brochantite forming thin rounded bladed crystals up to 0.6 mm long and seen to be replacing native copper and cuprite. Exceptional crystals up to 1 mm long encrust hemimorphite (Mason & Rust 1997); brochantite is found with cerussite, chalcopyrite, chrysocolla, galena, linarite, malachite and wulfenite on the dumps of Pen-y-bwlch shaft on the Glog-Fawr lode, SE of the village of Ysbyty Ystwyth (Ryback & Saville 1967).

Brochantite is also reported on dump material at Bonville's shaft, Vale of Towy mine, near Carmarthen (Bevins 1994), and from the Lodge Park copper trial, Tre'r-ddôl (Mason & Green 1996).

Gwynedd Brochantite is associated with chalcoalumite, lanthanite-(Ce), malachite and posnjakite at Britannia mine, Snowdonia (Bevins et al. 1985, Pollard et al. 1990). At Great Orme mine, Llandudno, aragonite, brochantite, calcite and malachite have been found cementing Bronze Age spoil material (Lewis 1990, Dutton et al. 1994, Ixer 2001). Brochantite is found as crystalline crusts, uncommon rounded thin tabular crystals up to 0.15 mm long, or rarely as star-like crystal clusters up to 0.6 mm across at a trial level, 2 km NW of Bontddu, near Dolgellau (Saich & Rust 1987). Brochantite is one of 25 post-mining secondary minerals recorded from the copper deposit at Parys Mountain, Anglesey (Jenkins et al. 2000).

Powys *Central Wales orefield:* Brochantite encrusts chalcopyrite-rich veinstone at Dyfngwm mine, Machynlleth (Rust & Rust 1987) – actually the Pen Dylife section of Dylife mine. It is one of many supergene species found on the dumps of Dylife mine, N of Hafren Forest (Jones & Moreton 1977, Moreton 1982). Bevins (1994) notes brochantite at Geufron mine, near Llanidloes. Elsewhere, brochantite has been found as dark-green radiating aggregates at Dolyhir quarry, Old Radnor, near Presteigne (T. F. Cotterell pers. comm. 2007).

Scotland

Dumfries & Galloway Emerald-green minutely crystallized films of brochantite and malachite occur on cleavage planes and fracture surfaces of a Silurian shale at a copper trial in Auchencat Burn, below Hartfell Spa, near Moffat (Moreton & Aspen 1992). A copper-rich (bornite and chalcopyrite) stockwork on the Solway coast near Needle's Eye, SW of Caulkerbush, is mostly oxidized and has produced blue and green coatings of botallackite, brochantite, connellite, gypsum and langite (Braithwaite & Knight 1990). The dumps of West Blackcraig lead–zinc mine (near Newton Stewart) contain a supergene assemblage consisting of aurichalcite, brochantite, cerussite, hemimorphite, hydrozincite, langite, linarite, malachite, posnjakite and wroewolfeite (Livingstone et al. 1976). Anthropogenic brochantite is noted on dump material from the Meadowfoot smelter, Wanlockhead (Green 1987a).

Highland A copper vein in a crush plane within Durness Limestone at Rassal, N of Loch Kishorn, contains pockets of bornite, brochantite and malachite (Heddle 1901, Berridge 1969).

Strathclyde Traces of brochantite, covellite and malachite are associated with molybdenite-bearing veins in a stream boulder from Allt Hallater (S of Ben Starav) in the Central Starav granite, a part of the Etive granite complex (Haslam et al. 1985).

Ireland

Cork Equant crystals of brochantite are associated with cuprite in cavities within goethite at Aghatubrid Beg mine, Glandore, and at Allihies on the Bearhaven (Berehaven) Peninsula, attractive encrustations of well formed langite and brochantite have been collected from mine debris (Ryback & Moreton 1991).

Kerry At Leamnamucka, to the N of Ballybunnion, mixed brochantite and malachite form green stains on pyritiferous shale (Moreton et al. 1995). Brochantite forms a bright-green microcrystalline crust on limestone at Muckross mine (Killarney National Park) and probably developed as a post-mining oxidation product of nearby banded chalcopyrite (and other sulphide) ore. The rare species monohydrocalcite, phaunouxite, rauenthalite and smolianinovite are also reported from this location (Ryback & Moreton 1993, Moreton et al. 1999).

Tipperary Brochantite forms green microcrystalline crusts at Gortdrum mine, near Tipperary town (Braithwaite & Wilson 2003).

Waterford Microcrystals of brochantite are associated with connellite, erythrite, langite and lavendulan on dump material derived from copper workings at Bunmahon, Knockmahon, SW of Waterford town (Ryback & Moreton 1991). Fine crystalline flowstone crusts of emerald-green brochantite, connellite and langite have been collected from Tankardstown mine, Knockmahon (Wolfe 1997, 1998, Green & Moreton 2001, Moreton 2005a).

brockite

hexagonal — RHABDOPHANE GROUP — $(Ca,Th,Ce)(PO_4)\cdot H_2O$

A very rare hydrated calcium–thorium–cerium phosphate found as an accessory mineral in the Dartmoor granite at East Wray Cleave quarry, Lustleigh, Devon. A single grain of brockite was found as an inclusion in apatite. Other accessory minerals in the granite are allanite, apatite, cordierite, epidote, fluorite, garnet, ilmenite, magnetite, monazite, pyrite, uraninite, xenotime and zircon (Ward et al. 1992).

Brockite was possibly found in a panned concentrate collected from Beinn Trilleachan (near Loch Etive) in the Etive granite complex, Strathclyde, Scotland, but confirmation is required (Haslam et al. 1985).

bromlite

TL Brownley (Bromley) Hill, Nenthead, near Alston Cumbria, England

A synonym of alstonite.

bronzite

A synonym of enstatite, pyroxene group.

brookite

orthorhombic — TiO_2

TL Twll maen grisial, Fron-oleu, Prenteg, Tremadog, Gwynedd, Wales?

Named after Henry James Brooke (1771–1857), crystallographer and co-author with W. H. Miller of *An elementary introduction to mineralogy* (1852); born in Exeter, Devon. Brookite is a widely distributed titanium oxide, trimorphous with anatase and rutile, and is found as an accessory mineral in alpine-type veins formed as a result of low-grade regional metamorphism, in contact-metamorphic zones and as a detrital phase. The type locality has been variously referred to as "Tremadog", "between Beddgelert and Tremadog", "Fronolen", or "near Snowdon". Dauphiné, Isère, France is considered to be a second type locality for brookite, but this is based on the record of an 1822 manuscript that is now lost (Dunn & Mandarino 1988, Starkey & Robinson 1992, Bevins 1994).

Considering the transparency, richness of colour, large size, high lustre, perfection of form and complex morphologies, crystals of brookite from Prenteg are considered by many to be among the world's finest (Bancroft 1973). Perhaps the best-known specimen, a fine single 2.5×3 cm crystal sitting on quartz, is at the NHM. Albite, anatase and fluorapatite are the other species in this alpine-type vein assemblage, which occurs in anastomosing quartz veins and fissures in altered dolerite intruded into Ordovician slaty mudstones (Greg & Lettsom 1858, Starkey & Robinson 1992). At Tanygrisiau, 2 km SW of Blaenau Ffestiniog, Gwynedd, brown tabular crystals of brookite were rarely found with albite, anatase, apatite, clinochlore, rutile

Tabular brookite, 2.5 mm across, and bipyramidal anatase (left) from Hendre quarry, near Glyn Ceiriog, Clwyd, Wales. (© National Museum of Wales, Cardiff; photographer M. P. Cooper)

and synchysite-(Ce) in alpine-type veins exposed in 1976 during construction work on the Ffestiniog railway (Green & Middleton 1996). Elsewhere in Wales, brookite is reported from: Arenig Station quarry, Blaenau Ffestiniog, Gwynedd (Bevins 1994); Hendre Henydd, Penrhyndeudraeth, near Porthmadog (R. E. Starkey pers. comm. 2005); Hendre quarry, Pandy, near Glyn Ceiriog, Clwyd (Starkey et al. 1991); Bwlch y Cywion, central Snowdonia, Gwynedd (Williams 1930a); Gimlet quarry, near Pwllheli, Lleyn Peninsula, Gwynedd (Harrison 1894, 1897); and Manod quarry, near Blaenau Ffestiniog, Gwynedd (Bevins 1994, Green & Middleton 1996).

Brookite is a common species in miarolitic cavities within granite at the Shap granite quarry, Cumbria. It forms well crystallized brown striated blades up to 2 mm across, associated with anatase and rutile (Young 1987a, Wirth & Green 2002). Two other Cumbrian granites also contain accessory brookite: the Skiddaw granite and the Threlkeld microgranite (Rastall & Wilcockson 1915).

In Devon, anatase and brookite are described from reddened Dartmoor granite that appears to have been pneumatolyzed during a period of tourmaline crystallization. Boulders on Hamel Down also contain cassiterite, topaz and zircon (Brammall & Harwood 1923, 1925, 1927). Brookite is noted as microscopic crystals in siderite associated with anatase and chlorite from Virtuous Lady mine, Buckland Monachorum, Devon (Greg & Lettsom 1858, Russell 1913a). Three Cornish locations for brookite are listed by Golley & Williams (1995): Lanterdan quarry, Tintagel; Prince of Wales quarry, Trebarwith; and Hingston Down quarry, Calstock.

Extremely varied morphology is a characteristic of brookite found in the Millstone Grit of west and southwest Yorkshire. At Crimsworth Dean, the Lower Kinderscout Grit contains dumb-bell (yellow tabular crystals, up to 0.15 mm across), fringe and lattice types; a bundle type is also reported from Buckden Pike (Butterfield 1939). Heavy-mineral concentrates from chamositic oolitic sandstones in the Hollins Basin, and Rosedale Abbey Bank, Rosedale, Yorkshire, contain abundant anatase and brookite (Rastall & Hemingway 1949). The Cleveland Ironstone contains thick tabular plates and extremely thin plates of deep reddish-brown brookite (Lindsey 1905). Other oolitic ironstones at Beckhole, Limber Hill and Ugglebarnby village, all in the valley of the Esk, near Whitby, Yorkshire, contain detrital brookite as thin yellowish-brown rectangular plates up to 0.2 mm across (Phillips 1932). Similar concentrates separated from the Triassic St Bees Sandstone at Dufton Mill, Dufton, and from the Permian Hilton Plant Beds at Hilton Beck, Hilton, both in Cumbria, also contain brookite (Versey 1939).

In Scotland, brookite occurs as a pale-yellow glassy accessory mineral in the microgranite of Ailsa Craig, Firth of Clyde, Strathclyde (Harrison et al. 1987), and as reddish-brown plates in calcite from Killin, Tayside Region (Macpherson & Livingstone 1982). Brookite occurs with anatase, andalusite, apatite, monazite and tourmaline in a heavy-mineral concentrate from pegmatite veins near Colehill farm, N of Ellon, Grampian (Mackie 1925).

Brookite is one of many accessory minerals recorded from the Leinster granite and is also reported from Kingstown and Killiney, Co. Dublin, Ireland (Smithson 1932). Minute crystals of brookite isolated from clay in joints in quartzite from Shankill, Co. Dublin, are now at the NHM.

A pegmatite at Sorel Point, Jersey, contains abundant reddish-brown plates of brookite up to 1 mm across associated with anatase, chlorite and plagioclase. The brookite is thought to be derived from the alteration of titanite (Mourant 1961).

brown hematite

A synonym of limonite.

brown linnets

A Derbyshire term used to describe mimetite.

brownlite

TL Brownley (Bromley) Hill, Nenthead, near Alston Cumbria, England

A synonym of alstonite.

brown ochre

A synonym of limonite.

brown spar

A synonym of magnesite or siderite (both calcite group) or iron-rich dolomite.

BRUCITE GROUP

Trigonal hydroxides with the general formula $M^{2+}(OH)_2$, where M^{2+} = Fe, Mg, Mn, Ni. Brucite, portlandite, pyrochroite and theophrastite are members of the group found in Britain or Ireland.

brucite

BRUCITE GROUP

trigonal $Mg(OH)_2$

Named after Archibald Bruce (1777–1919), physician, professor at Yale University, and founder of the *American Mineralogical Journal* in 1810; born in New York City of a British father. Brucite is a relatively uncommon magnesium hydroxide that forms during serpentinization of dunite and as a low-temperature hydrothermal-vein mineral in metamorphic limestones. It is a common alteration product of periclase.

Brucite occurs in the Mendip Hills, at Merehead (Torr Works) quarry, Cranmore, Somerset. Here Lower Carboniferous Limestone is veined by a varied assemblage of rare copper and lead minerals within pods of manganese oxides. The pods contain many small cavities in which calcite nodules have formed and it is within two cavities in calcite that brucite has been found. Compact masses of transparent brownish-green or pale-green prismatic–acicular crystals of brucite are arranged in radiating groups that grow out from the cavity walls. Alteration to hydromagnesite and later crystallization of blixite in one nodule is noted (Alabaster 1977).

Bright-green scales in a chlorite–magnetite–talc schist associated with serpentinite on the islet of Rhyd-y-Bont Creek, Holy Island, Anglesey, Gwynedd, Wales, have been optically identified as brucite (Greenly 1919, Bevins 1994).

On mainland Scotland, brucite is a significant component (modally up to 28% in places) of a Cambro-Ordovician marble outcropping in the Ledbeg River, N of Loch Borralan, Assynt, Highland Region, where it forms a rock that is white or yellowish-white mottled with grey. It has been produced by the thermal metamorphism of dolomitic limestone by the intrusion of the nearby Cnoc-na-Sroine syenite (Peach et al. 1907, Newman 1971, MacGregor & Phemister 1972, Smith et al. 1992). Brucite occurs at two other locations in Highland Region: in thermally metamorphosed Cambrian dolomites at Loubcroy, N of the main Lairg–Lochinver road (Newman 1971), and in metamorphosed siliceous dolomites within the Ballachulish granite aureole, near Kentallen (Berridge 1969, Ferry 1996). At Balmaha, close to the south-east corner of Loch Lomond, Central Region, brucite is interlayered with antigorite in metamorphosed Highland Border Suite serpentinites, and both species appear to have grown at the expense of lizardite (Ikin & Harmon 1983).

Small laths of brucite are found in a Lewisian metamorphosed limestone at Bay Steinigie, near the Finsbay (Fionnsabhagh) to Borve paragneiss belt on South Harris, Western Isles. Calcite and dolomite are the main constituents of the rock, although chondrodite, forsterite, humite, pargasite, phlogopite and minor graphite and titanite are also noted (Dearnley 1961). In a zone of thermal metamorphism around the Tertiary Beinn an Dubhaich granite, near Broadford, Isle of Skye, Highland Region, Cambrian dolomitic limestone has been converted to a brucite marble. At Cnoc Dubh, between Kilbride and Torrin, the marble contains appreciable forsterite. Brucite content is very variable (Harker 1904, Newman 1971). Some of the brucite marbles are characterized by a veined texture in which brucite-bearing white marble is veined by a spotted black or turbid brown brucite–calcite assemblage (Tilley 1951). At Camas Mòr, on the small island of Muck, Highland Region, a Tertiary dyke of olivine gabbro has thermally metamorphosed Jurassic limestones to produce a suite of marbles in which brucite, calcite, monticellite, periclase and spinel are found (Tilley 1947).

Brucite is found in veins cutting serpentinite at Swinna Ness, Unst, Shetland. It forms translucent, foliated plates of a silvery-white colour and also botryoidal

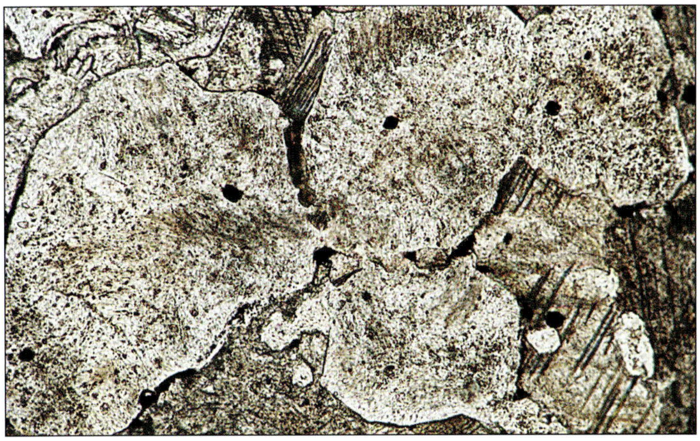

Irregular masses of brucite with interstitial calcite and minor magnetite. Brucite marble, Ledbeg, north of Loch Borralan, Assynt, Highland Region, Scotland; width 1.3 mm; viewed in PPL. (© K. Gardner)

masses (Greg & Lettsom 1858, Heddle 1878a). A vein at North Cross Geo, Haroldswick, Unst, contains fibres of pale-green brucite up to 38 mm long, associated with magnesite, magnetite and talc. At Haaf Gruney, a small island off the southeast coast of Unst, brucite is found with chromite and pyroaurite (Heddle 1901). Brucite is also noted from Hill of Clibberswick, Unst, and East Neap, Fetlar, by Flinn & Pentecost (1995). They report screes on high serpentinite cliffs internally cemented by aragonite, brucite and hydromagnesite, which they argue were formed after the most recent glaciation. The minerals were deposited from solutions originating as rain (contaminated by sea spray), which fell on the screes and dissolved magnesium and minor amounts of calcium from the finer components of the scree.

In Ireland, brucite forms pseudomorphs after periclase in thermally metamorphosed limestones from the Grange Cross area of Carlingford District, Co. Louth. Associated species are forsterite and monticellite (Nockolds & Vincent 1947).

brugnatellite

hexagonal $Mg_6Fe^{3+}(CO_3)(OH)_{13}\cdot 4H_2O$

A rare hydrated magnesium–iron carbonate hydroxide found on a brucite-bearing vein in serpentinite at Swinna Ness, Unst, Shetland, Scotland. Brugnatellite forms pearly-white to pale-brown efflorescences (Macpherson & Livingstone 1982).

bruiachite

TL Loch Bruicheach, SW of Beauly, Highland Region, Scotland

A synonym of fluorite; see Heddle (1889b).

brunsvigite

Chloritic material with a composition close to that of magnesium-rich chamosite.

brushite

monoclinic $CaHPO_4\cdot 2H_2O$

A rare hydrated calcium–hydrogen phosphate found lining the interior of a human skull from a cave near Kilkenny, Co. Kilkenny, Ireland. Brushite forms a thin crust of minute colourless platy crystals (Ryback et al. 1988).

Brushite is also found at Wheal Cock, near Botallack, St Just, Cornwall, as aggregates of small white to off-white crystals associated with siderite (Wolloxall 1989).

buergerite?

TOURMALINE GROUP

trigonal $NaFe^{3+}_3Al_6(BO_3)_3Si_6O_{18}(OH)_3F$

An uncommon sodium–iron–aluminium borosilicate hydroxide fluoride, whose composition was defined in the tourmaline classification by Hawthorne & Henry (1999). Clark (2007) describes the problems in calculating the structural formulae of tourmaline analyses obtained using EMP and points out that the common assumption that all iron is ferrous can result in the misidentification of buergerite as fluor-schorl. See fluor-schorl.

bukovskýite

triclinic $Fe^{3+}_2(AsO_4)(SO_4)(OH)\cdot 7H_2O$

A rare hydrated iron arsenate sulphate hydroxide found on a garden wall made up of blocks of arsenopyrite- and chalcopyrite-bearing mine waste derived from Devon

Blades of bukovskýite forming a pale-yellow crust on weathered arsenopyrite-rich mine waste from Devon Great Consols mine, Devon (near Gunnislake in Cornwall); width 70 µm; SEM image. (© M. R. Power, Camborne School of Mines)

Great Consols mine, Devon (which is near to Gunnislake, Cornwall). Bukovskýite is a secondary growth derived from weathering of the primary sulphides and it forms fine, almost moss-like pale-yellow crusts associated with gypsum and scorodite (M. R. Power pers. comm. 2004).

buntkupfererz

A synonym of bornite.

bustamite

triclinic $(Mn^{2+},Ca)_3Si_3O_9$

A generally uncommon manganese–calcium silicate, but an important constituent of the manganese ores of Treburland mine, 2 km SE of Fivelanes, Altarnun, Cornwall, where it is found associated with manganese-rich hedenbergite, rhodonite, spessartine and tephroite. The ore lies within, or in contact with, a highly calcareous calc-silicate rock, originally an impure limestone that has been thermally metamorphosed by the adjacent Bodmin Moor granite pluton. The manganese ore was probably deposited as rhodochrosite prior to metamorphism. Bustamite has a pale flesh-pink colouration and it forms fibrous crystals 25 mm or more in length (Russell 1946d, Tilley 1946). Brown lensoid masses of bustamite and hausmannite occur with rhodonite at Greystone quarry, Lezant, near Launceston, Cornwall. These species occur in a structurally complex metasedimentary sequence unrelated to the Cu–Pb–Zn vein mineralization for which this quarry is famous (Elton et al. 2004). Light-brown bustamite (pale flesh-pink when first exposed) from a similar paragenesis to Treburland is reported from the aureole of the Dartmoor granite at Meldon (British Railways quarry complex), near Okehampton, Devon (Howie 1965).

Minor bustamite with braunite, manganese-rich calcite, hematite, ilmenite, rhodochrosite, pyrophanite and possibly jacobsite form a complex stratiform manganese ore up to 4 m thick at Dalroy Burn, near Culloden, Highland Region, Scotland. The ore has a geothermal origin best explained by deposition from hot springs in a sub-lacustrine environment during Devonian times (Nicholson 1987b, 1990).

buttgenbachite

hexagonal $Cu_{19}Cl_4(NO_3)_2(OH)_{32}\cdot 2H_2O$

A rare hydrated copper chlorine hydroxyl nitrate found at Botallack mine, St Just, Cornwall (Golley & Williams 1995). Buttgenbachite is XRD-confirmed, but no further details are given. A Cornish specimen now in the NHM (originally thought to be connellite and to have been collected at Wheal Providence, St Day) is confirmed to be buttgenbachite (Leverett & Williams 2007).

byssolite

Both actinolite and tremolite (amphibole group) have been referred to as byssolite.

bytownite

A common triclinic calcium–sodium aluminosilicate with an intermediate (An_{70-90} mole %) composition between sodium-rich albite (An_{0-10}) and calcium-rich anorthite (An_{90-100}) end members. It is part of the plagioclase-feldspar solid-solution series that also includes oligoclase (An_{10-30}), andesine (An_{30-50}) and labradorite (An_{50-70}).

C

cacholong

An opaque variety of opal, said to be bluish white, white or cream. The name has also been used as a synonym of actinolite (var. nephrite), amphibole group.

cacoxenite

hexagonal $AlFe^{3+}_{24}(PO_4)_{17}O_6(OH)_{12}\cdot{\sim}75H_2O$

An uncommon (although widespread in small amounts) hydrated aluminium–iron phosphate hydroxide found in the southeastern part of the Bunny mine system at Gunheath china-clay pit, Carthew, St Austell, Cornwall. Cacoxenite is a secondary mineral from the oxidation zone and it occurs as yellowish-brown to orange-brown radiating fibres in a limonite vein. It forms aggregates up to 0.5 mm across on yellow brown stained fissure surfaces that cut the hydrothermally altered granite. Other species in the oxidation zone include arthurite, brochantite, cassiterite (varlamoffite), chalcosiderite, cyrilovite, dufrénite, leucophosphite, libethenite, scorodite, torbernite, turquoise, variscite and wavellite (Weiss 1994). Cacoxenite was one of several supergene species that were found with well crystallized wolframite when, in 1997, the summit car-park at Kit Hill subsided into the abandoned workings of Kit Hill Consols, Stoke Climsland, Cornwall. Stabilization work involved a large excavation of the old spoilheaps where cacoxenite was found later (Wolfe 1998). Ryback et al. (1988) note cacoxenite on specimens at the NHM from: Castle-an-Dinas mine and Wheal Phoenix, both in Cornwall, and Burdell Gill, Caldbeck Fells, Cumbria. It is not stated if these specimens are from the Kingsbury collection (and therefore potentially of dubious provenance – see Ryback et al. 1998, 2001), but Cooper & Stanley (1990) confirm this for the last locality. The Burdell Gill occurrence is also reported by Hartley (1984) and Young (1987a). Golley & Williams (1995) also list cacoxenite in Cornwall from Greystone quarry, Lezant, near Launceston, and from Wheal Jane, Kea, near Truro (said to be doubtful).

Cacoxenite fibres to 1 mm from Ballycormick, near Shanagolden, Co. Limerick, Ireland. (© M. P. Cooper)

In Wales, yellow, gold or orange cacoxenite occurs at Pwlldu beach, Bishopston, Gower Peninsula, West Glamorgan, in close association with other phosphate minerals – crandallite, variscite and wavellite. It is found on joint surfaces in chert (probably derived from the disused limestone quarry at Pwlldu Head) as small (<0.3 mm) tufted spheres and hemispheres with a silky lustre, and it is easily overlooked despite being relatively common (Dean & Cotterell 2003, Green 2005a).

Excellent specimens of cacoxenite in the form of golden-yellow fibrous 1 mm spherules and bunches of needles, associated with silky white acicular wavellite are reported from Ballycormick, near Shanagolden, Co. Limerick, Ireland. Samples were collected from the stream bank opposite the old fort of Lismeenagh in a quartzose gossan that also contains crandallite on greenish-blue crusts of turquoise (Ryback et al. 1988, Ryback & Moreton 1991).

cairngorm

TL Cairngorm Mountains, Grampian, Scotland

A clear transparent orange-brown to deep-brown gem variety of smoky quartz.

calamine

In the late 18th century it was discovered that the substance termed "calamine" actually consisted of two different minerals: a zinc silicate and a zinc carbonate (hemimorphite and smithsonite). Calamine is also a synonym of hydrozincite.

calaverite?

monoclinic $AuTe_2$

A very rare gold telluride tentatively identified from Afon Wen, Dolgellau goldbelt, Gwynedd, Wales, where a 2–4 µm-diameter grain of calaverite associated with petzite has been found included within an alluvial gold grain (Leake et al. 1993). Bevins (1994) notes a further occurrence in the same area, possibly from Panorama mine.

calcioancylite-(Ce)?

monoclinic $(Ce,Ca,Sr)CO_3(OH,H_2O)$

A rare hydrated cerium–calcium–strontium carbonate hydroxide thought to occur at Guisachan quarry, near Tomich, Strathglass, Highland Region, Scotland. It forms faintly pink bipyramidal crystals, up to 1 mm across, but further analysis is required to confirm its identity (D. I. Green and N. Hubbard pers. comms 2005).

calciopalygorskite

TL Strontian, Highland Region, Scotland

An uncertain composition; also known as mountain leather.

calciostrontianite

A calcium-rich variety of strontianite.

calciovolborthite

orthorhombic $CaCu(VO_4)(OH)$

A rare calcium–copper vanadate hydroxide found encrusting the surface of spherical vanadium-rich nodules and lenses of gypsum within the Triassic Keuper Marl at the Butterley and Blaby brick pit, Glen Parva, Leicestershire. Calciovolborthite forms microscopic, pale bluish-green, sponge-like reniform masses and is rimmed with crusts of yellowish-green volborthite. Etching of primary gypsum by meteoric water produced calcium sulphate in solution, and this, in combination with vanadium, leached from the nodules enabled the two vanadates to precipitate (King & Dixon 1971).

An occurrence of calciovolborthite associated with quartz and chalcocite from

Penrhyn quarry, Bethesda, Gwynedd, Wales, is briefly reported by Bevins (1994).

Burke (2006) reporting on an IMA-CNMNC-approved mass discreditation of species states that calciovolborthite is formally discredited and compositions are either tangeite of vésigniéite. British reports of calciovolborthite now require further confirmation.

CALCITE GROUP

Trigonal carbonates with the general formula $A^{2+}(CO_3)$, where A^{2+} = Ca, Cd, Co, Fe, Mg, Mn, Ni, Zn. Calcite, magnesite, otavite, rhodochrosite, siderite and smithsonite are members of the group found in Britain or Ireland.

calcite

CALCITE GROUP

trigonal $CaCO_3$

An extremely common calcium carbonate that is trimorphous with aragonite and vaterite, and which forms a series with rhodochrosite. Calcite is a major rock-forming mineral in chalk, limestone and marble; it forms stalagmites and stalactites (speleothems) in caves, and it cements many clastic sedimentary rocks. It is a common gangue mineral in hydrothermal veins and, less commonly, it occurs in alkalic-to-mafic igneous rocks such as carbonatite. In a work of this nature only representative examples of calcite can be presented. For Scotland, many more occurrences and 224 crystal drawings of calcite are presented by Heddle (1901).

England

Avon Well formed colourless-to-white scalenohedral crystals up to 2 cm long and partially covered with fluorite and goethite occur at Hampstead Farm quarry, Chipping Sodbury (Alabaster 1990b, Brooke 1994b). Scalenohedral calcite is noted from Southfields quarry, 2 km N of Chipping Sodbury (Curtis 1981), and at Wesley Theological College (Wesley mine), Westbury-on-Trym, calcite veins are host to chalcopyrite, galena, pyrite and a suite of rare lead-oxychloride species (Alabaster 1989b).

Cheshire Calcite is noted from Engine Vein mine, Alderley Edge ridge (Carlon 1979, Braithwaite 1994).

Cornwall The copper–tin mines of the St Just district (including Botallack, Levant, Wheal Cock, Wheal Owles, Wheal Edward) were celebrated for the beauty and diversity of their crystalline calcite, most notably pale-pink or orange roses, sometimes referred to as Levant paper, schiefer spar or slate spar (Embrey & Symes 1987, Wolloxall 1987, Bancroft & Weller 1993, Balzer et al. 2003). Rhombohedra and scalenohedra of colourless, white or pale-yellowish calcite associated with galena, pyrite, sphalerite and widenmannite are noted from North Zawn, near Botallack (Elton & Hooper 1996). Fine examples of calcite from Wheal Wrey, St Ive, near Liskeard, are now in collections at the NHM and the Royal Cornwall Museum, Truro. Crystals reach 10 cm in length, but perhaps the most magnificent specimen has 5 cm-long L-shape twinned acute rhombohedra on a groundmass of smaller calcite prisms encrusting galena (Rudler 1905, Rust 1983a, Embrey & Symes 1987, Bancroft & Weller 1993). Splendid clear rhombohedral calcite occurs in a zeolite assemblage veining gabbro at Dean quarry, St Keverne, Lizard Peninsula. Some crystals have brilliantly reflecting faces and reach 1 cm in length. Pale-brown rhombohedral calcite associated with prehnite is also present (Seager 1969, 1971, Wolters & Wolters 1986, Elton et al. 1997, 1998). The hydrothermal event producing this calcite also produced calcite at other locations on the Lizard Peninsula, including: Countybridge quarry, Goonhilly Downs (Holyer 1996); Kennack Sands (Bevins et al. 1987, Power et al. 1997); and Porthkerris Point, 3 km N of Dean quarry (Seager 1978b). Calcite is also noted from: Burraton Coombe quarry, St Stephens by Saltash (Russell 1925); Chywoon quarry, Longdowns – thin platy crystals in pegmatite (Moulding & Hooper 2005); Delabole slate quarry; Wheal Mary Ann, Menheniot (both Golley & Williams 1995); Greystone quarry, Lezant, near Launceston (Elton et al. 2004); Herodsfoot mine, near Liskeard (Rudler 1905); Lanterdan quarry, Tintagel (Bowman 1900); North Roskear mine, Camborne; Wheal Penrose, near Helston; West Wheal Grambler and Wheal Buller, near Redruth (all Collins 1892); Penberthy Croft mine, St Hilary (Betterton et al. 2000); Port Quin mine, St Endellion – very unusual stacked scalenohedra to 10 mm forming a post-mining crust on the mine walls (N. Carruth pers. comm. 2005); Rosevanion quarry, St Columb (Hosking 1951); South Crofty mine, Camborne (LeBoutillier 1996); South Tolgus shaft, Redruth (Davey 1948); and Treburland mine, 2 km SE of Fivelanes, Altarnun, Bodmin Moor – manganocalcite (Russell 1946d).

Cumbria *Caldbeck Fells:* Calcite is an abundant gangue mineral occurring in massive form and as attractive milky nailhead crystals in veinlets cross-cutting the Harding vein at Carrock mine, Grainsgill (Davidson & Thomson 1951, Green & Briscoe 2002). Small deeply etched lenticular crystals and crystals, with bright-red

Rosettes of paper calcite to 15 mm from the Levant section of Geevor mine, near Pendeen, St Just, Cornwall. (© D. I. Green)

Twinned calcite from Wheal Wrey, St Ive, near Liskeard, Cornwall; crystal length to 30 mm. (© Royal Cornwall Museum, Royal Institution of Cornwall; photographer D. I. Green)

fluorescence under UV light, occur at Red Gill mine (Briscoe 1986a, Cooper & Stanley 1990). Other occurrences are from: Driggith mine (Thimmaiah 1956, Stanley 1979); Hay Gill and Potts Gill mines (Hartley 1984); Ingray Gill (Cooper & Stanley 1990); Roughton Gill mine (Dewey & Eastwood 1925, Davidson & Thomson 1951). *West Cumbria:* Crystal-lined cavities were common in the hematite mines in this area and from them came thousands of specimens of calcite, barite, fluorite, hematite and quartz. Fine specimens of nailhead calcite were collected from Beckermet mine, SE of Egremont, and from the Banana Slide fault in this mine came both nailhead and scalenohedral crystal clusters stained bright red by included

CALCITE

Butterfly twin calcite, 50 mm across, from Bigrigg mine, near Egremont, Cumbria. (© D. I. Green)

hematite (Davidson & Thomson 1951, Green & Briscoe 2002). Spheroidal white calcite on specular hematite is noted from the neighbouring Florence mine – Ullcoat No. 1 pit (Green 1995a, Green & Briscoe 2002). However, classic calcite from this region is exemplified by that found at Bigrigg mine, near Egremont – well known for its butterfly twinned (Young 1987a, Balzer et al. 2003) and jackstraw calcite (Bancroft 1973), and at Wyndham mine, Bigrigg – beautiful glass-clear scalenohedral calcite filling large vugs (Davidson & Thomson 1951). Miers (1889a) describes the following varieties of calcite from the Egremont area: prismatic, pyramidal, trigonal, twinned, butterfly and two types of heart-shape crystals. Calcite is also reported from: Frizington (Tresise 1991, Balzer et al. 2003); Gillfoot and Haile Moor mines, Egremont and Distington quarries, Distington (Davidson & Thomson 1951, Balzer et al. 2003); Goose Green mine, Parkside, Frizington (Green & Briscoe 2002); Hodbarrow mine, Millom, near Haverigg (Smith 1924); Mossbay mine, Bigrigg (Young 1987a); Pallaflat mine, Egremont – excellent specimens including a 14.5 cm cluster of orange-yellow twin crystals (Davidson & Thomson 1951, Young 1987a, Green & Briscoe 2002, Balzer et al. 2003); boreholes in the Sellafield area – strongly fluorescent, zoned nailhead crystals (Milodowski et al. 1998); and Stank mine, near Barrow-in-Furness – very fine specimens (Rudler 1905, Young 1987a). *Elsewhere in the Lake District:* Finely crystallized nailhead and dogtooth (scalenohedral) calcite occur at Greenside mine, near Glenridding, Patterdale, and is notable for its deep salmon fluorescence under UV light (Goodchild 1884, Davidson & Thomson 1951, Young 1987a). Calcite is common as a late-stage infilling in veins and miarolitic cavities in the Shap granite (Wirth 1995, Wirth & Green 2002). Other locations include: Baron Cross quarry, Stainton, near Penrith; Cleator Moor area; Flusco quarry, Greystoke, near Penrith; Force Crag mine, Braithwaite, near Keswick; Shap Blue quarry (all Davidson & Thomson 1951); Barrow mine, Braithwaite (Eastwood 1921); Bonser mine, Coniston (Stanley & Vaughan 1982b); Christcliff and South Cumberland mines, near Boot, Eskdale (Young 1985e); Dale Head mine, Newlands Valley (Stanley 1979); Graphite mine, Seathwaite (Strens 1965); and Threlkeld quarry, Threlkeld (Rastall 1941). *Northern Pennine orefield:* Well crystallized colourless-to-white, pink, yellow or brown crystals of calcite up to 3 cm long and with nailhead terminations occur at Brownley Hill mine, Nenthead, near Alston (Wood 1991, Green et al. 2000). Well formed and often colourless crystals of calcite were formerly common at Tynebottom mine, NW of Garrigill, near Alston (Davidson & Thomson 1951, Dunham 1990, Green & Briscoe 2002), and post-mining hollow-straw stalactites, coloured pink with included erythrite, are also reported (Bridges & Young 1998). Other locations are: Capelcleugh and Smallcleugh mines, Nenthead, near Alston (Green & Briscoe 2002); Dow Scar level, Hilton mine, Scordale, near Appleby (Green & Briscoe 2002, Bridges & Green 2005b); Hagg's and Handsome Mea mines, Nenthead and Silverband mine, Milburn, near Appleby (Davidson & Thomson 1951); High Stennerskeugh level, Ravenstonedale (Young 1987b); Rampgill mine, Nenthead (Bridges & Green 2006b); Rotherhope mine, Leadgate, near Alston (Davidson & Thomson 1951, Dunham 1990, Green 1998b); and Silver Top quarry, Hallbankgate, Brampton (Young et al. 1989).

Derbyshire Calcite occurs in almost every cave and lead mine in the county, although in many cases is not of notable quality. The most attractive form is the scalenohedron that was found in particularly large crystals up to 30 cm long in the Masson Hill mines (Royal mine and Cumberland Cavern). Other superb examples,

Calcite scalenohedron from Millclose mine, Wensley, near Matlock, Derbyshire; length of crystal 90 mm. (© National Museum of Wales, Cardiff; photographer M. P. Cooper)

came from Millclose mine. Rusty or grey scalenohedra up to 16 cm long have been found at Calton Hill and further large examples are recorded from Golconda mine, Smalldale pipe at Bradwell, and Blende vein at Magpie mine. Coloured varieties of calcite include: fine amber to honey-colour scalenohedra at Magpie mine, apple green at Brightgate, near Winster, and Moorend, near Bonsall; brown (containing dispersed hydrocarbons) at High Rake, Little Longstone; orange at Speedwell mine, and red banded with hematite in the Bincliff mine levels in the Manifold Valley. Beautiful twinned crystals are reported from Bonsall Moor (axehead or butterfly twins) and Ladywash mine, Eyam (Derbyshire twins). The Peak District caves contain a range of speleothems that include a cream to amber variety at Treak Cliff Cavern and an orange "poached-egg" variety at Poole's Cavern, Buxton. Crushed calcite has been used in stucco pebble-dash from mines at Monsal Dale, near Chelmorton; Moss rake, Bradwell, and near Peak Forest (all Ford et al. 1993). Further descriptions are provided by the following authors: calcite–fluorite veins at Arrock Black Marble mine, Ashford-in-the-Water, near Bakewell (Orme & Ford 1970); very fine crystals at Ball Eye mine, Bonsall and Water Groove mine, Middleton Dale (Greg & Lettsom 1858); barite–calcite–fluorite–galena–sphalerite veins at Blackhole mine, Eyam; Dirtlow Rake, Castleton; Dirty Rake, Stoney Middleton; Hard Rake, near Sheldon; High Rake, Hucklow; Moss Rake on Bradwell Moor; Raper mine, Long Rake, near Middleton-by-Youlgreave; Tideslow Rake, Tideswell (all Robinson & Ineson 1979); calcite strongly associated with sphalerite at Dovegang mines, near Cromford; Edge Rake, near Tideswell; Field Grove mine, Sheldon; High Rake, Hucklow; Knotlow mine, Monyash; Ladywash mine, Eyam; Lathkilldale mine, S of Over Haddon; Moury mine, near Tideswell; Newall Nook mine, Bradwell, near Castleton; Odin mine, Castleton; Shaw's quarry, near Wirksworth; Shuttle Rake, near Bradwell (all Mostaghel 1985); beautiful 4 cm-long scalenohedra at Fall Hill quarry, Ashover (Smith 1982, Quirk 1993); spectacular transparent-to-translucent scalenohedra on fluorite at Gregory mine, Ashover (Cooper 2005); beautiful sharp transparent scalenohedra to 25 mm on fluorite and 7 cm-wide twinned crystals at Ladywash mine, Eyam (Braithwaite 1983, Balzer et al. 2003); calcite and sphalerite-lined open cavities in the Blende vein, Magpie mine, Sheldon (Worley 1976, Robinson & Ineson 1979, Mostaghel 1985); coarsely crystalline calcite in neptunian dykes at Manystones quarry, near Brassington (King 1966); massive, banded and dogtooth varieties, often of considerable size, at Millclose mine, Wensley, near Matlock (Sarjeant 1956b, Robinson & Ineson 1979, Mostaghel 1985); transparent scalenohedral crystals up to 60 mm long at Mill Dam mine, near Great Hucklow (Wolfe 2006); massive and white scalenohedra up to 40 mm long at Milltown quarry, near Ashover (Rust 1991c); scalenohedra in multiple groups up to 20 cm across in solution cavities in limestone at Pike Lane opencut, near Winster (Uttley 1989); massive, dogtooth, nailhead and stalactitic varieties in the Pindale area, Castleton (Sarjeant 1956b); 50 mm-long scalenohedra sprinkled with smithsonite, and transparent greenish calcite speckled with bornite and native copper at Riber mine, Matlock (Varvill 1959, Braithwaite 1983, Balzer et al. 2003); pink and red calcite veins containing small translucent scalenohedral and rhombohedral crystals at Snelston mine, near Ashbourne (Jackman 1996); pale honey-brown or colourless-turbid-white to pale-brown scalenohedra and combined rhombohedral–scalenohedral at Wapping mine, Cromford (Ince 1991); colourless and pink, massive and dogtooth varieties in the doleritic Waterswallows sill, NE of Buxton (Sarjeant 1956b, Ineson et al. 1983); and well developed scalenohedra to 10 cm in the roof of Witches' Cave, Treak Cliff Cavern (Ford 1955).

Devon Crystalline calcite is imprecisely described from: Bere Alston mines, near Bere Ferrers; Combe Martin; Venn limestone quarry, near Barnstaple; limestone quarries at Plymouth and Torquay; and Wheal Friendship, near Tavistock (all Hall 1868b). Calcite is associated with celestine at Peak Hill, near Sidmouth (Rudler 1905). Rough scalenohedral crystals of calcite are occasionally found in the gold- and selenium–gold-rich carbonate veins at Hope's Nose, Torquay (Harrison & Fuller 1987, Stanley et al. 1990a). Finely developed crystals of albite, calcite, datolite, fluorite, prehnite, pyrrhotite and quartz form veins at the Meldon aplite quarries, near Okehampton (Dearman 1960).

Dorset Small veins in chalk at Corfe Castle contain clear and colourless crystals of calcite up to 25 mm long. Some show internal ghosting or interpenetrant twinning (Bowman 1914). Yellow crystals of calcite are reported from the Isle of Portland (Greg & Lettsom 1858).

Durham Cambokeels mine, Eastgate, Weardale, is well known for a remarkable variety of calcite forms, including paper-thin hexagonal plates, simple rhombohedra, stacked hexagonal groups, pseudo-cubic and nailhead crystals (Green & Briscoe 2002). Nailhead is the dominant form of calcite at the nearby Boltsburn mine, Rookhope, where remarkable stalactites composed of these crystals have been found in cavities in the mineralized flats (King 1982b, Dunham 1990). Elsewhere in Weardale, calcite showing diverse habits is reported from: Blackdene mine, near Ireshopeburn; Frazer's Hush mine, SE of Allenheads; Heights quarry and flawless prismatic crystals at Stanhopeburn mine, N of Stanhope (Green & Briscoe 2002). Vugs in evaporite rocks of the Permian Lower Magnesian Limestone from the Stainton area (examined in borehole material) contain calcite rhombs up to 2.5 mm across and occasionally dogtooth calcite (Raymond 1962). Other occurrences in Magnesian Limestone include: in boreholes at Fishburn and Rushyford (Fowler 1943); colourless-to-white rhombohedral calcite lining cavities at Raisby cutting, E of Coxhoe (Young 2001); and calcite with a variety of forms in cavities at Rough Furze quarry, E of Ferryhill (Hirst & Smith 1974). The tufa variety of calcite occurs at Greenfoot quarry, Stanhope, and High Hurth Edge quarry, Langdon Beck, Teesdale (Bridges & Young 1998).

Hampshire Brightly luminescent calcite is a feature of early cementation in the Sherwood Sandstone from the Marchwood borehole N of Southampton and at other locations in the Wessex Basin (Strong & Milodowski 1987).

Hereford & Worcester Calcite is noted from the Everton quarries, near Ledbury (Greg & Lettsom 1858).

Humberside Brown to clear calcite occurs in a variety of habits in septarian nodules from Hilston Beaches, Holderness (Green 1993).

Kent Like most other chalk, Upper Cretaceous chalk from a borehole at Sprattling Court farm, NW of Ramsgate, is a very pure micritic limestone produced by planktonic algae, and at high magnification is shown to consist of micron-size plates and circular groupings known as coccoliths (Spears 1979). London Clay at Warden Point, Isle of Sheppey, contains septarian nodules in which calcite occurs as yellow or green linings to cracks or tiny crystals in drusy patches (Monk 1982).

Leicestershire Calcite occurs in various habits in mineralized veins within Precambrian rocks at Newhurst quarry, Shepshed. It displays pink or bright-red fluorescence under UV light and forms colourless scalenohedral crystals in veins of dolomite within Triassic sediments above a major unconformity (Ince 2005). Calcite is a secondary accessory mineral in altered gabbroic and granodioritic rocks in the Mountsorrel district (Taylor 1934). Other reports include: sharply defined scalenohedra from mines near Ashby-de-la-Zouch (Rudler 1905); coarsely crystalline calcite and minute scalenohedra growing into anhydrite nodules within the Carboniferous Hathern Anhydrite Series in the Hathern No. 1 borehole, NW of Loughborough (Llewellyn & Stabbins 1970); associated with sphalerite in septarian nodules – coarsely crystalline yellow calcite containing embedded plates of sphalerite at Lowesby railway cutting; two generations of calcite (pale-brown and white scalenohedra) have been found at a cutting close to the Grimston tunnel, near Old Dalby; and clear colourless calcite is noted from Old Parks coal pit, NE of Ashby-de-la-Zouch (King 1993).

Calcite with terminations selectively overgrown with radiating stripes of iridescent pyrite; Cambokeels mine, Weardale, Co. Durham; width 35 mm. (© M. P. Cooper)

Northumberland Crystalline barite and calcite invariably line cavities in the witherite deposit at Settlingstones mine, near Hexham (Trestrail 1931). Calcite is associated with witherite at Allenheads and West Allendale (Greg & Lettsom 1858), and it occurs with pectolite and stevensite in amygdale-like structures within the Whin Sill at Barrasford quarry, near Gunnerton (Randall 1959).

Nottinghamshire Translucent white calcite crystals up to 25 mm long line cavities in Permian mudstones and limestones at Vale Road quarry, Mansfield Woodhouse (Green & Spence 1988).

Oxfordshire Dogtooth crystals to 8 mm have been found in septarian nodules within Oxford Clay at Cassington and Yarnton (Atkins 1995).

Shropshire At Rorrington (West Snailbeach) mine, S of Meadowtown, calcite forms tapering terminated hexagonal prisms, simple rhombs and complex translucent crystals up to 4 mm long and sometimes contains inclusions of galena (Starkey et al. 1994). Pyramidal scalenohedral calcite, coated by prehnite and rare edingtonite, occurs with a later generation of calcite forming small rhombohedra on prehnite at Squilver quarry, Disgwylfa Hill (Hubbard & Enamy 1993). Calcite was of economic importance at the following Ba–Pb–Zn mines in the West Shropshire orefield around

Shelve: Cliffdale; East and South Roman Gravels, Grit and Tankerville (Pearce 1994a). Tankerville mine at one time produced some very fine penetration-twinned pale-lilac crystals of calcite with primitive rhombohedral form (Solly 1885). Greg & Lettsom (1858) note calcite from the Snailbeach mines and fine tabular crystals from Benthall.

Somerset Calcite veins in the Carboniferous Limestone of the Mendip Hills are not uncommon. Reports include large rough scalenohedra with chalcopyrite inclusions in a quarry to the S of Bryant's quarry, N of Cheddar, brown cuboid rhombohedra at Sandford Hill, and well formed crystals at Dulcote Hill quarries, 3 km SE of Wells (Kingsbury 1941). Calcite lines cavities in iron–manganese oxide-bearing veins at: Coleman's quarry, Holwell, near Frome (Alabaster 1990a): Higher Pitts mine, Priddy – 5 cm-long scalenohedra (Burr 1996); Merehead (Torr Works) quarry, Cranmore – associated with lead-oxychloride species (Alabaster 1975, Symes 1977, Welch et al. 1998); and Whatley quarry, near Frome – dogtooth calcite covered in golden goethite, also flowstone calcite (Alabaster 1976, House 2002).

Staffordshire Tiny colourless-to-white rhombohedra associated with aurichalcite and coarsely crystalline white to pale-pink scalenohedra containing inclusions, and attached acicular crystals of millerite occur at Waterbank mine, Ecton Hill (Starkey 1983, Rust 1995c). Fluid-inclusion studies of calcite and fluorite from Ecton Hill confirm that low-temperature Ca–Na–K brines were crucial for the development of the extensive carbonate-hosted copper mineralization in the area (Masheder & Rankin 1988).

Sussex Columnar calcite is noted from Plumpton Plains on the South Downs (Moore & Pearce 1997) and in the chalk at Beachy Head, Alfriston and Hamsey (Greg & Lettsom 1858).

Warwickshire Colourless-to-pink scalenohedra up to 1.5 cm long are generally doubly terminated at Judkins quarry, Nuneaton. Barite, bornite, chalcopyrite, mottramite and sphalerite are associated species (Ince et al. 1991).

Yorkshire Reports of calcite (often from lead veins) from the Northern Pennine orefield and adjacent areas include: Brandy Bottle Incline, Swaledale – pale-brown rhombohedra (Young 1987b); Coldstones quarry, 5 km W of Pateley Bridge – massive and rare butterfly twins (Ryde 1987, Young et al. 1989); Dam Rigg level, Arkengarthdale – excellent prismatic crystals (Wolfe 1996a); Duck Street quarry, SE of the summit of Greenhow Hill, near Pateley Bridge – vein calcite, part of a fluid-inclusion investigation (Rankin et al. 1987); Janet's Foss, Gordale Beck, near Malham – tufa (Mundy & Arthurton 1994); Great Sled Dale, 10 km SE of Kirkby Stephen – small transparent nailhead crystals (Briscoe & Green 1989); Priscilla level, Gunnerside Gill mine, Swaledale – calcite associated with strontianite and witherite in fissures (Wood 1993); Scaleber Bridge, 3 km SSE of Settle – coarse-grain calcite crystals with sulphide inclusions (Ixer & Stanley 1998); Wet Groves mines, Aysgarth, Wensleydale – white scalenohedral crystals (Young 1987b); and Yordas cave, near Ingleton (Greg & Lettsom 1858). Prismatic crystals up to 2 cm, milky rhombs and nailheads up to 1 cm long are also recorded from borehole material within an evaporite sequence in the Lower Magnesian Limestone in the Whitby area, near Robin Hood's Bay (Raymond 1962).

Wales

Clwyd Fine specimens of calcite have been collected from various parts of the Carboniferous Limestone in the Halkyn Mountain area, including aggregates of yellow rhombohedral crystals from Halkyn mine, dogtooth crystals from Rhosesmor (Halkyn East) mine, near Mold, and highly vitreous nailhead crystals from Pant-y-buarth mine, near Mold (Bevins 1994). Other good examples of calcite are from: Bryn-gwiog mine, near Mold – a massive yellowish variety with unusual form (Mountain 1924); Hendre mines, northwestern slopes of Moelfre – microscopic cruciform interpenetration twins of manganocalcite scalenohedra in calcite–chlorite–muscovite mudstone (Woodland 1938a); Pen-y-Bryn mine, near Holywell – of great purity and transparency (Rudler 1905); and Maes-Whitford – scalenohedra (Campbell Smith 1913).

Dyfed Calcite is widely developed as a gangue mineral in the lead veins of the Central Wales orefield, although notable occurrences are few. These include euhedral crystals from Hendrefelen mine, Ysbyty Ystwyth (Clark & Rust 1993), nailhead crystals from Logaulas (Greg & Lettsom 1858, Bevins 1994), and excellent flat rhombohedral crystals from Penrhiw mine, Ystumtuen (Jones & Moreton 1977). At Eaglebrook (Nantycagl) mine, Ceulanymaesmawr, calcite forms clear-to-white crystalline masses and small simple rhombs covering ankerite (Jones 1983).

Gwent Fine scalenohedral calcite crystals have been found at Caerwent quarry, near Caldicote (Bevins 1994).

Gwynedd At least three generations of calcite interbanded with botryoidal limonite, malachite and manganese oxides have been identified using cathodoluminescence techniques on copper-bearing ore samples from the Great Orme mines, Llandudno (Ixer & Davies 1996, Ixer & Stanley 1996, Ixer 2001). Attractive rhombohedral calcite coated with pyrite, or containing inclusions of marcasite, occur at Parc and Trecastell mines, near Llanrwst, N of Betws-y-coed (Bevins 1994). Other examples of calcite are from: Bwlch-y-Plwm, Penrhyndeudraeth – well formed nailhead crystals (Bevins 1994); Fron-oleu, Prenteg, Tremadog – opaque, white rhombohedral crystals to 3 cm (Starkey & Robinson 1992); and Gwaith mine, Llandegai – large crystalline plates intergrown with quartz (Dewey & Eastwood 1925).

Mid and South Glamorgan Excellent rhombohedral and scalenohedral calcite has been recovered from the Carboniferous Limestone of the South Wales coalfield, including: Castell Coch quarry; Creigiau quarry; Little Garth iron mine; Mwyndy mine, Llantrisant area (North 1916); and Taff's Well quarry, near Cardiff (Balzer et al. 2003). Crystals reach 30 cm in length at the last locality (Bevins 1994), and, at the nearby Ton Mawr quarry at Pentyrch, scalenohedra up to 20 cm long are reported (Wolfe 2000). Other occurrences of calcite include: Tŷ Coch mine, near Porthcawl – lenticular masses of varicoloured calcite, manganocalcite and iron-rich calcite associated with a manganese ore (Criddle & Symes 1977); Llanharry mine – fine colourless, transparent clusters of nailhead crystals (Bevins 1994); Machen quarry, 7 km ENE of Caerphilly – small sharp transparent honey-colour rhombohedra associated with barite, rarely etched scalenohedra up to 5 cm, and orange iron-stained blocky prisms up to 10 cm long overgrown with snow-white calcite (Plant & Jones 1995); and Vaynor quarry, Cefn Coed, Merthyr Tydfil – very good nailhead crystals and scalenohedra to 10 mm (Billington 1995, 1997a).

Powys Low rhombohedral crystals are recorded from Nantiago mine, near Hafren Forest (Jones & Moreton 1977), and well formed yellow rhombohedral calcite occurs at Weston mine, near Middletown (Bevins 1994). Stalactites of calcite are reported from Llanymynech Hill, near Pant, S of Oswestry (Campbell Smith 1913). Calcite is a common vein mineral at Dolyhir quarry, Old Radnor, near Presteigne. Good large scalenohedral and rhombohedral crystals and spherical aggregates occur in veins cutting Yat Wood Formation mudstones (T. F. Cotterell pers. comm. 2007).

Scotland

Central Pink calcite occurs in brecciated Dalradian quartzite and lead–zinc-enriched quartz sulphide veins at Tyndrum and is related to a phase of uranium mineralization (Pattrick 1985). White calcite from Earlsburn (Earlston) reservoir, Touch Hills, SW of Stirling, is associated with analcime, barite, harmotome, natrolite and stilbite (Green & Todd 2001).

Dumfries & Galloway Small amounts of dogtooth calcite occur in smoky quartz-lined vugs within uranium-rich veins along the Solway coast, mostly between Balcary Point and Caulkerbush (Miller & Taylor 1966). Veins of calcite in Carboniferous Limestone at Kelhead limestone quarry contain dogtooth crystals up to 25 mm long, whereas, at Burrow Head, nailhead calcite is noted in association with barite. At Glencrieff mine, Wanlockhead, dogtooth calcite is often twinned at the base and overgrown with nailhead calcite (Williams 1965). Doubly terminated scalenohedra up to 17.5 cm long from this location represent some of Scotland's most exquisite specimens (Livingstone 2002). A bluish-pink manganese-rich variety of calcite is recorded from Glencrieff mine and a lead-rich variety (plumbocalcite) has been found at the High Pirn vein, Wanlockhead (Heddle 1901, Livingstone 2002). The lead content of calcite from Bay and Belton Grain mines, Wanlockhead, was examined by Thomson (1887).

Fife White free-standing scalenohedra and transparent modified rhombohedra, both up to 10 mm long, occur in the exceptional pyrite locality of Goat quarry, 2 km NW of Aberdour (Ingram et al. 1992). Joints and fractures within Carboniferous basalt at Orrock quarry, near Burntisland, contain veins and stockworks of quartz and calcite, the latter with large colourless transparent cleavages up to 3 cm long and forming a few well developed twinned rhombohedral crystals (Wirth 2003, Moreton 2006). Langside quarry, near Kennoway, exposes similar basalt and, in the early 1990s, cross-cutting veins lined with scalenohedral white calcite crystals to several cm sprinkled with bipyramidal colourless quartz crystals were found (Moreton 2006). Heddle (1901) notes calcite in the central druses of agates in andesitic lavas at Heather Hill, Luthrie.

Highland Many fine calcite specimens have originated from the Strontian mines, where it is an important gangue mineral in the barite–galena veins of the area. These include: translucent brown rhombohedral calcite up to 5 cm long, overgrown with harmotome, from the Clashgorm section (Green & Todd 2001); very beautiful scalenohedra terminated with rhombohedra and coated in harmotome from Corrantee mine (Scott 1871); colour-zoned rhombohedra to 3 cm across from Middleshop mine (Livingstone 2002); red-tipped and white dogtooth crystals up to 2 cm long at Whitesmith mine (Savage 1997b), also scalenohedral crystals associated with brewsterite and kainosite-(Y) (Green & McCallum 2005); and bright-red fluorescent and non-fluorescent crystals, with zonation in iron and manganese, from the Strontian area (Long & Agrell 1965). Other occurrences include: beautiful water-clear symmetrical stout hexagonal prisms of calcite and combination prisms and scalenohedral crystals associated with barite, harmotome or sphalerite at Allt Tigh Cumhaig Burn, 3 km S of Struy, near Inverness (Russell 1946b); fine-grain calcite, manganese-rich calcite and rhodochrosite in a lacustrine hot-spring manganese deposit at Dalroy Burn, near Culloden (Nicholson 1987b, 1989c, 1990); calcite with an unusual intermediate form (between scalenohedral and nailhead) from Heilim, Loch Erribol (Heddle 1887b); pale-blue calcite in the Glen Dessary marble, W of Loch Arkaig (Livingstone 2002); magmatic strontium-rich calcite in a carbonatite associated with the Loch Borralan syenite, Assynt (Young et al. 1994b); massive calcite in veins with nailhead crystals in cavities at Loch an Arbhair, Lochinver–Ullapool road, SW of River Kirkaig (Waterston 1953); and well crystallized stubby prismatic doubly terminated crystals in fractures within Moine schists near the Great Glen fault at Moniack Burn, S of Kirkhill (R. E. Starkey pers. comm. 2005).

On the Isle of Skye, calcite is reported from a diopside–forsterite–tremolite skarn assemblage produced during thermal metamorphism of Durness Limestone by the Beinn an Dubhaich granite, near Broadford (Hoersch 1981). Septarian concretions in Jurassic mudstones at Flodigarry contain crack linings of brown fibrous and yellow sparry calcite up to 1 cm thick; barite and siderite are associated species (Hendry et al. 2006). Other occurrences are within Tertiary basalts: exceptional translucent pale-brown twinned rhombohedra up to 4 cm long on drusy white calcite at Biod a' Mhurain, both Duirinish Peninsula (Wolfe 1997, 2000, Green & Todd 2001); superb translucent rhombohedra to 4 cm on mesolite at Glen Ollisdal (Wolfe 1996b); simple rhombs varying from pale yellow to honey brown (or rarely as nailhead crystals) associated with a range of zeolite species at Moonen Bay, N of Dunvegan (Green & Todd 1996, 2001); well crystallized honey-brown rhombohedra to 8 mm at Talisker Bay, Minginish (Pearce 1996, Green & Todd 2001).

Strathclyde Carboniferous Clyde Plateau Lavas contain a diverse secondary-mineral assemblage dominated by zeolite species and are also an important source of well crystallized calcite specimens. Occurrences include: nailhead crystals with greenockite, harmotome and prehnite at the Glasgow–Greenock railway tunnel, Bishopton, and equant crystals associated with chalcopyrite, millerite and pyrite at Dockra quarry, near Beith (Heddle 1901); calcite with complex form and containing inclusions of native copper was found at Boyleston quarry, SW of Barrhead (Rudler 1905) – Heddle (1901) describes exquisite pellucid crystals (in nine different forms) from this locality; large colourless-to-white calcite specimens with chalcopyrite and pyrite at Branscroft quarry, near Kilbarchan, W of Glasgow (Green & Todd 2001); rare calcite in fissures and joints forming translucent colourless-to-white rhombic masses to 20 mm, associated with natrolite, prehnite and thomsonite at Hartfield Moss, 9 km WSW of Barrhead (Meikle 1990); superb cabinet specimens of white-to-orange scalenohedral calcite associated with prehnite, also transparent dogtooth and nailhead crystals to 35 mm from Loanhead quarry, near Beath (Meikle 1989b, Green & Todd 2001); fist-size scalenohedral calcite associated with colourful chalcedony in collapsed geodes to the N of Loch Carron (Wolfe 2006); and well formed colourless-to-white scalenohedral or prismatic crystals up to 50 mm long, accompanied by analcime, barite or quartz at Reilly quarry, Houston (Wolfe 1997, Meikle 2003).

Elsewhere, calcite occurrences include: scalenohedral crystals in cavities within fossil brachiopods at Trearne quarry, near Beith (Todd 1989); pink calcite associated with barite, fluorite and selenite occurs in druses within a trachyte sill at Craigmuschat quarry, Gourock (Heddle 1901, Nicholson & Durant 1986); major amounts of calcite gangue in Gasswater barite mine, near Muirkirk (Scott 1967), and Mannoch Hill mine, SW of Lesmahagow (Todd & McMullen 1991); large crystals of calcite infill analcime-lined cavities in the dolerite (teschenite) sill at Lugar, Glenmuir Water (Phillips 1968); more than 30 varieties of calcite are recorded from the Leadhills–Wanlockhead orefield, some producing exquisite specimens (Heddle 1901, Brown 1919, Temple 1955, Livingstone 2002); pearly-grey rhombohedra up to 1 cm long, sometimes pinkish because of finely divided hematite, occasional dogtooth crystals, fibrous and white or mauve cleavable calcite at Muirshiel barite mine, Queenside Muir, NE of Largs (Moore 1979, Todd & Laurence 1989); Tertiary basalts near the explosive vent in Ardnamurchan, and extending towards Morvern and towards Tobermory on the Isle of Mull, contain enough secondary calcite in vesicles for it to have been used as an index mineral to map the hydrothermal aureole of the Mull central volcano (Walker 1971, Green & Todd 2001); calcite forms stalactites in Nuns' Cave, near Carsaig on Mull (Heddle 1901); calcite, epidote, prehnite and zeolite species are found infilling fractures in endoskarns within the collapsed caldera of the Central ring complex at Creag an Fheidh, Isle of Arran (Cressey 1987); translucent-to-white calcite crystals up to 10 mm across occur in sulphide veins at Rubha a' Mhill, northeast Isle of Islay (Green 1988); and pink, minutely crystalline calcite occurs with subordinate dolomite in marble from Balephretrish Bay on Tiree. The marble also contains mineral aggregates composed of well formed crystals of amphibole, chlorapatite, diopside, forsterite, mica, orthoclase, meionite (scapolite), spinel and titanite (Coomaraswamy 1903).

Nail-head calcite crystals to 10 mm; Leadhills, Strathclyde, Scotland. (© The Trustees of the National Museums of Scotland)

Tayside A cluster of nailhead crystals 6 cm across have been recovered from Bleaton quarry, Kirkmichael, near Blairgowrie (Livingstone 2002). Amygdaloidal andesitic lavas at Ardownie quarry, Monifieth, well known for exceptional agates, are also host to massive dogtooth and nailhead calcite crystals up to 30 mm long (Ingram 1994). Brown scalenohedral calcite up to 35 mm long occurs with agates in vesicles in the nearby Ethiebeaton quarry, Broughty Ferry, W of Ardownie (Green & Todd 2001). Manganese-rich calcite is associated with groutite and woodruffite in a stratiform sulphide deposit at Ben Eagach, near Aberfeldy (Nicholson 1989b); in a similar deposit at Foss, near Aberfeldy, the carbonate assemblage includes barytocalcite, manganese-rich calcite, dolomite, norsethite, siderite and witherite (Moles 1985).

Fair Isle Coarsely crystalline and finely granular calcite occurs with scapolite in veins within Old Red Sandstone sediments in the S and W (Mykura & Young 1969).

Orkney Scalenohedra and prismatic crystals of calcite occur in veins on Hoy, together with stalactitic varieties in the SW of the island (Heddle 1901, Livingstone 2002).

Shetland On Mainland, calcite has been found: in fluorite–epidote-bearing veins at Hillswick promontory, North Quin Geo, and in amygdaloidal claystone in druses also lined with barite, chalcedony, fluorite and quartz at the Kirksands end of Papa Stour (Heddle 1901); and in massive, pale-pink veins at Horse of the Knabb, Lerwick (R. E. Starkey pers. comm. 2005). On Unst, calcite occurs at: Swinna Ness – with brucite in serpentine; Soul Hill – with aragonite in serpentine from chromite quarries; Hagdale chromite quarry; and Sandlodge (Heddle 1901).

Ireland

Antrim Calcite is widespread in amygdales and veins in Tertiary Antrim basalts. It is associated with zeolite species and it occurs in various habits, including steep scalenohedra, rhombohedra and unique barrel-shape forms (the last at Baird's Brae, Roughfort). Calcite is abundant near to fault zones, for example at Island Magee and along the N of the Garron Plateau, where it develops forms different from elsewhere. Notable scalenohedra have been found at Craigahulliar quarry, near Portrush, and in quarries either side of White Head, Lower Belfast. A large quarry on Carnmoney Hill, N of Belfast, has produced abundant calcite with steep scalenohedral and pseudo-cubic form, and a dull-grey plumose variety occurs in a basalt quarry to the E of Ballymoney (Herbert Smith et al. 1916, Walker 1960b).

Donegal Metasomatic alteration of limestone (calcite) inclusions in granite at Ballynacarrick has produced skarn assemblages containing (in different zones) grossular–vesuvianite, clinozoisite, plagioclase–pyroxene and hornblende–plagioclase–titanite nearest to the granite (Gindy 1953). A similar alteration assemblage is also reported from S of Bunbeg quay, E of Loughlin Island, where pegmatite dykes cut Dalradian calc-silicate hornfels (Whitten 1951).

Down Calcite occurs in late cavities or as vein infill in mineralized breccias from the lead-mining district between Conlig and Newtownards, where it forms flattened rhombohedra, nailhead crystals, equant crystals with pentagonal faces, and barrel-shape crystals (Moles & Nawaz 1996). A single brownish-yellow mass of calcite, fluorescing orange under short- and long-wavelength ultra-violet light, was noted in miarolitic granite from Lindsay's Leap quarry, Mourne Mountains (Green et al. 2005a).

Galway Blocky calcite crystals occur with well formed fluorite and quartz at Glengowla mine, near Oughterard (Ryback & Moreton 1992, Wolfe 2000, Green & Moreton 2001). Calcite, sometimes forming dogtooth crystals, was the dominant carbonate mineral in the stratiform Pb–Zn–(Ag–Cu) deposit at Tynagh mine, near Killimor (Morrissey et al. 1971, Clifford et al. 1986).

Kerry Some of the finest crystals of nailhead calcite from Ireland were discovered in 1997 at Ballyegan quarry, near Castleisland. They occurred in a large cavity within a dolomite vein and reached 10 cm in length (Wolfe 1998, Green & Moreton 2001).

Kildare The Harberton Bridge breccia-hosted Fe–Pb–Zn deposit, 40 km WSW of Dublin, is considered to be an example of MVT mineralization and has been studied using cathodoluminescence techniques. This reveals yellow–orange luminescent calcite with locally very well developed fine oscillatory zoning. Multiple phases of growth and dissolution are apparent (Trude & Wilkinson 2001).

Limerick Cathodoluminescence microscopy has identified three major luminescent zones within the equant calcite cement found in Carboniferous Limestone hosting the stratiform Pb–Zn–(Ag) sulphide deposit at Courtbrown, 25 km W of Limerick. An epigenetic origin is proposed (Grennan 1986, Reed & Wallace 2001).

Louth A pale-blue variety of calcite is noted in limestone xenoliths metasomatized by a syenite intrusion at Barnavave, Carlingford. Diopside, grossular and wollastonite are associated species (Osborne 1932b, Nockolds 1937).

Mayo Attractive specimens of colourless-to-white calcite containing inclusions of acicular millerite have been found at Sheefry Pb–(Ag) mine, 11 km S of Croagh Patrick (Moreton & Jackson 1991, Ryback & Moreton 1992).

Meath The black colouration of calcite in the world-class stratiform lead–zinc deposit at Navan mine is caused by inclusions of boulangerite (Ryback & Moreton 1993). Zoned crystals and ooliths of low-magnesium calcite are common in dolomitized micrite from the deposit, and these are associated with dolomite, minrecordite, pyrite and sphalerite (Kucha & Wieczorek 1984). Cathodoluminescence of calcite from Navan is discussed by Peace et al. (2003) and Wilkinson (2003), and the isotopic composition is described by Walshaw et al. (2006).

Sligo Some of the finest calcite ever found in Ireland occurred in a vuggy pipe-like structure close to a lead–zinc sulphide deposit at Abbeytown quarry (site of old Abbeytown mine), Ballysadare. Vugs were lined with crystals up to 8 cm long, which were dull and grey if facing upwards, or lustrous and transparent yellow if pointing downwards (Hitzman 1986, Ryback & Moreton 1992, Green & Moreton 2001). Curious yellow calcite crystals, up to several cm across, with curved crystal faces and a greasy feel, have been found at Aughamore quarry, near Lough Gill (Green & Moreton 2001).

Tipperary Bright luminescent calcite is a feature of the equant calcite cement found in Carboniferous Limestone hosting the stratiform lead–zinc sulphide deposit at Silvermines (Reed & Wallace 2004).

Offshore territories

Guernsey Large specimens of dogtooth calcite have been found in an open fissure at Mont Crevet, St Sampsons (Mourant 1939).

Jersey Calcite rarely produces good crystals here, but several varieties are recorded: massive in a stalactitic cave at Belle Hougie Point; nailhead crystals in veins at the German underground hospital, St Peter's Valley, St Lawrence; hexagonal prisms in geodes within the granite at L'Etacq and Ronez; calcite gangue in sulphide veins to the S of L'Ouaisné Bay; and short hexagonal prisms capped by flat rhombohedra in calcite veins at Val Rouget (Mourant 1961).

calcium catapleiite/calciocatapleiite

hexagonal $CaZrSi_3O_9{\cdot}2H_2O$

A very rare calcium–zirconium silicate that forms a series with catapleiite and is found in the Cnoc Rhaonastil dolerite intrusion on Islay, Strathclyde, Scotland. The intrusion represents a single body of alkali olivine-basalt magma that fractionated *in situ* to produce late-stage volatile-rich alkali syenite differentiates. Calcium catapleiite occurs as lath-shape inclusions within both albite and pyroxene. It is associated with other minor phases: aenigmatite, zirconium-rich aegirine, fluorapatite, monazite and zirconolite (Preston et al. 1998a, 2000a,b).

calc spar

A synonym of calcite.

Laths of calcium catapleiite (white) of variable size, associated with albite (black) and aegirine (dark grey) in a nepheline syenite from the Cnoc Rhaonastil intrusion, Isle of Islay, Strathclyde, Scotland; width 105 μm. BSE image. (© J. Still, University of Aberdeen)

caledonite

orthorhombic $Pb_5Cu_2(CO_3)(SO_4)_3(OH)_6$

TL Leadhills, Strathclyde, Scotland

An uncommon lead–copper carbonate sulphate hydroxide found in the oxidized zone of copper–lead deposits.

Caledonite has been found on a few specimens from Penberthy Croft mine, St Hilary, Cornwall, where it forms light bluish-green microcrystalline aggregates of divergent prismatic crystals to 0.5 mm within cavities in quartz veinstone or in altered galena. Associated species are anglesite, bayldonite, cerussite, linarite, malachite and members of the mimetite–pyromorphite series (Camm & Merry 1991a,b, Betterton et al. 2000). Caledonite is noted without description from Penberthy by Kingsbury (1954) and Golley & Williams (1995). Mineralized Cu–Pb–Zn veins at Greystone quarry, Lezant, near Launceston, contain caledonite as one of many supergene species. It forms light greenish-blue striated prisms up to 1 mm long associated with leadhillite, or as rosettiform aggregates with linarite (Elton et al. 2004).

Caledonite is one of the most beautiful and sought-after species from the Caldbeck Fells, Cumbria. At Red Gill mine, the colour ranges from light-blue to light bluish-green, and gem-like crystals to 1.5 mm have been found in association with

Pale-blue caledonite crystals to 0.7 mm on linarite; Red Gill, Caldbeck Fells, Cumbria. (© D. I. Green)

cerussite, leadhillite and linarite (Greg & Lettsom 1858, Wolloxall 1981, Cooper & Stanley 1990, Cooper 1991a). Very rarely caledonite from Red Gill is associated with queitite (Braithwaite et al. 1989a), native silver (Wirth 1989) and possibly native gold (Sweet 1960). Brilliant prismatic crystals coloured bluish-green, sky blue and sage green are noted by Davidson & Thomson (1951), and fine specimens containing bright-blue crystals up to about 3 mm long were recovered relatively recently (Wolfe 1998). Elsewhere in the Caldbeck Fells, caledonite is described from: eastern flank of Balliway Rigg, higher Roughton Gill (Green 1989, Green & Briscoe 2002, Green et al. 2006a); dumps of the 30-fathom level at Driggith mine, as well formed crystals (Green & Hubbard 1996, Green & Briscoe 2002); Roughton Gill (Cooper et al. 1988, Cooper & Stanley 1990); dumps of an old trial at Short Grain (Cooper et al. 1988, Cooper & Stanley 1990, Green & Briscoe 2002, Green et al. 2006a); Sandbed lead mine (Davidson & Thomson 1951), including probably genuine relatively large crystals to 3 mm on quartz in the Kingsbury collection at the NHM (Green et al. 2006a); Silver Gill (Cooper & Stanley 1990, Green et al. 2005b); and Thief Gills (Davidson & Thomson 1951). Caledonite from Brae Fell and Brandy Gill mines (Hartley 1984) and Wet Smale Gill (Young 1987a) requires further confirmation as the provenance is based on Kingsbury samples. Bridges et al. (2006) confirm caledonite as rare, small (< 1 mm) prismatic blue crystals, minute crusts and radiating needles at Brae Fell mine, adding that the Kingsbury specimen is probably fraudulent.

Still in Cumbria, caledonite is associated with beaverite, cerussite, leadhillite and linarite in a vein exposed at Rigg Head quarry, Borrowdale (Green & Briscoe 2002, Green et al. 2006c). At Closehouse mine, SW of Middleton-in-Teesdale, Lunedale, caledonite forms tiny acicular turquoise-blue crystals up to 0.5 mm long, lining a cavity in oxidized barite-galena veinstone (Young et al. 2005).

A single specimen containing poorly formed caledonite, anglesite, galena and linarite is noted from Snelston mine, near Ashbourne, Derbyshire (Jackman 1996), and a doubtful record of the species is described from Alderley Edge ridge, Cheshire (Carlon 1979, Braithwaite 1994).

In the Central Wales orefield of Dyfed (and Powys), caledonite is reported from abandoned lead mines, mostly in a post-mining dump environment: Brynyrafr mine (Bevins 1994); *in situ* bluish-green prismatic crystals up to 0.4 mm long associated with susannite at Bwlchglas mine, near Tal-y-bont (Rust 1996c); Bwlchrhennaid mine, near Goginan, and Darren mine, Pen-bont Rhydybeddau (Bevins 1994); extremely rare, light greenish-blue to bluish-green divergent prismatic crystals and botryoidal masses up to 1.7 mm across associated with anglesite, cerussite and leadhillite at Dyfngwm mine (actually the Pen Dylife section of Dylife mine), Machynlleth (Rust & Rust 1987); associated with bechererite, hydrocerussite, langite, lautenthalite, namuwite, schulenbergite, susannite at Dylife mine, Powys (Rust et al. 2003); light-blue to rich greenish-blue prismatic crystals up to 0.7 mm across or rarely as yellowish-green crystals at Esgair Hir mine, Ceulanymaesmawr (Rust & Mason 1988); Esgairmwyn mine (Bevins 1994); groups of pale-blue to greenish-blue crystals to 2 mm across, drusy crusts and pale-green crystal groups up to 1.5 mm across (possibly the finest in the Central Wales orefield) at Frongoch mine, between Devil's Bridge and Pont-rhyd-y-groes (Green et al. 1996b, Rust et al. 2003, Wolfe 2004); Glogfawr mine, Devil's Bridge (Bevins 1994); Hendrefelen mine, Ysbyty Ystwyth (Clark & Rust 1993); Llechweddhelyg mine, near Clawddmewlyn, between Bont-goch and Cwmsymlog (Bevins 1994); and South Nantycar mine, S of Rhayader, Powys (Bevins 1994).

Early reports of caledonite from the type locality describe it forming attached crystals (< 12 mm long) in the interior of geodes of cerussite and in hollows formed by the decomposition of galena. Lanarkite and leadhillite are associated species (Beudant 1832, Greg & Lettsom 1858). More precise locations are given by Heddle (1901), who notes it from the Susanna and Leadhills Dod mine, Leadhills, and in the High Pirn mine (Belton Grain vein) and Bay veins in nearby Wanlockhead, Dumfries & Galloway. Additional locations include: the Cove, Crawfurd's, East Stayvoyage, Laverock Hill, Marchbank, Old Glencrieff and Scar veins (Temple 1954a, Livingstone 2002). Exceptionally fine deep bluish-green crystals of caledonite up to 7 mm long from Susanna mine are in the Heddle collection at the NMS. Scottish caledonite is discussed further by Davies (1877), Rudler (1905), Brown (1919, 1927), Temple (1955), Williams (1965), Gillanders (1981), Livingstone & Sarp (1984), Morgan et al. (1986), Paar et al. (1986) and Livingstone et al. (1987). Green (1987a) notes caledonite on slag material from the Meadowfoot smelter at Wanlockhead.

Only a single cluster of caledonite crystals (< 1 mm) appears to have been found in Ireland – at Ballinafunshoge lead mine, Glenmalur, Co. Wicklow. Linarite and cerussite are associated species (Russell 1907a). Ryback & Moreton (1993) found cerussite, linarite, muscovite and pyromorphite on the mine dumps, but did not find

A doubly terminated, 4 mm long, caledonite crystal; Leadhills, Strathclyde, Scotland. (© M. P. Cooper)

calomel?

tetragonal Hg_2Cl_2

A rare mercury chloride that forms a series with kuzminite and is reported from Chatsworth mine, Grassington, Yorkshire (Kingsbury 1969). Ryback et al. (1998, 2001) have serious doubts about the provenance of many samples reported by Kingsbury (and in the Kingsbury collection, now in the NHM) and, until further discoveries are made, calomel cannot be considered to be a British or Irish species.

campylite

TL Dry Gill, Caldbeck Fells, Cumbria, England

A barrel-shape variety of mimetite.

CANCRINITE GROUP

The cancrinite group are hexagonal silicates with the general formula:

$$A_{6-9}(Si,Al_{12})O_{24}[(SO_4),(CO_3),Cl_2,(OH)]_{2-4}\cdot nH_2O$$

where A = Na, Ca, K. Cancrinite and vishnevite are members of the group found in Britain or Ireland.

cancrinite

CANCRINITE GROUP

hexagonal $Na_6Ca_2Al_6Si_6O_{24}(CO_3)_2$

A sodium–calcium aluminosilicate carbonate found as pseudomorphs after nosean, in phonolite from the Wolf Rock, 15 km from Gwennap Head, Land's End, Cornwall. Cancrinite and analcime were probably formed after prolonged heating of the phonolite, perhaps isothermally, and are not weathering products. Associated species are aegirine, nepheline and sanidine, with minor apatite, fluorite, pyrite and tourmaline also present (Harrison et al. 1977).

Sulphatic cancrinite is reported from pegmatites associated with the Borralan syenite complex at Allt a' Mhuilinn, Loch Borralan, Assynt, Highland Region, Scotland (Stewart 1941, Woolley 1973), but this is actually vishnevite (Livingstone 1993).

canfieldite

orthorhombic (pseudo-cubic) Ag_8SnS_6

An uncommon silver–tin sulphide that forms a series with argyrodite and was found in a quartz–sulphide lode at Wheal Jane mine, Kea, near Truro, Cornwall. Canfieldite occurs as a 5–10 μm-wide boundary phase between massive sulphides, especially chalcopyrite and stannite (Holl & Bromley 1988).

cann

A Cornish mining term for fluorite.

cannilloite?

AMPHIBOLE GROUP
monoclinic $CaCa_2(Mg_4Al)(Si_5Al_3)O_{22}(OH)_2$

A rare calcium–magnesium–iron aluminosilicate hydroxide that has been recognized when reprocessing wet chemical analyses using the AMPH-IMA04 software (Mogessie et al. 2004), which names amphiboles using the Leake (1997a) classification of the group. In all cases further confirmation is desirable.

Data originally presented by Heddle (1878c) have been renamed as a result of reprocessing, with two analyses classifying as "cannilloite" and one as "parvo-cannilloite". All data are from Scotland and were originally described as hornblende: from a diorite at Portsoy, Grampian; from a hornblendic gneiss on the slope of Ben Spinna, Kyle of Durness, Highland Region; and from a tuff and dykes at Elie, Fife. In all cases the author mentions impurities, and the data obviously represent contaminated compositions. A single wet chemical analysis of tremolite from serpentinized rocks associated with a layered alpine-type peridotite in the Langavat Valley, South Harris, Western Isles (Livingstone 1976a) is identified as parvo-cannilloite.

A partial wet chemical analysis of hornblende from syenite at Carlingford, Co. Louth, Ireland (Haughton 1856), is identified as aluminocannilloite.

cantonite

A covellite pseudomorph after galena.

carbonate-cyanotrichite/carbonatecyanotrichite

orthorhombic $Cu^{2+}_4Al_2(CO_3,SO_4)(OH)_{12}{\cdot}2H_2O$

A very rare hydrated copper–aluminium carbonate sulphate hydroxide found *in situ* underground at South Wheal Crebor (New East Wheal Russell), Tavistock Hamlets, Devon. Bright-blue matted acicular crystals of carbonate-cyanotrichite form crusts with other supergene species in a grey killas–quartz matrix associated with a chalcopyrite–pyrite orebody. Carbonate-cyanotrichite has grown directly on native copper or forms coatings on childrenite and cuprite. Chalcoalumite is a common associate (Starkey 1986b).

A single specimen of carbonate-cyanotrichite, consisting of sky-blue spherules up to 0.5 mm across and associated with malachite on barite, has been found at Lunehead mine, near Grains o' th' Beck, Lunedale, Co. Durham (Bridges & Young 1998).

In Wales, carbonate-cyanotrichite has been found at Wedding Cave mine, in the gorge of the Afon Nant-y-Ffrith, Bwlchgwyn, near Wrexham, Clwyd. Galena–chalcopyrite mineralization in narrow veins cuts highly brecciated veins in the mine, and carbonate-cyanotrichite appears to be a supergene alteration product of the chalcopyrite. It forms sky-blue crusts made up of interlocking spheroidal aggregates of minute feathery crystals. Anglesite, cerussite and linarite are associated species (Green 1995b).

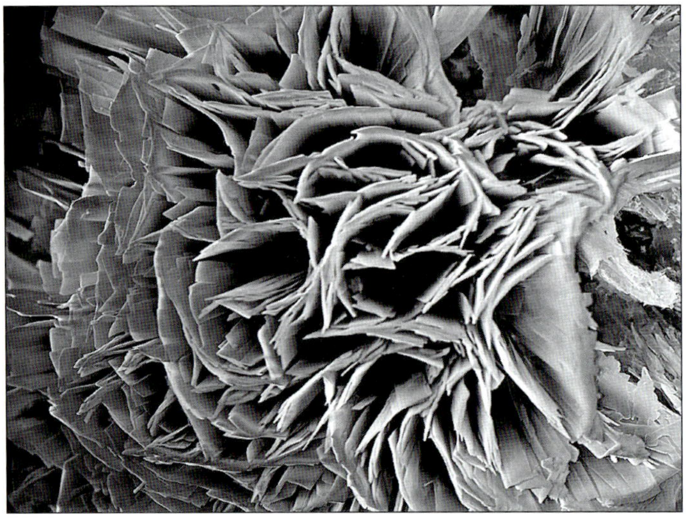

Rosette of curved carbonate-cyanotrichite crystals. Wedding Cave mine, Bwlchgwyn, near Wrexham, Clwyd, Wales; width 95 μm; SEM image. (© D. I. Green)

carbonate-fluorapatite

APATITE GROUP
hexagonal $Ca_5(PO_4,CO_3)_3(F,OH)$
TL Wheal Franco, Walkhampton, near Tavistock, Devon, England?

An uncommon calcium phosphate carbonate fluoride, whose type locality is unknown, although the type locality of francolite (a synonym for carbonate-fluorapatite) is Wheal Franco. In early literature the name francolite is used exclusively. At Fowey Consols, Tywardreath, St Blazey, Cornwall, carbonate-fluorapatite forms small, white and translucent, stalactitic crystallized masses and thin plates associated with quartz and chalcopyrite (Greg & Lettsom 1858, Butler 1887, Sandell et al. 1939). At Clitters United mine (also known as Gunnislake Clitters), Calstock, Cornwall, colourless crystals of carbonate-fluorapatite were found lining cavities in quartz veinstone on specimens collected from the mine dumps. Sage-green spherules of roscherite occurred liberally scattered on the carbonate-fluorapatite (Clark et al. 1983). Francolite is most commonly found in direct association with carbonates, particularly siderite and frequently around the margins of granite intrusions (Braithwaite & Cooper 1982). Elsewhere in Cornwall, other occurrences of carbonate-fluorapatite include: Crinnis and Carlyon mine, St Austell; East Gunnislake mine, Calstock; Holmbush mine, Stoke Climsland; Wheal Cock and Wheal Edward, St Just (all Golley & Williams 1995); Levant mine, St Just (Rudler 1905, Collins 1911a); Old Gunnislake mine, Calstock (Braithwaite & Cooper 1982); Wheal Pendarves (Tryphena lode), Camborne (Barstow & Cooper 1982). Also, in Devon: George and Charlotte mine, Tavistock Hamlets (Embrey & Fejer 1969); Lady Bertha mine, Buckland Monachorum, near Tavistock; and Tavistock canal tunnel, near Tavistock. Also, six locations around Tavistock Hamlets: Devon Great Consols mine; Ding Dong mine; East Wheal Russell; Russell United mine; South Bedford mine; Tavy Consols mine (all Braithwaite & Cooper 1982).

Being based on work by A. W. G. Kingsbury (Ryback et al. 1998, 2001), the occurrences of carbonate-fluorapatite from the following two Cornish locations require further verification: Cargol mine, Newlyn East (Kingsbury 1954); and Wheal Maudlin, 3 km NW of Lostwithiel (Kingsbury 1961).

A very different occurrence of carbonate-fluorapatite is noted from the Coprolite Bed phosphorite deposit (Lower Cretaceous) at Speeton, Yorkshire, where it occurs as pebbles (72–85% carbonate-fluorapatite) intimately mixed with microcrystalline

Interlocking hexagonal plates of carbonate-fluorapatite, up to 2 mm across, from South Wheal Basset, Carnkie, SE of Camborne, Cornwall. (© D. I. Green)

quartz, clay minerals, pyrite and organic matter. Some of the pebbles owe their shape to the internal chambers of ammonites and bivalves. The deposit is the result of three phases of phosphatization that have obliterated many of the original structures (Scott et al. 1987). Elsewhere in Yorkshire, carbonate-fluorapatite (francolite) forms hexagonal plates up to 0.5 mm across in Carboniferous oolitic ironstone at Robin Hood quarry, Thorpe on the Hill, S of Leeds. Together with calcite, kaolinite, pyrite, quartz and sphalerite, it crystallized within ooliths during diagenesis (Deans & Vincent 1938).

XRD analysis of collophane (an amorphous form of apatite) found in nodules within Ordovician mudstones exposed across North Wales has shown it to be carbonate-fluorapatite (Trythall 1988, Bevins 1994). Well crystallized carbonate-fluorapatite is also reported from the Coal Measures of South Wales where it is crystallized on millerite or siderite: Aber-pergwm colliery tips, Gly-neath; Abercynon colliery, Cilfnydd tips, Abercynon; Cambrian colliery, Clydach Vale, Rhondda; Gelli colliery, Rhondda – one specimen composed of large (up to 5 mm across) interlocking hexagonal plates; Parc colliery, Cwm Parc, Rhondda; Park Slip West opencast coalmine – very rich specimens of carbonate-fluorapatite in ironstone (septarian) nodules, including euhedral hexagonal plates up to 2 mm across (Plant & Evans 2005).

carbonate-hydroxylapatite

hexagonal — APATITE GROUP — $Ca_5(PO_4,CO_3)_3(OH,F)$

A rare calcium phosphate carbonate hydroxide found in black oolite from the Yorkshire Dogger (marine beds at the base of the Inferior Oolite) at Blakey mines, Farndale, Yorkshire. Flat plates of carbonate-hydroxylapatite (dahllite) have formed as a recrystallization product of mudstone pebbles in an oolitic limestone with a coarse-grain siderite matrix (Rastall & Hemingway 1949). Presumed carbonate-hydroxylapatite is also described from the calculi of Silurian polyzoa, where it takes the form of pearl-like bodies (spheroliths) that are considered to be in many respects comparable to gallstones or urinary calculi. These spheroliths have been noted in Wenlock Limestone and the Wenlock Shales at: Much Wenlock and Buildwas, Shropshire; Walsall and Dudley, near Birmingham, West Midlands; and Ty Mawr, Rumney, South Glamorgan, Wales (Oakley 1934).

cardenite

TL Carden Wood, Grampian, Scotland

A mixed alteration product of biotite.

carminite

orthorhombic — $PbFe_2^{3+}(AsO_4)_2(OH)_2$

A rare lead–iron arsenate hydroxide found as an alteration product of arsenopyrite in some lead-bearing mineral deposits.

Beautiful tufts of minute carmine-red or reddish-brown needles of carminite have been recovered from the 65-fathom level of Hingston Down Consols mine, 2 km W of Gunnislake, Calstock, Cornwall. These tufts were found on bluish scorodite or quartz, or as delicate star-like aggregates on prisms of mimetite. Other specimens contain fine velvety coatings on chalcopyrite and sphalerite, or, rarely, carminite forms minute stalactites with a drusy surface (Russell 1910a). At Penberthy Croft mine, St Hilary, carmine-red drusy crystalline aggregates of carminite occur in cavities within quartz veinstone associated with a suite of arsenate minerals. Acute rhombohedral crystals are translucent, sharp and up to 0.5 mm long. Flattened botryoidal groups of crystals also occur and a pseudo-cubic crystal form is also noted. Associated minerals include arsenopyrite, bayldonite, beudantite–segnitite, cerussite, chalcocite, chalcopyrite, goethite, hidalgoite, mimetite, pharmacosiderite and scorodite (Kingsbury 1954, Camm & Merry 1991b, Betterton et al. 2000). Carminite is also present at the nearby Croft Gothal mine, where it forms rich crystalline crusts, to 2 mm across, lining quartz cavities closely associated with yellow crystalline beudantite and minor pharmacosiderite (J. Betterton pers. comm. 2006). At Wheal Gorland, St Day, 2 km E of Redruth, where drusy botryoidal and small needles of carminite occur in quartz veinstone, the colour is noticeably browner than at Penberthy Croft. Here it is associated with mimetite, pharmacosiderite, plumbogummite, zeunerite and a yellow mineral near beudantite in composition (Kingsbury 1954, Rust 1981b). Old Treburgett, St Teath and Gravel Hill mine, Perranzabuloe, are also listed as in Cornwall where carminite has been found (Golley & Williams 1995).

Carminite has been documented from localities in the Caldbeck Fells, Cumbria, perhaps most notably at and around Brandy Gill lead mine, Carrock Fell, where As–Cu–Pb vein exposures have yielded a few rich specimens of carminite as minute crystals and drusy crusts on quartz associated with beudantite (Kingsbury & Hartley 1960, Cooper & Stanley 1990) – according to Green & Briscoe (2002); these are some of the best carminite specimens ever found in the area. To the west, at Iron Crag

Carminite crystals to 0.2 mm replacing mimetite; Penberthy Croft mine, St Hilary, Cornwall; width 6 mm. (© D. I. Green)

in Roughton Gill, micromount specimens of carminite, corkite and scorodite are reported (Wolfe 1998) and, near by in Burdell Gill, very rare carminite occurs as minute red acicular crystals associated with rhombs of beudantite on limonite, or as granular massive impregnations in quartz (Neall & Leppington 2003). Additionally, Cooper & Stanley (1990) report carminite with beudantite and cerussite from an exposure near Sandbed mine (probably the 60-fathom level, according to Green et al. 2006a) and from an old dump on Deer Hills. The Deer Hills quartz vein is host to arsenopyrite–(galena–pyrite) mineralization, and cavities left by oxidized sulphides contain locally abundant secondary carminite partially replaced by beudantite. Other supergene minerals in the vein include barite, mimetite, pharmacosiderite, scorodite and very rare schultenite (Symes et al. 1990). Carminite is also present as richly crystalline crusts overgrowing segnitite from an arsenate-rich gossan in the opencut between Driggith and Sandbed mines (Green et al. 2006a).

Further occurrences of carminite are noted by Kingsbury & Hartley (1960), but these occurrences remain dubious until further evidence is accumulated (Ryback et al. 1998, 2001). The localities in question are: Pengenna (Trewethen) mine, St Kew; Treore mine, Port Gaverne, near Port Isaac, St Teath; Wheal Carpenter, Gwinear, near Hayle, all in Cornwall; and from Driggith mine, Ingray Gill mine, Nether Row Brow (Top level, near Dumpy Stone) and Old Potts Gill barite mine, all in the Caldbeck Fells, Cumbria. These occurrences are further mentioned by Hartley (1984), Young (1987a), Cooper & Stanley (1990) and Green et al. (2006a). It is uncertain whether the record of carminite from Wanthwaite mine, St John's in the Vale, Cumbria (Anthony et al. 2000), is based on Kingsbury material.

carnallite

orthorhombic — $KMgCl_3{\cdot}6H_2O$

A generally uncommon but locally abundant hydrated potassium–magnesium chloride found in Permian evaporites that underlie much of North Yorkshire and extend far beneath the North Sea. Stewart (1965) considers carnallite to be a widespread minor constituent of the Upper and Middle Evaporite sequences and a major constituent of the Carnallitic Marl. In the Middle Evaporite sequence of Aislaby No. 2 borehole, Eskdale, near Whitby, carnallite forms single grains or (rarely) aggregates of small irregular grains that are normally light yellow but occasionally black. Carnallite grains reach 5 mm in length and are very deliquescent. In thin section, carnallite often shows polysynthetic twinning, similar to that found in plagioclase feldspars (Stewart 1951a). Evidence collected from four more boreholes in Yorkshire – at Robin Hood's Bay, Staintondale, Little Beck and Hawkster – suggested to Stewart (1956) that carnallite was an important original constituent of the Middle Evaporites, but that it was affected by complex replacements – an earlier series involving anhydrite, halite and sylvite, and a later series involving carnallite, halite, rinneite and sylvite. The Carnallitic Marl in the Whitby area is 12–17 m thick and it consists of anhydrite, carnallite, halite, hematite, quartz, clay minerals and marl fragments. In the Upgang (Eskdale No. 6) borehole, the Carnallitic Marl is a consolidated salt clay full of tiny ramifying veinlets seldom exceeding 0.1 mm in width. These are cut by oblique veins up to 3 cm wide carrying carnallite, halite, sylvite and rinneite, with occasional anhydrite, magnesite and quartz (Armstrong et al. 1951). Massive carnallite is also described from a thin bed at or near the base of the Boulby Potash (a potassium-rich subdivision of the Boulby Halite unit that lies beneath the Carnallitic Marl) in the N and W of Boulby potash mine, Loftus, Cleveland. Boracite nodules are commonly found in the Boulby carnallite (Green & Freier 1996). However, Woods (1979) states that carnallite is rarely found in the Boulby area, although it is present in the boreholes to the south (described above).

carnelian

An orange-to-red fine-grain gem variety of chalcedony (quartz ± moganite).

carnotite?

monoclinic $K_2(UO_2)_2(VO_4)_2{\cdot}3H_2O$

A bright-yellow earthy hydrated potassium uranyl vanadate reported from South Terras mine, St Stephen-in-Brannel, Cornwall (Gramaccioli 1955). The identification was by qualitative chemical tests, and Ryback & Tandy (1992) hint that another uranyl vanadate, francevillite (confirmed at this locality), may have been misidentified as carnotite.

carpholite?

orthorhombic $Mn^{2+}Al_2Si_2O_6(OH)_4$

A manganese aluminosilicate hydroxide that forms a series with ferrocarpholite and was supposedly found by Kingsbury & Hartley (1957d) at Carrock mine, Caldbeck Fells, Cumbria, and three localities in Cornwall: Hingston Down Consols mine, 2 km W of Gunnislake, Calstock; Kit Hill Consols, Stoke Climsland; and Stenna Gwyn mine, St Stephen-in-Brannel, St Austell. Kingsbury's collection is now held at the NHM and later detailed work on the material strongly suggests that the carpholite came from Slavok, Bohemia, or Wippra, Germany (Ryback et al. 1998, 2001). Unless further discoveries are made, this species cannot be considered native to Britain or Ireland. Hartley (1984), Young (1987a), Cooper & Stanley (1990) and Weiss (1994) make further reference to these occurrences.

carphosiderite

A synonym of hydronium jarosite; however, much carphosiderite is jarosite or natrojarosite.

caryopilite

monoclinic $(Mn^{2+},Mg)_3Si_2O_5(OH)_4$

A rare manganese–magnesium silicate hydroxide found as thin (<0.25 mm-wide) felt-like orange-to-brown veinlets cutting hard dense tephroite-rich manganese ore from Nant mine, Nant-y-Gadwen, Llanfaelrhys, Lleyn Peninsula, Gwynedd, Wales (Cotterell 2006c).

cassedanneite

monoclinic $Pb_5(VO_4)_2(CrO_4)_2{\cdot}H_2O$

A very rare hydrated lead vanadate chromate that forms a series with embreyite and is found on the dumps of Lady Anne Hopetoun shaft on the Brow–Hopeful vein system, Leadhills, Strathclyde, Scotland. Cassedanneite occurs as tiny lemon-yellow to intense orange-yellow acute rhombic crystals, exceptionally up to 0.2 mm long, thinly scattered over pyromorphite or lead-rich apatite. Mottramite is an associated species with which cassedanneite can easily be confused (Neall 2005).

cassiterite

RUTILE GROUP

tetragonal SnO_2

A common tin oxide found in medium- to high-temperature hydrothermal veins, greisen and stockworks. Less commonly it occurs in granites, granite pegmatites, rhyolites and contact-metamorphic deposits. Cassiterite is the main ore species of tin. In medieval times (and earlier) most cassiterite was extracted from alluvial deposits by the process of tin streaming. Since that time there have been many tin mines operating in Cornwall and Devon, mostly producing cassiterite from hydrothermal veins associated with granitic intrusions of Hercynian age. Where tin-bearing hydrothermal fluids from the granites have interacted with country rocks, cassiterite-rich subhorizontal sheet-like horizons (tin floors) or carbonas have developed (Jackson 1974, Dominy et al. 1996a, Camm & Dominy 1997, Selwood et al. 1998). Because of the abundance of cassiterite in Cornwall, emphasis here is on the different varieties and geochemically interesting associations, rather than on completeness. Dines (1956) provides comprehensive information on all the metalliferous mines in Cornwall and Devon, including details of Dolcoath mine, Camborne – the greatest of them all, which operated for 170 years and produced over 80 000 tonnes of cassiterite (black tin) in addition to 350 000 tonnes of copper ore. The most up-to-date history of tin mining in Cornwall is that of Buckley (2005), and the archaeological investigations of tin-mining sites by Gerrard (2000) also has much to recommend it.

England

Cornwall Good specimens of cassiterite are rarely reported, despite the abundance of the species in this county. However, Russell (1913b) reports many fine specimens of cassiterite including black and (very rarely) yellowish-red translucent twinned crystals from Kit Hill Great Consols mine, NW of Callington. Large lustrous black twinned crystals of cassiterite associated with topaz were once found at Trevaunance mine, St Agnes (Russell 1924), and, more recently, localized euhedral crystals up to 10 mm across have been noted from Carharrack mine (Whiteworks shaft) (Dominy et al. 1995). Good specimens of cassiterite are noted from Wheal Coates, St Agnes (Green 1984); large crystals have been found at the Imperial china-clay pit, St Austell (Grigg 1986); dark-brown, lustrous, faceted, transparent crystals up to 5 mm across occur in vugs within the lodes at Bunny mine, now part of Gunheath china-clay pit, St Austell (Weiss 1994); bright straw-yellow to dark clove-brown crystals, typically 1–2 mm long (rarely 5–6 mm), are described from Mulberry openwork, Lanivet (Starkey 1986a); very fine lustrous black crystals of cassiterite up to 15 mm across are associated with quartz, fluorite and pyrite at Hingston Down quarry, 2 km W of Gunnislake, Calstock (Wolfe 1996b, 2000). In general, crystalline cassiterite from the St Just mining district is inferior in both size and quality to examples from the St Agnes district (Bancroft & Weller 1993); and yet, 100 years earlier, fine crystals of cassiterite were found at Botallack, Wheal Owles and other St Just mines (Collins 1892). Greg & Lettsom (1858) report the most curious and beautiful crystals of cassiterite, remarkable for their perfection of form, from Wherry mine, Penzance (see sparable tin below).

The work of Bray & Spooner (1983) on mineralization at Goonbarrow china-clay pit, near Bugle, St Austell, provides a good summary of a tin-bearing hydrothermal system (stockwork), which they divide into three categories:

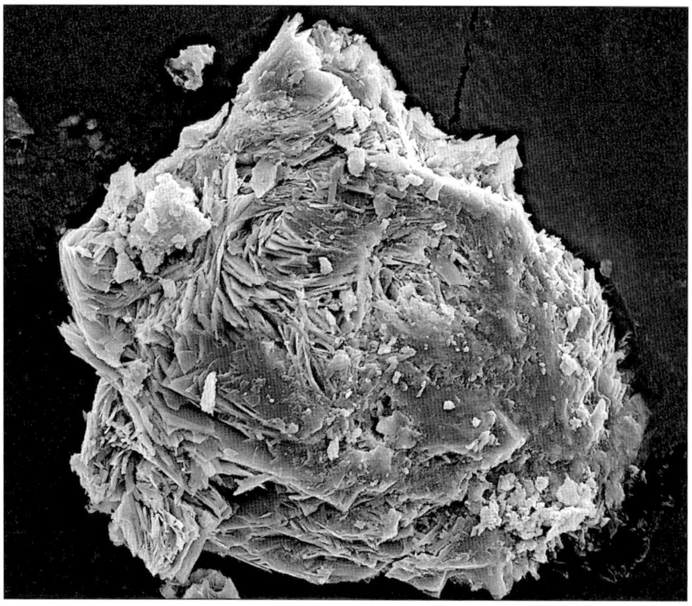

Detached fragment of platy caryopilite from Nant-y-Gadwen, near Rhiw, Lleyn Peninsula, Gwynedd, Wales; width 100 µm; SEM image. (© National Museum of Wales, Cardiff; photographers T. F. Cotterell & M. P. Lambert)

Cassiterite crystal, 5.5 cm across, on topaz from Trevaunance mine, St Agnes, Cornwall. (© The Natural History Museum, London)

- multi-mineral veins – thin (0.1–1.0 cm wide) veins containing apatite, arsenopyrite, cassiterite, fluorite, löllingite, topaz, tourmaline, wolframite and minor pyrite
- greisen-bordered quartz–tourmaline ± cassiterite ± wolframite sheeted veins containing muscovite, lepidolite or zinnwaldite
- barren quartz–hematite veins.

Other examples of cassiterite in stockworks include: Redmoor, N of Callington (Newall 1994); Balleswidden–Bostraze (associated with the Land's End granite); Great Wheal Fortune, NE of Breage (Tregonning–Godolphin granite); Mulberry, Lanivet (St Austell granite); Whiteworks mine, E of Carharrack (Carn Marth granite); Cligga Head, Perranzabuloe; St Michael's Mount, Marazion (all Dominy et al. 1995). Data from the last two localities show cassiterite from this paragenesis to have near-end-member composition, with little substitution of niobium or tantalum, but a slight increase in uranium content (trace amount) in the darker bands of zoned crystals (Moore & Howie 1979, Swart & Moore 1982). Some authors (e.g. Dunham et al. 1978) make a distinction between greisen-bordered vein swarms (subparallel veinlets with walls of greisen usually in granite) and other stockworks (narrow ramifying mineralized structures occurring in any country rock), but both are included above. A simpler style of mineralization and a more important source of cassiterite is the lode mineralization exemplified by blue tourmaline–cassiterite–quartz veins associated with the Carn Brea granite at South Crofty mine, near Camborne (Dominy & Camm 1996, 1997, LeBoutillier 1996). Mineralized lodes at Mount Wellington mine, Twelveheads, 7 km W of Truro, contain bipyramidal twinned cassiterite crystals with rutile inclusions that are closely associated with quartz and tourmaline (Kettaneh & Badham 1978). Cassiterite has been found associated with albite, garnet, muscovite and orthoclase in a pegmatite at Dolcoath mine, Camborne (Davison 1922, Hosking 1951, 1954b), and with albite, fluorite, pyrite, sphalerite, stokesite and tourmaline at Kessel Downs (Halvasso) quarry, Longdowns (Jackson 1978). A granite specimen from Beam mine, St Austell, containing black cassiterite crystals up to 12 mm across, bounded by short prism faces and associated with beryl and tourmaline (Hallimond 1939), probably represents a third pegmatite occurrence. An unusual disseminated cassiterite–sulphide association is reported from the lithium-mica–topaz granite of Hensbarrow stock, SSE of Roche, St Austell. Dark metallic spots, up to 2 cm across, in this granite consist of bornite, digenite, mawsonite, stannite, stannoidite and tennantite, with individual grains reaching 1 mm in diameter (Manning 1983). Finally, cassiterite is present as an accessory mineral in fine-grain tourmaline granite from the St Austell area. The granite also contains albite, quartz, topaz, zinnwaldite and accessory amounts of apatite, arsenopyrite, gorceixite, monazite, rutile and zircon (Manning et al. 1996).

Sparable tin is a variety of cassiterite peculiar to Cornwall, named because of its so-called resemblance to the sparable nails of the cobbler (probably a corruption of "sparrow-bill"). Fine sharply pointed jet-black to dark-brown crystals have been found at Dolcoath mine (Penhallurick 1997), Wheal Harriett (adjoining Dolcoath on the south) and Wheal Nancy, all near Camborne. Other examples are from: Carn Brea mines, near Redruth; Gaverigan stream works at St Columb Major, and Polgooth mine, both near St Austell; Great Wheal Vor, near Helston; Tregurtha Downs mine, Goldsithney; Wheal Grenville and Wheal Nancy, Camborne; Wheal Maudlin, Lostwithiel; Wheal Tremayne, Gwinear, near Hayle; Wheal Owles, St Just (Solly 1891, Embrey & Symes 1987). Russell (1949a) describes the largest examples of sparable tin from Wherry mine, Penzance, with pyramidal crystals reaching 35 mm in length.

Wood tin is a variety of cassiterite showing repeated light- and dark-brown bands, somewhat similar to that found in many varieties of agate – the combination of concentric banding and radial crystallization giving a resemblance to the structure of wood. It was originally found in alluvial (placer) deposits, but has since been found in *in situ* deposits in granites and metasediments, most spectacularly in strongly brecciated parts of certain lodes in non-calcareous metasediments, particularly at Wheal Metal, Breage and in the St Agnes area. Commonly, the deposition of wood tin is preceded by quartz, tourmaline and massive colour-zoned cassiterite and a magmatic–hydrothermal origin is favoured. Sometimes siderite and hematite are deposited subsequent to the development of wood tin. Notable locations include: Balleswidden mine; Bojewan placers and Garth mine, both Sancreed; Carbis Bay beach, near St Ives; Nancothan mine; Pendeen Consols (all associated with the Land's End granite); Whiteworks mine (Carn Brea granite); Pendarves mine (Carn Marth granite); Carnon Valley placers (Carnmenellis granite); Wheal Metal and Wheal Vor (Tregonning–Godolphin granite); Blue Hills, Penhalls, Polberro and West Kitty mines (St Agnes granite); Fowey Consols; Gaverigan mine and placers; Goss Moor placers, St Dennis; Prideaux Wood mine, Luxulyan; Rocks and Rostowrack china-clay pits; and Wheal Providence, near St Columb (St Austell granite) (Collins 1880, 1892, 1911b, Rudler 1905, Embrey & Symes 1987, Hosking et al. 1987). Toad's eye tin and shot tin are related varieties.

Dnepnoskite is a fine granular form of cassiterite (proto-wood tin) forming radiating clusters of acicular crystals at Gaverigan mine, St Columb Major, near St Austell (Camm & Dominy 1997).

Varlamoffite is a partially hydrated form of cassiterite that can be derived from the surficial alteration of stannite (Alderton 1991). It has the appearance of a clay-like colloidal mineral and it varies in colour from cream, ivory-yellow, brownish-yellow, greenish-yellow to primrose yellow. Varlamoffite is reported from: Bunny mine, St Austell; Cligga Head, Perranzabuloe; Kit Hill Consols, Stoke Climsland (Russell & Vincent 1952, Hosking et al. 1987); Gunheath china-clay pit (Hensbarrow section), Carthew, St Austell (Manning 1983, Hawkes et al. 1987, Elton 1998); Penberthy Croft mine, St Hilary (Hosking et al. 1987, Camm & Merry 1991b, Betterton et al. 2000); and St Michael's Mount, Marazion, and Wheal Rock, St Agnes (Hosking et al. 1987). An investigation of greisen from the Hensbarrow

Lustrous euhedral cassiterite from St Agnes, Cornwall; crystals to 14 mm. (© Royal Cornwall Museum, Royal Institution of Cornwall; photographer D. I. Green)

Cassiterite (variety wood-tin) from West Kitty mine, St Agnes, Cornwall; polished slice is 9 cm long. (© The Natural History Museum, London)

topaz granite has shown that topaz contains a varied suite of fluid and mineral inclusions, including columbite, ilmenorutile, qitianlingite and varlamoffite, which provide evidence for the formation of colloids in high-temperature hydrothermal systems (Williamson et al. 1997, 2002).

Other varieties include: rosin tin, a reddish or yellowish transparent variety (Collins 1892); bird's-eye tin – Redmoor placers, Lanlivery; dogtooth tin – Garth mine, Sancreed; dough tin – Nancegollan mine, Crowan; quilted tin – Sea Hole mine, St Agnes; ruby tin – West Wheal Kitty, St Agnes; and sealing-wax tin – Rosehill mine, St Ives (all Golley & Williams 1995).

Stream-tin or alluvial tin is found as pebbles, rolled masses, or as grains in gravels and sands. Additionally, angular debris above the china-clay regions and stanniferous estuarine sediments fall into this category. Native gold is an interesting but commercially insignificant byproduct of tin streaming, the process by which the cassiterite is concentrated. Carnon Valley at the head of Restronguet Creek was a rich source of stream tin ultimately derived from hydrothermal lodes associated with the Carn Marth and Carnmenellis granite intrusions (Collins 1880, 1892, Embrey & Symes 1987). Other locations include: Boscarne, near Bodmin; Breney Common, near Lanlivery; Castle Park, near Restormel Castle, Lostwithiel; Crow Hill and Nantellan, near Grampound; Milltown, near Cardinham; Pensagilles, near St Ewe; Porth River, below Treloy, near Newquay; Portreath beach; Trelower farm, St Mewan; Tresillian River, New Mills; Trewintmarsh, near Trewint, E of Bodmin Moor; Trewoder, E of Truro (all Camm 1995); Buttern Hill, near Bowithick, northern edge of Bodmin Moor, and Trewint Marsh, near Trewint, E of Bodmin Moor (Polkinghorne 1954); Camel and Gannel estuaries (Pirrie et al. 2000b); between Carbis and Relubbus, St Erth (Earl 1994); Colliford and Minzies Downs (Gerrard 2000); Conce Moor, Luxulyan, near Lostwithiel (Russell 1924); Fowey Estuary (Pirrie et al. 2002); Goss Moor, Creggan Moor and Breney Common, all North St Austell district (Camm & Hosking 1985); and Wendron and the Cobey Valley (Brooke 1994a).

Spectacular examples of cassiterite pseudomorphs have been found at Wheal Coates, St Agnes, where fine-grain cassiterite faithfully preserves original features of orthoclase such as Carlsbad twinning and feldspar cleavage. Crystals reach 75 mm in length and are actually casts rather than true pseudomorphs (Embrey & Symes 1987). Other examples are from Balleswidden mine, St Just, and Carn Brecon, St Mewan (Collins 1880, Cooper 1987), and Cameron quarry, St Agnes (Hosking & Camm 1985).

Cumbria A single grain of cassiterite in a Triassic basement breccia from Dufton Park Brow, Dufton, is noted by Versey (1939). Cassiterite is reported as a trace constituent of hübnerite–quartz veins at Carrock mine, Grainsgill, Caldbeck Fells (Hartley 1984, Ball et al. 1985a, Young 1987a, Cooper & Stanley 1990); however, it is possible that the identification is incorrect or based on unreliably provenanced Kingsbury samples (Ryback et al. 1998, 2001). Analysis of material from Carrock mine thought to be cassiterite proved to be rutile (T. Neall pers. comm. 2005). Further confirmation is desirable.

Devon Cassiterite is mostly found in or around the borders of the Dartmoor granite, where it occurs in quartz lodes and alluvial deposits. At Bedford United mines, on the eastern bank of the River Tamar, near Tavistock, fissure veins show evidence of polyphase mineralization as early cassiterite–hübernite–quartz lodes appear to have reopened to allow later crystallization of chalcopyrite, chlorite, quartz and then fluorite (Dunham et al. 1978). Cassiterite is intimately mixed with micaceous hematite and tourmaline in lodes cutting the Dartmoor granite at Chagford (Broughton 1967) and at the Birch Tor and Vitifer mines (MacAlister 1909, Dunham et al. 1978). Tourmaline-rich granite boulders on Hamel Down, Dartmoor, contain rare grains of cassiterite, probably introduced during pneumatolitic alteration of the granite (Brammall & Harwood 1925). A mineralized granite dyke at Hemerdon Bal, Plympton, containing a wolframite–cassiterite–(arsenopyrite)-bearing stockwork (greisen) was developed during the 1914–18 war to increase supplies of tungsten (Terrell 1920, Cameron 1951, Polkinghorne 1954, Ball et al. 1985b). Other locations include: Ashburton mine; Birch Tor mine, near North Bovey; Huey Sydney, near Plympton; Rix Hill, near Tavistock; Wheal Franco, near Buckland Monachorum; Yeoland Consols, near Buckland Monachorum (all Hall 1868b); Littleham Cove area, Budleigh Salterton (Harrison 1975b); North Devon United mine, Peter Tavy (Rumsey & Savage 2005); Shaptor and Hennock, east Dartmoor (Henson 1956); Virtuous Lady mine, Buckland Monachorum (Russell 1913a); Wheal Friendship (Devon Friendship), Mary Tavy (Dewey 1925, Braithwaite 1982b); and Whitstone quarry, Bovey Tracey (Hodge 1989). Additionally, many tin mines are described by Dines (1956).

Sparable tin is found at Wheal Mary Hutchings mine, Plympton (Solly 1891).

Wood tin is found at Brookwood on Dartmoor and Hemerdon Bal, Plympton (Hosking et al. 1987).

Alluvial tin is found at: Colston, Buckfastleigh (Scrivener & Walbeoffe-Wilson 1975); Golden Dagger mine, near Vitifer; Kester Brook, Bickington; and Sheepstor, Dartmoor (all Camm 1995).

Cassiterite is a trace constituent of the Tertiary granite on Lundy island, off the north Devon coast (Dollar 1941, Bromley & Holl 1986, Stone 1990, Thorpe et al. 1990).

Durham Microscopic amounts of cassiterite have been found in the ore from Groverake mine, W of Rookhope, Weardale (Young 1995). The source appears to be bismuthinite-bearing quartz veins that formed earlier than the main barite–fluorite–galena–sphalerite mineralization of the Northern Pennine orefield. The cassiterite is notably tungsten rich (Ixer et al. 1996).

Wales

Gwynedd Small grains of cassiterite have been found in hematite–magnetite–pyrite–quartz veins cutting the Ordovician Snowdon Volcanic Group at Shadow Gully, Cwm Tregalan, S of Snowdon (Colman & Appleby 1991, Reedman et al. 1985).

Scotland

Grampian Cassiterite forms colour-zoned (mainly attributable to variable iron and niobium contents) equidimensional crystals (up to 10 mm long) and irregular masses in a quartz–wolframite vein, in a zinnwaldite vein and in silicified alteration zones within zinnwaldite granite from the Glen Gairn granite complex, Gairnshiel Bridge, W of Ballater (Tindle & Webb 1989, Webb et al. 1992, Green & Todd 2001). The

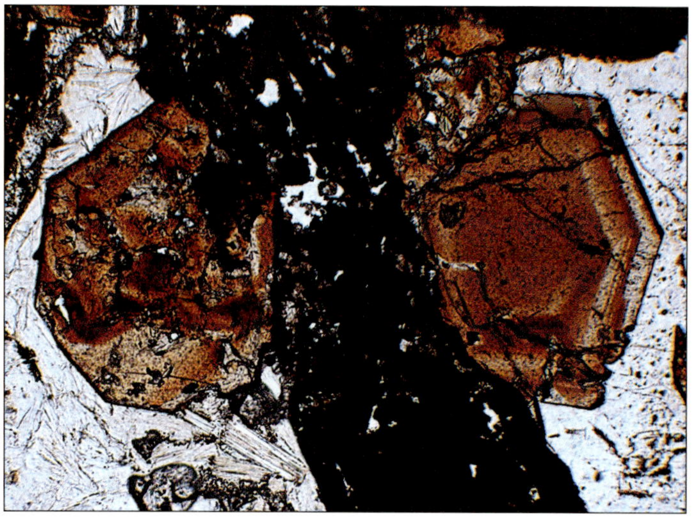

Euhedral zoned cassiterite crystals separated by tabular (black) hübnerite. From a hübnerite–quartz vein at Gairnshiel Bridge, Glen Gairn, Grampian Region, Scotland; width 1.2 mm; viewed in PPL. (© A. G. Tindle)

large tourmaline-bearing pegmatite at East Head, 1 km NE of Portsoy, contains accessory amounts of cassiterite associated with fluorapatite, manganocolumbite and spessartine (J. R. Holbrook specimen, Tindle unpub. EMP data). Cassiterite is one of many heavy minerals identified in panned concentrates of alluvium from the southeastern margin of the Cairngorm granite in Luibeg Burn, Derry Cairngorm (Beer & Bennett 1988, Haynes 1993).

Highland Cassiterite and magnetite are constituents of a garnetiferous albite gneiss belonging to the pre-Caledonian Carn Chuinneag igneous complex at Diebidale (Rudler 1905, Berridge 1969, Gallagher et al. 1971, Dunham et al. 1978). The association of cassiterite (pinkish-brown prismatic to acicular sprays, individual crystals up to 1 mm long) with magnetite is also reported from a skarn deposit at the margin of the Beinn an Dubhaich granite, near Kilchrist (A. Banthorpe pers. comm. 2005).

Strathclyde Stream-sediment samples from the Glen Sannox River on the Isle of Arran contain cassiterite thought to be derived from the Northern Arran granite pluton (Emeleus & Bell 2005).

Tayside Very pure cassiterite has been identified in a few thin sections of a barite-rich rock from the stratiform barite–base-metal deposit, near Aberfeldy (Moles 1985).

Ireland

Carlow Cassiterite is found in accessory amounts in a belt of lithium-rich pegmatites extending from Borris to Stranakelly in Co. Wicklow, and associated with the Leinster batholith. At Aclare, near Myshall, spodumene-bearing aplite–pegmatite and lepidolite greisen contain dark-brown irregular grains of cassiterite associated with manganese-rich apatite, beryl, columbite–tantalite, lithiophilite, pyrite, spessartine and tourmaline. The cassiterite is rich in tantalum. Other pegmatites are located at Rathnagerragh Castle, SW of Myshall and Seskinnamadra, W of Tomduff Hill (Steiger & von Knorring 1975, Scoon 1978, McArdle & Kennan 1989).

Down Cassiterite is responsible for anomalously high tin contents in alluvial samples derived from the Mourne Mountains granites. Additionally, W of Hare's Gap, in a SE-flowing tributary of the Trassey River, cassiterite has been identified in a thin quartz vein cutting granite (Clifford & Earls 1981).

Wicklow At Stranakelly, near Shillelagh, a pink and purple lepidolite greisen contains irregular aggregates of minute euhedral black cassiterite crystals associated with manganese-rich apatite, fluorite, microlite and spessartine (Steiger & von Knorring 1975, Scoon 1978). A cassiterite-bearing pegmatite like those in Co. Carlow occurs at Moylisha, near Shillelagh (Steiger & von Knorring 1975, Luecke 1981). At Ballingen, between Tinahely and Aughrim, cassiterite (sometimes corroded and replaced by stannite) occurs with arsenopyrite, chalcopyrite, fluorite, pyrrhotite, scheelite and sphalerite in altered microtonalite and associated quartz veins (Steiger & Bowden 1982, Gallagher 1989). Cassiterite, native gold, native platinum, blue corundum (sapphire) and wolframite have all been found in alluvium on the northeastern side of Croghan Mountain (Croghan Kinshela) (Ryback & Moreton 1993).

The abundance of cassiterite in Co. Wicklow and other parts of Ireland is thought to be too low for it to be the source of tin in the relatively common Early Bronze Age artefacts found in the area. It is now questionable if tin-bronze metallurgy developed in Ireland, and importation of the artefacts seems more likely (Budd et al. 1994).

Offshore territories

Jersey Cassiterite is an accessory mineral in copper-mineralized aplogranite found in abandoned quarries at La Moye (Ixer 1988).

cathkinite

TL Cathkin quarry, near Glasgow, Strathclyde, Scotland

An aluminosilicate of magnesium and iron, probably related to saponite. Also considered to be an iron-rich saponite.

cawk or caulk

A Derbyshire mining term for a dirty-white earthy and massive variety of barite.

cebollite

orthorhombic $Ca_5Al_2(SiO_4)_3(OH)_4$

A rare calcium–aluminium silicate hydroxide found in dolerite close to the contact with chalk at Scawt Hill, Larne, Co. Antrim, Ireland. Assimilation of chalk into the dolerite has modified its composition to a titanaugite–melilite rock containing plagioclase, nepheline and wollastonite, and with appreciable thomsonite in vesicles. Cebollite forms fine brownish fibres along cleavages and cracks in the melilite, which it sometimes completely replaces (Tilley & Harwood 1931). According to the IMA, the status of cebollite is questionable.

celadonite

MICA GROUP

monoclinic $K,Mg,Fe^{3+}Si_4O_{10}(OH)_2$

A common potassium–magnesium–iron silicate hydroxide widely found in altered basaltic rocks as a secondary mineral, and especially in vesicles and vugs. The difficulty of celadonite identification was encountered by Loveland & Bendelow (1984) in their examination of a weathered basalt from near Wood Hall, Bridekirk, Lake District, Cumbria. Zoned dark-green to pale yellowish-green celadonite forms sub-rounded grains (<50 µm to 3 mm in diameter) and occasionally is stained brown by alteration to iron oxides. Analysis shows that the mineral is most probably a celadonite/muscovite or celadonite/illite mixture, although a celadonite/phengite composition was not entirely discounted.

In Wales, a thin green crust on altered Ordovician tuffs at Llanelwedd quarry, near Builth Wells, Powys, and similar coatings on Silurian lavas near the Mew Stone, on Skomer Island, Dyfed, have both been confirmed as celadonite (Bevins 1994). Cope & Bevins (1993) also report celadonite in Precambrian volcaniclastic siltstones of the Coomb Volcanic Formation from the Llangynog inlier around Gelli and Trehyrn in Dyfed.

Dark-green celadonite occurs in great abundance, entirely filling most of the smaller druses in lavas, tuffs and agglomerate on Maol nan Damh on the shoulder of Ben More, Isle of Mull, Strathclyde, Scotland. However, its composition requires chemical analysis (Currie 1898). At Sgòrr Mòr, on the Isle of Rum, amygdaloidal basalt contains 2–3 cm-thick bands of apple-green celadonite apparently derived from flattened cavities, and thin layers of celadonite also coat the surface of agates (Heddle 1879f). Lustrous dark-green celadonite coats the outside of amygdales and occurs as filaments encased in silica to form moss agate at Ardownie quarry (unquestionably the most important Scottish agate locality of the 20th century), Monifieth, Tayside Region (Ingram 1994). Distinguishing between celadonite and glauconite is not easy, as was shown by French et al. (1977) in their examination of spherulitic and felted green-to-brown minerals found in various amygdaloidal volcanic rocks of the Western Ochils, Central Region. These authors also recorded compositions ranging from celadonite to vermiculite, noting celadonite spherules to 50 µm across in basalt and zoned celadonite–quartz-infilled vesicles in andesite. In brecciated rocks, celadonite forms thin veins, and in all rock types bright-green celadonite replaces clinopyroxene. Celadonite formed during a hydrothermal event and it suggests the presence of a potassium-rich hydrous fluid at a late stage in the volcanic history of the Ochils lavas. Further occurrences of celadonite reported by Heddle (1901) include:

DUMFRIES & GALLOWAY Whilk, near Lendalfoot. FIFE Scurr Hill, near Balmerino; E of Tay Bridge; Tayport slip; railway cutting W of Tayport. GRAMPIAN St Cyrus, N of Montrose. TAYSIDE Binn Hill, Glen Farg; Blue Hole, near Usan, Montrose; Hare Craig, near Broughty Ferry; Lunan railway cutting; Path of Condie. ORKNEY Sands Geo, Hoy.

Reports of celadonite in Ireland are limited to four occurrences: a brilliant dark apple-green variety filling druses and coating chalcedony in close association with magnesium-rich chamosite (chlorophaeite) in basalt to the S of the Giant's Causeway, Co. Antrim (Heddle 1879f); as a product of zeolite-facies metamorphism in Tertiary lavas from Co. Antrim (Walker 1960a); as a secondary mineral in low-grade metavolcanic greywackes and spilites in Co. Cavan (Oliver 1978); as overgrowths on muscovite from the Ougherard granite, Connemara, Co. Galway (Dempster et al. 1994).

celestine

BARITE GROUP

orthorhombic $SrSO_4$

A common strontium sulphate that forms a series with barite and is found in hydrothermal veins or as cavity and fissure-fillings from the precipitation of migrating groundwaters or basinal brines in carbonate rocks. Celestine can also have a primary sedimentary origin or can form diagenetically.

At Binner Downs mine, N of Helston, Cornwall, celestine occurred with cassiterite, chlorite, galena and sphalerite in a hydrothermal lode (Greg & Lettsom 1858, Collins 1892, Hosking 1951). However, its appearance in cavities in Triassic red marl at Peak Hill, near Sidmouth, Devon (Perceval 1880, Rudler 1905), has been called into question. A large cliff-fall from the W of Peak Hill revealed remarkable pale-blue to honey-brown crystals in cavities in calcite nodules that proved to be barite, not celestine (Corrie & Green 2003).

In economic terms, it is the Northavon district of Avon, to the NE of Bristol, that is Britain's most important celestine resource. Between 1875 and 1968 it produced 50–70% of world production from occurrences in the Keuper Marl and Lower Coal Measures. Within the Keuper Marl, celestine is found disseminated, in nodules, as thin stringers and veins, and as vuggy infillings, particularly in the Severnside

Evaporite Bed. The deposits are thought to be diagenetic after gypsum (or anhydrite) and the primary minerals were probably formed in a supratidal mudflat environment (sabkha), similar to the Persian Gulf today. A review of this mineralization by Nickless et al. (1975, 1976) produced the following locations for celestine: Awkley; Beanwood farm, Wapley; Charfield brick pit; Cowship Cottages, Wickwar; Cowship Lane, Cromhall; Grovesend railway cutting; Hall End, Wickwar; Hunter's Hall, Cromhall; Itchington; Leechpool, near Yate; Limekiln Plantation, Wickwar; Lodge farm, Siston; Lower Court farm, Almondsbury; Moorleaze; Oldland Common; Perrinpit farm, near Frampton Cotterell; Poundhouse farm, Charfield; Redfield Hill farm, Bitton; Sargeant's farm, Yate; Sheephouse farm, Latteridge; St Anne's Well, Siston; Stock Hill, Aust; Talbot's End, Cromhall; Thornbury Industrial Estate; Tytherington; Upper Culleyhall farm, Bitton; Wapley Common, Wapley; and West End farm, Wickwar. This area has previously produced celestine specimens of remarkable size and perfection of form, in particular from Wickwar and Yate (Rudler 1905). Clifton (Bristol) is an additional marl location that provided material for the analysis of celestine – the first analysis ever published by the fledgling *Mineralogical Magazine* (Stoddart 1876). A barium-rich variety of celestine (barytocelestine) is also reported from this locality (Collie 1879). Very beautiful, perfect pale-blue and yellowish transparent crystals, up to 25 mm long, were found during the making of a railway cutting at Pyle Hill, Bristol (Greg & Lettsom 1858). Triassic marl was also host to fine fibrous pale-blue celestine at Coley Mill, Hinton Blewett, Mendip Hills, Somerset (Kingsbury 1941).

The association of celestine with evaporite deposits is also noted in boreholes from: Aislaby (Eskdale No. 2), Eskdale, near Whitby, Yorkshire (Stewart 1949); Fordon, Humberside (Stewart 1963); and in Hathern No. 12, NW of Loughborough, Leicestershire (Llewellyn & Stabbins 1970). All these occurrences are minor. Other English occurrences include the following:

DORSET Common in the Lower Purbeck Beds at Durlston Head, Swanage, where it replaces anhydrite and calciostrontianite (Salter & West 1965). DURHAM Forming colourless crystals in a galena-bearing vein at Daddry Shield, near St John's Chapel, Weardale (Smythe 1922). LEICESTERSHIRE Colourless crystals to 1.5 mm associated with strontianite from Asquith Boulevard, Leicester (Cooper 1991b); in a dolomitic layer associated with gypsum beds within calcareous mudstones at Gipsy Lane brick pit, Leicester (Faithfull & Hubbard 1988); forming fans of barium-rich crystals in a siderite-mudstone nodule at Lowesby brickyard, N of Billesdon (King 1993). NORTHUMBERLAND Pale-blue radiating crystals at Burton (Greg & Lettsom 1858). OXFORDSHIRE Celestine is the major fracture filling sulphate in septarian nodules found in Oxford Clay at two locations between Cassington and Yarnton (Atkins 1995). SOMERSET Associated with analcime, dolomite, quartz and solid hydrocarbons that replace aragonitic shell debris in Lower Liassic rocks on the coast from Blue Anchor Point to Stolford Rocks (Alabaster 1980); in lenticular masses in limestone at Watchet (Greg & Lettsom 1858); as pale-blue crystals in a calcite vein at Merehead (Torr Works) quarry, Cranmore (R. E. Starkey pers. comm. 2005, Turner 2007). STAFFORDSHIRE Beautiful pale-blue crystals up to 12 mm long, sometimes associated with calcite and millerite at Clayton mine (Waterbank end), Ecton Hill, Wetton (Braithwaite et al. 1964a, Robey & Porter 1972, Braithwaite 1983, Starkey 1983, Ford et al. 1993). SUSSEX Colourless tabular crystals in a cavity within a gypsum seam mined at Mountfield mine, Battle (Young et al. 1985c). YORKSHIRE In Magnesian Limestone at Knaresborough and Ripon (Sedgwick 1835, Greg & Lettsom 1858, Fowler 1956); forming colourless crystals in a galena-bearing vein at Old Gang mine, E of Gunnerside Gill, Swaledale (Smythe 1922); white and opaque at Scotton Moor (Greg & Lettsom 1858).

Wales

Small, pale-blue, transparent, tabular crystals of celestine associated with strontianite and calcite are reported from rockfalls along the coast near Penarth, South Glamorgan (Wolfe 1996a,b) and as milky-white crystals in the nearby Penarth quarries (North 1916). Celestine (barytocelestine) had earlier been found during the excavation of Barry Dock and from thin veins cutting Carboniferous Limestone in quarries across Nell's Point on Barry Island and at Bendick Rock, near Barry (Connybeare & Phillips 1822, Howard 1895). A report of massive granular celestine in geodes and nodules within Triassic conglomerate collected from roadcuts made for the Llantrisant bypass, Mid Glamorgan (Thomas 1968), could not be confirmed by Bowler & Kingston (1970) or Alabaster (1990a).

Scotland

Fife Celestine forms blue fan-shape radiating crystals to 4 cm in dolerite near St Andrews (Heddle 1901) and has been found close to St David's, Inverkeithing (Livingstone 2002).

Grampian Fine-grain orange masses of celestine are associated with barite in fault breccia at Allt Chernie, Ballater area (Goodman et al. 1990).

Highland Celestine occurs as a vein mineral forming blue, foliated and, rarely, crystalline masses at Clachnaharry, W of Inverness (Heddle 1901). It is also found at the Fee Donald mines, 8 km from Strontian jetty, where it is associated with barite, calcite, galena, harmotome, pyrite, quartz and strontianite (Heddle 1901, Berridge 1969, Winrow 1986b), and within the Glenelg–Ratagain (diorite, syenite and quartz monzonite) intrusion, 10 km E of Glenelg (Nicholls 1951b).

Lothian Celestine forms pale-blue veins in trap basalt near North Berwick and small foliated masses in basalt at Calton Hill, Edinburgh (Greg & Lettsom 1858, Livingstone 2002).

Strathclyde Pale-blue to white massive celestine has been found in the disused Muirshiel barite mine, Queenside Muir, NE of Largs (Todd & Laurence 1989). It also occurs as bright-blue bladed crystals up to 40 mm long with strontianite in a single limestone block from Trearne quarry, near Beith (Wolfe 1996a) and filling a cavity in a small dyke near Lot's Wife, Auchenreoch Glen, Dumbarton (Livingstone 2002).

Tayside Deep-blue celestine occurs on a slickensided calc-silicate rock at Shierglas quarry, Blair Atholl (Livingstone 2002).

Orkney Minor barite and celestine occur with galena and sphalerite in a Middle Devonian algal breccia (part of an evaporitic lake system) at Hesta Geo, near Birsay, Mainland (Parnell & Janaway 1990).

Ireland

In Co. Leitrim, macroscopic pseudomorphs of celestine after euhedral gypsum reach 3 cm across in an evaporite facies of the Visean (Lower Carboniferous) at Carraun, near Kiltyclogher. A fibrous variety is also noted from nearby Glenkeel and from Briscloonagh, W of Glenfarne (West et al. 1968). Earlier reports of celestine from Newtownards lead mine, Co. Down, and from Seven Churches, Glendalough, Co. Wicklow (Hall 1868a), are unconfirmed.

celsian

FELDSPAR GROUP

monoclinic

$BaAl_2Si_2O_8$

A relatively uncommon, but locally abundant, barium aluminosilicate, dimorphous with paracelsian, that forms a series with hyalophane and orthoclase.

Celsian is found in lenses of manganese ore within Ordovician mudstones at Benallt mine, Rhiw, Lleyn Peninsula, Gwynedd, Wales, where it forms beautiful, well formed, colourless, transparent crystals accompanied by paracelsian. When found in cavities, celsian forms short stout prismatic crystals up to 1.5 cm long, or slender-to-acicular needles up to 5 mm long. The larger crystals enclose films of iron and manganese oxides. Barite is an associated species (Russell 1911b, Spencer 1942, Gay 1965, Gay & Roy 1968).

Celsian is reported from a stratabound zone of barium, lead and zinc mineralization in Middle Dalradian metasedimentary rocks N of Aberfeldy, Tayside Region,

Celsian crystal, 10 mm across, from Benallt mine, Rhiw, Lleyn Peninsula, Gwynedd, Wales. (© National Museum of Wales, Cardiff; photographer M. P. Cooper)

Scotland. The mineralization is contained in graphitic schists of the Ben Eagach Schist Formation and consists of interbanded barite rock, a cherty quartz–celsian rock and sulphide-bearing carbonate rocks. In the celsian–quartz rock, celsian occurs as platy grains, the precursor of which may have been cymrite. Celsian also occurs as porphyroblasts in marble and in brecciated carbonate-rich rocks. At one locality, coarse euhedral celsian occurs in sulphide-rich veinlets within a massive dolomitic marble, where it is particularly notable for its bottle-green colour. Other barium silicates recorded at Aberfeldy include cymrite, hyalophane and a barium-rich muscovite (Coats et al. 1980, Fortey & Beddoe-Stephens 1982, Willan & Coleman 1983). Analysis of barium feldspars from the Ben Eagach deposit and nearby Foss mine by Moles (1985) indicate a complete range of compositions between nearly pure celsian and hyalophane. In the area around Beinn Heasgarnich, S of Loch Lyon and 50 km SW of Aberfeldy, stratigraphically similar schists contain barium mineralization in which barite, hyalophane and barium-enriched muscovite are the main ore species, and celsian is of very restricted occurrence (Coats et al. 1984).

Lower Carboniferous shallow-water oolitic limestones at Moyvoughly, near Moate, 12 km N of Clara, Co. Westmeath, Ireland, have acted as a host for lead–zinc mineralization and in the process have been extensively replaced. In sphalerite-bearing samples, one such replacement is by celsian and dolomite, in which the composition of the celsian varies towards that of hyalophane (Kucha 1988).

centrallassite

A synonym of gyrolite.

cerargyrite

A synonym of chlorargyrite.

cerasine

A synonym of matlockite and phosgenite.

černýite?

tetragonal — STANNITE GROUP — Cu_2CdSnS_4

A copper–cadmium–tin sulphide found on samples in the Kingsbury collection at the NHM. Originally described as having been collected from Treore mine, St Endellion, Cornwall, this provenance is now considered dubious (Ryback et al. 1998, 2001). Unless further discoveries are made, this species cannot be considered native to Britain or Ireland.

ceruleite

triclinic — $Cu_2Al_7(AsO_4)_4(OH)_{13}\cdot 12H_2O$

A rare hydrated copper–aluminium arsenate hydroxide found as sky-blue botryoidal crystals on liroconite from Wheal Gorland, St Day, 2 km E of Redruth, Cornwall (Russell & Claringbull 1956, Spencer 1958). Many fine specimens were collected in 1977 from the dumps around Bawden's shaft (Rust 1981b, Fisher 1988). Ceruleite was also found 1 km away on a temporary dump derived from the Western Decline, an inclined shaft to Whiteworks lode in the Wheal Maid section of United Downs mine (a western extension of Wheal Jane). Here it forms sky-blue radiating spherules (0.5 mm across) on joint planes in tourmalinized killas associated with cassiterite, posnjakite, scorodite and either zeunerite or meta-zeunerite (Braithwaite 1987). Ceruleite was reported without description from Penberthy Croft mine, St Hilary, Cornwall, by Camm & Merry (1991b), but was not substantiated in the exhaustive study of the mine mineralogy by Betterton et al. (2000).

Pale-blue spherules of ceruleite, up to 0.5 mm across, associated with green plates of torbernite; Wheal Maid section of United Downs mine, near Chasewater, Cornwall. (© D. I. Green)

cerussite

orthorhombic — ARAGONITE GROUP — $PbCO_3$

A common lead carbonate found in the oxidized zones of lead deposits, often as an alteration product of galena. Sometimes its abundance makes it a viable ore species.

England

Avon Along the beach near the Pier hotel at Clevedon, 16 km west of Bristol, Triassic Dolomitic Conglomerate is faulted and mineralized with barite–(chalcopyrite–galena). Cavities up to 12 cm across in the barite contain secondary covellite (replacing tennantite) and other supergene species including aragonite, aurichalcite, azurite, beudantite, brochantite, cerussite, native copper, cotunnite, cuprite, phosgenite and native sulphur (Starkey 1984, Ixer et al. 1993, Bridges 2003). At Coombe Farm quarry, Westbury-on-Trym, low-temperature iron–manganese mineralization forms small podiform replacement deposits in Carboniferous Limestone. Many cavities in the pods contain barite, cerussite and wulfenite. Some cerussite is derived from hydrocerussite (Alabaster 1978). Massive opaque white cerussite, small euhedral twinned crystals to 5 mm, and pseudomorphs after hydrocerussite occur in another iron–manganese deposit at Wesley Theological College (Wesley mine), Westbury-on-Trym. Cerussite is also noted from Longwood quarry, near Long Ashton, and from the dumps of Providence iron mine, Bristol district (Alabaster 1989b). At Stancombe quarry, near Flax Bourton, Bristol, cerussite occurs with anglesite, calamine, native sulphur and rare greenockite in fractured galena pods within barite–calcite veins cutting a Carboniferous Limestone cave system (Alabaster 1983).

Cheshire Most cerussite occurrences are restricted to the copper and lead mines on Alderley Edge ridge. Extensive secondary alteration of the primary sulphides has produced an assemblage in which anglesite, cerussite (both replacing galena) and smithsonite (replacing sphalerite) are the major non-sulphide species (Ixer & Vaughan 1982). At Engine Vein mine, cerussite forms good crystals of various habits, including tabular, quartz-like prismatic and silky rod-like crystals, occasionally to 7 mm. Other locations include the Stormy Point area and Mottram, New Venture, Opencast, West and Wood mines (Carlon 1979, Braithwaite 1994). In the general area of Alderley Edge (including West mine) and Clive (Shropshire) copper mining areas, cerussite is reported with azurite, chrysocolla, dioptase, hydrocerussite, malachite, mimetite, pyromorphite and smithsonite in a secondary As–Co–Cu–Ni–P–Pb carbonate, hydrated silicate, oxide–hydroxide assemblage (Plant et al. 1999).

Cornwall Beautiful specimens of thin flat translucent twinned crystals of cerussite up to 8 mm long were once found in limonite geodes on the dumps of Wheal Rose, Sithney (Russell & Hutchinson 1927, Russell 1944). Crusts of cerussite coating joint surfaces are associated with an unknown lead uranyl carbonate, kasolite, malachite and widenmannite at Loe Warren Zawn, Botallack, St Just. The species forms white-to-colourless acicular crystals, stouter prismatic crystals, tabular crystals and cream anhedral lumps. Cerussite is also noted from the nearby North Zawn (Elton & Hooper 1995a,c, 1996). At Penberthy Croft mine, St Hilary, cerussite occurs in quartz veins as translucent colourless-to-white prismatic to tabular crystals (to 18 mm, but more commonly 1 mm) in cavities formed by the decomposition of galena (Camm & Merry 1991b, Betterton et al. 2000). In the antimony–lead mining district of north Cornwall, cerussite is a widespread but not particularly abundant secondary species replacing galena. At Bounds Cliff, E of Port Isaac, it is associated with barstowite, chalcopyrite, dolomite, jamesonite, phosgenite, pyrite, quartz and sphalerite (Stanley et al. 1991). Other occurrences in the area include: Pengenna mine and Treore mine, Port Gaverne, near Port Isaac, both St Teath (Russell 1949b); very fine (unrivalled) examples, some finer than hair, silvery white and up to 23 cm long from Pentire (Glaze) mine, near New Polzeath, St Minver (Greg & Lettsom 1858, Collins 1904b, Rudler 1905, Embrey & Symes 1987, Clayton et al. 1990, Penhallurick 1997); Port Quin mine, near Port Quin (Russell 1944, 1949b); beautiful growths at Trevinnick mine, St Kew (Dewey 1921, 1925, Russell 1949b, Fisk et al. 1982).

White to pale-yellow blocky, tabular, prismatic and acicular cerussite up to 15 mm long has been found in gossanous or galena-rich quartz veinstone at Greystone quarry, Lezant, near Launceston; dundasite, hemimorphite, sphalerite and wulfenite are associated species (Lloyd & Barstow 1983, Grigg 1996, Elton et al. 2004). Elsewhere, cerussite has been found at: Gunver Head, near Padstow Consols mine (Braithwaite & Merry 1996, Schlomann & Weiss 1998); Newporth beach, 2 km SW of Falmouth (Dean 1982, 1983, Dean et al. 1983); Park Mathews, S of St Austell;

Wheal Alfred, Phillack, near Hayle; Wheal Ann, Perranzabuloe; Wheal Confidence, near Newquay; Wheal Penrose, Sithney, near Helston; Wheal Primrose (West Kitty), near St Agnes (all Greg & Lettsom 1858); Polzeath Consols, St Minver and Wheal Rose, Newlyn (Golley & Williams 1995); Wheal Golden, Perranzabuloe and Wheal Mary Ann, Menheniot (Merry & Braithwaite 1995); Wheal Gorland, St Day, 2 km E of Redruth, (Rust 1981b); Wheal Trelawney, Menheniot (Dewey 1925).

Cumbria *Caldbeck Fells:* Cerussite is a common species in this area but it appears to have been of commercial value as an ore only at Mexico mine in Todd Gill, where it was crystallized and acicular (Greg & Lettsom 1858, Cooper & Stanley 1990, 1991b). At Red Gill mine, cerussite was abundant in small crystals, with some reaching 25 mm (Davidson & Thomson 1951, Wolloxall 1981), whereas others were replaced by psilomelane (Briscoe 1986a). Perfect 2.5 mm-long transparent twinned crystals of cerussite on goethite, particularly attractive crystals coated with a thin skin of malachite (Cooper & Stanley 1990), and massive cerussite containing cavities infilled with caledonite, leadhillite, queitite and susannite (Braithwaite et al. 1989a), are also noted from Red Gill mine. Good specimens of cerussite were obtained from Roughton Gill mine, including sprays of white crystals (<10 mm) scattered on acicular malachite and tabular crystals (10–20 mm) with an almost metallic lustre (Davidson & Thomson 1951, Cooper & Stanley 1990). Other occurrences include: colourless, white and pale-blue cerussite at Arm o' Grain (Green et al. 2006b); small crystals from the eastern flank of Balliway Rigg, near Thief Gills, higher Roughton Gill (Green & Briscoe 2002); tiny crystals at Bannerdale mine, Mungrisdale (Ziemba 1975); white acicular crystals (20 mm), including twinned crystals and the jackstraw variety at Brae Fell mine (Greg & Lettsom 1858, Cooper & Stanley 1990, Bridges et al. 2006); six-point star-shape crystals irregularly lining the centres of corroded galena at Brandy Gill (Goodchild 1884); Carrock End mine; from an old cross cut near Carrock mine; Ingray Gill mine (all Hartley 1984); tuft-like groups of slender prisms (Goodchild 1884), small greenish crystals (Cooper & Stanley 1990) and acicular-to-prismatic or tabular colourless-to-white crystals to 10 mm (Green et al. 2006a) at Driggith mine – paragenesis given by Stanley & Vaughan (1981); lustrous twinned crystals up to 20 mm at Dry Gill mine (Green & Briscoe 2002); Poddy Gill (Davidson & Thomson 1951, Hartley 1984); platy twinned crystals to 0.75 mm from the 60-fathom level dumps of Sandbed mine (Green et al. 2006a); cerussite associated with anglesite, caledonite, chenite, elyite, leadhillite, linarite, macphersonite, mattheddleite, scotlandite and susannite at Short Grain lead mine (Cooper et al. 1988, Green & Briscoe 2002); elongated white jackstraw crystals and small attractive striated prisms associated with acicular malachite at Silver Gill (Greg & Lettsom 1858, Wirth 1989, Cooper & Stanley 1990, Rothwell & Mason 1992, Green et al. 2005b); coated with pyromorphite at Thief Gills (Davidson & Thomson 1951). Cerussite from Nether Row Brow (Kingsbury & Hartley 1956a, Hartley 1984) and possibly the occurrences at Carrock End, Ingray Gill and Sandbed mines, and from an old cross cut near Carrock mine (Hartley 1984) deserve further confirmation, given the unreliable provenance of many Kingsbury specimens. One specimen claimed to have been collected at Driggith mine almost certainly came from Zellerfield, Germany (Ryback et al. 1998, 2001). Lustrous transparent platy-to-prismatic crystals up to a few mm long and translucent blocky crystals to 15 mm have also been found at Saddleback Old mine, near Mungrisdale, together with black cerussite containing inclusions of supergene galena (D. I. Green pers. comm. 2005). *Elsewhere in the Lake District and West Cumbria:* At Force Crag mine, Braithwaite, near Keswick, cerussite forms acicular jackstraw crystals up to 4 cm long in open fractures between brecciated wallrock clasts of the main barite vein (Green et al. 1997, Green & Briscoe 2002). Other descriptions of cerussite include: good crystals of cerussite in cavities (Davidson & Thomson 1951) and jackstraw cerussite to 20 mm long (Green & Briscoe 2002) at Barrow mine, Braithwaite; Brandlehow mine, Derwent Water (Davidson & Thomson 1951, Strens 1962); Brundholme (Blencathra) mine, Glenderaterra Valley (Eastwood 1921, Shackleton 1966); Castlenook mine, Newlands Valley (Eastwood 1921); worked at Glenderaterra mine, Glenderaterra Valley and Threlkeld mine, Threlkeld (Postlethwaite 1913, Firman 1978); Goldscope mine, Newlands Valley, near Keswick (Greg & Lettsom 1858, Goodchild 1884, Davidson & Thomson 1951, Young 1987a); Graphite mine, Seathwaite, Borrowdale (Young 1987a); Greenside mine, near Glenridding, Patterdale (Goodchild 1884, Davidson & Thomson 1951); common at Kinniside mine, Cleator Moor; jackstraw cerussite associated with wulfenite at Myers Head mine, Hartsop, Patterdale; cerussite associated with beaverite, caledonite, leadhillite and linarite at Rigg Head quarry, Borrowdale (all Green & Briscoe 2002); with anglesite in shale breccia at Millrigg mine, Staveley (Lawrence et al. 1986); Ruthwaite Lodge vein, Grisedale, SW of Glenridding, Ullswater (Young 1986); rare white translucent blocky crystals at the Shap granite quarry (Wirth & Green 2002); great abundance at Thornthwaite mine, Thornthwaite (Ward 1876, Goodchild 1884, Eastwood 1921, Davidson & Thomson 1951); fine acicular crystals (Greg & Lettsom 1858) and bundles of fine crystals, tennis-ball size, at Yewthwaite mine, Newlands Valley (Davidson & Thomson 1951). *Northern Pennine orefield:* White acicular cerussite is associated with bottinoite and native sulphur at Brownley Hill mine, Nenthead, near Alston (Clark & Rust 1993); compact cerussite and chalcopyrite are intimately mixed at Clouds mine, near Kirkby Stephen, and white cerussite with calamine occurs at Dufton mine, near Appleby (Davidson & Thomson 1951); cerussite forms sprays of acicular white crystals (<5 mm long), partly coating galena in association with barite and fluorite at Hilton mine, Scordale, near Appleby (Davidson & Thomson 1951, Bridges 1982, Winrow 1986a); specimens of cerussite, rarely of collector quality, have been found at Silverband mine, Great Dun Fell (Green & Briscoe 2002); cerussite occurs with linarite, malachite and pyromorphite at Smittergill Head mine, Melmerby Fell, and is noted as a grey alteration product of well crystallized galena on the Windy Brow vein, Tynehead, near Garrigill (Young 1985a); cerussite replaces partially oxidized galena at Hiddenholes mine, Tynehead; cerussite is noted at Hudgill Burn mine, Alston, and in old workings on Haggs vein, Hagg Bank, Nentsberry mine, Nenthead (Dunham et al. 2001). Significant quantities of cerussite have been found only at Hudgill Burn mine, where it was worked as an ore along with galena (Dunham 1990, Bridges & Young 1998).

Derbyshire Cerussite is the most common lead supergene mineral in the county, usually forming massive white coatings or small greyish-white or brownish-white crystals on galena. Crystals up to 5 cm were found in earlier centuries, at a time when small localized deposits of massive cerussite were economic to mine. Such deposits were known as white ore, wheatstone or white rake, of which there are examples at Bradwell, Brassington, Eyam, Flagg, Tideswell, Wardlow and Winster (Braithwaite 1983, Ford et al. 1993). Other localities include: Ashley Hay, near Wirksworth; Deep Rake, Longstone Edge; Magpie mine, Sheldon; Masson Hill, Matlock (all Ford & Sarjeant 1964a); Ashover, Matlock, Eyam and Stoney Middleton (Woodward & Mello 1881); Bage mine, Bole Hill, Wirksworth (Burr 1992); hexagonal twinned crystals (up to 2 mm long) with pyramidal terminations from the opencast above Ball Eye quarry, Bonsall (Bridges 1983); cerussite pseudomorphing anglesite in a replacement flat-type deposit of galena within Carboniferous dolomitized limestone immediately on top of a basic lava at Ball Eye opencast, north side of Via Gellia, near Cromford (Bridges & Green 2005a); Bole Hill, Sheldon (Rieuwerts 1963); Butts and Fall Hill quarries, and Gregory mine, all near Ashover (Smith 1982); good crystals in spaces between barite, fluorite and galena at Crich quarry (Wilson et al. 1995); very finely crystallized cerussite at Cromford Level mine, near Matlock; Hubberdale

Prismatic crystals of colourless cerussite to 4 mm, associated with green malachite; Red Gill mine, Caldbeck Fells, Cumbria. (© M. P. Cooper)

mine, Flagg, near Bakewell; Plackett mine, near Winster; Rent Tor, near Wirksworth (all Greg & Lettsom 1858); Golconda mine, Brassington (Dunham & Dines 1945); cerussite associated with pyromorphite at Green Linnets mine, Brassington (Rudler 1905); small amounts of cerussite in mines around the Heights of Abraham, Matlock Bath (Braithwaite & Ryback 1963a); small grey, white or brown cerussite on octahedral galena and honey-colour crystals in fluorite-dominated cavities at High Rake, Longstone Edge, north of Bakewell (Jeffrey 2001); cerussite mantling galena and associated with calcite, smithsonite and sphalerite at Ladywash mine, Eyam (Mostaghel 1985); colourless twinned crystals at Millclose mine, Wensley, near Matlock (Russell 1912b, Traill 1939); fairly common as prismatic or tabular crystals in vein exposures at Milltown quarry, near Ashover (Clark et al. 1984, Rust 1983b, 1991c); Mogshaw Rake, Sheldon; Royal mine, Matlock; Tearsall Rough, Matlock (all Ford et al. 1993); radiating sprays of clear-to-white acicular or bladed crystals to 10 mm at Pike Lane opencut, near Winster (Uttley 1989); Putty Hill, Monsal Dale; Wester Hollow and White Rake, Brassington (all Green & Strahan 1887); well developed cerussite at the boundary between fluorite and galena at Raper mine, Long Rake, near Middleton-by-Youlgreave (Ineson & Al-Kufaishi 1970); cerussite associated with barite, calcite, galena, pyromorphite and wulfenite in cavities within blocks of fluorite-rich limestone at Sacheveral farm, Winster (Rothwell & Mason 1992); microcrystals of cerussite coat almost every piece of galena except the veinlets in dolomitized limestone at Snelston mine, near Ashbourne (Dewey & Eastwood 1925, Jackman 1996); tabular and bipyramidal pale grey-yellowish crystals and lath-like pale-brown crystals up to 5 mm at Wapping mine, Cromford (Ince 1991); stumpy translucent crystals to 1 cm in cavities within barite–fluorite–galena veinstone at White Rake, Tideswell (Rudler 1905, Ward 1983a).

Devon Fine crystals associated with anglesite in decomposed galena at East Tamar mine, Bere Ferrers and forming small acicular crystals at Hennock, near Chudleigh (Greg & Lettsom 1858, Hall 1868b); associated with silver-rich galena at Frankmills mine, Christow, near Exeter (Rudler 1905, Dewey 1921); cerussite replaces clausthalite extensively in gold- and selenide-bearing carbonate veins at Hope's Nose, Torquay (Stanley et al. 1990a, Paar et al. 1998); limited development of anglesite, cerussite and malachite in voids within gold-bearing quartz–carbonate veins at Loddiswell Ag–Cu–Pb mine, 3 km N of Loddiswell (Stanley et al. 1990b); colourless microcrystals and crusts associated with anglesite and galena at Wheal Friendship (Devon Friendship), Mary Tavy (Brooke 1986).

Durham Cerussite is common throughout the Northern Pennine orefield as a normal supergene alteration product of galena, with white to pale-grey or buff crystalline coatings on oxidized galena reported from almost every lead mine in the orefield (Bridges & Young 1998). However, significant quantities occur only on the north-eastern slopes of Middle Fells, near Alston (Dunham 1990). Early descriptions of the species from Burne mine (Rashleigh 1797, Greg & Lettsom 1858) are probably referencing Stanhopeburn mine, N of Stanhope, where beautiful acicular jackstraw aggregates have been found. Similar material has also been recovered from other Weardale locations: Burtree Pasture mine, Cowshill; Redburn mine, near Rookhope; Stotfield Burn mine, Rookhope (Dunham 1990, Bridges & Young 1998, Green & Briscoe 2002). Abundant cerussite forming reticulated crusts of thin tabular white crystals is described from cavities in oxidized barite–galena veinstone at Closehouse mine, Lunedale (Young et al. 1994a, Dunham et al. 2001). Cerussite is also noted from: Dry Pry mine, Wolsingham; Grasshill mines (Highfield Hushes), Harwood, Teesdale; Groverake vein, Weardale (all Dunham et al. 2001); Eggles Hope Beck, Middleton Common, Teesdale (Bridges 1997b); Fairhill (Greg & Lettsom 1858) – thought by Bridges & Young (1998) to be Middle Fell, Alston; Frazer's Hush mine, Weardale (Green & Briscoe 2002); Grassfield mine, Nenthead and Stotfield Burn mine, Rookhope, Weardale (Bridges & Young 1998); Grasshill mine, Teesdale and Whitfield Brow mines, near Stanhope, Weardale (Young 1985a); and Rogerley mine, near Frosterley, Weardale (Ixer 2003).

Leicestershire Well developed crystals of cerussite (averaging 0.4 mm in length), galena and wulfenite have been found in red clay within an ancient (possibly Triassic) cave at Cloud Hill (Breedon) quarry, Breedon on the Hill (King 1991, Rothwell & Mason 1992). Euhedral, acicular single crystals or twinned equant crystals of cerussite had previously been found close to the Carboniferous/Triassic unconformity at this locality (King 1983). King (1973) notes minor cerussite associated with galena in a sandstone bed above the basal breccia at Newhurst quarry, Shepshed, and, since then, colourless-to-white or grey tabular crystals (to 2 mm) and colourless-to-white acicular crystals (to 3 mm) have been found there (Ince 2005). White-to-grey tabular and prismatic twinned crystals of cerussite occur rarely with mottramite and pyromorphite on malachite in a native copper and cuprite deposit near the Precambrian/Triassic unconformity at New Cliffe Hill quarry, Stanton under Bardon (Hubbard et al. 2005).

Northumberland Greg & Lettsom (1858) note cerussite without description, from Allenheads and Fallowfield, near Hexham.

Nottinghamshire The top 50 mm of the Lower Magnesian Limestone is feebly but persistently mineralized along its outcrop from Nottingham to Doncaster, predominantly with uranium-rich asphaltite, cerussite and galena and occasionally wulfenite. Localities for cerussite include the Bulwell district of Nottingham, and Papplewick Hall, N of Bulwell (Deans 1961).

Shropshire Oxidation of chalcopyrite–galena–sphalerite ore in calcite–quartz veins at Ladywell mine, Shelve, produced a secondary assemblage that includes white acicular cerussite together with anglesite, brochantite, devilline, goethite, jarosite, langite, linarite, malachite, pyromorphite and wroewolfeite (Braithwaite 1982b). Fine acicular crystals of cerussite are reported from Snailbeach mine, S of Minsterley (Greg & Lettsom 1858, Rudler 1905). The nearby Burgam mine has also produced beautiful transparent glassy colourless-to-creamy crystals (Starkey 1982). A study of fracture-controlled lead–zinc vein mineralization in Ordovician and Precambrian rocks of the Shelve inlier (belonging to the West Shropshire orefield) identified a secondary assemblage consisting of anglesite, cerussite, covellite, goethite and smithsonite (Pattrick & Bowell 1991). Samples studied came from mine dumps and museums, mainly from Cothercott, Pennerley, Roman Gravels, Snailbeach, Tankerville and Wrentnall mines.

Somerset Cerussite is often reported from veins within Carboniferous Limestone in the Mendip Hills: plentiful (but usually massive) cerussite in a calcite vein is associated with hydrocerussite and pockets of wad at Batts Combe quarry, near Cheddar (Eastwood 1983, Green 1986); cerussite occurs in a small hausmannite–manganite–pyrolusite vein at Coleman's quarry, Holwell, near Frome (Alabaster 1990a); large twinned crystals of cerussite have been found with pyromorphite and wad at Green Hill, near Charterhouse (Spencer & Mountain 1923); compact masses or loose rough crystals and nodules of cerussite are associated with crednerite, hydrocerussite and mendipite, all filling cavities in manganese ore at Higher Pitts farm, near Priddy (Spencer & Mountain 1923, Kingsbury 1941, Symes & Embrey 1977, Alabaster 1989b, Burr 1996); cerussite is a common species in the oxidized zone of a Cu–Mn–Pb deposit within Carboniferous Limestone at Merehead (Torr Works) quarry, Cranmore, from where hydrocerussite and rare oxychloride species have also been found, including blixite, chloroxiphite, diaboléite, mendipite, mereheadite, parkinsonite and symesite (Alabaster 1975, 1985b, 1989b, Symes 1977, Symes & Embrey 1977, Symes et al. 1994, Welch et al. 1998, 2000).

Staffordshire Cu–Pb–Zn mineralized fissure veins and replacement deposits have been found in Carboniferous Limestone on Ecton Hill. At Waterbank mine, 2 km NNW of Wetton village, on the eastern flank of Ecton Hill, cerussite is part of a supergene assemblage that commonly coats oxidized galena and forms small white-to-colourless striated crystals associated with malachite. A pale-blue variety is possibly copper-rich (Robey & Porter 1972, Rust 1992, 1995c).

Warwickshire Colourless to pale-brown crystals of cerussite to 2 mm and creamy-white twinned crystals to 3 mm have been found sparingly encrusting relict galena at Judkins quarry, Nuneaton (Ince et al. 1991).

Yorkshire The Carboniferous Limestone at Coldstones quarry, 5 km W of Pateley Bridge, hosts several veins containing barite, fluorite, galena and sphalerite. Strong oxidation of the sulphides has produced abundant cerussite, hemimorphite and smithsonite commonly lining cavities (which may represent original pockets of galena and sphalerite). Cerussite forms multiple twinned clear-to-white crystals up to 3 mm long in galena (Ryde 1987, Young et al. 1997). Other reports include: finely crystallized cerussite at Connonly mine, near Skipton (Greg & Lettsom 1858); cerussite associated with galena in Coal Measures sandstone at Ewden, Sheffield (Smith 1958, 1961); cerussite on the dumps of the Old Gang mines, E of Gunnerside Gill, Swaledale (Green 2003); cerussite replacing galena at Pikedaw mine, between Malham and Settle (Green et al. 2006c); cerussite associated with corroded galena, as crystalline crusts or rarely as single crystals at Wetgrooves mine, Aysgarth (Rust 1987b).

Wales

Clwyd Cerussite was first recorded in Wales in 1778 by Thomas Pennant, who described brown lead ore (caulk) from four localities: Claudh mine, Holywell; Crecras; Silver Rake on Halkyn Mountain; and Talacre mine, near Prestatyn. The only remaining specimen in Pennant's collection shows well crystallized cerussite coating galena, associated with pyromorphite (Campbell Smith 1913). Other localities for cerussite in this general area include: Jamaica mine, Mold (Greg & Lettsom 1858); Caeau mine, near Holywell; North Hendre mine, near Llangollen; Rhosesmor (Halkyn East) mine, near Mold (all Bevins 1994).

Dyfed Cerussite is widespread in the Central Wales orefield. Magnificent translucent tabular crystals were once found at Logaulas mine, near Ysbyty Ystwyth

Transparent crystal of cerussite, 5.5 mm tall, from Brandy Bottle Incline, Flincher Gill, Old Gang mines, Swaledale, Yorkshire. (© D. I. Green)

(Greg & Lettsom 1858, Otty 1982, Mason 2004), and these rank among the best cerussite specimens from Britain or Ireland. At Frongoch mine, between Devil's Bridge and Pont-rhyd-y-groes, large crystalline masses up to several cm across, coarse tabular crystals, twinned crystals to more than 3 cm, reticulate groups covering large hand specimens and an association with bechererite or brianyoungite are all reported (Greg & Lettsom 1858, Otty 1982, Bevins 1994, Green et al. 1996b, Cotterell et al. 2003, Rust et al. 2003, Mason 2004). Massive cerussite (occasionally with a pink tint) replaces silver-rich galena in bleached mudstone at Esgair Hir mine, Ceulanymaesmawr. White, green and blue dump-formed cerussite is also noted (Rust & Mason 1988, Rust et al. 2003, Mason 2004). Other occurrences include: Bodcoll mine; Brynyrafr mine; Bwlchrhennaid mine, near Goginan; Esgairmwyn mine; Florida mine; well formed colourless crystals to 2 mm at Rhydfach (Penpontbren) mine, near Tal-y-bont (all Bevins 1994); cerussite associated with anglesite, brochantite, linarite, schulenbergite and langite-group minerals at Brynarian mine, near Tal-y-bont (Rust 1992); platy crystals, well crystallized stalactitic crystals and clear separate crystals associated with pyromorphite and wulfenite at Bwlchglas mine, near Tal-y-bont, also pyromorphite epimorphs after cerussite (Braithwaite 1982a, Green 1986, Rothwell & Mason 1992, Rust 1996c, Mason 2004); Copa Hill mine, Cwmystwyth; Glogfawr mine, Devil's Bridge; Penycefn mine, near Tal-y-bont (all Jones & Moreton 1977); masses of acicular or jackstraw crystals at Eaglebrook (Nantycagl) mine, Ceulanymaesmawr (Braithwaite 1982b, Jones 1983, Mason 2004); tabular cerussite overgrowing pyromorphite at Cwmystwyth mine; tightly packed sheaves of acicular crystals forming grey-white pseudomorphous replacements of galena at Elgar mine, between Tal-y-bont and Penrhyncoch; most galena is partly altered to cerussite at Pandy mine, Tre'r-ddôl (both Mason 2004); associated with wulfenite at Darren mine, Pen-bont Rhydybeddau and Henfwlch mine, between Cerrig yr Hafan and Nant-y-moch reservoir, E of Tal-y-bont (both Rothwell & Mason 1992); attractive twinned crystals, and small colourless elongated and striated crystals sometimes coated with limonite or linarite, at Llechweddhelyg mine, near Clawddmelyn, between Bont-goch and Cwmsymlog – a classic locality for cerussite, malachite and pyromorphite (Jones 1987, Rothwell & Mason 1992, Mason 2004); cerussite associated with bindheimite, bottinoite, chrysocolla, hemimorphite, malachite, pyromorphite and wulfenite at Mynyddgorddu mine, W of Bont-goch, Ceulanymaesmawr (Clark & Rust 1993); crusts of lustrous transparent crystals to 5 mm across on gossany matrix at Penpompren mine (Wolfe 1997); small tabular crystals in cavities in galena, occasionally overgrown by linarite and hemimorphite at Penrhiw mine, Ystumtuen, Central Wales orefield (Mason & Green 1995); Ysbyty Ystwyth (Ryback & Saville 1967); greyish-white crystals up to 2 mm and massive veinlets up to 15 mm long associated with cuprite and malachite at Ystrad Einion mine, near Tal-y-bont (Mason & Rust 1997). Excellent acicular cerussite has also been found at Nant-y-Mwyn mine, Rhandirmwyn, W of Llanwrtyd Wells, and the Vale of Towy (Nant) mine, near Carmarthen (Bevins 1994).

Gwent Cerussite is noted from Ochrwyth quarry, near Risca (Bevins 1994).

Gwynedd Cerussite occurs in vein and vein–breccia deposits at mines in the Llanrwst lead–zinc district of Snowdonia: prismatic crystals enveloped by dundasite at Cyffty mine, between Betws-y-coed and Llanrwst (Otty 1986, Bevins 1994, Haggerty 1995); Gorlan mine, near Trefriw (Russell 1944); long prismatic crystals at New Pandora (Welsh Foxdale) mine, near Trefriw (Prior 1906); Penyrallt mine, between Betws-y-coed and Llanrwst (Haggerty 1995). Other reports are from: Great Orme mine, Llandudno (Ixer & Stanley 1996, Ixer 2001), and Penrhyn (Penrhyn Du) mine, near Abersoch, Lleyn Peninsula (Hubbard 1991).

Mid Glamorgan Partially oxidized galena contains anglesite, bindheimite and cerussite in a mineralized fault within Carboniferous Limestone at Machen quarry, 7 km ENE of Caerphilly. Here, cerussite forms lustrous, colourless, transparent to milky-white, blocky crystals up to 2 mm across. Reticulated masses up to 1 cm across are also noted (Bridges 1990, Goulding & Price 1995, Plant & Jones 1995, Wolfe 1996a).

Powys Reports of cerussite from the Powys section of the Central Wales orefield include: Aberdaunant and Geufron mines, near Llanidloes and Snowbrook (Nantyreira) mine, between Llanidloes and Tal-y-bont (Jones & Moreton 1977); Dyfngwm mine, Machynlleth (Rust & Rust 1987) – actually the Pen Dylife section of Dylife mine (Bevins 1994); Dylife mine, N of Hafren Forest (Moreton 1982, Bevins 1994); and South Nantycar mine, S of Rhayader (Bevins 1994). Cerussite has also been found at Craig-y-Mwyn mine, near Llangynog (Dunham et al. 1978), and Dolyhir quarry, Old Radnor, near Presteigne (Wolfe 2000).

Scotland

Dumfries & Galloway (and Strathclyde) The Wanlockhead–Leadhills orefield is well known for it galena–sphalerite vein mineralization and it is in the oxidation zone of these deposits that cerussite is widely distributed. It is usually white or colourless, but black, brown, grey, yellow and, rarely, blue, green and red varieties have all been found. Anglesite, pyromorphite and other secondary lead and copper minerals are associated species. Cerussite was so abundant at Susanna mine, Leadhills, that it was once worked as "white lead ore" (Wilson & Flett 1921). Very perfect and brilliant acicular crystals and rosettes are reported from the area (Greg & Lettsom 1858), with Rudler (1905) commenting that the Leadhills district was at one time remarkable for the beauty and variety of its crystallized cerussite. Pseudomorphs of cerussite after galena, lanarkite, leadhillite and pyromorphite are all noted by Brown (1927). Occurrences near Wanlockhead include: Margaret's or Straitstep vein (Brown 1919); silky white crystals up to 2 mm long on globular malachite, and jackstraw cerussite on limonite matrix, from Belton Grain vein (Gillanders 1981, Livingstone 2002). At High Pirn mine on the Belton Grain vein, cerussite crystals to 3.8 cm long were found in

Interlocking bunches of fibrous jackstraw cerussite. Leadhills, Strathclyde, Scotland; width of specimen 6 cm. (© M. P. Cooper)

cavities within crystalline quartz (Temple 1954a). Green (1987a) notes white to pale-blue anthropogenic cerussite on the dumps of the Meadowfoot smelter, Wanlockhead. At Leadhills, Strathclyde, cerussite is reported from: Hopeful vein (Temple 1955); Horner's vein (Rothwell & Mason 1992); Lady Anne vein (Brown 1919); Leadhills Dod (Brown 1919, Livingstone & Sarp 1984); and Susanna vein (Abdul-Samad et al. 1982, Rothwell & Mason 1992, Livingstone 2002). Additionally, cerussite has been found on the dumps of West Blackcraig lead–zinc mine, near Newton Stewart (Livingstone et al. 1976).

Grampian Cerussite forms coatings and euhedral crystals, including interpenetrant twins up to 11 mm and acicular bladed crystals up to 8 mm long in a galena–quartz vein cutting a Triassic breccia termed the Cherty Rock at Stotfield, Lossiemouth, near Elgin. Phosgenite, pyromorphite and sphalerite are associated species (Greg & Lettsom 1858, Heddle 1901, Starkey 1988a, Green & Todd 2001). Small white prismatic and tabular crystals of cerussite are reported from cavities in galena and fluorite at Corrybeg mine, Abergairn, near Ballater (Russell 1936a).

Highland Cerussite occurs as small patches throughout barite–galena veinstone from the Loch na Mèine workings of the Struy lead mines, Strath Glass, near Inverness (Russell 1946b).

Tayside Heddle (1901) notes cerussite associated with galena from Craigsoales, near Tarfside in Glen Esk.

Ireland

Clare Cerussite is associated with fluorite, hydrozincite, rosasite and smithsonite (calamine) at Sheshodonnell East mine (Cole 1922, Ryback & Moreton 1993).

Donegal The weathered dumps of Keeldrum mine, near Bedlam, have produced good specimens of cerussite, galena and pyromorphite (Ryback & Moreton 1992).

Down Cerussite, together with azurite, gypsum, limonite, malachite and pyromorphite, are the oxidation products of barite–galena mineralization found as breccias and vein infills within Ordovician and Silurian greywackes and shales in the lead-mining district of Conlig–Newtownards (Moles & Nawaz 1996, Moles et al. 2004).

Dublin The dumps of Ballycorus mine, near Carrickgollogan, have produced acicular cerussite, linarite and other copper and lead secondary minerals from a galena–quartz lode cutting granite (Ryback & Moreton 1993). Other occurrences include: galena-bearing veinstone on the beach at Killiney (Green 1998c); and Scalp mine, near Howth (Greg & Lettsom 1858).

Galway The limestone-hosted Cu–Pb–Zn–(Ag) deposit at Tynagh mine, near Killimor, once the largest lead–zinc mine in Europe, produced over a hundred species, including attractive crystals of cerussite up to 8 mm long (Schultz 1971, Clifford et al. 1986, Ryback & Moreton 1992, Green & Moreton 2001). A little acicular cerussite has also been found at Clements mine, Carrowgarriff, Maam (Russell 1912b).

Longford Cerussite is the dominant supergene mineral, replacing the margins of galena in the sphalerite-dominated deposit within faulted Carboniferous clastic sediments at Keel, Ardagh (Patterson 1970).

Tipperary Cerussite (cat-tooth ore) was abundant as masses of white crystals, sometimes stained with azurite or malachite at Gortnadyne (Gorteenadiha) mine, near Silvermines (Russell 1907b). Cerussite is also noted from other mines in the Silvermines district: Ballygown South (Russell 1907b, Rhoden 1959), Knockanroe (Rhoden 1959), Mogul (Ryback & Moreton 1991), and Shallee East (Russell 1907b).

Wexford Small cerussite crystals occur in chalcopyrite–galena ore on the dumps of Ballyhighland mine, W of Enniscorthy (Ryback & Moreton 1993).

Wicklow Good crystals of cerussite seated on galena have been found at the Luganure mines, Glendalough (Rudler 1905, Ryback & Moreton 1993). Other reports are from: Ballinafunshoge mine, Glenmalur and Van Dieman's Land mine, Glendalough (Ryback & Moreton 1993); and Seven Churches – fine white, reddish-brown and black specimens (Greg & Lettsom 1858).

Offshore territories

Isle of Man Cerussite is noted from galena-bearing quartz veins at the Bradda mines, N of Port Erin Bay (Smythe 1879–1884, Dewey & Eastwood 1925, Starkey 1998b).

Jersey Cerussite partially replaces galena in carbonate veins within granite at L'Ouaisné Bay, southwest Jersey (Ixer 1980), and forms crystals in cavities in galena–pyrite–sphalerite-bearing quartz veins at Le Pulec on the northwest coast (Stanley & Ixer 1982).

Sark The alteration assemblage at Sark's Hope Pb–(Ag) mine, Little Sark, includes cerussite (associated with galena) together with anglesite, azurite, chalcocite, covellite, blaubleibender covellite, limonite and malachite (Greg & Lettsom 1858, Mourant & Warren 1934, Ixer & Stanley 1983, Laffoley 1998).

cervantite

orthorhombic $Sb^{3+}Sb^{5+}O_4$

A rare antimony oxide found in an antimony-rich vein at Louisa mine, Glendinning, Glenshanna Burn, Dumfries & Galloway, Scotland. Cervantite and valentinite are secondary species that occur encrusting stibnite (the main ore species) in association with chalcopyrite, galena, pyrite, sphalerite, and rare jamesonite and semseyite (Williams 1965, 1973). Cervantite is also noted pseudomorphing stibnite on oxidized dump material from Knipe (Fountainhead) mine, Hare Hill, SE of New Cumnock, Dumfries & Galloway (Williams 1965), where fine specimens were once found in cavities in stibnite (antimony-glance) (Greg & Lettsom 1858).

Cervantite and stibnite are briefly reported from Robin Hood mine, Bassenthwaite, Cumbria (Davidson & Thomson 1951), and in Cornwall cervantite has been found at: Wheal Boys, Trewetha, St Endellion; Wheal Kine (Polberro mine), St Agnes; Wheal Lea (Leigh), 2 km N of Landrake; near St Minver; and near Port Isaac (Greg & Lettsom 1858).

cervelleite

cubic Ag_4TeS

A rare silver tellurium sulphide found in a brecciated gold-bearing quartz–pyrite vein (Halliday's vein), Tyndrum, Central Region, Scotland, where it occurs as irregular ≤30 μm-diameter inclusions in galena with hessite, petzite, sylvanite and an unidentified Ag–Te–S phase. These gold and silver minerals are associated with native gold (electrum with high silver contents), chalcopyrite and sphalerite. The gold–silver mineralization is thought to be attributable to hydrothermal activity related to copper–molybdenum mineralization associated with Late Caledonian granites (Pattrick et al. 1988). Cervelleite was originally referred to as "phase C", as the species had not been formally recognized at this time.

cesàrolite

hexagonal $PbMn^{4+}_3O_6(OH)_2$

Glossy, black, botryoidal cesàrolite from Eaglebrook (Nantycagl) mine, Ceulanymaesmawr, Dyfed, Wales; width 10 mm. (© National Museum of Wales, Cardiff; photographer T. F. Cotterell)

A rare lead-manganese oxide hydroxide found as a secondary mineral derived from the alteration of galena. Alteration of ankerite (a commonly associated gangue species) is thought to have contributed the manganese. Cesàrolite was identified on a specimen labelled "psilomelane" in the NMW collection in Cardiff and originally found at Eaglebrook (Nantycagl) mine, Ceulanymaesmawr, Dyfed, Wales. The specimen consists of a rusty, porous matrix, mostly coated with soft, glossy to dull black, botryoidal cèsarolite forming the outer part, and a mixture of cèsarolite and gibbsite forming the soft powdery interior (Cotterell 2006b). Cesàrolite has since been found associated with pyromorphite at Frongoch mine, between Devil's Bridge and Pont-rhyd-y-groes, Dyfed (Cotterell 2007).

ceylonite

A green iron-rich variety of spinel.

chabazite (series)

ZEOLITE GROUP
monoclinic $CaAl_2Si_4O_{12}·6H_2O$

A relatively common hydrated calcium aluminosilicate, whose name gained series status in the zeolite classification of Coombs et al. (1997). There are now four recognized chabazite species: -Ca, -K, -Na and -Sr. The entries below list the occurrences of chabazite for which the species name is not known.

Perfectly transparent and colourless crystals of chabazite up to 3 mm long have been found in cavities in pegmatitic granite at Colcerrow quarry, Luxulyan, near Lostwithiel, Cornwall, and excellent specimens were also found at Ramsley mine, South Tawton, near Okehampton, Devon (Russell 1910b). A second Cornish occurrence of chabazite is that from Dean quarry, St Keverne, Lizard Peninsula, where pale-brown or colourless transparent-to-translucent crystals up to 2 mm across occur in hydrothermal veins cutting gabbro (Seager 1971, Wolters & Wolters 1986).

In Co. Durham, chabazite occurrences are related to the final hydrothermal stages of the Carboniferous Whin Sill quartz dolerite. They include: Cambokeels (Cammock Eals) mine, Eastgate, Weardale – rare white chabazite, forming interpenetrant twinned rhombohedra to 3 mm with apophyllite in joints; Force Garth quarry, Teesdale – colourless, poorly formed rhombohedra to 0.5 mm on joint surfaces of dolerite; and High Force quarry, Upper Teesdale – colourless rhombohedra to 0.75 mm, exhibiting penetration twinning with stilbite on dolerite pegmatite (Young et al. 1991a).

Chabazite is one of many zeolite species found in vesicles within the Tertiary lavas on the Isle of Skye, Highland Region, Scotland. At Moonen Bay, N of Dunvegan, small crystals (< 2 mm) are transparent and colourless, but larger crystals may be brown, orange, pink or white. Exceptionally, chabazite rhombs reach 20 mm in length (Green & Todd 1996, 2001). Other occurrences include: mouth of Allt Preshal Beg, S of Talisker; foot of cliff, Beinn nan Cuithean; Beinn Edra; Geodha Tuill, Loch Eynort; Lyndale (fine crystals); Orbost, north shore of Loch Bharcasaig (Varkasaig); N of Sgùrr Mòr, Quiraing, N of the Snizort Peninsula; Rubha nan Clach, Loch Bracadale (all Heddle 1901); Sgùrr nam Boc, between Loch Brittle and Loch Eynort (Greg & Lettsom 1858, Green & Wood 1996); Sgùrr nam Fiadh (Heddle 1901, Wolfe 1997); the Storr, N of Portree (Greg & Lettsom 1858, Heddle 1901, Tschernich 1992, Livingstone 2002); and Talisker Bay, Minginish (Greg & Lettsom 1858, Heddle 1901, Tschernich 1992, Green & Wood 1996, Pearce 1996a, Green & Todd 2001). Chabazite also occurs as a late hydrothermal mineral in Tertiary rocks on other Scottish islands: lavas on An Gearna and Maol nan Damh, Ben More, Isle of Mull, Strathclyde (McLintock 1910b), and Mullach Sgar, St Kilda (Hirta), Western Isles (Harding et al. 1984).

On mainland Scotland, chabazite is a rare constituent of the secondary assemblage in Carboniferous basalts belonging to the Clyde Plateau Lavas. At Boyleston quarry, SW of Barrhead, Strathclyde, it forms clear transparent single or penetration-twinned crystals to 0.4 mm across associated with malachite-coated calcite crystals (Meikle 1989a), and at Loanhead quarry, near Beith, it occurs as opaque pseudocubes to 4 mm across associated with garnet, natrolite or thomsonite (Meikle 1989b, Green & Todd 2001). Other reports of chabazite in Scotland include: Coatsgate quarry, near Moffat, Dumfries & Galloway (Green & Todd 2001); Glen Farg, 16 km S of Perth, Tayside Region (Heddle 1901); Hushinish, North Harris, Highland Region (Dyer et al. 1996); seven localities in Strathclyde: in the banks of Rye Water, N of Baidland quarry; Creag an Fheidh, Isle of Arran (Cressey 1987); Grainger's quarry, N of Alton House, Kilmalcolm – very fine crystals, said to be the largest in Scotland (Greg & Lettsom 1858, Heddle 1901, Tschernich 1992); Kingscross Point, Isle of Arran (Dyer et al. 1996); Kilmalcolm, W of Bishopton (Livingstone 2002); Milngavie, N of Glasgow (Evans 1987, Livingstone 2002); hills behind Port Glasgow – glottalite (Hey 1960).

In Ireland, chabazite appears to be mainly restricted to the Tertiary basalt regions of Co. Antrim, where it is considered to be by far the most common amygdale species, and in places it forms crystals up to 12 mm across. Six different habits are recognized (Walker 1951, 1960a). Localities include: Ballintoy, between Dunseverick and Carnduff – fine white translucent crystals; Portrush – considerable size and transparency; the Ball in Rathlin Island; Island Magee; Salough Braes, near Larne (all Greg & Lettsom 1858); Ballymena quarries (Green & Moreton 2001); Ballynulto quarry, Broughshane; Giant's Causeway; Parkgate quarry, Templepatrick; Two Mouth Cave, S of Portmuck, Island Magee (all Tschernich 1992); Berk Hill, W of Glenarm (Walker 1960a); Black Cave, N of Larne (Walker 1959a); Catcairn basalt quarry, Legoniel, northwest Belfast (Nawaz 1978a); Coulters Hill quarry, Ballyclare (Nawaz 1978b); Craigahulliar, near Portrush; Killyflugh, N of Ballymena, Lower Toome; Whitehead quarries, Lower Belfast (Herbert Smith 1916); Garron Plateau (Walker 1951, 1960a); and Gobbins Cliff, Island Magee, near Larne (Tomkeieff 1934). Adair (1987) concludes that chabazite (mostly chabazite-Ca) is the single most abundant zeolite species in the Antrim basalts and notes a magnesium-rich variety associated with magnesium-rich heulandite and gonnardite from the southern and central part of Island Magee. Additionally, chabazite is reported from Co. Londonderry at: Benevene (Tschernich 1992); Castle Rocks, E of Magilligan (phacolite variety); and Portstewart (Greg & Lettsom 1858).

chabazite-Ca

ZEOLITE GROUP
trigonal $Ca_2(Al_4Si_8O_{24})·12H_2O$

A common hydrated calcium aluminosilicate whose name was established in the zeolite classification of Coombs et al. (1997). King (1976) reports chabazite-Ca from three Tertiary basalts from the Isle of Skye, Highland Region, Scotland, at: Beinn na Greine, near Portree; western slopes of Ben Horneval, near Dunvegan; and Sgùrr nam Boc, between Loch Brittle and Loch Eynort, on the west coast of Skye. Other locations on the Isle of Skye include Moonen Bay, Oisgill Bay, Sgùrr nam Fiadh, the Storr, and Talisker Bay (D. I. Green specimens and XRD, Tindle unpub. EMP data).

One analysis of chabazite-Ca is reported from amygdaloidal basalt of the Lower Basalt Formation, exposed along the road between Dunluce Castle and Portrush, near the Giant's Causeway, Co. Antrim, Ireland. Elongated vesicles in the basalt contain rims of saponite–chlorite, with analcime, calcite and thomsonite filling the remaining voids. Chabazite-Ca appears to be restricted to later veinlets crosscutting the earlier-formed zeolite species (Robert 2001). Chabazite-Ca is also confirmed from a few basalt locations in Co. Antrim: Craigs quarry, between Ballymena and Larne; Ladyhill quarry, Tardree; the Gobbins; Old quarry, Green Road, Ballyclare; Port-na-Boe, Giant's Causeway; and Aughrim quarry, Kilkeel, Co. Down (H. J. Foy and D. I. Green specimens and XRD, Tindle unpub. EMP data). Many more occurrences are documented by Adair (1987):

Chabazite-Ca exhibiting complex zonation in an amygdaloidal basalt; the Gobbins, Co. Antrim, Ireland; width 5.5 mm; viewed in PPL (upper) and XPL (lower). (© A. G. Tindle)

- South Antrim Escarpment to the W and SW of Belfast – Black Hill quarry, Castle Robin Primary School quarry, Hightown quarry, Cranny River section, McCourts quarry, White Mountain quarry
- Galboly River section, near Garron Point
- Mid-Antrim – Killyflugh quarry, Pioneer quarry
- Giant's Causeway area – Benanouran Head, Craigahulliar quarry, Plaiskin Head, Portnaweelan, Portninish
- Island Magee area – Black Cave, Glenarm cliff section, Gobbins Path.

At Lindsay's Leap quarry in the Mourne Mountains, Co. Down, drusy cavities and veins on joint planes in granite contain rare colourless blocky rhombs of chabazite-Ca up to 1.5 mm on edge, intergrown with heulandite-Ca and stilbite-Ca (Green et al. 2005a). The metamorphic aureole of the Omey granite hosts an amphibolite skarn at Barnahallia Lough, Claddaghduff, west Connemara, Co. Galway. The skarn contains cavities in which off-white slightly corroded rhombs of chabazite-Ca, up to 3 mm across, are associated with albite, epidote, grossular, heulandite-Ca and stilbite-Ca (Moreton & Green 2006a).

Chabazite-Ca is also present in hydrothermally altered gabbroic rocks at Dean quarry, St Keverne, Lizard Peninsula, Cornwall (D. I. Green specimen and XRD, Tindle unpub. EMP data).

chabazite-K?

ZEOLITE GROUP

trigonal $K_2NaCa_{0.5}(Al_4Si_8O_{24})\cdot 11.5H_2O$

An uncommon hydrated potassium-dominant aluminosilicate that has been found as milky-white rhombs, up to 6 mm across, on altered basalt from Lynedale, Loch Snizort, Isle of Skye, Highland Region, Scotland (Livingstone 2002) – see chabazite-Na.

chabazite-Na

ZEOLITE GROUP

trigonal $Na_{3.5}K(Al_{4.5}Si_{7.5}O_{24})\cdot 11.5H_2O$

An uncommon hydrated sodium-dominant aluminosilicate whose name was established in the zeolite classification of Coombs et al. (1997) and consequently the name does not appear in older publications. Instead the now-discredited name "herschelite" had been used, but unfortunately not all herschelite appears to be sodium dominant. For instance, Ryback et al. (1988) describe platy and rhombohedral crystals of "herschelite" in cavities in basalt, from Benanouran Head and Island Magee, Co. Antrim, Ireland, where some samples have K > Na, and Walker (1960a) describes herschelite from 1 km S of Drumnagreagh House, southeast Glenarm, Co. Antrim without any discussion of sodium dominance. Chabazite-Na and -Ca compositions have been determined on a single crystal from Old quarry, Green Road, Ballyclare, Co. Antrim (H. J. Foy specimen, Tindle unpub. EMP data). Other Antrim basalts containing chabazite-Na (some described as potassium herschelite) are reported by Adair (1987):

- Giant's Causeway area – W of Dunseverick Slip, Feigh Carrick, Portnachornan, Portneweelan, Portninish, Port Noffer, Doon Point on Rathlin Island
- Island Magee area – N of Black Cave; Castleshiels, Portmuck; Gobbins Path.

Livingstone (1993) describes milky-white rhombs of "herschelite" on altered basalt from Lynedale, Loch Snizort, Isle of Skye, Highland Region, Scotland, as an alkali-dominant chabazite, but B. Jackson (pers. comm. 2004) suggests that it is chabazite-Ca. Part of a single crystal of chabazite from vesicular basalt at Oisgill Bay on the Isle of Skye is chabazite-Na; the remainder is chabazite-Ca (D. I. Green specimen, Tindle unpub. EMP data).

CHALCANTHITE GROUP

Triclinic sulphates with the general formula $A^{2+}(SO_4)\cdot 5H_2O$, where A^{2+} = Cu, Fe, Mg, Mn. Chalcanthite and siderotil are members of the group found in Britain or Ireland.

chalcanthite

CHALCANTHITE GROUP

triclinic $Cu^{2+}SO_4\cdot 5H_2O$

A relatively uncommon hydrated blue copper sulphate found in a post-mining environment in the tin mines of Wheal Edward, Wheal Hazard, Wheal Cock and Botallack, all St Just, Cornwall (Bancroft & Weller 1993). At Wheal Jane, Kea, near Truro, Cornwall, chalcanthite occurs as large stalactitic formations to over 25 cm long and as crystalline crusts up to 10 cm thick (Rust 1982d). Well defined crystals up to 25 mm long have been found at Ting Tang mine, Carharrack, Cornwall (Greg & Lettsom 1858). Chalcanthite occurs rarely at South Crofty mine, near Camborne, where it occurs coating mine walls on the 290 and 315-fathom levels of the Dolcoath South lode (LeBoutillier 1996). Joint surfaces of the North Tincroft lode in New Cook's Kitchen mine, Camborne, contain occasional streaks of chalcanthite associated with limonite (LeBoutillier et al. 2000). Other reports from Cornwall include: Trevarthen, near Marazion; Gunnislake, Callington; Consolidated mines, near St Day; South Francis, near Redruth (Greg & Lettsom 1858, Collins 1892); Geevor mine, near Pendeen, St Just; St Ives Consols; Phoenix United mines, Linkinhorne (Golley & Williams 1995); and Pentire (Glaze) mine, near New Polzeath, St Minver (Clayton et al. 1990).

Chalcanthite is so soluble that it is not expected to form in the wet environments of the Peak District; however, rare crystals are reported from the Ecton mines, Staffordshire, and the species is also noted in Derbyshire from Dene Hollow, Cromford, and on Bonsall Leys (Robey & Porter 1972, Ford et al. 1993).

In the Northern Pennine orefield, crystals of chalcanthite associated with native copper and possible ferricopiapite are briefly described from the New Firestone level of Groverake mine, near Rookhope, Weardale, Co. Durham (Smith 1973a, Bridges & Young 1998). Greenish-blue crystals of chalcanthite are also noted from Beckermet mine (Winscales No. 1 pit), near Egremont, Cumbria (Young 1987a).

In Wales, chalcanthite (blue vitriol) is one of the post-mining precipitates recorded at Parys Mountain copper mine, Anglesey, where it is found on the walls of open pits and on the dumps (Greg & Lettsom 1858, Greenly 1919, Southwood & Bevins 1995).

Efflorescent chalcanthite from Lauchentyre mine, W of Gatehouse of Fleet, Dumfries & Galloway, Scotland, forms pale-blue curved crystals (< 1 mm) on a specimen now in the collection of the NMS. It is associated with chalcopyrite, pyrite and quartz (Macpherson & Livingstone 1982).

Chalcanthite occurs as bright sky-blue crystalline encrustations coating the walls

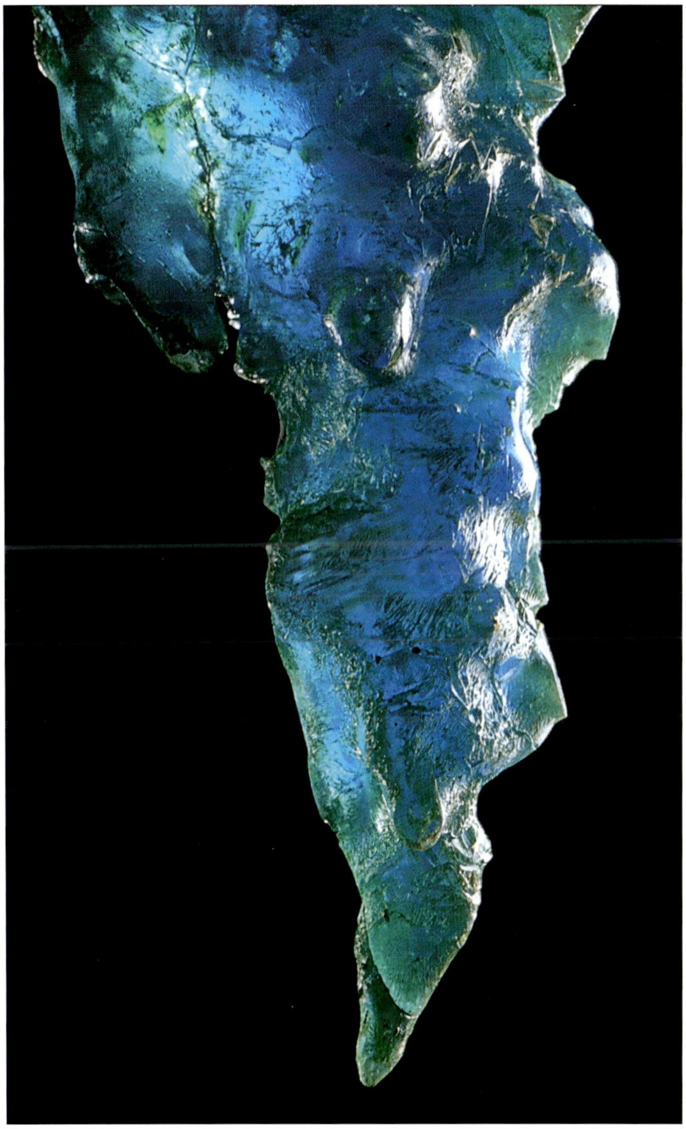

Chalcanthite and melanterite stalactite, 115 mm long; Wheal Jane, Kea, near Truro, Cornwall. (© M. P. Cooper)

and ceiling of an adit in Dooneen copper mine, Allihies, Co. Cork, Ireland, where a chalcopyrite–quartz vein was once worked. It is associated with botallackite, langite, a copper-silicate gel and an unidentified green copper chloride (Moreton & Aspen 1993). At Tankardstown mine, Knockmahon, 22 km SW of Waterford town, Co. Waterford, bright-blue chalcanthite encrusts wallrock containing decomposing chalcopyrite (Moreton 1991, 2005a), and at Cronebane, Avoca, Co. Wicklow, small bluish-green rhomboidal crystals of chalcanthite were once found in the mud at the bottom of an old shaft (Greg & Lettsom 1858).

chalcedony

A microcrystalline variety of quartz ±moganite; see also haytorite and quartz.

chalcoalumite

monoclinic $Cu^{2+}Al_4(SO_4)(OH)_{12}\cdot 3H_2O$

A rare hydrated copper–aluminium sulphate hydroxide found in a secondary alteration assemblage of primary chalcopyrite–pyrite ore at South Wheal Crebor (New East Wheal Russell), Tavistock Hamlets, Devon. Pale-blue to almost colourless radiating crystals of chalcoalumite form botryoidal crusts and overgrowths associated with brochantite, cuprite and native copper on earlier crystallized childrenite (Braithwaite & Cooper 1982, Starkey 1986, Ryback & Tandy 1992).

In Cornwall, chalcoalumite is described from Penberthy Croft mine, St Hilary, where it forms white spherules (≤0.05 mm) covering quartz crystals, with brochantite, chrysocolla, malachite and mimetite (Camm & Merry 1991b, Ryback & Tandy 1992).

At Sneyd's level in Britannia mine, Snowdonia, Gwynedd, Wales, rare chalcoalumite occurs in association with brochantite, lanthanite-(Ce), malachite and posnjakite (Bevins et al. 1985). Extremely rare chalcoalumite has also been found at a trial level, 2 km NW of Bontddu, near Dolgellau, Gwynedd. Here it replaces laths of devilline, malachite and a tyrolite-like mineral, on joints in veinstone (Saich & Rust 1987).

chalcocite

monoclinic Cu_2S

A common copper sulphide often found as a secondary mineral in or below the zone of oxidation in hydrothermal veins and low-grade porphyry deposits. Secondary enrichment often follows the sequence of chalcopyrite being converted to bornite, bornite to covellite, and covellite to chalcocite. Less commonly, chalcocite occurs as a primary hydrothermal mineral. Chalcocite is sometimes confused with djurleite.

England

Avon A chalcocite-bearing calcite veinlet is reported from small opencast workings in Fe–Mn–Pb-mineralized Carboniferous Limestone at Wesley mine, in the grounds of Wesley Theological College, Westbury-on-Trym (Alabaster 1989b).

Cheshire Uncommon chalcocite occurs in Cu–Pb–Zn mineralized veins at Engine Vein mine, West and Wood mines, Alderley Edge ridge (Carlon 1979, Braithwaite 1994, Plant et al. 1999).

Cornwall The St Just mining district on the margin of the Land's End granite has produced many fine specimens of chalcocite. Several lodes in the Botallack sett (Wheal Cock, Wheal Owles and Wheal Edward) produced chalcocite with a wide variety of habits, including prisms, unusual hexagonal lenticular crystals and rare twinned forms (Bancroft & Weller 1993). A 13 mm-long crystal associated with djurleite was recovered from the Corpus Christi lode in 1890 (Wolloxall 1989). At Loe Warren Zawn, Botallack, chalcocite is the major copper-bearing mineral and it occurs disseminated throughout an area of Cu–Pb–U mineralization and as swarms of tiny veinlets. Occasionally, narrow vugs have allowed small hexagonal, tabular, rarely prismatic, crystals to form (Elton & Hooper 1995a,c). A splendid twinned crystal with a somewhat sooty surface (4.5 cm long) from Levant mine is now in the collection of the NHM and is one of the world's finest examples of the species (Embrey & Symes 1987, Wolloxall 1987, Bancroft 1973). Chalcocite is noted with covellite and with atacamite, brochantite and sodium-zippeite from Geevor mine, near Pendeen, St Just (Ball et al. 1982, Elton & Hooper 1993, Green & Betz 2006). Other occurrences include: Bunny mine, St Austell (Russell & Vincent 1952); Carn Brea mines, near Redruth (Robson 1945a); Cligga Head, Perranzabuloe (Polkinghorne 1954); very fine crystals from Cook's Kitchen mine, Redruth (Rust 1982c, Watson 1987a, Penhallurick 1997); irridescent at Dolcoath mine, Camborne; fine elongated crystals at Wheal Fanny and North Wheal Basset, Illogan; as nailheaded copper ore and forming thin hexagonal plates at Wheal Abraham, Crowan (all Greg & Lettsom 1858); Spearn Moor mine, St Just; Tresavean mine, Gwennap (Golley & Williams 1995); in a granite-hosted cassiterite-bearing replacement body (carbona) at East Wheal Lovell (Fatwork section), Wendron (Dominy et al. 1996); Fowey Consols mine, Tywardreath, St Blazey (Lewis 1997); Great Work, Breage; South Wheal Frances, Illogan; Tincroft mine, Illogan; Trannack mine, Helston; Wheal Buller, near Redruth; Wheal Providence, near Carbis Bay, St Ives; Wheal Vor, Tregonning area (all Dewey 1925); in veins with cassiterite, fluorapatite and hydroxyl-chlorapatite at Gunheath china-clay pit, Carthew, St Austell (Elton 1997); Marke Valley mine, near Liskeard (Collins 1911a); in vugs within the North Tincroft lode, New Cook's Kitchen mine, Redruth (LeBoutillier et al. 2000); forming aggregates of hexagonal crystals up to 50 µm across at Penberthy Croft mine, St Hilary (Camm & Merry 1991a,b), although doubted by Betterton et al. (2000); small scattered crystals in massive chalcocite at Pendarves mine, Camborne; hexagonal crystals in parallel growth at West Wheal Basset, Illogan, near Redruth (both Rust 1982c); South and West Caradon mines, St Cleer and West Sharptor mine, near Cheesewring, N of St Cleer, near Liskeard (Shambrook 1982, Lawrence 1990); very rare at South Crofty mine, Camborne (LeBoutillier 1996); most of the copper ore at St Ives Consols, St Ives, was chalcocite, some producing fine crystals (Robson 1945b, Rust 1982c, Watson 1988); chalcocite and chalcopyrite were the major ores at Ting Tang mine, Carharrack (Stephens 1940a, Ball et al. 1982, Bruce & Aubrey-Jones 1998); Wheal Basset, Illogan, near Redruth (Zwaan et al. 1989); fine pseudo-hexagonal crystals to 8 mm from Wheal Crenver, Crowan (Greg & Lettsom 1858, Watson 1988); in veins, sometimes intergrown with bornite and chalcopyrite at Wheal Gorland, St Day, 2 km E of Redruth (Rust 1981b); Wheal Speed, St Ives (Ball et al. 1982); and Wheal Trenwith, St Ives (Hosking 1951).

Cumbria An unusual association of chalcocite with hematite occurred in a fault-bounded vein at Anty Cross mine, near Dalton-in-Furness (Smith 1924, Dewey & Eastwood 1925, Rose & Dunham 1977). Chalcocite and bornite replace chalcopyrite at Birk Fell mine, Tilberthwaite (Stanley 1979). Chalcocite also occurs with malachite at both Black Scar workings, Levers Water, Coniston, and Sea Wood mine, Bardsea, Ulverston (Young 1987a), and is a minor component of the copper-rich veinstone at Cornriggs, Alston Moor (Bridges & Young 1998). At Rigg Head quarry, Borrowdale, chalcocite occurs with covellite, digenite and yarrowite as alteration rims around chalcopyrite, galena and sphalerite (Green et al. 2006c). Minor amounts of chalcocite, cubanite, digenite and sphalerite occur in irregular masses of chalcopyrite within quartz veins at Long Work in the Vale of Newlands. Arsenopyrite and pyrite are associated species (Millward et al. 1999). Chalcocite is reported from Carrock and Driggith mines in the Caldbeck Fells (Davidson & Thomson 1951,

Chalcocite from Cook's Kitchen mine, Redruth, Cornwall; crystal height 30 mm. (© Royal Cornwall Museum, Royal Institution of Cornwall)

Fine twinned crystal of chalcocite, 4.5 cm long, with smaller crystals and an intergrowth of white platy calcite crystals (schieferspar); Levant mine, St Just, Cornwall. (© The Natural History Museum, London)

Thimmaiah 1956), but other locations deserve further confirmation: Brandy Gill, Driggith and Potts Gill mines (Hartley 1984), and Red Gill and Roughton Gill mines (Kendall 1884). All black copper sulphides found at Driggith and Sandbed mines have been shown to be either digenite or djurleite (Green et al. 2006a).

Derbyshire Substantial pieces of chalcocite from Snelston mine, near Ashbourne (now at the NHM), indicate it was an important part of the copper ore (Jackman 1996). Chalcocite is also noted from Golconda mine, Brassington (Ford & Sarjeant 1964a, Ford et al. 1993).

Devon Sporadic and often rich patches of bornite, chalcocite and chalcopyrite were once mined from discontinuous hematite–limonite–quartz veins in Upper Devonian slates and sandstones at Bampfylde mine, near Heasley Mill, North Molton. A little chalcocite was also mined from arsenopyrite–cassiterite–wolframite greisen-bordered vein swarms at Hemerdon Bal, Plympton (Dunham et al. 1978). In the Littleham Cove area, near Budleigh Salterton, chalcocite occurs in Cu–U–V-enriched concretions and also associated with sheets of native copper within Permian redbeds of the Littleham Mudstone Formation. The sheets display a complex sequence of alteration and subsequent mineral growth that includes early cuprite and tenorite, and later chalcocite and a suite of nickel arsenides (Harrison 1975, Milodowski et al. 2000, 2002). Sooty grey chalcocite is an alteration product of chalcopyrite at New East Wheal Russell (South Wheal Crebor), Tavistock (Starkey 1986b), whereas at Wheal Emily, Wembury, chalcocite is itself altered to malachite (Russell 1949b).

Durham Silvery-grey masses of chalcocite up to 10 mm across are associated with cuprite, malachite and minor native copper in brecciated limestones of the Permian Raisby Hill Formation in Raisby Hill quarry, Coxhoe, Ferryhill area (Hirst & Smith 1974), and in an old railway cutting near by (Bridges & Young 1998). Davidson & Thomson (1951) report chalcocite, chalcopyrite, malachite and olivenite from Tynehead mine in the Alston area.

Leicestershire Sand-filled pipe-like bodies near the Carboniferous Limestone/Triassic unconformity at Cloud Hill quarry, Breedon on the Hill, are lined with chalcocite, together with barite, calcite, galena, marcasite and copper carbonates (King 1966). Small black masses of chalcocite intergrown with malachite are noted from the sulphide assemblage associated with a native copper and cuprite deposit near the Precambrian/Triassic unconformity at New Cliffe Hill quarry, Stanton under Bardon (Hubbard et al. 2005). Chalcocite has also been described from Newhurst quarry, Shepshed (King 1973, 1983, King & Wilson 1976), but recent XRD analysis of greyish-black copper sulphides (thought to be chalcocite) found in carbonate-cemented breccia was confirmed as either anilite or djurleite (Ince 2005). However, this does not prove that chalcocite is absent at this locality.

Shropshire Strings of grey chalcocite, partly converted to malachite, are associated with barite in brecciated Precambrian rocks at Westcott mine, Westcott (Dewey & Eastwood 1925).

Somerset At Merehead (Torr Works) quarry, Cranmore, secondary enrichment of chalcopyrite–galena veins cutting Carboniferous Limestone has produced an assemblage of azurite, chalcocite, native copper, cuprite and malachite (Symes & Embrey 1977). Joint planes and pyroxene andesite immediately adjoining the joints at Moons Hill quarry, NE of Shepton Mallet, were once coated or veined with small amounts of chalcocite, chrysocolla, covellite, cuprite, malachite and melaconite, but the occurrence is now mined out (Kingsbury 1941). Burr (1996) suggests that the so-called grey copper at Higher Pitts mine, Priddy, may be chalcocite.

Staffordshire The principal copper ore minerals in the Ecton Hill copper–lead mines were bornite, chalcocite and chalcopyrite (mainly) (Critchley 1979, Robey & Porter 1972).

Warwickshire Chalcocite, often forming euhedral crystals, was an important primary phase at Judkins quarry, Nuneaton, although much has now undergone partial-to-complete alteration to bornite. The mineralization is dominated by a barite–calcite–copper mineral assemblage with minor Pb–V–Zn components, and is concentrated along the Precambrian/Cambrian unconformity (Ince et al. 1991).

West Midlands In the Knowle borehole, Solihull, radioactive nodules, enclosed in ferric-oxide-reduced haloes within redbeds of the Triassic Mercia Mudstone Group, contain coffinite cores with margins of chalcocite and nickeline (Harrison et al. 1983).

Yorkshire Secondary chalcocite is associated with bornite, chalcopyrite, covellite, native copper, cuprite, digenite and djurleite on dump material from the Middleton Tyas mines, near Scotch Corner (Dunham & Wilson 1985, Green 1987b). At Merrybent mine, a few km to the NW of Middleton Tyas, chalcocite has been found in an assemblage of bornite, chalcopyrite, covellite and digenite, all within galena-rich calcite veins (Small 1977, Dunham & Wilson 1985). At Forcett quarry, East Layton, 12 km N of Richmond, alteration of a fault-bounded barite–calcite–chalcopyrite–witherite vein produced secondary bornite, chalcocite and covellite (Dunham & Wilson 1985).

Wales

Dyfed In the Central Wales orefield, chalcocite is an alteration product of chalcopyrite and is itself altered to chrysocolla or malachite at: Camdwrbach mine, near Bryniau Rhyddion, NW of Nant-y-moch reservoir; Dolwen mine, near Devil's Bridge; and Llechweddhelyg mine, near Clawddmelyn, between Bont-goch and Cwmsymlog (Bevins 1994). Other occurrences in the orefield include: Cwmsymlog mine; Eaglebrook (Nantycagl) and Esgair Hir mines, Ceulanymaesmawr (Jones & Moreton 1977, Jones 1983); Frongoch mine, between Devil's Bridge and Pont-rhyd-y-groes, and Llettyhên mine, SE of Tal-y-bont (Bevins 1994).

Gwynedd Shallower core material from the Parys Mountain copper deposit on Anglesey contains secondary bornite, chalcocite and cuprite after chalcopyrite (Pointon & Ixer 1980). Other workers indicate that chalcocite, chalcopyrite, galena, pyrite and sphalerite were all abundant in this deposit (Greenly 1919, Dewey & Eastwood 1925, Ridgway 1980). Chalcocite has also been found: in pillow lavas at Cwm Tregalan, S of Snowdon (Williams 1927); in the porphyry copper deposit at Coed-y-Brenin, near Dolgellau (Rice & Sharp 1976); at Great Orme's Head, Llandudno; at Penrhyn quarry, near Bethesda; and at Port Ysgo, near Aberdaron, Lleyn Peninsula (Bevins 1994).

Mid Glamorgan Chalcocite is reported with digenite around galena from the Llantrisant area (Bowler & Kingston 1971).

Powys Bornite and chalcocite are important supergene species replacing chalcopyrite in Precambrian clastic rocks at Dolyhir quarry, Old Radnor, near Presteigne (Bevins & Mason 1997).

Scotland

Borders Reports of chalcocite from Brunta Burn (below Dodds' Mill), Crowheel and near Keelstone Pool (Heddle 1901) deserve further investigation. However, the species is confirmed from Elsford, near Duns (Livingstone 2002).

Central The bonanza silver deposit, hosted by Devonian andesitic volcanics at Alva mine, Silver Glen, Ochil Hills, is dominated by native silver, but also contains sulphides of copper and iron. These include bornite, chalcocite, chalcopyrite, covellite, digenite, djurleite, malachite, marcasite/pyrite and pyrrhotite (Heddle 1901, Moreton

1996, Moreton et al. 1998, Green & Todd 2001). Chalcocite has also been found at the nearby Blairlogie and Burn of Sorrow mines (Dickie & Forster 1974). At Tyndrum, early lead–zinc mineralization is followed by uranium-rich mineralization in which uraninite is associated with calcite, chalcopyrite, galena, pyrite, quartz and small (< 15 μm) grains of acanthite (argentite), chalcocite, native copper and native silver (Pattrick 1985).
Dumfries & Galloway Traces of chalcocite occur in the Black Stockarton Moor complex, NE of Kirkcudbright – a Caledonian subvolcanic complex that hosts disseminated and veinlet pyrite and copper mineralization comparable in all but scale to porphyry copper deposits (Leake & Brown 1979, Fortey 1980). Chalcocite is a common alteration product of chalcopyrite in the Leadhills–Wanlockhead orefield (Temple 1955).
Grampian Chalcocite occurs as inclusions in alluvial gold from Back Burn, W of Auchenblae (Chapman et al. 2000c).
Highland A sulphide-bearing horizon in Lewisian amphibolite, sandwiched between two marble bands, occurs at Druim na Fearna, Dubh Loch, near Gairloch, and contains disseminated bornite, chalcocite, chalcopyrite and malachite (Jones et al. 1987). Ferruginous chalcocite is reported from nodules in Lower Torridonian sandstone at Clachtoll, 7 km NW of Lochinver (Fermor 1951). On the Isle of Rum, chalcocite is one of the alteration products of chalcopyrite and pentlandite in the Eastern Layered Series of the Rum layered intrusion. The sulphides occur interstitially to chromite (Butcher et al. 1999, Power et al. 2000a, 2003) or form droplets in the upper part of chromitite units (Dunham & Wilkinson 1985). Boron–fluorine ore skarns produced by metasomatism of Durness Limestone in the aureole of the Beinn an Dubhaich granite, Broadford area, Isle of Skye, contain the copper minerals azurite, bornite, chalcocite, chalcopyrite, covellite and malachite (Tilley 1948b, 1951).
Lothian Chalcocite has been found in veins at Fasney Burn, Lammermuir Hills (Greg & Lettsom 1858).
Strathclyde Chalcocite is an uncommon species found in hydrothermally altered copper-enriched tuffaceous rocks of the Carboniferous Clyde Plateau Lava Series at Boyleston quarry, SW of Barrhead. It is either massive or it forms small (< 1 mm) hemihedral striated dull-to-lustrous equant crystals associated with native copper, cuprite and malachite (Meikle 1989a). A related investigation at Loanhead quarry, near Beith identified chalcocite as lustrous lead-grey masses surrounded by malachite staining (Meikle 1989b). Houston (1912) described chalcocite in barite–calcite–quartz veins cutting basaltic rocks at the Kaim mines, Renfrewshire Hills, near Lochwinnoch, but more recent examinations suggest that the main assemblage is malachite with bornite and chalcopyrite (Stephenson & Coats 1983). During a study of molybdenite–pyrite mineralization around the headwaters of Allt Choire Dhuibh, Central Starav granite, Etive granite complex to the E and NE of Ben Starav (on the Highland/Strathclyde border), a boulder was found containing abundant chalcocite and covellite (Haslam et al. 1985). Chalcocite and other copper sulphides are noted from a metasomatic replacement-fissure vein deposit at Kilfinan, near Lochgilphead (Berridge 1969, Willan 1980). Shining black chalcocite occurs in cavities and veins in limestone at Rubha a' Mhill, northeast Isle of Islay (Green 1988), and is also noted in sulphide veins from other locations on Islay: Ballygrant, Kilslevan, Mulreesh and North Ardachie (Baron & Parnell 2005).
Tayside Chalcocite occurs with native copper, datolite and prehnite in lavas at Glenfarg (Heddle 1901, Livingstone 2002). Chalcocite replaces chalcopyrite in the Foss stratiform barite–base-metal deposit, near Aberfeldy (Moles 1985). (Moles 1985).
Shetland Chalcocite borders heazlewoodite in PGM-enriched dunite pods within harzburgite from the Shetland ophiolite complex at Cliff in the Baltasound–Haroldswick area of Unst (Maynard 1993, Prichard et al. 1994).

Ireland

Cork All reports of chalcocite are from deposits hosted by Devonian redbeds in the West Carbery mining district. At Dhurode, thick quartz veins contain a late assemblage of native bismuth, bornite, chalcocite, djurleite, molybdenite and rare wittichenite, whereas, at Derrycarhoon, chalcocite is disseminated with bornite in a stratiform deposit. Other locations include: Brow Head, Cappaghglass, Castle Point, Kilkilleen, Letter and Mount Gabriel (Ni Wen et al. 1996, 1999, Kinnaird et al. 2002).
Limerick Bornite, chalcocite and chalcopyrite are disseminated in limestones at Aherlow, near Galbally (Schultz 1971).
Longford Pyrite-bearing quartz veins and veinlets in hornfelsed turbidites at Aghagah also contain chalcocite, chalcopyrite, molybdenite, pyrrhotite and possibly bornite (Morris 1984).
Tipperary Dolomitized intrusive breccias associated with the Cu–(Ag–Hg) deposit at Gortdrum mine, near Tipperary town, contain varying amounts of bornite, chalcocite, chalcopyrite, coffinite, stromeyerite, tennantite–tetrahedrite and uraninite (pitchblende). Economically important bornite and chalcocite are intimately intergrown in the lower part of the orebody (Morrissey et al. 1971, Halliday & Mitchell 1983, Duane et al. 1986, Steed 1986, Duane 1988). It was in a small bornite–chalcocite–chalcopyrite–cinnabar-bearing iron-rich barite–dolomite vein from this deposit that gortdrumite was first identified (Steed 1983). Chalcocite is listed as one of the many secondary metalliferous minerals found in the Silvermines district (Moreton 1999).
Tyrone Supergene alteration of sulphides (mainly pyrite) has produced a secondary assemblage of chalcocite, covellite, goethite and hematite in the quartz-vein hosted Lack gold deposit, WSW of Omagh (Cliff & Wolfenden 1992).

Offshore territories

Jersey Mourant (1961) reports rare chalcocite from a rhyolite quarry N of Archirondel, in granite from Gigoulande and with chalcopyrite in a calcite vein exposed during excavations for the dam at Val de la Mare.
Sark Samples now in the Lukis collection (Guernsey Museum and Art Gallery) from the copper lode at Le Pot, Little Sark, contain chalcocite, djurleite and a little bornite (Ixer 1992). Chalcocite is one of the alteration products of primary ore minerals in the nearby Pb–(Ag) lode (Ixer & Stanley 1983).

chalcodite

A synonym of stilpnomelane.

chalcolite

A synonym of torbernite, autunite group.

chalcomenite

orthorhombic $Cu^{2+}Se^{4+}O_3{\cdot}2H_2O$

A rare hydrated copper selenite, dimorphous with clinochalcomenite and found in a highly contorted shale bed at the base of coastal cliffs in a small bay NW of Doon East, 1 km N of Ballybunnion, Co. Kerry, Ireland. The shales contain pyritiferous and carbonaceous bands and in the local area are known to contain arsenopyrite, chalcanthite, chalcopyrite, melanterite, pyrite and wavellite. Chalcomenite forms bright-blue transparent elongated crystals up to 0.4×0.15 mm (generally smaller), scattered singly or in radiating clusters on joint surfaces of black shale. Rarely, chalcomenite forms doubly terminated prisms (Francis & Ryback 1987). Moreton et al. (1995) suggest that Cunnihish Cove was the actual location where chalcomenite had been found, but they were not able to confirm its presence at this locality. However, they did find at six locations in this area: secondary sulphates – brochantite, felsöbányaite (syn. basaluminite), gypsum, hydronium jarosite, metasideronatrite, sideronatrite; secondary phosphates – crandallite, turquoise, wavellite; and also malachite.

Bright-blue crystals of chalcomenite up to 0.4 mm across from Wheal Cock, near Botallack, St Just, Cornwall, won Bridget Belson the Maurice Grigg trophy for the best micromineral at the 1996 British Micromount Society symposium (Pearce 1996b). The Wheal Hen section of Botallack mine is another nearby location for chalcomenite (Golley & Williams 1995).

chalconatronite?

monoclinic $Na_2Cu(CO_3)_2{\cdot}3H_2O$

A rare hydrated sodium–copper carbonate found as a turquoise-blue powdery coating on a concretion of shells and pebbles from Trearddur Bay, Holy Island, Anglesey, Gwynedd, Wales. The chalconatronite was recovered from a shipwreck and possibly has an anthropogenic origin. A specimen has been donated to the National Museums of Scotland by Robert Talbot Gardner, an enthusiastic Scottish mineral collector (Livingstone 2002).

chalcophanite

trigonal $(Zn,Fe^{2+},Mn^{2+})Mn^{4+}_3O_7{\cdot}3H_2O$

An uncommon hydrated zinc–iron–manganese oxide found in patchy exposures of iron and manganese oxides in the Lecht area, near Tomintoul, Grampian, Scotland, especially at the site of the old ironstone mine, where a porous hydrothermal eruption breccia occurs. The deposit is mainly composed of goethite and cryptomelane, with occasional patches of hematite. Minor chalcophanite occurs as a fine-grain black constituent of the deposit, along with other manganese-rich species, pyrolusite, ramsdellite and woodruffite. The Lecht mineralization is derived from weathering of the local Dalradian sequence, and ponding of the resulting Fe–Mn–Zn solutions produced a seepage-bog type of deposit in which the permeable breccia acted as a

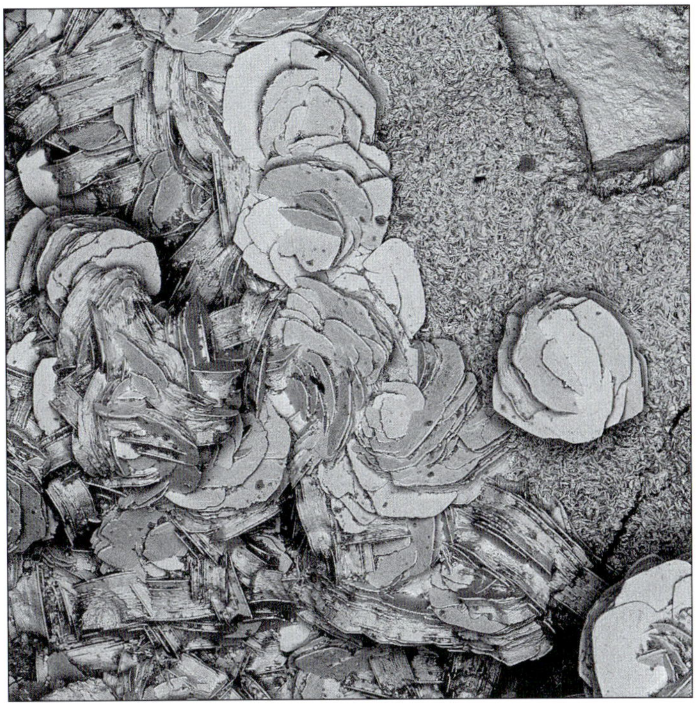

Platy rosettes of chalcophanite to 0.4 mm on manganese-rich matrix; Drosgol mine, Ponterwyd, Powys, Wales; SEM–BSE image. (© National Museum of Wales, Cardiff; photographers T. F. Cotterell & M. P. Lambert)

channel-way for these supergene solutions. Diagnostic lithogeochemical anomalies in the area also indicate the existence of a concealed and related Zn–Pb–(Ag)–(Au?) exhalative deposit (Nicholson 1987a, 1989a,b, 1992, Nicholson & Anderton 1989).

Dark-grey to black masses of chalcophanite in veins within bleached pinkish brecciated mudstones of Ashgill age are reported from Drosgol mine, Ponterwyd, Dyfed. The chalcophanite is remarkable for its abundance and the size of specimens (up to 5 cm). Minutely crystalline chalcophanite occurs as botryoidal masses that are relatively soft and have a brown streak. When examined under a microscope, masses of well formed platy crystals up to 0.4 mm in diameter can be seen. The mineralization is interpreted as a highly oxidized supergene remnant of original hypogene mineralization, with manganese supplied by wallrock leaching and zinc by the oxidation of sphalerite (Cotterell & Mason 2003).

chalcophyllite

trigonal $Cu_9Al(AsO_4)_2(SO_4)_{1.5}(OH)_{12}·18H_2O$

TL Wheal Tamar, Bere Ferrers, Devon, England?

A rare hydrated copper–aluminium arsenate sulphate hydroxide whose type locality is unknown. However, tamarite, a synonym of chalcophyllite, is from Wheal Tamar. Chalcophyllite is a secondary species formed in the oxidized zone of arsenic–copper deposits.

At Wheal Gorland, St Day, 2 km E of Redruth, Cornwall, chalcophyllite forms bright sharp dark-green hexagonal platy crystals, up to 10 mm across, on gossanous veinstone. It also occurs in sheaves lining cavities with chrysocolla, cuprite and malachite (Rust 1981b, Embrey & Symes 1987, Fisher 1988). Chalcophyllite also occurs at nearby Wheal Unity, St Day, Gwennap, forming distinctive bluish-green hexagonal plates up to 5 mm in diameter. Some crystals are pale and cloudy and are pseudomorphously altered to chrysocolla (Embrey & Symes 1987, Bancroft & Weller 1993). Small translucent light greenish-blue crystals of chalcophyllite are described from Penberthy Croft mine, St Hilary, where it forms rosettes up to 1.2 mm across, associated with arsenopyrite, chalcocite, chrysocolla, pharmacosiderite and scorodite (Penhallurick 1997, Betterton et al. 2000). Attractive crystals of chalcophyllite forming rosettes to 1.2 mm across lining cavities in quartz with pharmacosiderite and altered arsenopyrite have also been found at the nearby Croft Gothal mine (J. Betterton pers. comm. 2006). At Mulberry openwork (a tin-rich stockwork), Lanivet, St Austell, chalcophyllite occurs with brochantite and cyanotrichite on iron-stained quartz. It is also developed in cavities in altered chalcopyrite–stannite matrix as transparent green hexagonal crystals up to 1 mm across, infrequently associated with chalcopyrite and cyanotrichite (Starkey 1986a). Other Cornish occurrences of

Large pseudo-hexagonal crystal of chalcophyllite, 3.5 mm across, from South Tolcarne mine, Camborne, Cornwall. (© D. I. Green)

chalcophyllite include: Gunnislake, Callington; Ting Tang mine (Morcom's shaft), Carharrack; Wheal Muttrell, Gwennap (all Greg & Lettsom 1858); Hingston Down Consols mine, 2 km W of Gunnislake, Calstock; South Wheal Basset, Illogan; West Caradon mine, St Cleer (all Golley & Williams 1995); Marke Valley mine, near Liskeard; and Wheal Tremayne, near Hayle, Gwinear (Rust 1982b). Other descriptions of chalcophyllite from England are very dubious: the species cannot be confirmed in Derbyshire (Ford et al. 1993), despite it being described by Mello (1876) and Woodward & Mello (1881); and Green & Briscoe (2002) express concern about its claimed discovery by A. W. G. Kingsbury at Potts Gill mine, Caldbeck Fells, Cumbria (referenced in Hartley 1984, Young 1987a).

Cavities in a native copper–cuprite deposit near the Precambrian/Triassic unconformity at New Cliffe Hill quarry, Stanton under Bardon, Leicestershire, contain rare mm-size tabular hexagonal green chalcophyllite. Larger crystals up to 2.5 mm across also occur with azurite in fractures in the surrounding diorite (Hubbard et al. 2005).

In Wales, chalcophyllite has been found at two locations in the Central Wales orefield: at Dyfngwm mine, Machynlleth, Powys (actually the Pen Dylife section of Dylife mine), where it forms minute green crystals up to 0.2 mm across, associated with brochantite, langite and linarite (Rust & Rust 1987); and at Gwaithyrafon mine, SE of Tal-y-bont, Dyfed, where it forms minute (<0.5 mm) emerald-green hexagonal plates aggregated together in small cavities in quartz, associated with tyrolite (Rust & Mason 1994). A further occurrence is from a trial level on a 1 m-wide quartz vein, 2 km NW of Bontddu, near Dolgellau, Gwynedd, where chalcophyllite forms intergrown hexagonal green platy crystals in masses to 2.5 mm, as scattered single crystals, and as rosettes associated with devilline, malachite and tyrolite (Saich & Rust 1987).

Chalcophyllite is one of over a hundred species found in the residual orebody at Tynagh mine, near Killimor, Co. Galway, Ireland, once the largest lead–zinc mine in Europe (Ryback 1988a, Ryback & Moreton 1992).

CHALCOPYRITE GROUP

Tetragonal sulphides with the general formula $CuBX_2$, where B = Fe, Ga, In; and X = S, Se. Chalcopyrite and roquesite are members of the group found in Britain or Ireland.

chalcopyrite

CHALCOPYRITE GROUP

tetragonal $CuFeS_2$

A very common copper sulphide that forms a series with eskebornite and is found in hydrothermal veins, stockworks, disseminations and massive replacements. It is the most important copper ore species.

England

Avon Chalcopyrite (intergrown with sphalerite and tennantite) occurs in sulphide-rich bands within a barite vein in faulted Triassic Dolomitic Conglomerate along the beach near the Pier hotel, at Clevedon, 16 km W of Bristol. Chalcopyrite–bornite and chalcopyrite–tennantite, intergrown with an idaite-like phase, are also noted. Associated species (all <2 mm) include aragonite, aurichalcite, azurite, brochantite, cerussite, native copper, covellite, cuprite, galena, malachite, phosgenite and native sulphur (Starkey 1984, 1986c, Ixer 1986, Ixer et al. 1993). Emplacement of manganese orebodies into Carboniferous Limestone at Coombe Farm quarry and Wesley Theological College (Wesley mine), both Westbury-on-Trym, Bristol, was followed by chalcopyrite–galena–(pyrite ±sphalerite) veining (Alabaster 1978, 1989b).

Cheshire Vein and disseminated Cu–Pb–Zn redbed-type mineralization in coarse-grain clastic sediments of Triassic age on Alderley Edge ridge, contains a primary assemblage of common chalcopyrite, together with galena, pyrite (including bravoite), sphalerite and tetrahedrite (Ixer & Vaughan 1982). Tennantite-bearing chalcopyrite is also noted (Ixer & Budd 1998). Specific localities include: Stormy Point area, Engine Vein, West and Wood mines (Eagar & Broadhurst 1959, Warrington 1965, Carlon 1979, Braithwaite 1994, Plant et al. 1999). At Bickerton, also in the Cheshire Basin, chalcopyrite occurs with widespread barite in redbed-type mineralization closely associated with marginal basin faults. The paragenetic sequence is: early pyrite (bravoite and nickeloan pyrite) occasionally with Cu–Fe–Ni sulpharsenide rims, followed by bornite and tennantite (the most abundant sulphides), in which chalcopyrite is found as lamellae. Covellite is a secondary replacement species (Naylor et al. 1989).

Cornwall The richness of chalcopyrite-bearing veins and associated secondary enrichment zones in and around major granite intrusions was the reason many mines in Devon and Cornwall (notably those in the Gwennap, Camborne, Redruth and St Just areas) were able to produce more than 40% of the world's output of copper in the first half of the 19th century. The longevity of many mines was down to the fact that, as mining went deeper, copper was replaced by tin (cassiterite) in vertically zoned lodes (Hosking 1951, Dines 1956, Embrey & Symes 1987). *St Just mining district:* Some chalcopyrite occurrences are associated with uranium mineralization: Loe Warren Zawn, Botallack – minor chalcopyrite forming anhedral patches in chalcocite–pyrite veinlets near where widenmannite has been found (Elton & Hooper 1995a,c); North Zawn, Botallack – minor chalcopyrite in pyrite-bearing quartz–calcite boulders on which widenmannite has been found (Elton & Hooper 1996); Geevor mine, near Pendeen, St Just – spots of chalcopyrite in sodium-zippeite-bearing quartz veins and altered killas (Ball et al. 1982, Elton & Hooper 1993); Wheal Edward – chalcopyrite, compreignacite, siderite and pitchblende in dark hornfels, and chalcopyrite, coffinite, pyrite, sphalerite and uraninite (pitchblende) in quartz lodes (Hosking 1954b, Ball et al. 1982, Elton et al. 1994b); Wheal Owles, Penwith – minor chalcopyrite in tyuyamunite–zeunerite-bearing quartz from the Cargodna section of the mine (Elton & Hooper 1994). Other occurrences include: Botallack sett – most old-time specimens are of the blister-copper variety (Bancroft & Weller 1993); Gryll's Bunny and the Crowns, Botallack – accessory chalcopyrite in garnet-bearing skarn (van Marcke de Lummen 1985); Levant mine – well formed crystals up to 2 cm across and early reports of chalcopyrite pseudomorphs after tetrahedrite (Wolloxall 1987); Wheal Cock Zawn, Roscommon Cliffs – threads of chalcopyrite in a discordant axinite–grossular-bearing skarn (Couper & Barstow 1977). *St Ives–Helston mining district:* Carrick Du mine, near Porthmeor – pockets of axinite, calcite, chalcopyrite and epidote in metamorphosed epidiorite (Russell 1910b, Robson 1945b, 1949a); East Wheal Lovell (Fatwork section) and South Wendron mine, Wendron – granite replacement deposits (carbonas), where chalcopyrite occurs with cassiterite and pyrite ±bornite–chalcocite (Foster 1878c, Dominy et al. 1996a); Penberthy Croft mine, St Hilary – chalcopyrite was the main primary sulphide in the upper parts of the mine. It forms massive vein-fillings in quartz lodes and small distorted sphenoidal crystals to 0.5 mm occur in cavities (Barstow 1983, Betterton 1989, 1998, Rust 1996a, Betterton et al. 2000); Priest's Cove, near Cape Cornwall – droplets of chalcopyrite associated with galena and pyrite forming inclusions in quartz within siderophyllite- and zinnwaldite-bearing granite (Müller et al. 2006); Relistian mine, Gwinear – chalcopyrite with arsenopyrite, cassiterite and chalcocite in intrusion breccia (Dunham et al. 1978, Goode & Taylor 1980); St Ives Consolidated mines – minor chalcopyrite in the Great Carbona and good blister copper from a very large vug (in another carbona) on Daniel's lode (Gilbert 1878, Hosking 1951, Dominy et al. 1996a); St Michael's Mount, Marazion – chalcopyrite and stannite are confined to a few cassiterite–wolframite-bearing quartz veins associated with greisen (Davison 1920, Hosking 1954c, Jackson & Rankin 1976, Moore & Moore 1979); Trannack mine, Helston – chalcopyrite, covellite, galena, pyrite and sphalerite in late veins (Ball et al. 1982); Trevascus mine, near Hayle, Gwinear – chalcopyrite with arsenopyrite and cassiterite (±galena or sphalerite) in a quartz lode and intrusion breccia (Hosking 1951, Goode & Taylor 1980); West Wheal Darlington, Marazion – chalcopyrite, galena, pyrite and native silver in a quartz lode (Rust 1982a); Wheal Alfred, Phillack, near Hayle – traces of chalcopyrite and sphalerite in intrusion breccia (Goode & Taylor 1980), also "very perfect rhombic dodecahedrons" (Greg & Lettsom 1858); Wheal Metal, Breage – small quantities of chalcopyrite and sphalerite in a cassiterite-bearing lode (Russell 1913b); Wheal Providence, near Carbis Bay, St Ives – chalcopyrite replacing arsenopyrite and pyrite in greenstone (Ball et al. 1982); Wheal Rose, Sithney – the order of deposition was arsenopyrite, chalcopyrite, sphalerite, quartz, galena, siderite and dolomite (Russell & Hutchinson 1927, Hosking 1951); Wheal Trewavas, Trewavas Head, Breage – the paragenetic sequence at this mine was cassiterite, uraninite (pitchblende), arsenopyrite–pyrite, galena–sphalerite and finally chalcopyrite. The rare uranium-rich species tristramite was first found on a sample of chalcopyrite-bearing quartz vein from the dumps (Ball et al. 1982, Atkin et al. 1983); Wherry mine, Mount's Bay, Penzance – the mine worked the cassiterite-rich portion of an elvan dyke in which chalcopyrite was found at depth (Russell 1949a). *Camborne–Illogan–Redruth mining district (and southwards into the Carnmenellis granite and its environs):* Bosworgey granite borehole, near Fraddam – low-grade aggregates and euhedral crystals of chalcopyrite associated with arsenopyrite in the upper part of the borehole (Beer et al. 1975, Ball & Basham 1984); Carn Brea mines, near Redruth – sphalerite forms stringers in a porphyry dyke, and both mammillated and well crystallized (on cubes of fluorite) varieties are noted (Garby 1848, Greg & Lettsom 1858, Robson 1945a); Cook's Kitchen mine, Redruth – very fine blister copper (Greg & Lettsom 1858); Dolcoath mine, Camborne – the second most productive copper mine in Cornwall (350 000 tonnes of copper ore). It was first a copper and then a tin mine in which chalcopyrite (including attractive blister copper) and cassiterite occurred with other (Ag–As–Bi–Co–Fe–Pb–Sb–U–W–Zn) ore species (Dewey 1923, Hosking 1951, Dines 1956, Penhallurick 1997); East Pool mine, Illogan, near Redruth – enormous quantities of copper ore, including perfect tetrahedra of chalcopyrite associated with siderite (Garby 1848, Greg & Lettsom 1858, Dewey 1925); Holman's quarry, near Camborne – small chalcopyrite–fluorite–pyrite vein; and Pendarves Consols (South Roskear mine), Camborne – cassiterite, chalcopyrite and hematite in a chlorite–fluorite–quartz assemblage (Alderton et al. 1980); Kerriack Cove, between Porthtowan and Portreath, near Redruth – chalcopyrite–galena–sphalerite lode once worked principally for zinc as part of the Wheal Lushington sett (Dines 1956, Elton & Hooper 1995); Mount Wellington mine, Twelveheads, 7 km W of Truro – complex copper–tin lodes in which chalcopyrite is subsidiary to arsenopyrite and pyrite. Chalcopyrite contains stellate exsolution blebs of sphalerite, which in turn contains exsolved chalcopyrite and pyrrhotite (Kettaneh & Badham 1978); New Cook's Kitchen mine (North Tincroft lode), Redruth – complex polymetallic assemblage that includes anhedral masses of chalcopyrite up to 50 µm across, often intergrown with arsenopyrite and sphalerite (LeBoutillier et al. 2000); Newporth beach, 2 km SW of Falmouth – primary bournonite, chalcopyrite (minor), galena and tetrahedrite in a vein containing supergene cumengéite (Dean 1982, Dean et al. 1983); Parabola borehole, between Leedstown and Camborne – chalcopyrite disseminated in greenstone (Beer et al. 1975); Parson's beach, Mawnan – veinlets of chalcopyrite and pyrite in siderite (Camm 1995); deep mines in the neighbourhood of Redruth – remarkable composite crystals of chalcopyrite, which resemble ill developed pentagonal dodecahedra of pyrite (Lewis & Hall 1900); Seton mine, Camborne – chalcopyrite–pyrite–sphalerite lode (Hosking 1951); South Crofty mine, Camborne – well developed crystals in quartz, sometimes twinning in the form of arrowheads, also in veinlets and in massive form surrounding pyrite. Arsenopyrite, cassiterite,

Chalcopyrite crystals to 7 mm on siderite; Carn Brea mine, near Redruth, Cornwall. (© D. I. Green)

CHALCOPYRITE

fluorite, uraninite (pitchblende) and wolframite are associated species (Hosking 1951, 1954b, Polkinghorne 1954, Ball et al. 1982, Wolters & Wolters 1987, LeBoutillier 1996); South Tolgus shaft, Redruth – chalcopyrite–hematite–pyrrhotite–quartz stringers in biotite hornfels and arsenopyrite–cassiterite–chalcopyrite–siderite in heavily mineralized greenstone (Davey 1948); Tincroft mine, Illogan – considerable chalcopyrite and chalcocite, especially in killas, also exceptional specimens of blister copper (Dewey 1925, Penhallurick 1997); Ting Tang mine, Carharrack – immense quantities of chalcopyrite mined, and on the dumps, small golden-yellow masses of chalcopyrite (in vein quartz), partially altered to chalcocite (Stephens 1940a, Bruce & Aubrey-Jones 1998); Tolgus, near Redruth – obtuse rhomboids of chalcopyrite (Greg & Lettsom 1858); Tresavean mine, Gwennap – chalcopyrite–schorl-bearing quartz vein lode (Stephens 1940a); United Mines (Gwennap United), St Day – the most productive copper mines in Cornwall (935 000 tonnes of copper ore), also mined arsenopyrite, cassiterite, pyrite and sphalerite (Dewey 1923, Dines 1956); Wheal Basset, Illogan, near Redruth – chalcopyrite occurs:

- beneath a zone of oxidation
- with chalcocite and cassiterite (plus lesser amounts of arsenopyrite, bassetite, pyrite, vochtenite and wolframite) at the contact of the Carnmenellis granite with Devonian sandstones (Dewey 1925, Zwaan et al. 1989)
- in a quartz lode containing chalcopyrite, pyrite and uraninite (pitchblende) Ball et al. 1982).

Wheal Buller, near Redruth – similar to Wheal Basset, but associated with metatorbernite, ningyoite and uraninite (pitchblende) (Dewey 1925, Ball et al. 1982), also "very perfect tetrahedrons an inch and a half long on the edges" (Greg & Lettsom 1858); Wheal Damsel, St Day – chalcopyrite associated with ningyoite, pyrite, sphalerite and uraninite (pitchblende) (Ball et al. 1982); Wheal Gorland, St Day, 2 km E of Redruth – chalcopyrite is the main ore species in the deeper parts of the mine and it forms massive vein infillings and disseminations (Rust 1981b, Fisher 1988); Wheal Jane, Kea, near Truro – sharp sphenoidal and complex rounded crystals of chalcopyrite scattered on veinstone, also common large masses of chalcopyrite in the lode and remarkable exsolution of chalcopyrite in sphalerite (Rayment et al. 1971, Dunham et al. 1978, Rust 1982d,f, Holl & Bromley 1988); Wheal Unity, St Day, Gwennap – chalcopyrite associated with arsenopyrite, fluorite, jasper, molybdenite, pyrite and sphalerite (Hosking 1951); Wheal Virgin, part of Consolidated mines, Gwennap – the enormously copper-rich Taylor's lode included a suite of oxidized copper ores and masses of chalcopyrite and pyrite (Stephens 1940a). *St Agnes mining district:* Cligga Head, Perranzabuloe – greisen with central quartz veins containing arsenopyrite, cassiterite, chalcopyrite, tourmaline and wolframite. Chalcopyrite and pyrite are also associated with normal faulting and chloritic alteration (Rudler 1905, Jackson et al. 1977, Chick 1987); New Kitty mine – large sphenoidal crystals up to 37 mm on edge associated with quartz on a remarkable specimen now at the NHM (Embrey & Symes 1987); Perran Silver mine – arsenopyrite, chalcopyrite, pyrite and sphalerite have all been noted in the silver-bearing Sedgman lode (Wheal Sedgman lode), Perranuthnoe (Butler 1907); Wheal Golden, Perranzabuloe – cerussite, chalcopyrite, dolomite, galena, linarite, sphalerite, tetrahedrite and wulfenite on the dumps (Merry & Braithwaite 1995); Wheal Hope, ENE of Perranporth – sphalerite with chalcopyrite exsolution, also chalcopyrite included in supergene galena (Hosking & Camm 1981); Wheal Primrose (West Kitty), St Agnes – excellent crystals of chalcopyrite to 2.5 cm with well formed quartz (Rust 1982f); Wheal Towan, near St Agnes – fine crystals of chalcopyrite (originally described as towanite) measure 3 cm across and form interpenetrant symmetric twins with dodecahedral faces (Prior 1902, Rudler 1905, Hosking 1951). *Wadebridge–St Endellion mining district:* Stratabound mineralization is confined to the Devonian–Carboniferous volcano-sedimentary basinal succession and contains chalcopyrite in an assemblage that formed later than arsenopyrite–marcasite–pyrite and earlier than the main Pb–Sb–(Ag–Au) mineralization. Locations include: Bounds Cliff, Port Gaverne, near Port Isaac, St Teath (Forbes 1869, Rust 1982a, Clayton et al. 1990, Stanley et al. 1991, Camm 1995); Gilson's Cove mine, Lundy Bay, 5 km W of Port Gaverne – good specimens of chalcopyrite and galena (Shrimpton 1993); Gunver Head, near Padstow Consols mine – quartz vein containing chalcopyrite, galena, sphalerite and supergene cumengéite and diaboléite (Braithwaite & Merry 1996, Schlomann & Weiss 1998); Pentire (Glaze) mine, near New Polzeath, St Minver, and Reedy Cliff, near Port Quin (Clayton et al. 1990); Port Quin mine (Russell 1944, 1949b, Clayton et al. 1990); Treore mine, Port Gaverne, near Port Isaac, St Teath (Hosking 1951, Kingsbury & Hartley 1960); Wheal Thomas, St Endellion (Russell 1949b). *St Austell mining district:* Crinnis mine, St Austell – chalcopyrite crystals to 2.5 cm associated with native silver and also forming fine irridescent masses termed peacock copper (Greg & Lettsom 1858, Rust 1982a,f); Fowey Consols mine, Tywardreath, St Blazey – from 1827, and for many years after, this mine had the second highest production of copper in Cornwall (after United Mines) and in total produced 320 000 tonnes of copper ore. Bismuthinite was found in druses within William's (chalcopyrite–quartz–siderite) lode (Dewey 1923, Hosking 1951, Dines 1956, Lewis 1997); Gunheath and Hensbarrow china-clay pits, Carthew – chalcopyrite occurs in pegmatitic vugs in the roof zone of the granite, in greisen zones, in stockwork mineralization and in quartz veins (Alderton 1991, Weiss 1994); Mulberry mine, Lanivet – minor arsenopyrite, chalcopyrite and wolframite in a tin (cassiterite) stockwork (Russell 1924, Starkey 1986a); Silver Vein mine, Lostwithiel – calcite–quartz–siderite lode containing chalcopyrite, chlorargyrite, galena, polytelite, pyrargyrite, pyrite, sphalerite and possibly native silver (Rust 1982a); South Terras mine, St Stephen-in-Brannel – magnetite ore contained some arsenopyrite, cassiterite, chalcopyrite and sphalerite, whereas the assemblage in the Radium lode was arsenopyrite, autunite, chalcopyrite, galena, pyrite, torbernite, uraninite (pitchblende) and zippeite (Collins 1912, Robertson & Dines 1929, Dines 1956, Dunham et al. 1978, Ball et al. 1982, Rust 1982a); Treburland mine, 2 km SE of Fivelanes, Altarnun, Bodmin Moor – chalcopyrite is just one of many species found on the dumps from this manganese-rich contact-metasomatic deposit (Russell 1946d); Tremore quarries, Lanivet, SW of Bodmin – massive chalcopyrite in a chlorite–fluorite–sphalerite–talc-bearing lode (Barrow & Thomas 1908); Wheal Betsy and Wheal Mary, Lanivet, near Bodmin, and Wheal Maudlin, Lostwithiel – chalcopyrite is mentioned in the possibly dubious descriptions of danalite and helvine from these mines (Kingsbury 1961). *East Cornwall mining district:* Extends from the eastern margin of the Bodmin Moor granite, past the granite cupolas of Kit Hill and Hingston Down, towards the western margin of the Dartmoor granite in Devon. Chalcopyrite localities include: Cannaframe mine, Bodmin Moor – pegmatite vein carrying arsenopyrite, cassiterite, chalcopyrite, fluorite and wolframite (Brooks 2001); Caradon and Phoenix mining district, near Liskeard – quartz–fluorite lodes contain a suite of well formed and massive primary and secondary copper minerals, including azurite, bornite, chalcocite, chalcopyrite, native copper, cuprite, malachite, tenorite and tennantite (Rust 1982f, Shambrook 1982, Lawrence 1990, Turner & Bannon 1992). An NHM museum specimen with well crystallized quartz and fluorite on a chalcopyrite base came from West Caradon mine (Bancroft & Weller 1993); Greystone quarry, Lezant, near Launceston – chalcopyrite is the most abundant primary copper mineral and it forms complex interpenetrant sphenoidal golden-yellow to metallic-blue crystals, the largest reaching 18 mm on edge (Lloyd & Barstow 1983, Elton et al. 2004); Herodsfoot mine, Liskeard – chalcopyrite encrusting museum specimens of cog-wheel bournonite (Embrey & Symes 1987, Anon. 1993) and tetrahedrite (Green & Betz 2006); Hingston Down Consols mine, 2 km W of Gunnislake, Calstock – chalcopyrite and sphalerite coated with velvety carminite (Russell 1910a, Collins 1911a); Hingston Down quarry – chalcopyrite occurs with bertrandite in cavities in greisenized granite and with arsenopyrite, cassiterite, ferrokësterite and jeanbandyite–natanite in quartz veins (Ward 1983b, Betterton et al. 1998); Redmoor–Kit Hill, including Holmbush, Kelly Bray and Redmoor mines, Kelly Bray, Callington – sulphide veins in killas contain arsenopyrite, chalcopyrite, pyrite, pyrrhotite and sphalerite, along with cassiterite and wolframite (Newall & Newall 1989, Newall 1994); Silver Valley mine, S of Kit Hill, near Luckett – chalcopyrite–quartz lode (Polkinghorne 1954); Tolpetherwin quarry, Two Bridges, South Petherwin – quartz veins in greenstone containing bindheimite, chalcopyrite, jamesonite, pyrite and tetrahedrite (Russell 1949b); Wheal Bray, Altarnun – chalcopyrite and rammelsbergite mantling uraninite (pitchblende) (Ball et al. 1982); Wheal Langford (also known by other names), Callington – quartz lode containing arsenopyrite, chalcopyrite, galena, pyrite, sphalerite and a suite of silver-rich species (Rust 1982a); Wheal Mary Ann, Menheniot – chalcopyrite in a fluorite–galena–(sphalerite) cross-course vein (Gleeson et al. 2000); Wheal Trelawney, Menheniot – chalcopyrite with silver at depth (Dewey 1925); Wheal Wrey, St Ive, near Liskeard – widely scattered masses and thin veins of chalcopyrite forming a minor part of the sulphide mineralization. Rare highly twinned crystals to 6 mm, associated with pyrite in quartz vugs (Rust 1983a). *Other locations:* Carrick Luz, western part of the Lizard complex – lherzolites and dunites containing chalcopyrite and pentlandite associated with laurite, potarite and sperrylite (Gunn & Styles 2002); Countybridge quarry, Goonhilly Downs, Lizard Peninsula – a single specimen of banded chalcopyrite possibly replacing native copper (Holyer 1996); Fowey Estuary – detrital grains in intertidal sediments (Pirrie et al. 2002); Lill Cove mine, Trebarwith Strand, near Tintagel – vein containing arsenopyrite, chalcopyrite, malachite and pyrite (Sharpe 1990); and Parc Bean Cove, Mullion, Lizard Peninsula – calcite, chalcopyrite, datolite and natrolite in veins and geodes at the junction of hornblende schist and serpentine (Collins 1911a).

Cumbria Chalcopyrite is an extremely common mineral in a variety of parageneses

Tarnished chalcopyrite crystal 11 mm across, from Smallcleugh mine, Nenthead, near Alston, Cumbria. (© D. I. Green)

throughout the area and is by far the most common copper ore species. Perhaps the most important is the Lower Devonian (Caledonian) arsenopyrite–chalcopyrite–pyrite mineralization, which defines the earlier of two major mineralization types in the Lake District and is linked to the intrusion of a Lower Devonian composite granite batholith (with resulting de-watering of the overlying Borrowdale Volcanic Group). Much of the mineralization forms vein deposits cutting Ordovician Skiddaw Slates, although the mines near Coniston are an exception as they cut Ordovician Borrowdale Volcanic Group rhyolites and tuffs. Chalcopyrite is also present in other mineralization types, such as the hematite deposits of the West Cumbrian orefield. It is generally accepted that copper mining began in earnest when German miners arrived in 1563 during the reign of Queen Elizabeth I (Firman 1978, Stanley & Vaughan 1982a, Fleming 2000).

Chalcopyrite is reported from several distinct types of vein in the Caldbeck Fells:

- Lower Devonian scheelite–wolframite veins (e.g. Carrock mine)
- Lower Devonian arsenopyrite–chalcopyrite–pyrite veins (e.g. Carrock End, Hay Gill and Potts Gill copper mine – the main copper mines)
- Upper Carboniferous to Permian galena–sphalerite veins (e.g. Potts Gill barite, Driggith and Sandbed mines).

The copper mines were virtually exhausted by the end of the 17th century (Firman 1978, Cooper & Stanley 1990). Details include: Bannerdale mine, Mungrisdale – massive barite, chalcopyrite, galena and sphalerite on the dumps (Davidson & Thomson 1951); Brae Fell mine – small masses in quartz (Hartley 1984, Cooper & Stanley 1990, Bridges et al. 2006); Brandy Gill mine; Carrock End mine; Silver Gill mine, near Roughton Gill; Sandbed mine (all Davidson & Thomson 1951, Hartley 1984); Carrock mine, Grainsgill – massive chalcopyrite in small amounts, also chalcopyrite inclusions in sphalerite (Davidson & Thomson 1951, Eastwood 1959, Ewart 1962, Eastwood et al. 1968, Shepherd et al. 1976, Dunham et al. 1978, Fortey 1980, Hartley 1984, Ball et al. 1985a, Poitrasson et al. 1996); Driggith mine and vein – chalcopyrite and sphalerite enclosed by galena (Kingsbury & Hartley 1957e, Eastwood et al. 1968, Stanley & Vaughan 1981, Hartley 1984, Cooper & Stanley 1991b); Hay Gill mine (Greg & Lettsom 1858, Hartley 1984); Ingray Gill mine (Hartley 1984); Low Pike, near Potts Gill mine and Short Grain (Cooper & Stanley 1990); Nether Row Brow – traces of chalcopyrite in an arsenopyrite–galena–pyrite vein (Kingsbury & Hartley 1960); Potts Gill barite and copper mines – main sulphide is chalcopyrite at both mines (Kingsbury & Hartley 1960, Hartley 1984, Neall & Leppington 1994); Red Gill mine – principal ore minerals were chalcopyrite and galena (Davidson & Thomson 1951, Hartley 1984, Briscoe 1986a, Wirth 1989, Cooper & Stanley 1991b); Roughton Gill mine – active during the reign of Queen Elizabeth I; common chalcopyrite on the mine dumps (Greg & Lettsom 1858, Davidson & Thomson 1951, Eastwood 1959, Eastwood et al. 1968, Hartley 1984); Sandbed mine – rare, well formed sphenoidal crystals to 2.2 mm on edge in cavities in quartz veinstone (Green et al. 2006a). *West Cumbria:* Chalcopyrite is reported infrequently in the limestone-replacement hematite deposits of this area: Anty Cross mine, near Dalton in Furness – the only location where sulphides (chalcopyrite and chalcocite enclosed in hematite) occur in more than trace amounts, although a small copper vein was found in the Urswick Limestone at Sea Wood, Bardsea (Rose & Dunham 1977); Crossgill mine, Frizington – chalcopyrite on the dumps; Eskett mine, Rowrah, Frizington – chalcopyrite associated with barite on the mine dumps (Eastwood et al. 1931, Trotter 1944); Crowgarth mine, Cleator Moor – chalcopyrite (massive and tiny crystals) and pyrite earlier than the main hematite mineralization and also intimately associated with hematite elsewhere in the mine (Trotter 1944, 1945, Davidson & Thomson 1951); Florence mine, near Egremont – small amounts of barite, chalcopyrite, fluorite, galena, manganite and pyrite (Dunham & Hollingworth 1947); Sellafield area boreholes – traces of arsenopyrite, chalcopyrite, marcasite, galena, pyrrhotite, sphalerite, Ag–Bi–Se sulphosalts in uncommon pyrite–quartz veins, also traces of barite, fluorite, galena, hematite, pyrite, sphalerite and Ag-, Ag–Bi- and Bi-selenides in later calcite veins (Milodowski et al. 1998). *Elsewhere in the Lake District:* Applethwaite quarry, Troutbeck, Windermere – small tetrahedra up to 4 mm across in quartz–dolomite veins cutting Silurian rocks (Lawrence et al. 1986); Birkside Gill, Wythburn, Thirlmere; Bramcrag quarry, St John's in the Vale – small chalcopyrite tetrahedra; Holehouse Gill, Ulpha – massive chalcopyrite with arsenopyrite; Robin Hood mine, Bassenthwaite – chalcopyrite with dolomite, native gold and pyrite; Thirlspot mine, Thirlspot – chalcopyrite altering to goethite; Whitecombe Beck, Black Combe – massive arsenopyrite and chalcopyrite (all Young et al. 1987a); Black Craig vein, Manesty, Borrowdale – chalcopyrite associated with pyrite (Strens 1962); Old Brandlehow mine, Derwent Water – chalcopyrite, fluorite, galena and sphalerite on the mine dumps; Threlkeld mine – barite, cerussite, chalcopyrite, galena, malachite, pyrite and sphalerite all noted (all Davidson & Thomson 1951); Buckbarrow Beck, Corney Fell, Waberthwaite – quartz veins cutting the Eskdale granodiorite are chalcopyrite-rich in parts and also contain ferberite, scheelite, minor pyrite and the supergene species bismutite, bismutoferrite, brochantite, chrysocolla, covellite, cuprotungstite, eulytite, goethite, malachite, mixite, namibite and russellite (Young et al. 1986, 1991b, Neall et al. 1993); Castlenook mine, Newlands Valley – chalcopyrite–pyrite ore (Ineson & Mitchell 1974a); Coniston copper mines including Bonser mine, Coniston – massive chalcopyrite–pyrite with minor galena and sphalerite in quartz veins, although at depth arsenopyrite and then magnetite become more important (Russell 1925, Wheatley 1971a, Ineson & Mitchell 1974a, Dagger 1977, Holland 1981, Stanley & Vaughan 1982b, Millward et al. 1999, Green & Briscoe 2002); Copper Plate (Borrowdale) mine, Borrowdale; Dubhow copper mine, Patterdale; Glenderaterra mine, Glenderaterra Valley, near Keswick (all Dewey & Eastwood 1925); Dale Head mine, Newlands Valley – massive chalcopyrite and other sulphides in a pyrrhotite-dominated assemblage, some chalcopyrite has star-shape inclusions of sphalerite (Davidson & Thomson 1951, Stanley & Vaughan 1980, Millward et al. 1999); Damkirk Beck – chalcopyrite–galena–quartz veins, and Gate Crag, Bootle – actinolite–chalcopyrite–pyrrhotite veins, both associated with the Eskdale granite (Young 1985e, Young et al. 1986); Eagle Crag mine, Patterdale – chalcopyrite, galena and tetrahedrite in quartz veins (Stanley 1979, Green & Briscoe 2002); Gatesgarthdale Beck, W of the summit of Honister Pass, Wasdale – rare bornite, chalcopyrite and cubanite in hematite–pyrite-bearing quartz veins (Millward et al. 1999). Goldscope mine, Newlands Valley, near Keswick – worked for copper as early as the 13th century and one of the most productive locations for chalcopyrite. Anglesite, chalcopyrite, cerussite, chrysocolla, galena, linarite and pyromorphite have been found on the mine dumps (Greg & Lettsom 1858, Postlethwaite 1913, Davidson & Thomson 1951, Fleming 2000). Chalcopyrite from here was said to be gold rich by Greg & Lettsom (1858); Graphite mine, Seathwaite – chalcopyrite, pyrite and sphalerite in quartz veins associated with the graphite deposit (Strens 1965); Greenside lead mine, near Glenridding, Patterdale – the abundance of chalcopyrite and sphalerite increased with depth (Gough 1965, Firman 1978, Naylor 1981); Hartsop Hall lead mine, Hartsop, Patterdale – chalcopyrite associated with barite, fluorite, galena and supergene wulfenite (Russell 1944); Oughterside opencast coal site, Aspatria – septarian nodules in mudstones between coal seams contain chalcopyrite, millerite, pyrite and rarely sphalerite (Young & Nancarrow 1988a); Papcastle, near Cockermouth – economic chalcopyrite in hematite veins (Firman 1977); Piers Gill, 1 km N of Scafell summit, Wasdale – chalcopyrite occurs with arsenopyrite, galena, pyrite and sphalerite in a 60 cm-thick polyphase quartz vein cemented by ankerite, within the Scafell dacite. Rare associated species include acanthite (argentite), native gold (electrum), tetrahedrite, uytenbogaardtite and an unknown selenium–silver mineral (Aldridge 2005); Rigg Head quarry, Borrowdale (Green et al. 2006c); Ruthwaite Lodge vein, Grisedale, SW of Glenridding, Ullswater – minor chalcopyrite and tetrahedrite in a galena-bearing quartz vein (Young 1986); Sandbeds Gill mine, near Bassenthwaite – chalcopyrite–(galena)-bearing quartz vein with secondary annabergite, erythrite and symplesite (Thomson 1997b); Seathwaite (Tarnhead vein), near Coniston – arsenopyrite–chalcopyrite–pyrite-bearing quartz veins containing a suite of other copper sulphides including wittichenite (Stanley & Criddle 1979); Shap Blue quarry – chalcopyrite in metasomatized Borrowdale Volcanic Group rocks (Davidson & Thomson 1951, Young et al. 1988); Shap granite quarry – chalcopyrite occurs with bismuthinite, pyrite and sphalerite in calcite–quartz-dominated joint coatings. A single crystal with sphenoidal habit and a surface coating of covellite is also reported (Russell 1925, Firman 1978, Fortey 1980, Wirth & Green 2002); Simon's Nick, Coniston – brecciated chalcopyrite–pyrite ore cemented by siderite (Dagger 1977); Spothow Gill – chalcopyrite-bearing quartz veins in hornfelsed andesite within the aureole of the Eskdale granite (Firman 1977, Young 1985e, Young & Johnson 1985, Young et al. 1986); Ulpha copper mine – chalcopyrite–pyrite-bearing quartz veins (Firman 1977); Wanthwaite mine, St John's in the Vale, near Threlkeld – traces of chalcopyrite in a sulphide-rich quartz rib within fault breccia (Kingsbury & Hartley 1960); Wetherlam mine, Tilberthwaite (Shaw 1970). *Northern Pennine orefield:* Aisgill Moor, Blea Gill, Clouds Gill mine, Hazel Gill, and Turner Hill, all in the Ingleton–Dent area – chalcopyrite-bearing veins (all Bateson & Johnson 1984); Brownley Hill lead mine, Nenthead, near Alston – chalcopyrite forms complex brassy-yellow crystals exceptionally to 3 mm on edge, and is associated with ankerite, barite, calcite, fluorite, marcasite and pyrite, all in galena–sphalerite-rich ore. (Livingstone 1991, Green & McCallum 1993, Green et al. 2000); Hartley mine, near Kirkby Stephen (Davidson & Thomson 1951); Fell End Clouds (High Stennerskeugh level), Ravenstonedale, 10 km SSW of Kirkby Stephen – chalcopyrite–galena–tetrahedrite in quartz veins cutting Carboniferous Limestone (Small 1982, Young 1987b); Crossgill and Tynehead, S of Garrigill – quartz veinstone on the mine dumps contains chalcocite, chalcopyrite, malachite, olivenite and iron sulphides (Smythe 1924, Davidson & Thomson 1951, Dunham 1990); Great Sulphur vein, Alston Moor – moderate amounts of chalcopyrite in a pyrite–marcasite-dominated vein (Smythe 1924, Vaughan & Ixer 1980); Hanging Lund Scar, Mallerstang, 11 km S of Kirkby Stephen – chalcopyrite–tetrahedrite in quartz veins cutting Carboniferous Limestone (Small 1982); Higher Longrigg mine, S of Cote Garth, Hartley Birkett, near Kirkby Stephen – oxidation of small barite–calcite–chalcopyrite–fluorite–galena veins has produced conspicuous amounts of azurite and malachite (Dunham & Wilson 1985, Braithwaite 1988); Hilton lead mine, Scordale, near Appleby – uncommon massive chalcopyrite and malachite in fluorite–galena–quartz veins cutting Carboniferous Limestone (Davidson & Thomson 1951, Winrow 1986a); Nenthead mines, near Alston – chalcopyrite and galena in brecciated limestone on the mine dumps (Vaughan & Ixer 1980); Rotherhope mine, Leadgate, near Alston – chalcopyrite associated with bismuth-rich minerals (Ixer et al. 1996); Tynebottom mine, NW of Garrigill, near Alston – silicified ankerite-bearing limestone contains chalcopyrite crystals 1–2 cm across, intergrown with galena and a lesser amount of sphalerite (Vaughan & Ixer 1980, Ixer 1986) (the description of chalcopyrite by Ixer & Stanley [1987] on specimens said to come from Tynebottom mine was based on material from the Kingsbury collection at the NHM, and the provenance is now uncertain [Ryback et al. 1998, 2001]); Tynehead (Windy Brow vein), near Garrigill – specks of chalcopyrite in quartz veinstone (Young et al. 1985a). Occurrences of chalcopyrite are further described from Clouds, Hartley Birkett and Long Rigg, all near Kirkby Stephen (Dunham & Wilson 1985).

Derbyshire Chalcopyrite is widespread in small quantities, but was of greater importance in the adjacent Staffordshire section (Ecton Hill) of the Southern Pennine orefield. A general paragenetic sequence for the mineralization (vein, metasomatic replacement and cavern infill deposits) includes nickel-poor and -rich (bravoite) pyrite and marcasite, chalcopyrite, galena and finally sphalerite with barite, calcite, fluorite and quartz being the important gangue species (Ixer & Townley 1979, Ixer 1986). *Castleton–Bradwell area:* Chalcopyrite is noted without detail from Pindale, Castleton (Ford & Sarjeant 1964a), and is described from Odin mine, Castleton, where it forms twinned anhedral crystals to 0.5 cm mantling framboidal and spongy pyrite (Ixer & Townley 1979). *Hucklow-Eyam–Stoney Middleton area:* Chalcopyrite forms zoned inclusions in fluorite at Ladywash mine, Eyam (Braithwaite 1983), and is noted without detail from New Engine mine, Eyam, and Watersaw Rake, Longstone Edge (Ford & Sarjeant 1964a). *Ashford–Bakewell–Sheldon area:* An unusual mixed precipitate of chalcopyrite, galena and sphalerite is described from Blende vein at Magpie mine, Sheldon (Ford et al. 1993). *Bonsall–Matlock Bath–Cromford–Wirksworth area:* Cromford Rock mine, Matlock, and Dene quarry, Cromford (Ford et al. 1993); Cumberland and Rutland Caverns, Masson Hill, Matlock Bath, and Riber mine, Matlock – small scattered crystals of chalcopyrite (Lecornu 1897, Flindall & Hayes 1976, Ford et al. 1993); Golconda mine, Brassington (Green & Strahan 1887, Ford & Sarjeant 1964a); Ible quarry on the northern flank of Via Gellia – chalcopyrite in thin quartz veins within dolerite (Ford et al. 1993); Jug Holes mine, Masson Hill; Masson Cavern and opencast, Heights of Abraham, Matlock Bath and Oxclose mine, Snitterton, W of Matlock – traces of anhedral chalcopyrite rimmed with chalcocite, covellite, cuprite and malachite occur in mixed sulphide (galena, marcasite, pyrite, sphalerite) layers within void-infill and limestone-replacement deposits (Braithwaite & Ryback 1963a, Ixer 1974b); Millclose mine, Wensley – radiating nodules of pyrite (bravoite) are mantled by compact masses of chalcopyrite intergrown with galena and tetrahedrite in calcite veins at the contact of an altered toadstone, also pyritic inclusions in both calcite and fluorite contain some chalcopyrite (Bannister 1940, Mueller 1954b, Ford et al. 1993). *Ashover inlier:* Fall Hill quarry, Ashover – minor chalcopyrite in toadstone (Smith 1982); Tansley borehole, Ashover – chalcopyrite associated with millerite, pyrite, pyrrhotite and sphalerite in veinlets intersecting a toadstone (Ramsbottom et al. 1962, Ixer & Townley 1979, Smith 1982, Ford et al. 1993). *Ashbourne–Snelston area:* Snelston mine (Birchwoodpark), near Ashbourne – chalcopyrite forms isolated altered crystals up to 4 mm across and is found in cerussite–chalcopyrite–galena nodules coated with malachite in calcareous sandstone. It also occurs in association with massive chalcocite (Dewey & Eastwood 1925, Jackman 1996, Cornwell et al. 2001).

Elsewhere, chalcopyrite has been found in Coal Measures ironstone nodules near Chesterfield (Mello 1876, Ford & Sarjeant 1964a).

Devon Bampfylde mine, North Molton – largest copper mine in North Devon with almost 5000 tonnes of copper ore produced (Dunham et al. 1978, Camm 1995); Bedford United mines, Tavistock – early cassiterite–wolframite-bearing quartz veins were reopened and infilled with cores of arsenopyrite, chalcopyrite, chlorite and quartz, then by fluorite (Dewey 1925, Dunham et al. 1978); Belstone Consols mine, near Okehampton – chalcopyrite mixed with pyrite in an axinite–garnet skarn (Smith 1878, Solly 1885, Rudler 1905, Beer et al. 1989); Britannia mine, near North Molton – barite, chalcopyrite, siderite and tetrahedrite on the dumps (Rottenbury 1974, Camm 1995); Combe Martin mines - quartz–siderite veins containing chalcopyrite, galena (main ore), native silver, sphalerite, copper secondaries, and (both rare here) stibnite (antimonite) and millerite (Hall 1868b, Dunham et al. 1978); Copper Hill mine, near Okehampton; Devon and Courtenay mine; Fursdon Manor mine, near Okehampton; Great Devon Consols mine; Wheal Robert, Sanford Spinney; Willsworthy mine, near Tavistock – large chalcopyrite tetrahedra; and Wheal Franco, near Buckland Monachorum – massive and disseminated chalcopyrite; Molland mine, near North Molton; North Molton mine (all Hall 1868b); Hemerdon Bal, Plympton – minor chalcopyrite in a cassiterite–wolframite stockwork (Dunham et al. 1978); Hope's Nose, Torquay – accessory amounts of chalcopyrite and pyrite in a gold–selenide carbonate vein (Stanley et al. 1990); Ivy Tor mine, Belstone, Meldon area, near Okehampton – arsenopyrite, axinite, native bismuth, bismuthinite, chalcopyrite, marcasite, phaunouxite, rauenthalite and talmessite all noted (Braithwaite et al. 1996); Littleham Cove area, Budleigh Salterton – vanadium-rich nodules in Permian redbeds contain chalcopyrite among a suite of arsenic, cobalt, copper, nickel, silver and uranium minerals (Harrison 1975b, Milodowski et al. 2000); Meldon (British Railways quarry complex), near Okehampton – arsenopyrite, chalcopyrite and molybdenite in small veins within axinite–datolite skarns (Dearman 1966b); New East Wheal Russell (South Wheal Crebor), Tavistock – primary chalcopyrite–pyrite mineralization (Starkey 1986b); Red-a-ven mine, Meldon, near Okehampton – disseminated chalcopyrite in pyrrhotite-rich andradite–hedenbergite–ferrobustamite (iron wollastonite) skarns (Kingsbury 1961, El Sharkawi & Dearman 1966); St Mary's Bay, Brixham – uncommon chalcopyrite occurs with millerite and pyrite in quartz bedding-plane pods within Devonian shales outcropping on the beach (Braithwaite & Wilson 2001a); Teign Valley, near Hennock – fissure-filling veins containing chalcopyrite, galena, pyrite, sphalerite and tetrahedrite in barite–calcite–fluorite–siderite gangue (Beer & Ball 1977); Virtuous Lady mine, Buckland Monachorum – chalcopyrite was the principal ore and it formed large chalcopyrite tetrahedra associated with pyrite and siderite, also bisphenoids associated with quartz in a box-like cavity of siderite (Greg & Lettsom 1858, Hall 1868b, Rudler 1905, Russell 1913a, Rust 1982f, Embrey & Symes 1987); Wheal Eliza, Exmoor – disseminated chalcopyrite in an iron ore vein (Rudler 1905); Wheal Friendship (Devon Friendship), Mary Tavy – chalcopyrite occurs (with arsenopyrite, galena and sphalerite) in disseminated grains, massive veins and as large tetrahedra (Hall 1868b, Dewey 1925, Braithwaite 1982b, Barstow 1983, Brooke 1986); Wheal Tamar, Bere Ferrers (Phillips et al. 1852); Lundy island – chalcopyrite in minor sulphide veins (Hall 1868b, Bromley & Holl 1986, Thorpe et al. 1990).

Durham *Weardale:* Minerals in the veins and replacement flats of the Weardale area are, in descending order of abundance: galena, fluorite, quartz, siderite, ankerite, calcite, sphalerite, pyrite and chalcopyrite. Dunham (1948) mapped out five concentric zones over the orefield with a central fluorite–chalcopyrite zone. Appreciable quantities of chalcopyrite came from mines along Red vein (Groverake, Redburn and

Stotfield Burn); from a few locations to the NW and SE of Rookhope, in Sedling vein, near Cowshill; and in the Barbary, Slitt and Merkiel workings (King 1982b, Dunham 1990). Some of these veins are bismuthinite bearing and are close to the Rookhope cupola of the buried Weardale granite. They are earlier than the main barite–fluorite–galena–sphalerite mineralization of the orefield and they involved rhythmic crystallization of quartz, chalcopyrite and minor pyrite, followed by fluorite–quartz–chalcopyrite–minor sphalerite–altered pyrrhotite mineralization. Early chalcopyrite is tin bearing and it encloses argentopentlandite, bismuthinite, cosalite, cubanite, mackinawite, pyrrhotite and synchysite (Ixer et al. 1996). Other descriptions include: Blackdene mine, near Ireshopeburn – lustrous chalcopyrite or pyrite scattered over the surfaces of fluorite cubes (Green & Briscoe 2002); Boltsburn mine, Rookhope – minute tetrahedral crystals of chalcopyrite dispersed on fluorite and quartz from the Boltsburn vein (King 1982b); Cambokeels (Cammock Eals) mine, Eastgate – chalcopyrite and bismuth-rich minerals found on Slitt vein (Ixer et al. 1996); Frazer's Hush mine – chalcopyrite-rich quartz veinstone containing cavities partially infilled with bismuthinite (Wolfe 1998, Green & Briscoe 2002); Groverake mine, near Rookhope – chalcopyrite found with bismuthinite and cassiterite (Small 1982, Young 1995, Ixer et al. 1996); Redburn mine, near Rookhope – selvedges of early-phase chalcopyrite and quartz occur on the 40-fathom level, and on the 17-fathom level chalcopyrite gives way to marcasite and pyrite. Some chalcopyrite is tin bearing (Greenwood 1977, Ixer 1986); Rookhope borehole – chalcopyrite occurs in quartz veins within the Weardale granite and above it possibly from the footwall of Red vein (Dunham 1967, Ixer 1986, Brown et al. 1987a); Rookhope cupola (above the Weardale granite) – chalcopyrite with abundant bismuthinite inclusions (Ixer 1986); West Pasture mine, Stanhope – traces of chalcopyrite with sphalerite and the supergene species goethite, gypsum, malachite and rosasite (Young & Francis 1989). *Teesdale area:* Closehouse mine, Lunedale – chalcopyrite forms inclusions in sphalerite. Inclusions of chalcopyrite and sphalerite in barite have suffered secondary oxidation and are replaced by rosasite (Vaughan & Ixer 1980, Young et al. 1985a); Forcegarth quarry, High Force – irregular-to-anhedral chalcopyrite is intergrown with pyrrhotite in an ilvaite-bearing pegmatitic facies of the Whin Sill (Ixer et al. 2005b); Greenhurth mine, Upper Teesdale – small quantities of chalcopyrite (Young et al. 1985a); Lady's Rake mine, 6 km NW of Langdon Beck – minor chalcopyrite occurs in an unusual magnetite–nickeline skarn assemblage related to the intrusion of the Whin Sill (Young et al. 1985c, Green & Briscoe 2002); Teesdale Valley including Coldberry, Eggleshope mine, Flushie Mere mine, Middleton Common and Newbiggin Common – vein infilling and metasomatic ores containing predominantly chalcopyrite, galena, pyrite and sphalerite (Vaughan & Ixer 1980). *Durham coalfield:* New Brancepeth colliery, near Durham – specks of chalcopyrite and sphalerite in a massive barite vein (Spencer 1910, Smythe 1922, Ixer 1986); Raisby Hill quarry, 5 km ENE of Ferryhill – epigenetic copper-enriched breccias, hosted within Upper Permian limestones and associated with the Butterknowle fault contain azurite, chalcocite, chalcopyrite, native copper and malachite (Hirst & Smith 1974, Young 2001).

Lancashire Chalcopyrite, galena and sphalerite occur in a witherite lode on Anglezark Moor, near Chorley (Rudler 1905).

Leicestershire Chalcopyrite-rich quartz–carbonate veins and a supergene botryoidal blister-like form of chalcopyrite have been found at Newhurst quarry, near Shepshed. The quarry works Precambrian diorite close to the Triassic unconformity, and the contact of the two units played an important role in focusing much of the mineralization (King & Ford 1969, King 1983, 1998b, Ince 2005). A few small specimens of chalcopyrite (including some blister copper) have been found in rocks with similar paragenesis at New Cliffe Hill quarry, Stanton under Bardon (Pearson & Jeffrey 1997, Hubbard et al. 2005). Other occurrences include: Ashby-de-la-Zouch (Rudler 1905); Breedon Cloud quarry, S of Breedon on the Hill; Bluebell Wood quarry, near Grosby; Merry Lees Drifts, near Desford; Roaster Coal, near Heath – in septarian nodules; Sheethedges Wood quarry (all King 1998b); Breedon Cloud quarry – sharp tetrahedral crystals up to 3 mm long, on, and included in, colourless scalenohedral calcite (R. E. Starkey pers. comm. 2006); Croft quarry, SW of Leicester – tiny inclusions of chalcopyrite in analcime, within calcite–zeolite veins (Pearson & Jeffrey 1997); Mountsorrel (King 1959); Earl Ferrers mines, Staunton Harold – well developed single or twinned tetrahedral crystals (Sylvester-Bradley & King 1963, King 1998b).

Northumberland Chalcopyrite occurs with gersdorffite, nickeline, pyrite, sphalerite and ullmannite in a witherite vein close to the Whin Sill at Settlingstones mine, near Hexham (Young et al. 1985b). A chalcopyrite–malachite vein is noted from Langlee, Harthope Burn, Cheviots (Haslam 1975).

Nottinghamshire Chalcopyrite occurs with cerussite, galena and malachite in Permian dolomite marl at Papplewick Hall, N of Bulwell (Deans 1961). Similar rocks at Vale Road quarry, Mansfield Woodhouse, are mineralized in cavities and along joint planes, and contain uncommon chalcopyrite as brassy-yellow sphenoidal crystals up to 0.1 mm on edge on sphalerite (Green & Spence 1988).

Shropshire Veins cutting Ordovician sediments at Ladywell mine, Shelve, are mineralized with chalcopyrite, galena and sphalerite (Braithwaite 1982b). In general, Ordovician-hosted veins from the West Shropshire orefield of the Shelve inlier (Cothercott, Pennerley, Roman Gravels, Snailbeach, Tankerville and Wrentnall mines) have a paragenesis that began with the precipitation of sphalerite, followed by barite–galena mineralization, with chalcopyrite occurring late in the mineralizing episode (Jones 1924, Pattrick & Bowell 1991).

Somerset Good crystals of chalcopyrite and galena have been found associated with barite in manganese oxide veins within Carboniferous Limestone at Merehead (Torr Works) quarry, Cranmore (Symes & Embrey 1977). An unconfirmed report of chalcopyrite is noted from Higher Pitts mine, Priddy, Mendip Hills (Burr 1996).

Staffordshire Chalcopyrite was the principal ore species in the Ecton copper mines, from where many thousands of tonnes were raised between 1723 and 1891. At one time these were amongst the richest copper mines in Europe. Rich deposits of chalcopyrite were found lining large cavities in metasomatic pipes within the Carboniferous Limestone, but now the mineral is found only as minute crystals. Chalcopyrite was generally found at a lower level than galena and was associated with bornite (erubescite), chalcocite and malachite (Greg & Lettsom 1858, Rudler 1905, Dewey & Eastwood 1925, Robey & Porter 1972, Critchley 1979, Mostaghel 1985, Masheder & Rankin 1988, Ford et al. 1993, Porter 2004, Cooper 2005). Chalcopyrite is also noted from the Mixon mines (Ford et al. 1993).

Warwickshire Chalcopyrite forms compound tetragonal crystals in a mineralized nodule from Newdigate colliery, Bedworth, SW of Nuneaton. The nodule also

Tarnished chalcopyrite replacing bundles of acicular chalcocite crystals (to 7 mm) on calcite; Judkins quarry, Nuneaton, Warwickshire. (© M. P. Cooper)

contains millerite, galena, siderite and the rare species retgersite (Walker 1988). Ba–Ca–Cu mineralization occurs close to a well defined unconformity between igneous Precambrian rocks and overlying Cambrian sediments at Judkins quarry, Nuneaton. The mineralization includes chalcopyrite as small brassy metallic intergrown sphenoidal crystal groups up to 1 cm across, greenish-yellow granular masses, and an acicular slightly nodular variety forming sheaves up to 2 cm long (Ince et al. 1991, Hawthorne et al. 1999, Ixer et al. 2005a).

Yorkshire Minor chalcopyrite occurs with galena, molybdenite, pyrite and sphalerite in the buried Wensleydale granite at Raydale (Brown et al. 1987a). Bornite, chalcocite, chalcopyrite, native copper, covellite and cuprite have all been reported from the Middleton Tyas mines, near Scotch Corner. The area is grouped with the Richmond copper mines, which are described in a 1454 charter of Edward IV. Unfortunately, the exact location of the mines from this time cannot be established. Copper-rich veins at Billy Bank, East Layton, Feldom, Forcett, Kneeton Hall and Merrybent are also close to Richmond (Raistrick 1975, Small 1977, 1982, Dunham & Wilson 1985, Green 1987b). Chalcopyrite forms sphenoidal brassy crystals with striated faces to 3 mm on edge in fractured siltstone at Hickleton colliery, WNW of Doncaster (Uttley 1988). A regional geochemical survey of the Dent–Ingleton area identified chalcopyrite, often associated with barite, fluorite and galena, at the following locations: Bardale Head, Beckermonds, Fossdale Gill mine, Hull Pot, Hush Gutter mine, W of Ingleborough, Kingsdale Head, Mossdale Head and Tow Scar (Bateson & Johnson 1984). Chalcopyrite is also noted from: the upper reaches of Great Sleddale, SW of Keld (Bateson & Johnson 1984, Dunham & Wilson 1985); Pikedaw mine, between Malham and Settle (Green et al. 2006c); Scaleber Bridge, 3 km SSE of Settle (Ixer & Stanley 1998); and Old Gang mine, E of Gunnerside Gill, Swaledale (Green 2003).

Wales

Chalcopyrite is widely developed in Wales and has been exploited from mines in North and mid-Wales, often from quartz ± carbonate veins, where it is associated with galena and sphalerite. The most notable exception is the massive sulphide deposit at Parys Mountain on Anglesey (Dewey & Eastwood 1925).

Clwyd Chalcopyrite is noted from: Bryn-nantllech mine, N of Llansannan and West Holloway mine, Holywell (Dewey & Eastwood 1925); Bwlch-y-Plwm lead mine, Penrhyndeudraeth – a little chalcopyrite in galena–sphalerite ore; Nant-y-Plwm mine, Llansannan – chalcopyrite and galena in a calcite–quartz lode (both Archer 1959); Dyserth mine – minor chalcopyrite and cobaltite associated with pyrite in hematite fissure deposits cutting Carboniferous Limestone (Bolton 1996); Hendre quarry, near Pandy, Glyn Ceiriog – small isolated inclusions of chalcopyrite in calcite within veins cutting a large basic sill (Starkey et al. 1991); Minera mining district – traces of chalcopyrite and pyrite in galena–sphalerite dominated ore (Earp 1958); Pen-y-Bryn mine, Holywell, near Halkyn (Campbell Smith 1913); Wedding Cave mine, Bwlchgwyn, near Wrexham – chalcopyrite in galena-rich veins cutting brecciated Carboniferous sandstones (Green 1995).

Dyfed *Central Wales orefield:* Primary Cu–Pb–Zn mineralization occurs in veins along faults and breccia zones in Ordovician and Silurian sediments and includes chalcopyrite, galena, marcasite, pyrite, sphalerite, and minor arsenopyrite and cobaltite (Raybould 1974). These veins have been divided into two groups of assemblages (A1 early complex and A2 late simple) and include an A1-*a* iron-rich dolomite–chalcopyrite–quartz assemblage, and an A1-*b* early sphalerite assemblage in which quartz-cemented breccias contain abundant coarse-grain dark-brown sphalerite associated with chlorite, traces of chalcopyrite, iron-rich dolomite and pyrite. The sphalerite is affected by "chalcopyrite disease". A1 assemblages are also found at Darren mine, Pen-bont Rhydybeddau; Erglodd mine, N of Tal-y-bont; Esgair Hir mine, Ceulanymaesmawr; Esgair Fraith mine, near Tal-y-bont; and Eaglebrook (Nantycagl) mine, Ceulanymaesmawr. A2 assemblages are found at: Ystumtuen – A2-*b* chalcopyrite–galena–quartz assemblage, and Nantymwyn mine – A2-*d* coarsely crystalline quartz assemblage in which chalcopyrite occurs as inclusions within quartz (Mason 1997). Other reports of chalcopyrite include: Blaenceulan mine, near Tal-y-bont; Dolwen mine, near Devil's Bridge; Mynach Vale mine, near Devil's Bridge; Penrhiw mine, Ystumtuen (all Jones & Moreton 1977); Bronfloyd mine, ESE of Penrhyn-coch; Camdwrbach mine, nr Bryniau Rhyddion, NW of Nant-y-moch Reservoir; Castell mine, Ponterwyd, near Aberystwyth (all Raybould 1974); Bwlchglas mine, near Tal-y-bont – brecciated galena–sphalerite veinstone with minor chalcopyrite (Rust 1996c); Copa Hill–Cwmystwyth mine – chalcopyrite associated with galena in Comet lode, a Bronze Age copper mining site, blister copper also reported (Dewey & Eastwood 1925, Raybould 1974, Jones & Moreton 1977, Timberlake 1993, Mighall et al. 2002); Darren mine, Pen-bont Rhydybeddau (Jones & Moreton 1977); Eaglebrook (Nantycagl) mine, Ceulanymaesmawr – deeply oxidized chalcopyrite–galena–(sphalerite) assemblage in a quartz–dolomite gangue (Jones & Moreton 1977, Braithwaite 1982b, Jones 1983, Mason 2004); Erglodd mine, N of Tal-y-bont – chalcopyrite associated with galena, siegenite and sphalerite (Rust 1991); Esgair Hir mine, Ceulanymaesmawr – chalcopyrite was a major ore species and it occurs as disseminated spots, masses up to 10 cm across and as rare sphenoidal crystals (Jones & Moreton 1977, Rust & Mason 1988); Esgair Fraith mine, near Tal-y-bont (Jones & Moreton 1977, Bevins 1994); Hendrefelen mine, Ysbyty Ystwyth – chalcopyrite associated with ullmannite in cavities within quartz (Rust 1990); Llechweddhelyg mine, near Clawddmelyn, between Bont-goch and Cwmsymlog – primary chalcopyrite–galena-dominated assemblage now substantially oxidized. Some chalcopyrite is disseminated in quartz, or forms small walnut-size masses coated with brochantite or linarite (Jones 1987, Mason 2004); Lodge Park copper trial, Tre'r-ddôl – 2–5 cm-thick layers of chalcopyrite in a small quartz–(iron-rich dolomite) lode (Mason & Green 1996); Mynyddgorddu mine, W of Bont-goch, Ceulanymaesmawr – a bottinoite specimen encrusts ullmannite enclosed in quartz, with chalcopyrite and galena (Jones & Moreton 1977, Clark & Rust 1993); South Darren mine, Pen-bont Rhydybeddau – considerable chalcopyrite in galena–(sphalerite) ore (Dewey & Eastwood 1925, Hughes 1959, Bevins 1994); Ystrad Einion mine, near Tal-y-bont – polymetallic assemblage of mainly chalcopyrite–galena with accessory siegenite (Jones & Moreton 1977, Mason & Rust 1997).

Elsewhere, chalcopyrite is reported from: Mynyddygarreg district, near Kidwelly (Rudler 1905, Strahan et al. 1909); Cystanog mine, near Carmarthen (Carruthers et al. 1915); Dolocauthi mine, near Pumsaint (Annels & Roberts 1989).

Gwynedd Chalcopyrite was the principal ore species at the great Parys Mountain opencast, Amlwch, Anglesey – the site of a large volcanogenic massive sulphide deposit. Mining began in the Bronze Age, and by the mid-18th century it had become Europe's largest copper mine. The Great lode (the largest of the Parys Mountain lodes) exhibited zonation from a pyrite-rich base, through a chalcopyrite-dominated central zone to a galena–sphalerite-rich (bluestone) zone at the top. Chalcopyrite has been found both in massive form and as a constituent of bluestone, where it occurs as exsolution blebs (chalcopyrite disease) in sphalerite (Greg & Lettsom 1858, Greenly 1919, Dewey & Eastwood 1925, Manning 1959, Pointon & Ixer 1980, Ridgway 1980, Southwood & Bevins 1995). A resource potential has been identified for 15.6 to 18.6 million tonnes of massive sulphide mineralization at the base and for 31.5 Mt of disseminated sulphides at the top of the succession. If realized by future exploration, this represents a world-class massive sulphide deposit (Tyler 2004).

The Great Orme mines at Llandudno have a long history of producing chalcopyrite-rich ore from veins cutting Carboniferous Limestone. Bronze Age miners probably worked major veins and open voids within dolomitized limestone infilled by saddle dolomite and calcite–chalcopyrite–malachite–pyrite. Well formed tetrahedral crystals are noted (Greg & Lettsom 1858, Dewey & Eastwood 1925, Smith 1988, Bevins 1994, Dutton et al. 1994, Williams 1995, Ixer & Davies 1996, Ixer 2001). At Ty Gwyn mine on Great Orme's Head, chalcopyrite occurs with malachite and hydrocarbon botryoids containing nickeline and uraninite (Parnell 1988c). *Snowdonia:* Massive chalcopyrite was once worked from: vein deposits cutting Ordovician sedimentary and volcanic rocks in central Snowdonia (e.g. Britannia and Sygun mines and Hafod y Porth, SE of Snowdon); to the W in the Nantlle Valley (e.g. Simdde Dylluan and Drws-y-Coed mines); and in the SW in Cwm Pennant. Chalcopyrite occurs in four distinct assemblages – quartz–pyrite–sphalerite, quartz–pyrite ± pyrrhotite–sphalerite ± galena, galena–sphalerite ± chalcopyrite ± pyrite and calcite–marcasite ± chalcopyrite ± galena ± hematite ± sphalerite (Reedman et al. 1985, Bevins 1994). All the mines are described by Bick (1982). Other occurrences include: Castell Carn Dochan mine, near Bala – chalcopyrite–pyrite lode with gold (Archer 1959); Cwm Cynfal mine and Manod-bach mine, near Ffestiniog, Ceunant and Gwaith mines, Llandegai, S of Bethesda, and Tal-y-Sarn mine, Llanllyfni (Dewey & Eastwood 1925); Cwm Tregalan and Shadow Gully, both S of Snowdon; Lliwedd, SE of Snowdon (Colman & Appleby 1991); Gamallt mine, Careg y Fran, near Ffestiniog – chalcopyrite and sphalerite are the most common ore species (Archer 1959, Lynas 1973); Graig Wen and Llyn Bowydd, both near Ffestiniog (Lynas 1973); Llanrwst orefield, N of Betws-y-coed – lead–zinc veins with traces of chalcopyrite (Haggerty 1995); Mountain Lode mine, S of Tal-y-Sarn mine, Llanllyfni – chalcopyrite, galena and sphalerite in two quartz lodes and Pant-y-wrach mine, between Llanfrothen and Penryndeudraeth – quartz–calcite vein containing chalcopyrite and some galena and Prysor mine, Trawsfynydd – chalcopyrite–pyrite ore with some gold (Dewey & Eastwood 1925, Archer 1959). *Dolgellau goldbelt:* Chalcopyrite is an important constituent of many mesothermal gold-bearing veins in the belt: Bwlch-coch-uchaf, Caegwian and Hendreforion lodes (Andrew 1910); Clogau mine, Bontddu, near Dolgellau – chalcopyrite is the most abundant ore species, with arsenopyrite, erythrite, galena, pyrrhotite, pyrite, sphalerite, tetradymite and a suite of Bi–Pb–Sb tellurides also noted (Andrew 1910, Dewey & Eastwood 1925, Allen & Jackson 1985, Mason et al. 2002); Dolfrwynog and Ffridd-Goch mines, Llanfachreth – chalcopyrite and pyrite mainly, high-grade gold–silver mineralization and a highly variable telluride assemblage also noted from the latter (Dewey & Eastwood 1925, Mason et al. 2002); Friog Undercliff, Fairbourne, SW of Dolgellau – pyrrhotite dominantly, galena–sphalerite locally common, arsenopyrite–chalcopyrite–pyrite infrequent (Mason et al. 1999); Gwynfynydd mine, near Dolgellau – galena–pyrite–sphalerite dominated, with minor acanthite, arsenopyrite, chalcopyrite, hessite, marcasite, pyrrhotite and tetrahedrite. Chalcopyrite also occurs as exsolution blebs in sphalerite (Leake et al. 1993, Allen & Jackson 1985, Dominy et al. 1996c, Mason et al. 2002); Panorama mine, Llanaber, E of Barmouth – a great deal of chalcopyrite has been found in parts of the lode, where it is associated with pyrrhotite. Cobaltite, galena and sphalerite are noted elsewhere (Andrew 1910, Dewey & Eastwood 1925, Mason et al. 2002): Prince of Wales mine, near Dolgellau (Greg & Lettsom 1858); Tyddyn Gwladys Pb–(Ag) mine, N of Dolgellau – arsenopyrite, chalcopyrite, galena, freibergite–tetrahedrite (polytelite), sphalerite and both native silver and gold are reported (Forbes 1868, Mason et al. 2002); Vigra mine, Llanaber – mainly chalcopyrite, with galena and sphalerite near the walls of the lode (Andrew 1910, Dewey & Eastwood 1925, Mason et al. 2002). Chalcopyrite also occurs as low-grade disseminations in the Coed-y-Brenin porphyry copper deposit (Rice & Sharp 1976, Dunham et al. 1978) and with bornite, native gold and pyrite in a mineralized breccia pipe worked at Glasdir mine, Llanfachreth, near Dolgellau (Dewey & Eastwood 1925, Archer 1959, Allen & Easterbrook 1978, Allen & Jackson 1985). Chalcopyrite inclusions occur in alluvial gold from the Afon Wen (Leake et al. 1993).

Mid Glamorgan Composite crystals of chalcopyrite and sphalerite occur with galena in calcite-lined vugs within dolomitized Carboniferous Limestone at Bute quarry, Llantrisant, and both chalcopyrite and sphalerite are noted from similar rocks in the Locks Common borehole, Porthcawl (Fletcher & Young 1988). Other chalcopyrite-bearing veins cutting Carboniferous Limestone occur in the Mynyddygarreg district (Strahan et al. 1909) and at Machen quarry, 7 km ENE of Caerphilly (Plant & Jones 1995). Triassic rocks outcropping in the Llantrisant bypass roadcut contain calcite–witherite-lined vugs with some chalcopyrite (Alabaster 1990a). Coal Measures rocks containing chalcopyrite in clay–ironstone nodules are documented from: Caerau colliery, Maesteg – attractive twinned crystals (Bevins 1994); and Wyndham colliery, Ogmore Vale (Bevins & Horák 1985b). North (1916) notes chalcopyrite from Mwyndy mine in the Llantrisant area.

Powys *Central Wales orefield:* Cyfarthfa mine; Geufron mine, near Llanidloes and Snowbrook (Nantyreira) mine – blister copper (all Jones & Moreton 1977); Dyfngwm mine, Machynlleth (actually the Pen Dylife section of Dylife mine) – relatively uncommon chalcopyrite, forming disseminated spots and thin veins, occasionally developing crystal faces (Rust & Rust 1987); Dylife mine – chalcopyrite associated with galena, pyrrhotite and pyrite (Jones & Moreton 1977, Moreton 1982); Nantyricket, near Llanidloes – wide calcite–chalcopyrite-bearing lode in conglomerate (Dewey & Eastwood 1925).

Other locations include: Dolyhir quarry, Old Radnor – barite–chalcopyrite–galena–tennantite vein (Wolfe 2000); Llangynog mine, south flank of Berwyn Dome – galena–sphalerite lode with a little chalcopyrite (Archer 1959, Dunham et al. 1978).

Scotland

Borders Chalcopyrite inclusions in alluvial gold have been found at Craigierigg Burn, Cramalt Burn and Glengaber Burn, SW of Peebles (Leake et al. 1993, Chapman et al. 2000a,c). Heddle (1901) notes chalcopyrite from Stoneshiel Hill, NW of Preston, near Duns and Stonypath, Peebleshire.

Central Ag–Ba–Co–Cu mineralization has been exploited from mines between Bridge of Allan and Dollar, in the Ochil Hills, just N of the Highland Boundary fault (Dickie & Forster 1974). Chalcopyrite probably occurs in all of them, but is documented from: Airthrey Hill mine, Bridge of Allan – said to be the best chalcopyrite in Britain (Dickie & Forster 1974, Wolfe 2004); Alva mine, Silver Glen – chalcopyrite occurs in veins with acanthite (argentite), arsenopyrite, erythrite, galena, malachite, native silver and sphalerite, and in veinlets of bornite, dolomite and grey copper sulphides cut by clinosafflorite (Greg & Lettsom 1858, Heddle 1901, Dunham et al. 1978, Jassim et al. 1983, Moreton et al. 1998); Blairlogie mine and Burn of Sorrow mine, near Dollar (Dickie & Forster 1974); Borland Glen and surrounding area – chalcopyrite inclusions in alluvial gold (Chapman et al. 2000c, 2005). *Tyndrum area:* Ag–gold-bearing quartz–sulphide veins in a fault zone within Middle Dalradian metasediments at Cononish, 4 km SW of Tyndrum, contain (in decreasing order of abundance) pyrite, galena, chalcopyrite and sphalerite. Chalcopyrite inclusions in native gold are also noted (Parker et al. 1989, Earls et al. 1992, Curtis et al. 1993, Chapman et al. 2000c). Similar Ag–Au–chalcopyrite–sphalerite-bearing veins in the area contain cervelleite (listed as phase C: see cervelleite), electrum, hessite, petzite and sylvanite (Pattrick et al. 1988). Chalcopyrite also occurs in veins and vein breccias once worked for galena–(sphalerite) and other veins dominated by uranium-rich and other species including acanthite (argentite), chalcocite, kasolite, marcasite, pyrargyrite, pyrite, safflorite, tetrahedrite and uraninite (pitchblende) (Darnley et al. 1960, 1962b, Ineson & Mitchell 1974b, Pattrick 1978, 1985, Ridgway 1980). Pattrick & Hall (1984) describe a sulphide-rich boulder containing chalcopyrite exsolution blebs in sphalerite from the Crom Allt River, near Tyndrum. At Creag Bhocan, SSW of Tyndrum, copper–zinc stratabound mineralization hosted by chloritic schists and albitic quartzites within the Ben Challum Quartzite (Easdale Subgroup of the Argyll Group Dalradian) is characterized by chalcopyrite, pyrite and sphalerite (Smith et al. 1984, Fortey & Smith 1986, Scott et al. 1988). Chalcopyrite is noted with barite, galena and sphalerite from Clifton mine, Tyndrum (Greg & Lettsom 1858, Heddle 1901).

Dumfries & Galloway A diorite-hosted lensoidal mass of As–Cu–Ni mineralization at Talnotry (Palnure Burn), near Newton Stewart, contains an assemblage of chalcopyrite–pyrite–pyrrhotite with minor cobalt-rich gersdorffite in the lowermost 3 m and a chalcopyrite–gersdorffite–nickeline assemblage in the uppermost 1 m. Additionally, chalcopyrite forms stringers and lenticular masses in unmineralized diorite (Heddle 1901, Gregory & Leitch 1928, Dunham et al. 1978, Stanley et al. 1987, Power & Pirrie 2001, Gunn & Styles 2002, Power et al. 2004). Near by, arsenopyrite-rich pods containing pyrite, pyrrhotite and a little chalcopyrite and galena are reported from the marginal rocks around the Cairnsmore of Fleet granite, and in the granite itself. Orchars vein contains chalcopyrite (partially altered to covellite) with hematite, pyrite and colloform jasper (Lowry et al. 1997). A few km to the N is Penkiln Burn, where rare chalcopyrite has been found with pyrite–sphalerite-(pyrrhotite) disseminated in black Ordovician mudstone (Stone & Leake 1984). Accessory amounts of chalcopyrite, ilmenite and pyrite are also present in diorite from Loch Dee, at the southern end of the Loch Doon granite complex (Tindle 1982). Around Glendinning (Louisa) mine, Glendinning, 13 km NNW of Langholm, chalcopyrite is reported from drill-core sampling stratiform and disseminated arsenopyrite–pyrite mineralization, in which bournonite, galena, semseyite, sphalerite, stibnite, tennantite and tetrahedrite have been found (Gallagher et al. 1983). Glendinning mine worked a stibnite-bearing quartz vein (a turbidite-hosted As–Au–Sb deposit) containing minor chalcopyrite, galena and sphalerite (Williams 1973, Duller et al. 1997).

Other locations include: Balcary mine, Auchencairn Bay, E of Kirkcudbright – chalcopyrite with barite, malachite and quartz (Heddle 1901, Williams 1965); Barncorkrie farm, Drummore – chalcopyrite in a quartz–(calcite) vein cutting the Portencorkie diorite (Cameron & Lawson 1979); Beeswing, Potterland and Sandyhills, three localities all associated with the Criffell granite complex – uranium-rich veins containing chalcopyrite and uraninite (pitchblende), and also native bismuth at Sandyhills (Gallagher et al. 1971); Black Stockarton Moor – disseminated porphyry-style mineralization in diorite and granodiorite contains an assemblage of bornite–

Chalcopyrite on dolomite matrix from Wanlockhead, Dumfries & Galloway, Scotland; crystals to 1.2 cm. (© The Trustees of the National Museums of Scotland)

chalcocite–chalcopyrite–molybdenite–pyrite–tennantite–tetrahedrite (Leake & Brown 1979, Fortey 1980, Lowry et al. 1997); Cairngarroch Bay – chalcopyrite and pyrite intergrown in pods and disseminated in microtonalite of the Cairngarroch Bay intrusive complex, also chalcopyrite in a complex Bi–Cu–Fe–Te–Zn assemblage within adjacent greywacke-hosted quartz veins (Lowry et al. 1997); Cairnsmore of Fleet granite (Laour et al. 1990); Callisterhall, near Langholm – chalcopyrite occurs in galena and sphalerite in veins and disseminations in basal Carboniferous sediments underlying the Birrenswark Lavas (Smith et al. 1978); Cally mine, Gatehouse; Castle Douglas; Door of Cairnsmore; Kells; Kings Laggan; Knipe (Fountainhead) mine, Hare Hill – chalcopyrite occurs in quartz–carbonate veins, in stibnite veinlets, as inclusions in arsenopyrite and as disseminations in altered wallrocks adjacent to sulphide veins, all in a late Caledonian granodiorite intrusion (Naden & Caulfield 1989, Boast et al. 1990); Lauchentyre copper mine, W of Gatehouse of Fleet; Pibble mine, Creetown; West Grove mine, Wanlockhead (all Heddle 1901); Colvend copper mine, near Rockcliffe – chalcopyrite, connellite and malachite have been found in quartz–calcite gangue and crush rock (Heddle 1901, Wilson & Flett 1921, Moreton & Aspen 1992); East & West Blackcraig mines, SE of Newton Stewart – shattered barite–calcite–dolomite vein containing strings and patches of chalcopyrite, galena and sphalerite (Heddle 1901, Wilson & Flett 1921, Livingstone et al. 1976); Moorbrock Hill, Carsphairn, close to the margin of a granodiorite intrusion – chalcopyrite, minor marcasite, sphalerite and tetrahedrite in a gold prospect hosted by greywacke shale (Naden & Caulfield 1989); Needle's Eye area, SW of Caulkerbush – uranium-rich veins associated with the Criffell granite complex contain a primary ore assemblage of native bismuth, chalcopyrite, pyrite, uraninite and minor löllingite, nickeline and sphalerite (Miller & Taylor 1966, Basham et al. 1989, Braithwaite & Knight 1990, Green & Todd 2001); Stennies Water – chalcopyrite inclusions in alluvial gold (Leake et al. 1993); Wanlockhead lead–zinc orefield – widespread massive chalcopyrite, often as inclusions in galena and sphalerite (Brown 1927, Temple 1955, Livingstone 2002).

Fife Chalcopyrite occurs at: Burntisland on the Firth of Forth – cm-wide patches of chalcopyrite grains surrounded by malachite in non-marine algal limestones (Parnell 1983); Goat quarry, 2 km NW of Aberdour – tarnished bisphenoidal crystals to 4 mm on edge in a calcite–quartz vein (Ingram et al. 1992); Magus Muir limestone quarry (Heddle 1901); Orrock quarry, near Burntisland – small 1 mm tetrahedra associated with hematite encrusting calcite–pectolite in fissure fillings within Carboniferous basalt (Wirth 2003).

Grampian Chalcopyrite is commonly reported from cumulate units of the large gabbroic intrusions of the area: Arthrath intrusion, near Ellon – mafic–ultramafic cumulates contain magmatic sulphide mineralization consisting of pyrrhotite (mainly), chalcopyrite, pentlandite and traces of pyrite, often in monomineralic blebs (Fletcher et al. 1997, Gunn & Styles 2002, McKervey et al. 2007); Bin quarry and environs, near Huntly – cumulate units of the Huntly mafic–ultramafic intrusion contain disseminated chalcopyrite, sometimes displaying a net texture (Gunn & Shaw 1992); Littlemill and Auchencrieve, near Huntly – magmatic sulphide deposit associated with the Huntly–Knock gabbroic intrusion and consisting of pyrrhotite (mainly), chalcopyrite and pentlandite (Fletcher & Rice 1989, Gunn & Styles 2002, McKervey et al. 2007). Other occurrences of chalcopyrite include: Abergairn (Corrybeg) lead mine, near Ballater – small spots of chalcopyrite associated with galena and sphalerite, forming inclusions in fluorite (Russell 1936a); Back Burn, W of Auchenblae – chalcopyrite inclusions in alluvial gold (Chapman et al. 2000c); Coire Loch Kander, near Braemar – chalcopyrite is a minor constituent of a Dalradian armenite-bearing calc-silicate-rich cherty sulphidic quartzite (Fortey et al. 1991); Dobston quarry, Inverurie, and Lossiemouth, near Elgin (Heddle 1901); Kemnay granite, near Huntly (Laour et al. 1990); Water of Tanar Limestone Formation, Ballater area – disseminated sulphide mineralization, possibly derived from hydrothermal brines, consists of pyrite and pyrrhotite with minor bornite and chalcopyrite (Goodman et al. 1990).

Highland *Assynt:* The Allt Cathair Bhan section of the Loch Ailsh alkaline complex and the ultramafic section of the Loch Borralan alkaline complex contain post-magmatic sulphides, including chalcopyrite, galena, pyrite and sphalerite (Shaw et al. 1994, Styles et al. 2004). Chalcopyrite and pyrite are widespread accessory species in a melanite pyroxenite at Bad na h'Achlaise, SW of Ledmore (Stumpfl 1961), and in other pyroxenites and pyroxene syenites of the Borralan complex (Notholt et al. 1985, Gunn & Styles 2002). A carbonatite intrusion to the SW of Loch Borralan also contains minor chalcopyrite in white sövite close to limestone xenoliths (Young et al. 1994b). At Loch Bad nam Ban, 3 km SW of Lochinver, accessory chalcopyrite occurs with pyrite–pyrrhotite in biotite diorite (Stumpfl 1961). *Grudie and Loch Shin granite areas:* Low-grade sulphide mineralization is exposed in Moinian and Lewisian rocks intruded by the Caledonian Grudie and Shin granites to the W of Lairg. Accessory amounts of chalcopyrite, fluorite, scheelite and bismuth-rich minerals occur in thin molybdenite–pyrite-bearing quartz veins and breccia zones contain chalcopyrite and galena with barite or fluorite (Gallagher et al. 1974, Gallagher & Smith 1975, Dunham et al. 1978, Fortey 1980). Chalcopyrite is also present in bismuth sulphosalt-bearing quartz–sulphide veins cutting the Shin granite (Lowry et al. 1994) and near by it occurs in veins cutting syenite on Ord Hill (Heddle 1901). *Helmsdale area:* At Ceann Ousdale, chalcopyrite, uraninite and minor sphalerite occur in a thin arkosic horizon close to the eastern margin of the Helmsdale granite (Gallagher et al. 1971). Within the granite, chalcopyrite is found as vein and joint coatings with molybdenite, pyrite and quartz in Ord Burn (Tweedie 1979) and as inclusions in alluvial gold in Kildonan Burn (Leake et al. 1993, Chapman et al. 2000c). *Gairloch area:* The Lower Proterozoic Kerry Road stratiform Au–Cu–Zn sulphide deposit near Gairloch is hosted by pyritic quartz–carbonate schists and contains chalcopyrite in a pyrite–pyrrhotite-dominated ore. Disseminated chalcocite–bornite–chalcopyrite–malachite with magnetite–pyrite–pyrrhotite in massive amphibolite and chalcopyrite–pyrite–pyrrhotite in chloritic hornblende schist at Teangadh Bhuide Mhòr are related deposits (Jones et al. 1983, 1987). *Isle of Rum:* The layered basic and ultrabasic intrusion on Rum contains oxide–sulphide droplets within and above chromitite units. Chalcopyrite, magnetite and pentlandite are the most frequently occurring species, followed by bornite, chalcocite, native copper, cubanite, digenite, electrum and pyrrhotite. These rocks also host Au–Pd–Pt mineralization (Henderson & Suddaby 1971, Dunham & Wilkinson 1985, Butcher et al. 1999, Power et al. 2000a, 2003). Disseminated sulphides are abundant in sulphide-rich (mainly pyrrhotite) dykes within the adjacent West Sgaorishal ultramafic plug (Power et al. 2003). *Other areas:* Ardsheal peninsula, near Kentallen – high-grade chalcopyrite float associated with a Lower Devonian appinite pipe (Rice & Davies 1979); Gleann a' Chaolais, S of Ballachulish in the Ballachulish igneous complex – stockwork mineralization in granite contains fracture coatings and disseminations of chalcopyrite, molybdenite and pyrite (Evans 1977, Fortey 1980, Haslam & Kimbell 1981); Bridge of Westfield – mineralized breccia in the Loch Cander crush zone contain chalcopyrite, galena, marcasite, pyrite, smithsonite, sphalerite and uranium-rich hydrocarbons in a calcite–siderite gangue; Carn Chuinneag igneous complex – chalcopyrite and molybdenite in a thick quartz-breccia vein; Murkle Bay, near Castletown – calcite vein containing chalcopyrite, pyrite and uranium-rich hydrocarbons (all Gallagher et al. 1971); Loch Claise na Feàrna, Scourie Estate – galena and sphalerite mineralization with minor chalcopyrite in hornblendite and hornblende schist (Anon. 1974); western side of Loch Duich, along road to Glenelg and east side of Loch Duich, along road to Kyle of Lochalsh – abundant chalcopyrite occurs in veins with galena, gersdorffite, molybdenite, rutile, sphalerite and minor aikinite, gold (electrum), hessite, marcasite and tennantite associated with the Ratagain (diorite, syenite and quartz monzonite) intrusion (Alderton 1986, 1988b); S of Rhiconich – chalcopyrite in a vein within gneiss (Heddle 1901); Kilchrist, Broadford area, Isle of Skye – bornite, chalcopyrite and pyrite occur in a magnetite-rich skarn close to a dolerite dyke and within the aureole of the Beinn an Dubhaich granite (Tilley 1951, Groves 1952, Slater & Highley 1976); Meall na Cuilce, Isle of Skye – chalcopyrite and pentlandite are accessory species in chromitite belonging to the Peridotite Series of the Cuillin Complex (Pirrie et al. 2000a).

Lothian Chalcopyrite, galena and sphalerite are present in a stratabound sulphide deposit in Carboniferous non-marine algal limestone at South Queensferry (Parnell 1983). At Stobshiel, S of Haddington, galena and sphalerite occur with minor chalcopyrite and tetrahedrite in a greywacke–shale-hosted vein gold prospect (Naden & Caulfield 1989). Chalcopyrite is noted with minimal description from: Addiewell, West Calder; Corstorphine Hill; Elgin colliery; Hilderston, Bathgate Hills; and North Silver mine limestone quarry (all Heddle 1901).

Strathclyde Abhainn Srathain mines, S of Meall Mòr, South Knapdale – worked a chalcopyrite-rich stratiform deposit within Dalradian rocks (Smith et al. 1978, Willan 1980); Bishopton (bypass borrow pit), between the M8 motorway and the Glasgow–Greenock railway line (the type locality for greenockite) – rare bornite, chalcopyrite and galena near cavities in altered Carboniferous Clyde Plateau Lavas (Meikle 1992b); Boyleston quarry, SW of Barrhead – rare chalcopyrite forms groups of complex twinned crystals or coats prehnite in cavities within Clyde Plateau Lavas (Meikle 1989a); Branscroft quarry, near Kilbarchan, Glasgow area – chalcopyrite and pyrite on large calcite specimens (Green & Todd 2001); Cantyre, SW of Inverneil; Dockra quarry, near Beith; Erins, near Tarbert; Kilbrandon; Kilchattan; St Catherines, Loch Fyne; Stair (all Heddle 1901); Castletown mine, Lochgilphead – quartz-vein mineralization containing copper-rich pyrite with a little chalcopyrite and galena (Wilson & Flett 1921, Starkey & Hubbard 1998); Coille-bhraghad and Craignure mines, near Inveraray – stratabound metasomatic replacement deposit consisting of chalcopyrite, pyrite, pyrrhotite and pentlandite with associated gold

(Forbes 1868, Wilson & Flett 1921, Berridge 1969, Dunham et al. 1978, Willan 1980); Coire Buidhe, Glen Creran (within the Etive granite complex to the E and NE of Ben Starav) – massive chalcopyrite occurs with azurite, malachite and pyrite at the classic "Heddle" molybdenite–quartz vein locality (Heddle 1901, Ingram et al. 1993); Easedale Island, near Oban – chalcopyrite forms inclusions in pyrite and siderite within sulphide-rich slate of Dalradian age (Hall et al. 1988); East and West Kaim, Muirshiels and Swinlees mines, all Renfrewshire Hills – fissure-infill veins in Clyde Plateau Lavas typically contain bornite, chalcopyrite and malachite in a barite gangue, although often the veins are devoid of sulphides (Stephenson & Coats 1983); Etive granite complex to the E and NE of Ben Starav (on the Strathclyde/Highland Region border) – widespread chalcopyrite, molybdenite and pyrite occur in quartz veinlets cutting granite (Haslam et al. 1985); Fore Burn, Tairlow plantation, near Straiton – gold-rich chalcopyrite-rich veins and disseminated arsenopyrite–chalcopyrite stockwork mineralization in diorite and an associated volcanic suite, also chalcopyrite inclusions in alluvial gold (Allen et al. 1982, Charley et al. 1989, Chapman et al. 2000c); Garbh Achadh, near Inveraray – rafts of sulphide-rich hornfelsed quartzite adjacent to a stock-like porphyrite intrusion contain minor chalcopyrite associated with covellite, molybdenite, pyrite (major) and sphalerite (Ellis et al. 1978, Fortey 1980); Isle of Islay – the lead mines of Ballygrant, Gartness, Kilslevan (copper–lead), Mulreesh and North Ardachie contain chalcopyrite with chalcocite, galena, pyrite and sphalerite in veins and pockets within Dalradian limestone (Berridge 1969, Edmond 1983, Callender & Macaulay 1984, Nicholson 1989b, Baron & Parnell 2005). Sulphide mineralization at Rubha a' Mhill on Islay has a similar paragenesis, although chalcocite, galena and sphalerite were not found (Green 1988); Kilfinan (Murder lode), near Lochgilphead – metasomatic replacement sulphide ores containing bornite, chalcocite, chalcopyrite, covellite and pyrite (Wilson & Flett 1921, Berridge 1969, Willan 1980); Kilmartin copper mine, Upper Lorgie to Old Poltalloch – chalcopyrite and manganese oxides in an epidiorite sill (Heddle 1901, Wilson & Flett 1921, Nicholson 1989b); Kilmelford area, including Carn Dearg and Lagalochan – copper-rich intrusive suite containing disseminated chalcopyrite, molybdenite and pyrite, with galena and sphalerite in isolated veins peripheral to the main mineralized area (Fortey 1980, Zhou 1987, Harris et al. 1988); Leadhills – chalcopyrite associated with galena and sphalerite, also chalcopyrite inclusions in alluvial gold (Heddle 1901, Brown 1927, Temple 1955, Leake et al. 1998); Loanhead quarry, near Beith – chalcopyrite forms thin crusts lining cavities in Clyde Plateau Lavas near to a Tertiary dyke (Meikle 1989b); Mannoch Hill mine, SW of Lesmahagow – grains and stringers of chalcopyrite (with azurite and malachite) in barite veins (Todd & McMullen 1991); Poundland Burn, Ballantrae ophiolite complex – chalcopyrite, galena and native bismuth inclusions (+ less common awaruite, heazlewoodite, pentlandite) in chrome spinel within chromitite (Power & Pirrie 2004); Stronchullin, N of Tarbert – veins in Dalradian rocks, once worked for their galena content, also contain chalcopyrite and sphalerite (Ridgway 1980); Trearne quarry, near Beith – minor faulting in Carboniferous limestone and shale has created fissures in which crystals of chalcopyrite to 2 mm have grown (Todd 1989); Creag an Fheidh, Isle of Arran – chalcopyrite, pyrite and sphalerite occur in veinlets cutting metasomatic grossular–pyroxene endoskarns in agglomerate within the collapsed caldera of the Central ring complex (Cressey 1987).

Tayside *Aberfeldy area:* Chalcopyrite is a minor constituent of stratabound mineralization (principally barite, galena, pyrite, pyrrhotite and sphalerite) in Dalradian graphitic, calcareous and dolomitic schists from the Ben Eagach area, N of Aberfeldy (Coats et al. 1980, 1984, Fortey & Beddoe-Stephens 1982). Trace amounts of chalcopyrite are also noted in a similar deposit at Foss to the NW of Aberfeldy (Moles 1983). At Calliachar Burn, Aberfeldy, 14 steeply dipping quartz–sulphide ± gold veins, cutting Upper Dalradian gritty quartzites, quartz–mica schists and amphibolites, contain polyphase hypogene mineralization consisting of pyrite (most abundant), galena and sphalerite, plus minor arsenopyrite and chalcopyrite, and trace gold (electrum). In the nearby Urlar Burn, similar veins contain sporadic patches of galena ± chalcopyrite ± pyrite, with lesser amounts of arsenopyrite and sphalerite (Ixer et al. 1997). Chalcopyrite has also been noted as inclusions within *in situ* and alluvial gold from Calliachar Burn (Leake et al. 1997c, Chapman et al. 2000c). *Other areas:* Corrie Buie, Meall nan Oighcreag, Loch Tay – a stockwork cut by galena-bearing quartz veins contains minor amounts of native bismuth, chalcopyrite, electrum (36–65% gold), pyrrhotite, schirmerite and sphalerite phases belonging to the gustavite–lillianite series (Pattrick 1984); Glen Lednock, near Comrie and Glengarry; West Water, NW of Brigend – chalcopyrite inclusions in alluvial gold (all Chapman et al. 2000c); north end of Lochan na Lairige – the contact of the Ben Eagach Schist with the Ben Lawers Schist contains chalcopyrite, galena, pyrite and pyrrhotite (Green & Todd 2001); Tomnadashan, Loch Tay – chalcopyrite, molybdenite, pyrite and tetrahedrite occur as disseminations and in irregular veinlets associated with a late Caledonian alkali-diorite intrusion (Heddle 1901, Dunham et al. 1978, Fortey 1980, Ridgway 1980, Pattrick 1984). Heddle (1901) adds other localities with little description: Ardeonaig; Ardtalanaig; Arndrum Hill, Aberfoil; Ben Ledi; Craigie Barns, Dunkeld; Forgandelly; Glen Tilt; Kenmore plantation, Taymouth; Newtyle quarry; and Taymouth Park.

Western Isles On South Harris, disseminated chalcopyrite and pyrite occur in a layered alpine-type peridotite exposed between Rubha Sgeir nan Sgarbh and Loch Meurach in Langavat Valley (Livingstone 1976a). Minute quantities of chalcopyrite occur in the Western gabbro and the Cambir dolerite on St Kilda (Hirta) (Harding et al. 1984).

Fair Isle North Naaversgill (formerly Copper Geo), NW of Fair Isle – lenticular masses of chalcocite, chalcopyrite and malachite in a calcite–scapolite vein complex within a crush belt (Mykura & Young 1969); Dunrossness (Greg & Lettsom 1858, Heddle 1901).

Orkney Malachite and other secondary minerals are thought to have replaced chalcopyrite at Stromness lead mine, Mainland (Wilson & Flett 1921).

Shetland *Mainland:* Cunningsburgh and Garths Ness, near Bay of Quendale (Heddle 1901); Sandlodge mine – chalcopyrite and siderite below a gossan (Greg & Lettsom 1858, Heddle 1901, Dunham et al. 1978); Vidlin Ness – massive sulphide rocks in Dalradian hornblende schists contain chalcopyrite, marcasite, pyrite, pyrrhotite, and local galena and sphalerite (Garson & May 1976). *Unst:* Chalcopyrite is one of many sulphide species found in ultramafic rocks at the base of the Shetland ophiolite complex: Cliff area, Baltasound–Haroldswick area – palladium–platinum-rich sulphide-bearing dunite (Gunn & Styles 2002); W of Hagdale, N and E of Sandwick Bay, and Vord Hill, W of Nuda – chalcopyrite in gabbros and pyroxenites (Maynard 1993, Maynard et al. 1997); Mu Ness Peninsula – massive chalcopyrite–pyrite mineralization hosted by gabbro (Maynard 1993); Ordale – chalcopyrite associated with magnetite, pyrrhotite, sperrylite and unidentified iron–platinum phases in wehrlite (Prichard et al. 1994). Queyhouse talc quarry, W of Baltasound, represents the sheared base of the ophiolite complex and contains chalcopyrite in molybdenite-bearing chlorite–siderite nodules (Gass et al. 1982, Neary & Prichard 1985).

Ireland

Carlow Chalcopyrite is a rare species in the spodumene pegmatites at Aclare, near Myshall (Scoon 1978). At Coolasnaghta, in the same general area, arsenopyrite, chalcopyrite and pyrite are common accessory species in tourmalinites (quartz–tourmaline and muscovite–tourmaline schists) close to the southern margin of the Leinster granite batholith (McArdle et al. 1989).

Cavan Two minor chalcopyrite occurrences and thin chalcopyrite–molybdenite-bearing quartz veinlets are noted in the Crossdoney granodiorite at Drumcarban (Morris 1984).

Clare Disseminated and fracture-filled sulphide mineralization in Carboniferous Limestone within the crestal region of a faulted pericline at Ballyvergin is dominated by an assemblage of arsenopyrite, chalcopyrite, galena and pyrite (O'Brien 1959, Schultz 1971, Halliday & Mitchell 1983, Andrew 1986). Other occurrences include: Carrahan and Crowhill lead–zinc mines, Lassana township – chalcopyrite covered by cobalt bloom (O'Brien 1959); Castletown lead mine – vein deposit containing calcite, chalcopyrite, fluorite, galena, quartz and sphalerite (Cole 1922, O'Connor et al. 1993); Maghera Cross copper prospect, Maghera Townland – chalcopyrite–calcite veinlets and stringers in limestone with subordinate pyrite, minor arsenopyrite and galena (Schultz 1971).

Cork *West Carbery mining district:* Chalcopyrite occurs in both polymetallic major veins and stratiform-disseminated deposits in non-marine Devonian redbeds. The stratiform mineralization has a diagenetic bacteriogenic origin, and mobilization of metals from this source (in metamorphic fluids) during the Hercynian orogeny led to the major vein mineralization (Ni Wen et al. 1996). Localities include: Ballycummisk copper mine, Ballydehob – chalcopyrite in quartz veins associated with bornite, molybdenite, pyrite and tetrahedrite. Later quartz–carbonate veins contain minor chalcopyrite and tetrahedrite (O'Brien 1959, Bartlett 1980, Ni Wen et al. 1996, 1999, Kinnaird et al. 2002); Brow Head copper mine, Barley Cove, near Mizen Head – minor veins and disseminations containing bornite–chalcopyrite–(chalcocite) (Ni Wen et al. 1999); Cappagh (Cappaghglass) copper mine, near Ballycummisk – mainly bornite–chalcopyrite–galena in thin quartz veins within a stratiform deposit (all Ni Wen et al. 1996, 1999, Kinnaird et al. 2002); Crookhaven mine; Glanalin mine, between Killeen and Carravileen; Kilkilleen; Mount Gabriel mine – chalcopyrite (± arsenopyrite ± bornite ± galena ± pyrite ± tetrahedrite) in stratiform-minor vein deposits (Ni Wen et al. 1996, 1999); Dhurode (Carrigacat) mine, Dunmanus Bay – arsenopyrite–bournonite–chalcopyrite–galena–meneghinite–tetrahedrite in a stratiform-minor vein deposit (Ni Wen et al. 1991, 1996, 1999, Kinnaird et al. 2002); Letter; Rooska West; Skeagh - chalcopyrite ± (arsenopyrite ± bornite) in minor veins and disseminations (Ni Wen et al. 1999) (these mines are all noted by Cole [1922] and many are also reviewed by Reilly [1986]). *Elsewhere:* Chalcopyrite is reported from the Berehaven (Bearhaven) mines, Allihies, where an estimated 1 million tonnes of ore yielding 25 600 tonnes of copper was mined from chalcopyrite–(pyrite–tetrahedrite)-bearing quartz veins (Rudler 1905, Cole 1922, O'Brien 1959, Williams 1991, Kinnaird et al. 2002). At Dooneen mine, Allihies, chalcopyrite occurs in a quartz vein together with the copper secondary species botallackite, chalcanthite and langite (Halliday & Mitchell 1983, Moreton & Aspen 1993). Both barite and chalcopyrite were mined from Mountain mine, near Cloan, Allihies. Pyrite and tennantite–tetrahedrite are minor associated species (Cole 1922, Schultz 1971, Halliday & Mitchell 1983). A predominantly bornite–chalcopyrite ore at Tullacondra, NNW of Mallow is estimated to contain 3.6 million tonnes of 0.7% copper and 27.5 ppm silver. The deposit is part stratabound and part vein-hosted (Wilbur & Carter 1986). Cole (1922) documents a further 33 copper mines in Co. Cork.

Donegal Haughton (1862) mentions moderately large chalcopyrite and molybdenite crystals in granite veins at Castlecaldwell.

Down Breccias in Silurian greywackes and shales near a dolerite dyke at Conlig–Newtownards host galena – (chalcopyrite–pyrite–sphalerite) vein mineralization. Chalcopyrite typically occurs as well formed 2–8 mm crystals perched on saddle dolomite (Cole 1922, Dunham et al. 1978, Halliday & Mitchell 1983, Moles & Nawaz 1996). Molybdenite-bearing veins at Croreagh quarry in the Newry granodiorite also contain lesser amounts of chalcopyrite, magnetite and pyrite (Meighan & Ixer 1980). Chalcopyrite, galena and sphalerite are all found in altered limestone at Abbeytown, Ballysadare (Halliday & Mitchell 1983).

Fermanagh Chalcopyrite occurs with azurite and malachite in hematite ore at Magheramenagh mine, near Belleek (Cole 1922, Fowler 1959).

Galway Chalcopyrite is a subordinate primary ore species in the major limestone-hosted Pb–Zn–(Ag–Cu) deposit that formed close to the North Tynagh fault at Tynagh mine, near Killimor. Concentrates containing about 12 500 tonnes of copper were produced between 1966 and 1970. Chalcopyrite forms rims in association with galena and tennantite in massive ore, and as isolated grains it replaces pyrite and sphalerite (Schultz 1966b, 1971, Morrissey et al. 1971, Clifford et al. 1986). In Connemara, minor chalcopyrite is present in various associations within Dalradian rocks (McArdle et al. 1986a), in the Galway granite at Costelloe quarry, Murvey (O'Connor et al. 1993) and Mace Head (Talbot & Max 1984, Max & Talbot 1986, Derham & Feely 1988), and in the Oughterard granite (Laour et al. 1990).

Kerry Carbonate-hosted copper-polymetallic deposits occur at: Ardtully (Cloontoo) mine, near Kenmare – copper-dolomite vein mineralization contains bornite, chalcopyrite and tennantite (Cole 1922, Halliday & Mitchell 1983, Kinnaird et al. 2002); Crow Island, Lough Leane, near Muckross, Killarney – sphalerite is the main ore species and this occurs with chalcopyrite–galena and minor tennantite as 20–50 µm-diameter polygonal crystals forming abundant mosaics. Chalcopyrite exsolution textures in sphalerite occur in a later post-deformational assemblage (Cole 1922, Kinnaird et al. 2002, Ixer 2004); Muckross mine, Killarney – a stratabound Cu–Pb–Zn deposit with minor silver, cobalt and molybdenum, containing chalcopyrite–pyrite, chalcopyrite–galena–sphalerite and arsenopyrite–chalcopyrite–tennantite ores (Greg & Lettsom 1858, Campbell Smith 1913, Cole 1922, O'Brien 1959, Ryback & Moreton 1991, Ixer & Budd 1998, Moreton et al. 1999, Kinnaird et al. 2002); Kenmare – chalcopyrite–galena ore (O'Brien 1959); Ross Island mine, Killarney – an important Early Bronze Age copper mine (2400–2200 BC) in which chalcopyrite occurs with bornite and tetrahedrite in pockets, irregular patches, strings and epigenetic veins (Blue Hole is a large open pit that once worked the locally named bluestone, a fine-grain intimate mixture of galena, pyrite, sphalerite and minor chalcopyrite) (Cole 1922, Russell 1927b, Ixer & Budd 1998, Kinnaird et al. 2002, Nex et al. 2003). Additionally, disseminated bornite, chalcopyrite and djurleite mineralization infilling pore spaces in a Devonian redbed-hosted deposit is reported from Derrynane (Kinnaird et al. 2002). Chalcopyrite is also reported from a tennantite-bearing quartz vein cutting Devonian sandstones at Leaghillaun, Bearhaven (Berehaven) Peninsula (Moreton & Green 2006b).

Kilkenny Epigenetic chalcopyrite–tennantite–tetrahedrite vein mineralization has been found in boreholes cutting lead–zinc-mineralized Carboniferous Limestone at Galmoy, near Kilkenny (Doyle et al. 1992).

Limerick A potential Cu–(Ag) orebody of some 6 million tonnes grading 0.9% copper and 1.3oz/ton silver has been found in Carboniferous Limestone at Aherlow, near Galbally. Here, disseminated bornite, chalcopyrite and chalcocite occur with minor galena, native silver, sphalerite and stromeyerite (Cameron & Romer 1970, Schultz 1971). At Courtbrown, on the southern shore of the River Shannon, near Ballysteen, an epigenetic base-metal (lead–zinc) orebody hosted by dolomitized Carboniferous Limestone contains small quantities of chalcopyrite and tennantite in irregular veins and veinlets (Grennan 1986, Reed & Wallace 2001). Chalcopyrite and galena in a barite gangue were raised from the Oola (Oolahills) mines (Cole 1922).

Longford Hornfelsed turbidites in the aureole of the Amey granodiorite at Aghagah are host to pyrite-bearing quartz veinlets containing minor chalcocite, chalcopyrite, molybdenite and pyrrhotite (Morris 1984). The Keel prospect, Ardagh, 11 km S of Longford is a sphalerite-dominated Cd–Pb–Zn deposit formed in faulted, brecciated and dolomitized Carboniferous rocks. Chalcopyrite is mainly found as lath-like inclusions in sphalerite (Patterson 1970, Morrissey et al. 1971, Slowey 1986).

Louth Chalcopyrite and galena are minor constituents of a sphalerite-dominated breccia and quartz–carbonate vein deposit at Salterstown. Trials for both copper and lead were made in the mid-1820s on a nearby deposit, but all that remains now is a minor trace of chalcopyrite and sphalerite in vuggy quartz veins (Cole 1922, Morris 1984).

Mayo An Ordovician altered acidic porphyry 8 km to the S of Charlestown hosts disseminated copper mineralization dominated by chalcopyrite and pyrite, but with bornite in restricted high-grade areas (O'Connor & Poustie 1986, Poustie 1982). Native gold-bearing quartz veins at Cushinyen, in the Cregganbaun shear zone, contain mm-size aggregates of sulphides, including bournonite, chalcopyrite, galena, pyrite, sphalerite and tetrahedrite (Thompson et al. 1992). Alluvial and *in situ* grains of native gold with inclusions of chalcopyrite have been reported from: Bunowen River; Carrownisky River (Glenkeen Bridge); Croagh Patrick; Lugbeg River; Srahroosky (Leake et al. 1993, Chapman et al. 2000c). Rare chalcopyrite, nickeline and pyrrhotite occur with pyrite in gold-rich quartz veins cutting Silurian rocks in the Lecanvey prospect, Clew Bay. This occurrence is related to other chalcopyrite-bearing gold-rich quartz-vein deposits in the Sheefry Hills, S of Lecanvey (Aherne et al. 1989, 1992). At Sheefry Pb–(Ag) mine, 11 km S of Croagh Patrick, a quartz–carbonate vein in a mineralized fault contains chalcopyrite, galena, millerite and a suite of secondary species. Chalcopyrite forms euhedral crystals to 5 mm and irregular masses up to 1 cm across (Moreton & Jackson 1991). Chalcopyrite is also noted with galena and pyrite at Bolinglanna (Benderg) mine, near Corraun (Cole 1922).

Meath The world-class carbonate-hosted epigenetic syn-sedimentary lead–zinc replacement deposit at Navan contains traces of chalcopyrite in cavities within massive sulphide ore, and the species is also found in veinlets and as disseminations in surrounding rocks (Kucha & Wieczorek 1984, Andrew & Ashton 1985, Ashton et al. 1986, Wilkinson 2003).

Monaghan Very minor chalcopyrite–galena–sphalerite veinlets crosscut the main arsenopyrite–gold–pyrite–stibnite mineralization found in quartz veins and breccias at Clontibret mine, Tullybuck (Morris 1984, Morris et al. 1986, Ryback & Moreton 1993). Alluvial gold from Clontibret contains chalcopyrite inclusions (Chapman et al. 2000c). Quartz–dolomite veins containing minor chalcopyrite and sphalerite are noted from Annaglogh and Coolartragh (Bond) lead mines (Morris 1984).

Sligo Abbeytown lead–zinc mine, near Ballysadare, worked a dolomitized limestone-replacement deposit in which elongate laths of chalcopyrite occurred as

Tarnished sphenoidal chalcopyrite, 2 mm across, from Maye (Aughamore) quarry, near Lough Gill, Co. Sligo, Ireland. (© D. I. Green)

inclusions in dark reddish-brown sphalerite (Fernandez 1957, Halliday & Mitchell 1983, Hitzman 1986). The nearby Maye (Aughamore) quarries, near Lough Gill, contain brassy sphenoidal crystals of chalcopyrite to 5 mm scattered over calcite and dolomite (Ryback & Moreton 1992, Green & Moreton 2001).

Tipperary The Silvermines district is notable for its large stratiform sedimentary-exhalative Pb–Zn–barite deposit. Prior to its discovery, the area produced copper from mineralized fault breccias and fracture fillings close to a major fault system. Chalcopyrite is reported from the following mines: Ballynoe mine – lode contains barite, chalcopyrite and galena, also minor chalcopyrite in a mainly massive pyrite cap above a large barite orebody (Russell 1907b, Cole 1922, Mullane & Kinnaird 1998); Gortnadyne and Shallee mines – chalcopyrite partially replaced by tetrahedrite (Russell 1907b, Greg & Lettsom 1858, Cole 1922, Rhoden 1959); Mogul mine – subordinate chalcopyrite (Schultz 1971). Chalcopyrite is discussed more generally by other authors – Cole (1922), O'Brien (1959), Morrissey et al. (1971), Taylor & Andrew (1978), Samson & Russell (1983). Minor chalcopyrite is also noted from another carbonate-hosted deposit (in three stratabound sulphidic lenses) at Lisheen, between Thurles and Rathdowney, where it is found in a galena–marcasite–pyrite–sphalerite-dominated ore. Sphalerite and galena, and locally chalcopyrite and tennantite, overgrow and replace earlier colloform marcasite–pyrite (Strashimirov et al. 1999, Hitzman et al. 2002, Ixer 2002, Fusciardi et al. 2004, Wilkinson et al. 2005).

At Gortdrum mine, near Tipperary town, epigenetic mineralization related to the intrusion of two high-level basic intrusions has produced an orebody of chalcopyrite and mercury-rich tennantite. The mine dumps have produced good chalcopyrite crystals up to 3 mm on edge from cavities in veinlets of iron-rich dolomite within limestone. Sphenoidal chalcopyrite associated with cinnabar and tennantite is also noted (Morrissey et al. 1971, Schultz 1971, Steed 1983, 1986, Duane et al. 1986, Ryback & Moreton 1991, Green & Moreton 2001, Braithwaite & Wilson 2003).

Tyrone Gold-bearing quartz–carbonate veins at Curraghinalt, near Gortin, Sperrin Mountains, contain pyrite and subsidiary arsenopyrite, bornite, chalcopyrite, native copper, galena, sphalerite and tetrahedrite–tennantite (Clifford 1986, Earls et al. 1989, Clifford et al. 1990). Further investigations of vein mineralization at Curraghinalt and at Cavanacaw (near Omagh) confirm these observations and note common chalcopyrite and galena lining vugs and fractures in quartz. Chalcopyrite and other sulphides occur in crosscutting fractures and veinlets (Cliff & Wolfenden 1992, Parnell et al. 2000) and as inclusions in alluvial gold at Attagh Burn and Cavanacaw (Leake et al. 1993, Chapman et al. 2000c). Disseminations and veinlets of chalcopyrite and pyrite are also reported from nearby Cashel and Sleivemenagh (Leyshon & Cazalet 1976).

Waterford Steeply folded Ordovician sediments and volcanics at Bunmahon, Knockmahon, 22 km SW of Waterford town, are the host to important 19th-century copper mines. Chalcopyrite occurs in quartz veins virtually devoid of accompanying sulphides (Cole 1922, O'Brien 1959, Halliday & Mitchell 1983). Similar chalcopyrite-bearing veins at Tankardstown mine, near Knockmahon, were the richest in the area (Cole 1922, Moreton 1991, Braithwaite & Wilson 2001b). Rich lumps of chalcopyrite have recently been described from a small tip in the valley floor at Kilduane (Moreton 2005a).

Westmeath Chalcopyrite occurs as fine-grain crystals in vugs within dolomite in a deposit dominated by pyrite, sphalerite and minor galena (Patterson 1970, Jones & Brand 1986).

Wexford Considerable chalcopyrite and fine-grain intergrowths of galena and sphalerite have been found on the mine dumps of Barrystown mine, Bannow Estuary (O'Brien 1959). Decomposing chalcopyrite is associated with galena on the dumps of Ballyhighland lead mine, W of Enniscorthy (Ryback & Moreton 1993).

Wicklow The orebody at Avoca mine is a volcanogenic exhalative sulphide deposit containing considerable chalcopyrite and pyrite hosted in Ordovician volcanic rocks associated with black chloritic schists. Mined output amounted to 16 million tonnes at a grade of 0.6% copper, mainly from stratabound ore. Disseminated and vein mineralization also occurs. Sphalerite and galena are sub-economic constituents of the ore (Cole 1922, O'Brien 1959, Schultz 1971, Wheatley 1971, Badham 1978, Downes & Platt 1978, Sheppard 1980, Williams et al. 1986, McArdle 1989, 1993). Ryback & Moreton (1993) suggest that the main ore species was a copper-rich pyrite rather than chalcopyrite. A fine-grain purple ore, composed of a mixture of galena and sphalerite (and minor chalcopyrite), at the northeastern end of the Avoca district was named kilmacooite (after the local townland) before its composite nature was revealed (Cole 1922, Milner & McArdle 1992). A related Cu–Pb–Zn–(Bi–Au) deposit in the Goldmine River valley (10 km along strike from Avoca) within metamorphosed BIF rocks (mainly magnetite) contains chalcopyrite, as coarse-grain aggregates of polygonal crystals up to 1 cm across, or as intergrowths with galena, sphalerite and Bi–Cu–Pb sulphosalt species. In places the chalcopyrite contains native gold inclusions (McArdle & Warren 1987, McArdle 1989, Ixer et al. 1990). Along the southeastern margin of the Leinster batholith at Ballingen, between Tinahely and Aughrim, minor chalcopyrite and sphalerite (with inclusions of chalcopyrite) occur in cassiterite–scheelite-bearing quartz veins and in an altered microtonalite intrusion (Steiger & Bowden 1982, 1986, McArdle & Kennan 1988, Gallagher 1989). A mineralized lode at Ballycoog mine in the townland of Ballycoog Upper contains chalcopyrite, hematite, pyrite and sphalerite (Cole 1922). Forbes (1869) notes cassiterite, chalcopyrite, corundum (sapphire), galena, garnet, hematite, ilmenite (titanoferrite), magnetite, molybdenite, native platinum, pyrite, wolframite and zircon associated with large nuggets of alluvial gold from the northeastern side of Croghan Kinshela.

Offshore territories

Isle of Man The Bradda mines on the north side of Port Erin Bay worked veins cutting metamorphosed slates and grits of the Cambro-Ordovician Manx Group rocks and produced silver-rich galena and chalcopyrite. Attractive specimens of crystallized chalcopyrite on dolomite and sphalerite are noted (Rudler 1905, Dewey & Eastwood 1925, Starkey 1998). Minor chalcopyrite and malachite have also been found in galena–sphalerite-bearing veins at the Foxdale and Laxey mines (Dewey & Eastwood 1925, Dunham et al. 1978). At Laxey, a uranium-rich hydrocarbon (anthracite) from a copper-rich zone is reported to contain chalcopyrite inclusions (Davidson & Bowie 1951). Shortly after 1875, 11 tonnes of chalcopyrite were produced from the Langness mines, Langness Point (Dewey & Eastwood 1925).

Guernsey Sulphide veinlets in the Northern igneous complex at Baubigny, near St Sampson, contain chalcopyrite, magnetite, molybdenite, pyrite (abundant), pyrrhotite and sphalerite, whereas those at Bouette quarry, St Peter Port, contain chalcopyrite, galena, magnetite, pyrite (abundant), pyrrhotite, sphalerite and traces of pentlandite. Traces of chalcopyrite also occur on a specimen of galena from Baubigny, near St Sampson, now in the Lukis collection in the Guernsey Museum and Art Gallery (Ixer 1992).

Herm The Great Copper lode is a quartz vein containing hematite, magnetite and minor chalcopyrite (Ixer 1992).

Jersey Galena–sphalerite-bearing carbonate veins cutting Brioverian sediments at Le Pulec contain minor chalcopyrite as inclusions in sphalerite, as a replacement of tetrahedrite, and as relatively large (600 μm across) anhedral grains intergrown with galena, late sphalerite and quartz (Mourant 1961, Ixer & Stanley 1980, Stanley & Ixer 1982, Ixer 1992). Granite of the Northwest granite complex hosts several mineral veins: Le Doüet de la Mer, St Ouën – minor chalcopyrite in arsenopyrite–galena–pyrite–sphalerite-bearing carbonate veins; Gigoulande quarry, St Mary – chalcopyrite and pyrite are the main sulphides in quartz veins that also carry native antimony, bournonite, molybdenite and sphalerite (Ixer 1980, 1992); L'Etacq quarry – chalcopyrite–molybdenite–pyrite–sphalerite-bearing calcite–quartz veins produced some of Jersey's best specimens of chalcopyrite (Mourant 1961, Ixer 1980, 1992).

Chalcopyrite has also been found at: Bouley Bay, Trinity and La Crête quarry, Anne Port on the northeast coast of Jersey – chalcopyrite forming anhedral aggregates of polygonal crystals up to 2 mm on edge in copper-enriched flow-banded rhyolite (Mourant 1961, Ixer 1980, 1992); German underground hospital, St Peter's Valley, St Lawrence – calcite vein containing uncommon chalcopyrite inclusions in galena and sphalerite (Ixer 1980), and dominant chalcopyrite in a quartz–sulphide vein (Ixer 1990b); east of Côtil Point, Homestead and Westmount – chalcopyrite in andesite (Mourant 1961); La Corbière, southwest Jersey – chalcopyrite, galena and sphalerite in pyrite–quartz veins (Ixer 1980); La Moye – chalcopyrite forming inclusions in bornite within mineralized aplogranite (Ixer 1988); L'Ouaisné Bay, southwest Jersey – veins containing calcite, chalcopyrite, fluorite, galena, hematite, magnetite and sphalerite (Mourant 1961, Ixer 1980).

Sark Chalcopyrite is intimately associated with lead-rich tennantite in complex veinlets cutting pyrite and quartz on a specimen from Sark's Hope mine, Little Sark (Bishop et al. 1977). The primary ore from this mine consists predominantly of galena and pyrite, with local concentrations of chalcopyrite and tennantite–tetrahedrite (Mourant & Warren 1934, Ixer & Stanley 1983, Ixer 1992). Chalcopyrite also occurs with silver-rich pyrite in three parallel lodes at Port es Saies (Henwood 1871).

chalcosiderite

TURQUOISE GROUP

triclinic $Cu^{2+}Fe^{3+}_6(PO_4)_4(OH)_8 \cdot 4H_2O$

TL Wheal Phoenix mine, Linkinhorne, near Liskeard, Cornwall, England

A generally rare hydrated copper–iron phosphate hydroxide that forms a series with turquoise and was found in considerable quantities at the type locality in association

Green chalcosiderite associated with yellow cyrilovite on hematite-coated quartz. Gunheath china-clay pit, St Austell, Cornwall; crystal groups to 0.7 mm across. (© D. I. Green)

with dufrénite, andrewsite and turquoise (henwoodite). It occurs in light to dark-green sheaves or globular masses, with a radiating structure (Maskelyne 1875, Collins 1911a, Braithwaite 1981). Later work discredited andrewsite, showing it to be a mixture of hentschelite and rockbridgeite with minor chalcosiderite (Dunn et al. 1990).

Chalcosiderite has also been found in the Turquoise vein area of Gunheath china-clay pit, Carthew, St Austell, Cornwall. The species is associated with cyrilovite and it occurs both on the faces of small fractures in altered granite and on a layer of milky quartz, associated with a small hematite–quartz vein. The chalcosiderite forms dark-green crystals and crystalline aggregates to 2 mm (Ince 1986, Watson 1987). Additionally, pale-green chalcosiderite and goyazite are associated with wolframite in massive quartz veins within the Hensbarrow section of this pit (Elton & Grigg 1996), and powdery deposits of pale-green chalcosiderite are associated with variscite from the same location (Elton 1996). As the kaolin workings extended to Bunny mine, chalcosiderite forming bladed or lozenge-shape crystals, or botryoidal crystal aggregates, was commonly found in the Gossan vein in association with cyrilovite and dufrénite. Chalcosiderite appears to be part of a secondary assemblage from the oxidation zone at Gunheath, along with arthurite, brochantite, cacoxenite, cassiterite (varlamoffite), cyrilovite, dufrénite, leucophosphite, libethenite, morinite, scorodite, torbernite, turquoise, variscite and wavellite (Weiss 1994, Elton et al. 1996).

chalcotrichite

A variety of cuprite forming fine carmine-red capillary-shape crystals.

chalilite

TL Donegore Mountains, Sandy Brae, Co. Antrim, Ireland

An uncertain substance, although reported to be a variety of thomsonite, zeolite group (Greg & Lettsom 1858).

chalybite

A synonym of siderite, calcite group.

chamosite

CHLORITE GROUP

monoclinic $(Fe,Al,Mg)_6(Si,Al)_4O_{10}(OH)_8$

A common iron–aluminium–magnesium aluminosilicate hydroxide, dimorphous with orthochamosite, that forms a series with clinochlore. It is found in sedimentary ironstones, where it forms authigenically under reducing conditions in the presence of decomposed organic matter. Chamosite is also widely developed as a secondary mineral in altered igneous rocks, replacing minerals such as olivine or pyroxene. There is confusion in the literature regarding chamosite descriptions, as an XRD trace of berthierine with a small admixture of clinochlore can easily be mistaken for chamosite (Brindley 1982). Brindley (1982) suggests that the following descriptions refer to compositions close to berthierine: Frodingham, Lincolnshire; Thingdon mine, Finedon, Northamptonshire; Gateside farm, Ayrshire coalfield, 3 km E of Ayr, Strathclyde, Scotland; Blackburn, 3 km NNW of Carnlough, Co. Antrim, Ireland; Knowehead farm, Knowehead, 5 km SW of Ballyclare, Co. Antrim (Brindley 1951); Stanion Lane pit, Corby, Northamptonshire (Brindley & Youell 1953). An additional complication is that delessite is described by Ford et al. (1993) as a green magnesium-rich variety of chamosite for occurrences in Derbyshire (Calton Hill, Taddington; Hanging Flats mine, Stoney Middleton Dale; Lathkill Dale, S of Over Haddon; Mill Dam mine, Hucklow), but Bayliss (1975, 2000) considers delessite to be iron-rich clinochlore. To add further confusion, Clarke (1993) gives delessite species status, and de Fourestier (1999) considers it to be either magnesian chamosite or ferrous clinochlore. Bayliss (1975) recommends that the following mineral names should be discarded, as they have all been used to describe varieties of chamosite: brunsvigite, daphnite, diabantite, leptochlorite, orthochamosite, orthochlorite, pseudothuringite, ripidolite and thuringite. Aphrosiderite is considered to be a magnesium-rich chamosite (Bayliss 2000). The following records of chamosite should therefore be read with caution.

England

Avon Wickwar, N of Yate (Anthony et al. 1995).

Cornwall Geevor mine, near Pendeen, St Just (thuringite); Penzance (daphnite); Phoenix United mines, Linkinhorne (all Golley & Williams 1995); vein in chloritized slate at Tolgus mine, near Redruth (dark-green daphnite) (Guppy 1956).

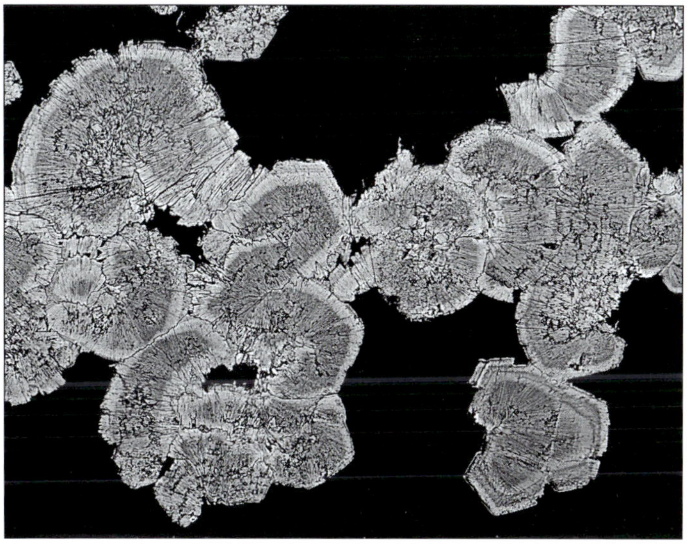

Zoned platelets of bright-green chamosite in quartz; Trevallas Porth beach, near Blue Hills, St Agnes, Cornwall; width 490 μm; BSE image. (© E. Condliffe, University of Leeds)

Lincolnshire Limonite ooliths fringed with chamosite in ironstone at Frodingham (Hey 1954, Guppy 1956, Slater & Highley 1976).

Northamptonshire Chamositic oolite at Irthlingborough mine, S of Irthlingborough church (Guppy 1956); chamosite–siderite oolite at Pitsford ironstone quarry (Hollingworth & Bannister 1950).

Yorkshire Chamositic oolite at Blakey mines, Farndale; Kitching's deposit, High Hollins farm, Rosedale; Rosedale Abbey Bank, Rosedale; Rosedale East mine; Sheriff's pit, Rosedale (all five localities noted by Rastall & Hemingway 1949); chamosite-oolite at Straight West Working, Longacres mine, Skelton (Guppy 1956).

Wales

Gwynedd Oolitic chamosite–magnetite rock at Betws Garmon, Snowdonia (Pulfrey 1933); pseudo-hexagonal crystals of a silvery-white aluminium-rich chamosite at Fron-oleu, Prenteg, Tremadog (Starkey & Robinson 1992); Penrhyn quarry, Llandegai, near Bangor (aphrosiderite–daphnite–thuringite) (Greg & Lettsom 1858, Hey

1954); oolitic iron ore consisting of concentric layers of magnetite set in a chamositic matrix at Pen-yr-allt iron mine, near Llanfrothen, Penrhyndeudraeth (Hallimond 1924); chamosite-bearing oolitic limestone at Tyllau Mwn, near Dolgellau (Pulfrey 1933, Matthews & Scoon 1964).
Mid Glamorgan Fforchaman colliery, Cwmaman, near Aberdare (Bevins 1994).

Scotland
Highland Light-green scaly masses embedded in quartz boulders at Dirty Burn, Bishop's Hill, Dunoon (aphrosiderite) (Heddle 1879f); ooliths in Raasay Ironstone, Raasay (Lee 1920, Slater & Highley 1976, Macpherson & Livingstone 1982, Draper & Draper 1992); botryoidal zinc-rich chamosite associated with kainosite-(Y) and hemimorphite from Whitesmith mine, Strontian (Green & McCallum 2005).
Lothian Campbell & Lunn (1925) describe dark olive-green to black chlorophaeite (possibly chamosite) as an alteration product infilling vesicles in a Carboniferous tholeiitic dolerite sill at Dalmahoy Hill and Kaimes Hills.

Ireland
Antrim Murlough Bay, Ballycastle (diabantite) (Hey 1954).
Down Mourne Mountains (brunsvigite) (Hey 1954); Lindsay's Leap quarry, Mourne Mountains (Green et al. 2005a).
Tipperary Shallee mine, Silvermines district (brunsvigite) (Rhoden 1959).
Wicklow Iron-rich chamosite in a pyrite-rich wallrock alteration assemblage adjacent to lenticular ore zones at Avoca Cu–Fe–Pb–Zn mine, Avoca (Wheatley 1971a).

chenevixite
monoclinic $Cu^{2+}_2Fe^{3+}_2(AsO_4)_2(OH)_4{\cdot}H_2O$
TL Wheal Gorland, St Day, 2 km E of Redruth, Cornwall, England

Named after Richard Chenevix FRS (1774–1830), an Irish–French chemist and mineralogist; born in Ballycommon, near Dublin, Ireland. Chenevixite is an uncommon copper–iron arsenate first described from an unknown locality in Cornwall, probably Wheal Gorland (Pisani 1866); a later description suggests Wheal Phoenix (Robson 1949b). Rust (1981) describes chenevixite from Wheal Gorland, where it occurs as dark blackish-green resinous masses in gossany veinstone associated with olivenite. Fisher (1988) describes it as a very common species at this site, but that was before the mine dumps were totally removed. At Ting Tang mine, also Gwennap, Cornwall, chenevixite occurs as green powdery masses in cavities within vein quartz. It appears to have formed by the oxidation of primary arsenopyrite–pyrite (Bruce & Aubrey-Jones 1998). Two additional localities in Cornwall are Wheal Unity, Gwennap, and Old Gunnislake mine, Calstock (Anthony et al. 2000).

Because of a Kingsbury connection, the chenevixite described by Hartley (1984) and Young (1987a) from Brandy Gill, Caldbeck Fells, Cumbria, requires further confirmation.

chenite
triclinic $Pb_4Cu^{2+}(SO_4)_2(OH)_6$
TL Susanna mine, Leadhills, Strathclyde, Scotland

A rare lead–copper sulphate hydroxide that at the type locality forms minute, singly terminated, transparent-to-translucent sky-blue crystals 0.1–1.0 mm long. Chenite is a secondary mineral that occurs on oxidized galena with minor chalcopyrite and is associated with the species caledonite, hydrocerussite, lanarkite, leadhillite, linarite and susannite. Near by, at the Glengaber smelting mill at Meadowfoot, Wanlockhead, chenite has been found on an old slag dump associated with elyite, lanarkite and other oxidation products (Paar et al. 1986). Light greenish-to-blue bladed crystals (<0.5 mm) of chenite are also noted from the dumps of Horner's vein, Leadhills (Rust 1994).

In the Central Wales orefield, translucent pale-blue chenite occurs at Llechweddhelyg mine, near Clawddmelyn, between Bont-goch and Cwmsymlog, Dyfed. It forms slightly rounded, subparallel plates, in groups to about 1.5 mm and was found in two cavities within a thin corroded galena veinlet (Rust 1994). At Esgair Hir mine, Ceulanymaesmawr, Dyfed, chenite forms equant sky-blue crystals up to 0.4 mm across in association with susannite, apparently with a post-mining origin (Rust 1997).

Small sky-blue crystals of chenite associated with elyite and caledonite are reported from dumps at the 12-fathom level of Driggith mine, Caldbeck Fells, Cumbria, and further noted from the nearby Red Gill and Short Grain mines, less than 3 km away (Green & Hubbard 1996, Green & Briscoe 2002, Green et al. 2006a).

chenocoprolite
A mixture of chlorargyrite and scorodite.

chernovite-(Y)?
tetragonal $YAsO_4$

A rare yttrium arsenate that forms a series with xenotime-(Y) and may be a constituent of black globules (<5 mm) on massive pink gypsum at Gypsy Lane, Leicester, Leicestershire (Embrey 1978).

chert
Fine-grain massive quartz ± moganite.

chessylite
A synonym of azurite.

chesterite?
orthorhombic $(Mg,Fe^{2+})_{17}Si_{20}O_{54}(OH)_6$

A rare magnesium–iron silicate hydroxide reported in Lewisian amphibolite-facies ultramafic rocks SW of Achmelvich, An Fharaid Mhòr, near Lochinver, Highland Region, Scotland. Chesterite (and probably the monoclinic polytype of chesterite) occurs with two other triple-chain pyriboles (clinojimthompsonite and jimthompsonite) and within a single crystal (<3 cm) these are often intergrown with actinolite, anthophyllite and cummingtonite. The pyriboles do not pseudomorph the amphibole species and probably grew during prograde Inverian metamorphism, possibly as the result of infiltration of CO_2-rich fluid. The many chemically similar pyriboles and their disposition within compound prisms suggests that crystallization was kinetically controlled. Chesterite is not confirmed as a British species because, with the small amount of material available and the very fine intergrowths, it is not possible to prove it to be orthorhombic. However, chemical analysis is consistent with its being chesterite (Droop 1994, G. T. R. Droop pers. comm. 2003).

chevkinite-(Ce)
monoclinic $Ce_4(Ti,Fe^{2+},Fe^{3+})_5O_8(Si_2O_7)_2$

A cerium–titanium–iron silicate, closely related to perrierite-(Ce) and found in accessory amounts in Tertiary epigranites (Beinn Dearg Mhòr, Coire Uaigneich, Loch Ainort and Maol na Gainmhlich) on the Isle of Skye, Highland Region, Scotland. Initially, the identity of the mineral was not recognized and crystals were described as "the most REE-rich sphenes yet reported" (Exley 1980). Chevkinite-(Ce) is petrographically similar to allanite, being dark reddish-brown and strongly pleochroic in thin section. Allanite, apatite, titanite and zircon are associated species.

Chevkinite is the most common of the three accessory minerals described from Tertiary granitic rocks on St Kilda (Hirta), Western Isles, the other species being allanite and zirconolite (zirkelite). Chevkinite is found in the Conachair, Glen Bay, and Na h-Eagan granites, where it forms elongate needles or stumpy prisms up to 200 µm long, enclosed in quartz or alkali feldspar (Harding et al. 1982, 1984).

Barritt (1983) reports chevkinite-(Ce) as an accessory species in granodiorite from the Southern Cruachan unit of the Etive granite complex to the E and NE of Ben

Chenite crystal, 0.6 mm long, from Driggith mine, Caldbeck Fells, Cumbria. (© D. I. Green)

Starav, Strathclyde. It occurs with other accessory species (apatite, titanite, thorite and zircon), but its composition is perrierite-(Ce) according to the chevkinite–perrierite discriminant diagram proposed by Macdonald & Belkin (2002).

chiastolite

A variety of andalusite containing carbonaceous inclusions arranged so that in cross section they form a dark cross.

childrenite

monoclinic $Fe^{2+}Al(PO_4)(OH)_2 \cdot H_2O$

TL near Tavistock, Devon, England

Named after John George Children FRS (1777–1852), a pioneer mineral chemist, astronomer, zoologist, keeper of natural history at the British Museum (1822–1840) and one-time secretary of the Royal Society (Gunther 1978); born in Tonbridge, Kent. Childrenite is a rare hydrated iron–aluminium phosphate hydroxide that was first found during the cutting of a canal near Tavistock and probably also during excavation of the Tavistock Canal tunnel. It was found as minute crystals with siderite and pyrite, and at Wheal Crebor, near Tavistock, it occurred as distinct crystals on a greyish-green chloritic matrix (Brooke 1823, Greg & Lettsom 1858, Russell 1926). Perhaps the most important discovery of childrenite was at George and Charlotte mine, Tavistock Hamlets, Devon, where fine lustrous light-brown striated crystals up to 2 cm across were found with pyrite, siderite and quartz on chlorite-encrusted slate (Greg & Lettsom 1858, Collins 1892, Thurgood 1982, Embrey & Symes 1987, Wolfe 2005).

In Cornwall, minute (up to 1 mm) brilliant light-brown to almost colourless crystals of childrenite have been found at Crinnis mine, near St Austell, where they occur on quartz and killas (Collins 1911a, Drugman 1915). Since then, honey-brown crystals of childrenite up to 1 mm across have been uncovered during engineering work at nearby Carlyon Bay (also known as Crinnis Cliff) and at Wheal Ramoth on the sand dunes near Perranporth, where it occurs with cassiterite and vein agate (Wolfe 2006). A very thorough review by Braithwaite & Cooper (1982) concluded that childrenite always occurs close to the margins of granite intrusions. The authors documented museum specimens and made many more additional discoveries in Cornwall and Devon, as follows.

Cornwall Drakewalls mine, Calstock – pale-brown bipyramidal crystals up to 1.5 mm long; Lady Ashburton mine, Callington – very pale creamy-brown bipyramidal crystals up to 0.25 mm; Okel Tor mine, Calstock – clear sharp very pale creamy-brown bipyramidal crystals up to 0.3 mm; Prince of Wales mine, Calstock, near Gunnislake – pale-brown bipyramidal crystals mostly 0.5–1.0 mm, occasionally up to 1.3 mm in small vugs; Wheal Jane, Kea, near Truro – greyish-brown crystals with tabular habit up to 0.5 mm.

Devon East Wheal Russell, Tavistock Hamlets – no details given; Furzehill (Furzehill Wood) mine, Walkhampton, near Tavistock – pale yellowish-brown bipyramidal crystals up to 0.5 mm long; Gawton mine, Tavistock Hamlets – deep reddish-brown sharp bipyramidal crystals up to 0.2 mm; Little Duke (North Tavy) mine, Tavistock Hamlets – very pale-brown bipyramidal crystals up to 0.2 mm; Sortridge Consols mine, Whitchurch – small, clear, very pale-brown, heavily iron-stained bipyramidal crystals up to 0.25 mm; South Wheal Crebor (New East Wheal Russell) mine, Tavistock Hamlets – almost colourless prismatic and bipyramidal crystals up to 0.5–1.0 mm; Wheal Anna Maria, Tavistock Hamlets – small, very pale-brown bipyramidal crystals up to 0.2 mm; Wheal Betsy, near Tavistock – no details given; William and Mary mine, Tavistock Hamlets – clear sharp pale-brown bipyramidal crystals up to 0.2–0.5 mm, occasionally in honey-brown clusters up to 1.5 mm across.

Childrenite from George and Charlotte mine, Tavistock, Devon; crystal 4 mm long. (© Royal Cornwall Museum, Royal Institution of Cornwall)

Kingsbury & Hartley (1957) describe childrenite from an unusual quartz–chlorite vein a little to the west of Causey Pike in Cumbria, but this is regarded as very dubious (Green & Briscoe 2002; see Ryback et al. 1998, 2001). However, a discovery of childrenite on the W side of the Vale of Newlands confirms childrenite in Cumbria (Braithwaite & Lamb 2007).

china clay

A synonym of kaolinite.

chloanthite

An arsenic-deficient variety of nickel-skutterudite.

chlorapatite/apatite-(CaCl)

APATITE GROUP

monoclinic $Ca_5(PO_4)_3Cl$

A rare calcium phosphate chloride found in iron-stained fracture veins within partially kaolinized granite at Gunheath china-clay pit, Carthew, St Austell, Cornwall. Analysis shows that the mineral is actually hydroxyl-chlorapatite and it occurs as a pale brown–cream horn-like encrustation (to about 1 cm thick) on a complex matrix of mixed phosphates. The bulk of the matrix is black barbosalite, but intermediate members of the frondelite–rockbridgeite and chalcosiderite–turquoise series, and also leucophosphite, are also present. The matrix is veined with cassiterite, chalcocite, fluorapatite and quartz (Elton 1997).

Apatite with a significant chlorine content, possibly chlorapatite, has been found in open fractures in ironstone (septarian) nodules within Coal Measures sediments close to Beggars Wood, near Gateshead, Tyne & Wear, and at Smithywood opencast site, Chapeltown, near Sheffield, Yorkshire. At Beggars Wood, the apatite forms tiny colourless-to-white tabular hexagonal crystals up to 0.5 mm across and is associated with dickite, quartz, siderite and rare chalcopyrite (Neall 2004).

Crystalline marble from Balephretrish Bay on the Isle of Tiree, Strathclyde, Scotland, contains aggregates of plentiful pale-blue chlorapatite, diopside, meionite (scapolite) and titanite. The marble also contains well formed crystals of amphibole, forsterite, mica, orthoclase and spinel (Coomaraswamy 1903).

chlorargyrite

cubic AgCl

A rare silver chloride that has occasionally been found in mineable quantities, such as in the Sedgman quartz lode at Perran mine, Perranuthnoe, near Marazion, Cornwall, where it occurred as distorted small cubes and cubo-octahedra, or in massive form with specimens approaching 1 kg. Chlorargyrite (cerargyrite) was found most abundantly in the gossanous part of the lode, particularly where limonite was abundant. Acanthite (argentite), arsenopyrite, chalcopyrite, galena, pyrite, siderite and sphalerite were also found in the lode (Rudler 1905, Butler 1907). A bromian variety of chlorargyrite occurs in a selenide assemblage within a gold-bearing carbonate vein cutting Middle Devonian limestones at Hope's Nose, Torquay, Devon. Associated species include chrisstanleyite, clausthalite, eucairite, fischesserite, klockmannite, naumannite, penroseite, tiemannite, trüstedtite, tyrrellite and umangite (Paar et al. 1998). Chlorargyrite was claimed to occur at Penberthy Croft mine, St Hilary, Cornwall (Kingsbury 1954), but this is disputed by Betterton et al. (2000). Robson (1954) and Camm & Merry (1991b) report chlorargyrite from this mine on the basis of the Kingsbury samples and, similarly, Hartley (1984) reports the species from Driggith mine, Caldbeck Fells, Cumbria; both occurrences must be considered very doubtful, particularly as there are no specimens from Driggith mine in the Kingsbury collection at the NHM (Green et al. 2006a). Other occurrences of chlorargyrite (which is referred to as kerate in the early references) include the following.

England

Cornwall Dolcoath mine, Camborne; Duchy Peru, near Cubert; Great Wheal Charlotte, St Agnes; Silver Vein mine, Lostwithiel; Wheal Alfred, Phillack; Wheal Ann, Phillack; Wheal Langford, Callington (all Rust 1982a); Levant mine, St Just; North Dolcoath mine, Camborne; Wheal Cock, Botallack, St Just; Wheal Herland, Gwinear (all Collins 1892, Rust 1982a); Wheal Basset, Perranzabuloe (Collins 1892); Wheal Brothers and Silver Valley mine, near Calstock; Wheal Duchy, Phillack; Wheal St Vincent, Calstock (all Greg & Lettsom 1858); Wheal Mexico, Calstock (Golley & Williams 1995); Wheal Mexico, near Perranporth, Perranzabuloe (Rust 1982a, Golley & Williams 1995).

Ireland

Dublin Ballycorus, near Carrickgollogan (Greg & Lettsom 1858).

Offshore territories

Sark Sark's Hope mine, Little Sark (Greg & Lettsom 1858, Mourant & Warren 1934, Ixer & Stanley 1983, Laffoley 1998).

CHLORITE GROUP

Monoclinic or triclinic silicates with the general formula $A_{4-6}Z_4O_{10}(O,OH)_8$, where A=Al, Fe^{2+}, Fe^{3+}, Li, Mg, Mn^{2+}, Ni, Zn; and Z=Al, B, Fe^{3+}, Si. Chamosite, clinochlore, cookeite, orthochamosite and pennantite are members of the group found in Britain or Ireland. More often than not, the species is not confirmed, particularly in petrogenetic studies, where green alteration products of mafic silicates are simply named "chlorite".

chloritoid

monoclinic or triclinic $(Fe^{2+},Mg,Mn)_2Al_4Si_2O_{10}(OH)_4$

An iron–magnesium–manganese aluminosilicate hydroxide that forms series with magnesiochloritoid and with ottrélite, and is found as a rock-forming mineral typically in low- and medium-grade regionally metamorphosed mudstones. Early examples were found in a coastal quarry close to Tintagel church in Cornwall. Since then, the British Geological Survey (Reid et al. 1910) have mapped chloritoid-bearing phyllites over a wider area. In thin section, chloritoid appears as plates with irregular terminations. Commonly, twinning and an hourglass structure defined by inclusion trails are observed. Chlorite, ilmenite, muscovite and rutile are associated species (Tilley 1925). Metamorphosed Ordovician slates in the outer aureole of the Skiddaw granite, Caldbeck Fells, Cumbria, contain rod-like crystals of chloritoid, and substantial spongy porphyroblasts are noted from one locality (Eastwood et al. 1968).

In Snowdonia, Gwynedd, Wales, chloritoid is widely distributed, being reported from Ordovician rhyolitic lavas (Harker 1889) and from Ordovician slates in the aureole of the Aber–Drosgl intrusion (Evans 1968). At Rhyd-Ddu, also in Snowdonia, a manganese-rich variety has been found in black slates of the Nant Ffracon Formation, where it occurs as small tabular porphyroblasts within a matrix of chlorite, muscovite and quartz, The chloritoid is 30–100 µm long and some crystals show classic fantail morphology (Brearley 1988). In a study of low-grade metamorphism of slate belonging to the Nantmor Group, Cwm Pennant area of Snowdonia, Roberts et al. (1989) were able to show chloritoid porphyroblasts grew pre-, syn- and post-tectonically.

Chloritoid–kyanite schists form the hills of Housi Field, Sothers Field, Saxa Vord and Ward of Norwick, all in the NE of Unst, Shetland, Scotland (Read 1932), and at Vannlip, Hillswick, on Mainland, Shetland, Heddle (1879f, 1901) reports quartz veins in talc schists containing dark-green chloritoid with inclusions of titanite (sphene) associated with kyanite and muscovite. Large green crystalline masses and brown foliated plates associated with rutile are further noted. Read (1934b) and Snelling (1957) suggest that the Saxa Vord schists formed during retrogressive metamorphism of an earlier andalusite–staurolite schist and, afterwards, progressive metamorphism introduced garnet as an additional phase. Porphyroblasts of chloritoid are randomly orientated and up to 1 mm long. Elsewhere in the Shetland Isles, at Loch of Watlee, Gunnister, Snarra Voe and Winner Houll, all on Unst, large and abundant chloritoid crystals occur in non-biotitic schists, clearly derived from the metamorphic recrystallization of kyanite and staurolite (Read 1934b). Chloritoid is further noted from a schist at Ness of Snabrough, Dale of Oddsta, northeast Fetlar (Guppy 1956).

Prominent porphyroblasts of chloritoid, up to 5 mm long, occur in Dalradian schists along the coastal section between Stonehaven and Aberdeen, Grampian. In thin section, most crystals are twinned and often their form is ragged and full of quartz inclusions. Alteration is common, and both chloritoid and chlorite are replaced by kaolinite, limonite and siderite (Snelling 1957). Two chlorite–chloritoid schists from along the coast N of Stonehaven also contain paragonite (Baltatzis & Wood 1977). Other chloritoid occurrences along this coastal region are documented by Barrow (1898), Williamson (1953), Phemister et al. (1960), Chinner (1967), Porteus (1973) and Atherton & Brotherton (1974). Inland from these outcrops, chloritoid-bearing rocks are described from pelites at five localities in the Macduff Slates, close to the northern margins of the Insch and Boganclough gabbroic intrusions. Small (< 1 mm) randomly orientated prismatic crystals of chloritoid overgrow the cleavage and show lamellar twinning, high relief and distinctive bluish-green pleochroism in thin section. Muscovite, quartz and opaque oxides are associated species (Leslie 1988).

At Grandtully, near Pitlochry, Tayside Region, zoned garnets in mica schists contain inclusions of choritoid showing compositional changes that have been used to establish a metamorphic history of garnet grown during regional metamorphism (Sivaprakash 1981).

Post-tectonic lamellar twinned chloritoid (< 1.5 mm long) occurs with garnet porphyroblasts containing orientated ilmenite inclusion trails in chlorite–muscovite–quartz schists from N of Bunnahowna, Renvyle Point, Connemara, Co. Galway, Ireland (Harvey & Ferguson 1973). Chloritoid also occurs in heavy-mineral concentrates with andalusite, corundum, garnet, kyanite, sillimanite and staurolite collected from the Lower Llanvirn Upper Derrylea Formation in the South Mayo Trough, Co. Galway, and together they signal the unroofing of the Dalradian metamorphic complex during the Ordovician collision of the Laurentian continental margin with a supra-subduction zone ophiolite/oceanic arc (Mange et al. 2003). Rare prismatic crystals of chloritoid (0.4 mm long) replace altered andalusite porphyroblasts in quartz–feldspar schists from the aureole of the Ardara granite, Co. Donegal (Atherton 1980).

Porphyroblast of twinned chloritoid in a muscovite (colourless), chlorite (green) schist; Unst, Shetland, Scotland; width 2.25 mm; viewed in PPL (upper) and XPL (lower). (© K. Gardner)

chloropal

A synonym of nontronite, smectite group.

chlorophaeite

TL Isle of Rum, Highland Region, Scotland

An alteration product of olivine, possibly a chlorite; also said to be talc (de Fourestier 1999) or chamosite (Bayliss 2000).

chlorophane

A green thermoluminescent variety of fluorite.

chlorophyllite

A green foliated pseudomorph after cordierite.

chlorospinel

A green variety of spinel, rich in ferric iron.

chloroxiphite

monoclinic $Pb_3Cu^{2+}Cl_2(OH)_2O_2$

TL Higher Pitts farm, Priddy, Mendip Hills, Somerset, England

A rare lead–copper oxychloride found at the type locality in calcite-rich veins, fissures and pockets of iron–manganese mineralization cutting Carboniferous Limestone, where it occurs as dull olive-green blade-like crystals embedded in mendipite. The mineral is very brittle and friable, and it forms crystals up to 3 cm long (Spencer & Mountain 1923). Chloroxiphite occurs as corroded crystals in cerussite and hydrocerussite (Kingsbury 1941, Burr 1996).

At Merehead (Torr Works) quarry, Cranmore, Somerset, mineralized veins, fissures and pockets similar to those at Higher Pitts farm contain chloroxiphite in subparallel groups of olive-green bladed crystals that are often curved and up to 4 cm long. Diaboléite is often associated with chloroxiphite, apparently as an alteration product. Other secondary lead and copper minerals associated with chloroxiphite include blixite, cerussite, cotunnite, hydrocerussite, mendipite, mereheadite, parkinsonite, symesite and wulfenite, all occurring in the cavernous parts of the manganese oxide ore (Symes 1977, Symes & Embrey 1977, Symes et al. 1994, Welch et al. 1998, 2000, Turner 2007). Alabaster (1989b) reports a specimen where chloroxiphite is intergrown with laurionite and paralaurionite, and adds that malachite is a common alteration product.

Elsewhere in the Mendip Hills, chloroxiphite is described with other lead oxychloride and manganese-rich species from Coleman's quarry, Holwell, near Frome, Somerset (Alabaster 1989b). Lead oxychlorides in the Mendips are thought to have formed by interaction of primary galena with low-temperature metalliferous brines along fractures cutting original Cu–Fe–Mn–Pb mineralization (Symes & Embrey 1977, Humphreys et al. 1980), although it is possible that they precipitated directly from a cooling brine (Welch et al. 2000). Alabaster (1989b) also reports chloroxiphite in association with parkinsonite from Wesley Theological College (Wesley mine), Westbury-on-Trym, Avon.

Complex intergrowth of dark-green bladed chloroxiphite (15 mm long), orange mereheadite, fibrous pinkish mendipite, thin purplish-red parkinsonite and white hydrocerussite; Merehead (Torr Works) quarry, Cranmore, Mendip Hills, Somerset. (© M. P. Cooper)

chlorspath

A synonym of mendipite.

chondrodite

HUMITE GROUP

monoclinic $(Mg,Fe^{2+})_5(SiO_4)_2(F,OH)_2$

An uncommon but locally abundant magnesium–iron silicate fluoride hydroxide described in the early literature from England (Cornwall), Scotland (near Loch Ness) and Ireland (Co. Donegal). All the occurrences are considered dubious by Read & Double (1935). These authors confirm chondrodite in limestone (now forsterite marble) from Allt Eas Mòr Chùl an Dùin, a tributary joining the Abhainn a' Ghlinne Bhig, Glenelg, Highland Region, Scotland. Other associated species are calcite, chlorite, diopside, phlogopite and iron ore. Chondrodite typically forms small shapeless granules around large forsterite (or forsterite–serpentinite) masses, or occurs as inclusions in many of the other species. It probably formed as an alteration product of forsterite during the passage of fluorine- and hydroxyl-bearing fluids. In thin section, chondrodite is pleochroic in shades of yellow and sometimes shows excellent lamellar twinning. Read & Double (1935) also report chondrodite from nearby locations in the Glenelg area at Balvraid, Gleann Beag and Allt Easan Mhic Garraidh.

Chondrodite is also found near Broadford, Isle of Skye, Highland Region, where the Beinn an Dubhaich granite has intruded dolomitic limestone. Along the contact, a multiple series of discrete skarn zones have developed, and at Kilbride, close to the northwestern tip of the granite, chondrodite is found associated with clinohumite, diopside and forsterite in a magnetite-rich skarn. In some other zones, chondrodite is found in a clinohumite–cuspidine–diopside–monticellite skarn and also in a ludwigite–magnetite skarn containing accessory fluoborite and szaibelyite. Another skarn containing chondrodite, clinopyroxene and magnetite occurs to the W at Kilchrist (No. 1 prospect). Expulsion of boron- and fluorine-rich fluids from the granite metasomatized Durness limestone near to the granite contact led to the crystallization of these skarn assemblages (Tilley 1948b, 1949, 1951).

At Bay Steinigie, near Finsbay (Fionnsabhagh), South Harris, Western Isles, Lewisian metamorphosed limestone is found in a long narrow xenolith within the South Harris igneous complex. Small yellow grains of chondrodite and humite (about 0.3 mm across) occur in an isolated outcrop of rock consisting mainly of calcite, dolomite and forsterite. Other minerals present include brucite, graphite, pargasite, phlogopite, titanite and opaque iron oxides (Dearnley 1961).

Intrusion of a carbonatite magma into Durness limestones and dolomites near Loch Borralan, Assynt, Highland Region, has produced metasomatic bands of chondrodite–diopside–phlogopite rock close to the contact. Within the carbonatite (a porphyritic layered white sövite), chondrodite forms vertical platelets reminiscent of comb-layering with diopside and phlogopite in a calcite matrix. Locally, equant crystals of chondrodite are also associated with pale-blue richterite needles in this rock. Chondrodite is a major component of the clasts in a sövite breccia, where it forms translucent red equant grains up to 7 mm across (Young et al. 1994b).

chonicrite

A mixture of unidentified minerals.

chrisstanleyite

orthorhombic $Ag_2Pd_3Se_4$

TL Hope's Nose, Torquay, Devon, England

Named after Christopher John Stanley (b. 1954), deputy head of mineralogy at the NHM, for his contributions to ore mineralogy; born in Nottingham, England. Chrisstanleyite is a rare palladium–silver selenide found in gold-bearing carbonate veins in Middle Devonian limestones. It is associated with palladium-rich and silver-rich gold, bromian chlorargyrite, clausthalite, eucairite, fischesserite, tiemannite, umangite, and possibly mertieite-II. Chrisstanleyite forms composite grains of anhedral crystals ranging in size from a few µm to several hundred µm (Paar et al. 1998).

christianite

A synonym of phillipsite, zeolite group.

christophite

A black iron-rich variety of sphalerite.

Composite grains of chrisstanleyite. Hope's Nose, Torquay, Devon; width 450 µm; viewed in reflected light (PPL upper and XPL lower). (© C. J. Stanley, The Natural History Museum, London)

chrome diopside

A chromium-rich variety of diopside, pyroxene group.

chrome ochre

A green chromium-rich clay.

chrome spinel

A synonym of magnesiochromite and chromium-rich varieties of spinel and hercynite, all spinel group.

chromic iron

A synonym of chromite, spinel group.

chromite

SPINEL GROUP

cubic $Fe^{2+}Cr_2O_4$

A common iron–chromium oxide that forms series with magnesiochromite and with hercynite and is dimorphous with donathite. Chromite is often found as a cumulus mineral in the ultramafic portion of layered mafic intrusions and is less commonly found in other mafic and ultramafic intrusions. The name is sometimes used to describe chrome spinel, an intermediate composition that can be closer to end-member spinel or hercynite than to chromite (e.g. Power et al. 2000a).

An unusual occurrence of magnesium-poor chromite and an unnamed chromium–titanium-rich oxide mineral are reported from a calc-alkaline augite–hornblende-rich porphyritic microdiorite intrusion at Scawgill Bridge, near Whinlatter Pass, 4 km W of Braithwaite, Cumbria. Chromite forms grains 10–20 µm across, accompanied by sulphide minerals of a similar size (chalcopyrite, pyrite and As–Co–Ni–S minerals) and a few 200 µm-diameter grains of the unnamed species (Fortey & Nancarrow 1990, Fortey et al. 1994). Chromite was one of the less common heavy minerals recovered from alluvial sediments in the Fowey Estuary during an investigation of siltation caused by past mining activities in this area of Cornwall (Pirrie et al. 2002). The source of the chromite is not discussed, although the relatively close proximity of the Lizard ophiolite complex and the identification of chromite in serpentinite lenses at Yellow Carn, Kynance, Lizard Peninsula (Robson 1949b), makes this probable.

At various localities within the Precambrian Mona Complex, such as Treflesg, near Llanfair-yn-Neubwll, close to the strait between Anglesey and Holy Island, Gwynedd, Wales, chromite occurs as an accessory component of serpentinized peridotite. It is associated with spinel (picotite), ilmenite and zircon (Greenly 1919, Maltman 1977). Ordovician sedimentary rocks at Cwm Tregalan, S of Snowdon, are enriched in magnetite and contain chromite and ilmenite in placer bands up to 3 mm thick (Colman & Appleby 1991). Chromite is also noted in peridotite nodules from the Great House diatreme, near Usk, Gwent (Boulton 1911).

Perhaps the greatest development of chromite in Britain or Ireland is associated with the ophiolite complex on Unst, Shetland, Scotland, where 50 000 tonnes of chromite ore were mined between 1823 and 1944. The chromite is concentrated in pods up to 1 m thick and 75 m long in serpentinized dunite near Baltasound (Dunham et al. 1978). Greg & Lettsom (1858) report octahedral crystals, up to 1 cm across, from Haroldswick, although most chromite is massive. These authors consider Buness House, Loch of Cliff and west of Hagdale to be further important localities for chromite on Unst. Heddle (1901) adds Hagdale quarry, between Heog and Keen Hills. Because of the relatively high abundance the PGMs associated with chromite in these rocks, considerable academic and commercial interest has built up (Prichard et al. 1981, Gass et al. 1982, Gunn & Leake 1985, Prichard & Lord 1988, Prichard & Tarkian 1988, Lord et al. 1994). Drilling of the ophiolite complex west of Hagdale produced serpentinite, which on analysis was shown to contain chrome spinel (intermediate compositions between chromite and spinel) rather than chromite *senso stricto* (Maynard 1993, Lord & Prichard 1997). Elsewhere in Shetland, chromite occurs at Hestaness on Mainland, Vord Hill on Fetlar, and on the small island of Haaf Gruney (Heddle 1901).

At Corrycharmaig, Glen Lochay, near Killin, Tayside Region, a small body of metamorphosed and altered ultramafic rocks contains chromite-rich lithologies in which accessory amounts of galena, gersdorffite, heazlewoodite, nickeline, native silver, and a nickel–iron alloy (possibly awaruite) occur. This assemblage is closely associated with relatively rare PGMs. It is most likely that the intrusion represents a fragment of oceanic crust (Hawson & Hall 1987, Power & Pirrie 2000). Elsewhere in Scotland, chromite is reported from: ultramafic rocks on the Isle of Rum, Highland Region (Henderson & Suddaby 1971, Dunham & Wilkinson 1985, Butcher et al. 1999, Power et al. 2000a, Upton et al. 2002); in the Peridotite Series of the Cuillin Complex, Isle of Skye (Gunn & Styles 2002); in serpentinized harzburgite from Alyth, near Blairgowrie, Tayside Region (Ikin & Harmon 1983); in the Ben Lui Schist, a semi-pelitic sequence of turbiditic origin at Tyndrum, Central Region (Fortey & Smith 1986); in altered ultramafic rocks near Balmaha, Loch Lomond, Central Region – close to the Highland Boundary fault (Henderson et al. 1983).

In Strathclyde, chromite has been reported with serpentinite, between Craig Hill and Knockdaw Hill, NW of Colmonell (Heddle 1901). The locality is on Poundland Moor near Craig Hill (half way between Pinwherry and Colmonell), but limited analytical data indicate that the mineral is magnesiochromite (D. G. Anderson pers. comm. 2006). A larger spinel-group dataset from this complex, including samples from Pinbain Bridge and the Poundland locality, exhibits a wide range of intermediate compositions best described as chrome spinel (Stone et al. 1986).

An examination of whole-rock and mineral-separates from BIF rocks at Ballycoog, Goldmine River valley, near Avoca, Co. Wicklow, Ireland, revealed chromite to be a trace component of an essentially sulphide-dominated assemblage. It occurs as crystals > 200 µm across, enclosed in thin iron-rich rims and wider magnetite overgrowths. Associated primary species in the concentrates include aikinite, bornite, chalcopyrite, galena, gold–silver alloy, hammarite, hematite, ilmenite, krupkaite, magnetite, pyrite and sphalerite. An origin of the chromite from Devonian basic dykes is proposed (Ixer et al. 1990). In the Sheefry Hills, S of Lecanvey, Co. Mayo, Lower Ordovician turbiditic slates, mudstones and sandstones of the Lower Ordovician Sheefry Formation contain detrital chromite thought to have crystallized in an ophiolite complex to the north (Aherne et al. 1992, Thompson et al. 1992).

chrysoberyl

orthorhombic $BeAl_2O_4$

A rare beryllium–aluminium oxide found in panned concentrates of alluvium from the River Lui, which drains the southeastern part of the Cairngorm granite, 8 km W of Braemar, Grampian, Scotland. Chrysoberyl was identified as a minor component of three concentrates. Associated species included anatase, cassiterite, fergusonite, ilmenite, magnetite, monazite, rutile, xenotime and zircon (Haynes 1974, 1993, Beer & Bennett 1988). No *in situ* source of the mineralization has ever been found, although it is likely to have been an evolved granite or pegmatite.

In Ireland, chrysoberyl is noted with albite, beryl, fluorite, gadolinite, olivine (peridot), orthoclase, smoky quartz and topaz in vugs within the Tertiary Mourne Mountains granite, Co. Down (Haughton 1856, Greg 1860, Richey 1927, Burke et al. 1964). The occurrence requires verification.

Chrysoberyl is also reported from the Meldon aplite quarry, near Okehampton, Devon, together with a list of other new species to the region (Kingsbury 1964b, 1965b). Unfortunately, many specimens in the Kingsbury collection (at the NHM) have been show to have incorrect provenance (Ryback et al. 1998, 2001). Later work on the aplite did not confirm the species (von Knorring & Condliffe 1984).

chrysocolla

orthorhombic $(Cu,Al)_2H_2Si_2O_5(OH)_4 \cdot nH_2O$

A common hydrated copper–aluminium–hydrogen silicate hydroxide found in the oxidized portions of many copper deposits. At Penberthy Croft mine, St Hilary, Cornwall, chrysocolla forms massive earthy cellular-to-botryoidal crusts up to 4 cm acorss covering other species in cavities within quartz veinstone. Its colour is very variable, but is often either blue, bluish-green, green or light-purple to black. Chrysocolla is difficult to distinguish from woodwardite (Camm & Merry 1991b, Betterton et al. 2000). Veins of massive blue to bluish-green chrysocolla occur at Ting Tang mine, Carharrack. Azurite, chalcocite, chalcopyrite, native copper, cuprite, malachite, melaconite, uraninite (pitchblende) and uranite (autunite or torbernite) are associated species at this mine (Dines 1956, Bruce & Aubrey-Jones 1998). Green-to-blue porcellanous masses of chrysocolla have been found in quartz gossan matrix from Wheal Gorland, St Day, 2 km E of Redruth (Rust 1981b). Blue to bluish-green crusts of chrysocolla on gossany aurichalcite and hemimorphite-bearing matrix are also noted from Greystone quarry, Lezant, near Launceston (Elton et al. 2004). Elsewhere in Cornwall, chrysocolla is noted from: Craddock Moor mine, St Cleer; Sharptor mine, Linkinhorne; Ghostcroft mine, Mullion (all Golley & Williams 1995); the Carn Brea mining field, near Redruth (Robson 1945a, Dines 1956); Old Gunnislake mine, Calstock, near Gunnislake (Braithwaite & Ryback 1994); Phoenix United mines, Linkinhorne (Dines 1956); Prince George mine, Gwinear and Wheal Edward, St Just (Greg & Lettsom 1858); West Caradon mine, St Cleer, Liskeard (Lawrence 1990); West Wheal Jewel, St Day (Collins 1892); and pseudomorphing chalcophyllite at Wheal Unity, Gwennap (Embrey & Symes 1987).

Chrysocolla commonly forms green crusts at Engine Vein, West and Wood mines, and Stormy Point, all close to Alderley Edge ridge, Cheshire. At Wood mine it forms magnificent bright-green rippled crusts coating the walls of an underground level (Carlon 1979, Braithwaite 1994, Plant et al. 1999).

In the Southern Pennine orefield, chrysocolla is rare, being briefly noted at: Calton Hill quarry, Taddington, and Dene quarry, Cromford (Braithwaite 1983); Long Tor mine, Matlock Bath (Ford et al. 1993), Snelston (Isolation) mine, near Ashbourne (Jackman 1996), all Derbyshire; and in the Ecton mines, Staffordshire (Ford & Sarjeant 1964a). At Waterbank mine, Ecton, chrysocolla forms bluish masses up to several mm across in a highly oxidized chalcopyrite matrix (Rust 1995c).

Small pale-green to bluish-green masses of chrysocolla associated with cuprite and malachite are noted from Newhurst quarry, Shepshed, Leicestershire (King 1973), and from a native copper–cuprite deposit near the Precambrian/Triassic unconformity at New Cliffe Hill quarry, Stanton under Bardon, Leicestershire (Hubbard et al. 2005). Copper-rich mineralization near the same unconformity, but at Bardon Hill, Coalville, has produced deep turquoise through to bluish-green, pale-blue and bluish-grey chrysocolla in association with azurite, calcite, native copper, cuprite, malachite, palygorskite and vésigniéite (Ince 2007).

Isolated occurrences of chrysocolla are noted from: Merehead (Torr Works) quarry, Cranmore, Mendip Hills, Somerset (M. S. Rumsey & R. W. Turner pers. comms 2006); Moons Hill quarry, NE of Shepton Mallet, Mendip Hills (Kingsbury 1941); Great Sled Dale, 10 km SE of Kirkby Stephen, Yorkshire (Briscoe & Green 1989); Westernhope Old mine, S of Westgate, SW of Stanhope, Weardale, Co. Durham (Bridges & Young 1998). The last two occurrences are within the Northern Pennine orefield, as are two occurrences in Cumbria: Murton mines, Scordale, Appleby (Bridges & Young 1998) and Tynebottom mine, NW of Garrigill, near Alston (Davidson & Thomson 1951, Braithwaite 1982b, Dunham 1990, Bridges & Young 1998). Elsewhere in Cumbria, chrysocolla is reported from locations in the Caldbeck Fells: eastern flank of Balliway Rigg, higher Roughton Gill (Kingsbury & Hartley 1958, 1960, Young 1987a, Green & Briscoe 2002); Dry Smale Gill (Shaw 1970); Carrock End mine; Carrock mine, Grainsgill; Driggith mine; Potts Gill mine (all Davidson & Thomson 1951, Hartley 1984); Red Gill (Greg & Lettsom 1858, Davidson & Thomson 1951, Wolloxall 1981, Hartley 1984, Wirth 1989); Roughton Gill (Greg & Lettsom 1858, Rudler 1905, Davidson & Thomson 1951). Chrysocolla from Brae Fell, Brandy Gill, Hay Gill, Ingray Gill, Sandbed and Silver Gill mines (Hartley 1984) deserves further investigation, as the identification could be based on material unreliably provenanced by A. W. G. Kingsbury (see Ryback et al. 1998, 2001). However, Green et al. (2005b) note that chrysocolla is widespread in the Caldbeck Fells and they confirm its occurrence at Silver Gill as blue to bluish-green spherules. At Driggith and Sandbed mines, chrysocolla occurs as blue to bluish-green spherules filling cavities in copper-rich veinstone (Green et al. 2006a). Other Cumbrian occurrences of chrysocolla include: Black Scar workings, Levers Water, Coniston (Stanley 1979); Brundholme mine, Glenderaterra Valley (Goodchild 1885, Eastwood 1921); Buckbarrow Beck, Corney Fell, Waberthwaite (Young et al. 1991b, 1986); Coniston copper mines; Goldscope mine, Newlands Valley; Shap Blue quarry (all Davidson & Thomson 1951); Dale Head mine, Newlands Valley (Greg & Lettsom 1858, Davidson & Thomson 1951); Longlands Fell mine, Uldale (Young 1987a); Millrigg mine, Staveley (Lawrence et al. 1986); Rigg Head quarry, Borrowdale (Green et al. 2006c); and South Cumberland mine, near Boot, Eskdale (Young 1985e).

Chrysocolla is not particularly common in Wales, although it is noted from a few localities in the Central Wales orefield (mainly in Dyfed): sky blue with an earthy habit at Esgair Fraith mine, near Tal-y-bont, and at Eaglebrook (Nantycagl) mine, Ceulanymaesmawr (Jones & Moreton 1977, Jones 1983); bluish-green crusts at

Botryoidal growths of pale-blue chrysocolla, up to 2 mm across, associated with an earlier generation of apple green pyromorphite; Roughton Gill, Caldbeck Fells, Cumbria. (© D. I. Green)

Esgair Hir mine, Ceulanymaesmawr (Rust & Mason 1988); bluish-green masses to 3 mm replacing chalcopyrite and green octahedral pseudomorphs up to 7 mm across after cuprite at Frongoch mine, between Devil's Bridge and Pont-rhyd-y-groes (Green et al. 1996b); shades of brown through green and blue replacing chalcopyrite and as a late phase (possibly post-mining) replacing malachite and tyrolite at Gwaith-yrafon mine, SE of Tal-y-bont (Rust & Mason 1994); forming bright-blue to bluish-green and green alteration products after malachite at Llechweddhelyg mine, near Clawddmelyn, between Bont-goch and Cwmsymlog (Jones 1987); Llety-Ifan-Hên mine (Bevins 1994); infilling voids in mudstone breccia at Mynyddgorddu mine, W of Bont-goch, Ceulanymaesmawr (Clark & Rust 1993); associated with brochantite, cerussite, chalcopyrite, galena, linarite, malachite, pyromorphite and wulfenite on mine dumps near the Glog-Fawr lode, Ysbyty Ystwyth (Ryback & Saville 1967); botryoidal crusts at Dylife mine, N of Hafren Forest, Powys (Jones & Moreton 1977, Moreton 1982). Elsewhere in Wales, chrysocolla occurs at a trial level, 2 km NW of Bontddu, near Dolgellau (Saich & Rust 1987), and at Dinorwic quarry, near Llanberis, Snowdonia, both Gwynedd (Bevins 1994).

In Scotland, chrysocolla occurs in widely scattered locations and with a more varied paragenesis than elsewhere:

CENTRAL In the Ochil Hills: Airthrie (Greg & Lettsom 1858); with barite at Alva (Heddle 1901); green and earthy at Blairlogie mine (Dickie & Forster 1974); in a chalcopyrite–galena–pyrite-bearing barite vein at Burn of Sorrow mine, near Dollar (Dickie & Forster 1974). DUMFRIES & GALLOWAY Castle Douglas (Greg & Lettsom 1858); pale-blue pseudomorphs partially replacing veszelyite on the dumps of Straitstep vein, Wanlockhead (Green 1990, Green & Todd 2001); emerald-green and finely mammillated on quartz at Whyte's Cleuch (Williams 1965). HIGHLAND In a chalcopyrite-bearing vein cutting gneiss S of Rhiconich (Heddle 1901). STRATHCLYDE Associated with greenockite, natrolite and prehnite at Boyleston quarry, SW of Barrhead (Heddle 1901); associated with native copper on a single specimen, where it forms small greenish-blue platy crystals or bluish-green to olive-green blocky crystals at Hartfield Moss, 9 km WSW of Barrhead (Meikle 1990). TAYSIDE With datolite and prehnite at Glen Farg, 16 km S of Perth (Heddle 1901); the old workings at Tomnadashan, Loch Tay, are encrusted with green malachite, chrysocolla and calcite (Wilson & Flett 1921).

In Ireland, thin coatings and 1–2 mm-diameter hemispheres of light-green chrysocolla have been found at Leaghillaun, Bearhaven (Berehaven) Peninsula, Co. Kerry, within a chalcopyrite–tennantite-bearing quartz vein hosted by Devonian sandstones. The suite of supergene minerals also includes azurite, clinoclase, cornwallite, malachite, tyrolite and zálesíite (Moreton & Green 2006b). Chrysocolla is also noted from Audley mine, Co. Cork, and from the Brathay Head mines on the Isle of Man (Greg & Lettsom 1858). A specimen of chrysocolla, labelled "Knockmahon mine" (22 km SW of Waterford town), Co. Waterford, is in the Russell collection at the NHM (Moreton 2005a).

chrysolite

A variety of olivine, but the name has also been applied to prehnite and apatite.

chrysoprase

A fine-grain green variety of quartz ± moganite.

chrysotile

KAOLINITE–SERPENTINE GROUP

monoclinic or orthorhombic — $Mg_3Si_2O_5(OH)_4$

A common asbestiform magnesium silicate hydroxide that did not have species status until clinochrysotile, orthochrysotile and parachrysotile were all discredited as part of a mass discreditation of species approved by the IMA-CNMNC (Burke 2006). The three discredited species are now considered to be polytypes (crystallographic variants) of the species chrysotile, and all are found in Britain or Ireland. Chrysotile often forms mats of fibres in serpentinite (hydrothermally altered peridotite) and is reported from the Lizard ophiolite complex, Cornwall, as a pseudofibrous mix of chrysotile and lizardite in vein serpentinite (Midgley 1951, Green 1964, Power et al. 1997). More specifically, chrysotile, talc and tremolite are described from veins cutting serpentinite in a quarry to the N of Church Cove, Landewednack (Hall & Jackson 1975), and chlorite, chrysotile, magnesite, magnetite, talc and tremolite replace serpentinized peridotite at the Rill, Kynance Cove (Clark 1967).

An interbedded mixture of fibrous pale-green chrysotile and nakauriite is noted in specimens of chromitite from Hagdale quarry, Unst, Shetland, Scotland (Braithwaite & Pritchard 1983b, Livingstone 2002).

churchillite

TL Churchill, Mendip Hills, Avon, England

A synonym of mendipite.

churchite-(Y)

monoclinic — $YPO_4{\cdot}2H_2O$

TL Tretoil mine, Lanivet, 8 km NW of Lostwithiel, Cornwall, England?

Named after Arthur Herbert Church FRS (1834–1915), professor of chemistry at the Royal Agricultural College at Cirencester, Gloucestershire, and later at the Royal Academy of Arts, London; born in London. Churchite-(Y) is a very rare hydrated yttrium phosphate, first found by Richard Talling (1820–1883) in a copper-rich lode near Lostwithiel, Cornwall, although the exact locality was never revealed. Churchite-(Y) was thought to be a cerium phosphate until its true identity was established (Claringbull & Hey 1953). It consists of pale smoke-grey (with a faint tint of flesh red) subparallel aggregates of platy crystals that form a crust on a quartz matrix and is generally similar to wavellite. Under the microscope it is transparent to translucent (Church 1865).

Sir Arthur Russell long suspected that Tretoil mine, Lanivet, 8 km NW of Lostwithiel, was the source of Talling's specimens, and both he and Arthur Kingsbury searched long and hard to rediscover the species. This effort was apparently rewarded when churchite-(Y) was found in 1955 (Kingsbury 1956, 1964). Ryback et al. (2001) re-examined the new specimen and concluded that it was probably an old Talling specimen that had been misrepresented. Tretoil mine cannot therefore be confirmed as the type locality of churchite-(Y).

Since then, churchite-(Y) has been found in the Tryphena lode, at Wheal Pendarves, Camborne, Cornwall. Churchite-(Y) lines cavities in gossany veinstone composed of quartz and limonite, and forms radiating tufts of silky white crystals up to 6 mm long, sometimes on barite. One spray shows churchite-(Y) needles stained brownish-red by iron oxides. Other minerals in the oxidized part of the Tryphena lode include connellite, native copper, cuprite, malachite, olivenite, pharmacosiderite and scorodite (Barstow & Cooper 1982).

cimolite

A mixture of alunite and the 7Å polymorph of halloysite, kaolinite–serpentinite group.

cinnabar

trigonal — HgS

A rare mercury sulphide, trimorphous with metacinnabar and hypercinnabar, that is usually found in low-temperature hydrothermal veins.

In southwest England, a single grain of cinnabar has been found in panned sediments from Portgaverne beach, near Port Isaac, St Teath, Cornwall, and occasional grains have been recovered from panned concentrates at unspecified sites along the Crediton Trough to the N of Dartmoor, Devon (Camm 1995). Cinnibar also forms a scarlet powder lining 1–2 mm-diameter cavities within barite at Stancombe quarry, near Flax Bourton, Bristol, Avon, where calcite–barite veins (some with minor galena, sphalerite or pyrite) cut Lower Carboniferous limestones. The origin of the cinnabar is problematical, but is probably related to the oxidation of sphalerite or galena containing small amounts of mercury (Zachariades 1993).

Mineralized fractures in silicified barite–calcite-rich fault rock associated with the West mine boundary fault at Alderley Edge, Cheshire, contain relict traces of primary acanthite (argentite), chalcopyrite and cinnabar, and the secondary species chrysocolla, dioptase, goethite, malachite and Mn-(Fe) oxyhydroxides (Plant et al. 1999).

In the Caldbeck Fells, Cumbria, cinnabar occurs as bright-red earthy coatings in 2–5 mm-diameter voids within coarsely granular quartz from Dry Gill mine. Residual sphalerite and an earthy iron–manganese oxide are associated with the cinnabar, and the quartz is often partly encrusted with the campylite variety of mimetite. At Potts Gill barite mine, cinnabar forms thin bright-crimson to brick-red, earthy or glistening, granular coatings in small cavities in highly cellular quartz. The cinnabar is associated with a suite of supergene minerals that includes adamite, hemimorphite and philipsburgite most abundantly, but with agardite and olivenite locally common (Neall & Leppington 1994). At Driggith mine, oxidation of sphalerite has produced red-brown crusts of cinnabar (Green et al. 2006a) and, at Silver Gill, rare pale orange-to-red powdery coatings of cinnabar coat cavities in quartz veinstone (Green et al. 2005b).

Small quantities of cinnabar have been discovered in the calcite–fluorite–galena orebodies of Rutland Cavern, Heights of Abraham, Matlock Bath, Derbyshire. Cinnibar occurs in cavities, mostly in fluorite, in association with smithsonite

(Braithwaite et al. 1963b, Braithwaite 1983). Cinnabar is also noted from the spoil-heaps of Magpie mine, Sheldon, Derbyshire, almost certainly from Blende vein (Ford & Sarjeant 1964a, Ford et al. 1993).

In Leicestershire, cinnabar occurs in dolomitized Carboniferous Limestone at Staunton Harold and at Cloud Hill quarry, near Breedon on the Hill (King 1973). The species is also present in panned concentrates of alluvium from Grace Dieu Brook, Thringstone, N of Coalville (Leake 1997b).

At a Mineralogical Society general meeting held in 1968, A. W. G. Kingsbury exhibited cinnabar, calomel, metacinnabarite and native mercury from mines in the Greenhow Hill and Grassington areas of northwest Yorkshire (Kingsbury 1969). The localities were listed as Moss and Turf Pits mines, Grassington; Cockhill, Craven Cross and Peat Pasture mines, Greenhow; and in Greenhow Rake at Greenhow quarry (Dunham & Wilson 1985). However, there is some doubt about these occurrences. A preliminary re-examination of the specimens (now in the Kingsbury collection at the NHM) suggests that the majority are genuine finds, but the Craven Cross specimen is now thought to have originated in the Berehove area, Ukraine. Small amounts of supergene cinnabar, associated with F–Pb–Zn mineralization, have since been independently confirmed throughout the Pennines (Ryback et al. 2001).

At Coldstones quarry, 5 km W of Pateley Bridge, Yorkshire, cinnabar occurs in the Garnet vein as continuous crusts up to 20 mm across, in cavities of irregular shape and lined with calcite, hemimorphite and smithsonite. The rare species doyleite, otavite and prosopite are also noted from Garnet vein. Dump material from the Friarfold–Surrender and Old Rake veins, NE of Gunnerside, Swaledale, contains cinnabar in oxidized veinstone. The veins carry galena with a little sphalerite and chalcopyrite, in a barite–fluorite–(strontianite)–witherite gangue. Localities include Bunton level, Gunnerside Gill; Friarfold vein, Flincher Gill; and Brandy Bottle Incline, also in Flincher Gill (Ineson & Young 1994, Young et al. 1989, 1997).

In Cumbria, orange-red earthy patches of cinnabar have been found on a few blocks of brown cellular dry-bone smithsonite at Natrass mines, Alston, and Silver Top quarry, Hallbankgate, Brampton. At the latter locality, cinnabar forms earthy masses and coatings filling cavities up to 1 mm across, in some instances encrusting aurichalcite (Young et al. 1989, 1997, Ineson & Young 1994, Bridges & Young 1998). Drainage sampling around Broughton Moor, S of Maryport, revealed the presence of cinnabar, chalcopyrite and isolated grains of native gold (Cooper et al. 1991, Leake et al. 1997b).

In Wales, brick-red powdery spots of cinnabar are reported in limonite–quartz gossan from Bwlchglas lead mine, near Tal-y-bont, Central Wales orefield, Dyfed (Braithwaite 1982a), and at Machen quarry, 7 km ENE of Caerphilly, Mid Glamorgan, cinnabar forms brick-red earthy coatings in small cavities in galena within a fault-bounded barite–cerussite vein. Minute orange-red crystals of cinnabar associated with galena and hemimorphite are noted from a 2 mm cavity in barite at Machen quarry (Bridges 1990, Plant & Jones 1995).

Panned concentrates from a small stream at Hartside and from Woodend Burn (both tributaries of the River Clyde), Devonshaw Hill, 3 km SW of Lamington, Strathclyde, Scotland, contain cinnabar (Dawson et al. 1979). Detrital cinnabar is also identified in panned concentrates of alluvium around the Louisa mine, Glendinning, Glenshanna Burn, 13 km NNW of Langholm, Dumfries & Galloway. At the mine itself, cinnabar has been identified in thin section as inclusions within other sulphide species found in mineralized veins dominated by stibnite and sphalerite in an ankerite–quartz matrix. Other species in the veins include arsenopyrite, bournonite, calcite, cervantite, chalcopyrite, galena, pyrite, semseyite, tennantite and tetrahedrite (Gallagher et al. 1983, 1989b, Duller et al. 1997). A doubtful record of cinnabar from Knipe (Fountainhead) antimony mine, SE of New Cumnock, Dumfries & Galloway, is noted by Williams (1973), who suggests that kermesite has been misidentified.

Bright cochineal-red clusters (<0.18 mm) of cinnabar are distributed randomly in annabergite-coated galena and nickeline crystal aggregates at Hilderston mine, Bathgate Hills, Lothian. The locality is notable for its silver mineralization found in barite-(calcite)-dominated veins adjacent to Late Carboniferous basalt dykes and a dolerite sill. Other minerals reported from Hilderston include acanthite, amalgam, barytocalcite, erythrite, harmotome, magnetite, maucherite, native mercury, native silver, sphalerite and witherite (Meikle 1994, Moreton et al. 1998). Cinnabar has also been identified in chalcopyrite-rich dykes associated with the West Sgaorishal ultramafic plug, Isle of Rum, Highland Region. The dyke hosts a range of accessory minerals, which include abundant galena, possible hessite and lesser quantities of froodite, gold (electrum), keithconnite, kotulskite, paolovite, native silver, sperrylite and telluropalladinite. Grainsize of the rare species is 1–40 μm (Power et al. 2003).

The unusual Cu–(Ag–Hg) deposit at Gortdrum mine, near Tipperary, Co. Tipperary, Ireland, is one of the few localities where well crystallized cinnabar occurs.

Cinnabar crystal, 1.5 mm across, from Gortdrum mine, Co. Tipperary, Ireland. (© D. I. Green)

It is found in carbonate veinlets as lustrous red complex crystals exceptionally up to 3 mm long, scattered on dolomite (Ryback & Moreton 1991, Green & Moreton 2001). The mine predominantly worked chalcopyrite and mercury-rich tennantite ore (bornite and chalcocite at greater depth) that occurred as disseminations and veinlets within shattered massive bioclastic limestones of Carboniferous age. Small amounts of native amalgam, aragonite, azurite, cobaltite, devilline, erythrite, gortdrumite, gypsum, lavendulan, malachite, posnjakite, pyrite, sphalerite, stromeyerite, tyrolite and wittichenite are also reported from this deposit (Steed 1983, Braithwaite & Wilson 2003).

cinnamonstone

A synonym of grossular, garnet group.

citrine

A translucent yellow variety of quartz.

claudetite

monoclinic As_2O_3

A rare arsenate, dimorphous with arsenolite and found in an antimony-rich quartz vein at Wet Swine Gill, Caldbeck Fells, Cumbria. The massive white quartz vein contains an early suite of As–Fe–Sb species (arsenopyrite, berthierite and stibnite) and a later suite of antimony–lead species (fülöppite, semseyite and zinkenite). Lustrous claudetite forms colourless-to-white thin tabular crystals up to 3 mm long and also foliated masses up to 5 mm across in cavities within the quartz. Parasymplesite occurs in the same matrix as the claudetite. Compositionally, the species is an antimony-rich claudetite and its origin is attributable to supergene oxidation of primary antimony–arsenic-bearing ore minerals (Leppington & Green 1998, Neall & Green 2001, Green & Briscoe 2002).

An earlier report of claudetite from Huel Sparnon, Redruth, Cornwall (Hintze 1904) requires further confirmation.

clausthalite

cubic PbSe

A rare lead selenide that forms a series with galena and is found in a gold-bearing carbonate vein cutting Middle Devonian limestones at Hope's Nose, Torquay, Devon. In a calcite-dominated gangue, the selenide assemblage includes eucairite, fischesserite, klockmannite, naumannite, penroseite, tiemannite, trüstedtite, tyrrellite and umangite. Clausthalite forms irregular patches 1–2 mm across, contains

Prismatic claudetite crystals to 2.5 mm; Wet Swine Gill, Caldbeck Fells, Cumbria. (© D. I. Green)

inclusions of penroseite and tyrrellite, and is extensively replaced by klockmannite and cerussite (Stanley et al. 1990a, Camm 1995). In another vein from this locality, clausthalite occurs with chrisstanleyite, fischesserite, silver-rich and palladium-rich gold, and two unknown selenides (Paar et al. 1998). Clausthalite, guanajuatite, tiemannite and unidentified selenide species are also found as inclusions in alluvial gold grains from the South Hams district of Devon. The selenides were deposited during a late phase of gold growth from a distinct type of solution that was saline, oxidizing and relatively low temperature (Leake et al. 1991, 1993). Vanadium-rich nodules from the Permian redbeds at Littleham Cove, Budleigh Salterton, Devon, contain clausthalite among a suite of arsenic, cobalt, copper, nickel, silver and uranium minerals (Harrison 1975, Stanley et al. 1990a, Milodowski et al. 2000). Clausthalite occurs in very minor amounts as tiny grains ($<5\,\mu m$) disseminated in a clay matrix and is most abundant in regions where uranium-silicate mineralization is developed (Milodowski et al. 2002).

In Scotland, clausthalite is briefly described from the Smid Hope Burn tributary of the River Tweed, near Tweedhope, Borders, where it has been found as 1–5 μm inclusions associated with possible palladseite, in two alluvial gold grains (Leake et al. 1993). Out of a total of 1794 alluvial gold grains examined by Leake et al. (1998) from the Leadhills area, along the Strathclyde/Dumfries & Galloway border, four grains are reported to contain clausthalite inclusions. Styles et al. (2004) report clausthalite in one sample from the Loch Ailsh alkaline complex, Assynt, Highland Region. Clausthalite with cubic morphology closely accompanies uraninite (pitchblende) in a uranium-rich carbonate vein at the margin of the Criffell granodiorite, Needle's Eye, SW of Caulkerbush, Dumfries & Galloway (Basham et al. 1989).

Clausthalite inclusions have been identified in one alluvial gold grain from Croagh Patrick, Co. Mayo, Ireland (Chapman et al. 2000).

clayite

A synonym of kaolin, and also used to describe sulpharsenite and sulphantimonite of copper and lead.

cleavelandite

A bladed variety of albite, feldspar group.

cleiophane

A colourless or white variety of sphalerite.

cliftonite

TL Clifton, near Bristol, Avon, England

Impure celestine, barite group (Greg & Lettsom 1858).

clinoatacamite

monoclinic $Cu^{2+}_2(OH)_3Cl$

A rare copper hydroxychloride, polymorphous with atacamite, botallackite and paratacamite. At New Cliffe Hill quarry, Stanton under Bardon, Leicestershire, clinoatacamite has been found on a single specimen from a native copper–cuprite deposit near the Precambrian/Triassic unconformity. It forms pseudo-octagonal emerald-green crystals up to 0.4 mm across, associated with connellite in a cavity in cuprite (Hubbard et al. 2005).

In Scotland, clinoatacamite is found associated with a copper-rich pyrite–quartz vein at Castletown mine, eastern side of Loch Gilp, Ballimore, near Lochgilphead, Strathclyde. It forms dark-green botryoidal crystal crusts and gem-like translucent bright-green euhedral crystals, and crystal aggregates up to 0.2 mm across. Associated species include anilite, botallackite and connellite. This suite of supergene copper chlorides probably formed by reaction between sea water and primary copper sulphides from the vein (Starkey & Hubbard 1998, Green & Todd 2001). Two other Scottish occurrences of clinoatacamite have been noted: Wha Taing, Burray, Orkney Isles (Starkey & Hubbard 1998), and Rubha a' Mhill, Isle of Islay, Strathclyde (Livingstone 2002).

Green crusts from the roof of a sea-level adit of the Stage lode complex in Cuan Cove, a part of Tankardstown mine, Knockmahon, 22 km SW of Waterford town, Co. Waterford, Ireland, have been shown to be a mixture of atacamite, clinoatacamite and botallackite (Braithwaite & Wilson 2001b).

clinochlore

CHLORITE GROUP

monoclinic $(Mg,Al)_6(Si,Al)_4O_{10}(OH)_8$

A very common magnesium–aluminium silicate hydroxide that forms a series with chamosite and is found in vein systems generated by low-grade regional metamorphism of basic rocks. Chlorite-group minerals have had a long and complex classification history, and Bayliss (1975) recommends that the following mineral names should be discarded as they have all been used to describe what is now known as clinochlore: delessite, leuchtenbergite, pennine (or penninite), pycnochlorite, ripidolite and sheridanite.

At Penberthy Croft mine, St Hilary, Cornwall, massive micaceous patches of iron-rich clinochlore occur in quartz veinstone with arsenopyrite, bismuthinite, dolomite, galena, siderite and minor chalcopyrite. Distinct crystals are rare and small, as clinochlore normally forms foliated granular-to-earthy aggregates (Betterton 1998, Betterton et al. 2000). Pale- to dark-green masses of clinochlore occur with galena, quartz and an amorphous black lead–manganese oxide at Greystone quarry, Lezant, near Launceston (Elton et al. 2004). Clinochlore is listed from three other Cornish locations by Golley & Williams (1995): the Lizard District, Penlee quarry (Newlyn) and Wheal Jane (Kea).

In Cumbria, well crystallized clinochlore (penninite) is reported from Coniston copper mine (Davidson & Thomson 1951) and the species is also found at the Shap granite quarry, where it occurs as aggregates of dark-green to black spherulitic masses in miarolitic cavities (Wirth & Green 2002). Minute colourless crudely hexagonal aggregates of clinochlore are common in sulphide-rich vein quartz in the opencut above Sandbed mine (Green et al. 2006a). Young (1987a) lists clinochlore in quartz veins on Crag Fell, Ennerdale and other possible clinochlore (penninite) occurrences in the Lake District. Clinochlore (penninite) pseudomorphs cordierite in a quartz–muscovite rock from Grainsgill, Caldbeck Fells (Roberts 1983).

Clinochlore is widely developed in Wales, particularly in altered basic igneous rocks. It replaces olivine and pyroxene, and infills veins and vesicles. The varieties

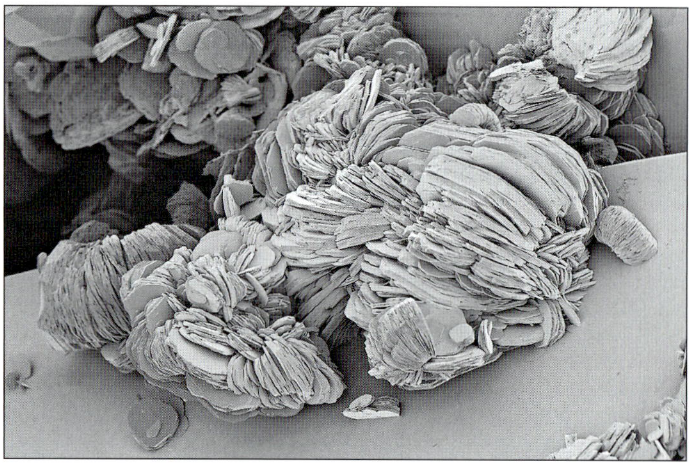

Cluster of platy clinochlore crystals on quartz. Croft Gothal mine, St Hilary, Cornwall; width 2.5 mm; SEM image. (© E. Condliffe, University of Leedsl)

pycnochlorite and ripidolite are noted by Bevins & Rowbotham (1983) and Bevins (1994). At Tanygrisiau, Blaenau Ffestiniog, Gwynedd, clinochlore commonly forms interlocking subspherical aggregates of platy crystals to 0.5 mm in alpine-type veins. The species also occurs as phantoms included in quartz (Green & Middleton 1996). Near the summit of Hanter Hill, Powys, pale-green platy or radiating clinochlore is found in altered gabbros accompanied by actinolite, albite, epidote and tourmaline (Holgate 1977). Clinochlore is an important component of a dolerite dyke cutting Precambrian basement at Dolyhir quarry, Old Radnor, near Presteigne, Powys (T. F. Cotterell pers. comm. 2007).

Pink, massive to indistinctly crystallized chromium-rich clinochlore (kammererite), hydromagnesite, pyroaurite and theophrastite are noted as colourful alteration products of pentlandite- and heazlewoodite-bearing chromitite from Hagdale quarry, Unst, Shetland, Scotland (Braithwaite & Pritchard 1983b). Clinochlore (pennine) forming small but "highly beautiful and perfect" apple-green crystals also occurs in chromitite from Hagdale (Heddle 1878a). A review of Scottish chloritic minerals, including varieties of clinochlore, is presented by Heddle (1879f). More recent descriptions include: an occurrence of dark-green to black clinochlore in small spherical vesicles within olivine basalt, associated with epistilbite, at Talisker Point, Isle of Skye (Green & Wood 1996); thin veins of clinochlore (penninite) and epidote cutting greywackes close to lead–zinc mineralization at Penkiln Burn, N of Newton Stewart, Dumfries & Galloway (Stone & Leake 1984); and a magnesium-rich variety of clinochlore occurs in Scourian ultramafic granulites from Drumbeg, Assynt (Sills 1982).

In Ireland, clinochlore (penninite) replaces biotite in small pegmatite dykes cutting Dalradian calc-silicate hornfels host rocks (now skarn) to the S of Bunbeg quay, E of Loughlin Island, Co. Donegal (Whitten 1951). Clinochlore is also found as colourless to light-brown hexagonal plates up to 1 cm across in veins and joints within ophicalcite (Connemara marble), which was quarried at Recess, Co. Galway (Hutchinson & Campbell Smith 1912). At Streamstown, Co. Galway, veins of colourless lamellar clinochlore are easily mistaken for mica (Ryback & Moreton 1992). Lower Palaeozoic schists close to lithium-bearing pegmatites at Aclare, near Myshall, Co. Carlow, contain clinochlore (penninite) together with andalusite, biotite, cordierite, garnet, hypersthene, muscovite, pyrite, quartz, staurolite and tourmaline (McArdle 1981, Whitworth 1992). Clinochlore encrusts anatase in quartz-lined cavities within a beach pebble collected below cliffs between Knockmahon and Bunmahon, Co. Waterford (Taylor 1996).

Clinochlore (penninite) replaces biotite in the contaminated granite of Bibette Head, Alderney, Channel Isles (Nockolds 1932).

clinochrysotile

A relatively common asbestiform magnesium silicate hydroxide found on the Lizard Peninsula, Cornwall, and in Glen Urquhart, W of Loch Ness, near Drumnadrochit, Highland Region, Scotland (Gaines et al. 1997). Clinochrysotile also forms fibrous and platy masses in altered ultramafic rocks near Balmaha, Loch Lomond, Central Region – close to the Highland Boundary fault (Henderson et al. 1983).

Burke (2006), reporting on an IMA-CNMNC-approved mass discreditation of species, states that clinochrysotile is formally discredited and that all samples are polytypes (crystallographic variants) of the species chrysotile.

clinoclase

monoclinic $Cu^{2+}_3(AsO_4)(OH)_3$

TL Wheal Gorland, St Day, 2 km E of Redruth, Cornwall, England

A rare copper arsenate hydroxide, dimorphous with gilmarite. The type locality is given as Wheal Gorland by Golley & Williams (1995) and Mandarino & Back (2004), but simply as Cornwall by Clark et al. (1995). Clinoclase is considered to be one of the most famous species from Wheal Gorland, where it occurs as deep-blue spherical crystal aggregates to over 2.5 cm in diameter. It is found in cavities within a gossan matrix, often with well developed crystals of liroconite and olivenite, and is said to be the best found anywhere in the world (Rust 1981b). There are also early reports of clinoclase from two other nearby copper mines; Wheal Unity and Ting Tang mine, in the vicinity of Carharrack (Greg & Lettsom 1858). At the latter mine in 1997 a systematic dig of the dumps believed to be related to Morcom's shaft produced a fine suite of supergene minerals (especially liroconite, olivenite, parnauite, pharmacosiderite and strashimirite) and one fan-shape thumbnail-size specimen of dark royal-blue clinoclase (Bruce & Aubrey-Jones 1998). Additionally in Cornwall, Golley & Williams (1995) list clinoclase at Wheal Muttrell (Gwennap), Providence mine (Lelant) and Wheal Hazard (St Just).

Clinoclase is briefly reported from Bedford United mines, near Tavistock, Devon, where it forms hemispherical and reniform masses (Hall 1868b).

Globular masses of deep blue clinoclase to 9 mm, showing radiating structure where broken; Bedford United mine, west of Tavistock, Devon. (© M. P. Cooper)

In the Royal Institution's former collection, now at the University of Liverpool, there is a specimen containing rosettes of clinoclase, cornubite and cornwallite on a slab of quartz. It is labelled "Fitfall Head", Shetland, but the source is probably Quendale Bay, south Mainland, Shetland (Rushton 1972).

Specimens of slate labelled "Ogwen Valley, Gwynedd, Wales" (now in the NMW) contain bright-blue friable coatings of clinoclase. The actual location is possibly Tan-y-Garth mine, near Bethesda, where a complex ore of arsenopyrite (+ pyrite, chalcopyrite and pyrrhotite) was once worked (Bevins 1994). Clinoclase occurs with aragonite, brochantite, erythrite, gypsum and jarosite in abandoned passageways of the Great Orme mine, Llandudno, Gwynedd. They are the oxidized products of primary copper mineralization (Jenkins & Johnson 1993, Dutton et al. 1994).

In Ireland, clear and pellucid clinoclase with beautiful chromatic striation of twin lamellae (seen in thin section) occurs in rhyolite from Tardree Mountain, Co. Antrim. Apatite, biotite, epidote, magnetite, sanidine, quartz and tridymite are associated species (von Lasaulx 1879). Confusingly, the name clinoclase was also used to describe triclinic feldspar at this time and it is plagioclase feldspar (andesine) that von Lasaulx is referring to. At Leaghillaun, Bearhaven (Berehaven) Peninsula, Co. Kerry, a chalcopyrite–tennantite-bearing quartz vein hosted by Devonian sandstones has produced an oxidized suite of supergene minerals that includes clinoclase as foliated to fan-shape crystal groups of individual bluish-black crystals up to 1 mm long associated with azurite, chrysocolla, cornwallite, malachite, tyrolite and zálesíite (Moreton & Green 2006b).

clinoenstatite?

PYROXENE GROUP

monoclinic $Mg_2Si_2O_6$

A rare magnesium silicate, dimorphous with enstatite, that forms a series with clinoferrosilite and is said to have been found in a xenolithic dyke on the south side of the peaks Streap and Streap Comlaidh, 7 km NW of Glennfinnan railway station, Highland Region, Scotland. The dyke is one of a swarm of camptonite and monchiquite dykes, probably of Permian age, that cut Moinian rocks in the area. Clinoenstatite (originally described as clinohypersthene with a relatively low iron content) is described as a constituent of one of the pyroxene-gneiss xenoliths (Walker & Ross 1954). This optical identification requires further confirmation.

clinohumite

HUMITE GROUP

monoclinic

$(Mg,Fe^{2+})_9(SiO_4)_4(F,OH)_2$

Twinned clinohumite crystal in clinohumite–monticellite–grossular skarn. Beinn an Dubhaich granite aureole, Kilbride, near Broadford, Isle of Skye, Highland Region, Scotland; width 9 mm; viewed in PPL (upper) and XPL (lower). (© K. Gardner)

A relatively common magnesium–iron silicate fluoride hydroxide found near Broadford, Isle of Skye, Highland Region, Scotland, where the Beinn an Dubhaich granite has intruded dolomitic limestone. Clinohumite is associated in a magnetite-rich skarn with chondrodite, cuspidine, forsterite, grossular, humite and monticellite. Small idioblastic crystals of clinohumite have also been found in forsterite–cuspidine marble near Camas Malag at the western end of this granite outcrop. Expulsion of boron- and fluorine-rich fluids from the granite-metasomatized nearby Durness Limestone and resulted in the crystallization of this skarn assemblage (Tilley 1948b, 1951).

Clinohumite, associated with diopside, spinel, phlogopite, quartz and troilite, has been found in impure marble within the Lewisian inlier at Sgiath Bheinn, S of Còsag, Glenelg, Highland Region, Scotland (Davidson 1995).

clinohypersthene

An obsolete term for clinoenstatite or clinoferrosilite, both pyroxene group.

clinojimthompsonite

monoclinic

$(Mg,Fe^{2+})_5Si_6O_{16}(OH)_2$

A very rare magnesium–iron silicate hydroxide, dimorphous with jimthompsonite and found in Lewisian amphibolite-facies ultramafic rocks from SW of Achmelvich, An Fharaid Mhòr, near Lochinver, Highland Region, Scotland. Clinojimthompsonite is the most common triple-chain pyribole at this location and it occurs with

Euhedral prisms of clinojimthompsonite from south west of Achmelvich, An Fharaid Mhòr, near Lochinver, Highland Region, Scotland. Lightest grey crystals and lamellae are cummingtonite; width 1.1 mm; BSE image. (© A. G. Tindle)

jimthompsonite and probable chesterite in complex intergrowths with actinolite, anthophyllite and cummingtonite. Individual crystals can reach 3 cm in length. The pyriboles do not pseudomorph the amphibole species and probably grew during prograde Inverian metamorphism, possibly as the result of infiltration of CO_2-rich fluid. The many chemically similar pyriboles and their disposition within compound prisms suggests that crystallization was kinetically controlled (Droop 1994).

clinoptilolite-Ca

ZEOLITE GROUP

monoclinic

$Ca_3(Si_{30}Al_6)O_{72}·20H_2O$

A relatively common hydrated calcium-dominant aluminosilicate that forms a series with members of the heulandite group and whose status changed from a species name to a series name during the zeolite reclassification of Coombs et al. (1997). Most references to clinoptilolite currently in the British and Irish literature do not give the new species names. The three exceptions are: Robert (2001), who describes clinoptilolite-Ca occurring with heulandite within vesicles in altered basalt at Craigahulliar quarry, near Portrush, Co. Antrim, Ireland; Adair (1987), who found it farther along

Lustrous clinoptilolite, 0.8 mm across, from Wheal Hazard, Botallack, St Just, Cornwall. (© D. I. Green)

the north coast of Antrim at Benanouran Head and Port Noffer; and Dyer et al. (2006), who discovered that some of the heulandite from the well known "heulandite" locality at Wheal Hazard, St Just, Cornwall, was actually clinoptilolite-Ca.

Transparent flat prisms of clinoptilolite up to 5 mm long have been found within cavities in tholeiitic basalt from Port Noffer, Giant's Causeway, Co. Antrim (Ryback et al. 1988), and, at the nearby National Trust reception centre, clinoptilolite may have been found in a fallen boulder of tholeiitic basalt that also contains erionite, paulingite and scarce phillipsite and heulandite (Foy 1992).

Livingstone (2002) reports clinoptilolite in veins cutting Palaeozoic lavas of the Midland Valley of Scotland (Strathclyde) from about 3 km W of Kilmacolm, where it is found with heulandite, calcite and quartz; and from 4 km NE of Dumbarton, where it is found with laumontite and stilbite. Earlier reports of clinoptilolite by Livingstone (1993) from lavas in the Campsie Hills, Central Region, and from amygdales in basalt from the Storr, N of Portree, Isle of Skye, are in error (A. Livingstone pers. comm. 2003).

In southeast England, zeolites of the clinoptilolite–heulandite series are reported from Jurassic, Cretaceous and Paleocene sediments, mainly in 1–20 µm fractions. They are associated with abundant montmorillonite and often with glauconite, amorphous silica or low-temperature disordered cristobalite–tridymite. Some have grown inside the chambers of foraminifera. An authigenic origin is proposed. Locations include: Brockram, Hog's Back, Merstham and Marden Castle near Godstone, all Surrey; Cholsey, Sinodun Hills, Sparsholt and Wantage, all Berkshire; Ellesborough, Halton and Shabbington Wood, all Buckinghamshire; Lower Froyle, Selborne and West Worldham, all Hampshire; Pegwell Bay and Upnor, both Kent; and Shillingford Bridge, Oxfordshire (Brown et al. 1969).

Clinoptilolite-K is the new name for the original species (clinoptilolite), but without supporting data the name cannot be generally applied to the above examples.

clinopyroxene

A general term for monoclinic pyroxene species such as augite and diopside.

clinosafflorite

monoclinic (Co,Fe,Ni)As_2

A rare cobalt–iron–nickel arsenate, dimorphous with safflorite and found in veinstone on the old dumps at Silver Glen, Alva, Ochil Hills, Central Region, Scotland. Disseminated and massive black clinosafflorite, sometimes with colloform or stellate habits, occurs in barite–dolomite gangue. Clinosafflorite commonly encrusts native silver and may itself be coated by the supergene arsenates annabergite, erythyrite and tyrolite, and rarely by conichalcite or picropharmacolite (Moreton 1996, Moreton et al. 1998, Green & Todd 2001).

Black clinosafflorite encrusting dendrites of native silver in a barite–dolomite gangue; Alva, Ochil Hills, Central Region, Scotland; width 7 mm. (© D. I. Green)

clinotyrolite

Clinotyrolite is described from Tynagh mine, near Killimor, Co. Galway, Ireland, where it occurs in an unconsolidated partially oxidized cap enriched in other supergene minerals above a primary sulphide orebody consisting predominantly of galena, pyrite and sphalerite (Ryback 1988a, Ryback & Moreton 1992). Work on the crystal chemistry of tyrolite confirms the mineral has monoclinic symmetry and two polytypes, -1*M* and -2*M* (Krivovichev et al. 2006). Tyrolite-1*M* is thought to be identical to clinotyrolite.

clinozoisite

EPIDOTE GROUP

monoclinic $Ca_2Al_3Si_3O_{12}(OH)$

A common calcium aluminosilicate hydroxide that forms a series with epidote and is dimorphous with zoisite. Clinozoisite is found in low- to medium-grade regionally metamorphosed igneous rocks and contact-metamorphosed calcium-rich sediments. It is often produced during the alteration of plagioclase feldspar (saussuritization).

Banded metabasite hornfels adjacent to subhorizontal sheets of cassiterite–tourmaline mineralization related to the Land's End granite at Gryll's Bunny, Botallack, St Just, Cornwall, contain an assemblage composed of clinozoisite, calcite, diopside, grossular and titanite (sphene), alternating with bands of magnetite (Jackson 1974). Clinozoisite and epidote are widely distributed in the andesitic lavas and tuffs of the Borrowdale Volcanic Series in Cumbria, where they often form in vesicles (e.g. at Eel Crags) or in veins such as the Seathwaite graphite vein, where clinozoisite forms needles up to 800 µm long (Strens 1964, 1965). The Carrock Fell gabbroic complex, Grainsgill, Caldbeck Fells, Cumbria, contains clinozoisite and rutile associated with ilmenite (Ewart 1962) and at Scawgill Bridge, 4 km W of Braithwaite, also in Cumbria, clinozoisite is found in an augite–phyric microdiorite, as an alteration product of plagioclase (Fortey et al. 1994). In Leicestershire, clinozoisite is reported in accessory amounts in granodiorite from the Mountsorrel area (e.g. Brazil Wood, Kinchley, the Main Mountsorrel quarry and Swithland reservoir). It is a secondary species that formed with calcite, chlorite, epidote, limonite, white mica, pyrite and zoisite (Taylor 1934).

At Bwlch y Ddeufaen, Conwy Valley, Snowdonia, Gwynedd, clinozoisite has been found with calcite, chlorite, prehnite, pumpellyite and zoisite in vesicles, and as an alteration product of plagioclase feldspar in the gabbroid and marginal portions of a dolerite intrusion (Jenkins & Ball 1964). Bevins (1994) reports clinozoisite in metabasites from the Myndd Preseli area of Dyfed and suggests that clinozoisite is much more common in Wales than the literature implies, as much of the so-called epidote is in fact clinozoisite.

In Scotland, clinozoisite is reported from a diverse range of occurrences. At Cnoc an t-Sidhean, near Stoer, Highland Region, it is found in leucocratic epidote–feldspar–hornblende–quartz gneisses containing biotite–scapolite augen (Cartwright et al. 1985) and at Meall Mheinnidh, to the NE of Loch Maree, Highland Region, bladed clinozoisite, chlorite and margarite occur in ellipsoidal bodies of plagioclase (anorthosite xenoliths) within metabasalts of Lewisian age (Cattell & Williams 1988). Symplectite-bearing nodules in the Ardgour marble at Coire Dubh, near Corran ferry, Highland Region, are noted to contain clinozoisite and 23 other minerals, including anorthite, augite, meionite, prehnite, vesuvianite and zoisite (Drever 1936). Clinozoisite in hornblendic psammites of Moinian age at Allt Bhlaraidh, Glen Moriston, Highland Region, forms subhedral prisms and vermicular intergrowths with feldspar and quartz (Rock 1984). At Corrie Loch Kander, near Braemar, Grampian, clinozoisite is a minor constituent of sulphidic calc-silicate-rich quartz rocks of Dalradian age and is associated with armenite, barite, chalcopyrite, galena, hyalophane, pyrite, pyrrhotite, salitic pyroxene, titanite and tremolite–actinolite (Fortey et al. 1991). In the Dorback granite area, Braes of Abernethy, 9 km W of Tomintoul, Grampian, clinozoisite occurs with epidote, diopside, grossular, brown mica, vesuvianite (idocrase), wollastonite and zoisite in metamorphosed limestones at the margins of the intrusion (Walker & Davidson 1935). At Kinnaird Head, Fraserburgh,

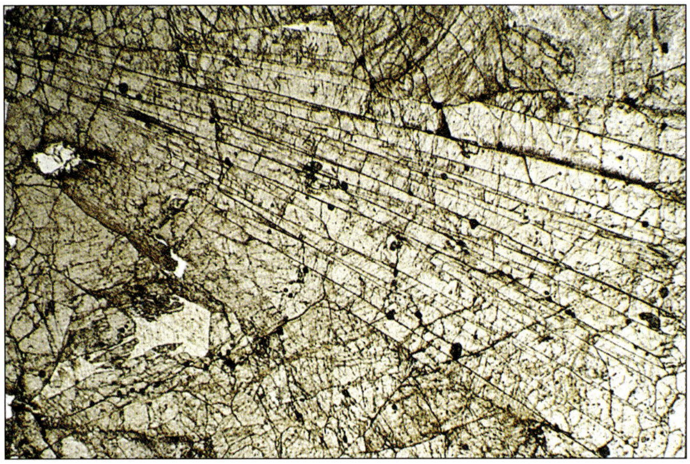

Radiating spray of clinozoisite from Pitscurrie quarry, Pitcaple, north west of Inverurie, Grampian Region, Scotland; width 7.5 mm; viewed in PPL (upper) and XPL (lower). (© K. Gardner)

Grampian, high abundances of clinozoisite, diopside ±calcic amphibole occur in a metamorphosed calc-silicate assemblage adjacent to garnetiferous aplites and pegmatites (Johnson et al. 2000). Quartz-bearing graphitic schists of Dalradian age near the River Tilt at Blair Atholl, Tayside Region, contain clinozoisite together with andesine-labradorite, biotite, chlorite, garnet and margarite (Chinner 1974) and adjacent to the Foss stratiform barite–base-metal deposit, near Aberfeldy, Tayside, Dalradian metasediments contain clinozoisite and epidote in solid solution within calcareous mica schists (Moles 1985). Highly altered basaltic lavas of Carboniferous age at Loanhead quarry, near Beith, Strathclyde, contain clinozoisite and epidote in vesicles. The clinozoisite forms groups and sprays of fine, transparent, well terminated yellowish-green crystals up to 0.4 mm long associated with laumontite or calcite (Meikle 1989b). A specimen from Dulnain Bridge, SW of Grantown-on-Spey, Highland Region, is in the Heddle collection at the NMS and comprises well formed prisms of brown clinozoisite up to 4 cm long in a quartz matrix (Livingstone 2002).

Good crystals of greyish clinozoisite up to 5 mm long have been reported from 2.5 cm-wide veins traversing a metamorphosed basic intrusion (now amphibolite) at the top of Camaderry Mountain, between Glendasan and Glendalough, Co. Wicklow, Ireland. Additionally, the veins carry albite, prehnite, titanite (sphene) and tremolite (Johnston 1949, Ryback & Moreton 1993). The amphibolite described from Glendalough, Co. Wicklow, is probably a related occurrence, but here the clinozoisite is said to occur as an accessory constituent of the amphibolite and as a constituent of veins, where the presence of cavities has allowed minute terminated crystals to form (Thomson 1908). In the area around Barnavave, Carlingford, Co. Louth, good crystals of clinozoisite are found with calcite and hornblende in impure thermally metamorphosed limestones and limestone xenoliths enclosed in granite (Osborne 1932b, Nockolds 1944). At Ballynacarrick, Co. Donegal, a similar metamorphic process has produced a thin clinozoisite zone around limestone inclusions within granite (Gindy 1953) and, in the northern aureole of the Ardara granite complex, Co. Donegal, prisms of clinozoisite occur with apatite, biotite and iron oxides in a grey micaceous schist, again the result of thermal metamorphism (Akaad 1956).

Clinozoisite is a minor constituent of a granite pegmatite found S of Bunbeg quay, E of Loughlin Island, Co. Donegal (Whitten 1951).

clintonite

monoclinic

MICA GROUP

$CaMg_2AlAl_3SiO_{10}(OH)_2$

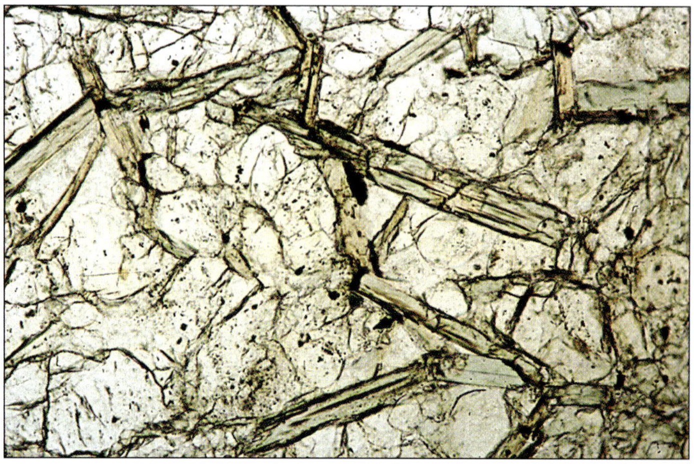

Blades of clintonite in a clintonite marble. Beinn an Dubhaich granite aureole, Camas Malag, near Broadford, Isle of Skye, Highland Region, Scotland; width 1.1 mm; viewed in PPL (upper) and XPL (lower). (© K. Gardner)

An uncommon calcium–magnesium aluminosilicate hydroxide found at Kilchrist and Camas Malag, near Broadford, on the Isle of Skye, Highland Region, Scotland, where the Beinn an Dubhaich granite has intruded dolomitic limestone and produced a diverse skarn assemblage. At Kilchrist (No. 1 prospect), the granite/marble contact contains a thin zone of clinopyroxene with dark patches rich in clintonite (xanthophyllite), grossular and spinel, separated from unaltered granite by a grossular–wollastonite assemblage (Tilley 1948b, 1949, 1951).

Reports of clintonite from hornfelses SW of Llanfaelog and graphite schists N of Bryn Ala, both Anglesey, Gwynedd, Wales, by Greenly (1919) have not been verified in the course of more recent studies (Bevins 1994).

cloustonite

TL Inganess, Mainland, Orkney, Scotland

Named after Revd Charles Clouston (1800–1884), born in Sandwick Parish, Orkney Isles, Scotland. Cloustonite is not a true mineral, but a poorly characterized jet-black organic substance.

Cl-tyretskite

TL Boulby mine, Cleveland, England

A synonym of hilgardite-1*Tc*.

cluthalite

TL Kilpatrick Hills, Dumbarton, Strathclyde, Scotland

"Clutha" is the name by which the valley of the Clyde is sometimes called. Cluthalite is a variety of analcime, zeolite group.

coalingite

trigonal $Mg_{10}Fe_2^{3+}(CO_3)(OH)_{24}\cdot 2H_2O$

A rare magnesium–iron carbonate hydroxide found on chromitite from Hagdale, Unst, Shetland, Scotland. Coalingite and pyroaurite form golden-brown micaceous flakes and plates on the chromitite (Livingstone 1993).

cobalt bloom

A synonym of erythrite, vivianite group.

COBALTITE GROUP

Cubic or pseudo-cubic sulphides with the general formula ABS, where A = Co, In, Ir, Ni, Pd, Pt, Rh, Ru; and B = As, Sb, Bi. Cobaltite, gersdorffite, hollingworthite, irarsite, micherenite, platarsite and ullmannite are members of the group found in Britain or Ireland.

cobaltite

COBALTITE GROUP

orthorhombic CoAsS

A relatively common cobalt sulpharsenide found in high-temperature hydrothermal veins, as disseminations, and as veins in contact-metamorphosed rocks. Cobaltite is very similar in composition to gersdorffite and may in some cases have been misidentified, particularly in early descriptions.

Cobaltite is described from Botallack mine, St Just, Cornwall, where it formed small particles interspersed in reddish quartz and chlorite (Greg & Lettsom 1858). In Cornwall, it has also been confusingly noted from: Dolcoath mine, Camborne; St Austell Consols; Wheal Sparnon, near Redruth; and Wherry mine, Penzance (Collins 1892, Robson 1949b, Golley & Williams 1995). Unfortunately, this is confusing because these occurrences appear to be abstracted from Greg & Lettsom (1858), who are describing skutterudite (smaltite) and not cobaltite. The situation is further complicated as Kingsbury (1964) claimed to have confirmed cobaltite from three Cornish locations: Wherry mine; Wheal Cock and near the Crowns engine houses at Botallack, both St Just; and from Wheal Friendship (Devon Friendship), Mary Tavy, Devon. These occurrences require further confirmation. The cobaltite noted from Botallack by Criddle & Stanley (1993) is based on a sample in the Kingsbury collection at the NHM and is now thought to be dubious (C. J. Stanley pers. comm. 2004). Wolloxall (1989) and Bancroft & Weller (1993) refer to the Kingsbury reference in their descriptions of the Botallack mine sett and it is possible that reference to cobaltite from Botallack and Dolcoath mines (Hosking 1951) and from Wheal Friendship (Braithwaite 1982b) are also based on the problematical earlier work. Two occurrences of cobaltite in Cornwall that do not appear to be contentious are from Fowey Consols mine, Tywardreath, St Blazey (Lewis 1997), and South Terras mine, St Stephen-in-Brannel (Gregory 1945). Subhedral to euhedral grains of cobaltite have also been found in fluorite–quartz veinstone on the dumps of North Devon United (East Wheal Friendship) mine, Peter Tavy, Devon, within the metamorphic aureole of the Dartmoor granite. It contains inclusions of native bismuth, bismuthinite, probable cosalite, gersdorffite and an unnamed bismuth telluride, and is associated with aikinite, arsenopyrite, bismite, cassiterite, erythrite, ikunolite, parkerite, sphalerite, stannite and ullmannite (Rumsey & Savage 2005).

In Cumbria, cobaltite occurs in the Scar Crag apatite–chlorite–quartz vein on Causey Pike as small euhedra ($<50\,\mu$m) at the outer margin of alloclasite, at the boundary between alloclasite and cobalt-rich arsenopyrite, or along fractures in these two minerals. Other associated species are glaucodot and skutterudite (Ixer et al. 1979a). Although these authors worked on some material from the Kingsbury collection at the NHM, which is now thought to have dubious provenance (Ryback et al. 1998, 2001), cobaltite was actually found on material collected from the mine dumps (R. A. Ixer pers. comm. 2004). There are also serious concerns about the validity of Kingsbury material (including cobaltite) said to have been collected at Tynebottom mine, NW of Garrigill, near Alston (Ixer & Stanley 1987, Ryback et al. 1998, 2001, Dunham et al. 2001). In the Vale of Newlands, 4 km SW of Keswick, cobaltite has been found in the Dale Head North vein, an arsenopyrite–chalcopyrite–pyrite-rich quartz–chlorite vein. Cobaltite is commonly euhedral and it occurs in an assemblage also containing native bismuth, bismuthinite, galena, marcasite, pyrrhotite, rutile, sphalerite and tennantite (Stanley & Vaughan 1980). These authors examined at least one Kingsbury specimen, but most specimens were either collected *in situ* or came from mine dumps (C. J. Stanley pers. comm. 2005). A classic hydrothermal brecciated sulphide-rich chlorite–quartz vein at Seathwaite mine, Seathwaite Tarn, Coniston, contains early-formed arsenopyrite and euhedral to subhedral cobaltite (<0.5 mm) in an assemblage with a paragenetic sequence of pyrite, digenite, djurleite, wittichenite, gold, bornite, chalcopyrite, spionkopite and yarrowite (the last two species together described as blaubleibender covellite), and covellite (Stanley & Criddle 1979).

An iron-poor cobalt–nickel sulpharsenide, similar in composition to nickel-rich cobaltite, is noted by Ixer & Vaughan (1982) from the copper–lead deposits of Alderley Edge ridge, Cheshire. However, Braithwaite (1994) considers this to be an unconfirmed occurrence of cobaltite.

The multiphase mesothermal gold veins of the Dolgellau goldbelt, Gwynedd, Wales, host early arsenopyrite–cobaltite–marcasite–pyrite mineralization in which cobaltite is relatively conspicuous. At Vigra mine it forms localized ribs 1–2 cm thick, coated with bright-pink erythrite and small (<0.5 mm), euhedral, modified cubes or pyritohedra are characteristic. Chalcopyrite–cobaltite intergrowths, in which cobaltite aggregates are enclosed in a matrix of later chalcopyrite, occur commonly at Panorama and Vigra mines (Mason et al. 2002). Elsewhere in the region, cobaltite is recorded from Wnion, Ffridd-goch, Afon Gain, Cefndeuddwr, Cefn Coch and Clogau mines and trials (Gilbey 1968, Naden 1988). A description of cobaltite from Hyddgen mine, near Nant-y-moch reservoir, Central Wales orefield, Dyfed (Raybould 1974), is disputed by Bevins (1994), who considers it to be cobalt-rich pyrite. Elsewhere in Wales, cobaltite is reported from: Great Orme mine, Llandudno, Gwynedd (Ixer 2001); from Dyserth quarry, Moel Hiraddug, near Prestatyn, Clwyd (Le Neve Foster 1882, Bolton 1996); and as an inclusion species in alluvial gold from Afon Mawddach, Snowdonia (Leake et al. 1993).

In Scotland, rare cobaltite inclusions have been found in alluvial gold from: the Head of Tweed River, Borders (Leake et al. 1993); Glen Lednock, near Comrie, Tayside Region, and Shortcleugh Water, Leadhills, Strathclyde (Chapman et al. 2000c); the Mauchline Basin, NW of Cumnock, Strathclyde (Leake et al. 1997b); and Calliachar Burn, 4 km from Aberfeldy, Tayside Region (Leake et al. 1997c, Chapman et al. 2000c). At Urlar Burn, 5 km from Aberfeldy, cobaltite has been found *in situ* within quartz–sulphide ±gold veins cutting Upper Dalradian gritty quartzites, quartz–mica schists and amphibolites. A bismuth-rich variety of galena commonly occurs in the veins, together with chalcopyrite, pyrite and lesser amounts of arsenopyrite and sphalerite. Gold (electrum) and some telluride species (including altaite, benleonardite, coloradoite, melonite, tsumoite and volynskite) form inclusions in the galena (Ixer et al. 1997). On Unst, Shetland, cobaltite forms solitary euhedral grains in altered pyroxenites from Vord Hill, W of Nuda, and intergrowths with pentlandite in talc–carbonate rocks to the E of Loch of Cliff, close to the basal thrust of the ophiolite complex (Maynard 1993). Gunn & Styles (2002) indicate that at Cliff and at Harold's Grave, also on Unst, cobaltite is one of several base-metal sulphides, arsenides and antimonides with which PGMs are associated.

Heddle (1901) notes cobaltite in association with annabergite and erythrite at Alva mine, Silver Glen, Ochil Hills, Central Region. Brown (1927) describes it as a rarely found species at New Glencrieff mine, Wanlockhead, Dumfries & Galloway, and this is repeated by Temple (1955). However, according to Macpherson & Livingstone (1982), these descriptions are unsubstantiated. The first confirmed occurrence in Scotland is therefore from the Fore Burn igneous complex, Tairlow plantation, near Straiton, Strathclyde, where rare cobaltite is found in <30 cm-wide veins associated with arsenopyrite, chalcopyrite, pyrite and tourmaline (Allen et al. 1982, Livingstone 1993).

Considerable quantities of cobaltite, associated with erythrite, were reported from the polymetallic sulphide mineralization hosted in Lower Carboniferous Limestone at Muckross, Killarney, Co. Kerry, Ireland (Greg & Lettsom 1858, Campbell Smith 1913), but it was thought to be skutterudite (var. smaltite) by Cole (1922). Ixer (2004) clarifies the situation at Muckross mine by describing an early arsenic–cobalt phase of mineralization consisting of abundant cobaltite with cobalt-rich arsenopyrite, glaucodot and skutterudite. Cobaltite is also noted from Ross Island mine approximately 2 km to the N of Muckross. Limestone-hosted epigenetic veins at Ardtully, Co. Kerry, Ireland, contain traces of cobaltite in a bornite–chalcopyrite–idaite–tennantite ore (Kinnaird et al. 2002).

Accessory amounts of cobaltite occur in the Gortdrum Cu–(Ag–Hg) mine, near Tipperary, Co. Tipperary, Ireland, in a mainly chalcopyrite–tennantite (bornite and chalcocite at depth) orebody, comprising disseminations, veinlets, infillings and replacements in massive dolomitized and silicified Carboniferous limestones. Pyrite is common and other accessory minerals present include native amalgam, arsenopyrite, cinnabar, gortdrumite, stromeyerite and wittichenite (Morrissey et al. 1971, Steed 1983, 1986).

A cluster of sulphides in veinstone from Tankardstown mine, Bunmahon, Co.

Waterford, has been found to contain small (<60 µm) euhedral prisms and rhombs of a cobalt arsenide mineral that has the correct chemical composition for cobaltite. It replaces and grows around earlier pyrite, and is also present as inclusions in chalcopyrite. Tetrahedrite is an associated species (Moreton 2005a). However, the mineral could conceivably be low-iron glaucodot or alloclasite, and XRD confirmation is therefore required.

cobalt pentlandite/cobaltpentlandite PENTLANDITE GROUP

cubic $(Co,Fe,Ni)_9S_8$

An uncommon cobalt–iron–nickel sulphide that forms a series with pentlandite and is found at Erglodd and Loveden mines, N of Tal-y-bont, Central Wales orefield, Dyfed, Wales. Cobalt pentlandite is a minor phase in late cavity-filling chalcopyrite–galena vein-type mineralization, comprising mineralized breccia and crustiform fissure fillings, emplaced along an ENE-trending fracture system. It is invariably associated with siegenite, with which it forms distinctive trellis-like intergrowths (Mason 1997, 1998). At Esgair Hir mine, Ceulanymaesmawr, also in the Central Wales orefield, cobalt pentlandite crystallized relatively early in the paragenetic sequence and, together with bournonite, linnaeite, tetrahedrite and ullmannite, it is found as inclusions in galena (Rust & Mason 1988). Coarse siegenite crystals from Erglodd often reveal weathering patterns that are visible to the naked eye, consistent with decomposition of cobalt pentlandite lamellae. Further localities for cobalt pentlandite in the Central Wales orefield include the following mines: Brynyrarian, Esgair Fraith, Nantycagl and Ystrad Einion (J. S. Mason pers. comm. 2004).

Cobalt pentlandite is one of the less common sulphides found in gabbros belonging to the ophiolite on Unst, Shetland, Scotland (for instance in the SE, on the Mu Ness Peninsula), where it is intergrown with chalcopyrite (5–40 µm across), pyrite and pyrrhotite. From a less precisely described location on Unst, cobalt pentlandite has been noted intergrown with bornite, chalcopyrite, magnetite and pentlandite in a cumulate pyroxenite (Maynard 1993).

cobalt pyrites

A synonym of linnaeite.

coccolite

An iron-rich variety of diopside, pyroxene group.

coeruleolactite

Burke (2006), reporting on an IMA-CNMNC-approved mass discreditation of species, states that coeruleolactite is formally discredited, and compositions are actually a mixture of planerite + variscite + wavellite. Milky-white to light-blue coeruleolactite had been reported from Wheal Phoenix, Linkinhorne, Cornwall (Roberts et al. 1990).

coffinite

tetragonal $U[(SiO_4(OH)_4]$

A rare uranium silicate hydroxide often found in hydrothermal vein deposits and in unoxidized uranium–vanadium deposits, where it replaces organic material in sandstone.

In Cornwall, coffinite is reported in ore samples from Roskrow United mine, Ponsanooth, where it occurs as a replacement of colloform pitchblende (Taylor & Harrison 1958, Harrison & Taylor 1966). At Geevor mine, near Pendeen, St Just, in the aureole to the NW of the Land's End granite, three primary uranium phases are present: uraninite, pitchblende (a variety of uraninite) and coffinite. Often, the latter two minerals occur together as late-stage vein fillings in the sulphide-rich parts of lodes and particularly where the lodes are brecciated or vuggy. Colloform coffinite is found with minor chalcopyrite, but mainly occurs as crystalline overgrowths on pyrite, and in polished section is distinguishable from pitchblende only by its lower reflectivity (Darnley et al. 1965). Near by at Wheal Edward, uranium enrichment of a brecciated and hematized cassiterite-bearing vein has produced coffinite and pitchblende in chalcopyrite–pyrite-bearing veinlets with a calcite–quartz gangue (Ball et al. 1982). A coffinite–pitchblende-bearing quartz vein is also noted from Wheal Owles mine (Pockley 1964, Jackson et al. 1982). To the east, in the aureole of the Carn Brea granite, coffinite and pitchblende are noted from South Crofty tin mine, Camborne (Ball et al. 1982, LeBoutillier 1996). At South Terras mine, St Stephen-in-Brannel, in the aureole of the St Austell granite, coffinite replaces smaller scattered pitchblende botryoids associated with finely divided sulphides (Pockley 1964). In 1958 a small piece of coffinite was found at South Tresavean mine, Falmouth (Rust 1982a).

At Kingswood mine, near Buckfastleigh, Devon, a milky quartz lode was worked for copper and, in brecciated parts of the vein, arsenolite, native bismuth, chalcopyrite, galena, cobalt-rich rammelsbergite, and uraninite (pitchblende) were deposited in the fractures, with coffinite forming much of the matrix (Darnley et al. 1965). From a completely different paragenesis, coffinite occurs with algodonite, native copper, lavendulan, malachite, maucherite, modderite, nickeline, rammelsbergite and native silver, all in Permian nodular concretions from the Littleham Cove area, Budleigh Salterton, Devon (Harrison 1973, Harrison et al. 1975b, Durrance & George 1976). De-watering of Permo-Triassic strata during burial compaction is favoured as the mechanism for producing metalliferous pore fluids capable of transporting uranium and associated Ag–As–Co–Cu–Ni–V metals (Milodowski et al. 2000, 2002).

Coffinite has been found as a major constituent of black spheroids (<8 mm in diameter) scattered through gypsum beds in the Triassic Mercia Mudstone Group (Keuper Marl) at Gipsy Lane brick pit, Leicester, Leicestershire. Many of the spheroids consist of soft-grey coffinite, others also contain unidentified sulphides and arsenides (Faithfull & Hubbard 1988). The Mercia Mudstone Group has also been examined in the Knowle borehole, Solihull, West Midlands, where radioactive nodules containing coffinite cores with outer margins of chalcocite and nickeline are found (Harrison et al. 1983).

Coffinite is tentatively identified as an accessory component of iron–manganese-rich calcite-filled veins within fractured Borrowdale Volcanic Group rocks and Carboniferous Limestone examined in borehole material from the Sellafield area, Cumbria (Milodowski et al. 1998).

In Scotland, coffinite is the only important uranium mineral in a fault-related occurrence in Old Red Sandstone sediments overlying the Helmsdale granite at Ousdale, Caithness, Highland Region. Coffinite forming submicroscopic grains up to 500 µm across is epigenetic and late, rimming sulphides and hydrocarbons. It is associated with a TiO_2 mineral (probably anatase), calcite, fluorite and quartz (Pointer et al. 1989). A uranium silicate mineral was discovered in a uraninite (pitchblende)-bearing carbonate vein close to the margin of the Criffell granodiorite at Needle's Eye, SW of Caulkerbush, Dumfries & Galloway. The authors were unable to identify the species positively and they suggest that it could be either coffinite or soddyite (Basham et al. 1989).

The cores of reduction spots in red mudrocks of Permo-Triassic age from the Avoniel borehole, Belfast, contain minor uranium–vanadium mineralization, in which coffinite, roscoelite, a vanadium-rich detrital mica and a vanadium-oxide mineral have all been found. Mineralization is thought to have occurred by the mixing of meteoric and saline connate fluids during early diagenesis of the sediments (Parnell 1988). Coffinite is also thought to occur as a minor constituent at Gortdrum mine, near Tipperary, Co. Tipperary, where it occurs with brannerite and uraninite (pitchblende) in an assemblage that formed prior to the main Cu–(Ag–Hg) mineralization (Duane et al. 1986, Duane 1988).

cog-wheel ore

A synonym of bournonite.

collieite

TL Leadhills, Strathclyde, Scotland

Named after Professor John Norman Collie FRS (1859–1942), a chemist who first analyzed the mineral in 1889; born at Alderley Edge, Cheshire. Collieite had been described as a calcium-, vanadium-rich variety of pyromorphite, apatite group, but examination of a black hemispherical sample from Belton Grain vein, Wanlockhead, shows it to be mottramite (Livingstone 2002).

collophane or collophanite

A synonym of carbonate-fluorapatite, apatite group.

collyrite

A mixture of gibbsite and allophane.

colophonite

A yellowish calcium–aluminium–iron garnet.

coloradoite SPHALERITE GROUP

cubic HgTe

A rare mercury telluride found as inclusions (<100 µm) in galena within quartz–sulphide ± gold veins cutting Upper Dalradian gritty quartzites, quartz–mica schists and amphibolites along Urlar Burn, Aberfeldy, Tayside Region, Scotland. Inclusions in galena are often complex and include altaite–coloradoite–tsumoite. Other inclusion species are benleonardite, chalcopyrite, gold (electrum), hessite, melonite and volynskite. In places galena is extensively and variably altered to cerussite and limonite, together with minor acanthite, bournonite and coloradoite (Livingstone 1993, Ixer et al. 1997). Coloradoite inclusions in alluvial gold are also noted from Balquandy Burn in the Ochil Hills, Central Region (Chapman et al. 2005).

columbite

See ferrocolumbite and manganocolumbite.

compreignacite

orthorhombic $K_2(UO_2)_6O_4(OH)_6\cdot 7H_2O$

A rare hydrated potassium uranyl oxide hydroxide found at Wheal Edward and West Wheal Owles, St Just, Cornwall. The mines were mainly exploiting tin- and copper-mineralized fissure veins, but some veins also carried uranium. At Wheal Edward, compreignacite forms yellow translucent spherules up to 0.1 mm across on dark mineralized hornfels containing chalcopyrite, siderite, uraninite (pitchblende) and potassium feldspar. Small amounts of brochantite, calcite, connellite and cuprosklodowskite are also present. Compreignacite from West Wheal Owles most commonly forms thin crusts of transparent bright-yellow or golden-yellow highly reflective crystals. The crystals are small (≤ 1 mm), short, prismatic, terminated by basal pinacoids and often intergrown. Associated species include cuprosklodowskite and uranophane. All compreignacite at West Wheal Owles occurs on open faces, possibly from *in situ* alteration of uraninite, but also because of the leaching of K^+ ions by percolating meteoric waters (Elton & Hooper 1994, Elton et al. 1994b).

Compreignacite with blocky habit from West Wheal Owles, St Just, Cornwall; width 130 µm; SEM image. (© N. J. Elton)

comptonite

Named after Spencer Joshua Alwyne Compton FRS (1790–1851), second marquess of Northampton, patron of science and the arts, and member of parliament for Northampton; born at Erlestoke Park, Wiltshire. Comptonite is a variety of thomsonite (zeolite group) first found near Vesuvius, Italy, and later reported from Kilmalcolm and Port Glasgow, Strathclyde, Scotland (Greg & Lettsom 1858).

condurrite

TL Condurrow mine, Camborne, Cornwall, England

A mixture of domeykite or ß-domeykite with cuprite or melaconite.

congolite

trigonal $(Fe^{2+},Mg,Mn)_3B_7O_{13}Cl$

A rare iron–magnesium–manganese borate with chloride, dimorphous with ericaite and described from Boulby potash mine, Loftus, Cleveland, where it occurs in nodules within anhydritic shale overlying halite–sylvite rock of the third Zechstein evaporite cycle. Milne et al. (1977) refer to it as iron-boracite. Isolated, euhedral, brown, subtranslucent, blocky crystals of the mineral, exceptionally to 2 cm across, in a pink sylvinite matrix are perhaps the world's best and are referred to as ericaite by Green & Freier (1994, 1996). Later work has shown that ericaite is unstable at room temperatures, and the mineral is now considered to represent a solid-solution series between congolite and trembathite (Burns & Carpenter 1996, Schindler & Hawthorne 1998, Green 2005b).

conichalcite

ADELITE GROUP

orthorhombic $CaCu^{2+}(AsO_4)(OH)$

A rare calcium–copper arsenate hydroxide that forms three series, with austinite, calciovolborthite and cobaltaustinite, and is found in the oxidized zone of some copper deposits.

Conichalcite is found as an apple-green stain very thinly covering a pale-buff rock associated with calcite at Braids Hill, Edinburgh, Scotland (Livingstone 1993). Dendritic silver from Alva mine, Silver Glen, Ochil Hills, Central Region, is often encrusted by clinosafflorite, which in turn may be coated by the supergene arsenates, annabergite, erythrite, tyrolite and, rarely, conichalcite and picropharmacolite (Moreton 1996, Green & Todd 2001).

Conichalcite from Wheal Kendall, St Hilary and Hingston Down quarry, 2 km W of Gunnislake, Calstock, both in Cornwall (Golley & Williams 1995) and from Dolyhir quarry, Old Radnor, near Presteigne, Powys, Wales (T. F. Cotterell pers. comm. 2003) requires further verification.

Other reports of conichalcite are all problematic, as they are based on descriptions by A. W. G. Kingsbury or from specimens in the Kingsbury collection at the NHM. These occurrences are noted in various publications, but have doubtful provenance: Bedford United mines, Tavistock, Devon (Kingsbury 1954, Robson 1954); three localities in the Caldbeck Fells, namely between Driggith and Sandbed mines, Potts Gill mine and Brandy Gill (Hartley 1984, Young 1987a, Cooper & Stanley 1990) – they are possibly all part of the Kingsbury provenancing problem outlined by Ryback et al. (1998, 2001). Conichalcite is one of several species noted from Potts Gill mine that Green & Briscoe (2002) suggest need careful examination. Green et al. (2006a) report that Kingsbury specimens said to be from between Driggith and Sandbed mines are untrustworthy.

conistonite

TL Coniston, Cumbria, England

An artificial compound, $CaC_2O_4\cdot 7H_2O$, once thought to be a mineral.

connellite

hexagonal $Cu^{2+}_{19}Cl_4(SO_4)(OH)_{32}\cdot 3H_2O$

TL Wheal Providence, St Day, Gwennap, Cornwall, England

Named after Arthur Connell FRS (1794–1863), a Scottish chemist at the University of St Andrews, who first examined the mineral; born in Edinburgh. Connellite is an uncommon hydrated copper chloride sulphate hydroxide found in the oxidized parts of copper deposits. Some authors quote the type locality for connellite simply as Wheal Providence, Cornwall (Clark et al. 1995, Blackburn & Dennen 1998). Although details are scant, the TL does not appear to be Wheal Providence, Lelant (Kingsbury 1964b), and most probably the actual TL is the Wheal Providence that, along with Wheal Muttrell and Wheal Pink, was incorporated into Wheal Gorland, St Day, 2 km E of Redruth (Rust 1981a). To make matters more confusing, Kingsbury (1954) reports connellite with brochantite and chalcophyllite from another mine called Wheal Providence, near Leedstown, Gwinear. Old specimens labelled tallingite are in fact connellite (Bannister et al. 1950, Kingsbury 1954).

Many confirmed occurrences of connellite are from Cornwall. Wheal Muttrell produced fine specimens of deep-blue connellite as velvety crusts of blue needles up to 3 mm long, associated with cuprite in cavities within vein quartz (Embrey & Symes 1987). The Muttrell lode was also worked at Wheal Gorland, where beautiful azure-blue acicular crystals to 8 mm were found in massive cuprite ore, on which chalcophyllite and cornubite also occur (Rust 1981b, Embrey & Symes 1987). At Penberthy Croft mine, St Hilary, connellite forms azure-blue spherules of radiating acicular crystals up to 1 mm long and these have been found both underground and on dump material. In weathered blocks of partially cemented slate-rich quartz veinstone, connellite is associated with aragonite, birnessite, brochantite, clinochlore, devilline, dolomite, goethite, langite, siderite, woodwardite and wroewolfeite (Kingsbury 1954, Rust 1982b, Betterton 1989, Betterton et al. 2000). Connellite is also reported from locations around Botallack, St Just: Loe Warren Zawn (Elton & Hooper 1995a); Botallack mine (Church 1865a, Bannister et al. 1950, Rust 1982b, Pollard et al. 1989, Wolloxall 1989); Botallack sett, which includes Wheal Cock, Wheal Owles and Wheal Edward (Wolloxall 1988, Bancroft & Weller 1993, Elton et al. 1994b); Geevor mine, near Pendeen (Golley & Williams 1995); Levant mine (Golley & Williams 1995); Wheal Cock (Kingsbury 1954, Rust 1982b); Wheal Hen (Golley & Williams 1995). Elsewhere in Cornwall, connellite is reported from: Carharrack, St Day (Greg & Lettsom 1858, Rudler 1905); Cligga Head, Perranzabuloe (Kingsbury 1954, Rust 1982b); Countybridge quarry, Goonhilly Downs, Lizard Peninsula (Holyer 1996); Mark Valley mine, near Minions, 9 km N of Liskeard (Miers 1885, Collins 1911a, Rust 1982b, Embrey & Symes 1987); New Cook's

Radiating spray of connellite needles, 1.25 mm across; New Cliffe Hill quarry, Stanton under Bardon, Leicestershire. (© D. I. Green)

Kitchen mine (North Tincroft lode), Redruth (LeBoutillier et al. 2000); Old Gunnislake mine, Calstock (Golley & Williams 1995); Perran St George mine (Droskyn lode), near Perranporth (Rust 1982b, Pollard et al. 1990); South Carn Brea mine, near Redruth (Rust 1982b); Ting Tang mine (Morcom's shaft), Carharrack (Bruce & Aubrey-Jones 1998); Tolcarne and South Tolcarne mines, Camborne (Robson 1949b, Rust 1982b, Green 2005a); United mines, St Day, Gwennap (Bancroft & Weller 1993); Wheal Coates, St Agnes (Kingsbury 1954, Rust 1982b); Wheal Damsel, NW of Carharrack (Semmons 1885, Collins 1892); Wheal Pendarves (Tryphena lode), Camborne (Barstow & Cooper 1982); Wheal Tremayne, Gwinear, near Hayle (Kingsbury 1954, Rust 1982b); Wheal Unity, NE of St Day (Collins 1892).

Connellite is associated with brochantite and langite at Bedford United mine, near Tavistock, Devon (Watson 1987), but requires further confirmation from Engine Vein mine, Alderley Edge ridge, Cheshire (Braithwaite 1994), and from Silver Gill, Caldbeck Fells, Cumbria (Green et al. 2005b). Connellite collected at Driggith mine, Caldbeck Fells, may in fact be Cornish material left by a thoughtless collector (Green et al. 2006a). Cooper & Stanley (1990) report connellite from Old Potts Gill mine in the Caldbeck Fells, but, as this is based on a description by A. W. G. Kingsbury, the provenance of the samples warrants further investigation (see Ryback et al. 1998, 2001).

Radiating acicular groups of bright-blue connellite to 3 mm are reported from cavities in cuprite within a native copper–cuprite deposit near the Precambrian/Triassic unconformity at New Cliffe Hill quarry, Stanton under Bardon, Leicestershire (Green 2005a, Hubbard et al. 2005).

In Wales, an unusual carbon-rich variety of connellite forms crusts of flattened blue needles associated with malachite and georgeite at Sneyd's level, Britannia mine, Snowdonia, Gwynedd (Pollard et al. 1989, 1990, 1991). The richest connellite yet discovered from Wales is from Lodge Park copper trial, Tre'r-ddôl, Dyfed, where it occurs as mid- to deep-blue acicular crystals, which often form aggregates to 0.5 mm on partly corroded cuprite. A post-mining origin is proposed (Mason & Green 1996). Other mines in the Central Wales orefield reported to contain connellite include: Gwaithyrafon mine, near Cwmsymlog, Dyfed (Rust & Mason 1994); Geufron mine, near Llanidloes, Powys (Bevins 1994); and possibly Llettyhên mine, SE of Tal-y-bont, Dyfed (Bevins 1994).

Connellite occurs at three locations in Dumfries & Galloway, Scotland: as a deep-blue secondary mineral (radiating spherules to 0.05 mm in diameter, associated with atacamite) in a copper-rich stockwork and uranium-enriched veins within the aureole of the Criffell granodiorite, at Southwick cliffs, near Dalbeattie (Miller & Taylor 1966, Braithwaite & Knight 1990); as dark-blue crystals forming thin encrustations with a rippled appearance in the entrance to a sea-level adit at Colvend copper mine, near Rockcliffe, to the W of the uranium-rich veins (Seawater action on the old mine-workings is assumed to have been involved in this post-mining occurrence [Moreton & Aspen 1992].); in paratacamite-bearing vein material collected near Barncorkrie farm, Drummore, Portencorkie area (Cameron & Lawson 1979).

At Castletown mine, 4 km SSE of Lochgilphead, Strathclyde, connellite forms deep-blue spherical crystal aggregates up to 0.1 mm in diameter, associated with anilite, botallackite, clinoatacamite and vésigniéite (Starkey & Hubbard 1998, Hubbard & Green 2001). Native copper and supergene copper minerals including connellite have been found in prehnite-bearing Tertiary basalts collected from the seacliffs of Sgùrr nan Cearcall, Cuillins, Isle of Skye (Wolfe 1998).

Particularly fine brochantite–connellite–langite flowstone crusts are reported from Tankardstown copper mine, Knockmahon, 22 km SW of Waterford town, Co. Waterford, Ireland (Wolfe 1997, 1998), although only langite and brochantite are noted by Moreton (1991) and (Green & Moreton 2001). Connellite does occur in the area, but in a seacliff adit at Trawnamoe Bay, just W of Bunmahon, where it forms light-blue to cobalt-blue mammillated coatings. Brochantite, langite and lavendulan are associated species (Moreton 1991, 2005a, Ryback & Moreton 1991, Braithwaite & Wilson 2001b, Green & Moreton 2001). Occurrences in Co. Cork include: deep-blue spherules and crusts of fibrous connellite with malachite at Dhurode mine; and connellite, brochantite, devilline and langite form greenish-blue crusts on mine debris at Allihies, Bearhaven (Berehaven) Peninsula (Ryback & Moreton 1991).

cookeite

CHLORITE GROUP

monoclinic $LiAl_4(Si_3Al)O_{10}(OH)_8$

A rare hydrous lithium aluminosilicate hydroxide most often found in granite pegmatites, but in Britain found in small cavities along shear and fracture planes traversing both gold-bearing ore and host Silurian and Ordovician shales at Ogofau mine, near Pumsaint, Dyfed, Wales. It forms patchy films or encrustations on other lode minerals and may be associated with hydromuscovite (Brammall et al. 1937, Steed et al. 1976, Annels & Roberts 1989, Bevins 1994).

At Redburn mine, near Rookhope, Weardale, Co. Durham, a creamy-colour clay in siderite-lined pockets was identified by XRD as "chlorite-Ia of the cookeite type". The mineral occurs in an area of wallrock alteration adjacent to a minor fluorite–quartz–galena vein, near a dolerite dyke and is associated with other clay minerals – berthierine–chamosite, dickite and kaolinite (Smith & Hardy 1981).

cooperite

tetragonal (Pt,Pd,Ni)S

A very rare platinum–palladium–nickel sulphide, dimorphous with braggite and found as inclusions in garnet and clinopyroxene within an ultramafic (biotite–magnetite pyroxenite) intrusion at Loch Borralan, Assynt, Highland Region, Scotland. Cooperite forms rounded grains $\leq 5\ \mu m$ in diameter clustered with possible braggite. Associated veins contain sperrylite, isomertieite and stibiopalladinite, which may be intergrown with chalcopyrite, galena and hessite (Gunn & Styles 2002, Styles et al. 2004).

Cooperite has also been qualitatively identified in a layered mafic–ultramafic intrusion on the Isle of Rum. Within one of the chromitite layers of the Eastern Layered Series, cooperite is found in association with species positively or tentatively identified as: braggite, hongshiite, irarsite, isoferroplatinum, keithconnite, kotulskite, laurite, moncheite, native palladium, palladoarsenide, platarsite, native platinum, sperrylite, telargpalite and telluropalladinite. Most of these grains are between 0.2–2.0 µm in diameter, although grains >15 µm are also present. The majority are in contact with sulphide minerals such as bornite, chalcopyrite or pentlandite. The chromitite layer was formed during magmatic fractionation and crystal settling of an open-system magma chamber, modified by magma mixing and liquid immiscibility of sulphide and silicate melts (Power et al. 2000a).

COPIAPITE GROUP

Triclinic sulphates with the general formula $A^{2+}Fe_4^{3+}(SO_4)_6(OH)_2\cdot18\text{–}20H_2O$ or $B_{2/3}^{3+}Fe_4^{3+}(SO_4)_6(OH)_2\cdot20H_2O$, where A^{2+}=Ca, Cu, Fe, Mg, Zn; and B^{3+}=Al, Fe. Aluminocopiapite, copiapite and ferricopiapite are members of the group found in Britain or Ireland.

copiapite

COPIAPITE GROUP

triclinic $Fe^{2+}Fe_4^{3+}(SO_4)_6(OH)_2\cdot20H_2O$

An uncommon hydrated iron sulphate hydroxide found as a post-mining secondary mineral, where it is associated with a 1 m-thick band of pyrite found along passageways in Cae Coch mine, near Trefriw, Conwy Valley, Llanrwst, Gwynedd, Wales.

Copiapite on a decomposed pyrite nodule from the Carboniferous Coal Measures sequence, Edinburgh, Scotland; width 15 mm. (© The Trustees of the National Museums of Scotland)

Copiapite is a trace component associated with copious mounds of fibroferrite, melanterite and jarosite (Johnson et al. 1979, Jenkins & Johnson 1993, Bevins 1994). Elsewhere in Wales, small yellow copiapite crystals (formed after mining ceased) have been discovered in the Great Opencast at Parys Mountain, Anglesey (Southwood & Bevins 1995, Todhunter 2002).

At the Oughterside opencast coal site, Aspatria, 10 km N of Cockermouth, Cumbria, copiapite occurs with epsomite, gypsum, jarosite, pickeringite and rozenite as encrustations and efflorescences on joint surfaces and the walls of old workings of pyrite-rich coal seams. In the Ten Quarters seam, copiapite forms vivid-yellow efflorescences up to 1 mm thick. Globular crusts up to 40 mm across, composed of a mixture of copiapite, epsomite and gypsum, are said to be abundant. In the Slaty Seam, copiapite forms bright-yellow nodular crusts up to 2 mm thick and 10 mm across associated with pickeringite (Young & Nancarrow 1988b). Copiapite was also thought to be a component of some scrapings of white spots taken from a clear-green stalactite at Ushaw Moor colliery, Durham (Smythe 1933). An unidentified member of the copiapite group, forming pale sulphur-yellow crystals in cavities within a highly cavernous mass of skeletal quartz, is reported from Mount Wellington mine, Bissoe, Gwennap, Cornwall (Brooke 1984, Golley & Williams 1995). This is likely to be copiapite, as the NMH has a specimen of copiapite in its collection from this locality (Young & Nancarrow 1988b).

Powdery yellow efflorescences of copiapite on altered pyrite have been found at Wanlockhead, Dumfries & Galloway, Scotland, and yellow granular aggregates of copiapite with coquimbite, rozenite and voltaite infill cracks in a decomposing pyrite nodule from Slateford (Macpherson & Livingstone 1982).

copper

cubic Cu

Copper (or native copper) is a common metal found as a secondary mineral in the oxidized zone of copper deposits.

England

Avon Native copper is one of the supergene species found in cavities in barite that has formed in faulted Triassic Dolomitic Conglomerate along the beach near the Pier hotel at Clevedon, 16 km W of Bristol. Associated species (all <2 mm) include aragonite, aurichalcite, azurite, brochantite, cerussite, chalcopyrite, covellite, cuprite, galena, malachite, phosgenite and native sulphur (Starkey 1984).

Cheshire Clusters of arborescent native copper, up to 3 mm across, are reported from a barite gossan at Engine Vein mine, Alderley Edge ridge (Braithwaite 1994).

Cornwall Native copper was abundant in all parts of the Botallack sett (includes Wheal Cock, Wheal Owles and Wheal Edward), St Just, where an extensive supergene zone of copper enrichment occurred. Mossy, arborescent and sheet-like masses, and pseudomorphs after cuprite, were all found (Wolloxall 1989, Bancroft & Weller 1993). Anhedral flakes of native copper are associated with boltwoodite, cuprite and zeunerite in a supergene Cu–Pb–U assemblage at Loe Warren Zawn, near Botallack (Elton & Hooper 1995a). Native copper was particularly abundant at the Carn Brea mines, near Redruth (as dull-brown octahedra), and Wheal Music, near Porthtowan (Garby 1848; see also Robson 1945a, Rust 1982b). Sheared serpentinite belonging

Native copper associated with pale-green malachite. South Caradon mine, St Cleer, Cornwall; width 5 cm. (© Royal Cornwall Museum, Royal Institution of Cornwall)

to the Lizard ophiolite complex contains thin leaves and arborescent native copper, sometimes partially altered to cuprite at Countybridge quarry, Goonhilly Downs, Lizard Peninsula (Holyer 1996), and at the junction of serpentinite with hornblende schist at Ghostcroft mine, Mullion, very large masses were once recovered (Rudler 1905, Selwood et al. 1998). A 2.25 m-long specimen from here (weighing 635 kg) is illustrated by Embrey & Symes (1987). Power et al. (1997) report dendritic masses of native copper containing small amounts of native silver from Trevassack quarry on the Lizard Peninsula. Distorted subhedral cubic crystals of native copper, up to 1.2 mm across, have been recovered from cavities in arsenopyrite–quartz dump material at Penberthy Croft mine, St Hilary (Betterton et al. 2000). Other occurrences include: Castle-an-Dinas mine, St Columb Major (Davison 1920, Hey et al. 1938); Condurrow mine, Camborne; Par Consols, near St Blazey; Relistian mine, Gwinear; Wheal Buller, Redruth; Wheal Virgin, near St Day (all Greg & Lettsom 1858); Fowey Consols mine, Tywardreath, St Blazey (Lewis 1997); Greystone quarry, Lezant, near Launceston (Elton et al. 2004); Gunheath china-clay pit, Carthew, St Austell (Weiss 1994); Levant mine, St Just (Wolloxall 1987); Phillips Rock, Goonhilly Downs, Lizard Peninsula (Robson 1954); Phoenix United mine, Linkinhorne; Tresavean mine and Wheal Unity, both Gwennap (all Golley & Williams 1995); Pol Cornick, near Mullion, Lizard Peninsula (Clark 1970a); in the shallower levels of Ting Tang mine, Carharrack (Stephens 1940a, Ball et al. 1982, Bruce & Aubrey-Jones 1998); West Caradon mine, St Cleer, Liskeard (Lawrence 1990); Wheal Buller, near Redruth (Cooper 1991a); very common at Wheal Gorland, St Day, 2 km E of Redruth (Greg & Lettsom 1858, Rust 1981a, Fisher 1988); Wheal Pendarves (Tryphena lode), Camborne (Barstow & Cooper 1982); Wheal Providence, near Carbis Bay, St Ives (Starkey 1989); and Wheal Prudence, by Hanover Cove, St Agnes (Hosking 1951). Dines (1956) documents many other occurrences in Cornwall.

Cumbria On Alston Moor, supergene native copper has been found as irregular masses up to 2 mm across, in chalcocite–quartz veinstone on the spoilheaps at Cornriggs (Dunham 1990, Bridges & Young 1998), and as thin coatings on chalcopyrite–quartz veinstone at the outcrop of the Great Sulphur vein at Cross Gill (Bridges & Young 1998). Native copper is also noted from Sir John's mine, Tynehead, 4 km S of Garrigill (Vaughan & Ixer 1980, Dunham 1990) and two locations in the Caldbeck Fells – Driggith mine (Stanley & Vaughan 1981) and Silver Gill (Green et al. 2005b).

Derbyshire Very rare native copper forms inclusions in calcite at Riber mine, Matlock (Varvill 1959, Braithwaite 1983). Blebs of native copper associated with cuprite occur at Snelston mine, near Ashbourne, where downward-percolating mineralizing fluids precipitated copper and lead minerals above a dolomitized limestone horizon (Jackman 1996).

Devon Very pure sheets of native copper, up to 160 mm in diameter, have been found in mudstone 10 cm below the Triassic Budleigh Salterton Pebble Beds at Littleham Cove, near Budleigh Salterton. Native silver occurs as inclusions in the copper. The sheets display a complex sequence of alteration and subsequent mineral growth that includes early cuprite and tenorite, and later chalcocite and nickel arsenides (algodonite, maucherite, nickeline, rammelsbergite and possibly modderite). The mineralization is closely associated with the formation of uranium- and vanadium-rich concretions that formed during burial diagenesis (Harrison 1973, Harrison et al. 1975b, Milodowski et al. 2000, 2002). Other occurrences include: massive, mossy and dendritic forms at: Devon and Courtenay mine, Great Devon Consols mine, and Wheal Crebor, all near Tavistock (Hall 1868b); and New East Wheal Russell (South Wheal Crebor), Tavistock (Starkey 1986b).

Durham In Weardale, supergene native copper occurs as platy dendritic growths associated with ferricopiapite on the surface of fluorite at the New Firestone level of Groverake mine, near Rookhope, and as coatings of dendrites on veinstone at Westernhope Old mine, S of Westgate, SW of Stanhope (Smith 1973a, Bridges & Young 1998).

Native copper has been described from a vein within limestones of the Raisby Formation in Raisby quarry, Coxhoe (Dunham 1990), but Bridges & Young (1998) could not trace any specimens. However, they did find localized native copper in brecciated limestone in the old railway cutting S of the quarry, as irregular masses (<1 mm) with cuprite (Young 2001).

Hereford & Worcester Inclusions of native copper have been found in mineralized specimens of cuprite from Triassic Mercia Mudstone Group sandstones to the W of Moorfield farm, N of Redditch (Old et al. 1991).

Humberside Beach pebbles derived from the chalk of Speeton Cliff, near Flamborough Head, contain diffuse spots of irregular shape up to 1 cm across, consisting of native copper intergrown with cuprite, mantled by calcite intergrown with malachite and an unidentified copper chloride (Roydhouse & Green 2003).

Leicestershire Native copper is one of the oxidation products of chalcopyrite that occurs as relict skins on galena-bearing nodules found in swallow holes in intensely dolomitized limestone at the base of the Triassic Bromsgrove Sandstone in Cloud Hill quarry, Breedon on the Hill (King 1983). "A couple of hundred tons" of (mainly) native copper and cuprite were discovered in 1990/1991 in veins cutting diorite close to the Precambrian/Triassic unconformity at New Cliffe Hill quarry, Stanton under Bardon. Most of it was buried on the dumps. Algodonite, azurite, bornite, chalcocite, covelline, domeykite, malachite, tenorite, yarrowite and a large suite of secondary copper-oxysalt minerals (bobkingite is one of them) are associated species (Pearson & Jeffrey 1997, Hawthorne et al. 2002, Hubbard et al. 2005, Ixer 2005a). Copper-rich mineralization near the same unconformity, but at Bardon Hill, Coalville, has produced arborescent masses of native copper in association with azurite, chrysocolla, cuprite, dolomite, malachite and pyrolusite, and also thin veins or strings of native copper in cuprite (Ince 2007). An impressive occurrence of copper mineralization was also found in 1937 at Newhurst quarry, Shepshed, below the unconformity between Triassic rocks and the North Charnwood diorite. Dendrites, filigree masses and a 12×9 m area covered by thin sheets (largest individual sheet being 3 m long) of native copper are reported (King 1967, 1973, Ince 2005).

Somerset Native copper was found (*c.* 1920) in a limestone quarry to the E of Pepper Hill, at Lower Aisolt in the Quantock Hills (Hamilton & Lawrence 1970). Leaching of primary copper sulphides in copper–lead-enriched manganese-oxide bodies at Merehead (Torr Works) quarry, Cranmore, has produced a secondary zone of enrichment in which azurite, chalcocite, native copper, cuprite and malachite have all been found (Symes & Embrey 1977).

Staffordshire Small quantities of native copper have been found at Ecton Hill (Ford & Sarjeant 1964a, Braithwaite 1983).

Yorkshire Minute masses of native copper and cuprite have been found in sulphidic matrix on the dumps of old copper mines at Middleton Tyas, near Scotch Corner (Raistrick 1936a, Green 1987b).

Wales

Dyfed Native copper is occasionally found in the enriched zones of copper orebodies in the Central Wales orefield: dull pointed crystals up to 1 mm long and rare anhedral masses enclosed in chrysocolla, cuprite and malachite at Frongoch mine, between Devil's Bridge and Pont-rhyd-y-groes (Green et al. 1996b); forming metallic particles within cuprite masses up to 3 mm across at Gwaithyrafon mine, SE of Tal-y-bont (Rust & Mason 1994); thin dendritic aggregates, and filiform and capillary crystals intimately intergrown with cuprite, at the Lodge Park copper trial, Tre'r-ddôl (Mason & Green 1996); dendritic growths up to 2 cm long in cavities within quartz at Ystrad Einion mine, near Tal-y-bont (Mason & Rust 1997); Cwmsymlog mine (requires verification) and Eaglebrook (Nantycagl) mine, Ceulanymaesmawr (Jones & Moreton 1977, Jones 1983); and Glogfawr mine (Bevins 1994).

Gwynedd Native copper forming leaf- and moss-like aggregates is reported from the gossan associated with the Great Lode at Parys Mountain (Greenly 1919, Dewey & Eastwood 1925), and arborescent masses up to 6 cm long are noted from Rhôs-mynach mine (Bevins 1994), both on Anglesey. Copper ores at Great Orme mine, Llandudno, are the result of simple, primary chalcopyrite–pyrite mineralization in veins containing vuggy dolomite. This was followed by multiple generations of supergene enrichment that produced azurite, native copper, covellite, digenite, malachite, spionkopite and tenorite (Ixer & Davies 1996, Ixer & Stanley 1996, Ixer 2001). Arborescent native copper was once found at Llandudno mine on Great Orme's Head (Vivian 1859). Native copper is a constituent of a copper-impregnated peat deposit termed "turf copper", near Coed-y-Brenin, Dolgellau (Archer 1959, Rice & Sharp 1976, Dunham et al. 1978, Allen & Jackson 1985). Other reports include Benallt mine, Rhiw, Lleyn Peninsula (Campbell Smith 1947), and a trial level, 2 km NW of Bontddu, near Dolgellau (Saich & Rust 1987).

Powys Native copper is uncommon in the eastern part of the Central Wales orefield, being reported only from Dylife mine (Jones & Moreton 1977, Moreton 1982) and Geufron mine, near Llanidloes (Mason & Hughes 1990, Rothwell & Mason 1992). It also occurs (at micromount size) in a barite–chalcopyrite–galena–tennantite vein at Dolyhir quarry, Old Radnor, near Presteigne (Wolfe 2000).

Scotland

Central At Tyndrum, early lead–zinc mineralization was followed by uranium-rich mineralization, in which uraninite is associated with calcite, chalcopyrite, galena, pyrite, quartz and small (<15 µm) grains of acanthite (argentite), chalcocite, native copper and native silver (Pattrick 1985).

Dumfries & Galloway Native copper occurs in the Wanlockhead and Leadhills (Strathclyde) orefield (Brown 1927). Rare anthropogenic native copper is also present on dump material from the Meadowfoot smelter, Wanlockhead (Green 1987a).

Highland A complex stratiform manganese ore at Dalroy Burn, near Culloden, contains calcite and rhodochrosite hosting inclusions of native copper up to 50 µm across. The ore has a geothermal origin best explained by deposition from hot springs in a sub-lacustrine environment during Devonian times (Nicholson 1987b, 1990). An investigation of sulphide droplets, in one of the chromite bands of the Eastern Layered Series in the Rum layered intrusion, identified common chalcopyrite, magnetite and pentlandite, with lesser amounts of bornite, chalcocite, native copper, cubanite, digenite, electrum and pyrrhotite (Dunham & Wilkinson 1985, Power et al. 2000a, 2003). The basaltic seacliffs of Sgùrr nan Cearcall, marginal to Soay Sound, Isle of Skye, are in the metamorphic aureole of the Cuillin intrusive complex and this has allowed the formation of a high-temperature secondary mineral assemblage containing masses of prehnite with calcite, connellite, native copper, epidote, laumontite, pumpellyite, quartz and unidentified copper minerals (Wolfe 1998, Green & Todd 2001). At the north end of Moonen Bay, also on the Isle of Skye, native copper occurs as bright metallic laths included in chabazite within Tertiary basalt (Green & Todd 1996).

Strathclyde Boyleston quarry, SW of Barrhead, is perhaps Scotland's most notable locality for native copper, with sheets up to 16 cm across, nuggets, leaves, stringers, irregular masses and, more rarely, crystals in cavities within Clyde Plateau basalts. Hydrothermal remobilization of disseminated chalcopyrite in the basalt is the probable mechanism of formation (Heddle 1887a, 1901, Rudler 1905, Hall et al. 1989, Green & Todd 2001, Livingstone 2002). Similar occurrences are described from Hartfield Moss, 9 km WSW of Barrhead (Meikle 1990), and Loanhead quarry, near Beith, the latter producing highly aesthetic specimens containing well formed cubo-octahedral crystals of native copper in transparent prehnite (Meikle 1989b). Rare oxidized coarsely dendritic masses of native copper occur in fractures within limestone at Rubha a' Mhill, northeast Isle of Islay (Green 1988). Plates of native copper occur on basalt at Gleniffer Hills, near Paisley, and on fracture surfaces and bedding planes of the New Red Sandstone at Ballochmyle (Livingstone 2002).

Tayside Filaments of native copper occur within prehnite in lavas at Glen Farg (Heddle 1901, Livingstone 2002).

Orkney Heddle (1901) reports native copper from the west end of Burray Island.

Shetland Accessory amounts of native copper intergrown with awaruite and magnetite occur in cumulate dunite from W of Hagdale and in dunite pods from Cliff quarry, both within the Shetland ophiolite complex on Unst (Maynard 1993). Because of the associated PGMs, these rocks have been extensively studied (Prichard et al. 1988, 1994, Prichard & Tarkian 1988, Lord & Prichard 1997, Maynard et al. 1997,). Heddle (1901) reports native copper from the southwestern corner of Haroldswick on Unst and from Sandlodge mine on Mainland.

Ireland

Antrim Native copper occurs in minute amounts in amygdaloidal lavas in the Glenarm to Island Magee area (Walker 1959a). A quarry on the north side of Whitehead at the extremity of the headland is one such location (Herbert Smith et al. 1916). The largest samples have been found at Woodburn Glen, near Carrickfergus, as anhedral or euhedral dendrites, films, or wires up to 3 cm long. Associated species include analcime, calcite, gmelinite, malachite, montmorillonite, natrolite and thomsonite. Gismondine is an associate at Brushlee quarry, Ballyclare. Magheramorne quarry, near Larne, is a further locality, from where many specimens have been recovered (Francis 1972).

Galway Native copper is one of many supergene species found in the residual orebody above the primary sulphide Cu–Pb–Zn–(Ag) orebody at Tynagh mine, near Killimor (Schultz 1971).

Kerry Campbell Smith (1913) documents Thomas Pennant's 1754 visit to Muckross copper mine, Killarney, where chalcopyrite, fahlerz (could be either tennantite or tetrahedrite) and limestone impregnated with native copper were found. Spike-like sprigs of native copper implanted with cuprite octahedra are described from Ross Island mine, Killarney (Russell 1927b).

Limerick Occasional flakes of azurite, native copper and malachite have been found in dolomitized limestone within the zone of weathering of the Glen of Aherlow Cu–(Ag) deposit, near Galbally (Cameron & Romer 1970).

Mayo Rare native copper is present in a secondary oxidation zone associated with a Silurian unconformity, 8 km to the S of Charlestown. Chalcocite is the major sulphide in this zone, with both chalcopyrite and pyrite as important primary minerals (Poustie 1982).

Tipperary Gortdrum mine, near Tipperary town, is an important fault-controlled Cu–(Ag–Hg) deposit found in basal Carboniferous rocks. It is a bornite–chalcocite–chalcopyrite–tennantite orebody with an oxidation zone containing azurite, covellite, digenite, malachite and rare native copper (Steed 1986).

Tyrone Gold-rich quartz veins at Curraghinalt Burn, near Gortin, Sperrin Mountains, host a pyrite-dominated assemblage in which native copper occurs with arsenopyrite, chalcopyrite, galena, sphalerite and tennantite–tetrahedrite (Clifford 1986, Earls et al. 1989, Clifford et al. 1990, McCaffrey & Johnston 1996).

Waterford Small amounts of native copper have been found in the upper parts of chalcopyrite-enriched quartz veins in the Knockmahon area, 22 km SW of Waterford town (Moreton 1991). Traces of native copper are also noted in this area from the Stage lode complex (to east of Cuan Cove) of Tankardstown mine (Braithwaite & Wilson 2001b). Excellent specimens of native copper, labelled "Waterford" (doubtless from the Bunmahon district), are in the collection of the National Museum of Ireland, Dublin (Moreton 2005a). Kilduane mine is also known to have produced crystalline and dendritic masses of native copper (Smyth et al. 1864).

Wicklow A few fine specimens of native copper were once found in the oxidized zone of the chalcopyrite–pyrite orebody in the Vale of Avoca (Ryback & Moreton 1993). Crystallized and massive native copper was once found at Cronbane (Greg & Lettsom 1858).

Offshore territories

Isle of Man A little native copper is reported from the upper levels of the Bradda mines, N of Port Erin Bay, where chalcopyrite, silver-rich galena, pyrite and sphalerite were once worked from quartz–carbonate veins cutting Carboniferous Limestone (Lamplugh 1903, Dewey & Eastwood 1925, Starkey 1998b).

copperas

A synonym of melanterite.

copper glance

A synonym of chalcocite.

copper mica

A synonym of chalcophyllite.

copper nickel

A synonym of nickeline.

copper pyrites

A synonym of chalcopyrite.

copper uranite

A synonym of torbernite, autunite group.

copper vitriol

A synonym of chalcanthite.

coquimbite

trigonal $Fe^{3+}_2(SO_4)_3{\cdot}9H_2O$

A rare hydrated iron sulphate, dimorphous with paracoquimbite, and found in an old railway cutting, W of Stanely, near Paisley, Strathclyde, Scotland. Coquimbite forms small clusters of clear glassy crystals associated with halotrichite, pyrite, römerite and voltaite (Macpherson & Livingstone 1982). Coquimbite is one of 25 post-mining secondary minerals from the copper deposit at Parys Mountain, Anglesey, Gwynedd, Wales (Jenkins et al. 2000).

cordierite

orthorhombic $Mg_2Al_4Si_5O_{18}$

A common magnesium–aluminium silicate that forms a series with sekaninaite and is dimorphous with indialite. Cordierite is usually found in thermally metamorphosed argillaceous sediments and high-grade metamorphosed schists, gneisses and granulites, but can also occur in mafic igneous rocks and granites. It appears that some authors do not make the distinction between cordierite (magnesium dominant) and sekaninaite (iron dominant). Sekaninaite is more likely to occur in granite than in metamorphosed sediments. See sekaninaite for examples.

Coarsely crystallized cordierite, cummingtonite and anthophyllite occur in hornfelsed dolerite (greenstone) quarries at the margin of the Land's End granite, near Kenidjack Castle, and North Zawn, NE of Cape Cornwall, St Just, Cornwall. Cordierite forms clear patches containing irregular clumps of green spinel and is associated with minor andalusite, biotite, diaspore, ilmenite and sillimanite. At nearby Priest's Cove, cordierite is found in an andalusite–cordierite–corundum (sapphire) hornfels (Collins 1911a, Tilley & Flett 1930, Chinner & Fox 1974). These rocks are discussed more generally by Khan (1972) and Mitropoulos (1984). Cordierite is also noted in fine- and coarse-grain biotite granite from Cornwall and the Isles of Scilly, where it occurs as a minor magmatic component or, more abundantly, as a constituent of entrained non-igneous xenoliths. The granite occurrences include: Bodmin Moor granite at Cheesewring quarry, SW of Henwood, Linkinhorne, and De Lank quarry on Bodmin Moor (Stimac et al. 1995); Carnmenellis granite at Carnsew quarry, near Penrhyn, and Chywoon quarry, near Longdowns (Stimac et al. 1995); Land's End granite at a small quarry near Knill's monument, Carbis Bay, St Ives (Robson 1949a), at Castle-an-Dinas and Lamorna Cove (Greg & Lettsom 1858, Stimac et al. 1995), at Zennor (Robson 1954), and once extremely plentiful in the now-flooded Castle Inch pegmatite quarry (Robson 1949a,b, Hosking 1954b); St Austell granite at Gready quarry, near Luxulyan, and at Golden Point quarry, near Pentruff (Collins 1911a), at Luxulyan quarry, near Lostwithiel (Stimac et al. 1995), and at many other locations (Richardson 1923); and Scilly Isles granite at Penninis Cove, St Mary's (Stimac et al. 1995). Xenolith occurrences include: in the Carnmenellis granite at Carnsew, and at both Chywoon and Kessel Down (Halvasso) quarries, near Longdowns; in the Bodmin Moor granite at De Lank quarry; in the St Austell granite at Luxulyan quarry (all Stimac et al. 1995). Cordierite is also present in anthophyllite–kyanite-bearing granulite of the Old Lizard Head Series, N of the main Lizard gabbro in the vicinity of Trelease Mill and Polkernogo, W of St Keverne, Lizard Peninsula (Tilley 1937b).

Sector-twinned cordierite is noted in garnetiferous granite from Sweltor, Dartmoor, Devon (Brammall & Rama Rao 1936). Elsewhere on Dartmoor, cordierite occurs at: Burrator (Brammall & Harwood 1932); Pewtor (Stimac et al. 1995); and Pithill quarry, Ivybridge, where alteration to pinite is recorded (Brammall & Harwood 1932, Brammall & Rama Rao 1936). A peculiar purplish layer of cordierite occurs at the junction of micaceous xenoliths with the Meldon aplite, near Okehampton, Devon (McLintock 1923).

Cordierite is possibly present in a heavy-mineral concentrate collected from a Millstone Grit sample at the Strid, Wharfedale, Yorkshire. Anatase, garnet, ilmenite, monazite and rutile were also found (Gilligan 1919).

In the Caldbeck Fells of Cumbria, intrusion of the Skiddaw granite has produced a thermal aureole in which black slates and siltstones have been converted to andalusite–cordierite hornfels and biotite–cordierite hornfels respectively. As the contact of the related Grainsgill greisen is approached, cordierite in the hornfels is replaced by chlorite (Roberts 1983). Elsewhere in Cumbria, cordierite is noted from Force Crag mine, Braithwaite, near Keswick; Thornthwaite mine, Thornthwaite; and Threlkeld quarry, Threlkeld (Rastall 1941, 1942); and Sinen Gill, Saddleback, and Swineside, Mosedale (Rastall 1910).

Reports of cordierite in Wales are few, although ovoid spots of chlorite (+ mica) in metasediments around the Tanygrisiau microgranite, Gwynedd (Bromley 1969), and the St David's Head intrusion, Carn Llidi, Dyfed (Roach 1969), are probably pseudomorphs after cordierite. Detrital cordierite has been identified in Carboniferous sandstones to the E of the River Taff in South Wales (Heard 1922).

Cordierite is very commonly reported from Scotland. Most reports are from the metamorphic aureoles of gabbroic or granitic intrusions, although magmatic cordierite forming lustrous deep-green nodules, up to 1 cm across, has been found in the central granitic unit of the Loch Doon granite complex in Pulskaig Burn, southeastern slopes of Mullwharchar, 20 km S of Dalmellington, on the border between Strathclyde and Dumfries & Galloway (Tindle 1982). The metamorphic occurrences include the following.

Scotland

Dumfries & Galloway Penkiln Burn, N of Newton Stewart (Stone & Leake 1984).
Grampian Around the Aberhirder, Huntly and Knock intrusions, and including cordierite-norite and cordierite-granitoid migmatites (Ashworth 1976, Ashworth & Chinner 1978); Black Hill, Lochnagar (Ashworth & Chinner 1978); Cowhyte Head, 3 km E of Portsoy (Gillen 1987a); Creag nam Ban, near Ballater, Deeside (Smith & Phillips 2002); Dorback granite aureole, Braes of Abernethy, 9 km W of Tomintoul (Walker & Davidson 1935); southeastern margins of the Lochnagar granite complex (Goodman & Lappin 1996); Morven–Cabrach gabbro aureole, around Strathdon (Allen 1970); Sparcraigs, Belhelvie (Stewart 1942, Leake 1960); NE of Wood of Schivas, xenoliths in Haddo House gabbroic complex, 32 km N of Aberdeen (Herd et al. 1984); and between Rothienorman and Methlick, Ythan Valley (Read 1952).

Sector-twinned cordierite; Broadland, 3km north west of Huntly, Grampian Region, Scotland; width 2.25 mm; viewed in XPL. (© K. Gardner)

Highland Ardgour district, W of Loch Linnhe (Drever 1940); Ballachulish aureole (Pattison 1987); Coire nam Muc, Glen Scaddle, near Fort William, and Sròn na Saobhaidh, near Strontian (Ashworth & Chinner 1978); Kinnaird Head, Fraserburgh (Johnson et al. 2000); Strontian aureole, Loch Sunart (Tyler & Ashworth 1981).
Strathclyde Buchite (partially melted rock) xenolith in dolerite at Bagh an Tigh-Stoir, Craignish Peninsula (Allison 1936); ragged poikiloblasts of cordierite in Dalradian pelite and semi-pelite from the southern part of the Etive granite aureole, near Loch Awe (Moazzen et al. 2001); shale/dolerite-dyke contact McInroy's Point, SW of Gourock, Firth of Clyde (Herriot 1971); Ross of Mull granite aureole, includes sekaninaite (Wheeler et al. 2004); around the Sidhean Sluaigh dolerite intrusion, 3 km SW of Strachur, Loch Fyne (Worden et al. 1987, Smith 1965).
Tayside Comrie diorite aureole, includes sekaninaite (Tilley 1924); Glen Clova, Angus (Leake 1960).
Western Isles Cordierite is a common constituent, often forming symplectites with enstatite, in sapphirine-bearing garnet–kyanite/sillimanite–orthopyroxene granulites to the S and SW of Roineabhal, South Harris (Baba 1999, 2003, Baba et al. 2000).

Ireland

Antrim Cordierite–enstatite–hematite–magnetite–mullite–tridymite buchite (porcellanite) at the contact of an olivine-dolerite plug at Tievebulliagh, near Cushendall (Agrell & Langley 1958, Chinner & Dixon 1973); euhedral colourless prisms of cordierite in a diopside–enstatite–feldspar–hematite–ilmenite–mullite–buchite at the contact of a doleritic intrusion with Old Red Sandstone tuffaceous sedimentary rocks at Tieveragh, near Cushendall (Kitchen 1984); and Rathlin Island (Greg & Lettsom 1858).
Donegal Ardara granite aureole between Naran and Ardlougher (Akaad 1956), and 2 km SE of Portnoo harbour, near Naran (Atkin 1978).
Dublin Pinitized crystals up to 37 mm long, near Dalkey, close to the River Dodder (Greg & Lettsom 1858).
Galway Commonly pinitized cordierite in the Galway granite aureole, between Clifden and Oughterard, Connemara (Barber & Yardley 1985); Cashel gabbro aureole, N and NW of Cashel Hill, Connemara (Treloar 1981, Ahmed-Said & Leake 1990); Currywongaun and Doughruagh, Connemara (Angus & Middleton 1985); Claggan roadcut (Ryback & Moreton 1992).
Mayo Large euhedral sector-zoned porphyroblasts of cordierite, crowded with quartz inclusions, occur in Lower Palaeozoic rocks within the metamorphic aureole of the Corvock granite, SW of Croagh Patrick (Dewey 1967).

corkite

BEUDANTITE GROUP
trigonal $PbFe^{3+}_3(SO_4)(PO_4)(OH)_6$
TL Glandore iron mine, Co. Cork, Ireland

A rare lead–iron sulphate phosphate hydroxide that was initially confused with pharmacosiderite and described as beudantite – from which it differs in having PO_4 in place of AsO_4 (Spencer 1958). At the type locality, corkite forms brilliant but small translucent brown crystals on a quartz and gossan matrix (Greg & Lettsom 1858). Ryback & Moreton (1991) describe corkite from Glandore as forming aggregated minute yellow rhombs in cavities within massive goethite and also note its presence at nearby Aghatubrid Beg mine.

Minute yellowish-green acute rhombohedral (lozenge-shape) crystals of corkite forming crusts and associated with beudantite are found at Penberthy Croft mine, St Hilary, Cornwall. The species also occurs lining iron-stained cavities in slate with parnauite and malachite (Camm & Merry 1991b, Betterton et al. 2000). A second Cornish occurrence of corkite is at Wheal Ludcott, St Ive, near Liskeard, where two lead veins were exploited. Brown drusy corkite coats an iron-stained quartz matrix in association with wulfenite and pyromorphite (Green & Merry 1997). Corkite is also briefly noted from Gravel Hill mine, Perranzabuloe, and Wheal Unity, Gwennap (Golley & Williams 1995).

In Cumbria, corkite is found in micromount-size specimens from Iron Crag, near Roughton Gill mine, Caldbeck Fells (Wolfe 1998), and from debris near the High level of Mexico mine, Todd Gill. At the latter site, corkite forms dark-brown to almost black pseudo-cubic crystals up to 1.5 mm across, in association with pyromorphite (Cooper & Stanley 1990).

Very rare corkite has been discovered at Frongoch mine, between Devil's Bridge and Pont-rhyd-y-groes, Central Wales orefield, Dyfed, Wales, where it is part of the *in situ* supergene secondary assemblage. It occurs as yellow-green microcrystalline

Aggregate of acutely rhombohedral corkite. Penberthy Croft mine, St Hilary, Cornwall; width 220 μm; SEM image. (© M. Rothwell)

coatings overgrowing coarse mauve-brown pyromorphite and quartz, and also as microcrystals on thin layers of goethite encrusting pyromorphite and quartz veinstone. Near transparent corkite with a characteristic rhombic form and reaching 0.3 mm in length is also found on limonitic coatings, where the underlying colour makes it appear reddish brown (Cotterell & Todhunter 2007).

Dark-brown to yellow corkite forms small stalactitic growths to 2 mm, and thin crusts on quartz, on samples from Belton Grain vein, Wanlockhead, Dumfries & Galloway, Scotland. It is associated with goethite and pyromorphite (S. A. Rust pers. comm. 2006).

cornetite?

orthorhombic $Cu_3^{2+}(PO_4)(OH)_3$

A rare copper phosphate hydroxide that was supposedly found at Wheal Phoenix, Linkinhorne, Cornwall, by A. W. G. Kingsbury. This is one of many species considered by Ryback et al. (1998, 2001) to be of doubtful provenance.

cornubianite

A synonym of cornubite, but also applied to albite, orthoclase and a deep-brown variety of tourmaline.

cornubite

triclinic $Cu_5^{2+}(AsO_4)_2(OH)_4$

TL Wheal Carpenter, near Hayle, Gwinear, Cornwall, England

Named after Cornubia, the Roman name for Cornwall, cornubite is a rare copper-arsenate hydroxide, dimorphous with cornwallite, which is found as apple-green botryoidal patches associated with cornwallite on well crystallized quartz from the type locality. A specimen from Wheal Unity, Gwennap, Cornwall, now in the NHM collection, contains hexagonal chalcophyllite crystals on a base of dark-green cornubite and chrysocolla encrusting chalcopyrite, all lining a cavity in brecciated quartz (Embrey & Symes 1987). A further specimen in this collection (labelled chenevixite from St Day, Cornwall) carries light- to dark-green massive porcellanous cornubite with crystals of olivenite. A third museum specimen almost certainly came from Wheal Gorland, St Day, 2 km E of Redruth, and contains massive light apple-green cornubite with liroconite (Claringbull et al. 1959). These authors also identified cornubite on specimens given to them by A. W. G. Kingsbury, who stated they came from: Wheal Muttrell, Gwennap, Cornwall; Wheal Phoenix, Linkinhorne, Cornwall; Marquis shaft, Bedford United mines Tavistock, Devon; and Potts Gill barite mine, Caldbeck Fells, Cumbria. The last occurrence was also referred to by Hartley (1984), Young (1987a) and Cooper & Stanley (1990). The work of Ryback et al. (1998, 2001) indicates that the provenance of many Kingsbury samples is dubious and further investigation is desirable.

Cornubite has since been found (very rarely) at Penberthy Croft mine, St Hilary, Cornwall, where it occurs as small dull medium- to deep-green botryoidal crusts up to 1 mm across. Goethite with minor chrysocolla and olivenite are associated species (Betterton et al. 2000).

In Scotland, green globular cornubite is one of many supergene species occurring in uranium-bearing veins on the Solway coast in the Needle's Eye area, SW of Caulkerbush, Dumfries & Galloway. Other species include annabergite, boltwoodite, bismutite, connellite, erythrite, eulytite, kahlerite, lavendulan, millerite, mixite, nováčekite, pharmacosiderite, schoepite, tyrolite, uranophane, vandendriesscheite, walpurgite, wulfenite and zeunerite. Most form tiny crystals or encrustations that are of particular interest to micromounters (Braithwaite & Knight 1990, Green & Todd 2001). Rushton (1972) notes cornubite, cornwallite and clinoclase on a specimen in the University of Liverpool collection; the specimen probably came from Quendale Bay, South Mainland, Shetland.

cornwallite

monoclinic $Cu_5^{2+}(AsO_4)_2(OH)_4$

TL Wheal Gorland, St Day, 2 km E of Redruth, Cornwall, England

Named after the county of Cornwall, Cornwallite is a rare copper arsenate hydroxide, dimorphous with cornubite, found at the type locality as green cellular masses and coatings in a quartz-rich type of gossan veinstone, and rarely as botryoidal crusts coating fibrous olivenite (Rust 1981b). Cornwallite is also noted from Wheal Unity, Gwennap (Anthony et al. 2000). Near by, on the dumps of Morcom's shaft, Ting Tang mine, Carharrack, cornwallite forms dark-green botryoidal crusts (<1 mm thick) lining flat-sided cavities between quartz crystals (Bruce & Aubrey-Jones 1998). Dark-green fibrous cornwallite (with cornubite) is also described from a specimen in the NHM collection, which originally came from Wheal Carpenter, near Hayle, Gwinear (Claringbull et al. 1959). These authors also describe cornwallite on specimens given to them by A. W. G. Kingsbury, who stated that they came from Wheal Phoenix, Linkinhorne, Cornwall, and Potts Gill barite mine, Caldbeck Fells, Cumbria. The last occurrence was also referenced by Shaw (1970), Dunham et al. (1978), Hartley (1984), Young (1987a) and Cooper & Stanley (1990). Additionally, Kingsbury (1954) notes cornwallite from Old Gunnislake mine, Calstock, near Gunnislake, and Penberthy Croft mine, St Hilary (both in Cornwall), and Bedford United mines, Tavistock, Devon. The work of Ryback et al. (1998, 2001) indicates that the provenance of many Kingsbury samples is dubious and further investigation is desirable.

Cornwallite has since been found by others at Penberthy Croft, where it occurs as dark-green microcrystalline radiating spheroidal aggregates, or crusts up to 16 mm across consisting of poorly formed minute pointed crystals. Associated species are bayldonite, goethite, and rare philipsburgite and olivenite. Cornwallite is visually indistinguishable from pseudomalachite; chemical analysis is required to differentiate the species (Camm & Merry 1991b, Betterton et al. 2000).

Botryoidal cornubite lining a cavity in iron-stained quartz veinstone; Penberthy Croft mine, St Hilary, Cornwall; width 4 mm. (© D. I. Green)

Spherules of cornwallite up to 0.1 mm in diameter, from Bedford Consols mine, near Calstock, Cornwall. (© D. I. Green)

Turquoise-blue cornwallite forming drusy crusts up to 10 mm thick have been found in mine spoil (vein quartz) from a trial adit on Low Pike vein, near Potts Gill mine, Caldbeck Fells, Cumbria. Cornwallite is typically found in the oxidation zones of copper- and lead-bearing mineral deposits, and at Low Pike it is accompanied by the following supergene species: barite, bayldonite, beudantite, brochantite, cornwallite, duftite, gartrellite, malachite, mimetite, mottramite, philipsburgite and pseudomalachite (Cooper & Stanley 1990, Young et al. 1990a, Green & Briscoe 2002, Green et al. 2003b). Specimens from Sandbed mine, Caldbeck Fells, containing rich-green globular crusts of cornwallite overgrowing malachite on limonitic quartz, are in the R. J. King collection at the NMW (Green et al. 2006a).

Rushton (1972) describes cornubite, cornwallite and clinoclase on a specimen in the University of Liverpool collection. The specimen probably came from Quendale Bay, South Mainland, Shetland, Scotland. Possible cornwallite (erinite), associated with caledonite, cerussite and linarite, is also noted by Brown (1927) from the Susanna vein, Leadhills, Strathclyde, and Belton vein, Wanlockhead, Dumfries & Galloway.

At Leaghillaun, Bearhaven (Berehaven) Peninsula, Co. Kerry, Ireland, a chalcopyrite–tennantite–quartz vein hosted by Devonian sandstones has produced an oxidized suite of supergene minerals that includes cornwallite as glassy green crusts associated with azurite, chrysocolla, clinoclase, malachite, tyrolite and zálesíite (Moreton & Green 2006b).

coronadite

CRYPTOMELANE GROUP

monoclinic (pseudo-tetragonal) $Pb(Mn^{4+},Mn^{2+})_8O_{16}$

An uncommon but locally abundant lead–manganese oxide found on the upper crosscut dumps in Dry Gill (between Dry Gill and Sandbed mines), Caldbeck Fells, Cumbria. Coronadite is a silver-white metallic mineral that tarnishes to dull black and forms massive, fibrous or botryoidal clusters associated with mimetite (campylite), psilomelane and quartz. The fine grainsize of the fibres often gives coronadite the appearance of black velvet (Hartley 1959). Coronadite occurs in oxidized veinstone once worked for fluorite and galena at High Sedling mine, Cowshill, Weardale, Co. Durham. On the spoilheaps, blocks up to 20 cm across are composed of compact closely spaced aggregates of small-scale mammillated crusts often entirely of coronadite. Freshly broken surfaces are dark steel-grey. Goethite and limonite are common associates. Coronadite at Sedling and at Dry Gill is clearly of supergene origin and presumably results from the oxidation of manganese-rich ankerite or siderite in the presence of galena. Coronadite's rarity is more apparent than real, as it would probably be reported more widely if identified correctly (Young et al. 1996, Bridges & Young 1998). Coronadite and other manganese-oxide minerals accompany mendipite and a suite of rare oxychloride minerals at Merehead (Torr Works) quarry, Cranmore, Somerset (Welch et al. 2000).

In Scotland, coronadite and wad are major components of a manganese-rich vein found close to Arndilly House, Arndilly, near Craigellachie, 20 km SE of Elgin, Grampian. Hollandite, pyrolusite and possibly romanèchite are minor components of the ore. The deposit is considered to be a supergene oxide capping on stockwork-type mineralization related to the Ben Rinnes granite (Nicholson 1986, 1989b, 1992).

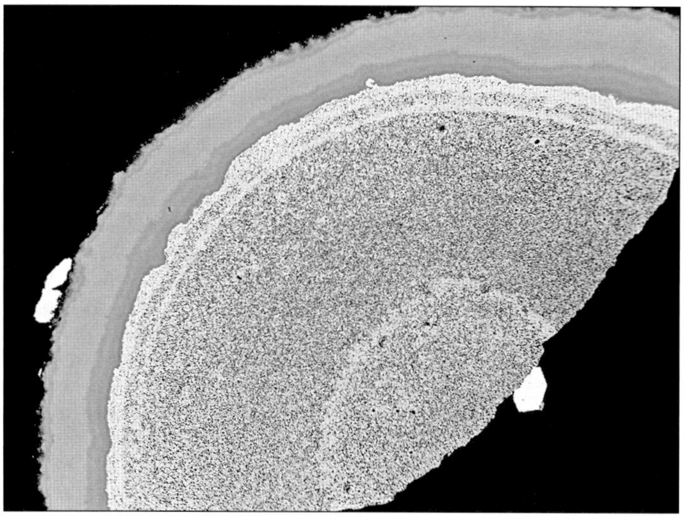

Coronadite forming a partially porous concentric growth; white crystals are pyromorphite. Old Luganure mine, Glendasan, Co. Wicklow, Ireland; width 1.2 mm; BSE image. (© A. G. Tindle)

Coronadite is one of the secondary metalliferous minerals recorded from the limonite-rich dumps of Ballygown South mine, near Silvermines, Co. Tipperary, Ireland. It forms dull dark crusts up to 1 mm thick, intermixed with hydrohetaerolite and associated with hemimorphite (Ryback & Moreton 1993, Moreton 1999). Coronadite has been found with romanèchite as mammillated and stalactitic brown-to-black coatings in a quartz vein hosted by the Leinster granite at Old Luganure mine, on the northeastern side of Camaderry, Glendasan, Co. Wicklow (Moreton et al. 2006).

corrensite

orthorhombic $(Ca,Na,K)_{1-x}(Mg,Fe,Al)_9(Si,Al)_8O_{20}(OH)_{10}\cdot nH_2O$

An uncommon hydrated calcium–sodium–potassium–magnesium–iron–aluminium aluminosilicate hydroxide clay mineral composed of a regular 1:1 interstratification of trioctahedral chlorite with either trioctahedral vermiculite or trioctahedral smectite. Corrensite is one of a few ordered interstratified clay minerals, others being brinrobertsite (pyrophyllite–smectite), rectorite (illite–smectite) and tosudite (chlorite–smectite). These minerals are generally the metastable intermediate products of a sequence of reactions involving end-member clay minerals (Dong et al. 2002).

Corrensite is present in Triassic (Rhaetic) Grey Marls at St Audries Bay, near Watchet, Somerset, and, with the mudcracks, ripplemarks and evaporites also found in these rocks, indicates formation under sabka conditions (Mayall 1979). Corrensite is noted in the clay fraction of Keuper Marl from the Wilkesley borehole, near Audlem, Cheshire (Poole & Whiteman 1966, Ryback & Tandy 1992). Corrensite is described more generally from the Cheshire Basin as a diagenetic phase in similar rocks renamed the Mercia Mudstone Group by Plant et al. (1999). Many other boreholes in England and Wales have sampled Keuper Marl and, although the locations of some of these are very imprecise, others include: Charnwood, Leicestershire; King's Norton, Birmingham; Solihull brickworks, Warwickshire; and West Burton, Nottinghamshire. In most cases, corrensite occurs in an assemblage that includes illite, carbonates, hematite and quartz (Davis 1967).

Both chlorite–vermiculite and chlorite–smectite varieties of corrensite occur in the $<2\,\mu m$ clay fraction separated from Arenig and Llanvirn pelites on the Lleyn Peninsula, Gwynedd, Wales, the former associated with the Rhiw basic intrusion and the latter in the vicinity of the more acidic Llanbedrog and Yr Eifl (near Llanaelhaern) intrusions (Merriman & Roberts 1985). These rocks had been subjected to contact metamorphism prior to regional metamorphism (Bevins 1994). In a similar study of Middle Cambrian metabentonites associated with Upper Caered and Nant Pig mudstones on St Tudwal's Peninsula, Lleyn Peninsula, Roberts & Merriman (1990) were able to identify corrensite, chlorite, kutnohorite, muscovite and smectite in the $<2\,\mu m$ clay fraction.

Corrensite fills veins in a dolerite sill at Hillhouse quarry, 7 km N of Prestwick, Strathclyde, Scotland (Nadeau et al. 1984). An examination of soil derived from a biotite-rich quartz gabbro in the Strathdon area, Grampian, by Wilson (1966) found that the natural weathering products of biotite are gibbsite, kaolinite and an aluminous chlorite–vermiculite that could possibly be corrensite.

corundophyllite

An aluminium–iron-rich variety of clinochlore, chlorite group.

corundum

HEMATITE GROUP

trigonal Al_2O_3

A relatively common aluminium oxide found in some quartz-free pegmatites and in regional or contact-metamorphosed aluminous rocks. Sapphire is the blue gem variety of corundum.

At Trencrom Hill, near St Ives, Cornwall, clots of corundum (sapphire) occur over a small area ($1 m^2$) within the Land's End granite. Corundum forms mid-blue euhedral plates, up to 4 mm across, arranged in a sheaf pattern within clusters of coarse-grain pink andalusite. The clots are thought to represent digested, very aluminous xenoliths, derived from the surrounding pelites (Wilson 1975). Small crystals of corundum (sapphire) have also been found in xenoliths and andalusite–cordierite hornfels country rocks at Priest's Cove, near Cape Cornwall, where the western margin of the Land's End granite is exposed (Tilley & Flett 1930). Corundum forms anhedral masses surrounded by cordierite in several non-igneous xenoliths collected from the St Austell granite (e.g. at Luxulyan quarry). Hercynite, ilmenite and sillimanite are associated species (Stimac et al. 1995). Similar xenoliths are also reported from the Carnmenellis granite (Jefferies 1985b). Other English occurrences include: Dartmoor granite, Cornwall (Brammall & Harwood 1925); Brandy Gill, Carrock Fell, Cumbria (Day 1928, Davidson & Thomson 1951); and possibly Shap granite, Cumbria (Rastall & Wilcockson 1915, Wirth & Green 2002).

Sheaf-like clusters of blue corundum (variety sapphire) in pale-pink to colourless andalusite. Land's End granite at Trencrom Hill, 2km south of Carbis Bay, near St Ives, Cornwall; width 5 mm; viewed in PPL. (© A. G. Tindle)

Euhedral corundum crystals (variety sapphire) in a buchite raft within basalt from Carsaig, Isle of Mull, Strathclyde, Scotland; width 3.6 mm. (© The Trustees of the National Museums of Scotland)

In Wales, royal-blue corundum crystals occur in Carboniferous sandstones and shales found to the E of the River Taff, Mid Glamorgan (Heard 1922), and the species has also been found in altered rhyolitic lavas and tuffs from a borehole near Treffgarne, Dyfed (Brown et al. 1987b).

In Scotland, corundum occurs in various igneous, xenolithic, hornfelsic and metamorphic environments. Centimetre-size crystals of corundum occurring with kyanite and staurolite in a white mica matrix are thought to represent residual material after an episode of partial melting. The assemblage is found in thin (<300 mm) quartz-free layers within a grey gneiss–paragneiss complex of Lewisian age at Stoer, NE of Achmelvich Bay, Highland Region (Cartwright et al. 1985). A 0.03 mm-diameter hexagonal crystal of corundum (sapphire) enclosed in red andalusite is described from a quartzose vein cutting schist at Clashnarae Hill, 6 km W of Lumsden, Grampian (Heddle 1891, Clough & Pollard 1899). Corundum is a constituent of a biotite–orthoclase–plagioclase–spinel hornfels from Glen Muick, Grampian (Leake 1960), and the species is particularly common in a hornfels from Tom a' Mhinn, near Crappich Hill, Glen Lednock, NW of Comrie, Tayside Region (Tilley 1924). To the NE of the Wood of Schivas, 32 km N of Aberdeen, corundum has been discovered in a zoned xenolith within the Caledonian Haddo House gabbro–norite complex. It occurs with biotite, cordierite and spinel in the core of the xenolith, surrounded by an assemblage of biotite, grandidierite, ilmenite, plagioclase, rutile, titanite and zircon (Read 1931a, Herd et al. 1984). On the north slopes of Craigmawhannel, 14 km S of Dalmellington, Strathclyde, a metasedimentary xenolith in the Loch Doon granite complex hosts corundum in a biotite-rich assemblage that also contains bottle-green and brown varieties of hercynite (Tindle 1982). The origin of the xenoliths is presumably similar to that proposed for a series of emery-like rocks that occur adjacent to a dolerite plug at Sidhean Sluaigh, Strachur, Loch Fyne, Strathclyde. They represent the aluminous residues of banded pelitic and psammitic phyllite host rocks after a granitic fraction has been selectively removed during contact metamorphism. Corundum at Sidhean Sluaigh is inclusion free, has a platy habit and is blue or bluish-green. Cordierite, magnetite, mullite, pseudobrookite, sanidine and spinel are associated species (Smith 1965). Precambrian metasediments caught up in the intrusion of lamprophyre dykes exposed on the shores of Loch Lomond contain occasional crystals of corundum, also thought to have been produced during contact metamorphism (Dempster & Bluck 1991). Perhaps the most famous corundum (sapphire) locality on mainland Scotland is Glebe Hill, 2 km NW of Kilchoan, Ardnamurchan, Highland Region, where the species occurs in colourless to deep-blue plates, up to 3 mm across, within a plagioclase–spinel hornfels that has been thermally metamorphosed by hypersthene gabbro (Richey & Thomas 1930, Wells 1951, Livingstone 2002). Corundum is also noted from the metamorphic aureoles of the following Scottish intrusions: Belhelvie ultrabasic complex, Grampian (Stewart 1946); Ben Cruachan granite, Strathclyde (Teall 1899); Coire nam Muc epidiorite complex, Ardgour, Highland Region (Drever 1940); and Lochnagar granite complex, near Braemar, Grampian (Goodman & Lappin 1996).

On the Isle of Mull, at locations such as Nuns' Pass, 2 km SW of Carsaig, and Seabank Villa, Loch Scridain, intense high-temperature metamorphism of phyllite incorporated into doleritic sills has produced xenoliths containing anorthite, corundum (sapphire), mullite and spinel. The original rocks have essentially fused to produce buchite glasses. Corundum commonly forms deep-blue tabular crystals from 1 mm up to (rarely) 15 mm across (Thomas 1922, Bowen et al. 1924, Livingstone 2002). However, the largest UK sapphire that has been cut (9.7 carats from a 39 carat crystal) came from a xenolithic monchiquite dyke at Cairisiadar (Carishader), Loch Roag, Isle of Lewis. Few crystals are less than 1 cm long and the average is 2–3 cm. Some examples show colour zoning or sector zoning, and euhedral crystals up to 2.5 cm across are reported (Upton et al. 1983, Jackson 1984, Offer 1995, Livingstone 2002).

In Ireland, corundum (sapphire) has been found with cassiterite, chalcopyrite, galena, garnet, native gold (sometimes in nuggets), hematite, ilmenite (titanoferrite), molybdenite, native platinum, pyrite, topaz, wolframite and zircon in alluvial deposits on the northeastern side of Croghan Mountain, Co. Wicklow (Forbes 1869, Ryback & Moreton 1993). In the thermal aureole of the late Caledonian Omey granite, Connemara, Co. Galway, corundum has grown within large muscovite crystals, but only in andalusite- and potassium-feldspar-bearing rocks (Ferguson & Al-Ameen 1985). Corundum occurs in thermally metamorphosed porcellanite inclusions (ferruginous bauxite and lithomarge, in which gibbsite, halloysite and hematite were the principal original constituents) within a dolerite plug at Tievebulliagh, near Cushendall, Co. Antrim. Corundum-bearing assemblages may also contain hematite, hercynite, ilmenite, magnetite, mullite or pseudobrookite (Agrell & Langley 1958). Corundum–mullite–spinel porcellanite inclusions occur close to the Brockley dolerite plugs on Rathlin Island, Co. Antrim, where corundum forms aggregates of colourless or, less commonly, deep-blue crystals up to 0.5 mm across (Wilson & Robbie 1966). Corundum also occurs in Dalradian pelites within 1 m from the contact, or in xenoliths within the Cashel gabbroic intrusion, near Toombeola, Connemara, Co. Galway. It forms fresh, euhedral crystals associated with biotite, cordierite, garnet, orthopyroxene, plagioclase and spinel (Ahmed-Said & Leake 1990).

cosalite

orthorhombic $Pb_2Bi_2S_5$

An uncommon lead–bismuth sulphide found in moderate-temperature hydrothermal deposits, in contact-metasomatic deposits and in pegmatites.

In Cumbria, cosalite has been found as anhedral <100 µm inclusions in galena in one sample from the Bonser vein, 3 km NW of Coniston, one of the most productive sources of copper in the English Lake District. Other species in the predominantly chlorite–quartz vein are arsenopyrite, native bismuth, bismuthinite, calcite, chalcopyrite, dolomite, joséite, laitakarite, magnetite, marcasite, pyrite, pyrrhotite, sphalerite and stilpnomelane (Stanley & Vaughan 1982b). Cosalite is probably present in bismuth-bearing assemblages from the Northern Pennine orefield at locations such as Redburn mine, Rookhope, Co. Durham, and Sir John's mine, 4 km S of Garrigill, Cumbria, but precise details are not recorded (Ixer et al. 1996). This is also the case for probable cosalite inclusions in galena, noted from the Teesdale Valley (Vaughan & Ixer 1980). Cosalite forming minute tufts in greisen at Carrock mine, Grainsgill, Caldbeck Fells, Cumbria, is reported by Kingsbury & Hartley (1956b). However, doubt has been expressed about the provenance of many Kingsbury specimens (Ryback et al. 1998, 2001). Reference to this occurrence is also found in Ewart (1962), Hartley (1984), Young (1987a) and Cooper & Stanley (1990). Fortunately, cosalite has since been independently confirmed from Carrock mine in a polished section, in association with bismuthinite, krupkaite and bismuth tellurides (Cooper & Stanley 1990). Since then, cosalite has been found in Cumbria as steel-grey metallic masses in joints within the Shap granite (Wirth & Green 2002).

Scattered small laths, rarely larger than 0.5×2 mm, and lamellar masses of cosalite, both embedded in galena and protruding from enclosing quartz, have been found at Braich-y-oen mine, Snowdon, Gwynedd, Wales (T. F. Cotterell pers. comm. 2003).

In Scotland, cosalite has been found in sulphide assemblages associated with the Gairnshiel zinnwaldite granite, Glen Gairn, W of Ballater, Grampian. A molybdenite-bearing assemblage in silicified wallrocks and zinnwaldite-rich veins adjacent to a wolframite-bearing quartz vein contains relict grains of silver-rich cosalite (5–25 µm across) within irregular masses of supergene molybdenum ochres (koechlinite and wulfenite have been identified). Narrow rims of acanthite mantle the cosalite, and the assemblage also contains chalcopyrite, galena, pyrite, native silver, sphalerite and an unnamed bromine–silver mineral. Intergrowths of cosalite, native bismuth and galena are also noted from a related sphalerite–(chalcopyrite–pyrite)-dominated assemblage (Tindle & Webb 1989, Webb et al. 1992).

Cosalite is noted as one of the rarest species found in a F–Sn–W mineralized microtonalite that was drilled near the southeastern margin of the Leinster granite at Ballinglen, between Tinahely and Aughrim, Co. Wicklow, Ireland. Arsenopyrite and scheelite are the dominant ore species occurring in both greisen and quartz veins within the microtonalite. Chalcopyrite, pyrrhotite and sphalerite with minor pyrite and stannite are also reported (Steiger & Bowden 1982, 1986, Gallagher 1989).

A dark-grey fibrous mineral found at l'Etacq quarry, Jersey, Channel Isles, was for many years thought to be boulangerite (plumosite) until analysis revealed it to be cosalite. However, boulangerite is an associated species (Mourant 1985). The cosalite probably came from calcite–quartz veins that fill the joints in the Northwest granite and would have been associated with bournonite, chalcopyrite, galena, molybdenite, pyrite, pyrrhotite, stannite and tetrahedrite (Ixer 1980, 1992).

cotterite

TL Rockforest, near Mallow, Co. Cork, Ireland

Named after Miss E. Cotter, who found the type specimen and gave it to Cork Geology Museum in 1876. Cotterite is a variety of quartz with a metallic pearly lustre resembling aluminium (Harkness 1878).

cotton stone

TL Isle of Skye, Highland Region, Scotland

A synonym of mesolite, zeolite group.

cotunnite

orthorhombic $PbCl_2$

A rare lead chloride that typically forms as an alteration product of galena in arid saline environments or as a volcanic sublimate in fumaroles.

Cotunnite was first described by Russell (1920b) from the wreck of the fireship *Firebrand* in Falmouth Harbour, Cornwall, where it formed on pieces of charcoal associated with metallic lead. The cotunnite has produced well defined and brilliant crystals with anglesite, leadhillite and galena (rare at this location). Cotunnite is colourless and transparent, up to 3 mm long and infrequently doubly terminated.

Cotunnite has also been found in a pebble of barite in the intertidal zone of Clevedon beach, 12 km W of Bristol, Avon. It formed in a cavity partly filled with corroded sphalerite and galena, and sealed from sea water by a rim of barite and crystalline masses of cerussite and galena (Bridges 2003). Rare cotunnite and diaboléite form a pale-blue alteration halo around chloroxiphite at Merehead (Torr Works) quarry, Cranmore, Mendip Hills, Somerset (Turner 2007).

Elongate blades of cotunnite from Clevedon beach, W of Bristol, Avon; crystals to 4 mm. (© D. I. Green)

Cotunnite is listed from Cromford, Derbyshire, by Ford & Sarjeant (1964a), but in a similar listing by the same authors (Ford et al. 1993) there is no mention of the species.

At Penrhyn Du mine, near Abersoch, Gwynedd, lead and silver were worked from veins cutting Ordovician mudstones and Cambrian sediments until 1870. Cotunnite has recently been found as whitish translucent tapering bladed crystals up to 4 mm long in a cavity in oxidized galena from one of the veins (Hubbard & Green 2005).

In Scotland, cotunnite is found with mimetite, plumbogummite and pyromorphite at High Pirn mine, Wanlockhead, Dumfries & Galloway, and with leadhillite at Leadhills (Hopeful vein), Strathclyde (Temple 1955).

coulsonite?

cubic SPINEL GROUP $Fe^{2+}V^{3+}_2O_4$

Named after Arthur Lennox Coulson (b. 1898), geologist in the Indian Geological Survey. Coulsonite is a rare iron vanadate thought to have been present in basal breccias of the Mercia Mudstone Group at Newhurst quarry, Shepshed, Leicestershire. It is now pseudomorphed by hematite, so cannot be confirmed. It was associated with the malachite and galena that sporadically cemented white sandstone in the south face of the quarry (King 1983). Later investigations of the material suggest that it may be hematite after magnetite (Ince 2005).

covellite

hexagonal CuS

A common copper sulphide most often found in the oxidation (secondary-enrichment) zone of copper deposits, but it can also have a primary hydrothermal origin. It frequently occurs as a tarnish on other sulphides. Covellite should not be confused with blaubleibender covellite, which has been shown to be a mixture of spionkopite and yarrowite.

England

Avon Along the beach near the Pier hotel at Clevedon, 16 km west of Bristol, Triassic Dolomitic Conglomerate is faulted and mineralized with barite–(chalcopyrite–galena). Cavities up to 12 cm across in the barite contain secondary covellite (replacing tennantite) and other supergene species including aragonite, aurichalcite, azurite, beudantite, brochantite, cerussite, native copper, cuprite, phosgenite and native sulphur (Starkey 1984, 1986c, Ixer et al. 1993).

Cheshire Vein and disseminated Cu–Pb–Zn redbed-type mineralization in coarse-grain clastic sediments of Triassic age on Alderley Edge ridge contains covellite and blaubleibender covellite as alteration products of all other sulphides. Covellite is also found as void infillings, where it forms acicular to plumose crystals collected into small rosettes (20 µm diameter) associated with malachite (Ixer & Vaughan 1982). Covellite also forms thin blue films on galena from Engine Vein mine on Alderley Edge (Braithwaite 1994). In the Cheshire Basin, at Bickerton, S of Warrington, covellite replaces bornite and tennantite in barite-dominated redbed-type mineralization closely associated with marginal basin faults (Naylor et al. 1989).

Cornwall Covellite replaces tetrahedrite in antimony–lead veins at Bounds Cliff, E of Port Isaac, and is noted to be a secondary species in the Pentire and Port Quin vein systems near by (Clayton et al. 1990). Pockets of uranium minerals in cassiterite lodes at Geevor mine, near Pendeen, St Just, contain paragenetically late covellite and chalcocite. Covellite is also noted from a similar paragenesis at: South Terras mine, St Stephen-in-Brannel; Trannack mine, Helston; Tregurtha Downs, Goldsithney; Wheal Bray, Altarnun and Wheal Speed, St Ives (Ball et al. 1982). Metallic-blue crusts and scaly hexagonal crystals of covellite occur with other supergene sulphides at Greystone quarry, Lezant, near Launceston. Some of the covellite has formed after collection (Elton et al. 2004). At Mount Wellington tin mine, Twelveheads, 7 km W of Truro, covellite, digenite and marcasite replace earlier-formed galena and pyrite (Kettaneh & Badham 1978). A tin-rich stockwork at Mulberry openwork, Lanivet, St Austell, contains scorodite partially replaced by covellite (Starkey 1986a). Covellite is intergrown with chalcocite as foliated to massive aggregates up to 4 mm across, at Penberthy Croft mine, St Hilary. It is often a post-mining replacement product after chalcopyrite (Betterton et al. 2000). Pseudomorphs of covellite after galena (cantonite) have been found at Perran St George mine, near Perranporth, and Wheal Falmouth, Kea (Davies 1877, Collins 1911a). Kaolinized tourmaline granite from the St Austell area contains xenoliths rimmed with arsenopyrite and covellite (Manning et al. 1996). Covellite has also been noted from: Carn Brea mines, near Redruth, Illogan (Robson 1945a); Hingston Down Consols mine, 2 km W of

Gunnislake, Calstock (Russell 1910a); Wheal Kine, St Agnes; and Wheal Maudlin, 3 km NW of Lostwithiel (both Greg & Lettsom 1858).

Cumbria *Caldbeck Fells:* Covellite replaces bornite and galena at Driggith mine, and chalcopyrite at Carrock mine, Grainsgill (Thimmaiah 1956, Stanley & Vaughan 1981). Reports of covellite on specimens from Brandy Gill mine, Old Potts Gill mine, Roughton Gill, Silver Gill mine, near Roughton Gill and Thief Gills (Hartley 1984, Cooper & Stanley 1990) require further confirmation, as they have a Kingsbury connection (see Ryback et al. 1998, 2001). However, covellite does occur at Silver Gill (and Sandbed mine) as thin rims around other sulphides and as (post-collecting) masses of minute metallic bluish-black plates on the surfaces of specimens collected from an old ore pile (Green et al. 2005b, Green et al. 2006a). It is also reported as irridescent purple-to-black coatings and crusts at Brae Fell mine, Dale Beck Valley (Bridges et al. 2006). The other locations also appear to be genuine, as covellite is a common species across the Caldbeck Fells (Green et al. 2006a, B. Young pers. comm. 2006). *Elsewhere in the Lake District:* The deeply oxidized portion of a chalcopyrite–ferberite–scheelite–quartz vein at Buckbarrow Beck, Corney Fell, Waberthwaite, contains a supergene assemblage that includes covellite and the rare species, bismutoferrite, cuprotungstite and russellite (Young et al. 1986, 1991b). Covellite is noted as a coating on chalcopyrite at the Shap granite quarry (Wirth & Green 2002) and in chalcopyrite–epidote–garnet veins within aureole rocks at Shap Blue quarry (Firman 1953b). At Seathwaite Tarn, Coniston, covellite has been collected from the dumps of an abandoned copper mine that worked two mineralized quartz veins cutting Ordovician Borrowdale Volcanic Series tuffs and agglomerates. Bornite, digenite and djurleite constitute the main sulphide assemblage, with minor arsenopyrite, chalcopyrite, cobaltite, blaubleibender covellite, digenite, pyrite and wittichenite also reported (Stanley & Criddle 1979). Additional reports include: Black Scar workings, Levers Water (Stanley & Criddle 1979); Dale Head mine, Newlands Valley, and Pave York mine, Greenburn Valley (Stanley 1979); Rigg Head quarry, Borrowdale (Green et al. 2006c). *Northern Pennine orefield:* Covellite replaces chalcopyrite in the Great Sulphur vein, Alston Moor (Vaughan & Ixer 1980).

Derbyshire Bornite, chalcocite, covellite, and cuprite replace chalcopyrite in galena–sphalerite veins at Oxclose mine, Snitterton, W of Matlock (Ixer & Townley 1979).

Devon Oxidized specimens from galena–quartz veins in the Combe Martin mines contain covellite associated with azurite and malachite (Rust 1982a). Covellite occurs with algodonite, bornite, chalcocite, chalcopyrite, coffinite, native copper, malachite, maucherite, nickeline, rammelsbergite and native silver in Cu–U–V-bearing nodules within Permian mudrocks at Littleham Cove, near Budleigh Salterton (Harrison et al. 1975b, Milodowski et al. 2000).

Durham Covellite is relatively uncommon, but is reported to replace galena at: Closehouse mine, Lunedale; at the Horse level horizon of Sedling mine, Cowshill, Weardale; and at Healeyfield mine, Castleside. It also forms coatings on anglesite–quartz matrix from Drypry vein, N of Wolsingham (Bridges & Young 1998).

Leicestershire Anhedral masses or, rarely, indigo-blue euhedral hexagonal plates to 0.6 mm of covellite occur in the basal breccia of the Triassic sequence at Newhurst quarry, Shepshed (King 1973, Ince 2005).

Shropshire Covellite, anglesite, cerussite, goethite and smithsonite are secondary minerals recognized in Ordovician-hosted Ba–Pb–Zn veins within the West Shropshire orefield (Pattrick & Bowell 1991).

Somerset At Moons Hill quarry, NE of Shepton Mallet, Mendip Hills, covellite is found on joint planes or in thin veinlets, but only in very small amounts (Kingsbury 1941).

Staffordshire Thin crystalline crusts of covellite growing on chalcopyrite almost certainly formed in the dump environment of Waterbank Cu–Pb–Zn mine, Ecton Hill (Rust 1995c).

Yorkshire Secondary covellite is associated with bornite, chalcocite, chalcopyrite, native copper, cuprite, digenite and djurleite on dump material from the Middleton Tyas mines, near Scotch Corner (Dunham & Wilson 1985, Green 1987b). At Merrybent mine, a few km to the NW of Middleton Tyas, covellite has been found in an assemblage of bornite, chalcocite, chalcopyrite and digenite, all within galena-rich calcite veins (Small 1977, Dunham & Wilson 1985). At Forcett quarry, East Layton, 12 km N of Richmond, alteration of a fault-bounded barite–calcite–chalcopyrite–witherite vein produced secondary bornite, chalcocite and covellite (Dunham & Wilson 1985).

Wales

Dyfed *Central Wales orefield:* Covellite forms as a result of weathering or alteration of copper sulphides (chalcocite or chalcopyrite). It sometimes coats galena and is often the first species to form in the post-mining environment. Reports include: Bwlchrhennaid mine, near Goginan (Bevins 1994); Eaglebrook (Nantycagl) mine, Ceulanymaesmawr (Jones & Moreton 1977, Jones 1983); Esgair Hir mine, Ceulanymaesmawr (Rust & Mason 1988); Frongoch mine, between Devil's Bridge and Pont-rhyd-y-groes (Green et al. 1996b); Llechweddhelyg mine, near Clawddmelyn, between Bont-goch and Cwmsymlog (Jones 1987); and Penrhiw mine, Ystumtuen (Mason & Green 1995).

Gwynedd Covellite is confirmed in the Dolgellau goldbelt, at: Caegwian mine (Gilbey 1968); Clogau mine, Bontddu, near Dolgellau (Naden 1988); Glasdir mine, Llanfachreth, near Dolgellau (Allen & Easterbrook 1978, Allen & Jackson 1985); and at a trial level, 2 km NW of Bontddu, near Dolgellau (Saich & Rust 1987). The species is also noted: with bornite, chalcocite, chalcopyrite, native copper, malachite, molybdenite and tetrahedrite in the Coed-y-Brenin porphyry copper deposit (Rice & Sharp 1976, Armstrong et al. 2003); replacing chalcopyrite in fracture-hosted hematite–magnetite–quartz veins within Ordovician rhyolitic tuffs at Shadow Gully, S of Snowdon (Colman & Appleby 1991); and with chalcopyrite and sphalerite (minor azurite, galena, pyrite) in a small lode at Gamallt mine, Careg y Fran, near Ffestiniog (Lynas 1973). At the Great Orme copper mine, Llandudno, covellite replaces chalcopyrite and also rare galena that initially shows alteration to digenite and djurleite, followed by covellite and spionkopite, and then to cerussite (Ixer & Davies 1996, Ixer & Stanley 1996, Ixer 2001). On the Isle of Anglesey, covellite is a relatively common replacement product after chalcopyrite in the Parys Mountain copper deposit (Pointon & Ixer 1980, Southwood & Bevins 1995).

Mid Glamorgan At Llantrisant, Bowler & Kingston (1971) report thin coatings of covellite attributable to oxidation of chalcopyrite by limonite.

Powys *Central Wales orefield:* Covellite forms as a blackish-blue alteration product on chalcopyrite and galena and as a thin plating on sphalerite at Dyfngwm mine (actually the Pen Dylife section of Dylife mine), Machynlleth (Rust & Rust 1987). These authors also report that covellite has been noted to grow on samples after collection. At South Nantycar mine, S of Rhayader, covellite forms an irridescent coating on chalcopyrite (Bevins 1994). Elsewhere, covellite forms iridescent-to-metallic purple hexagonal plates in cavities in a massive galena–tennantite vein at Dolyhir quarry, Old Radnor, near Presteigne. Some covellite-rich vein material contains inclusions of djurleite, roxbyite, spionkopite and yarrowite (T. F. Cotterell pers. comm. 2007).

Lustrous indigo-blue crystals of covellite to 3 mm associated with colourless barite; Dolyhir quarry, Old Radnor, near Presteigne, Powys, Wales. (© D. I. Green)

Scotland

Central At Alva silver mine in Silver Glen, Ochil Hills, covellite is intergrown with other sulphides and it occurs in the following paragenetic sequence: pyrrhotite, pyrite/marcasite, chalcopyrite, bornite, covellite and digenite in a gangue of barite and calcite (Jassim et al. 1983, Moreton et al. 1998, Green & Todd 2001).

Dumfries & Galloway The Black Stockarton Moor complex, to the NE of Kirkcudbright, is a Caledonian subvolcanic complex that hosts disseminated and veinlet pyrite and copper mineralization comparable in all but scale to porphyry copper deposits. Examination of borehole material has revealed significant bornite, chalcopyrite and tetrahedrite, and minor covellite and enargite (Leake & Brown 1979, Lowry et al. 1997). Covellite replaces chalcopyrite in greywacke-hosted quartz veins associated with the Cairngarroch Bay granodiorite, SE of Portpatrick and in the arsenopyrite–pyrite–pyrrhotite-bearing Orchars vein, SW of New Galloway, within the Cairnsmore of Fleet granite (Lowry et al. 1997). At Penkiln Burn, close to the margin of the Loch Doon granite pluton (nearest town Newton Stewart), black mudstones host a disseminated sulphide assemblage containing marcasite, pyrite, pyrrhotite and sphalerite, with rare chalcopyrite and covellite (Stone & Leake 1984). Complex As–Bi–Co–Cu–Fe–Pb–U-vein mineralization in the aureole of the Criffell granodiorite close to Needle's Eye, SW of Caulkerbush, hosts a supergene assemblage that includes covellite together with atacamite, bismoclite, boltwoodite, bornite, connellite, erythrite, mixite, uranophane and zeunerite (Miller & Taylor 1966, Braithwaite & Knight 1990).

Highland Weathering of chalcopyrite-bearing float close to a Lower Devonian appinite pipe on the SE of the Ardsheal Peninsula, near Kentallen, has produced covellite, malachite and hydrated iron oxides (Rice & Davies 1979). Small amounts of covellite, bornite and linnaeite occur with chalcopyrite, pyrite and titanomagnetite in an unusual garnet (melanite) pyroxenite at Bad na h'Achlaise, Ledmore, Assynt (Stumpfl 1961). Covellite occurs with malachite in a 4 cm-wide vein or fissure within sandstone at Kingsteps, NE of Nairn (Livingstone 2002). Quartz veins cutting Moinian psammites to the N of the Loch Shin monzogranite and W of Lairg contain fine-grain covellite together with chalcopyrite, galena, pyrite and sphalerite (Gallagher et al. 1974, Lowry et al. 1994). Sulphide minerals are associated with a diverse suite of PGMs and native gold in the layered basic and ultrabasic central complex on the Isle of Rum. Bornite, covellite, chalcocite and digenite all replace chalcopyrite (Power et al. 2000a). Similar primary and alteration assemblages are reported from the Sgaorishal ultramafic plug and associated dykes also on Rum (Power et al. 2003). Covellite occurs in boron–fluorine ore skarns in the metamorphic aureole around the Beinn an Dubhaich granite, Broadford area, Isle of Skye (Tilley 1948b, 1951).

Strathclyde Microscopic amounts of covellite occur in a chalcopyrite-rich stratiform deposit within Dalradian rocks at the Abhainn Srathain mines, S of Meall Mòr and Lochgilphead, South Knapdale (Smith et al. 1978a). Hydrothermal alteration of Clyde Plateau Lavas at Boyleston quarry, SW of Barrhead, has produced a diverse secondary assemblage that includes very rare covellite as small black to indigo-blue hexagonal crystals to 0.5 mm. Chrysocolla, malachite and prehnite are associated species (Meikle 1989a). Similar basalts at Hartfield Moss, 9 km WSW of Barrhead, contain covellite associated with diaspore, hematite, muscovite and native silver (Meikle 1990), whereas at Loanhead quarry, near Beith, covellite occurs within cavities in calcite associated with malachite (Meikle 1989b). Rare covellite replaces chalcopyrite in molybdenite–quartz veinlets within the Etive granite complex to the E and NE of Ben Starav (Haslam et al. 1985). Rafts of sulphide-rich hornsfelsed quartzite adjacent to a stock-like porphyrite intrusion at Garbh Achadh, W of Inveraray, contain minor covellite associated with chalcopyrite, molybdenite, pyrite (major) and sphalerite (Ellis et al. 1978). Covellite is common in oxidized chalcopyrite–galena ore within the Leadhills–Wanlockhead orefield (Temple 1955) and a single inclusion of covellite in an alluvial grain of native gold is also noted from a stream near Leadhills (Chapman et al. 2000c).

Tayside Sulphide mineralization dominated by galena and sphalerite (with pyrite and pyrrhotite), but also including minor covellite, is found in stratabound barite mineralization within Dalradian graphitic quartz–muscovite schists near Aberfeldy (Coats et al. 1981).

Shetland Covellite is noted in chromite-rich dunite lenses that formed near the base of the Shetland ophiolite complex at Cliff, Baltasound–Haroldswick area, Unst (Prichard et al. 1994). Maynard et al. (1997) identify covellite in pyroxenite from the same complex. At Matta Taing, West Mainland, covellite occurs with bornite, chalcopyrite, pyrite and sphalerite in nodules within Devonian laminated calcareous–carbonaceous-siltstone lake sediments known as the Upper Melmerby Fish Bed (Hall & Donovan 1978).

Ireland

Down Covellite replaces chalcopyrite in an occurrence of molybdenite–pyrite mineralization at Croreagh quarry, within the Newry granodiorite (Meighan & Ixer 1980).

Mayo Covellite is a supergene species replacing primary sulphides in gold-bearing quartz veins within Ordovician turbiditic slates, mudstones and sandstones of the Lower Ordovician Sheefry Formation along the line of the Cregganbaun shear zone, at Cushinyen (Thompson et al. 1992).

Tipperary Epigenetic mineralization related to the intrusion of two high-level basic intrusions at Gortdrum mine, near Tipperary town, has produced a dominantly chalcopyrite and mercury-rich tennantite orebody. The uppermost 15 m of the orebody is an oxidation zone containing covellite together with azurite, native copper, digenite, erythrite and malachite (Steed 1986). Covellite is listed as one of many secondary species found in the sedimentary-exhalative Ba–Pb–Zn deposit in the Silvermines district (Moreton 1999).

Tyrone Supergene alteration of sulphides (mainly pyrite) has produced a secondary assemblage of chalcocite, covellite, goethite and hematite in the quartz-vein hosted Lack gold deposit, WSW of Omagh (Cliff & Wolfenden 1992).

Wicklow Covellite is one of the minerals that replaces chalcopyrite in a Bi–Cu–Pb–Zn assemblage of sulphide and sulphosalt species in gold-bearing BIF rocks at Ballycoog, Goldmine River valley, near Avoca (Ixer et al. 1990).

Offshore territories

Jersey Extensive alteration of primary copper minerals to covellite is noted: in rhyolites at Bouley Bay, Trinity; in arsenopyrite–galena–pyrite–sphalerite-dominated carbonate veins within granite at Le Doüet de la Mer, St Ouën; in a chalcopyrite–pyrite-bearing quartz vein within granite at Gigoulande quarry, St Mary; and in chalcopyrite-dominated carbonate veins in the Jersey Shale formation at the German underground hospital, St Peter's Valley, St Lawrence (Mourant 1961, Ixer 1980, 1990, 1992).

Sark Examination of a sample of breccia from Sark's Hope mine on Little Sark showed it to contain chalcopyrite, pyrite and tennantite cemented together with the secondary species anglesite, covellite, scorodite and an unidentified phosphate (Bishop et al. 1977). A detailed examination of the complex ore mineralogy of the Sark's Hope Pb–(Ag) lode showed covellite to be a paragenetically late species replacing and pseudomorphing chalcopyrite, galena and sphalerite (Ixer & Stanley 1983).

cowlesite

orthorhombic — ZEOLITE GROUP — $Ca(Al_2Si_3)O_{10}·5–6H_2O$

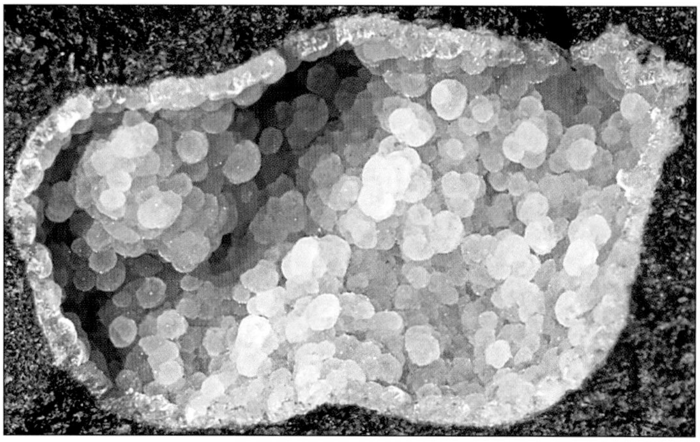

Cowlesite filling a vesicle in Tertiary basalt from Co. Antrim, Ireland; width 15 mm. (© The Trustees of the National Museums of Scotland)

An uncommon hydrated calcium aluminosilicate found as spherical balls up to 7 mm in diameter in Tertiary basalts from Coulters Hill quarry, Ballyclare, Co. Antrim, Ireland. Cowlesite is associated with calcite, chabazite, montmorillonite and offrétite. Another occurrence is at Dunseverick, near Giant's Causeway, Co. Antrim, where <2 mm-thick cavity linings of cowlesite are found in basalt (Nawaz 1978b, 1984). Soft globular linings of cowlesite inside vesicles and fractures of basaltic rocks from Co. Antrim are noted from: Galboly, Garron Point; Green Road quarry, Ballyclare; Kane's quarry, Boghill, Glengormley, near Belfast; Magheramorne quarry, near Larne; Parkgate and Tardree quarries; and also from Glover's quarry, Moneymore, Co. Londonderry. Broken hemispheres show the cowlesite crystals to be composed of thin radiating plates with a pearly lustre, sometimes associated with analcime or chabazite (Vezzalini et al. 1992). Cowlesite is also noted in basalt at other locations in Co. Antrim: Island Magee (Gottardi & Galli 1985); Benanouran Head and Port Noffer, both near the Giant's Causeway (Adair 1987); and Ladyhill quarry, Tardree, NE of Antrim town (H. J. Foy specimens, Tindle unpub. EMP data).

Cowlesite is quite rare in Scotland, but is found in cavities (to about 2 cm across) in water-worn boulders of Tertiary basalt at the northern end of Moonen Bay, N of Dunvegan, Isle of Skye. It forms white botryoidal spherules up to 1 mm across and, in other vesicles in this basalt type, erionite–offrétite, levyne and phillipsite are also found (Green & Todd 1996, 2001). At Oisgill Bay, a few km to the north of Moonen Bay, cowlesite occurs in vesicular basalt with chabazite, gyrolite, levyne, stilbite, thomsonite and excellent apophyllite (Robinson & Dossett 2002). Cowlesite in basalt is also noted from Kingsburgh, east side of Loch Snizort Beag, NW of Portree, Isle of Skye (Gottardi & Galli 1985).

craigtonite
TL Craigton, Hill of Fare, Grampian, Scotland
A mixture of iron and manganese oxides.

CRANDALLITE GROUP
Trigonal phosphates and arsenates with the general formula $AB_3(XO_4)_2(OH,F)_5$ or $AB_3(XO_4)_2(OH,F)_6$, where A = Ba, Bi, Ca, Ce, La, Nd, Pb, Sr, Th; B = Al, Fe^{3+}; and X = As, P, Si. Crandallite is a member of the group that occurs in Britain and Ireland.

crandallite
CRANDALLITE GROUP
trigonal $CaAl_3(PO_4)(PO_3OH)(OH)_6$

An uncommon calcium–aluminium phosphate hydroxide found in the Rowhurst Tonstein (a kaolinite-rich mudstone) at Red Street colliery, near Kidsgrove, Staffordshire. The tonstein consists mainly of elongate pale-brown and colourless pellety granules (0.01–0.02 mm), between which are illite, kaolinite, quartz and zircon particles. Apatite, crandallite and kaolinite are the constituents of the pellety granules (Wilson et al. 1966).

At Gunheath (Hensbarrow) china-clay pit, Carthew, St Austell, Cornwall, pegmatites cutting mineralized quartz and quartz–tourmaline veins contain several phosphate minerals in accessory amounts. These include apatite, crandallite, goyazite, mitridatite, rockbridgeite and waylandite (Hawkes et al. 1987). Crandallite has also been found in a large cavity (approximately 1.2×2×2 m) lined with green fluorite on amethystine quartz at Wheal Jane, Kea, near Truro. Pseudo-hexagonal prisms of crandallite occur as a thin brown crust on the surface of fluorite cubes in association with microscopic pyrite cubes (Ryde 1988).

In Wales, crandallite is described from Pwlldu beach, Bishopston, Gower Peninsula, West Glamorgan, where it occurs as a creamy-white powdery coating on limestone cherts and rottenstones. It is thought to be a weathering product of wavellite. Cacoxenite and variscite are associated species (Dean & Cotterell 2003).

Crandallite forms spiky microcrystals on greenish-blue crusts of turquoise coating vuggy white quartz at Ballycormick, near Shanagolden, Co. Limerick, Ireland, where it occurs in a brecciated slaty gossan in association with cacoxenite (Ryback & Moreton 1991). At Doon Bay, a cove less than 1 km N of Ballybunion, Co. Kerry, black and dark-grey shales containing crandallite, felsöbányaite (syn. basaluminite), turquoise and wavellite have been collected. Scarce crandallite is white and fibrous, and it occurs on fracture surfaces as radiating sprays up to 4 mm across and veins up to 4 mm thick (Ryback & Moreton 1993, Moreton et al. 1995).

creaseyite
orthorhombic $Pb_2Cu^{2+}_2Fe^{3+}_2Si_5O_{17}\cdot 6H_2O$

A rare hydrated lead–copper–iron silicate found in the oxidized zone of base-metal deposits. Creaseyite has been found in Whyte's Cleuch, Wanlockhead, Dumfries & Galloway, Scotland, in a mixture of pale-green to colourless pyromorphite veining light-blue chrysocolla, calcite and quartz, with minor undefined iron–manganese oxides, and is possibly derived from the Bay or Straitstep veins. Creaseyite forms pale- to light-green thickly packed minute lath-like crystals aggregated into 1.5 mm-diameter ball-like groups (S. A. Rust specimen, D. I. Green XRD and SEM).

crednerite
monoclinic $CuMnO_2$

An uncommon copper–manganese oxide first found in England on specimens at the NHM and originally from Higher Pitts farm, near Priddy, Somerset. Iron-black crednerite with a metallic lustre occurs on the outside of nodules of lead ore, between cerussite and wad, and forms radiating fan-like groups or hemispherical masses of thin plates. Single plates of crednerite penetrating hydrocerussite and a few small (1–2 mm) spherules consisting of a radial aggregate of plates embedded in mendipite are also reported (Spencer & Mountain 1923). The largest crystal of crednerite from High Pitts farm is 3.5×4×6 cm and at the time was thought to be one of the finest specimens of the mineral in existence (Kingsbury 1941).

At Merehead (Torr Works) quarry, Cranmore, veins of iron and manganese

Grey metallic plates of crednerite associated with green malachite; Merehead (Torr Works) quarry, Cranmore, Mendip Hills, Somerset; width 10 mm. (© D. I. Green)

oxides cutting Carboniferous Limestone contain crednerite crystals up to 2.5 cm long associated with calcite and cerussite or hydrocerussite. Margins of the crednerite are altered to tufts of malachite. Crednerite also occurs as small spherules in mendipite (Symes & Embrey 1977). This locality produced an iron–manganese-rich pod in 2006 containing bladed masses of crednerite associated with aragonite, calcite, cerussite, kentrolite, malachite, melanotekite, mimetite, amethystine quartz and wulfenite (Turner 2007). One Kingsbury specimen of crednerite from High Pitts and another from Merehead show an association with plattnerite (Alabaster 1985b).

Similar manganese-oxide deposits cutting Carboniferous Limestone at Coleman's quarry, Holwell, near Frome, and Wesley mine, Didsbury, near Bristol, contain crednerite, other manganese-rich species (goethite, hausmannite, kentrolite, manganite, rhodochrosite) and a lead oxychloride assemblage in crosscutting veins and small pockets (Alabaster 1989b, 1990a).

CRICHTONITE GROUP
Trigonal, or monoclinic, pseudo-trigonal oxides with the general formula $AB_{21}(O,OH,)_{38}$, where A = Ba, Ca, Ce, K, La, Na, Pb, Sr; and B = Cr^{3+}, Fe^{2+}, Fe^{3+}, Mg, Mn^{2+}, Ti, U, V^{3+} Zn, Zr. Lindsleyite is the only species of the group to be described from Britain or Ireland, but this requires confirmation (see also crichtonite).

crichtonite?
CRICHTONITE GROUP
trigonal $Sr(Mn,Y,U)Fe_2(Ti,Fe,Cr,V)_{18}(O,OH)_{38}$

Named after Alexander Crichton (1763–1856), a Scottish physician and mineral collector. Crichtonite itself has not been found in Britain or Ireland; however, it was thought to be synonymous with ilmenite until the two species were differentiated.

cristobalite
tetragonal SiO_2

A rare silicon oxide, polymorphous with coesite, quartz, stishovite and tridymite. It is stable above 1470°C, but can exist metastably below that temperature. Cristobalite is found in aegirine–riebeckite granite from the island of Rockall, North Atlantic Ocean, 440 km W of northern Skye, NW of Scotland. Some miarolitic cavities in the granite are infilled with an isotropic pale-brown material, identified as low-temperature α-cristobalite. Spherulitic growths of microcrystalline quartz mantle the α-cristobalite, and intergrowths with leucophosphite are also described. Other interstitial minerals include albite, bazirite, elpidite and monazite (Sabine 1960, Hawkes et al. 1975).

Cristobalite (or tridymite) occurs in thermally metamorphosed Tertiary basalts close to a dolerite plug at Tievebulliagh, near Cushendall, Co. Antrim, Ireland, where it is found in assemblages containing cordierite. Porcellanite inclusions (ferruginous bauxite and lithomarge, in which gibbsite, halloysite and hematite were the principal original constituents) in the dolerite also contain cristobalite in various assemblages that may also contain corundum, hercynite, ilmenite, mullite or magnetite (Agrell &

Langley 1958). At Glynn Hill quarry, 5 km S of Larne, cristobalite occurs with chlorophaeite within bands of lens-shape vesicles in a fine-grain quartz trachyte and also in pellets, up to 1 mm in diameter, on the walls of a fissure (Patterson 1951).

Uranium–vanadium-rich nodules in Permian marls and sandstones at Budleigh Salterton, Devon, contain α-cristobalite associated with metatyuyamunite and may indicate an association between silicification processes and the formation of radioactive minerals in the nodules (Durrance & George 1976).

Reports of cristobalite requiring further investigation include Cornwall (Mallard 1890) and Melbur china-clay pit, St Austell, Cornwall (Golley & Williams 1995).

crocalite

A variety of natrolite, zeolite group.

crocidolite

An asbestiform variety of magnesioriebeckite, amphibole group.

crocoisite

A synonym of crocoite.

crocoite

monoclinic $PbCrO_4$

A rare lead chromate found in small quantities with cerussite, leadhillite and pyromorphite on an overgrown spoilheap near Hopeful vein, W of Leadhills, Strathclyde, Scotland. Crocoite is "hyacynth" or aurora red (Brown 1927) and is also described as forming small blocky orange crystals (Green & Todd 2001).

Three claimed occurrences of crocoite in the Caldbeck Fells are documented by Cooper & Stanley (1990). In each case the occurrence is suspect: Brandy Gill – probably an incorrect provenance; Roughton Gill – a single specimen in Carlisle Museum is now thought to have its origin in Beresov, Siberia (see also Davidson & Thomson 1951); and Poddy Gill – a specimen in the Kingsbury collection at the NHM has probably been mislabelled. Reasons for these doubts are outlined in Ryback et al. (1998, 2001).

Crocoite is also reported from Cornwall, where suspected wulfenite from Greystone quarry, Lezant, near Launceston, proved on analysis to be crocoite (Rothwell & Mason 1992). Some fine small specimens of crocoite, wulfenite and sceptre quartz, as well as dolomite and pyrite, have been recovered from this site, where crocoite occurs as isolated orange-red flattened tabular crystals to about 0.5 mm (Wolfe 2000, Elton et al. 2004).

cromfordite

TL Cromford, Matlock, Derbyshire, England

A synonym of phosgenite.

cronstedtite

KAOLINITE–SERPENTINE GROUP

monoclinic or trigonal $Fe^{2+}_2Fe^{3+}(Si,Fe^{3+})O_5(OH)_4$

A rare iron silicate hydroxide found at Wheal Maudlin, 3 km NW of Lostwithiel, Cornwall, where it occurs on siderite and decomposing pyrite in association with hisingerite and vivianite. Crystals of cronstedtite form lustrous black three- or six-sided striated prisms or thin fibres (Greg & Lettsom 1858, Collins 1892). At Wheal Jane, Kea, near Truro, cronstedtite develops well formed microcrystals or black crystalline crusts and spherical masses on a pyrite-rich matrix and is associated with ludlamite and vivianite (Collins 1911a, Rust 1982d). On specimens from Wheal Jane, now in the Ludlam collection at the NHM, cronstedtite forms radiating groups of acicular crystals, rather suggestive of schorl, in association with siderite and a little limonite on pyrite (Rudler 1905). A vague reference to cronstedtite from the host rocks (killas) around the Hensbarrow granite area, St Austell, is given by Collins (1878b), and Golley & Williams (1995) also note the species at Penlee quarry, Newlyn.

crossite

A common species of amphibole formally discredited during the reclassification of this group (Leake et al. 1997a). Four amphibole species – ferroglaucophane, glaucophane, magnesioriebeckite and riebeckite – now overlap compositions that were referred to as crossite.

crucilite

TL Clonmell, Co. Waterford, Ireland

A pseudomorph of hematite after arsenopyrite.

crucite

A synonym of andalusite (var. chiastolite), but also used to describe pseudomorphs of hematite or limonite after arsenopyrite.

cryptolite

A synonym of monazite-(Ce).

CRYPTOMELANE GROUP

Monoclinic, tetragonal or pseudo-tetragonal oxides with the general formula AB_8O_{16}, where A = Ba, K, Mn^{4+}, Na, Pb, Sr; and B = Cr^{3+}, Fe^{3+}, Mg, Mn^{2+} Ti, V^{3+}, Zn, Zr. Coronadite, cryptomelane, hollandite and possibly manjiroite are members of the group found in Britain or Ireland.

cryptomelane

CRYPTOMELANE GROUP

monoclinic (pseudo-tetragonal) $K(Mn^{4+},Mn^{2+})_8O_{16}$

An uncommon potassium–manganese oxide found in oxidized manganese deposits as open-space fillings or replacing primary manganese-bearing minerals. It is commonly replaced by other manganese-bearing minerals.

At Pelyn Tor, 1 km E of Lanlivery, Cornwall, cryptomelane forms botryoids mantled by romanèchite in fissured and fragmented quartz veinstone (D. W. Williams specimen, Tindle unpub. EMP data). Cryptomelane is a common manganese-oxide species found in quartz veins cutting Ordovician greywackes and mudstones that have been thermally metamorphosed by the Skiddaw granite at Burdell Gill, Caldbeck Fells, Cumbria. It forms drusy black crusts less than 0.5 mm thick on pyrolusite or psilomelane, along with a wide range of other supergene species such as beudantite, carminite, lepidocrocite, mimetite, parasymplesite, pharmacosiderite, plumbogummite, rhodochrosite, scorodite, switzerite and vivianite (Neall & Leppington 2003).

At Merehead (Torr Works) quarry, Cranmore, Mendip Hills, Somerset, calcite veins contain lenticular areas composed of manganese oxides, surrounded by brown iron oxides (collectively termed manganese pods). Black manganese-oxide layers consist of manganite and pyrolusite, whereas brownish material is a mixture of cryptomelane and manganite (Turner 2007).

Cryptomelane occurs in patchy exposures of iron and manganese oxides in the Lecht Valley, near Tomintoul, Grampian, Scotland, especially at the site of the old ironstone mine, where a porous hydrothermal eruption breccia is exposed. The deposit is mainly composed of goethite, cryptomelane and occasional patches of hematite. Cryptomelane is usually bluish black, often with a metallic lustre, and may be intimately mixed with lithiophorite. Chalcophanite, pyrolusite, ramsdellite and woodruffite are other manganese-rich species found in minor amounts. The Lecht mineralization is derived from weathering of the local Dalradian sequence, and ponding of the resulting Fe–Mn–Zn solutions produced a seepage-bog type deposit in which the permeable breccia acted as a channel-way for these supergene solutions. Diagnostic lithogeochemical anomalies in the area also indicate the existence of a

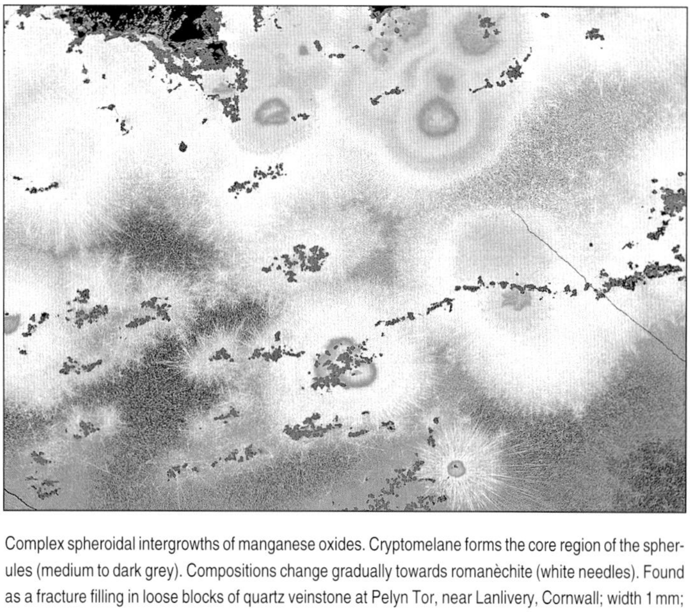

Complex spheroidal intergrowths of manganese oxides. Cryptomelane forms the core region of the spherules (medium to dark grey). Compositions change gradually towards romanèchite (white needles). Found as a fracture filling in loose blocks of quartz veinstone at Pelyn Tor, near Lanlivery, Cornwall; width 1 mm; BSE image. (© E. Condliffe, University of Leeds)

Botryoidal mass of cryptomelane and hollandite on a matrix of goethite; Cloghleigh mine, Blessington, Co. Wicklow, Ireland; width 75 mm. (© D. I. Green)

concealed and related Zn–Pb–(Ag)–(Au?) exhalative deposit (Wilson et al. 1970, Nicholson 1987a, 1989a,b, 1992, Nicholson & Anderton 1989).

Cryptomelane has been found with hollandite as intergrown botryoidal masses up to several cm across, associated with goethite in a quartz vein hosted by the Leinster granite at Cloghleagh mine, near Blessington, Co. Wicklow, Ireland (Moreton et al. 2006).

cubanite

orthorhombic $CuFe_2S_3$

An uncommon copper–iron sulphide, dimorphous with isocubanite and found in wolframite–scheelite–quartz veins related to the Skiddaw granite at Carrock mine, Grainsgill, Caldbeck Fells, Cumbria. Cubanite and chalcopyrite occur as inclusions in sphalerite within fractures and cavities in the veins and greisen zones. Bismuthinite, tungsten-rich columbite and native bismuth are associated species (Ball et al. 1985). Kingsbury & Hartley (1956b) also report cosalite and tetradymite, but, because many specimens in the Kingsbury collection (now at the NHM) have an incorrect provenance (Ryback et al. 1998, 2001), these species remain unproven. Hematite–pyrite-bearing quartz veins on the north bank of Gatesgarthdale Beck, W of the summit of Honister Pass, Wasdale, contain rare cubanite associated with chalcopyrite and bornite (Millward et al. 1999). Irregular masses of chalcopyrite containing minor amounts of chalcocite, cubanite, digenite and sphalerite occur in quartz veins at Long Work in the Vale of Newlands. Arsenopyrite and pyrite are associated species (Millward et al. 1999). In the Northern Pennine orefield, cubanite has been found in bismuth-bearing quartz veins related to the buried Weardale granite. Cubanite, argentopentlandite, bismuthinite, cosalite, pyrrhotite and synchysite form inclusions in chalcopyrite. In Co. Durham and Cumbria, 11 bismuth mineral localities were examined, but the precise locations where cubanite was found are not recorded (Ixer et al. 1996). Cubanite has also been described from Tynebottom mine, NW of Garrigill, near Alston, Cumbria (Ixer et al. 1987), but the specimens came from the Kingsbury collection at the NHM. Dunham et al. (2001) express serious doubt about the provenance of these specimens – a problem also discussed by Ryback et al. (1998, 2001).

At East Pool mine, Illogan, near Redruth, Cornwall, cubanite is described from a cassiterite-bearing vein (Brown 1934, Spencer 1958), but, as the species does not appear to have been analyzed and other occurrences have not been reported anywhere else in Cornwall, there is some doubt about the identification.

In Wales, cubanite occurs as thin ($<10\,\mu m$) subparallel lamellae within chalcopyrite at Cefndeuddwr mine, near Dolgellau, Gwynedd. It is part of a chalcopyrite–pyrrhotite association related to mesothermal gold lodes found in the Dolgellau area (Mason et al. 2002). Inclusions of cubanite in chalcopyrite are also noted from Hendreforion mine, 5 km W of Dolgellau (Gilbey 1968).

Cubanite is found in allivalite (a pyroxene-free gabbro) from Huntly, Grampian, Scotland, where it occurs as an exsolution product in chalcopyrite associated with mackinawite, pentlandite, pyrrhotite and valleriite (Babkine & Conquéré 1968). Later work revealed cubanite associated with chalcopyrite in ultramafic cumulates and picrites from this intrusion (Fletcher & Rice 1989). At Calliachar Burn, 4 km from Aberfeldy, Tayside Region, 14 steeply dipping quartz–sulphide ±gold veins, cutting Upper Dalradian gritty quartzites, quartz–mica schists and amphibolites, contain polyphase hypogene mineralization consisting of pyrite (most abundant), galena and sphalerite, plus minor arsenopyrite, chalcopyrite and trace gold (electrum). The veins also host benleonardite, cobaltite, cubanite, gersdorffite, hessite, mackinawite, marcasite, pentlandite, pyrrhotite, native tellurium and tetrahedrite. Again, there is a chalcopyrite–cubanite association (Ixer et al. 1997). Sulphide droplets in one of the chromite bands of the Eastern Layered Series, Isle of Rum, Highland Region, contain frequent chalcopyrite and pentlandite with minor bornite, chalcocite, native copper, cubanite, digenite, gold (electrum) and pyrrhotite (Dunham & Wilkinson 1986).

At Le Pulec on the northwest coast of Jersey, cubanite is described from quartz veins (<2 m wide) cutting Brioverian mudstones and siltstones that in the 1870s were worked for silver and lead. Sphalerite is the earliest and most abundant sulphide, followed by galena, chalcopyrite, and then arsenopyrite. Cubanite occurs as fine-grain intergrowths with chalcopyrite and galena, associated with minor native antimony, bournonite and tetrahedrite (Ixer & Stanley 1980). Near by, in the southeast quarry at L'Etacq, large anhedral crystals of chalcopyrite in excess of 1 cm in diameter contain many inclusions of the following species: arsenopyrite, bournonite, cubanite, galena, marcasite, pyrite, pyrrhotite, sphalerite (most common) and stannite. Elsewhere in the Channel Isles, a chalcopyrite–cubanite association is noted from Bouley Bay, north Jersey (Ixer 1980), Baubigny, near St Sampson, Guernsey (Ixer 1992), and from the Great Copper lode cutting across the southern end of Herm (Ixer 1992).

cube ore

A synonym of pharmacosiderite.

cumengéite/cumengeite

tetragonal $Pb_{21}Cu_{20}Cl_{42}(OH)_{40}\cdot 6H_2O$

A rare hydrated lead–copper chloride hydroxide found with other secondary lead and copper minerals within a primary Cu–Pb–Sb sulphide vein at Newporth beach, 2 km SW of Falmouth, Cornwall. Cumengéite forms pale-blue crystalline crusts on the rock matrix, and crystals up to 4 mm across, associated with bournonite, lining vugs in vein quartz. Other secondary minerals include anglesite, atacamite, bindheimite, cerussite, goethite, malachite and phosgenite. Weathering of the primary sulphides by sea water and atmospheric exposure produced the environment in which cumengéite formed (Dean 1982, 1983, Dean et al. 1983). Other coastal exposures in Cornwall

Well formed crystals of cumengéite from Gunver Head, near Padstow Consols mine, Cornwall; width 150 µm; SEM image. (© J. J. Hooper)

have also produced finds of cumengéite: Daymer beach, Polzeath, as deep-blue octahedral crystals associated with diaboléite and phosgenite in quartz veins (R. E. Starkey pers. comm. 2005); Gunver Head, near Padstow Consols mine, as small blue crystals associated with diaboléite in an oxidized galena–quartz veinlet (Braithwaite & Merry 1996, Schlomann & Weiss 1998); Loe Warren Zawn, Botallack, St Just, in a localized patch of Cu–Pb–U mineralization forming late in the paragenetic sequence with botallackite and connellite and after the rare uranium secondary species dewindtite and widenmannite (Elton & Hooper 1995c); Tolcarne beach, Great Western beach and Watergate Bay, all Newquay (Golley & Williams 1995); and Wheal Golden, Perranzabuloe, from lode outcrops in the cliffs in association with phosgenite (Merry & Braithwaite 1995). Cumengéite is also listed from Padstow Consols and Porthilly mine, St Minver (Golley & Williams 1995).

Cumengéite is found with atacamite and botallackite in fractures within pebbles of hydrothermally brecciated Ordovician siltstone containing bornite–chalcopyrite-bearing quartz veins, on the beach between Abersoch and Penrhyn Du, Gwynedd, Wales (Dossett & Green 1998).

cummingtonite

AMPHIBOLE GROUP

monoclinic $\square Mg_7Si_8O_{22}(OH)_2$

A common magnesium silicate hydroxide that forms a series with grunerite and is dimorphous with anthophyllite. Cummingtonite is found in medium-grade metamorphic rocks, metamorphosed iron formations and as a late-stage component of some gabbros. Where possible, identification is based on the 1997 classification of the amphibole group (Leake et al. 1997a, Mogessie et al. 2004).

Coarsely crystallized cummingtonite occurs with biotite, cordierite, ilmenite and plagioclase feldspar in thermally metamorphosed, hornfelsed dolerite (greenstone) found in quarries at the margin of the Land's End granite near Kenidjack Castle, and at North Zawn, St Just, Cornwall. Additionally, greyish-brown to clove-brown cummingtonite forms 12 mm-wide veinlets cutting through these rocks. Cummingtonite also forms radiating patches and veinlets in anthophyllite–cordierite hornfels at this location (Tilley & Flett 1930, Tilley 1935). Similar cummingtonite-bearing hornfels are noted from other locations near by (the Crowns, Wheal Edward Zawn, Zawn a Bal) (Mitropoulos 1984).

In Scotland, cummingtonite appears to have formed as an alteration product of orthopyroxene in paragneiss from Rubha Sgeir nan Sgarbh, South Harris, Western Isles (Livingstone et al. 1973), in a garnet–hornblende–pyroxene eclogite from Glen More, Glenelg, Highland Region (O'Hara 1960), and in quartz diorite from the Ben Nevis diorite–granite complex (Haslam 1968). The Cnoc an t-Sidhean supracrustals (part of the Lewisian complex), near Stoer, Highland Region, contain locally abundant cummingtonite in brown quartzo-feldspathic biotite–garnet gneiss (Cartwright et al. 1985). Elsewhere in Highland Region, cummingtonite is reported from a garnet schist, 1 km SW of the outlet of Loch Bad-na-Sgalaig, between Loch Maree and Gairloch (Teall 1907, Macpherson & Livingstone 1982), from certain schists and granulites in a migmatite complex at Strathy Point, between Armadale and Strathy (Collins 1942), from Scourian ultramafic granulites in the Assynt area (Sills 1982), and from Lewisian amphibolite-facies ultramafic rocks, SW of An Fharaid Mhòr, Achmelvich, near Lochinver (Droop 1994). Anthony et al. (1995) list cummingtonite from the Garabal Hill–Glen Fyne complex, W of the head of Loch Lomond, Strathclyde, and from Geodh Dubh (11 km E of Cape Wrath, Highland Region).

Cummingtonite is described from a Dalradian psammite–semipelite sequence at Creag nam Ban, near Ballater, Deeside, Grampian (Smith & Phillips 2002), but the data are grunerite according to the 1997 amphibole classification.

In Ireland, cummingtonite appears to occur only in Connemara, Co. Galway. The Currywongaun and Doughruagh masses represent sections through a syntectonic partially layered intrusion that was later metasomatized. The Doughruagh mass consists of quartz-rich gneisses in which colourless cummingtonite forms the cores of zoned crystals, with green tschermakitic hornblende forming the rims (Angus & Kanaris-Sotiriou 1982). Cummingtonite-bearing striped amphibolites (some garnetiferous) of Dalradian age are noted from Ballynaboleyglassa and Letterbreckaun (Leake 1972).

A Precambrian layered diorite–gabbro complex at Le Nez, southeast Jersey, Channel Isles, contains late composite megacrysts of cummingtonite in some mafic horizons, although actinolite and hornblende appear to be far more common (Bishop & Key 1983).

cupreous calamine

A synonym of aurichalcite.

cuprite

cubic Cu_2O

A common copper oxide found in the oxidized portion of many copper sulphide deposits and often associated with azurite, native copper, chalcocite and malachite. Chalcotrichite is a remarkable variety of cuprite that forms attractive capillary-shape crystals. Crystallized cuprite is sometimes found on ancient objects of copper or bronze that have lain in damp ground (Rudler 1905).

England

Avon Along the beach near the Pier hotel at Clevedon, 16 km west of Bristol, Avon, Triassic Dolomitic Conglomerate is faulted and mineralized with barite–(chalcopyrite–galena). Cavities up to 12 cm across in the barite contain cuprite and other supergene species including aragonite, aurichalcite, azurite, beudantite, brochantite, cerussite, native copper, covellite, phosgenite and native sulphur (Starkey 1984).

Cheshire Rare but beautiful crystals of cuprite, including octahedra up to 0.5 mm across, are reported with native copper from Engine Vein mine, Alderley Edge ridge (Carlon 1979, Braithwaite 1994).

Cleveland Very rare anhedral cuprite inclusions up to 0.25 mm across have been found in massive boracite at Boulby potash mine, Loftus (Green & Freier 1996).

Cornwall Many fine specimens of cuprite were once obtained from the shallower workings of mines in the Botallack sett (including Wheal Cock, Wheal Owles and Wheal Edward), St Just (Bancroft & Weller 1993). Excellent specimens from these mines are on display in the collection of the Royal Cornwall Museum, Truro (Wolloxall 1989). Minor cuprite is also noted from Loe Warren Zawn, near Botallack, where it is associated with boltwoodite and rutherfordine (Elton & Hooper 1995). Museum-quality specimens of cuprite (chalcotrichite), forming nests of fine capillary crystals in cavities within brecciated vein quartz, were once collected from Fowey Consols mine, Tywardreath, St Blazey (Embrey & Symes 1987, Lewis 1997). Lustrous wine-red octahedral crystals of cuprite were found in 2000 in loose boulders at Greystone quarry, Lezant, near Launceston (Elton et al. 2004). Gunheath china-clay pit, Carthew, St Austell, has produced sporadic cuprite as flattish grains and as chalcotrichite, in quartz veins belonging to the Bunny lodes (Grigg 1986,

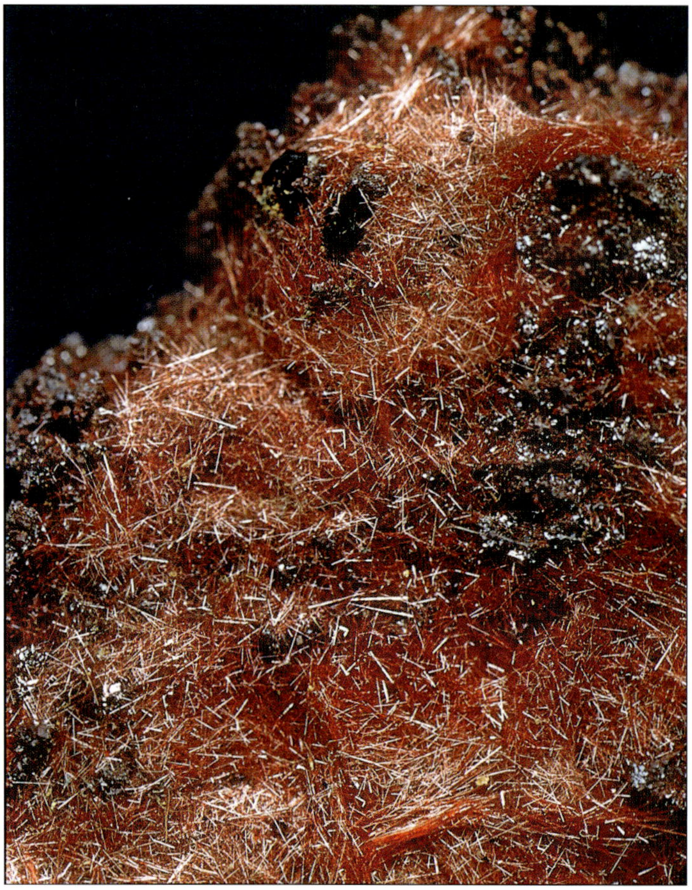

Cuprite (variety chalcotrichite) from an unknown location in Gwennap parish, Cornwall; width 25 mm. (© Royal Cornwall Museum, Royal Institution of Cornwall; photographer D. I. Green)

Weiss 1994). At Penberthy Croft mine, St Hilary, cuprite occurs widely on dump material as small cochineal-red cubic crystals up to 0.9 mm across, and as chalcotrichite. Bismutite, brochantite, clinochlore, connellite, native copper, goethite, langite and siderite are associated species (Camm & Merry 1991b, Rust 1995a, Betterton 1996e, Betterton et al. 2000). Near-surface oxidized copper-enriched lodes at South Caradon mine, St Cleer, near Liskeard, contain limonite associated with azurite (chessylite), native copper, cuprite, malachite and tenorite (melaconite). At lower levels, bornite, chalcocite and chalcopyrite appeared (Dewey 1925, Shambrook 1982). Like most other mines around St Day, Ting Tang mine, Carharrack, produced fine specimens of cuprite (and chalcotrichite). More recently, disappointing specimens containing octahedral crystals to 2 mm were recovered from the dumps around Morcom's shaft (Bruce & Aubrey-Jones 1998). Cuprite is also noted from Ting Tang mine by Stephens (1940a) and Ball et al. (1982). A cuprite specimen from Tremarcombe, Liskeard, containing a very unusual V-twin-shape crystal, won the British Micromount Society best micromount trophy in 1995 (Wolfe 1996a) and another superb micromount specimen of cuprite perched on a delicate branch of native copper from Wheal Buller, near Redruth, is described (and illustrated) by Cooper (1991a). At West Jewel mine, St Day, specimens of cuprite covered with many minute tabular crystals of oxide of uranium (possibly autunite) were once found (James 1947). Rich bunches of cuprite were encountered in the oxidized portion of Muttrell lode at Wheal Gorland, St Day, 2 km E of Redruth, where they formed solid vein sections and cellular masses, often associated with native copper. Well formed lustrous octahedral red crystals to 1 cm and chalcotrichite are both recorded (Greg & Lettsom 1858, Rudler 1905, Rust 1981b, Fisher 1988).

Cuprite is also recorded from: Carn Brea mines, near Redruth (Rust 1982b); Castle-an-Dinas, St Columb Major (Hey et al. 1938); Countybridge quarry, Goonhilly Downs, Lizard Peninsula (Holyer 1996); Dolcoath mine, Camborne; Old Gunnislake mine, Calstock; South Wheal Francis and Wheal Basset, both Illogan (all Golley & Williams 1995); Levant workings of Geevor mine, near Pendeen, St Just (Cooper & Green 1993); Marke Valley mine, near Liskeard (Rust 1982b); New Cook's Kitchen mine (North Tincroft lode), Redruth (LeBoutillier et al. 2000); Phillips Rock, Goonhilly Downs, Lizard Peninsula (Robson 1954); Phoenix United mines, Linkinhorne (Miers 1889b, Embrey & Symes 1987, Dunn 1990); West Caradon mine, St Cleer, Liskeard (Lawrence 1990); and Tryphena lode at Wheal Pendarves, Camborne (Barstow & Cooper 1982, Penhallurick 1997). Greg & Lettsom (1858), Collins (1892) and Dines (1956) list many more localities.

Cumbria *Caldbeck Fells:* Cuprite is very rare but occurs: with chrysocolla on three specimens from Brae Fell mine (Cooper & Stanley 1990, Bridges et al. 2006); as thin strings in masses of altered sulphides, as arborescent masses, presumably overgrowing native copper, and as minute euhedral octahedra up to 0.05 mm on edge in limonitic matrix at Silver Gill mine (Cooper & Stanley 1990, Green et al. 2000b); on Deer Hills, in barite workings to the E of Potts Gill barite mine and at Mexico mine in Todd Gill (Shaw 1970); at Red Gill mine (Briscoe 1986a). Reports of cuprite from Brandy Gill, Carrock End, Driggith, Hay Gill and Potts Gill (barite) mines (Hartley 1984, Cooper & Stanley 1990) deserve further investigation as the specimens all have a Kingsbury association and may have dubious provenance (see Ryback et al. 1998, 2001). The Driggith mine specimen is an excellent match for cuprite found at Wheal Prosper, St Hilary, Cornwall, although cuprite crystals have recently been found at both Driggith and Sandbed mines (Green et al. 2006a). *Elsewhere in the Lake District:* Cuprite occurs in the Coniston copper mines (Holland 1981); as wine-red pseudo-octahedral crystals up to 0.3 mm on edge at Force Crag mine, Braithwaite, near Keswick (Green et al. 1997); and at Tilberthwaite Gill (Dewey & Eastwood 1925). *Northern Pennine orefield:* Cuprite is noted from old workings on the Cornriggs vein, Alston Moor (Bridges & Young 1998); chalcopyrite has been altered to native copper, covellite and cuprite in the Great Sulphur vein, Alston Moor (Vaughan & Ixer 1980); and rare traces of cuprite occur at Stennerskeugh Clouds, Kirkby Stephen (Starkey 1987).

Derbyshire Reddish-brown submetallic-to-vitreous crystals are noted from: Hopton; Middleton-by-Wirksworth and Watersaw Rake on Longstone Edge (Ford & Sarjeant 1964a, Ford et al. 1993); and at Snelston mine, near Ashbourne, cuprite covers small areas on fracture surfaces, rarely as euhedral crystals and sometimes pseudomorphed by malachite (Jackman 1996).

Devon Crystallized cuprite is reported from Bedford United mines and Wheal Crebor, near Tavistock (Hall 1868b). A specimen of cornubite associated with cuprite, malachite and olivenite from Bedford United mines (collected by A. W. G. Kingsbury) was studied by Claringbull et al. (1959). Starkey (1986b) established a paragenetic sequence for copper mineralization at New East Wheal Russell (South Wheal Crebor), Tavistock, beginning with chalcopyrite that was replaced by chalcocite and then by cuprite and native copper. Detailed investigations of native copper in the form of sheets, found in mudstone 10 cm below the Triassic Budleigh Salterton Pebble Beds at Littleham Cove, near Budleigh Salterton, show cuprite to be the main alteration species. It is associated with hydrated copper arsenates thought to be either arhbarite or euchroite (Milodowski et al. 2000, 2002).

Durham Bright-red octahedral crystals of cuprite up to 0.1 mm on edge, and crusts associated with native copper and malachite, have been found on quartz at Westernhope Old mine, Weardale. Cuprite is also noted from Closehouse mine, Lunedale, and from a railway cutting S of Raisby quarry, Coxhoe, Ferryhill area (Bridges & Young 1998, Young 2001).

Hereford & Worcester Mineralized specimens of cuprite with inclusions of native copper and tenorite overgrowths, chalcocite, covellite and possibly chrysocolla have been found in Triassic Mercia Mudstone Group sandstones to the W of Moorfield farm, N of Redditch (Old et al. 1991).

Leicestershire Cuprite and native copper replace chalcopyrite in galena-rich spheroidal nodules formed at the Carboniferous/Triassic unconformity exposed in Cloud Hill quarry, Breedon on the Hill (King 1983). Massive cuprite intimately intergrown with native copper and forming blocks to over 100 kg is reported from a native copper–cuprite deposit near the Precambrian/Triassic unconformity at New Cliffe Hill quarry, Stanton under Bardon. Cavities in the deposit are occasionally lined with well formed cuprite, either as modified cubes or in prismatic-to-acicular habits tending towards chalcotrichite. Veinlets cutting cuprite contain algodonite or native silver (Pearson & Jeffrey 1997, Hawthorne et al. 2002, Hubbard et al. 2005, Ixer et al. 2005a). Copper-rich mineralization near the same unconformity but at Bardon Hill, Coalville, has produced blocks of massive cuprite and vivid to dark-red cuprite crystals, up to 1.1 mm across, associated with native copper, all coated with the oxidation products azurite, chrysocolla and malachite. Occasionally, ruby-red octahedral cuprite crystals have also been found in cavities in Triassic breccia (King 1973, Ince 2007). At Newhurst quarry, Shepshed, cuprite is associated with goethite and malachite as dark-red masses up to 1.5 mm across, and with chrysocolla and malachite in patches up to 5 mm across. In 1976 a thin vein of cuprite was discovered in a calcite-rich sandstone matrix containing voids in which chalcotrichite had formed (King 1973, Ince 2005).

Euhedral crystal of cuprite, 1.2 mm across, on malachite; New Cliffe Hill quarry, Stanton under Bardon, Leicestershire. (© D. I. Green)

Somerset *Mendip Hills:* Cuprite is one of many late-stage minerals found in calcite-lined cavities within Upper Triassic iron- and manganese-rich veins cutting Carboniferous Limestone at Coleman's quarry, Holwell, near Frome (Alabaster 1990a); leaching of primary copper sulphides in manganese-oxide deposits at Merehead (Torr Works) quarry, Cranmore, has produced cuprite together with azurite, chalcocite, native copper and malachite (Symes & Embrey 1977); small amounts of cuprite are noted from Moons Hill quarry, NE of Shepton Mallet (Kingsbury 1941).

Staffordshire Cuprite is one of a long list of copper minerals recorded from Ecton Hill (Sarjeant 1956b, Robey & Porter 1972, Ford et al. 1993).

Yorkshire Beach pebbles derived from the chalk of Speeton Cliff, near Flamborough Head, contain diffuse spots of irregular shape, up to 1 cm across, consisting of crudely octahedral cuprite intergrown with native copper mantled by calcite intergrown with malachite and an unidentified copper chloride (Roydhouse & Green 2003); minute masses of cuprite and native copper in a sulphide matrix are reported from the Middleton Tyas mines, near Scotch Corner (Green 1987b); and at Scaleber Bridge, 3 km SSE of Settle, minor As–Cu–Sb mineralization in dolomitized reef limestones contains chalcopyrite that is altered to 200 µm-wide aggregates of 1–2 µm-diameter spheres of cuprite, limonite and spionkopite. Enargite, famatinite, galena, luzonite, marcasite, pyrite and sphalerite are associated species (Ixer & Stanley 1998).

Wales

Dyfed *Central Wales orefield:* Cuprite is generally poorly formed and it probably occurs more widely that the literature suggests. It is found as deep-red crystalline masses and rarely as minute octahedral crystals associated with native copper and malachite at Eaglebrook (Nantycagl) mine, Ceulanymaesmawr (Jones & Moreton 1977, Jones 1983). Camdwrbach mine, nr Bryniau Rhyddion, NW of Nant-y-moch reservoir, and two mines near Devil's Bridge (Dolwen and Glogfawr) are noted by Bevins (1994). Very rare cuprite, forming veinlets intergrown with chalcocite and possible tenorite, and deep-red patches scattered through compact cerussite, occurs at Esgair Hir mine, Ceulanymaesmawr (Rust & Mason 1988). At Frongoch mine, between Devil's Bridge and Pont-rhyd-y-groes, octahedral crystals of cuprite up to 0.3 mm on edge have been found on schulenbergite (Green et al. 1996b), and at Gwaithyrafon mine, SE of Tal-y-bont, the species is intimately intergrown with native copper or develops drusy crusts on late-formed calcite (Rust & Mason 1994). Post-mining, lustrous transparent wine-red cuprite cubes, up to 0.5 mm across, on a mudstone matrix from the footwall of a lode at Lodge Park copper trial, Tre'r-ddôl, are perhaps the finest in the orefield (Bevins 1994, Mason & Green 1996). Cuprite is also noted from Penrhiw mine, Ystumtuen (Mason & Green 1995) and Ystrad Einion mine, near Tal-y-bont (Mason & Rust 1997).

Gwynedd Intense, widespread, polyphase supergene alteration of primary epigenetic copper-sulphide mineralization at the Great Orme mine, Llandudno, has produced secondary limonite and malachite (mainly) together with cuprite, native copper, covellite, digenite, spionkopite and tenorite (Ixer & Davies 1996, Ixer & Stanley 1996, Ixer 2001). Cuprite replaces chalcopyrite at the Parys Mountain copper deposit, Anglesey (Greenly 1919, Pointon & Ixer 1980, Southwood & Bevins 1995), and at Cwm Tregalan, S of Snowdon, it forms acicular crystals in vesicles within spilitic dolerite pillow lavas (Williams 1927). Saich & Rust (1987) note cuprite from a trial level on a chalcopyrite-bearing quartz vein, 2 km NW of Bontddu, near Dolgellau.

Powys Cuprite is recorded as an alteration product of chalcopyrite in the Central Wales orefield at Geufron mine, near Llanidloes, and Snowbrook (Nantyreira) mine, between Llanidloes and Tal-y-bont (Jones & Moreton 1977). It is also present at Dolyhir quarry, Old Radnor, where it is associated with a barite–chalcopyrite–galena–tennantite vein (Wolfe 2000, Bevins & Mason 1997, Green 2005a).

Scotland

Dumfries & Galloway Minute octahedra of cuprite (tile ore) associated with chalcopyrite are noted from Bay mine, Wanlockhead (Heddle 1901), and anthropogenic cuprite is reported on slag material from the Meadowfoot smelter, Wanlockhead (Green 1987a). Chalcotrichite and tile ore were once found in Balcary mine on Hestan Island in Auchencairn Bay, to the S of Almorness Point (Heddle 1901).

Strathclyde At Boyleston quarry, SW of Barrhead, cuprite occurs with analcime, calcite, native copper, malachite and prehnite, filling joints and cavities in fine-grain Carboniferous lavas. It forms cubo-octahedral crystals (<0.3 mm), elongate bladed crystals and hair-like masses of chalcotrichite. Cuprite formed during hydrothermal alteration of disseminated chalcopyrite in the lavas (Hall et al. 1989, Meikle 1989a). Barite–calcite veins cutting a Carboniferous intrusion exposed in Green Hill quarry, Kilbirnie, contain earthy cuprite associated with massive bornite and malachite (Wolfe 1997). Supergene alteration of sulphide-bearing veins and cavities in limestone at Rubha a' Mhill, northeast Isle of Islay, has produced dendrites of native copper in which minute lustrous cuprite crystals have grown (Green 1988). At Sannox Bay, Isle of Arran, stringers of cuprite have been found where Old Red Sandstone sediments have been baked by the injection of a basaltic dyke (Halsall 1980).

Ireland

Cork Massive cuprite associated with brochantite has been found in cavities within goethite in a fault-associated manganese-rich ore at Aghatubrid Beg mine, Glandore (Ryback & Moreton 1991). Crystallized cuprite and tile ore are recorded with azurite (chessylite), native copper and malachite from cavities in limestone at Ross Island mine, Killarney (Russell 1927).

Waterford Examination of the Stage lode of Tankardstown mine (to the east of Cuan Cove), Knockmahon, 22 km SW of Waterford town revealed massive cuprite with traces of native copper and botallackite in chalcopyrite–quartz veins (Braithwaite & Wilson 2001b). Specimens of cuprite on native copper from Kilduane mine are in the Russell collection at the NHM (Smyth et al. 1864, Moreton 2005a).

Wicklow Cuprite is one of the secondary species formed during oxidation of primary chalcopyrite–gold-bearing metamorphosed BIF rocks at Ballycoog, Goldmine River valley, near Avoca (Ixer et al. 1990).

Offshore territories

Isle of Man Minute octahedral crystals of cuprite are reported from the 40-fathom level of the South Bradda mines, N of Port Erin Bay (Smyth 1879–1884, Dewey & Eastwood 1925, Starkey 1998).

Jersey Alteration of bornite–chalcopyrite mineralization in ignimbrite from Bouley Bay, Trinity, produced a secondary assemblage containing cuprite together with azurite, covellite, spionkopite + yarrowite (blaubleibender covellite), limonite and malachite (Ixer 1980, 1992).

cuproadamite

A copper-rich variety of adamite. Cuproadamite was discredited as part of an IMA-CNMNC-approved mass discreditation of species (Burke 2006). The mineral should now be referred to as copper-bearing adamite. Since then, an IMA-approved formal description of zincolivenite has been published (Chukanov et al. 2007). It is an orthorhombic mineral with the composition $CuZn(AsO_4)(OH)$ and a compositional range of $Cu_{0.5}Zn_{1.5}$ to $Cu_{1.5}Zn_{0.5}$. Most cuproadamite is probably zincolivenite.

In Britain and Ireland, copper-bearing adamite has been found at: Penberthy Croft, St Hilary, Cornwall (D. Moulding specimen, D. I. Green XRD, Tindle unpub. EMP data); Driggith and Sandbed mines, Caldbeck Fells, Cumbria (Green et al. 2006a); Dolyhir quarry, Old Radnor, near Presteigne, Powys, Wales (D. I. Green XRD, Tindle unpub. EMP data); and Tynagh mine, near Killimor, Co. Galway, Ireland (Ryback 1988a, Ryback & Moreton 1992, D. I. Green XRD, Tindle unpub. EMP data). It is probable that this is all zincolivenite. See also adamite and zincolivenite.

cuprorivaite

tetragonal $CaCu^{2+}Si_4O_{10}$

A rare calcium–copper silicate briefly noted from Wheal Edward, St Just, Cornwall (Anthony et al. 1995, Gaines et al. 1997).

cuprosklodowskite

triclinic $Cu(UO_2)_2(SiO_3OH)_2·6H_2O$

A rare hydrated copper uranyl silicate hydroxide found as bright-green velvety botryoidal crusts and silky coatings composed of small radiating acicular crystals from the Cargodna section of West Wheal Owles, St Just, Cornwall (Ryback & Tandy 1992). Cuprosklodowskite is also reported from the Peeth lode, Geevor mine, near Pendeen, St Just, on a sample of chalcopyrite-spotted quartz on which a thin deposit of crystalline gypsum has been laid. It forms a thin pale-green crust of about 1 cm², composed of tiny spherulitic aggregates and divergent groups of acicular crystals up to about 0.01 mm long. Elsewhere along the Peeth lode, other secondary uranium minerals occur, including andersonite, johannite, schröckingerite and possibly becquerelite (Elton & Hooper 1992, Elton et al. 1992).

Spherulitic aggregates of cuprosklodowskite on gypsum from Geevor mine, Pendeen, St Just, Cornwall; width 180 µm; SEM image. (© N. J. Elton)

cuprotungstite

tetragonal $Cu_3^{2+}(WO_4)_2(OH)_2$

A rare copper tungstate hydroxide that forms part of a supergene assemblage associated with a chalcopyrite–ferberite–scheelite mineralized quartz vein cutting the Eskdale granodiorite at Buckbarrow Beck, Corney Fell, Waberthwaite, Cumbria. Cuprotungstite occurs as waxy bright-green crusts up to about 0.01 mm thick on fine granular scheelite. It is found with other supergene minerals such as bismutoferrite, brochantite, chrysocolla, covellite, goethite, malachite and russellite (Young et al. 1986, 1991b). Later investigations added bismutite, eulytite, namibite and mixite to this assemblage (Neall et al. 1993).

In Cornwall, cuprotungstite is reported as a greenish resinous alteration rind on wolframite from the dumps of Wheal Gorland, St Day, 2 km E of Redruth (Rust 1981b). Ryback & Tandy (1992) report that A. W. G. Kingsbury also found cuprotungstite here. Golley & Williams (1995) report cuprotungstite from Old Gunnislake mine, Calstock, near Gunnislake, and Gunheath china-clay pit, Carthew, St Austell.

cupro-uranite

A synonym of torbernite, autinite group.

CUSPIDINE GROUP

Monoclinic and triclinic silicates with the general formula $X_{16}(Si_2O_7)_4(O,OH,F)_8$, where X = Na, Ca, Zr, Ti, Nb, Mn^{2+}, Fe^{2+}, Y, REE. Baghdadite and cuspidine are members of the group found in Britain or Ireland.

cuspidine

CUSPIDINE GROUP

monoclinic $Ca_8(Si_2O_7)_2F_4$

An uncommon calcium silicate fluoride found in the limestone/gabbro contact zone in the Grange Irish neighbourhood close to Grange Cross, Carlingford, Co. Louth, Ireland. In hand-specimen, cuspidine is grey and it forms cleaved xenoblastic masses up to 9 mm in diameter that are difficult to distinguish from the associated minerals. It forms in the interstices between vesuvianite or diopside, or as a selvedge to phlogopite, or at the margins of calcite crystals. The species also occurs with apatite, edenite, grossular, magnetite, monticellite, pyrite, spinel and wollastonite, sometimes as large individual crystals. It is thought that this assemblage formed as a result of metasomatism of limestone by a hidden acid pegmatite (Osborne 1932a,b). Nockolds & Vincent (1947) report cuspidine from five localities in this area, three of which are in melilite-bearing rocks.

Central twinned crystal of cuspidine in a cuspidine–clinohumite–forsterite–calcite skarn. Beinn an Dubhaich granite aureole, Camas Malag, near Broadford, Isle of Skye, Highland Region, Scotland; width 4.5 mm; viewed in PPL and XPL. (© K. Gardner)

At Killala Bay, near Inishcrone, Co. Sligo, thermal metamorphism of limestone by basalt–dolerite dykes has produced spurrite-rich rocks that have later been hydrothermally altered. Cuspidine occurs in the primary assemblage as a subsidiary component in calcite–spurrite and spurrite–rankinite–tilleyite–wollastonite rocks, where it forms subhedral or rounded twinned grains. It is also one of the main alteration products and is found with afwillite, calcite, killalaite, scawtite, xonotlite and other hydrated calcium silicates (Nawaz 1974).

At Kilchrist and Camas Malag, near Broadford, on the Isle of Skye, Highland Region, Scotland, the Beinn an Dubhaich granite has intruded dolomitic limestone, and a mineralogically diverse skarn assemblage has developed. Expulsion of boron- and fluorine-rich fluids from the granite metasomatized Durness limestone near to the granite contact and led to the crystallization of this skarn assemblage. Cuspidine is found with chondrodite, clinohumite, forsterite, humite and monticellite in a magnetite-rich skarn at Kilchrist and in a clinohumite–forsterite–calcite skarn at Camas Malag (Tilley 1948b, 1951). Examination of polysynthetically twinned lanceolate crystals from Skye and Idaho has shown that cuspidine and custerite are in fact the same species (Tilley & Vincent 1947).

At Camas Mòr, Isle of Muck, Highland Region, an olivine-gabbro dyke intersecting Jurassic limestones has developed a metasomatic aureole dominated by an analcime–clinopyroxene–wollastonite skarn assemblage. Cuspidine forms interstitial grains mantling gehlenite and wollastonite, and also develops subhedral crystals enclosing calcite and gehlenite (Tilley 1947). In a similar environment at Kilchoan, Ardnamurchan, cuspidine–rustumite rocks occur sporadically and form part of a skarn assemblage that also includes åkermanite, grossular, kilchoanite, larnite, spurrite and wollastonite (Agrell 1965).

custerite

A synonym of cuspidine.

cyanite

A synonym of kyanite.

cyanosite

A synonym of chalcanthite.

cyanotrichite

orthorhombic $Cu_4^{2+}Al_2(SO_4)(OH)_{12}{\cdot}2H_2O$

A rare hydrated copper–aluminium sulphate hydroxide found as a secondary mineral in the oxidized portion of copper deposits.

The Mulberry tin stockwork, Lanivet, St Austell, Cornwall, contains fissure veins carrying coarsely crystallized cassiterite and traces of arsenopyrite, chalcopyrite, stannite and wolframite. Supergene alteration of this assemblage during oxidation of the primary sulphides has produced a suite of secondary minerals, including brochantite, chalcophyllite, covellite, cyanotrichite and scorodite. Cyanotrichite occurs in clusters (≤0.5 mm across) of bright-blue acicular crystals in cavities within iron-stained quartz (Starkey 1986a). At Penberthy Croft mine, St Hilary, minute bright-blue coatings or globules (≤0.4 mm across) of cyanotrichite occur rarely in vugs and as cavity infills to 1.7 mm across in goethite (Betterton et al. 2000). Hand specimens of azurite, cyanotrichite and torbernite were found at Wheal Basset, Illogan, near Redruth, during derelict-land reclamation works (Wolfe 1996a). Cyanotrichite is also reported from Phoenix United mines, Linkinhorne, Cornwall (Anthony et al. 2003), and from Leadhills, Dumfries & Galloway, Scotland (Gaines et al. 1997).

Other reports of cyanotrichite are all problematic: the lettsomite (a synonym of cyanotrichite) mentioned in Spencer (1898) cannot be traced (Macpherson 1983); cyanotrichite from Engine Vein mine, Alderley Edge ridge, Cheshire, is unconfirmed (Braithwaite 1994); and the provenance of cyanotrichite specimens in the Kingsbury collection at the NHM and described in various publications have dubious provenance: Wheal Gorland, St Day, 2 km E of Redruth, Cornwall (Kingsbury 1954, Rust 1981b); Potts Gill barite mine, Caldbeck Fells, Cumbria (Hartley 1984, Young 1987a, Cooper & Stanley 1990, Anthony et al. 2003) – they are probably all part of the Kingsbury problem outlined by Ryback et al. (1998, 2001).

Tiny spherules of cyanotrichite to 0.5 mm. Hingston Down, Calstock, Cornwall. (© M. P. Cooper)

cylindrite

triclinic $FePb_3Sn_4Sb_2S_{14}$

A very rare iron–lead–tin–antimony sulphide reported from the Navan lead–zinc orebody, Co. Meath, Ireland, where minor quantities have been identified in polished sections of the ore. Argyrodite, boulangerite, bournonite, franckeite, freibergite, jordanite, pyrargyrite and semseyite are also found as uncommon constituents in the predominantly galena–sphalerite ore (Andrew & Ashton 1985, Ashton et al. 1986).

cymrite

hexagonal, monoclinic or orthorhombic $Ba(Si,Al)_4(O,OH)_8{\cdot}H_2O$

TL Benallt mine, Rhiw, Lleyn Peninsula, Gwynedd, Wales

A rare hydrated barium aluminosilicate found in lenses of manganese ore within Ordovician mudstones at the type locality. Two occurrences are described from the mine: first, on the 60-foot level in the no. 1 orebody, where cymrite forms colourless plates up to 7 mm across and about 0.7 mm thick; and, secondly, in the no. 5 orebody, where it forms white crystals up to 1 mm long, with a very closely spaced cleavage, in association with ganophyllite (Campbell Smith et al. 1949).

In Scotland, cymrite is reported from a stratabound zone of barite, galena and sphalerite mineralization in Middle Dalradian metasedimentary rocks N of Aberfeldy, Tayside Region. The mineralization is contained in graphitic schists of the Ben Eagach Schist Formation and consists of interbanded barite rock, celsian–quartz rock and sulphide-bearing carbonate rocks. Cymrite was observed in a few specimens of schistose celsian–quartz rock. It was probably more abundant, but is thought to have been mainly replaced by celsian (Coats et al. 1980, Fortey & Beddoe-Stephens 1982, Willan & Coleman 1983). At Foss mine, 7 km NW of Aberfeldy, the Ben Eagach Schist formation contains widespread textural evidence for the partial-to-complete replacement of cymrite by coarser celsian, particularly in celcian-rich quartz–mica schist forming the upper mineralized horizons in the Foss East and Frenich Burn areas (Moles 1985).

cyrilovite

tetragonal $NaFe^{3+}_3(PO_4)_2(OH)_4{\cdot}2H_2O$

A rare hydrated sodium–iron phosphate hydroxide found associated with chalcosiderite and dufrénite in an area of copper mineralization at Gunheath china-clay pit,

Sheaf-like aggregate of cymrite associated with celsian. Benallt mine, Rhiw, Lleyn Peninsula, Gwynedd, Wales; width 780 µm. SEM–BSE image. (© National Museum of Wales, Cardiff; photographers T. F. Cotterell & M. P. Lambert)

Carthew, St Austell, Cornwall. These species occur on the faces of small fractures in kaolinized granite and on a layer of milky quartz, associated with a hematite–quartz vein. The area of the pit where the cyrilovite was found has been mined away, but was close to a prominent turquoise-rich vein that in addition contained cassiterite (including the variety varlamoffite), libethenite, torbernite and wolframite (Ince 1986). Cyrilovite forms clear "honey-yellow" or translucent "waxy-yellow" pseudo-octahedral crystals to 0.7 mm across and globular crystal aggregates to 0.5 mm across (Weiss 1994).

"Honey-yellow" crystals of cyrilovite to 0.3 mm with pale-yellow sprays of leucophosphite; Gunheath china-clay pit, St Austell, Cornwall. (© M. P. Cooper)

D

dahllite
A synonym of carbonate-hydroxylapatite, apatite group.

damourite
A term used to describe both muscovite and zincite.

danalite
cubic $Fe_8(Be_6Si_6O_{24})S_2$

A rare iron–beryllium silicate with sulphide, which forms two separate series, with genthelvite and with helvine. The first British description of danalite was on a specimen purchased in 1864 and whose provenance was given vaguely as Redruth, Cornwall. Danalite forms vitreous or resinous crystals varying in colour from columbine red to a brick red and is associated with arsenopyrite, calcite and quartz (Miers & Prior 1891). Resinous, reddish-brown masses of danalite, almost indistinguishable from the associated skarns, are reported from the cliffs of Chy Cornish at Botallack, St Just (Bancroft & Weller 1993), and in the nearby cliffs of Wheal Cock Zawn, Roscommon Cliffs (Wolloxall 1989). Other reports of danalite in Cornwall (Kingsbury 1961) are problematic, and, given the doubts about the provenance of other Kingsbury specimens (Ryback et al. 1998, 2001), further confirmation is required: Levant mine, St Just; New Tolgus shaft, Redruth; Tretoil mine, Lanivet; Wheal Clinton, near Trefusis Point, Mylor, Falmouth; and Wheal Maudlin, 3 km NW of Lostwithiel.

Danalite occurs as minute translucent brown tetrahedral crystals in miarolitic cavities within granite from Lindsay's Leap quarry, Mourne Mountains, Co. Down, Ireland (Green et al. 2005a).

danburite
orthorhombic $CaB_2(SiO_4)_2$

A rare calcium borosilicate supposedly found by A. W. G. Kingsbury in 1951–1952 at Cheesewring quarry, SW of Henwood, Linkinhorne, Cornwall. The specimen is now held by the NHM (Embrey 1978); this cannot be considered a confirmed occurrence (Ryback et al. 1998, 2001). Danburite is reported from Red-a-ven mine, Meldon, near Okehampton, Devon (El Sharkawi & Dearman 1966), and at Meldon quarry (British Rail quarry) it has been identified as greyish translucent grains within a datolite–grossular–wollastonite-rich rock (M. S. Rumsey pers. comm. 2006).

daphnite
TL Penzance, Cornwall, England

Magnesium-rich chamosite, chlorite group.

datholite
A synonym of datolite, gadolinite group.

datolite
GADOLINITE GROUP

monoclinic $CaBSiO_4(OH)$

A relatively common calcium borosilicate hydroxide found as a secondary mineral in mafic igneous rocks and in serpentinites, hornblende schists and skarns (within limestone).

Datolite is associated with chalcopyrite and natrolite in veins and geodes at the junction of serpentinite and hornblende schist at Parc Bean Cove, Mullion, Lizard Peninsula, Cornwall. It forms beautiful greyish-white crystals up to 2 cm long and also occurs in botryoidal form (McLintock 1910a, Collins 1911a). Datolite is also noted from a picrite on Clicker Tor, Menheniot, Cornwall (Robson 1949b), and from a sulphide-rich pegmatitic facies of the Whin Sill at Barrasford quarry, near Gunnerton, Northumberland (Randall 1959). The source of boron required to form datolite is not obvious in these occurrences, but in the Meldon aplite quarries, near Okehampton, Devon, datolite occurring in thin reaction veins cutting skarn has the nearby Dartmoor granite as an obvious source. In addition to datolite, the Meldon veins also contain albite, axinite, bavenite, calcite, fluorite, prehnite, pyrrhotite and quartz (Dearman & Claringbull 1960). Datolite is said to be abundant within the

Datolite crystal, 2.5 mm tall, from Parc Bean Cove, Lizard Peninsula, Cornwall. (© D. I. Green)

metamorphic aureole of the Dartmoor granite at Meldon, sometimes as well formed crystals. It occurs as a replacement of wollastonite in malayaite–wollastonite hornfels collected from the dumps of Red-a-ven mine (Dearman 1966b, El Sharkawi & Dearman 1966). In the Meldon railway quarry, calc-flinta rocks contain datolite in a grossular–vesuvianite–(fluorite) assemblage produced during metasomatic alteration of an original chert (El Sharkawi & Dearman 1967). Datolite is also known from Ramsley mine, South Tawton, near Okehampton (Symes 1977), and from the limestone of South Brent (McLintock 1910a).

In the Carboniferous Limestone quarry at Merehead (Torr Works), Cranmore, Somerset, a large complex manganese- and iron-rich vein, up to 4 m wide, was uncovered in 1975. Towards the bottom of the vein, in an area of 2×1 m, cavities in goethite were found to be lined with an assemblage of apophyllite, barite, calcite and milky-white fibrous radiating masses of datolite (Symes 1977, Symes & Embrey 1977).

A suite of boron-bearing minerals that includes datolite, fluoborite, harkerite, ludwigite and szaibelyite is found in the contact skarns surrounding the Beinn an Dubhaich granite in the Kilchrist area, near Broadford, Isle of Skye, Highland Region, Scotland. At the No.1 Prospect, Kilchrist, datolite appears to replace quartz in the granite at the actual contact with a grossular–wollastonite skarn and is associated with prehnite. A spherulitic botryolite variety is also noted here (Tilley 1948b, 1951). Elsewhere in Scotland, datolite is reported by Greg & Lettsom (1858) and Heddle (1901) from:

FIFE Isle of May, Firth of Forth (small crystals in prehnite, associated with scolecite). Datolite is also noted by Phemister & MacGregor (1942) from Chapel quarry, near Kirkcaldy, where it occurs with apophyllite, grossular, pectolite and talc in contact-altered limestone close to a thick sill of quartz dolerite. LOTHIAN Barnton Park railway cutting (fine crystals associated with analcime, pectolite, prehnite and natrolite); Corstorphine Hill, Edinburgh (colourless-to-white crystals); Salisbury Crag, Edinburgh (pale greenish-yellow, crystals up to 2 cm long). STRATHCLYDE Bishopton railway tunnel (large snow-white crystals associated with prehnite); Bowling quarry, Kilpatrick Hills, near Glasgow. TAYSIDE Glen Farg (pale-green associated with analcime).

davidsonite

TL Rubislaw granite quarry, Aberdeen, Grampian, Scotland

Named after James Davidson (1792–1841), professor of civil and natural history at Marischal College, Aberdeen, Grampian, Scotland. Davidsonite is a variety of beryl.

dechenite

A variety of descloizite.

delafossite

trigonal $Cu^{1+}Fe^{3+}O_2$

A rare copper iron oxide from Cornwall used by Buist et al. (1966) in their experimental work on the system Cu–Fe–O. Specimens of spherulitic delafossite used in the experiments came from the BM(NH). No further details are given. Delafossite is confirmed by XRD analysis from Tolvaddon mine, near Perranzabuloe, Cornwall (S. A. Rust pers. comm. 2004).

delessite

Delessite is described by Ford et al. (1993) as the magnesium-rich variety of chamosite, but Bayliss (2000) considers delessite to be iron-rich clinochlore. To add further confusion, Clarke (1993) gives delessite species status and de Fourestier (1999) considers it to be either magnesian chamosite or ferrous clinochlore.

dellaite

triclinic $Ca_6Si_3O_{11}(OH)_2$

TL Kilchoan, Ardnamurchan, Highland Region, Scotland

A rare calcium silicate hydroxide found at the type locality in nodular or concretionary limestone that has been thermally metamorphosed by dolerite and gabbroic intrusions. Dellaite forms irregular or bladed crystals in late-stage replacement veins with calcite and spurrite. Dellaite and foshagite, and dellaite and calcite (±tilleyite), replace kilchoanite and spurrite respectively (Agrell 1965).

demidoffite

A synonym of chrysocolla.

derbyshire spar

A synonym of fluorite.

DESCLOIZITE GROUP

Orthorhombic arsenates and vanadates with the general formula $PbM(XO_4)(OH)$, where $M = Cu^{2+}$, Fe^{2+}, Mn^{2+}, Zn; and $X = As^{5+}$, V^{5+}. Arsendescloizite, descloizite, mottramite and pyrobelonite are members of the group found in Britain or Ireland.

descloizite

DESCLOIZITE GROUP

orthorhombic $PbZn(VO_4)(OH)$

A lead–zinc vanadate hydroxide that forms a series with mottramite and typically occurs in the oxidized zone of lead–zinc deposits. Descloizite is described from New vein, Wanlockhead Dod, and High Pirn mine, Wanlock Dod, both Dumfries & Galloway, Scotland. Descloizite (dechenite) forms dark or reddish-yellow botryoidal crystals associated with hemimorphite (calamine), mottramite (pssitacinite), pyromorphite and vanadinite (Heddle 1901, Brown 1927). These occurrences are doubtful, but Bannister (1933) appeared to confirm descloizite (using XRD) as a thick black crust, yielding a canary-yellow powder, with vanadinite, on massive pyromorphite at Wanlockhead. Small brown spherules of vanadinite from this location had previously been identified as descloizite (Frenzel 1875). Recent examination of suspected descloizite from Leadhills confirms it to be mottramite, so the normal way of telling the two species apart (black crystals = mottramite and orange-red ones = descloizite) appears to fail. As the XRD pattern of descloizite is indistinguishable from other members of the group, the existence of this species in the Leadhills–Wanlockhead orefield is in doubt (D. I. Green XRD, Tindle unpub. EMP data).

Descloizite has been found in cavities in massive galena collected from a vein in the southeast corner of Greystone quarry, Lezant, near Launceston, Cornwall. It forms dark-brown translucent keel-shape crystals reaching 0.2 mm in length in association with orange vanadinite (Elton et al. 2004). Bannister (1933) describes descloizite forming minute black crystals and yielding an olive-green powder on sandstone from Pim Hill, near Harmerhill, 9 km N of Shrewsbury, Shropshire. Additionally, descloizite is mentioned as a rare species from small opencast workings in Carboniferous Limestone at Wesley Theological College (Wesley mine), Westbury-on-Trym, Avon (Alabaster 1989b), and with duftite from Old Potts Gill mine, Deer Hills, Caldbeck Fells, Cumbria (Cooper & Stanley 1990).

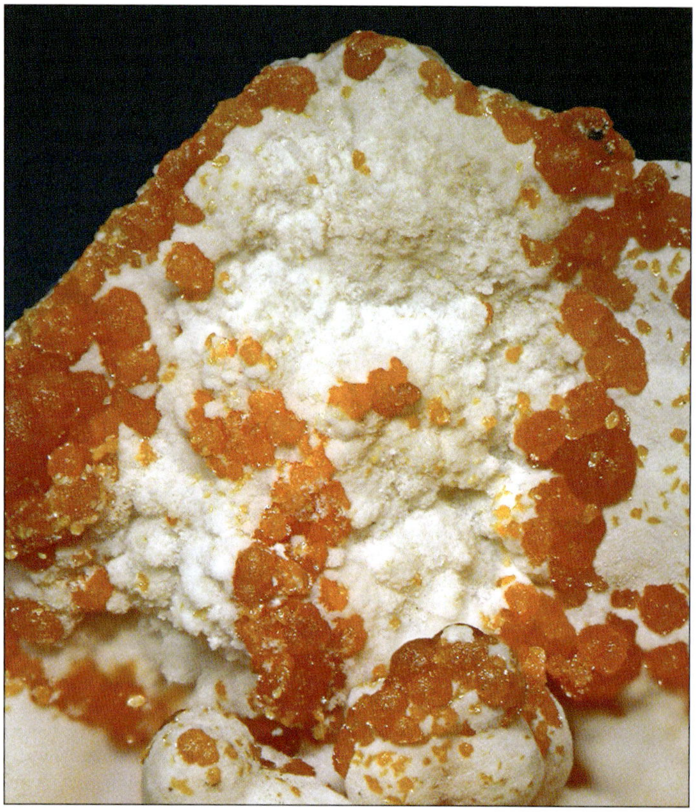

Orange descloizite clusters to 1 mm across with calcian pyromorphite (white) from Whitwell quarry, Whitwell, Derbyshire. (© D. I. Green)

Two other English occurrences of descloizite are less certain: the higher part of Brandy Gill, Caldbeck Fells, Cumbria (Kingsbury & Hartley 1956a; quoted by Hartley 1984, Young 1987a, Cooper & Stanley 1990), and Engine Vein mine, Alderley Edge, Cheshire (specimens in the Kingsbury collection at the NHM). Investigative work by Ryback et al. (1998) concludes that the specimens are probably from Arizona, USA and Germany, respectively.

In Wales, descloizite occurs with brochantite at Esgair Hir mine, Ceulanymaesmawr, Central Wales orefield, Dyfed (Bevins 1994).

desert rose

A synonym of petaline clusters (rosettes) of both barite and gypsum.

desmine

A synonym of stilbite, zeolite group.

devilline

monoclinic $CaCu^{2+}_4(SO_4)_2(OH)_6 \cdot 3H_2O$

TL Cornwall, England

A relatively rare hydrated calcium–copper sulphate hydroxide that forms in the oxidized zone of copper-sulphide deposits and can be of post-mining origin. Its early discovery from an unspecified Cornish locality (Pisani 1864) is further documented by Kingsbury (1954). Devilline is an important and locally widespread mineral at Penberthy Croft mine, St Hilary, Cornwall, where it is associated with birnessite, brochantite and connellite. It forms radiating spherules and rosettes, and tufted lath-like aggregates up to 2 mm across. It displays a range of colours from light to medium green, to bluish green, dirty yellow and silvery white (Betterton et al. 2000). Rust (1982b) also reports lightish-blue microbladed crystals of devilline from this locality. Other locations in Cornwall include: Botallack mine, St Just (Bancroft & Weller 1993); Ennis Wheal Virgin, St Hilary; Geevor mine, near Pendeen, St Just; Perran St George mine, Perranzabuloe (all Golley & Williams 1995); Fowey Consols mine, Tywardreath, St Blazey (Lewis 1997). Reports of devilline from Wheal Tremayne, near Hayle, Gwinear, and Wheal Providence, near Leedstown, Gwinear (Kingsbury 1954), require further confirmation in view of doubts about the provenance of many Kingsbury samples (Ryback et al. 1998, 2001).

Devilline is confirmed from Wheal Friendship (Devon Friendship mine), Mary

Tavy, Devon, where well developed microcrystals, sometimes pseudomorphing calcite, have been found associated with other secondary sulphate species – anglesite, brochantite, gypsum, jarosite, langite, linarite and serpierite (Brooke 1986).

In the Northern Pennine orefield, devilline occurs as post-mining encrustations on broken veinstone and wallrock, and locally on the walls of old workings at Tynebottom mine, NW of Garrigill, near Alston, Cumbria. It forms pale-blue to greenish-blue crusts of platy crystals, which are locally up to 1 mm thick and associated with brochantite, malachite and serpierite (Bridges 1987, Rust 1987a). Devilline is a minor component of similar crusts (dominated by serpierite) at Smallcleugh mine, E of Nenthead, near Alston, Cumbria (Bridges 1987). Green & Briscoe (2002) also report annabergite, brianyoungite, namuwite, schulenbergite, and an unnamed zinc analogue of ktenasite, forming post-mining supergene minerals from this mine.

Hartley (1984) reports devilline as green to bluish-green crusts of post-mining origin from Driggith and Sandbed mines in the Caldbeck Fells, Cumbria. This is corroborated by Green et al. (2006a), who found white to pale-blue or green crusts of devilline at these mines. Devilline is further confirmed from Ladywell mine, Shelve, Shropshire (Braithwaite 1982b).

In Wales, rare devilline has been found in a trial level on a 1 m-wide quartz vein, 2 km NW of Bontddu, near Dolgellau, Gwynedd, where it forms light bluish-green to rich bluish-green lath-like divergent crystals, sometimes pseudomorphed by chalcoalumite and iron oxide (Saich & Rust 1987). In the Central Wales orefield, devilline is reported from Eaglebrook (Nantycagl) mine, Ceulanymaesmawr, Dyfed, where primary chalcopyrite, galena and sphalerite were once worked. Green blades of devilline up to about 2 mm long coat fracture surfaces of massive dolomite associated with anglesite, brochantite, cerussite (abundant), langite, leadhillite, linarite, malachite, serpierite and wroewolfeite (Jones & Moreton 1977, Braithwaite 1982b, Jones 1983). Additionally, tentative identifications of devilline have been made on samples from Dyfngwm mine, Machynlleth, Powys (Rust & Rust 1987) (actually the Pen Dylife section of Dylife mine) and Gwaithyrafon mine, SE of Tal-y-bont, Dyfed (Rust & Mason 1994). In Powys, devilline occurs as small, pale-blue, wispy, feathery aggregates lining cavities in an area of copper mineralization at Dolyhir quarry, Old Radnor, near Presteigne. Djurleite and malachite are associated species (T. F. Cotterell pers. comm. 1007).

Devilline is common as bluish-green pearly crystals in cavities in slag from the dumps of the Meadowfoot smelter, Wanlockhead, Dumfries & Galloway, Scotland (Green 1987a).

At Muckross mine, Killarney National Park, Co. Kerry, Ireland, devilline occurs as pale-blue to green encrustations and minute poorly formed lath-like crystals on chalcopyrite. It is just one of many species formed by post-mining oxidation of the primary Ag–As–Co–Cu–Fe–Pb–Zn mineralization (Moreton et al. 1999). A second Irish occurrence is at Allihies, near the end of the Bearhaven (Berehaven) Peninsula, Co. Cork, where brochantite, connellite, devilline and langite occur in extensive secondary encrustations (Ryback & Moreton 1991).

devonite

TL Devon, England

A synonym of wavellite.

deweylite

A mixture of talc and serpentine.

dewindtite

orthorhombic $H_2Pb_3(UO_2)_6O_4(PO_4)_4{\cdot}12H_2O$

A very rare hydrated lead uranyl phosphate found in a localized patch of Cu–Pb–U mineralization near the base of the seacliff at Loe Warren Zawn, W of Botallack, St Just, Cornwall. Dewindtite has formed where surface water draining from rubble at the base of the cliff has trickled over veinlets carrying chalcocite, chalcopyrite and pyrite, and precipitated an assemblage of supergene carbonate and phosphate minerals on exposed surfaces and joint planes near a patch of uraninite mineralization. Dewindtite forming bright-yellow thick tapered laths (often doubly terminated) or radiating groups (spherules) of ragged crystals, is associated with widenmannite in some specimens, and together they appear to be the last formed of the uranium-bearing species at this site. Other associated species include cerussite, kasolite, malachite, and intermediate members of the torbernite–zeunerite series (Elton & Hooper 1995a,c).

diabantite

Iron-rich clinochlore, chlorite group.

Dewindtite with canoe-shape habit, from Loe Warren Zawn, St Just, Cornwall; width 120 μm; SEM image. (© N. J. Elton)

diaboléite/diaboleite

tetragonal $Pb_2Cu^{2+}Cl_2(OH)_4$

TL Higher Pitts farm, Priddy, Mendip Hills, Somerset, England

A rare lead–copper oxychloride found at the type locality in calcite-rich veins, fissures and pockets of iron–manganese mineralization cutting Carboniferous Limestone. It occurs as inclusions in mendipite in close association with chloroxiphite, with which it may also be embedded. Diaboléite forms bright sky-blue tabular crystals, with brilliant lustre and transparency, not unlike linarite or azurite in appearance. It also forms small compact finely granular aggregates with a dull lustre. Seldom occurring as well formed crystals, the largest diaboléite crystals are not more than 2 mm across (Spencer & Mountain 1923, Burr 1996).

At Merehead (Torr Works) quarry, Cranmore, Somerset, mineralized veins, fissures and pockets similar to those at Higher Pitts farm contain fine-grain pale-blue aggregates or broken, poorly formed prismatic crystals of diaboléite (up to 9 mm in length) associated with cerussite, cotunnite, hydrocerussite and wulfenite, and the lead oxychlorides chloroxiphite, mendipite and parkinsonite (Alabaster 1975, Symes & Embrey 1977, Symes et al. 1994, Turner 2007). Elsewhere in the Mendip Hills, diaboléite is also briefly described from Coleman's quarry, Holwell, near Frome (Alabaster 1990a), and possibly from Batts Combe quarry, near Cheddar (Eastwood 1983). Lead oxychlorides in the Mendips are thought to have formed by interaction of primary galena with low-temperature metalliferous brines along fractures cutting original Mn–Fe–Pb–Cu orebodies (Symes & Embrey 1977, Humphreys et al. 1980), although they may have precipitated directly from a cooling brine (Welch et al. 2000). See mereheadite for an alternative formation model.

Diaboléite is also described from a cliff exposure on Gunver Head, near Padstow Consols mine, Cornwall, where it forms sky-blue needles up to 0.4 mm long, on a matrix of chalcopyrite–galena–sphalerite-bearing quartz veinstone interspersed with slaty killas and closely associated with cumengéite (Braithwaite & Merry 1996, Schlomann & Weiss 1998). Starkey (1987a) reports diaboléite from Daymer Bay, Polzeath, Cornwall.

diaclasite

Serpentine pseudomorph after iron-rich enstatite, pyroxene group.

diackethyst

TL near the site of Black Jack's Castle, Craig, Montrose, Tayside, Scotland

Named after Revd Diack, minister of Craig parish for 27 years. Diackethyst is a local name for an inadequately described "translucent-wine" and amethystine-colour chalcedony – a variety of quartz (Diack 1979).

diadochite?

triclinic $Fe^{3+}_2(PO_4)(SO_4)(OH){\cdot}6H_2O$

A rare hydrated iron phosphate sulphate hydroxide formed by sulphate-rich solutions acting on earlier-formed phosphates. Diadochite is noted from Gravel Hill mine, Perranzabuloe, Cornwall, in a *South-West branch newsletter* of the Russell Society (Sparrow 1989, Golley & Williams 1995). Further details and analytical confirmation are desirable.

diallage
A vague term used to describe certain varieties of amphibole and pyroxene.

dialogite or diallogite
A synonym of rhodochrosite, calcite group.

diamond?
cubic C

A gem-quality diamond was reportedly found in the Colebrooke River, Co. Fermanagh, Ireland in 1816. It is referred to as the Brookeborough diamond and is now set in a ring of Irish (Co. Wicklow) gold. Despite considerable exploration interest, no likely diamond-bearing geological structures such as kimberlite pipes have been discovered and no diamonds have since been found (Ball 1887, Firsoff 1971, Bazley 1997, Arthurs & Earls 2004).

Another account of diamond is given by Smith (1895) in a description of a graphite mine at Craigman, New Cumnock, Strathclyde, Scotland. The whereabouts of the specimens are unknown and the occurrence is extremely implausible.

Diamond is also mentioned by Goodchild (1901) in describing a 19th-century specimen of aventurine (metaquartzite) collected by Professor Matthew Heddle from a lake NE of Ben Hope, Highland Region, Scotland. The specimen was said to contain red mica, red zircons, and either colourless garnets or diamonds. The specimen was rediscovered in 1998 in the Hunterian Museum, Glasgow, having been missing for most of the 20th century. Faithfull (2007) shows that it contains pale-yellow spessartine garnet rather than diamond, and piemontite instead of zircon.

diaspore
orthorhombic AlO(OH)

An uncommon aluminium oxyhydroxide, dimorphous with böhmite and found in a variety of basic (originally dolerite) hornfels types near the western margin of the Land's End granite, in the area around Kenidjack and Botallack, St Just, Cornwall. In an anthophyllite–cordierite–spinel (pleonaste) hornfels, the diaspore is intimately associated with spinel, and together they occur in clots of crystals up to 6 mm across, included in cordierite. Diaspore usually forms rhombic crystals at the core of these clots. Other varieties of hornfels contain, in differing combinations, biotite, cummingtonite, hornblende, plagioclase and quartz. Apatite, ilmenite and magnetite are minor components (Tilley & Flett 1930, Tilley 1935, Harker 1950). A re-examination of the rocks from the Kenidjack quarries and nearby Zawn a Bal showed that gedrite was also present, and an origin attributable to thermal metamorphism of existing rocks, rather than metasomatism, was suggested (Chinner & Fox 1974).

In Strathclyde, Scotland, at Dubbs quarry, Kilwinning, and Saltcoats beach, near Ardrossan, pisolitic bauxitic clays contain minute crystals (0.03 mm) of diaspore associated with böhmite. Both appear to have crystallized from a colloidal alumino-silica gel that was produced during alteration of basaltic rocks in a somewhat tropical climate (Lapparent 1934, Macpherson & Livingstone 1982). A very rare occurrence of diaspore is described by Meikle (1990) from a single highly mineralized basalt block found at Hartfield Moss, 9 km WSW of Barrhead, Strathclyde. Diaspore occurs in solution cavities in the matrix of the basalt, as creamy-yellow to off-white spherules up to 1.2 mm across, and also in 2 mm-wide open channels within masses of natrolite, prehnite and thomsonite, where it forms platy needle-like crystals up to 0.3 mm long. Hematite, muscovite, native silver and rare covellite are associated with the needle-like diaspore. Dissolution channels in basalt from Loanhead quarry, near Beith, Strathclyde, contain diaspore, again with a native silver–prehnite association (Meikle & Todd 1995).

Detrital minerals collected from a sandy boulder clay at Ludchurch, Dyfed, Wales, contain colourless plates of diaspore up to 0.4 mm across (Griffiths & Stuart 1940).

diatomite
Opaline silica; diatomaceous earth.

dichroite
A synonym of cordierite.

dickite
KAOLINITE–SERPENTINE GROUP

monoclinic $Al_2Si_2O_5(OH)_4$

TL Trwynbychan, near Amlwch, Anglesey, Gwynedd, Wales

Named after Allan Brugh Dick (1833–1926), a Scottish metallurgical chemist and designer of the Dick microscope, who described the mineral as kaolinite in 1888; born in Edinburgh. Subsequently, dickite, kaolinite and nacrite were all described as kaolinite, and it was not until 1930 that Ross & Kerr established dickite as a discrete species. Dickite is a relatively uncommon aluminosilicate hydroxide, polymorphous with halloysite, kaolinite and nacrite, which occurs along hydrothermally altered veins containing aluminosilicate minerals and is also found as an authigenic sedimentary mineral. It is uncertain if the kaolinite reported from Bolton Abbey in Yorkshire, Newcastle-upon-Tyne, and Congleton in Cheshire (Dick 1908) is actually dickite.

Russell (1949b) reports snow-white crystallized dickite often with ankerite or dolomite from four antimony mines in Cornwall: Dizzard (Duchy & Stuckey United) mine, Poundstock; Tredinnick mine, Landrake with St Erney; Trenute mine, near Lezant; and West Town Farm (mine), Botus Fleming. To this list, Golley & Williams (1995) add Millock beach, St Gennys, and Wheal Leigh, Pillaton.

Young (1987a) reports dickite in Cumbria from: Bank Wood quarry, Bassenthwaite; Embleton mine, Embleton, near Cockermouth (in mineralized breccia associated with dolomite and galena); and Robin Hood mine, Bassenthwaite (massive with stibnite). At the last locality, more recent excavations in a small quarry have confirmed the presence of dickite, sphalerite, stibiconite, stibnite and valentinite (Green & Briscoe 2002). White crystallized dickite was found in a large erratic boulder uncovered in 1895 during construction of the railway line between Troutbeck and Threlkeld. The boulder also contained quartz, stibiconite and stibnite (Russell 1949b). At Embleton quarry, near Cockermouth, dickite occurs in a quartz–carbonate vein associated with alunite, gypsum, thaumasite and zunyite (Embrey footnote in Knill 1960). This is a very dubious British occurrence requiring re-evaluation (Ryback et al. 1998). The occurrence is further noted by Young (1987a) and Young et al. (1998).

Cream flaky dickite has been found coating joints in Permian Magnesian Limestone sampled by the Hesleden Dene water borehole 2, ESE of Castle Eden, Co. Durham. It occurs as plates up to 0.5 mm in diameter, aggregated into sheaf-like or radial clusters, and is associated with barite, collophane, fluorite, galena, pyrite and sphalerite (Dunham et al. 1948, Fowler 1956, Guppy 1956). In Northumberland, a turquoise-blue variety of dickite fills cavities up to 1 cm across in calcite–galena–sphalerite veinstone on the dumps of Stonecroft mine, N of Haydon Bridge (Smith & Hardy 1981, Dunham 1990). At Beggars Wood, Tyne & Wear, dickite occurs fairly commonly in septarian ironstone nodules found in the Coal Measures, where it is associated with (rarely) apatite, chalcopyrite, quartz and siderite (Neall 2004). Apatite-bearing ironstone nodules collected from Coal Measures at Smithywood

Dickite from Trwynbychan, near Amlwch, Anglesey, Gwynedd, Wales; specimen height 11 cm. (© National Museum of Wales, Cardiff; photographer M. P. Cooper)

opencast site, Chapeltown, near Sheffield, Yorkshire, contain dickite and minor siderite (Neall 2004). Other occurrences of dickite in Yorkshire are in Carboniferous and Jurassic sandstones, and they include: Bondfield Gill, 10 km N of Helmsley; escarpment above Crag Pond, Farndale; 1 km W of Ellingstring, near Masham; and from a cliff location near West Pier, Whitby (Smithson & Brown 1957).

At the type locality in Wales, dickite occurs as a post-mining secondary mineral forming colourless-to-white hexagonal plates or stacks of plates up to 0.1 mm across on dolomite and quartz (Jenkins & Johnson 1993, Bevins 1994). The species is also found in Carboniferous sandstones at Bogdorgan Shore, Pont Malltraeth, Anglesey, and near Bangor, to the NE of the Menai suspension bridge (Smithson & Brown 1957). Additionally, it is noted from: Dylife mine, Central Wales orefield, Powys (Bevins 1994); the South Wales coalfield, West Glamorgan (Gill et al. 1977); and in the Aberfan borehole, Mid Glamorgan (Bevins 1994).

In Scotland, dickite has been found in a pipeline excavation near Hillend Roadhouse, Edinburgh, and in a Carboniferous sandstone from Taynuilt, Strathclyde (Macpherson & Livingstone 1982). Hubbard & Green (2001) report an unusual occurrence of dickite from the small abandoned copper mine at Castletown, Lochgilphead, Strathclyde, where it occurs with clinoatacamite, connellite, hematite, malachite, quartz and vésigniéite. Around Louisa mine, Glendinning, 13 km NNW of Langholm, Dumfries & Galloway, dickite is reported from drill-core sampling stratiform and disseminated arsenopyrite–pyrite mineralization, in which bournonite, chalcopyrite, galena, semseyite, sphalerite, stibnite, tennantite and tetrahedrite have been found (Gallagher et al. 1983).

At Lough Skillicore, Isle of Man, calcite, dickite and dolomite fill pore spaces and cement partially dolomitized Lower Carboniferous Limestone containing skeletal remains, pellets, intraclasts, and detrital quartz, illite and muscovite (Dickson & Barber 1977).

digenite

cubic, trigonal, pseudo-cubic Cu_9S_5

A relatively uncommon copper sulphide found in hydrothermal copper deposits, in mafic intrusions and in large exhalation deposits. Digenite cannot readily be distinguished from chalcocite, although it is distinctly blue in reflected light. It is probably more common than the literature would suggest.

Digenite is a constituent of a disseminated tin-sulphide assemblage in the Hensbarrow stock, a lithium-enriched stock within the St Austell granite, Cornwall, exposed in Gunheath china-clay pit. The crystallization sequence appears to be cassiterite, followed by stannite, stannoidite, bornite and mawsonite. Digenite occurs in veins cutting stannoidite and mawsonite. In turn, digenite may be replaced by tennantite (Manning 1983, Weiss 1994). In a study of the mineralization at Mount Wellington mine, Twelveheads, 7 km W of Truro, Cornwall, the hypothermal lode mineralization followed a similar sequence, from cassiterite, through base-metal sulphides, to metals and sulphosalts, as the source of the granite-related mineralizing fluid changed and was periodically tapped. Secondary digenite (neodigenite), covellite and marcasite occur as replacement species after galena and pyrite (Kettaneh & Badham 1978). A further occurrence of digenite in Cornwall is from Botallack mine, St Just (Embrey 1978, Bancroft & Weller 1993).

At Seathwaite Tarn, Coniston, Cumbria, digenite has been collected from the dumps of an abandoned copper mine that worked two mineralized quartz veins cutting Ordovician Borrowdale Volcanic Series tuffs and agglomerates. Bornite, digenite and djurleite comprise the main sulphide assemblage with smaller amounts of arsenopyrite, chalcopyrite, cobaltite, covellite, blaubleibender covellite (now recognized to be a mixture of spionkopite and yarrowite), pyrite and wittichenite. Digenite is extensively replaced by bornite and it develops graphic intergrowths with both bornite and wittichenite. Wittichenite and hematite also replace digenite along grain boundaries (Stanley & Criddle 1979). Irregular masses of chalcopyrite containing minor amounts of chalcocite, cubanite, digenite and sphalerite occur in quartz veins at Long Work in the Vale of Newlands. Arsenopyrite and pyrite are associated species (Millward et al. 1999). Digenite forms 5–20 µm-wide alteration rims surrounding chalcopyrite on the 60-fathom level dumps of Sandbed mine, Caldbeck Fells (Green et al. 2006a). At Rigg Head quarry, Borrowdale, digenite occurs with chalcocite, covellite and yarrowite as alteration rims around chalcopyrite, galena and sphalerite (Green et al. 2006c).

Digenite is noted without description from Littleham Cove, near Budleigh Salterton, Devon, where it occurs in a sulphide and arsenide assemblage containing native copper. It is associated with complex Ag–As–Co–Cu–Ni–U–V mineralization that developed in nodules as Permo-Triassic mudrocks were de-watered during burial compaction (Milodowski et al. 2000).

Additionally, digenite is reported from a few other locations in England: beach close to the Pier hotel at Clevedon, near Bristol, Avon (Ixer et al. 1993); Forcett quarry, 7 km NNW of Scotch Corner, Yorkshire (Dunham & Wilson 1985); Merrybent mine, NW of Middleton Tyas, Yorkshire (Small 1977, Dunham & Wilson 1985); Middleton Tyas mines, near Scotch Corner, Yorkshire (Dunham & Wilson 1985, Green 1987b); New Cliffe Hill quarry, Stanton under Bardon, Leicestershire (Hubbard et al. 2005); and Red Gill mine, Caldbeck Fells, Cumbria (Davidson & Thomson 1951, Briscoe 1986a).

Primary sulphide mineralization at the Great Orme copper mine, Llandudno, Gwynedd, Wales, was followed by widespread polyphase supergene alteration. Early chalcopyrite, accompanied by minor pyrite and marcasite, has altered mainly to limonite and malachite, but also to secondary copper sulphides and oxides including native copper, covellite, cuprite, digenite, spionkopite and tenorite. Less commonly, galena is altered initially along grain boundaries to 2–20 µm-wide rims of digenite or djurleite, followed by covellite and spionkopite, then to cerussite. Pyrite and marcasite also exhibit alteration rims of digenite at Great Orme (Ixer & Davies 1996, Ixer & Stanley 1996, Ixer 2001).

A mineralogical study of coarse-grain chalcocite from Alva silver mine, Ochil Hills, Central Region, Scotland, revealed intergrowths with digenite that indicate a temperature of formation of 100–300°C. An origin for the mineralization involving submarine faulting and the mixing of rising hydrothermal fluids with early Carboniferous sea water in feeder veins has been proposed (Jassim et al. 1983). Carbonate veins at the margin of the Criffell granodiorite, Needle's Eye, SW of Caulkerbush, Dumfries & Galloway, contain chalcopyrite and uraninite (pitchblende) associated with minor digenite, galena, pyrite and globular hydrocarbons (Basham et al. 1989). Sulphides are abundant in the Eastern Layered Series of the Tertiary layered ultrabasic intrusion on the Isle of Rum, Highland Region, particularly at the boundary between peridotite and the overlying leucotroctolite, where large blobs up to 1 cm occur. The primary sulphides chalcopyrite, pentlandite and minor pyrrhotite are altered to bornite, chalcocite, native copper, digenite, magnetite and iron hydroxides (Dunham & Wilkinson 1985, Power 2000a, Power et al. 2003). A specimen of digenite from the Falls of Rogie, 4 km W of Strathpeffer, Highland Region, is in the NHM collections (Embrey 1978). Crystals and irregular masses of digenite and pyrite are reported from calcite–scapolite veins at Copper Geo, Fair Isle, S of Shetland, in association with supergene covellite and goethite (Mykura 1972, Macpherson & Livingstone 1982). Pyroxenites belonging to the Shetland ophiolite complex on Unst contain primary sulphides such as chalcopyrite, pentlandite and pyrrhotite, all within serpentine that has formed interstitially to the pyroxene. The sulphide species follow a zonal sequence of alteration, starting with bornite, digenite and covellite, and ending with native copper (Maynard 1993).

At Dhurode copper mine, in the West Carbery mining district, Co. Cork, Ireland, digenite occurs in late-Hercynian mineralized veins containing a complex coarse-grain copper-rich assemblage of bornite, chalcopyrite, djurleite, tetrahedrite, wittichenite, and enargite-group minerals (Ni Wen et al. 1991). Digenite is a rare constituent of dominantly chalcocite mineralization developed within an Ordovician altered acidic porphyry located 8 km to the S of Charlestown, Co. Mayo (Poustie 1982), and it is also present in surficially oxidized material within the top 15 m of the Cu–(Ag–Hg) orebody at Gortdrum mine, near Tipperary town, Co. Tipperary (Steed 1986).

dimagnetite

Magnetite pseudomorph after ilvaite.

diopside

PYROXENE GROUP

monoclinic $CaMgSi_2O_6$

A very common calcium–magnesium silicate that forms separate series with hedenbergite and with johannsenite, and is found in skarn deposits, calcium–magnesium-rich gneisses and schists, and in metamorphosed siliceous pyroxene or epidote–amphibolite-facies hornfels. In igneous rocks, diopside occurs in peridotites, or less commonly in basalts or andesites. The definition of diopside now includes those pyroxenes described earlier as salite or sahlite (Morimoto 1989).

England

Cornwall Diopside, enstatite and olivine are common species in peridotite and serpentinite belonging to the Lizard ophiolite complex. To the N of Polbream Point, diopside crystals 3–4 cm long have been found in banded peridotite (Green 1964). Diopside is also reported from many other localities on the Lizard Peninsula: Carrick Luz, E of Kennack Sands (Bonney 1877); Coverack Cove (Teall 1888); Countybridge quarry, Goonhilly Downs; Coverack Cove; Gwenter and Kernewas, Goonhilly Downs; near Rill Cove, Kynance Cliff; N of Worvas; Perprean Cove, near

Coverack; the Bees, near Pedn Boar (all Green 1964); and Spernic Cove, Gwenter (Golley & Williams 1995). Analyses of diopside from Countybridge quarry, Gwenter and Kernewas are provided by Davies (1983). At Trevone Bay, Padstow area, diopside (salite) is one of the relict primary species found in altered Devonian amphibole-bearing alkali dolerites, termed greenstones (Floyd & Rowbotham 1982). Diopside also occurs with calcite, epidote, grossular, hornblende and magnetite in typical skarn rocks derived from metabasalts close to the margins of the Land's End granite at Carn Vellan, Gryll's Bunny, the Crowns and Wheal Cock Carn, all near Botallack, St Just (Jackson 1974, Alderton & Jackson 1978, van Marcke de Lummen 1985). Other skarn-related occurrences include: South Yolland farm, Caradon area (Barrow & Thomas 1909), and Treburland mine, 2 km SE of Fivelanes, Altarnun (Kingsbury 1961), both in the aureole of the Bodmin Moor granite.

Cumbria Augite and diopside phenocrysts up to 10 mm across occur in microdiorite at Scawgill Bridge, 4 km W of Braithwaite (Fortey et al. 1994). Diopside is common in dolerite dykes within or close to the margin of the Eskdale granite (Macdonald et al. 1988).

Derbyshire Carboniferous basalts from Calton Hill, Taddington, contain phenocrysts of augite and diopside, together with olivine, plagioclase and iron–titanium oxides (Hamad 1963, Macdonald et al. 1984). The lavas contain mantle-derived spinel lherzolite and harzburgite xenoliths, in which sub-equant anhedral crystals of chrome diopside up to 3 mm across have been found (Donaldson 1978).

Devon Metasomatism of Carboniferous cherts, limestones and shales in the aureole of the Dartmoor granite (and close to the Meldon aplite), near Okehampton, have produced skarn assemblages in which diopside occurs with axinite, grossular, quartz, tremolite and vesuvianite (El Sharkawi & Dearman 1966, 1967, von Knorring & Condliffe 1984). Diopside is associated with malayaite, vesuvianite and wollastonite at the nearby Red-a-ven mine (El Sharkawi & Dearman 1966). Diopside occurs with titanaugite in Tertiary dolerite dykes cutting the granite on Lundy island off the north coast of Devon (Thorpe & Tindle 1992, Tindle unpub. EMP data).

Merseyside Diopside is one of the most attractive species found in vesicles within altered gabbro (glacial erratics) on Thurstaston beach, Wirral. It forms light and dark-green terminated prismatic crystals reaching 3 mm in length (Rothwell & Dickinson 1989).

Northumberland Diopside and enstatite are widespread constituents in granitic rocks from the Cheviot complex, SW of Wooler (Haslam 1986).

Shropshire Microphenocrysts of diopside, olivine and plagioclase occur in Ordovician basalts from Sibdon Carwood (Leat & Thorpe 1986).

Wales

Dyfed The Ordovician Fishguard Volcanic Complex contains augite and diopside in both intrusive and extrusive rock types. Occasional pyroxene phenocrysts, together with dendrites and large purple skeletal crystals, are described from the basic lavas, whereas the intrusive rocks are noted for large ophitic crystals (Bevins 1982). Granophyric gabbro from the St David's Head layered intrusion also contains diopside (Bevins et al. 1994).

Gwent Emerald-green chrome diopside occurs in peridotite xenoliths within the Great House diatreme, near Usk (Bevins 1994).

Gwynedd Diopside occurs in the Lower Ultrabasic and Lower Gabbro Zones of the Rhiw layered intrusion, Lleyn Peninsula (Hawkins 1970, Cattermole 1976), in the Tal y Fan intrusion, SW of Conwy (Merriman et al. 1986), and in cumulo-phyric lavas of the Rhobell Volcanic Complex, Rhobell Fawr area, near Dolgellau (Allen & Jackson 1985, Kokelaar 1986). On the Isle of Anglesey, colourless anhedral crystals of diopside have been found in calc-silicate rocks in the Nebo inlier of the Mona Complex (Greenly 1919). Bevins (1994) notes diopside from the Bryn Fuches and Rhôs-mynach areas of northeast Anglesey. Dense, white rocks near fault zones at Rhyd-y-Bont on Holy Island (off the coast of Anglesey) contain diopside and represent altered serpentinites (Maltman 1977).

Scotland

Central A coarsely granular pyroxenite from NE of Arrochymore Point on the shores of Loch Lomond (close to the Highland Boundary fault) contains chromium-rich diopside with curved exsolution lamellae associated with chlorite (Henderson et al. 1983).

Dumfries & Galloway Diopside and augite are common species in pyroxene diorites found at the margins of the Loch Doon granite complex (Tindle 1982).

Fife Porphyroclastic spinel lherzolite xenoliths in volcanic vents and associated tuffs at Coalyard Hill and Ruddons Point, both near Elie, contain an assemblage of diopside, enstatite, forsterite and hercynite (Chapman 1976).

Grampian Diopside and augite are common species in the Insch intrusion, the largest of the Caledonian Newer Gabbro bodies. Diopside is particularly common in Middle Zone cumulates and fine-grain granular gabbros between Insch, Old Rayne, Oyne, Pitcaple and Daviot. Similarly, the Boganclogh intrusion to the W of Insch also contains diopside (Wadsworth 1986, 1988). Mafic bands associated with base-metal-enriched Dalradian metasediments at Corrie Loch Kander, near Braemar, contain diopside together with actinolite–tremolite, clinozoisite, hedenbergite, hyalophane, and the rare species armenite (Fortey et al. 1993). Dalradian calc-silicate hornfels at Pollagach Burn, E of Pannanich Hill, E of Ballater, were metasomatized by the Ballater granite and now contain diopside in skarn rocks dominated by grossular, wollastonite and vesuvianite (Crane 1987). A similar diopside-bearing assemblage is reported from Kinnaird Head, Fraserburgh, close to a suite of granitoid pegmatites (Johnson et al. 2000). A diopside-bearing calc-silicate gneiss at Ladyleys in the Ythan Valley (Read 1952) is also likely to be a skarn. At Shallowplough quarry, near Ellon, green diopside forms 5 mm-long prisms in a limestone-contaminated granitic pegmatite (Read 1925). Skarns at Michaelmuir quarry and Ardlethan, WSW of Ellon (Read 1925), and in the Dorback granite area, Braes of Abernethy, 9 km W of Tomintoul (Walker & Davidson 1935), also contain diopside. Heddle (1878c, 1901) reports diopside (referred to as coccolite, funkite, malacolite or sahlite) mostly in metasomatized limestone at: Alltcailleach, Coyle Hills, Glen Muick; Corntulloch; Delnabo quarry, Glen Gairn; Eslie, near Banchory; and Limehillock, near Grange.

Highland Cambrian dolomite in the burn N of Loch Sàil an Ruathair, Assynt, has been metasomatized by the Loch Ailsh syenite complex and replaced by diopside–tremolite rocks and forsterite marble (Macgregor & Phemister 1972). Diopside is also present in pyroxenite from this complex (Shaw et al. 1994). On the western side of Loch Ailsh, on Gruagach cliff, Seasgaich, Heddle (1878c) reports diopside (coccolite) from a gneissose rock associated with limestone. Abundant rounded grains of greenish-grey diopside have been found in a metamorphosed calc-silicate rock at Allt a' Mhuilinn, to the E of the Borralan syenite complex, Assynt (Shand 1910b), and, at Allt Bealach Choinnich, diopside is a constituent of metasomatized (fenitized) quartzites (Woolley et al. 1972). The Borralan complex itself contains pyroxenites in which diopside occurs with amphibole, garnet, mica and opaque minerals (Notholt et al. 1985). A carbonatite intrusion adjacent to the Borralan complex is responsible for producing metasomatic diopside in limestones and dolomites between Loch Borralan and Loch Urigill. Associated species are chondrodite and phlogopite (Young et al. 1994b). Pyroxene with the optical properties of aegirine-augite is described from the Glen Dessary syenite, W of Loch Arkaig, although chemically the mineral is diopside (Fowler 1992). Metamorphosed limestones containing diopside occur in Lewisian rocks of the Glenelg inlier at: Totaig, near the south shore of Loch Duich; on Beinn a' Chaoinich, between Glenelg and Totaig (Heddle 1878c); near the summit of Beinn a' Chapuill and Balvraid, Gleann Beag (Read & Double 1935). Diopside-bearing igneous rocks in the same area include: garnet-bearing olivine websterite from Beinn Fhada, W of Letterfearn, Loch Duich (Rawson et al. 2001); websterite and garnet–hornblende–pyroxene rock from the slopes of Glen More, Glenelg (O'Hara 1960, Mercy & O'Hara 1965); and pyroxene-rich xenoliths in diorite of the Ratagain complex, NNE of Suardalan (Nicholls 1951a).

Other occurrences of diopside include: a basalt dyke on the headland of Sròn Bheag, Ardnamurchan, W of Kilchoan (Wells et al. 1974); gabbro and tonalite from the Tertiary Centre III igneous complex, Ardnamurchan (Walsh 1975); quartzite xenoliths within granodiorite at Kentallen northern quarry, Ballachulish (Muir 1953); metamorphosed siliceous dolomite in the aureole of the Ballachulish granite complex (Ferry 1996); diorite and metastably in some granites of the Ballachulish granite complex (Weiss & Troll 1989); diorite and enclosed pyroxenite xenoliths from the Ben Nevis granite complex (Haslam 1968, 1970); biotite-bearing rock interbanded with semi-pelitic schists and granulites, SSW of Cuil, W of Allt Coire a' Bhoganaich, head of Loch Arkaig (Harry 1951); garnet pyroxenite xenoliths in a diatreme on the island of Eilean Shona (Upton et al. 2003); limestone at Shinness on the eastern shore of Loch Shin (Heddle 1878c) and Achnagonalin quarry, 1 km S of Grantown (Heddle 1901); Carboniferous–Permian lamprophyre dykes, N of the head of Loch Sunart and in a hornblende camptonite dyke at Allt Tarsuinn, Corrantee mine, 5 km NW of Strontian (Gallagher 1963); late-Caledonian minette at Glen Affric, E of Morvich (Canning et al. 1996); garnet–plagioclase amphibolite from the Scourie dyke (O'Hara 1961); strongly zoned crystals in a granulite-facies calc-silicate gneiss on Scourie More (Rollinson 1980).

On the Isle of Rum, diopside occurs in cumulate peridotites of the Eastern and Western Layered Series (e.g. on Hallival) of the Rum central complex (Heddle 1901, Brown 1954, Dunham & Wadsworth 1978, Tepley & Davidson 2003) and in satellite intrusions of peridotite (e.g. around Sgaorishal) intruding Torridonian country rocks around the main intrusion (Wadsworth 1994). On the Isle of Skye, diopside is found as a constituent of thermally metamorphosed chert-bearing Durness limestone and dolomite around the Beinn an Dubhaich granite (e.g. at Camas Malag and on the southern and northeastern slopes of Ben Suardal), W of Broadford. Grossular, hedenbergite and wollastonite are associated species (Harker 1904, Tilley & Vincent 1947, Tilley 1948b, 1949, 1951, Hoersch 1981, Holness 1992, Holness & Fallick 1994). Magnetite-rich endoskarns around this intrusion also contain diopside (Smith et al. 2002). Also on Skye, diopside occurs with augite in olivine tholeiite cone sheets that invade the Cuillin igneous complex (Bell et al. 1994) in Tertiary basalts (Thompson 1974); at Harker's Gully, Marsco area, Western Redhills, it is found in a mica pyroxenite lens within a Lewisian leucogneiss xenolith in ferrodiorite (Thompson 1981).

Strathclyde Diopside is a major constituent of a garnet–clinopyroxenite (eclogite) segregation in mantle harzburgite discovered between Knockormal and Carleton Mains farms (Bloxam & Allen 1959, Smellie & Stone 1984). The rocks form part of the Girvan–Ballantrae ophiolite complex (Smellie & Stone 1984). A hornblende-bearing garnet metapyroxenite containing diopside, exposed near Knocklaugh, is from the aureole of this complex (Treloar et al. 1980). The basanite vent at Black Rock, 2 km SSW of Fairlie, Hunterston Sands, contains diopside as zoned phenocrysts and as a constituent of lherzolite xenoliths (Alexander et al. 1986). Dark-green and pistachio-green diopside (coccolite) crystals up to 8 mm across occur in marble at the pink-marble quarry at Balephetrish, Isle of Tiree. The rock has discordant contacts with the adjacent Lewisian Gneiss and also contains amphibole, calcite, mica and scapolite (Heddle 1878c, Hallimond 1947, Guppy 1956). Adjoining the Tiree Marble is a rock containing rounded grains of green diopside (Davidson 1943). Diopside is also reported from: dolerite dykes cutting microgranite on Ailsa Craig in the Firth of Clyde (Harrison et al. 1987); garnet pyroxenite xenoliths in diatremes at Auchengarth, Heads of Ayr and Muirkirk (Upton et al. 2003); leucodolerite and nepheline syenite of the Cnoc Rhaonastil intrusion, Isle of Islay (Walker & Patterson 1959, Hole & Morrison 1992, Preston et al. 1998a); metasomatic grossular–pyroxene endoskarns in agglomerate within the collapsed caldera of the Central ring complex at Creag an Fheidh on the Isle of Arran (Cressey 1987); alkali dolerite (crinanite) of the Dippin Sill, southeast Isle of Arran (Gibb & Henderson 1978a); crinanite sills at Clauchlands Point and Kingscross Point, Lamlash Bay, Isle of Arran (Tindle unpub. EMP data); diorite and monzodiorite from the Quarry diorite unit at the southern end of the Etive granite complex to the E and NE of Ben Starav (Moazzen & Droop 2005); pargasite pyroxenite xenoliths in an olivine-nephelinite dyke cutting Moinian psammites at Gribun, Isle of Mull (Upton et al. 1998); basalts, hawaiites and mugearites belonging to the Tertiary lava succession on the Isle of Mull and Morvern (Kerr 1998); Devonian basalt, basaltic andesite and appinite from the Lorne Plateau (Groome & Hall 1974).

Tayside Diopside (syn. malacolite) is reported from the marble quarry, near Forest Lodge, Glen Tilt, E of Blair Atholl, and in limestone at Morenish, NW of Loch Tay (Heddle 1901).

Western Isles The Lewisian Bay Steinigie paragneisses (part of the South Harris igneous complex), near Finsbay (Fionnsabhagh), contain diopside amphibolite and diopside marble (Dearnley 1961), and diopside is also noted from an ultrabasic gneiss to the W of Loch Ossigary (Howie 1964). Diopside–hornblende–orthoclase rocks from the same complex are also noted from the Chaipaval (Ceapabhal) ridge, above the Liuri cliffs (Dearnley 1962). At Rodel on South Harris, diopside (coccolite and malacolite) has been found in limestone (Heddle 1901) and in calc-silicate gneiss (Davidson 1943). Diopside also occurs in an amphibolitized Lewisian intrusion on the shore E of Lochan nam Faoileann, Isle of Barra (Sobott 1979), and in alluvial deposits on St Kilda (Hirta) (Harding et al. 1984). The Tertiary Shiant Isles Main Sill is a differentiated basaltic intrusion that contains diopside, hedenbergite and rare aegirine-augite in a suite of rock types that includes crinanite, picrite, picrodolerite, olivine teschenite, and related pegmatites (Drever & Johnston 1965, Gibb & Henderson 1996, Henderson et al. 2000).

Shetland Diopside is present in wehrlitic dunite, harzburgite, cumulate pyroxenite and high-level pyroxenite from the Shetland ophiolite complex on Unst (Prichard 1985, Maynard 1993). Heddle (1901) reports diopside from Uriesetter on Fetlar (sahlite), and Colla Firth, Mainland (funkite).

Ireland

Antrim Stout black prisms of diopside (sahlite), usually 0.3–2.0 mm in length, but rarely 1 cm, some showing hourglass zoning, occur in pyroxenite associated with a chalk/dolerite contact zone at Carneal, Larne (Sabine & Young 1975). Similar occurrences are also noted at Tieveragh, near Cushendall (Kitchen 1984), and Scawt Hill, Larne (Tilley & Harwood 1931), where doleritic intrusions have interacted with surrounding country rocks.

Cork Diopside occurs in amphibole pyroxenite nodules within a megacryst-bearing pyroclastic pipe at Black Ball Head, near Cahermore (Pracht & Kinnaird 1995).

Donegal Diopside is a major constituent of zoned skarn xenoliths (originally limestone) in granite at Ballynacarrick (Gindy 1953). Chrome diopside is common in spinel lherzolite and harzburgite xenoliths hosted in an alkali basalt dyke near Inver (Shaw & Edgar 1997).

Other occurrences include: Aphort, Aran Island (Russell 1946a); in calc-silicate rocks around the Ardara granite complex (Akaad 1956); in Dalradian calc-silicate hornfels adjacent to pegmatitic dykes S of Bunbeg quay, E of Loughlin Island (Whitten 1951); in garnet–vesuvianite–wollastonite veins cutting limestone in the quarries of Lough Anure, Thorr area (Pitcher 1950); in leucodiorite from the Kilkenny School intrusion, Maas area (French 1976).

Down Granular, intersertal, ophitic and poikilitic varieties of augite and diopside are present in the dolerites and gabbros of the Tertiary Slieve Gullion central complex (Gamble 1982).

Galway The Fountain Hill skarn in west Connemara was produced by thermal metamorphism of impure limestone in the Omey granite aureole and it consists of calcite, diopside, grossular–andradite garnet, vesuvianite and wollastonite (Ahmed-Said & Leake 1996). A similar paragenesis is proposed for diopside from: Connemara – spectacularly zoned crystals (Yardley et al. 1991); Derreenagusfoor – in skarn rocks (McArdle et al. 1986a); and along the coast W of Grallagh (Ryback & Moreton 1992).

Limerick Carboniferous basalts from the Old Pallas area contain phenocrysts of diopside (Ashby 1946).

Louth Diopside is associated with calcite and rutile in limestone xenoliths enclosed in syenite at Barnavave, Carlingford (Nockolds 1937). Calc-silicate rocks surrounding the syenite contain xenoblastic diopside associated with grossular, vesuvianite and wollastonite (Osborne 1932b). A gabbro contaminated with limestone at Barnavave is also diopside bearing (Le Bas 1965).

Offshore territories

Jersey Diopside is a major constituent of Cadomian layered gabbros near Le Nez in southeast Jersey (Bishop & Key 1983) and in the gabbro dyke at Wolf's Cave, W of Frémont Point on the north coast (Rowbotham & Lees 1990). Bright-green euhedral diopside crystals are also noted from the South Hill lamprophyre near St Hélier (Smith 1936a, Mourant 1961) and from a mica lamprophyre at Plémont Point (Smith 1936b). Diopside is a constituent of an arfvedsonite-bearing minette at St Hélier (Wagner & Velde 1985).

dioptase

trigonal $Cu_6Si_6O_{18}{\cdot}6H_2O$

An hydrated copper silicate found in a silicified barite–calcite-rich fault rock associated with the West mine boundary fault at West mine, Alderley Edge ridge, Cheshire. Mineralized fractures contain relict traces of acanthite (argentite), chalcopyrite and cinnabar, and secondary dioptase, chrysocolla, goethite, malachite and Mn-(Fe) oxyhydroxides. Additionally, in the vicinity of Alderley Edge and Clive (Shropshire) copper mining areas, dioptase is reported with azurite, cerussite, chrysocolla, hydrocerussite, malachite, mimetite, pyromorphite and smithsonite (Plant et al. 1999). Dioptase was supposedly found by A. W. G. Kingsbury in 1953 at Old Potts Gill mine, Caldbeck Fells, Cumbria, with specimens now held in the Kingsbury collection at the NHM (Hartley 1984, Young 1987a, Ryback & Tandy 1992). Given the doubts about the provenance of other Kingsbury specimens (Ryback et al. 1998, 2001), further confirmation is required.

dipyre

A variety of marialite, scapolite group.

discachatae

A variety of agate (quartz), defined by the enclosure of discoidal patches of cacholong.

disthene

A synonym of kyanite.

djerfisherite

cubic $K_6(Fe,Cu,Ni)_{25}S_{26}Cl$

A rare potassium–iron–copper–nickel sulphochloride found in a high-temperature calcic skarn near the contact of the Cuillin gabbro complex at Camasunary Bay, Strathaird Peninsula, Isle of Skye, Highland Region, Scotland. Associated species include calcite, garnet (andradite–grossular), melilite, vesuvianite and wollastonite, together with minor amounts of strontium-rich apatite, magnetite, perovskite, pyrite, spurrite and tilleyite. Rare baghdadite, galena, kimzeyite and spinel are also noted (Beard & Drake 2007).

djurleite

monoclinic $Cu_{31}S_{16}$

A relatively uncommon copper sulphide found as a secondary mineral in the oxidized zones of copper deposits. Djurleite and digenite closely resemble chalcocite and are not easy to identify. Djurleite is probably more common than the literature would suggest.

In the St Just district of Cornwall, djurleite has been found at Wheal Owles, Penwith (Embrey 1978), and Geevor mine, near Pendeen (Bancroft & Weller 1993). Metallic blue masses of djurleite have been noted at Greystone quarry, Lezant, near Launceston (Weiss et al. 2001, Elton et al. 2004). Djurleite has also been found on the dumps of West Caradon and South Caradon mines, Liskeard, St Cleer. It occurs on a matrix of calcareous rock encrusted with tenorite (Lawrence 1990).

At Seathwaite Tarn, Coniston, Cumbria, djurleite has been collected from the dumps of an abandoned copper mine that worked two mineralized quartz veins cutting Ordovician Borrowdale Volcanic Series tuffs and agglomerates. Bornite, digenite and djurleite comprise the main sulphide assemblage, with smaller amounts of arsenopyrite, chalcopyrite, cobaltite, covellite, pyrite, wittichenite and blaubleibender covellite (now recognized to be a mixture of spionkopite and yarrowite). Djurleite is extensively replaced by bornite and wittichenite (Stanley & Criddle 1979).

Djurleite is the main secondary copper sulphide in vein mineralization around Alderley Edge ridge, Cheshire, where it forms mosaics of small (5 µm-diameter) crystals that replace chalcopyrite, galena and sphalerite. It also forms intergrowths with covellite, spionkopite and yarrowite (blaubleibender covellite) or copper and lead carbonates. In reflected light it is white to pale blue, and EMP analysis shows that it has a significant silver content (Ixer & Vaughan 1982).

Additionally, djurleite is reported from a few other locations in England: Cannington Park, near Bridgwater, Somerset (Embrey 1978); Dean quarry, St Keverne, Lizard Peninsula, Cornwall (Embrey 1978); Driggith mine, Caldbeck Fells, Cumbria (Stanley & Vaughan 1981); Gipsy Lane, Leicester (Embrey 1978); Judkins quarry, Nuneaton, Warwickshire, and New Cliffe Hill quarry, Stanton under Bardon, Leicestershire (Ixer et al. 2005a); Merehead (Torr Works) quarry, Cranmore, Mendip Hills, Somerset (Embrey 1978); Middleton Tyas mines, near Scotch Corner, Yorkshire (Green 1987b); and Raisby railway cutting, 1 km E of Coxhoe, Co. Durham (Young 2001).

At the Great Orme copper mine, Llandudno, Gwynedd, Wales, galena and chalcopyrite infill voids in a fault-joint plane cutting fine-grain dolostones. The galena is altered initially along grain boundaries to 2–20 µm-wide rims of digenite or djurleite, followed by spionkopite and covellite, then to cerussite. This phase of mineralization does not appear to be related to the much more abundant copper–dolomite mineralization for which Great Orme is famous (Ixer & Davies 1996, Ixer & Stanley 1996, Ixer 2001). Djurleite is also noted (but not described) from Port Ysgo, near Aberdaron, Lleyn Peninsula, Gwynedd (Bevins 1994). In Powys, djurleite occurs in pods and veinlets within fractured limestone at Dolyhir quarry, Old Radnor, near Presteigne (T. F. Cotterell pers. comm. 2007).

The bonanza-type silver deposit at Alva mine, Ochil Hills, Central Region, Scotland, has a sulphide assemblage associated with it. Djurleite forming bluish-black areas up to 3 mm across in a barite gangue is associated with bornite, chalcocite, chalcopyrite, covellite, digenite, galena (very rare), marcasite and pyrite (Moreton et al. 1998, Livingstone 2002). Large masses of bornite and djurleite were found in 1997 when excavations at Greenhill quarry, on the south side of Carwinning Hill, SW of Kilbirnie, Strathclyde, broke through into the old Swinlees mineworkings (Green & Todd 2001). The mine worked a copper-rich vein cutting the Clyde Plateau Volcanic Formation, in which azurite, barite, calcite, chalcopyrite, djurleite, malachite and quartz were found (Wilson & Flett 1921, Stephenson & Coats 1983).

At Dhurode copper mine, in the West Carbery mining district, Co. Cork, Ireland, djurleite occurs in late-Hercynian mineralized veins containing a complex coarse-grain copper-rich assemblage of bornite, chalcopyrite, digenite, tetrahedrite, wittichenite and enargite-group minerals (Ni Wen et al. 1991). Djurleite associated with bornite ± chalcopyrite has also been reported from stratiform disseminated sulphide deposits at Mount Gabriel, Co. Cork, and Derrynane, Co. Kerry (Kinnaird et al. 2002).

Old samples in the Lukis collection (Guernsey Museum and Art Gallery, St Peter Port) from the Le Pot copper lode that cuts foliated granite sheets at the southern end of Little Sark, Channel Isles, contain djurleite, chalcocite and a little bornite (Ixer 1992).

dnepnoskite

A fine granular form of cassiterite (proto-wood tin) forming radiating clusters of acicular crystals.

dogtooth spar

A scalenohedral variety of calcite.

dolianite

TL Knock Station, Grampian, Scotland

A mineral of uncertain composition, last examined in 1885 and of so little significance that further details are not warranted.

DOLOMITE GROUP

Trigonal carbonates with the general formula $AB(CO_3)_2$, where A = Ba, Ca; and B = Fe^{2+}, Mg, Mn^{2+}, Zn. All members of the group have been found in Britain or Ireland, namely ankerite, dolomite, kutnohorite, minrecordite and norsethite.

dolomite

DOLOMITE GROUP

trigonal $CaMg(CO_3)_2$

A very common calcium–magnesium carbonate that forms separate series with ankerite and with kutnohorite, and is found in limestones as a direct precipitate or as a replacement of calcite. It also occurs as a gangue mineral in hydrothermal veins and is a major component of some contact-metamorphic rocks and marbles.

England

Cleveland Boreholes in the Permian Lower Magnesian Limestone at Billingham and Wilton, near Guisborough, contain massive dolomite consisting of anhedral equant crystals up to 50 µm across (Raymond 1962).

Cornwall Old specimens of dolomite (pearl spar) are reported from the Botallack sett (includes Wheal Cock, Wheal Owles and Wheal Edward), St Just (Bancroft & Weller 1993). Other old specimens are from Garras mine, near Truro; Trevascus mine, near Hayle, Gwinear; North Roskear mine, Camborne; Polgooth mine, near St Austell; Wheal Castle, St Just (Greg & Lettsom 1858). Good crystals of dolomite to 7 mm, with typical curved faces, are common in cavities within veined blocks of metapelite (killas) on the dumps of Penberthy Croft mine, St Hilary. It is colourless-to-white, creamy yellow or light to dark brown, and is an important gangue mineral associated with the lead mineralization at this mine. Chalcopyrite, clinochlore, galena, millerite, quartz and siderite are associated species (Rust 1982b, Betterton et al. 2000). An interesting igneous occurrence of magnesium-rich dolomite is reported from a peralkaline minette at Pendennis Point, Falmouth, which contains 10% dolomite (sometimes compositionally zoned), together with apatite, biotite–phlogopite, magnesite, quartz, sanidine and accessory minerals (Hall 1982). Power et al. (1997) note that dolomite is the most common fracture-hosted vein mineral within the Lizard peridotite (e.g. at Pentreath beach), where it forms white-to-pink veins associated with calcite and quartz. Other occurrences include: Balleswidden mine, near St Just; Dolcoath mine, Camborne; Wheal Rose, Sithney, near Porthleven; Wheal Metal, Tregonning area (all Hosking 1951); Burraton Coombe quarry, St Stephens by Saltash (Russell 1925, 1944); Carleon Cove, near Poltesco, Lizard Peninsula; Geevor mine, near Pendeen, St Just; Eastern Cliff, S of Gwenter, Lizard Peninsula; Levant mine, St Just; West Chiverton mine, Perranzabuloe (all Golley & Williams 1995); Greystone quarry, Lezant, near Launceston (Wolfe 2000); New Rosewarne mine, Gwinear (Rudler 1905); South Crofty mine, Camborne (LeBoutillier 1996); South Tolgus shaft, Redruth (Davey 1948); Tredinnick mine, Landrake with St Erney; Trenute mine, near Lezant; Wheal Boys, Trewetha, St Endellion (all Russell 1949b); and Wheal Golden, Perranzabuloe (Merry & Braithwaite 1995).

Cumbria Carbonates are remarkably common in the northern part of the Northern Pennine orefield, and previous investigators (e.g. Greg & Lettsom 1858) have considered dolomite to be one of the commonest species. However, Dunham (1990) suggests that it is ankerite that is being described, and a detailed study of the mineralogy of Brownley Hill lead mine, Nenthead, near Alston (Green et al. 2000), would appear to confirm this. Iron-rich dolomite, ankerite and calcite are common in the quartz veins and are disseminated in wallrocks at Carrock mine, Grainsgill, Caldbeck Fells. Cream-to-brown massive dolomite-bearing veins also intersect and displace the quartz veins. It is likely that fluids carrying the carbonates were also responsible for the alteration of primary wolframite to scheelite in this deposit (Davidson & Thomson 1951, Eastwood 1959, Shepherd et al. 1976, Fortey 1978, Beddoe-Stephens & Fortey 1981, Ball et al. 1985a, Cooper & Stanley 1990). Elsewhere in the Caldbeck Fells, dolomite is abundant and it forms dark-brown crystals at Roughton Gill mine (Davidson & Thomson 1951) and Sandbed West barite mine (Shaw 1970). Green et

Dolomite rhombs to 2 mm and chalcopyrite from Sandbed mine, Caldbeck Fells, Cumbria. (© D. I. Green)

al. (2006a) describe dolomite from Driggith and Sandbed mines, and Hartley (1984) also notes it at Old Potts Gill barite mine.

Dolomite occurs in considerable quantity in the hematite orebodies of west and southwest Cumbria and Furness; there are fine examples in many collections (Goodchild 1884). Of these, the Florence, Ullcoats and Beckermet mines, near Egremont, were the most important mineralogically (Dunham & Hollingworth 1947, Davidson & Thomson 1951, Goldring & Greenwood 1977, 1990, Green & Briscoe 2002). Dolomite is also an important rock-forming mineral in the dolomites and dolomitic limestones of Carboniferous and Permian age, for example in the Distington quarries (Eastwood et al. 1931), and in the Vale of Eden (Arthurton et al. 1978).

Other localities include: Applethwaite quarry, Troutbeck (Lawrence et al. 1986); Barron Cross quarry, Stainton, Penrith; Barrow mine, Braithwaite; Embleton mine and quarry, Embleton, near Cockermouth; Robin Hood mine, Bassenthwaite (a Kingsbury specimen); Sir John pit, Bigrigg; South Cumberland mine, Eskdale (all Young 1987a); Bonser mine, Coniston (Stanley & Vaughan 1982b); Brandelhow mine, Derwent Water; Force Crag mine, Braithwaite, near Keswick; Handsome Mea mine, Nenthead (all Davidson & Thomson 1951); Driedly Gill, Stonethwaite Fell (Oliver 1961); Goose Green mine, Parkside, Frizington (Green & Briscoe 2002); Greenscoe mine, near Askham-in-Furness; and a pale-green variety at Hodbarrow mine, Millom, near Haverigg (Rose & Dunham 1977); Greenside mine, near Glenridding, Patterdale (Goodchild 1884, Gough 1963, Firman 1978); Nab Gill mine, Boot, Eskdale (Young 1984b); Paddy End mine, Coniston, Eskdale (Dewey & Eastwood 1925); Saltom Bay, St Bees Head (Arthurton & Hemingway 1972); Sandwith mine, Whitehaven (Harrison 1975); Sellafield area boreholes (Milodowski et al. 1998); Smallcleugh mine, Nenthead, near Alston (Livingstone 1991); Thornthwaite mine, Thornthwaite (Eastwood 1921, Ineson & Mitchell 1974a); Threlkeld quarry, Threlkeld (Rastall 1941); and the Windermere area (Bonney 1878).

Derbyshire Large areas of dolomitic limestone occur near Elton and Long Dale, at Brassington, Carsington and Wirksworth, and on Masson Hill. The passage of magnesium-rich brines through limestone has allowed dolomite to grow at the expense of calcite, destroying fossils and original textures in the process. Wherever vugs are found, dolomite forms small rhombic crystals (Bemrose 1904, Parsons 1922, Ford et al. 1993, Jackman 1996). Vein dolomite is uncommon in the Southern Pennine orefield, but is found associated with barite and galena at Golconda mine, Brassington (Robinson & Ineson 1979).

Devon Crystals of dolomite have been found with fluorite and siderite at the Bere Alston and South Hoo mines, near Bere Ferrers (Greg & Lettsom 1858, Hall 1868b, Rudler 1905). Buff iron-rich dolomite, containing inclusions of goethite and limonite, is found at the centre of gold and selenide-bearing veins at Hope's Nose, Torquay (Embrey & Symes 1987, Harrison & Fuller 1987, Stanley et al. 1990a). Manganese-rich dolomite and kutnohorite is reported from gold-bearing veins at Thorverton, Crediton (Shepherd et al. 2005). Dolomite is also noted from Wheal Friendship (Devon Friendship), Mary Tavy (Braithwaite 1982b) and from antimony–gold-mineralized quartz–carbonate veins at Loddiswell mine, 3 km N of Loddiswell (Stanley et al. 1990b).

Dorset Iron-bearing dolomite has infilled the pores of uncompacted mud during the diagenetic alteration of carbonate cementstones in the Kimmeridge Clay at Kimmeridge Bay (Irwin 1981). The porosity of the Triassic Sherwood Sandstone Group sandstones in the Wessex Basin is also affected by the precipitation of late iron-rich dolomite and iron–manganese-rich calcite (Strong & Milodowski 1987).

Durham The Lower Magnesian Limestone examined at Chilton and Rough Furze quarries, Ferryhill consists of soft buff-to-orange fine granular dolomite, pale-brown to pale-grey subporcellanous nodular and auto-brecciated dolomite, and thickly bedded dolomitic limestone (Hirst & Smith 1974). Dunham et al. (1948) found impure dolomite mudstone containing barite, collophane, dickite, fluorite, galena and sphalerite in Magnesian Limestone from a borehole at Hesleden Dene, near Castle Eden. Other boreholes where dolomite is described include those cutting Magnesian Limestone at Stainton, near Darlington (Raymond 1962), and others cutting Permian beds in the North Sea, 8 km east of Seaham harbour (Magraw 1978). Fine specimens of dolomite have been found at Frazer's Hush mine, near Rookhope, Weardale, together with fluorite, galena, pyrrhotite, quartz, siderite and sphalerite (Wolfe 1998).

Leicestershire Dolomite is a supergene species occupying joints and vein-like structures below the Precambrian/Triassic unconformity at Newhurst quarry, near Shepshed. It ranges in colour from white-to-greyish orange and forms rhombohedral crystals up to 4 mm across (King 1973, Ince 2005). At Breedon quarry, Breedon on the Hill, a cave passage system in Carboniferous Limestone is lined with well developed pale-brown dolomite rhombs (King 1991), and, at nearby Cloud Hill quarry, dolomite is described within spheroids of galena found near the Carboniferous/Triassic unconformity (King 1983).

Nottinghamshire A hard light-brown dolomite forms the top few cm of the Permian Magnesian Limestone at Bulwell, Nottingham, and, where fissured, contains rhombs up to 1 mm across, associated with asphaltite (a type of hydrocarbon), barite, galena, malachite and wulfenite. A similar assemblage also occurs at Kirkby in Ashfield and at Mansfield White Stone quarry (Deans 1961). Interbedded limestone and mudstone of Permian age at Vale Road quarry, Mansfield Woodhouse, contains minute transparent rhombs of dolomite lining solution cavities, often associated with sphalerite (Green & Spence 1988).

Somerset Authigenic dolomite occurs with analcime, barite, celestine, pyrite, quartz and sphalerite, and replaces aragonitic shell debris in a heavily faulted coastal exposure of Lower Jurassic marine sediments between Blue Anchor Point and Stolford Rocks (Alabaster 1980). Dolomite is uncommon in the iron–manganese ore at Higher Pitts farm, Priddy, Mendip Hills (Burr 1996). Kingsbury (1941) notes a pink manganese-rich dolomite from this locality.

Staffordshire Dolomite is one of the gangue minerals in the Ecton copper mines, Ecton Hill (Robey & Porter 1972).

Yorkshire Extensive replacement of dolomite by anhydrite has occurred in Permian evaporite rocks sampled in the Aislaby (Eskdale No. 2) borehole, Eskdale, near Whitby. Halite, magnesite and polyhalite are associated species in various sections of the core. Dolomite normally forms minute rounded granules up to 10 μm across (Stewart 1949, 1951a, Raymond 1962). Other boreholes sampling dolomite in evaporite beds include Fordon, on the Humberside/Yorkshire border (Stewart 1963), and Sleights (Eskdale No. 3), Eskdale, near Whitby (Armstrong et al. 1951). At Hickleton colliery, WNW of Doncaster, white transparent dolomite (massive and crystals to 2 mm) fills vugs and fractures in a faulted siltstone unit. Barite, chalcopyrite and kaolinite are associated species (Uttley 1988).

Wales

Dyfed Dolomite (or iron-rich dolomite) is a constituent of many mineral veins in the Central Wales orefield, with reports from: Cwmystwyth mine (Dewey & Eastwood 1925, Jones & Moreton 1977); Eaglebrook (Nantycagl) mine, Ceulanymaesmawr (Jones & Moreton 1977, Jones 1983); Darren mine, Pen-bont Rhydybeddau; Erglodd mine, N of Tal-y-bont; Esgair Fraith mine, near Tal-y-bont (all Mason 1997); Esgair Hir mine, Ceulanymaesmawr (Mason 1997, Rust & Mason 1988); Ystrad Einion mine, near Tal-y-bont (Mason & Rust 1997). However, it is only from Hafan mine that well formed crystals (5–10 mm across) are known (Bevins 1994). Excellent crystals of iron-rich dolomite have been found in a Coal Measures sandstone at Parc colliery, Tir-y-dail, near Ammanford (Rudler 1905, North 1916).

Gwynedd At Great Orme mine, Llandudno, copper mineralization is found within vuggy dolomitized limestone adjacent to a series of faults. Vugs are infilled with well formed saddle dolomite overgrown by chalcopyrite together with minor marcasite- and minor pyrite. This epigenetic mineralization belongs to the copper–dolomite class of deposit and has been exploited for at least four millennia (Dutton et al. 1994, Ixer & Davies 1996, Ixer & Stanley 1996, Ixer 2001). At the nearby Ty Gwyn mine on Great Orme's Head, saddle dolomite occurs with chalcopyrite, malachite and hydrocarbons containing nickeline inclusions (Parnell 1988c). Iron-rich dolomite, quartz and sericite are the principal gangue minerals at Parys Mountain, Anglesey – reasonable specimens of whitish-cream dolomite with gently curved rhombohedral faces were recovered from the dumps at Morfa Du (Southwood & Bevins 1995). Elsewhere on Anglesey, well formed rhombs of dolomite are noted from Pant-y-Gaseg mine, near Amlwch (Bevins 1994), and from Holy Island, where dolomite completely replaces serpentinite (Maltman 1977). Dolomite is a rare gangue species in the mineralized breccia pipe once worked at Glasdir copper mine, Llanfachreth, near Dolgellau (Allen & Easterbrook 1978).

Mid (South and West) Glamorgan Many of the Carboniferous Limestone units of South Wales are dolomitized and they contain cavernous spaces in which typically curved dolomite crystals have grown, for example at Machen quarry, 7 km ENE of Caerphilly (Plant & Jones 1995). Small-scale Fe–Pb–Zn–(Cu) mineralization with a predominantly barite–calcite–dolomite–witherite gangue occurs throughout the Triassic rocks (Dolomitic Conglomerate) to the S of Llantrisant, and at the Llantrisant bypass roadcut the walls of vugs and geodes are lined with saccharoidal dolomite intergrown with other species (Alabaster 1990). Bevins (1994) reports well formed dolomite in clay–ironstone nodules from Fforchaman colliery, Cwmaman, near Aberdare. In South Glamorgan, pink dolomite crystals have been found at Bendrick Rock, Barry (North 1916), and, at Penmain, Gower Peninsula, West Glamorgan, saddle dolomite has been shown to be strongly zoned on a microscopic scale (Searl 1989).

Scotland

Central Ag–Cu–Co vein mineralization occurs in a dolomite (later barite–calcite) gangue at Alva mine, Silver Glen, Ochil Hills (Moreton 1996, Moreton et al. 1998). Within the Dalradian Ben Lui Schist in the Tyndrum area is a soft green dolomitized schist containing bands of coarse-grain lenses, and pods of dolomite and magnesite (Fortey & Smith 1986).

Dumfries & Galloway Dolomite and quartz are the main gangue minerals in uranium-rich veins within the aureole of the Criffell granite complex around Needle's Eye, SW of Caulkerbush (Miller & Taylor 1966, Braithwaite & Knight 1990). Very fine lilac and brown twinned crystals of dolomite are reported from Blackcraig mine, Newton Stewart (Heddle 1901), and a brick-red variety is noted from Craiglebbock Rocks, near New Abbey (Williams 1965).

Fife Excellent oscillatory zoned (on a microscopic scale) saddle dolomite is described from the Lower Carboniferous rocks of Fife, although associated ankerite is even more impressive (Searl 1989, 1990). Heddle (1901) reports dolomite from Broomhall lime quarries; Chapel quarry, Kirkaldy; Charlestown limestone quarries; and Kinkell and Peanes quarry, Largo.

Grampian Opaque white dolomite is subsidiary to serpentine and talc in Damhead quarry, 2 km SSW of Portsoy. However, it is unusual in being found completely enclosed in lenses of serpentine and is thought to have formed by metasomatic processes operated by low-temperature solutions (Harry & Patterson 1949). Heddle (1901) notes dolomite at Abergairn lead mine, near Ballater; Sylavethy quarry, Alford; and Tyrebagger, Dyce.

Highland Dolomite is the dominant carbonate species in (Lewisian) Scourian ultramafic granulites subjected to Laxfordian shearing and is noted in Assynt: on An Fharaid Mhòr, SW of Achmelvich, near Lochinver (Droop 1994), at Drumbeg (between Stoer and Unapool), and at Gorm Loch, E of Scourie (Sills 1982). It is also a major constituent of the Cambro-Ordovician Durness Limestone near Loch Borralan, Assynt, where it is intruded by carbonatite (Young et al. 1994b). Siliceous dolomitic limestones (Appin Limestone), containing trace amounts of baddeleyite, geikielite or qandilite, have been used to understand mineral reactions and chemical processes that operated in the aureole of the Ballachulish granite (Ferry 1996). Similar studies of siliceous dolomites (dolostones) thermally metamorphosed by the Beinn an Dubhaich granite, Broadford area, Isle of Skye, suggest that calcite and tremolite are the earliest new phases formed by reaction of primary dolomite and quartz (Tilley 1948b). Nodular chert in this dolostone is mantled by talc, tremolite, diopside, and then forsterite, in a sequence of reaction rims as the granite is approached (Hoersch 1981). Additionally, these contact skarns contain a suite of relatively uncommon species derived by boron–fluorine metasomatism of original dolomite, including chondrodite, clinohumite, cuspidine, fluoborite, grossular, harkerite, ludwigite, monticellite, szaibelyite, wollastonite and xanthophyllite (Tilley & Vincent 1947, Tilley 1951). Zoned ultrabasic bodies (up to 1.5 m across) in Lewisian gneisses of the Moine Nappe at Ard Ghunel, Camas Baravaig and Cruard, in the SE of Skye, contain a talc–dolomite core surrounded by actinolite, and are thought to have formed by diffusion under regional metamorphic conditions (Matthews 1967).

Lothian Heddle (1901) reports dolomite from Addiewell (West Calder), Bathgate, Pumpherston and Whitburn, although in most cases the paragenesis is unclear.

Strathclyde Fine irridescent curved rhombohedra of dolomite were once found at Millhole quarry, near Ascog, Bute (Heddle 1901). Brow and Brown's veins near Leadhills contain galena-rich veinstone cemented with calcite and dolomite (Brown 1927). Drusy cavities or geodes in a Carboniferous trachyte sill at Craigmuschat quarry, Gourock, contain barite and quartz, minor dolomite and calcite, and rare fluorite (Nicholson & Durant 1986). Light-brown pearly saddle-shape dolomite rhombohedra occur in aggregates 5–10 mm across within fossil brachiopod shells at Trearne quarry, E of Beith (Todd 1989). White-to-brown intergrown dolomite rhombs up to 1 mm long, often coated with iron oxides, occur at Rubha a' Mhill, northeast Isle of Islay, where they line cavities in limestone (Green 1988).

Tayside Calcite and dolomite are common constituents of both the stratiform mineralization and the metasedimentary and metabasic rocks at Foss and Ben Eagach, near Aberfeldy, although no well crystallized examples are described. Ankerite, baotite, barytocalcite, celsian, cymrite, hyalophane, norsethite, siderite and witherite are associated species (Fortey & Beddoe-Stephens 1982, Moles 1985). Iron-rich dolomite and siderite are minor gangue minerals in gold- and sulphide-bearing vuggy quartz veins at Calliachar Burn, Aberfeldy (Ixer et al. 1997). The Corrycharmaig serpentinite intrusion at Glen Lochay, near Killin, is emplaced into Dalradian schists, and in places has been altered to produce areas of talc–carbonate (mainly magnesite) rock. Antigorite, chlorite, chromite, dolomite, magnetite and tremolite are associated species (Harrison 1985, Hawson & Hall 1987).

Western Isles Crystalline limestone lenses in Lewisian paragneisses at Bay Steinigie, near Finsbay (Fionnsabhagh), South Harris, contain calcite, dolomite and serpentinized forsterite, all with minor amounts of chondrodite and humite (Dearnley 1961). Snow-white dolomite in a vein cutting serpentinite is reported from near the lighthouse on Scalpay Island (Heddle 1901).

Orkney Abundant calcite and dolomite form a cement in Devonian sandstones of the Yesnaby Sandstone Group on the west coast of Mainland (at Ness, Quoyelsh and Yesnaby). They were examined by McReady & Parnell (1998) because they contain uranium-rich bitumen nodules.

Shetland Exceptionally fine talc is associated with highly translucent white dolomite, magnesite (breunnerite) and magnetite at Quin Geo, NE of Northwick Bay, Unst (Heddle 1878a). Dislocation metamorphism at the base of the Shetland ophiolite complex on Unst has produced recrystallized ultramafic rocks (now talc–carbonates), in which dolomite is subsidiary to magnesite (Maynard 1993).

Ireland

Clare Dolomite is a minor constituent of a disseminated and fracture-fill chalcopyrite deposit in Lower Carboniferous carbonate sediments at Ballyvergin (Schultz 1971).

Down Gangue minerals found in copper–lead mineralized breccia at Conlig–Newtownards include barite, calcite, chalcedony, dolomite and quartz (Moles & Nawaz 1996).

Fermanagh The most extensive dolomite deposit in Northern Ireland is on the south side of Belmore Mountain between Belcoo and Boho, and is confined to the basal division of the Lower Carboniferous (Smith et al. 1992).

Galway Four types of dolomite are related to Cu–Pb–Zn–(Ag) mineralization at Tynagh mine, near Killimor. Chemically it varies from iron-free dolomite to ankerite, and in places is replaced by sulphides (Morrissey et al. 1971, Schultz 1971, Gibson 1979, Clifford et al. 1986).

Kerry A cavity in a dolomite vein at Ballyegan quarry, near Castleisland, produced exceptional nailhead calcite crystals in 1997 (Wolfe 1998, Green & Moreton 2001).

Limerick The Courtbrown deposit on the southern shore of the River Shannon, near Ballysteen, is an epigenetic base-metal (lead–zinc) orebody hosted by dolomitized Carboniferous Limestone. Individual dolomite crystals reach 1.4 mm across and show zonation under cathodoluminescence imaging (Reed & Wallace 2001).

Longford Calcite and dolomite are major gangue species in the stratiform sphalerite–(galena–tetrahedrite) deposit at Keel, Ardagh (Morrissey et al. 1971, Ryback & Moreton 1993).

Meath Widespread (but minor) saddle dolomite was precipitated at about the same time as the sulphides (galena, pyrite–marcasite and sphalerite mainly) in the Tara orebody at Navan. It forms fine- to coarse-grain euhedral crystals filling primary and secondary porosity and fractures. In dolomite-cemented breccia it is white, but a distinctive black hydrothermal dolomite is developed as a halo around the sulphides (Kucha & Wieczorek 1984, Peace et al. 2003, Wilkinson 2003, Walshaw et al. 2006).

Monaghan Ankeritic dolomite is the only common carbonate in quartz veins and breccias hosting arsenopyrite–gold–pyrite–stibnite mineralization at Clontibret and the nearby Bryanlitter prospect (Steed & Morris 1997).

Sligo Fine colourless-to-creamy crystals of dolomite, up to 10 mm across, have been recovered from cavities in Carboniferous Limestone at Abbeytown lead–zinc mine, near Ballysadare (Ryback & Moreton 1992), but usually the dolomitized limestone (dolostone) contains only anhedral grains (with interlocking boundaries) included with carbonaceous material (Hitzman 1986). Aughamore (Maye) quarry, near Lough Gill, is a prolific source of well crystallized dolomite. Calcite and chalcopyrite are associated species (Ryback & Moreton 1992, Green & Moreton 2001).

Tipperary Carboniferous dolomitized limestones play host to the giant sedimentary-exhalative base-metal and barite deposit at Silvermines. Dolomite also occurs in the orebody itself mainly in massive form, although superb sharp crystals of dolomite associated with galena, pyrite and sphalerite have been recovered from one large cavity in the deposit (Ryback & Moreton 1991). Because of the importance of this deposit, dolomite has been discussed by many authors (Rhoden 1959, Schultz 1971, Taylor & Andrew 1978, Samson & Russell 1983, Mullane & Kinnaird 1998, Reed & Wallace 2004). Creamy iron-rich dolomite is a major gangue species in the epigenetic Cu–(Ag–Hg) orebody at Gortdrum mine, near Tipperary town, where it is closely associated with bornite, chalcocite, chalcopyrite, cinnabar and mercury-rich tennantite (Morrissey et al. 1971, Schultz 1971, Green & Moreton 2001, Braithwaite & Wilson 2003). Heavily dolomitized Carboniferous Limestone at Kedrah, NE of Cahir, is considered to be a high-grade resource of high-magnesium–low-iron dolomite suitable for use as an ingredient in glass-fibre manufacture (Jones & Fitzsimons 1992). Brecciated and veined dolomitic limestone contains a hydrothermal iron-rich dolomite infill in the carbonate-hosted massive sulphide Pb–Zn–(Ag) deposit at Lisheen, between Thurles and Rathdowney. Well formed dolomite occurs in cavities within the orebody, but only on a microscopic scale (Hitzman 1992, Hitzman et al. 2002, Wilkinson et al. 2005).

Tyrone Gold–sulphide-bearing quartz veins in metamorphosed Dalradian rocks at Curraghinalt, near Gortin in the Sperrin Mountains, contain a late carbonate assemblage that includes strongly zoned variably iron-rich calcite, dolomite and siderite, commonly with manganese substitution (Parnell et al. 2000).

Westmeath A pink crystalline vuggy variety of dolomite, often with chalcopyrite in the vugs, is distributed in minor amounts within sub-economic carbonate-hosted lead–zinc mineralization at Ballinalack, 15 km NW of Mullingar. The ore is dominated by barite, galena, pyrite–marcasite and sphalerite, all in a calcite gangue (Jones & Brand 1986). Dispersed dolomite rhombs, up to 15 μm across, occur with calcite as replacement species in bioclasts within carbonate-hosted lead–zinc mineralization at Moyvoughly, near Moate, 12 km N of Clara (Kucha 1988).

Offshore territories

Isle of Man A detailed chemical analysis of Carboniferous Limestone from Lough Skillicore showed that the original sediment, composed of skeletal remains, pellets and clasts, had been patchily replaced by dolomite, pyrite and quartz (Dickson & Barber 1977). Remarkably fine crystals of dolomite were found at the Foxdale lead mines (Greg & Lettsom 1858) and well defined rhombohedra of dolomite encrusted with chalcopyrite are noted from Laxey, E of Snaefell (Rudler 1905).

Jersey Precambrian sediments are cut by iron-rich dolomite–sphalerite veins at Le Pulec on the northwest coast of Jersey (Ixer & Stanley 1980); and at Petit Port, Trinity, iron-rich dolomite is identified in veins cutting andesite (Ixer 1980). Dolomite is also noted in shale excavated from the Beaumont tunnels (Mourant 1961).

domeykite

cubic Cu_3As

A rare copper arsenide found in Cornwall at Condurrow mine, near Camborne, and Wheal Druid, Carn Brea. Domeykite forms brownish-black nodular masses with a flat conchoidal fracture and metallic lustre. Fresh surfaces are tin white or slightly yellowish (Greg & Lettsom 1858). A sheet of native copper collected from loose material in a cliff talus debris flow at Littleham Cove, near Budleigh Salterton, Devon, also contains cuprite, domeykite and paramelaconite (all confirmed by XRD). Domeykite replaces cuprite and is intimately associated with two unnamed species whose compositions are Cu_4As and Cu_5As. The occurrence is associated with complex Ag–As–Co–Cu–Ni–U–V mineralization, which developed as the Permo-Triassic strata was de-watered during burial compaction (Milodowski et al. 2002).

Domeykite (probably domeykite-alpha) has been found in copper-rich void-infilling mineralization near the Precambrian/Triassic unconformity at New Cliffe Hill quarry, Stanton under Bardon, Leicestershire. It occurs as trace amounts in 10 μm-wide algodonite veinlets cutting native copper (Ixer et al. 2005a).

dopplerite

Calcium salt of a humic acid.

doranite

TL Carrickfergus, Co. Antrim, Ireland

Named after Patrick Doran (1779–1880), a mineral and fossil collector and dealer from Glassdrummond, Co. Down, Ireland, locally referred to as Diamond Pat. Doranite is a magnesium-rich analcime, zeolite group (Dyer et al. 1987).

doyleite

triclinic $Al(OH)_3$

A rare aluminium hydroxide, polymorphous with bayerite, gibbsite and nordstrandite, and found in veins cutting Lower Carboniferous limestones at Coldstones quarry, 5 km W of Pateley Bridge, Yorkshire. The veins carry fluorite, barite, galena and sphalerite, and in places the supergene minerals cerussite, hemimorphite and smithsonite are found commonly lining cavities. Fallen blocks from the Garnet vein also contain doyleite, otavite and prosopite. Doyleite occurs as a soft powdery white mineral filling small (up to 3 mm) cavities in greyish-white fluorite (Young et al. 1997).

dravite

TOURMALINE GROUP

trigonal $NaMg_3Al_6(BO_3)_3Si_6O_{18}(OH)_4$

A common sodium–magnesium–aluminium borosilicate hydroxide that forms series with schorl and with uvite, and is found in boron-metasomatized limestones and mafic igneous rocks, and in granite pegmatites and hydrothermal veins.

Dravite occurs in adinoles (quartz–albite rocks) at Dinas Head, a small promontory adjoining Trevose Head, 8 km W of Padstow, Cornwall. The adinoles have been produced by the complete metasomatism of Upper Devonian slates and cherts by an albite-dolerite intrusion. Minute needles of colourless to very pale-green dravite–uvite occur in minor amounts within pools of quartz and are associated with ankerite, calcite, chlorite, leucoxene and accessory graphite, prehnite, rutile and zoisite. In certain adinoles, bands with a tourmaline content up to 30% are noted. It is tentatively suggested that the boron required to form tourmaline came from a buried granite near by (Agrell 1939, 1941). There is no such doubt about the St Austell granite being the driving force behind a hydrothermal system that produced zoned dravite–magnesiofoitite–foitite–schorl crystals in a hydrothermal breccia at the southwestern margin of Roche Rock, 7 km N of St Austell, Cornwall. The presence of tin in certain zones within the tourmaline provides a valuable insight into the evolution of mineralizing fluids that ultimately produced tin (cassiterite) mineralization for which Cornwall is justly famous (Williamson et al. 2000). Dravite with a significant calcium content (i.e. towards uvite) has been found in another tourmaline breccia at Leigh Tor, Devon, within the aureole of the Dartmoor granite and also on the dumps of Botallack mine, St Just, Cornwall (Manning 1991). Dravite from two other Cornish localities requires further confirmation: Cligga Head mine, Perranzabuloe (Chick 1987), and Gunheath china-clay pit, St Austell (Golley & Williams 1995).

Also in Cornwall, quartz–tourmaline vein samples collected near St Ives along the shoreline at Carrick Du and Clodgy Point contain dravite. The samples were not *in situ*, but were probably derived from the nearby Carrick Du mine. At these localities, zoned dravite, showing pale-blue to pale-brown pleochroism in thin section, is the major tourmaline species present and this is sometimes rimmed with uvite. Spectacular oscillatory zonation in bright-blue (in thin section) tourmaline reflects alternating schorl and dravite compositions in a dump sample from Geevor mine, near Pendeen, St Just. Near by, at Priest's Cove close to Cape Cornwall, a tourmaline-rich

Dravite (medium grey) forms the cores of the acicular crystals and the triangular zone in the massive tourmaline (bottom right); light grey is schorl. From an aplite vein at Carn Moyle, near the Porthmeor Cove granite pegmatite, Cornwall; width 400 μm; BSE image. (© A. G. Tindle)

vein contains two discrete generations of tourmaline. The earliest generation is orange-brown (in thin section) schorl interpreted to be of magmatic origin; the second is bright-blue dravite (minor schorl), with a widely variable compositional range, and is interpreted to have a post-magmatic hydrothermal origin. At Gryll's Bunny, near Botallack, St Just, metasomatic alteration of amphibolite and actinolitic greenstone has produced coarsely granular (≤5 mm) tourmaline-rich rocks as part of a hydrothermal event induced by the emplacement of the Land's End granite. Optically blue to pale-brown tourmaline (black in hand specimen) is strongly zoned and mainly schorl or dravite in composition. Uvite and feruvite occur less commonly, either as discrete crystals or as rims to dravite–schorl crystals. In some cases, all four species are found within a single crystal. Perhaps the most complex occurrence of dravite is from a single blade of hydrothermal tourmaline found in a quartz vein cutting extensively kaolinized granite from Longstone china-clay pit (the site of the Trelawny Tigers speedway track), near Nanpean, within the St Austell granite complex. This single crystal is strongly zoned from core to rim with the following species of tourmaline: schorl, foitite, elbaite, foitite, dravite, magnesiofoitite, foitite, schorl, magnesiofoitite, foitite, dravite, schorl, dravite. Clearly, many changes occurred in the hydrothermal system during the formation of the crystal. Dravite–schorl–foitite compositions are also found in zoned tourmaline from the wolframite deposit at Hemerdon, Plympton, Devon. In thin section the radiating sprays of needle-shape bright-blue crystals are up to 2 mm in length (Tindle unpub. EMP data).

Diorite exposed in Newhurst quarry, Shepshed, Leicestershire, contains fault and joint surfaces coated with a bluish-grey striated microcrystalline dravite (Ince 2005).

Tourmaline is described from a wolframite-bearing vein and from loose blocks of vein quartz at Burdell Gill, near Coombe Height, Caldbeck Fells, Cumbria. It forms slender (1–4 mm long), pale grey-green to brown prisms with a triangular cross section, and its composition was predicted to be schorl (Neall & Leppington 2003). Analysis of this material revealed it to be mainly dravite, with minor schorl and magnesiofoitite. Another specimen from the same location, but containing smaller needles of tourmaline, contains foitite in addition to the other three species (Tindle unpub. EMP data).

Dravite is the major tourmaline group species in metasomatically altered metapelites in the aureole of the presumed Grasmoor granite beneath Crummock Water, Cumbria. Tourmalinization took place in wallrocks containing a series of faults and joints, and the resulting vein-like rocks contain nothing more than tourmaline and quartz. Tourmaline crystals have a grainsize of 5–10 µm, exhibit olive-green to brown colours in thin section, and show compositional variation to produce less common foitite and schorl. Another suite of tourmalinites exposed N of the Haweswater reservoir occur only in brecciated pipe-like bodies cut by tourmaline–quartz veins. Here, dravite is accompanied by minor magnesiofoitite (Eeckhout et al. 2004).

In Wales, tourmaline forms sporadic clots within altered gabbro, 300 m SW of the summit of Hanter Hill, Powys. The bluish-black clots are up to 15 mm across and consist of a single crystal or group of crystals in near-parallel orientation. Individual crystals exhibit well defined colour zones, becoming darker outwards, and the single chemical analysis is said to be of dravite composition with an unusually high calcium content (Holgate 1977). However, if the classification of Hawthorne & Henry (1999) is adopted, then the species is uvite. In view of the zonation and the fact that the analysis is based on a bulk mineral-separate (and hence prone to contamination), the species requires further investigation in Wales.

Fragments of dravite occur in quartz–mica schist at Loch Assapool, Ross of Mull, Isle of Mull, Strathclyde, Scotland (Livingstone 1993); and at the Foss barite–base-metal mine, near Aberfeldy, Tayside Region, dravite occurs in semi-pelites as pale-green to greenish-brown (in thin section) zoned crystals varying from dravite to schorl in composition (Moles 1985). A heterogeneous rock, best described as a chert but locally containing a high abundance of tourmaline, occurs in a sequence of Dalradian metasediments to the W of the Hill of Three Stones granite, to the S of Cabrach, Grampian. The tourmaline is zoned from olive-green cores to pale-brown rims (in thin section), but all compositions are dravite. Nearer to Cabrach, in Perkie Burn, a boulder of tourmaline pegmatite was also found to contain dravite. A fine-grain biotite hornfels in the aureole of the Loch Doon granite complex, near Loch Brecbowie in Carrick Forest (nearest village Straiton), Strathclyde, contains tiny green-to-brown zoned dravite in sub-mm veinlets (Tindle unpub. EMP data).

At the southwestern end of the Leinster batholith, near the village of Inistioge, Co. Kilkenny, metasomatic tourmaline belonging to the dravite–schorl series is found associated with disseminated arsenopyrite mineralization. Associated species include chalcopyrite, iron-rich chlorite, fluorite, manganese-rich ilmenite, molybdenite, muscovite, pyrite, pyrrhotite and rutile. The mineralization was produced during alteration of aluminous metasediments and is spatially related to pegmatitic quartz veins with greisen alteration (Appleby & Williams 1988). Dravite–schorl is reported to be common in the aureole of the Leinster granite, Co. Wicklow and Co. Carlow, southeast Ireland, particularly in coticules (quartz–tourmaline rocks) along the southern margin of the complex (Kennan 1997). Two analyses of dravite are presented by Gallagher (1988), from a coticule at Coolasnaghta and from a metasomatically altered schist adjacent to the Aclare lithium pegmatite, both in Co. Carlow. Additional occurrences related to the Leinster granite include: Glendasan, Co. Wicklow, where slender needles of tourmaline, up to 7 mm in length, are reported from a quartz vein cutting andalusite schist, within 10 m of the granite contact. The tourmaline is zoned from a core of blue foitite to a rim of brown dravite. A similar zonation pattern is developed in tourmaline within breccia-like zones of near-massive quartz–tourmaline rocks at Kilcarry Bridge, Co. Carlow, some 700 m from the granite contact. Dravite also occurs in wallrock phyllite adjacent to a microtonalite sheet complex and in an associated tin–tungsten mineralized quartz vein at Ballinglen, between Tinahely and Aughrim, Co. Wicklow (Gallagher & Kennan 1992). Additionally, dravite is noted in a schist xenolith within granite, collected at the top of the miner's path to Van Diemen's Land at the head of Glendalough, Co. Wicklow (S. Moreton sample, Tindle unpub. EMP data).

dreelite

Possibly a mixture of barite and gypsum, but also reported as a synonym of barite.

dudgeonite

TL Pibble mine, Creetown, Dumfries & Galloway, Scotland

Named after Patrick Dudgeon (1817–1895) of Cargen, near Newton Stewart, Dumfries & Galloway, Scotland; landowner, meteorologist, mineralogist and a founder member of the Mineralogical Society of Great Britain and Ireland; born at Marionville House, Restalrig, Edinburgh. Dudgeonite is a calcium-rich variety of annabergite, vivianite group (see Heddle 1889a).

dufrénite

monoclinic $Ca_{0.5}Fe^{2+}Fe^{3+}_5(PO_4)_4(OH)_6{\cdot}2H_2O$

An uncommon hydrated calcium–iron phosphate hydroxide found in association with chalcosiderite at Phoenix United mines (Stowe's section, western part of Wheal

Botryoidal encrustations of black dufrénite in aggregates to 1.5 mm wide, with dull yellow cyrilovite crystals; Gunheath china-clay pit, Carthew, near St Austell, Cornwall. (© M. P. Cooper)

Phoenix), Linkinhorne, Cornwall (Kingsbury 1954). Black to apple-green crystals of dufrénite with a vitreous lustre form confused groups or botryoidal masses (Kinch & Butler 1886) and dark olive-green dufrénite forms internally radiating globules on quartz (Kinch 1888). Collins (1911a) describes dufrénite as occurring rarely in crystals, usually fibrous or in radiating nodules with a silky lustre and amorphous green to brown. Braithwaite & Corke (1980) describe kidwellite pseudomorphing probable dufrénite in small cavities within a goethite–quartz gossan from the mine dumps at Wheal Phoenix and they suggest that many specimens of dufrénite in collections are, in fact goethite, to which dufrénite readily alters.

Dufrénite is also reported from small lenses full of iron staining and associated with chalcosiderite and cyrilovite at Gunheath china-clay pit, Carthew, St Austell, Cornwall (Grigg 1986). Watson (1987) reports a vein in kaolinized granite in the north face of this pit towards the Hensbarrow pit, containing dufrénite as dark-green, light-green and brown botryoidal spheres to 1 mm associated with cyrilovite, chalcosiderite, leucophosphite, libethenite and stannoidite. In the Hensbarrow section of Gunheath pit, iron-oxide-rich veins in kaolinized granite have been shown to contain dufrénite intimately associated with barbosalite, chalcosiderite–turquoise, fluorapatite, hentschelite, quartz and stannoidite (Elton 1998). Dufrénite is also described from Gossan vein in Bunny mine (also a part of Gunheath pit) by Weiss (1994), who divided the minerals of the pit into three paragenetic groups, describing dufrénite as a secondary mineral from the oxidation zone.

Because the dufrénite described from Burdell Gill, Caldbeck Fells, Cumbria (Hartley 1984), could be from a Kingsbury specimen, this occurrence requires further confirmation.

dufrénoysite

monoclinic $Pb_2As_2S_5$

A rare lead sulpharsenite found as tiny needles in a vug from Wheal Boys, Trewetha, St Endellion, Cornwall. Dufrénoysite is briefly described by Ryback in an editorial to Shrimpton (1994). Dufrénoysite has been found as inclusions in galena from the limestone-hosted Cu–Pb–Zn–(Ag) deposit at Tynagh mine, near Killimor, Co. Galway, Ireland (Boast et al. 1981a).

duftite

ADELITE GROUP

orthorhombic $PbCu(AsO_4)(OH)$

A rare lead–copper arsenate hydroxide found at Penberthy Croft mine, St Hilary, Cornwall, as euhedral, rhombic–bipyramidal, translucent, resinous, bright to dark olive-green crystals up to 1.2 mm across. It typically lines cavities in bayldonite and is associated with cerussite, mimetite and quartz. It has previously been confused with mottramite, with which it is indistinguishable by XRD analysis (Camm & Merry 1991b, Betterton et al. 2000). Duftite is also recorded from Mottram mine, Mottram St Andrew, Cheshire (D. I. Green specimen and XRD, Tindle unpub. EMP data).

Olive-green duftite to 0.5 mm with minor pale-green globular bayldonite. Penberthy Croft mine, St Hilary, Cornwall. (© D. I. Green)

At Brandy Gill, Caldbeck Fells, Cumbria, botryoidal duftite ocurs as a thin coating on lead minerals and has been found high up on the western slope, on a spoilheap from a small galena–chalcopyrite–quartz-bearing cross vein. Other associated species include, agardite–mixite, ankerite, cerussite, linarite, malachite, mimetite, pyromorphite and stolzite. The largest duftite crystals are 0.2 mm long (Claringbull 1951, Davidson & Thomson 1951, Cooper & Stanley 1990). Near by, a bright lemon-yellow crust of duftite intimately associated with bayldonite has been identified on two specimens from a mineralized quartz vein at Low Pike (Cooper & Stanley 1990, Young et al. 1990a, Green & Briscoe 2002, Green et al. 2003b) and from a specimen collected on Deer Hills (Cooper & Stanley 1990). Duftite is also confirmed as lustrous black crystals (up to 0.3 mm long and resembling mottramite), and greyish-green to sage-green crusts at Arm o' Grain, Caldbeck Fells (Green et al. 2006b).

Elsewhere in Cumbria, duftite is recorded from the dumps of an old shaft on Clouds vein at Stennerskeugh Clouds, 1 km SW of High Stennerskeugh level, Ravenstonedale, near Kirkby Stephen, where it is found in partly oxidized galena–tennantite ore. Duftite forms bright-green crusts up to 0.6 mm thick, composed of spherules up to 0.2 mm in diameter, and encrusts mimetite lining a cavity in barite. The duftite clearly results from supergene alteration of the arsenic-, copper- and lead-bearing ore (Starkey 1987b). Duftite described from Driggith mine and so-called calcioduftite from Higher Brandy Gill, both in the Caldbeck Fells, Cumbria (Hartley 1984, Young 1987a, Cooper & Stanley 1990) require further confirmation, as the identifications are based on samples supposedly collected by A. W. G. Kingsbury, and there is some doubt about their provenance (Ryback et al. 1998, 2001; A. Hart pers. comm. 2004). Green et al. (2006a) strongly suggest that the Driggith specimens are fraudulently provenanced, but confirm duftite as apple-green crusts on the nearby 60-fathom level dumps of Sandbed mine.

In Scotland, duftite is noted on specimens from an old dump at Tait's level, Whyte's Cleugh, Wanlockhead, Dumfries & Galloway (Macpherson & Livingstone 1982).

dumortierite?

orthorhombic $Al_7(BO_3)(SiO_4)_3O_3$

An uncommon aluminium borosilicate reported in the residue of heavy minerals separated from various samples in Grampian Region, Scotland: a paragneiss near Newburgh, S of Fraserburgh; small pegmatite veins cutting schists near Colehill farm, N of Ellon; and in the Cowhyte Gneiss and a foliated granite at Bogierow, both near Portsoy (Mackie 1925, 1928).

Dumortierite has also been described in heavy-mineral concentrates from England: Bagshot Beds, Bournemouth; Bridport Sands, Westport Bay; Upper Greensand, Lulworth Cove (all Dorset); Blackheath Beds, Shirley, Surrey (now Greater London); and Land's End granite, Trevella, and St Austell granite, Cornwall (Groves 1928).

In all cases the identification has been optical, on the basis of characteristic pleochroism from pale straw to an intense blue. However, it is possible to mistake dumortierite for andalusite, glaucophane, tourmaline or even lepidolite (Groves 1928), and confirmatory analytical data are therefore desirable.

Dumortierite is also noted on a specimen in the Kingsbury collection at the NHM. The sample is labelled as being from Ford Farm quarry, Devon, but Ryback et al. (2001) conclude that this is a falsification and suggest instead that the specimen is from Dehesa, California.

dundasite

orthorhombic $PbAl_2(CO_3)_2(OH)_4{\cdot}H_2O$

Dundasite is an uncommon hydrated lead–aluminium carbonate hydroxide found on a 19th-century specimen of galena from Wheal Rose, Sithney, Cornwall. Dundasite forms spheres and tufts of snow-white crystals on cerussite in a shallow cavity. Two specimens containing minute snow-white or slightly bluish dundasite, arranged in spherical aggregates of silky radiating acicular crystals, were also found on the dumps of Port Quin antimony mine, near Port Quin, St Endellion, Cornwall (Russell 1944, 1949b). More recently, large hand specimens richly encrusted in white botryoidal dundasite were recovered from the lead lode at Greystone quarry, Lezant, near Launceston, Cornwall. Near by, radiating crystals up to 3 mm across were found in cavities and fractures in gossanous matrix at the quarry. Dundasite is commonly associated with cerussite and sometimes with malachite and native copper. It is described as the finest in Cornwall (Elton et al. 2004). Snow-white radial aggregates of dundasite occur as secondary growths on limonite or galena at Gunver Head, near Padstow (Schlomann & Weiss 1998). Golley & Williams (1995) note dundasite from four other locations in Cornwall: Padstow Consols, Pengenna and Trevinnick mines (St Kew) and Wheal Penrose (Sithney). At Trevinnick mine, dundasite forms small pearly white free-standing acicular crystals and tufts on galena and surrounding matrix (Fisk et al. 1982).

At Hampstead Farm quarry, Chipping Sodbury, 16 km NE of Bristol, dundasite was found in two partially oxidized galena-bearing veins cutting Carboniferous Limestone close to the Rhaetic/Lower Carboniferous unconformity. Here it occurs with allophane, felsöbányaite (syn. basaluminite), gypsum, para-alumohydrocalcite, scarbroite and possibly alumohydrocalcite. Dundasite forms smooth white compact hemispheres, up to 0.2 mm diameter, composed of bladed compound crystals and it occurs in botryoidal crusts of bladed or acicular crystals, or silky mats of fibrous crystals. At this locality, the secondary aluminous minerals are thought to have been derived from the leaching of aluminous clays by strongly acidic solutions, resulting from the weathering and oxidation of pyrite. Dundasite is also noted on deeply oxidized smithsonite-bearing vein material associated with cavernous, mineralized Upper Triassic limestone breccias near East Harptree, N of the Mendip Hills, Avon (Alabaster 1990b), and associated with alumohydrocalcite in cavities within manganese oxides at Merehead (Torr Works) quarry, Cranmore, Mendip Hills, Somerset (M. S. Rumsey and R. W. Turner pers. comms 2006).

Minute white or yellow spheres of dundasite, up to 0.75 mm in diameter, the exterior smooth and silky, the interior radiating and silky, were once found at Millclose lead–zinc mine, Wensley, near Matlock, Derbyshire. Associated minerals are calamine, cerussite, fluorite, galena and sphalerite (Russell 1912b, Traill 1939).

The NHM holds samples of dundasite from two locations in Cumbria: Brandlehow mine, Derwentwater, and Wanthwaite mine, St John's in the Vale, near Threlkeld. Samples from the latter locality are in the NHM's Kingsbury collection, and have questionable provenance (Ryback et al. 1998, 2001).

In Wales, dundasite, forming small silky white tufts on cerussite, and spherical aggregates (or tufts), both consisting of a felted mass of minute radiating silky needles on cerussite (Prior 1906). A very good specimen of dundasite was also collected from the nearby Cyffty mine. It consists of acicular crystals of cerussite on which snow-white spheres of dundasite up to 1 mm in diameter have grown (Russell 1944). Another specimen collected at Gorlan mine, between Betws-y-coed and Trefriw, contains prismatic crystals of cerussite and bipyramidal crystals of quartz enveloped with a snow-white downy crust of very minute matted silky needles of dundasite. All three mines worked galena–marcasite–sphalerite lodes traversing slate (Russell 1944). Eaglebrook (Nantycagl) mine, Ceulanymaesmawr in the Central Wales orefield, Dyfed, is a fourth Welsh locality for dundasite. Here, it forms balls of radiating acicular crystals up to 0.5 mm in diameter, varying in colour from very pale-blue centres to white terminations (Bevins 1994). Dundasite is also confirmed (by XRD) from Dolyhir quarry, Old Radnor, near Presteigne, Powys (R. E. Starkey pers. comm. 2005).

Dundasite has been reported from Clements lead mine, Carrowgarriff, Maam, Co. Galway, Ireland. It normally forms minute snow-white silky balls or flattened spheres, but also occurs as snow-white loosely aggregated needles occupying cavities in quartz–calcite veinstone containing cerussite, galena, pyrite and sphalerite (Russell 1912b). Dundasite was just one of many supergene species collected at Tynagh mine, near Killimor, Co. Galway, Ireland, before it closed in 1982 (Green & Moreton 2001).

duporthite

TL Duporth, near St Austell, Cornwall, England

A mixture of talc and chlorite (Scott & Power 1995). See Collins (1877c).

durangite?

monoclinic $NaAl(AsO_4)F$

A sodium–aluminium arsenate fluoride that forms separate series with maxwellite and with tilasite. Durangite was supposedly found by A. W. G. Kingsbury in 1958 at Cheesewring quarry, SW of Henwood, Linkinhorne, Cornwall, and the specimens are now held in the Kingsbury collection at the NHM (Embrey 1978, Foord et al. 1985). Later workers have uncovered problems with the provenance of many Kingsbury samples (Ryback et al. 1998, 2001, Bridges et al. 2006, Green et al. 2006a) and, unless further discoveries are made, this species cannot be considered to be native to Britain or Ireland.

dypingite

monoclinic? $Mg_5(CO_3)_4(OH)_2{\cdot}5H_2O$

A rare hydrated magnesium carbonate hydroxide found at Hagdale, Unst, Shetland, Scotland. Dypingite forms a milky-white crust on chromite-bearing serpentinite (Livingstone 1993). It has also been identified by XRD as white spherulitic masses up to 0.3 mm across on witherite from Rampgill mine, Nenthead, Cumbria (Green & Young 2006).

dysclasite

A synonym of okenite.

dyscrasite?

orthorhombic Ag_3Sb

A rare silver antimonide briefly mentioned by Golley & Williams (1995) as occurring at Dolcoath mine, Camborne, Cornwall. Dyscrasite and a wide range of other sulphide species are described from Sark's Hope mine, Little Sark (Mourant & Warren 1934), but the species is not reported by later workers (e.g. Ixer & Stanley 1983). Further confirmation is desirable.

E

earthy cobalt

A synonym of asbolane.

eastonite

MICA GROUP

monoclinic $KMg_2^{2+}AlAl_2Si_2O_{10}(OH)_2$

A relatively uncommon potassium–magnesium aluminosilicate hydroxide most often found in regionally and thermally metamorphosed pelites. Eastonite is a rock-forming mineral that can be easily misidentified because of its similarity to biotite. Because the IMA classification of micas (Rieder et al. 1998) focuses only on end-member compositions, it is necessary to define here the compositional ranges for eastonite. They have been chosen to be Fe/(Fe+Mg)=0–0.5 and Al^{vi} = 0.5–1.5, where Fe is total Fe expressed as FeO. All data have been processed using the method for trioctahedral micas described in Tindle & Webb (1990). No data presented here are particularly close to the eastonite end member and in many cases the mineral is referred to as "biotite" by the original authors.

Eastonite occurs with chlorite, cordierite and ilmenite in thermally metamorphosed hornfels within the aureole of the Land's End granite at the Inner quarry, Kenidjack, St Just, Cornwall (Chinner & Fox 1974).

In Scotland, examples of eastonite have been reported from the following Dalradian assemblages: garnetiferous pelites between Dunkeld and Pitlochry, Tayside Region (Atherton 1968); exceptional barium-rich eastonite occurs with biotite and phlogopite in metasediments and metabasites associated with the Foss barite deposit, near Aberfeldy, Tayside Region (Moles 1985); graphite-bearing staurolite mica schists at the Craigs of Inchdowrie, SE of Clova village, Glen Clova, Tayside Region (Chinner 1965); thermally metamorphosed greenschist close to a dolerite feeder pipe near the summit of Sidhean Sluaigh, 3 km SW of Strachur, Strathclyde (Worden et al. 1987). Eastonite also occurs: in Moinian pelitic hornfels thermally metamorphosed by the Carn Chuinneag–Inchbae complex, Highland Region (Harker 1954b); in Lewisian epidote–garnet–hornblende gneisses at Camas Baravaig and Cruard, Isle of Skye (Matthews & Cheeney 1968); in graphite–kyanite–mica–pyrite schist, Glen Urquhart complex, W of Loch Ness, near Drumnadrochit (Rock et al. 1986b); with siderophyllite in xenoliths of gneiss within a Tertiary dolerite sill that outcrops on the south coast of the Ross of Mull, between Scoor and Uisken, Isle of Mull (Brearley 1986, 1987); and with phlogopite in sapphirine-bearing garnet–kyanite–orthopyroxene granulites, S and SW of Roineabhal, South Harris, Western Isles (Baba 2003). Thermally and regionally metamorphosed Moinian Supergroup metapelites in the aureole of the Ross of Mull granite, west Isle of Mull, Strathclyde, contain both eastonite and siderophyllite in assemblages in which combinations of almandine, andalusite, cordierite, garnet, ilmenite, kyanite, muscovite, orthoclase, plagioclase and sillimanite are also present (Wheeler et al. 2004). A few analyses of eastonite (and siderophyllite) have been collected together with many biotite analyses from the granodiorites and granites of the Criffell Main intrusion, Dumfries & Galloway, from the Peterhead granite, Grampian, and the Monadhliath granite, Highland Region. One sample of granite from Auldearn, near Nairn, Highland Region, also contains eastonite (Tindle unpub. EMP data).

In Ireland, eastonite occurs with siderophyllite in Dalradian metasediments from the aureole of the Ardara pluton at Clooney, close to the shore of Gweebara Bay, Co. Donegal (Kerrick 1987), and in blueschist-facies metabasites from central Achill, N of Clew Bay, Co. Mayo (Chew et al. 2003).

At Beaucette in north Guernsey, eastonite occurs with biotite and a suite of calcic amphiboles in meladiorite layers within veined diorite belonging to the late Precambrian Cadomian igneous complex (Bishop & French 1982).

ecandrewsite

ILMENITE GROUP

trigonal $ZnTiO_3$

A rare zinc–titanium oxide found in North Tincroft lode, New Cook's Kitchen mine, Redruth, Cornwall. The lode is a polymetallic fracture-hosted vein structure consisting mainly of arsenopyrite, cassiterite, chalcopyrite and sphalerite, all in a matrix of chlorite, fluorite and quartz. Paragenetically very early titanium oxides occur as zoned lath-like inclusions (up to 200 μm in length) in all the major phases. Compositionally these inclusions are mostly ilmenite, but ilmenorutile, pyrophanite and ecandrewsite are also found. Anatase, forming 50 μm-long laths, accompanies these species. A mesothermal origin (300–350°C) has been proposed (LeBoutillier et al. 2000).

ecdemite?

tetragonal $Pb_6As_2^{3+}O_7Cl_4$

A rare lead–arsenic oxychloride from Merehead (Torr Works) quarry, Cranmore, Somerset, very tentatively identified by the NHM (A. Hart pers. comm. 2003); requires further confirmation.

edenite

AMPHIBOLE GROUP

monoclinic $NaCa_2Mg_5(Si_7Al)O_{22}(OH)_2$

A relatively common sodium–calcium–magnesium aluminosilicate hydroxide that forms a series with ferro-edenite and is found in intermediate plutonic rocks and medium-grade metamorphic rocks such as amphibolites and marbles. Where possible, identification is based on the 1997 classification of the amphibole group (Leake et al. 1997a, Mogessie et al. 2004).

England

Cornwall Edenite occurs with magnesiohornblende in amphibolitic metabasites (Landewednack Schists) and in the mafic component of the Kennack Gneiss at Kennack Sands and its environs, Lizard Peninsula (Sandeman et al. 1995).

Cumbria Contact metamorphism of a dolerite dyke at the eastern margin of the Eskdale granite, near Stony Tarn, has produced edenite as a replacement product of pyroxene (Macdonald et al. 1988a).

Lancashire Erratic blocks of altered gabbro from a gravel pit at Burscough, SE of Southport, contain three species of amphibole: edenite, kaersutite and magnesiohornblende (D. I. Green specimen, Tindle unpub. EMP data). Cavities in the gabbro are infilled with radiating thomsonite-Ca similar to that found on Thurstaston beach, Wirral, Merseyside (Rothwell & Dickinson 1989).

Wales

Dyfed The xenolithic laminated olivine gabbro unit of the St David's Head layered intrusion contains edenite and pargasite that had originally been referred to as iron-rich pargasitic hornblende and magnesio-hastingsitic hornblende (Bevins et al. 1994).

Scotland

Central Well crystallized edenite is described from boulders in a stream bed to the SW of Trinafour, near Kinloch Rannoch (Green & Todd 2001), but later analysis could identify only tremolite (S. Moreton specimen, Tindle unpub. EMP data). So-called hornblende schists from Aberfoyle Forest, near Balmaha, Loch Lomond (close to the Highland Boundary fault), contain acicular green edenite, magnesiohornblende with lesser amounts of pargasite and tschermakite (Henderson et al. 1983).

Dumfries & Galloway The late Caledonian Newmains lamprophyre dyke, 7 km W of Dumfries, contains abundant euhedral prisms of edenite and pargasite, commonly cored by clinopyroxene (Macdonald et al. 1986).

Grampian A small Moinian amphibolitized metabasite intrusion, 4 km S of Tomatin, contains crystals of edenite up to 1 mm long, sieved by droplets of plagioclase and associated with ferropargasite, garnet and magnesiohornblende. The rock also contains diopside–plagioclase symplectites, suggesting that it may once have been an eclogite (Baker 1986). Edenite occurs with ferro-edenite, ferropargasite and pargasite in Middle Zone fine-grain granular gabbros in the Insch basic layered intrusion, to the E of Insch (Wadsworth 1988). Black edenite crystals reaching 50 cm in length are described from a diorite in Glenbuchat (Heddle 1878c). Edenite is a rock-forming mineral in diorite and granodiorite from the Hill of Fare granite complex at the Torphins railway cutting and in granodiorite from E of Little Craig, north Gairnshiel, Glen Gairn (Tindle unpub. EMP data).

Highland Edenite occurs in metasomatized (fenitized) quartzites at Four Burns, near Loch Borralan, Assynt. The nearby Borralan syenite complex was the source of the strongly alkaline fluids required to produce fenitization of these rocks. Green actinolite, edenite and magnesiohornblende developed in the early stages of metasomatism and this was followed by blue varieties that have magnesioriebeckite, richterite and winchite compositions (Woolley et al. 1972). Pale-green (in thin section) amphibole varying in composition from edenite to ferro-edenite occurs intimately associated with diopside (described as aegirine-augite) in syenite at Glen Dessary, W of Loch Arkaig (Richardson 1968, Fowler 1992). The rim of an amphibole-rich clot in a microgranular enclave within the Strontian granite complex is edenite. Although actinolite and magnesiohornblende compositions appear to be more common (Castro & Stephens 1992). Tonalite from the Morvern–Strontian complex (Nockolds & Mitchell 1946) and diorite from the Ballachulish granite complex (Weiss & Troll 1989) both contain edenite. Delicate fibres of black parvo-edenite, arranged in tufts and reaching 12 mm in length, are associated with calcite, magnesiohornblende and pyrrhotite in granular limestone from quarries above Milltown, Glen Urquhart (Heddle 1878c). At White Bridge, Glen Garry, W of the Great Glen, edenite (described as hastingsite) occurs with hastingsite and ferropargasite in hornblendic gneisses belonging to the Loch Eil division of the Moine (Rock 1984).
Strathclyde Green edenite, ferro-edenite and pargasite (originally called hornblende) are found in crossite–amphibolites associated with the Girvan–Ballantrae ophiolite complex to the W of Knockormal (Bloxam & Allen 1959; A. Livingstone unpub. EMP data). Euhedral brown crystals of edenite and pargasite occur in porphyrite dykes cutting the Loch Doon granite complex between Loch Dee and Craiglee (Tindle 1982).
Western Isles Edenite, magnesiohastingsite and pargasite occur in a metamorphosed spinel lherzolite lens within Rodel–Leverburgh metasediments at the head of Glen Rodel, South Harris. Associated species are augite, enstatite and accessory hercynite and magnetite (Livingstone 1976c). Amphibole-rich veins cutting gabbro at Mullach Bi, and around Claigeann Mòr on St Kilda (Hirta), are reported to contain actinolite, edenite and magnesiohornblende, although the data remain unpublished (Harding et al. 1984).

Ireland
Galway Edenite is reported from the Galway granite (Leake & Ahmed Said 1994, Leake 1998) and from Dalradian garnetiferous striped amphibolites in Connemara (Leake 1972).
Louth Metasomatically altered limestones in the Carlingford area are reported to contain edenite, although the data remain unpublished (Osborne 1932b).

Offshore territories
Guernsey At Beaucette in north Guernsey, brownish-green edenite occurs in a pegmatitic pocket in gabbro within veined diorite belonging to the late Precambrian Cadomian igneous complex. Amphiboles in the dioritic and gabbroic rocks include kaersutite, magnesiohornblende and pargasite (Bishop & French 1982).
Jersey Cadomian layered diorites associated with gabbro on the southeast coast at Le Nez contain a range of amphiboles considered to be secondary after pyroxene. They include actinolite, edenite, kaersutite, magnesiohastingsite, magnesiohornblende and pargasite (Bishop & Key 1983).

edingtonite
ZEOLITE GROUP
orthorhombic or tetragonal $Ba(Al_2Si_3)O_{10}\cdot 4H_2O$
TL Quarry 1 km NE of Old Kilpatrick, near Glasgow, Strathclyde, Scotland

Named after Thomas Edington (1783–1859), geologist, mineralogist and owner of the Phoenix foundry in Glasgow and, later, the Eagle foundry, Greenock. Edington was first to discover the mineral. Edingtonite is an hydrated barium aluminosilicate that at the type locality forms greyish-white translucent well formed crystals with curved faces in cavities within basalt (Greg & Lettsom 1858, Heddle 1901, Hey 1934, Livingstone 2002). Edingtonite is also noted from Bell's quarry and the adjacent Bowling quarry 2 km WNW of Old Kilpatrick and from unnamed quarries 1 km and 8 km N of Old Kilpatrick (Heddle 1901). At Loanhead quarry, near Beith, Strathclyde, similar Lower Carboniferous Clyde Plateau porphyritic basaltic lavas contain edingtonite as colourless, transparent, equant or rectangular, complex wedge-shape crystals (up to 3×3×1.5 mm) with sharp well defined faces, associated with barite, calcite, harmotome, hematite, natrolite, prehnite, thomsonite, and rarely diaspore, ilmenite and native silver (Meikle & Todd 1995).

In England, edingtonite is reported from joint planes and small brecciated zones in a dolerite intrusion at Squilver quarry, Disgwylfa Hill, 9 km N of Bishop's Castle, Shropshire. Colourless-to-white bipyramidal crystals of edingtonite cover areas up to 50 cm across, with individual crystals exceptionally reaching 20 mm. Prehnite and calcite are commonly associated species, and harmotome is infrequently found as inclusions in the edingtonite (Hubbard & Enamy 1993).

Wedge-shape crystal of edingtonite from Squilver quarry, Disgwylfa Hill, Shelve, Shropshire; width 9 mm. (© A. G. Tindle)

Etched crystals of edingtonite (100 µm across) on dolomite are reported from the Conlig–Whitespots mine site, Co. Down, Ireland (Moles 2007).

eggletonite
monoclinic $Na_2Mn_8(Si,Al)_{12}O_{29}(OH)_7\cdot 11H_2O$

A rare hydrated sodium–manganese aluminosilicate hydroxide found in lenses of manganese ore within Ordovician mudstones at Benallt mine, Rhiw, Lleyn Peninsula, Gwynedd, Wales. Eggletonite occurs within small compositionally zoned micaceous crystals in thin veins cutting the manganese ore. Parts of the crystals have eggletonite, ganophyllite or tamaite compositions (T. F. Cotterell pers. comm. 2006).

ehlite
A synonym of pseudomalachite.

eisenkiesel
A deep-red variety of quartz containing hematite inclusions.

eisen-nickelkies
A synonym of pentlandite.

elaeolite
A synonym of nepheline.

elbaite
TOURMALINE GROUP
trigonal $Na(Al_{1.5}Li_{1.5})Al_6(BO_3)_3Si_6O_{18}(OH)_4$

A rare sodium–aluminium–lithium borosilicate hydroxide that forms a series with schorl and is found in granites, granite pegmatites and high-temperature hydrothermal veins.

Transparent blue fibres of elbaite occur as inclusions in apatite or as discrete greenish-grey to almost colourless (achroite) needles in miarolitic cavities within pegmatites cutting the St Austell granite in Gunheath china-clay pit, near St Austell, Cornwall (Weiss 1994). A variation from schorl to elbaite composition is apparent in tourmaline-bearing granite from other china-clay pits in the St Austell district of Cornwall: Carclase; Goonbarrow; Longstone (the site of the Trelawny Tigers speedway track), near Nanpean; the West Gunheath section of Littlejohn pit; and Wheal Martin. An exceptional specimen of pale-green tourmaline from Dorothy china-clay pit proved to be zoned from elbaite to rossmanite in composition (N. Hubbard specimen, Tindle unpub. EMP data). Tourmaline (achroite), forming hair-like masses in pegmatitic cavities within the Carnmenellis granite at Chywoon quarry, near Longdowns, is possibly elbaite (Moulding & Hooper 2005), although analysis of similar material from Rocks china-clay pit, St Austell, proved to be foitite or schorl (Collins 1876c, E. Loye specimen, Tindle unpub. EMP data).

A sodium- and lithium-rich aplite at Meldon, Okehampton, Devon, 1 km NW of the main Dartmoor granite, is cut and metasomatized by successive generations of thin potassium- and lithium-rich veins containing two varieties of elbaite. In early magmatic veins and surrounding metasomatized aplite, green elbaite (verdelite) is found, but in later hydrothermal veins and associated aplite the colour changes to pink elbaite (rubellite). The green colour is attributed to a greater abundance of divalent iron in verdelite (Chaudry & Howie 1976), whereas the rubellite has a very low iron content and is perhaps pink because of its relatively high manganese content. Major species found in the aplite are albite, lepidolite, muscovite, orthoclase and quartz, with accessory amounts of apatite, native arsenic, beryllonite, elbaite, fluorite, herderite, manganocolumbite, microlite, montebrasite, petalite, pollucite, stibiconite and topaz (von Knorring & Condliffe 1984). If the tourmaline classification of Hawthorne & Henry (1999) is adopted, then both green and pink varieties of tourmaline consist of elbaite and fluor-elbaite on the basis of fluorine content (D. I. Green and R. Tayler specimens, Tindle unpub. EMP data).

Elbaite and fluor-elbaite have also been found enclosed in fluorite within granite on Lundy island, off the north coast of Devon (D. I. Green specimen, Tindle unpub. EMP data). They have a similar compositional range to those found at Meldon, with some data being very close to rossmanite in composition.

In Scotland, gem-quality elbaite, varying in colour from various shades of green to pink and mauve, occurs in a zoned granitic pegmatite intruded into Dalradian mica schists and phyllites at an isolated location approximately 1 km N of Peatfold farm, Glenbuchat, Grampian. The average size of the crystals is 8×3 mm, although some up to 4 cm in length are reported. Elbaite is found within a lithium-enriched zone of the pegmatite in association with albite (cleavelandite), lepidolite, muscovite, quartz and schorl, and accessory amounts of arrojadite, cassiterite, columbite–tantalite, galena, triplite and zircon (Jackson 1982). An investigation of tourmaline from this locality shows the very dark-bluish-black variety to be zoned from foitite cores to schorl rims and for a deep-blue variety to be elbaite and fluor-elbaite (I. Dossett and S. A. Rust specimens, Tindle unpub. EMP data). A single isolated tourmaline grain, 4 cm in cross section and colour zoned from a very dark-green core, to a pale-green intermediate zone and thin pink rim, was found in a lepidolite-bearing pegmatitic granite from East Head, at the base of Cowhythe Hill, near Portsoy, Grampian. All varieties are elbaite, but with significant solid solution towards schorl in the core of the specimen (Tindle unpub. EMP data).

Colour-zoned elbaite prisms to 12 mm from Glenbuchat, Grampian, Scotland. (© D. I. Green)

electrum

A silver-rich variety of gold.

ellonite

TL Ellon, Grampian, Scotland

A mixture of alunite, quartz and clay.

elpidite

orthorhombic $Na_2ZrSi_6O_{15}{\cdot}3H_2O$

A rare hydrated sodium–zirconium silicate found in aegirine–riebeckite granite from Rockall island, 440 km W of northern Skye, NW of Scotland. Elpidite is the most common late-stage interstitial mineral in the granite, filling voids commonly up to 5 mm across. It forms colourless or very pale-cream crystals, either as radiating sheaves of slender prisms up to 2 mm long or as needles and acutely terminated prisms intergrown with other interstitial minerals, such as albite, bazirite, cristobalite, leucophosphite, monazite and quartz (Sabine 1960, Hawkes et al. 1975, Young et al. 1978).

elyite

monoclinic $Pb_4Cu(SO_4)O_2(OH)_4{\cdot}H_2O$

A rare hydrated lead–copper sulphate hydroxide found in natural parageneses in the oxidation zones of lead–zinc deposits, generally by low-temperature post-mining oxidation processes. It can also form anthropogenically – as a slag mineral or where medieval and post-medieval fire-setting methods were used to extract ore.

Radiating acicular groups of purple elyite crystals up to 0.02 mm in length and derived from the Greystone lead lode are reported from Greystone quarry, Lezant, Cornwall (De Nul 2005). At Red Gill mine, Caldbeck Fells, Cumbria, elyite forms pale-purple crystals on galena (Briscoe 1986a) and, at the nearby Short Grain lead mine, elyite forms part of a supergene assemblage that also includes anglesite, caledonite, cerussite, chenite, leadhillite, linarite, macphersonite, scotlandite and susannite (Green & Briscoe 2002). Characteristic deep-purple acicular to lath-like crystals of elyite in small cavities in oxidized galena are also reported from the 12-fathom dumps of Driggith mine, Caldbeck Fells (Green et al. 2006a).

Elyite is reported from three locations in Dyfed part of the Central Wales orefield: Esgair Hir mine, Ceulanymaesmawr – violet fan-like sprays up to 0.2 mm across, scattered in a small cavity on altered galena in association with anglesite, caledonite,

Hair-like spray of elyite from Greystone quarry, Lezant, near Launceston, Cornwall; width 100 µm; SEM image. (© D. I. Green)

Fibrous sprays of lilac elyite, 1 mm across, from Dolyhir quarry, Old Radnor, near Presteigne, Powys, Wales. (© D. I. Green)

covellite and hydrocerussite (Rust & Mason 1988); Frongoch mine, between Devil's Bridge and Pont-rhyd-y-groes – minute pale-violet masses of lath-like crystals in small cavities in galena associated with cerussite and hydrocerussite. It is thought to have been formed by reaction between lime-rich mortar from collapsed masonry and a lead-rich mineral such as cerussite or galena (Bevins 1994, Green et al. 1996b); Llechweddhelyg mine, near Clawddmelyn, between Bont-goch and Cwmsymlog (Bevins 1994). Elswhere, rare elyite is reported from Dolyhir quarry, Old Radnor, near Presteigne, Powys (Bevins & Mason 1997).

Elyite forms sprays of lilac needles up to 0.4 mm across within cavities in a sample of cerussite, litharge and massicot from a stream bed at Leadhills, Strathclyde, Scotland. It is possible that the sample may be a dumped flue product (Livingstone 1993), particularly as elyite is also noted from the nearby Meadowfoot smelter at Wanlockhead, Dumfries & Galloway (Green 1987a). More recently, a natural occurrence of similar elyite, associated with cerussite, leadhillite, mattheddleite and possible chenite, has been found on corroded galena and quartz from the dumps of Susanna mine, Leadhills (S. A. Rust pers. comm. 2006).

embolite

A synonym of chlorargyrite or bromargyrite.

emerald

A green gem variety of beryl.

emerald nickel

A synonym of zaratite.

emery

An impure granular variety of corundum containing admixed spinel and iron oxides.

emmonite

A variety of strontianite.

emplectite

orthorhombic $CuBiS_2$

A rare copper sulphobismuthite found in a stockwork cut by galena-bearing quartz veins at Corrie Buie, Meall nan Oighcreag, Loch Tay, Tayside Region, Scotland. The veins also contain minor amounts of native bismuth, chalcopyrite, electrum (36–65% gold), pyrrhotite, sphalerite and schirmerite, and phases belonging to the gustavite–lillianite series. Mineralogy of the veins is typical of that associated with acid intrusions. A single grain of emplectite was found associated with dolomite in fractures within galena (Pattrick 1984).

enargite

orthorhombic Cu_3AsS_4

A relatively rare copper sulpharsenate, dimorphous with luzonite and found in hydrothermal veins formed at moderate or lower temperatures.

Specimens from Gregory mine, Ashover, Derbyshire, now in the Chatsworth House mineral collection, contain grey six-sided flat-top crystals of enargite in small crystal groups up to 2 mm across and they represent one of the few confirmed occurrences of arsenic-bearing minerals in Derbyshire (others being arsenopyrite and mimetite). Associated species are calcite, galena, fluorite, pyrite and possibly chalcopyrite (Cooper 1995, 2005). Close to the beach near the Pier hotel at Clevedon, near Bristol, barite–sulphide mineralization occurs in metre-wide fault-filling brecciated veins within Triassic Dolomitic Conglomerate and consists of banded barite with chalcopyrite, galena, sphalerite, minor amounts of hematite, marcasite, pyrite and tennantite, and trace amounts of bornite and enargite, together with their alteration products. Enargite crystallized late in the paragenetic sequence, as intergrowths with chalcopyrite and tennantite, and as 200 μm-wide rims around chalcopyrite (Ixer et al. 1993). Reports of enargite from Alderley Edge ridge, Cheshire, by Warrington (1965) and Carlyon (1979), appear to be in error (Braithwaite 1994), although a later discovery of optically zoned enargite-group minerals (collectively enargite, famatinite and luzonite), together with covellite and spionkopite, is reported by Ixer & Budd (1998). At Scaleber Bridge, 3 km SSE of Settle, Yorkshire, on the Askrigg Block, isolated brown-to-lilac (in thin section) untwinned crystals of enargite, famatinite and luzonite are found as inclusions up to 80 μm across in calcite within dolomitized and partially mineralized Carboniferous Limestone. Chalcopyrite, marcasite, pyrite and tennantite are associated inclusion species (Ixer & Stanley 1998).

Enargite has been noted in Wales (probably on the basis of an optical identification) from Great Orme's Head, Llandudno, Gwynedd (Bevins 1994), and from Dolyhir quarry, Old Radnor, near Presteigne, Powys (Ixer & Stanley 1998). At Dolyhir quarry, enargite forms metallic equant blocky crystals and minute bladed crystals (T. F. Cotterell pers. comm. 2007).

In Scotland, enargite and covellite are rare species found with bornite, chalcocite, chalcopyrite and pyrite, in an area of disseminated porphyry-style mineralization within the Black Stockarton Moor subvolcanic complex, near Kirkcudbright, Dumfries & Galloway (Leake & Brown 1979, Lowry 1997). One of the subvolcanic vents in this complex is a brecciated basaltic agglomerate mineralized with enargite, malachite and pyrite (Rock et al. 1986a). Livingstone (2002) reports trace amounts of enargite and tennantite forming rims separating galena and chalcopyrite in veins exposed in Calliachar and Urlar Burns, 4 km SW of Aberfeldy, Tayside Region.

At Tynagh mine, near Killimor, Co. Galway, Ireland, epigenetic Cu–Pb–Zn–(Ag) mineralization includes a suite of sulphide species that occur in trace amounts; native antimony, arsenopyrite, bournonite, enargite, geocronite, gratonite, jordanite, marcasite, millerite, pyrite, siegenite and tennantite, many occurring as inclusions in galena (Morrissey et al. 1971, Gibson 1979, Clifford et al. 1986). Additionally, Schultz (1966b) notes meneghinite, pyrargyrite, polybasite and stephanite occurring in minor amounts, together with galena and tennantite–tetrahedrite. Bronze Age miners are thought to have worked a stratabound copper ore containing trace amounts of enargite at Ross Island, Killarney, Co. Kerry (Ixer & Pattrick 2003). A possible occurrence of enargite was mentioned by Wen et al. (1991) from Dhurode mine, West Carbery mining district, Co. Cork.

In the Channel Isles, at Bouley Bay, Trinity, northeast coast of Jersey, grey flow-banded rhyolites with copper mineralization in microfractures, and similar in everything but size to porphyry-copper deposits, have a paragenetic sequence from pyrite, to chalcopyrite, to enargite and finally to the inadequately described species, idaite. Extensive alteration to covellite followed (Ixer 1980). Enargite is also described from a Pb–(Ag)-bearing brecciated vein cutting a foliated hornblende granite of Late Precambrian age at Sark's Hope mine, Little Sark. It occurs in a minor suite of silver-bearing minerals that also includes acanthite, pearceite, polybasite, pyrargyrite and tennantite–tetrahedrite (Ixer & Stanley 1983, Ixer 1992).

endellionite

TL Wheal Boys, Trewetha, St Endellion, Cornwall, England

A synonym of bournonite.

endellite

A synonym of halloysite-10Å

endiopside

A variety of enstatite–diopside, pyroxene group.

enstatite

PYROXENE GROUP

orthorhombic $Mg_2Si_2O_6$

A very common magnesium silicate that forms a series with ferrosilite and is dimorphous with clinoenstatite. Enstatite is found in pyroxenites, peridotites, dunites, mafic volcanics and granulite-facies metamorphic rocks. Intermediate compositions between enstatite and ferrosilite were once described as hypersthene, but this name was formally abandoned in the Morimoto (1989) classification of the pyroxene group, and enstatite now covers all compositions with Mg/(Mg+Fe)>0.5.

England

Cornwall Reports of enstatite are all from peridotite in the ophiolite complex on the Lizard Peninsula. A primary mineral assemblage consisting of anhedral diopside, enstatite, olivine and spinel is reported from Goonhilly Downs, near Gwenter, from near Worvas and from the Bees, near Pedn Boar. Serpentinization during diapiric rise of the peridotitic magma produced a mantle of cataclastic finely foliated recrystallized rocks (e.g. at Goonhilly Downs, near Kernewas and Rill Cove, Kynance Cliff), in which chromite, diopside, enstatite, olivine and plagioclase feldspar were the main alteration species (Green 1964). Enstatite is also described from peridotites at: the Balk, E of Lizard village; the Horse, 2 km NW of Kynance Cove (Bonney 1877); on Goonhilly Downs at Countybridge quarry and Gwenter; and Coverack (Davies 1983).
Derbyshire Nodules (spinel lherzolite and harzburgite xenoliths) in Carboniferous basalt at Calton Hill, Taddington, contain abundant forsterite with enstatite and minor chrome diopside and spinel (Hamad 1963, Donaldson 1978, Macdonald et al. 1984).
Northumberland Enstatite, diopside and an unusually titanium-rich biotite are reported from the Cheviot granite, SW of Wooler (Haslam 1986).

Two euhedral crystals of enstatite in an andesitic lava from the Cheviot Hills, Northumberland. The larger crystal contains a small hole where the enstatite has plucked out during sample preparation. Subsidiary microphenocrysts (top left) are plagioclase feldspar. Width 1.1 mm; viewed in PPL (upper) and XPL (lower). (© K. Gardner)

Wales

Dyfed A xenolithic laminated olivine-gabbro unit in the St David's Head intrusion on the northwestern extremity of the St David's Peninsula contains both pristine and altered enstatite in host magma and xenoliths (Bevins et al. 1994).
Gwent Enstatite occurs in peridotite xenoliths from the Great House diatreme, near Usk (Bevins 1994).
Gwynedd Enstatite has been described several times from the gabbroic layered intrusion at Penmaenmawr, near Conwy (Teall 1888b, Harker 1889, Sargent 1915, 1924). On the Lleyn Peninsula, enstatite occurs as phenocrysts in microtonalite from the Inner Garnfor intrusion (Croudace 1982).

Scotland

Fife Porphyroclastic spinel lherzolite xenoliths in volcanic vents and associated tuffs at Coalyard Hill and Ruddons Point, both near Elie, contain an assemblage of diopside, enstatite, forsterite and hercynite (referred to as picotite) or chrome spinel (Chapman 1976).
Dumfries & Galloway The zoned diorite–granodiorite–granite complex at Cairnsmore of Carsphairn contains hypersthene in marginal diorites (Deer 1938, Tindle et al. 1988), and analysis of a sample from Polsue Burn indicates that the mineral now classifies as enstatite (Tindle unpub. EMP data). Enstatite is a common constituent in a two-pyroxene diorite from the southern end of the Loch Doon granite complex around Loch Dee and in a similar diorite at the northwestern end of the complex, near Straiton, Strathclyde (Tindle 1982).
Grampian A comprehensive study of the Insch intrusion, the largest of the Caledonian Newer Gabbro bodies, has shown that orthopyroxene has a compositional range that varies from enstatite in the basic rocks to ferrosilite in more evolved rocks. Enstatite is particularly common in Middle Zone cumulates and associated gabbroic rocks between Insch, Old Rayne, Oyne, Pitcaple and Daviot. Similarly, the Bogancl ogh intrusion to the W of Insch also contains enstatite (Wadsworth 1986, 1988). Other Newer Gabbro bodies in the region are known to contain enstatite: Belhelvie, N of Aberdeen (Wadsworth 1991); Kennethmont, to the west of Insch (Busrewel et al. 1975); and Morven–Cabrach, around Strathdon (Allen 1970). Enstatite also occurs in Dalradian migmatite at Castle Bridge in the Huntly–Portsoy district (Ashworth & Chinner 1978). Heddle (1901) reports large dark-green crystals of enstatite, up to 25 mm in length, from the southwest side of the Hill of Milleath, near Huntly. Enstatite (hypersthene) is also a constituent of a plagioclase–spinel hornfels at Belhelvie (Howie 1964).
Highland Orthopyroxene phenocrysts in a pitchstone flow on Ardnamurchan are compositionally zoned from enstatite cores to ferrosilite rims (collectively described as hypersthene) and are occasionally mantled by augite or pigeonite. Associated species include ilmenite, plagioclase feldspar and titanomagnetite, all set in a light-brown perlitic glass (Preston et al. 1998b). The Sgùrr of Eigg pitchstone on the Isle of Eigg, near Rum, hosts similar zoned enstatite–ferrosilite phenocrysts (Tyrell 1928a, Carmichael 1960a, Ridley 1971, Preston et al. 1998b). Other occurrences of enstatite are reported from: diorite in the Ballachulish igneous complex (Weiss & Troll 1989); garnet-bearing olivine websterite at Beinn Fhada, near Letterfearn, Loch Duich (Rawson et al. 2001); quartz diorite and pyroxenite xenoliths in the Ben Nevis granite complex (Haslam 1968, 1970); websterite and garnet–hornblende–pyroxene rock from the slopes of Glen More, Glenelg (O'Hara 1960, Mercy & O'Hara 1965); picritic dykes cutting Lewisian Gneiss at Loch Feith an Leothaid, Lochinver area (Tarney 1969); charnockitic rock at Pairc a' Chladaich and pyroxenite to the NW of Scourie House, both near Scourie (Howie 1964); lherzolite and pyroxenite xenoliths in a monchiquite dyke at Streap, 7 km NW of Glenfinnan railway station (Walker & Ross 1954); mugearite lava on Bloodstone Hill, Isle of Rum (Ridley 1971); inclusions in chrome spinel at the Unit 7–8 boundary of the Eastern Layered Series of the Rum intrusion, Isle of Rum (Young 1984b); mica pyroxenite lens in Lewisian leucogneiss xenolith enclosed in ferrodiorite at Harker's Gully, Marsco area, Western Redhills, Isle of Skye (Thompson 1981).
Strathclyde Peridotite, pyroxenite, hypersthene gabbro and pyroxene–mica diorite of the Garabal Hill–Glen Fyne complex, W of the head of Loch Lomond, all contain enstatite, often in association with augite (Nockolds & Mitchell 1946). Enstatite is present in both diorite and monzodiorite from the Quarry Diorite unit at the southern end of the Etive diorite–granodiorite–granite complex (Moazzen & Droop 2005). Non-porphyritic pitchstones from Kilfinichen, Loch Scridain, Isle of Mull, contain elongate feathery crystallites of augite, enstatite and pigeonite, all forming sheaf-like aggregates (Preston et al. 1998b).
Tayside Serpentinized harzburgites from Alyth, near Blairgowrie, contain large porphyroclastic enstatite crystals in a finer- and equigranular-grain assemblage of enstatite, olivine and spinel. They probably represent residual mantle after primary

lherzolite had melted to produce the basaltic rocks of the Highland Border Series (Ikin & Harmon 1983).

Western Isles Enstatite is a ubiquitous species in sapphirine-bearing and sapphirine-free garnet–kyanite/sillimanite–orthopyroxene granulites to the S and SW of Roineabhal, South Harris (Baba 1999, 2003, Hollis et al. 2006). Ultrabasic gneiss W of the north end of Loch Ossigary, Rodel, South Harris, contains strongly pleochroic enstatite associated with amphibole, diopside, magnetite, green spinel and olivine, which apart from the olivine is reminiscent of charnockite. Enstatite data are also presented from a hornblende–plagioclase gneiss to the N of Rodel, from a plagioclase–pyroxene granulite to the N of Leverburgh (both on South Harris) and from a pegmatitic vein in pyroxene granulite at Caisteal Ard, Ardvourlie, North Harris (Howie 1964).

Shetland Enstatite is a common mineral in harzburgite from the Shetland ophiolite complex exposed on the islands of Unst and Fetlar (Prichard 1984, Maynard 1993). Lustrous pea-green crystals of enstatite, up to 12 mm across, at Swinna Ness on Unst are also from this complex (Heddle 1901).

Ireland

Antrim Partially melted rocks (buchites) at the contact of a doleritic intrusion with Old Red Sandstone tuffaceous sedimentary rocks at Tieveragh, near Cushendall, contain zoned enstatite crystals mantled by diopside (Kitchen 1984).

Donegal Large anhedral crystals of enstatite occur in peridotitic xenoliths within a 30 m-wide alkali basalt (dolerite) dyke that intrudes Lower Carboniferous limestones near Inver, 10 km W of Donegal. Xenoliths mainly consist of clinopyroxene, alkali feldspar, olivine, plagioclase and glass (Gallagher & Elsden 1990, Shaw & Edgar 1997).

Down Dolerites and gabbros from the Tertiary Slieve Gullion central complex contain augite, diopside and enstatite as groundmass phases varying in texture from poikilitic and ophitic to granular and intersertal (Gamble 1982).

Galway An orthopyroxenite within the Currywongaun and Doughruagh intrusion, Connemara, consists of subhedral grains of enstatite (hypersthene) and magnetite, both with minor ilmenite and spinel intimately intergrown with högbomite (Angus & Middleton 1985). A report of enstatite from Dawros (Anthony et al. 1995) is presumably from the peridotite intrusion, elsewhere described as the Inishdawros metaperidotite, Callow, Ballyconneely, Connemara (Ahmed & Leake 1978).

Sligo Enstatite crystals up to 3 cm long form a crust (or vein) on the northern margin of a serpentinite mass at Cooney to the W of Collooney (Lemon 1966).

enysite

TL St Agnes, Cornwall, England

Named after John Samuel Enys (1797–1872), owner of the quay at Trevaunance harbour, St Agnes, Cornwall; born in Worcestershire. Enysite was found in one of the caves at the old quay (Collins 1876a, Le Neve Foster 1876). It is a mixture of calcite, woodwardite and clay.

eosite

TL Leadhills, Strathclyde, Scotland

Status uncertain, although examples from the type locality (Heddle 1901) are in fact vanadinite (Livingstone 1992).

epichlorite

Status uncertain; possibly a chlorite.

EPIDOTE GROUP

Monoclinic and orthorhombic silicates with the general formula $A_2B_3(SiO_4)_3(OH)$ or $A_2B_3Si_3O_{11}(OH,F)_2$, where A = Ca, Ce, Pb, Sr; and B = Al, Fe^{3+}, Mg, Mn^{3+}, V^{3+}. A revised classification scheme for epidote-group minerals (Armbruster et al. 2006) divides this group into allanite, clinozoisite and dollaseite subgroups. Allanite-(Ce), allanite-(Y), clinozoisite, epidote, piemontite, piemontite-(Sr) and zoisite are members of the group found in Britain or Ireland.

epidote

EPIDOTE GROUP

monoclinic $Ca_2Al_2(Fe^{3+},Al)Si_3O_{12}(OH)$

A very common calcium–iron aluminosilicate hydroxide that forms a series with clinozoisite and is found in thermal metamorphic environments, where igneous rocks are in contact with calcareous metabasic or metasedimentary rocks. Epidote also forms in a variety of regionally metamorphosed rocks and can be produced during the alteration of plagioclase feldspar.

England

Cornwall Heterogeneous low-grade regionally metamorphosed basaltic hornfels in the aureole of the Land's End granite near Botallack, St Just, contains epidote together with actinolite, albite, calcite, chlorite, ilmenite–magnetite and quartz. One variety consists of a typical skarn assemblage in which epidote occurs with calcite, diopside, garnet and magnetite and can be found as sheets or veins at Botallack Head, Carn Du, Carn Vellan, Chycornish Carn, Crowns Rock, Gryll's Bunny, Trewellard North Cliff and Trewellard South Cliff. Similar material occurs on the dumps of Botallack mine and Wheal Edward (Jackson 1974, Hall & Jackson 1975, Alderton & Jackson 1978). Wheal Cock Zawn, Roscommon Cliffs, St Just, is a further location where epidote-bearing calc-silicate skarn occurs (Couper & Barstow 1977). Thin light-green radiating groups of epidote on dark hornblende rock were noted from Botallack Head by Collins (1892). Epidote from Crowns Rock and Gryll's Bunny is unusually tin rich (van Marcke de Lummen 1986b). Thermal metamorphism of host rocks around the Bodmin Moor granite has produced prisms and acicular epidote in some quantity at Tregullan quarry, S of Bodmin, and Tyland quarry, NE of Camelford. At the second location, epidote occurs within a garnet–vesuvianite skarn and also on joint surfaces, where it forms pale greenish-brown to greenish-purple prisms up to 4 cm in length (Barrow & Thomas 1908). Other Cornish locations for epidote include: Carrick Du mine, near Porthmeor (Robson 1945b); Church Cove, Landewednack, Lizard Peninsula (Robson 1949b); Lamorna Cove, Land's End, and South Tresavean, Gwennap, near Redruth (Collins 1892); Old Treburgett mine, E of Port Isaac, St Teath (Collins 1911a); Pen Olver and Porthkerris quarry, Lizard Peninsula (Golley & Williams 1995); South Tolgus shaft, Redruth (Davey 1948); St Minver (Kempe 1968); Tolpetherwin quarry, Two Bridges, South Petherwin (Russell 1949b); and Trenuth Mill and E of South Yolland farm, both in the Caradon area (Barrow & Thomas 1909).

Cumbria Epidote is an accessory species in the Skiddaw granite exposures in Sinen Gill and Grainsgill outcrops on the Caldbeck Fells (Hitchen 1934). Yellowish radiating prisms in quartz are noted from the Harding vein at Carrock mine in Grainsgill (Young 1987a). Miarolitic cavities within the Shap granite contain radiating groups of pistachio-green to brown prismatic epidote associated with bavenite, calcite, fluorite, muscovite, orthoclase, smoky quartz, scheelite or titanite (Russell 1925, Wirth 1995, Wirth & Green 2002). At Shap Blue quarry, alteration of plagioclase feldspar in hornfelsed andesite and rhyolite country rocks (Borrowdale Volcanics) within the aureole of the Shap granite has produced epidote, together with hornblende and garnet. Sometimes epidote–garnet veins are developed in which well terminated crystals of epidote have formed (Davidson & Thomson 1951, Firman 1957, Fortey 1980, Young et al. 1998). Strens (1964) describes various types of epidote in Borrowdale Volcanic Series andesitic lavas and tuffs, namely epidote–quartz vein fillings, so-called Shap-type veins, and metasomatized lavas and tuffs. Localities in Borrowdale include: Cat Gill, Fisher Crag, Fleetwith Pike, Grey Knotts, Hause Gill, Low Hows Wood, Seatoller Fell, Steel Fell and near Stonethwaite. The species also occurs at Dale Head mine in Newlands Valley and Eel Crags on Derwent Fells. Other occurrences include: in granophyre at Bowness Knott, Ennerdale and in dolerite at Burtness Coombe, Buttermere (Rastall 1906); in doleritic dykes from Eskdale (Macdonald et al. 1988a); terminated crystals at Falcon Crag, near Keswick (Greg & Lettsom 1858, Young 1987a); in andesite at Greenscoe quarry, Dalton-in-Furness (Rose & Dunham 1977); terminated crystals in Eycott Volcanics at Greystoke Forest, Greystoke, also at Park Brow quarry, Dockray (Young 1987a); in Borrowdale Volcanics on Harter Fell, Eskdale (Ward 1877); as a detrital mineral in sandstone at Langhorn quarry, Bigrigg, and at Overend quarry, near Whitehaven (Lewis 1931); in veins cutting porphyrite at Piers Gill, Wasdale (Oliver 1961); in microdiorite at Scawgill Bridge, near Whinlatter Pass, 4 km W of Braithwaite (Fortey & Nancarrow 1990, Fortey et al. 1994); in epidote–quartz veins discovered in boreholes in the Sellafield area (Milodowski et al. 1998); abundant at Skull Gill, Oxendale (Hartley 1932); as an accessory mineral in the Threlkeld microgranite (Rastall & Wilcockson 1915); abundant crystallized epidote at Walla Crag, Derwentwater, near Keswick (Greg & Lettsom 1858, Rudler 1905, Davidson & Thomson 1951).

Devon Epidote is widespread in accessory amounts in the Dartmoor granite and is noted between the Birch Tor and Vitifer iron–tin mines in central Dartmoor (Alderton 1988a) and at Blackingstone quarry, Doccombe, where it forms primary magmatic anhedral crystals enclosed in biotite (Ward et al. 1992). Epidote is found with albite, clinozoisite and titanite in chlorite schist at Start Point (Tilley 1923) and as a secondary accessory constituent of the Lundy granite near its contact with Devonian slates (Dollar 1941, Thorpe et al. 1990).

Leicestershire In the Mountsorrel area, bright yellowish-green epidote is one of the secondary accessory minerals found: in aplite in the Main Mountsorrel quarry; in diorite at Brazil Wood; in granodiorite at five locations – Cocklow quarry, Kinchley,

Rubbish Tip quarry, the Main Mountsorrel quarry and at Castle Hill (Taylor 1934). At New Cliffe Hill quarry, Stanton under Bardon, epidote has formed in the South Charnwood diorites as an alteration product of calcic amphibole, calcic plagioclase and quartz (Hawthorne et al. 2002). Small lustrous yellowish-green prisms of epidote have been found in alpine-type veins cutting volcaniclastic rocks at Bardon Hill quarry, Coalville (Ince 2007). Veins of epidote are also noted cutting diorite at Newhurst quarry, Shepshed (Ince 2005).

Shropshire Cavities in pyroxene andesite at Moons Hill quarry, NE of Shepton Mallet, Mendip Hills, contain brilliant crystals of epidote with quartz. Epidote also forms granular coatings on joint surfaces (Kingsbury 1941, van de Kamp 1969). Rare epidote occurs with edingtonite and prehnite on joint planes in dolerite at Squilver quarry, Disgwylfa Hill (Blyth 1942, Hubbard & Enamy 1993).

Wales

Clwyd Basaltic rocks subjected to prehnite–pumpellyite facies metamorphism have recrystallized to an albite–chlorite–clinopyroxene–epidote–prehnite–pumpellyite–titanite assemblage at Hendre quarry, Pandy, near Glyn Ceiriog (Bevins & Rowbotham 1983).

Dyfed Epidote is a widely developed replacement product of plagioclase feldspar and it forms vein or vesicle infillings associated with chlorite in metamorphosed Ordovician lavas and intrusions. Locations include: Garn Fawr, near Strumble Head; Mynydd Preseli (zoned crystals); and St David's Head (Bevins & Rowbotham 1983). Euhedral to subhedral epidote crystals are present in albitized plagioclase within Precambrian doleritic rocks belonging to the Coomb Volcanic Formation in the Llangynog district (Cope & Bevins 1993). Epidote, chlorite, prehnite, pumpellyite and titanite are secondary minerals found in regionally metamorphosed Ordovician lavas and intrusive rocks of the Fishguard Volcanic Complex (Bevins 1982, Metcalfe et al. 1994). Excellent crystalline epidote has been found at Marloes Bay, Pembroke (Bevins 1994).

Gwynedd Low-grade regional metamorphism of Ordovician basaltic rocks had produced secondary epidote at Cadair Idris and Cwm Eigiau, 6 km W of Trefriw (Bevins & Rowbotham 1983), and in the Tal y Fan intrusion, SW of Conwy (Bevins & Merriman 1988). On Holy Island, Anglesey, the intrusion of ultramafic bodies into Precambrian metasedimentary rocks has produced epidotic and hornfelsic rocks with up to 90% epidote content (Maltman 1977). At Hendrefor, Anglesey, epidote is a minor constituent of lawsonite blueschists of the Mona Complex (Gibbons & Mann 1983). Excellent crystalline epidote has been found at Daran Belydr, Ganllwyd, near Dolgellau, and at Dinorwic quarry, near Llanberis, Snowdonia (Bevins 1994).

Powys Pale-green epidote occurs with actinolite, albite, clinochlore and tourmaline in altered gabbros from near the summit of Hanter Hill (Holgate 1977). Altered dolerite at Dolyhir quarry, Old Radnor, near Presteigne, Powys, contains epidote as small yellowish-green crystalline masses (T. F. Cotterell pers. comm. 2007).

Scotland

Central Epidote occurs in Dalradian copper–lead mineralized schists at Creag Bhocan, SSW of Tyndrum (Scott et al. 1988).

Dumfries & Galloway Large pale-green dull radiating crystals of epidote occur with fluorite, pyrite and titanite in veins cutting granite at Beeswing granite quarry, and at Craignair granite quarry, Dalbeattie, epidote forms small bright-green crystals (Williams 1965). Fissure-type veinlets in hornfelsed country rocks on Black Stockarton Moor to the NE of Kirkcudbright (in the aureole of the Criffell granodiorite) contain actinolite (or hornblende), arsenopyrite, calcite, epidote and pyrite (Fortey 1980). At Coatsgate quarry, near Moffat, a dyke cutting Carboniferous Plateau basalts once yielded andradite, chabazite, green epidote and hematite (Green & Todd 2001). Epidote is a secondary species in the Late Caledonian Newmains lamprophyre dyke, 7 km W of Dumfries (Macdonald et al. 1986). Thin epidote–penninite veins occur in thermally metamorphosed greywacke near the margin of the Loch Doon granite complex at Penkiln Burn, N of Newton Stewart (Stone & Leake 1984). Heddle (1901) reports large dark-green epidote crystals in granite at Cassencarrie granite quarry.

Grampian Epidote occurs with a mineral described as piemontite in a calc-silicate rock from the Balnacraig Metabasite Formation, near Mill of Sterin, 3 km S of Ballater (Goodman et al. 1990). Feldspars are replaced by epidote in a fine-grain diopside-rich rock found in contact with limestone at the margin of the Dorback granite, Braes of Abernethy, 9 km W of Tomintoul (Walker & Davidson 1935). Epidote granulite, a rock type considered to be the product of localized metasomatism of associated metamorphic schists, is found at Glen Lui, in the Cairngorm Mountains (McLachan 1951). At Pitscurrie quarry, Pitcaple, 8 km NW of Inverurie, pistachio-green epidote occurs with calcite, pyrite and schorl in quartz-pegmatite veins cutting gabbro (Green & Todd 2001). Other occurrences of epidote include: radiating "sap-green" and pale-brown crystals at Delnabo quarry, Glen Gairn; in biotite gneiss at Dobston quarry, near Inverurie; in granitic veins between Girdleness and Torry; in a vein cutting diorite at the Old Battery, Portsoy; and acicular crystals in sandstone at Stonehaven Bay (Heddle 1901).

Plumose aggregates of epidote with minor actinolite, calcite and prehnite; Pitscurry quarry, Pitcaple, Inverurie, Grampian Region, Scotland; width 10 cm. (© The Trustees of the National Museums of Scotland)

Highland Accessory epidote occurs in a molybdenite-bearing amphibolite adjacent to a granite pegmatite, 2 km N of Achfary on the south shore of Loch Stack (von Knorring & Dearnley 1959a). Accessory epidote mantles allanite in the Inchbae hastingsite gneissic granite exposed in the Black Water, 3 km up stream of Inchbae Lodge (Harker & Leahy 2001). The Lewisian Cnoc an t-Sidhean supracrustal rocks, near Stoer, include leucocratic quartz–feldspar–hornblende–epidote–clinozoisite gneisses with biotite–scapolite augen (Cartwright et al. 1985). Epidote is not very common, but is richly developed in certain parts of a diopside-rich calc-silicate rock associated with syenite along Allt a' Mhuilinn, near Loch Borralan, Assynt (Shand 1910b). The Glen Dessary syenite contains an accessory mineral assemblage that includes allanite, apatite, calcite, epidote, magnetite, monazite, pyrite, quartz, strontianite, titanite, zircon and zirconolite (Fowler 1992). Small masses of metasomatized hornblendic rocks of Moinian age at White Bridge, near Invergarry, contain accessory epidote associated with allanite-(Ce), apatite, ilmenite, magnetite, prehnite, titanite and zircon (Rock 1984). In Coire Dhorrcail, near Barrisdale on the south side of Loch Hourne, dark-grey to greyish-brown epidote occurs in vein-like masses cutting muscovite–biotite gneiss. The largest well formed crystal was 15 cm across and is associated with actinolite, garnet and quartz (Thomas 1905). A brilliant-red variety of epidote (withamite) in trap (a type of volcanic rock) is described from Glencoe by Greg & Lettsom (1858). On the Isle of Skye, epidote is reported from Tertiary basalts thermally metamorphosed by the Cuillin intrusive complex at Sgùrr nan Cearcall (Green & Todd 2001) and as an accessory species in the Glamaig epigranite, Marsco area, Western Redhills (Thompson 1969). Many more occurrences of epidote are noted by Heddle (1901).

Strathclyde Altered and garnetized gabbros associated with serpentinite of the Girvan–Ballantrae ophiolite complex contain epidote, diopside, pectolite, prehnite and zoisite (Bloxam 1954). In the nearby Knockormal area, greenschists and glaucophane schists derived from spilitic lavas contain yellow and yellowish-brown granular epidote (Bloxam & Allen 1959). At Coire Buidhe, Glen Creran, Appin, a skarn deposit formed by the intrusion of the Cruachan granite into impure limestones contains lustrous, dark-green, euhedral epidote crystals up to 50 mm long, together with grossular, magnetite, molybdenite, powellite, scheelite, titanite and vesuvianite (Ingram et al. 1993, Green & Todd 2001). Epidote occurs alone or associated with laumontite in chlorite-lined cavities within Carboniferous Clyde Plateau Lavas adjacent to a dolerite dyke at Loanhead quarry, near Beith. It forms radial crystal groups or elongated, slender, yellowish-green to leek-green prisms up to 2 mm in length (Meikle 1989b). At Reilly quarry, Houston, W of Glasgow, alteration of a dolerite dyke has produced apple- to pistachio-green coatings and growths of epidote associated with prehnite (Wolfe 1997, Meikle 2003). Epidote is noted in a skarn deposit at the junction between vent pyroclastics and a xenolithic Cretaceous chalk block that subsided into the collapsed caldera of the Central ring complex at Creag an Fheidh on the Isle of Arran (Cressey 1987). Vesicles in Tertiary lavas on Ben More

Needle-like crystals of golden epidote forming spherulitic clusters (1 mm across) and associated with green pumpellyite; Creag Brimishgan, Isle of Mull, Strathclyde Region, Scotland. (© D. I. Green)

and at Loch na Keal, Isle of Mull, contain epidote varying in colour from blackish green, through various shades of yellowish green, to pale pink or almost colourless. Heulandite, prehnite, pumpellyite, scolecite and stilbite are associated species (Currie 1898, McLintock 1910, Green & Todd 2001). Other occurrences of epidote are noted by Heddle (1901).

Tayside Accessory amounts of clinozoisite–epidote (intermediate compositions) are recorded from Middle Dalradian barium-enriched (Ben Eagach) schists in the area NW of Aberfeldy, including at Foss mine (Coats et al. 1984, Moles 1985). Epidote occurs with almandine and ilmenite in Dalradian schists at the southern end of Lochan na Lairige, NNE of Killin (Green & Todd 2001).

Western Isles extensive hydrothermal alteration has produced epidote in Tertiary brecciated gabbros and dolerites on St Kilda (Hirta) (Harding et al. 1984). On South Harris the Loch a' Sgùrr pegmatite, near Finsbay (Fionnsabhagh), and small pegmatitic veins at Loch Vallarip (Balla-riop), Rodilpark area (Parc Roghadail), contain minor amounts of epidote (von Knorring & Dearnley 1960a).

Shetland Epidote and epidote–hematite encrustations have been found on joint planes in syenite at Bakkasetter quarry, southeastern flank of Noss Hill, South Mainland (Stephen 1954). Epidote lines cavities and forms veins in pegmatite associated with the Sandsting granite–diorite complex on the south shore of Taing of Koelkifield, Walls Peninsula, southwest Mainland (Mykura & Phemister 1976). High-level gabbros belonging to the Shetland ophiolite complex at Mu Ness, Unst, contain ubiquitous epidote as fine-grain disseminations and as euhedral crystals in crosscutting veinlets (Maynard 1993). Heddle (1878b, 1901) notes epidote at the mouth of Hamna Voe on Yell and at a few other Shetland locations.

Ireland

Antrim Epidote occurs with feldspar, fluorite and quartz in veins at Fair Head (Greg & Lettsom 1858).

Donegal Yellowish-green epidote forms irregular prisms with dusky-brown cores in arfvedsonite-bearing rhyolite intrusions at Carntreena, NW of Lough Agrougha, west Cloghaneely (Whitten 1954). Quartz lenses in a metadolerite intrusion within Dalradian quartzites on the roadside 3 km E of Bunbeg contain epidote associated with ferro-actinolite (or hornblende), grossular, scheelite and titanite (Ryback & Moreton 1992).

Down A fibrous stellate variety of epidote (rosstrevorite) is noted from Rosstrevor, near Warren Point (Greg & Lettsom 1858).

Galway Widespread secondary andradite–grossular–hydrogrossular garnet, chlorite, epidote, prehnite, sericite and titanite occur throughout the Galway granite and its satellite plutons; they were formed during the passage of post-magmatic fluids through the granite (Leake 1998). Epidote also occurs in axinite–plagioclase–tourmaline veins cutting amphibolite of the Balyconneely mass, N of Roundstone, and is attributed to hydrothermal fluids from the Galway granite (Benjamin 1968). Joint surfaces in granite at Toombeola quarry are coated with small crystals of epidote associated with chlorite and fluorite (Ryback & Moreton 1992). The metamorphic aureole of the Omey granite hosts an amphibolite skarn at Barnahallia Lough, Claddaghduff, west Connemara. The skarn contains cavities in which occasionally abundant dark-green euhedral crystals of epidote, up to 10 mm long, are associated with albite, chabazite-Ca, grossular, heulandite-Ca and stilbite-Ca. Additionally, epidote-rich material extends into the nearby granite (Moreton & Green 2006a).

Londonderry Epidote forms small acicular crystals and thin strings and veins in syenitic rocks of Slieve Gullion and Kildress (Greg & Lettsom 1858).

Wicklow The Aughrim portion of the Leinster batholith contains an almost colourless variety of epidote (Smithson 1932). Well formed pistachio-green epidote crystals to 3 mm from calcite–quartz veins cutting a pyroxene diorite intrusion were collected during quarrying operations at Parnell quarry, S of Arklow (Green & Moreton 2001).

Offshore territories

Alderney Granular epidote is associated with sericite derived from the alteration of plagioclase in the Bibette Head granite (Nockolds 1932).

Jersey Bright yellowish-green epidote, often forming masses of radiating prismatic crystals, occurs in granite pegmatites associated with diorites at Gigoulande, Grève d'Azette, the Hermitage, Icho, La Collette, La Rocque and Ronez. Well formed crystals of epidote occur in andesite at Westmount, and massive epidote is found with andradite garnet in veins and patches in andesite at Côtil Point. A dolerite dyke on the E of Portelet Bay contains epidote in vesicles (Mourant 1961).

epinatrolite

A synonym of natrolite, zeolite group.

epistilbite

ZEOLITE GROUP

monoclinic $Ca_3(Al_6Si_{18})O_{48}\cdot16H_2O$

A relatively rare hydrated calcium aluminosilicate that forms in amygdaloidal basalts at the beginning of zeolite crystallization.

At Talisker Bay, Isle of Skye, Highland Region, Scotland, epistilbite occurs in boulders of dark-green basalt 200 m south of the seastack at Talisker Point. Epistilbite forms aggregates of elongated prismatic crystals up to 1 mm long with dull flat terminations and it is often the only zeolite species found in the vesicles, although analcime, or rarely stilbite, may accompany it. In some vesicles, massive pinkish epistilbite contains cavities lined with drusy crystals (Greg & Lettsom 1858, Green & Wood 1996). At Sgùrr nam Boc, between Loch Brittle and Loch Eynort, on the west coast of Skye, epistilbite has been found in amygdaloidal basalt boulders on the shore below the cliffs. Epistilbite forms euhedral equant-to-prismatic pseudo-orthorhombic crystals up to 1.5 mm across and fan-shape crystal aggregates up to 5 mm across. Very rare tabular crystals up to 6 mm across are also found. Nearby vesicles

Prismatic epistilbite from a cavity in Tertiary basalt. Sgùrr nam Boc, Isle of Skye, Highland Region, Scotland; width 2.5 mm. (© The Trustees of the National Museums of Scotland)

in the same boulders contain other zeolite species, including chabazite, heulandite, mordenite and stilbite (Green & Wood 1996, Green & Todd 2001). Epistilbite is also recorded from two other locations on the Isle of Skye: white equant crystals up to about 2 mm across associated with calcite, epidote and quartz, from a quarry to the side of Loch Sligachan (King 1976), and, in association with analcime, from the shore below Beall (Heddle 1901). On the Isle of Mull, Strathclyde, epistilbite has been found on the beach below Creag Brimishgan, Loch na Keal, and from Dearg Sgeir, Loch Scridain (Livingstone 2002; D. I. Green pers. comm. 2004).

Small milky-white or clear colourless crystals of epistilbite are described from a network of irregular fissure veins in mudstone within a roadside quarry near the head of Dannemarche reservoir, St Lawrence valley, Jersey (Mourant 1961).

Other reports of epistilbite appear to be erroneous or in need of further confirmation: Park Brow quarry, Dockray, Cumbria (Young 1987a) – later shown to be laumontite (Green & Wood 1996); Rathlin and Portrush, Co. Antrim, Ireland (Bryce 1833, Phillips et al. 1852) – considered unproven by Nawaz (1988).

epsomite

orthorhombic $MgSO_4 \cdot 7H_2O$

TL Epsom, Surrey, England

A common hydrated magnesium sulphate found as efflorescences at the type locality and in a crystalline state in unspecified coalmines near Newcastle upon Tyne, Northumberland (Greg & Lettsom 1858). Large quantities of needle-shape epsomite are also reported from Hetton colliery, Co. Durham (Clapham & Dalglish 1863), and as efflorescent crusts of long slender snow-white crystals at New Brancepeth colliery, Durham (Smythe 1933). Outcrops of Permian dolomitic limestone on the north side of Hesledon Dene, Peterlee, Co. Durham, contain epsomite as granular efflorescences (Bridges & Young 1998).

In Cumbria, globular crusts of epsomite up to 40 mm across occur with copiapite, gypsum, jarosite, pickeringite and rozenite in pyrite-rich coal seams at the Oughterside opencast coal site, Aspatria. They are all supergene species formed during oxidation of pyrite-rich coals within 10 m of the present ground surface (Young & Nancarrow 1988b). Post-mining efflorescences of epsomite, gypsum, iron oxides and rare serpierite encrust the mine walls of Smallcleugh mine, Nenthead, near Alston. Epsomite is locally very abundant in parts of the workings, commonly forming colourless acicular crystals up to 4 cm long and <0.25 mm across (Livingstone et al. 1990, Bridges & Young 1998). At the nearby Brownley Hill mine, the mine walls are encrusted with fine hair-like crystals of epsomite, some reaching 10 cm in length and associated with gypsum (Green et al. 2000).

Epsomite, along with anhydrite, boracite, calcite, carnallite, dolomite, gypsum, magnesite, polyhalite and sylvite, are said to be the probable primary minerals in the Permian evaporite sequence beneath North Yorkshire and Cleveland (Stewart 1965). At the anhydrite mine under the ICI factory site at Billingham-on-Tees, Cleveland, epsomite was found with gypsum, halite, lepidocrocite, pyrite, sulphur, sylvite, talc and thenardite, as secondary efflorescences on mine walls after the mine had been abandoned (Raymond 1959). Epsomite is also briefly mentioned by Robson (1949b) from Dolcoath mine, Camborne, Cornwall.

Stalactitic masses of epsomite have been described from specimens in the collection of the NMW that were originally collected at Treharris, Mid Glamorgan, Wales (Bevins 1994). Heddle (1901) notes white, silky and capillary crystals of epsomite on the sides of mine walls cutting shale beds at Hurlet, near Paisley, Strathclyde, Scotland. An efflorescence of epsomite is further noted from Gigoulande quarry, Jersey, Channel Isles (Mourant 1961).

erdmannite

A synonym of gadolinite-(Ce), gadolinite-(Y), homilite + melanocerite-(Ce), and altered zircon.

ericaite?

orthorhombic $(Fe^{2+},Mg,Mn)_3B_7O_{13}Cl$

A rare iron–magnesium–manganese borate with halide, dimorphous with congolite, that forms a series with boracite and is described from Boulby potash mine, Loftus, Cleveland, where it is said to occur in nodules within anhydritic shale overlying halite–sylvite rock of the third Zechstein evaporite cycle. Milne et al. (1977) refer to it as iron-boracite. Isolated, euhedral, brown, subtranslucent, blocky crystals of "ericaite", exceptionally up to 2 cm long, in pink sylvinite matrix, are perhaps the world's best (Green & Freier 1994, 1996). Later work has shown that ericaite is unstable at room temperatures and the mineral is now considered to represent a solid-solution series between congolite and trembathite (Burns & Carpenter 1996, Schindler & Hawthorne 1998, Green 2005b).

erinite

TL Giant's Causeway, Co. Antrim, Ireland

Named after Ireland (Erin). Erinite has six synonyms: chalcophyllite, cornwallite, monazite-(Ce), nontronite (the mineral found at the Giant's Causeway) and yttrium-rich pyrope.

erionite (series)

ZEOLITE GROUP

hexagonal $(K_2,Ca,Na_2)_2Al_4Si_{14}O_{36} \cdot 15H_2O$

Erionite crystals to 0.5 mm from Edinbane, Isle of Skye, Highland Region, Scotland. (© D. I. Green)

A rare hydrated potassium–calcium–sodium aluminosilicate, whose status changed from a species name to a series name during the zeolite reclassification of Coombs et al. (1997). Erionite is difficult to identify and is easily confused with offrétite. Most descriptions of erionite do not give the species name, presumably because full chemical analyses were not available.

Creamy-colour erionite from the Storr, N of Portree, Isle of Skye, Highland Region, Scotland, is reported with garronite, as platy-fibrous aggregates (<0.2 mm) within a cavity in Tertiary vesicular basalt. Adjacent cavities contain analcime, apophyllite, chabazite, heulandite, levyne, mesolite, stilbite and thomsonite (Macpherson & Livingstone 1982, Livingstone 2002). Golden-brown erionite crystals, <1 mm in length, have been collected from a distinct horizon of vesicular basalt in a roadcut at Edinbane on the Isle of Skye. Nearby vesicles contain calcite, heulandite, stellerite and stilbite (Savage 1995, Green & Todd 2001). Erionite forming thin epitaxial overgrowths on levyne in vesicular basalt from Moonen Bay, N of Dunvegan, Isle of Skye, is visually identified by Green & Todd (1996).

In Ireland, hexagonal bundles of golden erionite crystals up to 3 mm long have been found at Kane's quarry, Boghill, Glengormley, near Belfast (Tschernich 1992, Savage 1995). Colourless to silky-white needles of erionite in basalts from the Giant's Causeway and Dunseverick reach 10 mm in length and form bow-tie groups and bush-like aggregates associated with heulandite and phillipsite. At Craigahulliar quarry, near Portrush, Co. Antrim, colourless-to-golden compound prisms of erionite reach 2 mm in length (Tschernich 1992). Anthony et al. (1995) note erionite from Parkgate quarry, Templepatrick, Co. Antrim.

erionite-Ca

ZEOLITE GROUP

hexagonal $Ca_4K_2(Al_{10}Si_{26})O_{72} \cdot 32H_2O$

In a review of the crystal chemistry of erionite and offrétite, Passaglia et al. (1998) present analyses of erionite-Ca from three localities in Co. Antrim, Ireland: Boghill Road quarry, Glengormley, near Belfast; Ballyclare; and Ladyhill quarry, Tardree, NE of Antrim town. No further details are given, except that erionite-Ca at the last two localities forms overgrowths on levyne. Presumably the samples came from amygdales within Tertiary basalts. A fourth analysis is quoted from a sample collected on the Isle of Skye, but the actual locality is not given.

erionite-K

hexagonal — ZEOLITE GROUP — $K_4Na_2Ca(Al_8Si_{28})O_{72}\cdot 32H_2O$

Rare erionite-K occurs with more common erionite-Na in a fibrous spray of crystals from a single vesicle in an amygdaloidal basalt from Port-na-Boe, Giant's Causeway, Co. Antrim. Identification is based on a single analysis (H. J. Foy specimen, D. I. Green XRD, Tindle unpub. EMP data). Adair (1987) documents erionite-K coating levyne in Antrim basalts from Portmuck on Island Magee.

erionite-Na

hexagonal — ZEOLITE GROUP — $Na_6K_2(Al_8Si_{28})O_{72}\cdot 25H_2O$

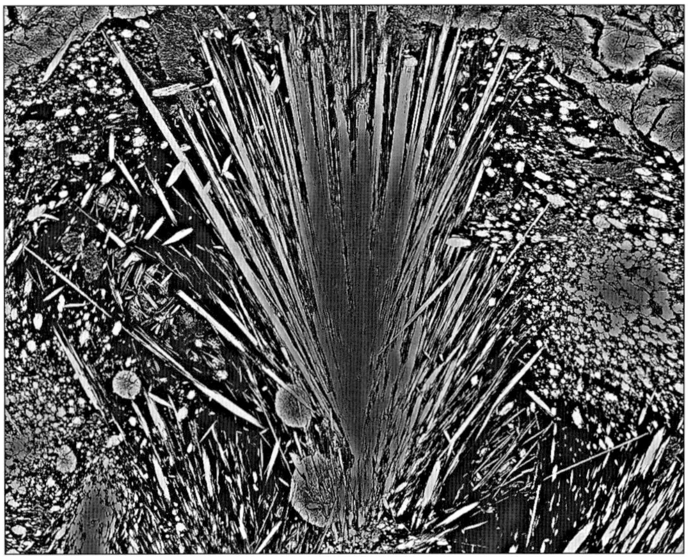

Radiating spray of erionite-Na in a vesicle within Tertiary basalt; Port-na-Boe, Giant's Causeway, Co. Antrim, Ireland; width 1.4 mm; BSE image. (© E. Condliffe, University of Leeds)

In a review of the crystal chemistry of erionite and offrétite, Passaglia et al. (1998) briefly describe erionite-Na occurring as an overgrowth on levyne from Dunseverick, Co. Antrim, Ireland. No further details are given, but presumably the sample came from an amygdaloidal Tertiary basalt. See also erionite-K.

Erionite-Na is confirmed from a Tertiary amygdaloidal basalt at Edinbane, Isle of Skye, Highland Region, Scotland (D. I. Green specimen and XRD, Tindle unpub. EMP data).

erlichmanite

cubic — PYRITE GROUP — OsS_2

A rare osmium sulphide that forms a series with laurite and is found in chromitites within a layered mafic–ultramafic intrusion on the Isle of Rum, Highland Region,

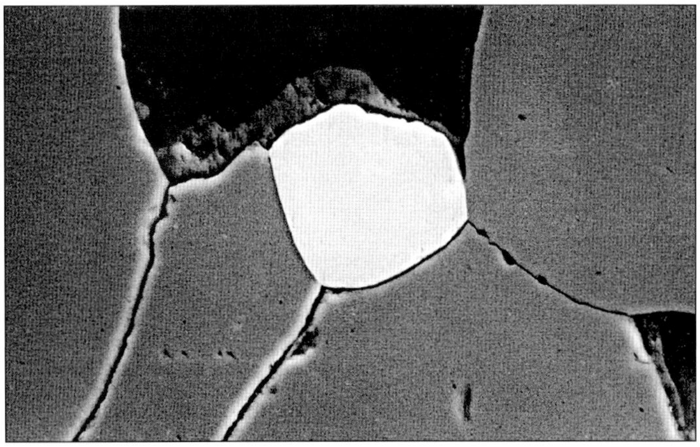

Erlichmanite surrounded by chrome spinel (grey) and plagioclase feldspar (black) in an olivine cumulate-hosted chromitite. Northern flanks of Ruinsival, Central intrusive complex, Isle of Rum, Strathclyde, Scotland; width 95 µm; BSE image. (© M. R. Power, Camborne School of Mines)

Scotland. A relatively large 30 µm cumulus grain of erlichmanite was found in the Central Series peridotites on the flanks of Ruinsival, together with cumulus chrome spinel and intercumulus plagioclase. Normally, grains of erlichmanite are < 10 µm in diameter and form inclusions in chrome spinel (M. R. Power pers. comm. 2004). Erlichmanite is also present at Pinbain Bridge as euhedral, < 10 µm-diameter inclusions in chrome spinel within chromitites belonging to the Ballantrae ophiolite complex, Strathclyde, Scotland. Laurite, irarsite, Ir–Os–Ru alloys and possibly kashinite are associated species (Power & Pirrie 2004).

erubescite

A synonym of bornite.

erythrite

monoclinic — VIVIANITE GROUP — $Co_3(AsO_4)_2\cdot 8H_2O$

A common hydrated cobalt arsenate that forms two series, with annabergite and with hörnesite, and is found in the oxidized zone of some As–Co–Ni-bearing mineral deposits. Erythrite is also known as cobalt bloom and often forms pink earthy masses and thin crusts associated with primary cobalt-bearing species.

England

Cornwall Massive cobaltite contains a powdery coating of erythrite at Botallack (the Crowns), St Just (Greg & Lettsom 1858, Bancroft & Weller 1993); on the mine dumps of Carrick Du mine, near Porthmeor (Green 1984); Dolcoath mine, Camborne; Wheal Polgooth, St Austell; Wheal Sparnon, near Redruth; Wheal Trenwyth, St Ives; Wheal Unity, St Day (all Greg & Lettsom 1858); East Pool mine, Illogan, near Redruth; South Terras mine, St Stephen-in-Brannel; Wheal Alfred, Phillack; Wheal Buller, near Redruth (all Robson 1949b); Ennys Wheal Virgin, St Hilary; Great Retallack mine, near St Agnes (both Green 1984); light-pink to crimson erythrite is the only supergene cobalt species found at Penberthy Croft mine, St Hilary, where it forms minor botryoidal and drusy aggregates up to 2 mm across, consisting of subhedral prismatic to acicular crystals (Betterton et al. 2000); rarely as blooms and occasional crystals at South Crofty mine, Camborne (LeBoutillier 1996); and crusts of erythrite on chlorite, associated with cassiterite and skutterudite (smaltite) at Wherry mine, Mount's Bay, Penzance (Russell 1949a, Kingsbury 1964b).
Devon Reports of erythrite include: Wheal Friendship (Devon Friendship), Mary Tavy (Braithwaite 1982b); and associated with native silver in a vein at Willsworthy mine, NE of Mary Tavy (Greg & Lettsom 1858, Rust 1982a).
Cheshire Erythrite is very rare at Engine Vein mine, Wood mine (small radiating aggregates of pink blades) and Stormy Point area, all Alderley Edge ridge (Carlon 1979, Braithwaite 1994).
Cumbria *Lake District:* Erythrite occurrences include: a single specimen of minute pink botryoids on shale at Force Crag mine, Braithwaite, near Keswick (Green et al. 1997); pale-pink encrustations at Holehouse Gill mine, Ulpha and Roughton Gill mine, Caldbeck Fells (Young 1987a); traces of erythrite on the dumps of Paddy End mine, Coniston, Eskdale (Russell 1925); pale-pink mm-size crystals, probably of post-mining origin, on the dumps of Sandbed mine, Caldbeck Fells and also found with annabergite and symplesite in the mine (Thomson 1997b, Green & Briscoe 2002); dull peach-red erythrite associated with skutterudite (smaltite) at Scar Crag cobalt mine, Causey Pike (Postlethwaite 1913, Davidson & Thomson 1951); associated with chlorite, hematite, nacrite and pyrite in joints and fault breccia at Shap Blue quarry in the aureole of the Shap granite (Firman 1953b, Fortey 1980). Erythrite is also reported with arseniosiderite, pharmacosiderite and scorodite from Ingray Gill, Caldbeck Fells (Kingsbury & Hartley 1957b), but further confirmation is desirable. Green et al. (2006a) note that visual identification of erythrite specimens from Driggith and Sandbed mines, Caldbeck Fells, is not reliable, as later analysis has shown them to be mainly annabergite or köttigite. *Northern Pennine orefield:* Reports of erythrite include: minute bright-pink spherules associated with ankerite, quartz and sphalerite at Brownley Hill mine, Nenthead, near Alston (Green et al. 2000); on quartz veinstone at Cross Gill Copper vein, Alston Moor and associated with annabergite at Hilton mine, Scordale, near Appleby (Bridges & Young 1998); rare traces of erythrite on witherite at Nentsberry Haggs mine, Nenthead, Alston Moor (Davidson & Thomson 1951); abundant erythrite in a block of mineralized shale breccia from the River Nent, near Smallcleugh mine, Nenthead (Young et al. 2005); small encrustations at Rotherhope mine, Leadgate, near Alston (Davidson & Thomson 1951, Bridges & Young 1998); and abundant erythrite staining of post-mining origin on the level walls and the dumps of Tynebottom mine, NW of Garrigill, near Alston (Greg & Lettsom 1858, Davidson & Thomson 1951, Ixer et al. 1979a, Rust 1987a, Bridges & Young 1998, Green & Briscoe 2002, Green et al. 2003a).

Well formed crystals of erythrite to 0.1 mm from Clogau mine, Bontddu, near Dolgellau, Gwynedd, Wales. (© D. I. Green)

Leicestershire Efflorescences of erythrite, malachite and lavendulan encrust coffinite-bearing subspherical nodules in calcareous mudstones of the Mercia Mudstone Group (Keuper Marl) at Gipsy Lane brick pit, Leicester (Faithfull & Hubbard 1988).

Wales

Dyfed Minor amounts of erythrite have been found at Cwmerfin, Cwmsymlog and Darren mines in the Central Wales orefield (Mason 1998). Pink coatings on quartz and breccia clasts at Gwaithyrafon mine, SE of Tal-y-bont, are tentatively identified as erythrite (Rust & Mason 1994). A visual identification of erythrite on a specimen of shale from Marloes Sands, Pembroke, is given by Starkey (1989).

Gwynedd *Dolgellau goldbelt* – occurrences of erythrite include: Clogau mine, Bontddu; Panorama mine, Llanaber, E of Barmouth; common on cobaltite-bearing vein material at Vigra mine, near Dolgellau (Dewey & Eastwood 1925, Gilbey 1968, Bevins 1994); pink to deep reddish-pink drusy botryoidal crusts, forming scattered balls to 0.6 mm, and as rare masses of minute divergent crystals associated with gypsum at a trial level, 2 km NW of Bontddu, near Dolgellau (Saich & Rust 1987).

Erythrite also forms post-mining efflorescences at Great Orme mine, Llandudno, where 10–20 µm-diameter spheres infill void spaces within mudstones at Roman shaft (Jenkins & Johnson 1993, Dutton et al. 1994, Ixer 2001).

Scotland

Central Native silver encrusted with clinosafflorite may be coated with a suite of post-mining arsenates, including peach-blossom or purple erythrite, together with annabergite, conichalcite, picropharmacolite and tyrolite at Alva mine, Silver Glen, Ochil Hills (Heddle 1901, Dickie & Forster 1974, Moreton 1996, Moreton et al. 1998, Green & Todd 2001). Acanthite (argentite), arsenopyrite, bornite, chalcopyrite, covellite, digenite, galena, malachite, marcasite or pyrite (or both), pyrrhotite and sphalerite in a barite–calcite gangue are also reported from this fissure-filled vein deposit cutting Lower Old Red Sandstone volcanics (Dunham et al. 1978, Jassim et al. 1983). Small quantities of silver-bearing erythrite are described from galena-bearing veins at Tyndrum lead–zinc mine. A dump specimen of uraninite (pitchblende) was found to contain secondary erythrite with kasolite, liebigite, uranospinite and zeunerite, associated with chalcopyrite–galena–marcasite–pyrite–safflorite mineralization (Ridgway 1980, Pattrick 1985). Heddle (1901) notes erythrite from Clifton lead mine, Tyndrum.

Dumfries & Galloway Uranium-rich dolomite–quartz veins on the Solway coast around Needle's Eye, SW of Caulkerbush, contain secondary erythrite together with annabergite, atacamite, bismoclite, boltwoodite, bornite, connellite, covellite, eulytite, kahlerite, millerite, mixite, nováčekite, pharmacosiderite, schoepite, tyrolite, uranophane, vandendriesscheite, walpurgite, wulfenite and zeunerite. They are alteration products of native bismuth–chalcopyrite–hematite–pyrite–uraninite mineralization (Miller & Taylor 1966, Green & Todd 2001). Well crystallized pink radiating blades of erythrite to 1 mm, associated with lavendulan occur in a copper-rich stockwork at this location (Braithwaite & Knight 1990). Erythrite also occurs with calcite, chalcopyrite, dolomite and pyrite at Blackcraig mine, Newton Stewart (Heddle 1901), and is one of several secondary species found at Glencrieff mine, Wanlockhead (Temple 1955).

Lothian Barite–calcite veins host Ag–Co–Ni mineralization at Hilderston mine, Bathgate Hills, and contain pink to whitish spherules and coatings of erythrite, together with annabergite, barytocalcite, cinnabar, galena, harmotome, maucherite, nickeline, native silver and sphalerite (Dunham et al. 1978, Ridgway 1980, Meikle 1994). Heddle (1901) notes erythrite from Broughton quarry, Edinburgh.

Ireland

Kerry Vivid pink to crimson mammillated and minutely crystalline crusts of erythrite, found on waste material from the dumps of Muckross mine, Killarney National Park, have a post-mining origin (Cole 1922, Ryback & Moreton 1991, Moreton et al. 1999, Green & Moreton 2001).

Leitrim The NHM contains specimens of erythrite from Pollboy mine (Ryback & Moreton 1992).

Mayo One specimen of bright-pink erythrite associated with annabergite has been found at the small Sheefry Pb–(Ag) mine, 11 km S of Croagh Patrick (Moreton & Jackson 1991).

Tipperary Well formed crystals of erythrite have been found with mercury-rich native silver and lavendulan in the Cu–(Ag–Hg) orebody at Gortdrum mine, near Tipperary town. The crystals form individual transparent well terminated blades up to 2 mm long and radiating clusters of blades that are considered to be the best examples ever recorded from Britain or Ireland (Steed 1986, Braithwaite & Wilson 2003). Erythrite is one of many secondary species listed from the Silvermines district (Moreton 1999).

Waterford Erythrite forms pink encrustations and spherules associated with lavendulan and occasionally brochantite, connellite or langite at Bunmahon, in the Knockmahon copper mining area, 22 km SW of Waterford town (Moreton 1991, Ryback & Moreton 1991). Erythrite has been collected with lavendulan from a level on Seven Dial lode, Knockmahon (Braithwaite & Wilson 2003). A rubbish-filled pit near the Tankardstown engine house has produced pink stains of erythrite associated with disseminated grains of a cobalt arsenide mineral (Moreton 2005a).

erythrite (of Thomson)

TL Kilpatrick Hills, near Glasgow, Strathclyde, Scotland

A red variety of orthoclase, feldspar group.

erythrosiderite

orthorhombic $K_2Fe^{3+}Cl_5 \cdot H_2O$

A rare hydrated potassium–iron chloride found on samples from the Permian salt deposits of the Aislaby No. 2 borehole, Eskdale, near Whitby, Yorkshire. When rinneite is dried in a desiccator, small deep reddish-brown crystals of erythrosiderite form on the surface. Erythrosiderite also occurs as a constituent of the altered crust of rinneite, and small irregular yellow crystals have been found enclosed in halite (Stewart 1951).

eskolaite

HEMATITE GROUP

trigonal Cr_2O_3

A rare chromium oxide found in boulder clay at Callowhill Upper, near the village of Newtown Mount Kennedy, Co. Wicklow, Ireland. Eskolaite forms submicroscopic grains in pebbles up to 2.5 cm long and does not appear to have been derived locally. It is premature to speculate on its origin (Oppenheim et al. 1977).

essonite

A variety of grossular, garnet group (and a synonym of hessonite).

ETTRINGITE GROUP

Hexagonal sulphates with the general formula $Ca_6X_2Y(O,OH)_{12} \cdot 24\text{–}26H_2O$, where X = Al, Cr^{3+}, Fe^{3+}, Mn^{2+}. Mn^{4+}, Si; and $Y = (SO_4,CO_3)_3$ or $(SO_4)_2B(OH)_4$. Ettringite and thaumasite are members of the group found in Britain or Ireland.

ettringite

ETTRINGITE GROUP

hexagonal $Ca_6Al_2(SO_4)_3(OH)_{12} \cdot 26H_2O$

A rare hydrated calcium–aluminium sulphate hydroxide found at Milltown quarry, near Ashover, Derbyshire, where fluorite veins cut limestone and tuffs. Ettringite forms colourless-to-white hexagonal prisms and capillary crystals to 1 mm associated

Delicate sprays of ettringite to 1 mm; Milltown quarry, Ashover, Derbyshire. (© M. P. Cooper)

with weddellite and less commonly with ashoverite, cerussite, hydrocerussite, litharge or sweetite (Rust 1991c).

Ettringite forms fibrous crystals in late-stage fissure veins (<0.5 mm wide) in limestone thermally metamorphosed by dolerite and gabbro at Kilchoan, Ardnamurchan, Highland Region, Scotland (Agrell 1965, Livingstone 2002).

At Scawt Hill, Larne, Co. Antrim, Ireland, ettringite has been found in vugs within an altered crust to a spurrite rock from the chalk/dolerite contact zone; afwillite and gyrolite are also found in the vugs (Tilley & Harwood 1931). Further work in this area revealed afwillite, calcite, ettringite, hydrocalumite and portlandite infilling vugs in larnite- and melilite-bearing rocks (Tilley 1933, 1934). Bannister (1936) reports that the largest crystal found at Scawt Hill was 1.1×0.8×0.5 mm.

eucairite
orthorhombic — CuAgSe

A very rare copper–silver selenide found in a gold-bearing carbonate vein cutting Middle Devonian limestones at Hope's Nose, Torquay, Devon. The selenide assemblage includes clausthalite, fischesserite, klockmannite, naumannite, penroseite, tiemannite, trüstedtite, tyrrellite and umangite in a gangue mainly consisting of calcite. Eucairite forms relatively large (to 300 μm) equant anhedral grains, mostly at or along calcite grain boundaries. It also forms dendritic to skeletal grains that project from these grain boundaries (Stanley et al. 1990a). Eucairite is tentatively identified as one of the inclusion species in chalcedonic silica within fault rocks at Clive mine, 10 km N of Shrewsbury, Shropshire. Other inclusion species positively or tentatively identified include barite, chalcocite or digenite, chalcopyrite, cinnabar, galena, naumannite and pyrite (Plant et al. 1999).

euchroite?
orthorhombic — $Cu^{2+}_2(AsO_4)(OH)\cdot 3H_2O$

A rare hydrated copper arsenate hydroxide, described from an altered sheet of native copper enclosed in relict mudstone from 10 cm below the Triassic Budleigh Salterton Pebble Beds at Littleham Cove, near Budleigh Salterton, Devon. Cuprite is the main alteration species within which hydrated copper arsenates have been found. EMP analysis indicates that these are either arhbarite or euchroite (Milodowski et al. 2002).

euclase?
monoclinic — $BeAlSiO_4(OH)$

A rare beryllium aluminosilicate said to have been found at Cligga Head, Perranzabuloe, Cornwall, and at Meldon aplite quarry, near Okehampton, Devon (Kingsbury 1958). Ryback et al. (1998, 2001) have serious doubts about the provenance of many samples in the Kingsbury collection (at the NHM) and, until further discoveries are made, euclase cannot be considered to be a British or Irish species.

eucolite
A softer altered variety (possibly ferrokentbrooksite) of eudialyte.

eucryptite?
trigonal — $LiAlSiO_4$

A rare lithium aluminosilicate described from the Meldon aplite quarry, near Okehampton, Devon (Drysdale 1991). Eucryptite was found on specimens in the Kingsbury collection at the NHM. Because this collection contains samples that have been incorrectly provenanced (Ryback et al. 1998, 2001), without further investigation it is impossible to confirm eucryptite as a British or Irish species.

EUDIALYTE GROUP
Trigonal silicates with the general formula:

$$A_{20-21}B_{2-6}C_{3-4}Si_{25}O_{73}(O,OH,H_2O)_3(OH,Cl)_2 \text{ or } A_{21}BC_4Si_{25}O_{74}F\cdot 2H_2O$$

where A=Na, Ca, Sr, REE; B = Fe^{2+}, Fe^{3+}, Mn^{2+}; and C= Nb, W, Zr. Eudialyte is the only member of the group found in Britain or Ireland.

eudialyte
EUDIALYTE GROUP

trigonal — $Na_{15}Ca_6(Fe^{2+},Mn^{2+})_3Zr_3(Si,Nb)(Si_{25}O_{73})(O,OH,H_2O)_3(OH,Cl)_2$

A rare complex sodium-rich zirconium silicate found in syenite veins cutting limestone to the east of Barnavave, Carlingford, Co. Louth, Ireland. Pale- to deep-pink eudialyte forms small segregations in the syenite, mostly in the neighbourhood of limestone. In thin section, eudialyte is colourless and has a tabular habit with a hexagonal cross section. Eudialyte replaces zircon. Associated species in the syenite are aegirine, potassium feldspar and quartz, accompanied by minor pectolite and wollastonite. Contamination of the syenite magma by digestion of the limestone produced unusual melt conditions from which these species crystallized (Nockolds 1950).

Sabine (1960) reported eudialyte (var. eucolite) in crushate from the aegirine–riebeckite granite of Rockall island, 440 km W of northern Skye, NW of Scotland. The occurrence was not confirmed by Hawkes et al. (1975).

The wide range in compositions of eudialyte from many world localities has required a revision of nomenclature, and a eudialyte group with six members has been proposed (Johnsen & Grice 1999, Johnsen et al. 2003). It is not known how the British and Irish examples fit into this classification.

eudidymite?
monoclinic — $NaBeSi_3O_7(OH)$

A rare sodium–beryllium silicate hydroxide, dimorphous with epididymite and reported from the Meldon aplite quarry, near Okehampton, Devon (Kingsbury 1964b, Embrey 1978). Eudidymite was found on specimens in the Kingsbury collection at the NHM. Because this collection contains samples that have been incorrectly provenanced (Ryback et al. 1998, 2001), without further investigation it is impossible to confirm eucryptite as a British or Irish species.

eudnophite
A cloudy variety of analcime, zeolite group.

eulite
A variety of pyroxene with a 70–90% ferrosilite component.

eulytite
cubic — $Bi_4(SiO_4)_3$

A rare bismuth silicate found in a native bismuth-rich vein close to Needle's Eye, SW of Caulkerbush, Dumfries & Galloway, Scotland. This and other similar veins are in the aureole of the Criffell granodiorite. Eulytite forms pale-green translucent radiating spherules to 0.25 mm on and around altered native bismuth. Other supergene minerals from this vicinity, produced during the alteration of primary As–Bi–Co–Cu–Fe–Pb–U minerals, include kahlerite, lavendulan, metakahlerite, schoepite, vandendriesscheite and walpurgite (Braithwaite & Knight 1990). A less precise location of "Southwick Cliffs area" is given by Macpherson & Livingstone (1982) for small green transparent-to-translucent, spherulitic masses of eulytite to 0.25 mm.

A narrow chalcopyrite–ferberite–scheelite-bearing quartz vein, which cuts the Eskdale granodiorite at Buckbarrow Beck, Corney Fell, Cumbria, contains colourless to honey-yellow eulytite tetrahedra up to 0.5 mm across and colour-banded aggregates up to 2 mm across. It is associated with other rare supergene species: bismutoferrite, cuprotungstite, mixite, namibite and russellite (Neall et al. 1993, Green & Briscoe 2002).

Reports of eulytite in Cornwall by Collins (1881) from Restormel (Royal) iron mine, N of Lostwithiel, and Wheal Coates, St Agnes, have been shown to be waylandite and bismutite respectively (Clark et al. 1986). A reported eulytite from

Eulytite cluster, 1 mm across, associated with creamy-yellow russellite and a pale-blue amorphous aluminium mineral; Buckbarrow Beck, Corney Fell, Waberthwaite, Cumbria. (© D. I. Green)

Lanlivery, a small village near Lostwithiel (Anthony et al. 1995), is probably based on the Collins publication.

euxenite-(Y)

orthorhombic $(Y,Ca,Ce,U,Th)(Nb,Ta,Ti)_2O_6$

A rare complex yttrium–niobium oxide found in pegmatites near Kinlochbervie, 18 km S of Cape Wrath, Highland Region, Scotland. Euxenite-(Y) is associated with allanite and thorite (Gallagher et al. 1971).

evansite?

amorphous $Al_3(PO_4)(OH)_6{\cdot}6H_2O$

Named after Brooke Evans (1797–1862), a Birmingham-born metallurgist, who brought the first specimens from Hungary. Evansite is an uncommon hydrated aluminium phosphate hydroxide reported from the Yoredale rocks at Ratcliff Wood, near Macclesfield, Cheshire. Intimately associated with evansite are brown iron ore, oxide of manganese, and calcite. "Beautiful perfect" crystals of pyrite, abundant ankerite or dolomite (pearl spar), and sphalerite occur near by (Woodward 1884). Mandarino & Back (2004) do not consider evansite to be a true species, probably because of its amorphous nature.

ewaldite

hexagonal $Ba(Ca,Y,Na,K)(CO_3)_2$

A rare barium–calcium–yttrium–sodium–potassium carbonate found with harmotome and calcite in thin veins within Precambrian clastic sediments, sandstones and siltstones of the Yat Wood and Strinds Formations at Dolyhir quarry, Old Radnor, near Presteigne, Powys. Ewaldite forms striated pale yellowish-brown, translucent, hexagonal prismatic-to-pyramidal crystals up to 3 mm in length (Green et al. 2005d).

Pale-brown barrel-shape ewaldite with colourless harmotome; Dolyhir quarry, Old Radnor, near Presteigne, Powys, Wales; width 5 mm. (© D. I. Green)

F

fahlerz

A name given to the tennantite–tetrahedrite solid-solution series.

FAIRFIELDITE GROUP

Triclinic arsenates and phosphates with the general formula $Ca_2B(XO_4)_2{\cdot}2H_2O$, where B = Co, Fe^{2+}, Mg, Mn^{2+}, Ni, Zn; and X = As, P. Talmessite and possibly fairfieldite are members of the group found in Britain or Ireland.

fairfieldite?

FAIRFIELDITE GROUP
triclinic $Ca_2(Mn^{2+},Fe^{2+})(PO_4)_2{\cdot}2H_2O$

An hydrated calcium–manganese–iron phosphate found on samples in the Kingsbury collection at the NHM and reportedly from Meldon railway quarry, near Okehampton, Devon. Because this collection contains samples that have been incorrectly provenanced (Ryback et al. 1998, 2001), without further investigation it is not possible to confirm fairfieldite as a British or Irish species.

famatinite

STANNITE GROUP
tetragonal Cu_3SbS_4

A rare copper–antimony sulphide that forms a series with luzonite and is a minor component of the sulphide assemblage on Melbecks Moor, N of Gunnerside, Yorkshire, in the Askrigg Block area of the Northern Pennine orefield. Famatinite is mantled by covellite and carbonate (all enclosed within galena) and it appears to form relatively late in the sequence of crystallization (Vaughan & Ixer 1980). Also on the Askrigg Block, at Scaleber Bridge, 3 km SSE of Settle, Yorkshire, famatinite forms brown-to-lilac (in reflected light) untwinned but highly anisotropic crystals up to 80 μm in length that are intergrown with chalcopyrite. Enargite and luzonite are similarly found as inclusions. The chalcopyrite (as well as pyrite and marcasite) itself occurs as inclusions in heavily malachite-stained coarse-grain vuggy calcite crystals. Enhanced silver, arsenic, copper and antimony concentrations in stream-sediment samples perhaps indicate that famatinite belongs to a type of mineralization different from the main barite–fluorite–galena–sphalerite (MVT) mineralization present within the Askrigg Block (Ixer & Stanley 1998).

An optical identification of famatinite associated with tennantite from Tynagh mine, near Killimor, Co. Galway, Ireland, has been made by Schultz (1968).

fargite

TL Glen Farg, Tayside Region, Scotland

A calcium-rich variety of natrolite, zeolite group.

faröelite

A radiating variety of thomsonite, zeolite group.

faserkeisel

A synonym of fibrolite, a variety of sillimanite.

fassaite

A synonym of aluminium–iron-rich diopside or augite, both pyroxene group, but also a term applied to an unidentified zeolite, probably stilbite or arduinite (syn. mordenite).

faujasite? (series)

ZEOLITE GROUP
cubic $(Na_2,Ca)Al_2Si_4O_{12}{\cdot}8H_2O$

A rare hydrated sodium–calcium aluminosilicate whose name has been given series status in the zeolite classification of Coombs et al. (1997). There are now three recognized faujasite species: -Na, -Ca and -Mg. Wilson et al. (2003) list faujasite from the Storr, Trotternish, Isle of Skye, Scotland, based on a note in Anderson & Dunham (1966), who suggest that faujasite and levyne may be present with analcime, chabazite and thomsonite in the mesostasis of lavas from this locality.

faustite

TURQUOISE GROUP
triclinic $(Zn,Cu^{2+})Al_6(PO_4)_4(OH)_8{\cdot}4H_2O$

A very rare hydrated zinc–copper–aluminium phosphate hydroxide found at Wheal Phoenix, Linkinhorne, Cornwall. In general, faustite is apple green with a waxy to dull lustre. No further details are given (Roberts et al. 1990).

fayalite

OLIVINE GROUP
orthorhombic $Fe^{2+}_2SiO_4$

A common iron silicate that forms series with forsterite and with tephroite, and is found in ultramafic volcanic and plutonic rocks or, less commonly, in felsic plutonic rocks. Examples are also found in metamorphosed iron-rich sediments and impure carbonate rocks. Compositions are often intermediate between fayalite and forsterite end members, and the ratio of molecular iron to magnesium is then used to ascribe a name to the mineral (e.g. 75% Mg or Fo_{75} is forsterite, and any Fo value <50 is fayalite). The two species can also be discriminated optically.

Fayalite is found in the Caledonian Newer Gabbro bodies in the Grampian region of Scotland, including the Morven–Cabrach intrusion, which stretches from the hamlet of Cabrach for about 24 km S towards Deeside. Fayalite (Fo_{43-45}) is one of the major constituents of dense black ferrogabbro that mainly occurs within areas of norite in the southern part of the intrusion. Associated species are augite, enstatite and plagioclase feldspar, with lesser amounts of apatite, biotite and iron oxides (Allan 1970). A few km to the NE, ferrogabbro containing fayalite (Fo_{5-40}) is reported from the Upper Zone cumulates of the Insch Layered Series, a part of the Insch layered intrusion (Clarke & Wadsworth 1970). Wadsworth (1986) reinvestigated these rocks, measured fayalite compositions ranging between $Fo_{5.5}$ and $Fo_{46.7}$, and was able to confirm a simple fractionation sequence for the cumulates.

Optical examination of olivine from the crinanite–dolerite–picrite Garbh Eilean sill, Shiant Isles, 8 km E of Lewis, Western Isles, was also used to confirm a simple fractionation sequence, with compositions varying continuously throughout the intrusion from forsterite (Fo_{79}) in picrite at the base of the intrusion to fayalite (Fo_{26}) in crinanite at the top (Johnston 1953). Chemical analyses of olivine from the sill show a wider range of compositions extending from forsterite (Fo_{83}) in lower units to fayalite (Fo_6) in upper units (Gibb & Henderson 1996). Fo_{39} in teschenite from an analcime-bearing leucodolerite sill at Cnoc Rhaonastil, Isle of Islay (Hole & Morrison 1992), has a similar paragenesis to that on Garbh Eilean.

Zoned crystals with forsterite cores (Fo_{60}) and fayalite (Fo_{30}) rims occur in the Glen Bay gabbro on St Kilda (Hirta), Western Isles. The assemblage is essentially the same as that of the Grampian gabbros, although olivine is less abundant (1–2%) and small amounts of alkali feldspar and quartz are present (Harding et al. 1984).

On the Isle of Skye, Highland Region, yellow fayalite (Fo_{22} hortonolite) occurs in ferrodiorite from Harker's Gully on the west side of Marsco, S of Sligachan (Wager & Vincent 1962), but it is in the nearby Tertiary epigranites of Skye that fayalite approaching the end-member composition occurs: Beinn Dearg Mhòr, Glen Sligachan, Loch Ainort, Marsco (Fo_3), Southern Porphyritic (Fo_{0-5}) and Meall Dearg (almost pure fayalite). The epigranites are predominantly equigranular (2–8 mm grainsize) alkali feldspar–plagioclase–quartz-rich rocks containing lesser amounts of amphibole, augite (ferrohedenbergite), biotite, ilmenite, magnetite, and accessory allanite, apatite, calcite, epidote, fluorite, prehnite, titanite (sphene) and zeolites. Local aenigmatite and almandine are additional accessory species found in the Southern Porphyritic epigranite (Thompson 1969, Exley 1980). Tertiary granites found in the Mourne Mountains, Co. Down, Ireland, have many similarities with the Scottish epigranites, including fayalite reported from Slievecarrach (Greg & Lettsom 1858) and large euhedral crystals at Eagle Rock, Slieve Donard (Richey 1927).

Many of the Tertiary pitchstones in Scotland contain fayalite: Isle of Arran, Strathclyde – porphyritic types show diverse phenocryst assemblages that include augite, hedenbergite, hypersthene, pigeonite, plagioclase, quartz and sanidine. Examples are found at Brodick school (Fo_{14}), Glen Shurig (pale brown, Fo_{12-14}) and near Blackwaterfoot, where two composite dykes, referred to as Judd dykes No. 1 and 9 contain rare fractured phenocrysts (Fo_{10-15}); Raasay, near the Isle of Skye –

fayalite (Fo_7) in pitchstone (Tyrell 1928a, Preston et al. 1998b); Rubh'an Tangaird, Isle of Eigg – fayalite in two thin pitchstone dykes (Ridley 1971). Fayalite in the allanite-bearing obsidian found in the Sandy Braes vent (a part of the Tardree rhyolite complex), Co. Antrim, Ireland, has a paragenesis similar to that of the Scottish examples (Brooks et al. 1981).

A different paragenesis for fayalite is recorded from the Scawt Hill chalk/dolerite contact zone, Co. Antrim, Ireland, where it occurs in 25 mm-wide segregation veins cutting dolerite near the edge of a hybrid zone. Apatite, alkali amphibole (hastingsite), augite, magnetite, plagioclase (labradorite) and thomsonite are associated minerals. Its origin is in the assimilation of limestone by basaltic magma (Tilley & Harwood 1931). At Carneal, Larne, Co. Antrim, further chalk–dolerite interaction has occurred, where blocks of Cretaceous chalk have been brought up in a dolerite plug that intrudes Tertiary lavas. The plug contains manganese-rich fayalite (Fo_{45} hortonolite) in pyroxenites adjacent to calc-silicate rocks (Sabine & Young 1975).

feather alum

A synonym of halotrichite and of alunogen.

feitknechtite

hexagonal or tetragonal — $\beta\text{-}Mn^{3+}O(OH)$

A rare manganese oxyhydroxide, trimorphous with groutite and manganite, and found on a NMW specimen from the northern end of 84 stope, Benallt mine, Rhiw, Lleyn Peninsula, Gwynedd, Wales. The specimen consists of coarse brown blades of hausmannite intergrown with feitknechtite. Visual distinction between the two species is impossible (T. F. Cotterell pers. comm. 2007).

FELDSPAR GROUP

Silicates with the general formula XZ_4O_8, where X = Ba, Ca, K, Na, NH_4, Sr; and Z = Al, B, Si. Albite, anorthite, anorthoclase, banalsite, celsian, hyalophane, microcline, orthoclase, paracelsian and sanidine are members of the group found in Britain or Ireland.

felsöbányaite/felsőbányaite

monoclinic — $Al_4SO_4(OH)_{10}{\cdot}4H_2O$

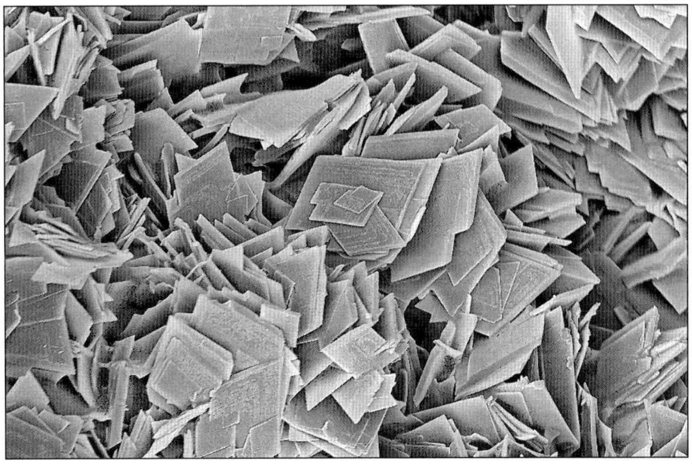

Felsöbányaite from Chickerell, near Weymouth, Dorset; width 20 µm. (© E. Condliffe, University of Leeds)

A rare hydrated aluminium sulphate hydroxide found as coatings in siderite deposits at Irchester, Northamptonshire, and Clifton Hill, Brighton, Sussex (Bernard & Hyršl 2004). This occurrence and others had been described as basaluminite – a species recommended for extinction in the IMA-CNMNC approved mass discreditation outlined by Burke (2006). See basaluminite.

ferberite

monoclinic — $Fe^{2+}WO_4$

A common iron tungstate that forms a series with hübnerite and is found in high-temperature hydrothermal veins, greisens, granitic pegmatites and some alluvial deposits. Ferberite is more common than the literature suggests, as the requisite chemical analysis required to identify the species is often lacking. Without an analysis, the species is usually referred to as wolframite. Ferberite and hübnerite are important ore species of tungsten.

Ferberite has been found in many mineralized quartz veins in Cornwall. At Penhale tin mine, St Cleer, it occurs in a brecciated quartz vein associated with an elvan dyke. Sheaves and bunches of black ferberite associated with chalcopyrite have also been found in three parallel quartz lodes exposed in an old railway cutting near Calstock Consols mine, near Gunnislake, and at nearby Wheal Arthur the same assemblage was encountered again; both mines exploited sulphide ores (Dines 1956). EMP analysis of wolframite from a greisen-bordered vein near the eastern contact of the St Michael's Mount granite, Marazion, has shown it to be oscillatory zoned with individual zones of intermediate composition between ferberite and hübnerite end members. This seems to be a common phenomenon, as the following instances of wolframite show the same variation: from a pegmatite vein near the southern contact of the Cligga Head granite, Perranzabuloe; from Chasewater, near Redruth; and from South Crofty, Camborne. Rapid changes in the composition of hydrothermal fluids, as a result of reaction between fluids and host rocks, is a plausible mechanism that could explain this variation (Moore & Howie 1978). Sharp prismatic crystals of wolframite (2–3 mm in length) associated with clinochlore and muscovite in vein quartz from Croft Gothal mine, St Hilary, have a more restricted compositional range within the ferberite compositional field (J. Betterton specimen, Tindle unpub. EMP data). Wolframite from the stockwork deposit at Hemerdon, Plympton, Devon, is more systematically zoned; from a hübnerite composition at the core to almost end-member ferberite composition at the rim (R. C. Scrivener specimen, Tindle unpub. EMP data). Elsewhere in Cornwall, Golley & Williams (1995) note ferberite from: Gunheath china-clay pit, St Austell; Hingston Down quarry, 2 km W of Gunnislake, Calstock; and North Herodsfoot mine, Lanreath. Individual crystals of wolframite from Bunny mine, St Austell (now exposed in Gunheath china-clay pit), have intermediate compositions either side of the ferberite/hübnerite compositional boundary (Tindle unpub. EMP data).

In Cumbria, ferberite is reported from Buckbarrow Beck, Corney Fell, Waberthwaite, where metalliferous mineralization occurs at the southwestern margin of the Eskdale granite–granodiorite intrusion. Quartz veins host a chalcopyrite–ferberite–scheelite assemblage in which ferberite is a minor (and relatively unzoned) constituent. Bismutoferrite, cuprotungstite, eulytite, mixite, namibite and russellite are some of the supergene species that accompany the primary mineralization (Young et al. 1986, 1991b, Green & Briscoe 2002). At Carrock mine, Caldbeck Fells, wolframite and scheelite occur in subvertical quartz veins that cut the Grainsgill granite – a greisenized cupola developed on the northern flank of the Skiddaw granite. Wolframite forms elongate bladed crystals, often several cm long, and typically clustered at the edges of veins. The composition is intermediate between ferberite and hübnerite end members (Ball et al. 1985a).

A very unusual niobium-rich variety of hübnerite occurs in a wolframite-bearing quartz vein cutting zinnwaldite granite near Gairnshiel Bridge, Glen Gairn, Grampian, Scotland. Surrounding the vein is a zone of silicified topaz-bearing greisen that contains niobium-rich wolframite, zoned between ferberite and hübnerite compositions. Russellite and stolzite are found in the greisen as alteration products of the wolframite (Tindle & Webb 1989, Webb et al. 1992).

fergusonite-(Y)

tetragonal — $YNbO_4$

Named after Robert Ferguson FRS (1767–1840), advocate, landowner, mineral collector, politician (Lord Lieutenant of the County of Fife and member of parliament), and fellow of the royal societies of London and Edinburgh, who resided at Raith House, near Kirkcaldy, Fife. He discovered the mineral in 1806. Fergusonite-(Y) is a rare yttrium niobate, dimorphous with fergusonite-beta-(Y), that forms a series with formanite-(Y) and is found in accessory amounts in miarolitic cavities within the Northern Arran granite, Isle of Arran, Strathclyde, Scotland. It is one of several REE minerals found principally in granites, pegmatites, alpine-type fissures and hydrothermal veins. Fergusonite-(Y) forms prismatic acicular crystals up to 2 mm in length. Most crystals are dull, although lustrous, submetallic and vitreous examples also occur. Allanite, beryl, gadolinite-(Y), topaz and zircon are additional accessory species found in the miarolitic cavities. This is the first account of the species from an *in situ* location (Hyslop et al. 1999). Other Scottish reports of fergusonite are from alluvial finds in stream-sediment panned concentrates and they include: Luibeg Burn, near Ben Macdui, Cairngorm Mountains, Grampian (Haynes 1974, 1993, Beer & Bennett 1988); and St Kilda (Hirta) (Harding et al. 1984).

feroxyhyte?

hexagonal — $Fe^{3+}O(OH)$

A ferrous hydroxide, trimorphous with goethite and lepidocrocite, and possibly found in a 1–3 mm-thick iron-pan (soil) formed from basalt at Killiechronan, SW of

Elongate crystal of fergusonite-(Y) in a miarolitic cavity within the Northern Arran granite, Strathclyde, Scotland; height of crystal 2 mm. (IPR/100-52C British Geological Survey © NERC. All rights reserved.)

Salen, Isle of Mull, Strathclyde, Scotland. Ferrihydrite is confirmed from this locality, but perhaps because of poor crystallinity, akaganéite and feroxyhyte could not be completely characterized (Goodman et al. 1988, Livingstone 1993).

ferri-actinolite*

AMPHIBOLE GROUP
monoclinic — no balanced formula

A rare calcium–magnesium–iron silicate hydroxide that forms series with tremolite, actinolite and ferro-actinolite, but for which a simple balancing formula cannot be written, although it approaches $\square Ca_2(Mg,Fe^{2+},Fe^{3+})_5Si_8O_{22}(OH)_2$. Ferri-actinolite is not listed as a species by Mandarino & Back (2004), suggesting that it is a named amphibole (Burke & Leake 2005).

A single analysis of colourless ferri-actinolite is reported from the Precambrian layered diorite–gabbro complex at Le Nez, southeast Jersey, Channel Isles. Other amphiboles in this complex are: cummingtonite, edenite, kaersutite, magnesiohastingsite, magnesiohornblende and pargasite (Bishop & Key 1983).

ferribarroisite?

AMPHIBOLE GROUP
monoclinic — $\square(CaNa)Mg_3Fe_2^{3+}(Si_7Al)O_{22}(OH)_2$

Small grains in rocks from the Ordovician Falahill Formation, near Heriot, N of Galashiels, Borders, Scotland, are described as ferribarroisite by Livingstone (2002). Identification is based on a single EMP analysis with a low analytical total and is given the name parvo-ferro-edenite in the amphibole classification of Leake et al. (1997a); see also Mogessie et al. (2004). Further investigation is desirable. Ferribarroisite does not appear to have been formally described (Mandarino & Back 2004).

ferric-nyböite*/ferric-nybøite*

AMPHIBOLE GROUP
monoclinic — $NaNa_2(Mg_3Fe_2^{3+})(Si_7Al)O_{22}(OH)_2$

A rare sodium–magnesium–iron aluminosilicate hydroxide that forms series with nyböite and with ferric-ferronyböite. Ferric-nyböite is a sodic subgroup end member and as such it is mentioned in the 1997 classification of the amphibole group (Leake et al. 1997a), but is it not listed as a species by Mandarino & Back (2004), suggesting that it has never been formally described. It is included here as a named amphibole (Burke & Leake 2005).

A single analysis of magnesio-arfvedsonite from a minette at St Hélier, Jersey, Channel Isles (Velde 1971a), has a ferric-nyböite composition when classified using the AMPH–IMA04 software (Mogessie et al. 2004).

ferricopiapite

COPIAPITE GROUP
triclinic — $(Fe_{2/3}^{3+}\square_{1/3})Fe_4^{3+}(SO_4)_6(OH)_2{\cdot}20H_2O$

A rare hydrated iron sulphate hydroxide forming part of a supergene assemblage at Groverake mine (New Firestone level), near Rookhope, Weardale, Co. Durham. Ferricopiapite occurs as very small pale-yellow botryoidal growths on native copper-bearing specimens (Smith 1973a).

At Hawthorn Hive, a wide bay on the Co. Durham coast (at and around Hive Point and Shippersea Bay), ferricopiapite has been found in mine waste mainly from Dawdon colliery. Ferricopiapite forms pale-yellow cauliflower-like efflorescent crusts, up to 30 cm across, on the upper surfaces of the spoil, but is washed away by prolonged rain during most of the winter months. Aluminocopiapite, apjohnite, calcite, gypsum, natrojarosite and sideronatrite are associated species (Young et al. 2008).

ferrierite-K

ZEOLITE GROUP
orthorhombic — $K_2NaMg(Al_5Si_{31})O_{72}{\cdot}nH_2O$

An uncommon hydrated potassium–sodium–magnesium aluminosilicate found at many localities in Carboniferous basalts and tuffs, such as at Burntisland, Fife, in the Midland Valley of Scotland. Ferrierite-K occurs as a replacement species in volcanic rocks, within amygdales in basalt and also in veins within tuff (Livingstone 1993).

ferri-ferrowinchite*

AMPHIBOLE GROUP
monoclinic — no balanced formula

A rare calcium–sodium–iron silicate hydroxide that forms a series with ferrowinchite, but for which a simple balancing formula cannot be written, although logically the formula approaches $\square(CaNa)(Fe_4^{2+}Fe_2^{3+})Si_8O_{22}(OH)_2$. It is not listed as a species by Mandarino & Back (2004), suggesting that it is a named amphibole (Burke & Leake 2005).

Ferri-ferrowinchite is found with ferrowinchite in miarolitic cavities in a Western Redhills Tertiary epigranite from Loch Ainort, Isle of Skye. The two amphiboles are blue to green (in thin section) and form zoned crystals with ferro-actinolite, ferro-edenite, ferrorichterite, katophorite and riebeckite. Other cavity species include hedenbergite, monazite, titanite and zircon (S. A. Rust specimen, Tindle unpub. EMP data).

ferrihydrite

trigonal — $5Fe_2O_3{\cdot}9H_2O$

An hydrated iron oxide found associated with goethite in >1 m thick ochre sediments formed from acid-mine drainage near West Chiverton (Cu–Pb–Zn) mine, Zelah, Perranzabuloe, Cornwall (Singh et al. 1999). Ferrihydrite is also reported in a 1–3 mm-thick iron pan formed from altered basalt at Killiechronan, SW of Salen, Isle of Mull, Strathclyde, Scotland. Here ferrihydrite is possibly accompanied by akaganéite and feroxyhyte (Goodman et al. 1988, Livingstone 1993).

ferri-magnesiohastingsite*

AMPHIBOLE GROUP
monoclinic — no balanced formula

A rare sodium–calcium–magnesium–iron aluminosilicate hydroxide that forms series with hastingsite and with magnesiohastingsite, but for which a simple balancing formula cannot be written, although logically the formula approaches $NaCa_2(Mg_3Fe_2^{3+})(Si_6Al_2)O_{22}(OH)_2$. It is not listed as a species by Mandarino & Back (2004), suggesting that it is a named amphibole (Burke & Leake 2005).

One analysis of an amphibole phenocryst in a pargasite–plagioclase–pyroxene phyric basalt lava from the Eglwys Rhobell Formation, Rhobell Fawr area, near Dolgellau, Gwynedd, Wales (Kokelaar 1986), has a ferri-magnesiohastingsite composition, when reprocessed using the amphibole classification of Leake et al. (1997a); see also Mogessie et al. (2004).

ferri-magnesiohornblende*

AMPHIBOLE GROUP
monoclinic — no balanced formula

A rare calcium–magnesium–iron aluminosilicate hydroxide that forms series with ferrohornblende and with magnesiohornblende, but for which a simple balancing formula cannot be written, although logically the formula approaches $\square Ca_2(Mg_3Fe_2^{3+})(Si_7Al)O_{22}(OH)_2$. It is not listed as a species by Mandarino & Back (2004), suggesting that it is a named amphibole (Burke & Leake 2005).

One partial analysis of an amphibole (described as hornblende) from an amphibole-rich reaction border in orthopyroxenite metasomatized by acid pegmatite veins in the Currywongaun–Doughruagh ultrabasic intrusion, Connemara, Co. Galway, Ireland (Kanaris-Sotiriou & Angus 1979), has a ferri-magnesiohornblende composition when reprocessed using the amphibole classification of Leake et al. (1997a); see also Mogessie et al. (2004).

ferrimolybdite

orthorhombic? $Fe_2^{3+}(Mo^{6+}O_4)_3 \cdot 8H_2O$

A relatively rare, but locally common, hydrated iron molybdate, found as a yellow oxidation product on molybdenite from a quartz vein cutting granite at Screel Hill, near Kirkmirran, N of Auchencairn, Dumfries & Galloway, Scotland (Williams 1973). The classic Heddle locality for molybdenite at Coire Buidhe, Glen Creran, Appin, Strathclyde, has recently been rediscovered by Ingram et al. (1993), who report rare yellow earthy coatings of probable ferrimolybdite in masses up to 0.5 mm between individual plates of molybdenite. Ferrimolybdite was also found coating molybdenite in a stream boulder in Allt Choire Dhuibh, draining the northern part of the Etive granite complex to the E and NE of Ben Starav, on the Dumfries & Galloway/Strathclyde border (Haslam et al. 1985).

In Ireland, massive molybdenite with a scaly yellow crust of ferrimolybdite is noted from Ballycummisk mine, near Rossbrin Cove, Co. Cork (Ryback & Moreton 1991). Disseminated molybdenite mineralization in vuggy quartz veins at Mountain mine, Allihies, Bearhaven (Berehaven) Peninsula, Co. Cork, contains ferrimolybdite (described as ferromolybdate) with both azurite and malachite as surface oxidation species on old stopes (Reilly 1986). Yellow encrustations of presumed ferrimolybdite on molybdenite are further described from Mace Head and Murvey, Connemara, Co. Galway (Ryback & Moreton 1992).

English descriptions of ferrimolybdite include: Carrock Fell mine, Grainsgill, Caldbeck Fells, Cumbria (Hartley 1984); Kit Hill Consols, Stoke Climsland, and Hingston Down quarry, 2 km W of Gunnislake, Calstock, both in Cornwall (Golley & Williams 1995). Unfortunately, these descriptions are based on samples supposedly collected by A. W. G. Kingsbury and their provenance is in doubt (Ryback et al. 1998, 2001). Positive identifications of ferrimolybdite and scorodite have been made at Water Crag, near Devoke Water, Cumbria, where they occur as supergene species associated with arsenopyrite–bismuthinite–fluorite–molybdenite mineralization in topaz greisen at the margin of the Eskdale granite (Young 1985d,e).

ferritchromit

Chromite, spinel group enriched in Fe^{2+} and Fe^{3+}.

ferrite

TL between Gleniffar and Boyleston, Strathclyde, Scotland

An iron-rich pseudomorph after olivine (smectite mixture). The name has also been used to describe brownmillerite, native iron and limonite.

ferritungstite

cubic $(W,Fe^{3+})_2(O,OH)_6 \cdot n(H_2O,K,Ca,Na)$

A rare hydrated potassium–calcium–sodium iron tungstate hydroxide found in cavities within bladed wolframite at Hingston Down quarry, 2 km W of Gunnislake, Calstock, Cornwall. Mpororoite, scorodite and stolzite are also found in the cavities, all probably formed because of the action of supergene oxidation processes (Golley & Williams 1995, Green et al. 1996a). Ferritungstite also occurs as sparkling bright-yellow crusts of octahedral crystals on weathered ferberite specimens found on the flanks of Coombe Height, Caldbeck Fells, Cumbria (D. I. Green pers. comm. 2006).

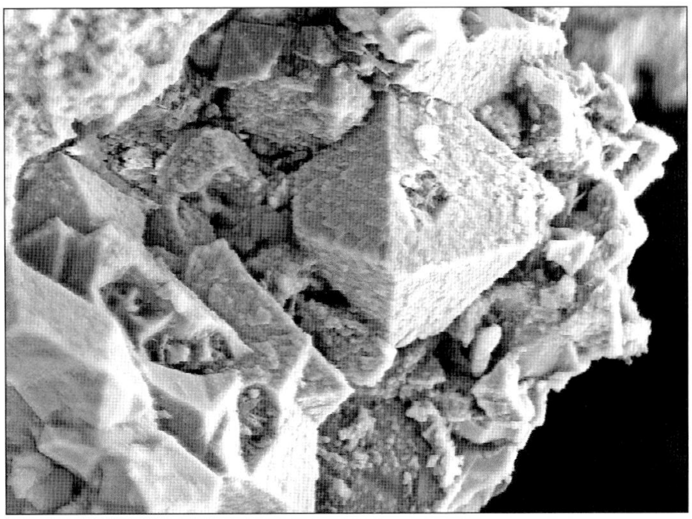

Ferritungstite from Coombe Height, Caldbeck Fells, Cumbria; width 75 μm; SEM image. (© D. I. Green)

ferriwinchite

AMPHIBOLE GROUP

monoclinic $\square Na,Ca(Mg_4Fe^{3+})Si_8O_{22}(OH)_2$

A rare sodium–calcium–magnesium–iron aluminosilicate hydroxide that forms a series with winchite and is found at Glen Lui (2 km SSE of Derry Lodge), Cairngorms, Grampian, Scotland, where it was originally described as "riebeckite–crossite". It is a pale-blue sodic amphibole that replaces pyroxene in aegirine granulite, a rock type considered to be the product of localized metasomatism of associated metamorphic schists. The granulite also contains aegirine-augite, albite, almandine, andradite, calcite, magnetite, orthoclase, titanite and zircon (McLachan 1951).

Ferriwinchite also occurs in a metasomatized (fenitized) quartzite at Loyne, near Loch Borralan, Assynt, Highland Region. The nearby Borralan syenite complex was the source of the strongly alkaline fluids required to produce this type of alteration. The petrography of these rocks shows green actinolite, edenite and magnesiohornblende developed in the early stages of metasomatism and this was later replaced by blue varieties that have magnesioriebeckite, richterite and winchite compositions (Woolley et al. 1972). Additional ferriwinchite and richterite data from the fenites at Loch Borralan are reported by Rock (1977).

ferro-actinolite

AMPHIBOLE GROUP

monoclinic $\square Ca_2Fe_5^{2+}Si_8O_{22}(OH)_2$

A relatively uncommon calcium–iron silicate hydroxide that forms a series with actinolite and tremolite, and is found in an appinitic microdiorite at Hope Beck, SSW of Scawgill Bridge, 4 km W of Braithwaite, Cumbria. Associated species include, augite, chromite, magnesiohornblende, olivine, pargasite, potassium feldspar and quartz (Fortey et al. 1994).

Ferro-actinolite has been described from the Tal y Fan olivine dolerite intrusion, SW of Conwy, Gwynedd, Wales (Bevins & Merriman 1988, Bevins 1994), but this appears to be on the basis of a single analysis that plots in the ferro-actinolitic hornblende field of Leake et al. (1978). Recent changes to nomenclature (Leake et al. 1997a) indicate that this composition should now be referred to as ferrohornblende.

In Scotland, the Tertiary granite on Ben More, Isle of Mull, Strathclyde, contains miarolitic cavities in which ferro-actinolite has formed as delicate hair-like crystals with euhedral diamond-shape cross sections. A similar granite at Loch Ainort, Isle of Skye, Highland Region, contains miarolitic cavities in which subhedral prismatic amphibole grains, some with feathery overgrowths, are found with titanite and zircon. The amphibole crystals are zoned, with ferro-actinolite, ferro-edenite, ferro-richterite, ferro-winchite, katophorite and riebeckite (S. A. Rust specimen, Tindle unpub. EMP data).

Ferro-actinolite forming a thin green skin on ferro-edenite in the Galway granite, Co. Galway, Ireland, developed under post-magmatic conditions when late hydrothermal fluids passed through the granite (Leake 1998).

ferro-augite

A synonym of augite, pyroxene group.

ferro-axinite/axinite-(Fe)

AXINITE GROUP

triclinic $Ca_2Fe^{2+}Al_2BO(OH)(Si_2O_7)_2$

A common calcium–iron–aluminium borosilicate hydroxide that forms a series with manganaxinite and is found in contact-metamorphic skarns.

Ferro-axinite has been found adjacent to a basic doleritic dyke in the Meldon aplite quarries, near Okehampton, Devon (Chaudry & Howie 1969), and with vesuvianite in the metamorphic aureole of the Land's End granite at Wheal Cock, near Botallack, St Just, Cornwall (Tindle unpub. EMP data). Dunn et al. (1980) notes ferro-axinite from Botallack mine and Roscommon Cliff (both near Wheal Cock), and from Liskeard, all in Cornwall.

Ferro-axinite is confirmed from four locations in Gwynedd, Wales, on specimens in the collection of the NMW: mauve-brown ferro-axinite in microdiorite from Penmaenmawr quarry, Penmaenmawr; brown platy ferro-axinite in microdiorite from Graiglwyd quarry, Penmaenmawr; brown ferro-axinite from veins cutting Gwna Group, Monian Supergroup pillow lavas from Carreg Ddu and Porth Dinllaen, both near Morfa Nefyn, Lleyn Peninsula (NMW specimens, Tindle unpub. EMP data).

Ferro-axinite is described from an epidote–plagioclase–quartz–tourmaline vein cutting amphibolite of the Ballyconneely mass to the N of Roundstone, Co. Galway, Ireland (Benjamin 1968).

Ferro-axinite is much more common than the references above suggest. Prior to the establishment of the axinite group, axinite was a species name and consequently many references never described ferro-axinite. It is highly likely that many of the occurrences listed under axinite are in fact ferro-axinite, although chemical analyses will be required to confirm this.

ferrobarroisite

AMPHIBOLE GROUP

monoclinic $\square(NaCa)Fe^{2+}_3AlFe^{3+}(Si_7Al)O_{22}(OH)_2$

An uncommon sodium–calcium–iron aluminosilicate hydroxide that forms a series with barroisite and is found in blueschist-facies metamorphic rocks. Green cores to blue amphiboles from the Mona Complex at the Marquis of Anglesey's monument, near Llanfairpwllgwyngyll, Anglesey, Gwynedd, Wales, were described as "hornblende" by Greenly (1919). However, analysis of similar material from there and from temporary roadcuttings made during construction of the Llanfairpwllgwyngyll bypass showed that the green cores consist of barroisite, ferrobarroisite and winchite, whereas the rims are pale-blue glaucophane (Gibbons & Gyopari 1986, Bevins 1994). Actinolite, alumino-ferrowinchite, ferroglaucophane, ferrowinchite and katophorite are also found in these rocks (Gyopari 1984).

ferrobrucite

An iron-rich variety of brucite.

ferrobustamite

triclinic $Ca_{1.67}Fe^{2+}_{0.33}Si_2O_6$

TL Camas Malag, near Broadford, Isle of Skye, Highland Region, Scotland

A rare calcium–iron silicate found in a contact skarn at the type locality, where the Beinn an Dubhaich granite has intruded dolomite. Ferrobustamite encases residual chert nodules in the dolomite and is associated with hedenbergite and grossular–andradite. Ferrobustamite was originally described as iron–wollastonite (Tilley 1948a). Iron–wollastonite, a synonym of ferrobustamite, was also mentioned when describing the skarn minerals associated with basic intrusions at Camas Mòr, Isle of Muck, Highland Region, Scotland (Tilley 1947), and Scawt Hill, Larne, Co. Antrim, Ireland (Tilley 1937).

ferrocalcite

The iron-rich analogue of dolomite.

ferrocolumbite/columbite-(Fe)

orthorhombic $Fe^{2+}Nb_2O_6$

A rare iron–niobium oxide that forms series with ferrotantalite and with manganocolumbite, and is found in evolved granites and granite pegmatites.

In Cornwall, ferrocolumbite is found as an accessory component of pegmatite veins cutting very evolved granite in Gunheath china-clay pit, near St Austell. Ferrocolumbite forms small subhedral platy crystals up to 5 mm across, either singly or in radiating clusters with microcline, quartz and zinnwaldite, and unusually as larger crystals to 2 cm in a vertical lens consisting of apatite, mica (var. gilbertite), quartz and rare torbernite (Hodkinson & Clark 1977). Extraction of china clay from

Stack of ferrocolumbite crystals to 0.5 mm from Chywoon granite quarry, near Longdowns, Cornwall. (© D. I. Green)

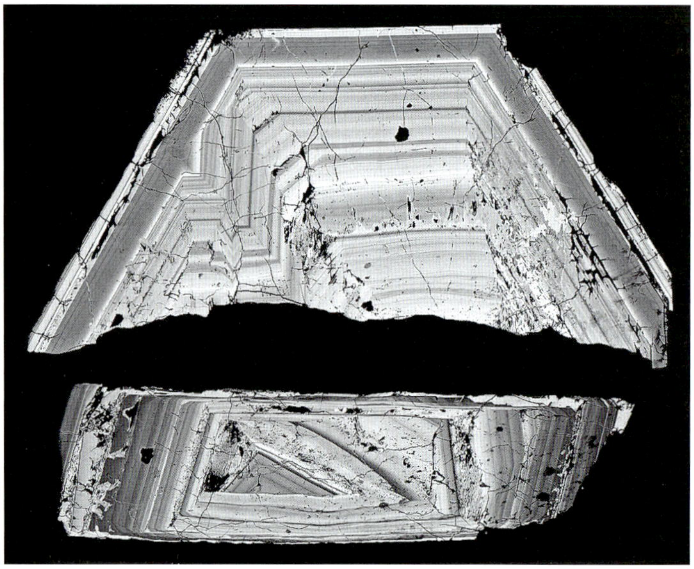

Oscillatory zoned ferrocolumbite from a granite pegmatite at Chywoon granite quarry, near Longdowns, Cornwall; width of large crystal 550 µm; BSE image. (© E. Condliffe, University of Leeds)

the St Austell granite has left behind waste tips and tailings lagoons that have concentrated accessory minerals from the granite and associated pegmatites. In the area W of Nanpean, a topaz-rich unit has been worked in the Goonvean and Rostrowrack china-clay pits, and sampling of the underflow waste from the pits has enabled the following species to be identified: apatite, cassiterite, columbite–tantalite and ilmenorutile, with lesser amounts of arsenopyrite, iron oxides, manganese-rich ilmenite, monazite, pyrite, sphalerite and zircon. Analysis of columbite–tantalite minerals shows that ferrocolumbite occurs most commonly, although ferrotantalite and manganocolumbite are also reported (Scott et al. 1998). Elsewhere in Cornwall, spectacular oscillatory zoned ferrocolumbite has been identified at Chywoon quarry, near Longdowns, where the Carnmenellis granite contains miarolitic cavities from a few cm to 2 m across. One cavity contained ferrocolumbite as euhedral tabular crystals on the prism faces of potassium feldspar. Albite, apatite, bertrandite and quartz are associated species (Moulding & Tindle 2005). Other cavities produced similar ferrocolumbite or manganocolumbite (Moulding & Hooper 2005). Ferrocolumbite is noted in the Land's End granite as an accessory component of a zinnwaldite-bearing variety of granite at Porth Ledden, near Cape Cornwall (Müller et al. 2006).

At Meldon aplite quarry, near Okehampton, Devon, a single ferrocolumbite grain (50 µm across) has been identified in a lithium- and fluorine-rich aplite dyke emplaced into the metamorphic aureole of the Dartmoor granite (Tindle unpub. EMP data). The aplite crystallized from an extremely evolved granitic magma and is host to many rare accessory species. Oscillatory zoned manganocolumbite is the major niobium–tantalum species found (with fersmite and microlite) in pegmatitic veinlets rich in elbaite, fluorite and lithium mica (von Knorring & Condliffe 1984).

In Scotland, ferrocolumbite occurs in accessory amounts (and occasionally as crystals up to 37 mm long) in the Lewisian granitic pegmatite at Chaipaval (Ceapabhal), near Northton (Taobh Tuath), South Harris, Western Isles. It is associated with minor amounts of the following species: allanite, beryl, gahnite, monazite, spessartine, thorite, tourmaline, uraninite and zircon (von Knorring & Dearnley 1960a). Ferrocolumbite is also found as a rare accessory mineral in the evolved Cairngorm Main granite, Grampian (Barritt 1983). Columbite has been identified in panned concentrates from alluvial deposits in the Glen Lui River, draining the Cairngorm Mountains, but the actual species was never determined (Beer & Bennett 1988).

Ferrocolumbite is just one of the rare species found in a borehole sample and a panned concentrate collected from alluvium in the grounds of Aclare House, near Myshall, Co. Carlow, Ireland. The samples also contain common cassiterite, ferrotapiolite, manganocolumbite, manganotantalite, and less common ferrotantalite, monazite and wodginite. Such an assemblage is typical of lithium-enriched granite pegmatites, and the source of the grains is most likely to be the nearby Aclare spodumene pegmatite (R. A. Ixer panned concentrate, Tindle unpub. EMP data).

ferrocordierite

A synonym of sekaninaite.

ferro-edenite

monoclinic AMPHIBOLE GROUP $NaCa_2Fe^{2+}_5(Si_7Al)O_{22}(OH)_2$

A relatively uncommon sodium–calcium–iron aluminosilicate hydroxide that forms a series with edenite and is found in medium-grade amphibolites and gneisses, and in plutonic igneous rocks, where it often occurs as an early or deuteric phase replacing pyroxene. Recognition is based on the amphibole classification of Leake et al. (1997a); see also Mogessie et al. (2004).

Scotland

Grampian Ferro-edenite forms interstitial grains with edenite, ferropargasite and pargasite in Middle Zone late fractionates of the Insch basic layered intrusion, to the E of Insch, around Daviot, Old Rayne, Oyne and Pitcaple (Wadsworth 1986).

Highland Amphiboles with a wide compositional range have been found in miarolitic cavities in one of the Western Redhills Tertiary epigranites at Loch Ainort, Isle of Skye. Crystals are blue to green (in thin section), sometimes euhedral and zoned. The following species are recognized: ferro-actinolite, ferro-edenite, ferrorichterite, ferrowinchite, katophorite and riebeckite. Other cavity species include: hedenbergite, monazite, titanite and zircon. The nearby miarolitic Glamaig epigranite at Eas a' Bradain, near Loch Ainort (between Broadford & Sligachan hotel), contains ferro-edenite and ferrohornblende, together with fluorapatite, hedenbergite, ilmenite and titanomagnetite (S. A. Rust specimens, Tindle unpub. EMP data). Pale-green (in thin section) amphibole, varying in composition from edenite to ferro-edenite, occurs intimately associated with diopside (described as aegirine-augite) in syenite at Glen Dessary, W of Loch Arkaig (Richardson 1968, Fowler 1992).

Strathclyde Green edenite, ferro-edenite and pargasite (originally called hornblende) are found in crossite amphibolites associated with the Girvan–Ballantrae ophiolite complex to the W of Knockormal (Bloxam & Allen 1959, A. Livingstone unpub. EMP data). Black, lustrous, prismatic crystals, up to 7.5 cm in length, originally described as barkevikite and occurring in veins of lugarite cutting picrite within the Lugar Sill, Glenmuir Water (Scott 1914b), are ferro-edenite. Unusual zirconium-rich ferro-edenite occurs within coarse and fine syenites in the differentiated alkali-olivine dolerite intrusion at Cnoc Rhaonastil, Isle of Islay, where arfvedsonite, ferrokaersutite, ferropargasite, hastingsite, kaersutite and magnesiokatophorite are associated amphibole species (Preston et al. 2000b). Ferro-edenite forms delicate crystallites and microphenocrysts in Tertiary pitchstone dykes at Corriegills on the east coast and near Blackwaterfoot (Judd dyke 1) on the west coast of the Isle of Arran. A further Judd dyke (9) contains ferro-edenite associated with almost pure end-member ferrosilite (Preston et al. 1998b, 2002). A single analysis of ferro-edenite was collected from the arfvedsonite-bearing microgranite on Ailsa Craig (BGS specimen, Tindle unpub. EMP data).

Western Isles The Glen Bay granite, gabbro and an associated gabbroic pegmatite on St Kilda (Hirta) are said to contain ferro-edenite, although the data remain unpublished (Harding et al. 1982, 1984).

Ireland

Galway Slightly zoned bluish-green magmatic ferro-edenite occurs in the centre of the Galway granite (Leake 1998).

Mayo Some Dalradian Supergroup blueschist-facies garnetiferous metabasites from central and south Achill, N of Clew Bay, contain ferro-edenite and ferrohornblende. Others also contain glaucophane or ferroglaucophane (Chew et al. 2003).

Offshore territories

Jersey The interior of the Wolf's Cave alkali-olivine dolerite dyke, near Frémont Point, contains poikilitic zoned amphiboles with ferro-edenite cores and ferropargasite rims (Rowbotham & Lees 1990).

ferroglaucophane

monoclinic AMPHIBOLE GROUP $\square Na_2(Fe^{2+}_3Al_2)Si_8O_{22}(OH)_2$

A relatively uncommon sodium–iron aluminosilicate hydroxide that forms a series with glaucophane and is found in high-pressure low-temperature blueschist-facies metamorphic rocks derived from either siliceous sedimentary, alkali basaltic or felsic volcanic predecessors.

Blue amphiboles from the Mona Complex at the Marquis of Anglesey's monument, near Llanfairpwllgwyngyll, Anglesey, Gwynedd, Wales, and from similar material from temporary roadcuttings made during construction of the Llanfairpwllgwyngyll bypass, have compositions described as crossite, ferroglaucophane and glaucophane (Gibbons & Gyopari 1986, Bevins 1994). Actinolite, alumino-ferrowinchite, aluminowinchite, ferrobarroisite, ferrowinchite, katophorite and winchite are also found in these rocks (Gyopari 1984).

Glaucophane schists associated with the Girvan–Ballantrae ophiolite complex and exposed on the crags near the Lendalfoot road, N of Knockormal, Strathclyde, Scotland, show many similarities with the Anglesey blueschists (Bloxam & Allen 1959). Analysis shows that both glaucophane and ferroglaucophane compositions are present (A. Livingstone unpub. EMP data) using the 1997 classification of the amphibole group (Leake et al. 1997a, Mogessie et al. 2004). Rare ferroglaucophane also occurs in Ordovician rocks with ophiolite affinities from the Falahill Formation, near Heriot, Galashiels area, Borders (Kassi & Weir 1993, Livingstone 1993).

At central and south Achill, N of Clew Bay, Co. Mayo, Ireland, blueschist-facies garnetiferous metabasites, whose origin is ophiolite related, belong to the Dalradian Supergroup and contain ferroglaucophane associated with ferro-edenite, ferrohornblende and glaucophane (Chew et al. 2003).

ferrohastingsite

A synonym of hastingsite, amphibole group.

ferrohedenbergite

A synonym of augite, pyroxene group.

ferrohornblende

monoclinic AMPHIBOLE GROUP $\square Ca_2[Fe^{2+}_4(Al,Fe^{3+})](Si_7Al)O_{22}(OH)_2$

A common calcium–iron aluminosilicate hydroxide that forms a series with magnesiohornblende and is found in granites, granodiorites, amphibolites, metabasalts and schists. Wherever possible, authors' original data have been reprocessed using the amphibole classification of Leake et al. (1997a) and Mogessie et al. (2004).

Wales

Gwynedd Actinolite, ferrohornblende and aluminowinchite replace primary olivine and clinopyroxene in the Tal y Fan olivine dolerite intrusion, SW of Conwy (Bevins & Merriman 1988).

Scotland

Dumfries & Galloway Actinolite and ferrohornblende are present in rocks from the propylitic alteration zone of the Black Stockarton Moor subvolcanic complex, a porphyry-copper style deposit near the Criffell granite complex (Leake et al. 1981).

Grampian Ferrohornblende is reported from Dalradian high-grade regionally metamorphosed amphibolites close to the margin of the Lochnagar granite complex, where it occurs with augite and ferropargasite (Goodman & Lappin 1996). Ferrohornblende also occurs in a foliated biotite granodiorite exposed in Keirn Burn, Hill of Three Stones, N of Glenbuchat (Tindle unpub. EMP data).

Highland The miarolitic Glamaig epigranite at Eas a' Bradain, near Loch Ainort (between Broadford and the Sligachan hotel, Isle of Skye) contains ferro-edenite and ferrohornblende, often forming clusters with fluorapatite, hedenbergite, ilmenite and titanomagnetite (S. A. Rust specimen, Tindle unpub. EMP data).

Strathclyde Post-obduction dyke-like doleritic–gabbroic (metadolerite) intrusions associated with the Girvan–Ballantrae ophiolite complex contain actinolite, ferrohornblende and magnesiohornblende that grew during low-grade metamorphism of these rocks (Holub et al. 1984). Parvo-ferrohornblende occurs with ferro-edenite and parvo-ferro-edenite as microcrystallites in the Corriegills pitchstone, N of Lamlash Bay, Isle of Arran (Preston et al. 2002).

Tayside An investigation of amphibole compositions from Dalradian Argyll and Southern Highland Group metabasites around and to the NE of Aberfeldy (centred on the classic area of Barrovian metamorphic zones that developed during the Grampian orogeny) shows that ferrohornblende is restricted to the chlorite zone (Zenk & Schulz 2004).

Ireland

Down A fibrous dark olive-green amphibole, probably from Dunmore Head, 7 km S of Newcastle (first described as kirwanite), was re-analyzed and named ferri-ferro-hornblende by Nawaz & Ryback (1989). Its new name is ferrohornblende.

Galway Green ferrohornblende is reported from Dalradian garnetiferous striped amphibolites at Faul in Connemara (Leake 1972).

Mayo Some Dalradian Supergroup blueschist-facies garnetiferous metabasites from central and south Achill, N of Clew Bay, contain ferro-edenite and ferrohornblende. Others also contain glaucophane or ferroglaucophane (Chew et al. 2003).

Offshore territories

Jersey Ferrohornblende is one of many amphibole species reported from an igneous complex (olivine-gabbro to aplogranite compositions) at Sorel Point in north Jersey (Salmon 1991), but the actual data are unpublished.

ferrohortonolite

A magnesium-rich variety of fayalite, olivine group.

ferrohypersthene

A synonym of ferrosilite, pyroxene group.

ferrokaersutite

AMPHIBOLE GROUP
monoclinic $NaCa_2(Fe^{2+}_4Ti)(Si_6Al_2)O_{23}(OH)$

An uncommon sodium–calcium–iron–titanium aluminosilicate hydroxide that forms a series with kaersutite and is found as dark-brown crystals in a sill of proterobase (altered basalt) belonging to a Devonian spilitic suite at St Minver, Cornwall (Kempe 1968). The mineral was originally named kaersutite and renamed ferrokaersutite according to the amphibole classification of Leake et al. (1997a); see also Mogessie et al. (2004).

Strongly pleochroic reddish-brown ferrokaersutite is found in coarse syenite within the differentiated alkali olivine-dolerite intrusion at Cnoc Rhaonastil, Isle of Islay, Strathclyde, Scotland (Preston et al. 2000b).

ferrokësterite

tetragonal $Cu_2(Fe,Zn)SnS_4$

TL Cligga mine, Perranzabuloe, Cornwall, England

A copper–iron–zinc–tin sulphide, dimorphous with stannite and found in quartz–sulphide veins in granite at the type locality. It is the iron-rich analog of kësterite. Ferrokësterite is steel grey, with a metallic lustre and it occurs with common arsenopyrite, cassiterite, löllingite, sphalerite and wolframite, and less common native bismuth, chalcopyrite and pyrite. The tin-bearing sulphides kësterite, stannite and stannoidite are also present, and distinguishing the first two of these species from ferrokësterite is difficult in hand specimen, if not impossible (Moore 1980, Kissin & Owens 1989, Alderton 1991). At Hingston Down quarry, 2 km W of Gunnislake, Calstock, also in Cornwall, an E–W-trending hydrothermal As–Sn–W vein contains arsenopyrite, cassiterite, chalcopyrite, ferrokësterite and quartz as the major primary minerals. Rare jeanbandyite and natanite are late-forming species in this vein (Betterton et al. 1998).

ferropargasite

AMPHIBOLE GROUP
monoclinic $NaCa_2(Fe^{2+}_4Al)(Si_6Al_2)O_{22}(OH)_2$

A common sodium–calcium–iron aluminosilicate hydroxide that forms a series with pargasite and is found in a range of igneous and metamorphic environments. Where possible, identification is based on the amphibole classification of Leake et al. (1997a); see also Mogessie et al. (2004).

At the Crowns, near Botallack, St Just, Cornwall, skarn assemblages in the aureole of the Land's End granite contain an unusual tin-rich variety of ferropargasite associated with cassiterite, malayaite and tin-bearing epidote (van Marcke de Lummen 1986b). Ferropargasite (originally called ferroan pargasitic hornblende), actinolite and edenite, with chlorite and epidote, are metamorphic replacement products of pyroxene in dolerite dykes cutting (or cut by) the Eskdale granite in the southwest Lake District, Cumbria. Macdonald et al. (1988a) interpret this as a contact-metamorphic effect.

In the Grampian Region of Scotland, ferropargasite is reported from Dalradian high-grade regionally metamorphosed amphibolites close to the margin of the Lochnagar granite complex, where it occurs with augite and ferrohornblende (Goodman & Lappin 1996). Ferropargasite has also been identified in foliated biotite granodiorite from Keirn Burn and from the summit of Hill of Three Stones, both N of Glenbuchat, and in biotite granite and diorite surrounding the lepidolite-bearing pegmatite locality at Peatfold, Glenbuchat (Tindle unpub. EMP data). Other occurrences include: as a corona around garnet in eclogitic amphibolite of the Grampian Moines at Tomatin (Baker 1986); as interstitial grains in the Insch basic layered intrusion (Wadsworth 1986, 1988).

To the SW of Fort Augustus, Highland Region, ferropargasite occurs in quartzose amphibolite masses found as enclaves within a sequence of Moinian argillaceous and arenaceous sedimentary rocks (now a granitic gneiss). The amphibole-rich rocks represent metamorphosed igneous rocks, with ferropargasite occurring as mostly fresh, olive- to bluish-green pleochroic grains up to 5 mm in length, sieved with inclusions of apatite, quartz and titanite (Highton 1994). The mineral was originally described as potassian hastingsite. At Allt Bhlaraidh, Glen Moriston, W of the Great Glen, Highland Region, ferropargasite (originally described as ferroan pargasitic hornblende) occurs with edenite and hastingsite in hornblendic gneisses belonging to the Loch Eil division of the Moine (Rock 1984). A single analysis of ferropargasite from iron-rich silicate–magnetite–quartz rocks (BIF) of Lewisian age from Flowerdale, near Gairloch, Highland Region (Williams 1986, Anthony et al. 1995), now classifies as alumino-ferropargasite.

An unusual zirconium-rich ferropargasite occurs within fine syenite in the differentiated alkali olivine-dolerite intrusion at Cnoc Rhaonastil, Isle of Islay, Strathclyde, where arfvedsonite, ferro-edenite, ferrokaersutite, hastingsite, kaersutite and magnesiokatophorite are associated amphibole group members (Preston et al. 2000b).

In the Western Isles, ferropargasite has been found in an altered basic dyke of Lewisian age at Garry-a-siar (Gearraidh a Siar) on the west coast of Benbecula (Beinn na Faoghla) (Williams 1987).

The interior of the Wolf's Cave alkali olivine-dolerite dyke, near Frémont Point, Jersey, Channel Isles, contains poikilitic zoned amphiboles with ferropargasite rims and ferro-edenite cores (Rowbotham & Lees 1990).

ferropericlase

TL Broadford area, Isle of Skye, Highland Region, Scotland

A synonym of magnesio-wüstite, an iron-rich variety of periclase.

ferropigeonite

An iron-rich variety of pigeonite, pyroxene group.

ferropyrophanite

An iron-rich variety of pyrophanite.

ferrorichterite

AMPHIBOLE GROUP
monoclinic $Na(CaNa)Fe^{2+}_5Si_8O_{22}(OH)_2$

An uncommon sodium–calcium–iron silicate hydroxide that forms a series with richterite and is found in granites with alkaline or peralkaline affinities.

In Wales, amphibole from the Mynydd Mawr microgranite in Gwynedd, which was originally described as hornblende (Harker 1888a) and then riebeckite (Harker 1888b, Bevins 1994), actually contains arfvedsonite, ferrorichterite and riebeckite (NMW specimen, Tindle unpub. EMP data) when the amphibole classification of Leake et al. (1997a) is applied; see also Mogessie et al. (2004).

The Southern Porphyritic epigranite, 80 m N of Fiaclan Dearg on the west face of Marsco, and the Maol na Gainmhich epigranite, Allt Darach, between Maol na Gainmhich and Meall Buide, both in the Western Redhills, Isle of Skye, Highland Region, Scotland, contain anhedral crystals of ferrorichterite. In one sample, crystals reach 1 mm across and occur with aenigmatite, perthitic alkali feldspar, iron–titanium oxides, sodic pyroxene, quartz and accessory allanite. In thin section, ferrorichterite exhibits pleochroism in shades of pale brown, greenish blue and dark blue (Thompson 1976). Ferrorichterite and parvo-ferrorichterite also occur in miarolitic cavities in a Western Redhills epigranite at Loch Ainort. They are blue to green (in thin section) and sometimes euhedral, and they form zoned crystals with ferro-actinolite, ferro-edenite, ferrowinchite, katophorite and riebeckite. Other cavity species include hedenbergite, monazite, titanite and zircon (S. A. Rust specimen, Tindle unpub. EMP data).

ferrosalite

A synonym of hedenbergite, pyroxene group.

ferrosilite

PYROXENE GROUP
orthorhombic $(Fe^{2+},Mg)_2Si_2O_6$

A common iron–magnesium silicate that forms a series with enstatite and is dimorphous with clinoferrosilite. Ferrosilite is found in basic and ultrabasic intrusions, pitchstone dykes and some iron-rich metamorphic rocks. Changes to the nomenclature of the pyroxene group (Morimoto 1988) have resulted in the widening of the definition of ferrosilite. The result of this change was the abandonment of hypersthene as a species name, and minerals previously given this name should now be referred to as enstatite if $Mg/(Mg+Fe) > 0.5$ or ferrosilite if $Mg/(Mg+Fe) < 0.5$.

Ferrosilite appears to have been found only in Scotland. It occurs with augite and ferrohornblende in Dalradian high-grade regionally metamorphosed amphibolites close to the margin of the Lochnagar granite complex, Grampian (Goodman & Lappin 1996). At Glenbuchat, ferrosilite occurs in a biotite-rich norite as brown crystals up to 4 mm long. Ferrosilite is also found in a series of late-stage fractionates (ferrogabbros and ferrodiorites, grading into syenodiorites and quartz syenites) belonging to the Insch gabbroic intrusion (Wadsworth 1986). Ferrosilite also occurs in Dalradian migmatite at Black Hill, near Lochnagar (Ashworth & Chinner 1978). At Druideag Lodge, Loch Duich, Highland Region, it occurs as green crystals accompanied by grunerite and garnet in an iron–magnesium-rich metamorphic rock, termed eulysite (Henry 1935). The ferrohypersthene described by O'Hara (1961) from the Scourie dyke (an altered iron-rich quartz dolerite) at Pol Eorna, Highland Region, is actually ferrosilite. A pyroxene–quartz-bearing ironstone in the first inlet W of the pier on the north shore of Scourie Bay contains ferrosilite associated with augite and pigeonite (Barnicoat & O'Hara 1979).

In Strathclyde, crystallites in one of the Tertiary Judd pitchstone dykes, N of Blackwaterfoot on the west coast of Arran, have an almost end-member ferrosilite composition, thought to be the most iron-rich magmatic orthopyroxene yet described. Related pitchstone dykes in this area and others at Glen Shurig also contain ferrosilite, the latter as phenocrysts associated with augite, fayalite, ilmenite, plagioclase, quartz and titanomagnetite in a light-brown glass (Preston et al. 1998b). Other occurrences are: in xenoliths within a Permian olivine nephelinite dyke at Gribun, Isle of Mull (Upton et al. 1998); in Dalradian greenschists thermally metamorphosed by a dolerite feeder pipe near the summit of Sidhean Sluaigh, 3 km SW of Strachur (Worden et al. 1987); and in a pyroxene–mica diorite at Carn Chois, part of the Garabal Hill–Glen Fyne igneous complex (Nockolds & Mitchell 1946).

ferrostrunzite?

triclinic $Fe^{2+}Fe^{3+}_2(PO_4)_2(OH)_2{\cdot}6H_2O$

A rare hydrated iron phosphate reported from along the Perran Iron lode at Gravel Hill mine, near Perranporth, Perranzabuloe, Cornwall. Massive limonite ore contains drusy cavities up to 8 cm across, on which shiny black manganese oxides, fine green needles of beraunite, and straw to golden-yellow radially arranged crystals of ferrostrunzite have grown. Individual clusters of ferrostrunzite are typically 3–6 mm across, but can reach 1 cm. Rockbridgeite and siderite are also found in the druses (Weiss 1989). This mineral has also been reported as strunzite (Ryback & Tandy 1992), a related species with similar XRD properties. EMP analysis reveals a high iron content (Fe/Mn = 15–20), so the mineral is not strunzite (Tindle unpub. data). However, further confirmation is required, as this type of analysis cannot discriminate between Fe^{2+} and Fe^{3+}, leaving ferristrunzite and ferrostrunzite as possibilities.

ferrotantalite/tantalite-(Fe)

orthorhombic $Fe^{2+}Ta_2O_6$

A rare iron–tantalum oxide that forms two series, with manganotantalite and ferrocolumbite, and is dimorphous with ferrotapiolite. Ferrotantalite is reported by Scott et al. (1998) from the Rostowrack china-clay pit, W of Nanpean, St Austell, Cornwall, where it has been found in a mineral concentrate (grainsize < 100 µm) produced as waste from kaolinite extraction. The source of the waste is a topaz granite unit that is the most chemically evolved part of the St Austell granite complex. Other accessory minerals in the concentrate are: cassiterite, ferrocolumbite and rutile, with minor apatite, zircon and monazite. Amblygonite, fluorite, ilmenorutile and uraninite are other accessory species found in the bulk rock (Manning et al. 1996).

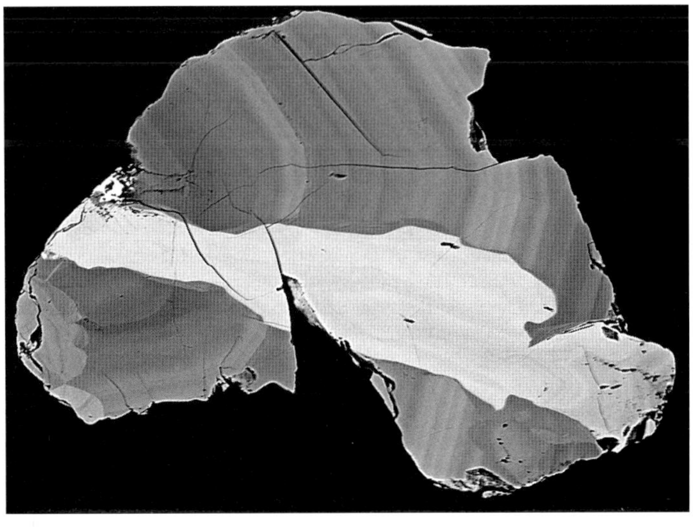

Oscillatory zoned ferrotantalite, partially replaced by a later generation of the same species with a higher tantalum content. From a panned concentrate collected in the grounds of Aclare House, Aclare, Co. Carlow, Ireland; width 300 µm; BSE image. (© A. G. Tindle)

FERROTAPIOLITE GROUP

Tetragonal oxides with the general formula $A^{2+}B^{5+}_2O_6$, where A^{2+} = Fe, Mg, Mn, Zn; and B^{5+} = Nb, Sb, Ta. Ferrotapiolite is reported from only Wales and Ireland.

ferrotapiolite/tapiolite-(Fe) FERROTAPIOLITE GROUP

tetragonal $Fe^{2}Ta_2O_6$

A rare iron–tantalum oxide that forms a series with manganotapiolite. Ferrotapiolite has been found in a panned concentrate of alluvium collected in the grounds of Aclare House, near Myshall, Co. Carlow, Ireland. It forms infrequent angular grains up to 185 µm in length, associated with minor wodginite, in a concentrate that also contains common cassiterite, ferrocolumbite, manganocolumbite and manganotantalite, and less common ferrotantalite and monazite. Such an assemblage is typical of lithium-enriched granite pegmatites, and the source of the grains is most likely to be the Aclare spodumene pegmatite near by. Further grains of ferrotapiolite have been identified in panned concentrates from: Gibbett Hill, E of Aclare; Ow River, Avoca; and Avoca Tributary, close to Avoca mine, Co. Wicklow (R. A. Ixer panned concentrates, Tindle unpub. EMP data).

Tapiolite (probably ferrotapiolite) is briefly reported from the vaguely expressed "near Tremadoc", Gwynedd, Wales (Bevins 1994).

ferrotschermakite AMPHIBOLE GROUP

monoclinic $\square Ca_2(Fe^{2+}_3AlFe^{3+})(Si_6Al_2)O_{22}(OH)_2$

A relatively uncommon calcium–iron aluminosilicate hydroxide that forms a series with tschermakite and is found in Lewisian hornblende (±garnet) gneiss from Sithean Mòr and Beasdale rail bridge in south Morar, Highland Region, Scotland (Lambert 1963), and in garnetiferous striped amphibolites of Dalradian age from Ballynaboleyglassa and Letterbreckaun in Connemara, Co. Galway, Ireland (Leake 1972). Ferrotschermakite is identified using the amphibole classification of Leake et al. (1997a); see also Mogessie et al. (2004).

ferrowinchite AMPHIBOLE GROUP

monoclinic $\square(CaNa)Fe^{2+}_4(Al,Fe^{3+})Si_8O_{22}(OH)_2$

A relatively common calcium–sodium–iron aluminosilicate hydroxide that forms a series with winchite and is found in blueschists belonging to the Mona Complex from the Penmynydd/Gwna boundary exposed in a roadcut near Llanfairpwllgwyngyll, southeast Anglesey, Gwynedd, Wales. Actinolite, alumino-ferrowinchite, aluminowinchite, ferrobarroisite, ferroglaucophane, glaucophane, katophorite and winchite are associated species (Gyopari 1984).

Ferrowinchite is also found with ferri-ferrowinchite in miarolitic cavities in a Western Redhills Tertiary epigranite at Loch Ainort, Isle of Skye, Highland Region, Scotland. Crystals are bluish green (in thin section) and sometimes euhedral, and are zoned towards ferro-actinolite, ferro-edenite, ferrorichterite, katophorite and riebeckite compositions. Other cavity species include hedenbergite, monazite, titanite and zircon (S. A. Rust specimen, Tindle unpub. EMP data).

ferrowollastonite

An iron-rich variety of wollastonite.

fersmite

orthorhombic $(Ca,Ce,Na)(Nb,Ta,Ti)_2(O,OH,F)_6$

A rare calcium–cerium–sodium niobate found in the Meldon aplite quarry, near Okehampton, Devon. This lithium- and fluorine-rich aplite dyke is emplaced into the metamorphic aureole of the Dartmoor granite. The aplite crystallized from an extremely evolved granitic magma and is host to many rare accessory species. Fersmite occurs in pegmatitic veinlets rich in elbaite, fluorite and lithium mica (where it partially replaces microlite) and is also found in the aplite and narrow axinite-bearing veinlets (von Knorring & Condliffe 1984).

feruvite TOURMALINE GROUP

trigonal $Ca(Fe,Mg)_3Al_6(BO_3)_3(Si_6O_{18})(OH,O)_4$

A rare calcium–iron–magnesium–aluminium borosilicate hydroxide found in a hydrothermal breccia within country rocks in the aureole of the St Austell granite at Gaverigan, Cornwall (Manning 1991). At Gryll's Bunny, near Botallack, St Just, Cornwall, metasomatic alteration of amphibolite and actinolitic greenstone has produced coarsely granular (≤5 mm) tourmaline-rich rocks as part of a hydrothermal event induced by the emplacement of the Land's End granite. In thin section, the

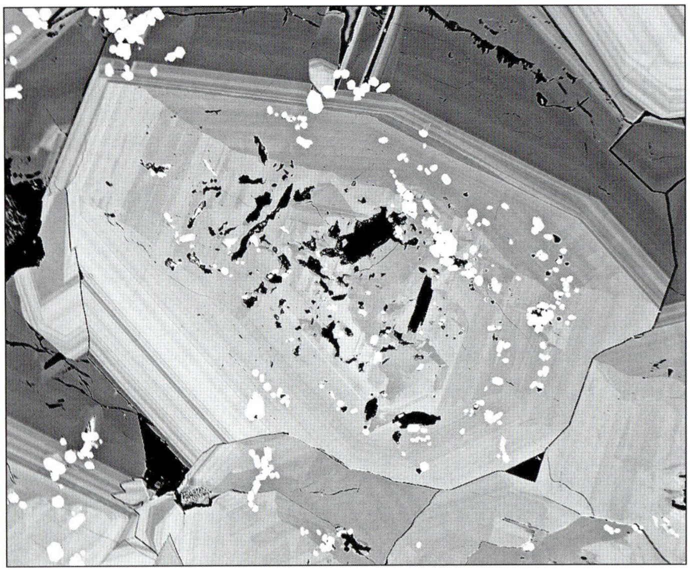

Oscillatory zoned feruvite containing fluorapatite inclusions (white), in a massive tourmalinite skarn from Gryll's Bunny, Botallack, St Just, Cornwall; width 1 mm; BSE image. (© A. G. Tindle)

tourmaline is blue to pale brown (black in hand-specimen), shows strong zonation, and is mainly dravite or schorl in composition. Feruvite and uvite occur less commonly, either as discrete crystals or as rims to dravite–schorl crystals (Tindle unpub. EMP data).

In Scotland, small pale- to medium-brown (in thin section) interstitial feruvite and schorl occur in a single sample of granodiorite belonging to the first magmatic pulse in the NW of the Loch Doon granite complex, near Straiton, Strathclyde. Since other tourmaline-bearing granites from this complex do not exhibit any calcium enrichment, it is thought that the feruvite may have formed during emplacement as a result of local contamination of the granitic magma by marginal metasediments (Tindle 1982 and unpub. EMP data).

fibroferrite

monoclinic $Fe^{3+}(SO_4)(OH)\cdot 5H_2O$

An hydrated iron sulphate hydroxide hydrate found in copious mounds along passageways in Cae Coch mine, near Trefriw, Conwy Valley, Gwynedd, Wales. Fibroferrite is a post-mining secondary mineral associated with a 1 m thick band of pyrite and is accompanied by jarosite, melanterite and traces of copiapite. Interestingly this acidic environment provides a remarkable setting for microbial activity in the form of pyrite-oxidizing bacteria (Johnson et al. 1979, Jenkins & Johnson 1993, Bevins 1994). Fibroferrite is one of 25 post-mining secondary minerals recorded from the copper deposit at Parys Mountain, Anglesey, also in Gwynedd (Jenkins et al. 2000).

fibrolite

A fibrous variety of sillimanite.

fichtelite

monoclinic $C_{19}H_{34}$

A rare hydrocarbon found in pine logs in moss at Handforth, Cheshire (Gaines et al. 1997, Bernard & Hyršl 2004) – the only British or Irish hydrocarbon that has been given species status by the IMA.

fire-opal

A variety of opal showing internal reflections of a fire-red colour.

fischesserite

cubic Ag_3AuSe_2

A very rare silver–gold selenide found in a gold-bearing carbonate vein cutting Middle Devonian limestones at Hope's Nose, Torquay, Devon. The selenide assemblage includes clausthalite, eucairite, fischesserite, klockmannite, naumannite, penroseite, tiemannite, trüstedtite, tyrrellite and umangite, all occurring in a calcite-dominated gangue. Fischesserite is the least common member of this assemblage and it occurs as anhedral grains (up to 200 μm across) associated with naumannite and eucairite. It is replaced by naumannite, eucairite and klockmannite (Stanley et al. 1990a). Near

Irregular mass of fischesserite (8 mm in height) associated with native gold; Hope's Nose, Torquay, Devon. (© D. I. Green)

by, another calcite vein, 4–6 cm wide, contains fischesserite and chrisstanleyite (a rare palladium–silver selenide) associated with cerussite, bromian chlorargyrite, clausthalite, eucairite, palladium-rich and silver-rich gold, tiemannite and umangite (Paar et al. 1998).

flêches d'amour

Dark smoky quartz containing inclusions of acicular rutile. The name has also been used to describe goethite and dark-grey aluminium-rich quartz.

flint

Cryptocrystalline dull greyish-black quartz ±moganite, often as nodules in Chalk.

flokite

A synonym of mordenite, zeolite group.

florencite-(Ce)

trigonal $CeAl_3(PO_4)_2(OH)_6$

A cerium–aluminium phosphate hydroxide found in a granitic pegmatite in the Hensbarrow section of Gunheath china-clay pit, Carthew, St Austell, Cornwall. Florencite-(Ce) was found with fluorapatite crystals in a vug within the pegmatite. Other associated species in the vug were feldspar, fluorite, schorl, torbernite and quartz. Pale-brown, through to dark-brown and near-black, florencite-(Ce) occurs as translucent rhombohedral crystals (≤3 mm long) with schorl inclusions. Late-stage radiogenically driven convective circulation of meteoric water is thought to have been responsible for the main kaolinization of the Cornish granites and it is likely that the late-stage pegmatitic phosphates formed at this time. However, earlier and contemporary stages of mineralization could have liberated the various elements from pre-existing species to allow the formation of florencite (Elton & Grigg 1996). A single zoned pseudo-cubic crystal of florencite-(Ce), 500 μm on edge, has been found in a small cavity within a massive tourmalinite vein from the West Gunheath section of Littlejohns china-clay pit, near St Austell (Tindle unpub. EMP data).

Qualitative analysis of detrital grains from mudstones of Ordovician age from the Corris district of Gwynedd, Wales, suggests the presence of florencite-(Ce) (Bevins 1994).

flos ferri

A stalactitic variety of aragonite.

fluellite

orthorhombic $Al_2(PO_4)F_2(OH)\cdot7H_2O$

TL Stenna Gwyn mine, N of St Stephen-in-Brannel, near St Austell, Cornwall, England

A rare hydrated aluminium fluorphosphate hydroxide found at the type locality, where it occurs in colourless or white minute crystals with pyramidal habit (up to 2.5 mm), associated with cassiterite, fluorite, muscovite, uranite (autunite or torbernite) and wavellite on a grey quartz rock. It occurs in the St Austell granite, but not in adjacent host rocks (Greg & Lettsom 1858, Collins 1878b, Rudler 1905, Weiss 1994). Fluellite described from the nearby Goonvean china-clay pit, SE of Treviscoe, St Austell (Kingsbury 1954, 1959b, 1964) requires further confirmation. Ryback et al. (1998, 2001) have shown that some samples in the Kingsbury collection (at the NHM) have been incorrectly provenanced and the status of fluellite is yet to be resolved.

Well formed glassy crystals of fluellite up to 1.8 mm across, from Stenna Gwyn mine, north of St Stephen-in-Brannel, near St Austell, Cornwall. (© D. I. Green)

fluoborite

hexagonal $Mg_3(BO_3)(F,OH)_3$

A rare magnesium borate fluorhydroxide found near Broadford, Isle of Skye, Highland Region, Scotland, in a contact skarn between the Beinn an Dubhaich granite and dolomitic limestone. Fluoborite is one of many skarn minerals that crystallized as granite-derived boron- and fluorine-rich fluids metasomatized Durness limestone near to the granite contact. Other boron-bearing species in the skarns are datolite, harkerite, ludwigite and szaibelyite. Fluoborite is best developed in ludwigite-bearing skarns, but is also found in accessory amounts in magnetite–chondrodite skarns, where it is intimately associated with chondrodite in the form of hexagonal prisms (Tilley 1948b, 1951).

fluor

A synonym of fluorite.

fluorapatite/apatite-(CaF)

APATITE GROUP

hexagonal $Ca_5(PO_4)_3F$

An extremely common calcium phosphate fluoride found as an accessory component in many igneous rocks and frequently in marbles, skarns, alpine-type veins and hydrothermal tin veins. In Britain and Ireland it is most often referred to as apatite, but, as these occurrences are missing the essential fluorine analysis required to confirm the species, they are not included here.

Fluorapatite is described from aplite–pegmatite sheets in the roof complex of the Tregonning–Godolphin granite at Megiliggar Rocks, near Porthleven, Cornwall. These granitic rocks are unusual in being enriched in elements such as cesium, fluorine, lithium and rubidium, and they contain albite, lithium micas (lepidolite and zinnwaldite) and topaz. Manganese-rich fluorapatite is just one of many phosphate minerals found at the locality, others being alluaudite, amblygonite, graftonite, lithiophilite, ludlamite, natromontebrasite, triphylite, triplite and vivianite (George et al. 1981, Stone & George 1983, Stone 1992).

Within the Carnmenellis granite at Trolvis quarry, near Longdowns, Cornwall, tabular fluorapatite occurs in pegmatite lenses within the granite. The fluorapatite is colour-zoned from various shades of greyish green to bluish green, separated by colourless bands. Bertrandite, Carlsbad-twinned feldspar, fluorite, muscovite, quartz, topaz and tourmaline are associated species (Hosking 1954a). Elsewhere in the granite, similar assemblages occur in the nearby Chywoon, Kessel Downs (Halvasso) and Maen quarries, where exceptionally fine fluorapatite has been found. At Chywoon, equant strongly coloured bluish-green crystals (up to 20×12 mm), intergrown equant-to-tabular lustrous indigo crystals (up to 7×12 mm), colourless-to-green strongly zoned prisms (to 5 mm), equant pale-blue to colourless crystals (to 2.3 mm), and violet prismatic crystals to 3 mm, are all reported (Moulding & Hooper 2005, Tindle unpub. EMP data).

A wide range of phosphate species have been found at Gunheath china-clay pit, Carthew, St Austell, within pegmatites, stockwork mineralization, greisen zones and quartz veins. Fluorapatite, arthurite, barbosalite, cacoxenite, chalcosiderite, chlorapatite, crandallite, cyrilovite, dufrénite, florencite-(Ce), frondelite, goyazite, hentschelite, hydroxylapatite, leucophosphite, libethenite, morinite, torbernite, variscite, wavellite and waylandite have all been found. Some of the fluorapatite forms beautiful hexagonal prisms of lilac-to-violet crystals, up to 5 mm in length, showing partial alteration to hydroxylapatite (Hawkes et al. 1987, Weiss 1994, Elton & Grigg 1996, Elton et al. 1996, Elton 1998).

Elsewhere in Cornwall, fluorapatite occurs in calc-silicate bodies at Wheal Cock, St Just (Alderton & Jackson 1978), and at Penberthy Croft mine, St Hilary, it forms tabular-to-short prismatic crystals, up to 0.3 mm across and ranging in colour from colourless-to-white, to yellow and orange (Betterton et al. 2000). At Botallack mine, St Just, fluorapatite forms small prismatic semi-transparent crystals ranging from white through creamy to yellowish green (Bancroft & Weller 1993). Although fluorapatite is noted from Fowey Consols mine, Tywardreath, St Blazey (Lewis 1997), it is said to be the variety francolite, so is actually carbonate-fluorapatite. Fluorapatite is also a common accessory species in the Land's End granite, where it is associated with monazite-(Ce), monazite-(Nd) and zircon (Müller et al. 2006).

Very fine celandon-green crystals (bluish-green with grey), nearly transparent and sometimes measuring 25 mm across, from Brandy Gill (probably the tungsten-rich veins around Carrock mine), Caldbeck Fells (Greg & Lettsom 1858, Cooper & Stanley 1990), are most likely to be fluorapatite. An analysis of apatite from Carrock mine shows it to be fluorapatite (Finlayson 1910). At Shap granite quarry, Cumbria, colourless, hexagonal, prismatic fluorapatite commonly occurs in miarolitic cavities. It is visually indistinguishable from the rare mineral milarite, which is also present (Wirth & Green 2002).

Transparent crystal of fluorapatite, 5 mm across, from the granitic pegmatite at Megiliggar Rocks, near Breage, Cornwall. (© D. I. Green)

Fluorapatite, 3 mm across, from Chywoon granite quarry, near Longdowns, Cornwall. (© D. I. Green)

Fine cream translucent crystals of apatite, up to 50 mm in length, and associated with tourmaline, from Bovey Tracey, near Chudleigh (Greg & Lettsom 1858), are almost certainly fluorapatite, but require confirmation.

Transparent colourless crystals of fluorapatite are associated with albite, brookite and quartz in alpine-type fissure veins associated with dolerite at Fron-oleu, Prenteg, Tremadog, Gwynedd, Wales. The crystals (0.5–1.5 mm in length) appear spherical because of their unusually complex morphology (Starkey & Robinson 1992). An extremely pure fluorapatite is reported from stilpnomelane-bearing calcite veins cutting Ordovician sedimentary iron ore at Tyllau Mwn mine, Drws-y-nant, near Dolgellau, Gwynedd. The fluorapatite varies from 1 mm to 12 mm in diameter and occurs as small flat crystals, floating free in calcite or attached to the vein walls (Matthews & Scoon 1964). Bevins (1994) notes fluorapatite in altered gabbros of the Pen Caer area, near Fishguard, Dyfed, Wales.

In Scotland, fluorapatite is noted from five very different locations. It forms rare squat prisms and radiating sheaves of hollow prismatic needles in carbonatites and silicate xenoliths discovered near the shore of Loch Urigill, approximately 400 m outside the contact of the Loch Borralan alkaline complex, Assynt, Highland Region (Young et al. 1994b). The Tertiary Shiant Isles Main Sill, Western Isles, is a differentiated basaltic intrusion that contains accessory amounts of fluorapatite in all rock types, including crinanite, olivine teschenite, picrite, picrodolerite, and related pegmatites (Gibb & Henderson 1996). At Cnoc Rhaonastil, Isle of Islay, Strathclyde, a Tertiary intrusion comprising dolerite, teschenite and minor nepheline syenite, contains fluorapatite that approaches the end-member composition but which can be enriched in REEs and exhibit complex zonation. In the syenite, zirconium-rich aegirine, aenigmatite, zirconium-rich arfvedsonite, calcium catapleiite, monazite and zircon are associated species (Preston et al. 2000a). At Creag an Fheidh, Isle of Arran, Strathclyde, a skarn assemblage occurs at the junction between vent pyroclastics and a xenolithic Cretaceous block that subsided into the collapsed caldera of the Central ring complex. Fluorapatite is a late-stage mineral that formed in fractures within the chalk during the passage of skarn-forming fluorine-rich metasomatic fluids (Cressey 1987). Livingstone (1994) reports an unusual white chalky carbonate–lead-rich fluorapatite associated with vanadinite from Belton Grain vein, Wanlockhead, Dumfries & Galloway. A large tourmaline-rich pegmatite at East Head, 1 km NE of Portsoy, Grampian, contains minor pale-blue manganese-rich fluorapatite associated with cassiterite, manganocolumbite and spessartine (Tindle unpub. EMP data).

Manganese-rich fluorapatite is an accessory constituent in spodumene pegmatites and greisen associated with the Leinster granite batholith in Co. Carlow and Co. Wicklow, particularly at: Aclare, near Myshall; Moylisha, near Shillelagh; Rathnagerragh Castle, SW of Myshall; Seskinnamadra, W of Tomduff Hill; and Stranakelly, 3 km N of Shillelagh. Other associated species include albite, almandine–spessartine garnet, beryl, bertrandite, cassiterite, columbite–tantalite, fluorite, lepidolite, lithiophilite, microlite, phosphosiderite, pyrite, quartz, tourmaline and zircon, although not all are present at every locality (Steiger & von Knorring 1975, Scoon 1978). Fluorapatite, autunite and torbernite are also noted by O'Connor (1986) from uranium-rich quartz breccias, veins and greisen in the Tullow pluton (a part of the Leinster batholith).

fluorapophyllite/apophyllite-(KF) — APOPHYLLITE GROUP

tetragonal or orthorhombic — $KCa_4Si_8O_{20}(F,OH)\cdot 8H_2O$

A rarely described hydrated (fluoride-dominant) potassium–calcium silicate fluorhydroxide that forms a series with hydroxyapophyllite and whose name came into existence when the apophyllite group was established (Dunn & Wilson 1978). Before this, apophyllite was the species name for all members of the group.

Well crystallized, colourless or slightly turbid fluorapophyllite encrusts analcime within calcite–chlorite–pectolite–quartz veins in the late Carboniferous quartz dolerite of the Whin Sill at Cambokeels (Cammock Eals) fluorite mine, Eastgate, Weardale, Co. Durham (Young et al. 1991).

At Moonen Bay, Isle of Skye, Highland Region, Scotland, altered amygdaloidal basalt contains small (up to 300 µm across) interstitial crystals of fluorapophyllite, hydroxyapophyllite (S. A. Rust and D. I. Green specimens, Tindle unpub. EMP data) and fibrous mesolite in cavities within amygdales that are mostly filled with massive fibrous gyrolite. Sgùrr nam Boc, between Loch Brittle and Loch Eynort, Isle of Skye, and Orrock quarry, 3 km NW of Burntisland, Fife, are other basalt-hosted fluorapophyllite localities (D. I. Green specimens, Tindle unpub. EMP data).

Fluorapophyllite is also confirmed from basaltic rocks in Co. Antrim, Ireland (Tindle unpub. EMP data): Craigs quarry, between Ballymena and Larne, Co. Antrim (D. I. Green specimen); Masllusk quarry, Newtownabbey (S. Moreton specimen); and Maxwell's quarry, Scotstown Road area, 8 km from Ballymena, off the main road to Larne (H. J. Foy specimen).

fluor-elbaite* — TOURMALINE GROUP

trigonal — $Na(Al_{1.5}Li_{1.5})Al_6(BO_3)_3Si_6O_{18}(OH)_3F$

A rare sodium–aluminium–lithium borosilicate hydroxide fluoride that forms a series with elbaite and whose composition was defined in the tourmaline classification of Hawthorne & Henry (1999). The classification is widely used, but fluor-elbaite has not been formally described and therefore the name does not have the approval of the IMA. It is considered to be a named tourmaline in much the same way as certain members of the amphibole group are (Burke & Leake 2005).

In Cornwall, fluor-elbaite is reported from a topaz granite forming part of the lithium-enriched Nanpean stock at Kernick quarry, St Austell (Manning 1991), and two locations associated with the Tregonning–Godolphin granite; in a lepidolite-bearing leucogranite at Megiliggar Rocks (pale-yellow unzoned grains) and a zinnwaldite-bearing aplite at Trequean Cliff (pale unzoned skeletal grains), both near Porthleven (Stone 1992).

Fluor-elbaite is much more common than the literature suggests, although still rare in comparison to other tourmaline species such as schorl and dravite. It is confirmed from the following locations (Tindle unpub. EMP data).

England

Cornwall Carclase china-clay pit, St Austell, where bluish- to greenish-grey radiating prisms of tourmaline consist of elbaite, fluor-elbaite and fluor-schorl within individual crystals (E. Loye specimen); and Goonbarrow china-clay pit, St Austell, where a quartz–tourmaline vein cutting kaolinized granite contains fluor-schorl, schorl, foitite, elbaite and fluor-elbaite, often with complex zonation patterns.

Devon Meldon, near Okehampton, where a lithium-enriched aplite in the aureole of the Dartmoor granite contains green and pink varieties of tourmaline, both with elbaite and fluor-elbaite compositions (D. I. Green and R. Tayler specimens); and Lundy island, off the coast of Devon, where a black elbaite and fluor-elbaite grain was found enclosed in fluorite within pegmatitic granite (D. I. Green specimen).

Yellowish-brown fluor-elbaite overgrowing orange-brown schorl in granite from the Tregonning–Godolphin granite complex, north west of Trewavas Head, near Porthleven, Cornwall; width 1.2 mm; viewed in PPL. (© A. G. Tindle)

Scotland

Grampian East Head, at the base of Cowhythe Hill, near Portsoy, in a lepidolite-bearing pegmatitic granite containing a single isolated tourmaline grain, 4 cm in cross section, colour zoned from a very dark-green core to a pale-green intermediate zone and thin pink rim, all colours having elbaite and fluor-elbaite compositions. Zoned fluor-elbaite–schorl crystals and foitite–schorl crystals also occur in tourmaline-rich pegmatite from this locality (J. R. Holbrook specimens); and at an isolated location approximately 1 km N of Peatfold farm, Glenbuchat, where fluor-elbaite occurs with schorl, fluor-schorl and elbaite in a zoned granitic pegmatite intruded into Dalradian mica schists and phyllites – blue gem-quality tourmaline from Glenbuchat is also fluor-elbaite (I. Dossett & S. Moreton specimens).

fluorite

cubic CaF_2

A very common calcium fluoride found in low- to high-temperature hydrothermal veins and stratabound deposits, also as an accessory mineral in granite and granite pegmatite. Fluorite frequently occurs in well developed cubes, and this, combined with its wide distribution and the unusually diverse range of colours it can develop, makes it a particularly popular species with collectors. Possible reasons why fluorite has variable colour (and fluorescence), such as trapped organic matter, lattice distortions and variable REE contents, are discussed by many authors (Dunham 1937, Bill et al. 1967, Dickson 1980, Howie et al. 1982, Ford et al. 1993, Bau et al. 2003, Ixer 2003). Fluorite is far too common to list all known localities, but extensive summaries are provided by Dunham (1952) and Notholt & Highley (1971).

England

Avon Fluorite has been found in a single barite–calcite vein cutting Carboniferous Limestone at Hampstead Farm quarry, Chipping Sodbury. Individual crystals reach 3 mm across and are colourless or deep golden yellow, pale yellow, deep purple or pale mauve. Cubes, bevelled-edge cubes and a rare tetrahexahedral form are all found, and some show interpenetrant twinning or colour zoning (Brooke 1994b).

Cleveland A trace of fluorite is noted from the main anhydrite seam at Billingham mine (under the ICI factory site) (Stewart 1947, Dunham 1960). Pale-grey fluorite and barite are abundant in borehole core passing through Magnesian Limestone at Middleton Road, West Hartlepool (Fowler 1956).

Cornwall Fluorite occurs as an accessory mineral in granite and as a gangue mineral in hydrothermal veins associated with the Cornubian batholith. Fluid inclusions in fluorite and quartz from many locations have been used to confirm different origins for fluids responsible for tin–tungsten greisen and lode mineralization from that of later cross-course lead–zinc mineralization (Alderton 1978, Alderton & Harmon 1991, Gleeson et al. 2000).

Occurrences associated with the Land's End granite include: fluorite associated with Bi–Cu–Sn–W mineralization at Balleswidden mine, near St Just, and with As–Cu–Fe–Zn mineralization at Wheal Unity, St Day, Gwennap (Hosking 1951); infrequent in mines of the Botallack sett (Wheal Cock, Wheal Owles and Wheal Edward), St Just (Bancroft & Weller 1993); deep-purple cubo-octahedral roughly crystallized fluorite to 50 mm on edge with bright-green cores at Geevor mine, near Pendeen, St Just (Harker 1973); pale-green to almost colourless fluorite in cubes with dodecahedral faces to 1 cm, associated with phenakite at St Day United mines, Gwennap (Russell 1911a); and fluorite in the pipe-like Bi–Cu–Sn–U replacement deposit (carbona) at St Ives Consols, St Ives (Hosking 1951, Robson 1945b, Dominy et al. 1996a), including spectacular sea-green to clear zoned crystals to 22 cm with sulphide inclusions (Jones 2006).

A deep-purple variety of fluorite is recorded from the Tregonning–Godolphin granite at Tolmennor (Hall 1930), and accessory amounts of fluorite occur in granite and associated topaz-bearing tin–tungsten veins from the nearby cupola of St Michael's Mount (Davison 1920b, Hosking 1951, 1954c).

Moving eastwards, the Camborne–Illogan–Redruth mining district hosts fluorite in granite, granite pegmatite and hydrothermal veins. Pegmatite occurrences include three near Longdowns in the Carnmenellis granite: deep-purple cubes and cubo-octahedral crystals up to 18 mm on edge and colour-zoned (green/purple/white) octahedra attached to large orthoclase crystals at Chywoon quarry, near Longdowns (Moulding & Hooper 2005); modified cubes of pale-purple fluorite at Kessel Downs (Halvasso) quarry (Couper & Barstow 1977); beautiful purple cubes, 12 mm across, with bevelled edges and modified corners, also smaller colour-zoned (green–purple) and entirely purple crystals at Trolvis quarry (Hosking 1954a, Harker 1973). Within the Carn Marth granite, pegmatite contains cubes of purple fluorite at Harvey's quarry (Davison 1921), and both purple and green fluorite have been found in veinstone on the dumps of St Aubyn United mine, near Redruth (Dunham 1952). At South Crofty mine, Camborne, fluorite is present in cassiterite-bearing quartz–chlorite veins and in later cross-course quartz–(hematite–pyrite) veins (mainly within the Carn Brea granite). Fluorite is usually massive, but it occasionally forms light-green cubes, sometimes associated with bismuthinite (Dewey 1952, Polkinghorne 1954, Harker 1973, Scrivener et al. 1986, Wolters & Wolters 1987, Dominy et al. 1994, Dominy & Camm 1996, LeBoutillier 1996). Dewey (1925) reports copper ore from South Crofty containing appreciable fluorite and pyrite. Very good crystals of fluorite to 1 cm on edge and larger apple-green crystals to 5 cm on edge have been found at Wheal Jane, Kea, near Truro. Bevel-edged crystals are common and one variety has 30 crystal faces. Colour varies from colourless-to-white, pale-green to blue, and less commonly pink. Fluorite is most common in the late cassiterite–chlorite–sulphide assemblage found at the mine (Rust 1982d, Holl & Bromley 1988). Other reports include: accessory green fluorite in granite and aplite veins and brown fluorite in greisen within the Carn Brea intrusion (Robson 1945a), also pale-green cubes to 6 cm from Carn Brea mine (Embrey & Symes 1987); fluorite (var. chlorophane) at East Pool mine, Illogan, and Pednandrea mine, Redruth (Robson

Deep-purple twinned fluorite cube, 0.7 mm on edge, with pale-blue transparent fluorapatite prisms; Chywoon granite quarry, near Longdowns, Cornwall. (© D. I. Green)

1949b), also 6.2 cm-wide twinned zoned bluish-green cubes from East Pool mine (Jones 2006); light to pale-green fluorite in an arsenopyrite–chalcopyrite-bearing chlorite–quartz lode at New Cook's Kitchen mine (North Tincroft lode), Redruth (LeBoutillier et al. 2000); octahedral crystals at North Downs, Scorrier, and Wheal Sparnon, near Redruth (Garby 1848b); pale-green to whitish and purple opaque masses and veins, rare yellow – good crystals, some with bevelled edges to 2.5 cm on edge (Rust 1981b) and pale-pink crystals to 5.5 cm on edge (Jones 2006) at Wheal Gorland, St Day, 2 km E of Redruth; pale-green fluorite at Wheal Seton, near Tuckingmills (Dewey 1952).

Later-stage magmatic and hydrothermal processes operating on the St Austell granite complex are responsible for many fluorite occurrences in and around this major region of china-clay production. Alteration effects in the granite include fluorite haloes around potassium-feldspar megacrysts, and fluorite along cleavages and in the cores of lithium micas. Hydrothermally altered topaz granite contains an abundance of fluorite (Richardson 1923, Bristow & Exley 1994, Manning et al. 1996). A large fissure zone in such fluorite granite at Great Wheal Prosper china-clay pit produced transparent colour-zoned (water blue passing into light green, grass green and dark violet) cubes to 4.2 cm (Weiss 1994). Descriptions from other china-clay pits include: purple-brown cubes in arsenopyrite–cassiterite–wolframite veins at Goonbarrow, near Bugle (Bray & Spooner 1983); dark-violet cubes to 2 cm in quartz veins at Gunheath, Carthew (Weiss 1994); purple cubes associated with turquoise and wavellite in kaolinized granite at Hensbarrow (Braithwaite 1981); and transparent, colourless-to-purple cubes to 0.8 mm at Wheal Remfry, near Indian Queens (Green et al. 2005d). Elsewhere in the St Austell area, fluorite occurs: with bismuthinite at Great Dowgas mine, near Sticker (Hosking 1951); at Stenna Gwyn mine, St Stephen-in-Brannel (Weiss 1994); as fine specimens including deep-purple crystals to 1.5 cm in pegmatite vugs at Colcerrow quarry, Luxulyan, near Lostwithiel (Russell 1910b, Embrey & Symes 1987, Weiss 1994); considerable amounts of bluish fluorite in calc-silicate rocks within the aureole of the St Austell granite near Tremore, SW of Bodmin (Barrow & Thomas 1908, Kingsbury 1961); and large semi-transparent octahedra and separately associated with topaz at Wheal Maudlin, Lanlivery (Greg & Lettsom 1858, Russell 1924, Kingsbury 1961).

The next major orefield extends from the eastern margin of the Bodmin Moor granite, past the granite cupolas of Kit Hill and Hingston Down towards the western margin of the Dartmoor granite in Devon. To the south it includes the important mining districts of Liskeard and Menheniot. Dines (1956) describes colourless, green and pink fluorite from Wheal Ludcott, St Ive, near Liskeard, and purple fluorite from neighbouring Wheal Wrey. A later investigation of fluorite from these mines confirmed colour-zoned crystals to 2.5 cm on edge (colourless to pale green), together with yellow, purple and rare pale-blue crystals (Rust 1983a). Other occurrences of fluorite include: in a pegmatite vein carrying arsenopyrite, cassiterite and chalcopyrite at Cannaframe mine, Bodmin Moor (Brooks 2001); small pale-violet octahedra and larger crystals to 9 cm associated with bertrandite and phenakite in joint planes within granite at Cheesewring quarry, SW of Henwood, Linkinhorne (Bowman 1911, Russell 1911a); beautiful transparent pale-yellow cubes to 2 cm attached to terminated quartz prisms on a specimen from West Caradon mine, St Cleer, Liskeard (Bancroft & Weller 1993), also fine cubes, octahedra and colourless elongated cubes at South Caradon mine (Greg & Lettsom 1858); deep-violet crystals associated with bertrandite, orthoclase and smoky quartz in cavities within granite at Gold Diggings quarry, near Linkinghorne (Russell 1913b, Hosking 1954b); fluorite in the wallrock of tin–tungsten greisen-bordered veins in the Redmoor–Kit Hill sheeted vein deposit (Newall & Newall 1989, Newall 1994); pale-violet or greenish, etched crystals associated with anglesite, carminite, pharmacosiderite and scorodite in a mineralized lode within granite at Hingston Down Consols mine, 2 km W of Gunnislake, Calstock (Russell 1910a), also fine pale-green fluorite in greisenized granite at Hingston Down quarry (Ward 1983b, Wolfe 2000); beautiful transparent colour-zoned (colourless to azure blue) cubes at Holmbush mine, Callington (Solly 1886); abundant pale-blue and purple fluorite in pegmatitic veins within granite at New Consols mine, Stoke Climsland (Embrey 1953); yellow fluorite at Wheal Trehane, near Liskeard (Howie et al. 1982); abundant fluorite including octahedral crystals in Cu–Pb–Zn lodes and extraordinary 30 cm deep-blue cubes at Wheal Mary Ann, Menheniot, also honey yellow, purple and green varieties (Greg & Lettsom 1858, Le Neve Foster 1878b, Rudler 1905, Dewey 1925, Hosking 1951, Dunham 1952, Harker 1973, Merry & Braithwaite 1995, Jones 2006); deep-blue bevelled cubes to 10 cm at Wheal Trelawney, near Liskeard (Greg & Lettsom 1858, Dines 1956). A fluid-inclusion and REE investigation of fluorite from polymetallic veins in the Kit Hill–Gunnislake area (arsenic, copper and tin) and the Tamar valley (silver, lead and zinc) shows a hydrothermal continuum from higher (post-magmatic) to lower (basinal brine) temperatures of formation (Shepherd & Scrivener 1987).

Elsewhere, fluorite is also noted at: Cligga Head, Perranzabuloe (Hall 1971); Wheal Golden, Perranzabuloe (Rust 1982a, Merry & Braithwaite 1995); East Wheal Lovell, Wendron (Dunham et al. 1978); Wheal Roots and South Wendron mine, Wendron (Dominy et al. 1996a); Polberro (Pell) mine, St Agnes – dark-purple to brownish cubes, a few cm across (Harker 1973), and, elsewhere in the Polberro sett, green fluorite at Bungay (Dunham 1952); St Agnes (Green & Betz 2006); Treburland mine, 2 km SE of Fivelanes, Altarnun, Bodmin Moor (Russell 1946d); Trevaunance mine and Wheal Devonshire, St Agnes – large translucent purple crystals and beautiful octahedrons made up of minute cubes (Greg & Lettsom 1858, Collins 1892, Embrey & Symes 1987, Jones 2006); Tyland quarry, NE of Camelford (Barrow & Thomas 1908); Wheal Kine, St Agnes – pale-green stepped octahedral crystals to 3 cm and lilac to purple tetrahexahedral crystals (Jones 2006); and Wolf Rock, SW of Land's End (Harrison et al. 1977).

Cumbria *Caldbeck Fells:* Rare fluorite occurs at Carrock mine, Grainsgill, where it has been found in wolframite-bearing quartz veins as small mauve and purple crystals (Davidson & Thomson 1951, Ball et al. 1985a), sharp colour-zoned (green to purple) octahedral crystals to 11 mm (Watson 1988, Cooper & Stanley 1990) and as pale bluish-green cubes in vein breccia (Stanley 1979). Fluorite also occurs as green crystals on joint faces in the surrounding greisen (Rastall 1942). *West Cumbria:* Fluorite is a widespread and significant, but minor, gangue component of the hematite replacement ore at Beckermet mine, near Egremont, where it forms dark-red cubes to 0.2 mm, mainly within compact hematite. It forms colourless or pale-green crystals in vugs associated with botryoidal and specular hematite. Fluorite also replaces calcite and is itself replaced by quartz (Goldring & Greenwood 1977, 1990). Davidson & Thomson (1951) note a little blue fluorite from this mine. At the neighbouring Florence mine, fluorite occurs as unusually large (40 mm across) blue cubes on dolomite (Green & Briscoe 2002) and as colourless, purple, pale-blue and sky-blue crystals (Davidson & Thomson 1951, Smith 1973b, Howie et al. 1982, Dunham 1984, Jones 2006). Other occurrences include: a fluorite vein at David Lawn mine, Bigrigg, and well developed yellow, green, pale- and deep-purple or colourless cubes in the Florence–Winscales orebody (Trotter 1945); colourless cubes on limestone from Kelton quarry, Kirkland, and, without detail, from Ulverston area and Wyndham pit, Bigrigg (all Young 1987a); fluorite replacing calcite cement and as thin purple films in joints within Carboniferous sandstone at Langhorn quarry, Bigrigg (Lewis 1931), also at this locality pale-yellow cubes with barite, calcite and dolomite (Young 1987a); purple fluorite at Sandwith anhydrite mine, Whitehaven (Harrison 1975); and minor fluorite and barite mineralization in boreholes through Carboniferous Limestone and Brockram rocks in the Sellafield area (Milodowski et al. 1998). *Elsewhere in the Lake District:* Ribs of amber fluorite up to 30 cm wide at Brandlehow mine, Derwent Water (Dunham 1952, Strens 1962, Firman 1978), and pale-yellow fluorite with chalcopyrite, galena and sphalerite on the dumps of Old Brandlehow mine (Davidson & Thomson 1951); rare pale-yellow crystals to 1 cm at Force Crag Ba–Pb–Zn mine, Braithwaite, near Keswick (Davidson & Thomson 1951, Young 1987a, Green et al. 1997, Green & Briscoe 2002); small pale-yellow cubes of fluorite at Hartsop Hall mine, Hartsop, Patterdale (Russell 1944,

Transparent, sky-blue fluorite crystals, up to 2 cm on edge, with a cluster of pale-pink dolomite crystals on one side. There is a thin coating of red hematite on the back, which shows through the fluorite. Florence mine, Egremont, Cumbria. (© J. Fisher)

Fluorite crystals with curved faces and up to 2 cm on edge. Hilton mine, Scordale, near Appleby, Cumbria. (© J. Fisher)

Davidson & Thomson 1951, Shaw 1970, Green 1986); colourless, emerald-green, violet and zoned green-to-purple cubo-octahedral crystals in cavities and joints at the Shap granite quarry (Harker & Marr 1891, Russell 1925, Grantham 1928, Davidson & Thomson 1951, Firman 1978, Wirth et al. 2001, Wirth & Green 2002); small colourless and yellow cubes with barite in the Shap Blue quarry; white, pale-green and pale-purple masses at Whitecombe Beck, Black Combe (all Young 1987a); deep-purple fluorite on joint surfaces in quartz–topaz greisen associated with the Eskdale granite at Water Crag, near Devoke Water (Young 1985d). *Northern Pennine orefield:* Cavities in stratiform metasomatic replacement lead–zinc deposits (flats) within Carboniferous Limestone have yielded some of the finest specimens of calcite and fluorite, both here and in Co. Durham (Rudler 1905). As much as 2.1 million tonnes of fluorite was mined from the orefield (Dunham 1990). Fluorite is distributed so widely as a gangue mineral that it is used as a zone species to outline the central area of the orefield, which has been shown to have a higher formation temperature on the basis of quartz and fluorite fluid-inclusion thermometry (Sawkins 1966, Vaughan & Ixer 1980, Ixer 1986, Dunham 1990, Green et al. 2000, Cann & Banks 2001). Superb specimens have been found at Hilton mine, Scordale, near Appleby, in every shade of amber-yellow and as colourless cubes. Barite is a common associate (Davidson & Thomson 1951, Guppy 1956, Bridges 1982, Winrow 1986a, Bouch et al. 2006, Jones 2006). Yellow (amber) twinned crystals to 2.8 cm are particularly attractive (Green & Briscoe 2002, Fisher 2004). Translucent amber fluorite associated with ankerite has been found in cavities several metres deep in the Jug vein at Brownley Hill mine, Nenthead, near Alston (Green et al. 2000, Green & Briscoe 2002), and fine crystals (colourless, yellow, green, almost every shade of blue, some colour banded or etched) have also been found at Rotherhope mine, Leadgate, near Alston (Sweet 1930, Davidson & Thomson 1951, Guppy 1956, Jones 2006). An exceptional, mainly mauve, penetration-twinned fluorite crystal, 5.5 cm on edge, from Rotherhope mine is illustrated in Fisher (2004). Fluorite is also present at the following locations: Clouds Gill mine (Bateson & Johnson 1984); Dowgang, 2 km south of Nenthead (Cann & Banks 2001); Dufton mine, near Appleby, and Silverband mine, Milburn, near Appleby (Davidson & Thomson 1951); Fell End vein, Clouds, 10 km SSW of Kirkby Stephen (Smith 1973b); Great Sulphur vein, Alston Moor (Ixer 2003); Hagg's mine, Nenthead (Howie et al. 1982); Hartley Birkett, Kirkby Stephen (Davidson & Thomson 1951, Rogers 1978); High Stennerskeugh level, Ravenstonedale (Starkey 1987, Young 1987b); Higher Longrigg mine, S of Cote Garth, Hartley Birkett, near Kirkby Stephen (Braithwaite 1988); Murton mine, near Appleby (Bunting 1994); Rampgill, Nenthead (Wolfe 2004); Smallcleugh mine, Nenthead, near Alston (Green & Briscoe 2002); and Tynebottom mine, NW of Garrigill, near Alston (Braithwaite 1982b, Ixer 2003, Bouch et al. 2006).

Derbyshire Fluorite (locally called Derbyshire spar) is a common gangue mineral in fracture-infill veins (rakes and scrins), dissolution cavities and zones of metasomatic replacement (flats and pipes) within Carboniferous Limestone, where it is associated with barite, calcite, galena (the main ore species) and sphalerite. It has been mined since as early as the 16th century. Fluorite tends to occur as well formed cubic crystals ranging from a few mm to several cm, although fine-grain metasomatic fluorite also occurs. Economic quantities of fluorite were restricted to a 5 km-wide zone along the eastern margin of the limestone, extending below surface towards the limestone inliers of Ashover and Crich (Dunham 1952, Mueller 1954b, Ford 1955, Ford & Ineson 1971, Ixer 1986). Fluorite occurrences are divided into local areas, beginning in the north and working southwards. *Castleton–Bradwell area:* Blue John is a world-famous blue/white banded variety of fluorite found lining open joints, veins and ancient caves around Treak Cliff, near Castleton. According to Ford (1955), Blue John owes its colour to bituminous inclusions, but this is disputed by Mackenzie & Green (1971) and Braithwaite et al. (1973), who suggest colloidal calcium is the chief colouring agent. In turn this is disputed by Calas (1972), Holgate (1973) and Galwey et al. (1979) – the last authors explaining the banding as zones of radiation damage during the intermittent but temporary deposition of radioactive material at the growing faces of fluorite crystals. Blue John forms spheroidal nodular masses with radiating crystalline structure, and contains blue bands of varying intensity separated by colourless, yellow or pale-blue bands. The centres of Blue John lined cavities are usually open and other minerals have been able to form on the interpenetrating cubes of fluorite. Well developed scalenohedra of calcite up to 10 cm long are notable on examples from Witches' Cave in Treak Cliff Cavern. At this locality (and Treak Cliff opencast), Blue John was deposited in the voids between limestone boulders in conglomerate before a major episode of vein (rake) mineralization. Yellow fluorite, barite, galena, pyrite and quartz are associated species (Ford 1955, 1969, Simpson & Broadhurst 1969). Further information on Blue John from Treak Cliff can be found in Guppy (1956), Sarjeant (1956b), Ford & Sarjeant (1964a), Ford & Ineson (1971), Rogers (1977), Howie et al. (1982), Ford et al. (1993) and Bau et al. (2003). Very similar fluorite also occurs at the neighbouring Blue John mine (Rudler 1905, Braithwaite et al. 1973, Rogers 1977, Braithwaite 1983). At Odin mine, near Mam Tor, major amounts of fluorite remain *in situ* and on spoilheaps, as it was only galena that was being sought when the mine was working. Fluorite is either pale bluish-green and finely granular or it forms saccharoidal aggregates of small, clear or pale-blue cubes (Ford 1955, Ford & Ineson 1971, Smith 1973b, Rogers 1977). Other fluorite occurrences include: colourless, white and purple crystals and blue/white banded masses (similar to Blue John), some with

Section through a fluorite (variety Blue John) slab showing growth banding; Blue John mine, Castleton, Derbyshire; field of view is 12 × 7 cm. (© National Museum of Wales, Cardiff)

pyrite (bravoite) inclusions, along Dirtlow Rake, a major economic source of fluorite (Ixer 1974a, Rogers 1977, Butcher & Hedges 1987, Ford et al. 1993, Jeffrey 1997, Ford & Rieuwerts 2000, Bau et al. 2003); purple banded fluorite at Old Tor vein, Castleton; colourless crystals at Faucet Rake and Nether Water mine, near Castleton; colourless crystals at Earl Rake, Hazlebadge Hall and Newall Nook, near Bradwell (all Rogers 1977); blue and amber crystals at Pindale quarry, Castleton (Rogers 1977, Ford & Rieuwerts 2000, Bau et al. 2003); Netherwater mine, Bradwell – redeveloped for fluorspar in the 1940s (Ford & Ineson 1971); grey and black fluorite with micro-inclusions of hematite and blue fluorite at Smalldale, Bradwell (Ford & Sarjeant 1964a, Butcher & Hedges 1987, Ford et al. 1993); dark-purple interpenetrant cubes to 1 cm from the far reaches of Speedwell Cavern and mine, Castleton (Quirk 1985); colourless crystals with the edges picked out in dark blue, and massive purple fluorite at Windy Knoll, Castleton (Rogers 1977, Ford et al. 1993). The use of Blue John as a gemstone and for ornamental purposes, together with the history of its mining, is discussed by Ford (1979, 2000). *Hucklow–Eyam–Stoney Middleton area:* Broadlow mine, Eyam – 6 m-wide vein mainly of fluorspar (Ford & Ineson 1971); Deep Rake, Longstone Edge – attractive colourless fluorite with purple edging (Goulding 1997b); purple fluorite at Dirty Rake, Stoney Middleton (Braithwaite et al. 1973); Glebe mine, Eyam – an 18th-century lead mine that was reconditioned for fluorspar mining in 1938–1939 and became one of Derbyshire's most important producers (Varvill 1959, Ford & Ineson 1971); High Rake, Longstone Edge – fluorite replacement of limestone (and galena) in an unusual galena-rich flat deposit (Jeffrey 2001); Hucklow Edge Rake – fluorite–smithsonite matrix containing chalcopyrite, galena and pyrite (bravoite) (Ixer 1974a); Ladywash mine, Eyam – yellow or red fluorite with limonite inclusions, blue fluorite associated with barite, purple cubes on massive fluorite, also colourless cubes with pyritic inclusions (Ford & Sarjeant 1964a, Braithwaite et al. 1973, Rogers 1977, Howie et al. 1982, Green 1986, Bau et al. 2003); Sallet Hole mine, Longstone Edge (massive colourless and banded fluorite) and Mill Dam mine, near Great Hucklow – major fluorspar producers (Ford & Ineson 1971, Rogers 1977, Ford & Rieuwerts 2000); and Red Rake on Longstone Edge and Slater's mine, near Eyam – massive colourless fluorite (Rogers 1977). *Ashford–Bakewell–Sheldon area:* Arrock Black Marble mine, Ashford-in-the-Water – fluorite-bearing calcite vein cutting chert (Orme & Ford 1970); Blyth mine, Alport – amber fluorite (Rogers 1977); Kirkdale on Dirtlow Rake (not the rake at Castleton with the same name) and Mogshaw Rake, E of Sheldon – important fluorspar opencuts, red fluorite at the latter (Ford & Ineson 1971, Ford et al. 1993); Magpie mine (Blende vein), Sheldon – purple fluorite associated with calcite and sphalerite in cavities within fluoritized silicified limestone (Worley 1976, Ford & Rieuwerts 2000); Raper mine and opencast, Long Rake, near Youlgreave – five generations of fluorite in fissure-infill veins and replacement zones, including pale-brown, violet and euhedral yellow crystals, all associated with barite (Ineson & Al-Kufaishi 1970, Ford & Ineson 1971); and Ricklow quarry, Monyash – blue fluorite filling an ancient cave (Ford et al. 1993). *Millclose mine, Darley Dale, near Wensley:* This mine is best known for its galena, which was mined for over 300 years until it closed in 1940. However, reprocessing of the spoilheaps in the mid-1970s returned about 250 000 tonnes of fluorspar and barite. Millclose produced spectacular specimens of crystalline fluorite, including colourless or slightly yellowish cubes, often showing included layers of marcasite or pyrite (bravoite). Amber and purple varieties were also found (Russell 1912b, Traill 1939, Bannister 1940, Ford & Ineson 1971, Ixer 1974a, Rogers 1977, Willies et al. 1989, Ford & Rieuwerts 2000). Attractive colourless cubes to 3 mm with purple edges were found at the associated Old Millclose mine (Ford et al. 1993). *Bonsall–Matlock Bath–Cromford–Wirksworth area:* Ball Eye mine, Bonsall – blue fluorite (Ford & Ineson 1971, Ford et al. 1993); Cobbler mine, Bonsall – deep-purple cubes (Braithwaite et al. 1973); Bage mine, Bole Hill, Wirksworth – colourless, non-fluorescent cubes associated with matlockite or phosgenite (Burr 1992); High Tor, Matlock – yellow or red fluorite with limonite inclusions (Ford & Sarjeant 1964a); Masson Hill, Matlock – a large replacement deposit containing common cubic crystals of fluorite, normally colourless to pale yellow, less commonly blue, green or purple, the last colour notable at the crystal edges, also known to contain inclusions of pyrite (bravoite), chalcopyrite and marcasite (Varvill 1959, Ford & Ineson 1971, Ixer 1974a,b, Rogers 1977, Dunham et al. 1978, Ford et al. 1993); Masson and Rutland Caverns, Heights of Abraham, Matlock Bath – colourless and rare purple fluorite, some containing cinnabar in cavities (Braithwaite & Ryback 1963a, Braithwaite et al. 1963b), pale-yellowish, greenish and faintly mauve varieties are noted by Dunham (1952) from the deeper workings of Masson mine; Oxclose mine, Snitterton – pyrite (bravoite) inclusions in fluorite (Ixer 1974a,b); Pike Lane opencut, near Winster – friable purple aggregates and gem-like purple cubes to 5 mm associated with calcite (Uttley 1989); and Wapping mine, Cromford – small colourless cubes, large (to 1 cm on edge) colourless cubes with spots and edges coloured purple hosting zoned inclusions, and large (to 2 cm on edge) colourless to pale greenish-yellow cubes (Ince 1991). *Ashover inlier:* Clay Cross mine, Milltown – blue/white banded fluorite similar to Blue John (Ford et al. 1993); Fall Hill (Gate) quarry, Ashover – extensive fluorite, some forming grey cubic crystals to 7.5 cm on edge, others with zoned marcasite inclusions (Smith 1982), also massive dark-blue and clear cubes (Spence 1982); Gregory mine – pale-mauve fluorite, colourless cubes to 4 cm invested with minute chalcopyrite crystals, and on other crystals with attached enargite crystal groups (Dunham 1952, Rogers 1977, Ford et al. 1993, Cooper 1995); Milltown quarry, near Ashover – large masses of fluorite up to 5 m wide in Blackwell's rake contained cavities in which corroded 7.5 cm cubes were found; colourless cubes associated with anglesite, calcite, cerussite, galena and sweetite, also dark-purple or yellowish cubes (Clark et al. 1984, Rust 1991c); and Starrs Wood – dull brick-red variety of fluorite (Ford et al. 1993). More generally, pale-yellow to grey transparent fluorite forming cubes to 20 mm on edge, some with pyrite inclusions, is noted from the mines and quarries around Ashover (Wolfe 2006). *Crich inlier:* The large limestone quarry at Crich has swallowed up several old lead mines, including the Glory mine from where colourless and blue fluorite was once obtained. Old End mine produced massive banded colourless fluorite. A yellowish frosted variety of fluorite with clear edges from this area (locally called Crich spar) was used for making vases and ornaments. Considerable fluorite remains unworked below the water table (Ford & Ineson 1971, Rogers 1977, Ford et al. 1993, Wilson et al. 1995, Braithwaite & Wilson 2007). *Ashbourne–Snelston area:* Calcite veins containing colourless cubes of fluorite to 8 mm associated with barite have been found at Snelston mine, near Ashbourne (Jackman 1996).

Devon The Pb–(Ag) mines at Bere Alston were a source of fine (and large) specimens of fluorite, some forming cubes, others with the form of octahedra, made up of aggregated cubes. The finest crystals were colourless or pale sea green internally and smalt-blue at the edges (Greg & Lettsom 1858, Hall 1868b, Rudler 1905). Commercial quantities of fluorspar were once recovered from Tamar mines, near Bere Ferrers, and Bedford United mine, near Tavistock (Hall 1868b, Le Neve Foster 1877, Vipan 1959), with green, reddish-yellow and white varieties noted at the latter (Dunham 1952). Many other locations are reported from within or around the Dartmoor granite, usually with little detail: between Birch Tor mine and Vitifer iron–tin mine, Dartmoor (Alderton 1988a); Butspill mine, S of Calstock – colourless and green fluorite (Durrance & Laming 1982); Crownhill Tor, southwest Dartmoor and Lee Moor china-clay pit, near Wotter, southwest Dartmoor – accessory amounts in granite (Knox & Jackson 1990); Furzehill (Furzehill Wood) mine at Walkhampton, North Hooe mine and South Tamar mine, all near Tavistock (all Alderton 1978); Dartmoor granite – accessory amounts (Brammall & Harwood 1925, Ward et al. 1992); Hemerdon Bal, Plympton (Keats 1981); Meldon aplite quarry, near Okehampton – in aplite, pegmatite and as dark-blue, violet or occasionally green crystals on joint surfaces (Worth 1920, McLintock 1923, von Knorring 1951, Dearman & Claringbull 1960, Chaudhry & Howie 1973b, Chaudhry & Mahmood 1979, Howie et al. 1982, von Knorring & Condliffe 1984, Wirth et al. 2001); Red-a-ven mine, Meldon, near Okehampton – in calc-silicate skarn (Kingsbury 1961, El Sharkawi & Dearman 1966); Teign Valley, near Hennock – fissure veins containing Cu–Fe–Pb–Zn sulphides in a barite–calcite–fluorite–siderite gangue (Beer & Ball 1977); Trowlesworthy Tor, Dartmoor – deep-violet fluorite replacing quartz in granite (Rudler 1905); Virtuous Lady mine, Buckland Monachorum – small yellowish cubes (Hall 1868b, Russell 1913a); Wheal Franco, near Buckland Monachorum (Hall 1868b); and Wheal Friendship (Devon Friendship), Mary Tavy (Hall 1868b, Braithwaite 1982b). On Lundy, off the north Devon coast, violet fluorite occurs in drusy cavities and as an accessory mineral in granite (McLintock & Hall 1912, Dollar 1941, Edmonds et al. 1979, Thorpe et al. 1990).

Durham *Northern Pennine orefield:* The Durham section of the orefield is by far the most important, both for the amount of fluorite mined and for world-class specimens. A temperature-controlled zonal arrangement of fluorite in the Alston Block part of the orefield has been noted. The central zone is dominated by purple and occasionally green varieties, whereas the outer zone is characterized by amber or yellow fluorite, usually accompanied by barite (Dunham 1937, King 1982b). Initially, fluorite was an unwanted byproduct of lead mining, but with the collapse of the lead-mining industry in the 1880s there came a demand for fluorspar (commercial-grade fluorite) for the steelmaking and chemical industries, and the mines continued to operate for another century. Sadly, the last mine (Frazer's Hush, near Allenheads) closed in 1998 and, since then, specimen mines, such as Rogerley mine, near Stanhope, are all that remain. Exquisite specimens from this and other mines in the region came from vugs in mineralized veins and from cavities in stratiform metasomatic replacement lead deposits (flats) within Carboniferous Limestone. Penetration-twinned cubes are common, and crystals to 15 cm on edge are not uncommon in the

Ice-blue fluorite cubes to 5 cm; Cambokeels mine, Eastgate, Weardale, Co. Durham. (© D. I. Green)

flats (Dunham 1990, Wolfe 1998, Fisher 2004). Many important fluorite mines are in Weardale and East Allendale, centred above the buried Weardale granite at Rookhope (Greenwood 1977, Dunham 1959a, 1990, Brown et al. 1987a, Ixer & Vaughan 1993, Fairbairn 1996, Fisher 2004, Jones 2006). They include: Barbary mine, Ireshopeburn; Billing Hills and Cumnock Isle mines, Eastgate; Rigg mine, Westgate (Sweet 1930); Blackdene mine, near Ireshopeburn – purple cubes with attached chalcopyrite, lustrous purple twinned crystals to 2.2 cm, twinned yellow cubes to 2.5 cm on edge, pale-green cubes with pale-purple zonation and overgrowths of calcite, and a remarkable mauve elongate crystal of fluorite 60 mm high (Green 1991, Tresise 1991, Green & Briscoe 2002, Fisher 2004); Blue Circle cement quarry, near Eastgate – pale-green fluorite twin 14 mm on edge, deeper-green twinned crystal 2 cm on edge and others with purple colour zones towards the centre of the crystals (Green & Briscoe 2002, Fisher 2004, Bouch et al. 2006, Jones 2006, Wolfe 2006); Boltsburn mine, Rookhope – predominantly purple fluorite including a pale-purple cube to 18 cm on edge, less commonly green, also some optical-grade fluorite (Smythe 1921, Dunham 1952, Guppy 1956, Rankin 1977, Dunham et al. 1978, King 1982b, Young 1995, Jones 2006); Burtree Pasture mine (Greenwood 1977); California mine, S of Stanhope – brown fluorite and white fluorite with brilliant bluish fluorescence (Guppy 1956); Cambokeels (Cammock Eals) mine, Eastgate (the mine is noted to have been one of the most productive sources of fluorite in Weardale) – colourless and ice-blue cubes to 3 cm and 50 mm on edge respectively, purple twinned crystals to 2.6 cm on edge, fine twinned gem-like lavender fluorite and green fluorite, some associated with pyrrhotite (Greenwood 1977, Dunham et al. 1978, Rankin & Graham 1988, Wolfe 1998, Green & Briscoe 2002, Fisher 2004); Crawley Spar mine, Stanhope – one of the most productive fluorite mines (Smythe 1921); Frazer's Hush mine – superb large museum pieces of lustrous deep-purple twinned crystals up to 5 cm on edge, poorer-quality crystals to 30 cm on edge, distorted crystals, and, in the upper workings, green and purple cubes together with cerussite, galena and sphalerite (Green & Briscoe 2002, Bau et al. 2003, Fisher 2004, Jones 2006); Frazer's Grove – a name given to Frazer's Hush and Groverake mines when their underground workings were connected in the 1990s (Younger 2003); Greenlaws mine, S of Daddry Shield – a specimen mine producing non-twinned purple cubes to 10 cm on edge and rarely amber cubes (Green & Briscoe 2002, Fisher 2004); also at Greenlaws mine – mud-filled fluorite-lined vugs in mineralized flats, including amber cubes with purple outer zones and other crystals to 12 cm on edge (Moreton 2005b); Groverake mine, near Rookhope (an important, and the last, commercial fluorspar producer) – massive purple- and green-banded fluorite, also some pastel-green fluorite, and fluorite with platy dendritic growths of native copper (Dunham 1959a, 1978, Smith 1973a, Greenwood 1977, Rankin & Graham 1988, Young 1995, Fisher 2004); Heights mine and quarry, between Eastgate and Westgate – a source of magnificent emerald-green fluorite crystals up to 4 cm on edge, opaque deep-green crystals up to 12 cm on edge, and well formed pale-purple to colourless fluorite (Dunham 1952, 1959a, Rankin 1977, Howie et al. 1982, Briscoe 1986c, Rankin & Graham 1988, Green & Briscoe 2002, Fisher 2004, Jones 2006); Killhope mine, E of Nenthead – sparse purple fluorite (Dunham 1992); Lanehead veins, near Stanhope – lustrous watery-green to bluish-violet fluorite twins to 20 mm on edge (Wolfe 2006); Middlehope mine – good purple fluorite (Dunham 1952) (Fisher [2006] suggests that White's level in Middlehope Burn was the source of many 19th- and early 20th-century specimens of fluorite labelled "Alston Moor" and "Cumberland"); Old Fall mine, Boltsburn – perfect fluorite cubes to 7 mm and calcite–fluorite–quartz stalactites (Watson 1900, Smythe 1921, Bridges & Young 1998); Redburn mine, near Rookhope – mostly white fluorite, also green, lilac, purple and yellow varieties (Greenwood 1977, Dunham et al. 1978, Day 1979, Rankin & Graham 1988, Green & Briscoe 2002, Fisher 2004); Rogerley mine, between Frosterley and Stanhope – a specimen mine producing exceptionally fine emerald-green and bluish-green fluorite up to 3 cm on edge, smaller twinned crystals, also deep inky-purple crystals (Wolfe 2000, Green & Briscoe 2002, Fisher & Greenbank 2003, Fisher 2004, Bouch et al. 2006); Sedling mine, Cowshill – pale-mauve twinned cubes to 8 mm (Seager 1953, Guppy 1956, Dunham 1959a); Stanhopeburn mine, N of Stanhope – massive purple fluorite and occasionally cubes (Smythe 1921, Dunham 1959a, Greenwood 1977, Rankin & Graham 1988, Green & Briscoe 2002); Stotfieldburn mine, SE of Rookhope – attractive specimens, including lustrous twinned crystals of purple and purple-to-green fluorite (Guppy 1956, Dunham 1959a, Houston 1964, Greenwood 1977, Fisher 2004); West Pastures mine, Stanhope – attractive untwinned apple-green cubes to 2 cm that fade to grey with exposure to light, also yellow and dark-purple cubes (Green 1998b, Green & Briscoe 2002, Fisher 2004); Whiteheaps mine, N of Rookhope – green–purple massive fluorite and purple-to-lavender twinned crystals to several cm (Greenwood 1977, Rankin & Graham 1988, Fisher 2004).

Purple twinned fluorite crystal, 2.2 cm on edge, sitting on a cluster of opaque, pale-purple fluorite; Blackdene mine, Irehopesburn, Weardale, Co. Durham. (© J. Fisher)

Deep-green penetration-twinned fluorite crystals, up to 2 cm across, covering silicified limestone; Rogerley mine, near Frosterley, Weardale, Co. Durham. (© J. Fisher)

Fairburn (2003) is an excellent source of information on the fluorspar mining industry of the northern Pennines.

Elsewhere, fluorite occurs as water-clear, pale-yellow, amber, violet or violet-brown, dark-brown to black, or cream crystals in Permian Magnesian Limestone examined in borehole cores to the W of the Northern Pennine orefield, most notably in the Ferryhill area. Barite, galena and sphalerite are associated species (Fowler 1943, 1956, Hirst & Smith 1974). Dunham (1990) reports fine examples of amber fluorite with barite at Chilton quarry, Ferryhill.

Lancashire Clear fluorite with pyrite (bravoite) inclusions is reported from Bellman Park quarry, Clitheroe (Ixer 1974a).

Northumberland Some of the finest fluorite ever found in the Pennines was recovered from Beaumont mine, Allenheads, in the 1970s; it included lustrous pale-lavender crystals associated with siderite and sphalerite (Greenwood 1977, Green & Briscoe 2002, Fisher 2004). St Peter's mine, between Allenheads and Allendale town produced good specimens of bright apple-green fluorite in the 1930s, but recent activity at the mine failed to find further examples. However, amber crystals and lustrous purple-to-grey untwinned cubes to 3 cm, with overgrowths of siderite, were found instead (Wolfe 1996b, Green & Briscoe 2002, Fisher 2004, Jones 2006). Bright-lilac disseminated fluorite has been found in a xenolithic raft of limestone (now marble) metasomatized by the Whin Sill at Barrasford dolerite quarry. A few medium- to deep-purple cubes to 0.5 mm (some colour-zoned) were recovered from small cavities in the marble (Young et al. 2005).

Fluorite forming lavender twinned crystals, up to 1.8 cm, associated with sphalerite and creamy siderite from Beaumont (Allenheads) mine, Diana vein, Allenheads, East Allendale, Northumberland. (© J. Fisher)

Shropshire Colourless or green crystalline masses and cubes of fluorite to 10 mm on edge occur with calcite, barite and witherite at Rorrington lead mine, S of Meadowtown (Starkey et al. 1994). Fluorspar is also noted from the neighbouring Snailbeach mine, where it was exploited commercially from 1874 to 1913 (Dines 1958, Pearce 1994a).

Somerset Bryant's quarry, SSW of Shipham, Mendip Hills – thin veinlets and cubes of fluorite to 3 mm on edge, in rounded nodules of fine-grain granular calcite, some colour zoned from colourless to dark purple (Kingsbury 1941); Coleman's quarry, Holwell, near Frome, Mendip Hills – fluorite in calcite-lined cavities within manganese-rich veins (Alabaster 1990a); Whatley quarry, near Frome, Mendip Hills – deep-purple fluorite (Goulding 1997a, House 2002); Wick quarry, near Kingswood, between Bath and Bristol – minor amounts of pale-purple fluorite (Cowdry 2002b).

Staffordshire "Water-coloured" and light-blue fluorite has been described from Ecton Hill mine (Mawe 1802, Rudler 1905, Robey & Porter 1972, Masheder & Rankin 1988).

Yorkshire Many occurrences of fluorite are described from galena-bearing veins and flats in Carboniferous rocks within the Askrigg Block – the southern part of the Northern Pennine orefield. The entire area is underlain by the Wensleydale granite, which played an enigmatic part in the development of the orefield (Dunham & Wilson 1985, Brown et al. 1987a). Fluid inclusions in fluorite from barite–calcite–galena–sphalerite veins at many localities in the Askrigg Block have been studied by Rogers (1978). *North Swaledale mineral belt:* Arkendale and Old Gang mines – dumps reprocessed for barite and fluorite (Dunham et al. 1978); Brandy Bottle dumps (colourless or yellowish-white fluorite) and Hard level dumps (colourless fluorite), both NE of Gunnerside (Young 1987b, Ineson & Young 1994); East Arn Gill (Adelaide level) – amber and greenish-yellow twinned crystals of fluorite (Dunham & Wilson 1985); Sir Francis and Priscilla levels on Old Rake vein, Gunnerside Gill – pale-yellow crystals (< 1 cm) with pyrite inclusions, some associated with strontianite (Wood 1993); and Great Sled Dale, 10 km SE of Kirkby Stephen – amber groups of cubic crystals (Briscoe & Green 1989). *Wensleydale:* Keld Heads mine, near Wensley – fluorite with pyrite (bravoite) inclusions (Rogers 1974). Warton Hall level produced crystals of fluorite with amber and purple banding (Dunham & Wilson 1985). Fluorite is also reported from: Cloud End Fell, near Sedbergh; Wet Grooves and Worton mines, near Aysgarth (all Rogers 1978); Seata mine, near Aysgarth (Smith 1973b, Rogers 1978). *Wharfedale & Nidderdale:* Coldstones quarry, 5 km W of Pateley Bridge – blue-tinted crystals to 25 mm, also colourless, yellow and dark-blue crystals, some gem-like. Garnet and Sun veins in this quarry are mainly composed of fluorite (± barite–calcite). Supergene prosopite occurs on 1–3 cm cubes of fluorite (Ryde 1987, Watson 1987, Young et al. 1989, 1997, Bridges 1997a, 2002). Duck Street quarry, SE of summit of Greenhow Hill, near Pateley Bridge exposes Greenhow Rake – a major calcite–fluorite–galena vein; the largest crystals of fluorite (7.6 cm on edge) in the Askrigg Block are from this rake. Elsewhere in the Greenhow area, fluorite is colourless, very pale yellow or pale mauve (Dunham & Stubblefield 1944, Guppy 1956, Dunham & Wilson 1985, Rankin et al. 1987). Gill Heads vein, near Skyreholme, occasionally produced colourless cubes of fluorite to 25 mm, whereas Lolly Scar and Old Providence veins produced pale-yellow cubes to 25 mm (Dunham & Wilson 1985). Deep-purple fluorite is associated with calcite and bitumen in veins cutting a reef knoll at Swinden Limeworks quarry, near Threshfield, where small zoned cubes of blue fluorite have also been found (Dunham 1952, Braithwaite et al. 1973). Fluorite with pyrite (bravoite) inclusions is noted from Wharfedale mine, Kettlewell (Rogers 1974). Large matrixless specimens of intergrown frosted fluorite crystals to 40 mm with blue edges are reported from Elbolton Hill mine, near Thorpe, Lower Wharfdale (Wolfe 2006).

Minor amounts of fluorite are also reported from Permian evaporite deposits discovered in boreholes near Whitby (Stewart 1954, 1965, Raymond 1962).

Wales

Clwyd Fluorite occurs in the lead–zinc orefield of the Halkyn mining district, with particularly fine crystals noted from Bryn-gwiog mine, near Halkyn, and Rhosesmor mine (Halkyn East), near Mold (Carruthers et al. 1916). Translucent purple cubes to 3 cm with colourless or yellow centres, and colourless cubes with purple faces, are reported from Halkyn mine (Jones 2006). Other examples of fluorite in the district include: massive colourless crystals at East Pant Du mine, near Llanarmon; massive purple crystals at three localities – Bryn-gwiog mine, near Halkyn; Ffrith quarry, near Minera; Pant-y-gwlanod vein, near Llanarmon (Smith 1973b); Maes-y-safn mine, near Mold, and Llanarmon-yn-Ial quarry, near Tan-y-graig mine (Strahan 1890, Bevins 1994); yellow cleavage fragments and purple cubes from Moel-y-crio, Halkyn Mountain (Campbell Smith 1913). Additionally, fluorite has been identified at Llandudno mine, Llandudno (Bevins 1994).

Dyfed Fluorite is noted from Alltycrib mine, Tal-y-bont, in the Central Wales orefield (Jones & Moreton 1977), and from Brynambor, Llandewi Brefi (Hall 1993c). Both occurrences are considered dubious (Bevins 1994).

Gwynedd Fluorite has been found in magnetite–pyrite-rich rocks on spoilheaps near Llyn Cwellyn, SE of Mynydd Mawr, 10 km NW of Cwm Tregalan, and in the nearby Mynydd Mawr microgranite (Colman & Appleby 1991). Other reports are from: the contact aureole of the Bwlch y Cywion intrusion, Y Llymllwyd (Williams 1927); near the cromlech (prehistoric stone circle) at Benellech on Anglesey; and Llandudno mine, Llandudno (both Bevins 1994).

Mid (and South) Glamorgan Colour-zoned cubes of fluorite to 25 mm on edge occur in cavities in Carboniferous Limestone at Vaynor quarry, Cefn Coed, Merthyr Tydfil. Colours range from purple to pale blue, yellow, green and colourless (Bevins 1994, Billington 1995). Galena-rich veins at Risca also contain fluorite (Howard 1899). In South Glamorgan, fluorite occurs as colourless crystals in thin veins within Carboniferous Limestone on Sully Island (Bevins 1994) and as pale cubes in a boring for water in Working Street, Cardiff (Storrie 1885).

Scotland

Dumfries & Galloway Colourless and honey-yellow crystals of fluorite to 1 mm and a massive pale lilac-pink variety have been found at Beeswing granite quarry (Williams 1965). Fluorite is noted from the lead-mining district of Wanlockhead at

Glencrieff, High Pirn mine (large muddy-white cubes) and Whyte's Cleuch mines (Heddle 1901, Dunham 1952).

Fife Bluish-grey crystals of fluorite have been found in veins within zeolitic albitized basalt at Glen Farg railway cutting (Heddle 1901, Dunham 1952).

Grampian The Corrybeg lead–zinc mine at Abergairn, near Ballater, worked a calcite–fluorite vein cutting Dalradian schist. Pale-yellow cubic crystals to 25 mm on edge and bluish-grey, green, red, yellow and violet varieties were once found (Heddle 1901, Russell 1936a). The occurrence is possibly related to the nearby fluorine–lithium-rich zinnwaldite granite at Gairnshiel Bridge in Glen Gairn, where fluorite is an accessory species (Hall & Walsh 1972, Webb et al. 1992). Fluorite is also present along joint planes and in miarolitic cavities within the Ballater granite, at the Pass of Ballater, where it is associated with beryl, bertrandite and kaolinite (Crane 1987, Green & Todd 2001). Early reports of green and purple cubes of fluorite and purple veins associated with galena from "Balater House" and "near Monaltrie House and the Pass of Ballater" (Greg & Lettsom 1858, Heddle 1901) were re-examined by Russell (1936a), who describes the locality as Creag an t-Seabhaig on the north side of the Pass of Ballater. He found four types of fluorite cubes: colourless, pale-green, violet-to-grey to 2 cm, and colour-zoned violet-to-green to 2 mm. In the Ballater area, Russell also found fluorite at Tomnakiest granite quarry, Cambus O'May, and Neil's granite quarry, Rinabaich. Other Heddle (1901) locations include: Burn of Baharm, Lower Ardoch, W of Keith – purple cubes; Findrassie quarry, near Elgin – yellow and red cubes; W of Middleton of Balquhain – purple and pale-green isolated octahedra; Muir limestone quarry, Deeside – deep-purple octahedra; Murdoch's Cairn quarry, near Peterhead – blue crystals of exceptional quality. Russell (1936a) visited other Heddle localities, but failed to find fluorite at: Coilacriech, north bank of the River Dee, 3 km W of Abergairn (see also Dunham 1952); Crathie limestone quarry at Creag a' Chlamhain; Maisley limestone quarry, near Keith; Midstrath limestone quarry, Aboyne; Na Tri Chaochain, a small stream feeding into Strath Avon between Tomintoul and Inchrory, 2 km S of Wester Gaulrig (where pale-green cubes 25 cm on edge had been reported). However, at Na Tri Chaochain, a locality at the contact of the Glen Avon granite with mica schists, a fluorite-rich breccia was found containing dark violet-to-grey to blackish violet-grey octahedra to 4.5 cm on edge and pinkish cubes to 2 mm. A galena-bearing silicified fault zone 2 km E of Burghead, N of Elgin, contains pale-purple cubes of fluorite (Gillen 1987b) and at Shierglas quarry, Blair Atholl, octahedral crystals of pale-green fluorite up to 1 cm across have been found in calcite veins cutting calc-schists and limestones (R. E. Starkey pers. comm. 2005).

Highland Purple-to-colourless fluorite occurs in galena- and molybdenite-bearing veins crosscutting Moine schists close to the Grudie granite and as disseminations through the granite (Gallagher 1970, Gallagher et al. 1971). Quartz veinlets and scattered grains of fluorite also occur with molybdenite in and around the nearby Shin granite (Gallagher et al. 1974, Gallagher & Smith 1975, Lowry et al. 1994). At Caen and Ousdale Burns, in the Helmsdale granite, and Ceann Ousdale in the metamorphic aureole, fluorite occurs in uranium-enriched shear zones associated with galena and pyrite (Heddle 1901, Gallagher et al. 1971, Dunham et al. 1978, Tweedie 1979). Pointer et al. (1989) found fluorite with calcite, coffinite, hydrocarbons and sulphides in a borehole through faulted Old Red Sandstone sediments near the south-west margin of the Helmsdale granite. Small strings of fluorite occur in a barite vein cutting Old Red Sandstone conglomerate at Loch Bruicheach, 12 km SW of Beauly – greyish-yellow, white-to-colourless and rare bluish-lavender varieties are noted (Heddle 1901, Russell 1946c). Beautifully crystalline fluorite from this locality had previously been described as a new species (bruiachite) by Macadam (1886a). Other localities include (Heddle 1901 unless otherwise stated): Abriachan granite quarry, N of Drumnadrochit – blue cubes; in a giant erratic boulder at Beinn Bhreac, 2 km E of Tongue – pale purple (Heddle 1883); Carn Chuinneag – accessory species in orthogneiss (Gallagher et al. 1971); Loch Borralan igneous complex, Assynt – minor purple fluorite as a fracture fill in carbonatite, also colourless or purple varieties in melanite syenite at Cnoc-na-Glas-Choille (Shand 1910b, Young et al. 1994b); Ratagain granite complex, Loch Duich, near Glenelg – fluorite in Ag–Au–Te-bearing calcite–quartz veins (Alderton 1988b); and Whitesmith opencast (part of the Strontian mines), Loch Sunart – small colourless to faintly green or purple cubes on wallrock (Green & Todd 2001).

On the Isle of Skye, minor amounts of fluorite have been found in the aureole of the Beinn an Dubhaich granite, near Broadford (Tilley 1951, Smith et al. 2004), and as fine transparent crystals with scheelite and zircon at several imprecise locations within miarolitic Red Cuillin granite (Wolfe 2000, Green & Todd 2001). Thompson (1969) notes accessory fluorite in the Glamaig and Marsco epigranites, in the Western Redhills of Skye.

Strathclyde Fluorite is noted from the lead mining district of Leadhills – on the spoilheaps at Rakehead washery, Susanna mine and Wembley shaft (Dunham 1952). Drusy cavities in a trachyte sill at Craigmuschat quarry, Gourock, are well known for the fluorite, calcite and quartz that they contain. Fluorite crystals (cubes and octahedra) are usually purple or emerald green, although colourless, honey-yellow and pink varieties are also described (Heddle 1901, Rudler 1905, Nicholson & Durant 1986). Amber cubes of fluorite up to 1 cm on edge, and smaller colourless cubes associated with strontianite, occur in fossil brachiopod shells at Trearne quarry, near Beith (Todd 1989, Green & Todd 2001). Other reports include: abundant fluorite inclusions in amphibole and pyroxene within the alkali granite on Ailsa Craig in the Firth of Clyde (Harding 1983, Harrison et al. 1987); Hillhead limestone quarry, SW of Beith – pale-yellow cubes (Heddle 1901); Waterland limestone quarry, Lugton, Dunlop (Heddle 1901, Anderson & Lamont 1935); Isle of Arran – accessory amounts in the groundmass of the Northern Arran granite (England 1992, Hyslop et al. 1999).

Tayside Pale-red cubes of fluorite are noted from the foot of Beinn a' Ghlo, Glen Tilt, E of Blair Atholl (Heddle 1901).

Western Isles The Conachair granite, St Kilda (Hirta), contains drusy cavities in which fluorite has been found (Harding et al. 1984).

Shetland Heddle (1901) localities include: Kirkavoe (probably Mid Yell Voe, Yell) – colourless to deep-purple colour-zoned cubes; western shore of Sandwick Bay, Hillswick, Mainland – pale pea-green fluorite; North Quin Geo – violet fluorite in calcite vein; and Papa Stour (a small island off the west coast of Mainland) – pale-violet and dark-blue cubes in druses within amygdaloidal claystone (basalt).

Ireland

Antrim Greg & Lettsom (1858) report small green octahedra of fluorite in the granite of Slieve Carne.

Carlow Fluorite is a rare accessory species in spodumene pegmatites near the southern margin of the Leinster granite batholith at Aclare, near Myshall (Scoon 1978).

Clare Purple fluorite (rarely well crystallized) has been found with botryoidal smithsonite on the dumps of Sheshodonnell East lead mine, the Burren (Green & Moreton 2001). Fluorite from this locality, Ballyganner South, Castletown, Gleann na Manach and Skancullin mines has been the subject of a fluid-inclusion and geochemical investigation that suggested an origin under hydrothermal conditions with a source in the Galway leucogranites, but activated by later continental rifting in Triassic or later times (O'Connor et al. 1993).

Down Rare pale-green and pale-purple octahedral crystals of fluorite (some colour zoned and to 10 mm on edge) are described from miarolitic cavities in granite at Lindsay's Leap quarry, Mourne Mountains (Green et al. 2005a). Similar occurrences in the Mourne granites and associated greisen veins are also noted by Haughton (1856), Richey (1927) and Nockolds & Richey (1939).

Dublin Greg & Lettsom (1858) report fluorite in cavities in granite at Dalkey Island, Dublin Bay.

Galway Glengowla mine, near Oughterard, was once a source of very fine large specimens of pale-green and lilac octahedra of fluorite (Greg & Lettsom 1858, Ryback & Moreton 1992). Inverin copper–lead mine also produced fine octahedra of fluorite (Greg & Lettsom 1858), although in more recent times only small (2 mm) purple crystals could be found (Ryback & Moreton 1992). Elsewhere, fluorite occurs at: Costelloe granite quarry, Murvey – colourless pale-green and purple fluorite, sometimes colour zoned, in 2 cm-wide calcite–quartz veins (O'Connor et al. 1993, Menuge et al. 1997); Kinvarra, Connemara – dark octahedral purple fluorite to 20 mm on edge associated with barite, chlorite, galena and sphalerite (Wolfe 2006); Toombeola quarry – lilac fluorite along joints in granite (Ryback & Moreton 1992); Tynagh mine, near Killimor – a minor gangue species in the primary lead–zinc orebody (Clifford et al. 1986).

Kerry An early description of fluorite (Ainsworth 1834) could not be confirmed by Moreton et al. (1995) during their investigation of sulphide mineralization N of Ballybunnion.

Kilkenny At the southwestern end of the Leinster batholith, near the village of Inistioge, fluorite veinlets cut metasomatic tourmaline-rich rocks containing disseminated arsenopyrite associated with chalcopyrite, iron-rich chlorite, manganese-rich ilmenite, molybdenite, muscovite, pyrite, pyrrhotite and rutile (Appleby & Williams 1988).

Sligo A large limestone quarry near Abbeytown lead–zinc mine, near Ballysadare, produced pale-purple fluorite, but it was not found in the mine itself (Hitzman 1986, Ryback & Moreton 1992).

Wicklow Fluorite is present in pyrrhotite–scheelite-bearing quartz veins associated with a fluorine-rich microtonalite sheet at Ballingen, between Tinahely and Aughrim (Steiger & Bowden 1982, 1986, Kennan et al. 1986, McArdle & Kennan 1988, Gallagher 1989, Gallagher & Kennan 1992). Accessory amounts of fluorite occur in

Octahedral fluorite, 8 mm on edge, from Glengowla mine, Oughterard, Co. Galway, Ireland. (© D. I. Green)

a lepidolite greisen at Stranakelly, near Shillelagh (Steiger & von Knorring 1975, Scoon 1978). Fluorite is associated with barite, harmotome and strontianite in sulphide (mainly galena) veins at the Luganure mines, Glendalough (Rudler 1905).

Offshore territories

Jersey Fluorite occurs in galena–sphalerite-bearing calcite–chlorite veinlets cutting granite at L'Ouaisné Bay, southwest Jersey (Ixer 1980). Other locations are documented by Mourant (1961): in cavities within granite at Corbière (La Rosière), La Moye, L'Etacq and Mont Mado (good cubes); in rhyolite at Jeffrey's Leap; in veins cutting granite at La Carrière quarry and South Hill; and in a chlorite–hematite vein to the S of L'Ouaisné Bay.

fluor-schorl*

TOURMALINE GROUP

trigonal $NaFe^{2+}_3Al_6(BO_3)_3Si_6O_{18}(OH)_3F$

An uncommon sodium–iron–aluminium borosilicate hydroxide fluoride that forms a series with schorl and whose composition was defined in the tourmaline classification of Hawthorne & Henry (1999). Fluor-schorl is discussed further by Ertl et al. (2006). The classification is widely used, but fluor-schorl has not been formally described and therefore the name does not have the approval of the IMA. It is considered to be a named tourmaline in much the same way as certain members of the amphibole group are (Burke & Leake 2005). Clark (2007) describes the problems in calculating the structural formulae of tourmaline analyses obtained using EMP and she points out that the common assumption that all iron is ferrous can result in the misidentification of buergerite as fluor-schorl. At the present time, it is not possible to confirm if any of the fluor-schorl described below is actually buergerite.

In Cornwall, fluor-schorl is found in tourmaline aplite at Hantergantick quarry on Bodmin Moor, in greisen at Old Beam mine, Goonbarrow china-clay pit, St Austell, in tourmaline breccia from Wheal Remfry, near Indian Queens, St Austell, and in tourmaline hornfels at Gryll's Bunny, Botallack, St Just (Manning 1991). In most cases the tourmaline is strongly zoned, but the limited data presented do not allow the compositional ranges of individual samples to be assessed, although dravite–schorl substitution is most common. Fluor-schorl is considered to occur at the following localities (Tindle unpub. EMP data):

England

Cornwall Longstone china-clay pit (the site of the Trelawny Tigers speedway track), near Nanpean, St Austell, where a quartz–tourmaline vein cutting kaolinized granite contains fluor-schorl, schorl, foitite, magnesiofoitite and elbaite, often with complex zonation patterns; Priest's Cove, close to Cape Cornwall, where a tourmaline-rich vein derived from the Land's End granite contains an early-formed magmatic generation of tourmaline with fluor-schorl and schorl compositions; Megiliggar Rocks, where an aplite–pegmatite complex at the margin of the Tregonning–Godolphin granite, near Porthleven, contains discrete fluor-schorl crystals, zoned schorl–fluor-schorl and fluor-schorl–fluor-elbaite crystals.
Cumbria Upper Poddy Gill, Caldbeck Fells, where zoned dravite–fluor-schorl crystals occur in Skiddaw Slates hydrothermally modified by the nearby Skiddaw granite (D. I. Green specimen).

Scotland

Grampian At an isolated location approximately 1 km N of Peatfold farm, Glenbuchat, where fluor-schorl occurs with schorl, fluor-elbaite and elbaite in a zoned granitic pegmatite intruded into Dalradian mica schists and phyllites (D. I. Green specimen).

fluorspar

A synonym of fluorite. Fluorspar is also used to describe fluorite-rich ore containing admixed quartz, carbonates, sulphides and REE-bearing minerals.

foitite

TOURMALINE GROUP

trigonal $\square[(Fe^{2+}_2(Al,Fe^{3+})]Al_6(BO_3)_3Si_6O_{18}(OH)_4$

An uncommon iron–aluminium borosilicate hydroxide found in zoned dravite–magnesiofoitite–foitite–schorl crystals from a hydrothermal breccia at the southwestern margin of Roche Rock, 7 km N of St Austell, Cornwall. The presence of relatively high tin content in the foitite and magnesiofoitite zones within the tourmaline provides a valuable insight into the evolution of mineralizing fluids that ultimately produced the tin (cassiterite) mineralization for which Cornwall is justly famous (Williamson et al. 2000). Anhedral brown tourmaline with a foitite composition occurs in a medium-grain feldspar–phyric muscovite granite from Goonbarrow china-clay pit, St Austell, Cornwall (Wilson & Young 1983), but its identity was not fully established until foitite was formally described (MacDonald et al. 1993). In a study of quartz-cored tourmaline from pegmatitic veins, sheets and joint fillings in Cornwall, Lister (1979) presents six analyses of foitite from: Priest's Cove, Cape Cornwall (Land's End granite); Megiliggar Rocks, near Porthleven (Tregonning–Godolphin granite); Goonbarrow china-clay pit, near St Austell, and Tregarden, Luxulyan (both St Austell granite). In all cases the tourmaline is compositionally zoned, with schorl being the other species present. Foitite compositions do not appear to be correlated with colour (two are brown, two are blue) or position within a crystal (three are core compositions and one a rim composition).

Foitite is accompanied by dravite and schorl in metasomatically altered metapelites in the aureole of the presumed Grasmoor granite beneath Crummock Water, Cumbria. Tourmalinization took place in wallrocks containing a series of faults and joints and the resulting vein-like rocks contain nothing more than quartz and tourmaline. Tourmaline crystals have a grainsize of 5–10 μm and exhibit olive-green to brown colours in thin section (Eeckhout et al. 2004). A few analyses of foitite and magnesiofoitite have been collected from tourmaline collected in Burdell Gill, Caldbeck Fells, although the samples are predominantly of dravite composition (D. I. Green and C. M. Leppington specimens, Tindle unpub. EMP data).

In Ireland, tourmaline is encountered along the margin of the Leinster granite batholith, where it occurs in various environments. At Aclare, Co. Carlow, it has been found in an altered tonalite sheet associated with lithium-rich pegmatites. At Glendasan, Co. Wicklow, slender needles of tourmaline up to 7 mm in length are reported from a quartz vein cutting andalusite schist, within 10 m of the granite contact. The tourmaline is zoned from a core of blue foitite to a rim of brown dravite. A similar zonation pattern is developed in tourmaline within breccia-like zones of massive quartz–tourmaline rocks at Kilcarry Bridge, Co. Carlow, some 700 m from the granite contact. Foitite also occurs in a tin–tungsten-mineralized quartz vein, at Ballinglen, between Tinahely and Aughrim, Co. Wicklow, where it forms fine acicular crystals that are intergrown with scheelite (all Gallagher & Kennan 1992).

Foitite has been confirmed (Tindle unpub. EMP data) in many samples from

Cornwall and other regions, where it usually forms zoned foitite–schorl crystals. In most cases it is foitite that is first to crystallize and in thin section the species is typically brown or orange-brown (black in hand specimen). Where known, the likely source of the fluids thought to be responsible for tourmaline formation is given in brackets.

England

Cornwall Cligga Head, Perranzabuloe (Cligga Head granite); Mousehole, near Newlyn; Porth Ledden, near Cape Cornwall (both Land's End granite); Goonbarrow china-clay pit, St Austell; Longstone china-clay pit (the site of the Trelawny Tigers speedway track), near Nanpean; Wheal Martyn china-clay pit, St Austell; Rocks china-clay pit, Bugle, near St Austell – bright-blue (in thin section) needles (achroite) in a tourmaline veinlet cutting through a well formed quartz crystal; Wheal Remfry, near Indian Queens (all St Austell granite); Megiliggar Rocks, near Porthleven; E end of Praa Sands beach; NW of Trewavas Head, near Rinsey (all Tregonning–Godolphin granite); Phoenix United mines, Linkinhorne (Bodmin Moor granite).
Cumbria Burdell Gill, near Coombe Height, Caldbeck Fells (Skiddaw granite).
Devon Hemerdon, near Plympton; Meldon, near Okehampton (Dartmoor granite).

Scotland

Grampian Glenbuchat, near Peatfold (evolved lithium-rich pegmatite); East Head, at the base of Cowhythe Hill, near Portsoy (lepidolite-bearing pegmatitic granite and pegmatite).

fool's gold

A synonym of pyrite.

fornacite

monoclinic $Pb_2Cu(AsO_4)(CrO_4)(OH)$

A rare lead–copper arsenate chromate hydroxide found at Merehead (Torr Works) quarry, Cranmore, Somerset. Fornacite forms unusual golden-yellow to yellowish-green platy crystals on cerussite and was one of the last minerals to form in manganese-rich pods at this location. In one such pod, found in 2005, fornacite occurred with kentrolite, mereheadite, mimetite and symesite (Wolfe 2006, Turner 2007).

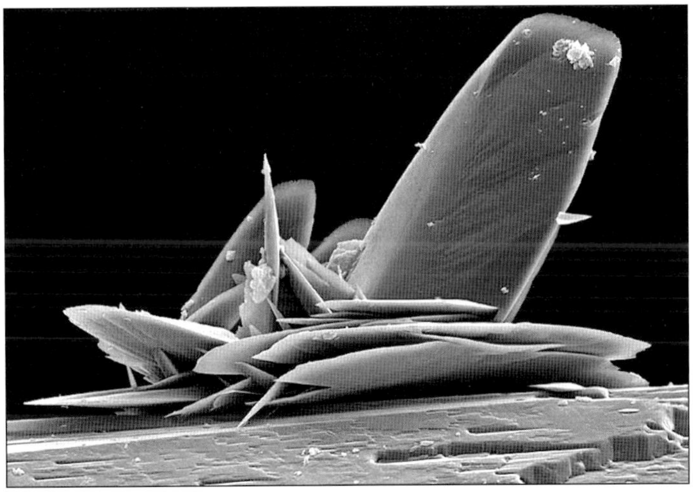

Petaline fornacite from Merehead (Torr Works) quarry, Cranmore, Mendip Hills, Somerset; width 780 µm; SEM image. (© D. I. Green)

forsterite

OLIVINE GROUP
orthorhombic Mg_2SiO_4

Named after Adolarius Jacob Forster (1739–1806), a London-based German mineral collector and dealer (Frondel 1972, Whitehead 1973, Cooper 2001, 2007). Forsterite is a very common magnesium silicate that forms a series with fayalite and is trimorphous with ringwoodite and wadsleyite. It is found in many mafic and ultramafic rocks and in thermally metamorphosed impure dolomitic limestones. Compositions are frequently intermediate between forsterite and fayalite end members, and the ratio of molecular iron to magnesium is then used to ascribe a name to the mineral (e.g. 75% Mg or Fo_{75} is forsterite, and any Fo value <50 is fayalite). Optical methods can also be employed to discriminate between these two species.

Perhaps the largest area of forsterite-bearing rocks in England is on the Lizard Peninsula, Cornwall, where a suite of gabbros and peridotites form part of an ancient ophiolite complex. Peridotite from four localities (near Rill Cove, Kynance Cliff; N of Worvas; S of Kernewas; and W of Gwenter) contains forsterite Fo_{90-91} in association with enstatite, diopside and a green chrome spinel. A later troctolite unit containing forsterite (Fo_{83}) intrudes the peridotite at Perprean Cove, Coverack (Green 1964). At Holmead farm, Loxbeare, Devon, forsterite (Fo_{82}) is a component of an unusual analcime-bearing lamproite dyke of Permian age (Velde 1971b). On Lundy island, Devon, forsterite (Fo_{66-82}) occurs with augite (salite), plagioclase feldspar (labradorite) and titaniferous magnetite in dolerite dykes cutting the Tertiary granite (Thorpe & Tindle 1992). In the southern Pennines, basaltic lavas of Carboniferous age commonly outcrop in the area around Matlock and between Buxton and Tideswell, Derbyshire. At Calton Hill, the lavas contain phenocrysts of forsterite (Fo_{80-85}), augite, plagioclase (labradorite) and titanium oxides, and, in addition, contain forsterite (Fo_{88-92}) in larger xenocrysts derived from digested xenoliths. Alteration of forsterite to combinations of calcite, chlorite, serpentine, iddingsite and iron–titanium oxides is common (Hamad 1963, Donaldson 1978, Macdonald et al. 1984).

In Wales, forsterite (Fo_{79-86}) is reported from many of the layered units in the Rhiw intrusion, Trwyn Talfarach, Lleyn Peninsula, Gwynedd (Hawkins 1970, Cattermole 1976). Peridotite nodules in the Great House diatreme, near Usk, Gwent, contain forsterite (Fo_{90}) (Bevins 1994). Reports of forsterite in crystalline limestones of the Mona Complex, Anglesey, Gwynedd (Greenly 1919) are unsubstantiated (Horák 1993).

Forsterite is widespread in Scotland. Representative examples include:

FIFE Fo_{81-89} in clinopyroxenite and lherzolite inclusions in Carboniferous–Permian vents along the coast near Elie (Chapman 1976). GRAMPIAN Fo_{77-87} in ultramafic, troctolitic and gabbroic rocks belonging to the Belhelvie layered intrusion, W of Balmedie, near Aberdeen (Wadsworth 1991); Fo_{79-81} in the Arthrath mafic–ultramafic intrusion near Ellon, 25 km N of Aberdeen (McKervey et al. 2007). HIGHLAND Abundant (optically identified) forsterite in a chondrodite-bearing limestone with minor chlorite, diopside and phlogopite on the west side of the Allt Eas Mòr Chùil an Dùin, a tributary joining the Amhainn a' Ghlinne Bhig, Glenelg (Read & Double 1935): Fo_{99} in contact-metamorphosed siliceous dolomite in the northeast corner of the metamorphic aureole of the Ballachulish granite (Ferry 1996); Fo_{73-75} in a garnet-bearing websterite rock within the Eastern Glenelg Lewisian at Beinn Fhada, 2 km W of Letterfearn, near Loch Duich (Rawson et al. 2001); in thermally metamorphosed dolomitic limestone near the Cnoc-na-Sroine syenite, Loch Borralan (Peach et al. 1907); partially or totally serpentinized forsterite in a black speckled metamorphosed dolomitic limestone at Coire nam Muc, 4 km WNW of Corran ferry, Ardgour, SW of Fort William (Drever 1939); forsterite–spinel marble at Glenmore of Glenelg, 1 km W of Sgiath Bheinn (Clough & Pollard 1899); Fo_{90} producing well formed but slightly corroded phenocrysts, up to 1 mm across, in a monchiquite dyke at Streap, 7 km NW of Glennfinnan station (Walker & Ross 1954); forsterite marble (contact-altered dolomite) in the burn N of Loch Sàil an Ruathair, N of Loch Ailsh (Macgregor & Phemister 1972); Fo_{83} in a rhythmically layered picritic body at Suisnish Point, Isle of Skye (Parslow 1976); Fo_{87-93} in a small Tertiary ultrabasic sill at Sgùrr Dearg, Cuillin Hills, Isle of Skye (Gibb 1972); Fo_{88-94} in layered peridotite from the Cuillin igneous complex (Bell & Claydon (1992); Fo_{85-87} in olivine tholeiite cone sheets that invade the Cuillin igneous complex (Bell et al. 1994); Fo_{74} in a highly amygdaloidal Tertiary basalt lava at Allt Coir' a' Ghobhainn, near Drynoch, Isle of Skye (Livingstone 1988); in Tertiary basaltic lavas from Achtaslean, near Portree, Fo_{84}, and to the S of Fairy Bridge Fo_{67} (Thompson 1974); forsterite in a brucite marble (thermally metamorphosed dolomitic limestone) along the margins of the Beinn an Dubhaich granite at Cnoc Dubh, between Kilbride and Torrin, Isle of Skye (Harker 1904, Newman 1971), and Fo_{95-99} in a related dolomite–marble–tremolite also close to the contact (Hoersch 1981, Holness 1992); Fo_{78-86} in layered peridotite and allivalite from the Eastern Layered Series of the Rum central complex, Isle of Rum (Palacz & Tait 1985); Fo_{79-88} in layered peridotite and allivalite from the Western Layered Series of the Rum central complex (Dunham & Wadsworth 1978); Fo_{83-90} in minor intrusions of peridotite to the N of the Rum central complex at West Sgaorishal (Wadsworth 1994); Fo_{81-93} forming phenocrysts up to 8 mm in diameter, or Fo_{75-77} in skeletal quench crystals within late picritic dykes belonging to the Rum central complex (Upton et al. 2002). STRATHCLYDE Fo_{39-69} (some is fayalite) in an analcime-bearing leucodolerite sill at Cnoc Rhaonastil, Isle of Islay (Walker & Patterson 1959, Hole & Morrison 1992); Fo_{44-66} in crinanite within the Dippin Sill, southeast Isle of Arran (Gibb & Henderson 1978b); Fo_{86} in wehrlite belonging to an ultrabasic group of the Girvan–Ballantrae ophiolite complex, near Knockormal (Bloxam & Allen 1959); Fo_{73} in a Devonian appinitic rock (possibly a feeder for the basic Lorne lavas) from the northern part of the Lorne Plateau, to the E of Oban (Groome & Hall 1974); Fo_{68-89} in magnesium-rich basalt varying to Fo_{59-64} in evolved basaltic hawaiite lavas from many locations on the Isle of Mull (Kerr 1998); forsterite in crystalline marble from Balephretrish Bay on the Isle of Tiree – the marble also contains mineral aggregates composed of well formed crystals of amphibole, chlorapatite, diopside, mica, orthoclase, meionite (scapolite), spinel and titanite (Coomaraswamy 1903). TAYSIDE Fo_{91-92} in Highland Border Suite serpentinized ultramafic rocks (dunite, harzburgite and orthopyroxenite) at Alyth, near Blairgowrie (Ikin & Harmon 1983). WESTERN ISLES Fo_{50-83} in the basal part (remainder of the intrusion contains fayalite) of a crinanite–dolerite–picrite layered intrusion at Garbh Eilean, Shiant Isles, 8 km E of the Isle of Lewis (Johnston 1953, Drever & Johnston 1965, Gibb & Henderson 1996, 2006, Henderson et al. 2000); forsterite in chondrodite–humite-bearing marble at Bay Steinigie, near Finsbay (Fionnsabhagh), South Harris (Dearnley 1961); Fo_{85} in ultrabasic gneiss W of Loch Ossigary, Rodel, South Harris (Howie 1964); Fo_{84} in garnet peridotite on the eastern shore of Loch Rodel, W of St Clement's church, South Harris (Livingstone 1967); Fo_{53-63} in the Cambir dolerite, Fo_{60} in the Glen Bay gabbro, Fo_{70-78} in the Western gabbro, all on St Kilda (Hirta) (Harding et al. 1984). SHETLAND Fo_{87-92} in dunite, harzburgite and wehrlite from the Shetland ophiolite complex (Maynard 1993).

Some descriptions of forsterite from Ireland relate to chalk/dolerite contact zones where forsterite has formed as a primary magmatic component of the dolerite or during the thermal metamorphism of dolomitic limestone. Examples include: unaltered dolerite at Scawt Hill, Larne, Co. Antrim, containing anhedral forsterite (Fo_{71-78}) intergrown with augite and plagioclase and intruded into Cretaceous Chalk and Tertiary basaltic lavas containing partially serpentinized forsterite (Fo_{65-81}) phenocrysts (Kitchen 1985); dolerite plug containing forsterite (Fo_{77-87}) at Tieveragh, near Cushendall, Co. Antrim (Tomkeieff 1940); and thermally metamorphosed limestone converted to brucite–forsterite–spinel marble close to a syenite intrusion in the area of Barnavave, Carlingford, Co. Louth (Osborne 1932b, Nockolds 1947).

Still in Ireland, the Tertiary Antrim basalts of the Giant's Causeway area contain partially serpentinized forsterite, usually as a constituent of the groundmass or as zoned microphenocrysts (Fo_{60-74}), where chemical variation into fayalite compositions (Fo_{44-63}) occurs (Lyle & Preston 1993). The Slieve Gullion central complex in Co. Down represents a deeply denuded shield volcano and consists of a ring dyke within which a range of dolerites, gabbros, microgranites and granophyres occur. Forsterite forms phenocrysts (Fo_{89}) and it also occurs in the groundmass (Fo_{56-74}) in the dolerites and gabbros (Gamble 1982). A highly xenolithic dolerite dyke at Inver, Co. Donegal, contains forsterite (Fo_{91}) within spinel–lherzolite xenoliths (Gallagher & Elsdon 1990). The hyalosiderite variety of forsterite (Fo_{50-70}) is described from an olivine-poor dolerite dyke on the northeastern shore of Lower Lough Erne, Co. Fermanagh (Preston 1967), and from Jurassic xenoliths in the Portrush dolerite sill, Castle Island, Portrush, Co. Antrim (Hawkes & Wilson 1975). The Inishdawros metaperidotite intrusion at Callow, Ballyconneely, Connemara, Co. Galway, contains near-euhedral forsterite crystals (Fo_{84-89}) included in pyroxene or hornblende (Ahmed & Leake 1978).

In the Channel Isles, the Wolf's Cave gabbro dyke on the north Jersey coast, E of Frémont Point, is an 8 m-wide dolerite intrusion that contains forsterite (Fo_{75-79}) microphenocrysts in the chilled margin, and optically zoned forsterite (Fo_{60-80}) forming crystals up to 2 mm across in the interior of the dyke. Augite, biotite, iron oxides and plagioclase feldspar complete the primary assemblage. Vesicle-like patches in the dolerite contain a post-magmatic–deuteric assemblage of green amphibole, calcite, chlorite, prehnite and pumpellyite (Rowbotham & Lees 1990).

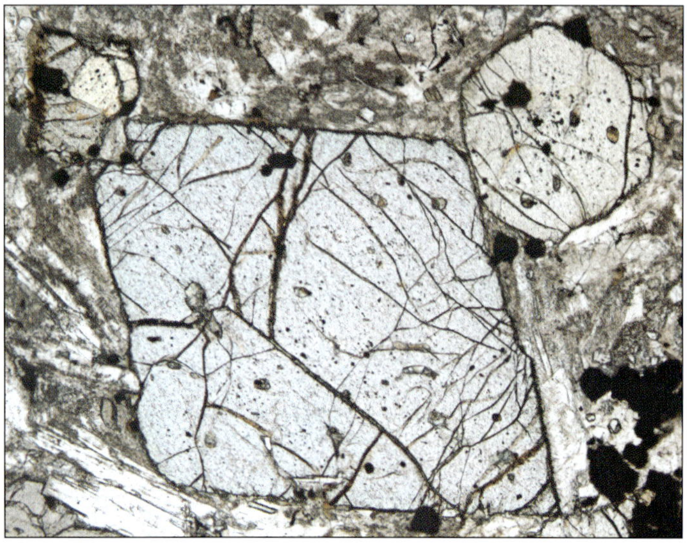

Colourful forsterite with plagioclase feldspar laths and augite (bottom left) in essexite from Crawfordjohn, near Douglas, Dumfries & Galloway, Scotland; width 2 mm; viewed in PPL (upper) and XPL (lower). (© A. G. Tindle)

foshagite

triclinic $Ca_4Si_3O_9(OH)_2$

A rare calcium silicate hydroxide found in late-stage veins cutting Jurassic limestones thermally metamorphosed by gabbro at Kilchoan, Ardnamurchan, Highland Region, Scotland. Foshagite is primarily derived from the replacement of kilchoanite or rustumite. It forms colourless fibrous aggregates, commonly perpendicular to the margins of thin (<2 mm) veins, with sharp contacts against kilchoanite, larnite or merwinite (Agrell 1965, Livingstone 2002).

fraiponite

KAOLINITE–SERPENTINE GROUP

monoclinic $(Zn,Al)_3(Si,Al)_2O_5(OH)_4$

A rare zinc aluminosilicate hydroxide found with gibbsite in dump material derived from workings on the Virgin Moss vein, East Bolton Moor, Wensleydale, Yorkshire. Fraiponite is dull and apple green to whitish green, and is present as pockets and discontinuous bands up to 5 cm across within ochrous brown or cellular limonite. Most specimens have an enamel-like texture, and colloform banding is common. The mineral is considered to be the product of a supergene reaction between sphalerite and aluminium and silica-rich groundwater derived from shale and chert wallrocks (Young et al. 1992c). A soft white botryoidal crust of fraiponite with smithsonite and hemimorphite from the Copperthwaite vein, Swaledale, Yorkshire, is mentioned by Ryback & Tandy (1992).

Fraiponite has been found as soft eggshell-blue crusts up to 0.1 mm thick, associated with brochantite, cerussite and linarite in quartz veinstone from Silver Gill, Caldbeck Fells, Cumbria (Green et al. 2005b).

At Machen quarry, 7 km ENE of Caerphilly, Mid Glamorgan, Wales, hexagonal or pseudo-hexagonal crystals of fraiponite with barite and hemimorphite (±anglesite and cerussite) are found in brecciated galena–sphalerite-bearing calcite veins cutting Carboniferous Limestone. Pearly pale-green fraiponite occurs in two cavities within vein material as crystals up to 0.4 mm across (Goulding & Price 1995). Waxy creamy-white fractured botryoids of fraiponite have also been identified here, although their X-ray patterns were poor (Bevins 1994, Plant & Jones 1995).

francevillite

orthorhombic $Ba(UO_2)_2V_2O_8{\cdot}5H_2O$

A rare hydrated barium uranyl vanadate that forms a series with curienite and is described from Dalveen, S of Leadhills, Dumfries & Galloway, Scotland. Francevillite occurs beneath the Carboniferous unconformity and is associated with calcite veinlets in greywacke (Gallagher et al. 1971).

In England, francevillite has been found in a supergene environment at South Terras mine, St Stephen-in-Brannel, Cornwall, where it forms a yellow earthy crust on a pitchblende nodule (Embrey 1978, Elton & Hooper 1994).

franckeite

triclinic $Pb_6Sb_2FeSn_2S_{14}$

A very rare lead–antimony–iron–tin sulphide reported from the Navan lead–zinc orebody, Co. Meath, Ireland, where a single grain has been found in a polished section of the ore. Minor quantities of argyrodite, boulangerite, bournonite, cylindrite, freibergite, jordanite, pyrargyrite and semseyite are also described from this deposit (Andrew & Ashton 1985, Ashton et al. 1986).

francolite

TL Wheal Franco, Walkhampton, near Tavistock, Devon, England

A synonym of carbonate-fluorapatite, apatite group.

freibergite

cubic TETRAHEDRITE GROUP $Ag_6Cu_4Fe_2Sb_4S_{13}$

A rare silver–copper–iron sulphantimonite that forms series with argentotennantite and with tetrahedrite, and is reported from a galena–pyrite–quartz–siderite lode at Silver Vein mine (Wheal Fortescue), near Lostwithiel, Cornwall (Davies 1867b, Dines 1956). At Pentire (Glaze) mine, near New Polzeath, St Minver, galena contains intergrown inclusions of freibergite, polybasite and rare bournonite in an antimony–lead vein system that also contains arsenopyrite, chalcopyrite, jamesonite, pyrite, sphalerite and tetrahedrite (Clayton et al. 1990).

Freibergite and pyrargyrite are both reported as inclusion species in galena from Tyddyn Gwladys mine, Dolgellau goldbelt, Gwynedd, Wales (Gilbey 1968), and probable freibergite is also noted from the Darren mines, Pen-bont Rhydybeddau, Central Wales orefield, Dyfed (Mason & Hughes 1990).

In Scotland, freibergite and polybasite are the main carriers of silver in a hydrothermal vein deposit at Calliachar Burn, 4 km from Aberfeldy, Tayside Region. 14 steeply dipping quartz–sulphide ± gold veins, cutting Upper Dalradian gritty quartzites, quartz–mica schists and amphibolites, contain polyphase hypogene mineralization consisting of pyrite (most abundant), galena and sphalerite, plus minor chalcopyrite, arsenopyrite and trace gold (electrum). The veins also host benleonardite, cobaltite, cubanite, gersdorffite, hessite, mackinawite, marcasite, pentlandite, pyrrhotite, native tellurium and tetrahedrite (Ixer et al. 1997). Cadmium-rich tetrahedrite inclusions up to 100 μm across in galena are reported from the Tyndrum mining district (particularly from Hard vein), Central Region, and some with over 20 wt% silver are considered to be freibergite (Pattrick 1978). At Lagalochan, near Kilmelford, Strathclyde, a Caledonian diatreme–porphyry-breccia complex contains "crush" or shear zones of irregular shape are rich in freibergite and native silver (both occurring as inclusions in galena). This style of mineralization is associated with a distinctive galena–sphalerite, arsenopyrite–(chalcopyrite)–hessite–pyrite–tetrahedrite suite of veins and disseminations in gouge and rockflour matrix. The Lagalochan complex is thought to represent the lower portions of a now eroded, potentially productive, epithermal precious-metal system (Harris et al. 1988).

Microscopic amounts of boulangerite, cylindrite, freibergite, pyrargyrite and semseyite have been noted in the galena–sphalerite of the Navan orebody, Co. Meath, Ireland (Andrew & Ashton 1985, Ashton et al. 1986, 2004). Another zinc deposit at Keel, Co. Longford, also contains galena and small amounts of the sulphosalts freibergite, jalpaite and pyrargyrite (Ryback & Moreton 1993). A microscopic and EMP study of Ag–Pb–Zn ores from the B-zone, Silvermines, Co. Tipperary, has shown that pyrite (or marcasite) and sphalerite were the first species to crystallize, followed by galena and geocronite, then tetrahedrite-group minerals and proustite–pyragyrite. The tetrahedrite-group minerals have intermediate compositions and some could be considered to be arsenic-rich freibergite (Zakrzewski 1989).

freieslebenite?

monoclinic $AgPbSbS_3$

A silver–lead sulphantimonite reported from the Isle of Sark, Channel Isles, and described by Mourant & Warren (1934) from Sark's Hope mine. Analysis showed that specimens presented to the Geological Museum of the Institute of Geological Sciences in London were in fact galena with a trace of antimony (Macpherson 1983). Freieslebenite requires further confirmation.

freirinite

A synonym of lavendulan.

friedrichite

orthorhombic $Cu_5Pb_5Bi_7S_{18}$

A rare copper–lead sulphobismuthite found in quartz veins cutting monzogranite at Loch Shin, 2 km W of Lairg, Highland Region, Scotland. Friedrichite is accompanied by aikinite, gladite, hammarite, krupkaite, lindströmite and pekoite. Compositions are said to approach friedrichite (Livingstone 2002).

fritzscheite

tetragonal AUTUNITE GROUP $Mn^{2+}(UO_2)_2(PO_4,VO_4)_2\cdot 4H_2O$

A rare hydrated manganese uranyl phosphate vanadate vaguely reported from Cornwall (Clark 1993).

frondelite

orthorhombic $Mn^{2+}Fe^{3+}_4(PO_4)_3(OH)_5$

A rare manganese–iron phosphate hydroxide that forms a series with rockbridgeite and is found in N- or NE-trending iron-oxide veins in kaolinized granite at Gunheath china-clay pit (Hensbarrow section), Carthew, St Austell, Cornwall. Frondelite is intimately mixed with barbosalite, and both probably crystallized from phosphatic solutions as they interacted with early-formed iron-rich veins during the later stages of kaolinization (Elton 1998).

froodite

monoclinic $PdBi_2$

A very rare palladium bismuthide found in the Ben Buie peridotite intrusion at Craig, Isle of Mull, Strathclyde, Scotland. Froodite occurs in, or adjacent to, chromitite layers in association with laurite, ruarsite, sperrylite and minerals of the hollingworthite–irarsite–platarsite series. Grainsize is 1–3 μm (Pirrie et al. 2000a, Power et al. 2003).

Froodite has also been identified in the West Sgaorishal ultramafic plug, Isle of Rum, Highland Region. Discrete PGMs, especially kotulskite and sperrylite, occur within disseminated sulphide (chalcopyrite–pentlandite) mineralization. Less commonly, froodite, gold (electrum), sperrylite and unidentified Pt–Bi–Te and Pd–Sb–Te phases are also found. Grainsize is up to 10 μm, but more often about 2 μm. Froodite is also noted in associated pyrrhotite-rich dykes, where it is clustered together with kotulskite, pentlandite and an unidentified Ag–Pd–Te phase. In related chalcopyrite-rich dykes, froodite is found with gold (electrum), keithconnite, kotulskite, paolovite, native silver, sperrylite and telluropalladinite (Power et al. 2003).

At Talnotry, near Newton Stewart, Dumfries & Galloway, froodite occurs as one of the subordinate species of an arsenic-rich magmatic sulphide assemblage that has developed at the base of an altered diorite sheet. The principal ore minerals are chalcopyrite, gersdorffite, nickeline, pentlandite and pyrrhotite, and, within the mineralized zone, small grains, less than 10 μm across, of sperrylite, irarsite, gold (electrum) and an unnamed bismuth–palladium telluride have been identified (Power & Pirrie 2001, Gunn & Styles 2002). Later work confirms gold (electrum), irarsite and sperrylite to be the dominant PGMs present, with hollingworthite, michenerite and merenskyite being the other subordinate species. Annabergite, arsenopyrite, bismuthinite, galena and pyrite are also noted (Power et al. 2004). Trace amounts of native bismuth, froodite and native gold are reported from a borehole cutting sulphide-rich gabbro at Littlemill farm, near Huntly, Grampian Region (McKervey et al. 2007).

fuchsite

A green chromium-rich variety of muscovite, mica group.

fuller's earth

A calcium-rich variety of montmorillonite, smectite group.

fülöppite

monoclinic $Pb_3Sb_8S_{15}$

A rare lead sulphantimonite found in an antimony-rich quartz vein in the headwaters of Wet Swine Gill, Coombe Height, Caldbeck Fells, Cumbria. The massive white vein contains an early suite of As–Fe–Sb species (arsenopyrite, berthierite and stibnite) and a later suite of Pb–Sb species (fülöppite, semseyite and zinkenite). Fülöppite forms mm-size equant submetallic subhedral crystals with a deep cherry-red colour on thin edges (Fortey et al. 1984, Green & Briscoe 2002, Green et al. 2005e).

funkite

A synonym of hedenbergite, pyroxene group.

G

GADOLINITE GROUP

Monoclinic silicates with the general formula $W_{2-3}X(B,Be)_2(Si,B)_2(O,OH)_{10}$, where W = Ca, Ce, Y, Yb; and X = Fe^{2+}, Mg. Datolite and gadolinite-(Y) are members of the group found in Britain or Ireland.

gadolinite-(Y)

GADOLINITE GROUP
monoclinic $Y_2Fe^{2+}Be_2Si_2O_{10}$

A rare yttrium–iron–beryllium silicate found in accessory amounts in miarolitic cavities within the Northern Arran granite, Isle of Arran, Strathclyde, Scotland. Gadolinite-(Y) is one of many REE minerals found principally in granites, pegmatites, alpine-type fissures and hydrothermal veins. This is the first account of the species from an *in situ* location. Brown to dark-green gadolinite-(Y) forms vitreous prismatic crystals up to 2.5 mm long, associated with and growing on albite and orthoclase. Allanite, beryl, fergusonite-(Y), topaz and zircon are additional accessory species found in the miarolitic cavities (Hyslop et al. 1999). Gadolinite-(Y) and other REE species, such as kainosite-(Y) have also been found at several localities in the granites of the Red Cuillin, Isle of Skye, Highland Region, associated with well crystallized fluorite and zircon (Green & Todd 2001).

References to gadolinite in Ireland lack detail and deserve further investigation: Co. Galway (Greg & Lettsom 1858 and probably Seymour 1902); Blue Lough and Diamond Rocks, Mourne Mountains granite, Co. Down (Burke et al. 1964).

Prismatic gadolinite-(Y) in a miarolitic cavity within the Northern Arran granite, Strathclyde, Scotland; crystal is 2.5 mm across. (IPR/100-52C British Geological Survey © NERC. All rights reserved.)

gahnite

SPINEL GROUP
cubic $ZnAl_2O_4$

An uncommon zinc–aluminium oxide that forms a series with spinel and with hercynite, and is found in accessory amounts in the Lewisian granitic pegmatite at Chaipaval (Ceapabhal), near Northton (Taobh Tuath), South Harris, Western Isles, Scotland. Gahnite is associated with minor amounts of the following species: allanite, beryl, columbite, monazite, spessartine, thorite, tourmaline, uraninite and zircon (von Knorring & Dearnley 1959b, 1960). Robinson & Dossett (2002) report that gahnite at Chaipaval is greenish-blue in colour and 1–3 mm in length. Garnet–staurolite–sillimanite gneiss of the Strathy complex, at Strathy lighthouse, near Melvich, Highland Region, contains accessory amounts of gahnite, together with apatite, ilmenite, rutile and zircon (Burns et al. 2004).

At Creag Bhocan, SSW of Tyndrum, Central Region, gahnite is one of 21 species found as ≤ 10 µm inclusions in the cores of pyrite crystals. They are thought to be primary phases that were trapped during hydrothermal pyrite growth. The locality is the site of copper–zinc stratabound mineralization within the Ben Challum Quartzite (Easdale Subgroup of the Argyll Group Dalradian), characterized by pyrite, chalcopyrite and sphalerite hosted by chloritic schists and albitic quartzites (Scott et al. 1988). Drainage sampling of the Mauchline Basin, NW of Cumnock, Strathclyde, produced gahnite in an isolated sample (Leake et al. 1997b).

At Botallack (the Crowns), St Just, Cornwall, a spinel phase of gahnite–hercynite–spinel composition occurs locally in garnet–magnetite skarns that are found as veins or massive replacements of various rocks within the Land's End granite aureole (van Marcke de Lummen 1985).

galactite

TL Glen Farg, Tayside Region, Scotland
A milky variety of natrolite, zeolite group.

galena

cubic PbS

A very common lead sulphide that forms a series with clausthalite and is found in many geological environments. It is commonly found associated with sphalerite in hydrothermal veins and limestone-replacement deposits. Galena is the main ore species of lead, and silver is often an important minor constituent. However, some occurrences of so-called silver-rich galena may actually be silver-poor galena containing exsolved tetrahedrite inclusions (Ixer & Pattrick 2003).

England

Avon Galena is the dominant sulphide species in a barite vein within faulted Triassic Dolomitic Conglomerate found along the beach near the Pier hotel at Clevedon, 16 km W of Bristol. It forms euhedral to subhedral crystals 3–5 mm across. Associated species (all < 2 mm) include aragonite, aurichalcite, azurite, brochantite, cerussite, chalcopyrite, native copper, covellite, cuprite, malachite, phosgenite, sphalerite and native sulphur (Starkey 1984, 1986c, Ixer 1986, Ixer et al. 1993). Emplacement of manganese orebodies into jointed Carboniferous Limestone at Coombe Farm quarry and Wesley Theological College (Wesley mine), both Westbury-on-Trym, Bristol, was followed by chalcopyrite–galena–(pyrite ± sphalerite) veining (Alabaster 1978, 1989b). Other reports of galena in Carboniferous Limestone include: Churchill – galena associated with mendipite (Rudler 1905); Corporation quarry, Upper Knoll, Brentry, Bristol; Durdham Downs, near Bristol – curious flattened octahedral crystals (both Greg & Lettsom 1858); Leigh Woods, E of Abbots Leigh, Bristol; Wick, near Doynton church, Abston (all Guppy 1956); Hampstead Farm quarry, Chipping Sodbury – with barite–calcite–marcasite–pyrite–sphalerite along joints, with sphalerite in a cave system, and rarely forming cubes, cubo-octahedra and octahedra in barite–pyrite stringers (Alabaster 1990b, Brooke 1994b, Sharp 1997); Lulsgate quarry, SW of Bristol – coralloid or cauliflower-like galena (Cowdry 1995); Southfields quarry, 2 km N of Chipping Sodbury – galena–pyrite–sphalerite in barite–calcite veins within joints, also disseminated galena forming cubes to 4 mm in Rhaetic shales (Curtis 1981); Stancombe quarry, near Flax Bourton, Bristol – minor galena, sphalerite and pyrite in barite–calcite veins (Alabaster 1983, Zachariades 1993); Woodleaze quarry, Tytherington – galena–pyrite–sphalerite in calcite veins within joints (Alabaster 1989a). At East Harptree, N of Mendip Hills, minor galena occurs with sphalerite in deeply oxidized veins within Triassic breccias (Alabaster 1990b).

Cheshire Vein and disseminated Cu–Pb–Zn redbed-type mineralization in coarse-grain clastic sediments of Triassic age on Alderley Edge ridge contains a primary assemblage of chalcopyrite, galena, pyrite (including bravoite), sphalerite and tetrahedrite. Galena is the main sulphide cement enclosing all other sulphides and it shows extensive alteration to anglesite, beaverite, cerussite, covellite, blaubleibender covellite (spionkopite and yarrowite) and djurleite (Ixer & Vaughan 1982). Specific localities include: Stormy Point area, Engine Vein, New Venture, Opencast, West and Wood mines (Eagar & Broadhurst 1959, Warrington 1965, Thompson 1970, Carlon 1979, Braithwaite 1994, Lawton 1994, Plant et al. 1999).

Cornwall Galena and sphalerite occur mainly in cross-course veins that developed after the main episode of copper–tin mineralization. Evidence from fluid inclusions suggests that the base-metal mineralizing fluids were derived from Permo-Triassic basinal sequences (Gleeson et al. 2000). Galena is the chief source of silver in Cornwall, and the content of this metal is said to have been highest in the upper levels of the mines. In medieval times, the galena lodes were exploited mainly for their silver content, and, between 1845 and 1886, 250 000 tonnes of lead ore containing 5.7 million oz. of silver were raised (Dines 1956). *St Just mining district:* Balleswidden mine, near St Just – lode containing cassiterite, galena, pyrite and sphalerite (Hosking 1951); North Zawn, Botallack – quartz–sulphide boulders containing pyrite spherulites with galena and sphalerite cores (Elton & Hooper 1996); Wheal Owles, Penwith – silver-rich galena associated with native bismuth and uraninite (pitchblende) (Ball et al. 1982). *St Ives–Helston mining district:* Binner Downs mine, Helston – minor galena in a complex lode containing cassiterite, native copper, iron oxide, pyrite and sphalerite; Trevaskis mine, near Hayle – arsenopyrite–cassiterite–chalcopyrite–galena–sphalerite lode (both Hosking 1951); Penberthy Croft mine, St Hilary – cross-course MVT Cu–Pb–Zn veins contain chalcopyrite, silver-rich galena, pyrite and sphalerite in a quartz–carbonate gangue. Galena occurs both massive and as crystals with partial cubic faces, and when altered sometimes forms laurionite (Camm & Merry 1991b, Wallace 1998, Betterton et al. 2000); Perran Silver mine (Wheal Sedgman lode), Perranuthnoe – minor galena in a lode also containing acanthite (argentite), chlorargyrite and native silver, all associated with iron-ore deposits (Butler 1907, Rust 1982a); Porthleven – minor galena in pyrite-rich quartz veins (Gleeson et al. 2000); St Aubyn & Grylls, Breage – cross-course lode containing arsenopyrite, chalcopyrite, galena, pyrite, sphalerite and uraninite (pitchblende); Trannack mine, Helston – small amounts of chalcopyrite, covellite, galena, pyrite and sphalerite in a cross-course lode; Wheal Trenwith, St Ives – late-forming species in a cassiterite–quartz lode include native bismuth, galena, molybdenite, sphalerite and uraninite (pitchblende) (all Ball et al. 1982); Wheal Herland, Gwinear – silver-rich galena with acanthite (argentite), arsenopyrite, native bismuth, native silver, skutterudite (smaltite), stibnite and uraninite (pitchblende) in a silver-rich cross-course lode (Rudler 1905, Ball et al. 1982, Rust 1982a); Wheal Penrose, Sithney, near Helston – possibly worked for lead by the Romans, galena, siderite and sphalerite on the dumps (Vipan 1959, Hall & Jackson 1975); Wheal Rose, Sithney – the sequence of deposition was arsenopyrite, chalcopyrite, sphalerite, quartz, galena, siderite and dolomite (Russell & Hutchinson 1927, Russell 1944, Hosking 1951); Wheal Tremayne, Gwinear, near Hayle – native silver in silver-rich galena (Rust 1982a); Wheal Trewavas, Trewavas Head, Breage – arsenopyrite, chalcopyrite, galena, marcasite, pyrite, sphalerite, tristramite and uraninite (pitchblende) on the dumps (Atkin et al. 1983, Ball et al. 1982); Wherry mine, Mount's Bay, Penzance – small quantities of galena associated with chalcopyrite (Russell 1949a). *Camborne–Illogan–Redruth mining district (and southwards into the Carnmenellis granite and its environs):* Dolcoath mine, Camborne – silver-rich galena was one of the many ore species found in the polymetallic (Ag–As–Bi–Co–Cu–Fe–Pb–Sn–U–W–Zn) Main lode (Hosking 1951, Rust 1982a); Kerriack Cove, between Porthtowan and Portreath, near Redruth – minor chalcopyrite and galena in a pyrite–sphalerite-rich quartz vein (Elton & Hooper 1995b); Mount Wellington mine, Gwennap – minor galena and native gold were deposited in a phase of sphalerite-dominated mineralization after the main cassiterite–(wolframite) mineralization (Kettaneh & Badham 1978); New Cook's Kitchen mine (North Tincroft lode), Redruth – complex polymetallic assemblage that includes occasional inclusions of galena in sphalerite (LeBoutillier et al. 2000); Newporth beach, 2 km SW of Falmouth – primary bournonite, chalcopyrite (minor), galena and tetrahedrite in a vein containing supergene cumengéite (Dean 1982, Dean et al. 1983); Pennance mine, Maenporth, near Falmouth – galena-rich ore on dumps and Roscarrack quarry, near Falmouth – galena, pyrite and sphalerite associated with native gold (Camm 1995); Roskrow United mine, Ponsantooth – galena enclosing uraninite (pitchblende) and associated with chalcopyrite, löllingite, pyrite, safflorite and skutterudite (Ball et al. 1982); South Tresavean mine, Gwennap, near Redruth – silver-rich galena with kupfernickel (could be gersdorffite or nickeline), native silver and uraninite (pitchblende) (Hosking 1951, Rust 1982a), also large cubes noted from Tresavean (Greg & Lettsom 1858); Wheal Clinton, near Trefusis Point, Mylor, Falmouth – chalcopyrite–galena–sphalerite lode (Kingsbury 1961); Wheal Gorland, St Day, 2 km E of Redruth – very rare galena (Rust 1981b, Fisher 1988); Wheal Jane, Kea, near Truro – cross-course veins contain galena (silver-rich in places) and pyrite; galena forms bright crystals to 1 cm in modified cubes (Rayment et al. 1971, Rust 1982d). *St Agnes mining district:* East Wheal Rose, near Newlyn East, is considered to be Cornwall's most important lead mine. Between 1845 and 1885 it produced 48 000 tonnes of 62% lead ore, 210 000 oz. of silver, pyrite and sphalerite from two parallel lodes within killas (Dines 1956, Douch 1964). West Chiverton mine, Perranzabuloe, was another important lead mine, with an output between 1859 and 1886 of 45 000 tonnes of 73% lead ore, 1.2 million oz. silver and 22 000 tonnes of zinc ore from galena–sphalerite–quartz lodes (Dewey 1925). Galena was also mined at Lambriggan mine, near Perranzabuloe, from a galena–pyrite–sphalerite lode (Hosking 1951). At Wheal Hope, ENE of Perranporth, galena with chalcopyrite inclusions and long irregular six-sided galena pseudomorphs after pyromorphite are both noted (Greg & Lettsom 1858, Rudler 1905, Hosking & Camm 1981). *Wadebridge–St Endellion mining district:* Polymetallic (antimony–lead mainly) stratabound mineralization has been mined from the Middle Devonian to Lower Carboniferous volcano-sedimentary basinal succession of this area. In addition to antimony, these mines produced subordinate quantities of galena and gold–silver ores, and included: Bounds Cliff, E of Port Isaac – early arsenopyrite–marcasite–pyrite followed by chalcopyrite–galena–sphalerite–lead sulphosalt mineralization (Forbes 1869, Rust 1982a, Clayton et al. 1990, Stanley et al. 1991, Camm 1995); Pengenna (Trewethen) mine, St Kew – arsenopyrite–galena–jamesonite–pyrite lode, antimony-rich becoming lead-rich at depth (Russell 1949b, Kingsbury & Hartley 1960); Pentire (Glaze) mine, near New Polzeath, St Minver – brecciated zones and vein infillings contain galena–carbonate mineralization that formed after chalcopyrite and sphalerite. The silver-rich species freibergite, polybasite and rare bournonite are also present (Clayton et al. 1990); Port Quin mine and Reedy Cliff, both near Port Quin – galena contains inclusions of bournonite and tetrahedrite, and forms intergrowths with other antimony–lead sulphosalts. These and other species found on the dumps include: anglesite, bindheimite, cerussite, chalcopyrite, dundasite, jamesonite (abundant), nadorite, senarmontite, sphalerite and tetrahedrite (Russell 1944, 1949b, Clayton et al. 1990); shear zones within brecciated and carbonatized (metasomatized) basaltic volcanics near Port Quin contain galena with inclusions of bournonite, tetrahedrite and antimony–silver sulphosalts (Clayton 1993); Poltreworgy mine, near Port Isaac – jamesonite–stibnite with bindheimite, boulangerite, galena, pyrite, sphalerite and tetrahedrite (Russell 1949b); Treore mine, Port Gaverne, near Port Isaac, St Teath – arsenopyrite–pyrite-rich lode with associated bindheimite, boulangerite, bournonite, chalcopyrite, galena, native gold, jamesonite, scheelite, sphalerite, stibnite and tetrahedrite (Russell 1949b, Hosking 1951, Kingsbury & Hartley 1960); Trevinnick mine, St Kew – fine-grain masses of galena associated with jamesonite and stibnite (Dewey 1925, Russell 1949b, Fisk et al. 1982); Wheal Boys, Endellion – galena in a polymetallic lode similar to that at Treore mine. Jamesonite and stibnite occurred in the mine at higher levels than galena (Hosking 1951, Russell 1949b).

Mining other than for antimony included the lead workings at: Duchy Peru mine on the Perran Iron lode – worked a silver-rich galena-bearing cross-course vein (Rudler 1905, Vipan 1959, Henley 1971, Slater & Highley 1976, Rust 1982a); Gilson's Cove mine, Lundy Bay, 5 km W of Port Gaverne – good specimens of chalcopyrite and galena (Shrimpton 1993); Great Retallack mine, NW of St Agnes – galena developed large crystals in a lead–zinc lode that became copper rich at depth (Dewey 1925); Gunver Head, near Padstow Consols mine – chalcopyrite–galena–sphalerite quartz-bearing vein (Braithwaite & Merry 1996, Schlomann & Weiss 1998); Penhale mine, Penhale Sands, near Gravel Hill – galena–tetrahedrite ore (Russell 1949b); Wheal Golden, Perranzabuloe – silver-rich galena associated with chalcopyrite, fluorite, native silver and tetrahedrite (Rust 1982a, Merry & Braithwaite 1995); Treburgett mine, St Teath, E of Port Isaac – galena and sphalerite in a cross-course vein, good specimens of ankerite, galena and pyrite on the dumps (Kingsbury & Hartley 1960, Shrimpton 1993). *St Austell mining district:* Silver Vein mine, Lostwithiel – silver-rich galena with chalcopyrite, chlorargyrite, polytelite, pyrite and sphalerite in a quartz–calcite–siderite lode (Rust 1982a); St Austell Consols mine – finely disseminated galena occurs with zoned uraninite (pitchblende) in a cross-course lode (Ball et al. 1982); Terrace (Terras) Hill quarry, Lostwithiel – small masses of galena, pyrite and sphalerite are noted on specimens of apophyllite–axinite–garnet (Russell 1910b); Wheal Mary (Wheal Ding section), Lanivet, near Bodmin – minor chalcopyrite and galena occur with appreciable sphalerite in a garnet–pyroxene–wollastonite skarn assemblage; Wheal Maudlin, Lostwithiel – galena in a polymetallic assemblage (both Kingsbury 1961); South Terras mine, St Stephen-in-Brannel – rare arsenopyrite, chalcopyrite, galena and more common pyrite occur in a pitchblende-bearing quartz lode (Rudler 1905, Robertson & Dines 1929, James 1947, Dines 1956, Ball et al. 1982); Treburland mine, 2 km SE of Fivelanes, Altarnun, Bodmin Moor – trace of galena on the dumps from this manganese-rich contact-metasomatic deposit (Russell 1946d, Kingsbury 1961). *East Cornwall mining district:* Extending from the eastern margin of the Bodmin Moor granite, past the granite cupolas of Kit Hill and Hingston Down towards the western margin of the Dartmoor granite in Devon: Greystone quarry, Lezant, near Launceston – small perfect octahedral and cubo-octahedral crystals of galena to 25 mm and occasional rounded crystals to 5 cm have been found in the Greystone lead lode (Lloyd & Barstow 1983, Wolfe 1997, Elton et al. 2004); Herodsfoot mine, near Liskeard – a Pb–(Ag) mine that worked silver-rich galena from a pyrite–sphalerite-bearing quartz vein (Rudler 1905); New Consols mine, Stoke Climsland – cross-course lode carrying galena and siderite (Hosking 1951); Penharget mine (Caradon Great Consols), Penharget, 7 km NE of Liskeard, and Wheal Caradon (Caradon–Slade), NE of Pensilva, 6 km NE of Liskeard – spots and traces of galena on the dumps of these mines (Shambrook 1982); Redmoor–Kit Hill, Kelly Bray, Callington – silver-rich galena occurs with arsenopyrite, pyrite and sphalerite in quartz–carbonate cross-course veins (Newall & Newall 1989, Newall 1994); West Town Farm (mine), Botus Fleming – chalcopyrite–galena–tetrahedrite–(pyrite–sphalerite) ore (Russell 1949b); Wheal Baron (St George's mine), near Tregeare – gold-rich galena–pyrite lode (Camm 1995); Wheal Langford (had other names too), Callington – silver-rich galena with chalcopyrite, chlorargyrite, pyrargyrite, pyrite, native silver and sphalerite (Rust 1982a); Wheal Mary Ann, Menheniot – Cornwall's deepest lead mine worked a cross-course quartz–carbonate lode containing barite, chalcopyrite, fluorite, galena, pyrite and sphalerite. It produced "very perfect" crystals of galena, combining cube and octahedral forms (Greg & Lettsom 1858, Le Neve Foster 1878b, Dewey 1925, Hosking 1951, Dines 1956, Gleeson et al. 2000); Wheal Trelawney, Menheniot – a vertically zoned lode with galena at the top, followed by pyrite, galena–sphalerite and, at depth, chalcopyrite and native silver (Dewey 1925, Hosking 1951); Wheal Wrey and Ludcott United, St Ive, near Liskeard – cross-course Pb–(Ag) veins contain large ore shoots, veinlets, ribs and masses of galena. Uncommon crystals are cubo-octahedral and about 1.5 cm across, but rarely reach 5 cm (Dines 1956, Rust 1983a). *Other areas:* Camel and Gannel estuaries – galena in intertidal sediments (Pirrie et al. 2000b); Dean quarry, St Keverne, Lizard Peninsula (Wolters & Wolters 1986); wreck of the fireship *Firebrand* in Falmouth harbour – an anthropogenic occurrence of sharp cubo-octahedra of galena to 1 mm on nodules of fused lead (Russell 1920b).

Cumbria Galena is extremely common in the metalliferous veins of the central Lake District and it was the main ore species in many of the mines. It is most abundant in Lower Carboniferous barite–chalcedony–quartz–sphalerite-bearing veins or locally in Upper Carboniferous barite-dominated veins. Greenside mine, near Glenridding, Patterdale, is an example of the former (and the third richest lead mine in Britain, producing 200 000 tonnes of lead concentrate) and Potts Gill barite mine, Caldbeck Fells, an example of the latter. In common with most British galena, it is silver rich, particularly in the presence of barite (Eastwood 1921, Firman 1978, Stanley and Vaughan 1982a). *Caldbeck Fells:* Arm o' Grain – very rare on quartz veinstone (Green et al. 2006b); eastern flank of Balliway Rigg, higher Roughton Gill – highly oxidized galena with cavities containing scotlandite (Green 1989); Bannerdale mine, Mungrisdale – barite, chalcopyrite, galena and sphalerite on the dumps; Thief Gills – solid galena mantled by pyromorphite (both Davidson & Thomson 1951); Brae Fell mine – barite, chalcopyrite, galena and sphalerite on the dumps (Shaw 1970, Hartley 1984, Bridges et al. 2006); Brandy Gill lead mine – uncertain reference to galena; Poddy Gill – barite, chalcopyrite, galena and sphalerite on the dumps (both Davidson & Thomson 1951, Hartley 1984, Cooper & Stanley 1990); Carrock mine, Grainsgill – galena is common in a series of ankerite–quartz veins crosscutting earlier wolframite veins. The wolframite veins themselves also host a minor late-stage assemblage of galena, pyrite and sphalerite (Davidson & Thomson 1951, Eastwood 1959, Ewart 1962, Moorbath 1962, Hartley 1984, Ball et al. 1985a, Neall & Green 2001a); Carrock End mine; Ingray Gill mine and Silver Gill mine, near Roughton Gill (all Hartley 1984); Deer Hills vein – minor galena in an arsenopyrite–(pyrite)-dominated vein (Symes et al. 1990); Driggith mine – over 5000 tonnes of dressed lead ore (includes galena) were mined between 1850 to 1870 from a barite–(chalcopyrite–sphalerite) vein. Other primary species occurring in the vein in lesser amounts include native antimony, arsenopyrite, bournonite and tetrahedrite (Davidson & Thomson 1951, Eastwood 1959, Ineson & Mitchell 1974a, Stanley & Vaughan 1981, Hartley 1984, Cooper & Stanley 1991b, Green et al. 2006a); Dry Gill mine – scarce galena (Eastwood 1921, Shaw 1970); Dry Smale Gill – galena-rich quartz veinstone containing cavities lined with mattheddleite (Green & Briscoe 2002); Nether Row Brow – fine-grain disseminated arsenopyrite–galena–pyrite–sphalerite in a barite-bearing quartz vein (Kingsbury & Hartley 1960, Hartley 1984); Potts Gill barite mine – minor galena in a chalcopyrite-rich vein with arsenopyrite, löllingite, pyrite and sphalerite in small veinlets and stringers (Kingsbury & Hartley 1960, Hartley 1984); Mexico mine, Todd Gill (Shaw 1970); Red Gill mine – massive galena, sometimes intergrown with chalcopyrite in two (out of three) barite–quartz veins (Davidson & Thomson 1951, Hartley 1984, Briscoe 1986a, Wirth 1989, Cooper & Stanley 1991b); Roughton Gill – active as a copper mine during the reign of Elizabeth I. Afterwards galena became the main ore species and, with sphalerite, it was more abundant than chalcopyrite at depth (Davidson & Thomson 1951, Eastwood 1959, Eastwood et al. 1968, Hartley 1984); Sandbed lead mine – massive chalcopyrite, galena and sphalerite on the dumps (Davidson & Thomson 1951, Eastwood 1959, Hartley 1984, Green et al. 2006a); Short Grain – small inclusions in massive barite; Swinburn Gill – mostly massive galena with some cubo-octahedral crystals almost entirely altered to cerussite (both Cooper & Stanley 1990). *West Cumbria:* Minor quantities of galena occur locally in the hematite deposits found in Carboniferous Limestone: Crossgill mine, Frizington – stringers of galena on the dumps (Eastwood et al. 1931, Trotter 1944); Florence mine, near Egremont – small amounts of barite, chalcopyrite, fluorite, galena, manganite and pyrite (Dunham & Hollingworth 1947); Hensingham boreholes, Whitehaven – galena replacing Carboniferous Limestone (Moorbath 1962); Kinniside mine, Cleator Moor (Eastwood et al. 1931); Sellafield area boreholes – traces of arsenopyrite, chalcopyrite, galena, marcasite, pyrrhotite, sphalerite, and Ag–Bi–Se sulphosalts in uncommon pyrite–quartz veins, also traces of barite, fluorite, galena, hematite, pyrite, sphalerite and Ag-, Ag–Bi- and Bi-selenides in later calcite veins (Milodowski et al. 1998); Whitecombe Beck, Black Combe – fluorite, galena and sphalerite on the dumps (Young 1987a). *Elsewhere in the Lake District:* In and around Newlands Valley, galena is the main sulphide in N–S trending lead–zinc veins, and in earlier E–W-trending copper veins it is a subsidiary constituent of chalcopyrite-rich ore. Important localities include: Barrow mine, Braithwaite – native antimony, boulangerite, bournonite and ullmannite occur as inclusions in galena (Davidson & Thomson 1951, Eastwood 1959, Stanley & Vaughan 1981, Young 1987a, Criddle & Stanley 1993); Castlenook mine, W of Grange (Eastwood 1921, Strens 1962, Ineson & Mitchell 1974a); Dale Head mine – galena is the last species formed during ore mineralization of the Dale Head North vein (Stanley & Vaughan 1980); Goldscope mine – one mass from the richest part is said to have contained the equivalent of 50 m^3 of solid galena (Eastwood 1921, Davidson & Thomson 1951, Stanley & Vaughan 1981); Old Brandlehow mine, Derwent Water – cerussite, chalcopyrite, fluorite, galena, pyromorphite and sphalerite on the dumps (Davidson & Thomson 1951); Stoneycroft mine, Braithwaite (Eastwood 1921, Strens 1962); Yewthwaite mine – galena with sphalerite in ribs and nests (Eastwood 1921, 1959, Davidson & Thomson 1951). *Other locations:* Blackbeck Tarn, Buttermere – galena associated with barite, chalcopyrite and pyrite in the Buttermere vein (Strens 1962, 1965, Moorbath 1962); Bonser mine, Coniston – galena is the last sulphide to form and contains inclusions of native bismuth and cosalite (Stanley & Vaughan 1982b); Brundholme mine, Glenderaterra Valley and Loweswater mine, Loweswater (Eastwood 1921); Burtness Wood, Buttermere (Clark 1963); Cat Gill, Borrowdale (Davidson & Thomson 1951, Strens 1964); Coniston copper mines – galena and sphalerite are subordinate to chalcopyrite and pyrite (Ineson & Mitchell 1974a, Dagger 1977, Young 1987a); Damkirk Beck – chalcopyrite–galena–quartz veins associated with the Eskdale granite (Young et al. 1986); Eagle Crag mine, Patterdale – galena (hosting native antimony and bournonite inclusions) and sphalerite were the ore species (Davidson & Thomson 1951, Stanley & Vaughan 1981, Green & Briscoe 2002); Embleton mine, Embleton, near Cockermouth – galena with chalcopyrite, dickite and dolomite in vein breccia; Fleetwith Pike, Buttermere Fells – galena in several small veins; Goat Crag, Borrowdale – steel galena; Honister House and Low Wax Knott, Buttermere – small veins; Manesty copper vein, Grange, Buttermere (all Strens 1962); Gate Gill mine, Threlkeld (part of Threlkeld mine) – worked a galena-rich vein; Greenside mine, Glenridding – good crystals, some with fine irridescent tarnish; Longlands Fell mine, Uldale – galena associated with chalcopyrite; Woodend mine, Threlkeld – corroded galena (all Young 1987a); Force Crag mine, Braithwaite, near Keswick – produced barite, galena and sphalerite from an E–W-trending vein (Davidson & Thomson 1951, Eastwood 1959, Ineson & Mitchell 1974a, Stanley & Vaughan 1981, Green et al. 1997, Tyler 2005); Graphite mine, Seathwaite – calcite–galena veinstone on the dumps (Strens 1962, 1965); Greenside lead mine, near Glenridding, Patterdale – an extremely rich galena-bearing lode that at one time produced attractive crystals. Associated species are pyrite and sphalerite (Goodchild 1884, Davidson & Thomson

1951, Eastwood 1959, Firman 1978, Naylor 1981, Stanley & Vaughan 1981); Hartsop Hall lead mine, Hartsop, Patterdale – dates back to Elizabethan times and contains cubo-octahedral crystals of galena associated with barite, calcite, chalcopyrite, fluorite and sphalerite (Russell 1944, Davidson & Thomson 1951, Shaw 1959, Stanley & Vaughan 1981, Rothwell & Mason 1992); Hellen's mine, Tilberthwaite – galena with arsenopyrite, chalcopyrite, pyrite and sphalerite in a chlorite–quartz gangue (Stanley 1979); Helvellyn (Wythburn) mine, Thirlmere – disseminated galena (some highly silver rich) in a quartz–barite gangue (Ward 1876, Shaw 1970); Knipe Tarn, Staveley – barite–galena vein (Aveline & Hughes 1888, Eastwood 1921); Millrigg mine, Staveley (Lawrence et al. 1986); Leady Moss, near Sosgill (Eastwood et al. 1931); Myers Head mine, Hartsop – galena–sphalerite with a little chalcopyrite, galena (containing native antimony and bournonite inclusions) and sphalerite with minor chalcopyrite (Shaw 1970, Stanley & Vaughan 1981); Piers Gill, 1 km N of Scafell summit, Wasdale, galena occurs in a 60 cm-thick polyphase quartz vein cemented by ankerite, within the Scafell dacite, and is associated with arsenopyrite, chalcopyrite, pyrite and sphalerite, and rare accessory species including acanthite (argentite), native gold (electrum), tetrahedrite, uytenbogaardtite and an unknown selenium–silver mineral (Aldridge 2005); Ruthwaite Lodge vein, Grisedale, SW of Glenridding, Ullswater – deeply weathered cubo-octahedral galena with minor chalcopyrite and tetrahedrite in a quartz vein (Young 1986); Saddleback Old mine, below Foule Crag, Glenderamackin Valley – galena in masses 1 cm across in vein quartz or barite, also supergene galena on cerussite (Tyler 2003, Green et al. 2005c); Sandbeds Gill mine, near Bassenthwaite – traces of galena in a chalcopyrite-bearing quartz vein (Thomson 1997b); Shap Blue quarry – massive galena in veins cutting earlier garnet-bearing veins (Davidson & Thomson 1951, Firman 1953a, 1978, Fortey 1980, Young et al. 1988); Shap granite quarry – galena is widespread but uncommon as anhedral masses and rare as subhedral to euhedral crystals up to 1 mm across in miarolitic granite (Bradshaw 1967, Wirth & Green 2002); Thornthwaite mine, Thornthwaite – cerussite–galena–pyrite–sphalerite on the dumps (Davidson & Thomson 1951); Threlkeld mine, Threlkeld – attractive groups of galena and sphalerite (Davidson & Thomson 1951, Eastwood 1921, 1959); Tilberthwaite Gill, Tilberthwaite – galena–(sphalerite) in lumps and strings within a brecciated vein (Eastwood 1921, Stanley 1979); Wanthwaite mine, St John's in the Vale, near Threlkeld – brecciated vein containing arsenopyrite–pyrite, minor galena–sphalerite–stibnite and traces of chalcopyrite (Kingsbury & Hartley 1958, 1960); Whiteoak Beck, High Nook, Loweswater (Eastwood 1921, Eastwood et al. 1931). *Northern Pennine orefield:* Lead mineralization can be classified into two basic types: vein ore-shoots along near-vertical fissures, and metasomatic flats, formed by replacement of flat-lying beds of limestone. All workable mineral deposits are in country rocks of Carboniferous age, with the majority in the Carboniferous Limestone (formerly called the Lead Measures). Galena occurs in coarse crystalline bands within veins, and in the flats it may be disseminated through silicified, ankeritized or fluoritized limestone as tiny cubes, crystallized bands or masses. Although not rich in silver, galena from this orefield (includes parts of Co. Durham and Northumberland) produced 5.5 million oz. of silver from 1.5 million tonnes of lead concentrate in the period 1729–1870. Galena with an exceptionally high silver content has been found along Sir John's vein and its branch, Stow Crag vein, in the neighbourhood of Tynehead. Mineral deposits in the northern Pennines are not uniform in composition and a zonal sequence with a central fluorite zone and a barite- or witherite-rich fringe has been recognized. The most important galena–(sphalerite)-rich deposits occur at the outer margin of the fluorite zone (Dunham 1934, 1944, 1990, Raistrick 1936b, Dunham et al. 2001). The history of the Alston Moor mines is discussed in detail by Fairbairn (1993). Galena is reported from many localities, including: Blagill mine, near Alston – barite, galena and witherite were the ore species; Hartley mine, Kirkby Stephen – barite, chalcopyrite, fluorite, galena and linarite on the dumps (both Davidson & Thomson 1951); Brownley Hill mine, Nenthead, near Alston – distorted cube and octahedral forms of galena, with crystals exceptionally to 10 cm once found. Sphalerite with minor chalcopyrite, marcasite and pyrite complete the primary sulphide assemblage (Greg & Lettsom 1858, Young et al. 1990b, Clark & Rust 1993, Green & McCallum 1993, Green et al. 2000); Clouds Gill mine – chalcopyrite and galena with calcite and fluorite (Bateson & Johnson 1984); Browngill vein, River South Tyne, Garrigill; Cowslitts vein, Nenthead; Farnberry mine, Alston; Garrigill–Nenthead road (Browngill vein); Hiddenholes mine, Tynehead; Stow Crag mine, Tynehead – galena from these localities exhibits a wide range of silver content (all Dunham et al. 2001); Dufton mine, near Appleby – old records speak of large rough octahedral crystals of galena (Greg & Lettsom 1858, Bramall 1921, Davidson & Thomson 1951); Fell End Clouds, 10 km SSW of Kirkby Stephen (Dunham & Wilson 1985); Great Sulphur vein, Alston Moor – minor galena, pyrrhotite and sphalerite in a marcasite–pyrite–(chalcopyrite)-dominated quartz–carbonate vein (Vaughan & Ixer 1980); Higher Longrigg mine, S of Cote Garth, Hartley Birkett, near Kirkby Stephen – chalcopyrite–galena in a barite gangue (Braithwaite 1988); High Stennerskeugh level, Ravenstonedale – limestone abundantly veined by calcite and fluorite, and containing minor chalcopyrite, galena and tennantite on the dumps (Small 1982, Young 1987b); Hilton mine, Scordale, near Appleby – galena forms moderate-size modified cube and cubo-octahedral crystals associated with barite, fluorite and witherite. Euhedral cubo-octahedral crystals of galena on gersdorffite and galena intergrown with nickeline were found in an area of nickel mineralization close to the Whin Sill (Dunham & Hollingworth 1947, Davidson & Thomson 1951, Solomon et al. 1971, Bridges 1982, Ixer 1986, Winrow 1986a, Bouch et al. 2006); Hudgill Burn mine, Alston – worked for both cerussite and galena

Galena cubes to 12 mm on ankerite; Old Fan flats, Smallcleugh mine, Nenthead, near Alston, Cumbria. (© D. I. Green)

Cubo-octahedral galena to 25 mm on ankerite; Hansome Mea flats, Smallcleugh mine, Nenthead, near Alston, Cumbria. (© D. I. Green)

(Bridges & Young 1998, Dunham et al. 2001); Nentsberry (Hagg's) mine, Nenthead – notable barite, galena, sphalerite and witherite (Davidson & Thomson 1951, Solomon et al. 1971, Dunham et al. 2001); Rampgill, Nenthead – fluorite, galena, gypsum and witherite in well crystallized and sometimes spectacular specimens (Wolfe 2004); Rotherhope mine, Leadgate, near Alston – crystallized galena, marcasite, pyrite and siderite, also abundant fluorite (Dunham 1947, 1959a, 1990, Davidson & Thomson 1951, Solomon et al. 1971); Silverband mine, Milburn, near Appleby – after galena deposits were exhausted, the mine became an important barite producer (Dunham & Hollingworth 1947, Davidson & Thomson 1951, Solomon et al. 1971); Silver Top quarry, Hallbankgate, Brampton – specks of galena in a barite-bearing fissure vein (Young et al. 1989); Smallcleugh mine (Hansome Mea Flats), Nenthead, near Alston – crystallized calcite, galena, gypsum (selenite) and sphalerite. Galena forms cubic or cubo-octahedral crystals often associated with ankerite, and in 1990 unusual elongated cubo-octahedral crystals were collected from Middlecleugh vein (Davidson & Thomson 1951, More et al. 1991, Green & Briscoe 2002, Bouch et al. 2006); Staveley Head, Staveley (Kendall 1884); Stennerskeugh Clouds, Kirkby Stephen (Starkey 1987b); Threlkeld Side level, Dufton – worked for lead and produced galena from large pockets and nodules surrounded by barite (Bramall 1921); Tynebottom mine, NW of Garrigill, near Alston – the order of crystallization of the ore species was pyrite, pyrrhotite, marcasite, glaucodot, gersdorffite, ullmannite, chalcopyrite, galena and sphalerite (Vaughan & Ixer 1980, Bouch et al. 2006); Windy Brow vein, Tynehead, near Garrigill – well crystallized galena partially altered to cerussite (Young 1985a).

Derbyshire In the Southern Pennine orefield, galena has been found in over 2000 named veins, where it is associated with minor sphalerite and the gangue species barite, calcite and fluorite. Euhedral crystals, usually combination cubes and octahedra, are uncommon and often show a degree of alteration because of surface oxidation. In extreme cases, oxidation produces a pseudomorph of cerussite after galena. Lustrous cleavable masses, granular steel ore and a radiate form are described, and, in some examples, inclusions of pyrite occur. Derbyshire galena in general does not have a high silver content, although Ball Eye mine, Bonsall, is an exception. The main host rock for lead mineralization is Carboniferous Limestone, but a few examples are within Millstone Grit or basaltic volcanic rocks (toadstones). Galena has been worked from major veins (rakes), minor veins (scrins), metasomatic limestone-replacement deposits, and open cavity infills (flats and pipes). It is probable that mining started during the Early Bronze Age and continued through the Roman occupation into Anglo-Saxon and Medieval times. The 18th and 19th centuries saw the industry reach the peak of production, with major technical advances in the form of gunpowder blasting and of steam engines used to pump water out of the mines and to stamp the ore. It is estimated that 3–6 million tonnes of lead concentrates have been raised from over 270 mines since mining began (Schnellmann & Willson 1947, Ford et al. 1993, Ford & Rieuwerts 2000). Details of the mines are noted by Carruthers & Strahan (1923). *Castleton–Bradwell area:* Bradwell Moor, including Dirtlow Rake – cubes to 5 mm in calcite (Robinson & Ineson 1979, Jeffrey 1997); Odin mine, Castleton – galena encloses native silver (Ford & Ineson 1971, Ixer & Townley 1979, Mostaghel 1985); Pindale, Castleton – galena in bands and as freely grown crystals (Sarjeant 1956b, Robinson & Ineson 1979, Ford et al. 1993); Speedwell mine, Castleton – massive galena (Robinson & Ineson 1979); Treak Cliff, Castleton – galena with barite, calcite, fluorite and pyrite in pipe veins, some forming blocks up to 36 kg (Ford 1955, Sarjeant 1956b). *Hucklow–Eyam–Stoney Middleton area:* Blackhole mine, Eyam – galena in calcite or fluorite; Dirty Rake, Stony Middleton – massive galena (Robinson & Ineson 1979); Glebe mine, Eyam – originally a lead mine, later an important fluorite producer (Varvill 1959); High Rake, Longstone Edge, N of Bakewell – galena-rich replacement flat with a well developed supergene assemblage (Robinson & Ineson 1979, Jeffrey 2001); Ladywash mine, Eyam – overgrowths of pyrite on galena and sphalerite (Robinson & Ineson 1979, Mostaghel 1985); Mill Dam mine, near Great Hucklow – common galena and sphalerite, also cerussite and fluorite with overgrowths of supergene galena (Green et al. 2005c). *Ashford–Bakewell–Sheldon area:* Hard Rake, near Sheldon – galena cubes to 7 mm in calcite (Robinson & Ineson 1979); Hubberdale mine, Flagg, near Bakewell – one of the richest mines in Derbyshire in the 18th century, where a 10-tonne lump of galena was found in 1768 (Green & Strahan 1887, Varvill 1959, Robinson & Ineson 1979); Long Rake, near Middleton-by-Youlgreave – galena–pyrite–sphalerite in a barite–fluorite gangue, Raper mine is one of the workings along it (Ineson & Al-Kufaishi 1970, Robinson & Ineson 1979, Ball et al. 1985b); Magpie mine, Sheldon – a pipe deposit in reef-facies limestones containing galena with fluorite, pyrite and abundant hydrocarbons (Worley 1976, Robinson & Ineson 1979); Mogshaw mine, SE of Sheldon – galena commonly enclosed in banded barite (Ford & Sarjeant 1964b). *Bonsall–Matlock Bath–Cromford–Wirksworth area:* Ashley Hay, near Wirksworth – galena hosted by Millstone Grit (Lacey 1862, Ford & Sarjeant 1964a); Bage mine, Bole Hill, Wirksworth – water-worn galena associated with matlockite and phosgenite (Burr 1992); Ball Eye opencast, north side of Via Gellia, near Cromford – 11 cm-thick flat-type deposit of galena in Carboniferous dolomitized limestone immediately on top of a basic lava (Bridges & Green 2005a); Devonshire and Tearsall mines, Matlock (Robinson & Ineson 1979); Bonsall Moor area – galena in bands or massive (Sarjeant 1956b); Golconda mine, Brassington – massive and banded galena (Ford & King 1965, Robinson & Ineson 1979, Ford et al. 1993); Green Linnets mine, Brassington – galena encrusted with cerussite and pyromorphite (Rudler 1905); Jughole mine, Masson Hill, Matlock – galena occurs with barite, chalcopyrite, pyrite (bravoite) and sphalerite in a fluorite–smithsonite matrix (Ixer 1974a); Masson and Rutland Caverns, and Nestus mine, all Heights of Abraham, Matlock Bath – galena has mostly been mined out and the residue forms strings, pockets and individual grains (Braithwaite & Ryback 1963a); Masson Hill, Matlock – large subhedral to euhedral cubes of galena in a fluorite-rich flat, also some octahedral crystals (Ixer 1974b, Ford et al. 1993); Millclose mine, Wensley, near Matlock – Derbyshire's most productive lead mine (500 000 tonnes of lead concentrate between 1865 and 1940) contained cubes of galena up to 5 cm on edge, associated with large crystals of calcite (Rudler 1905, Russell 1912b, Traill 1939, Bannister 1940, Ford & Ineson 1971, Mostaghel 1985, Ixer 1986, Willies et al. 1989); Oxclose mine, Snitterton, W of Matlock (Ixer 1974a,b); Pike Lane opencut, near Winster – massive galena in calcite and fluorite, also occasional dull-grey rough crystals surrounded by pyromorphite and cerussite (Uttley 1989); Riber mine, Matlock – band of galena associated with calcite (Varvill 1959, Robinson & Ineson 1979); Wapping mine, Cromford - small amounts of galena are scattered throughout Moletrap rake, some lining vugs and forming crystals to 3 cm, also cubo-octahedral crystals to 1.5 cm, cubes to 1 cm, and a large mass of galena (40 kg) found in 1965. (Robinson & Ineson 1979, Ince 1991, Ford et al. 1993). *Ashover and Crich inliers:* Fall Hill (Gate) quarry, Ashover – silver-rich galena was mined from Ashover Old vein in the 18th and 19th centuries. Small cubo-octahedra of galena are associated with fluorite, and cerussite is a frequent alteration product (Robinson & Ineson 1979, Smith 1982, Spence 1982); Gregory mine, Ashover – a complex cubo-octahedral crystal of galena (5 cm across) on fluorite on a specimen in the Devonshire collection, Chatsworth House (Ford et al. 1993, Cooper 1995); Milltown quarry – scattered masses of galena or rarely octahedral or cubic crystals to 15 mm, also one specimen of galena associated with barite, cerussite and weddellite, and others associated with anglesite, cerussite, hydrocerussite or sweetite (Robinson & Ineson 1979, Rust 1983b, 1991c, Clark et al. 1984); Crich quarry – a thick pod of galena with cerussite in barite–fluorite gangue is locally dusted with well formed crystals of wulfenite (Wilson et al. 1995). *Ashbourne–Snelston area:* Snelston mine, near Ashbourne – small nodules and fragments of galena are nearly always encrusted with cerussite, but mainly unaltered crystals are found filling spaces between calcite scalenohedra in dolomitized limestone (Dewey & Eastwood 1925, Jackman 1996). This locality is described as Birchwoodpark, Snelston, by Cornwell et al. (2001), who also note nodules of cerussite, chalcopyrite and galena in a malachite-cemented calcareous sandstone at nearby Limestone Hill, between Ashbourne and Stanton. *Elsewhere:* Axe Edge, SW of Buxton and Whaley Bridge, N of Buxton – galena found in Millstone Grit (Ford & Sarjeant 1964a); Bincliffe mine, Manifold Valley – galena with barite (Robinson & Ineson 1979); Whitwell, 19 km SE of Sheffield – barite–galena vein with extensive metasomatic replacement of Permian Magnesian Limestone, also supergene galena coating cerussite, lanarkite and leadhillite (Ineson et al. 1972, Green et al. 2005c).

Devon Silver-rich galena has been worked (intermittently) from quartz–siderite veins and lenses at the Combe Martin mines from 1294 (during the reign of Edward I) until 1875. Sphalerite and minor amounts of chalcopyrite, millerite, pyrite, stibnite and tetrahedrite are also reported from these mines. The orebodies are irregular reef-like masses with base-metal sulphides developed in the cores of near-isoclinal overturned folds in Carboniferous slates (Hall 1868b, 1877, Dunham et al. 1978, Scrivener & Bennett 1980, Durrance & Laming 1982, Rust 1982a). Apart from a brief note of galena among a suite of arsenic, cobalt, copper, nickel, silver and uranium minerals in vanadium-rich nodules in Permian redbeds from the Littleham Cove area, Budleigh Salterton (Milodowski et al. 2000), all other occurrences are related to the intrusion of the Dartmoor granite: Bere Alston mines – many fine specimens of fluorite and galena came from these mines (Hall 1868b, Rudler 1905, Ball et al. 1985b); Bridford barite mine, Teign Valley, Christow, 10 km SW of Exeter – large quantities of galena (±pyrite–sphalerite) were extracted before the mine became an important barite producer (Polkinghorne 1951); Devon and Courtenay mine; East Tamar mine, Bere Ferrers; Hennock, near Chudleigh Hall; Hennock, near Chudleigh; Holestock Hall; Okehampton Consols; South Hoo mine, near Bere Alston; Tamar mines, near Bere Ferrers; Wheal Betsy, near Tavistock (all Hall 1868b); Frankmills mine and Wheal Exmouth, east Dartmoor – the largest Pb–(Ag) producers (Durrance & Laming 1982, Embrey & Symes 1987); Lockridge (Goldstreet) mine, SW of Bere Alston – galena associated with tennantite, tetrahedrite and intergrown sphalerite–wurtzite (Grguric & Nickel 2006); Red-a-ven mine, Meldon, near Okehampton – galena is associated with disseminated löllingite and pyrrhotite in a mineralized skarn (Kingsbury 1961); South Tamar mine, near Tavistock – galena particularly rich in silver, averaging 37 oz. per tonne of lead concentrates (Vipan 1959); Teign Valley, near Hennock – silver-rich galena with chalcopyrite, pyrite, sphalerite and tetrahedrite in a barite–calcite–fluorite–siderite gangue (Beer & Ball 1977); Wheal Emily, Wembury – rare galena in a semseyite-dominated ore, accompanied by jamesonite, pyrite and rare bournonite (Russell 1949b); Wheal Friendship (Devon Friendship), Mary Tavy – galena in a cross-course vein cutting earlier polymetallic ore (Hall 1868b, Braithwaite 1982b, Barstow 1983, Brooke 1986).

Durham The mining of galena may have begun in Roman times, although direct evidence is lacking. It is likely that the bishops of Durham owned the mineral rights of the county from 1072, but it was not until the 18th century that major development of lead mining and smelting occurred. This continued well into the 20th century, when Boltsburn mine in Weardale became the most successful and profitable lead mine in the northern Pennines. Between 1818 and 1940 it produced 128 000 tonnes of lead concentrates (70% between 1901 and 1931). As with lead mines elsewhere in the orefield, Boltsburn worked ore-shoots in fissure veins and metasomatic replacement flats in Carboniferous Limestone (Dunham 1959a, 1990, King 1982b). The history of the Weardale mines is discussed further by Fairbairn (1996). *Weardale:* All the Weardale mines are discussed by Dunham (1990) and further details are given by other authors: Barbary mine (Dunham 1959a, Solomon et al. 1971); Blackdene mine, near Ireshopeburn – a particularly fine specimen of galena is noted (Moreton 1998); Boltsburn mine, Rookhope – large crystals of galena up to 2.75 cm across were found in cavities within metasomatic flats, usually associated with siderite (Smythe 1921, Solomon et al. 1971, King 1982b); Burtree Pasture mine, Cowshill – fluorite–galena veins (Smith & Hardy 1981); Daddry Shield, near St John's Chapel – galena associated with celestine (Smythe 1922); Frazer's Hush mine (Wolfe 1998, Green & Briscoe 2002); Greenlaws mine, S of Daddry Shield – corroded cubes with octahedral modifications in mud-filled fluorite-lined vugs within mineralized flats (Moreton 2005b); Groverake mine, near Rookhope – worked initially for galena, it later became the largest single producer of fluorite in the orefield (Dunham 1959a, Greenwood 1977, Young 1995, Dunham et al. 2001); Heights quarry – dull but sharp cubo-octahedral galena associated with remarkable fluorite (Green & Briscoe 2002); Killhope mine, between Cowshill and Nenthead – fluorite, galena, siderite and sphalerite in veins (Solomon et al. 1971, Dunham 1992, Young 1995, Dunham et al. 2001); Park level mine, Killhope – cavities in a block of galena contain minute supergene galena cubes associated with ankerite (or siderite), cerussite and fluorite (Green et al. 2000f); Redburn mine, near Rookhope – galena in fluorite (Day 1979); Rogerley mine, near Frosterley – galena forms subhedral to euhedral

Cube-octahedral galena, up to 2.5 cm across, intermixed with creamy-yellow siderite; Boltsburn mine, Rookhope, Weardale, Co. Durham. (© J. Fisher)

crystals or is intergrown with fluorite. Cerussite replaces the cores of some crystals (Ixer 2003); Rookhope borehole – galena occurs with arsenopyrite, chalcopyrite, molybdenite, pyrite, sphalerite and a tetrahedrite-group mineral in minor amounts within the buried Weardale granite and, possibly from the footwall of Red vein, with fluorite, pyrrhotite and sphalerite (Bradshaw 1967, Dunham 1967, Brown et al. 1987a); Sedling mine, Cowshill and Wolfcleugh mine, near Rookhope – mineralized with fluorite, galena and siderite, but not of economic grade (Dunham 1959a); Stanhopeburn mine, N of Stanhope and Stotfield Burn mine, near Rookhope – both worked Red vein, 5000 tonnes of galena mined from the latter (Dunham 1959a, Solomon et al. 1971); Whitfield Brow mine, near Stanhope – galena partially replaced by cerussite with a thin crust of pyromorphite (Young 1985a). *Teesdale:* Closehouse mine, Lunedale – lead mining in the district dates from the 13th century. Small amounts of massive galena are associated with a fault-bounded vein and metasomatic replacement barite deposit. Less commonly, galena occurs as 10–30 μm inclusions in sphalerite (Hill & Dunham 1968, Solomon et al. 1971, Vaughan & Ixer 1980, Young et al. 1994a, Dunham et al. 2001); Coldberry mine, 4 km north of Middleton-in-Teesdale (Dunham 1959a); Cow Green mine, Harwood-in-Teesdale and Lunehead mine, near Grains o' th' Beck, Lunedale (Solomon et al. 1971); Grasshill mine, near Harwood – concentric bands of pyromorphite surrounding cores of galena (Young 1985a, Dunham et al. 2001); Lady's Rake mine, 6 km NW of Langdon Beck – galena associated with chalcopyrite, pyrite, pyrrhotite and sphalerite in a magnetite–nickeline ore found close to the Whin Sill (Young et al. 1985b, Green & Briscoe 2002); Teesdale Valley – complex assemblage comprising chalcopyrite, galena, pyrite and sphalerite, with minor arsenopyrite, gersdorffite, tetrahedrite and ullmannite; Whiteheaps mine, near Blanchland – minor galena in a fluorite vein (both Vaughan & Ixer 1980). *Other areas:* Cubes of galena replace dolomite on joints within fossiliferous Permian Magnesian Limestone in the Hesleden Dene boreholes, near Castle Eden, NW of West Hartlepool. Galena is also noted in similar rocks from Chilton quarry, near Ferryhill. The main associated species are barite, calcite, fluorite and sphalerite (Dunham et al. 1948, Fowler 1956, Hirst & Smith 1974). Boreholes reaching the base of the Permian at Fishburn contain minor galena lining joints and as thin films in limestone or marl (Fowler 1943). At New Brancepeth colliery, minor galena occurs with alstonite, chalcopyrite, sphalerite, ullmannite and witherite in a major barite vein cutting faulted Coal Measures rocks (Spencer 1910, Smythe 1922, Ixer 1986). Galena is also noted from: Dry Pry mine, Wolsingham – includes cerussite coated with supergene galena (Dunham et al. 2001, Green et al. 2005c); Healeyfield mine, Castleside – well crystallized cubes to 3 mm on edge (Solomon et al. 1971, Young 1985a); and Pike Law workings, Newbiggin (Solomon et al. 1971).

Gloucestershire Galena occurs with pyrite, sphalerite and probably chalcopyrite in thin cracks and fissures within Wenlock strata at Glasshouse Brook, near Birches, 4 km S of Newent (Bick 1986a).

Lancashire Galena is noted with chalcopyrite, sphalerite and witherite from a lode on Anglezark Moor, near Chorley (Rudler 1905). The mining history of this district is covered by Gill (1987).

Leicestershire A major review of galena in Leicestershire is provided by King

(1983), who considers almost all the deposits to be epigenetic, but notes a lack of data to allow a chronology to be established. These descriptions are supplemented by other published works: Cloud Hill quarry, Breedon on the Hill – occasionally quite spectacular developments of galena, including beautiful very high-lustre material. Spheroids of galena form masses weighing up to 29 kg; Earl Ferrers lead mine, Staunton Harold, Dimminsdale – an ancient mine noted for its outstanding finely crystallized galena, including cubo-octahedral crystals up to 26 mm (average 18 mm) and a second generation forming bizarre skeletal crystals on calcite or the earlier galena; Gipsy Lane brick pit, Leicester – minor amounts of galena on fracture planes in calcareous mudstones of the Mercia Mudstone (Keuper Marl) (Faithfull & Hubbard 1988); Newhurst quarry, near Shepshed – strings of galena in quartz–carbonate veins cutting North Charnwood diorite and fine-grain galena replacing calcite in Triassic sediments (Ince 2005); opencast coal site opposite the lane to Old Parks farm, Ashby-de-la-Zouch – galena in sphalerite-rich septarian siderite mudstone nodules; Tickow Lane mine, Blackbrook, nr Shepshed galena was the main ore species, occurring with cerussite, pyromorphite and wulfenite (King & Ludlam 1969), and fine crystals were found at "Ticknil Hall" (Greg & Lettsom 1858), which is probably the same location. Galena is also noted from Ashby-de-la-Zouch (Rudler 1905), and from Breedon quarry, Breedon on the Hill, where a Triassic cave system in Carboniferous Limestone produced euhedral crystals of galena to 60 mm on edge entirely coated with crystallized cerussite (King 1991).

Northumberland Most occurrences of galena are from the Allendale, Derwent and Tynedale districts, whose mining history is documented by Fairbairn (2000): Allenhead (Beaumont) mine – the largest galena producer in the Northern Pennine orefield (>260 000 tonnes of lead concentrate) worked large fluorite-rich ore zones (Dunham 1959a, Solomon et al. 1971, Greenwood 1977); Barrasford quarry, 9 km N of Hexham – joints in the Whin Sill carry barite, galena and sphalerite in fissure veins (Money et al. 1995); Fallowfield mine – mining may have begun as early as Roman times, although the earliest documentary record is from 1611. Galena occurs in an ore-bearing fault zone with alstonite, barite, barytocalcite, calcite and witherite (the chief constituent of the vein). A coal seam containing an intimate mixture of alstonite, coal, galena and witherite is also noted (Spencer 1910, Briscoe 1986b); Langley Barony mine, near Hexham – peak production of 4700 tonnes of galena in 1882 (Trestrail 1931, Harley 1999); Settlingstones mine, near Hexham – sandstones and shales close to the Whin Sill contain minor galena intimately associated with nickeline and sphalerite. Chalcopyrite, gersdorffite, marcasite, millerite, pyrite (bravoite), rammelsbergite, vaesite and violarite are associated species (Russell 1927, Trestrail 1931, Solomon et al. 1971, Ixer 1986); Shildon mine, Blanchland and St Peter's mine, Allenheads (Solomon et al. 1971); Sipton vein, N of Spartylea – spots of galena in a fluorite vein (Dunham 1990).

Nottinghamshire The top 5 cm of the Lower Magnesian Limestone is feebly but persistently mineralized along its outcrop from Nottingham to Doncaster. Galena predominates and both cerussite and uranium-rich asphaltite are associates. Additionally, wulfenite is present at Bulwell, Nottingham and Kirkby in Ashfield. Other galena localities include: Clipstone colliery; Limekiln Wood, Newstead, near Mansfield; and Papplewick Hall, N of Bulwell (all Deans 1961). Galena forming hopper-faced cubo-octahedral crystals has been found coating fractures in Permian limestones (Cadeby Formation) at Vale Road quarry, Mansfield Woodhouse. Aurichalcite, barite, calcite, chalcopyrite, dolomite, malachite, melanterite, pyrite and sphalerite are associated species (Green & Spence 1988).

Shropshire Galena was the major species mined in the West Shropshire orefield from veins mainly cutting Ordovician sediments and volcanics. Barite, calcite and sphalerite were mined much later. Some of the mines date back to Roman times and these (and others) flourished at various times between the 12th and 18th centuries. Dines (1958) provides a summary of the mines, most within a 10 km radius of Shelve: Bog, Callow Hill, the Grits, Huglith (Precambrian host rocks), Ladywell, Pennerley, Perkins Beach, Ritton Castle, Roman Gravels, Rorrington, Shelve, Snailbeach, Tankerville and Wotherton. Samples from many of these have been used to show that multiple sources and complex mineralizing fluids were involved in the genesis of the deposits (Pattrick & Bowell 1991, Haggerty et al. 1996b). Good (but rare) crystals of galena are noted from Burgam mine, SSW of Pontesbury (Starkey 1982).

Somerset Galena was mined in large quantities from veins cutting Carboniferous Limestone in the Mendip Hills. In places the galena occurred as matrix-free nodules, and at Lamb Bottom, near West Harptree, and Greendown, near Litton, a stalactitic variety has been found (Cowdry 1995). At Grebe Swallet mine, Charterhouse, nodules of galena up to 28 kg have been recovered from clay- and rubble-filled fissures (Stanton 1991). Other occurrences include: W of Eaker Hill farm, WSW of Chewton Mendip church – one small specimen of dendritic galena (Kingsbury 1941); Higher Pitts mine, Priddy (Burr 1996); Lamb Lair Cavern, Gibbet's Brow, near Compton Martin – stalactites of galena are mentioned in old documents (Cowdry 1996); Merehead (Torr Works) quarry, Cranmore – good crystals of chalcopyrite and galena associated with barite in manganese-oxide-rich veins (Symes & Embrey 1977); Whatley quarry, near Frome – barite–galena–smithsonite-bearing veinlets and solution cavities (Alabaster 1976). A review of the Mendip mines is provided by Gough (1930).

Staffordshire Although better known for the mining of copper, Ecton Hill produced considerable quantities of galena from small veins in the vicinity of major open cavity-infill mineralization. Galena was often found at high levels on the hill, whereas copper sulphides (chiefly chalcopyrite) predominated deeper in the hill (Rudler 1905, Dewey & Eastwood 1925, Robey & Porter 1972, Mostaghel 1985, Masheder & Rankin 1988, Ford et al. 1993). Dewey & Eastwood (1925) also document other occurrences: galena associated with copper mineralization in conglomerate at Shore Hill, N of Cannock, and in boreholes at Fair Oak colliery, NE of Huntington gravel pit; and Mixon mine, Butterton – galena in copper ore.

Warwickshire Galena occurs with millerite in septarian nodules at Newdigate colliery, Bedworth, SW of Nuneaton (Walker 1988), and in rocks of the Caldecote Volcanic Series beneath the Precambrian/Cambrian unconformity at Judkins quarry, Nuneaton (Ince et al. 1991). At Judkins quarry, galena forms sporadic single cubo-octahedral crystals to 2 mm, occasionally with a skeletal appearance on some faces, and tabular twinned crystals with hexagonal cross section up to 3 mm across.

Yorkshire Apart from fringe areas in the N and NW, the entire southern half of the Northern Pennine orefield is in Yorkshire (the Askrigg Block), where a major (indirect) influence on ore formation has been the buried Wensleydale granite intrusion centred at Raydale. Galena has been mined in Yorkshire for two millennia, often from remote and inaccessible locations. The earliest evidence of mining was during the Roman occupation and is based on the discovery of pigs of lead found at Heyshaw Bank, 4 km SE of Greenhow, and dated AD 81. Documentary evidence exists of mining activity throughout the ages, but it was only in the 18th century that the collection of mining statistics began. Since that time (up to 1939) an estimated 1 million tonnes of lead concentrates was produced (Dunham & Wilson 1985). Regional histories of lead mining include: Arkengarthdale (Tyson 1995), Grassington (Gill 1993), Greenhow (Gill 1998), Grinton (Tyson et al. 1995), Marrick in Swaledale (Tyson 1989), South Craven (Gill 1987), Swaledale (Gill 2004) and Wharfedale (Gill 1994). *North Swaledale mineral belt:* Friarfold–Surrender vein, between Keld and Langthwaite – the most continuously mineralized vein in the belt, contains galena associated with barite, fluorite, a little sphalerite and traces of chalcopyrite (Dunham & Wilson 1985, Young et al. 1989); Gunnerside Gill mines (Dunham & Wilson 1985, Wood 1993); Lover Gill, W of Muker – galena with high silver content associated with barite, calcite and fluorite (Bateson & Johnson 1984, Dunham & Wilson 1985); Old Rake vein (includes Old Gang mines), NE of Gunnerside Gill (Dunham & Wilson 1985, Young et al. 1989, Ineson & Young 1994, Green 2003); Watersykes vein, Victoria level, NE of Gunnerside (Ineson & Young 1994); and Whaw, Arkengarthdale (Rodgers 1975). *Wensleydale:* East Bolton Moor, near Askrigg – intergrown galena–sphalerite associated with barite, calcite and locally abundant witherite (Vaughan & Ixer 1980, Young et al. 1992c); Raydale borehole – samples of Wensleydale granite from the base of the borehole contain minor amounts of chalcopyrite, galena, molybdenite, pyrite and sphalerite (Brown et al. 1987a); Wet Grooves mine, Aysgarth and Worton Hall level, near Askrigg – both areas are within a central fluorite zone (Dunham & Wilson 1985). *Wharfedale & Nidderdale:* Coldstones quarry, 5 km W of Pateley Bridge – strongly oxidized veins in Carboniferous Limestone containing large massive lumps of galena (with occasional cubic faces) associated with barite, fluorite and sphalerite (Ryde 1987, Young et al. 1989, 1997, Bridges 1997a, 2002); Duck Street quarry, SE of summit of Greenhow Hill, near Pateley Bridge – exposure of Greenhow Rake, a major calcite–fluorite–galena vein, also coarse crystals in other veins (Dunham & Stubblefield 1944, Rankin et al. 1987); Grassington Moor; Greenhow Hill; and Providence mine, near Thwaite (Dunham & Wilson 1985).

Galena is also noted at: Bitholmes mine, Ewden Valley, 9 km NNW of Sheffield (Moore 1988); Pikedaw Gill, NW of Malham (Raistrick 1983, Mundy & Arthurton 1994); Merrybent mine, NW of Middleton Tyas (Small 1977, Dunham & Wilson 1985); and Scaleber Bridge, 3 km SSE of Settle (Ixer & Stanley 1998).

Wales

Clwyd Galena occurs with sphalerite in many veins cutting Carboniferous Limestone from Minera in the S to Abergele in the N (i.e. along the entire 30 km-long outcrop of the limestone). Evidence from the many ingots found in Clwyd dates the earliest smelting of galena as Roman (AD 74–96). This area was the most important lead producer in Wales, and between 1845 and 1938 over 658 000 tonnes of lead concentrates were extracted from 245 mines (Earp 1958, Ixer & Vaughan 1993, Bevins 1994, Bennett 1995). A selection of occurrences include: Bwlch-y-Plwm mine, Penrhyndeudraeth; Nant-y-Plwm mine, Llansannan; Pennant mine, near St Asaph (all Archer 1959); Clawdd-ffordd, near Holywell and Crecas, Whitford; Silver Rake, Halkyn Mountain; Talacre mine, near Prestatyn; Trelogan mine (all Campbell Smith 1913); Halkyn mine, and Hendre, Pandy, near Glyn Ceiriog (Fletcher et al. 1993); Talargoch mine, near Dyserth – the abundance of galena relative to sphalerite decreased with depth. Drift gravels above the veins produced significant galena in water-worn lumps termed gravel ore (Rudler 1905, Campbell Smith 1913, Dunham et al. 1978); Wedding Cave mine, Bwlchgwyn, near Wrexham – galena associated with chalcopyrite and pyrite, but with very little sphalerite (Green 1995b).

Dyfed *Central Wales orefield:* Galena is a major component of fracture-hosted mineral veins in this area, particularly those in the W, with sphalerite being more common towards the E. A paragenetic study indicates a sequence of crystallization in the veins of pyrite–ankerite–chalcopyrite–sphalerite–galena, coupled with prolonged quartz deposition (Raybould 1974, Bevins 1994). The veins have been divided into two groups of assemblages (A1 – early complex; A2 – late simple) and include an economically important silver-rich A1-*c* polymetallic assemblage, in which quartz-cemented breccias host chalcopyrite and galena, often in fine-grain intergrowths. Boulangerite, bournonite, cobalt pentlandite, gold (electrum), gersdorffite, millerite, pyrite, pyrrhotite, tetrahedrite, tučekite and ullmannite are minor associated species. A1 assemblages are found at: Erglodd mine, N of Tal-y-bont; Esgair Hir mine, Ceulanymaesmawr; Esgair Fraith mine, near Tal-y-bont; Eaglebrook (Nantycagl) mine, Ceulanymaesmawr; and Darren mine, Pen-bont Rhydybeddau. Of the A2 assemblages, the A2-*a* sphalerite–galena quartz assemblage was the most important economic source of lead and zinc, and has been found at: Brynyrafr mine; Cefngwyn mine, near Aberystwyth; Dolclettwr mine, Tre'r-dol; Frongoch mine, between Devil's Bridge and Pont-rhyd-y-groes; Nantymwyn mine; and Ystumtuen mine (all Mason 1997). Mining histories include: Bwlchglas (Hughes 1989c), Cwmystwyth (Hughes 1981), Darren (Hughes 1989a), Frongoch (Bick 1986b), Goginan (Hughes 1988), Llywernog (Hughes 1990), Ystrad Einion (Hughes 1989b), and other authors provide more general accounts (Lewis 1967, Bick 1988, 1990, 1991, 1992).

Occurrences of galena are reported from: Alltycrib and Leri Valley mines, Tal-y-bont and Nantycreiau mine (Mason 1998); Bronfloyd mine, ESE of Penrhyn-coch and Camdwrbach mine, near Bryniau Rhyddion, NW of Nant-y-moch Reservoir (Raybould 1974); Brynyrafr mine and Logaulas, near Ysbyty Ystwyth (Fletcher et al. 1993); Bwlchglas mine, near Tal-y-bont – brecciated galena–sphalerite-bearing quartz veinstone subjected to extensive supergene oxidation (Rust 1996c, Mason 2004); Caegynon mine, near Ponterwyd – small octahedral crystals; Goginan mine – fibrous steel ore; Penrhiw mine, Ystumtuen – cubo-octahedral crystals (all Jones & Moreton 1977); Copa Hill mine, Cwmystwyth – galena was mined from Comet lode during the Bronze Age and near by is a site of medieval lead smelting (Timberlake 1993, Mighall et al. 2002); Cwmsymlog mine – highly silver-rich galena (Ridgway 1980, Fletcher et al. 1993); Cwmystwyth mine – chalcopyrite–galena–sphalerite in a calcite–dolomite–quartz gangue (Dewey & Eastwood 1925, Raybould 1974, Fletcher et al. 1993); Darren mine, Pen-bont Rhydybeddau – boldly crystallized and granular galena, both highly silver rich (Rudler 1905, Ridgway 1980); Eaglebrook (Nantycagl) mine, Ceulanymaesmawr – worked a deeply oxidized chalcopyrite–galena lode, notable for its supergene assemblage. Some galena is the fine-grain steel-ore variety, and partial octahedral crystals to 1 cm on edge are also recorded (Braithwaite 1982b, Jones 1983, Mason 2004); Elgar mine, 3 km S of Tal-y-bont – distinctive quartz–chalcopyrite–ullmannite–galena sequence, and Pandy mine, Tre'r-ddôl – galena partially altered to cerussite (Mason 2004); Erglodd mine, N of Tal-y-bont – galena intergrown with siegenite (Rust 1991a); Esgair Hir mine, Ceulanymaesmawr – on the dumps, cleavage fragments to a few cm, steel-ore and rare rounded cubo-octahedral crystals to 3 mm (Rust & Mason 1988, Fletcher et al. 1993); Frongoch mine, between Devil's Bridge and Pont-rhyd-y-groes (Fletcher et al. 1993, Green et al. 1996b); Hendrefelen mine, Ysbyty Ystwyth – disseminated calcite–chalcopyrite–galena–ullmannite occasionally form euhedral crystals in cavities within quartz veinstone (Clark & Rust 1993); Ysbyty Ystwyth area – fine specimens of galena (Bevins 1994); Llechweddhelyg mine, near Clawddmelyn, between Bont-goch and Cwmsymlog – galena and chalcopyrite disseminated in quartz veinstone and rare cubo-octahedral crystals of galena with dulled faces (Jones 1987, Rust 1994, Mason 2004); South Darren mine, Pen-bont Rhydybeddau – silver-rich galena associated with considerable chalcopyrite and some sphalerite (Dewey & Eastwood 1925, Hughes 1959); Ystrad Einion mine, near Tal-y-bont – polymetallic assemblage of chalcopyrite and galena with accessory siegenite (Mason & Rust 1997).

Elsewhere, a significant quantity of galena was mined near Llanfyrnach, SSE of Cardigan (Bevins 1994), and, at the Dolaucothi (Ogofau) mines, near Pumsaint, minor amounts of galena occur with sphalerite in gold-rich arsenopyrite–pyrite veins cutting pyritiferous shale bands (Brammall et al. 1937, Steed et al. 1976, Annels & Roberts 1989). Galena is also reported from the Carboniferous Limestone in the Mynyddygarreg district, near Kidwelly (Rudler 1905, Bevins 1994), and from veins in the Carmarthen area, for example at the Vale of Towy (Nant) mine (Bevins 1994).

Gwent Thin stringers and veins of galena occur in barite–calcite veins cutting Carboniferous Limestone at Dan-y-graig quarry, near Risca (Alabaster 1983).

Gwynedd Galena and sphalerite are intimately intergrown in a minor ore known locally as bluestone at Mona and Parys mines, Parys Mountain, Amlwch, Anglesey. They occur with the main chalcopyrite-rich ore in a major volcanogenic massive sulphide deposit that was worked from Bronze Age times. Silver-rich galena is reported, but this may be attributable to inclusions of tennantite–tetrahedrite – species also found as rims on chalcopyrite and galena, as small veinlets in sphalerite, and as anhedral isolated grains (Greenly 1919, Dewey & Eastwood 1925, Pointon & Ixer 1980, Ridgway 1980, Timberlake 1988, Fletcher et al. 1993, Southwood & Bevins 1995).

The Great Orme mines, Llandudno, are another site of Bronze Age mining, where galena occurs in a chalcopyrite-dominated ore. Mineralization is within dolomitized Carboniferous Limestone. Galena forms euhedral-to-anhedral crystals up to 1 cm across that enclose gersdorffite, Co–Fe–Ni sulphides, rare tennantite and sphalerite. Rarely, inclusion-free galena is found in veins along joint planes separate from the copper ores (Dutton et al. 1994, Ixer & Davies 1996, Ixer & Stanley 1996, Joel et al. 1996, Ixer 2001). A few km to the S, at Bwlch mine, near Deganwy, Conwy, galena, marcasite and sphalerite are minor species found in antimony–lead (jamesonite–plagionite–semseyite–stibnite–zinkenite) mineralization within nodular Ordovician ignimbrites (Russell 1949b, Bevins et al. 1988, Ryback & Francis 2001). *Eastern Snowdonia:* The Llanrwst orefield (between Betws-y-coed and Trefriw) produced significant galena, sphalerite and minor chalcopyrite from epigenetic veins and vein breccias in an Ordovician volcano-sedimentary sequence. Total production of all the mines was 13 000 tonnes of lead concentrates and 14 000 tonnes of zinc ore between 1848 and 1938 (Otty 1986). Occurrences include: Cilenus – sphalerite-dominated ore with galena and pyrite in a quartz–ankerite cemented breccia; Klondyke mine – galena–pyrite–sphalerite ore with minor chalcopyrite; New Pandora mine – sphalerite–galena–marcasite–pyrite ore, lead–zinc–quartz rich in the upper levels and iron-calcite rich at depth (all Haggerty 1995); Parc mine – the largest and most successful mine, worked galena–marcasite–sphalerite ore in an ankerite–calcite–dolomite–quartz gangue (Archer & Elliott 1965, Otty 1986, Haggerty 1995). Lead isotopic data for these and other mines in the orefield (Cyffty, Hafna and Penyrallt) are provided by Haggerty et al. (1996a). *Elsewhere in Snowdonia and the Lleyn Peninsula:* Benallt mine, Rhiw, Lleyn – two galena–sphalerite-bearing quartz lodes (Archer 1959); Britannia mine, Snowdon; Cwm Llan, SE of Snowdon; Lliwedd, SE of Snowdon – the Snowdon Volcanic Group hosts Cu–Pb–Zn vein mineralization in which chalcopyrite, pyrite and sphalerite dominate. Silver-rich galena is locally abundant (all Reedman et al. 1985); Drws-y-Coed mine, near Nantlle – principally chalcopyrite with galena, pyrite, pyrrhotite and sphalerite in a quartz gangue (Dewey & Eastwood 1925, Braithwaite 1982b); Gwaith mine, Llandegai – chalcopyrite–galena–pyrite ore; Manod-bach mine, near Ffestiniog – chalcopyrite–galena–pyrite–sphalerite ore in a quartz gangue; Pan-ty-wrach mine, between Llanfrothen and Penryndeudraeth; Pen-y-groes – mixed sulphide ore of chalcopyrite, galena and sphalerite (all Dewey & Eastwood 1925); Fron-oleu, Prenteg, Tremadog – galena is a rare accessory mineral associated with brookite in some calcite–quartz veins (Starkey & Robinson 1992); Gamallt mine, Carreg y Fran, near Ffestiniog – chalcopyrite–galena–sphalerite ore (Dewey & Eastwood 1925, Archer 1959, Lynas 1973); Gorlan mine, near Trefriw – specimens of dundasite forming crusts on galena, marcasite, sphalerite and quartz (Russell 1944); Graig Wen, near Ffestiniog – chalcopyrite–galena–pyrite–sphalerite-bearing lode; Llyn Bowydd, near Ffestiniog – large quantities of fine-grain sphalerite with patches of chalcopyrite and intergrown cubes of galena (both Lynas 1973); Penrhyn (Penrhyn Du) mine, Lleyn – barite–galena–sphalerite in a quartz vein cutting Ordovician mudstones with phosgenite found in cavities within the galena (Hubbard 1991); Tal-y-Sarn mine, Llanllyfni – chalcopyrite ore passing downwards into mixed chalcopyrite–galena–sphalerite ore (Dewey & Eastwood 1925, Archer 1959); Trecastell mine, near Conwy – especially fine specimens of galena (Archer 1959, Bevins 1994). A mining history of the Llanegan orefield is provided by Bennett & Vernon (2002). *Dolgellau goldbelt:* The complex shear-zone-hosted mesothermal gold-lode mineralization has a four-stage paragenetic sequence, the last of which is a lead–zinc assemblage. The assemblage is more common to the east, particularly between Foel Ispri and Gwynfynydd mines. Galena was worked prior to the gold-mining era at East Cwmheisian mine (Mason et al. 2002). Other reports of galena include: Clogau mine, Bontddu – many species reported, including arsenopyrite, chalcopyrite (abundant), erythrite, galena, native gold, pyrite, pyrrhotite, sphalerite and tetradymite. More recent work has identified an Ag–Au–Bi–Sb–Te bonanza gold–silver assemblage, in which galena occurs in symplectite-like intergrowths with tellurobismuthite and tetradymite (Andrew 1910, Dewey & Eastwood 1925, Naden 1988, Bevins & Stanley 1990, Mason et al. 2002); Friog Undercliff, Fairbourne, SW of Dolgellau – fine-grain sulphide veins dominated by pyrrhotite, with local concentrations of galena and sphalerite (Mason et al. 1999); Gwynfynydd mine – galena–pyrite–sphalerite-dominated veins in which some galena and native gold is replaced by sphalerite. Galena contains inclusions of pyrargyrite and tetrahedrite (Allen & Jackson 1985, Dominy et al. 1996, Platten & Dominy 1999, Mason et al. 2002); Panorama lode, Llanaber, E of Barmouth and Vigra lodes – native gold-bearing chalcopyrite–galena–sphalerite lodes (Andrew 1910, Dewey & Eastwood 1925); Prince Edward mine – native gold-bearing ribbon-quartz veins carrying arsenopyrite and pyrite, followed by chalcopyrite and pyrrhotite, and finally galena and sphalerite (Mason & Williamson 2001); Tyddyn Gwladys mine – electrum with a high silver content intergrown with pyrargyrite and silver-rich tetrahedrite, overgrown by chalcopyrite and enclosed by galena and sphalerite (Forbes 1868, Mason et al. 2002). Additionally, galena occurs as inclusions in alluvial gold from Afon Mawddach and Afon Wen (Leake et al. 1993).

Mid Glamorgan Galena-bearing veins cut Carboniferous Limestone around the margins of the South Wales coalfield and include: Bute quarry, Llantrisant – dolomitized limestone containing galena cubes to 1 cm and composite crystals of chalcopyrite and sphalerite (Fletcher & Young 1988); Groes Faen, near Miskin, Llantrisant area, and between Machen and Rudry – barite–galena-bearing veins (North 1916); Machen quarry, 7 km ENE of Caerphilly – partially oxidized galena in a mineralized fault zone hosts anglesite, cerussite and cinnabar in cavities. Rare crystals of galena form groups of dull, rough cubes and octahedra up to 10 mm across (Bridges 1990, Bevins 1994, Goulding & Price 1995, Plant & Jones 1995).

Galena is also found in clay–ironstone nodules in the coalfield, sometimes associated with chalcopyrite, millerite or sphalerite: Senghenydd pit, Abertridwr (Moore & Pearce 1997); International colliery, Blaengarw, near Maesteg, and Park Slip West opencast coalmine, Aberkenfig (both Plant 1997); Wyndham colliery, Ogmore Vale – delicate octahedral galena perched on millerite needles (Cooper 1991a, Fletcher et al. 1993, Bevins 1994). Thin galena-bearing veins are further described from Jurassic rocks at Ogmore-by-Sea and Porthcawl (Fletcher 1988, Billington 1997b).

Powys *Central Wales orefield:* Bryntail and Nantiago mines – examples of A2-type "late simple" assemblages (see Dyfed), excellent crystals of octahedral galena in the latter (Billington 1997c, Mason 1997); Dyfngwm mine (actually the Pen Dylife section of Dylife mine), Machynlleth – small masses of galena to 5 cm, commonly found embedded in sphalerite, also rarely as cubo-octahedral crystals to 4 mm (Rust & Rust 1987); Dylife mine – abundant galena in the Llechwedd Ddu lode (Moreton 1982); Pen-y-Clun mine, NW of Llanidloes – weathered masses and crystals of galena on the dumps (Morgan & Starkey 1991); Van mine, Llanidloes – excellent cubo-octahedral crystals of galena associated with sphalerite in vein breccia (Rudler 1905, Jones & Moreton 1977, Hughes 1991).

Elsewhere galena is noted from: Bulthy mine, Middletown, Breidden Hills (Dines 1958); Craig-y-Mwyn mine (Dunham et al. 1978); Cwm Orog mine, Llangynog (Archer 1959); Dolyhir quarry, Old Radnor, near Presteigne – galena–tennantite vein (Wolfe 2004); Llangynog mine, southern flank of Berwyn Dome (Archer 1959, Dunham et al. 1978); Llanymynech Hill, near Oswestry (Campbell Smith 1913); Rhandirmwyn mine (Bevins 1994).

South Glamorgan Galena was worked in levels between St Hilary and New Beaupre, both near Cowbridge, where Triassic breccias rest unconformably on Carboniferous Limestone. Also in the Cowbridge area at Langland farm, near Brocastle, W of Cowbridge, and Pentre, near Llantrithyd, galena is restricted to Carboniferous rocks (North 1916). Bevins (1994) notes that galena was exploited from Llangan mine, near Cowbridge.

Scotland

Borders Galena occurs as inclusions in alluvial gold at: Craigierigg Burn; Cramalt Burn; Glengaber Burn, SW of Peebles; head of River Tweed (Leake et al. 1993, Chapman et al. 2000a,c). Heddle (1901) notes galena from the Vaults, near Dunbar.

Central *Ochil Hills:* Alva mine – minor galena is associated with acanthite (argentite), bornite, chalcocite, chalcopyrite, covellite, digenite, erythrite, malachite, maucherite, pyrite, pyrrhotite, native silver and sphalerite, all in a barite–calcite gangue within veins cutting Old Red Sandstone volcanics (Heddle 1901, Dunham et al. 1978, Jassim et al. 1983, Moreton et al. 1998); Burn of Sorrow mine, near Dollar – chalcopyrite–galena–pyrite vein (Dickie & Forster 1974); Westerton, near Alva – cubes of galena (Heddle 1901). Galena is also reported as inclusions in alluvial gold (Chapman et al. 2005). *Tyndrum mining district:* Galena and sphalerite were mined between 1745 and 1925 from mineralized veins and vein breccias cutting Dalradian metasediments. The galena contains small inclusions (< 100 µm) of marcasite, pyrargyrite, pyrite and tetrahedrite. So-called "steel ore", a fine-grain massive variety of galena with inclusions of chalcopyrite and fragments of earlier quartz and sphalerite, on rare occasions produced 1.2 kg silver per tonne (but the silver was probably in tetrahedrite inclusions). Tyndrum Main mine worked Hard vein (a major galena–sphalerite–quartz lode) and this is crosscut by minor barite–calcite–chalcopyrite–uraninite veinlets, in which galena has been found as inclusions in the latter two species. Galena is also present in gold–silver veins, where it replaces pyrite and contains inclusions of gold (electrum) and hessite (Heddle 1901, Wilson & Flett 1921, Darnley et al. 1960, Darnley 1962b, Ineson & Mitchell 1974b, Pattrick 1978, 1985, Ridgway 1980, Pattrick et al. 1988, Curtis et al. 1993). Another gold–silver vein deposit occurs 4 km SW of Tyndrum at Cononish, where galena occurs intergrown with native gold in microfractures within pyrite (Parker et al. 1989, Earls et al. 1992, Curtis et al. 1993). Alluvial gold from Cononish contains inclusions of galena (Chapman 2000c). Heddle (1901) notes galena from Clifton mine, and down stream from the Tyndrum mineralization in the Crom Allt River, a galena-bearing mineralized boulder has been found (Pattrick & Hall 1984). Galena is also found with sphalerite in Dalradian stratabound deposits (interpreted as relict hydrothermal rocks) near Ben Challum to the E of Tyndrum (Smith et al. 1984, Fortey & Smith 1986), and with chalcopyrite and pyrite in chloritic schists at Creag Bhocan, SSW of Tyndrum (Scott et al. 1988).

Dumfries & Galloway The mining of galena in this part of the Leadhills–Wanlockhead orefield dates back to 1680, during the reign of James VI (although more probably they were rediscovered at this time). Early mining was from Straitstep vein under Dod Hill. By 1921 the total amount of Pb + Zn ore mined approached 500 000 tonnes. Galena, forming massive coarse-grain crystalline aggregates or fine-grain

Complex galena crystals with white scalenohedral calcite on a quartz matrix (85 × 114 mm); New Glencrieff mine, Wanlockhead, Dumfries & Galloway, Scotland. (© M. P. Cooper)

"steel ore" was the main ore species, with sphalerite abundant in some veins. Chalcopyrite, jamesonite and pyrite were the other primary ore species. Details of all the mines are summarized by Wilson & Flett (1921). Other reports include Wanlockhead (Brown 1927, Temple 1955), and Glencrieff mine, Wanlockhead (Temple 1954b, Nicholson & Banks 1987).

At Palnure Burn, Talnotry, 5 km NE of Newton Stewart, galena containing segregations of matildite occurs in an arsenopyrite-bearing quartz vein at the contact of the Cairnsmore of Fleet granite with hornfelsed greywackes (Lowry 1993, Lowry et al. 1997). Galena is also present with bismuthinite (perhaps also altaite and tellurobismuthite) in a zone of arsenic-rich mineralization within a diorite intrusion at Talnotry. Additionally, chalcopyrite, gersdorffite and nickeline are present in parts of the orebody (Power et al. 2004).

Around Glendinning (Louisa) mine, 13 km to the NNW of Langholm, galena is reported from drill-core sampling stratiform and disseminated arsenopyrite–pyrite mineralization, in which bournonite, chalcopyrite, semseyite, sphalerite, stibnite, tennantite and tetrahedrite were found (Gallagher et al. 1983). Glendinning mine worked a stibnite-bearing quartz vein (a turbidite-hosted As–Au–Sb deposit) containing minor chalcopyrite, galena and sphalerite (Williams 1965, 1973, Ball et al. 1985b, Duller et al. 1997).

Other occurrences of galena include: Callisterhall, near Langholm – galena with chalcopyrite and sphalerite in veins and disseminations in basal Carboniferous sediments underlying the Birrenswark Lavas (Smith et al. 1978); Carsphairn lead mines – massive and occasionally crystalline galena with chalcopyrite and sphalerite on the dumps (Williams 1965); Dalmellington; Pibble mine, Creetown; Woodhead, near Garryhorn House, Carsphairn (all Heddle 1901); East and West Blackcraig mines, SE of Newton Stewart – shattered barite–calcite–dolomite–(quartz) vein containing strings and patches of galena, sphalerite and chalcopyrite (Heddle 1901, Wilson & Flett 1921, Livingstone et al. 1976); Glenhead Burn, Loch Dee – minor amounts of galena and sphalerite in carbonate veinlets following native gold and arsenopyrite–pyrite veining in a Silurian–Ordovician turbidite sequence close to the Loch Doon granite complex (Leake et al. 1981). Knipe (Fountainhead) mine, Hare Hill, 5 km SE of New Cumnock – galena and sphalerite occur in association with stibnite in late-stage veins cutting a Caledonian granodiorite (Boast et al. 1990);

Fife Fine octahedral crystals of galena in veins partly cutting dolerite, partly cutting Carboniferous sandstone at Inverkeithing, Castlelandhill (Greg & Lettsom 1858, Heddle 1901).

Grampian Galena, sphalerite and barium-rich minerals (armenite, barite and hyalophane) occur in a thick band of Dalradian cherty sulphidic quartzite exposed in cliff faces SE of Coire Loch Kander, near Braemar (Gallagher et al. 1989a, Fortey et al. 1991, 1993). Silver-rich galena was mined between 1870 to 1875 from a fault-bounded fissure vein cutting Dalradian schist at Abergairn (Corrybeg) lead mine, near Ballater. It forms small dull-grey cubes and octahedra associated with calcite (schiefer spar), fluorite, psilomelane and sphalerite (Heddle 1901, Russell 1936a, Dunham 1952, Hall & Walsh 1972, Nicholson 1989b). Other occurrences include: Back Burn, W of Auchenblae – galena inclusions in alluvial gold (Chapman et al. 2000c); coastal section 2 km E of Burghead, 12 km W of Lossiemouth – mineralized fault containing cubes of fluorite and galena (Gillen 1987b); Stotfield, West Beach, Lossiemouth – disseminated galena associated with phosgenite, pyrite, pyromorphite and sphalerite in a Triassic breccia termed the Cherty Rock (Wilson & Flett 1921, Starkey 1988a, Green & Todd 2001).

Highland The Allt Cathair Bhan section of the Loch Ailsh alkaline complex and the ultramafic section of the Loch Borralan alkaline complex, Assynt, contain post-magmatic sulphides including chalcopyrite, galena, pyrite and sphalerite. In places, galena is intergrown with PGMs (e.g. cooperite, sperrylite), hessite and bismuth tellurides (Shaw et al. 1994, Gunn & Styles 2002, Styles et al. 2004). Heddle (1901) reports small specks of galena associated with babingtonite, lepidocrocite (rubinglimmer), magnetite, microcline (amazonite), thorite, titanite and topaz, probably from pegmatitic veins cutting syenite in the cliffs of Sgòr Chaonasaid, Ben Loyal.

Galena has been described from barite–fluorite–hematite–quartz veins cutting the eastern margin of the Grudie granite and adjacent Moinian schists, near Lairg, and from aikinite–bismuthinite–chalcopyrite–fluorite-bearing quartz veins to the N of the Loch Shin granite (Gallagher 1970, Gallagher et al. 1974, Gallagher & Smith 1975, Fortey 1980, Lowry et al. 1994).

Silver-rich galena was the primary ore mineral of the Strontian mines (Bellsgrove, Corrantee, Fee Donald, Middleshop and Whitesmith) and was found with jamesonite, pyrite and sphalerite in a barite–calcite–quartz gangue at the northern end of the Strontian granite complex. The mines worked intermittently from 1722 to 1872, and demand for barite in the 1980s led to a reworking of the veins. Fine specimens of cubo-octahedral galena crystals to 4 mm were recovered from the mine dumps as recently as 2000 (Scott 1871, Heddle 1901, Wilson & Flett 1921, Bradshaw 1967, Winrow 1986b, Green & Todd 2001). Small amounts of galena have been found in barite veins to the E of the Strontian granite complex at Garbh Bheinn, Ardgour (Gallagher 1964).

Other occurrences include: Achanarras farm, near Halkirk – barite–galena veins in a mineralized crush zone within Old Red Sandstone rocks (Berridge 1969, Dunham et al. 1978); Beinn a' Chaisil, Glen Sanda – galena–pyrite–sphalerite vein and nearby at Rubha a' Chamais Bhàin – galena–magnesite–millerite–sphalerite vein; Brawl and Skinnet, near Halkirk; Dulnain Bridge, 3 km SW of Grantown-on-Spey – diopside–galena–pyrite–sphalerite–zoisite association; Camasunary Bay, Strathaird Peninsula, Isle of Skye – rare galena associated with djerfisherite, garnet, melilite, perovskite, spurrite, tilleyite, vesuvianite and wollastonite in a high-temperature calcic skarn near the contact of the Cuillin gabbro complex; Gie-uisg Geo, SW of Crosskirk – asphalt–calcite–galena–pyrite–sphalerite vein; Glen Nevis – two galena-bearing veins; in a burn on road between Jeantown and Loch Kishorn – fluorite–galena (all Heddle 1901); Bridge of Westfield in the Loch Cander crush zone – mineralized breccia with chalcopyrite, galena, marcasite, pyrite, smithsonite, sphalerite and uranium-rich hydrocarbons in a calcite–siderite gangue; Broubster – galena–sphalerite in a calcite–dolomite vein cutting Old Red Sandstone rocks; Carn Chuinneag granite complex – galena in a sheared lamprophyre dyke to the north of the complex; Helmsdale, including Caen Burn, Ceann Ousdale and Longwell Water – galena in uranium-rich veinlets cutting granite and aureole rocks; Shurrery, Caithness – disseminated galena in loose blocks of arkose (all Gallagher et al. 1971); Gairloch schist belt (Proterozoic) including the Kerry Road sulphide (pyrite–pyrrhotite mainly) deposit – minor species include chalcopyrite, galena, native gold, magnetite, marcasite and sphalerite (Jones et al. 1983, 1987); Kildonan, Suisgill and Torrish Burns, NW of Helmsdale – galena inclusions in alluvial gold (Leake et al. 1993, Chapman et al. 2000c); Loch Claise na Feàrna, Scourie Estate – galena and sphalerite mineralization with minor chalcopyrite in hornblendite and hornblende schist (Anon. 1974); McPhun's Cairn, Loch Fyne – stratiform sulphide mineralization in Cambrian phyllites is dominated by pyrite, but also contains appreciable galena and sphalerite (Willan 1980, Willan & Hall 1980); Ratagain complex, Loch Duich, near Glenelg – chalcopyrite, galena, matildite (only at Invershiel), molybdenite, pyrite, rutile and sphalerite in calcite–fluorite–quartz veins cutting quartz monzonite (Alderton 1986, 1988b, Lowry 1993); Struy mines, Loch na Mèine, mid-way between Strath Glass and Strath Farrar, 3 km SW of Struy – embedded masses and coarse cleavage cubes of silver-rich galena with sphalerite in barite veins; Allt Tigh Cumhaig Burn, 3 km S of Struy – barite vein containing more sphalerite than galena (both Russell 1946b, Berridge 1969); Sgaorishal ultramafic plug, Isle of Rum – small blebs (<20 μm) of galena in sulphide-rich (pyrrhotite- or chalcopyrite-dominated) dykes crosscutting the plug (Power et al. 2003); Beinn an Dubhaich, Broadford area, Isle of Skye – galena in boron–fluorine-rich ore skarns at the margins of the granite (Tilley 1948b, 1951).

Lothian Galena occurs with dolomite, pyrite and sphalerite in seams of coal at Bathgate and with barite and scleretinite (a pitchy resin) at the "silver mine" at the north-east foot of Cairnpapple Hill, in the Bathgate Hills (Heddle 1901). This locality is referred to as Hilderston mine by Gallagher (1964) and also by Meikle (1994), who describes often dull cubo-octahedral crystals of galena to 10 mm on edge and smaller cubes associated with acanthite, cinnabar and nickeline in a vein close to an altered dolerite dyke. At South Queensferry, Carboniferous stromatolites in a non-marine algal limestone have been found with layers containing galena and sphalerite in aggregates up to 1 cm across (Parnell 1983). At Stobshiel, S of Haddington, galena and sphalerite occur with minor chalcopyrite and tetrahedrite in a greywacke–shale-hosted vein gold prospect (Naden & Caulfield 1989). Galena is also noted from Granton quarry and Little Vantage, W of Balerno, both near Edinburgh (Heddle 1901).

Strathclyde The mining of galena in the Strathclyde part of the Leadhills–Wanlockhead orefield dates back at least to 1239, when the monks of Newbattle worked a lead mine at Glengonnar. It was here, along Brown's vein, that cubo-octahedral crystals of galena up to 4 cm were found. About 40 veins have been mined for galena. Supergene alteration of this galena (especially at Susanna mine) produced a wide range of secondary lead minerals, including cerussite, pyromorphite (both relatively common), anglesite, caledonite, chenite, lanarkite, leadhillite, linarite, macphersonite, mattheddleite, scotlandite and susannite (Wilson & Flett 1921, Brown 1927, Temple 1955, Dunham et al. 1978, Gillanders 1981, Livingstone & Sarp 1984, Paar et al. 1986). On the dumps of Lady Anne Hopetoun shaft, Leadhills, supergene galena cubes (<0.1 mm) have been found in a small cavity within barite, quartz and deeply corroded galena matrix (Green et al. 2005c). Galena is also one of the most common species found as inclusions in alluvial gold from Shortcleugh

Water, Leadhills mining district (Leake et al. 1998, Chapman et al. 2000c).

Callender & Macaulay (1984) describe the mining history of many lead mines on the Isle of Islay. The mineralization (chalcopyrite–galena–pyrite–sphalerite) occurs in veins and breccia zones hosted in Ordovician–Silurian metamorphic (mainly limestone) rocks (Wilson & Flett 1921, Barnett 1959, Edmonds 1983, Nicholson 1989b). Results of fluid-inclusion microthermometric analysis of calcite or quartz from four of the mines (Ballygrant, Kilslevan, Mulreesh, and North Ardachie mine, near Ballygrant) are presented by Baron & Parnell (2005), who propose an origin involving multiple fluid-flow events.

Galena is also reported from: Abhainn Srathain copper mines, S of Meall Mòr, South Knapdale – uncommon galena in quartz–feldspar veins, also in boreholes near the mines (Smith et al. 1978a); Arduaine, near Kilmelford – Dalradian breccia close to the An Chap sheeted porphyrite contains veinlets rich in bornite, galena, pyrite and sphalerite (Zhou 1987); Bishopton railway tunnel and bypass borrow pit – rare galena close to or within altered Clyde Plateau Lavas adjacent to cavities and joints (Heddle 1901, Meikle 1992b); Cantyre, SW of Inverneil – galena–quartz–siderite veins cutting mica slate and Garngad Road, leading to Provanmill – cubes of galena associated with millerite in fireclay within Millstone Grit (Heddle 1901); Castletown mine, Lochgilphead – quartz vein containing copper-rich pyrite with a little chalcopyrite and galena (Wilson & Flett 1921, Starkey & Hubbard 1998); Clachan Beag, NE of Inveraray – silver-rich galena, pyrite and sphalerite in a Dalradian stratabound deposit (Willan 1980); Easedale Island, near Oban – micro-inclusions (< 20 μm) of chalcopyrite, galena, pyrrhotite and sphalerite in pyrite cubes within Dalradian slates (Hall et al. 1988); Etive granite complex to the E and NE of Ben Starav – coarse-grain galena in panned concentrates from W of Ben Starav (Haslam et al. 1985); Kilmelford – galena and sphalerite occur in isolated veins peripheral to porphyry-copper style mineralization within a dacitic porphyrite (Fortey 1980); Lagalochan, near Kilmelford – sulphide veinlets in a porphyrite–breccia complex contain pyrite with lesser amounts of chalcopyrite, galena, molybdenite and sphalerite. Galena contains inclusions of freibergite and native silver (Zhou 1987, Harris et al. 1988); Loanhead quarry, near Beith – rare lustrous isolated distorted cubes up to 5 mm across in calcite and prehnite, sometimes associated with greenockite, in Clyde Plateau Lavas (Meikle 1989b); Lochy River, W of Tyndrum – galena inclusions in alluvial gold (Chapman et al. 2000c); Mannoch Hill mine, SW of Lesmahagow – many barite–galena breccia veins within 1 km of the mine (Todd & McMullen 1991); Muirshiels barite mine, Renfrewshire Hills, W of Paisley – small specks of galena are reported from borehole material sampling the Muirshiel barite vein (MacGregor 1944, Stephenson & Coats 1983); Stronchullin, N of Tarbert – veins in Dalradian rocks, once worked for their galena content, also contain chalcopyrite and sphalerite (Berridge 1969, Ridgway 1980).

Tayside Galena and sphalerite occur in thin quartz bands associated with stratiform barium–zinc mineralization within Middle Dalradian schists in the Ben Eagach area, N of Aberfeldy (Coats et al. 1980, 1981, Fortey & Beddoe-Stephens 1982, Smith et al. 1984). Similar galena–pyrrhotite–sphalerite mineralization has been found in the Foss deposit, along strike from Ben Eagach (Moles 1983, 1985). Massive quartz containing galena cubes to 3 cm, and associated with chalcopyrite, pyrite and sphalerite, have been found on the dumps of Corrie Buie lead mine, Meall nan Oighcreag, Loch Tay. The galena is silver rich (Ridgway 1980, Pattrick 1984, Plant et al. 1989).

Other locations for galena include: Calliachar and Urlar Burns, 4 km from Aberfeldy – gold-bearing quartz–sulphide veins containing chalcopyrite, galena and sphalerite (Ixer et al. 1997), also alluvial gold from Calliachar Burn containing inclusions of galena (Leake et al. 1993, 1997c, Chapman et al. 2000c); Corrycharmaig, Glen Lochay, NW of Killin – accessory amounts of galena, gersdorffite, heazlewoodite, nickeline, an iron–nickel alloy (possibly awaruite) and native silver occur in chromite-rich lithologies within a small body of metamorphosed and altered ultramafic rocks (Power & Pirrie 2000); Glen Lednock, near Comrie; Glengarry; West Water, NW of Brigend – inclusions of galena in alluvial gold (Chapman et al. 2000c); northern end of Lochan na Lairige, NNE of Killin – the contact of the Ben Eagach Schist with the Ben Lawers Schist contains chalcopyrite, galena, pyrite and pyrrhotite (Green & Todd 2001); Tomnadashan copper deposit, Loch Tay – galena occurs with native bismuth, chalcopyrite and tetrahedrite in fractures within pyrite from a minor diorite–granite intrusion (Heddle 1901, Pattrick 1984, Plant et al. 1989).

Orkney Coarse-grain galena and a little sphalerite were mined at Stromness lead mine, Mainland (Wilson & Flett 1921). Barite–galena–(sphalerite) replacement veins and breccia zones have been found in Basal Breccia and Sandstone members of the Old Red Sandstone near the Moine unconformity at Blackhall and Mill of Cairston, Stromness area (Gallagher et al. 1971, Michie 1972). At Warbeth in the Stromness district, barite, celestine, galena and sphalerite replace stromatolites and adjacent calcareous siltstones (Muir & Ridgway 1975), and abundant barite, celestine and galena occur similarly in a Middle Devonian algal breccia (part of an evaporitic lake system) at Hesta Geo, near Birsay, Mainland (Parnell & Janaway 1990). Silver-rich galena was mined from Hoy before 1529, although most workings were in the 18th century (Wilson & Flett 1921, Ridgway 1980). Galena is also noted from Graemsay, Scabra Head on Rousay, and Sanday (Heddle 1901, Wilson & Flett 1921).

Shetland Localized galena and sphalerite occur at the margins of Dalradian amphibolites and massive sulphide rocks (mainly pyrite and chalcopyrite) at Vidlin Ness, Mainland (Garson & May 1976). Primary galena, pyrite and sphalerite occur in sulphide nodules within a carbonaceous–calcareous–siltstone facies of the Devonian Upper Melby Fish Bed at Matta Taing, Walls and Sandness, West Mainland (Hall & Donovan 1978).

Ireland

Armagh College (Carryhugh) mine, SW of Keady and the nearby Derrynoose mine (where galena occurs with pyrite and sphalerite) are the two main lead mines in Armagh. Other mines in the vicinity include Aughnagurgan mine, near Tullynawood Lake, and Clay (Carryhugh) mine, near Derrynoose. Silver-rich galena is associated with barite, pyrite (mundic) and quartz at Creggan mine (Cole 1922, Fowler 1959).

Cavan Cloghstuckagh, near Cootehill – a lead mine worked in 1750 and again in about 1832; Cornaurney – monomineralic patches and veinlets of galena in greywacke, some associated barite and sphalerite; Muff – barite–sphalerite vein with subordinate amounts of galena and tetrahedrite (all Morris 1984).

Clare Rudler (1905) reports curiously striped coarsely crystalline galena from Ballyhickey mine, near Tulla. Ballyvergin mine, also near Tulla, worked a disseminated and fracture-fill arsenopyrite–chalcopyrite–galena–pyrite deposit, with minor bornite, sphalerite and tennantite (Cole 1922, Schultz 1971, Andrew 1986). The Maghera Cross copper prospect, Maghera Townland, is a similar deposit with dominant chalcopyrite, subordinate pyrite, and minor galena and sphalerite (Schultz 1971). Castletown lead mine worked a calcite–chalcopyrite–fluorite–galena–quartz–sphalerite vein deposit (O'Connor et al. 1993, Cole 1922). Cole (1922) notes ten other lead mines within the Carboniferous Limestone, and comments that the ore is usually silver rich and often associated with sphalerite. Additionally, galena occurs at Sheshodonnell mine, once worked for smithsonite (calamine).

Cork West Carbery mining district: Cappagh (Cappaghglass) copper mine, near Ballycummisk – mainly bornite–chalcopyrite–galena in thin quartz veins (Cole 1922, Ni Wen et al. 1996, 1999, Kinnaird et al. 2002); Dhurode (Carrigacat) mine, Dunmanus Bay – arsenopyrite–bournonite–chalcopyrite–galena–meneghinite–tetrahedrite in a stratiform-minor vein deposit (Ni Wen et al. 1991, 1996, 1999, Kinnaird et al. 2002); Glanalin mine, between Killeen and Carravileen – chalcopyrite–galena–pyrite; Kilkilleen – arsenopyrite–chalcopyrite–bornite–galena–tetrahedrite, both in stratiform-minor vein deposits (both Ni Wen et al. 1996, 1999); Rooska West – arsenopyrite–chalcopyrite–(galena–tennantite–tetrahedrite) in minor veins and disseminations (Ni Wen et al. 1999). Cole (1922) reports a further ten locations where galena occurs: some were worked for lead alone (e.g. Gortacloona mine, near Bantry, and Ringabella mine, Ringabella Bay), whereas others were mainly copper mines (e.g. Ballydehob mine, near Bantry, and Bouleysallagh mine, near Crookhaven) or barium–copper mines (e.g. Duneen Bay mine, west side of Clonakilty Bay).

Donegal Glenaboghil mine, NNW of Fintown, worked a fine-grain stratiform deposit of galena and sphalerite (Cole 1922, Ryback & Moreton 1992). The ore at Glentogher (Carrowmore) mine, near Carndonagh was galena and sphalerite (Cole 1922, O'Brien 1959). South of the Donegal main granite, Dalradian impure limestones host finely intergrown galena–sphalerite and a little barite in veinlets and breccias (O'Brien 1959, McArdle et al. 1986a). Cerussite, galena and pyromorphite can be found on the mine dumps of Keeldrum mine, near Bedlam (Ryback & Moreton 1992). Cole (1922) documents a further seven lead mines in Co. Donegal. Galena has been found as inclusions in alluvial gold at Balwoges (Chapman et al. 2000c).

Down Breccias in Silurian greywackes and shales near a dolerite dyke at Conlig–Newtownards host galena (chalcopyrite–pyrite–sphalerite) vein mineralization. Galena occurs as crystalline masses and well formed cubic crystals in vugs (Cole 1922, Fowler 1959, Dunham et al. 1978, Halliday & Mitchell 1983, Moles & Nawaz 1996). Galena occurs with chalcopyrite and sphalerite in altered limestone at Abbeytown, Ballysadare (Schultz 1971, Halliday & Mitchell 1983).

Dublin The only substantial lead mine was Ballycorus mine, near Carrickgollogan, where a galena–quartz lode cutting granite was worked (Ryback & Moreton 1993). A further seven lead mines are noted by Cole (1922). Galena pebbles containing crystals of phosgenite have been found at Killiney beach (Green & Moreton 2001).

Galway Galena is an important primary ore species in the major limestone-hosted Cu–Pb–Zn–(Ag) deposit that formed close to the North Tynagh fault at Tynagh mine, near Killimor. Concentrates containing nearly 280 000 tonnes of lead were produced between 1966 and 1970, and at one time it was the largest lead–zinc mine in Europe. Galena contains inclusions of bournonite, enargite, geocronite, gratonite, tennantite and rarely native antimony, and itself forms inclusions in chalcopyrite, sphalerite and tennantite (Schultz 1966b, 1971, Morrissey et al. 1971, Gibson 1979, Boast et al. 1981a,b, Halliday & Mitchell 1983, Clifford et al. 1986, Ryback & Moreton 1992, Banks et al. 2002). Russell (1912b) notes a calcite–galena–sphalerite–quartz lode containing secondary allophane, cerussite and dundasite at Clements mine, Carrowgarriff, Maam. The small dumps of Glengowla mine, near Oughterard, are rich in calcite, galena, pyrite and quartz (Ryback & Moreton 1992). Additional lead mines in Co. Galway include: five in the Oughterard district, two in southern Galway and four others (Cole 1922).

Kerry Crow Island, Lough Leane, near Muckross, Killarney – chalcopyrite, galena, sphalerite (the main ore species) and minor tennantite occur as 20–50 μm-diameter polygonal crystals forming abundant mosaics. These species replace arsenopyrite, pyrite, pyrrhotite and sphalerite from an earlier metasomatic replacement deposit (Kinnaird et al. 2002, Ixer 2004); Muckross mine, Killarney – a stratabound Cu–Pb–Zn deposit with minor arsenic, cobalt and molybdenum, containing chalcopyrite–pyrite, chalcopyrite–galena–sphalerite and arsenopyrite–chalcopyrite–tennantite ores (Moreton et al. 1999, Kinnaird et al. 2002); Kenmare – chalcopyrite–galena ore (O'Brien 1959); Ross Island mine, Killarney – an exceptionally rich bornite–chalcopyrite–tetrahedrite deposit that in parts contain a fine-grain intimate mixture of galena–pyrite–sphalerite with a little chalcopyrite (Russell 1927b, Kinnaird et al. 2002); Shanagarry (Caher West) mine, near Cleady – a stratiform galena–pyrite–sphalerite deposit (Cole 1922, Halliday & Mitchell 1983). Cole (1992) reports a further four lead mines in Co. Kerry.

Kildare The small breccia-hosted Fe–Pb–Zn deposit at Harberton Bridge, 40 km WSW of Dublin, is widely considered to be an example of MVT mineralization. It consists of galena–marcasite–sphalerite mineralization in a calcite gangue (Emo 1986, Gallagher et al. 1992, Trude & Wilkinson 2001).

Kilkenny The Galmoy orebody, near Kilkenny, is a stratiform lead–zinc replacement deposit that developed in dolomitized and brecciated Carboniferous Limestone. Galena occurs as fine-grain disseminations enclosed within sphalerite and as coarse intergrowths with sphalerite (Doyle et al. 1992, Lowther et al. 2004). Knockadrina (Flood Hall) is an ancient lead mine known for its very rich silver ore (Cole 1922).

Leitrim Silver-rich galena and native silver are reported from veins in Carboniferous Limestone at Twigspark mine, near Manor Hamilton (Cole 1922, Halliday & Mitchell 1983).

Limerick Galena, pyrite and sphalerite are the major sulphides in a complex limestone-hosted replacement deposit at Courtbrown, on the southern shore of the River Shannon, near Ballysteen. Galena (associated with sphalerite) forms coarse crystalline masses or disseminated euhedral grains. Finer-grain galena and sphalerite occur in crosscutting calcite veins (Grennan 1986, Reed & Wallace 2001). Other reports of galena include: Aherlow, near Galbally – disseminated copper deposit in a Carboniferous limestone–shale sequence with galena and sphalerite often in calcite veins peripheral to the main mineralization (Cameron & Romer 1970, Schultz 1971); Carricklittle lead–zinc prospect, Carricklittle Townland – sphalerite, pyrite and galena in a fracture-controlled deposit within Carboniferous Limestone intruded by feldspar porphyry (Schultz 1971). Cole (1922) notes a further six lead mines in Co. Limerick.

Longford The Keel prospect, Ardagh, 11 km S of Longford, is a sphalerite-dominated Cd–Pb–Zn deposit formed in faulted, brecciated and dolomitized Carboniferous rocks. Galena is of minor overall importance in the deposit, although where it contains silver-rich tetrahedrite inclusions it has economic importance. In places, tetrahedrite also replaces galena (Patterson 1970, Morrissey et al. 1971, Schultz 1971, Halliday & Mitchell 1983, Slowey 1986, Ryback & Moreton 1993). The Newtown Cashel prospect is a fracture-controlled sphalerite–(barite–galena) deposit to the W of Keel that contains minor amounts of galena (Crowe 1986).

Louth Galena and chalcopyrite are minor constituents of a sphalerite-dominated breccia and quartz–carbonate vein deposit at Salterstown (Morris 1984).

Mayo Barite–galena–sphalerite mineralization occurs in the basal zone of a brecciated and altered acidic porphyry 8 km to the S of Charlestown. It is overlain by chalcopyrite- and pyrite-rich zones (Poustie 1982, O'Connor & Poustie 1982). Native gold-bearing quartz veins at Cushinyen, in the Cregganbaun shear zone, contain mm-size aggregates of sulphides, including bournonite, chalcopyrite, galena, pyrite, sphalerite and tetrahedrite (Thompson et al. 1992). Silver-rich galena was mined at Sheeffry (Tawnycrower) mine, near Westport from a mineralized fault zone. The vein assemblage is dominated by galena (subhedral crystals and masses to 3 cm across) accompanied by chalcopyrite and millerite in a calcite gangue (Cole 1922, Moreton & Jackson 1991, Aherne et al. 1992). Alluvial and *in situ* grains of native gold, with inclusions of galena, have been found at: Carrownisky River (Glenkeen Bridge), Cregganbaun (E and W), Croagh Patrick, Cushinyen, Lough Alisheen, Owannadornaun River, and Srahroosky (Leake et al. 1993, Chapman et al. 2000c).

Meath The ore at Navan mine comprises fine layers of sphalerite and coarser-grain crystalline galena, with varying amounts of pyrite, marcasite and barite. It is a world-class carbonate-hosted, epigenetic, syn-sedimentary replacement deposit. In 1985 it had pre-production ore reserves of 69 million tonnes grading 10.1% zinc and 2.6% lead, and in the 1990s it was the largest zinc producer in Europe. By 2004, production was 57.7 million tonnes grading 8.3% zinc and 2.1% lead (Ashton et al. 2004, Ashton 2005). Some of the galena layers show crustification textures and dendritic reticulate growths. Elsewhere, bush-like skeletal to cubic growths of galena on laminated sphalerite occur. Three generations of galena are recognized in the U-lens (the south-western extension of the orebody) – dendrites of anhedral crystals, discontinuous bands of euhedral crystals, and replacements of earlier sulphides (Schultz 1971, Boast et al. 1981b, Boyce et al. 1983, Andrew & Ashton 1985, Ashton et al. 1986, Ryback & Moreton 1993, Anderson et al. 1998, Blakeman et al. 2002, Peace et al. 2003, Wilkinson 2003). The Tatestown–Scallanstown deposit (3.6 million tonnes grading 6.9% Zn+Pb), near Navan, is regarded as a satellite to the major Navan orebody. Sphalerite, galena and pyrite–marcasite dominate the mineral assemblage (Andrew & Poustie 1986). Beds stratigraphically equivalent to those at Navan host a sub-economic barite–galena–sphalerite vein deposit at Drumlerry, near Oldcastle (Brand & Emo 1986).

Monaghan Galena occurs with chalcopyrite and sphalerite in carbonate veinlets crosscutting arsenopyrite–stibnite-bearing veins once worked for antimony at Clontibret mine, Tullybuck. Galena–sphalerite-bearing quartz veins also occur in the local area (Morris 1984, Morris et al. 1986, Ryback & Moreton 1993, Steed & Morris 1997). Twelve other lead mines are recorded by Cole (1922) and Morris (1984), but only the loose isolated euhedral crystals of galena to 2 cm from Annaglogh mine, near Tassan, are notable. Alluvial gold containing inclusions of galena has been found at Clontibret (Chapman et al. 2000c).

Sligo Abbeytown lead–zinc mine, near Ballysadare, worked a dolomitized limestone-replacement deposit in which galena was less abundant than sphalerite. Galena is generally euhedral, cubic, silver rich and contains inclusions of boulangerite (Cole 1922, Fernandez 1957, O'Brien 1959, Schultz 1971, Halliday & Mitchell 1983, Hitzman 1986, Ryback & Moreton 1992). A thin string of galena has been reported from Deadman's Point, opposite Coney Island, Sligo harbour (Cole 1922).

Tipperary The Silvermines district is notable for its large stratiform sedimentary-exhalative Ba–Pb–Zn deposit (estimated in 1978 to contain 4.6 million tonnes of ore with 6.5% zinc, 2.5% lead and 30 ppm silver), but considerable early mining activity (records date back to 1600) was focused on mineralized fault breccias and fracture fillings close to a major fault system. It was, at least in part, because of the abundance of silver-rich galena that the district acquired its name. Galena is described from the following mines: Ballygown South mine – galena with exceptionally high silver content in primary and oxidized ores (Russell 1907b, Rhoden 1959, Boland et al. 1992); Ballynoe mine – lode containing barite, chalcopyrite and galena, in which a

Large twinned crystal of galena, 3 cm across, on a creamy dolomite matrix; Mogul mine, Silvermines, Co. Tipperary, Ireland. (© D. I. Green)

large vug containing magnificent galena was discovered in the late 1990s (Russell 1907b, Wolfe 1996a); Ballynoe (Magcobar) barite open pit – minor sphalerite and galena (Schultz 1971, Mullane & Kinnaird 1998); Gortnadyne mine – galena containing rare inclusions (Russell 1907b, Rhoden 1959); Mogul mine – stratiform ore zones consisting of variable abundances of galena, pyrite and sphalerite. One remarkable 3 m-long cavity found in 1978 produced exceptional crystals of bournonite, galena and sphalerite, including magnificent cubo-octahedral galena crystals to 2.5 cm, often in intergrown groups (Schultz 1971, Zakrzewski 1989, Ryback & Moreton 1991, Wolfe 1996a, Moreton 1999, Green & Moreton 2001); Shallee mine – galena is the most abundant sulphide and contains a multitude of inclusions of boulangerite, bournonite and tetrahedrite (Greg & Lettsom 1858, Rhoden 1959, Halliday & Mitchell 1983). The Silvermines district is discussed more generally by other authors – Cole (1922), O'Brien (1959), Morrissey et al. (1971), Taylor & Andrew (1978, 1979), Boast et al. (1981b), Samson & Russell (1983), Taylor (1984), Russell (1990), Reed & Wallace (2004). Another carbonate-hosted deposit (22 million tonnes of ore with 11.5% zinc, 1.9% lead and 26 ppm silver) occurs at Lisheen, between Thurles and Rathdowney, where galena is found with marcasite, pyrite, sphalerite and other minor sulphides in three stratabound sulphidic lenses. The majority of the sulphides replace black matrix breccia and, where cavities have developed, these are infilled with colloform sulphide masses intergrown with calcite or dolomite (Hitzman 1992, Strashimirov et al. 1999, Hitzman et al. 2002, Fusciardi et al. 2004, Wilkinson et al. 2005).

At Gortdrum mine, near Tipperary town, epigenetic mineralization related to the intrusion of two high-level basic intrusions has produced a dominantly chalcopyrite–Hg-rich tennantite orebody in which galena occurs as an accessory species (Schultz 1971, Morrissey et al. 1971, Duane et al. 1986).

Tyrone Gold-bearing quartz–carbonate veins at Curraghinalt, near Gortin, Sperrin Mountains, contain pyrite and subsidiary arsenopyrite, bornite, chalcopyrite, native copper, galena, sphalerite and tennantite–tetrahedrite (Clifford 1986, Earls et al. 1989, Clifford et al. 1990). Similar vein mineralization at nearby Cavanacaw, W of Omagh, contains galena frequently intergrown with chalcopyrite and tetrahedrite, infilling vugs in quartz veins and fractures in arsenopyrite and pyrite (Cliff & Wolfenden 1992, Parnell et al. 2000). Galena has also been found in the Sperrin Mountains as inclusions in alluvial gold from Attagh Burn, Curraghinalt Burn and Glencurry Burn (Chapman et al. 2000c). Minor galena and sphalerite are reported from quartz veins at Cashel (Leyshon & Cazalet 1976).

Waterford Sphalerite and galena occur infrequently in copper-rich quartz veins cutting Ordovician rocks at Knockmahon, 22 km SW of Waterford town (Holdsworth 1834, Moreton 1991).

Westmeath At Ballinalack, 15 km NW of Mullingar, a carbonate-hosted lead–zinc deposit has formed as discrete pods within a series of reef knolls. It is spatially related to the Ballinalack fault and has a style of mineralization similar to that at Tynagh in Co. Galway. Galena is most common in the basal massive sulphide part of the mineralization, where extensive replacement of carbonate by sulphide has occurred. It forms discrete cubes and intergrowths with sphalerite. Barite and marcasite–pyrite in a calcite–(dolomite) gangue complete the main assemblage (Schultz 1971, Patterson 1970, Halliday & Mitchell 1983, Jones & Brand 1986). Shallow-water carbonates host another lead–zinc deposit at Moyvoughly, near Moate, 12 km N of Clara. Massive sulphides formed by replacement of limestone and consist of sphalerite and pyrite, with minor galena making up the high-grade sections of the orebody (Patterson 1970, Poustie & Koucha 1986, Kucha 1988).

Wexford The miners of old found it difficult to separate fine-grain galena from intergrown sphalerite in quartz veins at Barrystown mine, close to the Bannow Estuary, and Caim, near Enniscorthy (O'Brien 1959, Halliday & Mitchell 1983). At Ballyhighland mine, W of Enniscorthy, galena occurs with cerussite, sphalerite and decomposing chalcopyrite (Cole 1922, Ryback & Moreton 1993). Barrystown mine, Bannow Estuary dates back to Tudor times and worked galena–sphalerite from a siderite–quartz lode (Cole 1922).

Wicklow The orebody at Avoca mine is a volcanogenic exhalative sulphide deposit with considerable chalcopyrite and pyrite hosted in Ordovician volcanic rocks associated with black chloritic schists. Mineralization is also found in stringer ore. Galena and sphalerite occur in sub-economic amounts in both ore types at the margins and top of the mineralized areas (Cole 1922, O'Brien 1959, Schultz 1971, Wheatley 1971, Badham 1978, Downes & Platt 1978, Sheppard 1980, Halliday & Mitchell 1983, McArdle 1989, 1993, Ryback & Moreton 1993). A fine-grain purple ore, composed of a mixture of galena and sphalerite–(chalcopyrite) at the northeastern end of the Avoca district was named kilmacooite (after the local townland) before its composite nature was revealed (Cole 1922, Milner & McArdle 1992). A related Cu–Pb–Zn–(Au–Bi) deposit in the Goldmine River valley, 10 km along strike from Avoca, is hosted by metamorphosed BIF rocks and contains minor galena subsidiary to (in sequence of abundance) magnetite, chalcopyrite, pyrite, hematite and sphalerite (Ixer et al. 1990). Other occurrences along the southeastern margin of the Leinster batholith are at: Ballingen, between Tinahely and Aughrim – minor galena in altered microtonalite associated with cassiterite–scheelite mineralization (Steiger & Bowden 1982, 1986, McArdle & Kennan 1988, Gallagher 1989) and Luganure mines, Glendalough – produced over half of Ireland's lead output in the 19th century (Rudler 1905, Cole 1922, O'Brien 1959, Halliday & Mitchell 1983, Ryback & Moreton 1993). Cole (1922) notes a further 11 lead mines in Co. Wicklow.

Hexagonal spinel-twin crystals of galena, up to 11 mm across, from North Hero Shallow Adit tip, Glendasan, Co. Wicklow, Ireland. (© D. I. Green)

Offshore territories

Isle of Man Galena–sphalerite-bearing veins cutting metamorphosed slates and grits of the Cambro-Ordovician Manx Group were exploited at the Foxdale and Laxey mines, where a combined total of 500 000 tonnes of lead and zinc concentrates were produced between 1845 and 1938. The galena was silver rich, sometimes forming aggregated cubo-octahedral crystals up to 25 cm on edge. Minor chalcopyrite and malachite are associated species (Greg & Lettsom 1858, Rudler 1905, Dewey & Eastwood 1925, Russell 1949b, Dunham et al. 1978, Alderton et al. 1995, Scarffe 2004). The old Bradda mines on the southwest coast worked veins cutting Manx Group rocks and produced silver-rich galena and chalcopyrite (Rudler 1905, Dewey & Eastwood 1925, Starkey 1998). At Snaefell mine, galena and sphalerite occur with minor chalcopyrite, pyrite and pyrrhotite in a calcite–dolomite–quartz gangue (Lamplugh 1903).

Guernsey Sulphide veinlets in the Northern igneous complex at Bouette quarry, St Peter Port, contain chalcopyrite, galena, magnetite, pyrite (abundant), pyrrhotite, sphalerite and traces of pentlandite. The Lukis collection in the Guernsey Muesum and Art Gallery) holds specimens of galena from Baubigny, near St Sampson (Ixer 1992), and from Bouet and Moulin Huet, below Mount Durand (Mourant 1939).

Jersey Galena cements and forms veinlets in sphalerite within carbonate veins cutting Brioverian sediments at Le Pulec. It occurs as subhedral crystals to 800 µm across and replaces arsenopyrite, bournonite, marcasite, pyrite and tetrahedrite (Mourant & Warren 1934, Mourant 1961, Ixer & Stanley 1980, Stanley & Ixer 1982, Ixer 1992). Granite of the Northwest granite complex hosts many mineral veins: Le Doüet de la Mer, St Ouën – small quantities of arsenopyrite, galena, pyrite and sphalerite in a small carbonate vein. Galena replaces and cements arsenopyrite and pyrite, or forms intergrowths with chalcopyrite or sphalerite (Robinson 1934, Mourant 1961, Ixer 1980, 1992); Gigoulande quarry, St Mary – bournonite, galena and sphalerite cementing pyrite veinlets within chalcopyrite grains (Ixer 1980, 1992); L'Etacq quarry – minor galena in calcite–quartz veins containing chalcopyrite, galena, molybdenite, pyrite and sphalerite (Mourant 1961, Ixer 1980, 1992); Les Handois, Mont Mado and Samarès (Mourant 1961).

Galena has also been found at: Bouley Bay, Trinity, northeast coast of Jersey – traces of galena and sphalerite in copper-enriched flow-banded rhyolite (Ixer 1980); German underground hospital, St Peter's Valley, St Lawrence – galena and sphalerite in a chalcopyrite–pyrite–quartz vein, and galena more abundant than sphalerite or pyrrhotite in a calcite vein (Robinson 1947, Ixer 1980, 1990b); La Corbière, southwest Jersey – chalcopyrite, galena and sphalerite in pyrite–quartz veins (Ixer 1980); L'Ouaisné Bay, southwest Jersey – veins containing calcite, chalcopyrite, fluorite, galena, hematite, magnetite and sphalerite (Mourant 1961, Ixer 1980); near Tesson Mill – minor galena in veins (Mourant 1961).

Sark The primary ore from Sark's Hope Pb–(Ag) mine, Little Sark, is mainly galena and pyrite with local concentrations of chalcopyrite and tennantite–tetrahedrite. Galena contains inclusions of acanthite, enargite, pearceite, polybasite, pyrite (bravoite) and tennantite (Mourant & Warren 1934, Ixer & Stanley 1983, Ixer 1992, Laffoley 1998). Another galena-rich lode occurs at Le Pot on Little Sark (Ixer 1992).

galenite

A synonym of galena.

galenobismutite

orthorhombic $PbBi_2S_4$

A rare lead sulphobismuthite found in Cu–Pb–Zn mineralization at Parys Mountain, 3 km S of Amlwch, Anglesey, Wales, where it occurs as fine-grain (5–20 µm) intergrowths associated with native bismuth, bismuthinite, bournonite and kobellite. Galenobismutite also occurs as inclusions within sphalerite, as a cement to framboidal pyrite, and as rims around chalcopyrite, galena and tetrahedrite-group minerals. The deposit is considered to be the result of sedimentary–exhalative mineralization accompanying minor rhyolitic volcanism (Sivaprakash 1977, Pointon & Ixer 1980, Bevins 1994, Southwood & Bevins 1995).

Brief reports of galenobismutite from Gupworthy mine, Brendon Hills, Somerset, and Wheal Penrose, Sithney, near Helston, Cornwall, are given by Criddle & Stanley (1993) and Golley & Williams (1995) respectively.

In Scotland, galenobismutite occurs as (<0.5 cm) plates in galena from the old mine dumps at Corrie Buie, Meall nan Oighcreag, S of Loch Tay, Tayside Region, and at Borland Glen, Ochil Hills, Central Region, it is reported as inclusions in two alluvial gold grains (Leake et al. 1993, Chapman 2000c).

galmei

A synonym of hydrozincite and smithsonite (calamine).

ganophyllite

monoclinic $(K,Na)_6(Mn,Al,Mg)_{24}(Si_{32.5}Al_{7.5})O_{96}(OH)_{16}\cdot21H_2O$

A rare hydrated potassium–sodium–manganese–magnesium aluminosilicate hydroxide found in lenses of manganese ore within Ordovician mudstones at Benallt mine, Rhiw, Lleyn Peninsula, Gwynedd, Wales. Ganophyllite occurs as small fibrous patches (varying in colour from pale pink to brown) in thin veins cutting the manganese ore in the Ty Canol incline. Six-sided crystals up to 4 mm across have a perfect basal cleavage and a brilliant lustre ranging from pearly to metallic or micaceous. Ganophyllite forms on fracture surfaces and is associated with calcite (Campbell Smith & Bannister 1948). Some of the material that was originally described as ganophyllite was shown to be bannisterite by Smith & Frondel (1948). Additional ganophyllite data are provided by Dunn et al. (1983). calcium-rich ganophyllite from Benallt mine is mistaken for tamaite by Kato (1980) and Matsubara et al. (2000). An earlier (and very brief) description of ganophyllite associated with rhodonite from the nearby Nant mine, Llanfaelrhys, is given by Russell (1946d). Recent EMP investigations of ganophyllite from Benallt mine indicate it is compositionally zoned, with parts of crystals having eggletonite, ganophyllite or tamaite compositions. A single EMP analysis of barium-rich ganophyllite from Nant mine suggests there is a fourth ganophyllite-like species, but this is as yet unpublished (T. F. Cotterell pers. comm. 2006).

GARNET GROUP

Cubic silicates, with the general formula $A_3B_2(SiO_4)_3$ (and for hibschite and katoite, $A_3B_2(SiO_4)_{3-x}(OH)_{4x}$), where A = Ca, Fe^{2+}, Mg, Mn^{2+}; and B = Al, Cr^{3+}, Fe^{3+}, Mn^{3+}, Si, Ti, V^{3+}, Zr. Si is partially replaced by Al, Fe^{3+}. Almandine, andradite, goldmanite, grossular, hibschite, pyrope, schorlomite and spessartine are members of the group found in Britain or Ireland. Kimzeyite and uvarovite have also been reported. Kimzeyite and uvarovite have also been reported.

garnierite

A general term for hydrated nickel silicates; was genthite.

garronite

ZEOLITE GROUP

orthorhombic (pseudo-tetragonal) $NaCa_{2.5}(Al_6Si_{10})O_{32}\cdot14H_2O$

TL Western side of Glenariff Valley, Garron Plateau, Co. Antrim, Ireland

A relatively rare, but locally abundant, hydrated sodium–calcium aluminosilicate found in less than 1% of amygdales within the Tertiary basalts of Co. Antrim examined by Walker (1960a). A formal description of the species followed and nine additional locations (many along the eastern coastal strip of Antrim) were documented: S of Black Burn, N of Carnlough; W of the Ticloy dolerite plug; E of Glenarm; NW of Kilwaughter Castle, Larne; SW of Kilwaughter crossroads, S of Kilwaughter Castle; Langford Lodge borehole; northeast slope of Squire's Hill; roadside quarry W of Whitehead village and quarry on the north side of White Head (Walker 1962a). Garronite forms radiating aggregates completely filling amygdales up to 2 cm across (crystal faces have not been observed) next to other amygdales containing analcime, chabazite, levyne, natrolite, phillipsite or thomsonite (faröelite). It is white or colourless with a vitreous to greasy lustre. In the Langford Lodge borehole, garronite is associated with chabazite, levyne, calcite, saponite and thomsonite (Walker 1962a). Additionally, Nawaz (1982) presents data for garronite from Black Cave, near Larne, Co. Antrim. Radiating aggregates of garronite in rubbly basalt are reported from the Galboly River section, near Garron Point, Co. Antrim, and Moneymore quarry, near Magherafelt, Lough Neagh, Co. Londonderry (Adair 1987).

In Scotland, garronite is found in amygdales within Tertiary basalts from the eastern side of the Storr, N of Portree, Isle of Skye, Highland Region, by Macpherson & Livingstone (1982). Garronite from Moonen Bay, also on Skye, forms whitish-grey fibrous compact masses up to 1.5 cm across (Livingstone 2002). Green & Todd (2001) record massive translucent cavity infills of garronite from Moonen Bay.

gartrellite

TSUMCORITE GROUP

triclinic $PbCuFe^{3+}(AsO_4)_2(OH)H_2O_2$

A rare hydrated lead–copper–iron arsenate hydroxide found in mine spoil from a trial adit on Low Pike vein, near Potts Gill mine, Caldbeck Fells, Cumbria. Yellow crusts of gartrellite coat fractures in quartz veinstone and cover areas up to several cm square, sometimes forming crude spherules up to 0.1 mm across. Gartrellite is typically found in the oxidation zones of lead- and copper-bearing mineral deposits, and in Cumbria it is accompanied by the following supergene species: bayldonite, beudantite, brochantite, cornwallite, duftite, malachite, mimetite, mottramite, philipsburgite and pseudomalachite (Green et al. 2003b).

Gartrellite has been confirmed more recently from Dolyhir quarry, Old Radnor, near Presteigne, Powys, Wales, where it occurs on Precambrian mud–siltstone as a patchy thin pale-yellow powdery coating associated with bayldonite (T. F. Cotterell pers. comm. 2005).

gastaldite

A synonym of glaucophane, amphibole group.

gaylussite

monoclinic $Na_2Ca(CO_3)_2\cdot5H_2O$

A rare hydrated sodium–calcium carbonate found in clays at Jarrow, Tyne & Wear, and from near Glasgow (Bernard & Hyršl 2004).

gearksutite

monoclinic $CaAl(OH)F_4\cdot H_2O$

A rare hydrated calcium–aluminium hydroxide fluoride formed by late-stage low-temperature hydrothermal alteration of fluorine-bearing minerals at the Old Gang lead mines, E of Gunnerside Gill, Swaledale, Yorkshire. Abundant gypsum and minor quantities of blue and green copper minerals (ktenasite and serpierite) are found on the mine walls of Lucky String, the western section of Alderson's vein. On top of these minerals are overgrowths of white spherules up to 0.5 mm across aggregated into spherulitic masses to 5 mm across and containing allophane, gearksutite and gypsum. Gearksutite is a product of post-mining oxidation and probably crystallized over the past hundred years at a temperature of approximately 10°C (Green 2003, Wolfe 2004). Gearksutite has also been found at the Ball Eye opencast, north side of Via Gellia, near Cromford, Derbyshire, in a replacement flat-type deposit within Carboniferous dolomitized limestone and basic lavas. It forms white masses on anglesite in a supergene assemblage replacing the primary ore species fluorite and galena (Bridges & Green 2005a).

Gearksutite spherules to 0.5 mm from Hard level, Old Gang mine, near Gunnerside, Yorkshire. (© D. I. Green)

gedrite

monoclinic AMPHIBOLE GROUP $\square Mg_5Al_2(Si_6Al_2)O_{22}(OH)_2$

An uncommonly reported magnesium aluminosilicate hydroxide that forms a series with ferrogedrite and is found in high-grade metamorphic rocks and metasomatized contact-metamorphosed rocks.

Gedrite, mantled by spinel and diaspore, occurs in thermally metamorphosed hornfels within the aureole of the Land's End granite at the Inner quarry, Kenidjack, St Just, Cornwall. Biotite, chlorite, cordierite and ilmenite are also present (Chinner & Fox 1974).

In Scotland, gedrite is a common constituent of a Moine Series migmatite complex near Strathy, Highland Region. At Uamh nan Soithichean, to the NW of Armadale, it occurs in a biotite–garnet–oligoclase–quartz schist and in thin section displays yellowish-brownish and smoke-grey pleochroism (Collins 1942, Livingstone 2002). At Glen Urquhart, near Milton, Highland Region (a locality described as "near the Free Church of Milton" by Heddle), clove-brown gedrite forms radiating fans with prisms up to 6 cm in length, in a monomineralic rock. Occasionally the rock contains giant hornblende crystals reaching 11 cm. Hydrobiotite, rutile, serpentine and talc are associated species. The occurrence is in a zoned hydrothermal body within a large serpentinite mass (Francis 1955). The gedrite is considered to be an unusual hydrous variety; however, the published data are based on material contaminated by chlorite (Law 1982).

geerite

pseudo-cubic Cu_8S_5

A rare copper sulphide found in void-infilling carbonate-associated copper mineralization near the Precambrian/Triassic unconformity at Judkins quarry, Nuneaton, Warwickshire. It occurs in trace amounts with possible digenite as 20–30 μm-long veinlets cutting bornite and chalcopyrite. Geerite is usually replaced, partially or totally, by spionkopite (Ixer et al. 2005a).

At Dolyhir quarry, Old Radnor, near Presteigne, Powys, Wales, geerite is tentatively identified together with djurleite and roxbyite, all as minute lamellae intergrown with chalcopyrite (T. F. Cotterell pers. comm. 2007).

Geerite is also found in low-grade stratiform disseminated copper mineralization within the Castlehaven Formation of the Munster Basin at Letter and Mount Gabriel, both near Ballydehob, West Carbery mining district, Co. Cork, Ireland. Bornite and djurleite are the main sulphide phases accompanied by trace amounts of chalcocite, chalcopyrite, pyrite and wittichenite. Supergene alteration of sulphide-rich zones has modified the primary assemblage to either bornite, altered bornite with chalcopyrite and covellite lamellae, or covellite, djurleite, geerite, spionkopite and yarrowite (Ni Wen et al. 1999).

gehlenite

tetragonal MELILITE GROUP $Ca_2Al(AlSi)O_7$

A rare but locally abundant calcium aluminosilicate that forms a series with åkermanite and is found in the chalk/dolerite contact zone at Scawt Hill, Larne, Co. Antrim, Ireland. Intense metasomatism of the chalk has produced a rock containing 30 wt% gehlenite with major amounts of merwinite, spinel, spurrite and minor perovskite (Tilley & Harwood 1931). Referred to as melilite–gehlenite by Tilley (1929b), the mineral forms euhedral tabular crystals with a prominent basal cleavage. Later work at Scawt Hill revealed the new mineral bredigite in a gehlenite–larnite–spurrite assemblage from the contact zone (Tilley & Vincent 1948). At Carneal, Larne, Co. Antrim, further chalk–dolerite interaction has occurred where blocks of Cretaceous chalk have been brought up in a dolerite plug that intrudes Tertiary lavas. Gehlenite is a major constituent, making up about 20% of the rock in a larnite–magnetite assemblage that replaces the original chalk. It forms coarse tablets and prisms up to 1 mm long in an aggregate of larnite that is partially enclosed in, and altered to, spurrite. Bredigite is an associated species and it indicates a very high-temperature, low-pressure crystallization history (Sabine et al. 1982, 1985). Gehlenite is also found at Killala Bay, near Inishcrone, Co. Sligo, in thermally metamorphosed limestone close to the contact with many basalt–dolerite dykes. It is a constituent of two primary metasomatic assemblages; gehlenite–larnite–(bredigite) ± spurrite and larnite–(bredigite)–spurrite ± gehlenite (Nawaz 1974).

In Scotland, gehlenite has developed in a gabbro/limestone contact zone at Camas Mòr, Isle of Muck, Highland Region. In the exocontact zone, Jurassic limestones have been converted to rock types in which the following pairs of minerals are dominant: gehlenite–monticellite, gehlenite–wollastonite, monticellite–merwinite, monticellite–periclase and spurrite–gehlenite. Gehlenite forms tabular crystals of variable size 0.1–0.5 mm across. Other associated species include cuspidine, larnite, merwinite, rankinite and tilleyite (Tilley 1947). Near Camphouse, Ardnamurchan, Highland Region, extreme thermal metamorphism of Liassic limestones by a gabbroic intrusion has produced rock types in which there is extensive development of bredigite, calcite, gehlenite, larnite, rankinite and spurrite (Agrell 1951).

geikielite

trigonal ILMENITE GROUP $MgTiO_3$

Named after Sir Archibald Geikie FRS (1835–1924), a geologist and director-general of the Geological Survey of Great Britain (1882–1901); born in Edinburgh. Geikielite is a very rare magnesium–titanium oxide that forms a series with ilmenite and has been reported in trace amounts from the siliceous dolomites of the Allt Guibhsachain area, near Ballachulish, S of Fort William, Highland Region, Scotland. It forms isolated euhedral crystals associated with baddeleyite, qandilite, rutile and zircon (Ferry 1996, Livingstone 2002).

genkinite

tetragonal $(Pt,Pd)_4Sb_3$

A very rare platinum–palladium antimonide found as two irregular crystals, 4 × 12 μm and 2 × 8 μm, at Harold's Grave, Baltasound–Haroldswick area, Unst, Shetland, Scotland. Genkinite forms composite grains with hollingworthite, hongshiite, irarsite, sperrylite and unnamed Pt–Pd–Cu and copper–nickel alloys in the altered silicate matrix of chromite-rich dunite lenses that formed near the base of the Shetland ophiolite complex (Prichard & Tarkian 1988, Prichard et al. 1994).

genthelvite

cubic $Be_3Zn_4(SiO_4)_3S$

A very rare beryllium–zinc silicate with sulphide that forms series with danalite and with helvine, and is found in miarolitic cavities in adamellite at Coire an Lochain in the Cairngorm Mountains, Grampian, Scotland. Superficially yellowish-brown tetrahedral crystals of genthelvite (dark green on fracture surfaces) up to 1.3 cm in length occur in cavities that also contain quartz (varieties cairngorm and rock crystal), microcline and plagioclase (oligoclase). Bertrandite emplanted on genthelvite was found in one cavity (Morgan 1967). Further investigation of the genthelvite revealed oscillatory zonation, with some zones low in zinc and more enriched in both iron and manganese (Clark & Fejer 1976).

Genthelvite is also reported from Treburland mine, 2 km SE of Fivelanes, Altarnun, Cornwall (Kingsbury 1961); however, given the doubts about the provenance of other Kingsbury specimens (Ryback et al. 1998, 2001), further confirmation is desirable.

genthite

Probably a garnierite; a general term for hydrated nickel silicates.

geocronite

orthorhombic $Pb_{14}(Sb,As)_6S_{23}$

A rare lead–antimony–arsenic sulphide that forms a series with jordanite and is found at Kilbricken (Monanoe) lead mine, between Ennis and Quin, Co. Clare, Ireland. Geocronite (syn. kilbrickenite) forms lead-grey prismatic crystals with a metallic lustre and is found with bournonite (Greg & Lettsom 1858, Prior 1902, Russell 1949b). Elsewhere in Co. Clare, at Ballyvergin, disseminated and fracture-filled sulphide mineralization occurring in the crestal region of a faulted pericline contains trace amounts of geocronite, bournonite and native silver in an assemblage dominated by arsenopyrite, chalcopyrite, galena and pyrite (Brown 1967, Andrew 1986). At Tynagh mine, near Killimor, Co. Galway, Ireland, epigenetic Cu–Pb–Zn–(Ag) mineralization includes a suite of sulphide species that occur in trace amounts; native antimony, arsenopyrite, bournonite, enargite, geocronite, gratonite, jordanite, marcasite, millerite, pyrite, siegenite and tennantite, many occurring as inclusions in galena (Morrissey et al. 1971, Gibson 1979, Clifford et al. 1986). A report of geocronite from Silvermines (B-zone), Co. Tipperary, Ireland (Zakrzewski 1989), gives analyses with As > Sb, so the species is actually jordanite.

Criddle & Stanley (1993) note geocronite from Treore mine, Port Gaverne, near Port Isaac, St Teath, Cornwall, but this is based on a sample in the Kingsbury collection at the NHM and its provenance is now considered dubious (C. J. Stanley pers. comm. 2004).

georgeite

amorphous $Cu_2CO_3(OH)_2$

A very rare copper carbonate hydroxide found in vein material on Sneyd's level, Britannia mine, Snowdonia, Gwynedd, Wales. Georgeite is a late-forming oxidation product that forms pale-blue amorphous masses overlain by carbonatian connellite, goethite and malachite (Pollard et al. 1989). Synthetic studies of georgeite have shown that it is an amorphous analogue of malachite and that previously quoted chemical formulae for the species were in error, probably because of admixed chalconatronite in the material used for analysis (Pollard et al. 1991).

gersdorffite

COBALTITE GROUP

cubic NiAsS

A rare nickel sulpharsenide found in moderate-temperature hydrothermal deposits.

Gersdorffite has been found in quartz–fluorite veinstone on the dumps of North Devon United (East Wheal Friendship) mine, Peter Tavy, Devon (within the metamorphic aureole of the Dartmoor granite). It occurs as inclusions in cobaltite and is associated with arsenopyrite, cassiterite, cosalite, erythrite, parkerite, sphalerite, stannite and ullmannite, and a variety of bismuth-rich minerals including aikinite, bismite, native bismuth, bismuthinite, ikunolite and an unnamed bismuth telluride (Rumsey & Savage 2005). At Greystone, Lezant, near Launceston, cobalt-rich gersdorffite has been found with chalcopyrite, galena, jamesonite, pentlandite, pyrite and pyrrhotite in carbonate-rich altered dolerite (Power & Scott 1995). Gersdorffite is reported from South Terras mine, St Stephen-in-Brannel, Cornwall (Davis et al. 1965, Golley & Williams 1995), but the occurrence requires further confirmation, given the doubts about the provenance of other Kingsbury samples (Ryback et al. 1998, 2001). Another possible Cornish occurrence of gersdorffite, from St Austell Consols mine, near Sticker (Robertson & Dines 1929), also requires confirmation because the mineral is described as kupfernickel and this is a synonym for both gersdorffite and nickeline. Gersdorffite occurs as inclusions in tetrahedrite in antimony–gold mineralized quartz–carbonate veins at Loddiswell mine, 3 km N of Loddiswell, Devon (Stanley et al. 1990b).

On Alderley Edge ridge, Cheshire, trace amounts of gersdorffite are found in Cu–Pb–Zn redbed-type mineralization, which forms veins and disseminations in coarse-grain clastic sediments of Triassic age. Gersdorffite is optically zoned and forms crystals 5–10 µm across, enclosing pyrite (bravoite). Nickel-rich cobaltite, and possibly pararammelsbergite, are associates (Ixer & Vaughan 1982, Braithwaite 1994).

At Tynebottom mine, NW of Garrigill, near Alston, Cumbria, gersdorffite and glaucodot form overgrowths up to 40 µm wide on arsenical marcasite within an ankerite–calcite–quartz gangue. Galena, pyrite and sphalerite are the main ore species. Material collected from the Great Sulphur vein and from workings at Nenthead contains small (5–10 µm), euhedral gersdorffite crystals in quartz, associated with anhedral chalcopyrite and pyrite (Ixer et al. 1979, Vaughan & Ixer 1980, Ixer 1986, Ixer & Stanley 1987). These investigations were based on samples collected by the authors and from the Kingsbury collection at the NHM. Elsewhere in Cumbria, gersdorffite has been found in a small deposit of nickel minerals on the Dow Scar level of Hilton lead mine, Scordale, near Appleby. It occurs as a brilliant silvery microcrystalline crust over nickeline and as thin bands in nickeline. The deposit itself occupied a brecciated area in which cavities were infilled mainly with nickeline (Bridges 1982, Winrow 1986a). Primary gersdorffite associated with annabergite and erythrite has also been found on the 60-fathom-level dumps of Sandbed mine, Caldbeck Fells (Green & Briscoe 2002, Green et al. 2006a).

Nickeline and other nickel-bearing species, including gersdorffite and ullmannite, are described from magnetite-rich ore at Lady's Rake mine, along Harwood Beck, Upper Teesdale, Co. Durham, and at Settlingstones mine, near Hexham, Northumberland. Fluids from the Whin Sill are thought to have circulated through the Teesside fault system to produce the mineralization (Young et al. 1985b).

In the Central Wales orefield of Dyfed, gersdorffite is a component of a post-Caledonian (probably early Devonian) polymetallic vein assemblage within quartz-cemented breccias at the following mines: Darren mine, Pen-bont Rhydybeddau; Eaglebrook (Nantycagl) mine, Ceulanymaesmawr; Erglodd mine, N of Tal-y-bont; Esgair Hir mine, Ceulanymaesmawr; Esgair Fraith mine, near Tal-y-bont. Associated minor species include bournonite, millerite, siegenite, tetrahedrite and ullmannite, together with rare cobalt pentlandite, electrum, pyrite, pyrrhotite and tučekite (Mason 1997). Gersdorffite also forms rare inclusions (< 100 µm across) in galena within a complex polymetallic assemblage at Gwaithyrafon mine, SE of Tal-y-bont (Bevins 1994, Mason 1998). Elsewhere in Wales, gersdorffite is reported from Great Orme mine, Llandudno, Gwynedd (Ixer & Stanley 1996, Ixer 2001), where it occurs as a trace component of an assemblage dominated by chalcopyrite, galena, marcasite, pyrite and siegenite. It has also been found as an inclusion species in alluvial gold from Afon Mawddach, Snowdonia, Gwynedd (Leake et al. 1993).

In Scotland, gersdorffite has been found in a talc–carbonate rock at the base of the Shetland ophiolite complex, to the E of Loch of Cliff, Baltasound–Haroldswick area, Unst. The assemblage includes chalcopyrite, pentlandite, pyrrhotite, chromium–iron-rich spinel (ferritchromit) and violarite (Maynard et al. 1997). On the west side of Loch Duich (along the road to Glenelg) and on the east side of Loch Duich (along the road to Kyle of Lochalsh), Highland Region, sporadic but locally abundant gersdorffite occurs with abundant chalcopyrite, galena, molybdenite, rutile, sphalerite and minor aikinite, gold (electrum), hessite, marcasite and tennantite in veins associated with the Ratagain (diorite, syenite and quartz monzonite) intrusion. The gersdorffite forms small (10 µm) inclusions associated with chalcopyrite in galena and sphalerite (Alderton 1986, 1988b). At a mine between Coillebhragdad and Eas a' Chosain, 2 km W of Inveraray, Strathclyde, gersdorffite occurs in crosscutting veins in quartzose bands within schist, where it is found with chalcopyrite and pentlandite (Livingstone 2002). Small strings of nearly pure gersdorffite up to 50 mm wide and chalcopyrite were also found at Craignure nickel mine, 12 km SW of Inveraray (Forbes 1868, Heddle 1901, Anthony et al. 1990, Criddle & Stanley 1993). Quartz–sulphide ± gold veins cutting Upper Dalradian gritty quartzites, quartz–mica schists and amphibolites along the Calliachar and Urlar Burns, Aberfeldy, Tayside Region, contain rare gersdorffite rimmed by arsenopyrite, and thin rims of gersdorffite have formed around chalcopyrite, galena and pyrite (Ixer et al. 1997). At Corrycharmaig, Glen Lochay, NW of Killin, Tayside Region, a small body of metamorphosed and altered ultramafic rocks contains chromite-rich lithologies in which accessory amounts of galena, gersdorffite, heazlewoodite, nickeline, an iron–nickel alloy (possibly awaruite) and native silver occur. This assemblage is closely associated with relatively rare PGMs, including irarsite, laurite, the native metals iridium, osmium and ruthenium, and unnamed species containing variable amounts of antimony, arsenic, osmium, platinum and ruthenium. The chemistry and association of the PGMs is strongly indicative of an ophiolite assemblage and it is most likely that the intrusion represents a fragment of oceanic crust (Power & Pirrie 2000). A diorite-hosted lensoidal mass of As–Cu–Ni mineralization at Talnotry, near Newton Stewart, Dumfries & Galloway, contains an assemblage of chalcopyrite–pyrite–pyrrhotite with minor cobalt-rich gersdorffite (perfectly euhedral crystals up to 25 µm across, some with sperrylite cores) in the lowermost 3 m and a chalcopyrite–gersdorffite–nickeline assemblage in the upper 1 m (Power & Pirrie 2001, Power et al. 2004). Gersdorffite forming inclusions in alluvial gold is reported from Balquandy, Glenclach, Hodyclach Lamb and Westplace Burns in the Ochil Hills, Central Region (Chapman et al. 2005).

In Ireland, gersdorffite was one of the many species found in the B-zone of Mogul mine, Silvermines, Co. Tipperary, within a remarkable lens of barite that hosted many silver-bearing minerals. The species include acanthite (argentite), argentotennantite, argyrodite, freibergite, miargyrite, proustite, pyrargyrite, native silver, smithite, tennantite, tetrahedrite and xanthoconite, often in complex intergrowths. Gersdorffite occurs as euhedral crystals up to 20 µm across, associated with proustite–pyrargyrite and galena, but in most cases as fine-grain intergrowths with geocronite and galena. The deposit formed in a classic sedimentary-exhalative environment (Taylor 1974, Zakrzewski 1989, Moreton 1999). At Lisheen, between

Thurles and Rathdowney, Co. Tipperary, a carbonate-hosted deposit, consisting of three stratabound sulphidic lenses dominated by galena, marcasite, pyrite and sphalerite, contain minor arsenopyrite, bornite, bournonite, chalcopyrite, gersdorffite, gratonite, jordanite, nickeline, pararammelsbergite, seligmannite and tennantite (Strashimirov et al. 1999, Hitzman et al. 2002, Ixer 2002, Fusciardi et al. 2004). The large carbonate-hosted syn-sedimentary Cu–Pb–Zn–(Ag) deposit at Tynagh mine, near Killimor, Co. Galway, contains gersdorffite as inclusions in galena (Boast et al. 1981a). Scattered grains of gersdorffite and arsenopyrite occur in Ordovician turbiditic slates, mudstones and sandstones of the Lower Ordovician Sheefry Formation along the line of the Cregganbaun shear zone, Co. Mayo (Thompson et al. 1992). The Lack gold deposit, 5 km WSW of Omagh, Co. Tyrone, consists of a swarm of economically significant gold-bearing quartz veins containing trace amounts of gersdorffite, marcasite and pyrrhotite. Pyrite dominates the sulphide assemblage, accompanied by accessory arsenopyrite, chalcopyrite and galena, and minor sphalerite and tetrahedrite (Cliff & Wolfenden 1992).

Other Scottish and Irish occurrences of gersdorffite are as inclusions in alluvial gold, as follow:

BORDERS Craigierigg Burn; Cramalt Burn; Glengaber Burn, SW of Peebles (all Chapman et al. 2000c). STRATHCLYDE Fore Burn, Tairlow plantation, near Straiton (Chapman et al. 2000c); Leadhills–Wanlockhead area (Leake et al. 1998); Lochy Trib (Eas Daimh), W of Tyndrum and Shortcleugh Water, Leadhills (Chapman et al. 2000c). TAYSIDE Calliachar and Urlar Burns, Aberfeldy (Leake et al. 1997c, Chapman et al. 2000c); Glengarry and West Water, NW of Brigend (Chapman et al. 2000c)

CO. MAYO Glenkeen bridge on the Carrownisky River; Croagh Patrick; Lugbeg River (all Chapman et al. 2000c). CO. TYRONE Lack, Omagh (Leake et al. 1993).

geversite

cubic — PYRITE GROUP — $Pt(Sb,Bi)_2$

A very rare platinum antimonide bismuthide found in chromite-rich dunite lenses that formed near the base of the Shetland ophiolite complex at Cliff, Baltasound–Haroldswick area, Unst, Shetland, Scotland. The main sulphides at this locality are pentlandite, heazlewoodite and millerite, which are accompanied by trace amounts of spionkopite and sperrylite, and the native metals copper, gold and silver (Prichard et al. 1994). Geversite is also reported with native platinum in a chromite-rich sample from Jimmie's quarry, S of Hagdale, Unst (Prichard et al. 1988).

gibbsite

monoclinic — $Al(OH)_3$

A common aluminium hydroxide, polymorphous with bayerite, doyleite and nordstrandite. Gibbsite is a typical product of weathering of aluminous minerals and also forms in low-temperature hydrothermal and metamorphic envoronments.

Gibbsite occurs with allophane in white compact nodules from Clifton Hill, Brighton, Sussex (Hollingworth & Bannister 1950, Macpherson 1983). Similar nodules are described from collapsed Tertiary sediments filling solution pipes in the Upper Chalk near the coastguard station at Newhaven, Sussex, where gibbsite forms very friable orange masses, up to 20 cm across, that grade into the surrounding clay-rich sand. It occurs with other aluminium-rich species such as aluminite, bayerite, felsöbányaite (syn. basaluminite) and nordstrandite. All appear to have formed by the reaction of acid groundwater with Tertiary clays and the Chalk (Wilmot & Young 1985). Gibbsite is also noted from collapsed Tertiary sediments, infilling solution hollows in sands and clays along the approach road to Portobello outfall works, Telscombe, near Newhaven, and in temporary exposures created during the construction of the Falmer bypass, both in Sussex (Wilmot & Young 1985). The occurrence of gibbsite in fissures within Cave Oolite of Middle Jurassic age in the Cockle Pits, near Brough church, Cumbria (Smith 1967), would appear to have an origin similar to that of the occurrences in Sussex.

Gibbsite is sparingly developed at three localities in Avon. In Upper Triassic limestone breccias near East Harptree, Mendip Hills, it occurs as crystalline, monomineralic, white powdery-to-porcellanous cavity fillings up to 3 cm in diameter, associated with deeply oxidized smithsonite-bearing veins, formerly worked for zinc. Associated species include allophane, dundasite, halloysite, kaolinite, scarbroite and a hydrotalcite-like mineral. At the south end of Hampstead Farm quarry, Chipping Sodbury, 16 km NE of Bristol, gibbsite is found in heavily oxidized pyritiferous veins cutting Carboniferous Limestone close to the Rhaetic/Lower Carboniferous unconformity. Here it occurs with allophane, felsöbányaite (syn. basaluminite), gypsum, para-alumohydrocalcite, scarbroite and possibly alumohydrocalcite. At this locality, the secondary aluminous minerals are thought to have been derived from the leaching of aluminous clays by strongly acidic solutions, resulting from the weathering and oxidation of pyrite (Alabaster 1990b). Gibbsite also occurs with allophane, alumohydrocalcite, and a white carbonate-containing aluminous phase that is amorphous to X-rays, in cavities within a sandstone filling in a sinkhole in the Carboniferous Limestone at Woodleaze quarry, near Tytherington, Bristol (Alabaster 1989a).

Gibbsite is reported from a few other locations: two in Cornwall – weathered Carnmenellis granite at Bosahan quarry, Constantine (Rice 1973), and Wheal Gorland, St Day, 2 km E of Redruth (Macpherson 1983); one in Derbyshire – Bees Nest sand pit, Brassington (Ford 1967); and in weathered granite from several locations on Dartmoor, Devon (Green & Eden 1971).

In Wales, gibbsite is the only crystalline component of the clay fraction of a <25 cm-thick soil horizon developed over the Bwlch y Cywion granite, on Y Llymllwyd, W of Nant Ffrancon, Snowdonia (Ball 1964). It has also been found on mine dumps at Glasdir, near Dolgellau, Gwynedd, where it forms white globules up to 0.2 mm in diameter on clear, glassy, colourless to pale-blue allophane (Bevins 1994).

At Penkiln Burn, close to the margin of the Loch Doon granite pluton (nearest town Newton Stewart), Dumfries & Galloway, Scotland, intergrowths of gibbsite, goethite, illite, kaolinite and limonite occur in highly altered igneous and sedimentary rocks, particularly at the margins of minor dioritic intrusions containing disseminated sulphides. The low-temperature assemblage forms a gossan that has also affected sphalerite and galena-bearing quartz veinlets. Beudantite, malachite and plumbogummite are other low-temperature species noted in the gossan (Stone & Leake 1984). An examination of soil derived from a biotite-rich quartz gabbro in the Strathdon area, Grampian, found that gibbsite and kaolinite, together with aluminous vermiculite–chlorite, are the natural weathering products of biotite (Wilson 1966).

Cream friable gibbsite occurs in Upper Carboniferous phosphate-rich black shales at Grouse Lodge quarry, 5 km S of Shanagolden, Co. Limerick, Ireland, sometimes in intimate association with planerite. Other species found at this site include kaolinite, koninckite, turquoise and wavellite (Moreton & Green 2005). Bauxite, consisting of a mixture of gibbsite and diaspore, was mined at the following locations in Co. Antrim, Ireland: Cargan mine; Clegnagh mine, Ballintoy; Correen mine, near Broughshane; Cullinane mine, near Carnlough; Essathohan mine; Evishacrow mine; Lemnagh More mine, Ballintoy; Solomon's Drift mine, Skerry Water Valley, above Newtown Crommelin; Straid (Irish Hill) mine; Tuftarney mine, between Newtown Crommelin & Cargan; Urbanreagh mine. Bauxite was also worked at Killygreen mine, Co. Londonderry. Weathering of basalt and some rhyolite under tropical conditions (laterization) allowed the bauxite to form (Coles 1922). At Lyle Hill mine, W of Newtonabbey, Co. Antrim, weathered basalt is veined with 5 cm-wide gibbsite-rich veins (Hill et al. 2000).

gibsonite

TL Hartfield Moss, 9 km WSW of Barrhead, Strathclyde, Scotland

Named after Alexander Gibson (1800–1867), a botanist who published on Indian organic natural resources from 1839 to 1853; born in Laurencekirk, Kincardineshire (part of Grampian), Scotland. Gibsonite is a pink fibrous variety of thomsonite, zeolite group.

gigantolite

A mixture of muscovite and biotite.

gilbertite

TL Stenna Gwyn, St Stephen-in-Brannel, near St Austell, Cornwall, England

Named after Davies Gilbert FRS (1767–1839), a scientist, applied mathematician, antiquary, and member of parliament for Helston and Bodmin; born in St Erth, Cornwall. Gilbertite has been described as a fluorine-rich variety of muscovite, mica group. It often occurs as an alteration product of feldspar, and pseudomorphs of gilbertite after orthoclase feldspar have been found in Cornwall.

gilpinite

An iron-rich variety of johannite.

girasol

A synonym of opal-A.

gismondine

monoclinic (pseudo-tetragonal) — ZEOLITE GROUP — $Ca_2(Si_4Al_4)O_{16}\cdot 8H_2O$

A rare hydrated calcium aluminosilicate found in fewer than 1% of localities (out of a total of 670) by Walker (1960a) in a study of the regional distribution of amygdale minerals in Tertiary lavas of Co. Antrim, Ireland. However, later work showed it to

Spray of pink gismondine on chabazite; Coulters Hill quarry, Ballyclare, Co. Antrim, Ireland; width 3.5 mm. (© D. I. Green)

Pseudo-tetragonal gismondine from the Storr, north of Portree, Isle of Skye, Highland Region, Scotland; width 200 μm; SEM image. (© D. I. Green)

be more widely distributed. Some of the best gismondine occurs at Brushlee quarry, near Ballyclare, Co. Antrim, where it is found mostly as chalky-white radiating aggregates lining cavities in vesicles up to 5 cm across. A nearly colourless variety, forming octahedron-like crystals up to 8 mm across, has also been encountered. Gismondine is associated with apophyllite, gyrolite, saponite, thomsonite (faröelite) and pre-zeolite calcite. In a highly amygdaloidal basalt at another quarry 2 km away, gismondine is more abundant and closely associated with chabazite and saponite. In this quarry, apophyllite, calcite, gyrolite, levyne, mesolite, phillipsite and thomsonite (var. faröelite) are also found. Rubbly Antrim basalts containing chalky aggregates of gismondine are noted from Garron Point (Adair 1987). Other localities in Co. Antrim are within a 10 km radius and include: Call Hill farm, near Cookstown junction, Randalstown; near the summit of Carnearney Hill; Carnmoney Hill, N of Belfast; Catcairn Hill, Legoniel, Belfast; Collinward, near Belfast; southeastern side of Craig Hill, Ballyclare; quarry SW of Kilwaughter crossroads; Nelson's quarry, Ballyhenry; Shane's Hill, near Larne; Skerrywhirry, E of Ballymena; Slemish; Upton Park borehole, Templepatrick; Whitewell, N of Belfast; Wolf Hill, Legoniel, Belfast (Walker 1962b). Nawaz (1982) showed that gismondine is also present in amygdaloidal basalt at the Giant's Causeway, where it has crystallized on early-formed phillipsite. At Carneal, close to the contact of a Tertiary volcanic plug with chalk, gismondine forms white rims in amygdales, up to 1 cm across, in olivine basalt (Sabine & Young 1975). In Co. Londonderry, gismondine occurs in basalts at Castlerock (Walker 1962b) and Moneymore quarry, near Magherafelt, Lough Neagh (Adair 1987).

Gismondine is briefly described from a lopolith at Clauchland's Point, 3 km NE of Lamlash, Isle of Arran, Strathclyde, Scotland (Dyer et al. 1996), and a pseudo-tetragonal variety is noted in basalt from the Storr, N of Portree, Isle of Skye (D. I. Green pers. comm. 2006).

A barium-rich variety of gismondine is reported from the weathering of slags at three lead smelting mills in North Yorkshire – Marrick Low mill, Marrick; Surrender mill, Reeth, Whashton lead mill, Whashton – and one locality in Derbyshire, in the bed of the River Derwent at Cromford. Gismondine forms colourless, transparent, bipyramidal crystals, often displaying interpenetrant twins (Braithwaite et al. 2001).

gladite

orthorhombic $CuPbBi_5S_9$

A rare copper–lead sulphobismuthite found in quartz veins cutting monzogranite at Loch Shin, 2 km W of Lairg, Highland Region, Scotland. Elsewhere in the monzogranite, mineralization is disseminated and mainly restricted to fluorite, molybdenite and pyrite. However, in the veins, molybdenite does not occur and the other species are accompanied by native bismuth, chalcopyrite, covellite, galena, sphalerite and a suite of bismuth sulphosalts: aikinite, berryite, gladite, hammarite, lindströmite, krupkaite, pekoite and tetradymite. Gladite forms coarse fibrous crystals and is a member of the aikinite–bismuthinite series (Livingstone 1993, Lowry et al. 1994).

Sulphosalts belonging to the aikinite–bismuthinite series have been found with quartz and other sulphides on joints in granite and in enclaves from the Shap granite quarry, Cumbria. On the basis of XRD analysis, one occurrence is considered to be an intimate intergrowth of bismuthinite and gladite, another a mixture of gladite and krupkaite. Further analytical work is desirable to confirm these species (Wirth & Green 2002).

glauberite

monoclinic $Na_2Ca(SO_4)_2$

A rare sodium–calcium sulphate found in the anhydrite–polyhalite subzone of the Permian Lower Evaporites in the Fordon F.3 borehole, Humberside (13 km S of Scarborough, Yorkshire). The Permian evaporite sequence underlies much of North Yorkshire and Humberside and extends beneath the North Sea. At Fordon the sequence is about 430 m thick. Well formed or irregular grains of glauberite up to 7 mm across occur in cavities within anhydrite–polyhalite rocks and make up 2% by volume of the rock. Glauberite replaces both anhydrite and polyhalite, and is itself replaced by halite (Stewart 1963, 1965).

glaucocerinite?

hexagonal $(Zn,Cu^{2+})_{10}Al_6(SO_4)_3(OH)_{32}\cdot 18H_2O$

A rare hydrated zinc–copper–aluminium sulphate hydroxide said to form pale-blue botryoidal coatings at Ffridd-Goch mine, Llanfachreth, and Simdde Dyllhuan, Drwys-y-coed, near Nantlle, both in Gwynedd, Wales. Its composition is dominated by copper almost to the exclusion of zinc and has been described as the copper analogue of glaucocerinite (Raade et al. 1985). The samples came from the collection of the NMW and were labelled as "woodwardite coating grass" and "woodwardite coating bracken". Bevins (1994) reports similar material (under woodwardite) from Drwys-y-coed mine. The Simdde Dyllhuan material had previously been examined by Nickel (1976), who concluded it was not woodwardite. The uncertainty was finally resolved when Witske (1999) showed that the material was the newly described species, hydrowoodwardite.

glaucodot

ARSENOPYRITE GROUP

orthorhombic (pseudo-cubic) $(Co,Fe)AsS$

An uncommon cobalt–iron arsenate, dimorphous with alloclasite and found in mineralized vein samples from the dumps of Tynebottom mine, NW of Garrigill, near Alston, Cumbria. Glaucodot is associated with gersdorffite, and both species form overgrowths (up to 40 μm in width) on arsenic-rich marcasite within a quartz, calcite, ankerite gangue (Ixer et al. 1979a, Vaughan & Ixer 1980). Although Ixer et al. (1979b) worked on some material from the Kingsbury collection now at the NHM that is thought to have dubious provenance (Ryback et al. 1998, 2001), glaucodot was actually found on material from the mine dumps (R. A. Ixer pers. comm. 2004).

At Scar Crag cobalt mine, Causey Pike, Cumbria, an apatite–chlorite–quartz vein contains markedly zoned glaucodot and alloclasite overgrowths on arsenopyrite in association with minor amounts of cobaltite and skutterudite. Native bismuth, bismuthinite, marcasite, molybdenite and pyrite are the other primary phases noted from this locality (Ixer et al. 1979a). Glaucodot also occurs with cobalt-rich arsenopyrite and cobaltite at Paddy End mine, Coniston, Eskdale, Cumbria (Young 1987a).

glauconite (series)

MICA GROUP

monoclinic $(K,Na)(Fe^{3+},Al,Mg)_2(Si,Al)_4O_{10}(OH)_2$

A compositional range of dioctahedral interlayer-deficient micas. During a reclassification of the mica group, it was decided that, as glauconite does not define an end-member composition (and is therefore not a true species), it should be given series status to reflect more accurately its current usage in describing a relatively large volume of compositional space (Rieder et al. 1998). Glauconite forms mainly during diagenesis in sedimentary rocks accumulated under shallow-water marine conditions (especially in loosely consolidated sandstone, impure limestone and siltstone).

Glauconite imparts the colour, and is particularly abundant in, the greensands that are well represented in rocks of Cretaceous age. Large glauconite grains are found in the Upper Greensand in Wiltshire, Dorset, Devon and the Isle of Wight within coarse quartz sands containing fragments of feldspar and mica. Similar glauconitic sandstones are found at Carsaig, Isle of Mull, Strathclyde, Scotland, and in Co. Antrim, Ireland, where glauconite grains are unusually large. The Cambrian Lower Comley Sandstone (Comley, Shropshire) and Tertiary Bracklesham Beds (East Wittering, Sussex coast) also contain glauconite (Harker 1954a). Rare zoned glauconite grains (125–150 μm across) exhibiting green rims surrounded by brown centres are described from the Upper Greensand in the Vale of Pewsey, Wiltshire (Loveland 1980), and a similar "corona" structure was observed in glauconite from the Lower Glauconitic Marl at Rocken End, Isle of Wight (Zumpe 1971). An analysis of glauconite from the Bracklesham Beds on Oystershell Hill, north slopes of Staple Hill, Chobham Common, Surrey, is presented by Guppy (1956). Buckley et al. (1978) analyzed glauconite in Cretaceous and Tertiary sediments from many other locations: Barrington, near Cambridge; Brook, King Garn Gutter and Shepherds Gutter, all Hampshire; Abbots Cliff, Copt Point, Oxney and Sandling, all near Folkestone, Kent (also Dover); Tolgate borehole, near Sandwich, Kent; Glyndebourne, Sussex. Hallimond (1922) describes a borehole containing dark-green glauconite within Cretaceous Greensand at Iford Manor, near Lewes, Sussex.

At Speeton in Yorkshire, the Coprolite Bed, a Lower Cretaceous phosphorite deposit, contains glauconite that impregnates the compactly accumulated phosphatic (carbonate-fluorapatite) pebbles and also occurs with detrital quartz in the calcareous matrix of the rock. Abundant gypsum (var. selenite) coats the surface of the bed and all cracks around the pebbles (Scott et al. 1987).

In a study of the impact of mining on sedimentation in the Camel and Gannel estuaries, Cornwall, Pirrie et al. (2000b) found glauconite in coarse-grain sands together with amphibole, biotite, muscovite, quartz, tourmaline and other materials such as molluscan skeletal carbonates, foraminifera and lithic metamorphic rock fragments. These are associated with a suite of diagenetic minerals that includes magnesium-rich calcite, galena, framboidal pyrite, siderite, iron monosulphide, copper–iron sulphides, zinc sulphide and possibly zinc sulphate.

An unusual occurrence of iron-poor glauconite associated with solid hydrocarbons and calcite is reported from Cave Dale and Odin vein near Castleton, Derbyshire (Young et al. 1968). It occurs in veinlets, pockets and irregular coatings in rubbly apron-reef limestones, in proximity to barite–calcite–fluorite–galena–sphalerite scrins and veins.

A totally different paragenesis is reported for a glauconitic mineral found in the Tertiary alkali microgranite of Ailsa Craig, Strathclyde, Scotland. Clusters of intergrown plates of the brown or green fibrous mineral forms interstitial patches up to 1 mm across, surrounded by amphibole, but could not be positively identified because not enough material could be collected for XRD work (Harding 1983, Harrison et al. 1987). Elsewhere in Scotland, glauconite occurs as dark-green flaky material on micaceous sandstone from Salisbury Crags, Edinburgh (Livingstone 2002). Heddle (1901) notes glauconite from Ashgrove quarry, near Elgin, Grampian, and between Elie and St Monance, Fife.

Glauconite is noted from Wales – North (1916) describes it in the Sutton Stone, Bridgend area, Mid Glamorgan; and from Ireland – Seymour (1902) lists it from an unspecified location. Greg & Lettsom (1858) report it from Scotland and Co. Antrim, Ireland, but at the time there was confusion between the compositions of glauconite, green earth and kirwanite, so the details are not presented here.

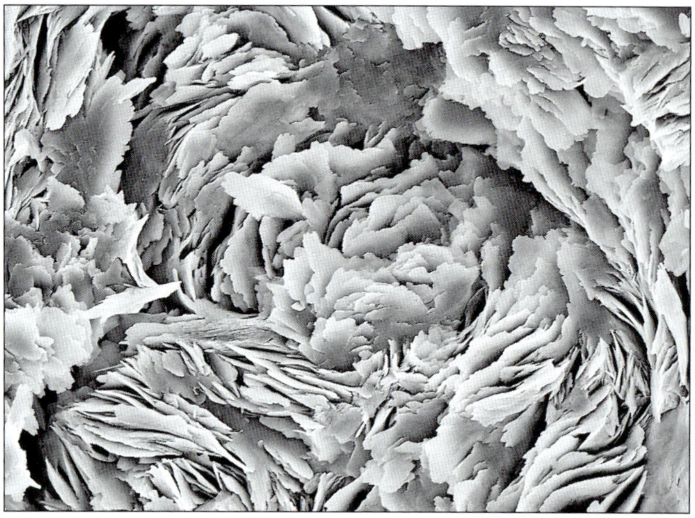

Nodule of glauconite from the Upper Cretaceous Chalk of Co. Antrim, Ireland; width 10 μm; SEM image. (© E. Condliffe, University of Leeds)

glaucophane

AMPHIBOLE GROUP

monoclinic $\square Na_2(Mg_3Al_2)Si_8O_{22}(OH)_2$

An uncommon sodium–magnesium–aluminosilicate hydroxide that forms a series with ferroglaucophane and is found in basic igneous rocks that have suffered characteristic low-pressure high-temperature metamorphism. Glaucophane is typically blue (bright blue in thin section), giving rise to the term "blueschist-facies"

Blue glaucophane in a glaucophane–epidote schist. Marquis of Anglesey monument, Anglesey, Gwynedd, Wales; width 1.1 mm; viewed in PPL (upper) and XPL (upper). (© K. Gardner)

metamorphism. Where possible, identification is based on the 1997 classification of the amphibole group (Leake et al. 1997a, Mogessie et al. 2004).

In Wales, optical identifications of glaucophane from a strongly foliated Mona Complex blueschist exposed near the Marquis of Anglesey's monument, near Llanfairpwllgwyngyll, Anglesey, Gwynedd, were made by Blake (1888) and Greenly (1919), the latter author describing zoned crystals with green cores and blue rims. Because of the fine grainsize, analysis proved difficult, but was attempted by Holgate (1951), who reported a crossite composition, presumably because his sample consisted mainly of green cores, with the blue rims being lost during sample preparation. This composition was reclassified as magnesio-arfvedsonite by Macpherson (1983) and is here given the name magnesiokatophorite. Reliable glaucophane data were later collected by Horák & Gibbons (1986), who describe green barroisite cores with blue-crossite–glaucophane–ferroglaucophane rims. The current nomenclature suggests a katophorite core and a rim consisting of aluminowinchite and glaucophane. A detailed examination of amphibole compositions from blueschists belonging to the Mona Complex from the Penmynydd/Gwna boundary exposed in a roadcut near Llanfairpwllgwyngyll indicates actinolite, alumino-ferrowinchite, aluminowinchite, ferrobarroisite, ferroglaucophane, ferrowinchite, katophorite and winchite are associated species (Gyopari 1984). The Mona Complex is also exposed on the Lleyn Peninsula at Penrhyn Nefyn, and here blue amphiboles in deformed basic volcanic rocks have been described as glaucophanic amphibole (Gibbons 1981), but appear to be true glaucophane.

Pale-blue glaucophane was optically identified in a tough bluish rock exposed on the crags near the Lendalfoot road, N of Knockormal, Strathclyde, Scotland, close to the margin of the Girvan–Ballantrae ophiolite complex (Bloxam & Allen 1959). The species was later confirmed by EMP analysis (Livingstone 1993). Previously, glaucophane had been described (but not confirmed) from the Ben Ledi schists in Perthshire, Tayside Region (Boswell 1926).

At Central and South Achill, N of Clew Bay, Co. Mayo, Ireland, blueschist-facies garnetiferous metabasites, whose origin is ophiolite-related, belong to the Dalradian Supergroup and contain ferro-edenite, ferroglaucophane, ferrohornblende and glaucophane (Chew et al. 2003).

glimmer

A synonym of mica.

glockerite

A synonym of lepidocrocite.

glottalite

TL Port Glasgow, Strathclyde, Scotland

Named after "Glotta", a name for the River Clyde; a synonym of chabazite, zeolite group.

glushinskite

monoclinic $Mg(C_2O_4)\cdot 2H_2O$

A rare magnesium oxalate dihydrate found at the lichen/rock interface on serpentinite colonized by *Lecanora atra* at Mill of Johnston, 6 km WSW of Insch, Grampian, Scotland. Glushinskite occurs in a creamy-white layer intermingled with the hyphae of the lichen fungus and consists of crystals up to 5 µm across, showing a distorted pyramidal form, often with curved and striated faces. It is associated with whewellite and traces of serpentine and quartz (Wilson et al. 1980). Gaines et al. (1997) note glushinskite without detail from the Isle of Rum, Highland Region.

gmelinite (series)

ZEOLITE GROUP

hexagonal $(Ca_{0.5},Na,K)_4(Si_8Al_4)O_{24}\cdot 11H_2O$

TL Little Deer Park, Glenarm, County Antrim, Ireland (co-type with Vicenza, Italy)

An uncommon, but locally abundant hydrated calcium–potassium–sodium aluminosilicate, whose name has been given series status in the zeolite classification of Coombs et al. (1997). There are now three gmelinite species – -Ca, -K and -Na – but these names do not appear in older publications and, unless chemical analyses become available, their species status cannot be established. At Little Deer Park, gmelinite forms flattened salmon-colour crystals up to 1 cm in diameter, intergrown with chabazite. Crystals often have milky-white to colourless cores and opaque, red or orange exteriors (Tschernich 1992). Walker (1959a) mapped the distribution of zeolite species in the Tertiary basalts of Co. Antrim and concluded that gmelinite is confined to a narrow zone along the eastern seaboard of Antrim. The zone comprises less than 1% of the area of the basalts, but in it gmelinite is extremely abundant. The

Orange-pink gmelinite crystals to 4 mm associated with colourless natrolite prisms from Little Deer Park quarry, Glenarm, Island Magee, Co. Antrim, Ireland. (© D. I. Green)

Gmelinite cluster, 15 mm across, in vesicular basalt from Little Deer Park quarry, Glenarm, Island Magee, Co. Antrim, Ireland. (© D. I. Green)

best exposures of gmelinite-bearing basalts are on the peninsula of Island Magee. Whitehead, Black Cave and E and SE of Glenarm are other notable locations near by. Large crystals of gmelinite were found by Tomkeieff (1934) along the shore S of the Gobbins, Island Magee. Greg & Lettsom (1858) describe large flesh-colour and nearly transparent crystals of gmelinite, sometimes embedded in chabazite, at Black Head, Island Magee, and large, nearly opaque crystals of greenish-white gmelinite at Portrush on the north coast of Co. Antrim. The last locality is considered doubtful by Walker (1959a), as are others in Co. Londonderry, and it is thought that chabazite (var. phacolite) was misidentified as gmelinite. Gmelinite with a platy habit is associated with gobbinsite in roughly spherical vugs, up to 5 mm across at Hills Port, S of the Gobbins area. Chabazite and gonnardite (or tetranatrolite) occur near by in other vugs (Nawaz & Malone 1982). Hexagonal salmon-pink crystals of gmelinite, up to 10 mm across, have been found at Madman's Window, near Little Deer Park, and at Magheramorne quarry, near Larne, water-clear sceptre gmelinite crystals occur in small cavities in the basalt (Green & Moreton 2001).

In Scotland, gmelinite appears to be restricted to two locations on the Isle of Skye, Highland Region: Talisker Bay, where it forms large colourless twinned crystals

(Greg & Lettsom 1858, Heddle 1901), and Echoing Crag, Quiraing, where imperfect glassy crystals up to 2 mm across line cavities in levyne-bearing amygdaloidal basalt (Livingstone 2002). Gmelinite has also been found in a dyke at Breadlebane on the Isle of Arran, Strathclyde (Dyer et al. 1996). These authors suggest that the Talisker gmelinite may actually be chabazite (phacolite), as good examples of this species can be collected from the seastack at Talisker.

gmelinite-Ca

ZEOLITE GROUP
hexagonal $Ca_2(Si_8Al_4)O_{24}{\cdot}11H_2O$

An uncommon hydrated calcium aluminosilicate originally described as gmelinite and found in a basalt quarry on the south side of Whitehead, immediately to the S of Island Magee, Co. Antrim, Ireland. Gmelinite-Ca occurs as very pale pinkish-yellow crystals in one of the upper flows of basalt exposed in the quarry (Herbert Smith et al. 1916). Gmelinite-Ca has also been confirmed from amygdales in basalt at Magheramorne quarry, near Larne, Co. Antrim (H. J. Foy specimen, Tindle unpub. EMP data).

gmelinite-Na

ZEOLITE GROUP
hexagonal $Na_4(Si_8Al_4)O_{24}{\cdot}11H_2O$

An uncommon hydrated sodium aluminosilicate originally described as gmelinite and found in a quarry on the north side of Whitehead, immediately to the S of Island Magee, Co. Antrim, Ireland. Gmelinite-Na occurs in dense yellow, white or opaque crusts lining irregular-shape cavities in highly decomposed Tertiary basalt, sometimes accompanied by aggregates of analcime, calcite or natrolite. Occasionally, two generations of gmelinite are present on the same specimen, with larger crystals lying on smaller ones. The smaller crystals are often almost water clear, and twinned crystals are far more common that simple crystals (Herbert Smith et al. 1916). The types of crystals analyzed are not stated. Gmelinite-Na has also been confirmed from amygdales in basalt from Magheramorne quarry, near Larne, and from Fair Head, near Ballycastle (Passaglia et al. 1978b), both Co. Antrim. Other gmelinite-Na specimens have been collected from rubbly basalts on the east coast of Co. Antrim on Island Magee at Brown's Bay, Castleshiels, Portmuck and Gobbins Path, and also from a few km to the N at Glenarm (Adair 1987).

gobbinsite

ZEOLITE GROUP
orthorhombic $Na_5(Si_{11}Al_5)O_{32}{\cdot}11H_2O$

TL Coastal escarpment near Hills Port, S of the Gobbins, Island Magee, Co. Antrim, Ireland

An extremely rare hydrated sodium aluminosilicate forming chalky-white fibrous crystals in the amygdales of Tertiary basalts at the type locality. Gobbinsite occurs in roughly spherical vugs, up to 5 mm in diameter, either alone or in association with gmelinite. Chabazite and gonnardite (or tetranatrolite) occur near by in other vugs (Nawaz & Malone 1982, Nawaz 1983b, Green & Moreton 2001). At Magheramorne quarry, near Larne, Co. Antrim, rare gobbinsite fills vugs and vesicles in amygdaloidal basalts. It is chalky-white or slightly pink and each vug also contains phillipsite.

Radiating gobbinsite infilling a cavity in Tertiary basalt from near Black Cave, Island Magee, Co. Antrim, Ireland; width 14 mm. (© The Natural History Museum, London)

Other cavities in the altered basalts contain calcite and the zeolite species analcime, cowlesite, garronite, gmelinite, gonnardite, natrolite or thomsonite (Artioli & Foy 1994). Fibrous aggregates of gonnardite are noted in rubbly basalt from Black Cave and Gobbins Path, both on Island Magee (Adair 1987).

Gobbinsite occurs in basaltic dykes of suggested Tertiary age, along the foreshore to the W of the Bunnahabhain distillery, 4 km N of Port Askaig, northeast Isle of Islay, Strathclyde, Scotland. Associated species include analcime, natrolite and thomsonite (Dyer et al. 1996).

godlevskite

orthorhombic $(Ni,Fe)_9S_8$

A rare nickel–iron sulphide found in a disused chromitite quarry at Cliff, Baltasound–Haroldswick area, Unst, Shetland, Scotland. At this locality, sulphide-bearing dunite lenses in harzburgite contain accessory amounts of godlevskite, heazlewoodite, maucherite, millerite and orcelite, as well as the more common sulphides bornite, chalcopyrite, covellite, digenite, djurleite, nickeline, pentlandite and the rare PGE species geversite, hongshiite, sperrylite and stibiopalladinite. Godlevskite is found as 1–15 µm-wide parallel lamellae replacing heazlewoodite (Ixer 1990a, Maynard 1993, Prichard et al. 1994). Gunn et al. (1985) also report a cobalt-rich variety of godlevskite from Cliff.

goethite

orthorhombic $Fe^{3+}O(OH)$

A very common iron oxide hydroxide, trimorphous with feroxyhyte and lepidocrocite, and found as a weathering product of many iron-bearing minerals. Goethite can also precipitate in hydrothermal, marine and bog environments during oxidation of reduced iron-bearing waters.

England

Avon Black spherules of goethite up to 1 mm across and occasional radiating fans of needles up to 1 mm occur attached to calcite and fluorite in veins cutting Carboniferous Limestone at Hampstead Farm quarry, Chipping Sodbury. Barite and pyrite are associated species (Brooke 1994b). Clusters of spherulitic goethite are also noted on strontianite in celestine deposits found in Lower Coal Measures sandstones and shales near Leechpool farm, Yate (Faithfull 1988). Earthy goethite found at Wesley Theological College (Wesley mine), Westbury-on-Trym, may be an alteration product of melanotekite (Alabaster 1985a). At the Providence iron mines, Bristol district, goethite was associated with the finest amethystine quartz in England (Greg & Lettsom 1858).

Cornwall Goethite is an abundant and commonly well crystallized species in all the mines of the Botallack sett, St Just, and other closely associated mines such as Wheal Cock, Wheal Edward and Wheal Owles. In particular, the Parknoweth section produced fine lustrous jet-black botryoidal groups. Small radiating golden sprays of acicular goethite and goethite pseudomorphs after siderite are also found in this region (Rudler 1905, Bancroft & Weller 1993). Beautiful brown sprays of goethite, up to 4 mm in length, on clear quartz have been recovered from the dumps of Wheal

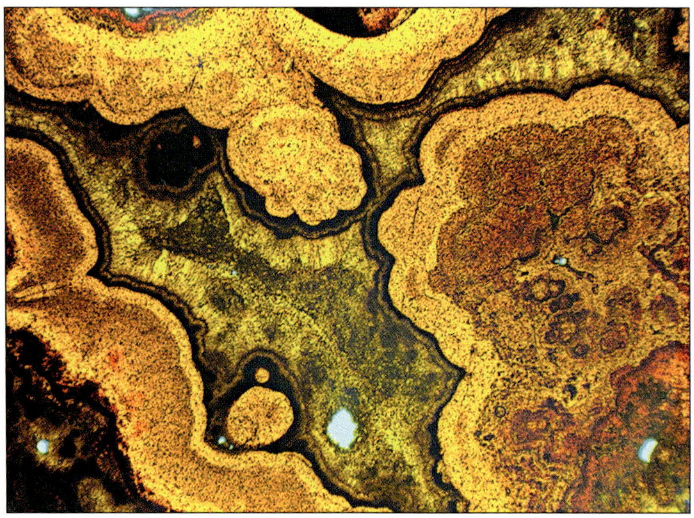

Botryoidal growths of goethite from Wheal Phoenix, Linkinhorne, near Liskeard, Cornwall; width 2.8 mm; viewed in PPL. (© A. G. Tindle)

GOETHITE

Cock (Wolloxall 1989) and Wheal Edward (Wolloxall 1988). Additionally, the nearby Levant mine has produced large groups of botryoidal goethite (Luff 1996). Remarkable specimens of concentrically zoned, banded quartz, limonite, goethite and hematite, and stout black composite prisms up to 5 cm long on earlier-formed goethite and quartz, have been recovered from cavities in veinstone at Restormel (Royal) iron mine, near Lostwithiel, to the SW of the Bodmin Moor granite mass (Greg & Lettsom 1858, Rudler 1905, Embrey & Symes 1987). Goethite showing botryoidal, earthy, dendritic, massive, stalactitic and compact forms occurs at Penberthy Croft mine, St Hilary. Specimens reach 11 cm across and are coloured with shades of black, brown, orange, red and yellow. Rare pseudomorphs after cubic pyrite are also recorded (Betterton 1989, 1997, Betterton et al. 2000).

Other goethite occurrences include: Carn Brea mine, Wheal Beauchamp, Wheal Buller and Wheal Druid, all near Redruth (Greg & Lettsom 1858, Collins 1892); Greystone quarry, Lezant, near Launceston (Elton et al. 2004); Kernick Iron lode, St Austell, and Polzeath Consols, St Minver (Golley & Williams 1995); New Cook's Kitchen mine (North Tincroft lode), Redruth (LeBoutillier et al. 2000); Pentire (Glaze) mine, near New Polzeath, St Minver (Clayton et al. 1990); Perran Iron lode, near Perranporth (Slater & Highley 1976); Phoenix United mines, Linkinhorne (Braithwaite & Corke 1980, Braithwaite 1981, Embrey & Symes 1987, Dunn 1990).

Cumbria *Caldbeck Fells:* Goethite is commonly found in many veins as thin crusts and masses, notably from Brae Fell mine, Burdell Gill, Driggith mine, Harestone umber mine, Old Potts Gill mine, Roughton Gill, Sandbed mine and Thief Gills. At Mexico mine, black botryoidal goethite is associated with corkite (Davidson & Thomson 1951, Stanley & Vaughan 1981, Young 1987a, Cooper & Stanley 1990, Green et al. 2006a). Some of these occurrences are also documented by Kingsbury & Hartley (1958, 1960) and Hartley (1984). Goethite forms abundant friable limonitic crusts in cavities in vein quartz and hard black encrustations, thin botryoidal films, and rarely minute euhedral crystals at Arm o' Grain (Green et al. 2006b). Green et al. (2005b) report abundant goethite as the major component of pale-brown friable limonitic crusts and as hard black encrustations on vein quartz from Silver Gill. *Elsewhere in the Lake District and West Cumbria:* Goethite is common as irregular masses and coatings of earthy limonite in the oxidized part of a chalcopyrite–ferberite–quartz–scheelite vein cutting the Eskdale granodiorite at Buckbarrow Beck, Corney Fell, Waberthwaite (Young et al. 1986, 1991b); goethite pseudomorphs pyrite at Longlands Fell mine, Uldale, takes a stalactitic form at Lowling, near Wigton, and is also noted from Haile Moor mine, Egremont (Young 1987a). Other localities include: Dale Head mine, Newlands Valley (Stanley 1979); Sellafield area boreholes (Milodowski et al. 1998); Shap Blue quarry (Young et al. 1988); Shap granite quarry, and in brecciated veins within the metamorphic aureole of the Shap granite (Firman 1978, Wirth & Green 2002). *Northern Pennine orefield:* Goethite occurs with brianyoungite and secondary copper and zinc minerals in a post-mining assemblage at Brownley Hill mine, Nenthead, near Alston (Livingstone & Champness 1993, Green et al. 2000); goethite pseudomorphs nailhead calcite at Hard Rigg Edge, Melmerby Fell (Young 1985a); oxidation of a stockwork in the Great Scar Limestone at High Stennerskeugh level, Ravenstonedale, near Kirkby Stephen, has produced common goethite and malachite (Young 1987b); pink powdery goethite and hematite are noted from Silverband lead mine, Milburn, near Appleby (Davidson & Thomson 1951); fissures in limestone at Silver Top quarry, Hallbankgate, Brampton, are replaced for up to 1.5 m by dark-brown earthy goethite, some pseudomorphing calcite. Aurichalcite, barite, cinnabar and galena are associated species (Young et al. 1989); partially oxidized galena–tennantite at Stennerskeugh Clouds, Kirkby Stephen, has produced a supergene assemblage consisting of azurite, cuprite, duftite, goethite and malachite (Starkey 1987b).

Spherules of velvet-like goethite from Driggith mine, Caldbeck Fells, Cumbria; width 8 mm. (© D. I. Green)

Derbyshire Goethite is not uncommon in this county, although is sometimes referred to as limonite or ochre. It has been found forming hemispheres and needles in quartz crystals within volcanics on Calton Hill, Taddington, and as small bundles of stalactitic rods in a small cavity at Masson quarry, Matlock (Braithwaite 1983). At Masson quarry (also at the nearby Jug Holes mine and Oxclose mine, Snitterton) it occurs as a supergene replacement product after pyrite (var. bravoite) in metasomatic replacement flats dominated by fluorite and galena (Ixer 1974b). Nodular, mammillary concretions of dark-brown goethite (probably pseudomorphs after pyrite) up to 20 cm across occur with barite, calcite and galena gangue in a fault-brecciated zone in Carboniferous Limestone at Hindlow quarry, near Earl Sterndale church (Harrison et al. 1975a). At Snelston mine, near Ashbourne, goethite pseudomorphs both chalcopyrite and pyrite, and good botryoidal goethite was also found in North quarry (Jackman 1996). Coatings and crusts of yellowish-brown material on earlier-formed barite, calcite and fluorite has been shown to be poorly crystalline goethite and the species also forms brown crystals to 2 cm pseudomorphing marcasite at Wapping mine, Cromford (Ince 1991). Many other occurrences are summarized by Ford et al. (1993): Bakewell Chert mine; Bonsall Moor; Bradwell Moor; Brassington; Crich; Gank Hole and Mandale mines, Lathkill Dale; Hucklow, near Eyam; Millclose mine, Wensley, near Matlock; Moor Jarnett Lane, above Taddington; Odin mine, Castleton; Portway mine, W of Castleton; and Rutland Cavern, Matlock Bath.

Devon Goethite and limonite pseudomorphs after chalcopyrite and pyrite are noted in the gold-bearing selenide vein at Hope's Nose, Torquay (Stanley et al. 1990a). Goethite and limonite are the oxidation products that replace calcite and iron-rich dolomite in antimony–gold-bearing quartz–carbonate veins at Loddiswell mine, 3 km N of Loddiswell (Stanley et al. 1990b).

Durham Friable brown earthy goethite is associated with aurichalcite, chalcopyrite, smithsonite and sphalerite on the dumps of Greenhurth mine, Upper Teesdale (Young et al. 1985a). Supergene goethite of probable post-mining origin is noted with gypsum and malachite from West Pasture mine, Stanhope, Weardale (Young & Francis 1989).

Gloucestershire Fine specimens of coarsely banded and radiate yellow and brown goethite can be found in the fields near Holder's farm, 3 km NW of Newent. At Walden Court, 4 km N of Newent, massive goethite is associated with blocks of quarried sandstone capping a garden wall, and, at Birches, 4 km S of Newent, iron-grey goethite has been recovered from ploughed fields. These occurrences are related to the extraction of iron ores from rocks of Silurian to Permo-Triassic age that took place between medieval times until the latter half of the 19th century (Bick 1986a).

Leicestershire Oxidation of chalcopyrite concentrated at the line of unconformity between Carboniferous Limestone and overlying Triassic rocks at Cloud Hill quarry, Breedon on the Hill, has produced a secondary assemblage consisting of native copper, cuprite, goethite and malachite. Goethite also replaces marcasite (producing fine rosettes up to 7 mm across) and pyrite (King 1973, 1983). Small sprays of golden-brown acicular goethite and massive dark-brown goethite occur in fractured wall-rock associated with a native copper–cuprite deposit near the Precambrian/Triassic unconformity at New Cliffe Hill quarry, Stanton under Bardon (Hubbard et al. 2005). At Newhurst quarry, Shepshed, detrital goethite occurs in the basal beds of Triassic sediments, and the species is also found as a common oxidation product in the supergene assemblage associated with sulphide mineralization at the quarry (King 1973,

Ince 2005). Other goethite occurrences are described by King (1973): quartz–carbonate veins cutting tuffs in the dam quarry at Blackbrook; abundant goethite pseudomorphs after pyrite in a volcanoclastic sequence SE of Rise Rocks farm; excellent goethite pseudomorphs after pyrite from the cuttings of Markfield bypass; tabular plates of goethite after hematite at Bardon Hill quarry; goethite (including excellent pseudomorphs after pyrite) in dolomite–hematite–quartz–sulphide veins at Sheethedges Wood quarry, Grosby; microscopic spheroids of goethite associated with calcite and quartz at Croft quarry.

Northamptonshire The lower part of fissures in Northamptonshire Ironstone (Inferior Oolite) are infilled with banded goethite mud showing small-scale mammillation at Pitsford ironstone quarry, Irchester, 3 km S of Wellingborough (Hollingworth & Bannister 1950).

Shropshire In the upper levels of Ladywell mine, Shelve, oxidation of chalcopyrite–galena–pyrite–sphalerite-bearing veins produced a supergene assemblage consisting of anglesite, brochantite, cerussite, devilline, goethite (limonite), jarosite, langite, linarite, malachite, pyromorphite and wroewolfeite (Braithwaite 1982b).

Somerset Small pockets and veins of iron–manganese oxides cut Carboniferous Limestone at Coleman's quarry, Holwell, near Frome, Mendip Hills. Goethite and hematite are a minor part of the hausmannite–manganite–pyrolusite-dominated ore, occurring in discrete pockets of massive ore and as thin discontinuous selvages along vein margins (Alabaster 1990a). Goethite and hematite appear to be more common at Higher Pitts mine, Priddy, Mendip Hills, where an iron–manganese oxide deposit formed in dolomitic conglomerate within a hollow in Carboniferous Limestone (Kingsbury 1941, Burr 1996). Perhaps the best goethite in the Mendips is that described by Kingsbury (1941) from a limestone quarry, N of Lamb Lair Cavern, near West Harptree, where iron oxide-infilled fissures were reported to contain cavities lined with crusts of brilliant velvety goethite needles. Iron–manganese oxide-filled fissures, veins and pods in Carboniferous Limestone are also reported from Merehead (Torr Works) quarry, Cranmore. Some pods of mottled bluish-grey MnO_2 are rimmed by goethite and contain cavities in which the rare oxychloride species blixite, chloroxiphite, diaboléite, mendipite and paralaurionite have crystallized (Alabaster 1975). Turner (2007) reports euhedral crystals of goethite, manganite and pyrolusite within calcite-lined cavities at this location. In the lower levels of one particularly large vein, cavities in goethite were encrusted with apophyllite, barite, calcite and datolite (Symes 1977).

Staffordshire Goethite is noted from Ecton Hill in the Southern Pennine orefield (Sarjeant 1956a).

Yorkshire Dark-brown goethite pseudomorphs after characteristic spear-shape crystals of marcasite are abundant on the dumps of Virgin Moss vein on East Bolton Moor in Wensleydale (Young et al. 1992c). Platy boxworks of goethite after marcasite occur at Faggergill mine, Arkengarthdale, and the species is considered to be widely present in the oxidized ores of the Askrigg Block section of the Northern Pennine orefield (Dunham & Wilson 1985).

Wales

Dyfed *Central Wales orefield:* Dark-brown to black goethite, forming resinous masses and nodules infilling cavities in quartz, occur at Eaglebrook (Nantycagl) mine, Ceulanymaesmawr (Jones 1983). Goethite commonly forms thin yellowish-to-black crusts that are often overgrown with cerussite or pyromorphite at Frongoch mine, between Devil's Bridge and Pont-rhyd-y-groes (Green et al. 1996b). At Llechweddhelyg mine, near Clawddmelyn, between Bont-goch and Cwmsymlog, goethite is very common and it forms pulverescent brown masses or deep-brown resinous masses, especially infilling cavities within galena or quartz (Jones 1987). Two cavities in galena were found to contain patches of yellowish goethite (probably replacing chalcopyrite) and chenite (Rust 1994). A substantial zone of supergene mineralization on the Camdwr lode at Llechweddhelyg mine is derived from sulphides that have been completely converted to a carbonate-dominated goethite-hosted assemblage (Mason 2004). Black-to-brown botryoidal goethite coats shale adjacent to areas mineralized with brochantite and malachite at Lodge Park copper trial, Tre'r-ddôl (Mason & Green 1996).

Gwynedd Small quantities of goethite and other ferric hydroxides are associated with carbonatian connellite, georgeite and malachite at Britannia mine, Snowdonia (Pollard et al. 1989). Goethite–limonite pseudomorphs after cockscomb marcasite are noted with anglesite, cerussite, malachite and smithsonite in the oxidized zone of primary galena–sphalerite veins and vein breccias at Cyffty and Penyrallt mines, between Betws-y-coed and Llanrwst (Haggerty 1995). At Great Orme mine, Llandudno, small goethite crystals, dolomite and malachite line 20–25 cm-wide veins infilled with a black, highly friable ore called "copper dhu" or "black copper" (Ixer & Davies 1996, Ixer 2001). Thinner veins (1–2 cm wide) containing a dark-brown massive mixture of malachite and goethite enclosing chalcopyrite also occur in N–S joints and faults within Carboniferous Limestone at Great Orme (Dutton et al. 1994). Bevins (1994) reports goethite pseudomorphs after cuprite from Great Orme's Head. Impressive sprays of goethite needles, up to 4 mm in length, have been collected from the Deep Adit level of the Chidlaw lode at Gwynfynydd mine, near Dolgellau. Spectacular rosettes of goethite crystals, up to 5 mm in length, are found in association with marcasite at Gwynfynydd mine (Bevins 1994). Goethite is noted without description from Parys Mountain, Anglesey (Southwood & Bevins 1995).

Mid Glamorgan Goethite forms pseudomorphs after cubic or octahedral pyrite at: Blaengwynlais quarry, near Rhiwbina; Miskin quarry, near Llantrisant; and Machen quarry, 7 km ENE of Caerphilly (Bevins 1994, Plant & Jones 1995). At Llanharry mine, Llanharry, goethite–hematite deposits within Carboniferous Limestone form roughly cylindrical orebodies. Goethite is the dominant ore species and it occurs in layers or bands (1 mm to 1 m thick) with colloform, crystalline, spheroidal or stalactitic form (Gayer & Criddle 1970, Dunham et al. 1978, Rankin & Criddle 1985, Bevins 1994). Garth iron mine, near Taff's Well, worked a similar goethite–hematite deposit in the same orefield (Rankin & Criddle 1985). At Tŷ Coch, near Porthcawl, botryoidal and compound crystallites of goethite are interlayered with romanèchite in a limestone-hosted manganese-rich ore also known to contain braunite, hausmannite, psilomelane, pyrolusite and pyrobelonite (Criddle & Symes 1977). Perhaps the best goethite in South Wales was collected from Ton-mawr quarry, Pentyrch, Cardiff, where rich aggregates of acicular crystals up to 2 mm long on a matrix of massive goethite were recovered during the construction of a new haul road (Wolfe 2000).

Powys Lustrous black goethite coats galena and is found with cerussite and pyromorphite at Aberdaunant mine, near Llanidloes, in the Central Wales orefield (Jones & Moreton 1977).

Scotland

Borders Ferromagnesian sediments at Noblehouse, near West Linton, contain goethite in addition to birnessite, hematite and todorokite (Nicholson 1989a).

Dumfries & Galloway A gossan zone consisting of beudantite, goethite, limonite, malachite and plumbogummite occurs above lead–zinc mineralized siliceous mudstones, siltstones and chert at Penkiln Burn, N of Newton Stewart (Stone & Leake 1984).

Fife At Magus Muir quarry, 5km W of St Andrews, limestone contains deep-brown goethite forming brushes of acicular crystals rising in tufts from druses lined with quartz (rock crystal) (Heddle 1901). Sprays of goethite needles to 1 mm have been found on calcite and quartz within joints and fractures in basalt at Orrock quarry, 3 km NW of Burntisland (Wirth 2003).

Grampian A soil clay-mineral study of an iron–manganese pan in brown forest soil from the Blairmore Estate, Glass, near Huntly, showed two oxyhydroxide species, goethite and lepidocrocite (dominant), to be present (Russell et al. 1984a). In the Lecht valley, near Tomintoul, manganese-rich ironstone occurs in several outcrops within deeply weathered Dalradian metasediments. The deposit is composed principally of yellowish-brown, soft, porous masses and harder, more compact, varieties of goethite, together with cryptomelane. Minor chalcophanite, hematite, lithiophorite, pyrolusite, ramsdellite and woodruffite are also noted (Nicholson 1987a, Nicholson & Anderton 1989). Lead–zinc mineralization in a Triassic breccia termed the Cherty Rock (Triassic) at Stotfield, W of Lossiemouth, near Elgin, contains fibrous botryoidal crusts of goethite mantling galena and quartz, and pyrite much altered to goethite (Starkey 1988a).

Highland Goethite and malachite are the main secondary species in chalcopyrite-, molybdenite-, pyrite-bearing veins cutting the Ballachulish granite in Gleann a' Chaolais (Haslam & Kimbell 1981). Mammillated goethite is overgrown by quartz (var. rock crystal) that contains rutile inclusions at Burn of Isauld, Achvarasdal, near Reay (Heddle 1901).

Lothian Fine radiating specimens of goethite were once found in a quarry at the north end of Salisbury Crags, Edinburgh (Heddle 1901). A mammillated variety, associated with red quartz, was recovered from veins exposed near the Hopetoun monument in the Garleton Hills (Heddle 1901).

Strathclyde Veins, coatings and discrete masses of mixed iron–manganese oxides occur in the Dalradian Jura Quartzite at Dùn Athad on the southwest coast of the Isle of Islay. Goethite and hematite are the main iron-bearing species, whereas hollandite is the principal manganese-bearing species (Nicholson 1987b, 1988a). Elsewhere on Islay, small black lustrous stalactitic growths of goethite on dolomite are reported from copper-mineralized veins at Rubha a' Mhill, on the northeast coast (Green 1988). Goethite is associated with jarosite and native sulphur in amygdaloidal basalt at Craignish on the Isle of Mull (Livingstone & Baird 1991). Dark golden-brown

Compact, banded goethite with yellow quartz. Garleton Hills, Lothian Region, Scotland; width of specimen 11 cm. (© The Trustees of the National Museums of Scotland)

acicular crystals of goethite up to 2.5 mm long are associated with botallackite and hematite on quartz veinstone from Castletown mine, Lochgilphead (Starkey & Hubbard 1998). At Craigmuschat quarry, Gourock, geodes, variably composed of ankerite, barite, calcite, chalcedony, dolomite, fluorite, goethite, gypsum, hematite, prehnite, quartz, romanèchite and wad, occur in a trachyte sill (Nicholson & Durant 1986). Mineralization related to the intrusion of a Tertiary dyke cutting Carboniferous basaltic lavas at Loanhead quarry, near Beith, includes goethite as a partial alteration product of chalcopyrite and as stalactitic growths with a golden-yellow fibrous internal structure on and replacing hematite (Meikle 1989b). Mannoch Hill mine, SW of Lesmahagow, worked barite veins in which goethite occurred as lustrous black radiating groups to 6 mm and as pseudomorphs after chalcopyrite and pyrite (Todd & McMullen 1991). Heddle (1901) reports goethite from Boyleston quarry, near Barrhead; Bishopton railway tunnel; Gourock quarry; and Kilmalcolm.

Tayside Golden-brown to black needles of goethite up to 8 mm long are either included in, or penetrate, quartz crystals in geodes at Scotlands' premier agate locality – Ardownie quarry, Monifieth (Ingram 1994).

Orkney Large mammillated specimens of goethite were once found with quartz (var. rock crystal) in the chasm of the Burn of the Sale, the Bring, Hoy (Heddle 1901).

Ireland

Cork At Aghatubrid Beg manganese mine, Glandore, massive goethite contains cavities in which rhombs of corkite have grown (Ryback & Moreton 1991).

Galway Leaching of copper, lead and zinc from a residual orebody (gossan) at Tynagh mine, near Killimor, left behind a silver-rich goethite–limonite deposit (Clifford et al. 1986).

Mayo Gold-mineralized quartz veins of the Lecanvey prospect, Clew Bay, contain associated minor sulphides and both goethite and hematite (Aherne et al. 1992).

Tipperary A residual deposit representing a zone of deep weathering at Knockanroe and Ballygown South mines, Silvermines (where acid groundwaters have passed through dolomitized and mineralized bedrock), contains goethite, hemimorphite, siderite and smithsonite (Rhoden 1959, Boland et al. 1992).

Tyrone Supergene alteration of primary sulphides to chalcocite, covellite, goethite and hematite has occurred in gold-bearing quartz veins of the Lack deposit, WSW of Omagh (Cliff & Wolfenden 1992).

Waterford At Drumslig, between Youghal and Dungarvon, goethite and hematite form veins and replacements in Devonian shales and sandstones (Halliday & Mitchell 1983).

Offshore territories

Jersey Radiating fibrous goethite occurs within narrow veins in crushed shales at Mont Sohier, St Brelade (Mourant 1961).

gold

cubic Au

A rare metal, often referred to as native gold, which forms a series with silver-3C. The name electrum is commonly applied to intermediate compositions where there is some substitution of gold by silver. Native gold (abbreviated below to gold) is found in very small quantities in a wide range of rock types. It is often reported from epithermal quartz veins with pyrite and other sulphides and tellurides, and from alluvial (placer) deposits. Reviews of gold in Cornwall and Devon, and in the south of Scotland, are provided by Camm (1995) and Dudgeon (1876) respectively.

England

Cheshire Mineralized fractures in silicified barite–calcite-rich fault rock associated with the West mine boundary fault at West mine, Alderley Edge ridge, contain relict traces of chalcopyrite with acanthite (argentite), cinnabar and native gold (<2 μm) inclusions (Plant et al. 1999).

Cornwall Sulphide-bearing veins belonging to a phase of antimony–lead mineralization in north Cornwall contain gold associated with pyrite and various sulphosalt species (boulangerite, bournonite, jamesonite, plagionite and tetrahedrite) at Bounds Cliff, E of Port Isaac, St Endellion (Forbes 1869, Clayton et al. 1990, Stanley et al. 1990b). At nearby Treore mine, Port Gaverne, near Port Isaac, St Teath, similar veins contain gold together with arsenopyrite, boulangerite, bournonite, chalcopyrite, galena, jamesonite, pyrite, sphalerite, stibnite and tetrahedrite (Dewey 1925, Russell 1949b, Hosking 1951, Criddle & Stanley 1993, Camm 1995). Other veins in the area, at Port Quin and nearby Reedy Cliff, are also gold bearing (Clayton et al. 1990, Stanley et al. 1990b). In nearby quartz veins at Tregildern quarry, near Pendoggett, St Teath, gold has been found in fractures within tetrahedrite and is associated with copper and silver at Wheal Sampson, near Bounds Cliff (Camm 1995). Elsewhere in Cornwall, *in situ* gold occurrences include: associated with russellite in jig concentrates from Castle-an-Dinas wolframite mine, St Columb Major (Russell 1944, Hosking 1973); East United mine, Gwennap and Phoenix United mine, Linkinhorne (Camm 1995); Great Dowglas mine, Sticker, near St Austell (Rudler 1905, Camm 1995); Great Roughtor Consols mine, Bodmin Moor (Pattison 1854); Great Work mine, near Breage (Camm 1995); Halloon mine, near St Columb Road railway station (Hamilton Jenkin 1964, Camm 1995); Inny Consols, near Polyphant, and Wheal Baron (St George's mine), near Tregeare (Hamilton Jenkin 1969, Camm 1995); Mount Wellington mine, Twelveheads, 7 km W of Truro (Kettaneh & Badham 1978); Nangiles mine, Kea (Greg & Lettsom 185, Forbes 1869); Pennance mine, Maenporth, near Falmouth, and Wheal Anna Maria, Mawnan (Hamilton Jenkin 1967, Camm 1995); in a schorl rock at Porthmeor Cove, near Zennor (Robson 1954); Swanpool mine, near Falmouth and Wheal Clinton, Flushing, near Falmouth (Stephens 1899, Camm 1995); Trevinnick mine, St Kew, and Ino iron mine (Brogden Iron), near Porth Navas (Dines 1956); West Tolgus mine, Redruth (Hambly 1897); Wheal Gorland, St Day, 2 km E of Redruth (Rust 1981b, Camm 1995); Wheal Jane, Kea, near Truro (Rust 1982d); Wheal Sparnon, near Redruth (Greg & Lettsom 1858, Forbes 1869, Rudler 1905, Camm 1995). Gold from Cligga Head, Perranzabuloe (Rogers 1968, Camm 1995), deserves further investigation.

Many gold occurrences in Cornwall are alluvial, often being discovered during tin-streaming activities. These are admirably documented by Camm (1995) and other authors: Boscarne, near Bodmin; Breney Common, near Lanlivery, and Cober Valley, near Wendron (Henwood 1874); Carnon Valley, near Gwennap – largest nugget 56 g (Greg & Lettsom 1858, Forbes 1869, Collins 1904a, Embrey & Symes 1987, Penhallurick 1997); Castle Park, near Restormel Castle, Lostwithiel; Crow Hill, Trewarda; Ladock, near Grampound; Lanlivery, near Lostwithiel (all Greg & Lettsom 1858); Durra Stream, near Manaccan, Lizard Peninsula; Glynn, near Bodmin Road railway station; Godrevy beach, near Gwithian; Laity Moor, near Redruth (all Garby 1846); Lavrean, near Luxulyan; around Manaccan, Lizard Peninsula; Nantellan, near Grampound; Pensagilles, near St Ewe; Port Gaverne beach, near Port Isaac, St Teath; Port Isaac Road station, St Teath; Porthcurnick beach, Porth River, below Treloy, near Newquay; Portreath beach (all Hambly 1897); Restronguet Creek, near Devoran – largest nugget approximately 59 g (Henwood 1855); Rosemullion Point, Falmouth Bay; St Austell River on St Austell Moor (both Forbes 1869); Trebartha Lemarne mine, northeast Bodmin Moor (Henwood 1874, Hamilton Jenkin 1969); Trelower farm, St Mewan; Tresillian River, Ladock; Trewiggett farm, near Pendoggett, St Teath; and Trewoder, E of Truro.

A specimen of gold in the Kingsbury collection at the NHM, labelled Porthcurnick beach, Portscatho, and described in Criddle & Stanley (1993) and Camm (1995), appears to have been given an incorrect provenance. It almost certainly came from Kalgoorlie, Australia (Ryback et al. 1998).

Cumbria *Caldbeck Fells:* Gold disseminated in small specks on lead ore is recorded from an unspecified location in this area (Calvert 1853), and a specimen labelled "Cumberland" is possibly from Red Gill mine (Sweet 1960a). More precisely, gold is noted from Carrock mine, Grainsgill, where it accompanies native bismuth (Dewey & Dines 1923, Shaw 1970). Young (1987a) reports gold with arsenopyrite in a sample of quartz vein in greisen from the large dumps near

Grainsgill Beck. *Elsewhere in the Lake District:* Copper-rich sulphide veins in Newlands Valley, near Keswick are noted to contain gold at Dale Head (Stanley & Vaughan 1980) and Goldscope (Calvert 1853, Postlethwaite 1913) mines. Gold is associated with a bismuth telluride at Paddy End mine, Coniston, Eskdale (Stanley 1979), and one 20 µm grain is reported from the Upper level dump of Seathwaite mine, Seathwaite Tarn, Coniston (Stanley & Criddle 1979). At Piers Gill, 1 km N of Scafell summit, Wasdale, a 60 cm-thick polyphase quartz vein cemented by ankerite, within the Scafell dacite, contains microscopic inclusions of gold (electrum) in an Ag–As–Au–Cu–Pb–Zn assemblage similar in many respects to that found in a porphyry-copper system. Associated species include acanthite (argentite), pyrargyrite, tetrahedrite, uytenbogaardtite and an unknown selenium–silver mineral (Aldridge 2005).

Other gold specimens supposedly collected by A. W. G. Kingsbury are documented by Young (1987a) and Cooper & Stanley (1990), but their provenance is now considered dubious. Localities include: Birkside Gill mine, Wythburn, Thirlmere; Embleton quarry, near Cockermouth; Robin Hood mine, Bassenthwaite; Slate Fell, Cockermouth; Wanthwaite mine, St John's in the Vale, near Threlkeld; four localities in the Caldbeck Fells – Ingray Gill mine; Mexico mine, Todd Gill; Roughton Gill mine; Swinburn Gill. It is uncertain if the gold described from Brandlehow mine, Derwent Water (Goodchild 1882, Postlethwaite 1913), is native gold or a minor constituent of a sulphide phase.

Derbyshire Gold (native gold?) is confirmed only from a toadstone at Temple mine, Matlock Bath. Other reports list Lathkill Dale, S of Over Haddon and Ible quarry on the north flank of Via Gellia. However, limited evidence suggests that these were financial swindles based on pyrite or planted gold (Ford et al. 1993).

Devon Particularly attractive delicate dendritic growths of gold (ferns up to 3 cm in length) have been discovered totally embedded in calcite within carbonate veins cutting Devonian limestones at Hope's Nose, Torquay. The gold is remarkable for its silver and palladium content, and for its associated selenide assemblage, which includes chrisstanleyite, clausthalite, eucairite, fischesserite, klockmannite, naumannite, oosterboschite, penroseite, tiemannite, trüstedtite, tyrrellite and umangite (Gordon 1922, Russell 1929, Clark & Criddle 1982, Scrivener et al. 1982, Harrison & Fuller 1987, Stanley et al. 1990a, Leake et al. 1991, Criddle & Stanley 1993, Penhallurick 1997, Paar et al. 1998, Bachmann et al. 2003, Shepherd et al. 2005). Similar gold with dendritic form and palladium enrichment is also known from Brownstone in the South Hams district and was collected as panned concentrates in drainage and overburden samples derived from Devonian sediments and volcanics. Native gold is widespread in stream sediments in this district and at locations such as Whympston (near Modbury) and Challon's Combe, several varieties are reported. These include gold enriched in silver, gold containing inclusions of selenide species (clausthalite, guanajuatite, tiemannite and unnamed varieties), and gold associated with potarite. Compositional zonation is common (Leake et al. 1989, 1991, 1993, Camm 1995, Chapman et al. 2000b). Alluvial gold has also been found in the Crediton Trough at Becka Moor, Deckport, Exbourne, NW of Hatherleigh, Hatherleigh Moor, Honeychurch, Hopebrook, Smallbrook, Solland and Venn Lake (Camm 1995). *In situ* gold-bearing carbonate veins have been discovered: in boreholes at Thorverton and Upton Pyne in the Crediton Trough (Leake et al. 1994, Shepherd et al. 2005) and at surface, cutting Devonian slates, 3 km N of Loddiswell, where associated species are acanthite, native bismuth, bournonite, chalcopyrite, covellite, galena, gersdorffite, jamesonite, sphalerite, tetrahedrite and yarrowite (Stanley et al. 1990b, Camm 1995). In the Exmoor area at North Molton, gold is reported from a copper-rich gossan at Poltimore (Bampfylde) mine (Rudler 1905) and from Prince Regent (Britannia) mine (Hall 1868b). Camm (1995) notes that both were the cause of a minor gold rush and fraud, although gold undoubtedly was found, perhaps as early as Roman times. Other hard-rock occurrences include: Bampfylde (Poltimore) mine, North Molton (Pattison 1855, Durrance & Laming 1982); in an amphibole pegmatite at Bittleford Down, Widecombe (Brammall & Harwood 1924); Daddy Hole, Torquay (Camm 1995); a claimed (doubted) report from a gossan at Drewsteignton, ESE of Okehampton (Dines 1956); in Dartmoor granite at North Hessary Tor, Princetown; Vixen Tor quarry, Sampford Spinney; Welstor Cross quarry, near Buckland Beacon (Brammall 1926); in a gossan at Okehampton Consols (Hamilton Jenkin 1981); and in veins at South Molton (Robson 1949b). Alluvial gold reports include: Colston, Buckfastleigh (Scrivener & Walbeoffe-Wilson 1975); Golden Dagger mine, near Vitifer (Baring-Gould 1888); Kester Brook, Bickington (Camm 1995); and Sheepstor, Dartmoor (English 1851, Greg & Lettsom 1858).

Gloucestershire Old Red Sandstone conglomerates near Micheldean in the Forest

Fern-like dendritic growths of native gold etched out of calcite from Hope's Nose, Torquay, Devon; field of view is *c.* 18×30 mm (© D. I. Green).

Fern-like dendritic growth of native gold (1 cm long) on calcite. Hope's Nose, Torquay, Devon. (© M. P. Cooper)

of Dean have been shown to contain gold (Hart 1944), although native gold does not appear to have been confirmed.

Leicestershire Gold has been found in a small outcrop of gossan on the north face of Bardon Hill. It forms flakes 0.8 mm across in cavernous quartz within a fault-bounded shatterbelt (King 1967, 1973). A specimen containing minuscule flakes of gold on a limonite-quartz matrix labelled "Springhill, Whitwick" (now Peldar Tor quarry) is in the NMW, and may originate from the same swarm of quartz veins on Peldar Tor described by Hill & Bonney (1880).

Yorkshire The report of gold from Jamie vein, Greenhow, near Pateley Bridge, remains unconfirmed (Rodgers 1975, Dunham & Wilson 1985, Ixer & Stanley 1998).

Wales

Dyfed Gold is present in arsenopyrite–pyrite–(galena)-bearing quartz veins within strongly folded Ordovician–Silurian turbiditic shales and sandstones at the Dolocauthi mines near Pumsaint. Ogofau mine is perhaps the best known of these, where gold has been exploited since Roman times. Often the gold occurs as inclusions (15–400 µm across) in the sulphides (Greg & Lettsom 1858, Rudler 1905, Brammall et al. 1937, Hughes 1959, Steed et al. 1976, Dunham et al. 1978, Annels & Roberts 1989, Bevins 1994). In the Central Wales orefield, minor amounts of gold have been found in certain veins, including those at: Caegynon mine, near Ponterwyd (doubtful) and Copa Hill mine, Cwmystwyth (Jones & Moreton 1977); Eaglebrook (Nantycagl) mine, Ceulanymaesmawr; Erglodd mine, N of Tal-y-bont; Llechweddhelyg mine, near Clawddmelyn, between Bont-goch and Cwmsymlog (all Bevins 1994). Mason (1997) notes gold (electrum) inclusions (<100 µm across) in galena within the chalcopyrite–galena-dominated A1-*c* polymetallic assemblage at Erglodd mine, N of Tal-y-bont.

Gwynedd Gold has been worked from steeply dipping anastomosing quartz veins at over 30 mines in the Dolgellau goldbelt, which is a multiphase, mesothermal gold-lode province around the southern and eastern margins of the Harlech Dome. It extends from Barmouth in the SW to Cwm Prysor near Trawsfynydd in the NE. The veins are dominated by an arsenopyrite–chalcopyrite–pyrite–pyrrhotite assemblage in the SW and a galena–pyrite–sphalerite assemblage in the NE. Gold typically occurs disseminated in quartz or is associated with sulphides, particularly sphalerite, although at Clogau–St David's mine, Bontddu, near Dolgellau, it is associated with aleksite, galena and tellurobismuthite (Naden 1988, Bevins & Stanley 1990). Telluride species are also noted from the Caegwian, Cefn Coch, Ffridd-Goch and Panorama mines (Mason et al. 2002). Particularly attractive leaf-gold aggregates to 4 cm have been found at Gwynfynydd mine, near Dolgellau, where high-grade veins cut Cambrian black shales. Silver-rich galena and sphalerite are associated species (Andrew 1910, Allen & Jackson 1985, Bevins 1994, Dominy et al. 1996b,c, Dominy & Phelps 1996, Platten & Dominy 1999, Chapman et al. 2000b, Mason et al. 2002). The following localities are summarized by Andrew (1910) and other authors: Berthlŵyd mine, near Tynygroes; Cae-gwian lode, near Pont Glandwr; Cae-Mawr mine; Cambrian mine; Cwm Hesian mine; Farchynys lode; Garth-gell mine, near Bontddu; Prince of Wales mine (all Andrew 1910); Cefn Coch mine, Tynygroes (Platten & Dominy 1999); Dolfrwynog mine, Llanfachreth (Greg & Lettsom 1858, Dewey & Eastwood 1925); Ffridd-Goch mine, Llanfachreth (Mason et al. 2002); Glasdir mine, near Llanfachreth; Panorama mine, Llanaber, E of Barmouth, and Vigra mine (all Dewey & Eastwood 1925); Tyddyn-Gwladys mine, near Dolgellau (Forbes 1868, Dewey & Eastwood 1925, Mason et al. 2002). Alluvial gold is also noted from the Afon Las, Afon Mawddach and Afon Wen (Ward 1982, Leake et al. 1993, Green 1998b, Chapman et al. 2000b). Elsewhere, gold is noted from quartz veins at Castell Carn Dochan mine, near Bala (Readwin 1864, Archer 1959, Bevins 1994); Parys Mountain, Anglesey, and in arsenopyrite-, bismuthinite-, chalcopyrite-, pyrite- and tetradymite-bearing quartz veins at Rhôs-mynach-fawr mine, Anglesey (both Greenly 1919); Prince Edward (Mod Croesau) mine, Trawsfynydd (Leake et al. 1993, Mason & Williamson 2001).

South Glamorgan Grains of gold associated with fluorite are reported from a borehole in Triassic rocks at Working Street, Cardiff (Storrie 1895).

Scotland

Borders Grains of alluvial gold, often with inclusions of other species (e.g. chalcopyrite, galena, gersdorffite, pyrite, sphalerite, tetrahedrite), are reported from: Craigierigg Burn; Cramalt Burn; Glengaber Burn, SW of Peebles (coarse grains, largest 8 mm long, weighing 0.7 g); Glenwhappen Burn, Tweedsmuir; Hawkshaw Burn, Tweed headwaters, Tweedsmuir; Lower Tweed, Tweed headwaters; Smid Hope Burn, Tweed headwaters; Talla Water, SE of Tweedsmuir; Tweed Head, N of Moffat; Upper Tweed, SW of Ettrickbridge (Leake et al. 1993, 1997c, Chapman et al. 2000a,b,c).

Central Quartz–sulphide veins at Cononish, 4 km SW of Tyndrum, are both silver- and gold bearing. They occur in a fault zone within Middle Dalradian metasediments and (in decreasing order of abundance) contain pyrite, galena, chalcopyrite and sphalerite. Gold mainly forms 5–100 µm grains, although 1–2 mm grains are also noted (Parker et al. 1989, Earls et al. 1992, Curtis et al. 1993, Treagus et al. 1999). Other fault-related veins in the region are also gold-bearing and one example near Tyndrum (Halliday's vein) has been shown to contain gold (electrum 22–61 wt% Au), hessite, petzite, sylvanite and unnamed telluride species (Pattrick et al. 1988). One of these tellurides is probably cervelleite. Alluvial gold is noted from: Cononish River (Chapman et al. 2000a,b); Crom Allt River, near Tyndrum (Green & Todd 2001); Chapel Burn and Cloan Burn, near Glen Devon Youth Hostel, Ochil Hills (Davidson 1995b, Green & Todd 2001); Borland Glen and Fanny Burn, Ochil Hills (Leake et al. 1993, 1997c, Chapman et al. 2000c).

Dumfries & Galloway Gold (electrum 76–77 wt% Au) occurs as small blebs and veins within gersdorffite, often associated with chalcopyrite veinlets, within the arsenic-rich magmatic sulphide deposit at Talnotry (Palnure Burn), 5 km NE of Newton Stewart (Power & Pirrie 2001, Gunn & Styles 2002, Power et al. 2004). Minor gold (grains to 20 µm) is also noted in quartz–stibnite veins at Glendinning mine, Glenshanna Burn (Duller et al. 1997), in an arsenopyrite–quartz vein at Glenhead Burn, Loch Dee (Leake et al. 1981), and in chalcopyrite–tetrahedrite-bearing quartz veins at Moorbrock Hill, Carsphairn (Naden & Caulfield 1989). Rare native gold, silver and bismuth accompany chalcopyrite and pyrite in quartz and carbonate veinlets within the aureole of the Criffell granodiorite at Needle's Eye, SW of Caulkerbush (Basham et al. 1989). Alluvial gold is reported from: Wanlockhead (Brown 1927); Glengap Burn, SW of Moffat (Leake et al. 1997c, Chapman et al. 2000c); Mennock Water, Wanlockhead (Chapman et al. 2000b, Green & Todd 2001, Chapman & Leake 2005); and Stennies Water (Leake et al. 1993, 1997c).

Grampian Submicron particles of native gold occur in a Devonian hot-spring system associated with a hydrothermally altered fault zone at Rhynie (Rice et al. 1995). Native gold forms clusters of irregular grains, up to 20 µm across, associated with native bismuth and froodite, in a borehole cutting sulphide-rich gabbro at Littlemill farm, near Huntly (McKervey et al. 2007). Alluvial gold is reported from Back Burn, W of Auchenblae (Leake et al. 1997c, Chapman et al. 2000c), and Greenknowle Burn, Clatterin Brig, 5 km N of Fettercairn, produced relatively large nuggets up to 2 g (Green & Todd 2001).

Highland The Kerry Road stratiform sulphide deposit near Gairloch is hosted by pyritic quartz–carbonate schists and it contains rare gold in a pyrite–pyrrhotite-dominated ore (Jones et al. 1983, 1987). On the west side of Loch Duich (along the road to Glenelg) and on the east side of Loch Duich (along the road to Kyle of Lochalsh) veins associated with the Ratagain granitic intrusion contain sporadically distributed gold (electrum with 67 wt% Au) associated with dominant chalcopyrite, galena, molybdenite, pyrite, rutile, sphalerite and locally abundant aikinite, hessite, marcasite and tennantite (Alderton 1986, 1988b). A possible hot-spring gold deposit occurs at Meall Odhar, near Lairg, although gold has so far been found only in quartz-cemented breccia float samples and panned concentrates (Crummy et al. 1997). The Kildonan alluvial goldfield is perhaps the best known and richest deposit in Britain and was the centre of a minor goldrush in 1868–1870 (Berridge 1969). Gold is reported from Kildonan Burn (largest nugget weighed 28 g); Suisgill Burn, N of Kildonan (largest nugget weighed 60 g), and Torrish, NW of Helmsdale (Heddle 1901, Rudler 1905, Leake et al. 1993, 1997c, Chapman et al. 2000c). Rare gold (electrum with 46–76 wt% Au) occurs in disseminated sulphide mineralization within the mafic–ultramafic layered intrusion and in associated minor intrusions on the Isle of Rum (Dunham & Wilkinson 1985, Power et al. 2000a, 2001, 2003).

Strathclyde Gold is associated with chalcopyrite, freibergite, galena, pyrite, sphalerite and tennantite in a Caledonian diatreme–porphyry-breccia complex at Lagalochan farm, near Kilmelford. Grainsize reaches 100 µm (Harris et al. 1988). Gold mineralization at Hare Hill, 5 km SE of New Cumnock, is associated with zones of disseminated arsenopyrite and hosted by hydrothermally altered and quartz-veined granodioritic intrusives. Grainsize is assumed to be submicroscopic, although two samples produced grains 1–10 µm across (Boast et al. 1990). Gold (electrum with 50–78 wt% Au) is present in arsenopyrite–pyrite-rich quartz veins associated with the Fore Burn igneous complex, Tairlow plantation, near Straiton (Charley et al. 1989). At Coille-bhraghad mine, near Inveraray, gold has been found in fine-grain Dalradian stratabound chalcopyrite–pentlandite–pyrite–pyrrhotite ore (Dunham et al. 1978). Microscopic gold is also noted from the Ailsa Craig microgranite (Heddle 1901). Alluvial gold is noted from: Fore Burn (Chapman et al. 2000a,b); Leadhills–Wanlockhead region (Brown 1927, Heddle 1901, Leake et al. 1998, Chapman et al. 2000a,b); Lochy River, W of Tyndrum (Leake et al. 1997c, Chapman et al. 2000c); and Lochy Trib (Eas Daimh), W of Tyndrum (Chapman et al. 2000c). By far the largest alluvial find has been the 4 cm-long 32 g gold–quartz nugget discovered at the head of the Shortcleugh Water, Leadhills (Russell 1944, Livingstone 2002).

Tayside Gold (electrum 50–65 wt% Au) has been found in quartz–sulphide veins cutting Dalradian metasediments and metabasic rocks at Calliachar Burn, 4 km SW of Aberfeldy. It is primarily associated with galena and pyrite, and it forms blebs 10–20 μm across, locally concentrated into 100–200 μm patches. In the nearby Urlar Burn, similar veins contain altaite–hessite–tsumoite and mercury-poor electrum inclusions in pyrite, and in one alteration zone mercury-rich electrum is noted (Ixer et al. 1997). Traces of gold (electrum 36–65 wt% Au), native bismuth and bismuth–copper sulphide minerals occur in a stockwork of galena-rich quartz veins at Corrie Buie, Meall nan Oighcreag, Loch Tay (Pattrick 1984). One grain of gold is reported from a small body of metamorphosed and altered ultramafic rocks at Corrycharmaig, Glen Lochay, NW of Killin (Power & Pirrie 2000). Alluvial gold is noted from: Calliachar Burn, Aberfeldy (Leake et al. 1993, 1997c, Chapman et al. 2000a,b); Glen Garry, near Dalnacardoch; Glen Lednock, near Comrie; and West Water, NW of Brigend (all Leake et al. 1997c, Chapman et al. 2000c). A 3 cm-long 65 g gold nugget was found at Turreich, near Amulree (Livingstone 2002), although Heddle (1901) places doubt on the provenance.

Shetland A chromite-rich sulphide-bearing dunite belonging to the Shetland ophiolite complex at Cliff on Unst contains minor gold associated with a diverse PGE assemblage. Sulphide-bearing dunite pods, cumulate dunite, sulphide-bearing wehrlite and pyroxenite from other unspecified locations in the complex also contain gold (Prichard et al. 1986, 1994).

Ireland

Carlow Alluvial gold recovered from Aclare, near Myshall, and Coolasnaghta was potentially derived from tourmaline-rich schists close to the margins of the Leinster granite (McArdle et al. 1987, 1989).

Donegal Alluvial gold occurs at: Balwoges; Cabry River, Inishowen Peninsula; and Crownarad (Chapman et al. 2000a,b).

Down Alluvial gold is noted from many locations in the Mourne Mountains (Greg & Lettsom 1858, Moreton 1993, Smith et al. 1996, Chapman et al. 2000c).

Kerry Rare electrum, intergrown with chalcopyrite and overgrowing pyrite, occurs in a polymetallic carbonate-hosted deposit on Crow Island, near Muckross, Killarney (Ixer 2004).

Mayo Isoclinally folded turbiditic sediments and tuffs in the Cregganbaun shear zone contain gold-rich quartz veins. Overlying drainage samples are also gold bearing (Leake et al. 1993). At nearby Croagh Patrick, gold occurs in quartz veins within Silurian quartzites and in associated placer deposits (McArdle 1989, Ryback & Moreton 1992, Leake et al. 1993, Chapman et al. 2000a,b). At Kilgeever, 7 km to the W of Croagh Patrick, visible but fine-grain gold has been found in quartzite (Moreton 1993). Visible gold is said to be common in quartz veins of the Lecanvey prospect, Clew Bay, where it forms 10 μm to 3 mm grains (rarely to 8 mm) associated with arsenopyrite and pyrite (Aherne et al. 1989, 1992, Green & Moreton 2001). Other *in situ* gold occurrences include those at: Doo Lough (Cregganbaun prospect) and Inishturk, off the Mayo coast (Ryback & Moreton 1992, Moreton 1993); Lough Alisheen (Leake et al. 1993); and the Sheefry Hills, S of Lecanvey (Aherne et al. 1992). Alluvial gold has been found in the following rivers: Bunowen, Carrownisky, Glenlaur, Lugatoran, Lugbeg and Owannadornaun (Chapman et al. 2000a,b).

Monaghan Gold is present both in native form (insignificant) and in solid solution within arsenopyrite or pyrite within ankerite–quartz veins at Clontibret mine, Tullybuck (Ireland's only antimony mine). Stibnite is an associated species (Morris 1984, Morris et al. 1986, Ryback & Moreton 1993, Steed & Morris 1997, Chapman et al. 2000c).

Tyrone Extensive panning, soil geochemistry and mineralized boulder mapping resulted in the discovery of gold-bearing quartz veins in Dalradian metasediments at Curraghinalt, near Gortin, Sperrin Mountains. Arsenopyrite, bornite, chalcopyrite, native copper, galena, pyrite (dominant), sphalerite and tennantite–tetrahedrite are associated species. Prior to this discovery, alluvial gold had been reported from the Sperrin Mountains for centuries (Clifford 1986, Earls et al. 1989, Clifford et al. 1990, Moreton 1993, McCaffrey & Johnston 1996, Chapman et al. 2000a,b). Exploration of areas with high alluvial-gold contents led to the discovery of gold-bearing pyrite–quartz veins in shear zones at Golan Burn, near Strabane (Woodham et al. 1989, Moreton 1993). *In situ* gold has also been found at Cavanacaw, near Omagh, in the Sperrin Mountains (Parnell et al. 2000), and both gold and electrum occur at Lack, WSW of Omagh (Cliff & Wolfenden 1992, Leake et al. 1993).

Wicklow Production from the placer deposit in the Goldmine River valley, near Avoca, amounts to almost 300 kg of gold and appears to have been released to form native gold during the breakdown of gold-rich chalcopyrite within Ordovician BIF rocks in Tertiary–Quaternary times. The largest nugget weighed about 750 g. Ballycoog, Croghan Kinshela and Woodenbridge are related alluvial deposits (Greg & Lettsom 1858, Forbes 1869, Rudler 1905, Cole 1922, O'Brien 1959, McArdle & Warren 1987, McArdle 1989, 1993, Ixer et al. 1990, McArdle & Kennan 1992, Ryback & Moreton 1993). A 682 g gold nugget (the Wicklow nugget) from the stream in Ballin Valley, found in 1795 and once the property of George III, is in the collection of the NHM. At Kilmacoo, towards the northeast end of the Avoca district, gold is closely associated with chalcopyrite and pyrite, and it occurs in quartz veins within Ordovician silicified chloritic tuff (Milner & McArdle 1992). Elsewhere, alluvial gold is noted down stream of tourmalinites at Aghowle (McArdle et al. 1989).

Gold nugget, 27 mm in height, from Croghan Kinshela, Co. Wicklow, Ireland. (© Royal Cornwall Museum, Royal Institution of Cornwall; photographer D. I. Green)

goldmanite

GARNET GROUP

cubic $CaV^{3+}_2(SiO_4)_3$

A rare calcium–vanadium silicate (the vanadium analogue of andradite) found as a detrital grain in a sidewall core from the Amoco oil exploration well 9/27-1 in the UK sector of the North Sea, East Shetland platform, at a depth of 1902 m and within the Palaeocene Montrose Group sandstone. The sub-rounded grass-green goldmanite grain was discovered during routine heavy-mineral analysis. It is the purest goldmanite analyzed to date (Hallsworth et al. 1992).

gonnardite

ZEOLITE GROUP

tetragonal $(Na,Ca)_2(SiAl)_5O_{10}\cdot 3H_2O$

A rare hydrated sodium–calcium aluminosilicate found in vesicles in Tertiary basalt at Allt Ribhein, Fiskavaig Bay, on the south side of Loch Bracadale, Isle of Skye, Highland Region, Scotland. The vesicles are up to 2 cm in diameter and are filled with white zeolite species (heulandite, stilbite, thomsonite) or with saponite. Two samples contained gonnardite with later saponite infills (Mackenzie 1957). Vugs in a dyke from Ard Fernal, Isle of Jura, Strathclyde, contain off-white crystals of gonnardite associated with mesolite, natrolite and scolecite (Dyer et al. 1996), and at Milngavie, N of Glasgow, gonnardite is found in amygdales with analcime, chabazite, cristobalite, laumontite, natrolite, thomsonite and wairakite (Evans 1987, Livingstone 2002).

Nawaz (1988) reviewed all published data of gonnardite, including that found in Tertiary basalts at Hills Port, S of the Gobbins, Island Magee, Co. Antrim, Ireland,

Gonnardite crystals with diamond-shape cross sections, associated with needles of a calcium-rich phase (possibly mesolite). South of Gobbins cliff, south of Island Magee, Co. Antrim, Ireland; width 300 µm; BSE image. (© E. Condliffe, University of Leeds)

Prisms of gortdrumite, up to 1 mm long, coated with red cinnabar and associated with black tennantite. Gortdrum mine, Co. Tipperary, Ireland. (© D. I. Green)

and concluded that their compositions are distinct from tetranatrolite. However, Alberti et al. (1995) have shown that there is a very close structural relationship between the two species, so it is possible that some previous records may be erroneous. At Hills Port, gonnardite forms salmon-red to white spherules, up to 3 cm in diameter, in association with chabazite or heulandite, or both (Nawaz 1988). Tetranatrolite has since been discredited and the mineral confirmed to be gonnardite (Gottiardi & Galli 1999). Rubbly Antrim basalts containing fibrous aggregates of gonnardite are noted from the Galboly River section, near Garron Point, and to the N of Black Cave, Island Magee (Adair 1987).

goose-dung ore

A synonym of ganomatite (probably an impure scorodite). Not to be confused with ganomalite.

gorceixite

monoclinic (pseudo-trigonal) $BaAl_3(PO_4)(PO_3OH)(OH)_6$

A rare barium–aluminium phosphate hydroxide found in a vein at Hospital quarry, Elgin, Grampian, Scotland, where it forms microscopic greenish six-sided platy crystals with halloysite (Macpherson & Livingstone 1982).

In Cornwall, gorceixite is a minor constituent that occurs in fine-grain tourmaline–albite granite at St Austell, Cornwall. The granite is topaz- and zinnwaldite-bearing and also contains accessory apatite, arsenopyrite, cassiterite, monazite, rutile and zircon (Manning et al. 1996).

gorlandite

TL Wheal Gorland, St Day, 2 km E of Redruth, Cornwall, England

A synonym of mimetite, apatite group.

gortdrumite

orthorhombic $(Cu,Fe)_6Hg_2S_5$

TL Gortdrum mine, near Tipperary, Co. Tipperary, Ireland

A rare copper–iron mercury sulphide found in an iron-rich dolomite–barite vein cutting dolomitized limestones at the type locality. The mineralization is epigenetic and related to the intrusion of two high-level basic intrusions, and has produced a dominantly chalcopyrite and mercury-rich tennantite orebody, beneath which bornite–chalcocite mineralization occurs. Accessory minerals in the deposit include cinnabar, cobaltite, gortdrumite, sphalerite, stromeyerite, wittichenite and native amalgam (var. kongsbergite). Gortdrumite superficially resembles chalcocite and forms grains (less than 50×200 mm) either in isolation or intergrown with bornite. The temperature of deposition of the gortdrumite is considered to be 100–200°C or less (Steed 1983, 1986).

goslarite

orthorhombic $ZnSO_4{\cdot}7H_2O$

A rare hydrated zinc sulphate found in the Leadhills and Wanlockhead lead–zinc mines, along the Strathclyde/Dumfries & Galloway border, Scotland, where it occurs as feathery growths attached to the sides and timbers of some of the upper and dry levels (Wilson & Flett 1921).

A nickel-bearing variety of goslarite, intimately associated with epsomite, is reported from the barite–witherite workings of Ushaw Moor colliery, Durham. The goslarite forms efflorescent bundles of long slender snow-white crystals (Smythe 1933, Bridges & Young 1998). Nickel-bearing and -free varieties of efflorescent goslarite were also collected at New Brancepeth colliery, near Durham, on a Mineralogical Society excursion (Dunham 1946).

In Cornwall, goslarite encrusts zinc ore at Wheal Castle, St Just, and Tresavean and Trethellan mines, both Gwennap, near Redruth (Greg & Lettsom 1858).

A very early description of white goslarite from Holywell, Clwyd, is given by Sowerby (1811).

göthite

A synonym of goethite.

goyazite

trigonal $SrAl_3(PO_4)_2(OH,H_2O)_6$

A rare hydrated strontium–aluminium phosphate hydroxide found in the clay fraction of a soil derived from Middle Old Red Sandstone rocks in the Culbokie area, northwest flank of the Black Isle, Highland Region, Scotland (Macpherson & Livingstone 1982). Goyazite is also a minor component in titanium–uranium-rich bitumen nodules found in Lower and Middle Devonian sandstones and conglomerates at Ness, Quoyelsh and Yesnaby, all on the west coast of Mainland Orkney. Other subordinate species found in the nodules are barite, calcite, dolomite, pyrite, quartz and uraninite, as well as unnamed titanium oxides. The predominant titanium–uranium phase is brannerite – a diagenetic species formed by the adsorption of uranium onto titanium (McReady & Parnell 1998). A single grain of goyazite is reported from stream sediments collected at An Lag on St Kilda (Hirta), Western Isles. Other exotic species from this location are andalusite, corundum (sapphire), fergusonite, garnet (almandine, andradite and spessartine), kyanite, staurolite and tourmaline (Harding et al. 1984).

In England, euhedral calcium-rich goyazite and florencite-(Ce) have been identified on pegmatite matrix from the Hensbarrow section of Gunheath china-clay pit, Carthew, St Austell, Cornwall. Goyazite forms pale-green to colourless crystals up to 0.2 mm across in a vug within massive quartz and is associated with masses of acicular tourmaline, wolframite and a little chalcosiderite. Its composition is intermediate between goyazite and crandallite (Elton & Grigg 1996). Goyazite, crandallite and waylandite are also noted, without detail, in phosphate-rich pegmatites in the Hensbarrow granite (Hawkes et al. 1987).

graftonite

monoclinic $(Fe^{2+},Mn^{2+},Ca)_3(PO_4)_2$

An iron–manganese calcium phosphate found in two small metamorphic segregation pods 15 and 30 cm long, within striped kyanite–sillimanite-bearing Moine gneiss from Gleann Còsaidh, Loch Quoich, Knoydart, Highland Region, Scotland. The segregation pods contain apatite, garnet, muscovite, plagioclase and quartz together with rare, very small clusters of graftonite, jahnsite, johnsomervilleite, mitridatite, phosphosiderite, rockbridgeite and vivianite. Graftonite forms well cleaved, grey to fleshy-pink greasy grains up to 4 mm long. It contains many minute colourless rod-shape inclusions of an unidentified phase close to $MgFe^{2+}_2(PO_4)_2$ (Livingstone 1980).

gramenite

A grass-green variety of nontronite, smectite group; see Collins (1877a).

grandidierite

orthorhombic $MgAl_3O_2(BO_3)SiO_4$

A rare magnesium–aluminium borosilicate found in the outer margin of a biotite-rich pelitic xenolith in norite, close to the contact of a tourmaline-bearing granite in the Haddo House gabbro–norite complex, NW of the Wood of Shivas, Grampian, Scotland. Boron for the grandidierite was introduced via granite-derived metasomatic fluids. Grandidierite is a colourless to pale-blue mineral that sometimes develops well formed prism faces and is found included in plagioclase and biotite, in association with corundum, spinel and locally cordierite. Grains reach 1.2 mm in length and have good cleavage (Herd et al. 1984). Livingstone (2002) briefly reports bluish-green plates of grandidierite from the inner contact aureole of the Comrie granodiorite, Grampian.

grandite

A garnet with grossular–andradite composition.

graphite

hexagonal or trigonal C

A common carbon species, polymorphous with chaotite and diamond, and most often found in metamorphosed carbon-rich sedimentary rocks.

Graphite from Cornwall is not well described. It is noted in elvan dykes in the vicinity of Tuckingmill and in a granite quarry on Kergiliack Estate, near Penryn (Garby 1848, repeated by Greg & Lettsom 1858). At Dinas Head, near Padstow, a series of albite–quartz rocks, termed adinoles, contain small amounts of graphite together with ankerite, calcite, chlorite, dravite, leucoxene, rutile and zoisite (Agrell 1939). Robson (1949b) adds to these locations: Wheal Alfred, Phillack, near Hayle; Wherry mine, Penzance; and either the village or the parish of St Erth. Golley & Williams (1995) add Wheal Jane, Kea, and Botallack, St Just. Graphite also occurs at Millclose mine, Wensley, Derbyshire, where it is found as inclusions in calcite (Mueller 1951, Braithwaite 1983, Ford et al. 1993).

Perhaps the most important graphite deposits in Britain or Ireland were discovered in Cumbria, possibly in the mid-16th century. The species was mined for use in the making of Cumberland™ pencils. It occurred in pipes, strings and irregular masses called "sops", associated with altered diorite dykes or sills and with diabase intrusions cutting Borrowdale Volcanic Group rocks (Rudler 1905, Eastwood 1959). Specimens with either a massive or nodular structure (some nodules up to 60 cm in length) are reported from the graphite mine at Seathwaite, Borrowdale, where the vein system consisted of eight fault-controlled veins in diabase. A magmatic origin for the carbon was proposed by Strens (1965). However, later isotopic work concluded that the carbon was organic and possibly derived from Lower Palaeozoic mudstones underlying the volcanic rocks (Weiss et al. 1981, Parnell 1982). Graphite has also been found in the Grisedale Pike area, 8 km NNW of Seathwaite, but this may be related to the smuggling of graphite when it was a profitable local industry (Strens 1965). At Bannerdale graphite mine, Mungrisdale, graphite occurs in clots up to 25 mm in diameter, along the sides of a quartz vein containing chlorite, manganite, muscovite and traces of iron sulphides (Greg & Lettsom 1858, Davidson & Thomson 1951, Ziemba 1975). Elsewhere in Cumbria, graphite is reported from: Bowscale Tarn, Caldew Valley and the Shap granite quarry (Young 1987a); Dale Head mine, Newlands Valley (Stanley & Vaughan 1980); and Wanthwaite mine, St John's in the Vale, near Threlkeld (Kingsbury & Hartley 1958). It is uncertain whether graphite actually occurs at Wanthwaite mine, as the provenance of other species described from this mine by Kingsbury has been called into question (Ryback et al. 1998).

Minor graphite is widely distributed in Lower Palaeozoic rocks of Wales and it occurs at Nant mine, Llanfaelrhys, Rhiw, Lleyn Peninsula, Gwynedd (Bevins 1994).

In Scotland and Ireland, graphite occurrences are many and varied, but the species is rarely described in any great detail.

Scotland

Grampian Kyanite–labradorite–margarite schist at Sròn Dubh, Glen Clunie (Chinner 1974); Easdale Subgroup graphitic and sulphidic metasediments of Dalradian age near the margins of the Huntly and Knock mafic–ultramafic intrusions at Littlemill and Auchencrieve, near Huntly (Fletcher & Rice 1989); in a cutting near Rothiemay railway station, Knock (Macadam 1885); Au–Pt–Pd-rich graphite and sulphide-bearing orthopyroxene-rich pegmatite at Bin quarry, Cairnie, Huntly (Gunn & Shaw 1992, Gunn & Styles 2002).

Highland A peculiar gneissose pyrrhotite–tremolite-bearing rock containing abundant distinct scales of graphite at Glenelg (Clough 1899); graphite–kyanite–mica–pyrite schist, Glen Urquhart complex, W of Loch Ness, near Drumnadrochit (Rock et al. 1986b); graphite once mined from a vein cutting garnetiferous muscovite–biotite gneiss below the foot of Loch a' Mhuillidh on the north side of the River Farrar, Glen Strathfarrar (Heddle 1901).

Strathclyde Massive and columnar graphite at Craigman coalmine, near New Cumnock (Greg & Lettsom 1858); massive granular lumps of graphite near Pennycross, Killiemore Estate, southern shore of Loch Scridain, Isle of Mull (Heddle 1901); minor graphite in sulphide-bearing slate of Dalradian age at Easdale Island, near Oban (Hall et al. 1988); graphite-rich xenoliths in a late-Caledonian lamprophyre dyke exposed on the east and west shores of Loch Lomond (Dempster & Bluck 1991).

Tayside Brilliant crystals of graphite, together with calcite and hornblende, S of the Spittal of Glen Shee, 21 km N of Blairgowrie (Heddle 1901); garnet–graphite–pyrrhotite-bearing mica schist at Beinn Heasgarnich, Ben Eagach area, near Aberfeldy (Coats et al. 1984); calcareous graphitic pelite containing andesine, chlorite, clinozoisite, garnet, margarite and quartz, River Tilt, Blair Atholl (Chinner 1974).

Western Isles Graphite-bearing metamorphosed limestone paragneiss containing brucite, calcite, chondrodite, dolomite, humite, pargasite, phlogopite and titanite (sphene) at Bay Steinigie, near Finsbay (Fionnsabhagh), South Harris (Dearnley 1961); diopside–hornblende–orthoclase paragneiss at Liuri Cliffs, Chaipaval (Ceapabhal), South Harris (Dearnley 1962).

Shetland Graphite schist at Wick of Belmont, Unst (Read 1934b).

Ireland

Cork Graphite is associated with Devonian redbed-hosted copper mineralization at Ballycummisk, Cappagh, Kileen North, Glanalin and Letter, all in the West Carbery mining district (Ni Wen et al. 1999).

Donegal Dalradian graphite–mica–pyrite-bearing limestone in the quarries of Lough Anure, Thorr area (Pitcher 1950); Dalradian graphite-bearing pelite containing cordierite, hercynite, staurolite and sillimanite from the aureole of the Ardara granite on Inishkeel island and 2 km SE of Portnoo harbour, near Naran (Akaad 1956, Atkin 1978).

Tyrone Dalradian graphitic schist in the Lack inlier, WSW of Omagh (Cliff & Wolfenden 1992).

Offshore territories

Isle of Man Rudler (1905) reports graphite, in very limited quantity, in the Snaefell lode and from Beary Mountain.

grastite

A synonym of clinochlore, chlorite group.

gratonite

trigonal $Pb_9As_4S_{15}$

A rare lead sulpharsenite found in waste material on the dumps of North Laxey mine, by the side of Ballaglass stream, Isle of Man. The mine worked a mineralized N–S striking quartz vein. Well formed gratonite crystals up to 2 mm across occur with arsenopyrite, galena, pyrite, pyrrhotite and sphalerite. The samples are coated with a white material that appears to be zinc-rich melanterite (Alderton et al. 1995).

At Tynagh mine, near Killimor, Co. Galway, Ireland, gratonite occurs as inclusions in galena with bournonite, enargite, geocronite, tennantite and rare native antimony. The galena itself forms inclusions in chalcopyrite, sphalerite and tennantite (Gibson 1979, Clifford et al. 1986). Gratonite is also noted in three stratabound sulphidic lenses dominated by galena, marcasite, pyrite and sphalerite, with minor arsenopyrite, bornite, chalcopyrite, gersdorffite, jordanite, seligmannite and tennantite from the carbonate-hosted Lisheen deposit, between Thurles and Rathdowney, Co. Tipperary (Hitzman et al. 2002).

Well formed trigonal crystals of gratonite coated with powdery zincian melanterite. North Laxey mine, Isle of Man; width of largest crystal is 500 µm; SEM image. (© D. H. M. Alderton)

graubraunstein

A synonym of pyrolusite.

greenalite

KAOLINITE–SERPENTINE GROUP
monoclinic $(Fe^{2+},Fe^{3+})_{2-3}Si_2O_5(OH)_4$

A rare iron silicate hydroxide found in an Ordovician chert at Glenluce, Dumfries & Galloway, Scotland, where it occurs in association with spilitic pillow lavas and gabbroic intrusions. Greenalite forms rounded, ellipsoidal or irregular granules, sometimes partially or totally replaced by spherulitic aggregates of dusty quartz, embedded in a matrix of cherty quartz. In thin section it can be dark and opaque, but may also be olive green or brownish green (Kennedy 1936, Sabine & Young 1971).

green calamine

A synonym of aurichalcite.

green earth

A synonym of celadonite and glauconite, mica group.

greenite

A synonym of chlorite.

green john

A green variety of fluorite.

greenockite

WURTZITE GROUP
hexagonal CdS
TL Bishopton railway tunnel, 7 km WNW of Renfrew, Strathclyde, Scotland

Named after Charles Murray Cathcart (1783–1859), soldier and geologist, the second Earl of Cathcart, who was earlier known as Lord Greenock; born in Walton, Essex. Greenockite was discovered in material excavated during construction of the Glasgow–Greenock railway and was described as a new species by Jameson (1840). The first crystal was found in about 1810 by Thomas Brown (1774–1853) of Lanfine and was mistaken for sphalerite. Greenockite is a rare cadmium sulphide, dimorphous with hawleyite, and is typically found as a supergene species replacing sphalerite.

At Bishopton tunnel, greenockite with a beautiful yellow lustre occurs in pockets and fractures within hydrothermally altered Clyde Plateau Lavas. Generally, greenockite produces well formed hexagonal pyramidal crystals up to 5 mm long. Prehnite and less abundant analcime, calcite, harmotome, natrolite, saponite and thomsonite are associated species. In lava adjacent to cavities and joints, the species bornite, chalcopyrite, galena and pyrite are also found (Greg & Lettsom 1858, Heddle 1901, Meikle 1992b). The history of the discovery is intriguingly described by Meikle (1992b) and Livingstone (2002). In 1974–1975, during construction of the nearby Bishopton bypass, euhedral yellow greenockite crystals to 5 mm, intimately associated with prehnite, were recovered from basaltic rubble in a borrow pit next to the road. Some crystals were coated by the rare species otavite (Faithfull & Hubbard 2001, Green & Todd 2001).

Greenockite has been found in altered basaltic lavas at Loanhead quarry, near Beith, Strathclyde. It is well crystallized and it forms distinct yellow to reddish-orange crystals up to 7 mm long, either in prehnite or at the margins of prehnite-bearing fissures (Meikle 1989). Other Scottish occurrences of greenockite include: Bowling quarry, Kilpatrick Hills, near Glasgow, Strathclyde – crystals up to 12 mm long, associated with prehnite (Heddle 1901); Boyleston quarry, SW of Barrhead, Strathclyde – greenockite with chrysocolla, natrolite and prehnite (Heddle 1887a, 1901); and Wanlockhead, Dumfries & Galloway – citron-yellow earthy coatings on sphalerite (Brown 1927, Livingstone 2002).

Alabaster (1983) and Zachariades (1993) note greenockite from Stancombe limestone quarry, near Flax Bourton, Bristol, Avon, where mineralization consists of a few narrow subvertical barite–calcite veins containing galena, pyrite and sphalerite. Greenockite occurred with these primary minerals, native sulphur and an unidentified antimony-rich powder in an infilled mineralized cave.

In Derbyshire, a yellow earthy variety of greenockite is reported from Millclose lead–zinc mine, Wensley, near Matlock (Russell 1912b, Traill 1939), and Braithwaite (1983) suggests that hawleyite may also be present. At Wapping mine, Cromford, an amorphous bright canary-yellow material on hemimorphite is presumed to be greenockite or hawleyite (Ince 1991). The two species are confirmed as powdery canary-yellow coatings on fluorite or sphalerite in a galena-rich vein (the Ashover Old vein) worked for fluorite at Fall Hill (Gate) quarry, Ashover. Marcasite, pyrite and smithsonite are associated species (Smith 1982, Spence 1982). Flecks of greenockite (5 µm) and relics of sphalerite occur with smithsonite at Oxclose mine, Snitterton, W of Matlock (Ixer & Townley 1979). Bradwell Moor is another location mentioned by Ford & Sarjeant (1964a).

In the northern Pennines, vivid-yellow earthy supergene cadmium-sulphide encrustations on sphalerite have been collected from several mines in the orefield, and greenockite is confirmed from some of them: Blagill mine, Nenthead, near Alston, Cumbria; Ashgillhead, Greenhurth, Reddycombe, Willyhole and Trough mines, all in Upper Teesdale, Co. Durham (Dunham 1948, 1990, Young et al. 1987, Bridges & Young 1998).

Greenockite is further noted as crusts on sphalerite at Greenside mine, near Glenridding, Patterdale, Cumbria (Davidson & Thomson 1951), and Wet Grooves mine, Aysgarth, Yorkshire (Rust 1987b).

At Dylife mine, Central Wales orefield, Powys, where copper, lead and zinc mineralization was worked from three lodes, greenockite commonly occurs as canary-yellow earthy globular coatings in cavities in sphalerite and as fine powdery yellow coatings on hemimorphite (Jones & Moreton 1977, Moreton 1982). The species has also been noted at Bwlchglas lead mine, near Tal-y-bont, Dyfed, where it occurs as thin yellow films on sphalerite (Braithwaite 1982a). Bevins (1994) reports further occurrences of greenockite at Cwmystwyth and Frongoch mines, both in Dyfed and Pen-y-Bryn shaft, Halkyn, Clwyd. Galena–tennantite vein material from Dolyhir quarry, Old Radnor, near Presteigne, Powys, contains inclusions of greenockite (Bevins & Mason 1997, T. F. Cotterell pers. comm. 2007).

Hemimorphic greenockite crystal, 3 mm in length, on basalt; Bishopton railway tunnel, Strathclyde, Scotland. (© M. P. Cooper)

Sheshodonnell East mine, the Burren, Co. Clare, Ireland, is perhaps best known for its botryoidal smithsonite, whose bright-yellow colouration is attributable to minute inclusions of greenockite (Green & Moreton 2001). Traces of bright-yellow powdery greenockite occur with crusts of hydrozincite, smithsonite and striated crystals of sphalerite at Glenaboghil mine, NNW of Fintown, Co. Donegal, Ireland (Ryback & Moreton 1992).

greenovite

Named after George Bellas Greenough FRS (1778–1855), London-born geologist, mapmaker, one of the founder members of the Geological Society of London and author of *A critical examination of the first principles of geology* (1819). Greenovite is a manganese-rich variety of titanite.

green vitriol

A synonym of melanterite.

gregorite

Named after Revd William Gregor (1761–1817), vicar of Creed and discoverer of the element titanium; born in Trewarthenick, near Truro, Cornwall. Gregorite is a synonym of manaccanite (an iron-rich variety of ilmenite), and also of bismutite.

greifensteinite

monoclinic or triclinic $Ca_2Be_4(Fe^{2+},Mn^{2+})_5(PO_4)_6(OH)_4 \cdot 6H_2O$

A rare hydrated calcium–beryllium–iron–manganese phosphate hydroxide that is reported from Clitters United mine (Gunnislake Clitters mine), Calstock, near Gunnislake, Cornwall (Chukanov et al. 2003). The mineral was originally described as an iron-rich roscherite and it commonly consists of dark sage-green spherules with a silky crystalline exterior, ranging up to 5 mm in diameter, associated with colourless carbonate-fluorapatite lined cavities in quartz. Flat radiating aggregates of roscherite (to 1 cm) are also found on joint surfaces of iron-stained quartz (Clark et al. 1983).

greigite

LINNAEITE GROUP

cubic $Fe^{2+}Fe^{3+}_2S_4$

A rare magnetic iron sulphide found in authigenic siderite-cemented concretions within intertidal sediments at Warham Marshes, Norfolk. When collected, the concretions could not have been more than 30 years old and must have formed during rapid sedimentation and burial of organic matter during the early stages of marsh growth. Concretions are up to 40 cm in diameter, with quartz constituting the main framework grains, siderite the main constituent of the cement, and greigite (another constituent of the cement) probably responsible for the black colouration (Pye 1981, 1984, Al-Agha et al. 1995). Greigite is also noted from Treore mine, Port Gaverne, near Port Isaac, St Teath, Cornwall (Criddle & Stanley 1993), but the specimen is from the Kingsbury collection at the NHM and its provenance is now considered dubious (C. J. Stanley pers. comm. 2004).

Livingstone (1993) reports greigite forming a soft black corrosion product on a garnet–gold sword pommel unearthed from soil in Fife, Scotland.

grenatite

A synonym of leucite and staurolite.

grey antimony

A synonym of stibnite.

grey copper

A synonym of tennantite and tetrahedrite.

groddeckite

A synonym of gmelinite, zeolite group.

grossular

GARNET GROUP

cubic $Ca_3Al_2(SiO_4)_3$

A common calcium aluminosilicate that forms three series, with andradite, with hibschite and katoite, and with uvarovite. Grossular is found in contact or regionally metamorphosed calcareous rocks and in some schists and serpentinites. Like all other species of garnet, end-member compositions are rare and most examples show some degree of solid solution towards other members of the group. Where possible, the proportions of the two most abundant end members are listed below (al = almandine, an = andradite, gr = grossular).

England

Cornwall Discordant garnet-bearing calc-silicate bodies produced during metasomatic alteration of metabasalt, in the aureole of the Land's End granite contain grossular (gr_{65-72}, an_{22-28}) in a typical skarn assemblage. These sheet-like horizons occur over a small area near Botallack, St Just, with larger examples at Carn Vellan and the Crowns, and smaller horizons at Botallack Head, Carn Du, Chycornish Carn, Gryll's Bunny, Stamps and Jowl Zawn, Trewelland Cliffs and Wheal Cock Carn (Alderton & Jackson 1978). At Gryll's Bunny the associated species are calcite, clinozoisite, diopside and titanite, alternating with bands of magnetite (Jackson 1974). Small grossular-bearing cavities in axinite pods within the calc-silicates at Wheal Cock Zawn, Botallack, also contain the rare species stokesite (Couper & Barstow 1977) and wickmanite (Hubbard 1989). Calcium metasomatism of gabbroic dykes within serpentinite at East Cliff, Kennack Sands, Lizard Peninsula, has produced a garnet–vesuvianite assemblage in which grossular occurs as colourless-to-white crystals (Hooper 1997).

Cumbria Grossular occurs in talcose schist from Carrock Fell and Swinside, Newlands, Keswick (Young 1987a), and in calc-silicate rocks from the aureole of the Shap granite (Firman 1978). Garnet resembling grossular is also noted in altered Ordovician limestones at Wasdale Head, Shap (Harker & Marr 1891).

Devon Grossular-rich selvedges in Carboniferous cherts and limestones in the metamorphic aureole of the Dartmoor granite have been found at Meldon railway quarry, near Okehampton, and Red-a-ven mine near Meldon. Datolite, heulandite and vesuvianite are associated species at Red-a-ven mine, where grossular forms euhedral reddish-brown crystals in veins cutting cherts (Dearman 1965, 1966b, El Sharkawi & Dearman 1966, 1967).

Wales

Gwynedd Grossular (gr_{88-90}, an_{13}) forms clusters up to 2 mm wide in altered gabbros at Dinas-bach, Holy Island, and Mynachdy, near Llanfairynghornwy, northwest Anglesey (Maltman 1977). Other examples are from the Bryn Fuches and Rhôs-mynach areas of northeast Anglesey, where grossular occurs in meta-limestones and meta-calc-pelites from the Nebo Gneisses (Horák 1993).

Scotland

Grampian Xenoliths of metamorphosed calcareous rocks (now mainly calcite–wollastonite) in the Dorback granite, in the Braes of Abernethy, 9 km W of Tomintoul, contain grossular porphyroblasts ranging in size from 1 mm to 2.5 cm across. Clinozoisite, diopside, epidote, brown mica, vesuvianite and zoisite are associated species (Walker & Davidson 1935). In Deeside, other reports include: grossular (gr_{48}, an_{42}) at Crathie limestone quarry, Creag Mhòr, NW of Balmoral (Heddle 1901); grossular or cinnamonstone (gr_{85}, al_{10}) at Delnabo quarry, Glen Gairn (Heddle 1901, Crane 1987); grossular (not analyzed) at Pollagach Burn, E of Pannanich Hill, E of Ballater (Crane 1987). All appear to be from skarn deposits at the margins of major granitic intrusions. Wollastonite-bearing marbles associated with a suite of granitic pegmatites at Kinnaird Head, Fraserburgh contain grossular (gr_{84}, an_{12}) in another typical skarn assemblage. Grossular is one of many heavy minerals identified in panned concentrates from the southeastern margin of the Cairngorm granite in Luibeg Burn, Derry Cairngorm (Beer & Bennett 1988, Haynes 1993).

Highland At Camasunary Bay, Strathaird Peninsula, Isle of Skye, end-member grossular occurs with andradite (an_{47-67}, gr_{14-33}) and schorlomite in calcic skarns found at the margin of the Cuillin gabbro complex. Associated species include calcite, melilite, spurrite, tilleyite, vesuvianite and wollastonite, together with minor amounts of strontium-rich apatite, magnetite and pyrite. Rare baghdadite, djerfisherite, kalsilite, kimzeyite, perovskite and spinel complete the assemblage (Beard & Drake 2007). Grossular–andradite, diopside–hedenbergite, plagioclase and wollastonite are the most abundant minerals in primary skarns formed in similar metasomatically altered Durness limestone surrounding the Beinn an Dubhaich granite, near Broadford, Isle of Skye (Tilley 1949, 1951); however, later analysis has shown the garnet to be close in composition to end-member andradite (Smith et al. 2004). With this in mind, the grossular described (but not analyzed) from skarn deposits at the gabbro/limestone contact at Camas Mòr, Isle of Muck (Tilley 1947) deserves further investigation (see hibschite). At Lòn Ostatoin, near Duntulm, on the Trotternish Peninsula, Isle of Skye, subhedral and euhedral grossular (gr_{72-75} and $_{24-28}$) crystals (5–10 µm across) and saponite occur as alteration products of Jurassic limestone metasomatised by a doleritic sill of Tertiary age (Kemp et al. 2005).

Strathclyde At Coire Buidhe, Glen Creran, abundant brown grossular crystals up to 8 mm across, and massive grossular associated with epidote, occur in a skarn deposit (metamorphosed impure limestones of Dalradian age) related to the intrusion of the Cruachan granite (Ingram et al. 1993). At Creag an Fheidh, near Glenloig Bridge on

the Isle of Arran, fine-grain grossular (gr_{54}, an_{44}) intergrown with diopside–hedenbergite (partially replaced by epidote) occurs in a skarn assemblage at the junction between vent pyroclastics and a xenolithic Cretaceous chalk block that subsided into the collapsed caldera of the Central ring complex (Cressey 1987). White, cream and clear grossular (gr_{80-92}, an_{8-19}) and andradite garnet crystals up to 5 mm across occur as sparkling crystalline crusts lining vesicles in Clyde Plateau Lavas at Loanhead quarry, near Beith, where they form part of a low-temperature hydrothermal assemblage that also includes fibrous amphibole, analcime, calcite, epidote, hematite, prehnite, pumpellyite and thomsonite (Livingstone 1989a, Meikle 1989b, Green & Todd 2001). Vesicles in Tertiary lavas on An Gearna, a spur of Ben More, Isle of Mull, contain grossular (gr_{87}, an_7) associated with albite, chabazite, epidote, prehnite, scolecite and thomsonite. Well formed grossular crystals up to 0.5 cm across occur fairly frequently and vary in colour from pale yellow or colourless, to deep brown with a beautiful "wine-yellow" tint being commonest (McLintock 1910b).
Western Isles Panned concentrates of alluvium from Abhainn Mhòr on St Kilda (Hirta) contain grossular (Harding et al. 1984).

Ireland

Antrim Dolerite contaminated by chalk at Scawt Hill, Larne, has produced grossular as an alteration product of melilite in a typical skarn assemblage (Tilley & Harwood 1931). However, as the authors did not analyze the mineral, further confirmation is desirable, particularly as a similar assemblage at Carneal (also near Larne) contains hydrogrossular (i.e. hibschite) instead of grossular (Henmi et al. 1973).
Galway Attractive crystals of grossular, together with diopside, vesuvianite and wollastonite, have been found in skarns on the west coast of Grallagh (Ryback & Moreton 1992). The metamorphic aureole of the Omey granite hosts an amphibolite skarn at Barnahallia Lough, Claddaghduff, west Connemara. The skarn contains cavities in which euhedral dark-brown grossular crystals, up to a few mm across, are associated with albite, chabazite-Ca, epidote, heulandite-Ca and stilbite-Ca (Moreton & Green 2006a).
Donegal Grossular occurs with epidote, ferro-actinolite (or hornblende), scheelite and titanite in quartz lenses in a metadolerite body within Dalradian quartzites by the roadside 3 km E of Bunbeg (Ryback & Moreton 1992). Calc-silicate rocks forming bands in sillimanite hornfels within the aureole of the Ardara granite complex at Clooney, near Gweebara Bay, contain grossular associated with diopside, epidote, vesuvianite and zoisite (Akaad 1956). Limestone inclusions in granite at Ballynacarrick show a skarn armour made up of definite mineral zones. Nearest to the limestone is a grossular–vesuvianite–wollastonite zone (Gindy 1953). Other skarn-related occurrences of grossular are reported from the limestone quarries of Lough Anure, Thorr area (Pitcher 1950), and from S of Bunbeg quay, E of Loughlin Island (Whitten 1951).
Louth Thermal metamorphism of calcareous aureole rocks by an unusual syenite intrusion in the area around Barnavave, Carlingford, has produced a suite of rocks in which grossular occurs with diopside, vesuvianite and wollastonite at the junction of impure limestone, and with diopside and wollastonite in calcareous silt and altered limestone xenoliths. Grossular is generally xenoblastic and either colourless or pale yellow (Osborne 1932b, Nockolds 1937).
Mayo Blueschist-facies metabasites belonging to the Dalradian Supergroup from Central Achill, N of Clew Bay, contain a few zoned garnets where the grossular component is dominant (gr_{38-40}, al_{31-33}). Most garnets in these rocks have almandine-dominant compositions (Chew et al. 2003).
Sligo Some thermally metamorphosed limestones close to two basalt–dolerite dykes at Killala Bay, near Inishcrone, contain a calcite–grossular assemblage (Nawaz 1974).

grothite

A synonym of titanite.

groutite

orthorhombic $Mn^{3+}O(OH)$

An uncommon manganese oxide hydroxide, trimorphous with manganite and feitknechtite, and found in veins rich in manganese and iron oxides within the Dalradian Jura Quartzite at Dùn Athad, Mull of Oa peninsula, Isle of Islay, Strathclyde, Scotland. Groutite was found as a black surface coating on two specimens dominated by hollandite and minor pyrolusite. The oxides were deposited from groundwater that penetrated the Jura Quartzite via a local fault system, probably in Permo-Triassic time (Nicholson 1988a, 1989a,b, 1992).

Groutite also occurs in association with woodruffite at Ben Eagach, near Aberfeldy, Tayside Region, where it forms a weathered oxide crust on a manganese-rich Dalradian carbonate horizon (Nicholson 1989b, 1992).

grovesite

TL Benallt mine, Rhiw, Lleyn Peninsula, Gwynedd, Wales

Named after Arthur William Groves (1903–1956), geologist in the Mineral Resources Division of the Colonial Geological Surveys; born in Sutton, Surrey. It was thanks to him that the BM(NH) received the material in which the new minerals banalsite, cymrite and pennantite were discovered. Grovesite is a synonym of pennantite – the manganese-rich analogue of berthierine (Bannister et al. 1955). Burke (2006) formally discredits grovesite in an IMA-CNMNC-approved mass discreditation of species.

grünerde

A synonym of celadonite and glauconite, both mica group.

grunerite

AMPHIBOLE GROUP
monoclinic $\square Fe^{2+}_7Si_8O_{22}(OH)_2$

A relatively common iron silicate hydroxide that forms a series with cummingtonite and is dimorphous with ferro-anthophyllite. Grunerite is found in medium- to high-grade metamorphosed iron formations, in contact-metamorphic environments, and in some blueschist-facies metaquartzites.

At Zawn a Bal, Botallack, St Just, Cornwall, metabasic rocks in the aureole of the Land's End granite have been metasomatically altered to produce a biotite–grunerite–quartz hornfels. Well formed almandine garnet and cordierite are associated species. In hand specimen, grunerite is greenish brown, but in thin section it is almost colourless (Alderman 1935, Tilley 1935).

Grunerite is also described from Lewisian rocks of the Glenelg area, Highland Region, Scotland. To the SW of Druideig Lodge, Loch Duich, it occurs in discrete bands clearly replacing both hypersthene and fayalite. A fibrous habit and multiple twinning can be observed in thin section. Garnet and ilmenite are associated species (Tilley 1936). Near by, at Gleann Beag, Glenelg, grunerite occurs in a manganese-rich schist containing almandine–spessartine garnet, magnetite and pyroxmangite, together with accessory hedenbergite, hypersthene and pyrrhotite (Tilley 1937a).

The above descriptions are based on optical identifications of grunerite, and further work is desirable to confirm that these compositions still classify as grunerite (Leake et al. 1997a). However, grunerite is confirmed in silicate–magnetite–quartz rocks (BIF) of Lewisian age from Flowerdale, near Gairloch, Highland Region, where it is the most abundant silicate present. It is very pale brown in thin section and it forms prismatic crystals up to 10 mm long, often in radiating aggregates or spectacular spherulitic structures (Williams 1986). Reprocessing of data using the amphibole classification of Leake et al. (1997a) and the software described in Mogessie et al. (2004) suggests that samples from a Dalradian psammite–semipelite sequence at Creag nam Ban, near Ballater, Deeside, Grampian (Smith & Phillips 2002), which had originally been described as cummingtonite, are actually grunerite.

grünlingite

TL probably Carrock Fell, Cumbria, England

A mixture of joséite and bismuthinite.

guanajuatite?

orthorhombic Bi_2Se_3

A very rare bismuth selenide thought to be present as inclusions in alluvial gold grains from the South Hams district of Devon. Other selenide inclusions are possibly clausthalite and tiemannite. Unidentified palladium, mercury and copper selenides are also reported. Unfortunately, the inclusions are too small to be defined accurately with the EMP. The selenides were deposited during a late phase of gold growth from a distinct type of low-temperature solution that was both oxidizing and saline (Leake et al. 1991).

gudmundite

ARSENOPYRITE GROUP
monoclinic FeSbS

A rare iron–antimony sulphide found in the giant sedimentary-exhalative base-metal and barite deposit at Gortnadyne and Shallee mines, Silvermines, Co. Tipperary, Ireland. Gudmundite has been identified only in polished sections; but, although it closely resembles arsenopyrite, its strong rose-pink pleochroism makes it very distinctive. In some of the brecciated ore, conversion of pyrrhotite to marcasite has been shown to take place by loss of iron, resulting in the formation of narrow shrinkage cracks. These were later infilled with chalcopyrite, tetrahedrite, a thin layer of gudmundite, and finally quartz, together producing a very unusual angular texture. Gudmundite is one of 17 primary sulphides found at Silvermines (Rhoden 1959, Moreton 1999).

Gudmundite is also reported from Esgair Hir mine, Ceulanymaesmawr, Central Wales orefield, Dyfed (Rust & Mason 1988), but this is a misidentification of tučekite (J. S. Mason pers. comm. 2004).

guitarmanite

An arsenic–lead sulphide, possibly a mixture.

gunningite

KIESERITE GROUP

monoclinic $ZnSO_4 \cdot H_2O$

A rare hydrated zinc sulphate found at Parys Mountain, Anglesey, Wales, as a sparse white encrustation on a sphalerite-rich vein within opencast workings. Identification is by XRD (T. F. Cotterell pers. comm. 2003).

gurolite

A synonym of gyrolite and the original name of the species (Anderson 1851).

gustavite

orthorhombic $AgPbBi_3S_6$

A rare silver–lead sulphobismuthite that forms a series with lillianite and is found in a stockwork cut by galena-bearing quartz veins at Corrie Buie, Meall nan Oighcreag, Loch Tay, Tayside Region, Scotland. The veins also contain minor amounts of native bismuth, chalcopyrite, electrum (36–65% gold), emplectite (only one grain found), pyrrhotite, sphalerite, schirmerite and phases belonging to the gustavite–lillianite series. The mineral assemblage of the veins is typical of that associated with acid intrusions. Gustavite, and its close associate ourayite, form bladed aggregates (up to 0.5 cm across) associated with pyrrhotite and galena, and also develop as exsolution blades (up to 20 µm long) in galena (Pattrick 1984).

gyps

A synonym of gypsum.

gypsum

monoclinic $CaSO_4 \cdot 2H_2O$

A very common hydrated calcium sulphate found in sedimentary rocks, particularly marine salt deposits. It can form directly by evaporation, or later by hydration of anhydrite. Gypsum is commonly encountered as efflorescences in mines as a result of oxidation of primary sulphide species.

England

Avon Gypsum forms free-grown crystals up to 6 cm long within clay gouge along a fault plane at Hampstead Farm quarry, Chipping Sodbury, where heavily oxidized pyritiferous veins are found in Carboniferous Limestone close to the Rhaetic/Lower Carboniferous unconformity. It also replaces calcite and locally encrusts unaltered Fe–Pb–Zn sulphides (Alabaster 1990b). At several points along the beach at Aust Cliff, gypsum occurs as small rosettes forming along joint planes in grey shale, probably because of the dissolution and reprecipitation of the nodules (M. R. Stolworthy pers. comm. 2007).

Cambridgeshire Water-clear gypsum (var. selenite) forming sharply developed individual and twinned crystals up to 8 cm long occurs in the clays of the Ely area (particularly the Jurassic Ampthill Clay). Crystals up to 5 cm long have been collected from near the Old Bedford River, NE of Mepal Bridge (Grguric 1998).

Cheshire Efflorescences and coatings of gypsum are found on mine walls in Engine Vein mine and Wood mine (Carlon 1979, Braithwaite 1994).

Cleveland Gypsum replaces shell fragments, occurs as efflorescences and fibrous veins, and is said to form inclusions, sometimes 30 cm long, at Billingham anhydrite mine (under the ICI factory site) (Raymond 1959, 1962). At Boulby potash mine, Loftus, gypsum is an uncommon post-mining mineral resulting from the dissolution of anhydrite and is found as translucent prismatic crystals exceptionally up to 3 cm long (Woods 1973, Green & Freier 1996).

Cornwall Gypsum is occasionally found in uranium-rich veins at North Zawn, Botallack (Elton & Hooper 1996), and Geevor mine, near Pendeen, both at St Just (Elton & Hooper 1992, 1993, Elton et al. 1992). At Penberthy Croft mine, St Hilary, gypsum occurs in the basic copper sulphate assemblage, forming subhedral prismatic colourless crystals up to 1 mm long and distorted aggregates intergrown with other minerals (Betterton et al. 2000). Gypsum forms beautiful clear crystals to 0.5 cm associated with bismuthinite at South Crofty mine, Camborne (Wolters & Wolters 1987). It is also noted from: Levant mine, St Just (Golley & Williams 1995); Wheal Busy, St Day; Wheal Penrose, Sithney, near Helston (both Robson 1949b); and Wheal Mary Ann, Menheniot (Merry & Braithwaite 1995).

Cumbria Gypsum is a major constituent of the Upper Permian evaporites in the Vale of Eden, the Carlisle Plain, and the St Bees Shales of west and south Cumbria: Acorn Bank mine, Temple Sowerby (Greg & Lettsom 1858); Barrowmouth alabaster mine, Whitehaven – in seams 8.7 m thick (Eastwood et al. 1931); Culgaith, Temple Sowerby

Rosette of colourless gypsum, 2 mm across, on bedding plane in shale; Aust Cliff, Avon. (© D. I. Green)

Spray of bladed gypsum (variety selenite) from Smallcleugh mine, Nenthead, near Alston, Cumbria; blades are 12 mm long. (© D. I. Green)

(Meyer 1965); Kirkby Thore quarry (Trechmann 1901); Knott Hill or Boaterby mine, Cotehill (Trotter & Hollingworth 1932); Little Salkeld, near Penrith (Postlethwaite 1913); Long Meg mine, Lazonby, and Thistle Plaster quarry and mine, Kirkby Thore (Sherlock & Hollingworth 1938). The satin spar variety and stellate groups of pink bladed gypsum in bands 30 cm thick have been found at New Stamp Hill mine, Kirkby Thore (Dunham & Hollingworth 1947, Davidson & Thomson 1951, Worley 2005).

Gypsum also forms efflorescences in many mines in the Northern Pennine orefield, wherever sulphate-rich groundwaters are actively attacking calcareous wallrocks or vein minerals, and good examples have been found at Brownley Hill and Smallcleugh mines, Nenthead, and Tynebottom mine, NW of Garrigill, all near Alston (Bridges 1987, Rust 1987a, Bridges & Young 1998). Gypsum is noted in limonite near oxidizing pyrite at Hilton mine, Scordale, near Appleby (Dunham 1990), and as colourless selenite crystals up to 15 mm across at: Long Fell mine, near Brough; Handsome Mea (Smallcleugh) and Tynebottom mines (Davidson & Thomson 1951, Bridges & Young 1998).

Other occurrences in the northern Pennines and elsewhere include: Birks Head mine, Long Marton, Appleby; Cotehill, near Carlisle; Flusco quarry, Greystoke (Davidson & Thomson 1951); Carrock mine, Grainsgill, Caldbeck Fells (Young 1987a); Cocklakes and Acrehead mine, Cumwhinton, near Carlisle (Trotter & Hollingworth 1932); Embleton quarry, near Cockermouth (Embrey, in footnote to Knill 1960); Goose Green mine, Parkside, Frizington (Green & Briscoe 2002); Nentsberry Haggs mine (Horse level), Nenthead (Thomson 1997a); Oughterside opencast coal site, Aspatria (Young & Nancarrow 1988b); Paddy End mine, Coniston, Eskdale (Young & Johnson 1985); Rampgill mine, Nenthead (Wolfe 2004); boreholes in the Sellafield area (Milodowski et al. 1998); Shap granite quarry (Wirth & Green 2002); and Smallcleugh mine (Rust 1991b). A substantial review of gypsum occurrences, including some not mentioned here, is provided by Young (1987a).

Derbyshire Gypsum (alabaster) has been worked for ornamental purposes since the 13th century at Chellaston, S of Derby (Dunham et al. 1978, Young 1990), and is widespread in the Triassic formations surrounding the Peak District. Almost any mine close to a limestone/shale contact, where pyrite has suffered oxidation, is likely to have small crusts of selenite crystals on its walls. Selenite is well developed at Odin mine, Castleton; Glebe and Ladywash mines, Eyam; Masson Hill mines, Matlock; Old Millclose mine, Darnley Dale; and the mines at Ashover and Crich (Ford et al. 1993). The famous Selenite Chamber in Cumberland Cavern, Masson Hill, Matlock Bath (Ford et al. 1993), described as part of Wapping mine (Ince 1991), produced very large white specimens from the walls and roof, although only colourless bladed crystals, white acicular crystals to 4 cm, or "rams horns" rosettes of colourless or white crystals to 6 cm in diameter, can now be found. Masses of colourless-to-white gypsum fibres fill cavities on a light-brown clay within fluorite-lined vugs at Milltown quarry, near Ashover (Rust 1991c), and at Fall Hill (Gate) quarry, Ashover, clear single crystals of selenite up to 35 mm long were recovered from mud-filled cavities (Spence 1982). Snelston mine, near Ashbourne, produced transparent flat plates of gypsum 1 mm across on a sandstone matrix (Jackman 1996).

Devon The Dunscombe and Branscombe Mudstone Formations at Weston Mouth, and W towards Sidmouth, contain veins and nodules of gypsum (Corrie & Green 2003). At Wheal Friendship (Devon Friendship), Mary Tavy, gypsum is closely associated with devilline and it forms colourless crystals up to 5 mm long on shale and quartz (Brooke 1986).

Dorset Gypsum occurs with felsöbányaite (syn. basaluminite) and hydrobasaluminite surrounding concretions within the weathering zone of the Oxford Clay at Crook Hill brickyard, Chickerell, near Weymouth (Clayton 1980). Gypsum was once worked in the Upper Jurassic beds of Durlston Bay, Isle of Purbeck (Rudler 1905).

Durham Raymond (1962) describes irregular inclusions of gypsum up to 7 cm long in the Stainton area, and offshore boreholes in the Lower Magnesian Limestone have been shown to contain abundant acicular gypsum associated with dolomite (Magraw 1978). Unusual transparent gypsum crystals etched into complex shapes have been recovered from West Pasture mine, Stanhope, Weardale (Young & Francis 1989, Green & Briscoe 2002), and efflorescences of gypsum occur at Boltsburn mine, Rookhope (Bridges & Young 1998). A glacial erratic boulder of gypsum from Shippersea Bay, near Easington colliery, is described by Young (2005).

Essex Clear, well formed, twinned crystals of gypsum (var. selenite), up to 10 cm long, were found in temporary exposures of London Clay at Brentwood during excavations for the M25 motorway (Lucy 1999).

Gloucestershire Gypsum occurs with thenardite as a white coating on Lower Jurassic mudstones in a gravel pit at Wingmoor farm, Bishop's Cleeve (King 1997). Red mudstones of the Mercia Mudstone Group (the Keuper Marl) contain massive nodules of gypsum.

Hereford & Worcester Locally abundant gypsum was formerly mined from the Triassic Mercia Mudstone Group at Spernall, SE of Redditch. A borehole at Saleway encountered abundant thin gypsum veins and anhydrite–gypsum nodules in the lower part of the Mercia Mudstone Group (Old et al. 1991).

Kent Rudler (1905) reports a group of lenticular crystals of gypsum from the Oldhaven Bed exposed at Reculver Cliff; and, at Warden Point, Isle of Sheppey, the variety selenite (some crystals twinned) occurs in London Clay with barite, calcite and pyrite (Monk 1982). Shelford quarry, near Canterbury, produced plentiful selenite and gypsum sand roses in the 1980s (Kingsmill 1982).

Leicestershire In a study of vanadium-rich nodules at the Butterley and Blaby clay pit, Glen Parva, King & Dixon (1971) note a large lens of pink gypsum consisting of single crystals and veinlets of fibres in Upper Triassic Mercia Mudstone Group (Keuper Marl). At the Gipsy Lane brick pit, Leicester, at least three fairly continuous gypsum beds have been found in Triassic calcareous mudstones. Gypsum is also present in thinner discontinuous nodular layers (Faithfull & Hubbard 1988). Minor amounts of gypsum occur at the margins of some anhydrite nodules from the Carboniferous Hathern Anhydrite Series examined in the Hathern No. 1 borehole, NW of Loughborough (Llewellyn & Stabbins 1970).

Northumberland Gypsum forms thick coral-like crusts of minute crystals in the Great Limestone at Crindledykes quarry, near Bardon Mill, and crystals of selenite occur in cavities in supergene barite on the dumps of Leadbitter shaft, Langley Barony mine, 2 km NW of Haydon Bridge (Bridges & Young 1998).

Nottinghamshire Veins of fibrous gypsum (satin spar) up to 25 cm wide reached their greatest perfection at East Bridgford, between Nottingham and Newark, where they were mined for the manufacture of ornaments (Richardson 1920). Satin spar is also noted from East Bridgford (Rudler 1905). A petrographic description of Nottinghamshire gypsum and anhydrite is provided by Richardson (1921). Economic concentrations of gypsum have been worked at East Leake mine (Ford et al. 1993).

Oxfordshire Greg & Lettsom (1858) report very large perfect crystals of gypsum that are detached, transparent or greyish white in the clay at Shotover Hill, whereas much less spectacular dirty-grey crusts up to 3 cm thick occur in nodules within the Oxford Clay at Cassington (Atkins 1995). Fine crystals of selenite are described from the Kimmeridge Clay at Headington, on the outskirts of Oxford (Rudler 1905).

Somerset Gypsum occurs with halite in a clay-mineral–evaporite sequence within Triassic marls at St Audries Bay, near Watchet (Mayall 1979).

Staffordshire Gypsum forms colourless-to-white drusy crusts on dump material from Waterbank mine, Ecton Hill (Rust 1995c), and is briefly noted to be of economic importance at Tutbury mine (Ford et al. 1993).

Sussex Colourless single crystals, simple twins and stellate aggregates of selenite occur in the topmost clays in the cliffs below the coastguard station at Newhaven (Wilmot & Young 1985). Rudler (1905) describes Upper Jurassic gypsum in a borehole at Mountfield, near Netherfield, comparable in age to that reported from Brightling mine (Holliday 1973).

Tyne & Wear Greg & Lettsom (1858) report delicate white acicular crystals of gypsum and elongate twin crystals of a brownish-yellow colour from Felling colliery. Gypsum is associated with barite and sideronatrite in cavities within the Brockwell coal seam at the Rising Sun colliery, Backworth (Randall & Jones 1966).

West Midlands Anhydrite–gypsum nodules and secondary fibrous gypsum occur at Knowle, Solihull, in borehole material sampling the Triassic Mercia Mudstone Group (Old et al. 1991).

Wiltshire Greg & Lettsom (1858) report fine detached crystals of gypsum from Telsford (thought to be Tellisford on the Wiltshire/Somerset border) and, more recently, superb transparent, euhedral crystals up to 6 cm across were found during construction of the Semington bypass (Wolfe 2004).

Yorkshire Permian evaporites underlie much of North Yorkshire and extend considerably beneath the North Sea. In the Aislaby (Eskdale No. 2) borehole, Eskdale, near Whitby, anhydrite and halite are the major constituents of the sequence, but major changes, caused by replacement and recrystallization since original precipitation, have occurred and textural evidence suggests that gypsum was also a primary phase. Evidence of early gypsum is found in polyhalite-bearing rocks, where anhydrite, halite and polyhalite form pseudomorphs after gypsum (Stewart 1949, 1951a, 1965). Thick encrustations of gypsum occur with allophane, gearksutite, ktenasite and serpierite on the mine walls of Old Gang mine, E of Gunnerside Gill, Swaledale (Green 2003). Crystalline crusts of gypsum also occur with greenockite at Wetgrooves mine, Aysgarth (Rust 1987b). A 6 m-thick bed of gypsum in Permian Upper Marls has been worked at Sherburn-in-Elmet, near Leeds (Dunham et al. 1978).

Wales

Dyfed Gypsum (selenite) occurs as thin films on Ordovician black shales at Abereiddi and as efflorescences on mine walls at the Lodge Park copper trial, Tre'r-ddôl (Mason & Green 1996). At Ogofau mine, near Pumsaint, and Cwmrheidol mine, near Devil's Bridge, gypsum probably results from the breakdown of pyrite (Bevins 1994). Gypsum forms colourless prismatic crystals on oxidized marcasite, pyrite and sphalerite at Frongoch mine, between Devil's Bridge and Pont-rhyd-y-groes (Green et al. 1996b), and flat radiating groups of crystals (selenite) on bedding planes and joints at Goginan mine (Jones & Moreton 1977).
Gwent Gypsum on the walls of caves in the Carboniferous Limestone escarpment of Mynydd Llangattwg at Agen Allwed and Ogof y Daren Cilau was produced from solutions passing by diagenetic pyrite in the overlying Millstone Grit (Bottrell 1991).
Gwynedd Abundant gypsum occurs in the weathered zone of the Cae Coch volcanogenic massive sulphide deposit, near Trefriw, Llanrwst (Ball & Bland 1985, Ball et al. 1985b, Otty 1986). It is also noted from old mine walls and spoilheaps at Britannia mine, Snowdonia, and Castell Carn Dochan mine, near Bala (Bevins 1994). Selenite needles are found rarely at the Great Orme mine, Llandudno (Dutton et al. 1994). Other occurrences include Parys Mountain, Anglesey (Southwood & Bevins 1995, Jenkins et al. 2000), and, at a trial level, 2 km NW of Bontddu, near Dolgellau (Saich & Rust 1987).
Powys Dylife mine, in the Central Wales orefield, contains gypsum (selenite) on bedding planes and joints in shale, on spoilheaps and as transparent prisms associated with greenockite in vugs in sphalerite (Jones & Moreton 1977).
South Glamorgan Veins of fibrous gypsum (satin spar) and nodules up to 60 cm across of white to pale-pink gypsum (alabaster) occur in the Triassic Mercia Mudstone on the coast between Barry and Penarth (North 1916, Waters & Lawrence 1987). Fibrous gypsum is noted from the Rhaetic Series at Penarth, and at Lavernock and SE of Cog, the variety alabaster was once worked (North 1916, Wolfe 1996b).

Scotland

Borders Heddle (1901) records veins of gypsum up to 5 cm thick in clay beds at Lochton, near Birgham, and 2.5 cm-long transparent crystals in cliffs 2 km above Tweed Mill. Gypsum crystals form beds, concretions and joint fillings in Carboniferous Cementstone Group shales around Hutton Castle, S of Eyemouth (Greigg 1988). Gypsum also forms inclusions in native gold within alluvial deposits from Craigierig Burn, near Megget reservoir (Chapman et al. 2000c).
Dumfries & Galloway Well formed gypsum crystals up to 1 cm across coat the roof and walls of Colvend copper mine, near Rockcliffe (Moreton & Aspen 1992). Near Needle's Eye, SW of Caulkerbush, small crystals of gypsum have been found with langite in the oxidized part of a bornite–chalcopyrite stockwork within the aureole of the Criffell granite complex (Braithwaite & Knight 1990). Gypsum is noted without description from the Leadhills–Wanlockhead orefield (Brown 1927) and more specifically from Glencrieff mine, Wanlockhead (Temple 1955). Anthropogenic gypsum is identified as a relative rare species on the dumps of the Meadowfoot smelter, Wanlockhead (Green 1987a).
Fife The exceptional pyrite occurrence in Carboniferous shales at Goat quarry, 2 km NW of Aberdour, contains colourless-to-orange gypsum in radiating sprays up to 1 cm across (Ingram et al. 1992).
Highland Alluvial gold from Kildonan Burn contains inclusion species including gypsum (Chapman et al. 2000c). A sea-washed cavity in vesicular basalt from Moonen Bay, N of Dunvegan, Isle of Skye, was found to contain colourless prismatic crystals of gypsum associated with cowlesite (Green & Todd 1996).

Gypsum (variety selenite) from roadworks on the Semington bypass, Wiltshire; width of crystal 40 mm. (© D. I. Green)

Lothian Gypsum has been found in stellate groups in joints in Carboniferous shales in the Dalkeith coalfield on the shore of South Queensferry (Heddle 1901), and the selenite variety is noted (with anhydrite) in veins, vugs and druses within the Traprain Law phonolite, near Haddington (MacGregor 1960).
Strathclyde Sugary white microcrystals of gypsum have been found in fossil cavities in Carboniferous Limestone from Trearne quarry, near Beith (Todd 1989), and a fibrous variety occurs in calcite veins at Southerness and Arbigland, near Kirkbean (Williams 1965). On Ailsa Craig, Firth of Clyde, gypsum forms off-white botryoidal masses up to 1 cm thick at the foot of Mains Craigs (Harrison et al. 1987).

Ireland

Antrim A large quantity of gypsum was once recovered from brickyards working Triassic clays on the west side of Belfast (Cole 1922) and the species is also reported from a Permo-Triassic sequence encountered in the Belfast Harbour borehole No. 1 (Smith 1986). Thin bands of gypsum were discovered in red marls above a salt dome at Duncrue mine, near Carrickfergus (Cole 1922). Greg & Lettsom (1858) report large, transparent and aggregated crystals over a salt bed at Kilroot, near Carrickfergus.
Cavan Reddish gypsum has been mined from Triassic formations at Drumgill mine, near Kingscourt (Cole 1922, Visscher 1971), and a potentially economic evaporite deposit of Carboniferous age has been discovered in boreholes at Glangevlin. Here, the gypsum is variegated and greyish white in colour (Grennan 1992).
Down Anhydrite, calcite and gypsum occur as a cement and in veins within Triassic sediments next to a Tertiary dolerite sill encountered in a borehole at Ballyalton, Newtownards (Knill 1960). Gypsum is one of the secondary alteration products found in copper–lead mineralized breccia at Conlig–Newtownards (Moles & Nawaz 1996).
Galway Abundant well formed crusts of gypsum occur at Clements mine, near Maum (Ryback & Moreton 1992).
Kerry Oxidation of sulphides in Carboniferous shales along the coast to the N of Ballybunnion has produced a sulphate assemblage in which gypsum occurs with brochantite, felsőbányaite (syn. basaluminite), hydronium jarosite, metasideronatrite and sideronatrite (Moreton et al. 1995). A similar, but post-mining, sulphate assemblage has also been found at Muckross copper mine, Killarney National Park, where the primary sulphides (chalcopyrite, galena, pyrite and sphalerite) and sulpharsenides (alloclasite, arsenopyrite and tennantite) have oxidized (Moreton et al. 1999).
Leitrim Gypsum occurs along the western bank of Lough Allen, near Gubb, Lecarrow, and at Drummans Lower Township, in a Visean (Lower Carboniferous) evaporite sequence (Kincaid 1857, West et al. 1968).
Mayo Greenish or grey crystalline encrustations of gypsum line the walls of Sheefry Pb–(Ag) mine, 11 km S of Croagh Patrick (Moreton & Jackson 1991).
Monaghan Keuper Marls contain gypsum of a "fair" white quality at Coagh, along the banks of Ballinderry River, and at Knocknacran East, near Carrickmacross, a reddish variety stained by iron hydroxides was worked from beds up to 18 m thick (Cole 1922).
Tipperary Crusts of gypsum are noted from Gortdrum mine, near Tipperary town (Braithwaite & Wilson 2003).
Wicklow A 1–3 mm-thick encrustation of gypsum contaminated with soot occurs on the walls and roof of railway tunnels, S of Rathdrum, and is thought to have originally been ejected from steam locomotives as a mixture of anhydrite, ash and soot derived from impure coal (Strogen & Elsdon 1974). Efflorescences of gypsum are associated with limonite, morenosite and native sulphur at St Patrick's copper mine, Avoca (King & Evans 1964).

Offshore territories

Jersey Efflorescences of gypsum are recorded from a cave on the east side of Devil's Hole. Elsewhere, it has been found on joints in granite at Les Handois and Mont Mado, in cleavage cracks in fluorite at Mont Mado, and on fluorite at the southeast quarry at L'Etacq (Mourant 1961).

gyrolite

triclinic (pseudo-hexagonal) $NaCa_{16}(Si_{23}Al)O_{60}(OH)_8 \cdot 14H_2O$

TL The Storr, 10 km N of Portree, Isle of Skye, Highland Region, Scotland

A relatively common hydrated sodium–calcium aluminosilicate hydroxide predominantly found infilling vesicles in basaltic rocks.

In Scotland, gyrolite is a pearly-white to creamy-white amygdale mineral, often associated with zeolite species and reported from many Tertiary basalt localities on the Isle of Skye, Highland Region, and the Isle of Mull, Strathclyde (Greg & Lettsom 1858, Heddle 1901). At the type locality, gyrolite is associated with apophyllite in an amygdaloidal basalt. It forms small spherical concretions composed of thin plates

Coarse white platy aggregates of gyrolite to 15 mm associated with botryoidal thomsonite and minor saponite; in vesicular basalt from Moonen Bay, Isle of Skye, Highland Region, Scotland. (© D. I. Green)

radiating from a centre, and the external surface of each concretion has an exceedingly beautiful striated appearance (Anderson 1851). The finest gyrolite and largest crystals of apophyllite were said to be found up a small valley, close to the road between Uig and the Quiraing, which is 15 km N of the type locality (Heddle 1879f). Less than 2 km N of Portree, gyrolite occurs in a small vesicular olivine-dolerite intrusion cutting vesicular basalts, where it is associated with calcite, mesolite, saponite, tacharanite, thomsonite, tobermorite and xonotlite. Amygdales containing gyrolite may be intergrown with mesolite or calcite, and rimmed by mesolite, thomsonite or tobermorite (Sweet 1961). A detailed investigation of the minerals from a weathered vein cutting basalt, 2 km N of Uig village, proved that uigite is an intimate mixture of gyrolite and the faröelite variety of thomsonite (Sweet 1959). Gyrolite is also noted from: Beinnan Lochain, NE of Healabhal Bheag; E of Roskill, Ben Vatten, S of Portree; Rubh' an Dúnain, S of Loch Brittle; Lòn na h Atha, W of Portree; and the Storr, N of Portree (King 1976). Moonen Bay, on the west coast of the Duirinish Peninsula, Isle of Skye, is a well known locality for zeolites, and here the gyrolite–apophyllite association is again found. Gyrolite forms subspherical aggregates of thin white pearly plates (Green & Todd 1996). Gyrolite is also found with apophyllite, together with mesolite and some of the best calcite ever collected from the basalts of Skye, in a remote and inaccessible coastal area SE of Moonen Bay, at Biod a' Mhurain, near Glen Ollisdal (Wolfe 1996b, Green & Todd 2001).

At 'S Airde Beinn, 5 km SW of Tobermory, Isle of Mull, Strathclyde, Tertiary basalts in the aureole of a volcanic plug have been thermally metamorphosed. One type of amygdale found to the N of the plug contains radiating aggregates of colourless plates of gyrolite, with smaller amounts of analcime, natrolite and thomsonite, but metamorphism has replaced the gyrolite with reyerite, and this in turn has been replaced by wollastonite (Cann 1965a). Gyrolite has also been found on the adjacent mainland at Ardtornish Bay, Morvern Peninsula, Highland Region, where it fills vesicles in hard basaltic rocks (Currie 1905). Small white crystals of gyrolite associated with analcime are further noted in interstitial and discrete patches in a large basic alkaline (teschenite) sill at Dippin on the Isle of Arran, Strathclyde (Gibb & Henderson 1978b).

Gyrolite is also noted from veins in a serpentinite from a quarry on the north side of Colla Firth, north Mainland, Shetland, where it occurs in association with pectolite, thaumasite and xonotlite (Butler et al. 1960), At Binhill quarry, Cairnie, Huntly, Grampian, white micaceous and pale-blue gyrolite is found with apatite, laumontite, prehnite and xonotlite within tourmaline-bearing granitic pegmatites cutting the Huntly gabbro (Livingstone 1974).

Gyrolite is noted from eight localities on the Isle of Skye by Heddle (1901) – Beinn Edra, 5 km E of Uig, Trotternish Peninsula; Beinn nan Cuithean, 1 km S of Talisker Point; Carn Liath, 2 km N of the Storr, N of Portree; Coire Faoin, SE of the Storr, N of Portree; southeast shore of Loch Varkasaig, west coast of Loch Bracadale; Rubha Thearna Sgùrr, N of Loch Brittle; Rubha na h-Airde Glaise, 3 km NE of Portree; and Sgùrr nam Fiadh, 3 km S of Talisker Bay – and also western shore of Isle of Canna; Uamh Fhraing, south end of Isle of Eigg; and Isle of Eilean Chathastail, off the Isle of Eigg.

Heddle (1901) also reports six localities for gyrolite on the Isle of Mull – Beinn na Croise, E of head of Loch Scridain; Bloody Bay, NW of Tobermory; Carsaig Arches, near Malcolm's Point; Carrachan Mòr, SW of Bearraich, Ardmeanach; Sgùrr Dearg, 9km W of Duart Point; and Killimore House, Kilfinichen, Loch Scridain – and also five of the Treshnish Isles, near the Isle of Mull: Cairn na Burgh Beg, Cairn na Burgh More, Fladda, Lunga and Sgeir a' Chaisteil.

In Ireland, gyrolite and apophyllite are described in Tertiary basalts from 30 localities, mostly in the SE of Co. Antrim, especially near Belfast: Catcairn quarry, Legoniel; Wolf Hill quarry; north and west side of Carnmoney Hill; quarries near Ballyhenry; and in the White Mountains area: quarries near Castlerobin and in a quarry NW of the summit of Collin Mountain. Gyrolite occurs in cavities in very tough massive black basalt lava flows. The cavities are lined with compact radiating fibrous thomsonite (faröelite), on which the apophyllite sits. Gyrolite formed later than thomsonite and occurs as rosettes of pearly-white leaves. With few exceptions, the gyrolite and apophyllite are found only in quarries, suggesting that weathering may have destroyed many other natural occurrences (Fleischermann 1910, Walker 1960a). The finest examples today are from the Wolf Hill, Hightown and White Mountain quarries to the SW of Belfast (Adair 1987). Nawaz (1978) notes that gyrolite is associated with apophyllite, chabazite and okenite at Catcairn quarry. Gyrolite is also found with apophyllite, chabazite, mesolite and thomsonite in basalt quarries around Ballymena, Co. Antrim (Green & Moreton 2001). One of these quarries is Maxwell's quarry, Scotstown Road area, 5 km from Ballymena, off the main road to Larne, where radiating sheaves of gyrolite are found with tiny sprays of thomsonite and euhedral hydroxyapophyllite (H. J. Foy specimen, Tindle unpub. EMP data).

At Scawt Hill, Larne, Co. Antrim, gyrolite occurs in cavities in the crust of a spurrite contact rock (contaminated chalk) close to a chalk/dolerite contact. It occurs as flat plates up to 1 mm across with a perfect basal cleavage and appears to be an alteration product of spurrite. Associated species include afwillite, calcite, ettringite, gehlenite, larnite, merwinite, perovskite, spinel, wollastonite and xonotlite (Tilley & Harwood 1931). Gyrolite is one of the secondary minerals found in another chalk/dolerite contact zone associated with a Tertiary volcanic plug at Carneal, Larne. Hydration and alteration following assimilation of the chalk have produced complex assemblages that also contain allophane, calcite, hibschite, portlandite, tacharanite, thomsonite, tobermorite and xonotlite (Henmi et al. 1973).

Gyrolite forms densely packed pearly-white micaceous flakes lining the walls of amygdales in Tertiary basalt at Sounding Hill quarry, NNW of Magherafelt, Co. Londonderry (Old 1972).

An aluminium-free variety of gyrolite is reported from a fracture vein in gabbro at Dean quarry, St Keverne, Lizard Peninsula, Cornwall. It forms aggregates of colourless to very pale-ochrous-brown platy crystals to 6 mm and is embedded in calcite (Elton et al. 1998).

Gyrolite spherules associated with thomsonite from Parkgate quarry, Templepatrick, Co. Antrim, Ireland; width 12 mm. (© D. I. Green)

H

haarkies
A synonym of millerite and of capillary pyrite.

haema-ovoid-agate
A reddish variety of agate (quartz + moganite) with ovoidal patches of cacholong.

haematite
Alternative (but not IMA-approved) spelling for hematite.

haidingerite?
orthorhombic $Ca(AsO_3OH){\cdot}H_2O$

A rare hydrated calcium arsenate hydroxide found coating brickwork and plasterwork on an arsenic calciner at Roseworthy, near Camborne, Cornwall. Haidingerite occurs with gypsum and the arsenic-bearing species arsenolite, pharmacolite and weilite, and together they form large undulating sheets several metres wide and up to 5 cm thick. Confirmation is by XRD (M. R. Power pers. comm. 2004). In this case, haidingerite is technically not a mineral and, more correctly, it should be regarded as an anthropogenic substance or synthetic equivalent (Nickel 1995).

Haidingerite has also been found on samples in the Kingsbury collection at the NHM. The specimens were originally said to have been collected from Wanthwaite mine, St John's in the Vale, near Threlkeld, Cumbria; however, given the doubts about the provenance of other Kingsbury specimens (Ryback et al. 1998, 2001), further confirmation is desirable.

Haidingerite is not to be confused with haidingerite (of Berthier) – a synonym of berthierite.

halite
cubic NaCl

A very common sodium chloride that occurs as a major component of sedimentary evaporite deposits. Permian evaporites underlie much of North Yorkshire and extend considerably beneath the North Sea. Halite and anhydrite are the major constituents of an evaporite sequence, consisting of three distinct beds, discovered in the Aislaby (Eskdale No. 2) borehole, Eskdale, near Whitby, Yorkshire. Large-scale changes through replacement and recrystallization since original precipitation have occurred and textural evidence suggests that some anhydrite and gypsum has been replaced by halite, and that halite has itself been replaced by polyhalite. The top 30 m of the evaporite sequence is dominated by halite and it is also plentiful in other horizons, associated with anhydrite, polyhalite and talc. In the Lower Evaporite Bed, halite is colourless and the equidimensional grains vary in size from a few mm to 25 mm across (Stewart 1949). The top 60 m of the Middle Evaporite Bed contain abundant halite, forming grains of irregular shape up to 50 mm across. Colour varies from colourless-to-grey, faint to light brown, pink, brick red and blood red, the last attributable to hematite inclusions (Stewart 1951a). The Upper Evaporite Bed also contains an abundance of halite in its upper part, with pale-pink granular crystals up to 10 mm across intergrown with coarser-grain colourless or pink crystals up to 75 mm across. Anhydrite, rinneite and sylvite are associated species (Stewart 1951b). Later borings proved abundant halite, with individual crystals commonly reaching 25 mm at other locations in Eskdale (No. 3 Sleights, No. 4 Sneaton and No. 6 Upgang), and confirmed the extent of the evaporite sequence in the Whitby area (Armstrong et al. 1951). Another borehole (F.3) at Fordon in Humberside contained a 430 m-thick evaporite sequence in which halite is a major component, with one subzone consisting of 64% colourless halite. Anhydrite, dolomite, kieserite and polyhalite are the other major constituents, with minor amounts of aphthitalite, celestine, glauberite, magnesite, pyrite, sulphoborite, sulphur and talc also noted (Stewart 1963, 1965).

The Permian evaporite sequence extends into Cleveland, where anhedral halite crystals up to 10 mm across have been found associated with anhydrite in the M3 borehole at Wilton, near Guisborough (Raymond 1962). Halite is the dominant species in the Boulby Halite and with sylvite, a major component of the Boulby Potash encountered at Boulby potash mine, near Loftus. Well crystallized specimens are not known, although samples are found with unusual gneissic textures and white hopper-faced crystals formed by the evaporation of temporary pools of brine (Green & Freier 1994, 1996). An evenly bedded sequence of pink halite occurs near the top of the Boulby Halite. Above this the halite is of variable colour, from brown to grey, and is intermixed with anhydrite and an interstitial clay (Woods 1979). A Permian halite bed (7 m thick) has also been discovered in a borehole to the E of Kirkby Thore, Cumbria (Worley 2005). At the Rising Sun colliery, Backworth, Tyne & Wear, an unusual occurrence of halite is reported close to two sulphide-lined cavities within coal seams. The halite formed stalactites up to 38 mm in diameter and thin-walled hemispherical shells on the walls and timbers of the mine roadway (Randall & Jones 1966). Another discovery of halite from a coalmine was at Hickleton colliery, WNW of Doncaster, Yorkshire, where one of the best recent examples of British halite was reported to be on a 175×100×50 mm specimen containing multiple cubic twins up to 50 mm across (Uttley 1986).

Another large halite (salt) field is within the Mercia Mudstone Group (Keuper Marl) of the Cheshire–Shropshire Basin, which extends over several hundred km^2 and contains more than 70% halite in beds totalling more than 580 m in thickness. Greg & Lettsom (1858) list Middlewich, Nantwich, Northwich and Winsford, all in Cheshire, as locations for halite. In 1978, Meadowbank mine, Winsford, was producing 20% of Britain's salt requirements (mainly to combat ice formation on roads in winter) from a 24 m-thick bed of halite (Dunham et al. 1978). A smaller saltfield reaching a thickness of 180 m exists beneath Preesall in Lancashire, and in Somerset a 50 m-thick sequence of saliferous beds are conservatively estimated to contain 13 000 million tonnes of halite (Dunham et al. 1978). Halite is further described from St Audries Bay, near Watchet, Somerset (Mayall 1979), and from a series of boreholes sampling Mercia Mudstone Group evaporites at Burton Row in Somerset to

Lustrous, elongate crystals of halite, up to 5 mm on edge, from Kilroot, Co. Antrim, Ireland. (© D. I. Green)

Marshwood, near Lyme Regis, and Cranbourne, both in Dorset (Gallois 2003).

In Cornwall, halite has been noted as micron-size cubic crystals associated with copper sulphides at Penberthy Croft mine, St Hilary, Cornwall (Camm & Merry 1991a, Betterton et al. 2000). Halite crystals of similar size also occur as fluid inclusions in quartz from a tourmaline-rich fault breccia at Roche Rock, near St Austell (Williamson et al. 2000).

Cavities in oil shales contain halite in association with barite and calcite at Midcalder, Pumpherston, Lothian, Scotland (Livingstone 2002). Halite, sylvite and an opaque phase (probably magnetite) are noted as daughter crystals in fluid inclusions within clinopyroxene from a skarn at the margin of the Beinn an Dubhaich granite, near Broadford, Isle of Skye (Smith et al. 2004).

Commercial deposits of halite occur in Triassic Keuper Marl between Carrickfergus and Larne, Co. Antrim. Rudler (1905) reports clear, colourless, crystalline halite from Maidenmount mine, near Carrickfergus, and other occurrences in Co. Antrim are listed by Cole (1922): Ballyedward borehole; Ballylig; Burleigh Hill mine; Downshire mine, near Eden; Duncrue mine, near Carrickfergus; Eden mine; French Park mine; Magheramorne boreholes; New mine, near Eden; Redhall mine, near Ballycarry. In 1971 the only mine working in Northern Ireland was the Kilroot salt mine, Co. Antrim (Wilson 1971), and this is still in operation today. An indication of the abundance of halite in Ireland can be gauged by the Larne No. 2 borehole, where a 113 m-thick bed of Permian halite (part of the Belfast Group) and three Triassic halite beds (part of the Mercia Mudstone Group) totalling 400 m in thickness were encountered (Mitchell 2004). Hopper-face clastic casts of halite up to 6 mm across have been found in a Visean (Lower Carboniferous) evaporite sequence along the shore of Lough Allen, near Lecarrow, Co. Leitrim (West et al. 1968).

halloysite

KAOLINITE–SERPENTINE GROUP

monoclinic or hexagonal $Al_2Si_2O_5(OH)_4$

An uncommon aluminosilicate hydroxide, polymorphous with dickite, kaolinite and nacrite, and formed during surface weathering of aluminosilicate rocks or as a hydrothermal alteration product.

An early description of possible halloysite is from Cook's Kitchen mine, Redruth, Cornwall, where it was referred to as lithomarge. Yellowish-white lithomarge formed a vein 22 cm wide and initially was very soft, but soon after became much harder (Garby 1848). Because the name "lithomarge" was also applied to a variety of clay-like minerals such as kaolinite, halloysite is not positively confirmed at Cook's Kitchen or at Tincroft mine, Illogan (Greg & Lettsom 1858). There are also doubts about the Cornish halloysite mentioned by Robson (1949b) from altered granite in the St Austell area (Hensbarrow–St Stephens) and from granites at St Just, Germoe, St Day, Hawk's Tor (Bodmin) and Lee Moor. Golley & Williams (1995) add a further unsubstantiated Cornish locality at Phoenix United mines, Linkinhorne. Betterton et al. (2000) report an XRD confirmation of halloysite occurring with quartz at Penberthy Croft mine, St Hilary.

There are actually two forms of halloysite, one form (7Å) being less hydrous than the other (10Å). At Kirkham's and Spencer's sandpits, W of Brassington, Derbyshire, the 7Å form occurs as hard waxy white-to-cream pieces up to 25 mm across, which are enclosed by a soft white powdery form, both occurring in nodules within silica sand-pocket deposits. A light-bluish variety of halloysite associated with wad and a red or buff clay is noted in a trial hole at Stevenson's spar workings on Gallows Knoll, above Wirksworth, Derbyshire (Ford 1963, Ford & Sarjeant 1964a, Ford et al. 1993).

Halloysite is one of several secondary aluminium-rich species associated with deeply oxidized smithsonite-bearing veins in Upper Triassic limestone breccias, near East Harptree, Mendip Hills, Avon. It occurs as relatively pure well crystallized monomineralic cavity infillings up to 4 cm in diameter and also in thin veinlets in mineralized breccia. Two types are recognized: a waxy-looking, hard, brittle, translucent, brown-to-white variety, which breaks with a conchoidal fracture, and a compact chalky-looking variety of similar colour. Halloysite is associated with allophane, hydroscarbroite, kaolinite and scarbroite (Alabaster 1990b).

Other occurrences in England include: as a replacement of analcime that occurs growing into and replacing aragonitic shell debris in a heavily faulted coastal exposure of Lower Jurassic marine sediments, in particular at Quantoxhead, west Somerset (Alabaster 1980); as encrustations on fluorite in barite–fluorite–galena–sphalerite veins cutting Lower Carboniferous limestones at Coldstones quarry, 5 km W of Pateley Bridge, Yorkshire. Halloysite (10Å) is associated with prosopite, smithsonite and several uncharacterized aluminium hydroxides or silicate phases (or both) (Young et al. 1997).

At Twll maen grisial ("cave of the crystals"), near Prenteg, Tremadog, Gwynedd, Wales, anastomosing quartz veins cutting a dolerite intrusion were the source of some world-class brookite crystals (probably collected between 1800 and 1840). In a much later study, Starkey & Robinson (1992) found halloysite to be a white late-stage clay mineral common in many of the veins.

In Scotland, halloysite (7Å) has been found in a vein at Hospital quarry, Elgin, Grampian (Heddle 1882a, Macpherson & Livingstone 1982), and tubular particles of halloysite (7Å and 10Å) are noted in the clay fraction of soils derived from gabbro, around Insch, Grampian (Wilson & Tait 1977).

HALOTRICHITE GROUP

Monoclinic sulphates with the general formula $AB_2(SO_4)_4{\cdot}22H_2O$, where $A = Fe^{2+}$, Mg, Mn^{2+}, Ni, Zn, and B = Al, Cr^{3+}, Fe^{3+}. Apjohnite, halotrichite and pickeringite are members of the group found in Britain or Ireland.

halotrichite

HALOTRICHITE GROUP

monoclinic $Fe^{2+}Al_2(SO_4)_4{\cdot}22H_2O$

A relatively common hydrated iron–aluminium sulphate that forms a series with pickeringite and typically develops as a result of alteration of pyrite. Halotrichite is often found growing on mine walls in association with other sulphate minerals. Bevins (1994) reports halotrichite in Wales, from coalmines in Mid Glamorgan – Cymmer colliery, Porth; Deep Navigation colliery, Treharris – and from three locations in Gwynedd – Gwynfynydd mine, near Dolgellau; Trwyn Cae Iago, near Porthmadog; and the copper deposit at Parys Mountain, Anglesey. Samples from the coalmines had previously been identified as alunogen, but XRD analysis proved them to be halotrichite.

Halotrichite was said to be abundant in the shales of exhausted coal beds at Hurlet and Campsie, near Paisley, Strathclyde, Scotland (Greg & Lettsom 1858), and at Hurlet it is described as forming yellowish-white silky fibrous crystals with an inky astringent taste (Heddle 1901). Halotrichite is also noted from a railway cutting W of Stanely, near Paisley (Livingstone 2002), and from Smallcleugh mine, Nenthead, near Alston, Cumbria, England (Uttley 1987b).

A. W. G. Kingsbury claimed to have found a nodular sample of halotrichite at Mealbank quarry, Ingleton, Yorkshire, but Ryback et al. (2001) suggest that it is an old specimen from a foreign location.

Fibrous halotrichite from Parys Mountain, Anglesey, Gwynedd, Wales; width of specimen 65 mm. (© National Museum of Wales, Cardiff; photographer M. P. Cooper)

hammarite

orthorhombic $Cu_2Pb_2Bi_4S_9$

A rare copper–lead sulphobismuthite found in quartz veins cutting monzogranite at Loch Shin, 2 km W of Lairg, Highland Region, Scotland. Elsewhere in the monzogranite, mineralization is disseminated and mainly restricted to fluorite, molybdenite and pyrite. However, in the veins, molybdenite does not occur and the other species are accompanied by native bismuth, chalcopyrite, covellite, galena and sphalerite, and a suite of bismuth sulphosalts: aikinite, berryite, gladite, hammarite, lindströmite, krupkaite, pekoite and tetradymite. Hammarite forms coarse fibrous crystals and is a member of the aikinite–bismuthinite series (Livingstone 1993, Lowry et al. 1994).

In Ireland, at Ballycoog, Goldmine River valley, near Avoca, Co. Wicklow, chalcopyrite–magnetite–pyrite–siderite (± galena, hematite, sphalerite) mineralization is reported from metamorphosed BIF rocks. Sulphide content of the BIF is low, with chalcopyrite hosting gold inclusions. Bismuth sulphosalts are a minor component, consisting of members of the aikinite–bismuthinite series (aikinite, krupkaite and hammarite) with traces of bornite, chromite, silver-rich gold, mackinawite and pyrrhotite being observed microscopically. The sulphosalts form coarse-grain crystals, 50–200 µm across, intergrown with chalcopyrite. The BIF was deposited in a volcanic environment during the late Ordovician (Ixer et al. 1990).

harkerite

cubic $Ca_{12}Mg_4Al(CO_3)_5(BO_3)_3(SiO_4)_4{\cdot}H_2O$

TL Camas Malag, near Broadford, Isle of Skye, Highland Region, Scotland

Named after Alfred Harker FRS (1859–1939), an eminent English petrologist of St John's College, University of Cambridge, and president of the Geological Society, 1916–1918; author of *Petrology for students* (1895), *Natural history of igneous rocks* (1909) and *Metamorphism* (1932); born in Kingston upon Hull, Humberside. Harkerite is a very rare hydrated calcium–magnesium–aluminium carbonate borosilicate found in harkerite–monticellite and pure harkerite rocks at the type locality, within a contact skarn where granite has intruded dolomite. It is colourless, it forms simple octahedra, and it readily alters to calcite. Accessory amounts of bornite, chalcocite, diopside and magnetite accompany harkerite and other boron-bearing minerals occur locally, for example datolite, fluoborite, ludwigite and szaibelyite (Tilley 1948b, 1951, Grew et al. 1999). Starkey (1988b) lists 39 species from this skarn area.

Interlocking crystals of colourless harkerite with compositionally zoned calcite replacing core regions from the Beinn an Dubhaich granite aureole, Camas Malag, near Broadford, Isle of Skye, Highland Region, Scotland; width 1.75 mm; viewed in PPL. (© K. Gardner)

harmotome

ZEOLITE GROUP

monoclinic $Ba_2(Na,K,Ca_{0.5})(Al_5Si_{11})O_{32}{\cdot}12H_2O$

A relatively uncommon hydrated barium–sodium–potassium–calcium aluminosilicate that forms a series with phillipsite-Ca and is mainly found in late-stage hydrothermal Pb–(Ag) veins and also in volcanic rocks, where it is derived from the alteration of barium-rich feldspar.

In the Northumberland part of the Northern Pennine orefield, harmotome has been found close to the Whin Sill along a witherite vein system at Settlingstones and Stonecroft mines, near Hexham. Originally discovered, and described at a Mineralogical Society meeting in 1933, by Sir Arthur Russell, at the first location the species forms 1.5 mm-long colourless doubly terminated cruciform-twinned crystals or perfectly formed colourless crystals. Similar, but slightly larger (to 5 mm), crystals occur at the second location (Ineson 1972, Young & Bridges 1984, Dyer & Wilson 1988, Dunham 1990). The dumps of Maughan's shaft at Greysides mine, near Newbrough, contain harmotome crystals up to 1.5 mm across associated with barytocalcite, in vugs within barytocalcite–calcite veinstone. A single specimen containing cruciform twins of harmotome to 1 mm long has been found on the dumps near Wellhope shaft, Nentsberry Haggs mine (Young & Bridges 1984). Harmotome is a very late-stage mineral occurring in open vugs at all these localities. A fifth locality in the northern Pennines is Brownley Hill mine, Nenthead, near Alston, Cumbria, where harmotome has been found as colourless-to-white elongate prismatic twinned crystals up to 1.5 mm long in a large quartz-lined cavity within metasomatized limestone. Unlike the other occurrences it appears to be a primary mineral unrelated to the emplacement of the Whin Sill (Green & Nudds 1998, Green et al. 2000). Harmotome has also been confirmed at Ruthwaite Lodge, Grisedale, SW of Glenridding, Ullswater, Cumbria, where it forms prismatic 2 mm-long core-bit twin crystals on galena and quartz. It is overgrown by mimetite and wulfenite (Todhunter 2007).

Colourless-to-white prismatic crystals of harmotome up to 7 mm across are associated with edingtonite and occur on the joint planes of a dolerite intrusion quarried for roadstone at Squilver quarry on Disgwylfa Hill, 9 km N of Bishop's Castle, Shropshire (Hubbard & Enamy 1993).

On the dumps of Pen-y-Clun lead mine, NW of Llanidloes, Central Wales orefield, Powys, common-to-abundant harmotome forms transparent colourless cruciform-twinned euhedral crystals up to 2 mm long (Morgan & Starkey 1991). Similar crystals were also recovered from Cwm Orog mine, near Llangynog, Powys, in a barite vein dominated by witherite (Robinson & King 1993). Harmotome is further noted from Gorn mine, near Pen-y-Clun, and Craig Rhiwarth mine, near Cwm Orog (Bevins 1994). A few small crystals of harmotome were recovered from Dolyhir quarry, Old Radnor, near Presteigne, Powys, close to a barite–chalcopyrite–galena–tennantite vein (Bevins & Mason 1997, Wolfe 2000, 2004). Cruciform-twinned crystals of harmotome associated with alstonite, calcite, ewaldite and witherite have also been found (Green et al. 2005d). Dolyhir quarry is now the premier Welsh locality for harmotome (T. F. Cotterell pers. comm. 2007). Harmotome is briefly described from the manganese workings (orebodies 2 and 5) of Benallt mine, Rhiw, Lleyn Peninsula, Gwynedd, where it occurs as minute crystals up to 0.03 mm across coating banalsite in association with barite (Campbell Smith et al. 1944, 1945).

The lead mines (e.g. Bellsgrove, Corrantee, Fee Donald, Middleshop and Whitesmith mines) located about 8 km to the north of Strontian, near Loch Sunart, Highland Region, are perhaps the most famous locations for Scottish harmotome. Here it occurs with brewsterite and strontianite in open spaces within ore veins of barite, calcite, galena and sphalerite that traverse the Strontian granite along a near-vertical contact with gneiss of Moinian age. Exceptional specimens of colourless-to-white harmotome were found from 1722 to 1871, when the mines were worked for lead. Reworking of the area between Bellsgrove and Whitesmith mines for barite in the 1980s produced many more remarkable specimens. Most of the specimens of exceptional quality came from Bellsgrove mine, including crystals up to 2.5 cm long on calcite from the Clashgorn stope. Brilliant harmotome twins up to 8 mm in length on

Colourless, 9 mm long, twinned harmotome crystal, on pale-brown calcite; Corrantee mine, Strontian, Highland Region, Scotland. (© D. I. Green)

Harmotome cluster, 95 mm tall, with crystals to 12 mm; Strontian mines, Highland Region, Scotland. (© D. I. Green)

amber calcite from Corrantee mine are also notable (Greg & Lettsom 1858, Heddle 1901, Winrow 1986b, Tschernich 1992, Green & Todd 2001). Harmotome always forms complex interpenetrant twins and these can be divided into four types, each producing distinctive crystals: Marburg, Morvenite, Perier and Stempel. Morvenite (a variety of harmotome showing morvenite twinning) was named after the mountainous area near Strontian and has mistakenly been quoted as the type locality for harmotome (Nickel & Nichols 1991). A summary of the early work on harmotome (pre-1858) is given by Livingstone (2002). Elsewhere in Scotland, harmotome is reported from the following places:

CENTRAL Burn of Sorrow, Dollar Glen (Livingstone 2002); Campsie Fells, N of Strathblane (Heddle 1901, Tschernich 1992); Earlsburn (Earlston) reservoir, Touch Hills, SW of Stirling (Heddle 1901, Green & Todd 2001); Gargunnock, near Stirling (Livingstone 2002). DUMFRIES & GALLOWAY Coatsgate quarry, near Moffat (D. I. Green specimen and XRD, Tindle unpub. EMP data). HIGHLAND In a sphalerite-bearing barite–calcite vein in Allt Tigh Cumhaig Burn, 3 km S of Struy, near Inverness (Russell 1946b); water-clear lustrous cruciform twinned harmotome coating fissures in a hematite-rich breccia zone from a quarry near Loch an Arbhair, Lochinver–Ullapool road, SW of the River Kirkaig (Waterston 1953, Tschernich 1992); in miarolitic cavities within syenite near the northeast contact of the Ben Loyal igneous complex (Gallagher et al. 1971); Guisachan quarry, Cannick, near Drumnadrochit; and Struy mines, Loch na Mèine, midway between Strath Glass and Strath Farrar, 3 km SW of Struy, near Inverness (D. I. Green specimens and XRD, Tindle unpub. EMP data). LOTHIAN Barnton Brae quarry, near Edinburgh (Tschernich 1992); Corstorphine Hill, Edinburgh (Heddle 1901); Hilderston mine, Bathgate Hills (Meikle 1994). STRATHCLYDE Bishopton railway tunnel, near Glasgow (Heddle 1901, Tschernich 1992); colourless harmotome to 3 mm associated with edingtonite and thomsonite in basaltic andesite at Bowling quarry, 2 km WNW of Old Kilpatrick, near Glasgow (Heddle 1901, Waterston 1953, Tschernich 1992); Craignaught quarry, Dunlop (Livingstone 2002); Glen Arbuck, Old Kilpatrick (Tschernich 1992); rare transparent, colourless, well terminated harmotome up to 1.5 mm associated with barite and prehnite in cavities in Carboniferous basalt at Hartfield Moss, 9 km WSW of Barrhead (Meikle 1990); transparent colourless cruciform-twinned harmotome up to 3 mm long in Carboniferous basalt at Loanhead quarry, near Beith (Meikle 1989b, Meikle & Todd 1995); Scoat Head and Touch Forest, Touch Hills (D. I. Green specimen and XRD, Tindle unpub. EMP data).

In Ireland, harmotome has been found forming cruciform twins up to 10 mm long in the Luganure mines and on the dumps extending from Glendalough to Glendasan, Co. Wicklow (Joly 1886, Tschernich 1992). Green et al. (2005g) note its occurrence on the dumps of Glendalough mine, Foxrock mine and at Deep Adit, Old Luganure mine, Glendasan, where it was found with brewsterite-Ba. Harmotome also occurs in spoil from abandoned lead mines between Newtownwards and Conlig, Co. Down. Here barite–chalcopyrite–galena vein mineralization within silica- and dolomite-matrix breccias was intruded by a Tertiary dolerite dyke, and harmotome probably formed during a phase of hydrothermal activity following on from this. Clusters of transparent prismatic cruciform-twinned harmotome up to 5 mm in length occur within cavities. The species also forms equant 100–200 µm crystals dusting the surfaces of vugs (Moles & Nawaz 1996, Moles 2007). Harmotome is also confirmed from Tara mine, Navan, Co. Meath, where it occurs as mm-size crystals with calcite in late-stage veins cutting the main lead–zinc orebody (D. I. Green specimen and XRD, Tindle unpub. EMP data). Harmotome described by Greg & Lettsom (1858) from the Giant's Causeway, Co. Antrim, is actually phillipsite.

harringtonite

TL Portrush and the Skerries, Co. Antrim, Ireland

A tough snow-white mixture of mesolite and thomsonite (both members of the zeolite group). Harringtonite is probably named after Robert Harrington (1751–1837) of Carlisle, Cumbria, a surgeon and writer on science, but this is not confirmed.

hastingsite

AMPHIBOLE GROUP

monoclinic $NaCa_2(Fe^{2+}_4Fe^{3+})(Si_6Al_2)O_{22}(OH)_2$

A relatively common sodium–calcium–iron aluminosilicate hydroxide that forms a series with tschermakite and is found in various igneous (granite, syenite) and metamorphic (amphibolite, gneiss, schist) rocks. Wherever possible, authors' original data have been reprocessed using the amphibole classification of Leake et al. (1997a) and Mogessie et al. (2004).

England

A specimen of hastingsite intimately intergrown with magnetite from the dumps of Haytor iron mine, Ilsington, Devon, is in the R. W. Barstow collection at Plymouth City Museum.

Wales

Gabbros belonging to the St David's Head intrusion, Dyfed, and the Rhiw intrusion, Lleyn Peninsula, Gwynedd, are noted to contain hastingsite (Bevins 1994), although supporting data are unpublished.

Scotland

Highland Highly pleochroic hastingsite (optical identification) is closely associated with aegirine-augite in marginal syenites belonging to the Cnoc nan Cuilean intrusion, SE of Ben Loyal (King 1942). Anhedral crystals of hastingsite, many poikilitic and up to 2 mm across, occur with arfvedsonite in a gneissic granite belonging to the Carn Chuinneag–Inchbae complex in the Black Water, 3 km up stream from Inchbae Lodge, Easter Ross (Harker & Leahy 2001). W of the Great Glen, hastingsite is widespread in amphibole-rich quartzites, psammites, semipelites, mafic rocks and granite gneisses belonging to the Loch Eil and Glenfinnan divisions of the Moine (Rock 1984, Livingstone 1993). Associated species include allanite-(Ce), anorthite, britholite-(Y), diopside, hedenbergite and magnetite (Smith et al. 2002).

Strathclyde Gabbroic pegmatites, teschenites and coarse syenites within the differentiated alkali olivine dolerite intrusion at Cnoc Rhaonastil, Isle of Islay, contain unusual zirconium-rich hastingsite (Preston et al. 2000b). The Carboniferous Lugar Sill exposed at Glenmuir Water and at Barshaw, near Paisley, contains black lustrous prismatic crystals of hastingsite up to 75 mm in length, which had earlier been described as barkevikite (Scott 1914b, Howie 1963). Endoskarn associated with the Beinn an Dubhaich granite, Isle of Skye, contains hastingsite as discrete grains or replacing pyroxene. Associated species include allanite-(Ce), anorthite, britholite-(Y), diopside, hedenbergite and magnetite (Smith et al. 2002).

Ireland

Antrim Hastingsite is one of many species noted from the contact of a dolerite contaminated by surrounding chalk at Scawt Hill, Larne (Tilley & Harwood 1931), although no supporting data have been published.

Down Hastingsite occurs in a hornblende granite at Slieve Donard in the Mourne Mountains (Deer et al. 1997).

Offshore territories

Jersey Hastingsite is one of many amphibole species described from an igneous complex (olivine-gabbro to aplogranite compositions) at Sorel Point in north Jersey (Salmon 1991), but the actual data are unpublished.

hatchettite

TL Merthyr Tydfil, Mid Glamorgan, Wales

Named after Charles Hatchett (1765–1847), a chemist and mineralogist who toured Cornwall with W. G. Maton in 1794 and 1796, author of *Analysis of magnetical pyrites* (1804), and who discovered the element columbium (now called niobium); born in Long Acre, Covent Garden, London. Hatchettite in not a true mineral but a natural paraffin wax with a composition near $C_{39}H_{78}$.

hatterel

A banded, clear and yellow variety of fluorite.

HAUCHECORNITE GROUP

Tetragonal sulphosalts with the general formula A_9BCS_8, where A = Co, Ni, B = As, Bi, Sb, and C = Bi, Sb, Te. Hauchecornite and tučekite are members of the group found in Britain or Ireland.

hauchecornite

HAUCHECORNITE GROUP

tetragonal Ni_9BiSbS_8

A rare nickel–bismuth–antimony sulphide found as inclusions in hydrocarbons from copper-mineralized veins at Laxey, E of Snaefell, Isle of Man. Hauchecornite is associated with pyrite (bravoite), siegenite and an unidentified bismuth sulphide (Parnell 1988c).

haughtonite

Named after Samuel Haughton FRS (1821–1897), geologist and physiologist, professor of geology at Trinity College, Dublin, and an associate of M. F. Heddle; born in Carlow, Ireland. A variety of biotite, mica group (see Heddle 1879d, 1881).

hausmannite

tetragonal $Mn^{2+}Mn^{3+}_2O_4$

A locally abundant manganese oxide, found in predominantly pyrolusite ore at Wesley Theological College (Wesley mine), Westbury-on-Trym, Avon. Hausmannite occurs in massive form and as crusts of sub-mm octahedral crystals lining cavities in the massive ore. Manganite and rhodochrosite are associated species. The mineralization is associated with Triassic rubble-filled fissures in Carboniferous Limestone. Hausmannite with a very similar paragenesis is noted from Coombe Farm quarry, Westbury-on-Trym, Avon, and two locations in the Mendip Hills, Somerset – in the large limestone quarry at Merehead (Torr Works), Cranmore, and at Coleman's quarry, Holwell, near Frome. At the latter location, small pockets and veins of primary pyrolusite and manganite ore are accompanied by a diverse late-stage mineral assemblage that includes aragonite, barytocalcite, calcite, cerussite, chloroxiphite, crednerite, cuprite, diaboléite, fluorite, goethite, hausmannite, hydrocerussite, kentrolite, malachite, manganite, melanotekite, mendipite, mimetite, paralaurionite, pyromorphite, quartz and rhodochrosite (Alabaster 1985b, 1989b, 1990). Dump material from the Providence iron mine, Long Ashton, Bristol, also contains hausmannite (with cerussite, pyrolusite and rhodochrosite), but its paragenesis is not described (Alabaster 1989b).

Brown lensoid masses of bustamite and hausmannite occur with rhodonite at Greystone quarry, Lezant, near Launceston, Cornwall. These species occur in a structurally complex metasedimentary sequence unrelated to the Cu–Pb–Zn vein mineralization for which this quarry is famous (Elton et al. 2004).

In Cumbria, a vague reference to hausmannite from the Cleator area is given by Goodchild (1875), and Young (1987a) describes pyramidal hausmannite on hematite from the Bigrigg area. Commercially mined hausmannite from Wyndham mine, Bigrigg, near Egremont, is noted by Davidson & Thomson (1951). Examination of a small amount of dump material from Wyndham mine revealed that the hausmannite occurs as compact dark-grey crystalline masses up to 8 cm across associated with calcite and rare barite. Some of the hausmannite formed in cavities within the ore and produced well formed bipyramidal crystals (Young & Nancarrow 1990).

On dumps at an abandoned 19th-century manganese mine at Tŷ Coch, near Porthcawl, Mid Glamorgan, Wales, massive, fine-grain, compact, very hard, dark-grey manganese ore contains euhedral to subhedral braunite and hausmannite in equal amounts, with interstitial and vug filling barite, calcite, iron-rich calcite, dolomite and manganocalcite. Minor pyrobelonite and vanadinite are also present (Criddle & Symes 1977). Hausmannite is also confirmed from Benallt mine, Rhiw, Lleyn Peninsula, Gwynedd, Wales, where it occurs as brown irridescent plates associated with altered pyrochroite (Bevins 1994).

Spray of hausmannite in calcite from Tŷ Coch, Nottage, near Porthcawl, Mid Glamorgan, Wales; width 13 mm. (© National Museum of Wales, Cardiff)

Lustrous brownish-black crystals of hausmannite, up to 2 mm long, and associated with barite, have been found in a small area of stratiform manganese ore at Dalroy Burn, near Culloden (10 km E of Inverness), Highland Region, Scotland (Livingstone 1993).

haüyne?

SODALITE GROUP

cubic $Na_6Ca_2Al_6Si_6O_{24}(SO_4)_2$

A sodium–calcium aluminosilicate with sulphate, thought to occur in the phonolite at Wolf Rock, 15 km from Gwennap Head, off Land's End, Cornwall. Minute cloudy crystals of haüyne (pronounced "haween") are mentioned by Ayde (1907), but this is probably an error of identification. Most likely it is nosean, partially altered to analcime and cancrinite, which is being described (Harrison et al. 1977).

hawleyite

SPHALERITE GROUP

cubic CdS

A rare cadmium sulphide, dimorphous with greenockite and found at Fall Hill (Gate) quarry, Ashover, Derbyshire, where it forms a powdery coat on fluorite or, less commonly, sphalerite. Hawleyite is slightly yellower than the greenockite found at this location. Latterly the quarry worked fluorite from an 8 m-wide vein cutting limestone, but, prior to that, galena was the principal ore mineral. Other associated species found in the quarry are barite, goethite, marcasite, millerite, pyrite and smithsonite (Smith 1982). The occurrence is also noted by Spence (1982), Braithwaite (1983), Young et al. (1987) and Ford et al. (1993).

In Ireland, hawleyite is reported from two lead–zinc deposits: Tynagh mine, near Killimor, Co. Galway (Morrissey 1970, Ryback et al. 1988), and tentatively from Ballinalack, 15 km NW of Mullingar, Co. Westmeath (Patterson 1970).

haytorite

TL Hay Tor, Devon, England

Chalcedony pseudomorphs after datolite, gadolinite group.

heavy spar

A synonym of barite.

heazlewoodite

trigonal Ni_3S_2

An uncommon nickel sulphide found associated with pentlandite in chromitite at Hagdale quarry, Baltasound–Haroldswick area, Unst, Shetland, Scotland (Livingstone & Bish 1982, Braithwaite & Pritchard 1983b). At Cliff (a disused chromite quarry) in the same area, an arsenic-rich assemblage within a dunite lens contains intergrown heazlewoodite and millerite, with associated native copper, geversite, native gold, native silver, sperrylite, spionkopite and palladium antimonides, all in accessory amounts. In a sulphide-bearing dunite at Cliff, pentlandite is observed altering to awaruite, heazlewoodite and magnetite, and, in fractures crosscutting

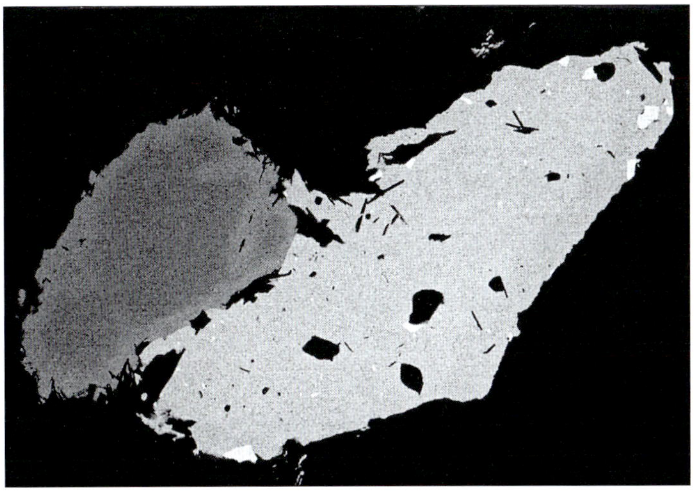

Elongate heazlewoodite containing many blebs of nickeline (white) and tiny needles of antigorite (black); the rounded medium-grey crystal to the left is chromite. From a peridotite at Corrycharmaig, NW of Killin, Tayside Region, Scotland; width 700 µm; BSE image. (© M. R. Power, Camborne School of Mines)

chromite, heazlewoodite occurs with iron–nickel alloy and palladium antimonides. Heazlewoodite (with a grainsize of > 100 µm) is also noted to contain inclusions of breithauptite (Prichard et al. 1994). Maynard et al. (1997) describes blocks of dunite to the west of Hagdale, containing heazlewoodite–pentlandite intergrowths up to 2 mm. Gunn et al. (1985) and Lord et al. (1994) report heazlewoodite from the Cliff area of Unst and suggest that some of the nickel sulphides associated with PGMs are related to hydrothermal alteration of the Shetland ophiolite, rather than being primary magmatic phases.

At Corrycharmaig, Glen Lochay, NW of Killin, Tayside Region, Scotland, a small body of metamorphosed and altered ultramafic rocks contains chromite-rich lithologies in which accessory amounts of galena, gersdorffite, heazlewoodite, nickeline, native silver and an iron–nickel alloy (possibly awaruite) occur. Heazlewoodite is the most abundant and it forms small (< 100 µm) anhedral to subhedral grains that commonly enclose elongate crystals of antigorite. This assemblage is closely associated with relatively rare PGMs, including irarsite, laurite, the native metals ruthenium, iridium and osmium, and unnamed species containing variable amounts of antimony, arsenic, osmium, platinum and ruthenium. The chemistry and association of the PGMs is strongly indicative of an ophiolite assemblage and it is most likely that the intrusion represents a fragment of oceanic crust (Power & Pirrie 2000).

Rare heazlewoodite has been found at Poundland Burn (E of Colmonell), within the Ballantrae ophiolite complex, Strathclyde, Scotland. It occurs with awaruite and pentlandite as small blebs (< 10 µm diameter) in serpentinite associated with nodular chromitites. In the nearby Pinbain Bridge area, chromitite contains abundant chrome spinel with inclusions of erlichmanite, irarsite, laurite, Ir–Os–Ru alloys and possibly kashinite (Power & Pirrie 2004).

heddlite

Named after Matthew Forster Heddle (1828–1897), Scottish mineralogist, physician, and professor of chemistry at the University of St Andrews (Fife) and author of *The mineralogy of Scotland*; born at Melsetter, Hoy, Orkney Isles (Livingstone 1990a). Heddlite is an artificial potassium oxalate that was originally thought to be natural.

hedenbergite

PYROXENE GROUP
monoclinic $CaFe^{2+}Si_2O_6$

A common calcium–iron silicate that forms series with diopside and with johannsenite, and is found in alkali granites, syenites, iron–manganese skarns and metamorphosed iron formations. For many examples given below, identification is optical and confirmatory data are lacking.

England

Cornwall Minute prismatic crystals of hedenbergite occur with actinolite, axinite, calcite, epidote and garnet in thermally metamorphosed calc-silicate rocks surrounding the Bodmin Moor granite at Tregullan quarry, Bodmin (Barrow & Thomas 1908). Within the same aureole, a manganese-rich variety of hedenbergite was later described with bustamite and tephroite in manganese ores at Treburland mine, 2 km

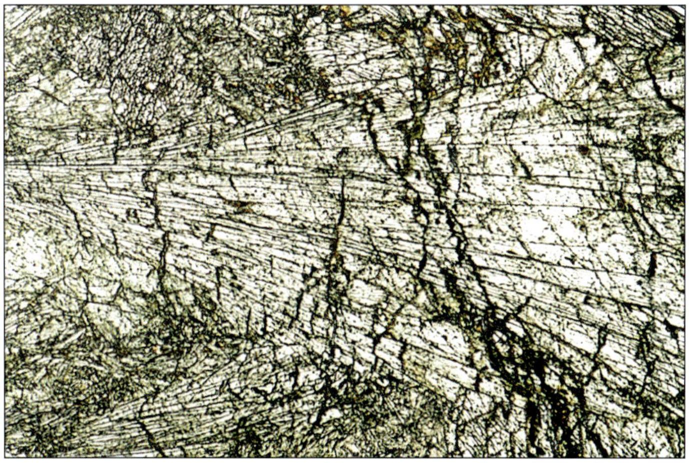

Radiating spray of hedenbergite in a calc-silicate rock from the aureole of the Dartmoor granite at Aish, near South Brent, Devon; width 5.2 mm; viewed in PPL (upper) and XPL (lower). (© K. Gardner)

SE of Fivelanes, Altarnun, Bodmin Moor (Tilley 1946). Metasomatism of basic rocks cutting sphalerite-enriched breccia in the Perran Iron lode is proposed for the origin of garnet–hedenbergite veins at Great Retallack mine, near St Agnes. Hedenbergite is bottle green, with very coarse grain and of radiating to columnar habit (Reid & Scrivenor 1906, Dines 1956, Henley 1971). Skarns at Botallack (Gryll's Bunny and the Crowns), St Just, in the aureole of the Land's End granite, contain a pyroxene zone in which compositions intermediate between diopside and hedenbergite occur (van Marcke de Lummen 1985).

Cumbria The Carrock Fell igneous complex consists of a series of gabbroic rocks with hedenbergite found in both granophyre and coarse pegmatitic gabbro (Eastwood et al. 1968).

Devon Some limestones in the aureole of the Dartmoor granite at Red-a-ven mine, Meldon, near Okehampton have been converted into andradite–ferrobustamite–hedenbergite (iron-wollastonite) skarns containing abundant massive pyrrhotite and other sulphides (El Sharkawi & Dearman 1966).

Wales

Gwynedd Hedenbergite occurs in Precambrian calc-silicate marbles and clinopyroxene–feldspar gneisses of the Coedana Complex, Nebo inlier, Anglesey (Horák 1993).

Scotland

Dumfries & Galloway A pale-green mineral belonging to the diopside–hedenbergite series is reported from calcite veins close to a porphyritic intrusion in a quarry at the top of Heughs, Caulkerbush (Williams 1965).

Grampian At Corrie Loch Kander, near Braemar, hedenbergite occurs with actinolite, diopside, hyalophane, pyrrhotite, scapolite, tremolite and zoisite in schistose bands, within a rock interpreted as a Dalradian metamorphosed exhalative sulphidic dolomitic chert (Gallagher et al. 1989a, Fortey et al. 1991, 1993). The Insch layered gabbro contains hedenbergite in the most chemically evolved Upper Zone cumulate

rocks at locations such as Bridge of Johnston and Hill of Newleslie (Wadsworth 1986).
Highland Lewisian rocks at Druideig Lodge, in the vicinity of Totaig on Loch Duich (between Totaig and Letterfearn), Glenelg area, include iron-rich eulysites (fayalite-rich rocks) and garnet–hedenbergite gneisses. In the latter, green hedenbergite forms porphyroblasts up to 4 mm across (Tilley 1936). Near by, at Gleann Beag, accessory hedenbergite occurs in a manganese-rich schist containing almandine–spessartine garnet, grunerite, magnetite and pyroxmangite (Tilley 1937a). Multiple bands of rudely foliated hornblendic rocks within psammite of Moinian age at Allt Bhlàraidh, Glen Moriston, NW of Fort Augustus, contain pale-green hedenbergite (ferrosalite) associated with albite–oligoclase, clinozoisite, pargasite and quartz (Rock 1984). Dark-green hedenbergite (funkite) is reported with diopside (pale sap-green sahlite and milky malacolite) in contact rocks between limestone and granitic gneiss at Shinness on the eastern shore of Loch Shin (Heddle 1882c). At Rubh' an Tangaird on the Isle of Eigg, hedenbergite (described as ferro-augite), occurs with alkali feldspar and titanomagnetite in a deep-blue glassy pitchstone dyke (Ridley 1971, Preston et al. 1998b). Scattered euhedral phenocrysts of hedenbergite and sodic plagioclase occur in another pitchstone dyke, on the Isle of Raasay near the Isle of Skye. Crystallites of hedenbergite forming comb-like aggregates are also found in the glassy matrix (Preston et al. 1998b). Hedenbergite is abundant in group 1 skarns (at the immediate contact) within the aureole of the Beinn an Dubhaich granite, Broadford area, Isle of Skye (Tilley 1948b, 1951, Smith et al. 2002). It is also reported from the Beinn Dearg Mhòr and Loch Ainort fayalite-bearing epigranites, all in the Western Red Hills, Isle of Skye (Exley 1980), and from the Meall Dearg granophyre (1955). However, Anwar's two analyses now classify as diopside. Anhedral crystals of hedenbergite have been found in the Glamaig epigranite at Eas a' Bradain, near Loch Ainort, between Broadford and the Sligachan hotel, and from the Beinn an Dubhaich granite at Camas Malag, Loch Slapin, both on the Isle of Skye (S. A. Rust specimen, Tindle unpub. EMP data).
Strathclyde Hedenbergite occurs in trachyte rocks within the Tertiary lava succession on the Isle of Mull, to the W of the summit of Ben More (Kerr 1998). It has also been found mantling magnetite and zircon in epigranite from Ben More (S. A. Rust specimen, Tindle unpub. EMP data) and with augite in trachyte from Holy Island, Lamlash Bay, Isle of Arran (Tindle unpub. EMP data). Hedenbergite is present in crystallite form in the Corriegills pitchstone on the Clauchland shore, Lamlash Bay (Preston et al. 1998b, 2002), and is also recorded as deep-green or yellowish-green subhedral or interstitial crystals, varying in composition towards aegirine, in microgranite on Ailsa Craig in the Firth of Clyde (Harding 1983, Harrison et al. 1987).
Western Isles Hedenbergite occurs in crinanite and pegmatites from the Shiant Isles Main Sill, a classic composite alkaline basic sill of Tertiary age (Gibb & Henderson 1996).

Ireland

Antrim Titanaugite from a melilite-rich contact skarn rock at the margins of a dolerite plug at Carneal, Larne (Sabine & Young 1975), now classifies as hedenbergite.
Louth Abundant crystals of andradite, hedenbergite and wollastonite occur in pegmatites contaminated by skarn material in the Carlingford district (Osborne 1932b).

hedleyite

trigonal Bi_7Te_3

A rare bismuth telluride found in greisen from the Carrock mine, Grainsgill, Caldbeck Fells, Cumbria. Hedleyite forms grains up to 200 µm across and is associated with arsenopyrite, native bismuth, bismuthinite, chalcopyrite, joséite-B and molybdenite (Cooper & Stanley 1990).

Naden (1988) reports hedleyite from Clogau mine, Bontddu, near Dolgellau, Gwynedd, Wales, where it occurs as 20 µm inclusions in a galena specimen from the Kingsbury collection at the NHM. Given the doubts about the provenance of other Kingsbury specimens (Ryback et al. 1998, 2001), further confirmation is desirable.

hedyphane

APATITE GROUP

hexagonal $Pb_3Ca_2(AsO_4)_3Cl$

A rare lead–calcium arsenate chloride found in the top west level of Brandy Gill mine, Caldbeck Fells, Cumbria. Hedyphane forms radial to club-shape aggregates of creamy prismatic crystals up to 2 mm long with duftite on quartz (Cooper & Stanley 1990, Ryback & Tandy 1992).

Hedyphane is also reported from Tynagh mine, near Killimor, Co. Galway, Ireland, where it occurs as white-to-greenish tufts with malachite (Ryback & Tandy 1992).

heliotrope

A green variety of chalcedony (quartz) with spots of red jasper.

helminthe

A variety of iron-rich clinochlore (chlorite group) that forms worm-like aggregates.

helvine?

cubic $Mn_4^{2+}Be_3(SiO_4)_3S$

A manganese–beryllium silicate with sulphide, which forms series with danalite and genthelvite. Helvine was supposedly found by A. W. G. Kingsbury at Red-a-ven mine, Meldon, near Okehampton, Devon, and at three localities in Cornwall: Wheal Cock, near Botallack, St Just; Wheal Betsy, Tremore, W of Lanivet; Wheal Mary (Wheal Ding section), between Lanivet and Bodmin (Kingsbury 1961). Ryback et al. (1998, 2001) and other authors have expressed serious doubts about the provenance of many samples in the Kingsbury collection (now at the NHM) and, until further discoveries are made, helvine cannot be considered to be a British or Irish species.

helvite

A widely used synonym of helvine.

HEMATITE GROUP

Trigonal oxides with the general formula A_2O_3, where A = Al, Cr^{3+}, Fe^{3+}, V^{3+}. Corundum, eskolaite and hematite are members of the group found in Britain or Ireland.

hematite

HEMATITE GROUP

trigonal Fe_2O_3

A very common iron oxide, dimorphous with maghemite, found in metamorphosed rocks and in rocks subjected to contact metamorphism. It also occurs as an accessory mineral in felsic igneous rocks, hydrothermal veins and some sandstones. Hematite occurs in various forms, including kidney or fibrous red iron ore (botryoidal or reniform with a radiating and sometimes concentric lamellar structure), martite (pseudomorphous after magnetite), micaceous (red or dark-grey plates), pencil (conical masses derived from broken kidney ore), red (compact and massive), reddle (or smit – a soft variety of hematite), ring (or stalactitic), specularite (well crystallized with iron-black colour and brilliant lustre) and turgite (or hydrohematite).

England

Avon Minor hematite, marcasite, pyrite and tennantite accompany chalcopyrite, galena and sphalerite in vein-style base-metal–barite mineralization within dolomitic limestones close to the beach near the Pier hotel, Clevedon (Ixer et al. 1993). Brooke (1994b) lists hematite from Hampstead Farm quarry, Chipping Sodbury.
Cheshire Sediments of Triassic age on Alderley Edge ridge, contain detrital hematite (Ixer & Vaughan 1982).
Cleveland Hematite is widespread as a minor constituent of the Permian Boulby Potash and has been found in borate nodules and in lenses of anhydrite and halite at Boulby potash mine, Loftus (Woods 1979, Green & Freier 1996).
Cornwall Hematite has been found at Botallack sett, a series of mines in the aureole of the Land's End granite near St Just, and including Botallack mine, Carnyorth mine, Parknoweth mine, Stamps and Jowl Zawn, Wheal Cock, Wheal Edward and

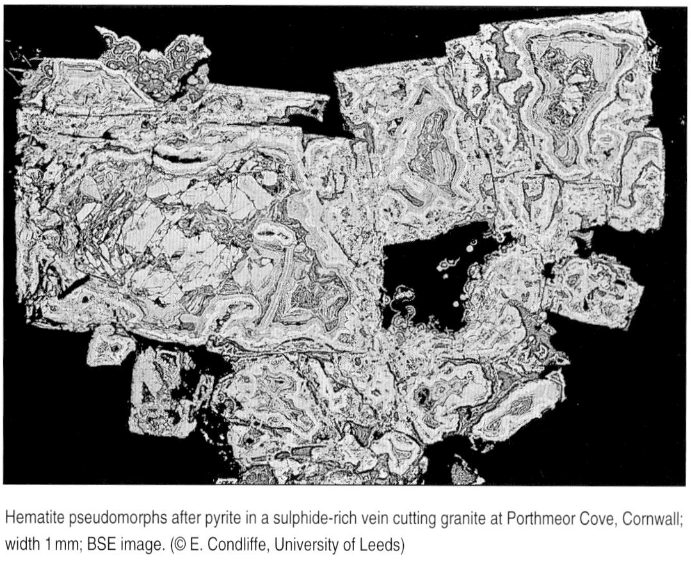

Hematite pseudomorphs after pyrite in a sulphide-rich vein cutting granite at Porthmeor Cove, Cornwall; width 1 mm; BSE image. (© E. Condliffe, University of Leeds)

Wheal Owles. In addition to specular hematite and kidney ore, fine specimens of hematite forming nailhead crystals have been found at Botallack mine (Greg & Lettsom 1858, Rudler 1905, Couper & Barstow 1977, Green 1984, Wolloxall 1989, Bancroft & Weller 1993). Hematite was an important ore species in iron lodes in and around the St Austell granite and extending into the Wadebridge and St Agnes districts. It occurs in Devonian slates and within the granite near the centre of St Austell Moor, as at Ruby, Knightor and Trebyn mines. Restormel iron mine, N of Lostwithiel, was a leading producer among several mines, which included Westdowns, Retire Nanstallon, Ruthvoes, Indian Queens and Treffry Consols (Rudler 1905, Slater & Highley 1976). Other locations include: Cligga Head, Perranzabuloe (Jackson et al. 1977); massive and micaceous at various Carn Brea mines, near Redruth (Greg & Lettsom 1858, Robson 1945a); botryoidal masses at Countybridge quarry, Goonhilly Downs, Lizard Peninsula (Holyer 1996); in veins cutting peridotite at Eastern Cliff, Kennack Sands, Lizard Peninsula (Bevins et al. 1987); earthy hematite at Geevor mine, near Pendeen, St Just; in quartz veins containing hematite, marcasite, pyrite and uraninite (pitchblende) at South Terras mine, St Stephen-in-Brannel; in pitchblende–sulphide veins at Trannack mine, Helston; hematite intimately associated with bornite chalcopyrite, covellite, pitchblende and smoky quartz at Wheal Bray, Altarnun; in a selenium-bearing pitchblende–sulphide lode at Wheal James, Withiel (all Ball et al. 1982); red hematite associated with amethystine quartz in veins at Goonbarrow china-clay pit, near Bugle, St Austell (Bray & Spooner 1983); Great Retallack mine, near St Agnes (Dewey 1925); Greystone quarry, Lezant, near Launceston (Weiss et al. 2001); specular hematite associated with cassiterite and pyrite at Halabezack farm, near Porkellis, and specular hematite at Wheal Roots (part of Wendron Consols mine), both near Wendron (Dominy et al. 1996); an earthy variety at Davidstow, near Camelford; reniform masses at Ladock iron mine, near Grampound; Tincroft mine, Illogan; Wheal Beauchamp, near Redruth (all Garby 1848); Levant mine, St Just (Garby 1848, Wolloxall 1987); Pawton mine, near Wadebridge (Collins 1911a); Shallow Water mine on Shallow Water Down, Bodmin Moor, and East Pool mine, Illogan (Golley & Williams 1995); specular hematite lines drusy cavities and occurs in massive earthy form at South Crofty mine, Camborne (Scrivener et al. 1986, Dominy et al. 1994, Dominy & Camm 1996, LeBoutillier 1996); minute plates and crystalline masses in an axinite–pyroxene skarn rock at Tregullan quarry, Bodmin Moor (Barrow & Thomas 1908); and Tresavean mine, Gwennap, near Redruth (Robson 1949b). Many of these mines and others are also documented by Dines (1956).

Cumbria Hematite was once of great commercial importance in the S and W of Cumbria, surrounding the domal structure of the Lake District. Mining is known to have occurred in Roman times, but peak production was between 1870 and 1890, with nearly 102 million tonnes of ore being produced in west Cumbria between 1855 and 1974. These deposits are found in fractured Carboniferous Limestone – in solution hollows and pockets (sops), as irregular or tabular masses (flats), or as steeply dipping lodes. Arguments have been proposed to suggest that the ores are Cretaceous or Tertiary in age, being formed by downward percolation of warm solutions from Triassic sandstones resting on a former land surface (Slater & Highley 1976, Rose & Dunham 1977, 1984, Shepherd & Goldring 1993). However, palaeomagnetic dating suggests a Permian age for the mineralization (Evans & El-Nikhely 1982). The reality may be even more complex, as fracture mineralization examined in borehole cores from the Sellafield area indicates nine episodes of mineralization, some very characteristic of the hematite–(dolomite–fluorite) mineralization of the West Cumbrian orefield (Milodowski et al. 1998).

Botryoidal hematite from Florence mine, Egremont, Cumbria; width of specimen 100 mm. (© National Museum of Wales, Cardiff; photographer M. P. Cooper)

Large replacement orebodies, known as sops, occupy solution hollows in Lower Carboniferous limestones and occur only in the Furness area (South Cumbrian orefield). The largest (Park sop, near Askham-in-Furness) was 420 m long, 20–180 m wide and up to 180 m deep. In the Askam–Park–Roanhead district, other sops include California (three sops), Nigel, North, Paddy, Rita, Sandscale, Sandscale–Peggy and Woodhead; whereas, in the Dalton–High Crossgates district, many more (but smaller) sops were mined, including Crossgates, Elliscales, Eure Green, Haume and Mouzell (Wurzburger 1872, Rose & Dunham 1977). Additional deposits were worked from flats or veins in south and west Cumbria in the areas of Cleator Moor, Egremont, Frizington, Furness, Lamplugh and Millom. Details of individual mines can be found in Smith (1924), Hibbert et al. (1940) and Rose & Dunham (1977). A select few occurrences with mineralogical interest follow: small lenses of chalcocite–chalcopyrite are enclosed in hematite at Anty Cross, near Dalton-in-Furness (Rose & Dunham 1977); massive, compact, bluish ore; botryoidal ore (including kidney and pencil varieties); smit, stalactitic and well crystallized specular hematite in a gangue notable for its fluorite content at Beckermet mine, near Egremont (Davidson & Thomson 1951, Goldring & Greenwood 1977, 1990); good specimens of kidney and pencil ore at many mines around Bigrigg and Cleator Moor (Smith 1924); lenses of barite, chalcopyrite and pyrite are associated with hematite at Crowgarth mine, Cleator Moor (Trotter 1944, 1945); fine specimens of kidney ore encrusted with specular hematite, dolomite and quartz are a feature of Florence mine, near Egremont; in one cavity, masses of kidney ore over 2 m wide were found (Smith 1924, Davidson & Thomson 1951); Florence mine also produced very attractive specimens of specular hematite on which unusual spheroidal calcite has grown and, exceptionally, rosettes of lustrous cm-size jet-black specular hematite were found on cabinet-size specimens – the finest examples of the species in Britain (Green & Briscoe 2002); kidney, specular and stalactitic ore at Gillfoot Park and Helder mines, Egremont (Smith 1924); Hodbarrow mine, Millom, near Haverigg, produced large specimens of kidney and pencil ore, highly prized for hematite jewellery, also hematite pseudomorphous after fluorite or pyrite (Smith 1924, Green & Briscoe 2002); layers of kidney ore with compact ore between Kelton Fell, Kirkland and Knockmurton mines, Kirkland (Smith 1924); dark-red kidney ore, pencil ore and fossils partly or wholly converted to hematite at Lindal Moor mine, Lindal-in-Furness (Schneider 1884, Smith 1924, McFadzean 1988, 1989); kidney ore at Margaret mine, Frizington; North Newton and North Stank mines, Dalton-on-Furness; Park mines, Askam-in-Furness; Postlethwaite's Moor Row mine, Moor Row (all Smith 1924); pseudomorphing fluorite at Stank mine, Barrow-in-Furness (Miers 1897, Young 1987a); hematite pseudomorphing barite is recorded from Ulverston (Rudler 1905); and specularite coats barite at Winscales mine (Trotter 1945).

Hematite also occurs in other locations apart from the main South and West Cumbrian orefields, including: mines working veins in Ennerdale and Eskdale, especially Birker Moor, Blea Tarn and Nab Gill mines (all near Boot), and the Brant Rake mines, S of Eskdale Green, all in Eskdale. Particularly fine kidney ore has been recovered from Nab Gill mine (Young 1984b, 1985e, 1987a), and specular hematite occurs at South Cumberland mine and in a greisen to the W of Devoke Water (Young 1985e), all in Eskdale; in the Caldbeck Fells, where hematite is noted from Carrock Fell crosscut and the mines of Brandy Gill, Carrock, Potts Gill and Sandbed (all Hartley 1984), also Roughton Gill mine (Davidson & Thomson 1951); dark maroon-red earthy films at Burdell Gill (Neall & Leppington 2003); kidney ore at Ewer Gap mine, Red Tarn mine on Wrynose Fell and Tongue Gill mine, Grasmere (Davidson & Thomson 1951); minor hematite associated with calcite and chalcopyrite in epidote–quartz veins at Fleetwith Pike, Buttermere Fells (Strens 1964); as authigenic coatings on detrital specularite in the St Bees Sandstone (e.g. at Fleswick Bay, St Bees Head), also pseudomorphing cubic and framboidal pyrite (Turner & Ixer 1977, Ixer et al. 1979b); curved platy crystals of hematite associated with pyrite in quartz veins in Gatesgarthdale Beck, W of the summit of Honister Pass, Wasdale, and at the foot of Water Crag, Wasdale (Millward et al. 1999); hematite forms the cores of graphite nodules at Graphite mine, Seathwaite (Strens 1965); specularite in altered sandy mudstone of the Permian Eden Shales adjacent to the Armathwaite tholeiite dyke (Arthurton & Wadge 1981); intimately associated with magnetite in the Long Crag vein at Long Crag mine, Greenburn Valley, Little Langdale (Stanley 1979); pseudomorphing galena at Longlands Fell, Uldale (Young 1987a); microcrystalline and botryoidal varieties in copper sulphide-bearing quartz veins at Seathwaite mine, Seathwaite Tarn, Coniston (Stanley & Criddle 1979); massive in fault breccia, shatter belts and chlorite–quartz veins at Shap Blue quarry in the aureole of the Shap granite (Firman 1953a,b, Young et al. 1988, 1998); in late-stage barite–calcite–fluorite–quartz veins cutting the Shap granite (Firman 1978, Loughlin 1999).

Derbyshire Hematite is widespread in the Peak District as an earthy granular material or as a red stain on the surface of other minerals. Red earthy hematite was mined near Hartington railway station from calcite veins, and small quantities were also worked in Hand Dale, N of Hartington and S of Newhaven House at Friden. Other trials were in Narrow Dale and near Winster (Cantrill et al. 1919). Hematite has been found in fault breccia at Bincliff, near Wetton, and in Alsop Moor limeworks quarry. An earthy form of hematite, associated with manganese-rich wad, occurred in lead mines around Elton and in silica sand pockets around Brassington and Doglow Wood, Carsington. Red ochrous hematite has been found at Glory mine, Crich. Specular hematite, forming small tabular crystals associated with calcite and quartz, occurs in veinlets cutting basalt at Calton Hill quarry, Taddington (Ford et al. 1993). Other occurrences include: in Ashover tuff (Mello 1876); in limestone at Chelmorton (Sarjeant 1957a); Masson Hill and Bonsall Moor (Sarjeant 1956b); and at Rutland mine, Matlock (Workman 1960).

Devon From at least 1839 to 1969, specular hematite was mined in the central and northeast part of the Dartmoor granite. Birch Tor and Vitifer mine in the centre of Dartmoor was one of several tin mines that worked cassiterite intimately associated with specular hematite (Dines 1956, Dunham et al. 1978). Unfortunately, this by-product was commercially useless, as it was too hard to grind as a pigment and not suitable for smelting (MacAlister 1909). In the Hennock and Lustleigh areas, hematite has much finer grain (micaceous hematite) and was mined extensively for use in the paint and other industries. By the end of the 19th century, several thousand tonnes had been raised from Hawksmoor, Kelly, Plumley, Shaptor and Shuttamor mines. The Great Rock mine at Hennock, near Chudleigh, produced extremely fine-grain ore from very irregular lodes between 1902 and 1969. The ore is considerably brecciated and it contains kaolinized feldspar, altered biotite, tourmaline, occasional cassiterite and some pyrite (Edge 1920, Henson 1956, Slater & Highley 1976, Durrance & Laming 1982, Nicholas & Scrivener 2000, Brooks 2004). Further occurrences include: in copper-sulphide-bearing quartz–siderite veins at Bampfylde (Poltimore) mine, North Molton (Dunham et al. 1978, Camm 1995); Bideford; Bratton Fleming, Shirwell, East Down; Buckfastleigh; Buckland Brewer, near Bideford; Combe Martin; Georgeham and Viveham, near Barnstaple; Ilfracombe; Lynton; North Molton; West Down; Wheal Forest, Meldon Gorge, near Okehampton (all Hall 1868b); Cornham Ford, Exmoor (Durrance & Laming 1982); micaceous hematite at Great Weeks Consols, Chagford, Dartmoor (Broughton 1967); Hope's Nose, Torquay (Stanley et al. 1990a); Merrivale quarry, Dartmoor (Ball et al. 1982); pseudomorphs of hematite after pyrite near the Triassic/Devonian unconformity at Saltern Cove, Torquay (Solly 1889); and unusually large crystals of specular hematite up to 7 mm across associated with cassiterite at Waterhill mine, W of Birch Tor (MacAlister 1909, Dines 1956).

Durham Hematite occurs in minor veins cutting the Weardale granite sampled via the Rookhope borehole (Bradshaw 1967) and as minute anhedral flakes in Permian carbonate and evaporite rocks from boreholes around Stainton, near Darlington (Raymond 1962).

Gloucestershire Small amounts of kidney ore were found in purplish-red mineralized sandstones at Hillend Green mine, 3 km NW of Newent (Bick 1986a).

Leicestershire At Newhurst quarry, Shepshed, where a Precambrian/Triassic unconformity is exposed, hematite occurs: as specularite associated with dolomite; as reniform hematite associated with barite and dolomite; as dark reddish-brown coatings on joint faces; and as radiating groups of plates (eisenrosen) up to 4 mm across in steeply dipping veins. Hematite pseudomorphs (probably after magnetite) are also found in Triassic breccias from the quarry (King 1973, Ince 2005). Secondary finely divided hematite is widespread in the late Precambrian porphyritic dacite at Springhill quarry, Peldar Tor, Whitwick (Faithfull & Ince 1992).

Somerset Mining for iron and manganese was carried out in the Mendip Hills, possibly as early as Roman times. Hematite was the main iron ore species at Higher Pitts mine, Priddy, Mendip Hills, and this was mined with pyrolusite (wad) from a localized replacement deposit in Triassic Dolomitic Conglomerate, probably before 1752. Limonite and psilomelane are lesser constituents of the ore, together with the species crednerite, goethite and manganite (Kingsbury 1941, Evans & Evans 1977, Burr 1996). Other locations for hematite in the Mendip Hills include: Lamb's Lair, SSW of West Harptree (Evans & Evans 1977); Mells quarry, near Frome (Bradshaw & Phillips 1967); Merehead (Torr Works) quarry, Cranmore (Welch et al. 2000); and Moons Hill quarry, N of Shepton Mallet (van de Kamp 1969).

Staffordshire Hematite is noted in fault breccia in and near Cauldon Low quarry, Waterhouses (Ford et al. 1993).

Warwickshire Earthy-to-massive dark reddish-brown hematite has been collected from the north face of Judkins quarry, Nuneaton (Ince et al. 1991).

Yorkshire Permian evaporite deposits forming part of the Yorkshire potash field have been examined in boreholes near Whitby at Aislaby and Sleights, and shown to contain dolomite, hematite, magnesite, pyrite, quartz, native sulphur and talc, in addition to a range of evaporite species. Hematite is described as a widespread minor constituent, but a major constituent in certain beds. It forms inclusions in much of the sylvite and less commonly in boracite, halite and quartz (Lees & Taitt 1946, Armstrong et al. 1951, Stewart 1951a,b, 1965).

Wales

Clwyd Hematite is associated with cobalt and nickel mineralization in Carboniferous Limestone at Moel Hiraddug, Cwm, near Dyserth, and also occurs at the nearby Cwm iron mine, near Tyddyn-y-cyll, at Henfryn mine, near Dyserth, and at Bodfari mine, NE of Denbigh (Le Neve Foster 1882, Strahan 1885, Warren et al. 1984). Dunham et al. (1978) reports a replacement deposit in limestone at Nant Uchaf mine, SE of Abergele, with hematite at the bottom and manganese carbonate above.

Gwynedd Hematite is a constituent of magnetite–pyrite–quartz veins cutting the Ordovician Snowdon Volcanic Group at Cwm Tregalan and Shadow Gully, to the S of Snowdon. At Cwm Tregalan, hematite is the most abundant vein mineral (swirling bands associated with magnetite) and also forms thin euhedral acicular crystals up to 1 cm long in vesicles within the host pillow basalts (Reedman et al. 1985, Colman & Appleby 1991). Elsewhere in Snowdonia, large hematite crystals up to 30 mm across have been found in chlorite–quartz veins cutting slates at Penrhyn quarry, near Bethesda (Bevins 1994). Other occurrences of hematite include: earthy hematite in a marcasite-rich carbonate vein at Gwynfynydd mine, near Dolgellau (Dominy et al. 1996c); minor hematite in Ordovician oolitic iron ores at Llandegai, near Bangor (Pulfrey 1933); hematite is important in manganese ores at Mynydd Rhiw and Nant-y-Gadwen on the Lleyn Peninsula (Woodland 1939); Parys Mountain, Anglesey (Southwood & Bevins 1995); and in marcasite–quartz veins unrelated to the gold-bearing veins at Prince Edward mine, Dolgellau goldbelt (Mason & Williamson 2001).

Mid Glamorgan Important limestone-hosted goethite–hematite ore deposits occur across the southern edge of the South Wales coalfield near Llantrisant at Llanharry, Mwyndy, Bute and Trecastell mines; near Pentyrch at Garth mine; Fforest Fawr on the opposite side of the River Taff; and Rudry mine, near Caerphilly. The deposits are believed to represent both replacements of the limestone and cavity infillings (Vivian 1885, North 1916, Dunham et al. 1978, Rankin & Criddle 1985). Hematite, associated with botryoidal goethite, is a minor constituent of cavity infill and replacement manganese ore at Tŷ Coch, near Porthcawl (Criddle & Symes 1977).

South Glamorgan Hematite replaces crinoids and other fossils in limestone at Rhiwbina (Strahan 1899).

Scotland

Borders A ferromanganese sediment of exhalative origin at Noblehouse, near West Linton consists of hematite associated with birnessite, goethite and todorokite (Nicholson 1989a).

Central Heddle (1901) reports specular hematite associated with columnar and corded pyrite in clay slate at Birnam Hill quarries, near Dunkeld.

Dumfries & Galloway Beautiful samples of black and red botryoidal hematite, sometimes banded with yellow limonite and a little specularite, have been found on the dumps of Auchenleck hematite mine, Auchencairn (Williams 1965). Both botryoidal and specular hematite are reported from Auchenlosh quarry, near Dalbeattie (Green & Todd 2001). It is unclear if these two occurrences are uranium-rich, like the hematite–quartz veins described from Beeswing, Needle's Eye, Sandyhills, Potterland and Craiglebbock Rocks, near New Abbey (Williams 1965, Miller & Taylor 1966, Gallagher et al. 1971), but all are marginal to the Criffell granite complex. Other occurrences of hematite include: thin red-to-black bladed crystals associated with andradite, epidote and chabazite in a basaltic dyke at Coatsgate quarry, near Moffat (Green & Todd 2001); in stratiform and disseminated arsenopyrite–pyrite mineralization examined in boreholes around Glendinning (Louisa) mine, 13 km NNW of Langholm (Gallagher et al. 1983); associated with colloform jasper, chalcopyrite and pyrite in the Ochars vein cutting the Cairnsmore of Fleet granite, SW of New Galloway (Lowry et al. 1997); in botryoidal masses in the oxidized tops of veins (e.g. Belton Grain vein), Wanlockhead (Wilson & Flett 1921); magnetite on the mine dumps at Whyte's Cleuch, Wanlockhead, contains lamellae of hematite (Nicholson & Banks 1987).

Fife Small rosettes of black scaly hematite up to 1 mm across encrust quartz in veins at Orrock quarry, near Burntisland (Wirth 2003).

Grampian Hematite occurs with chalcophanite, lithiophorite, pyrolusite, ramsdellite and woodruffite as a minor constituent of a cryptomelane-goethite orebody in deeply weathered Dalradian metasediments at Lecht mine, near Tomintoul (Wilson et al. 1970, Nicholson 1987). Dalradian quartzite hosts brecciated goethite–hematite veins at Arndilly house, Arndilly (Nicholson 1986). Vuggy vein quartz, hematite and psilomelane occur along mineralized faults at Quartz Cliff, within the Kincardine (Mount Battock) granite, NW of Mount Keen (Goodman et al. 1990). Heddle (1901) reports twisted blackish-brown plates of hematite in a feldspar quarry at Pitfichie Hill, near Monymusk.

Highland Hematite is a constituent of minor chalcocite ore found in small nodules within Torridonian rocks in the bay at Clachtoll, 7 km NW of Lochinver (Fermor 1951). Irregular pockets of hematite and manganese oxides occur in a sub-lacustrine hot-spring deposit near the unconformable base of the Old Red Sandstone at Dalroy, near Culloden (Berridge 1969, Nicholson 1990). Hematite is also noted: as specularite in a giant erratic boulder at Beinn Bhreac, 2 km E of Tongue (Heddle 1883); in calcite–fluorite–galena–quartz veins at the contact of the Grudie granite with Moine Schists in Allt Bad na h'Achaise, 5 km SW of Lairg (Gallagher 1970, Gallagher et al. 1971); in uranium-rich veinlets associated with alteration, fracturing and shearing in the Helmsdale granite (Tweedie 1979); in weathered Tertiary basalt at Killundine, Morvern Peninsula (Bain et al. 1980); within mineralized Lewisian Gneiss breccia at Loch an Arbhair, Lochinver–Ullapool road, SW of the River Kirkaig, where it is either earthy and amorphous or it forms mammillated surfaces in cavities and fissures (Waterston 1953). Heddle (1879d, 1901) reports hematite with biotite (haughtonite) and orthoclase in gneiss on the western slope of Foinaven (Foinne Bhein).

Lothian Pseudomorphs after olivine in Markle basalt from Holyrood Park, Edinburgh consist of an intimate mixture of chlorite and hematite (Smith 1959). Heddle (1901) reports flat nailhead crystals of hematite with quartz in cavities in goethite from Salisbury Crags, Edinburgh.

Strathclyde Hematite is one of many secondary minerals found in the Carboniferous Clyde Plateau Lavas. It forms minute micaceous flakes as a coating covering analcime and prehnite, and forms dull metallic greyish-brown botryoids or granular-to-earthy coatings in cavities in tuffaceous rocks at Boyleston quarry, SW of Barrhead (Hall et al. 1989, Meikle 1989a). At Hartfield Moss, 9 km WSW of Barrhead, hematite forms aggregates of tiny plates in vesicles in the lavas or brilliant golden-red rosettes on analcime in larger cavities (Meikle 1990). Unusual golden hematite rosettes up to 3 mm across, associated with andradite and grossular, have been found at Loanhead quarry, near Beith, where a Tertiary dyke cuts the lavas (Meikle 1989b, Green & Todd 2001). A similar secondary suite is also found near another dyke at Reilly quarry, Houston, although the hematite is earthy or of the specular variety (Wolfe 1997, Meikle 2003). Other reports of hematite include: well developed as finely divided grains disseminated through barite in veins at Pockmuir Burn, Auchinstilloch, SW of Lesmahagow (Evans & El-Nikhely 1982); dark-red platy crystals and crystal rosettes up to 0.5 mm across on quartz at Castletown mine, Lochgilphead (Starkey & Hubbard 1998, Hubbard & Green 2001); coating joint planes and fractures in trachyte at Craigmuschat quarry, Gourock (Nicholson & Durant 1986, Nicholson 1988c); fine botryoidal hematite and specularite in barite veins cutting Silurian and Devonian rocks on Mannoch Hill, SW of Lesmahagow (Todd & McMullen 1991; see also Evans & El-Nikhely 1982); and in iron–manganese-rich veins within the Dalradian Jura Quartzite at Dùn Athad, Mull of Oa peninsula, Isle of Islay (Nicholson 1987b, 1988a). Heddle (1901) reports reddish-brown striated cubic crystals of hematite (turgite) in phyllite on the Isle of Kerrera, which are interpreted to be pseudomorphs after pyrite.

Tayside Hematite occurs infrequently as reddish-brown botryoidal layers in some geodes within an andesite lava flow (well known for its agates) at Ardownie quarry, Monifieth (Ingram 1994).

Orkney Scaly patches of earthy hematite occur in the Old Red Sandstone at Walls, near Tor Ness, and, near by on the eastern shore of Aith Hope, hollow stalactites have been found. Granular hematite occurs in veins at the Manse, north Hoy (Heddle 1901).

Shetland Syenite at Bakkasetter quarry, southeastern flank of Noss Hill, South Mainland, contains hematite–quartz veins and epidote–hematite encrustations along joint planes (Stephen 1954). Heddle (1901) reports specular hematite in quartz veins cutting mica schist at Vanlup, Hillswick, Mainland.

Ireland

Antrim Hematite occurs in thermally metamorphosed cordierite-bearing assemblages close to an olivine dolerite plug cutting Tertiary basalts at Tievebulliagh, near Cushendall (Agrell & Langley 1958).

Cavan The Claragh deposit, 4 km SE of Redhills, is a stratabound lenticular orebody containing films, layers and veinlets of hematite (locally specular), limonite and manganese oxides (Morris 1984).

Cork Micaceous hematite and occasional specular crystals in small cavities within quartz have been found on the dumps of Ballycummisk mine, near Rossbrin Cove, where emplacement of early barite–hematite–quartz veins was followed by copper-polymetallic sulphide mineralization (O'Brien 1959, Bartlett 1980, Ryback & Moreton 1991, Ni Wen et al. 1999, Kinnaird et al. 2002). Letter and Skeagh are nearby mines where barite–hematite veins have also been found (Ni Wen et al. 1999). Micaceous hematite is a minor constituent of the copper-ore assemblage at the Berehaven (Bearhaven) mines, Allihies (Moreton & Aspen 1993). Much hematite and manganese ore is noted from Aghaturbrid mine, Glandore district (Cole 1922).

Down In the Mourne Mountains, tiny scales of hematite line drusy cavities in granite, ENE of the summit of Slieve Muck (Richey 1927), and, at Silent Valley quarry near Newcastle, hematite (some specular) is found with limonite, manganite, psilomelane and pyrolusite in veins cutting granite (Nockolds & Richey 1939). Hematite also occurs in altered Silurian slate near a granite margin at Deehommed mine, W of Bing's End (Cole 1922).

Fermanagh Hematite and copper ore were worked from deposits in Carboniferous limestone at Castle Caldwell, Rossbeg and Magheramenagh mine, E of Belleek (Cole 1922).

Galway Hematite and chert are the most abundant ironstone minerals in the Carboniferous Tynagh Ironstone Formation, near Killimor. The original sediment is thought to have been composed of colloidal ferric oxide and silica, accompanied by carbonate mud that may have been precipitated by algae (Schultz 1966a). Russell (1975) suggests a sedimentary-exhalative origin for the deposit.

Kerry Hexagonal plates of hematite are reported with prolific crystals of quartz in ribbon-like tension gashes within Devonian sandstone at Smerwick Harbour, near Slea Head, Dingle Peninsula (Wolfe 2006).

Hexagonal plate of black hematite, 5 mm across, marginally altered and associated with green clinochlore and colourless quartz, Ardnamore, Smerwick harbour, Dingle, Co. Kerry, Ireland. (© D. I. Green)

Mayo Minor hematite occurs in gold-bearing quartz veins from the Lecanvey prospect, Clew Bay (Aherne et al. 1992). A pyrite-bearing vein in Precambrian quartzite containing hematite (of extraordinary richness) is described from Carrignahelty, Bolinglanna (Cole 1922).

Meath Chalcopyrite veinlets containing minor barite and hematite (specularite) occur in a syenite intrusion near the Navan lead–zinc orebody (Andrew & Ashton 1985).

Monaghan Bluish-black to red hematite occurs within shales in a small quarry on Logwood Hill, Calliagh, 8 km SW of Monaghan. Pyrolusite and rhodochrosite are associated species (Morris 1984, Ryback & Moreton 1993).

Sligo Hematite occurs in Carboniferous rocks at Ballynakill (Cole 1922).

Tipperary The Ballynoe deposit, 2 km W of Silvermines village, is a stratabound body of barite containing variable amounts of pyrite and minor galena, hematite and sphalerite (Mullane & Kinnaird 1998). The nearby Magcobar barite deposit similarly contains hematite (Taylor & Andrew 1978).

Waterford Hematite and goethite occur in the joints of shaly beds of white Devonian sandstone at Drumslig, between Youghal and Dungarvon. The deposit was apparently worked in 1600 by Sir Walter Raleigh, who was clearly involved in activities other than his well known expeditionary and colonial exploits (Cole 1922, Halliday & Mitchell 1983).

Wicklow Both primary and secondary hematite (after magnetite) are present in BIF rocks at Ballycoog, Goldmine River valley, near Avoca. The primary species, in order of abundance, are magnetite, chalcopyrite, pyrite, hematite, sphalerite, galena with native gold appearing in a later hydrothermal event (Ixer et al. 1990).

Offshore territories

Isle of Man Hematite was worked from veins cutting slaty rocks in the neighbourhood of Manghold Head (Rudler 1905).

Jersey Sulphide-bearing carbonate veins cutting granite at Le Doüet de la Mer, St Ouën and L'Ouaisné Bay in the SW, contain minor hematite as small (5–10 µm) laths or basal crystals, or aggregates up to 300 µm across (Ixer 1980, 1992). Magnetite in quartz veins cutting granite near La Corbière Point, southwest Jersey, shows extreme alteration to hematite (Ixer 1980). Other reports of hematite are from: bornite–cassiterite-bearing aplogranite at La Moye (Ixer 1988); in veins cutting rhyolite at Archirondel and La Crête; in granite near Noirmont and Île Percée; in andesite at Le Mont à l'Abbé; in a quartz vein at L'Oeillière; in a fluorite vein S of L'Ouaisné Bay; and as inclusions in quartz at Mont Mado and Mont Orgueil (Mourant 1961).

hemimorphite

orthorhombic $Zn_4Si_2O_7(OH)_2 \cdot H_2O$

A common hydrated zinc silicate hydroxide found as a secondary mineral in the supergene zone of zinc-bearing orebodies, wherever sphalerite has been oxidized in the presence of siliceous solutions. Paler varieties of hemimorphite can be confused with smithsonite. The name "calamine" has been applied to both species.

England

Cheshire Hemimorphite is a very rare species on Alderley Edge ridge, being noted as small radiating sheaves of blades at Engine Vein mine and as colourless sheaves associated with aurichalcite at Wood mine (Braithwaite 1994).

Cornwall Colourless, white or pale-blue fan-shape radiating aggregates of hemimorphite up to 2 mm across occur in gossanous quartz veinstone at Greystone quarry, Lezant, near Launceston. Aurichalcite, dundasite, malachite, sphalerite and wulfenite are associated species (Lloyd & Barstow 1983, Grigg 1996, Elton et al. 2004). Hemimorphite is also noted from Fowey Consols mine, Tywardreath, St Blazey, and Wheal Herland, Gwinear (Golley & Williams 1995).

Cumbria *Caldbeck Fells:* Roughton Gill is a classic locality for hemimorphite, where it was once found in large quantities cementing vein breccia. It forms beautiful pale- to sky-blue crusts up to 30 cm across on 19th-century specimens, and more recently it has been found *in situ* in an outcrop of the South lode on Balliway Rigg in higher Roughton Gill. Associated species on the dumps here include chrysocolla, hydrocerussite, malachite and pyromorphite (Greg & Lettsom 1858, Goodchild 1884, Davidson & Thomson 1951, Cooper & Stanley 1990, Green & Briscoe 2002). Hemimorphite is also recorded from Brae Fell mine (Cooper & Stanley 1990, Bridges et al. 2006), Potts Gill mine (Neall & Leppington 1994), Thief Gills (Davidson & Thomson 1951) and probably from Red Gill (Sweet 1960a). Other occurrences noted by Hartley (1984) from Driggith, Ingray Gill, Sandbed and Silver Gill mines require further confirmation, as they may be based on samples incorrectly provenanced by A. W. G. Kingsbury (see Ryback et al. 1998, 2001). Hemimorphite is noted from Sandbed mine by Young (1987a) on the basis of a specimen in the Kingsbury collection now at the NHM. Widespread but inconspicuous hemimorphite does occur at Driggith and Sandbed mines, and many Kingsbury specimens from these locations appear to be genuine; however, one specimen has more in common with material from Greenside mine, Patterdale (Green et al. 2006a). Genuine hemimorphite also occurs at Silver Gill, as white to pale-blue crusts in cavities within partly oxidized veinstone, and as radiating fans of colourless euhedral bladed crystals, up to 0.8 mm long, often associated with aurichalcite (Green et al. 2005b). *Elsewhere in the Lake District:* Specimens of bright-blue mammillated hemimorphite, comparable in quality with those from Roughton Gill, were once found on the dumps near the Glencoyne adit of Greenside mine, Glenridding, Patterdale (Davidson & Thomson 1951). Hemimorphite associated with wulfenite is also recorded from two other Patterdale localities: Eagle Crag mine, E of Grisdale Beck, and Myers Head mine, near Hartsop (Rothwell & Mason 1992, Green & Briscoe 2002). Supergene alteration of chalcopyrite-rich veins at Rigg Head quarry, Borrowdale, has produced an assemblage of white-to-brown crystalline hemimorphite associated with caledonite, cerussite, chrysocolla, leadhillite and linarite (Green et al. 2006c). *Northern Pennine orefield:* Hemimorphite is locally associated with much more common smithsonite and it forms colourless-to-brown radiating crystal aggregates at: Bayle Hill mine, Alston; Nenthead Field and Nenthead mines, Nenthead; and Swathbeck mine, Great Dun Fell (Dunham 1990). Hemimorphite is tentatively identified from Hilton mine, Scordale, near Appleby (Winrow 1986a).

Hemimorphite from Roughton Gill, Caldbeck Fells, Cumbria; width of specimen 90 mm. (© National Museum of Wales, Cardiff)

Derbyshire Hemimorphite forms small lustrous-to-translucent brownish crystals at many locations in Derbyshire (Ford et al. 1993). It is well known from Golconda mine, Brassington (Ford & King 1965), and it forms radiating groups of colourless to light-brown blades and dark-brown prisms at Wapping mine, Cromford (Ince 1991). It is also reported from: Bage mine, Bole Hill, Wirksworth (Jones 1982, Burr 1992); Ball Eye mine, Bonsall; Devonshire Cavern, Matlock, and Royal Cavern, Matlock Bath (all Ford et al. 1993); Black Hole mine, Eyam; Crimbo Engine mine, Monyash; Ladywash mine, Eyam; Masson openworks, Matlock; Shining Cloud mine, Hopton (all Ford & Sarjeant 1964a); Chance mine, Hopton (Carruthers & Strahan 1923); Cumberland and Rutland mines, Matlock Bath (Adam 1846, Ford et al. 1993); Millclose mine, Wensley, near Matlock (Sarjeant 1957); Milltown quarry, near Ashover (Rust 1991c); Pike Lane opencut, near Winster (Uttley 1989); and Sacheveral farm, Winster (Rothwell & Mason 1992).

Durham Hemimorphite occurs as colourless-to-white radiating crystals encrusting fluorite at West End Hushes, Pike Law mines, Newbiggin Common, Teesdale, and as pale-brown crystalline masses on the dumps of Grasshill mine along Harwood Beck, Teesdale (Bridges & Young 1998). It is also noted from Ashgillhead mine, Upper Teesdale, and Foster's Hush, Lunedale (Young et al. 1985a, Dunham 1990, Bridges & Young 1998).

Leicestershire Small radiating aggregates of colourless-to-white blades up to 1 mm long and pale greenish-brown crusts of hemimorphite have been found at Newhurst quarry, Shepshed (Ince 2005).

Staffordshire Hemimorphite occurs with aurichalcite, chalcopyrite, fluorite, gypsum, serpierite and smithsonite in barite gangue from Clayton adit, Ecton Hill, Wetton (Braithwaite & Knight 1968).

Yorkshire Blocks of cavernous barite contain small sprays of colourless hemimorphite on the dumps of Brandy Bottle Incline, Friarfold veins, NE of Gunnerside, Swaledale. Aurichalcite, calcite, cinnabar, fluorite, strontianite and witherite are associated species (Young 1987b, Young et al. 1989, Ineson & Young 1994). Crystals and mammillated growths of white, yellow and brown hemimorphite are noted from Copperthwaite mine, Marrick Moor, Swaledale (Dunham & Wilson 1985). Blocky colourless crystals of hemimorphite up to 5 mm long, richly encrusting cavities in a limonitic matrix, occur underground at Buckden Gavel mine, Buckden (R. E. Starkey pers. comm. 2005). Oxidation of sphalerite to form abundant clear to light-brown hemimorphite and smithsonite has occurred in barite–fluorite–galena veins at Coldstones quarry, 5 km W of Pateley Bridge. Aurichalcite, cerussite, cinnabar, doyleite, otavite, prosopite and rosasite are associated species (Ryde 1987, Young et al. 1989, 1997). Hemimorphite is also noted from: Virgin Moss vein and level, East Bolton Moor, Wensleydale (Young et al. 1992c); Castaway–Ripley vein, Grassington Moor (Young 1987b); and Wetgrooves mine, Aysgarth (Rust 1987b).

Wales

Clwyd Hemimorphite is briefly noted from the Halkyn and Minera mining districts. At the latter it forms 3–5 mm sheaves on quartz and calcite (Bevins 1994).

Dyfed All localities are in the Central Wales orefield, where hemimorphite is widely developed: white crystalline crusts on veinstone at Bwlchrhennaid mine, near Goginan (Bevins 1994); colourless-to-white crusts of drusy bladed crystals on botryoidal brown to green hemimorphite at Esgair Hir mine, Ceulanymaesmawr (Rust & Mason 1988); drusy bladed crystals lining cavities and radiating crystal groups to 3 mm on goethite at Frongoch mine, between Devil's Bridge and Pont-rhyd-y-groes (Green et al. 1996b); tufts of clear transparent tabular crystals on sphalerite at Eaglebrook (Nantycagl) mine, Ceulanymaesmawr (Jones & Moreton 1977, Jones 1983); radiating spherical aggregates to 2.5 mm of acicular hemimorphite at Cwmystwyth mine; sprays 2–3 mm across on sphalerite and iron-rich dolomite at Henfwlch mine, between Cerrig yr Hafan and Nant-y-moch reservoir, E of Tal-y-bont; most attractive white to pale-blue botryoids up to 5 cm across at Llettyhên mine, SE of Tal-y-bont (all Bevins 1994); flat crystals on joints in sphalerite at Mynyddgorddu mine, W of Bont-goch, Ceulanymaesmawr (Clark & Rust 1993, Bevins 1994); sprays of well terminated crystals up to 0.5 mm long lining cavities in sulphidic breccias at Penrhiw mine, Ystumtuen (Mason & Green 1995). Hemimorphite is also recorded from: Brynyrafr, Esgairlle West, Nantycreiau and Nantycreiau West mines, all near Ponterwyd (all Bevins 1994); Bwlchglas mine, near Tal-y-bont(Braithwaite 1982a); Castell mine, Ponterwyd, near Aberystwyth (Rust 1990); and Ystrad Einion mine, near Tal-y-bont (Mason & Rust 1997).

Mid Glamorgan Hemimorphite forms radiating aggregates of colourless acicular crystals up to 5 mm long at Cwmleyshon quarry, near Draethen (Bevins 1994), and is relatively common as aggregates of colourless to honey-colour radiating tabular crystals up to 4 mm across at Machen quarry, 7 km ENE of Caerphilly (Goulding & Price 1995, Plant & Jones 1995). Hemimorphite also occurs at Hendy quarry, Miskin, near Llantrisant (Bevins 1994).

Powys Hemimorphite forms tufts of clear transparent tabular crystals on sphalerite at Dylife mine in the Central Wales orefield (Jones & Moreton 1977) and is also noted from Llangynog and Craig Rhiwarth mines, Llangynog (Bevins 1994).

Scotland

Dumfries & Galloway Hemimorphite is associated with wulfenite in a vein exposed in a cutting below the dam at Clatteringshaws reservoir, New Galloway (Rothwell & Mason 1992). Hemimorphite is also found on the dumps of West Blackcraig mine, near Newton Stewart, where it is one of many supergene species replacing a primary chalcopyrite–galena–sphalerite assemblage (Livingstone et al. 1976). Widely distributed hemimorphite is found around Wanlockhead, forming glassy crystals or botryoidal coatings on oxidized vein material: Bay mine (Thomson 1883, Rudler 1905); Belton vein (Brown 1919); fine crystals lining cavities, associated with cerussite, galena, pyromorphite and sphalerite – also variegated white, yellow, orange and brown stalactitic and botryoidal masses (Wilson & Flett 1921, Gillanders 1981); green hemimorphite associated with vanadinite at Cove vein; white to eggshell-blue crusts or sprays of euhedral transparent blades up to a few mm in length and associated with chrysocolla and veszelyite on the dumps of Waygate shaft on Straitstep vein; and fine specimens of botryoidal white, green and blue hemimorphite in thick crusts from the dumps of Whyte's Cleuch (all Green & Todd 2001).

Ireland

Tipperary Hemimorphite is described only from the Silvermines district, where cavities in limonite were lined with brilliant sharp stout crystals at Ballygown South mine. This is the only location where hemimorphite was worked in any quantity (Russell 1907b, Ryback & Moreton 1991, 1993, Boland et al. 1992, Moreton 1999). Rhoden (1959) suggests that at Ballygown South and Knockanroe mines, hemimorphite and goethite occur in the lower parts of the oxidation zone, and above them siderite and smithsonite were the dominant species. Hemimorphite is also reported from Gortnadyne mine. Later discoveries at Mogul mine include clusters of transparent pale-yellowish hemimorphite crystals on sphalerite (Ryback & Moreton 1991, Moreton 1999).

Wicklow Small crystals of hemimorphite occur on the dumps of the Luganure mines, Glendasan (Ryback & Moreton 1993).

hemloite?

triclinic $(As^{3+},Sb^{3+})_2(Ti,V^{3+},Fe^{2+},Fe^{3+})_{12}O_{23}(OH)$

A rare arsenic–antimony–titanium–vanadium–iron oxide hydroxide tentatively identified as tiny (<10 µm) inclusions within an unidentified aluminium–arsenic sulphate found growing as a drusy crust on an arsenic calciner wall at Tolgus, near Redruth, Cornwall. Secondary growths of potassium alum, dendritic gypsum and scorodite are also found at this site (M. R. Power pers. comm. 2004). These growths are all considered to be anthropogenic.

hentschelite

LAZULITE GROUP

monoclinic $Cu^{2+}Fe^{3+}_2(PO_4)_2(OH)_2$

A rare copper–iron phosphate hydroxide found on the type specimen of andrewsite from Stowe's section of the Phoenix United mines, Linkinhorne, Cornwall. Hentschelite and rockbridgeite were discovered during a re-investigation of the sample, and andrewsite was formally discredited (Dunn 1990). Both species form pale-green crystals on the outer shell of fibrous, radial spherules up to 4 mm across. Barbosalite and goethite form the core of the spherules.

Hentschelite is reported from vesicles in black barbosalite specimens from iron-stained fracture veins within partially kaolinized granite at Gunheath china-clay pit, Carthew, St Austell, Cornwall. The vesicles also contain chalcosiderite–turquoise compositions, dull-green fibrous dufrénite, fluorapatite, colourless-to-yellowish leucophosphite, bronzy-brown stannoidite and greenish-yellow cassiterite (var. varlamoffite). Hentschelite and the other phosphates probably crystallized from phosphatic solutions as they interacted with early-formed iron-rich veins during the later stages of kaolinization (Elton 1998).

henwoodite

TL West Phoenix mine, Liskeard, Cornwall, England

Named after the Cornish mining surveyor, William Jory Henwood FRS (1805–1875); born at Perran Wharf, Cornwall. Henwoodite is a synonym of turquoise (see Collins 1876a, Le Neve Foster 1876).

hepatic pyrite

An impure brown variety of pyrite.

hepatite

An impure earthy variety of barite, which emits a foetid odour when rubbed or hit.

hercynite

SPINEL GROUP

cubic $Fe^{2+}Al_2O_4$

A common iron–aluminium oxide that forms three series (with spinel, gahnite and chromite), and is typically found in mafic and ultramafic igneous rocks and high-grade metamorphosed ferruginous argillaceous sediments.

Small grains of hercynite (referred to as picotite) occur in serpentinite on the Lizard Peninsula, Cornwall, at: Goonhilly Downs; Lower Predannack quarry; S of Mullion (in thin section; dark-brown and greenish semi-transparent grains); Mullion Cove; Rill quarry; W of Kynance Cove and the Balk, E of Lizard village (Bonney 1877). A few pelitic xenoliths in the Carnmenellis granite contain deep-green porphyroblasts of zinc-rich hercynite associated with andalusite, corundum, ilmenite and sillimanite (Jefferies 1985b).

In Scotland, picotite forms inclusions within olivine in a remarkable xenolithic monchiquite dyke on the south side of Streap and Streap Comlaidh, NE of Glenfinnan, Highland Region (Walker & Ross 1954). A deep-brown variety of picotite mantles green spinel in ultrabasic wehrlite from the Girvan–Ballantrae ophiolite complex at Knocknormal, Strathclyde (Bloxam & Allen 1959). In all these cases, identification

Hercynite (green and brown varieties) in a metasedimentary xenolith within the Loch Doon granite, Craigmawhannel, near Dalmellington, Dumfries & Galloway, Scotland; width 0.4 mm; viewed in PPL. (© A. G. Tindle)

is by optical means and, because of the extensive solid solution possible between members of the spinel group, confirmation of hercynite is considered to be tentative.

At Sidhean Sluaigh, Strachur, Loch Fyne, Strathclyde, a series of emery-like rocks containing hercynite have formed within the metamorphic aureole of a dolerite plug. The rocks were originally albite–chlorite–muscovite–quartz phyllites and now represent the aluminous residues after a granitic fraction has been selectively removed. Hercynite occurs at the margins of magnetite grains or, together, they form complex runic intergrowths. Hercynite also occurs in phengite pseudomorphs. Cordierite, corundum, mullite, pseudobrookite and sanidine are associated species (Smith 1965, Worden et al. 1987). Desilication of a metasedimentary source has also been noted in the Loch Doon granite complex, on the north slopes of Craigmawhannel, 14 km S of Dalmellington, Strathclyde, where a metasedimentary xenolith containing a biotite–corundum residue is peppered with two types of hercynite: a bottle-green variety, and a brown variety exhibiting limited substitution towards qandilite (Tindle 1982). Contact metamorphism of Dalradian metapelites in the thermal aureole of the Lochnagar granite complex, SW of Ballater, Grampian, has produced a high-grade assemblage containing hercynite with andalusite, cordierite, corundum, plagioclase and potassium feldspars, and sillimanite. The assemblage is particularly well developed in the southern and eastern margins of the complex, where marginal quartz diorites occur (Goodman & Lappin 1996). Similar hercynite also occurs in hornfels within the aureole of the Belhelvie granite at Sparcraigs, near Whitecairns, to the N of Aberdeen (Stewart 1942). Analysis of spinel-group minerals from a xenolith of pelitic gneiss collected from a Tertiary dolerite sill on the southeast coast of the Ross of Mull between Scoor and Uisken, Isle of Mull, Strathclyde, shows them to be either pure magnetite or a hercynite-rich spinel (pleonaste) with variable gahnite, galaxite, spinel and ulvöspinel components. They appear to have formed during the metamorphic breakdown of muscovite or aluminium–iron biotite. Alkali feldspar, corundum and mullite are associated species (Brearley 1986, 1987). Dark-green hercynite forms (< 0.05 mm) irregular granules in pelitic hornfels surrounding the southern sector of the Etive granite complex, Strathclyde (Droop & Moazzen 2007). An intermediate hercynite–spinel composition is noted from a pargasite pyroxenite xenolith of lower crustal or possible mantle origin in a Permian olivine-bearing nephelinite dyke at Gribun, on the west coast of the Isle of Mull (Upton et al. 1998). Hercynite occurs in xenoliths in the Haddo House gabbroic complex, NE of Wood of Schivas, 32 km N of Aberdeen, although the authors describe the mineral as spinel (Herd et al. 1984).

In Ireland, a zinc-rich variety of hercynite is present in pelites within the aureole of the Ardara granite pluton, just W of Naran crossroads, 2 km SE of Portnoo harbour, Co. Donegal. It is always mantled by almandine, cordierite or feldspar, and it appears on textural and chemical evidence to have formed during the breakdown of zinc-rich staurolite. Hercynite occurs as clusters of dark-green subhedral grains enclosed in cordierite or in complex intergrowths with cordierite, sillimanite and staurolite (Atkin 1978). At Scawt Hill, Larne, Co. Antrim, a Tertiary dolerite plug has intruded Cretaceous chalk and Tertiary basaltic lava flows. On the basalt side of the dolerite/basalt contact, total recrystallization has occurred and a magnesium-rich hercynitic spinel has formed in association with lenses and streaks of devitrified basalt melt, dendritic olivine, bronzite–hypersthene, skeletal magnetite and micro-crystalline plagioclase (Kitchen 1985). Chrome spinel and a green aluminous spinel-group mineral, with an intermediate composition between hercynite and spinel, are noted from the Dawros peridotite, Connemara, Co. Galway. Both appear to have a magmatic origin (Rothstein 1972).

Other descriptions of hercynite (where analytical data are not presented) deserve further investigation: a Lewisian biotite-diorite dyke at Bad na h'Achlaise, Ledmore, Assynt, Highland Region (Stumpfl 1961); xenoliths of hornfels in hypersthene gabbro at Glebe Hill, near Kilchoan, Ardnamurchan, Highland Region, contain corundum (sapphire), hercynite and magnetite (Richey et al. 1930, Wells 1951); porcellanite inclusions (ferruginous bauxite and lithomarge, in which halloysite, gibbsite and hematite were the principal original constituents) within the Tertiary dolerite plug at Tievebulliagh, near Cushendall, Co. Antrim, containing cordierite, corundum, cristobalite (or tridymite), hercynite, ilmenite, mullite and magnetite in various assemblages (Agrell & Langley 1958).

herderite

orthorhombic $CaBePO_4F$

A rare calcium–beryllium fluorophosphate that forms a series with hydroxylherderite. Mandarino & Back (2004) consider herderite to be a doubtful species, as they claim there are no chemical data with F > OH, although Anthony et al. (2000) suggest that data from Brazil are valid.

Herderite is reported from the Meldon aplite quarries, near Okehampton, Devon. The Meldon aplite is a lithium–fluorine-rich dyke emplaced into the metamorphic aureole of the Dartmoor granite. Herderite forms intergrowths with beryllonite in an aplitic matrix close to a pegmatite veinlet. Bavenite and beryl are associated beryllium-bearing species (Dearman & Claringbull 1960, von Knorring & Condliffe 1984). In view of the discreditation of many rare British minerals described by Kingsbury (Ryback 1998, 2001), other beryllium species, namely chrysoberyl, eudidymite and rhodizite, that have been claimed to occur here (Embrey 1978, Kingsbury 1964b) require reinvestigation. The reports of herderite in Cornwall from Colcerrow quarry, Luxulyan, near Lostwithiel, and Wheal Cock, near Botallack, St Just (Kingsbury 1958, 1961), are problematic. Given the doubts about the provenance of other Kingsbury specimens (Ryback et al. 1998, 2001), further confirmation of herderite from these locations is desirable. It is unclear whether the report of herderite from Wheal Cock by Bancroft & Weller (1993) represents an additional discovery or is based entirely on specimens in the Kingsbury collection at the NHM.

herrengrundite

A synonym of devilline.

herschelite

Named after Sir John Frederick William Herschel FRS (1792–1871), astronomer, mathematician and mineral collector, who was president of the Royal Astronomical Society and author of the *Treatise on astronomy* (1833) and *Outlines of astronomy* (1849); born in Slough, Berkshire. Herschelite is a synonym of chabazite-Na, zeolite group.

herzenbergite?

orthorhombic SnS

A rare tin sulphide, briefly described with stokesite from Kessel Downs (Halvasso) quarry, near Longdowns, Cornwall (Golley & Williams 1995). Requires confirmation.

hessite

monoclinic Ag_2Te

A rare silver telluride, identified in ultramafic rocks (heterogeneous biotite–magnetite pyroxenite and garnet pyroxenite) at Loch Borralan, Assynt, Highland Region, Scotland. Hessite occurs in late-stage low-temperature calcite veins containing isomertieite, sperrylite and stibiopalladinite, all intergrown with chalcopyrite, galena and bismuth tellurides (Gunn & Styles 2002, Styles et al. 2004).

Hessite, with gold (electrum) and silver-rich galena, forms a distinctive assemblage in calcite–fluorite–quartz veins cutting quartz monzonite in the Ratagain diorite–granite complex, Loch Duich, near Glenelg, Highland Region. Hessite is a relatively common species in sulphide-rich parts of the veins, forming small (0.4 mm) irregular inclusions in galena and chalcopyrite (Alderton 1986, 1988b).

A further occurrence of hessite is in quartz–sulphide ± gold veins cutting Upper Dalradian gritty quartzites, quartz–mica schists and amphibolites along the Calliachar and Urlar burns, Aberfeldy, Tayside Region. Here, galena is one of the main sulphide species and it contains inclusions of hessite as discrete grains or forming intergrowths with other species such as altaite, benleonardite, chalcopyrite, coloradoite and gold (electrum) (Ixer et al. 1997, Chapman et al. 2000).

Hessite is also found in a brecciated gold-bearing pyrite–quartz vein (Halliday's vein), Tyndrum, Central Region, where it occurs as ≤ 20 μm-diameter inclusions in galena with petzite, sylvanite, cervelleite (described as phase C: see cervelleite) and an unidentified Ag–Te–S phase. These gold and silver minerals are associated with native gold (electrum with high silver content), chalcopyrite and sphalerite. The Au–(Ag) mineralization is thought to be attributable to hydrothermal activity related to copper–molybdenum mineralization associated with Late Caledonian granites (Pattrick et al. 1988, Curtis et al. 1993, Treagus et al. 1999).

At Lagalochan, near Kilmelford, Strathclyde, a Caledonian diatreme–porphyry breccia complex contains hessite and a bismuth–lead telluride, in a zoned shear-related phase of Ag–As–Au–Pb–Sb–Zn mineralization. The Lagalochan complex is thought to represent the lower portions of a now-eroded potentially productive epithermal precious-metal system (Harris et al. 1988). In sulphide-rich zones (dominated by chalcopyrite, pentlandite and pyrrhotite) within the Arthrath mafic–ultramafic intrusion near Ellon, 25 km N of Aberdeen, Grampian, accessory amounts of hessite occur with native bismuth, native gold, melonite, merenskyite and michenerite (McKervey et al. 2007).

Other Scottish reports of hessite are as inclusions in alluvial gold grains from: Borland Glen, Ochil Hills, Central Region; Kildonan Burn, Highland Region; Eas Daimh, Lochy Trib, W of Tyndrum, Strathclyde; Leadhills–Wanlockhead, Strathclyde; Glengarry, Tayside Region; West Water, NW of Bridgend, 13 km from

Brechin, Tayside Region (Leake et al. 1993, 1998, Chapman et al. 2000, 2005).

Hessite has been reported from Clogau gold mine, Bontddu, near Dolgellau, Gwynedd, Wales. It occurs as 30–40 µm-diameter inclusions in tellurobismuthite or galena, associated with tetradymite, altaite and gold (Gilbey 1968, Naden 1988, Bevins 1994). At Gwynfynydd mine, near Dolgellau, Gwynedd, hessite and tetrahedrite are minor components in gold-bearing samples, associated with chalcopyrite, galena and sphalerite. Other species recorded are acanthite, arsenopyrite, marcasite, pyrrhotite, pyrite, and bismuth–silver tellurides (Gilbey 1968, Pryor 1988, Leake et al. 1993).

A further association of hessite in quartz–sulphide ± gold vein mineralization is from the Sperrin Mountains of Ireland; it includes micron-scale crystals in pyrite at Cavanacaw, near Omagh, and Curraghinalt, near Gortin, both Co. Tyrone. Associated phases include altaite, bismuth telluride and mercury telluride (Parnell et al. 2000). Hessite, also forms inclusions in alluvial gold grains from the same area (Chapman 2000c).

hessonite

A synonym of grossular, garnet group.

hetaerolite

tetragonal $ZnMn^{3+}_2O_4$

A rare zinc–manganese oxide found closely associated with hematite in secondary mineral veins (no more than 4 cm wide) cutting the serpentinized peridotites of the Lizard ophiolite complex in the vicinity of Eastern Cliff, Kennack Sands, Lizard Peninsula, Cornwall. Hetaerolite forms small black octahedral crystals generally less than 0.75 mm across, but occasionally crystal aggregates form lenses up to 4 mm across. Because of their resistance to weathering relative to their lizardite ± calcite host, these crystals stand out on exposed surfaces. Individual crystals show well developed faces, have a submetallic lustre and, in the aggregates, commonly show twinning (Bevins et al. 1987).

heterogenite?

trigonal or hexagonal $Co^{3+}O(OH)$

A rare cobalt oxide hydroxide claimed by A. W. G. Kingsbury to have been found at Engine Vein mine, Alderley Edge ridge, Cheshire (Ryback & Tandy 1992, Braithwaite 1994). Ryback et al. (1998) have shown this to be incorrect and they suggest that the material came from Zaire.

heteromorphite?

monoclinic $Pb_7Sb_8S_{19}$

A rare lead sulpharsenite briefly described from Wheal Boys, Trewetha, St Endellion, Cornwall (Russell 1949b, Embrey 1978). This requires confirmation.

heulandite (series)

ZEOLITE GROUP

monoclinic $(Na,Ca)_{4-6}(Si,Al)_{36}O_{72}\cdot 24H_2O$

Named after John Henry Heuland (1778–1856), a mineral dealer and collector, based in London, who was also secretary of the Geological Society of London; born "Johann Heinrich" in Bayreuth, Germany (Frondel 1972, Whitehead 1973, Cooper 2001, 2007). Heulandite is a relatively common hydrated sodium–calcium aluminosilicate, whose name has been given series status in the zeolite classification of Coombs et al. (1997). There are now five recognized heulandite species: -Ba, -Ca, -K, -Na and -Sr. In most cases the new species names do not appear in the literature and, until chemical data become available, their status in Britain and Ireland cannot be established. Heulandite is one of the more common zeolite species and is commonly found in vesicles within volcanic rocks, as a devitrification product of volcanic glasses and tuffs, and in some sedimentary rocks containing volcanic material.

A large block of rock consisting almost entirely of oligoclase feldspar, found on the dumps of Wheal Forest, Meldon Gorge, near Okehampton, Devon, was shown by Russell (1910b) to be partially covered with a thick crust of pearly heulandite. The species was also found as sharp and brilliant crystals (the largest 2 mm long) in cavities in the same rock. Near by, on the dumps of Red-a-ven mine, Meldon, heulandite occurs with datolite and tin-rich grossular in veins cutting calc-silicate hornfels (calc-flinta). Both occurrences are within the metamorphic aureole of the Dartmoor granite (Dearman 1966b). Bright and fairly well defined crystals of colourless heulandite, barely 1 mm long, were found on a single mass of shale on the dumps of Ramsley mine, South Tawton, near Okehampton (Russell 1910b).

Abundant heulandite forms small colourless to lemon-yellow coffin-shape crystals coating an altered hornfels matrix in the Wheal Hazard section of Botallack mine, St Just, Cornwall (Wolloxall 1989, Bancroft & Weller 1993). Crusts of yellowish-brown or sometimes colourless heulandite, forming crystals up to 1 mm long, cover joint surfaces of fine-grain dolerite on the dumps of Carrick Du mine, near Porthmeor beach, St Ives (Russell 1910b). Heulandite is one of several zeolite species (others being analcime, chabazite, natrolite and stilbite) found in veins cutting gabbro at Dean quarry, 2 km SE of St Keverne, Lizard Peninsula. It is colourless and transparent, and it forms single tabular crystals up to 0.75 mm long. Crusts of heulandite growing on analcime, calcite, natrolite and prehnite confirm it to be one of the last zeolites to crystallize (Seager 1971, Wolters & Wolters 1986). Additionally, Golley & Williams (1995) note heulandite from Levant mine, St Just, but give no further details. Analysis of "heulandite" from Wheal Hazard has revealed that most of it is actually clinoptilolite-Ca (Dyer et al. 2006). Unless XRD and full chemical analysis are available, these species cannot be reliably identified at species level.

Brown et al. (1969) report zeolites with intermediate clinoptilolite–heulandite composition from soil clays derived from the basal Thanet Beds at Upnor, Kent, and from a Harwell series soil at Halton, Buckinghamshire, the latter in association with amorphous silica and disordered cristobalite–tridymite.

Heulandite has not been verified in Wales, although Bevins (1994) reports Sir Arthur Russell exhibiting specimens of heulandite and prehnite from Wales at a general meeting of the Mineralogical Society in London on 14 March 1914.

Tertiary basalts at 'S Airde Beinn, 5 km SW of Tobermory, Isle of Mull, Strathclyde, Scotland, occasionally contain vesicles infilled with heulandite or stilbite, although analcime, gyrolite, natrolite and thomsonite appear to be more common (Cann 1965a). At Maol nan Damh, on the shoulder of Ben More, on the Isle of Mull, milky-white heulandite forms minute crystals lining cavities (probably in basalt), often coating earlier crystallized tufts of epidote. Rarely, heulandite forms stalactitic growths encrusting calcite (Currie 1898).

At Allt Ribhein, Fiskavaig Bay on the Isle of Skye, Highland Region, heulandite occurs in vesicles up to 2 cm across in Tertiary lavas, together with other white zeolites (stilbite and thomsonite) and saponite (Mackenzie 1957). Crystals of heulandite to 2 mm are noted from a distinct layer of vesicular basalt in a roadside cutting at Edinbane, on Skye, where they are associated with calcite, epistilbite, erionite, stellerite and stilbite (Savage 1995, Green & Wood 1996). Perhaps the most spectacular heulandite comes from Sgùrr nam Boc, between Loch Brittle and Loch Eynort, where fine white and pink crystals were found with apophyllite, calcite, laumontite, mesolite, quartz and stilbite by Pearce & Pearce (1991) during their search for an old Heddle locality. Their success was such that Green & Todd (2001) refer to the place as the premier zeolite locality in Scotland. These authors also describe lustrous curved translucent crystals of heulandite from another great Tertiary basalt zeolite locality at Moonen Bay on the Duirinish Peninsula, in the NW of Skye. Here heulandite often forms crystals up to 10 mm long (exceptionally to 25 mm) that are commonly associated with chabazite and mesolite. Moonen Bay is also noted for a wide range of other species, including analcime, apophyllite, cowlesite, garronite, laumontite, levyne, offrétite, phillipsite, stellerite, stilbite and thomsonite (Green & Todd 1996). To the N of Moonen Bay is Oisgill Bay, where vesicular basalts contain glassy coffin-shape heulandite with natrolite and stilbite. The best Scottish levyne is also found here (Green & Todd 2001). The relative scarcity of heulandite on Skye is highlighted by King (1976), who examined zeolites from 404 localities and found it in only 1% of them.

On mainland Scotland, heulandite occurs in Carboniferous basalts belonging to the Clyde Plateau Lavas. At Hartfield Moss, 7 km NE of Beith, SW of Glasgow, colourless or reddish-brown tabular heulandite crystals up to 0.5 mm long commonly occur in vesicles. Analcime, harmotome, laumontite, natrolite, stilbite and thomsonite are other zeolite species found at this locality (Meikle 1990). Altered basaltic lavas at Loanhead quarry, near Beith, contain extremely small colourless tooth-shape crystals of heulandite that form compact coatings in small vugs. Analcime, greenockite, harmotome, laumontite, mottramite, natrolite, prehnite, native silver, thomsonite and vanadinite are also found in fractures and vesicles in the lavas at Loanhead (Meikle 1989b). Excellent brick-red heulandite up to 10 mm long and stilbite, both associated with analcime, barite, calcite, harmotome, natrolite and quartz; have recently been collected from exposures along the shore of Earlsburn (Earlston) reservoir in the Touch Hills, SW of Stirling, Central Region (Green & Todd 2001). Heulandite from Todhead Point, 9 km S of Stonehaven, Grampian (Heddle 1901), was not substantiated by Dyer et al. (1993).

Other locations in Scotland mentioned by Heddle (1901) and other authors include the following:

CENTRAL Balhennan Hill, near Flintry; Campsie Fells, N of Strathblane (Gottardi & Galli 1985) – but doubted by Heddle (1893a); Carbeth, 3 km W of Strathblane, N of Glasgow. FIFE Kincraig, 3 km W of Elie; Scurr Hill, Balmerino, SW of Newport-on-Tay. GRAMPIAN Off the Inverbervie–Stonehaven road, near the turning to Kinneff church (Heddle 1879f); quarry at the cliff of Slainges, 10 km S of

Heulandite with dark-green chlorite on basalt; Kilmalcolm, Strathclyde, Scotland; crystals to 1 cm. (© M. P. Cooper)

Stonehaven; Thornyhive, 3 km S of Stonehaven; Tremuda Bay, 2.8 km S of Stonehaven. HIGHLAND Mouth of Allt Mòr (or Allt Leth-uilt), between Loch Brittle and Loch Eynort (Heddle 1893b); north shore of Loch Bharcasaig (Varkasaig), west coast of Loch Bracadale; Sgùrr an Duine, between Loch Brittle and Loch Eynort; the Storr, N of Portree – all Isle of Skye; Isle of Canna. STRATHCLYDE Bac Mòr, Treshnish Isles, off the west coast of Mull; Ballygroggan, Machrihanish Bay, Kintyre Peninsula; Boyleston quarry, SW of Barrhead (Hall et al. 1989); Carsaig Arches, Isle of Mull; Kilmalcolm, 10 km SE of Greenock; Little Cumbrae island, off the coast near West Kilbride; Lang Craigs and Loch Humphrey, both Kilpatrick Hills, above Dumbarton (Heddle 1893a, Tschernich 1992, Wolfe 2000). SHETLAND Papa Stour, a small island off the west coast of Mainland.

In Ireland, heulandite is a scarce species found in cavities within a boulder of phillipsite-rich tholeiitic basalt, near the National Trust reception centre, Giant's Causeway, Co. Antrim. It is accompanied by erionite, paulingite and possibly clinoptilolite (Foy 1992). Walker (1960a) describes well shaped crystals of white or colourless heulandite up to 5 mm long from the basalts of Co. Antrim, but apart from noting an occurrence at the Giant's Causeway does not give other location details. Heulandite is found as colourless transparent crystals at Gibons quarry, White Mountain, SW of Belfast (Tschernich 1992), and is particularly abundant at the nearby Sales Corner quarry (Adair 1987). Heulandite also occurs with analcime and thomsonite in the contact aureole of a dolerite dyke at Blind Rock, SW of Mount Charles, Co. Donegal (Nawaz 1979).

heulandite-Ba

ZEOLITE GROUP

monoclinic $NaBa_4(Si_{27}Al_9)O_{72}\cdot 24H_2O$

A very rare hydrated sodium–barium aluminosilicate found associated with strontianite on specimens collected from Hugh's shaft at Whitesmith mine, Strontian, Highland Region, Scotland. Heulandite-Ba forms colourless coffin-shape crystals up to 5 mm long and exhibits oscillatory zonation with individual zones either dominated by strontium or, much less frequently, by barium (D. I. Green specimen and XRD of a "bulk" sample, Tindle unpub. EMP data).

heulandite-Ca

ZEOLITE GROUP

monoclinic $NaCa_4(Si_{27}Al_9)O_{72}\cdot 24H_2O$

Heulandite-Ca is the new name for the most common species of the heulandite series (Coombs et al. 1997). It is an hydrated sodium–calcium aluminosilicate found in the vesicles of hydrothermally altered Tertiary basaltic lavas at the Giant's Causeway, Co. Antrim, Ireland. The first stage of alteration is the formation of saponite, chlorite–smectite and dioctahedral smectite on vesicle walls. This is followed by the crystallization of zoned heulandite-Ca, heulandite-K, clinoptilolite and finally calcite. These secondary minerals develop as pseudomorphs of primary phases, replacing groundmass minerals and glass, and coating or infilling veins and vesicles (Robert 2001). At Lindsay's Leap quarry in the Mourne Mountains, drusy cavities and veins on joint planes in granite contain a few wedge-shape crystals of heulandite-Ca up to 2 mm across and always associated with stilbite-Ca (Green et al. 2005a). The metamorphic aureole of the Omey granite hosts an amphibolite skarn at Barnahallia Lough, Claddaghduff, west Connemara, Co. Galway. The skarn contains cavities in which common mm-size creamy crystals of heulandite-Ca occur associated with albite, chabazite-Ca, epidote, grossular and stilbite-Ca (Moreton & Green 2006a).

Heulandite-Ca also occurs in England: Carrick Du quarry, near Porthmeor, St Ives and Wheal Hazard section of Botallack mine, St Just, both in Cornwall; and High Force quarry, Upper Teesdale, Co. Durham. In Scotland, it is found: in amygdaloidal basalts from Edinbane, Sgùrr nam Boc, and the Hoe on the Isle of Skye, Highland Region; at Loch Fada on the Isle of Mull; at Lang Craigs and Old Kilpatrick, Kilpatrick Hills, near Dumbarton, all Strathclyde; also from Todd Head Point, Grampian (D. I. Green specimens and XRD, Tindle unpub. EMP data).

heulandite-K

ZEOLITE GROUP

monoclinic $KCa_4(Si_{27}Al_9)O_{72}\cdot 24H_2O$

An hydrated potassium–calcium aluminosilicate found in the vesicles of hydrothermally altered Tertiary basaltic lavas at the Giant's Causeway, Co. Antrim, Ireland. The first stage of alteration is the formation of saponite, chlorite–smectite and dioctahedral smectite on vesicle walls. This is followed by the crystallization of zoned heulandite-Ca, heulandite-K and clinoptilolite, and finally calcite. These secondary minerals develop as pseudomorphs of primary phases, replacing groundmass minerals and glass, and coating or infilling veins and vesicles (Robert 2001).

heulandite-Sr

ZEOLITE GROUP

monoclinic $NaSr_4(Si_{27}Al_9)O_{72}\cdot 24H_2O$

A very rare hydrated sodium–strontium aluminosilicate found as a late-stage mineral at Whitesmith mine, Strontian, Highland Region, Scotland. Heulandite-Sr forms colourless coffin-shape crystals up to 10 mm long. Some crystals show oscillatory zonation, with individual zones either dominated by strontium or less frequently by barium, and are associated with strontianite; others are strontium dominant and are associated with calcite (D. I. Green specimen and XRD, Tindle unpub. EMP data).

Complex, compositionally zoned crystal with a heulandite-Sr core and a heulandite-Ba rim; Strontian mine, Highland Region, Scotland; width 1.25 mm; BSE image. (© A. G. Tindle)

HEXAHYDRITE GROUP

Monoclinic sulphates with the general formula $A^{2+}SO_4 \cdot 6H_2O$, where A^{2+} = Co, Fe, Mg, Mn, Ni, Zn. Hexahydrite is the only species of the group confirmed in Britain or Ireland, although nickelhexahydrite is also reported.

hexahydrite

HEXAHYDRITE GROUP
monoclinic $MgSO_4 \cdot 6H_2O$

A rare hydrated magnesium sulphate hydroxide found on exposed serpentinite drill core from North Ballaird bore No. 3, Balsalloch farm, near Ballantrae, Strathclyde, Scotland. Hexahydrite forms a secondary pale-green surface coating on the core (Macpherson & Livingstone 1982). XRD analysis of specimens in the NMW labelled "epsomite" from Treharris, Mid Glamorgan, Wales, revealed the presence of hexahydrite. This is most likely to be the result of dehydration of epsomite in the museum environment (Bevins 1994).

hibbertite

TL Nikka Vord and Hagdale, Unst, Shetland, Scotland

Named after Samuel Hibbert-Ware (1782–1848), the discoverer of chromite on Unst, Shetland, Scotland; born in Manchester. Hibbertite is described as mixture of calcite, dolomite and pyroaurite (or sjögrenite), and also of brucite and calcite.

hibschite

GARNET GROUP
monoclinic $Ca_3Al_2(SiO_4)_{3-x}(OH)_{4x}$ (x=0.2–1.5)

An uncommon calcium aluminosilicate hydroxide that forms two series, with grossular and with katoite. Although hibschite is an IMA-approved species, some authorities suggest that it should be recognized as an intermediate composition of the grossular–katoite series.

Hibschite occurs close to the Whin Sill, on samples from the dumps near Cadger Well level, Harwood Beck, Teesdale, Co. Durham. Hibschite (syn. hydrogrossular) is associated with yellow garnets of unknown composition (described as ugrandite type), which form granular aggregates and subhedral crystals, up to 0.5 mm across, in garnet-rich selvages between magnetite- and calcite-rich rocks (Young et al. 1985).

At Camas Mòr, Isle of Muck, Highland Region, Scotland, an olivine gabbro dyke intersecting Jurassic limestones has developed a high-temperature contact assemblage of gehlenite, larnite, rankinite and spinel. Gehlenite is frequently replaced by dense aggregates of isotropic grains – possibly a member of the hydrogrossular series (Tilley 1947). Livingstone (2002) considers them to be hibschite.

Hibschite (hydrogrossular) is also reported from the Girvan–Ballantrae complex, Strathclyde, Scotland at Byne Hill, S of Girvan (Bloxam 1954), and at the Sailors monument, N of Lendalfoot (Bloxam 1964). It occurs in considerably altered gabbro dykes cutting serpentinite, which also now contain chlorite, pectolite and prehnite.

Hibschite is found in a chalk/dolerite contact zone associated with a Tertiary volcanic plug at Carneal, Larne, Co. Antrim, Ireland. Assimilation of the chalk has produced complex metasomatic assemblages containing hibschite (hydrogrossular). These include larnite–spinel and scawtite–tobermorite–xonotlite rocks. Well formed hibschite occurs as equigranular crystals, up to 0.5 mm across, in a fine-grain grey rock with a groundmass of calcite, tobermorite and other minerals. Co-existing flint has been converted to an assemblage containing hibschite, plagioclase, quartz and wollastonite. Hibschite is also noted in turbid brown material within pyroxene-rich dolerite and in a metasomatic merwinite-rich vein cutting dolerite (Sabine et al. 1968, Sabine & Young 1975).

hidalgoite

BEUDANTITE GROUP
trigonal $PbAl_3(As,S)O_4(OH,H_2O)_6$

A relatively rare, although locally abundant hydrated lead–aluminium arsenate sulphate hydroxide found on the waste dumps of Penberthy Croft mine, St Hilary, Cornwall, where it forms minute colourless to pale-yellow crystals encrusting arsenopyrite–quartz veinstone (Rust 1996d). The appearance of hidalgoite is quite variable and acute rhombohedral crystals forming rosettes are also noted. Additionally, drusy aggregates up to several mm across of bayldonite, beudantite, hidalgoite and mimetite are also found on a goethite–quartz matrix (Betterton et al. 2000). An hidalgoite-like mineral is reported from Hingston Down Consols mine, 2 km W of Gunnislake, Calstock, Cornwall (Davis & Hey 1964), and a specimen of bluish-grey so-called plumbogummite from Roughton Gill, Caldbeck Fells, Cumbria is considered to be a carbonate-bearing mix-crystal of plumbogummite, hinsdalite and hidalgoite (Förtsch 1967). Both occurrences require confirmation.

Hidalgoite occurring with beudantite as a secondary mineral at Tynagh mine, near Killimor, Co. Galway, Ireland, also requires confirmation (Morrissey 1970, Ryback et al. 1988).

Curved aggregate of yellowish-green hidalgoite with colourless prismatic mimetite; Penberthy Croft mine, St Hilary, Cornwall; width 3 mm. (© D. I. Green)

hiddenite

A green chromium-rich gem variety of spodumene.

hilgardite

monoclinic or triclinic $Ca_2B_5O_9Cl \cdot H_2O$

A very rare hydrated calcium borate with chloride, found in Permian evaporite deposits at Boulby Potash mine, Loftus, Cleveland. Hilgardite can form three polymorphs and it is hilgardite-1*Tc* that is present in various forms at Boulby. Pinkish monomineralic crystals of hilgardite are found in nodules and it is also found interstitially within boracite lattice-works as paper-thin crystals. The brick-red colouration of some of the nodules is caused by admixed hematite. Crystalline nodules reach

Cluster of hilgardite crystals measuring 5 mm across; Boulby mine, Loftus, Cleveland. (© D. I. Green)

25 cm in diameter and individual crystals reach 1 cm, perhaps representing the finest development of the species anywhere in the world (Green & Freier 1994, 1996).

hillebrandite

monoclinic $Ca_2SiO_3(OH)_2$

A rare calcium silicate found in thermally metamorphosed limestones to the south of Barnavave, Carlingford, Co. Louth, Ireland. Some of the metamorphosed limestones containing grossular, spurrite, tilleyite, vesuvianite and wollastonite have suffered retrograde changes; spurrite, tilleyite and wollastonite are replaced by varying amounts of hillebrandite. In thin section, hillebrandite is colourless and it forms fibrous aggregates that, between crossed polars, show anomalous ultra-blue interference colours (Nockolds & Vincent 1947).

hinsdalite

BEUDANTITE GROUP

trigonal $PbAl_3(SO_4)(PO_4)(OH)_6$

A rare lead–aluminium phosphate sulphate hydroxide found on gossan breccia at Penkiln Burn, 13 km NNE of Newton Stewart, Dumfries & Galloway, Scotland. Hinsdalite forms small ivory blebs and coatings on the breccia (Livingstone 1993).

Hinsdalite commonly occurs as white to yellowish-brown encrustations at an old trial on a small manganese-rich vein to the NW of Saddleback Old mine at the head of the Glenderamackin Valley. Hinsdalite also forms epimorphs and pseudomorphs after elongate hexagonal pyromorphite crystals at this locality. A record of hinsdalite from Potts Gill barite mine (Old Potts Gill mine), Caldbeck Fells, Cumbria (Hartley 1984, Young 1987a), is in need of careful re-examination, as it is based on Kingsbury material of uncertain provenance (Ryback et al. 1998, 2001, Green & Briscoe 2002). Hinsdalite mixed with hidalgoite and plumbogummite, described from Roughton Gill in the Caldbeck Fells (Förtsch 1967), also requires confirmation.

In Wales, hinsdalite occurs as crusts of minute spiky crystals coating and partially replacing earlier coarse pyromorphite crystals at Frongoch mine, between Devil's Bridge and Pont-ryd-y-groes, Dyfed (Cotterell & Todhunter 2007).

Well formed crystals of hinsdalite coating pyromorphite (white) from Frongoch mine, between Devil's Bridge and Pont-rhyd-y-groes, Dyfed, Wales; width 900 µm; SEM–BSE image. (© National Museum of Wales, Cardiff; photographers T. F. Cotterell and M. P. Lambert)

Pseudomorphs of hinsdalite (to 3 mm) after pyromorphite from an old trial near Saddleback Old mine, near Mungrisdale, Cumbria. (© D. I. Green)

hisingerite

monoclinic or amorphous $Fe^{3+}_2Si_2O_5(OH)_4{\cdot}2H_2O$

A rare hydrated iron silicate hydroxide found at Wheal Jane, Kea, near Truro, Cornwall. Hisingerite occurs in scaly and brittle masses of a dark-brown colour, intimately associated with pentlandite and vivianite, and was originally thought to be beraunite (Greg 1860). The occurrence is noted by Collins (1892) and further described by Rust (1982d), who found finely crystallized masses of hisingerite in pyrite-rich veinstone. Dark-brown amorphous hisingerite is thought to have been found on a specimen of tamarite from Wheal Gorland, St Day, 2 km E of Redruth (Church 1870). Hisingerite is noted without comment from Penlee quarry, Newlyn, Cornwall (Golley & Williams 1995). Specimens of hisingerite were supposedly collected from Wheal Gorland and Phoenix United mines by Kingsbury (1954). However, because many specimens in the Kingsbury collection (now at the NHM) have questionable provenance (Ryback et al. 1998, 2001), these occurrences must be doubtful. The occurrence of hisingerite at Brandy Gill mine, Caldbeck Fells, Cumbria (Hartley 1984), is also in doubt, as this too could be a Kingsbury specimen.

hislopite

A mixture of calcite and a glauconite-like mineral.

hitchcockite

A synonym of plumbogummite (Hartley 1900).

hodrušhite?

monoclinic $Cu_8Bi_{12}S_{22}$

A rare copper sulphobismuthite reported from a stockwork cut by galena-bearing quartz veins at Corrie Buie, Meall nan Oighcreag, Loch Tay, Tayside Region, Scotland. The veins also contain minor amounts of native bismuth, chalcopyrite, electrum (36–65% Au), pyrrhotite, schirmerite and sphalerite, and phases belonging to the gustavite–lillianite series. The mineral assemblage of the veins is typical of that associated with acid intrusions. Pattrick (1984) describes a phase where substitution of iron and lead for bismuth has occurred, and which is interpreted as a variety of hodrušhite.

högauite

A synonym of natrolite, zeolite group.

högbomite

hexagonal or trigonal $(Mg,Fe^{2+})_2(Al,Ti)_5O_{10}$

A rare magnesium–iron–aluminium–titanium oxide found in a chlorite–corundum–ilmenite–magnetite xenolith in the Cashel basic intrusion, near Toombeola, Co. Galway, Ireland. Anhedral, medium- to light-brown högbomite and green spinel are intergrown with magnetite in a corundum matrix traversed by pale chloritic veins. Some högbomite grains have spinel cores. Three polytypes of högbomite are present: *6H*, *4H* and *15R* (McKie 1963 – but note revised classification by Armbruster 2002).

Within the Currywongaun and Doughruagh intrusion, Connemara, also in Co. Galway, högbomite has been found in two distinct assemblages. Both are within 30 m of the edge of the intrusion, which is emplaced into a series of Middle Dalradian semi-pelitic and psammitic schists. The first assemblage is in a cordierite-rich (90% by volume) hornfelsed pelitic xenolith and the second an orthopyroxenite of magmatic origin. Högbomite forms less that 0.5% of the xenolith and is associated with accessory amounts of ilmenite and magnetite. Anthophyllite, biotite, spinel and possibly corundum are the other minor components. The second assemblage is in an enstatite–magnetite orthopyroxenite, containing minor amounts of högbomite intimately intergrown with ilmenite and spinel. The högbomite also forms euhedral crystals with rectangular or hexagonal outlines. An origin involving regional metamorphism of Middle Dalradian sediments is proposed (Angus & Middleton 1985).

hollandite

CRYPTOMELANE GROUP

monoclinic (pseudo-trigonal) $Ba(Mn^{4+},Mn^{2+})_8O_{16}$

Named after Thomas Henry Holland FRS (1868–1947), a director of the Geological Survey of India and principal of the University of Edinburgh; born in Helston, Cornwall. Hollandite is an uncommon barium–manganese oxide that has been found in fracture-mineralized Permo-Triassic St Bees Sandstone examined in borehole material from the Sellafield area, Cumbria, where it forms surface impregnations and stains, thin films and dendritic coatings on fracture surfaces and bedding fissures, together with other manganese oxides (goethite, romanèchite and todorokite). It is derived mainly from the dissolution and oxidative decomposition of iron–manganese-rich calcite host-rock cements and vein fillings (Milodowski et al. 1998). At Croft quarry, SW of Leicester, joint surfaces of diorite and Caledonian basement conglomerates contain minor hollandite associated with palygorskite, pyrolusite and unidentified manganese dendrites (Pearson & Jeffrey 1997).

Manganese-mineralized yellowish tuff of the Llyn Conwy Formation at Mochowgryn and in the vicinity of Nant yr Helfa, Migneint area, Snowdonia, Wales, contains hollandite. The mineralization is related to N–S vertical faults and contains pyrolusite in addition to hollandite (Lynas 1973).

Hollandite and pyrolusite (and possibly romanèchite) are minor components of a manganese vein composed primarily of coronadite and amorphous wad found close to Arndilly House, Arndilly, near Craigellachie, 20 km SE of Elgin, Grampian, Scotland. The deposit is considered to be supergene oxide capping on stockwork-type mineralization related to the Ben Rinnes granite (Nicholson 1986). At Dùn Athad, Mull of Oa Peninsula, Isle of Islay, Strathclyde, veins, coatings and discrete masses of mixed iron and manganese oxides are found within the Jura Quartzite (Dalradian Supergroup) of Precambrian–Cambrian age. Hollandite is the principal manganese mineral, together with pyrolusite and minor groutite, and these species are postdated by calcite, manganese-rich calcite and rhodochrosite. The crystal form of hollandite is quite variable, with radiating fibres, colloform textures intergrown with hematite, and oolitic textures produced by radiating hollandite fibres centred on grains of quartzite. The manganese enrichment is thought to be derived from Old Red or New Red Sandstone sediments and deposited from freshwater (non-hydrothermal) solutions (Nicholson 1987b, 1988a, 1989b, 1992).

Hollandite has been found with cryptomelane as intergrown botryoidal masses up to several cm across associated with goethite in a quartz vein hosted by the Leinster granite at Cloghleagh mine, near Blessington, Co. Wicklow, Ireland (Moreton et al. 2006).

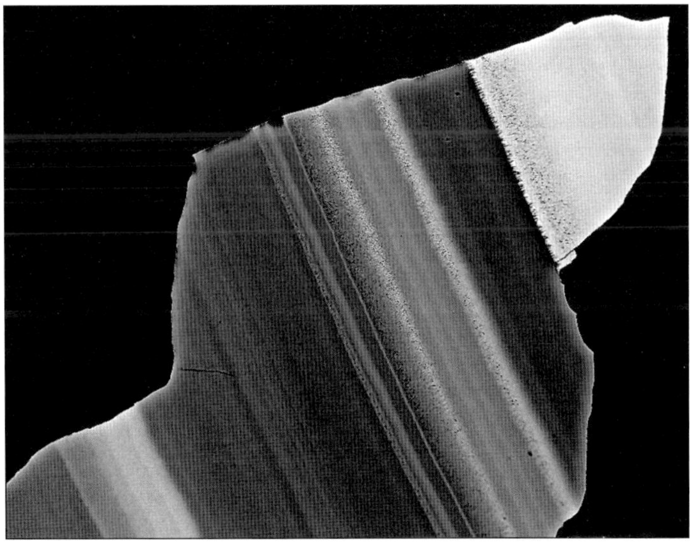

Finely banded cryptomelane (dark grey) and hollandite (light grey) from a goethite-bearing vein cutting granite at Clogea mine, Blessington, Co. Wicklow, Ireland; width 2.2 mm; BSE image. (© A. G. Tindle)

hollingworthite

COBALTITE GROUP

cubic $(Rh,Pt,Pd)AsS$

Named after Sydney Ewart Hollingworth (1899–1966) Yates–Goldsmid Professor of Geology, University College, London; born at Floore, Northamptonshire. Hollingworthite is a rare rhodium–platinum–palladium arsenide that forms a series with irarsite and is found at Cliff and at Harold's Grave (two disused chromite quarries), Baltasound–Haroldswick area, Unst, Shetland, Scotland. It forms euhedral tabular or hexagonal equant crystals, often less than 5 μm across, the composition of which is intermediate between hollingworthite and irarsite. Composite grains of hollingworthite with genkinite, hongshiite, irarsite, sperrylite, stibiopalladinite and unnamed Cu–Pd–Pt and copper–nickel alloys are found in the altered silicate matrix of chromite-rich dunite lenses that formed near the base of the Shetland ophiolite complex (Prichard et al. 1986, 1994, Tarkian & Prichard 1987, Prichard & Tarkian 1988). Near by, at the junction of chromite with silicate matrix at Nikkavord North, Prichard et al. (1988) note a small composite grain containing hollingworthite, irarsite, native osmium, native platinum, a nickel–rhodium antimonide and a ruthenium-rich Co–Fe–Ni sulphide. Very small grains of hollingworthite are also reported from the same area by Gunn et al. (1985), although precise details are not quoted.

At Talnotry, near Newton Stewart, Dumfries & Galloway, hollingworthite occurs as a rare species in the nickeline–gersdorffite zone of an arsenic-rich magmatic sulphide assemblage that developed near the base of an altered diorite sheet. It is associated with gold (electrum), irarsite and sperrylite. Other mineralized zones in the diorite are dominated by chalcopyrite, pentlandite and pyrrhotite, with froodite, merenskyite and michenerite being other subordinate species. Arsenopyrite, annabergite, bismuthinite, galena and pyrite are also noted (Power et al. 2004).

Prout & Andersen (2001) report rare hollingworthite with laurite, michenerite and sperrylite from the Ben Buie ultramafic intrusion on the Isle of Mull, Strathclyde, Scotland.

holmbushite

TL Holmbush Mine, Callington, Cornwall, England

An acicular microcrystalline variety of carbonate-fluorapatite, apatite group.

honey blende

A golden-yellow variety of sphalerite.

honessite

trigonal $Ni_6Fe^{3+}_2(SO_4)(OH)_{16}{\cdot}4H_2O$

A rare hydrated nickel–iron sulphate, related to the hydrotalcite group and found on chromitite at Hagdale quarry, Unst, Shetland, Scotland, where it forms fine-grain yellow encrustations intimately mixed with reevesite and associated with theophrastite. The specimens also contain aragonite, chlorite, goethite and lesser amounts of heazlewoodite and pentlandite (Bish & Livingstone 1981, Livingstone & Bish 1982, Livingstone 2002). Elsewhere in Scotland, at the Craignure mine, 12 km SW of Inveraray, Strathclyde, aluminium–copper analogues of honessite and hydrohonessite form bluish-green coatings associated with reevesite on dump material consisting mainly of chlorite schist with narrow quartz veins and calcite-rich bands (Livingstone 1990b).

hongshiite

trigonal $(Pt,Fe)Cu$

A very rare platinum–iron–copper alloy found at Cliff and at Harold's Grave, Baltasound–Haroldswick area, Unst, Shetland, Scotland. Hongshiite forms composite grains with genkinite, hollingworthite, irarsite, sperrylite and unnamed Cu–Ni and Cu–Pd–Pt alloys in the altered silicate matrix of chromite-rich dunite lenses that formed near the base of the Shetland ophiolite complex (Prichard & Tarkian 1988, Prichard et al. 1994).

In a layered mafic–ultramafic intrusion on the Isle of Rum, Highland Region, within the chromitite layer between units 7 and 8 of the Eastern Layered Series, a copper–platinum alloy (possibly hongshiite) is found in association with species, positively or tentatively identified as braggite, cooperite, irarsite, isoferroplatinum, keithconnite, kotulskite, laurite, moncheite, palladoarsenide, platarsite, platiniridium, sperrylite, telargpalite, telluropalladinite and the native species palladium and platinum. Most of these grains are 0.2–2 μm in diameter, although grains $>$ 15 μm are also present (Butcher et al. 1999, Power et al. 2000a).

hornblende

A name commonly used to describe brown or dark-green to black calcic amphiboles in many volcanic and intrusive rocks. It is not a species name, although ferrohornblende and magnesiohornblende are. However, not all hornblende descriptions refer to these two species, and other members of the amphibole group are possible (e.g. hastingsite, kaersutite, pargasite or tschermakite). Since the introduction of complex classification schemes for amphibole nomenclature (Leake et al. 1978, 1997), it is not usually possible to identify amphibole species positively without a chemical analysis. Hornblende is still an acceptable colloquial name for the hand specimen or optical identification of calcic amphibole.

horn silver

A synonym of chlorargyrite.

hornstone

A red fine-grain massive variety of quartz ±moganite.

horseflesh ore

A synonym of bornite; a term used by Cornish miners, referring to the purplish-red colour of the ore.

hortonolite

A manganese-rich variety of fayalite, olivine group.

hovite (near collyrite)

TL Hove, Brighton, Sussex, England

Possibly an impure scarbroite.

hübnerite

monoclinic $Mn^{2+}WO_4$

A common manganese tungstate that forms a series with ferberite and is found in high-temperature hydrothermal veins, greisens, granitic pegmatites and some alluvial deposits. Hübnerite is more common than the literature suggests, as the chemical analysis required to identify the species is often not quoted. Without an analysis, the species is usually referred to as wolframite.

EMP analysis of wolframite from a greisen-bordered vein, near the eastern contact of the St Michael's Mount granite, Marazion, has shown it to be oscillatory zoned, with individual zones of intermediate composition between ferberite and hübnerite end members. This seems to be a common phenomenon, as wolframite from a pegmatite vein near the southern contact of the Cligga Head granite, near Perranzabuloe, from Chasewater, near Redruth, and from South Crofty, Camborne, show the same feature. Because of reaction between fluids and host rocks, rapid changes in the composition of hydrothermal fluids is a plausible mechanism that could explain this variation (Moore & Howie 1978). Individual crystals of wolframite from Bunny mine, St Austell (now exposed in Gunheath china-clay pit), have intermediate compositions that classify as either ferberite or hübnerite (Tindle unpub. EMP data). Wolframite from the stockwork deposit at Hemerdon, Plympton, Devon, is more systematically zoned, from a hübnerite composition at the core, to almost end-member ferberite composition at the rim (R. C. Scrivener specimen, Tindle unpub. EMP data). Golley & Williams (1995) note hübnerite from Hingston Down quarry, 2 km W of Gunnislake, Calstock and Wheal Gorland, St Day, 2 km E of Redruth, both in Cornwall, but these descriptions appear to be based on samples in the Kingsbury collection at the NHM and they may have an unreliable provenance (Ryback et al. 1998, 2001).

At Carrock tungsten mine, Caldbeck Fells, Cumbria, wolframite and scheelite occur in subvertical quartz veins cutting the Skiddaw (Grainsgill) granite, a greisenized cupola developed on the north flank of the Skiddaw granite. It forms elongate bladed crystals, often several cm long, typically in clusters at the edges of veins. Individual crystals show chemical zonation with intermediate compositions that are closest to the hübnerite end member. Where hübnerite occurs adjacent to vugs in the vein quartz, it is fractured and partially replaced by scheelite, arsenopyrite, pyrite and carbonate minerals and rarely by quartz (Beddoe-Stephens & Fortey 1981). Ball et al. (1985a) quote a larger range of wolframite compositions from Carrock mine, which indicates that ferberite is also present, but it is unclear whether this represents a different phase of mineralization. An earlier report of hübnerite from Carrock mine was given by Kingsbury (1959a). Young (1987a) describes reddish-brown crystalline hübnerite from open workings on the Combside vein on the north flank of Coombe Height, but this is a Kingsbury specimen whose provenance requires confirmation.

A very unusual niobium-rich variety of wolframite occurs in a wolframite-bearing quartz vein cutting zinnwaldite granite near Gairnshiel Bridge, Glen Gairn, Grampian, Scotland. The wolframite is strongly zoned, with very variable niobium contents and compositions that approach pure end-member hübnerite. Blades of hübnerite reaching 2–3 cm in length are partially replaced by scheelite along discrete zones. Cassiterite is an associated species. Surrounding the vein is a zone of silicified topaz-bearing greisen that contains niobium wolframite, zoned between ferberite and hübnerite compositions. Russellite and stolzite are found in the greisen as alteration products of the wolframite (Tindle & Webb 1989, Webb et al. 1992, Livingstone 1993).

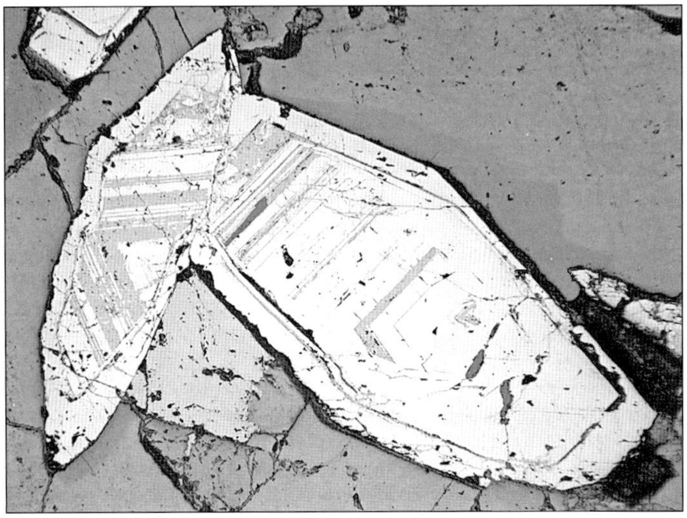

Zoned crystals of hübnerite from a quartz vein at Gairnshiel Bridge, Glen Gairn, Grampian Region, Scotland. The medium-grey crystal between the hübnerite crystals, and the darker zones in the hübnerite, are scheelite; width 1.0 mm; viewed in reflected light. (© A. G. Tindle)

hudsonite

A synonym of hastingsite, amphibole group.

hullite

TL Carnmoney Hill, near Belfast, Co. Antrim, Ireland

Named after Edward Hull FRS (1829–1917), stratigrapher and economic geologist, who was director of the Irish Geological Survey, Dublin, and a professor at the Dublin Royal College of Sciences; born in Antrim town, Co. Antrim, Ireland. Hullite is possibly identical to chlorophaeite and has also been described as magnesium-rich chamosite, chlorite group.

humboldtilite

An intermediate composition between åkermanite and gehlenite, melilite group, or any member of the melilite group.

humboldtine?

monoclinic $Fe^{2+}C_2O_4{\cdot}2H_2O$

A rare iron oxalate forming a yellow microcrystalline crust associated with crystals of cassiterite, quartz and tourmaline on a sample from Pendarves mine, Camborne, Cornwall (Ryback & Tandy 1992). The Pendarves sample, and another from Wheal Jane near by, came from the deep levels of the mines and may be artificial (Green 2004). An organic mineral such as humboldtine is not expected in this environment, but if iron minerals being cleaned in oxalic acid to remove iron staining are left too long in the acid, the result will be a precipitate of iron oxalate – humboldtine. It is quite likely, therefore, that the reported humboldtine could be anthropogenic.

HUMITE GROUP

Orthorhombic or monoclinic silicates with compositions A_2SiO_4, $A_2SiO_4{\cdot}A(OH)_2$, $2A_2SiO_4{\cdot}A(OH)_2$, $3A_2SiO_4{\cdot}A(OH)_2$, $4A_2SiO_4{\cdot}A(OH)_2$, where $A = Fe^{2+}$, Mg, Mn^{2+}; (OH) is partially replaced by F. Alleghanyite, chondrodite, clinohumite, humite, manganhumite, norbergite and sonolite are members of the group found in Britain or Ireland.

humite

HUMITE GROUP

orthorhombic $Mg_7(SiO_4)_3(F,OH)_2$

Named after Sir Abraham Hume FRS (1749–1838), English connoisseur and collector of works of art, precious stones and minerals; born in Berkeley Square, London. Humite is a rare magnesium silicate fluorhydroxide found at Kilbride and Kilchrist, near Broadford, on the Isle of Skye, Highland Region, Scotland, where the Beinn an Dubhaich granite has intruded dolomitic limestone and produced a diverse skarn assemblage. Humite, with chondrodite, clinohumite, cuspidine, forsterite and monticellite, is characteristic of the ore (mainly magnetite) skarns on the limestone side of the contact zone. At Kilbride, humite is present in parallel growth with clinohumite and with chondrodite at Kilchrist (No. 1 Prospect). This assemblage belongs to the group II skarn zone and is associated with boron–fluorine metasomatism (Tilley 1948b, 1951). In a study of contact metamorphism around the Beinn an

Dubhaich granite, Hoersch (1981) notes humite-group minerals occurring with flake-like form or in shreddy aggregates from a zone of Strath Suardal dolomite to the S and E of Ben Suardal.

At Bay Steinigie, near Finsbay (Fionnsabhagh), South Harris, Western Isles, Scotland, Lewisian metamorphosed limestone is found in a long narrow xenolith within the South Harris igneous complex. Small yellow grains of humite and chondrodite (about 0.3 mm across) occur in an isolated outcrop of rock consisting mainly of calcite, dolomite and forsterite. Other minerals present include brucite, graphite, pargasite, phlogopite, titanite and opaque iron oxides (Dearnley 1961).

hureaulite?

monoclinic $Mn_5^{2+}(PO_4)_2[PO_3(OH)]_2 \cdot 4H_2O$

A rare hydrated manganese phosphate hydroxide reported from a fine-grain spodumene pegmatite at Moylisha, near Shillelagh, Co. Wicklow, Ireland. A salmon-pink alteration coating on the pegmatite is described as "salmonsite" – a mixture of hureaulite and jahnsite. A similar occurrence, on boulders of spodumene pegmatite, is also noted at Seskinnamadra, W of Tomduff Hill, Co. Carlow, Ireland (Scoon 1978).

hyacinth

A name used primarily to describe reddish gem-quality zircon, but also to describe corundum, grossular or vesuvianite.

hyaline quartz

A claret variety of chalcedony (quartz).

hyalite

A colourless variety of opal found as globular concretions. The name has also been used as a synonym of axinite.

hyalophane

FELDSPAR GROUP

monoclinic $(K,Ba)Al(Si,Al)_3O_8$

An uncommon potassium–barium aluminosilicate, intermediate in composition between orthoclase and celsian. Hyalophane is reported from a stratabound zone of barium and lead–zinc mineralization in Middle Dalradian metasedimentary rocks N of Aberfeldy, Tayside Region, Scotland. The mineralization is contained in graphitic schists of the Ben Eagach Schist Formation and it consists of interbanded barite rock, celsian–quartz rock and sulphide-bearing carbonate rocks. Hyalophane, together with cymrite, magnetite, fuchsite mica and rutile, is a minor constituent of an ore dominated by barite, celsian, pyrite, pyrrhotite and sphalerite (Coats et al. 1980, Willan & Coleman 1983). Analysis of barium feldspars from the Ben Eagach deposit and nearby Foss mine by Moles (1985) indicate a complete range of compositions between hyalophane and nearly pure celsian. In one specimen, composite porphyroblasts with a complex internal structure consist of co-existing hyalophane and celsian in a muscovite–quartz schist (Fortey & Beddoe-Stephens 1982).

In the area around Beinn Heasgarnich, S of Loch Lyon and 50 km SW of Aberfeldy, stratigraphically similar schists contain barium mineralization in which barite, hyalophane, and barium-enriched muscovite are the main ore species. Hyalophane forms a granoblastic mosaic (50–500 µm grainsize) with barite, calcite and quartz, alternating with bands of undeformed muscovite and interstitial hyalophane. Partial replacement of muscovite by hyalophane is also noted (Coats et al. 1984).

At Corrie Loch Kander, near Braemar, Grampian, Middle Dalradian rocks of the Argyll Group contain a unit best described as a porphyroblastic hyalophane–armenite-bearing calc-silicate-rich cherty sulphidic quartzite. The rock, which may originally have been a dolomitic marl, also contains diopside, galena, pyrite, pyrrhotite, sphalerite and tremolite–actinolite, with minor chalcopyrite, clinozoisite, titanite and rare barite. Hyalophane is locally mantled and partially replaced by armenite. The mineralization appears to be an eastern extension of the Aberfeldy and Beinn Heasgarnich assemblages and is interpreted as the result of seafloor exhalative activity related to subsidence and lateral basin extension during the deposition of the Middle Dalradian (Gallagher et al. 1989a, Fortey et al. 1991, Fortey et al. 1993).

Lower Carboniferous shallow-water oolitic limestones at Moyvoughly, near Moate, 12 km N of Clara, Co. Westmeath, Ireland, have acted as host for lead–zinc mineralization and, in the process, have been extensively replaced. In sphalerite-bearing samples, one such replacement is by celsian and dolomite, in which celsian shows a degree of solid solution towards the hyalophane end-member (Kucha 1988).

In Gwynedd, Wales, hyalophane is noted from Benallt mine, Rhiw, Lleyn Peninsula and Cyffty mine, between Betws-y-coed and Llanrwst (T. F. Cotterell pers. comm. 2003).

hyalosiderite

A glassy variety of olivine with a composition intermediate between fayalite and forsterite.

hydrargillite

A synonym of aluminite, gibbsite, turquoise and wavellite.

hydrated labradorite

A misidentification of calcium-rich marialite, scapolite group.

hydro-andradrite

An hydrated variety of andradrite, garnet group.

hydrobasaluminite

monoclinic $Al_4(SO_4)(OH)_{10} \cdot 12H_2O$

TL Lodge Pit, Irchester, near Wellingborough, Northamptonshire, England

A rare hydrated aluminium sulphate hydroxide that coats joint surfaces and fills fissures up to 30 cm wide in Northampton Ironstone (Inferior Oolite) at the type locality. Hydrobasaluminite is a white plastic clay-like mineral that, on drying, forms a white powder of felsöbányaite (syn. basaluminite) composition and structure. It is associated with allophane, aragonite, gypsum, goethite and halloysite in discrete layers. The fissures also contain a brecciated oolite containing chamosite and siderite (chalybite), above which is a siderite mudstone. These fissures probably acted as traps for the soluble products of weathering, from which the alumina-rich minerals precipitated (Bannister & Hollingworth 1948, Hollingworth & Bannister 1950).

Hydrobasaluminite and felsöbányaite (syn. basaluminite) have also been found in the weathering zone of the Oxford Clay at Crook Hill brickyard, Chickerell, Dorset. They occur in reaction rims surrounding carbonate concretions, although the latter species is found only on those concretions that have fallen to the floor of the pit, suggesting that it formed as a dehydration product of hydrobasaluminite (Clayton 1980).

hydrobiotite

A 1:1 regular interstratification of biotite and vermiculite. Hydrobiotite was listed as a Welsh species by Bevins (1994) on the basis of its occurrence in highly fractured phyllitic rock associated with the Coed-y-Brenin copper mineralization, NNE of Dolgellau, Gwynedd (Rice & Sharp 1976).

hydrocalcite

A synonym of monohydrocalcite.

hydrocalumite

monoclinic $Ca_4Al_2(OH)_{12}(Cl,CO_3,OH)_{2-x} \cdot 4H_2O$

TL Scawt Hill, Larne, Co. Antrim, Ireland

A rare hydrated calcium–aluminium hydroxide chloride carbonate found in vugs within larnite rock carrying melilite–wollastonite aggregates from the chalk/dolerite contact zone at Scawt Hill. Hydrocalumite forms light-greenish crystals showing a perfect cleavage and is intimately associated with afwillite. Ettringite and portlandite are also found with the hydrocalumite (Tilley 1934).

hydrocarbons

Hydrocarbon compounds are usually not considered to be mineral species, as they have non-crystalline structures and organic origins (fichtelite is the only British exception). To avoid confusion with mineral names, some examples are listed here: adipocerite (TL Wales), albafite, albertite, alyphite, amber, anthracite, argillo-bitumen, argilole, aromite, asphaltite, asphaltum, bathvillite (TL Scotland), bernalite (named after J. D. Bernal, p. 80), bitumen, bitumenite, bituminous flag, black amber, bog butter, boghead coal (TL Scotland), brewsterlinite, butyrellite, butyrite, cannels, caoutchouc (TL England), carbonite, carboresinite, cloustonite (TL Scotland; named after C. Clouston, p. 154), coal, copalite (TL England), cryptoline, cyclite, cyclo-resinite, elaterite (TL England), fichtelite, foxite, fusain, gilsonite, grahamite, guay-aquillite, hartite, hatchettite (TL Wales; named after C. Hatchett, p. 266), highgate resin (TL England), humacite, impsonite, jet, lignite, metabituminite, middletonite (TL England), mineral charcoal, mountain tallow (TL Scotland), mutabilite, naptha, nigrite, olefinite, oxycarbonite, oxicyclite, oxy-olefinite, ozocerite, paraffinite, parrot coal (TL Scotland), peat, peat jet, pellite, petroleum, pigotite (TL England), polyelaterite, pyrobitumen, retinasphalt, retinite, schraufite, scleretinite (TL England), settlingite (TL England), splint coal, torbanite (TL Scotland), uintahite and urpethite (TL England).

hydrocerussite

trigonal $Pb_3(CO_3)_2(OH)_2$

An uncommon, but locally abundant, lead carbonate hydroxide found in the oxidized zone of lead deposits.

Hydrocerussite is apparently quite rare in Cornwall, although it has been reported from Greystone quarry, Lezant, near Launceston. It was found as pearly white flakes and radiating aggregates up to 1 mm across in the Greystone lead lode (Elton et al. 2004). In Somerset, hydrocerussite appears to be much more common. It is most often described from Merehead (Torr Works) quarry, Cranmore, where it occurs in massive manganese oxide bodies associated with cerussite, mendipite and other rare secondary lead minerals (blixite, chloroxiphite, diaboléite, mereheadite, paralaurionite and parkinsonite). The mineralization is found within joints and faults in Carboniferous Limestone (Alabaster 1975, 1977, Symes 1977, Symes & Embrey 1977, Symes et al. 1994, Welch et al. 2000). Hydrocerussite occurs in a very similar paragenesis at Higher Pitts farm, near Priddy, Mendip Hills. Manganese and iron oxides and small concentrations of lead ore occur within pockets in dolomitic conglomerate at the base of the Carboniferous Limestone. Hydrocerussite occurs as a zone of alteration around mendipite or as isolated patches associated with cerussite. Crystals are enormous when compared to those from other locations known at the time and are either thick tabular up to 2–3 cm across or lenticular and up to 6 cm across (Spencer & Mountain 1923, Kingsbury 1941, Burr 1996). Hydrocerussite has also been found at other locations in the Mendip Hills: Batts Combe quarry, near Cheddar (Eastwood 1983); Coleman's quarry, Holwell, near Frome (Alabaster 1990a); Halls quarry, Felton, S of Bristol; Avon Ham Wood, near Shepton Mallet; Mendip quarry, Leighton, W of Shepton Mallet (all Symes & Embrey 1977); and Wesley Theological College (Wesley mine), Westbury-on-Trym, Avon (Alabaster 1989b).

In the general area of Alderley Edge (including West mine) and Clive (Shropshire) copper mining areas, Cheshire, hydrocerussite is reported with azurite, cerussite, chrysocolla, dioptase, malachite, mimetite, pyromorphite and smithsonite in a secondary As–Co–Cu–Ni–P–Pb carbonate, hydrated silicate, oxide–hydroxide assemblage (Plant et al. 1999).

At Milltown limestone quarry, near Ashover, Derbyshire, cream to very pale-orange hydrocerussite occurs: as hexagonal plates up to 0.5 mm across sometimes associated with calcite, gypsum and weddellite; as plates up to 0.3 mm across in cavities in galena; and as flaky cellular masses to 12 mm across. In the zinc hydroxide zone in this quarry (where the rare species ashoverite, sweetite, wülfingite and an unnamed zinc hydroxide have been found), hydrocerussite replaces cerussite. The possible presence of a lime kiln in the quarry could have a bearing on the formation of the zinc hydroxides at this location (Clark et al. 1984, Rust 1991c). Ford et al. (1993) note hydrocerussite from Long Tor mine, Matlock Bath, Derbyshire.

Because inaccuracies have been discovered with the provenance of A. W. G. Kingsbury specimens (Ryback et al. 1998, 2001), occurrences of hydrocerussite from the following locations in the Caldbeck Fells, Cumbria (as reported by Hartley 1984, Briscoe 1986, Cooper & Stanley 1990, Anthony et al. 2003) require further confirmation: Balliway Rigg, Brae Fell mine, Carrock Fell crosscut at Grainsgill, Driggith mine, Poddy Gill, Red Gill mine, Roughton Gill mine, Sandbed mine and Swinburn Gill. Independent verification of hydrocerussite is available only for Balliway Rigg (Green & Briscoe 2002) and Driggith mine (Rust et al. 2003). Green et al. (2006a) corroborate the Driggith and Sandbed occurrences, adding there is no reason to doubt Hartley (1984), particularly as no specimens have been found in the Kingsbury collection. Hydrocerussite from Burtness Wood, Buttermere, Cumbria (Rothwell & Mason 1992), and Wet Swale Gill, Caldbeck Fells (Young 1987a), do appear to be genuine. A single specimen of hydrocerussite, forming white to pale-green pearly crusts on partly oxidized sulphides, is noted from Silver Gill, Caldbeck Fells (Green et al. 2005b). Bridges et al. (2006) report that rare hydrocerussite occurs at Brae Fell mine, adding that Kingsbury specimens listed from this locality probably came from the Leadhills–Wanlockhead orefield on the Dumfries & Galloway/ Strathclyde border, Scotland.

Hydrocerussite is confirmed as a supergene species from several Cu–Pb–Zn mine sites in the Central Wales orefield of Dyfed and Powys, Wales. At Esgair Hir mine, Ceulanymaesmawr, Dyfed, it occurs in four forms: as white-to-greenish divergent fan-like crystal sprays up to 0.4 mm across, as flattened hexagonal crystals up to 0.4 mm across with shallow pyramidal faces, as colourless-to-white zoned triangular-sectioned prismatic crystals up to 0.5 mm long and, most commonly, as a pearly white-to-creamy encrustations. Bechererite, cerussite, hemimorphite, hydrozincite, leadhillite, linarite, malachite and schulenbergite are associated species (Rust & Mason 1988, Rust et al. 2003). Hydrocerussite forms rare groups of creamy-white acicular crystals up to 0.7 mm long and tiny tabular crudely hexagonal crystals up to 0.4 mm across at Frongoch mine, between Devil's Bridge and Pont-rhyd-y-groes, Dyfed, where it is associated with bechererite, caledonite and susannite (Green et al. 1996b, Rust et al. 2003). Hydrocerussite forms encrustations of colourless-to-white platy crystals associated with anglesite, caledonite and mattheddleite at Darren mine, Pen-bont Rhydybeddau, Dyfed (Rust 1990), and is also noted at Llechweddhelyg mine, near Clawddmelyn, between Bont-goch and Cwmsymlog (Bevins 1994). In Powys, hydrocerussite has been found at two localities in the Dylife sett: on the dumps below the Star inn and on a dump S of Penycrocbren Hill related to the Dylife–Dyfngwm sett, near Machynlleth, where it occurs as cream-to-white colourless zoned prismatic crystals or divergent crystals in masses up to 0.8 mm across. Associated species include bechererite, caledonite, lautenthalite, langite, namuwite, schulenbergite and susannite (Rust & Rust 1987, Rust et al. 2003).

Intergrown thin plates of hydrocerussite to 1.25 mm with minor pale-green pyromorphite; Higher Roughton Gill, Caldbeck Fells, Cumbria. (© M. P. Cooper)

In Scotland, hydrocerussite is a supergene alteration product restricted to the lead–zinc veins of Leadhills and Wanlockhead. It forms white coatings in cavities in galena, associated with cerussite and plumbocalcite in the Belton Grain vein (Heddle 1901, Wilson & Flett 1921, Livingstone 2002). In the New Glencrieff vein it forms thin silvery crystals associated with aurichalcite and cerussite (Brown 1919). Hydrocerussite also occurs on a specimen of chenite from Susanna mine, with caledonite, leadhillite, linarite, lanarkite and susannite (Paar et al. 1986).

hydrodolomite

A term used to describe hydromagnesite ± calcite.

hydrogarnet

An OH-rich grossular or katoite. See hydrogrossular.

hydroglockerite

TL Parys Mountain, Anglesey, Wales

An indefinite composition found to contain more loosely bound water than glockerite does; also said to be hydrated lepidocrocite.

hydrogrossular

A name originally used to describe a range of hydrous garnets between grossular and hibschite compositions. With the discovery of katoite $Ca_3Al_2(SiO_4)_{3-x}$ (x = 1.5–3), hydrogrossular is considered to be a group name for hydrated garnets in the series

$Ca_3Al_2(SiO_4)_3$ to $Ca_3Al_2(OH)_{12}$ (Clark 1993). This grouping is not recognized by Mandarino & Back (2004). *MBI* discusses all occurrences of hydrogrossular as hibschite, as there are no references to katoite in the British or Irish literature.

hydrohematite
An hydrated variety of hematite.

hydrohetaerolite
tetragonal $Zn_2Mn^{3+}_4O_8 \cdot H_2O$

A rare hydrated zinc–manganese oxide found on calcite in oxidized veinstone from Copperthwaite vein, Swaledale, Yorkshire (Ryback & Tandy 1992). Hydrohetaerolite (in very minor amounts) is also one of the secondary metalliferous species found at Ballygown South mine, near Silvermines, Co. Tipperary, Ireland, where it occurs with coronadite as thin black crusts on limonite containing hemimorphite (Ryback & Tandy 1992, Moreton 1999).

hydrohonessite
hexagonal $Ni_6Fe^{3+}_2(SO_4)(OH)_{16} \cdot 7H_2O$

A rare hydrated nickel–iron sulphate hydroxide found on chromitite at Hagdale quarry, Unst, Shetland, Scotland, where it forms fine-grain yellow encrustations associated with honessite, reevesite and theophrastite. The specimens also contain aragonite, chlorite and goethite, and lesser amounts of heazlewoodite and pentlandite (Bish & Livingstone 1981, Livingstone & Bish 1982, Livingstone 2002). At Craignure mine, 12km SW of Inveraray, Strathclyde, Scotland, dump material consisting mainly of chlorite schist with narrow quartz veins and calcite-rich bands, contain aluminium-copper analogues of honessite and hydrohonessite as bluish-green coatings associated with reevesite (Livingstone 1990b).

hydrolite
A synonym of gmelinite (zeolite group) and siliceous sinter.

hydromagnesite
monoclinic $Mg_5(CO_3)_4(OH)_2 \cdot 4H_2O$

A rare hydrated magnesium carbonate hydroxide found encrusting and infilling fractures in altered ultramafic rocks and serpentinites. Hydromagnesite also occurs in hydrothermally altered limestones.

At Merehead (Torr Works) quarry, Cranmore, Somerset, hydromagnesite is an alteration product of brucite that has developed in the central cavities of two calcite nodules. These nodules were found in iron- and manganese-rich mineralized faults and fissures that transect Lower Carboniferous limestones. Blixite is an associated species in one nodule (Alabaster 1977). In the northern Pennines, at Brownley Hill mine, Nenthead, near Alston, Cumbria, hydromagnesite spherules to 1 mm in diameter richly encrust alstonite crystals found in cavities within the sulphide-rich vein filling Holmes's rise on High Cross vein (Green et al. 2000, Green & Briscoe 2002, Green & Young 2006). Hydromagnesite also occurs as white spherules on alstonite at Fallowfield mine, Hexham, Northumberland (Green & Young 2006).

Durness Limestone horizons beyond the skarn zones, but still within the aureole of the Beinn an Dubhaich granite, Broadford area, Isle of Skye, Highland Region, Scotland, contain hydromagnesite. Associated species are brucite, chlorite, diopside, forsterite, grossular, periclase, phlogopite, serpentine, spinel, talc, tremolite, vesuvianite and rare wollastonite (Tilley 1948b, 1951).

At Hagdale quarry, Unst, Shetland, Scotland, hydromagnesite forms white, sometimes radiating, crusts on chromitite in association with clinochlore (kämmererite), nakauriite, theophrastite and material with a composition close to pyroaurite (Braithwaite & Pritchard 1983b). On Unst and on the nearby island of Fetlar, screes on high serpentinite seacliffs have been cemented by deposits (up to 1 cm thick) of brucite, hydromagnesite and minor aragonite. These travertine deposits are best developed at Hill of Clibberwick on Unst and East Neap on Fetlar. The source of water for the deposits was rain (occasionally contaminated by sea spray). Calcium and magnesium were dissolved from the serpentinite and later precipitated as the calcium- or magnesium-rich species aragonite, brucite and hydromagnesite (Flinn & Pentecost 1995). Hydromagnesite is also reported from cavities in brucite-rich veins associated with serpentinite at Swinna Ness, Unst. Hydromagnesite forms small (<3 mm) lustrous crystals and an earthy form is also noted to coat brucite (Greg & Lettsom 1858, Heddle 1878a, 1901). Unfortunately, this occurrence is unsubstantiated (Macpherson & Livingstone 1982). Rare rosettes of hydromagnesite crystals on serpentinite from Portsoy, Grampian (Heddle 1901) are also doubtful.

Hydromagnesite is also noted to encrust, as stellate aggregates, a specimen of serpentinite collected to the north of Louisburgh, Co. Mayo, Ireland (Nawaz 1978b).

Hydromagnesite balls to 0.4 mm from Brownley Hill mine, Nenthead, Cumbria. (© D. I. Green)

hydromica
An obsolete group name for altered micas, low in alkali and high in water.

hydromuscovite
TL Ogofau, near Pumsaint, Dyfed, Wales

A variety of muscovite or hydrated muscovite grading into illite. Hydromuscovite has also been used as a group name for altered muscovite (damourite, gilbertite, margarodite and sericite).

hydronephelite
Probably a mixture of nepheline or natrolite, or both, with diaspore or gibbsite.

hydronium jarosite/hydroniumjarosite
ALUNITE GROUP

trigonal $(H_3O)Fe^{3+}_3(SO_4)_2(OH)_6$

An uncommon hydronium iron sulphate hydroxide formed by oxidation of sulphides in Carboniferous shales on the coast at Ballybunnion (near the castle ruin), Co. Kerry, Ireland. Additionally, this process has produced brochantite, felsőbányaite (syn. basaluminite), gypsum, metasideronatrite and sideronatrite in nearby coves. Hydronium jarosite forms dull earthy pale-yellow encrustations on joint surfaces of a small exposure of black shale some 10–20 m above sea level (Ryback & Moreton 1993, Moreton et al. 1995).

Hydronium jarosite is also reported from Muckross mine, Killarney National Park, Co. Kerry, Ireland, where it commonly forms brown powdery masses up to 6 cm across on material from the mine dump. It is one of 17 secondary species formed by post-mining oxidation of mine waste. Such a range of secondary species were able to form because of a varied primary assemblage that includes arsenic, cobalt, copper, iron, lead and zinc mineralization (Moreton et al. 1999).

In Wales, hydronium jarosite is confirmed from Brynycastell (Cross Foxes) mine, Dolgellau, Gwynedd, where it occurs in bright sulphur-yellow efflorescences over a 35 cm-wide area on pyrite-rich pisolitic iron ore. On close inspection the efflorescences proved to be composed of delicate acicular sprays of gypsum intimately intergrown with hydronium jarosite (Cotterell 2006a). Post-mining hydronium jarosite is also reported as extensive deposits in the underground passages of Parys

Mountain, Anglesey (Jenkins & Johnson 1993). However, XRD patterns of many samples do not conform to those of known minerals, and Bevins (1994) suggests that further verification is required.

hydrophane

A semi-opaque variety of opal that becomes translucent or transparent when it is immersed in water.

hydrophilite

A synonym of antarcticite.

hydroplumbite

A very doubtful species, probably hydrocerussite. Said to have come from Cumbria, but suspected to be from Leadhills, Strathclyde, Scotland (Heddle 1889a).

hydroscarbroite

triclinic $Al_{14}(CO_3)_3(OH)_{36}\cdot nH_2O$

TL White Nab, South Bay, Scarborough, Yorkshire, England

A rare hydrated aluminocarbonate hydroxide found in vertical fissures within Jurassic sandstone in a small cove just south of South Bay, Scarborough. White, compact or powdery hydroscarbroite and scarbroite line fissures, together with dark-brown or grey calcite, gibbsite, kaolinite, and varying amounts of quartz. Hydroscarbroite spontaneously dehydrates to scarbroite when the material is equilibrated at room temperature; the transformation is irreversible (Duffin & Goodyear 1960, King 1982).

Hydroscarbroite and scarbroite form massive compact chalky-looking white microgranular cavity infillings up to 4 cm in diameter and thin veinlets associated with deeply oxidized smithsonite-bearing veins in Upper Triassic limestone breccias, near East Harptree, Mendip Hills, Avon. Allophane, halloysite and kaolinite are associated species (Alabaster 1990b).

HYDROTALCITE GROUP

Trigonal carbonates with the general formula $A_6B_2(CO_3)(OH)_{16}\cdot 4H_2O$, where A = Mg, Ni, and B = Al, Cr^{3+}, Fe^{3+}, Mn^{2+}, Co^{3+}. Hydrotalcite, hydrowoodwardite, pyroaurite, reevesite and possibly stichtite are members of the group found in Britain or Ireland.

hydrotalcite

HYDROTALCITE GROUP

trigonal $Mg_6Al_2(CO_3)(OH)_{16}\cdot 4H_2O$

A rare hydrated magnesium–aluminium carbonate hydroxide, dimorphous with manasseite and found in chromitite from the southwestern headwaters of Poundland Burn, Colmonell, Carrick, Strathclyde, Scotland (Livingstone 2002).

Colourless botryoidal crusts of what was thought to be allophane, which lined a smithsonite cavity at East Harptree, Mendip Hills, Avon, gave an XRD pattern that closely matched hydrotalcite. The crusts consist of a jumbled mass of slender prismatic crystals up to 20 µm long composed of three equal prism faces with large pinacoid terminations. Larger, triangular or hexagonal prisms up to 50 µm long are sparingly distributed throughout the crusts. Allophane invariably underlies the hydrotalcite crusts (Alabaster 1990b).

hydrotungstite

monoclinic $H_2WO_4\cdot H_2O$

A rare hydrated tungsten oxide found as a bright-yellow powdery coating on a niobium-rich wolframite (hübnerite) crystal in a wolframite-bearing quartz vein cutting zinnwaldite granite near Gairnshiel Bridge, Glen Gairn, Grampian, Scotland (Tindle & Webb 1989, Livingstone 1993).

hydrowoodwardite

HYDROTALCITE GROUP

trigonal $(Cu,Al)_9(SO_4)_2(OH)_{18}\cdot nH_2O$

Named after Samuel Pickworth Woodward (1821–1865), an English naturalist and geologist; born in Norwich, Norfolk. Hydrowoodwardite is a rare hydrated copper–aluminium sulphate hydroxide that forms pale-blue botryoidal coatings at Ffridd-Goch mine, Llanfachreth, and Simdde Dyllhuan, Drwys-y-coed, near Nantlle, both in Gwynedd, Wales. The samples are in the collection of the NMW, where they are labelled as "woodwardite coating grass" and "woodwardite coating bracken". However, the Simdde Dyllhuan material has been examined by Nickel (1976), who concluded it was not woodwardite, and by Raade et al. (1985), who describe it as the copper analogue of glaucocerinite. The confusion was ended when Witske (1999) showed that the material was the newly described species, hydrowoodwardite.

hydroxyapophyllite/apophyllite-(KOH)

tetragonal $KCa_4Si_8O_{20}(OH,F)\cdot 8H_2O$

An uncommon hydrated potassium–calcium silicate hydroxide silicate that forms a series with fluorapophyllite and whose name came into existence when the apophyllite group was established (Dunn & Wilson 1978). Before this, apophyllite was the species name for all members of the group.

Colourless crystals of hydroxyapophyllite up to 5 mm long are found within calcite–chlorite–pectolite–quartz veins in the late Carboniferous quartz dolerite of the Whin Sill at Copt Hill quarry, Cowshill, Weardale, Co. Durham (Young et al. 1991).

Hydroxyapophyllite is found lining vugs in Tertiary basalts on the Isle of Skye, Highland Region, Scotland, at Dunvegan, Loch Bracadale and Quiraing (Livingstone 1993). Hydroxyapophyllite is also confirmed in amygdaloidal basalts from two other localities on the Isle of Skye: at the Storr, N of Portree, subtransparent colourless to milky-white hydroxyapophyllite forms pseudo-hexagonal crystals (up to 4 mm across) growing on beds of earlier-formed fibrous mesolite within partially filled amygdales; at Moonen Bay, on the Duirinish Peninsula, an altered amygdaloidal basalt contains small (up to 300 µm) interstitial crystals of fluorapophyllite, hydroxyapophyllite and fibrous mesolite in cavities within amygdales mostly filled with massive fibrous gyrolite. Biod a' Mhurian on the Duirinish Peninsula, Isle of Skye, is a further location for hydroxyapophyllite (S. A. Rust and D. I. Green specimens, Tindle unpub. EMP data).

Hydroxyapophyllite is confirmed from vesicles in Tertiary basalt at Maxwell's quarry, Scotstown Road area, 8 km from Ballymena, Co. Antrim, Ireland (H. J. Foy specimen, Tindle unpub. EMP data). Hydroxyapophyllite has also been found at Clinty and Cross Moorfields quarries, Ballymena, Co. Antrim, and at Budore quarry, 12 km W of Belfast (S. Moreton specimens, Tindle unpub. EMP data).

Euhedral hydroxyapophyllite associated with fibrous thomsonite-Ca (dark grey) and gyrolite (light grey) in vesicular basalt from Maxwell's quarry, near Ballymena, Co. Antrim, Ireland; width 1 mm; BSE image. (© E. Condliffe, University of Leeds)

hydroxylapatite/apatite-(CaOH)

APATITE GROUP

hexagonal $Ca_5(PO_4)_3(OH)$

An uncommon calcium phosphate hydroxide found in the vuggy parts of quartz veins cutting the main Bunny lodes at Gunheath china-clay pit, Carthew, St Austell, Cornwall. Beautiful lilac-to-violet fluorapatite crystals are partially altered to a milky-white hydroxylapatite. Pastel-blue to silky-white hydroxylapatite also occurs capping pencil-shape dark greenish-blue apatite prisms in pegmatite from the northern face of Gunheath pit (Weiss 1994).

In Wales, carbonaceous dirt bands in anthracite containing 40% hydroxylapatite and 10% kaolinite are recorded from the Lower Hornlo (Red vein) coal seam in the Brynamman district of the South Wales coalfield, West Glamorgan (Baqri & Hodson 1979). The identification requires confirmation (Bevins 1994).

An investigation of a chalky-white mineral investing exquisite vanadinite specimens in the Sutcliffe–Greenbank collection, now in the NMS, has shown it to be a very unusual lead-rich variety of hydroxylapatite, with a significant carbonate-hydroxylapatite component. The specimens are thought to have originated from the High Pirn mine, Wanlockhead, Dumfries & Galloway, Scotland (Livingstone 1994b). A similar apatite from Whyte's Cleuch, Wanlockhead, was described as lead hydroxylapatite by Temple (1955) and erroneously thought by Livingstone et al. (1987) to be identical to mattheddleite.

hydrozincite

monoclinic $Zn_5(CO_3)_2(OH)_6$

A common zinc carbonate hydroxide found as a supergene oxidation product in zinc-bearing ore deposits, usually from the alteration of sphalerite or smithsonite. Hydrozincite also forms as encrustations or stalagmites in old mineworkings and caves.

In Cornwall, hydrozincite has been visually identified at Greystone quarry, Lezant, near Launceston, and Old Treburgett mine, St Teath (Golley & Williams 1995, Elton et al. 2004).

At Milltown quarry, near Ashover, Derbyshire, cream and white hydrozincite forms thin coatings on other zinc-rich minerals or forms divergent groups of lath-like crystals up to 0.5 mm long (Rust 1991c). At Waterbank Cu–Pb–Zn mine, Ecton Hill, Staffordshire, hydrozincite is a late-forming species, developing botryoidal crusts up to 8 mm thick on large areas of vein material (Rust 1992, 1995c). Hydrozincite is noted from other localities in the Southern Pennine orefield (mostly in Derbyshire): Fall Hill (Fall Gate) quarry, Ashover (Spence 1982); Golconda mine, Brassington (Ford & Sarjeant 1964a); Long Tor mine, Matlock Bath (Ford et al. 1993); Millclose mine, Wensley, near Matlock (Traill 1939); Netherwater mine, Hazlebadge, near Castleton (Braithwaite & Ryback 1963b); and Ecton Hill mines, Staffordshire (Braithwaite et al. 1964b, Robey & Porter 1972).

Hydrozincite is common throughout the Northern Pennine orefield, where it often forms white crusts of post-mining origin in mineworkings and on sphalerite-rich dumps. A massive white variety is noted from Hudgill Burn mine, between Alston and Nenthead, Cumbria (Dunham 1990). Particularly good examples form thick stalagmitic crusts around Nenthead (Brownley Hill mine, Carr's level and Smallcleugh mine) and at Scraithole mine, West Allendale, Northumberland. At Breckonsike mine, Cowshill, Weardale, Co. Durham, crusts of hydrozincite cement material on the dumps to form a breccia (Bridges & Young 1998). A supergene assemblage that includes aragonite, brochantite, chrysocolla, erythrite, gypsum, hydrozincite, linarite, malachite and wroewolfeite occurs at Tynebottom mine, NW of Garrigill, near Alston, Cumbria, where it replaces the primary species galena, sphalerite, minor chalcopyrite, marcasite and pyrite (Braithwaite 1982b, Bridges 1987).

Two reports of hydrozincite from Cumbria are based on specimens in the Kingsbury collection at the NHM and possibly have a dubious provenance: Old Potts Gill mine, Caldbeck Fells (Cooper & Stanley 1990), and Wanthwaite mine, St John's in the Vale, near Threlkeld (Young 1987a). Other reports from Cumbria are genuine: white and pale bluish-white crusts at Eagle Crag mine, Patterdale, and pale-blue crusts at Sandbed mine, Caldbeck Fells (Young 1987a; B. Young pers. comm. 2005). Green et al. (2006a) confirm hydrozincite at Sandbed mine, as thin crusts of a bright to dirty-white colour, and as globular aggregates up to 1 mm across, composed of minute lath-like crystals with a pearly lustre. The species is also noted from Barrow mine, Braithwaite, and Hartsop Hall mine, Hartsop, Patterdale (Davidson & Thomson 1951), and from Hilton mine, Scordale (Bridges & Green 2005b).

Two isolated occurrences of hydrozincite are from the Gunnerside Gill mines, Swaledale, Yorkshire (Wood 1993), and Rorrington mine (Engine shaft), to the S of Meadowtown, Shropshire (Starkey et al. 1994).

Hydrozincite is relatively abundant in Wales, particularly in the Central Wales orefield, where it typically forms white earthy coatings on sphalerite or rock matrix. In Dyfed it is noted from: Castell mine, Ponterwyd, near Aberystwyth (Jones & Moreton 1977); Cwmystwyth mine (Bevins 1994); Eaglebrook (Nantycagl) mine, Ceulanymaesmawr (Jones 1983); Esgair Hir mine, Ceulanymaesmawr (Rust & Mason 1988, Rust et al. 2003); Frongoch mine, between Devil's Bridge and Pont-rhyd-y-groes (Green et al. 1996b); Llechweddhelyg mine, near Clawddmelyn, between Bont-goch and Cwmsymlog (Jones 1987, Rust 1990); Penrhiw mine, Ystumtuen (Mason & Green 1995); and Plynlimon mine (Jones & Moreton 1977). In Powys, hydrozincite occurs at Dyfngwm mine (actually the Pen Dylife section of Dylife mine), near Machynlleth (Rust & Rust 1987) and Dylife mine (Bevins 1994). Spectacular specimens of hydrozincite with attractive rippled surfaces have been collected from Van mine, Llanidloes, Powys (Bevins 1994).

Apple- and ice-green nickel-rich varieties of hydrozincite have been found at Parc mine, near Llanrwst, Gwynedd (Alwan & Williams 1979, Bevins et al. 1982a), and, in the search for further material in the collection of the NMW, Bevins et al. (1982b) found a specimen labelled hydrozincite from Aberllyn mine, 2 km N of Betws-y-coed,

Thin coating of nickeloan hydrozincite from Parc mine, Llanrwst, Gwynedd, Wales; width 20 mm. (© National Museum of Wales, Cardiff; photographer M. P. Cooper)

Gwydyr Forest, Snowdonia, on which the new species namuwite was later identified. Post-mining hydrozincite can be found at this mine as a white mud, up to 0.5 m thick, in some of the lower levels of the mine (Jenkins & Johnson 1993). Hydrozincite is also recorded from Castell Carn Dochan mine, near Bala, Gwynedd, and Trecastell mine, near Llanrwst, N of Betws-y-coed, Gwynedd (Bevins 1994). Haggerty (1995) reports extensive hydrozincite in the old workings of Cyffty mine, between Betws-y-coed and Llanrwst, Gwynedd.

Elsewhere in Wales, the species is noted from: Machen quarry, 7 km ENE of Caerphilly, Mid Glamorgan (Plant & Jones 1995); Cwmleyshon quarry, near Draethen, Mid Glamorgan, and Llanfyrnach mine, Dyfed (Bevins 1994); and Dan-y-graig quarry, near Risca, Gwent (Alabaster 1983).

In Scotland, hydrozincite forms widespread white encrustations on sphalerite in the Leadhills–Wanlockhead area, on the Dumfries & Galloway/Strathclyde border (Wilson & Flett 1921, Temple 1995). It is also found on the dumps of West Blackcraig mine, near Newton Stewart, Dumfries & Galloway, where it is one of many supergene species replacing a primary chalcopyrite–galena–sphalerite assemblage (Livingstone et al. 1976).

Abundant thin (sometimes botryoidal) coatings of hydrozincite occur on sphalerite-bearing sulphide ores at Muckross mine, Killarney National Park, Co. Kerry, Ireland (Ryback & Moreton 1991, Moreton et al. 1999). At Glenaboghil mine, NNW of Fintown, Co. Donegal, supergene alteration of a stratiform deposit of galena and sphalerite has produced fibrous crusts of hydrozincite, together with smithsonite and greenockite (Ryback & Moreton 1992). At Sheshodonnell East mine, Co. Clare, hydrozincite occurs with cerussite and rosasite (Ryback & Moreton 1993).

hypersthene

An intermediate member of the enstatite–ferrosilite series, pyroxene group.

hypostilbite

A silica-poor variety of stilbite, zeolite group.

hystatite

A mixture of ilmenite with hematite or magnetite, possibly an exsolution mixture.

I

Ice globules coating grass outside Saddleback mine, near Mungrisdale, Cumbria (© D. I. Green)

ice

hexagonal H_2O

An extremely common species found at one time or another everywhere in Britain and Ireland. There are at least eight crystallized phases of ice, and a glassy supercooled phase is known, but only ice-I appears to occur naturally (Clark 1993). In the form of snow, it produces small acicular crystals with the shapes of stars, feathers or dendrites (Greg & Lettsom 1858). Libbrecht (2006) reports that ice crystals can also form when water vapour condenses from the air to form ice crystals on the ground. These tiny crystals produce what is commonly called frost, but if larger crystals are produced, faceting and branching becomes apparent and the name changes to hoar frost. This author describes 35 different types of snowflake based on studies in Canada and USA, noting that the most dramatic, large and photogenic crystals form at low humidity around –15°C.

iceland spar

A colourless and transparent variety of calcite with strong double-refracting properties.

idaite?

hexagonal Cu_3FeS_4

A relatively common copper–iron sulphide (with a questionable chemical formula – Cu_5FeS_6 has also been suggested) that forms as a decomposition product of bornite, commonly associated with fine spindles of chalcopyrite. Although a recognized species, it is inadequately defined. Idaite is described as a pale-brown to yellow iron-rich bornite that characteristically carries crystallographically orientated 1 µm-wide chalcopyrite laths sheathed within thinner covellite. This results in a crazed texture suggestive of progressive alteration (R. A. Ixer pers. comm. 2006).

On Alderley Edge ridge, Cheshire, veins and disseminations of redbed-type mineralization occurring in sediments of Triassic age contain an early diagenetic assemblage, followed by main-stage Cu–Pb–(Fe–Zn) mineralization accompanied by marcasite, Co–Fe–Ni sulpharsenides and tetrahedrite, all cementing clastic grains. Idaite replacing chalcopyrite belongs to a third phase of mineralization that includes a supergene assemblage of copper, lead and zinc carbonates, sulphates and limonite (Ixer & Vaughan 1982).

Close to the beach near the Pier hotel at Clevedon, near Bristol, barite–sulphide mineralization occurs in 1 m-wide fault-filling brecciated veins within Triassic Dolomitic Conglomerate, where idaite again replaces chalcopyrite in association with other secondary sulphides such as bornite, covellite, blaubleibender covellite (spionkopite and yarrowite), digenite, tennantite and limonite. So-called idaite-like material shows variation in surface colour from brownish orange to yellowish orange, and forms 40 µm-wide rims mantling the outer zones of chalcopyrite, which it replaces in a rhythmical sequence (Ixer 1986, Ixer et al. 1993).

In Co. Kerry, Ireland, idaite is described from limestone-hosted copper-polymetallic veins at Ross Island and Ardtully. Other major associated species are bornite, chalcopyrite and tennantite (Kinnaird et al. 2002). Copper mineralization associated with Devonian redbeds in the West Carbery district of Co. Cork contains a primary bornite–djurleite assemblage in which yellow bornite (idaite) occurs as a supergene alteration product. The composition of this phase is $Cu_{4.3}FeS_4$ (Ni Wen et al. 1999).

At Bouley Bay, Trinity, northeast coast of Jersey, Channel Isles, grey flow-banded rhyolites with copper mineralization in microfractures, and similar in everything but size to porphyry-copper deposits, have a paragenetic sequence from pyrite, to chalcopyrite, to enargite, and finally to idaite. Extensive alteration to covellite followed (Ixer 1980). Idaite is also described from crosscutting quartz veinlets in the La Moye aplogranite, southwest Jersey, where it is found with bornite, chalcopyrite and tetrahedrite (Ixer 1988).

iddingsite

A mixture of silicates formed by alteration of olivine.

idocrase

A synonym of vesuvianite.

igelströmite

A synonym of pyroaurite, hydrotalcite group.

ikaite?

monoclinic $CaCO_3{\cdot}6H_2O$

An uncommon hydrated calcium carbonate that itself has not been found in Britain or Ireland. However, it is the parent mineral of the calcite pseudomorphs, called jarrowite or pseudogaylussite, found during dredging of the River Tyne at Jarrow, Tyne & Wear, and the River Clyde at Cardross, Strathclyde, Scotland (Shearman & Smith 1985). The pseudomorphs are elongated prismatic and almost square to rhomb shape in cross section, with undulose faces. Ikaite transforms rapidly to calcite and vaterite at ambient temperatures (Shaikh 1990, Mikkelsen et al. 1999), preventing long-term preservation in the geological record. The presence of ikaite (now calcite) reflects very cold periglacial conditions during their formation (Livingstone 2002).

Lustrous pseudomorph of calcite after ikaite, dredged from the River Clyde muds near Cardross, Strathclyde, Scotland; length of crystal is 20 cm. (© The Trustees of the National Museums of Scotland)

ikunolite

trigonal $Bi_4(S,Se)_3$

A rare bismuth sulphide selenide found in fluorite–quartz veinstone on the dumps of North Devon United (East Wheal Friendship) mine, Peter Tavy, Devon. The mine is situated in the metamorphic aureole of the Dartmoor granite. The main sulphide present in the only specimen examined is arsenopyrite, which is associated with cobaltite and a suite of bismuth-rich minerals, including aikinite, native bismuth, bismuthinite and ikunolite. The mine is also noted for parkerite, which occurs as small (<50 μm) grains clustered together with native bismuth, bismite, cassiterite, gersdorffite, sphalerite, ullmannite and an unknown copper–lead sulphide selenide within a large grain of stannite, all partially enclosed in fluorite (Rumsey & Savage 2005).

illite (series)

MICA GROUP

monoclinic $K_{0.65}Al_{2.0}\square Al_{0.65}Si_{3.35}O_{10}(OH)_2$

A potassium aluminosilicate hydroxide that represents a compositional range of dioctahedral interlayer-deficient micas. During a reclassification of the mica group, it was decided that, as illite does not define an end-member composition (and is therefore not a true species), it should be given series status to reflect its current usage more accurately (Rieder et al. 1998).

The Carboniferous Rowhurst tonstein, a dark slightly purplish-grey mudstone encountered at Red Street colliery, near Kidsgrove, Staffordshire, contains minor illite together with apatite, crandallite, kaolinite, pyrite, quartz and carbonaceous material (Wilson et al. 1966). At Littleham Cove, near Budleigh Salterton, Devon, vanadium-rich illite or vanadium-rich mica is reported to be intergrown with native copper in a uranium–vanadium mineralized section of the Permian Littleham Mudstone Formation. Illite is the major component of the <2 μm clay fraction in these rocks (Milodowski et al. 2000, 2002). Illite is also a major component of the clay mineral content (<2 μm size fraction) of the Triassic Grey Marls sequence from St Audries Bay, near Watchet, Somerset, where it is associated with chlorite, corrensite and illite–smectite; excluding corrensite, a similar sequence also occurs near Seaton, Devon (Mayall 1979). The Upper Jurassic Kimmeridge Clay Formation from the North Sea and on shore has been shown to contain mixed-layer illite–smectite with variable amounts of kaolinite and minor chlorite (Scotchman 1987). A study of the clay fraction of Upper Cretaceous Chalk from a borehole at Sprattling Court farm, NW of Ramsgate, Kent, has shown smectite to be more abundant than illite (Spears 1979). Illite and illite–chlorite are the major constituents of a detrital clay-mineral suite in Eocene–Oligocene palaeosols at Whitecliff Bay on the Isle of Wight (Huggett & Cuadros 2005). Fibrous illite and hematite have crystallized in fault gouges and clusters of slip planes closely associated with faults in borehole material cutting Permo-Triassic St Bees Sandstone from the Sellafield area, Cumbria (Milodowski et al. 1998).

A study of the sources of dust in South Wales coalmines carried out on behalf of the Pulmonary Disease Committee of the Medical Research Council identified illite in shales associated with coal seams (Nagelschmidt & Hicks 1943).

Using hydromuscovite from Ogofau, near Pumsaint, Dyfed, Wales, Mackenzie & Milne (1953) were able to show that dry grinding has a profound effect on the crystal structure of micas and they produced a form of illite by the end of the experiment. Illite crystallinity determinations on Palaeozoic pelitic rocks from the Fishguard area of Dyfed were used by Robinson et al. (1980) to investigate Caledonian and Variscan metamorphic episodes.

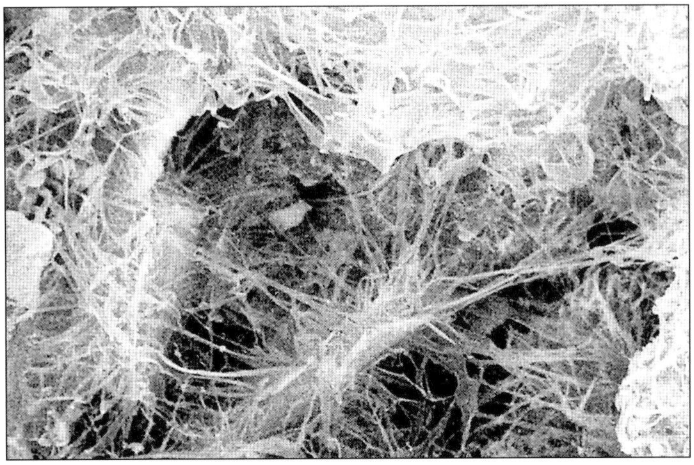

Authigenic illite forming a tangled web of filaments in Sherwood Sandstone; from a borehole at Cleethorpes, Lincolnshire; width 50 μm; SEM image. (IPR/100-52C British Geological Survey © NERC.)

In Scotland, illite has been found as a light greenish-yellow clay-like material penetrating and coating altered granite along a vein exposed during roadworks, 2 km NE of Ballater, Grampian. The illite has a platy character and, under the microscope, is nearly colourless (Mackenzie et al. 1949). In Penkiln Burn, to the S of the Loch Doon granite complex, near Newton Stewart, Dumfries & Galloway, illite, gibbsite, goethite, kaolinite and limonite form a complex intergrowth at the margins of highly altered dioritic minor intrusions. Beudantite, malachite and plumbogummite are associated species, and all are related to a distinctive low-temperature episode of disseminated sulphide mineralization (Stone & Leake 1984).

A major investigation into the timing of mineralization of Irish orebodies used the K/Ar dating technique on clay concentrates removed from close to ore minerals (mainly chalcopyrite, galena, pyrite, sphalerite) and showed in most cases that illite is a major constituent (Halliday & Mitchell 1983). Locations include:

ARMAGH Coolartragh; Lisdrumgormley. CLARE Carrahin. CORK Ballydehob; Cappaghglass; Coosheen; Crookhaven; Dereennalomane; Duneen Bay; Dooneen, Allihies; Glandore; Mountain mine, Allihies; Mount Gabriel. DOWN Conlig–Newtownards. GALWAY Tynagh. KERRY Shanagarry. LEITRIM Twigspark. LIMERICK Aherlow; Oola Hills. LONGFORD Keel. MONAGHAN Clontibret. MEATH Navan. TIPPERARY Ballynoe; Gortdrum; Shallee West (also Silvermines – Moreton 1999). WATERFORD Bunmahon; Drumslig. WEXFORD Barrystown; Caim, near Enniscorthy. WICKLOW Avoca; Ballinafunshoge; Glendalough; Luganure.

At Sark's Hope mine, Sark, Channel Islands, illite occurs with calcite and hematitic quartz as a gangue mineral in a Pb–(Ag) lode whose primary ore species include arsenopyrite, chalcopyrite, galena, marcasite, pyrite (including bravoite), pyrrhotite, tennantite and tetrahedrite, and the silver-bearing minerals acanthite, pearceite, polybasite and pyrargyrite (Ixer & Stanley 1983).

ILMENITE GROUP

Trigonal oxides with the general formula $M^{2+}TiO_3$, where M^{2+} = Fe, Mg, Mn, Zn. All members of the group are found in Britain or Ireland: ecandrewsite, geikielite, ilmenite and pyrophanite.

ilmenite

ILMENITE GROUP

trigonal $Fe^{2+}TiO_3$

A very common iron–titanium oxide that forms two series, with geikielite and with pyrophanite, and is found as a widespread accessory mineral in igneous rocks (gabbros, granites, granite pegmatites), high-grade metamorphic rocks and alpine-type veins. Ilmenite also forms alluvial deposits.

England

Alluvial deposits include: in the Durra stream, below Manaccan, Lizard Peninsula, Cornwall, where it is found with native gold (Camm 1995); Gwendra, near Coverack and Porthallow, both Lizard Peninsula (Robson 1949b). At Buttern Hill, near Bowithick on the northern edge of Bodmin Moor, cassiterite, ilmenite and wolframite are important alluvial species derived from the weathering of granite and associated vein mineralization. Between 1916 and 1918, British Tungsten Mines was set up to exploit the alluvial deposits around West Moor, near Buttern Hill, but work was abandoned when ilmenite was found to have been mistaken for wolframite (Polkinghorne 1954, Brooke 2001). Ilmenite is a minor component of the heavy-mineral assemblage collected from the Camel and Gannel estuaries, north Cornwall (Pirrie et al. 2000b), and the Fowey Estuary, south Cornwall (Pirrie et al. 2002). The release of waste from tin streaming or hard-rock mining is responsible for the anomalous trace-element content of these deposits, as cassiterite, galena, sphalerite and wolframite are the associated species. A heavy-mineral concentrate from the waste of the Goonvean and Rostrowrack china-clay pits, where a topaz-rich unit of the St Austell granite is worked, contains a manganese-rich variety of ilmenite, together with apatite, cassiterite, columbite–tantalite, ilmenorutile and lesser amounts of arsenopyrite, iron oxides, monazite, pyrite, sphalerite and zircon (Scott et al. 1998).

Hard-rock examples of ilmenite in Cornwall include two near the margin of the Land's End granite: in greenstone hornfels at Kenidjack, St Just, where streaks of ilmenite are partially replaced by titanite (Tilley 1935, Harker 1950, Chinner & Fox 1974), and in skarn deposits formed by the replacement of various rocks, mostly metapelites and metabasalts, such as at Gryll's Bunny, Botallack, where tourmaline-rich tin floors are found (van Marcke de Lummen 1985). Large (2–5 mm) irregular concentrations of ilmenite occur in white garnet–serpentine veins cutting peridotite at Enys Head, near Cadgwith, Lizard Peninsula (Hall 1979). A review of igneous and non-igneous xenoliths within the granites of southwest England lists ilmenite in all samples examined, albeit in minor amounts (the first eight being in Cornwall): Carnsew quarry, near Penrhyn, Carnmenellis; Castle-an-Dinas, Land's End (not the same as the location N of St Austell); Chywoon and Kessel Downs (Halvasso) quarries, near Longdowns, Carnmenellis; De Lank quarry, Bodmin Moor; Lamorna Cove, Land's End; Luxulyan quarry, near Lostwithiel; Sennen Cove, Land's End; also Birchtor and Pewtor on Dartmoor, Devon; and Peninnis Cove, St Mary's, Isles of Scilly (Stimac et al. 1995). Ilmenite, anatase and monazite are said to be significant in the North Tincroft lode worked at New Cook's Kitchen mine, Redruth. The mineralization is polymetallic, with arsenopyrite, cassiterite, chalcopyrite and sphalerite all found in a dense non-brecciated chlorite–fluorite–quartz matrix (LeBoutillier et al. 2000). Additionally, Golley & Williams (1995) list, but do not describe, three other Cornish locations for ilmenite: Dinas Head, near Padstow; Downgate mine, Callington; and Wheal Vincent, Altarnun.

Ilmenite is the commonest iron-bearing species in conglomerates, grits, sandstones and shales belonging to the Carboniferous Millstone Grit in Yorkshire. It is found in 32 heavy-mineral separates from the Grassington, Horsforth, Ilkley, Knarsborough, Leeds and Wharfedale areas of the county (Gilligan 1919). A skeletal variety of ilmenite is reported from a teschenite sill at Landywood, near Cannock, Staffordshire (Barton 1963). Rounded blebs of ilmenite are found in a 10–30 cm-thick mineralized vein of dioritic pegmatite within the Whin Sill, at the Forcegarth quarry, 16 km NW of Middleton-in-Teesdale and 2 km W of High Force waterfall, Co. Durham. It is closely associated with ilvaite, sphalerite and titanomagnetite, but chalcopyrite, marcasite, molybdenite, pyrite, pyrrhotite, rutile and titanite are also present (Ixer et al. 2005b). Rastall & Wilcockson (1915) report ilmenite (locally altered to leucoxene) as a rare accessory mineral in the Shap granite, Cumbria.

Wales

In Wales, ilmenite is present in many basic igneous rocks, most spectacularly in: Ordovician tuffs in central Snowdonia near Bedgellert; Cwm Meirch; around Lliwedd; in the crags of Gallt-y-llyn and Coed; Pant Agored; and near the summit of Snowdon. Crystals are manganese rich and up to 0.3 mm in length (Williams 1927, Bevins 1994). Ilmenite also occurs in the St David's Head layered gabbroic intrusion, where it reaches 12% of the mode in quartz ferrogabbros (Bevins 1994). In the Tal y Fan olivine dolerite intrusion, SW of Conwy, Gwynedd, ilmenite is replaced by titanite (Merriman et al. 1986). Plagioclase mesocumulates belonging to the Rhiw layered intrusion on the Lleyn Peninsula, Gwynedd, contain ilmenite that is mostly replaced (Cattermole 1976). A single specimen of ilmenite and titanite has been found with fine brookite crystals in an alpine-type vein assemblage from Fron-oleu, Prenteg, Tremadog, Gwynedd.

Scotland

Dumfries & Galloway Ilmenite is a common accessory mineral throughout the Loch Doon granite complex, S of Dalmellington, and is manganese-rich in cordierite microgranite on Dungeon Hill, near Merrick (Tindle 1982).

Grampian Rounded grains of ilmenite and magnetite occur in discontinuous lenticular sulphide bodies associated with olivine cumulates, contaminated gabbros and metasediments, found on the southeast margin of the Knock gabbroic intrusion, near the farms of Littlemill and Auchencrieve, 2 km NW of Rothiemay, N of Huntly (Fletcher & Rice 1989). Ilmenite is a major component of panned concentrates derived from the Cairngorm granite in the Glen Lui drainage area, Derry Cairngorm, N of Braemar (Beer & Bennett 1988). Minor ilmenite, grandidierite, rutile, titanite and zircon occur in a biotite-rich xenolith within norite close to the contact of a tourmaline-bearing granite in the Haddo House gabbro–norite to the NE of the Wood of Schivas, Haddo House, 32 km N of Aberdeen (Herd et al. 1984).

Highland Ilmenite, magnetite, spinel and a range of sulphide species form droplets within allivalite–chromitite–peridotite from the Eastern Layered Series, Isle of Rum ultramafic complex (Dunham & Wilkinson 1985); ilmenite is intergrown with magnetite within picritic dykes, Isle of Rum ultramafic complex (Upton et al. 2002); rutile is replaced by ilmenite in a basaltic dyke from Sròn Bheag headland, southwest Ardnamurchan (Wells et al. 1974); ilmenite forms glomeroporphyritic clots with pyroxene and plagioclase in a pitchstone flow to the SE of Ben Hiant, Ardnamurchan (Preston et al. 1998b); ilmenite coexists with titanomagnetite in monzodiorite from the Ballachulish igneous complex (Weiss & Troll 1989); manganese-rich ilmenite, apatite and garnet form a minor component of a stratiform manganese-rich ore (mainly braunite and bustamite) in Dalroy Burn, Dalroy (Nicholson 1990); ilmenite–magnetite intergrowths are associated with chalcopyrite, pentlandite and pyrrhotite in a Lewisian biotite-diorite dyke at Loch Bad nam Ban, 3 km SW of Lochinver (Stumpfl 1961); ilmenite forms flattened crystals, up to 4 mm in length, with an 8% geikielite component in white sövite (carbonatite) from the Loch Borralan igneous complex, Assynt (Young et al. 1994b); complex exsolution of ilmenite and magnetite has been observed in trondhjemite and granulite-facies calcareous gneiss from the Lewisian Scourian complex, near Upper Badcall and Scourie More respectively (Rollinson 1979, 1980); ilmenite is a minor component of an aphyric Tertiary basaltic lava collected to the S of Fairy Bridge (Thompson 1974).

Strathclyde Irregularly shaped skeletal masses of ilmenite up to 1 mm in diameter occur in teschenite from the Lugar Sill, Glenmuir Water (Phillips 1968); magnesium-rich (giesekitic) ilmenite is present in pargasite pyroxenite and pyroxene granulite gneiss xenoliths in an olivine-nephelinite dyke at Gribun, Isle of Mull (Upton et al. 1998); ilmenite forms microphenocrysts in porphyritic pitchstone from Glen Shurig, Isle of Arran (Preston et al. 1998b); isolated euhedral prisms of ilmenite and titaniferous magnetite up to 0.25 mm across have been found in composite pitchstone sheets (Judd dykes), near Blackwaterfoot, Isle of Arran (Preston et al. 1998b).

Tayside Andalusite-bearing hornfels from the thermal aureole of the Ben Vuirich granite, 10 km NE of Pitlochry, contains manganese- and vanadium-rich ilmenite (Ahmed-Said & Tanner 2000); ilmenite is associated with almandine and epidote at the contact of the Ben Eagach Schist and Ben Lawers Schist near the southern end of Lochan na Lairige, NNE of Killin (Green & Todd 2001).

Plates of ilmenite in quartz. Loch na Lairige, Killin, Tayside Region, Scotland; width of specimen 10 cm. (© The Trustees of the National Museums of Scotland)

Western Isles Manganese-rich ilmenite forms inclusions in hematite within metasomatically altered Lewisian Gneiss at Garry-a-siar (Gearraidh a Siar), west coast of Benbecula (Williams 1987); manganese-rich ilmenite is associated with allanite, apatite, magnetite, titanite and zirkelite in pegmatite cutting the Glen Bay gabbro, St Kilda (Hirta) (Harding, et al. 1982); skeletal ilmenite crystals up to 2.5 mm across coexist with spinel in crinanite and picrodolerite from the Shiant Isles alkali basaltic sill (Gibb & Henderson 1996, Henderson et al. 2000).

Ireland

Antrim Ilmenite is a groundmass species in basalt near the dolerite plug at Scawt Hill, Larne (Kitchen 1985).

Donegal Ilmenite and hercynite form clusters of grains in pelites adjacent to the Ardara granite pluton at Narin, 2 km SE of Portnoo harbour (Atkin 1978).

Down Ilmenite inclusions in titanite and molybdenite occur in the Newry granodiorite at Croreagh quarry (Meighan & Ixer 1980); dolerite and gabbro from the Slieve Gullion central complex contain coarse granular exsolution products of ilmenite in magnetite (Gamble 1982).

Dublin Manganese-rich ilmenite is reported from a biotite-quartz diorite belonging to the Leinster granite batholith (Elsdon 1975).

Galway Ilmenite forms trains of inclusions in well formed garnet crystals within chlorite–chloritoid–muscovite–quartz schist at Bunnahowna, Renvyle Point, Connemara (Harvey & Ferguson 1973); ilmenite occurs in the Currywongaun–Doughruagh intrusion, Connemara, where it is associated with anthophyllite, högbomite and magnetite in a cordierite-rich xenolith (Angus & Middleton 1985).

Wicklow Ilmenite–magnetite, ilmenite–hematite and chromite are minor constituents of gold-bearing mineral separates within metamorphosed iron ores at Ballycoog, Goldmine River valley, near Avoca (Ixer et al. 1990).

Offshore territories

Jersey Ilmenite forms small brilliant plates with tourmaline in pegmatite veins at Frémont Point and is noted in diorite at Ronez Mont, and in granite at Mont Mado (Mourant 1961); interstitial ilmenite occurs in the coarse-grain Wolf's Cave dolerite dyke, near Frémont Point (Rowbotham & Lees 1990); an arfvedsonite-bearing minette at St Hélier contains accessory amounts of manganese-rich ilmenite (Wagner & Velde 1985).

ilmenorutile

tetragonal $(Ti,Nb,Fe)O_2$

A rare titanium–niobium–iron oxide, found as inclusions up to 20 µm across in lithium-mica within the Hensbarrow topaz granite stock, near St Austell, Cornwall. In earlier works, this type of granite was termed non-megacrystic lithium-mica granite. Albite, orthoclase, quartz and zinnwaldite are the essential silicate components of this granite (Manning 1983, Manning & Hill 1990). In studies of fluid and mineral inclusions within the topaz from Hensbarrow, Williamson et al. (1997, 2002) were able to identify albite, cassiterite (varlamoffite), columbite, ilmenorutile, potassium-feldspar, muscovite, pyrrhotite, qitianlingite, quartz, stannite and zircon. A single vug in a pegmatite from Gunheath china-clay pit, which cuts the Hensbarrow stock, was found to contain stout greyish-black submetallic ilmenorutile crystals 1–3 mm across (Weiss 1994). Elsewhere in the St Austell granite, to the W of Nanpean, a similar topaz-rich unit has been worked in the Goonvean and Rostrowrack china-clay pits. Ilmenorutile has been found in the waste from these pits, together with apatite, cassiterite and columbite–tantalite, and lesser amounts of arsenopyrite, iron oxides, manganese-rich ilmenite, monazite, pyrite, sphalerite and zircon (Scott et al. 1998). A general review of the St Austell granites, their associated mineralization and alteration is given by Hawkes et al. (1987) and Manning et al. (1996). Ilmenorutile was also found during an investigation of the North Tincroft lode, New Cook's Kitchen mine, between Camborne and Pool, Cornwall. This polymetallic fissure-fill vein structure occurs in the metamorphic aureole of the Carnmenellis granite and contains a telescoped assemblage of arsenopyrite, cassiterite, chalcopyrite and sphalerite in a dense, non-brecciated chlorite–fluorite–quartz matrix. In total, 28 species have been indentified and, of these, four are titanium oxides: ilmenite (most common), ecandrewsite, ilmeonorutile and pyrophanite; all form lath-like crystals up to 200 µm long and occur as inclusions in all the other major phases (LeBoutillier et al. 2000).

In Scotland, ilmenorutile is a minor component of three panned concentrates of alluvium from the River Lui, which drains the southeastern part of the Cairngorm granite, 8 km W of Braemar, Grampian. Associated species included anatase, cassiterite, chrysoberyl, columbite, fergusonite, ilmenite, magnetite, monazite, rutile, topaz, xenotime and zircon (Beer & Bennett 1988, Haynes 1974, 1993). No *in situ* source of the mineralization has ever been found, although it is likely to have been an evolved granite or pegmatite. Corundum megacrysts in a xenolithic monchiquite dyke at Cairisiadar (Carishader), Loch Roag, Isle of Lewis, Western Isles, are occasionally intergrown with ilmenorutile, sanidine or zircon (Upton et al. 1983).

ilsemannite?

amorphous $Mo_3O_8 \cdot nH_2O$?

An uncommon hydrated molybdenum oxide probably present as a bright-blue post-mining efflorescence on limestone and anhydrite in the mine under the ICI factory site at Billingham-on-Tees, Cleveland. The source of the molybdenum is unknown, although it is thought to have precipitated out of the Upper Magnesian Limestone. Other minerals confirmed in efflorescences at the mine include epsomite, gypsum, halite, lepidocrocite, pyrite, sulphur, sylvite, talc, thenardite and trona. Apthitalite, arsenopyrite, molybdenite, tachyhydrite and tamarugite may also be present (Raymond 1959, Stewart 1965). Specimens of black sand, found at Tickow Lane lead mine near Blackbrook and stored under centrally heated conditions, developed dark-blue to bluish-black surface efflorescences that are thought to be ilsemannite (King 1973).

ilvaite

monoclinic or orthorhombic $Ca Fe^{3+}(Fe^{2+})_2O(Si_2O_7)(OH)$

A rare calcium–iron silicate hydroxide supposedly found by A. W. G. Kingsbury and now in the Kingsbury collection at the NHM. Locations given are Wheal Messer (Tretoil mine), Lanivet, Cornwall; Ramsley mine, South Tawton, near Okehampton, and Smallacombe iron mine, Ilsington, both in Devon (Embrey 1978). Unfortunately, the provenance of many samples in the Kingsbury collection is suspect (Ryback et al. 1998, 2001) and therefore ilvaite remains to be proven from these localities.

Ilvaite is confirmed from a mineralized vein of dioritic pegmatite, 10–30 cm thick, within the Whin Sill, at Forcegarth quarry, 16 km NW of Middleton-in-Teesdale and 2 km W of High Force waterfall, Co. Durham. It is closely associated with ilmenite, titanomagnetite and rare sphalerite, but chalcopyrite, marcasite, molybdenite, pyrite, pyrrhotite, rutile and titanite are also present. In thin section, ilvaite is an opaque species, but in reflected light it is bright blue with intense pleochroism. It occurs as interstitial patches mantling rounded grains of ilmenite, or as fracture infillings in altered pyroxene (Ixer et al. 2005b).

Irregular, pale-blue ilvaite, partially enclosing ilmenite (white) in a dolerite pegmatite facies of the Whin Sill. Forcegarth quarry, near High Force, Teesside, Co. Durham; width 0.3 mm; viewed in reflected light. (© A. G. Tindle)

imogolite

amorphous $Al_2SiO_3(OH)_4$

An aluminosilicate hydroxide also described as a tubular paracrystalline mineral, a tubular aluminium silicate polymer and a gel variety of allophane. Imogolite is not accorded species status by Mandarino & Back (2004), although it was approved by the IMA in 1986 (imogolite had previously been rejected by the IMA in 1967). It forms coatings on Scottish soils at many unspecified locations (Livingstone 2002).

indicolite

A blue variety of elbaite, tourmaline group. Not all blue tourmaline is elbaite.

indigo copper

A synonym of covellite.

ingersonite (of Bridge)

The name ingersonite was proposed by T. E. Bridge for a calcium silicate discovered at Scawt Hill, Co. Antrim, Ireland. Unfortunately, he died before the description could be published, and the name was then re-used by other authors for a calcium–manganese antimonate from Långban, Värmland, Sweden.

ingodite

trigonal Bi_2TeS

TL Caldbeck Fells, Cumbria, England

A rare bismuth sulphotelluride discovered on a museum specimen originating from Brandy Gill, Carrock Fell, Caldbeck Fells, Cumbria, and on specimens from the Verkhne–Ingodinskoye tin deposit, Ingoda River, central Transbaikal region, Russia, with which it shares type specimen status (Zav'yalov & Begizov 1981a). Later publications suggest that Carrock mine is the actual location (Young 1987a, Cooper & Stanley 1990). Ingodite is silvery white with a metallic lustre and perfect cleavage, and was discovered during a study of the dubious mineral grünlingite (now known to be a mixture of bismuthinite and joséite-A).

inverarite

TL Inveraray, Strathclyde, Scotland

A mixture of pentlandite, pyrrhotite and pyrite.

iolite

A synonym of cordierite.

irarsite

cubic COBALTITE GROUP (Ir,Ru,Rh,Pt)AsS

A rare iridium-dominant arsenide that forms a series with the rhodium-rich species hollingworthite. Both are found in chromite-rich dunite at Cliff and at Harold's Grave, Baltasound–Haroldswick area, Unst, Shetland, Scotland. Irarsite is commonly rimmed by hollingworthite and sometimes by sperrylite (Tarkian & Prichard 1987, Prichard & Tarkian 1988, Prichard et al. 1994). A composite grain of irarsite, platinum and an unnamed Rh–Sb–S phase has been found in a chromite-rich sample from Jimmie's quarry, S of Hagdale, also on Unst (Prichard et al. 1988). Nikkavord North is a further location on Unst, where irarsite occurs with hollingworthite, native osmium and unnamed phases (Prichard et al. 1986).

Irarsite has been identified in a layered mafic–ultramafic intrusion on the Isle of Rum, Highland Region, Scotland. Within certain chromitite layers of the Eastern Layered Series, irarsite is found in association with species positively or tentatively identified as braggite, cooperite, hongshiite, isoferroplatinum, keithconnite, kotulskite, laurite, moncheite, palladoarsenide, platarsite, platiniridium, sperrylite, telargpalite and telluropalladinite, and the native species palladium and platinum. Most of these grains are 0.2–2 μm in diameter, although grains > 15 μm are also present. The majority are in contact with sulphide minerals such as chalcopyrite, bornite or pentlandite. The chromitite layer was formed during magmatic fractionation and crystal settling of an open-system magma chamber, modified by magma mixing and liquid immiscibility of sulphide and silicate melts (Butcher et al. 1999, Power et al. 2000a). A restricted PGM assemblage within chromitites of the mafic–ultramafic Ben Buie intrusion on Mull (specifically the western extremity at Caigeann Doire nan Cuilean and Craig) includes irarsite, laurite, sperrylite and unidentified Bi–Pb–Te and Bi–Pd–Sb phases. The PGMs occur as either euhedral crystals within chrome spinel grains or as clusters of minute inclusions (< 2 μm) within intercumulus plagioclase and clinopyroxene (Pirrie et al. 2000, Power et al. 2003).

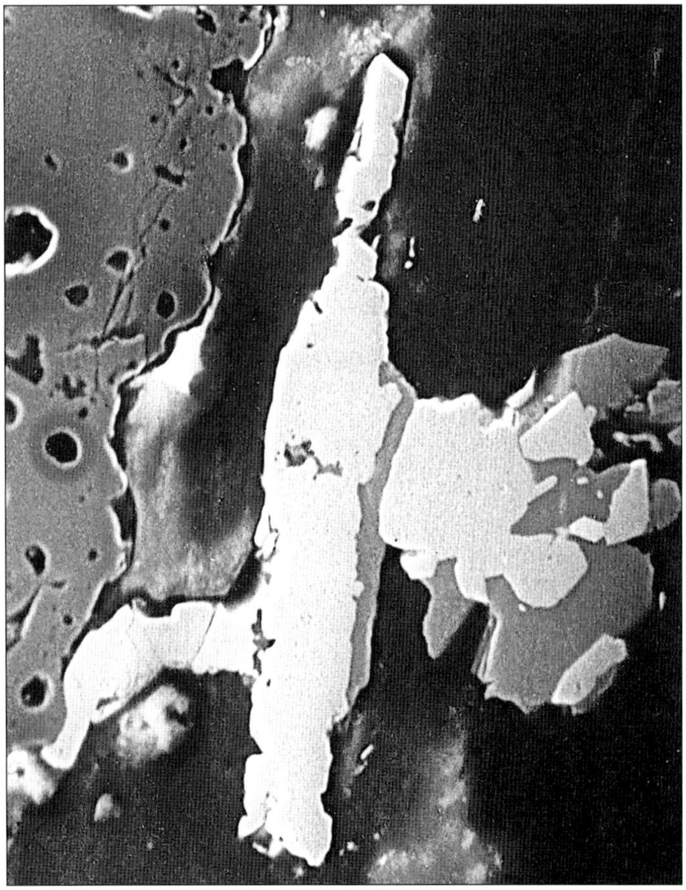

Elongate blade and rounded grain of irarsite (light grey) partially surrounded (on the right) by ruthenian pentlandite (darker grey). The small diamond-shape grain (middle right) is native osmium. The light-grey appended grain on left of irarsite blade is sperrylite. Cliff, west of Baltasound, Unst, Shetland, Scotland; width 70 μm; BSE image. (© H. M. Prichard)

At Corrycharmaig, Glen Lochay, NW of Killin, Tayside Region, a small body of metamorphosed and altered ultramafic rocks contains small irarsite and laurite inclusions in chromite. Other PGMs found in these rocks include the native species iridium, osmium and ruthenium, and unnamed alloys of antimony, arsenic, osmium, platinum and ruthenium. The chemistry and association of the PGMs is strongly indicative of an ophiolite assemblage and it is most likely that the intrusion represents a fragment of oceanic crust (Power & Pirrie 2000).

Farther south, at Talnotry, near Newton Stewart, Dumfries & Galloway, a small body of copper–nickel mineralization has developed at the base of an altered diorite sheet. Its principal ore minerals are chalcopyrite, gersdorffite, nickeline, pentlandite and pyrrhotite, and these are associated with a PGM assemblage of gold (electrum), irarsite, sperrylite and an unnamed bismuth–palladium telluride (Power & Pirrie 2001, Gunn & Styles 2002). Later work by Power et al. (2004) confirms irarsite and sperrylite to be the dominant PGMs, with froodite, hollingworthite, merenskyite and michenerite being the other subordinate species. Arsenopyrite, annabergite, bismuthinite, galena and pyrite are also noted.

In Ireland, irarsite occurs with laurite as inclusions in chromite within the Dawros More peridotite emplaced into the Dalradian rocks of Connemara, Co. Galway. Other associated sulphides are bornite, chalcocite, chalcopyrite, covellite, heazlewoodite and nickel-rich pentlandite (McElduff & Johnston 1989).

iridium

cubic (Ir,Os,Ru) with Ir dominant

Platiniridium, a rare platinum-rich variety of iridium, is described from the Leadhills area, along the Strathclyde/Dumfries & Galloway border, Scotland, where it occurs at the core of a complex alluvial grain with projections of silver-rich native gold. Accompanying species found as inclusions in alluvial gold grains include isoferroplatinum, tulameenite and unnamed copper–platinum and copper-oxide compounds (Gunn & Styles 2002).

Platiniridium has also been qualitatively identified in a layered mafic–ultramafic intrusion on the Isle of Rum, Highland Region, Scotland. Within certain chromitite layers of the Eastern Layered Series, platiniridium is found in association with species positively or tentatively identified as braggite, cooperite, hongshiite, irarsite, isoferroplatinum, keithconnite, kotulskite, laurite, moncheite, palladoarsenide, platarsite, sperrylite, telargpalite and telluropalladinite, and the native species palladium and platinum. Most of these grains are 0.2–2 μm in diameter, although grains > 15 μm are common; the majority are in contact with sulphide minerals such as bornite, chalcopyrite or pentlandite. The chromitite layer was formed during magmatic fractionation and crystal settling of an open-system magma chamber, modified by magma mixing and liquid immiscibility of sulphide and silicate melts (Butcher et al. 1999, Power et al. 2000a).

iridosmine

A synonym of osmium.

iron

cubic Fe

A very rare native metal found in two inclusion-rich gold grains that also contain arsenic–iron alloy inclusions. The gold grains are 4–10 μm across and were collected from an alluvial sample in Afon Mawddach, Dolgellau goldbelt, Snowdonia, Gwynedd, Wales. Other gold grains at this location contain inclusions of cobaltite, galena, gersdorffite, pyrrhotite and sphalerite. The assemblage suggests that there may be hidden sources of gold in the area distinct from the Gwynfynydd gold mineralization already documented (Leake et al. 1993). Early records of iron, such as those from the basaltic rocks of Co. Antrim, Ireland (Greg & Lettsom 1858), Unst, Shetland, and various locations in the Highland Region of Scotland (Heddle 1901), are almost certainly misidentifications.

Minerals with extremely high iron contents are much more commonly found in meteorites, two examples being the Barwell meteorite, Leicestershire, England (Jobbins et al. 1966), and the Crumlin meteorite, Co. Antrim, Ireland (Fletcher 1921). Here the iron is nickel rich and it forms the species kamacite and taenite.

iron alum

A synonym of halotrichite.

iron-boracite

A synonym of ericaite (but see congolite–trembathite).

iron-cordierite

A synonym of sekaninaite.

iron-corundum

An iron-rich variety of corundum.

iron-glance

A synonym of specular hematite.

iron mullite

An iron-rich variety of mullite.

iron pyrites

A synonym of pyrite.

iron-vitriol

A synonym of melanterite.

iron-wollastonite

A synonym of ferrobustamite.

iserine

A supposed cubic variety of ilmenite.

isoferroplatinum

cubic Pt_3Fe

A very rare platinum–iron alloy, identified in a layered mafic–ultramafic intrusion on the Isle of Rum, Highland Region, Scotland. Within certain chromitite layers of the Eastern Layered Series, isoferroplatinum is found in association with species positively or tentatively identified as kotulskite, laurite, michenerite, moncheite, palladoarsenide, members of the hollingworthite–irarsite–platarsite series and sperrylite, and the native species gold and silver. Most of these grains are 0.2–2 μm in diameter, although, for some species, grains > 15 μm are also found. The majority are in contact with sulphide minerals, such as bornite, chalcocite, chalcopyrite, pentlandite or pyrrhotite. The chromitite layer was formed during magmatic fractionation and crystal settling of an open-system magma chamber, modified by magma mixing and liquid immiscibility of sulphide and silicate melts (Power et al. 2000a).

Two alluvial gold grains from the Leadhills–Wanlockhead orefield, along the Strathclyde/Dumfries & Galloway border, Scotland, contain isoferroplatinum inclusions (Leake et al. 1998).

isomertieite

cubic $Pd_{11}Sb_2As_2$

A very rare palladium antimonide arsenide, dimorphous with mertieite-I and found in a gold-bearing carbonate vein cutting Middle Devonian limestones at Hope's Nose, Torquay, Devon. Isomertieite, and a mineral close to mertieite-II in composition, form inclusions close to, or on the outer surface of, a few palladium-rich gold dendrites etched out from the vein (Clark & Criddle 1982). Inclusions of isomertieite have also been found in one alluvial gold grain from Whympston, near Modbury, South Hams area of Devon (Leake et al. 1991).

In the Loch Ailsh alkaline complex, along the Allt Cathair Bhan Valley, Highland Region, Scotland, there is widespread PGE enrichment in the pyroxenites, pyroxene mylonites and pyroxenite skarns. Two isomertieite grains from an area of 20 μm^2 are reported with sperrylite and complex tellurides of bismuth, lead, palladium and silver. A magmatic origin associated with late magmatic volatile-rich fluids deficient in sulphur has been suggested for the origin of these rocks (Shaw et al. 1994, Gunn & Styles 2002, Livingstone 2002). In the Loch Borralan alkaline pyroxenite–syenite complex, Assynt, probable isomertieite, sperrylite and bismuth–palladium tellurides mainly occur in late brittle fractures and along grain boundaries, often with chalcopyrite, galena and hessite (Styles et al. 2004). Isomertieite has also been qualitatively identified in a layered mafic–ultramafic intrusion on the Isle of Rum, Highland Region (Power et al. 2000a).

isopyre

TL St Just, Cornwall, England

An impure variety of opal.

isostannite

TL Cligga Head, Perranzabuloe, Cornwall, England

A synonym of kësterite; shown to be a member of the kësterite–ferrokësterite series.

ixiolite

monoclinic (Ta,Mn,Nb)O_2

A rare tantalum–manganese–niobium oxide found in a granite pegmatite at Megiliggar Rocks, near Breage, Cornwall. Initial identification was by XRD (Ryback & Tandy 1992). Tungsten-rich ixiolite (a single 100 μm-long black lath) was confirmed more recently on another sample from this locality. A single (< 100 μm) grain of an Mn–Nb–Ta oxide with an even higher tungsten content, from the Meldon aplite, near Okehampton, Devon, is also thought to be a variety of ixiolite (Tindle unpub. EMP data).

izoklakeite?

orthorhombic $Pb_{27}(Cu,Fe)_2(Sb,Bi)_{19}S_{57}$

A lead–copper–iron sulphantimonite found on samples from the Kingsbury collection at the NHM and whose provenance is described as Wheal Cock, St Just, Cornwall. The work of Ryback et al. (1998, 2001) indicates that the provenance of many Kingsbury samples is dubious, and further investigation of this species is therefore desirable.

J

jacobsite

cubic SPINEL GROUP $(Mn^{2+},Fe^{2+},Mg)(Fe^{3+},Mn^{3+})_2O_4$

A rare manganese–iron–magnesium oxide that forms a series with magnetite and is found at Benallt manganese mine, Rhiw, Lleyn Peninsula, Wales. Jacobsite, ganophyllite and pyrophanite are subsidiary manganese- and iron-rich species in the deposit. The main manganese ore consists of a complex mixture of manganese silicates and includes alleghanyite, pennantite, spessartine and tephroite. Jacobsite is strongly magnetic and it forms black angular patches up to 1 cm across in banalsite-bearing veins, where the sequence of crystallization is alleghanyite, tephroite, jacobsite, biotite, barite and calcite (Campbell Smith et al. 1944). Oolitic structures in the ore suggest that it is a metasomatically altered chamositic mudstone (Dunham et al. 1978).

Jacobsite is possibly present with braunite, bustamite, manganese-rich calcite, manganese-rich ilmenite, hematite, pyrophanite and rhodochrosite in a complex (up to 4 m thick) stratiform manganese ore at Dalroy Burn, near Culloden, Highland Region, Scotland. The ore has a geothermal origin best explained by deposition from hot springs in a sub-lacustrine environment during Devonian times (Nicholson 1987b, 1990).

jade tenace

A type of jade, later called saussurite (= zoisite or epidote + albite).

jahnsite-(CaMnFe)

monoclinic WHITEITE GROUP $CaMn^{2+}Fe^{2+}_2Fe^{3+}_2(PO_4)_4(OH)_2 \cdot 8H_2O$

An hydrated calcium–manganese–iron phosphate hydroxide found in two small metamorphic segregation pods, 15 cm and 30 cm long, occurring in striped kyanite–sillimanite-bearing Moine gneiss from Gleann Còsaidh, Loch Quoich, Knoydart, Highland Region, Scotland. The segregation pods contain apatite, garnet, muscovite, plagioclase and quartz, together with very small clusters of jahnsite-(CaMnFe), graftonite, johnsomervilleite, mitridaite, phosphosiderite, rockbridgeite and vivianite. Jahnsite-(CaMnFe) is nut brown in colour, has a greasy lustre, and forms anhedral grains up to 1 mm across (Livingstone 1980).

A salmon-pink alteration coating on spodumene pegmatite from Moylisha, near Shillelagh, Co. Wicklow, Ireland, is named "salmonsite" – a mixture of jahnsite and hureaulite. A similar occurrence, on boulders of spodumene pegmatites, is also noted at Seskinnamadra, W of Tomduff Hill, Co. Carlow, Ireland (Scoon 1978).

jalpaite

tetragonal Ag_3CuS_2

A rare silver–copper sulphide found in an epigenetic cadmium–zinc deposit within Lower Carboniferous clastic sediments at Keel, Ardagh, Co. Longford, Ireland. Jalpaite has a black sooty appearance and it occurs in complex intergrowths with pyrargyrite. Cadmium-rich sphalerite, galena and tetrahedrite are the major ore species. They occur with minor amounts of barite, bournonite, chalcopyrite, freibergite, marcasite and pyrite in a calcite, dolomite and quartz gangue (Patterson 1970, Morrissey et al. 1971, Slowey 1986, Ryback & Moreton 1993).

jamborite

hexagonal $(Ni^{2+},Ni^{3+},Fe)(OH)_2(OH,S,H_2O)$?

An uncommon hydrated nickel–iron hydroxide found in Mid Glamorgan, Wales, at Cymmer colliery, Porth, and possibly at Bedwas colliery, Trethomas (Bevins 1994). Jamborite forms as an alteration product of primary nickel-sulphide minerals.

jamesonite

monoclinic $Pb_4FeSb_6S_{14}$

TL St Endellion, Cornwall, England

Named after Robert Jameson FRS (1774–1854), a geologist, mineralogist and professor of natural history at Edinburgh University; born in Leith, near Edinburgh. Jamesonite is a relatively rare, but locally common, lead–iron sulphantimonite that forms a series with benavidesite and is dimorphous with parajamesonite. It is found as a late-stage hydrothermal mineral in low- to moderate-temperature Ag–Pb–Zn veins.

Silvery-grey jamesonite and yellow bindheimite in a quartz gangue; Treore mine, St Endellion, Cornwall; width of specimen 6 cm. (© D. I. Green)

In Cornwall, steel-grey or black jamesonite, with a metallic lustre, has been reported from mines in the parish of St Endellion, where it occurs as acicular crystals or fibrous masses in antimony–lead-rich quartz veins cutting Upper Devonian slates and intercalated spilitic pillow lavas. The spatial relationship of antimony mineralization with the lavas suggests a genetic linkage between the two (Edwards 1976). The principal district where antimony ores are found extends into the neighbouring parishes of St Minver, St Kew and St Teath (Dewey 1925). At Bodannon mine, near Port Isaac, St Endellion, jamesonite was the main ore species, although bindheimite, senarmontite, nadorite and valentinite have also been noted (Russell 1927a, 1949b). At Trevinnick mine, St Kew, jamesonite is accompanied by bindheimite, stibnite and stibiconite (antimony ochre), whereas at Treore mine, Port Gaverne, St Teath, it is found partially or totally altered to bindheimite in association with boulangerite, bournonite and tetrahedrite (Dewey 1925, Russell 1949b, Fisk et al. 1982). Up stream of Treore mine, large masses of jamesonite, together with cassiterite, cerussite, native gold, pyrite and wolframite, have been recovered from alluvial deposits (Camm 1995). A jamesonite-bearing vein on the foreshore at Bounds Cliff, St Endellion, produced specimens from which the new species, barstowite, was discovered (Stanley et al. 1991). Other Cornish locations for jamesonite include: Endellion (Berry 1940); Greystone, Lezant, near Launceston (Power & Scott 1995); Linnick, South Petherwin (Russell 1949b); Pendoggett mine, near Port Isaac, St Kew (Edwards 1976); Pentire (Glaze) mine, near New Polzeath, St Minver (Clayton et al. 1990); Poltreworgy mine, near Port Isaac, St Endellion (Hosking 1951); Port Quin mine, St Endellion (Russell 1944, 1949b); Reedy Cliff, near Port Quin (Clayton et al. 1990); Tolpetherwin quarry, Two Bridges, South Petherwin (Russell 1949b); Tresungers (Rose or Wheal Arthur) mine, near Port Isaac, St Endellion (Russell 1949b, Hosking 1951, Edwards 1976, Criddle & Stanley 1993); Trewane mine, near Pendoggett, St Kew (Greg & Lettsom 1858, Edwards 1976); Wheal Boys, Trewetha, St Endellion (Collins 1904b, Rudler 1905, Russell 1949b); also Wheal Emily, Wembury (Russell 1949b), and Loddiswell mine, 3 km N of Loddiswell (Stanley et al. 1990b), both in Devon.

Jamesonite forms straight or curved hair-like capillary crystals 60–250 μm long in an antimony-bearing quartz vein on Coombe Height, at the head of Wet Swine

Gill, Caldbeck Fells, Cumbria. It is a minor component of a primary assemblage that includes berthierite, stibnite, zinkenite and lesser native antimony, arsenopyrite and sphalerite. Supergene alteration has added beudantite, bindheimite, claudetite, füllöppite, parasymplesite, pharmacosiderite, scorodite, semseyite, senarmontite, stibiconite and native sulphur to this assemblage. A second jamesonite-bearing vein is also noted 500 m to the E of Wet Swine Gill (Fortey et al. 1984, Neall & Green 2001, Green & Briscoe 2002). Reports of jamesonite from Carrock mine, Grainsgill, Caldbeck Fells, Cumbria (Kingsbury & Hartley 1956b, Hartley 1984), and a specimen at the NHM from Wanthwaite mine, St John's in the Vale, near Threlkeld, Cumbria (Young 1987a), require further confirmation, as the provenance may be incorrect (see Ryback et al. 1998, 2001).

At Bwlch mine, near Deganwy, 3 km S of Llandudno, Gwynedd, Wales, antimony-enriched quartz veins containing galena, marcasite and sphalerite cut nodular ignimbrite. The antimony-bearing mineral assemblage includes capillary crystals of jamesonite associated with bindheimite, plagionite, rosiaite, semseyite, stibiconite, stibnite and zinkenite, and possibly boulangerite, kermesite and robinsonite (Russell 1944, Bevins et al. 1988, Ryback & Francis 2001).

Jamesonite is reported from Glendinning mine, Glenshanna Burn, Dumfries & Galloway, Scotland, by Ball et al. (1985b), but not by Gallagher et al. (1983). The species is also noted from Knipe mine, SE of New Cumnock (Williams 1973), and from the Leadhills–Wanlockhead orefield, on the Strathclyde/Dumfries & Galloway border (Brown 1927, Williams 1965, 1973). However, Livingstone (2002) suggests that these occurrences require confirmation.

Minor jamesonite altering to bindheimite has been found in a wulfenite-bearing aplite vein in the Northwest quarry, L'Etacq, Jersey, Channel Isles (Mourant 1961). Jamesonite is described with bindheimite, chalcopyrite, galena, molybdenite, pyrite, pyrrhotite and wulfenite from sulphide-rich quartz veins at Le Pulec, northwest Jersey, and with bindheimite, chalcopyrite, galena, molybdenite, pyrite, pyrrhotite, wulfenite and a wide range of other sulphide species from Sark's Hope mine, Little Sark (Mourant & Warren 1934).

jarosite

ALUNITE GROUP

trigonal $KFe_3(SO_4)_2(OH)_6$

A relatively common potassium–iron sulphate hydroxide that forms as a low-temperature hydrothermal mineral or as a secondary mineral in the oxidized parts of sulphide-bearing rocks.

Jarosite is found at Penberthy Croft mine, St Hilary, Cornwall, where it forms yellow earthy microcrystalline patches within goethite cavities in iron-stained quartz (Golley & Williams 1995, Betterton et al. 2000), and at Greystone quarry, Lezant, near Launceston, where it forms brown amorphous crusts, presumably because of oxidation of pyrite (Elton et al. 2004). Jarosite (requiring further confirmation) is noted from the dumps of the abandoned As–Cu–Sn–W mine Wheal Friendship (Devon Friendship), Mary Tavy, Devon (Brooke 1986).

Compact, massive, yellowish bands of jarosite, associated with gypsum (selenite), commonly occur in Tertiary London Clay near the coastguard station at Newhaven, Sussex (Wilmot & Young 1985). At Alum Bay and Whitecliffe Bay, Isle of Wight, similar Tertiary clays (Barton Beds and Bracklesham Beds) contain finely crystalline yellow jarosite coating quartz grains closely associated with glauconite, gypsum and limonite (Wright & Wilson 1970). The latter authors suggest that jarosite is the product of alteration of limonite, which can result from the initial weathering of pyrite.

Reddish-brown to dark-brown microcrystalline crusts and masses of jarosite associated with azurite and malachite are noted from Newhurst quarry, Shepshed, Leicestershire (King 1973).

Jarosite, forming yellow powdery films on the surface of black shales, was initially thought to be sulphur. Copious quantities of earthy or powdery jarosite occur at Mealbank quarry, Ingleton, Yorkshire. It is a decomposition product of pyrite and it forms crystals < 1 μm across in fireclay (Cosgrove & Hodson 1963, Ryback et al. 2001). Buff jarosite from Ladywell Pb–Zn–Cu mine, Shelve, Shropshire, forms part of a supergene assemblage that includes anglesite, brochantite, cerussite, langite, linarite, malachite, pyromorphite and wroewolfeite (Braithwaite 1982b), and in the Engine Vein mine rift, Alderley Edge ridge, Cheshire, yellow-to-brownish microcrystalline or powdery jarosite is found in conglomerate associated with galena (Braithwaite 1994).

In the northern Pennines, jarosite occurs as a dull yellow crust associated with cerussite and malachite at Yew Tree mine, Stanhope, Co. Durham (Bridges & Young 1998), and at Groverake mine, Rookhope, Weardale, it forms thin pale-yellow crusts on fracture surfaces of shale (Young et al. 2005). At Brownley Hill mine, Nenthead, near Alston, Cumbria, jarosite has crystallized on harmotome crystals within a large

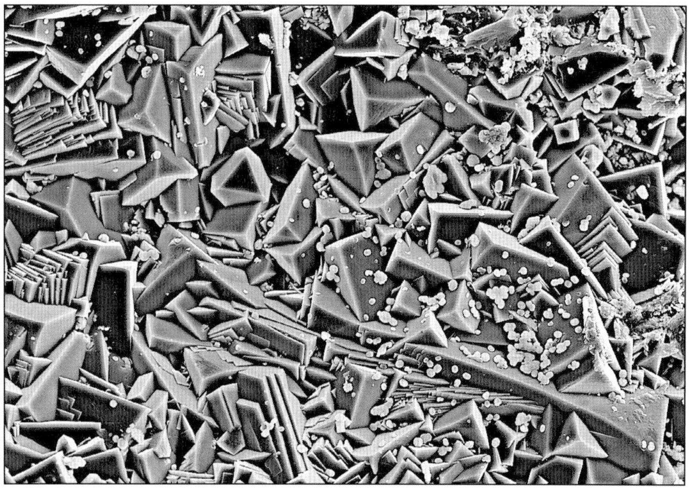

Jarosite rhombs associated with possible goethite from Tynebottom mine, NW of Garrigill, near Alston, Cumbria; width 120 μm; SEM image. (© D. I. Green)

quartz-lined cavity in metasomatized limestone (Green & Nudds 1998, Green et al. 2000). Jarosite is one of several species encrusting pyrite-rich coal seams at Oughterside opencast coal site, Aspatria, Cumbria (Young & Nancarrow 1988b).

Descriptions of jarosite from the Caldbeck Fells, Cumbria, are mostly problematic, as they are either by A. W. G. Kingsbury or based on specimens in the Kingsbury collection at the NHM. The work of Ryback et al. (1998, 2001) indicates that the provenance of many Kingsbury samples is dubious and further investigation of jarosite from these locations is therefore desirable: Potts Gill barite mine (Kingsbury & Hartley 1958); Balliway Rigg in higher Roughton Gill; Driggith mine (both Hartley 1984); Brandy Gill mine; Burdell Gill; Carrock Fell mine, Grainsgill; Ingray Gill mine; Nether Row Brow; higher Roughton Gill (all Kingsbury & Hartley 1958, Hartley 1984). Wanthwaite mine, St John's in the Vale, near Threlkeld, is a further location outside the Caldbeck Fells noted by Kingsbury & Hartley (1958). Most of these occurrences are also mentioned by Young (1987a) and Cooper & Stanley (1990). Fortunately, jarosite has since been confirmed at Burdell Gill (Neall & Leppington 2003) and Silver Gill (Green et al. 2005b), and unspecified jarosite-group minerals are reported from Carrock mine by Green & Briscoe (2002). Jarosite has been confirmed as pale yellowish-brown earthy crusts at Driggith and Sandbed mines (Green et al. 2006a).

In Wales, reports of jarosite are few, but together with aragonite, brochantite, clinoclase, erythrite and gypsum, it forms as an oxidation product in the copper-rich orebody at Parys Mountain, Anglesey (Bevins 1994, Dutton et al. 1994). It occurs with copiapite and fibroferrite in post-mining deposits formed in the abandoned pyrite mine at Cae Coch, near Trefriw, Llanrwst, Gwynedd (Johnson et al. 1979). Jarosite requires confirmation at Gwaithyrafon mine, SE of Tal-y-bont, Central Wales orefield, Dyfed (Rust & Mason 1994).

Yellow earthy patches, or minute pale-yellow platy crystals, of jarosite have been found in sulphide-rich gneisses at Loch Garbhaig, Letterewe Forest, Loch Maree, Highland Region, Scotland (Macpherson & Livingstone 1982), and it occurs with goethite and gypsum on granite matrix near decomposing pyrite at Bellsgrove mine, on the Strontian Main vein, Strontian, Highland Region (Livingstone 2002).

jarrowite

TL Jarrow, Tyne & Wear, England

A synonym of thinolite; a pseudomorph of calcite after ikaite. See Browell (1861).

jaskólskiite?

orthorhombic $Pb_{2.2}Cu_{0.2}(Sb,Bi)_{1.8}S_5$

A lead–copper–antimony–bismuth sulphide found on samples in the Kingsbury collection at the NHM. Originally thought to have been collected from Wheal Cock, St Just, Cornwall, this provenance is now considered dubious (Ryback et al. 1998, 2001). Unless further discoveries are made, this species cannot be considered native to Britain or Ireland.

jasper

An impure variety of quartz (chalcedony) with hematite inclusions.

jasper opal
A deep-red, almost opaque, variety of opal.

jeanbandyite
STOTTITE GROUP
tetragonal (pseudo-cubic) $(Fe^{3+},Mn^{2+})Sn^{4+}(OH,O)_6$

A very rare iron–manganese tin hydroxide. At Penberthy Croft, Cornwall, jeanbandyite occurs in an intimate mixture with natanite as lustrous yellow transparent-to-translucent pseudo-octahedral crystals up to 0.4 mm across. The crystals occur in high-temperature sulphide veins on an arsenopyrite–quartz matrix with associated clinochlore, galena and sphalerite, and occasionally anglesite and cerussite (Betterton et al. 1998, Betterton et al. 2000). At Hingston Down quarry, 2 km W of Gunnislake, Calstock, Cornwall, jeanbandyite–natanite intergrowths were found along an E–W-trending vein in two areas: at the first, forming lustrous pseudo-octahedral crystals to 1.2 mm; and, at the second, forming opaque yellowish-brown octahedral crystals to 2 mm. Jeanbandyite crystallized later than cassiterite in the vein paragenesis and it is therefore later than all the major primary minerals; an origin from residual tin-rich hydrothermal fluid is proposed (Betterton et al. 1998). The larger crystals are perhaps the largest examples of the species in the world (Wolfe 1997). Jeanbandyite has also been found as isolated minute yellow octahedra to about 0.3 mm across in small cavities on partly altered galena, pyrite, sphalerite and chalcopyrite at Killham mine, St Neot, Cornwall (J. Betterton pers. comm. 2006). Possible jeanbandyite, forming transparent pale-orange octahedra <0.5 mm across on arsenopyrite, has also been found at Wheal Charlotte, Charlotte United Mines, Perranuthnoe, near Penzance, but this awaits confirmation (M. R. Stolworthy pers. comm. 2007).

Pyramidal mixed crystal of jeanbandyite–natanite, 0.4 mm across. Hingston Down quarry, Calstock, near Gunnislake, Cornwall. (© D. I. Green)

jelletite
A green variety of andradite, garnet group.

jenaite
A synonym of ilvaite.

jenkinsite
A variety of antigorite, kaolinite–serpentine group.

jimthompsonite
orthorhombic $(Mg,Fe^{2+})_5Si_6O_{16}(OH)_2$

A very rare magnesium–iron silicate hydroxide, dimorphous with clinojimthompsonite and found in Lewisian amphibolite-facies mafic–ultramafic rocks SW of Achmelvich, in the vicinity of An Fharaid Mhòr, near Lochinver, Highland Region, Scotland. Jimthompsonite and two other triple-chain pyriboles (clinojimthompsonite and probable chesterite) occur in complex intergrowths with actinolite, anthophyllite and cummingtonite. Individual crystals can reach 3 cm in length. The pyriboles do not pseudomorph the amphibole species and they probably grew during prograde Inverian metamorphism, possibly as the result of infiltration of CO_2-rich fluid. The many chemically similar pyriboles and their disposition within compound prisms suggests that crystallization was kinetically controlled (Droop 1994).

johannite
triclinic $Cu(UO_2)_2(SO_4)_2(OH)_2{\cdot}8H_2O$

A rare hydrated copper uranyl sulphate hydroxide briefly described from the Peeth lode, Geevor mine, near Pendeen, St Just, Cornwall. Along the Peeth lode, other secondary uranium minerals include andersonite, cuprosklodowskite, schröckingerite and possibly becquerelite (Elton & Hooper 1992, Elton et al. 1992). In the No. 2 Branch of Geevor mine, johannite underlays sodium-zippeite and is associated with antlerite, atacamite, brochantite, chalcocite and gypsum (Elton & Hooper 1993). Johannite also occurs at three other Cornish locations: Wheal Basset, Illogan, near Redruth; South Terras mine, St Stephen-in-Brannel (Robson 1949b); and South Crofty mine, Camborne (LeBoutillier 1996).

johannsenite?
PYROXENE GROUP
monoclinic $CaMn^{2+}Si_2O_6$

A rare calcium–manganese silicate that forms two series, with diopside and with hedenbergite, and is thought to occur in an Early Devonian hydrothermal manganese deposit overlying Dalradian Grampian Group schists at Dalroy Burn, near Culloden Moor, Highland Region, Scotland (Nicholson 1989b). Confirmation is required.

johnsomervilleite
trigonal $Na_2Ca(Mg,Fe^{2+},Mn)_7(PO_4)_6$
TL Gleann Còsaidh, Loch Quoich, Knoydart, Highland Region, Scotland

Named after John M. Somerville (1908–1978), who discovered the mineral in Gleann Còsaidh, Loch Quioch, Scotland; born in Rutherglen, Glasgow. Johnsomervilleite is a transition metal phosphate found in two small metamorphic segregation pods (15 cm and 30 cm long) occurring in striped kyanite–sillimanite-bearing Moine gneiss from the type locality. The segregation pods contain apatite, garnet, muscovite, plagioclase and quartz, together with very- mall clusters of johnsomervilleite associated with graftonite, jahnsite, mitridaite, phosphosiderite, rockbridgeite and vivianite. Johnsomervilleite forms dark-brown anhedral grains or pools measuring up to 1.5 mm across. It has a vitreous lustre and usually is in close association with graftonite or vivianite (Livingstone 1980).

johnstonite (of Haidinger)
TL Dufton, Cumbria, England

Named after James Finlay Weir Johnston (1796–1855), chemist, mineralogist, reader in chemistry and mineralogy at the University of Durham and author of *Elements of agricultural chemistry and geology* (1842) and *Catechism of agricultural chemistry and geology* (1844); born in Paisley, Strathclyde, Scotland. Johnstonite is probably a mixture of anglesite, galena and sulphur. It requires confirmation.

johnstonotite
An intermediate composition: grossular–spessartine, garnet group.

johnstrupite
A synonym of rinkite, but also said to be a variety of mosandrite.

jordanite
monoclinic $Pb_{14}As_6S_{23}$

A rare lead sulpharsenite that forms a series with geocronite and is reported from Penberthy Croft mine, St Hilary, Cornwall, by Robson (1949b), Kingsbury (1954), Camm & Merry (1991b) and Golley & Williams (1995). However, more recent work has failed to confirm its occurrence (Betterton et al. 2000).

In Wales, a rare lead sulpharsenide with a composition close to jordanite has been reported from Parys Mountain, Anglesey (Sivaprakash 1977, Bevins 1994, Southwood & Bevins 1995).

Jordanite occurs in some of the large stratabound lead–zinc ore deposits in Ireland. At Tynagh mine, near Killimor, Co. Galway, epigenetic Cu–Pb–Zn–(Ag) mineralization contains a suite of sulphide species that occur in trace amounts: native antimony, arsenopyrite, bournonite, enargite, geocronite, gratonite, jordanite, marcasite, millerite, pyrite, siegenite and tennantite, many occurring as inclusions in galena (Morrissey et al. 1971, Gibson 1979, Clifford et al. 1986). The Navan lead–zinc orebody in Co. Meath has a simple mineral assemblage consisting of galena and sphalerite, with minor marcasite and pyrite, although less commonly boulangerite, bournonite, cylindrite, freibergite and pyrargyrite are also found. Single grains of argyrodite, franckeite and jordanite are also recorded (Boast 1978, Andrew & Ashton 1985, Ashton et al. 1986). In the Silvermines area, Co. Tipperary, the syn-sedimentary and syn-diagenetic Upper G orebody is predominantly composed of marcasite and pyrite (making up 75% of the sulphide content – with the remainder being galena and sphalerite). Jordanite is a minor component of the ore, along with boulangerite, bournonite, chalcopyrite, pyrrhotite and tennantite (Taylor & Andrew 1978, Moreton 1999). A report of geocronite forming grains up to 200 µm across from Silvermines B zone (Zakrzewski 1989) gives analyses with As > Sb, so the species is actually jordanite. Jordanite is also noted in three stratabound sulphidic lenses dominated by galena, marcasite, pyrite and sphalerite, with minor arsenopyrite, bornite, chalcopyrite, gersdorffite, gratonite, seligmannite and tennantite in the carbonate-hosted Lisheen deposit, between Thurles and Rathdowney, Co. Tipperary (Strashimirov et al. 1999, Hitzman et al. 2002, Fusciardi et al. 2004, Wilkinson et al. 2005).

jordisite?

amorphous MoS_2

A rare molybdenum sulphide, dimorphous with molybdenite and described from a small lead working at Tickow Lane mine, Shepshed, Leicestershire. Patches of a dark-brown to black uncemented sand found in the calcite-cemented walls of the mine were shown to contain an amorphous material enriched in molybdenum and it was concluded that this was jordisite (King 1973). Further confirmation is desirable.

joséite-A

trigonal Bi_4TeS_2

A rare bismuth–tellurium sulphide, sometimes simply referred to as joséite, found at Carrock mine, Grainsgill, Caldbeck Fells, Cumbria, where it is usually intergrown with bismuthinite. It also crystallizes as resplendent round plates, with a blue tint on fresh fracture surfaces (Davidson & Thomson 1951). There has been great confusion over the identification of bismuth tellurides, and many specimens labelled as grünlingite, joséite or tetradymite appear to have been misidentified (Cooper & Stanley 1990). With a redefinition of joséite into the two species joséite-A and joséite-B (Peacock 1941) and the discovery of ingodite (Zav'yalov & Begizov 1981a,b), identification problems seem to have been resolved. Samples labelled as "grünlingite" from Carrock Fell have now been shown to contain joséite-A, joséite-B, bismuthinite and minor ingodite (Cooper & Stanley 1990). Zav'yalov & Begizov (1981a,b) confirm that "grünlingite" from Brandy Gill mine, Caldbeck Fells, was a mixture of joséite-A and bismuthinite, but Cooper & Stanley (1990) suggest that this too is material from the Carrock mine site. Young (1987a) notes "joséite including joséite-A" from Brandy Gill on the basis of a specimen in the Kingsbury collection now at the NHM, and another specimen from this collection is listed by Golley & Williams (1995) – joséite-A from Penlee quarry, Newlyn, Cornwall. It is uncertain if they are mislabelled specimens (see Ryback et al. 1998).

joséite-B

trigonal Bi_4Te_2S

A rare bismuth–tellurium sulphide that was defined by Peacock (1941) during a re-evaluation of the ill defined species, joséite (see joséite-A above). A selenium-rich variety of joséite, nearer in composition to joséite-B than joséite-A and closely associated with laitakarite (< 10 µm), occurs in the Bonser vein, 3 km NW of Coniston, Cumbria. This was one of the most productive sources of copper in the Lake District and it contains a mineral assemblage that also includes arsenopyrite, native bismuth, bismuthinite, calcite, chalcopyrite, chlorite, cosalite, dolomite, galena, magnetite, marcasite, pyrite, pyrrhotite, sphalerite and stilpnomelane, all in a quartz-dominated gangue (Stanley & Vaughan 1982b). Museum specimens labelled as tetradymite from Carrock mine, Grainsgill, Caldbeck Fells, Cumbria, have been shown to contain a mixture of joséite-B and ingodite (Cooper & Stanley 1990). Joséite-B noted by Golley & Williams (1995) from Penlee quarry, Newlyn, Cornwall, was identified on a specimen in the Kingsbury collection at the NHM; it is uncertain whether this is one of the mislabelled specimens in the collection (Ryback et al. 1998).

10 mm-long mass of joséite-A mantled by a thin rim of golden joséite-B from Carrock mine, Caldbeck Fells, Cumbria. (© D. I. Green)

In Scotland, joséite-B is described from quartz veins, up to 2 cm wide, cutting greywackes adjacent to the Cairngarroch Bay porphyry complex, S of Portpatrick, Dumfries & Galloway. Joséite-B and an unnamed bismuth–lead telluride occur as inclusions in bismuthinite in association with native bismuth, chalcopyrite, pyrite, pyrrhotite and sphalerite (Livingstone 1993, Lowry et al. 1997).

julgoldite-(Fe^{2+})

PUMPELLYITE GROUP

monoclinic $(Ca_2,Fe^{2+})(Fe^{3+})_2(Si_2O_7)(SiO_4)(OH)_2{\cdot}H_2O$

A rare hydrated calcium–iron silicate hydroxide that forms series with pumpellyite-(Fe^{2+}) and with pumpellyite-(Mg). Julgoldite-(Fe^{2+}), forming small black crystals in pectolite, has been found in Scotland at Ratho quarry, near Edinburgh, and at Auchinstarry quarry, Kilsyth, near Glasgow. At the latter locality, julgoldite-(Fe^{2+}) is highly lustrous and it forms plumose clusters up to 2 mm long. Both occurrences are from rare pockets in quartz dolerite and probably result from highly localized late-stage alteration processes acting on the quartz dolerite. Julgoldite-(Fe^{2+}) had previously been mistakenly described as aenigmatite or goethite (Heddle 1901, Livingstone 1976b).

junkerite

A synonym of siderite.

jurinite

A synonym of brookite.

K

kaersutite

AMPHIBOLE GROUP
monoclinic $NaCa_2(Mg_4Ti)(Si_6Al_2)O_{23}(OH)$

A common sodium–calcium–magnesium–titanium aluminosilicate hydroxide that forms a series with ferrokaersutite and is found in alkali volcanic rocks, alkali dolerites and syenites. Where possible, identification is based on the 1997 classification of the amphibole group (Leake et al. 1997a, Mogessie et al. 2004).

Discrete grains of brown kaersutite and rims of earlier-formed clinopyroxene are described from metadolerites (greenstones) metamorphosed to prehnite–pumpellyite facies at Trevone Bay, W of Padstow, Cornwall (Floyd & Rowbotham 1982). Erratic blocks of altered gabbro from a gravel pit at Burscough, SE of Southport, Lancashire, contain three species of amphibole: edenite, kaersutite and magnesiohornblende (D. I. Green specimen, Tindle unpub. EMP data).

Kaersutite is described from two locations in Wales; in the Tal y Fan dolerite intrusion, SW of Conwy, Gwynedd (Merriman et al. 1986), and in doleritic and gabbroic intrusions such as the Llanwnda intrusion, W of Fishguard, Dyfed (Bevins 1994).

In Scotland, black rounded megacrysts of kaersutite up to 3 cm in length occur in a basanite vent at Black Rock, 2 km SSW of Fairlie, Hunterston Sands, Strathclyde (Alexander et al. 1986). In another vent at Waterside, near Muirkirk, Strathclyde, kaersutite is a minor component of garnet pyroxenite xenoliths found in a Carboniferous basanitic intrusion (Upton et al. 2003). Scarce black rounded kaersutite megacrysts up to 3 cm across have been noted in the xenolithic olivine-nephelinite dyke at Gribun, Isle of Mull, Strathclyde (Upton et al. 1998). Certain pyroxenite nodules found in Carboniferous–Permian vents near Elie in Fife contain kaersutite associated with augite, forsterite and, in one variety, interstitial oligoclase feldspar. Anorthoclase, augite, kaersutite and pyrope megacrysts are also reported in the tuff of the Elie Ness vent (Chapman 1976). Other Scottish kaersutite localities are associated with alkali dolerite intrusions: as late-stage interstitial grains in picrite within the Shiant Isles alkali dolerite sill, Western Isles (Gibb & Henderson 1996); within fine syenite in the differentiated alkali olivine-dolerite intrusion at Cnoc Rhaonastil, Isle of Islay, Strathclyde, where arfvedsonite, ferro-edenite and magnesiokatophorite are associated amphibole-group members (Preston et al. 1998a, 2000b); in veins of lugarite cutting picrite within the Lugar Sill, Glenmuir Water, Strathclyde (Howie 1963, Mason 1968). A monchiquite dyke at Wart Holm, Copinsay, Orkney, contains large phenocrysts of an amphibole originally described as hornblende (Flett 1900), which, because of its high titanium content, Macpherson & Livingstone (1982) describe as kaersutite. It now classifies as potassicpargasite.

A unique pyroclastic pipe at Black Ball Head, near Cahermore, Co. Cork, Ireland, contains megacrysts of kaersutite, phlogopite and titanium-rich magnetite, accompanied by ultramafic nodules of amphibole pyroxenite in a completely altered chlorite–clay matrix. Kaersutite is black with a bright-green rim and it forms embayed anhedral to subhedral crystals up to 4 cm in length (Pracht & Kinnaird 1995).

Amphibole with a kaersutite core and a pargasite rim occurs in an acicular gabbro in the Cadomian igneous complex at Beaucette, north Guernsey (Bishop & French 1982). Deep reddish-brown kaersutite occurs in appinitic diorites at Le Nez, south-east Jersey (Bishop & Key 1983). Kaersutite is one of many amphibole species described from an igneous complex (olivine-gabbro to aplogranite compositions) at Sorel Point in north Jersey (Salmon 1991), but the data are unpublished.

Euhedral kaersutite from the Lugar Sill, Strathclyde, Scotland; width 2.75 mm; viewed in PPL (upper) and XPL (lower). (© K. Gardner)

kahlerite

AUTUNITE GROUP
tetragonal $Fe^{2+}(UO_2)_2(AsO_4)_2 \cdot 10\text{–}12H_2O$

A rare hydrated iron uranyl arsenate found on joint surfaces of shaly to slaty country rocks half-way up the cliff, adjacent to a highly uranium-rich fissure vein close to Needle's Eye, SW of Caulkerbush, Dumfries & Galloway, Scotland. This and other similar veins are in the aureole of the Criffell granodiorite. Well crystallized encrustations of kahlerite form pale creamy-yellow plates (<1 mm) in association with metakahlerite. Other supergene minerals from this vicinity, produced during the alteration of primary As–Bi–Co–Cu–Fe–Pb–U minerals include cornubite, eulytite, lavendulan, malachite, meta-autunite, metazeunerite, mixite, schoepite, tyrolite, vandendriesscheite, walpurgite, wulfenite and zeunerite (Braithwaite & Knight 1990).

Translucent yellow platy crystals of kahlerite found near Needle's Eye, SW of Caulkerbush, Dumfries & Galloway, Scotland; crystals up to 0.4 mm in length. (© D. I. Green)

kainosite-(Y)
orthorhombic $Ca_2(Y,Ce)_2Si_4O_{12}(CO_3)\cdot H_2O$
A rare hydrated calcium–yttrium–cerium silicate carbonate found at several localities within miarolitic cavities in Tertiary epigranites of the Red Cuillin on the Isle of Skye, Highland Region, Scotland. Other species in the cavities include fluorite, gadolinite and zircon (Green & Todd 2001). One such locality is near Eas a Bhradain, where colourless, blocky-to-prismatic crystals of kainosite-(Y) have been found. Kainosite-(Y) also occurs as a rare, late-stage primary mineral at Corrantee and Whitesmith mines, and at Strontian barite mine, N of the village of Strontian, Highland Region. It forms white-to-yellow translucent-to-opaque radiating aggregates of acicular, prismatic or platy crystals, up to 1 mm in length, and is associated with ancylite-(Ce), brewsterite and harmotome (Green & McCallum 2005).

White and yellow hemispheres of kainosite-(Y), up to 0.5 mm across, on brown chamosite; (lower) one hemisphere of kainosite-(Y) is perched on quartz; Whitesmith mine, near Strontian, Highland Region, Scotland. (© D. I. Green)

kakochlor
An iron–aluminium-rich variety of asbolane.

kalcglimmer
A synonym of margarite.

kalinite?
monoclinic? $KAl(SO_4)_2\cdot 11H_2O$
An uncertain hydrated potassium–aluminium sulphate species considered synonymous with alum (Heddle 1901). Unfortunately, "alum" is also used as a general term for hydrous sulphates of aluminium and an alkali, thus making problematic the confirmation of kalinite.

kalk or kalkspath
A synonym of calcite.

kalk-uranite
A synonym of autunite.

kalsilite
hexagonal $KAlSiO_4$
A rare potassium aluminosilicate, polymorphous with kaliophilite, panunzite and trikalsilite. Kalsilite occurs in a high-temperature calcic skarn (the rock types include carbonates and a cherty nodule) near the contact of the Cuillin gabbro complex at Camasunary Bay, Strathaird Peninsula, Isle of Skye, Highland Region, Scotland. Associated species include calcite, garnet (andradite–grossular), melilite, vesuvianite and wollastonite, together with minor amounts of magnetite, monticellite, perovskite and pyrite (Beard & Drake 2007).

kämmererite
A chromium-rich variety of clinochlore, chlorite group.

kampylite
A synonym of phosphorus-rich mimetite.

kankite/kaňkite
monoclinic $Fe^{3+}AsO_4\cdot 3.5H_2O$
A rare hydrated iron arsenate forming green botryoidal crusts with zeunerite at King's Wood mine, Buckfastleigh, Devon, and pale-green crusts at South Terras mine, St Stephen-in-Brannel, Cornwall (Ryback & Tandy 1992).

kaolin
A name, generally much like the term "china clay", which carries no reference to the exact composition of the material. It is now used to describe the three polytypes dickite, kaolinite and nacrite. Older literature undoubtedly refers to mixtures of these species.

KAOLINITE–SERPENTINE GROUP
Triclinic, monoclinic, orthorhombic, trigonal and hexagonal silicates with the general formula $M_{2-3}Z_2O_5(OH)_4\cdot nH_2O$, where M = Al, Fe^{3+}, Fe^{2+}, Mg, Mn^{2+}, Ni, Zn, and Z = Al, Fe^{2+}, Si. In Britain and Ireland, the group is represented by amesite, antigorite, berthierine, chrysotile, cronstedtite, dickite, fraiponite, greenalite, halloysite, kaolinite, lizardite and nacrite. Burke (2006), reporting on an IMA-CNMNC-approved mass discreditation of species, states that clinochrysotile, orthochrysotile and parachrysotile are formally discredited and that all samples are polytypes (crystallographic variants) of the species chrysotile. All three variants are found in Britain or Ireland. It could be argued that dickite, kaolinite and nacrite are polytypes of a single species, the polytypes being -1*A* (kaolinite), -1*M* (nacrite), and -2*M* (dickite).

kaolinite KAOLINITE–SERPENTINE GROUP
trigonal $Al_2Si_2O_5(OH)_4$
A very common aluminosilicate hydroxide, polymorphous with dickite, halloysite and nacrite, and formed during hydrothermal alteration and weathering, particularly of alkali feldspars and other aluminium-bearing minerals. Kaolinite is a common constituent of the clay-size fraction of sediments. It is an important industrial mineral, used as a filler in paper and in the manufacture of ceramics.

England
Avon Deeply oxidized smithsonite-bearing vein material associated with cavernous mineralized Upper Triassic limestone breccias near East Harptree, N of the Mendip Hills, contains relatively pure monomineralic off-white chalky cavity linings consisting of allophane and kaolinite. Dundasite, gibbsite, scarbroite and a hydrotalcite-like mineral are associated with secondary aluminium species at this location (Alabaster 1990b).
Cornwall Current production of kaolinite is from hydrothermally altered granite worked in 19 china-clay pits in Cornwall and a further 5 in Devon. The major production centre for kaolinite is within the St Austell granite, central Cornwall, where the most intense kaolinization has occurred and where over 200 abandoned clay works are also to be found. A china-clay museum at Wheal Martyn, Carthew, preserves and records the history of china-clay production in the St Austell area. Mineralogical reports list the following china-clay pits around St Austell, although most discuss the process of kaolinization rather than kaolinite itself: Blackpool; Bodelva; Great Carclaze; Gunheath, Carthew; Hensbarrow; Littlejohn; Wheal Prosper; Wheal Remfry; Whitemoor–Dorothy (all Weiss 1994); Goonbarrow, near Bugle (Bray & Spooner 1983, Grigg 1986, Weiss 1994); Goonvean and Rostowrack, W of Nanpean

(Weiss 1994, Scott et al. 1998); Melbur, near Treviscoe (Weiss 1994, Psyrillos et al. 1998); Trelavour Downs (Alderton et al. 1980); Trethosa, Carthew (Psyrillos et al. 1998); unnamed clay pit, St Austell (Hogg et al. 1975). More generally, kaolinization in the St Austell granite area is discussed by Collins (1887), Manning & Exley (1984) and Manning et al. (1996).

Within the Land's End granite, kaolinite has been extracted from the Bostraze–Balleswidden china-clay pit, 2 km E of St Just. The kaolinized zone is centred around a stockwork or sheeted-vein swarm in poorly megacrystic biotite granite. In one vein type, layers of kaolinite occur within growth zones of clear, smoky or amethystine quartz crystals. The highly altered granite is composed of coarse (>10 μm) kaolinite books set in a finer-grain matrix. XRD studies indicate that the clay fraction of this granite contains 75–90% kaolinite; dickite and illite are the other components (Jackson et al. 1982, Dominy & Camm 1998). A study of a deep weathering profile on the Carnmenellis granite, showed that gibbsite and kaolinite occur in all size fractions and in all horizons. The size of some of the grains (>200 μm) indicates that they are pseudomorphs after feldspar (Rice 1973). Kaolinite is an extensive white earthy mineral filling cavities in tin lodes at Wheal Jane, Kea, near Truro (Rust 1982d). Elsewhere in Cornwall, kaolinization is noted over parts of the Bodmin and Tregonning–Godolphin granites (Thurlow 2001).

Cumbria Kaolinite is noted from: China Clay mine (Harestones Umber mine), Caldbeck Fells (Young 1987a); High Pike, Caldbeck Fells (Davidson & Thomson 1951); Oughterside opencast coal site, Aspatria (Young & Nancarrow 1988a): Park farm, near Askham-in-Furness (Rose & Dunham 1977); Scar Crag cobalt mine, Causey Pike (Strens 1962); Sellafield area boreholes (Milodowski et al. 1998); Shap granite quarry (Firman 1972, Wirth & Green 2002); Wabberthwaite quarry (Dwerryhouse 1909).

Derbyshire Sandpits in the Brassington and Friden areas (Ford et al. 1993).

Devon Bovey and Plympton (Hall 1868b). Ball clay, a type of ultra-fine highly plastic clay containing a large proportion of kaolinite, and used for white pottery, including sanitary ware, tableware, floor tiles and electrical insulators, is produced from the Bovey and Petrockstowe basins to the E and N of the Dartmoor respectively (Ball Clay Heritage Society 2003).

Dorset Kimmeridge Bay and Portesham (Scotchman 1987). Wareham basin is well known for its kaolinite-bearing ball-clay deposits (Ball Clay Heritage Society 2003).

Durham Groverake mine (Firestone sublevel), near Rookhope, Weardale (Smith & Hardy 1981, Dunham 1990).

Kent Shore Hill, Otford (Brindley & Robinson 1946).

Leicestershire Old Parks coal pit, NE of Ashby-de-la-Zouch (King 1993); Newhurst quarry, Shepshed (Ince 2005).

Lincolnshire Cleethorpes borehole (Milodowski et al. 1987); Donnington-on-Bain (Scotchman 1987).

Norfolk Denver Sluice, North Runcton and North Wootton (Scotchman 1987).

Northamptonshire Hartwell (Scotchman 1987).

Northumberland Settlingstones mine, near Hexham (Ineson 1972).

Staffordshire Rowhurst Tonstein at Red Street colliery, near Kidsgrove (Wilson et al. 1966).

Surrey Warlingham (Scotchman 1987); Westcott, Dorking (Brindley & Robinson 1946).

Warwickshire Near Mancetter church, NE of Bentley church (Guppy 1956).

Wiltshire Swindon, Tisbury and West Lavington (Scotchman 1987).

Yorkshire Exceptionally well crystallized kaolinite is reported from inclusions in ferruginous deposits on the Yorkshire coast, just N of Scalby, near Scarborough. Crystallite size is 30–50 μm (Goodyear & Duffin 1961). Kaolinite is also reported from: Hickleton colliery, WNW of Doncaster (Uttley 1988); Marton and Reighton (Scotchman 1987); Robin Hood quarry, Thorpe on the Hill, S of Leeds (Deans & Vincent 1938); and West Gill, Northdale (Rastall & Hemingway 1949).

Wales

In Wales, kaolinite takes the form of a white powder consisting of small plates; it is present in many Carboniferous rocks in the South Wales coalfield, especially in fault zones, such as at Cwm Fforch-wen in the Garw Valley, Mid Glamorgan. It is probably derived from the decomposition of feldspars that formed part of the Pennant Sandstone and Grits (North 1916, Bevins 1994). An investigation of clay minerals in carbonate and terrigenous rocks from the South Wales coalfield revealed the widespread presence of kaolinite, chlorite, illite and illite–montmorillonite mixed-layer clay. Increased metamorphism in the western part of the coalfield (the anthracite area) has allowed vermicular allevardite, dickite, kaolinite and pyrophyllite to form (Gill et al. 1977). Carboniferous dirt bands in the Hornlo seam examined in borehole material and at the Pengasto opencast site in the Brynammon district of the South Wales coalfield contain 10% kaolinite and 40% hydroxylapatite. A biogenic source is proposed (Baqri & Hodson 1979). Kaolinite is also reported from Carboniferous rocks in southwest Dyfed (Robinson et al. 1980) and from heavily altered lavas and tuffs in a borehole near Treffgarne (Brown et al. 1987b).

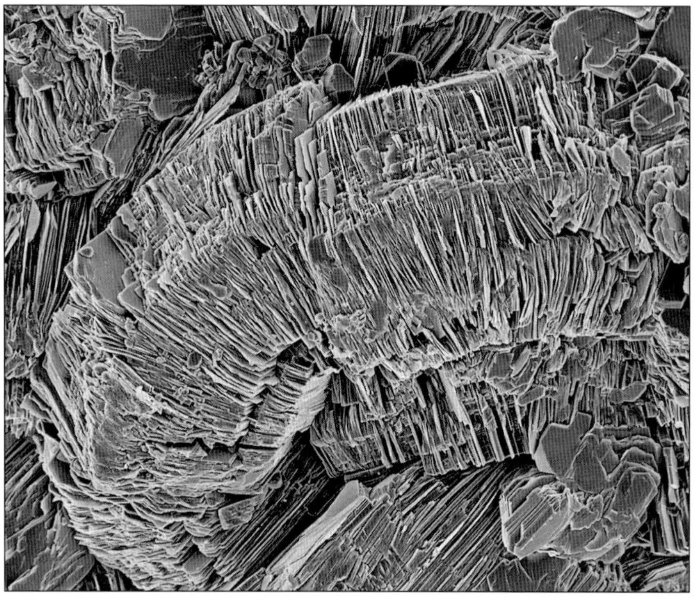

Curved book of kaolinite crystals from a joint infilling; South Bay, Scarborough, Yorkshire; width 75 μm; SEM image. (© E. Condliffe, University of Leeds)

Scotland

Kaolinite is a major component of an illite–macaulayite-bearing clay derived from a deeply weathered granite near Bennachie, 7 km WNW of Inverurie, Grampian (Russell et al. 1984, Wilson et al. 1984). It also occurs intergrown with gibbsite, goethite, illite and limonite in highly altered igneous and sedimentary rocks at the margins of minor lead–zinc mineralized dioritic intrusions around Penkiln Burn, close to the margin of the Loch Doon granite complex, N of Newton Stewart, Dumfries & Galloway (Stone & Leake 1984). An extensive zone of kaolinitic alteration up to 30 m thick is developed in Dalradian quartz–feldspar granulite at Tresta on Mainland, Shetland, but it is kaolin rather than kaolinite that is being described (Berridge 1969, Dunham et al. 1978). Other reports of kaolinite from Scotland, all of a minor nature, include: Alva mine, Silver Glen, Ochil Hills, Central Region (Moreton et al. 1998); 500 m NNW of Barncorkrie farm, Drummore, Portencorkie area, Dumfries & Galloway (Cameron & Lawson 1979); Saltcoats beach, near Ardrossan, Strathclyde (de Lapparent 1934); Lamb Hoga, Fetlar, Shetland (Livingstone 2002).

Ireland

In Ireland, kaolinite has been identified in many clay concentrates derived from samples associated with mineral deposits: Coolartragh and Lisdrumgormley, Co. Armagh; Cappaghglass; Dereenalomane; Dooneen and Mountain mine, Allihies; Duneen Bay, near Clonakilty; Mount Gabriel (all Co. Cork); Conlig–Newtownards, Co. Down; Clontibret, Co. Monaghan; Tynagh, near Killimor, Co. Galway. K/Ar dating of these concentrates indicates that most were formed during Carboniferous times or earlier (Halliday & Mitchell 1983). Killyflugh quarry, N of Ballymena, Lower Toome, Co. Antrim (Herbert Smith 1916), and Grouse Lodge quarry, near Shanagolden, Co. Limerick (Moreton & Green 2005), are two other Irish kaolinite localities.

karstenite

A synonym of anhydrite.

kashinite?

orthorhombic $(Ir,Rh)_2S_3$

A rare iridium–rhodium sulphide thought to have been found (but too small to analyze) at Pinbain Bridge, Strathclyde, Scotland. Kashinite forms euhedral <10 μm-diameter inclusions in chrome spinel within chromitites belonging to the Ballantrae ophiolite complex. Erlichmanite, laurite (mainly), irarsite and Ir–Os–Ru alloys are associated species (Power & Pirrie 2004).

kasolite

monoclinic $Pb(UO_2)SiO_4 \cdot H_2O$

Spherical aggregates of kasolite from Loe Warren Zawn, St Just, Cornwall; width 90 μm; SEM image. (© N. J. Elton)

A rare hydrated lead uranyl silicate found at Loe Warren Zawn, W of Botallack, St Just, Cornwall, where surface water draining from rubble at the base of the cliff has trickled over veinlets carrying chalcocite, chalcopyrite and pyrite, and precipitated an assemblage of supergene carbonate and phosphate minerals on exposed surfaces and joint planes near a patch of uraninite mineralization. Pale whitish-yellow through canary-yellow to dark-yellow and orangey-yellow kasolite occurs with cerussite, dewindtite, malachite, intermediate torbernite–zeunerite compositions, and rare cumengéite. It forms compact botryoidal crusts, fibrous crusts, and areas of discrete and intergrown compact or open spherules of radiating lath-like or petaline crystals (Elton & Hooper 1995a,c).

In Scotland, kasolite is found in the Lewisian granitic pegmatite at Chaipaval (Ceapabhal), 3 km NW of Northton (Taobh Tuath), South Harris, Western Isles. It is an accessory constituent of the pegmatite, along with allanite, beryl, columbite, gahnite, magnetite, monazite, pyrite, pyrochlore, spessartine, thorite, topaz, tourmaline, uraninite and zircon (von Knorring & Dearnley 1959b, 1960). Kasolite was found in small fractures containing fluorite, galena and pyrite on erratic boulders of siliceous adamellite related to the Helmsdale granite, at Caen Burn, near Helmsdale, Highland Region (Bowie et al. 1970, Gallagher et al. 1971, Tweedie 1979). Kasolite also occurs in late uranium-rich mineralization hosted by Dalradian metasediments at Tyndrum Main mine, Central Region. The assemblage includes barite, chalcopyrite, galena, marcasite, pyrite, safflorite and uraninite (pitchblende), associated with secondary erythrite and the uranium minerals liebigite, kasolite, uranospinite and zeunerite. The secondary species probably formed during post-depositional oxidation of the deposit (Darnley 1962a, Ridgway 1980, Pattrick 1985).

katharite

A synonym of alunogen.

katophorite

AMPHIBOLE GROUP

monoclinic $Na(CaNa)Fe^{2+}_4(Al,Fe^{3+})(Si_7Al)O_{22}(OH)_2$

A rare sodic–calcic amphibole that forms a series with magnesiokatophorite. See also aluminokatophorite.

Katophorite forms the cores of zoned amphibole crystals in blueschists belonging to the Mona Complex from the Penmynydd/Gwna boundary exposed in a roadcut near Llanfairpwllgwyngyll, southeast Anglesey, Gwynedd, Wales. A wide range of amphibole species occur in these rocks, including actinolite, alumino-ferrowinchite, aluminowinchite, ferrobarroisite, ferroglaucophane, ferrowinchite, glaucophane and winchite (Gyopari 1984).

Katophorite occurs in miarolitic cavities in a Western Redhills Tertiary epigranite at Loch Ainort, Isle of Skye, Highland Region, Scotland. It is bluish green (in thin section), sometimes euhedral, and zoned to ferro-actinolite, ferro-edenite, ferro-richterite, ferrowinchite and riebeckite compositions. Other cavity species include: hedenbergite, monazite, titanite and zircon (S. A. Rust specimen, Tindle unpub. EMP data).

kegelite

monoclinic (pseudo-hexagonal) $Pb_4Al_2Si_4O_{10}(SO_4)(CO_3)_2(OH)_4$

A very rare lead aluminosilicate sulphate carbonate hydroxide found on the waste tips of Glencrieff mine at the foot of Whyte's Cleuch, Wanlockhead, Dumfries & Galloway, Scotland. Kegelite occurs on a single specimen as small patches up to 5 mm across, each consisting of many tiny (0.2–0.3 mm) coalescing rosettes of radiating pale greenish-yellow fibres. Kegelite is embedded in leadhillite and is intimately associated with cerussite. Other species found on the sample include descloizite, galena, quartz and vanadinite (Neall & Green 2005).

keilhauite

An iron- and yttrium-rich variety of titanite.

keithconnite

trigonal $Pd_{20}Te_7$

A very rare palladium telluride identified in a layered mafic–ultramafic intrusion, on the Isle of Rum, Highland Region, Scotland. In certain chromitite layers of the Eastern Layered Series, keithconnite is found in association with species positively or tentatively identified as braggite, cooperite, hongshiite, irarsite, isoferroplatinum, kotulskite, laurite, michenerite, moncheite, palladoarsenide, members of the hollingworthite–irarsite–platarsite series, sperrylite, telargpalite, telluropalladinite and the native species gold, silver, palladium and platinum. Most of these grains are 0.2–2 μm in diameter, although, for some species, grains >15 μm are also present. The majority are in contact with sulphide minerals such as bornite, chalcocite, chalcopyrite, pentlandite or pyrrhotite. The chromitite layers were formed during magmatic fractionation and crystal settling of an open-system magma chamber, modified by magma mixing and liquid immiscibility of sulphide and silicate melts (Power et al. 2000a). Keithconnite has also been identified in chalcopyrite-rich dykes associated with the West Sgaorishal ultramafic plug on Rum. It occurs with froodite, kotulskite, paolovite, sperrylite, telluropalladinite, gold (electrum), native silver and Ag–Bi–Pd–Te phases, usually as rounded anhedral blebs enclosed within chalcopyrite or pentlandite. Grainsize varies from <1 μm to 40 μm, although the majority of grains are <5 μm (Power et al. 2003).

kentrolite

orthorhombic $Pb_2Mn^{3+}_2Si_2O_9$

A rare lead–manganese silicate that forms a series with melanotekite and is found in iron and manganese-oxide mineralized fractures within Carboniferous Limestone at Coleman's quarry, Holwell, near Frome, Mendip Hills, Somerset. Kentrolite forms part of a late-stage mineral assemblage and is mostly developed in calcite-lined cavities within manganese oxides. Other secondary species found in veins at this location include aragonite, barite, barytocalcite, calcite, cerussite, chloroxiphite, crednerite, cuprite, diaboléite, fluorite, goethite, hausmannite, hydrocerussite, malachite, melanotekite, mendipite, mimetite, paralaurionite, pyromorphite, quartz and rhodochrosite (Alabaster 1990a). At Merehead (Torr Works) quarry, Cranmore, Mendip Hills, kentrolite was originally found only in minute quantities in manganese-rich pods. However, discoveries in 2005 and 2006 produced large quantities in other pods, including jet-black euhedral crystals up to 5 mm long (originally misidentified as goethite) and brownish-black blades replacing crednerite (Turner 2007).

References to kentrolite from Higher Pitts farm, Priddy, Mendip Hills (Alabaster 1985a, Ryback & Tandy 1992, Burr 1996) are all based on a sample from the Kingsbury collection at the NHM. It is uncertain whether this is one of many mislabelled specimens in the collection (Ryback et al. 1998), although its appearance suggests that it is genuine.

kerargyrite

A synonym of chlorargyrite.

kerasine

A synonym of mendipite and phosgenite.

kerate

A synonym of chlorargyrite.

kermes

A synonym of kermesite.

kermesite?
triclinic (pseudo-monoclinic) Sb_2S_2O

A rare antimony oxysulphide that forms as a red oxidation product of stibnite and is described from Wanthwaite mine, St John's in the Vale, near Threlkeld, Cumbria, in association with beudantite, dundasite, hydrozincite, löllingite, senarmontite, sphalerite, stibnite and symplesite (Young 1987a). The sample is from the Kingsbury collection at the NHM and its provenance is doubtful (Ryback et al. 1998, 2001, A. Hart pers. comm. 2004). Films of a red mineral on quartz grain boundaries and around primary antimony minerals in a vein at Wet Swine Gill, Caldbeck Fells, Cumbria, may also be kermesite, but this could not be confirmed (Fortey et al. 1984). Cherry-red coatings on stibnite in quartz veins from Bwlch mine, near Deganwy, Gwynedd, Wales, are thought to be kermesite, but, despite XRD analysis, its identity could not be confirmed (Bevins et al. 1988). Other reports of kermesite appear to be on the basis of visual identification alone and further confirmation is required.

England

Cornwall Cutlinwith farm, Landrake with St Erney (Russell 1949b); Poltreworgey mine, St Endellion; Port Quin beach, St Endellion; Trebullet mine, Lezant; Tredinninck mine, St Erney; Wheal Leigh, Pillaton; Wheal Prosper, St Ewe (all Golley & Williams 1995).

Scotland

Dumfries & Galloway Knipe mine or, more generally, Hare Hill, 5 km SE of New Cumnock (Greg & Lettsom 1858, Heddle 1901, Russell 1949b, Williams 1965, 1973).

kerolite
A variety of talc or serpentine with variable composition.

kësterite
tetragonal $Cu_2(Zn,Fe)SnS_4$

A rare copper–zinc–iron–tin sulphide, related to the stannite group and found in greisen-bordered sulphide veins in granite at Cligga mine, Perranzabuloe, Cornwall. A reinvestigation of the holotype isostannite specimen from Cligga showed it to consist mainly of massive (but polycrystalline) grains of kësterite and stannite that display complex interrelationships, together with quartz and chalcopyrite. Ferrokësterite is an associated species (Kissin & Owens 1989). Apparently, stannite is the most abundant tin sulphide at Cligga Head, and kësterite, containing exsolution bodies of sphalerite, has been detected in only one thin section (out of about a hundred), where it displays a dark-grey colour in reflected light (Moore & Howie 1984). Distinguishing stannoidite (with a yellowish-bronze appearance) from kësterite and stannite (which are grey) is relatively straightforward, but distinguishing kësterite from stannite is much more difficult, if not impossible, in hand specimen. Kësterite is much more common in greisen-bordered veins associated with another small granite pluton, at St Michael's Mount, Marazion, Cornwall, where it is commonly found associated with stannoidite. Interestingly, kësterite forms exsolution lamellae in stannoidite, and stannoidite forms exsolution lamellae in kësterite, even in the same sample. Stannite at St Michael's Mount is relatively rare (Moore & Howie 1984). A paragenetic sequence for the mineralization at St Michael's Mount shows early cassiterite and wolframite, followed by arsenopyrite and löllingite, kësterite–stannite, and, finally chalcopyrite–pyrite, as the circulating hydrothermal system evolved from high-temperature oxide dominant to lower temperature sulphide dominant (Moore & Moore 1979). Analytical data for all the tin-sulphide species from Cornwall are summarized by Alderton (1991).

In Scotland, trace amounts of stannite-group minerals are reported as fine, symplectite-like intergrowths within chalcopyrite and sphalerite from silicified greisen at Gairnshiel Bridge, Glen Gairn, Grampian. Although of intermediate composition, analyses are enriched in zinc relative to iron so are closer to kësterite than stannite. The occurrence is related to vein-hosted cassiterite–wolframite mineralization within a zinnwaldite-bearing granite pluton of Caledonian age (Webb et al. 1992).

khmaralite?
monoclinic $(Mg,Al,Fe^{2+},Fe^{3+})_4(Al,Si,Be)_3O_{10}$

A rare magnesium–aluminium–iron–beryllium silicate reported from Lewisian Gneiss (containing potassium feldspar, kyanite, orthopyroxene and quartz), 0.5 km S of Roineabhal summit, South Harris, Western Isles, Scotland. These rocks contain accessory amounts of sapphirine, and a silica-rich variety of sapphirine containing inclusions of surinamite, which is presumed to be khmaralite (Baba 1999, 2004). Further confirmation is required.

kidney ore
A botryoidal variety of hematite.

kidwellite
monoclinic $NaFe^{3+}_9(PO_4)_6(OH)_{10}{\cdot}5H_2O$

Kidwellite from Wheal Phoenix, Linkinhorne, Cornwall; spherules are up to 0.5 mm across. (© D. I. Green)

A rare hydrated sodium–iron phosphate hydroxide found on a single specimen of quartz–goethite gossan at Phoenix United mines (Wheal Phoenix), Linkinhorne, Cornwall. Kidwellite forms greenish-yellow fibres grouped in radiating globular aggregates (< 1 mm in diameter) in cavities within the gossan, and is apparently pseudomorphous after dufrénite. Chalcosiderite is an associated species (Braithwaite & Corke 1980).

kielselmalachite
A synonym of chrysocolla.

kieselzinkerz
A synonym of hemimorphite.

KIESERITE GROUP
Monoclinic sulphates with the general formula $X^{2+}SO_4{\cdot}H_2O$, where X^{2+} = Cu, Fe, Mg, Mn, Ni, Zn. Gunningite, kieserite and szomolnokite are members of the group found in Britain or Ireland.

kieserite
KIESERITE GROUP
monoclinic $MgSO_4{\cdot}H_2O$

A relatively uncommon hydrated magnesium sulphate found in the Permian Lower Evaporite bed of the Yorkshire potash field sampled in the Eskdale No. 3 borehole at Sleights, Eskdale, near Whitby, Yorkshire. Kieserite is a highly soluble species that forms irregular twinned crystals up to 2.5 cm across, enclosing and replacing polyhalite, which itself replaced anhydrite (Armstrong et al. 1951). It also occurs in the Middle Evaporite beds in the Sleights borehole and in the Lower Evaporite beds in the Fordon borehole, Humberside (Stewart 1954, 1963). Summarizing many years of work on the Yorkshire potash field, Stewart (1965) considers kieserite to be a major constituent in the anhydrite–halite–kieserite zone of the Lower Evaporites, a widespread minor constituent in two other zones of the Lower Evaporites and a scarce minor constituent in two Middle Evaporite zones and one Lower Evaporite zone.

kilbrickenite
TL Monanoe (Kilbricken) mine, Quin, Co. Clare, Ireland

A synonym of geocronite (Prior 1902).

kilchoanite

orthorhombic $Ca_6(SiO_4)(Si_3O_{10})$

TL Kilchoan, Ardnamurchan, Highland Region, Scotland

A rare calcium silicate, dimorphous with rankinite and found at the type locality in limestones thermally metamorphosed by gabbro. Kilchoanite forms in metamorphic assemblages that essentially include melilite, rankinite and spurrite or wollastonite. These give way to assemblages that also include cuspidine, grossular, tilleyite and vesuvianite. Kilchoanite forms coarse (1–2 mm) irregular interlocking crystals with common inclusions of grossular, spurrite and vesuvianite, and rare inclusions of larnite and merwinite. Kilchoanite is always found as a replacement of rankinite, of which it is a polymorph (Agrell & Gay 1961, Agrell 1965), and it produces bright orange-yellow cathodoluminescence (Long & Agrell 1965).

Kilchoanite is briefly reported from Carlingford, Co. Louth, Ireland, where it again forms in limestones thermally metamorphosed by gabbro (Agrell & Gay 1961). Ryback & Moreton (1993) add that it does not occur as free-growing crystals in cavities.

killalaite

hexagonal $Ca_3Si_2O_7{\cdot}H_2O$

TL Killala Bay, near Inishcrone, Co. Sligo, Ireland

An hydrated calcium silicate found in thermally metamorphosed limestone that has interacted with two basalt–dolerite dykes at Killala Bay. Killalaite is a secondary mineral found in cavities and veins with calcite or afwillite in spurrite–wollastonite rocks. It occurs as well formed colourless crystals (<2 mm long), with a characteristic bow-tie twinning when seen in cross section, and probably formed in a CO_2-deficient hydrothermal environment at 350–550°C and 0.5–3.5 kbar (Nawaz 1974, Taylor 1977).

Nawaz (1977) reports a second occurrence of killalaite from Carneal, Larne, Co. Antrim, Ireland. Here, in a chalk/dolerite contact zone associated with a Tertiary volcanic plug, a secondary assemblage of afwillite, calcite, killalaite, portlandite and plombièrite forms veinlets and cavities in larnite-rich rocks. Killalaite produces crystals up to 0.3×0.1 mm, but is otherwise identical to material from the type locality.

killinite

TL Killiney Bay, Co. Dublin, Ireland

A pseudomorph after spodumene; shown to be dioctahedral hydromuscovite.

kilmacooite

TL Magpie and Connary mines, Avoca, Co. Wicklow, Ireland

A local name for a lead–zinc sulphide; probably a mixture.

kimzeyite?

GARNET GROUP

cubic $Ca_3(Zr,Ti)_2(Si,Al,Fe^{3+})_3O_{12}$

A rare calcium–zirconium–titanium–iron aluminosilicate reported from a high-temperature calcic skarn near the contact between the Cuillin gabbro complex at Camasunary Bay, Strathaird Peninsula, Isle of Skye, Highland Region, Scotland. Associated species include calcite, garnet (andradite–grossular), melilite, vesuvianite and wollastonite, together with minor amounts of strontium-rich apatite, magnetite, monticellite, perovskite, pyrite and tilleyite. Rare baghdadite, djerfisherite and spinel complete the assemblage. The occurrence of zirconium-rich minerals suggests that the skarn underwent a degree of metasomatism with the introduction of fluids rich in aluminium, potassium, titanium and zirconium, possibly from the Coire Uaigneich granophyre (Beard & Drake 2007). EMP data indicate the mineral is actually zirconium-rich andradite with a 23–28 mol% kimseyite component.

kinoshitalite

MICA GROUP

monoclinic $BaMg_3Al_2Si_2O_{10}(OH)_2$

A rare barium–magnesium aluminosilicate hydroxide found in contact-metamorphosed siliceous dolomite in the Allt Guibhsachain area, near Ballachulish, S of Fort William, Highland Region, Scotland. Most mica in the dolomite is phlogopite, but, 2 m from the contact with the Ballachulish granite, kinoshitalite has been found instead. It is thought that barium was metasomatically transferred from the adjacent pluton (Ferry 1996).

kintoreite

ALUNITE GROUP

trigonal $PbFe^{3+}_3(PO_4)_2(OH,H_2O)_6$

Kintoreite is an hydrated lead–iron phosphate hydroxide with a structure and XRD pattern closely similar to beudantite and other lead-bearing members of the alunite group (Pring et al. 1995). An investigation of supposedly XRD-confirmed beudantite crystals from Burdell Gill, Caldbeck Fells, Cumbria, found that none were beudantite. Instead, they are mostly the closely related species kintoreite or segnitite, commonly intergrown. Several of the investigated specimens were from the collection of D. R. Middleton, which are described as beudantite by Cooper & Stanley (1990) and also repeated by Neall & Leppington (2003). This casts doubt an all claimed XRD-confirmed identifications of British "beudantite". Other crystals have variable compositions, and parts of individual crystals fall within the kintoreite, philipsbornite, plumbogummite and segnitite compositional fields (D. I. Green XRD, Tindle unpub. EMP data).

Pyramidal kintoreite–segnitite crystals to 0.3 mm on iron- and manganese-rich veinstone; Burdell Gill, Caldbeck Fells, Cumbria. (© D. I. Green)

kipushite?

monoclinic $Cu_6(PO_4)_2(OH)_6{\cdot}H_2O$

A rare hydrated copper phosphate hydroxide that may have been found in an old trial on Low Pike, Caldbeck Fells, Cumbria. It is associated with bayldonite, cornwallite, mimetite and quartz. On the basis of its infrared spectrum it is either philipsburgite or a highly arsenatian kipushite (Braithwaite & Ryback 1988).

kirwanite

Named after Richard Kirwan FRS (1733–1812), chemist, mineralogist, inspector of His Majesty's Mines in Ireland, who was a fellow of the royal societies of London and of Edinburgh and author of *Elements of mineralogy* (1784); born at Cloghballymore House, near Kinvara, Co. Galway, Ireland. Kirwanite is a chlorite-like aluminosilicate of calcium and ferric iron found on the northeast coast of Ireland. It has also been described as an impure amphibole. Greg & Lettsom (1858) consider it to be a synonym of glauconite and green earth. New data presented by Nawaz & Ryback (1989), when reprocessed using the amphibole classification of Leake et al. (1997a) and software provided by Mogessie et al. (2004), show it to be ferrohornblende.

klaprothite

Thought to be identical to cuprobismutite.

klockmannite

tetragonal CuSe

A very rare copper selenide found in a gold-bearing carbonate vein cutting Middle Devonian limestones at Hope's Nose, Torquay, Devon. The selenide assemblage includes clausthalite, eucairite, fischesserite, naumannite, penroseite, tiemannite, trüstedtite, tyrrellite and umangite, all in a gangue of dominantly calcite. Klockmannite and cerussite both extensively replace other selenides (Stanley et al. 1990a).

knebelite

A manganese-rich variety of fayalite, olivine group.

kobellite
orthorhombic $Pb_{11}(Cu,Fe)_2(Bi,Sb)_{15}S_{35}$

A rare lead–copper–iron–bismuth–antimony sulphide that forms a series with tintinaite and is found as a relatively late phase within major Cu–(Pb–Zn) mineralization at Parys Mountain, Anglesey, Wales. Kobellite forms fine-grain intergrowths (5–20 µm) associated with native bismuth, bismuthinite, bournonite and galenobismutite, all around chalcopyrite, galena and tetrahedrite-group minerals. It also forms as a cement to framboidal pyrite and as inclusions in sphalerite (Sivaprakash 1977, Pointon & Ixer 1980, Bevins 1994, Southwood & Bevins 1995).

At Kilmacoo, Avoco district, Co. Wicklow, Ireland, kobellite is found in chalcopyrite-bearing quartz veins and silicified chloritic tuffs within an area of Ordovician chalcopyrite–pyrite mineralization. Kobellite occurs in fractures within pyrite in association with native bismuth, chalcopyrite, galena and gold. The gold mineralization is considered to have formed when fluid movements during the Caledonian orogeny remobilized and enhanced low-grade gold contents from an Ordovician volcanogenic massive sulphide deposit (Milner & McArdle 1992).

koechlinite
orthorhombic Bi_2MoO_6

A rare bismuth molybdate found in the silicified wallrocks of a quartz–wolframite (hübnerite) vein cutting zinnwaldite granite near Gairnshiel Bridge, Glen Gairn, Grampian, Scotland. Koechlinite is a supergene phase associated with acanthite, cassiterite, cosalite and molybdenite, and it forms interstitial patches up to 70 µm across (Tindle & Webb 1989, Webb et al. 1992). Soft yellow koechlinite is also found in a molybdenite-rich greisen from the same locality (Livingstone 1993).

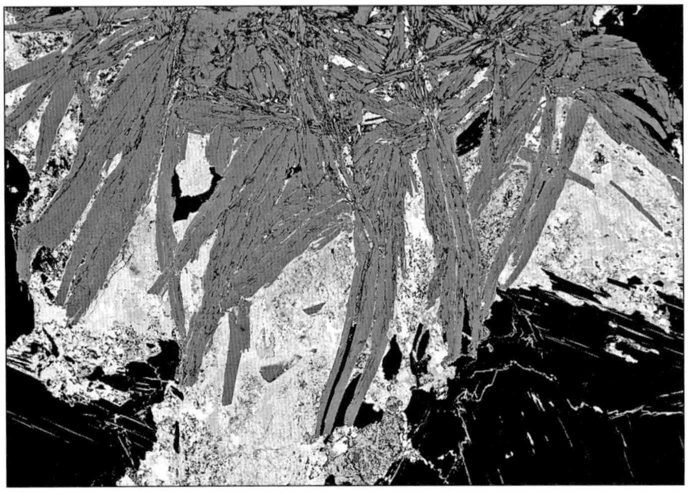

Interstitial patches of koechlinite (light grey) between blades of molybdenite (medium grey); from a zinnwaldite-rich vein at Gairnshiel, Glen Gairn, Grampian, Scotland; width 1.5 mm; BSE image. (© E. Condliffe, University of Leeds)

kombatite
monoclinic $Pb_{14}(VO_4)_2O_9Cl_4$

A very rare lead–vanadium oxychloride found at Wesley mine, Westbury-on-Trym, Avon. Kombatite occurs as bright-orange isolated cleavage plates up to 2 mm across within calcite related to manganese pod mineralization. In hand specimen it is similar to wulfenite, but has a more pronounced perfect cleavage and thus a more reflective surface (M. S. Rumsey pers. comm. 2008).

kongsbergite
A variety of native amalgam; mercury-rich silver containing about 5% mercury.

konilite
A powdery variety of quartz found in zeolitic cavities.

koninckite
tetragonal $Fe^{3+}PO_4{\cdot}3H_2O$

A very rare and poorly characterized hydrated iron phosphate found at Grouse Lodge quarry, 5 km S of Shanagolden, Co. Limerick, Ireland. Koninckite is a hard colloform iron-stained white material, forming masses several cm across in Upper Carboniferous phosphate-rich black shales. An SEM was needed to resolve the radiating masses of tiny lath-like crystals, which are up to 10 µm long but only 1 µm thick. Associated species are gibbsite, kaolinite, planerite, turquoise and wavellite (Moreton & Green 2005).

koodilite
TL Co. Down, Ireland

Impure thomsonite, zeolite group.

köttigite
VIVIANITE GROUP

monoclinic $Zn_3(AsO_4)_2{\cdot}8H_2O$

A rare hydrated zinc arsenate that forms a series with parasymplesite and is found in the oxidized zone of some zinc deposits. Köttigite occurs as colourless and transparent crystals, up to 1 mm across, at Hilton mine, Scordale, Cumbria. It is associated with annabergite on a galena–gersdorffite specimen from a nickel-rich pod (Bridges & Green 2006a). Pale-pink feathery crystals of köttigite (previously identified as erythrite) have also been found at Sandbed mine, Caldbeck Fells, Cumbria (Green et al. 2006a). Additionally, köttigite is noted from Moll Doyle mine, Co. Wicklow, Ireland (Bridges & Green 2006a).

kotulskite
hexagonal Pd(Te,Bi)

A rare palladium telluride–bismuthide, identified in a layered mafic–ultramafic intrusion on the Isle of Rum, Highland Region, Scotland. In certain chromitite layers of the Eastern Layered Series, kotulskite is found in association with species positively or tentatively identified as braggite, cooperite, hongshiite, irarsite, isoferroplatinum, keithconnite, laurite, michenerite, moncheite, palladoarsenide, sperrylite, telargpalite, telluropalladinite and members of the hollingworthite–irarsite–platarsite series, and the native species gold, silver, palladium and platinum. Most of these grains are 0.2–2 µm in diameter, although for some species grains > 15 µm are also present. The majority are in contact with sulphide minerals such as bornite, chalcocite, chalcopyrite, pentlandite or pyrrhotite. The chromitite layers were formed during magmatic fractionation and crystal settling of an open-system magma chamber, modified by magma mixing and liquid immiscibility of sulphide and silicate melts (Butcher et al. 1999, Power et al. 2000a, 2003). Kotulskite has also been identified in the West Sgaorishal ultramafic plug on the Isle of Rum. Discrete PGMs, especially kotulskite and sperrylite, occur within disseminated sulphide (chalcopyrite–pentlandite) mineralization. Less commonly, froodite, gold (electrum) and unidentified Bi–Pt–Te and Pd–Sb–Te phases are also found. Grainsize is up to 10 µm, but more often about 2 µm. In associated pyrrhotite-rich dykes, kotulskite (typically < 5 µm) clusters around froodite, pentlandite and an unidentified Ag–Pd–Te phase. Related chalcopyrite-rich dykes host kotulskite with froodite, keithconnite, paolovite, sperrylite, telluropalladinite, gold (electrum) and native silver (Power et al. 2001, 2003).

kröhnkite?
monoclinic $Na_2Cu^{2+}(SO_4)_2{\cdot}2H_2O$

A rare hydrated sodium–copper sulphate that forms in the oxidized zone of copper deposits, typically in arid environments. Bowell (1992) describes kröhnkite in a unique supergene assemblage from the Wheal Hazard section of Botallack mine, St Just, Cornwall. Although part of the assemblage (gold) is in the collection of the NHM, no specimens (or XRD data) of kröhnkite exist, making confirmation difficult. Kröhnkite does not normally occur in wet acid environments, as would have occurred in this mine.

krupkaite
orthorhombic $PbCuBi_3S_6$

A rare lead–copper sulphobismuthite found in quartz veins cutting monzogranite at Loch Shin, 2 km W of Lairg, Highland Region, Scotland. Elsewhere in the monzogranite, mineralization is disseminated and mainly restricted to fluorite, molybdenite and pyrite. However, in the veins, molybdenite does not occur and the other species are accompanied by bismuth, chalcopyrite, covellite, galena and sphalerite, and a suite of bismuth sulphosalts – aikinite, berryite, gladite, hammarite, krupkaite, lindströmite, pekoite and tetradymite. Krupkaite is a member of the aikinite–bismuthinite series (Lowry et al. 1994).

In Ireland, at Ballycoog, Goldmine River valley, near Avoca, Co. Wicklow, magnetite–siderite–(chalcopyrite–pyrite ± galena, hematite, sphalerite) mineralization is reported from metamorphosed BIF rocks. Sulphide content of the BIF is low, with chalcopyrite hosting gold inclusions. Bismuth sulphosalts are a minor component, consisting of members of the aikinite–bismuthinite series (aikinite, krupkaite and hammarite), with traces of bornite, chromite, silver-rich gold, mackinawite and pyrrhotite being observed microscopically. The sulphosalts form coarse-grain crystals, 50–200 µm across, intergrown with chalcopyrite. The BIF was deposited in a volcanic environment during the late Ordovician (Ixer et al. 1990).

In England, krupkaite forms minute 30–60 µm grains as a replacement product of bismuthinite at Carrock mine, Grainsgill, Caldbeck Fells, Cumbria (Cooper & Stanley 1990). Sulphosalts belonging to the aikinite–bismuthinite series have been found with quartz and other sulphides on joints in granite and in enclaves from the Shap granite quarry, Cumbria. On the basis of XRD analysis, one occurrence is considered to be an intimate intergrowth of bismuthinite and gladite, another a mixture of gladite and krupkaite. Further analytical work is desirable to confirm these species (Wirth & Green 2002).

krutovite?

cubic — PYRITE GROUP — $NiAs_2$

A rare nickel arsenide reported from a sheet of native copper collected from loose material in a cliff-talus debris flow of Littleham Mudstone Formation at Littleham Cove, near Budleigh Salterton, Devon. The sample of native copper also contains cuprite, domeykite and paramelaconite (all confirmed by XRD), and krutovite that was identified only provisionally. Krutovite is close in composition to pararammelsbergite, which may have previously been misidentified. The occurrence is associated with complex Ag–As–Co–Cu–Ni–U–V mineralization that developed as the Permo-Triassic mudrocks were de-watered during burial compaction (Milodowski et al. 2002).

ktenasite

monoclinic — $(Cu^{2+},Zn)_5(SO_4)_2(OH)_6{\cdot}6H_2O$

A rare hydrated copper–zinc sulphate hydroxide found encrusting the walls of a stope on Middlecleugh First Sun vein at Smallcleugh mine, Nenthead, near Alston, Cumbria. Ktenasite (the k is not pronounced) occurs in association with gypsum and copper minerals (serpierite and schulenbergite) of post-mining origin. It occurs as aggregates of tiny bluish-green tabular crystals to 0.4 mm and was identified solely by XRD (Rust 1991b). EMP analysis indicates that the species has a high zinc content, and Livingstone (1991) suggests it should more correctly be referred to as an unnamed zinc analogue of ktenasite. Additionally, Livingstone (1991) notes its presence at Brownley Hill mine, Nenthead, where it forms blue to bluish-green grains up to 0.3 mm in length.

Ktenasite is also reported from gypsum-encrusted walls in Old Gang mine, E of Gunnerside Gill, Swaledale, Yorkshire, where it occurs with serpierite. Ktenasite forms pale-green or blue aggregates of lath-like crystals up to 1 mm in length and 0.05 mm in thickness. White spherules on the gypsum and copper minerals are identified as a mixture of allophane, gypsum and gearksutite. All species appear to have formed by post-mining oxidation as acid metal-rich solutions percolated through the overlying lead–zinc orebody (Green 2003, Wolfe 2004).

In Wales, ktenasite is described from another copper–zinc deposit – at Ystrad Einion mine, near Tal-y-bont, Central Wales orefield, Dyfed – where it forms parallel groups of thin pale emerald-green bladed crystals up to about 3 mm long, associated with gypsum and serpierite. Corroded lenticular crystals to 0.7 mm are also noted on joints in silty sandstone. Again a post-mining origin is proposed (Bevins 1994, Mason & Rust 1997).

kullerudite

orthorhombic — MARCASITE GROUP — $NiSe_2$

A rare nickel selenide, dimorphous with penroseite. A single inclusion was recorded in one of 38 placer gold grains from Smallbrook, near Crediton, Devon. The single 5 µm-diameter bleb is hosted in native gold with a composition of 99.8% gold and 0.2% silver. Associated species, also found as inclusions, include klockmannite, naumannite, tiemannite, umangite and several unnamed selenide and telluride species (R. J. Chapman pers. comm. 2007).

kunzite

Light-violet or pink gem variety of spodumene.

kupferblau

A mixture of azurite and chrysocolla.

kupferglimmer

A synonym of chalcophyllite.

kupferkies

A synonym of chalcopyrite.

kupfernickel

A synonym of gersdorffite or nickeline.

kupferuranite

A synonym of torbernite or zeunerite.

kupfervitriol

A synonym of chalcanthite.

kuramite

tetragonal — STANNITE GROUP — Cu_2CuSnS_4

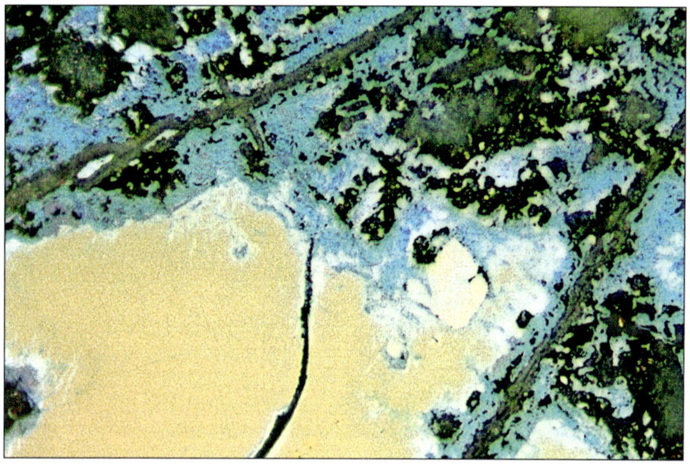

Yellowish-orange kuramite with marginal alteration (white) to an unnamed phase, Cu_6SnS_8, and bornite (blue). Gunheath china-clay pit, St Austell, Cornwall; width 150 µm; viewed in reflected light. (© C. J. Stanley, The Natural History Museum, London)

A rare copper–tin sulphide found in granite from Gunheath china-clay pit, Carthew, St Austell, Cornwall. Kuramite forms grey granular metallic veins with possible covellite and stannoidite (Ryback & Tandy 1992). Much of the "stannite" at Gunheath is actually kuramite and stannoidite, with only minor stannite. The kuramite is altered to Cu_6SnS_8, an unnamed copper analogue of canfieldite (Alderton 1991).

kutnohorite

trigonal — $Ca(Mn^{2+},Mg,Fe^{2+})(CO_3)_2$

An uncommon calcium–manganese–magnesium–iron carbonate that forms two series, with dolomite and with ankerite.

In the Crediton Trough, Devon, kutnohorite is associated with manganese-rich calcite and dolomite in narrow veinlets cutting Permian basalts at Thorverton, and at Upton Pyne it is found associated with extensive manganese mineralization in veins hosted by Permian breccias. At both locations native gold is also present (Shepherd et al. 2005).

Kutnohorite occurs in Cambrian manganese-rich sediments belonging to the Hafotty Formation in the Barmouth area, Harlech Dome, Gwynedd, Wales. It is the main carbonate species in redbed ore horizons, where it is associated with aluminosilicates, ferropyrophanite, hematite, magnetite and quartz. The mineralization is thought to be partly syngenetic and partly syndiagenetic, with manganese probably supplied by submarine exhalative hydrothermal solutions (Bennett 1987b). Kutnohorite is also found in Middle Cambrian metabentonites associated with Upper Caered mudstones on St Tudwal's Peninsula, Lleyn, Gwynedd, where it has been identified in the <2 µm fraction of the clay (Roberts & Merriman 1990). Bennett (1987a) reports kutnohorite from a concretion in a thick-bedded fine-grain altered tuff at Port Ceiriad on St Tudwal's Peninsula.

kyanite

triclinic — Al_2SiO_5

A common aluminosilicate, trimorphous with andalusite and sillimanite, and found in gneisses, schists and quartz veins, from moderately high-pressure regional metamorphism of pelitic rocks.

Kyanite is present as isolated grains and as fine intergrowths with sillimanite in anthophyllite–cordierite-bearing granulites of the Old Lizard Head Series, N of the main Lizard gabbro in the vicinity of Trelease Mill and Polkernogo, W of St Keverne, Lizard Peninsula, Cornwall (Tilley 1937b).

In the Grampian Region of Scotland, kyanite is a fairly common species in Dalradian pelitic schists W of the Portsoy–Duchray Hill lineament. Bright-blue blades of kyanite up to 6 cm in length occur in a quartz vein associated with muscovite at Botriphnie, 9.75 km SW of Keith (Greg & Lettsom 1858, Wallace 1884, Heddle 1901, Livingstone 2002); pseudomorphs of kyanite after andalusite have been found in graphitic schists at Glen Fiddich, S of Auchindoun Castle, and in chiastolite schist at Portsoy (Beddoe-Stephens 1990); coarse-grain Moinian gneiss containing large porphyroblasts of kyanite occurs with tremolite limestone, garnetiferous mica schist and aegirine granulite at Glen Lui, 2 km SSE of Derry Lodge, Cairngorm Mountains (McLachlan 1951); margarite-bearing black schist containing biotite, garnet, graphite, oligoclase, quartz and tourmaline occurs at Ailnack Bhuilg, near Tomintoul (Chinner 1974).

A range of kyanite occurrences are reported from Highland Region: at the entrance to Gleann Còsaidh, near Loch Quoich, kyanite–sillimanite gneiss of Moinian age hosts small metamorphic segregation pods containing almandine–spessartine garnet, apatite, muscovite, quartz and plagioclase, and the rare species graftonite, jahnsite, johnsomervilleite, mitridaite, phosphosiderite, rockbridgeite and vivianite

Blades of kyanite to 6 cm in quartz; Botriphnie, Grampian, Scotland. (© The Trustees of the National Museums of Scotland)

(Livingstone 1980); an unusual kyanite-bearing amphibolite derived from eclogite is described from near Beinn a' Chapuill, Glenelg (Tilley 1937c); kyanite occurs as pseudomorphs after andalusite and cordierite in the Moinian pelitic hornfels thermally metamorphosed by the Carn Chuinneag–Inchbae complex (Tilley 1935a, Harker 1954b); prismatic porphyroblasts of kyanite up to 20 mm long are found in a matrix of muscovite, together with corundum and staurolite, in Lewisian Gneiss near Stoer, NW of Lochinver (Barnicoat et al. 1987); large isolated crystals and sheaf-like aggregates of kyanite occur in gneiss at Glen Urquhart above Milltown, near Loch Ness (Heddle 1901); cm-size crystals of kyanite occurring with corundum and staurolite in a white mica matrix are thought to represent residual material after an episode of partial melting. The assemblage is found in thin (<300 mm) quartz-free layers within a grey gneiss–paragneiss complex of Lewisian age at Stoer, NE of Achmelvich Bay (Cartwright et al. 1985).

Reports of kyanite from Tayside Region include: in garnet–muscovite schist at Gleann Feàrnach, 5 km N of Kirkmichael (Harker 1954a); in graphite-bearing staurolite–mica schists at the Craigs of Inchdowrie, SE of Clova village, Glen Clova (Chinner 1965); in Dalradian hornfels surrounding the Ben Vuirich granite, 10 km NE of Pitlochry (Ahmed-Said & Tanner 2000); in attractive kyanite–quartz veins, where bluish-white kyanite prisms up to 10 cm in length and hematite–kyanite–muscovite schist occur above Craigoshina Cottage, 10 km E of Tarfside, Glen Esk (Harte 1987).

Extremely rare ultra-high-pressure sapphirine-bearing (and sapphirine-free) garnet–orthopyroxene–quartz granulites of Lewisian age from S and SW of Roineabhal, South Harris, Western Isles, contain kyanite intergrown with enstatite (Baba 1999, 2003, Baba et al. 2000, Hollis et al. 2006).

Kyanite is well developed in the Shetland Isles, with masses of broadly aligned blades up to 13 cm long described from 2 km SW of Scalla Field summit, on Mainland (Livingstone 2002), and large brushes and plumose groupings of tile-red crystals in massive quartz are reported from Cunningburgh, also on Mainland (Heddle 1901). A rock composed of rosettes of pale-blue kyanite, up to 5 cm in length and set in a base of glassy quartz, is said to be one of the most beautiful on Unst and it occurs in vein-like bodies; three important localities are Dale Burn, between Sothers Fields and Ward of Norwick, and the upper part of the Burn of Shaw. The country rocks to this type are chloritoid–kyanite schists, which are widely distributed in the NE of Unst, forming the hills of Housi Field, Sothers Field, Ward of Norwick and Saxa Vord. Andalusite, biotite, garnet, magnetite and staurolite are commonly associated species in the schist (Heddle 1901, Read 1932, 1934b). Other locations for kyanite on Shetland include: distinctly blue and terminated crystals at Hillswickness Point, north Mainland (Greg & Lettsom 1858); white and grey kyanite near Levenwick, eastern side of the Ward of Skousburgh, south Mainland (Heddle 1901); and on the hillside between Woodwick and Loch of Cliff, Unst (Heddle 1878a).

Traces of kyanite are reported from hornfelsed pelites surrounding the Kilkenny School leucodiorite intrusion, Maas area, Co. Donegal, Ireland (French 1976), and more generally the species occurs in the metamorphic aureoles of both the Ardara and Main Donegal granites (Atherton et al. 1975). Kyanite is noted without description from Altnapaste, Co. Donegal (Greg & Lettsom 1858).

Kyanite also occurs as a detrital mineral and has been reported in heavy-mineral concentrates derived from alluvial, drift or bedrock deposits: Holden's Wood, Brentwood, Essex; Hayes Common, Kent; Kennel Wood, Shirley, Surrey (now Greater London); Northwood, Middlesex (now Greater London); Plumstead, Kent (now Greater London) (all Davies 1915); 2 km W of Wetherby, Yorkshire (Crampton 1959); Castell-toch, near Llansadurnen, Dyfed, Wales (Strahan et al. 1909); Ludchurch, Dyfed, Wales (Griffiths & Stuart 1940); St Kilda (Hirta), Western Isles, Scotland (Harding et al. 1984); Hunt Hill to the N of Rothes and Bennachie, 10 km W of Inverurie, both Grampian, and Inchbae, N of Garve, Highland Region, Scotland (Mackie 1928); South Mayo Trough, Co. Mayo, Ireland (Mange et al. 2003).

L

labradorite

A very common triclinic sodium–calcium aluminosilicate with an intermediate (An_{50-70} mole %) composition between sodium-rich albite (An_{0-10}) and calcium-rich anorthite (An_{90-100}) end members. It is part of the plagioclase feldspar continuous solid-solution series that also includes oligoclase (An_{10-30}), andesine (An_{30-50}) and bytownite (An_{70-90}).

lacroixite?

monoclinic $NaAl(PO_4)F$

A rare sodium–aluminium phosphate fluoride related to tilasite and similar in composition to natromontebrasite. A study of the composition and structure of natromontebrasite resulted in its discreditation and it was concluded that it is a mixture of OH-rich amblygonite with lacroixite and subordinate wardite (Fransolet et al. 2007). Lacroixite is not confirmed in Britain or Ireland, but natromontebrasite has been described from complex granitic pegmatites in the layered roof zone of the Tregonning granite at Megiliggar Rocks, near Breage, Cornwall (Stone & George 1983).

lady's slipper

TL Virtuous Lady mine, Buckland Monachorum, Devon, England

An epimorph of pyrite and siderite after barite. Specimens form tapering casts up to 26 cm in length that were found in quantity at the type locality in 1832–1833 (Embrey & Symes 1987).

laitakarite

trigonal Bi_4Se_2S

A rare bismuth sulphoselenide found in Bonser vein, 3 km NW of Coniston, Cumbria. The vein was the source of one of the most productive copper deposits in the area and was particularly rich where it crosscuts Ordovician Borrowdale Volcanic Group rhyolites. Excluding gangue species, the paragenetic sequence is: magnetite, arsenopyrite, pyrrhotite, sphalerite, chalcopyrite, pyrite, marcasite, bismuthinite, bismuth, laitakarite, joséite, galena and cosalite. Laitakarite forms grains $<50\,\mu m$ across and is associated with bismuth and bismuthinite (Stanley & Vaughan 1982b, Ryback & Tandy 1992).

lanarkite

monoclinic $Pb_2O(SO_4)$

TL Susanna mine, Leadhills, Strathclyde, Scotland

A rare lead oxysulphate formed during the oxidation of galena and found at the type locality in divergent sprays of slender crystals 2.5–5 cm long associated with caledonite and susannite. Lanarkite forms transparent-to-translucent yellowish-white or greenish crystals with an adamantine-to-resinous lustre (Greg & Lettsom 1858, Heddle 1901). As well as occurring in the Susanna vein, lanarkite has also been found on spoilheaps from Margaret's vein (with leadhillite), at the dressing floors of the Crawford vein, as very fine crystals from the New Glencrieff vein, from Cove vein, and from Straitstep vein. These occurrences are all within 3 km of the Susanna vein, but closer to the neighbouring village of Wanlockhead, Dumfries & Galloway (Brown 1919, Temple 1954a, Livingstone 2002). Conditions of formation of lanarkite from Susanna mine were investigated by Abdul-Samad et al. (1982). Yellow, brown or pale-blue sprays of anthropogenic lanarkite are also common in the interior of metal-rich slag pieces found near the Meadowfoot smelter, close to New Glencrieff mine (Green 1987a).

In the Caldbeck Fells, Cumbria, lanarkite is a very rare species, but has it been reported from Red Gill mine, where it forms sprays of pale-blue tapering crystals on altered galena (Cooper & Stanley 1990), and from the South vein workings on the flank of Balliway Rigg in higher Roughton Gill, where it occurs as interlocking masses of white bladed crystals up to about 5 mm long, commonly associated with drusy crusts of prismatic mattheddleite and scotlandite, in pods of altered galena (Green 1989, Cooper & Stanley 1990, Green & Briscoe 2002). A single specimen from Silver Gill contains acicular blades of lanarkite (<0.2 mm long) in a relict

Divergent spray of lanarkite crystals, up to 5 cm long; Susanna mine, Leadhills, Strathclyde, Scotland. (© M. P. Cooper)

galena-bearing quartz-lined cavity (Green et al. 2005b), and two specimens of corroded galena containing sprays of very pale-blue bladed crystals of lanarkite up to 0.7 mm long are recorded from Driggith mine by Green et al. (2006a). A fifth occurrence at Brae Fell mine (Hartley 1984, Young 1987a, Cooper et al. 1988, Cooper & Stanley 1990) is based on samples in the Kingsbury collection, now at the NHM. However, the work of Ryback et al. (1998, 2001) indicates that the provenance of many Kingsbury samples is dubious and further investigation of these specimens is therefore desirable. Bridges et al. (2006) state that lanarkite probably does occur at Brae Fell (the single specimen is too small for confirmatory analysis) and that "several specimens claimed to have been collected (from this locality) by A. W. G. Kingsbury are probably fraudulent" (they are typical of the Leadhills district).

Green et al. (2005c) describe lanarkite coated with supergene galena from Whitwell quarry, Derbyshire.

Lanarkite forms radiating sprays of bladed crystals up to 2.5 mm across and subparallel groups of prismatic crystals up to 2 mm long on two specimens from Frongoch mine, between Devil's Bridge and Pont-rhyd-y-groes, Central Wales orefield, Dyfed, Wales. The lanarkite is crystallized on a covellite crust, in association with caledonite and susannite (Green et al. 1996). In Powys, very rare lanarkite forms white elongated prismatic crystals associated with elyite in the oxidized portion of a chalcopyrite–galena–tennantite vein at Dolyhir quarry, Old Radnor, near Presteigne (T. F. Cotterell pers. comm. 2007).

langbeinite?

cubic $K_2Mg_2(SO_4)_3$

A rare potassium–magnesium sulphate thought to have been identified in Permian Lower Evaporite beds of the Yorkshire potash field from the Eskdale No. 3 borehole at Sleights, near Whitby, Yorkshire. In the anhydrite–polyhalite zone, a meagre quantity of a pale-brown isotropic mineral interlayered with lenticular wisps of magnesite could possibly be langbeinite (Armstrong et al. 1951, Stewart 1965).

langite

monoclinic $Cu_4^{2+}(SO_4)(OH)_6{\cdot}2H_2O$

TL St Blazey and St Just, Cornwall, England

A common hydrated copper sulphate hydroxide, dimorphous with wroewolfeite and found in the oxidized zone of copper sulphide deposits, sometimes with a post-mining origin. The type localities noted above are parishes, and the actual locations do not appear to have been recorded. However, langite is found at Fowey Consols mine, Tywardreath, St Blazey (Braithwaite 1982b, Lewis 1997), and at Botallack mine, St Just (Anthony et al. 2003). Langite is widespread at Penberthy Croft mine, St Hilary, Cornwall, where it forms euhedral-to-subhedral, sky-blue to greenish-blue, transparent-to-translucent prismatic crystals up to 1 mm across. It is often associated with birnessite, brochantite, connellite, devilline, dolomite, goethite, woodwardite and wroewolfeite in slate-rich quartz veinstone (Camm & Merry 1991b, Betterton et al. 2000). Collins (1892) reports langite from Copper Hill mine and Wheal Basset, both near Redruth, and East Pool and Dolcoath mines, both near Camborne, where it was observed as brilliant blue crystals of a beautiful deep-blue colour, or as rippled crusts of a greenish-blue tint. Elsewhere in Cornwall, langite is noted from: Geevor mine, near Pendeen, St Just; Phoenix United and Sharptor mines, Linkinhorne (all Golley & Williams 1995); New Cook's Kitchen mine (North Tincroft lode), Redruth (LeBoutillier et al. 2000); and Wheal Grenville, Camborne (Robson 1949b). Langite occurs as thin bluish-green crusts of acicular crystals and deep-blue crystalline aggregates at the As–Cu–Sn–W mine of Wheal Friendship (Devon Friendship), Mary Tavy, Devon (Brooke 1986). At Bedford United mine, near Tavistock, Devon, large crystals of langite, the equal of any similar world occurrence, and occasionally associated with brochantite and connellite, are noted by Watson (1987) and Anthony et al. (2003).

In the Caldbeck Fells, Cumbria, langite has been found as: minute crystalline aggregates on a single specimen from Brae Fell mine, Dale Beck Valley (Cooper & Stanley 1990, Bridges et al. 2006); as small pseudo-hexagonal crystals at Mexico mine, Todd Gill; as deep prussian-blue crystals with caledonite at Red Gill mine (both Cooper & Stanley 1990); as pseudo-hexagonal crystals to 0.25 mm at Silver Gill, near Roughton Gill (Cooper & Stanley 1990, Green et al. 2005b); and as a widespread component of bluish-green crusts surrounding fragments of copper-rich ore on the dumps of Driggith and Sandbed mines (Green et al. 2001, 2006a). Elsewhere in Cumbria, langite is a common component of copper-rich dripstones on the walls of Force Crag mine, Braithwaite, 8 km W of Keswick. Less commonly, it occurs (rarely) in open fractures between brecciated wallrock clasts, as pseudo-hexagonal blue crystals up to 0.8 mm long, associated with brochantite, namuwite and serpierite (Green et al. 1997, Green & Briscoe 2002), Vivid turquoise-blue microcrystalline crusts of langite and posnjakite have been found on heavily weathered chalcopyrite–quartz veinstone on the dumps of a small copper mine at Spothow Gill, Eskdale, and the assemblage brochantite–gypsum–langite–malachite is noted on the walls of a stope in Paddy End mine, Coniston, Eskdale (Young & Johnson 1985). Some, if not all, of these occurrences have a post-mining origin.

Lustrous blue pseudo-hexagonal langite crystal group, 1.6 mm across. Silver Gill, Caldbeck Fells, Cumbria. (© D. I. Green)

Isolated occurrences of langite include: deep bluish-green transparent equant twins and twin clusters of blades up to 3 mm long in barite gossan at Engine Vein mine, Alderley Edge ridge, Cheshire (Carlon 1979, Braithwaite 1994); Snelston mine (Isolation mine), near Ashbourne, Derbyshire (Ford et al. 1993); and Ladywell mine, Shelve, Shropshire (Braithwaite 1982b).

Langite is a widely distributed species on the mine dumps of the Central Wales orefield of Dyfed (mainly) and Powys. At Frongoch mine, between Devil's Bridge and Pont-rhyd-y-groes, langite is occasionally found as euhedral pseudo-hexagonal bluish-green crystals to 1 mm, as etched blocky crystals to 0.5 mm, and as fibrous silky blue crusts on cerussite, where it is associated with native copper, cuprite, malachite and schulenbergite (Jones & Moreton 1977, Green et al. 1996b). A few specimens of deep bluish-green langite have been recovered from the copper–lead mine at Llechweddhelyg, near Bont-goch, Tiryminach, associated with anglesite, brochantite, cerussite and malachite (Jones 1987). Langite occurs both above and below ground at Ystrad Einion mine, near Tal-y-bont, forming dark-to-light bluish-green crystals up to 0.6 mm long with blocky pseudo-hexagonal form (Mason & Rust 1997), and at Esgair Hir mine, Ceulanymaesmawr, it forms bluish-green blocky crystals up to 0.5 mm long (Rust & Mason 1988). Langite develops various crystal habits at Dyfngwm mine (actually the Pen Dylife section of Dylife mine), Machynlleth, Powys, including pseudo-hexagonal crystals up to 0.6 mm long, distorted octahedra up to 0.3 mm long, and occasionally with the form of a rice grain (Rust & Rust 1987). Langite or wroewolfeite may also be present at Gwaithyrafon mine, SE of Tal-y-bont (Rust & Mason 1994). Other occurrences in the district include: Cwmystwyth mine (Bevins 1994); Eaglebrook (Nantycagl) mine, Ceulanymaesmawr (Braithwaite 1982b, Jones 1983, Bevins 1994); Esgair Fraith mine, near Tal-y-bont (Jones & Moreton 1977, Bevins 1994); Glogfawr mine, Devil's Bridge (Bevins 1994); Lodge Park copper trial level, near Loveden mine, Tre'r-ddôl (Bevins 1994, Mason & Green 1996); and Dylife and Geufron mines, near Llanidloes, both Powys (Bevins 1994, Rust et al. 2003).

Elsewhere in Wales, rare flattened or euhedral pseudo-cubic crystals of langite, associated with brochantite, devilline and malachite are reported from a trial level, 2 km NW of Bontddu, near Dolgellau, Gwynedd (Saich & Rust 1987), and from Vigra mine, near Dolgellau (Dewey & Eastwood 1925).

Supergene alteration of a suite of primary sulphides (chalcopyrite, galena and sphalerite) has produced blue scales and tiny (0.3 mm) clear-blue tabular crystals of langite associated with brochantite, posnjakite, serpierite and wroewolfeite on two specimens collected from the dumps of West Blackcraig mine, near Newton Stewart, Dumfries & Galloway, Scotland (Livingstone et al. 1976). Near Needle's Eye, SW of Caulkerbush, Dumfries & Galloway, langite laths up to 0.5 mm long occur with botallackite, brochantite, connellite and gypsum in a bornite–chalcopyrite stockwork (Braithwaite & Knight 1990). Rare langite and azurite have been found with barite, chalcopyrite and malachite at Airthrie Hill mine, Mine Wood, Ochils, Central Region (Wolfe 2004). Anthropogenic langite is relatively common on metal-rich slag on the dumps of the Meadowfoot smelter, Wanlockhead, Dumfries & Galloway. It is usually white, but can be stained yellow or brown; it occurs in a variety of different habits (Green 1987a).

In Ireland, langite occurs on the walls of an adit near sea level at Dooneen copper mine, Allihies, Co. Cork, where it forms deep-blue crystals up to 2 mm across. It is associated with botallackite, chalcanthite and an amorphous copper-silicate gel (Moreton & Aspen 1993). Old mineworkings (including Mountain mine) on the mountainside above the Allihies copper mines, Bearhaven (Berehaven) Peninsula, Co. Cork, contain well formed langite and brochantite crystals up to 2 mm long in encrustations carrying connellite and devilline (Ryback & Moreton 1991, Green & Moreton 2001). Dump material from Sheefry mine, 11 km S of Croagh Patrick, Co. Mayo, contains a secondary assemblage of anglesite, annabergite, calcite, erythrite, gypsum, langite, linarite and quartz derived from the alteration of primary chalcopyrite–galena–millerite (Moreton & Jackson 1991). Particularly fine cabinet-size specimens of langite, exhibiting rippled crystalline crusts mixed with brochantite, connellite and malachite, have been found at Tankardstown mine, Knockmahon, 22 km SW of Waterford town, Co. Waterford (Wolfe 1997, 1998, Green & Moreton 2001). Bright-blue powdery encrustations on the walls of an adit on the Bonhiva lode, near Cassaunnagreana Rock, Knockmahon, contain mixed connellite and langite (Moreton 1991, Ryback & Moreton 1991). Small crystals of langite, cerussite, hemimorphite, linarite, pyromorphite and sphalerite can be found on the dumps of the Luganure mines at Glendasan, Co. Wicklow (Ryback & Moreton 1993).

lanthanite-(Ce)

orthorhombic $(Ce,La,Nd)_2(CO_3)_3{\cdot}8H_2O$

TL Britannia mine, Snowdonia, Gwynedd, Wales

A rare hydrated cerium–lanthanum–neodymium carbonate found on the walls and the back of the adit on Sneyd's level at the type locality, where it occurs as a secondary mineral in oxidized copper ore. Lanthanite-(Ce) forms well crystallized colourless transparent plates (up to 5×5×0.5 mm) covered by radiating tufts of malachite. It is associated with brochantite, chalcoalumite, malachite and posnjakite. It crystallized as REE-bearing groundwaters percolated through the old mine. The source of the REEs was probably the nearby acid pyroclastic rocks of the Snowdon Volcanic Group (Bevins et al. 1985).

lapis ollaris

A synonym of potstone or steatite (talc).

larnite

monoclinic Ca_2SiO_4

TL Scawt Hill, Larne, Co. Antrim, Ireland

An uncommon but locally important calcium silicate found in the chalk/dolerite contact zone at Scawt Hill, where it occurs in a reaction zone within the chalk. Larnite is next in importance to spurrite and together they produce fine-grain granular grey rocks that also contain calcite, merwinite and spinel. Often, larnite forms drop-like grains included in spurrite. Exceedingly fine polysynthetic twinning is characteristic in thin section (Tilley 1929b). Larnite is strictly stable only above 1420°C and probably survives only in an unstable equilibrium (Tilley & Harwood 1931). In the hybrid zone at the contact, larnite rocks carrying melilite–wollastonite aggregates contain vugs infilled with afwillite, ettringite, hydrocalumite and portlandite (Tilley 1934). Larnite is also found in association with melilite, rankinite and wollastonite in melilite-rich rocks and at the contact of flint nodules in metasomatized chalk (Tilley 1942). The chalk/dolerite contact at Ballycraigy, near Scawt Hill also contains larnite assemblages (McConnell 1954, 1955, Long & McConnell 1959).

Braided larnite from Scawt Hill, near Larne, Co. Antrim, Ireland; width 1.3 mm; viewed in PPL (upper) and XPL (lower). (© K. Gardner)

The Tertiary dolerite plug at Carneal, 9 km NNW of Carrickfergus, Co. Antrim, cuts basalt lavas and incorporates blocks of chalk and flint from underlying Cretaceous rocks. Assimilation of the limestone and flint took place under the very rare highest-temperature low-pressure conditions as must have been the case at Scawt Hill, and larnite occurs in larnite–spinel hybrid rocks derived from the metasomatic alteration of limestone and flint. Larnite generally consists of a granular aggregate of crystals 0.05–0.15 mm across, with prominent cleavage and polysynthetic twinning commonly observed in thin section (Sabine & Young 1975). In one specific type of hybrid, aggregates of larnite occur with coarse-grain tablets and prisms of gehlenite and magnetite (Sabine et al. 1982). In another hybrid, a dense dark-grey rock has been produced containing bredigite, a little calcite, gehlenite, larnite, magnetite, perovskite and spurrite (to which larnite may be altered) (Sabine et al. 1985).

Larnite has also formed in limestone thermally metamorphosed by two basalt–dolerite dykes at Killala Bay, near Inishcrone, Co. Sligo. It is part of a primary assemblage that changes away from the dolerite contact and also contains calcite, cuspidine, gehlenite, grossular, rankinite, spurrite, tilleyite and wollastonite (Nawaz 1974).

At Camas Mòr, Isle of Muck, Highland Region, Scotland, an olivine gabbro dyke intersecting Jurassic limestones has developed a high-temperature contact assemblage of gehlenite, larnite, rankinite and spinel. Larnite is an important constituent of assemblages on the western contact of the gabbro, where it forms isolated grains and drop-like grains in larger crystals of rankinite (Tilley 1947). In a similar environment at Kilchoan, Ardnamurchan, Highland Region, extreme metamorphism of limestone adjacent to a gabbroic intrusion has produced assemblages that include åkermanite, bredigite, calcite, cuspidine, gehlenite, grossular, kilchoanite, larnite, rustumite and spurrite (Agrell 1951, 1965).

Larnite occurs in a somewhat different environment within thermally metamorphosed amygdales in Tertiary basalts in the aureole of the volcanic plug at 'S Airde Beinn, 5 km SW of Tobermory, Isle of Mull, Strathclyde. One type of amygdale originally containing calcite is now represented by the anhydrous calcium silicates, larnite, rankinite and wollastonite (zoned from core to rim in this sequence). Late reaction has led to the replacement of both basalt and calcium silicates by melilite. Aegirine, equant melilite crystals, perovskite and iron oxides are found at the rims of these amygdales (Cann 1965a).

larsenite?

orthorhombic $PbZnSiO_4$

A rare lead–zinc silicate reported from in altered Ordovician rhyolites in central Snowdonia, Wales. Larsenite–fayalite of presumed intermediate composition is one of many secondary (hydrothermal) accessory species found in these rocks using BSE imagery, others being aeschynite, lanthanite, parisite, synchisite, thorite and possibly gadolinite (or spencite). These occur as anhedral crystals and intergrowths, in veins and as secondary rims. Associated primary magmatic accessory species include apatite, monazite, titanite, xenotime and zircon. Allanite in these rocks can be either primary or secondary (Howells et al. 1991).

latrobite

A synonym of anorthite, feldspar group.

laumontite

ZEOLITE GROUP

monoclinic $Ca(Si_4Al_2)O_{12}{\cdot}4H_2O$

An uncommon but locally abundant hydrated calcium aluminosilicate with a hydrothermal origin, often found lining cavities in igneous rocks.

At Chywoon quarry, near Longdowns, Cornwall, laumontite occurs as lath-like fragments up to 2 cm in length associated with a white clay in pegmatitic cavities within the Carnmenellis granite (Moulding & Hooper 2005). An XRD trace of a mineral close to laumontite is reported from Geevor mine, near Pendeen, St Just, Cornwall, by Golley & Williams (1995). Laumontite also occurs with apophyllite, pectolite and thaumasite in a late hydrothermal assemblage at Shap Blue quarry, Cumbria. Host rocks to the mineralization are metasomatized Borrowdale Volcanic Group rocks within the aureole of the Shap granite (Young et al. 1988, 1998). Pink crystals of laumontite are noted by Young (1987a) from Park Brow quarry, Dockray, Cumbria. Pink to brownish-red laumontite also occurs in calcite–quartz veins and fractures within brecciated Silurian andesite at Moons Hill quarry, N of Shepton Mallet, Mendip Hills, Somerset (Kingsbury 1941, van de Kamp 1969). Salmon-pink masses, and prismatic laths up to 18 cm long intergrown with calcite rhombs, were once found in calcite–zeolite veins cutting diorite at Croft quarry, SW of Leicester (Dyer & Wilson 1988, Pearson & Jeffrey 1997).

In Wales, massive pale-pink laumontite (originally thought to be heulandite) has been identified in thin (20–30 mm-wide) calcite-rich veins cutting altered Ordovician basic lavas at Llanelwedd quarry, Builth Wells, Powys. Laumontite occurs with prehnite and pumpellyite in rocks subjected to low-grade regional metamorphism, but as it is found in veins it is possible that it may have formed as a result of submarine hydrothermal alteration of the lavas (Bevins & Horák 1985a).

Laumontite is one of many zeolite species found in amygdaloidal Tertiary basalts on the Isle of Skye, Highland Region, Scotland. Anderson (1851) reports abundant and beautiful apophyllite, stilbite and laumontite from the Storr, N of Portree, and at a roadside quarry near Flodigarry, Quiraing, fine hand specimens of stilbite and white prismatic laumontite have been found (Green & Todd 2001). To the west, on the Duirinish Peninsula, laumontite occurs at Moonen Bay, where it forms colourless-to-white prismatic crystals up to about 10 mm long in association with analcime and stilbite (Green & Todd 1996). At nearby Sgùrr nam Boc, a very inaccessible locality, outstanding specimens of apophyllite, calcite, heulandite, laumontite, mesolite and quartz have been collected (Pearce & Pearce 1991). White or cream transparent-to-translucent prismatic crystals of laumontite up to 0.6 cm long are noted from amygdales throughout the lower lavas and especially in veins. Specific localities include: Beinn na Greine, near Portree; Tianavaig Bay, 6 km SE of Portree; and Rubh an Dùnain, Loch Brittle (King 1976). Laumontite is also found at Sgùrr nan Cearcall, in thermally modified basalt within the aureole of the Cuillin intrusive complex, where a higher-temperature mineral assemblage has been produced in which laumontite is accompanied by calcite, epidote, prehnite, pumpellyite, quartz and some unidentified copper-bearing minerals (Green & Todd 2001). Other localities on Skye, noted by Heddle (1901), include: Beinn nan Cuithean, S of Bioda

Terminated prism of laumontite, 40 mm long, from a cavity in basalt at Beinn na h-Iolare on the Isle of Mull, Strathclyde, Scotland. (© D. I. Green)

Ruahd, between Talisker and Loch Einort; between Loch Einort and Loch Brittle; Leacan Fhionn, Quiraing; Sgùrr an Duine, S of Loch Eynort (36 mm-long doubly terminated lustrous crystals); Sgùrr nam Fiadh, between Talisker and Loch Eynort (very fine pink crystals up to 25 mm long); and in 75 mm-thick veins associated with stilbite at Snizort. Because of its ion-exchange properties, laumontite from Bernisdale on Skye has been examined as a candidate material for aqueous nuclear-waste treatment (Dyer et al. 1991). Laumontite (leonhardite) is also noted as an accessory species in the Tertiary Glamaig epigranite in the Marsco area of Skye (Thompson 1969).

On the Isle of Mull, Strathclyde, and the adjacent mainland of the Morvern Peninsula, Highland Region, laumontite occurs widely in amygdaloidal basalts. Based on a study of zeolite assemblages from 520 localities, Walker (1971) was able to map out a distinct laumontite zone: at the top of lava flows, laumontite occurs with analcime, mesolite and thomsonite, together with sporadic chabazite, heulandite and stilbite; in the rather sparse amygdales in the interior of flows, laumontite occurs with gyrolite, mesolite and thomsonite (sporadic apophyllite and tobermorite). Using evidence from the zonation pattern within the basalts, Walker (1971) was able to reconstruct the original thickness of the Mull lava pile. Heddle (1901) reports laumontite from Mull: near Beinn na Croise, between the head of Loch Scridain and Loch Buy (well crystallized); Bloody Bay, near Tobermory; Carsaig Arches; and on the shore near the cliffs to the N of Aros Castle.

Two very different occurrences of laumontite are found on the Isle of Arran, Strathclyde: in leucocratic veinlets and lenses cutting the Dippin Sill, a Tertiary basic alkaline intrusion composed of coarse analcime–olivine dolerite (Gibb & Henderson 1978b); in late-stage veinlets cutting skarn rocks that have developed at the junction of vent pyroclastics and a xenolithic chalk block that subsided into the collapsed caldera of the Central ring complex at Creag an Fheidh (Cressey 1987).

Laumontite is one of the secondary minerals found in the Carboniferous Clyde Plateau Lavas to the W and SW of Glasgow. At Hartfield Moss, 7 km NE of Beith, laumontite is common in small vesicles and cavities, where it forms white-to-transparent colourless (sometimes terminated) prisms up to 6 mm long, and as intergrown crystals completely filling the cavities. Analcime, harmotome, natrolite, stilbite and thomsonite are other zeolite species found in association (Heddle 1901, Meikle 1990). Similar cavities in altered basaltic lavas at Loanhead quarry, near Beith, contain rare laumontite as small clear-to-opaque (and white) equant or elongated prisms (Meikle 1989). Analcime, barite, calcite, laumontite and thomsonite have been found at Reilly quarry, Houston, in a similar paragenesis (Wolfe 1997). Some of the finest laumontite in Scotland (single crystals up to 5 cm long) occurs in a lava flow at the base of a crag high up on Beinn na Sreine, Isle of Mull, Strathclyde (R. E. Starkey pers. comm. 2005).

In the Grampian Region of Scotland, laumontite has been found in hornblende schist exposed in a stream near Bowman Hillock, 4 km W of Huntly. It appears that the laumontite formed via hydrothermal alteration of sodic plagioclase feldspar, and an advanced stage of this process locally reduced the laumontite to an isotropic mass of clay (Wilson 1974). Narrow fracture-filled veins and stringers of gyrolite, laumontite, prehnite, scawtite, tacharanite and xonotlite cut the Morven–Cabrach gabbro at Binhill quarry, Cairnie, Huntly, possibly as a result of uralitization of the gabbro (Livingstone 1974). Heddle (1901) describes 50 mm-long flesh-red crystals of laumontite within amygdales in Devonian olivine basalt at Todhead Point, 9 km S of Stonehaven, and notes laumontite with quartz and saponite in druses at Thornyhive, 3 km S of Stonehaven. Dyer et al. (1993) confirmed laumontite and also discovered mordenite and stellerite at Todhead Point.

Other Scottish localities noted by Heddle (1901) include:

CENTRAL Carbeth, 3 km W of Strathblane, N of Glasgow; Campsie Fells, N of Strathblane; Earlsburn reservoir, Touchadam Muir, 13 km SW of Stirling; Mugdock tunnel, S of Strathblane. STRATHCLYDE Bishopton railway tunnel; Bowling quarry, 2 km WNW of Old Kilpatrick; Boyleston quarry, SW of Barrhead; Cloak, near Lochwinnoch, 16 km E of Largs; Kilmalcolm, 10 km SE of Greenock. TAYSIDE Glen Farg, 16 km S of Perth.

The Tertiary basalts of Co. Antrim, Ireland, were extensively examined by Walker (1960a), but laumontite, forming prismatic crystals 5 mm long associated with chabazite, heulandite and stilbite, was found at only one locality – Belshaw's quarry, 2 km SW of White Mountain. However, laumontite is also noted from Ballintoy, in Co. Down (Greg & Lettsom 1858), and three locations in Co. Antrim: the Ballymacilroy borehole, where it occurs in a vein cutting the Portrush Sill (Adair 1987), Larne and Portrush.

laurionite

orthorhombic PbCl(OH)

An uncommon lead oxychloride, dimorphous with paralaurionite and found on a specimen from Wheal Rose, Sithney, Cornwall, purchased by Sir Arthur Russell in 1905. Colourless and translucent laurionite forms blades up to 4 mm long in a small geode (7×5 cm) lined with limonite, ferruginous quartz, siderite and minor galena. Anglesite and phosgenite are also noted to have crystallized on this lining. An origin involving supergene alteration of galena-bearing veins in the presence of sea water was proposed (Russell & Hutchinson 1922, 1927). Laurionite is also noted from the dumps of Daw's shaft, Penberthy Croft mine, St Hilary, Cornwall, where it forms colourless transparent blades up to 1.25 mm long with oblique terminations and bevelled edges, and as transparent-to-translucent flat hexagonal crystals up to 1.5 mm long. It forms small groups with phosgenite and native sulphur (Wallace 1998, Betterton et al. 2000).

At Wesley Theological College (Wesley mine), Westbury-on-Trym, Avon, laurionite, and especially paralaurionite, are fairly abundant, occurring together as coarse- to fine-grain intergrowths or as isolated corroded grains on cerussite. Transparent-to-translucent laurionite varies in colour from dark brown through to pale brown or pale grey, and the largest grains measure 3×1 mm. Laurionite and other lead compounds (cerussite, hydrocerussite, mendipite, paralaurionite, plattnerite, wulfenite) form part of a secondary mineral suite associated with replacement deposits of iron and manganese oxides of Triassic age (Alabaster 1985b, 1989b).

In Wales, laurionite and altered galena are reported from the dumps of Eaglebrook (Nantycagl) mine, Ceulanymaesmawr, near Tal-y-bont, Central Wales orefield, Dyfed. Laurionite forms colourless-to-white bladed-to-prismatic crystals up to 0.75 mm long in cavities within oxidized dolomite associated with (Rust 1995b).

laurite

PYRITE GROUP

cubic RuS_2

A rare ruthenium sulphide that forms a series with erlichmanite and is found in trace amounts near Carrick Luz on the Lizard Peninsula, Cornwall. Laurite occurs with potarite, sperrylite and various alloys in lherzolites and dunites belonging to the Lizard ophiolite complex. It is associated with altered pentlandite and chalcopyrite or secondary magnetite, and probably has a magmatic origin modified by local hydrothermal redistribution (Hutchinson 2001, Gunn & Styles 2002).

Laurite is abundant (a relative term) in the unit 7–8 boundary chromitite of the Eastern Layered Series in the layered mafic–ultramafic intrusion on the Isle of Rum, Highland Region, Scotland. It is found in association with species positively or tentatively identified as braggite, cooperite, isoferroplatinum, kotulskite, michenerite, moncheite, members of the hollingworthite–irarsite–platarsite series, sperrylite and telluropalladinite, and the native species gold, silver, palladium and platinum. Most of these grains are 0.2–2 µm in diameter, although for some species grains >15 µm are common. The majority are in contact with sulphide minerals such as bornite, chalcocite, chalcopyrite, pentlandite or pyrrhotite. The chromitite layers were formed during magmatic fractionation and crystal settling of an open-system magma chamber, modified by magma mixing and liquid immiscibility of sulphide and silicate melts (Butcher et al. 1999, Power et al. 2000a, 2003).

Laurite is also the dominant PGM at Pinbain Bridge, Strathclyde, where it occurs as euhedral <10 µm-diameter inclusions in chrome spinel within chromitites belonging to the Ballantrae ophiolite complex. Erlichmanite, irarsite, Ir–Os–Ru alloys and possibly kashinite are associated species (Power & Pirrie 2004).

In the Outer Layered Series of the Cuillin complex (part of a layered ultramafic intrusion) on the Isle of Skye, Highland Region, laurite is the most common PGM encountered, occurring as 1–7 µm-diameter inclusions in chromite with sperrylite, ruarsite or in more common sulphides such as chalcocite or pentlandite (Pirrie et al. 2000a, Power et al. 2003). In a similar setting, the Ben Buie peridotite intrusion at Craig, Isle of Mull, Strathclyde, laurite occurs in, or adjacent to, chromitite layers in association with froodite, ruarsite, sperrylite and minerals of the hollingworthite–irarsite–platirasite series. The PGMs occur as either euhedral crystals within chrome spinel grains or clusters of minute inclusions (<2 µm) within intercumulus plagioclase and clinopyroxene (Pirrie et al. 2000a, Power et al. 2003). Additionally, Prout & Andersen (2001) report michenerite from Ben Buie.

Laurite containing inclusions of native osmium has been found within chromitite horizons in a small body of mineralized, metamorphosed and altered ultramafic rocks at Corrycharmaig, Glen Lochay, NW of Killin, Tayside Region. Accessory amounts of galena, gersdorffite, irarsite and laurite, and the native metals gold, iridium, osmium, ruthenium and silver, and unnamed alloys of As, Co, Fe, Ni, Os, Pd, Pt, Ru, Sb and Sn occur either as inclusions (up to 40 µm across) in chromite or are associated with the nickel sulphides heazlewoodite and nickeline (usually ≤2 µm across). The chemistry and association of the PGMs is strongly indicative of an ophiolite assemblage and it is most likely that the intrusion represents a fragment of oceanic crust (Power & Pirrie 2000).

Chromite-rich dunite lenses that formed near the base of the Shetland ophiolite complex at Cliff and at Harold's Grave, Baltasound–Haroldswick area, Unst, Shetland, contain two types of laurite: an osmium-rich (up to 22% Os) variety, entirely enclosed in chromite, and an osmium-free variety located in the silicate matrix interstitial to the chromite. Osmium-free laurite growing near the rims of chromite grains contains tiny inclusions of native osmium. Laurite grains up to 0.4 mm across are reported (Tarkian & Prichard 1987, Prichard & Tarkian 1988, Prichard et al. 1994). Gunn et al. (1985) report laurite from Cliff as small grains up to 10 µm across in a chromitite sample, where it is associated with sperrylite and other unnamed species. Elsewhere on Unst, laurite is reported from: Jimmie's quarry, S of Hagdale; Keen of Hamar; Nikkavord (N and E); and Quoys. On Fetlar, chromitites associated with dunite, harzburgite and wehrlite also contain laurite (Prichard et al. 1986).

In Ireland, laurite occurs with irarsite as inclusions in chromite within the Dawros More peridotite emplaced into the Dalradian sequence of Connemara, Co. Galway. Other associated sulphides are bornite, chalcocite, chalcopyrite, covellite, heazlewoodite and nickel-pentlandite (McElduff & Johnston 1989).

lautenthalite

monoclinic $PbCu_4^{2+}(OH)_6(SO_4)_2{\cdot}3H_2O$

A rare hydrated lead–copper sulphate hydroxide found in the oxidized zone of copper–lead deposits.

Lautenthalite has been tentatively identified on a single specimen from the Force Crag mine, Braithwaite, near Keswick, Cumbria, where it forms minute (<0.3 mm long) green plates forming epitaxial overgrowths about a langite-group mineral (Green et al. 1997).

At Dyfngwm mine, Machynlleth, Central Wales orefield, Powys, Wales (actually the Pen Dylife section of Dylife mine), lautenthalite occurs in a mineralized vein as bladed emerald-green crystals up to 0.7 mm long and as tabular crystals up to 0.5 mm long in association with anglesite, cerussite and langite (Rust & Rust 1987). Along the same vein at Dylife mine, the dump south of Penycrocbren Hill contains lautenthalite with bechererite, caledonite, langite, hydrocerussite, namuwite, schulenbergite and susannite (Rust et al. 2003).

In the Dyfed section of the Central Wales orefield lautenthalite occurs: on a few specimens as minute emerald-green crystals associated with brochantite, linarite, schulenbergite and less commonly as an epitaxial overgrowth on wroewolfeite at Penrhiw mine, Ystumtuen (Mason & Green 1995); very rare emerald-green tabular

Crystals of lautenthalite to 0.6 mm from Eaglebrook (Nantycagl) mine, Ceulanymaesmawr, Dyfed, Wales. (© D. I. Green)

and bladed crystals (to 0.7 mm) of lautenthalite occur at Frongoch mine, between Devil's Bridge and Pont-rhyd-y-groes, in association with brochantite and linarite on cerussite overgrowing goethite-coated quartz (Green et al. 1996b); and lautenthalite is also noted from Hendrefelen mine, Ysbyty Ystwyth (Bevins 1994).

In Scotland, an anthropogenic occurrence of lautenthalite is reported from the dumps of the Meadowfoot smelter, Wanlockhead, Dumfries & Galloway (Green 1987a).

lavendulan

orthorhombic $NaCaCu_5^{2+}(AsO_4)_4Cl{\cdot}5H_2O$

A rare hydrated sodium–calcium–copper arsenate chloride found at Loe Warren Zawn, W of Botallack, St Just, Cornwall, where surface water draining from rubble at the base of the cliff has trickled over veinlets carrying chalcocite, chalcopyrite and pyrite, and precipitated an assemblage of supergene carbonate and phosphate minerals on exposed surfaces and joint planes near a patch of uraninite mineralization. Lavendulan has been identified on one specimen, where it forms sky-blue sheafy aggregates of thin squarish plates, typically 20 mm across. It is associated with connellite, zeunerite and an unidentified amorphous copper silicate (Elton & Hooper 1995a,c). At Gunver Head, near Padstow, lavendulan forms tiny sky-blue hemispherical aggregates up to 0.1 mm long (Schlomann & Weiss 1998). Golley & Williams (1995) list lavendulan from two other locations in Cornwall: Cligga mine, Perranzabuloe, and Pentire Glaze mine, St Minver.

Hemispherical aggregates of a sky-blue mineral from the 60-fathom-level dumps of Sandbed mine, Caldbeck Fells, Cumbria, are considered to be lavendulan by Green et al. (2006a), although further analytical confirmation is desirable.

Sky-blue powdery lavendulan is one of the species found in Permian nodular concretions from the Littleham Cove area, Budleigh Salterton, Devon, where dewatering of Permian strata during burial compaction produced metalliferous pore fluids capable of transporting As–Cu–Ni and associated Ag–Co–U–V mineralization. A wide range of species were precipitated from these fluids, including algodonite, clausthalite, coffinite, native copper, lavendulan, malachite, maucherite, modderite, montroseite, nickeline, rammelsbergite, roscoelite, safflorite and native silver (Harrison 1973, Harrison et al. 1975b, Milodowski et al. 2000).

Lavendulan is a minor constituent, occurring with more common malachite and erythrite, as a coating on coffinite spheroids scattered through gypsum beds in the Triassic Mercia Mudstone Group (Keuper Marl) at Gipsy Lane brick pit, Leicester, Leicestershire (Faithfull & Hubbard 1988). An unconfirmed report of lavendulan from Wood mine, Alderley Edge ridge, Cheshire, is noted by Braithwaite (1994).

In the area around the Needle's Eye, SW of Caulkerbush, Dumfries & Galloway, Scotland, pale-blue spherical aggregates (to 0.1 mm) of lavendulan occur in vugs within a stockwork of copper-rich veins and mineralized fissure veins in the aureole of the Criffell granodiorite. Lavendulan and other associated rare species are the products of oxidative supergene alteration of primary As–Bi–Co–Cu–Fe–Pb–U species. At one location, beautiful spherical aggregates of pale-blue lavendulan to 0.25 mm occur with cornubite, kahlerite, malachite, metakahlerite, metazeunerite, mixite, tyrolite and zeunerite (Braithwaite & Knight 1990, Green & Todd 2001).

In Ireland, thin sky-blue crusts of lavendulan are closely associated with erythyrite on a quartz-rich matrix at Gortdrum mine, near Tipperary town (Braithwaite & Wilson 2003). The same authors also found lavendulan with erythrite from a level of the Seven Dial lode, Knockmahon, 22 km SW of Waterford town. Near by, pale-turquoise fibrous spherules (0.07 mm) of lavendulan, with brochantite, connellite, erythrite and langite, occur on dumps close to Bunmahon village (Ryback & Moreton 1991, Moreton 2005a). At Muckross mine, Killarney National Park, Co. Kerry, lavendulan forms blue mammillated crusts of thin lath-like crystals – just one of many species formed by post-mining oxidation of the primary Ag–As–Co–Cu–Fe–Pb–Zn mineralization (Moreton et al. 1999, Green & Moreton 2001).

lawsonite

orthorhombic $CaAl_2Si_2O_7(OH)_2{\cdot}H_2O$

Named after Andrew Cowper Lawson (1861–1952), professor of geology and anthropology at the University of California, Berkeley; born in Anstruther, Fife, Scotland. Lawsonite is an uncommon hydrated calcium aluminosilicate hydroxide, dimorphous with parthéite, that is found in blueschists of the Mona Complex on Anglesey, Gwynedd, Wales, where it occurs within the southeastern margin of the Aethwy Unit at localities on the northern shore of the Menai Straits and N of Hendrefor. Lawsonite forms tabular crystals up to 1 mm long (generally 0.2×0.4 mm) in narrow bands interleaved with crossite-bearing basic schists. It is a low-temperature moderate- to high-pressure mineral that formed during dynamic metamorphism of the area (Nataraj 1967, Gibbons & Mann 1983, Bevins 1994).

LAZULITE GROUP

Monoclinic phosphates with the general formula $A^{2+}B_2^{3+}(PO_4)_2(OH)_2$, where A^{2+} = Cu, Fe, Mg; and B^{3+} = Al, Fe. Barbosalite, hentschelite and lazulite are members of the group found in Britain or Ireland.

lazulite

LAZULITE GROUP

monoclinic $MgAl_2(PO_4)_2(OH)_2$

A rare manganese–aluminium phosphate hydroxide that forms a series with scorzalite and is found in white Dalradian quartzite at Achill Island, Co. Mayo, Ireland. Lazulite forms aggregates of blue anhedral grains up to 0.5 mm across (Ryback et al. 1988). Previous descriptions of lazulite proved on analysis to be grandidierite (Herd et al. 1984).

lead?

cubic Pb

A very rare native metal species that can be found in the oxidized zone of lead-bearing mineral deposits, but in Britain is an extremely dubious species. Greg & Lettsom (1858) report globules of native lead from Alston Moor, Cumbria, explaining that the Romans used to extract lead from orebodies by unearthing a lode, covering it with combustible materials and burning it to cause fracturing as it cooled. The ore appears to have released globules of lead as the primary galena was roasted. Native lead has also been reported from: Hope's Nose, Torquay (Camm 1995); Golch Rake, near Holywell, Clwyd, Wales (Sowerby 1817); and Leadhills–Wanlockhead area, on the Strathclyde/Dumfries & Galloway border, Scotland (Brown 1919, 1927). Native lead is not reported from Ireland.

lead chlorocarbonate

A synonym of phosgenite.

lead glance

A synonym of galena.

leadhillite

monoclinic (pseudo-trigonal) $Pb_4(SO_4)(CO_3)_2(OH)_2$

TL Leadhills, Strathclyde, Scotland

A rare lead sulphate carbonate hydroxide, trimorphous with macphersonite and susannite, and usually found as a secondary mineral in the oxidation zone of lead deposits.

Leadhillite has been found on slag-like specimens from the wreck of the fireship *Firebrand*, which in 1780 was burned in Falmouth Harbour, Cornwall. It forms small crystals on fused metallic lead together with anglesite, cotunnite and rare galena (Russell 1920b). Colourless, tabular, hexagonal, thin plates of leadhillite up to 0.3 mm across have been identified within galena-bearing quartz-lined cavities at Penberthy Croft mine, St Hilary, Cornwall (Betterton et al. 2000), and at Greystone quarry, Lezant, near Launceston, the species forms colourless-to-white pseudo-hexagonal plates and blocky crystals up to about 5 mm across in galena-bearing veins (Elton et al. 2004).

In the Southern Pennine orefield, traces of leadhillite have been found in Derbyshire in the Masson mines, Matlock, and Rutland mine, Matlock Bath (Flindall & Hayes 1976). The mineral had earlier been described from Matlock by Bertrand (1878). Green et al. (2005c) note leadhillite coated with supergene galena from Whitwell quarry, Derbyshire. Leadhillite has also been identified at Waterbank mine, Ecton Hill, Staffordshire (Rust 1983c).

Colourless to very-pale greenish leadhillite forms short prismatic crystals up to 0.4 mm across at Engine Vein mine, Alderley Edge ridge, Cheshire, where it is associated with anglesite, cerussite and pyromorphite in altered galena (Braithwaite 1994). Another isolated occurrence is from Kingston, near Taunton, Somerset, where a small pit sunk for galena was found to contain small brilliant crystals of leadhillite (Greg & Lettsom 1858).

Leadhillite is noted from locations in the Caldbeck Fells, Cumbria: in cavities within highly oxidized galena, associated with caledonite, lanarkite, mattheddleite and scotlandite on the eastern flank of Balliway Rigg (near Thief Gills), higher Roughton Gill (Young 1987a, Green 1989, Green & Briscoe 2002); uncommon but in good specimens as small sharp equant colourless-to-greenish crystals (<2 mm) with cerussite, lanarkite, mattheddleite and pyromorphite at Brae Fell mine, Dale Beck valley (Hartley 1984, Cooper et al. 1988, Cooper & Stanley 1990, Bridges et al. 2006); poorly crystalline brownish aggregates on quartz at Brandy Gill mine (Cooper & Stanley 1990); with chenite, caledonite and linarite on galena–sphalerite-bearing quartz veinstone on the dumps of Driggith mine (Green & Hubbard 1996), also noted

Tabular pseudohexagonal leadhillite crystal 6.5 mm across; Brown's vein, Glengonnar, Leadhills, Strathclyde, Scotland. (© The Trustees of the National Museums of Scotland)

A large platy crystal of leadhillite, 25 mm across, associated with blue caledonite; Susanna vein, Leadhills, Strathclyde, Scotland. (© D. I. Green)

to be widespread but uncommon on the dumps of both Driggith and Sandbed mines (Green et al. 2006a); the finest leadhillite in England – well formed and brilliant white crystals up to 1 cm across, in cavities within quartz or on galena at Red Gill mine (Greg & Lettsom 1858, Davidson & Thomson 1951, Wolloxall 1981, Hartley 1984, Briscoe 1986a, Cooper et al. 1988, Braithwaite et al. 1989a, Cooper & Stanley 1990, Green & Briscoe 2002); with caledonite, cerussite and susannite in cavities within barite, once filled with galena at Short Grain barite–galena trial, near Hay Gill (Cooper et al. 1988, Green & Briscoe 2002); and white thick tabular pseudo-hexagonal crystals to 1 mm associated with anglesite and caledonite at Silver Gill (Green et al. 2005b). Specimens of leadhillite from Sandbed mine (Hartley 1984), now in the Kingsbury collection at the NHM, require further investigation. Green et al. (2006a) state that most Kingsbury specimens labelled Driggith or Sandbed are genuine, but a few are "clearly fraudulent" and are from the classic Leadhills locality. Elsewhere in Cumbria, leadhillite occurs at Kinniside mine, Cleator Moor, and at Rigg Head quarry, Borrowdale (Green & Briscoe 2002, Green et al. 2006c).

Lustrous white-to-colourless hexagonal prisms of leadhillite up to 1 mm long occur with cerussite in cavities within oxidized barite–galena veinstone at Closehouse mine, SW of Middleton-in-Teesdale, Lunedale (Young et al. 1994a, Bridges & Young 1998), and from workings on the Middlegrove vein, Killhope, Weardale, both in Co. Durham (Bridges & Young 1998).

In the Central Wales orefield of Dyfed (and at one locality in Powys) leadhillite is reported from the following lead mines: as a late dump-formed variably coloured species, forming tabular pseudo-hexagonal crystals up to 0.7 mm long or rare prismatic crystals at Dyfngwm mine, Machynlleth, Powys (Rust & Rust 1987) (actually the Pen Dylife section of Dylife mine); as microscopic clear-to-yellowish hexagonal tablets at Eaglebrook (Nantycagl) mine, Ceulanymaesmawr (Braithwaite 1982b, Jones 1983); as uncommon tabular pseudo-hexagonal crystals up to 1 mm long, usually colourless-to-white, although occasionally colour zoned, at Esgair Hir mine, Ceulanymaesmawr (Rust & Mason 1988); as very rare transparent-to-translucent colourless blocky pseudo-hexagonal crystals up to 0.4 mm long, intergrown with anglesite at Penrhiw mine, Ystumtuen (Mason & Green 1995); also briefly noted from Cwmystwyth, Darren and Llety-Ifan-Hên mines (Bevins 1994). The leadhillite reported from Frongoch mine, between Devil's Bridge and Pont-rhyd-y-groes, Dyfed (Bevins 1994), may actually be susannite (Green et al. 1996b).

Very fine specimens of leadhillite were once collected from lead–zinc veins at Leadhills, Strathclyde, Scotland (Greg & Lettsom 1858). The exact location of the type locality is not recorded, but the early characterization of leadhillite is well documented by Livingstone (2002). Leadhillite is widely dispersed in all the Leadhills veins, although never in any abundance. Crystals are yellowish-white and rarely reach 2.5 cm across (Gillanders 1981). Leadhillite is noted from Margaret's (Straitstep) vein (Brown 1919), and from Susanna and Hopeful veins (Brown 1927), and Temple (1955) reports it from a further ten veins, including some from the adjacent Wanlockhead part of the orefield (in Dumfries & Galloway). Many examples of leadhillite (often with cerussite or lanarkite) are in the collection of the NMS, including specimens from Brown's vein, Glengonnar; Susanna mine, Leadhills; and Wanlock Dod, Whyte's Cleuch, Wanlockhead (Livingstone 2002). The new species macphersonite (Livingstone & Sarp 1984), chenite (Paar et al. 1986) and mattheddleite (Livingstone et al. 1987) are all associates of leadhillite, and they occur on oxidized galena or quartz veinstone from this orefield. Caledonite, cerussite, hydrocerussite, lanarkite, pyromorphite and susannite are other associated species (particularly cerussite or lanarkite). Moderately rare anthropogenic leadhillite is also noted on slag material from the dumps of the Meadowfoot smelter at Wanlockhead (Green 1987a). Russell (1936c) records a single specimen of leadhillite from Drumruck mine, N of Gatehouse of Fleet, Dumfries & Galloway.

A single specimen containing well formed platy crystals of leadhillite associated with anglesite and corroded galena was found at the Luganure mines, Glendasan, Co. Wicklow, Ireland (Ryback & Moreton 1993). A further report is from an old silver mine on Sark, Channel Isles (Greg & Lettsom 1858), presumably from the Pb–(Ag)-bearing brecciated vein cutting foliated hornblende granite at Sark's Hope mine, Little Sark. The occurrence is also noted by Mourant & Warren (1934), who also describe a wide range of sulphide species from this location.

lead spar

A synonym of cerussite, aragonite group.

lead vitriol

A synonym of anglesite, barite group.

ledikite

TL Ledikin, Grampian, Scotland

The trioctahedral equivalent of illite, mica group.

leedsite

TL Leeds, Yorkshire, England

A mixture of barite and anhydrite.

lehuntite

TL Glenarm, Co. Antrim, Ireland

Named after Captain Lehunt; lehuntite is a synonym of natrolite, zeolite group.

leonhardite

A (partially dehydrated?) variety of laumontite, zeolite group.

lepidocrocite

orthorhombic $Fe^{3+}O(OH)$

An iron oxide hydroxide, trimorphous with feroxyhyte and goethite, more common than its frequency in the literature implies, and found in the anhydrite mine under the ICI factory site at Billingham-on-Tees, Cleveland. Lepidocrocite developed with epsomite, gypsum, halite, pyrite, sulphur, sylvite, talc and thenardite as secondary efflorescences on mine walls once the mine had been abandoned (Raymond 1959). Minute scattered or drusy orange-red crystals of lepidocrocite have been found in small (<5 mm) cavities in manganese-rich quartz veins at Burdell Gill, Caldbeck Fells, Cumbria, where it is associated with a pharmacosiderite assemblage. Bright orange-red lepidocrocite is found in wolframite-bearing quartz veins at this locality and it also occurs in friable mudstone breccias in which cavities may be lined with either lustrous dark reddish-brown to almost black botryoidal lepidocrocite or lustrous drusy tabular brownish-red crystals up to 1 mm in length (Neall & Leppington 2003). Additionally, lepidocrocite is noted from: three locations in Cornwall – Great Retallack and Gravel Hill mines, Perranzabuloe, and Trewethen mine, St Kew (Golley & Williams 1995); Embleton quarry, Cumbria (Young 1987a); and Low mine, near Bonsall, Derbyshire (Mostaghel 1984, Ford et al. 1993).

Bevins (1994) suggests that lepidocrocite is widely developed in Wales, although specific localities where it has been found are not documented. Hydrated lepidocrocite (hydroglockerite) is noted from Parys Mountain, Anglesey, by Greenly (1919) and Bor (1950). A single specimen of lepidocrocite, forming reddish-brown metallic platy crystals and lining a small cavity in oxidized galena–tennantite is reported from Dolyhir quarry, Old Radnor, near Presteigne, Powys (T. F. Cotterell pers. comm. 2007).

Lepidocrocite has been found with biotite, chlorite, goethite, kaolinite and vernadite in a manganese-enriched fraction of clay material from an iron–manganese pan in poorly drained brown forest soil from the Blairmore Estate, near Huntly, Grampian, Scotland (Russell et al. 1984a). Earlier reports of lepidocrocite (rubinglimmer) in Scotland (Heddle 1901) are unsubstantiated, according to Macpherson & Livingstone (1982).

Radially arranged platelets of lepidocrocite from Tynagh mine, near Killimor, Co. Galway, Ireland; width 290 µm; SEM image. (© D. I. Green)

lepidolite (series)

MICA GROUP

A compositional range of lithium-rich trioctahedral micas on, or close to, the trilithionite–polylithionite join. Lepidolite does not define an end-member composition and is therefore not a true species, but it was given series status in the IMA classification of the mica group (Rieder et al. 1998). Its distinctive purple colouration makes it easy to identify. Lepidolite is typically found in granite pegmatites, where it often occurs as a metasomatic replacement of muscovite. It is also found in lithium-enriched granite, greisen and high-temperature quartz veins.

At Chywoon quarry, near Longdowns, Cornwall, dark slabs of compacted mica associated with albite and muscovite (gilbertite) in pegmatitic cavities within the Carnmenellis granite have a lepidolite composition (Moulding & Hooper 2005). Pegmatites in the roof zone of the Tregonning–Godolphin granite near Porthleven, Cornwall, contain mica compositions transitional between zinnwaldite and lepidolite (Stone et al. 1988). Lepidolite has also been found in fluor-elbaite–schorl-bearing granite N of Trewavas Head, in tourmaline (fluor-elbaite) aplite and tourmaline (fluor-elbaite and schorl) pegmatite at Megiliggar Rocks, and in cassiterite-bearing granite from Tregonning Hill, all in this same granite complex (Tindle unpub. EMP data).

Parts of the St Austell granite, Cornwall, are lithium enriched and contain units such as the Hensbarrow stock (a non-megacrystic lithium-mica granite) and fluorite granite, in which lepidolite occurs in interstitial patches (<2 mm) with topaz and apatite. The Hensbarrow stock is exposed at Gunheath china-clay pit, where petrographic textures suggest that the lepidolite is primary. A range of hydrothermal alteration effects (autometasomatism) have overprinted this unit to produce the fluorite granite (Henderson et al. 1989). Manning & Hill (1990) note that the late-stage volatile-rich granitic stock of Hensbarrow and the nearby Nanpean stock contain primary albite, amblygonite, lepidolite, orthoclase, topaz, tourmaline and zinnwaldite, and consider them to have evolved from a unique source of magma. At Gunheath china-clay pit, a light-brown to silver-grey mica resembles muscovite but is iron-rich lepidolite (Weiss 1994). Other Cornish occurrences of lepidolite include: Carn Brea (Robson 1945a); Cligga Head, Perranzabuloe, and Castle-an-Dinas, St Columb Major (Robson 1949b); Goonbarrow china-clay pit, St Austell (Bray & Spooner 1983); Lanarth, Redruth (Golley & Williams 1995); St Michael's Mount, near Marazion (Greg & Lettsom 1858, Collins 1892); and Treburland manganese mine, 2 km SE of Fivelanes, Altarnun (Russell 1946d).

Lepidolite occurs in aplite and associated crosscutting pegmatitic veins at Meldon, near Okehampton, Devon. This lithium–fluorine-rich aplite dyke crystallized from an extremely evolved granitic magma and is emplaced into the metamorphic aureole of the Dartmoor granite. Its main constituents are albite, orthoclase, lepidolite and quartz, with apatite, elbaite, fluorite and petalite occurring in accessory amounts (Chaudry & Howie 1973a). Lepidolite is closely associated with beryllonite, elbaite, orthoclase, petalite and pollucite in the pegmatite veins (von Knorring & Condliffe 1984).

In Wales, pale-blue to pink pleochroic crystals of lepidolite are reported from glacial deposits exposed in the area between the River Neath, West Glamorgan, and the River Tywi, Dyfed (Griffiths 1939).

Fine lilac scaly aggregates of lepidolite (probably the finest in Britain; scales <3 mm across) occur in lithium-enriched tourmaline pegmatite boulders on the moors near Peatfold, 5 km N of Kirkton of Glenbuchat, Grampian, Scotland (Macpherson & Livingstone 1982). Lepidolite has also been found in a tourmaline (foitite–schorl) pegmatite cutting gneiss on Cowhythe Hill, near Portsoy, Grampian (Tindle unpub. EMP data).

In Ireland, lepidolite occurs in albite–spodumene pegmatites found in boulders along the southeastern flank of the late Caledonian Leinster granite batholith. At Aclare, Co. Carlow, one such pegmatite is thought to contain lepidolite in association with albite, microcline, muscovite, quartz and spodumene, and minor amounts of manganese-rich apatite, beryl, biotite, cassiterite, garnet, manganotantalite, tourmaline and iron oxides (Steiger & von Knorring 1975). However, Scoon (1978) suggests that lepidolite does not occur at Aclare. Boulders of micaceous greisen and albite–amblygonite–quartz–spodumene pegmatite at Stranakelly, 3 km N of Shillelagh, Co. Wicklow, both contain pink and purple lepidolite (Scoon 1978). In the greisen, lepidolite is an alteration product of spodumene (O'Connor et al. 1991).

lepidomelane

An iron-rich variety of biotite, mica group.

leptochlorite

Chlorite with fine scales or of indistinct fibrous form.

lettsomite

Named after William Garrow Lettsom (1805–1887), co-author with Robert Philips Greg of the *Manual of the mineralogy of Great Britain and Ireland*, published in 1858; born at Fatham, Middlesex. Lettsomite (Percy 1850) is a synonym of cyanotrichite. Curiously, a specimen of lettsomite at the NHM, mentioned in Spencer (1898), cannot be traced (Macpherson 1983). The locality where the specimen was originally found is also in doubt.

leuchtenbergite

A synonym of clinochlore, chlorite group.

leucite

ZEOLITE GROUP

tetragonal (pseudo-cubic) $KAlSi_2O_6$

A relatively uncommon potassium aluminosilicate found in a 30 m-wide alkali basalt dyke, containing peridotitic xenoliths, that intrudes Lower Carboniferous limestones near Inver, 10 km W of Donegal, Co. Donegal, Ireland. Leucite, analcime and titanium-rich magnetite are trace components in a rock consisting mainly of olivine, plagioclase, clinopyroxene, alkali feldspar and glass (Shaw & Edgar 1997). When examined under a microscope, museum specimens of fine-grain black basalt at the NHM collected from Rathjordan, Co. Limerick, Ireland, were found to contain small round sections of isotropic material containing inclusions that resembled leucite. However, chemical analysis showed the material to be analcime. Prior (1910) suggested that the rock could be an altered leucite basalt.

White and pink pseudo-leucite crystals up to 2.5 cm across occur in a type of alkali syenite termed "borolanite", which is found in the Loch Borralan alkaline igneous complex, Assynt, Highland Region, Scotland. Aegirine, andradite (melanite) and biotite are associated species (Teall 1900, Woolley 1973). Pseudo-leucite is a pseudomorph of orthoclase and nepheline after leucite.

At Holmead farm, Loxbeare, near Tiverton, Devon, leucite has been described from loose boulders of minette (a biotite–orthoclase-rich lamprophyre), where it has a characteristic polygonal form, often containing trains of inclusions (Tidmarsh 1932). This rock is mentioned again by Knill (1969) in a thorough review of the Permian lavas in Devon. Two leucite-bearing rocks are described: an orendite, containing microphenocrysts of phlogopite and euhedral leucite (size 0.04 mm), in a biotite–orthoclase–plagioclase groundmass; and an olivine–leucitite, where abundant euhedral pinkish-brown and occasionally zoned leucite reaches 0.6 mm in diameter. Augite, hematite and orthoclase are associated species. A subsequent study (Velde 1971b) has shown that there is no leucite in the so-called leucite minette, but instead feldspar pseudomorphs, with regular octagonal outlines that may (or may not) represent original leucite crystals. The study also identified analcime and concluded that "minette" is an inappropriate term to describe the rock, the author instead preferring to call it a lamproite.

leucochalcite

A white variety of olivenite.

leucophanite?

triclinic (pseudo-orthorhombic) $NaCaBeSi_2O_6F$

A sodium–calcium–beryllium silicate with fluoride found on samples in the Kingsbury collection at the NHM. Although listed from Meldon railway quarry, near Okehampton, Devon, this locality is now doubted. Ryback et al. (1998, 2001) and other authors have expressed serious doubts about the provenance of many samples in the Kingsbury collection and, until further discoveries are made, leucophanite cannot be considered to be a British or Irish species.

leucophosphite

monoclinic $KFe^{3+}_2(PO_4)_2(OH)\cdot 2H_2O$

An hydrated potassium–iron phosphate hydroxide found in the Gossan vein cutting kaolinized granite in the Gunheath china-clay pit (Hensbarrow section), Carthew, St Austell, Cornwall. Leucophosphite forms transparent light yellowish-green spheres and wedge-shape crystals to 1 mm and pale-yellow crystal aggregates to 2 mm, and is associated with chalcosiderite, cyrilovite, dufrénite, libethenite and stannoidite (Watson 1987, Weiss 1994, Green 2005a). Elton (1998) reports libethenite and leucophosphite lining open cavities within vein material from this location, uncommonly accompanied by libethenite and torbernite. Pale-brown sheaves of leucophosphite with dark-green dufrénite are reported from Gunheath china-clay pit, from loose boulders on which a chalcosiderite–turquoise series mineral, fluorapatite, fluorite, morinite, muscovite and torbernite were also found (Elton 1996). Leucophosphite probably crystallized from phosphatic solutions as they interacted with early-formed iron veins during the later stages of kaolinization.

Waxy orange interstitial grains of leucophosphite are also found in the aegirine–riebeckite granite of Rockall island, 440 km W of northern Skye, NW of Scotland. In thin section, it appears as a yellow colloform material moulded upon or enclosing other minerals, in both the framework and the interstices of the rock. It also occurs in radiating groups of fibrous, colourless to pale-yellow crystals replacing pyroxene, feldspar and elpidite. Other associated species are albite, bazirite, cristobalite, monazite and quartz (Hawkes et al. 1975).

Rosette of colourless leucophosphite associated with green chalcosiderite; Gunheath china-clay pit, St Austell, Cornwall; spheroidal crystal aggregate 1.1 mm across. (© D. I. Green)

leucopyrite

A synonym of löllingite.

leucoxene

A general term for alteration products of ilmenite.

levyne (series)/lévyne (series)

ZEOLITE GROUP

trigonal $[Ca_3 \text{ or } Na_6](Si_{12}Al_6)O_{36}\cdot 18H_2O$

A relatively common hydrated calcium–sodium aluminosilicate, whose name has been given series status in the zeolite classification of Coombs et al. (1997). There are now two levyne species – -Ca and -Na – but these names do not appear in older publications, and only when chemical data become available will it be possible to establish the true species. Walker (1951, 1960a) mapped the distribution of zeolite species in the Tertiary basalts of Co. Antrim and concluded that levyne occurs in both analcime–natrolite and chabazite–thomsonite assemblages over a very wide area of the Antrim Plateau. It is considered to be one of the earliest zeolites to crystallize and nearly always forms thin tabular crystals that are invariably twinned. Sheaf-like aggregates are also found. Levyne tends to occur alone in a cavity and is often pink (Walker 1959a). Many dolerite intrusions associated with the Antrim lavas contain zeolites and, in the Parkgate plug, Templepatrick, abundant levyne occurs with chabazite and phillipsite. Rare levyne is also found in the Ticloy plug, at Ticloy Water, SW of Carnlough (Walker 1960a). Greg & Lettsom (1858) note levyne from Little Deer Park, Glenarm (very distinct and perfect translucent white crystals), and Island Magee, near Larne (crystals up to 12 mm across), both Co. Antrim, and Banevenagh; Cabragh; Cluntygeeragh, near Dungiven; Magilligan, all Co. Londonderry. Other descriptions of levyne in Co. Antrim include: hexagonal plates up to 3 mm across in the upper basalt flows at Audley's Head, Glenarm; colourless-to-white crystals, up to 7 mm across, covered with a thin overgrowth of erionite–offrétite and commonly associated with cowlesite at Green Road quarry, Ballyclare (both Tschernich 1992); fairly abundant crystals in vesicles within basalt at Brushlee quarry, near Ballyclare; and with chabazite, garronite, phillipsite and thomsonite in cavities in the dolerite plug at Ticloy (both Walker 1962b). Levyne is also found at Glover's quarry, Moneymore, Co. Londonderry (Tschernich 1992).

Levyne is an uncommon occupant of vesicles in the Tertiary basalts at Moonen Bay, Duirinish Peninsula, Isle of Skye, Highland Region, Scotland. It occurs as thin opaque hexagonal plates (exceptionally reaching 10 mm across) with an epitaxial overgrowth of erionite–offrétite, and as thick transparent hexagonal crystals possessing large pinacoid faces. A particularly characteristic assemblage at the north end of the bay is an assemblage of analcime, chabazite, cowlesite, levyne and phillipsite, the last three species often occurring as the only occupants of the small cavities in which they crystallized (Green & Todd 1996). Immediately to the north, at Oisgill Bay, the basalts contain cavities hosting cm-size white hexagonal plates of levyne, which are among the best yet collected in Scotland (Green & Todd 2001).

Levyne of similar size is also found with superb stilbite crystals at a beach locality at the northern lip of Loch Eynort, on the west coast of Skye (Green & Todd 2001). On the northeast coast of Skye, levyne is found with a boxwork habit in vesicles up to 1.5 cm across within basalt at Quiraing, near Staffin Bay (Livingstone & Macpherson 1983), and is also noted from the Storr, N of Portree (Heddle 1901, Tschernich 1992). Flesh-red opaque levyne is also noted in basalts from Hartfield Moss, 9 km WSW of Barrhead, near Glasgow, Strathclyde (Greg & Lettsom 1858).

levyne-Ca/lévyne-Ca

ZEOLITE GROUP

trigonal $Ca_3(Si_{12}Al_6)O_{36}\cdot 18H_2O$

A relatively common hydrated calcium–potassium–sodium aluminosilicate found in vesicles within Tertiary basalts in Ireland and Scotland, but data are sparse and, quite likely, grossly underestimate its abundance. In Ireland levyne-Ca is recorded from an unspecified location in Co. Londonderry and three locations in Co. Antrim: Garron Point; Parkgate quarry, Templepatrick; and S of Audley's Head, Little Deer Park, Lower Glenarm (Galli et al. 1981). Other locations in Co. Antrim include: Ladyhill quarry, Tardree, NE of Antrim town, and S of Portmuck, Island Magee (H. J. Foy specimens, Tindle unpub. EMP data). Adair (1987) also documents levyne-Ca from: South Antrim Escarpment, W and SW of Belfast – Aughrim Plantation quarry and McCourts quarry; Galboly River section, near Garron Point; Mid-Antrim – Brushlee quarry and Parkgate quarry; Giant's Causeway area – Portninish and Port Noffer.

On the Isle of Skye, Highland Region, Scotland, white or colourless levyne-Ca forms thin tabular crystals at Beinn na Greine, near Portree (King 1976), and in addition is confirmed from Tertiary amygdaloidal basalts at: Echoing Crag, Quiraing; Loch Einort; Moonen Bay; Oisgill Bay; the Storr (specimens and XRD confirmation by D. I. Green and NMS, Tindle unpub. EMP data).

levyne-Na/lévyne-Na

ZEOLITE GROUP

trigonal $Na_6(Si_{12}Al_6)O_{36}\cdot 18H_2O$

An hydrated sodium–calcium–potassium aluminosilicate, confirmed in Tertiary amygdaloidal basalts from Dunseverick and Island Magee, near Larne, both Co. Antrim, Ireland. At Dunseverick it forms transparent colourless hexagonal plates covered with a thick silky white-to-yellowish overgrowth of erionite–offrétite and is associated with cowlesite (Galli et al. 1981). Adair (1987) presents many analyses of levyne-Na from rubbly basalts on Island Magee from Black Cave, Brown's Bay, and Castleshiels near Portmuck.

In Scotland, levyne-Na is confirmed from Tertiary amygdaloidal basalts at: Echoing Crag, Quiraing; Loch Einort; Moonen Bay; Oisgill Bay; the Storr (D. I. Green specimens and XRD, Tindle unpub. EMP data).

Thin hexagonal plates of levyne (up to 10 mm across) in vesicular basalt; Moonen Bay, Isle of Skye, Highland Region, Scotland. (© D. I. Green)

levynite

A synonym of levyne, zeolite group.

libethenite

orthorhombic $Cu^{2+}_2(PO_4)(OH)$

A rare copper phosphate hydroxide found at Gunheath china-clay pit, Carthew, St Austell, Cornwall, where it occurs as a supergene species in association with a prominent turquoise-rich quartz vein cutting kaolinized granite (Ince 1986, Hawkes et al. 1987). Dark-green crusts of libethenite, forming pseudo-tetragonal truncated bipyramids up to 2 mm long are also noted from fissures in pegmatites and in the topaz granite of Half Stope at Gunheath. Arthurite, cacoxenite, chalcosiderite, cyrilovite, dufrénite, leucophosphite, torbernite, variscite and wavellite are other secondary phosphates found at this site (Weiss 1994). Typical olive-green highly lustrous libethenite is reported from a single specimen collected at Cligga Head, Perranzabuloe, where it infills a vug in a massive wolframite- and scorodite-bearing quartz vein (Chick 1987). A green crust of arsenic-rich libethenite has been found on the dumps of East Gunnislake mine, Calstock, near Gunnislake (Braithwaite & Ryback 1994). Elsewhere in Cornwall, libethenite is noted from: Basset mines, Illogan; Old Gunnislake mine, Calstock; Phoenix United mine, Linkinhorne; Pentire Glaze mine, St Minver (all Golley & Williams 1995); South Wheal Francis, Illogan (Collins 1892, Braithwaite & Ryback 1994); Tolcarne mine, Camborne (Braithwaite & Ryback 1994). Some of these occurrences deserve further investigation. Earlier references to dark-green crystalline aggregates of libethenite occurring with pseudomalachite from Penberthy Croft mine, St Hilary, are probably in error, as Betterton et al. (2000) have shown similar crystals to be olivenite.

Dark greenish-grey striated crystals of libethenite (largest 8.3 mm in height) on botryoidal dufrénite; Phoenix mine, Linkinhorne, Cornwall. (© The Natural History Museum, London)

Libethenite forms sprays of blocky to prismatic crystals up to 0.2 mm long and equant crystals on crusts of chrysocolla and malachite in fractured diorite associated with a native copper–cuprite deposit near the Precambrian/Triassic unconformity at New Cliffe Hill quarry, Stanton under Bardon, Leicestershire (Hubbard et al. 2005).

A report of libethenite from Engine Vein mine, Alderley Edge ridge, Cheshire (Greenwood 1919, Carlon 1979, Braithwaite 1994), is doubtful and requires confirmation. Further doubts have also been expressed about the occurrence of libethenite at Old Potts Gill mine, Caldbeck Fells, Cumbria (Hartley 1984, Young 1987a), as they are based on specimens of doubtful provenance in the Kingsbury collection at the NHM (Ryback et al. 1998, 2001, Green & Briscoe 2002).

liebigite

orthorhombic $Ca_2(UO_2)(CO_3)_3 \cdot 11H_2O$

A rare hydrated calcium uranyl carbonate found on a specimen erroneously labelled "bassetite" and originally from Wheal Basset, Illogan, near Redruth, Cornwall (Evans & Frondel 1950). Liebigite in association with pink calcite, uraninite and sulphides in vuggy cavities, is a component of late uranium-rich mineralization hosted by Dalradian metasediments at Tyndrum Main mine, Central Region, Scotland. The assemblage includes chalcopyrite, galena, marcasite, pyrite and safflorite associated with secondary erythrite and the uranium minerals liebigite, kasolite, uranospinite and zeunerite. The secondary species probably formed during post-depositional oxidation of the deposit (Darnley 1962b, Pattrick 1985).

A pitchblende-bearing carbonate vein close to the margin of the Criffell granodiorite at Needle's Eye, SW of Caulkerbush, Dumfries & Galloway, Scotland, contains hydrocarbons in which peculiar uranium-rich filaments (apparently of organic origin) have been found. The walls and cells of the filaments are lined with a tabular calcium–uranium-enriched mineral that may be liebigite. Further confirmation is required (Basham et al. 1989).

liebnerite

Muscovite pseudomorph after nepheline.

lievrite

A synonym of ilvaite.

likasite?

orthorhombic $Cu^{2+}_3(NO_3)(OH)_5 \cdot 2H_2O$

A rare hydrated copper nitrate hydroxide that is questionable as a true British species. Likasite has been found at Hingston Down Consols mine, 2 km W of Gunnislake, Calstock, Cornwall, where it is thought to have formed from the reaction of copper with nitrates in blasting explosives (Pearce 1994b). It is clearly anthropogenic.

lime?

cubic CaO

A calcium oxide described from Hadley in Hereford & Worcester, where it is said to occur as minute crystals associated with barite in a copper-stained calcareous conglomerate (Roberts 1860, Dewey & Eastwood 1925). This is a highly unlikely occurrence, because lime readily alters to calcium hydroxide under normal atmospheric conditions. It is more likely that the authors were describing carbonate of lime (i.e. calcite), as this was a term in common usage at the time of the first description (see Greg & Lettsom 1858).

lime-uranite

A synonym of autunite.

limnite

An impure variety of limonite containing much water.

limonite

Hydrated iron oxide, mostly goethite. In current usage, limonite is used to describe any unidentified hydroxides or oxides of iron.

linarite

monoclinic $PbCu^{2+}(SO_4)(OH)_2$

An uncommon lead–copper sulphate hydroxide found in the oxidized zone of copper–lead deposits.

Fine specimens of linarite were recovered in the 1990s from the underground orebody at Penberthy Croft mine, St Hilary, Cornwall. Dark-blue crystal laths up to 4 mm long were found in 3 cm-long cavities with cerussite, brochantite, devilline and langite. Some radiating crystal groups show alteration to caledonite. Rare transparent doubly twinned terminated crystals have also been collected on dump material (Camm & Merry 1991b, Betterton et al. 2000). Deep-blue, transparent, flat prismatic-to-acicular crystals, exceptionally up to 3 mm long, have been found in oxidized veinstone within the Greystone lead lode at Greystone quarry, Lezant, near Launceston (Elton et al. 2004). Brief references to other linarite occurrences in Cornwall include: Gunnislake Clitters mine, Calstock (Green 1984); Pentire (Glaze) mine, near New Polzeath, St Minver (Clayton et al. 1990); traces of dump-grown linarite and brochantite at the Prince of Wales mine, Calstock, near Gunnislake (Braithwaite & Cooper 1982); perfectly formed, brilliant indigo-blue terminated crystals up to 1 mm long at Trevinnick mine, St Kew (Fisk et al. 1982); on the dumps

Beautifully formed 1 mm long blade of linarite; Penberthy Croft mine, St Hilary, Cornwall. (© M. P. Cooper)

at Wheal Golden, Perranzabuloe, and Wheal Mary Ann, Menheniot (Merry & Braithwaite 1995); tiny crystals at Wheal Penrose, Sithney, near Helston (Collins 1892); Garras mine, St Erme; Penhale mine, Perranzabuloe; Wheal Carpenter, Gwinear; Polzeath Consols, St Minver (all Golley & Williams 1995).

Undoubtedly, it is in Cumbria that linarite is best represented, with Red Gill mine, Caldbeck Fells, perhaps the most famous locality. Here, thick deep-blue tabular crystals of linarite (quite brilliant and perfect), to nearly 25 mm, were found lining cavities in a matrix of quartz and sulphides, and at the time were the world's best (Greg & Lettsom 1858, Eastwood et al. 1968, Wolloxall 1981, Cooper & Stanley 1990). Bryce Wright (in Jenkinson 1875) describe them as "truly one of the gems of the great national collection". Unfortunately, linarite of this calibre can no longer be found, but the species has often been collected from the mine dumps (Davidson & Thomson 1951, Hartley 1984, Briscoe 1986a, Wirth 1989, Cooper 1991a, Wolfe 1998, Green & Briscoe 2002). Many other locations for linarite are reported from the Caldbeck Fells: Brandy Gill mine (Davidson & Thomson 1951, Hartley 1984); Driggith mine (Hartley 1984, Green & Hubbard 1996, Green et al. 2001, 2006a, Green & Briscoe 2002); Dry Smale Gill (Shaw 1970); Mexico mine in Todd Gill (Greg & Lettsom 1858, Cooper & Stanley 1990); Sandbed mine (Davidson & Thomson 1951, Hartley 1984, Green et al. 2006a); Short Grain mine (Cooper & Stanley 1990, Green & Briscoe 2002); Silver Gill mine, near Roughton Gill (Davidson & Thomson 1951, Hartley 1984, Green et al. 2005b); and Thief Gills (Davidson & Thomson 1951). Reports of linarite from Brae Fell, Carrock End, Ingray Gill and Old Potts Gill mines, higher Roughton Gill; Poddy Gill; an old cross cut at the foot of Brandy Gill, Grainsgill and Swinburn Gill (Hartley 1984, Cooper & Stanley 1990) requires further confirmation, as the material was probably provenanced by A. W. G. Kingsbury. Green et al. (2006a) suggest that Kingsbury's "most spectacular fake" from the Caldbeck Fells is a specimen of linarite labelled Driggith mine, which is much more likely to be from Llaillay, Chile, or more locally from Red Gill or Roughton Gill. A Kingsbury specimen at the NHM, labelled Sandbed mine, probably originated from Red Gill mine. XRD confirmation of linarite has been made on material from Brae Fell mine, Dry Gill mine, Ingray Gill and Swinburn Gill (D. I. Green pers. comm. 2004). Bridges et al. (2006) note that they could not find a specimen of linarite from Brae Fell mine in the Kingsbury collection. The provenance of linarite from Roughton Gill is questioned by Davidson & Thomson (1951) despite there being fine 19th-century museum specimens labelled as such.

Elsewhere in the Lake District, linarite occurs at: Eagle Crag mine, Patterdale – encrustations of minute sparkling azure crystals (Green & Briscoe 2002); Force Crag mine, Braithwaite, near Keswick – radiating sprays of deep-blue crystals up to 1.5 mm long (Green et al. 1997, Green & Briscoe 2002); Goldscope mine, Newlands Valley, near Keswick (Davidson & Thomson 1951); and Rigg Head quarry, Borrowdale (Green & Briscoe 2002, Green et al. 2006c).

In the Cumbrian part of the Northern Pennine orefield, linarite is widespread (but only in trace amounts) and in many cases is described as having a post-mining origin: Clouds mine and Hartley mine, both near Kirkby Stephen; top dump of Rotherhope mine, Leadgate, near Alston (all Davidson & Thomson 1951); Smittergill Head mine, Melmerby Fell, Alston Moor (Young 1985a); as small deep-blue brilliant crystals up to 0.5 mm long at Tynebottom mine, NW of Garrigill (Davidson & Thomson 1951, Braithwaite 1982b, Bridges 1987, Rust 1987a, Bridges & Young 1998). In the eastern part of the orefield, in Co. Durham, linarite is noted at: Closehouse mine,

Middleton-in-Teesdale (Young et al. 1994a); at a trial on Drypry vein, N of Wolsingham; as minute crystals and crusts at Healeyfield mine, Castleside; and as crystals up to 1.5 mm long at Westernhope Old mine, S of Westgate, SW of Stanhope, Weardale (all Bridges & Young 1998).

Microscopic specks of linarite in granular quartz have been recovered from dump material belonging to the As–Cu–Pb–Sn deposit at Wheal Friendship (Devon Friendship), Mary Tavy, Devon (Brooke 1986).

Linarite is a rare species in Derbyshire, being reported as small bright-blue crystals from Masson and Rutland Caverns, Heights of Abraham, Matlock Bath (Braithwaite 1983, Ford et al. 1993). At Snelston (Isolation) mine, 4 km SW of Ashbourne, linarite forms acicular or flattened microcrystals on joint surfaces and is associated with cerussite and galena (Braithwaite & Greenland 1966, Jackman 1996). Over the border on Ecton Hill in Staffordshire, linarite occurs with serpierite at Clayton mine (Braithwaite 1983); and at Waterbank mine, where scattered azure-blue crystals to 1.5 mm (some forming beautiful fan-shape divergent groups) are associated with other post-mining supergene minerals (Rust 1995c).

Isolated occurrences of linarite are described from Engine Vein mine and Wood mine, Alderley Edge ridge, Cheshire (Braithwaite 1994), from Ladywell mine, Shelve, Shropshire (Braithwaite 1982b), and from Wet Grooves mine, Aysgarth, Yorkshire (Rust 1987b).

Linarite is described many times from mine dumps in the Central Wales orefield. In Dyfed, occurrences include: Brynarian mine, near Tal-y-bont (Rust 1992); Brynyrafr mine, near Nant-y-moch reservoir; Bwlchrhennaid mine, near Goginan; Dolwen-Mynach Vale mines, near Devil's Bridge (all Bevins 1994); beautiful blue blades to 2 mm at Eaglebrook (Nantycagl) mine, Ceulanymaesmawr (Jones & Moreton 1977, Braithwaite 1982b, Jones 1983); fine bladed prismatic crystals to 2 mm and superb rich divergent crystal groups encrusting malachite crusts at Esgair Hir mine, Ceulanymaesmawr (Rust & Mason 1988); Esgair Fraith mine, near Tal-y-bont, and Glogfawr mine, Devil's Bridge (Jones & Moreton 1977); blue bladed crystals to 2 mm and intimately associated with brochantite at Frongoch mine, between Devil's Bridge and Pont-rhyd-y-groes (Jones & Moreton 1977, Green et al. 1996b); poorly formed blades adjacent to galena and associated with cerussite, brochantite and malachite at Gwaithyrafon mine, SE of Tal-y-bont (Rust & Mason 1994); light-blue weathered crusts on quartz veinstone, and blue spherules up to 1 mm across with a composition midway between linarite and schmiederite at Llechweddhelyg mine, near Clawddmelyn, between Bont-goch and Cwmsymlog (Jones 1987, Rust 1990c); rich deep-blue prismatic crystals on shaly matrix, also intimately intergrown with ramsbeckite, posnjakite and schulenbergite at Penrhiw mine, Ystumtuen (Bevins 1994, Mason & Green 1995); poorly formed blades at Ystrad Einion mine, near Tal-y-bont (Mason & Rust 1997). In Powys, linarite occurs at: Dyfngwm mine, Machynlleth (Jones & Moreton 1977, Rust & Rust 1987); Dylife mine (Jones & Moreton 1977, Moreton 1982); and Geufron mine, near Llanidloes (Jones & Moreton 1977).

Sprays of linarite crystals, up to 3 mm across; Esgair Hir mine, Ceulanymaesmawr, Dyfed, Wales. (© National Museum of Wales, Cardiff; photographer M. P. Cooper)

Other Welsh occurrences of linarite include: as minute blue bladed crystals at Dolyhir quarry, Old Radnor, near Presteigne, Powys (T. F. Cotterell pers. comm. 2007); as bright-blue crystals in copper–lead veins with barite gangue at Machen quarry, 7 km ENE of Caerphilly, Mid Glamorgan – originally thought to be schmiederite (Plant & Jones 1995); associated with brochantite and cerussite on dump material at the Vale of Towy (Nant) mine, near Carmarthen, Dyfed (Bevins 1994); with anglesite, carbonate-cyanotrichite, cerussite and an unidentified species at Wedding Cave mine, Bwlchgwyn, near Wrexham, Clwyd (Green 1995b).

Excellent examples of linarite have been recovered from Hopeful vein and East Stayvoyage vein (Brown 1927), Lady Anne vein, Brown's vein and Susanna mine (Livingstone 2002), all Leadhills, Strathclyde, Scotland. At Susanna mine, associated species are caledonite, chenite, hydrocerussite, lanarkite, leadhillite and susannite (Paar et al. 1986). Additionally, in this area, linarite is reported from Leadhills Dod, High Pirn and Belton Grain mines (Heddle 1901), and Margaret's vein (Brown 1919), all Wanlockhead, Dumfries & Galloway. Linarite is also known from Pibble lead mine, Creetown (Heddle 1901), and the dumps of West Blackcraig lead–zinc mine, near Newton Stewart (Livingstone et al. 1976), all Dumfries & Galloway. Linarite is further reported from the Burn of Sorrow mine, near Dollar, Ochil Hills, Central Region (Dickie & Forster 1974). An anthropogenic occurrence of linarite on slag material from the dumps of the Meadowfoot smelter, Wanlockhead, Dumfries & Galloway is noted by Green (1987a).

Linarite is one of a vast array of supergene species found in the oxidized part of the Cu–Pb–Zn–(Ag) ore deposit at Tynagh mine, near Killimor, Co. Galway, Ireland, and recently it has been found together with adamite, bayldonite, mixite, olivenite, philipsbornite, native silver, strashimirite, tyrolite and wulfenite (Ryback & Moreton 1992, Green & Moreton 2001). In Co. Kerry it has been found at Ross Island and Muckross copper mines, Killarney National Park (Ryback & Moreton 1991, Moreton et al. 1999, Green & Moreton 2001), and in Co. Mayo it occurs with wroewolfeite in a Pb–(Ag) orebody at Sheefry mine, 11 km S of Croagh Patrick (Moreton & Jackson 1991, Ryback & Moreton 1992). Other Irish occurrences of linarite are at: Ballycorus mine, near Carrickgollogan, Co. Dublin (Ryback & Moreton 1993); Mogul mine, Silvermines, Co. Tipperary (Ryback & Moreton 1991); Ballinafunshoge mine, Glenmalure, Co. Wicklow (Russell 1907a); and the Luganure mines, Glendasan, also in Co. Wicklow (Ryback & Moreton 1993).

lindackerite?

monoclinic $Cu_5(AsO_3OH)_2(AsO_4)_2 \cdot 10H_2O$

A rare hydrated copper arsenate hydroxide found as a secondary mineral in copper-smelting slags in the area around Aberavon, Port Talbot, West Glamorgan, Wales. The material is not recognized as a mineral as it is derived from manmade materials (Nickel 1995). It occurs as radiating aggregates of fine needles showing colour zoning from apple green to almost colourless and it encrusts cavities within a copper-rich slag (Plant 2003).

lindbergite

monoclinic $Mn(C_2O_4) \cdot 2H_2O$

A rare hydrated manganese oxalate found as a secondary alteration product of manganese-rich ore at Lecht mine, near Tomintoul, Grampian Region, Scotland. Lindbergite occurs on the mine dumps in cryptomelane-rich ore covered with unusually thick growths of the lichen Pertusaria corallina (Wilson & Jones 1984, Atencio et al. 2004).

lindgrenite?

monoclinic $Cu^{2+}_3(MoO_4)_2(OH)_2$

A rare hydrated copper molybdate, said to occur at Brandy Gill mine, Caldbeck Fells, Cumbria (Kingsbury & Hartley 1955). The occurrence was later reported by Hartley (1984), Young (1987a) and Cooper & Stanley (1990, 1991b). Ryback et al. (2001) consider the provenance to be doubtful and suggest that the material came from

Chuquicamata, Chile. Unless further discoveries are made, this species cannot be considered to be native to Britain or Ireland.

lindsleyite

trigonal CRICHTONITE GROUP $(Ba,Sr)(Zr,Ca)(Mg,Fe)_2(Ti,Cr,Fe)_{18}O_{38}$

A rare barium–zirconium–magnesium–titanium oxide found in sulphidic quartz–celsian rock and sulphide-banded barite rock at Foss, near Aberfeldy, Tayside Region, Scotland. Quantitative data indicate variable compositions within the crichtonite group, but, with notable enrichment in barium and chromium, lindsleyite is the preferred species name. Unusually, some data also show enrichment in lead, uranium and vanadium. A sedimentary exhalative origin has been proposed for the mineralization (Moles 1985).

lindströmite

orthorhombic $Cu_3Pb_3Bi_7S_{15}$

A rare copper–lead sulphobismuthite found in quartz veins cutting monzogranite at Loch Shin, 2 km W of Lairg, Highland Region, Scotland. Elsewhere in the monzogranite, mineralization is disseminated and mainly restricted to fluorite, molybdenite and pyrite. However, in the veins, molybdenite does not occur and the other species are accompanied by bismuth, chalcopyrite, covellite, galena, sphalerite and a suite of bismuth sulphosalts: aikinite, berryite, gladite, hammarite, lindströmite, krupkaite, pekoite and tetradymite. Lindströmite is a member of the aikinite–bismuthinite series (Lowry et al. 1994).

LINNAEITE GROUP

Cubic sulphides with the general formula $A^{2+}B_2^{3+}X_4$, where A^{2+} = Co, Cu, Fe, Ni, Zn; B^{3+} = Co, Cr, Fe, In, Ni, Sb; and X = S, Se. Greigite, siegenite, trüstedtite, tyrrellite and violarite are members of the group found in Britain or Ireland. Linnaeite is also reported, but without positive identification.

linnaeite?

cubic LINNAEITE GROUP $Co^{2+}Co_2^{3+}S_4$

An uncommon cobalt sulphide that forms a series with polydymite and is described from Bad na h-Achlaise, Ledmore, Assynt, Highland Region, Scotland, where it is reported in a melanite pyroxenite of unusual character that is genetically related to the Loch Borralan alkaline complex. Ilmenite-poor titanomagnetite is the dominant ore species, accompanied by common chalcopyrite, pyrite and small amounts of bornite, covellite and linnaeite (Stumpfl 1961).

In Wales, linnaeite has been reported from "Rhonda Valley, Glamorganshire" (Des Cloizeaux 1880, North 1916), Ferndale No. 1 Pit, Mid Glamorgan (North & Howarth 1928), Wyndham colliery, Ogmore Vale, Mid Glamorgan (Bevins & Horák 1985b), Esgair Hir mine, Ceulanymaesmawr, Central Wales orefield, Dyfed (Rust & Mason 1988), and Great Orme's Head, Llandudno, Gwynedd (Bevins 1994, Ixer & Stanley 1996). However, XRD and EMP analysis of all grains from Wales looking like linnaeite are actually siegenite (Mason 1998) and on this evidence the Scottish occurrence deserves further attention.

linsenerz

A synonym of liroconite.

liparite

A synonym of chrysocolla, fluorite and jerntalk (ferrugeneous talc).

liroconite

monoclinic $Cu_2^{2+}Al(AsO_4)(OH)_4{\cdot}4H_2O$

TL Wheal Gorland, St Day, 2 km E of Redruth, Cornwall, England

A rare hydrated copper–aluminium arsenate hydroxide described as "a beautiful variety of arseniate of copper octahedral arseniate" by Garby (1848). Exceptional specimens, including the largest and finest crystal (2.1 cm across) known from any locality, are in the Rashleigh collection, now in the Royal Cornwall Museum, Truro (Penhallurick 1997). Liroconite is distinguished from all other copper arsenates by its sky-blue colour and sharply defined crystal form. Wheal Gorland is remarkable for its wealth of supergene species, many occurring in very fine crystallized form, and both a sky-blue and a dark-green variety of liroconite associated with clinoclase and olivenite are recorded (Rust 1981b). The occurrence of liroconite in Cornwall is virtually restricted to the mines of the St Day area, and at Ting Tang mine, Carharrack, Gwennap, it occurs as typical blue to bluish-green euhedral crystals up to 4 mm long, associated with parnauite in narrow cavities in quartz veins (Bruce & Aubrey-Jones 1998). At Wheal Muttrell, Gwennap, liroconite occurs as deep-blue pyramidal crystals up to 2.5 mm across, closely associated with olivenite on a limonite–quartz gossan (Embrey & Symes 1987). It is also reported from Wheal Gorland, St Day, 2 km E of Redruth (Greg & Lettsom 1858, Rudler 1905), and from Great Hewas mine, Sticker, near St Austell (Robson 1949b, Green 1984). Liroconite is also noted from Gunnislake, near Calstock, Cornwall, by Robson (1949b), but there are many mines in the area (e.g. Gunnislake Clitters, Old Gunnislake, South Bedford and Russell United) and the location is therefore uncertain.

The world's largest known crystal of liroconite on a base of pale-green strashimirite; 35 mm wide at base. Wheal Gorland, Gwennap, Cornwall; specimen is on display in the Royal Cornwall Museum, Truro, Cornwall. (© The Natural History Museum, London)

Doubtful occurrences of liroconite include: Alderley Edge ridge, Cheshire (Warrington 1965, Carlyon 1979, Braithwaite 1994), and Old Potts Gill mine, Caldbeck Fells, Cumbria, the latter claimed by A. W. G. Kingsbury (Green & Briscoe 2002).

liskeardite

monoclinic or orthorhombic $(Al,Fe^{3+})_3(AsO_4)(OH)_6{\cdot}5H_2O$

TL Marke Valley mine, near Liskeard, Cornwall, England

A rare hydrated aluminium–iron arsenate hydroxide found at the type locality as an encrustation associated with arsenopyrite, chalcopyrite, chlorite, pyrite, quartz and scorodite. Liskeardite is a fibrous white crystalline material, with a slight tint of blue

Soft, fibrous crystals of liskeardite forming a pale-green botryoidal crust on a matrix of chlorite; Marke Valley mine, Liskeard, Cornwall; width 70 mm. (© The Natural History Museum, London)

Delicate coating of liskeardite in a cavity within decomposing hornfels; Penberthy Croft mine, St Hilary, Cornwall; width 650 µm; SEM image. (© E. Condliffe, University of Leeds)

or greenish blue and it has proved to be unsuited to X-ray structural analysis (Maskelyne 1878, Flight 1883, Rudler 1905, Embrey & Symes 1987). At Penberthy Croft mine, St Hilary, Cornwall, liskeardite is locally abundant in arsenate-rich assemblages and is found in open cavities, where it forms white (with a greenish tint) botryoidal crusts with a radial or fibrous structure up to about 1.2 mm thick. Associated species are arsenopyrite, clinochlore, goethite, pharmacosiderite, quartz, scorodite and rare mansfieldite (Robson 1949b, Camm & Merry 1991b, Betterton et al. 2000).

litharge

tetragonal PbO

An uncommon lead oxide, dimorphous with massicot and found in an oxidized area of a vein exposure in Milltown quarry, near Ashover, Derbyshire. Litharge occurs as cubo-octahedral pseudomorphs replacing lead hydrates in association with anglesite, cerussite, hydrocerussite and sweetite (Clark et al. 1984, Rust 1991c). Minute opaque orange-red to scarlet crystals of litharge on galena are also noted from Millclose mine, Wensley, near Matlock, Derbyshire (Ford et al. 1993).

Litharge pseudomorphs probably after an unnamed hydrated lead oxide; Milltown quarry, Ashover, Derbyshire; width 2.5 mm. (© M. P. Cooper)

Litharge is reported from Frongoch mine, between Devil's Bridge and Pont-rhyd-y-groes, Central Wales orefield, Dyfed, Wales, where it occurs in scattered blebs, amorphous masses (to 0.4 mm) and occasionally as octahedra (to 0.5 mm) on cerussite. It is thought to have formed by reaction between lime-rich mortar from collapsed masonry and cerussite or galena (Bevins 1994, Green et al. 1996b).

Red plates (up to 1 mm) of litharge associated with cerussite, elyite and massicot have been found in a sample from a stream bed at Leadhills, Strathclyde, Scotland. It is possible that the sample may be a dumped flue product (Livingstone 1993), particularly as litharge is also noted from the nearby Meadowfoot smelter at Wanlockhead, Dumfries & Galloway (Green 1987a).

lithionglimmer

A synonym of lepidolite, mica group.

lithiophilite

orthorhombic $LiMn^{2+}PO_4$

An uncommon lithium–manganese phosphate that forms a series with triphylite and is found in accessory amounts in spodumene-bearing pegmatite at Aclare, near Myshall, and Seskinnamadra, W of Tomduff Hill, both Co. Carlow, Ireland. Lithiophilite is associated with apatite, beryl, bertrandite, columbite–tantalite, phosphosiderite, uranium-rich microlite, pyrite, spessartine and tourmaline (Steiger & von Knorring 1975, Ryback & Moreton 1993).

lithiophorite

trigonal $(Al,Li)Mn^{4+}O_2(OH)_2$

A rare aluminium–lithium–manganese oxyhydroxide found in a brecciated fracture zone of manganese-rich ore at Lecht mine, near Tomintoul, Grampian, Scotland. Lithiophorite usually occurs as massive black fine-grain aggregates, occasionally showing botryoidal form, and is associated with cryptomelane, goethite, hematite and quartz (Wilson et al. 1970). The Lecht mineralization is derived from weathering of the local Dalradian sequence and ponding of the resulting Fe–Mn–Zn solutions produced a seepage-bog type of deposit in which the permeable breccia acted as a channel-way for these supergene solutions (Nicholson 1987a, 1989a,b, 1992, Nicholson & Anderton 1989). In a study of manganese oxides cementing and coating raised-beach deposits between Auchenmaig and Craignarget Burn, Luce Bay, Dumfries & Galloway, Scotland, Nicholson (1988b, 1989a,b) was able to identify birnessite, lithiophorite, quenselite and amorphous wad and concluded the oxides were deposited from local groundwaters.

Black films of lithiophorite, rarely exceeding 1 mm in thickness, are described coating joint surfaces of granophyre brecciated by the Red Gill fault at Clews Gill, Ennerdale, Cumbria (Clark 1963, Young 1987a).

XRD analysis showed lithiophorite to be present in massive to minutely botryoidal bluish manganese-rich material at Drosgol, Ponterwyd, Dyfed, Wales (T. F. Cotterell pers. comm. 2003).

lithomarge

A variety of kaolinite and of halloysite, also called tuesite. It is a very early name for several soft clay-like minerals.

livesite

Named after Sir George Thomas Livesey (1834–1908), gas engineer and industrialist, born in Islington, London. Livesite is an intermediate composition between halloysite and kaolinite (both kaolinite–serpentine group) found in the fireclays of Yorkshire.

lizardite

KAOLINITE–SERPENTINE GROUP

trigonal or hexagonal $Mg_3Si_2O_5(OH)_4$

TL Kennack Cove, Lizard Peninsula, Cornwall, England

A relatively common magnesium silicate hydroxide, polymorphous with antigorite, clinochrysotile, orthochrysotile and parachrysotile, that forms a series with népouite and is found in a vein cutting serpentinized peridotite at the type locality. Lizardite is a white soapy mineral that was initially thought to be saponite and later to be a variety of antigorite. Under high magnification, lizardite has a plate-like habit (Midgley 1951). XRD analysis enabled Whittaker & Zussman to declare it a new species in 1956 and to note a further occurrence in serpentinite at the ICI quarry on the Lizard Peninsula. These authors state that most of the light- to yellowish-green massive rock specimens examined (and commonly called serpentine) are in fact

Botryoids of lizardite from Kennack Sands, Lizard Peninsula, Cornwall; width 4 mm. (© D. I. Green)

lizardite. Soft white fine-grain to massive lizardite also occurs with calcite and hetaerolite in parallel-sided or anastomosing veins cutting serpentinized peridotite from an area of Eastern Cliffs, near Kennack Sands (Bevins et al. 1987). Peridotite at Enys Head, near Cadgwith, hosts thin white veins in which lizardite is found with calcite, chrysotile, garnet, ilmenite and zoisite (Hall 1979). More generally, pale- to dark-green pseudo-fibrous mixtures of lizardite and chrysotile are reported in fractures within the Lizard peridotite (Power et al. 1997).

Lizardite is reported from the Rhiw intrusion, Trwyn Talfarach, Lleyn Peninsula, Gwynedd, Wales (Bevins 1994), and less precisely from serpentinized peridotite of the Mona Complex, Holy Island, Anglesey. At the second location it occurs associated with antigorite and replaces primary olivine and orthopyroxene. It produces rock textures in the serpentinites, variably described as mesh, ribbon, bladed mat and disturbed, and little now remains of the original peridotite (Maltman 1977).

In Scotland, lizardite is found in serpentinite replacing ultramafic rocks at Alyth, near Blairgowrie, and Carity Burn, Kirriemuir, both Tayside Region. These serpentinites are composed almost exclusively of lizardite, although some relict forsterite, enstatite and spinel are preserved. Three ultramafic intrusions in Strathclyde, at Balmaha, Innellan and Scalpsie Bay, also contain lizardite, but replacement by brucite–antigorite has occurred during prograde metamorphism (Ikin & Harmon 1983). In the Shetland ophiolite complex on Unst, two undeformed serpentinized cumulate dunites from north Baltasound have been shown to contain mesh- and hourglass-textured lizardite replacing olivine intergrown with minor amounts of chlorite, carbonate and pentlandite (Maynard 1993). Another ophiolite complex is exposed at Pinbain Bridge and Poundland Burn, N of Ballantrae, Strathclyde, where lizardite occurs with chrysotile in serpentinized ultramafic rocks (Power & Pirrie 2004). Lizardite is also present in altered serpentinite from Glen Urquhart, Highland Region (Livingstone 2002), and with antigorite and chrysotile in serpentinite at Arrochymore Point on the shores of Loch Lomond, Central Region, close to the Highland Boundary fault (Henderson et al. 1983).

EMP analysis suggests that lizardite is also present in serpentinites from Old Head, near Louisburg, Co. Mayo, Ireland (H. J. Foy specimen, Tindle unpub. data). Because lizardite was described only relatively recently, many older works do not refer to it, and it may well be much more common than the literature suggests. Almost certainly, some of the descriptions of serpentine by Greg & Lettsom (1858), Collins (1892) and Heddle (1901) are actually of lizardite.

loganite

Named after Sir William Edmond Logan FRS (1798–1875), educated at the University of Edinburgh, and later to be the first director of the Geological Survey of Canada; born in Montreal, Canada, of Scottish parents. Loganite is a pseudomorph of pennine (now clinochlore, chlorite group) after amphibole, found at Lisoughter, Co. Galway, Ireland (Harkness 1866).

LÖLLINGITE GROUP

Orthorhombic arsenides and antimonides with the general formula AB_2, where A = Co, Fe, Ni, and B = As, S, Sb. Löllingite, rammelsbergite and safflorite are members of the group found in Britain or Ireland.

löllingite LÖLLINGITE GROUP

orthorhombic $FeAs_2$

A relatively common iron arsenide that can easily be mistaken for arsenopyrite and is found at many locations in Cornwall. Considerable quantities of löllingite are reported from Castle-an-Dinas mine, St Columb Major, where it occurs in vuggy quartz veins as a major component of wolframite ore. Fluorite, russellite, scorodite, topaz and wavellite are minor species also found in the lode. The mineralization is associated with a small granite cupola to the NW of the St Austell granite (Hosking 1949, Polkinghorne 1954, Dines 1956, Hosking & Trounson 1959, Ball et al. 1985b). Löllingite is associated with triplite in pegmatite and leucogranite near the top of the main granite sheet at Megiliggar Rocks, near Breage (George et al. 1981). The pegmatite is related to the evolved lithium-enriched Tregonning–Godolphin granite. Early-formed flat-lying "pegmatitic lodes" containing arsenopyrite, löllingite, wolframite and rare stannite are described from South Crofty mine, Camborne, and Wheal Pendarves, SW of Camborne (Scrivener et al. 1986, Dominy et al. 1994). Greisen-bordered veins at Cligga Head, Perranzabuloe, contain löllingite in an ore assemblage consisting of arsenopyrite, cassiterite, chalcopyrite, ferrokësterite, kësterite, stannite and stannoidite (Alderton 1991). Similar löllingite-bearing greisen also occurs at Goonbarrow china-clay pit, near Bugle, St Austell (Bray & Spooner 1983), in the Redmoor–Kit Hill area, Stoke Climsland (Newall & Newall 1989, Newall 1994), at St Michael's Mount, Marazion (Moore & Moore 1979, Alderton 1991), and at Wheal Jane, Kea, near Truro (Holl & Bromley 1988), at the latter site in a wolframite, cassiterite, native bismuth, muscovite and quartz assemblage. A single specimen of finely granular löllingite intergrown with arsenopyrite and associated with chalcopyrite and pyrite is reported from New Consols mine, Stoke Climsland (Embrey 1953). Other Cornish occurrences of löllingite include: East Pool mine, Illogan, near Redruth, and Dolcoath mine, Camborne (Collins 1892); Pednandrea mine, Redruth (Le Neve Foster 1877); Penlee quarry, Newlyn (Golley & Williams 1995); Roskrow United mine, Ponsanooth (Ball et al. 1982); St Austell Consols mine, NW of Sticker (Ball et al. 1982); Tregargus china-clay pit, St Stephen-in-Brannel (Embrey 1953); Wheal Alice, Grampound Road (Ball et al. 1982); and Wheal Sparnon, near Redruth (Embrey 1953).

Arsenic-rich species are encountered only rarely in the highly mineralized Meldon aplite, near Okehampton, Devon, but löllingite, arsenic-rich native antimony and native arsenic have all been found. Crosscutting pegmatitic veins contain native arsenic and arsenolite in addition to other rare species such as columbite–tantalite, petalite and pollucite (von Knorring & Condliffe 1984). The Meldon aplite is intruded into the thermal aureole of the Dartmoor granite and, near by at Red-a-ven mine, this thermal event has metasomatized limestones to produce andradite–hedenbergite–ferrobustamite (iron-wollastonite) skarns containing abundant massive pyrrhotite with lesser amounts of disseminated arsenopyrite, bornite, chalcopyrite, löllingite and sphalerite (El Sharkawi & Dearman 1966). Boreholes in the Bovey Tracey area of east Dartmoor produced samples containing disseminated and thin discontinuous stratabound veinlets of sulphides (mainly galena and sphalerite) associated with minor arsenopyrite, chalcopyrite and löllingite (Beer et al. 1992, Benham et al. 2004).

Löllingite from the following locations requires further confirmation, as the descriptions were either made by A. W. G. Kingsbury or were based on samples in the Kingsbury collection at the NHM: Huckworthy Bridge mine, Sampford Spiney, Devon (Kingsbury 1964b); Paddy End mine, Coniston, Eskdale, and Wanthwaite mine, St John's in the Vale, near Threlkeld, both Cumbria (Young 1987a); Potts Gill barite mine, Caldbeck Fells, Cumbria (Kingsbury & Hartley 1960, Hartley 1984); Red-a-ven mine, Meldon, near Okehampton, Devon (Kingsbury 1961); Tynebottom mine, NW of Garrigill, near Alston, Cumbria (Ixer & Stanley 1987). Ryback et al. (1998, 2001) have shown that some samples in the Kingsbury collection have been incorrectly provenanced and the status of the löllingite samples is yet to be resolved.

At Inverneil, S of Adrishaig, Knapdale, Strathclyde, Scotland, Dalradian-hosted lead–zinc mineralization includes uranium-rich hydrocarbons containing löllingite and uraninite (Parnell 1988). Uranium-bearing veins cutting the aureole of the Criffell granodiorite at Southwick, near Dalbeattie, Dumfries & Galloway, contain löllingite hosted in dolomite, with nickeline and sphalerite as associated species. Many arsenic-, bismuth-, copper-, cobalt- and uranium-rich species occur in the veins exposed along the coastline in this region (Miller & Taylor 1966, Braithwaite & Knight 1990). A single alluvial gold grain containing löllingite inclusions is noted from the Leadhills area, along the Strathclyde/Dumfries & Galloway border (Leake et al. 1998).

At Gortdrum mine, near Tipperary, Co. Tipperary, Ireland, epigenetic mineralization related to the intrusion of two high-level basic intrusions has produced an orebody of chalcopyrite and mercury-rich tennantite, in which löllingite is one of the minor constituents. Löllingite also occurs in the lead–zinc orebody at Silvermines, Co. Tipperary, where it is reported to be one of the minor and trace species of the upper G orebody (Morrissey et al. 1971, Moreton 1999).

lonchidite
A variety of marcasite containing a minor amount of arsenic.

lorettoite
Shown to be manmade material (slag-formed $Pb_7O_6Cl_2$).

ludlamite
monoclinic $Fe^{2+}_3(PO_4)_2{\cdot}4H_2O$
TL Wheal Jane, Kea, near Truro, Cornwall, England

Ludlamite crystals to 7 mm on quartz; Wheal Jane, Kea, near Truro, Cornwall. (© M. P. Cooper)

Named after Henry Ludlam (1824–1880), a London-based hosier and mineral collector (Cooper 2004). Ludlamite is a rare hydrated iron phosphate that in Britain and Ireland has only been found at the type locality. It is a beautiful pale- to dark-green species that forms sharp but composite crystals up to 2–3 cm across or small transparent crystals richly encrusting pyrite matrix. It occurs as a secondary mineral in the oxidized zone of the mine, sometimes associated with cronstedtite, quartz, siderite or vivianite (Collins 1911a, Rust 1982d,g, Embrey & Symes 1987, Bancroft & Weller 1993, Cooper & Green 1993). Ludlamite (requiring further investigation) is described by Stone & George (1983) from the complex granite pegmatite associated with the Tregonning granite, at Megiliggar Rocks, near Breage, Cornwall.

LUDWIGITE GROUP
Orthorhombic borates with the general formula $X_2Y(BO_5)$, where $X = Fe^{2+}$, Mg, Ni, and $Y = Al, Fe^{3+}, Mg, Mn^{3+}$, Sb, Ti. Ludwigite is the only species of the group found in Britain or Ireland.

ludwigite
orthorhombic $Mg_2(Fe^{3+})O_2(BO_3)$
A rare magnesium–iron borate that forms a series with vonsenite and is found in a contact skarn where the Tertiary Beinn an Dubhaich granite has intruded dolomite at Kilbride, near Broadford, Isle of Skye, Highland Region, Scotland. Ludwigite occurs in a carbonate–chondrodite–ludwigite–magnetite skarn in which fluoborite

Dense black fibrous mass of ludwigite in metasomatically altered limestone from Kilchrist, near Broadford, Isle of Skye, Highland Region, Scotland; width of specimen 7 cm. (© A. G. Tindle)

and szaibelyite are accessory phases. At this locality, ludwigite appears as radiating clusters of velvety black needles up to 1 mm long. Traces of ludwigite, in skarn at Kilchrist and in marble at Camas Malag, are also reported from the same area (Tilley 1948b, 1951).

lunnite
A synonym of pseudomalachite.

lussatite
A fibrous variety of cristobalite.

lutecite
A synonym of moganite, a fibrous, microcrystalline polymorph of quartz.

luzonite
STANNITE GROUP
tetragonal Cu_3AsS_4
A copper sulpharsenate, dimorphous with enargite and found as inclusions in malachite-stained vuggy calcite crystals in limestone from Scaleber Bridge, 3 km SSE of Settle, North Yorkshire. Chalcopyrite, marcasite and pyrite are the main inclusion species, with trace amounts of luzonite, famatinite and tennantite also present. Luzonite (As > Sb) and famatinite (Sb > As) show a range of intermediate compositions and form crystals up to 80 μm across, with multiple colours varying from brown to lilac (Ixer & Stanley 1998).

In Wales, luzonite is reported from temporary exposures at Dolyhir quarry, Old Radnor, near Presteigne, Powys. It occurs in barite–calcite boulders, where it forms colloform to nodular masses, to 17 mm across, and veinlets with a characteristic pinkish metallic hue (T. F. Cotterell pers. comm. 2007).

lydian stone
A black jasper; a synonym of basanite (not bassanite) and touchstone.

lyellite
Named after Sir Charles Lyell FRS (1797–1875), one of the most influential British geologists of the 19th century, professor of geology at King's College, London, and author of *Principles of geology* (1830); born at Kinnordy, Forfarshire (part of Tayside Region), Scotland. Lyellite is reported from Cornwall and is a synonym of devilline.

M

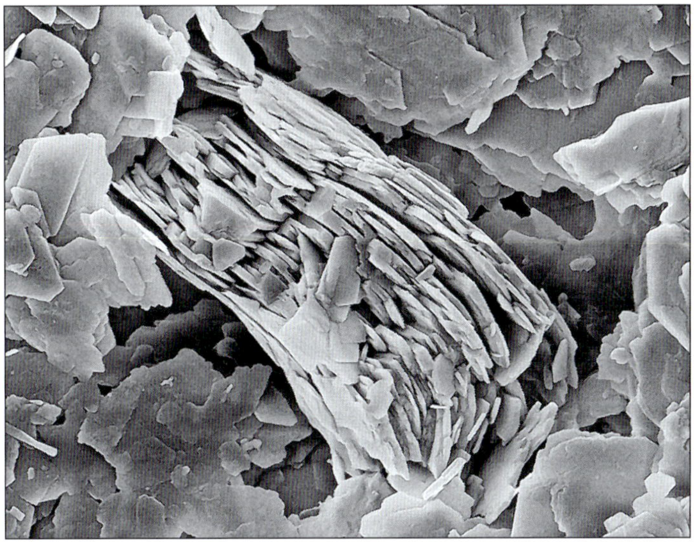

Platelets of macaulayite in deeply weathered granite from Bennachie, near Inverurie, Grampian Region, Scotland; width 3.6 µm; SEM image. (© E. Condliffe, University of Leeds)

macaulayite

monoclinic $(Fe^{3+},Al)_{24}(Si_4O_{43})(OH)_2$

TL Bennachie car-park, 7 km WNW of Inverurie, Grampian, Scotland

Named after the Macaulay Institute for Soil Research, Craigiebuckler, Aberdeen, Scotland, which was founded by Thomas Bassett Macaulay (1860–1942) – the parent body of the authors. Macaulayite is a rare iron aluminium silicate hydroxide found intimately associated with illite and kaolinite in an outcrop of reddened deeply weathered granite. It is earthy, blood red and with very fine grain, and has been described as a "swelling hematite/layer-silicate mineral" (Wilson et al. 1981, 1984, Russell et al. 1984a).

mackinawite

tetragonal FeS

An uncommon iron sulphide found in the serpentinized peridotite of the Lizard, Cornwall, in the vicinity of the Rill, Kynance Cliff. Mackinawite forms a reticulate network of minute flakes averaging 20–30 µm in diameter within chlorite–magnetite patches, and probably formed during the late-stage serpentinization of the peridotite after the breakdown of primary chrome spinel. In reflected light, mackinawite has a diagnostic pinkish-grey colour and it exhibits marked anisotropism from bluish white to dark grey (Clark 1967). At New Brancepeth colliery, near Durham, a galena–ullmannite vein cutting Coal Measures rocks contains small framboidal pyrite grains mantled by skutterudite, gersdorffite and ullmannite, and pyrite and marcasite surrounded by mackinawite (Ixer 1986). Mackinawite has also been described from Tynebottom mine, NW of Garrigill, near Alston, Cumbria (Ixer 1986), but the specimens came from the Kingsbury collection at the BM(NH). Dunham et al. (2001) express serious doubt about the provenance of these specimens.

In Scotland, mackinawite is reported from a hydrothermal quartz–sulphide ± gold vein deposit cutting Upper Dalradian gritty quartzites, quartz–mica schists and amphibolites at Calliachar Burn, 4 km from Aberfeldy, Tayside Region. Pyrite is the most abundant sulphide, forming discrete crystals in aggregates ≤ 1 cm across and hosting inclusions ≤ 50 µm across. Mackinawite is one of the inclusion species that occurs with mixed chalcopyrite–pyrrhotite, cubanite, galena, hematite, pentlandite, pyrrhotite, sphalerite and TiO_2 minerals. At the nearby Urlar Burn, pyrite encloses large (up to 200 µm in diameter) intergrowths of altaite–hessite–tsumoite and minor quantities of mercury-poor electrum, or it contains inclusions of chalcopyrite–mackinawite, chalcopyrite–pyrrhotite, chalcopyrite–cubanite, chalcopyrite–galena–sphalerite and sphalerite (Ixer et al. 1997). Mackinawite is found in allivalite (a pyroxene-free gabbro) from Huntly, Grampian, where it occurs as anhedral inclusions (50–100 µm across) in pentlandite or, less commonly, chalcopyrite, in association with cubanite and valleriite (Babkine & Conquéré 1968). On Unst, Shetland, serpentinites from the basal thrust of the ophiolite complex contain a sulphide assemblage that includes chalcopyrite–mackinawite–pyrrhotite intergrowths and mackinawite associated with pyrrhotite and pentlandite. Cobaltite, marcasite, nickeline, orcelite, pyrite and valleriite are other associated sulphide species (Maynard 1993).

At Carn Menyn in the Mynydd Preseli, Dyfed, Wales, mixed chalcopyrite–mackinawite–pyrrhotite grains occur in an altered basic intrusion (Thorpe et al. 1991), and, in the Dolgellau goldbelt of Gwynedd, mackinawite occurs as inclusions in chalcopyrite or as vermicular growths with chalcopyrite associated with pyrrhotite (Gilbey 1968). Bevins (1994) notes mackinawite from Cefndeuddwr mine, near Dolgellau, and from a lode in the Foel Ispri area, and it suggests that the species may be more common in the ultrabasic and basic rocks of Wales than the literature suggests. It has not been reported more often because studies of opaque minerals in these rocks are few.

In the Ballinglen area, between Tinahely and Aughrim, Co. Wicklow, Ireland, arsenopyrite–fluorite–pyrrhotite–scheelite mineralization occurs along the south-eastern margin of the Leinster granite in a swarm of narrow microgranite sheets cut by greisen-bordered quartz veinlets. A diverse assemblage of minor mineral phases (mainly within the veinlets) includes apatite, native bismuth, cassiterite, chalcopyrite, cosalite, dravite, galena, mackinawite, molybdenite, proustite, rutile, schorl, sphalerite, stannite and titanite. It is uncertain if the mineralization represents the deep-seated portion of a subvolcanic system (perhaps related to a tungsten porphyry) or is related to a granite pegmatite system (Steiger & Bowden 1982, 1986, McArdle & Kennan 1988).

macphersonite

orthorhombic $Pb_4(SO_4)(CO_3)_2(OH)_2$

TL Leadhills Dod, Leadhills, Strathclyde, Scotland

Named after Harry Gordon Macpherson (1925–2001), keeper of minerals at the NMS; born in Aberdeen. Macphersonite is an exceedingly rare lead sulphate carbonate hydroxide, trimorphous with leadhillite and susannite, and was first described

Macphersonite crystal group; Leadhills, Strathclyde, Scotland; width of specimen 2 cm. (© The Trustees of the National Museums of Scotland)

from a cavity within a small quartz specimen in the Heddle collection at the NMS. It is pale amber to tan, resinous to adamantine, and often polysynthetically twinned. Macphersonite is associated with caledonite, cerussite, leadhillite, pyromorphite, susannite and an unnamed "lead-hydroxyapatite" mineral (later named mattheddleite). Leadhillite and macphersonite can mimic each other, and an X-ray powder photograph or an infrared spectrum is required to describe these species positively (Livingstone & Sarp 1984). An infrared investigation of the three polymorphs is provided by Russell et al. (1984b) and the crystal structure of macphersonite has been examined by Steele et al. (1998). Macphersonite is a secondary mineral that forms during oxidation of galena (Livingstone 2002).

Small colourless prismatic crystals of macphersonite (triangular in cross section), associated with caledonite, leadhillite, mattheddleite and pyromorphite, are briefly described from Red Gill mine, Caldbeck Fells, Cumbria (Cooper et al. 1988, Cooper & Stanley 1990, Green & Briscoe 2002), and from the nearby Short Grain mine (Green & Briscoe 2002).

maghemite
cubic $Fe_{2.67}O_4$

An iron oxide, dimorphous with hematite, briefly described from the Tertiary basalts of the isles of Rum, Eigg, Canna and Muck, Highland Region, Scotland (Ridley 1971). Traces of maghemite occur in red ferruginous bauxite from Lyles Hill mine, Templepatrick, Co. Antrim, Ireland, associated with anatase, böhmite, corundum, hematite and magnetite (Eyles 1952). The identification is questioned by Embrey (1978).

magnesio-anthophyllite
The name "magnesio-anthophyllite" was formally abandoned when the IMA-CNMNC reclassified the amphibole group (Leake et al. 1997a). Minerals of this composition should now be referred to as anthophyllite. Magnesio-anthophyllite had been described from a metamorphosed layered alpine-type peridotite in the Langvat Valley, South Harris, Western Isles, Scotland (Guppy 1956, Livingstone 1976a).

magnesio-arfvedsonite
AMPHIBOLE GROUP
monoclinic $NaNa_2(Mg_4Fe^{3+})Si_8O_{22}(OH)_2$

An uncommon sodium–magnesium–iron silicate hydroxide that forms a series with arfvedsonite and is found in the Pendennis peralkaline minette, a minor intrusion cutting Devonian slates on the foreshore of Pendennis Point, Falmouth, Cornwall. Magnesio-arfvedsonite occurs with potassic magnesio-arfvedsonite, potassic-ferro-eckermannite and magnesioriebeckite, although together they are only a minor constituent of the rock. All crystals are compositionally heterogeneous, with a short prismatic habit and some with fibrous core regions (Hall 1982).

A peralkaline minette at St Hélier, Jersey, contains elongated 200 μm-long crystals of magnesio-arfvedsonite associated with biotite, phlogopite, altered olivine and alkali feldspar (Wagner & Velde 1985).

magnesiochromite
SPINEL GROUP
cubic $MgCr_2O_4$

An uncommon magnesium–chromium oxide that forms two series, with chromite and with spinel, and is found as an accessory species in mafic and ultramafic igneous rocks (e.g. basalts and peridotites).

Some chrome spinel data (with $Fe^{2+}/(Fe^{2+}+Mg)<0.5$ and $Cr/(Cr+Al)>0.5$) from Tertiary picritic basalts on the northeastern side of An Cruachan and from Sgùrr Beag, Isle of Skye, Highland Region, Scotland, indicate that magnesiochromite is present, although the data should perhaps be considered intermediate between magnesiochromite and spinel compositions (Bell & Williamson 1994). Similar data have also been collected from ultramafic rocks belonging to the Shetland ophiolite complex (Gass et al. 1982) and from the layered intrusion on the Isle of Rum (Henderson & Suddaby 1971, Power et al. 2000b). Picritic dykes associated with the Rum layered intrusion contain accessory amounts of magnesiochromite and chromite (Upton et al. 2002).

In Strathclyde, magnesiochromite has been found as cm-size rounded crystals within dunite on Poundland Moor near Craig Hill (half way between Pinwherry and Colmonell). The occurrence is within a sequence of layered harzburgites, dunites, werhlites and olivine gabbros that form part of the Girvan–Ballantrae ophiolite complex (D. G. Anderson pers. comm. 2006). A larger spinel-group dataset from this complex, including samples from Pinbain Bridge and the Poundland locality, exhibit a wide range of intermediate compositions best described as chrome spinel or picotite (Stone et al. 1986).

Altered ultramafic rocks near Balmaha, Loch Lomond, Central Region (close to the Highland Boundary fault), contain deep brownish-red translucent crystals of magnesium-rich chromite and iron-rich magnesiochromite (Henderson et al. 1983).

magnesiocummingtonite
A synonym of cummingtonite. Magnesiocummingtonite was established as a species in an early classification of the amphibole group (Leake et al. 1978), but was formally abandoned in the later revision (Leake et al. 1997a). The amphibole described as cummingtonite by Droop (1994) from Lewisian amphibolite-facies ultramafic rocks at An Fharaid Mhòr, SW of Achmelvich, near Lochinver, Highland Region, Scotland, would have been called magnesiocummingtonite had the 1978 amphibole classification been used.

magnesiofoitite
TOURMALINE GROUP
trigonal $\square(Mg_2,Al)Al_6(BO_3)_3Si_6O_{18}(OH)_4$

An uncommon X-site vacant magnesium–aluminium borosilicate hydroxide found in zoned dravite–foitite–magnesiofoitite–schorl crystals from a hydrothermal breccia at the southwestern margin of Roche Rock, 7 km N of St Austell, Cornwall. The presence of relatively high tin content in the magnesiofoitite and foitite zones within the tourmaline provides a valuable insight into the evolution of mineralizing fluids that ultimately produced tin (cassiterite) mineralization, for which Cornwall is justly famous (Williamson et al. 2000). Composite grains of colourless, brown, yellow and dark-green tourmaline, with compositions varying from magnesiofoitite to schorl, occur in a quartz–tourmaline rock from the Providence mines area, near Carbis Bay, St Ives, Cornwall (Wilson & Young 1983), but their identity was not fully established until magnesiofoitite was first described (Hawthorne et al. 1999b).

At Porth Ledden, to the N of Cape Cornwall, tourmaline-rich rocks occur at the contact of the Land's End granite with pelitic hornfels. The tourmaline is predominantly schorl, but, in narrow brecciated veinlets cutting through the earlier tourmaline-rich rocks, magnesiofoitite is found in blue (in thin section) crystals of irregular shape (Tindle unpub. EMP data).

A complex occurrence of magnesiofoitite is from a single blade of hydrothermal tourmaline found in a quartz vein cutting extensively kaolinized granite from Longstone china-clay pit (the site of the Trelawny Tigers speedway track), near Nanpean, within the St Austell granite complex, Cornwall. This single crystal is strongly zoned from core to rim with the following species of tourmaline: schorl, foitite, elbaite, foitite, dravite, magnesiofoitite, foitite, schorl, magnesiofoitite, foitite, dravite, schorl, dravite (Tindle unpub. EMP data).

Tourmaline is described from a wolframite-bearing vein and from loose blocks of vein quartz at Burdell Gill, near Coombe Height, Caldbeck Fells, Cumbria. It forms slender (1–4 mm long) pale greyish-green to brown prisms with a triangular cross section, and its composition was predicted to be schorl (Neall & Leppington 2003). EMP analysis of this material revealed it to be mainly dravite, with minor schorl and magnesiofoitite also present. Another specimen from the same location, but containing smaller needles of tourmaline, contains foitite in addition to the other three species (D. I. Green & C. M. Leppington specimens, Tindle unpub. EMP data). Minor magnesiofoitite accompanies dravite in tourmalinites exposed N of the Haweswater reservoir, Cumbria. The tourmalinites form brecciated pipe-like bodies cut by tourmaline-bearing quartz veins (Eeckhout et al. 2004).

magnesiohastingsite
AMPHIBOLE GROUP
monoclinic $NaCa_2(Mg_4Fe^{3+})(Si_6Al_2)O_{22}(OH)_2$

A common sodium–calcium–magnesium–iron aluminosilicate hydroxide that forms a series with hastingsite and is found in alkalic basalts, andesites and dioritic rocks. Where possible, identification is based on the 1997 classification of the amphibole group (Leake et al. 1997a, Mogessie et al. 2004).

Wales
Gwynedd Some amphibole phenocrysts in pargasite–pyroxene–plagioclase-phyric basalt lavas of the Eglwys Rhobell Formation, Rhobell Volcanic Complex, Rhobell Fawr area, near Dolgellau, have magnesiohastingsite or ferri-magnesiohastingsite compositions (Kokelaar 1986).

Scotland
Fife Titanium-rich magnesiohastingsite, previously identified as kaersutite, occurs in pyroxenite nodules within Carboniferous–Permian volcanic vents near Elie (Chapman 1976).

Highland Pyroxenite and hornblendite rocks from the southwestern part of the Loch Borralan syenite complex contain magnesiohastingsite, although the authors indicate a range extending into pargasite compositions (Notholt 1985). Biotite pyroxenite from the Loch Ailsh alkaline igneous complex contains large plates of brown amphibole with a magnesiohastingsite or pargasite composition (Shaw et al. 1994). Euhedral (but strongly resorbed) brown magnesiohastingsite occurs in rhyodacite lavas at the centre of the Ben Nevis diorite–granite complex, near Fort William (Haslam 1968). Magnesiohastingsite is also found: as deep-green crystals in diorite from the Glenelg–Ratagain igneous complex (Nicholls 1951b); as green crystals in granodiorite from the Morvern–Strontian igneous complex (Nockolds & Mitchell 1946); and with pargasite as inclusions in spinel within the ultramafic portion of the Rum intrusion, Isle of Rum (Young 1984b).

Strathclyde A Devonian andesite from the northern Lorne Plateau, to the E of Oban, contains magnesiohastingsite (Groome & Hall 1974). Magnesiohastingsite is brown (in a hornblendite) or brownish-green (in an appinitic diorite) within the Garabal Hill–Glen Fyne igneous complex (Nockolds & Mitchell 1946).

Tayside Typical diorite from the Glen Tilt igneous complex, E of Blair Atholl, contains magnesiohastingsite. Amphiboles in associated appinites, diorites, schists and xenoliths are either magnesiohornblende or tschermakite (Deer 1938).

Western Isles Magnesiohastingsite occurs in a metamorphosed spinel lherzolite lens within Rodel–Leverburgh metasediments at the head of Glen Rodel, South Harris. Associated species are augite, edenite, enstatite, pargasite and accessory hercynite and magnetite (Livingstone 1976c). One analysis of magnesiohastingsite had previously been referred to as magnesiotaramite (Macpherson & Livingstone 1982).

Ireland

Cork Some kaersutite described from an amphibole pyroxenite nodule within a megacryst-bearing pyroclastic pipe at Black Ball Head, near Cahermore (Pracht & Kinnaird 1995), is actually titanium-rich magnesiohastingsite. It is black with a bright-green rim and it forms embayed anhedral-to-subhedral crystals up to 4 cm long.

Offshore territories

Jersey Late Precambrian Cadomian layered diorites associated with gabbro on the southeast coast at Le Nez contain a range of amphiboles considered to be secondary after pyroxene; they include actinolite, edenite, kaersutite, magnesiohastingsite, magnesiohornblende and pargasite (Bishop & Key 1983).

magnesiohornblende

monoclinic — AMPHIBOLE GROUP — $\square Ca_2[Mg_4(Al,Fe^{3+})](Si_7Al)O_{22}(OH)_2$

A very common calcium–magnesium–iron aluminosilicate hydroxide that forms a series with ferrohornblende and is found in a wide range of igneous and metamorphic rocks, including gabbros, diorites, granodiorites, granites, amphibolites and schists. Where possible, identification is based on the 1997 classification of the amphibole group (Leake et al. 1997a, Mogessie et al. 2004).

England

Cornwall Magnesiohornblende occurs with edenite in amphibolitic metabasites (Landewednack Schists) and in the mafic part of the Kennack Gneiss at Kennack Sands and its environs, Lizard Peninsula (Sandeman et al. 1995).

Cumbria Greenish-brown magnesiohornblende is a hydrothermal or auto-metasomatic alteration product of pyroxene in microdiorite at Hope Beck, SSW of Scawgill Bridge, 4 km W of Braithwaite (Fortey et al. 1994).

Lancashire Erratic blocks of altered gabbro from a gravel pit at Burscough, SE of Southport, contain three species of amphibole: edenite, kaersutite and magnesiohornblende (D. I. Green specimen, Tindle unpub. EMP data).

Wales

Dyfed Green needle-like crystals of magnesiohornblende are reported from the Llech Dafad intrusion to the NE of Aberbach (Bevins 1994), but the data are unpublished. Magnesiohornblende from gabbros of the St David's Head intrusion, mentioned in Bevins (1994) and published in Bevins et al. (1994), reclassify as edenite and pargasite.

Gwynedd Magnesiohornblende occurs in a cumulo-phyric basalt lava from the Eglwys Rhobell Formation, Rhobell Volcanic Complex, Rhobell Fawr area, near Dolgellau (Kokelaar 1986), and is also noted (on the basis of unpublished data) from basic rocks of the Sarn igneous complex on the Lleyn Peninsula (Bevins 1994).

Powys Tourmaline-bearing amphibolitized gabbro from the Stanner–Hanter Hill igneous complex contains interstitial patches of brown magnesiohornblende associated with actinolite (NMW specimen, Tindle unpub. EMP data).

Scotland

Central So-called hornblende schists from Aberfoyle Forest, near Balmaha, Loch Lomond (close to the Highland Boundary fault), contain acicular green edenite, magnesiohornblende, and lesser amounts of pargasite and tschermakite (Henderson et al. 1983).

Dumfries & Galloway Slender prismatic green amphibole, sometimes reaching 35 mm in length, from a pegmatitic patch of basic rock hybridized by granite at Cairnsmore Gairy, Carsphairn (Deer 1937), is magnesiohornblende. Within the Loch Doon granite complex (which extends across the border into Strathclyde), green magnesiohornblende is perhaps the most common and widespread amphibole species and is found in appinite, diorite, quartz diorite, granodiorite, tonalite, granite and microgranular enclaves (Tindle 1982, Tindle & Pearce 1983). Boreholes in aureole rocks to the SW of the Loch Doon complex at Glenhead Burn, near Loch Dee, pass through metasomatic zones in which green magnesiohornblende (described as actinolite) is developed (Leake et al. 1981).

Grampian A small Moinian eclogitic amphibolite, 4 km S of Tomatin, contains clinopyroxene–plagioclase symplectites, the latter partially replaced by magnesiohornblende (Baker 1986). Amphibolite-facies regional metamorphism of metabasalts and ultrabasic rocks from the Dalradian Blackwater Formation, near Cabrach, has produced actinolite, magnesiohornblende and tremolite as alteration products of primary clinopyroxene (Macdonald et al. 2005). Magnesiohornblende has also been found in several intrusions in this region: Boganclough gabbro (Wadsworth 1988), Glen Gairn quartz diorite and granodiorite, Mount Battock granodiorite, Lochnagar granodiorite, Hill of Fare diorite and granodiorite, and a pyroxene diorite near Peatfold, Glenbuchat (Tindle unpub. EMP data).

Highland The petrography of metasomatized (fenitized) quartzites at Four Burns and Loyne, near Loch Borralan, Assynt, shows green actinolite, edenite and magnesiohornblende developed in the early stages of metasomatism and this was later replaced by blue varieties that have magnesioriebeckite, richterite and winchite compositions. The nearby Borralan syenite complex was the source of the strongly alkaline fluids required to produce fenitization of these rocks (Woolley et al. 1972). In the Ballachulish granite complex, late-magmatic pyroxene-consuming reactions have produced diorites containing actinolite, edenite and magnesiohornblende (Weiss & Troll 1989). Within the same complex, a granodiorite from Beinn a' Bheithir and an amphibole-rich corona around a quartzite xenolith both contain magnesiohornblende (Muir 1953). Green crystals of magnesiohornblende occur in adamellite, diorite and syenite from the Glenelg–Ratagain igneous complex (Nicholls 1951b). Magnesiohornblende has also been found in several other intrusions in this region: in appinite and diorite from the Ben Nevis diorite–granite complex (Haslam 1968, 1970); in adamellite, quartz diorite and syenitic hybrids from the Glenelg–Ratagain complex (Nicholls 1951b); and in tonalite and microgranular enclaves within the Strontian granite complex (Castro & Stephens 1992). Green magnesiohornblende occurs with biotite, epidote, plagioclase feldspar and garnet in intermediate gneisses of Lewisian age at Camas Baravaig on the Isle of Skye (Matthews & Cheeney 1968) and in a Lewisian biotite–hornblende gneiss at Mallaigmore in North Morar (Lambert 1963).

Foliated amphibolite xenoliths at Meall Mheinnidh, to the NE of Loch Maree, contain magnesiohornblende and represent metamorphosed and deformed inclusions of anorthosite from a body of magma cogenetic with Lewisian metabasalts of the Loch Maree Group (Cattell & Williams 1988). Delicate fibres of pale-green magnesiohornblende, arranged in tufts and reaching 12 mm in length, are associated with calcite, edenite and pyrrhotite in granular limestone from quarries above Milltown, Glen Urquhart (Heddle 1878c).

Strathclyde Post-obduction dyke-like doleritic–gabbroic (metadolerite) intrusions associated with the Girvan–Ballantrae ophiolite complex contain actinolite, ferrohornblende and magnesiohornblende that grew during low-grade metamorphism of these rocks (Holub et al. 1984). Within the Kilmelford igneous complex, magnesiohornblende is reported from four separate diorite intrusions at An Cnap, Ceann Mòr, Meall Mòr and Tom Soilleir. Elsewhere in the complex, actinolite and pargasite are also found (Zhou 1987). Pale-green magnesiohornblende is also reported from the gabbro of the Garabal Hill–Glen Fyne igneous complex, W of the head of Loch Lomond (Nockolds & Mitchell 1946). Amphibole in the Etive diorite–granodiorite–granite complex is all magnesiohornblende (Moazzen & Droop 2005).

Tayside An investigation of amphibole compositions from Dalradian Argyll and Southern Highland Group metabasites around and to the NE of Aberfeldy (centred on the classic area of Barrovian metamorphic zones that developed during the Grampian orogeny) shows that magnesiohornblende is distributed throughout the biotite, garnet, kyanite, sillimanite and andalusite zones (Zenk & Schulz 2004). Appinites, and schist xenoliths derived from Dalradian epidiorite country rocks, that are

Two prismatic sections and a twinned basal section of magnesiohornblende in granodiorite from the Loch Doon granite complex. Carrick Lane, south west of Loch Doon, Strathclyde, Scotland; width 1.2 mm; viewed in PPL (upper) and XPL (lower). (© A. G. Tindle)

associated with diorite of the Glen Tilt igneous complex, E of Blair Atholl contain magnesiohornblende (Deer 1938, 1953).

Western Isles Within the Lewisian complex of South Harris, magnesiohornblende occurs in a plagioclase amphibolite near Loch Allityer (Dearnley 1963), and parvo-magnesiohornblende occurs in a layered alpine-type peridotite from the Langavat Valley (Livingstone 1976a). Magnesiohornblende forms zoned crystals with actinolite and edenite in veins cutting the Western gabbro on St Kilda (Hirta) (Harding et al. 1984), but the data are unpublished.

Shetland Gabbros to the N of Sandwick Bay (on the east coast of Unst) belonging to the Shetland ophiolite complex, contain actinolite and magnesiohornblende formed by the replacement of clinopyroxene (Maynard 1993). Green fibrous, and brown, magnesiohornblende in gabbro from the same area are also described by Prichard (1985).

Ireland

Galway Magnesiohornblende is reported from the Galway granite (Leake & Ahmed Said 1994) and from Dalradian garnetiferous striped amphibolites in Connemara (Leake 1972). The predominantly ultrabasic Currywongaun–Doughruagh intrusion in Connemara contains anthophyllite- and magnesiohornblende-rich reaction borders in orthopyroxenite metasomatized by acid pegmatite veins (Kanaris-Sotiriou & Angus 1979). The Doughruagh mass consists of quartz-rich gneisses in which magnesiohornblende or cummingtonite and tschermakite are found as alteration products of primary igneous species (Angus & Kanaris-Sotiriou 1982). Amphibolite from the Ballyconneely mass to the N of Roundstone also contains magnesiohornblende (Benjamin 1968).

Offshore territories

Guernsey Injection of granitic pipe intrusions into dioritic host magma within the late Precambrian Northern igneous complex at Port Soif, northeast Guernsey, provides evidence for magma mingling. Around the margins of the pipes, well formed zoned crystals of green to pale-green actinolite, magnesiohornblende and pargasite grew in response to sudden and marked changes in magma conditions during this event (D'Lemos 1992). Within the same complex, Bishop & French (1982) report green-to-brown magnesiohornblende in pegmatitic pockets in gabbro and associated diorites at Beaucette, and in the Chouet granodiorite.

Jersey Late Precambrian Cadomian layered diorites, associated with gabbro on the southeast coast at Le Nez, contain a range of amphiboles considered to be secondary after pyroxene and they include actinolite, edenite, kaersutite, magnesiohastingsite, magnesiohornblende and pargasite (Bishop & Key 1983). Magnesiohornblende is one of many amphibole species described from an igneous complex (olivine-gabbro to aplogranite compositions) at Sorel Point in north Jersey (Salmon 1991), but the actual data are unpublished.

magnesiokatophorite

AMPHIBOLE GROUP

monoclinic $Na(CaNa)(Mg_4(Al,Fe^{3+})(Si_7Al)O_{22}(OH)_2$

An uncommon sodium–calcium–magnesium–iron aluminosilicate hydroxide that forms a series with katophorite and is found in alkali igneous rocks. Zirconium-rich magnesiokatophorite occurs within fine syenite in the differentiated alkali olivine–dolerite intrusion at Cnoc Rhaonastil, Isle of Islay, Strathclyde, where alumino-ferropargasite, arfvedsonite, ferro-edenite, ferrokaersutite, hastingsite and kaersutite are associated amphibole group members (Preston et al. 2000b).

Magnesiokatophorite is the new name given to an amphibole from early Cambrian blueschists at the Marquis of Anglesey's monument, just E of Llanfairpwllgwyngyll, Anglesey, Wales (Leake et al. 1997a, Mogessie et al. 2004). The single analysis had been referred to as crossite by Holgate (1951). The mineral was renamed calcio-anophorite by Horák & Gibbons (1986), who suggested that it possibly represents a contaminated analysis.

magnesioriebeckite

AMPHIBOLE GROUP

monoclinic $\square Na_2(Mg_3,Fe^{3+}_2)Si_8O_{22}(OH)_2$

A relatively common sodium–magnesium–iron silicate hydroxide that forms a series with riebeckite and is found in the Pendennis peralkaline minette, a minor intrusion cutting Devonian slates on the foreshore of Pendennis Point, Falmouth, Cornwall. Amphibole is only a minor constituent of the rock, and crystals are compositionally heterogeneous with a short prismatic habit and some have fibrous core regions (Hall 1982). Applying the latest amphibole classification to Hall's data indicates that magnesioriebeckite occurs with magnesio-arfvedsonite, potassic-magnesio-arfvedsonite and potassic-ferro-eckermannite (Leake et al. 1997a, Mogessie et al. 2004).

In Scotland, magnesioriebeckite occurs in metasomatized (fenitized) quartzites at Four Burns and Loyne, near Loch Borralan, Assynt, Highland Region. The nearby Borralan syenite complex was the source of the strongly alkaline fluids required to produce fenitization of the surrounding rocks. The petrography of these rocks shows green actinolite, edenite and magnesiohornblende developed in the early stages of metasomatism, and this was later replaced by blue varieties that have magnesioriebeckite, richterite and winchite compositions (Woolley et al. 1972). Magnesioriebeckite in a Lower Carboniferous trachyte from the Cuff Hill Plantation, 3 km NE of Beith, Strathclyde (Herriot 1968), requires confirmation.

The magnesioriebeckite forming blue felted aggregates with aegirine and hematite in joints and shears in the Inverness area, Highland Region (Macpherson & Livingstone 1982), was described first as abriachanite (Heddle 1879c) and then as crocidolite (Heddle 1901). Fenitization of Moinian metasediments and part of the Abriachan granite, near Loch Ness, and of metasediments around Dochfour, SW of Inverness, has produced blue fibrous magnesioriebeckite as overgrowths on hornblende or in clusters within quartz (Garson et al. 1984).

magnesiosadanagaite

AMPHIBOLE GROUP

monoclinic $NaCa_2[Mg_3(Fe^{3+},Al)_2]Si_5Al_3O_{22}(OH)_2$

A rare sodium–calcium–magnesium–iron aluminosilicate hydroxide that forms a series with sadanagaite and is found in skarn deposits near the western margin of the Land's End granite at the Crowns and Gryll's Bunny, Botallack, St Just, Cornwall.

Amphibole-rich veins cutting metapelite and metabasalt host rocks contain amphibole with compositions transitional between pargasite and sadanagaite. Two analyses plot in the magnesiosadanagaite field (van Marcke de Lummen 1985). Magnesiosadanagaite is considered to be a potassium-rich species by Anthony et al. (1995), but a sodium-rich species by Leake et al. (1997a).

magnesiotarmite

AMPHIBOLE GROUP

monoclinic $Na(CaNa)[Mg_3(Al,Fe^{3+})](Si_6Al_2)O_{22}(OH)_2$

An uncommon sodium–calcium–magnesium–iron aluminosilicate hydroxide that forms a series with taramite. A pyroxenite nodule found in a Carboniferous–Permian volcanic vent near Elie, Fife, Scotland, contains rare green magnesiotaramite previously identified as edenite–ferroan pargasite (Chapman 1976).

Magnesiotaramite has also been described from a metamorphosed spinel-lherzolite lens within Rodel–Leverburgh metasediments at the head of Glen Rodel, South Harris, Western Isles. Associated species are augite and enstatite, with accessory hercynite and magnetite (Livingstone 1976c, Macpherson & Livingstone 1982). According to the amphibole classification of Leake et al. (1997a) and Mogessie et al. (2004), this composition should be referred to as magnesiohastingsite.

magnesite

trigonal $MgCO_3$

A common magnesium carbonate that forms series with gaspéite and with siderite, and is found in marine evaporite deposits, in serpentinized ultramafic rocks and as a replacement of dolomite in contact-metamorphosed limestones.

Magnesite is a widespread major or minor primary constituent in the Permian evaporites of the Yorkshire potash field. It is the normal carbonate of the Upper and Middle Evaporite beds, and the only carbonate of the zones carrying polyhalite and kieserite in the Lower Evaporite bed (Stewart 1965). In the Lower Evaporite bed of the Eskdale No. 2 borehole at Aislaby, Eskdale, near Whitby, Yorkshire, magnesite forms platy crystals that reach 1 cm in width in coarse anhydrite–halite rocks. The plates often have ragged edges and sometimes form radiating bunches. Magnesite is always associated with anhydrite and it replaces it in coarse anhydrite–halite–magnesite rocks (Stewart 1949). In the Middle Evaporite bed encountered in this borehole, magnesite forms small tabular crystals up to 0.5 mm wide. It is mainly found in interstitial areas, but some isolated crystals or crystal groups occur enclosed in halite and it is particularly abundant near the top of the halite zone, where it is associated with anhydritic clay (Stewart 1951a). In the Upper Evaporite bed of the Eskdale No. 2 borehole, magnesite forms tiny rounded granules up to 0.01 mm in diameter, aggregated together or disseminated through anhydrite (Stewart 1951b). Magnesite is also reported from the Lower, Middle and Upper Evaporite beds and from the Carnallitic Marl from the Eskdale No. 3 borehole at Sleights, Eskdale (Armstrong et al. 1951). In the Fordon F.3 borehole, Humberside (13 km S of Scarborough, Yorkshire), the Lower Evaporites contain a bed consisting of 66% anhydrite and 33% polyhalite, with minor amounts of glauberite, halite, magnesite, pyrite, talc and bituminous material (Stewart 1963). At Boulby potash mine, Loftus, Cleveland (but still part of the Yorkshire potash field), colourless magnesite occurs at the base of the Boulby Potash in a band known locally as the borate nodule bed. It is associated with spectacular sky-blue boracite and reddish hilgardite (Green & Freier 1994, 1996). Upper Permian evaporites also occur in Cumbria, where interstitial magnesite is reported from the Sandwith Anhydrite unit of the St Bees Evaporites. Dolomite and anhydrite are the major associated species (Arthurton & Hemingway 1972).

Magnesite is a minor but important species contributing to the formation of spherulitic ironstones in certain beds of the Carboniferous Millstone Grit and Coal Measures. The greatest development is at Snail Green borehole, Rombalds Moor, Kinderscout, Yorkshire, where dark-grey shaly fireclay–mudstone contains patches and stringers of siderite–(calcite–magnesite) nodules that appear to have grown from siderite-rich gel during diagenesis (Deans 1934).

In Cornwall, magnesite occurs in the pre-Devonian peridotite of the Lizard Peninsula in the vicinity of the Rill, Kynance Cliff. Serpentinization of original clinopyroxene, enstatite, olivine, plagioclase and spinel rocks has resulted in the formation of chlorite, chrysotile, mackinawite, magnesite, magnetite, talc and tremolite (Clark 1967). On the foreshore of Pendennis Point, Falmouth, magnesite occurs in a peralkaline minette dyke cutting the Devonian slates of the Gramscatho Beds. The rock is rich in primary igneous carbonate, with both dolomite and magnesite present in moderate amounts. Magnesite forms large irregular aggregates with no euhedral boundaries, and its composition is highly ferroan. Apatite, biotite, quartz and sanidine are the other major species present, with arfvedsonite, hedenbergite, hematite, ilmenite, pyrite, riebeckite, rutile and unidentified copper sulphides forming accessory components (Hall 1982).

Magnesite is thought to occur in some of the carbonates in serpentinites and calc-silicate rocks of the Mona Complex on Anglesey, Wales (Greenly 1919), but this requires verification. Magnesite is also noted in association with barite and smithsonite from Ochrwyth, near Risca, Gwent (Bevins 1994), but no further details are given.

At Creag Bhocan, SSW of Tyndrum, Central Region, Scotland, the basal 50 m of the Dalradian Ben Lui Schist contains chloritic or talcose bands up to 4 m thick, and locally coarse-grain lenses and pods of dolomite and magnesite. The rocks also contain minor amounts of chalcopyrite, chromite, magnetite, chromium-rich mica, pentlandite and pyrrhotite, and probably represent altered ultramafic rocks (Fortey & Smith 1986). Scott et al. (1988) note magnesite in chloritic schists from the Ben Challum Horizon at Creag Bhocan.

At Corrycharmaig, Glen Lochay, NW of Killin, Tayside Region, a small body of Middle Dalradian ultramafic rocks has been metamorphosed and altered to produce a magnetite–talc rock, in which magnesite (breunnerite) forms coarse granular interlocking crystals containing abundant antigorite flakes (Guppy 1956, Hawson & Hall 1987). Some samples from Corrycharmaig contain over 60% magnesite (Harrison 1985). Within the ultramafic rocks, accessory amounts of galena, gersdorffite, irarsite, laurite, five native metals (gold, iridium, osmium, ruthenium, silver) and unnamed alloys of As, Co, Fe, Ni, Os, Pd, Pt, Ru, Sb and Sn occur either as inclusions (up to 40 μm across) in chromite or associated with the nickel sulphides heazlewoodite and nickeline (usually ≤ 2 μm). It is most likely that the intrusion represents a fragment of oceanic crust (Power & Pirrie 2000).

In the Lewisian sequence found in the Moine Nappe, near Ard Ghunel and Camas Baravaig on the south coast of the Isle of Skye, Highland Region, a serpentinite body contains large brown spots of magnesite (breunnerite) up to 2.5 cm across, associated with chlorite, magnetite and talc (Matthews 1967).

Pale-brown pearly rhombohedra of magnesite occur in a vein with brucite (nemalite), chlorite, magnetite and talc at North Cross Geo, Haroldswick, Unst, Shetland (Heddle 1901). This is possibly the same location (Taing of Norwick) noted for breunnerite by Greg & Lettsom (1858). Heddle (1878a) notes magnesite from Quin Geo on the NE of Norwick Bay. At Queyhouse quarry to the N of Baltasound, Unst, a talc–magnesite deposit at the northern margin of the Unst serpentinite mass was a major producer of talc (Dunham et al. 1978, Smith et al. 1992). Talc–carbonate rocks on Unst are found all along the basal thrust of the Shetland ophiolite and are thought to have been harzburgites prior to serpentinization. They contain accessory amounts of many sulphide and oxide species, including chalcopyrite, cobaltite, ferritchromit, mackinawite, magnetite, marcasite, nickeline, pentlandite, pyrite, pyrrhotite, valleriite and violarite. At Cliff, in the Baltasound–Haroldswick area, magnesite contains inclusions of pyrrhotite, and rhombs of magnesite up to 50 μm across are intergrown with cm-wide veins of antigorite and fine-grain magnetite that crosscut talc–carbonate rocks (Maynard 1993). Extensive talc–magnesite deposits occur in a belt of metamorphosed basic and ultrabasic rocks on Mainland, Shetland, at Cunningsburgh, 13 km SSW of Lerwick. Here, magnesite is fairly equigranular, often forming grains 0.15 mm across, but occasionally up to 5 mm (Bain et al. 1971, Smith et al. 1992).

magnesium-diopside

A variety of clinoenstatite, pyroxene group; also used to describe pigeonite.

magnetite

SPINEL GROUP

monoclinic $Fe^{2+}Fe^{3+}_2O_4$

A very common iron oxide that forms series with jacobsite and with magnesioferrite, and is found as an accessory mineral in many igneous and metamorphic rocks. In these rocks, magmatic segregation or contact metamorphism can produce economic deposits. Magnetite is found extensively in BIF rocks and in detrital deposits.

England

Cornwall Massive crystallized magnetite occurs in bands and veins in the cliffs around Botallack and in the Crowns Rock, St Just, often associated with axinite, epidote and garnet (Collins 1892, Bancroft & Weller 1993). At Gryll's Bunny, near Botallack, flat-lying tourmaline-rich skarn deposits are overlain by garnet–magnetite hornfels derived from basic igneous rocks metasomatized by the Land's End granite (Jackson 1974, Hall & Jackson 1975, van Marcke de Lummen 1985). At South Terras mine, St Stephen-in-Brannel, magnetite was recovered from a lode that also contained arsenopyrite, chalcopyrite, sphalerite, tourmaline and unnamed silver-bearing minerals (Dewey 1925, Dines 1956, Slater & Highley 1976). Lustrous black octahedral crystals of magnetite up to 0.5 mm on edge occur in rhodonite-bearing matrix at Greystone quarry, Lezant, near Launceston (Elton et al. 2004). Magnetite has been identified in serpentinized peridotite belonging to the Lizard ophiolite complex at Pol Cornick, near Mullion (Clark 1970a), and the Rill, Kynance Cove (Clark 1967).

Other occurrences include: Barras Nose, Tintagel (Basta 1957); Dean quarry, St Keverne; Polyphant and Trewen quarries, South Petherwin; Tregurtha Downs mine, St Hilary; Tretoil mine, Lanivet (all Golley & Williams 1995); Fowey Consols mine, Tywardreath, St Blazey (Lewis 1997); detrital in the Fowey Estuary (Pirrie et al. 2002); Ponsantooth, near Penryn, and Rosudgeon mine, near Perranuthnoe (Robson 1949b); Treburland mine, 2 km SE of Fivelanes, Altarnun, Bodmin Moor (Russell 1946d); Treluswell, 2 km NW of Penryn and Wheal Harmony, Redruth (Greg & Lettsom 1858); Wheal Jane, Kea, near Truro (Collins 1892); and in accessory amounts within phonolite at Wolf Rock, 15 km from Gwennap Head, Land's End (Allport 1874, Harrison et al. 1977).

Cumbria The Coniston mines are well known for their chalcopyrite–pyrite–(galena–sphalerite) ore found in veins cutting Borrowdale Volcanic Series rocks; however, the primary mineralization appears to be zoned and at Bonser vein an increasing amount of magnetite was found at depth. Here, magnetite is either massive or it forms bladed or radiating grains up to 2 mm across, associated with native bismuth, bismuthinite, joséite and laitakarite (Kendall 1884, Dagger 1977, Stanley & Vaughan 1982b). Similar magnetite occurs at Long Crag mine in the Greenburn Valley, 3 km N of Bonser vein (Stanley 1979, Stanley & Vaughan 1982b). Massive magnetite has also been found at Birker Force, Eskdale (Young 1987a). Other occurrences are mostly in accessory amounts and they include: Armathwaite tholeiite dyke, Brampton district (Dixon et al. 1926); in dolerite at Burtness Combe, Buttermere (Rastall 1906); Graphite mine, Seathwaite, Borrowdale (Ward 1876, Strens 1962, Mitchell & Ineson 1975); euhedral crystals in a lamprophyre dyke at Grassgarth Height, Haverthwaite (Rose & Dunham 1977); detrital in sandstone at Langhorn quarry, Bigrigg, and Overend quarry, near Whitehaven (Lewis 1931); in metasomatized Borrowdale Volcanic Group rocks at Shap Blue quarry (Fortey 1980, Young et al. 1988, Young et al. 1998); in the Shap granite (Harker & Marr 1891, Rastall & Wilcockson 1915, Firman 1978, Wirth & Green 2002); in slates and gabbros within the aureole of the Skiddaw granite (Eastwood et al. 1968); and in felsite on Watch Hill, Cockermouth (Green 1917).

Derbyshire Magnetite is mainly found as an accessory constituent in basaltic rocks: in toadstones at Ashover, Matlock, Miller's Dale and Winster (Woodward & Mello 1881); in the Ashover tuff (Mello 1876); in dolerites and basalts at Calton Hill, Taddington (Ford et al. 1993); in xenoliths within the Great Rocks Dale dyke, near Buxton (Cope 1933); in olivine dolerite of the Ible Sill, near Matlock (Garnett 1923a); between limestone and shale at Mam Tor, near Castleton (Stokes 1879); and at Millclose mine, Wensley, near Matlock (Garnett 1923b, Ixer 1986).

Devon Hay Tor mine, Ilsington, Dartmoor, worked a fine-grain magnetite ore associated with axinite, garnet, hornblende and pyrite, all interbedded with Carboniferous sandstones and shales (Hall 1868b, Rudler 1905, MacAlister 1909, Dines 1956, Slater & Highley 1976, Durrance & Laming 1982, Scrivener et al. 1987). Other occurrences include: Buckland-in-the-Moor, near Ashburton, and veins on Lundy island (Hall 1868b); in the Dartmoor granite (Brammall & Harwood 1925); and in an analcime-bearing lamproite at Holmead farm, Loxbeare (Velde 1971b).

Durham Fine-grain dull masses of magnetite–nickeline ore containing chalcopyrite, galena, gersdorffite, pyrite, pyrrhotite, sphalerite and ullmannite occur at Lady's Rake mine, 6 km NW of Langdon Beck, Teesdale. They are probably part of a skarn deposit related to the intrusion of the Whin Sill (Young et al. 1985a, Green & Briscoe 2002). Traces of titanium-rich magnetite (titanomagnetite), ilmenite and ilvaite are reported from a pegmatitic facies of the Whin Sill at Forcegarth quarry, High Force, Teesdale (Ixer et al. 2005b). Magnetite is one of the alteration products of pyrrhotite in quartz veins cutting the Weardale granite at Rookhope (Brown et al. 1987a).

Gloucestershire Alluvial magnetite (possibly associated with gold) has been found in Glasshouse Brook, Rocks Wood, 4 km S of Newent (Bick 1986a).

Leicestershire An investigation of the Mountsorrel granodiorite and associated igneous rocks found accessory magnetite at Brazil Wood, Cocklow quarry, Kinchley, the Main Mountsorrel quarry, Rubbish Tip quarry and Swithland reservoir (Taylor 1934).

Northumberland Magnetite and ilmenite are accessory minerals in the Cheviot granite (Haslam 1986). Magnetite–garnet–pyrrhotite–sulphide-bearing calc-silicate rocks have been found close to the Whin Sill at Ratcheugh quarry, SE of Alnwick (Ixer & Stanley 1987).

Yorkshire Magnetite is one of many heavy minerals found in: the Millstone Grit at Clayton Wood, Crag Hill and Whitehall quarries, Horsforth; Fountains Fell; Grassington Moor; Great Whernside, Kettlewell; Netherwood Plantation quarry, Silsden; Old Cote Moor, Arncliffe; Penyghent; and Steeton Moor (Gilligan 1919). Magnetite veins cut Arenig rocks in the Beckermonds Scar borehole, between Hawes and Kettlewell (Dunham & Wilson 1985). Rare magnetite associated with pyrite and enclosed in halite has been tentatively identified in evaporite deposits sampled in the Eskdale No. 2 borehole at Aislaby (Stewart 1951a). A Jurassic magnetite-bearing oolite was once worked at the West mines, Rosedale; the magnetite is diagenetic and it forms rims on chamosite grains or concentric zones with chamosite, all in a siderite cement (Rudler 1905, Rastall & Hemingway 1949, Hemingway & Riddler 1982).

Wales

Gwynedd Octahedral magnetite crystals up to 5 mm across have been described from chlorite schists of the Precambrian Mona Complex exposed near Rhoscolyn, Holy Island, Anglesey (Greenly 1919, Maltman 1977). Related serpentinites on Holy Island contain up to 3% magnetite and are associated with chromite, ilmenite and zircon (Maltman 1977). Magnetite is abundant in parts of the Rhiw layered intrusion, Lleyn Peninsula (Hawkins 1970, Cattermole 1976), and is also reported from the Parys Mountain copper deposit on Anglesey (Southwood & Bevins 1995). In Snowdonia, veins cutting Ordovician Snowdon Volcanic Group basalts and rhyolitic tuffs at Cwm Tregalan and Shadow Gully (S of Snowdon) contain small euhedral crystals of magnetite associated with hematite, pyrite, quartz and rare cassiterite (Colman & Appleby 1991, Reedman et al. 1985). Magnetite is found with fluorite and pyrite on spoil tips near Llyn Cwellyn, SE of Mynydd Mawr, 10 km NW of Cwm Tregalan (Colman & Appleby 1991). Elsewhere in Snowdonia, magnetite is a constituent of Ordovician magnetite–spessartine schists at Tan-yr-allt, near Nant Peris (Williams 1927, 1930a), and is a minor species in oolitic chamosite mudstones at Pen-yr-allt iron mine, near Llanfrothen, Penrhyndeudraeth (Hallimond 1924). Pulfrey (1933) notes magnetite in chlorite–magnetite oolites at Betws Garmon, Cross Foxes (Cadair Idris) and Pen yr-allt. Similar chamosite pisolitic mudstones from Tyllau Mwn, Drws-y-nant, near Dolgellau, contain magnetite octahedra up to 0.1 mm across (Matthews & Scoon 1964). Additional magnetite-bearing oolites in Snowdonia are noted from Bwlch y Cywion (a saddle on Foel Goch between the Ogwen and Llanberis valleys), Cadair Idris, Llanberis Valley and Rhyd (Trythall 1988).

Scotland

Central A pelitic schist from the base of the Dalradian Ben Lui Schist at Ben Challum, Tyndrum, contains almandine crystals in which exsolved magnetite and inclusion trails of chloritoid, quartz, staurolite and white mica have been found (Brearley & Champness 1986). Magnetite is a minor constituent of a stratabound copper–zinc deposit within chloritic schists and albitic quartzites of Dalradian age at Creag Bhocan, SSW of Tyndrum (Scott et al. 1988).

Dumfries & Galloway Reddish-brown blocks of vuggy magnetite lined with well crystallized quartz have been found on the dumps near Whyte's Cleuch, Wanlockhead (Nicholson & Banks 1987).

Grampian The Dalradian Longshank Gneiss Formation, to the SW of Lochnagar, is a metasedimentary unit of pelites and psammitic gneisses, in which abundant magnetite and localized tourmaline occur. Both are considered to be detrital (Goodman et al. 1990). Primary magmatic ilmenite and magnetite are reported from mineralized sulphide zones at the margins of the Huntly–Knock gabbroic intrusion at Auchencrieve and Littlemill, N of Huntly. Magnetite is also an alteration product of pyrrhotite and pentlandite in the mineralized horizons (Fletcher & Rice 1989).

Highland A pyroxene–garnet (melanite) rock containing approximately 20% magnetite (titanomagnetite) was found in the Allt Bad na h'Aclaise stream between Loch Urigill and Loch Borralan, Assynt (close to the Loch Borralan syenite complex). Drilling to the S of Loch Urigill revealed related pyroxenites and syenite veins with a magnetite content 5–90%; grainsize is variable from 20 μm to 2 mm (Newman 1971, Notholt et al. 1985, Gunn & Styles 2002). Elsewhere in Assynt, magnetite is described (in lower abundances) from: pyroxenite at Allt a' Mhuilinn, E of Loch Borralan (Shand 1910b); feldspathic granitoid (syenite) at Cnoc na Sròne, Loch Borralan (Heddle 1901); pyroxenite in the Allt Cathair Bhan valley, NE of Loch Ailsh (Shaw et al. 1994, Gunn & Styles 2002); pyroxenite at Bad na h'Achlaise, Ledmore, and biotite diorite from Loch Bad nam Ban, southern end of Loch Inver, 3 km SW of Lochinver (Stumpfl 1961); and ultramafic rocks from An Fharaid Mhòr, SW of Achmelvich, near Lochinver (Droop 1994). High concentrations of magnetite (~20%) are reported from a pyroxene–quartz-bearing ironstone in the first inlet W of the pier on the north shore of Scourie Bay. Magnetite is rimmed with garnet and it occurs in small localized lenses or thin bands up to 0.5 cm wide (Barnicoat & O'Hara 1979). Trondhjemites from the Scourian Complex at nearby Badcall Bay contain magnetite, formed when original titanomagnetite exsolved into large-scale (up to 50 μm wide) ilmenite–magnetite lamellae (Rollinson 1979). Similar ilmenite–magnetite pairs are also described from granulite-facies calc-gneisses at Scourie (Rollinson 1980). On the north coast at Rispol, Loch Erribol, bluish-black magnetite (the finest massive magnetite in Scotland) had been found in a granitic vein. Smaller granitic veins at nearby Kean na Bin, opposite the western end of Eilean Hoan, contain well formed octahedral crystals of magnetite (Heddle 1881). A garnetiferous albite gneiss containing cassiterite–magnetite-rich bands is reported from Carn Chuinneag (Rudler 1905, Gallagher et al. 1971, Dunham et al. 1978). At Easter Ross, magnetite occurs in a Lewisian BIF (a quartz–magnetite schist) around Flowerdale, near Gairloch (Jones et al. 1983, 1987, Williams 1986). Within the Lewisian Glenelg inlier, magnetite approaches 7% by volume in a garnet websterite from Beinn Fhada, near Letterfearn, Loch Duich; it forms small lozenge-shape inclusions within pyroxene (Rawson et al. 2001). On the Isle of Rum, magnetite is one of the common species occurring in oxide–sulphide droplets within chromitites in the Eastern Layered Series of the Rum ultramafic complex (Dunham & Wilkinson 1985). These rocks and associated dykes have been studied in great detail, not least because of associated PGM assemblages they also contain (Butcher et al. 1999, Upton et al. 2002, Power et al. 2000a, 2003). Bodies of magnetite accompanied by diopside–garnet skarn rocks occur for about 3 km along the contact of the Beinn an Dubhaich granite with Durness Limestone, W of Broadford, Isle of Skye. Metasomatism of limestone by the granite produced the sub-economic deposits that also contain bornite, chalcopyrite and pyrite. The best examples are at the No. 1 Prospect, Kilchrist (Cill Chriosd), and at Camas Malag (Harker 1904, Tilley 1949, 1951, Groves 1952, Slater & Highley 1976). Similar irregular magnetite bodies are also known from the Strath district of Skye (Berridge 1969). Other occurrences of magnetite are from: pitchstone on Ardnamurchan (Preston et al. 1998b); the Ballachulish granite complex (Weiss & Troll 1989); granitic veins in a giant erratic boulder at Beinn Bhreac, 2 km E of Tongue (Heddle 1883); garnet–grunerite–pyroxmangite schist at Gleann Beag, Glenelg (Tilley 1937a); dolerite and lamprophyre dykes at the head of Loch Sunart (Gallagher 1963); pegmatite at Raven Rock, 3 km NW of Strathpeffer, and granitic veins in the cliffs of Sgòrr Chaonasaid, Ben Loyal (Heddle 1901); picrite at Suisnish Point, Isle of Skye (Parsons 1976); and titanomagnetite occurs in olivine tholeiite cone sheets, which invade the Cuillin igneous complex, Isle of Skye (Bell et al. 1994).

Lothian Fresh, well formed, lustrous black octahedra of magnetite up to 1.5 mm across are a major component of quartz dolerite at Hilderston silver mine, Bathgate Hills (Meikle 1994). Heddle (1901) reports crystals of magnetite in cavities within Carboniferous Limestone at Kirkton quarry, Bathgate.

Strathclyde Diopside megacrysts containing abundant inclusions of magnetite occur in the Black Rock basanite vent, 2 km SSW of Fairlie, Hunterston Sands (Alexander et al. 1986). Altered Carboniferous tuff at Loanhead quarry, near Beith, contains small masses of magnetite octahedra associated with calcite, prehnite, mottramite and vanadinite (Meikle 1989b). Magnetite has been found as euhedral to subhedral crystals up to 5 mm across associated with epidote in cavities at the molybdenite–quartz vein and skarn locality of Coire Buidhe, Glen Creran (Ingram et al. 1993). At Creag an Fheidh on the Isle of Arran, a zone of magnetite occurs between andradite exoskarn and metasomatized agglomerate (endoskarn) related to the collapsed caldera of the Central ring complex (Cressey 1987). Near the summit of Sidhean Sluaigh, 3 km SW of Strachur, a series of Dalradian chlorite–muscovite (phengite) phyllites have been thermally metamorphosed adjacent to a dolerite plug. The resulting emery-like rocks contain magnetite together with cordierite, corundum, hercynite, mullite, pseudobrookite and sanidine (Smith 1965, Worden et al. 1987). Magnetite is an exsolved phase in garnets within Moinian pelitic gneisses on the Ross of Mull (Brearley & Champness 1986). Other occurrences include: in gneiss at Cnoc an Each, Ben Vorlich, near Loch Lomond (Heddle 1901); in quartz veinlets cutting the Carn Dearg porphyrite, NW of Kilbride (Zhou 1987); large skeletal crystals in a dolerite (crinanite) dyke near Deucheran Hill, Carradale Glen, Kintyre Peninsula (McCallien 1932); chromium-rich magnetite in ultrabasic rocks belonging to the Girvan–Ballantrae ophiolite complex, near Knockormal (Bloxam & Allen 1959); titanomagnetite in the Tertiary Dippin (crinanite) Sill, southeast Isle of Arran (Gibb & Henderson 1978b); titanomagnetite in pitchstone at Glen Shurig and in the Judd dykes, near Blackwaterfoot, Isle of Arran (Preston et al. 1998b); in granulite gneiss and pyroxenite xenoliths within a Permian nephelinite dyke at Gribun, Isle of Mull (Upton et al. 1998); titanomagnetite in the Ailsa Craig granite, Firth of Clyde (Harrison et al. 1987); and in a magnetite-rich pyroxene gneiss of Lewisian age between lochs a' Phuill and Bhasapoll, Isle of Tiree (Groves 1952).

Tayside Magnetite is a minor component of stratabound barium–zinc mineralization in Dalradian schists in the Ben Eagach area, near Aberfeldy (Coats et al. 1980). Magnetite occurs with a chromium-rich variety of magnetite and chromite in a talc-carbonate rock that is considered to be an altered part of the Corrycharmaig serpentinite, Glen Lochay, near Killin. These minerals form serrate, rounded subhedral-to-anhedral crystals 0.5–2 mm across and aggregates of crystals up to 1 cm across (Guppy 1956, Harrison 1985).

Western Isles The pod-like Loch a' Sgùrr pegmatite, near Finsbay (Fionnsabhagh), South Harris, is a biotite–magnetite variety typical of many lenses throughout the Langavat Valley (von Knorring & Dearnley 1960a). Magnetite is noted in various units of the Shiant Isles Main Sill, a classic composite alkaline basic sill of Tertiary age (Heddle 1901, Gibb & Henderson 1996), and in the Glen Bay gabbro on St Kilda (Hirta) (Harding et al. 1982, 1984). Heddle (1901) reports magnetite as: flattened octahedra in quartz veins at Macphee's Hill on Mingulay; coarse crystals and "lumps the size of goose eggs" in a granitic vein on Stromay island, near Northton (Taobh Tuath), in the Sound of Harris; small crystals in a vein cutting serpentinite at Klibberness on Scalpay island; and tarnished cubes of magnetite in gneiss at Suenish, near Newtonferry, North Uist.

Shetland A massive band of compact magnetite, surrounded by skarn rocks composed of garnet, epidote and pyroxene at Clothister Hill, Sullom, Mainland, is thought to be a metasomatic deposit associated with a nearby Devonian granite (Guppy 1956, Berridge 1969, Mykura & Phemister 1976, Slater & Highley 1976). It makes an excellent ore and is considered to be the most important occurrence of magnetite in Scotland (Groves 1952, Dunham et al. 1978). Magnetite mantles chromite in magnesite–talc schist at Cunningsburgh, Mainland (Bain et al. 1971). Chlorite schist from Pundy Geo, Point of Fethaland, Mainland, contains octahedra of magnetite up to 1.5 cm on edge (Green & Todd 2001, Livingstone 2002). Magnetite is a common (but accessory) constituent of ultramafic rocks belonging to the Shetland ophiolite complex on the Isles of Fetlar (Oddsta on the northwest coast), Haaf Gruney (a small island off southeast Unst) and, most significantly, on Unst itself, where octahedral crystals up to 25 mm across have been found. Unst localities include: Hill of Saxavord; Kleber Geo, Fiedland Peninsula; Loch of Cliff hillside towards Woodwick; Quin Geo, NE of Northwick Bay; and Sandwick Bay (Greg & Lettsom 1858, Heddle 1878a, 1901, Maynard 1993, Maynard et al. 1997).

Ireland

Antrim Magnetite is a constituent of skarn deposits at the contact between chalk and dolerite at two localities near Larne. At Carneal it occurs with wollastonite in marble and in bredigite–larnite–spinel assemblages (Sabine et al. 1968, 1985, Sabine & Young 1975), and at Scawt Hill it is found in larnite- and melilite-rich rocks, and in nearby metasomatized basalt (Tilley & Harwood 1931, Tilley & Alderman 1934, Kitchen 1985). A similar origin is ascribed to magnetite granules and octahedra found in metamorphosed lithomarge (porcellanite) within a plug of olivine dolerite at Tievebulliagh, near Cushendall (Agrell & Langley 1958). Magnetite is also found at Tieveragh, where cordierite–enstatite–hematite–osumilite–tridymite buchite has been formed where another dolerite plug has fused ashy Old Red Sandstone rocks (Chinner & Dixon 1973). At Brockley, on Rathlin Island, magnetite is found in calcite veinlets cutting agglomerate and as coatings on some joint surfaces (Wilson & Robbie 1966). Rudler (1905) reports magnetite octahedra on a specimen of basalt from the Isle of Muck, near Island Magee, off the east coast of Antrim, and even larger octahedra are noted on a specimen from Island Magee itself.

Cork Round black titanomagnetite grains up to 0.8 cm across form megacrysts in a

Euhedral, partially skeletal crystal of magnetite (light grey) in basalt, showing exsolution lamellae of ilmenite (darker grey); the Gobbins, Co. Antrim, Ireland; width 420 µm; BSE image. (© E. Condliffe, University of Leeds)

pyroclastic pipe at Black Ball Head, near Cahermore. Kaersutite and phlogopite are associated species (Pracht & Kinnaird 1995).

Down Slightly zoned euhedral crystals of magnetite occur in molybdenite–pyrite-bearing granodiorite from Croreagh quarry, Newry (Meighan & Ixer 1980). Opaque minerals (magnetite mostly together with ilmenite and sulphides) are abundant (modally up to 10%) in dolerites and gabbros of the Tertiary central complex of Slieve Gullion (Gamble 1982).

Galway Magnetite occurs with högbomite in a chlorite–corundum–ilmenite xenolith in the Cashel basic intrusion, near Toombeola (McKie 1963). An association of magnetite with högbomite is also recorded from an orthopyroxenite and a cordierite-rich hornfelsed pelitic xenolith in the Currywongaun ultrabasic intrusion, Connemara (Angus & Middleton 1985).

Mayo (and Sligo) Magnetite replaces almandine in oxidized pelite from the aureole of the Easky Lake adamellite, N of Tobercurry, Ox Mountains (Yardley & Long 1981).

Tyrone Porphyroblasts of magnetite up to 2 mm across were discovered in schists during an investigation of the Lack gold deposit, WSW of Omagh (Cliff & Wolfenden 1992).

Wicklow Magnetite zones are reported from the margins of an Ordovician volcanogenic copper–iron deposit at Avoca mine, Avoca (Wheatley 1971a). Ballycoog mine at Ballycoog Upper once worked a chalcopyrite–magnetite lode (Rudler 1905, Cole 1922). McArdle (1993) notes a BIF consisting of laminated magnetite and siderite at Ballycoog. A remarkable lode of magnetite is described from Ballard (Ballycapple) mine, SW of Wicklow town (Cole 1922), and probably represents another BIF. Magnetite and hematite were the main ore species found in green chloritic schist at Ballymoneen and Knocknamohill mines (West Avoca mines), near Ballygahan, whereas at Ballycoog–Moneyteige, near Avoca, magnetite occurs with copper ore and possibly gold (Cole 1922). Primary gold certainly does occur in chalcopyrite–magnetite–pyrite-rich bedrock samples within Ordovician metavolcanics close to the Goldmine River, near Avoca. Again, this is a BIF in which magnetite is the most abundant opaque phase. Magnetite forms aggregates of 10–50 µm-diameter crystals in cm-thick bands or forms the cores of 50–100 µm-long radiating laths intergrown with hematite (Ixer et al. 1990).

Offshore territories

Guernsey Ixer (1992) reports magnetite with chalcopyrite and pyrite in sulphide veinlets cutting the Northern igneous complex at Baubigny, near Sampson, and Bouette quarry, St Peter Port.

Jersey Pegmatites at Le Hocq, Le Nez Point, Portelet and near La Moye House contain crystals of magnetite up to 2 cm across. Granite around La Moye and Corbière Point in southwest Jersey contain magnetite-bearing veins and, where these cut pegmatite, the junction is marked by fine-grain magnetite ore shoots that form masses up 1 kg (Mourant & Warren 1934, Mourant 1961). Vein samples from N of Corbière Point show three generations of magnetite: colour-zoned tabular crystals up to 400 µm long, polygonal crystals up to 150 µm across, and small (5–10 µm) octahedral crystals (Ixer 1980). Magnetite is also present: in diorite at Mourier Bay; in jasper nodules in andesite to the W of Belle Hougue Point; in metamorphic rocks at Côtil Point (Mourant 1961); and in carbonate veins cutting granite at L'Ouaisné Bay and at Le Doüet de la Mer, St Ouën (Ixer 1980, 1992).

mahlmoodite/malhmoodite

monoclinic $Fe^{2+}Zr(PO_4)_2 \cdot 4H_2O$

A very rare hydrated iron–zirconium phosphate found in a coastal exposure of zinc–iron mineralization at Kerriack Cove, between Porthtowan and Portreath, near Redruth, Cornwall. The quartz lode was worked principally for zinc (as part of the Wheal Lushington sett), with chalcopyrite, galena and sphalerite being the common ore species. Near the middle of the cove, a fissure vein (up to 4 cm wide) contains many small vugs in which occasional spherules of mahlmoodite are found. The natural colour of the spherules is creamy-white, although many are brownish from iron-staining. The spherules are up to 0.15 mm in diameter and are concentrically banded, with individual crystals being lath-like, sometimes in short stacks. Mahlmoodite is most commonly associated with sphalerite, rarely with anatase, and may easily be overlooked (Elton & Hooper 1995b).

malachite

monoclinic $Cu_2^{2+}(CO_3)(OH)_2$

A very common copper carbonate hydroxide found in the oxidation zone of many copper deposits, sometimes in such abundance as to become an ore.

England

Avon Malachite is noted with azurite and wulfenite in sulphide veins associated with a podiform replacement deposit of iron and manganese oxides at Coombe Farm quarry, Westbury-on-Trym (Alabaster 1978). It also occurs in barite pebbles on the beach near the Pier hotel, Clevedon (Starkey 1984).

Cheshire Malachite is ubiquitous in copper–lead deposits within the Trias at Alderley Edge ridge, where it forms green disseminations, mainly following bedding planes or botryoidal masses and fibrous crystals growing into voids. Localities include: Engine Vein mine (good specimens), Mottram mine, Stormy Point area, West mine and Wood mine (Greg & Lettsom 1858, Carlon 1979, Ixer & Vaughan 1982, Braithwaite 1994, Lawton 1994, Plant et al. 1999).

Cornwall Malachite has been found at Botallack sett, a series of mines in the aureole of the Land's End granite, near St Just, and including Botallack mine, Wheal Cock, Wheal Edward and Wheal Owles (Greg & Lettsom 1858, Bancroft & Weller 1993). It forms patches and clumps of acicular crystals associated with uranium–lead mineralization at Loe Warren Zawn, Botallack (Elton & Hooper 1995a,c). Malachite is a common species at Penberthy Croft mine, St Hilary, where it forms attractive deep-green botryoidal crusts up to 10 mm across associated with cerussite, chrysocolla, goethite, linarite, mimetite–pyromorphite, pseudomalachite, quartz and occasionally cyanotrichite (Camm & Merry 1991b, Betterton et al. 2000). The association of malachite with rosasite and native silver from Penberthy Croft (Kingsbury & Hartley 1957c) requires further confirmation. At Countybridge quarry, Goonhilly Downs, Lizard Peninsula, light to dark-green malachite is associated with cuprite and also appears to form lustrous green cubic crystals – but these are actually an intimate mixture of malachite and lizardite pseudomorphing cuprite (Holyer 1996). Other occurrences of malachite include: Binner Downs, near Leedstown; Perran St George United mine, SW of Perranporth; Phoenix United mines, Linkinhorne; Providence mine, near St Ives; Treveddoe, NW of Warleggon; West Godolphin mine, St Hilary; Wheal Music, near Porth Towan; Wheal Sisters, St Neot (all Dines 1956); Carharrack mine, Camborne; Gunheath china-clay pit, St Austell; Old Gunnislake mine, Calstock; Polzeath Consols, St Minver (all Golley & Williams 1995); Carn Brea mines, near Redruth (Robson 1945a, Rust 1982b); Dolcoath mine, Camborne; Wheal Buller and Wheal Harmony, Redruth (all Collins 1892); Fowey Consols mine, Tywardreath, St Blazey (Lewis 1997); Greystone quarry, Lezant, near Launceston (Lloyd & Barstow 1983, Elton et al. 2004); Lill Cove mine, Trebarwith Strand, near Tintagel (Sharpe 1990); Mark Valley mine, near Liskeard (Collins 1911a, Rust 1982b); Newporth beach, 2 km SW of Falmouth (Dean 1982, Dean et al. 1983); Pentire (Glaze) mine, near New Polzeath, St Minver (Collins 1892, Clayton et al. 1990); Port Quin mine (Russell 1944, 1949b); Prince George mine, Gwinear; Wheal Hewas, St Austell; Wheal Unity, Gwennap, near Redruth (Greg & Lettsom 1858); South Caradon mine, St Cleer, near Liskeard (Rudler 1905, Dewey 1925, Dines 1956, Shambrook 1982, Lawrence 1990); South Wheal Frances, Illogan and Wheal Carpenter, Gwinear, near Hayle (Braithwaite & Ryback 1994); Ting Tang mine (Morcom's shaft), Carharrack (Collins 1892, Stephens 1940a, Dines 1956, Ball et al. 1982, Bruce & Aubrey-Jones 1998); Tolpetherwin quarry, Two Bridges, South Petherwin (Russell 1949b); Treburland mine, 2 km SE of Fivelanes, Altarnun, Bodmin Moor (Russell 1946d); Trevinnick mine, St Kew (Fisk et al.

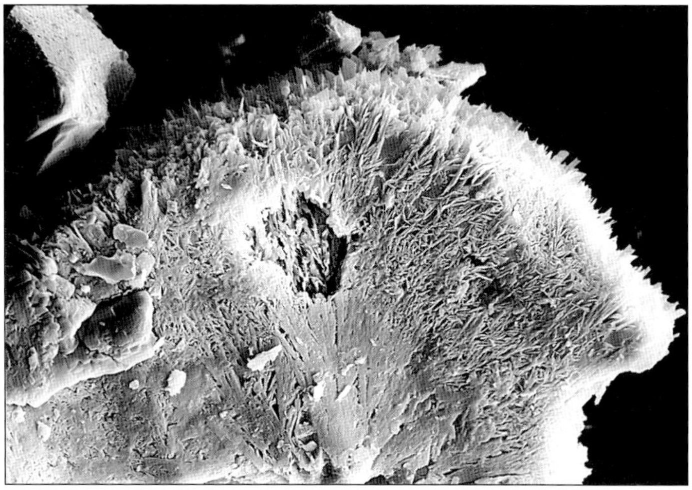

Cross section through a mahlmoodite spherule from Kerriack Cove, between Porthtowan and Portreath, Cornwall; width 160 µm; SEM image. (© N. J. Elton)

Interlocking sprays of malachite to 0.7 mm from Saddleback Old mine, below Foule Crag, Glenderamackin Valley, Cumbria. (© D. I. Green)

1982); Wheal Alfred, Phillack (Rust 1982a); Wheal Basset, Illogan and Wheal Jewel, Gwennap (Embrey & Symes 1987); Wheal Gorland, St Day, 2 km E of Redruth (Greg & Lettsom 1858, Fisher 1988, Rust 1981b); Wheal Leisure, near Perranzabuloe (Rudler 1905); Wheal Muttrell, Gwennap (Claringbull et al. 1959); and Wheal Pendarves (Tryphena lode), Camborne (Barstow & Cooper 1982).

Cumbria *Caldbeck Fells:* Malachite is widespread in small amounts, with fine specimens once found at Roughton Gill, where it formed masses of acicular emerald-green crystals interspersed with needles of cerussite. Rare well formed crystals and botryoidal encrustations are also noted (Greg & Lettsom 1858, Goodchild 1884, Davidson & Thomson 1951, Eastwood 1959, Shaw 1970, Cooper & Stanley 1990). Specimens of malachite from Potts Gill barite mine reported by Kingsbury & Hartley (1957e), and repeated by others (Hartley 1984, Young 1987a), and from Dry Gill and Ingray Gill mines (Hartley 1984), are problematical, as they could potentially be part of the incorrectly provenanced material described by Ryback et al. (1998, 2001). However, malachite does occur at this locality (Davidson & Thomson 1951, Braithwaite & Ryback 1988). Green et al. (2006a) note that, although most malachite specimens in the Kingsbury collection from Driggith and Sandbed mines appear to be genuine, there are two that are most probably fraudulent. Malachite also occurs at: Arm o' Grain (Green et al. 2006b); Balliway Rigg, higher Roughton Gill (Green & Briscoe 2002); Bannerdale mine, Mungrisdale (Ziemba 1975); Brae Fell mine (Eastwood 1921, Hartley 1984, Bridges et al. 2006); Brandy Gill (Davidson & Thomson 1951, Hartley 1984); Carrock End and Hay Gill mines (Greg & Lettsom 1858, Davidson & Thomson 1951, Hartley 1984); Carrock mine, Grainsgill (Davidson & Thomson 1951, Hartley 1984, Young 1987a); Driggith mine (Eastwood 1921, Davidson & Thomson 1951, Hartley 1984, Stanley & Vaughan 1981, Green et al. 2006a); Low Pike, near Potts Gill mine (Young et al. 1990a, Green et al. 2003b); Mexico mine, Todd Gill (Greg & Lettsom 1858, Goodchild 1884, Hartley 1984); Red Gill mine (Greg & Lettsom 1858, Davidson & Thomson 1951, Wolloxall 1981, Hartley 1984, Briscoe 1986a); Sandbed mine (Davidson & Thomson 1951, Hartley 1984, Green et al. 2006a); Silver Gill mine, near Roughton Gill (Davidson & Thomson 1951, Cooper & Stanley 1990, Green et al. 2005b); Thief Gills (Young 1987a); and Wet Smale Gill (Shaw 1970). *Elsewhere in the Lake District and West Cumbria:* Malachite occurs in an oxidation assemblage with azurite, bornite, covellite and chrysocolla cementing fractured quartz at Black Scar workings, Levers Water, Coniston (Firman 1978, Stanley 1979). Crusts and pockets of brochantite, chrysocolla and malachite occur in chalcopyrite–ferberite–scheelite-bearing quartz veins in granodiorite at Buckbarrow Beck, Corney Fell, Waberthwaite (Young et al. 1986, 1991b); Other occurrences include: Anty Cross mine, near Dalton-in-Furness (Smith 1924); Braithwaite, near Keswick (Rudler 1905); Brundholme mine, Glenderaterra Valley (Eastwood 1921); Burtness Wood, Buttermere (Clark 1963); Coniston and Threlkeld mines (Davidson & Thomson 1951); Crossgill mine, Frizington (Shepherd 1973); Crowgarth mine, Cleator Moor (Trotter 1945, Davidson & Thomson 1951); Dale Head mine, Newlands Valley (Goodchild 1884, Davidson & Thomson 1951); Easy Gully on Dow Crags, 2 km SW of Paddy End mine, Coniston (Fleming 2000); Millrigg mine, Staveley Head Fell, Staveley (Lawrence et al. 1986); Paddy End mine, Coniston, Eskdale (Young & Johnson 1985); Parkhouse quarry, Johnby (Green & Briscoe 2002); Rigg Head quarry, Borrowdale (Green et al. 2006c); Seathwaite mine, Seathwaite Tarn, Coniston (Stanley & Criddle 1979); Sea Wood mine, Bardsea, Ulverston (Young 1987a); Shap Blue quarry (Young et al. 1988); Shap granite quarry (Harker & Marr 1891, Russell 1925, Wirth & Green 2002); Staveley, near Windermere (Greg & Lettsom 1858, Goodchild 1884); and Yewthwaite mine, Newlands Valley (Dewey & Eastwood 1925). A report of malachite from Brandlehow mine, Derwent Water (Young 1987a), is based on a specimen in the Kingsbury collection at the NHM and its provenance requires confirmation. *Northern Pennine orefield:* Malachite, like azurite is widespread throughout the orefield, but generally in small amounts. However, it is locally abundant: as earthy and crystalline crusts in the copper-rich Cornriggs vein, Alston Moor; on the Copper Slitts vein, Windy Brow and at nearby Allen's Cleugh, SSE of Garrigill (Young et al. 1985a, Dunham 1990, Bridges & Young 1998). Malachite is also noted from: Brownley Hill mine, Nenthead, near Alston (Livingstone 1991, Green et al. 2000); Higher Longrigg mine, S of Cote Garth, Hartley Birkett, near Kirkby Stephen (Braithwaite 1988); High Stennerskeugh level, Ravenstonedale (Young 1987b); Hilton mine, Scordale, near Appleby (Davidson & Thomson 1951, Winrow 1986a, Bridges & Young 1998); Smallcleugh mine, Nenthead, near Alston (Livingstone 1991); Smittergill Head mine, Melmerby Fell (Young 1985a); Stennerskeugh Clouds, Kirkby Stephen (Starkey 1987b); Tynebottom mine, NW of Garrigill, near Alston (Davidson & Thomson 1951, Braithwaite 1982b, Bridges 1987, Rust 1987a); and Torpenhow, NE of Cockermouth (Eastwood et al. 1968).

Derbyshire Scattered crusts and stains of malachite occur wherever traces of copper ore are found, or where fluorite and calcite have chalcopyrite inclusions. At Snelston mine, near Ashbourne (also referred to as Birchwoodpark), malachite cements calcareous sandstone and forms coatings on chalcopyrite nodules. Beautiful specimens of malachite forming delicate dark-green mammillated drusy crusts were recovered from this mine prior to 1917 (Dewey & Eastwood 1925, Jackman 1996, Cornwell et al. 2001). Other records of malachite include: Goodluck mine, Via Gellia; Jug Holes mine, Masson Hill, Matlock; Long Rake, near Middleton-by-Youlgreave (all Ford et al. 1993); Gregory mine, Ashover (Smith 1982); Hassop Station quarry (Mueller 1951); Matlock and Middleton (Woodward & Mello 1881); Millclose mine, Wensley, near Matlock (Sarjeant 1957); Red Rake, Calver; Treak Cliff Cavern, Castleton; Watersaw Rake, Longstone Edge (all Ford & Sarjeant 1964a); Rutland Cavern, Heights of Abraham, Matlock Bath (Workman 1960, Braithwaite & Ryback 1963a); Whitwell, 19 km SE of Sheffield (Ineson et al. 1972).

Devon Anglesite, cerussite, malachite and quartz infill voids in a gold-bearing carbonate vein at Loddiswell mine, 3 km N of Loddiswell (Stanley et al. 1990b), and, at Hope's Nose, Torquay, malachite replaces klockmannite in a gold-bearing selenide vein (Stanley et al. 1990a, Camm 1995). Near-surface weathering of native copper and earlier diagenetic copper minerals in the Permian Littleham Mudstone Formation at Littleham Cove, Budleigh Salterton, has produced secondary malachite and a range of copper sulphates, oxides, chloride, copper–uranium arsenate and uranium vanadates (Harrison 1975b, Milodowski et al. 2000, 2002). Other occurrences include: Bampfylde mine, NNE of South Molton (Dines 1956, Camm 1995); Bedford United mines (Marquis shaft), Tavistock (Claringbull et al. 1959) – requires confirmation as this is a Kingsbury specimen; Buckfastleigh and Hennock, near Chudleigh (Hall 1868b); Combe Martin mines (Hall 1868b, Dines 1956, Rust 1982a); New East Wheal Russell (South Wheal Crebor), Tavistock (Starkey 1986b); St Mary's Bay, Brixham (Braithwaite & Wilson 2001a); Wheal Emily, Wembury (Russell 1949b).

Durham Malachite is widespread in the Durham section of the Northern Pennine orefield, although only in trace amounts. The small copper veins at Crossgill and Tynehead are exceptions (Davidson & Thomson 1951, Dunham 1990). Relatively common crusts of malachite composed of minute acicular crystals lining cavities in limonitic veinstone are noted from Westernhope Old mine, S of Westgate, SW of Stanhope, Weardale (Bridges & Young 1998). Other occurrences include: Closehouse mine, Middleton-in-Teesdale, Lunedale (Young et al. 1994a); Raisby Hill quarry and railway cutting, 5 km ENE of Ferryhill (Smythe 1924, Hirst & Smith 1974, Bridges & Young 1998, Young 2001); and West Pasture mine, Stanhope, Weardale (Young & Francis 1989).

Leicestershire At Newhurst quarry, Shepshed (where Precambrian diorites are worked for roadstone), malachite is common in the neighbourhood of any veins containing bornite, chalcocite or chalcopyrite (Ince 2005). Another quarry working similar rock types is at New Cliffe Hill, Stanton under Bardon, where copper-rich sulphide veins contain azurite, malachite and the rare species bobkingite, all replacing cuprite (Hawthorne et al. 2002, Hubbard et al. 2005, Ixer et al. 2005a). Malachite is also a common supergene species at Bardon Hill quarry, Coalville, where it forms pastel-green encrustations, small botryoidal masses and dark green tufts of acicular crystals in association with azurite, chrysocolla, native copper and cuprite. Wires of native copper coated with acicular malachite crystals are particularly notable (King 1973, Ince 2007). Cloud Hill quarry, near Breedon on the Hill (King 1983), and Gipsy Lane brick pit, Leicester (Faithfull & Hubbard 1988), are other localities where malachite has been found.

Northumberland A copper vein with malachite and a small amount of chalcopyrite is recorded near Langlee, crossing Harthope Burn, Cheviots (Haslam 1975).

Nottinghamshire Specks of malachite associated with asphaltite (a hydrocarbon), barite (rare), cerussite, galena and wulfenite occur in fissured dolomite at the top of the Permian Lower Magnesian Limestone near Bulwell, Nottingham. Similar assemblages occur at Kirkby in Ashfield and Papplewick Hall, N of Bulwell (Deans 1961). Rare spheroidal malachite is also noted from cavities formed from the dissolution of evaporites and on joint planes in interbedded Permian mudstones and limestones (Cadeby Formation) at Vale Road quarry, Mansfield Woodhouse. Aurichalcite, barite, calcite, chalcopyrite, dolomite, galena, melanterite, pyrite and sphalerite are associated species in the mineralized horizon (Green & Spence 1988).

Shropshire Malachite is one of several secondary species derived from the oxidation of a chalcopyrite–galena–sphalerite ore at Ladywell mine, Shelve (Braithwaite 1982b). Malachite replaces chalcocite at Westcott copper mine, Westcott Hill, 5 km S of Pontesbury, and is also found at Wilderley copper mine, SSW of Castle Pulverbatch (Dewey & Eastwood 1925). Greg & Lettsom (1858) note malachite from Llanymynich Hill, SW of Oswestry. A small boulder of almost solid malachite (drusy crusts of sparkling well formed crystals up to 3 mm long) and chalcopyrite, possibly from a localized copper pod in dolomitic limestone, has been discovered at Llynclys quarry, near Oswestry (Wolfe 2006).

Somerset *Mendip Hills:* Faults and joints in Lower Carboniferous limestone at Coleman's quarry, Holwell, near Frome, contain small pockets and veins of iron and manganese oxides and sulphide veins that have oxidized to produce many secondary species, including malachite (Alabaster 1990a). Similar veins are also found at Batt's Combe quarry, near Cheddar (Kingsbury 1941, Eastwood 1983), and Higher Pitts mine, Priddy (Spencer & Mountain 1923, Burr 1996). Malachite is reported from both localities (silky tufts and small clear prisms at the latter), but only in small amounts. Additionally, malachite occurs with azurite, chalcocite, native copper, cuprite and galena in veins associated with massive manganese oxide bodies cutting limestone at Merehead (Torr Works) quarry, Cranmore (Symes & Embrey 1977). Kingsbury (1941) describes two further (minor) localities for malachite: Moons Hill quarry, NE of Shepton Mallet, and the western end of Sandford Hill. *Quantock Hills:* Mammillated and fibrous malachite, associated with azurite, was once mined as a copper ore at Dodington mine, between Dodington and Nether Stowey (Rudler 1905). Other descriptions are from copper mines and showings at: Buckingham mine, near Bridgwater (Greg & Lettsom 1858); Cannington Park (Dines 1956); Crowcombe church; Crowcombe Court; Great Wood, E of Great Holwell farm; Merridge mine, between Merridge and Courtway (fine specimens); Jacob's Pond quarry, near Capton; Pepper Hill, near Over Stowey; Smith's Combe quarry, near Perry Hill, West Quantoxhead (all Hamilton & Lawrence 1970).

Curly rams-horn crystals of malachite with minor blue azurite on dolomitized limestone. Cannington Park quarry, Bridgwater, Somerset; width 6 mm. (© R. E. Starkey)

Staffordshire Tufts of fibrous malachite occur in the chimney-pipe workings of Clayton mine, Ecton (Ford et al. 1993). At Waterbank mine, Ecton Hill, green botryoidal coatings of malachite are common, but rarely well crystallized (Rust 1995c). An impressive specimen (9×13 cm) of chalcopyrite and malachite on barite from Ecton mine is in the Devonshire mineral collection at Chatsworth House, Derbyshire (Cooper 2005). Dewey & Eastwood (1925) report malachite from Fair Oak colliery, NE of Huntington gravel pits.

Warwickshire Malachite is a secondary alteration product of barite–calcite–copper mineralization found close to the Precambrian/Cambrian unconformity at Judkin's quarry, Nuneaton. It forms small mid- to dark-green velvety crystal aggregates and botryoids up to 2 mm in diameter (Cummins 1969, Ince et al. 1991).

Yorkshire Pale-green malachite forms an alteration halo around native copper- and cuprite-bearing pebbles derived from the chalk of Speeton Cliff, near Flamborough Head (Roydhouse & Green 2003). Beautiful specimens of botryoidal malachite have been found at Old Gang mine, E of Gunnerside Gill, Swaledale (Rodgers 1975). The copper trials of Great Sled Dale, 10 km SE of Kirkby Stephen, have produced richly coated specimens of malachite associated with azurite (Briscoe & Green 1989, Wolfe 2006). Rhombic and lath-shape crystals of malachite overgrow limonite pseudomorphs after chalcopyrite at Scaleber Bridge, 3 km SSE of Settle (Ixer & Stanley 1998). Other reports of malachite are from: Castaway–Ripley vein, Grassington Moor (Young 1987b); Merrybent mine, NW of Middleton Tyas (Small 1977, Dunham & Wilson 1985); Middleton Tyas mines, near Scotch Corner (Raistrick 1936a, Green 1987b); S of Newton Kyme, near Tadcaster (Leake et al. 1997b); Pikedaw Gill, NW of Malham (Raistrick 1983, Mundy & Arthurton 1994, Green et al. 2006c); Richmond mines, Richmond (Dewey & Eastwood 1925); and Wetgrooves mine, Aysgarth (Rust 1987b).

Wales

Clwyd A vein of chalcopyrite and malachite was once mined at Bryn-nantllech mine, N of Llansannan (Dewey & Eastwood 1925). Minor amounts of malachite are also known from mines in the vicinity of Llanfair Talhaiarn, 10 km NW of Denbigh (Bevins 1994).

Dyfed *Central Wales orefield:* Malachite is widespread and it forms thin or botryoidal coatings at: Bwlchglas mine, near Tal-y-bont (Rust 1996c); Bwlchrhennaid mine, near Goginan (Bevins 1994); Castell mine, Ponterwyd, near Aberystwyth (Rust 1990d); Copa Hill mine, Cwmystwyth; Darren mine, Pen-bont Rhydybeddau; Esgair Fraith mine, near Tal-y-bont; Llety-Ifan-Hên (Vaughan) mine; Penrhiw mine, Ystumtuen (all Jones & Moreton 1977); Eaglebrook (Nantycagl) mine, Ceulanymaesmawr (Jones & Moreton 1977, Jones 1983); Gwaithyrafon mine, SE of Tal-y-bont (Rust & Mason 1994); Mynyddgorddu mine, W of Bont-goch, Ceulanymaesmawr (Jones & Moreton 1977, Clark & Rust 1993); Penrhiw mine, Ystumtuen (Mason & Green 1995); Ystrad Einion mine, near Tal-y-bont (Mason & Rust 1997). At Frongoch mine, between Devil's Bridge and Pont-rhyd-y-groes, malachite is found in a normal supergene assemblage, associated with cerussite and hemimorphite, and in a post-mining suite associated with brochantite, cerussite and langite (Green et al. 1996b). Rare post-mining malachite is also noted on langite–wroewolfeite at Esgair Hir mine, Ceulanymaesmawr (Rust & Mason 1988). A chalcopyrite–galena lode containing much quartz-cemented breccia at Llechweddhelyg mine, near Clawddmelyn, between Bont-goch and Cwmsymlog, is traversed by many open joints that have allowed meteoric waters to convert the massive sulphide assemblage into a suite of spectacular supergene species. Malachite of excellent quality (the best in Wales) has been produced, including large bright-green fibrous and banded sprays on specimens, up to 4.5 cm across, and acicular and largely confused crystal groups showing good terminations (Jones 1987, Mason 2004). Elsewhere in Dyfed, malachite is reported from: Lodge Park copper trial, Tre'r-ddol (Mason & Green 1996); Mynyddygarreg, near Cross Hands (Strahan et al. 1909); Vale of Towy (Nant) mine, near Carmarthen (Bevins 1994).

Gwynedd On Anglesey, malachite is noted from Pant-y-Gaseg mine, near Amlwch (Bevins 1994), and from the Parys Mountain volcanogenic sulphide deposit (Greenly 1919). Malachite, in places accompanied by azurite, encrusts joint surfaces and lines vugs within epigenetic copper veins cutting Carboniferous Limestone at Great Orme and Ty-Gwyn mines, on Great Orme's Head, Llandudno (Parnell 1988c, Bevins 1994, Dutton et al. 1994, Ixer & Davies 1996, Ixer & Stanley 1996, Ixer 2001). At Britannia mine, Snowdon, spherical aggregates of radiating malachite needles are associated with carbonatian connellite, georgeite and lanthanite-(Ce)

(Bevins et al. 1985, Pollard et al. 1989). Other occurrences are from: Turf copper mine, Coed-y-Brenin, Snowdonia (Rice & Sharp 1976, Armstrong et al. 2003); Cyffty and Penyrallt mines, between Betws-y-coed and Llanrwst (Haggerty 1995); Drws-y-Coed mine, near Nantlle, Snowdonia (Bevins 1994); Gamallt mine, Careg y Fran, near Ffestiniog, Snowdonia (Lynas 1973); and a trial level, 2 km NW of Bontddu, near Dolgellau (Saich & Rust 1987).

Mid (and South) Glamorgan Oxidation of thin chalcopyrite-bearing veins in Carboniferous Limestone near Cardiff has produced malachite at: Creigiau quarry; Cwmleyshon quarry, near Draethen; Cwrt-yr-Ala Park, Michaelston-le-Pit, South Glamorgan; Goldsland Wood, Wenvoe, South Glamorgan (all Bevins 1994); Machen quarry, 7 km ENE of Caerphilly (Plant & Jones 1995). Malachite is found with gypsum in Triassic rocks on the Penarth coast, South Glamorgan (Wolfe 1996b). At Bute quarry, Llantrisant, zinc-rich malachite is associated with aurichalcite and rosasite. A similar assemblage is also noted from the Locks Common borehole, Porthcawl (Fletcher & Young 1988). Hendy quarry, Miskin, near Llantrisant is a further location where malachite has been found (Bevins 1994).

Powys Some of the finest malachite ever found in Wales was collected in 2001 from veins and joints in limestone at Dolyhir quarry, Old Radnor, and included specimens to cabinet size, displaying brilliant-green velvety crystalline crusts on highly oxidized copper sulphide mineralization (T. F. Cotterell pers. comm. 2005). Earlier reports include: Dolyhir quarry, Old Radnor (Wolfe 2000); Old Radnor (Hall 1868a, Bevins 1994); mines in the vicinity of Llanymynech (Bevins 1994); many veins in the Central Wales orefield, in particular, the Dyfngwm section of Dylife mine at Machynlleth (Rust & Rust 1987); Dylife mine (Jones & Moreton 1977, Moreton 1982); Geufron mine near Llanidloes, which is well known for massive malachite, native copper and cuprite (Jones & Moreton 1977, Rothwell & Mason 1992); and Snowbrook (Nantyreira) mine, between Llanidloes and Tal-y-bont (Jones & Moreton 1977).

Scotland

Borders Livingstone (2002) reports malachite associated with chalcocite from Elmsford, near Duns.

Central Annabergite, erythrite, malachite and tyrolite are post-mining oxidation products found as efflorescences on the dumps of Alva mine, Silver Glen, Ochil Hills (Dunham et al. 1978, Moreton 1996, Green & Todd 2001). Malachite has also been reported from: Airthrey Hill mine, Mine Wood, Bridge of Allan (Wolfe 2004, Ineson & Mitchell 1974b); above Blairlogie on Dumyat (Heddle 1901); and Burn of Sorrow mine, near Dollar (Dickie & Forster 1974), all in the Ochil Hills. Galena–sphalerite–(chalcopyrite–pyrite)-veins in the Tyndrum district contain malachite, pyromorphite, smithsonite (calamine) and native sulphur in the secondary assemblage (Ridgway 1980).

Dumfries & Galloway Connellite, gypsum, malachite and iron oxides undoubtedly formed on the walls of Colvend mine, near Rockcliffe, after it had been abandoned. A similar origin is likely for the bright-green finely mammillated films and coatings of malachite associated with brochantite, found impregnating shales at the entrance to a copper trial at Auchencat Burn, below Hartfell Spa, near Moffat (Moreton & Aspen 1992). Other occurrences include: minor amounts at Balcary mine, S of Dalbeattie; Carsphairn; Skipper's Bridge, S of Langholm (Williams 1965); Kings Laggan, N of the Cairnsmore of Fleet pluton (Heddle 1901); Knockskelly, Black Stockarton Moor (Rock et al. 1986a); Needle's Eye, SW of Caulkerbush (Braithwaite et al. 1990); Penkiln Burn, at the southwestern margin of the Loch Doon granite complex (Stone & Leake 1984); Pibble mine, near Creetown, and Straitstep vein dump, Wanlockhead (Livingstone 2002); Wanlockhead (Brown 1927, Gillanders 1981); West Blackcraig mine, near Newton Stewart (Livingstone et al. 1976); also anthropogenic malachite on the dumps of the Meadowfoot smelter, Wanlockhead (Green 1987a).

Fife Centimetre-wide patches containing chalcopyrite grains surrounded by malachite occur in non-marine algal limestones at Burntisland on the Firth of Forth (Parnell 1983).

Highland Weathering of chalcopyrite-bearing float close to a Lower Devonian appinite pipe on the SE of the Ardsheal Peninsula, near Kentallen, has produced covellite, malachite and hydrated iron oxides (Rice & Davies 1979). A stratiform sulphide deposit in Lewisian Loch Maree Group amphibolites at Druim na Fearna, Dubh Loch, near Gairloch, contains bornite, chalcocite, chalcopyrite and malachite (Jones et al. 1987). Malachite is also thought to occur in copper-rich Lewisian paragneiss to the SSW of Cuil, W of Allt Coire a' Bhoganaich, at the head of Loch Arkaig (Harry 1951). Other reports include: in chalcopyrite–molybdenite–pyrite-bearing quartz veinlets at Gleann a' Chaolais, in the Ballachulish granite (Evans 1977, Haslam & Kimbell 1981); in veins cutting Durness Limestone at Rassal, 4 km N of Loch Kishorn (Heddle 1901, Berridge 1969); in shales of the Ousdale mudstones, near Helmsdale and to the N at Shurrery (Gallagher et al. 1971); and in skarns adjacent to a dolerite dyke within the aureole of the Beinn an Dubhaich granite, Kilchrist (Cill Chriosd), Broadford area, Isle of Skye (Tilley 1948b, 1951, Groves 1952).

Strathclyde Malachite occurs as green stains or botryoidal encrustations in barite-bearing calcite–quartz veins: in the Renfrewshire Hills at Calderbank Bleachfield trial, up stream from Lochwinnoch; Drumshantie mine, SE of Gourock; East and West Kaim mines, near Lochwinnoch; Swinlees mine, Dalry (Heddle 1901, Stephenson & Coats 1983). Carboniferous Clyde Plateau Lavas at Loanhead quarry, near Beith, contain uncommon malachite as well developed free-standing crystals, fibrous sprays or coatings, and stains in calcite and prehnite (Meikle 1989b). At the contact of a stock-like porphyrite intrusion at Garbh Achadh, W of Inveraray, malachite occurs in small dilational quartz veins (Ellis et al. 1978). Malachite is also recorded from: Abhainn Srathain mines, S of Meall Mòr, South Knapdale (Smith et al. 1978a); Boyleston quarry, SW of Barrhead, and Brown's vein, Glengonnar, Leadhills (Livingstone 2002); Castletown mine, Lochgilphead (Hubbard & Green 2001); Coire Buidhe, Glen Creran (Ingram et al. 1993); Etive granite complex to the E and NE of Ben Starav (Haslam et al. 1985); Green Hill quarry, Kilbirnie (Wolfe 1997); Hartfield Moss, 9 km WSW of Barrhead (Meikle 1990); Kilfinan, near Lochgilphead (Berridge 1969, Willan 1980); Leadhills Dod, Leadhills; Macringans Point, near Campbeltown, Kintyre Peninsula (both Heddle 1901); Mannoch Hill mine, SW of Lesmahagow (Todd & McMullen 1991); Murder lode, on eastern side of Loch Fyne, S of Otter Ferry, near Kilfinan (Wilson & Flett 1921); and Rubha a' Mhill, northeast Isle of Islay (Green 1988).

Tayside At Tomnadashan, near Ardtalanaig, Loch Tay, the sides of old copper (chalcopyrite) workings are lined with calcite, chrysocolla and malachite (Wilson & Flett 1921). Heddle (1901) reports malachite from Glen Farg; near Glentully; and from Kenmore plantation, Taymouth.

Orkney Malachite occurs in a barite vein near Hullion, on Rousay (Heddle 1901).

Shetland and Fair Isle Malachite occurs with chalcocite and chalcopyrite in a vein complex at Copper Geo, Fair Isle (Mykura & Young 1969). The species is also noted from Dunrossness, Fair Isle, and in fine acicular crystals at Sandlodge mine, Mainland, Shetland (Greg & Lettsom 1858, Heddle 1901).

Ireland

Cork The Mizen Head Peninsula contains many copper mines where malachite has been found, including those at Coosheen and Dhurode (Ryback & Moreton 1991), Streek Head and, most notably, Cappagh, near Ballydehob, where surely the most beautiful of Irish mineral specimens have been found – brilliant violet bornite in a matrix of malachite and chloritized phyllite (Bartlett 1980). Malachite has also been found at the Coom and Irhin (Urhin) mines (Moreton & Aspen 1993) and at Mountain mine, Allihies (Reilly 1986), all on the Bearhaven (Berehaven) Peninsula.

Down Malachite is a weathering product of chalcopyrite–galena–pyrite–sphalerite ore found in veins and silica–dolomite-cemented breccias at Conlig–Newtownards (Moles & Nawaz 1996).

Dublin Minor amounts of malachite occur in a barite–galena–phosgenite–sphalerite-bearing vein at Killiney (Green 1998c).

Fermanagh Veins containing azurite, chalcopyrite and malachite occur near Belleek (Fowler 1959).

Galway Tynagh mine, near Killimor, is best known to collectors for its azurite and

Botryoidal malachite from Tynagh mine, near Killimor, Co. Galway, Ireland; width of specimen 50 mm. (© D. I. Green)

malachite, which were important ore species in the upper part of the mine (Schultz 1971, Clifford et al. 1986, Ryback & Moreton 1992, Green & Moreton 2001).

Kerry Rare azurite, native copper, cuprite and malachite are noted in bornite–chalcopyrite–tetrahedrite ore from Ross Island mine, Killarney (Russell 1927). Green stains in pyritiferous shales at Leamnamucka, Ballybunnion, proved to be a mixture of brochantite and malachite (Moreton et al. 1995). At Leaghillaun, Bearhaven (Berehaven) Peninsula, a chalcopyrite–tennantite-bearing quartz vein hosted by Devonian sandstones has produced an oxidized suite of supergene minerals, including deep-green velvety hemispheres up to 3 mm across of malachite, associated with azurite, chrysocolla, clinoclase, cornwallite, tyrolite and zálesíite (Moreton & Green 2006b).

Limerick The bornite–chalcocite–chalcopyrite orebody at Glen of Aherlow, near Galbally, contains minor azurite and malachite (Cameron & Romer 1970).

Louth Minor amounts of chalcopyrite and malachite are noted from a small copper deposit at Salterstown (Morris 1984).

Mayo Cunningham (1950) reported malachite from Sheefry mine, 11 km S of Croagh Patrick, but this was not confirmed by Moreton & Jackson (1991).

Sligo Velvety spherules to 2 mm and bunches of needles up to 1.5 mm long have been found in cavities in barite at Tormore mine, between Gleniff and Glencar (Ryback & Moreton 1992).

Tipperary Fractures in sandstones and limestones derived from the upper part of the Cu–(Ag–Hg) deposit at Gortdrum mine, near Tipperary town, contain azurite, cinnabar and botryoidal malachite (Steed 1986, Ryback & Moreton 1991, Braithwaite & Wilson 2003). Malachite is also known from the Silvermines area (Greg & Lettsom 1858, Russell 1907b, Ryback & Moreton 1991, Moreton 1999) and from Lackamore mine, near Newport (Greg & Lettsom 1858).

Waterford Rippled crystalline crusts of mixed brochantite, connellite, langite and malachite have been found at Tankardstown mine, Knockmahon, 22 km SW of Waterford town (Wolfe 1998). Spherules of malachite up to 5 mm across can be found in quartz from North Tankardstown mine (Moreton 2005a). Malachite also occurs in fractured chalcopyrite-bearing quartz veins in this area (Moreton 1991, Braithwaite & Wilson 2001).

Wicklow Greg & Lettsom (1858) report malachite from Audley mine.

Offshore territories

Isle of Man Malachite associated with chalcopyrite, galena and sphalerite were the main ore species at Laxey mines, E of Snaefell (Dewey & Eastwood 1925). Malachite is also noted from the Bradda mines, N of Port Erin Bay (Smyth 1879–1884, Dewey & Eastwood 1925, Starkey 1998b).

Jersey Malachite replaces chalcopyrite in net-veined ignimbrite at Bouley Bay, Trinity, northeast coast of Jersey (Mourant 1961, Ixer 1980, 1992). The species is also noted from: Archirondel; Beaumont tunnels; Gigoulande; Meadowbank; Westmount (Mourant 1961); the tunnels of the German underground hospital, St Peter's Valley, St Lawrence (Ixer 1990b); and La Moye (Mourant & Warren 1934, Mourant 1961, Ixer 1988).

Sark Azurite and malachite replace chalcopyrite at Sark's Hope mine, Little Sark (Mourant & Warren 1934, Ixer & Stanley 1983).

malacolite

A synonym of diopside, pyroxene group.

malaconite

A synonym of cyrtolite, which is a variety of zircon.

malayaite

monoclinic $CaSnSiO_5$

An uncommon calcium–tin silicate found in wollastonite hornfels within the contact-metamorphic aureole of the Dartmoor granite at Red-a-ven mine, Meldon, near Okehampton, Devon. The rocks were originally siliceous limestones, but have been converted to 5 cm-wide white bands of wollastonite interspersed between green bands of diopside, löllingite, vesuvianite and minor datolite. Yellowish-brown malayaite up to 4 mm in diameter is found with fibrous wollastonite. In another assemblage, malayaite occurs with greenish brown garnet (El Sharkawi & Dearman 1966).

In the aureole of the Land's End granite at the Crowns, Botallack, St Just, Cornwall, hornfels has been transformed by intense metasomatism to produce skarns in which malayaite has formed. It is associated with cassiterite and three silicate species enriched in tin: epidote, alumino-ferropargasite and titanite (van Marcke de Lummen 1985, 1986b). At Treliver mine, St Columb Major, near St Austell, malayaite is a minor component of the tin mineralization, where it is hosted by calc-silicates (Camm & Dominy 1997).

mallardite?

MELANTERITE GROUP

monoclinic $MnSO_4{\cdot}7H_2O$

A rare hydrated manganese sulphate, thought to occur in an isomorphous mixture with epsomite, melanterite, goslarite and morenosite in a clear green stalactite of presumed post-mining origin from Ushaw Moor colliery, Durham (Smythe 1933). Mallardite requires further confirmation.

manaccanite

TL Tregonwell Mill, near Manaccan, Cornwall, England

An iron-rich variety of ilmenite.

MANASSEITE GROUP

Hexagonal carbonates with the general formula $Mg_6B_2(CO_3)(OH)_{16}{\cdot}4H_2O$, where $B = Al$, Cr^{3+}, Fe^{3+}. In Britain and Ireland, the group is possibly represented by barbertonite.

manganalmandine

An intermediate composition between almandine and spessartine, garnet group.

manganapatite

A manganese-rich variety of fluorapatite, apatite group.

manganaxinite/axinite-(Mn)

AXINITE GROUP

triclinic $Ca_2Mn^{2+}Al_2BSi_4O_{15}(OH)$

An uncommon calcium–manganese–aluminium borosilicate hydroxide that forms series with ferro-axinite and with tinzenite. Manganaxinite is reported from an iron–manganese deposit at Longwood Hill, Calliagh, 8 km SW of Monaghan, Co. Monaghan, Ireland. It forms veinlets, up to 2 cm across, of lustrous yellowish-brown fan-shape masses in quartz. Associated species are hematite, pyrolusite and rhodochrosite (Ryback et al. 1988, Ryback & Moreton 1993). Manganaxinite had previously been misidentified as idocrase (O'Reilly 1897).

manganhedenbergite

A manganese-rich variety of hedenbergite, pyroxene group.

manganhumite

HUMITE GROUP

orthorhombic $(Mn^{2+},Mg)_7(SiO_4)_3(OH)_2$

A rare manganese–magnesium silicate hydroxide found in a microconcretion and in carbonate-rich red laminated manganese ore from the Cambrian Hafotty Formation at Llyn Dywarchen, near Moel Ysgyfarnogod, Gwynedd, Wales. Manganhumite is orange-yellow in colour and is associated with barite and sonolite (Bennett 1987a).

manganilmenite

A manganese-rich variety of ilmenite.

manganite

monoclinic (pseudo-orthorhombic) $Mn^{3+}O(OH)$

A common manganese oxide hydroxide, trimorphous with feitknechtite and groutite, and found in low-temperature hydrothermal or hot-spring manganese deposits, as a replacement of other manganese minerals in sedimentary deposits, and as a component in some clay deposits and laterites.

Many late-forming iron lodes in and near the southern part of the St Austell granite, Cornwall, contain oxides of manganese and iron. Pyrolusite is the most common species, but acicular manganite and psilomelane also occur at Restormel (Royal) iron mine, N of Lostwithiel. However, manganite is not found in any quantity in Cornwall (Greg & Lettsom 1858, Hosking 1951). Brilliant crystals of manganite from Botallack mine, St Just, are reported by Collins (1892) and it is recorded by Bancroft & Weller (1993) from dump material in the same area (Botallack, Parknoweth and Wheal Cock). Elsewhere in Cornwall, massive manganite is recorded from: Veryan; Trebartha, near North Hill; Indian Queen (Ruthers?) mine, near St Columb (all Greg & Lettsom 1858); and well crystallized varieties are noted from St Minver and Wheal Bucketts, Redruth (Greg & Lettsom 1858, Collins 1892). Golley & Williams (1995) note the occurrence of manganite at Lidcott mine, Laneast, and Trevena mine, Tintagel.

Fine prismatic crystals of manganite have been found at Upton Pyne, NW of Exeter, Devon (Greg & Lettsom 1858). Shepherd et al. (2005) note that oxidation of interbanded psilomelane, kutnohorite and rhodochrosite produced the massive manganite–pyrolusite ore at Upton Pyne. At Pound Living mine, Upton Pyne, the water-filled workings, known locally as Black Pit, contain nodular masses of manganite

Cluster of well formed manganite crystals from Gerran's Bay, Porthscatho, Cornwall; width 220 µm; SEM image. (© D. I. Green)

and rhodochrosite, with some alteration to earthy pyrolusite (Scrivener & Edwards 1990). Manganite is also noted from Doddiscombleigh, near Chudleigh, Devon (Hall 1868b).

Manganite occurs in massive iron–manganese oxide bodies within Carboniferous Limestone in the Mendip Hills, Somerset. It is reported as a late-stage mineral in manganese-rich veins at Coleman's quarry, Holwell, near Frome (Alabaster 1990a), from Higher Pitts mine, Priddy (Burr 1996), and from Merehead (Torr Works) quarry, Cranmore (Symes & Embrey 1977, Turner 2007). Crednerite, cryptomelane, goethite and pyrolusite are associated species at these locations. A related occurrence at Durnford quarry, Long Ashton, near Bristol, has produced pseudo-hexagonal plates of manganite associated with barite, calcite and wulfenite in rhodochrosite-lined cavities within manganese-rich pods (A. Napier pers. comm. 2007).

Hexagonal plates of lustrous manganite, 3.5 mm across, on pink rhodochrosite; Durnford quarry, Long Ashton, near Bristol, Avon. (© D. I. Green)

Manganite has been worked on a commercial basis in the iron ore mines around Bigrigg, Cumbria. At Wyndham pit it occurs within and on the fringe of a hematite body, forming lenses up to 2 m in length. Braunite, hausmannite, pyrolusite and rhodochrosite are associated species (Trotter 1945, Davidson & Thomson 1951, Young 1987a). At nearby Florence mine, hematite mineralization, in fault veins and replacement orebodies in Carboniferous Limestone, is accompanied by small amounts of barite, chalcopyrite, fluorite, galena, manganite and pyrite (Dunham & Hollingworth 1947). Manganese ore was also worked at the Woodhead Sop, Roanhead mine, Dalton-in-Furness, Cumbria, and the ore possibly included manganite and hausmannite (Rose & Dunham 1977). Other Cumbrian occurrences of manganite include: Bannerdale graphite mine, Mungrisdale, Caldbeck Fells (Ziemba 1975); Burdell Gill, Caldbeck Fells (Greg & Lettsom 1858); Force Crag mine, Braithwaite, near Keswick (Greg & Lettsom 1858); Saddleback (Blencathra), near Keswick (Greg & Lettsom 1858); Wiley Gill, Carrock Fell (Kendall 1884); and possibly Thief Gills, Caldbeck Fells (Davidson & Thomson 1951).

Descriptions of manganite from Wales are brief. Reported occurrences are from: Benallt manganese mine, Rhiw, Lleyn Peninsula, and Gyrndu, near Clynog, Gwynedd (both Bevins 1994); and the Gower Peninsula, West Glamorgan (Hall 1868a). The species is also noted in tuffs of Ordovician age in the vicinity of Lliwedd, Snowdonia, Gwynedd (Williams 1927).

In Scotland, 79 manganese oxide occurrences have been recorded (Nicholson 1992), although only nine contain manganite. Nicholson (1989b) summarizes these occurrences as follows:

DUMFRIES & GALLOWAY Kinharvie, SW of New Abbey – crystallized in psilomelane (also noted by Heddle 1901 and McMullen & Todd 1990). GRAMPIAN Bridge of Don, N of Aberdeen – loose crystalline blocks in till contain sharp lustrous black manganite crystals up to 0.2 mm long within cavities throughout massive pyrolusite (also noted by Chew 1978, Green & Todd 2001 and Livingstone 2002); Corrybeg lead mine, Abergairn, near Ballater (Russell 1936a); Laverock Braes, Grandholm, Aberdeen – in veins with barite, hematite and pyrolusite (last two localities also noted by Heddle 1901). LOTHIAN Fenton Tower quarry, near North Berwick – elongate crystals associated with pyrolusite and saponite (also noted by Heddle 1901). STRATHCLYDE Near Stranny Point, Ailsa Craig, Firth of Clyde – minor manganite in hematite cobbles on shore; Dùn Athad, Mull of Oa Peninsula, Isle of Islay – veins in Dalradian Jura Quartzite (also noted by Berridge 1969); Queenside barite mine – black crystals, scattered through barite; Ruchburn, Lochwinnoch.

At Silent Valley quarry, near Newcastle, Mourne Mountains, Co. Down, Ireland, manganite occurs with hematite, pyrolusite, psilomelane and soft brown limonite in black greisen-like replacement veins within granite (Nockolds & Richey 1939). The reports of manganite by Greg & Lettsom (1858) deserve further investigation: Slieve-an-oir, Co. Clare; around Ross, Leap, Noharval and Castleventry, all Co. Cork; and near Howth, Co. Dublin.

manganocalcite

A manganese-rich variety of rhodochrosite, calcite group.

manganocolumbite/columbite-(Mn)

orthorhombic $(Mn^{2+},Fe^{2+})(Nb,Ta)_2O_6$

A rare manganese–iron–niobium–tantalum oxide that forms series with manganotantalite and with ferrocolumbite, and is usually found in accessory amounts in rare-element granitic pegmatites and some evolved granites.

At Carthew, Goonvean and Rostowrack china-clay pits, W of Nanpean, St Austell, Cornwall, manganocolumbite has been found in mineral concentrates (grainsize < 100 µm) produced as waste from kaolinite extraction. The source of the waste is a topaz granite unit that is the most chemically evolved part of the St Austell granite complex. Other accessory minerals found in the concentrate include cassiterite, ferrocolumbite, manganotantalite, pyrochlore and rutile, with minor amounts of apatite, zircon and monazite (Scott et al. 1998). Manganocolumbite is also noted from Gunheath china-clay pit near by (Golley & Williams 1995).

Columbite–tantalite minerals occur in pegmatitic veinlets cutting the lithium–fluorine-rich Meldon aplite dyke, which is emplaced into the metamorphic aureole of the Dartmoor granite near Okehampton, Devon. Manganocolumbite occurs as perfectly formed rectangular or pointed prismatic crystals (0.02–1.0 mm long), which are often black but occasionally translucent brown or reddish. An EMP investigation

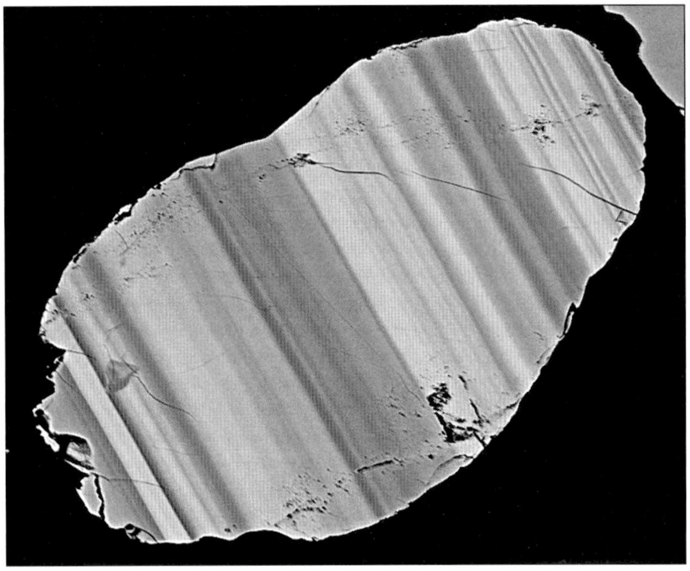

Complex zonation in a water-worn grain of manganocolumbite, collected in a panned concentrate from the grounds of Aclare House, Aclare, Co. Carlow, Ireland; width 210 µm; BSE image. (© A. G. Tindle)

revealed extensive oscillatory zonation within the compositional field of manganocolumbite, and one manganotantalite analysis was also reported. Associated species include amblygonite, beryl, elbaite, fluorite, lepidolite, microlite, petalite and pollucite (von Knorring & Condliffe 1984).

An unusual tungsten-rich manganocolumbite occurs as minute (generally <30 µm across, but up to 180 µm) bladed black crystals in wolframite-bearing quartz veins at Carrock mine, Grainsgill, Caldbeck Fells, Cumbria. It is spatially associated with ankerite and it forms inclusions in arsenopyrite and scheelite. The manganocolumbite appears to have formed during metasomatic alteration of the wolframite (Beddoe-Stephens & Fortey 1981).

In Scotland, very small (<20 µm-diameter) grains of manganocolumbite are reported from an evolved zinnwaldite-bearing granite and associated greisen at Gairnshiel, Glen Gairn, near Ballater, Grampian. The greisen is associated with a hübnerite–quartz vein. Acanthite, barite, bismuthinite, cassiterite, chalcopyrite, cosalite, fluorite, koechlinite, molybdenite, russellite, scheelite, sphalerite, stolzite, stannite–kësterite, topaz and wulfenite have all been found in close proximity, although only in trace amounts (Tindle & Webb 1989, Webb et al. 1992). EMP data indicate extreme enrichment of manganese in manganocolumbite and manganotantalite from a lepidolite–tourmaline pegmatite at Peatfold, Glenbuchat (Tindle unpub. EMP data). The large tourmaline-bearing pegmatite at East Head, 1 km NE of Portsoy, contains accessory amounts of manganocolumbite with tantalum-enriched rims, associated with cassiterite, fluorapatite and spessartine (J. R. Holbrook specimen, Tindle unpub. EMP data).

Manganocolumbite has been identified in spodumene pegmatite sampled in a borehole and also in a panned concentrate of alluvium collected in the grounds of Aclare House, near Myshall, Co. Carlow, Ireland. In the concentrate, many grains are >500 µm in diameter and show remarkable zonation patterns, indicating a very complex crystallization history. The assemblage is dominated by tantalum-bearing species, including common cassiterite, ferrocolumbite, manganotantalite and less-common ferrotantalite, ferrotapiolite and wodginite (R. A. Ixer panned concentrate, Tindle unpub. EMP data). The occurrence is related to the nearby Leinster granite batholith. Scoon (1978) presents a single manganocolumbite analysis from the Aclare pegmatite and he notes that the species forms tiny reddish-black grains to 1 mm across.

manganomelane

A synonym of romanèchite.

manganosite

PERICLASE GROUP

cubic — $Mn^{2+}O$

A rare manganese oxide found with hausmannite and pyrochroite at Benallt manganese mine, Rhiw, Lleyn Peninsula, Gwynedd, Wales (Bevins 1994). Black crystalline material from Roughton Gill, Caldbeck Fells, Cumbria, also contains manganosite (Ryback & Tandy 1992). A possible third occurrence of manganosite on a specimen of chrysoberyl from Ford Farm quarry, Sticklepath, Devon, is less certain, as its identification in unconfirmed (Ryback & Tandy 1992) and because it was identified on a specimen in the Kingsbury collection at the NHM – a collection in which the provenance of many specimens is in doubt (Ryback et al. 1998, 2001).

manganotantalite/tantalite-(Mn)

orthorhombic — $Mn^{2+}Ta_2O_6$

A rare manganese–tantalum oxide, dimorphous with manganotapiolite that forms series with manganocolumbite and with ferrotantalite, and is found in accessory amounts in extremely evolved granitic pegmatites and some granites.

At Carthew, Goonvean and Rostowrack china-clay pits, W of Nanpean, St Austell, Cornwall, manganotantalite has been found in mineral concentrates (grain-size <100 µm) produced as waste from kaolinite extraction. The source of the waste is a topaz granite unit that is the most chemically evolved part of the St Austell granite complex. Other accessory minerals found in the concentrate include cassiterite, ferrocolumbite, manganocolumbite, pyrochlore and rutile, with minor amounts of apatite, zircon and monazite (Scott et al. 1998). Manganotantalite has also been recognized in another lithium-enriched granite – the Tregonning–Godolphin granite, near Porthleven, Cornwall (Hawkes et al. 1987).

Columbite–tantalite minerals occur in pegmatitic veinlets cutting the lithium–fluorine-rich Meldon aplite dyke, which is emplaced into the metamorphic aureole of the Dartmoor granite near Okehampton, Devon. Manganocolumbite is the dominant niobium–tantalum species, although one analysis of manganotantalite is reported. Associated species include amblygonite, beryl, elbaite, fluorite, lepidolite, microlite, petalite and pollucite (von Knorring & Condliffe 1984). Manganotantalite associated with wolframite is noted from the Meldon aplite by Hawkes et al. (1987).

In Scotland, irridescent manganotantalite is reported from a lepidolite–tourmaline pegmatite at Glenbuchat, Grampian, where it forms very dark-brown grains, <1 mm long (Livingstone 1993). Additional data indicate extreme enrichment of manganese in both manganocolumbite and manganotantalite from this pegmatite (Tindle unpub. EMP data).

Manganotantalite has been identified by EMP in spodumene pegmatite sampled in a borehole and also in a panned concentrate of alluvium collected in the grounds of Aclare House, near Myshall, Co. Carlow, Ireland. In the concentrate, many grains are >500 µm in diameter and show remarkable zonation patterns, indicating a very complex crystallization history. The assemblage is dominated by tantalum-rich species and also includes common cassiterite, ferrocolumbite and manganocolumbite, and less-common ferrotantalite, ferrotapiolite, wodginite and monazite (R. A. Ixer panned concentrate, Tindle unpub. EMP data). The occurrence is related to the nearby Leinster granite batholith. Another panned concentrate from a tributary of the Avoca River, close to Avoca mine, Co. Wicklow, has a similar paragenesis and contains manganotantalite, together with cassiterite, ferrotantalite, ferrotapiolite and microlite (R. A. Ixer sample, Tindle unpub. EMP data). In a related occurrence, lithium pegmatite boulders containing cassiterite, manganotantalite and microlite have been found 16 km to the NE of Aclare at Stranakelly, near Shillelagh, Co. Wicklow (Steiger & von Knorring 1975, McArdle & Kennan 1988, McArdle & Kennan 1989).

manganpyrosmalite/pyrosmalite-(Mn)

trigonal — $(Mn^{2+})_8Si_6O_{15}(OH,Cl)_{10}$

A rare manganese silicate hydroxide chloride that forms a series with ferropyrosmalite and is dimorphous with brokenhillite. Manganpyrosmalite is found at Treburland mine, Altarnun, Cornwall (Anthony et al. 1995).

manjiroite?

CRYPTOMELANE GROUP

tetragonal — $(Na,K)(Mn^{4+},Mn^{2+})_8O_{16}\cdot nH_2O$

A rare sodium–potassium–manganese oxide found in the oxidized zone of bedded manganese deposits and noted as an unconfirmed species from Glandore, Co. Cork, Ireland (Ryback 1992). It forms a black mammillated coating on a pale slaty rock. XRD indicates a hollandite-group manganese oxide, possibly manjiroite, but a full quantitative analysis is required to confirm this (S. Moreton pers. comm. 2006).

mansfieldite

VARISCITE GROUP

orthorhombic — $AlAsO_4\cdot 2H_2O$

A rare hydrated aluminium arsenate that forms a series with scorodite and is found on a few of the old spoilheaps at Penberthy Croft mine, St Hilary, Cornwall. Mansfieldite forms small white to light-grey botryoidal crusts and cavity infills in vein quartz, in association with arsenopyrite and possibly scorodite. Bright-yellow glassy mansfieldite has also been found in nearby host rocks (Golley & Williams 1995, Betterton et al. 2000).

Partially iron-stained mansfieldite spherules, up to 0.4 mm across. Penberthy Croft mine, St Hilary, Cornwall. (© D. I. Green)

MARCASITE GROUP

Orthorhombic sulphides, selenides and tellurides with the general formula AX_2, where A = Co, Fe, Ni; and X = S, Se, Te. Kullerudite and marcasite are members of the group found in Britain or Ireland.

marcasite

MARCASITE GROUP
orthorhombic FeS_2

A very common iron sulphide, dimorphous with pyrite and found in hydrothermal veins and sedimentary environments (shales, limestones and coals). It forms under low-temperature highly acidic conditions.

England

Avon Lead–zinc veins in Carboniferous Limestone at Hampstead Farm quarry, Chipping Sodbury, have undergone deep supergene alteration resulting in the replacement of marcasite and pyrite by hydrated iron oxides (Alabaster 1990b, Brooke 1994b). Barite–sulphide veins cutting Triassic Dolomitic Conglomerate near the Pier hotel, Clevedon, contain galena surrounded sequentially by partial or complete rims of pyrite and marcasite, followed by chalcopyrite, tennantite, sphalerite and, finally, enargite (Ixer 1986, Starkey 1986c, Ixer et al. 1993).

Cheshire Very rare marcasite has been found in copper–lead ores within Permo-Triassic redbeds at Engine Vein mine and in the Stormy Point area on Alderley Edge ridge. It forms rims up to 10 µm wide around nickel-rich pyrite (bravoite) cubes (Ixer & Vaughan 1982, Braithwaite 1994).

Cornwall At Mount Wellington mine, Twelveheads, 7 km W of Truro, in host rocks around the Carnmenellis granite, polymetallic Sn–(As–Cu–Fe) mineralization developed at the footwall of an elvan dyke and was controlled by faulting and brecciation. Quartz veins contain marcasite with covellite and digenite (neodigenite) in a secondary assemblage replacing galena and pyrite. Botryoidal and stalactitic marcasite is particularly common in fault-controlled cross courses (Kettaneh & Badham 1978). At nearby Wheal Jane, Kea, marcasite occurs as well formed crystals, rarely up to 1 cm across, or encrusts quartz veinstone in large masses. Pseudomorphs after pyrrhotite and stalactitic marcasite are also noted (Rust 1982d). Marcasite is a constituent of pitchblende-bearing quartz veins from Wheal Trewavas, Trewavas Head, Breage, where it occurs with chalcopyrite, galena, pyrite, sphalerite and tristramite (Atkin et al. 1983). Serpentinization of peridotite belonging to the Lizard ophiolite complex at the Rill, Kynance Cove, Lizard Peninsula, produced minor marcasite, pyrite, pyrrhotite and valleriite as secondary species (Clark 1967). Marcasite has also been noted from: Dean quarry, St Keverne, Lizard Peninsula (Wolters & Wolters 1986); Dolcoath mine, Camborne (Hosking 1951); Fowey Consols mine, Tywardreath, St Blazey (Collins 1892, Lewis 1997); Greystone quarry, Lezant, near Launceston, as lenticular crystals (Elton et al. 2004); Mount Wellington mine, Twelveheads, 7 km W of Truro (Rust 1982e); New Cook's Kitchen mine (North Tincroft lode), Redruth (LeBoutillier et al. 2000); South Crofty mine, Camborne (LeBoutillier 1996); South Terras mine, St Stephen-in-Brannel; Treburland mine, 2 km SE of Fivelanes, Altarnun, Bodmin Moor, rare as a thin crust on rhodonite (Russell 1946d); Wheal Alice, Grampound Road and Wheal James, Withiel (Ball et al. 1982); Wheal Clifford and Wheal Unity, Gwennap; Hingston Down quarry, Calstock (all Golley & Williams 1995); and Wheal Mary Ann, Menheniot (Gleeson et al. 2000).

Cumbria Bonser vein, Coniston, was one of the most productive sources of copper in the Lake District and it contains marcasite with many other species in a (mainly) chlorite–quartz gangue. Marcasite is a primary species and it occurs in the following paragenetic sequence: magnetite, arsenopyrite, pyrrhotite, pyrite, marcasite, native bismuth, bismuthinite, laitakarite, joséite, arsenopyrite, chalcopyrite, sphalerite, galena and cosalite (Stanley & Vaughan 1982b). At Dale Head mine, Newlands Valley, copper-rich veins contain marcasite replacing chalcopyrite and pyrrhotite. It forms aggregates of small grains arranged in lamellar fashion with crosscutting euhedral pyrite (Stanley & Vaughan 1980). Marcasite occurs with galena, pyrite and sphalerite in iron-rich dolomite veins cutting the main quartz–wolframite veins at Carrock mine, Grainsgill, Caldbeck Fells (Davidson & Thomson 1951, Ball et al. 1985a, Young 1987a). Fan-shape aggregates of blade-like marcasite were once found at Potts Gill mine, Caldbeck Fells, but these spoilheaps are now covered by waste from the later barite workings (Davidson & Thomson 1951). The opencut between Driggith and Sandbed mines, also in the Caldbeck Fells, contains microscopic amounts of marcasite associated with gersdorffite or pyrite (Green et al. 2006a). Cavities in sulphide-rich vein material at Brownley Hill lead mine, Nenthead, near Alston, contain marcasite crystals up to 1 mm long. It is one of the minor primary sulphides (together with chalcopyrite and pyrite) in an ore dominated by galena and sphalerite (Green & McCallum 1993, Green et al. 2000). Other occurrences of marcasite include: Garrigill, near Alston (Greg & Lettsom 1858); Nenthead and Great Sulphur vein, Alston Moor (Vaughan & Ixer 1980, Dunham 1990); Greenside mine, near Glenridding, Patterdale and Rotherhope mine, Leadgate, near Alston (Davidson & Thomson 1951); Mowbray mine, Frizington (Green & Briscoe 2002); Sellafield area boreholes (Milodowski et al. 1998); Shap Blue quarry, Shap (Firman 1978, Young et al. 1988, Young et al. 1998); Woodend mine, Threlkeld (Kendall 1884). Minor marcasite undoubtedly occurs at Tynebottom mine, NW of Garrigill, near Alston, where it is found as a primary mineral (with galena and sphalerite + minor chalcopyrite and pyrite) in a series of mineralized flats in Carboniferous Limestone (Davidson & Thomson 1951, Braithwaite 1982b). Here, marcasite replaces pyrite extensively; EMP analysis indicates that it has a high aresenic content (Vaughan & Ixer 1980). Note that some material thought to come from Tynebottom mine and described in Ixer et al. (1979b), Ixer (1986) and Ixer & Stanley (1987) was supposedly collected by A. W. G. Kingsbury and may have a false provenance (see Ryback et al. 1998, 2001); however, marcasite does appear to be genuine (C. J. Stanley pers. comm. 2005).

Derbyshire Marcasite occurs in small quantities in many lead veins in the Southern Pennine orefield, often on joint faces of dolomitized limestone. Massive layers of marcasite occur interlayered with fluorite where one such vein cuts toadstone at Oxclose mine, Snitterton, W of Matlock (Ford et al. 1993). Ixer & Townley (1979) note radiating euhedral and bladed crystals of marcasite closely associated with bravoite (nickel-rich pyrite), chalcopyrite and pyrite from this locality. Marcasite is one of the most interesting minerals found at Fall Hill quarry, Ashover, where it is often altered to goethite, but in some vugs in the Ashover Old vein it forms bright unaltered crystals up to 12 mm long associated with fluorite (Smith 1982). At Wapping mine, Cromford, marcasite occurs as inclusions in fluorite (Ince 1991), whereas on nearby Masson Hill it is included in barite, calcite and fluorite (Ixer & Townley 1979). Marcasite also forms large tabular crystals in non-dolomitized areas, such as in open workings at Moor Jarnett Lane, above Taddington and Hubberdale mine, Flagg, near Bakewell, where it is pseudomorphed by goethite and associated with barite (Worley et al. 1978, Ford et al. 1993). Burr (1992) notes marcasite (associated with barite, cerussite, fluorite, galena, pyrite and sphalerite) on fine old-time specimens of matlockite from Bage mine, Bole Hill, Wirksworth (Burr 1992). Ford & Sarjeant (1964a) suggest that many old records have confused marcasite with pyrite, despite their having different crystal form. Other localities include: Cawdor quarry, Matlock (Robinson & Ineson 1979); Dirtlow Rake, Castleton (Ixer 1978); Ladywash mine, Eyam, and Mawstone mine, Youlgreave (Ford & Sarjeant 1964a); Millclose mine,

Wensley, near Matlock, forming large balls with a bright platy exterior (Russell 1912b, Sarjeant 1957a, Ixer 1986); Odin mine, Castleton (Smith 1962, Ixer 1978, Ixer & Townley 1979); Tansley borehole, Ashover (Ramsbottom et al. 1962, Smith 1982); Windy Knoll, Castleton (Russell 1912b); and Winster (Ixer 1978).

Devon Fine crystallized specimens of marcasite, and marcasite pseudomorphs after calcite, are reported from the Tamar lead–zinc mines, near Bere Ferrers, and, at Virtuous Lady mine, Buckland Monachorum, large aggregated crystals on quartz were once found. Marcasite has also been recovered from Wheal Crebor and Wheal Crowndale, both near Tavistock (Greg & Lettsom 1858, Hall 1868b). Collins (1892) suggests that the arsenic-rich variety lonchidite occurs at Wheal Crebor. Braithwaite et al. (1996) lists marcasite as one of the species found at Ivy Tor mine, Belstone, Meldon area, near Okehampton.

Durham Late barite–fluorite–galena–sphalerite mineralization in the Teesdale Valley, and in the central cupola of the Weardale granite at Rookhope, are characterized by bladed marcasite replacing pyrrhotite; for example, at Closehouse mine, SW of Middleton-in-Teesdale, Lunedale, quartz–sulphide veins cutting autometasomatized Whin Sill dolerite carry pyrite and bladed marcasite pseudomorphs after pyrrhotite. The veins are extensively shattered and cemented by galena (Vaughan & Ixer 1980, Ixer 1986). Large bladed specimens of marcasite, some partially altered to pyrite, were recovered from the lower levels of Frazer's Hush mine, Weardale (Green & Briscoe 2002), and, at nearby Groverake mine, spherules of marcasite were found on euhedral fluorite crystals (Smith & Hardy 1981). Other locations include: Boltsburn mine, Rookhope (King 1982b); Harehope Gill mine, Weardale (Dunham 1990); New Brancepeth colliery, near Durham (Ixer 1986); Redburn mine, near Rookhope (Greenwood 1977); Rogerley mine, near Frosterley, Weardale (Ixer 2003); Rookhope borehole into Weardale granite (Brown et al. 1987a); and Whiteheaps fluorite mine, near Blanchland (Vaughan & Ixer 1980, Ixer et al. 1996).

Kent Marcasite occurs in boulders on the wave-cut chalk platform (accessible only at low tide) at Samphire Hoe, Dover, where it forms remarkable isolated arrowhead crystals up to 23 mm long and sharp bright clusters of intergrown crystals up to 10 cm across (R. E. Starkey pers. comm. 2005). Marcasite also occurs in the Lower Chalk at Abbots Cliff, Folkestone. It forms twinned arrowhead crystals, both isolated and in crystal aggregates up to about 4 mm long (J. Betterton pers. comm. 2006).

Leicestershire At the Carboniferous/Triassic unconformity seen at Cloud Hill quarry, Breedon on the Hill, sand-filled pipe-like bodies contain an assemblage of barite (caulk), calcite, chalcocite, galena, marcasite and copper carbonates (King 1966). Relics of tabular crystals, up to 7 mm long, have also been found in dolomite-bearing oxidized supergene veins at Newhurst quarry, Shepshed (King 1973, Ince 2005).

Northumberland A 24 m-long cavity in a dolerite dyke at West Sleekburn colliery, Bedlington, was found to contain magnesium-rich ankerite, ankeritic calcite, marcasite, pyrite and a large volume of (probably) methane gas. The sulphides formed botryoidal masses and, near the top of the cavity, stalactitic growths (Randall 1953).

Shropshire Fan-like aggregates of spear-shape marcasite up to 2.5 mm long on coarse barite are noted from Rorrington (West Snailbeach) lead–zinc mine, S of Meadowtown. Barite, barytocalcite, calcite, galena, sphalerite and witherite are associated species (Starkey et al. 1994).

Tyne & Wear Two large (one 12 m long) and unusual cavities were found in the Rising Sun colliery (375 m below sea level in the Brockwell coal seam), Backworth. Botryoidal marcasite (mainly) and pyrite line the cavities (Randall & Jones 1966).

Yorkshire Marcasite (now replaced by goethite pseudomorphs) was a primary species in barite-, calcite-, galena- and witherite-bearing veins on East Bolton Moor (Virgin Moss vein and level), Wensleydale (Young et al. 1992c). Chalcopyrite, marcasite and pyrite are present as discrete grains or are intergrown with each other and with trace amounts of enargite-group minerals at Scaleber Bridge, 3 km SSE of Settle. The occurrence is in dolomitized reef limestone (Ixer & Stanley 1998). Nickel-rich and Ni-poor marcasite have both been found associated with pyrite in barite–calcite–quartz-bearing veins at Trollers Gill, ESE of Grassington; all are subsidiary to the main galena–sphalerite mineralization (Vaughan & Ixer 1980). Other locations include: Greenhow Hill, and High Faggergill, Faggergill Moor, Arkengarthdale (Vaughan & Ixer 1980), and at Raydale in Wensleydale (Brown et al. 1987a).

Wales

Dyfed Marcasite is widely developed in the Central Wales orefield, and with pyrite it forms coarse crystals in the A2-*f* assemblage – the final event of primary mineralization in this county. Examples include net-veined marcasite–pyrite replacing iron-rich dolomite at Esgair Fraith mine, near Tal-y-bont; coarse bladed marcasite coating minute pyrite cubes in alternating layers with fine-grain quartz–clay minerals mixture; and coarse crystalline quartz at Brynyrafr mine. Ribs of solid marcasite (>2 m thick) still remain underground at Ystumtuen mine, near Devil's Bridge (Mason 1997). Minute tabular crystals of marcasite forming a late deposit on breccia surfaces are noted from Bwlchgwyn, Penrhiw and Ystumtuen mines, which all worked the Castell lode (Raybould 1974). Fine crystals have been found at both Brynyrafr and Plynlimon mines (Jones & Moreton 1977). Elsewhere in the orefield, marcasite is reported from: Alltycrib mine, Tal-y-bont; Blaenceulan mine, near Tal-y-bont; Hafan mine; Tan-yr-allt mine, near Aberystwyth (Bevins 1994); Clara mine (Jones & Moreton 1977); Eaglebrook (Nantycagl) mine, Ceulanymaesmawr (Jones & Moreton 1977, Jones 1983); Esgair Hir mine, Ceulanymaesmawr (Rust & Mason 1988).

Gwynedd In the Llanrwst lead–zinc orefield, excellent marcasite crystals have been found in association with calcite at Parc and Trecastell mines, near Llanrwst, and the species has also been found at Cyffty, Hafna, Klondyke and New Pandora (Welsh Foxdale) mines, all between Betws-y-coed and Trefriw (Bevins 1994). Russell (1944) notes that Gorlan mine, like Cyffty and Welsh Foxdale mines, worked a galena–marcasite–sphalerite lode traversing slate. As with the Central Wales orefield, marcasite formed after the main lead–zinc ore assemblage had crystallized (Haggerty 1995). Marcasite also occurs in the Dolgellau goldbelt, with reports from: Gwynfynydd mine (Allen & Jackson 1985, Leake et al. 1993, Bevins 1994, Dominy et al. 1996c); Prince Edward mine, Trawsfynydd (Mason & Williamson 2001); and West Cwmheisian mine (Mason et al. 2002). Glasdir mine, Llanfachreth, near Dolgellau, which worked a chalcopyrite-rich orebody within an intrusive breccia pipe, contained, in addition, arsenopyrite, marcasite, pyrite and sphalerite (Allen & Easterbrook 1978, Allen & Jackson 1985). Chalcopyrite, galena, marcasite, pyrite and siegenite are the main sulphides in a polymetallic ore within dolostones and dolomitic limestones at Great Orme mine, Llandudno. The assemblage is considered to represent a separate phase of mineralization from the main copper ore at this site (Ixer & Davies 1996, Ixer & Stanley 1996, Ixer 2001). Other occurrences include:

Marcasite forming cog-wheels in association with acicular goethite; Gwynfynydd mine, near Dolgellau, Gwynedd, Wales; width 10 mm. (© M. P. Cooper)

Britannia mine, Snowdonia – forming impressive spheroids (Reedman et al. 1985); Bwlch antimony mine, near Deganwy – associated with jamesonite, galena, plagionite, semseyite, sphalerite, stibnite and zinckenite in quartz veins cutting ignimbrite (Bevins et al. 1988, Ryback & Francis 2001); Parys Mountain copper deposit, Anglesey – minor amounts (Pointon & Ixer 1980); and the railway tunnel between Blaenau Ffestiniog and Roman Bridge – crystalline cockscombs (Bevins 1994).

Powys Pointed metallic crystals of marcasite occur as crystals within corals at Dolyhir quarry, Old Radnor, near Presteigne (T. F. Cotterell pers. comm. 2007).

Scotland

Central Barite–As–Co–Cu–Pb ± Ag vein mineralization in Old Red Sandstone rocks from Alva mine, Silver Glen, Ochil Hills, has the following paragenetic sequence: pyrrhotite, marcasite–pyrite, chalcopyrite, bornite, covellite, digenite and chalcocite (Jassim et al. 1983, Moreton et al. 1998). Rare marcasite is noted from galena- and pitchblende-bearing veins at Tyndrum (Darnley 1962b, Pattrick 1985).

Dumfries & Galloway Secondary blades of marcasite are commonly associated with chalcopyrite or pyrite in granitoid-related greywacke-hosted quartz veins at Cairngarroch Bay (Lowry et al. 1997). Mineralization with a similar paragenesis on Moorbrock Hill, Carsphairn, includes marcasite in chalcopyrite–gold-bearing quartz veins (Naden & Caulfield 1989), whereas at Glenhead Burn, near Loch Dee (at the southwestern margin of the Loch Doon granite complex), marcasite occurs in arsenopyrite–gold-bearing quartz veins (Leake et al. 1981). Marcasite also occurs with pyrite, pyrrhotite and sphalerite in disseminated mineralization within Ordovician black mudstones at Penkiln Burn, only 2 km from Glenhead Burn (Stone & Leake 1984). At Talnotry, 5 km NE of Newton Stewart, a lenticular Co–Cu–Ni deposit near the base of a diorite intrusion contains nickel-rich marcasite, but only as a late alteration product (Stanley et al. 1987). Rare secondary marcasite has also been found in chalcopyrite–galena–sphalerite veinstone on the nearby dumps of West Blackcraig mine, 3 km SE of Newton Stewart (Livingstone et al. 1976). Massive marcasite, associated with ankerite, chalcopyrite, pyrite and quartz, is noted from the Leadhills–Wanlockhead lead–zinc mining district (Temple 1955).

Fife Marcasite has been tentatively identified in a stockwork of veins consisting primarily of calcite and quartz at Orrock quarry, near Burntisland (Wirth 2003). Sphaerocrystalline marcasite associated with coal is noted from Balbirnie by Heddle (1901).

Grampian Cockscomb marcasite is reported on psammite from Netherglen quarry, 10 km S of Elgin (Livingstone 2002). Shear zones in amphibolite at the southeastern margin of the Knock gabbroic intrusion, near the farms of Littlemill and Auchencrieve, 2 km NW of Rothiemay, N of Huntly, contain hydrothermal fracture fillings with minor amounts of mackinawite, magnetite, marcasite and pyrite (Fletcher & Rice 1989).

Highland A mineralized breccia at Bridge of Westfield in the Loch Cander crush zone contains chalcopyrite, galena, marcasite, pyrite, smithsonite, sphalerite and uranium-rich hydrocarbons in a calcite–siderite gangue (Gallagher et al. 1971). The Kerry Road sulphide deposit near Gairloch is hosted by pyritic quartz–carbonate schists and contains minor marcasite in a pyrite–pyrrhotite-dominated ore (Jones et al. 1987). On the western side of Loch Duich (along the road to Glenelg) and on the eastern side of Loch Duich (along the road to Kyle of Lochalsh), veins associated with the Ratagain granitic intrusion contain common sphalerite, less common chalcopyrite, galena, molybdenite, pyrite and rutile, and, in some examples, locally abundant aikinite, hessite, marcasite and tennantite (Alderton 1986, 1988b). Marcasite and pyrrhotite are notable accessory species in ferrodiorite from Harker's Gully, Marsco, Isle of Skye (Wager & Vincent 1962). At Gie-uisg Geo, 2 km W of Crosskirk, near Thurso, cockscomb marcasite has been found with asphalt (a hydrocarbon), calcite, galena and sphalerite (Heddle 1901).

Strathclyde Marcasite and epidote occur in gneiss at Small Isles Bay on the Isle of Jura (Heddle 1901).

Tayside Hydrothermal quartz–sulphide ± gold veins at Calliachar Burn and Urlar Burn, near Aberfeldy, contain minor amounts of marcasite, sometimes as inclusions in pyrite (Ixer et al. 1997). The Foss stratiform barite–base-metal deposit near Aberfeldy primarily consists of barite, Ba-muscovite and Ba-feldspar (celsian) in cherts, schists and carbonate rocks, with disseminated-to-massive galena, pyrite and sphalerite. However, in a few samples, pentlandite and marcasite replacing pyrrhotite have been noted (Moles 1983, 1985).

Orkney Marcasite coats sandstone in an old lead mine W of the Point of Ness, Stromness (Heddle 1901).

Shetland At Vidlin Ness on Mainland, marcasite and pyrite form stockworks and develop colloform fillings in cavities within a stratiform sulphide horizon in Dalradian calc-silicate granulites. Chalcopyrite, pyrrhotite and sphalerite are the main sulphides present (Garson & May 1976). Pyrrhotite, followed by pyrite, chalcopyrite, marcasite and cobalt pentlandite, are the main sulphide species in the mafic portion of the ophiolite on the Isle of Unst (Maynard 1993).

Ireland

Antrim Marcasite occurs with magnetite, melilite, plagioclase, thomsonite, titanaugite and tobermorite in dolerite contaminated by Cretaceous chalk and flint within the Tertiary plug at Carneal, 9 km NNW of Carrickfergus (Sabine & Young 1975).

Clare Greg & Lettsom (1858) report lance-shape marcasite from Kilkee.

Galway Marcasite was a primary but subsidiary mineral in the stratiform Cu–Pb–Zn–(Ag) deposit at Tynagh mine, near Killimor, although much has been replaced by pyrite. It also forms colloform and crustified cavity linings and fillings (Schultz 1966b, 1971, Morrissey et al. 1971).

Kerry Trace amounts of marcasite are reported from epigenetic chalcopyrite–tennantite veinlets at Western mine on Ross Island (Kinnaird et al. 2002).

Kildare The Harberton Bridge breccia-hosted Fe–Pb–Zn orebody, 40 km WSW of Dublin, is an example of an MVT deposit. It contains minor amounts of marcasite in microcrystalline aggregates that progress outwards into a mass of lamellar hedgehog-like crystals, all surrounded by coarsely crystalline marcasite. Often, marcasite has nucleated on sphalerite (Trude & Wilkinson 2001).

Kilkenny Galena, marcasite–pyrite and sphalerite are the main sulphides found in the stratiform replacement orebody in Carboniferous Limestone at Galmoy, near Kilkenny. A 4 m-thick marcasite–pyrite gossan caps the deposit (Doyle et al. 1992, Lowther et al. 2004).

Longford Marcasite is a minor species in epigenetic sphalerite-dominated mineralization found in breccia zones and fracture fillings within Carboniferous sandstones and conglomerates at Keel, Ardagh (Patterson 1970, Morrissey et al. 1971, Slowey 1986).

Meath Marcasite occurs in variable amounts in the large stratiform carbonate-hosted galena–sphalerite–(marcasite–pyrite) deposit worked at Navan mine. Two marcasite and four pyrite generations are recognized, with the former developing botryoidal and stalactitic growths, 2–3 cm long, on the roofs and floors of cavities (Boyce et al. 1983, Andrew & Ashton 1985, Ashton et al. 1986, 2004, Anderson et al. 1998, Peace et al. 2003). A satellite lead–zinc deposit at Tatestown, 3 km to the SSE of Navan, is dominated by galena–sphalerite–(barite–marcasite–pyrite), which occur as rhythmic colloform infill to fractures and voids in dolomitized Carboniferous Limestone (Andrew & Poustie 1986).

Monaghan Marcasite is a very minor constituent of carbonate veins associated with the arsenopyrite–gold–stibnite-bearing quartz vein deposit at Clontibret mine (Lisglassan–Tullybuck deposit), Tullybuck (Morris et al. 1986).

Sligo Minor pyrrhotite, precipitated within dolostone as intragranular grains, was replaced by marcasite and then by pyrite during dolomitization of the Carboniferous Limestone at what is now Abbeytown lead–zinc mine, near Ballysadare (Hitzman 1986).

Tipperary Marcasite is a minor species in the sedimentary-exhalative barite–pyrite deposit at Ballynoe mine, near Silvermines (Mullane & Kinnaird 1998). Textural evidence suggests that sphalerite and marcasite–pyrite were the first sulphides to precipitate in the B zone of the Ag–Pb–Zn deposit at Silvermines (Zakrzewski 1989), and Rhoden (1959) describes marcasite as an alteration product of pyrrhotite at Gortnadyne and Shallee mines, Silvermines. Marcasite at Silvermines is also noted by Morrissey et al. (1971), Schultz (1971), Taylor & Andrew (1978), Samson & Russell (1983) and Moreton (1999). Galena, marcasite, pyrite and sphalerite are the dominant sulphides, and all replace brecciated carbonate and infill dissolution cavities in the carbonate-hosted Lisheen deposit, between Thurles and Rathdowney. In places, early colloform arsenic-rich pyrite–marcasite, with a radial cockscomb texture, is particularly well developed (Hitzman 1992, Hitzman et al. 2002, Fusciardi et al. 2004). Marcasite is also noted from Gortdrum mine, near Tipperary town (Morrissey et al. 1971).

Tyrone Traces of gersdorffite, marcasite and pyrrhotite have been found in pyrite-dominated gold–quartz veins of the Lack deposit, WSW of Omagh (Cliff & Wolfenden 1992).

Westmeath Carbonate-hosted lead–zinc mineralization at Ballinalack, 15 km NW of Mullingar, contains minor millerite associated with pyrite (bravoite) in an ore dominated by barite, galena, marcasite–pyrite and sphalerite, all in a calcite gangue (Jones & Brand 1986).

Offshore territories

Jersey Marcasite is a minor constituent of arsenopyrite–galena–pyrite–sphalerite-bearing carbonate veins cutting granite at Le Doüet de la Mer, St Ouën (Ixer 1980, 1992). Lead–zinc veins cutting Precambrian Brioverian sediments at Le Pulec contain paragenetically early arsenopyrite–marcasite–pyrite. Marcasite forms tabular clusters up to 700 μm long that enclose both pyrite and pyrrhotite grains (Ixer & Stanley 1980, Stanley & Ixer 1982). At the German underground hospital, St Peter's Valley, St Lawrence, carbonate veins cutting shales are dominated by galena and sphalerite, but contain minor marcasite as radiating aggregates up to 400 μm across with calcite, or as 10–20 μm laths in galena. Euhedral rhombs of marcasite are also found in the shales (Ixer 1980, 1990b). Marcasite replaces pyrite in a chalcopyrite–pyrite–quartz vein at Gigoulande quarry, St Mary (Ixer 1980, 1992), and is also reported from chalcopyrite–molybdenite–pyrite–sphalerite veins in the southeast granite quarry at L'Etacq (Mourant 1961, Ixer 1980) and from Mont Cochon (Mourant 1961).

Sark Marcasite is one of the subsidiary sulphides found at Sark's Hope mine, Little Sark, where a galena–pyrite lode cutting Precambrian granite was once worked for its lead and silver content (Mourant & Warren 1934, Ixer & Stanley 1983).

margarite

MICA GROUP

monoclinic $CaAl_2\square Al_2Si_2O_{10}(OH)_2$

A rare but locally abundant calcium aluminosilicate hydroxide found in Dalradian quartz-bearing graphitic schists, both as a primary phase and as an alteration product of kyanite. Blair Atholl series calcareous graphitic pelites exposed in the River Tilt, near Blair Atholl, Tayside Region, Scotland, contain a garnet-zone assemblage consisting of andesine/labradorite–biotite–chlorite–clinozoisite–garnet–margarite and Ben Eagach black schists in Tayside Region and Grampian contain biotite–garnet–graphite–kyanite–oligoclase–quartz–tourmaline assemblages in which the kyanite is partially replaced by margarite. In Grampian, Dubh (Glen Clunie) and Ailnack Bhuilg (near Tomintoul) are two specific locations where the kyanite–margarite reaction has been noted (Chinner 1974). Later work identified margarite pseudomorphs after kyanite in a Dalradian graphitic gneiss collected 8 km NW of Fettercairn, Glen Esk, Tayside Region (Baltatzis & Katagas 1981). Margarite is noted as inclusions in garnet from Dalradian mica schists at Grandtully, near Pitlochry, Tayside Region (Sivaprakash 1981).

Margarite is also found in ellipsoidal bodies of plagioclase (lenticles) up to 1 m long that occur in metabasalts of the Loch Maree Group, a supracrustal succession within the Precambrian Lewisian Complex in Highland Region. The lenticles occur within amphibolites that have a greenschist-facies overprint comprising saussuritized plagioclase, chlorite, clinozoisite and margarite, and are thought to represent metamorphosed and deformed inclusions of anorthosite from a body of magma cogenetic with the erupted basalts (Cattell & Williams 1988).

There is a suggestion that margarite occurs in certain quartzites belonging to the Mona Complex on Anglesey, Wales (Greenly 1919), but this requires confirmation.

margarodite

A magnesium-rich muscovite, mica group.

marialite

SCAPOLITE GROUP

tetragonal $Na_4Al_3Si_9O_{24}Cl$

An uncommon sodium–aluminium silicate chloride whose occurrence is probably more widespread than the literature suggests. There are many scapolite-group mineral occurrences in the literature, but a lack of chemical data often means that the authors cannot confirm the actual species. However, marialite (dipyre) has been found as a minor constituent of garnet amphibolites at Beinn Chapuill, Glenelg, Highland Region, Scotland (Tilley 1937c, Livingstone 2002), and is also described from veins and replacement deposits in the Walls Peninsula, Mainland, Shetland, and on Fair Isle. The marialite on Fair Isle forms vein complexes associated with both basic and acidic dykes, and with WNW-trending shear belts. It generally forms subtranslucent white crystals with a marked bluish tint. On weathered surfaces it is chalky or waxy white to ochre-brown. Marialite from shear zones on Mainland, Shetland, is cherty in texture and greyish-white or reddish, whereas that in veins is creamy white, bluish pink or pale brown (Mykura & Young 1969). Scapolite from Ward of Silwick, Mainland, has a calcium-rich marialite composition and contains many tiny inclusions of zircon and thorite (S. Moreton sample, Tindle unpub. EMP data). Gallagher et al. (1971) suggests that hydrothermal alteration associated with the intrusion of Devonian post-orogenic granites is responsible for the formation of scapolite in the shear zones and veins on Shetland.

In England, marialite is reported from the New Aplite quarry, Meldon, Devon, where it forms white fibrous compact aggregates associated with tourmaline (Ryback & Tandy 1992).

marmatite

A black iron-rich variety of sphalerite.

marmolite

A variety of chrysotile, kaolinite–serpentine group.

martite

Hematite pseudomorphous after magnetite.

mascagnite

orthorhombic $(NH_4)_2SO_4$

A rare hydrated ammonium sulphate briefly described from ignited coal beds at Bradley in Staffordshire (Greg & Lettsom 1858). Mascagnite is also found in fibrous and mealy encrustations in shale at Emily coal pit, Arniston, Lothian, Scotland (Shand 1910a).

massicot

orthorhombic PbO

An uncommon lead oxide, dimorphous with litharge and found in a stream bed at Leadhills, Strathclyde, Scotland. Soft creamy-yellow massicot is associated with cerussite, elyite and litharge, and may be a dumped flue product (Livingstone 1993).

In England, the occurrence of massicot is less certain. Minute light-yellow crystals associated with litharge from Milltown quarry, near Ashover, Derbyshire (Rust 1991c), and descriptions of massicot from Engine Vein mine, Alderley Edge ridge, Cheshire (Greenwood 1919, Carlon 1979, Braithwaite 1994), require confirmation.

Massicot has been noted (visual identification) at Frongoch mine, between Devil's Bridge and Pont-rhyd-y-groes, Central Wales orefield, Dyfed, Wales, by Rust & Mason (1994), but not by later workers (Green et al. 1996b).

matildite

trigonal $AgBiS_2$

A rare silver sulphobismuthite found in quartz veins at the northwestern contact between the Cairnsmore of Fleet granite and hornfelsed greywackes at Palnure Burn, Talnotry, 5 km NE of Newton Stewart, Dumfries & Galloway, Scotland. Matildite forms a complex intergrown with galena, and both are paragenetically late. Native bismuth is an associated species found as inclusions in the galena. A temperature of formation below 235°C is proposed (Lowry 1991, 1993, Lowry et al. 1997). In a roadcut at Invershiel, Loch Duich area, 10 km E of Glenelg, Highland Region, matildite has been found in a quartz vein at the contact of the Ratagain monzonite and Moinian metasediments. Euhedral crystals of matildite, occurring within a mass of bismuth- and silver-bearing galena more than 3 mm across, have undergone almost complete breakdown to an unnamed composition close to eskimoite (Lowry 1993). Matildite is also reported as an inclusion species in galena from a stockwork cut by quartz veins at Corrie Buie, Meall nan Oighcreag, Loch Tay, Tayside Region. The veins also contain minor amounts of native bismuth, chalcopyrite, electrum (36–65% gold), emplectite (one grain), pyrrhotite, schirmerite and sphalerite, and phases belonging to the gustavite–lillianite series. The mineral assemblage found in the veins is typical of that associated with acidic intrusions (Pattrick 1984).

At Kildonan Burn, Strath of Kildonan, near Helmsdale, Highland Region, matildite has been found with galena as 2–6 μm inclusions in three alluvial gold grains. A further alluvial gold grain from Torrish Burn, Strath of Kildonan, also contains an inclusion of matildite (Leake et al. 1993, Chapman et al. 2000c).

A single small 30–40 μm grain of matildite included in galena was found in a complex assemblage of altaite–gold–hessite–tellurobismuthite–tetradymite from Clogau mine, Bontddu, near Dolgellau, Gwynedd, Wales (Naden 1988, Bevins 1994), and a possible occurrence of matildite in cassiterite–proustite–mackinawite–tourmaline-bearing quartz veins from Ballinglen, between Tinahely and Aughrim, Co. Wicklow, Ireland, is mentioned by Steiger & Bowden (1982).

matlockite

tetragonal $PbFCl$

TL Cromford, near Matlock, Derbyshire, England

A rare lead fluorchloride found in decomposed galena associated with phosgenite (cromfordite) in one of the old air shafts of an old level near Cromford, about 3 km S of Matlock. Matlockite is yellowish, sometimes with a greenish tint, transparent-to-translucent, tabular, and seldom forms crystals larger than 50 mm across (Greg & Lettsom 1858). A more detailed description by Bannister (1934) indicates four chief types of matlockite: single crystals, mostly square, tabular and colourless to pale amber; flat aggregates of platy crystals, in subparallel growth and sometimes quite colourless; hemispherical rosette-like aggregates of platy crystals; and stumpy bipyramidal crystals with small prism faces. Bannister's suggestion that a specimen of matlockite at the NHM (where a fine collection of specimens has been assembled)

Lustrous matlockite crystal 22×25 mm associated with galena; Bage mine, Bole Hill, Wirksworth, Derbyshire. (© D. I. Green)

came from Rutland mine, Matlock Bath, is corrected by Braithwaite & Ryback (1963a), who indicate that it came instead from Arkwright mine, Cromford. This locality is also in dispute, as there is no record of it in the Barmaster's Records, but it may be supposed that it refers to Cromford Sough. Bridges & Smith (1983) conclude that Bage mine on Bole Hill, between Cromford and Wirksworth, and Wallclose mine, Cromford, are the only confirmed localities for matlockite. The history of matlockite is well documented by Burr (1992). Matlockite is also described by Rudler (1905), Ford & Sarjeant (1964a), Braithwaite (1983) and Ford et al. (1993).

matlockite (of Chapman)

TL Cromford, Matlock, Derbyshire, England

A synonym of phosgenite.

mattheddleite

hexagonal $Pb_{20}(SiO_4)_7(SO_4)_4Cl_4$

TL Leadhills Dod, Leadhills, Strathclyde, Scotland

Named after Matthew Forster Heddle (1828–1897), chemist, mineralogist, physician, professor of chemistry at the University of St Andrews, Fife, and author of *The mineralogy of Scotland* (1901); born at Melsetter, Hoy, Orkney Isles (Livingstone 1990a). Mattheddleite is a rare lead silicate sulphate chloride, isostructural with minerals of the apatite group, that was originally found as a creamy-white lining to a drusy cavity in quartz. Associated species include a characteristic Leadhills suite of secondary minerals: caledonite, cerussite, leadhillite, macphersonite, pyromorphite and susannite. Terminated hexagonal prisms of mattheddleite are up to 100 μm long and form radiating rosettiform aggregates some 0.2 mm in diameter (Livingstone et al. 1987). Since then, mattheddleite has been found at Horner's vein, Leadhills, with cerussite, pyromorphite, queitite and susannite (Jackson 1990), and with chenite (Paar et al. 1996). It has also been found on the dumps of Cove vein and Wilson's vein in nearby Wanlockhead (Livingstone 2002). Analytical work on material from Leadhills has identified remarkable chemical zonation in mattheddleite, which is attributed to rapid changes in fluid chemistry during late-stage hydrothermal processes (Essene et al. 2006).

In England, mattheddleite occurs at Penberthy Croft mine, St Hilary, Cornwall, where it forms white prismatic crystals up to 0.2 mm long in cavities within vein quartz. It is associated with anglesite, clinochlore and sphalerite (Betterton 1996c, Betterton et al. 2000). A further occurrence is from Whitwell quarry, Derbyshire, where mattheddleite is found with lanarkite and leadhillite as typical radiating sprays of minute pencil-shape crystals up to 0.3 mm long (D. I. Green pers. comm. 2004). Mattheddleite is also reported from several localities in the Caldbeck Fells, Cumbria: Brae Fell mine (Cooper et al. 1988, Cooper & Stanley 1990, 1991b, Bridges et al. 2006); eastern flank of Balliway Rigg, higher Roughton Gill (Cooper et al. 1988, Green 1989, Cooper & Stanley 1990, 1991b, Green & Briscoe 2002); Dry Smale Gill (Green & Briscoe 2002); Red Gill mine (Cooper et al. 1988, Wirth 1989, Cooper & Stanley 1990, 1991b, Green & Briscoe 2002), Short Grain mine (Cooper et al. 1988, Cooper & Stanley 1991b, Green & Briscoe 2002). At Red Gill mine, mattheddleite forms radiating sprays of minute (<0.25 mm) acicular crystals on caledonite in a cavity in a galena–quartz matrix, and at Short Grain it forms deeply etched colourless prisms lining small cavities in barite (Cooper & Stanley 1990). Only two specimens are known from Brae Fell mine and one of these is in the Kingsbury collection at the NHM. Because many specimens in this collection have an incorrect provenance (Ryback et al. 1998, 2001), the mattheddilite specimen requires further investigation. The second specimen does appear to be genuine. A specimen now in Manchester Museum, labelled mattheddleite from Driggith mine, has been shown to contain anglesite and lanarkite, but no mattheddleite (Green et al. 2006a).

Spray of mattheddleite crystals, 0.2 mm across, on quartz; Red Gill, Caldbeck Fells, Cumbria. (© D. I. Green)

At Esgair Hir copper–lead mine, Ceulanymaesmawr, Central Wales orefield, Dyfed, Wales, mattheddleite is one of many supergene minerals found in cavities in oxidized sulphides. It forms colourless acicular crystals up to 0.2 mm long and

Cluster of mattheddleite crystals from Whyte's Cleuch, Wanlockhead, Dumfries & Galloway, Scotland; width 170 μm; SEM image. (© D. I. Green)

encrusts caledonite in association with leadhillite (Rust & Mason 1988). A few km away at Darren mine, Pen-bont Rhydybeddau, 9 km E of Aberystwyth, mattheddleite has been found in oxidized ore on the dump at the mouth of Darren Fach adit. Colourless acicular crystals of mattheddleite are 0.2–1.5 mm long and occur with caledonite, hydrocerussite and rare anglesite (Rust 1990b). An unconfirmed occurrence of mattheddleite associated with anglesite and scotlandite on oxidized galena from Machen quarry, 7 km ENE of Caerphilly, Mid Glamorgan, is noted by Plant & Jones (1995).

maucherite

tetragonal $Ni_{11}As_8$

A rare nickel arsenide found with algodonite, coffinite, native copper, lavendulan, malachite, modderite, nickeline, rammelsbergite and native silver in Permian nodular concretions from the Littleham Cove area, Budleigh Salterton, Devon (Harrison 1973, Harrison et al. 1975b). Unusual sheet-like plates of native copper occurring with the concretions show a complex sequence of alteration and subsequent mineral growth on their surfaces, which includes the species algodonite, maucherite and nickeline. De-watering of Permo-Triassic strata during burial compaction is favoured as the mechanism for producing metalliferous pore fluids capable of transporting the As–Cu–Ni and associated Ag–Co–U–V mineralization (Milodowski et al. 2000).

On Unst, Shetland, Scotland, the nickel arsenides heazlewoodite, godlevskite, maucherite and orcelite occur in many serpentinized peridotites that are associated with PGE mineralization (Gunn et al. 1985). Two such locations for maucherite in the Baltasound–Haroldswick area of Unst are Cliff in the NW and Ordale in the SE. At the former, maucherite forms 5–50 μm anhedral crystals with nickeline interstitial to, and along cracks in, chrome spinel, in a serpentinized matrix within chromitite. Sheared and serpentinized pyroxenites are also noted to contain antimony-rich maucherite, orcelite and composite orcelite–pentlandite grains (Maynard 1993).

Elsewhere in Scotland, maucherite and rammelsbergite are the only primary nickel arsenides at Alva mine, Silver Glen, Ochil Hills, Central Region. Only two specimens (1 cm across) containing maucherite (with galena and native silver) have been found (Moreton et al. 1998). Hilderston mine, Bathgate Hills, Lothian, is a further locality where pale copper-colour maucherite has been found with annabergite and galena, or as individual crystals (up to 1 mm across). One specimen contains threads of maucherite up to 0.09 mm long, associated with native silver (Meikle 1994). In common with Alva, the nickel arsenides at Hilderston seem to be mainly the product of post-mining oxidation (Green & Todd 2001).

mawbyite

TSUMCORITE GROUP

monoclinic $Pb(Fe^{3+})_2(AsO_4)_2(OH)_2$

Named after Sir Maurice Alan Edgar Mawby (1904–1977), for his contribution to the Australian mining industry and his knowledge of the minerals of Broken Hill, New South Wales; born at Broken Hill of an English father. Mawbyite is a very rare lead–iron arsenate hydroxide, dimorphous with carminite and found as a component of a brownish-red to blood-red crystalline crust from an arsenate-rich area above Sandbed 60-fathom level, Caldbeck Fells, Cumbria. Both mawbyite and carminite are confirmed by XRD, although it was not physically possible to separate the two species on the specimen (Green et al. 2006a).

mawsonite

tetragonal $Cu_6Fe^{3+}_2Sn^{4+}S_8$

Named after Sir Douglas Mawson FRS (1882–1958), an English–Australian geologist and Antarctic explorer; born in Shipley, near Bradford, Yorkshire. Mawsonite is a rare copper–iron–tin sulphide found in the Hensbarrow stock at Gunheath china-clay pit, Carthew, St Austell, Cornwall, where it is disseminated within non-megacrystic lithium-mica granite. Mawsonite with a dark metallic lustre, together with two other tin-bearing sulphides (stannite and stannoidite) occurs as spots up to 2 cm across mixed with silicates. Mawsonite characteristically occurs as irregular rims between bornite and stannoidite, and also partially replaces stannoidite. Common associates are albite, lepidolite, topaz and zinnwaldite, which are found with the less common species bornite, cassiterite, digenite, ilmenorutile and tennantite. The disseminated tin-sulphide mineralization is thought to be related to large-scale post-magmatic processes that involved meteoric water as well as fluid–rock interaction with the granite and with country rocks (Manning 1983, Weiss 1994). A comprehensive review of tin sulphides in southwest England is provided by Alderton (1991).

meionite

SCAPOLITE GROUP

tetragonal $Ca_4Al_6Si_6O_{24}CO_3$

An uncommon calcium–aluminium silicate carbonate, whose occurrence is probably more widespread than the literature suggests. There are many scapolite-group mineral occurrences in the literature, but a lack of chemical data often means that the authors cannot confirm the actual species. However, anhedral crystals of meionite have been found enclosing other minerals in symplectite-bearing nodules in the Ardgour marble, Coire Dubh, 3 km NW of Corran, Ardgour, Highland Region, Scotland (Drever 1936, Livingstone 2002). Minutely crystalline marble from Balephretrish Bay on the Isle of Tiree, Strathclyde, contains crystals of meionite in mineral aggregates associated with biotite, chlorapatite, diopside and titanite (Coomaraswamy 1903).

melaconite

A synonym of tenorite.

melanite

A titanium-rich variety of andradite, garnet group.

melanotekite

orthorhombic $Pb_2Fe^{3+}_2Si_2O_9$

A rare lead–iron silicate that forms a series with kentrolite and is found at three localities near Bristol, where it forms a minor constituent of the secondary mineral assemblage associated with Carboniferous Limestone-hosted replacements of iron and manganese oxides. At Wesley Theological College (Wesley mine), Westbury-on-Trym, Avon, melanotekite forms surface films and small pockets (max. 5×10 mm) of sulphur-yellow to blackish-yellow powder encrusting hausmannite and iron-rich wad. It is associated with earthy goethite and hematite. One specimen of melanotekite, found at Merehead (Torr Works) quarry, Cranmore, Somerset, is similar in appearance to that at Wesley mine and it coats a fracture surface on a block of mixed iron and manganese oxides. The yellow powder rims goethite, suggesting that the latter may be an alteration product. Additionally, some of the melanotekite from Merehead is crystalline and forms rhomb-shape and compound crystals, occasionally with curved crystal faces. Broken surfaces reveal an amorphous, granular or spongy internal structure. Further melanotekite specimens have been found at Merehead quarry as coatings and small granular masses of a distinctive greenish-yellow colour in the iron-rich outer parts of manganese-rich pods (Turner 2007). The third locality is Coombe Farm quarry, Westbury-on-Trym, Avon, where melanotekite is commonly associated with earthy hematite and amorphous masses of kaolinite–serpentine-group minerals. Powdery and crystalline melanotekite occurs in pockets within serpentine or in thin layers parallel to the margins of the pockets. The crystalline material is more compact than at Merehead, although some crumbly pale-yellow material remains. A few euhedral crystals were found in drusy cavities, where they form tabular rhombs or complex crystals that are stepped or concertina like (Alabaster 1985a). Coleman's quarry, Holwell, near Frome, Somerset, is a fourth locality where melanotekite is briefly noted (Alabaster 1990a).

MELANTERITE GROUP

Monoclinic sulphates with the general formula $A^{2+}(SO_4)\cdot 7H_2O$, where A^{2+} = Co, Cu, Fe, Mn, Zn. Melanterite and possibly bieberite and mallardite are members of the group found in Britain or Ireland.

melanterite

MELANTERITE GROUP

monoclinic $FeSO_4\cdot 7H_2O$

A common hydrated iron sulphate that forms in old coal and metalliferous mines as a post-mining product of oxidation of marcasite and pyrite. Distinct and well formed crystals rarely occur in nature. Melanterite is said to be found in most mining districts of Cornwall, in particular: St Agnes; Ting Tang mine, Carharrack; Fowey Consols, Tywardreath, St Blazey; United mines, St Day, Gwennap; and Wheal Maudlin, 3 km NW of Lostwithiel (Collins 1892). At Wheal Jane, Kea, near Truro, melanterite (syn. pisanite) forms large stalactitic masses of a greenish-blue colour (Rust 1982d). In South Crofty mine, Camborne, melanterite forms stalactitic masses in iron-rich lodes, particularly on the 380-fathom Cooks–Robinson crosscut (LeBoutillier 1996).

Melanterite forms a bright-green aggregate of fibrous crystals up to 15 mm in length at New Brancepeth colliery, near Durham, where it was found on the surface of pyrite-rich shale (Spencer 1910). A clear green stalactite covered in white spots mainly of melanterite has been described from the barite–witherite workings of Ushaw Moor colliery, Durham (Smythe 1933) and large quantities of melanterite (syn. copperas) were also noted at Hetton colliery, Co. Durham (Clapham & Dalglish

Irregular growths of melanterite, up to 10 mm in height, from the Parys Mountain copper deposit, Anglesey, Gwynedd, Wales. (© D. I. Green)

1863). Across in Cumbria pale-blue masses of curved capillary crystals of melanterite are associated with gypsum and jarosite at Brownley Hill lead mine, Nenthead, near Alston (Green et al. 2000). At Smallcleugh mine, Nenthead, melanterite forms bluish-green stalactitic aggregates or fibrous growths up to 2 cm long, particularly at the flat above Wheel flat, Hetheringtons crosscut and the roadway on First Sun vein. It is associated with epsomite, gypsum, halotrichite, limonite and pyrite (Uttley 1987b, Dunham 1990).

Elsewhere in England, blue-green efflorescences of melanterite (pisanite) occur uncommonly at Engine Vein mine, Alderley Edge ridge, Cheshire (Eagar & Broadhurst 1959, Carlon 1979, Braithwaite 1994), and at Odin mine, Castleton, Derbyshire (Hall 1868a, Braithwaite 1983). At Copperas farm, near Ringinglow, Derbyshire, pale-green encrustations of melanterite (copperas) were collected as a source of sulphuric acid (Ford et al. 1993). White vermicular efflorescences of melanterite occur on the outer surface of pyrite nodules collected from Fe–Pb–Zn mineralization in Permian carbonates at Vale Road quarry, Mansfield Woodhouse, Nottinghamshire. These may have developed after the samples had been collected (Green & Spence 1988). Fine specimens of melanterite are noted from the fuller's earth pits at Widcomb Hill, near Bath, Avon (Greg & Lettsom 1858). Upper Eocene rocks (Bagshot Formation) of the Thames Basin are also melanterite bearing (Irving 1891).

Melanterite is probably widely distributed in Wales, although reports are few. It is noted from Parys Mountain, Anglesey, where it forms fine green stalactites (Greg & Lettsom 1858, Hall 1868a, Greenly 1919, Southwood & Bevins 1995). Bor (1950) reports pale-blue acicular crystals of pisanite, the copper-rich variety of melanterite, from this location. In the abandoned underground workings of Cae Coch mine, near Trefriw, Llanrwst, Gwynedd, seams of green melanterite occur in shales associated with pyrite mineralization. Passageways in this mine contain copious amounts of "acid streamers" of bacterial origin consisting of fibroferrite accompanied by copiapite and melanterite (Johnson et al. 1979, Jenkins & Johnson 1993). Melanterite is noted from Abertysswg colliery, Rhymney, Mid Glamorgan, by Bevins (1994).

Heddle (1901) suggests that melanterite occurs at many locations in Scotland, although quotes only Hurlet, in Paisley, Strathclyde, where it forms green efflorescences and fibres on artificially exposed aluminous shale containing pyrite (see also Greg & Lettsom 1858). Livingstone (2002) reports yellowish-green stalactitic growths of melanterite and rozenite from West Mains colliery, West Calder, Lothian.

In Ireland, small crystals of melanterite up to 5 mm long have been found on the walls of an old coalmine NW of Arigna in Co. Roscommon (Ryback & Moreton 1992). Melanterite has been reported from Ballybunnion, Co. Kerry (Ainsworth 1834), but was not confirmed by Moreton et al. (1995). Stewart (1999) describes melaterite from the lead–zinc mine at Lisheen, Co. Tipperary.

Melanterite is noted from Laxey mine on the Isle of Man (Lamplugh 1903).

MELILITE GROUP

Tetragonal silicates with the general formula A_2BZSiO_7, where A = Na, Ca; B = Al, Be, Mg, Zn; and Z = Al, Si. Åkermanite, gehlenite and melilite are members of the group found in Britain or Ireland.

melilite

tetragonal — MELILITE GROUP — $(Ca,Na)_2(Al,Mg)(Si,Al)_2O_7$

An uncommon (but locally abundant) calcium–sodium–aluminium–magnesium aluminosilicate that belongs to the series åkermanite–gehlenite. Its composition is intermediate between these two species and therefore some authorities (i.e. Anthony et al. 1995) do not recognize it as a distinct species. In the literature, the term is common, as it can be applied on the basis of optical data only. Because of this, it is not always possible to confirm which species of the melilite group is actually being described. Another complication is that major substitution of iron for magnesium or aluminium can produce compositions markedly different from any of the defined species in the group, and the terms ferro-åkermanite and ferri-gehlenite are to be found in the literature. Sodium melilite is another variety not given species status.

Abundant melilite is found in the chalk/dolerite contact zone at Scawt Hill, Larne, Co. Antrim, Ireland, both in contaminated dolerite and in a reaction zone within the chalk. In the latter it is associated with calcite, larnite, merwinite, spinel and spurrite. In the contaminated dolerite, brown to honey-yellow vitreous to resinous melilite forms crystals up to 25 mm in diameter associated with titaniferous augite, nepheline and varying amounts of aegirine, apatite, iron oxides, labradorite, perovskite and wollastonite (Tilley 1929a). One hybrid rock type, found in small lenses and patches surrounded by metasomatized chalk, contains abundant melilite with aegirine and perovskite, and no titaniferous augite or nepheline. Melilite in this hybrid rock is very susceptible to alteration and replacement by cebollite or scawtite (Tilley 1930a, 1938b, Tilley & Harwood 1931). Another melilite-bearing assemblage was produced during the metasomatism of flint nodules within the chalk. Replacement of melilite by magnetite is a characteristic feature of this hybrid, and aegirine, pectolite, wollastonite and xonotlite are common associates (Tilley & Alderman 1934). Later investigations show that rankinite is also found with larnite, melilite and wollastonite in certain melilite hybrids and at the contact of flint nodules with metasomatized chalk (Tilley 1942).

The Tertiary dolerite plug at Carneal, 9 km NNW of Carrickfergus, Co. Antrim, cuts basalt lavas and incorporates blocks of chalk and flint from underlying Cretaceous rocks. Assimilation of the limestone and flint took place under very rare highest-temperature low-pressure conditions (as must have been the case at Scawt Hill) and, once again, melilite occurs in both endomorphic (derived from dolerite) and exomorphic (derived from limestone and flint) hybrid assemblages. To emphasize the difficulty of nomenclature relating to melilite, it is referred to as being gehlenitic (or rarely åkermanitic) in the exomorphic hybrid rocks, and an analysis from an endomorphic melilite–titanaugite rock is said to have a 65% åkermanite component (Sabine & Young 1975). However, an analysis of melilite from an exomorphic larnite–magnetite rock is one of the most highly gehlenitic melilites ever recorded, with a 94% gehlenite component (Sabine et al. 1982). Analysis of dark-green elongate laths of "åkermanite" collected from Carneal show a melilite XRD pattern and exceptional zoning from magnesium-rich cores to iron-rich rims (H. J. Foy specimen, D. I. Green XRD, Tindle unpub. EMP data).

Similar occurrences of melilite occur in thermally metamorphosed limestones around Barnavave, Carlingford, Co. Louth. Melilite is a subordinate constituent in rocks containing spurrite and tilleyite, where it forms small and rather rounded

Euhedral crystals of strongly zoned melilite from the chalk/dolerite contact zone at Carneal, Larne, Co. Antrim, Ireland; width 580 μm; BSE image. (© E. Condliffe, University of Leeds)

grains. At other localities it is abundant and some rocks are nearly pure melilite. Both gehlenitic and åkermanitic melilite is reported (Nockolds & Vincent 1947).

In Scotland, melilite occurs in thermally metamorphosed amygdales within Tertiary basalts in the aureole of the volcanic plug at 'S Airde Beinn, 5 km SW of Tobermory, Isle of Mull, Strathclyde. One type of amygdale originally containing calcite is now represented by the anhydrous calcium silicates, larnite, rankinite and wollastonite (zoned from core to rim in this sequence). Late reaction has led to the replacement of both basalt and calcium silicates by melilite. Within the rims of these amygdales, aegirine and equant melilite crystals are found with perovskite and iron oxides (Cann 1965a). At Kilchoan, Ardnamurchan, Highland Region, melilite, rankinite, spurrite or wollastonite are essential components of limestones thermally metamorphosed by gabbro. Cuspidine, kilchoanite and tilleyite are found with melilite in other assemblages within the contact zone (Agrell & Gay 1961). Melilite is also reported from a high-temperature calcic skarn near the contact of the Cuillin gabbro complex at Camasunary Bay, Strathaird Peninsula, Isle of Skye. Associated species include calcite, garnet (andradite–grossular), spurrite, tilleyite, vesuvianite and wollastonite, together with minor amounts of magnetite, monticellite, perovskite and pyrite. Rare baghdadite, djerfisherite, galena, kalsilite, kimzeyite and titanite complete the assemblage. The occurrence of zirconium-rich minerals suggests that the skarn underwent a degree of metasomatism, with the introduction of fluids rich in aluminium, potassium, titanium and zirconium, possibly from the Coire Uaigneich granophyre (Beard & Drake 2007). On Mainland, Orkney, melilite is described as a late crystallizing phase in alkaline dykes (alnöite and monchiquite) cutting Old Red Sandstone sediments at a few localities: Long Geo, Galtick, near Holm; Naversdale, Orphir; and Rennibuster Point, near Kirkwall (Flett 1900). Confirmation of all the Scottish occurrences of melilite is desirable, particularly as the Long Geo occurrence has since been proved to be apatite (Macpherson & Livingstone 1982).

melnikovite

A synonym of greigite, but also used informally to describe poorly crystalline botryoidal brown pyrite.

melnikovite-marcasite

A colloform variety of marcasite; analogous to melnikovite–pyrite.

MELONITE GROUP

Trigonal sulphides and tellurides with the general formula AB_2, where A = Ni, Pd, Pt, Sn; and B = Bi, S, Se, Te. Melonite, merenskyite and moncheite are members of the group found in Britain or Ireland.

melonite

MELONITE GROUP

trigonal — $NiTe_2$

A very rare nickel telluride that forms a series with merenskyite and is found at Urlar Burn, 5 km from Aberfeldy, Tayside Region, Scotland, in quartz–sulphide ±gold veins cutting Upper Dalradian gritty quartzites, quartz–mica schists and amphibolites. Melonite is one of several telluride species forming inclusions in bismuth-bearing galena, others being altaite, benleonardite, coloradoite, hessite, tsumoite and volynskite (Ixer et al. 1997). Along the Strathclyde/Dumfries & Galloway border, in the Leadhills area, a single alluvial gold grain is reported to contain melonite inclusions (Leake et al. 1998). A single grain of melonite is noted from the Arthrath mafic–ultramafic intrusion near Ellon, 25 km N of Aberdeen, Grampian. The assemblage also contains trace amounts of native bismuth, native gold, hessite, merenskyite and michenerite (McKervey et al. 2007).

At Croagh Patrick, Co. Mayo. Ireland, melonite occurs as inclusions in gold grains thought to be derived from a mesothermal quartz vein or stockwork (Leake et al. 1993, Chapman et al. 2000c).

menaccanite

An iron-rich variety of ilmenite (see manaccanite).

mendipite

orthorhombic — $Pb_3Cl_2O_2$

TL Churchill, Mendip Hills, Avon, England

A rare lead oxychloride, described from the type locality together with galena, black earthy manganese and hematite (Greg & Lettsom 1858), and from workings of the Somerset Manganese and Iron Company at Higher Pitts farm, near Priddy, Somerset, where it was found with cerussite, a little copper ore, and oxide of manganese (Rudler 1905). At Higher Pitts, mendipite with perfect prismatic cleavage occurs as nodules, with irregular corroded surfaces, embedded in manganese ore. Dirty-white or grey, but sometimes yellow or delicate pink crystals up to 16 cm long, fibrous, splintery and with a pearly or silky lustre are reported from old museum collections and from field excavations. Partial alteration to cerussite or hydrocerussite (or possibly both) is common (Spencer & Mountain 1923, Kingsbury 1941, Burr 1996).

A pinkish mass of mendipite with minor cerussite from Merehead (Torr Works) quarry, Cranmore, Somerset; width 3.7 cm. (© R. E. Starkey)

In the large-scale Carboniferous Limestone quarry at Merehead, Cranmore, Somerset, mendipite is the commonest oxychloride and it forms irregular lumps, cavity fillings and thin veinlets in black manganese oxide ore. Fibrous-looking aggregates are white to creamy yellow, or pinkish purple, and contain inclusions of chloroxiphite, diaboléite, or, less commonly, crednerite and parkinsonite. No free-growing crystals have been found. Partial alteration to hydrocerussite and cerussite is common, and alteration to paralaurionite is also reported (Alabaster 1975, Symes & Embrey 1977, Turner 2007).

At Wesley Theological College (Wesley mine), Westbury-on-Trym, Avon, veins of manganese oxides with many calcite-lined cavities cut the Carboniferous Limestone. Secondary lead and manganese oxide species are found as fracture and cavity infillings in the original manganese oxides or as drusy encrustations, discrete crystals or veinlets in calcite nodules and geodes. Mendipite occurs rarely as compact cleavable masses of lustrous off-white translucent sprays of brittle prismatic–acicular crystals in some of the cavities. Associated species are barite, chloroxiphite, crednerite, manganite, mereheadite, parkinsonite and wulfenite. The largest specimen is 3×6 cm (Alabaster 1989b).

The lead oxychlorides are thought to have formed by interaction of primary galena with low-temperature metalliferous brines along fractures cutting the original Fe–Mn–(Cu–Pb) orebody (Symes & Embrey 1977, Humphreys et al. 1980), although it is possible that they precipitated directly from a cooling brine (Welch et al. 2000). See also Turner (2007).

meneghinite

orthorhombic — $Pb_{13}CuSb_7S_{24}$

A rare lead–copper sulphantimonite found in barite–sulphide ore at Tynagh mine, near Killimor, Co. Galway, Ireland. Meneghinite and other sulphosalt minerals (bournonite, enargite, pyrargyrite, polybasite and stephanite) occur in microscopic amounts mainly associated with galena and tennantite–tetrahedrite (Schultz 1966b). At Dhurode, West Carbery mining district, Co. Cork, vein material from the abandoned copper mine contains meneghinite partially replaced by a symplectite of galena and bournonite. This replacement is thought to have occurred because of the addition of copper and sulphur from vein-forming fluids. Meneghinite forms 50–

200 μm-long euhedral-to-subhedral crystals mantled by bournonite. Common associates are galena and native antimony (Ni Wen et al. 1991, Kinnaird et al. 2002).

Meneghinite has also been reported from Shallowford Bridge, near South Molton, Devon, and Pengenna (Trewethen) mine, St Kew, Cornwall (Kingsbury 1965c, Embrey 1978). However, because many specimens in the Kingsbury collection (at the NHM) have been shown to have an incorrect provenance (Ryback et al. 1998, 2001), the meneghinite samples deserve further investigation. Golley & Williams (1995) note meneghinite in Cornwall from the Pentire vein system, St Minver, and from the Port Quin vein system, St Endellion.

mercury

a liquid — Hg

The only native metal that is a liquid at room temperature. Mercury, forming μm-size particles, was found during an SEM investigation of acanthite from Hilderston silver mine, Bathgate Hills, Lothian, Scotland. Vein-hosted mineralization at Hilderston occurs adjacent to an extensively hydrothermally altered quartz-dolerite dyke (Meikle 1994, Moreton et al. 1998). Native silver from Alva mine, Ochil Hills, Central Region, Scotland, contains up to 16 wt% mercury in association with tiemannite (the mercury selenide), but, so far, no native mercury is reported from this location (Moreton et al. 1998).

Mercury has also been reported from Chatsworth mine, Grassington Moor, Yorkshire (Kingsbury 1968, Embrey 1978, Dunham & Wilson 1985). The occurrence requires further confirmation, as Ryback et al. (1998, 2001) have shown that some samples in the Kingsbury collection (at the NHM) have been incorrectly provenanced and the status of mercury is therefore in doubt.

mereheadite

monoclinic — $Pb_2O(OH)Cl$

TL Merehead (Torr Works) quarry, Cranmore, Somerset, England

A very rare lead oxychloride related to litharge, found in lenses and cavities in veins of manganese and iron oxide minerals cutting Lower Carboniferous dolomitic limestone at the type locality. Mereheadite forms anhedral transparent or translucent, pale-yellow to reddish-orange grains, and is part of a secondary lead and copper oxychloride assemblage. It is intimately associated with blixite, calcite, cerussite, hydrocerussite and mendipite. Other minerals found in the cavities include chloroxiphite, datolite, paralaurionite, parkinsonite and symesite. The composition and physical characteristics of mereheadite and blixite are very similar (Welch et al. 1998, 2000). The lead oxychlorides are thought to have formed by interaction of primary galena with low-temperature metalliferous brines along fractures cutting the original Fe–Mn–(Cu–Pb) orebody (Symes & Embrey 1977, Humphreys et al. 1980), although it is possible that they precipitated directly from a cooling brine (Welch et al. 2000). Turner (2007) reports a discovery in 2005 of a manganese-rich pod from this locality producing an assemblage of fornacite, kentrolite, mereheadite, mimetite and symesite, and suggests a complex origin involving: formation of galena-rich veins in the late Permian or Triassic period; regional uplift and subsequent erosion, with exposure of the deposit to Jurassic seas and the triggering of manganate mineral formation; conversion of manganate minerals to manganese oxides during a hydrothermal event; and late-stage deposition of the rare oxychloride species.

Irregular patch of orange mereheadite (*c.* 8 mm across) in a vein cutting manganese-rich pods at Merehead (Torr Works) quarry, Cranmore, Mendip Hills, Somerset. (© D. I. Green)

merenskyite

MELONITE GROUP

trigonal — $PdTe_2$

A very rare palladium telluride that forms a series with melonite and has been qualitatively identified in a layered mafic–ultramafic intrusion on the Isle of Rum, Highland Region, Scotland. Associated species positively or tentatively identified include braggite, cooperite, hongshiite, irarsite, isoferroplatinum, isomertieite, keithconnite, kotulskite, laurite, michenerite, moncheite, palladoarsenide, platarsite–irarsite, platiniridium, potarite, sperrylite, telargpalite and telluropalladinite, in addition to the native species gold, silver, palladium and platinum, and unnamed alloys, arsenides, antimonides and sulphides. Most of these grains are 0.2–2 μm in diameter, although, for some species, grains > 15 μm are also present. The majority are in contact with sulphide minerals such as chalcopyrite, bornite or pentlandite, which were concentrated in chromitite layers during magmatic fractionation and crystal settling of an open-system magma chamber, modified by magma mixing and liquid immiscibility of sulphide and silicate melts (Power et al. 2000a, 2003). An analysis of merenskyite from a chalcopyrite-rich dyke associated with the West Sgaorishal ultramafic plug on the Isle of Rum is presented by Power et al. (2003), and grains too small to analyze are also noted in the plug itself (M. R. Power pers. comm. 2004).

At Talnotry, near Newton Stewart, Dumfries & Galloway, merenskyite occurs in the chalcopyrite–pyrrhotite–(pentlandite) zone of a small body of copper–nickel mineralization that developed at the base of an altered diorite sheet. It is associated with minor bismuthinite, galena, cobalt-rich gersdorffite, gold (electrum), a Bi–Pb–Re sulphide, silver and bismuth tellurides (possibly hessite and tellurobismuthite respectively), and the PGMs froodite, irarsite, michenerite and sperrylite (Power et al. 2004). Merenskyite is reported as inclusions in pentlandite or pyrrhotite, in sulphide-rich zones within the Arthrath mafic–ultramafic intrusion, near Ellon, Grampian. Associated accessory species include native bismuth, native gold, hessite, melonite and michenerite (McKervey et al. 2007).

merlinoite?

ZEOLITE GROUP

orthorhombic — $K_5Ca_2(Al_9Si_{23})O_{64}{\cdot}24H_2O$

A rare hydrated potassium–calcium aluminosilicate listed (but not described) from Tertiary basalts in Co. Antrim, Ireland (Adair 1987). A merlinoite-like mineral (intergrown with gobbinsite and garronite) had previously been noted in basalt found near Hills Port, S of the Gobbins, Island Magee, Co. Antrim (Nawaz & Malone 1982). Further confirmation is desirable.

meroxene

A variety of biotite (mica group), with an optic axial plane parallel to (010).

mertieite-II

trigonal — $Pd_8(Sb,As)_3$

A very rare palladium antimonide arsenide, close to mertieite-II in composition, has been found in a gold-bearing carbonate vein cutting Middle Devonian limestones at Hope's Nose, Torquay, Devon. It forms inclusions with isomertieite, close to, or on the outer surface of, a few palladium-rich gold dendrites etched out from the vein (Clark & Criddle 1982).

In Scotland, a chromite-rich dunite at Cliff, Baltasound–Haroldswick area, Unst, Shetland, is described as containing palladium antimonides of irregular shape that, on the basis of available chemical data, are either stibiopalladinite, mertieite-I or mertieite-II (Prichard & Tarkian 1988). However, Prichard et al. (1994) would appear to confirm them to be stibiopalladinite.

merwinite

monoclinic — $Ca_3Mg(SiO_4)_2$

A rare calcium–magnesium silicate found in fine-grain granular grey rocks near a chalk/dolerite contact at Scawt Hill, Larne, Co. Antrim, Ireland, where it occurs in a reaction zone within metasomatized chalk. Merwinite forms twinned tabular crystals up to 3 mm long associated with spurrite, both minerals enclosing many drop-like grains of larnite. Calcite and spinel are also present in this rock. In another assemblage, merwinite occurs with calcite, gehlenite, perovskite and spinel (Tilley 1929b). These unusual assemblages are thought to have developed because of the relatively high temperature prevailing during metamorphism, and elsewhere are represented by rocks containing calcite, diopside, grossular, vesuvianite and wollastonite (Tilley & Harwood 1931).

The Tertiary dolerite plug at Carneal, 9 km NNW of Carrickfergus, Co. Antrim,

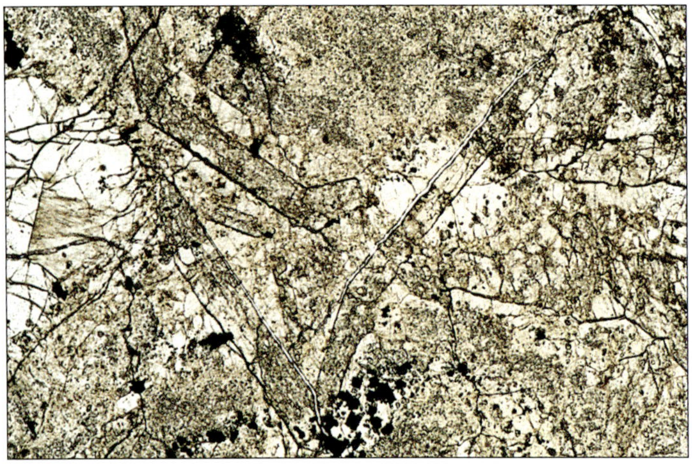

Elongate prisms of merwinite, some bent, others showing twinning in a matrix of spurrite, both minerals enclosing many drop-like grains of larnite. From the chalk/dolerite contact zone at Scawt Hill, near Larne, Co. Antrim, Ireland; width 4.5 mm; viewed in PPL and XPL. (© K. Gardner)

cuts basalt lavas and incorporates blocks of chalk and flint from underlying Cretaceous rocks. Assimilation of the limestone and flint took place at temperatures similar to those at Scawt Hill, and the chalk was transformed into several different assemblages, including one containing calcite, merwinite, spurrite and spinel, with minor hibschite (syn. hydrogrossular) and scawtite. A metasomatic vein containing very fine granules of merwinite and melilite is also noted (Sabine & Young 1975).

At Camas Mòr, Isle of Muck, Highland Region, Scotland, an olivine-gabbro dyke intersecting Jurassic limestones has developed high-temperature contact assemblages in which merwinite occurs with calcite, gehlenite and monticellite in a metasomatized limestone. In thin section, merwinite exhibits characteristic interpenetrant twinned laths 0.25 mm long set in calcite, sometimes mantled by a thin rim of spurrite (Tilley 1947). In a similar environment at Kilchoan, Ardnamurchan, nodular or concretionary limestones have been thermally metamorphosed by intrusions of dolerite and gabbro. Merwinite occurs in four metamorphic assemblages, together with combinations of the following species: åkermanite, cuspidine, grossular, kilchoanite, larnite, monticellite, rustumite, spurrite and vesuvianite (Agrell 1965).

mesitite or mesitine spar

A variety of magnesite, calcite group.

mesole

A synonym of faröelite – a radiating variety of thomsonite, zeolite group.

mesolite

ZEOLITE GROUP

orthorhombic $Na_2Ca_2(Al_6Si_9)O_{30}{\cdot}8H_2O$

An hydrated sodium–calcium aluminosilicate found most commonly in Tertiary amygdaloidal basalts from the Isle of Skye, Highland Region, Scotland, and Co. Antrim, Ireland. At Moonen Bay, on the W coast of Skye, white mesolite forms fibrous aggregates with individual fibres reaching 30 mm in length, sometimes with

Hair-like mesolite up to 12 mm long spearing chabazite; Moonen Bay, Isle of Skye, Highland Region, Scotland. (© D. I. Green)

apophyllite or chabazite crystals perched on top of them (Green & Todd 1996). Mesolite, along with apophyllite, calcite, heulandite, laumontite and stilbite are the main species filling vesicles in basalt at Sgùrr nam Boc, between Loch Brittle and Loch Eynort, a Heddle locality rediscovered in the mid-1980s. Size and quality are said to be outstanding (Pearce & Pearce 1991). Heddle (1857) describes mesolite from three localities on Skye: delicate feathery tufts of white crystals at the Storr, N of Portree; delicate interlacing white crystals at Talisker Bay; and, at Kilmore, yellowish-white radiating crystals of mesolite, forming solid masses. A red variety of mesolite is also noted by Heddle (1901) from the first of these localities and from Quiraing in the N of the Snizort Peninsula. Radiating acicular mesolite sprays are also found in vesicles in an olivine dolerite 1 km N of Portree, Isle of Skye. It is just one of several species filling vesicles in the area, others being calcite, gyrolite, saponite, tacharanite, thomsonite, tobermorite and xonotlite. Often, two or more species are found in the same vesicle (Sweet 1961). The distribution of mesolite on Skye is perhaps best illustrated by King (1976) and Wilson et al. (2003), who record 44 and 69 separate localities respectively for the species, with many around Dunvegan, Portree and S of Loch Bracadale. Mesolite is associated with some of the best calcite

Hair-like mesolite with attached crystals of calcite, infilling a vesicle in Tertiary basalt from Talisker Bay, Isle of Skye, Highland Region, Scotland; width 12 mm. (© A. G. Tindle)

ever collected from the basalts of Skye, in a remote and inaccessible coastal area about 7 km SE of Moonen Bay, at Biod a Mhurain, near Glen Ollisdal (Wolfe 1996b, Green & Todd 2001).

On the Isle of Mull, Strathclyde, pale-blue mesolite occurs mantling tobermorite in vesicular basalt to the N of the pier at Tobermory (Heddle 1880b). In a study of zeolite distribution patterns on Mull, Walker (1971) showed that mesolite occurs widely over much of the island, being found in his mesolite, laumontite and prehnite zones. Other associated species include analcime, chabazite, gyrolite, heulandite, stilbite, thomsonite and, less commonly, apophyllite and tobermorite. The outpourings of the volcano that produced these lavas also extended to the Morvern Peninsula on the mainland, where Walker (1971) mapped two mesolite-bearing zones. Currie (1905) notes analcime and tobermorite on mesolite in poorly amygdaloidal lava from Ardtornish Bay on Morvern.

A large alkali dolerite (crinanite) sill exposed at Clauchlands Point, Lamlash Bay, Isle of Arran, Strathclyde, contains accessory amounts of analcime, mesolite and thomsonite-Ca, infilling interstices between earlier-formed pyroxene and plagioclase-feldspar crystals (Tindle unpub. EMP data). Other Scottish occurrences mentioned in Heddle (1901) include the following:

HIGHLAND Mostly on the Isle of Skye – Dunvegan; southern shore of Loch Eynort; Orbost, 5 km S of Dunvegan, near Loch Bracadale; Rubha Thearna Sgùrr, N of Loch Brittle; Rubha na h-Airde Glaise, 3 km NE of Portree; Rubha nan Clach, N of Talisker Bay, Loch Bracadale; and Sgùrr nam Fiadh, between Talisker and Loch Eynort. Also Isle of Canna and Uamh Fhraing on the Isle of Eigg. LOTHIAN Crossgreen, near Linlithgow. STRATHCLYDE Bac Mòr and Fladda, Treshnish Isles, near Mull; seacliffs S of Beinn Chreagach, Brolass, Isle of Mull; Calgary Pier, W of Tobermory, Isle of Mull; SW of Carsaig Bay, Isle of Mull; Dumbarton; Hartfield Moss, 9 km WSW of Barrhead; Kilmalcolm, 10 km SE of Greenock; Pap Craig, Tinto, 12 km SE of Lanark; and Philipshill quarry, East Kilbride. WESTERN ISLES Garbh Eilean and Eilean Mhuire, both Shiant Isles.

In the Tertiary basalts of the Garron plateau area, Co. Antrim, Ireland, mesolite very commonly occurs with natrolite and scolecite as tufts of delicate hair-like crystals. Most frequently, these species are found at the bottom of the lava series (Walker 1951). Mesolite (antrimolite and harringtonite), chabazite, phillipsite, stilbite and thomsonite are the dominant zeolite species found in vesicular olivine basalt at the Gobbins, Island Magee, near Larne, Co. Antrim (Tomkeieff 1934). Nawaz (1988a) presents data for mesolite from Dunseverick, near the Giant's Causeway, and the species is also noted with analcime and chabazite-Ca (phacolite) from Killyflugh quarry, N of Ballymena, Lower Toome (Herbert Smith 1916, Hey 1933), and from White Rocks, near Portrush (Robert 2001). In the basalt quarries around Ballymena, mesolite occurs with chabazite, gyrolite and thomsonite, together with fine cabinet specimens of apophyllite (Green & Moreton 2001). Mesolite and heulandite are particularly abundant at Sales Corner quarry, near White Mountain, SW of Belfast. Other reports of mesolite, often as delicate tufts, are from: Benanouran Head; Portnaweelan, near Dunseverick; Portrush (all along the Giant's Causeway section); Castle Robin Primary School quarry, between Collin and White Mountains; Dog quarry, NE of White Mountain and Aughrim Plantation quarry, SW of Collin Mountain (all W or SW of Belfast); Moneymore quarry, near Magherafelt, Lough Neagh, Co. Londonderry (Adair 1987). Greg & Lettsom (1858) report mesolite from Down Hill, Co. Londonderry.

Mesolite is said to have been found along the cliffs between Botallack and Wheal Cock, St Just, Cornwall (Hall 1868a, Russell 1910b, Robson 1949b, Golley & Williams 1995), but this requires confirmation.

Mesolite occurs as white fibres at the Bouet, Guernsey (Mourant 1939).

mesolitine

TL Newtown Crumlin, Clough, Co. Antrim, Ireland

A synonym of thomsonite, zeolite group.

mesotype

A subfamily consisting of mesolite, natrolite, scolecite and thomsonite – zeolite group.

meta-aluminite

monoclinic $Al_2(SO_4)(OH)_4{\cdot}5H_2O$

A rare hydrated aluminium sulphate hydroxide found at Newhaven, Sussex (Anthony et al. 2003). See aluminite.

META-AUTUNITE GROUP

Tetragonal or orthorhombic uranyl phosphates and arsenates with the general formula $A(UO_2)_2(XO_4)_2{\cdot}nH_2O$, where A = Ba, Ca, Co, Cu^{2+}, Fe^{2+}, $(H_3O)_2$, K_2, $(NH_4)_2$, Zn; and X = As^{5+}, P^{5+}. Bassetite, meta-autunite, metakahlerite, metatorbernite, metazeunerite and possibly meta-uranocircite are members of the group found in Britain or Ireland. Metanováčekite is also discussed.

meta-autunite

META-AUTUNITE GROUP

tetragonal $Ca(UO_2)_2(PO_4)_2{\cdot}2\text{–}6H_2O$

A rare hydrated calcium uranyl phosphate found in boulders believed to be derived from an adjacent silicified fracture zone near Ousdale Burn, 2 km N of Helmsdale, Highland Region, Scotland. Meta-autunite is found with a variety of secondary uranium minerals, including metatorbernite, in association with quartz and barite. The boulders consist of Devonian Old Red Sandstone arkose formed by erosion of the Helmsdale granite, modified by later fracturing and hydrothermal alteration (Gallagher et al. 1971, Dunham et al. 1978, Tweedie 1979).

metabasaluminite

An artificial mineral created by heating hydrobasaluminite at 150°C.

metacinnabar

SPHALERITE GROUP

cubic HgS

A rare mercury sulphide, trimorphous with cinnabar and hypercinnabar, and briefly noted with mercury-rich sphalerite as inclusions in cinnabar from a unique vein of massive cinnabar and dolomite at Gortdrum mine, near Tipperary, Co. Tipperary, Ireland. The orebody consists of four principal species – bornite, chalcocite, chalcopyrite and mercury-rich tennantite – and is thought to have formed as a consequence of convective water circulation, mainly above the unconformity between Devonian and Silurian rocks, because of high geothermal gradients related to Carboniferous volcanic activity in the area (Steed 1986).

Metacinnabar is also described from Cockhill mine, Bewerley, near Pateley Bridge, Yorkshire (Kingsbury 1969, Embrey 1978). This occurrence requires further confirmation, as Ryback et al. (1998, 2001) have shown that some samples in the Kingsbury collection (at the NHM) have been incorrectly provenanced and the status of metacinnabar is therefore in doubt.

metagreenalite

A synonym of greenalite.

metahalloysite

Given species status by some authorities (Clark 1993), but considered by others to be a synonym of halloysite (Bayliss 2000).

metakahlerite

META-AUTUNITE GROUP

tetragonal $Fe^{2+}(UO_2)_2(AsO_4)_2{\cdot}8H_2O$

A rare hydrated iron uranyl arsenate found on joint surfaces of shaly to slaty country rocks half way up the cliff, adjacent to a highly uranium-rich fissure vein close to Needle's Eye, SW of Caulkerbush, Dumfries & Galloway, Scotland. This and other similar veins are in the aureole of the Criffell granodiorite. Metakahlerite forms yellow scales and is associated with kahlerite. Other supergene minerals from this vicinity, produced during the alteration of primary As–Bi–Co–Cu–Fe–Pb–U minerals, include cornubite, eulytite, lavendulan, malachite, meta-autunite, metazeunerite, mixite, schoepite, tyrolite, vandendriesscheite, walpurgite, wulfenite and zeunerite (Macpherson & Livingstone 1982, Braithwaite & Knight 1990).

metaköttigite?

triclinic $(Zn,Fe^{3+})(Zn,Fe^{3+},Fe^{2+})_3(AsO_4)_2{\cdot}8(H_2O,OH)$

A rare hydrated zinc–iron arsenate hydroxide, possibly present in cavities in quartz at Sandbeds Gill, Bassenthwaite, Cumbria, where it forms minute yellowish-brown acicular crystals. XRD analysis could not distinguish between metaköttigite and a metaköttigite–symplesite intermediate composition, hence the doubtful status (Ryback & Tandy 1992).

metanatrolite

A synonym of natrolite, zeolite group.

metanováčekite

tetragonal — META-AUTUNITE GROUP — $Mg(UO_2)_2(AsO_4)_2 \cdot 4–8H_2O$

An uncommon hydrated magnesium uranyl arsenate found in disused mineworkings of the Cargodna section of Wheal Owles, St Just, Cornwall. Metanováčekite is the lower hydrate of nováčekite, with which it is intimately mixed. In warm and dry conditions, dehydration produces the metahydrate, but the process is reversible. Metanováčekite and nováčekite are supergene species that lined a small vug in quartz–sulphide veins containing arsenopyrite, chalcopyrite and sphalerite. Magnesium necessary for the formation of these species most likely came from the decomposition of the mafic hornfels host rock (Elton et al. 1994a).

metasideronatrite

orthorhombic — $Na_2Fe^{3+}(SO_4)_2(OH) \cdot H_2O$

A rare hydrated sodium–iron sulphate hydroxide found as an alteration product of pyrite, typically in seashore environments.

Metasideronatrite is associated with gypsum and pyrite at Trerubies Cove, near Delabole, Cornwall. Metasideronatrite forms yellow botryoidal crusts mixed with sideronatrite (Ryback & Tandy 1992). Barton-on-Sea, near Lymington, Hampshire, is another locality for metasideronatrite (Anthony et al. 2003).

Oxidation of sulphides in Carboniferous shales on the coast at Doon Bay, N of Ballybunnion, Co. Kerry, Ireland, has produced the secondary sulphate minerals metasideronatrite and sideronatrite, as well as felsöbányaite (syn. basaluminite), brochantite, gypsum and hydronium jarosite in nearby coves. Metasideronatrite is intimately mixed with sideronatrite and forms pale-yellow botryoidal crusts 1–2 mm thick associated with tiny gypsum crystals. Sea water is assumed to be the source of Na^+ (Moreton et al. 1995).

metastibnite

amorphous — Sb_2S_3

A rare antimony sulphide, dimorphous with stibnite, that has been found in a sulphide-bearing quartz vein from the headwaters of Wet Swine Gill, at Coombe Height, Caldbeck Fells, Cumbria. Metastibnite occurs as a bright red encrustation a few tens of microns thick on senarmontite and stibnite, and in thin fractures in vein quartz (D. I. Green pers. comm. 2007).

Deep-red coating of metastibnite (up to 0.2 mm thick) in a quartz-rich vein from Wet Swine Gill, Coombe Height, Caldbeck Fells, Cumbria. (© D. I. Green)

metastrengite

A synonym of phosphosiderite.

metatorbernite

tetragonal — META-AUTUNITE GROUP — $Cu^{2+}(UO_2)_2(PO_4)_2 \cdot 8H_2O$

A rare hydrated copper uranyl phosphate, very similar in composition to torbernite, with the main difference being a lower water content in metatorbernite (Hallimond 1920). The finest metatorbernite known in Cornwall was found at Old Gunnislake mine, Calstock, near Gunnislake (Philips 1816, Embrey & Symes 1987). Tiny dispersed crystals of metatorbernite and torbernite are reported from altered tourmaline granite wallrock from the Godaver tin mine, St Cleer, in the heart of the Bodmin Moor granite (Ball et al. 1982). In Cornwall, metatorbernite is associated with ningyoite, torbernite and uraninite (pitchblende) at Wheal Alice, Grampound Road, and with ningyoite and uraninite at Wheal Buller, near Redruth; the two occurrences are associated with chalcopyrite-dominated vein mineralization (Ball et al. 1982). Additionally, metatorbernite is reported from Wheal Alice by Harrison & Taylor (1966), who also describe weathered uraninite (var. pitchblende) and metatorbernite from Wheal Bray, Altarnun. Metatorbernite (and torbernite) occurs as pale-green small thin platy crystals on gossan veinstone, sometimes with olivenite and scorodite, at Wheal Gorland, St Day, 2 km E of Redruth (Rust 1981b), and as distinct grass-green platelets up to 3 mm across, coating limonitic quartz at Cligga Head, near St Agnes (Chick 1987). At Ting Tang mine (Morcom's shaft), Carharrack, Gwennap, Bruce & Aubrey-Jones (1998) report visually indistinguishable torbernite and zeunerite that have partially dehydrated to produce translucent-to-opaque metatorbernite and metazeunerite.

In Scotland, metatorbernite is found with a variety of secondary uranium minerals, including meta-autunite, in association with quartz and barite in boulders believed to be derived from an adjacent silicified fracture zone near Ousdale Burn, 2 km N of Helmsdale, Highland Region. The boulders consist of Devonian Old Red Sandstone arkose formed by erosion of the Helmsdale granite and modified by later fracturing and hydrothermal alteration (Gallagher et al. 1971, Dunham et al. 1978, Tweedie 1979). At Blackhall and Mill of Cairston, Stromness district, Mainland, Orkney, secondary metatorbernite occurs in thin coatings on minor fractures within dolomite, phosphatic siltstones and phosphatic sandstones of the Old Red Sandstone Flagstone series. Both locations are near the Moine/Devonian unconformity and adjacent to a major fault (Michie 1972).

Metatorbernite crystals to 1.5 mm from Goonbarrow china-clay pit, St Austell, Cornwall. (© D. I. Green)

metatyuyamunite?

orthorhombic — $Ca(UO_2)_2(VO_4)_2 \cdot 3H_2O$

A rare hydrated calcium uranyl vanadate said to form a yellow efflorescence on vanadium-rich nodules of Permian age, found at Budleigh Salterton, Devon (Wyley 1962, Durrance & George 1976). It is considered to have formed by the spontaneous dehydration of tyuyamunite, which in turn formed from an original coffinite and thucolite assemblage. However, Embrey (1978) and King & Dixon (1971) suggest that metatyuyamunite is probably pascoite. In a report on sheet-like plates of native copper occurring with uranium–vanadium concretions in the Littleham Cove area, near Budleigh Salterton, Milodowski et al. (2000) refer to the earlier literature and list metatyuyamunite as one of many species present in black mineralized clays and nodules. The occurrence requires confirmation.

meta-uranocircite?/metauranocircite? META-AUTUNITE GRP
tetragonal $Ba(UO_2)_2(PO_4)_2{\cdot}8H_2O$

A rare hydrated barium uranyl phosphate described from Merrivale granite quarry, near Princetown, Devon, and identified by XRD. Meta-uranocircite is mentioned in field-trip notes compiled by the South-West branch of the Russell Society. Further confirmation is desirable.

meta-uranopilite?/metauranopilite?
orthorhombic? $(UO_2)_6(SO_4)(OH)_{10}{\cdot}5H_2O$

A rare hydrated uranyl sulphate hydroxide, reported as green or yellow needles and vaguely described as occurring in Cornwall (O'Donoghue 1994). Requires confirmation.

metavariscite
monoclinic $AlPO_4{\cdot}2H_2O$

A rare hydrated aluminium phosphate, dimorphous with variscite and found at Gunheath china-clay pit, Carthew, St Austell, Cornwall. Anhedral metavariscite crystals are found as friable aggregates with minor kaolinite in the cores of variscite spherules that grow to 2 mm in diameter. A cryptocrystalline crust on one spherule is nearly pure metavariscite. Variscite and metavariscite typically occur on a matrix of granite pegmatite or coarse quartz–tourmaline vein material. Cassiterite occurs in the matrix of several specimens, and wavellite is a common and intimate associate (Elton 1996).

metavivianite
triclinic $(Fe^{2+},Fe^{3+})_3(PO_4)_2{\cdot}8H_2O$

A rare hydrated iron phosphate thought to have been found during an investigation of the original materials used in medieval wall paintings in Winchester Cathedral, Hampshire. Vivianite was extensively used as a blue pigment and, over time, it oxidized to produce a yellow alteration product. Because of the small sample size, it proved very difficult to separate enough of the yellow material for XRD analysis; however, analysis of one mixed sample indicated that metavivianite was present (Howard 1996). Metavivianite appears to be confirmed from Wheal Jane, Kea, near Truro, Cornwall, although no details of the occurrence are given (Anthony et al. 2000).

metavoltine
hexagonal $K_2Na_6Fe^{2+}Fe^{3+}_6(SO_4)_{12}O_2{\cdot}18H_2O$

An uncommon hydrated potassium–sodium–iron sulphate found along the cliffs near the sea-level adit to Wheal Edward, St Just, Cornwall, where it forms sulphur-yellow botryoidal crusts (Ryback & Tandy 1992). Metavoltine also occurs as a locally abundant supergene mineral encrusting weathered surfaces of sandstone in Brigham quarries, Cockermouth, Cumbria. Here it produces patchy vivid-yellow efflorescent crusts coating areas up to 1.5 m across, made up of minute platy crystals. Metavoltine probably formed very recently because of oxidation of pyrite in nearby mudstone, together with diffusion of sulphate-rich fluids through the porous sandstone (Young et al. 1992a).

metazeunerite META-AUTUNITE GROUP
tetragonal $Cu^{2+}(UO_2)_2(AsO_4)_2{\cdot}8H_2O$

An uncommon hydrated copper uranyl arsenate found in mineralized fissure veins at the contact of the Land's End granite with hornfels at Wheal Edward, St Just, Cornwall. Metazeunerite forms dark-green tabular crystals up to 15 mm across and also globular and plate-like clusters. Associated uranium-rich species include autunite, compreignacite, torbernite and uraninite (Collins 1871, Wolloxall 1988, Elton et al. 1994b). Phosphuranylite and saléeite may also be present (Ryback & Tandy 1992). Metazeunerite is also noted from Wheal Owles, St Just (Anthony et al. 2000). Near by, at Geevor mine, near Pendeen, St Just, rare metazeunerite and rare scorodite, with antlerite, atacamite, brochantite, chalcanthite, gypsum and sodium-zippeite, were found on the 17 level, Levant North lode, associated with a chalcopyrite-rich vein (Elton & Hooper 1993). Metazeunerite is mentioned as a dehydration product of zeunerite on the dumps of Morcom's shaft, Ting Tang mine, Carharrack, associated with sulphide-rich (arsenopyrite, chalcopyrite, pyrite) veins cutting hornfels, near the margins of the Carn Marth granite (Bruce & Aubrey-Jones 1998). Metazeunerite from Wheal Gorland, St Day, 2 km E of Redruth, described by Kingsbury (1954), requires further confirmation given the doubts about the provenance of other Kingsbury specimens (Ryback et al. 1998, 2001).

In Scotland, metazeunerite is found in a highly uranium-rich fissure vein close to Needle's Eye, SW of Caulkerbush, Dumfries & Galloway. This and other similar veins are in the aureole of the Criffell granodiorite. Metazeunerite and zeunerite form green platy crystals up to 0.2 mm across associated with lavendulan on joint and cavity surfaces. Other supergene minerals from this vicinity, produced during the alteration of primary As–Bi–Co–Cu–Fe–Pb–U minerals, include cornubite, eulytite, kahlerite, malachite, metakahlerite, mixite, schoepite, tyrolite, uranophane, vandendriesscheite, walpurgite and wulfenite (Macpherson & Livingstone 1982, Braithwaite & Knight 1990).

miargyrite
monoclinic $AgSbS_2$

A rare silver sulphantimonite found in a very high-grade pod of polymetallic silver mineralization cutting pyrite mineralization in the B zone of the Silvermines orebody, Co. Tipperary, Ireland. The silver mineralization included acanthite, argyrodite, miargyrite, proustite, smithite and xanthoconite, together with barite, galena and sphalerite (Gasparrini 1978, Taylor & Andrew 1979). Taylor (1984) also reports geocronite–jordanite, gersdorffite and pyrargyrite from the same zone. A comprehensive review of the mineralization in the Silvermines district is provided by Moreton (1999), who suggests that the mineralization has a sedimentary-exhalative origin related to a system of WNW-trending faults. Miargyrite and tennantite are also reported from Tynagh mine, near Killimor, Co. Galway, Ireland (Schultz 1968).

MICA GROUP

A large group of pseudo-hexagonal monoclinic layer silicates with the general formula $IM_{2-3}\square_{0-1}T_4O_{10}A_2$, where *I*, *M*, *T* and *A* indicate elements having similar ionic radii and being capable of replacing each other in the structure. *I* = K, Na, Ca, Cs, NH_4, Rb, Ba; *M* = Li, Fe^{2+}, Fe^{3+}, Mg, Al, Ti, Mn^{2+}, Mn^{3+}, Zn, Cr, V; *T* = Al, Fe^{3+}, Si, Be, B; and *A* = F, OH, Cl, O (oxymicas), S. Mica-group minerals are classified into three different types: true mica, brittle mica and interlayer-deficient mica (all can be either dioctahedral or trioctahedral). A revised nomenclature for micas, which elevates biotite, brammallite, glauconite, illite, lepidolite, phengite and zinnwaldite to series status, is provided by Rieder et al. (1998). A further (but not IMA approved) classification is provided by Tischendorf et al. (2007). Because these classifications do not clearly define compositional boundaries between all species, a simple approach based on aluminium, iron and magnesium contents is adopted here (see annite, biotite, eastonite, siderophyllite and phlogopite for further details). Annite, celadonite, clintonite, eastonite, kinoshitalite, margarite, muscovite, paragonite, phlogopite, polylithionite, roscoelite and siderophyllite are members of the group found in Britain or Ireland.

michenerite COBALTITE GROUP
cubic PdBiTe

A very rare palladium–bismuth telluride, identified in a layered mafic–ultramafic intrusion on the Isle of Rum, Highland Region, Scotland. Within certain chromitite units of the Eastern Layered Series, michenerite is found in association with species positively or tentatively identified as isoferroplatinum, kotulskite, moncheite, members of the hollingworthite–irarsite–platarsite series, sperrylite, telluropalladinite and the native species gold, silver, palladium and platinum. Most of these grains are 0.2–2 µm in diameter, although for some species grains > 15 µm are common. The majority are in contact with sulphide minerals such as bornite, chalcocite, chalcopyrite, pentlandite or pyrrhotite. The chromitite layers were formed during magmatic fractionation and crystal settling of an open-system magma chamber, modified by magma mixing and liquid immiscibility of sulphide and silicate melts (Power et al. 2000a, 2003).

The Ben Buie ultramafic intrusion on the Isle of Mull, Strathclyde, contains michenerite with hollingworthite, laurite and sperrylite associated with chromitite layers, and it represents one of the earliest products of the Mull layered complex (Prout & Andersen 2001).

At Talnotry, near Newton Stewart, Dumfries & Galloway, michenerite forming rounded blebs, less than 10 µm across, occurs in the chalcopyrite–pyrrhotite–(pentlandite) zone of a small body of copper–nickel mineralization that developed at the base of an altered diorite sheet. It is associated with minor cobalt-rich gersdorffite, bismuthinite, galena, gold (electrum), silver telluride and bismuth telluride (possibly hessite and tellurobismuthite respectively), a Bi–Pb–Re sulphide and the PGMs froodite, irarsite, merenskyite and sperrylite (Power et al. 2004). A single grain of michererite has been found with native bismuth, froodite and native gold, in a borehole cutting sulphide-rich gabbro at Littlemill farm, near Huntly, Grampian Region (McKervey et al. 2007).

microcline

FELDSPAR GROUP
triclinic $KAlSi_3O_8$

A common potassium aluminosilicate, dimorphous with orthoclase and found in granites, granite pegmatites, syenites, some metamorphic rocks and also hydrothermal veins formed at medium to low temperatures, Microcline is often intimately intergrown with quartz (myrmekite) or albite (perthite). Amazonite is a pale-green variety of microcline.

The alkali feldspars in Cornish granites are micro- and cryptoperthites in which the potassic phase may be microcline or orthoclase. Microcline appears to be more common in the siderophyllite granites of Godolphin, Bodmin Moor, Dartmoor and the lithium-mica granites of Tregonning and St Austell, but, as the inversion of monoclinic orthoclase to triclinic microcline is a complex process involving low-temperature fluids, it is expected that both occurred during the subsolidus recrystallization of all the granites (Exley & Edmondson 1993).

The porphyritic granite at Bosahan quarry, near Constantine, Cornwall, is a typical Cornish granite; it contains large and abundant phenocrysts of microcline perthite associated with biotite, muscovite, plagioclase (oligoclase) and quartz, and accessory amounts of andalusite, apatite, ilmenite, monazite, pyrite, tourmaline and zircon. Hydrothermal alteration has replaced some of the microcline with muscovite (Rice 1973). Microcline, quartz and zinnwaldite are major components of a pegmatite vein in Gunheath china-clay pit, St Austell. The vein is notable for its accessory assemblage of apatite, arsenopyrite, ferrocolumbite, opal, topaz and tourmaline and turquoise (Hodkinson & Clark 1977). Barstow (1982) adds that cassiterite, orthoclase, stannite, torbernite and wolframite also occur in these veins. A pink specimen labelled rhodonite, from the lead lode at Greystone quarry, Lezant, has been shown to be microcline (Elton et al. 2004).

At Arm o' Grain, Caldbeck Fells, Cumbria, microcline is associated with grey chalcedonic quartz and it forms euhedral white-to-colourless crystals, up to a few mm across, in cavities within earlier-formed vein quartz (Green et al. 2006b).

In Wales, rare microcline is reported from the St David's granophyre, SW of St David's, Dyfed (Bloxam & Dirk 1988), and from various units of the Precambrian on Anglesey, including granitic rocks of the Coedana complex and calc-silicate gneisses in the Nebo inlier (Horák 1993).

The Lewisian pegmatite at Sletteval, South Harris, Western Isles, Scotland, contains a central zone rich in microcline and biotite, and, together with another nearby pegmatite at Chaipaval (Ceapabhal), was once worked for its potash feldspar (Robertson 1945, von Knorring & Dearnley 1960a). Difficulty in the analysis of microcline because of incipient perthitic exsolution is noted from Dalradian muscovite granitic migmatites around the Aberhirder, Huntly and Knock gabbroic intrusions in Grampian (Ashworth 1976). Quartz veins cutting the Cruachan granite contain euhedral, yellow or pink crystals of microcline, up to 5 mm across, at the rediscovered classic Heddle locality for molybdenite at Coire Buidhe, Glen Creran, Appin, Strathclyde (Ingram et al. 1993).

Microcline in granitic and syenitic rocks is documented from several intrusions in the Highland Region of Scotland, including the Loch Ailsh syenite complex in Assynt, three syenite intrusions around Loch Loyal, near Tongue, and the Foyers tonalite–granite complex, 30 km S of Inverness (Parsons & Boyd 1971). These authors suggest that microcline becomes more abundant relative to orthoclase in the later more fractionated members of these intrusive sequences. Small masses of metasomatized hornblendic rocks of Moinian age at White Bridge, near Invergarry, contain barium-rich microcline forming homogeneous crystals, microperthite and myrmekite-like intergrowths with oligoclase feldspar (Rock 1984). Remarkable crystals of beryl-green amazonite (amazonstone) up to 50 cm long were once found in a giant erratic boulder at Beinn Bhreac, 2 km E of Tongue. They were thought to be derived from the Ben Loyal (Beinn Laoghal) syenite complex (Heddle 1877). *In situ* occurrences of bright-green amazonite, forming crystals up to 25 mm long, were later reported in pegmatitic veins cutting syenite in the cliff of Sgòrr Chaonasaid, Ben Loyal (Heddle 1901). Many more occurrences of microcline are also listed by Heddle (1901). Microcline occurs with aegirine-augite, apatite and richterite in fenite associated with the Loch Borralan syenite complex, Assynt. It is interstitial, shows no visible exsolution and in thin section sometimes exhibits cross-hatched twinning (Rock 1977).

In Ireland, microcline is found in pegmatites along the southern flank of the Leinster batholith. Microcline data are reported from two spodumene pegmatites at Borris, Co. Carlow, and Moylisha, Co. Wicklow (Luecke 1981), and the species is also noted from amblygonite- and petalite-bearing pegmatites in the same area (Steiger & von Knorring 1975). In Co. Donegal, orthoclase microperthite and microcline are both present in the Ardara granite pluton, between Ardara and Maas (Hall 1966). A structural study of potassium-feldspar from the Galway granite, Co. Galway, has shown a transition from orthoclase to microcline as a result of late deformation (Wilson & Coats 1972).

Microcline (var. amazonite), 50 mm across, from the giant erratic boulder (Heddle's boulder) on Beinn Bhreac, near Tongue, Highland Region, Scotland. (© D. I. Green)

microlite

PYROCHLORE GROUP
cubic $(Ca,Na_2)Ta_2(O,OH,F)_7$

A rare calcium–sodium–tantalum oxide hydroxide that forms a series with pyrochlore and is found in lithium-rich rare-element pegmatites, where it can be a primary phase or a replacement of earlier-formed tantalum-rich species such as manganotantalite.

Minute yellow, grey and brownish octahedral crystals of microlite have been found in aplite and associated crosscutting pegmatitic veins at Meldon, near Okehampton, Devon. This lithium–fluorine-rich aplite dyke is emplaced into the metamorphic aureole of the Dartmoor granite. The largest crystals of microlite are 0.8 mm across and are partially intergrown with elbaite, lithium mica and quartz. One grain of microlite was partially replaced by fersmite. The aplite crystallized from an extremely evolved granitic magma and is host to many rare accessory species, including native antimony, native arsenic, arsenolite, autunite, herderite, manganocolumbite and stibiconite (von Knorring & Condliffe 1984). Further analysis of microlite from aplite at this locality reveals it to be tungsten rich and, in one sample, so enriched in uranium that it approaches uranmicrolite in composition (Tindle unpub. EMP data).

Grey, yellowish and brown microlite, occasionally in small octahedral crystals, and with a significant uranium content, occurs in lithium-mica pegmatite boulders containing cassiterite and manganotantalite at Stranakelly, 3 km N of Shillelagh, Co. Wicklow, Ireland. Albite (cleavelandite), manganese-rich apatite, fluorite, quartz and spessartine are associated species (Steiger & von Knorring 1975, Luecke 1981, McArdle & Kennan 1989). Irregular crystals of brown microlite up to 3 mm across were separated from lepidolite greisen at Stranakelly and shown to be intergrown with bariopyrochlore (pandaite). Bariopyrochlore is considered to be an exsolved phase rather than a replacement product of microlite (Scoon 1978). The occurrence is related to the nearby Leinster granite batholith.

Panned concentrates of alluvium from a tributary of the Avoca River, close to Avoca mine, Co. Wicklow, and two localities in Co. Carlow (Gibbett Hill and Tombreen Bridge, both E of Aclare) contain assemblages indicating that they have sampled rare-element pegmatites adjacent to the Leinster granite. Crystals of manganotantalite up to 500 μm long are partially replaced in a very irregular manner by microlite. Cassiterite, ferrotantalite and ferrotapiolite are other tantalum-bearing species noted in the panned concentrates (R. A. Ixer sample, Tindle unpub. EMP data).

milarite

OSUMILITE GROUP
hexagonal $KCa_2AlBe_2Si_{12}O_{30}\cdot 0.5H_2O$

A rare hydrated potassium–calcium–aluminium–beryllium aluminosilicate found in late-stage miarolitic cavities within granite at the Shap granite quarry, Cumbria. Milarite forms colourless transparent prismatic crystals up to 2 mm long and is one of three late-stage beryllium species (others: bavenite, bertrandite) that probably grew as beryllium was concentrated during the final stages of crystallization of the

Milarite crystals to 1.8 mm in a miarolitic cavity within granite. Shap granite quarry, Cumbria. (© D. I. Green)

granitic magma. Associated cavity species are anatase, apatite, biotite, fluorite, pyrite and titanite (Wirth et al. 2001, Wirth & Green 2002).

Milarite was also reported from Cheesewring quarry, SW of Henwood, Linkinhorne, Cornwall, and Meldon aplite quarry, near Okehampton, Devon (Kingsbury 1964b, Embrey 1978). These occurrences require further confirmation, as Ryback et al. (1998, 2001) have shown that some samples in the Kingsbury collection (at the NHM) have been incorrectly provenanced. Wirth et al. (2001) have since reported milarite from Meldon, but their specimen has a matrix different from that of the Kingsbury sample and, as other beryllium species are found at this locality, there is no reason to doubt that this milarite is correctly located.

milk opal

A synonym of white opal-CT.

milky quartz

A translucent-to-opaque white variety of quartz containing abundant fluid inclusions.

millerite

trigonal NiS

Named after William Hallowes Miller FRS (1801–1880), a mineralogist, distinguished mathematician, crystallographer, professor of mineralogy at the University of Cambridge, and author of *Treatise on crystallography* (1838); born at Felindry, near Llandovery, Dyfed, Wales. Millerite is a common nickel sulphide often found as a diagenetic species in sedimentary rocks associated with Coal Measures. It also occurs in low-temperature hydrothermal nickel-bearing sulphide veins.

England

Cleveland Young et al. (1985b) notes millerite from an underground borehole at Blackhall colliery.

Cornwall Millerite has been found in a cross-course vein associated with kupfernickel (either gersdorffite or nickeline) and other nickel ore at Fowey Consols mine, Tywardreath, St Blazey (Collins 1892, Dines 1956, Lewis 1997), and in a cavity lined with pink calcite within a vein cutting black shales at Gerrans Bay, Porthscatho (Collins 1911a). Late-crystallizing brassy-yellow acicular-to-prismatic crystals of millerite (to 0.6 mm) occur in tiny cavities within a sulphide-rich matrix on the dumps at Penberthy Croft mine, St Hilary (Rust 1996a,d, Betterton et al. 2000). At Greystone quarry, Lezant, near Launceston, millerite forms bright metallic acicular crystals up to 20 mm long in quartz–carbonate veins (Elton et al. 2004). Other occurrences of millerite include: Pengelly mine, near St Ewe, 4 km SSE of St Stephen (Collins 1892, Dines 1956); in capillary and divergent fibres of a dull yellow colour at Wheal Sparnon, near Redruth; and at Wheal Chance (South Roskear mine), St Austell (all Collins 1892).

Cumbria Millerite is rare in the Northern Pennine orefield, but at Brownley Hill mine, Nenthead, near Alston, it forms bright brassy-yellow acicular crystals up to 25 mm long filling cavities lined with ankerite, calcite and siderite. It is associated with galena and pyrite, and is sometimes covered by an unidentified green oxidation product (Green & McCallum 1993, Green et al. 2000, Green & Briscoe 2002). The species has also been found in a small gersdorffite–nickeline orebody on the Dow Scar level of Hilton lead mine, Scordale, near Appleby, where it forms 2 mm-long needles in gersdorffite-coated cavities within massive nickeline (Bridges 1982, Ixer 1986, Winrow 1986a). One bed of clay–ironstone nodules in Coal Measures at the Oughterside opencast coal site, Aspatria, contains siderite-lined cracks in which millerite is locally common as brass-colour capillary crystals up to 10 mm long (Young & Nancarrow 1988a).

Derbyshire Millerite forms rare needle-like crystals in sulphide-rich veins from the Tansley borehole, S of Ashover (Ramsbottom et al. 1962, Ixer & Townley 1979, Ixer 1986), and forms filamentous inclusions in a calcite vein within olivine dolerite (toadstone) at Fall Hill (Gate) quarry, Ashover (Smith 1982, Spence 1982, Braithwaite 1983).

Devon Spectacular millerite crystals, up to 1 cm long, have been found in quartz pods in a blackish Middle Devonian shale on the beach at St Mary's Bay, Brixham. They are sometimes fresh and bright golden, or tarnished and altered to a translucent green mixture of variable composition, and are associated with pyrite or, less commonly, chalcopyrite, galena and sphalerite. Millerite has also been noted at Brookwood mine, Dartmoor (Braithwaite & Wilson 2001a). Vague reports of rare millerite from the Combe Martin mines describe minute filaments lining cavities and dispersed among crystals of galena (Greg & Lettsom 1858, Hall 1868b, Collins 1892, Dines 1956, Dunham et al. 1978, Rust 1982a, Braithwaite & Wilson 2001a).

Durham Specimens of millerite from Boltsburn mine, Rookhope, are in the Russell collection at the NHM. The species has also been found at Cowgreen mine, Harwood-in-Teesdale (Dearman & Jones 1967, Dunham 1990).

Gloucestershire Needle-like crystals of millerite were once found in the casts of fossil shells within Wenlock limestones and shales at Gorsley Common, W of Newent (Symonds 1872, Bick 1986a).

Northumberland Minor amounts of chalcopyrite, marcasite, millerite and pyrite (including bravoite) occur in galena–nickeline–sphalerite–ullmannite-bearing calcite veins close to the Whin Sill at Settlingstones mine, near Hexham (Ixer 1986).

Staffordshire Golden acicular crystals of millerite occur either encrusting, or as inclusions within, calcite and celestine at the Waterbank end of Clayton adit, Ecton Hill. A bright-green alteration product covering some of the millerite is described as morenosite (Starkey 1983, Ford et al. 1993).

Tyne & Wear Millerite is reported from Boldon colliery (between South Shields and Sunderland) and Walker colliery, near Wallsend (Dearman & Jones 1967, Young et al. 1985b). Forming brassy-colour tufts of slender acicular crystals, millerite was also found in open joints in Coal Measures sandstones from the Green's Farm borehole, 5 km SW of Newcastle-upon-Tyne (Anderson & Smythe 1942, Young et al. 1985b).

Warwickshire Millerite forming acicular or capillary crystals up to 24 mm long, and associated with galena and retgersite, was discovered in a Coal Measures clay–ironstone septarian nodule from Newdigate colliery, Bedworth, SW of Nuneaton (Walker 1988).

Wales

Dyfed Many occurrences of millerite have been documented from the Central Wales orefield: it occurs as very long slender capillary crystals intergrown with quartz at Eaglebrook (Nantycagl) mine, Ceulanymaesmawr (Jones & Moreton 1977, Jones 1983), and richly at Brynyrafr mine, where tangled masses of acicular crystals up to 20 mm long fill quartz cavities. It is invariably associated with chalcopyrite, and other cobalt–nickel minerals may be present (Bevins 1994, Mason 1998). Other localities include: Brynarian mine, near Tal-y-bont; Camdwrbach mine, near Bryniau Rhyddion, NW of Nant-y-moch reservoir; Erglodd mine, N of Tal-y-bont; Esgair Hir mine, Ceulanymaesmawr; Esgair Fraith mine, near Tal-y-bont; Henfwlch mine, between Cerrig yr Hafan and Nant-y-moch reservoir, E of Tal-y-bont; Hyddgen mine, near Nant-y-moch reservoir (Bevins 1994, Mason 1997, 1998). Alteration of thiospinel to millerite is said to occur at Esgair Hir mine (Rust & Mason 1988).

Gwent Reports of millerite in the South Wales coalfield (usually in clay–ironstone

Acicular millerite crystals, up to 4 cm long, from the Seven Feet seam, Treharris, Mid Glamorgan, Wales. (© National Museum of Wales, Cardiff; photographer M. P. Cooper)

nodules) include: Ebbw Vale (Greg & Lettsom 1858); Bedwellty, near Tredegar; Blaenavon; Nantyglo (North 1916); Bedwellty pits; Blaina; Newbridge (North & Howarth 1928); Beynon colliery, Brynmawr; Big colliery, Blaenavon; Cwm-felinfach colliery, Cwmfelinfach; Marine colliery, Ebbw Vale; Nine Mile Point colliery, Wattsville; Tirpentwys colliery, Pontypool; Ty Trist, Tredegar (Firth 1971); and the following four collieries – Blackwood, Blaenrhondda; Celynen South, Newbridge; Tower, Hirwaun (Bevins 1994); and Oakdale (Wolfe 1997).

Gwynedd Rarely, millerite forms inclusions (<60 μm) in galena within the Great Orme copper deposits hosted by Carboniferous Limestone, Llandudno (Ixer & Stanley 1996, Ixer 2001). The species is also noted from Drws-y-Coed mine, near Nantlle, Snowdonia (Starkey 1983).

Mid Glamorgan Twisted acicular crystals of millerite up to 4 cm long, forming sprays in cavities within ironstone nodules from the Seven Feet seam, Treharris, are now in the NMW (Bevins 1994). When the tip was being moved, many fine specimens of millerite were recovered from the classic millerite locality of Coed Ely colliery, Llantrisant, as sprays up to 30 mm long (Wolfe 1996b), and from Deep Navigation colliery, Treharris as sprays to 25 mm (Wolfe 1996a). A spectacular association of acicular millerite apparently spearing a euhedral galena crystal from Wyndham colliery, Ogmore Vale, is documented by Cooper (1991a). Millerite is also associated with siegenite at this locality (Bevins & Horák 1985b). Other occurrences from this section of the South Wales coalfield include: Llanbradach colliery (North 1916); Brynna pits, Pencoed; Cymmer colliery, Porth; Ferndale (No. 1 Pit), Rhondda Fach; Hopkinstown; Ogmore Vale (North & Howarth 1928); Bedwas colliery, Trethomas; Bryn colliery, Maesteg; Cwm colliery, Llantrisant; Cwm Dare colliery, Aberdare; Maerdy colliery, Rhondda Fach; Parc colliery, Cwm Parc; Penallta colliery, Ystrad Mynach; Royal Arms opencast site, Merthyr Tydfil; St John's colliery, Maesteg; Tylorstown colliery; Wern Tarw, Pencoed; Windsor colliery, Abertridwr, near Bedwas (all Firth 1971); Cwm Cynon colliery, Mountain Ash; Dowlais Cardiff colliery, Abercynon; Ffaldau colliery, Pontycymmer; Fforchaman colliery, Cwmaman, near Aberdare; International colliery, Blaengarw, near Maesteg; Lewis Merthyr colliery, Trehafod; Ocean colliery, Ferndale; Pentre colliery, Pentre; Scotch colliery, Gilfach Goch; Universal colliery, Senghenydd (all Bevins 1994); and Senghenydd pit, Abertridwr (Moore & Pearce 1997). A review of millerite in the South Wales coalfield is provided by Jones (1981).

West Glamorgan Occurrences of millerite in this section of the South Wales coalfield (usually in clay–ironstone nodules) include: Avan colliery, Abergwynfi (Firth 1971); Eagle Brickworks quarry, Cwmavon, Port Talbot, and Wern Fawr colliery, North Gower, near Swansea (Bevins 1994).

Scotland

Dumfries & Galloway Millerite is one of many supergene sulphide species found in uranium-bearing veins on the Solway coast around the Needle's Eye area, SW of Caulkerbush (Green & Todd 2001). The species is also one of the minor sulphides found on the dumps of West Blackcraig lead–zinc mine, near Newton Stewart (Livingstone et al. 1976).

Fife Heddle (1901) reports millerite from Broomhall lime quarries, from Lathalmond, N of Dunfermline, and from a quarry to the W of Chapel quarry, Kirkcaldy.

Highland Disseminated millerite occurs with chalcopyrite and pyrite as minor sulphides within pyroxenite and pyroxene syenite belonging to the Loch Borralan syenite complex, Assynt (Notholt et al. 1985). Golden metallic sprays up to 20 mm across of crystalline millerite were collected from the dumps of the Middleshop section of Strontian barite mine, Loch Sunart (Green & Todd 2001). An older discovery was at Rubha a' Chamais Bhàin, near Glen Sanda Castle, where tufts of millerite up to 25 mm in length were found in magnesite (var. mesitite) with barite, calcite, galena, morenosite, quartz and sphalerite (Heddle 1901).

Strathclyde Rare millerite is reported as 5 mm-long needle-shape crystals at the Trearne limestone quarry (Todd 1989, Green & Todd 2001) and as brilliant crystals up to 50 mm long in cavities within fossils or in fissures associated with calcite and pearl spar (ankerite or dolomite) at Dockra quarry (Heddle 1901, see also Livingstone 2002), both near Beith. Other Heddle (1901) localities include: in clay–ironstone nodules at Blochairn fireclay quarry, NE of Glasgow; near Dunoon; in fireclay on Garngad Road leading to Provanmill, Glasgow; and in limestone at Pollokshields, Glasgow.

Shetland Millerite is one of the alteration products of pentlandite in the ultramafic rocks of the Shetland ophiolite complex on Unst and it formed as a result of serpentinization of a precursor magmatic sulphide assemblage. Reports focus on a chromium–arsenic-rich sulphide-bearing dunite at Cliff, Baltasound–Haroldswick area, although the species is also known to occur in pyroxenites (Prichard et al. 1994, Maynard 1993, Maynard et al. 1997, Gunn & Styles 2002).

Ireland

Galway Epigenetic millerite is a minor constituent of chalcopyrite–galena–sphalerite ore at Tynagh mine, near Killimor (Morrissey et al. 1971).

Mayo Greg & Lettsom (1858) vaguely report millerite from the foot of Croagh Patrick. At Sheefry Pb–(Ag) mine, 11 km S of Croagh Patrick, millerite forms locally abundant golden acicular crystals up to 10 mm long in cavities in calcite and in dense masses between galena crystals (Moreton & Jackson 1991, Ryback & Moreton 1992, Green & Moreton 2001). Millerite has also been found as inclusions in alluvial gold at Glenkeen bridge on the Carrownisky River (Chapman et al. 2000c) and at Creganbaun (Leake et al. 1993).

Westmeath Sub-economic carbonate-hosted lead–zinc mineralization at Ballinalack, 15 km NW of Mullingar, contains minor millerite associated with pyrite (bravoite) in an ore dominated by barite, galena, marcasite–pyrite and sphalerite, all in a calcite gangue (Jones & Brand 1986).

mimetite

APATITE GROUP

hexagonal $Pb_5(AsO_4)_3Cl$

A relatively common lead arsenate chloride, dimorphous with clinomimetite and found in the oxidized zone of arsenic-bearing lead deposits. Mimetite is one of the most abundant arsenate species at Penberthy Croft mine, St Hilary, Cornwall, where it commonly forms subhedral-to-euhedral short-to-long prismatic transparent-to-opaque crystals up to 10 mm long. Other forms include: elongated hexagonal prisms, rounded barrel-shape crystals, subparallel aggregates tapering to a point, and near-acicular crystals. Colour varies from colourless to white, cream, greenish-blue, yellow, yellowish-green and brown. Differentiation of mimetite from pyromorphite ($Pb_5(PO_4)_3Cl$) on the basis of habit or colour is not reliable (Collins 1892, Kingsbury 1954, Camm & Merry 1991b, Betterton et al. 2000). Thick yellowish to light-brown translucent crystals of mimetite up to 2 cm long were once found in the copper-rich vein deposit at Wheal Unity, St Day, Gwennap (Greg & Lettsom 1858, Embrey & Symes 1987, Penhallurick 1997), and as fine yellow or small brown prismatic crystals on gossany veinstone at Wheal Gorland, St Day, 2 km E of Redruth (Collins 1892, Kingsbury 1954, Kingsbury & Hartley 1960, Rust 1981b). The variety of mimetite termed "gorlandite" was first collected from the latter location (Clark 1993). Other Cornish occurrences include: orange-brown barrel-shape crystals at Greystone quarry, Lezant, near Launceston (Elton et al. 2004); slender colourless crystals and stout cream prisms at Hingston Down Consols mine, 2 km W of Gunnislake, Calstock (Russell 1910a); Pentire Glaze mine, near New Polzeath, St Minver (Clayton et al. 1990); Trevinnick mine, St Kew (Fisk et al. 1982); Wheal Boys, Endellion and Wheal Rose, Sithney, near Helston (Collins 1892); Wheal Penrose, Sithney and Wheal Golden, near St Agnes (Greg & Lettsom 1858); Croft Gothal mine, St Hilary; Pentire (Glaze) mine, St Minver; and Wheal Carpenter, Gwinear (all Golley & Williams 1995). Localities deserving further investigation include: Old Treburgett mine, St Teath, E of Port Isaac; Pengenna (Trewethen) mine, St Kew; Treore mine, Port Gaverne, near Port Isaac, St Teath (all Kingsbury & Hartley 1960);

Mimetite (variety campylite) from the upper level of Dry Gill mine, Caldbeck Fells, Cumbria; specimen is 120 mm wide. (© D. I. Green)

and Wheal Alfred, Phillack, near Hayle, Gwinear (Bowman 1903). The Bere Alston mines, near Bere Ferrers, Devon, are also known to have produced grey mimetite with other lead-bearing species (Hall 1868b).

Without doubt, it is from the Caldbeck Fells of Cumbria that the best and most well known mimetite has been found. It is widespread but rarely of end-member composition, grading imperceptibly into pyromorphite. Dry Gill mine is world renowned for a remarkable phosphorus-rich variety of mimetite (campylite), which forms highly lustrous barrel-shape crystals up to 30 mm in diameter. Matrix specimens up to 20 cm across and stalactites up to 50 mm long were recovered when the mine was working. Colour zoning and chemical zoning is common. The colour is usually a shade of pale to dark orange-brown, but red, yellow, greenish-yellow, and lime- to olive-green varieties have also been found at this locality, Cinnabar and earthy iron–manganese oxides are closely associated species (Greg & Lettsom 1858, Davidson & Thomson 1951, Cooper & Stanley 1990, 1991b, Neall & Leppington 1994, Green & Briscoe 2002). Fine specimens of pale-green mimetite, forming globular aggregates and twisted stalactitic growths, have been collected from Roughton Gill (Cooper & Stanley 1990). A rich golden-yellow variety is described from the nearby Mexico mine (Greg & Lettsom 1858, Cooper & Stanley 1990) and a translucent white variety forming rosettes up to 1 mm across is known from Potts Gill mine (Cooper & Stanley 1991b). Other occurrences include: buff, pale-brown, yellow, greyish-brown, and orange-brown resinous barrel-shape to platy crystals, up to 1 mm across, at Arm o' Grain (Cooper & Stanley 1990, Green et al. 2006b); grey to pale-green tabular-to-prismatic crystals at Brandy Gill mine (Cooper & Stanley 1990, 1991b); tiny white pointed crystals at Burdell Gill (Cooper & Stanley 1990, Neall & Leppington 2003); Carrock mine, Grainsgill (Fortey 1980); associated with other supergene species, barite, beudantite, carminite, pharmacosiderite, schultenite and scorodite in the Deer Hills vein (Symes et al. 1990, Green & Briscoe 2002); green lenticular crystals at Driggith mine (Young 1987a, Cooper & Stanley 1991b, Green & Briscoe 2002); also, end-member composition mimetite forming colourless highly elongated prismatic crystals, up to a few mm in length, from an arsenate-rich gossan in the Driggith–Sandbed opencut (Green et al. 2006a); yellow to yellowish-green crusts and small white hexagonal crystals at Low Pike, near Potts Gill mine (Young et al. 1990a, Green et al. 2003b); globular sprays at Red Gill mine (Cooper & Stanley 1990); pale-yellow mimetite overgrows and replaces cerussite at Sandbed mine (Green et al. 2006a); and white elongated prismatic crystals up to 3 mm long associated with beudantite and yellow to orange-yellow crusts at Silver Gill (Green et al. 2005b). Hartley (1984) lists many of these occurrences and a few others as pyromorphite/mimetite, indicating that further confirmation is required. Mimetite is also recorded on partially oxidized galena–tennantite ore from Stennerskeugh Clouds, 1 km SW of High Stennerskeugh level, Ravenstonedale, near Kirkby Stephen, Cumbria (Young 1987b, Starkey 1987b), and as inconspicuous white encrustations in cavities within a limonite-encrusted quartz–fluorite specimen containing minor galena–gersdorffite–pyrite from the nickel deposit in the Dow Scar level flats of Hilton mine, Scordale (Bridges & Green 2005b).

Mimetite and pyromorphite are relatively common associates in ancient lead workings on the Mendip Hills, Somerset. At Merehead (Torr Works) quarry, Cranmore, mimetite forms sulphur-yellow microcrystalline coatings on fractures or joints in manganese-oxide ore (Alabaster 1975, Symes & Embrey 1977). Manganese-rich pods found at this locality in 2005 and 2006 produced mimetite and kentrolite, together with a range of other rare species including crednerite, fornacite, melanotekite, mereheadite, symesite and wulfenite (Turner 2007). At Higher Pitts farm, Priddy, mimetite forms orange-yellow or orange-red crystalline crusts with a brilliant lustre on calcite, cerussite or hydrocerussite (Spencer & Mountain 1923, Kingsbury 1941, Burr 1996). Mimetite also occurs in manganese-rich veins associated

Well formed hexagonal colour-zoned mimetite crystals, up to 4 mm long, on romanèchite; Dry Gill, Caldbeck Fells, Cumbria. (© D. I. Green)

with galena–pyrite–sphalerite mineralization within Carboniferous Limestone from Coleman's quarry, Holwell, near Frome (Alabaster 1990a).

In Derbyshire, a light-brown variety of mimetite (once known as brown linnets) is reported from Moorhigh (Maury) mine above Miller's Dale, and Wheals Rake, Youlgreave (Ford & Sarjeant 1964, Braithwaite 1983, Ford et al. 1993). Minerals of the mimetite–pyromorphite solid-solution series are also noted from Alderley Edge ridge, Cheshire (Braithwaite 1994). In the vicinity of Alderley Edge (including West mine) and Clive (Shropshire) copper mining areas, mimetite is reported with azurite, cerussite, chrysocolla, dioptase, hydrocerussite, malachite, pyromorphite and smithsonite in a secondary As–Co–Cu–Ni–P–Pb carbonate, hydrated silicate, oxide–hydroxide assemblage (Plant et al. 1999).

At Gwaithyrafon mine, SE of Tal-y-bont, Central Wales orefield, Dyfed, mimetite forms colourless-to-yellow or pale-green prismatic-to-blocky crystals coating joint surfaces on quartz veinstone. Crystals are sometimes zoned, up to 1.5 mm long, and mainly associated with cerussite and malachite (Rust & Mason 1994). Mimetite has also been confirmed from: the nearby Darren mine, Pen-bont Rhydybeddau; Dylife mine, N of Hafren Forest; Hendrefelen mine, Ysbyty Ystwyth (Clark & Rust 1993, Bevins 1994); and in Powys at Dolyhir quary, Old Radnor, near Presteigne (Bevins & Mason 1997).

In Scotland, mimetite is reported from two locations in the lead-mining area of Wanlockhead, Dumfries & Galloway. At Belton Grain vein it forms orange-red encrustations on quartz (Livingstone 2002) and at High Pirn mine it is associated with cotunnite, plumbogummite and pyromorphite (Temple 1955). Anthony et al. (2000) note mimetite from Leadhills, Strathclyde, but this is presumably a general reference to the Leadhills–Wanlockhead orefield.

Clear to translucent hexagonal crystals of mimetite up to 5 mm long are described from a single specimen of massive sulphide ore collected from the dumps of Muckross mine, Killarney National Park, Co. Kerry, Ireland (Ryback & Moreton 1993, Moreton et al. 1999). Attractive specimens of azurite, cerussite, dundasite, malachite and mimetite (forming acicular crystals up to 2 mm long) were collected at Tynagh lead–zinc mine, near Killimor, Co. Galway, prior to its closure in 1982 (Ryback 1988a, Ryback & Moreton 1992, Green & Moreton 2001).

minium

tetragonal $Pb_2^{2+}Pb^{4+}O_4$

A rare secondary lead oxide recorded with little description from Engine Vein mine, Alderley Edge ridge, Cheshire (Thompson 1970, Carlon 1979, Braithwaite 1994), and from the Castaway–Ripley vein, near Keld, Grassington Moor, Yorkshire (Raistrick 1973, Young 1987b). Greg & Lettsom (1858) note minium from Grasshill mine, Teesdale, Co. Durham, but, in an extensive review of supergene minerals from the Northern Pennine orefield, the species could not be confirmed (Bridges & Young 1998).

Mimetite (var. campylite) on psilomelane from Dry Gill mine, Caldbeck Fells, Cumbria; crystals to 7 mm. (© M. P. Cooper)

Fragments of vivid orange-red minium are confirmed among lead slag from Nenthead smelt mill, Nenthead, Cumbria (Bridges & Young 1998), and there are unconfirmed records of the species in Derbyshire from Masson Cavern, Matlock Bath, and from Old Millclose mine, Darnley Dale (Ford et al. 1993), but these too may be related to smelting operations.

Dunham & Wilson (1985) report minium from Turf Pits mine, Grassington Moor, Yorkshire, and Old Moss mine, Killhope Burn, Weardale, Co. Durham, based on samples supposedly collected by A. W. G. Kingsbury. The work of Ryback et al. (1998, 2001) indicates that the provenance of many Kingsbury samples is dubious and further investigation is desirable.

In Wales, minium is recorded from Parys Mountain, Anglesey (Hall 1868), and from lead-bearing veins cutting Carboniferous Limestone to the N of Cefn Onn, South Glamorgan (Howard 1899, Bevins 1994).

In Scotland, exceedingly rare bright-red minium is briefly noted from Susanna vein, Leadhills, Strathclyde. Plattnerite is an associated species in this vein (Heddle 1901, Temple 1955).

minnesotaite

monoclinic $(Fe^{2+},Mg)_3Si_4O_{10}(OH)_2$

A rare iron–magnesium silicate hydroxide found in veinlets cutting peridotite at Pol Cornick, near Mullion, Lizard Peninsula, Cornwall. Minnesotaite is associated with calcite, talc, valleriite and an unnamed fibrous iron–magnesium sulphide (Clark 1970a).

Minor amounts of minnesotaite, associated with stilpnomelane, are described from Lower Carboniferous cherty ironstones from Tynagh, near Killimor, Co. Galway, Ireland. It occurs in irregular fibrous aggregates and in bundles or sheaves of radiating fibres or curved needles; the maximum length of individual crystals is about 150 µm, but the average is much smaller. Minnesotaite replaces chert and dolomite. An origin for the ironstone involving deposition of iron-rich sediments in a restricted marine basin bordered by low-lying land undergoing deep chemical weathering under tropical or subtropical conditions has been proposed (Schultz 1966a).

minrecordite

DOLOMITE GROUP
trigonal $CaZn(CO_3)_2$

A rare calcium–zinc carbonate found in dolomitized micrite in the Navan (Tara) lead–zinc orebody, Co. Meath, Ireland. Minrecordite occurs with iron- and zinc-bearing low-magnesium calcite, dolomite, iron dolomite, sphalerite and pyrite. This assemblage probably formed in an intertidal environment into which exhalation of base metals occurred. The magnesium- and zinc-rich species were the precursors to low-grade sulphide-ore deposition, with some being converted to sphalerite during H_2S metasomatism of the deposit (Kucha & Wieczorek 1984).

mirabilite

monoclinic $Na_2SO_4 \cdot 10H_2O$

A rare hydrated sodium sulphate found attached to a gypsum rock from a quarry at Kirkby Thore, Cumbria. Perfectly clear, pellucid and colourless mirabilite formed a layer 1.5 cm thick and 5 cm long attached to the parent rock and probably represented a portion of a lenticular cavity (Trechmann 1901). Mirabilite is also noted from Petit Portelet, Jersey, Channel Isles, where it was found as an incrustation (Mourant 1961). A report of mirabilite from the Hurlet Limestone, near Glasgow (Glen & Young 1876), requires confirmation.

mispickel

A synonym of arsenopyrite.

mitridatite

monoclinic $Ca_2Fe^{3+}_3(PO_4)_3O_2 \cdot 3H_2O$

A rare hydrated calcium–iron oxyphosphate found at Gunheath (Hensbarrow) china-clay pit, Carthew, St Austell, Cornwall. At this location, pegmatites cut by mineralized quartz and quartz–tourmaline veins contain several phosphate minerals in accessory amounts, including apatite, crandallite, goyazite, mitridatite, rockbridgeite and waylandite (Hawkes et al. 1987). However, it is the iron oxide-rich fracture veins at Gunheath that apparently contain the rare phosphates according to Elton (1998), who describes complex and intimate mixtures of phosphates and sulphides from a rib of rock that originally separated the Gunheath and Hensbarrow pits. Cacoxenite, chalcosiderite, cyrilovite, leucophosphite and mitridatite all contain trivalent iron and indicate oxidizing conditions during formation. Other phosphates in the assemblage include barbosalite, dufrénite, fluorapatite, frondelite, hentschelite, hydroxyl-chlorapatite, rockbridgeite and turquoise (Elton 1998).

Mitridatite has been found in two small segregation pods, 15 and 30 cm long, in striped kyanite–sillimanite-bearing Moine gneiss at the entrance to Gleann Còsaidh, Loch Quoich, Knoydart, Highland Region, Scotland. The pods contain clusters of an iron–manganese phosphate assemblage that includes graftonite, jahnsite, mitridatite, phosphosiderite, rockbridgeite and vivianite, in addition to apatite, garnet (almandine–spessartine), muscovite, plagioclase and quartz. Mitridatite forms earthy yellowish to khaki-green coatings and compact masses in the phosphate clusters (Livingstone 1980).

MIXITE GROUP

Hexagonal arsenates or phosphates with the formula $ACu_6(XO_4)_3(OH)_6 \cdot 3H_2O$, where A = Al, Bi, Ca, Ce, La, Nd, Y; and X = As, P. Agardite-(La), agardite-(Y), mixite and zálesíite are members of the group found in Britain or Ireland.

mixite

MIXITE GROUP
hexagonal $BiCu^{2+}_6(AsO_4)_3(OH)_6 \cdot 3H_2O$

A relatively rare hydrated bismuth–copper arsenate hydroxide found as a supergene species in an oxidized goethite and manganese-oxide gossan assemblage dominated by arsenate species at Penberthy Croft mine, St Hilary, Cornwall. Mixite forms beautiful divergent acicular groups of brilliant emerald-green, bluish-green to whitish hexagonal crystals, occasionally up to 6.5 mm long. Tufted crystal sprays and fibrous veinlets also occur. Common associates include bayldonite, olivenite and parnauite, and rarely cassiterite and cornwallite (Camm & Merry 1991b, Betterton et al. 2000). Reports of mixite from Wheal Owles, Penwith, St Just, Cornwall, referenced by Embrey (1978) and Wolloxall (1988), are based on samples supposedly collected by A. W. G. Kingsbury. The work of Ryback et al. (1998, 2001) indicates that the provenance of many Kingsbury samples is dubious and further confirmation is desirable. Four other Cornish occurrences of mixite are noted by Golley & Williams (1995):

Acicular sprays of green mixite to 6.5 mm on a gossanous quartz matrix. Penberthy Croft mine, St Hilary, Cornwall. (© D. I. Green)

South Terras mine, St Stephen-in-Brannel; Old Gunnislake mine, Calstock, near Gunnislake; Wheal Carpenter, Gwinear, near Hayle; and Wheal Edward, St Just.

Minute acicular sprays of pale turquoise to almost white acicular crystals of mixite were found associated with eulytite and namibite at Buckbarrow Beck, Corney Fell, Waberthwaite, Cumbria. The mixite forms encrustations or fills cavities in the oxidized part of a chalcopyrite–ferberite–quartz–scheelite-bearing vein cutting the Eskdale granodiorite. The vein is remarkable for its rare supergene mineral assemblage, which also includes bismutoferrite, cuprotungstite and russellite (Neall et al. 1993, Green & Briscoe 2002).

Mixite is a supergene species that forms grass-green terminated prismatic crystals as an alteration product of native bismuth and it also occurs as emerald-green mammillary growths in the Needle's Eye area, SW of Caulkerbush, Dumfries & Galloway, Scotland. The primary ore minerals are native bismuth, chalcopyrite, hematite, pyrite and uraninite, with minor nickeline, löllingite and sphalerite in a quartz and dolomite gangue. Other supergene minerals from this vicinity include annabergite, boltwoodite, bismutite, connellite, cornubite, erythrite, eulytite, kahlerite, lavendulan, millerite, metakahlerite, nováčekite, pharmacosiderite, schoepite, tyrolite, uranophane, vandendriesscheite, walpurgite, wulfenite and zeunerite (Miller & Taylor 1966, Braithwaite & Knight 1990, Green & Todd 2001).

Green & Todd (2001) report mixite as one of the unusual and usually minute supergene species found at Tynagh lead–zinc mine, near Killimor, Co. Galway, Ireland. Additionally, adamite, anglesite, bayldonite, cerussite, dundasite, linarite, malachite, mimetite, olivenite, philipsbornite, native silver, strashimirite, tyrolite and wulfenite are also noted.

mizzonite

An intermediate member of the marialite–meionite series, scapolite group.

mocha-stone

Fine-grain banded quartz and pyrolusite.

mock lead

A synonym of sphalerite – an English miners' term derived from its resemblance to galena.

modderite?

orthorhombic CoAs

A rare cobalt–iron arsenide reported with algodonite, coffinite, native copper, lavendulan, malachite, maucherite, nickeline, rammelsbergite and native silver in Permian nodular concretions from the Littleham Cove area, Budleigh Salterton, Devon. EMP data were obtained that indicated a close match with modderite, but confirmation by XRD was not possible (Harrison et al. 1975b). Modderite is also listed but not discussed by Milodowski et al. (2000), who suggest that de-watering of the Permo-Triassic strata during burial compaction could have produced the metalliferous pore fluids that led to the mineralization.

moganite/mogánite

monoclinic SiO_2

A recently described polymorph of quartz (Flörke et al. 1984), whose status as a species is not universally recognized. It appears in Bayliss (2000) and Deer et al. (2004), but not in Anthony et al. (1995) or Mandarino & Back (2004). De Fourestier (1999) describes it as an impure quartz (lutecite). Moganite is a microcrystalline fibrous species, often found intergrown with chalcedony and quartz, that has been found in the majority of microcrystalline quartz-rich silica deposits (chalcedony, chert, flint) examined worldwide (Heaney & Post 1992). It occurs as a metastable phase that will ultimately transform to quartz, given sufficient time or a change in ambient conditions (Rodgers & Cressey 2001). The only confirmed occurrence in Britain or Ireland is in flint from the Cretaceous Upper Chalk at Margate, Kent, which contains about 20% moganite (G. Cressey pers. comm. 2006). However, moganite is also likely to be present in agate, beekite, carnelian, chrysoprase, hornstone, sard and sardonyx.

molybdene

A synonym of ferrimolybdite.

molybdenite

hexagonal or trigonal MoS_2

A common molybdenum sulphide, dimorphous with jordisite and found in high-temperature hydrothermal veins, in disseminated deposits of porphyry type, in greisen, granites, aplites and pegmatites, and also in contact-metamorphic skarn deposits in limestones.

England

Cornwall Molybdenite occurs in greisen at Carn Brea, Illogan, near Redruth, and Cligga Head, Perranzabuloe, bordering veins that carry cassiterite and wolframite (Hosking 1951, Robson 1949b). A few scales of molybdenite were once found in a quartz vein near the granite country-rock (killas) contact at Hanover Bay, between St Agnes and Cligga Head (Le Neve Foster 1877). Tourmaline stringers cutting granite pegmatite at Hawks Wood mine, North Hill, contain a little molybdenite (Hosking 1954b). Quartz veins containing sporadic molybdenite are associated with wolframite-bearing pegmatites at Kit Hill Consols, Stoke Climsland (Dewey 1925). Minute very-bright flakes of molybdenite, included in rhodonite and pyroxene within a belt of calcareous sediments and greenstones near the margin of the Bodmin Moor granite, are reported from Treburland manganese mine, 2 km SE of Fivelanes, Altarnun (Russell 1946d, Kingsbury 1961). Although molybdenite is reported from St Michael's Mount, Marazion (Davison 1920, Hosking 1954c), this may be an erroneous reference to Wheal Friendship, which is near by (Greg & Lettsom 1858, Collins 1892). Molybdenite is also noted from: Camborne Vean; Levant mine, St Just; Penlee, Newlyn (all Robson 1954); Chywoon quarry, near Longdowns (Moulding & Hooper 2005); Drakewalls mine, Calstock, and Wheal Trenwith, St Ives (Dines 1956); Wheal Gorland, St Day, 2 km E of Redruth; Wheal Mary, Lelant; Wheal Unity, St Day (all Greg & Lettsom 1858, Green 1984); Hingston Down quarry, 2 km W of Gunnislake, Calstock (Jewson 1997); Killifreth mine, Kenwyn, near Redruth (Bawden 1929); and Prince of Wales mine, Calstock (Golley & Williams 1995).

Cumbria A series of polymetallic quartz veins cutting greisenized Skiddaw granite and the Carrock Fell gabbro complex were once worked for wolframite at Carrock mine, Grainsgill, Caldbeck Fells. Associated species include arsenopyrite, native bismuth, bismuthinite, cassiterite, chalcopyrite, molybdenite (particularly in one W-poor vein), pyrite, scheelite, sphalerite and tetradymite (Finlayson 1910, Hitchen 1934, Davidson & Thomson 1951, Eastwood 1959, Eastwood et al. 1968, Shepherd et al. 1976, Fortey 1978, 1980, Ball et al. 1985a, Cooper & Stanley 1990). Greg & Lettsom (1858) list five separate localities for this one occurrence. Abundant molybdenite occurs on joint planes (anhedral masses) and miarolitic cavities (hexagonal or complex curved crystals up to 5 mm across) in granite from the Shap granite quarry (Russell 1925, Davidson & Thomson 1951, Wirth & Green 2002). At Shap Blue quarry, metasomatized Borrowdale Volcanic Group rocks in the aureole of the granite contain early-formed garnet–epidote-bearing veins followed by assemblages containing molybdenite and other common sulphides (Young et al. 1988). Molybdenite is also noted from: Brandy Gill mine, Caldbeck Fells (Davidson & Thomson 1951); Scar Crag cobalt mine, Causey Pike (Ixer et al. 1979a); and Water Crag, near Devoke Water (Young 1985d,e).

Devon Pegmatite veins cutting Tertiary granite on Lundy island contain occasional leaves of molybdenite associated with apatite and lithium mica (Dollar 1941, Thorpe

Metallic plate of molybdenite, 4 mm across, in granite from the Shap granite quarry, Cumbria. (© D. I. Green)

et al. 1990). Molybdenite is also noted from: Dartmoor granite (Brammall & Harwood 1925); Hemerdon Bal, Plympton (Keats 1981); Littleham Cove, near Budleigh Salterton (Milodowski et al. 2000); British Railways quarry complex, Meldon, near Okehampton (Dearman 1965, 1966b); Red-a-ven mine, Meldon, near Okehampton (El Sharkawi & Dearman 1966); and Wheal Friendship (Devon Friendship), Mary Tavy (Braithwaite 1982b).

Durham Minor amounts of arsenopyrite, molybdenite, pyrite, sphalerite and tetrahedrite occur in chalcopyrite–galena-bearing veinlets within the buried Weardale granite, discovered during a drilling program at Rookhope, Weardale (Ixer 1986, Brown et al. 1987a).

Leicestershire Molybdenite plates are reported on the walls of joints in granodiorite at the Main Mountsorrel quarry, Mountsorrel district, near Loughborough (Rudler 1905, Taylor 1934).

Shropshire Molybdenite is identified as inclusions in quartz cement within a silicified fault rock at Clive mine, 10 km N of Shrewsbury, Shropshire. It is associated with barite, chalcocite or digenite, chalcopyrite, covellite, galena and rare acanthite (Plant et al. 1999).

Yorkshire Molybdenite is said to occur in a chalcopyrite, galena and pyrite ore association related to the buried Wensleydale granite sampled in the Raydale borehole, 7 km SSE of Hawes (Brown et al. 1987a). However, these authors also state that no molybdenum mineralization was found.

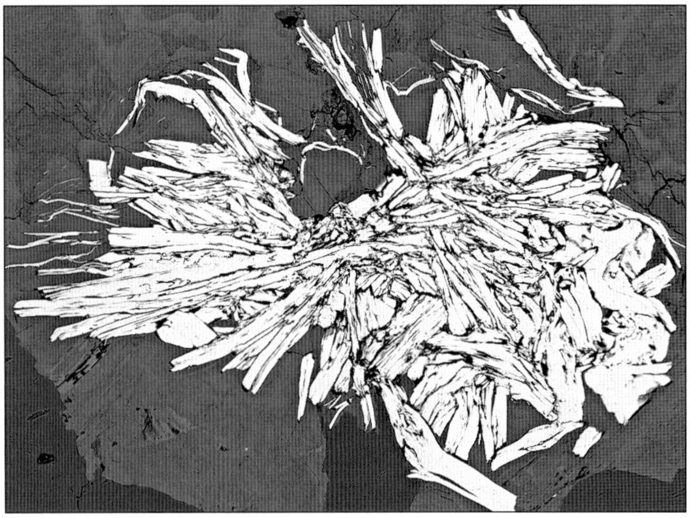

Cluster of molybdenite laths in a quartz vein cutting zinnwaldite-bearing granite; Gairnshiel, Glen Gairn, Grampian, Scotland; width 1.2 mm; BSE image. (© E. Condliffe, University of Leeds)

Wales

Gwynedd The earliest report of molybdenite from Wales is from Blaen-y-nant, Nant Ffrancon area, Snowdonia (Williams 1802, Dewey & Eastwood 1925). Traces of molybdenite (sometimes associated with native gold) occur along joints, fractures and locally in quartz–calcite veinlets within the central part of the Coed-y-Brenin porphyry copper deposit, in southern Snowdonia (Rice & Sharp 1976, Allen & Jackson 1985). Near by, at Clogau mine, Bontddu, near Dolgellau, molybdenite forms small isolated laths (10–20 μm) and clumps of crystals associated with galena (Bevins 1994). Molybdenite has also been found: in the volcanogenic massive sulphide (mainly pyrite) deposit at Cae Coch mine, near Trefriw, Llanrwst (Ball & Bland 1985); in the Ffestiniog granite quarry, Cefn Bychan, near Ffestiniog (Bevins 1994); associated with allanite, pyrophyllite and quartz in pipes cutting the Tanygrisiau granite (Roberts 1979).

Scotland

Central Rare molybdenite occurs in Dalradian stratabound exhalative copper–zinc mineralized horizons at Allt Auchtertyre, to the E of Tyndrum (Fortey & Smith 1986). Alluvial gold from the central Ochil Hills contains rare molybdenite as an inclusion species (Chapman et al. 2005).

Dumfries & Galloway Some of the finest Scottish molybdenite has been found in quartz veins cutting granite at the margins of the Criffell granite complex (around Dalbeattie), particularly at Almorness Point, E of Auchencairn, and Screel Hill, near Kirkmirran, N of Auchencairn. Additionally, this complex contains molybdenite on Hestan Island in Auchencairn Bay, bismuthinite–molybdenite–pyrite veins at Lotus Hill, near Beeswing, and in surrounding hornfelsed aureole rocks at Lower Porterbelly quarry, Kirkgunzeon, molybdenite occurs as a surface coating near the granite contact. In the Needle's Eye area, SW of Caulkerbush, molybdenite also occurs as a minor constituent of uranium-rich veins cutting this aureole (Heddle 1901, Williams 1973, Green & Todd 2001, Livingstone 2002). To the W of the Criffell granite complex at Black Stockarton Moor, molybdenite occurs in a copper–molybdenum assemblage associated with sill-like granodioritic intrusions and swarms of porphyrite dykes, interpreted to be part of a porphyry copper deposit (Fortey 1980).

Grampian At Gairnshiel Bridge, Glen Gairn, molybdenite occurs in quartz veins, in greisen associated with hübnerite, koechlinite and wulfenite, and within zinnwaldite-bearing granite (Tindle & Webb 1989, Livingstone 1992, Webb et al. 1992). Spectacular clusters of molybdenite rosettes up to 10 mm across have been found in quartz veins and greisen associated with the small Middleton granite intrusion near Chapel of Gairloch, to the W of Inverurie. Additionally, Dalradian schists at the margins of this granite contain molybdenite along joint surfaces (Heddle 1901, Colman et al. 1989). At Souter Head, 6 km S of Aberdeen, an explosion breccia pipe intruding Dalradian gneiss contains cavity infillings of quartz and calcite, in which pyrite and subordinate molybdenite have been found (Porteous 1973, Kneller & Gillen 1987, Nicholson 1989c).

Highland Low-grade sulphide mineralization is exposed in Moinian and Lewisian rocks intruded by the Caledonian Grudie and Shin granites to the W of Lairg. Molybdenite occurs with pyrite and accessory aikinite–bismuthinite, chalcopyrite and fluorite in thin quartz veins and as coatings to joints and foliation planes in the schists and gneisses (Gallagher 1970, Gallagher et al. 1971, 1974, Gallagher & Smith 1975, Fortey 1980). Molybdenite occurs in Lewisian amphibolites close to a granitic pegmatite in a roadcut 2 km N of Achfary on the south shore of Loch Stack (von Knorring & Dearnley 1959a) and in quartz veins SW of the house of Badnabay, 8 km WNW of Loch Stack (Nutter 1953). Other reports of molybdenite include: in quartz veinlets and disseminated in a stockwork-type deposit within the Ballachulish granite at Gleann a' Chaolais, Ballachulish (Evans 1977, Haslam & Kimbell 1981); in a Middle Old Red Sandstone black-shale horizon at Blar Cnoc na Gaoith, near Achvarasdal (Gallagher et al. 1971); with chalcopyrite in a quartz-breccia vein in pelitic hornfels at the northern edge of the Carn Chuinneag (Inchbae) granite gneiss (Gallagher et al. 1971, Dunham et al. 1978); in calcite–fluorite–quartz veins associated with the Ratagain complex, Loch Duich, near Glenelg (Alderton 1986, 1988b); in chlorite schist at Glenelg, in augen gneiss at Inchbae and in a granitic vein at Dochfour Burn at the northern end of Loch Ness (Heddle 1901); in veins and as joint coatings in a brecciated area of Helmsdale granite in the valley of the Ord Burn (Tweedie 1979); and very rare in a limestone contact belt at Shinness on the eastern shore of Loch Shin (Heddle 1878c).

Strathclyde Fine molybdenite, forming six-sided crystals up to 10 cm across, was once found at Coire Buidhe, Glen Creran, Appin (Greg & Lettsom 1858, Heddle 1901). The locality was rediscovered by Ingram et al. (1993), who note molybdenite in veins associated with granitic intrusions belonging to the Etive complex, as disseminations in acidic porphyritic rocks, and in epidote–garnet skarn rocks. Elsewhere in the Etive complex, molybdenite is recorded from Barrs quarry, Loch Etive, and the Ben Cruachan granite quarries (Heddle 1901), and at the head of Loch Etive, particularly near Allt Coire Dubh (on the Highland/Strathclyde border) and two tributaries of Allt nam Meirleach (Haslam et al. 1985). A Caledonian a brecciated-porphyry diatreme complex at Lagalochan, near Kilmelford, contains bornite, molybdenite and chalcopyrite in a hydrothermally altered hypogene assemblage typically found on or adjacent to quartzose veinlets (Fortey 1980, Zhou 1987, Harris et al. 1988). A stock-like porphyrite intrusion at Garbh Achadh, W of Inveraray, contains minor molybdenite, chalcopyrite and pyrite in schlieren-like pockets, and these species are accompanied by covellite and sphalerite in rafts of sulphide-rich hornsfelsed quartzite (Ellis et al. 1978, Fortey 1980). Molybdenite also occurs in pyrite–quartz veins on the margin of the Moor of Rannoch granite at Blackmount, near Bridge of Orchy (Smith & Marsden 1977, Curtis et al. 1993).

Tayside At Tomnadashan, on the south shore of Loch Tay, chalcopyrite–molybdenite–pyrite–tetrahedrite mineralization is associated with a late Caledonian alkali diorite intrusion. Molybdenite occurs in both calcite and quartz veins, being particularly abundant in samples rich in pyrite (Heddle 1901, Wilson & Flett 1921, Fortey 1980, Ridgway 1980, Pattrick 1984, Plant et al. 1989). Inclusions of molybdenite in alluvial gold are reported from Glen Lednock, near Comrie (Chapman et al. 2000c).

Shetland Molybdenite-bearing chlorite veins associated with siderite nodules occur in the serpentinized basal part of the Shetland ophiolite complex at Queyhouse talc quarry, W of Baltasound, Unst (Gass et al. 1982, Neary & Prichard 1985).

Ireland

Cavan Chalcopyrite–molybdenite veins and veinlets cut the Crossdoney granodiorite at Drumcarban (Morris 1984).

Cork In the West Carbery mining district, molybdenite occurs: in Devonian redbeds; altering to ferrimolybdite and associated with bornite, chalcopyrite and tetrahedrite in quartz veins at Ballycummisk mine, near Rossbrin Cove; in two copper-rich vein assemblages at Dhurode mine, Mizen Head Peninsula; in veinlets associated with bornite and chalcocite mineralization at Letter; with sphalerite in calcrete at Mount Gabriel (Ryback & Moreton 1991, Ni Wen et al. 1999, Kinnaird et al. 2002). At Allihies on the Bearhaven (Berehaven) Peninsula, minor molybdenite has been found in chalcopyrite-rich ore at the Berehaven (Bearhaven) mines (Moreton & Aspen 1993) including Mountain mine (Schultz 1971).

Donegal Molybdenite is briefly recorded from granitic rocks at Garvary Wood, Castlecaldwell (Haughton 1862), at Lough Lara, near Glenties, and in quartz veins cutting Precambrian limestone and quartzite at Inishdooey, near Tory Island (Cole 1922). Occasional flakes of molybdenite are seen in quartz veins in granite, pegmatite and garnet skarns at Lough Anure, near Dunglow (O'Brien 1959). Within the Rosses granite ring complex, molybdenite occurs with pyrite in beryl greisen and quartz veins (Burke et al. 1964).

Down Primary chalcopyrite, molybdenite and pyrite occur with minor magnetite at Croreagh quarry in the Newry granodiorite (Meighan 1979, Meighan & Ixer 1980).

Galway Within the Galway granite batholith, a molybdenite-bearing quartz vein is reported from the margins of the Murvey granite at Mace Head, around Murvey, and near Roundstone (O'Brien 1959, Max & Talbot 1986, Derham & Feely 1988, Ryback & Moreton 1992). The nearby Carna granodiorite also hosts similar quartz veins (Talbot & Max 1984).

Kerry Minor molybdenite is reported in stratabound Cu–Pb–Zn ores at Muckross, Killarney, and in epigenetic limestone-hosted copper-polymetallic veins at Ross Island (Kinnaird et al. 2002).

Longford Minor molybdenite occurs in quartz–pyrite veins and veinlets cutting hornfelsed turbidite in the aureole of the Arva granodiorite at Aghagah (Morris 1984).

Tyrone Molybdenite occurs on shear surfaces of chalcopyrite–pyrite-bearing quartz veins at Cashel, near Greencastle (Leyshon & Cazalet 1976).

Wicklow A trace of molybdenite has been found in a sulphide-rich fluoritized microadamellite at Ballinglen, between Tinahely and Aughrim (Steiger & Bowden 1986). Molybdenite is one of many species found with alluvial gold on the northeastern side of Crogan Kinshela (Forbes 1869).

Offshore territories

Isle of Man Nockolds (1931) reports minor molybdenite in certain parts of the Dhoon (Isle of Man) granite.

Guernsey Minor molybdenite occurs in pyrite-rich veinlets cutting diorite of the Northern igneous complex at Baubigny, near St Sampson (Ixer 1992).

Jersey Molybdenite has been extracted from calcite–quartz veins in joints within granite at L'Etacq in northwest Jersey. Arsenopyrite, boulangerite, bournonite, chalcopyrite, cosalite, galena, marcasite, pyrite, pyrrhotite, tetrahedrite and wulfenite are associated species (Mourant 1985, Ixer 1992). Within the northwestern granite mass, molybdenite also occurs: in granite pegmatites at Gigoulande quarry, near La Hague Mill, and at Côtil Point, La Moye, Mont Mado and Ronez; in veins at La Saline; and in granite at Handois and Ronez Point (Mourant 1961). Ixer (1988) reports molybdenite in a quartz vein from Vatcher's quarry, Handois (now Handois reservoir).

molybdic ochre

A synonym of molybdite.

molybdite

orthorhombic MoO_3

A rare molybdenum oxide that forms as a weathering product of molybdenite and has been described from Carrock Fell mine, Grainsgill, Caldbeck Fells, Cumbria (Davidson & Thomson 1951, see also Young 1987a). Greg & Lettsom (1858) report it in Scotland from Mount Coryby (Coire Buidhe), near Loch Creran, Strathclyde, and East Tulloch, S of Loch Tay, Tayside Region. Unfortunately, these occurrences have not been confirmed by later workers (Cooper & Stanley 1990, Macpherson & Livingstone 1982, Livingstone 2002). At Middleton, 3 km SE of Chapel of Garioch, near Inverurie, Grampian, molybdenite-bearing quartz-vein material occurs in the field hedges. Some specimens show fractures with irregular coatings of a yellow secondary mineral, proved to be an iron-rich molybdite (Colman et al. 1989).

MONAZITE GROUP

Monoclinic arsenates, phosphates and silicates with the general formula ABO_4, where A = Bi, Ca, Ce, La, Nd, Th; and B = As^{5+}, P^{5+}, Si^{4+}. Monazite-(Ce), monazite-(La), monazite-(Nd) and possibly rooseveltite are members of the group found in Britain or Ireland.

monazite-(Ce)

monoclinic — MONAZITE GROUP — $(Ce,La,Nd)PO_4$

A widely distributed cerium-dominant phosphate found in accessory amounts in granites, syenites and their pegmatites. The species can also occur in fissure veins, high-grade gneissic metamorphic rocks and commonly in alluvial or beach sands as a detrital mineral. It is one of the main mineral hosts for REEs and thorium in the continental crust, and can be used for Sm–Nd and U–Th–Pb geochronology. The descriptions that follow include examples only where the dominant ion is quoted; there are many other descriptions that do not.

Monazite-(Ce) occurs in alpine-type veins within Devonian metasediments at Croft Gothal mine, 2 km E of Goldsithney, St Hilary, Cornwall. It is found associated with albite (later shown to be orthoclase, var. adularia – J. Betterton pers. comm. 2005) in cavities in quartz, as acicular-to-prismatic crystals up to 0.6 mm long and, more commonly, as euhedral pale- to mid-brown opaque-to-translucent tabular crystals up to 2 mm long (Betterton 1996d). At Porth Nanven, near Cape Cornwall, monazite-(Ce) is associated with thorite in a siderophyllite granite from the Land's End granite pluton (Müller et al. 2006). Monazite-(Ce) is also noted from the Bodmin Moor granite by Holmes & Smales (1948). In a study of the distribution of REEs in the Carnmenellis granite, SE of Camborne, Cornwall, Jefferies (1985a) was able to confirm the presence of accessory amounts of monazite-(Ce) together with apatite, xenotime and zircon. Charoy (1986) found apatite, monazite-(Ce) and zircon in Carnmenellis granite from Chywoon quarry, near Longdowns. Hydrothermally altered monazite-(Ce) from the Carnmenellis granite was also used by Poitrasson et al. (1996) to examine the effects of fluid–rock interaction. These authors also examined hydrothermally altered monazite-(Ce) from the Grainsgill cupola (a greisenized facies of the Skiddaw granite) in the Caldbeck Fells, Cumbria, and were able to show distinct differences in the mechanisms of alteration at the two sites. Golley & Williams (1995) list three other Cornish locations for monazite-(Ce): Hingston Down quarry, 2 km W of Gunnislake, Calstock; Lanterdan quarry, Tintagel; and Prince of Wales quarry, Tintagel. In Devon, monazite-(Ce) has been found in accessory amounts in the Tertiary Lundy granite, together with almandine, tungsten-rich columbite, thorium-rich uraninite and xenotime (Thorpe et al. 1990, 1995).

A study of fluorite ores from the Northern Pennine orefield revealed that they contain accessory amounts of monazite-(Ce), xenotime, and the fluorcarbonate synchysite. Monazite-(Ce) is noted from Rotherhope mine, Leadgate, near Alston, Cumbria, but is more common at Groverake, Redburn, Stotfield Burn and Whiteheaps mines, all near Rookhope, Weardale, Co. Durham. It occurs as discrete equant crystals up to 40 μm in diameter, or in clusters up to 150 μm across, and encloses very small (< 1 μm) bismuthinite, hematite and uraninite inclusions. It is itself included in synchysite and xenotime (Ixer 2003). Webb & Brown (1984a) found accessory amounts of monazite-(Ce) in the Skiddaw granite, Caldbeck Fells, Cumbria.

At Fron-oleu, Prenteg, Tremadog, Gwynedd, Wales, monazite-(Ce) is found in Ordovician slaty mudstones adjacent to an alpine-type vein assemblage that has

Pseudo-hexagonal monazite-(Ce) crystals from a sphalerite-bearing quartz vein cutting slate at Kerriack Cove, near Porthtowan, Truro, Cornwall; width 1 mm. (© J. J. Hooper)

produced some of the finest brookite crystals anywhere in the world. Monazite-(Ce) occurs sparingly as transparent honey-colour rhombic crystals up to 1.5 mm across, associated with albite, fluorite, goethite and possibly cookeite (Starkey & Robinson 1992). Mineralized joints at Hendre quarry, Pandy, near Glyn Ceiriog, Clwyd, contain pinkish-to-yellow rhombic crystals of monazite-(Ce) up to 0.5 mm across, associated with anatase and xenotime-(Y) (Starkey et al. 1991). Grey ellipsoidal nodular monazite-(Ce), 0.05–2 mm in diameter, occurs in Lower Palaeozoic sedimentary rocks from the following areas of Wales: Afon Gwesyn catchment, Dyfed; Berwyn Dome, Powys; Harlech Dome, Gwynedd; and Mynydd Preseli, Dyfed. The nodules probably developed by *in situ* recrystallization and authigenic growth of detrital monazite (Cooper et al. 1983, Read et al. 1987).

Livingstone (2002) notes crystals of monazite-(Ce) to 0.6 mm from acid pegmatites of Lewisian age on South Harris, Western Isles, Scotland. Monazite-(Ce) is also reported in accessory amounts from several Caledonian granite complexes throughout Scotland: Cairngorm granite, Grampian (Barritt 1983, Webb & Brown 1984b); Ballater and Mount Battock granites, Grampian (Webb & Brown 1984b); Central Starav granite, Etive, Strathclyde (Barritt 1983); quartz diorite, granodiorite, microgranite and aplites from the Loch Doon granite complex, which straddles the Strathclyde/Dumfries & Galloway boundary (Tindle & Pearce 1981, Tindle 1982).

At Lindsay's Leap quarry in the Mourne Mountains, Co. Down, Ireland, drusy cavities in granite contain very rare honey-brown, transparent, euhedral, blocky-to-prismatic crystals of monazite-(Ce) up to 0.5 mm long, on white albite (Green et al. 2005a).

monazite-(La)

monoclinic — MONAZITE GROUP — $(La,Ce,Nd)PO_4$

A rare lanthanum-dominant phosphate found in fissure veins, granites and gneissic metamorphic rocks. At Penberthy Croft mine, St Hilary, Cornwall, monazite-(La) forms translucent, buff to pale-brown, thin curved platy subhedral crystal groups up to 0.5 mm across, associated with anatase, clinochlore, minor siderite and quartz (Betterton 1996a, Betterton et al. 2000). Monazite-(La) is probably more common than the literature suggests, but frequently the dominant ion is not quoted.

monazite-(Nd)

monoclinic — MONAZITE GROUP — $(Nd,Ce,La)PO_4$

A rare neodymium-dominant phosphate found in accessory amounts in a zinnwaldite-bearing variety of the Land's End granite at Priest's Cove, near Cape Cornwall, Cornwall. Monazite-(Nd) forms domains with monazite-(Ce) crystals (Müller et al. 2006).

moncheite

trigonal — MELONITE GROUP — $Pt(Te,Bi)_2$

A rare platinum–bismuth telluride, identified in a layered mafic–ultramafic intrusion on the Isle of Rum, Highland Region, Scotland. Within certain chromitite layers of the Eastern Layered Series, moncheite is found in association with species positively or tentatively identified as braggite, cooperite, isoferroplatinum, kotulskite, laurite, michenerite, members of the hollingworthite–irarsite–platarsite series, sperrylite, telluropalladinite, and the native species gold, silver, palladium and platinum. Most of these grains are 0.2–2 µm in diameter, although for some species grains > 15 µm are common. The majority are in contact with sulphide minerals such as bornite, chalcocite, chalcopyrite, pentlandite or pyrrhotite. The chromitite layers were formed during magmatic fractionation and crystal settling of an open-system magma chamber, modified by magma mixing and liquid immiscibility of sulphide and silicate melts (Butcher et al. 1999, Power et al. 2000a, 2003).

monohydrocalcite

hexagonal — $CaCO_3{\cdot}H_2O$

An uncommon hydrated calcium carbonate found on lumps of primary sulphide ore at Muckross mine, Killarney National Park, Co. Kerry, Ireland. Monohydrocalcite is one of the secondary minerals formed during post-mining oxidation of the spoil-heap. It forms chalky white spheroidal masses, 1–1.5 mm across, and yellowish mammillated encrustations, associated with aragonite, phaunouxite and rauenthalite (Moreton et al. 1999, Green & Moreton 2001).

In Wales, monohydrocalcite is confirmed from a trial level, 2 km NW of Bontddu, near Dolgellau, Gwynedd. The trial is on a 1 m-wide quartz vein spotted with calcite, chlorite and sulphides, and was possibly investigated for gold in the mid-19th century. Monohydrocalcite forms minute colourless pyramidal triangular sections in a vug within chalcopyrite–quartz veinstone. It is associated with aragonite and calcite (Saich & Rust 1987).

montebrasite

triclinic — AMBLYGONITE GROUP — $(Li,Na)Al(PO_4)(OH,F)$

An hydrated lithium–sodium–aluminium phosphate fluorhydroxide that forms a series with amblygonite and is found in lithium-rich granites and granitic pegmatites. Montebrasite was found with albite, amblygonite, apatite, elbaite, fluorite, lithium mica and topaz, and occasionally petalite, in a pegmatite vein in the old aplite quarry at Meldon, Okehampton, Devon – but possibly this is doubtful, as it was reported as being collected by A. W. G. Kingsbury, whose collection (now at the NHM) is known to contain incorrectly provenanced samples (Kingsbury 1964b, Moss et al. 1969, Ryback et al. 1998, 2001). Further confirmation is desirable. However, von Knorring & Condliffe (1984) found montebrasite from the same locality, describing it as being less widespread than amblygonite and associated with grey lithian mica and orthoclase, and with microcline- and green tourmaline-bearing veinlets. Montebrasite is also associated with amblygonite and a secondary mica in a sample of fine-grain aplite from Meldon (Tindle unpub. EMP data).

monticellite

orthorhombic — $CaMgSiO_4$

An uncommon calcium–magnesium silicate that forms a series with kirschsteinite and is found in the limestone/gabbro contact zone in the Grange Irish neighbourhood close to Grange Cross, Carlingford, Co. Louth, Ireland. Monticellite forms irregular crystals associated with apatite, calcite, cuspidine, diopside, edenite, grossular, magnetite, phlogopite, spinel (var. pleonaste) and wollastonite. It is thought that this assemblage formed through metasomatism of limestone by a hidden acid pegmatite (Osborne 1932a,b). Monticellite is also noted from the chalk/dolerite contact zone at Scawt Hill, Larne, Co. Antrim (Anthony et al. 1995).

Monticellite–magnetite skarns have been produced during metasomatic alteration of dolomitic limestone by boron- and fluorine-rich fluids from the Beinn an Dubhaich granite, near Broadford, Isle of Skye, Highland Region, Scotland. In addition, the skarns contain chondrodite, clinohumite, cuspidine, forsterite and humite.

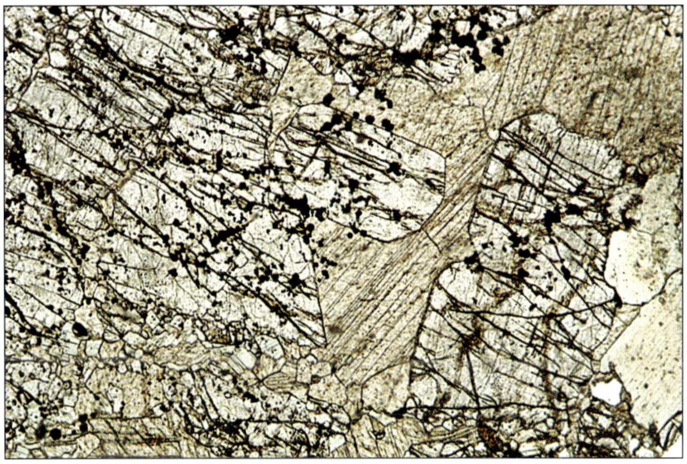

Colourful euhedral and subhedral monticellite with interstitial calcite and phlogopite (pink in XPL). Carlingford, Co. Louth, Ireland; width 3.4 mm; viewed in PPL (upper) and XPL (lower). (© K. Gardner)

Monticellite is also one of the most important silicate species in marbles adjacent to the skarns, particularly near Camas Malag at the western end of the granite outcrop. The colour of monticellite varies from light brown to greyish white away from magnetite-rich areas (Tilley 1948b, 1951). Minor amounts of monticellite are also reported from the carbonate-rich part of a high-temperature calcic skarn near the contact of the Cuillin gabbro complex at Camasunary Bay, Strathaird Peninsula, Isle of Skye. It occurs with calcite, garnet (andradite–grossular), melilite, vesuvianite and wollastonite, together with minor amounts of magnetite, perovskite and pyrite. Rare djerfisherite and kimzeyite complete the assemblage (Beard & Drake 2007).

At Camas Mòr, Isle of Muck, Highland Region, an olivine gabbro dyke intersecting Jurassic limestones has developed high-temperature contact assemblages in which monticellite occurs in marbles, together with periclase or brucite, and as a constituent of carbonate-poor assemblages with gehlenite. In one rock adjacent to the gabbro, monticellite forms coarse grains up to 6 mm in diameter in a gehlenite–spinel assemblage also carrying magnetite, perovskite and pyrrhotite (Tilley 1947).

montmorillonite

SMECTITE GROUP

monoclinic $(Na,Ca)_{0.3}(Al,Mg)_2Si_4O_{10}(OH)_2{\cdot}nH_2O$

A common hydrated sodium–calcium–magnesium aluminosilicate hydroxide that forms in pegmatite dykes and wallrocks bordering hydrothermal mineral deposits. It is also an alteration product of volcanic ash and tuff, forming bentonite beds.

Abundant montmorillonite is reported as a bluish-grey alteration product of analcime and natrolite in gabbro from Dean quarry, St Keverne, Lizard Peninsula, Cornwall. Thin warped yellowish flakes and massive material have been found and, additionally, white spheres of montmorillonite, <0.1 mm in diameter, occur singly or in chains on the surfaces of calcite crystals at this locality. Some spheres are enclosed by a late generation of colourless calcite (Seager 1969, 1978a, Wolters & Wolters 1986). Montmorillonite is briefly described with decomposing pyrite at Great Retallack mine, Perranzabuloe, Cornwall (Collins 1878b), and in a hydrothermal replacement deposit (carbona) at Levant mine, St Just (Jackson et al. 1982). A pink variety of montmorillonite is described from the lithium–fluorine-rich aplite dyke at Meldon, Okehampton, Devon, where it is said to be an alteration product of petalite (Worth 1920, McLintock 1923). The occurrence does not appear to have been subjected to modern analytical confirmation.

Calcium montmorillonite (fuller's earth) of Middle Jurassic or Lower Cretaceous age occurs in economic concentrations (some now mined out) at: Woburn Sands and Clophill, Bedfordshire; Coombe Hay, S of Bath, Avon; Redhill and Nutfield, Surrey; and Baulking, Vale of the White Horse, between Abingdon, Oxfordshire, and Swindon, Wiltshire. The unweathered clays are waxy and greenish blue, becoming yellowish on oxidation. When dry, they are hard and brittle, but they disintegrate into a clay slurry when placed in water (Newton 1937, Brammall & Leech 1940, Poole & Kelk 1971, Dunham et al. 1978). In the past, various clays have been referred to as fuller's earth, so early descriptions such as those in Greg & Lettsom (1858) and Heddle (1901) are not confirmed occurrences of montmorillonite.

In Scotland, calcium montmorillonite is found in the Ardtun Leaf Beds, an interbasaltic horizon near the base of the lava pile, at Slochd an Uruisge and Bremanoir, cliff sections respectively to the NW and N of Ardtun, southwest Isle of Mull, Strathclyde. It occurs in a brown or grey, unctuous, laminated argillaceous rock that also contains feldspar, smectite, kaolinite, quartz, a clay mica and lesser amounts of analcime, calcite and a member of the heulandite–clinoptilolite series (Sabine et al. 1972). On the Morvern Peninsula, adjacent to Mull, red weathered basalt at Killundine contains an interstratified illite–montmorillonite clay together with hematite and anatase. The montmorillonite is an alteration product of labradorite feldspar, which in the most altered specimens is totally replaced by clay minerals (Bain et al. 1980). Clay-mineral analysis of the Jurassic Lower Broadford Beds on Mull, Ardnamurchan and the Isle of Skye, Highland Region, has shown them to contain up to 30% montmorillonite, perhaps derived from the weathering of basic igneous rocks exposed in the sediment source in Late Triassic times (Amiri-Garroussi 1977). Montmorillonite is also noted at the northeast contact of the Cnoc nan Cuilean syenite, where it occurs with harmotome and stilbite (Gallagher et al. 1971), and in clay-filled vugs at the margins of fissure-filled veins almost entirely composed of barite, at Muirshiels mine, Renfrewshire Hills, W of Paisley, Strathclyde (Stephenson & Coats 1983).

Montmorillonite is associated with calcite, cowlesite, chabazite and offrétite in amygdaloidal Tertiary basalt from Coulters Hill quarry, Ballyclare, Co. Antrim, Ireland (Nawaz 1978a).

montroseite

orthorhombic $(V^{3+},Fe^{2+},V^{4+})O(OH)$

A rare vanadium–iron oxyhydroxide found in Permian uranium- and vanadium-rich nodular concretions from the Littleham Mudstone Formation at Littleham Cove area, Budleigh Salterton, Devon. The cores of the concretions have been extensively replaced by vanadium-rich illite or roscoelite and montroseite. The nodules also contain a complex assemblage of copper, cobalt and nickel arsenides, chalcocite, clausthalite, coffinite and uraninite. It is thought that the nodules developed before maximum burial and compaction had taken place (Nancarrow 1985, Milodowski et al. 2000, 2002).

moolooite

orthorhombic $Cu^{2+}(C_2O_4){\cdot}nH_2O$ $(0 \leq n < 1)$

A rare hydrated copper oxalate found within the medulla of four lichen species growing on copper-bearing rocks from the Coniston copper mines, Coniston, Cumbria. Moolooite forms vivid blue inclusions (up to 0.5 mm across) in white weddellite. It seems most probable that reaction of groundwater or surface water containing copper with oxalic acid excreted by the lichen lead to the precipitation of moolooite (Chisholm et al. 1987).

moonstone

A variety of orthoclase and of plagioclase (both feldspar group), which shows a silvery or bluish iridescence.

mooraboolite

An impure variety of natrolite, zeolite group.

mordenite

ZEOLITE GROUP

orthorhombic $(Na_2,Ca,K_2)_4(Al_8Si_{40})O_{96}{\cdot}28H_2O$

An uncommon hydrated sodium–calcium–potassium aluminosilicate found in cavities within a trachyte at Dumbarton, Strathclyde, Scotland. The walls of the cavities are lined with quartz and heulandite, upon which a layer of mordenite (ptilolite), calcite and quartz has developed. Mordenite is brick red and it forms radiating groups of needle-shape crystals (Bøggild 1922). Devonian olivine basalts at Todhead Point, 9 km S of Stonehaven, Grampian, contain a suite of zeolite species that includes mordenite, laumontite and stellerite (Dyer et al. 1993). Mordenite is also found in Tertiary basalts at Scotland's premier zeolite locality, Sgùrr nam Boc, between Loch Brittle and Loch Eynort, Isle of Skye, Highland Region, where it forms fibrous masses of white acicular crystals up to 2 cm in length. Aragonite, apophyllite, calcite, chabazite, heulandite, laumontite, quartz, stilbite and rare epistilbite are associated species in nearby vesicles in the same boulders (Green & Wood 1996, Green & Todd 2001). At Aros, Isle of Mull, Strathclyde, fibrous crystals of mordenite are noted by Waymouth et al. (1938).

In Ireland, mordenite is one of the rarest zeolite species found infilling vesicles within the Tertiary basalts of Co. Antrim, being found as aggregates of delicate fibres at only two localities: in a shore exposure at Black Cave, near Larne, and from an unnamed quarry, 0.5 km NNE of the summit of White Mountain. Walker (1960a)

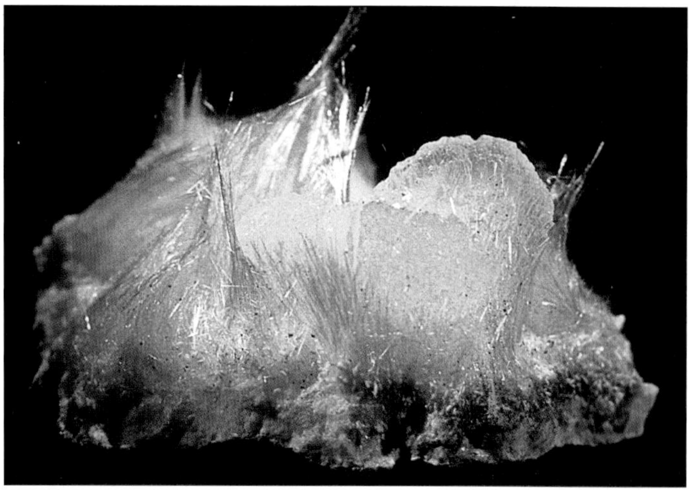

Mordenite tufts to 2 cm associated with stilbite. Sgùrr nam Boc, Isle of Skye, Highland Region, Scotland. (© The Trustees of the National Museums of Scotland)

suggests that mordenite may be more common than this suggests. Fine white fibres of mordenite have also been found in basalts from Boghill Road quarry, Glengormley, near Belfast (H. J. Foy sample, D. I. Green XRD, Tindle unpub. EMP data), from Portnaboe, near the Giant's Causeway, and in a basaltic andesite dyke at Slievenagravery quarry, SW of Hannahstown (Adair 1987).

morenosite

orthorhombic $NiSO_4{\cdot}7H_2O$

A rare hydrated nickel sulphate described from St Patrick's copper mine, Avoca, Co. Wicklow, Ireland, where it occurs as a greenish-white efflorescence in cavities and joints under overhanging rocks, protected from rain. A fine-grain white efflorescence from an old adit in the mine is also morenosite. Tiny prismatic crystals of morenosite up to 0.12 mm in length are readily soluble in water and have a strong astringent taste (King & Evans 1964, Ryback & Moreton 1993).

In Wales, reports of green coatings of morenosite on millerite are probably misidentifications: Elled vein, Bedwellty, Gwent (North & Howarth 1928); Tylorstown colliery, Mid Glamorgan (Firth 1971); and Gelli colliery, Rhondda, Mid Glamorgan (Bevins 1994). EMP analysis of specimens in the collection of the NMW, failed to identify any sulphur and it is suggested that the mineral could instead be a nickel carbonate, oxide or hydroxide (J. M. Horák pers. comm. 2004). An unpublished report suggests that the species is nickelhexahydrite (T. F. Cotterell pers. comm. 2007). At Brownley Hill mine, Nenthead, Cumbria, gypsum has been confirmed as a green coating on millerite (Green & McCallum 1993).

Morenosite is also reported from the Waterbank end of Clayton adit, a part of Clayton copper mine, Ecton, Staffordshire, where it occurs as a bright-green coating on acicular crystals of millerite. The millerite is free-standing or encrusts cavities in celestine and is associated with microcrystalline pyrite cubes (Starkey 1983). The identification was visual and it requires verification.

Heddle (1901) briefly reports morenosite from Highland Region in Scotland – at Rubha a' Chamais Bhàin, near Glen Sanda Castle, and Morvern, Strontian, but this was not substantiated by Macpherson & Livingstone (1982).

morinite

monoclinic $NaCa_2Al_2(PO_4)_2(OH,F)_4{\cdot}2H_2O$

Intergrown crystals of morinite from Gunheath china-clay pit, St Austell, Cornwall; width 600 μm; SEM image. (© N. J. Elton)

An hydrated sodium–calcium–aluminium phosphate fluorhydroxide found on loose boulders in the Gunheath china-clay pit (Hensbarrow section), Carthew, St Austell, Cornwall. Morinite forms crusts of intergrown crystals coating the surface of a thin druse of quartz traversing partially kaolinized granite. The morinite crusts are pale bluish grey to grey and are made up of colourless transparent morinite at the surface and white opaque-to-translucent massive morinite below. Occasional tabular torbernite crystals occur on the morinite, and fluorapatite, fluorite and muscovite are found in the granitic matrix. Morinite is a low-temperature hydrothermal phosphate found in granite–pegmatite environments (Elton 1996).

morion

An almost black variety of quartz.

mornite

TL Mourne Mountains, Co. Down, Ireland

A synonym of labradorite, a member of the plagioclase series, feldspar group.

morvenite

TL Strontian, Highland Region, Scotland

Morvenite is named after Morven (now known as Morvern), a district that lies to the south of the Strontian mines (Thomson 1836). It is a transparent spindle-shape variety of harmotome (zeolite group).

moss agate

A green and white variety of chalcedonic quartz containing arborescent growths of pyrolusite.

mossottite

A strontium-rich variety of aragonite.

mottramite

DESCLOIZITE GROUP

orthorhombic $PbCuVO_4(OH)$

TL Mottram St Andrew, near Alderley Edge, Cheshire, England

A rare lead–copper vanadate hydroxide that forms a series with descloizite and is reported as resinous black crystalline encrustations, sometimes with a velvety appearance, on Keuper sandstone. Thin particles are subtranslucent and they transmit yellow light (Roscoe 1876). The type material could have been collected from a processing plant at the type locality, but the specimens may instead have originated from Pim Hill, near Harmerhill, 9 km N of Shrewsbury, Shropshire, where mottramite has been found *in situ* (Bannister 1933, Kingsbury & Hartley 1956a, Braithwaite 1994). To complicate matters further, arsenatian mottramite has since been found *in situ* underground, in the northern part of Mottram mine, Mottram St Andrew. This material forms thick brown botryoidal crusts on sandstone and does not resemble the type material (Braithwaite 1994).

At Greystone quarry, Lezant, near Launceston, Cornwall, mottramite occurs as black rhombic bipyramidal crystals and crystalline crusts with green arsenic-rich pyromorphite on quartz veinstone derived from the Greystone lead lode (Elton et al. 2004). Mottramite–duftite described by Camm & Merry (1991b) from Penberthy Croft mine, St Hilary, Cornwall, has been shown to be on the duftite side of the solid-solution series and should be labelled vanadium-rich duftite (Betterton et al. 2000).

Kingsbury & Hartley (1956a, 1960) report mottramite from Brandy Gill mine and Nether Row Brow, both Caldbeck Fells, Cumbria; however, evidence has been collected to indicate that the provenance of some of Kingsbury's samples is incorrect (Ryback et al. 1998, 2001). Hartley (1984) repeated these claims, unaware that they could be dubious. Mottramite has since been found as minute sharp pale-green bipyramidal crystals at Brandy Gill mine by Cooper & Stanley (1990), who also note other occurrences of mottramite in the Caldbeck Fells: minute, brownish-black rice-grain crystals associated with mimetite–pyromorphite from Arm o' Grain; minute translucent pistachio-green rounded crystals on goethite and quartz from

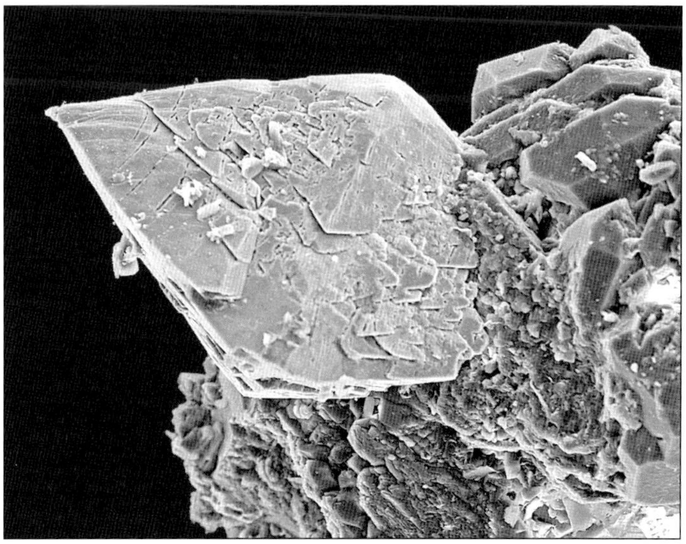

Rhombohedral mottramite from Pim Hill mine, Shropshire; width 225 μm; SEM image. (© D. I. Green)

Roughton Gill mine (debris below Iron Crag); and thin greenish-brown encrustations with bayldonite from Low Pike, near Potts Gill mine. Green et al. (2006b) describe mottramite from Arm o' Grain as black elongated boat-shape crystals up to 1 mm long. At Low Pike, mottramite is one of the supergene species occurring with bayldonite, beudantite, brochantite, cornwallite, duftite, gartrellite, malachite, mimetite, philipsburgite and pseudomalachite, which occur in thin fractures in quartz veinstone and surrounding wallrocks (Green et al. 2003b). A yellow-brown crust from the 60-fathom level dumps of Sandbed mine has been identified as mottramite (Green et al. 2006a), and dark-brown to buff, minute crystals (< 0.1 mm) of rice-grain shape covering areas several centimetres across are reported from Brae Fell mine, Dale Beck valley (Bridges et al. 2006).

At Judkins quarry, Nuneaton, Warwickshire, mottramite occurs in a minor Pb–V–Zn assemblage associated with Ba–Ca–Cu mineralization, close to a well defined unconformity between igneous Precambrian rocks and overlying Cambrian sediments. Mottramite forms yellow to olive-green or brown crystalline crusts coating barite, calcite and vanadinite. Rounded olive-green rhombohedral crystals up to 0.2 mm long, and overgrowths on, or replacing, galena, are also reported (Ince et al. 1991). At Springhill quarry, Peldar Tor, Whitwick, Leicestershire, mottramite was found in a 20 cm-wide quartz vein cutting Precambrian porphyritic dacite (a part of the Whitwick complex). Various habits are reported, including: small tabular olive-green to black vitreous crystals (0.1–1 mm across) with curved faces, occurring individually or in groups, and larger dark-brown to black crystals, up to 7 mm across, forming globular interpenetrant clusters, cog-wheels, elongate blades and spear-like bundles. It is thought that the mottramite is associated with pre-Triassic low-grade metamorphism (Faithfull & Ince 1992). More recently, yellowish-brown to green botryoidal drusy crusts and lustrous black pyramidal clusters of mottramite on dolomite have been confirmed from a native copper–cuprite deposit at New Cliffe Hill quarry, Stanton under Bardon, Leicestershire (Hubbard et al. 2005). At Newhurst quarry, Shepshed, Leicestershire, mottramite has been found in various forms: light to dark-green aggregates of small globules up to 1 mm across, dark-green to greenish-brown crusts of microcrystals, small light to dark-green and brown crystals and yellowish- to dark-green crusts in Triassic breccia within weathered fractures on the S side of the quarry (Ince 2005).

In Scotland, mottramite has been briefly noted from the Belton vein, near the head of Whyte's Cleuch, Wanlock Dod, Wanlockhead, Dumfries & Galloway, and Leadhills Dod, Leadhills, Strathclyde (Brown 1927). Mottramite is also found on the dumps of Lady Anne Hopetoun shaft on the Brow–Hopeful vein system, Leadhills, as reddish-brown resinous globular aggregates or amber-yellow rosettes, commonly overgrown with cassedanneite (Neall 2005).

At Loanhead quarry, near Beith, Strathclyde, mottramite is found in fractures and vesicles within Clyde Plateau Lavas of Carboniferous age that have been cut by a dolerite dyke. It is a very rare species and it occurs as lemon to lime-yellow and clove-brown crystals up to 0.5 mm across, or as coatings with vanadinite and greenockite on prehnite. Other species from this environment include analcime, andradite, barite, bornite, calcite, chabazite, chalcocite, edingtonite, native copper, covellite, galena, grossular, harmotome, hematite, heulandite, laumontite, magnetite, malachite, muscovite, natrolite, quartz, native silver and thomsonite (Meikle 1989b, Green & Todd 2001).

Mottramite from Peldar Tor, Whitwick, Leicestershire; 3 mm crystal aggregate. (© M. P. Cooper)

mountain cork or mountain leather

A synonym of asbestos; variously used to describe actinolite (amphibole group), palygorskite and sepiolite (smectite group).

mountain silk

A synonym of asbestos; a fibrous variety of amphibole.

mountain soap

A synonym of saponite, smectite group, and of halloysite, kaolinite–serpentine group.

mountain wood

A synonym of asbestos; fibrous amphibole or chrysotile. Also applied to palygorskite and saponite (smectite group).

mpororoite

triclinic? $Al_2O(WO_4)_2{\cdot}6H_2O$

A rare hydrated aluminium tungstate found at Hingston Down Consols mine, 2 km W of Gunnislake, Calstock, Cornwall, where it is associated with ferritungstite and scorodite. All three species occur in cavities within bladed wolframite (Golley & Williams 1995, Green et al. 1996a). It is pronounced "muh-pororo-ite".

mullite

orthorhombic $Al_{4+2x}Si_{2-2x}O_{10-x}$ ($x \approx 0.4$)

TL Seabank Villa, Isle of Mull, Strathclyde, Scotland

A rare aluminosilicate found in fused argillaceous sediments (the so-called sillimanite buchites) that occur rather frequently as inclusions (xenoliths) in Tertiary dolerite intrusions. The buchites consist principally of glass in which are embedded rare plates of corundum and abundant minute prisms of colourless to delicate lilac-pink mullite, looking like sillimanite. Also on the Isle of Mull, occurrences of mullite are reported from a buchite at Nuns' Pass, 2 km SW of Carsaig (Bowen et al. 1924), and, from an unnamed paragenesis at Loch Scridain, pink masses of mullite have been found in blue corundum crystals up to 3 mm long (Robinson et al. 1992). Mullite-bearing xenoliths, collected from the upper surface of a dolerite sill outcropping between Scoor and Uisken on the south coast of the Ross of Mull, were used by Brearley (1986) in an electron optical study of muscovite–phengite breakdown in pelitic xenoliths during pyrometamorphism (high-temperature contact metamorphism).

On mainland Scotland, but only 20 km from the Isle of Mull, an olivine dolerite dyke at Bàgh an Tigh-Stòir, Craignish Peninsula, Strathclyde, contains two contact-metamorphosed xenoliths of Craignish phyllite, each 60 cm across. The xenoliths consist of globular clots of crystalline quartz, a little cordierite, pink mullite needles, possible spinel and glass (Allison 1936). Contact metamorphism is also responsible for the occurrence of buchite glasses containing possible mullite needles after muscovite in partially melted Moine arkoses adjacent to a dolerite plug at Glenmore, Ardnamurchan, Highland Region, and to banded cordierite–magnetite–mullite–pseudobrookite–spinel rocks replacing Dalradian phyllites near a dolerite plug at Sidhean Sluaigh, Strachur, Loch Fyne, Strathclyde. Close to the dolerite contact, bright-blue corundum (var. sapphire) is also present (Smith 1965, Worden et al. 1987).

In Ireland, two varieties of mullite are found in porcellanite inclusions within the Tertiary dolerite plug at Tievebulliagh, near Cushendall, Co. Antrim. The porcellanites correspond to ferruginous bauxite and lithomarges in which halloysite, gibbsite and hematite were the principal original constituents. Mullite is colourless or very pale pink in thin section and is found in the immediate vicinity of the igneous contact. It usually has a grainsize of 0.05–0.1 mm, and sporadic nests of crystals up to

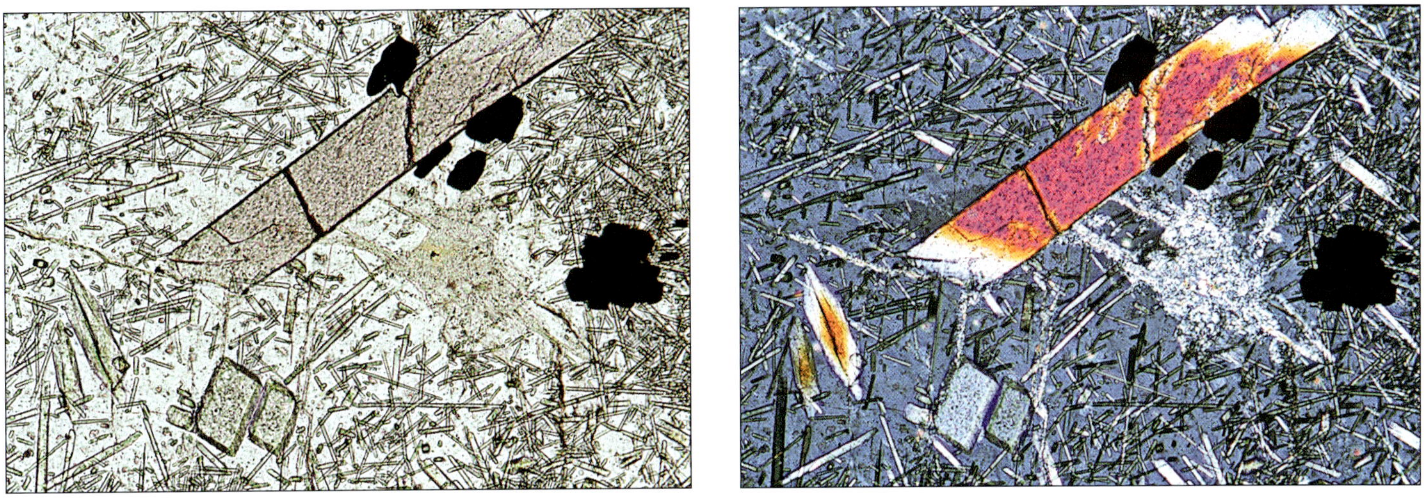

Large and small needles of mullite embedded in anorthite from the Rubh' a' Chromain sill, Carsaig, Isle of Mull, Strathclyde, Scotland; width 1.3 mm; viewed in PPL (left) and XPL (right). (© K. Gardner)

0.75 mm in length also occur. A second, colourless or pale-yellow, variety of mullite is iron rich and is found as a fine-grain felt or mesh of fibres, with individual crystals barely distinguishable and ranging from 0.01 mm to 0.05 mm in length. Rounded patches of mullite, up to 0.5 cm across, associated with cristobalite or tridymite, probably represent original halloysite segregations in the original lithomarge. Colourless to pale-blue corundum is a common associate, together with cordierite, hematite, magnetite, ilmenite, pseudobrookite and green spinel (Agrell & Langley 1958). These authors also report mullite in corundum–magnetite–mullite porcellanites at the contact of the Brockley dolerite plug on Rathlin Island and from thermally metamorphosed shales and sandstones (now mullite–cordierite hornfels) adjacent to the Carrickmore dolerite dyke, 8 km E of Ballycastle, both in Co. Antrim. The Brockley occurrence is further documented by Wilson & Robbie (1966), who note an association of mullite with sekaninaite (iron cordierite). Partially melted rocks (buchites) at the contact of doleritic intrusion with Old Red Sandstone tuffaceous sedimentary rocks at Tieveragh, near Cushendall, also contain mullite (Kitchen 1984).

mundic

A Cornish mining term for massive pyrite.

murchisonite

TL Dawlish and Heavitree, Devon, England

Named after Sir Roderick Impey Murchison FRS (1792–1871), director general of the Geological Survey of Great Britain and author of *The Silurian System* (1839); born in Tarradale, Ross-shire (part of Highland Region), Scotland. Murchisonite is a flesh-red variety of orthoclase, feldspar group. In Devon, fine masses have been found in rolled pebbles within Triassic conglomerates at Heavitree, near Exeter, and on the beach between Teignmouth and Dawlish (Hall 1868b, Collins 1892). Heddle (1901) notes murchisonite in granite from the east side of Goat Fell, Isle of Arran, Strathclyde, Scotland.

muscovite

MICA GROUP

monoclinic

$KAl_2\square AlSi_3O_{10}(OH)_2$

A very common potassium aluminosilicate hydroxide found in granites, granite pegmatites, aplites, phyllites, schists and gneisses. It also forms detrital or authigenic grains in sediments. Muscovite compositions with significant lithium trend towards the lepidolite series, whereas those with substantial iron trend towards the phengite series. Unless supporting data are available, it is not always possible to confirm the identity of these essentially colourless (or pale) micas.

England

Cornwall Muscovite is an important, but minor, constituent of many of the granites of the southwest England batholith, with reports from the Bodmin Moor granite at De Lank quarry, the Carnmenellis granite at Carnsew and Chywoon quarries, the Land's End granite at Castle-an-Dinas and Lamorna Cove, the St Austell granite at Luxulyan quarry, and from Penninis Cove in the Isles of Scilly. Biotite is a common associate, and accessory cordierite, sillimanite or tourmaline are often present. Enclaves within these granites can also contain appreciable muscovite (Stimac et al. 1995). Large crystals of a pinkish-lilac phengite variety are noted from the Carn Marth granite, near Redruth (Charoy 1986). A full review of mica chemistry in these intrusions, with particular emphasis on the lithium enrichment, is given by Stone (1984), Stone et al. (1988) and Henderson et al. (1989). A study of two small granite cupolas at Cligga Head (Perranzabuloe) and St Michael's Mount has shown them to be evolved granites containing zinnwaldite (protolithionite) and muscovite, whose texture is mainly a result of metasomatism and recrystallization during the final stages of emplacement. These metasomatic (greisenization) changes include the replacement of biotite and zinnwaldite by tin-rich muscovite in a process that also produced disseminated tin–tungsten stockwork mineralization (Alderton & Moore 1981). Greisenization at Cligga Head has produced rocks with over 40% iron-rich muscovite and no zinnwaldite near to mineralized veins (Hall 1971). A coarse muscovite–quartz greisen containing minor chlorite, topaz and tourmaline has also developed near a quartz–wolframite vein at Hawkswood mine, on the eastern margin of the Bodmin Moor granite (Alderton et al. 1980), and an arsenopyrite–cassiterite-bearing greisen is reported from Wheal Jane, Kea, near Truro (Dominy & Camm 1997). Newall (1994) describes muscovite-bearing granite and greisen in a report of the sheeted tin–tungsten vein deposit at Redmoor, 2 km N of Callington. Greisenization and extensive hydrothermal alteration of granite has left behind large quantities of commercial kaolin, together with muscovite (including the varieties gilbertite and sericite) and quartz, in the china-clay pits of Cornwall, particularly in the St Austell district (Bray & Spooner 1983, Manning & Exley 1984, Williamson et al. 1997, Thurlow 2001). Muscovite is also found in pegmatites and aplites associated

Muscovite rosette (variety gilbertite) from Kessel quarry, Longdowns, Cornwall; width 18 mm. (© Royal Cornwall Museum, Royal Institution of Cornwall; photographer D. I. Green)

with the granitic intrusions. It is common in the pegmatitic roof zone of the Tregonning granite, between Rinsey and Megiliggar Rocks, near Porthleven (Stone 1992), and forms large flakes in a pegmatite at Priest's Cove, Cape Cornwall, and stellate aggregates (gilbertite) in pegmatite at Luxulyan quarry, near Lostwithiel (Hosking 1954b). At Trolvis quarry, near Longdowns, pegmatites in the Carnmenellis granite have crystallized muscovite on one side of earlier-formed feldspar crystals, thereby indicating the direction of flow of hydrothermal fluids (Hosking 1954a). Chywoon quarry, near Longdowns, also works Carnmenellis granite, and here the associated pegmatites contain perfect hexagonal crystals of muscovite up to 2 mm across and books of brown muscovite plates up to 6 mm across, sometimes intergrown with albite (Moulding & Hooper 2005). Other reports of muscovite include: Balleswidden mine, E of St Just (Robson 1949b); in weathered Carnmenellis granite at Bosahan quarry, near Constantine (Rice 1973); in pegmatites at Carn Marth and Kessel Downs (Halvasso) quarry, near Longdowns (Jackson 1978); in cavities in vein quartz at Croft Gothal mine, St Hilary (Betterton 1996d); in a carbona at East Wheal Lovell (Fatwork section), Wendron (Dominy et al. 1996a); in intertidal sediments containing particulate mine waste in the Fowey and Gannel estuaries (Pirrie et al. 2000b, 2002); crude hexagonal flakes at Greystone quarry, Lezant, near Launceston (Elton et al. 2004); in schists at Lizard Head (Robson 1949b); in granite and luxullianite veins at Luxulyan quarry, near Lostwithiel (Lister 1978); in granite at Pendarves mine, Camborne (Alderton et al. 1980); in a greisen at South Crofty mine within the Carn Brea granite (Webb 1946); and the variety gilbertite occurs in granite at Stenna Gwyn mine, St Stephen-in-Brannel, near St Austell (Rudler 1905, Kingsbury 1959).

Cumbria The Grainsgill section of the Skiddaw granite has been metasomatized and is now composed of quartz, feldspar (partially altered to muscovite) and biotite (mostly altered to chlorite). Metasomatism has also affected surrounding biotite–cordierite hornfels, leaving behind a rock composed almost entirely of quartz and muscovite, the latter forming large crystals and radiating fan-like clusters. Muscovite-bearing quartz veins also occur at this locality (Roberts 1983). At the nearby Carrock mine, muscovite (including gilbertite) is said to be abundant (Davidson & Thomson 1951), although Shepherd et al. (1976) consider it to be a minor component of the quartz–scheelite–wolframite veins. Other reports of muscovite include: at Bannerdale graphite mine, Mungrisdale, Caldbeck Fells (Ziemba 1975); in sandstones at Langhorn quarry, Bigrigg, and Overend quarry, near Whitehaven (Lewis 1931); in the finer sediments of the Silurian Coniston Grits to the E of the River Lune (Furness 1965); in the copper-bearing quartz vein at Dale Head (North vein), Newlands Valley (Stanley & Vaughan 1980); in the Eskdale granite (Simpson 1934) and related chalcopyrite–ferberite–scheelite veins at Buckbarrow Beck, Corney Fell, Waberthwaite (Young et al. 1986); and in aplites and miarolitic cavities within the Shap granite (Russell 1925, Wirth & Green 2002).

Derbyshire Muscovite is reported from xenoliths in the Great Rocks Dale dyke, near Buxton (Cope 1933, Ford et al. 1993).

Devon Muscovite is considered to be a primary magmatic species in the Crownhill and Lee Moor intrusions in the southwestern lobe of the Dartmoor granite (Knox & Jackson 1990). Elsewhere in the Dartmoor granite, secondary muscovite (including the pinite variety after sillimanite or cordierite) is described from Haytor, Pithill and Sweltor quarries, and pegmatitic muscovite (gilbertite) is noted from Merrivale quarry (Brammall & Harwood 1932). Pervasive hydrothermal alteration, including sericitization, chloritization, kaolinization and tourmalinization, locally affects all Dartmoor granite types, but is particularly advanced in the china-clay areas. Muscovite is noted in Dartmoor granite from Birch Tor and Pewtor (Stimac et al. 1995); between Birch Tor mine and Vitifer iron–tin mine (Alderton 1988); from Lee Moor (Hogg & Meads 1970); and more generally by Brammall & Harwood (1925) and Ward et al. (1992). Metasomatic pale yellowish-green secondary muscovite (gilbertite) occurs along joints in the Meldon aplite, a fluorine–lithium-rich aplite dyke located 1 km NW of the main Dartmoor granite (Chaudhry & Howie 1973b). However, in the aplite itself, members of the lepidolite series are found instead (Chaudhry & Howie 1973a); these had previously been referred to as "white mica" (McLintock 1923). On Lundy island, phengitic muscovite is an essential minor component of the Tertiary granite and it forms intergrowths with siderophyllite (Stone 1990, Thorpe et al. 1990). Pale-green talc-like material associated with millerite in fractures and cavities in quartz pods within Middle Devonian shale on the beach at St Mary's Bay, Brixham, proved to be muscovite (Braithwaite & Wilson 2001a).

Durham Compact pale yellowish-green muscovite commonly occurs in Weardale as a breccia matrix or veinlet infilling at Burtree Pasture mine, Cowshill; and, near Rookhope, at Frazer's Hush and Redburn mine. Muscovite may be a useful indicator mineral of fluorite–galena vein mineralization (Smith & Hardy 1981). The Weardale granite also contains muscovite (Tindle unpub. EMP data).

Leicestershire At Newhurst quarry, Shepshed, colourless to greenish-white plates of sericitic muscovite occur in veinlets and on joint surfaces associated with shatter belts and aplites (King 1973, Ince 2005).

Wales

Dyfed Muscovite is widespread as an alteration product of primary plagioclase feldspar in Ordovician metabasites, such as those in the St David's Head intrusion and Garn Turne Rocks, E of Wolf's Castle (Bevins & Rowbotham 1983). A soft silvery-white substance described as hydromuscovite from Ogofau, near Pumsaint (Brammall et al. 1937, Steed et al. 1976), is actually illite.

Gwynedd Muscovite (including the iron-rich phengitic variety) occurs in Ordovician metabasites from Cadair Idris and Cwm Eigiau, 6 km W of Trefriw (Bevins & Rowbotham 1983). Colourless to pale-green sericitic muscovite, forming talc-like aggregates of very small grains, has been found in open joints and veins in quartz strings associated with a dolerite sill at Tan-y-Bwlch (Hutchinson & Campbell Smith 1912). The Precambrian Coedana granite on Anglesey, and the surrounding hornfels, both contain muscovite, as do higher-grade gneisses in the Holland Arms area (Horák 1993).

Scotland

Central Muscovite is one of many inclusion types (commonly ≤ 10 µm in diameter) found in pyrite from a stratabound copper–zinc deposit within chloritic schists and albitic quartzites of Dalradian age at Creag Bhocan, SSW of Tyndrum (Scott et al. 1988). The Dalradian Ben Lui Schist in the Auchtertyre–Challum area, to the E of Tyndrum is a semipelitic sequence of turbiditic origin that hosts chalcopyrite, chromite, chromium-rich magnetite and muscovite, and pyrrhotite. The chromium-rich horizons extend intermittently for 20 km towards Glen Lyon (Fortey & Smith 1986).

Dumfries & Galloway Muscovite is a common primary constituent in the peraluminous Cairnsmore of Fleet granite, where it is associated with siderophyllite (Tindle unpub. EMP data). Near Clatteringshaws Loch, pegmatite and aplitic veins at the margin of Fleet pluton contain plates of muscovite up to 3 cm across, associated with sillimanite and almandine–spessartine garnet (Williams 1965, Macleod 1992). Small amounts of muscovite are also noted from a cordierite microgranite at the centre of the Loch Doon granite complex and from associated aplite dykes (Tindle 1982).

Grampian The Caledonian Glen Gairn granite complex, to the NW of Ballater, contains lithium-enriched granites that at Gairnshiel Bridge host Ag–Bi–Mo–Sn–W mineralization. Colourless and pale-green muscovite (including sericitic and phengitic varieties) occur in the granites, greisen and a wolframite-bearing quartz vein (Tindle & Webb 1989, Webb et al. 1992, Tindle unpub. EMP data). Both primary and secondary (after siderophyllite) phengitic muscovite is present in the Cairngorm granite and the associated Glen Avon pluton, although never in abundance (Harrison 1990, Tindle unpub. EMP data). Other Caledonian granites containing muscovite, albeit not in all samples examined, include: Aberchirder, Aberdeen, Ardlethan, Ballater, Bennachie, Cove Bay, Hill of Three Stones, Keith, Mount Battock, Peterhead, Skene, Strichen and Windyhill. Muscovite is also present in the lepidolite pegmatite at Peatfold in Glenbuchat (Tindle unpub. EMP data). Fine specimens of a pinkish-brown muscovite were found in the granite quarry at Rubislaw, Aberdeen, and at Auchindoir, 3 km SW of Rhynie (Greg & Lettsom 1858).

Muscovite is associated with an unusual sodium-rich siderophyllite in Dalradian staurolite-bearing schists at Craig of Finnoch, Glen Lethnot (Dempster & Jackson 1996), and frequently occurs in garnet–mica schists along the Portsoy–Duchray Hill lineament between Tomintoul and Portsoy (Beddoe-Stephens 1990). Muscovite-bearing granitoid migmatites are also a feature along this lineament, particularly around the Huntly, Knock and Portsoy gabbroic intrusions (Ashworth 1976). North of Stonehaven, muscovite occurs with paragonite in chlorite–chloritoid schists and has been used to establish crystallization temperatures in these Dalradian rocks (Baltatzis & Wood 1977).

Highland Muscovite was an economic commodity used for a wide range of radio, electrical and scientific purposes during the Second World War and was quarried from Moinian muscovite-rich pegmatites on Sgùrr Coire nan Gobhar on the Tom an Neoil ridge, near Loch Nevis, Knoydart Peninsula. The largest book reached 60 cm in diameter, and crystals of 30 cm or more were fairly common. Albite, apatite, beryl, garnet and quartz are associated species. Similar muscovite was also quarried at Little Scatwell in the Garve–Strathpeffer district, and occurrences at nearby Carn Gorm and Brae Tollie were also investigated (Kennedy & Lawrie 1943, Turner & Harris 1971, Starkey 1987c, Green & Todd 2001, McIntosh 2001). Muscovite is a minor constituent in the granite intrusions of Ardclach, Grantown, Moy (Tindle unpub. EMP data), and Mullach nan Coirean, near Ben Nevis (Haslam 1971). Other occurrences include: 30 cm-diameter books of muscovite in a quartz-rich pegmatite at Ardarie, near Strontian, and in a Moinian mica schist near Dalilea House, between the summit of Ceann Loch Uachdrach and Lochan a' Mhuilinn (Kennedy & Lawrie 1943); a manganese-rich variety in boulders of Moinian pink mica quartzite from near Kinloch, S of Tongue (Faithfull 2007); bright-green chromium-rich muscovite (fuchsite) in an ankerite–pyrite–quartz rock associated with a fault zone at An Leth Allt, north slopes of Sgùrr an Airgid, Loch Duich (Clifford 1959); in Dalradian pelites within the aureole of the Ballachulish granite complex (Pattison 1987); chromium-rich muscovite (fuchsite) associated with actinolite in Lewisian ultrabasic pods at Camas Baravaig, southeast Isle of Skye (Matthews 1967, Matthews & Cheeney 1968); intergrown with siderophyllite in Moinian garnet–mica schist at Glenfinnan (Rieder 1995); in pelitic schists and gneisses within the Glen Urquhart serpentinite metamorphic complex, W of Loch Ness, near Drumnadrochit (Rock et al. 1986b); in bismuth-bearing quartz veins within the Loch Shin monzogranite, 2 km W of Lairg (Lowry et al. 1994). Heddle (1901) reports 23 cm-diameter plates of muscovite in boulders from the head of Glen Shiel and 25 cm-diameter plates in boulders from the foot of Loch Glass, N of Dingwall. Fine rose-tinted crystals, sometimes 37 cm across, are also noted from Glensgaich railway cutting, Strathpeffer.

Strathclyde Muscovite (phengite) forms abundant spherules up to 150 µm in diameter in mudstones of the Dalradian Bonahaven Formation, at Loch Giùr-Bheinn on the Isle of Islay. The spherules probably represent recrystallized glauconite-infilled microfossils (Fairchild 1977). Other occurrences include: in vugs, associated with thomsonite and other zeolites within Carboniferous basalts at Hartfield Moss, 9 km WSW of Barrhead, and Loanhead quarry, near Beith (Meikle 1989b, 1990); fine-grain phengite in Dalradian greenschists thermally metamorphosed by the Sidhean Sluaigh dolerite intrusion, 3 km SW of Strachur, Loch Fyne (Worden et al. 1987); in thermally and regionally metamorphosed sediments in the aureole of the Ross of Mull granite (Wheeler et al. 2004); in a Moinian pelitic xenolith within a dolerite sill between Scoor and Uisken, Isle of Mull (Brearley 1986).

Tayside Dalradian sulphidic quartzose cherts from the Foss stratiform barite–base-metal deposit, near Aberfeldy, contain very unusual barium–chromium-rich muscovite that in parts has the distinctive green coloration of fuchsite. A murky olive-brown to translucent pale-green variety proved to be a barium–vanadium-rich muscovite (Moles 1985). Fuchsite and barium-rich muscovite have also been described from mineralized horizons in the Ben Eagach Schists around Aberfeldy and SW towards Tyndrum, Central Region. Celsian, cymrite, hyalophane, pyrite and quartz are associated species (Coats et al. 1980, 1984, Fortey & Beddoe-Stephens 1982). Muscovite is also present with zinc-rich staurolite in Dalradian gneisses from the southern contact aureole of the Glen Doll diorite (Schumacher 1985) and with garnet, biotite and kyanite in hornfels in the aureole of the Ben Vuirich granite, 10 km NE of Pitlochry (Ahmed-Said & Tanner 2000). A Dalradian graphitic gneiss from 8 km NW of Fettercairn, Glen Esk, contains muscovite coexisting with margarite (Baltatzis & Katagas 1981).

Western Isles Phengitic muscovite is a minor constituent of the Lewisian Chaipaval (Ceapabhal) pegmatite, 3 km NW of Northton (Taobh Tuath), South Harris (von Knorring & Dearnley 1959b, 1960a). Heddle (1879c) reports this occurrence as muscovite with a peculiar green tint from the great vein of Ben Capval on Harris. At the southern end of Taransay (an island off the west coast of South Harris), muscovite crystals up to 6 cm across have been found in granitic veins (Heddle 1901).

Shetland Colourless to almost white margarodite from a quartz vein in micaceous gneiss from Vannlip, Hillswick, Mainland (Heddle 1879e), could be muscovite.

Ireland

Carlow Muscovite is an important constituent of spodumene pegmatites found along the southeastern margin of the Leinster granite batholith. Perhaps the most important of the pegmatites is that found in boulders and in borehole core, in the grounds of Aclare House, near Myshall, where both ragged green muscovite phenocrysts up to 4 cm in diameter and secondary muscovite after spodumene occur. Pale-green muscovite is also found in associated aplites (Steiger & von Knorring 1975, Scoon 1978, McArdle & Kennan 1987, 1988, Gallagher 1988, Whitworth & Rankin 1989, Whitworth 1992). Other muscovite-bearing pegmatites (and some quartz–tourmaline rocks) are located at Aghlowe, at Coolasnaghta, at Kilcarry Bridge, at Seskinnamadra, W of Tomduff Hill, and northeastwards into Co. Wicklow and Co. Dublin. Muscovite is also present in granite at Borris and in various schistose host rocks (Luecke 1981, Gallagher 1988, O'Connor et al. 1991, Ryback & Moreton 1993).

Cork Lower Carboniferous limestones from the Kilmaclenine area, near Mallow, contain detrital zoned magmatic muscovite thought to be derived from a buried granite in southwest Ireland (Clipstone & Roycroft 1992).

Donegal Within the aureole of the Ardara granite complex, muscovite occurs in minor amounts in the Lower Portnoo Limestone (now marble), in biotite–chlorite phyllonites and as thick books in andalusite–quartz veins (Akaad 1956). Muscovite data from the Donegal granite complex at Castlecaldwell and near Ballygihen in Dooish Mountain are presented by Haughton (1862). Muscovite commonly occurs in association with spessartine in both granite and associated pegmatites in the Rosses ring complex and Trawenagh Bay granite (Hall 1965b). Sericitized plagioclase in the Rosses complex is enriched in ammonium, introduced during pervasive hydrothermal alteration (Hall 1993a).

Down Muscovite (white mica) is a minor constituent of siderophyllite–topaz-bearing replacement veins of greisen type within granite at Newcastle, Mourne Mountains (Nockolds & Richey 1939). Greg & Lettsom (1858) note "very perfect" crystals of muscovite associated with albite, beryl, feldspar and topaz from the Mourne Mountains, but do not give a locality.

Dublin Muscovite is considered to be an essential mineral in the northern portion of the Leinster granite and it forms large hexagonal plates on Three Rock Mountain (Haughton 1856, O'Reilly 1897, Smithson 1932). These crystals are termed "megacrysts" by Roycroft & Kennan (2000), who report that they are up to 12 mm across and contain core regions filled with sillimanite needles. Exceptional zoned muscovite crystals (> 25 mm across), some displaying in excess of a hundred discrete zones, have been collected from a quarry at Stepaside (Roycroft 1989). At the margin of the granite, muscovite also occurs in tourmaline-bearing mica schist at Rathmichael and Piperstown Hill, and in small tourmaline-bearing aplite and quartz bodies at Killiney (Gallagher 1988).

Galway Thermal metamorphism has caused corundum to grow at the expense of muscovite in Dalradian country rocks within the Omey granite aureole, Connemara (Ferguson & Al-Ameen 1985). The Oughterard granite in Connemara contains chemically zoned muscovite with celadonite rims (Dempster et al. 1994).

Mayo Chromium-rich muscovite (phengite) occurs in association with other chromium-rich minerals in a peculiar schist that lies along a complex shear zone within the pre-Caledonian Erris complex at Scotchport Bay, NW of Belmullet (Max et al. 1983). Chromium-rich muscovite (fuchsite) mantles detrital chromite in Ordovician turbidites within the Cregganbaun shear zone (Thompson et al. 1992) and is a constituent of the alteration assemblage associated with gold-vein mineralization in tuffs in the Sheefry Hills, S of Lecanvey (Aherne et al. 1992). Other occurrences include: in contact-metamorphosed hornfels around the Easky adamellite, Ox Mountains (Yardley & Long 1981); in blueschist-facies metabasites belonging to the Dalradian Supergroup from Central Achill, N of Clew Bay (Chew et al. 2003).

Tipperary Chromium-rich muscovite (fuchsite) is reported from altered basic dykes and fault breccias near the Cu–(Ag–Hg) deposit at Gortdrum mine, near Tipperary (Duane 1988). Sericitic muscovite is one of many gangue minerals in the Silvermines sedimentary-exhalative base metal and barite deposit (Moreton 1999).

Westmeath Lead–zinc sulphide mineralization at Moyvoughly, near Moate, 12 km N of Clara, is hosted within Lower Carboniferous carbonates and contains replacement muscovite (phengite) enriched in magnesium and zinc if found embedded in sphalerite or enriched in barium and lead if found in galena (Kucha 1988).

Wicklow Muscovite-bearing pegmatites are found at Monaughrim, Moylisha and Stranakelly, all around Shillelagh, where albite, potassium feldspar, quartz, ± spodumene and accessory cassiterite, garnet and tantalum oxides are associated species (Steiger & Bowen 1975, Luecke 1981, Whitworth & Rankin 1989). At Moylisha the muscovite is lithium poor, but at Stranakelly analyses of a purplish-pink lithium-manganese-rich variety trends to lepidolite compositions, particularly in associated greisen (Scoon 1978, Luecke 1981, Gallagher 1988). Elsewhere, an attractive pale-yellowish muscovite is reported from the dumps of Ballinafunshoge lead mine, Glenmalur (Ryback & Moreton 1993), and a grey silvery phengitic muscovite (margarodite) is noted from granite in the Glendalough Valley (Haughton 1856).

Offshore territories

Isle of Man Clipstone & Roycroft (1992) note euhedral magmatic muscovite from the Foxdale granite.

Jersey Muscovite is a constituent of pegmatites at Côtil Point, Echéhous, L'Etacq, Le Pulec, Maisons and Pipettes (both Minquiers), and Paternoster, and also occurs in the joints of metamorphosed rhyolite at Côtil Point (Mourant 1961).

N

nacrite

monoclinic — KAOLINITE–SERPENTINE GROUP — $Al_2Si_2O_5(OH)_4$

An uncommon aluminosilicate hydroxide, polymorphous with dickite, halloysite and kaolinite, and found on slickensided syenite with quartz veins in a shatter zone of the main fault at Sheet Hedges Wood quarry, Groby, Leicestershire. Nacrite is a creamy-white soapy or powdery mineral that occurs in small cavities within quartz (Claringbull 1952). During an investigation of the hydrothermal mineralization associated with the Shap granite, nacrite was found in the Shap Blue quarry, Cumbria, in metamorphosed Borrowdale Volcanic Group lavas. It coats shear joints in fault breccias and shatter belts, and is associated with chlorite, hematite, pyrite and traces of erythrite, and often with late barite, calcite and quartz veins. Nacrite is pale apple-green to white and has a greasy feel (Firman 1953b, Young et al. 1988). However, XRD examination of a nacrite specimen from Shap Blue quarry, now at the NHM, show it to be glauconite (Young 1987a). Nacrite (described as margarodite) from Cornwall (Greg & Lettsom 1858) is actually muscovite (Spencer 1958).

Work on the A74 near Abington, Strathclyde, Scotland, produced a temporary exposure just to the NW of the Southern Uplands fault. The outcrop contained orange and white clay gouges in which nacrite was identified (Livingstone 1993).

Bevins (1994) provides confirmation of nacrite in Wales from the Trimsaran borehole No. 2, Kidwelly, Dyfed, and the Aberfan borehole, Mid Glamorgan.

nadorite

orthorhombic — $PbSb^{3+}O_2Cl$

A rare lead–antimony oxychloride found at the Bodannon antimony mine, Port Gaverne, St Endellion, Cornwall. Smoky brownish-yellow to reddish-brown nadorite was found in a 2.5×1.5 cm cavity within jamesonite, where it forms aggregates of parallel, nearly square, exceedingly thin, transparent plates, the largest of which is 1.5 mm across. The remainder of the cavity is filled with quartz, lemon-yellow bindheimite and a few small needles of jamesonite. A second specimen contains an aggregate of dark reddish-brown nadorite crystals 1 cm across within a cavity in jamesonite and quartz. Anglesite, bindheimite and senarmontite are associated species (Russell 1927a). Near by, on the dumps of the Port Quin antimony mine, St Minver Highlands, nadorite occurs along with allophane, anglesite, dundasite, cerussite, chalcopyrite, galena, malachite, pyrite, senarmontite and sphalerite. Neighbouring roadside walls also contain jamesonite-rich veinstone superficially altered to bindheimite (Russell 1949b). Golley & Williams (1995) briefly report a few other Cornish localities for nadorite: Wheal Leigh, Pillaton; Treore mine, Port Gaverne, near Port Isaac, St Teath; and Trevinnick mine, St Kew. The last site is listed because of the work of Kingsbury (1957b) and should be considered doubtful until further confirmation is available.

nagyágite

orthorhombic — $(Au,Te)Pb(Pb,Sb)S_2$

A very gold–tellurium–lead sulphide reported from the Clogau gold mine, Bontddu, near Dolgellau, Gwynedd, Wales. It occurs at the contact between gold and telluride minerals in a galena–gold–tellurobismuthite–tetradymite assemblage (Gilbey 1968, Bevins 1994).

nailhead spar

A variety of calcite with flat rhombohedral terminations.

nakauriite

orthorhombic — $Cu_8^{2+}(SO_4)_4(CO_3)(OH)_6{\cdot}48H_2O$

A rare hydrated copper sulphate carbonate hydroxide found on chromitite from Hagdale quarry, Unst, Shetland, Scotland. The chromitite contains occasional heazlewoodite and pentlandite grains embedded in pale-green chrysotile (olivine), some with colourful alteration products. These include clinochlore, hydromagnesite, pyroaurite, theophrastite and pale-blue fibrous nakauriite (Braithwaite & Pritchard 1983).

namibite

monoclinic — $Cu^{2+}(BiO)_2V^{5+}O_4(OH)$

A very rare copper–bismuth vanadate hydroxide found in a narrow quartz–chalcopyrite–ferberite–scheelite vein that cuts the Eskdale granodiorite at Buckbarrow Beck, Corney Fell, Cumbria. Namibite is very dark green and it forms roughly spherical masses up to 1.5 mm across, sometimes with a sheaf-like platy appearance. It occurs in isolation and in small cavities within quartz veinstone associated with bismutite, bismutoferrite, eulytite and mixite (Neall et al. 1993, Green & Briscoe 2002).

namuwite

trigonal — $(Zn,Cu^{2+})_4(SO_4)(OH)_6{\cdot}4H_2O$

TL Aberllyn lead mine, 2 km N of Betws-y-coed, Gwydyr Forest, Gwynedd, Wales

Named after the National Museum of Wales, in which collection the type specimen was found. Namuwite is an hydrated zinc–copper sulphate hydroxide that was found on a single specimen labelled "hydrozincite". The matrix of the specimen is composed of a breccia of fragments of black pyritiferous slate from the Crafnant Volcanic Group, which is cemented by ankerite, calcite, quartz and sphalerite. Pale sea-green namuwite is amorphous to the naked eye, but at very high magnification can be seen to consist of aggregates of euhedral to subhedral hexagonal plates up to 60 µm across (Bevins et al. 1982b). In the lower levels of Aberllyn mine, post-mining hydrozincite has accumulated in abandoned passages as a white mud up to 0.5 m thick, and in a few places pale sea-green namuwite is an associated species (Jenkins & Johnson 1993). Further details of the primary and secondary mineral assemblages from the Llanrwst orefield are provided by Haggerty (1995).

Encrustation of bluish-green namuwite from Aberllyn mine, Llanrwst, Gwynedd, Wales; width 3 cm. (© National Museum of Wales, Cardiff)

Very rarely, namuwite occurs as pale bluish-green pseudo-hexagonal platy crystals up to 0.5 mm across and sometimes grouped into rosettes, on galena-bearing quartz veinstone at Dylife mine, Central Wales orefield, Powys (Green et al. 1996b). Namuwite has also been noted from the nearby Dyfngwm mine (Livingstone et al. 1990) and from Frongoch mine, between Devil's Bridge and Pont-rhyd-y-groes, Dyfed (Anthony et al. 2003).

Post-mining groundwater percolating through Smallcleugh mine, Nenthead, near Alston, Cumbria, has oxidized primary lead–zinc ore minerals and left behind encrustations on the mine walls that contain epsomite, gypsum, namuwite, schulenbergite and serpierite. Namuwite forms pearly sea-green, flat triangular plates, up to 200 µm across, that are often curved and grouped together as sheaves (Livingstone et al. 1990). Ankerite, beaverite, dolomite, smithsonite and the zinc analogue of ktenasite are also noted from the Middlecleugh First Sun vein running through this mine (Livingstone 1991). At Force Crag (Ba–Pb–Zn) mine, Braithwaite, near Keswick, Cumbria, a small area of supergene mineralization was discovered in the lower workings of the mine in 1990. It contained unusually large pyromorphite, common cerussite and lesser amounts of brochantite, langite, linarite, namuwite, serpierite and native silver (forming delicate dendrites) (Green & Briscoe 2002). Rust (1995c) reports namuwite as a rare supergene species forming whitish to watery-green translucent crystals to 0.3 mm at Waterbank mine, Ecton Hill, Staffordshire. It usually overgrows hydrozincite or smithsonite, or develops as a parallel overgrowth on schulenbergite.

Pseudo-cubic natrojarosite from a small seacliff by Cudmore Grove, Mersea Island, Essex; width 8 µm; SEM image. (© E. Condliffe, University of Leeds)

nantokite?

cubic CuCl

A rare copper chloride reported from Levant mine, St Just, Cornwall (Golley & Williams 1995). It is unclear how the species was identified.

nasonite

hexagonal $Pb_6Ca_4Si_6O_{21}Cl_2$

A rare lead–calcium silicate chloride found at Merehead (Torr Works) quarry, Cranmore, Mendip Hills, Somerset. A patch of blue material, originally identified by the NHM as ganomalite, has been shown to be nasonite (A. Hart pers. comm. 2003). A second specimen containing pale-blue nasonite has more recently been found associated with apophyllite, ardennite-(As), barite, datolite and goethite. Nasonite is associated with a late-stage hydrothermal event and not the manganese-rich mineralization at this locality (Turner 2007).

natanite

SCHOENFLIESITE GROUP

cubic $Fe^{2+}Sn^{4+}(OH)_6$

A very rare iron–tin hydroxide found in high-temperature veins in Cornwall at Penberthy Croft, St Hilary, and Hingston Down quarry, 2 km W of Gunnislake, Calstock. At Penberthy Croft, natanite occurs in intimate intergrowths with jeanbandyite as lustrous pale-yellow transparent-to-translucent octahedral crystals up to 0.4 mm across. At Hingston Down quarry, jeanbandyite–natanite intergrowths were found along an E–W-trending vein in two areas, at the first forming lustrous pseudo-octahedral crystals to 1.2 mm across and, at the second, opaque yellowish-brown octahedral crystals to 2 mm. A late-stage origin from residual tin-rich hydrothermal fluid is proposed (Wolfe 1997, Betterton et al. 1998, Betterton et al. 2000).

native species

See antimony, arsenic, copper, gold, iridium, iron, mercury, osmium, palladium, platinum, ruthenium, silver, sulphur and tellurium. Despite references to lead and tin in the literature, native lead and native tin do not appear to be British or Irish species. Native zinc requires further confirmation.

natrojarosite

ALUNITE GROUP

trigonal $NaFe_3(SO_4)_2(OH)_6$

A rare sodium–iron sulphate hydroxide found in a temporary exposure of Middle Jurassic Oxford Clay at Cassington, 8 km NW of central Oxford. Natrojarosite formed lemon-yellow films and nodular aggregates in the clay and in associated gypsum clusters. Crystals are up to 4 µm across, with either a hexagonal platy or pseudo-cubic habit (Atkins 1995). Pseudo-cubic natrojarosite is also reported from a small seacliff by Cudmore Grove, Mersea Island, Essex (R. Tayler pers. comm. 2005).

Kingsbury & Hartley (1958) report natrojarosite from Wanthwaite mine, St John's in the Vale, near Threlkeld, and Embleton quarry, near Cockermouth, both in Cumbria. The work of Ryback et al. (1998, 2001) indicates that the provenance of many Kingsbury samples is dubious and, therefore, further investigation is desirable.

At Hawthorn Hive, a wide bay on the Co. Durham coast (at and around Hive Point and Shippersea Bay), natrojarosite has been found in mine waste derived mainly from Dawdon colliery. Natrojarosite is the most abundant supergene species in the spoil and it forms soft pale lemon-yellow crusts associated with aluminocopiapite, apjohnite, calcite, ferricopiapite, gypsum and sideronatrite (Young et al. 2008).

natrolite

ZEOLITE GROUP

orthorhombic $Na_2(Al_2Si_{13})O_{10}{\cdot}2H_2O$

A common hydrated sodium aluminosilicate often found in amygdaloidal basalts as a late-stage hydrothermal cavity infill species.

Colourless and completely transparent, strongly prismatic or acicular natrolite, forming crystals up to 7 cm long, occurs with analcime, calcite and prehnite in secondary hydrothermal veins (infilled fault fissures) cutting gabbro at Dean quarry, St Keverne, Lizard Peninsula, Cornwall. It alters to blue-grey montmorillonite (Seager 1969, 1971, Wolters & Wolters 1986, Dyer & Wilson 1988). Other occurrences of natrolite on the Lizard Peninsula include: Chynhalls Point, St Keverne, where it formed needle-like crystals radiating from many centres, in fissures within serpentinite (Lowe 1900, Russell 1910b); George's Cove (Golley & Williams 1995); Henscath, near Mullion (Robson 1949b); Parc Bean Cove, Mullion, where it is associated with datolite or prehnite (Russell 1910b, 1912a); Porthkerris Point, 3 km N of

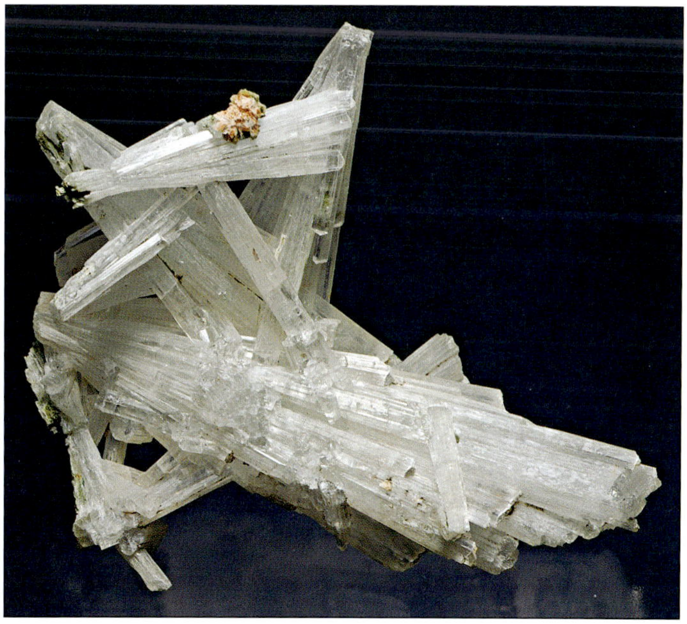

Blades of natrolite, up to 70 mm in length, from Dean quarry, St Keverne, Lizard Peninsula, Cornwall. (© D. I. Green)

Dean quarry, where it occurs associated with analcime, calcite, orthoclase, prehnite and quartz, in hydrothermal veins cutting amphibolite and granulite (Seager 1978a); Potstone Point (Golley & Williams 1995). Elsewhere in Cornwall, natrolite is reported with prehnite between Botallack and Wheal Cock, St Just; from Wheal Carn, St Just; and from Stenna Gwyn, St Austell (all Greg & Lettsom 1858).

Isolated occurrences of natrolite are reported: from vesicles in dolerite dykes cutting granite on Lundy island, off the north Devon coast (Thorpe & Tindle 1992); from vesicles in altered gabbro (glacial erratics) on Thurstaston beach, Wirral, Merseyside (Rothwell & Dickinson 1989); from vesicles in analcime basalt at Calton Hill quarry, Taddington, Derbyshire. At the last locality, natrolite occurs as pale salmon-pink compact fibrous crystalline masses completely filling vesicles up to 10 mm across (Dyer et al. 2001).

In Wales, excellent specimens of water-clear to opaque-white natrolite prisms up to 5 cm long have been found at Benallt manganese mine, Rhiw, Lleyn Peninsula, Gwynedd (Dewey & Bromehead 1915, Hey 1932b, Campbell Smith 1944). The species is noted in a basic mass within gneisses of the Precambrian Mona Complex, at Henblas, Anglesey, Gwynedd (Greenly 1919), and in olivine dolerite dykes between Nant Peris and Nant Ffrancon in central Snowdonia, Gwynedd (Williams 1930b). Natrolite is also reported from a monchiquitic basanite in the Great House diatreme, near Usk, Gwent (Boulton 1911), and in a monchiquite dyke in nearby Glen Court (Haslett 1992). Some of these identifications are tentative and do not appear to be supported by analytical data.

A group of Devonian alkaline basic igneous rocks in the eastern part of the Orkney Isles, Scotland, are hosts to minor late-stage zeolite mineralization. Fibrous natrolite, analcime, apatite, chlorite and anorthoclase–sanidine are found in the interstices between earlier-formed primary minerals at Muckle Castle, SE of Stonehall, and at Point of Ayre, both Deerness, on Mainland, and at the Burg, Shapinsay (Kellock 1969).

At Eilean an Tighe and Garbh Eilean in the Shiant Isles, Western Isles, natrolite and analcime fill interspaces between earlier-formed plagioclase and pyroxene within olivine dolerite (crinanite) sills (Walker 1930).

In Highland Region, natrolite forms abundant needles in orthoclase within a ledmorite dyke at Loch Roe, An Fharaid Mhòr, near Achmelvich, NW of Lochinver (Sabine 1952). Another ledmorite dyke on Rubha Mòr, Coigeach, W of Inverpolly National Nature Reserve, contains fine-grain natrolite as an alteration product of nepheline (Sabine 1953). At both locations aegirine, andradite (melanite) and orthoclase are essential primary constituents of these syenitic dykes. At Oisgill Bay, N of Moonen Bay, Isle of Skye, small stellate natrolite sprays are found with stilbite and heulandite infilling vesicles within Tertiary basalt lava flows (Green & Todd 2001). Natrolite covers epidote in gabbro from Sgùrr Dearg, 3 km E of Glen Brittle youth hostel, Cuillins, Isle of Skye (Heddle 1901). Natrolite is an uncommon amygdale species in Tertiary basalts on the Isle of Skye, but has been reported as tufts of radiating acicular colourless crystals from Allt Mòr Sgùrr a Ghobhaim, W of Loch Brittle, and Osdale River, between Macleod's Tables (Healabhal Mhòr and Healabhal Bheag) (King 1976).

Amygdales in Tertiary basalts within the aureole of the volcanic plug at 'S Airde Beinn, 5 km SW of Tobermory, Isle of Mull, Strathclyde, contain radiating finely fibrous natrolite, analcime and thomsonite, which, during metamorphism, recrystallized to form plagioclase feldspar (Cann 1965a). Natrolite is also reported from Bloody Bay, 3 km NW of Tobermory (Heddle 1901). The Dippin Sill, an alkali dolerite (crinanite) intrusion on the SE of the Isle of Arran, contains interstitial and discrete patches of white analcime, natrolite, chlorite and clay minerals (Gibb & Henderson 1978b). Natrolite also occurs in vesicular Tertiary basaltic dykes at Ard Fernal, Isle of Jura, with gonnardite, mesolite and scolecite; and at Bunnahabhain, northeast Isle of Islay, with analcime, gobbinsite and thomsonite (Dyer et al. 1996).

Elsewhere in Strathclyde (on the mainland), natrolite occurs within vesicles of Carboniferous Clyde Plateau basalts. At Hartfield Moss, 9 km WSW of Barrhead, natrolite forms well crystallized, colourless, transparent, slightly tapering prisms up to 30 mm in length, intimately associated with prehnite and thomsonite (Meikle 1990). Well crystallized, although usually small, examples of natrolite occur at Loanhead quarry, near Beith, in a complex assemblage that includes analcime, barite, calcite, edingtonite, low-temperature garnet, harmotome, prehnite and thomsonite, and rarely diaspore, ilmenite and native silver. Natrolite forms slender, transparent, colourless to pale-blue elongated prisms up to 15 mm long in cavities that also contain analcime, calcite or thomsonite. Emplacement of a Tertiary dyke into the lava pile probably gave rise to a localized hydrothermal system, from which many of these species formed (Meikle 1989b, Meikle & Todd 1995, Green & Todd 2001). Other examples of natrolite in Clyde Plateau basalts include: pearly, acicular and interlacing crystals in the Glasgow–Greenock railway tunnel at Bishopton (Heddle 1901); spheres of radiating white crystals at Bowling quarry, Kilpatrick Hills, near Glasgow (Heddle 1901); Boyleston quarry, SW of Barrhead, where doubly terminated crystals to 8 cm long were once found (Heddle 1887a, 1901, Hall et al. 1989, Livingstone 2002); Reilly quarry, Houston (Meikle 2003). Additionally, Green & Todd (2001) note natrolite in basalts from Earlsburn (Earlston) reservoir, Touch Hills, SW of Stirling, Central Region.

At the southern end of Craignish Peninsula, SW of Kilmelford, Strathclyde, natrolite occurs with analcime and thomsonite in Tertiary crinanite dykes (Allinson 1936), and in the teschenite of the Lugar Sill, Glenmuir Water, Strathclyde, natrolite forms spheroidal masses generally ≤0.3 mm across, associated with the other late-forming minerals – analcime, calcite, chlorite, mica and nepheline (Phillips 1968).

At Glen Farg, 16 km S of Perth, Tayside Region, natrolite is intergrown with mesolite; it forms fine, transparent, colourless or "flesh-red" crystals (Greg & Lettsom 1858, Heddle 1901, Hey 1932b). At Orrock quarry, near Burntisland, Fife, natrolite forms aggregates of parallel crystals, up to 20 mm long, associated with apophyllite and pectolite in fracture-filled veins cutting Carboniferous basalts (Wirth 2003).

Natrolite is abundant in the Tertiary basalts of Ireland, being one of the last zeolite species to crystallize. It occurs with analcime around the edge of the plateau basalts and along fault zones. When one species is abundant, the other is not, and natrolite is predominant at most eastern localities in Co. Antrim from Carnlogh to Belfast and along the north coast near Whitepark Bay, and in Co. Londonderry at Downhill, Carntogher and Craignashoke (Greg & Lettsom 1858, Walker 1960a). Sounding Hill quarry, NNW of Magherafelt, is a further location in Co. Londonderry where fibrous natrolite (and thomsonite) have been reported (Old 1972). On the face of the cliffs at Cregganboy, near Kinbane Castle, Co. Antrim, colourless or pure-white natrolite occurs in veins and cavities, where it forms finely fibrous to acicular radiating masses of needle-like crystals up to 1.5 cm long (Ashcroft 1916). Natrolite, mesolite and scolecite commonly occur together as tufts of delicate hair-like crystals in basalts from the Garron Plateau, Co. Antrim (Walker 1951). Cavities in basalt at Little Deer Park and along the coastal cliffs, near Glenarm, contain stout colourless crystals of natrolite up to 1 cm long, forming loose aggregates or radiating ball-like groups with white centres on analcime and gmelinite (Tschernich 1992, Green 2005a). Analcime, chabazite, garronite, gismondine and thomsonite are associates of natrolite in basalts near Kilwaughter Castle, Larne, Co. Antrim (Walker 1962a). Fine specimens of colourless natrolite needles up to 4 cm long are found in veins and cavities in basalt at Magheramorne quarry, near Larne (Tschernich 1992). Orange natrolite is also reported from Magheramorne (Nawaz 1988a). At Hills Port on Island Magee, Co. Antrim, natrolite occurs with chabazite, gonnardite, heulandite and the debatable species paranatrolite and tetranatrolite (Nawaz 1988a), and exceptional coarse-grain 2 cm-long needles of natrolite line large cavities in basalt at Whitehead (Herbert Smith 1916, Walker 1959a, Tschernich 1992). Fine specimens of natrolite are also found at Black Cave, near Larne, and Hogg's quarry, Magherafelt, Co. Londonderry (Tschernich 1992). Other occurrences in Antrim basalts are documented by Adair (1987): White Mountain quarry, SW of Belfast; Galboly, near Garron Point; Port Noffer to the E of the Giant's Causeway; and fine examples on Island Magee – Brown's Bay; Castleshiels, near Portmuck; and Glenarm Magee.

Two Tertiary dolerite intrusions in Ireland contain natrolite and other zeolites infilling vesicles: Scawt Hill, Larne, Co. Antrim (Tilley & Harwood 1931), and Ballyalton, Newtownards, Co. Down (Knill 1960).

Natrolite crystals to 3 mm from Little Deer Park, Co. Antrim, Ireland. (© D. I. Green)

Handsome groups of radiating natrolite crystals have been found in veins at Ronez quarry, Jersey, Channel Isles, and even finer specimens occur in veins within diorite on Guernsey (Mourant 1961). These are possibly the prisms of natrolite encrusted with colourless analcime described from the Bouet, Guernsey (Mourant 1939).

natromontebrasite

triclinic $(Na,Li)Al(PO_4)(OH,F)$

A rare sodium–lithium–aluminium phosphate fluorhydroxide found in complex granitic pegmatites in the layered roof zone of the Tregonning granite at Megiliggar Rocks, near Breage, Cornwall. It is a replacement product after amblygonite (Stone & George 1983). However, Fransolet et al. (2007) have published an IMA-CNMNC approved discreditation of natromontebrasite and they confirm it to be a mixture of OH-rich amblygonite with lacroixite and subordinate wardite.

natron?

monoclinic $Na_2(CO_3)\cdot 10H_2O$

A rare hydrated sodium carbonate formed on the margins around soda-rich lakes or as an efflorescence on lavas. Natron has been reported from Botallack mine, St Just, Cornwall (Wolloxall 1989, Bancroft & Weller 1993), but this could not be substantiated (Ryback & Tandy 1992). Bowell (1992) notes natron in a unique supergene assemblage from the Wheal Hazard section of Botallack mine. Although part of the assemblage (gold) is in the collection of the NHM, no specimens (or XRD data) of kröhnkite exist, making confirmation difficult. Natron does not normally occur in wet acid environments, as would have been the case in this mine.

naumannite

orthorhombic (pseudo-cubic) Ag_2Se

A very rare silver selenide found in a gold-bearing carbonate vein cutting Middle Devonian limestones at Hope's Nose, Torquay, Devon. The selenide assemblage includes clausthalite, eucairite, fischesserite, klockmannite, penroseite, tiemannite, trüstedtite, tyrrellite and umangite in a gangue of dominantly calcite. Naumannite forms moderately common isolated grains (up to 100 μm) in calcite or clausthalite and is extensively replaced by klockmannite at eucairite/naumannite boundaries (Stanley et al. 1990a). Naumannite is tentatively identified as the most abundant inclusion species in chalcedonic silica within fault rocks at Clive mine, 10 km N of Shrewsbury, Shropshire. Other inclusion species positively or tentatively identified include barite, chalcocite or digenite, chalcopyrite, cinnabar, eucairite, galena and pyrite (Plant et al. 1999).

necronite

A variety of orthoclase, feldspar group.

needle iron ore

A fibrous variety of goethite.

needlestone

A term used to describe quartz + rutile (hairstone), but more commonly as a general term for acicular species of the zeolite group (natrolite, mesolite or scolecite).

nemalite

A thread-like fibrous variety of brucite.

neodigenite

A synonym of digenite.

neotocite

amorphous or monoclinic $(Mn^{2+},Fe^{2+})SiO_3\cdot H_2O$?

An amorphous hydrated manganese–iron silicate found at Wheal Edward and Wheal Owles, Penwith, St Just, Cornwall, where it was described as penwithite by Collins (1878a, 1879a, 1911a). At these localities, neotocite is translucent to opaque and dark amber to blackish-brown. Clark et al. (1978) suggests neotocite represents a series of X-ray amorphous compositions that are fairly common in manganese deposits of the volcano-sedimentary type. They are thought to have been formed from gelatinous siliceous suspensions deposited in cavities at low temperature. These authors also note the occurrence of neotocite at Geevor mine, near Pendeen, St Just, Cornwall, and Nant mine (No. 2 level), Llanfaelrhys, Rhiw, Lleyn Peninsula, Gwynedd, Wales.

neotype

A synonym of baricalcite (barium-rich calcite) related to the aragonite group.

nepheline

hexagonal $(Na,K)AlSiO_4$

A relatively common sodium–potassium aluminosilicate found in a nosean-bearing phonolite at Wolf Rock, 15 km from Gwennap Head, Land's End, Cornwall. In thin section, nepheline forms a major part of the rock, occurring as water-clear phenocrysts and as a component of the fine-grain groundmass of the phonolite. Nepheline forms simple and complex twins, and contains inclusions of sanidine laths, nosean particles and acicular needles of titanite. Basal sections form perfect hexagonal outlines. Associated species are aegirine, nosean (now mostly altered) and sanidine (the major phenocryst phase), with minor apatite, fluorite, pyrite and tourmaline also present (Allport 1871, 1874, Adye 1907, Harrison et al. 1977). Nepheline has also been noted as an accessory species in the Carboniferous Bonsall dolerite sill, near Matlock, Derbyshire (Smith et al. 1967, Macdonald et al. 1984).

Rectangular crystals of nepheline associated with aegirine (green), sanidine and isotropic nosean; Wolf Rock phonolite, south west of Land's End, Cornwall; width 0.65 mm; viewed in PPL (upper) and XPL (lower). (© K. Gardner)

Perhaps the best known nepheline occurrence in Britain or Ireland is that found in the Loch Borralan alkaline igneous complex, Assynt, Highland Region, Scotland. Along the northern contact of the intrusion, in tributaries of the Ledbeg River, unaltered nepheline is found with andradite (melanite) and biotite in a nepheline syenite, and with aegirine, albite, biotite, sodalite and titanite in another variety of syenite (Peach et al. 1907, Macgregor & Phemister 1972). Crystals of nepheline 12 mm across, and with a dull-green waxy appearance when altered, are noted by Teall (1900). Nepheline is further described from this complex by Shand (1910b), Sabine (1953), Tilley (1958) and Woolley (1973). In the same area, minor intrusions at Loyne, Sgonnan Mòr and several related ledmorite dykes to the west at Rubha Mòr, Cul Mòr, and Cnoc Breac near Elphin also contain nepheline (Sabine 1953).

Contamination of an olivine dolerite dyke by Jurassic limestones at Camas Mòr,

Isle of Muck, Highland Region, has produced a gabbro containing analcime, melilite, nepheline, titanaugite and iron-rich wollastonite (Tilley 1947).

Elsewhere in Scotland, clear and colourless nepheline with an ophitic texture towards co-existing feldspar occurs in the Traprain Law sodalite-bearing phonolite, near Haddington, Lothian (MacGregor & Ennos 1922, MacGregor 1960), and mottled grey nepheline, clouded with abundant minute inclusions of calcite and sericite, is found, with kaersutite, titanaugite and labradorite, in teschenite of the Lugar Sill, Strathclyde (Scott 1914b, Phillips 1968, Livingstone 2002). Small grains of nepheline are further noted in the interstices between plagioclase laths in a large basic alkaline (teschenite) sill at Dippin on the Isle of Arran, Strathclyde. Replacement by analcime is common (Henderson & Gibb 1977, Gibb & Henderson 1978b). There are also reports of hexagonal crystals of nepheline occurring in a dolerite at Eilan Mhuire, Shiant Isles, Western Isles, by Heddle (1901), and from an analcime–nepheline syenite unit of the alkali dolerite intrusion of Cnoc Rhaonastil, Isle of Islay, Strathclyde (Walker 1930, Walker & Patterson 1959).

In Ireland, nepheline is found in dark contaminated igneous rocks produced when chalk was assimilated by a doleritic magma at Scawt Hill, Larne, Co. Antrim. The main components of the rock are titaniferous augite and gehlenite (melilite), with varying amounts of aegirine, calcite, labradorite, perovskite and wollastonite also present. Nepheline has grown at the expense of plagioclase, which is resorbed, giving ragged irregular grains mantled by nepheline. Nepheline also occurs in graphic intergrowths with pyroxene, sometimes as outgrowths from the boundaries of large titaniferous augite grains. Although still displaying well formed hexagonal and rectangular outlines, much of the nepheline is converted to the poorly defined phase, hydronephelite (Tilley 1929a, Tilley & Harwood 1931). Nepheline is also found in a Tertiary olivine dolerite volcanic plug contaminated with chalk at Carneal, Larne, Co. Antrim, where it occurs in an aegirine–apatite–augite–nepheline assemblage (Sabine & Young 1975).

nephelite

A synonym of nepheline.

nephrite

A tough fine-grain variety of actinolite, amphibole group.

neptunite

monoclinic $KNa_2Li(Fe^{2+},Mn^{2+})_2Ti_2Si_8O_{24}$

A rare complex titanium silicate that forms a series with mangan-neptunite and is found in syenite veins cutting limestone to the east of Barnavave, Carlingford, Co. Louth, Ireland. Deep-red neptunite is a late-forming mineral and it occurs as small interstitial flakes or partially replacing titanite. Associated species in the syenite are aegirine, potassium feldspar and quartz, with minor eudialyte, pectolite and wollastonite. Contamination of the syenite magma by the digestion of limestone produced unusual melt conditions from which these species crystallized (Nockolds 1950).

niccolite

A synonym of nickeline.

nickel bloom

A synonym of annabergite, vivianite group.

nickel glance

A synonym of gersdorffite, cobaltite group.

nickelhexahydrite?

HEXAHYDRITE GROUP

monoclinic $NiSO_4{\cdot}6H_2O$

An hydrated nickel sulphate, dimorphous with retgersite, which has been reported as a bright-green surface alteration on brassy acicular millerite crystals found within clay ironstone nodules from Llanbradach colliery, near Caerphilly, Mid Glamorgan, Wales (T. F. Cotterell pers. comm. 2007). Further confirmation is desirable.

NICKELINE GROUP

Hexagonal antimonides, arsenides, bismuthides, selenides, stannides and tellurides, with the general formula *AX*, where *A* = Co, Ni, Pd, Pt; and *X* = As, Bi, Sb, Se, Sn, Te. Breithauptite and nickeline are members of the group found in Britain or Ireland.

nickeline

NICKELINE GROUP

hexagonal NiAs

A common nickel arsenide, usually found in high-temperature hydrothermal veins or as a magmatic component of norites and peridotites. In some reports, nickeline is called kupfernickel (a term also used to describe gersdorffite), niccolite, or nickelite.

In Cornwall, nickeline has been noted from cross-course veins in several mines: with millerite at Fowey Consols mine, Tywardreath, St Blazey (Greg & Lettsom 1858, Dines 1956, Lewis 1997); Killifreth mine, Kenwyn (Robson 1949b); massive and in fine fibres at Pengelly mine, St Ewe (Greg & Lettsom 1858); associated with coffinite, rammelsbergite and uraninite (pitchblende) at Roskrow United mine, Ponsantooth (Ball et al. 1982); St Austell Consols mine, near Sticker (Robertson & Dines 1929); with millerite at Wheal Chance, St Austell (Greg & Lettsom 1858, Collins 1892); Wheal Jane, Kea, near Truro (Robson 1949b). Nickeline has also been recorded from South Tresavean mine, Falmouth (Pearce 1878, Hosking 1951), but not confirmed during further investigations (Rust 1982a).

At Littleham Cove, near Budleigh Salterton, Devon, nickeline is identified with algodonite and maucherite, and possibly coffinite, as alteration products at the margins of native copper "plates" found associated with uranium–vanadium-rich Permian nodular concretions that have formed in localized horizons within the Littleham Mudstone Formation (Harrison et al. 1975b, Milodowski et al. 2000, 2002). Hall (1868b) briefly mentions nickeline associated with rhodonite and psilomelane from Black Down, near Tavistock, Devon.

Characteristic pale copper-red nickeline, associated with chalcopyrite, galena, millerite, rammelsbergite, sphalerite and ullmannite, has been found at Settlingstones lead mine, Fourstones, near Hexham, Northumberland, on the Whin Sill dolerite side of a pipe of galena in a witherite vein (Russell 1927, Ineson 1972, Young et al. 1985b, Ixer 1986). A magnetite-rich ore containing nickeline with chalcopyrite, galena, gersdorffite, pyrite, pyrrhotite, sphalerite and ullmannite at Lady's Rake mine, along Harwood Beck, Upper Teesdale, Co. Durham, is described as a skarn deposit, probably related to the intrusion of the Whin Sill (Young et al. 1985b). The Whin Sill may also have been the source of nickel that produced a small nickeline–gersdorffite–millerite orebody on the Dow Scar level of Hilton barite–lead mine, Scordale, near Appleby. The deposit occupied a brecciated area in which cavities were infilled mainly with nickeline (Bridges 1982). Nickeline is associated with rammelsbergite, safflorite and skutterudite (smaltite) at Paddy End copper mine, Coniston, Eskdale, Cumbria (Russell 1925, Young 1987a), and an unconfirmed report of the species is noted from Newlands Valley, Cumbria (Davidson & Thomson 1951, Young 1987a). Nickeline identified on material from Tynebottom mine, NW of Garrigill, near Alston, Cumbria (Ixer & Stanley 1987), came from specimens in the Kingsbury collection now at the NHM. Dunham et al. (2001) express serious doubt about the provenance of the museum specimens, and evidence compiled by Ryback et al. (1998, 2001) on other specimens in this collection suggests that further confirmation is needed.

Cobalt-rich nickeline has been found in uraninite-bearing hydrocarbons from the Ty Gwyn mine, Great Orme's Head, Llandudno, Gwynedd, Wales (Parnell 1988).

At Palnure Burn and in its tributary, Glen of the Bar, near Talnotry, 5 km NE of Newton Stewart, Dumfries & Galloway, Scotland, nickeline is one of the principal ore minerals in a lenticular Co–Cu–Ni deposit found near the base of a diorite intrusion. Small irregularly distributed blebs of nickeline, closely associated with gersdorffite (referred to as cobaltite in the early works), form an assemblage close to the diorite with chalcopyrite-rich and pyrrhotite-rich assemblages occurring farther away (Wilson & Flett 1921, Gregory & Leitch 1928, Stanley et al. 1987). More recently, the deposit has been shown to contain a wide range of PGMs (Power & Pirrie 2001, Gunn & Styles 2002, Power et al. 2004). Elsewhere in Dumfries & Galloway, nickeline is reported from: Menimuir Burn, near Cassencarrie (Temple 1954b); Moneypool Burn, W of Pibble mine, near Creetown (Heddle 1901); Needle's Eye, SW of Caulkerbush (Miller & Taylor 1966); and Pibble mine, near Creetown (Livingstone 2002). At New Glencrieff lead mine, nickeline (rare at this locality) occurs with rammelsbergite (Brown 1927, Temple 1955) and West Grove lead mine (Heddle 1901), both near Wanlockhead, and at Leadhills, in Strathclyde (Heddle 1901). A heavy-mineral concentrate from Raggengill Burn, in the Abington–Biggar–Moffat area of Strathclyde, was found to contain nickeline (Dawson et al. 1979).

An Ag–Co–Ni assemblage found in a barite vein has been worked at Hilderston mine, Bathgate Hills, Lothian. The mineralization is thought to be related to the intrusion of nearby basaltic dykes and a quartz-dolerite sill, and consists of annabergite, chalcopyrite, galena, nickeline, pyrite and sphalerite. Away from the igneous rocks the mineralization does not contain Ag–Co–Ni-bearing minerals (Heddle 1901, Wilson & Flett 1921, Gallagher 1964, Ridgway 1980, Meikle 1994). Another mine famous for its silver, and only 25 km away from Hilderston, is Alva mine, Silver Glen, Ochil Hills, Central Region, where minor nickeline occurs with rammelsbergite. A hydrothermal system related to a local quartz-dolerite dyke swarm played a dominant role in the mineralization episode (Moreton et al. 1998). Heddle (1901) briefly notes nickeline from Ballagan Glen, Campsie Hills, Central Region.

Nickel mineralization is common within chromite-rich lithologies of the Corrycharmaig ultramafic intrusion, Glen Lochay, NW of Killin, Tayside Region. Heazlewoodite is the main nickel-bearing species, but both nickeline and gersdorffite are associated with it. Nickeline occurs as small (< 10 µm) blebs enclosed within heazlewoodite and also as larger (> 20 µm) subhedral grains. The intrusion is also host to PGE mineralization, suggesting an ophiolitic origin for the intrusion (Power & Pirrie 2000).

On Unst, Shetland, disseminated pink nickeline grains (10–300 µm grains) occurs in podiform chromitite and pyroxenite belonging to the Shetland ophiolite complex at Cliff, Baltasound–Haroldswick area, and on Vord Hill, W of Nuda, in southwest Unst. Associated species include chrome spinel, cobaltite, magnetite, maucherite and composite pentlandite–pyrrhotite grains (Maynard 1993, Gunn & Styles 2002).

A minor sulphide and oxide assemblage in gold-bearing quartz veins has been discovered in the Lecanvey prospect, Clew Bay, Co. Mayo, Ireland; it consists of goethite, hematite and pyrite, and rare chalcopyrite, nickeline and pyrrhotite (Aherne et al. 1992). The carbonate-hosted Lisheen Pb–Zn–(Ag) deposit, between Thurles and Rathdowney, contains an As–Co–Cu–Ni assemblage near major faults. The mineralization is dominated by bournonite, cobaltite–gersdorffite, nickeline, pararammelsbergite, pyrite (var. bravoite) and tennantite (Ixer 2002, Fusciardi et al. 2004).

nickel ochre

A synonym of annabergite, vivianite group.

nickel-skutterudite/nickelskutterudite

cubic $(Ni,Co,Fe)As_{2-3}$

An uncommon nickel–cobalt–iron arsenide that forms a series with skutterudite and is found at Kingswood mine, near Buckfastleigh, Devon. Nickel-skutterudite (var. chloanthite) is associated with uraninite (var. pitchblende) and native bismuth (Russell 1922). Four Cornish occurrences require confirmation: Dolcoath mine, Camborne; Fowey Consols, Tywardreath, St Blazey; St Austell Consols; and Wheal Ann, Phillack (Golley & Williams 1995).

The nickel-skutterudite identified on material from Tynebottom mine, NW of Garrigill, near Alston, Cumbria (Ixer & Stanley 1987), came from specimens in the Kingsbury collection at the BM(NH). Dunham et al. (2001) express serious doubt about the provenance of the museum specimens, and evidence compiled by Ryback et al. (1998, 2001) on the other specimens in this collection suggests that further confirmation is needed.

nickel vitriol

A synonym of morenosite.

nigrine

A black variety of rutile.

ningyoite

RHABDOPHANE GROUP

orthorhombic (pseudo-hexagonal) $(U,Ca,Ce)_2(PO_4)_2{\cdot}1{-}2H_2O$

A rare hydrated uranium–calcium–cerium phosphate found in uranium-rich veins at three localities in Cornwall: Wheal Alice, Grampound Road – ningyoite associated with metatorbernite, torbernite and uraninite (var. pitchblende) in killas; Wheal Buller, near Redruth – metatorbernite, ningyoite and pitchblende in granite overlain to the north by killas; and Wheal Damsel, St Day – ningyoite and pitchblende in granite overlain to the east by killas. The uranium assemblages are mainly associated with chalcopyrite–sphalerite mineralization (Ball et al. 1982).

nisaite

An arsenic-rich variety of phurcalite.

nitre

A synonym of natron.

nitrocalcite?

monoclinic $Ca(NO_3)_2{\cdot}4H_2O$

A rare hydrated calcium nitrate that forms white or grey efflorescent silken tufts and coatings and is reported from unspecified locations – on old walls, in limestone caves or on spots of earth where the soil is highly calcareous (Greg & Lettsom 1858). Nitrocalcite requires further confirmation.

nontronite

SMECTITE GROUP

monoclinic $Na_{0.3}Fe^{3+}_2(Si,Al)_4O_{10}(OH)_2{\cdot}nH_2O$

An uncommon but locally abundant hydrated sodium–iron aluminosilicate hydroxide, described as the gramenite variety of chloropal and found with fluorite in fissures within a granite close to Wheal Ludcott at Carclaze china-clay pit, St Austell, Cornwall (Church 1866b, Collins 1892). At the nearby Gunheath china-clay pit, nontronite occurs with elbaite–schorl, ferrocolumbite, fluorite, ilmenorutile and zinnwaldite in miarolitic granite pegmatites (Weiss 1994). Nontronite (gramenite) is also vaguely noted from granite in the Hensbarrow granite area, St Austell (Collins 1878b), and two other Cornish localities near Breage: Carn Bray and Tremearne (Collins 1892). In Devon, nontronite (gramenite) is reported from Smallacombe, near Bovey Tracey (Collins 1877a, 1911a), and, in Derbyshire, alteration of the Waterswallows dolerite sill, 3 km to the NE of Buxton, has produced smectite-infilled veins and olivine pseudomorphs with nontronite and saponite compositions (Ineson et al. 1983).

Nontronite is a pale-green alteration product of actinolite in syenite from the Bakkasetter quarries, southeastern flank of Noss Hill, South Mainland, Shetland, Scotland. It contains epidote inclusions and is often associated with granular aggregates of anatase and hematite. Palygorskite is also found at this site (Stephen 1954).

In Ireland, a dull waxy mineral with a starchy fracture from Carnmoney Hill, 8 km N of Belfast, Co. Antrim, has been shown by analysis to be nontronite (chloropal). It occurs with calcite in a cavity in a volcanic agglomerate adjacent to a dolerite plug (Tomkeieff 1934). Nontronite–saponite is also recorded from Scawt Hill, Larne, Co. Antrim, as a constituent (with calcite and zeolite species) of amygdales in Tertiary basalts that have been partially melted adjacent to another dolerite plug (Kitchen 1985).

norbergite

HUMITE GROUP

orthorhombic $Mg_3(SiO_4)(F,OH)_2$

A rare magnesium silicate fluorhydroxide found in crystalline limestone at Loch Ness, Highland Region, Scotland. Norbergite forms waxy yellowish-brown crystals up to 12 mm long, associated with arsenopyrite (Macpherson & Livingstone 1982).

nordstrandite

triclinic $Al(OH)_3$

An aluminium hydroxide, polymorphous with bayerite, doyleite and gibbsite, and found in collapsed Tertiary sediments filling solution pipes in the Upper Chalk near the coastguard station at Newhaven, Sussex. Nordstrandite occurs in impure nodules in white or pale-cream bands together with black manganese-rich matter. It is soft, friable and has fine grain, with an earthy texture and sometimes a vitreous lustre. Other aluminium-rich species discovered in the pipes include aluminite, bayerite, felsöbányaite (syn. basaluminite) and gibbsite. All appear to have formed by the reaction of acid groundwater with Tertiary clays and the Chalk (Wilmot & Young 1985).

norsethite

DOLOMITE GROUP

trigonal $BaMg(CO_3)_2$

A rare barium–magnesium carbonate found in sulphide, barite and carbonate rock at Foss, near Aberfeldy, Tayside Region, Scotland. The Foss stratiform barite–base-metal deposit consists of barite, Ba-muscovite, Ba-feldspar (celsian) in cherts, schists and carbonate rocks, with galena, disseminated-to-massive pyrite and sphalerite. Norsethite and barytocalcite were discovered as inclusions in coarse pyrite. Calcite, dolomite, siderite, witherite and an Fe–Mn–Zn carbonate are also present in this deposit. A sedimentary-exhalative origin has been proposed for the mineralization (Moles 1985).

northupite?

cubic $Na_3Mg(CO_3)_2Cl$

A rare sodium–magnesium carbonate chloride, possibly found in the Permian Lower Magnesian Limestone of the Aislaby No. 2 borehole, Eskdale, near Whitby, Yorkshire. Northupite is thought to occur as inclusions in secondary fluorite within a dolomite vein (Raymond 1962, Stewart 1965).

A nosean pseudomorph showing unusual pentagonal outline and interior altered to analcime and cancrinite; Wolf Rock phonolite, south west of Land's End, Cornwall; width 1.3mm; viewed in PPL. (© K. Gardner)

nosean

SODALITE GROUP

cubic $Na_8Al_6Si_6O_{24}(SO_4)\cdot H_2O$

A sodium aluminosilicate with sulphate found in phonolite from the Wolf Rock, 15km from Gwennap Head, Land's End, Cornwall. Dull blackish turbid nosean occurs as microphenocrysts and in the groundmass of the phonolite, and is characterized by complex zonation and abundant inclusions. The microphenocrysts show simple or compound hexagonal outlines, commonly with corroded margins. Analcime and cancrinite pseudomorph much of the nosean. Associated species are aegirine, nepheline and sanidine, with minor apatite, fluorite, pyrite and tourmaline also present (Allport 1874, Adye 1907, Harrison et al. 1977).

Nosean reported from 1 km E of Ledbeg Cottage, Assynt, Highland Region, Scotland, requires confirmation (Tilley 1958, Macpherson & Livingstone 1982).

nováčekite

AUTUNITE GROUP

tetragonal $Mg(UO_2)_2(AsO_4)_2\cdot 12H_2O$

An uncommon hydrated magnesium uranyl arsenate found in disused mineworkings of the Cargodna section of Wheal Owles, St Just, Cornwall. Nováčekite forms bright greenish-yellow botryoidal crusts of interlocking plates and scales with an open structure like an egg box. Individual crystals may form euhedral plates and are typically 10–20 µm across, but may reach up to 0.1 mm. Fluorescence is unusually bright. Nováčekite is a supergene species that lined a small vug in quartz–sulphide veins containing arsenopyrite, chalcopyrite and sphalerite. Magnesium necessary for the formation of nováčekite most probably came from the decomposition of the mafic hornfels host rock (Elton et al. 1994a). Transparent yellow rectangular plates of nováčekite up to 0.5 mm long on altering arsenopyrite are also noted from Roughtor Great Consols, Cornwall (N. Hubbard pers. comm. 2006). Nováčekite is noted without detail from Wheal Edward, St Just (Anthony et al. 2000).

Delicate plates of yellow nováčekite associated with pale-green zeunerite; Wheal Edward, St Just, Cornwall; crystal groups to 0.8mm across. (© D. I. Green)

In the Needle's Eye area, SW of Caulkerbush, Dumfries & Galloway, Scotland, nováčekite has been identified within dolomite-rich fissure veins in hornfels, where it occurs as greenish-yellow botryoidal crusts and flakes on uraninite (var. pitchblende). Rich specimens of this and other colourful uranium secondary species, such as boltwoodite and uranophane, are common in the highly radioactive veins along this coastline (Livingstone 1993, Green & Todd 2001). Additionally, Basham et al. (1989) confirm zeunerite and suggest that coffinite, becquerelite and uranospinite may also be present at this location.

O

oakstone
A pseudostalactitic variety of barite (Ford & Sarjeant 1964b, Sarjeant & Ford 1993).

ochre
Fine-grain goethite or hematite.

octahedrite
A synonym of anatase.

offrétite/offretite
ZEOLITE GROUP
hexagonal $CaKMg(Al_5Si_{13})O_{36}·15H_2O$

A rare hydrated calcium–potassium–magnesium aluminosilicate found in basalt at Coulters Hill quarry, Ballyclare, Co. Antrim, Ireland, where it forms intergrowths with levyne. Calcite, chabazite, cowlesite and montmorillonite are associated species (Nawaz 1978b). Later work by Adair (1987) confirms offrétite coating levyne in other Antrim basalts: South Antrim escarpment to the W and SW of Belfast – Aughrim Plantation quarry; McCourts quarry; Galboly River section, near Garron Point.

Offrétite is more commonly reported from Scotland, typically from vesicular basalts. Fibrous outgrowths of offrétite, less than 0.1 mm thick, on glassy levyne occur in cavities up to 1.5 cm across, at Quiraing, Isle of Skye, Highland Region (Livingstone 2002), and at Moonen Bay, N of Dunvegan, also on the Isle of Skye, minute acicular crystals of erionite–offrétite form epitaxial outgrowths on levyne (Green & Todd 1996, 2001). Rare greenish-grey needle-like bladed crystals of possible offrétite are noted from two exposures of vesicular basalt at Hartfield Moss, 9 km WSW of Barrhead, Strathclyde. In one location, offrétite occurs as a coating on titanite; in the second location, it forms on the chlorite lining of vesicles as tufted aggregates up to 0.2 mm long, along with velvety-surfaced spherulitic aggregates up to 0.45 mm across (Meikle 1990). At Loanhead quarry, near Beith, Strathclyde, Scotland, offrétite resembling shredded coconut lines vugs in basalt, associated with clusters of pink albite (Robinson & King 1993).

Amygdale filled with bladed offrétite in a Tertiary basalt from Co. Antrim, Ireland; width 9.3 mm. (© The Trustees of the National Museums of Scotland)

okenite
triclinic $Ca_5Si_9O_{23}·9H_2O$

A rare hydrated calcium silicate found with gyrolite, pectolite, reyerite, tacharanite, tobermorite and xonotlite in amygdales within Tertiary olivine basalts on the Isle of Mull, Strathclyde, and across the Sound of Mull on the Morvern Peninsula, Highland Region, Scotland (Walker 1971). Yellowish to bluish-white fibrous radiating crystals of okenite are also noted by Heddle (1901) from the mouth of the An Leth Allt, near Loch Brittle, and Dùnan Earr an Sgùirr (coast between Loch Einort and Loch Brittle), both on the Isle of Skye. The okenite mentioned by Macpherson (1983) from Carleton Castle, Strathclyde, requires confirmation.

In Ireland, okenite occurs in metasomatically altered flint nodules in a chalk/dolerite contact zone at Scawt Hill, Larne, Co. Antrim. Colourless fibres of okenite are found in small areas or nests among a felt (i.e. resembling velvet in having a soft surface) of xonotlite, wollastonite and rare diopside (Tilley & Alderman 1934). At Catcairn basalt quarry, Legoniel, Belfast, Co. Antrim, white okenite develops fibres up to 1 mm in length, or is massive and chalky. Apophyllite, chabazite and gyrolite are associated species (Nawaz 1978a).

oligoclase
A very common triclinic sodium–calcium aluminosilicate with an intermediate (An_{10-30} mole %) composition between sodium-rich albite (An_{0-10}) and calcium-rich anorthite (An_{90-100}) end members. It is part of the plagioclase feldspar continuous solid-solution series that also includes andesine (An_{30-50}), labradorite (An_{50-70}) and bytownite (An_{70-90}).

olivenite
monoclinic $Cu_2^{2+}(AsO_4)(OH)$
TL Carharrack mine, Gwennap, Cornwall, England

A relatively common copper arsenate hydroxide that forms a series with adamite and is found in the oxidized zone of hydrothermal copper deposits.

Remarkably fine olivenite was found at several mines near St Day, Cornwall, particularly Wheal Gorland, Wheal Unity and Wheal Unity Wood (Embrey & Symes 1987). Specimens from Wheal Gorland have been divided into three types: well formed crystals up to 6 mm long, bright green to almost black; white to light-green fibrous crystals forming crusts up to 2.5 cm thick; and as "wood copper", a dense fibrous form varying from white to yellow and green to near black. Clinoclase is an associated species (Rust 1981b). The infrared properties of olivenite from Wheal Gorland are described by Braithwaite & Pritchard (1983a) and the sad removal of the mine dumps is documented by Fisher (1988). The type locality is given as Carharrack mine by Mandarino & Back (2004), but it is said to be Wheal Gorland by Golley & Williams (1995); other authorities note it simply as Cornwall (Clark et al. 1995). Specimens of olivenite occurring with cyanotrichite are described from Wheal Gorland by Kingsbury (1954); however, evidence compiled by Ryback et al. (1998, 2001) on other specimens reported by Kingsbury suggests that further confirmation is needed. At Wheal Unity, Gwennap, olivenite forms dark-green needle-like prisms up to 2 mm in length associated with hematite on a milky quartz matrix (Greg & Lettsom 1858, Rudler 1905, Embrey & Symes 1987). At Penberthy Croft mine, St Hilary, olivenite is found in a wide range of euhedral translucent crystal forms. It is usually dark green, but the white variety leucochalcite is also noted as a dense network of fibres infilling cavities in vein quartz. Associated species are bayldonite, cornwallite, goethite and infrequently mixite (Betterton et al. 2000). Very fine crystals of olivenite were found at the Marke Valley mine, near Liskeard, and well formed crystals associated with pharmacosiderite and scorodite are noted from South Terras mine, St Stephen-in-Brannel (Collins 1876b, 1911a). Other Cornish occurrences of olivenite include: Beam mine and North Goonbarrow china-clay pit, St Austell (Collins 1892, Weiss 1994); Pednandrea mine, Redruth (Greg & Lettsom 1858, Collins 1892); Pentire Glaze mine, St Minver (Golley & Williams 1995); Tincroft mine, Illogan (Greg & Lettsom 1858, Collins 1892); fine specimens from Ting Tang mine, Carharrack (Greg & Lettsom 1858, Bruce & Aubrey-Jones 1998); Wheal Pendarves (Tryphena lode), Camborne (Barstow & Cooper 1982); Old Gunnislake mine, Calstock (Kingsbury 1954, Golley & Williams 1995); Phoenix United mines, Linkinhorne (Kingsbury 1954, Golley & Williams 1995); and Wheal Cock, near Botallack, St Just (Kingsbury 1954). Further confirmation is desirable for the last three occurrences.

Irregular sprays of prismatic olivenite to 2 mm on vein quartz; Ting Tang mine, Carharrack, Gwennap, Cornwall. (© Royal Cornwall Museum, Royal Institution of Cornwall; photographer D. I. Green)

At Bedford United mines, near Tavistock, Devon, very beautiful specimens of olivenite were once found, forming olive-green prismatic crystals and fibrous or acicular masses (Greg & Lettsom 1858, Hall 1868b, Kingsbury 1954, Anthony et al. 2000). Olivenite is also reported from Engine Vein mine and Wood mine, Alderley Edge ridge, Cheshire, where it is associated with erythrite and tyrolite (Carlon 1979, Braithwaite 1994).

Until recently, the only specimens of olivenite from the Caldbeck Fells, Cumbria, were supposedly collected by A. W. G. Kingsbury and their provenance is questionable: Carrock mine, Grainsgill; Dry Gill mine; Ingray Gill; Potts Gill barite mine (referred to in Kingsbury & Hartley 1960, Hartley 1984, Young 1987a, Cooper & Stanley 1990). Locally abundant olivenite has since been found at the last locality with the supergene species adamite, agardite, hemimorphite and philipsburgite (Neall & Leppington 1994), and less commonly at Sandbed mine, where it forms drusy crusts of olive-green acicular crystals associated with parnauite (Green et al. 2006a). Olivenite is also noted from Goldscope mine, Newlands Valley, near Keswick, Cumbria (Goodchild 1885), but this and another occurrence at Tynehead mine, Co. Durham, does not appear to be confirmed by Davidson & Thomson (1951).

Olivenite is confirmed as green-to-brown acicular crystals in vugs within a native copper–cuprite matrix and as matted pale-green to white capillary crystals (leucochalcite) in a copper deposit near the Precambrian/Triassic unconformity at New Cliffe Hill quarry, Stanton under Bardon, Leicestershire (Hubbard et al. 2005).

The leucochalcite variety of olivenite is a rare constituent of an unusual supergene arsenate assemblage found at Gwaithyrafon mine, 6 km SE of Tal-y-bont, Central Wales orefield, Dyfed. It occurs as aggregated silky-white radial crystal groups up to 0.75 mm across on goethite-coated quartz (Bevins 1994, Rust & Mason 1994). Olivenite is also noted with anglesite, azurite, cerussite, native copper, cuprite, harmotome, malachite and tyrolite from Dolyhir quarry, Old Radnor, Powys (Wolfe 2000).

Minute tufted groups of olivenite were once found with chrysocolla and leadhillite at Brown's vein, Glengonnar, Leadhills, Strathclyde, Scotland (Heddle 1901).

In Ireland, reports of olivenite include: green-to-brown drusy crusts of minute blocky crystals associated with erythrite and copper sulphides at Muckross mine, Killarney National Park, Co. Kerry (Moreton et al. 1999, Green & Moreton 2001), and Tynagh mine, near Killimor, Co. Galway (Ryback 1988a, Ryback & Moreton 1992, Green & Moreton 2001).

OLIVINE GROUP

Orthorhombic silicates with the general formula $A_2^{2+}SiO_4$, where A^{2+} = Fe, Mg, Mn, Ni. Fayalite, forsterite and tephroite are members of the group found in Britain or Ireland.

olsacherite

orthorhombic $Pb_2(SeO_4)(SO_4)$

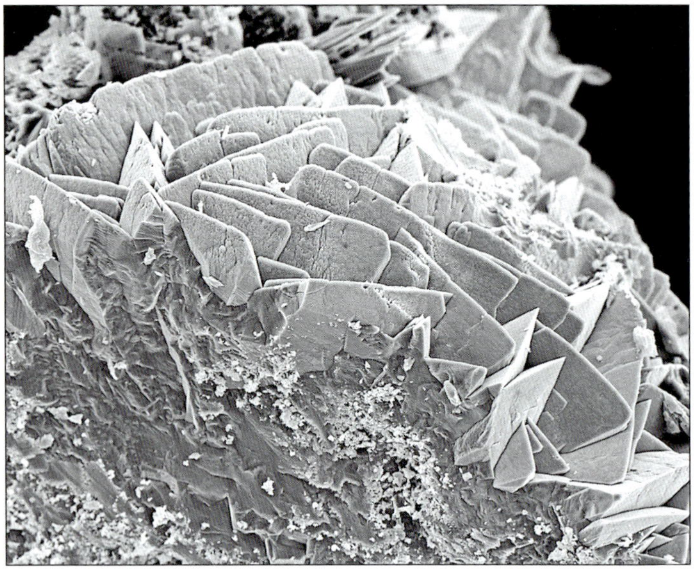

Bladed crystals of olsacherite; Waterbank mine, near Ecton, Staffordshire; width 225 μm; SEM image. (© D. I. Green)

A very rare lead selenate sulphate found on a single specimen from the abandoned Cu–Pb–Zn Waterbank mine, 2 km NNW of Wetton village, near Ecton, Staffordshire, where it occurred as a supergene species in a drusy cavity within carbonate veinstone. Olsacherite formed a pale-brown spheroidal aggregate about 0.25 mm across and, when examined with an SEM, shows slightly curved blades up to 0.1 mm long with sharp steep terminations. Unfortunately, the specimen was destroyed in preparation for the XRD analysis used to confirm its identity (Rust & Green 2005).

omphacite

PYROXENE GROUP

monoclinic $(Ca,Na)(Mg,Fe,Al)Si_2O_6$

An uncommon calcium–sodium–magnesium–iron–aluminosilicate that occurs as a major component of eclogite. It has a compositional range between aegirine (0–25%), augite (25–75%) and jadeite (25–75%). A dull-green eclogite found within Lewisian Gneiss at Totaig Ferry, Loch Duich, 9 km NE of Kirkton of Glenelg, Highland Region, Scotland, contains almandine garnet and a pale-green anhedral pyroxene, shown to be omphacite. Hornblende, plagioclase feldspar and rutile are the other primary species in the assemblage (Teall 1891a). Later work confirmed that eclogitic rocks occur in some abundance in the Lewisian of the Glenelg area, such as on the hills W of the Postman's Path to Ardintoul, Glenelg, on the north slopes of Glen More, and NNE of the summit of Beinn a' Chapuill. These rocks are thought to have formed as a result of extreme metamorphism of basic igneous rocks (Alderman 1936, Tilley 1937c, Harker 1950, O'Hara 1960, Mercy & O'Hara 1968).

An eclogitic-facies garnet–amphibole pyroxenite from the west side of Glen Strondeval and 0.4 km SW of Leverburgh post office, South Harris, Western Isles, contains a pale-green clinopyroxene with a relatively sodium-poor composition (Livingstone 1967) considered to be omphacite by Edgar et al. (1969). Eclogite (containing essential omphacite) is said to be widely distributed throughout the Archaean basic and feldspathic orthogneisses of South Harris, including those at Rodelpark (Pairc Roghadail), Mullach an Stughadh, Beinn na h-Aire, Roineabhal, Sletteval and especially on Beinn Tharsuinn (Davidson 1943). Omphacite is also described from a plagioclase eclogite at Ballyshannon, Co. Donegal, Ireland (Edgar et al. 1969).

onegite
An acicular variety of goethite.

onyx
A variety of evenly banded agate.

onyx stone
A brown pseudostalactitic variety of barite.

oosterboschite?
orthorhombic $(Pd,Cu)_7Se_5$ or $Cu_2Pd_3Se_4$

A rare palladium–copper selenide tentatively described from gold-bearing carbonate veins in Middle Devonian limestones at Hope's Nose, Torquay, Devon. An irregularly shaped phase (<10 μm across), similar in composition to oosterboschite, and two additional unknown phases, were found as inclusions in a single gold grain from a selenide assemblage that also contains bromian chlorargyrite, chrisstanleyite, clausthalite, eucairite, fischesserite, tiemannite and umangite, and possibly mertieite-II (Paar et al. 1998).

opacite
A name given to black unidentified grains.

opal
amorphous $SiO_2 \cdot nH_2O$

A fairly common variety of silica that contains water and which can be divided into three structural categories: opal-C (well ordered high-temperature cristobalite); opal-CT (disordered α-cristobalite and low-temperature tridymite); and opal-A (highly disordered, amorphous). In most cases, the type of opal is unknown. Opal occurs in many varieties: cacholong (bluish-white or cream variety), common opal (no play of colours), ferruginous opal, fire-opal (semi-transparent, red, yellow, sometimes irridescent), hyalite (transparent or semi-transparent), hydrophane (becomes more transparent when immersed in water), semi-opal (dull and opaque) and wood opal (silicified wood).

Opal is found in pegmatitic lenses and quartz–tourmaline veins cutting kaolinized granite at Gunheath and the adjacent Hensbarrow china-clay pits, Carthew, near St Austell, Cornwall. The pegmatites also contain arsenopyrite, cassiterite (varlamoffite), ferrocolumbite, fluorapatite, gilbertite mica, microcline, orthoclase, quartz, stannite, topaz, torbernite, tourmaline, turquoise, variscite, wolframite and zinnwaldite (Barstow 1982, Hodkinson & Clark 1977). Weiss (1994) subdivides the mineralization in Gunheath pit into three parageneses, and places opal in the category "stockwork, greisen zones and quartz veins". Towards the end of the 1970s, at the bottom of Gunheath pit a chlorite-rich greisen zone, containing abundant (water) opal, was found with occasional smoky quartz. The material is of variable colour, but predominantly yellowish, greyish-blue and milky white. Transparent, colourless, glassy and rose varieties were less common. Rudler (1905) reports a milky-white and pale-blue variety of opal associated with cassiterite from Trumpet Consols, near Helston, Cornwall.

Reports of opal from Derbyshire at Glebe mine, Eyam (Mueller 1951), and in pockets in Bakewell Chert (Ford & Sarjeant 1964a), are considered dubious by Ford et al. (1993).

In Scotland, opal has been described from a thin skin on an agate from the Isle of Mull, Strathclyde, and as botryoids on calcite from near Trotternish, Isle of Skye (Livingstone 2002). The variety cacholong is described from vesicles in basalt at: Quiraing, near Portree, Isle of Skye; Usan, near Montrose, Tayside Region; and Balmeadowside, Fife (Heddle 1901).

A thin vein of opal associated with chalcedony from Ronez quarry, Jersey, Channel Isles, is reported by Mourant (1961). The opal is transparent and deep orange (fire-opal) or nearly opaque and bright red (jasper opal).

Greg & Lettsom (1858) and Collins (1892) report many other occurrences of opal from which a selection are noted here.

England

Cornwall Botallack mine, Trewellard mine, Wheal London, Wheal Spearn, Wheal Stennack, all St Just; Wheal Stennach and Wheal Spinster, St Day; North Roskear, Camborne; Wheal Rosewarne, Gwinear; Wheal Poligine, Wheal Buller and Wheal Damsel, all near St Day and Redruth; Stenna Gwyn mine, St Stephen-in-Brannel, St Austell (all common opal, many also have semi-opal); Wheal Gorland, St Day, 2 km E of Redruth (fire-opal); Wheal Buller, near Redruth (semi-opal). Botallack mine also has fire opal and cachalong.
Devon Blackdown Hills (wood opal); Hay Tor (common opal); Okehampton (semi-opal).

Ireland

Antrim Giant's Causeway and Rathlin Island (hydrophane); Crossreagh, Ballywillin (hydrophane); Sandy Braes (jasper opal).
Down Donald's Hill (hyalite); Quarus, S of Newcastle (hyalite).

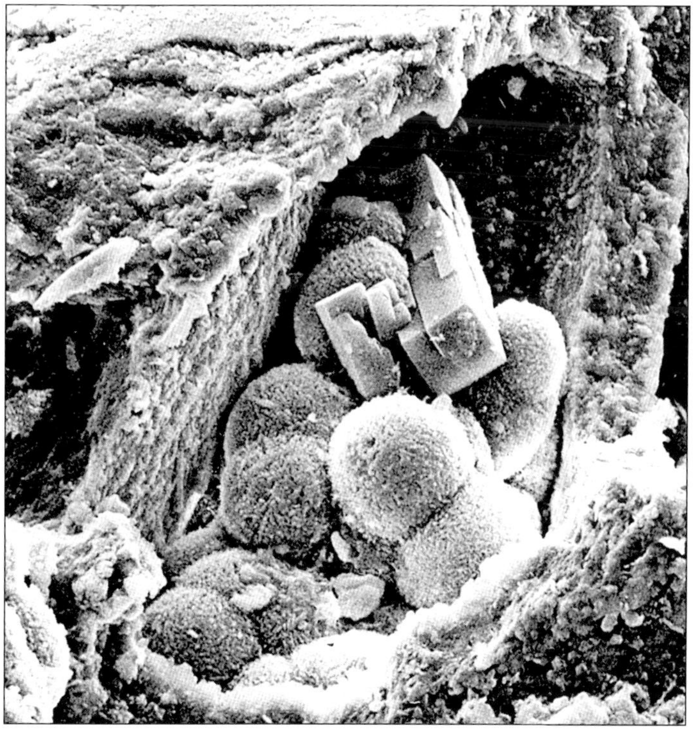

Spherules of opal-CT with blocky clinoptilolite crystals, lining a cavity in a calcareous microfossil shell within siliceous mudstone from the Gault Clay. Harwell borehole H3, Oxfordshire; width 95 μm; SEM image. (IPR/100-52C British Geological Survey © NERC. All rights reserved. A. E. Milodowski)

orangite
An hydrated variety of thorite.

orcelite
hexagonal $Ni_{5-x}As_2$

A rare nickel arsenide found in a disused chromitite quarry at Cliff, Baltasound–Haroldswick area, Unst, Shetland, Scotland. At this locality, sulphide-bearing dunite lenses in harzburgite contain accessory amounts of orcelite, godlevskite, heazlewoodite, maucherite and millerite, as well as the more common sulphides bornite, chalcopyrite, covellite, digenite, djurleite, nickeline, pentlandite and the rare PGE species geversite, hongshiite, sperrylite and stibiopalladinite (Gunn et al. 1985, Ixer 1990a, Maynard 1993, Prichard et al. 1994, Gunn & Styles 2002).

orpiment
monoclinic As_2S_3

An uncommon arsenic sulphide said to have been found near a barite–chalcopyrite–galena lode cutting Silurian and Carboniferous limestones at Gortshaneroe (Ballynoe) mine, Silvermines district, Co. Tipperary, Ireland (Gages 1860, Russell 1907b, Cole 1922). However, later investigations suggest that this was a misidentification, possibly of an alteration product of cadmium-rich sphalerite (Russell 1944).

A second problematical occurrence of orpiment is noted from the Clevedon Brick and Tile works, near Clevedon, Avon. Small lemon-yellow foliated masses of pure orpiment from this locality were held by the Bristol Museum until they were destroyed by bombing during World War II. Only a piece held by Weston-super-Mare Museum and another held by Sir Arthur Russell survived, and subsequent fieldwork at Clevedon failed to find further specimens (Russell 1944). Orpiment is listed without description from Perranarworthal, N of Penryn, and Burraton Coombe quarry, St Stephens by Saltash, both in Cornwall (Golley & Williams 1995).

In Wales, orpiment, pyromorphite and mimetite are listed from Gwynfynydd mine, near Dolgellau, Gwynedd (Andrew 1910, Allen & Jackson 1985), but further confirmation is desirable. Cowdry (2002a) reports orpiment and realgar (and a trace of pyrite) in a large fresh exposure of shale at Dolyhir quarry, Old Radnor, near Presteigne, Powys.

orthite

A synonym of allanite, epidote group.

ortho-antigorite

TL Unst, Shetland, Scotland

A variety of lizardite, kaolinite–serpentine group, but once thought to be an orthorhombic polymorph of antigorite.

orthochamosite

CHLORITE GROUP

orthorhombic $(Fe,Al,Mg,Mn)_6(Si,Al)_4O_{10}(OH)_8$

An uncommon iron–aluminium–magnesium–manganese aluminosilicate hydroxide, dimorphous with chamosite, that is briefly noted from the coalfields of Ayrshire (part of Strathclyde), Scotland (Anthony et al. 1995).

orthochrysotile

A relatively common asbestiform magnesium silicate hydroxide, found in veinlets cutting serpentinite at Glen Urquhart, W of Loch Ness, near Drumnadrochit, Highland Region, and Unst, Shetland, Scotland (Gaines et al. 1997).

Burke (2006) reporting on an IMA-CNMNC-approved mass discreditation of species states that orthochrysotile is formally discredited and that all samples are polytypes (crystallographic variants) of the species chrysotile.

orthoclase

FELDSPAR GROUP

monoclinic $KAlSi_3O_8$

An extremely common potassium aluminosilicate that forms two series, with celsian and with hyalophane, and is dimorphous with microcline. Orthoclase is commonly found in granites, granite pegmatites and syenites. It also occurs in high-grade metamorphic rocks and can form during potassic hydrothermal alteration. Adularia is a more completely ordered low-temperature variety of orthoclase, and murchisonite is a red variety found in Devon.

The alkali feldspars in Cornish granites are micro- and cryptoperthites in which the potassic phase may be orthoclase or microcline. Orthoclase appears to be more common in the siderophyllite granites of the Isles of Scilly, Land's End, Carnmenellis, St Austell and Hingston Down; however, as the inversion of monoclinic orthoclase to triclinic microcline is a complex process involving low-temperature fluids, it is expected that both occurred during the subsolidus recrystallization of the granites (Exley & Edmondson 1993). Laser ablation analyses of potassium-feldspar megacrysts (as these feldspars are commonly called) from the Land's End granite at Porth Ledden and Porth Loe show relatively high levels of lead, rubidium, tantalum and tin, indicating crystallization from a highly differentiated magma (Müller et al. 2006).

Well formed crystals of orthoclase are uncommon, but at Bodelva china-clay pit, St Blazey Gate, E of St Austell, tabular crystals have been found weathering out of soft kaolinized granite (Weiss 1994); however, collecting is no longer possible as the site has been converted into a tourist attraction (the Eden Project). Alteration of orthoclase to kaolinite or scaly aggregates of muscovite is often encountered in the St Austell granite complex, where intense hydrothermal activity has been focused (Richardson 1923). Elsewhere in this complex, a different type of hydrothermal alteration has produced the distinctive rock type, termed luxullianite, comprising pink orthoclase in a black groundmass of tourmaline and quartz. The rock forms a series of discontinuous veins in a granite quarry near Luxulyan, to the E of St Austell. Adjacent unaltered granite contains phenocrysts of white orthoclase microperthite up to 3 cm long (Lister 1978).

Orthoclase from one of the smaller granite plutons in Cornwall, at Cligga Head, Perranzabuloe, is described by Hall (1971), who states that the potassic feldspar occurs mainly as phenocrysts (4–20 mm), but also as smaller crystals (0.5–4 mm), and appears in thin section to be an orthoclase microperthite. It is everywhere sericitized and kaolinized, particularly near greisen. Cornish granite containing much larger phenocrysts (or megacrysts), up to 10 cm in length and often with a zonal pattern of inclusion trains mainly consisting of dark mica (siderophyllite), can be found in many towns and cities around the UK, as it is a popular building stone. Excellent examples of zoned orthoclase in granite of unknown provenance (but probably derived from the local St Austell granite complex) can be found on the steps leading down to the picturesque harbour of Fowey, 10 km E of St Austell. More Cornish locations are documented by Collins (1892).

Iron-stained orthoclase with small attached muscovite crystals; Chywoon granite quarry, Longdowns, Cornwall; width of large crystal is 30 mm. (© D. I. Green)

Large and well formed crystals of orthoclase are noted from various granite pegmatites in Cornwall, such as at Megiliggar Rocks, Breage (crystals up to 10 cm in length), associated with the Tregonning–Godolphin granite and on Trelavour Downs and at Tresayes quarry, Roche, near St Austell, both associated with the St Austell granite (Robson 1949b). Exceptional, perfectly formed orthoclase (or possibly microcline) crystals up to 13 cm in length associated with albite, ferrocolumbite, muscovite, quartz and schorl have recently been recovered from pegmatitic cavities in the Carnmenellis granite at Chywoon quarry, near Longdowns (Moulding & Hooper 2005). Similar orthoclase found in pegmatitic cavities within granite in the nearby Trolvis quarry shows growth of apatite, bertrandite, fluorite and muscovite (gilbertite) on one side of the crystals. The orthoclase crystals appear to have acted as baffles, inhibiting the flow of hydrothermal fluids and causing precipitation of orthoclase in pressure shadows (Hosking 1954a,b). Euhedral crystals of orthoclase showing Baveno twinning and reaching 15 cm in length have also been collected from Brill No. 1 quarry, Constantine, in pegmatite at the southern margin of the Carnmenellis granite (Embrey & Symes 1987).

Minute pink crystals of adularia associated with chlorite and tourmaline from drusy cavities within an elvan dyke are reported from spoil on the site of Wherry mine, Penzance, Cornwall (Russell 1949a). The formation of adularia is widespread in the Lizard ophiolite complex of southwest Cornwall. Exceedingly fresh, completely colourless and transparent adularia is reported from hydrothermal veins within amphibolites near Porthkerris Point, 3 km N of Dean quarry, Lizard Peninsula, and, at Dean quarry, flesh-pink adularia occurs with stilbite in hydrothermal veins cutting gabbro (Seager 1978a,b, Seager et al. 1978). Other adularia localities on the Lizard are documented by Halliday & Mitchell (1976).

Zoned orthoclase crystal containing inclusions of siderophyllite; from the steps leading down into Fowey, Cornwall (possibly from Colcerrow quarry, 6 km to the NW); width 8 cm. (© A. G. Tindle)

Orthoclase is one of the major constituents in pegmatitic veins cutting the highly mineralized lithium–fluorine-rich aplite at Meldon, near Okehampton, Devon. It forms distinct, often twinned, crystals up to 5 cm long (von Knorring & Condliffe 1984). These authors state that orthoclase is exceptionally rare in granite pegmatites, where the principal feldspar is commonly microcline. Unfortunately, the majority of examples quoted here do not have confirmatory structural data and, as with the case of the Chywoon orthoclase, identification is probably based on the lack of fine-scale tartan twinning often seen in thin sections of microcline (Deer et al. 2001). The complexity of identification of microcline and orthoclase is highlighted by Chaudry (1971), who, despite having XRD data for a suite of potassic feldspars from the Meldon aplite, prefers to call the mineral potassium feldspar. The name potassium feldspar is also used by Shelley (1966) in an extremely detailed discussion of granophyric and myrmekitic textures found in porphyritic granite from Lundy island, off the north coast of Devon.

Perhaps the most distinctive granite found as polished slabs on many UK city streets is the Shap granite from Cumbria. The most striking feature of the rock is the occurrence of flesh-colour phenocrysts of orthoclase, often 25 mm but up to 50 mm in length in the granite, and more perplexingly in rounded grey igneous xenoliths scattered throughout the pluton (Harker & Marr 1891). Later work has shown that these phenocrysts comprise a fine-scale mixture of crypto- and microperthites, with altered micropore-rich feldspar with irregular microstructures. A tweed variety of orthoclase is noted to be partially replaced by irregular microcline during deuteric alteration (Lee et al. 1995). Microporous and micropermeable textures in alkali feldspar from Shap, and some Cornish and Scottish granites, are discussed in detail by Walker et al. (1995), who conclude that most examples are penetratively altered.

In Wales, very rare anhedral orthoclase is reported from the St David's granophyre, SW of St David's, Dyfed (Bloxam & Dirk 1988), and from Station quarry, Arenig, Gwynedd (Bevins 1994).

Orthoclase is reported many times from the granites of Scotland, although descriptions are sparse. The Ballachulish complex, S of Fort William, Highland Region, contains interstitial alkali feldspar in early-formed diorites and strongly zoned microperthitic megacrysts in later-formed granite. The megacrysts do not show tartan twinning and are considered to be low microcline, but structural studies indicate, in addition, the metastable preservation of orthoclase or intermediate microcline (Weiss & Troll 1989). Other descriptions of orthoclase include: microperthite phenocrysts, commonly euhedral, but showing marginal outgrowths into the matrix in the Ben Nevis granite, near Fort William (Haslam 1968); microperthite in andalusite-bearing granite on Mullach nan Coirean, S of Fort William (Haslam 1971); orthoclase and microcline in the Foyers tonalite–granite complex, 30 km S of Inverness, Highland Region (Parsons & Boyd 1971); microperthite in the Central Starav granite of the Etive complex, Strathclyde (Haslam et al. 1985); reddish flesh-colour twinned crystals up to 21 cm long in granitic veins in the Rubislaw granite quarry, Aberdeen, Grampian (Heddle 1877); rhombic prisms of orthoclase up to 1 cm across in galena–sphalerite-bearing veins at Corrybeg lead mine, Abergairn, near Ballater, Grampian (Russell 1936a). Two granite-related occurrences of orthoclase from classic intrusions are documented by Heddle (1877): the colourless to slightly white translucent orthoclase forming crystals up to 1 cm in the Corriegills pitchstone, near Lamlash, Isle of Arran, Strathclyde, and the well formed brick-red and ochre-yellow to cream orthoclase that reaches 25 mm in length in the Canisp porphyry, Assynt, Highland Region.

Syenites and related rocks are also sources of orthoclase, and the species has been documented from intrusions in the Highland Region of Scotland: in borolanite from the Loch Borralan alkaline igneous complex, Assynt (Horne & Teall 1892, Teall 1900, Shand 1910b), and in metasomatized quartzites around it (Woolley et al. 1972); in syenite at Loch Ailsh, Assynt, and three syenite intrusions around Loch Loyal, near Tongue (Parsons & Boyd 1971); in syenite at Glen Dessary, W of Loch Arkaig (Fowler & Williams 1986); and in a ledmorite dyke at Achmelvich, near Lochinver (Sabine 1952). The Newmains dyke, a mantle-derived lamprophyre–syenite, exposed at several locations to the W of Dumfries in Dumfries & Galloway, contains fresh tabular grains of almost end-member orthoclase as well as other feldspar variants (Macdonald et al. 1986).

In the Mourne Mountains, Co. Down, Ireland, orthoclase is a common constituent of the granitic rocks. At Diamond Rocks, it has developed well formed white crystals in miarolitic cavities, associated with beryl, mica, smoky quartz and topaz (Green & Moreton 2001). At Lindsay's Leap quarry, creamy-white blocky orthoclase is one of the earliest species to crystallize in drusy cavities (Green et al. 2005a), and crystals up to 1 cm long have been found with smoky quartz in miarolitic granite in a quarry near Left Hand Buttress on Pigeon Rock (S. Moreton pers. comm. 2005). Beautiful crystals have also been found at Slieve Corragh (Haughton 1856).

Curved crystals of orthoclase from a miarolitic cavity in granite on the A' Chir ridge, Glen Rosa, Isle of Arran, Strathclyde Region, Scotland; the main crystal is 45 mm across. (© D. I. Green)

In Co. Donegal, orthoclase microperthite and microcline are present in the Ardara granite pluton, between Ardara and Maas (Hall 1966), and, in roadside exposures at Dunlewy, orthoclase (adularia) occurs in quartz veins cutting schists (Ryback & Moreton 1992). A structural study of potassium feldspar from the Galway granite, Co. Galway, has shown a transition from orthoclase to microcline as a result of late deformation (Wilson & Coats 1972). Large crystals of pure orthoclase are reported from locations within the Leinster granite in the counties of Carlow, Dublin, Wexford and Wicklow (Haughton 1856).

Pink-to-red orthoclase is ubiquitous in the granites of Jersey, with particularly porphyritic varieties containing orthoclase phenocrysts found in the Dicq granite and around Seymour Tower. Baveno twinned crystals have been found in miarolitic cavities and pegmatitic patches at Ecréhou, Gigoulande, La Moye, L'Etacq, Mont Mado, Paternoster and Portelet Bay; and, at Les Pipettes, Minquiers, pale translucent crystals up to 30 cm long have been found. Orthoclase is also common in porphyries and porphyritic rhyolites on the island and tiny white crystals of adularia occur in veins in the German quarry, S of L'Etacq (Mourant 1961). Following up claims that prismatic crystals up to 1 mm long of colourless-to-milky epistilbite had been found in Waterworks Valley, St Lawrence, Jersey, Bishop et al. (1982) were able to show that this was a misidentification of adularia.

orthoferosilite

A synonym of ferrosilite, pyroxene group.

orthopyroxene

A general term for orthorhombic pyroxene species such as enstatite and ferrosilite.

orthoserpierite

orthorhombic $Ca_2(Cu^{2+},Zn)_4(SO_4)_2(OH)_6 \cdot 3H_2O$

An hydrated calcium–copper–zinc sulphate, dimorphous with serpierite and found at Tynagh mine, near Killimor, Co. Galway, Ireland. Orthoserpierite is found in an unconsolidated partially oxidized cap enriched in other supergene minerals above a primary sulphide orebody consisting predominantly of galena, pyrite and sphalerite. Altogether 104 species have been described from this locality, including some new to Ireland (bayldonite, beudantite, chalcophyllite, hedyphane, olivenite, parnauite and strashimirite) and some new to Britain and Ireland (clinotyrolite, orthoserpierite and theisite) (Ryback & Moreton 1992).

osarizawaite

ALUNITE GROUP

trigonal $Pb(Al,Cu)_3(SO_4)_2(OH)_6$

A very rare lead–aluminium–copper sulphate hydroxide found as a yellowish-green crust and impregnation on sandstone from an unspecified locality on Alderley Edge ridge, Cheshire (Embrey 1978, Braithwaite 1994).

osmiridium

A synonym of iridium.

Orthoserpierite associated with green spherules of malachite and colourless gypsum; Foxrock mine, Glendasan, Co. Wicklow, Ireland; width 6 mm. (© D. I. Green)

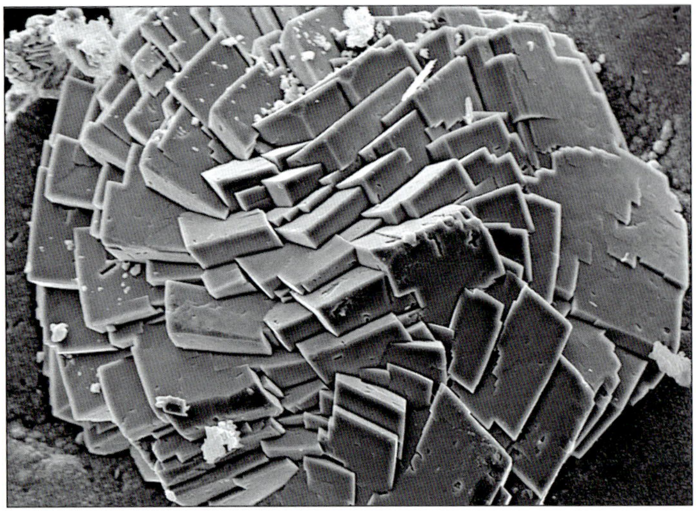

Otavite spheroid showing rhombohedral form. Dolyhir quarry, Old Radnor, near Presteigne, Powys, Wales; width 100 µm; SEM image. (© M. Rothwell)

osmium

hexagonal (Os,Ir,Ru) with Os dominant

A very rare native metal found in the altered silicate matrix of chromite-rich dunite lenses that formed near the base of the Shetland ophiolite complex at Cliff and at Harold's Grave, Baltasound–Haroldswick area, Unst, Shetland, Scotland. Native osmium and osmium-poor laurite form composite inclusions (many <5 µm across) within chromite grains. Irarsite, pentlandite and sperrylite are other associated species (Prichard et al. 1986, 1994, Tarkian & Prichard 1987, Prichard & Tarkian 1988). Nikkavord North on Unst is a further location, where native osmium has been found in harzburgite-hosted chromitite (Prichard et al. 1986).

Native osmium has also been found within chromitite horizons at Corrycharmaig, Glen Lochay, NW of Killin, Tayside Region, in a small body of mineralized, metamorphosed and altered ultramafic rocks. Accessory amounts of irarsite, laurite, galena and gersdorffite, the native metals osmium, iridium, ruthenium, silver and gold, and unnamed alloys of As, Co, Fe, Ni, Os, Pd, Pt, Ru, Sb and Sn, occur either as inclusions (up to 40 µm across) in chromite or are associated with the nickel sulphides heazlewoodite and nickeline (usually ≤2 µm). The chemistry and association of the PGMs is strongly indicative of an ophiolite assemblage and it is most likely that the intrusion represents a fragment of oceanic crust (Power & Pirrie 2000).

A report of alluvial iridosmine (osmium) from the Dolgellau goldbelt, Gwynedd, Wales, requires verification (Bevins 1994).

OSUMILITE GROUP

The osumilite group are hexagonal and orthorhombic silicates, with the general formula $A_{1-2}B_{2-3}C_3Z_{12}O_{30}\cdot nH_2O$, where A = Ba, Ca, K, Na; B = Fe^{2+}, Li, Mg, Mn^{2+}, Na, Sn, Ti, Y, Zn, Zr; C = Al, B, Be, Fe^{2+}, Fe^{3+}, Li, Mg, Zn; and Z = Al, Si. Armenite, milarite and osumilite-(Mg) are members of the group found in Britain or Ireland.

osumilite-(Mg)

OSUMILITE GROUP

hexagonal $KMg_2(Al,Fe)_3(Si,Al)_{12}O_{30}$

TL Tieveragh, near Cushedall, Co. Antrim, Ireland

A potassium–magnesium aluminosilicate found in fused sandstone (buchite) at the contact of the Tertiary dolerite plug of Tieveragh. Osumilite-(Mg) forming tabular crystals up to 100 µm across occurs as a common to irregularly distributed constituent, usually as masses or trails of euhedral and anhedral crystals, set in a colourless glass. Associated species include cordierite, orthopyroxene (hypersthene), sanidine, sodic plagioclase and tridymite in the form of marginal inversion needles bordering relict quartz grains. At Tieveragh, osumilite-(Mg) is described as being close to the magnesium–potassium end-member composition (Chinner & Dixon 1973). Osumilite-(Mg) is not approved by the IMA-CNMNC.

otavite

CALCITE GROUP

trigonal $CdCO_3$

A very rare cadmium carbonate found in veins cutting Lower Carboniferous limestones at Coldstones quarry, 5 km W of Pateley Bridge, Yorkshire. The veins carry barite, fluorite, galena and sphalerite, and in places the supergene minerals cerussite, hemimorphite and smithsonite occur, commonly lining cavities. Fallen blocks from the Garnet vein also contain otavite, doyleite and prosopite. One cavity in fluorite veinstone contains otavite occurring as small pure-white nodular masses up to 2 mm across, some of which exhibit crude twinned rhombohedral outlines. Otavite perches on fluorite, hemimorphite or smithsonite and is clearly the last mineral to form in the cavity. Oxidation of sphalerite, to form abundant hemimorphite and smithsonite within the veins, no doubt released the cadmium to form otavite (Young et al. 1997).

Small quantities of otavite are reported from scree below the railway cutting leading to the tunnel at Bishopton, Strathclyde, Scotland. Otavite forms creamy-white crusts coating greenockite and develops hemispheres (up to 0.5 mm across) or botryoidal crusts on prehnite near greenockite grains. The spheroidal and botryoidal otavite consists of clusters of tiny rhombohedra. The occurrence here is unusual, as cadmium is normally mobilized only under extreme conditions resulting from the oxidation of sulphide orebodies (Faithfull & Hubbard 2001).

Minute buff rhombohedral crystal masses with individual crystallites up to 0.1 mm long from Dolyhir quarry, Old Radnor, near Presteigne, Powys, Wales, have been identified as otavite by XRD, EMP and wet chemical techniques (T. F. Cotterell pers. comm. 2003).

ottrélite?

monoclinic or triclinic $(Mn^{2+},Fe^{2+},Mg)_2Al_4Si_2O_{10}(OH)_4$

A rare manganese-iron-magnesium aluminosilicate hydroxide reported from "Snowdon", Gwynedd, Wales (Hall 1868a), although this could equally well be a broad reference to Snowdonia rather than the mountain peak. Ottrélite has not been confirmed in later studies and it may have been mistaken for chloritoid, a similar mineral found in the same district (Bevins 1994). Ottrélite is also described from slate quarried in the cliffs near to Tintagel, Cornwall (Hutchings 1889), and from black slates in the aureole of the Skiddaw granite, Caldbeck Fells, Cumbria (Rastall 1910), but, as these are optical identifications, further confirmation is desirable.

ourayite

orthorhombic $Ag_3Pb_4Bi_5S_{13}$

A rare silver–lead sulphobismuthite found in a stockwork cut by galena-bearing quartz veins at Corrie Buie, Meall nan Oighcreag, Loch Tay, Tayside Region, Scotland. The veins also contain minor amounts of native bismuth, chalcopyrite, electrum (36–65% gold), emplectite (one grain), pyrrhotite, schirmerite and sphalerite, and phases belonging to the gustavite–lillianite series. The mineral assemblage of the veins is typical of that associated with acid intrusions. Ourayite and its close associate gustavite form bladed aggregates (up to 0.5 cm long) associated with pyrrhotite and galena, and also develop as exsolution blades (up to 20 µm long) in galena (Pattrick 1984).

P

palagonite

A basaltic tufa, formerly considered to be a mineral species.

palladium

cubic Pd or (Pd,Hg)

Native palladium has been identified in a layered mafic–ultramafic intrusion on the Isle of Rum, Highland Region, Scotland. Within the intra-unit 14 chromitite of the Eastern Layered Series, palladium is found in association with species positively or tentatively identified as michenerite, moncheite, sperrylite and the native species gold and platinum. Most of these grains are 0.2–2 μm in diameter, although grains >15 μm are also present for some species. The majority are in contact with sulphide minerals such as bornite, chalcocite, chalcopyrite, pentlandite or pyrrhotite. The chromitite layers were formed during magmatic fractionation and crystal settling of an open-system magma chamber, modified by magma mixing and liquid immiscibility of sulphide and silicate melts (Butcher et al. 1999, Power et al. 2000a). Palladium is also briefly described from the Smid Hope Burn tributary of the River Tweed, near Tweedhope, Borders, where it has been found as 2 μm-diameter inclusions associated with berzelianite in two alluvial gold grains (Leake et al. 1993, Livingstone 2002).

palladoarsenide

monoclinic Pd_2As

A very rare palladium arsenide, identified in a layered mafic–ultramafic intrusion on the Isle of Rum, Highland Region, Scotland. Within the unit 7–8 inter-unit chromitite layer of the Eastern Layered Series, palladoarsenide is found in association with species positively or tentatively identified as braggite, cooperite, hongshiite, irarsite, isoferroplatinum, keithconnite, kotulskite, laurite, moncheite, platarsite, sperrylite, telargpalite and telluropalladinite, and the native species palladium and platinum. Most of these grains are 0.2–2 μm in diameter, although grains >15 μm are also present for some species. The majority are in contact with sulphide minerals such as bornite, chalcopyrite or pentlandite. The chromitite layer was formed during magmatic fractionation and crystal settling of an open-system magma chamber, modified by magma mixing and liquid immiscibility of sulphide and silicate melts (Power et al. 2000a).

palladseite?

cubic $Pd_{17}Se_{15}$

A very rare palladium selenide, qualitatively identified from the Smid Hope Burn tributary of the River Tweed, near Tweedhope, Borders, Scotland, where it has been found as a 1 μm-diameter inclusion associated with clausthalite in an alluvial gold grain (Leake et al. 1993).

palygorskite

monoclinic or orthorhombic $(Mg,Al)_2Si_4O_{10}(OH)\cdot 4H_2O$

An uncommon hydrated magnesium aluminosilicate hydroxide found as an alteration product of magnesium-rich silicates in soils and sediments, also in carbonate and mafic igneous rocks.

Palygorskite is found in the zinc-hydroxide zone of Milltown limestone quarry, near Ashover, Derbyshire. This zone is thought to have been formed in a highly alkaline oxidizing environment, perhaps by leaching of calcined limestone, itself the result of lime burning, or from the pre-gunpowder technique of fire setting. Palygorskite forms whitish coatings on joint surfaces and is sometimes found in cavities, where it has a spheroidal form and is partially coated with calcite. The rare species ashoverite, sweetite and wülfingite, and an unnamed zinc hydroxide, are found with hemimorphite, hydrocerussite and hydrozincite at this locality (Rust 1991c). Clark et al. (1984) describe joints in fluorite veins from Milltown quarry as being encrusted in white attapulgite (palygorskite). In Leicestershire, at Newhurst quarry, Shepshed, white coatings on joints in diorite have been tentatively identifed as palygorskite (Ince 2005), whereas at Bardon Hill quarry, Coalville, and New Cliffe Hill quarry, Stanton under Bardon (Evans & King 1962), Croft quarry (Pearson & Jeffrey 1997), and at Warren quarry, Enderby (Anthony et al. 1995), the species appears to be confirmed. Densely packed microscopic fibres of palygorskite occur in voids within Keuper Marl from Bindon and Haven Cliffs on the South Devon coast. Light-grey to white flexible veinlets cutting mudstone at nearby Branscombe West Cliff proved to contain a mixture of calcite, celestine and palygorskite (Jeans 1978).

Palygorskite (var. mountain leather) from Mwyndy mine, near Llantrisant, Mid Glamorgan, Wales; width of specimen 11.5 cm. (© National Museum of Wales, Cardiff; photographer M. P. Cooper)

"Mountain leather" is a poorly defined term often used in old literature to describe either palygorskite or fibrous amphibole. The term was used by North (1916) to describe an asbestiform mineral found with iron ore from Mwyndy mine, near Llantrisant, Mid Glamorgan, Wales, and this was later confirmed as palygorskite by Criddle & Symes (1977). Subsequently, fibrous grey and white palygorskite has been found on strongly sheared and slickensided surfaces of fine-grain compact very-hard dark-grey manganese ore at Tŷ Coch farm, near South Cornelly, Mid Glamorgan (Criddle & Symes 1977).

Greyish-white compact palygorskite (attapulgite) is found associated with montmorillonitized syenite in the Bakkasetter quarries, southeastern flank of Noss Hill, South Mainland, Shetland, Scotland. The palygorskite occurs sporadically in irregular pocket-like masses and thin veinlets, and in thin section it displays an "intimate felted texture". An XRD investigation showed that this mineral is identical to pilolite from Burn of Daugh, Cabrach, Grampian, and "mountain leather" from Nawton, Yorkshire, and Seaton, Devon (Stephen 1954). Palygorskite (referred to as pilolite, mountain leather and mountain cork) is briefly described by Heddle (1879a, 1901) from many other locations throughout Scotland. Those with supporting analytical data are: Fife – Parton Craig, Tayport; Grampian – Mouth of the Burn of Boyne; near Portsoy; Todhead Point, 9 km S of Stonehaven; Highland – Strontian; Strathclyde – Leadhills.

pandaite

A synonym of bariopyrochlore, pyrochlore group.

paolovite

orthorhombic Pd_2Sn

A very rare palladium–tin alloy identified in chalcopyrite-rich dykes associated with the West Sgaorishal ultramafic plug, on the Isle of Rum, Highland Region, Scotland. Paolovite forms euhedral or skeletal grains almost exclusively within chalcopyrite, bornite or covellite. Associated species, usually occurring as rounded anhedral blebs enclosed within chalcopyrite or pentlandite, include froodite, gold (electrum), keithconnite, kotulskite, native silver, sperrylite, telluropalladinite and Ag–Bi–Pd–Te phases. Grainsize varies between 1 μm and 40 μm, although the majority of grains are <5 μm. Concentration of the PGEs in an immiscible sulphide liquid within a mantle magma reservoir must have occurred prior to the emplacement of the dykes (Power et al. 2001, 2003).

para-alumohydrocalcite

monoclinic or triclinic $CaAl_2(CO_3)_2(OH)_4{\cdot}6H_2O$

A rare secondary hydrated calcium–aluminium carbonate hydroxide found in a localized series of interconnected white vuggy veinlets up to 1 mm thick, filling allophane-lined fractures in a sandy iron-stained and partially decalcified limestone at the southern end of Hampstead Farm quarry, Chipping Sodbury, Avon. The veinlets are associated with heavily oxidized pyritiferous veins in Carboniferous Limestone close to the Rhaetic/Lower Carboniferous unconformity. Para-alumohydrocalcite forms continuous encrustations in the fractures or, where space permits, divergent or hemispherical sprays of acicular crystals up to 50 μm long and with euhedral terminations. The secondary aluminous minerals at this locality (allophane, basaluminite, dundasite, gibbsite and scarbroite) are thought to have been derived from the leaching of aluminous clays by strongly acidic solutions, resulting from the weathering and oxidation of pyrite (Alabaster 1989a, 1990b). Para-alumohydrocalcite from an 8 mm-wide vein in ferruginous sandstone is also reported from a quarry at Boughton, near Northampton (Ryback & Tandy 1992).

paracelsian

FELDSPAR GROUP

monoclinic $BaAl_2Si_2O_8$

A rare barium aluminosilicate, dimorphous with celsian and found at Benallt manganese mine, Rhiw, Lleyn Peninsula, Gwynedd, Wales. Paracelsian forms either colourless and transparent, or white and cloudy, prismatic crystals up to 5 cm long and 1 cm wide, in association with celsian. Both species are locally abundant in a band of shales and sandstones associated with Ordovician lavas and tuffs, and with beds of black manganese ore. Paracelsian probably crystallized during a metamorphic event caused by the intrusion of a nearby dolerite sill (Spencer 1942).

Paracelsian crystals, up to 25 mm in length; Benallt mine, Rhiw, Lleyn Peninsula, Gwynedd, Wales. (© National Museum of Wales, Cardiff; photographer M. P. Cooper)

parachrysotile

KAOLINITE–SERPENTINE GROUP

orthorhombic $Mg_3Si_2O_5(OH)_4$

An asbestiform magnesium silicate hydroxide found as a minor component of many serpentinites. It is probably more common than the literature would suggest, because it is difficult to characterize. Parachrysotile has been found at Glen Urquhart, W of Loch Ness, near Drumnadrochit, Highland Region, Scotland (Gaines et al. 1997).

Burke (2006), reporting on an IMA-CNMNC approved mass discreditation of species, states that parachrysotile is formally discredited and that all samples are polytypes (crystallographic variants) of the species chrysotile.

paracostibite?

orthorhombic CoSbS

A rare cobalt sulphantimonite, dimorphous with costibite and found on specimens in the Kingsbury collection at the NHM; the specimens are thought to have been collected from Wheal Cock, near Botallack, St Just, Cornwall (Criddle & Stanley 1993). Paracostibite requires further confirmation, as the provenance of the museum specimens is dubious (Ryback & Tandy 1992, Ryback et al. 1998, 2001; C. J. Stanley pers. comm. 2004).

paragonite

MICA GROUP

monoclinic $NaAl_2\square AlSi_3O_{10}(OH)_2$

A common sodium aluminosilicate hydroxide found in glaucophane schists and hornblende schists at Plasberw, Anglesey, Gwynedd, Wales (Greenly 1919), and as a major constituent of thin veins (<0.5 mm wide) associated with analcime, banalsite and pennantite from the Ty Canol incline, Benallt manganese mine, Rhiw, Lleyn Peninsula, Gwynedd (Campbell Smith et al. 1946). Paragonite is also a common constituent of Lower Palaeozoic pelites in Wales, being found in roofing slates of the Cambrian Llanberis Slate Formation from Bethesda and Nantlle, Snowdonia (Bevins 1994). Pelites of Caradoc age between Cwm Pennant and Penmachno, and of Arenig and Llanvern age extending across Gwynedd, contain abundant paragonite (Bevins 1994). From a detailed study of the <2 μm fraction of over 300 pelite samples collected on the Lleyn Peninsula and Snowdonia, Merriman & Roberts (1985) were able to show that paragonite is dominant in stage III metapelites, corresponding to the clinozoisite and biotite zones of greenschist-facies metamorphism. Paragonite forms intergrowths with muscovite and a metastable intermediate potassium–sodium mica at the margin of Ordovician Caradoc metabasites intruded into a sequence of metapelitic rocks at Tremadog (Jiang & Peacor 1993).

In Scotland, paragonite is found as small fibrous crystals and as intergrowths with muscovite in Dalradian chlorite–chloritoid–mica schists from the N of Stonehaven, Grampian (Baltatzis & Wood 1977). At Creag Bhocan, SSW of Tyndrum, Central Region, paragonite is one of many inclusion types (commonly ≤10 μm in diameter) found in pyrite from a stratabound copper–zinc deposit within albitic quartzites and chloritic schists of Dalradian age (Scott et al. 1988). Paragonite is also an inclusion species within zoned metamorphic garnets in Dalradian metapelites from Grandtully, near Pitlochry, Tayside Region, and it occurs together with chlorite, chloritoid, margarite, rutile and staurolite, often in the same garnet host (Sivaprakesh 1981). Minor paragonite is noted in Strathclyde from pyritiferous black and grey slates belonging to the Easdale Slate Formation of the Argyll Group in the Middle Dalradian of the late Precambrian, particularly at Easdale Island and Cuan Ferry (Hall et al. 1988), and from red Dalradian phyllites also containing chlorite, hematite, muscovite, quartz and rutile from near Dunoon (McNamara 1963).

Paragonite is noted from the Great Rocks Dale dyke, near Buxton, Derbyshire (Cope 1933, Ford et al. 1993), and in bleached and indurated siltstones and mudstones of the Skiddaw Group within the Crummock Water aureole, near Buttermere, Cumbria (Cooper et al. 1988).

parahilgardite

A synonym of hilgardite-3*A*.

parahopeite?

triclinic $Zn_3(PO_4)_2{\cdot}4H_2O$

A very rare hydrated zinc phosphate, dimorphous with hopeite and described from Roughton Gill, Caldbeck Fells, Cumbria (Hartley 1984), and two localities in Yorkshire: Cockhill mine, Bewerley, and Turf Pits mine, Grassington Moor (Dunham & Wilson 1985). These descriptions are based on specimens supposedly collected by A. W. G. Kingsbury. However, the work of Ryback et al. (1998, 2001) indicates that the provenance of many Kingsbury samples is doubtful and further investigation is therefore desirable. The specimens are also reported in compilations by Young (1987a), Cooper & Stanley (1990), Ryback & Tandy (1992) and Anthony et al. (2000).

paralaurionite

monoclinic $PbCl(OH)$

An uncommon lead oxychloride, dimorphous with laurionite and found in a small geode lined with limonite from Wheal Confidence, near Newquay, Cornwall (now in the Brooke collection of the Sedgwick Museum, University of Cambridge). Paralaurionite forms long narrow platy crystals and is associated with well defined crystals of phosgenite (Russell & Hutchinson 1922, 1927). Elsewhere in Cornwall, paralaurionite is noted without description from Daymer Bay, Padstow (Golley & Williams 1995), and Wheal Rose, near Sithney (Gaines et al. 1997).

Paralaurionite is a very rare species at Merehead (Torr Works) quarry, Cranmore, Somerset, and it occurs in veins cutting Carboniferous Limestone, where it forms translucent, sulphur-yellow or orange-yellow, corroded crystalline masses within cerussite–hydrocerussite-lined cavities in brownish-black manganese ore. It is closely associated with blixite, cerussite, chloroxiphite, coronadite, hematite, hydrocerussite, mendipite, mereheadite, parkinsonite, pyrolusite and symesite. The lead oxychlorides are thought to have formed by interaction of primary galena with low-temperature metalliferous brines along fractures cutting the original Fe–Mn–(Cu–Pb) orebody (Alabaster 1975, Symes & Embrey 1977, Welch et al. 2000, Turner 2007). At Coleman's quarry, Holwell, near Frome, paralaurionite is a late-stage mineral found in similar manganese-rich veins cutting Lower Carboniferous limestones (Alabaster 1990a).

At Wesley mine (Wesley Theological College), Westbury-on-Trym, Avon, laurionite, and especially paralaurionite, are fairly abundant, occurring together as coarse- to fine-grain intergrowths or as isolated corroded grains on cerussite. Samples of paralaurionite up to 5×1.5 cm are always encrusted with cerussite and, when broken open, reveal usually colourless (or rarely pale-brown) transparent-to-translucent crystals with a brilliant lustre. Paralaurionite and other lead compounds (cerussite, hydrocerussite, laurionite, mendipite, plattnerite, wulfenite) form part of a secondary mineral suite associated with replacement deposits of iron and manganese oxides of Triassic age (Alabaster 1985b, 1989b).

Green (1987a) notes isolated transparent lath-like crystals of anthropogenic paralaurionite (several mm long but rarely more than 0.05 mm thick) associated with lanarkite from the dumps of the Meadowfoot smelter, 2 km NW of Wanlockhead, Dumfries & Galloway, Scotland.

paralstonite

trigonal $BaCa(CO_3)_2$

A rare barium–calcium carbonate, trimorphous with alstonite and barytocalcite, and found at Dolyhir quarry, Old Radnor, near Presteigne, Powys, Wales. A white, fibrous-to-powdery mineral associated with harmotome produced an XRD pattern with both alstonite and paralstonite present (Cotterell & Dean 2007). Dolyhir is the site of a barite–chalcopyrite–galena–tennantite vein, in which anglesite, azurite, cerussite, native copper, cuprite, malachite, olivenite and tyrolite are also found (Wolfe 2000).

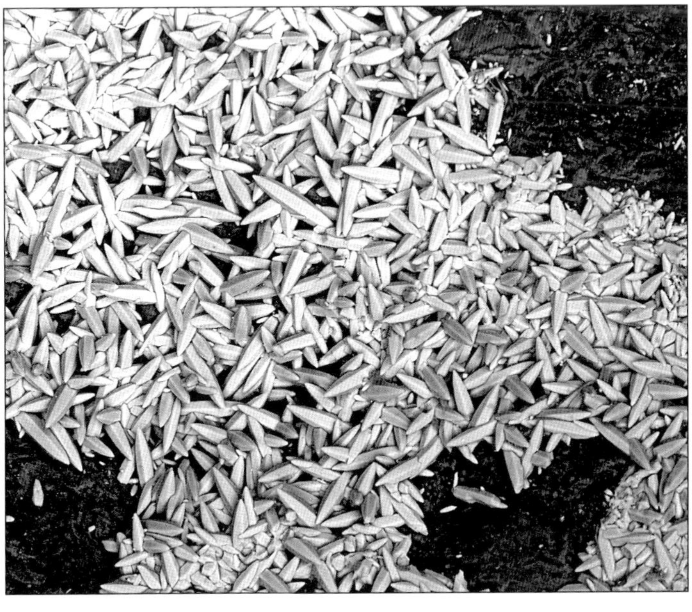

Elongated bipyramidal crystals of white paralstonite from Dolyhir quarry, Old Radnor, near Presteigne, Powys, Wales; width 740 µm; SEM image. (© National Museum of Wales, Cardiff; photographers T. F. Cotterell & M. P. Lambert)

paramelaconite

tetragonal $Cu^{1+}_2Cu^{2+}_2O_3$

A rare copper oxide found on a specimen of native copper collected from loose material in a cliff talus debris flow at Littleham Cove, near Budleigh Salterton, Devon. Paramelaconite was identified by XRD but was not seen during petrographic analysis. Associated minerals include cuprite, domeykite, krutovite (provisional identification) and two unnamed species whose compositions are Cu_4As and Cu_5As. The occurrence is associated with complex Ag–As–Co–Cu–Ni–U–V mineralization that developed as Permo-Triassic strata were de-watered during burial compaction (Milodowski et al. 2002).

paranatrolite

monoclinic (pseudo-hexagonal) $Na_2Al_2Si_3O_{10}\cdot3H_2O$

An hydrated sodium aluminosilicate found at Hills Port, S of the Gobbins, Island Magee, Co. Antrim, Ireland. Throughout the district, chabazite, gonnardite, heulandite, natrolite, paranatrolite, tetranatrolite, thomsonite and other zeolite minerals infill vesicles in Tertiary basalts (Nawaz 1988a). Paranatrolite is not a well studied mineral and its species status is debatable. It could be regarded as overhydrated natrolite, tetranatrolite or gonnardite (Coombs et al. 1997).

pararammelsbergite

orthorhombic $NiAs_2$

A rare nickel arsenide, trimorphous with rammelsbergite and krutovite, and described from Alderley Edge ridge, Cheshire, where it forms inclusions in nickel-rich cobaltite. Although the inclusions have the optical properties of pararammelsbergite, the small grainsize (5–10 µm) and complex zoning and intergrowths prevent quantitative identification (Ixer & Vaughan 1982, Braithwaite 1994). Milodowski et al. (2002) suggest that pararammelsbergite can be mistaken for krutovite, as the two species are almost identical in composition.

Pararammelsbergite appears to be confirmed in the carbonate-hosted Lisheen Pb–Zn–(Ag) deposit, between Thurles and Rathdowney, Co. Tipperary, Ireland, where an As–Co–Cu–Ni assemblage has been found near major faults. Associated species are bournonite, cobaltite–gersdorffite, nickeline, pyrite (bravoite) and tennantite, and traces of arsenopyrite, barite, chalcopyrite, hematite and ullmannite (Ixer 2002, Fusciardi et al. 2004).

parastilbite

TL Kilkeel, Co. Down, Ireland

A synonym of epistilbite or stilbite, zeolite group.

parasymplesite

VIVIANITE GROUP

monoclinic $Fe^{2+}_3(AsO_4)_2\cdot8H_2O$

A rare hydrated iron arsenate, dimorphous with symplesite, that forms a series with köttigite. It is the arsenate analogue of vivianite and is found in the oxidation zone of some arsenic-rich hydrothermal base-metal deposits.

Parasymplesite occurs in an antimony-rich quartz vein at Wet Swine Gill, Caldbeck Fells, Cumbria. The massive white quartz vein contains an early suite of As–Fe–Sb species (arsenopyrite, berthierite and stibnite) and a later suite of Pb–Sb species (fülöppite, semseyite and zinkenite). Parasymplesite forms radiating aggregates of pale- to dark-blue or green crystals up to 2 mm across. Clusters of green fibres also develop radiating rosettes up to 3 mm across. A close association with arsenopyrite and claudetite is common and it is likely that a combination of saturation from groundwater and arsenopyrite oxidation was responsible for its formation (Neall & Green 2001, Green & Briscoe 2002). Less than 1 km away at Burdell Gill, gossanous quartz containing dark-green to dark indigo-blue sheaves of vivianite, with variable arsenic substitution to a parasymplesite composition, have been described. Both parasymplesite and vivianite appear to be confined to isolated blocks containing rhodochrosite and switzerite.(Neall & Leppington 2003).

The only other occurrence of parasymplesite in Britain or Ireland is at Muckross mine, Killarney National Park, Co. Kerry, Ireland, where clear prismatic crystals 0.1–0.5 mm long have been found. Post-mining oxidation of mine waste produced secondary carbonates, sulphates and arsenates (of which parasymplesite was just one) from the primary Ag–As–Co–Cu–Fe–Pb–Zn mineralization (Moreton et al. 1999).

paratacamite

trigonal $Cu_2^{2+}Cl(OH)_3$

A relatively uncommon copper hydroxychloride, trimorphous with atacamite and botallackite, and found at Botallack mine, St Just, Cornwall (Bannister et al. 1950). Good specimens up to 1.5 cm long are said to have been found in the mine's submarine lodes (Bancroft & Weller 1993). At the adjacent Geevor mine, near Pendeen, paratacamite forms small dark-green crystals and crystalline masses up to 4 mm across (Wolloxall 1987). Specimens containing paratacamite from Penberthy Croft mine, St Hilary, Cornwall, described by Kingsbury (1954), and now in the Kingsbury collection at the NHM, have since been shown to consist of brochantite, langite and devilline (Betterton et al. 2000). This occurrence is also noted by Camm & Merry (1991b) and Golley & Williams (1995). Until independent verification is assembled, other Cornish occurrences of paratacamite noted by Kingsbury (West Wheal Owles, St Just; Wheal Cock, St Just; and Cligga Head, Perranzabuloe) must be considered doubtful. Golley & Williams (1995) list paratacamite from Levant mine, St Just, but later analysis has shown it to be clinoatacamite (M. P. Cooper pers. comm. 2007).

The identification of paratacamite is further complicated since Jambor et al. (1996) have suggested that the commonly used Powder Diffraction File standard for paratacamite (card 25–1427) is that of a new monoclinic polymorph, named clinoatacamite. So-called paratacamite from Rubha a' Mhill, northeast Isle of Islay, Strathclyde, Scotland (Green 1988), and from near Barncorkrie farm, Drummore, Portencorkie area, Dumfries & Galloway (Cameron & Lawson 1979, Livingstone 2002), is in fact clinoatacamite (Starkey & Hubbard 1998). It is uncertain whether the paratacamite noted from Dererach, Loch Scridain, Isle of Mull, Strathclyde, by Livingstone (2002) is actually clinoatacamite.

pargasite

monoclinic AMPHIBOLE GROUP $NaCa_2(Mg_4,Al)(Si_6Al_2)O_{22}(OH)_2$

Common sodium–calcium–magnesium aluminosilicate hydroxide that forms a series with ferropargasite and is found in basalts, gabbros, altered ultramafic rocks, schists, amphibolites and metamorphosed siliceous-limestone skarn deposits. Where possible, identification is based on the 1997 classification of the amphibole group (Leake et al. 1997a, Mogessie et al. 2004).

England

Cornwall A finely banded plagioclase lherzolite from the Lizard ophiolite complex at Countybridge quarry, Goonhilly Downs, Lizard Peninsula, contains both pargasite and tschermakite (Davies 1983). Colourless pargasite is also described from an olivine–plagioclase–spinel assemblage in contact areas of the Lizard peridotite (Green 1964). Tin-enriched pargasite occurs in a calc-silicate body (representing metasomatically altered metabasalt) in the aureole of the Land's End granite at Chycornish Carn, St Just (Alderton & Jackson 1978). An occurrence of bladed dark-green pargasite in skarn rocks from the same aureole at Wheal Cock Zawn, Botallack, St Just (Couper & Barstow 1977, Hubbard 1989), requires confirmation.
Cumbria Brown pargasite occurs with actinolite and magnesiohornblende as hydrothermal or auto-metasomatic alteration products of pyroxene in microdiorites from Scawgill Bridge and Hope Beck, SSW of Scawgill Bridge, 4 km W of Braithwaite (Fortey et al. 1994).

Wales

Dyfed The xenolithic laminated olivine gabbro unit of the St David's Head layered intrusion contains edenite and pargasite that had originally been referred to as ferroan pargasitic hornblende and magnesio-hastingsitic hornblende (Bevins et al. 1994).
Gwynedd Pargasite is the major amphibole species in basalt lavas from the Ordovician Rhobell complex, Rhobell Fawr area, near Dolgellau. It forms dark-green subhedral-to-anhedral crystals up to 4 cm long, which in thin section can be seen to be zoned (Kokelaar 1986). Unpublished data suggest that pargasite is also present in gabbros of the Rhiw intrusion, Lleyn Peninsula (Bevins 1994).

Scotland

Central So-called hornblende schists from Aberfoyle Forest, near Balmaha, Loch Lomond (close to the Highland Boundary fault), contain acicular green edenite and magnesiohornblende, and minor amounts of both pargasite and tschermakite (Henderson et al. 1983).
Dumfries & Galloway The late Caledonian Newmains lamprophyre dyke, 7 km W of Dumfries, contains abundant, euhedral prisms of edenite and pargasite, commonly cored by clinopyroxene (Macdonald et al. 1986). Euhedral brown crystals of edenite and pargasite occur in porphyrite dykes cutting the Loch Doon granite complex between Loch Dee and Craiglee (Tindle 1982).

Sector-zoned pargasite in a porphyrite dyke cutting the Loch Doon granite complex; Craiglee, north of Loch Dee, Dumfries & Galloway, Scotland; width 1.2 mm; viewed in PPL (upper) and XPL (lower). (© A. G. Tindle)

Fife Certain pyroxenite nodules found in a Carboniferous–Permian vent at Elie Ness contain green pargasite associated with albite and biotite. Other nodules contain kaersutite, magnesiohastingsite or, rarely, magnesiotaramite (Chapman 1976).
Grampian Gabbros in the Insch basic layered intrusion contain pargasite, with ferro-edenite and ferropargasite in the Upper Zone cumulate sequence (Wadsworth 1986) to the W and SW of Insch, and with edenite, ferro-edenite and ferropargasite in Middle Zone fine-grain granular gabbros to the E of Insch (Wadsworth 1988).
Highland Large plates of brown amphibole in a biotite pyroxenite associated with the Loch Ailsh alkaline igneous complex have magnesiohastingsite or pargasite compositions (Shaw et al. 1994). Ultramafic granulites at Drumbeg and Gorm Loch, NE of Scourie, Assynt, contain granulite-facies pargasite that exsolved opaque oxides and recrystallized as actinolite during retrograde metamorphism (Sills 1982). A garnet–pyroxene eclogite from Glen More, Glenelg, contains very dark-green strongly pleochroic pargasite (O'Hara 1960). Pargasite is said to replace omphacite and garnet in eclogite at Letterfearn, to the NE of Glen More (Brewer et al. 2003), although no data are quoted. Aluminopargasite is also recorded from Glenelg, in a garnet–kyanite amphibolite (Tilley 1937c), but admixed kyanite may have contaminated the analysis. Problems with analysis (Leake 1968) may also explain the unusual aluminopargasite composition from a Lewisian epidote–garnet–hornblende gneiss at Camas Baravaig, southeast Isle of Skye (Matthews & Cheeney 1968). The amphibole described as hornblende from a camptonite dyke at Allt Tarsuinn, Strontian, Argyll (Gallagher 1963), is actually brown pargasite. Pargasite and magnesiohastingsite occur as inclusions in spinel within the ultramafic portion of the Rum intrusion, Isle of Rum (Young 1984b). On the Isle of Muck, pargasite occurs in a gabbro dyke with biotite, clinopyroxene, olivine, plagioclase and opaque oxides (Ridley 1971). A diatreme associated with an alkaline dolerite dyke on the Isle of Eilean Shona, N of Ardnamurchan, contains xenoliths of garnet-bearing spinel pyroxenite in which pargasite is a subordinate species (Upton et al. 2003).

Strathclyde Within the Kilmelford igneous complex, pargasite occurs in the diorite intrusions at Kilbride and Tom Soilleir. Elsewhere in the complex, actinolite and magnesiohornblende are also found (Zhou 1987). Green edenite, ferro-edenite and pargasite (originally called hornblende) are found in crossite-bearing amphibolites associated with the Girvan–Ballantrae ophiolite complex to the W of Knockormal (Bloxam & Allen 1959, A. Livingstone unpub. EMP data). A hornblende-bearing garnet metapyroxenite containing pargasite and tschermakite, exposed near Knocklaugh, is from the aureole of this complex (Treloar et al. 1980). At Black Rock, a Late Carboniferous basanitic vent on the west coast of Scotland, on the Hunterston Sands 2 km SSW of Fairlie, pargasite occurs in lherzolite inclusions (Alexander et al. 1986). The analcime dolerite (crinanite) at Clauchlands Point, near Lamlash on the Isle of Arran, contains minor brown pargasite (Tindle unpub. EMP data). Brown pargasite, commonly poikilitically enclosing pyroxene, occurs in a pyroxenite xenolith of lower-crustal or possible mantle origin in a Permian olivine-bearing nephelinite dyke at Gribun, on the west coast of the Isle of Mull, Strathclyde (Upton et al. 1998). Gneiss-like inclusions in pink Tiree Marble from a quarry at Balephetrish, Isle of Tiree, contain pargasite together with apatite, biotite, calcite, diopside and scapolite (Hallimond 1947).

Western Isles Edenite, magnesiohastingsite and pargasite occur in a metamorphosed spinel-bearing lherzolite lens within Leverburgh–Rodel metasediments at the head of Glen Rodel, South Harris. Associated species are augite, enstatite and accessory hercynite and magnetite (Livingstone 1976c). Pargasite in a garnet–amphibole pyroxenite, from SW of Leverburgh post office, is a related occurrence (Livingstone 1967). "Honey-brown" pargasite is reported from a metamorphosed limestone at Stuaidh, and a greenish-brown variety occurs in a diopside–pargasite rock near St Clement's church, both at Rodel (Davidson 1943). Elsewhere on South Harris, pargasite occurs in metagabbro at Bolaval–Meaval col (Dearnley 1963) and replacing cummingtonite in paragneiss at Rubha Sgeir nan Sgarbh (Livingstone et al. 1973). Pargasite is also reported from a Lewisian chondrodite–humite-bearing metamorphosed limestone at Bay Steinigie, near Finsbay (Fionnsabhagh) (Dearnley 1961), but the data are unpublished.

Offshore territories

Guernsey Injection of granitic pipe intrusions into dioritic host magma within the late Precambrian Northern igneous complex at Port Soif, northeast Guernsey, provides evidence for magma mingling. Around the margins of the pipes, well formed zoned crystals of green to pale-green actinolite, magnesiohornblende and pargasite grew in response to sudden and marked changes in magma conditions during this event (D'Lemos 1992). Amphiboles in the dioritic and gabbroic rocks of this complex at Beaucette include kaersutite, magnesiohornblende and pargasite (Bishop & French 1982).

Jersey Late Precambrian Cadomian layered diorites associated with gabbro on the southeast coast at Le Nez contain a range of amphiboles considered to be secondary after pyroxene and include actinolite, edenite, kaersutite, magnesiohastingsite, magnesiohornblende and pargasite (Bishop & Key 1983). Pargasite is one of many amphibole species described from an igneous complex (olivine-gabbro to aplogranite compositions) at Sorel Point in north Jersey (Salmon 1991), but the actual data are unpublished. The suite of amphibole species ranges from coarse-grain, strongly zoned, sub-ophitic crystals with clinopyroxene cores, to fine-grain subhedral homogeneous crystals.

parisite

trigonal $Ca(Ce,La,Nd)_2(CO_3)_3F_2$

A rare calcium–cerium–lanthanum–neodymium carbonate fluoride found in accessory amounts in altered Ordovician rhyolites from central Snowdonia, Wales. It is not known if the species is parisite-(Ce) or parisite-(Nd). Parisite is one of many secondary (hydrothermal) accessory species found in these rocks, others being aeschynite, lanthanite, larsenite–fayalite, synchysite, thorite, and possibly gadolinite. They occur as anhedral crystals and intergrowths, in veins and as secondary rims. Associated primary magmatic accessory species include apatite, monazite, titanite, xenotime and zircon. Allanite in these rocks can be either primary or secondary (Howells et al. 1991). Parisite-(Ce) is apparently confirmed from around Gallt y Wenallt, Gwynedd (Anthony et al. 2003), but no details of the occurrence are given.

In Ireland, a single 7 mm-long pseudo-hexagonal pinkish-brown crystal of probable parisite-(Ce) has been found associated with quartz, both lining a cavity in a chalcopyrite–galena-bearing fluorite vein from Martin Bowen's quarry, Inveran, Co. Galway. An energy-dispersive detector fitted to a variable-pressure SEM was used to make the identification (N. Peters pers. comm. 2007).

Pinkish-brown crystal of probable parisite-(Ce) associated with quartz, both lining a cavity in a chalcopyrite–galena-bearing fluorite vein from Martin Bowen's quarry, Inveran, Co. Galway, Ireland. (© D. Whipp)

parkerite

monoclinic $Ni_3(Bi,Pb)_2S_2$

A rare nickel–bismuth–lead sulphide found in quartz–fluorite veinstone on the dumps of North Devon United (East Wheal Friendship) mine, Peter Tavy, Devon. The mine is situated in the metamorphic aureole of the Dartmoor granite. Parkerite occurs as small (<50 µm) grains within a large grain of stannite, clustered together with bismite, native bismuth (also found as inclusions in parkerite), cassiterite, gersdorffite, sphalerite, ullmannite and an unknown copper–lead sulphide selenide, all partially enclosed in fluorite. The dominant sulphide present in the only specimen examined is arsenopyrite, which is associated with cobaltite and a range of bismuth minerals, including aikinite, native bismuth, bismuthinite and ikunolite (Rumsey & Savage 2005).

parkinsonite

tetragonal $(Pb_7Mo,\square)_8O_8Cl_2$

TL Merehead (Torr Works) quarry, Cranmore, Somerset, England

Named after Frank Reginald Derek Parkinson (1928–1993), a mineral collector and dealer, who discovered the mineral (Morse 2002); birth registered at Wandsworth, London. Parkinsonite is a very rare lead–molybdenum oxychloride found at the type locality in fractures and cavities in carbonate vugs within veins of manganese and iron oxide and hydroxide minerals that cut Lower Carboniferous dolomitic limestone. It occurs as compact clusters or patches (<3 mm) of red to purplish-red bladed crystals with an adamantine lustre and is associated with cerussite, chloroxiphite, diaboléite, hydrocerussite, mendipite and wulfenite (Symes et al. 1994). A detailed discussion of the manganese mineralization at Merehead quarry is provided by

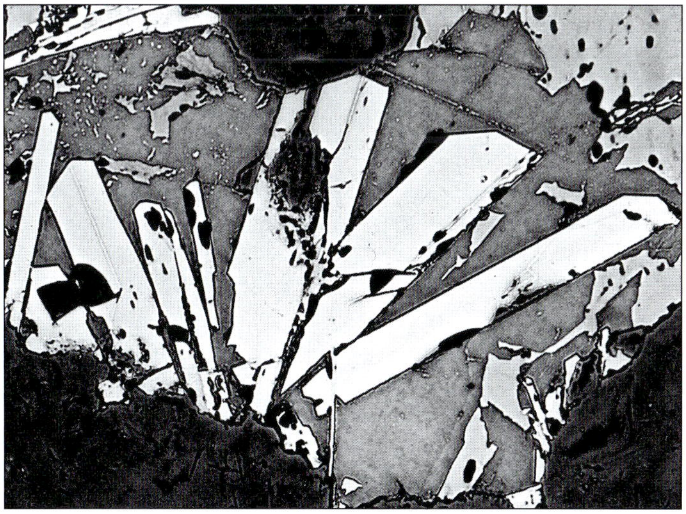

Prismatic crystals of parkinsonite projecting into cerussite and associated with wulfenite (dark grey). Merehead (Torr Works) quarry, Cranmore, Mendip Hills, Somerset; width 1 mm; viewed in reflected light. (© C. J. Stanley, The Natural History Museum, London)

Turner (2007) – see mereheadite. Parkinsonite has also been found at Wesley mine, Westbury on Trym, northwest Bristol, where it again forms part of a lead oxychloride assemblage associated with manganese oxide veins in Carboniferous Limestone (Alabaster 1989b). The lead oxychlorides are thought to have formed by interaction of primary galena with low-temperature metalliferous brines along fractures cutting the original Fe–Mn–(Cu–Pb) orebody (Symes & Embrey 1977, Humphreys et al. 1980), although it is possible that they precipitated directly from a cooling brine (Welch et al. 2000).

parnauite

orthorhombic $Cu_9^{2+}(AsO_4)_2(SO_4)(OH)_{10}\cdot 7H_2O$

A rare hydrated copper arsenate sulphate hydroxide found in arsenate-rich assemblages at Penberthy Croft mine, St Hilary, Cornwall, where it forms micaceous pale- to mid-green aggregates, fan-like sheaves, small spherules and rosette-like groups up to 0.6 mm across. Parnauite is commonly altered and bleached, and it occurs with cassiterite, clinochlore, quartz and occasionally mixite (Camm & Merry 1991b, Betterton et al. 2000). At nearby Croft Gothal mine, parnauite occurs as scattered or isolated green square-to-elongated platy crystals up to 0.5 mm long on iron-stained quartz associated with minor pharmacosiderite; it is a rare species at this locality (J.jBetterton pers. comm. 2006). The finest parnauite in Cornwall appears to be from Morcom's shaft, Ting Tang mine, Carharrack, where dark-green botryoidal aggregates composed of fan-like sheaves of micaceous crystals occur with azurite, liroconite and malachite (Bruce & Aubrey-Jones 1998). Bright-green crystalline parnauite has also been identified on a specimen of spangolite from Wheal Gorland, St Day, 2 km E of Redruth, and now in the Russell collection at the NHM (Ryback & Tandy 1992). Other Cornish locations for parnauite include: Wheal Unity, Gwennap; South Tolcarne mine, Camborne; Mulberry stockworks, Lanivet, and Marke Valley mine, Linkinhorne (all Golley & Williams 1995).

Aggregates of green parnauite laths up to 0.7 mm long, associated with olivenite, are reported on a single specimen of iron-stained quartz from the 60-fathom level dumps of Sandbed mine, Caldbeck Fells, Cumbria (Green et al. 2006a).

Parnauite is a rare species at New Cliffe Hill quarry, Stanton under Bardon, Leicestershire, where it forms deep-green to bluish-green curved sheaf-like aggregates composed of platy crystals up to 0.5 mm across in a native copper–cuprite deposit near the Precambrian/Triassic unconformity (Hubbard et al. 2005).

At Gwaithyrafon mine, SE of Tal-y-bont, Central Wales orefield, Dyfed, Wales, parnauite forms pale bluish-green thin drusy crusts lining small (<3 mm cavities) in quartz and as thin crusts on corroded chalcopyrite. It is associated with brochantite, chrysocolla, connellite, malachite and tyrolite (Rust & Mason 1994).

Tynagh lead–zinc mine, near Killimor, Co. Galway, is the only location in Ireland where parnauite has been found. Here, it is associated with clinotyrolite, tennantite and turquoise (Ryback & Moreton 1992, Ryback & Tandy 1992).

Parnauite stalactites, up to 4 mm long; Mulberry openwork, Lanivet, Cornwall. (© D. I. Green)

parvo-amphibole

In 2004 the IMA revised the classification of the amphibole group and introduced the Na–Ca–Mg–Fe–Mn–Li subgroup after the discovery of new species (Leake et al. 2004). To minimize the number of new species names in the subgroup, the prefix "parvo" was introduced to represent those compositions where the B site contains >0.5 apfu and lithium ≥0.5 apfu. There have been no discoveries of any of these species (parvo-mangano-winchite, parvo-manganotremolite and parvowinchite) in Britain or Ireland, but a few analyses when processed using the Hypercard program AMPH-IMA04 (Mogessie et al. 2004) produced named amphiboles (see Burke & Leake 2005), with parvo prefixes: parvo-actinolite, parvo-cannilloite, parvo-edenite, parvo-ferro-edenite, parvo-magnesiohornblende, parvo-ferrorichterite and parvo-tschermakite. These minerals are all described under their root names (e.g. for parvo-actinolite see actinolite).

parsettensite

monoclinic (pseudo-hexagonal)

$(K,Na,Ca)_{7.5}(Mn,Mg)_{49}(Si,Al)_{72}O_{168}(OH)_{50}\cdot nH_2O$, where $n=24–48$

A rare hydrated potassium–sodium–calcium–manganese–magnesium alumino-silicate hydroxide hydrate briefly reported from Cwm Nantcol, Harlech Dome, Gwynedd, Wales (Bevins 1994). It is usually found, in or peripheral to, manganese ore deposits.

pascoite?

monoclinic $Ca_3V_{10}^{5+}O_{28}\cdot 17H_2O$

A rare hydrated calcium vanadate that is described as a yellow efflorescence on vanadium-rich nodules of Permian age, at Budleigh Salterton, Devon (Wyley 1962). The mineral was originally thought to be metatyuyamunite, but Embrey (1978) and King & Dixon (1971) suggest that it was probably pascoite.

passyite

An impure silica, probably a siliceous sinter.

paulingite (series)

ZEOLITE GROUP

cubic $(Ca_{0.5},K,Na)_{10}(Al_{10}Si_{32})O_{84}\cdot 27–44H_2O$?

A rare hydrated potassium–calcium–sodium aluminosilicate that forms in cavities within a phillipsite-rich tholeiitic basalt near the National Trust reception centre, Giant's Causeway, Co. Antrim, Ireland. Paulingite is accompanied by erionite and forms colourless transparent-or-translucent rhombic dodecahedral crystals confirmed up to 4 mm across (although 10 mm crystals have also been reported); they are undoubtedly the world's largest. Other associated species are heulandite and possibly clinoptilolite (Ryback et al. 1988, Foy 1992). It is not known if the species is paulingite-Ca or paulingite-K (but see below). Paulingite is also noted from Port Ganny and Lacada Point in the Giant's Causeway (Tschernich 1992) and from a derelict quarry at Craigahulliar, Portrush, Co. Antrim (Foy 1992).

paulingite-K

ZEOLITE GROUP

cubic $(K,Ca_{0.5},Na)_{10}(Al_{10}Si_{32})O_{84}\cdot 27–44H_2O$

Paulingite-K has been found in cavities within a Tertiary basalt from the Giant's Causeway, Co. Antrim, Ireland (H. J. Foy and D. I. Green specimens, D. I. Green XRD, Tindle unpub. EMP data). Interestingly, a few analyses are sodium dominant and they point to the possibility of a new species, paulingite-Na. Sodium dominance in paulingite is also noted by Passaglia et al. (2001) in samples from Port Ganny, on the Giant's Causeway. Paulingite-K is also reported from massive basalt flows at Portnaboe, close to the Giant's Causeway (Adair 1987).

paulite

A synonym of hypersthene, a discredited species of the pyroxene group.

peach

A Cornish miners' term for massive chlorite in cassiterite-bearing veins.

peacock ore

A synonym of bornite.

pearceite

monoclinic $Cu(Ag,Cu)_6Ag_9As_2S_{11}$

A rare copper–silver sulpharsenite that forms a series with polybasite (see also antimonpearceite) and is found in quartz veins cutting monzonite of the Ratagain complex, Loch Duich area, Highland Region, Scotland. Pearceite forms <100 µm inclusions in tennantite and galena (Livingstone 1993).

Pearceite occurs as inclusions in galena from the limestone-hosted Cu–Pb–Zn–(Ag) deposit at Tynagh mine, near Killimor, Co. Galway (Boast et al. 1981a).

In the Channel Isles, a Pb–(Ag)-bearing brecciated vein cutting a foliated hornblende granite of Late Precambrian age at Sark's Hope mine, Little Sark, contains a minor suite of silver-bearing minerals that includes acanthite, enargite, pearceite, polybasite, pyrargyrite and tennantite–tetrahedrite. Pearceite either replaces enargite or forms random isolated inclusions in galena. It may itself contain inclusions of acanthite (Ixer & Stanley 1983, Ixer 1992) See antimonpearceite.

pearl spar

A synonym of dolomite and of ankerite, dolomite group, but originally applied to a variety of calcite.

pectolite

triclinic $NaCa_2Si_3O_8(OH)$

A relatively common sodium–calcium silicate hydroxide that forms a series with sérandite and is found as a hydrothermal mineral in cavities in basalt and dolerite.

Pectolite occurs as 1 mm–3 cm-wide veins in gabbro at Dean quarry, 3 km NE of Coverack, St Keverne, Lizard Peninsula, Cornwall. It forms pure-white or creamy-yellow radiating acicular aggregates associated with calcite, prehnite, pyrite and a little chalcopyrite (Seager 1971). At Porthoustock quarry, 1 km N of Dean quarry, pectolite has been found not only in veins but also within the gabbro (Holyer 1997). Analcime and calcite are associated species.

The Shap Blue quarry, Cumbria, is famous for its range of minerals present in metasomatized Borrowdale Volcanic Group rocks within the aureole of the Shap granite. Laumontite and pectolite are species formed in a late hydrothermal assemblage. White crystalline pectolite coats fracture surfaces and encrusts apophyllite in rare epidote–garnet–prehnite–thaumasite-bearing veins (Young et al. 1988, 1998). Other Cumbrian occurrences include: Embleton quarry (Young 1987a); Mosedale on Carrock Fell (Melmore 1920); Caw Burn, near Haltwhistle – spheroids up to 10 cm across in the Whin Sill (Tomkeieff 1929); and the Armathwaite tholeiite dyke, Brampton district (Trotter & Hollingworth 1932).

Pectolite-, calcite-, chlorite- and quartz-bearing veins and joint coatings are found in the Late Carboniferous quartz dolerite of the Whin Sill in the northern Pennines (within County Durham). The mineralization represents the final hydrothermal stages of the sill's development and is noted from: Cambokeels (Cammock Eals) mine, Eastgate, Weardale, where a little white fibrous pectolite encrusts apophyllite; Copt Hill quarry, Cowshill, Weardale, where pale-buff fibrous pectolite is associated with apophyllite; Force Garth quarry, Teesdale, where pale pinkish-buff radiating pectolite forms veins associated with later-formed prehnite (Young et al. 1991a). Pectolite and zeolite species were also collected from High Force quarry, Upper Teesdale, on a Mineralogical Society field excursion (Dunham 1946). Stevensite is reported as an alteration product of pectolite in amygdaloidal structures within the Whin Sill at Barrasford quarry, near Gunnerton, Northumberland (Randall 1959).

A specimen of pectolite, forming crusts and botryoidal masses of off-white radiating crystals associated with apophyllite in calcite-filled joints, is noted from Hendre quarry, Pandy, near Glyn Ceiriog, Clwyd, Wales (Starkey et al. 1991, Bevins 1994). Pectolite has also been found at Gimlet quarry, near Pwllheli, Lleyn Peninsula, Gwynedd (Bevins 1994), and is reported from axinite veins within the Penmaenmawr intrusion in the Graig Lwyd area, Gwynedd (Sargent 1925). The latter occurrence requires further confirmation, as a museum specimen of supposed pectolite from this intrusion has since been shown to be prehnite (Bevins 1994).

Pectolite appears to be most abundant in Scotland. It is accompanied by gyrolite, thaumasite and xonotlite in veins cutting serpentinite in a quarry on the north side of Colla Firth, North Mainland, Shetland, Scotland (Butler et al. 1960, Young et al. 1998). At Orrock quarry, near Burntisland, Fife, where basaltic rocks are quarried for roadstone, pectolite commonly develops in joints and as fissure fillings. It forms botryoidal masses or compact aggregates associated with calcite or prehnite, sometimes in a layered sequence (Green & Todd 2001, Wirth 2003).

Pectolite is noted from a few Highland Region localities in Scotland: in metamorphosed limestones of the Moine Series, Ardgour district, north side of Loch Linnhe (Drever 1939); and intergrown with aegirine in syenite dykes from Rhu More and Camas Eilean Ghlais, Coigach (Tilley 1958). On the Isle of Skye, pectolite occurs in large orbicular masses of pale greyish-green colour within druses in Tertiary basalt at Prince Charlie's Cave and Talisker Bay (Greg & Lettsom 1858, Heddle 1893b, Heddle 1901), in reyerite-bearing amygdales also within Tertiary basalt at Allt Coir' a' Ghobhainn, near Drynoch (Livingstone 1988), and in Durness Limestone horizons within the metamorphic aureole of the Beinn an Dubhaich granite, Broadford area (Tilley 1948b, 1951).

Botryoidal pectolite crystals up to 15mm long from Hendry quarry, Glyn Ceiriog, Clwyd, Wales. (© National Museum of Wales, Cardiff; photographer M. P. Cooper)

In Strathclyde, pectolite occurs on Byne Hill, S of Girvan, within the Girvan–Ballantrae ophiolite complex. It is found in dyke-like doleritic–gabbroic intrusions that have been subjected to low-temperature hydrothermal alteration, and is associated with secondary calcite, chlorite (pennine), diopside, epidote, hibschite (hydrogrossular), prehnite and zoisite (Bloxam 1954, 1964, Holub et al. 1984). After an extensive study of amygdale minerals in the Tertiary basalts of Mull and Morvern, Walker (1971) was able to establish pectolite as one of the relatively high-temperature secondary species in the basalts, in common with okenite, reyerite, tacharanite, tobermorite and xonotlite. Sgùrr Dearg, 9 km W of Duart Point, is one of the localities on Mull where pectolite occurs (Heddle 1893b). At Castle Hill, 3 km WSW of Kilbirnie, pectolite, prehnite and tobermorite are secondary species in a basaltic plug of Carboniferous age (Webb 1971). Interestingly, it was on specimens of colourless transparent crystals of pectolite from Auchinstarry quarrry, Kilsyth, that Livingstone (1976b) found the second occurrence in the world of julgoldite in pockets within a quartz-dolerite intrusion.

White earthy-looking masses several cm long and disseminations of pectolite occur in contact-altered limestone close to a thick sill of quartz dolerite at Chapel quarry, near Kirkcaldy, Fife. Associated species include apophyllite, datolite, grossular and talc (Phemister & MacGregor 1942).

In the Edinburgh area, well formed greenish-grey crystals of pectolite (ratholite)

Pectolite vein from a beach locality on the south side of Lendalfoot Bay, between Ballantrae and Girvan, Strathclyde, Scotland; width 180 mm. (© D. G. Anderson)

were found in large vesicles in an intrusive sheet of Lower Carboniferous dolerite at Ratho quarries (Greg & Lettsom 1858, Heddle 1901, Rudler 1905) and from a similar setting on Corstorphine Hill the variety walkerite (probably a mixture of pectolite and stevensite) was first described (Heddle 1880b).

In Ireland, pectolite is reported from limestone xenoliths within a eudialyte–neptunite–quartz syenite at Barnavave, Carlingford, Co. Louth (Nockolds 1944, 1950), and from wollastonite–xonotlite-bearing flint nodules within chalk that have been metasomatized by a dolerite intrusion at Scawt Hill, Larne, Co. Antrim (Tilley & Alderman 1934). In both cases, pectolite replaces wollastonite.

pekoite

orthorhombic $PbCuBi_{11}S_{18}$

A rare lead–copper sulphobismuthite found in quartz veins cutting monzogranite at Loch Shin, 2 km W of Lairg, Highland Region, Scotland. Elsewhere in the monzogranite, mineralization is disseminated and mainly restricted to fluorite, molybdenite and pyrite. However, in the veins, molybdenite does not occur and the other species are accompanied by native bismuth, chalcopyrite, covellite, galena, sphalerite and a suite of bismuth sulphosalts: aikinite, berryite, gladite, hammarite, lindströmite, krupkaite, pekoite and tetradymite (Lowry et al. 1994).

pencil ore

A variety of hematite taking the form of elongate conical masses derived from broken botryoidal kidney ore. The name has also been used as a synonym of graphite, a species once used as a raw material in the making of Cumberland pencils.

pennantite

CHLORITE GROUP

monoclinic $Mn^{2+}_5Al(Si_3Al)O_{10}(OH)_8$

TL Benallt mine, Rhiw, Lleyn Peninsula, Gwynedd, Wales

Pennantite in manganese ore; Benallt mine, Rhiw, Lleyn Peninsula, Gwynedd, Wales; width 2.25 mm; viewed in PPL. (© K. Gardner)

Named after the Welsh traveller, zoologist and mineralogist, Thomas Pennant FRS (1726–1798); author of *Native fossils* (1757); born at Whitford, near Holywell, Clwyd. Pennantite is a rare manganese aluminosilicate hydroxide that pervades the manganese ores at Benallt mine. It forms poorly crystallized orange patches and small specks in the matrix of the ore. Less commonly, it occurs in thin veinlets with alleghanyite, analcime, banalsite, cymrite, ganophyllite, jacobsite, pyrochroite, pyrophanite, tephroite and a colourless mica (Campbell Smith et al. 1946, Dunham et al. 1978). Pennantite also occurs in manganese ore of the Hafotty Formation, Harlech Dome, Gwynedd (Bennett 1987b), and possibly at Penrhyn slate quarry, Bethesda, Gwynedd (Bevins 1994). The "grovesite" described by Bannister et al. (1955) is in fact pennantite.

pennine or penninite

A pseudo-trigonal variety of clinochlore, chlorite group.

pennite

A variety of hydrodolomite (a mixture of hydromagnesite and calcite).

penroseite

PYRITE GROUP

cubic $(Ni,Co,Cu)Se_2$

A very rare nickel–cobalt–copper selenide found in a gold-bearing carbonate vein cutting Middle Devonian limestones at Hope's Nose, Torquay, Devon. The selenide assemblage includes clausthalite, eucairite, fischesserite, klockmannite, naumannite, tiemannite, trüstedtite, tyrrellite and umangite in a calcite dominated gangue. Penroseite forms abundant inclusions in clausthalite (grains up to 100-200 μm), or as a complex intergrowths (grains up to 50 μm) in clausthalite (Stanley et al. 1990a).

PENTLANDITE GROUP

Cubic sulphides with the general formula AB_8X_8, where A = Ag, Cd, Co, Fe, Mn, Ni, Pb; B = Co, Cu, Fe, Ni; and X = S, Se. Argentopentlandite, cobalt pentlandite and pentlandite are members of the group found in Britain or Ireland.

pentlandite

PENTLANDITE GROUP

cubic $(Fe,Ni)_9S_8$

TL Craignure mine, near Inveraray, Strathclyde, Scotland

Named after Joseph Barclay Pentland (1797–1873), an Irish diplomat, natural historian and traveller, who first noted the mineral in nickel ores at Sudbury, Ontario, Canada; born at Ballybofey, Co. Donegal. Pentland trained as a geologist in Paris and was the British consul-general in Lima, Peru (1826–1828), and in La Paz, Bolivia (1836–1839) (Cooper 2007). Pentlandite is a common iron–nickel sulphide that forms a series with cobalt pentlandite and is usually found intimately associated with pyrrhotite in ultramafic rocks.

In Cornwall, pentlandite was found in some quantity, associated with vivianite, at Wheal Jane, Kea (Collins 1892, Rudler 1905, Criddle & Stanley 1993). Dunite and lherzolite rocks from near Carrick Luz in the eastern part of the Lizard complex, Lizard Peninsula, contain altered pentlandite and chalcopyrite associated with rare laurite, potarite and sperrylite (Gunn & Styles 2002). At Greystone quarry, Lezant, near Launceston, pentlandite occurs with chalcopyrite, galena, gersdorffite, jamesonite, pyrite and pyrrhotite in carbonate-rich altered dolerite (Power & Scott 1995).

Pentlandite is apparently uncommon in Wales (but see cobalt pentlandite), with only two confirmed occurrences: single elongate lamellae of pentlandite in pyrrhotite from Cesailgwm and Clogau, Bontddu, near Dolgellau, Gwynedd (Gilbey 1968); and rare exsolution flames of pentlandite in pyrrhotite within dolerite from Carn Menyn, Mynydd Preseli, Dyfed (Thorpe et al. 1991).

On Unst, Shetland, Scotland, pentlandite occurs in a wide range of rocks related to the Shetland ophiolite complex. Prichard et al. (1994) report it in dolomite–magnesite–serpentinite–talc rocks of the basal thrust and in dunite pods, chromitite, harzburgite, pyroxenite and wehrlite. A ruthenium-rich variety of pentlandite (crystals up to 30 μm across), associated with irarsite or laurite and rarely containing inclusions of native osmium, occurs at Cliff and at Harold's Grave, two localities well known for chromite-rich samples containing an unusual PGM assemblage (Prichard & Tarkian 1988, Prichard et al. 1988, Gunn & Styles 2002). It forms grains 0.01–1 mm across in samples from the Cliff area and also occurs in rocks with no apparent PGM enrichment (Gunn et al. 1985). Pentlandite from a dunite to the W of Hagdale has altered to heazlewoodite during serpentinization (Maynard et al. 1997). A compilation of pentlandite data from Unst is provided by Maynard (1993).

In the Highland Region of Scotland, accessory amounts of pentlandite are reported from Loch Bad nam Ban, 3 km SW of Lochinver, where it forms flame-like intergrowths with pyrrhotite in a biotite-diorite dyke cutting Lewisian rocks (Stumpfl 1961), and from Meall na Cuilce, to the W of Loch Coruisk, Isle of Skye, where it occurs with chalcopyrite in chromitites from the base of the Peridotite Series of the Cuillin complex (Pirrie et al. 2000a). Pentlandite (10–60 μm across), bornite and chalcopyrite are the major sulphide species in chromite-rich rocks from the layered central complex on the Isle of Rum. They are associated with a very varied and diverse PGM assemblage that includes gold (electrum), hollingworthite, platarsite, sperrylite and several unnamed phases (Dunham & Wilkinson 1985, Butcher et al. 1999, Power et al. 2000a). Dykes containing 40% sulphides (of which, 65% pyrrhotite, 30% chalcopyrite, 5% pentlandite) from the nearby West Sgaorishal plug, Isle of Rum, are also associated with a range of PGM species (Power et al. 2003).

On the Isle of Mull, pentlandite forms grains < 100 μm across with chalcopyrite and arsenic–platinum sulphides in chromitite within the Ben Buie gabbro at Caigeann Doire nan Cuilean (Pirrie et al. 2000a). Elsewhere in Strathclyde, pentlandite occurs at: Coille-bhraghad mine, near Inveraray (Berridge 1969, Willan 1980); Craignure mine, 12 km SW of Inveraray (Dufrénoy 1856, Berridge 1969, Livingstone 1990b); Leadhills–Wanlockhead area – inclusions in alluvial gold (Leake et al. 1998); Pinbain Bridge – rare blebs associated with awaruite and heazlewoodite in chromitite, belonging to the Girvan–Ballantrae ophiolite complex (Power & Pirrie 2004).

Pentlandite is a minor constituent of a hydrothermal quartz–sulphide ±gold vein deposit at Calliachar Burn, 4 km from Aberfeldy, Tayside Region, where it forms flame structures intergrown with pyrrhotite (Ixer et al. 1997). Barite- and barium-silicate-bearing calcareous schist near the top of the Dalradian Ben Eagach Schist at Foss, 7 km NW of Aberfeldy, Tayside Region, contains minor pentlandite in a sulphide assemblage dominated by galena, pyrite, pyrrhotite and sphalerite (Moles 1985). Traces of pentlandite have also been found with chalcopyrite, chromite, chromium-rich magnetite, chromium-rich mica and pyrrhotite in a varied suite of rocks belonging to the Dalradian Ben Lui Schist in the Auchtertyre–Challum area, to the E of Tyndrum, Central Region (Fortey & Smith 1986). Mineralization at the last two locations is considered to be of the sedimentary-exhalative type.

In the Grampian Region of Scotland, pentlandite is found in mafic–ultramafic cumulates within the Ordovician Arthrath intrusion, near Ellon, 25 km N of Aberdeen. The disseminated sulphide assemblage is predominantly pyrrhotite with subordinate chalcopyrite and pentlandite (Fletcher et al. 1997, Gunn & Styles 2002, McKervey et al. 2007). On the southeastern margin of the Knock gabbroic intrusion, near the farms of Littlemill and Auchencrieve, 2 km NW of Rothiemay, N of Huntly, pentlandite occurs with chalcopyrite and pyrrhotite in discontinuous lenticular sulphide bodies associated with olivine cumulates, contaminated gabbros and metasediments (Fletcher & Rice 1989, Gunn & Styles 2002). Irregular discordant dykes of graphite- and sulphide-bearing orthopyroxene-rich pegmatites intruding cumulate rocks at Bin quarry and its environs, W of Huntly, contain a sulphide assemblage of pyrrhotite (90–100%) with minor chalcopyrite and pentlandite (Gunn & Shaw 1992).

A small deposit of chalcopyrite–nickeline–pentlandite–pyrrhotite ore within the base of a sill-like diorite body at the western edge of the Cairnsmore of Fleet granite in Palnure Burn, Talnotry, 5 km NE of Newton Stewart, Dumfries & Galloway, contains pentlandite intimately associated with pyrrhotite, either along grain boundaries or as flame-like intergrowths. Associated species include bismuthinite, galena, cobalt-rich gersdorffite, gold (electrum), Bi–Pb–Re sulphide, silver and bismuth tellurides, and the PGM species froodite, irarsite, merenskyite, michenerite and sperrylite (Dunham et al. 1978, Stanley et al. 1987, Power et al. 2004). At Glencrieff Burn, Wanlockhead, Dumfries & Galloway, flame-like exsolution lamellae of pentlandite within pyrrhotite have been found in samples of greywacke adjacent to several small quartz veins (Nicholson & Banks 1987).

Reports of pentlandite in Ireland are from surficial deposits, where it has been found as inclusions in alluvial gold at Glenkeen bridge, Carrownisky River, in the Cregganbaun shear zone, Co. Mayo (Leake et al. 1993, Chapman et al. 2000c).

Sulphide veinlets cutting diorite of the Northern igneous complex on Guernsey, Channel Islands, contain trace amounts of pentlandite. Chalcopyrite, cubanite, galena, ilmenite, magnetite, molybdenite, pyrite, pyrrhotite and sphalerite are all noted from these veins (Ixer 1992).

penwithite

TL Wheal Owles, Penwith, Cornwall, England

A synonym of neotocite. See Collins (1878a, 1879).

PERICLASE GROUP

Cubic oxides with the general formula A2+O, where A2+O = Cd, Fe, Mg, Mn, Ni. Manganosite and periclase are members of the group found in Britain or Ireland.

periclase

PERICLASE GROUP

cubic

MgO

An uncommon magnesium oxide found in thermally metamorphosed limestones at Grange Irish, S of Barnavave, Carlingford, Co. Louth, Ireland. Limestone has recrystallized to produce a brucite–forsterite–spinel marble. Pre-existing cubo-octahedral outlines of periclase still remain, but these are now distorted by volume change following hydration of the species as it was replaced by fibrous and acicular brucite. Apatite, cuspidine (custerite), melilite, monticellite and phlogopite are associated species (Osborne 1932b, Nockolds & Vincent 1947). It is unclear whether any primary periclase still remains.

Periclase is preserved at Camas Mòr, Isle of Muck, Highland Region, Scotland, where an olivine-gabbro dyke has thermally metamorphosed Jurassic limestones. Some of the resulting carbonate rocks contain rounded yellow grains of periclase along with calcite, brucite, monticellite and spinel. Partial replacement of periclase to a brown ferrobrucite and opaque iron ore is noted (Tilley 1947).

At Kilbride, near Broadford, Isle of Skye, Highland Region, contact skarns have been produced where the Tertiary Beinn an Dubhaich granite has intruded dolomitic limestones. In the metamorphic aureole surrounding the skarn zones are forsterite–periclase–spinel marbles that, near the contact with skarn, become depleted in periclase. Fluorite, fluoborite and ludwidgite occur in accessory amounts in these marbles (Tilley 1948b, 1951, Holness 1992).

peridot

A green gem variety of forsterite, olivine group, but the name has also been used to describe a greenish-yellow variety of tourmaline.

PEROVSKITE GROUP

Orthorhombic or monoclinic (pseudo-cubic) oxides with the general formula ABO_3, where A = Ca, Ce, Na, Sr; and B = Nb, Ti, Fe. Perovskite is the only member of the group found in Britain or Ireland.

perovskite

PEROVSKITE GROUP

orthorhombic (pseudo-cubic)

$CaTiO_3$

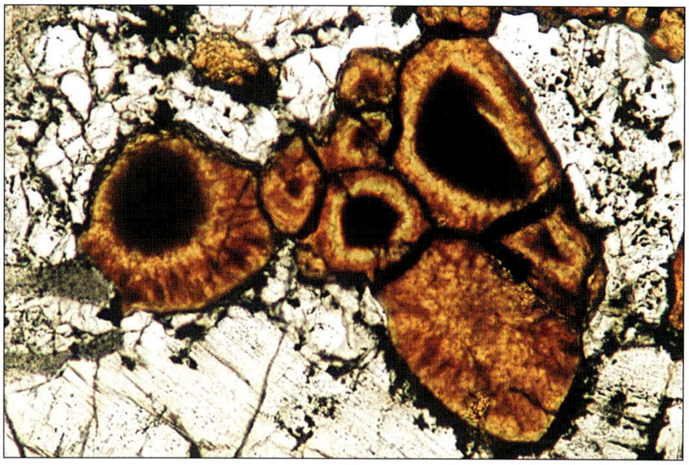

Cluster of zoned perovskite in a bredigite–larnite rock. Scawt Hill, near Larne, Co. Antrim, Ireland; width 1.3 mm; viewed in PPL. (© K. Gardner)

An uncommon calcium–titanium oxide found in the chalk/dolerite contact zone at Scawt Hill, Larne, Co. Antrim, Ireland. Perovskite is an accessory mineral found in coarsely crystalline hybrid rocks, in which the main components are titaniferous augite and gehlenite (melilite). Calcite, larnite, merwinite, spurrite and spinel are associated species. These pseudo-igneous rocks were produced when chalk was assimilated by doleritic magma (Tilley 1929a,b). Elsewhere in the contact zone, rocks consisting mainly of aegirine, melilite and wollastonite contain perovskite in much greater abundance (Tilley & Harwood 1931).

A rare rock type, containing accessory amounts of perovskite and representing an extreme example of silica metasomatism of chalk, is a larnite–spinel rock found in another chalk/dolerite contact zone associated with a Tertiary volcanic plug at Carneal, Larne, Co. Antrim. The spinel is ferrian chlorospinel and magnetite, and minor amounts of calcite, hibschite, melilite and spurrite are also present (Sabine et al. 1968, Sabine & Young 1975). In thin section, perovskite is deep yellow to yellowish brown and it forms sub-rounded grains, often with compositional zoning.

Perovskite is also found as an accessory component of brucite–forsterite–spinel marbles and in a rock type whose essential components are calcite, diopside, phlogopite, spinel (pleonaste) and vesuvianite. These rocks occur in a limestone/gabbro contact zone close to Grange Cross, Carlingford, Co. Louth (Osborne 1932b).

In Scotland, accessory amounts of perovskite are found in a marble immediately adjacent to a gabbroic intrusion at Camas Mòr, Isle of Muck, Highland Region. The rock is dominated by gehlenite, monticellite and spinel, and minor magnetite and pyrrhotite are also present (Tilley 1947). Perovskite is also reported from a high-temperature calcic skarn near the contact of the Cuillin gabbro complex at Camasunary Bay, Strathaird Peninsula, Isle of Syke (Beard & Drake 2007).

Perovskite possibly occurs in: alnöite monchiquite dykes from Naversdale, 3 km N of Orphir, and Rennibuster Point, near Kirkwall, both on Mainland, Orkney Isles, Scotland (Flett 1900); olivine minette boulders from Holmead farm, Loxbeare, near Tiverton, Devon, England (Tidmarsh 1932); dolerite dykes from the Llyn Padarn area, Gwynedd, Wales (Elsden 1904); pyroxenites and gabbros from the Mona Complex, Holy Island, Anglesey, Gwynedd (Greenly 1919); and altered palagonite tuffs from Afon Glaslyn, Cwm Dyli, Snowdon, Gwynedd (Williams 1927). The last three occurrences at least are probably misidentifications of titanite (Bevins 1994).

perrierite-(Ce)

monoclinic $(Ce,La,Ca,Sr)_4(Fe^{2+},Mg,Mn^{2+})(Ti,Fe^{3+})_4Si_4O_{22}$

An REE–calcium–strontium–iron–magnesium–manganese–titanium silicate, closely related to chevkinite and found in accessory amounts in granodiorite from the Southern Cruachan unit of the Etive granite complex, near Bonawe quarry, Strathclyde, Scotland. It occurs with other accessory species – apatite, titanite, thorite and zircon – and had been described as chevkinite (Barritt 1983). According to the chevkinite–perrierite discriminant diagram proposed by Macdonald & Belkin (2002), the composition is clearly that of perrierite-(Ce).

perthite

An intimate mixture of alkali and plagioclase series feldspars.

petalite

monoclinic $LiAlSi_4O_{10}$

A rare lithium aluminosilicate found in lithium, fluorine and rare- element (Cs–Rb–Nb–Ta)-enriched aplites, pegmatites and granitic rocks. Petalite occurs in aplite and crosscutting pegmatite veins (up to 5 cm wide) at Meldon, Okehampton, Devon. The locality is 1 km NW of the main Dartmoor granite. Petalite has been found where the aplite is cut by the Red-a-ven mine and in the old aplite quarry. Major associated species are albite, muscovite, lepidolite, orthoclase and quartz; accessory species are apatite, native arsenic, beryllonite, elbaite, fluorite, herderite, manganocolumbite, microlite, montebrasite, pollucite, stibiconite and topaz. Petalite decomposes to form montmorillonite (McLintock 1923, von Knorring 1951, Chaudry & Howie 1970, 1973a, von Knorring & Condliffe 1984). Descriptions of petalite, eucryptite, pollucite and spodumene by Drysdale (1985, 1991) are perhaps less reliable, as they are based on samples from the Kingsbury collection at the NHM. This collection is known to contain specimens with dubious provenance (Ryback et al. 1998, 2001).

petitjeanite

triclinic $Bi_3O(OH)(PO_4)_2$

A very rare bismuth phosphate hydroxide that forms a series with preisingerite and schumacherite and is found in cavities within quartz veinstone on the dumps of Carrock mine, Caldbeck Fells, Cumbria. The species occurs as minutely radiating deep-brown sub-mm crystals associated with a pale yellowish-green russellite-like phase, bismite and primary bismuthinite. It probably forms from the secondary alteration of the primary bismuth minerals at the site. Currently only a single specimen is known, which resides at the NHM (Rumsey et al. 2007).

petuntze

Green microcline feldspar and quartz.

petzite

cubic Ag_3AuTe_2

Petzite is a very rare silver–gold telluride found in a brecciated gold-bearing quartz–pyrite vein (Halliday's vein), Tyndrum, Central Region, Scotland, where it occurs as ≤20 μm-diameter inclusions in galena with cervelleite (described as phase C: see cervelleite), hessite, sylvanite and an unidentified Ag–Te–S phase. These gold and silver minerals are associated with chalcopyrite, native gold (electrum with high silver content) and sphalerite. The silver–gold mineralization is thought to be attributable to hydrothermal activity related to copper–molybdenum mineralization associated with Late Caledonian granites (Pattrick et al. 1988, Curtis et al. 1993).

Petzite is also tentatively identified from Afon Wen, Dolgellau goldbelt, Gwynedd, Wales, where an 8 μm-diameter grain of petzite associated with possible calaverite has been found included within an alluvial gold grain (Leake et al. 1993).

phacolite

A synonym of chabazite-Ca, zeolite group. The term refers to a crystal habit of lens shape rather than the normal cubic chabazite habit, and arises from twinning around the *c*-axis.

pharmacolite?

monoclinic $CaHAsO_4{\cdot}2H_2O$

A rare hydrated calcium–hydrogen arsenate found coating brickwork on an arsenic calciner at Devon Great Consols mine, near Gunnislake, Devon. Pharmacolite is white, with a colloform habit. It is a secondary species that has precipitated from arsenic-rich fluids derived from the leaching of arsenic-impregnated masonry and is associated with gypsum and the arsenic-bearing species, arsenolite and scorodite. Colloform growths of pharmacolite also coat brickwork and plasterwork on an

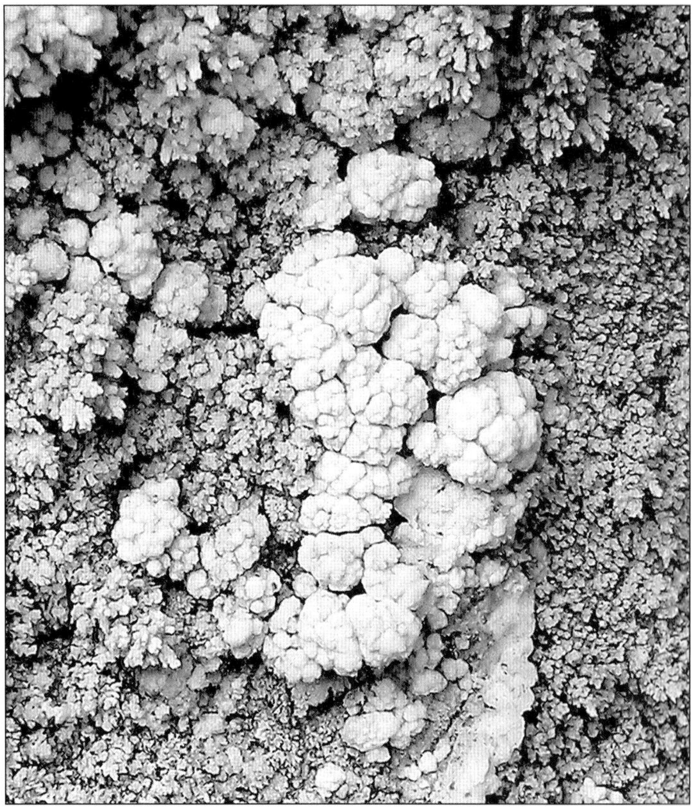

Colloform pharmacolite surrounded by finer-grain gypsum, coating brickwork on an arsenic calciner at Devon Great Consols mine, near Gunnislake, Devon; width 20 cm. (© M. R. Power, Camborne School of Mines)

arsenic calciner at Roseworthy, near Camborne, Cornwall, where the species forms large undulating sheets several metres wide and up to 5 cm thick, associated with gypsum and the arsenic-bearing species arsenolite, haidingerite and weilite. The colloform growths typically comprise concentric alternations of pharmacolite and gypsum, which probably represent seasonal growth, with the pharmacolite precipitating during relatively dry periods (M. R. Power pers. comm. 2004). In these cases, pharmacolite is technically not a mineral and it should more correctly be regarded as an anthropogenic substance or synthetic equivalent (Nickel 1995).

Pharmacolite is also described with picropharmacolite, weilite and wapplerite from Wanthwaite mine, St John's in the Vale, near Threlkeld, Cumbria (Young 1987a). However, this and other rare species described from Wanthwaite mine are from the Kingsbury collection at the NHM and their provenance is in doubt (Ryback et al. 1998, 2001, Green & Briscoe 2002).

pharmacosiderite

cubic $KFe^{3+}_4(AsO_4)_3(OH)_4{\cdot}6\text{–}7H_2O$

TL Carharrack mine, Gwennap, and Tincroft mine, Pool, Illogan, both Cornwall, England

An uncommon hydrated potassium–iron arsenate hydroxide found as an oxidation product of arsenic-bearing sulphides. In Cornwall, pharmacosiderite is green and it forms sharply defined cubic crystals, from which the discredited name "cube ore" is derived (Rudler 1905). At Penberthy Croft, St Hilary, pharmacosiderite commonly forms individual or intergrown groups of cubic crystals up to 1 mm on edge. Crystals with olive-green to yellowish-brown colouration are translucent-to-transparent and lustrous (Camm & Merry 1991b, Betterton et al. 2000). At Ting Tang mine, Carharrack, Gwennap, pharmacosiderite was widespread as apple-green to brownish or yellowish cubes up to 3 mm on edge (Bruce & Aubrey-Jones 1998). Fine crystals associated with olivenite and scorodite have been found at South Terras mine, St Stephen-in-Brannel (Collins 1876b, 1911a, Dines 1956). At Cligga Head mine, Perranzabuloe, it occurs in vugs within massive quartz and in association with massive arsenopyrite. Crystals are mint green and up to 2 mm across (Chick 1987). Other occurrences of pharmacosiderite include, in Cornwall: Botallack mine, St Just – uncommon apple-green to yellow cubes usually associated with arsenopyrite (Greg & Lettsom 1858, Bancroft & Weller 1993); Carn Brea, near Redruth (Greg &

Cubic crystals of pharmacosiderite to 3 mm; almost certainly from Wheal Muttrell, Gwennap, Cornwall. (© D. I. Green)

Lettsom 1858); Carharrack mine, Gwennap (Collins 1892, Penhallurick 1997); Hingston Down Consols mine, 2 km W of Gunnislake, Calstock (Russell 1910a); Old Beam mine, SW of Bugle, near St Austell – associated with olivenite and wavellite in the oxidized part of cassiterite-rich veins (Dines 1956, Weiss 1994); Wheal Falmouth, Kea (Collins 1892); Wheal Gorland, St Day, 2 km E of Redruth (Greg & Lettsom 1858, Carruth 1981, Rust 1981b); Wheal Jane, Kea, near Truro (Greg & Lettsom 1858, Rust 1982d); Wheal Pendarves (Tryphena lode), Camborne – in an oxidized quartz lode associated with barite, churchite, connellite, native copper, cuprite, malachite, olivenite and scorodite (Barstow & Cooper 1982); Wheal Unity, St Day, Gwennap (Collins 1892). At Hemerdon Bal, Plympton, Devon, small but exemplary specimens of brilliant apple-green pharmacosiderite associated with scorodite have been found (Robson 1949b, Allen 2002).

In the Caldbeck Fells, Cumbria, pharmacosiderite is described from a few locations: as minute pale-green cubes on spongy quartz from the dry tributary gill to the E of Brandy Gill (Cooper & Stanley 1990); as sparkling crusts of small crystals on spongy quartz impregnated with iron and manganese oxides at Burdell Gill (Greg & Lettsom 1858, Carruth 1981, Cooper & Stanley 1990, Green & Briscoe 2002, Neall & Leppington 2003); on the Emerson vein dumps of Carrock mine (Cooper & Stanley 1990); as pale-green masses and small transparent cubes in altered veinstone from an old working on the Deer Hills quartz vein (Cooper & Stanley 1990, Symes et al. 1990); Dry Gill (Davidson & Thomson 1951); as pale-green crystals on massive scorodite from Nether Row Brow (Cooper & Stanley 1990); and Wet Swine Gill (Neall & Green 2001b, Green & Briscoe 2002). Further confirmation is required for the reported occurrences of pharmacosiderite in the Caldbeck Fells from Ingray Gill and Potts Gill barite mine, and from Wanthwaite mine, St John's in the Vale, near Threlkeld, Cumbria (Kingsbury & Hartley 1960, Hartley 1984, Young 1987a), as the descriptions are based on samples from the Kingsbury collection now at the NHM. The provenance of many samples in this collection has been called into question by Ryback et al. (1998, 2001), Green & Briscoe (2002) and Green et al. (2006a,c). However, a specimen of pharmacosiderite from Potts Gill mine, now in the NMW, does appear to be genuine (T. F. Cotterell pers. comm. 2006). New analyses of samples labelled "pharmacosiderite" from the opencut above Sandbed mine have proved to be barium-pharmacosiderite; however, a few pale-green pseudomorphs after an unknown phase are pharmacosiderite intergrown with scorodite (Green et al. 2006a).

Pharmacosiderite is also found in barite gossan from Engine Vein mine, Alderley Edge ridge, Cheshire, where it forms orange cubes up to 2 mm across associated with antlerite, native copper, cuprite, langite and posnjakite (Braithwaite 1994).

Additional occurrences include: Gwaithyrafon mine, SE of Tal-y-bont, Central Wales orefield, Dyfed, Wales, forming small colourless to pale-green or yellow cubic crystals on goethite and quartz (Bevins 1994, Rust & Mason 1994); Needle's Eye area, SW of Caulkerbush, Dumfries & Galloway, Scotland – in uranium-rich veins containing a wide range of supergene species (Green & Todd 2001); Tynagh mine, near Killimor, Co. Galway, Ireland (Ryback 1988a); South Bradda mines, north side of Port Erin Bay, Isle of Man – small green transparent cubic crystals up to 0.2 mm on edge, with scorodite on fracture surfaces and cavities in quartz veinstone (Starkey 1998b).

phaunouxite

triclinic $Ca_3(AsO_4)_2{\cdot}11H_2O$

An exceedingly rare hydrated calcium arsenate from the disused Cu–(Co) mine at Muckross, Killarney National Park, Co. Kerry, Ireland. Post-mining oxidation in the mine waste has given rise to secondary arsenates, carbonates and sulphates, six species of which proved new to Ireland (alloclasite, monohydrocalcite, parasymplesite, phaunouxite, rauenthalite and smolyaninovite). Phaunouxite forms groups of white, translucent, acicular crystals associated with erythrite, rauenthalite and monohydrocalcite; it also occurs as 0.5–2 mm sprays (Ryback & Moreton 1993, Moreton et al. 1999, Green & Moreton 2001).

Phaunouxite (or rauenthalite) has also been described from the metamorphic aureole of the Dartmoor granite at Ivy Tor mine, Belstone, Meldon area, near Okehampton, Devon. A piece of veinstone from the mine dump, when broken open, revealed pink dolomite gangue and a primary löllingite ore containing a cavity lined with white radiating spherules of talmessite, often mixed with rauenthalite or phaunouxite. Rauenthalite can arise by the reversible dehydration at room temperature of phaunouxite, depending upon the humidity of the surrounding air, so the original phase could have been either of these species. Other minerals described from the mine include arsenopyrite, axinite, native bismuth, bismuthinite, chalcopyrite, chlorite garnet, marcasite, quartz, scheelite, and a little cobalt ore (Braithwaite et al. 1996).

phenakite

trigonal Be_2SiO_4

A rare beryllium silicate found in pegmatites, greisens and hydrothermal veins.

Russell (1911a) describes examples of phenakite from Cornwall: Cheesewring quarry, SW of Henwood, Linkinhorne – three crystals of colourless prismatic phenakite (largest 1.5×2.5 mm), closely associated with fluorite, doubly terminated quartz and rare bertrandite, were found on a dark-green chloritic crust from a joint plane in a coarse-grain muscovite–tourmaline granite; South Phoenix mine, Linkinhorne – prismatic crystals to 4.5 mm on cassiterite–tourmaline-bearing quartz veinstone, either perfectly colourless or yellowish from a film of iron oxide; St Agnes – wine-yellow striated prismatic crystals to 7×2 mm on a quartz specimen, the crystals becoming perfectly colourless and transparent on being rubbed between the fingers, the colour evidently attributable to an external film; South Crofty mine, Camborne – a remarkable specimen with white radiating prismatic crystals up to 2.5 cm long from the mine dumps, the matrix consisting of dark-green chlorite, containing arsenopyrite and cassiterite with minor quartz and fluorite; St Day United mines, Gwennap – two specimens from a wolframite lode, now in the collection of the NHM and labelled apatite, were recognized to be exceedingly fine specimens of phenakite, one specimen consisting of a mass of long quartz crystals, coated with globular chlorite and studded with beautiful crystals of phenakite, the other consisting of a mass of pale-green to colourless cubes of fluorite, on which are sprinkled many exceedingly beautiful crystals of colourless and transparent or slightly turbid phenakite; Wheal Gorland, St Day, 2 km E of Redruth – a single well defined specimen of phenakite, showing prismatic crystals up to 9 mm long, associated with chalcopyrite, chlorite, pale-green fluorite and quartz, and sphalerite containing plentiful small crystals of arsenopyrite.

Phenakite is also noted from Trolvis quarry, near Longdowns, Carnmenellis, Cornwall, from a pegmatitic pod in granite (Harker 1973) and from a very inaccessible rocky inlet known as Stamps and Jowl Zawn, at the foot of Roscommon Cliff, under Wheal Cock, St Just, Cornwall (Russell 1920a). A further description of phenakite in drusy cavities within granite from Cheesewring quarry is given by Green et al. (1996a). Other reports of phenakite in Cornwall are based on samples supposedly collected by A. W. G. Kingsbury, and their provenance requires further confirmation. Ryback et al. (1998, 2001) provide convincing evidence that some so-called British samples in the Kingsbury collection (now at the NHM) came from as far afield as Chile, Mexico, Germany and Greece. Such records include Wheal

Cock and the Wheal Edward section of Wheal Owles, both near Botallack, St Just (Kingsbury 1961, 1964), and Wheal Kitty, St Agnes (Kingsbury 1954).

In Scotland, pegmatitic veins cutting syenite at Sgòrr Chaonasaid, Ben Loyal, Highland Region, contain colourless crystals of phenakite up to 1.5 cm long. Albite, magnetite, microcline and tourmaline are associated species (Macpherson & Livingstone 1982). Phenakite is also noted in a stream-sediment concentrate from the southwestern flank of Cam Eas, Cairngorm Mountains, Grampian, and from the northeastern slopes of Ben Macdui, also in the Cairngorms. A 1 cm-long phenakite crystal group from the latter location is in the private collection of Gordon Sutherland (Livingstone 2002).

phengite (series)

MICA GROUP

A compositional range of dioctahedral micas between, or close to, the joins muscovite–aluminoceladonite and muscovite–celadonite. Phengite is an iron-rich variety of muscovite that does not define an end-member composition and is therefore not a true species. However, it has been elevated to series status in the reclassification of the micas (Rieder et al. 1998) and as such is included in *Fleischer's glossary of mineral species* (Mandarino & Back 2004). Fleet (2003) indicates that phengite often contains 3.1 to 3.3 atoms per formula unit of silicon – compositions typically associated with low-grade, low- and medium-*P–T* metamorphic rocks or in high *P–T* and ultra-high *P* metamorphic rocks the silicon ranges from 3.2 to 3.6 atoms per formula unit.

Phengite has been identified in Moine mylonites in the Dundonnell and Knockan parts of the Assynt–Ullapool section of the Moine thrust zone, Highland Region, Scotland. The phengite grew during Moine thrust shearing and is associated with albite and potassium feldspar (Freeman et al. 1998). In a study of thermal metamorphism of Dalradian greenschists caused by the intrusion of the Sidhean Sluaigh intrusion, 3 km SW of Strachur, Loch Fyne, Strathclyde (a dolerite feeder pipe), Worden et al. (1987) were able to show that phengite decomposed to produce pseudomorphs of biotite, mullite, sanidine and spinel. Because of the fine grainsize of the reaction products, a transmission electron microscope was required for the analysis. Phengite breakdown reactions have also been examined in pelitic xenoliths in a Tertiary dolerite sill that outcrops on the south coast of the Ross of Mull, between Scoor and Uisken, Isle of Mull, Strathclyde (Brearley 1986). Around Loch Katrine in the Aberfoyle district, Central Region, very pale-green to pink phengite occurs in regionally metamorphosed Dalradian metapelites and, together with data from chlorite, microcline and siderophyllite, has been used to model the reactions involved in increasing metamorphic grades (Mather 1970). The Foss stratiform barite–base-metal deposit near Aberfeldy, Tayside Region, consists of barite, Ba-muscovite and Ba-feldspar (celsian) associated with disseminated-to-massive pyrite, sphalerite and galena in cherts, schists and carbonate rocks. The muscovite is said to show a broad range of phengitic substitutions (Moles 1985). Zoned and recrystallized phengite is common in Dalradian metasediments between Aberfeldy, Central Region, and Stonehaven, Grampian (Dempster 1992).

Chrome phengite is described with other chromium-rich minerals from a peculiar schist that lies along a complex shear zone within the Precambrian Erris Complex at Scotchport Bay, NW of Belmullet, northwest Co. Mayo, Ireland. The schist contains the mineral assemblage garnet, kyanite, Cr-mica, microcline, plagioclase, rutile, serpentine, sillimanite and talc, and is interpreted to have a complex metamorphic and metasomatic origin wherein the chemical constituents of ultrabasic-to-acid rocks have been recombined, almost certainly under structural control (Max et al. 1983).

philipsbornite

trigonal $PbAl_3(AsO_4)_2(OH,H_2O)_6$

A rare hydrated lead–aluminium arsenate hydroxide found as a pale-brown drusy crust of minute rhombohedral crystals on an old-time specimen from Wheal Unity, Gwennap, Cornwall, where steeply dipping veins on the eastern flank of the Carn Marth granite were worked first for copper and later for tin. Philipsbornite produces sharp trigonal crystals up to 0.05 mm long (Green et al. 2005f).

Philipsbornite has also been described with other supergene species at Tynagh Pb–Zn–(Cu) mine, near Killimor, Co. Galway, Ireland. It was found on the old mine dumps in association with adamite, bayldonite, linarite, olivenite, mixite, native silver, strashimirite, tyrolite and wulfenite (Green & Moreton 2001). However, analysis suggests that the mineral is pyromorphite (D. I. Green specimen, Tindle unpub. EMP data).

Sharp trigonal philipsbornite crystals from Wheal Unity, Gwennap, Cornwall; width 210 µm; SEM image. (© D. I. Green)

philipsburgite

monoclinic $(Cu,Zn)_6(AsO_4)_2(OH)_6{\cdot}H_2O$

An uncommon hydrated copper–zinc arsenate hydroxide found in one boulder of vein quartz with barite from the dumps of Potts Gill barite mine, Caldbeck Fells, Cumbria. Philipsburgite is a supergene species that forms translucent pale-green spherules that are spiky rather than fibrous. It coats terminated quartz crystals in drusy cavities (Braithwaite & Ryback 1988). In a few cases it is associated with adamite, agardite, cinnabar, hemimorphite and olivenite (Neall & Leppington 1994). Young et al. (1990a) report philipsburgite from nearby Low Pike. It exhibits several different habits and it forms crusts of radiating acicular pale turquoise-blue crystals in spherulitic masses up to 2 mm across on quartz. Globular aggregates, up to 1 mm across, of radiating pale turquoise-blue bladed crystals and, less commonly, bluish-green to apple-green bladed crystals lining cavities in quartz, are also found. At Driggith mine, philipsburgite occurs with adamite, bechererite, caledonite, chenite, leadhillite, linarite, schulenbergite and ramsbeckite (Green & Briscoe 2002). It is common on the 30-fathom level dumps of Driggith mine and on the 60-fathom dumps of nearby Sandbed mine, where it forms rich-green crusts that can easily be mistaken for malachite (Green et al. 2006a).

In Cornwall, philipsburgite is found at Penberthy Croft mine, St Hilary, where it

Botryoidal philipsburgite from Sandbed mine, Caldbeck Fells, Cumbria; width 6 mm. (© D. I. Green)

is relatively uncommon and forms druses of bright-green crystals with chisel-shape terminations. Samples of goethite on the dumps contain emerald-green to dark-green druses of philipsburgite up to 7 mm in diameter (Betterton et al. 2000). Ryback & Tandy (1992) report a further occurrence in Cornwall at Wheal Carpenter, Gwinear, near Hayle, where philipburgite forms botryoidal crusts in small cavities in quartz.

phillipsite (series)

monoclinic — ZEOLITE GROUP — $(K,Na,Ca_{0.5})_6(Si_{10},Al_6)O_{32}\cdot10H_2O$

Named after William Phillips FRS (1775–1828), a London-born printer, bookseller, mineralogist, founder member of the Geological Society of London, and author of *An elementary introduction to mineralogy* (1816). Phillipsite is a common hydrated potassium–sodium–calcium aluminosilicate, whose name attained series status with the zeolite classification of Coombs et al. (1997). There are now three phillipsite species: -Ca, -K and -Na, but these names do not appear in older publications and, until chemical data become available, species status cannot be established.

In Co. Antrim, Ireland, phillipsite is found in amygdales within Tertiary basalts at the Giant's Causeway. This is perhaps the best British locality for the species, and crystals up to 1.5 cm long have been reported (Walker 1960a). A phillipsite-rich tholeiitic basalt near the National Trust reception centre at the Giant's Causeway contains the rare zeolite paulingite and, in nearby scree, erionite, heulandite and possibly clinoptilolite have also been found (Foy 1992). Greg & Lettsom (1858) note phillipsite from basalt at Pleaskins, along the Giant's Causeway, and in similar basalt to the E near Dunseverick it has been found as curved white crystals up to 8 mm across, associated with cowlesite, gismondine, levyne–offrétite and thomsonite (Tschernich 1992). Phillipsite, often with a red colouration, forms radiating aggregates, especially in the gmelinite-bearing lavas around Portmuck, Island Magee, near Larne. Dolerite intrusions associated with the lavas usually contain zeolites in accessory amounts. The Ticloy dolerite plug contains chabazite, phillipsite, rare levyne and possibly analcime (Walker 1959a, 1960a). The rare zeolite mineral gobbinsite has been found intimately associated with phillipsite in vugs and vesicles of altered amygdaloidal basalts at Magheramorne quarry, near Larne (Artioli & Foy 1994). In the Glenariff Valley, at the western edge of the Garron Plateau, zeolites are abundant in thin olivine basalt flows. Phillipsite develops well formed crystals lining amygdales up to 2 cm across. Associated species are analcime, chabazite, garronite, levyne, natrolite and thomsonite (var. faröelite). Occasionally, in incompletely filled amygdales, phillipsite coats the surface of garronite aggregates (Walker 1962a).

Reports of phillipsite in Scotland are sparse and it is said to be one of the rare zeolites on the Isle of Skye. Blocky white to pale salmon-pink phillipsite crystals up to 20 mm long have been found in Tertiary basalts at Moonen Bay, Duirinish Peninsula, Isle of Skye, Highland Region. Additionally, it forms drusy white to orange-brown crystalline crusts in cavities up to 5 cm across within brown basaltic rocks that also contain cowlesite and levyne (Green & Todd 1996, 2001). Phillipsite has also been found at nearby Oisgill Bay (Dyer et al. 1996). Phillipsite is briefly noted from Cairisiadar (Carishader), Loch Roag, Isle of Lewis, Western Isles, where it occurs as an inclusion in apatite and as a crust on a sapphire-bearing monchiquite dyke (Livingstone 1993, Dyer et al. 1996). Calcic phillipsite is described as a deuteric alteration product that formed post-magmatically in amygdaloidal pegmatitic alkali dolerite (crinanite) within the Shiant Isles Main Sill, Western Isles, particularly at Eilean an Tighe near the top of the sill (Gibb & Henderson 1996). A vesicular basaltic dyke at Bennan Head, on the Isle of Arran, Strathclyde, contains amygdales filled with analcime, phillipsite and thomsonite (Dyer et al. 1996). Heddle (1901) notes phillipsite from the Bishopton railway tunnel, near Glasgow, and from Corstorphine Hill, Edinburgh.

Amygdale of phillipsite in Tertiary basalt from Co. Antrim, Ireland; width 15 mm. (© The Trustees of the National Museums of Scotland)

Barium-rich phillipsite, forming a solid-solution series with harmotome, is described from a cavity in metasomatized Great Limestone adjacent to the Brownley Hill vein, Brownley Hill mine, Nenthead, near Alston, Cumbria (Green & Nudds 1998).

phillipsite-Ca

monoclinic (pseudo-orthorhombic) — ZEOLITE GROUP — $Ca_3(Si_{10}Al_6)O_{32}\cdot12H_2O$

An hydrated calcium-dominant aluminosilicate that forms a series with harmotome and is found in amygdales within Tertiary basalts from the Giant's Causeway, Co. Antrim, Ireland (Nawaz 1982). Radiating aggregates (locally abundant) of phillipsite-Ca are also present in rubbly basalt from Black Cave, Island Magee, near Larne (Adair 1987). Additional data confirm the species on the Isle of Skye, Highland Region, Scotland, at Lealt, N of Portree, and Oisgill Bay, Duirinish Peninsula, where phillipsite-Ca is found within amygdales in Tertiary basalt (D. I. Green samples and XRD, Tindle unpub. EMP data).

phillipsite-K

monoclinic (pseudo-orthorhombic) — ZEOLITE GROUP — $K_6(Si_{10}Al_6)O_{32}\cdot12H_2O$

An hydrated potassium-dominant aluminosilicate, confirmed on the basis of a single EMP analysis from an amygdale in a Tertiary basalt found at the Giant's Causeway, Co. Antrim, Ireland (Nawaz 1982). Adair (1987) reports phillipsite-K from rubbly Causeway basalts at Port na Callian and Benanouran Head, and from similar basalts in the Galboly River section near Garron Point. Phillipsite-K is confirmed from two other locations in Co. Antrim: Port-na-Boe, Giant's Causeway, and Stookon, Dunseverick. It is also confirmed from Oisgill Bay, Isle of Skye, Highland Region, Scotland (D. I. Green samples and XRD, Tindle unpub. EMP data).

phillipsite-Na

monoclinic (pseudo-orthorhombic) — ZEOLITE GROUP — $Na_6(Si_{10}Al_6)O_{32}\cdot12H_2O$

An hydrated sodium-dominant aluminosilicate, found in amygdales within Tertiary basalts from the Giant's Causeway and Dunseverick, Co. Antrim, Ireland (Nawaz 1982), and from nearby basaltic lava flows at Benanouran Head and Port Noffer (Adair 1987). Phillipsite-Na is also confirmed from Black Cave and Brown's Bay, Island Magee, near Larne (Adair 1987); and from Lealt, N of Portree, Isle of Skye, Highland Region, Scotland (D. I. Green samples and XRD, Tindle unpub. EMP data).

phlogopite

monoclinic — MICA GROUP — $KMg_3AlSi_3O_{10}(OH)_2$

A common potassium–magnesium aluminosilicate hydroxide found in ultramafic rocks such as peridotites, lamproites and peridotites, and also in metamorphosed impure limestones. The end-member composition of phlogopite is defined in the IMA classification of the mica group (Rieder et al. 1998), but unfortunately the classification does not establish a compositional range for the species. For *MBI*, phlogopite boundaries are defined as Fe/(Fe+Mg) = 0–0.25 and Al^{vi} = 0–0.5 (molecular proportions). The other complication is that compositional data for phlogopite are often lacking, with identification based on optical characteristics alone. Phlogopite is superficially very similar to siderophyllite and biotite (now a series name and defined as trioctahedral micas between, or close to, the annite–phlogopite and eastonite–siderophyllite joins; all are dark micas without lithium).

Analytical data are available from: the foreshore of Pendennis Point, Falmouth, Cornwall – a peralkaline minette, a minor intrusion cutting Devonian slates, contains brown mica flakes, strongly zoned from pale phlogopite centres to dark biotite rims (Hall 1982); Holmead farm, Loxbeare, near Tiverton, Devon – light-brown phlogopite occurs in a highly potassic Permian lava (Tidmarsh 1932, Knill 1969, Thorpe et al. 1986); Foss, near Aberfeldy, Central Region, Scotland – syngenetic barium–zinc mineralization occurs in Middle Dalradian graphitic schists where metasediments contain a wide range of mica species (many barium enriched), including eastonite, biotite, muscovite and phlogopite (Moles 1985); Ballachulish, Highland Region – almost end-member phlogopite occurs in siliceous dolomitic limestones in the aureole of the Ballachulish granite (Ferry 1996); E of Laggan Inn, Highland Region (possibly between Loch Laggan and Newtownmore, although another Laggan is located between Loch Lochy and Loch Ness) – bronzy phlogopite (described as biotite) occurs in a limestone quarry (Heddle 1879e); shore of Loch Urigill near

Loch Borralan, Assynt, Highland Region – phlogopite occurs in phlogopite sövite, a type of carbonatite (Threadgould et al. 1994, Young et al. 1994b); Loyne, near Loch Borralan – phlogopite is associated with more common biotite in metasomatized quartzite (Woolley et al. 1972); Harker's Gully, Marsco area, Western Redhills, Isle of Skye, Highland Region – phlogopite occurs in a pyroxenite lens within a Lewisian leucogneiss xenolith enclosed in ferrodiorite (Thompson 1981); Beinn an Dubhaich granite aureole, Isle of Skye, Highland Region – minor amounts of phlogopite occur in thermally metamorphosed dolomitic limestone; the unit 11/12 chromite band of the Eastern Layered Series, Isle of Rum, Highland Region – contains droplets of phlogopite and a range of oxide and sulphide species (Dunham & Wilkinson 1985), and at the unit 7–8 boundary phlogopite occurs as inclusions in chrome spinel (Young 1984b); on the island of Eilean Shona, NW of Ardnamurchan, Highland Region – phlogopite occurs in spinel-bearing pyroxenite xenoliths within a Permo-Carboniferous diatreme associated with a dolerite dyke (Upton et al. 2003). Similar lower-crust or mantle-derived xenoliths (pargasite pyroxenite and pyroxene granulite gneiss) containing phlogopite ±biotite have also been found in an olivine-bearing nephelinite dyke at Gribun, Isle of Mull, Strathclyde (Upton et al. 1998); Kintail–Glen Affric, Highland Region, and Criffell, Dumfries & Galloway – biotite and phlogopite are ubiquitous phenocryst phases in Caledonian minettes (Canning et al. 1996); Beinn Lora, near Benderloch, Strathclyde – phlogopite (described as biotite) occurs in a Devonian (Lorne Plateau) basalt lava (Groome & Hall 1974); Loch Dubh Sletteval, Langavat Valley, South Harris, Western Isles – phlogopite occurs in a mica peridotite lens associated with a large ultrabasic intrusion (Livingstone 1976a); S and SW of Roineabhal, South Harris, Western Isles – phlogopite and eastonite (both described as biotite) occur in sapphirine-bearing garnet–kyanite–orthopyroxene granulites (Baba 2003); chromium-rich phlogopite and biotite occur in an arfvedsonite-bearing minette at St Hélier, Jersey (Wagner & Velde 1985).

Localities where phlogopite data are lacking, or are better described as biotite, include: Castle-an-Dinas mine, St Columb Major, Cornwall (Dines 1956); in Wales at Arenig Station quarry, near Blaenau Ffestiniog, Gwynedd, and Great House, near Usk, Gwent (Bevins 1994); and Esgair Hir mine, Ceulanymaesmawr, Central Wales orefield, Dyfed (Rust & Mason 1988); six localities in the Highland Region of Scotland: SW of Achmelvich, An Fharaid Mhòr, near Lochinver (Droop 1994); Ardgour district (Drever 1939, 1940); Glenelg district (Clough & Pollard 1899, Read & Double 1935, Davidson 1995a); Loch Ailsh, Assynt (Gunn & Styles 2002); Loch Borralan, Assynt (Woolley et al. 1972, Notholt et al. 1985); Loch Scye, 8 km W of Scotcalder (Smith Aitken 1965); aureole of the Beinn an Dubhaich granite, Broadford area, Isle of Skye (Tilley 1951); Black Rock vent, 2 km SSW of Fairlie, Hunterston Sands, Strathclyde (Alexander et al. 1986); and Bay Steinigie, near Finsbay (Fionnsabhagh), South Harris, Western Isles (Dearnley 1961); and three localities in Ireland: Black Ball Head, near Cahermore, Co. Cork (Pracht & Kinnaird 1995); Barnavave, Carlingford, Co. Louth (Osborne 1932b, Nockolds & Vincent 1947); and Cregganbaun shear zone, Co. Mayo (Thompson et al. 1992).

phoenicochroite

monoclinic $Pb_2O(CrO_4)$

A rare lead chromate from the Hopeful vein, Leadhills, Strathclyde, Scotland. Phoenicochroite is a secondary mineral associated with cerussite, leadhillite and pyromorphite (Temple 1955, Macpherson & Livingstone 1982).

phosgenite

tetragonal $Pb_2(CO_3)Cl_2$

TL Bage mine, Bole Hill, between Cromford and Wirksworth, Derbyshire, England

A rare lead carbonate chloride formed by the alteration of galena in hydrothermal lead deposits or by the reaction of lead minerals with sea water.

Phosgenite is associated with anglesite and matlockite in decomposed galena from one of the old air shafts at the type locality. Colourless, white passing into grey, yellow, greenish and brown crystals of phosgenite (cromfordite), with the form of a right square prism, are described by Greg & Lettsom (1858). Some of these crystals are exceedingly fine and up to 9 cm long, an excellent example from the nearby Wallclose mine, Cromford, being illustrated in James Sowerby's *British mineralogy* (1811). The history of phosgenite is well documented by Burr (1992), who also notes that it has a strong yellow fluorescence under short ultraviolet light. Phosgenite from Derbyshire is also described by Mawe (1802), Rudler (1905), Ford & Sarjeant (1964a), Braithwaite (1983), Bridges & Smith (1983) and Ford et al. (1993), the last four references disagreeing about reputed occurrences at Millclose mine, Darley Dale, near Wensley, 5 km NW of Matlock, and at Meerbrook Sough mine in the Bole Hill area.

Phosgenite from Wheal Confidence, near Newquay, Cornwall (Greg & Lettsom 1858), is similarly disputed by Burr (1992), who suggests that it may be a Derbyshire specimen, and by Kingsbury (1957c), who suggests it is from Wheal Rose, near Porthleven, Sithney, Cornwall. Undoubtedly, phosgenite does occur at Wheal Rose, where sea water has entered lead lodes, and exceptionally beautiful crystals associated with laurionite and paralaurionite have formed. One 6 mm-long specimen of phosgenite is transparent, has a high lustre, and sits on a matrix of black limonite (Russell & Hutchinson 1927, Russell 1944, Dines 1956, Burr 1992). Kingsbury (1954, 1957c) found two small crystals of phosgenite associated with anglesite on decomposed galena in a cavity within "gossany" galena-bearing quartz veinstone from Penberthy Croft mine, St Hilary, Cornwall. The specimen cannot now be located, but others have been reported more recently, forming thick tabular crystals to about 0.75 mm long associated with altered galena, laurionite and native sulphur (Wallace 1998, Betterton et al. 2000). At nearby Croft Gothal mine, rare phosgenite typically forms transparent, occasionally striated, tetragonal prisms up to 1 mm long, in cavities within galena–quartz veinstone (J. Betterton pers. comm. 2006). Newporth beach, 2 km SW of Falmouth is another Cornish location where phosgenite has been found, occurring with cumengéite in a hypogene Cu–Pb–Sb sulphide vein (Dean et al. 1983). At the western end of Bounds Cliff, St Endellion, phosgenite occurs in an antimony–lead vein in association with barstowite, cerussite, chalcopyrite, dolomite, pyrite, quartz and sphalerite (Stanley et al. 1991). Phosgenite has also been found in an oxidized chalcopyrite–galena–sphalerite-bearing quartz veinlet from a cliff exposure at Gunver Head, near Padstow Consols mine, where it is associated with anglesite, cerussite, cumengéite, diaboléite and native sulphur (Braithwaite & Merry 1996, Schlomann & Weiss 1998). Additionally, Golley & Williams (1995) note two other locations in Cornwall: Daymer beach, Padstow (a

Phosgenite (35×10 mm) from Bage mine, Bole Hill, Cromford, Derbyshire. (© M. P. Cooper)

well formed crystal is illustrated in Green & Betz 2006), and Tolcarne beach, Newquay; and Merry & Braithwaite (1995) report phosgenite and cumengéite from lode outcrops in the coastal cliffs near Wheal Golden, Perranzabuloe. Excellent euhedral colourless-to-pink blocky crystals of phosgenite up to 6 mm across have been found in a quartz vein on the shore of Treubies Cove, Tregardock (R. E. Starkey pers. comm. 2005).

Along the beach near the Pier hotel at Clevedon, 16 km west of Bristol, Avon, Triassic Dolomitic Conglomerate is faulted and mineralized with barite–(chalcopyrite–galena). Cavities up to 12 cm across in the barite contain several supergene species, including aragonite, aurichalcite, azurite, beudantite, brochantite, cerussite, native copper, covellite, cuprite, phosgenite and native sulphur. Good euhedral crystals of phosgenite are common but no longer than 2 mm (Starkey 1984, 1986c). A single pebble from Clevedon beach has been found to contain phosgenite and the rare species cotunnite (Bridges 2003). At Wesley Theological College (Wesley mine), Westbury-on-Trym, Avon, manganese oxide mineralization in the Triassic Dolomitic Conglomerate was exposed during foundation excavations in 1983–1984. Alabaster (1989b) reports phosgenite as a supergene species on nine out of over a thousand individual samples collected. Phosgenite is also reported from a small lead working at Tickow Lane mine, Shepshed, Leicestershire (King 1973), and from inside an ancient lead pot deep inside Wookey Hole, Somerset (Midgley 1958).

Colourless phosgenite from the main Driggith–Sandbed vein (Driggith mine), Caldbeck Fells, Cumbria, described by Kingsbury (1957c) and repeated by Young (1987a), is very dubious. Given the doubts about the provenance of other Kingsbury specimens (Ryback et al. 1998, 2001; A. Hart pers. comm. 2004), further confirmation is desirable. The classic phosgenite locality at Monteponi in Sardinia is the likely source of the specimen (Green et al. 2006a).

At Penrhyn (Penrhyn Du) mine, near Abersoch, Lleyn Peninsula, Wales, loose blocks of galena-bearing quartz-vein material contain white-to-pink striated prisms and colourless tabular crystals of phosgenite up to 1.5 mm long. It is associated with anglesite or cerussite, and probably formed by the action of sea water on galena (Hubbard 1991). Phosgenite is also reported from the Pen-y-Bryn shaft, Halkyn, Clwyd (Bevins 1994).

A similar origin is proposed for phosgenite found on the coast at Stotfield to the W of Lossiemouth, near Elgin, Grampian, Scotland, where a coastal outcrop of Pb–(Zn) mineralization occurs in a Triassic breccia known as Cherty Rock. Phosgenite forms brilliant colourless-to-yellow transparent euhedral crystals, generally between 0.2–1.0 mm long, but rarely to 4×5 mm. It encrusts galena and quartz as single or interpenetrant groups. Cerussite, pyrite, pyromorphite and sphalerite are associated species (Heddle 1901, Starkey 1988a, Green & Todd 2001).

Phosgenite has also formed in lead-bearing veins along the coast at Killiney, Co. Dublin, Ireland, where transparent-to-translucent tabular crystals up to 0.8 mm across occur in barite–quartz veinstone near oxidized galena (Green 1998c).

phosphohedyphane

APATITE GROUP

hexagonal $Pb_3Ca_2(PO_4)_3Cl$

A lead–calcium phosphate chloride found at many localities worldwide and previously referred to as "calcian pyromorphite". It is probably a more common species than the examples here suggest. Phosphohedyphane occurs as drusy crusts of white to pale-green hexagonal crystals formed by the supergene oxidation of a galena vein in Permian dolomitic limestone of the Cadeby Formation at Whitwell quarry, near Whitwell, Derbyshire. The crystals, which rarely exceed 1 mm in length, can cover specimens to small hand size and sometimes form pseudomorphs after cerussite. In Scotland in the Leadhills–Wanlockhead orefield, on the Strathclyde/Dumfries & Galloway border, relatively common phosphohedyphane formed as white to pale-green cellular masses and crusts during extensive oxidation of lead veins hosted by Ordovician greywacke (D. I. Green specimens and XRD, Tindle unpub. EMP data).

Phosphohedyphane pseudomorphs after cerussite, 12 mm tall, from Whitwell quarry, Whitwell, Derbyshire. (© D. I. Green)

phosphorite

A fibrous concretionary form of carbonate-hydroxylapatite or carbonate-fluorapatite, apatite group.

phosphosiderite

monoclinic $Fe^{3+}PO_4{\cdot}2H_2O$

A rare hydrated iron phosphate, dimorphous with strengite and found in small metamorphic segregation pods, occurring in striped kyanite–sillimanite-bearing Moine gneiss from Gleann Còsaidh, Loch Quoich, Knoydart, Highland Region, Scotland. The segregation pods contain major amounts of apatite, garnet, muscovite, plagioclase and quartz, and very small rare clusters of graftonite, jahnsite, johnsomervilleite, mitridaite, phosphosiderite, rockbridgeite and vivianite. Pink-to-white phosphosiderite is the rarest of the transition-metal phosphates in the assemblage and occurs in one small patch (Livingstone 1980).

In Ireland, blue phosphosiderite (metastrengite) is reported from albite–spodumene pegmatite boulders along the southeastern flank of the Leinster granite at Seskinnamadra, W of Tomduff Hill, Co. Carlow, and appears to be a secondary alteration product of lithiophilite. Other accessory minerals in the pegmatites include cassiterite, columbite–tantalite, manganese-rich apatite, spessartine and a yellow alteration product described as "salmonsite" (Steiger & von Knorring 1975).

Phosphosiderite is also noted from Gravel Hill mine, Perranzabuloe, Cornwall (Golley & Williams 1995).

phosphuranylite?

orthorhombic $KCa(H_3O)_3(UO_2)_7(PO_4)_4O_4{\cdot}8H_2O$

A rare hydrated potassium–calcium hydronium uranyl phosphate oxide, tentatively identified from Wheal Edward, St Just, Cornwall. Phosphuranylite is reported as bright-yellow crystalline crusts on shale (Ryback & Tandy 1992).

phurcalite

orthorhombic $Ca_2(UO_2)_3O_2(PO_4)_2{\cdot}7H_2O$

A rare hydrated calcium–uranium phosphate found on joint surfaces in granite at Merrivale quarry, Dartmoor, Devon. Deep-yellow phurcalite, up to 0.2 mm long, forms microscopic bladed crystals, often in clusters and associated with yellow autunite (Braithwaite et al. 1989b).

Rosettes and needle-shape crystals of yellow phurcalite from Merrivale quarry, Dartmoor, Devon; the sphere on the right is 0.5 mm across. (© D. I. Green)

pickeringite

monoclinic HALOTRICHITE GROUP $MgAl_2(SO_4)_4{\cdot}22H_2O$

An uncommon hydrated magnesium–aluminium sulphate that forms a series with halotrichite and is found in a slate quarry at Dunkeld, Tayside Region, Scotland, where it forms aggregates of pale-yellow fibrous crystals (Macpherson & Livingstone 1982). Pickeringite is also noted on a specimen from Newtyle, Tayside (Macpherson 1983).

In England, pickeringite, together with copiapite, epsomite, gypsum, jarosite and rozenite, is reported from the Oughterside opencast coal site, Aspatria, Cumbria, where it occurs as encrustations and efflorescences on joint surfaces and on the walls of old workings in coal seams. Pickeringite forms stellate clusters of white acicular crystals up to 1.5 mm across and probably formed by oxidation of pyrite-rich coals within 10 m of the present ground surface (Young & Nancarrow 1988b); these authors also mention a specimen from Bishop Auckland, Co. Durham, that is now at the NHM.

Other museum specimens containing pickeringite are recorded from Gwynfynydd mine, near Dolgellau, and Parys Mountain, Anglesey, both in Gwynedd, Wales (Bevins 1994, Southwood & Bevins 1995).

picotite

A chromium–magnesium-rich variety of hercynite, spinel group.

picrochromite

Hypothetical molecule of the chromite–spinel series.

picrolite

A variety of antigorite, kaolinite–serpentine group (Whittaker & Zussman 1956).

PICROMERITE GROUP

Monoclinic sulphates with the general formula $A_2B^{2+}(SO_4)_2{\cdot}6H_2O$, where A = K, (NH_4); and B^{2+} = Cu, Fe, Mg, Ni. Picromerite is the only member of the group found in Britain or Ireland.

picromerite

monoclinic PICROMERITE GROUP $K_2Mg(SO_4)_2{\cdot}6H_2O$

A rare hydrated potassium–magnesium sulphate found in the potash subzone of the Permian Middle Evaporites in the Fordon F.3 borehole, Humberside (13 km S of Scarborough, Yorkshire). Other rare species in this subzone include apthitalite, boracite, kieserite and pyrite. More common species include carnallite, hematite, halite, magnesite, quartz, rinneite and sylvite. The Permian evaporite sequence underlies much of North Yorkshire and extends considerably beneath the North Sea. At Fordon the sequence is 430 m thick (Stewart 1954, 1965). Anthony et al. (2003) list picromerite from Whitby, Yorkshire, but give no details.

picropharmacolite

triclinic $Ca_4Mg(AsO_3OH)_2(AsO_4)_2{\cdot}11H_2O$

A rare hydrated calcium–magnesium arsenate hydroxide, said to be found as crusts of white satiny fibres and massive with pharmacolite, weilite and wapplerite at Wanthwaite mine, St John's in the Vale, near Threlkeld, Cumbria (Young 1987a). The identification of picropharmacolite is based on specimens from the Kingsbury collection now at the NHM, and their provenance is in doubt (see Ryback et al. 1998, 2001, Green & Briscoe 2002).

In Scotland, post-mining oxidation of the old dumps at Silver Glen, Alva, Ochil Hills, Central Region, has produced small white spheres of picropharmacolite on a barite matrix. Stains and efflorescences are also found of annabergite, erythrite, malachite, tyrolite and, rarely, conichalcite. The richest bonanza of native silver ever found in Britain was at the Alva silver mine (Livingstone 1993, Moreton 1996).

piedmontite

A synonym of piemontite, epidote group.

piemontite

monoclinic EPIDOTE GROUP $Ca_2(Al,Mn^{3+},Fe^{3+})_3Si_3O_{12}(OH)$

An uncommon calcium–aluminium–manganese–iron aluminosilicate hydroxide found in pink psammite boulders in glacial drift near Kinloch, S of Tongue, Highland Region, Scotland. The boulders are almost certainly derived from the local Moine rocks. Braunite, manganese-rich hematite, manganese-rich muscovite and spessartine are associated species (Faithfull 2007). The locality had previously been noted as "from a lake NE of Ben Hope" by Goodchild (in Heddle 1901), and a specimen

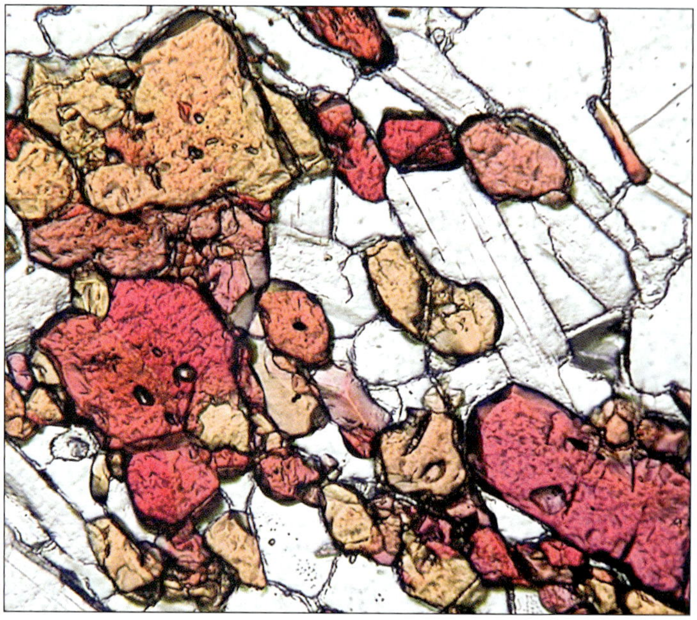

Piemontite showing pink and orange pleochroic colours, with muscovite and quartz from a psammite boulder in glacial drift, near Kinloch, south of Tongue, Highland Region, Scotland; width 0.75 mm; viewed in PPL. (© J. W. Faithfull)

of aventurine (meta-quartzite) containing either colourless garnet or diamond, red mica and red zircon was described. New analyses have confirmed the zircon to be piemontite, and other red zircons described by Heddle from a quartz schist at Ben Hope, E of An Gorm Loch, are probably also piemontite (Faithfull 2007).

Piemontite has been reported in veinlets and infilling vesicles in an altered andesite approximately 9 km W of King's House, Glencoe, Highland Region. The vesicles vary from 0.3 mm to 70 mm and the larger ones may also contain calcite. The mineral is always separated from the surrounding rock by a thin layer of milky-white epidote, between which there is a complete and gradual colour variation. In thin section, the mineral is strongly pleochroic, with colour variations from deep reddish pink to more subtle shades of pink and light apple green, and is described as withamite by Greg & Lettsom (1858) and Hutton (1938). However, it has a low manganese content and should more correctly be referred to as manganese-rich epidote (Faithfull 2007). A strongly coloured mineral in calc-silicate rocks of the Dalradian Balnacraig Metabasite Formation at the southwest margin of the Ballater granite, Grampian, has been identified from its optical properties as piemontite (Goodman et al. 1990). An analysis is desirable to confirm the identification, as strong colouration is not sufficient to identify this species.

In Wales, two thin sections of green flaggy or laminated mica schists from the New Harbour Group, near Holyhead, Anglesey, contain presumed piemontite in an albite–quartz-dominated assemblage with minor apatite, epidote, ilmenite, titanite and zoisite (Greenly 1919). Small greenish-brown to reddish-brown prismatic crystals from Benallt mine, Rhiw, Lleyn Peninsula, are confirmed to be piemontite (Bevins 1994). Rose-pink strongly pleochroic so-called epidote from the Llanberis Pass area, Snowdonia (Williams 1927), is also presumed to be piemontite.

piemontite-(Sr)

monoclinic EPIDOTE GROUP $CaSrMn^{3+}Al_2(Si_2O_7)(SiO_4)O(OH)$

A rare hydrated calcium–strontium–manganese–aluminium silicate hydroxide found in lenses of manganese ore within metamorphosed Ordovician mudstones at Benallt mine, Rhiw, Lleyn Peninsula, Gwynedd, Wales. Piemontite-(Sr) occurs as small deep-red acicular crystals within cavities in crystalline celsian and is associated with pyrophanite. A chemical variation towards a strontium-rich epidote composition is noted (T. F. Cotterell pers. comm. 2007). This species was originally called strontiopiemontite, but was renamed as part of the reclassification of the epidote group (Armbruster et al. 2006).

pigeonite

monoclinic PYROXENE GROUP $(Mg,Fe^{2+},Ca)(Mg,Fe^{2+})Si_2O_6$

A relatively common magnesium–iron–calcium silicate found in rapidly cooled basaltic volcanic rocks and which is stable only at high temperatures. Depending on magnesium and iron content, pigeonite inverts to orthopyroxene between 1140°C and 980°C, but can survive in the metastable condition if the temperature falls quickly. With slower cooling, as is the case in gabbroic and some doleritic intrusions, it inverts to an orthorhombic orthopyroxene. Such "inverted pigeonite" can often be recognized in thin section, although it is quite similar to augite.

At Dinas Head, adjoining Trevose Head, 8 km W of Padstow, Cornwall, pigeonite forms colourless relict ophitic plates in albite–quartz rocks termed "adinoles", which were formed when an intrusion of dolerite metasomatically altered slate (Agrell 1939). In Derbyshire, magnesium-rich pigeonite occurs with diopside (salite) in Carboniferous basalt at Lower Miller's Dale (Macdonald et al. 1984). Pigeonite is also noted from Ordovician Borrowdale Volcanic Group andesites in the Cockermouth district of Cumbria (Eastwood et al. 1968) and, from the Cleveland dyke, a basaltic andesite intrusion exposed at many places across southern Scotland and Cumbria, and extending through Co. Durham (e.g. Cockfield) to Cleveland (Macdonald et al. 1988b). Pigeonite was previously referred to as uniaxial augite (Heslop 1912).

The St David's Head intrusion, on the northwest extremity of the St David's Peninsula, Dyfed, Wales, is a tholeiitic sill comprising a suite of mafic gabbros, which are in part layered. EMP data indicate that pigeonite occurs with augite and enstatite in xenolithic laminated olivine gabbros that are restricted to an outcrop at Carn Llidi (Bevins et al. 1994).

Scotland

Central Pigeonite occurs with augite and plagioclase feldspar (labradorite) in a tholeiite dyke at Kinkell, 11 km N of Glasgow (Walker et al. 1952). Pigeonite is also said to be common in a sill complex extending from Stirling and Denny to Kilsyth in neighbouring Strathclyde (Walker 1935).

Dumfries & Galloway A tholeiitic basalt from the summit of Castle Hill, 16 km NW of Langholm, Eskdalemuir, Snar Valley, contains hypersthene rimmed with equal amounts of monoclinic pyroxene and pigeonite, together with small granules of augite and pigeonite. The pigeonite forms strongly zoned prisms (Elliot 1956).

Highland A pigeonite–(augite–ferrosilite)-bearing meta-ironstone set in basic gneisses occurs in the first inlet W of the pier on the north shore of Scourie Bay. In thin section, some of the pyroxenes are highly exsolved and, in places, fibrous amphibole overgrows or replaces them (Barnicoat & O'Hara 1979). On Ardnamurchan, pigeonite (probably inverted pigeonite) is one of the phenocryst phases occurring in variable assemblages with fayalite, ferro-augite, ferrohedenbergite, orthopyroxene, plagioclase, quartz and sanidine in pitchstone flows (Preston et al. 1998b). A pitchstone on Ben Hiant, Ardnamurchan, contains pigeonite as discrete phenocrysts, as cores to augite phenocrysts, and itself forming cores of hypersthene crystals. Pigeonite phenocrysts also occur in porphyritic felsites at west Minishal and Beinn nan Stac, and in granophyre at Loch a' Ghillie Reamhra, all on the Isle of Rum (Emeleus et al. 1971).

Lothian A review of Carboniferous quartz dolerite and tholeiite dykes and sills in Scotland describes elongate prisms of pigeonite associated with rhombic pyroxene in the intrusions around Edinburgh and the Bathgate Hills (Walker 1935).

Strathclyde Pigeonite forms phenocrysts in a Tertiary tholeiitic dyke to the NE of Kilmartin and forms acicular crystals in another dyke at Rubh' an Tighe Loisgte, Loch Melfort, both in the Craignish area (Allison 1936). On the Isle of Mull, pigeonite is noted from two locations: at Pennyghael, described as uniaxial augite forming minute (<0.5 mm) rounded phenocrysts in a dark glassy magnetite-bearing andesite cutting Tertiary basalts (Hallimond 1914, Virgo & Ross 1973); and at Burg on the north side of Loch Scridain, Ardmeanach, as rounded crystals (<2 mm across) in pitchstone (Preston et al. 1998b). Walker (1935) briefly notes pigeonite in tholeiitic dykes on the Isle of Jura.

Western Isles The Tertiary Glen Bay granite on St Kilda (Hirta) contains ferropigeonite together with apatite, calcic amphibole (ferrohornblende or ferro-edenite), chevkinite, ferro-augite, alkali feldspar, ilmenite, magnetite, plagioclase (oligoclase), quartz and zircon (Harding et al. 1984).

Ireland

Antrim Pigeonite occurs in a Tertiary dolerite plug at Tieveragh, near Cushedall (Tomkeieff 1940), and in mineral-filled cavities within partially melted basalts adjacent to a dolerite intrusion at Scawt Hill, Larne (Kitchen 1985). The tholeiitic basalts of the Giant's Causeway area also contain zoned crystals of pigeonite (Lyle & Preston 1993).

Down Dolerites and gabbros from the Tertiary central complex of Slieve Gullion contain pigeonite (now inverted) with complex graphic exsolution textures, coexisting with augite and hypersthene (Gamble 1982).

pihlite

A white micaceous mineral, probably a mixture of feldspar and mica. Also described as a mixture of albite and a muscovite pseudomorph after spodumene.

pilolite

A synonym of palygorskite.

pilsenite

trigonal Bi_4Te_3

A rare bismuth telluride found in galena-dominated gold-bearing veins at Clogau mine, Bontddu, near Dolgellau, Gwynedd, Wales (Bevins 1994). The original description was of "wehrlite" (Naden 1988), but this is a mixture of pilsenite and hessite. Pilsenite is one of the minor species in Ag–Au–Bi–Sb–Te "bonanza gold" mineralization at Clogau, where it is associated with aleksite, bismuth, bismuthinite, boulangerite, bournonite, hedleyite and nagyágite (Mason et al. 2002).

In Scotland, one gold grain from a panned concentrate collected in the Ochil Hills, Central Region, contains inclusions with a composition intermediate between pilsenite and tsumoite (Leake et al. 1993), and at Boreland Glen, Caulkerbush, Dumfries & Galloway, panned alluvial gold grains contain inclusions of chalcopyrite, galenobismutite, pilsenite, pyrite, tennantite, tetrahedrite and wittichenite (Chapman et al. 2000c).

pimelite

A synonym of népolite and pecoraite, but also applied to members of the smectite group.

pinguite

Probably a synonym of nontronite, smectite group.

pinite

A group of pseudomorphs, mostly of muscovite after cordierite.

pinnoite?

tetragonal $MgB_2O(OH)_6$

A rare magnesium borate hydroxide described from the Eskdale No. 3 borehole, Sleights, near Whitby, Yorkshire. In the Lower Halite unit of the Lower Evaporite Bed, hair-like needle-shape inclusions in halite could possibly be pinnoite, although their refractive index appears to be too high. Further investigation is required (Armstrong et al. 1951).

pirssonite?

orthorhombic $Na_2Ca(CO_3)_2{\cdot}2H_2O$

A rare hydrated sodium–calcium carbonate described from a cave within the Great Opencast of Parys Mountain, Anglesey, Gwynedd, Wales, on a Russell Society field trip in 2002. In places, pirssonite was said to be "richly encrusted" (Todhunter 2002). Unfortunately, this is a mistake and the mineral being described is actually melanterite, var. pisanite (T. F. Cotterell pers. comm. 2004).

pisanite

A copper-rich variety of melanterite.

pistacite

A synonym of epidote.

pistomesite

A magnesium-rich variety of siderite, calcite group.

pitchblende

A colloidal variety of uraninite.

pitchy copper ore

A synonym of chrysocolla.

pitticite?

amorphous $Fe^{3+}(AsO_4)(SO_4)H_2O$?

An uncommon hydrous ferric arsenate–sulphate referred to in vague terms by various authors: from England (O'Donoghue 1994); Botallack, St Just, Cornwall (Bancroft & Weller 1993); Dolcoath mine, Camborne (Anthony et al. 2000); Penberthy Croft mine, St Hilary, Cornwall (Golley & Williams 1995, Betterton et al.

2000); probably from Wheal Sparnon, Redruth, Cornwall (Greg & Lettsom 1858); and Brandy Gill mine, Caldbeck Fells, Cumbria (Hartley 1984). All require further confirmation.

plagioclase

Plagioclase is a term used to describe a very commonly occurring series of triclinic sodium- to calcium-rich feldspars found in many igneous, metamorphic and sedimentary rocks. The end members have species status, with albite (An_{0-10} mole %) at the sodium-rich end and anorthite (An_{90-100}) at the calcium-rich end of a continuous solid-solution series. Intermediate compositions are not species, but have been named oligoclase (An_{10-30}), andesine (An_{30-50}), labradorite (An_{50-70}) or bytownite (An_{70-90}), depending on their calcium–sodium content.

plagionite

monoclinic $Pb_5Sb_8S_{17}$

A rare lead sulphantimonite found in stratabound quartz veins confined to a Middle Devonian to Lower Carboniferous volcano-sedimentary basinal succession at Bounds Cliff, E of Port Isaac, Cornwall. The mineral assemblage is antimony–lead sulphosalt dominated with a minor gold association. Plagionite forms at a late stage in the paragenesis along intergrain fractures and at vein intersections, and is intergrown with bournonite, boulangerite, jamesonite and tetrahedrite (Clayton et al. 1990).

Epithermal quartz veins cutting nodular ignimbrite from Bwlch mine, near Deganwy, S of Llandudno, Gwynedd, Wales, also contain rare plagionite. Associated species are jamesonite, semseyite, stibnite and zinkenite, and possibly boulangerite and robinsonite. The assemblage occurs in irregular patches, and in thin veins and stringers up to 2 cm in width (Bevins et al. 1988).

At Le Pulec, northwest coast of Jersey, Channel Isles, plagionite is described from quartz veins (<2 m wide) cutting Brioverian mudstones and siltstones that in the 1870s were worked for lead and silver. Sphalerite is the earliest and most abundant sulphide, followed by galena, chalcopyrite and then arsenopyrite. Plagionite forms small euhedral rhombs, 5–20 μm long, overgrowing the last three species. It is part of a late sulphosalt-rich assemblage that contains native antimony, bournonite, cubanite, stibnite and tetrahedrite (Ixer & Stanley 1980).

planchéite?/plancheite?

orthorhombic $Cu_8(Si_4O_{11})_2(OH)_4 \cdot H_2O$

A rare hydrated copper silicate hydroxide found on three specimens in the Kingsbury collection at the NHM. The localities are: Driggith mine, Caldbeck Fells, Cumbria; Alderley Edge ridge, Cheshire; and Gunheath china-clay pit, Carthew, St Austell, Cornwall. The specimens are reported in compilations by Carlyon (1979), Young (1987a), Cooper & Stanley (1990), Ryback & Tandy (1992), Braithwaite (1994), Weiss (1994), Anthony et al. (1995), Golley & Williams (1995) and Green et al. (2006a). Unfortunately, the provenance of these specimens is considered very dubious and the samples probably originated from Shaba (Katanga), Congo, or Mammoth District, Tiger, Arizona (Ryback et al. 1998, 2001). An additional matrix-free specimen of sky-blue planchéite, 3 mm long, was found in the southeastern end of the Engine Vein mine rift, Alderley Edge ridge, but this does not appear to have been analyzed (Braithwaite 1994).

planerite

TURQUOISE GROUP

triclinic $\square Al_6(PO_4)_2(PO_3OH)_2(OH)_8 \cdot 4H_2O$

A rare hydrated aluminium phosphate hydroxide found at Grouse Lodge quarry, 5 km S of Shanagolden, Co. Limerick, Ireland. Planerite is a component of buff granular material that occurs in veins within Upper Carboniferous phosphate-rich black shales. It is intimately associated with gibbsite and kaolinite. Turquoise, wavellite and the rare species koninckite are also reported (Moreton & Green 2005).

plasma

A bright-green variety of chalcedony (quartz).

platarsite

COBALTITE GROUP

cubic $(Pt,Rh,Ru)AsS$

A very rare platinum–rhodium–ruthenium sulpharsenite, qualitatively identified in a layered mafic–ultramafic intrusion on the Isle of Rum, Highland Region, Scotland. Within the unit 7–8 inter-unit chromitite layer of the Eastern Layered Series, platarsite is found in association with species positively or tentatively identified as braggite, cooperite, hongshiite, irarsite, isoferroplatinum, keithconnite, kotulskite, laurite, moncheite, palladoarsenide, sperrylite, telargpalite, telluropalladinite, and the native species palladium and platinum. Most of these grains are 0.2–2 μm in diameter, although grains >15 μm are also present for some species. The majority are in contact with sulphide minerals such as bornite, chalcopyrite or pentlandite. The chromitite layer was formed during magmatic fractionation and crystal settling of an open-system magma chamber, modified by magma mixing and liquid immiscibility of sulphide and silicate melts (Butcher et al. 1999, Power et al. 2000a).

platiniridium

A platinum-rich variety of iridium.

platinum

cubic Pt (commonly containing Pd, Ir, Fe, Ni)

A composite grain of platinum, irarsite and an unnamed Rh–Sb–S phase has been found in a chromite-rich sample from Jimmie's quarry, S of Hagdale, Unst, Shetland, Scotland (Prichard et al. 1988). In a layered mafic–ultramafic intrusion on the Isle of Rum, Highland Region (within the Eastern Layered Series unit 5/6 boundary and unit 7–8 inter-unit chromitite layers), platinum is found in association with species positively or tentatively identified as braggite, cooperite, hongshiite, irarsite, isoferroplatinum, keithconnite, kotulskite, laurite, moncheite, palladoarsenide, platarsite, sperrylite, telargpalite, telluropalladinite, and the native species gold and palladium. Most of these grains are 0.2–2 μm in diameter, although grains >15 μm are also present for some species (Butcher et al. 1999, Power et al. 2000a).

An iridium-rich variety of platinum is one of many heavy minerals panned from the Gold Mines River on the northeastern side of Croghan Mountain, Co. Wicklow, Ireland. Other species include blue corundum (sapphire), cassiterite, gold and wolframite (Forbes 1869, Ryback & Moreton 1993). A 48 g platinum nugget reported from Curran, Larne, Co. Antrim, is thought to be of Russian origin (Doughty et al. 1982).

Records of platinum in England and Wales require further confirmation (Readwin 1888, McPherson & Lamb 1921, Bevins 1994, Gunn & Styles 2002).

plattnerite

RUTILE GROUP

tetragonal PbO_2

TL Leadhills, Strathclyde, Scotland (probably)

A rare lead oxide, dimorphous with scrutinyite and found as a minor late-stage replacement of oxychlorides and, in one case, hydrocerussite at three locations in the Bristol District: Merehead (Torr Works) quarry, Cranmore, Somerset; Higher Pitts mine, Priddy, Mendip Hills, Somerset; and Wesley Theological College (Wesley mine), Westbury-on-Trym, Avon. In all cases, plattnerite is part of a secondary assemblage associated with low-temperature replacement deposits of iron and manganese oxides of Triassic age. Generally, plattnerite occurs as irregular grains (up to 4 mm in diameter), stringers, films, crystalline crusts and clumps of distorted crystals. The colour varies from chestnut brown to jet black. Red internal reflections may be observed under a binocular microscope (Alabaster 1985b). At Merehead (Torr Works) quarry, plattnerite occurs as dark-brown resinous masses (up to 1 cm across) and groups (up to 5 mm across) of bright-black sub-mm distorted crystals, and replaces chloroxiphite, diaboléite or mendipite (Burr 1996).

Early records of plattnerite in Scotland are vague, being restricted to the probable type locality (Greg & Lettsom 1858, Heddle 1901, Brown 1927), although Temple (1955) is more precise, noting its occurrence in the Raik and Susanna veins. Black botryoidal masses of plattnerite are associated with yellow and green pyromorphite, and cerussite, in a 1.5 × 1.5 cm cavity in quartz from the Belton Grain vein, in nearby Wanlockhead, Dumfries & Galloway (Heddle 1889a, Livingstone 2002).

pleonaste

An iron-rich variety of spinel.

plinthite (or plynthite)

TL Co. Antrim, Ireland

The type specimen is a mixture of analcime, hematite and montmorillonite, but other specimens are mixtures of thomsonite and chabazite, or thomsonite and hematite (Sweet 1960b).

plombièrite

orthorhombic $Ca_5Si_6O_{16}(OH)_2 \cdot 7H_2O$

A rare hydrated calcium silicate hydroxide found in thermally metamorphosed limestone at Kilchoan, Ardnamurchan, Highland Region, Scotland. Plombièrite forms in late-stage replacement veins as a breakdown product of larnite or merwinite (Agrell 1965, Livingstone 2002).

In the chalk/dolerite contact zone at Ballycraigy, Larne, Co. Antrim, Ireland, a plombièrite gel replaces both larnite and bredigite in larnite-rich rocks produced by the magmatic silicification of chalk (McConnell 1954). At Scawt Hill, 8 km away, in another chalk/dolerite contact zone, metasomatic reaction of flint nodules with surrounding chalk has produced larnite nodules mantled by a plombièrite gel. The pink to reddish-brown gel is transparent to translucent and it breaks with a conchoidal fracture (McConnell 1955). In a chalk/dolerite contact zone associated with a Tertiary volcanic plug at Carneal, Larne, Co. Antrim, larnite-rich rocks contain many narrow veinlets and cavities filled with secondary minerals, including plombièrite gels (three varieties), afwillite, calcite, killalaite and portlandite (Nawaz 1977). From this same locality, Sabine & Young (1975) report a scawtite–xonotlite rock containing a silicate hydrate identified as plombièrite, and a similar hydrate associated with tobermorite in an amygdale in basalt.

plumbago

A synonym of graphite and of molybdenite.

plumbo-aragonite

TL Leadhills, Strathclyde, Scotland

A synonym of tarnowitzite – a mixture of aragonite and cerussite.

plumbocalcite

TL Wanlockhead, Dumfries & Galloway, Scotland

A variety of calcite, also considered to be a mixture of calcite and cerussite.

plumbogummite

trigonal $PbAl_3(PO_4)_2(OH,H_2O)_6$

An uncommon hydrated lead–aluminium phosphate found in the oxidized zone of lead deposits. At Roughton Gill mine, Caldbeck Fells, Cumbria, exceptional specimens of plumbogummite have been found that are probably the most beautiful in the world. Most prized is bright cobalt-blue minutely botryoidal plumbogummite on a quartz matrix, coated with yellowish globular pyromorphite. Other varieties of plumbogummite found here are almost colourless or bluish-grey to lavender, and occasionally the species forms sugary white epimorphs after pyromorphite (Rudler 1905, Davidson & Thomson 1951, Cooper & Stanley 1990). At Dry Gill mine, less than 2 km away, dark turquoise-blue plumbogummite occurs in association with manganese oxides and canary-yellow pyromorphite. A very pale-blue to colourless variety, forming drusy encrustations on pyromorphite–mimetite, is also noted (Cooper & Stanley 1990, Green & Briscoe 2002). At Arm o' Grain, plumbogummite has been identified as colourless to pale-brown or rarely pale-blue rhombohedral crystals (a few μm across) on pyromorphite or quartz (Green et al. 2006b).

Blue botryoidal plumbogummite with green pyromorphite crystals up to 1.5 mm across. Roughton Gill, Caldbeck Fells, Cumbria. (© National Museum of Wales, Cardiff; photographer M. P. Cooper)

Plumbogummite forms minute colourless drusy crystals (< 50 μm across) encrusting goethite-coated quartz at Burdell Gill, Caldbeck Fells. Additionally, a single specimen is described where galena has been replaced by a hollow boxwork of plumbogummite (Neall & Leppington 2003). Early descriptions confuse plumbogummite with hemimorphite and smithsonite, and the species has also been described as plumboresinite (Miers 1897) and hitchcockite (Hartley 1900).

Descriptions of plumbogummite from other locations in the Caldbeck Fells are problematic, as they are either by A. W. G. Kingsbury or based on specimens in the Kingsbury collection at the NHM. The problems of inaccurate provenancing of specimens in this collection are outlined by Ryback et al. (1998, 2001). These occurrences are noted in various publications, but have doubtful provenance: Brae Fell mine, Dale Beck Valley (Young 1987a, Cooper & Stanley 1990, Bridges et al. 2006); Mexico mine in Todd Gill, and (Old) Potts Gill mine (Young 1987a, Cooper & Stanley 1990); Brandy Gill, Carrock Fell (Kingsbury & Hartley 1955, Hartley 1984, Young 1987a, Cooper & Stanley 1990); Nether Row Brow (Hartley 1984, Young 1987a, Cooper & Stanley 1990); and Sandbed mine (Cooper & Stanley 1990). Green et al. (2006a) conclude that the plumbogummite sample supposedly from Sandbed mine is likely to have been incorrectly provenanced. Fortunately, plumbogummite has since been confirmed at one of these localities: brown plumbogummite associated with campylite is noted from the Todd Gill dumps of Higher Mexico mine (Green & Briscoe 2002).

Fine specimens of light-bluish to greenish plumbogummite have been found in the underground workings of Penberthy Croft mine, St Hilary, Cornwall. They occur as botryoidal crusts up to several cm^2 on a slaty iron-stained matrix with goethite, pyromorphite and quartz. Surface specimens are more variable in colour (light blue or more commonly grey, brown, yellowish or greenish) and their habit may be botryoidal, globular or stalactitic (Betterton et al. 2000). The occurrence has also been noted by Kingsbury (1954), Camm & Merry (1991b) and Golley & Williams (1995). At Wheal Gorland, St Day, 2 km E of Redruth, Cornwall, plumbogummite occurs as small (up to 4 mm across) bluish-grey patches with carminite, mimetite and zeunerite on quartz gossan (Rust 1981b). Golley & Williams (1995) note plumbogummite from Wheal Alfred, Phillack, near Hayle. Other Cornish occurrences are alluvial and include Buttern Hill, Altarnun (Robson 1954), and possibly the Camel and Gannel estuaries on the north coast (Pirrie et al. 2000b).

In Scotland, plumbogummite occurs with cotunnite, mimetite and pyromorphite at High Pirn mine, Wanlockhead, Dumfries & Galloway, and with plattnerite from the dumps of Raik vein, Leadhills, Strathclyde. At both locations the mineral is rare (Temple 1955). At the headwaters of Penkiln Burn, N of Newton Stewart, Dumfries & Galloway, plumbogummite was discovered in alluvial samples and was traced to outcrops in a fault-bounded mineralized gossan zone, close to the Loch Doon granite complex. Beudantite, goethite, limonite and malachite were also found and they represent the oxidized products of disseminated chalcopyrite, covellite, galena, marcasite, pyrite, pyrrhotite and sphalerite mineralization hosted in black mudstones, microdiorite intrusions and quartz veins (Stone & Leake 1984). Fine-grain glauconitic calcareous sandstones of Jurassic age are noted to contain plumbogummite-group minerals at Gribun and Loch Don on the Isle of Mull, Strathclyde, and on the adjacent mainland in the Poll Luachrain area, Morvern Peninsula, Highland Region (Bain 1970).

plumbojarosite

trigonal JAROSITE GROUP $PbFe^{3+}_6(SO_4)_4(OH)_{12}$

A rare lead–iron sulphate hydroxide found at the following locations in England.

Cornwall Penberthy Croft mine, St Hilary (Kingsbury 1954, Kingsbury & Hartley 1960, Camm & Merry 1991b, Golley & Williams 1995, Betterton et al. 2000); Wheal Carpenter, Gwinear, near Hayle (Kingsbury & Hartley 1960, Golley & Williams 1995).

Cumbria Balliway Rigg, higher Roughton Gill, Caldbeck Fells (Kingsbury & Hartley 1960, Hartley 1984, Young 1987a, Cooper & Stanley 1990); Brandy Gill mine, Caldbeck Fells (Young 1987a, Cooper & Stanley 1990); Driggith mine, Caldbeck Fells (Kingsbury & Hartley 1957f, Hartley 1984, Young 1987a, Cooper & Stanley 1990); Nether Row Brow (Lower level), Caldbeck Fells (Kingsbury & Hartley 1957f, 1960, Young 1987a, Cooper & Stanley 1990); Wanthwaite mine, St John's in the Vale, near Threlkeld (Kingsbury & Hartley 1960, Young 1987a).

Yorkshire Turf Pits mine, Grassington Moor, Yorkshire (Kingsbury 1965, Dunham & Wilson 1985).

A common theme throughout the descriptions is A. W. G. Kingsbury, who appears to have collected all the specimens. Other authors are predominantly referencing Kingsbury's work. Because the Kingsbury collection (now at the NHM) contains samples that have been incorrectly provenanced (Ryback et al. 1998, 2001,

Green & Briscoe 2002, Green et al. 2006a), it is not possible, without further investigation, to confirm the plumbojarosite occurrences. It is unclear whether the reference to plumbojarosite from Treore mine, St Teath, Cornwall (Anthony et al. 2003), is based on Kingsbury's work.

At Greystone quarry, Lezant, near Launceston, Cornwall, dark-brown equant crystals of plumbojarosite, up to 0.4 mm across, occur in a drusy crust found on a single specimen derived from the Greystone lead lode (Elton et al. 2004). Plumbojarosite is tentatively identified in association with beaverite from Ystrad Einion mine, near Tal-y-bont, Central Wales orefield, Dyfed (Bevins 1994).

plumbonacrite?

TL Wanlockhead, Dumfries & Galloway, Scotland

Plumbonacrite was first described by Heddle (1889a) and later considered to be a synonym of hydrocerussite. It is metastable and it transforms easily to a composition close to (but different from) hydrocerussite (Haacke & Williams 1981, Krivovichev & Burns 2000). Requires confirmation.

plumboresinite

A synonym of plumbogummite, crandallite group.

plumosite

A name applied to feathery antimony ore; usually boulangerite, but sometimes jamesonite, jaskólskiite or zinkenite.

podolite

A synonym of carbonate-hydroxylapatite, apatite group.

polianite

A pseudomorph after manganite and a synonym of pyrolusite, rutile group.

pollucite

ZEOLITE GROUP

cubic $Cs(Si_2Al)O_6{\cdot}nH_2O$

A rare hydrated cesium aluminosilicate that forms a series with analcime and is found in the lithium–fluorine-rich aplite dyke at Meldon aplite quarry, near Okehampton, Devon. The aplite was emplaced into the metamorphic aureole of the Dartmoor granite and hosts many rare accessory species. Saccharoidal pollucite, forming triangular outlines 1 cm across, is found in petalite-rich pegmatitic veins crosscutting the aplite. Associated species are beryllonite, lepidolite and orthoclase. Nodular pollucite, partially replaced by a clay mineral, fluorite, quartz and lithian mica, is more common, forming aggregates up to 2 cm across (von Knorring & Condliffe 1984).

Other descriptions of pollucite from Meldon (Kingsbury 1964b, Drysdale 1985, 1991) are based on samples supposedly collected by A. W. G. Kingsbury. Because the Kingsbury collection (now at the NHM) contains samples that have been incorrectly provenanced (Ryback et al. 1998, 2001), it is not possible, without further investigation, to confirm that these pollucite samples came from Meldon.

polybasite

pseudo-hexagonal $Cu(Ag,Cu)_6Ag_9Sb_2S_{11}$

A rare silver–copper sulphantimonite that forms a series with pearceite (see also arsenpolybasite) and is found in antimony–lead veins at Pentire (Glaze) mine, near New Polzeath, St Minver, Cornwall. Polybasite occurs as unorientated rounded or irregular-shape inclusions intergrown with bournonite or freibergite (silver-rich tetrahedrite) within galena. Together they make a large contribution to the high silver values reported in analyses of galena from this area. Polybasite is also noted without comment from the nearby antimony–lead vein system at Port Quin and Reedy Cliff (Clayton et al. 1990).

In the Dolgellau goldbelt, Gwynedd, Wales, polybasite has been noted as inclusions in galena in association with freibergite and pyrargyrite (Gilbey 1968).

Polybasite occurs in a hydrothermal vein deposit at Calliachar Burn, 4 km from Aberfeldy, Tayside Region, Scotland, where it is found in trace amounts in quartz–sulphide ±gold veins cutting Upper Dalradian gritty quartzites, quartz–mica schists and amphibolites. It forms 100 µm-diameter complex intergrowths with chalcopyrite, freibergite, gersdorffite, sphalerite and in association with arsenopyrite, benleonardite and hessite (Ixer et al. 1997).

At Tynagh mine, near Killimor, Co. Galway, Ireland, epigenetic Cu–Pb–Zn–(Ag) mineralization contains a minor sulphosalt assemblage that includes: bournonite, enargite, meneghinite, pyrargyrite, polybasite and stephanite. They occur with galena and tennantite–tetrahedrite (Schultz 1966b).

In the Channel Isles, a Pb–(Ag)-bearing brecciated vein cutting a foliated horn-

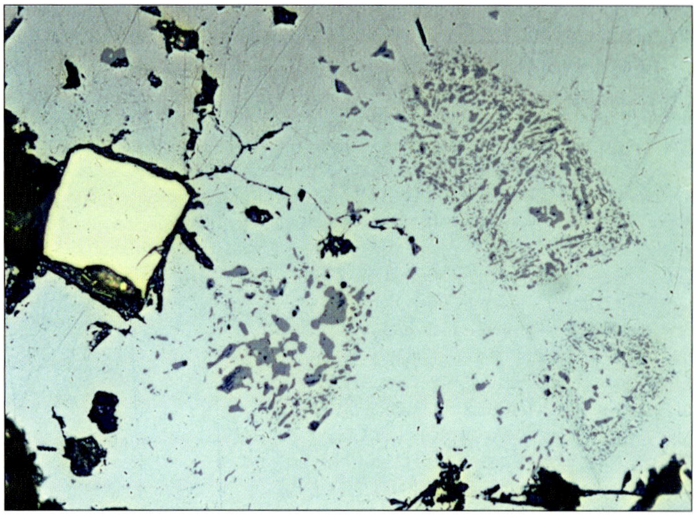

Dark grey intergrowths of polybasite in galena; the yellow cube is pyrite. Sark's Hope mine, Little Sark, Channel Isles; width 150 µm. (© C. J. Stanley, The Natural History Museum, London)

blende granite of Late Precambrian age at Sark's Hope mine, Little Sark, contains a minor suite of silver-bearing minerals that includes acanthite, enargite, pearceite, polybasite, pyrargyrite and tennantite–tetrahedrite. The primary ore is predominantly galena and pyrite (Ixer & Stanley 1983, Ixer 1992).

polyhalite

triclinic $K_2Ca_2Mg(SO_4)_4{\cdot}2H_2O$

A locally abundant hydrated potassium–calcium–magnesium sulphate found in the Lower Evaporites of the Yorkshire potash field from Eskdale No. 2 borehole at Aislaby, Eskdale, near Whitby, Yorkshire. Polyhalite occurs as fine-grain aggregates (0.001–0.2 mm across) of irregular crystals arranged in fibrous or spherulitic groups. Larger subhedral or euhedral grains up to 1.5 mm long are occasionally found associated with anhydrite, lining halite cavities, or enclosed in halite. In thin section, simple, lamellar or sector twinning is common. It is thought that polyhalite is a replacement product of anhydrite and halite, and the species also pseudomorphs gypsum (Tilley 1943, Stewart 1949). In the Lower Evaporites from the Eskdale No. 3 borehole at Sleights, Eskdale, polyhalite forms a mosaic of closely interlocking crystals and is partially replaced by kieserite (Armstrong et al. 1951). In the Fordon F.3 borehole, Humberside (13 km S of Scarborough, Yorkshire), the Lower Evaporites contain a bed consisting of 66% anhydrite and 33% polyhalite, both with minor amounts of glauberite, halite, magnesite, pyrite, talc and bituminous material (Stewart 1963). Small rounded aggregates of polyhalite partly coated in clay make up a minor component of the Carnallitic Marl in the Eskdale No. 3 borehole (Armstrong et al. 1951). Summaries of the Yorkshire evaporite deposits are given by Dunham (1960) and Stewart (1965). Additionally, polyhalite is noted from Boulby potash mine, Loftus, Cleveland (Anthony et al. 2003).

polylithionite?

MICA GROUP

monoclinic $KLi_2AlSi_4O_{10}F_2$

A rare potassium–lithium aluminosilicate fluoride found in peraluminous granites and granite pegmatites. Iron-rich polylithionite is described as a minor but widely distributed late- to post-magmatic mica that partially overgrows siderophyllite in the Land's End granite, Cornwall. Cordierite and muscovite are associated species (Müller et al. 2006). The authors use the classification scheme of Tischendorf et al. (2001), which in this case hides the fact that the compositions are far removed from the polylithionite end member. Lepidolite is a more appropriate name for these micas.

polyphant stone

TL Polyphant, Cornwall, England

A greyish-green pot-stone flecked with white and brown, and used as an ornamental stone in churches since Norman times.

polysphaerite

A globular variety of pyromorphite.

polytelite
A synonym of freibergite, tetrahedrite group, but also used to describe a variety of tetrahedrite.

portlandite
BRUCITE GROUP
hexagonal $Ca(OH)_2$
TL Scawt Hill, Larne, Co. Antrim, Ireland

A rare calcium hydroxide found in the chalk/dolerite contact zone at Scawt Hill, Larne, Co. Antrim, Ireland, and named from the fact that the mineral is a common product of the hydration of Portland cement. Portlandite forms colourless hexagonal platy crystals associated with afwillite, calcite and ettringite, all lining cavities in a larnite–spurrite rock. The manner of occurrence indicates that both portlandite and afwillite form by hydration of earlier-formed calcium orthosilicates (Tilley 1933). Further investigations of vug minerals from Scawt Hill revealed portlandite intimately associated with afwillite, ettringite and hydrocalumite in a larnite rock carrying melilite–wollastonite aggregates (Tilley 1934).

Also in Ireland, larnite rocks at Ballycraigy, Larne, Co. Antrim, from a similar environment to that at Scawt Hill, contain hydration cavities infilled with afwillite, portlandite and a calcium-silicate hydrogel (Long & McConnell 1959). At Carneal, Larne, Co. Antrim, in a chalk/dolerite contact zone associated with a Tertiary volcanic plug, a secondary assemblage of portlandite with afwillite, calcite, killalaite and plombièrite forms veinlets and cavities in larnite-rich rocks. Portlandite forms a pale-bluish coating on the surface of a larnite–spinel rock and has developed microscopic crystals (<0.2 mm) intergrown with afwillite, from which it is distinguished by slight turbidity, lower refractive index and a perfect basal cleavage (Sabine & Young 1975, Nawaz 1977).

posnjakite
monoclinic $Cu^{2+}_4(SO_4)(OH)_6{\cdot}H_2O$

A rare hydrated copper sulphate hydroxide found as crystalline crusts, associated with brochantite and langite, coating the slaty mudstone walls of a now-inaccessible stope in the disused workings of Drakewalls mine, Calstock, Cornwall. Posnjakite is a post-mining formation deposited by the steady trickling of copper-bearing waters down the mine walls. It also occurs with langite on specimens at the NHM, originally collected from Fowey Consols mine, Tywardreath, St Blazey, Cornwall (Knight & Barstow 1970, Braithwaite 1982b). It forms blue subhedral blocky crystals up to about 0.2 mm long as a component of bluish-green crusts at Greystone quarry, Lezant, Cornwall (Elton et al. 2004). At Littleham Cove, near Budleigh Salterton, Devon, posnjakite is identified as an alteration product at the margins of a native copper "plate" found associated with uranium–vanadium-rich Permian nodular concretions that have formed in localized horizons within the Littleham Mudstone Formation (Milodowski et al. 2002).

Subhedral blue grains of posnjakite are found with langite, ramsbeckite and schulenbergite in thin carbonate-rich veinlets within oxidized sphalerite at Driggith mine, Caldbeck Fells, Cumbria (Green et al. 2001, 2006a). On the old dumps at Silver Gill, also in the Caldbeck Fells, blocky turquoise-blue crystals of posnjakite occur with langite, linarite and wroewolfeite on altered sulphides (Cooper & Stanley 1990, Green et al. 2005b). Elsewhere in Cumbria, posnjakite occurs with langite at Paddy End mine, Coniston, Eskdale, where it forms turquoise-blue spherules up to 1 mm in diameter on tennantite, associated with brochantite and malachite, and at Spothow Gill, Eskdale, where it forms a turquoise-blue microcrystalline crust on heavily weathered chalcopyrite–quartz veinstone. At these localities the langite and posnjakite have a post-mining origin (Young & Johnson 1985).

Posnjakite forms thin blue crusts associated with antlerite, native copper, cuprite, langite and pharmacosiderite in a barite-rich gossan at the northwest end of Engine Vein mine, Alderley Edge ridge, Cheshire (Braithwaite 1994). At Waterbank mine, an abandoned Cu–Pb–Zn mine on Ecton Hill, 2 km NNW of Wetton village, Staffordshire, posnjakite is a fairly common species, forming tiny greenish-blue blocky pseudo-hexagonal crystals up to 0.4 mm long, as either crusts or scattered single crystals. It may be intergrown with serpierite or associated with linarite, malachite or ramsbeckite (Rust 1995).

Excavations in undisturbed dump material at Penrhiw mine, Ystumtuen, Central Wales orefield, Dyfed, revealed an unusual suite of secondary Cu–Pb–Zn minerals that included exceptional ramsbeckite as well as anglesite, brochantite, cerussite, covellite, cuprite, hemimorphite, hydrozincite, lautenthalite, leadhillite, linarite, malachite, posnjakite, queitite, schulenbergite, serpierite, native sulphur and wroewolfeite. Posnjakite forms poorly defined corroded tabular crystals to 1.5 mm and lines cavities in siltstone blocks. It can overgrow ramsbeckite and schulenbergite, or, in a different paragenesis, these two species may overgrow it (Mason & Green 1995).

Posnjakite as typically poorly formed intergrown crystals; Penrhiw mine, Ystumtuen, Dyfed, Wales; width 2 mm. (© D. I. Green)

A few km to the SE at Copa Hill mine, Cwmystwyth, blue tabular crystals of posnjakite, associated with brochantite and galena, are reported by Bevins (1994) and, at Sneyd's level, Britannia mine, Snowdonia, Gwynedd, it forms stalactitic growths with brochantite (Bevins et al. 1985).

An examination of the dumps of West Blackcraig mine, near Newton Stewart, Dumfries & Galloway, Scotland, revealed a suite of primary sulphides – chalcopyrite, galena and sphalerite – together with many characteristic secondary alteration products; aurichalcite, brochantite, cerussite, hemimorphite, hydrozincite, linarite and malachite. Langite, dark-blue posnjakite and wroewolfeite were found on two specimens (Livingstone et al. 1976). A visual identification of deep-blue glassy posnjakite has been made on material collected from the dumps of the Meadowfoot smelter, Wanlockhead, Dumfries & Galloway (Green 1987a).

In Ireland, posnjakite occurs as <0.2 mm crystals or blue crusts at Muckross mine, Killarney National Park, Co. Kerry, and was formed by post-mining oxidation within the mine waste (Ryback & Moreton 1993, Moreton et al. 1999). At Gortdrum mine, near Tipperary, sparse patches of posnjakite forming bluish-green blades up to about 1 mm long were found associated with erythrite and lavendulan on a greyish quartz-rich matrix from the mine dump (Braithwaite & Wilson 2003).

potarite
tetragonal PdHg

A very rare palladium mercury alloy found associated with alluvial gold grains from both Brownstone and Whympston, in the South Hams district of Devon. Potarite forms dendritic growths and rims on native gold and often there is a spectrum of intermediate compositions between the two species (Leake et al. 1991). In lherzolites and dunites of the Lizard complex at Carrick Luz, potarite is described as the dominant PGE species, occurring with laurite, sperrylite and various alloys. It is associated with altered chalcopyrite and pentlandite or secondary magnetite and probably has a magmatic origin modified by local hydrothermal redistribution (Hutchinson 2001, Gunn & Styles 2002).

Chromite containing a single inclusion of potarite has been described (and qualitatively analyzed) from the chromite quarry at Cliff, Baltasound–Haroldswick area, Unst, Shetland, Scotland (Prichard et al. 1986, 1988). Potarite is also possibly present in the unit 6–7 boundary chromitite of the Eastern Layered Series in the layered mafic–ultramafic intrusion on the Isle of Rum, Highland Region. Associated rare species positively or tentatively identified include native gold, irarsite, moncheite, platarsite, sperrylite and telluropalladinite (Power et al. 2000a).

potash alum
A synonym of potassium alum.

potash mica
A synonym of muscovite, mica group.

potassic-ferro-eckermannite*
AMPHIBOLE GROUP
monoclinic $(Na,K)Na_2(Fe^{2+}_4Al)Si_8O_{22}(OH)_2$

A rare sodium–potassium–iron aluminosilicate hydroxide that forms series with eckermannite and with ferro-eckermannite. Potassic-ferro-eckermannite occurs in the Pendennis peralkaline minette, a minor intrusion cutting Devonian slates on the foreshore of Pendennis Point, Falmouth, Cornwall. All crystals are compositionally heterogeneous with a short prismatic habit and some have fibrous core regions (Hall 1982). Associated amphibole species are magnesio-arfvedsonite, potassic-magnesio-arfvedsonite and magnesioriebeckite (using the classification of Leake et al. 1997a). Potassic-ferro-eckermannite is not listed as a species by Mandarino & Back (2004), suggesting it is a "named amphibole" (Burke & Leake 2005).

potassic-magnesio-arfvedsonite
AMPHIBOLE GROUP
monoclinic $(Na,K)Na_2(Mg_4Fe^{3+})Si_8O_{22}(F,OH)_2$

A rare sodium–potassium–magnesium–iron silicate hydroxide that forms series with arfvedsonite and with magnesio-arfvedsonite. Potassic-magnesio-arfvedsonite occurs in the Pendennis peralkaline minette, a minor intrusion cutting Devonian slates on the foreshore of Pendennis Point, Falmouth, Cornwall. All crystals are compositionally heterogeneous with a short prismatic habit and some have fibrous core regions (Hall 1982). The amphibole classification of Leake et al. (1997a), Mazdab (2003) and Mogessie et al. (2004) indicates that potassic-magnesio-arfvedsonite is associated with magnesio-arfvedsonite, potassic-ferro-eckermannite and magnesioriebeckite.

potassicpargasite
AMPHIBOLE GROUP
monoclinic $(Na,K)Ca_2(Mg,Fe,Al)_5(Si_6Al_2)_{22}(OH,F)_2$

A rare sodium–potassium–calcium–magnesium–iron aluminosilicate hydroxide that forms series with pargasite and with ferropargasite. Large phenocrysts of potassicpargasite (originally described as hornblende) were found in a monchiquite dyke at Wart Holm, Copinshay, Orkney Isles (Flett 1900). Potassicpargasite from a marble on the Isle of Tiree, Strathclyde, Scotland, is mentioned in Bernard & Hyršl (2004), but the data are unpublished. If the identification is based on data in Hallimond (1947), where K=0.48 apfu, the mineral could be referred to as potassian pargasite if an optional modifier is used.

potassium alum/alum-(K)
cubic $KAl(SO_4)_2{\cdot}12H_2O$

An uncommon hydrated potassium aluminosilicate found in Scotland as fibrous white efflorescences on shales at Hurlet, near Paisley, Strathclyde, and Moffat,

Potassium alum growing on the walls of an arsenic calciner at Tolgus, near Redruth, Cornwall; width 170 μm; SEM image. (© M. R. Power, Camborne School of Mines)

Dumfries & Galloway. At the first locality, potassium alum is associated with alunogen and tamarugite (Livingstone 2002). Potassium alum is said to be common at many other locations in Scotland where shales containing pyrite have been exposed to the action of the weather (Heddle 1901).

A record of potassium alum and gypsum by Hall (1868a) was not confirmed by Moreton et al. (1995) during their thorough investigation of sulphate mineralization in Carboniferous shales at Ballybunnion, Co. Kerry, Ireland.

Potassium alum is also briefly noted in slate at Newport, Dyfed, Wales (Greg & Lettsom 1858), in alum shales from near Whitby, Yorkshire, and in clay at Chudleigh, Devon. It is also found growing as a drusy crust on an arsenic calciner wall at Tolgus, near Redruth, Cornwall. Other secondary growths on the wall consist of dendritic gypsum, scorodite and an unidentified aluminium–arsenic sulphate containing tiny inclusions of possible hemloite (M. R. Power pers. comm. 2004).

potstone
A synonym of clinochlore and talc.

powellite
tetragonal $CaMoO_4$

An uncommon calcium molybdate that forms a series with scheelite and is found at Traprain Law, near Haddington, Lothian, Scotland. A single subhedral crystal of powellite was found, with analcime, apophyllite and calcite, in a geode within phonolite. It is greenish-black in colour, about 1 cm long, and is apparently a primary deuteric phase (Battey & Moss 1962). Powellite is very rare at the rediscovered classic "Heddle" locality for molybdenite at Coire Buidhe, Glen Creran, Appin, Strathclyde Region, where yellow squat octahedral bipyramids to 10 mm have been found in quartz veins cutting granite (Ingram et al. 1993). XRD analysis on the original specimens showed them to be scheelite (Starkey 1995), but yet further analysis identified powellite in addition to scheelite. Powellite appears to have formed during the weathering of molybdenite (Ingram et al. 1996).

In Wales, powellite has been found as colourless to pale-yellow pyramidal crystals, up to 0.4 mm across, in a cavity within altered dolerite at Benallt Mine, Rhiw, Lleyn Peninsula, Gwynedd (Dossett et al. 2007).

Colourless to pale-yellow blocky prisms of powellite up to 0.4 mm across in a small cavity within altered dolerite; Benallt mine, Rhiw, Lleyn Peninsula, Gwynedd, Wales. (© D. I. Green)

Powellite from Carrock mine, Grainsgill, Caldbeck Fells, Cumbria (Hartley 1984, Young 1987a, Cooper & Stanley 1990), is less certain, as the specimens are assumed to have a Kingsbury connection. The work of Ryback et al. (1998, 2001) indicates that the provenance of many Kingsbury samples is in doubt, and further investigation of powellite from this locality is therefore desirable.

prase
A translucent leek-green variety of quartz.

prasine
A synonym of pseudomalachite.

prasolite or prasilite

TL Kilpatrick Hills, near Glasgow, Strathclyde, Scotland

A soft leek-green fibrous iron-rich clinochlore, chlorite group. Also described as a synonym of cordierite and a variety of saponite, smectite group.

prehnite

orthorhombic $Ca_2Al_2Si_3O_{10}(OH)_2$

A common calcium aluminosilicate hydroxide found in hydrothermal veins and cavities in basaltic rocks. Prehnite is a typical product of low-grade metamorphism.

England

Cornwall At Dean quarry, NE of Coverack, Lizard Peninsula, Cornwall, several different forms and colours of prehnite occur in hydrothermal veins cutting schistose to pegmatitic gabbro. A green botryoidal variety grows on colourless calcite on which analcime, brown calcite, heulandite, natrolite and stilbite have all grown later. Other varieties include: colourless-to-white botryoidal; pale-green to white, thick tabular euhedral crystals up to 3 mm long; small white tabular euhedral crystals up to 1 mm long; and pale-brown tabular crystals and colourless, glassy euhedral crystals up to 2 mm long (Seager 1969, 1971, Wolters & Wolters 1986). Elsewhere on the Lizard Peninsula, prehnite occurs as: tabular and prismatic crystals of greenish-white, up to 1 mm across, encrusting hornblende schist at Parc Bean Cove, Mullion (Russell 1912a); colourless or pale-green botryoidal crystals associated with analcime and calcite in veins cutting hornblende schist on Porthkerris Point, 3 km N of Dean quarry (Seager 1978b); fine-grain colourless crystals replacing brecciform gabbro at Coverack Cove and in altered gabbroic dykes at Enys Head, near Cadgwith (Hall 1979). Additionally, prehnite is noted from: Porthoustock quarry, St Keverne; Ghostcroft mine, Mullion; Predannack Head and Ogo Dour (Golley & Williams 1995), and Meneage, E of Falmouth (Barnes & Andrews 1981). Other occurrences of prehnite include: as pale-green crystals in a vein cutting hornfels between Botallack mine and Wheal Cock, near Wheal Button (Greg & Lettsom 1858, Bancroft & Weller 1993); Clicker Tor, Menheniot (Robson 1949b); small amounts in metasomatized adinoles at Dinas Head, near Padstow (Agrell 1939); Levant mine, near Pendeen (Collins 1892, Wolloxall 1989); and replacing plagioclase in metadolerites (greenstones) at Trevone Bay, W of Padstow (Floyd & Rowbotham 1982).

Cumbria Prehnite is found: with albite, apatite, biotite, hedenbergite and hornblende in gabbroic pegmatite from Carrock Fell and replacing plagioclase feldspar in diorite from Seathwaite How (Eastwood et al. 1968); as creamish-white radiating botryoidal crystals at Potts Gill mine, Caldbeck Fells (Young 1987a); as a rare constituent of epidote–garnet veins cutting metasomatized Borrowdale Volcanic Group rocks within the aureole of the Shap granite at Shap Blue quarry (Young et al. 1988, 1998); and with calcite and quartz in narrow veins cutting granitic aplites of the Haweswater Complex in the Walla Crag area (Nutt 1979).

Devon Prehnite occurs with albite, axinite, calcite, datolite, fluorite, microcline, pyrrhotite and quartz in narrow veinlets cutting aplite at the Meldon aplite quarry, near Okehampton (McLintock 1923, Dearman & Claringbull 1960, von Knorring & Condliffe 1984).

Durham Apophyllite, analcime, chabazite, prehnite and stilbite fill joints in quartz dolerite of the Whin Sill at Copt Hill quarry, Cowshill, Weardale, and Force Garth quarry, Teesdale (Young et al. 1991a).

Leicestershire Alteration haloes around calcite–zeolite veins cutting diorite at Croft quarry, SW of Leicester, contain radiating sheaves of prehnite and iron-rich pumpellyite associated with analcime, chlorite, kaolinite, quartz and sericite (Pearson & Jeffrey 1997).

Shropshire Prehnite forms thin crusts of pale yellowish-green crystals associated with calcite, edingtonite and harmotome on joint planes within dolerite at Squilver quarry, Disgwylfa Hill, 9 km NW of Bishop's Castle (Blythe 1942, Hubbard & Enamy 1993). At Shelve, near the Welsh border, prehnite occurs with albite, calcite, chlorite, clinopyroxene, pumpellyite and titanite in an alteration assemblage within metabasic lavas of Ordovician age (Bevins & Rowbotham 1983).

Yorkshire Prehnite from Kilnsey Moor (Rodgers 1975) is unconfirmed.

Wales

Prehnite is widely distributed in Wales, where it usually occurs as an alteration mineral in Ordovician basic lavas and intrusions associated with albitized plagioclase, calcite, chlorite, epidote, pumpellyite, stilpnomelane and titanite. Its origin is attributed to low-grade burial metamorphism (Bevins & Rowbotham 1983). Localities include the following.

Clwyd Hendre quarry, Pandy, near Glyn Ceiriog (Bevins & Rowbotham 1983).

Dyfed Allt Tre-hyrn, Llangynog (Cope & Bevins 1993); Garn Folch, near Fishguard (Bevins et al. 1993); Garn Turne Rocks, E of Wolf's Castle, and Mynydd Preseli (Bevins & Rowbotham 1983); Pen Caer area, near Fishguard (Metcalfe et al. 1994); St David's Head intrusion, Porth Llong (Roach 1969, Bevins & Rowbotham 1983, Bevins et al. 1994); forming spots in acidic tuffs at Strumble Head, near Fishguard, and at Y Garn, near Pontiago (Bevins 1982, 1994, Bevins & Rowbotham 1983).

Gwynedd Aberdaron, Lleyn Peninsula, and Llanfachraeth, Anglesey (Bevins & Rowbotham 1983); in soils derived from local doleritic and gabbroic intrusions at Bwlch y Ddeufaen, Conwy Valley, Snowdonia (Jenkins & Ball 1964); Gimlet Rock, near Pwllheli (Harrison 1897); Penmaenmawr intrusion, near Conwy (Sargent 1916, 1925); Rhiw intrusion, southwest Lleyn Peninsula (Cattermole 1976); and Tal y Fan intrusion, SW of Conwy (Bevins & Merriman 1988).

Powys Breidden Hills and Builth Wells (Bevins & Rowbotham 1983); Llanelwedd quarry, near Builth Wells (Nicholls 1958, Bevins 1985, Metcalfe 1990).

Scotland

Prehnite is an important indicator mineral in Lower Palaeozoic prehnite–pumpellyite facies spilites and basic-clast greywackes across the Southern Uplands of Scotland, where it occurs in metamorphic assemblages that also contain albite, chlorite, epidote, phengite and titanite (Oliver & Leggett 1980). However, prehnite is perhaps best represented by occurrences in Carboniferous Clyde Plateau basalts farther north. At Bishopton, excavations for the railway tunnel on the Glasgow–Greenock line produced perfectly botryoidal yellowish-green translucent prehnite with individual botryoids up to 1.5 cm across (Heddle 1901, Livingstone 2002). In the 1970s prehnite associated with greenockite and otavite was found in basaltic rubble between the Bishopton railway track and the M8 motorway (Meikle 1992b, Faithfull & Hubbard 2001, Green & Todd 2001). Boyleston quarry, to the SW of Barrhead, worked vesicular basalt and produced some of the finest prehnite ever collected in Scotland. Meikle (1989a) describes a specimen of prehnite consisting of greyish-yellow botryoids up to 25 mm across on an opaque base of prehnite coloured deep pink by included native copper. Acanthite, analcime, calcite, cuprite, lepidocrocite (rubinglimmer), malachite, natrolite, native silver and thomsonite are other associated species found at this location (Heddle 1887a, 1901, Meikle 1992a). Some of the prehnite is transparent enough to facet small gems (Green & Todd 2001). Prehnite and other secondary species are thought to have formed because of active volcanogenic geothermal processes during Lower Carboniferous times (Hall et al. 1989). Prehnite is associated with natrolite and thomsonite in lavas on Hartfield Moss, 9 km WSW of Barrhead (Greg & Lettsom 1858, Meikle 1990), and, at Loanhead quarry, near Beith, superb cabinet specimens of dark apple-green botryoidal prehnite have been found together with calcite and thomsonite (Green & Todd 2001). Many other species are found infilling vesicles and fractures in the lavas and a crosscutting dolerite dyke at Loanhead (Livingstone 1989a, Meikle 1989b). Across the River Clyde from these localities is Bowling (basalt) quarry in the Kilpatrick Hills, near Glasgow, where Heddle (1901) found large quantities of prehnite in two great veins. More recently, translucent botryoidal masses have been found (Green & Todd 2001). This is probably the locality described as Frisky Hall by Greg & Lettsom (1858), where sulphur-yellow prehnite was found.

On the Isle of Mull, Strathclyde, prehnite is widely distributed in a band of lavas around the Mull central intrusive complex. It has been used as an index mineral to calculate the original thickness of the lava pile (Walker 1971). White orbicular crystalline aggregates of prehnite overlain with scolecite have been found on Maol nan Damh, a shoulder on Ben More, Mull (Currie 1898), where cubes of pyrite enclosed in prehnite are also reported (McLintock 1910b). Vesicles in lavas on Beinn Fhada to the N of Ben More contain prehnite associated with epidote, garnet and common albite (McLintock 1910b). Very unusual prehnite has been found during the upgrading of the A484 road S of Tobermory, where stalactites up to 10 mm long were found in cavities in basalt (Green & Todd 2001). Many more localities are described by Heddle (1901) and some of these (and other occurrences) are listed below.

Borders Quartz veins cutting a metasomatized Carboniferous basaltic sill at Haddenrig quarry, E of Kelso, contain small amounts of prehnite (Tomkeieff 1941).

Fife Prehnite forms white or pale-green globular masses with crystal terminations on the sides of pectolite blades, filling joints in Carboniferous lava at Orrock quarry, near Burntisland (Wirth 2003).

Grampian Prehnite is associated with gyrolite, laumontite and xonotlite in stringers and fracture-filled veins cutting gabbro at Binhill quarry, Cairnie, Huntly (Livingstone 1974); hydrothermally altered hornblende schist near Bowman Hillock, 4 km W of Huntly, now consists of halloysite (metahalloysite) and prehnite (Wilson 1974); pale-green to colourless prehnite occurs with anorthite, diopside (coccolite) and vesuvianite (idocrase) at Delnabo limestone quarry, Glen Gairn (Heddle 1901).

Highland On the mainland: minor amounts of prehnite, pyrrhotite, quartz, serpentine

Botryoidal prehnite; Bishopton railway tunnel, Strathclyde, Scotland; width of specimen is 7 cm. (© M. P. Cooper)

and titanite (sphene) occur in Moine Series impure marbles in the Ardgour district. Prehnite is considered to have a widespread distribution in metasomatically altered limestones, such as that at Coire Dubh, near the Corran ferry crossing, Ardgour (Drever 1939); biotite in diorite belonging to the Ben Nevis diorite–granite complex shows alteration to chlorite or prehnite (Haslam 1970); small masses of metasomatized hornblendic rocks of Moinian age contain reddened iron-rich prehnite in cm-wide veins at White Bridge, near Invergarry (Rock 1984); at Camphouse, Ardnamurchan, prehnite occurs with analcime, hydrogarnet, thomsonite and xonotlite in contaminated augite diorite close to the contact with Jurassic limestone (Agrell 1951). Isle of Skye: prehnite is one of many species found in primary skarns near the contact of Durness Limestone with the Beinn an Dubhaich granite, Broadford area (Tilley 1951); prehnite is an accessory mineral in the Glamaig and Marsco epigranites, Western Redhills, where it infills interstices and small drusy cavities in association with epidote and laumontite (Thompson 1969); fine greenish masses of prehnite line drusy cavities up to 5 cm across in Tertiary basalt at Sgùrr nan Cearcall within the Cuillin metamorphic aureole (Wolfe 1998, Green & Todd 2001).

Lothian Prehnite occurs with pectolite in basalt on Castle Hill and with analcime and datolite in basalt on Salisbury Crags, both Edinburgh (Greg & Lettsom 1858, Heddle 1901). Taylor (1884) also describes prehnite from Salisbury Crags and from similar rocks on Corstorphine Hill, Edinburgh.

Strathclyde Altered gabbroic dykes associated with serpentinite on Byne Hill, belonging to the Girvan–Ballantrae ophiolite complex, contain chlorite, diopside, epidote, hydrogrossular, pectolite, prehnite and zoisite (Bloxam 1954, 1964). A related garnet–hornblende clinopyroxenite exposed between Knockormal and Carleton Mains farms contains rare platy prehnite (Smellie & Stone 1984) and, more generally, prehnite is found with pectolite in veinlets formed because of secondary low-temperature alteration of the complex (Holub et al. 1984); vugs in a basaltic plug of probable Carboniferous age on Castle Hill, 3 km WSW of Kilbirnie, contain pectolite and prehnite and very rarely tobermorite (Webb 1971); prehnite is associated with ankerite, barite, calcite, fluorite, goethite, gypsum, hematite, quartz, romanèchite and tourmaline in geodes within trachyte at Craigmuschat quarry, Gourock (Nicholson & Durant 1986); prehnite and calcite occur as fracture infills in epidotized vent agglomerate close to a marble contact at Creag an Fheidh, near Glenloig, Isle of Arran (Cressey 1987).

Tayside Pale-green or pink, and sometimes crystalline, prehnite with chalcocite, native copper, datolite and malachite have been found at Glen Farg, 16 km S of Perth (Heddle 1901).

Western Isles Primary augite and labradorite in very coarse-grain gabbroic pegmatite cutting gabbro has been replaced by albite, chlorite, epidote and prehnite at the Cambir (An Campar), Mullach Bi and Mullach Sgar, all on St Kilda (Hirta) (Harding et al. 1984).

Ireland

Prehnite replaces biotite in appinitic rocks associated with the Ardara granite near the summit of Meenalargan Hill and SW of Kilrean post office, Co. Donegal (Hall 1965a). Minor prehnite is noted with albite, clinozoisite, titanite and tremolite in a metamorphosed basic intrusion (now amphibolite) at the top of Camaderry Mountain, between Glendasan and Glendalough, Co. Wicklow (Johnston 1949, Ryback & Moreton 1993). Widespread secondary (andradite–grossular–hydrogrossular) garnet, epidote, prehnite and titanite occur throughout the Galway granite, Co. Galway

(Leake 1998), and, in the Carlingford area around Barnavave quarry, Co. Louth, prehnite occurs as an accessory species in calc-silicate rocks metasomatized by a syenitic intrusion (Osborne 1932b). Nockolds (1938a) notes colourless prehnite enclosing needles of epidote in the interstices of a hybridized eucrite mass at Barnavave, Carlingford, Co. Louth.

Offshore territories

The Lukis and Island Museum catalogue, Guernsey, Channel Isles, notes imperfect rounded pale-greenish crystals of prehnite massed together in veins (the location is not given) and perfectly terminated crystals possibly from Fort Joli (Mourant 1939). In Jersey, a pale-green variety of prehnite with somewhat barrel-shape crystals has been found in pegmatitic cavities within diorite at Ronez quarry (Mourant 1961).

preisingerite

triclinic $Bi_3O(OH)(AsO_4)_2$

A very rare bismuth arsenate hydroxide found in fractures within quartz veinstone on an old NHM specimen from Carrock mine, Caldbeck Fells, Cumbria. Preisingerite occurs intergrown on a microscopic scale with another bismuth arsenate, rooseveltite, and together they form small glassy light-grey massive material infilling fractures up to 2 mm across. They occur with joséite and unidentified powdery yellow bismuth ochres (Rumsey 2007).

priceite?

triclinic $Ca_4B_{10}O_{19}\cdot 7H_2O$

A rare hydrated calcium borate said to have been collected by A. W. G. Kingsbury from a pegmatite in the Meldon aplite quarry, near Okehampton, Devon (Embrey 1978). However, Ryback et al. (1998, 2001) and other authors have expressed serious doubts about the provenance of many Kingsbury samples (now at the NHM) and, until further discoveries are made, priceite is not confirmed.

prochlorite

A synonym of ripidolite, a silica-poor chlorite-group mineral, now referred to as clinochlore.

prosopite

monoclinic $CaAl_2(F,OH)_8$

A rare calcium–aluminium fluorhydroxide found in veins cutting Lower Carboniferous limestones at Coldstones quarry, 5 km W of Pateley Bridge, Yorkshire. The veins carry fluorite, barite, galena and sphalerite, and in places the supergene minerals cerussite, hemimorphite and smithsonite occur, commonly lining cavities. Fallen blocks from the Garnet vein also contain prosopite, doyleite and otavite. Prosopite occurs as white-to-pale-blue-tinted chalky or soft earthy masses up to 10 mm across, and typically it encrusts fluorite or lines irregular open cavities within fluorite veinstone. Individual crystals are prismatic and 1–5 μm long (Young et al. 1997).

protolithionite

A variety of mica intermediate between annite and zinnwaldite.

proustite

trigonal Ag_3AsS_3

A rare silver sulpharsenide, dimorphous with xanthoconite and found in the oxidized and enriched zones of hydrothermal sulphide deposits.

Proustite is reported from brecciated quartz lodes (up to 1 m wide) at Wheal Ludcott, St Ive, near Liskeard, Cornwall. Mineralization was in four stages: 1. chalcopyrite, galena and sphalerite; 2. fluorite; 3. carbonate and iron sulphides; 4. silver mineralization. The deposit has produced some of the best examples in Britain or Ireland of argentite, stephanite and possibly proustite. Microprismatic proustite is intergrown with stephanite on quartz–siderite veinstone (Dines 1956, Rust 1983a). One of the lodes was also worked at nearby Wheal Wrey, where considerable amounts of capilliary silver and pyrargyrite were mined. Proustite and stephanite were also found (Dines 1956). Proustite is noted with a wide range of primary minerals in the Main lode of Dolcoath mine, Camborne, Cornwall, including argentite (acanthite), native arsenic, arsenopyrite, bismuthinite, cassiterite, chalcopyrite, cobaltite, galena, marcasite, pyrargyrite, pyrite, skutterudite (smaltite), sphalerite, uraninite (pitchblende) and wolframite (Hosking 1951).

In Wales, elongate prismatic mm-size crystals of deep-red proustite have been found at Dolyhir quarry, Old Radnor, Powys, on a single micromount specimen. This is the only silver mineral found at the quarry and it has been confirmed by XRD and EMP analysis (T. F. Cotterell pers. comm. 2003).

Proustite crystal, 1.0 mm in length, with barite and tennantite; Dolyhir quarry, Old Radnor, near Presteigne, Powys, Wales. (© D. I. Green)

In the Ballinglen area, between Tinahely and Aughrim, Co. Wicklow, Ireland, arsenopyrite–fluorite–pyrrhotite–scheelite mineralization occurs along the south-eastern margin of the Leinster granite in a swarm of narrow microgranite sheets cut by greisen-bordered quartz veinlets. Mainly within the veinlets, a diverse assemblage of minor mineral phases includes apatite, native bismuth, cassiterite, chalcopyrite, cosalite, dravite, galena, mackinawite, molybdenite, proustite, rutile, schorl, sphalerite, stannite and titanite. It is uncertain if the mineralization represents the deep-seated portion of a subvolcanic system (perhaps related to a tungsten porphyry) or is related to a granite pegmatite system (Steiger & Bowden 1982, 1986, McArdle & Kennan 1988).

Proustite was also found in a very high-grade pod of polymetallic silver mineralization cutting pyrite mineralization in the B zone of the Silvermines orebody, Co. Tipperary, Ireland. The silver mineralization included acanthite, argyrodite, miargyrite, proustite, smithite and xanthoconite, together with barite, galena and sphalerite (Gasparrini 1978, Taylor & Andrew 1979). Taylor (1984) also reports geocronite–jordanite, gersdorffite and pyrargyrite from the same zone. A comprehensive review of mineralization in the Silvermines District is provided by Moreton (1999). Chemical data for many of the ore species are provided by Zakrzewski (1989), who shows that the "proustite" is actually an intermediate composition between proustite and pyrargyrite, and that it forms veinlets and nestlets often in a myrmekitic intergrowth with galena.

przibramite

A cadmium-rich variety of sphalerite.

pseudoboléite/pseudoboleite

tetragonal $Pb_{31}Cu^{2+}_{24}Cl_{62}(OH)_{48}$

A lead–copper chloride hydroxide briefly described with boléite, cerussite, cumengéite, malachite and phosgenite from Tolcarne beach, Newquay, Cornwall (Ryback & Tandy 1992).

pseudobrookite

orthorhombic $(Fe^{3+},Fe^{2+})_2(Ti,Fe^{3+})O_5$

A rare iron–titanium oxide found in thermally metamorphosed basalts and bauxitic clays (now termed porcellanites) adjacent to an inclined plug of olivine dolerite at Tievebulliagh, near Cushendall, Co. Antrim, Ireland. In the altered basalts, pseudobrookite forms rods and thin blades (transparent and sienna brown) in cordierite and orthopyroxene-bearing assemblages. In the porcellanites, granular pseudobrookite is particularly concentrated in bands rich in cristobalite, iron-corundum and mullite (Agrell & Langley 1958).

At Sidhean Sluaigh, Strachur, by Loch Fyne, Strathclyde, Scotland, granular or lath-shape pseudobrookite forms honey-brown crystals, associated with cordierite, corundum (some sapphire blue), magnetite, mullite, sanidine and spinel in emery-like rocks adjacent to a dolerite plug. In thin section, pseudobrookite is deep russet in colour. The emery-like rocks represent the aluminous residues from banded pelitic, semipelitic and psammitic phyllites after a granitic fraction has been selectively remobilized (Smith 1965, Macpherson & Livingstone 1982).

pseudogaylussite

A pseudomorph of calcite after ikaite.

pseudo-hypersthene

A synonym of diallage, a type of pyroxene.

pseudoleucite

A nepheline and orthoclase pseudomorph after leucite.

pseudomalachite

monoclinic $Cu^{2+}_5(PO_4)_2(OH)_4$

An uncommon copper phosphate hydroxide, trimorphous with ludjibaite and reichenbachite, and found in the oxidized zone of hydrothermal copper deposits.

Pseudomalachite has been reported by Kingsbury (1954) from Old Gunnislake mine, Calstock, near Gunnislake, Cornwall; however, later analysis has shown that some specimens are of reichenbachite (Braithwaite & Ryback 1994). These authors were also able to confirm pseudomalachite from other Cornish locations: Penberthy Croft mine, St Hilary (bright-green minute spherules); Old Gunnislake and East Gunnislake mines, Calstock; South Wheal Frances, Illogan (with libethenite and malachite); Tolcarne mine, Camborne (arsenatian with libethenite); Wheal Carpenter, near Hayle, Gwinear (arsenatian); Wheal Phoenix and West Wheal Phoenix, Linkinhorne; and Wheal Unity, St Day. Betterton et al. (2000) add that the Penberthy Croft pseudomalachite is commonly found on heavily iron-stained quartz and rarely with bayldonite, chrysocolla, malachite and stilpnomelane. It forms pale crusts on fractured quartz and, in its spherulitic form, is coloured various shades of bluish-green, turquoise or darker green hues. Complete solid solution towards cornwallite also seems to occur. Other descriptions of pseudomalachite from southwest England include: Bampfylde mine, North Molton, Devon (Braithwaite & Ryback 1994); Bedford United mines, near Tavistock, Devon (Robson 1954); and, in Cornwall, Botallack, St Just (Rudler 1905); Gunheath china-clay pit, Carthew, St Austell (Golley & Williams 1995); Wheal Gorland, St Day, 2 km E of Redruth, – requires confirmation (Rust 1981b, Fisher 1988); Wheal Providence, near Leedstown, Gwinear (Kingsbury 1954) – requires confirmation.

Lustrous spheres of greenish-blue pseudomalachite, up to 0.2 mm in diameter; Penberthy Croft mine, St Hilary, Cornwall. (© D. I. Green)

In the Caldbeck Fells, Cumbria, arsenic-rich pseudomalachite is reported from: Low Pike, near Potts Gill mine – porcellanous pale turquoise-blue crusts on quartz with bayldonite; Potts Gill mine and Short Grain (Cooper & Stanley 1990, Young et al. 1990a, Braithwaite & Ryback 1994). It is possible that pseudomalachite also occurs at two other Caldbeck Fells locations – Driggith and Sandbed mines (Hartley 1984) – but these have not been subjected to rigorous analysis and, as the specimens were supposedly collected by A. W. G. Kingsbury, their provenance is in doubt (Ryback et al. 1998, 2001). Green et al. (2006a) go further, stating that the Driggith and Sandbed samples are clearly fraudulent and they conclude that some of the specimens are an exact match to material from Virneberg in the Rhineland, Germany. Other specimens examined by Green et al. (2006a) of suspect pseudomalachite from the Driggith–Sandbed mines proved to be other species, mainly rosasite. Pseudomalachite from Potts Gill could also be dubious, as it was identified on specimens in the Kingsbury collection at the NHM.

Olive-green to blackish-green partially vitreous pseudomalachite is confirmed from a 1 cm-long cavity in quartz from the barite mine dump at Auchencairn, Rerrick, Dumfries & Galloway, Scotland (Livingstone 1993, Braithwaite & Ryback 1994), and there are reports of the species from Cappagh and Coosheen mines, Co. Cork, Ireland (Ryback & Moreton 1991).

pseudomesolite

A zeolite-group mineral, chemically the same as mesolite but with different optical properties.

pseudophite

A compact variety of clinochlore, chlorite group.

psilomelane

An old name for several hard massive black manganese oxides, e.g. hollandite, romanèchite.

psittacinite

A synonym of mottramite, descloizite group.

ptilolite

A synonym of mordenite, zeolite group.

PUMPELLYITE GROUP

Monoclinic silicates with the general formula $Ca_2XY_2(SiO_4)(Si_2O_7)(OH)_2{\cdot}H_2O$, where X = Al, Fe^{2+}, Fe^{3+}, Mg, Mn^{2+}; and Y = Al, Fe^{3+}, Cr^{3+}, Mn^{3+}. Julgoldite-(Fe^{2+}) and pumpellyite-(Mg) are the two members of the group confirmed in Britain or Ireland, but, because of the lack of information concerning the oxidation state of iron when analysis is by EMP, it is not possible to confirm two other possible species from the group – pumpellyite-(Fe^{2+}) [$Ca_2Fe^{2+}Al_2(SiO_4)(SiO_7)(OH)_2{\cdot}H_2O$] and pumpellyite-($Fe^{3+}$) [$Ca_2(Fe^{3+},Mg)Al_2(SiO_4)(SiO_7)(OH,O)_2{\cdot}H_2O$]. For this reason, all iron-rich pumpellyite is described below under pumpellyite-(Fe)? and, for convenience, those identifications made only by optical methods are also grouped under this heading. In reality, some of the optical identifications may in fact refer to pumpellyite-(Mg), but, of course, confirmatory data are lacking.

pumpellyite-(Fe)?

PUMPELLYITE GROUP

trigonal $Ca_2Fe^{2+}Al_2(SiO_4)(Si_2O_7)(OH)_2{\cdot}H_2O$

A calcium–iron aluminosilicate hydroxide that develops in basic igneous rocks affected by low-grade metamorphism.

Pumpellyite (forming large colourless radiate sheaves), carbonate, epidote, muscovite and prehnite are secondary species formed during hydrous low-grade metamorphism of plagioclase feldspar in metadolerites (greenstones) at Trevone Bay, W of Padstow, Cornwall (Floyd & Rowbotham 1982). Alteration haloes around calcite–

zeolite veins cutting diorite at Croft quarry, SW of Leicester, contain radiating sheaves of prehnite and iron-rich pumpellyite associated with analcime, chlorite, kaolinite, quartz and sericite (Pearson & Jeffrey 1997)

In Wales, pumpellyite is widely distributed in Lower Palaeozoic basic lavas and intrusions, where it forms small prismatic crystals in albitized plagioclase feldspar, in association with chlorite, epidote and prehnite (Bevins 1994). At Llanelwedd quarry, near Builth Wells, Powys, pumpellyite occurs in altered Ordovician basic lavas that have been subjected to a submarine hydrothermal circulation system. Pumpellyite occurs in vesicles and vein-like alteration patches, and is dark green to colourless and strongly pleochroic in thin section. Calcite, chlorite, prehnite and titanite are associated species. Locally, the rock is intensively altered and is now composed almost entirely of pumpellyite (Nichols 1958, Bevins 1985). Ordovician lavas and dolerites belonging to the Fishguard Volcanic Group, between Fishguard and Strumble Head, Dyfed, have suffered low-grade metamorphism and contain a secondary mineral assemblage consisting of actinolite, calcite, chlorite, epidote, prehnite, pumpellyite, stilpnomelane and titanite. In alteration veins, pumpellyite occurs as stout prismatic crystals up to 0.4 mm long and as amygdale fillings, where it consists of long thin crystals in a radial form. Colourless to greenish-brown pleochroism typical of iron-rich pumpellyite is noted in thin section. Pumpellyite data are presented for Fishguard Volcanic Group dolerites outcropping at Treseissllt and Y Garn by Bevins (1978). The Tal y Fan olivine dolerite intrusion, SW of Conwy, Gwynedd, contains iron-rich pumpellyite in the outer 6–8 m of its central zone, in association with albite, chlorite, epidote, prehnite, quartz and titanite (Bevins & Merriman 1988). Elsewhere in Wales, pumpellyite is recorded from soils in the Conwy Valley district of Gwynedd, which Jenkins & Ball (1964) traced back to altered gabbros and dolerites in the vicinity of Bwlch y Ddeufaen. This pumpellyite has colourless to bluish-green pleochroism and it forms stellate and radiating groups of bladed crystals in the Bwlch y Ddeufaen dolerite. A second type of pumpellyite with a golden-yellow to very pale bluish-green pleochroic scheme could not be sourced. In addition, Bevins & Rowbotham (1983) report pumpellyite from metabasites belonging to the paratectonic Caledonides at many localities: Berwyn Hills, Clwyd; St David's Head and Mynydd Preseli, Dyfed; Snowdonia, Anglesey and the Lleyn Peninsula, Gwynedd; Breidden Hills, Powys; and Shelve, Shropshire. Rare pumpellyite is further noted in Precambrian doleritic rocks of the Llangynog inlier, SW of Carmarthen, Dyfed (Cope & Bevins 1993).

At Loanhead quarry, near Beith, Strathclyde, Scotland, highly altered basaltic lavas of Carboniferous age are cut by a Tertiary dyke and, close to the dyke, a distinctive secondary mineral assemblage has formed within vesicles in the lavas or on joint surfaces. Pumpellyite occurs sparingly as blackish-green fibrous blades, up to 1.5×0.2 mm, lying on or protruding from chlorite-rich surface linings. Albitic feldspar, analcime, calcite, chalcopyrite, clinozoisite, epidote, magnetite, prehnite and thomsonite are associated species (Meikle 1989b). Along the southern shore of Loch na Keal, Isle of Mull, Strathclyde, dark-green pumpellyite occurs in acicular crystal sprays with epidote, prehnite and quartz in small cavities within a minor doleritic intrusion (Green & Todd 2001). Basalts at Sgùrr nan Cearcall, Cuillins, Isle of Skye, Highland Region, on the edge of the Cuillin metamorphic aureole, contain pumpellyite as small well formed prismatic-to-acicular crystals. Here, calcite, epidote, laumontite, prehnite and quartz, as well as several uncharacterized secondary copper minerals, are the associated species (Green & Todd 2001). Pumpellyite is an important indicator mineral in Lower Palaeozoic prehnite–pumpellyite facies spilites and basic-clast greywackes across the Southern Uplands of Scotland, where it occurs in metamorphic assemblages that also contain albite, chlorite, epidote, phengite and titanite (Oliver & Leggett 1980).

pumpellyite-(Mg)

PUMPELLYITE GROUP

trigonal $Ca_2MgAl_2(SiO_4)(Si_2O_7)(OH)_2{\cdot}H_2O$

A calcium–magnesium aluminosilicate hydroxide that forms two series, with pumpellyite-(Fe^{2+}) and with julgoldite-(Fe^{2+}). Pumpellyite-(Mg) is probably much more common than the literature would suggest and is a widespread mineral in low-grade regional metamorphic rocks that form a melange of basic volcanics and greenstones on the Lizard Peninsula and in the Roseland–Gorran area of south Cornwall. Spongy aggregates and flecks of pumpellyite-(Mg) are found as very fine-grain (<8 µm) pale-green to colourless (in thin section) crystals. It also occurs in veins, where it forms acicular crystals up to 50 µm long, in association with albite, actinolite, carbonate, chlorite, quartz and, occasionally, prehnite. This assemblage is well developed at Great Perhaver Point, Nare Head, and inland around Caerhays, N of Portholland. At Porthallow on the Lizard Peninsula, pumpellyite-(Mg) occurs in the matrix feldspars of greenstones, where it occurs with actinolite and chlorite (Barnes & Andrews 1981). A single analysis of pumpellyite-(Mg) is presented by Floyd & Rowbotham (1982) from a metadolerite (greenstone) at Trevone Bay, W of Padstow, Cornwall.

Pumpellyite-(Mg) occurs around the margins and at the edges of spherules in this amygdale within basalt from Creag Brimishgan, Isle of Mull, Strathclyde, Scotland; later growth is iron dominant. Width 1.5 mm; viewed in PPL. (© A. G. Tindle)

Pumpellyite-(Mg) is confirmed from altered Ordovician spilitic lavas forming part of the Girvan–Ballantrae ophiolite complex near Knocknormal farm, 2 km E of Lendalfoot, Strathclyde, Scotland (particularly near outcrops of glaucophane schist and serpentinite). Pumpellyite-(Mg) is generally developed along the walls of quartz veins, where it forms needle-like clusters with a stellate habit, usually associated with albite and calcite. Pseudomorphs of calcite and pumpellyite-(Mg) after plagioclase feldspar indicate a low-temperature origin related to the albitization of plagioclase during low-grade metamorphism (Bloxam 1958, Bloxam & Allen 1959, Passaglia & Gottardi 1973). Pumpellyite containing more magnesium than that found at Knocknormal is recorded from Ordovician lavas at Llanelwedd quarry, near Builth Wells, Powys (Bevins 1985), but, because the data do not include information about the oxidation state of iron, it is not possible to confirm the species.

purple copper

A synonym of bornite.

pycnochlorite

An iron-rich clinochlore, chlorite group.

pyrargyrite

trigonal Ag_3SbS_3

A rare silver sulphantimonite, dimorphous with pyrostilpnite and usually found in hydrothermal veins as a primary late-stage low-temperature mineral, but the species can also form by secondary processes. It is described in older literature as red silver or ruby silver. Unfortunately, these names also refer to proustite, which had not been isolated from pyrargyrite at the time. Neither species can be satisfactorily confirmed if referred to by the older names, for instance, as ruby silver was from Wheal Duchy, near Callington, and Wheal Jewel, St Day, Gwennap, both in Cornwall (Collins 1904a, Rudler 1905). Fortunately, Greg & Lettsom (1858) note pyrargyrite from Wheal Duchy and this appears to be confirmed by Rust (1982a), who describes a large amount of native silver, acanthite (argentite) and pyrargyrite from Wheal Brothers (the new name given to Wheal Duchy when the mine reopened in 1833). Similarly, pyrargyrite is noted from Wheal Jewel by Collins (1892).

Pyrargyrite is associated with stephanite at Wheal Boys, Trewetha, St Endellion, Cornwall (Prior 1890), and, at Wheal Ludcott, St Ive, near Liskeard, the species occurred with acanthite (argentite), proustite, stephanite and large amounts of native capillary silver, where a lead lode was intersected by a later fissure (Hosking 1951, Dines 1956, Rust 1982a). Elsewhere in Cornwall, pyrargyrite is reported from: Dolcoath mine, Camborne; Harrowbarrow (Wheal Newton) mine, SW of Gunnislake; Herodsfoot mine, Lanreath; Holmbush mine, Stoke Climsland; North Dolcoath mine, Camborne; Silver Vein mine, Lostwithiel; Wheal Alfred, Phillack, near Hayle; Wheal Herland, Gwinear; and Wheal Mexico, Perranzabuloe (Collins 1892, Dines 1956, Rust 1982a, Golley & Williams 1995).

At Piers Gill, 1 km N of Scafell summit, Wasdale, Cumbria, a 60 cm-thick polyphase quartz vein cemented by ankerite, within the Scafell dacite, contains microscopic inclusions of pyrargyrite in an Ag–As–Au–Cu–Pb–Zn assemblage similar in many respects to that found in a porphyry-copper system. Associated species include acanthite (argentite), native gold (electrum), tetrahedrite, uytenbogaardtite and an unknown selenium–silver mineral (Aldridge 2005). The pyrargyrite described from Tynebottom mine, NW of Garrigill, near Alston, Cumbria, by Ixer & Stanley (1987), is most probably an error, as the samples came from the Kingsbury collection at the NHM. Ryback et al. (1998, 2001) and other authors have expressed serious doubts about the provenance of many samples in the Kingsbury collection and this could be one of them.

The only report of pyrargyrite in Wales is from Tyddyn Gwladys mine in the Dolgellau goldbelt, Gwynedd, where it is associated with freibergite and polybasite (Gilbey 1968, Bevins 1994).

In Scotland, pyrargyrite is found with tetrahedrite in fault-bounded quartz veins containing massive galena at Tyndrum, Central Region. Pyrargyrite forms discrete grains up to 100 µm across and also occurs as replacement rims around tetrahedrite (Pattrick 1978, Ridgway 1980, Pattrick 1985).

Pb–Zn–(Ag) ores from the Silvermines deposit, Co. Tipperary, Ireland, contain a wide range of silver-bearing species. Ruby silver is noted by Zakrzewski (1989), and other authors list acanthite (argentite), argentotennantite, argyrodite, freibergite, miargyrite, proustite, pyrargyrite, native silver, smithite, tennantite, tetrahedrite and xanthoconite, often in complex intergrowths. Many of the species were found in the B-zone of Mogul mine, within a remarkable lens of barite. The deposit formed in a classic sedimentary-exhalative environment (Taylor 1974, Moreton 1999). At Keel, Ardagh, Co. Longford, a disseminated, brecciated and fracture-filled Ba–Pb–Zn orebody has formed around the Keel fault system. Sphalerite is the main ore species, but it is in association with galena that the silver-rich species freibergite, jalpaite, pyrargyrite and tetrahedrite have been found (Slowey 1986, Ryback & Moreton 1993). Various sulphosalt species (bournonite, enargite, meneghinite, pyrargyrite, polybasite, stephanite) occur in microscopic amounts together with galena and tennantite–tetrahedrite at Tynagh mine, near Killimor, Co. Galway (Schultz 1966b), and very minor Ag–Cu–Pb sulphosalts (including argyrodite, boulangerite, bournonite, cylindrite, freibergite, jordanite and pyrargyrite) are noted from the Navan lead–zinc orebody, Co. Meath (Andrew & Ashton 1985, Ashton et al. 2004).

In the Channel Isles, minor pyrargyrite associated with chalcopyrite and tennantite–tetrahedrite has been found in a Pb–(Ag) lode. The vein is dominated by galena and pyrite, and it cuts foliated hornblende granite at Sark's Hope mine on Little Sark, 10 km E of Guernsey. Other primary species in the ore include acanthite, arsenopyrite, enargite, pearceite, polybasite and pyrrhotite (Greg & Lettsom 1858, Mourant & Warren 1934, Ixer & Stanley 1983, Ixer 1992).

pyribole

A collective term used to describe members of the pyroxene and amphibole groups.

PYRITE GROUP

Cubic sulphides, arsenides etc. with the general formula AXY or AX_2, where A = Au, Co, Cu, Fe, Mn, Ni, Os, Pd, Pt, Ru; and X and Y = As, Bi, S, Sb, Se, Te. Aurostibite, erlichmanite, geversite, laurite, penroseite, pyrite, sperrylite and vaesite are members of the group found in Britain or Ireland.

pyrite

PYRITE GROUP
cubic FeS_2

A very common iron sulphide that forms a series with cattierite and is dimorphous with marcasite. Pyrite is the most abundant sulphide species in the world and is found under a wide range of conditions: in hydrothermal veins, as magmatic segregations, as an accessory mineral in igneous rocks, in contact-metamorphic deposits, in metamorphic rocks, and as a diagenetic replacement species in sedimentary rocks.

England

Avon Joints in Carboniferous Limestone at Hampstead Farm quarry, Chipping Sodbury, and Woodleaze quarry, Tytherington, contain pyrite associated with barite, calcite, galena and sphalerite (Alabaster 1989a, 1990b). Southfields quarry, 2 km N of Chipping Sodbury, has produced isolated cubes and tubose pyrite associated with calcite in faulted mineralized (lead–zinc) limestone (Curtis 1981). Residual pyrite grains are also found in fractured iron-oxide-impregnated Upper Triassic breccias at East Harptree, N of the Mendip Hills (Alabaster 1990b). On the beach near the Pier hotel, Clevedon, minor pyrite has been found mantling grains of galena in a barite–sulphide vein deposit (Starkey 1986c, Ixer et al. 1993). Zachariades (1993) notes pyrite in other barite–sulphide veins at Stancombe quarry, near Flax Bourton, Bristol.
Cheshire Pyrite is a subsidiary species in chalcopyrite–galena-bearing veins and disseminations within Triassic sediments at Engine Vein and Wood mines, and in the Stormy Point area, Alderley Edge ridge. It forms cubes up to 100 µm across and is accompanied by nickel-rich pyrite (bravoite), which forms discrete single crystals (40–80 µm across) or small groups of crystals enclosed in later sulphides (Warrington 1965, Ixer 1974a, Carlon 1979, Ixer & Vaughan 1982, Braithwaite 1994). Pyrite, chalcopyrite and possibly chalcocite appear to have been the main ore minerals at West mine on Alderley Edge (Plant et al. 1999). Minute but beautifully perfect pyrite cubes are noted from Ratcliff Wood, near Macclesfield (Woodward 1883). At Bickerton, S of Warrington, in the Cheshire Basin, euhedral pyrite crystals up to 50 µm across are reported from barite-dominated redbed-type mineralization closely associated with marginal basin faults. The paragenetic sequence is: early pyrite (bravoite and nickeloan pyrite) occasionally with Cu–Fe–Ni sulpharsenide rims, followed by bornite and tennantite (the most abundant sulphides), in which chalcopyrite is found as lamellae (Naylor et al. 1989).
Cleveland Traces of pyrite have been noted in Permian evaporites encountered at Billingham mine (under the ICI factory site) (Raymond 1959, 1962) and Boulby potash mine, Loftus (Woods 1979, Green & Freier 1996).
Cornwall Pyrite (known locally as mundic) is widely distributed in many of the mineralized veins of the county, often occurring as a late arrival, and forming crystals encrusting vugs or as a constituent of late cross-course veins. In gossans it readily decomposes to give an earthy brown ore. Economic amounts have been recovered from most districts (apart from St Just) usually as a byproduct, but with large tonnages being mined from Wheal Jane and West Wheal Jane to the E of St Day (Dines 1956). *St Just mining district:* Balleswidden and Wheal Castle, near St Just – polymetallic vein containing cassiterite with minor galena, pyrite and sphalerite (Hosking 1951); Loe Warren Zawn, Botallack – minor pyrite and chalcopyrite in a Cu–Pb–U vein, very rarely as octahedra (Elton & Hooper 1995a,c); North Zawn, Botallack – spherulitic pyrite filling cavities in a quartz boulder (Elton & Hooper 1996); Geevor mine, near Pendeen – an exceptional vug lined with pyrite crystals up to 25 mm across, associated with cassiterite, fluorite and quartz (Harker 1973, Ball et al. 1982); Wheal Edward – pyrite, chalcopyrite and sphalerite associated with pitchblende (Ball et al. 1982). Fine specimens of pyrite have occasionally been found at Wheal Spearn, Botallack and Levant mines (Collins 1892). *St Ives–Helston mining district:* Penberthy Croft mine, St Hilary – euhedral-to-subhedral bright cubes of pyrite as individuals or groups up to 0.6 mm across (Camm & Merry 1991b, Betterton et al. 2000); Porthleven – minor pyrite, some as rims on siderite, in cross-course veins and disseminated pyrite in E–W veins (Gleeson et al. 2000); St Ives Consolidated mines, St Ives – pyrite is a minor constituent of the Great Carbona, a large cassiterite–tourmaline replacement deposit (Hosking 1951, Dominy et al. 1996a); St Michael's Mount, Marazion – pyrite is a minor component of cassiterite–löllingite–wolframite-bearing veins (Jackson & Rankin 1976, Moore & Moore 1979); Wheal Providence, near Carbis Bay, St Ives – pyrite and chalcopyrite occur as fine disseminations and small grains replacing pitchblende (Ball et al. 1982). *Camborne–Illogan–Redruth mining district (and southwards into the Carnmenellis granite and its environs):* Balmynheer mine; Halabezack farm, near Porkellis; East Wheal Lovell (Fatwork section) and Wheal Roots, all Wendron area – minor pyrite in cassiterite-dominated carbona deposits (Stephens 1940b, Dominy et al. 1996a); Bosworgey (granite) borehole, near Fraddam – euhedral-to-anhedral crystals of pyrite in accessory amounts (Beer et al. 1975, Ball & Basham 1984); Chywoon quarry, near Longdowns – tiny crystals dusting feldspars in granite pegmatite (Moulding & Hooper 2005); Dolcoath mine, Camborne (Hosking 1951); East Pool mine, Illogan, near Redruth – delicate stalactites of pyrite (Garby 1848); Flushing, near Falmouth – podiform pyrite with euhedral crystals and inclusions of native gold (Camm 1995); Kessel Downs (Halvasso) quarry, Carnmenellis – accessory pyrite in granite pegmatite (Couper & Barstow 1977, Jackson 1978); Mount Wellington mine, Twelveheads, 7 km W of Truro – large amounts of pyrite in cassiterite-bearing lodes and cross courses with excellent euhedral crystals up to 2 cm across (Kettaneh & Badham 1978, Rust 1982e); South Crofty mine, Camborne – fairly common, pyrite occurs massive, in cubes and rarely as pyritohedra (Dewey 1925, Ball et al. 1982, LeBoutillier 1996); South Tolgus shaft, Redruth – veinlets of pyrite in mica–hornfels and considerable amounts of pyrite, forming well defined cubes up to 12 mm on edge, associated with cassiterite in quartz veins (Davey 1948); Ting Tang mine (Morcom's shaft), Carharrack – pyrite fairly common as small metallic cleavages in vein quartz (Bruce & Aubrey-Jones 1998); Wheal Basset, near Carnkie, Illogan – minor pyrite in the copper–tin mineralized contact between the Carnmenellis granite and Devonian sandstones (Zwaan et al. 1989); Wheal Falmouth and Sperries – produced 48 000 tonnes of pyrite and lesser amounts of galena and copper ore (Stephens

1940a); Wheal Gorland, St Day, 2 km E of Redruth – small crystals lining cavities in vein quartz (Rust 1981b, Fisher 1988); Wheal Jane, Kea, near Truro – the B lode is mainly mineralized with cassiterite and tourmaline, but in some places massive pyrite and sphalerite occur; pyrite is also present in cross-course veins; pyrite forms excellent bright crystals to 1 cm, larger crystals to 4 cm and stalactitic, botryoidal forms to 10 cm; crystals form cubes mainly, but also octahedrons, pyritohedrons and other modifications (Rayment et al. 1971, Dunham et al. 1978, Rust 1982d, Holl & Bromley 1988, Dominy & Camm 1997); Wheal Virgin, part of Consolidated mines, Gwennap – large oreshoot of chalcopyrite, pyrite and quartz in slate breccia cemented by pyrite (Stephens 1940a). *St Agnes mining district:* Cligga Head, Perranzabuloe – pyrite and chalcopyrite associated with chloritic alteration in a late fracture system (Jackson et al. 1977); Duchy Peru mine, Perranzabuloe – minor galena–pyrite–sphalerite mineralization (Henley 1971); Wheal Hope, ENE of Perranporth – pyrite framboids in sediment partially replaced by supergene galena (Hosking & Camm 1981); Wheal Towan, near St Agnes – the Great Lode carries cassiterite, chalcopyrite, pyrite and sphalerite (Hosking 1951). *Wadebridge–St Endellion mining district:* This district is noted for its sulphide-bearing vein systems, especially galena–stibnite-bearing veins. However, an earlier assemblage is dominated by arsenopyrite–marcasite–pyrite mineralization (Clayton et al. 1990). Examples include: Bounds Cliff, E of Port Isaac (Clayton et al. 1990, Stanley et al. 1991); Pengenna mine, St Kew; Poltreworgy mine, near Port Isaac; Wheal Boys, Trewetha, St Endellion (all Russell 1949b); Pentire (Glaze) mine, near New Polzeath, St Minver, and Reedy Cliff, near Port Quin (Clayton et al. 1990); Port Quin mine (Russell 1949b, Clayton et al. 1990); Treore mine, Port Gaverne, near Port Isaac, St Teath – pyrite associated with gold (Russell 1949b, Kingsbury & Hartley 1960, Camm 1995). *St Austell mining district:* Fowey Consols, Tywardreath, St Blazey – finely crystallized pyrite (Greg & Lettsom 1858); Great Dowglas mine (Coffin lode), between Grampound and St Austell – pyrite associated with gold (Camm 1995); Mulberry mine, Lanivet, St Austell – polymetallic quartz veins containing arsenopyrite, cassiterite, pyrite, stannite and wolframite (Dominy et al. 1995); South Terras mine, St Stephen-in-Brannel – rare pyrite crystals in uranium ore (Robertson & Dines 1929, Dines 1956, Ball et al. 1982); Wheal Prosper (Lanhadron), St Ewe, SSW of St Austell (Russell 1949b). *East Cornwall mining district:* Extends from the eastern margin of the Bodmin Moor granite, past the granite cupolas of Kit Hill and Hingston Down towards the western margin of the Dartmoor granite in Devon. The Caradon and Phoenix copper mines all appear to have contained some pyrite and it was fairly abundant at Wheal Caradon (Caradon–Slade), NE of Pensilva, 6 km NE of Liskeard (Dines 1956, Shambrook 1982). Other localities include: Gold Diggings quarry, Linkinhorne – pyrite associated with bertrandite, fluorite, orthoclase and sphalerite on joint faces in granite (Russell 1913b, Hosking 1954b); Greystone quarry, Lezant, near Launceston – very common striated golden-yellow to metallic-blue cubic and pyritohedral crystals of pyrite up to 15 mm on edge (Lloyd & Barstow 1983, Wolfe 2000, Elton et al. 2004); Holmbush, Kit Hill and Redmoor (Callington United) mines – pyrite in polymetallic sulphide veins, highly pyritic in cassiterite-bearing veins (Dines 1956, Newall & Newall 1989, Newall 1994); Linnick, South Petherwin – jamesonite associated with considerable arsenopyrite and pyrite (Russell 1949b); New Consols mine, Stoke Climsland – granular pyrite associated with fluorite in a pegmatite vein (Embrey 1953); Trenute mine, near Lezant – considerable (late) pyrite associated with arsenopyrite, siderite, sphalerite and stibnite (Russell 1949b, Hosking 1951, Dines 1956); Wheal Baron (St George's mine), near Tregeare – pyrite and galena in a gold-bearing lode (Camm 1995); Wheal Trelawney, Menheniot – produced 920 tonnes of pyrite from the deeper levels of the mine (Dewey 1925, Dines 1956); Wheal Wrey, St Ive, near Liskeard – crystals to 1 cm, botryoidal masses, pyritohedra and pseudomorphs after fluorite on the dumps (Rust 1983a). *Tintagel district:* Lanterdan quarry (West quarry) – striated pyrite cubes embedded in slate, also combination octahedral–pyritohedral forms (Bowman 1900); Lill Cove mine, Trebarwith Strand, near Tintagel – large masses of country rock rich in pyrite (Sharpe 1990). *Other occurrences:* Camel, Fowey and Gannel estuaries – detrital pyrite in intertidal sediments (Pirrie et al. 2000b, Pirrie et al. 2002).

Cumbria Pyrite is an extremely common mineral in a variety of parageneses throughout the area, and in some sulphide veins it is locally abundant. Lower Devonian (Caledonian) arsenopyrite–chalcopyrite–pyrite mineralization defines the earlier of two major mineralization types in the Lake District and is linked to the intrusion of a Lower Devonian composite granite batholith, with resulting de-watering of the overlying Borrowdale Volcanic Group (Firman 1978, Stanley & Vaughan 1982a). *Caldbeck Fells:* Pyrite is widespread, usually massive and in small amounts. Occurrences are unremarkable: Brae Fell mine, Dale Beck Valley (Bridges et al. 2006); Brandy Gill, Hay Gill and Potts Gill mines (Davidson & Thomson 1951, Hartley 1984); Burdell Gill – 1–2 mm cubes of goethite after pyrite (Neall & Leppington 2003); Carrock mine, Grainsgill – localized pockets of pyrite in scheelite–wolframite-bearing veins (more arsenopyrite, pyrite and pyrrhotite northwards), pyrite also found in later iron-rich dolomite veins (Hitchen 1934, Davidson & Thomson 1951, Eastwood 1959, Ewart 1962, Shepherd et al. 1976, Pattrick 1980, Hartley 1984, Ball et al. 1985a, Neall & Green 2001a); Deer Hills vein – pyrite associated with arsenopyrite and galena (Symes et al. 1990); Driggith mine – pyrite is the earliest sulphide to form in this lead–zinc vein and it forms small cubic crystals (Stanley & Vaughan 1981, Green et al. 2006a); Nether Row Brow – finely disseminated pyrite and other sulphides (Kingsbury & Hartley 1960, Hartley 1984); Sandbed mine – small cubic and octahedral crystals (Hartley 1984, Green et al. 2006a); Silver Gill mine, near Roughton Gill (Greg & Lettsom 1858). *West Cumbria:* Concentrations of pyrite and chalcopyrite are noted from a few of the hematite deposits: Anty Cross mine, near Dalton in Furness (Smith 1924); Beckermet mine, near Egremont – tiny pyrite crystals lining irregular cavities in hematite ore (Shepherd 1973); Crowgarth mine, Cleator Moor (Trotter 1945); Florence mine, near Egremont (Dunham & Hollingworth 1947). Pyrite is also noted at: Eskett mine, Rowrah, Frizington – pyrite pseudomorphs after barite (Young 1987a); Saltom Bay, St Bees Head – the Saltom Siltstone, at the base of the St Bees evaporite sequence, is enriched in silica and pyrite (Arthurton & Hemingway 1972); Ulpha copper mine – pyrite in chalcopyrite-bearing quartz veins (Firman 1977). *Elsewhere in the Lake District:* Pyrite is an important constituent of chalcopyrite-bearing veins in the Coniston copper mines and is especially common in a cross vein known as the Great Sulphur vein (Dewey & Eastwood 1925). Dagger (1977) notes: the primary mineralization at Coniston is vertically zoned with chalcopyrite–pyrite ore near surface; some pyrite pseudomorphs pyrrhotite; and at Simon's Nick fragmented chalcopyrite–pyrite ore is cemented with siderite. At Bonser mine, Coniston, pyrite partially replaces magnetite and pyrrhotite (Stanley & Vaughan 1982b). Other occurrences include: Oughterside opencast coal site, Aspatria – small crystals of pyrite with kaolinite and siderite in septarian-structured clay–ironstone nodules (Young & Nancarrow 1988a); Barrow mine, Braithwaite (Davidson & Thomson 1951); Black Craig vein, Manesty, Borrowdale – pyrite with arsenopyrite and chalcopyrite (Strens 1962); Brandlehow mine, Derwent Water (Eastwood 1921); Buckbarrow Beck, Corney Fell, Waberthwaite – minor pyrite in ferberite–scheelite veins (Young et al. 1986); Castlenook mine, Newlands Valley – chalcopyrite–pyrite ore (Strens 1962, Ineson & Mitchell 1974a); Force Crag mine, Braithwaite, near Keswick (Davidson & Thomson 1951); Goldscope mine, Newlands Valley, near Keswick (Dewey & Eastwood 1925); Graphite mine, Seathwaite – pyrite forms cores and rims to many graphite nodules (Strens 1965); Dale Head mine, Newlands Valley – pyrite is the dominant ore mineral and often contains up to 1 wt% cobalt (Dewey & Eastwood 1925, Davidson & Thomson 1951, Stanley & Vaughan 1980); Dubhow Gill, Patterdale – nodules of pyrite (Dakyns et al. 1897); Greenside mine, near Glenridding, Patterdale (Eastwood 1921, Naylor 1981); Piers Gill, 1 km N of Scafell summit, Wasdale – pyrite occurs with arsenopyrite, chalcopyrite, galena and sphalerite in a 60 cm-thick polyphase quartz vein cemented by ankerite, within the Scafell dacite. Rare associated species include acanthite (argentite), native gold (electrum), tetrahedrite, uytenbogaardtite and an unknown selenium–silver mineral (Aldridge 2005); Robin Hood mine, Bassenthwaite – pyrite associated with native gold (Young 1987a); Scar Crag mine, Causey Pike – pyrite and molybdenite follow sulpharsenide mineralization in this cobalt–nickel enriched vein (Ixer et al. 1979a); Seathwaite mine, Seathwaite Tarn – pyrite crystallized with arsenopyrite and cobaltite prior to the main bismuth–copper mineralization (Stanley & Criddle 1979); Shap Blue quarry – pyrite in metasomatic garnet-bearing veins cutting Borrowdale Volcanic Group rocks (Davidson & Thomson 1951, Firman 1953a, 1957, 1978, Young et al. 1988); Shap granite quarry – occasional fine crystals of pyrite occur in joints and druses within the granite (Russell 1925, Davidson & Thomson 1951, Firman 1978, Wirth 1995, Wirth et al. 2001); Threlkeld microgranite – granular and euhedral pyrite associated with quartz haloes and veins (Caunt 1984); Thornthwaite mine – cavernous masses of pyrite associated with dolomite, fine crystals and pseudomorphs after dolomite (Eastwood 1921, Davidson & Thomson 1951, Young 1987a); Threlkeld mine, Threlkeld – considerable quantities of pyrite (Postlethwaite 1913, Davidson & Thomson 1951); Woodend mine, Threlkeld (Greg & Lettsom 1858, Kendall 1884); Yewthwaite mine, Newlands Valley (Davidson & Thomson 1951). *Northern Pennine orefield:* Pyrite is the earliest epigenetic sulphide found in galena-bearing veins at Tynebottom mine, in veins around Nenthead and in the Great Sulphur vein (all around the town of Alston) and it is also present in framboidal form in adjacent host rocks. Chalcopyrite, gersdorffite, marcasite, pyrrhotite, sphalerite and ullmannite are associated species (Vaughan & Ixer 1980). Other locations include: Brownley Hill mine, Nenthead, near Alston – widespread occurrence, tarnished crystals up to 1.5 cm across showing cube and pyritohedral forms, highly elongated crystals and crystals speared by millerite

(Clark & Rust 1993, Green & McCallum 1993, Green et al. 2000); Handsome Mea mine, Nenthead and Leadgate, near Alston (Davidson & Thomson 1951); Hilton mine, Scordale, near Appleby – spots of pyrite occur in an area of nickel mineralization (Bridges 1982, Ixer 1986); Silverband mine, Great Dun Fell, Milburn, near Appleby – limonite pseudomorphs after pyrite (Davidson & Thomson 1951, Green & Briscoe 2002); Smallcleugh mine, Nenthead, near Alston – alteration of pyrite has produced melanterite (Uttley 1987b).

Derbyshire Pyrite is widely distributed in small quantities, occurring as (all Ixer 1974a,b):

- a primary mineral in veins within limestone
- an alteration product at limestone-toadstone junctions – Jug Holes mine, Masson Hill, Matlock; Ravenstor in Miller's Dale
- in veinlets within basaltic toadstones – Black Rocks, above Wirksworth; Calton Hill, Taddington; Cavedale, Castleton; Magpie mine, Sheldon; Masson Hill, Matlock; Sallet Hole mine, Stoney Middleton
- dispersed in shales
- segregations in shale or limestone – Eldon Hill quarry, SW of Castleton and Nutseats quarry, Pickory Corner, near Bakewell
- in nodules and replacing fossils in shale – Mam Tor, Castleton (all Ford & Sarjeant 1964a, Ford et al. 1993)
- inclusions in fluorite (or calcite) – Ball Eye mine, Cromford; Dirtlow Rake, Castleton; Jug Holes mine, Masson Hill; Milltown quarry, near Ashover (Rust 1991c); Raper mine, Long Rake, near Middleton-by-Youlgreave (Ineson & Al-Kufaishi 1970); Treak Cliff, Castleton (Rogers 1974).

An exceptional vein occurrence of pyrite was at Millclose Pb–(Zn) mine, Wensley, near Matlock, where a rib of pyrite was found in the middle of the main galena-bearing vein (Parsons 1897). Some specimens of pyrite from here (found on toadstone) form radiating nodular layers with pyritohedral faces and weather to a pinkish tint. When sliced these nodules show pyrite cores and nickel-rich pyrite (bravoite) rims (Traill 1939, Bannister 1940). Early diagenetic framboidal pyrite, bravoite inclusions in fluorite, and zoned bravoite and nickel-rich pyrite crystals, have also been found at Millclose (Vaughan 1969, Ford & Ineson 1971, Rogers 1974, Ixer & Townley 1979). Other bravoite occurrences are from: Bole Hill, near Sheldon, and Treak Cliff, Castleton (Ford et al. 1993); Crich and near Winster (Ixer & Townley 1979); Hucklow Edge Rake, N of Eyam; Masson opencast, Masson Hill, Matlock; Oxclose mine, Snitterton, W of Matlock (Ixer 1974a,b); Riber mine, Matlock (Ford & Sarjeant 1964a); Tansley borehole, Ashover (Ramsbottom et al. 1962).

Other notable occurrences include: Fall Hill quarry, Ashover – small pyrite cubes in a massive calcite vein cutting toadstone (Smith 1982, Spence 1982); Gregory mine, Ashover – silver-rich pyrite (Lecornu 1879) requires confirmation; Magpie mine, Sheldon – cavities in Blende vein are lined with calcite, fluorite and sphalerite, all with minor pyrite and galena (Worley 1976); Odin mine, Castleton – early framboidal pyrite in shale or limestone, associated with spongy pyrite and enclosed by chalcopyrite (Ixer & Townley 1979).

Devon Pyrite is widely distributed in Devon but is usually unremarkable; however, large cubes to 4 cm were once found at: Virtuous Lady mine, Buckland Monachorum; Wheal Friendship (Devon Friendship), Mary Tavy; and Wheal Robert, Sanford Spinney (Greg & Lettsom 1858, Hall 1868b). Russell (1913a) describes large interpenetrant groups of pyrite, elongated and deeply striated cubes, and brilliant pentagonal dodecahedra of pyrite from Virtuous Lady mine, and ranks them among the best examples of the species in Britain. Remarkable fine-grain tapering casts of pyrite (after barite) reaching 26 cm in length and known as lady's slippers were also found at Virtuous Lady mine (Embrey & Symes 1987). Wheal Friendship (Devon Friendship), Mary Tavy, worked a vertically zoned copper–tin lode in which arsenopyrite, chalcopyrite and pyrite dominated the upper levels, with cassiterite occurring at depth (Dewey 1925). Other occurrences include: Belstone mine (Mid-Devon copper mine), near Okehampton – massive pyrite mixed with chalcopyrite (Solly 1886); Combe Martin – forming cubes (Hall 1868b); Bridford barite mine, Teign Valley, Christow, 10 km SW of Exeter – galena, pyrite and sphalerite on the dumps (Polkinghorne 1951); Dartmoor – pyrite is a rare accessory species in the granite (Brammall & Harwood 1925); Franco Consols mine (part known as Wheal Robert), Horrabridge, near Tavistock – pyrite in shale and associated with chalcopyrite in veinstone (Dines 1956); Hemerdon Bal, Plympton – pyrite in the inner metamorphic aureole of a mineralized granite dyke (Ball et al. 1985b); Hope's Nose, Torquay – goethite and limonite pseudomorphs after pyrite and chalcopyrite in a native gold–selenide-bearing vein (Stanley et al. 1990); Littleham Cove, near Budleigh Salterton – pyrite is one of many sulphide and arsenide species found in uranium- and vanadium-rich nodules in the Permian Littleham Mudstone Formation (Milodowski et al. 2000, 2002); Loddiswell mine, 3 km N of Loddiswell – intergrown aggregates of subhedral-to-anhedral zoned pyrite and arsenopyrite near the margins of a carbonate vein bearing native gold (Stanley et al. 1990b); Meldon quarry, near Okehampton – good specimens of pyrite (Robinson 1997a); New East Wheal Russell (South Wheal Crebor), Tavistock – primary chalcopyrite–pyrite mineralization (Starkey 1986b); St Mary's Bay, Brixham – pyrite cubes associated with millerite in cavities within quartz pods in Middle Devonian shales (Braithwaite & Wilson 2001).

Durham Pyrite is reported from many locations in Weardale including: Blackdene mine, near Ireshopeburn – lustrous crystals scattered over fluorite and unusual spherical aggregates (Green & Briscoe 2002); Boltsburn mine, Rookhope – several fine specimens forming lustrous cubic crystals to 3 mm and interpenetrant twins encrusting fluorite (King 1982b); Cambokeels (Cammock Eals) mine, Eastgate – fluorite–pyrite–pyrrhotite–quartz veins and common pyrite in cavities within a sphalerite-rich flat deposit (Solomon et al. 1971, Smith & Hardy 1981, Green & Briscoe 2002); Groverake mine, near Rookhope – abundant pyrite (minor fluorite, chalcopyrite and sphalerite) in bismuth-bearing quartz veinstone from the mine dumps (Ixer et al. 1996); Redburn mine, near Rookhope – minor pyrite and bornite in a fault-hosted fluorite deposit within limestone (Greenwood 1977, Day 1979); Rogerley mine, near Frosterley – quartz veins above the Weardale granite contain chalcopyrite, pyrite and REE minerals (Ixer 2003); Rookhope borehole – minor pyrite in pegmatite and in veinlets cutting the Weardale granite (Ixer 1986, Brown et al. 1987a).

Elsewhere, pyrite is subsidiary to galena and barite in the vein and replacement deposit at Closehouse mine, SW of Middleton-in-Teesdale, Lunedale. It crystallized early, forms massive interlocking grains to 1 mm, and contains inclusions of pyrrhotite and sphalerite (Hill & Dunham 1968, Vaughan & Ixer 1980, Ixer 1986, Young et al. 1994a). Other reports include: Lady's Rake mine, 6 km NW of Langdon Beck, Teesside – minor pyrite in nickeline–magnetite-rich ore (Young et al. 1985a); New Brancepeth colliery, near Durham – small cubes, cubo-octahedra and massive pyrite (also bravoite) occur in the barite–witherite vein at this mine (Spencer 1910, Ixer 1986); Stainton, near Darlington – borehole samples of Permian evaporites contain rounded grains of pyrite 35 µm across, at the edges of gypsum inclusions (Raymond 1962); Whiteheaps mine, near Blanchland – banded chalcedony–marcasite–pyrite–sphalerite veins associated with fluorite mineralization, also pyrite framboids in nearby silicified limestones (Vaughan & Ixer 1980).

Globular pyrite aggregates covering a 4 × 6 cm fluorite crystal; Fraser's Hush, Rookhope, Weardale, Co. Durham. (© M. P. Cooper)

Gloucestershire Thin limestone bands enclosing nodular pyrite and crystallized pyrite in clay–ironstone nodules occur at Wingmoor farm, Bishop's Cleeve (King 1997).

Kent Pyrite is common in London Clay at Warden Point, Isle of Sheppey, where it replaces fossils and forms microcrystalline masses (Monk 1982). Nodules of pyrite, some exhibiting small distorted octahedra, which might easily be mistaken for marcasite, are present in the Lower Chalk on the foreshore below the Warren, Folkestone (Bannister 1932). The Lower Chalk also hosts pyrite nodules in the Medway Valley at Great Culand quarry, Burham, near Chatham, and Holborough quarry, near Snodland (Dines et al. 1954, J. Betterton pers. comm. 2006).

Lancashire Pyrite (bravoite) occurs as inclusions in fluorite near Clitheroe at Bellman Park quarry and Skeleron mine (Rogers 1974, Ixer 1974a).

Leicestershire Much of the pyrite at Newhurst quarry, Shepshed, has been altered to goethite, although very small cubes to 0.5 mm have been found in a partially oxidized bornite-chalcopyrite vein (King 1973, Ince 2005).

Merseyside Pyrite, forming cubes, octahedra and pyritohedra, is the commonest sulphide species in glacial erratics from Thurstaston beach, Wirral (Rothwell & Dickinson 1989).

Northamptonshire Pyrite, berthierine and siderite are the major iron-rich phases in the Northampton Ironstone as seen at Cowthick quarry, Corby (Gehring 1990).

Northumberland Nodules of pyrite up to 15 cm across occur in shales close to the Whin Sill at Settlingstones mine, near Hexham (Trestrail 1931). At West Sleekburn colliery, Bedlington, reniform ankeritic calcite, marcasite and pyrite were found in a large cavity within a dolerite dyke (Randall 1953).

Nottinghamshire Attractive well crystallized pyrite, forming lustrous cubes with minor octahedral modification, have been found in Permian carbonates at Vale Road quarry, Mansfield Woodhouse. Calcite and sphalerite are associated species (Green & Spence 1988).

Oxfordshire Microcrystalline coatings of pyrite are noted in septarian nodules within Oxford Clay at Cassington. Similar rocks at Yarnton contain pyrite replacing ammonite shells (Atkins 1995).

Somerset Beautiful golden pyrite crystals suspended between barite and calcite are reported from Whatley quarry, near Frome, Mendip Hills (House 2002).

Staffordshire Pyrite and chalcopyrite (with lesser amounts of galena) are the dominant hypogene sulphides in the Ecton Hill copper deposits (Masheder & Rankin 1988), and both nickel-rich pyrite and bravoite are reported (Ixer 1978). Minute cubes of pyrite encrusting celestine crystals adjacent to millerite have been collected from Clayton mine, Ecton (Starkey 1983).

Surrey Compact radiating blade-like aggregates of pyrite form nodules in the Lower Chalk at Oxted (Bannister 1932).

Sussex Well crystallized pyrite nodules have been collected on Climping beach, near Littlehampton (Moore & Pearce 1997, Green 1998b). Small ellipsoidal nodules of suspected marcasite in a pebble of calcareous shale from Newhaven proved on analysis to be pyrite (Bannister 1932). Large pyrite nodules (> 14 cm across) have been recovered from the Lower Chalk at Beachy Head, W of Eastbourne (J. Betterton pers. comm. 2006).

Tyne & Wear Carboniferous rocks in the cliffs below Tynemouth Castle, around Cullercoats harbour and in a quarry at Clousden Hill, contain disseminated pyrite immediately beneath the Permian unconformity (Leake et al. 1997b). Pyrite and marcasite were discovered lining cavities in the Brockwell coal seam at Rising Sun colliery, Backworth (Randall & Jones 1966).

Yorkshire Small quantities of pyrite are present throughout the Permian evaporite sequence as sampled at Aislaby (Eskdale No. 2 borehole), Eskdale, near Whitby, and the Fordon boreholes, on the Humberside/Yorkshire border (Stewart 1951a, 1963, 1965, Beevers & Stewart 1960). Examples of pyrite in Carboniferous sediments include: uncommon small cubes and rounded grains of pyrite in the Millstone Grit at locations such as East Keswick, Bardsey; Snosty quarry, Bardsey; in the Tannery borehole, Meanwood, Leeds (Gilligan 1919); diagenetic pyrite develops framboidal and other textures in marine Coal Measures rocks at the Hepworth Iron Company workings, near Penistone (Love et al. 1983); siderite in spherulitic ironstone at High Birks pit, Thornton, W of Bradford, and Siddal, Halifax, has been replaced by pyrite (Deans 1934); pyrite is associated with barite and dolomite in Coal Measures sandstone at Silverwood colliery, SW of Doncaster (Uttley 1987a). The remaining examples are from the Northern Pennine orefield: bravoite at eight localities in the Askrigg Block, including the Arkengarthdale, Greenhow, Swaledale areas (Ixer 1978); Keld Heads mine, Wensleydale and Wharfedale mine, Kettlewell – bravoite inclusions in fluorite (Rogers 1974); accessory amounts of pyrite associated with chalcopyrite, galena and molybdenite in the buried Raydale granite, Wensleydale (Brown et al. 1987a); Scaleber Bridge, 3 km SSE of Settle – 10–100 μm pentagonal dodecahedral crystals of pyrite associated with chalcopyrite, marcasite and enargite-group minerals (Ixer & Stanley 1998); and Trollers Gill, ESE of Grassington – euhedral zoned inclusions of bravoite in pyrite (Vaughan & Ixer 1980).

Wales

Clwyd Bwlch-y-Plwm mine, Penrhyndeudraeth, worked a galena–pyrite–quartz lode cutting Ordovician sediments and volcanics (Archer 1959). Pyrite, chalcopyrite and cobaltite were the primary sulphides in the asbolane–hematite fissure deposit worked at what is now Dyserth quarry, Moel Hiraddug, near Prestatyn (Bolton 1996). At Wedding Cave mine, Bwlchgwyn, near Wrexham, pyrite occurs with galena in fault-related mineralization within Carboniferous sandstone (Green 1995b). Pyrite aggregates are reported from basic igneous rocks at Llechlwyd quarry, Towyn, near Rhyl (Bevins 1994). Traces of chalcopyrite and pyrite occur in galena–sphalerite-dominated ore from the Minera mining district (Earp 1958).

Dyfed *Central Wales orefield:* Pyrite is widely distributed, occurring in the Lower Palaeozoic Welsh Basin as pre-tectonic framboids, as nodules and in cubes. This early mineralization was followed by polymetallic vein mineralization in which pyrite was usually the first sulphide to form, followed by ankerite, chalcopyrite, sphalerite and then galena. Localities include: Bodcoll mine, E of Devil's Bridge (Bick 1992); Brynyrafr mine (Jones & Moreton 1977, Mason 1997); Castell mine, Ponterwyd, near Aberystwyth – cubes, cubo-octahedral and filiform crystals (Rust 1990d); Cefngwyn mine, near Aberystwyth; Dolclettwr mine, Tre'r-dol; Eaglebrook (Nantycagl) mine, Ceulanymaesmawr; Erglodd mine, N of Tal-y-bont; Esgair Fraith mine, near Tal-y-bont; Frongoch mine, between Devil's Bridge and Pont-rhyd-y-groes; Nantymwyn mine, NE of Llandovery (all Mason 1997); Cwmystwyth mine; Penrhiw mine, Ystumtuen; Plynlimon mine – excellent cubic crystals; Ystumtuen mine, near Devil's Bridge (all Jones & Moreton 1977); Darren mine, Pen-bont Rhydybeddau – crystal aggregates to 2 cm (Raybould 1974, Mason 1997); Eaglebrook (Nantycagl) mine, Ceulanymaesmawr – small crystalline masses and confused partial cubes (Jones 1983, Mason 1997); Esgair Hir mine, Ceulanymaesmawr – rare pyrite in masses of goethite (Rust & Mason 1988, Mason 1997); and Llechwedd-helyg mine, near Clawddmelyn, between Bont-goch and Cwmsymlog – sparse crystalline masses and veinlets (Jones 1987).

Elsewhere in Dyfed, pyrite is the principal metallic mineral in gold-bearing quartz veins cutting Ordovician shales at Dolaucothi (Ogofau mine), near Pumsaint (Rudler 1905, Brammall et al. 1937, Hughes 1959, Steed et al. 1976, Annels & Roberts 1989). Excellent pyritohedra up to 25 mm across are reported from Cwmgwili colliery, Llanelli (Bevins 1994). Abundant disseminated pyrite is present in the

Pyrite pyritohedron from black Coal Measures shale at Cwmgwili colliery, Llanelli, Dyfed, Wales; width 25 mm. (© National Museum of Wales, Cardiff; photographer M. P. Cooper)

Nant-y-Coy Formation, a sequence of sedimentary and pyroclastic rocks below Little Treffgarne Rocks (Brown et al. 1987b).

Gwynedd Pyrite is the most common sulphide in the polymetallic volcanogenic massive sulphide deposit at Parys Mountain, Amlwch, northwest Anglesey, and four different generations are recognized (Greenly 1919, Dewey & Eastwood 1925, Bor 1950, Manning 1959, Pointon 1980, Pointon & Ixer 1980, Ridgway 1980, Southwood & Bevins 1995, Tennant & Steed 1997). Another volcanogenic massive sulphide deposit, at Cae Coch, near Trefriw, Llanrwst, was mined for pyrite (the only pyrite mine in Wales) and contains features such as tubes of pyrite 5–10 mm across, suggesting a deep-sea black-smoker origin. However, a syn-diagenetic inhalative origin has also been proposed. The mine produced over 100 000 tonnes of ore (Sherlock 1918, Archer 1959, Dunham et al. 1978, Ball & Bland 1985, Ball et al. 1985b, Otty 1986, Bottrell & Morton 1992, Jenkins & Johnson 1993). Other examples of pyrite in Snowdonia and adjacent areas include: Cwm Tregalan, S of Snowdon; Cwm Llan, Hafod y Porth; Lliwedd, SE of Snowdon; Shadow Gully, S of Snowdon; Sygun mine, S of Snowdon – colloform, framboidal and cubic pyrite associated with chalcopyrite and sphalerite ±(galena–hematite–magnetite–pyrrhotite) in thin veins, narrow stockworks and disseminations in rocks of the Snowdon Volcanic Group (Reedman et al. 1985, Colman & Appleby 1991); Cwt-y-Bugail quarry, near Blaenau Ffestiniog – cubes and pyritohedra in slate (Savage 1997a); Cyffty mine, between Betws-y-coed and Llanrwst – abundant pyrite on the dumps (Otty 1986); Drws-y-Coed mine, near Nantlle – veins in Cambrian and Ordovician sediments contain pyrite associated with arsenopyrite, chalcopyrite, galena, pyrrhotite and sphalerite (Braithwaite 1982b); Fron-oleu, Prenteg, Tremadog – minute cubes to 1 mm in brookite-bearing veins (Starkey & Robinson 1992); Hafna mine, Llanrwst orefield, N of Betws-y-coed – attractive cockscomb pyrite on the dumps (Otty 1986); Llanrwst orefield, N of Betws-y-coed – pyrite in epigenetic lead–zinc veins, also framboidal and cubic pyrite in Ordovician host rocks (Haggerty 1995); Llyn Bochlwyd, Ogwen – single 5 cm cube now in the NMW (Bevins 1994); Oakley quarry, Blaenau Ffestiniog and Penrhyn quarry, Bethesda – cubes and modified cubes in Cambrian and Ordovician slates (Bevins 1994); Parc mine, near Llanrwst, N of Betws-y-coed – notable specimens from lead–zinc veins cutting Lower Palaeozoic rocks (Bevins 1994); Prysor mine, Trawsfynydd – chiefly a pyrite ore with 1–2% chalcopyrite (Dewey & Eastwood 1925, Archer 1959).

Pyrite forms discrete euhedral crystals (20–250 μm across) or aggregates with marcasite, enclosed in chalcopyrite within dolomitized Carboniferous Limestone at the Great Orme mines, Llandudno (Ixer & Davies 1996, Ixer & Stanley 1996, Ixer 2001).

The Dolgellau goldbelt contains pyrite in gold-rich quartz–sulphide veins cutting Cambrian sedimentary and intrusive rocks. The veins show regional variation from an arsenopyrite–chalcopyrite–cobaltite–pyrite–pyrrhotite assemblage in the SW to a galena–pyrite–sphalerite assemblage in the NE (Gilbey 1968). Pyrite often forms large (5–10 mm) euhedral crystals containing inclusions (sometimes of native gold) and has been noted at: Cefndeuddwr mine, near Dolgellau (Mason et al. 2002); Clogau and Vigra mines, Llanaber – abundant chalcopyrite, pyrite and pyrrhotite (Andrew 1910, Dewey & Eastwood 1925); Friog Undercliff, Fairbourne, SW of Dolgellau (Mason et al. 1999); Glasdir mine, Llanfachreth – ubiquitous pyrite in a breccia pipe (Dewey & Eastwood 1925, Allen & Easterbrook 1978, Allen & Jackson 1985); Gwynfynydd mine, near Dolgellau – arsenopyrite–chalcopyrite–galena–pyrite–sphalerite-bearing lodes in black shales (Allen & Jackson 1985, Dominy et al. 1996c, Mason et al. 2002); Prince Edward mine, near Trawsfynydd – ribbon quartz veins containing massive pyrite, with cavities infilled with remarkable filiform pyrite (Mason & Williamson 2001, Mason et al. 2002); Turf Copper mine, Coed-y-Brenin – pyrite forms a disjointed halo surrounding high-grade copper ore (Rice & Sharp 1976). Pyrite also occurs in mudstone associated with manganese carbonates at the Hafotty mines, near Barmouth (Glasby 1974), and in manganese-rich shales of the Harlech Grits at Cell-fechan, near Barmouth, and the Cwm-yr-Afon mines, Artro Valley, Llanbedr district (Woodland 1938b).

Pyrite in Cambrian slate from Penrhyn quarry, Bethesda, Gwynedd, Wales; 10×12 mm on edge. (© National Museum of Wales, Cardiff; photographer M. P. Cooper)

Mid Glamorgan Euhedral pyrite is associated with alstonite in dolomite-lined cavities within Triassic conglomerate at the Llantrisant bypass roadcut and at a nearby old mine site (Alabaster 1990a). Fracture-infilled veins within dolomitized Carboniferous Limestone at Machen quarry, 7 km ENE of Caerphilly, contain pyrite, but most of it has been pseudomorphed by goethite (Plant & Jones 1995). Pyrite is a minor component of iron ores once worked at Llanharry and Mwyndy mines, near Llantrisant (Bevins 1994).

Powys Good cubic crystals of pyrite have been found in the Central Wales orefield on the mine dumps of Dylife mine, N of Hafren Forest, and at Geufron mine, near Llandiloes (Jones & Moreton 1977). Elsewhere in the orefield, pyrite has been reported from: Bryntail and Nantiago mines (Mason 1997); Dyfngwm mine, Machynlleth (Rust & Rust 1987). Large euhedral pyrite cubes (to 40 mm on edge) have also been found in slate from Craig Dyfnant on the east side of the Claerwen reservoir, Rhayader (R. E. Starkey pers. comm. 2006). Pyrite is a minor primary mineral at Dolyhir quarry, Old Radnor, near Presteigne, where it forms small euhedral cubes, matted aggregates of hair-like crystals, and distorted cubes (Bevins & Mason 1997, T. F. Cotterell pers. comm. 2007).

Scotland

Borders Pyrite is the most common inclusion species in alluvial gold found at the following locations: Craigierigg Burn; Cramalt Burn; Glengaber Burn, SW of Peebles; head of the River Tweed; and Smid Hope Burn, Tweed headwaters, N of Moffat. Associated species in pyrite-bearing or other gold grains include arsenopyrite, berzelianite, chalcopyrite, clausthalite, galena (common), gersdorffite, native palladium, sphalerite and tetrahedrite (Leake et al. 1993, Chapman et al. 2000a,c).

Central Pyrite is found at Airthrie Hill mine, Bridge of Allan, and most other mines in the Ochil Hills. Barite, chalcopyrite and galena are the main associated species (Dickie & Forster 1974). Heddle (1901) notes pyrite in a vein cutting andesite at Westerton, near Alva, Ochil Hills. More generally, pyrite is a common inclusion species in alluvial gold from the central Ochil Hills (Chapman et al. 2005).

Pyrite–quartz veins cutting Dalradian metasediments and subsidiary metavolcanics are an important host for gold mineralization in the Tyndrum area and have attracted recent commercial interest at Cononish, 4 km SW of Tyndrum (Pattrick et al. 1988, Parker et al. 1989, Earls et al. 1992, Curtis et al. 1993, Treagus et al. 1999). Pyrite is also found in galena–sphalerite veins in the Tyndrum lead–zinc mining district (Pattrick 1978, Ridgway 1980, Pattrick 1985). Creag Bhocan, SSW of Tyndrum, is the location of a Dalradian stratiform sedimentary-exhalative deposit in which pyrite and chalcopyrite with lesser amounts of sphalerite are the main sulphides present. Some of this pyrite shows delicate growth zoning on a microscopic scale, characterized by variable cobalt content (Scott et al. 1988). To the NE of Tyndrum, low-grade stratiform sulphide (mainly pyrite) mineralization in the Dalradian Ben Challum Quartzite is also considered to be a sedimentary-exhalative deposit (Smith 1977, Fortey & Smith 1986). A related occurrence in the Ben Lawers Schist has produced elongated cubes of pyrite with curved faces up to 5 cm across from the ridge between Meall Tarmachan and Creag na Caillich, near Killin (Heddle 1901, Green & Todd 2001).

Dumfries & Galloway Copper–nickel mineralization at the base of a diorite intrusion near Talnotry, 5 km NE of Newton Stewart, contains zoned nickel-rich pyrite as euhedral modified cubes to 200 μm (Stanley et al. 1987). Near by, along Palnure Burn and close to the margin of the Cairnsmore of Fleet granite, pyrite occurs in sheared brecciated spaces within arsenopyrite-bearing quartz veins. Within this granite, euhedral pyrite, forming octahedral and multi-faceted crystals, is noted from the Orchars chalcopyrite-bearing quartz vein (Lowry et al. 1997). Elongate cubo-octahedral pyrite is reported from Blackcraig mine, SE of Newton Stewart (Heddle 1901). In this same general area, disseminated pyrite occurs in siliceous mudstone in a mineralized gossan zone at Penkiln Burn (Stone & Leake 1984) and is associated with arsenopyrite in metasomatically altered igneous rocks at Glenhead Burn, near Loch Dee (Leake et al. 1981). Both locations are close to the contact of the Loch Doon diorite–granite complex.

Early-formed pyrite occurs in the quartz–stibnite vein at Glendinning (Louisa) mine, Glenshanna Burn, 13 km NNW of Langholm (Heddle 1901, Herbert Smith 1919, Russell 1949b, Williams 1973). Gallagher et al. (1983) and Duller et al. (1997) also note disseminated pyrite in turbidite-hosted stratabound arsenopyrite–gold mineralization from this region.

Other reports of pyrite are from: Bail Hill, N of Sanquhar (Gallagher et al. 1971); Beeswing granite quarry (Williams 1965); Black Stockarton Moor (Leake & Brown 1979, Fortey 1980, Rock et al. 1986a, Lowry et al. 1997); Cairngarroch Bay (Lowry et al. 1997); Lotus Hill, near Beeswing, and Lower Porterbelly quarry, Kirkgunzeon (Williams 1973); Needle's Eye, SW of Caulkerbush (Miller & Taylor 1966, Basham et al. 1989); and the Wanlockhead lead–zinc orefield (Heddle 1901, Temple 1955).

Fife The finest source of specimen pyrite in Scotland is an occurrence, probably of syn-sedimentary origin, in Carboniferous mudstones at Goat quarry, 2 km NW of Aberdour. Here pyrite forms lens-shape crystalline masses and individual (distorted) cubic crystals to 4 cm (Ingram et al. 1992, Green & Todd 2001, Moreton 2006). Pyrite is also briefly noted from Orrock quarry, near Burntisland (Wirth 2003).

Grampian Base-metal enrichment, represented by anhedral patches of galena, pyrite and sphalerite, occurs in Dalradian cherty quartzite at Corrie Loch Kander, near Braemar, and is considered to be an exhalative deposit (Gallagher et al. 1989a, Fortey et al. 1991, 1993). Schistose xenoliths at the margins of the Dorback granite, Braes of Abernethy, 9 km W of Tomintoul, are exceptionally rich in pyrite (Walker & Davidson 1935). The Dalradian breccia pipe at Souter Head, Cove Bay, 6 km S of Aberdeen, contains pyrite and subordinate molybdenite in the matrix and associated quartz veins (Porteous 1973, Kneller & Gillen 1987). Pyrite forms small grains and rims on galena and sphalerite in a Triassic breccia termed the Cherty Rock at Stotfield, W of Lossiemouth, near Elgin (Starkey 1988a). Pyrite is also noted at: Abergairn (Corrybeg) lead mine, near Ballater (Russell 1936a); Pollagach Burn, E of Pannanich Hill, E of Ballater, Deeside, and southwards towards Glen Doll – disseminated pyrite in the Water of Tanar Limestone Formation (Goodman et al. 1990); Rhynie – spongy pyrite and zoned arsenic-rich cubes of pyrite in geyserite from a Devonian hot-spring deposit (Rice et al. 1995). Laouar et al. (1990) note pyrite in the Aberdeen granite.

Highland Faulted Old Red Sandstone sediments (arkose) at Ousdale and Ord Burns, near Helmsdale, are uranium-rich and contain pyrite and fluorite in mineralized zones (Gallagher et al. 1971, Dunham et al. 1978, Tweedie 1979). Borehole material from here revealed pyrite intergrown with coffinite (Pointer et al. 1989), and, in the nearby Kildonan and Suisgill Burns, and in the River Helmsdale at Torrish, alluvial gold containing inclusions of pyrite has been found (Leake et al. 1993, Chapman et al. 2000c).

Pyrite occurs in thin post-foliation quartz veins and on joint surfaces in Moinian and Lewisian schists and gneisses intruded by Caledonian granites W of Lairg. Chalcopyrite, fluorite, galena, molybdenite and bismuth minerals (aikinite–bismuthinite) are associated species. Fluorite and pyrite are also found disseminated throughout the Grudie granite (Gallagher 1970, Gallagher et al. 1974, Gallagher & Smith 1975, Fortey 1980).

Magmatic and introduced sulphides (chalcopyrite, pentlandite, pyrite and pyrrhotite) are present in pyroxenites, hornblendites and other rocks belonging to the Loch Borralan and Loch Ailsh syenite complexes (Shand 1910b, Stumpfl 1961, Notholt et al. 1985, Shaw et al. 1994, Styles et al. 2004). Both pyrite and pentlandite occur as accessory phases in an associated carbonatite at Loch Borralan (Young et al. 1994b).

Other occurrences of pyrite include: An Leth Allt, north slopes of Sgùrr an Airgid, Loch Duich – pyrite in a Moinian rock containing ankerite, fuchsite mica, and quartz (Clifford 1959); Ardsheal Peninsula, near Kentallen – pyrite disseminated in the wallrocks and within a porphyrite explosion breccia pipe (Rice & Davies 1979); Ballachulish – pyrite cubes to 12 mm on edge in slate (Neumann 1950) and pyrite associated with chalcopyrite and molybdenite in disseminated, veinlet and fracture infills within granite (Evans 1977, Haslam & Kimbell 1981); Broadford area, Isle of Skye – pyrite, chalcopyrite and bornite in magnetite skarn at the margin of the Beinn an Dubhaich granite (Tilley 1948b, 1951, Slater & Highley 1976); Bridge of Westfield and Murkle Bay, near Castletown – pyrite in uranium-rich veins and fault breccia within Old Red Sandstone sediments (Gallagher et al. 1971); Gairloch – the Lower Proterozoic Loch Maree Group contains stratiform sulphide deposits in which pyrite and pyrrhotite are the main ore species, and chalcopyrite, galena, native gold, marcasite and sphalerite are found in lesser abundance (Jones et al. 1983, 1987); Glen Urquhart complex, near Drumnadrochit – conspicuous pyrite in kyanite-rich mica schist enveloped by Moinian psammites and pelites (Rock et al. 1986b); Ratagain complex, Loch Duich – quartz monzonite containing Ag–Au–Te mineralization in quartz–(calcite–fluorite) veins dominated by galena, pyrite and sphalerite (Alderton 1986, 1988b); Strontian mines – pyrite in barite–galena veins (Heddle 1901, Winrow 1986b, Wolfe 1998). Laouar et al. (1990) note pyrite in granites from Ballachulish; Corran, near the Great Glen fault; Mullach nan Coirean, near Ben Nevis and Ratagain.

Lothian Pyrite cubes up to 0.1 mm across occur in a stratabound sulphide (chalcopyrite–galena–sphalerite) deposit in Carboniferous non-marine algal limestone at South Queensferry (Parnell 1983). Heddle (1901) notes fine cubes of pyrite from Skolie Burn, West Calder.

Strathclyde Stratabound metasomatic replacement deposits in quartzose schists were worked for copper and then nickel at Craignure and Coille-bhraghad mines, near Inveraray. The ore consisted of chalcopyrite, pyrite, pyrrhotite and pentlandite (Wilson & Flett 1921, Berridge 1969, Dunham et al. 1978, Willan 1980). At Garbh Achadh, near Inveraray, low-grade porphyry-type copper–molybdenum mineralization includes disseminated pyrite in a sericitic rock centred about a stock-like porphyrite intrusion (Ellis et al. 1978, Fortey 1980). Disseminated pyrite occurs with chalcopyrite, galena, molybdenite and sphalerite in other porphyry intrusions around Kilmelford (Ellis 1977, Fortey 1980, Zhou 1987, Harris et al. 1988). Laouar et al. (1990) note pyrite in granites from: Arrochar, Etive and the Ross of Mull.

Other occurrences of pyrite include: Abhainn Srathain mines, S of Meall Mòr, South Knapdale – Dalradian stratiform sulphide deposit containing a 10 km-long pyrite zone in the Erin Quartzite, and pyrite and chalcopyrite in epidiorite (Smith et al. 1978a); Ailsa Craig, Firth of Clyde – accessory amounts of polyframboidal-to-coarse fragments of pyrite in microgranite (Harrison et al. 1987); Balgit Burn, Dunoon – pyrite in manganese-mineralized quartz blocks (Nicholson 1989b); Ben More (Maol nan Damh), Isle of Mull – cavities in Tertiary basalts contain cubes of pyrite enclosed in prehnite (McLintock 1910b); Blackmount, on the margin of the Moor of Rannoch granite, near Bridge of Orchy – pyrite in molybdenite–quartz veins (Smith & Marsden 1977, Curtis et al. 1993); Branscroft quarry, near Kilbarchan, Glasgow area – pyrite and chalcopyrite on calcite crystals (Green & Todd 2001); Castletown mine, Lochgilphead – copper-rich pyrite with chalcopyrite and galena in a vein cutting Dalradian rocks (Wilson & Flett 1921, Starkey & Hubbard 1998); Clachan Beag, NE of Inveraray – silver-rich galena, pyrite and sphalerite in a Dalradian stratabound deposit (Willan 1980); Coire Buidhe, Glen Creran – skarns at the margin of the Etive granite complex contain massive pyrite and chalcopyrite, also pyrite crystals, up to 20 mm across, associated with epidote and quartz , all in cavities (Ingram et al. 1993); Cuan Ferry and Easedale Island, near Oban – skeletal crystals and cubes of pyrite in Middle Dalradian slates (Hall et al. 1988); Etive granite complex to the E and NE of Ben Starav – molybdenite–pyrite veinlets cutting granite (Haslam et al. 1985); Knipe (Fountainhead) mine, Hare Hill, 5 km SE of New Cumnock – arsenopyrite, pyrite and stibnite occur in carbonate–quartz veins and disseminated in a hydrothermally altered granodiorite intrusion (Boast et al. 1990); Fore Burn, Tairlow plantation, near Straiton – gold-rich sulphide–quartz veins containing arsenopyrite and pyrite hosting native gold inclusions (Allen et al. 1982, Charley et al. 1989); Isle of Islay – at Ballygrant, Kilslevan, Mulreesh and North Ardachie, pyrite occurs with chalcopyrite, galena and sphalerite in veins and breccia zones cutting Dalradian rocks (Wilson & Flett 1921, Baron & Parnell 2005), and is also noted in Dalradian dolomite on the northeast coast (Hall et al. 1994); Leadhills–Wanlockhead orefield – pyrite forms inclusions in chalcopyrite and sphalerite at Glencrieff mine and Gordon's vein respectively (Temple 1955), and inclusions in alluvial gold from Shortcleugh Water, near Leadhills (Leake et al. 1998, Chapman et al. 2000c); McPhun's Cairn, Loch Fyne – stratiform sulphide mineralization in Cambrian phyllite containing pyrite and pyrrhotite pods (Willan 1980, Willan & Hall 1980); Muirshiel barite mine, Queenside Muir, NE of Largs – cubes and pyritohedra to 2 mm studded over barite, also pyrite inclusions in calcite (Todd & Laurence 1989); Reilly quarry, Houston – well crystallized pyrite in a Permo-Carboniferous quartz-dolerite dyke (Meikle 2003); Small Isles Harbour, Isle of Jura – pyrite in gneiss (Heddle 1901); Trearne quarry, near Beith – pyrite forms nodules in shale bands, replaces the shells of fossil brachiopods, and develops bright cubes to 1 mm in fissures within limestone (Todd 1989).

Tayside Pyrite is the most abundant sulphide in gold-rich vuggy fractured quartz veins at Calliachar and Urlar Burns, near Aberfeldy. It forms discrete crystals or aggregates up to 1 cm across associated with chalcopyrite, galena and sphalerite (Ixer et al. 1997). Inclusions of pyrite in alluvial gold are noted from Calliachar Burn, Aberfeldy (Leake et al. 1993, 1997c, Chapman et al. 2000c). Pyrite, pyrrhotite and sphalerite are the main sulphides in a Dalradian barium-enriched sulphidic schist from the Ben Eagach area, near Aberfeldy (Coats et al. 1980, 1984, Fortey & Beddoe-Stephens 1982, Willan & Coleman 1983); the same formation has been worked for barite at Foss mine, 7 km NW of Aberfeldy (Moles 1983, 1985).

At Tomnadashan, Loch Tay, fine crystals of pyrite, up to 25 mm on edge, have been found in disseminated copper mineralization associated with granitic lenses intruded into a Caledonian diorite. Chalcopyrite, molybdenite and tetrahedrite are also noted (Heddle 1901, Wilson & Flett 1921, Fortey 1980, Pattrick 1984). A few km to the S, at Corrie Buie, Meall nan Oighcreag, minor pyrite occurs in galena-bearing quartz veins (Pattrick 1984). Approximately 15 km SE of Tomnadashan, the Comrie diorite complex hosts gold mineralization associated with pyrite (Plant et al. 1989).

Attractive tarnished cubes of pyrite to 2 cm in slate from Birnam quarry, Dunkeld are noted by Livingstone (2002) and slightly larger crystals (up to 5 cm) have been found in schist on the ridge between Meall nan Tarmachan and Meall Garbh, SW of Ben Lawers (R. E. Starkey pers. comm. 2007).

Western Isles Pyrite is briefly noted from: dolerite at Eilean Mhuire, Shiant Isles (Heddle 1901); granite pegmatite at Loch a' Sgùrr, near Finsbay (Fionnsabhagh), South Harris (von Knorring & Dearnley 1959b); layered alpine-type peridotite at the Sgeir nan Sgarbh promontory to Loch Meurach, Langavat Valley, South Harris (Livingstone 1976a); in the Cambir dolerite and Western gabbro, St Kilda (Hirta) (Harding et al. 1984).

Orkney Flett (1900) describes a porphyritic dyke on Scabra Head, Rousay, containing large crystals of pyrite.

Shetland A chlorite slate found in a creek between Muness harbour and Sandwick Bay on Unst contains such a high concentration of pyrite that the locality is named after its colour – Golden Geo (Heddle 1878a). At Vidlin Ness on Mainland, Dalradian rocks host a 10 m-thick sulphide zone of massive sulphides, sulphide-bearing tremolite schist and a sulphide-bearing quartz rock (possibly a recrystallized chert). This stratabound sulphide deposit contains chalcopyrite, pyrite, and local galena and sphalerite (Garson & May 1976). Almost pure layers and crystal-lined vugs containing pyrite are found with bornite, chalcopyrite, galena, sphalerite and tennantite in nodules within a laminated Devonian siltstone at Matta Taing, Sandness, West Mainland (Hall & Donovan 1978). Pyrite is a common accessory species in gabbro, amphibole pegmatite and plagiogranite from the Shetland ophiolite complex, and is noted from the Vord Hill area and N and E of Sandwick Bay, Unst. It often forms euhedral crystals intergrown with chalcopyrite or pyrrhotite (Maynard 1993, Maynard et al. 1997). Heddle (1901) reports massive pyrite from Quendale on Mainland.

Ireland

Armagh Galena, pyrite and sphalerite were found at Derrynoose lead mine (Cole 1922).

Carlow Accessory amounts of pyrite occur in a suite of lithium pegmatites on the southeastern flank of the Leinster granite batholith and the species has also been reported as poorly developed cubes to 3 mm at Aclare, near Myshall, and at Seskinnamadra, W of Tomduff Hill (Scoon 1978, McArdle & Kennan 1988, Whitworth 1992). At Coolasnaghta, in the same general area, tourmalinites (quartz–tourmaline and muscovite–tourmaline schist) also contain accessory pyrite (McArdle et al. 1989).

Clare A disseminated and fracture-fill sulphide deposit in Lower Carboniferous carbonate sediments at Ballyvergin is dominated by arsenopyrite, chalcopyrite, galena and pyrite mineralization. Pyrite forms cubes to 1 mm in laminae on bedding surfaces and as infill or replacement of bioclasts (Schultz 1971, Halliday & Mitchell 1983, Andrew 1986). The Maghera Cross copper prospect, Maghera Townland, is a similar deposit with dominant chalcopyrite, subordinate pyrite, and minor galena and sphalerite (Schultz 1971).

Cork Disseminated and vein-hosted pyrite is reported as a constituent of polymetallic ores from the West Carbery mining district: Ballycummisk mine (near Skull), Cappaghglass (Cappagh), Castle Point, Coosheen, Crookhaven, Dhurode, Glanalin, Kilkilleen, Letter and Mount Gabriel (Ni Wen et al. 1999, Kinnaird et al. 2002). Pyrite also occurs in a barite deposit at Duneen Bay mine, western side of Clonakilty Bay (Cole 1922), and in chalcopyrite veins at Dooneen and Mountain mines, Allihies (O'Brien 1959, Schultz 1971, Halliday & Mitchell 1983, Moreton & Aspen 1993).

Donegal McArdle et al. (1986a) note galena, pyrite and sphalerite as the main sulphides in stratiform and vein mineralization in Dalradian rocks close to (but not necessarily associated with) the Donegal granite complex, but give no further details.

Down Pyrite is subordinate to chalcopyrite and galena in a vein deposit at Conlig–Newtownards (Dunham et al. 1978, Griffith & Wilson 1982, Moles & Nawaz 1996). Euhedral or subhedral crystals of pyrite, some intergrown with chalcopyrite, have been found with molybdenite in the Newry granodiorite at Croreagh quarry (Meighan & Ixer 1980).

Galway Limestone-hosted Cu–Pb–Zn–(Ag) mineralization at Tynagh mine, near Killimor, includes pyrite as one of the primary ore species, although a substantial portion is derived from inversion of marcasite (Schultz 1966b, 1971, Morrissey et al. 1971, Russell 1975, Halliday & Mitchell 1983, Ryback & Moreton 1992, Green & Moreton 2001). An unusual polygonal quasi-framboidal variety of pyrite from Tynagh is illustrated by Morrissey (1972). Pyrite also occurs in minor amounts in a bedded ironstone unit found during drilling operations at Tynagh (Schultz 1966a). McArdle et al. (1986a) report several pyrite locations in Connemara, including pyrite–pyrrhotite-dominated sulphide lenses and disseminations, near the top of the Dalradian Lakes Marble Formation, that were once mined for their pyrite and copper content. Chalcopyrite, molybdenite and pyrite are the most common sulphides in a metasomatized ring structure at Mace Head, in the Carna granodiorite, a part of the Galway granite batholith (Talbot & Max 1984). Dodecahedral crystals of pyrite up to 2 mm across occur on the dumps of Glengowla mine, Oughterard (Ryback & Moreton 1992). Derrylea lead mine, near Clifden, is another notable locality, where pyrite appears to have been mistaken for native gold (Cole 1922). Laouar et al. (1990) note pyrite in Connemaran granite from Carna, Cashel, Costelloe Murvey, Errisbeg Townland, Inish, Omey, Oughterard and Roundstone.

Kerry Carboniferous Limestone hosts stratabound Cu–Fe–Pb–Zn mineralization at Crow Island (galena–sphalerite mainly), Ross Island (chalcopyrite–pyrite mainly) and Muckross (complex ore assemblages), all near Lough Leane, Killarney (Russell 1927b, Moreton et al. 1999, Kinnaird et al. 2002). Ixer (2004) notes pyrite replacing crinoidal limestone, stylolites and shale laminae, and pyrite itself being replaced by chalcopyrite and galena at Crow Island. Another stratiform limestone-replacement deposit containing pyrite occurs at Shanagarry (Caher West) mine, near Cleady (Halliday & Mitchell 1983).

Kilkenny A 4 m-thick marcasite–pyrite cap has formed above the Galmoy orebody, near Kilkenny – a stratiform lead–zinc replacement deposit that developed in dolomitized and brecciated limestone. Pyrite also forms fine crystals and occurs intimately associated with galena and sphalerite in the main orebody (Doyle et al. 1992, Lowther et al. 2004).

Limerick The Courtbrown limestone-hosted Pb–Zn replacement deposit, 25 km W of Limerick contains minor early-formed pyrite and later fine-grain or acicular pyrite associated with galena and sphalerite. The style of mineralization is clearly epigenetic (Grennan 1986, Reed & Wallace 2001).

Longford Minor pyrite occurs in a variety of forms, including spheroids, pyritohedra and nodular aggregates of euhedral crystals in the Keel prospect, 11 km S of Longford. The deposit is enriched in Cd–Pb–Zn (predominantly sphalerite) and it formed in faulted, brecciated and dolomitized Carboniferous rocks (Patterson 1970, Morrissey et al. 1971, Schultz 1971, Watling 1974, Slowey 1986). The Newtown Cashel prospect is a fracture-controlled sphalerite–(barite–galena) deposit to the W of Keel, in which pyrite was an early-crystallizing (probably authigenic) phase. Coarsely crystalline (recrystallized) pyrite has also been found in drill cores (Crowe 1986).

Mayo The Cregganbaun shear zone at Cushinyen contains pyrite as mm-size flecks associated with bournonite, chalcopyrite, galena, sphalerite and tetrahedrite in gold-bearing ribboned quartz veins within Ordovician tuffs (McArdle 1989, Thompson et al. 1992). Pyrite is disseminated throughout a silicified porphyry at the western end of the Curlew Mountains, S of Charlestown, where it is the main sulphide species in a low-grade copper (chalcopyrite) deposit (O'Connor & Poustie 1986, Poustie 1982). Pyrite is also noted at: Bolinglanna (Benderg) copper mine, near Corraun; Gubnabinniaboy mine, Clew Bay – a pyrite lode mined for sulphur; Srahmore copper mine (all Cole 1922); Lecanvey gold prospect, Clew Bay (Aherne et al. 1992); and Sheefry mine, 11 km S of Croagh Patrick (Moreton & Jackson 1991). Inclusions of pyrite in alluvial gold are reported from many locations along the shear zone (Leake et al. 1993, Chapman et al. 2000c).

Meath The world-class Carboniferous carbonate-hosted Pb–Zn replacement deposit at Navan is mineralogically simple, and is dominated by sphalerite and accompanied by barite, galena, marcasite and pyrite (although conglomeratic units and other localized areas are dominated by the iron sulphides). In places the mineralization is characterized by virtually complete replacement of the host carbonate by sphalerite and lesser amounts of pyrite. Colloform, stalactitic and botryoidal pyrite have all been found (Schultz 1971, Kucha & Wieczorek 1984, Andrew & Ashton 1985, Ashton et al. 1986, 1992, 2004, Boyce et al. 1994, Anderson et al. 1998, Blakeman et al. 2002, Peace et al. 2003, Wilkinson 2003). The Tatestown prospect, near Navan, is a satellite deposit to the main Navan orebody; it contains marcasite–pyrite in various textures and grainsizes, but is generally of minor significance (Andrew & Poustie 1986).

Monaghan Pyrite containing inclusions of native gold and associated with arsenopyrite occurs in an As–Au–Sb vein deposit at Clontibret mine, Tullybuck (Morris

1984, Morris et al. 1986, McArdle 1989, Steed & Morris 1997). Interestingly, alluvial gold from here contains inclusions of pyrite (Chapman et al. 2000c). Abundant disseminated arsenopyrite and pyrite occur in altered greywackes close to another gold-rich vein at Bryanlitter, near Clontibret (Steed & Morris 1997).

Sligo At Abbeytown lead–zinc mine, near Ballysadare, original pyrrhotite, which was precipitated within dolostone as intragranular grains, was replaced by marcasite and then pyrite during dolomitization in a limestone-replacement deposit (Schultz 1971, Hitzman 1986, Ryback & Moreton 1992). Pyrite, chalcopyrite and dolomite are noted from the Maye quarries, Lough Gill (Ryback & Moreton 1992). In the Benbulben Range, seams of pyrite crystals up to 1 mm across separate regularly distributed chert lenses from enclosing Carboniferous Limestone (Michel 1994).

Tipperary The large sedimentary-exhalative lead–zinc-barite deposits in the Silvermines district have been worked from several mines. Pyrite is described from the following: Ballygown South mine – alteration of pyrite in a residual (weathering-induced) deposit has produced limonitic boxworks (Boland et al. 1992); Ballynoe barite deposit – pyrite occurs as disseminations within barite, forms chimneys in the footwall, is found in stockworks and joint-controlled veins, and develops a 2 m-thick cap above the barite orebody (Russell 1907b, Schultz 1971, Larter et al. 1981, Boyce et al. 1983, Mullane & Kinnaird 1998, Moreton 1999); Gortnadyne mine – sphalerite partially replaced by pyrite, both subsequently replaced by galena (Rhoden 1959); Knockanroe mine – spherulites of radiating pyrite 5 cm across with relict marcasite cores (Rhoden 1959); Mogul mine – pyrite is present in two ore zones associated with galena and sphalerite ±(arsenopyrite, chalcopyrite, marcasite and tetrahedrite–tennantite), also pyrite cubes to 4 cm on edge associated with a fossil hydrothermal vent and irridescent pyrite from a large cavity (Schultz 1971, Boyce et al. 1983, Ryback & Moreton 1991, Moreton 1999); and more generally by other authors – O'Brien (1959), Morrissey et al. (1971), Taylor & Andrew (1978), Taylor (1984), Reed & Wallace (2004).

The carbonate-hosted Lisheen Pb–Zn–(Ag) deposit, between Thurles and Rathdowney is similar to that at Silvermines and consists of three stratabound sulphidic lenses dominated by galena, marcasite, pyrite and sphalerite mineralization. Pyrite, commonly containing alternating bands of marcasite, forms coarse concentrically zoned colloform masses (Hitzman 1992, Hitzman et al. 2002, Fusciardi et al. 2004, Wilkinson et al. 2005).

Pyrite is a minor species in the dominantly chalcopyrite–Hg-rich tennantite orebody at Gortdrum mine, near Tipperary town (Morrissey et al. 1971, Schultz 1971, Steed 1983, Duane et al. 1986).

Tyrone Gold-bearing quartz veins hosted by Dalradian metasediments contain pyrite with subsidiary arsenopyrite, bornite, chalcopyrite, native copper, galena, sphalerite, tennantite (near end member) and tetrahedrite at Curraghinalt, near Gortin, Sperrin Mountains. Here pyrite forms euhedral crystals containing inclusions of native gold (electrum), gold–silver telluride or bismuth telluride (Clifford 1986, Clifford et al. 1990, Earls et al. 1989, McCaffrey & Johnston 1996, Parnell et al. 2000). Pyrite is also one of the inclusion species found in alluvial gold at three locations in the Sperrin Mountains: Attagh Burn, Curraghinalt Burn and Glencurry Burn (Chapman et al. 2000c).

Pyrite and arsenopyrite, forming coarse (1–2 cm) aggregates, were the earliest sulphides to crystallize in gold-bearing quartz veins of the Lack deposit, WSW of Omagh (Cliff & Wolfenden 1992). Sub-economic veinlets and disseminations of chalcopyrite and pyrite are noted from Cashel and Sleivemenagh (Leyshon & Cazalet 1976). Other gold-bearing veins containing significant pyrite are reported from Cavanacaw, near Omagh (Parnell et al. 2000), and Golan Burn, near Strabane (Woodham et al. 1989).

Westmeath Barite, galena, marcasite, pyrite and sphalerite are the main ore species found in the limestone-replacement deposit at Ballinalack, 15 km NW of Mullingar. Some of the pyrite is nickel-rich bravoite (Patterson 1970, Schultz 1971, Halliday & Mitchell 1983, Jones & Brand 1986). After sphalerite, pyrite is the most abundant sulphide in a similar deposit at Moyvoughly, near Moate, 12 km N of Clara, where it is found in veinlets infilling stylolites, and as disseminations, blebs and clusters of small grains (Poustie & Kucha 1986, Kucha 1988).

Wicklow An Ordovician volcano-sedimentary sequence is host to a volcanogenic exhalative sulphide deposit at Avoca. Syn-sedimentary pyrite occurs with chalcopyrite in massive ore consisting of laminated pyritic sulphides hosted by chloritic schist and a more complex stringer ore hosted by siliceous tuffs. Since 1970 over 5.5 million tonnes of ore have been mined, but total production is much higher, as the pyrite was responsible for the development of the sulphur trade in Ireland in 1840. Notable mines include Ballygahan, Ballymurtagh, Connary, Cronebane and Tigroney (Greg & Lettsom 1858, Cole 1922, O'Brien 1959, Schultz 1971, Wheatley 1971a,b, Badham 1978, Sheppard 1980, Halliday & Mitchell 1983, Williams et al. 1986, McArdle 1993, Ryback & Moreton 1993). At Kilmacoo at the northeast end of the Avoca district, gold-bearing polymetallic (arsenopyrite–chalcopyrite–galena–pyrite–sphalerite–stibnite) vein and disseminated sulphide deposits have been found in extensively altered and sheared volcanic rocks (McArdle 1989, Milner & McArdle 1992). Pyrite also occurs in a gold-rich chalcopyrite–magnetite–siderite-rich BIF sequence in the Goldmine River, 7 km to the S of Avoca (McArdle & Warren 1987, Ixer et al. 1990). Greg & Lettsom (1858) describe large quantities of pyrite from Ballygahan mine in Glandore being sent to Liverpool for the manufacture of sulphuric acid.

Offshore territories

Isle of Man Pyrite occurs with uranium-rich hydrocarbons in low-temperature hydrothermal lead–zinc mineralized veins cutting Ordovician slates at the Laxey mines, E of Snaefell (Parnell 1988, Alderton et al. 1995). Greg & Lettsom (1858) report pyrite from the Foxdale lead mines and, much more recently, rich specimens of arsenopyrite–chalcopyrite–pyrite in blocks of quartz veinstone have been described from the South Bradda mines, N of Port Erin Bay (Starkey 1998b). Partially dolomitized limestone at Lough Skillicore contains large euhedral crystals and framboids of pyrite (Dickson & Barber 1977).

Guernsey Pyrite is present in sulphide veins cutting diorite of the Northern igneous complex at Bouette quarry, St Peter Port (Ixer 1992).

Jersey Pyrite is particularly notable in andesites from Les Rouaux and Westmount, and large masses were found on the site of the Grands Vaux reservoir dam. Molybdenite and pyrite occur in veins at L'Etacq and Les Handois. Pyritohedra have been found at Mont Mado and Gigoulande, and plate-like masses pseudomorphing pyrrhotite are recorded from L'Etacq and Gigoulande. Small cubes of pyrite occur in shales at St Ouën's Bay and Meadowbank tunnel (Mourant 1961). Ore-microscopy studies of mineralization within carbonate veins at the German underground hospital (Meadowbank tunnel), St Peter's Valley, St Lawrence, have shown that pyrite was the earliest, and locally the most abundant, sulphide to crystallize. It forms euhedral 80 µm-diameter grains associated with chalcopyrite, galena, pyrrhotite and sphalerite (Ixer 1980, 1990). Pyrite cubes are common in quartz veins cutting granite at La Corbière, southwest Jersey (Ixer 1980), and, at Le Pulec, on the northwest coast of Jersey, it forms euhedral crystals in lead–zinc veins cutting Precambrian sediments (Stanley & Ixer 1982). Zoned 20 µm-diameter crystals and slightly larger grains containing inclusions of chalcopyrite, enargite and galena are noted in rhyolitic rocks from Bouley Bay, Trinity, northeast coast of Jersey (Ixer 1980).

Sark A specimen of mineralized breccia from Sark's Hope Pb–(Ag) mine, Little Sark, contains tennantite and three different forms of pyrite: euhedral (dodecahedral) crystals, arsenic-rich collomorphic pyrite, and framboidal pyrite aggregates (Bishop et al. 1977). Ixer & Stanley (1983) report strongly colour-zoned nickel-rich pyrite (bravoite) from this mine, and Mourant & Warren (1934) also record nodular and radiating pyrite.

pyroaurite

trigonal — HYDROTALCITE GROUP — $Mg_6Fe^{3+}_2(CO_3)(OH)_{16}{\cdot}4H_2O$

An uncommon hydrated magnesium–iron carbonate hydroxide, dimorphous with sjögrenite and found at several localities on Unst, Shetland, Scotland: in serpentinite with brucite and chromite from the northeastern extremity of Haaf Grunay, a small island off the southeast coast of Unst (Heddle 1878b, 1901); in chromitite, occasionally with pentlandite and heazlewoodite in pale-green chrysotile from Hagdale quarry (Heddle 1901, Braithwaite & Pritchard 1983b); as small golden-yellow grains in veins within chromitite and serpentinite from quarries S of Nikka Vord (Heddle 1901); and in a green serpentinite at Cliff quarry, W of Baltasound, and from mid-Unst (Maynard 1993). A nickeloan pyroaurite has been found in serpentinite from 1 km W of Leslie, near Glenrothes, Grampian (Wilson et al. 1976).

pyrobelonite

orthorhombic — DESCLOIZITE GROUP — $PbMn^{2+}(VO_4)(OH)$

A rare lead–manganese vanadate hydroxide found in fine-grain, compact, very hard, dark-grey manganese ore at Tŷ Coch farm, near South Cornelly, Mid Glamorgan, Wales. Braunite and hausmannite are the main ore species occurring with interstitial and vug-filling barite, calcite and dolomite. Black to deep-red pyrobelonite, with an adamantine to submetallic lustre, is a minor constituent, forming euhedral to anhedral grains or intergrowths with vanadinite. Pyrobelonite also occurs as veinlets (<0.5 mm thick) that can be traced for up to 3 cm (Criddle & Symes 1977).

PYROCHLORE GROUP

Complex cubic oxides with the general formula $A_{1-2}B_2O_6(O,OH,F)\cdot nH_2O$, where A = Ba, Bi, Ca, Ce, Cs, K, Na, Pb, Sb^{3+}, Sn, Sr, Th, U, Y, Zr; and B = Fe, Nb, Ta, Ti, W. Three subgroups are also defined: microlite, pyrochlore and betafite (Hogarth 1977). Microlite and pyrochlore, and possibly betafite, are members of the group found in Britain or Ireland.

pyrochlore

PYROCHLORE GROUP

cubic $(Ca,Na)_2Nb_2O_6F$

A rare calcium–sodium niobium oxide that forms a series with microlite and is reported from the Goonvean and Rostowrack china-clay pits, W of Nanpean, St Austell, Cornwall, where it occurs in mineral concentrates (grainsize <100 µm) produced as waste from kaolinite extraction. The source of the waste is a topaz granite unit that is the most chemically evolved part of the St Austell granite complex. Other accessory minerals found in the concentrate include apatite, cassiterite, ferrocolumbite, monazite, rutile and zircon. Pyrochlore and uraninite form inclusions in ferrocolumbite grains (Scott et al. 1998).

Pyrochlore is possibly present in the Lewisian granitic pegmatite at Chaipaval (Ceapabhal), 3 km NW of Northton (Taobh Tuath), South Harris, Western Isles, Scotland. A metamict betafite-like mineral is described by von Knorring & Dearnley (1960), who provide XRD data that Macpherson & Livingstone (1982) interpret as probably being pyrochlore. The pyrochlore is an accessory constituent of the pegmatite, along with allanite, beryl, columbite, gahnite, kasolite, magnetite, monazite, pyrite, spessartine, thorite, topaz, tourmaline, uraninite and zircon.

Accessory pyrochlore is also possibly present in the aegirine–riebeckite granite on Rockall island, 440 km W of northern Skye, NW of Scotland. A tiny colourless octahedron in a single thin section was thought to be a member of the microlite–pyrochlore series (Sabine 1960), but the species could not be confirmed by Hawkes et al. (1975).

A report of microlite or pyrochlore (or both) with columbite from lithium-rich spodumene pegmatites at Stranakelly and Moylisha, both near Shillelagh, Co. Wicklow, Ireland (Luecke 1981), requires further confirmation.

pyrochroite

BRUCITE GROUP

trigonal $Mn^{2+}(OH)_2$

A rare manganese hydroxide found at the Benallt manganese mine, Rhiw, Lleyn Peninsula, Wales. The ore consists of a complex mixture of manganese silicates (alleghanyite, pennantite, spessartine and tephroite) and other subsidiary magnesium and iron species (ganophyllite, jacobsite and pyrophanite). Pyrochroite is a secondary oxidation product that occurs with hausmannite on one specimen from this locality now at the NHM (Dunham et al. 1978, Bevins 1994).

pyrolusite

RUTILE GROUP

tetragonal $Mn^{4+}O_2$

A common manganese oxide, trimorphous with akhtenskite and ramsdellite, and found in strongly oxidized manganese-bearing hydrothermal deposits, in bogs and lakes, under shallow marine conditions and frequently as an alteration product of manganite or other manganese-rich species.

Delicate and beautiful dendrites of pyrolusite have been found at Countybridge quarry, Goonhilly Downs, Lizard Peninsula, Cornwall, where brecciated serpentinite has been impregnated with the species (Holyer 1996). Earthy black pyrolusite coats the surface of birnessite on dump material from Treburland manganese mine, 2 km SE of Fivelanes, Altarnun, Bodmin Moor, Cornwall. At one time, pyrolusite was the main manganese ore at this mine (Young 1988). Russell (1946d) also reports scoriaceous masses and glossy black botryoidal aggregates from this location and suggests that pyrolusite is an alteration product of rhodonite. Other Cornish locations for pyrolusite include: Greystone Wood, Lezant; Indian Queens (Ruthvoes) mine, St Columb Major; Lidcott mine, Laneast; Pencrebar mine, Callington; Pendower, Veryan; Tremollet (Coads Green) mine, near North Hill; Wheal Truscott, St Stephens, Launceston (all Russell 1946d); Callington mine, near Liskeard, and Restormel (Royal) iron mine, N of Lostwithiel (Robson 1949b); Greystone quarry, Lezant and Tremore quarry, Lanivet (Golley & Williams 1995).

In Cumbria, pyrolusite is recorded many times: Argill mine (an unknown locality) and Bigrigg mine, both near Egremont; Kelton Fell mine, Kirkland (all Young 1987a); Arm o' Grain, Caldbeck Fells (Day 1928); Brandlehow mine, Derwent Water; Driggith mine, Caldbeck Fells (fine dendrites on barite); Thief Gills, Caldbeck Fells (all Davidson & Thomson 1951); hard crystalline black masses, or as tiny brilliant diamond-shape tablets at Burdell Gill, Caldbeck Fells (Davidson & Thomson 1951, Neall & Leppington 2003); Dry Gill, Caldbeck Fells (Cooper & Stanley 1990); in the hematite mines of the Cleator district (Goodchild 1875); Remmington's vein on Great Dunn Fell (Bridges & Young 1998); Silverband mine, Great Dun Fell (Bramall 1921); amorphous and in crystals at Wiley Gill, Carrock Fells (Kendall 1884); Wyndham mine, Bigrigg, near Egremont (Davidson & Thomson 1951, Young & Nancarrow 1990). Green et al. (2006a) note that suspected pyrolusite specimens from Driggith and Sandbed mines are in fact romanèchite, and Green et al. (2006b) suggest that pyrolusite does not occur at Arm o' Grain.

In Devon, manganese deposits were worked at several localities in the Permian sedimentary rocks close to the southern boundary fault of the Crediton Trough. One such locality is Pound Living mine (Black Pit), Upton Pyne, NW of Exeter, where pyrolusite is found as an earthy alteration product of manganite and rhodochrosite (Scrivener & Edwards 1990). Shepherd et al. (2005) note that oxidation of interbanded psilomelane, kutnohorite and rhodochrosite produced the massive manganite–pyrolusite ore at Upton Pyne. An extensive deposit of pyrolusite, associated with manganocalcite, psilomelane and rhodochrosite, was once worked at Chillaton and Hogstor mine, Milton Abbot (Russell 1946d, Dines 1956, Durrance & Laming 1982). Pyrolusite forms lustrous irregular crystals to several mm across on wad from Fullabrook mine, Ilfracombe (Robinson & King 1993). Pyrolusite is also described from: Allerford mine, Marystow; Sydenham and Lee Wood mine, Marystow, and Week, Milton Abbot (Russell 1946d); Brent Tor, 6 km N of Tavistock; Georgeham, NW of Barnstaple; and Viveham, N of Barnstaple (Hall 1868b).

Pyrolusite forms a black botryoidal crust on oxidized veinstone from the dumps of Grasshill mine, along Harwood Beck, Teesdale, Co. Durham (Young et al. 2005).

Quartz-rich veins cutting a late Precambrian porphyritic dacite at Springhill quarry, Peldar Tor, Whitwick, Leicestershire, contain pyrolusite as black films on chlorite aggregates, or on joints, as lustrous or sooty-black acicular crystals. Pyrolusite is often associated with barite, calcite, dolomite or mottramite (Faithfull & Ince 1992). Black acicular crystals of pyrolusite have been recovered from vugs in a carbonate–quartz vein near the Precambrian/Triassic unconformity at Bardon Hill quarry, Coalville, Leicestershire (Ince 2007). At Croft quarry, SW of Leicester, joint surfaces of diorite and Caledonian basement conglomerates contain pyrolusite associated with palygorskite, minor hollandite and unidentified manganese dendrites (Pearson & Jeffrey 1997). Greg & Lettsom (1858) report small well formed crystals of pyrolusite from Hartshill, Warwickshire.

Pyrolusite is a constituent of iron- and manganese-rich mineralized veins cutting Carboniferous Limestone at Merehead (Torr Works) quarry, Cranmore, Somerset, where it forms amorphous wad, botryoidal masses and silvery-grey metallic fibres up to 2 mm in length. It is associated with blixite, cerussite, chloroxiphite, coronadite, hematite, hydrocerussite, mereheadite, paralaurionite, parkinsonite and symesite. Polianite, a hard steel-grey variety of pyrolusite forming composite crystals up to 6 mm long, is also found at Merehead (Alabaster 1975, Welch et al. 2000). Turner (2007) reports euhedral crystals of goethite, manganite and pyrolusite within calcite-lined cavities at this location. Earthy, fibrous and granular pyrolusite was the main manganese ore at Higher Pitts farm, Priddy, Mendip Hills (Kingsbury 1941, Burr 1996). Kingsbury (1941) also reports dendritic pyrolusite from the western end of Sandford Hill, Mendip Hills, and a similar form is noted at Whitwell, 19 km SE of Sheffield, Derbyshire, where dendrites have formed in a barite–galena vein cutting Permian Magnesian Limestone (Ineson et al. 1972).

Dendritic pyrolusite on a joint plane in Lower Carboniferous sandstone; Llangothlan, Gwynedd, Wales; width 4 cm. (© K. Gardner)

Occurrences of pyrolusite in Wales are rare, although the species is confirmed from Benallt manganese mine, Rhiw, Lleyn Peninsula, Gwynedd, and from Tŷ Coch, near Porthcawl, Mid Glamorgan (Bevins 1994). Pockets of pyrolusite in shales adjacent to a Mn–Pb–Zn lode are described from Llety-Ifan-Hên (Vaughan) mine, Central Wales orefield, Dyfed (Jones & Moreton 1977). Disseminated pyrolusite and hematite occur in the Haffoty Formation, a cherty manganese-rich horizon exposed on St Tudwal's Peninsula, S of Llanengan, Lleyn Peninsula (Dunham et al. 1978).

In reviewing early literature for manganese minerals in Scotland, Nicholson (1989b) suggests that the name pyrolusite has often been used as a field term for bluish-black crystalline material with a metallic lustre, and the material being described should not be regarded as a precise species. Pyrolusite appears to be confirmed at the following locations, in all cases as a supergene species:

BORDERS In ferromanganese sediment of exhalative origin at Noblehouse, near West Linton (Nicholson 1989b). DUMFRIES & GALLOWAY Lining cavities in limestone at Closeburn (Nicholson 1989b). GRAMPIAN In a vein within brecciated Dalradian Ben Aigan Quartzite at Arndilly House, Arndilly, near Craigellachie (Nicholson 1986, 1989b); crystalline blocks of pyrolusite in till at Bridge of Don, Aberdeen (Chew 1978, Nicholson 1989b, 1992); Laverock Braes, Grandholm, Aberdeen (Nicholson 1989b); in seepage-bog type ironstone hosted by breccia at Lecht mine, near Tomintoul (Nicholson 1987a, 1989b, 1992, Nicholson & Anderton 1989). LOTHIAN Elongate crystals in felsite at Fenton Tower quarry, near North Berwick, and at Kingston, North Berwick (Nicholson 1989b). STRATHCLYDE Carngriffe quarry, Tinto (Nicholson 1989b); dendritic coatings on barite at Muirshiel mine, Queenside Muir, NE of Largs (Todd & Laurence 1989); in veins within Dalradian Jura Quartzite at Dùn Athad, Mull of Oa Peninsula, Isle of Islay (Nicholson 1987b, 1988a, 1992). ORKNEY Radiating crystals to 5 cm at Hoy (Livingstone 2002).

Despite reports of pyrolusite from Leadhills, Strathclyde, and Wanlockhead, Dumfries & Galloway (Brown 1927, Williams 1965), the mineral does not appear to have been confirmed.

In Ireland, pyrolusite occurs with hematite and rhodochrosite in a small quarry at Logwood Hill, Calliagh, Co. Monaghan (Ryback & Moreton 1993), and with hematite, limonite, manganite and psilomelane in greisen-like fissure veins within granite in Silent Valley quarry, near Newcastle, Mourne Mountains, Co. Down (Nockolds & Richey 1939).

A nodule of bluish-black divergent crystalline fibres embedded in wad in a joint in mudstone, in the quarry at the foot of Jubilee Hill, St Peter, Jersey, Channel Isles, is described as pyrolusite (Mourant 1961).

pyromorphite

APATITE GROUP

hexagonal $Pb_5(PO_4)_3Cl$

A common lead phosphate chloride found in the oxidized zone of lead deposits, where it often replaces galena.

England

Cheshire The copper mines of Alderley Edge ridge have produced mimetite–pyromorphite crystals with little correlation between appearance (short prismatic crystals, needles and cauliflower-like masses), colour (bright green, grey green, yellowish and colourless) and composition. Localities include Engine Vein, Mottram, West and Wood mines, and the Stormy Point opencast (Carlon 1979, Rothwell & Mason 1992, Braithwaite 1994, Plant et al. 1999).

Cornwall Remarkably fine specimens of lustrous apple-green prismatic pyromorphite were once found at Wheal Alfred, Phillack, near Hayle (Greg & Lettsom 1858, Rudler 1905, Watson 1988). Crystals of similar colour forming hexagonal prisms or barrel shapes, rarely up to 15 mm in length, have been reported from vugs and cavities in gossanous matrix at Penberthy Croft mine, St Hilary, although analysis indicates that many specimens thought to be pyromorphite are in fact mimetite (Dines 1956, Camm & Merry 1991b, Betterton et al. 2000). Wheal Golden, Perranzabuloe, is another classic locality, where long spiky crystals of pale-green pyromorphite have been shown to have end-member composition (Greg & Lettsom 1858, Merry & Braithwaite 1995, Green et al. 1996a). Acicular, elongate prismatic and barrel-shape crystals (<5 mm) are described from the Greystone lead lode exposed in Greystone quarry, Lezant, near Launceston (Elton et al. 2004). Other reports of pyromorphite are from: Garras mine, St Erme; Wheal Helston; Watergate Bay, Newquay; Whipsiderry beach adits, Newquay; Gunheath china-clay pit, St Austell (all Golley & Williams 1995); Pengenna mine, St Kew and Tolpetherwin quarry, Two Bridges, South Petherwin (Russell 1949b); Pentire (Glaze) mine, near New Polzeath, St Minver (Clayton et al. 1990); Perran Iron lode, near Perranporth, and Wheal Trelawny, E of Liskeard (Dines 1956); Trevinnick mine, St Kew (Dewey 1925, Russell 1949b, Fisk et al. 1982); Wheal Confidence, near Newquay (Russell & Hutchinson 1927); Wheal Hope, ENE of Perranporth (Hosking & Camm 1981); Wheal Ludcott, St Ive, near Liskeard (Green & Merry 1997); Wheal Mary Ann, Menheniot (Merry & Braithwaite 1995); Wheal Penrose, Sithney, near Helston (Greg & Lettsom 1858); Wheal Rose, Sithney (Russell 1944, Dines 1956); Wheal Saturn, near Breage (Collins 1904b); and Wheal Unity, St Day, Gwennap (Penhallurick 1997).

Pyromorphite from Wheal Penrose, Sithney, Cornwall; width 25 mm. (© Royal Cornwall Museum, Royal Institution of Cornwall; photographer D. I. Green)

Cumbria Some of the world's finest pyromorphite has come from lead lodes in the Caldbeck Fells, Roughton Gill mine being perhaps the most impressive source. Hundreds of tonnes were raised from here in the early days of mining. "Oily-green" pyromorphite on smalt-blue plumbogummite is unique to the locality. Bright yellow-green to oily-green terminated prismatic or tapering crystals, usually less than 10–15 mm in length, are characteristic. Brown, grey, honey-colour, orange and white crystals are also known, but here, as elsewhere, some may actually be mimetite (Rudler 1905, Postlethwaite 1913, Davidson & Thomson 1951, Förtsch 1967, Cockbain 1968, Hartley 1984, Wolfe 1998, Green & Briscoe 2002). Particularly good reviews of pyromorphite from Roughton Gill and other Caldbeck Fell locations are to be found in Cooper & Stanley (1990, 1991b); these include: green, elongated, prismatic

Fine, lustrous crystals of pale-green pyromorphite to 20 mm from Roughton Gill mine, Caldbeck Fells, Cumbria. (© M. P. Cooper)

to spiky crystals and crystalline crusts at Arm o' Grain (Rothwell & Mason 1992, Green et al. 2006b); eastern flank of Balliway Rigg, higher Roughton Gill (Young 1987a, Green & Briscoe 2002); Blea Gill (Young 1987a); olive- to emerald-green crystals at Brae Fell mine (Hartley 1984, Cooper et al. 1988, Bridges et al. 2006); Brandy Gill and Sandbed lead mines, and as yellow coatings and tiny green (occasionally yellow or blue) crystals, at Red Gill mine (Davidson & Thomson 1951, Wolloxall 1981, Hartley 1984, Green et al. 2006a); Carrock End mine (Hartley 1984); Carrock mine, Grainsgill (Dewey & Dines 1923, Hartley 1984); steeply pyramidal to acicular dark-green crystals along part of Roughton Gill South lode at Clints Gill (Green & Briscoe 2002); at Driggith mine, pyromorphite was mined as an ore and it forms rounded, prismatic, or seed-like crystals of milky green or khaki colour (Gregg & Lettsom 1858, Davidson & Thomson 1951, Stanley & Vaughan 1981, Rothwell & Mason 1992, Hartley 1984, Green & Briscoe 2002, Green et al. 2006a); abundant at Dry Gill mine, where arsenic-free bright canary-yellow tapering prisms have been found (Davidson & Thomson 1951, Hartley 1984, Green & Briscoe 2002); Higher Mexico mine, Todd Gill (Green & Briscoe 2002); Ingray Gill, Mexico and Potts Gill (barite) mines (Hartley 1984); bright-green rice-grain to acicular crystals up to 15 mm long at Iron Crag, near Roughton Gill mine (Wolfe 1998, Green & Briscoe 2002); minutely crystalline green crusts at Poddy Gill (Davidson & Thomson 1951, Green 1986, Rothwell & Mason 1992); pale-blue radiating crystals to 1.5 mm at Sandbed mine (Green et al. 2006a); crusts of curved oily-green crystals up to several mm long and yellow barrel-shape crystals at Silver Gill, near Roughton Gill (Hartley 1984, Rothwell & Mason 1992, Green et al. 2005b). Pyromorphite is also noted from Saddleback Old mine, below Foule Crag, Glenderamackin Valley, where it occurs in a supergene assemblage with cerussite, hinsdalite and pseudomalachite (Tyler 2003, Green et al. 2005c, Wolfe 2006).

In the northern Pennines, pyromorphite forms encrustations coating fracture surfaces of other vein minerals and in places surrounds cores of galena. Several locations are described by Young (1985a) and other workers: yellowish coatings at Clouds mine, near Kirkby Stephen (Davidson & Thomson 1951); small amounts at Gale Sike, Smittergill Head, Alston Moor (Bridges & Young 1998); clusters of green

Spheres of apple green pyromorphite on pale-blue plumbogummite; Roughton Gill, Caldbeck Fells, Cumbria; specimen is 30 mm tall. (© D. I. Green)

Stepped cluster of pale-green prismatic pyromorphite, 30 mm tall, from Roughton Gill, Caldbeck Fells, Cumbria. (© D. I. Green)

prismatic crystals at Hard Rigg Edge, Melmerby Fell (Dunham 1990, Green & Briscoe 2002) – this is the same location described as Aglionby Beck, Alston Moor, by Dunham (1948); Hilton mine, Scordale, near Appleby (Bridges & Young 1998); small amounts at Silverband mine, Great Dun Fell (Bridges & Young 1998, Green & Briscoe 2002); deep-green prisms and wart-like aggregates up to 2 mm across at Smittergill Head mine, Melmerby Fell; and dull-green crystalline patches from Windy Brow vein, Tynehead, near Garrigill.

Other descriptions of pyromorphite include: brown prismatic crystals at Barrow mine, Braithwaite (Davidson & Thomson 1951, Green & Briscoe 2002); grass-green crusts at Brandlehow mine, Derwent Water (Davidson & Thomson 1951, Strens 1963, Young 1987a); pale-brown to black prismatic crystals up to 20 mm long at Force Crag mine, Braithwaite, near Keswick (Wolfe 1996b, Green et al. 1997, Green & Briscoe 2002); Goldscope mine, Newlands Valley, near Keswick (Davidson & Thomson 1951, Strens 1963); Kinniside mine, Cleator Moor (Young 1987a, Green & Briscoe 2002); Millrigg mine, Staveley Head Fell, Staveley (Lawrence et al. 1986); green crusts on galena at Ruthwaite Lodge, Grisedale (Rothwell & Mason 1992); green crystals in cavities within granite at Shap granite quarry (Wirth & Green 2002); Threlkeld mine, Threlkeld (Young 1987a); Yewthwaite mine, Newlands Valley (Eastwood 1921, Strens 1963).

Derbyshire Pale apple-green to almost colourless crusts of pyromorphite on corroded galena, on fluorite and on barite are reported from Wapping mine, Cromford (Ince 1991). At Pike Lane opencut, near Winster, pyromorphite commonly forms yellowish-green powdery coatings on galena or, less commonly, dull to light-green acicular crystals up to 12 mm across (Utley 1989). Greg & Lettsom (1858) describe green and yellow crystals of pyromorphite from Bonsall Moor and at Brassington, and small green crystals from near Wirksworth. Other occurrences include: Ashley Hay, near Wirksworth (Ford & Sarjeant 1964a); Ballidon (Hall 1868a); Deep Rake, Longstone Edge; Bole Hill, Sheldon (Rieuwerts 1963); Fall Hill quarry, Ashover (Smith 1982); Golden Valley (Golconda) mine, Brassington (Lecornu 1879); Green Linnets mine, Brassington (Rudler 1905); Hazard mine, Dirtlow Rake, Castleton

Green botryoidal pyromorphite (30 cm across); Roughton Gill, South lode above New Mexico mine, Caldbeck Fells, Cumbria. (© D. I. Green)

(Ince 1991); Heyspot mine, Elton (Green & Strahan 1887); Masson Hill, Matlock, and Youlgreave (Braithwaite 1983); Red Rake, Calver (Rieuwerts 1960); Rutland mine, Matlock Bath (Ford et al. 1993); Sacheveral farm, Winster (Rothwell & Mason 1992); and Tideslow Rake, Tideswell (Cotter-Howells & Giddens 1990).

Devon Grey pyromorphite is noted from Bere Alston, near Bere Ferrers (Hall 1868b).

Durham Pyromorphite forms deep grass-green tapering prisms up to 3 mm long, encrusted on barite or altered dolerite wallrock at Closehouse mine, SW of Middleton-in-Teesdale, Lunedale (Dunham 1948, Young 1985a, Young et al. 1994a, Bridges & Young 1998). At Grasshill mine, along Harwood Beck, Teesdale, pyromorphite forms distinctive bright yellowish-green cellular masses up to 10 cm across associated with limonite, or stubby hexagonal prisms (Greg & Lettsom 1858, Young 1985a, Bridges & Young 1998). Other localities include: Healeyfield mine, Castleside; Whitfield Brow mine, Bollihope, Weardale; Yorkshire Silverband mine, near Cauldron Snout, Teesdale (all Young 1985a and Bridges & Young 1998) and Pikestone mine, Pikestone Fell, Weardale (Bridges & Young 1998).

Leicestershire Crusts of white to pale-green acicular pyromorphite are associated with cerussite and mottramite on a few specimens of malachite from a native copper–cuprite deposit near the Precambrian/Triassic unconformity at New Cliffe Hill quarry, Stanton under Bardon (Hubbard et al. 2005). Bright-yellow encrustations of pyromorphite associated with cerussite, galena and wulfenite are noted from Tickow Lane lead mine, Blackbrook, near Shepshed (King & Ludlam 1969).

Shropshire Uncommon green coatings of pyromorphite are derived from the oxidation of Cu–Pb–Zn ore at Ladywell mine, Shelve (Braithwaite 1982b). Excellent specimens of bright to yellowish-green acicular-to-prismatic crystals of pyromorphite, typically 2–3 mm across, have been found on the Day level of Burgam mine, Shelve (R. E. Starkey pers. comm. 2005).

Somerset In the Mendip Hills, pyromorphite appears to have been worked in considerable quantities (as green lead ore) as early as 1728. It formed yellowish-green crystalline encrustations and cellular botryoidal masses at Green Hill and Blagden Hill, both near Charterhouse. The village of Green Ore, near Priddy, is perhaps another location (Spencer & Mountain 1923). At Merehead (Torr Works) quarry, Cranmore, pyromorphite is found with mimetite as powdery coatings on manganese ore (Alabaster 1975, Symes & Embrey 1977). Kingsbury (1941) reports further (minor) occurrences at Batts Combe quarry, near Cheddar, and at Higher Pitts farm, Priddy, where both yellowish-green encrustations of prismatic crystals and warty crystalline aggregates have been found (Alabaster 1975).

Yorkshire Pyromorphite forms sprays of bright-green elongated hexagonal crystals up to 3 mm long at Coldstones quarry, 5 km W of Pateley Bridge (Wolfe 1998). It is also noted from Ashfold Side Beck (Merryfield vein) and from the Old Gang mines, Gunnerside Gill, Swaledale (Small 1982, Dunham & Wilson 1985, Ineson & Young 1994).

Wales

Clwyd Campbell Smith (1913) describes the mineral collection of Thomas Pennant (1726–1798), in which were specimens of pyromorphite from Gelly–Fowler Fields mine and Silver Rake, both on Halkyn Mountain.

Dyfed Pyromorphite is widely distributed in the Central Wales orefield and is characterized by its prismatic habit and bright grass-green colour. Some of the best pyromorphite in Britain or Ireland came from Bwlchglas mine, near Tal-y-bont, where it formed beautiful 1–2 mm-long (occasionally up to 6 mm long) crystals coating quartz fragments in a breccia pipe (Braithwaite 1982a, Hughes 1989c). At Eaglebrook (Nantycagl) mine, Ceulanymaesmawr, pyromorphite occurs as small translucent-to-frosty light-green crystals forming barrel-shape stubby-to-equant, or rarely acicular, hexagonal crystals (Jones 1983). Unusual prismatic, colourless, brown and pale mauve varieties (exceptionally to 2.5 cm) have been found at Frongoch mine, between Devil's Bridge and Pont-rhyd-y-groes (Greg & Lettsom 1858, Green et al. 1996b). Many other localities are described by Jones & Moreton (1977), Rothwell & Mason (1992), Bevins (1994) and (Mason 2004), including: Alltycrib mine, Tal-y-bont; Brynyrafr mine; Caegynon mine, near Ponterwyd; Copa Hill mine, Cwmystwyth; Cwmbrwyno mine, near Devil's Bridge; Cwmdarren mine, Trefeirig; Cwmystwyth mine; Darren mine, Pen-bont Rhydybeddau; East Glogfach mine, Ysbyty Yswyth; Esgair Hir mine, Ceulanymaesmawr (all Rust & Mason 1988); Florida mine; Glogfawr mine, Devil's Bridge; Henfwlch mine, between Cerrig yr Hafan and Nant-y-moch reservoir, E of Tal-y-bont; Llechweddhelyg mine, near Clawddmelyn, between Bont-goch and Cwmsymlog (all Jones 1987); Logaulas mine, near Ysbyty Ystwyth; Mynyddgorddu mine, W of Bont-goch, Ceulanymaesmawr (Clark & Rust 1993); Nant-y-Cae mine, near Ponterwyd; Pandy mine, Tre'r-ddol; Penybanc mine, Tal-y-bont; Penycefn mine, near Tal-y-bont; Ysbyty Ystwyth (Ryback & Saville 1967); Ystrad Einion mine, near Tal-y-bont (Mason & Rust 1997). Outside of the Central Wales orefield, pyromorphite has been found at Nant-y-Mwyn mine, Rhandirmwyn, W of Llanwrtyd Wells, and at the Vale of Towy (Nant) mine, near Carmarthen (Bevins 1994).

Gwynedd Pyromorphite is noted from Gwynfynydd mine, near Dolgellau (Andrew 1910, Allen & Jackson 1985); there is also evidence to suggest that the species was once found at Parys Mountain, Anglesey (Bevins 1994, Southwood & Bevins 1995).

Mid Glamorgan A single specimen of pyromorphite in the NMW is labelled Rhiwsaeson, Llantrisant, but no other material is known from this area (Bevins 1994).

Powys Mines in the Central Wales orefield where pyromorphite has been found include: Aberdaunant, near Llanidloes; Dyfngwm, Machynlleth; Dylife, N of Hafren

Purplish-brown translucent prisms of pyromorphite up to 2 mm in length; Frongoch mine, between Devil's Bridge and Pont-rhyd-y-groes, Dyfed, Wales. (© M. P. Cooper)

Forest; Llanerch-yr-aur; Snowbrook (Nantyreira); South Nantycar (Jones & Moreton 1977, Rust & Rust 1987, Rothwell & Mason 1992). Elsewhere, pyromorphite is recorded from Cwm Orog and Ochr Graig mines, Llangynog (Bevins 1994).

Scotland

Central Pyromorphite occurs with malachite, smithsonite (calamine) and native sulphur in veins, irregular masses and metasomatic replacements in the lead–zinc mining district of Tyndrum (Ridgway 1980).

Dumfries & Galloway Around Wanlockhead (and extending into Leadhills, Strathclyde), massive pyromorphite, pseudomorphing galena ranges in colour from black, through green, yellow and white, whereas crystalline material displays green, yellow, and red to orange. Localities where individual varieties occur are often vague (Davies 1877, Heddle 1901, Rudler 1905, Brown 1927), but include Belton Grain vein at High Pirn mine (Brown 1919, Temple 1955, Livingstone 1994, 2002), Margaret's or Straitstep vein (Brown 1919), Glencrieff shaft at Wanlockhead mine (Wilson & Flett 1921) and Whyte's Cleuch (Temple 1955).

Grampian A few specimens of drusy green pyromorphite have been found in a lead vein cutting a Triassic breccia termed the Cherty Rock on the west beach, Lossiemouth, near Elgin. Associated species include cerussite, galena, phosgenite, pyrite, quartz and sphalerite (Starkey 1988a, Green & Todd 2001).

Highland Abundant grass-green pyromorphite is associated with wulfenite in barite–galena–quartz vein material at the Struy mines, Loch na Mèine, midway between Strath Glass and Strath Farrar, 3 km SW of Struy, near Inverness (Russell 1946b, Green 1986, Rothwell & Mason 1992, Green & Todd 2001). At Ceann Ousdale, Helmsdale, pyromorphite occurs in barite–galena veins in deeply weathered Moine schists, W of the Helmsdale granite (Gallagher et al. 1971). Green and yellow pyromorphite is recorded from Strontian mine (Greg & Lettsom 1858, Hall 1868b, Heddle 1901), but no specimens are now known to exist (D. I. Green pers. comm. 2005).

Strathclyde Pyromorphite occurs widely on the dumps around Leadhills, including: green from East Stayvoyage vein (Green & Todd 2001); Hopeful vein (Temple 1955); Horner's vein; yellow at Raik vein, Broadlaw (both Rothwell & Mason 1992); olive-green and yellow at Leadhills Dod, Leadhills (Heddle 1901, Livingstone & Sarp 1984); orange and red at Susanna vein (Heddle 1901, Rothwell & Mason 1992).

Short prisms of brick-red pyromorphite to 2 mm in diameter; Leadhills, Strathclyde Region, Scotland. (© M. P. Cooper)

More generally, pyromorphite is noted by Heddle (1882a) and Gillanders (1981), and beautiful orange-red prismatic crystals with hexagonal cross sections (2 mm diameter) are illustrated by Livingstone (2002). See also Dumfries & Galloway. Elsewhere, yellow pyromorphite associated with wulfenite is reported from Gatehouse, Lauchentyre (Heddle 1901), and a green microcrystalline coating of pyromorphite on galena has been found at Mannoch Hill mine, SW of Lesmahagow (Todd & McMullen 1991).

Ireland

Donegal The dumps of Keeldrum lead mine, near Bedlam, have yielded specimens of cerussite, galena and pyromorphite (Ryback & Moreton 1992).

Down Secondary oxidation products from the Whitespots–Conlig lead vein to the N of Newtownwards include pyromorphite, together with azurite, cerussite, gypsum, limonite and malachite (Dunham et al. 1978, Moles & Nawaz 1996).

Monaghan A crystallized pyromorphite–(wulfenite) specimen from Lemgare mine, between Monaghan and Castleblaney, is in the Russell collection at the NHM (Ryback & Moreton 1993).

Tipperary Moreton (1999) lists pyromorphite as one of many secondary metalliferous minerals found in the Silvermines district.

Wicklow Yellow-to-green drusy pyromorphite, associated with wulfenite, is reported from mines in the Leinster granite, near Glendalough (Green & Moreton 2001). Pyromorphite forms pale-green prisms up to 2 mm long in vugs in quartz at Van Dieman's Land mine, Glendalough, and pale-yellowish transparent stubby prisms at Ballinafunshoge lead mine, Glenmalur. The Luganure lead mines in Glendasan are marked by many dumps, on which small crystals of pyromorphite can be found (Ryback & Moreton 1993).

Offshore territories

Isle of Man Pyromorphite has been found near Dumbell's shaft at Laxey mine (Lamplugh 1903).

Sark Pyromorphite is noted from Sark's Hope Pb–(Ag) mine, Little Sark (Greg & Lettsom 1858, Mourant & Warren 1934).

pyrope

GARNET GROUP

cubic

$Mg_3Al_2(SiO_4)_3$

A common magnesium aluminosilicate that forms series with almandine and with knorringite, and is found in ultramafic rocks such as eclogite, peridotite and serpentinite. Like all other species of garnet, end-member compositions are rare and most examples show some degree of solid solution towards other members of the group. Where possible, the proportions of the two most abundant end members are listed below (al=almandine, py=pyrope).

In Strathclyde, Scotland, pale-pink pyrope (py_{43-46}, al_{31-32}) has been found in garnet–hornblende clinopyroxenite (previously described as eclogite), within the Girvan–Ballantrae complex, between Knockormal and Carleton Mains farms (Bloxam & Allen 1959, Smellie & Stone 1984). On the eastern and western shores of Loch Lomond, pyrope occurs as bright-pink crystals (py_{46-53}, al_{42-47}), up to 4 mm across (and sometimes zoned), with almandine, amphibole and sodic feldspar, in xenoliths within a lamprophyre dyke (Dempster & Bluck 1991). Pyrope also occurs within pyroxenite xenoliths and as discrete crystals in Permo-Carboniferous alkali basalt diatremes in the Midland Valley of Scotland (part of Strathclyde). Localities include: Auchengarth (py_{67}, al_{19}) and Black Rocks (py_{51}, al_{32}), both near Largs; Dunaskin Glen, 3 km NW of Dalmellington (py_{57}, al_{28}); Heads of Ayr, immediately S of Ayr (py_{57}, al_{28}); Holmbyre, 2 km SW of Dalry (py_{51}, al_{34}); Muirkirk (py_{69}, al_{17}); Seamills, N of Saltcoats (py_{45}, al_{39}). Additionally, pyrope has been found in spinel pyroxenite xenoliths within a diatreme on the island of Eilean Shona, NW of Ardnamurchan, Highland Region (py_{59}, al_{26}), and at Elie Ness, Fife, it forms deep-red gem-quality sub-rounded grains (py_{65-66}, al_{20-21}), up to 2.5 cm across, in volcanic agglomerate and tuffs (Greg & Lettsom 1858, Heddle 1901, Colvine 1968, Chapman 1976, Upton et al. 2003). These grains are responsible for the names Elie rubies and Ruby Bay (Moreton 2006). Pyrope is further noted from the nearby Ardross Neck agglomerate (Cumming 1935) and from columnar basalt in Ruddons Point Neck, W of Kincraig (Heddle 1901).

Other occurrences are in the Glenelg district, Highland Region: (py_{41}, al_{38}) in eclogite (a pyrope–omphacite rock) from Beinn a' Chapuill, Glenelg (Alderman 1936); (py_{51-62}, al_{26-33}) in olivine websterite from Beinn Fhada, near Letterfearn, Loch Duich (Rawson et al. 2001); (py_{41}, al_{38}) and almandine in a garnet–two-pyroxene–hornblende rock from Glen More, Glenelg (O'Hara 1960). At Rodel, South Harris, Western Isles, pink pyrope (py_{45-60}, al_{22-37}) has been found in an amphibole–garnet pyroxenite said to be almost identical to the Glen More rock (Livingstone 1967, 1976c). Close by, to the S and SW of Roineabhal, South Harris, zoned pyrope (py_{46-58}, al_{31-46})–almandine (al_{47-54}, py_{39-47}) porphyroblasts occur in sapphirine-bearing and sapphirine-free kyanite/sillimanite–orthopyroxene granulites (Baba 1999, 2003, Hollis et al. 2006).

pyrophanite

trigonal — ILMENITE GROUP — $Mn^{2+}TiO_3$

A rare manganese–titanium oxide that forms a series with ilmenite and is found in metamorphosed manganese deposits or, less commonly, as an accessory mineral in granite, amphibolite or serpentinite.

Pyrophanite occurs at New Cook's Kitchen mine (North Tincroft lode), Redruth, Cornwall, as lath-like euhedral-to-subhedral crystals, some of which are intermediate in composition between ilmenite, ilmenorutile and pyrophanite end members. Some crystals have pyrophanite cores and ecandrewsite rims. The North Tincroft lode is an anatase–cassiterite–chlorite–ilmenite–monazite–sulphide assemblage, thought to have developed as a result of mixing between magmatic–hydrothermal and meteoric fluids during or after intrusion of the Carnmenellis granite (LeBoutillier et al. 2000).

Pyrophanite is present in a 5 mm-thick pennantite-bearing vein cutting massive manganese ore at Benallt mine, Rhiw, Lleyn Peninsula, Wales. Pyrophanite was observed in thin section as orange-yellow crystals (<0.25 mm diameter) with a tabular habit, high refractive index and very high birefringence. It is associated with analcime, banalsite, calcite, colourless mica, ganophyllite and probable tephroite (Campbell Smith et al. 1946, Campbell Smith & Claringbull 1947). An iron-rich variety of pyrophanite is noted from the Harlech manganese deposit (Hafotty Formation of Cambrian age), which, although only 40 cm thick, occurs over an area of 190 km^2 in North Wales. Pyrophanite is a minor component (along with apatite, barite, hematite, magnetite, pennantite and rhodonite) of the banded cherty ore, which consists of Ca–Fe–Mg–Mn carbonates, chlorite, quartz and spessartine (Bennett 1987, Bevins 1994).

Minor pyrophanite with braunite, bustamite, hematite, manganese-rich calcite, manganese-rich ilmenite, rhodochrosite and possibly jacobsite form a complex stratiform manganese ore up to 4 m thick at Dalroy Burn, near Culloden, Highland Region, Scotland. Pyrophanite is closely associated with apatite, garnet, manganese-rich ilmenite and quartz. The ore has a geothermal origin best explained by deposition from hot springs in a sub-lacustrine environment during Devonian times (Nicholson 1987b, 1989b, 1990, 1992).

pyrophyllite

monoclinic or triclinic — $Al_2Si_4O_{10}(OH)_2$

A relatively uncommon clay-like aluminosilicate hydroxide found in Middle and Upper Devonian slates along the north coast of Cornwall from Padstow to Tintagel. An XRD analysis of the <2 μm clay fraction of the slates shows pyrophyllite to be a minor component (Primmer 1985). Pyrophyllite is also recorded from Brookwood mine, Dartmoor, Devon, by Collins (1911a), who points out the remarkable exfoliation property of the species when strongly heated. An NHM specimen of pyrophyllite from the Lizard Peninsula, Cornwall, is noted without description by Golley & Williams (1995).

In Wales, XRD analysis of the <2 μm fraction of Ordovician pelitic rocks from the Lleyn Peninsula and Snowdonia was undertaken by Merriman & Roberts (1985) to try to establish the regional distribution of greenschist and prehnite–pumpellyite facies metamorphism. Pyrophyllite was recorded in samples from the aureoles of four acid–intermediate intrusions: Bwlch Mawr, Llanbedrog, Penmaenmawr and Yr Eifl. Using the same technique, Roberts & Merriman (1990) have also shown that pyrophyllite and rectorite (plus chlorite) are the main components of the clay fraction of Ordovician metabentonites, which they argue were derived from a sub-alkaline rhyolitic magma. Parc Penrhyn, near Bangor, is one location where the metabentonite is found. Elsewhere in Wales, pyrophyllite is reported from: certain sandstones and shales of Carboniferous age from the anthracite zone (i.e. western part) of the South Wales coalfield (Gill et al. 1977); sedimentary rocks of Silurian and Devonian age from the Marloes area of Dyfed (Nicholls 1979, Robinson et al. 1980); altered rhyolite and rhyolitic tuffs from a borehole near Treffgarne, Dyfed (Brown et al. 1987b); and from pod-like structures in the Ffestiniog granite quarry, Cefn Bychan, near Ffestiniog, Gwynedd (Roberts 1979).

Chloritoid–kyanite-bearing rocks within regionally metamorphosed Dunrossness phyllites, Dunrossness, South Mainland, Shetland, Scotland, also contain pyrophyllite (Livingstone 2002).

pyrosclerite

Pennine pseudomorphous after pyroxene.

pyrosmalite

Any member of the series ferropyrosmalite–manganpyrosmalite.

pyrostilpnite?

monoclinic — Ag_3SbS_3

A rare silver sulphantimonite, dimorphous with pyrargyrite, said to have been found at St Teath, Cornwall, by A. W. G. Kingsbury (Ryback & Tandy 1992). However, Ryback et al. (1998, 2001) and other authors have expressed serious doubts about the provenance of many samples in the Kingsbury collection (now at the NHM) and, until further discoveries are made, pyrostilpnite cannot be considered to be a British or Irish species.

PYROXENE GROUP

Orthorhombic or monoclinic silicates with the general formula ABZ_2O_6, where A = Ca, Fe^{2+}, Li, Mg, Mn^{2+}, Na, Zn; B = Al, Cr^{3+}, Fe^{2+}, Fe^{3+}, Mg, Mn^{2+}, Sc, Ti, V^{3+}; and Z = Al, Si. Aegirine, aegirine-augite, augite, clinoenstatite, diopside, enstatite, ferrosilite, hedenbergite, jadeite, omphacite, pigeonite, spodumene and possibly clinoenstatite and johannsenite are members of the group found in Britain or Ireland. A classification of the group is provided by Morimoto et al. (1988).

pyroxmangite

triclinic — $Mn^{2+}SiO_3$

A rare manganese silicate that forms a series with pyroxferroite and is found as a series of thin (<0.6 mm-wide) orange-brown veinlets cutting massive manganese ore dominated by rhodochrosite at Nant mine, Llanfaelrhys, Lleyn Peninsula, Wales. Broken veinlets reveal aggregates of flat prismatic crystals no more than a few mm in length (Cotterell 2006b).

Pyroxmangite is also reported at a single outcrop within the Lewisian inlier at Gleann Beag, Glenelg, Highland Region, Scotland. Pink grains of pyroxmangite occur in garnet–grunerite schist, where it forms anhedral crystals 0.5–0.75 mm in length. Exceptionally, it develops porphyroblasts up to 5 mm across. Associated species are ferrosilite, hedenbergite, magnetite and pyrrhotite (Perutz 1937, Tilley 1937a).

pyrrhosiderite

A synonym of lepidocrocite.

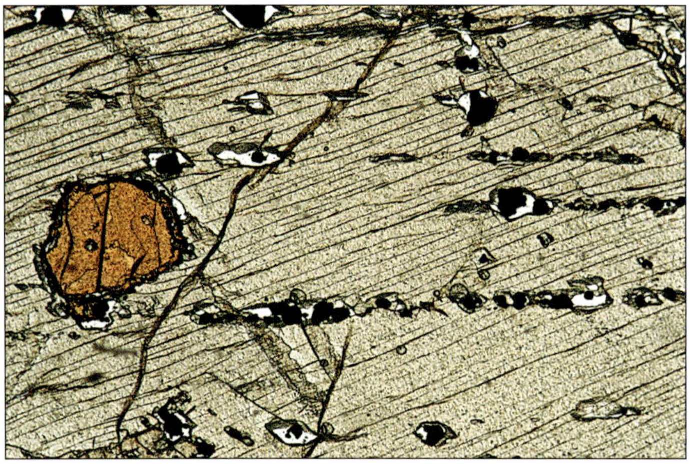

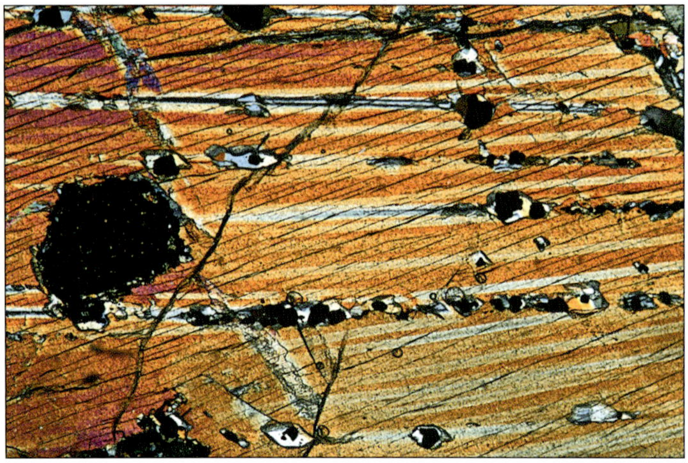

Well cleaved and twinned pyroxmangite containing trains of magnetite inclusions and a large orange-brown manganese-rich garnet; Glen Beag, Glenelg, Highland Region, Scotland; width 3.4 mm; viewed in PPL and XPL. (© K. Gardner)

pyrrhotite

monoclinic or hexagonal Fe_7S_8

A common iron sulphide found mainly in mafic igneous rocks, although it also occurs in high-temperature hydrothermal and replacement veins and in some sedimentary and metamorphic rocks.

England

Cornwall Pyrrhotite is listed as one of the primary minerals in the Crowns lode at Botallack mine, St Just, and it occurs in calc-flintas around the Bodmin Moor granite at Camelford (Hosking 1951). A contact-metamorphic deposit close to the margin of the Bodmin Moor granite at Treburland mine, 2 km SE of Fivelanes, Altarnun, Bodmin Moor, contains abundant magnetite and pyrrhotite associated with a bedded manganese ore (Russell 1946d, Kingsbury 1961). Large light-to-dark bronzy masses of pyrrhotite associated with other sulphides occur at Wheal Jane, Kea, near Truro (Dunham et al. 1978, Rust 1982d). The old mines of Redmoor, Kelly Bray and Holmbush to the W of Kit Hill worked sulphide veins in killas and all contain arsenopyrite, chalcopyrite, pyrite and pyrrhotite, along with cassiterite and wolframite (Newall & Newall 1989, Newall 1994). Minute tabular crystals of pyrrhotite have been found on millerite needles from quartz veins at Greystone quarry, Lezant, near Launceston (Elton et al. 2004). Other occurrences include: East Pool mine, Illogan; St Austell Consols; Great Onslow Consols, St Breward (all Golley & Williams 1995); Levant mine, St Just (Collins 1892, Wolloxall 1987); in arsenopyrite–pyrite-dominated sulphide veins at Mount Wellington mine, Twelveheads, 7 km W of Truro (Kettaneh & Badham 1978); in an anatase–cassiterite–chlorite–ilmenite–monazite–sulphide assemblage at New Cook's Kitchen mine (North Tincroft lode), Redruth (LeBoutillier et al. 2000); in greenstone at South Tolgus shaft, Redruth (Davey 1948); minor amounts of pyrrhotite are associated with bournonite, galena and tetrahedrite in a small quartz lens on Newporth beach, 2 km SW of Falmouth (Dean 1982, Dean et al. 1983); disseminated pyrrhotite occurs in a diabase dyke close to Wheal Sedgman lode of Perran silver mine, Perranuthnoe (Butler 1907); in peridotite at the Rill, Kynance Cove, Lizard Peninsula (Clark 1967); Wheal Kine, St Agnes (Collins 1892); cleavable and massive pyrrhotite at Wheal Maudlin, Lostwithiel, and in calc-flintas at Wheal Mary (Wheal Ding section), Lanivet, near Bodmin (Greg & Lettsom 1858, Kingsbury 1961).

Cumbria Pyrrhotite is plentiful in the Harding vein at Carrock mine, Grainsgill, Caldbeck Fells, where it is associated with arsenopyrite, pyrite, scheelite and wolframite. Bronze-colour hexagonal platy crystals of pyrrhotite up to 15 mm in length, in vugs or enclosed in massive calcite, are also known from this mine (Hitchen 1934, Davidson & Thomson 1951, Eastwood 1959, Shepherd et al. 1976, Dunham et al. 1978, Fortey 1980, Ball et al. 1985a, Watson 1988, Green & Briscoe 2002). Pyrrhotite was common in the arsenopyrite-rich assemblage found in the opencut above the 60-fathom level at Sandbed mine, Caldbeck Fells (Green et al. 2006a). In the Coniston area, pyrrhotite is a constituent of polymetallic ore from Bonser mine (Stanley & Vaughan 1982b) and it replaces pyrite and cobalt-rich arsenopyrite at Paddy End mine (Stanley 1979). It was a significant mineral in the ores, but is now represented by pseudomorphs of pyrite and marcasite (Dagger 1977). Occurrences of pyrrhotite in Newlands Valley include minute amounts, usually enclosed in arsenopyrite or pyrite, at Castlenook, Copper Plate (Borrowdale) and Goldscope mines, and abundant masses at Dale Head mine (Davidson & Thomson 1951, Stanley & Vaughan 1980, Young 1987a). Other occurrences in the Lake District include: Armboth dyke, Thirlmere (Rastall & Wilcockson 1915); associated with the Eskdale granite at Gate Crag and on Corney Fell (Young 1985e, Young et al. 1986); exceptionally plentiful at Packhorse Hill, Shap (Harker & Marr 1891); Sellafield area boreholes (Milodowski et al. 1998); Shap granite quarry (Wirth & Green 2002); Wanthwaite mine, St John's in the Vale, near Threlkeld (Stanley 1979). In the Northern Pennine orefield, pyrrhotite has been found at: Tynebottom mine, NW of Garrigill, near Alston (Vaughan & Ixer 1980), and at four localities on Alston Moor, to the S of Garrigill – along the Great Sulphur vein, Crossgill trials, Calvert mine and Sir John's mine (Dunham 1990).

Derbyshire Vein mineralization hosted within altered lavas at Millclose mine, Wensley, near Matlock, contains pyrite (some is nickel-rich bravoite) with inclusions of pyrrhotite (Ixer 1978, 1986, Ixer & Townley 1979). Pyrrhotite is also noted: from a sulphide vein in limestone intersected by a borehole at Tansley, near Ashover (Ramsbottom et al. 1962, Ford et al. 1993); from Deep Rake, Longstone Edge and Ladywash mine, Eyam (Ford & Sarjeant 1964a).

Devon Abundant massive disseminated löllingite–pyrrhotite–(arsenopyrite–bornite–chalcopyrite–sphalerite) has been found in andradite–hedenbergite skarns within the aureole of the Dartmoor granite at Red-a-ven mine, Meldon, near Okehampton (Kingsbury 1961, El Sharkawi & Dearman 1966). At the nearby Meldon (British Railways) quarry complex, near Okehampton, metasomatism of quartzite by the Dartmoor granite has produced a coarsely crystalline pyrrhotite–sphalerite rock and a massive pyrrhotite–rhodonite–tephroite rock (Dearman 1966a, Beer et al. 1989). Joints in wallrocks adjacent to the Meldon aplite contain pyrrhotite (Dearman & Claringbull 1960). Hall (1868b) notes pyrrhotite from Meldon quarry and the mines at Bere Alston. Pyrrhotite is also found as an accessory mineral in the Dartmoor granite (Brammall & Harwood 1925) and the Lundy granite (Rastall & Wilcockson 1915, Dollar 1932). A borehole at Honeymead farm on Exmoor exposed disseminated pyrrhotite and pyrite mineralization in sandstone of the Ilfracombe Slate Formation (Edmonds et al. 1985, Benham et al. 2004).

Durham Pyrrhotite occurs in an ilvaite-bearing dolerite–pegmatite facies of the Whin Sill at Forcegarth quarry, close to High Force waterfall, NW of Middleton-in-Teesdale (Ixer et al. 2005b), and in garnet–magnetite–sulphide-bearing calc-silicate rocks at Ratcheugh quarry, SE of Alnwick – which were formed immediately adjacent to the Whin Sill (Ixer & Stanley 1987). Roughly hexagonal crystals of pyrrhotite up to 10 cm across were found sporadically in fluorite–pyrite–quartz veins at Cambokeels (Cammock Eals) mine, Eastgate, Weardale (Smith & Hardy 1981, Dunham 1990, Green & Briscoe 2002). Bladed marcasite pseudomorphs after pyrrhotite, and inclusions of pyrrhotite in marcasite and pyrite, are noted from Whiteheaps mine, near Blanchland (Vaughan & Ixer 1980, Ixer et al. 1996). Small amounts of pyrrhotite occur in mineralized White Whin (metasomatically altered quartz dolerite) at Closehouse mine, SW of Middleton-in-Teesdale, Lunedale, where it forms lath-shape inclusions in pyrite (Vaughan & Ixer 1980, Ixer 1986). Pyrrhotite is found with nickeline in a magnetite-rich ore at Lady's Rake mine, 6 km NW of Langdon Beck, Teesdale (Young et al. 1985b, Green & Briscoe 2002). Other reports are from a borehole at Wynch Bridge, Teesdale (Dunham 1990), Frazer's Hush mine, Weardale (Wolfe 1998), and Groverake mine, near Rookhope, Weardale (Ixer et al. 1996). All these occurrences could be attributed to the heat of the Whin Sill, as has been suggested for the conversion of syngenetic pyrite in shale to pyrrhotite at Scoberry

PYRRHOTITE

Lustrous crystal of pyrrhotite 55×35 mm; Cambokeels mine, Weardale, Co. Durham. (© M. P. Cooper)

Bridge, Teesdale (Dunham 1990). Pyrrhotite–quartz–sulphide veins have also been found in a borehole cutting the Weardale granite at Rookhope (Dunham 1967, Solomon et al. 1971, Brown et al. 1987a).

Leicestershire Pyrrhotite is a ubiquitous accessory mineral in igneous rocks from the Mountsorrel district: in gabbro at Swithland reservoir; in diorite at Brazil Wood; in granodiorite at Cocklow quarry, the Main Mountsorrel quarry and Rubbish Tip quarry; and in granodiorite and a granophyre vein at Kinchley (Taylor 1934).

Northumberland A bronze-colour mineral forming inclusions in the Whin Sill at Snook Point, N of Dunstanburgh, is identified as pyrrhotite (Smythe 1924).

Wales

Dyfed In a review of polyphase and polymetallic vein mineralization in the Central Wales orefield, Mason (1997) notes widespread but rare pyrrhotite in a galena–chalcopyrite-dominated A1-*c* polymetallic assemblage from: Darren mine, Pen-bont Rhydybeddau; Eaglebrook (Nantycagl) and Esgair Hir mines, Ceulanymaesmawr; Erglodd mine, N of Tal-y-bont; and Esgair Fraith mine, near Tal-y-bont. Microscopic amounts of pyrrhotite have also been found with chalcopyrite and pyrite in altered dolerite from Carn Menyn, Mynydd Preseli (Thorpe et al. 1991).

Gwynedd Pyrrhotite was one of the earliest sulphides to form in the Parys Mountain copper deposit, Anglesey, where it has been found as rounded 10–30 µm inclusions (intimately associated with chalcopyrite) in pyrite and as aggregates up to 350 µm across replacing pyrite (Pointon & Ixer 1980). In northern Snowdonia, rare pyrrhotite is noted from: Cwmbychan, Betws Garmon and Llyn Aran, Cadair Idris – in pyritic oolites and mudstones (Pulfrey 1933); Cwmdwyfor mine, near Rhyd-ddu, and Pan-ty-wrach mine, between Llanfrothen and Penryhndeudraeth (Bevins 1994); Parc mine, near Llanrwst – in albite–diabase intrusions (Archer & Elliott 1965); Tan-y-Garth mine, Ogwen Valley, near Bethesda (Dewey 1920); between Lliwedd and Hafod y Porth, SE of Snowdon – in volcanogenic pyrite-dominated veins (Reedman et al. 1985). It is a major component of sulphide veins at Drws-y-Coed mine (Braithwaite 1982b) and Simdde Dyllhuan mine (Bevins 1994), both in the Nantlle Valley. In southern Snowdonia, pyrrhotite is widespread in gold-bearing veins in the Dolgellau goldbelt, often with chalcopyrite and pre-dating the major galena–sphalerite mineralization. In bonanza-gold associations, pyrrhotite overgrows gold and telluride species (Mason et al. 2002). Reports are from: Caegwian mine (Gilbey 1968); Clogau mine, Bontddu (Andrew 1910, Dewey & Eastwood 1925, Gilbey 1968, Naden 1988, Allen & Jackson 1985, Mason et al. 2002); Ffridd-Goch mine, Llanfachreth, and Gwynfynydd mine (Mason et al. 2002); Foel Ispri and Wnion mines, near Dolgellau (Gilbey 1968); Friog Undercliff, Fairbourne, SW of Dolgellau (Mason et al. 1999); Panorama lode, Llanaber, E of Barmouth (Andrew 1910, Dewey & Eastwood 1925, Gilbey 1968); Prince Edward mine (Gilbey 1968, Mason & Williamson 2001); St David's mine (Andrew 1910); trial level, 2 km NW of Bontddu, near Dolgellau (Saich & Rust 1987); Vigra mine (Andrew 1910, Gilbey 1968, Mason et al. 2002). At Tyllau Mwn mine, 3 km SW of Drws-y-nant station, near Dolgellau, pyrrhotite is abundant in thin stilpnomelane-bearing veins cutting through lenticular masses of sedimentary iron ore (Matthews & Scoon 1964). The Cae Coch stratiform massive pyrite deposit, near Trefriw, Llanrwst, contains rare poikilitic pyrrhotite, and patches of the mineral frequently forming stockworks are also noted in adjacent doleritic rocks (Ball & Bland 1985, Bottrell & Morton 1992).

Mid Glamorgan Hexagonal platy crystals of pyrrhotite occur in clay–ironstone in: the Coal Measures at Ferndale colliery (Llanwynno tips); Gelli colliery, Rhondda; and Parc colliery, Cwm Parc (Firth 1971, Bevins 1994).

Powys Superb short prismatic hexagonal pyrrhotite crystals up to 7 mm across, and associated with chalcopyrite, galena and pyrite, have been found at Dylife mine, N of Hafren Forest (Jones & Moreton 1977, Moreton 1982).

Scotland

Central Rare pyrrhotite is reported from Dalradian pyritic quartzite near Tyndrum (Fortey & Smith 1986) and from stratabound chalcopyrite–sphalerite deposits at Creag Bhocan, SSW of Tyndrum (Scott et al. 1988). Minor pyrrhotite is also noted from barite–As–Co–Cu±Ag±Pb veins at Alva mine, Ochil Hills (Jassim et al. 1983).

Dumfries & Galloway A small pyrrhotite–pentlandite–(chalcopyrite–nickeline–skutterudite) ore in a sill-like diorite body at the western end of the Cairnsmore of Fleet granite was once mined at Palnure Burn, Talnotry, 5 km NE of Newton Stewart (Heddle 1901, Wilson & Flett 1921, Gregory & Leitch 1928, Cook 1976, Dunham et al. 1978, Lowry et al. 1997, Gunn & Styles 2002, Power et al. 2004). Stanley et al. (1987) add that the skutterudite is probably misidentified gersdorffite. Pyrrhotite is uncommon: in greywacke-hosted quartz veins at Cairngarroch Bay (Lowry et al. 1997); in greywacke, close to sulphide-bearing quartz veins at Glencrieff Burn, Wanlockhead (Nicholson & Banks 1987); disseminated (with arsenopyrite and pyrite) in the margins of monzonitic intrusions near the southwestern margin of the Loch Doon granite complex, Glenhead Burn, near Loch Dee (Leake et al. 1981), and (with marcasite, pyrite and sphalerite) in black mudstones in nearby Penkiln Burn (Stone & Leake 1984).

Grampian Pyrrhotite is an important species in disseminated magmatic copper–nickel sulphide assemblages: in the Arthrath gabbroic intrusion, near Ellon, 10 km N of Huntly; at Bin quarry in the Huntly gabbro; in the Knock gabbro between Littlemill and Auchencrieve farms, near Huntly (Fletcher & Rice 1989, Gunn & Shaw 1992, Fletcher et al. 1997, Gunn & Styles 2002, McKervey et al. 2007). Other reports include: disseminated pyrrhotite and pyrite in the Dalradian Water of Tanar Limestone Formation, S of Ballater (Goodman et al. 1990); in metasomatized limestone at Eslie, near Banchory (Heddle 1878c); and in zinnwaldite granite at Gairnshiel Bridge, Glen Gairn (Hall & Walsh 1972). Heddle (1901) lists other localities, but does not describe their paragenesis: Corntulloch; Crathie; Delnabo quarry, Glen Gairn; Deskry; Forester Hill; Leac Gorm; Limehillock, near Grange; Muir, Deeside; Old Battery, Craigbuirach, Glenbuchat; Redhythe.

Highland Within the Lewisian inlier of Glenelg, pyrrhotite is associated with troilite in impure marble at Sgiath Bheinn, S of Còsag (Davidson 1995), and with garnet, grunerite and pyroxmangite in schist at Gleann Beag (Tilley 1937a). Pyrrhotite is a prominent sulphide in the Glenelg eclogite, on the southwest shore of Loch Duich, near Fern Villa (Mercy & O'Hara 1968). Less precisely, pyrrhotite is noted in graphite-bearing altered sediments from Glenmore of Glenelg (Clough & Pollard 1899) and with serpentine at Totaig, Loch Duich (Heddle 1901). The stratiform sulphide deposit at Kerry Road and Teangadh Bhuidhe Mhòr, near Gairloch, is hosted by Lewisian carbonate–quartz-bearing schists in which fine-grain pyrite and pyrrhotite are the dominant sulphide species. Near by, massive or banded pyrrhotite occurs in a black gossan within Flowerdale hornblende schists, to the NW of the summit of Sithean Mòr, and, at Druim na Fearna, Dubh Loch, disseminated pyrrhotite occurs with a suite of copper sulphides in amphibolites of the Loch Maree Group (Jones et al. 1983, 1987).

Other occurrences on the mainland include: in clinopyroxenite from the Loch Ailsh alkaline complex, Allt Cathair Bhan Valley, Assynt (Shaw et al. 1994); in carbonatite at Loch Borralan, Assynt (Young et al. 1994b); in a biotite-diorite dyke at Loch Bad nam Ban, 3 km SW of Lochinver (Stumpfl 1961); replacing pyrite in slate at North Ballachulish slate quarry, 2 km E of Onich (Neumann 1950); in metamorphosed siliceous dolomite at Ballachulish (Ferry 1996); in calc-silicate hornfels at Coire Dubh, near the Corran ferry crossing, Ardgour (Drever 1939); in granular limestone at Shinness, on the eastern shore of Loch Shin (Heddle 1878c); in granodiorite within St Adamnan's explosion breccia pipe on the Ardsheal Peninsula, near Kentallen (Rice & Davies 1979). Heddle (1901) adds further localities with little description: Achnagonalin, Grantown; Arscaig, Loch Shin; Dulnain Bridge, Laggan; and Glen Urquhart. A sulphur-isotope investigation of Caledonian granites used samples of pyrrhotite from intrusions at Ballachulish, Ratagain, Strontian and Upper Findhorn, near Inverness (Laouar et al. 1990).

On the Isle of Rum, pyrrhotite occurs with chalcopyrite, pentlandite and magnetite in oxide–sulphide droplets within the unit 11/12 chromite band, Eastern Layered Series, although the authors suggest that the pyrrhotite is best described as troilite (Dunham & Wilkinson 1985, Power et al. 2000a). A sulphide-rich dyke cutting the West Sgaorishal ultramafic plug on the Isle of Rum is dominated by pyrrhotite, with lesser quantities of chalcopyrite and pentlandite (Power et al. 2003). On the Isle of Skye, globular pyrrhotite occurs in ferrodiorite from Harker's Gully, Marsco (Wager & Vincent 1962, Thompson 1969).

Strathclyde Fine-grain Dalradian stratabound chalcopyrite–pentlandite–pyrite–pyrrhotite ore (with associated gold) has been mined for copper and then for nickel at Coille-bhraghad and Craignure mines, both near Inveraray (Wilson & Flett 1921, Berridge 1969, Dunham et al. 1978, Willen 1980, Livingstone 1990b). Stratiform pyrrhotite–(chalcopyrite–pyrite) form a massive ore between thin quartzitic horizons close to a stock-like porphyrite intrusion at Garbh Achadh, W of Inveraray (Ellis et al. 1978, Fortey 1980). At Cuan Ferry and Easedale island, near Oban, pyrrhotite formed during the main regional metamorphism of the Grampian orogeny in the late Precambrian Dalradian Easdale Slate Formation. It has since been retrogressed to form pyrite (Hall et al. 1988). Syn-sedimentary pyrrhotite is noted at McPhun's Cairn, Loch Fyne, where it occurs with pyrite in narrow layers and pods within the Lower Cambrian Ardrishaig Phyllite (Willan 1980, Willan & Hall 1980). Other occurrences include: in limestone at Balephetrish, Isle of Tiree (Heddle 1901); massive pyrrhotite in a vein close to the molybdenite locality at Coire Buidhe, Glen Creran (Ingram et al. 1993); rare pyrrhotite and chalcopyrite are disseminated in the Central Starav granite, Etive complex, head of Loch Etive (Haslam et al. 1985); pyrrhotite occurs in quartz–carbonate veins and disseminated in a granodiorite intrusion at Knipe (Fountainhead) mine, Hare Hill, 5 km SE of New Cumnock (Boast et al. 1990); in granite at Kilmelford (Laouar et al. 1990); pyrrhotite forms inclusions in alluvial gold grains from Shortcleugh Water, Leadhills (Leake et al. 1998, Chapman et al. 2000c).

Tayside Common pyrrhotite occurs with chalcopyrite, pyrite, galena and sphalerite in predominantly Ba–Pb–Zn-enriched Dalradian rocks at the top of the Ben Eagach Schist, near Loch Lyon. The mineralized sequence extends eastwards to the Foss deposit, NW of Aberfeldy, and the Ben Eagach deposit, N of Aberfeldy (Coats et al. 1980, 1984, Willan 1980, Fortey & Beddoe-Stephens 1982, Moles 1983, 1985, Willan & Coleman 1983). Chalcopyrite, galena, pyrite and pyrrhotite are all found at the contact of the Ben Eagach Schist with the Ben Lawers Schist at the north end of Lochan na Lairige, NNE of Killin (Green & Todd 2001). Pyrrhotite is a constituent of gold–sulphide-bearing quartz veins at Calliachar and Urlar Burns, near Aberfeldy (Ixer et al. 1997, Leake et al. 1997c, Chapman et al. 2000c). Minor amounts of pyrrhotite are noted from galena-bearing quartz veins at Corrie Buie, Meall nan Oighcreag, Loch Tay (Pattrick 1984), and from granite at Glen Doll (Laouar et al. 1990). Heddle (1901) reports pyrrhotite from: Edintian quarry, S of Tulloch Hill, Blair Atholl; Sawmill Burn (Allt a' Bhealaich), Taymouth; and on the western shoulder of Craigsoales, Tarfside. At Shierglas quarry, E of Tulloch Hill, Blair Atholl, pyrrhotite occurs as massive segrations in vein quartz and as platy inclusions in calcite (R. E. Starkey pers. comm. 2005).

Western Isles Pyrrhotite is present in the Shiant Isles Main Sill, particularly in the upper picrites, in pegmatite segregations in the upper part of the crinanite, and in the olivine teschenite margins of the sill (Gibb & Henderson 1996, Henderson et al. 2000). Heddle (1901) reports pyrrhotite in limestone from Rodil, South Harris.

Shetland At Vidlin Ness, Mainland, pyrrhotite occurs in a sulphide horizon within tremolite schists and amphibolite, all interbedded with calc-silicate rocks (Garson & May 1976, Dunham et al. 1978, Smith et al. 1984). Pyrrhotite occurs in three parageneses within the ophiolite complex on Unst: with chalcopyrite and pentlandite in pyroxenite; with chalcopyrite and pyrite in gabbro; and in talc–carbonate rocks at the base of the complex. Localities include: east of Loch of Cliff, Baltasound–Haroldswick area; W of Hagdale; Nuda; N of Sandwick Bay; Vord Hill, W of Nuda (Maynard 1993, Maynard et al. 1997).

Ireland

Antrim In the Larne area, pyrrhotite is a minor species in contaminated dolerites from Carneal (Sabine & Young 1975) and Scawt Hill (Tilley & Harwood 1931).

Carlow Disseminated arsenopyrite and pyrrhotite are found in granite sheets emplaced into the Ballybeg Pelitic Formation in the Tinahely area. Pyrrhotite may be intergrown with scheelite (McArdle & Kennan 1988).

Donegal Pyrrhotite occurs in graphite–hercynite–staurolite-bearing pelites in the aureole of the Ardara granite at Narin, 2 km SE of Portnoo harbour (Atkin 1978). Greg & Lettsom (1858) report bronze-colour pyrrhotite from Leahtown.

Down Trace amounts of pyrrhotite are noted in molybdenite–pyrite-bearing granodiorite from a quarry at Croreagh (Meighan & Ixer 1980).

Galway The Middle Dalradian Lakes Marble Formation in the Glanns area, NW of Oughterard, Connemara, contains pyrrhotite–(pyrite)-dominated sulphide lenses and disseminations in a stratiform deposit closely associated with metavolcanic amphibolites. Near by, at Teernakill South mine, a thinner sulphide deposit containing pyrrhotite–(chalcopyrite–pyrite) occurs adjacent to another amphibolite unit (Cole 1922, McArdle et al. 1986a). The Oughterard granite also contains pyrrhotite (Laouar et al. 1990).

Longford Pyrrhotite occurs with chalcocite, chalcopyrite and molybdenite in hornfelsed turbidites in the aureole of the Arva granodiorite at Aghagah (Morris 1984).

Mayo Gold-bearing quartz veins from the Lecanvey prospect, Clew Bay, contain minor pyrrhotite (Aherne et al. 1992). Gold grains containing inclusions of pyrrhotite have been found at Croagh Patrick and in alluvium at Glenkeen Bridge on the Carrownisky River (Leake et al. 1993, Chapman et al. 2000c).

Monaghan Sheaves and clusters of pyrrhotite occur with native antimony, arsenopyrite, pyrite and stibnite in gold-bearing quartz–carbonate veins at Bryanlitter, near Clontibret. Pyrrhotite is rare in the arsenopyrite–gold veins at Clontibret (Steed & Morris 1997).

Sligo At Abbeytown mine, near Ballysadare, original pyrrhotite, which was precipitated within dolostone as intragranular grains, was replaced by marcasite and then pyrite during dolomitization in a replacement-style MVT deposit (Hitzman 1986).

Tipperary Accessory chalcopyrite, galena and pyrrhotite occur in a sedimentary-exhalative barite–pyrite deposit hosted by Carboniferous limestone at Ballynoe mine, near Silvermines (Mullane & Kinnaird 1998). Pyrrhotite is also noted in syngenetic galena–sphalerite deposits in the Silvermines district by Morrissey et al. (1971), Taylor & Andrew (1978) and Moreton (1999).

Tyrone Traces of marcasite and pyrrhotite have been found in pyrite-dominated gold–quartz veins at the Lack deposit, WSW of Omagh (Cliff & Wolfenden 1992).

Wicklow Pyrrhotite is present in cassiterite–scheelite-mineralized microtonalite sheets at Ballingen, between Tinahely and Aughrim (Steiger & Bowden 1982, Kennan et al. 1986, Gallagher 1989). BIF rocks, consisting of magnetite–(chalcopyrite–pyrite–siderite), at Ballycoog on the NW of Goldmine River valley, near Avoca, contain trace amounts of pyrrhotite (Ixer et al. 1990).

Offshore territories

Isle of Man Pyrrhotite occurs with arsenopyrite, galena, gratonite, pyrite and sphalerite in quartz veins at North Laxey mine, by Ballaglass stream (Alderton et al. 1995). Pyrrhotite, chalcopyrite and pyrite are minor constituents of galena–sphalerite-bearing veins at Laxey and Snaefell mines (Lamplugh 1903).

Guernsey Sulphide veinlets cutting diorite of the Northern igneous complex contain pyrrhotite associated with chalcopyrite, cubanite, ilmenite, magnetite, molybdenite, pyrite (abundant) and sphalerite at Baubigny, near St Sampson, and with chalcopyrite, galena, magnetite, pyrite and sphalerite at Bouette quarry, St Peter Port (Ixer 1992).

Jersey Polymetallic mineralization associated with granites of the Northwest igneous complex includes pyrrhotite as one of the earliest sulphide species to form, with examples reported from: galena–sphalerite-bearing carbonate veins at Le Pulec (Ixer & Stanley 1980, Ixer 1992); chalcopyrite–molybdenite–pyrite–sphalerite veins at L'Etacq quarry (Mourant 1961, Ixer 1980, 1992); galena–sphalerite carbonate veins in the German underground hospital tunnels at Meadowbank (Mourant 1961, Ixer 1980, 1990b). Pyrrhotite also occurs in diorite at Gigoulande, in granite at Le Frêt Point, and in dolerite at Ronez (Mourant 1961).

Sark Pyrrhotite is one of the minor sulphide-ore species in the galena–pyrite lode of Sark's Hope mine, Little Sark (Ixer & Stanley 1983, Ixer 1992).

Q

qandilite

cubic — SPINEL GROUP — $(Mg,Fe^{2+})_2(Ti,Fe^{2+},Al)O_4$

A very rare magnesium–iron–titanium–aluminium oxide found in trace amounts in metamorphosed siliceous dolomite in the Allt Guibhsachain area, near Ballachulish, S of Fort William, Highland Region, Scotland. Qandilite forms isolated euhedral crystals associated with baddeleyite, geikielite, rutile and zircon (Ferry 1996, Livingstone 2002). The species was formally described only relatively recently (Al-Hermezi 1985) and it is quite likely that other occurrences will emerge.

qitianlingite

orthorhombic — $(Fe^{2+},Mn^{2+})_2(Nb,Ta)_2W^{6+}O_{10}$

Qitianlingite is a very rare iron–manganese–niobium–tantalum tungstate found only as inclusions in topaz within greisen at the Gunheath china-clay pit (Hensbarrow section), Carthew, St Austell, Cornwall. Other inclusion species within the topaz include albite, apatite, columbite–tantalite, ilmenorutile, muscovite, varlamoffite, zinnwaldite and zircon. Greisen formation along the margins of quartz–tourmaline and cassiterite, stannite, wolframite and copper mineralized veins was a result of flow along fractures of acid high-temperature post-magmatic fluids that reacted with granite wallrocks (Williamson et al. 1997, 2002).

quartz

trigonal — SiO_2

An extraordinarily common silicon oxide, polymorphous with coesite, cristobalite, moganite, stishovite and tridymite, and found in hydrothermal veins, including epithermal and alpine types. It is an essential mineral in granites and granite pegmatites, in sandstones and in quartzites, and is common in some carbonate and metamorphic rocks. It produces a variety of forms that can be divided into cryptocrystalline (agate, basanite, beekite, bloodstone, carnelian, chalcedony, chert, chrysoprase, flint, haytorite, heliotrope, hornstone, hyaline, jasper, lutecite, plasma, sard, sardonyx) and crystalline (amethyst, berg crystal, cairngorm, citrine, cotterite, eisenkiesel, milky, morion, prase, rock crystal, rose, smoky, tiger eye); high temperature (or beta-quartz) is also noted. Opal has chemistry similar to quartz, but is amorphous and is therefore listed separately. Only those occurrences with distinguishing features are listed below. Unless otherwise stated, authors are usually referring to transparent (rock crystal) or milky quartz.

England

Avon The finest amethystine quartz in England is said to have come from Providence iron mine, 2.5 km SW of Clifton, Bristol (Greg & Lettsom 1858).

Bedfordshire Flint is noted from Kensworth Chalk pit, near Dunstable (Mortimore et al. 2001).

Berkshire Flint (including a variety with lilac cortices) is common at Fognam quarry, near Lambourne, and at the Boxford Chalk pit, N of Donnington (Mortimore et al. 2001).

Cleveland Pyramidal crystals of quartz up to 0.5 mm long with hematite inclusions occur in the Boulby Potash at Boulby mine, Loftus (Green & Freier 1996).

Cornwall Quartz is the major gangue mineral in hydrothermal Cu–Pb–Sn–W veins associated with the Cornubian batholith and is widespread in its distribution. An investigation of inclusions in quartz from many locations in Cornwall (and Devon) concluded that, within restricted fracture systems, both magmatic fluids exsolved from the granites and basinal brines circulating contributed to the mineralization (Alderton & Harmon 1991). Investigations using argon isotope geochemistry on fluids extracted from fluid inclusions in quartz concluded that the mineralization persisted for up to 80 million years after the emplacement of the granite (Turner & Bannon 1992).

The St Just mining district at the western margin of the Land's End granite has produced elongated crystals up to 5 cm long at Levant mine (Wolloxall 1987) and quartz pseudomorphs after calcite rhombohedra at Botallack mine (Rudler 1905). Bancroft & Weller (1993) record splendid quartz crystals in vugs, some coloured yellow or amethystine, and finely textured jasper, ranging in colour from yellow through chocolate brown to black in the Botallack sett. Other reports include: amethyst, citrine, rock crystal and a stalactitic-form of quartz at Botallack mine; amethyst and prase at Wheal Bellon (both Collins 1892); rock crystal at Wheal Diamond (Greg & Lettsom 1858); and smoky quartz at Wheal Edward (Ball et al. 1982). At Bostraze–Balleswidden china-clay pit, near St Just, massive milky or clear granular quartz veins in kaolinized granite contain vugs with clear amethyst and smoky quartz crystals up to 3 cm long (Hall & Jackson 1975, Dominy & Camm 1998). From Boscaswell Downs, St Just, Collins (1877b) reports black quartz containing many inclusions of tourmaline from a vug in schorl-rich granite. Another orefield occurs at the northern margin of the Land's End granite and is centred on St Ives and Hayle, where quartz occurrences include: 5–15 cm-long crystals lining a very large vug (12 m long) at St Ives Consolidated mines (Gilbert 1878); rock crystal at Wheal Alfred, Phillack, near Hayle (Golley & Williams 1995); beautiful chalcedony in arborescent, dendritic and stalactitic forms at Trevascus mine, near Hayle, Gwinear (Garby 1848, Greg & Lettsom 1858, Miers 1929, Embrey & Symes 1987). Mineralized veins in the granite stock of St Michael's Mount, Marazion, to the SE of the Land's End granite, produced citrine, rock crystal and rose quartz (Collins 1892), whereas on Marazion beach it is poor-quality vein agate that has been found (Macpherson 1989). Both chalcedony and very large quartz crystals have been found at nearby Penlee quarry, Newlyn (Tuffnell 1997). Around the coast at Nangisel Cove, Sennen, near Land's End, it is amethyst that is noted (Greg & Lettsom 1858). A cathodoluminescence study of quartz from the Land's End granite revealed a complex crystallization history in which magmatic and hydrothermal varieties were identified (Müller et al. 2006).

In and around the Tregonning–Godolphin granite, the following quartz occurrences are recorded: babel quartz at Great Work mine, Breage (Russell 1905); vein agate at Looe Bar, near Porthleven (Macpherson 1989); massive lumps of amethyst and small crystals of milky quartz at Penberthy Croft mine, St Hilary (Betterton et al. 2000); acicular crystals (rare for quartz) up to about 2–4 mm in length and larger milky quartz crystals (of poor quality) displaying interesting oscillatory and sector zonations on a cm scale with smoky coloured areas at Croft Gothal mine, St Hilary (J. Betterton pers. comm. 2006). To the E, on the Lizard Peninsula, veins cutting serpentinite contain subtranslucent brown chalcedony at Countybridge quarry on Goonhilly Downs (Holyer 1996) and amethyst in vugs within massive quartz at Eastern Cliff, Kennack Sands (Hooper 1997). Both amethyst and white quartz occur in veins cutting gabbro at Dean quarry, St Keverne (Wolters & Wolters 1986). Tiger eye has been found at Kynance Cove (Golley & Williams 1995).

The next major group of quartz occurrences is from the Camborne–Ilogan–Redruth mining district, centred on the Carn Brea and Carn Marth granite intrusions, extended here to cover the Carnmenellis granite and its environs. *Near to Camborne:* Rock crystal from Dolcoath mine, and at North Roskear mine, blue quartz, cairngorm, haytorite, rock crystal and smoky quartz (Collins 1892, Golley & Williams 1995); crustiform, vuggy, chalcedonic and jasperoid quartz at South Croft mine (Dominy et al. 1994, Dominy & Camm 1996, Pearce 1996b). *Near to Redruth:* Yellow (but rarely transparent), amethyst and brown quartz at Carn Brea mines (Garby 1848, Robson 1945a); agate in quartzose rock on the southeastern part of Carn Brea Hill; large clear crystals, some with inclusions, at North Pool mine; rose quartz at Wheal Bucketts (all Garby 1848); large crystals with epimorphs after siderite and elongate fingers of chalcedony at Cook's Kitchen mine (Rudler 1905, Penhallurick 1997); smalt blue massive quartz and white chalcedony at Pednandrea mine (Garby 1848, Penhallurick 1997); lustrous crystals in groups up to 20 cm across at Ting Tang mine, Carharrack (Bruce & Aubrey-Jones 1998); amethyst, citrine and smoky quartz at Wheal Buller (Greg & Lettsom 1858, Golley & Williams 1995); rock crystal at Wheal Sparnon (Golley & Williams 1995); amethyst at Wheal Tolgus and Wheal Uny (Collins 1892, Rudler 1905, Embrey & Symes 1987). *E of Redruth:* Well developed crystals up to 1.5 cm long associated with barite and pyrite and quartz epimorphs after galena at Mount Wellington mine, Twelveheads, 7 km W of Truro (Rust 1982e); doubly terminated quartz prisms from St Day United mines,

Elongate growths of quartz (variety chalcedony) to 82 mm, probably from Cook's Kitchen mine, Redruth, Cornwall. (© Royal Cornwall Museum, Royal Institution of Cornwall)

Gwennap (Russell 1911a); jasper at Wheal Unity, Gwennap (Hosking 1951); a fibrous variety of quartz at Wheal Virgin, Scorrier (Greg & Lettsom 1858); well developed transparent-to-translucent crystals up to 5 cm long, doubly terminated crystals to 4 cm, hematite-tipped crystals, citrine and amethyst clusters looking like a dahlia head at Wheal Jane, Kea, near Truro (Rust 1982d, Golley & Williams 1995, Luff 1996). *In the Carnmenellis granite:* Druses in pegmatitic granite at Kessel Downs (Halvasso) quarry, near Longdowns, sometimes contain doubly terminated or smoky quartz crystals (Couper & Barstow 1977, Jackson 1978); Trolvis quarry, large crystals to 5 cm of alpha-quartz in pegmatite (Hosking 1954a); well formed clear, milky and pale smoky quartz crystals in pegmatite at Chywoon quarry, near Longdowns (Moulding & Hooper 2005).

The St Austell china-clay district is centred over the St Austell granite complex and is well known for its crystalline quartz. Large crystals were once found at Blackpool pit, good amethyst was collected from Longstone pit, beautiful clear crystals and black quartz occurred at Treviscoe pit, and probably the finest smoky quartz came from Rostowrack pit, SE of Treviscoe (Grigg 1986). Amethyst crystals up to 1.8 cm long and morion up to 5 mm long are noted from Bodelva pit (now the site of the Eden Project), whereas at Goonvean pit, SE of Treviscoe, beautiful stout crystals, colour zoned from smoky brown to black and reaching 3.5 cm in length, occurred in vertical quartz veins. Cap-shape morion crystals up to 2.5 cm long have been found at nearby Treviscoe pit. A greisen zone in Gunheath pit, Carthew, produced malformed smoky quartz crystals up to 10 mm long in an opal matrix (Weiss 1994). Hematite–quartz veins at Goonbarrow pit, near Bugle, produced fine- to coarse-grain slightly amethystine crystals containing hematite inclusions (Bray & Spooner 1983). Green (1984) notes amethyst at Hensbarrow, High Fraddon and Rostowrack pits, and rock crystal at Goonvean and Hendra pits. A cathodoluminescence study of quartz from late hydrothermal veins at Melbur pit revealed complex oscillatory and sector zonation on a micro-scale (Psyrillos et al. 1998). Away from the china-clay pits, quartz veins are common in and around the St Austell granite and have produced smoky quartz at Colcerrow quarry, Luxulyan, near Lostwithiel, and South Terras mine, St Stephen-in-Brannel, and citrine and a green variety of quartz at Restormel (Royal) iron mine, N of Lostwithiel (Golley & Williams 1995). The red variety of quartz (eisenkiesel), containing inclusions of hematite, is noted from Restormel mine (Rudler 1905), morion is recorded from Fowey Consols, Tywardreath (Collins 1892), and doubly terminated bipyramids of beta-quartz have been collected from the soils around Belowda Beacon, near Roche, and from a nearby abandoned china-clay pit (Drugman 1927).

Around Liskeard, quartz occurrences include: smoky quartz associated with bertrandite, fluorite and orthoclase in cavities within granite at Gold Diggings quarry, near Linkinghorne (Russell 1913b); hornstone at Herodsfoot mine, Liskeard (Greg & Lettsom 1858); attractive milky quartz, chalcopyrite and fluorite clusters at West Caradon mine, St Cleer (Bancroft & Weller 1993); doubly terminated bipyramids of beta-quartz at Wheal Coates (Drugman 1927); pseudomorphs of chalcedony after fluorite (Harker 1973) and fine rock crystal, some doubly terminated, at Wheal Mary Ann, Menheniot (Greg & Lettsom 1858, Dewey 1925). Penhale quarry, Common Moor, Liskeard was opened up for the production of quartz (Polkinghorne 1954). *Other quartz occurrences:* Colourless crystals to 5 cm associated with cassiterite at Cligga Head, Perranzabuloe (Chick 1987); rock crystal and rutilated quartz at Delabole slate quarry and in nearby cliffs (Greg & Lettsom 1858, Rudler 1905, Golley & Williams 1995); colourless-to-milky prismatic crystals (sceptre quartz) and jasper at Greystone quarry, Lezant, near Launceston (Wolfe 2000, Elton et al. 2004); rutilated quartz at Lanterdan quarry, Tintagel (Golley & Williams 1995); epimorphs up to 3 cm long after barite or orthoclase at Perran iron mine, near Perranporth (Wolfe 1998); smoky quartz at Wheal Bray, Altarnun, within the Bodmin Moor granite pluton (Ball et al. 1982); vein agate associated with cassiterite and childrenite at Wheal Ramoth on the sand dunes near Perranporth (Wolfe 2006).

Cumbria *Caldbeck Fells:* Quartz is the most common gangue mineral in this area, but rarely forms good crystals. The principal veins of Brandy Gill and its vicinity produced large crystals in cavities within massive quartz veins. Opaque, tarnished examples up to 35 cm long are noted by Day (1928) and are probably from Carrock mine, from where crystals 60 cm long and 20 cm in diameter (Dewey & Dines 1923) and 15 cm long (Davidson & Thomson 1951) are also reported. Large crystals from this locality are also noted by Eastwood (1959), Shepherd et al. (1976), Fortey (1978) and Young (1987a), the last author also noting pale-blue translucent masses of chalcedony. Very fine quartz crystals from Emerson vein on nearby Coombe Height are described by Shaw (1970). Other localities worthy of note include: quartz phantoms with unidentified green spheroidal inclusions at Arm o' Grain (Green et al. 2006b); rock crystal, and crystals with attenuated prism faces, at Brandy Gill mine (Greg & Lettsom 1858, Davidson & Thomson 1951); well grown pyramids up to 7.5 cm long in a vein outcrop high up on Iron Crag in Roughton Gill (Day 1928); rock crystal at Potts Gill and chalcedony at Silver Gill (Greg & Lettsom 1858); crystals at Dry Gill and Red Gill mines (Davidson & Thomson 1951); crystals and chalcedony at Roughton Gill mine (Greg & Lettsom 1858, Eastwood 1921, Davidson & Thomson 1951). *West Cumbria:* Excellent crystallized specimens of smoky quartz and other varieties of quartz are noted from hematite mines in the area, especially around Cleator Moor, Egremont and Frizington (Rudler 1905, Young 1987a). Specific examples of these and other occurrences include: dark-red quartz containing inclusions of hematite at Beckermet mine, near Egremont (Goldring & Greenwood 1990); translucent colourless terminated crystals associated with hematite and eisenkiesel at Florence mine, Egremont (Green & Briscoe 2002), also quartz pseudomorphs after barite (Davidson & Thomson 1951); agate in beach shingle at Fleswick Bay, Whitehaven (Macpherson 1989); red chalcedony in dolomite from Haile Moor mine, Egremont; pseudomorphs after barite at Parkside mine, Frizington; pseudomorphs after fluorite at Stank mine, Barrow-in-Furness, and Yarlside mines, Dalton-in-Furness (all Young 1987a); chalcedony in vesicles within Ordovician andesite at Hare Slack Hill, Askham-in-Furness (Rose & Dunham 1977); chalcedony concretions in an anhydrite seam at Sandwith mine, Whitehaven (Harrison 1975). *Elsewhere in the Lake District:* Bipyramidal beta-quartz in acid dykes including the Armboth dyke, Thirlmere (Strens 1962); bipyramidal beta-quartz in a lamprophyre dyke at Brant Rake, Eskdale; amethyst at Cat Gill, Derwent Water; crystal groups at Threlkeld mine, Threlkeld; compact masses of radiating crystals at Whiteoak Beck, High Nook, Loweswater (all Young 1987a); pseudomorphs of quartz after barite at Castlenook mine, Newlands Valley (Strens 1962); small quartz euhedra at Dale Head mine, Newlands Valley (Stanley & Vaughan 1980); agate, carnelian, jasper, prase and rock crystal at Falcon Crag, near Keswick (Greg & Lettsom 1858, Davidson & Thomson 1951); pale-purple amethyst and chalcedony at Force Crag mine, Braithwaite, near Keswick (Stanley 1979, Young 1987a); red quartz (eisenkiesel) associated with chalcopyrite at Graphite mine, Seathwaite, Borrowdale (Strens 1965); brilliant semi-transparent crystals and pseudomorphs after barite at Greenside mine, near Glenridding, Patterdale (Young 1987a, Davidson & Thomson 1951); milky-looking botryoidal chalcedony in limestone at Orton Scar (Davidson & Thomson 1951); large smoky quartz crystals (up to at least 9 cm long) in cavities within pegmatite and smaller colourless crystals in miarolitic cavities within the Shap granite (Russell 1925, Davidson & Thomson 1951, Wirth & Green 2002); bipyramidal beta-quartz in the Threlkeld microgranite (Hadfield & Whiteside 1936); jasper at Walla Crag, Derwentwater, near Keswick (Shackleton 1966); pseudomorphs after barite at Yewthwaite mine, Newlands Valley (Strens 1962, Firman 1978). *Northern Pennine orefield:* Colourless to milky-white pyramidal quartz up to 5 cm long (some richly covered specimens), smaller drusy crystals and minor chalcedony in vein and replacement flat deposits at Brownley Hill mine, Nenthead, near Alston (Clark & Rust 1993, Green & Nudds 1998, Green et al. 2000); peculiar quartz stalactites covered with dolomite, galena and pyrite at Handsome Mea mine, Nenthead, and carnelian in basement conglomerate from Soulby Fell, near Dacre (Davidson &

Thomson 1951); limonite-stained crystals at Hard Rigg Edge, Melmerby Fell (Young 1985a); good colourless euhedral microcrystals (often doubly terminated) in a fluorite matrix at Hilton mine, Scordale, near Appleby (Bridges 1982); epimorphs of quartz after fluorite at Rampgill Horse level, Nenthead (Bridges & Wilkinson 2003, 2005, Wolfe 2004); chalcedony and white pyramidal quartz (some up to 2 cm long and doubly terminated) at Tynebottom mine, NW of Garrigill, near Alston (Foy 1997, Green & Briscoe 2002). Additionally, techniques to examine the pore system of quartz grains and associated quartz cement in the Penrith Sandstone from Bongate Scar, near Appleby, have been presented by Cooper & Hunter (1995) and fluid inclusions in quartz have confirmed the existence of four hydrothermal fluids involved in ore formation over the Alston Block (Cann & Banks 2001).

Derbyshire Most quartz occurs in one of four distinct associations:

- *In the weathered zone of some volcanic rocks* Crystals to 2 cm, notably at Calton Hill, Taddington (colourless hexagonal prisms, amethyst, brown ferruginous quartz, jasper, rose quartz, rare smoky quartz); Ible, near Matlock; Miller's Dale and less commonly Cavedale, Castleton (Sarjeant 1956b, Ford et al. 1993); tuff – amethystine in tuff at Ashover (Ford et al. 1993); toadstone – crystals at Harpur Hill, Buxton; Magpie Sough; Tideswell Dale quarry and Waterswallows quarry near Buxton, the latter producing high-grade purple amethyst and some pink crystals (Sarjeant 1956b, Ford et al. 1993), also chalcedony filling vesicles in toadstones at Cavedale (Mawe 1802), High Tor (Adam 1846) and Hucklow Edge, near Eyam (Sarjeant 1957).
- *Authigenic crystals in limestone* Needle-like crystals up to 5 mm long in many limestones or rarely larger crystals up to 3 cm from Countess Cliff, S of Buxton, which were known as Buxton diamonds (Ford et al. 1993). Buxton diamonds are also described from the Ashover tuff, Calton Hill, Hucklow, Knotlow and Priestcliffe in Miller's Dale, Masson Hill and the Waterswallows Sill (Mueller 1954a, Ford & Sarjeant 1964a).
- *Metasomatic replacement* Massive quartz as a metasomatic replacement of limestone, particularly around Bonsall, and on Masson Hill, and also above Pindale, Castleton. Cavities in these rocks are often lined with pyramidal crystals such as at Dirtlow Rake, W of Bakewell (Ford et al. 1993), and Wapping mine, Cromford (Ince 1991).
- *Vugs in limestone* Euhedral crystals lining vugs in limestone adjacent to mineral veins such as those at Fall Hill quarry, Ashover (Smith 1982); the Masson mines, Matlock (Ford et al. 1993); Milltown quarry, near Ashover (Rust 1991); Snelston mine, near Ashbourne (Jackman 1996).

Other occurrences include: banded chert composed of silicified biomicrite and lutecite in the Arrock Black Marble mine, Ashford-in-the-Water, near Bakewell (Orme & Ford 1970); beekite on Bradwell Moor and in Pindale (Orme & Brown 1963); clear crystals lining cavities in hematite at Doglow Wood, Hopton, and as inclusions in fluorite at Mawstone mine, Youlgreave (Ford & Sarjeant 1964a); xenoliths in the Great Rocks Dale dyke, near Buxton (Cope 1933); chert at Holme Bank mine, Bakewell; dark smoky quartz groups from Jug Holes mine, Masson Hill, Matlock (R. E. Starkey pers. comm. 2005); reddish crystals with hematite inclusions at Kenslow Knoll, Friden; beekite at Longstone Edge, in Pindale; Masson Hill and Monsal Dale, near Chelmorton (all Ford et al. 1993).

Terminated quartz (variety amethyst) cluster, 20 mm tall, from Calton Hill quarry, Taddington, Derbyshire. (© D. I. Green)

Smoky quartz (40 mm in height) from a miarolitic cavity in granite on Lundy island, off the north Devon coast. (© D. I. Green)

Devon Hydrothermal veins associated with the Dartmoor granite have produced the best crystals, with the rock crystal from Wheal Friendship, near Tavistock, perhaps the finest (crystals up to 15 cm long). Notable rock crystal has also been found at: Wheal Betsy, near Tavistock; Gidleigh, near Moreton Hampstead; and Wheal Robert, Sampford Spinney (Greg & Lettsom 1858, Hall 1868b). Other related occurrences include: red jasper and grey hornstone at Belstone mine, near Okehampton (Solly 1886); jasper at Brent Tor, near Tavistock; Doddiscombsleigh, near Chudleigh, and Ivybridge; agate, haytorite and rock crystal at Hay Tor; pseudomorphs after calcite at Hay Tor iron mine, Islington; amethyst at Whitchurch Down, near Tavistock (all Hall 1868b); small crystals associated with albite, bavenite and fluorite in fissures at the Meldon aplite quarries, near Okehampton (Dearman & Claringbull 1960), also well crystallized examples in druses within the aplite (McLintock 1923); very beautiful specimens of quartz forming large milky-white crystals up to 15 cm in length, studded with siderite, also a capped variety occasionally reaching 14 cm at Virtuous Lady mine, Buckland Monachorum (Russell 1913a); and white and amethystine quartz at Willsworthy copper mine, NE of Mary Tavy (Collins 1904a).

The mineralized region of north Devon is another important source of quartz. Localities include: babel quartz and pseudomorphs after fluorite at the Bere Alston and South Hoo mines, near Bere Ferrers (Greg & Lettsom 1858, Hall 1868b, Rudler 1905, Miers 1929); rock crystal at Combe Martin; hornstone at East Tamar mine, Bere Ferrers; small but brilliant crystals embedded in hematite at Georgeham and Viveham, near Barnstaple (Hall 1868b); delicate pink agate from near Lynmouth (Pabian et al. 2006); large opaque reddish twinned crystals at North Bovey (Greg & Lettsom 1858).

Other occurrences include: abundant chert in greensands at Haldon, near Exeter, and at Orleigh Court, near Bideford; agate at Mary Church, near Torquay (Hall 1868b); flint, including the paramoudra type (an unusual variety of flint that forms large cylindrical growths around the trace fossil *Bathichnus paramoudrae*), occurs at Hooken Cliffs, between Branscombe East Cliff and Bee Head (Mortimore et al. 2001); crystals up to 2.5 cm long, varying from black through pale brown to almost colourless, in the granite on Lundy (Hall 1868b, McLintock & Hall 1912).

Dorset Flint occurs in chalk at Shillingstone quarry, near Blandford Forum (Mortimore et al. 2001).

Durham *Northern Pennine orefield:* Quartz and chalcedony in crustiform bands alternating with fluorite and galena, also well formed perfectly colourless and transparent pseudo-bipyramids up to 14 mm long at Boltsburn mine, Rookhope (King 1982b); fine white pyramidal crystals up to 5 cm long covering specimens to up 50 cm across at Cambokeels mine, Weardale; also noted from Blackdene mine, near Ireshopeburn, Weardale, and Whiteheaps mine, near Blanchland (Green & Briscoe 2002); abundant quartz and chalcedony at Groverake mine, W of Rookhope, Weardale (Young 1995); fine examples of quartz at Frazer's Hush mine, Weardale (Wolfe 1998).

Gloucestershire Very fine ribbon jasper is noted from Dressing Green, near Tortworth (Greg & Lettsom 1858).

Hampshire Flint, including the paramoudra type, occurs at Downend Chalk pit, Portsdown (Mortimore et al. 2001).

Isle of Wight Flint is very common in bands in the Upper Chalk from Compton Bay and Whitecliff (Mortimore et al. 2001).

Kent Flint is abundant and characteristically it forms thick bands in the Upper Chalk seacliffs from Folkestone and Dover to Kingsdown. The paramoudra type is common at specific stratigraphical horizons. Flint is also very common in the Chalk cliffs, particularly along the Thanet coast from Ramsgate, around Broadstairs, from Margate to Minis Bay, and E of Birchington (Mortimore et al. 2001). Flint is also common in the Upper Chalk at White pit, NE of Upper Halling, and at Blue Bell Hill quarry, Burham, Chatham (J. Betterton pers. comm. 2006).

Leicestershire Minor euhedral quartz is associated with hematite or muscovite in thin veins cutting diorite at Newhurst quarry, Shepshed, also amethystine quartz is found associated with sulphides in crush breccia (King 1973, Ince 2005).

Norfolk Flint, a tough grey to black variety of chalcedony that breaks with a conchoidal fracture leaving sharp edges, was well known to Stone Age people, who mined it extensively at Grimes Graves, near Thetford Forest (Naylor 1981). It was worked in Neolithic times between 2500 and 2200 BC, and represents some of the oldest known mining activity in Britain. Flint is also found at Caistor St Edmund Chalk pit, Norwich, and in cliff exposures from Overstrand to Trimingham Cliffs (Mortimore et al. 2001). Carnelian pebbles have been found on Cley beach on the north coast (Macpherson 1989).

Northumberland In the neighbourhood of Dunstanborough, water-clear quartz crystals are known as Dunstanborough diamonds. Excellent nests of amethyst are reported from cavities in a dyke at Haydon Bridge, near Hexham, whereas in another dyke at Hampeth Burn, rose quartz occurs. Other dykes show pockets of chalcedony, for example, the Bingfield and Bothal dykes (Smythe 1927b). The Cheviot Hills, particularly around the River Coquet, have produced good-quality agate (Pabian et al. 2006).

Shropshire Burgam mine, SSW of Pontesbury, has produced small terminated crystals up to 3–4 mm long in barite-bearing quartz veins (Starkey 1982).

Somerset Quartz occurrences include red agate (potato stone) from Dulcote quarry, near Wells (Pabian et al. 2006), and Sandford in the Mendip Hills (Moore & Pearce 1997), and doubly terminated crystals to 3 cm from a cave infill in an iron–manganese vein at Whatley quarry, near Frome, Mendip Hills (House 2002). Snowdon Hill quarry, Chard, is well known for its exposures of Whitecliff Chert – a silica-rich unit of the Upper Greensand (Mortimore et al. 2001).

Surrey Slightly amethystine quartz is reported from a fuller's earth pit at Nutfield, near Redhill (Rudler 1905). Flint is a common constituent of the Chalk at Artington pit, W of Guildford, and at Coombs pit at West Horsley (Dines & Edmunds 1929).

Sussex Flint occurs in the Chalk: along cliffs between Newhaven to Brighton; from Cuckmere to Seaford; at Southerham pit, Lewis (Mortimore et al. 2001).

Wiltshire Flint is noted from West Harnham chalk pit, Salisbury (Mortimore et al. 2001).

Yorkshire Agate (some with amethyst centres) and carnelian occur on the beaches of Cornelian Bay, near Scarborough, and Stoupe Beck, Robin Hood's Bay (Rodgers 1975, Pabian et al. 2006). Flint is found at Welton Wold quarry, Melton, near Welton. It is also very abundant in bands at the Enthorpe railway cutting, NE of Market Weighton, and along a 17 km stretch of chalk cliffs around Flamborough Head (Mortimore et al. 2001).

Wales

Clwyd Hendre quarry, Pandy, near Glyn Ceiriog, has produced fine colourless euhedral crystals of quartz up to 20 mm long from alpine-type veins (Wedd et al. 1929, Starkey et al. 1991).

Dyfed *Central Wales orefield:* Good quartz crystals are reported from Blaenceulan mine, near Tal-y-bont; Brynyrafr mine, near Nant-y-moch reservoir; Copa Hill mine, Cwmystwyth; Grogwynion mine, Pont-rhyd-y-groes; Mynyddgorddu mine, W of Bont-goch, Ceulanymaesmawr (all Jones & Moreton 1977). At Eaglebrook (Nantycagl) mine, Ceulanymaesmawr, quartz forms clear and well terminated crystals up to 1 cm long (Jones & Moreton 1977, Jones 1983). Attractive specimens are also noted from Cefngwyn mine, near Aberystwyth (Bevins 1994). Mason (1997) describes veins in the orefield from the "Group A1 – early complex assemblage" containing open vugs with characteristic long thin transparent milky-white crystals (to 15 mm) with water-clear terminations and from the "A2 – late simple assemblage" containing large, strongly growth zoned pyramidal crystals, the most spectacular from Nantymwyn mine, where crystals reach 15 cm. Other quartz occurrences include: equant crystals to 6 cm with bubble inclusions, from the Coal Measures near Saundersfoot (Green 1998b); clear crystals up to 8 cm long associated with

Sceptre quartz crystals (variety rock crystal). Saundersfoot, Dyfed, Wales; the specimen is 40 mm long. (© D. I. Green)

albite, goethite and coarsely crystalline prehnite veins from a locality in the St David's Head area (Wolfe 2000; T. F. Cotterell pers. comm. 2005); jasper associated with doleritic rocks in the Strumble Head area, near Fishguard (Bevins 1979).

Gwent Quartz crystals were recovered from the tips of Oakdale colliery during landscaping (Wolfe 1997).

Gwynedd Probably the finest and most spectacular quartz crystals in Britain come from the neighbourhood of Snowdon (Rudler 1905). Those from Mynydd Drws-y-Coed, near Nantlle, form clear faceted prisms up to 7 cm long with pyramidal terminations. Additionally, excellent clear prismatic quartz crystals occur in alpine-type veins at: Votty and Bowydd (Maen-offeren) and Manod quarries, near Blaenau Ffestiniog; Gimlet quarry, near Pwllheli, Lleyn Peninsula; Minffordd quarry, near Porthmadog; and Penrhyn quarry, Bethesda (Bevins 1994). Similar veins have produced exceptional crystals to 8 cm at Fron-oleu, Prenteg, Tremadog (Starkey & Robinson 1992), and flattened crystals containing rutile inclusions and water-clear terminated crystals up to 5 cm long, sometimes with clinochlore inclusions at Tanygrisiau, Blaenau Ffestiniog (Green & Middleton 1996). Other occurrences include: jasper at Aberdaron, Lleyn Peninsula (Matley 1928); Newborough, on Anglesey (Greenly 1902); the Strumble Head area, near Fishguard (Bevins 1979); excellent quartz specimens coated with marcasite and pyrite, and aggregates of terminated crystals on galena and sphalerite, at Parc and Trecastell mines, near Llanrwst, N of Betws-y-coed; agate in rhyolitic rocks at Pen-y-chain, Lleyn Peninsula (all Bevins 1994); and doubly terminated crystals coated in dickite (but referred to as kaolinite) from cavities in a porphyritic (elvan) dyke at Porth-yr-hwch, near Amlwch, Anglesey (Dick 1888). Agate nodules (thunder eggs) are reported from volcanic ash deposits at Pen-y-chain and Carig-y-defaid (Pabian et al. 2006).

Mid Glamorgan Excellent specimens of quartz coloured amber, buff, black, pink, red or yellow from inclusions of iron-rich minerals (also clear and occasionally amethystine) have been found to the NW of Cardiff in ore from the old iron mines, in particular Llanharry and Mwyndy mines (Rudler 1905, North 1916, Bevins 1994). Clear doubly terminated quartz crystals, which may reach 4 cm in length (called Merthyr diamonds), have been found associated with millerite in clay–ironstone nodules across much of the South Wales coalfield, for example Hafod pit, Rhondda (Wolfe 1996a), and International colliery, Blaengarw, near Maesteg (Bevins 1994, Moore & Pearce 1997). Quartz is also noted from Machen quarry, 7 km ENE of Caerphilly – rare doubly terminated clear crystals (Plant & Jones 1995); and the Porthcawl area beaches – potato stones, a variety of agate (Bevins 1994).

South (and West) Glamorgan Agate and chalcedony occur in Carboniferous Limestone at Coed Cefn Garw quarry, near Tongwynlais, and cavities in iron ores have produced clear colourless doubly terminated crystals at Taff's Well quarry, near Cardiff (North 1916). In West Glamorgan, clear gem-like quartz crystals coated with wavellite have been found at Pwlldu Bay on the Gower Peninsula (Plant & Jones 2001).

Powys The Central Wales orefield has produced: good quartz crystals from Dylife mine, N of Hafren Forest, and Snowbrook (Nantyreira) mine, between Llanidloes and Tal-y-bont (Jones & Moreton 1977); small pyramidal crystals up to 5 mm long, also doubly terminated up to 2 mm long in sphalerite vugs at Dyfngwm mine, Machynlleth (Rust & Rust 1987) (actually the Pen Dylife section of Dylife mine). Attractive specimens have also been collected from: the Upper Boat level tip of Rhandirmwyn mine – crystals up to 15 cm long; Llanelwedd quarry, near Builth Wells – agate nodules in lavas (both Bevins 1994); Dolyhir quarry, Old Radnor, near Presteigne – water-clear doubly terminated crystals (Cowdry 2002a), and pale smoky-colour doubly terminated crystals up to 3.4 cm long (T. F. Cotterell pers. comm. 2007).

Scotland

Readers with an interest in particular varieties of quartz (especially agate) will find the format (and breadth) of occurrences provided by Heddle (1901) most helpful. However, paragenetic relationships are obscured by Heddle's arrangement, so a regional approach is used here. Harry Macpherson's book on agates, with many illustrations of examples from Old Red Sandstone lavas, can also be recommended.

Borders Brown and (rarely) large red and white specimens of agate have been found at Carlop. Agate is also noted from gravels at Linhouse (Heddle 1901, Macpherson 1989).

Central In the Ochil Hills, the best locality for white quartz is said to be Airthrie Hill mine, Bridge of Allan. Beautiful delicate-purple amethyst lines geodes in volcanic rocks cut by Balquharn Burn, near Menstrie. Agate is widely distributed, with notable occurrences at Balquharn Glen, at Kirk Hill, near Tillycoultry, and at Myreton Hill, 2 km W of Alva. Jasper is also reported from Airthrie Hill mine and Myreton Hill (Dickie & Forster 1974). In a report on silver mineralization at Alva mine in Silver Glen, Moreton et al. (1998) describe cream, milky and grey chalcedonic quartz, sometimes containing clinosafflorite; colourless, doubly terminated crystals; smoky quartz; and amethystine quartz in amygdales in the surrounding lavas.

Black Hill, 8 km NW of Ballindean, has produced substantial amounts of large (up to 15 cm long) well banded reddish-brown and white agates. Rare agates are also noted from Carboniferous lavas on Campsie Hill, Stirling (Macpherson 1989). Well formed opaque white quartz occurs in gold–sulphide-bearing quartz veins at Cononish mine, near Tyndrum (Parker et al. 1989).

Dumfries & Galloway Quartz has been found at many locations in and around the Criffell granite complex, including: specimens of typical pyramidal amethyst crystals from Boreland Glen, Caulkerbush, and smoky quartz from Auchenlosh quarry, near Dalbeattie, represent some of Scotland's finest quartz. Amethystine quartz is reported from Screel Hill, near Dalbeattie (Green & Todd 2001). Clear euhedral amethystine and smoky quartz occurs in uranium-rich veins along the Solway coast, for example Lot's Wife, Marbruie Cove, and Needle's Eye, SW of Caulkerbush (Heddle 1901, Williams 1965, Miller & Taylor 1966). At Kinharvie Burn, 6 km W of New Abbey, agate, chalcedony, morion, smoky quartz and pale-pink hexagonal crystals up to 8 mm long encrusting vugs in kaolinized feldspar are all noted (Williams 1965, McMullen & Todd 1990). Smoky quartz is reported from Auchencairn hematite mine (Williams 1965).

Elsewhere, quartz occurrences include: chalcedony, jasper and quartz pseudomorphs after barite, calcite, galena and psilomelane from the Wanlockhead orefield (Brown 1927); red-to-yellow jasper at Redkirk Point, Gretna; high-quality colourless-to-white vein agate at Sowie's Pot and dirty-white agate at Skipper's Bridge, both near Langholm (Williams 1965).

Fife The farmland of north Fife is the most prolific agate-bearing area in Scotland. Many examples are strongly coloured and they produce attractive specimens when polished. Particularly notable are those from the Cupar area: grey-white fortification and stalactitic agates from Norman's Law; bluish-grey agates with red-tinted centres from Balmeadowside; sagenitic agate from Dunbog; bluish agates with milky-white bands from a large area around Luthrie; brilliant red-yellow or red-white agates (including onyx) and moss agate from Middlefield farm; red and yellow moss agate and plasma from Scurr Hill, Balmerino; carnelian agate from the Tay railway bridge (Heddle 1901, Field 1989, Macpherson 1989, Moreton 2006, Pabian et al. 2006).

Dolerite sills worked for roadstone have produced a variety of specimens: pale amethyst crystals line cavities in quartz veins at Goat quarry, 2 km NW of Aberdour (Ingram et al. 1992); milky quartz crystals up to 15 cm long in groups up to 40 cm across and pale- to medium-purple amethyst crystals up to 2 cm long at Orrock quarry, near Burntisland (Wirth 2003, Moreton 2006). Heddle (1901) notes salmon-colour doubly terminated crystals with blue barite on dolomite, in druses within tufa at Kinkell.

Grampian The light-to-dark brown variety of quartz is named cairngorm after the large mountainous area in this region. It is generally found as loose crystals weathered out of granitic veins and is associated with orthoclase, rare beryl and topaz (Heddle 1901). A 53 cm-long example of cairngorm from Beinn a' Bhuird, weighing 21 kg, is on display in Braemar Castle (Dunlop 2005). Specimens of cairngorm from Murdoch Head (crystals up to 20 cm long), Stirling Hill, Peterhead (15 cm crystal), and the Culblean and Loch Avon area, are in the collections of the NMS (Livingstone 2002). Large smoky quartz crystals (up to 15 cm long and 7 cm in diameter), associated with gem-quality beryl, from a weathered pegmatite have been found on the top of Beinn a' Bhuird, a high and precipitous summit in the Cairngorm Mountains (R. E. Starkey pers. comm. 2005).

Other examples of crystalline quartz include a fine specimen of smoky quartz from Banchory (Rudler 1905); colourless euhedral crystals, up to 5 mm across, in a Triassic breccia termed the Cherty Rock, at West Beach, Lossiemouth, near Elgin (Starkey 1988a); opaque or yellow doubly terminated crystals from a vein in gneiss to the E of Kinnaird lighthouse; magnificent doubly terminated rich-brown crystals up to 15 cm long from Sterling Hill quarry, S of Peterhead; yellow quartz near the Ladder, Mount Keen, and on the south side of Loch Avon (all Heddle 1901).

Occurrences of agate are concentrated in the SE of the area between Montrose and Stonehaven, where Old Red Sandstone lavas are exposed: Allardice, Arbuthnott; Barras quarry – fine onyx and blue/grey agate; Finella Den; Pulpit Rock, near Grange Burn, Kinneff; St Cyrus, 8 km N of Montrose – fine colours including red/white onyx agate (Heddle 1901, Macpherson 1989, Pabian et al. 2006).

Highland Cairngorm is reported from the northern cliff of Ben Loyal; from a giant erratic boulder at Beinn Bhreac, 2 km E of Tongue; and from near the summit of Quinag (Heddle 1901, Berridge 1969). Massive prase and rock crystal occur in thin veins on the southeastern shore at the head of Loch Torridon and on Bulgay Island, Loch Hourn (Heddle 1901). Well formed pale smoky quartz crystals up to 8 mm long on cockscomb barite represent a new find from the eastern end of the Clashgorm

Smoky quartz (var. cairngorm) crystal (10.5×5 cm). Ben Avon, Cairngorm Mountains, Grampian, Scotland. (© M. P. Cooper)

opencut, Strontian (Wolfe 2004). On Bloodstone Hill, Sgòrr Mhòr, Isle of Rum, the uppermost part of thin lava flows contain vesicles infilled with agate, chalcedony, heliotrope (bloodstone) and prase (Tomkeieff 1942). Blue/grey/white agates are noted from the beaches below Bloodstone Hill (Macpherson 1989), and a violet variety has been found on the Isle of Luinga Bheag. Onyx occurs at Quiraing and the Storr on the Isle of Skye (Heddle 1901).

Lothian Agate occurrences include: blood-red agate associated with lavender and violet chalcedony, carnelian and jasper in andesitic lavas from Blackford Hill, Edinburgh; violet and red-mottled agate at Dunglass beach; grey agate at Hillend, Pentland Hills (Heddle 1901).

Strathclyde Carboniferous Clyde Plateau Lavas contain quartz, usually in amygdales: very rare crystals up to 2.5 mm long on analcime and large epimorphs after heulandite at Loanhead quarry, near Beith (Meikle 1989b); rare crystals and pseudomorphs after prehnite at Hartfield Moss, 9 km WSW of Barrhead (Meikle 1990); Lang Craigs, Dumbarton (Wolfe 2000); stubby colourless, transparent crystals up to 3 mm long near a dolerite dyke at Reilly quarry, Houston (Meikle 2003). Blue, brown and white chalcedony coating calcite groups in collapsed geodes with these lavas are noted to the N of Loch Carron (Wolfe 2006). Both Carboniferous and Old Red Sandstone lavas occur around Ayr. Agates are found in Burn Anne, near Galston and along the coastal stretch from Heads of Ayr to Dunure. Burn Anne is famous for its beautiful jasp agates, a type of vein agate in which red, orange and yellow jasper is found associated with bluish-grey or violet chalcedony. Fortification agate is also found at this location. From Heads of Ayr have come scenic agate and onyx agate, and from Dunure both onyx agate and a finely banded brilliant red agate. Blue/grey/white agates reaching 3 cm across, and rarely jasp and moss agate, are found on the beaches of Iona. They are thought to be of Tertiary age, derived from lavas on the Isle of Mull (Macpherson 1989). An extraordinarily large agate (24×16 cm cut surface) has been found at Dunure (Livingstone 2002). Agate is also noted from the shore of Machrihanish Bay, near Campbeltown (Heddle 1901). The microstructure of agates from many locations in the Midland Valley of Scotland is discussed in great detail by Moxon (1996). A brown/white fortification agate from Culzean showing spherulitic growth partially replaced by a typical fortification pattern of growth is illustrated by Moxon (1991).

Smoky quartz occurs with allanite, fergusonite, gadolinite and zircon in miarolitic cavities within the Northern Arran granite, Isle of Arran (Hyslop et al. 1999), and amethystine and clear varieties are also recorded (Emeleus & Gyopari 1992). Large matrix-less smoky quartz crystals, some reaching 8 cm in length, have been found in this granite around Goat Fell (Wolfe 2004). Diack (1979) suggests that these crystals are often clearer and less flawed than typical examples from the Cairngorm Mountains. Other crystalline quartz occurrences include: large and fine specimens of amethyst from S of Campbeltown, near the summit of Killellan Hill and at Wigle (Heddle 1901); cairngorm at Ballirouline Hill and Ballygroggan, near Campbeltown (Berridge 1969); white quartz crystals up to 35 mm long in molybdenite-bearing quartz veins at Coire Buidhe, Glen Creran (Ingram et al. 1993); amethystine quartz with barite, bornite, calcite, chalcopyrite and malachite in veins at the Kaim mines, Renfrewshire Hills, near Lochwinnoch (Stephenson & Coats 1983); smoky-to-black crystals up to 0.5 mm long and pink (occasionally doubly terminated) crystals to 14 mm on barite from the dumps of Mannoch Hill mine, SW of Lesmahagow (Todd & McMullen 1991); small euhedral crystals, often doubly terminated, forming sugary coatings on barite at Muirshiel mine, Queenside Muir, NE of Largs (Todd & Laurence 1989); smoky-to-transparent euhedral crystals up to 10 mm long, some with tourmaline inclusions in cavities and veins within limestone at Rubha a' Mhill, northeast Isle of Islay (Green 1988); lustrous clear crystals up to 5 mm long coating or infilling brachiopods, also agate and chalcedony completely replacing the shells at Trearne quarry, near Beith (Todd 1989). Fluid inclusions in quartz are described from the Starav monzogranite – the central unit of the Etive granite complex (Batchelor et al. 1992), and from veins in the Stronchullin area, Loch Fyne (Anderson 2004).

Tayside The Old Red Sandstone lavas that produced fine agates in Fife continue into Tayside Region, where many more agates have been found. The relatively recent discovery of agate in an andesitic flow at Ardownie quarry, Monifieth, near Dundee, is the most important find for many years. In addition to an abundance of agates of exceptional size (up to 6 cm across) and quality, geodes lined with lustrous crystals of amethyst and smoky quartz are not uncommon. Most of the agates are blue/grey/black, but red/orange/yellow/brown examples have been found in weathered rock. Moss agate containing filaments of celadonite is also noted (Ingram 1994, Smith 1998, Green & Todd 2001, Field 2005). Smaller finely banded blue/brown agates

Quartz-cored agate from Ardownie quarry, Monifieth, near Dundee, Tayside Region, Scotland; the specimen is 105×50 mm. (© D. G. Anderson)

have been found at nearby Ethiebeaton quarry, Broughty Ferry (Green & Todd 2001). The most celebrated Scottish agate locality is, however, the Blue Hole at Usan, near Montrose. Brilliant inky-blue to white agates are characteristic of the locality, but some yellow and reddish-brown examples also occur. Smoky quartz lines the druses of some of these agates. Other important agate discoveries have been made around Montrose: East Balgay; near Ethie House; Ferryden; Lunan Bay railway cutting; Panmure Den; Rock of St Skae; and Scurdie Ness. In the Perth area, occurrences include: Agate Knowe; Ballindean; Binn Hill, Glen Farg; Black Hill; Gourdie; Inchture; Kinnoull Hill; Path of Condie; near Pitkeathly; Pitroddie Den; Rossie Ochil; and Tinkletop (Heddle 1901, Field 1989, Macpherson 1989, Pabian et al. 2006). "Translucent wine [*sic*] and amethystine coloured chalcedony" (described as diackethyst) is found near the site of Black Jack's Castle, in the old parish of Craig, Montrose (Diack 1979).

Western Isles On St Kilda (Hirta), stubby bipyramidal crystals of quartz (inverted beta-quartz) occur in the Conachair and Oiseval granite, and amethyst lines drusy cavities in the Mullach Sgar complex (Harding et al. 1984). Quartz veins on the island of North Rona mainly consist of massive white quartz, although pink and smoky varieties are also noted (Stewart 1932).

Orkney Brilliant doubly terminated crystals (rock crystal) were once found at Burn of the Sale, the Bring, on Hoy. Velvety-black basanite is noted from Point of Ness, Stromness, Mainland, and dendritic jasper is recorded from veins on North Ronaldsay (Heddle 1901).

Shetland Agates, often with barite cores and up to 10 cm across, are present in the vesicular tops of basalt flows on the south shore of Papa Stour, a small island off the west coast of Mainland (Mykura & Phemister 1976). Heddle (1901) describes: agate and amethyst in vesicular basalt from Northmaven, opposite Dore Holm on Mainland; rose quartz from North and South Quin, Hillswick, Mainland; and rock crystal from Gruting Bay, Hestaness on Fetlar. Large milky quartz crystals up to 10 cm long have been found on the island of Whalsay (R. E. Starkey pers. comm. 2005).

Ireland

Antrim Near Carmean, cavities in Cretaceous Chalk at Glover's pit and Carmean Limestone quarry contain minute doubly terminated quartz crystals trapped in fibrous mats of amorphous quartz (Preston 1977). Agate occurs on the coast near Ballycastle and along the shores of Lough Neagh (Greg & Lettsom 1858). Amygdaloidal Tertiary basalts at Craigahulliar quarry, 2.5 km SE of Portrush, contain well formed quartz crystals up to 50 mm long, which are associated with small stalactites of chalcedony in the larger cavities (Black 1976).

Carlow Fluid inclusions in quartz are described from barren and spodumene pegmatites at Aclare, near Myshall (Whitworth & Rankin 1989).

Cork Cotterite, a variety of quartz with a metallic pearly lustre resembling aluminium, was once found coating quartz crystals in a vein of rock crystal and calcite from Rockforest (Harkness 1878).

Donegal Ryback & Moreton (1992) describe slender milky crystals in quartz veins in schists at Dunlewy, and colourless quartz prisms in quartzite on Errigal Mountain.

Down Smoky quartz (exceptionally up to several cm long) is associated with feldspar, mica and topaz within miarolitic cavities in granite at Lindsay's Leap quarry, Mourne Mountains (Green & Moreton 2001, Green et al. 2005a). Greg & Lettsom (1858) report very beautiful quartz with beryl and topaz from the Mourne Mountains. Pale-green chalcedony (associated with fine-grain white quartz) is described from silicified breccia near a Tertiary dyke between Conlig and Newtownards (Moles & Nawaz 1996).

Galway Slender quartz prisms up to 2 cm long and transparent needle quartz occur at Glengowla mine, near Oughterard (Ryback & Moreton 1992, Wolfe 2000, Green & Moreton 2001).

Kerry Slender well formed quartz crystals up to 4 mm long occur in quartzite near Aghnanus Bridge, SE of Killarney (Ryback & Moreton 1991). Prolific crystals of quartz to 15 cm, including doubly terminated, phantom and faden quartz, have been found in ribbon-like tension gashes within Devonian sandstone at Dunquin, Dingle Peninsula. Near by at Smerwick harbour, near Slea Head, these varieties are accompanied by hexagonal plates of hematite (Wolfe 2006).

Limerick Vuggy white quartz is noted from Ballycormick, near Shanagolden (Ryback & Moreton 1991).

Londonderry Greg & Lettsom (1858) describe exceptionally fine rock crystal from Dungiven. Specimens were exhibited by Russell (1946a).

Mayo Attractive specimens (10 cm across) of well formed pale-purple amethyst coated with later white quartz are reported from Achill Island, where massive vein quartz was once quarried (Greg & Lettsom 1858, O'Brien 1959, Ryback & Moreton 1992, Wolfe 2004). Small bipyramidal crystals of beta-quartz occur in a copper-mineralized porphyry to the S of Charlestown (Poustie 1982). Small hexagonal pyramidal quartz crystals form drusy crusts in fault breccia at Sheefry mine, 11 km S of Croagh Patrick (Moreton & Jackson 1991). Rutilated blue detrital quartz is described from greywackes of the Clew Bay complex (Mange et al. 2003).

Mayo Pale-blue quartz is reported from near Navan (Greg & Lettsom 1858).

Sligo Colourless prismatic quartz crystals up to 40 mm long with unidentified inclusions occur at Abbeytown quarry (site of old Abbeytown mine), Ballysadare; and at Aughamore limestone quarry, near Lough Gill, transparent quartz replaces fossil corals (Green & Moreton 2001).

Tipperary Clear flattened prisms of quartz up to 3 cm long are noted from Mogul and Magcobar mines, Silvermines; and at Gortdrum mine, near Tipperary town, slender, doubly terminated quartz crystals have been found (Ryback & Moreton 1991).

Wicklow Pure quartz has been quarried from 6 m-wide veins near Tinahely (O'Brien 1959). Yellow crystals are noted from Glenmalure (Greg & Lettsom 1858).

Offshore territories

Jersey Chalcedony occurs in veins, often containing barite, in the diorite and granite from Les Mouriers to Ronez, and at Crabbé and Val Rouget. Chalcedony is present in veins cutting mica lamprophyre at Les Reuses, Mourier Bay, Perrique and Ronez, and in geodes within granite between Ronez and Sorel, and N of Noirmont Point. It also pseudomorphs scalenohedral calcite at La Plaine. Jasper is found: in rhyolite N of Anne Port; in andesite at Colombière Point, SW of Belle Hougue Point; at Westmount; and at Le Mont à l'Abbé. Well formed crystalline quartz, some smoky, occurs in geodes within granite at Les Handois, and amethyst is occasionally found at Mont Mado. The largest crystals are from granite pegmatites at La Rocque (up to 20 cm long) and from Gigoulande. Other pegmatitic occurrences are from Minquiers and the Pipettes (Mourant 1961). Cathodoluminscence techniques have revealed fine-scale growth zoning in euhedral bipyramidal quartz crystals within dacite from Bonne Nuit Bay (Watt et al. 1997).

Sark Agate, jasper, hornstone and crystalline quartz (hematitic, milky, rose and brown) are recorded from the Pb–(Ag) lode of Sark's Hope mine, Little Sark (Mourant & Warren 1934, Ixer & Stanley 1983).

quartzine

A mixture of fine-grain quartz ± moganite.

queitite

monoclinic $Pb_4Zn_2(SO_4)(SiO_4)(Si_2O_7)$

An extremely rare lead–zinc sulphate silicate found on a single specimen from the dumps of the Old Dutch level of Red Gill mine, Caldbeck Fells, Cumbria. Queitite was found in a cavity within massive cerussite (minor quartz), associated with caledonite, leadhillite and susannite. It forms thin white botryoidal crusts (0.5 mm thick) on leadhillite and susannite (Braithwaite et al. 1989a, Cooper & Stanley 1990).

A single specimen of queitite has been found at Penrhiw mine, Ystumtuen, Central Wales orefield, Dyfed, forming aggregates of minute lath-like crystals to about 50 μm long (Mason & Green 1995).

In Scotland, dark-grey hemispheres of queitite up to 1 mm across and with a fibrous radial structure have been found on dump material from Horner's vein, Leadhills, Strathclyde. Associated species are cerussite, galena, mattheddleite, pyromorphite and susannite (Jackson 1990, Livingstone 1993).

quenselite

monoclinic $PbMn^{3+}O_2(OH)$

A rare lead–manganese oxide hydroxide found in manganese-cemented Quaternary gravels that form a raised-beach deposit between Auchenmaig and Craignarget Burn, Luce Bay, Dumfries & Galloway, Scotland. The cement consists of dark-brown to black amorphous wad, in which are found harder grains, usually < 1 mm across, consisting of birnessite, lithiophorite and quenselite. It is referred to as a supergene groundwater soil deposit (Nicholson 1988b, 1989a,b, 1992).

R

radiolite

An impure variety of natrolite, zeolite group.

rammelsbergite

LÖLLINGITE GROUP
orthorhombic $NiAs_2$

A rare nickel arsenide, trimorphous with pararammelsbergite and krutovite, and found in uranium–vanadium-rich Permian nodular concretions that have formed in localized horizons within the Littleham Mudstone Formation at Littleham Cove, Budleigh Salterton, Devon. Rammelsbergite and three other nickel arsenides (maucherite, nickeline and skutterudite) sometimes form dense concentric zones within the cores of the nodules, with a diameter range of 1–10 cm. Other species found in the nodules include algodonite, clausthalite, coffinite, native copper, lavendulan, malachite, modderite, montroseite, roscoelite, safflorite, native silver and many common sulphide species. The origin and paragenesis of the nodules is complex and not fully understood, but the nodules are authigenic and the latest theory is that de-watering of Permo-Triassic strata during burial compaction produced pore fluids that were capable of transporting the Ag–As–Co–Cu–Ni–U–V metals needed for the mineralization (Harrison et al. 1975b, Milodowski et al. 2000).

At Wheal Bray, Altarnun, Cornwall, within the Bodmin Moor granite pluton, Co–Ni–U mineralization occurs in late cross-course veins. Uraninite (pitchblende) is mantled by rammelsbergite and chalcopyrite in an assemblage that also contains bornite, covellite and hematite (Ball et al. 1982). Rammelsbergite and nickeline associated with uraninite (pitchblende) are also noted by these authors from Roscrow United mine, Ponsanooth, Cornwall. Elsewhere in Cornwall, rammelsbergite has been reported from South Terras mine, St Stephen-in-Brannel (Davis et al. 1965). The samples were supposedly collected by A. W. G. Kingsbury and their provenance is now in doubt. This is also the case for rammelsbergite from Tynebottom mine, NW of Garrigill, near Alston, Cumbria, described by Ixer et al. (1987), in reports based on samples from the Kingsbury collection now at the NHM (Ryback et al. 1998, 2001).

Rammelsbergite is reported from the Coniston copper mines, Cumbria, where it occurs as well formed interpenetrating lamellar twins, which are sometimes developed in a plumose texture growing on native bismuth. In some specimens it replaces nickeline (Stanley 1979, Young 1987a). At Settlingstones mine, near Hexham, Northumberland, rammelsbergite has been visually identified replacing nickeline along grain edges and fractures. Other associated species in the Ni–Pb–Zn ore are galena, gersdorffite, marcasite, millerite, pyrite (var. bravoite), skutterudite, sphalerite, ullmannite, vaesite and violarite (Ixer 1986).

An unexpected (visually identified) occurrence of rammelsbergite, forming 50 µm-diameter inclusions in massive sulphide (probably galena–tennantite), is reported from Dolyhir quarry, Old Radnor, near Presteigne, Powys. No other nickel-bearing minerals are found at this site (T. F. Cotterell pers. comm. 2003).

In Scotland, rammelsbergite has been found: associated with nickeline in galena-bearing ankerite–quartz veinstone on the dumps of Glencrieff mine, Wanlockhead, Dumfries & Galloway; on specimens in the NMS collected from Menimuir Burn, near Cassencarrie, Dumfries & Galloway; and possibly also from Hilderston, Bathgate Hills, Lothian (Temple 1954b, 1955). Green & Todd (2001) report rammelsbergite (and many other species) from the uranium-bearing veins along the Solway coast around Needle's Eye, SW of Caulkerbush, Dumfries & Galloway. Rammelsbergite is the principal nickel-bearing species at the Alva silver mine, Silver Glen, Ochil Hills, Central Region, and forms small (> 1 cm) solid masses of a bright metallic-grey colour, occasionally accompanied by nickeline (Moreton 1996, Moreton et al. 1998). Disseminated sulphides forming 1–200 µm grains in serpentinites from the basal thrust of the Shetland ophiolite on Unst include nickeline with alteration rims of orcelite and rammelsbergite (Maynard 1993).

ramsbeckite

monoclinic $(Cu^{2+},Zn)_{15}(SO_4)_4(OH)_{22}\cdot 6H_2O$

A rare but locally abundant hydrated copper–zinc sulphate hydroxide found on dump material derived from the fissure veins and pipe-like orebodies once worked at Waterbank mine, Ecton Hill, about 2 km NNW of Wetton village, Staffordshire. Ramsbeckite forms pseudo-orthorhombic emerald-green crystals up to 0.6 mm across, sometimes aggregated together into small crusts, and is associated with brochantite, cerussite, hydrozincite, namuwite, posnjakite, schulenbergite, serpierite and smithsonite. It is a supergene species produced during post-mining oxidation of the primary Cu–Pb–Zn ore and is typically one of the last species to form. It can be difficult to distinguish visually between brochantite and ramsbeckite (Rust 1992, 1995c). Ramsbeckite has also been found on a single specimen from the Prince of Wales mine, Calstock, near Gunnislake, Cornwall, where it forms flattened emerald-green crystals up to 0.5 mm long within cavities in partially oxidized sphalerite. Schulenbergite is an associated species (Green et al. 2001). This association is found again on specimens from the dumps of the 30-fathom level of Driggith mine, Caldbeck Fells, Cumbria, where the two species occur in thin carbonate-bearing veinlets in oxidized sphalerite. At Driggith mine, ramsbeckite reaches 0.4 mm in length, but on the dumps of the nearby Sandbed mine (derived from the same vein), crystals to 2 mm were found (Green et al. 2001, 2006a). Ramsbeckite has also been found as a bluish-green crust and as pseudo-orthorhombic crystals up to 0.3 mm across associated with linarite and schulenbergite from Silver Gill, Caldbeck Fells (Green et al. 2005b).

In Wales, five occurrences of ramsbeckite from the Central Wales orefield, Dyfed, are reported:

- Exceptionally large (up to 3 mm across) diamond-shape crystals of dark-green ramsbeckite associated with schulenbergite and linarite were found in a small isolated mineralized pocket on the dumps of Penrhiw mine, Ystumtuen. Single flat-lying crystals and aggregates of star-like rosettes occur in cavities left by dissolution of carbonate veinlets in siltstone blocks and in sulphide blocks (Bevins 1994, Mason & Green 1995).
- At the nearby Frongoch mine, between Devil's Bridge and Pont-rhyd-y-groes, ramsbeckite is occasionally found on schulenbergite in association with caledonite, langite and wroewolfeite (Green et al. 1996b).
- Very rare ramsbeckite is found with anglesite, brochantite, cerussite, linarite, schulenbergite and langite-group minerals at Brynarian mine, near Tal-y-bont (Rust 1992).

Translucent green ramsbeckite crystal 0.3 mm across associated with schulenbergite; Penrhiw mine, Ystumtuen, Dyfed, Wales. (© D. I. Green)

- Tiny emerald-green blocky crystals of ramsbeckite, too small for XRD analysis, cover a botryoidal crust of schulenbergite and an associated langite-group mineral on a single specimen from Ystrad Einion mine, near Tal-y-bont (Mason & Rust 1997).
- A visual identification of ramsbeckite from Eaglebrook (Nantycagl) mine, Ceulanymaesmawr is noted by Bevins (1994).

At Dylife mine, also in the Central Wales orefield, but in Powys, typical pseudo-orthorhombic emerald-green crystals of ramsbeckite and tiny elongated rod-like prisms are found (Rust 1992). For all the Welsh occurrences, post-mining oxidation of Cu–Pb–Zn-rich mine waste is thought to be responsible for ramsbeckite crystallization.

ramsdellite

orthorhombic $Mn^{4+}O_2$

An uncommon manganese oxide, trimorphous with akhtenskite and pyrolusite, and found in patchy exposures of iron and manganese oxides in the Lecht Valley, near Tomintoul, Grampian, Scotland, especially at the site of the old ironstone mine, where a porous hydrothermal-eruption breccia occurs. The deposit is mainly composed of goethite and cryptomelane, both with occasional patches of hematite. Fine-grain ramsdellite occurs as a black coating on a cryptomelane–goethite sample. Chalcophanite, lithiophorite, pyrolusite and woodruffite are other manganese-rich species found in minor amounts. The Lecht mineralization is derived from weathering of the local Dalradian sequence, and ponding of the resulting Fe–Mn–Zn solutions produced a seepage-bog type of deposit in which the permeable breccia acted as a channel-way for these supergene solutions. Diagnostic lithogeochemical anomalies in the area also indicate the existence of a concealed and related Zn–Pb–(Ag)–(Au?) exhalative deposit (Nicholson 1987a, 1989a,b, 1992, Nicholson & Anderton 1989).

Metallic-grey prismatic crystals of ramsdellite (identified by XRD) forming layers with chalcophanite are reported from Drosgol mine, Ponterwyd, Dyfed (T. F. Cotterell pers. comm. 2003).

ranciéite

hexagonal $(Ca,Mn^{2+})Mn_4^{4+}O_9{\cdot}3H_2O$

A rare hydrated calcium–manganese oxide that forms a series with takanelite and is briefly described from a vague locality near Newbiggin, Teesdale, Co. Durham. Ranciéite forms a pink-brown powder (Ryback & Tandy 1992). In Wales, ranciéite occurs at Parys Mountain, Anglesey, Gwynedd, where it forms dark-brown to black, soft, porous, stalactitic post-mining aggregates up to several cm across, in the Joint (45-fathom) level accessed via Dyffryn Adda (T. F. Cotterell pers. comm. 2007).

Individual crystals of ranciéite to 0.1 mm forming velvety clusters on saponite; the Gobbins, Co. Antrim, Ireland; width 1.5 mm. (© D. I. Green)

rankinite

monoclinic $Ca_3Si_2O_7$

TL Scawt Hill, Larne, Co. Antrim, Ireland

A rare calcium silicate, dimorphous with kilchoanite and found in the chalk/dolerite contact zone at Scawt Hill. Rankinite is associated with the minerals larnite, melilite and wollastonite. The melilite rocks have been produced by localized intense soaking of solutions from the dolerite into the chalk, and here rankinite forms colourless rounded or irregular grains devoid of cleavage – in melilite rocks and at the contact of flint nodules in metasomatized chalk. Flint nodules in the chalk have been metasomatized to produce quartz–wollastonite aggregates encased by larnite–spurrite assemblages and, along the wollastonite borders, narrow alteration zones consisting of colourless rankinite have developed (Tilley 1942, Moody 1953).

At Camas Mòr, Isle of Muck, Highland Region, Scotland, an olivine-gabbro dyke intersecting Jurassic limestones has developed a high-temperature contact assemblage of gehlenite, larnite, rankinite and spinel. Rankinite forms rounded grains devoid of cleavage and is often the host for inclusions of gehlenite (Tilley 1947). In a similar environment at Kilchoan, Ardnamurchan, Highland Region, rankinite forms part of a metamorphic assemblage that also includes åkermanite, cuspidine, grossular, kilchoanite, rustumite and spurrite (Agrell 1965). Perhaps the most unusual occurrence of rankinite is from 'S Airde Beinn, 5 km SW of Tobermory, Isle of Mull, Strathclyde, where a volcanic plug has thermally metamorphosed amygdaloidal plateau basalts. Original calcite-filled amygdales have been replaced inwards by a sequence of aegirine-augite, melilite, perovskite, wollastonite, rankinite and larnite (Cann 1965a). Also in Strathclyde, rankinite is reported as a contact-metamorphic mineral, where a Tertiary dyke has intruded Carboniferous basalt of the Clyde Valley Plateau Lavas at Loanhead quarry, near Beith. Evans (1987) notes extensive replacement of the basalt by calcite and epidote, and vesicles infilled by prehnite and thomsonite.

Rankinite has also formed in a limestone that has been thermally metamorphosed by two basalt–dolerite dykes at Killala Bay, near Inishcrone, Co. Sligo, Ireland. It is part of a primary assemblage that changes away from the dolerite contacts and collectively contains calcite, cuspidine, gehlenite, grossular, rankinite, spurrite, tilleyite and wollastonite (Nawaz 1974).

rashleighite

TL Bunny mine, St Austell and Castle-an-Dinas mine, St Columb Major, Cornwall, England

Named after the mineral collector and author Philip Rashleigh FRS (1729–1811) of Menabilly, Fowey, Cornwall; born in the East End of London. Rashleighite is an intermediate composition between turquoise and chalcosiderite.

raspite

monoclinic $PbWO_4$

A very rare lead tungstate, dimorphous with stolzite and found in quartz veinstone from Carrock mine, Grainsgill, Caldbeck Fells, Cumbria, where it appears to have formed by supergene oxidation of lead- and tungsten-bearing minerals. Raspite forms small (up to 0.3 mm long) lustrous yellow prismatic crystals associated with stolzite (Neall & Green 2001a).

ratholite

TL Ratho quarries, Edinburgh, Lothian, Scotland

A synonym of pectolite, wollastonite group.

rauenthalite

monoclinic or triclinic $Ca_3(AsO_4)_2{\cdot}10H_2O$

An exceedingly rare hydrated calcium arsenate from the disused Cu–(Co) mine at Muckross, Killarney National Park, Co. Kerry, Ireland. Post-mining oxidation of the mine waste has given rise to secondary carbonates, sulphates and arsenates, six species of which proved new to Ireland (alloclasite, monohydrocalcite, parasymplesite, phaunouxite, rauenthalite and smolyaninovite). Rauenthalite forms chalky white spheroidal masses 1–1.5 mm across and is indistinguishable from monohydrocalcite. It also occurs as colourless acicular sprays of crystals up to 5 mm long (Ryback & Moreton 1993, Moreton et al. 1999, Green & Moreton 2001).

Rauenthalite (or phaunouxite) has also been described from within the metamorphic aureole of the Dartmoor granite at Ivy Tor mine, Belstone, Meldon area, near Okehampton, Devon. A piece of veinstone from the mine dump, when broken open, revealed dolomite gangue and a primary löllingite ore containing a cavity lined with white radiating spherules of talmessite, often mixed with rauenthalite or phaunouxite. Rauenthalite can arise by the reversible dehydration at room temperature of

Sprays of rauenthalite crystals from Muckross mine, Killarney, Co. Kerry, Ireland; width 0.7 mm. (© D. I. Green)

phaunouxite, depending upon the humidity of the surrounding air, so the "original" phase could have been either of these species. Other minerals described from the mine include arsenopyrite, axinite, native bismuth, bismuthinite, chalcopyrite, garnet, marcasite, quartz, scheelite and a little cobalt ore (Braithwaite et al. 1996).

realgar

monoclinic AsS

An uncommon sulpharsenide, dimorphous with pararealgar and found in Burraton Coombe quarry, St Stephens by Saltash, Cornwall. Realgar forms small "aurora-red" crystalline patches and long prismatic crystals up to 1.25 mm in length embedded in cleavage masses of calcite resting directly upon a reniform mass of native arsenic. Arsenolite, dolomite and quartz are associated species (Russell 1944).

In Wales, realgar has been found at Dolyhir quarry, Old Radnor, near Presteigne, Powys. Here, millimetre-thick veins of bright-red realgar occur in Precambrian mudstones near a major strike-slip fault. The veins show no relationship to the original bedding and do not appear to be syngenetic (Cowdry 2002a, T. F. Cotterell pers. comm. 2003).

Realgar crystal, 2 mm in height, from Dolyhir quarry, Old Radnor, near Presteigne, Powys, Wales. (© D. I. Green)

rectorite

monoclinic $(Na,Ca)Al_4(Si,Al)_8O_{20}(OH)_4{\cdot}2H_2O$

An uncommon hydrated sodium–calcium–aluminium aluminosilicate hydroxide clay mineral composed of a regular 1:1 interstratification of a dioctahedral mica and a dioctahedral smectite. Rectorite is found at Parc Penrhyn, Bangor, Gwynedd, Wales, in a metabentonite of Caradoc age, where it has been identified in the <2 µm fraction of the clay, in association with pyrophyllite (Roberts & Merriman 1990). In the valley of the Afon Seiont, 2 km SE of Caernarfon (and 13 km SW of Bangor), rectorite is also reported from the clay fraction of a pelite sample (Jiang et al. 1990). At Tremadog, Gwynedd, the <2 µm clay fraction of another pelite contains rectorite and pyrophyllite (Merriman & Roberts 1985). Less precisely, Gill et al. (1977) note rectorite (var. allevardite) and pyrophyllite in the <2 µm clay-mineral fraction of rocks from the anthracite zone of the South Wales coalfield (i.e. West Glamorgan).

In England, rectorite has been described from Boulby potash mine, Loftus, Cleveland, where it forms interstitially between boracite crystals as white triangular crystals up to 0.5 mm across with a perfect cleavage and pearly lustre, and as subspherical crystal groups to 2 mm (Green & Freier 1996).

red copper

A synonym of cuprite.

red lead

A synonym of minium.

red ochre

A soft variety of hematite.

reddle

A soft variety of hematite.

redgillite

monoclinic $Cu_6(SO_4)(OH)_{10}{\cdot}H_2O$

TL Golden Hugh level, Silver Gill vein, Caldbeck Fells, Cumbria, England

A rare hydrated copper sulphate hydroxide formed as a result of supergene oxidation of Cu–Pb–Zn mineralization. Redgillite was first described as an unknown species

Randomly arranged blades of redgillite from Silver Gill, Caldbeck Fells, Cumbria; width 200 µm; SEM image. (© D. I. Green)

Plumose spray of pale-green redgillite (1 mm tall) from Eaglebrook (Nantycagl) mine, Ceulanymaesmawr, Dyfed, Wales. (© D. I. Green)

from Red Gill mine, less than 1 km away from Silver Gill (Cooper & Stanley 1990). Since that report, the mineral has commonly been referred to as redgillite, despite the fact that it had not been approved by the IMA-CNMNC. Material from Red Gill was not of sufficient quality to characterize the species adequately and so Silver Gill material became the holotype. Redgillite from Silver Gill occurs as translucent to transparent grass-green bladed crystals up to 0.15 mm long with squared-off or tapering terminations, usually in radiating groups. The Silver Gill vein consists of primary chalcopyrite and galena, with lesser sphalerite, covellite and pyrite in a dominantly quartz gangue. Redgillite is found in thin fractures in partly oxidized sulphides, where it is associated with langite and more rarely with brochantite, connellite, cuprite and malachite. Other supergene minerals found in the vein include: anglesite, aurichalcite, bechererite, caledonite, cerussite, native copper, fraiponite, goethite, hemimorphite, hydrocerussite, leadhillite, linarite, posnjakite, pyromorphite, ramsbeckite, rosasite, schulenbergite, smithsonite, native sulphur, susannite, wroewolfeite and wulfenite (Green et al. 2005b, Pluth et al. 2005, Wolfe 2006). Pale-green redgillite forming feathery acicular crystals up to 0.3 mm long has also been found in sphalerite-rich veinstone at Driggith mine, Caldbeck Fells (Green et al. 2006a).

Material with an XRD pattern identical to that of redgillite was noted from Esgair Hir mine, Ceulanymaesmawr, near Tal-y-bont, Central Wales orefield, Dyfed, Wales (Cooper & Stanley 1990), and more recently redgillite has been found at other mines in the district: Bwlchrhennaid, near Goginan; Darren, Pen-bont Rhydybeddau, near Tal-y-bont; Frongoch, between Devil's Bridge and Pont-rhyd-y-groes, Dyfed; and Llechweddhelyg, near Clawddmelyn, between Bont-goch and Cwmsymlog (Pluth et al. 2005).

The unknown species PC1, which forms very rare bright grass-green individual or acicular radiating crystal groups, to 0.2 mm, on botryoidal chrysocolla and goethite at Penberthy Croft mine, St Hilary, Cornwall (Betterton et al. 2000), is also redgillite.

redondite

An identical composition to Messbach-type variscite.

redruthite

TL Redruth, Cornwall, England

A synonym of chalcocite.

red silver

A synonym of proustite and pyrargyrite, used before the two species had been isolated.

reevesite

HYDROTALCITE GROUP

trigonal $Ni_6Fe_2^{3+}(CO_3)(OH)_{16}\cdot 4H_2O$

A rare hydrated nickel–iron carbonate hydroxide found on chromitite at Hagdale quarry, Unst, Shetland, Scotland, where it forms fine-grain yellow encrustations intimately mixed with honessite and associated with theophrastite. The specimens also contain aragonite, chlorite and goethite, and lesser amounts of heazlewoodite and pentlandite (Bish & Livingstone 1981, Livingstone & Bish 1982, Livingstone 2002). Elsewhere in Scotland, at the Craignure mine, 12 km SW of Inveraray, Strathclyde, reevesite forms a fluffy citron-yellow coating associated with hydrohonessite and honessite on dump material consisting mainly of chlorite schist with narrow quartz veins and calcite-rich bands (Livingstone 1990b).

reichenbachite

monoclinic $Cu_5^{2+}(PO_4)_2(OH)_4$

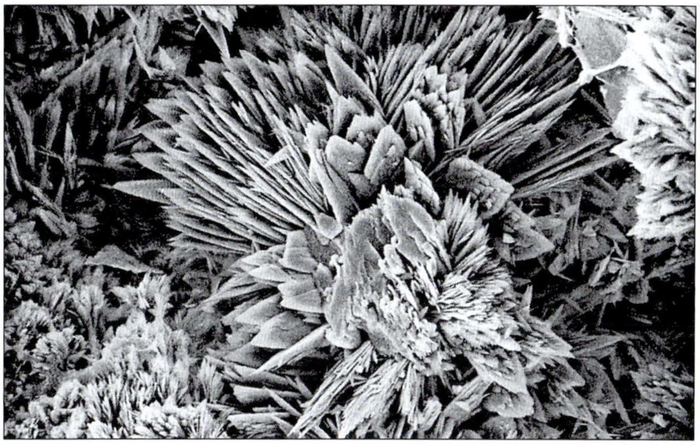

Spray of terminated reichenbachite crystals. Old Gunnislake mine, Calstock, Cornwall; width 160 µm; SEM image. (From Braithwaite & Ryback 1994, and reproduced with permission from the Mineralogical Society)

A copper phosphate hydroxide, trimorphous with ludjibaite and pseudomalachite, and commonly found on the dump near Michael's shaft, Old Gunnislake mine, Calstock, near Gunnislake, Cornwall. Reichenbachite is a late-generation supergene mineral, forming dark-green crusts to mammillary coatings, with a compact, indistinctly fibrous radiating habit. The coatings fill cracks and line cavities on a matrix of milky quartz occasionally in association with chrysocolla, malachite or metatorbernite. Well formed emerald-green platy crystals (0.02 mm across) are also reported. Reichenbachite is visually indistinguishable from pseudomalachite, which also occurs at this locality (Braithwaite & Ryback 1994).

reichite

TL Alston Moor, Cumbria, England

A synonym of calcite.

reinite

Ferberite pseudomorph after scheelite.

resin opal

A synonym of opal-CT.

restormelite

TL Restormel (Royal) iron mine, N of Lostwithiel, Cornwall, England

A doubtful or unidentified aluminosilicate mineral containing potassium and sodium.

retgersite

tetragonal $NiSO_4\cdot 6H_2O$

A rare hydrated nickel sulphate, dimorphous with nickelhexahydrite and found at Menimuir Burn, near Cassencarrie, Dumfries & Galloway, Scotland, in association with annabergite and nickeline. Retgersite forms pale-blue to bluish-green crusts and minute curved or twisted crystals (Macpherson & Livingstone 1982).

Retgersite is also reported from a clay–ironstone septarian nodule collected at Newdigate colliery, Bedworth, SW of Nuneaton, Warwickshire. Retgersite forms minute single crystals and groups of crystals coating acicular millerite. Other species in the nodule include chalcopyrite, galena, siderite and zaratite (Walker 1988).

reyerite

trigonal $Na_2Ca_{14}Al_2Si_{22}O_{58}(OH)_8{\cdot}6H_2O$

A rare hydrated sodium–calcium aluminosilicate hydroxide found with gyrolite, okenite, pectolite, tacharanite, tobermorite, xonotlite, and a wide range of zeolite species in amygdales within Tertiary olivine basalts on the Isle of Mull, Strathclyde, and across the Sound of Mull on the Morvern Peninsula, Highland Region, Scotland (Walker 1971). Near 'S Airde Beinn, 5 km SW of Tobermory, Isle of Mull, a volcanic plug intrudes amygdaloidal plateau basalts containing amygdales infilled with analcime, gyrolite, natrolite and thomsonite; and in the metamorphic aureole of the plug, gyrolite has been replaced by pale-green radiating aggregates of reyerite (Cann 1965a). At Allt Coir' a' Ghobhainn, near Drynoch, Isle of Skye, Highland Region, reyerite is again found in amygdaloidal basalts, where it has infilled vesicles along with tobermorite and a hard porcellanous intergrowth of calcium-rich analcime and feldspar (bytownite). Reyerite is micaceous and it infills central zones within the vesicles and is considered to have formed during zeolitization when calcium- and strontium-rich fluids were expelled from the deepest parts of the lava-pile (Livingstone 1988, 1989). Green & Todd (2001) report mesolite, reyerite and large blocky apophyllite crystals from Dunvegan quarry, also on the Isle of Skye.

RHABDOPHANE GROUP

Hexagonal or pseudo-hexagonal phosphates with the general formula XZO_4, where X = Ca, Ce, Fe^{3+}, La, Pb, Th; and Z = P, S. Brockite, ningyoite, rhabdophane-(Nd) and tristramite are members of the group found in Britain or Ireland.

rhabdophane-(Nd)

RHABDOPHANE GROUP

hexagonal $(Nd,Ce,La)PO_4{\cdot}H_2O$

TL Fowey Consols mine, Tywardreath, St Blazey, Cornwall, England?

A rare hydrated neodymium-dominant phosphate originally discovered by W. G. Lettsom, but labelled as "from Cornwall". Rhabdophane occurs in small irregular mammelated concretions and is garnet red (Lettsom 1878, Hartley 1882). In 1964, A. W. G. Kingsbury claimed to have rediscovered the location at Fowey Consols. However, Ryback et al. (1998, 2001) have shown that the provenance of many Kingsbury specimens is in doubt, so the type locality of rhabdophane remains unconfirmed. Lewis (1997) repeats Kingsbury's description of rhabdophane. Further analysis by Bowles & Morgan (1984) has shown the species to be rhabdophane-(Nd) and not rhabdophane-(Ce), as listed by Blackburn & Dennen (1997).

rhaetizite

A synonym of kyanite.

rhodalite

TL Co. Antrim, Ireland

An undefined clay mineral.

rhodizite?

cubic $(K,Cs)Al_4Be_4(B,Be)_{12}O_{28}$

A very rare potassium–cesium–aluminium–beryllium borate reported from Meldon aplite quarry, near Okehampton, Devon (Kingsbury 1964b, Embrey 1978). Rhodizite is also listed from Meldon by Anthony et al. (2003), presumably on the basis of Kingsbury's description. Ryback et al. (1998, 2001) have expressed serious doubts about the provenance of many Kingsbury samples (now at the NHM) and, without further confirmation, rhodizite cannot be considered to be a British or Irish species.

rhodochrome and rhodophyllite

Synonyms of kammererite; a chromium-rich variety of clinochlore, chlorite group.

rhodochrosite

CALCITE GROUP

trigonal $Mn^{2+}CO_3$

A relatively common manganese carbonate that forms series with calcite and with siderite, and is found in a contact-metamorphic deposit close to the margin of the Bodmin Moor granite at Treburland mine, 2 km SE of Fivelanes, Altarnun, Bodmin Moor, Cornwall. It forms pale-pink spheres up to 0.75 mm in diameter, built up of successive layers and forming rosettes in a bedded manganese ore dominated by pyrolusite, with lesser amounts of bustamite, rhodonite and tephroite (Russell 1946d).

Pink rhodochrosite with an irregular surface; crystals up to 15 mm across replace earlier calcite crystals; Durnford quarry, Long Ashton, near Bristol, Avon. (© D. I. Green)

Russell also reports pale-pink rhodochrosite associated with rhodonite in pyrolusite-rich ore that forms pockets and strings in hard chert at Lidcott mine, Laneast, Cornwall. In the parish of St Just in Cornwall, rhodochrosite is noted from: Botallack mine (Collins 1892, Robson 1949b, Bancroft & Weller 1993); Levant mine, in massive form and as small pink saddle-shape crystals usually no more than 3 mm across (Wolloxall 1987); Wheal Owles (Robson 1949b, Golley & Williams 1995). A single specimen of rhodochrosite and rhodonite was recovered from the dumps of Allerford mine, Marystow, Devon; it contained many minute blebs of glassy quartz and its exterior consisted of pyrolusite and a white chalky substance (Russell 1946d). Nodular masses of manganite and rhodochrosite with some alteration to earthy pyrolusite are described from the dumps of the Black Pit, Pound Living mine, Upton Pyne, Exeter, Devon (Scrivener & Edwards 1990). Shepherd et al. (2005) note that oxidation of interbanded psilomelane, kutnohorite and rhodochrosite produced the massive manganite–pyrolusite ore at Upton Pyne. Dearman (1966a) notes the restricted occurrence of rhodochrosite and rhodonite in the basal bed of the Calcareous Group of the Lower Culm Measure (Lower Carboniferous) at the British Railways quarry complex, Meldon, near Okehampton, Devon. Other reports of rhodochrosite from southwest England include: Simm's lode, Geevor mine, near Pendeen, St Just, and Trebarwith Strand, Tintagel, both in Cornwall (Golley & Williams 1995); and, in Devon, Stancombe mine, Bovey Tracey (Greg & Lettsom 1858, Russell 1946d), and Chillaton and Hogstor mine, Milton Abbot (Russell 1946d).

Pale rhodochrosite forms massive nodules, or occasionally botryoidal crusts, or rounded aggregates of crystals lining cavities at Higher Pitts farm, Priddy, Mendip Hills, Somerset (Kingsbury 1941). At Batts Combe quarry, near Cheddar, Mendip Hills, rhodochrosite and manganocalcite are found in thin veins cutting pockets of manganese-rich wad within a 0.5 m-wide calcite vein. Cerussite, hydrocerussite and wulfenite are associated species (Eastwood 1983). Rhodochrosite is developed only sporadically at Merehead (Torr Works) quarry, Cranmore, where it forms small white microcrystalline or granular cavity fillings in manganese oxide pods that occur as replacement deposits in the Carboniferous Limestone. It is more common as a major cavity filling in manganese-poor iron ores associated with the pods and it forms pink and white colour-banded botryoidal encrustations. A sequence of crystallization from calcite to rhodochrosite to lead carbonates is noted in some cavities (Alabaster 1975). Two other Mendip localities are near Frome: Whatley quarry – very small quantities of rhodochrosite; Coleman's quarry, Holwell – rhodochrosite and many late-stage minerals in manganese-rich veins (Alabaster 1990a). Dump material from Providence iron mine, Bristol area, contains rhodochrosite with cerussite, hausmannite and pyrolusite (Alabaster 1989b). At Durnford quarry, Long Ashton, near Bristol, encrustations of attractive pink rhodochrosite line cavities in manganese-rich pods in Carboniferous Limestone. Associated species include barite, calcite, manganite and wulfenite (A. Napier pers. comm. 2007).

Rare minute glassy crystals of rhodochrosite, usually less than 0.2 mm across and grey or pale pink, are scattered over joint faces and cavities in gossanous quartz-vein material at Burdell Gill, Caldbeck Fells, Cumbria. Crystals form rounded subhedral grains or sharp scalenohedra truncated by rhombohedral faces. Barium-pharmacosiderite is an associated species (Neall & Leppington 2003). Rhodochrosite is associated with braunite, hausmannite and pyrolusite at Wyndham mine, Bigrigg, near Egremont, Cumbria (Davidson & Thomson 1951). Rhodochrosite is also noted from Hubberdale mine, Flagg, near Bakewell, Derbyshire (Mostaghel 1983, Ford et al. 1993) and from "the Hon. Mr West's mine" (location now uncertain), near Oswestry, Shropshire (Greg & Lettsom 1858).

In Wales the principal occurrence of rhodochrosite is in the Lower and Middle Cambrian manganese deposits of the Harlech Dome, Gwynedd, that extend over an area of 190 km^2 from Harlech in the north to Barmouth in the south and Y Garn in the east. The ore has extremely fine grain, is compact and cherty in appearance, and consists of a mosaic of red laminated bands of rhodochrosite (dialogite) and colourless to pale-yellow spessartine. Rhodochrosite (dialogite) is also found with garnet and magnetite in associated mudstones (Woodland 1939). Later work by Bennett (1987b) indicates that kutnohorite is the dominant carbonate species in the red bands, although micritic calcium-rich rhodochrosite is found in micro-concretions that in places form continuous creamy-yellow bands. Apatite, barite, chlorite, ferropyrophanite, hematite, magnetite, quartz and rhodonite are noted as associated species. A diagenetic origin for the manganese ore has been proposed, probably in a shallow marine basin in which reducing conditions allowed manganese in the sediment column to be extensively remobilized and deposited at the sediment/water interface as rhodochrosite (Glasby 1974, Allen & Jackson 1985). Manganese-rich shales of the Mulfran Formation on the St Tudwal's Peninsula, Lleyn, Gwynedd, have been correlated with the Harlech deposits and contain both manganese-rich calcite and calcium-rich rhodochrosite (Bennett 1987b). Elsewhere in Wales, rhodochrosite has been noted from: the Dolgellau goldbelt (Readwin 1888); Benallt and Nant mines, near Rhiw, Lleyn Peninsula, Gwynedd (Woodland 1939, Dunham et al. 1978); and the Cambrian Slate quarry, Glyn Ceiriog, Clwyd (Bevins 1994). Hall (1993c) records rhodochrosite from Talachddu, near Brecon, Powys, but XRD analysis suggests that it is pink dolomite (Bevins 1994).

Rhodochrosite is a minor component of two manganese deposits in Scotland. It occurs as coatings and veinlets on and within goethite–hematite–hollandite–pyrolusite mineralization found in the Dalradian Jura Quartzite at Dùn Athad, Mull of Oa Peninsula, Isle of Islay, Strathclyde, and at Dalroy, near Culloden Moor, 10 km SE of Inverness, Highland Region. The species also occurs within cavity-infill structures in a Dalradian stratiform mixed oxide and silicate ore interpreted to be a lacustrine hot-spring deposit. Strontium-rich barite, braunite (the dominant ore species), bustamite, calcite, manganese-rich ilmenite and pyrophanite are associated species (Macpherson & Livingstone 1982, Nicholson 1987b, 1989c, 1990).

Rhodochrosite occurs with hematite and pyrolusite in an iron–manganese deposit exposed in a small quarry on Logwood Hill, Calliagh, Co. Monaghan, Ireland (Ryback & Moreton 1993). Greg & Lettsom (1858) report yellowish-grey rhodochrosite from Glendree, near Tulla, Co. Clare.

rhodonite

triclinic $CaMn_4Si_5O_{15}$

A relatively common calcium–manganese silicate that can form by hydrothermal, contact- and regional-metamorphic, and sedimentary processes.

Rhodonite is extremely abundant in a contact-metamorphic deposit close to the margin of the Bodmin Moor granite at Treburland mine, 2 km SE of Fivelanes, Altarnun, Bodmin Moor, Cornwall, where it forms extremely fine specimens together with bustamite and tephroite in a bedded manganese ore dominated by pyrolusite. It forms large very-hard crystalline blocks externally altered to a thin black crust of pyrolusite. Grainsize ranges from fine to coarse, with exceptional crystals reaching 6.5 cm in length, the colour varying from rose pink to brownish pink or colourless along altered margins. Large polished slabs of massive rhodonite, measuring up to 17×25 cm, form beautiful specimens. In some specimens, rhodonite is intergrown with fine-grain green pyroxene and in others small (up to 7 mm) part-cleaved and interlocking crystals occur as very acute rhombohedra. The latter crystals are pale pink and sometimes transparent. Inclusions of chalcopyrite, galena, malachite, molybdenite or spessartine are sometimes found embedded in the rhodonite (Russell 1946d). Rhodonite also occurs at Greystone quarry, Lezant, near Launceston, Cornwall, where low-grade metamorphism of a presumed rhodochrosite–quartz assemblage has produced rhodonite lenses and bands in deformed fine-grain metasediments. Polished slabs of pinkish rhodonite, tens of cm across, show thin quartz veins, fine-grain sedimentary matrix and reddish jasper (Elton et al. 2004).

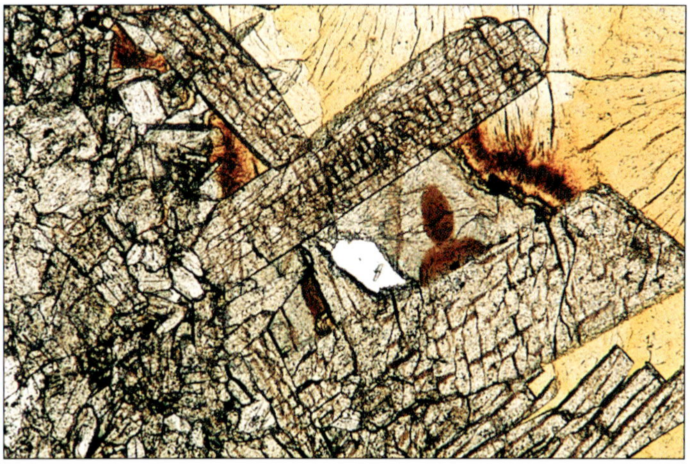

Prismatic rhodonite associated with yellow bementite. Treburland mine, Altarnun, Cornwall; width 1.75 mm; viewed in PPL (upper) and XPL (lower). (© K. Gardner)

In Devon, massive pink rhodonite has been formed by thermal metamorphism of Lower Carboniferous sediments within the thermal aureole of the Dartmoor granite at the British Railways quarry complex, Meldon, near Okehampton, Devon. Bustamite, spessartine and tephroite are associated species. The massive rhodonite contains druses in which well formed crystals of rhodonite up to 5 mm in length occur and on which small globules and sheaves of rhodochrosite, pyrite and pyrrhotite have later grown. In habit the rhodonite forms very acute rhombohedra (Howie 1965, Dearman 1966a). Other reports of rhodonite from southwest England include five localities in Cornwall – Atway mine, Launceston; Lidcott mine, Laneast; Greystone Wood mine, Lezant; Pencrebar mine, 2 km S of Callington (many beautiful specimens); Torr Wood mine, Pillaton (all Russell 1946d, Golley & Williams 1995) – and five in Devon – Allerford mine, Marystow; Sydenham and Lee Wood mine, Marystow; Week manganese mine, Milton Abbot; Black Down, Mary Tavy, near Tavistock; Upton Pyne manganese mine, Exeter (Greg & Lettsom 1858, Hall 1868b, Rudler 1905); (Russell 1946d). The two Devon occurrences are disputed by Russell (1946d), who also considers the following as unsubstantiated locations for rhodonite: Indian Queens iron–manganese mine, St Columb Major; Pendower, Veryan; Tremollet manganese mine, North Hill; West Down End manganese mine, Egloskerry; Lewannick Hill, Lewannick; St Cleer, all in Cornwall, and Chillaton and Hogstor manganese mine, Milton Abbot, Devon.

Lower Cambrian manganese ore in the Harlech Dome part of Gwynedd, Wales, was for many years thought to contain a major amount of rhodonite; however, Woodland (1935, 1939) was able to show that spessartine is in fact the major silicate phase. Rhodonite does occur in these rocks, but only in minor amounts. The manganese ore is thought to have a diagenetic origin. Manganese in the sediment column was deposited at the sediment/water interface as rhodochrosite and, during later regional metamorphism, this recrystallized to form rhodonite (Glasby 1974, Allen & Jackson 1985). At Vigra mine, Dolgellau goldbelt, Gwynedd, rhodonite is a minor component of a gold-bearing quartz lode (Andrew 1910) and the species is also noted from Benallt and Nant mines on the Lleyn Peninsula, Gwynedd (Russell 1946d, Dunham et al. 1978).

At Gleann Beag, in the Lewisian inlier of Glenelg, Highland Region, Scotland, rhodonite occurs in veins up to 2.5 cm wide, in a grunerite–magnetite–spessartine schist associated with pyroxmangite (Tilley 1937a).

rhoetizite

A white variety of kyanite.

rhyacolite

A synonym of sanidine, feldspar group.

richterite

AMPHIBOLE GROUP

monoclinic $Na(CaNa)Mg_5Si_8O_{22}(OH)_2$

An uncommon sodium–calcium–magnesium silicate hydroxide that forms a series with ferrorichterite and is found in metasomatized (fenitized) quartzite at Glas Choille, SE of Loch Borralan, Assynt, Highland Region, Scotland. Richterite is pale blue in thin section and it occurs with aegirine-augite, apatite, biotite, orthoclase and titanite in these rocks. The nearby Borralan syenite complex was the source of the strongly alkaline fluids required to produce fenitization of the surrounding rocks (Woolley et al. 1972). Fenite from another locality to the SW of the Borralan complex contains both richterite and ferriwinchite (Rock 1977). Carbonatite rocks (porphyritic white sövite) intruding Durness Limestone close to the contact of the Borralan complex contain richterite-bearing limestone xenoliths. The richterite forms radiating aggregates of acicular prisms (which in large clusters are pale blue) and irregular clusters of needles up to 10 mm across. Calcite is the major species of the sövite, with minor chondrodite, diopside, fluorapatite, ilmenite, phlogopite and pyrrhotite also present (Young et al. 1994b).

riebeckite

AMPHIBOLE GROUP

monoclinic $\square Na_2(Fe^{2+}_3Fe^{3+}_2)Si_8O_{22}(OH)_2$

An uncommon sodium–iron silicate hydroxide that forms a series with magnesio-riebeckite and is found in alkali granites and syenites or occasionally as crocidolite in BIF rocks. Where possible, identification is based on the 1997 classification of the amphibole group (Leake et al. 1997a, Mogessie et al. 2004).

The black and blue lustrous amphibole, forming crystals usually 6 mm long, in microgranite at Mynydd Mawr, 5 km W of Snowdon, Gwynedd, Wales, was first identified as riebeckite using optical techniques (Bonney 1888, Harker 1888a,b, Cole 1891). Subsequent investigations confirmed the species and identified arfvedsonite (most common) and ferrorichterite (Bevins 1994, Tindle unpub. EMP data).

In Scotland, riebeckite occurs, sometimes with aegirine, in a suite of blue and bluish-green granophyric felsite dykes exposed between Ronas Voe and Colla Firth, North Roe, Mainland, Shetland. One dyke is described from the western end of Beorgs of Uyea, another in a burn S of North Hill. The riebeckite-bearing dykes include a non-porphyritic homogeneous type and a porphyritic spherulitic type, the former beautiful and admirably suitable for curling stones, the latter handsome, with blue spherulites of riebeckite up to 3 mm across (Phemister 1950). Amphibole with a wide range of compositions has been found in miarolitic cavities in a Western Redhills Tertiary epigranite at Loch Ainort, Isle of Skye. Crystals are bluish green (in thin section), sometimes euhedral and zoned. The following species are recognized: ferro-actinolite, ferro-edenite, ferrorichterite, ferrowinchite, katophorite and riebeckite. Invariably, riebeckite is one of the last of these species to crystallize. Other cavity species include: hedenbergite, monazite, titanite and zircon (S. A. Rust sample, Tindle unpub. EMP data). Riebeckite (unconfirmed) is noted in a felsite intrusion at Dirrington Little Law, Lammermuir Hills, W of Duns, Borders (Irvine 1930).

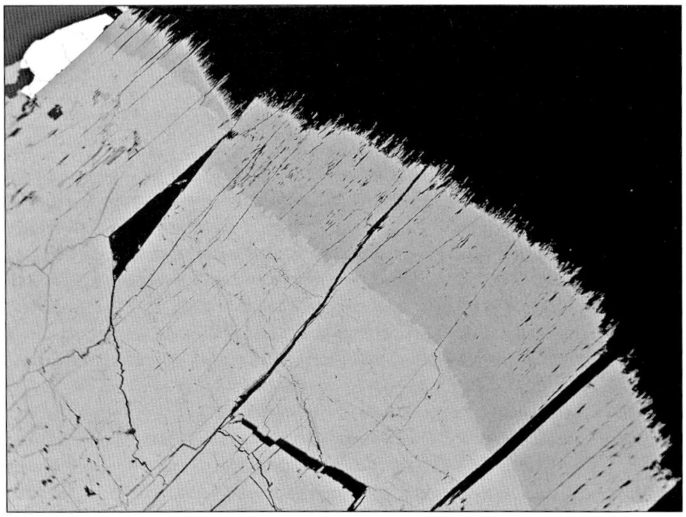

Compositionally zoned amphibole from a miarolitic cavity in Tertiary granite from Loch Ainort, Isle of Skye, Highland Region, Scotland; the crystal has a mainly ferro-edenite and ferrorichterite core (light grey) and a fringed riebeckite rim (darker grey); white is zircon; width 2 mm; BSE image. (© A. G. Tindle)

Surprisingly, riebeckite could not be confirmed in microgranites from either Ailsa Craig (Teall 1891b, Heddle 1901, Harding 1983, Harrison et al. 1987) or Holy Island, Lamlash Bay, Arran (Phemister 1950), both in Strathclyde, or the Rockall alkali granite (Sabine 1960, Hawkes et al. 1975), which instead contain blue amphibole of mainly arfvedsonite composition (Howie & Walsh 1981, Tindle unpub. EMP data). Other reports of riebeckite requiring confirmation include: rhyolite of Winkston Hill, Peebles, Borders (Thirlwall 1981); pegmatite associated with the analcime olivine dolerite Dippin Sill, Isle of Arran, Strathclyde (Henderson & Gibb 1977); syenite at Black Rock Burn, N of Benmore Lodge, near Loch Ailsh, Highland Region (Macgregor & Phemister 1972); and granophyre from Meall Dearg, a Western Redhills granitic intrusion on the Isle of Skye (Heddle 1901). Riebeckite described from gneiss of the Carn Chuinneag–Inchbae complex, Easter Ross, Highland Region (Harker 1962, Gallagher et al. 1971, Harker & Leahy 2001), is actually arfvedsonite.

Riebeckite is said to occur at Ross Strand, Killala Bay, Co. Mayo, Ireland, in an unusual drusy alkali-rich rock within gabbro, where it forms crystals in cavities associated with aegirine, arfvedsonite, quartz and accessory minerals (Ryback & Moreton 1992), but confirmation is required.

rinneite

trigonal $K_3NaFe^{2+}Cl_6$

An uncommon potassium–sodium–iron chloride found in Permian Upper Evaporite beds of the Yorkshire potash field from the Eskdale No. 2 borehole at Aislaby, Eskdale, near Whitby, Yorkshire. Rinneite is highly deliquescent and on exposure to air forms a brown liquid, from which sylvite and halite can crystallize. When fresh, rinneite is transparent and almost colourless, but sometimes has a delicate violet tint. Associated species are anhydrite, halite and sylvite (Stewart 1949, 1951b). Rinneite is a late-forming species in veins cutting the Carnallitic Marl in two other Eskdale boreholes: Sleights (Eskdale No. 3) and Upgang (Eskdale No. 6). It is associated with anhydrite, carnallite, gypsum, halite, hematite, magnesite, polyhalite, quartz and sylvite (Armstrong et al. 1951). Reporting on the Yorkshire potash field, Stewart (1965) considers rinneite to be a widespread (but scarce in certain units) minor constituent of the Upper Evaporites, Carnallitic Marl and Middle Evaporites.

ripidolite

An iron-rich variety of clinochlore, chlorite group.

riversideite?

pseudo-orthorhombic $Ca_5Si_6O_{16}(OH)_2$

A rare calcium silicate hydroxide noted from Ballycraigy, Larne, Co. Antrim, Ireland, by Anthony et al. (1995) and Gaines et al. (1997). These authors are probably referring to the earlier work of McConnell (1953), who describes calcium silicate hydrates (plombièrite and tobermorite) associated with scawtite in cavities within chalk/dolerite contact rocks at Ballycraigy. The hydrates have been preserved because of the compact and desiccating nature of enclosing larnite-bearing rocks. It appears that the Ballycraigy riversideite was produced only during dehydration experiments on the natural hydrates and so it remains a questionable species until it can be proved to exist in nature.

robinsonite?

monoclinic $Pb_4Sb_6S_{13}$

A rare lead sulphantimonite reported from epithermal quartz veins cutting nodular ignimbrite from Bwlch mine, near Deganwy, S of Llandudno, Gwynedd, Wales. A heat source to drive the hydrothermal system was probably a nearby contemporary high-level intrusion such as that at Penmaenmawr. Associated species are jamesonite, plagionite, semseyite, stibnite, zinkenite and possibly boulangerite. Quantitative data indicate an ideal robinsonite composition, but material was too sparse for XRD work, hence the species still requires verification (Bevins et al. 1988).

rockbridgeite

orthorhombic $Fe^{2+}Fe^{3+}_4(PO_4)_3(OH)_5$

A rare iron phosphate hydroxide that forms a series with frondelite and is found on the type specimen of andrewsite from Stowe's section of the Phoenix United mines, Linkinhorne, Cornwall. Barbosalite and goethite form the core of fibrous, radial spherules up to 4 mm across. The outer shell of these spherules was described as pale-green andrewsite, later shown to be a mixture of rockbridgeite and hentschelite (Dunn 1990).

Intermediate members of the frondelite–rockbridgeite series occur in iron-stained fracture veins within partially kaolinized granite at Gunheath china-clay pit, Carthew, St Austell, Cornwall. The mixed phosphate assemblage is complex and includes intermediate members of the chalcosiderite–turquoise series, barbosalite, chlorapatite and leucophosphite. The matrix is veined with chalcocite, quartz, fluorapatite and cassiterite (Elton 1997, 1998). In the Hensbarrow part of Gunheath china-clay pit, granitic pegmatites close to the biotite/granite contact and near to crosscutting mineralized quartz and quartz–tourmaline veins contain a suite of phosphate species that includes apatite, crandallite, goyazite, mitridatite, rockbridgeite and waylandite. The pegmatites are lithium-rich and contain accessory amounts of columbite–tantalite, fluorite, ilmenorutile, topaz and uraninite (Hawkes et al. 1987). At Gravel Hill mine, Perranzabuloe, Cornwall, rockbridgeite is found with beraunite, ferrostrunzite and siderite (Weiss 1989).

Two specimens of rockbridgeite that can be fitted together to make a single large sample are in the Kingsbury collection at the NHM. One specimen is labelled "Restormel mine, Lanlivery, Cornwall" and the other "East Wheal Russell, near Tavistock, Devon". A description of the latter specimen is given by Kingsbury (1957a). Ryback et al. (2001) have discredited the claimed localities and conclude that the specimens came from Lexington, Rockbridge County, Virginia, USA. Ryback et al. (2001) add that rockbridgeite has been independently verified from East Wheal Russell, where it forms black fibrous radiating crystals in vugs within iron-stained quartz. It is uncertain whether the specimens of rockbridgeite, claimed to have come from Wheal Phoenix (Phoenix United mines) by Kingsbury (1957a) and referenced by Braithwaite (1981), are correctly provenanced, but this may actually be an accurate description. However, serious doubt surrounds a specimen of rockbridgeite in the Kingsbury collection that is said to come from Burdell Gill, Caldbeck Fells, Cumbria, referenced by Young (1987a) and Cooper & Stanley (1990).

In Scotland, rockbridgeite has been found in small (15–30 cm) metamorphic segregation pods within striped kyanite–silliamanite-bearing Moine gneisses exposed near the entrance of Gleann Còsaidh, Loch Quoich, Highland Region. It is associated with apatite, garnet, graftonite, jahnsite, johnsomervilleite, mitridatite, phosphosiderite and vivianite (Livingstone 1980).

rock crystal

A colourless transparent gem variety of quartz.

rock milk

A soft efflorescent and fine-grain variety of calcite.

rock salt

A synonym of halite.

rock silk

TL Tayport, Fife, Scotland

A synonym of asbestos, a variety of amphibole.

rock soap

A synonym of montmorillonite, saponite and oropion (a clay similar in composition to halloysite).

romanèchite

monoclinic $(Ba,H_2O)_2(Mn^{4+},Mn^{3+})_5O_{10}$

A relatively uncommon hydrated barium–manganese oxide, often formed as a weathering product of manganese-bearing oxides, carbonates and silicates, and a principal component of psilomelane.

Both Anthony et al. (1997) and Gaines et al. (1997) note romanèchite from Restormel mine, N of Lostwithiel, Cornwall. Near by, at Pelyn Tor, 1 km E of Lanlivery, romanèchite has been found mantling botryoidal cryptomelane in fissured and fragmented quartz veinstone (D. W. Williams sample, Tindle unpub. EMP data).

Romanèchite is found in fracture-mineralized Permo-Triassic St Bees Sandstone

Delicate spray of romanèchite needles around a core of cryptomelane. Found as a fracture filling in loose blocks of quartz veinstone at Pelyn Tor, near Lanlivery, Cornwall; width 275 µm; BSE image. (© E. Condliffe, University of Leeds)

examined in borehole material from the Sellafield area, Cumbria, where it forms surface impregnations, and stains, thin films and dendritic coatings on fracture surfaces and bedding fissures. Romanèchite, and other manganese oxides (goethite, hollandite and todorokite) with which it is admixed, are derived mainly from the dissolution and oxidative decomposition of iron–manganese-rich cements and vein fillings in the calcite host rock (Milodowski et al. 1998). Romanèchite occurs at Remmington's veins on Great Dunn Fell, northern Pennines, Cumbria, where it had originally been called psilomelane. Pyrolusite forms a thin crystalline coating on the romanèchite (Young et al. 1996, Bridges & Young 1998). On the Caldbeck Fells of Cumbria, romanèchite occurs as a dull greyish-black dense-to-silky fibrous botryoidal species at Burdell Gill. Its lack of characteristic features makes it easy to overlook and it may be relatively abundant at this site (Neall & Leppington 2003). Romanèchite, forming a hard black coating on quartz, is reported from Brae Fell mine, Dale Beck Valley (Bridges et al. 2006). Many examples of black manganese minerals from Dry Gill mine, Caldbeck Fells, commonly labelled coronadite or psilomelane, have been confirmed to be romanèchite (Wolfe 2006). At Driggith mine, two specimens of black manganese oxide are confirmed to be romanèchite (Green et al. 2006a). Botryoidal romanèchite also occurs in dump material derived from a hematite vein at Nab Gill mine, Boot, Eskdale, Cumbria (Young 1984b, 1985e).

At the abandoned 19th-century manganese mine at Tŷ Coch, near South Cornelly, Mid Glamorgan, Wales, botryoidal and collform romanèchite and goethite have been found on dump material, where they infill cavities lined with white fibrous calcite. The ore is an interbedded sequence of iron- and manganese-rich layers resting conformably on Carboniferous Limestone and overlain by Keuper conglomerate. Other manganese-rich species found at Tŷ Coch include hausmannite, pyrobelonite and amorphous phases described as psilomelane and wad. Multi-stage deposition from solution and from a colloidal gel is suggested (Criddle & Symes 1977). Bevins (1994) reports romanèchite from Benallt mine, Rhiw, Lleyn Peninsula, Gwynedd.

In Scotland, romanèchite is briefly described from geodes within the Craigmuschat trachyte quarry, Gourock, Strathclyde. It is found in this environment with ankerite, barite, calcite, fluorite, goethite, gypsum, hematite, quartz and tourmaline (Nicholson & Durant 1986). At the Bridge of Don Industrial Estate, Aberdeen, Grampian, loose blocks containing black cryptocrystalline bands of romanèchite are found with manganite and pyrolusite in supraglacial till (Chew 1978, Nicholson 1989b), and at Pennan, near Gamrie, Grampian, romanèchite cements Old Red Sandstone sediments (Nicholson 1989b). Romanèchite has also been found in Kinharvie Burn and its tributaries, 6 km W of New Abbey, and near Criffell, Dumfries & Galloway. It occurs as a supergene species in manganese-rich veins cutting the Criffell granodiorite and it forms black botryoids, often coated in a thin brown layer of limonite associated with quartz. There is no doubt that the psilomelane described from this site by Heddle (1901) is actually romanèchite (McMullen & Todd 1990). An unconfirmed report of romanèchite from a manganese-rich vein found close to Arndilly House, Arndilly, near Craigellachie, 20 km SE of Elgin, Grampian, is given by Nicholson (1986, 1992). Coronadite and wad are major components of the ore, with minor hollandite, pyrolusite and possibly romanèchite also reported.

Botryoidal romanèchite encrusted with pyromorphite has been described from

Botryoidal romanèchite from near St Johns Head, Lead Geo, Hoy, Orkney, Scotland; specimen is 15×11 cm. (© The Trustees of the National Museums of Scotland)

the highest dump on the Luganure lode in Glendasan, Co. Wicklow, Ireland. The occurrence is associated with Cu–Pb–Zn mineralization within the Leinster granite (Ryback & Moreton 1993). More recently, coronadite has been found with romanèchite as mammillated and stalactitic brownish-black coatings in a quartz veinstone from the dumps of Old Luganure mine, on the northeastern side of Camaderry, Glendasan. Individual botryoids are compositionally zoned from coronadite cores to romanèchite rims (Moreton et al. 2006).

roméite?

cubic — STIBICONITE GROUP — $(Ca,Fe^{2+},Mn^{2+},Na)_2(Sb,Ti)_2O_6(O,OH,F)$

A rare complex antimonate briefly described from Hingston Down quarry, 2 km W of Gunnislake, Calstock, Cornwall (Golley & Williams 1995). Further confirmation is required.

römerite

triclinic — $Fe^{2+}Fe^{3+}_2(SO_4)_4 \cdot 14H_2O$

An hydrated iron sulphate reported from a railway cutting, W of Stanely, near Paisley, Strathclyde, Scotland. Römerite forms small clusters of tiny amber crystals associated with coquimbite, halotrichite, pyrite and voltaite (Macpherson & Livingstone 1982). Römerite is an oxidation product of pyritic ore.

In Wales, a chestnut-brown alteration mineral from Parys Mountain, Anglesey, was thought to be römerite (Bor 1950), but this requires confirmation.

rooseveltite

monoclinic — MONAZITE GROUP — $BiAsO_4$

A rare bismuth arsenate dimorphous with tetrarooseveltite and found in fractures within quartz veinstone on an old NHM specimen from Carrock mine, Caldbeck Fells, Cumbria. Rooseveltite forms grey glassy grains and is intergrown on a microscopic scale with another (less common) bismuth arsenate, preisingerite. The two species are associated with joséite and unidentified powdery yellow bismuth ochres (Rumsey 2007).

It has previously been reported that rooseveltite and zavaritskite may be present as supergene alteration products of arsenopyrite–bismuth–bismuthinite–fluorite–molybdenite mineralization in topaz greisen from the margin of the Eskdale granite at Water Crag, near Devoke Water, Cumbria (Young 1985d,e). Further confirmation is required.

roquesite

tetragonal — CHALCOPYRITE GROUP — $CuInS_2$

A copper–indium sulphide occurring as grains and aggregates, up to 100 μm across, in a bornite–chalcocite–wittichenite assemblage at Geevor mine, near Pendeen, St Just, Cornwall. It is not uncommon, especially in sulphide-rich material from the 16 level on Coronation lode (Criddle & Stanley 1993, C. J. Stanley pers. comm. 2003).

ROSASITE GROUP

Monoclinic or triclinic carbonates with the general formula $A_2(CO_3)(OH)_2$ or $AB(CO_3)(OH)_2$, where A and B = Co, Cu, Mg, Ni, Zn. Rosasite is the only member of the group found in Britain or Ireland.

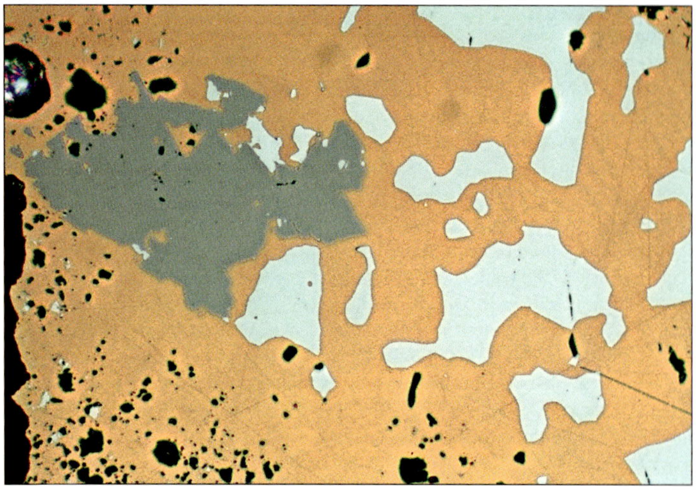

Dark-grey irregular inclusion of roquesite in bornite (orange-brown); light bluish-grey inclusions are chalcocite. Geevor mine, Pendeen, St Just, Cornwall; width 200 μm; viewed in reflected light. (© C. J. Stanley, The Natural History Museum, London)

rosasite

monoclinic — ROSASITE GROUP — $(Cu^{2+},Zn)_2(CO_3)(OH)_2$

A rare copper–zinc carbonate hydroxide found as a supergene mineral in the oxidized zone of copper–zinc deposits (sometimes with a post-mining origin).

Small green spherules of rosasite associated with aurichalcite and hemimorphite have been found in mineralized veins in the southeast corner of Greystone quarry, Lezant, near Launceston. It is unclear if rosasite is hosted in the Greystone lead lode, earlier-formed chalcopyrite–dolomite–quartz–sphalerite veins, or aragonite–sphalerite (rare bournonite) quartz veins (Elton et al. 2004). Rosasite has been described from Penberthy Croft mine, St Hilary, Cornwall, by Kingsbury & Hartley (1957c), but later analysis has shown it to be brochantite (Betterton et al. 2000).

In Staffordshire, rosasite has occasionally been found in fissure veins and replacement deposits at the small abandoned Waterbank Cu–Pb–Zn mine, Ecton Hill, Wetton, where it forms scattered bluish-green spherules or small botryoidal groups to about 5 mm^2, sometimes coated by aurichalcite or hydrozincite (Braithwaite & Ryback 1963a, Robey & Porter 1972, Rust 1995c). Rosasite is also recorded from the nearby Clayton mine on Ecton Hill (Ford & Sarjeant 1964a) and, elsewhere in the Southern Pennine orefield, the species occurs at six Derbyshire localities: Masson and Rutland Caverns and Nestus mine, Matlock Bath (at Nestus forming pale bluish-green to dark-green botryoidal crusts or globules up to 2 mm in diameter and as clusters of microneedles on fluorite or calcite); Red Rake mine, Calver (all Braithwaite & Ryback 1963a); Snelston mine, near Ashbourne (Jackman 1996); and Watersaw Rake, Longstone Edge (Ford & Sarjeant 1964a).

Rosasite is found at a few locations in the Northern Pennine orefield. At Coldstones quarry, 5 km W of Pateley Bridge, Yorkshire, traces of rosasite together with aurichalcite, hemimorphite and smithsonite (and less commonly cinnabar, doyleite, otavite and prosopite) occur in a supergene assemblage developed on Garnet vein, a 2 m-wide barite–calcite–fluorite–galena–sphalerite vein cutting Carboniferous Limestone (Young et al. 1989, 1997). Rosasite forms a green coating on barite, fluorite, goethite, pyrite and witherite at the Brandy Bottle incline on Friarfold vein, Swaledale, Yorkshire, where a post-mining origin is suspected (Dunham & Wilson 1985). Small turquoise-blue spherules of rosasite up to 1 mm across on barite have been found at Closehouse mine, SW of Middleton-in-Teesdale, Lunedale, Co. Durham, and appear to be the derived from the oxidation of chalcopyrite and sphalerite (Young et al. 1985a, Cooper 1991b, Bridges & Young 1998). Similar spherules of vivid turquoise-blue rosasite up to 0.2 mm across on quartz are noted from West Pasture mine, Stanhope, Weardale, Co. Durham (Young & Francis 1989). At Hilton mine, Scordale, Cumbria, rosasite occurs as blue-green crusts made up of spherulitic masses to 0.3 mm across on iron-stained fluorite matrix (Bridges & Green 2005b). Reported finds of rosasite on Grassington Moor, Yorkshire, from Coalgrove Head and Turf Pitts mine (Dunham & Wilson 1985), require further confirmation, as the discoveries were made by A. W. G. Kingsbury, who has a dubious reputation for provenancing samples (Ryback et al. 1998, 2001). For the same reason, the reports of rosasite in Cumbria (five localities in the Caldbeck Fells: Driggith mine, Potts Gill barite mine, Roughton Gill, Sandbed mine, and Silver Gill mine, near Roughton Gill; plus Grisedale mines, Patterdale, and Ruthwaite Lodge, Grisedale) by Kingsbury &

Spherules of rosasite up to 0.4 mm across, from Driggith mine, Caldbeck Fells, Cumbria. (© D. I. Green)

Hartley (1957c), and repeated by Hartley (1984) and Young (1987a), are questionable. Green et al. (2006a) state that some of Kingsbury's rosasite reported from Driggith mine is so unlike anything otherwise reported from the Caldbeck Fells that it is "likely to be fraudulently" provenanced. Rosasite noted from the Caldbeck Fells and Cornwall by Anthony et al. (2003) is probably based on Kingsbury's work. However, rosasite is widespread but uncommon at both Driggith and Sandbed mines, where it occurs in zinc-rich veinstone associated with adamite, hemimorphite and philipsburgite (Green et al. 2006a). Small turquoise-blue spherulitic masses of rosasite, up to 0.3 mm across, lining cavities in iron-stained quartz veinstone from Silver Gill have since been described by Green et al. (2005b).

Rosasite, aurichalcite and zinc-rich malachite are reported from calcite-lined cavities in Carboniferous Limestone at Bute quarry, Llantrisant, Mid Glamorgan, Wales, and subsequently the same assemblage was encountered at a depth of 35 m in a borehole at Locks Common, Porthcawl. The primary sulphide mineral assemblage at Bute quarry consists of galena and composite crystals of chalcopyrite and sphalerite, and it is the alteration of the latter two species that has allowed rosasite to form (Fletcher & Young 1988). At Hendy quarry, Miskin, near Llantrisant, less than 1 km from Bute quarry, rosasite is thought to have been found, but confirmation is required (Bevins 1994). Faults and joints within dolomitized Carboniferous Limestone are the host to galena and sphalerite (minor chalcopyrite and pyrite) mineralization at Machen quarry, 7 km ENE of Caerphilly, Mid Glamorgan. Extensive alteration has produced a wide range of secondary species, including bluish-green spheroids or botryoidal aggregates of rosasite up to 5 mm across in small cavities in barite gangue, sometimes overgrown by aurichalcite. Anglesite, bindheimite, cerussite, cinnabar, fraiponite, goethite, hematite, hemimorphite, hydrozincite, linarite, malachite, mattheddleite, scotlandite, schmiederite and smithsonite are other species found in the assemblage (Plant & Jones 1995). In the Dyfed part of the Central Wales orefield, bluish-green spherules of rosasite up to 5 mm across occur on hemimorphite, occasionally in association with lath-like aurichalcite at Frongoch mine (one of the most important lead–zinc mines), between Devil's Bridge and Pont-rhyd-y-groes. Rosasite was formed by supergene alteration of primary sulphides, but does not occur in the post-mining assemblage (Green et al. 1996b). Gwaithyrafon mine, SE of Tal-y-bont, Dyfed, hosts rosasite in chalcopyrite-rich quartz veins and breccias, in an unusual assemblage that also includes the following supergene species: agardite-(Y), bayldonite, brochantite, chrysocolla, connellite, native copper, cuprite, linarite, malachite, mimetite, olivenite, parnauite, pharmacosiderite, tyrolite and a few other unconfirmed species (Rust & Mason 1994).

In Ireland, rosasite forms bluish-green spherules up to 2 mm across associated with cerussite and hydrozincite at Sheshodonnell East mine, Co. Clare (Ryback & Moreton 1993).

roscherite?

monoclinic or triclinic $CaMn^{2+}_5Be_4(PO_4)_3(OH)_3{\cdot}2H_2O$

An iron-rich variety of this rare hydrated calcium–manganese beryllium phosphate hydroxide is reported from Clitters United mine (Gunnislake Clitters mine), Calstock, near Gunnislake, Cornwall. Roscherite commonly consists of dark sage-green spherules with a silky crystalline exterior ranging up to 5 mm in diameter, associated with colourless carbonate-fluorapatite-lined cavities in quartz. Flat radiating aggregates of roscherite (to 1 cm) are also reported on joint surfaces of iron-stained quartz (Clark et al. 1983). According to Chukanov et al. (2003) this is actually greifensteinite.

roscoelite

monoclinic MICA GROUP $KV_2\square AlSi_3O_{10}(OH)_2$

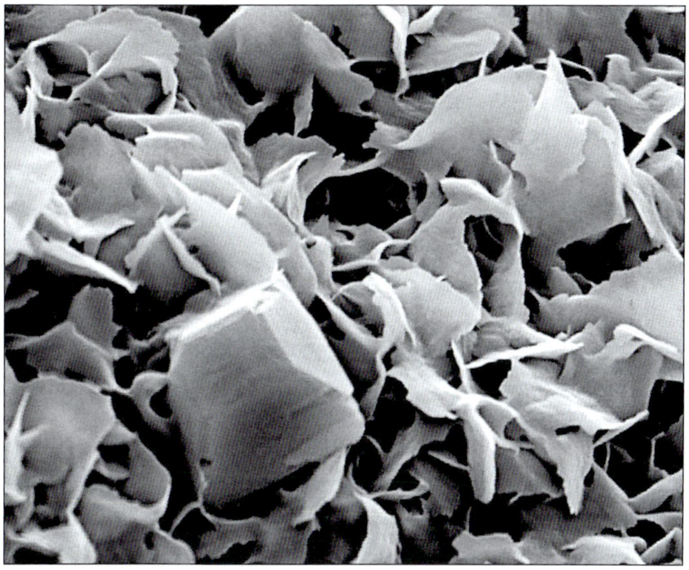

Curved plates of roscoelite associated with prismatic anatase (lower left), in a Devonian sandstone from Gamrie Bay, Moray Firth, Grampian, Scotland; width 18 µm; SEM image. (© J. Still, University of Aberdeen)

Named after Sir Henry Enfield Roscoe FRS (1833–1915), a London-born chemist who worked at the University of London and later at Owen's College, Manchester, and who was the first to prepare pure vanadium; born in London. Roscoelite is an uncommon potassium–vanadium aluminosilicate hydroxide found within white reduction spots in Devonian redbeds at Gamrie Bay, Gardenstown, 11 km E of Banff, Grampian, Scotland. It is brown, has extremely fine grain, and fills sandstone pores and the spaces where grains have decomposed. Additionally, roscoelite is seen to replace plagioclase in the altered rim of a granitic rock fragment and, as with all other occurrences, it is surrounded by a bleached zone. It is associated with feldspar, quartz and iron or iron–titanium minerals, together with detrital apatite, garnet, monazite, wakefieldite-(Ce), xenotime and zircon, and an unnamed vanadium oxide (van Panhuys-Sigler et al. 1996).

Roscoelite has been found in the Avoniel borehole, Belfast, Co. Antrim, Ireland, where it occurs as a constituent of minor uranium–vanadium mineralization in Permian–Triassic redbeds. Reduction spots in red mudrocks contain coffinite, roscoelite, an unnamed vanadium-rich detrital mica and a vanadium-rich oxide. Mineralization may have occurred by the mixing of meteoric and saline connate fluids during early diagenesis (Parnell 1988b).

A tentative identification of roscoelite, again in Permian redbeds, has been made in the Littleham Cove area, Budleigh Salterton, Devon. Here roscoelite occurs in concretions that may also contain algodonite, bornite, chalcocite, chalcopyrite, coffinite, native copper, covellite, lavendulan, malachite, maucherite, nickeline, rammelsbergite and native silver, and possibly modderite (Harrison et al. 1973, 1975b, Milodowski et al. 2000).

ROSELITE GROUP

Monoclinic arsenates with the general formula $Ca_2X^{2+}(AsO_4)_2{\cdot}2H_2O$ where X^{2+} = Co, Mg, Mn, Zn. Brandtite is the only member of the group described from Britain or Ireland – a discovery reported by A. W. G. Kingsbury. Later workers have shown that the provenance of many Kingsbury samples is now in doubt (Ryback et al. 1998, 2001) and further confirmation is therefore required.

rose quartz

A pink variety of quartz.

rosiaite
trigonal $PbSb^{5+}_2O_6$

A rare lead antimonate found in mineralized nodular ignimbrite at Bwlch mine, near Deganwy, S of Llandudno, Gwynedd, Wales. Rosiaite was found in a cavity a few mm across, lined with a thin light orange-brown crust coating stibnite and bindheimite. The surface of the crust consists of rounded platelets $<50\,\mu m$ across. The less common associated species confirmed at Bwlch mine are anglesite, jamesonite, plagionite, semseyite, stibnite and zinkenite, all occurring in quartz patches and veinlets (Ryback & Francis 2001).

rossmanite
trigonal TOURMALINE GROUP $\square(Li,Al_2)Al_6(BO_3)_3Si_6O_{18}(OH)_4$

Pale bluish-green rossmanite sprays with schorl compositions in darker areas; Dorothy china-clay pit, St Austell, Cornwall; width 4 cm. (© D. I. Green)

A rare sodium–lithium–aluminium borosilicate hydroxide, usually found in lithium-rich pegmatites, but in Britain found in a lithium-mica granite at Dorothy china-clay pit, St Austell, Cornwall. Rossmanite occurs in an unusual sample of tourmaline–quartz veinstone that proved to contain various tourmaline species. They represent the final crystallization products of a complex magmatic–hydrothermal crystallization/recrystallization sequence that saw foitite form early, followed by schorl, fluor-schorl, elbaite and rossmanite. Rossmanite forms sprays of pale-green crystals and compositionally is very unusual in being enriched in magnesium (N. Hubbard specimen, Tindle unpub. EMP data). Grigg (1986) reports green tourmaline from this locality collected in 1984, which may be similar material.

rosstrevorite
TL Rosstrevor, Co. Down, Ireland

A fibrous stellate variety of epidote (Greg & Lettsom 1858).

roxbyite
monoclinic Cu_9S_5

A rare copper sulphide found as an important component of colloform supergene copper-sulphide mineralization in Silurian limestone at Dolyhir quarry, Old Radnor, near Presteigne, Powys, Wales. Roxbyite occurs as: extremely fine lamellar intergrowths with chalcopyrite, covellite, djurleite and spionkopite; massive to bladed crystals containing bornite inclusions with feathery covellite; and intergrowths with covellite in djurleite-rich veinlets (T. F. Cotterell pers. comm. 2007).

ROZENITE GROUP

Monoclinic sulphates with the general formula $A^{2+}(SO_4)\cdot 4H_2O$, where A^{2+} = Co, Fe, Mg, Mn, Ni, Zn. Rozenite is the only member of the group found in Britain or Ireland.

rozenite
monoclinic ROZENITE GROUP $Fe^{2+}SO_4\cdot 4H_2O$

An uncommon hydrated iron sulphate found encrusting pyrite-rich coal seams in the Oughterside opencast coal site, near Aspatria, Cumbria. Thin white saccharoidal crusts of rozenite, <0.2 mm thick and 5 mm across, occur with copiapite, epsomite, gypsum, jarosite and pickeringite. The minerals occur as encrustations and efflorescences on joint surfaces and on the walls of old workings in coal seams near the outcrop (Young & Nancarrow 1988b).

Jenkins et al. (2000) records rozenite as one of 25 post-mining secondary species found at the Parys Mountain copper deposit, Anglesey, Gwynedd, Wales.

At the West Mains coalmine, West Calder, Lothian, Scotland, porcelain white patches of rozenite replace a greenish melanterite stalactite and, at Howcommon Limestone mine, Kilmarnock, Strathclyde, bluish-green crystals of rozenite (<6 mm), partly covered with white encrustations, are found in a laminated shale (Macpherson & Livingstone 1982). Additionally, rozenite is noted from Slateford, near Edinburgh (Anthony et al. 2003).

ruarsite
monoclinic ARSENOPYRITE GROUP RuAsS

A very rare ruthenium sulpharsenide found in the Ben Buie peridotite intrusion at Craig, Isle of Mull, Strathclyde, Scotland. Ruarsite occurs in, or adjacent to, chromitite layers in association with froodite, laurite and sperrylite, and minerals of the hollingworthite–irarsite–platarsite series. Grainsize of all species is 1–3 µm (Pirrie et al. 2000a). In the Outer Layered Series of the Cuillin complex, Isle of Skye, Highland Region, the layered peridotite contains ruarsite in an assemblage similar to that of Mull (although froodite is absent), but with the rare species occurring as inclusions in chromite or in more common sulphides such as chalcocite or pentlandite (Pirrie et al. 2000a, Power et al. 2003).

rubellite
A pink variety of elbaite, tourmaline group.

rubinglimmer
A synonym of lepidocrocite.

rubislite
TL Rubislaw, Aberdeen, Grampian, Scotland

A doubtful member of the chlorite group. XRD analysis indicates an iron-rich saponite, smectite group.

ruby blende
An ambiguous name used to describe deep-red sphalerite and zincite.

ruby silver
A synonym of proustite or pyrargyrite.

rucklidgeite
trigonal $(Bi,Pb)_3Te_4$

A very rare bismuth–lead telluride found in the Leadhills area, close to the Strathclyde/Dumfries & Galloway border, Scotland. A single alluvial gold grain is reported to contain rucklidgeite inclusions (Leake et al. 1998).

russellite
orthorhombic Bi_2WO_6

TL Castle-an-Dinas wolfram mine, St Columb Major, Cornwall, England

Named after Sir Arthur Edward Ian Montagu Russell (1878–1964), a mineral collector of Swallowfield Park, Reading, Berkshire, who discovered the mineral. Arthur Russell began mineral collecting at seven years of age and, during his lifetime, built up the finest and most comprehensive private collection of British minerals ever made. He served as president of the Mineralogical Society of Great Britain and Ireland, was made a Member of the British Empire, and was awarded an honorary DSc by the University of Oxford. On his death he bequeathed his superb collection (about 12 000 specimens) to the British Museum (Natural History) (Kingsbury 1966, Starkey 1988c). Russellite is a bismuth–tungsten oxide that has formed as an alteration product of native bismuth. It has fine grain, is compact, and forms pale-yellow rounded fragments (averaging 7 mm across), and because of its high density it was first seen in jig concentrates in the mine mill. The mine worked a wolframite-bearing quartz lode, and russellite was found associated with native bismuth, bismuthinite and wolframite from a single stope. Associated species in the lode and surrounding greisen include topaz, tourmaline and zinnwaldite, with minor arsenopyrite, cassiterite, native copper, cuprite, scorodite, turquoise, wavellite and rare native gold (Hey et al. 1938, Russell 1944).

In Cumbria, russellite and other rare supergene species (bismutoferrite, cuprotungstite, eulytite, mixite and namibite) have been found in chalcopyrite–ferberite–scheelite veins within the southeastern margin of the Silurian Eskdale granodiorite at Buckbarrow Beck, Corney Fell, Waberthwaite, Cumbria. Russellite is locally abundant in the veins and it forms thin (< 1 mm) pale-buff to greenish-yellow discontinuous crusts on the joint and fracture surfaces of massive quartz. In places, spherulitic masses up to 1 mm across are found, in which a radiating crystalline texture, sometimes with concentric colour banding, can be distinguished. In view of the presence of the rare species, a rescue collection of the site was made in 1987 and the top metre of the outcrop was removed for future researchers; the vein is still exposed *in situ* (Young et al. 1986, 1991b, Neall et al. 1993).

Webb et al. (1992) identified russellite as a secondary species forming irregularly shaped masses that occupy cavities in silicified wallrocks associated with greisen, near a hübnerite-bearing quartz vein cutting zinnwaldite granite at Gairnshiel Bridge, Glen Gairn, Grampian, Scotland. Russellite appears to have formed from the decomposition of cosalite and hübnerite.

Spherules of russellite to 0.7 mm on wolframite. Hemerdon Bal, Plympton, Devon. (© D. I. Green)

rustumite

monoclinic $Ca_{10}(Si_2O_7)_2(SiO_4)Cl_2(OH)_2$

TL Kilchoan, Ardnamurchan, Highland Region, Scotland

A rare calcium silicate with chloride and hydroxide, found at the type locality in nodular or concretionary limestones that have been thermally metamorphosed by dolerite and gabbroic intrusions. Rustumite occurs as tabular crystals up to 2 mm long in a zone between spurrite-dominant and kilchoanite-dominant assemblages. Crystals are colourless and, in thin section, rustumite is similar in appearance to cuspidine (Agrell 1965). Anthony et al. (1995) report a second occurrence of rustumite at Carlingford, Co. Louth, Ireland.

rutheniridosmine

hexagonal (Ir,Os,Ru), with Ir dominant

A rare iridium-rich metal found during an examination of alluvial gold grains from Chapel Burn and Cloan Burn, both about 5 km from the Glen Devon youth hostel, Ochil Hills, Tayside Region, Scotland. Three light-grey grains (< 0.2 mm) of rutheniridosmine and grains of mercury-rich silver were discovered, but unfortunately the source of these species could not be determined (Davidson 1995b).

ruthenium

hexagonal Ru

A very rare native alloy found at Corrycharmaig, Glen Lochay, NW of Killin, Tayside Region, Scotland, in a small body of metamorphosed and altered ultramafic rocks containing nickel-rich mineralization within chromitite horizons. Accessory amounts of the native species gold, iridium, osmium, ruthenium and silver, together with galena, gersdorffite, irarsite and laurite, and unnamed alloys of As, Co, Fe, Ni, Os, Pd, Pt, Ru, Sb and Sn occur either as inclusions (up to 40 μm across) in chromite or associated with the nickel sulphides heazlewoodite and nickeline (usually ≤ 2 μm). The chemistry and association of the PGMs is strongly indicative of an ophiolite assemblage and it is most likely that the intrusion represents a fragment of oceanic crust (Power & Pirrie 2000).

rutherfordine

orthorhombic $UO_2(CO_3)$

Named after Ernest Rutherford FRS (1871–1937), 1st Baron Rutherford of Nelson, an atomic physicist and prominent worker in radioactivity, who was Macdonald Research Professor of Physics at McGill University, Montreal, Canada; he held the Langworthy Chair of Physics at the University of Manchester, and later the Cavendish Chair of Physics at the University of Cambridge. He became a Nobel laureate for chemistry in 1908 and was author (or co-author) of several books, including *Radioactivity* (1904) and *Radiations from radioactive substances* (1906); born near Nelson, New Zealand.

Rutherfordine is a rare uranyl carbonate found at Loe Warren Zawn, W of Botallack, St Just, Cornwall, where surface water draining from rubble at the base of the cliff trickled over veinlets carrying chalcocite, chalcopyrite and pyrite, and precipitated an assemblage of supergene carbonate and phosphate minerals on exposed surfaces and joint planes near a patch of uraninite mineralization. Waxy rutherfordine forms dull green (or rarely brown) crusts, spherulitic aggregates and fan-shape groups of lath-like crystals up to 0.2 mm long associated with chalcocite and

Spray of rutherfordine crystals from Loe Warren Zawn, W of Botallack, St Just, Cornwall; width 120 μm; SEM image. (© N. J. Elton)

Bright-yellow spray of fibrous rutherfordine from South Terras mine, at St Stephen-in-Brannel, near St Austell, Cornwall; width 650 μm; SEM–BSE image (© J. C. Ø. Andersen, Camborne School of Mines)

uraninite, or less frequently with boltwoodite, cuprite, trögerite or vandendriesscheite (Elton & Hooper 1995). Further confirmation is desirable for reports of rutherfordine from South Terras mine, St Stephen-in-Brannel, near St Austell, Cornwall (James 1947, Robson 1949b), and from schists at Boherkyle, near Graiguenamanagh, Co. Kilkenny, Ireland (O'Connor 1986, Ryback & Moreton 1993).

RUTILE GROUP

Tetragonal oxides with the general formula $X^{4+}O_2$, where X^{4+} = Ge, Mn, Pb, Si, Sn, Te, Ti. Cassiterite, plattnerite, pyrolusite and rutile are members of the group found in Britain or Ireland (ilmenorutile in not a member of this group).

rutile — RUTILE GROUP

tetragonal — TiO_2

A very common titanium oxide, trimorphous with anatase and brookite, and found as a high-temperature accessory mineral in igneous rocks, in hydrothermally altered rocks, in schists, gneisses and contact-metamorphosed limestone. The occurrence of rutile is often related to the breakdown of primary iron–titanium oxide minerals such as ilmenite. It also occurs as a detrital mineral in sedimentary rocks and alluvial deposits. See also strüverite.

England

Heavy-mineral assemblages collected from the Camel, Fowey and Gannel estuaries in Cornwall contain small amounts of rutile in cassiterite-enriched intertidal sediments contaminated by detritus from the many nearby copper and tin mines (Pirrie et al. 2000b, 2002). The Millstone Grit in Yorkshire, particularly some shale units, contains abundant detrital rutile forming light-yellow, through deep-amber and brown, to "foxy-red" crystals. Anatase, barite, brookite, garnet, ilmenite, magnetite, monazite, pyrite, topaz, tourmaline, xenotime and zircon are associated species found in heavy-mineral residues from these rocks. Localities include: Bowland Knotts, near Slaidburn; Bramley Fall quarry, Newlay; Clayton Wood and Hawkesworth Wood quarries, Horsforth; Earl's Crag and Netherwood Plantation quarry, Silsden; East Keswick, Bardsey; Fountains Fell; Grassington Moor; Great Almescliffe Crags, near Weeton, Wharfedale; Great Whernside, Kettlewell; Grimbald's Crag and Plumpton Rocks, Knaresborough; Ingleborough; Keighley Moor; Old Cote Moor, Arncliffe; Penyghent; Rumbles Moor, above Ilkley; Snosty quarry, Bardsey; Steeton Moor; Tannery borehole, Meanwood, Leeds; the Strid, Bolton Woods, Wharfedale; and Whitehall quarry, Butcher Hill, Horsforth (Gilligan 1919). Detrital rutile is also found in Cretaceous and Eocene sands at: Faringdon, Berkshire; Holden's Wood, Brentwood, Essex; Hayes Common and Plumstead, Kent (now Greater London); Kennel Wood, Shirley, Surrey (now Greater London); Northwood, Middlesex (now Greater London); and Shotover Hill, Oxfordshire (Davies 1915). Gallagher (1974) reports detrital rutile in Northumbrian beach sands at: Alnmouth; Budle Bay, near Bamburgh; and Hauxley Haven, near Amble. Abundant authigenic and detrital rutile is found with anatase and hematite in Triassic redbeds (St Bees Sandstone) at St Bees Head, SSW of Whitehaven, Cumbria (Ixer et al. 1979b). Detrital rutile is also noted in sandstone at Langhorn quarry, Bigrigg, and Overend quarry, near Whitehaven (Lewis 1931). Rutile is a detrital species in redbeds adjacent to copper–lead mineralization on Alderley Edge ridge (Ixer & Vaughan 1982) in Cheshire.

Cornwall At Lanterdan quarry, near Tintagel, Cornwall, very fine hair-like needles of golden rutile up to 5 mm long have been found with albite, monazite and quartz along joint planes within slate (Bowman 1900, Rudler 1905, Cooper 1991b). Brilliant prismatic crystals of rutile embedded in transparent quartz were also found in joints at Delabole slate quarry, Cornwall (Garby 1848, Rudler 1905).

Rutile with a different paragenesis occurs at a few locations near St Austell, where it is associated with the Hercynian granite complex. A niobium–tantalum-rich variety of rutile transitional into ilmenorutile compositions has been recovered from the waste at Goonvean and Rostowrack china-clay pits, W of Nanpean (Scott et al. 1998). At Wheal Remfry china-clay pit, near Indian Queens, rutile occurred in cavities within a 1 m-wide vertical quartz–topaz–tourmaline vein. Crystals up to 2 mm long have a blocky to pseudo-hexagonal form and some groups develop a reticulated branching habit (Hacker 1997). Rutile, cassiterite and hematite are present in tin-rich calc-silicate rocks at the northwestern margin of the St Austell granite complex, near Gaverigan and Treliver (Camm & Dominy 1997). Rutile is also found in the granite itself (biotite-, lithium-mica- and tourmaline-rich units) as an accessory species, where it occurs with apatite, monazite, uraninite and zircon (Manning et al. 1996). Other locations for rutile include: with graphite, prehnite and zoisite in albite–quartz-(±dravite) adinoles at Dinas Head, near Padstow (Agrell 1939, 1941); brown needle-like crystals up to 2 mm long in malachite and pyromorphite-bearing veinstone at Greystone quarry, Lezant, near Launceston (Elton et al. 2004); minor occurrence in cassiterite-rich lodes at Mount Wellington mine, Twelveheads, 7 km W of Truro (Kettaneh & Badham 1978); minor occurrence in the sanidine-dominated minette at Pendennis Point, Falmouth (Hall 1982); in a cassiterite-rich tourmalinite breccia assemblage at Wheal Jane, Kea, near Truro (Holl & Bromley 1988).

Cumbria Small dark-brown anhedral grains up to 2 mm across in veins within hornfelsed slate at Bowness Knott and in quartz veins at Brown How, both Ennerdale (Young 1987a); extensively developed as an alteration product of ilmenite at the contact between gabbro and wolframite-bearing quartz veins at Carrock mine, Grainsgill, Caldbeck Fells (Fortey 1978, 1980); an accessory species in the Skiddaw granite at Sinen Gill (Rastall & Wilcockson 1915, Hitchen 1934) and associated greisen at Grainsgill (Eastwood et al. 1968); irregular grains and subhedra (<100 μm) in a pyrrhotite-rich vein and adjacent chloritized Skiddaw Slates, and <20 μm at Dale Head mine, Newlands Valley (Stanley & Vaughan 1980); irregular grains in fragments of altered country rocks within mineralized veins at Eagle Crag mine, Patterdale (Stanley 1979); twinned detrital grains in silicified argillaceous limestone close to the marcasite–pyrite-rich Great Sulphur vein, Alston Moor (Vaughan & Ixer 1980); radiating bunches of fine needles (<100 μm in length) or subhedral grains (<30 μm) in quartz within an apatite–chlorite–quartz vein at Scar Crag cobalt mine, Causey Pike (Ixer et al. 1979a); black to dark-brown reticulated mats, typically associated with anatase and brookite, in miarolitic cavities within granite at Shap granite quarry (Wirth & Green 2002); forming clusters of minute euhedra enclosed within quartz crystals in greisen associated with the Eskdale granite at Taw House (Young 1985d); in altered slate and along vein margins at Wanthwaite mine, St John's in the Vale, near Threlkeld (Stanley 1979); forming needles in metamorphosed Silurian flags at Wasdale Beck, Shap (Harker & Marr 1891).

Devon Rutile is a rare constituent in tourmaline nodules within the Dartmoor granite on Hamel Down (Brammall & Harwood 1925) and at Lee Moor china-clay pit, near Wotter, southwest Dartmoor, it is an accessory component in coarse-grain tourmaline-muscovite granite (Knox & Jackson 1990). Secondary rutile, with very rare anatase, is unusually abundant in severely pneumatolyzed tourmaline-bearing granite exposed in the quarry below the reservoir dam on Holne Moor, Dartmoor (Brammall & Harwood 1927). The Tertiary Lundy granite contains accessory amounts of rutile together with anatase, apatite, beryl, cassiterite, columbite, cordierite, epidote, fluorite, garnet, monazite-(Ce), pyrrhotite, titanite, topaz, tourmaline, uraninite, xenotime-(Y), zircon and zoisite (Thorpe et al. 1990).

Leicestershire In the Mountsorrel area, rutile is a widespread accessory species in granodiorite and gabbro, where it forms honey-yellow to reddish-brown prismatic crystals up to 0.15 mm long. Localities include: Cocklow quarry; Kinchley; Main Mountsorrel quarry; Rubbish Tip quarry; and Swithland reservoir (Taylor 1934).

Wales

Clwyd Rutile occurs in alpine-type veins (produced during low-grade regional metamorphism) at Hendre quarry, Pandy, near Glyn Ceiriog (Starkey et al. 1991).

Dyfed In the Central Wales orefield, rutile is a rare species, but has been noted from Brynyrafr mine, near Nant-y-moch reservoir, and Darren mine, Penbontrhydybeddau (Bevins 1994). Transparent-to-translucent golden prismatic crystals of rutile 1–2 mm long have been found with chlorite and quartz in thin veins cutting rhyolite at Treffgarne Rocks, Haverfordwest (R. E. Starkey pers. comm. 2005).

RUTILE

Gwynedd Rutile is the least abundant of the TiO_2 polymorphs at Fron Oleu, Prenteg, Tremadog, where it forms acicular silvery-yellow crystals associated with chamosite and quartz in veins cutting slaty mudstones of Ordovician age (Starkey & Robinson 1992). Hair-like rutile inclusions in quartz up to 20 mm long also occur in quartz veins at Tanygrisiau, Blaenau Ffestiniog (Green & Middleton 1996). Altered volcanic rocks adjacent to the Parys Mountain copper deposit, Anglesey, contain 10 µm box-works or laths of anatase and rutile replacing primary iron–titanium oxides. Additionally, rutile forms 40–60 µm twinned tabular crystals intergrown with or enclosed by chalcopyrite and pyrite (Pointon & Ixer 1980). The rare tantalum-rich variety of rutile (strüverite) is reported from an altered microgranite at Y Llymllwyd, Nant Ffrancon, where it is associated with brookite and gibbsite (Bevins 1994).

Scotland

Rutile occurrences in Scotland are widespread and of varied paragenesis. Many Caledonian granites throughout Scotland have been found to contain accessory amounts of rutile (Mackie 1928). Other occurrences include the following.

Borders Rutile inclusions in native gold from alluvial deposits in Craigierigg Burn (Chapman et al. 2000c).

Central Rutile inclusions in pyrite within Dalradian stratabound exhalative copper–zinc mineralized horizons and their unmineralized chloritic schist and albitic quartzite hosts at Creag Bhocan, SSW of Tyndrum (Fortey & Smith 1986, Scott et al. 1988); silicified and brecciated Dalradian quartzites E of Tyndrum are cut by gold–silver-bearing quartz veins and a network of pyrite veinlets containing rutile and sericite (Pattrick et al. 1988).

Dumfries & Galloway Rutile is an accessory species in argillaceous xenoliths entrained in the marginal units of the Cairnsmore of Carsphairn granite complex (Deer 1937b).

Grampian Rutile forms inclusions in native gold from alluvial deposits in Back Burn, W of Auchenblae (Chapman et al. 2000c). It is found with actinolite, biotite, margarodite, pyrite and pyrrhotite in limestone quarries between Glenbuchat and Glen Nochty; as bright-red crystals in crystalline limestone inland from Redhythe, W of Portsoy; in beryl-bearing granitic veins cutting gneiss at Torry, Aberdeen (all Heddle 1901). A niobium–tantalum-rich variety of rutile occurs in zinnwaldite granite at Gairnshiel Bridge, Glen Gairn (Webb et al. 1992); and it is associated with almandine, anatase, cassiterite, chrysoberyl, columbite, fergusonite, ilmenorutile, monazite, spessartine, topaz, xenotime and zircon in panned concentrates from Luibeg Burn, Glen Lui, Cairngorms (Beer & Bennett 1988, Haynes 1993).

Highland Rutile occurs with chalcopyrite, molybdenite and pyrite in stockwork mineralization at Gleann a' Chaolais, S of South Ballachulish in the Ballachulish igneous complex (Evans 1977, Fortey 1980), and less commonly in metamorphosed siliceous dolomites surrounding the pluton (Ferry 1996); rutile is a constant accessory species in Lewisian eclogitic rocks in the Glenelg district, such as at Totaig Ferry and on the hills W of the Postman's Path to Ardintoul (Teall 1891a, Alderman 1936); eclogite at Letterfairn and Sandaig, Glenelg, contains small anhedral rutile inclusions in garnet (Brewer et al. 2003); eclogite exposed NW of Fern Villa, southwest shore of Loch Duich, Glenelg, contains blebs of dark-brown rutile, often in lamellar intergrowth with ilmenite (Mercy & O'Hara 1968); on the west side of Loch Duich (along the road to Glenelg) and on the east side of Loch Duich (along the road to Kyle of Lochalsh), rutile occurs in electrum–hessite-bearing quartz veins associated with the Ratagain diorite–syenite intrusion (Alderton 1986, 1988b); in the heavy-mineral fraction of drill sludges from the Allt na Muic borehole, Helmsdale (Gallagher et al. 1971); as inclusions in native gold found in alluvial deposits in Kildonan Burn (Chapman et al. 2000c); calc-silicate pods in Moinian pelite contain accessory rutile included within titanite at Loch Eilt, near the Sound of Arisaig (Tanner & Evans 2003); rutile partially altered to, or overgrown by, ilmenite occurs in the carbonate-enriched margins of a basic dyke on Sròn Bheag headland, Ardnamurchan, W of Kilchoan (Wells et al. 1974); conspicuous rutile, graphite and pyrite occur in kyanite-rich mica schists in Glen Urquhart, W of Loch Ness, near Drumnadrochit (Rock et al. 1986b).

Strathclyde Dark-brown to yellow prisms, pyramids and twinned crystals of rutile occur in aegirine–arfvedsonite microgranite on Ailsa Craig, Firth of Clyde (Harrison et al. 1987); abundant minute needles of rutile occur with corundum and mullite in buchite from Nuns' Pass, 2 km SW of Carsaig, Isle of Mull (Bowen et al. 1924).

Tayside Superb specimens of rutile have been collected from the west shore of Lochan na Lairige, NNE of Killin, where the contact of the Dalradian Ben Eagach Schist with the Ben Lawers Schist is exposed. Sulphide-rich horizons contain lustrous red-to-pink crystals of rutile (<7 cm) in carbonate pods within schist and, in chlorite lenses, dark-red to black prisms are found (Green & Todd 2001, Livingstone 2002); rutile is a widespread but minor component of all rock types at Foss, near Aberfeldy, where syngenetic barite–sphalerite mineralization occurs in Middle Dalradian graphitic schists of the Ben Eagach Schist Formation (Coats et al. 1980, 1984, Fortey & Beddoe-Stephens 1982, Willan & Coleman 1983, Moles 1985); ilmenite, muscovite, quartz and rutile are common inclusion species in zoned garnet porphyroblasts within Dalradian pelites at Grandtully, near Pitlochry (Sivaprakash 1981).

Striated prisms of rutile on quartz; Lochan na Lairige, NNE Killin, Tayside Region, Scotland; width of specimen 8.5 cm. (© M. P. Cooper)

Shetland Rutile occurs rarely and with well crystallized faces in muscovite (margarodite) schist at Vanlup, Hillswick, Mainland, and also in schist on the Isle of Burra (Greg & Lettsom 1858, Heddle 1901).

Ireland

Prismatic brown crystals of rutile with striated faces occur with small black crystals of the variety nigrine within mica schist at Cush-cum-Curragh, near Borrishdale, Clew Bay, Co. Mayo, Ireland (Rudler 1905). Coarse-grain schist belonging to the pre-Caledonian Erris complex at Scotchport Bay, NW of Belmullet, Co. Mayo, also contains rutile together with biotite, garnet, kyanite, chromium-rich mica, microcline, plagioclase, quartz, serpentine, sillimanite, talc and zoisite (Max et al. 1983).

Accessory amounts of apatite, rutile and titanite occur in microgranite belonging to the Tullow Unit of the Leinster granite batholith at Ballingen, between Tinahely and Aughrim, Co. Wicklow. Locally this unit is host to greisen-type mineralization (Steiger & Bowden 1982, 1986). Elsewhere in the Leinster granite, rutile appears to be widely distributed (but rare), forming yellow or dull-brown grains (<0.2 mm long) and lath-shape or stumpy prisms (Smithson 1932). A single grain of tantalum-rich rutile (var. strüverite), 250 µm across and partially mantled by microlite, was found in a panned concentrate of alluvium collected at Tombreen Bridge, E of Aclare, Co. Carlow. Strüverite is often found in lithium-enriched granite pegmatites, and the source of the grain is most likely to be the nearby Aclare spodumene pegmatite (R. A. Ixer sample, Tindle unpub. EMP data).

Rutile is a minor component of the Upper Portnoo Limestone found along the northern margin of the Ardara granite complex, Co. Donegal (Akaad 1956), and is also found in cordierite–enstatite–hematite–magnetite–mullite–tridymite buchite (porcellanite) at the contact with an olivine-dolerite plug at Tievebulliagh, near Cushendall, Co. Antrim (Agrell & Langley 1958). Greg & Lettsom (1858) note large prisms of rutile from Malin Beg, and small crystals at Dunfanaghy, both Co. Donegal.

Offshore territories

In Jersey, minute black prisms of rutile have been found in a thin pegmatitic vein cutting rhyolite at Bonne Nuit Bay (Mourant 1961). The southeast quarry at L'Etacq in northwest Jersey hosts polymetallic mineralization in granite in which large rutile crystals and abundant zircon are recorded (Ixer 1980). At La Corbière in the S of the island, brown twinned rutile (30–70 µm diameter) is intergrown with magnetite in chlorite–quartz veins cutting granite (Ixer 1980), and, at La Moye, rutile is briefly noted from mineralized aplogranite (Ixer 1988).

S

sadanagaite?

AMPHIBOLE GROUP

monoclinic $NaCa_2[Fe^{2+}_3(Fe^{3+},Al)_2]Si_5Al_3O_{22}(OH)_2$

A rare sodium–calcium–iron aluminosilicate hydroxide that forms a series with magnesiosadanagaite and is reported from skarn deposits near the western margin of the Land's End granite at the Crowns and Gryll's Bunny, Botallack, St Just, Cornwall. Amphibole-rich veins cutting metapelite and metabasalt host rocks contain amphibole with compositions transitional between pargasite and sadanagaite (van Marcke de Lummen 1985). Sadanagaite is considered to be a potassium-rich species by Anthony et al. (1995) but a sodium-rich species by Leake et al. (1997a). The Cornish example is intermediate between the two, but, according to the reclassification of the amphiboles (Leake et al. 1997a, Mogessie et al. 2004), the composition should now be referred to more correctly as aluminosadanagaite.

saddlebackite

hexagonal $Pb_2Bi_2Te_2S_3$

A very rare lead–bismuth sulphotelluride found with aleksite, chalcopyrite, galena, hessite, pyrrhotite, tellurobismuthite, tetradymite and an unnamed phase (corresponding to $PbBi_4Te_4S_3$) in a calcite–chlorite–quartz matrix at Clogau mine, Bontddu, near Dolgellau, Gwynedd, Wales. The single grain of saddlebackite is intergrown with aleksite (Cook et al. 2007).

safflorite

LÖLLINGITE GROUP

orthorhombic $CoAs_2$

A rare cobalt arsenide, dimorphous with clinosafflorite. Safflorite has been reported from the Littleham Cove area, near Budleigh Salterton, Devon (Milodowski et al. 2000), at Roskrow United mine, Ponsanooth, Cornwall (Ball et al. 1982), at the Coniston copper mines, Coniston, Cumbria (Green & Briscoe 2002), and at Tynebottom mine, NW of Garrigill, near Alston, Cumbria (Ixer et al. 1987). The last description is based on samples from the Kingsbury collection at the NHM and the provenance requires further confirmation. Ryback et al. (1998, 2001) have shown that some samples in this collection have been incorrectly provenanced and their status is yet to be ascertained.

Safflorite has been reported from an old mine dump at Silver Glen, Alva, Ochil Hills, Central Region, Scotland, where it is described as tin-white patches associated with erythrite and native silver in barite gangue (Livingstone 1993, 2002). However, Moreton et al. (1998) point out that the clinosafflorite they have identified at Silver Glen is easily mistaken for safflorite, and other occurrences reported here may similarly be in doubt. The work of Moreton (2007) suggests that safflorite does not occur at Alva – it all appears to be clinosafflorite.

At Tyndrum, Central Region, a uranium-rich vein, postdating the main sulphide mineralization, contains cobalt-rich safflorite associated with chalcopyrite, galena, hematite, marcasite, sphalerite and uraninite (pitchblende), all in a pink calcite gangue. The sulphides occur mainly in vuggy cavities (Darnley 1962b, Pattrick 1985). Primary safflorite with chalcopyrite, native bismuth, bismuthinite, millerite, molybdenite, nickeline, rammelsbergite and uraninite are also noted in uranium-rich veins along the Solway coast at Needle's Eye, SW of Caulkerbush, Dumfries & Galloway (Green & Todd 2001). Out of a total of 1794 alluvial gold grains examined by Leake et al. (1998) from the Leadhills area along the Strathclyde/Dumfries & Galloway border, two contain safflorite inclusions.

sagenite

A complex reticulated twinned intergrowth of acicular rutile.

sahlite

A synonym of salite and diopside, both pyroxene group.

sal ammoniac/salammoniac

cubic NH_4Cl

An uncommon ammonium chloride forming a greyish-white scaly efflorescence on fused coal–shale fragments at Emily coal pit, Arniston, Lothian, Scotland (Livingstone 2002). Sal ammoniac is also mentioned by Heddle (1901) from a similar environment at two other Scottish localities: West Wemyss, Fife and Hurlet, near Paisley, Strathclyde. Greg & Lettsom (1858) and Gaines et al. (1997) note sal ammoniac in the vicinity of ignited coal seams at Newcastle-upon-Tyne.

saléeite

AUTUNITE GROUP

monoclinic (pseudo-tetragonal) $Mg(UO_2)_2(PO_4)_2{\cdot}10H_2O$

A rare hydrated magnesium uranyl phosphate identified at three locations in Cornwall: Wheal Edward (Wheal Owles sett), St Just; Wheal Basset, Redruth; and Roughtor Great Consols, St Clether (Ryback & Tandy 1992). At Wheal Edward, saléeite forms small pale-yellow sheaf-like crystals up to 4 mm long (Wolloxall 1988) and radiating masses of pale-yellow plates up to 0.5 mm long on edge in a chalcopyrite matrix (T. Neall specimen, D. I. Green XRD, Tindle unpub. EMP data). South Terras mine, St Stephen-in-Brannel, Cornwall, is a further location where saléeite is confirmed (Anthony et al. 2000).

Radiating masses of saléeite mantled by green torbernite plates; Wheal Edward, St Just, Cornwall; width 4 mm. (© M. P. Cooper)

salite

An iron-rich variety of diopside, pyroxene group.

salmiac

A synonym of sal ammoniac.

salmonsite

A mixture of hureaulite and jahnsite, whiteite group.

samarskite-(Y)?

monoclinic $(Y{+}REE,Fe^{2+},Fe^{3+},U,Th,Ca)(Nb,Ta,Ti)O_4$

A rare yttrium–REE–iron–uranium–thorium–calcium niobate that may be present in a corundum-bearing xenolithic dyke in the Loch Roag area of the Isle of Lewis, Western Isles, Scotland (Livingstone 2002).

sanidine

FELDSPAR GROUP

monoclinic $(K,Na)AlSi_3O_8$

A common potassium–sodium aluminosilicate often found as phenocrysts in high-temperature volcanic rocks such as rhyolite and trachyte, and as spherulites in volcanic glass. Sanidine is also found in high-temperature contact-metamorphic rocks and is characterized by simple Carlsbad twinning, unlike plagioclase-series feldspars that are lamellar twinned. The distinction of sanidine from orthoclase or microcline is not easily made in hand specimen.

Sanidine, forming elongated, columnar, colourless, euhedral, tabular crystals (to 13×8 mm) is the major phenocryst phase in phonolite from the Wolf Rock, 15 km from Gwennap Head, Land's End, Cornwall. Associated species are aegirine, nepheline and nosean (now mostly altered), with minor apatite, fluorite, pyrite and tourmaline also present (Allport 1871, 1874, Adye 1907, Harrison et al. 1977). Elsewhere in Cornwall, abundant sanidine has been found in the Pendennis peralkaline minette, a minor intrusion cutting Devonian slates of the Gramscatho Beds on the foreshore at Pendennis Point, Falmouth. The rock is unusual in having a primary igneous carbonate component (dolomite and iron-rich magnesite). Other constituents of the minette include biotite, apatite and alkali amphiboles described as riebeckite and potassium arfvedsonite (Hall 1982).

An arfvedsonite-bearing rhyolite dyke cutting the granite on Lundy island, off the north coast of Devon, contains phenocryst and groundmass alkali feldspar with a composition intermediate between sanidine and anorthoclase (Thorpe & Tindle 1992). A visual identification of porphyritic Carlsbad-twinned sanidine crystals, up to 50 mm long, is reported from a biotite–quartz porphyry dyke outcropping to the S of Wasdale Old Bridge, near the Shap granite, Cumbria (Harker & Marr 1891).

Davies & Bloxam (1990) report sanidine and albite phenocrysts with flow-aligned feldspar microphenocrysts in late Precambrian rhyolitic lavas exposed at Pen Pedol and Pen-y-Foel, to the W of St David's, Dyfed, Wales.

In Scotland, sanidine is reported as a groundmass constituent of fine-grain trachyte belonging to the Mull Plateau Group lava series, from the western slopes of Ben More, Isle of Mull, Strathclyde (Kerr 1998). Heddle (1901) notes sanidine from:

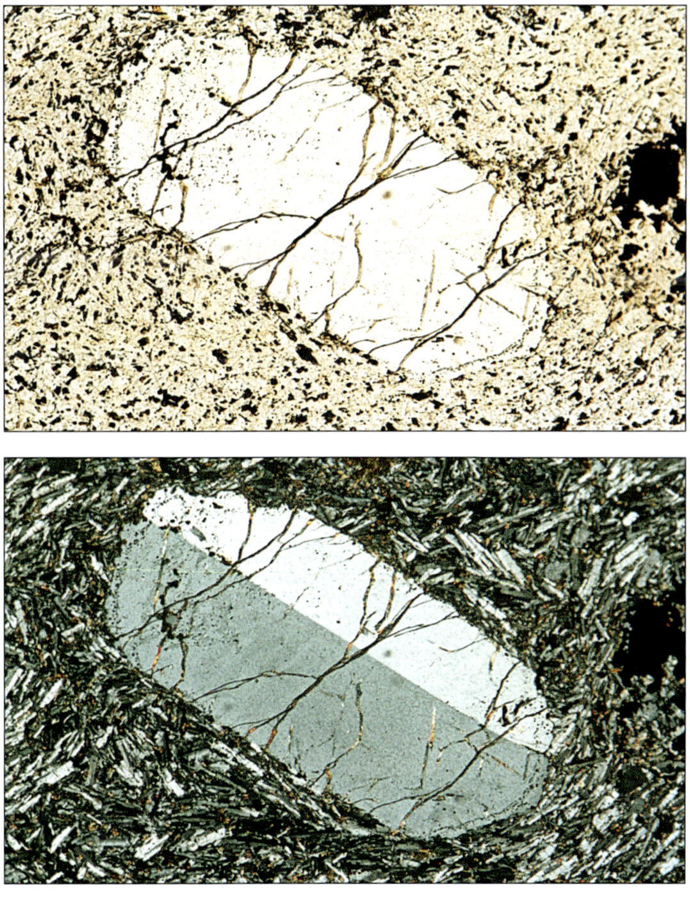

Carlsbad-twinned sanidine phenocryst in a groundmass of small sanidine crystals; from a trachyte on Garleton Hills, north of Haddington, Lothian, Scotland; width 3.4 mm; viewed in PPL (upper) and XPL (lower). (© K. Gardner)

pitchstones at Corriegills, N of Clauchland Point, near Lamlash, Isle of Arran, Strathclyde; Sgùrr of Eigg, Isle of Eigg, Highland Region; and on the tiny island of Oigh-sgeir to the SW of the Isle of Canna, Highland Region. The pitchstone on Eigg contains abundant phenocrysts of sanidine, with less common phenocrysts of augite and plagioclase in a dull-brown perlitically fractured microcrystalline glass (Carmichael 1960b). Other examples occur in lava flows on the Isle of Canna and in tuffs (250 μm euhedral crystals) on the Isle of Muck (Chambers et al. 2005). The mineral chemistry and petrology of Tertiary pitchstones of Scotland, including some of the occurrences listed above, is extensively reviewed by Preston et al. (1998b), although these authors often use the term alkali feldspar instead of sanidine. Sanidine is noted in a pyroxenite–granulite gneiss xenolith of lower crustal or possible mantle origin in a Permian olivine-nephelinite dyke at Gribun, on the west coast of the Isle of Mull, Strathclyde (Upton et al. 1998). Large glassy phenocrysts or xenocrysts of sanidine up to 6 cm in length are described from a trachybasalt dyke cutting lavas at Bangley quarry, near Haddington, Lothian, and the species is also found as megacrysts in the alkali olivine basalts of east Fife (Livingstone 2002). Upton et al. (1983) report sanidine megacrysts in basaltic hosts from Kellie Law, Fife, and also from Fidra Island, Lothian, and Loch Roag, Isle of Lewis, Western Isles. Large twinned crystals of sanidine up to 5 cm long occur in tuff at Kinkell, E of St Andrews, Fife (Heddle 1877). Phenocrysts of sanidine are further noted in Tweeddale peralkaline rhyolite lavas from Winkston Hill and Wrae Hill, near Peebles, Borders (Thirlwall 1981).

An example of sanidine occurring in high-temperature metamorphic rocks is provided by Butler (1961), who describes the metamorphism and partial fusion of Moine sediments adjacent to a small dolerite plug at Glenmore, Ardnamurchan, Highland Region. Reactions in pelitic rocks suggest sanidine and mullite (or spinel) are produced by the breakdown of mica (muscovite or aluminous biotite) and quartz. This process of pyrometamorphism is investigated in much greater detail by Worden et al. (1987) in a study of interbedded Dalradian psammites, pelites and greenschists from near the summit of Sidhean Sluaigh, 3 km SW of Strachur, Strathclyde. These authors show that breakdown of phengite and chlorite in these rocks was caused by the intrusion of a dolerite feeder pipe, and the reaction products included crystals of microscopic-scale biotite, mullite, spinel and sanidine.

Contact metamorphism of tuffaceous sandstone and conglomerate of Old Red Sandstone age has occurred around the Tertiary dolerite plug of Tieveragh, near Cushendall, Co. Antrim, Ireland, and has produced a pitchstone-like fused rock described as vitric cordierite–buchite. Patches of buchite are also found in the dolerite and these contain water-clear sanidine in the form of rectangular or rhombic prisms with hollow ends. These appear as forked terminations in thin section and are associated with plagioclase feldspar with similar morphology (Tomkeieff 1940). Later work on buchite from Tieveragh by Chinner & Dixon (1973) identified assemblages that include enstatite, cordierite, hematite, magnetite, osumilite, sodic plagioclase, sanidine and tridymite. Phenocrysts of sanidine (up to 3 mm long), oligoclase and quartz constitute 20% of the allanite-bearing obsidian found in the Sandy Braes vent that forms part of the Tardree rhyolite complex, Co. Antrim (Brooks et al. 1981).

saponite

SMECTITE GROUP

monoclinic $(Ca_{0.5},Na)_{0.3}(Mg,Fe^{2+})_3(Si,Al)_4O_{10}(OH)_2 \cdot 4H_2O$

TL Lizard Point, Cornwall, England

A relatively common hydrated calcium–sodium–magnesium–iron aluminosilicate hydroxide that is typically found in altered basic igneous rocks, but whose early description is somewhat confused. It has variously been referred to as bowlingite, cathkinite, mountain soap, rubislite, prasilite, seifenstein, steatite and thalite, names that have also been used to describe chlorite-group minerals, kaolinite–serpentine-group minerals and talc. Further confusion arises because saponite has been used in the past as a group name for the minerals delessite, chlorophaeite, hullite, saponite and celadonite (Heddle 1879f). The first three minerals have since lost species status and the last has been shown to belong to the mica group. The early descriptions therefore require confirmation with modern analytical techniques.

Saponite (soapstone) is soft and amorphous, and can be white, grey or yellowish. Greg & Lettsom (1858) report it from three localities in Cornwall: Cheesewring, SW of Henwood, Linkinhorne; in serpentinite near Lizard Point, Lizard Peninsula; and S of Mullion, Lizard Peninsula. Collins (1892) refers to it as "steatite" and adds that it is found at locations in the Lizard area including Kynance Cove, Pedn Boar, Cadgwith and Coverack. As steatite is a synonym for talc, these and other occurrences in Cornwall mentioned by Collins are doubtful, although Power et al. (1997) confirm its occurrence at the contact of the Lizard peridotite with Kennack Gneiss and where talc veins cut basic dykes within the peridotite. Similar doubts surround the saponite (prasilite) described with scolecite, lining cavities in basalt from Maol nan Damh, on the shoulder of Ben More, Isle of Mull, Strathclyde (Currie 1898), and at many other locations in Scotland (Heddle 1901). Bowlingite from Bowling quarry, Kilpatrick Hills, near Glasgow, is possibly saponite. It is deep green and it occurred in a thin vein cutting dolerite (Hannay 1877).

At Merehead (Torr Works) quarry, Cranmore, Mendip Hills, Somerset, saponite occurs as a white powder associated with magnesium-rich calcite and purple quartz, in a manganese-rich pod (M. S. Rumsey and R. W. Turner pers. comms 2006). Many irregular 1–5 cm-wide veins of dark-green fibrous saponite cut analcime basalt at Calton Hill quarry, Taddington, 7 km E of Buxton, Derbyshire (Curtis 1976). Fibrous saponite and nontronite pseudomorphs after olivine are found in the Waterswallows sill, Buxton, Derbyshire. The rock is an olivine microphyric dolerite that has suffered deuteric alteration. The fibrous clays also infill veins at the base of the sill (Ineson et al. 1983). This occurrence is essentially identical to that reported by Roberts (1981) for saponite-bearing metadolerite sills from near Pwllheli and Dinas on the Lleyn Peninsula, Gwynedd, Wales, and by Archer et al. (1965) from Parc mine (No. 3 level), near Llanrwst, Gwynedd. Saponite is also noted from Gimlet quarry, near Pwllheli, and at Llanelwedd quarry, near Builth Wells, Powys (Bevins 1994).

XRD and chemical data confirm as saponite the pale-blue, almost white, mineral found in a 5 cm-wide vein at a temporary exposure in Allt Ribhein, Fiskavaig Bay, Isle of Skye, Highland Region, Scotland. The vein occurred in vesicular basalt, in which amygdales were infilled with heulandite, stilbite and thomsonite (Mackenzie 1957). Chemically, the saponite is very similar in composition to that described from amygdales in basalt from along the road between Uig and Quiraing, northeast Isle of Skye (Heddle 1901). Saponite is also found in vesicles in an olivine dolerite exposed 1 km N of Portree, Isle of Skye. It is just one of eight species filling vesicles in the area, others being calcite, gyrolite, mesolite, tacharanite, thomsonite, tobermorite and xonotlite. Often, two or more species are found in the same amygdale – an example being one containing saponite and tobermorite, another with these two species and probable cristobalite (Sweet 1961). Farther north, at Lòn Ostatoin, near Duntulm, grossular and saponite occur as alteration products of Jurassic limestone metasomatized by a doleritic sill of Tertiary age (Kemp et al. 2005). At Sgùrr Beag, near Loch Eynort on Skye, saponite, yellowish-green chlorite and iron oxides are the alteration products of olivine in two flows of picritic basalt (Bell & Williamson 1994). In Dumfries & Galloway, saponite is found in quartz-vein samples associated with faulting cutting through diorite in the Portencorkie diorite–granite complex, 500 m NNW of Barncorkrie farm, Drummore. Other species found in the vein are calcite and chalcopyrite with minor arsenopyrite, chlorite, connellite, kaolinite and paratacamite (Cameron & Lawson 1979). At Fenton Tower quarry, near North Berwick, Lothian, saponite is noted with manganite and pyrolusite in a felsite (Nicholson 1989b).

A study of the alteration of Tertiary vesicular basalt at the Giant's Causeway, Co. Antrim, Ireland, has shown that saponite, chlorite–smectite and dioctahedral smectite form successive rims in the same vesicle. Basaltic glass, itself partially replaced by heulandite and clinoptilolite, fills the remaining voids (Robert 2001). Amygdaloidal basalts adjacent to a Tertiary dolerite plug at Scawt Hill, Larne, Co. Antrim, originally contained vesicles that were infilled with calcite, saponite–nontronite and zeolite species. Thermal metamorphism caused by the dolerite intrusion has melted the amygdale assemblage and it now consists of a quenched brown glass containing complexly zoned pyroxene, plagioclase and opaque oxides (Kitchen 1985). Greg & Lettsom (1858) report saponite from the amygdaloidal basalts of Antrim and at Magilligan in Co. Londonderry, where it is associated with levyne. At Lindsay's Leap quarry in the Mourne Mountains, Co. Down, joint planes in granite contain yellow amorphous-looking waxy masses of saponite associated with danalite and zeolite species (Green et al. 2005a).

sappare

TL Botriphnie, near Drummuir, SW of Keith, Grampian, Scotland

A synonym of kyanite.

sapphire

A blue gem variety of corundum, hematite group.

sapphirine

monoclinic or triclinic $(Mg,Al)_8(Al,Si)_6O_{20}$

A rare magnesium aluminosilicate found in association with spinel and orthopyroxene in a hornfels adjacent to an ultrabasic intrusion on Coire nam Muc, Ardgour, Highland Region, Scotland (Livingstone 2002). Rare sapphirine is also reported from garnet–kyanite/sillimanite–orthopyroxene granulites to the S and SW of Roineabhal, South Harris, Western Isles, where it forms pale-blue inclusions up to 0.1 mm across in sillimanite (Baba 1999, 2003, Baba et al. 2000).

Irregular-shaped pale-blue sapphirine and brown eastonite–phlogopite included in colourless garnet, in a pyroxene granulite from the South Harris igneous complex, Western Isles, Scotland; width 0.5 mm; viewed in PPL. (© S. Baba)

In Ireland, the cordierite–hercynite–mullite portion of a hornfels derived from metamorphism of a laterite at Tievebulliagh, near Cushendall, Co. Antrim, may contain sapphirine. On the basis of its optical properties and refractive index, small spots of a blue mineral in a single thin section from this rock are very likely to be sapphirine (Sørensen 1955).

sarcite

Possibly named after the River Sark on the England/Scotland border. Sarcite is a name used to describe analcime or leucite.

sard or sardonyx

Brown fine-grain gem quartz ± moganite. Sardonyx is banded.

satin spar

A name applied to fibrous varieties of calcite, aragonite or gypsum.

saussurite

A mixture of zoisite and plagioclase.

SCAPOLITE GROUP

A tetragonal series with a formula $Na_4Al_3Si_9O_{24}Cl$ to $Ca_4Al_6Si_6O_{24}(CO_3,SO_4)$. The group contains two species – marialite and meionite – both found in Britain or Ireland.

scarbroite

triclinic $Al_5(CO_3)(OH)_{13}\cdot 5H_2O$

TL White Nab, South Bay, Scarborough, Yorkshire, England

A relatively uncommon hydrated aluminium carbonate hydroxide found in vertical fissures within Jurassic sandstone in a small cove just S of South Bay, Scarborough. White, compact or powdery, scarbroite and hydroscarbroite line fissures together with dark-brown or grey calcite, gibbsite, kaolinite and varying amounts of quartz (Duffin & Goodyear 1960, King 1982a). White clay found in septarian ironstone nodules at this locality is confirmed to be dickite (King 1982a) and not scarbroite as Greg & Lettsom (1858) imply. Ryback (1988b) found that in Maidstone Museum some old specimens labelled as scarbroite from the type locality, were in fact alumohydrocalcite.

In Northamptonshire, scarbroite has been found in a temporary roadside exposure near Weston Favell, again in Jurassic sandstone. Subspheroidal masses of orange scarbroite up to 60 mm in diameter occur in a brown ferruginous sandy deposit within widened joints opened by periglacial weathering (known as gulls). At both localities, Pleistocene deposits overlie the Jurassic host lithologies and it seems likely that weathering processes involving downward-circulating water played an important role in the genesis of scarbroite and other aluminium-rich minerals such as basaluminite and hydrobasaluminite (King 1982a).

Scarbroite is one of several secondary aluminium-rich species associated with

Rhombs of scarbroite coated with minute crystals of an unidentified species. South Bay, Scarborough, Yorkshire; width 3.6 μm; SEM image. (© E. Condliffe, University of Leeds)

deeply oxidized smithsonite-bearing veins in Upper Triassic limestone breccias, near East Harptree, Mendip Hills, Avon, and it is also present at Hampstead Farm quarry, Chipping Sodbury, Avon, where heavily oxidized pyritiferous veins are found in Carboniferous Limestone close to the Rhaetic/Lower Carboniferous unconformity. At East Harptree, scarbroite occurs as massive, compact, chalk-like microgranular cavity infillings up to 4 cm in diameter and in thin veinlets in the mineralized breccia. It is associated with allophane, halloysite, hydroscarbroite and kaolinite. At Hampstead Farm quarry, scarbroite, forming 20–40 μm-diameter spheres composed of stacks of tabular rhombic microcrystals, is found with allophane as massive, earthy to compact, white cavity-fillings in the upper portions of the oxidized sulphide veins (Alabaster 1990b).

scawtite

monoclinic $Ca_7Si_6O_{18}(CO_3)\cdot 2H_2O$

TL Scawt Hill, Larne, Co. Antrim, Ireland

A rare hydrated calcium silicate carbonate found in hybrid rocks formed by assimilation of chalk by the dolerite intrusion at Scawt Hill. It is intimately associated with melilite-bearing hybrids and is a constituent of some of the late-formed vesicles in these rocks. Calcite and thomsonite are also found in the vesicles and may be accompanied by needles of aegirine that project from the melilite into the vesicles. Colourless crystals of scawtite form flattened plates up to 1.5 mm in length and in places form veinlets cutting through melilite (Tilley 1930b, Tilley & Harwood 1931). Scawtite also occurs in the exogenous contact zone (i.e. within the marbles derived from the chalk), where it appears as white turbid zones in clear glassy spurrite associated with calcite. All stages from partial to complete replacement of spurrite to scawtite can be traced, and typically the scawtite occurs as fibrous aggregates of radiating overlapping flakes 0.03–0.05 mm in length (Tilley 1938b). Newer material from Scawt Hill has been used to define the crystal structure of scawtite (Grice 2005). At Carneal, Larne, Co. Antrim, in another chalk/dolerite contact zone associated with a Tertiary volcanic plug, spherulitic scawtite nodules are found in a white plombièrite–tobermorite–xonotlite rock adjacent to a hybrid gabbro. The scawtite, dark grey in hand specimen, encloses abundant magnetite and is associated with hibschite (hydrogrossular). Minor amounts of scawtite are also noted in hibschite–quartz–wollastonite rocks produced by the metasomatic introduction of lime (CaO) into flint (Sabine & Young 1975). Scawtite is also noted from a third chalk/dolerite contact zone at Ballycraigy, Larne, Co. Antrim (McConnell 1954).

At Kilchoan, Ardnamurchan, Highland Region, Scotland, scawtite occurs in late-stage veins within nodular or concretionary limestones that have been thermally metamorphosed by dolerite and gabbroic intrusions. Scawtite is relatively uncommon and it replaces spurrite as scaly aggregates, at the margins of veins or ettringite-filled fractures. Other species found in the veins include calcite, dellaite, foshagite, kilchoanite, larnite, merwinite, plombièrite, rustumite, tilleyite and wollastonite (Agrell 1965). Scawtite also occurs sporadically in narrow fracture-filled veins in gabbro at Binhill quarry, Cairnie, Huntly, Grampian. Scawtite-rich areas appear sheared and contain tacharanite and xonotlite pools. Scawtite also forms the margins of a pink xonotlite vein (Livingstone 1974).

schalenblende

A colloform, banded variety of yellow sphalerite ± wurtzite ± pyrite ± galena.

scheelite

tetragonal $CaWO_4$

A common calcium tungstate that forms a series with powellite and is found in high-temperature hydrothermal veins and greisen. It has bright bluish-white (or, less commonly, cream) fluorescence under short-wavelength ultraviolet light.

Scheelite, said to be the best found in Cornwall, is reported from Stamps and Jowl Zawn, Wheal Cock, near Botallack, St Just. It forms brownish-yellow to pale "wax-yellow", partially translucent, simple bipyramidal crystals up to 72 mm long, attached to quartz crystals or embedded in loose chlorite. Cassiterite and phenakite are associated species (Russell 1920a, Wolloxall 1989, Bancroft & Weller 1993). It has also been found in the nearby Levant mine (Le Neve Foster 1877, Collins 1878b, Polkinghorne 1954) and in cliffs near Wheal Button (Bancroft & Weller 1993). Elsewhere in the aureole of the Land's End granite, scheelite is noted from a vein in dolerite at Carrick Du quarry, near Porthmeor, St Ives (Polkinghorne 1954). In the Camborne–Redruth mining district of Cornwall, scheelite has been found in wolframite lodes at New Roskear mine (Robson 1954) and in the Roskear complex of South Crofty mine (Hosking 1954b, LeBoutillier 1996, Wolfe 1998). A secondary pink variety replacing wolframite has been recovered from the deep levels of East Pool and New Dolcoath mines (Collins 1878b, Polkinghorne 1954). Other Cornish occurrences of scheelite include: with wolframite at Hawkswood mine, North Hill, Bodmin Moor (Polkinghorne 1954, Robson 1954); subhedral to euhedral pale-yellowish crystals up to 1 cm long in clinochlore, associated with minor cassiterite and quartz at Penberthy Croft mine, St Hilary (Camm & Merry 1991a,b, Betterton et al. 2000); in a tin lode at Pengelly Croft mine, Nancegollan, near Breage, and at Wheal Maudlin, Lostwithiel (Greg & Lettsom 1858, Collins 1878b, Polkinghorne 1954); Penhale mine, St Cleer; Old Gunnislake mine, Calstock; Hingston Down quarry, Calstock, and Treburland mine, Altarnun (all Golley & Williams 1995); Stenna Gwyn mine, St Stephen-in-Brannel, near St Austell (Polkinghorne 1954); associated with bindheimite, boulangerite, bournonite, cerussite, galena, native gold, jamesonite, pyrite and sphalerite at Treore antimony–arsenic mine, Port Gaverne, near Port Isaac (Russell 1949b); with wolframite at Trebartha Lemarne mine, North Hill (Robson 1954, Polkinghorne 1954, Brooks 2001).

In Devon, reports of scheelite include: Bedford United mine, near Tavistock (Collins 1911a); Devon United mine and Wheal Jewell, both near Mary Tavy, and Hemerdon Bal, Plympton (Polkinghorne 1954); rich-yellow octahedra of exceptional quality, up to 4 cm long, embedded in chlorite and associated with wolframite at Wheal Friendship (Devon Friendship), Mary Tavy (Greg & Lettsom 1858, Polkinghorne 1954, Braithwaite 1982b). There is some doubt about the scheelite occurrences described from: Ford mine, Sticklepath, near Okehampton; Ivy Tor mine, Belstone, Meldon area, near Okehampton; Limestone quarry, Meldon area; British Railways quarry complex at Meldon; Ramsley mine, South Tawton; Red-a-ven mine, Meldon, and Wheal Forest, Meldon Gorge (Kingsbury 1961, Kingsbury 1964, El Sharkawi & Dearman 1966) – they are based on material supposedly collected by A. W. G. Kingsbury. Ryback et al. (1998, 2001) have shown that some samples in this collection have been incorrectly provenanced.

Bipyramidal scheelite, 2 cm across, from Wheal Friendship, Mary Tavy, near Tavistock, Devon. (© The Natural History Museum, London)

At Carrock mine, Grainsgill, Caldbeck Fells, Cumbria, high-temperature wolframite-rich quartz veins, within greisenized Skiddaw granite and aureole rocks, have produced translucent sherry-brown bipyramidal crystals to several cm on edge. More typically, crystals are less than 30 mm across, although uncommon, poorly formed, heavily fractured crystals up to 20 cm are reported. Colourless, wine-yellow and clove-brown varieties are also noted. Partial replacement of wolframite by scheelite is common in these veins and it has been suggested that the scheelite has a metasomatic origin. Arsenopyrite, native bismuth, chalcopyrite, galena, molybdenite, pyrite, sphalerite and tetradymite are associated species (Finlayson 1910, Hitchen 1934, Davidson & Thomson 1951, Eastwood 1959, Ewart 1962, Eastwood et al. 1968, Appleton & Wadge 1976, Shepherd et al. 1976, Beddoe-Stephens & Fortey 1981, Ball et al. 1985a, Young 1987a, Watson 1988, Cooper & Stanley 1990, Green & Briscoe 2002). Scheelite is common in chalcopyrite–ferberite–quartz veins cutting granodiorite at the southeastern margin of the Eskdale granite complex at Buckbarrow Beck, Corney Fell, Waberthwaite. The locality is also notable for its supergene assemblage containing the rare species bismutoferrite, cuprotungstite, eulytite, mixite, namibite and russellite (Young et al. 1986, 1991b, Neall et al. 1993, Green & Briscoe 2002).

Other Cumbrian scheelite occurrences include: miarolitic cavities within the Shap granite (Cooper 1991b, Wirth 1995, Wirth & Green 2002); massive fawn scheelite in quartz veins in Higher Brandy Gill, Caldbeck Fells (Young 1987a); stream sediments from Wet Swine Gill, Caldbeck Fells (Appleton & Wadge 1976); and a dubious occurrence noted from Scar Crag, Causey Pike (Kingsbury & Hartley 1957a, Young et al. 1986).

Rare scheelite is reported from pyritic mudstones of the Cambrian Hafotty Formation, Harlech Dome, Gwynedd, Wales. Chlorite with rare chalcopyrite, ferroan pyrophanite and an yttrium-rich phosphate are associated species (Bennett 1987a, Bevins 1994).

Scotland

Dumfries & Galloway Pale bluish-green massive scheelite is reported on wolframite in granite at Craignair quarry, W of Dalbeattie, Dumfries & Galloway (Williams 1973, Livingstone 2002).

Grampian At Gairnshiel Bridge, Glen Gairn, scheelite replaces niobium-rich wolframite in a hübnerite-bearing quartz vein cutting zinnwaldite granite. The wolframite is strongly zoned with some niobium-rich zones preferentially replaced by scheelite. Scheelite also forms crosscutting microveinlets within the wolframite. Cassiterite and muscovite are associates (Tindle & Webb 1989, Livingstone 1992, Webb et al. 1992). Abundant scheelite has been found in vein quartz from a gas-pipeline trench near Middleton and Mill of Braco, both between Chapel of Garioch and Inverurie. The scheelite veins and nearby molybdenite-bearing veins and greisen are associated with the small Middleton granite intrusion (Colman et al. 1989).

Highland Whitish crystals of scheelite have been found in and near quartz veinlets from Gleann a' Chaolais within the Ballachulish granite. Chalcopyrite, molybdenite and pyrite are associated primary species (Haslam & Kimbell 1981). Trace amounts of bismuth-rich minerals, chalcopyrite, fluorite, scheelite and sphalerite have been found in molybdenite–pyrite-mineralized quartz veins cutting Precambrian schists and gneisses close to the Grudie and Shin granites to the SW of Lairg (Gallagher et al. 1974, Gallagher & Smith 1975, Fortey 1980).

Strathclyde At Coire Buidhe, Glen Creran, coarse bipyramidal scheelite (initially confused with powellite) occurs in cavities and quartz masses within skarn deposits related to the intrusion of the Cruachan granite into metamorphosed impure limestones of Dalradian age. The complex assemblage also contains andradite, epidote, magnetite, powellite, titanite and vesuvianite, some in well formed crystals (Ingram et al. 1993, 1996, Starkey 1995, Green & Todd 2001). Scheelite and chalcopyrite are said to be widespread in quartz veinlets associated with molybdenite–pyrite mineralization centred around the headwaters of Allt Choire Dhuibh, Central Starav granite, Etive granite complex to the E and NE of Ben Starav (on the Highland/Strathclyde border) (Haslam et al. 1985).

Ireland

Donegal A 10 mm-long crystal of scheelite in quartz is described from a roadside quarry, 3 km E of Bunbeg, where a metadolerite intrudes Dalradian quartzites (Ryback & Moreton 1992, Green & Moreton 2001).

Galway Late veins cutting stratiform pyrite–pyrrhotite-dominated sulphide lenses within the Middle Dalradian Lakes Marble Formation at Teernakill, Connemara, contain minor scheelite intergrown with sulphides. Granite-related skarn and quartz-vein-hosted chalcopyrite–pyrite–(scheelite) mineralization is noted in the formation from the Glann area and from around the Omey and Letterfrack granites, also in Connemara (McArdle et al. 1986a). One such skarn, in the aureole of the Omey granite, at Barnahallia Lough, Claddaghduff, is epidote-rich and it hosts white-to-cream anhedral-to-subhedral scheelite, occasionally up to 1 cm in length (Moreton & Green 2006a). Arsenopyrite, fluorite and scheelite are disseminated in the Errisbeg Townland granite (part of the Main Galway granite) on the western shore of Travore Bay (O'Raghallaigh et al. 1997).

Wicklow A tin–tungsten mineralized (greisen) microtonalite sheet complex centred on Ballingen, between Tinahely and Aughrim, contains arsenopyrite–scheelite-bearing fluorite–muscovite–quartz–tourmaline veinlets. Scheelite forms isolated grains and intergrowths with sulphides – pyrrhotite in particular (Steiger & Bowden 1982, Kennan et al. 1986, McArdle & Kennan 1988, 1992, Gallagher 1989, McArdle et al. 1989, Gallagher & Kennan 1992).

schiefer spar

A paper-thin lamellar and friable variety of calcite.

schiller spar

A mixture of chrysotile ±anthophyllite or lizardite or talc.

schirmerite

orthorhombic $Ag_4PbBi_4S_9$

A rare silver–lead–bismuth sulphide found in a stockwork cut by galena-bearing quartz veins at Corrie Buie, Meall nan Oighcreag, near Loch Tay, Tayside Region, in Scotland. The veins also contain minor amounts of bismuth, chalcopyrite, electrum (36–65% gold), emplectite, pyrrhotite, sphalerite, and phases belonging to the gustavite–lillianite series. Mineralogy of the veins is typical of that associated with acid intrusions. Schirmerite was found associated with dolomite in fractures within galena (Pattrick 1984).

Schirmerite in tourmaline-bearing microgranite dykes from Ballinglen, between Tinahely and Aughrim, Co. Wicklow, Ireland (Steiger & Bowden 1982), requires confirmation.

schmiederite

monoclinic $Pb_2Cu_2^{2+}(Se^{4+}O_3)(Se^{6+}O_4)(OH)_4$

A rare lead–copper selenate hydroxide found in copper-rich quartz veins at Greystone quarry, Lezant, Cornwall, where it occurs as powdery crusts and sprays of flat pale-blue radiating needles up to 3 mm long (Golley & Williams 1995, Plant & Williams 1995, Elton et al. 2004).

Schmiederite is also found in the oxidized zone of a copper–lead orebody at Llechweddhelyg mine, near Clawddmelyn, between Bont-goch and Cwmsymlog, Central Wales orefield, Dyfed, where it forms deep-blue spherules, sometimes aggregated into crusts. EMP analysis indicates a composition close to the linarite/schmiederite boundary (but selenium rich) and it is therefore a sulphatian schmiederite (Rust 1990c). Schmiederite is also reported from Eaglebrook (Nantycagl) and Esgair Hir mines, Ceulanymaesmawr, Dyfed, and Dyfngwm mine, Machynlleth, Powys, all in the Central Wales orefield, and from copper–lead veins in Machen quarry, 7 km ENE of Caerphilly, Mid Glamorgan (Bevins 1994). However, on the basis of further analytical data, Plant & Jones (1995) state that the schmiederite at Machen quarry is a misidentification of linarite.

Tiny light-blue lath-like crystals of schmiederite associated with anglesite, caledonite, linarite and serpierite are found in cavities in slag from the Meadowfoot smelter, Wanlockhead, Dumfries & Galloway, Scotland (Green 1987a).

SCHOENFLIESITE GROUP

Cubic hydroxides with the general formula $X^{4+}Sn^{4+}(OH)_2$, where X^{4+} = Ca, Cu, Fe, Mg, Mn, Zn. Natanite and wickmanite are members of the group found in Britain or Ireland.

schoepite

orthorhombic $(UO_2)_8O_2(OH)_{12}{\cdot}12(H_2O)$

A rare hydrated uranyl oxide hydroxide found half way up a cliff, in a highly uranium-rich fissure vein close to Needle's Eye, SW of Caulkerbush, Dumfries & Galloway, Scotland. This and other similar veins are in the aureole of the Criffell granodiorite. Schoepite forms brownish to orange-yellow short prismatic pseudo-hexagonal crystals up to 0.1 mm long and is associated with vandendriesscheite and uranophane. Other supergene minerals from this vicinity, produced during the alteration of primary As–Bi–Co–Cu–Fe–Pb–U minerals include eulytite, kahlerite, lavendulan, metakahlerite and walpurgite (Braithwaite & Knight 1990). A less precise location of "Southwick Cliffs area" is given by Macpherson & Livingstone (1982) for lemon-yellow schoepite forming minute crystals up to 0.1 mm long in aggregates or as fine-grain massive fissure-fillings and encrustations associated with uraninite (pitchblende).

schorl

TOURMALINE GROUP

trigonal

$NaFe^{3+}_3Al_6(BO_3)_3Si_6O_{18}(OH)_4$

A very common sodium–iron–aluminium borosilicate hydroxide that forms a series with dravite and is found in granites and granitic pegmatites, high-temperature hydrothermal veins and some metamorphic rocks. It is unquestionably widespread in Cornwall and Devon, although in most cases it is simply reported as tourmaline, with the species not identified; because of this uncertainty, many probable occurrences of schorl are not included here. Even where schorl is apparently confirmed, identification can be suspect, because tourmaline-group species are commonly zoned and this would not be apparent with early wet-chemical analyses. A further complication is that the definition of schorl has changed with time: Collins (1892) describes the species as a near-opaque dark variety of tourmaline with a large proportion of iron, whereas the tourmaline-group classification of Hawthorne & Henry (1999) requires a full chemical analysis before the species can be confirmed. It is not unusual for fluor-schorl, foitite or manganofoitite to be described as schorl; this is an irresolvable problem when analytical data are not quoted. An extensive dataset of new EMP analyses aims to address these problems.

Schorl is reported from a few locations in and around the Land's End granite, Cornwall. "Pleasant" specimens of schorl encapsulated in water-clear quartz have been collected from the dumps of Botallack mine, St Just (Bancroft & Weller 1993), and tourmalinized biotite schist from Crowns Rock near by also contains schorl (Wilson & Long 1983). At Gryll's Bunny, Botallack, within the aureole of the granite, a 60 cm-thick metasomatic tourmaline-rich skarn contains strongly zoned euhedral crystals of schorl (Jackson 1974, Wilson & Long 1983). Fluor-schorl is also noted by Manning (1991). EMP analysis confirms spectacular zonation on a microscopic scale and indicates that the tourmaline mainly belongs to an unusual dravite–schorl solid-solution series, with feruvite and uvite compositions also occurring (Tindle unpub. data). As part of an investigation of quartz-cored tourmaline, Lister (1979) examined a tourmaline-rich pegmatite at Priest's Cove, near Cape Cornwall, and found it to contain crystals ranging from 0.5 mm to 10 mm in diameter and up to 10 cm in length. In thin section the crystals are brown, pale blue or dark blue. Most of the published data are schorl, but two analyses of the blue tourmaline are foitite. Further work on tourmaline from Priest's Cove indicates that early-crystallized orange-brown tourmaline is schorl or fluor-schorl, and late-crystallized blue tourmaline is dravite or less commonly schorl (Tindle unpub. EMP data). Schorl is also noted from: Carrick Du, near Porthmeor beach, St Ives; the dumps of Geevor mine, near Pendeen, St Just; Mousehole; Porth Ledden, near Cape Cornwall; Sennen Cove, near Land's End; Trencrom Hill, S of St Ives (all Tindle unpub. EMP data); Pellitras Point, near Land's End (Müller et al. 2006); Wheal Owles, Penwith, St Just, and St Ives Consols, St Ives (Hosking 1951); and Providence mines area and Rosewall Hill, both near St Ives (Wilson & Long 1983).

The Tregonning–Godolphin pluton is an albite–lithium-rich mica–topaz granite with a well developed pegmatite–aplite roof complex exposed at Megiliggar Rocks, near Porthleven, Cornwall. Schorl and foitite occur in the pegmatite as elongated prismatic needles up to 10 cm long. They are brown in thin section and some have quartz cores (Lister 1979). New data adds fluor-schorl and foitite as additional species in the pegmatite (Tindle unpub. EMP data), whereas, in other parts of the complex, fluor-elbaite is also noted (Stone 1992, Tindle unpub. EMP data).

Schorl has been collected from many locations within the St Austell granite complex and in particular within the china-clay pits. Flat radiating aggregates of brownish-black schorl formed in pegmatite at Gunheath china-clay pit (Weiss 1994) and, within vugs in another pegmatite, needles of schorl occur with florencite, fluorapatite and fluorite (Elton & Grigg 1996). At Goonbarrow china-clay pit, pegmatitic veins contain pale-brown to brown (in thin section) schorl, sometimes forming quartz-cored crystals up to 5 cm long (Lister 1979). Quartz–tourmaline greisen at this locality contains hydrothermal schorl exhibiting remarkable compositional zonation (Manning 1991). Greisen at Carrigan, 3 km ENE of Roche, also contains schorl, although supporting data are lacking (Le Neve Foster 1878a). Complex zonation is apparent in hydrothermal fault breccias within massive quartz–tourmaline rocks at Roche, as multiple analyses of a single crystal reveal compositional differences leading to dravite, magnesiofoitite, foitite and then schorl crystallization (Williamson et al. 2000). Tourmaline compositions in a hydrothermal explosion breccia at Wheal Remfy china-clay pit are also extremely complex, with dravite, foitite, schorl and uvite occurring in strongly zoned crystals. In the West Gunheath section of Littlejohn's china-clay pit, near St Austell, hundreds of tiny pale-blue tourmaline crystals pseudomorph individual orthoclase phenocrysts in an elvan dyke. The crystals vary in composition, with schorl dominant, and foitite and elbaite also present. Massive tourmaline veins at this locality consist entirely of schorl, apart from very rare florencite-(Ce) crystals in microcavities. In hand specimen all the tourmaline looks black. Blackpool, Cocksbarrow and Wheal Martyn china-clay pits are other sites where schorl is confirmed (Tindle unpub. EMP data). Manning (1991) reports colourless-to-blue needles of fluor-schorl from this locality. Hydrothermal quartz veins cutting a small quartz–tourmaline stock at Roche Rock, N of St Austell, contain quartz-cored crystals of schorl varying from brown to blue in thin section (Müller et al. 2005). At Luxulyan (Tregarden) quarry, between St Austell and Lostwithiel, brown (in thin section) radiating crystals of schorl and foitite, a few cm long, occur in pegmatitic veins cutting granite, which itself contains schorl (Lister 1979, Manning 1991). Tourmaline in the well known luxullianite variety of granite is predominantly schorl (Tindle unpub. EMP data). It is unlikely in the near future that anyone will collect massive schorl or cassiterite from greisen bands up to 1 m thick in Bodelva china-clay pit, near St Blazey Gate (Weiss 1994), as this is now the site of the famous Eden Project.

Other Cornish occurrences of schorl include: Chywoon and Kessel Downs (Halvasso) granite quarries, near Longdowns (Tindle unpub. EMP data); Cheesewring quarry, SW of Henwood, Linkinhorne (Green et al. 1996a); Cligga Head, Perranzabuloe (Hall 1971, Chick 1987, Tindle unpub. EMP data); Hingston Down quarry, 2 km W of Gunnislake, Calstock (Ward 1983b); and Tresavean mine, Gwennap (Stephens 1940a).

Schorl is undoubtedly present in granite and granitic pegmatites in Devon, but

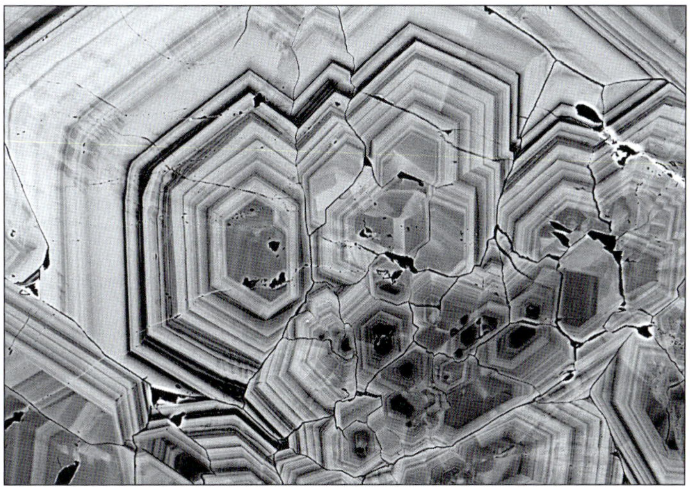

Exceptional zoning in schorl from Geevor mine, near Pendeen, St Just, Cornwall; darkest zones have dravite composition; width 300 μm; BSE image. (© E. Condliffe, University of Leeds)

Fine needles of schorl in an altered granite (luxullianite) from Luxulyan, near St Austell, Cornwall; width 1.0 mm; viewed in PPL. (© A. G. Tindle)

confirmed occurrences are sparse. It is mentioned by Power (1968) from granite at Lock's Gate Cross, Sherberton Common, near Pontsworthy, Dartmoor, and from a vein at Sheltor, Dartmoor. A dyke-like mass of greisen at the reservoir quarry of Paignton waterworks, 4 km NW of Holne, contains stout black acicular crystals of schorl, which in thin section show brown and blue colour zoning (Blyth 1949). Massive dark-brown to black, almost fibrous, tourmaline from hornfels at Meldon quarry, near Okehampton (in the aureole of the Dartmoor granite), has schorl or foitite composition, and tourmaline from the wolframite deposit at Hemerdon, Plympton, consists of these species together with dravite (Tindle unpub. EMP data). Rosettes of tourmaline associated with fluorite on joint surfaces in the Lundy granite are described as schorl by Dollar (1941), but recent analysis has found only elbaite and fluor-elbaite compositions (D. I. Green sample, Tindle unpub. EMP data). Very fine, large (up to 9 cm across and 8 cm long), black, doubly terminated tourmaline crystals, associated with apatite, from a granite pocket on Wolley farm, near Bovey Tracey (Greg & Lettsom 1858, Collins 1892, Embrey & Symes 1987, Gruguric 2005), are also considered to be schorl.

Schorl is described from quartz veins in Cumbria on Coombe Height, near Burdell Gill, and from Poddy Gill, Caldbeck Fells, by Davidson & Thomson (1951). However, recent work indicates that the tourmaline at these localities is mainly dravite, foitite or magnesiofoitite, and that compositional zoning is a common feature. Schorl has been identified at Burdell Gill, Poddy Gill and Upper Poddy Gill, but only very rarely. Fluor-schorl is also found at Upper Poddy Gill (D. I. Green sample, Tindle unpub. EMP data). Schorl is accompanied by dravite and foitite in metasomatically altered metapelites in the aureole of the presumed Grasmoor granite beneath Crummock Water, Cumbria. Tourmalinization took place in wallrocks containing a series of faults and joints, and the resulting vein-like rocks contain nothing more than tourmaline and quartz. Tourmaline crystals have a grainsize of 5–10 µm and exhibit olive-green to brown colours in thin section (Eeckhout et al. 2004).

A single wet-chemical analysis of bluish-black tourmaline occurring as clots in amphibolitized gabbro at Hanter Hill, Powys, is uvite (Holgate 1977), but recent work does not confirm this. Instead, a range of compositions either side of the compositional boundary between dravite and schorl was found (NMW specimen, Tindle unpub. EMP data). Examination of detrital tourmaline from Caradocian sandstone at Cwm Bychan, Beddgelert, Gwynedd, revealed schorl to be the major species, with some dravite and foitite also present (NMW specimen, Tindle unpub. EMP data).

In Scotland, schorl appears to be somewhat uncommon, although lack of analytical data may be the reason, rather than its actual absence. Deep bluish-black schorl occurs at Glenbuchat, Grampian, as crystals up to 3 cm long associated with elbaite in quartz–feldspar pegmatite (Livingstone 2002). EMP analysis indicates that bluish-black tourmaline at Glenbuchat is usually schorl or foitite (minor fluor-schorl), but that a more noticeably deep-blue variety has elbaite and fluor-elbaite compositions (I. Dossett and D. I. Green samples, Tindle unpub. EMP data). Fine prismatic black schorl occurs in pegmatites cutting gabbro at Pitscurrie quarry, Pitcaple, 8 km NW of Inverurie, Grampian (Green & Todd 2001, Tindle unpub. EMP data). The large granitic pegmatite at East Head, 1 km NE of Portsoy, Grampian, contains large black tourmaline crystals up to 10 cm long that have been described as schorl by Gillen (1987a), although confirmatory data are lacking. Later analysis indicates that these crystals are zoned, with both foitite–schorl and fluor-elbaite–schorl compositions present (J. R. Holbrook specimens, Tindle unpub. EMP data). This also appears to be the case for the rough schorl associated with beryl and garnet from the Carn Gorm mica prospect, Strathgarve, Highland Region (Green & Todd 2001). Occurrences of schorl are confirmed from the following additional locations in Scotland: granitic vein cutting gabbro on southwest coast of Uyea, Shetland; granitic pegmatite cutting the Belhelvie ultrabasic complex, Grampian; granitic pegmatite at Chaipaval (Ceapabhal), near Northton (Taobh Tuath), South Harris, Western Isles; aplite and garnetiferous pegmatite cutting the Loch Doon granite along the western shore of Loch Doon, S of Dalmellington, Strathclyde (all Tindle unpub. EMP data). Ferguson et al. (1998) report 1 mm-diameter radiating aggregates of black schorl in a sub-Cambrian palaeosol on the Proterozoic Lewisian unconformity in Highland Region.

Tourmaline is a common mineral in the aureole of the Leinster granite in southeast Ireland, where it occurs in: quartz–tourmaline rocks, country-rock schists and host tonalite intrusions in lithium pegmatite complexes; country-rock schists and quartz veins, with or without aplites and pegmatites, at the granite contact; stratiform quartz–tourmaline rocks (tourmalinites); wallrocks and mineralized veins in a tin–tungsten mineralized sheet. At the southwest end of the Leinster batholith, near the village of Inistioge, Co. Kilkenny, metasomatic tourmaline belonging to the dravite-schorl series is found associated with disseminated arsenopyrite mineralization. Associated species include chalcopyrite, iron-rich chlorite, fluorite, manganese-rich ilmenite, molybdenite, muscovite, pyrite, pyrrhotite and rutile. The mineralization was produced during alteration of aluminous metasediments and is spatially related to pegmatitic quartz veins with greisen alteration (Appleby & Williams 1988). A small published dataset confirms the occurrence of schorl in quartz–tourmaline rock and wallrock schist at Coolasnaghta, Co. Carlow, and in schist at Rathmichael, Co. Dublin. Other data are dravite or foitite (Gallagher & Kennan 1992). Tourmaline in altered granite hosting the lithium pegmatite at Aclare, Co. Carlow, is also schorl (Gallagher 1988). Elsewhere in Ireland, schorl is confirmed from: an axinite–epidote–tourmaline vein cutting amphibolite at Balyconneely, N of Roundstone, west Co. Galway (Benjamin 1968); pegmatitic segregation in gneiss from a roadside quarry, near Maam Cross, between Oughterard and Clifden, Connemara, Co. Galway, and in miarolitic cavities in granite at Pigeon Rock and Diamond Rocks, Mourne Mountains, Co. Down (S. Moreton & S. A. Rust samples, Tindle unpub. EMP data).

schorlite

A synonym of schorl, tourmaline group.

schorlomite

GARNET GROUP

cubic $Ca_3(Ti,Fe^{3+})_2[(Si,Fe)O4]_3$

An uncommon calcium–titanium–iron silicate that forms a series with andradite and has been found in a pegmatite near Camphouse, Ardnamurchan, Highland Region, Scotland. Schorlomite occurs with titanaugite and an iron-rich variety of wollastonite, and, although the rock type is not specified, it is likely to be extremely undersaturated (Howie & Woolley 1968). Schorlomite is also reported from a high-temperature calcic skarn near the contact of the Cuillin gabbro complex at Camasunary Bay, Strathaird Peninsula, Isle of Skye, Highland Region. Associated species include calcite, garnet (andradite–grossular), melilite and vesuvianite, together with minor amounts of strontium-rich apatite, magnetite, pyrite and wollastonite. Kimzeyite, perovskite, spinel and rare djerfisherite complete the assemblage (Beard & Drake 2007). However, EMP data indicate that the mineral is actually titanium-rich andradite with a 39 mol% schorlomite component.

schröckingerite

triclinic $NaCa_3(UO_2)(CO_3)_3(SO_4)F{\cdot}10H_2O$

A rare hydrated sodium–calcium uranyl carbonate sulphate fluoride found along 4 m of the Peeth lode, Geevor mine, near Pendeen, St Just, Cornwall. Schröckingerite occurs as small greenish-yellow spheroidal globules up to 4 mm in diameter. The globules are composed of aggregates of tiny euhedral pseudo-hexagonal plates, sometimes cemented by gypsum. Andersonite is closely associated with the schröckingerite. Elsewhere along the Peeth lode other secondary uranium minerals include cuprosklodowskite, johannite and possibly becquerelite (Elton & Hooper 1992, Elton et al. 1992).

schrötterite

An opaline variety of allophane.

schulenbergite

trigonal $(Cu^{2+},Zn)_7(SO_4,CO_3)_2(OH)_{10}{\cdot}3H_2O$

A rare hydrated copper–zinc sulphate carbonate hydroxide that typically forms by supergene oxidation processes during the post-mining alteration of primary copper and zinc sulphides.

Light-turquoise to greenish-blue schulenbergite forms botryoidal crusts up to 2 mm across on partially oxidized black sphalerite at Penberthy Croft mine, St Hilary, and is also noted from the Prince of Wales mine, Calstock, both in Cornwall (Betterton 1996b, Betterton et al. 2000). At the latter locality, schulenbergite forms abundant greenish-blue feathery aggregates (0.2–0.3 mm across) associated with ramsbeckite (Green et al. 2001).

In Cumbria, epsomite, gypsum, namuwite, schulenbergite, serpierite and various iron oxides form coatings on mine walls in the abandoned lead–zinc mine at Smallcleugh, Nenthead, near Alston. Specimens of schulenbergite collected from the Middlecleugh First Sun vein form pearly bluish-green crystals that exhibit sharp or rounded corners, together with a perfect basal cleavage (Livingstone et al. 1990). Further examination of material from this mine and the nearby Brownley Hill mine revealed that schulenbergite is associated with beaverite, gypsum, a zinc analogue of ktenasite, malachite, namuwite, serpierite and smithsonite (Livingstone 1991). Nentsberry Haggs Horse level on High Raise vein is a third location around Nenthead where schulenbergite has been reported. Here it forms a bluish-green encrustation in association with gypsum and bright-blue serpierite (Thomson 1997). Elsewhere in Cumbria, foliated aggregates of schulenbergite up to 0.3 mm across are

Schulenbergite (turquoise) aggregates up to 0.5 mm across, associated with linarite (dark blue) and tiny ramsbeckite (green) crystals; Penrhiw mine, near Ystumtuen, Dyfed, Wales. (© National Museum of Wales, Cardiff; photographer M. P. Cooper)

associated with posnjakite, ramsbeckite and serpierite in thin carbonate-bearing veinlets cutting oxidized sphalerite on the dumps of 30-fathom level at Driggith mine, Caldbeck Fells (Green et al. 2001, 2006a, Green & Briscoe 2002). At Silver Gill, Caldbeck Fells, schulenbergite forms crudely hexagonal foliated aggregates of pale-turquoise crystals up to 0.3 mm across, associated with linarite, posnjakite and ramsbeckite (Green et al. 2005b).

Another location where schulenbergite occurs is Waterbank mine, Wetton, Ecton Hill, Staffordshire, where pale greenish-blue platy masses of intergrown crystals up to 4 mm long are reported from mine spoil. Other species formed in the post-mining environment are brochantite, cerussite, hydrozincite, namuwite, serpierite and smithsonite (Rust 1992, Rust 1995c).

In Wales, schulenbergite occurs with anglesite, brochantite, cerussite and linarite at Brynarian mine, near Tal-y-bont, Central Wales orefield, Dyfed, all as supergene oxidation products of primary Cu–Pb–Zn sulphides (Rust 1992). Bevins (1994) reports schulenbergite in Dyfed from: Copa Hill mine, Cwmystwyth; Eaglebrook (Nantycagl) mine, Ceulanymaesmawr; and Penrhiw mine, Ystumtuen. Another location in the same general area is Ystrad Einion mine, near Tal-y-bont, where schulenbergite forms crusts of near-white hexagonal platy crystals up to 0.75 mm across with minor serpierite (Mason & Rust 1997). Schulenbergite also occurs with bechererite, caledonite, cerussite, hemimorphite, hydrocerussite, namuwite and susannite at Frongoch mine, between Devil's Bridge and Pont-rhyd-y-groes, Central Wales orefield, Dyfed (Rust et al. 2003), and at Dyfngwm mine, Pengroes, Powys (Anthony et al. 2003).

schultenite

monoclinic $PbHAsO_4$

A very rare lead–hydrogen arsenate found in a 1.5 m-wide quartz–sulphide vein at Deer Hills, Caldbeck Fells, Cumbria. Arsenopyrite is the main sulphide in the vein, although pyrite is common and galena is also present. Oxidation of the vein has resulted in the formation of cavities that are partially filled with yellowish-green beudantite, some of which replaces earlier carminite. Unaltered carminite and scorodite are locally abundant. Other supergene minerals include barite, mimetite and pharmacosiderite. Schultenite was found in one of these cavities as a single spray of colourless transparent blades up to 1 mm long with acute triangular terminations (Symes et al. 1990).

schwertmannite?

tetragonal $Fe^{3+}_{16}O_{16}(OH)_{12}(SO_4)_2$

Difficult to identify, but probably a common iron oxyhydroxide sulphate found in acid-mine drainage produced by the atmospheric oxidation of pyrite near metal-ore mining areas. A study of ochre sediments near West Chiverton mine, Zelah, Perranzabuloe, Cornwall, identified ferrihydrite and goethite as the main species, both associated with filamentous algae. In one sample, minor lepidocrocite or schwertmannite is also present (Singh et al. 1999).

scolecite

ZEOLITE GROUP

monoclinic $Ca(Al_2Si_3)O_{10}{\cdot}3H_2O$

A relatively common hydrated calcium aluminosilicate found infilling vesicles in the Tertiary basaltic lavas on Maol nan Damh, southwest slope of Ben More, Isle of Mull, Strathclyde, Scotland. Scolecite occurs in radiating groups of white fibrous crystals up to 10 cm long, sometimes associated with epidote or, more rarely, with celadonite or heulandite (Currie 1898). It is said to occur in profusion on An Gearna and to be common on Beinn Fhada, two areas on the northern slopes of Ben More (McLintock 1910b). Walker (1971) suggests that scolecite occurs sporadically around the central intrusive complex of Mull, within an amygdale assemblage described as the epidote zone. Off the coast of Mull, scolecite has been found on the Isle of Staffa, also in basaltic lavas (Greg & Lettsom 1858). Farther north, on the Isle of Skye, brilliant colourless acicular crystals of scolecite are found in Tertiary amygdaloidal basalts at Talisker, and at Sgùrr nam Fiadh, between Talisker and Loch Eynort, the species has been found in vapour cavities up to 1.5 m across (Heddle 1901).

The fragmented intrusions of dolerite (also microdiorite and granite) forming the Mullach Sgar complex, St Kilda (Hirta), Western Isles, contain small quantities of apophyllite, calcite, chabazite, epidote, prehnite, quartz (amethyst), scolecite and titanite lining drusy cavities (Harding et al. 1984), Dyer et al. (1996) note scolecite with chabazite and thomsonite from a boulder found at Kingscross Point, Isle of Arran, Strathclyde. Hall et al. (1989) report rare scolecite with heulandite, laumontite, natrolite, botryoidal prehnite and thomsonite, all from Lower Carboniferous basaltic lavas exposed in Boyleston quarry, SW of Barrhead, Strathclyde. Other locations in Scotland noted by Heddle (1901) include the following:

FIFE Kincraig, 3 km W of Elie; Isle of May, near Anstruther Easter, off the coast of Fife. HIGHLAND Isle of Skye: Beinn nan Cuithean, 1 km S of Talisker Point; Sgùrr an Duine, between Loch Brittle and Loch Eynort; west side of the Old Man of Storr, 10 km N of Portree; south side of Talisker Bay. STRATHCLYDE Four localities on the Isle of Mull: cliffs to SW of Bearraich, Ardmeanach; Beinn na Croise, E of head of Loch Scridain; Carsaig Arches, near Malcolm's Point; Nuns' Cave, below Beinn Chreagach. Four localities on the Treshnish Isles, near the Isle of

Radiating fan of scolecite; Maol nan Damh, Ben More, Isle of Mull, Strathclyde, Scotland; width 17 cm. (© The Trustees of the National Museums of Scotland)

Mull: Bac Mòr, Fladda, Lunga and Sgeir a' Chaisteil. Also on the Killdaloig shore, near Campbeltown, on the Mull of Kintyre. An analysis of scolecite from An Gearna, Ben More on the Isle of Mull, is presented by Hey (1936).

In Co. Antrim, Ireland, rare mesolite and scolecite occur with natrolite as tufts of delicate hair-like crystals in amygdaloidal Tertiary basalts from the Garron Plateau area (Walker 1951, 1959a). At Parkgate quarry, Templepatrick, scolecite forms a white base on which delicate needles of mesolite up to 5 mm long are found. Vesicular basalt from Dunseverick contains thin white radiating fibres of scolecite in masses up to 4 cm thick and, less commonly, it forms delicate 1 cm-long sprays on cowlesite and thomsonite (Tschernich 1992). Other Irish occurrences include: Down Hill, Co. Londonderry; Giant's Causeway, Co. Antrim (Wilson et al. 2003); Kane's quarry, Glengormley, near Belfast; Old Ballyclare quarry, Ballyclare; Two-Mouth Cave, S of Portmuck, Island Magee, all Co. Antrim (all Tschernich 1992); South Antrim Escarpment to the W of Belfast – Black Hill quarry, SE of Hannahstown; Hightown quarry, Collinward; Giant's Causeway area – Benanouran Head; Feigh Carrick; Portnabrock; Portnaweelan; Port Noffer, all Co. Antrim; and Moneymore quarry, near Magherafelt, Lough Neagh, Co. Londonderry (all Adair 1987).

Scolecite is listed without comment from Dean quarry, St Keverne, Lizard Peninsula, Cornwall (Golley & Williams 1995).

scorodite

VARISCITE GROUP

orthorhombic

$Fe^{3+}AsO_4{\cdot}2H_2O$

A locally common hydrated iron arsenate that forms a series with mansfieldite and is found as a secondary alteration product of arsenic-bearing sulphides, particularly arsenopyrite. Unusually, this mineral is named after its odour (when heated, it smells like garlic).

Scorodite has been reported from many locations in Cornwall. At Penberthy Croft mine, St Hilary, scorodite forms small lustrous transparent-to-opaque crystals up to 0.5 mm across and spherules to 0.6 mm lining or projecting into cavities or fractures. Its colour ranges from colourless through white, pale-green, bluish, and light to medium brown. Collectively, scorodite crystals form drusy crusts with a botryoidal, earthy, massive or stalactitic habit, and these are often found on heavily iron-stained quartz with arsenopyrite, bayldonite, beudantite, carminite, chalcopyrite, cyanotrichite (uncommon), goethite and pharmacosiderite (Betterton et al. 2000). Pale bluish-green transparent crystals of scorodite occur with pharmacosiderite in samples of the gossan from the old copper mine at Wheal Gorland, St Day, 2 km E of Redruth, and the species is also noted from Wheal Unity, St Day, Gwennap, and most of the mines near St Day (Garby 1848, Greg & Lettsom 1858, Rudler 1905). Later, it was found as attractive pale-blue to greenish radiating masses up to 6 mm across, as crystallized botryoidal crusts, and as well terminated lustrous greenish to pale-blue individual crystals on the dumps of Wheal Gorland (Rust 1981b). Sadly, these dumps were bulldozed away in 1977 (Fisher 1988). Other Cornish occurrences include: Carharrack mine, Gwennap; Crinnis mine, St Austell; Pednandrea mine, Redruth; Tincroft mine, Illogan; Wheal Muttrell, Gwennap (all Collins 1892); Castle-an-Dinas mine, St Columb Major (Hey et al. 1938, Polkinghorne 1954); Cligga Head, Perranzabuloe (Russell & Vincent 1952, Polkinghorne 1954, Chick 1987); Geevor mine, near Pendeen, St Just (Elton & Hooper 1993); Gunheath china-clay pit, Carthew, St Austell (Weiss 1994); Hingston Down Consols mine, 2 km W of Gunnislake, Calstock (Russell 1910a, Green et al. 1996a); Imperial china-clay pit, St Austell (Grigg 1986); Kit Hill Consols, Stoke Climsland (Russell & Vincent 1952); Mulberry mine, Lanivet, St Austell (Russell 1924, Starkey 1986a); Prince of Wales mine, Calstock, near Gunnislake (Braithwaite & Cooper 1982); South Terras mine, St Stephen-in-Brannel (Collins 1876b, 1911a, Dines 1956); Ting Tang mine, Carharrack (Bruce & Aubrey-Jones 1998); and Wheal Pendarves, Camborne (Barstow & Cooper 1982). Three other Cornish locations are noted by Kingsbury & Hartley (1960): Old Treburgett mine, St Teath, E of Port Isaac; Pengenna (Trewethen) mine, St Kew; and Treore mine, Port Gaverne, near Port Isaac, St Teath. However, Ryback et al. (1998, 2001) note that the provenance of many Kingsbury samples requires further investigation.

Isolated occurrences of scorodite are reported from: Hemerdon Bal, Plympton, Devon (Davison 1920a, Ball et al. 1985b, Bernard & Hyršl 2004); Virtuous Lady mine, Buckland Monachorum, Devon (Russell 1913a); Engine Vein mine, Alderley Edge ridge, Cheshire (Carlon 1979), but confirmation is required at this locality (Braithwaite 1994).

Scorodite is relatively common in the Caldbeck Fells, Cumbria. A sulphide-rich exposure on the west bank of Arm o' Grain Beck contains brown to yellowish-brown crusts of scorodite in arsenopyrite–pyrite-rich rock (Green et al. 2006b). At Brandy Gill mine it occurs in a quartz vein just down stream from the waterfall (Day 1928, Davidson & Thomson 1951). Colourless to light-grey drusy scorodite occurs associated with residual arsenopyrite in a wolframite-bearing quartz vein at Burdell Gill. Minute pale-yellow bladed aggregates (<0.1 mm) of scorodite are also noted near by (Neall & Leppington 2003). On the Deer Hills quartz lode, yellowish-brown scorodite forms glistening encrustations on quartz with pharmacosiderite, and a pale-green to bluish-green variety is also locally abundant (Cooper & Stanley 1990, Symes et al. 1990). The arsenate-rich outcrop workings between Driggith and Sandbed mines contain abundant brown to yellow-brown drusy encrustations of scorodite associated with barium-pharmacosiderite, beudantite, carminite and minor mimetite (Green et al. 2006a). Arsenopyrite has altered to yellowish-green massive scorodite in greisen associated with the Skiddaw granite at Grainsgill, Carrock Fell (Hitchen 1934, Davidson & Thomson 1951). On Coombe Height, among the headwaters of Wet Swine Gill, scorodite has been found with beudantite, bindheimite and pharmacosiderite in quartz veinstone (Neall & Green 2001b, Green & Briscoe 2002). Sadly, there is doubt over the provenance of specimens containing scorodite reportedly from Ingray Gill mine, Nether Row Brow and Old Potts Gill mine (Kingsbury & Hartley 1960, Hartley 1984). These occurrences are also noted by Young (1987a) and Cooper & Stanley (1990).

Elsewhere in Cumbria, scorodite occurs at Scar Crag cobalt mine, where it replaces arsenopyrite in quartz veins possibly related to a stock-like intrusion beneath Causey Pike (Ixer et al. 1979a). At Water Crag, near Devoke Water, ferrimolybdite and scorodite are supergene species associated with arsenopyrite–bismuthinite–fluorite–molybdenite mineralization in topaz greisen at the margin of the Eskdale granite (Young 1985d,e). Scorodite from Embleton quarry, near Cockermouth (Young 1987a), and Wanthwaite mine, St John's in the Vale, near Threlkeld (Kingsbury & Hartley 1960), have questionable provenance.

At Gerlan, Bethesda, Snowdonia, Wales, scorodite is confirmed as a post-mining secondary alteration product of arsenopyrite (Jenkins & Johnson 1993, Bevins 1994), and at Clogau mine, Bontddu, near Dolgellau, Gwynedd, it replaces arsenopyrite in veins also containing calcite, chlorite, goethite–lepidocrocite and quartz (Gilbey 1968, Bevins 1994). A grey to pale-yellow mineral forming scattered crystal aggregates and single crystals (<0.3 mm) from Gwaithyrafon mine, SE of Tal-y-bont, Central Wales orefield, Dyfed, has been visually identified as scorodite (Rust & Mason 1994).

Scorodite is noted from one locality in Scotland, as an alteration product of arsenopyrite in a chalcopyrite–pyrite–tourmaline vein associated with the Fore Burn

Unusual blue scorodite from Hemerdon Bal, Plympton, Devon; width 8 mm. (© D. I. Green)

diorite–granodiorite complex, Tairlow plantation, near Straiton, Strathclyde (Allen et al. 1982). It also forms a complex network of veinlets with other secondary species (anglesite, covellite and an unidentified phosphate) at Sark's Hope mine, Little Sark, Channel Isles (Bishop et al. 1977); and, at the South Bradda mines, N of Port Erin Bay, Isle of Man, it is found on fracture surfaces and cavities in arsenopyrite–chalcopyrite–pyrite-bearing quartz veinstone (Starkey 1998).

scotlandite

monoclinic $PbSO_3$

TL Susanna vein, Leadhills, Strathclyde, Scotland

A rare lead sulphite first found on an old specimen of massive barite from the type locality that contains a cavity displaying a variety of lead secondary minerals and which is labelled "lanarkite, anglesite, leadhillite/susannite". In addition, the specimen contains pale-yellow to greyish white and colourless, sometimes transparent, chisel-shape or bladed crystals (up to 1 mm long) and crystal aggregates of an unknown mineral. These crystals were formally described as scotlandite by Paar et al. (1984). A second example in the NMS contains grey scotlandite clusters up to 0.5 mm across, with susannite and mattheddleite, on a galena-bearing lanarkite-rich specimen, also from the type locality (Livingstone 2002).

Scotlandite was found *in situ* at an exposure of the Roughton Gill (South) vein, Caldbeck Fells, Cumbria, where subtranslucent-to-transparent pointed crystals with a tendency to form radiating clusters, or individual chisel-shape blades (exceptionally up to 0.6 mm long), were noted. The crystals are white, inclining towards grey, brown or colourless, and they occur in cavities in oxidized galena, commonly associated with caledonite, lanarkite, leadhillite and mattheddleite (Green 1989). Subsequently, scotlandite has been found at three other locations in the Caldbeck Fells: eastern flank of Balliway Rigg, near Thief Gills, higher Roughton Gill; Red Gill mine; and Short Grain mine (Green & Briscoe 2002). A single specimen of probable scotlandite has been found at Brae Fell mine, Dale Beck Valley, but confirmation is required (Bridges et al. 2006).

A study of the occurrence, morphology and paragenesis of scotlandite in England and Scotland discusses the above examples in some detail, and reports on additional discoveries: five localities in the Leadhills–Wanlockhead district (Broadlaw deep adit, Cove vein, New Cove vein, Horner's vein and Lady Anne Hopetoun shaft); Dry Smale Gill trial, Caldbeck Fells and Whitwell quarry, Derbyshire. At the last locality, scotlandite forms "honey-brown", white and colourless translucent crystals up to 1.2 mm long, and colourless-to-white crystals up to 2.5 mm long, associated with lanarkite and leadhillite (Neall et al. 2006a).

In Wales, scotlandite is reported associated with anglesite and mattheddleite on oxidized galena in a lead-rich vein exposed in 1980 at Machen quarry, 7 km ENE of Caerphilly, Mid Glamorgan. However, the occurrence requires verification (Plant & Jones 1995).

Cluster of scotlandite crystals from Horner's vein, Leadhills, Strathclyde, Scotland; width 245 μm; SEM image. (© D. I. Green)

Sprays of scotlandite crystals (individual crystals to 0.4 mm) from Lady Anne Hopetoun shaft, Leadhills, Strathclyde, Scotland. (© D. I. Green)

scoulerite

TL Portrush, Co. Antrim, Ireland

Named after John Scouler (1804–1871), a surgeon, botanist and naturalist who was professor of geology, natural history and mineralogy at the Andersonian University, Glasgow, and later professor of mineralogy to the Royal Society, Dublin; born in Glasgow. Scoulerite is a variety of thomsonite, zeolite group.

scyelite

TL Loch Scye, Caithness, Highland Region, Scotland

Variously described as a lizardite pseudomorph after enstatite (bastite) and as a mixture of chrysotile ± lizardite or talc or anthophyllite.

segnitite

trigonal $PbFe^{3+}_3H(AsO_4)_2(OH)_6$

A rare complex lead–iron–hydrogen arsenate hydroxide forming globular to powdery microcrystalline coatings in cavities within vein quartz at Penberthy Croft mine, St Hilary, Cornwall. Segnitite is associated with bayldonite, carminite and an unnamed species. It is bright yellow to olive green and it forms acute rhombohedral crystals similar to hidalgoite (Betterton et al. 2000).

The arsenate-rich outcrop workings between Driggith and Sandbed mines, Caldbeck Fells, Cumbria, contain yellow powdery crusts that are either segnitite or beudantite. The two species commonly encrust and replace mimetite, and are themselves overgrown by carminite (Green et al. 2006a).

seifenstein

A synonym of saponite, smectite group.

sekaninaite

orthorhombic $(Fe^{2+},Mg)_2Al_4Si_5O_{18}$

An apparently uncommon iron-dominant (magnesium) aluminosilicate that forms a series with cordierite and which has often been referred to as iron cordierite. Sekaninaite probably occurs more often that the literature suggests, as many identifications of "cordierite" are based on optical determinations, but chemical analysis is required to confirm iron or magnesium dominance, and hence also the species name. The problem is compounded because the name "sekaninaite" was not established until 1968, so the earlier literature does not refer to it at all.

In Cornwall, sekaninaite is confirmed in a sample of garnet–"cordierite"–biotite granite from the Dartmoor granite intrusion at Sweltor quarry, Dartmoor. In many cases "cordierite" in the granite can be attributed to contamination from hornfelsed shale xenoliths and marginal shales, but at Sweltor the "cordierite" appears to be magmatic (Brammall & Harwood 1932). Further work indicates that sekaninaite is transitional to cordierite in composition and that it is the pale lavender-grey and glass-clear lilac-blue varieties that have the highest iron content (Brammall & Rama Rao 1936). Analyses by Stone (1988) of euhedral cyclically twinned "cordierite" from Sweltor quarry are actually sekaninaite. Most analyses of "cordierite" from Cornish granites and entrained enclaves presented by Stimac et al. (1995) are of

Pale-yellow segnitite crust with minor red carminite; Penberthy Croft mine, St Hilary, Cornwall; width 5 mm. (© D. I. Green)

sekaninaite, including: non-igneous enclaves and Carnmenellis granite at Chywoon quarry, near Longdowns; Land's End granite at Sennen Cove; non-igneous enclaves and St Austell granite at Luxulyan quarry, near Lostwithiel.

At Gallowhill Brae, Arnage, Grampian, Scotland, sekaninaite constitutes 90% of a biotite–"cordierite"–spinel hornfels in which plagioclase feldspar is the only other component (Read 1929). In the thermal aureole of the Ross of Mull granite, Strathclyde, cordierite and sekaninaite occur in pelitic sediments previously metamorphosed to amphibolite facies. Coarse almandine, andalusite, biotite, feldspar (plagioclase and potassium feldspar), kyanite, muscovite, quartz and sillimanite are also present in these rocks (Wheeler et al. 2004). Similar pelites in the aureole of the Etive granite complex, Strathclyde, contain euhedral crystals of cordierite and sekaninaite (Droop & Moazzen 2007). It is likely that the ferrous-iron cordierite described by Tilley (1924) from the slopes of Creag na h-Iolaire, in the aureole of the Comrie diorite complex, Tayside Region, is also sekaninaite.

Thermally metamorphosed rock types (hornfels and porcellanite) adjacent to a dolerite plug at Brockley, Rathlin Island, Co. Antrim, Ireland, are said to contain two varieties of cordierite, both richer in iron than normal. One is moderately iron-rich and could be either cordierite or sekaninaite, but the second type is described as iron cordierite (i.e. sekaninaite) and this is found as equidimensional pseudo-hexagonal colourless crystals abundantly spread through a mullite groundmass in one of the porcellanite samples (Wilson & Robbie 1966).

selenite

A transparent variety of gypsum.

selenium?

trigonal Se

A rare native metal reported from a pitchblende-bearing carbonate vein close to the margin of the Criffell granodiorite at Needle's Eye, SW of Caulkerbush, Dumfries & Galloway, Scotland. The vein contains hydrocarbons in which peculiar uranium-rich filaments (apparently of organic origin) have been found. The walls and cells of the filaments are lined with possible native selenium and a calcium–uranium-enriched mineral that may be becquerelite or liebigite. Further confirmation is required (Basham et al. 1989).

seligmannite

orthorhombic $PbCuAsS_3$

A very rare lead–copper arsenic sulphide that forms a series with bournonite and is found in the carbonate-hosted Lisheen Pb–Zn–(Ag) deposit, between Thurles and Rathdowney, Co. Tipperary, Ireland. The deposit consists of three stratabound sulphidic lenses, dominated by galena, marcasite, pyrite and sphalerite, in which seligmannite occurs with arsenopyrite, bornite, chalcopyrite, gersdorffite, gratonite, jordanite and tennantite (Hitzman et al. 2002).

semseyite

monoclinic $Pb_9Sb_8S_{21}$

A rare lead sulphantimonite found in an antimony-rich quartz vein within the aureole of the Skiddaw granite, at Coombe Height, at the headwaters of Wet Swine Gill, Caldbeck Fells, Cumbria. The massive white quartz vein (0.5 m wide) contains an early suite of arsenic–antimony–iron species (arsenopyrite, berthierite and stibnite) and a later suite of antimony–lead species (fülöppite, semseyite and zinkenite). Semseyite forms coatings on what may be late-formed fractures and so is placed late in the sequence of crystallization (Fortey et al. 1984). The supergene species beudantite, bindheimite, claudetite, parasymplesite, pharmacosiderite, scorodite, senarmontite, stibiconite, native sulphur and valentinite are also found at this locality (Leppington & Green 1998, Neall & Green 2001, Green & Briscoe 2002). Semseyite has been noted, but not described, by Golley & Williams (1995) from Wheal Boys, Trewetha, St Endellion, Cornwall, an important producer of antimony in the 18th and 19th centuries. The authors refer to an unpublished manuscript on antimony mines in Great Britain and Ireland by Sir Arthur Russell (1949b), but semseyite is not mentioned. However, the same manuscript does describe massive and rarely crystallized semseyite ore, with jamesonite in a matrix of ankerite, siderite (chalybite), pyrite and quartz from Wheal Emily, Wembury, Devon. These occurrences deserve further investigation.

At Bwlch mine, near Deganwy, Gwynedd, Wales, a semseyite- and stibnite-bearing chalcedony–quartz lode consisting of irregular patches and thin veins and stringers (up to 2 cm wide) occurs in nodular Ordovician ignimbrite. Steel-grey to iron-black semseyite occurs both massive and as flat prismatic crystals up to 2 mm long. The latter are found lining cavities in quartz and are coated by an alteration crust probably of bindheimite. Associated species are anglesite, ankerite, barite, galena, marcasite, plagionite, pyrite, rosiaite, sphalerite, stibiconite and zinkenite, and possibly boulangerite, jamesonite, kermesite and robinsonite (Russell 1944, 1949b, Bevins et al. 1988, Ryback & Francis 2001).

In Scotland, at Louisa mine, Glendinning, Glenshanna Burn, 13 km NNW of Langholm, Dumfries & Galloway, a mineralized quartz vein, dominated by stibnite intimately mixed with galena, was found to contain drusy cavities lined with semseyite crystals up to 0.4 mm across. Associated species are ankerite, calcite, cervantite, pyrite, sphalerite and valentinite (Herbert Smith 1919, Williams 1973, Macpherson & Livingstone 1982). Investigation of drill-core material from four boreholes near the mine also revealed the following sulphides: arsenopyrite, bournonite, chalcopyrite, tennantite and tetrahedrite (Gallagher et al. 1983). A paragenetic model involving large-scale faulting and brecciation in the host Silurian greywackes, followed by disseminated arsenic–gold mineralization and later polymetallic quartz-stibnite veining, was proposed for the Glendinning deposit by Duller et al. (1997).

Semseyite is a minor component of the large stratiform sphalerite–galena–(marcasite–pyrite) deposit at Navan mine, Co. Meath, Ireland. It occurs with other sulphosalts such as argyrodite, boulangerite, bournonite, cylindrite, franckeite (a single grain), freibergite, jordanite and pyrargyrite (Andrew & Ashton 1985, Ashton et al. 1986, Ryback & Moreton 1993, Ashton et al. 2004).

senarmontite/sénarmontite

cubic Sb_2O_3

A rare antimony oxide, dimorphous with valentinite and found in an antimony-rich quartz vein within the aureole of the Skiddaw granite, at Coombe Height, at the headwaters of Wet Swine Gill, Caldbeck Fells, Cumbria. The massive white quartz vein (0.5 m wide) contains an early suite of As–Fe–Sb species (arsenopyrite, berthierite and stibnite) and a later suite of antimony–lead species (fülöppite, semseyite and zinkenite). Senarmontite is one of the less common supergene species that occurs as grey masses and minute colourless octahedra derived from the alteration of stibnite, and is found with beudantite, bindheimite, claudetite, parasymplesite, pharmacosiderite, scorodite, stibiconite, native sulphur and valentinite (Fortey et al. 1984, Cooper & Stanley 1990, Leppington & Green 1998, Neall & Green 2001, Green & Briscoe 2002). There is some doubt about the senarmontite described by Young (1987a) from Wanthwaite mine, St John's in the Vale, near Threlkeld, Cumbria, as this is based on material supposedly collected by A. W. G. Kingsbury, whose collection, now at the NHM, contains many mislabelled specimens (Ryback et al. 1998, 2001; A. Hart pers. comm. 2004).

A vaguely located specimen of senarmontite, forming regular octahedra and accompanied by cerussite, is described from near St Endellion, Cornwall (Davies 1867a). This is possibly Port Quin mine, St Minver Highlands, where the dumps yielded abundant jamesonite superficially altered to bindheimite with cavities containing small crystals of anglesite and minute octahedra of senarmontite (Russell 1944, Russell 1949b). Senarmontite was also found at Bodannon mine, St Endellion, with jamesonite (apparently the main ore species), anglesite, bindheimite, nadorite, valentinite and a single specimen of native silver (Russell 1949b). Three other Cornish occurrences of senarmontite are listed by Golley & Williams (1995): Bounds Cliff, Port Gaverne, near Port Isaac, St Endellion; East Wheal Leigh, Pillaton; and Treore mine, St Teath.

Senarmontite is briefly reported as rare minute (<0.5 mm) octahedral crystals in cavities in stibnite at Hare Hill, 5 km SE of New Cumnock, Strathclyde, Scotland (Macpherson & Livingstone 1982), and more vaguely from the Dolgellau goldbelt, Gwynedd, Wales (Readwin 1888).

sepiolite

orthorhombic $Mg_4Si_6O_{15}(OH)_2 \cdot 6H_2O$

An uncommon hydrated magnesium silicate hydroxide found in a coastal exposure of serpentinite, 500 m SE of Kynance Cove hotel, Lizard Peninsula, Cornwall. Fibrous sepiolite occurs in a 5 cm-wide calcite-rich veinlet and is thought to be an alteration product of the serpentinite (Caillère & Hénin 1949). Power et al. (1997) report that sepiolite is abundant in fractures within the Lizard peridotite and they note that it forms fibrous mats more than 3 mm thick at the eastern end of Kennack Sands. Sepiolite is also noted in the Keuper Marl from an undisclosed location in the Midlands of England. A sample representing 9 m of stratigraphical thickness contains up to 39% sepiolite, with the remainder comprising calcite, ferric oxide, mica, quartz and a trace of kaolinite (Keeling 1956). Microscopic fibres of sepiolite forming mats and tangled masses fill voids in Keuper Marl from Dunscombe and Weston Cliffs on the south Devon coast (Jeans 1978).

In Scotland, a talc–carbonate rock from Swinna Ness, Unst, Shetland, contains a soft white powdery mineral confirmed by XRD to be sepiolite (Macpherson & Livingstone 1982).

sericite

Fine-grain muscovite or paragonite, mica group; often found as a secondary alteration product of feldspar.

serpentine

A rock name, also used as part of a group name (kaolinite–serpentine group). Serpentine was once synonymous with chrysotile, and early works list it as a species name. Many examples are listed in these works, such as the plentiful quantity at Lizard Point, Cornwall, and the very fine varieties at Portsoy, Grampian, Scotland, and Ballynahinch, Co. Down, Ireland (Greg & Lettsom 1858); these all deserve further investigation. Later work confirms that some serpentine at Lizard Point is a mixture of chrysotile and lizardite (Midgley 1951, Green 1964, Power et al. 1997).

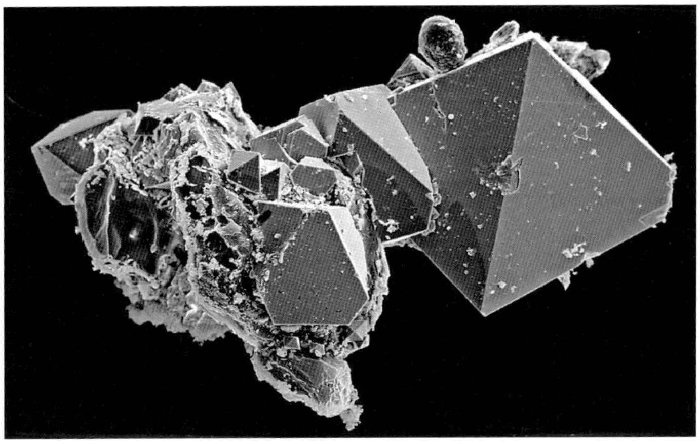

Minute octahedra of senarmontite replacing stibnite from the antimony-rich vein in Wet Swine Gill, Caldbeck Fells, Cumbria; width 450 µm; SEM image. (© D. I. Green)

A history of serpentine working in Cornwall is provided by Sagar-Fenton & Smith (2005).

serpierite

monoclinic $Ca(Cu^{2+},Zn)_4(SO_4)_2(OH)_6 \cdot 3H_2O$

A rare hydrated calcium–copper–zinc sulphate hydroxide, dimorphous with orthoserpierite and found as a secondary mineral (sometimes post-mining) in the oxidized zone of hydrothermal copper–zinc-bearing mineral deposits. Serpierite is noted from Penberthy Croft mine, St Hilary, Cornwall (Golley & Williams 1995), but the occurrence is not confirmed by Betterton et al. (2000), who suggest that confusion with devilline may have occurred. A visual identification of serpierite on a single specimen showing small irregular sprays of pale-blue acicular crystals, up to 2 mm in length and closely associated with devilline, is reported from Wheal Friendship (Devon Friendship), Mary Tavy, Devon (Brooke 1986).

In the Southern Pennine orefield, serpierite occurs rarely as crusts of beautiful blue microscopic blades closely associated with gypsum and chalcopyrite in the Clayton adit, Ecton Hill, Wetton, Staffordshire (Braithwaite & Knight 1968, Braithwaite 1983). At Waterbank mine, Ecton Hill, serpierite forms light-blue crusts and divergent groups of lath-like crystals to 2 mm in cavities in oxidized sulphide-rich material. Linarite, malachite, namuwite, posnjakite, ramsbeckite and schulenbergite are associated species (Rust 1992, 1995c). Serpierite is also reported from Isolation mine, Snelston, Derbyshire (Ford et al. 1993).

In the Northern Pennine orefield, pale-blue to greenish-blue crystalline crusts of serpierite and devilline were found on broken veinstone in old workings at Tynebottom mine, NW of Garrigill, near Alston, Cumbria. Serpierite occurs as radiating sprays of turquoise-blue acicular crystals, although devilline is the main component of the crusts, which are up to 3 mm across. Associated species are aragonite, brochantite and gypsum (Bridges 1987, Rust 1987a, Bridges & Young 1998). At Smallcleugh mine, Nenthead, near Alston, Cumbria, bright turquoise-blue serpierite forms acicular crystals up to 2 mm long in crusts of gypsum containing minor devilline (Bridges 1987, Bridges & Young 1998). Livingstone et al. (1990) found turquoise- to copper-blue serpierite at Smallcleugh mine, forming 1 mm-long bundles comprising wispy or stellate acicular groups, associated with late-forming namuwite and schulenbergite. The zinc analogue of ktenasite is a further associated species (Livingstone 1991, Rust 1991b). At both mines, serpierite, devilline and the other supergene minerals have a post-mining origin and appear to have formed where circulating groundwater has passed through limestone containing spots of sulphides. This origin is proposed for the blue lath-like crystals of serpierite associated with brochantite, gypsum and harmotome at Brownley Hill mine, Nenthead (Green et al. 2000), and the serpierite associated with gypsum and schulenbergite from Nentsberry Haggs mine, Nenthead (Thomson 1997a), presumably formed in the same way.

Elsewhere in Cumbria, serpierite has been found in a small area of supergene mineralization at Force Crag Ba–Pb–Zn mine, at the head of Coledale, 8 km W of Keswick, where it forms radiating aggregates of pale-blue feathery to lath-like

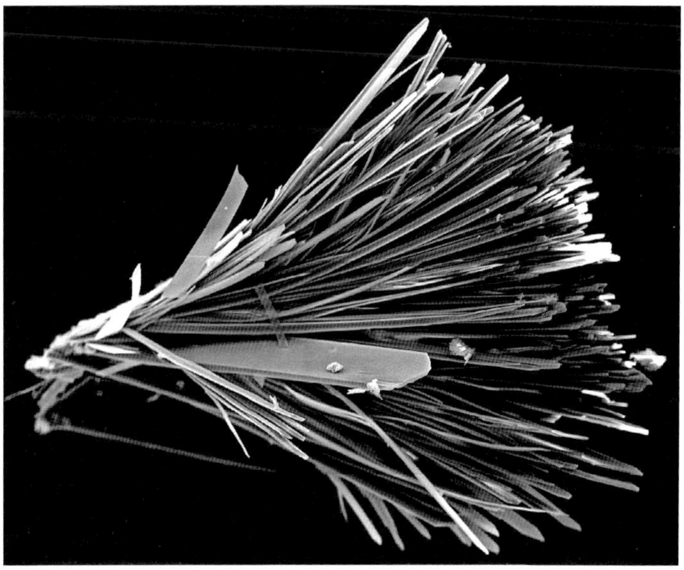

Divergent spray of serpierite crystals from Force Crag mine, Braithwaite, near Keswick, Cumbria; width 320 µm; SEM image. (© D. I. Green)

Radiating clusters of serpierite crystals, up to 1 mm long, from Ystrad Einion mine, near Tal-y-bont, Dyfed, Wales. (© National Museum of Wales, Cardiff)

crystals up to 1.5 mm long, commonly associated with linarite. Cerussite, brochantite, langite, namuwite, pyromorphite and native silver were also recovered from the same area (Green et al. 1997, Green & Todd 2002). Reports of serpierite in the Caldbeck Fells from Driggith and Old Potts Gill mines (Kingsbury & Hartley 1957e, Hartley 1984, Young 1987a) are problematical, as they could be part of the incorrectly provenanced material described by Ryback et al. (1998, 2001). Cooper & Stanley (1990) note that the Kingsbury samples from Driggith mine are typical of dump-formed serpierite, suggesting there is probably no reason to doubt their authenticity. Green et al. (2001, 2006a) have been able to provide independent confirmation of serpierite as pale-blue lath-like crystals from Driggith and Sandbed mines, and the species has also been found a few kilometres away at Red Gill mine (Cooper & Stanley 1990).

At the Old Gang lead mines, E of Gunnerside Gill, Swaledale, Yorkshire, minor quantities of ktenasite and serpierite are found with gypsum on the mine walls of Lucky String, the western section of Alderson's vein. Crystallizing on top of these minerals are overgrowths of the rare species gearksutite (Green 2003).

Serpierite has been noted from a few mines in the Central Wales orefield (mainly Dyfed): pale-blue microscopic radiating blades associated with brochantite, cerussite, native copper, cuprite, devilline, hemimorphite, leadhillite, linarite, malachite and wroewolfeite at Eaglebrook (Nantycagl) mine, Ceulanymaesmawr (Braithwaite 1982b, Jones 1983); light-blue intergrown contorted lath-like crystals filling cavities and associated with anglesite, brochantite, caledonite, langite, leadhillite and linarite at Esgair Hir mine, Ceulanymaesmawr (Rust & Mason 1988); very rare broad bluish-green plates up to 0.5 mm across and sky-blue lath-like crystals up to 0.7 mm long, associated with cerussite, hemimorphite and langite at Frongoch mine, between Devil's Bridge and Pont-rhyd-y-groes (Bevins 1994, Green et al. 1996b); pale-blue pearly lath-like crystals, up to 0.5 mm long, tentatively identified as devilline or serpierite, at Gwaithyrafon mine, SE of Tal-y-bont (Rust & Mason 1994); tangled blue fibres in minute amounts, as coatings on dolomite and overgrowths on posnjakite, schulenbergite and ramsbeckite at Penrhiw mine, Ystumtuen (Mason & Green 1995); turquoise-blue radiating sprays and rosettes up to 5 mm long, probably the finest yet discovered in Britain, at Ystrad Einion mine, near Tal-y-bont (Bevins 1994, Mason & Rust 1997); and in Powys as sky-blue tufted aggregates of lath-like crystals and also as botryoidal crusts associated with calcite or linarite at Dylife mine, N of Hafren Forest (Jones & Moreton 1977, Moreton 1982).

Pearly fibrous to lath-like aggregates of sky-blue serpierite are fairly common among a suite of secondary minerals in dump material from the West Blackcraig lead–zinc mine, 4 km SE of Newton Stewart, Dumfries & Galloway, Scotland. The dumps also contain the primary ore species chalcopyrite, galena and sphalerite, and minor sulphides marcasite, millerite, pyrite and siegenite, and the secondary species aurichalcite, brochantite, cerussite, hemimorphite, hydrozincite, langite, linarite, posnjakite and wroewolfeite (Livingstone 1976). Anthropogenic serpierite is also present in slag material from the Meadowfoot smelter, Wanlockhead, Dumfries & Galloway, Scotland (Green 1987a).

In Ireland, well defined crystals of serpierite were found on the dumps of an old copper mine at Ross Island, Killarney, Co. Kerry (Russell 1927), but only unattractive specimens were found by Ryback & Moreton (1991). Anglesite, aurichalcite, hydrozincite and linarite are associated species.

serpophite

A compact variety of serpentine, kaolinite–serpentine group.

shattuckite?

orthorhombic $Cu_5(SiO_3)_4(OH)_2$

A rare copper silicate hydroxide reported with ceruleite at the Wheal Maid decline, Gwennap, Cornwall (Golley & Williams 1995). Requires confirmation.

sheridanite

An aluminium-rich variety of clinochlore, chlorite group.

sicklerite

orthorhombic $Li(Mn^{2+},Fe^{3+})PO_4$

A rare lithium–manganese–iron phosphate that forms a series with ferrisicklerite. Sicklerite occurs in borehole material from the Aclare spodumene pegmatite, Co. Carlow, Ireland. A single grain of orange sicklerite, 600 µm across, has been found mantled by fluorapatite, adjacent to manganotantalite and a large altered grain of spodumene (Tindle unpub. EMP data).

siderite

CALCITE GROUP

trigonal $Fe^{2+}CO_3$

A very common iron carbonate that forms three series – with magnesite, rhodochrosite and smithsonite – and is found as a constituent of bedded sedimentary iron ores and metamorphic iron formations. It also occurs as a late-stage gangue mineral in (often low-temperature) hydrothermal veins, particularly in Cornwall and Devon.

England

Cornwall Pale lemon-yellow to bright brownish-red siderite is a common gangue species in tin-bearing lodes within the Botallack, Wheal Owles and Parknoweth sections of Botallack mine, St Just, where it forms simple rhombohedra, complex aggregations of offset crystals and botryoidal masses (Greg & Lettsom 1858, Hosking 1951, Bancroft & Weller 1993). However, it is from Wheal Maudlin, Lostwithiel, in the aureole of the St Austell granite, that some of the most impressive siderite (syn. chalybite) has been found. Here, well formed cream and brownish-grey short prismatic crystals, up to 13 mm across, with concentrically zoned pseudo-hexagonal cross sections, partially infill cavities in mineralized lodes (Greg & Lettsom 1858, Rudler 1905, Kingsbury 1961, Embrey & Symes 1987, Penhallurick 1997). Collins (1892) adds that Charlestown United and other mines near St Austell also produced "very perfect" and irridescent siderite. These include Crinnis mine, near Charlestown, and Fowey Consols, Tywardreath, St Blazey (Greg & Lettsom 1858, Kingsbury 1964b, Lewis 1997). Museum-quality specimens forming stout prisms up to 1.7 cm long have also been recovered from the little-known Great Onslow Consols copper mine, St Breward (Embrey & Symes 1987). At Penberthy Croft mine, St Hilary, siderite is a common vein mineral forming botryoidal and globular infillings to vugs and cavities within vein quartz. Tabular crystals up to 0.8 mm long, massive and earthy forms are also reported (Betterton et al. 2000). Siderite is the primary iron-bearing mineral in the Perran Iron lode, near Perranporth, although the main source of ore was the upper oxidized zone, where siderite has been altered to goethite and limonite (Slater & Highley 1976). The Restormel (Royal) iron mine, N of Lostwithiel, also worked an oxidized zone above a primary siderite deposit (Hosking 1951,

Short prismatic crystals of siderite exhibiting compositional zonation; Wheal Maudlin, Lanlivery, Cornwall; crystals are mainly about 13 mm across. (© The Natural History Museum, London)

Slater & Highley 1976). In the Camborne–Illogan–Redruth mining district, mineralization is related to the intrusion of the Carn Brea and Carn Marth granites, and it includes (often well formed) siderite at: Carn Brea mines, near Redruth (Robson 1945a); Cook's Kitchen mine, Illogan (Collins 1892); Dolcoath mine, Camborne (Hosking 1951); East Pool mine, Illogan (Rudler 1905); South Crofty mine, Camborne (Wolters & Wolters 1987, LeBoutillier 1996, Wolfe 1998); South Tolgus shaft, Redruth (Davey 1948); Tincroft mine, Illogan (Embrey & Symes 1987). The north Cornwall (Port Isaac–St Endellion) antimony–lead district produced siderite at: Bounds Cliff, E of Port Isaac; Reedy Cliff, near Port Quin, and Pentire (Glaze) mine, near New Polzeath (Clayton et al. 1990); Port Quin mine, and Treore mine, Port Gaverne (Russell 1949b, Hosking 1951); and Wheal Boys, St Endellion (Rudler 1905). Around Liskeard, at the southern margin of the Bodmin Moor granite, siderite is reported from: Herodsfoot mine (Dewey 1925); Wheal Ludcott and Wheal Wrey, St Ive – as pale creamy-brown well formed lenticular crystals up to 1 cm long and noted with rich silver ore (Hosking 1951, Dines 1956, Rust 1982a, 1983a). Siderite is also noted from: Bodannon mine, Port Gaverne (Russell 1927); mines in the neighbourhood of Camborne (Hutchinson 1903); Carrick Du mine, near Porthmeor (Russell 1910b); Duchy Peru mine, Perranzabuloe, and Great Retallack mine, near St Agnes (Henley 1971); East Wheal Lovell and Wheal Roots, both Wendron district – in cassiterite-bearing carbonas (Ball et al. 1982, Dominy et al. 1996a); Fowey Estuary – as detrital grains in intertidal sediments (Pirrie et al. 2002); Gravel Hill mine, Perranzabuloe (Weiss 1989); Greystone quarry, Lezant, near Launceston; Strangles beach, St Gennys; Old Treburgett mine, St Teath; Trewethen mine, St Kew, and Wheal Drea, St Just (all Golley & Williams 1995); Holmbush mine, Stoke Climsland (Russell 1913a); New Consols mine, Stoke Climsland, and Trenute mine, near Lezant (Hosking 1951); Parson's beach, below Mawnan church (Camm 1995); Perran Silver mine (Wheal Sedgman lode), Perranuthnoe (Butler 1907); Silver Vein mine, Lostwithiel; Wheal Newton (Harrowbarrow) mine, near Gunnislake (Lewis 1882, Rust 1982a); West Wheal Darlington, Marazion and Wheal Ann, Phillack (Rust 1982a); Redmoor–Kit Hill sheeted-vein complex, N of Callington (Newall & Newall 1989, Newall 1994); South Terras mine, St Stephen-in-Brannel (Ball et al. 1982); Wheal Betsy, Tremore, Lanivet (Ball et al. 1982); Wheal Golden, Perranzabuloe (Merry & Braithwaite 1995); Wheal Hope, ENE of Perranporth – impressive amounts laid down after galena had formed (Hosking & Camm 1981); Wheal Jane, Kea, near Truro (Rudler 1905); Wheal Mary Ann, Menheniot (Le Neve Foster 1878b, Merry & Braithwaite 1995); Wheal Penrose, Sithney (Hall & Jackson 1975); Wheal Rose, Sithney (Russell & Hutchinson 1927, Hosking 1951); and Wheal Trenwith, St Ives (Ball et al. 1982). Botryoidal masses of siderite (sphaerosiderite) resembling calamine are reported from Madron, near Penzance, and from Mount mine, near Marazion (Greg & Lettsom 1858, Collins 1892).

Cumbria In the Northern Pennine orefield, siderite forms interlocking crusts of opaque, dark-brown to black, dull to lustrous, crystals, commonly encrusting galena at Brownley Hill mine, Nenthead, near Alston (Green et al. 2000), and it occurs crystalline at Rotherhope mine, Leadgate, near Alston (Davidson & Thomson 1951). Minor siderite is found in quartz gangue at Smallcleugh mine and Handsome Mea flats, Nenthead, where it is associated with chalcopyrite, galena, gersdorffite, pyrite and sphalerite (More et al. 1991). In the Lake District, siderite is a minor gangue species in barite–galena–sphalerite veins at Force Crag mine, Braithwaite, near Keswick (Young & Cooper 1988, Green et al. 1997, Green & Briscoe 2002), and in arsenopyrite–chalcopyrite–pyrite veins at Seathwaite mine, Seathwaite Tarn, Coniston (Stanley & Criddle 1979). It is also reported from the Simon Nick area, Coniston, cementing brecciated chalcopyrite–pyrite ore in an altered rhyolite (Dagger 1977) and with little description from Barrow mine, Braithwaite; Brigham area, Keswick; Capel Gill, Seathwaite; Middle Tongue Gill, Grasmere (all Young 1987a). At Oughterside opencast coal site, Aspatria, siderite lines cracks in nodules within Coal Measures mudstone (Young & Nancarrow 1988a). Siderite is said to be absent from the hematite deposits of West Cumbria (Dunham et al. 1978), although Trotter (1945) describes it encrusting kidney ore at Crowgarth mine on Cleator Moor. The species has also been found in Ordovician Coniston Limestone at Greenscoe Crags, near Askham-in-Furness (Green 1913), and in Ordovician mudstones at Park farm, also near Askham-in-Furness (Rose & Dunham 1977). Additionally, it is reported from: Dalzells Moor Row mine, Moor Row and Salterhall mine, Arlecdon (Young 1987a); Postlethwaite's Eskett and Windergill mines, Arlecdon (Young 1984a); and Beckermet iron ore mine, near Egremont (Goldring & Greenwood 1990).

Derbyshire Siderite is rare in the Southern Pennine orefield, although it is present in nodules within the Carboniferous Edale Shales (e.g. Mam Tor, Castleton) and Coal Measures. It is also noted without description from Pindale, Castleton, and there are old records of its occurrence on Brassington Moor and at Cumberland mine, Matlock (Ford & Sarjeant 1964a, Ford et al. 1993).

Devon The Virtuous Lady copper mine NW of Buckland Monachorum, near Tavistock, has yielded many fine specimens of siderite, none more so than the hollow cubic shells (epimorphs after fluorite) of dark- to light-brown siderite forming translucent lenticular crystals that have aggregated into globular masses to produce "boxes" up to 10 cm across (Greg & Lettsom 1858, Hall 1868b, Rudler 1905, Embrey & Symes 1987). One example in the collection of the NHM has irridescent chalcopyrite and white prismatic quartz growing in the hollow and caused Russell (1913a) to remark "It is impossible to describe adequately the beauty of these specimens". This locality is also well known for a long flat tapering variety of rich-brown siderite, forming crystals up to 31 cm long and termed "lady's slippers". They are thought to be epimorphs after barite (Russell 1913a, Embrey & Symes 1987, Turner 1990). Fine crystals of siderite are also reported from other mines near Tavistock, especially Wheal Betsy, Wheal Crebor, Wheal Friendship and the Bedford United mines, whereas in the Bere Alston mines, near Bere Ferrers, siderite pseudomorphs after calcite occur (Hall 1868b). In the same general area, siderite is noted from: Franco Consols mine, Horrabridge, near Tavistock (Dines 1956), and George and Charlotte mine, Tavistock Hamlets (Embrey & Fejer 1969), as well as from: Bampfylde mine, North Molton and the Combe Martin mines (Dunham et al. 1978, Rust 1982a); Britannia mine, near North Molton (Camm 1995); Hemerdon Bal, Plympton (Keats 1981); British Railways quarry, Meldon, near Okehampton (Dearman 1965); Plympton railway cutting – "of a beautiful white" (Greg & Lettsom 1858, Hall 1868b); St Mary's Bay, Brixham (Braithwaite & Wilson 2001a); Teign Valley, near Hennock (Beer & Ball 1977); and Wheal Emily, Wembury (Russell 1949b).

Durham Very fine-grain siderite is the principal constituent of the primary (replacement) iron ore known as fine steel-grey ore that was worked from near-surface flats and veins in limestone at many iron mines in Weardale, such as Boltsburn and Rispley mines, Rookhope. At Boltsburn, cavities in metasomatized iron-rich carbonate rocks produced well formed (normally honey-yellow) siderite crystals with curved faces associated with fluorite or galena (King 1982b, Dunham 1990). Data from Carricks mine, Weardale, indicate siderite is both magnesium- and manganese-rich (Smythe & Dunham 1947). Frazer's Hush mine, Weardale, well known for its purple fluorite, has also produced attractive specimens of siderite (Wolfe 1998, Green & Briscoe 2002). Other localities include: Heights quarry, Weardale (Green & Briscoe 2002); Old Moss vein, Killhope Burn, Weardale; and West Rigg quarries, N of Westgate, near St John's Chapel (Young 1995).

Leicestershire At Old Parks coal pit, NE of Ashby-de-la-Zouch, well crystallized siderite has been found in Carboniferous septarian clay–ironstone nodules (King 1993). Other occurrences of well crystallized siderite, including the variety sphaerosiderite, are noted in mudstone nodules from a coal pit at Albert Village. Crystals are colourless or tinted light orange and have sharp rhombohedral form with slightly curved faces (King 1973).

Lincolnshire The Jurassic Frodingham Ironstone, worked in the Scunthorpe area, was a major source of iron until relatively recently. It is composed of chamosite ooliths (in parts replaced by siderite and calcite), sideritized shell fragments and pebbles of limestone usually containing more siderite that the surrounding matrix. Siderite is probably diagenetic. The Claxby Ironstone, a Cretaceous chamositic oolitic limestone with a sideritic and chamositic matrix, mined at Nettleton, between Audleby and Donnington-le-Bain, was mixed with the Frodingham ore during smelting (Slater & Highley 1976).

Norfolk Siderite concretions are common in intertidal marsh sediments on the coast of East Anglia. Samples from Wareham Marshes show that siderite forms spherulitic grains as a passive void-filling cement and is associated with calcite, potassium feldspar, greigite, illite, kaolinite, quartz and iron monosulphides (Pye 1981, 1984).

Northamptonshire The Northampton Ironstone is the most prominent Jurassic oolitic limestone in Britain. It contains berthierine, siderite and pyrite as the major iron-rich phases, and investigations at Cowthick quarry, Corby, suggest that they all formed in a marine environment during diagenesis (Gehring 1990). Siderite is also reported from the Northampton Ironstone at Lodge Pit, Irchester (Hollingworth & Bannister 1950), Stanion Lane pit, NE of Stanion Lodge, near Corby; Thingdon mine, E of Finedon church, and West Field Lodge, WSW of Irthlingborough church (Guppy 1956). Very fine-grain Middle Lias (Jurassic) siderite (chalybite) mudstone nodules containing septarian cracks infilled with barite and calcite have been recovered from the banks of the River Welland, near Rockingham (Taylor 1950).

Northumberland Some of the finest fluorite ever found in the Pennines is associated with pale-brown siderite and sphalerite, and was recovered from Beaumont mine, Allenheads, in the 1970s (Green & Briscoe 2002). St Peter's mine, Allenheads, has been worked for fluorite specimens and these contain coatings of brown siderite and occasional doubly terminated quartz crystals (Wolfe 1996b, 1997, Green & Briscoe 2002). Metasomatic alteration of quartz dolerite (White Whin) at Settlingstones mine, near Hexham, has produced a rock containing anatase, ankerite, calcite, illite, kaolinite, muscovite, pyrite and siderite adjacent to barite–witherite mineralization (Ineson 1972).

Staffordshire Siderite is an important gangue mineral in the Ecton copper mines, although it is usually altered to limonite (Sarjeant 1956a).

Tyne & Wear Lenticular siderite is reported from an apatite-bearing ironstone nodule in Coal Measures siltstone at Beggars Wood, near Gateshead (Neall et al. 2004).

Warwickshire A mineralized Coal Measures clay–ironstone septarian nodule from Newdigate colliery, Bedworth, SW of Nuneaton, was found to contain siderite-coated septa voids in which chalcopyrite, millerite, galena, retgersite, sphalerite and zaratite have grown (Walker 1988).

Yorkshire Millstone Grit and Coal Measures rocks in West Yorkshire include a series of sedimentary ironstones characterized by bedded or nodular spherulitic siderite (sphaerosiderite). Nodules are build up of narrow fibrous crystals radiating from the centre and they vary from 5 cm to 60 cm across, grading into lenticular masses up to 3 m long. Septarian structures are not developed. Localities include: Ashday, Elland; Beacon Hill and Siddal, Halifax; High Birks pit, Thornton, W of Bradford; Nab Scar, Oxenhope; Robin Hood quarries, near Wakefield; Snail Green borehole, Rombalds Moor, Kinderscout; Wamesley Scar, Mixenden (all Deans 1934). Siderite in Coal Measures concretions and sheets from Hazlehead (Hepworth Iron Company quarry), Penistone, has extremely fine grain and is intimately mixed with clay, organic matter and phosphates. A calcium-rich siderite and a magnesium-rich variety (pistomesite) are both present (Pearson 1974, Curtis et al. 1975). Chamositic oolite from the Jurassic Yorkshire Dogger at Rosedale East mine contains sideritized ooliths and a light- to dark-brown siderite matrix capable of producing 39% metallic iron (Rastall & Hemingway 1949). Siderite is also found in ironstones at: Blakey mines, Farndale; Rosedale Abbey Bank, Rosedale; Sheriff's pit, Rosedale; West Gill, Northdale (all Rastall & Hemingway 1949); North Cheek, Robin Hood's Bay; West Kilns, Rosedale (Hemingway & Riddler 1982). Dunham & Wilson (1985) note siderite at Old Providence, near Kettlewell, and in cavities in veinstone from the Copperthwaite vein, Swaledale.

Wales

Dyfed (Powys) Siderite is reported in the Central Wales orefield from: Siglenlas mine, near Devil's Bridge; Ystrad Einion mine, near Tal-y-bont; and Geufron mine, near Llanidloes, Powys (Jones & Moreton 1977). However, Bevins (1994) warns that Fe-rich dolomite may have been mistaken for siderite at these three locations. The report of siderite from Eaglebrook (Nantycagl) mine, Ceulanymaesmawr (Jones 1983) requires confirmation. Greg & Lettsom (1858) report simple rhombs of siderite on quartz from Goginan mine.

Gwynedd Ordovician ironstones containing siderite have been found at: Bryn Poeth, Anglesey; Garreg Fawr and Ystrad mines, near Betws Garmon, Snowdonia; Llandegai, near Bangor; Maes y Gaer, near Aber; and Tremadog, Snowdonia (Bevins 1994). The variety sphaerosiderite has been found at Pen y Gaer, Llangian, Lleyn Peninsula, where it forms radiating spherules up to 0.55 mm across (Pulfrey 1933). At Pen-yr-allt iron mine, near Llanfrothen, Penrhyndeudraeth, siderite infills fissure veins in an ironstone bed, close to an igneous intrusion (Hallimond 1924).

Mid Glamorgan Siderite is the principal carbonate component in clay–ironstone nodules of the South Wales coalfield (North 1916, Firth 1971) and has been reported as orange-brown six-point star-shape crystals, 1.5 mm across, from International colliery, Blaengarw, near Maesteg (Bevins 1994), and as white six-point stars up to 0.4 mm across from Wyndham colliery, Ogmore Vale (Anon. 1989a). Rare stellate groups of siderite up to 1 mm across were also found on the dumps of Hafod pit, Rhondda prior to tip reclamation and landscaping (Wolfe 1996a).

Scotland

Central Siderite has been identified as inclusions in pyrite in copper–zinc strata-bound mineralization within the Dalradian Ben Challum Quartzite at Creag Bhocan, SSW of Tyndrum (Scott et al. 1988). Pattrick & Hall (1984) found a lead–zinc mineralized boulder in the Crom Allt River, near Tyndrum, containing siderite as rhombohedra up to 5 mm across overgrowing acicular quartz.

Dumfries & Galloway Siderite (chalybite) is listed without detail from the Leadhills–Wanlockhead orefield (Brown 1927); Heddle (1901) notes siderite with ores of copper and lead from Pibble mine, 5 km NE of Creetown.

Highland A mineralized breccia at Bridge of Westfield in the Loch Calder crush zone contains chalcopyrite, galena, marcasite, pyrite, smithsonite, sphalerite and uranium-rich hydrocarbons in a calcite–siderite gangue (Gallagher et al. 1971). The marginal facies of a fine-grain basaltic dyke at Sròn Bheag headland, Ardnamurchan, W of Kilchoan, is unusual in being carbonated, with siderite occurring in microgranular aggregates, probably as a result of a build-up of CO_2 during magmatic crystallization (Wells et al. 1974). Jurassic ironstone on the Isle of Raasay (near the Isle of Skye) consists of oolitic chamosite, partially replaced by siderite (Berridge 1969, Slater & Highley 1976).

Siderite rhombs, up to 5 mm on edge, from Llantrisant, Mid Glamorgan, Wales. (© National Museum of Wales, Cardiff; photographer M. P. Cooper)

Strathclyde Translucent yellowish-brown microcrystals of siderite on calcite have been found in Carboniferous Limestone at Trearne quarry, near Beith (Todd 1989). Heddle (1901) reports crystallized siderite (chalybite) with chalcopyrite and galena in quartz veins cutting mica schist to the SW of Inverneil; and at Erines, near Tarbert, it forms simple rhombs associated with chalcopyrite and goethite; both localities are in Knapdale. Siderite also forms clear rhombs in a quartz vein cutting chlorite schist at the head of Loch Long, near Arrochar.

Tayside Iron-rich dolomite and siderite are minor gangue minerals in gold- and sulphide-bearing vuggy quartz veins at Calliachar Burn, Aberfeldy (Ixer et al. 1997). Siderite and calcite were found on dump samples from galena-bearing quartz-vein mineralization at Corrie Buie, Meall nan Oighcreag, and, at nearby Tomnadashan, Loch Tay, siderite is one of the gangue species in chalcopyrite–molybdenite–pyrite–tetrahedrite vein mineralization (Wilson & Flett 1921, Pattrick 1984). Dark siderite is a late metamorphic or supergene phase found occasionally in the stratiform sulphide deposit at Foss, near Aberfeldy (Moles 1985).

Shetland Sandlodge mine on Mainland was exploited unsuccessfully for its iron ore, which consisted of primary siderite (sideroplesite) and chalcopyrite, weathered to limonite and hematite near the surface. The deposit is in a fault-controlled 3–5 m-wide vein in Old Red Sandstone (Berridge 1969, Livingstone 2002). Heddle (1901) notes siderite in veins on Aithness Hill on Mainland. Siderite nodules are associated with molybdenite-bearing chlorite veins at Queyhouse quarry, Unst (Neary & Prichard 1985).

Ireland

Cork Siderite and quartz are the major gangue species in coarse-grain copper-rich veins at Dhurode in the West Carbery mining district (Ni Wen et al. 1991). In the Berehaven (Bearhaven) copper mines, Allihies, siderite appears to be less common, with quartz the dominant gangue species (O'Brien 1959, Moreton & Aspen 1993).

Monaghan Siderite is noted as a gangue species in the barite–galena vein deposit at Hope mine, Cornalough–Carrickagarvan (Morris 1984).

Tipperary Siderite is a minor species in the sedimentary-exhalative barite–pyrite deposit at Ballynoe mine, near Silvermines (Mullane & Kinnaird 1998). There are also references to siderite as a hydrothermal mineral in the lead–zinc deposit at Silvermines, although details are sparse (Rhoden 1959, Samson & Russell 1983, Taylor 1984, Moreton 1999).

Waterford Siderite was occasionally found in the rich copper deposits at Knockmahon, 22 km SW of Waterford city (Holdsworth 1834, Moreton 1991).

Wicklow A metamorphosed BIF consisting of chalcopyrite, magnetite, pyrite and siderite occurs in Ordovician volcanic rocks close to the Goldmine River valley, near Avoca, and has been shown to be a source of gold (McArdle & Warren 1987, Ixer et al. 1990, McArdle 1993).

sideronatrite

orthorhombic $Na_2Fe^{3+}(SO_4)_2(OH)\cdot 3H_2O$

An hydrated sodium–iron sulphate hydroxide found as yellow botryoidal crusts mixed with metasideronatrite in association with gypsum and pyrite at Trerubies Cove, near Delabole, Cornwall (Ryback & Tandy 1992).

At Hawthorn Hive, a wide bay on the Co. Durham coast (at and around Hive Point and Shippersea Bay), sideronatrite has been found in mine waste derived mainly from Dawdon colliery. Sideronatrite forms abundant bright deep-yellow efflorescent crystalline crusts about 2 m above beach level. Aluminocopiapite, apjohnite, calcite, ferricopiapite, gypsum and natrojarosite are associated species (Young et al. 2008). Sideronatrite is also reported from two unusual cavities (the largest 4.3 m long) at the Rising Sun colliery (375 m below sea level in the Brockwell coal seam), Backworth, Tyne & Wear. The walls of the cavities were partially lined with botryoidal marcasite and pyrite, and in places were coated with a fungoid-like growth of a lemon-yellow to orange material identified as sideronatrite. Individual crystals are prismatic and up to 1 mm long. The cavities also contained barite, calcite, gypsum and halite, and are thought to have developed because of transcurrent fault movements. Passage of saline groundwaters over the mineralized areas allowed the sideronatrite to form (Randall & Jones 1966).

Sideronatrite is briefly noted from coastal localities at Barton-on-Sea, near Lymington, Hampshire (Anthony et al. 2003), and South Bay, Scarborough, Yorkshire (Moreton et al. 1995).

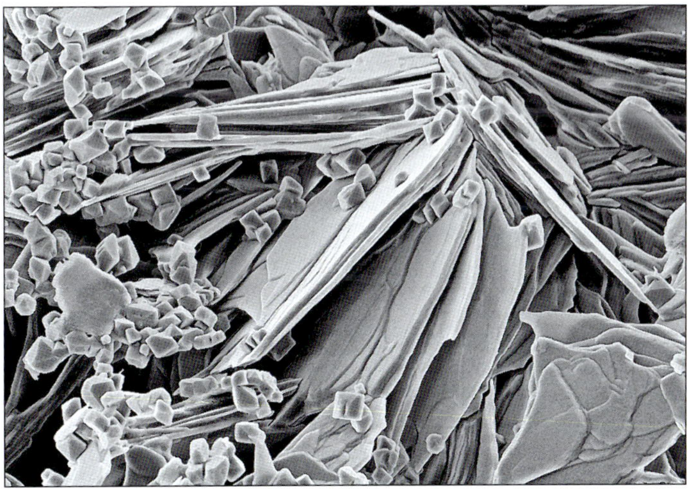

Irregular plates of lemon-yellow sideronatrite on a joint plane in Jurassic sandstone from Carnelian Bay, near Scarborough, Yorkshire; small bipyramidal crystals are thought to be alunite; width 11 µm; SEM image. (© E. Condliffe, University of Leeds)

Oxidation of sulphides in Carboniferous shales on the coast at Doon Bay, N of Ballybunnion, Co. Kerry, Ireland, has produced metasideronatrite and sideronatrite, as well as basaluminite, brochantite, gypsum and hydronium jarosite in nearby coves. Sideronatrite is intimately mixed with metasideronatrite and it forms pale-yellow botryoidal crusts 1–2 mm thick associated with tiny gypsum crystals. Near by, at Cunnihish Cove, soft botryoidal canary-yellow encrustations of sideronatrite occur in the middle of the bay from beach level to 10 m up the cliff. Sea water is assumed to be the source of Na^+ (Moreton et al. 1995).

siderophyllite

monoclinic MICA GROUP $KFe^{2+}_2AlAl_2Si_2O_{10}(OH)_2$

A probably quite common potassium–iron aluminosilicate hydroxide found in peraluminous granites, aplites and some metamorphic rocks. Siderophyllite is a rock-forming mineral that is often misidentified because of its similarity to biotite (biotite has lost its species status and it is now a series name intended to designate incompletely investigated trioctahedral micas for the benefit of field geologists and petrographers). Because the IMA classification of micas (Rieder et al. 1998) focuses only on end-member compositions, it is necessary to define here the compositional ranges for siderophyllite. They have been chosen to be Fe/(Fe + Mg) = 0.5–1.0 and Al^{vi} = 0.5–1.5, where Fe is total Fe expressed as FeO. All data have been processed using the structural-formula recalculation method for trioctahedral micas described in Tindle & Webb (1990).

In Cornwall, siderophyllite has been shown to follow a substitution scheme to zinnwaldite and then to lepidolite compositions as lithium contents increase during fractionation of granitic magma. Examples are presented from Bodelva and Goonbarrow china-clay pits, St Austell, and Luxulyan, near Lostwithiel, all from the St Austell granite complex (Henderson et al. 1989, Mueller et al. 1999). Manning et al. (1996) were able to show that 70% of the outcrop of this complex consists of a lithology termed "biotite granite", in which the dark mica varies in composition from annite to siderophyllite. Elsewhere in southwest England, siderophyllite is reported from the following Hercynian granites: Bodmin Moor (Stone et al. 1988, Stimac et al. 1995); Carnmenellis (Rice & Williams 1969, Jefferies 1985a, Stone et al. 1988, Stimac et al. 1995, Stone 1995, 2000, Tindle unpub. EMP data); Dartmoor (Brammall & Harwood 1932, Al-Saleh et al. 1977, Stimac et al. 1995, Stone 1995, 2000); Isles of Scilly (Stone 1995, 2000); Land's End (Stone et al. 1988, Müller et al. 2006); and from the Tertiary granite on Lundy island, off the north Devon coast (Stone 1990, Thorpe et al. 1990).

Other granite or granite-related occurrences of siderophyllite include: *England* – Skiddaw granite, Cumbria (Hitchen 1934); Weardale granite, Rookhope, Co. Durham

Reddish-brown siderophyllite containing many inclusions of zircon or monazite; De Lank granite quarry, Bodmin Moor, Cornwall; width 2.75 mm; viewed in PPL (upper) and XPL (lower). (© K. Gardner)

(Tindle unpub. EMP data). *Scotland* – granites in Grampian; Aberchirder, Aberdeen, Ardlethan, Ballater, Bennachie, Ben Rinnes, Boat of Garten, Cove Bay, Glenbuchat, Hill of Three Stones, Huntly, Keith, Kemnay, Mount Battock, Peterhead, Portsoy and Strichen (Tindle unpub. EMP data); Cairngorm granite, Grampian (Harrison 1990, Tindle unpub. EMP data); Ardclach, Grantown, Monadhliath and Moy granites, Highland Region (Tindle unpub. EMP data); Cairnsmore of Fleet and Criffell granites, Dumfries & Galloway (Tindle unpub. EMP data); Loch Doon cordierite microgranite, Dumfries & Galloway (Tindle 1982); Sletteval pegmatite, Isle of Harris, Western Isles (von Knorring & Dearnley 1960a). *Ireland* – coarse-grain miarolitic granite at Diamond Rocks, Mourne Mountains, Co. Down (S. A. Rust sample, Tindle unpub. EMP data); a bluish-green variety of siderophyllite is associated with quartz and topaz in greisen veins (within the Mourne Mountains granite complex) at a disused quarry near Newcastle (Nockolds & Richey 1939); Galway granite, Connemara, Co. Galway (Leake 1974).

Siderophyllite also occurs in metamorphic rocks, particularly in Scottish Dalradian metasediments. It is noted in both regionally and thermally metamorphosed andalusite–magnetite–staurolite gneisses in Tayside Region around the sio margin of the Glen Doll diorite between Glen Doll and Glen Clova (Schumacher 1985), and in garnetiferous pelites between Dunkeld and Pitlochry (Atherton 1968). Biotite and siderophyllite are present in Dalradian metapelites around Loch Katrine in the Aberfoyle district, Central Region (Mather 1970). In Grampian, siderophyllite occurs in: cordierite–orthoclase–plagioclase-bearing pelitic schists at Kinnaird Head, near Fraserburgh (Johnson et al. 2000); garnet–mica schists along the Stonehaven coast (Baltatzis 1979); cordierite–garnet–mica–spinel hornfels close to a mafic–ultramafic intrusion at Sparcraigs, Belhelvie (Stewart 1942); grunerite schists forming part of a psammite–semipelite sequence at Creag nam Ban, near Ballater, Deeside (Smith & Phillips 2002). A very unusual sodium-rich variety of siderophyllite occurs in lower amphibolite facies garnet–mica–staurolite schists near Craig of Finnoch, Glen Lethnot, Angus, and also in Grampian (Dempster & Jackson 1996). Other Scottish occurrences of siderophyllite include: Moinian migmatites from Coire nam Muc, Glen Scaddle, near Fort William, and Sròn na Saobhaidh, near Strontian, both Highland Region (Ashworth & Chinner 1978); regionally metamorphosed Moinian metasediments within the aureole of the Strontian granodiorite, Highland Region (Tyler & Ashworth 1981); Moinian "muscovite–biotite" schist from a roadcut near the Glenfinnan monument, W of Fort William, Highland Region (Rieder et al. 1995); thermally and regionally metamorphosed Moinian Supergroup metapelites in the aureole of the Ross of Mull granite, west Isle of Mull, Strathclyde (Wheeler et al. 2004); a body of BIF within amphibolite-facies rocks of the supracrustal Lewisian Loch Maree Group in Flowerdale, at a locality 1 km W of Gairloch pier, Highland Region (Williams 1986; the author refers to the mica as biotite with an 85% annite component). On this evidence Livingstone (1993) described the mineral as annite, but using the criteria discussed here, the mineral is siderophyllite; xenoliths of gneiss in a Tertiary dolerite sill that outcrops on the south coast of the Ross of Mull, between Scoor and Uisken, Isle of Mull, Strathclyde (Brearley 1986, 1987).

In Ireland, siderophyllite occurs in Dalradian metasediments, with examples reported from: the aureole of the Ardara pluton at Clooney, close to the shore of Gweebara Bay, Co. Donegal (Atkin 1978, Kerrick 1987); pelitic hornfels within the aureole of the Cashel ultrabasic intrusion, Connemara, Co. Galway (Trelour 1981); pelitic hornfels within the aureole of the Easky adamellite, N of Tobercurry, Ox Mountains, Co. Mayo and Co. Sligo (Yardley & Long 1981). Siderophyllite is also an important species in granites from the Caledonian batholiths of Leinster and Donegal, occurring with muscovite at Borris, Co. Carlow (Luecke 1981); and as 12 mm-diameter plates at Ballygihen (Haughton 1862), respectively.

sideroplesite

A magnesium-rich variety of siderite, calcite group.

siderotil

CHALCANTHITE GROUP

triclinic $Fe^{2+}SO_4 \cdot 5H_2O$

An uncommon hydrated iron sulphate found at Parys Mountain, Anglesey, Gwynedd, Wales. Siderotil (confirmed by XRD) is a blue earthy mineral that is associated with halotrichite (T. F. Cotterell pers. comm. 2003).

siegenite

LINNAEITE GROUP

cubic $CoNi_2S_4$

A uncommon cobalt–nickel sulphide found in low-temperature hydrothermal veins particularly in the Central Wales orefield, Dyfed, Wales. Siegenite is widespread in an early polymetallic phase of lead–zinc mineralization found in quartz-cemented breccias, especially at mines to the N and E of Tal-y-bont. These include: Brynyarian mine, near Tal-y-bont; Darren mine, Pen-bont Rhydybeddau; Eaglebrook (Nantycagl) mine, Ceulanymaesmawr; Erglodd mine, N of Tal-y-bont; Esgair Hir mine, Ceulanymaesmawr; Esgair Fraith mine, near Tal-y-bont; and Loveden and Ystrad Einion mines, near Tal-y-bont. It is locally abundant at Erglodd mine, where it forms cubo-octahedral crystals up to 2 mm in diameter. At other locations it is greyish white with a distinct pink tint and it forms finely granular intergrowths with chalcopyrite and galena or cobalt pentlandite (Mason 1997, 1998). At Erglodd mine it forms pinkish to bronzy metallic grains and masses up to 10 mm across, scattered in vein quartz, and is associated with chalcopyrite, galena, millerite and sphalerite (Rust 1991a).

Well formed crystal of siegenite from a clay–ironstone nodule in Coal Measures strata. Wyndham colliery, Ogmore Vale, Mid Glamorgan, Wales; width 640 µm; SEM-BSE image. (© National Museum of Wales, Cardiff; photographer M. P. Lambert)

At Gelli colliery, Rhondda and Wyndham colliery, Ogmore Vale, both Mid Glamorgan (in the South Wales coalfield), clay–ironstone nodules lined with siderite contain small (<0.75 mm) silver-colour metallic octahedral crystals of siegenite (Bevins & Horák 1985b, Bevins 1994). These authors add that similar material has previously been reported as linnaeite, but it is now thought highly doubtful that this species occurs in Wales. Linnaeite, now thought to be siegenite, is described from: "Rhondda Valley, Glamorganshire" (Des Cloizeaux 1880, North 1916); Ferndale No. 1 Pit, Mid Glamorgan (North & Howarth 1928); Esgair Hir mine, Ceulanymaesmawr (Rust & Mason 1988); and Great Orme's Head, Llandudno, Gwynedd (Bevins 1994, Ixer & Stanley 1996). At Great Orme, siegenite forms 20–250 µm grains within vuggy dolostones, together with chalcopyrite, dolomite, marcasite and strongly zoned pyrite in a polymetallic assemblage, possibly formed by fluids expelled during late-stage basin de-watering. Siegenite has also been found during landscaping of waste tips at Deep Navigation colliery, Treharris, Mid Glamorgan (Wolfe 1996a), and Oakdale colliery, Oakdale, Gwent (Wolfe 1997).

In Scotland, rare siegenite is reported from the dumps of West Blackcraig mine, near Newton Stewart, Dumfries & Galloway, where it forms fine-grain bronze-like patches associated with the primary sulphides chalcopyrite, galena and sphalerite, and minor marcasite, millerite and pyrite (Livingstone et al. 1976).

At Tynagh, near Killimor, Co. Galway, Ireland, a large Pb–Zn–(Ag–Cu) deposit is concentrated within a Lower Carboniferous carbonate sequence structurally controlled by faulting. Siegenite, together with arsenopyrite, bournonite, enargite, geocronite, jordanite, marcasite and millerite, are minor sulphides found in galena–sphalerite–(chalcopyrite and tennantite–tetrahedrite) ore (Morrissey et al. 1971, Boast et al. 1981a, Clifford et al. 1986). Another occurrence is in the carbonate-hosted stratabound lead–zinc deposit at Lisheen mine, Moyne, Co. Tipperary, where tennantite-rich ore contains exsolved blebs of siegenite up to 100 µm across (S.Moreton sample, Tindle unpub. EMP data)

Detailed analysis of complex Bi–Co–Ni–Sb inclusions in hydrocarbons from low-temperature hydrothermal veins at Laxey mine, E of Snaefell, Isle of Man, suggest the presence of hauchecornite, pyrite (bravoite), siegenite and an unidentified bismuth-rich sulphide (Parnell 1988).

silicite

TL Co. Antrim, Ireland

A variety of labradorite, a member of the plagioclase series, feldspar group.

silicomonazite

A silicon-rich variety of monazite.

sillimanite

orthorhombic Al_2SiO_5

A common aluminium silicate, trimorphous with andalusite and kyanite, and found in high-grade metamorphic schists, gneisses and hornfels or rarely in granitic pegmatites. Sillimanite also occurs as a detrital mineral and as such is reported with ilmenite, kyanite, rutile, staurolite, tourmaline and zircon in Eocene sands from Holden's Wood, Brentwood, Essex (Davies 1915), and in panned concentrates of stream sediments collected on St Kilda (Hirta), Western Isles, Scotland (Harding et al. 1984).

Sillimanite is an uncommon species in Cornwall, but is reported: as a constituent of hornfelsed xenoliths of slate in Bodmin Moor granite at Cheesewring quarry, SW of Henwood, Linkinhorne (Shambrook 1982); and from non-igneous xenoliths in Bodmin Moor granite at De Lank quarry, 2 km N of Blisland, and St Austell granite at Luxulyan quarry (Stimac et al. 1995). A few pelitic xenoliths in the Carnmenellis granite contain sillimanite associated with andalusite, corundum, hercynite and ilmenite (Jefferies 1985b). In Devon, sillimanite occurs as an accessory mineral in the Dartmoor granite (Brammall & Harwood 1925).

In Wales, sillimanite (often forming bundles of needles) is an abundant species in high-grade biotite–feldspar–quartz gneiss from the Coedana Complex on Anglesey, Gwynedd. It is particularly common in coastal regions of the Nebo inlier and in the valley SE of Llechcynfarwydd church (Greenly 1919).

Sillimanite is a constituent of many rock types in the Highland Region of Scotland. At Coire nam Muc, 4 km WNW of the Corran ferry, Ardgour, it forms large conspicuous crystals, exceptionally up to 50 mm long, in Moine Series sillimanite–garnet–muscovite–biotite gneiss bordering large masses of dioritic gneiss (Drever 1940). At Cnoc na Beiste, Learable Hill and Suisgill Burn, all 16 km NW of Helmsdale, near Kildonan, sillimanite (fibrolite) occurs in migmatitic Moine Series granulite gneiss that forms part of a high-grade injection complex. Sillimanite also forms thin veins and knots (faserkiesel) associated with biotite, muscovite and feldspar in pegmatitic and aplitic rocks (Read 1931b, Watson 1948). Rare sillimanite (fibrolite) occurs in a xenolithic dyke on the south side of the peaks Streap and Streap Comlaidh, 4 km NE of Glennfinnan. The dyke is one of a swarm of camptonite and monchiquite dykes of probable Permian age that cut Moinian rocks in the area (Walker & Ross 1954). Two metamorphic segregation pods are described in striped kyanite–sillimanite-bearing Moine gneiss from the entrance to Gleann Còsaidh, Loch Quoich. The pods contain a suite of phosphates, including graftonite, jahnsite, johnsomervilleite, mitridatite, phosphosiderite, rockbridgeite and vivianite (Livingstone 1980).

In Grampian, sillimanite occurrences include: fibrolite in thin veins in gneiss on the northwest side of the summit of Pressendye Hill, 4 km N of Tarland (Heddle 1901); beautiful sillimanite-bearing hornfels interbanded with anthophyllite and plagioclase in the aureole of the Dorback granite, Braes of Abernethy, 9 km W of Tomintoul (Walker & Davidson 1935); sporadic elongated colourless prisms up to 2 mm across from biotite–garnet gneiss exposed in a cutting during widening of the Huntly to Keith road, near Cairnie church, 7 km from Huntly (Stewart 1950); in the metamorphic aureole of the Morven–Cabrach basic intrusion, near Strathdon, in occasional outcrops of biotite–garnet schist, in gneissose rocks on the E of the mass and in boulders of silica-poor hornfels (with cordierite–garnet–plagioclase–spinel) close to the western margin of the intrusion in Glenbuchat (Allen 1970); with garnet and cordierite as widespread constituents of Ellon Series gneisses and Haddo gneisses to the W of Ellon, Ythan Valley (Read 1952); a 1–2 km-wide belt of sillimanite-bearing migmatites extends from near Portsoy on the Moray Firth inland towards the Aberchirder granite and on to the Huntly gabbroic intrusion (Ashworth 1976); fibrolite in migmatitic gneisses are infolded with a small metabasite body at Tomatin (Baker 1986); at Creag Dearg, 8 km S of Ballater, semi-pelitic and pelitic gneisses belonging to the Dalradian Queen's Hill Formation contain abundant fibrous bladed masses of sillimanite (probably pseudomorphing kyanite) and large zoned garnets (Goodman et al. 1990).

Regionally metamorphosed Dalradian gneisses contain sillimanite (fibrolite) and muscovite at the southern contact of the Glen Doll diorite, Tayside Region (Schumacher 1985). Pelitic gneisses in Glen Clova, Tayside, contain sillimanite that has grown at the expense of biotite (Chinner 1961). Precambrian metasediments caught up in the intrusion of lamprophyre dykes exposed on the shores of Loch Lomond, Strathclyde, contain occasional crystals of needle-like sillimanite up to 2 mm long, together with chlorite, corundum, magnetite and staurolite in essentially biotite–garnet–feldspar rocks (Dempster & Bluck 1991). In Dumfries & Galloway, a pegmatite vein cutting the Cairnsmore of Fleet granite at Clatteringshaws contains sillimanite, muscovite and garnet, and at Knocknairling, New Galloway, sillimanite is found with biotite, muscovite and garnet in contact-metamorphosed rocks (Williams 1965).

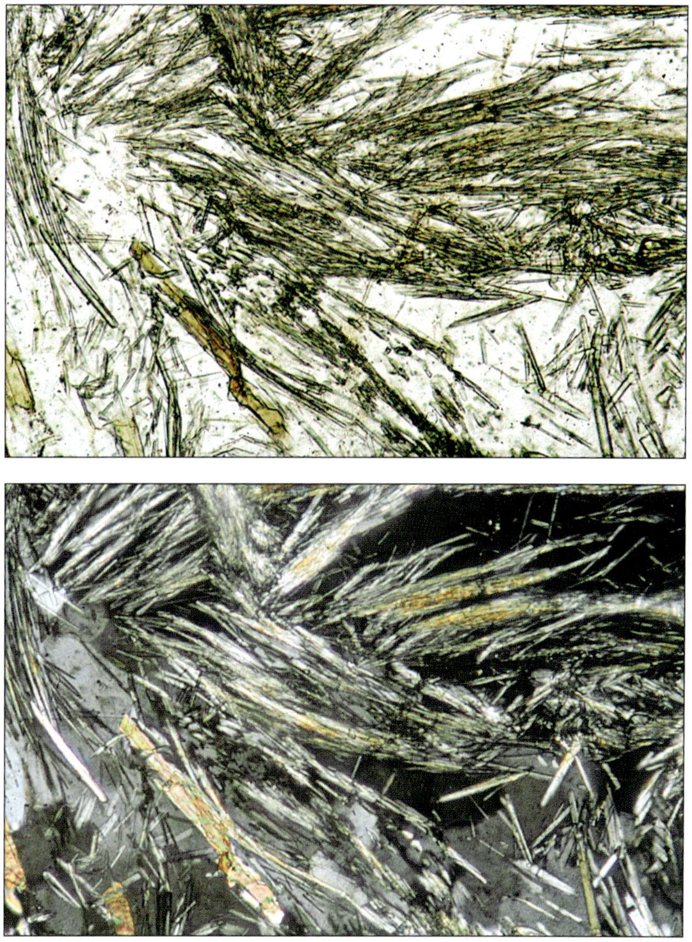

Mass of sillimanite needles (variety fibrolite) associated with biotite (brown). From a mica schist on the west side of Loch Brandy, Glen Clova, Tayside Region, Scotland; width 1.75 mm; viewed in PPL and XPL. (© K. Gardner)

In the Western Isles, sillimanite inclusions in enstatite porphyroblasts are reported from Lewisian sapphirine-bearing and sapphirine-free garnet–kyanite/sillimanite–orthopyroxene granulites to the S and SW of Roineabhal, South Harris (Baba 1999, 2003, Hollis et al. 2006).

In Co. Donegal, Ireland, regionally metamorphosed Dalradian rocks, thermally metamorphosed by the Ardara granite pluton, have developed an inner sillimanite-bearing aureole at Clooney and Naran, close to the shore of Gweebara Bay. Sillimanite is coarse and prismatic, or fibrolitic where it replaces biotite or garnet (Akaad 1956, Atkin 1978, Kerrick 1987). Near by, in the Maas area, semi-pelitic hornfels within the aureole of the Kilkenny School leucodiorite intrusion, one of the satellite intrusions of the Ardara pluton, contains sillimanite prisms up to 1 cm in length (French 1976). A pelitic horizon in the aureole of the Fanad pluton (another of the Donegal granites) on the eastern shore of Mulroy Bay contains sillimanite replacing andalusite, and fibrolite replacing biotite (Kerrick 1987).

Other Irish occurrences of sillimanite include: Dalradian biotite hornfels around the Leinster granite batholith (e.g. at Ballycorus), near Rathmichael, Co. Dublin (Brindley 1957); Dalradian pelitic hornfels around the Cashel basic and ultrabasic intrusion, Connemara, Co. Galway (Treloar 1981, Barber & Yardley 1985, Ahmed-Said & Leake 1990); as a detrital component in the Ordovician Upper Derrylea Formation to the S of Clew Bay, Co. Mayo (Mange et al. 2003).

Sillimanite is noted (with no details) from the La Moye aplogranite, southwest Jersey, Channel Islands (Ixer 1988).

silver

cubic or hexagonal Ag

An uncommon metal that forms a series with gold (intermediate compositions are often called electrum) and is found as a primary hydrothermal mineral, or more often as a supergene species in the oxidized zone of mineral deposits.

England

Avon Close to the beach near the Pier hotel at Clevedon, near Bristol, barite–sulphide mineralization occurs in metre-wide fault-filling brecciated veins within Triassic Dolomitic Conglomerate and consists of banded barite with chalcopyrite, galena (most abundant) and sphalerite, minor amounts of hematite, marcasite, pyrite and tennantite, and trace amounts of bornite and enargite, together with their alteration products. Secondary botryoidal galena is associated with native silver, native copper and pyrite (Ixer et al. 1993).

Cheshire Very rare native silver dendrites have been found in limonite gossan at Engine Vein mine, Alderley Edge ridge (Braithwaite 1994).

Cornwall In the St Just district, native silver, forming dendrites and elongated octahedral crystals, is reported from Levant mine (Garby 1848, Rust 1982a), and, at Wheal Cock, a piece of native silver "the size of a walnut" was once found (Collins 1904a). Fine specimens of native silver were found at Wheal Brothers, near Calstock (Garby 1848), and beautiful filiform specimens were recovered from: Wheal Basset, Illogan, near Redruth; Wheal Vincent, E of Callington, Calstock; and Wheal Herland, Gwinear (Greg & Lettsom 1858). Wheal Herland was probably the most prolific silver mine in Cornwall, producing 156 tonnes of the metal before 1814, from an acanthite (argentite) and native silver deposit found in a cross-course vein that intersected the main copper–tin lodes (Hosking 1951, Dines 1956, Penhallurick 1997). Considerable amounts of capillary native silver were also found at Wheal Ludcott, St Ive, near Liskeard, in a dominantly acanthite (argentite) ore, again from a cross-course lode. Three silver-rich species (proustite, pyrargyrite and stephanite), quartz and siderite are associated species (Collins 1911a, Dines 1956, Rust 1982a, 1983a). More recent discoveries include: arborescent growths associated with native copper and cuprite in serpentinite from Countybridge quarry, Goonhilly Downs (Holyer 1996), and small amounts of native silver in dendritic masses of native copper at Trevassack quarry (Power et al. 1997), both on the Lizard Peninsula; minute wires in a cavity within quartz from NE–SW veins associated with a lead lode at Greystone quarry, Lezant, near Launceston (Elton et al. 2004); minute < 1 μm-diameter grains in copper sulphides at Penberthy Croft mine, St Hilary (Camm & Merry 1991a,b, Betterton et al. 2000). A thorough review of silver minerals in Cornwall by Rust (1982a) describes native silver from the following additional locations: Bounds Cliff, E of Port Isaac; Crinnis mine, St Austell; Dolcoath and North Dolcoath mines, Camborne; Duchy Peru mine, Perranzabuloe; Fowey Consols, Tywardreath, St Blazey; Harrowbarrow (Wheal Newton) mine, near Gunnislake; Holmbush mine, Callington; Silver Vein mine, Lostwithiel; South Tresavean (Roskrow United) mine, Ponsanooth; West Wheal Darlington, Marazion; Wheal Alfred, Phillack, near Hayle; Wheal Ann, Phillack; Wheal Golden, Perranzabuloe; Wheal Jewel, near St Day; Wheal Langford (but worked under other names too), Callington; Wheal Mexico, near Perranporth, Perranzabuloe; Wheal Tremayne, near Hayle, Gwinear. Other authors list a few additional locations: Bodannon mine, Port Gaverne (Russell 1949b); Old Treburgett mine, E of Port Isaac, St Teath (Hosking 1951); Penhale mine, near Gravel Hill, SW of Newquay, and Retallack mine, Perranzabuloe (Robson 1949b); Perran Silver lode, Perranuthnoe (Butler 1907); Wheal Sparnon, Redruth and Wheal Maid Decline, Gwennap (Golley & Williams 1995).

Cumbria Supergene processes have produced delicate dendrites of native silver, restricted to cerussite-rich areas bounding the main barite vein at Force Crag mine, Braithwaite, near Keswick (Green et al. 1997, Green & Briscoe 2002). In the Caldbeck Fells, small amounts of native silver occur at Driggith, Red Gill and Short Grain mines. At Red Gill it forms arborescent crystalline aggregates, up to 0.25 mm across, associated with anglesite, caledonite, cerussite, chalcopyrite, chrysocolla, galena, leadhillite, linarite, mattheddleite, sphalerite and susannite. A supergene origin resulting from the alteration of silver-rich sulphosalts is proposed (Wirth 1989, Cooper & Stanley 1990, Green et al. 1997, 2006a). Subhedral masses of native silver associated with bindheimite have been found in a 10 mm cavity on a single specimen of quartz from Brae Fell mine, Dale Beck Valley (Bridges et al. 2006).

Derbyshire The Ible Sill, an altered olivine dolerite found near Matlock, contains rare native silver blebs along joint planes (Ford & Sarjeant 1964a, Ford et al. 1993). Rare native silver is enclosed by galena at Odin mine, Castleton (Ixer & Townley 1979).

Devon Quartz–siderite veins and lenses cutting Middle Devonian slates mainly produced galena, but native silver in filaments and capillary form was also found, together with chalcopyrite, millerite and rare stibnite (antimonite) at the Combe Martin mines (Hall 1868b, Dunham et al. 1978, Rust 1982a). Stratiform occurrences of sheet-like native copper are accompanied by traces of native silver (as small inclusions) in the Permian Littleham Mudstone Formation at Littleham Cove, near Budleigh Salterton (Harrison et al. 1975b, Milodowski et al. 2000, 2002). The most remarkable capillary native silver appears to have been found in arsenical cobalt ore within a copper lode in Carboniferous slate close to the margin of the Dartmoor granite at Willsworthy mine, NE of Mary Tavy (Hall 1868b, Collins 1904a, Dines 1956, Rust 1982a).

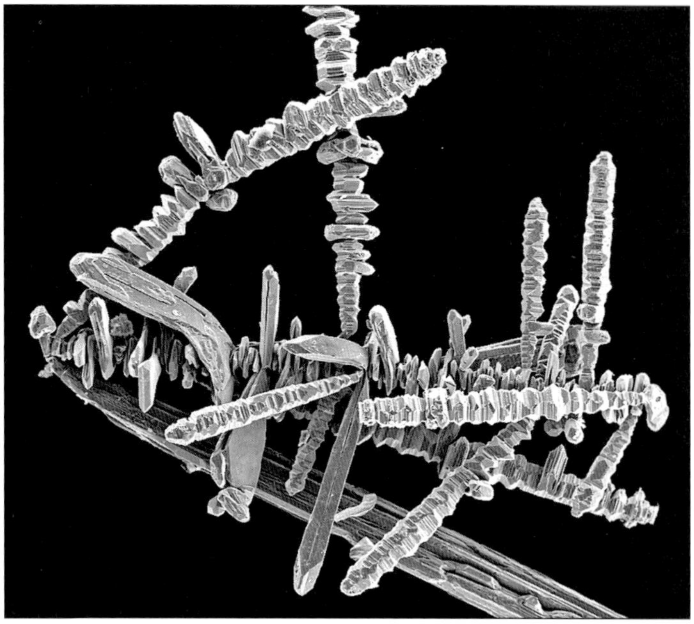

Distorted elongate dendritic crystals of native silver; Force Crag mine, Coledale, Cumbria; width 550 μm; SEM image. (© D. I. Green)

Leicestershire Small subhedral masses and occasional sharp single crystals of native silver have been found on malachite, and thin malleable sheets occur on fractured wallrock from a native copper–cuprite deposit near the Precambrian/Triassic unconformity at New Cliffe Hill quarry, Stanton under Bardon (Hubbard et al. 2005). Ixer et al. (2005a) report trace amounts of native silver in veinlets cutting cuprite and as 2–15 μm-diameter inclusions in azurite or tenorite from this location.

Wales

Dyfed A small dendritic mass of native silver in cerussite was found in 1992 at Frongoch mine, between Devil's Bridge and Pont-rhyd-y-groes, Central Wales orefield (Bevins 1994, Green et al. 1996).

Gwynedd Native silver is noted from the Parys Mountain copper deposit, Anglesey (Southwood & Bevins 1995).

Scotland

Central Faulted Lower Devonian andesitic lavas at Silver Glen mine, near Alva, Ochil Hills, once hosted an epigenetic hydrothermal bonanza-type silver deposit. The mine is famous for its exceptional native silver, which is mercury rich and occurs almost entirely as dendrites, occasionally 30 mm long. Cubic crystals and rough octahedral crystals with hopper faces are also noted. Native silver was the first major primary phase to be deposited and this was followed by cobalt–nickel arsenides, copper–iron sulphides and other species including: annabergite, bornite, chalcocite, chalcopyrite, clinosafflorite, conichalcite, covellite, djurleite, erythrite, galena, malachite, picropharmacolite, pyrite, rammelsbergite, safflorite, skutterudite and tyrolite in a dolomite–(barite–calcite) gangue (Heddle 1901, Ridgway 1980, Jassim et al. 1983, Parnell 1988a, Moreton 1996, 2007, Moreton et al. 1998, Green & Todd 2001). Acanthite is reported too, but is disputed by Livingstone (2002). Elsewhere in the Ochil Hills, minor silver occurrences are noted at Burn of Care, above Dollar, and at Daiglen Burn mine, near Tillycoultry (Heddle 1901, Moreton et al. 1998). Dalradian metasediments host lead–zinc mineralization in veins and vein breccia within fractures associated with the Tyndrum–Glen Fyne fault. Galena is the main ore species, but a later phase of ore-mineral formation is copper- and uranium-rich (chalcopyrite–uraninite) and it contains minor acanthite (argentite), chalcocite, native

Dendritic crystal aggregate of mercurian silver etched from enclosing carbonate matrix. Alva silver mine, Silver Glen, Ochil Hills, Central Region, Scotland; dendrites to 22 mm long. (© D. I. Green)

copper, safflorite, native silver and tetrahedrite in a calcite gangue (Pattrick 1985). Later work identified cervelleite (described as phase C: see cervelleite), electrum, hessite, petzite and sylvanite in these veins, and concluded that they were produced by hydrothermal activity related to copper–molybdenum mineralization associated with Late Caledonian granites (Pattrick et al. 1988).

Dumfries & Galloway Rare native silver, gold and bismuth accompany chalcopyrite and pyrite in quartz and carbonate veinlets within the aureole of the Criffell granodiorite at Needle's Eye, SW of Caulkerbush (Basham et al. 1989).

Grampian Native silver is locally present between laths of molybdenite and within masses of molybdenum ochres in a sulphide-bearing assemblage at Gairnshiel, Glen Gairn, W of Ballater. Acanthite, cosalite, koechlinite and wulfenite are associated species. The mineralization is associated with a zinnwaldite-bearing granite (Webb et al. 1992).

Highland On the Isle of Rum, sulphide-rich dykes near the West Sgaorishal ultramafic plug contain discrete PGMs, especially kotulskite and sperrylite, within disseminated sulphide (chalcopyrite–pentlandite) mineralization. Native silver is one of the less common associated species, occurring as $<5\ \mu m$ grains with cinnabar, froodite, galena, gold (electrum), sperrylite, tellurium and unidentified Bi–Pt–Te and Pd–Sb–Te phases (Power et al. 2003). Native silver also occurs in the Rum layered intrusion, where it is found in chromite-rich horizons associated with nickel- and copper-rich sulphides and PGMs. Many platinum tellurides from the unit 11/12 boundary chromitite contain native silver inclusions (Butcher et al. 1999, Power et al. 2000a). Heddle (1901) reports filiform native silver associated with native copper in a ferruginous gangue within gneiss at Bona, near Abriachan, N of Drumnadrochit, Loch Ness.

Lothian Hilderston in the Bathgate Hills became famous as early as 1606, when native silver was found in a local stream. The source was found to be a barite vein close to a hydrothermally altered dolerite dyke, and for many years afterwards both native silver and galena were successfully mined. There are two distinct metallogenic assemblages, one containing an Ag–Co–Ni suite in which filiform native silver occurs with annabergite, erythrite and nickeline, and a second dominated by galena with chalcopyrite, pyrite, sphalerite and solid hydrocarbons. A few small specimens of native silver and acanthite were recovered in the early 1990s, together with amalgam, cinnabar, maucherite and native mercury (Ridgway 1980, Stephenson et al. 1983, Meikle 1994, Moreton et al. 1998, Green & Todd 2001). Heddle (1901) refers to this locality as the "silver–lead mine at the northeastern foot of Cairn-naple (Cairnpapple) Hill, the highest of the Hilderston Hills".

Strathclyde Specimens of native silver in and on prehnite have been found in mineralized joints and cavities in Clyde Plateau basalts at Boyleston quarry, SW of Barrhead. Crystals are bright and lustrous (irridescent when tarnished) and exhibit a variety of forms, from simple or twinned cubes, to unequal bipyramidal forms. Acanthite platelets have subsequently grown on the surface of the silver (Meikle 1989a, 1992a). Similar occurrences of native silver are noted from Loanhead quarry, near Beith (Meikle 1989a, 1992a), and, without later acanthite growths, at Hartfield Moss, 9 km WSW of Barrhead (Meikle 1990, 1992a). The Fore Burn diorite–granodiorite complex within the Tairlow plantation, near Straiton, contains many fault-bounded quartz–carbonate veins. Arsenopyrite and pyrite are the main sulphide species found in the veins and these contain inclusions of native bismuth, bismuthinite, chalcopyrite and native gold. Native silver and silver-rich tennantite–tetrahedrite are minor associated species (Charley et al. 1989). Native silver forms well developed cubic crystals to 2.7 mm on edge on calcite and sphalerite within barite–calcite veins at Mannoch Hill mine, SW of Lesmahagow (Todd & McMullen 1991). Heddle (1901) suggests that native silver in capillary form was once found in veins cutting limestone on the Isle of Islay. The locality may have been Gartness lead mine, near Ballygrant, which was noted for its silver production (Callender & Macaulay 1984).

Tayside Silvery-grey mercury-rich native silver and rutheniridosmine have been found in alluvial gold samples from Chapel Burn and Cloan Burn near Glen Devon Youth Hostel, Ochil Hills (Davidson 1995). Gold-bearing quartz–sulphide veins in Urlar Burn, Aberfeldy, contain altered galena replaced by bournonite, coloradoite, copper sulphides or acanthite, or by trace amounts of native silver (Ixer et al. 1997). Heddle (1901) notes filamentous native silver on galena from Craigsoales lead mine, near Tarfside in Glen Esk.

Shetland Native silver occurs as submicroscopic inclusions associated with a wide range of PGMs in chromite-rich sulphide-bearing dunite from the Shetland ophiolite complex at Cliff on the island of Unst (Prichard et al. 1994).

Ireland

Clare A disseminated and fracture-fill sulphide deposit in Lower Carboniferous carbonate sediments at Ballyvergin is dominated by arsenopyrite, chalcopyrite, galena and pyrite. Native silver occurs with bournonite and geocronite in trace amounts (Brown 1967, Andrew 1986).

Cork The Tullacondra copper–silver deposit, NNW of Mallow, is hosted by Carboniferous limestones, shales and sandstones, and contains considerable amounts of very finely disseminated tennantite in a silver-rich zone. Silver occurs in its native form as discrete, occasionally visible grains (Wilbur & Carter 1986).

Dublin Native silver was once found in a galena-bearing quartz lode in granite at Ballycorus mine, near Carrickgollogan (Greg & Lettsom 1858, Ryback & Moreton 1993).

Galway Native silver forming dendrites has been found on the dumps of Tynagh Pb–Zn–(Cu–Ag) mine, near Killimor (Green & Moreton 2001, Green 2005a). It occurs in a supergene assemblage that also contains native copper and a wide range of carbonates, oxides, silicates, sulphates and sulphides (Schultz 1971).

Kerry Extremely rare primary native silver occurs in a limestone-hosted copper-polymetallic deposit at Muckross mine, Killarney National Park. The primary ore also contains alloclasite, arsenopyrite, chalcopyrite, galena, pyrite, skutterudite and sphalerite (Ryback & Moreton 1993, Moreton et al. 1999, Green & Moreton 2001).

Kilkenny A galena–marcasite–pyrite–sphalerite-dominated stratiform replacement orebody in Carboniferous Limestone at Galmoy, near Kilkenny, also contains a separate phase of fracture-fill copper–silver mineralization. Here native silver is associated with chalcopyrite and tetrahedrite–tennantite (Doyle et al. 1992).

Leitrim Specks of native silver associated with copper, lead and zinc ores are recorded from Twigspark mine, near Manor Hamilton (Cole 1922).

Limerick Disseminated sulphides (mainly bornite, chalcopyrite and chalcocite) in fractured Carboniferous Limestone are associated with minor native silver, galena, sphalerite and stromeyerite at Aherlow, near Galbally (Cameron & Romer 1970, Schultz 1971).

Tipperary Epigenetic mineralization related to the intrusion of two high-level basic intrusions at Gortdrum mine, near Tipperary town, has produced a dominantly chalcopyrite–Hg-rich tennantite-bearing orebody in which mercury-rich silver (native amalgam, var. kongsbergite) has been found as an accessory mineral, along with cinnabar, cobaltite, gortdrumite, sphalerite, stromeyerite and wittichenite (Morrissey et al. 1971, Steed 1986, Duane 1988, Braithwaite & Wilson 2003). The large sedimentary-exhalative base metal and barite deposit in the Silvermines district contains a wide range of sulphides, sulphosalts and secondary species, one of which is reputed to have been native silver (Kinahan 1886). Although likely to be present, native silver is not confirmed by later workers (Boland et al. 1992, Moreton 1999).

Wicklow Traces of native silver and gold are reported from chalcopyrite–pyrite-rich disseminated and vein mineralization associated with the banded volcanogenic Avoca copper deposit (McArdle 1993).

silver fahlerz

A synonym of freibergite, tetrahedrite group.

silver glance
A synonym of acanthite and argentite.

silver mundic
A synonym of arsenopyrite.

simonkolleite
hexagonal $Zn_5(OH)_8Cl_2{\cdot}H_2O$

A very rare hydrated zinc hydroxychloride formed in a diagenetic concretion that has grown since the late 1870s at Penpol in the Gannel Estuary, north Cornwall. The catchment area of the estuary includes the Ag–Pb–Zn mining district around Newland Downs. At Penpol quay, many wooden and metal artefacts were discarded in the estuary and now form nuclei for concretions. The simonkolleite-bearing concretion is complex, being dominated by siderite, calcite, iron monosulphide and zinc sulphide (or sulphate) cements, and it probably formed as a result of corrosion of a zinc-plated artefact within the estuary (Pirrie et al. 2000c).

skaergaardite?
cubic CuPd

A rare copper–palladium alloy found in composite microglobules within tholeiitic gabbro in the Skaergaard alkaline intrusion, Kangerdlugssuaq area, east Greenland (Rudashevsky et al. 2004). These authors suggest that an unnamed copper–palladium alloy, from the layered intrusion on the Isle of Rum, Highland Region, Scotland (Power et al. 2000a), may also be skaergaardite.

skiagite
TL Glen Skiag, Highland Region, Scotland

Hypothetical garnet molecule; a synonym iron-andradite.

skutterudite
cubic $CoAs_3$

A relatively uncommon cobalt arsenide that forms a series with nickel-skutterudite and is found in moderate- to high-temperature hydrothermal veins, often with other cobalt–nickel minerals.

At St Austell Consols mine, 1 km NW of Sticker, near St Austell, Cornwall, skutterudite is associated with löllingite and nickeline in cross-course veins that in places also contain uraninite (var. pitchblende) and possibly barite and galena (Robertson & Dines 1929, Dines 1956, Ball et al. 1982). Skutterudite (var. smaltite) occurs with cassiterite and erythrite in chlorite-rich rocks from Wherry mine, Penzance (Greg & Lettsom 1858, Russell 1949a). Skutterudite (smaltite) has also been reported from the following Cornish locations: Botallack mines, St Just (Greg & Lettsom 1858); Dolcoath mine, Camborne (Greg & Lettsom 1858, Hosking 1951); East Pool mine, Illogan; Roskrow United mine, Ponsanooth; Wheal Trugo, St Columb (all Golley & Williams 1995); Fowey Consols mine, Tywardreath, St Blazey (Lewis 1997); Killifreth mine, 2 km NE of St Day (Dines 1956); Polgooth mine, 2 km SSW of St Austell (Greg & Lettsom 1858, Dines 1956); South Terras mine, St Stephen-in-Brannel (Davis et al. 1965); Wheal Herland, Gwinear (Greg & Lettsom 1858, Dines 1956, Rust 1982a); Wheal Sparnon, near Redruth (Greg & Lettsom 1858, Collins 1892, Rudler 1905). In Devon, vanadium-rich nodules from the Permian redbeds at Littleham Cove, Budleigh Salterton, contain skutterudite among a wide-ranging suite of Ag, As, Co, Cu, Ni and U minerals (Milodowski et al. 2000).

Coniston copper mine, 2 km NW of Coniston, Cumbria, contains skutterudite (smaltite) and nickeline (niccolite) in chalcopyrite–quartz veins that in places are rich in sphalerite. Both the Bonser vein and the Paddy End section contain this assemblage. Eyes or small irregular patches of nickeline are surrounded by skutterudite in grey and white crystalline calcite encrusted with a pink, greenish or nearly white mixture of annabergite and erythrite. Other specimens consist of beautifully crystallized brilliant and tin-white skutterudite forming simple cubes with markedly curved faces. The largest cubes measure 9 mm on edge and are the best examples of the species in Britain (Russell 1925). Young (1987a) notes a specimen of skutterudite (smaltite) associated with bismuthinite and rammelsbergite from the Coniston mines (now in the Russell collection at the NHM). Skutterudite also occurs in an arsenopyrite-rich apatite–chlorite–quartz vein at the cobalt mine on Scar Crag, Causey Pike, Cumbria, where it forms grains up to 20 μm across enclosed by alloclasite. Native bismuth, bismuthinite, cobaltite, erythrite, glaucodot, marcasite, molybdenite, pyrite, rutile, scorodite and tourmaline are associated species (Postlethwaite 1913, Ixer et al. 1979a). A specimen of skutterudite from Wanthwaite mine, St John's in the Vale, near Threlkeld, Cumbria, now in the Kingsbury collection at the NHM (Young 1987a), requires further confirmation. Serious doubts about the provenance of other species from this locality in this collection have been expressed by Ryback et al. (1998, 2001). The skutterudite described by Ixer (1986) from Tynebottom mine, NW of Garrigill, near Alston, Cumbria, also came from the Kingsbury collection and it too has an uncertain provenance. Additionally, Ixer (1986) notes trace amounts of skutterudite and pyrite (bravoite) enclosed within nickeline from Settlingstones mine, near Hexham, Northumberland, and Greg & Lettsom (1858) report the species at Force Crag mine, Braithwaite, near Keswick, Cumbria.

In Scotland, skutterudite (smaltite) occurs in a small deposit of chalcopyrite–nickeline–pyrrhotite–pentlandite ore within a sill-like diorite body at the western edge of the Cairnsmore of Fleet granite in Palnure Burn, Talnotry, 5 km NE of Newton Stewart, Dumfries & Galloway. Skutterudite forms almost complete crystals surrounded by nickeline, and irregular patches and rims encircling nickeline (Gregory & Leitch 1928, Dunham et al. 1978). A small portion of a radioactive vein near Needle's Eye, SW of Caulkerbush, Dumfries & Galloway, contains skutterudite (smaltite) encrusted by erythrite and rare annabergite (Williams 1965). Earlier reports identify skutterudite (smaltite) from Essochossan Glen, 3 km from Inveraray, Strathclyde, and at Breeton, near Alva, Central Region (Greg & Lettsom 1858).

Fine-grain disseminations of alloclasite and skutterudite in calcite are described from the disused Cu–Pb–Zn–(Co) mine at Muckross, Killarney National Park, Co. Kerry, Ireland. Post-mining oxidation of these cobalt arsenides has produced erythrite and smolyaninovite (Moreton et al. 1999, Green & Moreton 2001).

slate-spar
A tabular variety of calcite.

sloanite
A name applied to laumontite, natrolite and thomsonite.

smaltite
An arsenic-deficient variety of skutterudite.

smaragdite
A synonym of actinolite, amphibole group.

SMECTITE GROUP
Monoclinic silicates with the general formula $X_{0.3}Y_{2\text{–}3}Z_4O_{10}(OH)_2{\cdot}nH_2O$, where X (exchangeable ions) = $Ca_{0.5}$, Li, Na; Y = Al, Cr^{3+}, Cu^{2+}, Fe^{2+}, Fe^{3+}, Li, Mg, Ni, Zn; and Z = Al, Si. Beidellite, montmorillonite, nontronite, saponite and stevensite are members of the group found in Britain or Ireland. Another member of the group, volkonskoite, requires confirmation.

smithite
monoclinic $AgAsS_2$

Named after George Frederick Herbert Smith (1872–1953), a crystallographer and keeper of minerals at the British Museum (Natural History), London, who discovered paratacamite, a species related to herbertsmithite; author of *Gemstones and their distinctive characters* (1912); born in Edgbaston, Warwickshire. Smithite is a rare silver sulpharsenide, dimorphous with trechmannite, that has been found in a very high-grade pod of polymetallic silver mineralization cutting pyrite mineralization in the B zone of the Silvermines orebody, Co. Tipperary, Ireland. The silver mineralization included acanthite, argyrodite, miargyrite, proustite, smithite and xanthoconite, together with barite, galena and sphalerite (Gasparrini 1978, Taylor & Andrew 1979). Taylor (1984) also reports geocronite–jordanite, gersdorffite and pyrargyrite from the same zone. A comprehensive review of mineralization in the Silvermines area is provided by Moreton (1999).

smithsonite
CALCITE GROUP

trigonal $ZnCO_3$

TL Derbyshire, and the Mendip Hills, Somerset, England

Named after James Lewis Smithson FRS (1765–1829), a mineralogist and postumously the founder of the Smithsonian Institute in Washington DC, USA. He was born in Paris as Jacques Louis Macie, the illegitimate son of the Duke of Northumberland. Smithsonite is a common zinc carbonate found as a secondary mineral in the oxidized zone of zinc-bearing deposits, sometimes in such abundance as to be an ore species. It often replaces sphalerite. In early works the term "calamine" is synonymous with smithsonite, but confusingly the term was also used to describe hemimorphite.

England

Avon Partially silicified Triassic Dolomitic Conglomerate at East Harptree, 20 km S of Bristol, and on the northern edge of the Central Mendip lead–zinc orefield, contains cavernous and veined regions impregnated with barite, iron oxides, smithsonite and occasionally galena (Rudler 1905, Alabaster 1990). Supergene alteration of sphalerite in As–Ba–Cu–Pb–Zn-mineralized veins cutting Dolomitic Conglomerate near the Pier hotel, Clevedon, has led to the formation of blaubleibender covellite, covellite and smithsonite (Ixer et al. 1993).

Cheshire Very rare smithsonite forms rounded zoned crystals replacing sphalerite in faulted Triassic copper–lead-enriched sandstones on Alderley Edge ridge (Ixer & Vaughan 1982, Braithwaite 1994). In the general area of the Alderley Edge (including West mine) and Clive (Shropshire) copper mining areas, smithsonite is reported with azurite, cerussite, chrysocolla, dioptase, hydrocerussite, malachite, mimetite and pyromorphite in a secondary As–Co–Cu–Ni–P–Pb carbonate, hydrated silicate, oxide–hydroxide assemblage (Plant et al. 1999).

Cornwall Sphalerite-rich veinstone from Greystone quarry, Lezant, near Launceston contains uncommon encrustations of pale-brown lenticular smithsonite crystals up to 2 mm long (Elton et al. 2004). Smithsonite is also noted from: Fowey Consols mine, Tywardreath, St Blazey; Perran St George mine, near Perranporth; and Wheal Mary, Lelant (all Robson 1949b). A single specimen of smithsonite has been found at Penberthy Croft mine, St Hilary (Betterton et al. 2000).

Cumbria *Caldbeck Fells:* The occurrence of smithsonite in this area is problematical, first because, at some locations (e.g. Roughton Gill), material originally thought to be smithsonite has later proved to be hemimorphite (Davidson & Thomson 1951); and, secondly, because it is possible that some reports (Hartley 1984, Young 1987a) are based on material supposedly collected by A. W. G. Kingsbury and may have dubious provenance. Locations include: Driggith mine, Potts Gill mine (old No. 2 cross cut) and Silver Gill mine, near Roughton Gill. Further confirmation is desirable. It does appear that smithsonite occurs at Sandbed lead mine (Davidson & Thomson 1951) and it has also been confirmed from Silver Gill (Green et al. 2005b). Green et al. (2006a) found yellowish-brown rice-grain crystal aggregates of smithsonite at Driggith and Sandbed mines, but add that Kingsbury specimens from these locations are "almost certainly fraudulent". *Northern Pennine orefield:* Smithsonite is the most widespread supergene zinc mineral in this area and is common at many localities. At Middle Fell, near Alston, it was particularly abundant and was extracted as calamine at several mines. A cellular brown dry-bone variety is most common, with excellent examples being found at the Bayle Hill mines and Hudgill Burn mine, both near Alston, and in old workings on the Cowslitts vein, Nenthead. Smithsonite pseudomorphs after sphalerite are reported from the latter location and from Lough vein, Blagill, NE of Alston (Bridges & Young 1998). Massive green and yellow smithsonite was once obtained from Farnberry and Holyfield mines, near Alston, where it formed beautiful mammillated masses, some colour-banded in shades of yellowish-green to brown (Davidson & Thomson 1951, Dunham 1990, Bridges & Young 1998). Smithsonite was noted (with brianyoungite) from Brownley Hill mine, Nenthead, near Alston, by Livingstone & Champness (1993), but was not found by Green et al. (2000). Other occurrences include: Dow Scar level flats of Hilton mine, Scordale (Bridges & Green 2005b); Haggs vein, Haggs Bank, Nentsberry mine, Nenthead (Davidson & Thomson 1951, Dunham et al. 2001); Hilton mine, Scordale (Bouch et al. 2006); Natrass mines, near Alston, and Silver Top quarry, Hallbankgate, Brampton (Young et al. 1989); Smallcleugh mine, Nenthead, near Alston (Livingstone 1991); and Windy Brow vein, Tynehead, near Garrigill (Young et al. 1985a).

Derbyshire Smithsonite occurs widely in the Southern Pennine orefield, particularly in the Matlock area. Crystals can be rhombohedral, generally <2 mm long, and well formed, but more often smithsonite forms cream mammillated crusts replacing oxidized sphalerite. The dry-bone variety is present in surface weathered zones of most sphalerite-bearing veins. Rounded colourless-to-white smithsonite rhombohedra up to 2 mm long and epimorphs after hemimorphite occur at Wapping mine, Cromford (Ince 1991), and pale amber to dark-brown varieties form botryoidal clusters encrusting fluorite at Fall Hill quarry, Ashover (Smith 1982). Mostaghel (1985) examined the supergene transition from sphalerite to smithsonite in great detail using material from: Ball Eye mine, Bonsall; Fall Hill quarry; Knotlow mine, Monyash; and Ladywash mine, Eyam. Other occurrences include: Bage mine, Bolehill, Cromford; Bonsall Moor; Butts quarry, Ashover; Cumberland and Rutland mines, Masson Hill, Matlock Bath; Golconda mine, Brassington; Hard Rake, near Sheldon; Old Millclose mine, Darnley Dale; Owlet Hole and Royal mines, Matlock; Red Seats vein, Castleton (all Ford et al. 1993); Millclose mine, Wensley, near Matlock (Russell 1912b); Milltown quarry, Ashover (Rust 1991c); Oxclose mine, Snitterton, (Ixer 1974b, Ixer & Townley 1979); Riber mine, Matlock (Braithwaite 1983).

Durham In the Northern Pennine orefield, alteration of cadmium-rich sphalerite has produced smithsonite associated with crusts of greenockite at Ashgillhead mine and Reddycombe (Willyhole) mine, near Harwood, both in Upper Teesdale (Young et al. 1987, Bridges & Young 1998). Good examples of cellular brown dry-bone smithsonite can be found at Greenhurth mine, Upper Teesdale (Bridges & Young 1998).

Leicestershire Small pale-green to brownish-green globules and crusts of smithsonite are noted from Newhurst quarry, Shepshed (Ince 2005).

Northumberland Dry-bone smithsonite occurs at Stonecroft mine, N of Haydon Bridge (Bridges & Young 1998).

Shropshire Fracture-controlled veins in Ordovician and Precambrian rocks of the Shelve inlier (belonging to the West Shropshire orefield) contain three generations of sphalerite together with barite, chalcopyrite and galena. Supergene alteration of the primary ore minerals has produced a secondary assemblage consisting of anglesite, cerussite, covellite, goethite and smithsonite. Samples studied came from mine dumps and museums, mainly from Cothercott, Pennerley, Roman Gravels, Snailbeach, Tankerville and Wrentnall mines (Pattrick & Bowell 1991).

Somerset Smithsonite forming a cellular, massive or earthy deposit occurred in some abundance in the Mendip Hills, particularly in Triassic conglomerates. The areas around Rowberrow and Shipham were centres of activity in the 18th century, and later attempts at mining followed near Chewton Mendip. In the Green Ore area, smithsonite (syn. calamine) is associated with barite and calcite (Rudler 1905, Kingsbury 1941, Green 1958). Warty or cellular, brown or pink, encrustations and thick crusts of pale-tan smithsonite pseudomorphing scalenohedral calcite have been found at Whatley quarry, near Frome (Alabaster 1976; R. E. Starkey pers. comm. 2005).

Staffordshire Smithsonite is very common at Waterbank mine, Ecton Hill, in a secondary assemblage also containing brochantite, cerussite, hydrozincite, namuwite, ramsbeckite, serpierite and schulenbergite. It forms translucent greyish crusts on sphalerite, and rounded white to dark-grey, pale-yellow to yellowish-brown and reddish-brown crystals exceptionally up to 1.5 mm long (Rust 1992, 1995c). A richly mineralized pocket in Clayton adit, Ecton Hill, Wetton, has yielded excellent specimens of aurichalcite in a friable barite gangue, with fluorite, gypsum, hemimorphite, smithsonite and decomposing chalcopyrite (Braithwaite & Knight 1968, Robey & Porter 1972).

Yorkshire Smithsonite (calamine) was worked in the Calamine mine to the NW of Malham – actually a Quaternary cavern system discovered in 1788 during copper mining. Up to 5000 tonnes of smithsonite, used mainly in brassmaking, was extracted before the mine closed in 1830 (Mundy & Arthurton 1994). Minor amounts of dry-bone smithsonite are also reported from nearby Pikedaw mine (Green et al. 2006c). Several barite–fluorite–galena–sphalerite veins cutting Carboniferous Limestone at Coldstones quarry, 5 km W of Pateley Bridge, contain smithsonite as an alteration product of sphalerite (Young et al. 1989, 1997). Smithsonite is a fairly common secondary mineral associated with hemimorphite in oxidized sphalerite veinstone at Wetgrooves mine, Aysgarth (Rust 1987b). It is also reported from: Buckden Gavel and Middlesmoor mines, both near Kettlewell (D. I. Green pers. comm. 2007); Castaway–Ripley vein, Grassington Moor (Young 1987b); Friarfold–Surrender and Old Rake veins, Swaledale (Young et al. 1989); Level House Bridge, Flincher Gill, NE of Gunnerside, Swaledale (Ineson & Young 1994); and Old Gang mine, E of Gunnerside Gill, Swaledale (Green 2003).

Botryoidal coating of lustrous yellow smithsonite, 60 mm across, from Smiddy shaft at Buckden Gavel mine, Buckden, near Kettlewell, Yorkshire. (© D. I. Green)

Greyish-white stalactitic smithsonite, 45 mm tall, cut to reveal internal greenish-yellow to grey colour zoning. Middlesmoor mine, near Kettlewell, Yorkshire. (© D. I. Green)

Wales

Clwyd Fine canary-yellow botryoidal smithsonite from Talargoch, near Dyserth, and epimorphs of smithsonite on cubes of fluorite and pseudomorphs after calcite, both from Moel-y-crio, Halkyn Mountain, are noted by Campbell Smith (1913) in describing the mineral collection of Thomas Pennant (1726–1798). Other samples of smithsoniten this collection are from: near Cilcain; near Holywell – Clawdd-ffordd, Coetia Butler, Coetia-mawr, Pen-y-Bryn; Talacre mine, near Gronant; and Trelogan. Botryoidal smithsonite has been found at Milwr mine in the Holywell district (Mountain 1924) and specimens now in the NMW indicate that smithsonite occurred at: Halkyn United mine; Rhosesmor (Halkyn East) mine, near Mold; and West Halkyn mine (Bevins 1994). Smithsonite is also known from Minera mine, Bwlchgwyn, near Wrexham (Bevins 1994).

Dyfed Smithsonite is recorded from a few locations in the Central Wales orefield: bright-yellow botryoidal crusts at Eaglebrook (Nantycagl) mine, Ceulanymaesmawr (Jones & Moreton 1977, Jones 1983); rare white to bluish-grey botryoidal masses up to 0.4 mm across at Esgair Hir mine, Ceulanymaesmawr (Rust & Mason 1988); colourless crystals and white botryoidal crusts at Esgair Fraith mine, near Tal-y-bont; also Brynyrafr mine and Copa Hill mine, Cwmystwyth (all Jones & Moreton 1977).

Gwent A specimen of smithsonite in the NHM mineral collection is thought to be from Ochrwyth, near Risca (Bevins 1994).

Gwynedd The Llanrwst lead–zinc orefield contains simple epigenetic galena–marcasite–pyrite–sphalerite veins and vein breccias in Ordovician volcano-sedimentary rocks through which meteoric waters have passed, leaving behind an oxidized zone containing anglesite, cerussite, limonite (or goethite), malachite and smithsonite. This is noted at Cyffty and Penyrallt mines, between Betws-y-coed and Llanrwst (Haggerty 1995).

Mid Glamorgan At Machen quarry, 7 km ENE of Caerphilly, some creamy-beige botryoidal coatings on galena are confirmed to be smithsonite, although others are hemimorphite. Excellent small translucent brown botryoidal groups of smithsonite up to 3 mm across on barite were found in a rich pod of galena here in 1994 (Plant & Jones 1995, Wolfe 1996a). Bevins (1994) notes smithsonite from Rhyd-y-Gwern mine, near Machen.

Powys Smithsonite is reported from: Craig Rhiwarth and Ochr Graig mines, Llangynog (Bevins 1994); Dyfngwm mine, Machynlleth (Rust & Rust 1987); and Llanymynech Hill, near Oswestry (Greg & Lettsom 1858).

Scotland

Dumfries & Galloway Very pale-green botryoidal smithsonite encrusting quartz has been found near the head of Whyte's Cleuch, Wanlockhead (Williams 1965).

Highland A mineralized fault breccia outcropping 2 km N of Lake Calder and near the Bridge of Westfield contains calcite, galena, marcasite, pyrite, siderite, smithsonite and uranium-rich hydrocarbons (Gallagher et al. 1971).

Ireland

Clare Very attractive smithsonite (perhaps the best Britain or Ireland), stained bright yellow by included greenockite and associated with fluorite has been collected from Sheshodonnell East mine, the Burren. The mineralization is related to the intrusion of the Galway granite (Cole 1922, Ryback & Moreton 1991, O'Connor et al. 1993, Green & Moreton 2001).

Donegal Dump specimens from the fine-grain stratiform deposit of galena and sphalerite at Glenaboghil mine, NNW of Fintown, contain white microbotryoidal smithsonite (Ryback & Moreton 1992).

Galway Smithsonite is one of many species found in the residual ore (secondary enriched zone) in the stratiform Cu–Pb–Zn–(Ag) deposit at Tynagh mine, near Killimor (Schultz 1971, Clifford et al. 1986).

Kerry Smithsonite containing spots of chalcopyrite and galena once occurred as brownish cellular and minutely stalactitic masses at Ross Island mine, Killarney. Cavities in the smithsonite contained serpierite (Russell 1927b).

Tipperary Smithsonite was exploited at Knockanroe and Ballygown South mine, near Silvermines, from a zinc-enriched zone, known as the Calamine Bed. The primary ore forms a sedimentary-exhalative deposit in Lower Carboniferous Limestone

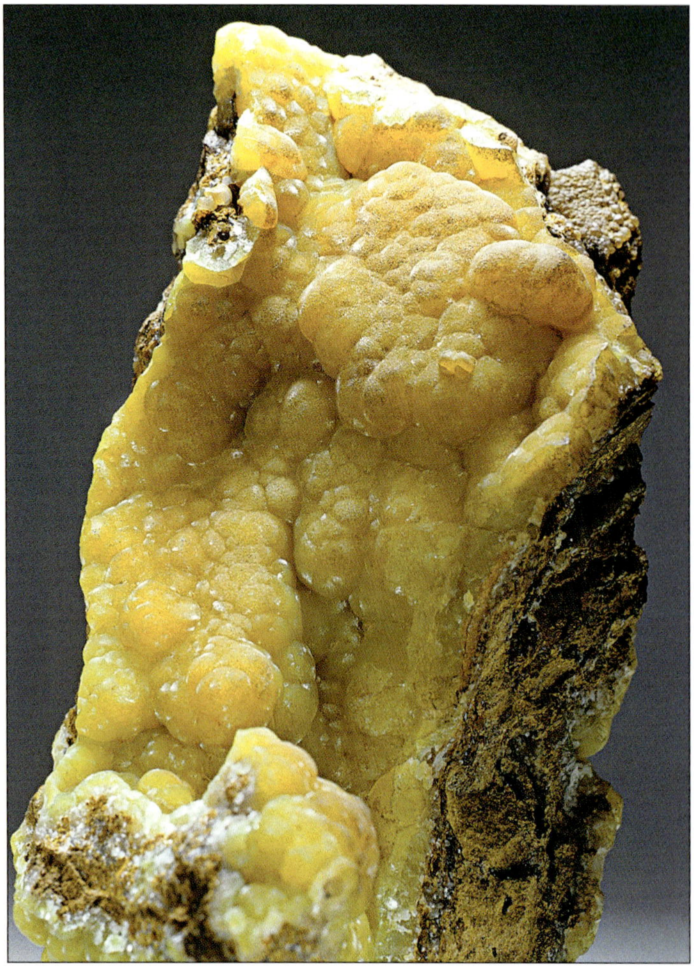

Botryoidal smithsonite from Halkyn Mountain, Clwyd, Wales; 8.5 cm tall. (© National Museum of Wales, Cardiff; photographer M. P. Cooper)

series dolomite (Rhoden 1959, Boland et al. 1992, Mullane & Kinnaird 1998, Moreton 1999).

Offshore territories

Jersey Lead–zinc veins cutting Brioverian sediments at Le Pulec on the northwest coast of Jersey contain smithsonite replacing sphalerite (Stanley & Ixer 1982). Probable smithsonite is also noted in sulphide-bearing carbonate veins at the German underground hospital, St Peter's Valley, St Lawrence (Ixer 1990).

smoky quartz

A translucent brown variety of quartz.

smolianinovite

orthorhombic $(Co)_3(Fe^{3+})_2(AsO_4)_4 \cdot 11H_2O$

An exceedingly rare cobalt–iron hydrated arsenate from the disused Cu–(Co) mine at Muckross, Killarney National Park, Co. Kerry, Ireland. Post-mining oxidation in the mine waste has given rise to secondary carbonates, sulphates and arsenates, six species of which proved new to Ireland (alloclasite, monohydrocalcite, parasymplesite, phaunouxite, rauenthalite and smolyaninovite). Smolianinovite forms brown, translucent, resinous, crumbly masses associated with erythrite (Ryback & Moreton 1993, Moreton et al. 1999, Green & Moreton 2001).

smrkovecite

monoclinic $Bi_2O(OH)(PO_4)$

A very rare bismuth oxide hydroxide phosphate found at the western end of the Hensbarrow granite (part of the St Austell granite complex) in Wheal Remfry china-clay pit, near Indian Queens, Cornwall. Smrkovecite was collected from a vertical 1 m-wide quartz–topaz–tourmaline vein previously reported to contain rutile (Hacker 1997) and the bismuth minerals bismutite and waylandite (Rust 1998). All the bismuth minerals were found at the bottom of a triangular topaz-rich pod in miarolitic cavities lined by muscovite and topaz crystals. A single almost-rounded grain of smrkovecite, 1 mm across, was found mantled by a thin rim of bismutite and partially altered to flaky waylandite (S. A. Rust sample, Tindle unpub. EMP data). Unlike the only other occurrence of smrkovecite, in the Czech Republic (Řídkošil et al. 1996), the Cornish occurrence does not appear to be a weathering product.

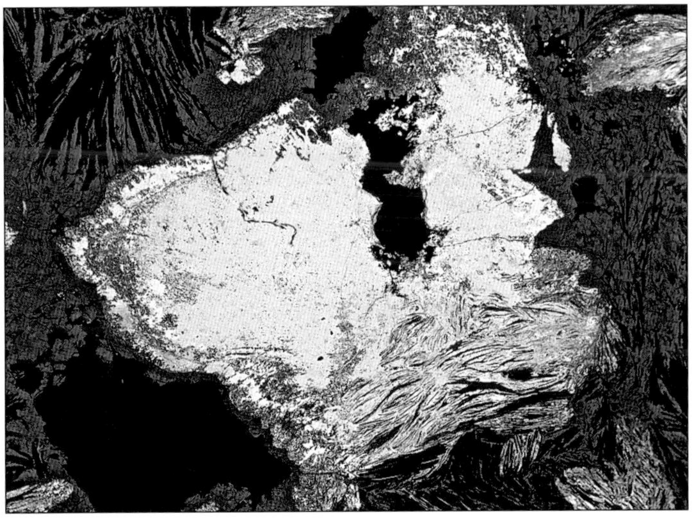

Irregular crystal of smrkovecite (medium grey) mantled by patches of bismutite (light grey) and altered to a waylandite-like phase (bottom right); the dark grey is waylandite. From a topaz-rich pod in Wheal Remfry china-clay pit, Indian Queens, near St Austell, Cornwall; width 1.8 mm; BSE image. (© E. Condliffe, University of Leeds)

soapstone

A synonym of talc or saponite, smectite group.

SODALITE GROUP

Cubic silicates with the formula $(Na,Ca)_{4-8}Al_6Si_6(O,S)_{24}(SO_4,Cl,(OH),S)_{1-2} \cdot nH_2O$. Nosean and sodalite are members of the group found in Britain or Ireland. Haüyne, another member of the group, requires confirmation.

sodalite

SODALITE GROUP

cubic $Na_8Al_6Si_6O_{24}Cl_2$

An uncommon sodium aluminosilicate with chloride, found in the Traprain Law phonolite, near Haddington, Lothian, Scotland. Sodalite forms small (<0.2 mm) interstitial patches in the groundmass, or larger (<0.5 mm) roundish patches with ophitic relations to the feldspars, in association with analcime, for which it can easily be mistaken. Nepheline, green augite and olivine (probably fayalite) are also present in the phonolite (MacGregor & Ennos 1922, MacGregor 1960, Hatch et al. 1972, Livingstone 1993).

To the E of Ledbeg and 1 km N of Ledmore Junction, along a tributary of the Ledbeg River, Assynt, Highland Region, sodalite enclosed by all the later species forms a "beautiful trachytic rock" (Shand 1910) referred to as assyntite, augite syenite or augite–sodalite syenite. Additionally, the rock contains albite, apatite, aegirine-augite, biotite, magnetite, nepheline, orthoclase and titanite. Sodalite associated with biotite, feldspar, nepheline and garnet (var. melanite) is also found in a pegmatitic vein near by (Shand 1910, MacGregor & Phemister 1972).

In Ireland, rhombdodecahedral crystals of sodalite are reported in irregular patches of pyroxenite pegmatite associated with a composite lava flow at McIlroy's Port, Island Magee, Co. Antrim. Aegirine-augite, augite (titanaugite), magnetite and plagioclase feldspar are associated species (Walker 1959b).

soddyite?

orthorhombic $(UO_2)_2SiO_4 \cdot 2H_2O$

Named after Frederick Soddy FRS (1877–1956), professor of chemistry at the universities of Aberdeen and Oxford, and winner of the Nobel Prize for chemistry in 1921 for his work in radiochemistry; born in Eastbourne, Sussex. Soddyite is a rare hydrated uranyl silicate reported from a pitchblende-bearing carbonate vein close to the margin of the Criffell granodiorite at Needle's Eye, SW of Caulkerbush, Dumfries & Galloway, Scotland (Basham et al. 1989). The authors were unable to identify the species positively and they suggest that the mineral could equally well be coffinite.

sodium illite

A synonym of brammallite.

sodium-uranospinite/natrouranospinite

AUTUNITE GROUP

tetragonal $(Na_2,Ca)(UO_2)_2(AsO_4)_2 \cdot 5H_2O$

A rare hydrated sodium–calcium uranyl arsenate found in a uranium-rich vein within the aureole of the Land's End granite at Wheal Edward, St Just, Cornwall. Mixed crystals of bright-green to yellowish-green sodium-uranospinite and uranospinite overgrow zeunerite. Sodium-uranospinite has a distinctive XRD pattern, so can be identified with confidence, but further analyses are required to determine the nature of the intergrowth with uranospinite (D. I. Green pers. comm. 2007).

sodium-zippeite/natrozippeite

ZIPPEITE GROUP

orthorhombic $Na_5(UO_2)_8(SO_4)_4O_5(OH)_3 \cdot 12H_2O$

A rare hydrated sodium uranyl sulphate hydroxide found at Geevor mine, near Pendeen, St Just, Cornwall. Sodium-zippeite has various habits and associations:

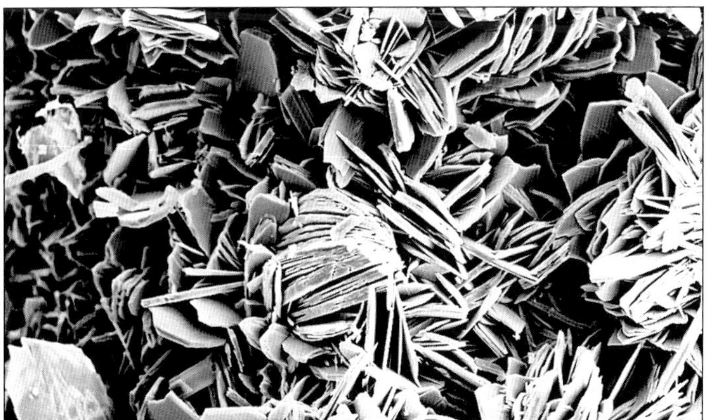

Intergrown tabular crystals of sodium-zippeite from Geevor mine, near Pendeen, St Just, Cornwall; width 70 μm; SEM image. (© N. J. Elton)

extremely thin, translucent, pale lemon-yellow leaves of indistinct shape associated with atacamite, brochantite, chalcocite and gypsum; minute tabular crystals occurring as a creamy-yellow efflorescence and associated with atacamite, chalcopyrite and gypsum; small pale-yellow globular aggregates of randomly orientated acicular crystals with indistinct shape and closely associated with johannite and gypsum; fine translucent petal-shape laths to about 1 mm long, forming booklets and desert-rose aggregates; sprays of very fine narrow laths or spindles and as stacks of stout plates. In the last two habits, sodium-zippeite frequently encrusts atacamite and, in the same area, antlerite, brochantite, chalcanthite, gypsum and, more rarely, metazeunerite and scorodite are also found. Sodium-zippeite forms along joint planes in granite and its host rocks, or in partially altered and veined granite (Elton & Hooper 1993). Tiny powdery aggregates of sodium-zippeite associated with cuprosklodowskite and gypsum are also briefly reported from Geevor mine (Elton et al. 1992).

Small crusts and blebs of bright-yellow microcrystalline sodium-zippeite occur sparingly on beach boulders at North Zawn, W of Botallack, St Just, in or near areas of altered galena and sometimes close to widenmannite. Both are very late-formed minerals and the matrix on which they are found is chiefly spherulitic pyrite and massive quartz, with minor calcite, chalcopyrite and galena. It is suggested that both widenmannite and sodium-zippeite formed by the action of sea spray and rainwater on pyrite, calcite and galena in the host rock, as uranium secondary minerals are not found on host rocks elsewhere (Elton & Hooper 1996).

sonnenstein

A synonym of sunstone (orthoclase, feldspar group).

sonolite

HUMITE GROUP

monoclinic $Mn^{2+}_9(SiO_4)_4(OH)_2$

A rare manganese silicate hydroxide, dimorphous with jerrygibbsite and found in manganese ore of the Haffoty Formation, Llyn Dywarchen, near Moel Ysgyfarnogod, Gwynedd, Wales. Sonolite forms orange-yellow aggregates with barite and manganhumite in a microconcretion and in adjacent carbonate-rich red laminated ore (Bennett 1987a).

spangolite

trigonal $Cu^{2+}_6Al(SO_4)(OH)_{12}Cl{\cdot}3H_2O$

A rare hydrated copper–aluminium sulphate hydroxide chloride found lining small cavities in quartz gossan at Wheal Gorland, St Day, 2 km E of Redruth, Cornwall. Spangolite occurs as well formed bluish-black tabular (hexagonal) crystals up to 3 mm long (Rust 1981b). Miers (1894) describes spangolite from an unidentified mine in the St Day district, near Redruth, where it formed brilliant translucent crystals of a deep emerald-green colour tinted with indigo and up to 2.5 mm in length. It is associated with massive cuprite that is partially replaced by azurite, chrysocolla, clinoclase, malachite and liroconite. Brief mentions of spangolite in Cornwall from Wheal Unity, St Day, Gwennap, and West Caradon mine, St Cleer, Liskeard, are given by Robson (1949b) and Golley & Williams (1995) respectively.

Spangolite is a rare species at New Cliffe Hill quarry, Stanton under Bardon, Leicestershire, where it forms groups of rounded tabular pale bluish-green crystals in cavities in cuprite or on fracture surfaces between wallrock and cuprite in a native copper–cuprite deposit near the Precambrian/Triassic unconformity. Spectacular hemimorphic triangular cone-shape crystals of spangolite up to 1.5 mm long on chrysocolla coat joints in diorite from this location (Hubbard et al. 2005).

A spangolite-type mineral is reported from Dylife mine, N of Hafren Forest, Central Wales orefield, Powys, Wales (T. F. Cotterell pers. comm. 2003).

Hemimorphic triangular crystals of spangolite up to 1.5 mm long on chrysocolla; New Cliffe Hill quarry, Stanton under Bardon, Leicestershire. (© D. I. Green)

sparable tin

A variety of cassiterite peculiar to Cornwall, named because of its so-called resemblance to "sparable nails" of the cobbler (probably a corruption of "sparrow bill") (Solly 1891).

spathic iron

A synonym of siderite, calcite group.

specularite

A variety of hematite with a brilliant metallic lustre.

spencerite?

monoclinic $Zn_4(PO_4)_2(OH)_2{\cdot}3H_2O$

Named after Leonard James Spencer FRS (1870–1959), keeper of minerals at the NHM and editor of *Mineralogical Magazine* for many years; born in Worcester. Spencerite is a rare hydrated zinc phosphate hydroxide reported from Turf Pits mine, Grassington Moor, Yorkshire, by A. W. G. Kingsbury (Embrey 1978). However, as many samples in the Kingsbury collection (now at the NHM) are incorrectly provenanced (Ryback et al. 1998, 2001), doubts remain about the occurrence of spencerite.

sperrylite

PYRITE GROUP

cubic $PtAs_2$

A rare platinum arsenide found in lherzolites and dunites of the Lizard complex at Carrick Luz, Cornwall, where it is found with laurite, potarite (dominant) and various alloys. It is associated with altered pentlandite and chalcopyrite or secondary magnetite, and probably has a magmatic origin modified by local hydrothermal redistribution (Hutchinson 2001, Gunn & Styles 2002).

In the Loch Ailsh alkaline complex, along the Allt Cathair Bhàn Valley, Highland Region, Scotland, there is widespread PGE enrichment in the pyroxenites, pyroxene mylonites and pyroxenite skarns, and here sperrylite, isomertieite and complex tellurides of bismuth, lead, palladium and silver are found in accessory amounts. Sperrylite is also one of the most abundant PGMs in the Borralan alkaline igneous complex near by, occurring with probable isomertieite and bismuth–palladium tellurides mainly in late brittle fractures and along grain boundaries. A magmatic origin associated with late magmatic volatile-rich fluids deficient in sulphur has been proposed (Shaw et al. 1994, Gunn & Styles 2002, Styles et al. 2004).

In the Eastern Layered Series of the layered mafic–ultramafic intrusion on the Isle of Rum, Highland Region, sperrylite represents the most common PGE species in some of the chromitite units and is variously accompanied by braggite, cooperite, hongshiite, irarsite, isoferroplatinum, keithconnite, kotulskite, laurite, moncheite,

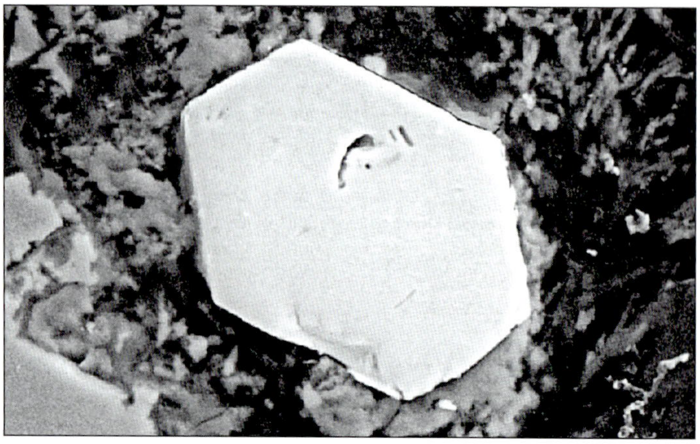

Euhedral sperrylite enclosed by plagioclase feldspar and altered pyrrhotite. From a gabbroic dyke close to the margin of the Sgaorishal ultramafic plug, Isle of Rum, Highland Region, Scotland; width 55 mm; BSE image. (© M. R. Power, Camborne School of Mines)

palladoarsenide, platarsite, sperrylite, telargpalite and telluropalladinite, and the native species gold, palladium, platinum and silver (Butcher et al. 1999, Power et al. 2000a, 2003). From a similar magmatic environment, the Ben Buie peridotite intrusion at Caigeann Doire nan Cuilean, Isle of Mull, Strathclyde, contains sperrylite in, or adjacent to, chromitite layers in association with froodite, laurite, ruarsite and minerals of the hollingworthite–irarsite–platirasite series. All species have a grainsize of 1–3 µm across (Pirrie et al. 2000a, Power et al. 2003). Additionally, Prout & Andersen (2001) report michenerite from Ben Buie. In the Outer Layered Series of the Cuillin complex (also part of a layered ultramafic intrusion) on the Isle of Skye, Highland Region, ruarsite, sperrylite, Fe–Pt–Rh alloy and Pd arsenides occur at the grain boundaries of silicate, sulphide and chrome spinel. All are subordinate to laurite, which is typically found in chromite as inclusions 1–7 µm in diameter (Pirrie et al. 2000a, Power et al. 2003).

A single alluvial grain of sperrylite was recovered by Gunn et al. (1990) from Cachnamuin Stripe, Garmuch Burn, near Bridgend, SW of Huntly, Grampian. The grain is cubic, 150 µm in length and very fresh in appearance.

At Talnotry, Newton Stewart, Dumfries & Galloway, a small body of copper–nickel mineralization has developed at the base of an altered diorite sheet. Its principal ore minerals are chalcopyrite, gersdorffite, nickeline, pentlandite and pyrrhotite, and, within the mineralized zone, small grains (< 10 µm across) of gold (electrum), irarsite, sperrylite and an unnamed bismuth–palladium telluride have been identified (Power & Pirrie 2001, Gunn & Styles 2002). Later work by Power et al. (2004) confirms irarsite, gold (electrum) and sperrylite to be the dominant PGMs present, with hollingworthite, merenskyite and michenerite being the other subordinate species. Arsenopyrite, annabergite, bismuthinite, galena and pyrite are also noted.

Sperrylite occurs in chromite-rich dunite at Cliff, Baltasound–Haroldswick area, Unst, Shetland, where it forms 10–20 µm grains (rarely to 100 µm across) in association with chalcopyrite, cobaltite, heazlewoodite, irarsite, millerite, nickeline, orcelite, native osmium, pentlandite, pyrrhotite and stibiopalladinite (Gunn et al. 1985, Prichard et al. 1986, Gunn & Styles 2002). Composite grains of sperrylite, hongshiite and a copper–nickel alloy have also been found in dunite at Cliff (Prichard & Tarkian 1988). Between Ordale and Keen of Hammar, on Unst, Prichard et al. (1994) report 1 µm-diameter grains of sperrylite associated with bornite, chalcopyrite and pyrrhotite, in a wehrlite and pyroxenite unit. All these occurrences are from the Shetland ophiolite complex.

spessartine

cubic — GARNET GROUP — $Mn^{2+}_3Al_2(SiO_4)_3$

A common manganese aluminosilicate that forms a series with almandine and is found in some granite pegmatites, granites and rhyolites. Spessartine also forms in metasomatic manganese-rich rocks adjacent to igneous intrusions. Like all other species of garnet, end-member compositions are rare and most examples show some degree of solid solution towards other members of the group. Where possible, the proportions of the two most abundant end members are listed below (al = almandine, gr = grossular, py = pyrope, sp = spessartine).

England

Cornwall At Treburland mine, 2 km SE of Fivelanes, Altarnun, Bodmin Moor, manganese-rich sediments thermally metamorphosed by the Bodmin Moor granite produced small clear "honey-yellow" aggregates of spessartine in crystalline rhodonite and dark yellowish-brown crystals covering altered pyroxene. Crystals are well formed and up to 1 mm across. Other associated species are andradite, bustamite, pyrrhotite, tephroite, vesuvianite and wollastonite (Russell 1946d, Tilley 1946).

Devon Contact metamorphism of calc-silicate rocks exposed in the British Railways quarry complex (sp_{79}, gr_{13}) and the nearby aplite quarry (sp_{71}, py_{14}) at Meldon, near Okehampton, has produced an assemblage similar to Treburland, with golden-yellow spessartine (2–3 mm across), associated with bustamite, pyrrhotite, rhodonite and tephroite. Garnet with an intermediate composition (sp_{40}, gr_{35}), and forming brownish-red crystals up to 5–6 mm across, has also been found in quartz veins in the quarry (Howie 1965, Dearman 1966a).

Wales

Dyfed Contact metamorphism of pelites adjacent to the St David's Head intrusion, Carn Llidi, has produced rocks containing spessartine (Roach 1969).

Gwynedd Spessartine is widely developed in the Harlech Dome, where it is an important component of bedded manganese ores of Cambrian age. Bennett (1987a) describes isolated euhedral crystals, 0.005–0.03 mm across, with variable composition (sp_{77-90}, al_{9-14}) and (sp_{85-90}, gr_{5-9}) in quartz–chlorite pods in the Gamlan Formation at Trawsfynydd. The Hafotty mines, above Llanaber, near Barmouth, once worked a bedded manganese-rich ore containing quartz, rhodochrosite and spessartine (Allen & Jackson 1985, Bennett 1987b). Woodland (1939) considered the spessartine to have a regional metamorphic origin, stating that the crystals are small (averaging 0.014 mm), reddish-brown and euhedral. Other accounts of these rocks are provided by Woodland (1935), Matley & Wilson (1946) and Glasby (1974). Spessartine is said to be a common species in bedded manganese ore from Benallt mine, Rhiw, and on St Tudwal's Peninsula, both on the Lleyn Peninsula (Dunham et al. 1978), although later work indicates that andradite is the major garnet species at Benallt (T. F. Cotterell pers. comm. 2007). At Gwastadnant, near Nant Peris, Snowdonia, coffee-brown, violet and brick-red spessartine occurs with magnetite in a schistose grit band associated with the Talgau lavas (Williams 1927, 1930b). More recently, spessartine was found in metamorphosed pelitic rocks associated with blueschists in a temporary roadcut near Llanfairpwllgwyngyll, Anglesey (Bevins 1994). Williams (1922) notes spessartine in Snowdonian lavas near Beddgelert and in a honestone quarry at Pen-y-Gwryd.

Scotland

Central Garnet in a mica schist associated with stratiform sulphide mineralization in the Dalradian Ben Challum Quartzite to the E of Tyndrum is mostly almandine, although one analysis (sp_{41}, al_{31}) is spessartine dominant (Fortey & Smith 1986).

Grampian Accessory magmatic yttrium-rich spessartine (sp_{53-54}, al_{44-45}) forms anhedral crystals up to 2.5 mm across in an aplite sheet cutting the Cairngorm granite pluton at Allt Ruadh (Harrison 1988). Alluvial spessartine in Luibeg Burn, Derry Cairngorm, is derived from this pluton (Beer & Bennett 1988, Haynes 1993). Small spessartine (sp_{59-64}, al_{34-39}) crystals have also been found in red aplite and granite 2 km SW of Gairnshiel Bridge in Glen Gairn (Tindle unpub. EMP data). The large tourmaline-bearing pegmatite at East Head, 1 km NE of Portsoy, contains accessory spessartine (sp_{58-68}, al_{31-41}) associated with cassiterite, fluorapatite and manganocolumbite (J. R. Holbrook specimen, Tindle unpub. EMP data).

Highland A so-called diamond-bearing sample in the Hunterian Museum, Glasgow, originally collected from a boulder in glacial drift near Kinloch, S of Tongue, has been shown instead to contain colourless to very pale-yellow spessartine (sp_{86}, gr_{13}). Such boulders were probably derived from local Moine rocks (Faithfull 2007).

Strathclyde A schorl-bearing pegmatite dyke cutting granodiorite at the southern end of Loch Doon (part of the Loch Doon granite complex) contains small (< 2 mm) pinkish-red spessartine (sp_{55-57}, al_{40-41}) crystals in accessory amounts (Tindle 1982). Dalradian albite schists near Loch Lomond contain zoned porphyroblasts with spessartine (sp_{41-44}, al_{36-43}) cores and almandine rims (Mathavan & Bowes 2005).

Western Isles Spessartine (sp_{46-50}, al_{46-49}) is locally abundant and forms masses up to 15 cm across in the large Chaipaval (Ceapabhal) pegmatite on South Harris (von Knorring & Dearnley 1960a, Tindle unpub. EMP data) and is also noted in the Loch a' Sgùrr pegmatite, near Finsbay (Fionnsabhagh) (von Knorring & Dearnley 1960a). Alluvial spessartine has been identified in panned concentrates from the An Lag area, St Kilda (Hirta) (Harding et al. 1984).

Ireland

Carlow (& Wicklow) Lithium-rich pegmatites and aplites related to the Leinster granite often carry spessartine as an accessory component. At Aclare, near Myshall, orange to dark-red or brown subhedral to euhedral crystals of spessartine (sp_{63-89}, al_{10-36}) up to 15 mm across occur in spodumene pegmatite and garnetiferous aplite (Scoon 1978, Whitworth 1992, Tindle unpub. EMP data). Other examples are reported from Rathnagerragh Castle, SW of Myshall (Scoon 1978), and near Shillelagh in Co. Wicklow from Moylisha (Scoon 1978) and Stranakelly (Steiger & von Knorring 1975).

Donegal Garnet is a common accessory mineral in muscovite granites, aplites and pegmatites of the Donegal granite batholith. One example in a pegmatite segregation in biotite granite within the Trawenagh Bay granite is spessartine (sp_{80}, al_{18}), whereas another from the Main Donegal granite is spessartine (sp_{49}, al_{47}) (Hall 1965b). Beautiful deep-red lustrous euhedral spessartine (sp_{63-72}, al_{25-32}) crystals up to 3.5 mm across have also been found in granitic pegmatite at Gweebarra Bay (Ryback & Moreton 1992, Green et al. 2005a, S. Moreton sample, Tindle unpub. EMP data).

Galway Spessartine (sp_{58-72}, al_{23-39}) forms zoned crystals with manganese-rich cores and iron-rich rims in the western part of the Murvey granite and associated aplites at Rick's Lough, Roundstone (Leake 1967). Spessartine in this granite was also investigated by Whitworth & Feely (1994), who describe cherry-red euhedral crystals of spessartine (sp_{86-87}, al_{11-12}) up to 5 mm across in pegmatite, and spessartine (sp_{65-77}, al_{20-34}) crystals up to 1 mm across in an aplite vein within the Murvey granite

Mayo Blueschist-facies metabasites belonging to the Dalradian Supergroup from

Euhedral spessartine, 3.5 mm across, in granitic pegmatite from Gweebarra, Co. Donegal, Ireland. (© D. I. Green)

central Achill Island, N of Clew Bay, contain a few zoned garnets where the spessartine component is dominant (sp_{35-54}, al_{30-36}). Most garnets in these rocks have almandine composition (Chew et al. 2003).

spessartite
A synonym of spessartine, garnet group.

sphaerodialogite
TL Gwynedd, Merionethshire (now part of Gwynedd), Wales
A variety of rhodochrosite, calcite group.

sphaerosiderite
A synonym of siderite, calcite group.

sphaerostilbite
A botryoidal fibrous variety of thomsonite, zeolite group.

SPHALERITE GROUP
Cubic sulphides, selenides and tellurides with the general formula *AX*, where *A* = Cd, Fe, Hg, Zn; and *X* = S, Se, Te. Coloradoite, hawleyite, metacinnabar, sphalerite and tiemannite are members of the group found in Britain or Ireland.

sphalerite
SPHALERITE GROUP
cubic
ZnS

A very common zinc sulphide, trimorphous with matraite and wurtzite, and found under a wide variety of hydrothermal conditions. Sphalerite is frequently found associated with galena in carbonate-hosted replacement deposits. It is the main ore species for zinc and in old reports is described as blende or black jack. Britain has five principal lead–zinc–(barite–fluorite) orefields hosted within Upper Palaeozoic limestones and dolomites and classified as MVT deposits. They are the Mendip orefield, the North Wales orefield, the Southern Pennine orefield (Derbyshire), and the Askrigg and Alston Block areas of the Northern Pennine orefield (Ixer 1986). The large Carboniferous carbonate-hosted lead–zinc deposits of Ireland have many characteristics of MVT deposits, but are sufficiently different to justify their own type (Hitzman & Large 1986).

England
Avon Very rare tiny transparent pale-brown gem-like crystals of sphalerite have been found in a major barite–calcite–galena vein within Carboniferous Limestone close to the Rhaetic/Lower Carboniferous unconformity. They are associated with galena in a cave system at Hampstead Farm quarry, near Chipping Sodbury (Alabaster 1990b, Brooke 1994b, Sharp 1997). The vein has a mineral assemblage similar to those at the adjacent Southfields quarry (Curtis 1981) and to other veins at Stancombe quarry, near Flax Bourton, Bristol (Zachariades 1993), and Woodleaze quarry, Tytherington (Alabaster 1989a). Close to the beach near the Pier hotel, Clevedon, near Bristol, vein-style mineralization cutting Upper Triassic dolomitic limestone breccia comprises banded barite with minor chalcopyrite, enargite, galena (most abundant), sphalerite and tennantite, together with their alteration products. Many generations of sphalerite are described, including a cadmium-rich variety (Ixer 1986, Ixer et al. 1993, Starkey 1986c).

Cheshire Sphalerite occurs in redbed-type mineralization within faulted Triassic sandstones at Engine Vein mine and Stormy Point, Alderley Edge ridge. It contains inclusions of chalcopyrite, pyrite (bravoite) and tetrahedrite, and is itself commonly enclosed in galena. Some of the sphalerite is cadmium rich (Carlon 1979, Ixer & Vaughan 1982, Braithwaite 1994).

Cornwall Sphalerite (and galena) mainly occur in cross-course veins that developed after the main episode of copper–tin mineralization in the region. Sphalerite has also been worked from the upper levels of a few copper lodes such as those N of Camborne at Roskear and Seton mines, and in some of the St Day mines. Thick masses of sphalerite lined the walls of the Perran Iron lode, near Perranporth, prior to major iron mineralization, and mines along this lode were some of the largest producers of zinc in Cornwall. The St Agnes district was the most important mining area for sphalerite, raising 75% of the total production of 85 000 tonnes (Dines 1956). Some of the most interesting and important occurrences are described as follows. *St Just mining district:* Balleswidden mine – cassiterite lode containing some galena, pyrite and sphalerite (Hosking 1951); North Zawn, Botallack – large blackish grains or yellowish crusts of sphalerite in galena–pyrite-bearing quartz boulders (Elton & Hooper 1996); Botallack mine (Crowns lode) – sphalerite in polymetallic lodes (Hosking 1951, Bancroft & Weller 1993); Wheal Edward – sphalerite in a quartz lode associated with chalcopyrite, pyrite and uraninite (Ball et al. 1982). *St Ives–Helston mining district:* Balmynheer mine, Wendron – granite-hosted cassiterite-bearing replacement deposit (carbona) containing considerable pyrite, sphalerite and minor wolframite (Stephens 1940b, Dominy et al. 1996a); Binner Downs mine, Helston – sphalerite associated with cassiterite, celestine, galena and pyrite (Hosking 1951); Carnmeal mine, E of Godolphin, and Trannack mine, Helston – uranium-rich cross-course veins containing sphalerite with chalcopyrite, galena and pyrite (Ball et al. 1982); Great Wheal Vor, near Helston – sphalerite in Woolf's cross course (Stephens 1940); Penberthy Croft mine, St Hilary – dark-brown anhedral masses to 2 cm across are widespread in the primary assemblage, with oxidation on the mine dumps producing schulenbergite as an alteration product (Betterton 1996b, Betterton et al. 2000); Relistian mine, Gwinear – buff-to-brown botryoidal sphalerite (Kingsbury 1964b); St Michael's Mount, Marazion – sphalerite in polymetallic veins and associated greisen bands within granite (Davison 1920b, Hosking 1954c); Trevascus mine, near Hayle, Gwinear – sphalerite in a polymetallic telescoped orebody including an intrusion breccia (Hosking 1951, Goode & Taylor 1980); Wheal Alfred, Phillack, near Hayle – mineralized lodes changed from being sphalerite-rich to copper-rich at depth (Dewey 1925); Wheal Fortune, near Helston – considerable quantities of sphalerite (Dewey 1925); Wheal Metal, Breage – sharp black tetrahedra of sphalerite up to 9 mm on edge associated with bertrandite and orthoclase, and with albite, cassiterite, dolomite and pyrite (Russell 1913b, Hosking 1951); Wheal Penrose, Sithney, near Helston – sphalerite associated with galena and siderite in mesothermal lodes (Robson 1949b, Hall & Jackson 1975); Wheal Rose, Sithney – the sequence of crystallization was arsenopyrite, chalcopyrite, sphalerite, quartz, galena, siderite and dolomite (Hosking 1951); Wheal Trewavas, Trewavas Head, Breage – sphalerite associated with chalcopyrite, galena, marcasite, pyrite and tristramite in uraninite-bearing quartz veins (Atkin et al. 1983). *Camborne–Illogan–Redruth mining district (and southwards into the Carnmenellis granite and its environs):* Bosworgey borehole, near Fraddam – sphalerite as an accessory mineral in an As–Bi–Cu–Sn–W-enriched part of the Carnmenellis granite (Ball & Basham 1984); Chywoon quarry, near Longdowns – rare golden-brown sphalerite crystals to 3 mm long on sericite covered pegmatite within the Carnmenellis granite (Moulding & Hooper 2005); Copper Bottom mine dump, near Camborne – common sphalerite and wolframite in an intrusion breccia (Goode & Taylor 1980); Dolcoath mine, Camborne – sphalerite in silver ore containing large amounts of chalcopyrite and pyrite (Hosking 1951, Rust 1982a); East Wheal Falmouth, St Day – produced copper, lead and 487 tonnes of sphalerite ore (Stephens 1940a); Kerriack Cove, between Porthtowan and Portreath, near Redruth – appreciable sphalerite in a chalcopyrite–galena lode once worked for zinc as part of the Wheal Lushington sett (Dines 1956, Elton & Hooper 1995b); Kessel Downs (Halvasso) quarry, near Longdowns, Carnmenellis – sharp clear pale-brown sphalerite associated with apatite, cassiterite, fluorite, pyrite and stokesite, all in pegmatite lenses within the Carnmenellis granite (Couper & Barstow 1977, Jackson 1978); Mount Wellington mine, Twelveheads, 7 km W of Truro – polymetallic veins containing sphalerite with abundant chalcopyrite or pyrrhotite inclusions, also later mineralization with chalcopyrite and pyrite with sphalerite inclusions (Kettaneh & Badham 1978); New Cook's Kitchen mine (North Tincroft lode), Redruth – fine-grain polymetallic lode in which sphalerite grains to 1 mm across occur in isolation or intimately intergrown with arsenopyrite and chalcopyrite (LeBoutillier et al. 2000); Seton and South Roskear mines – major copper mines, produced 19 and 6 tonnes of sphalerite respectively (Dines 1956); Tresavean mine, Gwennap, near Redruth – vertically zoned lode with sphalerite near surface, then chalcopyrite followed by cassiterite at depth (Stephens 1940a, Hosking 1951); Wheal Jane, Kea, near Truro – dominantly mined for cassiterite, but massive pyrite and sphalerite zones and caunter lodes containing narrow chalcopyrite and sphalerite are also present. Sphalerite overgrows cassiterite and contains inclusions of chalcopyrite and stannite. Attractive specimens consisting of sharp black crystals of sphalerite to 1 cm across on vein quartz have been collected. Other crystals reach 4 cm across, develop a plate-like form, or show polysynthetic twinning (Rayment et al. 1971, Rust 1982d, Holl & Bromley 1988, Dominy & Camm 1997); Wheal Unity, St Day, Gwennap – polymetallic ore, produced buff-to-brown botryoidal sphalerite (Collins 1892, Hosking 1951, Kingsbury 1964b). *St Agnes mining district:* Cligga Head, Perranzabuloe – sphalerite with chalcopyrite and stannite exsolution, also delicate exsolution lamellae of sphalerite in kësterite and zinc-rich stannite (Moore & Howie 1984); Duchy Peru and Great Retallack mines, near St Agnes – worked the Perran Iron lode and produced 20 000 and 11 500 tonnes, respectively, of fine-grain black sphalerite and late translucent red sphalerite (Dewey 1925, Dines 1956, Vipan 1959, Henley 1971); Lambriggan mine, near Perranzabuloe – unusual pale (yellowish-green) sphalerite (Robson 1949b, Hosking 1951, Embrey & Symes 1987); Trevaunance mine – octahedra–tetrahedra of brown-to-black sphalerite associated with fluorite (Embrey & Symes 1987); West Chiverton mine, Perranzabuloe – partially dissolved galena and sphalerite on the mine dumps (Dewey 1925); Wheal Hope, ENE of Perranporth – sphalerite containing inclusions of chalcopyrite and pyrite, also dark sphalerite rimmed with chalcopyrite in a galena-bearing lode (Hosking & Camm 1981); Wheal Kitty – fine crystals of dark-brown iron-rich sphalerite, variety christophite (Collins 1879b). *Wadebridge–St Endellion mining district:* This district is best known for its antimony–lead vein mineralization, which has been classified into:

- Type 1: arsenopyrite dominated
- Type 2: galena (+ chalcopyrite & sphalerite) dominated
- Type 3: galena–lead-sulphosalt dominated.

Examples are described from: Bounds Cliff, E of Port Isaac (Clayton et al. 1990, Stanley et al. 1991); Old Treburgett mine, St Teath, E of Port Isaac (Kingsbury & Hartley 1960); Pengenna mine, St Kew (Robson 1949b); Pentire (Glaze) mine, near New Polzeath, St Minver (Clayton et al. 1990); Poltreworgy mine, near Port Isaac (Russell 1949b); Port Quin mine (Russell 1944, 1949b, Clayton et al. 1990); Reedy Cliff, near Port Quin (Clayton et al. 1990); Treore mine, Port Gaverne, near Port Isaac, St Teath (Russell 1949b, Kingsbury & Hartley 1960); Wheal Boys, St Endellion (Rudler 1905, Russell 1949b). Additionally, sphalerite has also been found in chalcopyrite–galena veinstone at Gunver Head, near Padstow Consols mine (Braithwaite & Merry 1996, Schlomann & Weiss 1998). *St Austell mining district:* Fowey Consols, Tywardreath, St Blazey – sphalerite varieties include buff-to-brown botryoidal sphalerite, cadmium-rich sphalerite and "white, transparent" cleiophane (Garby 1848, Greg & Lettsom 1858, Kingsbury 1964b); Goonvean and Rostowrack china-clay pits, W of Nanpean, St Austell – sphalerite is one of the accessory minerals found in the waste from kaolinized topaz granite (Scott et al. 1998); Silver Vein mine, Lostwithiel – sphalerite in a silver-bearing polymetallic lode (Rust 1982a); South Terras mine, St Stephen-in-Brannel – sphalerite in a silver–uranium-bearing polymetallic lode (Dines 1956, Dunham et al. 1978, Rust 1982a); Terrace (Terras) Hill quarry, Lostwithiel – sphalerite associated with apophyllite, axinite, calcite, galena, garnet and pyrite in a metasomatic deposit (Russell 1910b); Tremore quarries, Lanivet, SW of Bodmin – sphalerite associated with actinolite, axinite, fluorite, garnet, pyroxene, talc, tourmaline and vesuvianite in a metasomatic (calc-flinta) deposit (Barrow & Thomas 1908). *East Cornwall mining district:* Cheesewring quarry, SW of Henwood, Linkinhorne, and Gold Diggings quarry, near Cheesewring quarry, St Cleer – red sphalerite with bertrandite, fluorite, orthoclase (adularia) and pyrite on joint faces in granite (Bowman 1911, Russell 1913b, Hosking 1954b); East Kit Hill mine, Stoke Climsland – a similar occurrence to Cheesewring (Russell 1913b); Greystone quarry, Lezant, near Launceston – a source of fine well crystallized specimens of sphalerite, which occur as complex translucent yellow, green to black crystals up to 23 mm long (Lloyd & Barstow 1983, Grigg 1996, Elton et al. 2004); Holmbush, Kelly Bray and Redmoor mines, Kelly Bray, Callington – all worked sphalerite-bearing sulphide veins cutting country rock (killas) adjacent to a minor granite intrusion. At Redmoor, sphalerite was also found in galena-bearing cross-course veins (Newall & Newall 1989, Newall 1994); Prince of Wales mine, Calstock, near Gunnislake – partially oxidized sphalerite contains cavities lined with ramsbeckite (Green et al. 2001); Trebullett and Trenute mines, near Lezant – sphalerite in a stibnite-bearing lode (galena-bearing at depth), light brown at Trenute (Russell 1949b, Hosking 1951, Dines 1956); Tredinnick mine, Landrake with St Erney – a little brown sphalerite in a stibnite lode (Russell 1949b); Wheal Langford (had other names too), Callington – sphalerite in a silver-bearing polymetallic lode (Rust 1982a); Wheal Ludcott and Wheal Wrey, St Ive, near Liskeard – scarce orange-brown microcrystals and translucent reddish-brown veinlets and masses of sphalerite in silver-bearing chalcopyrite–galena–quartz lodes (Dines 1956, Rust 1983a); Wheal Mary Ann, Menheniot – minor sphalerite in cross-course lead lodes (Gleeson et al. 2000). *Other occurrences:* Camel, Fowey and Gannel estuaries – detrital sphalerite in intertidal sediments (Pirrie et al. 2000b, Pirrie et al. 2002).

Cumbria Sphalerite is a very common mineral in metalliferous veins of the central Lake District and is most abundant in Lower Carboniferous galena–sphalerite veins formed along normal faults. It is also present in lesser amounts in Lower Devonian copper- and tungsten-bearing veins (Firman 1978, Stanley & Vaughan 1982a). In the early days of mining, sphalerite was a waste material. *Caldbeck Fells:* Although present in many veins in the area, sphalerite is usually massive or forms fine-grain aggregates. Euhedral crystals are uncommon and small. Localities include: Bannerdale mine, Mungrisdale (Eastwood 1921, Davidson & Thomson 1951); Brae Fell mine, Dale Beck Valley (Hartley 1984, Bridges et al. 2006); Brandy Gill and Silver Gill mines (Hartley 1984); Carrock mine, Grainsgill – sphalerite occurs in scheelite–wolframite veins and in later galena-bearing veins. In places, sphalerite replaces arsenopyrite and pyrite. Some sphalerite is iron rich (marmatite), and rare cubanite has been found with chalcopyrite as inclusions in sphalerite (Hitchen 1934, Davidson & Thomson 1951, Eastwood 1959, Ewart 1962, Eastwood et al. 1968, Shepherd et al. 1976, Dunham et al. 1978, Hartley 1984, Ball et al. 1985a, Young 1987a, Neall & Green 2001a); Driggith mine – sphalerite associated with barite, chalcopyrite and galena, also radiating veins of sphalerite in quartz (Eastwood 1921, 1959, Stanley & Vaughan 1981, Hartley 1984, Green et al. 2006a); Dry Gill mine – a little residual sphalerite associated with cinnabar (Neall & Leppington 1994); Nether Row Brow (Kingsbury & Hartley 1960, Hartley 1984); Potts Gill mine – sphalerite (marmatite) in a mainly chalcopyrite–galena-bearing barite vein, also cinnabar associated with residual sphalerite (Kingsbury & Hartley 1960, Shaw 1970, Hartley 1984, Neall & Leppington 1994); Red Gill mine – sphalerite associated with native silver (Wirth 1989); Roughton Gill mine – large amounts of massive sphalerite left on the dumps, generally light brown (Eastwood 1921, 1959, Davidson & Thomson 1951, Eastwood et al. 1968, Hartley 1984); Sandbed mine and along the Driggith–Sandbed vein at High Pike – the principal ore mineral was galena with appreciable (estimate 1100 tonnes) massive sphalerite remaining on the dumps (Greg & Lettsom 1858, Davidson & Thomson 1951, Eastwood 1959, Eastwood et al. 1968, Hartley 1984, Green et al. 2006a); Wet Swine Gill – sphalerite is one of the lesser species in a quartz–stibnite vein, where it forms isolated orange-red grains up to 50 µm across (Fortey et al. 1984, Neall & Green 2001b). *Elsewhere in the Lake District and West Cumbria:* Barrow mine, Braithwaite – occasional good specimens of sphalerite and galena on the dumps (Davidson & Thomson 1951); Bonser mine, Coniston (Stanley & Vaughan 1982b); Brandlehow mine, Derwent Water and Yewthwaite mine, Newlands Valley – galena is the principal ore mineral, although there is some sphalerite (Davidson & Thomson 1951, Eastwood 1959); Castlenook mine, Newlands Valley – sphalerite plentiful on the tips (Strens 1962, Ineson & Mitchell 1974a); Coniston copper mines, Coniston – lead–zinc vein, a good deal of sphalerite in places (Eastwood 1921, Russell 1925, Dagger 1977); Dale Head mine, Newlands Valley – unzoned iron-rich sphalerite in a pyrite-dominated quartz vein (Stanley & Vaughan 1980); Eagle Crag mine and Hartsop Hall mine, Hartsop, Patterdale – sphalerite and galena were the ore species (Davidson & Thomson 1951); Fleetwith Pike, Borrowdale – sphalerite in several small veins and at Goat Crag, Borrowdale – pale-yellow crystals, also Manesty Copper vein, Grange, Borrowdale (Strens 1962); Force Crag mine, Braithwaite, near Keswick – finely crystallized lustrous black specimens and colour-zoned sphalerite are reported. Sphalerite is ten times as common as galena and is found in the lower levels of the mine (Greg & Lettsom 1858, Postlethwaite 1913, Davidson & Thomson 1951, Eastwood 1959, Ineson & Mitchell 1974a, Dickinson & Pattrick 1987, Green et al. 1997, Green & Briscoe 2002, Tyler 2005); Goldscope mine, Newlands Valley (Eastwood 1921); Graphite mine, Seathwaite – dark sphalerite with chalcopyrite (Strens 1962, 1965); Greenburn mine, Little Langdale – sphalerite associated with arsenopyrite, chalcopyrite and pyrite (Stanley 1979); Pave York mine, Greenburn Valley – sphalerite in thin veins, locally replaced by bornite and chalcopyrite (Stanley 1979); Greenside mine, near Glenridding, Patterdale – sphalerite and chalcopyrite increased with depth (Gough 1965, Firman 1978);

Hellen's mine, Tilberthwaite – very dark sphalerite associated with arsenopyrite, chalcopyrite, pyrite and galena (Shaw 1970, Stanley 1979); Myers Head mine, Hartsop, Patterdale (Dakyns et al. 1897, Eastwood 1921); Oughterside opencast coal site, Aspatria – sphalerite in millerite-bearing clay–ironstone nodules (Young & Nancarrow 1988a); Piers Gill, 1 km N of Scafell summit, Wasdale – sphalerite occurs with arsenopyrite, chalcopyrite, galena and pyrite in a 60 cm-thick polyphase quartz vein cemented by ankerite, within the Scafell dacite. Rare associated species include acanthite (argentite), native gold (electrum), tetrahedrite, uytenbogaardtite and an unknown selenium–silver mineral (Aldridge 2005); Robin Hood mine, Bassenthwaite – small orange-brown grains in quartz (Russell 1949b, Young 1987a, Green & Briscoe 2002); Ruthwaite Lodge vein, Grisedale, SW of Glenridding, Ullswater – chalcopyrite–galena–sphalerite–tetrahedrite assemblage (Young 1986); Sellafield area boreholes – sphalerite in polymetallic veins found in boreholes (Milodowski et al. 1998); Shap Blue quarry – sphalerite in polymetallic veins cutting Borrowdale Volcanic Group rocks (Young et al. 1988); Shap granite quarry – sphalerite in miarolitic cavities (bright blue in transmitted light), in quartz veins, on joint coatings and in zones of sericitic alteration within granite (Grantham 1928, Firman 1978, Fortey 1980, Wirth & Green 2002); Thornthwaite mine – small amounts of sphalerite, galena and cerussite on the dumps (Postlethwaite 1913, Davidson & Thomson 1951); Threlkeld mine – attractive groups of sphalerite and galena (Postlethwaite 1913, Davidson & Thomson 1951, Eastwood 1959); Tilberthwaite Gill – sphalerite–galena vein (Eastwood 1921); Wanthwaite mine, St John's in the Vale, near Threlkeld – granular aggregate of arsenopyrite, galena, pyrite and sphalerite in quartz vein (Kingsbury & Hartley 1960); Whitecombe Beck, Whicham – sphalerite with fluorite, galena and quartz (Young 1987a). *Northern Pennine orefield:* Sphalerite is abundant in the Nent and West Allendale area, where it occurs in fissure veins and flats as brown to dark-brown crystalline bands associated with galena. In cavities, it forms highly lustrous near-black crystals. A sequence of mineral zoning has been established for the Alston Block, with sphalerite present in three of the zones and sphalerite-rich areas close to the inner fluorite/outer barite junction (Dunham 1990, Cann & Banks 2001). Localities include: Brownley Hill mine (includes Guddam Gill level), Nenthead, near Alston – lustrous black crystals up to 5 mm across, dull crystals to 2 cm across in cavities, and small transparent red to reddish-brown crystals (Smythe 1924, Young et al. 1987, Young et al. 1990b, Clark & Rust 1993, Green & McCallum 1993, Green et al. 2000); Dufton mine, near Appleby – galena–sphalerite veins with barite (Goodchild 1884, Davidson & Thomson 1951); Great Sulphur vein, Alston Moor – minor sphalerite in a marcasite–pyrite-dominated vein (Vaughan & Ixer 1980); Handsome Mea mine, Nenthead – fine crystals of sphalerite associated with gypsum, also examples with delicate growth zoning visible in transmitted light (Davidson & Thomson 1951, More et al. 1991); Hilton mine, Scordale, near Appleby (Ixer 1986); Nentsberry Haggs mine, Nenthead – brilliant crystals of sphalerite associated with barite, galena and witherite, also associated with alstonite (Davidson & Thomson 1951, Solomon et al. 1971, Green 1993); Rampgill mine, Nenthead – euhedral black sphalerite with complex form (Green & Briscoe 2002, Wolfe 2004); Rotherhope mine, Leadgate, near Alston – spherical nodules of sphalerite (Dunham 1947, Davidson & Thomson 1951); Smallcleugh mine, Nenthead, near Alston – spectacular sphalerite specimens from the Ballroom Flat area and massive specimens with delicate growth zoning visible when viewed in transmitted light (Young et al. 1987, More et al. 1991, Green & Briscoe 2002); Tynebottom mine, NW of Garrigill, near Alston – dark-brown crystalline sphalerite undoubtedly present, but some descriptions are based on A. W. G. Kingsbury samples with dubious provenance

Well formed crystals of sphalerite from Smallcleugh mine, Nenthead, Cumbria; width 9 mm. (© A. G. Tindle)

Cluster of well formed sphalerite crystals from Prouds Sump, Rampgill mine, Nenthead, near Alston, Cumbria; 70 mm tall. (© D. I. Green)

(Vaughan & Ixer 1980, Ixer 1986, Ixer & Stanley 1987, Young et al. 1987).

Derbyshire The distribution of sphalerite is more restricted than that of galena, in part because it is more easily leached or oxidized by weathering processes. In many cases it is coated or extensively replaced by oxidation products such as hemimorphite or smithsonite. It is mainly found along the eastern margin of the Southern Pennine orefield, commonly in association with calcite. Cadmium and copper are important trace elements in the sphalerite, but, unlike sphalerite from the Alston Block, iron content is low, indicating a low formation temperature. Inclusion species in sphalerite include calcite, chalcopyrite, galena, pyrite and, rarely, barite and fluorite (Mostagel 1985). Some mines produced sphalerite as a subsidiary ore to galena, with Millclose mine by far the largest producer at 100 000 tonnes of sphalerite concentrate produced (Ford et al. 1993). *Castleton–Bradwell area:* Newall Nook mine, Bradwell and Odin mine, Castleton (Mostaghel 1985). *Hucklow–Eyam–Stoney Middleton area:* Edge Rake and Moury mine, near Tideswell, and High Rake, Hucklow (Mostagel 1985); Glebe mine, Eyam (Ford et al. 1993); Ladywash mine, Eyam – good crystals usually scattered on fluorite (Braithwaite 1983, Mostaghel 1985, Ford et al. 1993). *Ashford–Bakewell–Sheldon area:* Blythe mine, Alport and Field Grove mine, Sheldon (Mostaghel 1985); Long Rake, near Middleton-by-Youlgreave (Ball et al. 1985b); Magpie mine, Sheldon – open cavities lined with calcite, fluorite and sphalerite (Ford & Sarjeant 1964a, Worley 1976, Robinson & Ineson 1979, Mostaghel 1985, Ford et al. 1993). *Bonsall–Matlock Bath–Cromford–Wirksworth area:* Ball Eye mine and quarry, Bonsall; Dovegang mines, near Cromford; Golconda mine, Brassington; Knotlow mine, Monyash; Lathkilldale mine, S of Over Haddon; Shaw's quarry, near Wirksworth; Shuttle Rake, near Bradwell (all Mostaghel 1985); Cawdor quarry, Matlock (Robinson & Ineson 1979); Masson Hill and Rutland Caverns, Heights of Abraham, near Matlock Bath – iron-poor "honey-colour" sphalerite extensively altered to smithsonite (Braithwaite & Ryback 1963a, Ixer 1974b); Millclose mine, Wensley, near Matlock – 538 tonnes of sphalerite concentrate raised in 1902, 30 700 tonnes raised between 1904 and 1938 from calcite–sulphide veins hosted within altered lavas. Large "honey-colour" to brown crystals and fine-grain black sphalerite are noted (Rudler 1905, Russell 1912b, Traill 1939, Varvill 1959, Ford & Ineson 1971, Robinson & Ineson 1979, Mostaghel 1985, Ixer

1986, Young et al. 1987, Willies et al. 1989); Riber mine, Matlock – calcite vein containing plentiful sphalerite and minor galena (Varvill 1959, Robinson & Ineson 1979). *Ashover inlier:* Fall Hill (Fall Gate) quarry – sphalerite crystals up to 6 mm across associated with fluorite (Smith 1982, Mostaghel 1985); Milltown quarry, near Ashover – dark-brown to translucent yellowish-brown crystals of sphalerite to 2.5 mm across scattered on fluorite (Rust 1991); Tansley borehole, Ashover – chalcopyrite–millerite–pyrite–pyrrhotite–sphalerite-bearing vein (Ramsbottom et al. 1962, Ixer & Townley 1979). *Ashbourne–Snelston area:* Fine-grain sphalerite and galena occur in clots and fracture infillings within borehole core from Limestone Hill, between Ashbourne and Stanton (Cornwell et al. 2001). Jackman (1996) notes sphalerite from Snelston mine, near Ashbourne.

Elsewhere, sphalerite has been found in Coal Measures Ironstone nodules near Chesterfield (Ford & Sarjeant 1964a) and is noted with barite and galena from Manifold Valley (Mostaghel 1985).

Devon Sphalerite is generally associated with galena in veins in and around the margins of the Dartmoor granite. Locations include: Bere Alston mines – minor sphalerite in fluorite–galena lodes (Hall 1868b, Ball et al. 1985b); Bridford barite mine and the Teign Valley, near Hennock – sphalerite associated with galena, pyrite and tetrahedrite, plus bournonite at depth (Hall 1868b, Polkinghorne 1951, Beer & Ball 1977); Lockridge (Goldstreet) mine, SW of Bere Alston – intergrown wurtzite–sphalerite associated with galena, tennantite and tetrahedrite (Grguric & Nickel 2006); Loddiswell mine, 3 km N of Loddiswell – sphalerite and other sulphides occur with native gold and jamesonite in an antimony–gold-rich carbonate vein cutting Devonian slates (Stanley et al. 1990); Meldon (British Railways quarry complex), near Okehampton, and Red-a-ven mine, Meldon – coarsely crystalline black sphalerite in metasomatically altered quartzite and calc-silicate hornfels (Kingsbury 1961, Dearman 1966a, El Sharkawi & Dearman 1966, Beer et al. 1989); North Devon United mine, Peter Tavy – primarily a copper mine where sphalerite occurs in a complex polymetallic assemblage that includes the rare species parkerite (Rumsey & Savage 2005); Wheal Betsy, near Tavistock (Hall 1868b); Wheal Friendship (Devon Friendship), Mary Tavy – sphalerite and a wide range of other oxide and sulphide species (Hall 1868b, Dines 1956, Braithwaite 1982b).

Sphalerite is also reported from: Combe Martin mines – small quantities of sphalerite in silver-rich galena-bearing quartz–siderite veins cutting Devonian slates (Hall 1868b, Hall 1877, Dunham et al. 1978, Scrivener & Bennett 1980, Rust 1982a); Landkey, near Barnstaple (Hall 1868b); Littleham Cove, near Budleigh Salterton – authigenic sphalerite occurs with copper, iron, lead and molybdenum sulphides and cobalt, copper and nickel arsenides within uranium–vanadium-rich nodules in Permian mudstones (Milodowski et al. 2000); St Mary's Bay, Brixham – sporadic very pale-brown translucent-to-transparent sphalerite tetrahedra to 3 mm associated with chalcopyrite, galena, millerite and pyrite in pods within Devonian shale (Braithwaite & Wilson 2001).

Durham *Teesdale area:* A review of the cadmium content of sphalerite concluded there was little correlation between supergene encrustations of yellow earthy crusts found on oxidized surfaces and the cadmium content of sphalerite, although the most cadmium-rich sphalerite in the area, from Reddycombe mine, near Harwood, Upper Teesdale, was the only one to be encrusted with supergene greenockite. Other samples were from: Dirt Pit mine, Teesdale and Lady's Rake mine, 6 km NW of Langdon Beck – dark-brown sphalerite with no cadmium encrustation; Willyhole mine, near Harwood, Upper Teesdale, and Trough mine, Teesdale – mid- to dark-brown cadmium-rich sphalerite with cadmium encrustations (all Young et al. 1987); Greenhurth mine, Upper Teesdale – abundant sphalerite (Young 1985a). Other locations are: Closehouse mine, SW of Middleton-in-Teesdale, Lunedale – large subhedral grains of iron-rich sphalerite with inclusions of arsenopyrite, chalcopyrite, galena, pyrrhotite and tetrahedrite-group minerals (Vaughan & Ixer 1980); Ettersgill and Wynch Bridge, Teesdale – workable but small deposits of sphalerite (Dunham 1990). *Weardale:* Boltsburn mine, Rookhope – sphalerite is relatively uncommon, but fine specimens were found sporadically, especially in High Flat (King 1982b, Solomon et al. 1971); Cambokeels mine – reddish-brown crystalline sphalerite (Young et al. 1987); Frazer's Hush mine (Wolfe 1998, Green & Briscoe 2002); Killhope mine, near Nenthead (Solomon et al. 1971, Dunham 1992); Old Moss vein, Killhope Burn (Young 1995); Rookhope borehole– minor amounts of sphalerite in the buried Weardale granite and associated pegmatites, also the colloform banded variety schalenblende in a replacement flat in the Jew Limestone (Dunham 1967, 1990, Solomon et al. 1971, Ixer 1986, Brown et al. 1987a); Sedling mine, Wearhead (Solomon et al. 1971); West Pasture mine, Stanhope – dark-brown sphalerite (Young & Francis 1989). *Other areas:* Chilton quarry, Ferryhill – specimen of dolomite enclosing pale-yellow sphalerite and Hesleden Dene boreholes, between Ferryhill and West Hartlepool – rare specks of brown resinous-looking sphalerite (Fowler 1956, Hirst & Smith 1974); New Brancepeth colliery, near Durham – specks of chalcopyrite and sphalerite in massive barite (Spencer 1910, Smythe 1922, Ixer 1986); Whiteheaps mine, near Blanchland – iron-rich sphalerite in a small chalcedony–marcasite–pyrite vein associated with extensive fluorite–galena mineralization (Vaughan & Ixer 1980).

Gloucestershire Sphalerite crystals, up to 3 mm across, occur with galena and pyrite in veins cutting Carboniferous Limestone near Birches, 4 km S of Newent (Bick 1986a).

Lancashire Sphalerite occurs with chalcopyrite, galena and witherite in a vein cutting Millstone Grit on Anglezark Moor, near Chorley (Rudler 1905).

Leicestershire Reviewing sphalerite occurrences in Leicestershire, King (1993) describes remarkable complex and multiple twinned sphalerite from Earl Ferrers mine, Staunton Harold, Ticknall, Ashby-de-la-Zouch. Multiple crystals take the form of rosettes up to 4.1 mm across and show colour zoning from greyish-orange cores through golden to black rims. In the Leicestershire coalfield, sphalerite is noted from: Albert Village coal pit; Netherseal colliery; Ravenstone opencast site and Old Parks coal pit, NE of Ashby-de-la-Zouch. Mesozoic rocks, particularly the Lias, have provided examples of sphalerite from septarian siderite–mudstone nodules at Catthorpe borrow pit (plates up to 8 mm across), Hallaton temporary British Gas trench (black lustrous plates up to 4 mm), Lowesby brickyard (black adamantine plates in attractive groups), Lowesby and Slawston cuttings (dark-brown plates up to 9 mm with strikingly high lustre), Rockingham Road bridge excavations at Market Harborough (black, lustrous plates associated with calcite), and Old Dalby railway cutting (highly lustrous dark-brown plates). Pale greenish-yellow aggregates of crystalline sphalerite and colourless to pale-brown sphalerite in galena-bearing veins have also been found at Newhurst quarry, Shepshed (Ince 2005).

Northumberland Sphalerite related to the Alston Block mineralization occurs at a few locations N of the Stublick fault system near Haydon Bridge and Hexham: Barrasford quarry, 9 km N of Hexham – barite–galena–sphalerite veins cutting the Whin Sill dolerite (Money et al. 1995); Fallowfield mine, near Hexham – barite–galena–witherite veins with subordinate sphalerite (Briscoe 1986b); Langley Barony, 2 km NW of Haydon Bridge – pale-brown sphalerite (Dunham 1990); Settlingstones mine, near Hexham – pale-brown sphalerite in places intimately associated with galena, nickeline, ullmannite and witherite, also a thin vein or stringer of sphalerite continuous throughout the mine (Russell 1927b, Trestrail 1931, Young et al. 1985a); Stonecroft–Greyside mines, near Settlingstones – brown-banded, pale-brown and dark-brown varieties of sphalerite have all been found, the latter in association with harmotome and witherite (Guppy 1956, Young & Bridges 1984, Young et al. 1987, Dunham 1990). Sphalerite is also reported from Beaumont mine, Allenheads – characteristic lustrous crystals (Green & Briscoe 2002) and from St Peter's and Swinhope mines, near Allenheads, the latter well mineralized with sphalerite (Solomon et al. 1971, Dunham 1990).

Nottinghamshire Cavities in Permian mudstones and limestones at Vale Road quarry, Mansfield Woodhouse, contain complex black sphalerite crystals with transparent overgrowths, often associated with dolomite and pyrite (Green & Spence 1988).

Oxfordshire Honey-brown to black platy crystals of sphalerite up to 2 mm across have been found in the cavities of ammonites in temporary exposures of Oxford Clay at Yarnton (Atkins 1995).

Shropshire The West Shropshire mining district contains Ba–Pb–Zn veins occupying fractures in Ordovician and Precambrian rocks of the Shelve inlier. The sequence of crystallization began with sphalerite and was followed by barite–galena, then chalcopyrite. Three generations of sphalerite are recognized, including distinct cadmium–iron-rich and indium-rich varieties (Pattrick & Bowell 1991). At Rorrington (West Snailbeach) mine, S of Meadowtown, sphalerite forms lustrous transparent yellow-orange crystals up to 2 mm across in cavities within quartz (Starkey et al. 1994).

Somerset Sphalerite was of secondary importance to galena in the Mendips orefield. Near the hamlet of Green Ore, SSW of Chewton Mendip, it occurs on the mine dumps with barite, galena and smithsonite, and in places bright canary-yellow patches indicate that it is cadmium rich. At the western end of Sandford Hill it forms small stringer-veins (Kingsbury 1941). Authigenic sphalerite engulfs shattered shell debris and forms euhedral crystals in cavities in ammonites within Jurassic marine sediments between Blue Anchor Point and Stolford Rocks on the Somerset coast (Alabaster 1980).

Staffordshire Sphalerite occurred in considerable quantities within galena-bearing veins on Ecton Hill (Robey & Porter 1972, Mostaghel 1985).

Warwickshire Grey-brown to dark-brown hemispherical aggregates, up to 4 mm across, of small transparent euhedral sphalerite crystals with a relatively unusual end-member composition occur in a Pb–V–Zn assemblage at Judkins quarry,

Nuneaton. The mineralization is close to the Precambrian/Cambrian unconformity (Ince et al. 1991).

Yorkshire Sphalerite is common in three sections of the Northern Pennine orefield, usually in galena-bearing veins. *North Swaledale mineral belt:* Minor amounts of sphalerite were exploited from galena-bearing veins and replacement flats in Carboniferous Limestone. Examples have been described from: Friarfold–Surrender and Old Rake veins (Young et al. 1989); Flincher Gill and Hard Level Gill, NE of Gunnerside (Ineson & Young 1994); Gunnerside Gill mines (Wood 1993); High Faggergill, Arkengarthdale Moor (Vaughan & Ixer 1980); Old Gang mines, E of Gunnerside Gill (Green 2003); Whitaside mine, SE of Gunnerside – a few tonnes of sphalerite mined (Dunham & Wilson 1985). *Wensleydale:* East Bolton Moor, near Askrigg – barite, galena and sphalerite on the dumps. Sphalerite from Worton Hall, near Askrigg, has very low iron content (Vaughan & Ixer 1980). Minor amounts of sphalerite are present in the buried Wensleydale granite sampled in a borehole at Raydale (Brown et al. 1987a). *Wharfedale and Nidderdale:* Carboniferous limestones are cut by several veins carrying barite, fluorite, galena and sphalerite at Coldstones quarry, 5 km W of Pateley Bridge (Young et al. 1997). Sphalerite from Greenhow, near Pateley Bridge, has very low iron content (Vaughan & Ixer 1980).

Other descriptions of sphalerite are from: North Cheek, Robin Hood's Bay and Rosedale Wyke – authigenic sphalerite in the chambers of Jurassic ammonites and within siderite mudstone concretions (Hemingway & Riddler 1982); Scaleber Bridge, 3 km SSE of Settle – discrete grains of galena and sphalerite, up to 100 μm across, in As–Cu–Sb mineralization within Carboniferous Limestone (Ixer & Stanley 1998).

Wales

Clwyd Sphalerite occurs widely in galena-bearing veins cutting Carboniferous Limestone, with excellent reddish-brown crystals found at Minera, near Wrexham, and small (2–3 mm) red ruby-blende crystals reported from the Halkyn area (Bevins 1994). The Minera lode was by far the most productive in the Clwyd area, and substantial reserves of sphalerite are likely to occur below the limits of existing workings (Earp 1958, Dunham et al. 1978). Other examples are from: Bwlch-y-Plwm mine, Penrhyndeudraeth (Archer 1959); Pennant mine, near St Asaph (Carruthers et al. 1915, Archer 1959); Talargoch mine, near Dyserth – reputed to have produced 3000 tonnes of sphalerite annually between 1845 and 1884 (Dunham et al. 1978); Wedding Cave mine, Bwlchgwyn, near Wrexham – rare sphalerite (Green 1995b); Trelogan mine – masses of reddish-brown sphalerite associated with fluorite (Bevins 1994).

Dyfed *Central Wales orefield:* Sphalerite is a major component of fracture-hosted mineral veins, particularly those in the east, with galena being more common towards the west (Bevins 1994). These veins have been divided into two groups of assemblages (A1 – early complex; A2 – late simple) and they include an A1-*b* early-sphalerite assemblage in which quartz-cemented breccias contain abundant coarse-grain dark-brown sphalerite associated with chlorite, traces of chalcopyrite, iron-rich dolomite and pyrite. The sphalerite is distinctive in being affected by chalcopyrite disease. A1 assemblages are found at: Darren mine, Pen-bont Rhydybeddau; Erglodd mine, N of Tal-y-bont; Esgair Hir mine, Ceulanymaesmawr; Esgair Fraith mine, near Tal-y-bont; and Eaglebrook (Nantycagl) mine, Ceulanymaesmawr. Of the A2 assemblages, the A2-*a* galena–sphalerite–(quartz) assemblage was the most important economic source of lead and zinc and has been found at: Brynyrafr mine; Cefngwyn mine, near Aberystwyth; Dolclettwr mine, Tre'r-dol; Frongoch mine, between Devil's Bridge and Pont-rhyd-y-groes; Nantymwyn mine; and Ystumtuen mine (Mason 1997). Other records of sphalerite include: Bronfloyd mine, ESE of Penrhyn-coch; Camdwrbach mine, near Bryniau Rhyddion, NW of Nant-y-moch Reservoir; and Cwmystwyth mine (all Raybould 1974); Bwlchrhennaid mine, near Goginan; Caegynon mine, near Ponterwyd; Castell mine, Ponterwyd, near Aberystwyth; Cwmbrwyno mine, near Devil's Bridge – highly complex form of flattened crystals; Penrhiw mine, Ystumtuen; Rhoswydol mine, near Machynlleth; Wemyss (Frongoch West) mine (all Jones & Moreton 1977); Cystanog mine, near Carmarthen (Carruthers et al. 1915); Dolwen mine, near Devil's Bridge – black sphalerite crystals (Bevins 1994); Eaglebrook (Nantycagl) mine, Ceulanymaesmawr – dark-brown to black crystalline masses and rough crystals, also light-brown masses associated with galena and quartz (Jones 1983); Erglodd mine, N of Tal-y-bont (Rust 1991a); Esgair Hir mine (Rust & Mason 1988); Frongoch mine – unusual pale-purple sphalerite and radiating aggregates of crystals forming spherules up to several cm across (Jones & Moreton 1977, Bevins 1994); Llechweddhelyg mine, near Clawddmelyn, between Bont-goch and Cwmsymlog – dark-brown corroded masses (Jones 1987); South Darren mine, Pen-bont Rhydybeddau (Dewey & Eastwood 1925, Hughes 1959); Ystrad Einion mine, near Tal-y-bont (Mason & Rust 1997). Elsewhere, sphalerite occurs in gold-bearing quartz–sulphide veins at the Dolaucothi mines, near Pumsaint (Brammall et al. 1937, Steed et al. 1976, Annels & Roberts 1989).

Gwynedd Sphalerite and galena are intimately intergrown in a minor ore, known locally as bluestone, at Mona and Parys mines, Parys Mountain, Amlwch, Anglesey. Chalcopyrite and pyrite are both important associates in this volcano-sedimentary deposit, although they form a separate ore. Both inclusion-free and "chalcopyrite-diseased" sphalerite replace and cement the earlier copper-rich ores (Greenly 1919, Dewey & Eastwood 1925, Pointon & Ixer 1980, Ridgway 1980, Southwood & Bevins 1995). Around Llandudno, minor amounts of sphalerite are reported from Bwlch antimony mine, near Deganwy (Russell 1944, 1949b, Bevins et al. 1988, Ryback & Francis 2001), and the Great Orme copper mines, Llandudno (Ixer & Davies 1996, Ixer & Stanley 1996, Ixer 2001). At Trecastell lead–zinc mine, near Conwy – 12 500 tonnes of sphalerite were produced between 1892 and 1913, with further mining activity occurring from 1953 to 1955 (Archer 1959). *Snowdonia and the Lleyn Peninsula:* Aberllyn lead mine, 2 km N of Betws-y-coed, Gwydyr Forest – ankerite–calcite–quartz–sphalerite breccia with later growth of hydrozincite and namuwite (Bevins et al. 1982b, Otty 1986); Benallt mine, Rhiw, Lleyn Peninsula – two quartz-rich lodes containing galena and sphalerite (Archer 1959); Lliwedd mine, SE of Snowdon and other mines in the Snowdon caldera – both colourless and dark-brown (iron-rich) sphalerite occur in subsidiary amounts to chalcopyrite and pyrite in thin quartz veins and stockworks within Snowdon Volcanic Group rocks (Reedman et al. 1985); Drws-y-Coed mine, near Nantlle – a copper mine that also produced some sphalerite (Dunham et al. 1978, Braithwaite 1982b); Gamallt mine, near Ffestiniog – mainly chalcopyrite–sphalerite in a fault-bounded lode; Graig Wen, near Ffestiniog – major quartz vein hosting chalcopyrite, galena, pyrite and sphalerite; and Llyn Bowydd, near Ffestiniog – large quantities of fine-grain sphalerite with patches of chalcopyrite and galena (all Lynas 1973); New Pandora mine, between Betws-y-coed and Llanrwst – fine-scale zoned sphalerite and crustiform banding, where galena precipitated prior to sphalerite in lead–zinc veins cutting Ordovician volcano-sedimentary rocks (Haggerty 1995); Parc mine, near Llanrwst – Principal lode consists of galena, brown sphalerite and marcasite in calcite–quartz gangue. The dumps were once a fine collecting area for sphalerite and hydrozincite (Archer 1959, Archer & Elliott 1965, Otty 1986); Tal-y-Sarn mine, Llanllyfni – copper mine in which galena and sphalerite increased with depth (Dewey & Eastwood 1925, Archer 1959, Dunham et al. 1978). *Dolgellau goldbelt:* Complex, shear-zone-hosted mesothermal gold-lode mineralization has a four-stage paragenetic sequence, the last of which is a lead–zinc assemblage. Galena and sphalerite are rare at Clogau mine, Bontddu, with sphalerite occurring as flower-shape inclusions in chalcopyrite. The assemblage is more common to the east, particularly between Foel Ispri and Gwynfynydd mines, where sphalerite often aggressively replaces chalcopyrite, galena, gold and pyrrhotite, leaving behind relict arsenopyrite, cobaltite and pyrite inclusions (Mason et al. 2002). Further descriptions include: Afon Mawddach – sphalerite inclusions in alluvial gold grains (Leake et al. 1993); Clogau and Vigra mines – sphalerite and galena in chalcopyrite-rich gold-bearing quartz lodes (Dewey & Eastwood 1925); Gwynfynydd mine – abundant galena and sphalerite in quartz veins (Readwin 1888, Gilbey 1968, Allen & Jackson 1985, Dominy et al. 1996, Platten & Dominy 1999); Friog Undercliff, Fairbourne, SW of Dolgellau – fine-grain sulphide veins dominated by pyrrhotite, with local concentrations of galena and sphalerite (Mason et al. 1999); Glasdir mine, Llanfachreth – blebs of exsolved sphalerite in chalcopyrite from an intrusion breccia related to the Ordovician Rhobell volcanic complex (Allen & Easterbrook 1978, Allen & Jackson 1985); West Cwmheisian mine – alternating bands of calcite, marcasite and sphalerite in crustiform banded Alpine-type fissure veins (Mason et al. 2002).

Mid Glamorgan Massive resinous brown sphalerite occurs in small patches in lead–zinc veins within Carboniferous dolomitized limestone at Machen quarry, 7 km ENE of Caerphilly. Rare fraiponite probably formed as a supergene alteration product of this sphalerite (Goulding & Price 1995, Plant & Jones 1995). Sphalerite occurs with chalcopyrite and galena in calcite-lined cavities within Carboniferous Limestone at Bute quarry, Llantrisant, and with chalcopyrite in calcite veins cutting similar rocks in the Locks Common borehole, Porthcawl (Fletcher & Young 1988).

Powys *Central Wales orefield:* Aberdaunant mine, near Llanidloes; Gorn mine; Pen-y-Clun mine, NW of Llanidloes (all Carruthers et al. 1915); Bryntail mine (Carruthers et al. 1915, Mason 1997); Cwmbyr mine – botryoidal sphalerite (Jones & Moreton 1977); Dyfngwm mine, Machynlleth (actually the Pen Dylife section of Dylife mine) – large masses to 15 cm across and small vugs commonly lined with dark brown crystals up to 4 mm across (Rust & Rust 1987); Dylife mine – botryoidal sphalerite with a radiating fibrous structure in hemispherical aggregates up to 10 cm across, and good crystals in vuggy calcite breccia (Raybould 1974, Jones & Moreton 1977, Moreton 1982, Bevins 1994); Nantiago mine – good crystals in vuggy calcite breccia (Jones & Moreton 1977, Mason 1997); Van mine, Llanidloes – small red crystals (Rudler 1905, Bevins 1994). *Elsewhere:* Galena–sphalerite-bearing fissure

Orange-brown sphalerite crystals to 0.2mm, scattered over needles of millerite; Coed Ely colliery, Llantrisant, Mid Glamorgan, Wales. (© M. P. Cooper)

veins (with barite–witherite gangue) occur in the Ordovician sediments and volcanics of the Llangynog area on the southern flank of the Berwyn Dome. The principal mines were Craig-y-Mwyn, Cwm Orog and Llangynog (Archer 1959, Dunham et al. 1978).

Scotland

Borders Sphalerite occurs as inclusions in alluvial gold from Craigierigg Burn and Glengaber Burn, SW of Peebles (Leake et al. 1993, Chapman et al. 2000a,c).

Central Sphalerite and galena are the two main ore species once worked in the lead–zinc mining district of Tyndrum. The sphalerite has low iron content and contains copper-rich zones forming a pervasive symplectite with digenite. Sphalerite formed early, mainly in vein breccias related to the Tyndrum–Glen Fyne fault (Greg & Lettsom 1858, Heddle 1901, Darnley et al. 1960, Ineson & Mitchell 1974b, Pattrick 1978, 1985, Ridgway 1980, Dickinson & Pattrick 1987). In the same area, sphalerite is a minor constituent of gold-bearing pyrite–quartz veins at Cononish, 4km SW of Tyndrum (Parker et al. 1989, Earls et al. 1992, Curtis et al. 1993). A boulder found in the Crom Allt River, near Tyndrum, consists of vuggy mineralized breccia in which sphalerite occurs as twinned crystals containing patches of exsolved chalcopyrite (Pattrick & Hall 1984). Low-grade stratabound sedimentary-exhalative sulphide (mainly pyrite) mineralization in Dalradian quartzite and chloritic schists near Tyndrum, at Ben Challum and Creag Bhocan, also contains minor sphalerite (Smith et al. 1984, Fortey & Smith 1986, Scott et al. 1988).

In the Alva area, N of the Ochil fault, sphalerite occurs with chalcopyrite and galena in silver-bearing veins cutting Old Red Sandstone volcanics (Dunham et al. 1978) and, more generally in the central Ochil Hills, as inclusions in alluvial gold (Chapman et al. 2005).

Dumfries & Galloway Only the Wanlockhead part of the Leadhills–Wanlockhead lead–zinc orefield produced sphalerite, and prior to 1880 it was allowed to run to waste. From then on, the area gradually increased production to 1258 tonnes in 1915 (Wilson & Flett 1921). Early quartz veins containing albite, gold, muscovite and pyrite were followed by major lead–zinc mineralization in which ankerite, barite, calcite and quartz are the gangue species (Temple 1955). Other locations include: Glencrieff Burn – sphalerite up to 32 µm across, forming inclusions in pyrrhotite (Nicholson & Banks 1987); Margaret's or Straitstep vein – abundant sphalerite on the mine dumps (Brown 1919, Mackay 1959); Wanlockhead mine, working the New Glencrieff vein – galena gave way to sphalerite at depth, also good samples of sphalerite noted on the dumps (Heddle 1901, Wilson & Flett 1921, Williams 1965). This and other lead–zinc mineralization in the area developed in a Lower Palaeozoic greywacke/shale sequence. In many cases mineralization is associated with Silurian to Devonian granitic intrusions and major strike-slip faulting.

Other locations include: Cairngarroch Bay – minor sphalerite in greywacke-hosted sulphide (mainly pyrite) veins (Lowry et al. 1997); Talnotry, Newton Stewart – globular, cadmium-rich sphalerite and late sphalerite with a chalcopyrite exsolution texture (chalcopyrite disease) in a complex sulphide vein at the margin of the Cairnsmore of Fleet granite (Stanley et al. 1987, Lowry et al. 1997); Glendinning mine, Glenshanna Burn – black jack (marmatite) and honey-blende varieties of sphalerite occur in a highly complex barite–galena–jamesonite–stibnite ore overprinting earlier arsenopyrite and pyrite (Heddle 1901, Dewey et al. 1920, Russell 1949b, Williams 1965, 1973, Gallagher et al. 1983, Ball et al. 1985b, Duller et al. 1997); Glenhead Burn, Loch Dee – a few grains of ruby-red sphalerite in carbonate veins cutting metasomatically altered sediments near the Loch Doon granite complex (Leake et al. 1981); Knipe (Fountainhead) mine, Hare Hill – sphalerite in an Ag–As–Cu–Sb–Pb–Zn assemblage hosted by a late-Caledonian diorite–granodiorite intrusion (Boast et al. 1990); Lauchentyre copper mines, W of Gatehouse of Fleet (Heddle 1901); Moorbrock Hill, Carsphairn – minor sphalerite in chalcopyrite-dominated gold–silver veins close to the margin of a granodiorite intrusion (Naden & Caulfield 1989); Needle's Eye, SW of Caulkerbush – minor sphalerite in uranium-rich fissure veins in the aureole of the Criffell granite complex (Miller & Taylor 1966, Basham et al. 1989); Penkiln Burn, N of Newton Stewart – disseminated marcasite, pyrite, pyrrhotite and sphalerite mineralization in a black mudstone sequence near the soutwestern margin of the Loch Doon granite complex (Stone & Leake 1984); Pibble mine, Creetown – black graphitic shale on the dumps containing small crystals of galena and sphalerite (Heddle 1901, Wilson & Flett 1921); West Blackcraig mine, near Newton Stewart – examples of chalcopyrite, galena and sphalerite on the mine dumps (Livingstone et al. 1976); Wood of Cree mine, near Newton Stewart – complex chalcopyrite–galena–pyrite–sphalerite ore with 117 tonnes of sphalerite produced between 1915 and 1919 (Wilson & Flett 1921); Woodhead mine, near Garryhorn House, Carsphairn – crystalline and fibrous sphalerite on the mine dumps (Heddle 1901, Wilson & Flett 1921, Williams 1965). At Callisterhall, near Langholm, sphalerite occurs with chalcopyrite and galena in veins and disseminations in Carboniferous sediments below the Birrenswark Lavas (Smith et al. 1978).

Fife Carboniferous non-marine algal limestones at Burntisland contain sphalerite and galena concentrated along certain laminae within stromatolites. Chalcopyrite is also present (Parnell 1983). Small crystals of brown sphalerite up to 1.2mm across have been found associated with pectolite in mineralized joints and fractures within basalt at Orrock quarry, near Burntisland (Wirth 2003).

Grampian Sphalerite occurs with galena in a calcite–fluorite vein cutting Dalradian schist at Abergairn (Corrybeg) mine, near Ballater. The sphalerite is yellowish-brown, tawny to honey-yellow in colour and forms cleavage masses and rough crystals to 1.5cm (Heddle 1901, Wilson & Flett 1921, Russell 1936a, Hall & Walsh 1972, Nicholson 1989b). Only a few km to the W, at Gairnshiel Bridge, Glen Gairn, sphalerite (with chalcopyrite disease) occurs in sulphide-rich pods within greisenized zinnwaldite granite and in a zinnwaldite vein (Webb et al. 1992). Other reports of sphalerite include: Back Burn, W of Auchenblae – sphalerite inclusions in alluvial gold (Chapman et al. 2000c); Corrie Loch Kander, near Braemar – stratabound barite, galena, pyrite, pyrrhotite and sphalerite mineralization in Dalradian calc-silicate-rich cherty quartzite. The sulphides form anhedral patches and grains intermingled with armenite, barite, hyalophane, salitic pyroxene and actinolite–tremolite (Gallagher et al. 1989a, Fortey et al. 1991, 1993); Stotfield, W of Lossiemouth, near Elgin – pale-yellow sphalerite intergrown with galena (Starkey 1988a).

Highland The Allt Cathair Bhàn section of the Loch Ailsh alkaline complex and the ultramafic section of the Loch Borralan alkaline complex, Assynt, contain post-magmatic sulphides including chalcopyrite, galena, pyrite and sphalerite (Shaw et al. 1994, Styles et al. 2004). Minor barite, galena and sphalerite mineralization occurs in Moinian schists along the eastern margin of the Grudie granite, near Lairg (Gallagher et al. 1974, Gallagher & Smith 1975, Fortey 1980, Lowry et al. 1994). In Caithness, sphalerite occurs in uranium-rich fracture zones at: Blar Cnoc na Gaoith; Bridge of Westfield; Broubster; Ceann Ousdale, Helmsdale (Gallagher et al. 1971); and at Suisgill Burn, N of Kildonan, it forms inclusions in alluvial gold (Leake et al. 1993, Chapman et al. 2000c). The Kerry Road stratiform sulphide deposit near Gairloch is hosted by Lower Proterozoic quartz–carbonate schists and contains minor

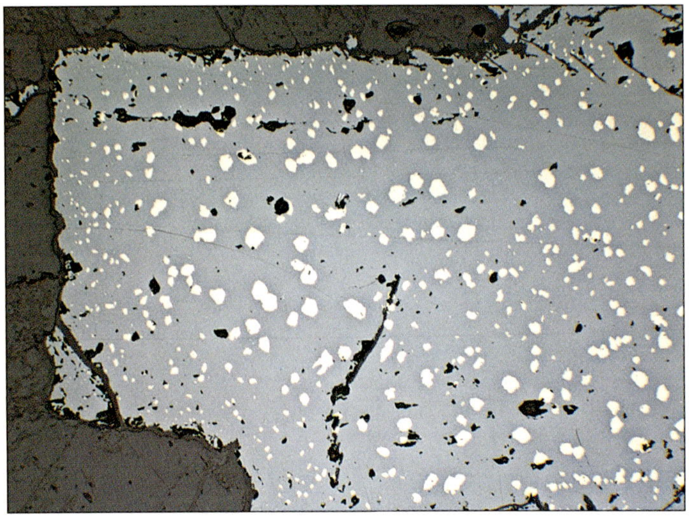

Sphalerite containing exsolved blebs of chalcopyrite, from a sulphide-rich pod in zinnwaldite granite, Gairnshiel Bridge, Glen Gairn, Grampian, Scotland; width 1.0 mm; viewed in PPL (upper) and reflected light (lower). (© A. G. Tindle)

chalcopyrite, galena, marcasite and sphalerite in a pyrite–pyrrhotite-dominated ore. Traces of sphalerite are also found in another mineralized area within chloritic hornblende schist at nearby Teangadh Bhuide Mhòr (Jones et al. 1983, 1987).

Other reports of sphalerite are from: Ballachulish – sphalerite is a rare constituent of metamorphosed siliceous dolomite at the margin of the Ballachulish granite (Ferry 1996); Corrantee mine, Allt Tarsuinn, NW of Strontian – considerable sphalerite in barite–galena veins (Heddle 1901, Winrow 1986b, Wolfe 2006); Loch Claise na Feàrna, Scourie Estate – galena and sphalerite with minor chalcopyrite in hornblendite and hornblende schist (Anon. 1974); Ratagain complex, Loch Duich, near Glenelg – chalcopyrite, galena, molybdenite, pyrite and sphalerite in quartz veins within faulted quartz monzonite (Alderton 1986, 1988b); lead mines, Loch na Mèine, midway between Strath Glass and Strath Farrar, and Allt Tigh Cumhaig Burn, S of Struy – minor black sphalerite in barite–calcite–galena veins (Russell 1946b, Berridge 1969); Beinn an Dubhaich, Broadford area, Isle of Skye – sphalerite in boron–fluorine-rich ore skarns at the margin of the granite (Tilley 1948b, 1951).

Lothian Low-grade stratabound chalcopyrite–galena–sphalerite mineralization has been found in Carboniferous non-marine algal limestone at South Queensferry (Parnell 1983). Sphalerite is also present as reddish-brown crystals up to 4 mm long in earthy annabergite spherules from Hilderston mine, Bathgate Hills (Meikle 1994). At Stobshiel, S of Haddington, galena and sphalerite occur with minor chalcopyrite and tetrahedrite in a greywacke–shale-hosted vein gold prospect (Naden & Caulfield 1989). Heddle (1901) notes reddish-yellow sphalerite associated with galena in sandstone from Granton quarry and honey-yellow transparent crystals in quartz veins from Ratho quarry, both near Edinburgh.

Strathclyde Abhainn Srathain mines, S of Meall Mòr, South Knapdale – borehole material from a stratiform chalcopyrite-rich deposit in Middle Dalradian rocks contains microscopic disseminated sphalerite (Smith et al. 1978a); Arduaine, near the southern shore of Loch Melfort, Kilmelford – sphalerite associated with bornite, galena and pyrite in veinlets within an explosion breccia of fractured Dalradian rocks (Zhou 1987); Bowling quarry, Kilpatrick Hills, near Glasgow – honey-yellow, and colourless transparent, sphalerite associated with prehnite, and Leadhills – uncommon sphalerite forming black crystals up to 5 cm across on calcite (Heddle 1901); Clachan Beag, NE of Inveraray – silver-rich galena, pyrite and sphalerite in a Dalradian stratabound deposit (Willan 1980); Easedale Island, near Oban – chalcopyrite, galena, pyrrhotite and sphalerite form micro-inclusions in pyrite within Middle Dalradian slates (Hall et al. 1988); Garbh Achadh, near Inveraray – sphalerite associated with molybdenite, chalcopyrite and (major) pyrite in rafts of sulphide-rich hornfelsed quartzite adjacent to a stock-like porphyrite intrusion (Fortey 1980); Isle of Islay: Ballygrant, Kilslevan, Mulreesh and North Ardachie mines – chalcopyrite, chalcocite, galena and pyrite and sphalerite in veins cutting Dalradian rocks (Baron & Parnell 2005); Lagalochan, near Kilmelford – sphalerite occurs in a phase of Ag–Au–As–Pb–Sb–Zn mineralization and in later carbonate-hosted lead–zinc veinlets within a Caledonian diatreme–porphyry–breccia complex (Harris et al. 1988, Zhou 1987); Mannoch Hill district, SW of Lesmahagow – barite veins contain black masses and lustrous crystals of sphalerite up to 2 mm across with native silver on Law Hill and honey to light-brown masses on Nutberry Hill (Todd & McMullen 1991); McPhun's Cairn, Loch Fyne – stratiform sulphide mineralization in Cambrian phyllites is dominated by pyrite, but also contains appreciable sphalerite and galena (Willan 1980, Willan & Hall 1980); Reilly quarry, Houston – multi-twinned spherical growths (up to 2 mm across) of semi-transparent pale-grey lustrous sphalerite occur with barite and calcite in Carboniferous Clyde Plateau Lavas close to a dolerite dyke (Meikle 2003); Shortcleugh Water, Leadhills – sphalerite inclusions in alluvial gold (Leake et al. 1998, Chapman et al. 2000c); Stronchullin, N of Tarbert – chalcopyrite, galena and sphalerite in gold-bearing veins cutting Dalradian rocks (Berridge 1969, Ridgway 1980); Creag an Fheidh, near Glenloig, Isle of Arran – chalcopyrite, laumontite, prehnite and sphalerite crystallized from late-stage fluids in endoskarns associated with the collapsed caldera of the Central ring complex (Cressey 1987).

Tayside Sphalerite occurs with chalcopyrite, galena, pyrite and pyrrhotite in stratiform barium–zinc mineralization within Middle Dalradian schists to the N of Aberfeldy (Fortey & Beddoe-Stephens 1982). Pyrite and sphalerite (including a manganese-rich variety) are the most common sulphides in related mineralization in the Ben Eagach area, near Aberfeldy (Coats et al. 1980, 1984, Willan & Coleman 1983, Smith et al. 1984). Sphalerite is common in mineralized rocks and host metasediments in the Foss deposit (along strike from Ben Eagach), where barite has been mined. It occurs as disseminated grains or in laminae parallel to lithological layering, and it forms equant and subhedral highly irregular embayed crystals up to 3 mm across. Zonation caused by variable iron and manganese contents has also been noted (Moles 1983, 1985). Minor sphalerite occurs in galena-bearing quartz at Corrie Buie, Meall nan Oighcreag, Loch Tay (Pattrick 1984). Gold-bearing quartz–sulphide veins at Urlar Burn, 4 km from Aberfeldy, contain patches of chalcopyrite–galena–pyrite and minor arsenopyrite–sphalerite (Ixer et al. 1997). A single grain of alluvial gold from Calliachar Burn, Aberfeldy, contains inclusions of sphalerite (Leake et al. 1993, 1997c, Chapman et al. 2000c).

Orkney Barite–galena–(sphalerite) replacement veins and breccia zones have been found in Basal Breccia and Sandstone members of the Old Red Sandstone near the Moine unconformity at Blackhall and Mill of Cairston, Stromness area, Mainland (Gallagher et al. 1971, Michie 1972). Sphalerite is the predominant ore mineral in Middle Devonian algal breccias (part of an evaporitic lake system) at Hesta Geo, near Birsay, and at both Ness and Warbeth, near Stromness, Mainland (Parnell & Janaway 1990). Heddle (1901) notes sphalerite in an old lead mine, W of the Point of Ness, near Stromness.

Shetland Low-iron sphalerite occurs with other sulphides in nodules within a carbonaceous–calcareous siltstone facies of the Devonian Upper Melby Fish Bed at Matta Taing, Walls, and Sandness, west Mainland (Hall & Donovan 1978). Sphalerite and galena occur at the margins of Dalradian amphibolites and massive sulphide rocks (pyrite and chalcopyrite) at Vidlin Ness, Mainland (Garson & May 1976).

Ireland

Armagh Galena, sphalerite and pyrite occur at Derrynoose lead mine (Cole 1922).

Cavan Common coarse-grain sphalerite occurs in monomineralic veins and veinlets in vuggy quartz at Cornaurney lead mine, and minor sphalerite occurs in a barite vein at Muff (Morris 1984).

Clare Sphalerite is noted from the following mines: Ballyhickey (Pb), near Tulla;

Ballyvergin (Cu–Pb–Zn), near Tulla; Castletown (Ag–Pb–Zn), near Mogouhy; Carrahan (Pb), Lassana Townland; Crowhill (Pb–Zn), Knockaphreagaun; Milltown, near Tulla (Pb–Zn) – sphalerite and galena in the Great Calcite lode (Cole 1922). At Ballyvergin, sphalerite, bornite and tennantite are minor species in an epigenetic copper–silver (arsenopyrite–chalcopyrite–galena–pyrite) ore within folded Carboniferous carbonate sediments (Andrew 1986).

Cork Rare sphalerite occurs in chalcopyrite-bearing quartz veins within Devonian redbeds at Dhurhode, West Carbery mining district (Ni Wen et al. 1999, Kinnaird et al. 2002). Calcretes associated with bornite–chalcocite mineralization at Mount Gabriel, 15 km to the E of Dhurode, also carry minor sphalerite (Ni Wen et al. 1999).

Donegal Glenaboghil mine, NNW of Fintown worked a fine-grain stratiform deposit of galena and sphalerite, the latter forming dark-brown striated crystals up to 1.5 mm across (Ryback & Moreton 1992). The ore at Glentogher (Carrowmore) mine, near Carndonagh, was galena and sphalerite (O'Brien 1959). South of the Donegal main granite, Dalradian impure limestones host finely intergrown galena–sphalerite and minor barite in veinlets and breccias (O'Brien 1959, McArdle et al. 1986a).

Down Breccias in Silurian greywackes and shales near a dolerite dyke at Conlig–Newtownards host galena–(chalcopyrite–pyrite–sphalerite) vein mineralization. Sphalerite forms dark-brown 2–10 mm crystals usually enclosed in barite (Greg & Lettsom 1858, Cole 1922, Dunham et al. 1978, Moles & Nawaz 1996). Minor sphalerite is noted with molybdenite and pyrite in the Newry granodiorite at Croreagh quarry (Meighan & Ixer 1980). Sphalerite also occurs at Castleward lead mine, on Dickson's Island, in the grounds of Castleward manor house, NE of Downpatrick (Cole 1922).

Dublin Sphalerite is present in a galena-bearing barite–quartz vein at Killiney (Green 1998c).

Galway Sphalerite (a low-Fe variety) is the main primary ore species in the major limestone-hosted Pb–Zn–(Ag–Cu) deposit that formed close to the North Tynagh fault at Tynagh mine, near Killimor. Concentrates containing over 140 000 tonnes of zinc were produced between 1966 and 1970 and at one time it was the largest lead–zinc mine in Europe. Colloform textures are well developed in parts of the deposit, with banded sphalerite noted to replace pyrite extensively (Schultz 1966b, 1971, Morrissey et al. 1971, Gibson 1979, Boast et al. 1981a, Halliday & Mitchell 1983, Clifford et al. 1986, Ryback & Moreton 1992, Banks et al. 2002). Cole (1922) notes sphalerite in barite–calcite gangue at Glengowla lead mine, Oughterard.

Kerry Sphalerite and galena mineralization occurs in a small stratabound deposit hosted by Lower Carboniferous limestones and dolomites on Crow Island, within Lough Leane, Killarney. Sphalerite (var. honey blende) is the main sulphide species and this is intergrown with coarse patches of galena and calcite, and lesser amounts of arsenopyrite, chalcopyrite and pyrite. There are at least two generations of sphalerite (Kinnaird et al. 2002, Ixer 2004). Sphalerite is also present in fine-grain stratiform Cu–Pb–Zn ore at the nearby Muckross mine (Moreton et al. 1999, Kinnaird et al. 2002). At Ross Island, in Lough Leane, sphalerite is a trace constituent of chalcopyrite–cobaltite–tennantite ore found in epigenetic veins (Russell 1927, Kinnaird et al. 2002).

Kildare The small breccia-hosted Fe–Pb–Zn deposit at Harberton Bridge, 40 km WSW of Dublin, is widely considered to be an example of MVT mineralization. It consists of galena–marcasite–sphalerite mineralization in a calcite gangue (Emo 1986, Gallagher et al. 1992, Trude & Wilkinson 2001).

Kilkenny The Galmoy orebody, near Kilkenny, is a stratiform lead–zinc replacement deposit that developed in dolomitized and brecciated Carboniferous Limestone. Sphalerite is the main sulphide phase and it occurs as fine-grain cream-to-buff crystals in fracture and void infill as well as a replacement species. Banded galena–sphalerite and dark or pale crystalline sphalerite are also found (Doyle et al. 1992, Lowther et al. 2004).

Limerick Galena, pyrite and sphalerite are the major sulphides in a complex limestone-hosted replacement deposit at Courtbrown, on the southern shore of the River Shannon, near Ballysteen. Sphalerite (associated with galena) forms coarse crystalline masses and is often spherulitic or colour banded. Finer-grain sphalerite and galena occur in crosscutting calcite veins (Grennan 1986, Reed & Wallace 2001). Other reports of sphalerite are from: Aherlow, near Galbally – disseminated copper deposit in a Carboniferous limestone/shale sequence with galena and sphalerite often in calcite veins peripheral to the main mineralization (Schultz 1971, Cameron & Romer 1970); Carricklittle lead–zinc prospect, Carricklittle Townland – galena, pyrite and sphalerite in a fracture-controlled deposit within Carboniferous Limestone intruded by feldspar porphyry (Schultz 1971); and Cloghatrida mine (Cole 1922).

Longford The Keel prospect, Ardagh, 11 km S of Longford, is a sphalerite-dominated Cd–Pb–Zn deposit formed in faulted, brecciated and dolomitized Carboniferous rocks. The sphalerite is yellow to light-brown in colour and, when found in cavities, it forms fine gem-like crystals up to 4 mm long. It is cadmium rich and it contains some mercury (Patterson 1970, Morrissey et al. 1971, Schultz 1971, Watling 1974, Halliday & Mitchell 1983, Slowey 1986, Ryback & Moreton 1993, Hitzman & Beaty 2004). The Newtown Cashel prospect is a fracture-controlled sphalerite–(barite–galena) deposit to the W of Keel, which consists principally of crystalline brown sphalerite with minor galena and pyrite. A later generation of orange sphalerite is thought to be cadmium rich (Crowe 1986).

Louth Considerable amounts of pale yellowish-brown sphalerite and lesser amounts of dark-brown sphalerite occur with chalcopyrite and galena in a breccia and quartz–carbonate vein deposit at Salterstown (Morris 1984).

Mayo Native gold-bearing quartz veins at Cushinyen, in the Cregganbaun shear zone, contain mm-size aggregates of sulphides, including bournonite, chalcopyrite, galena, pyrite, sphalerite and tetrahedrite (Thompson et al. 1992). An Ordovician, copper-mineralized, altered acidic porphyry, 8 km to the S of Charlestown, contains a late pulse of barite–galena–sphalerite mineralization at the base of the mineralized zone (Poustie 1982, O'Connor & Poustie 1986). Alluvial gold containing inclusions of sphalerite is reported from: Bunowen River, Carrownisky River (Glenkeen Bridge), Lugbeg River (Chapman et al. 2000c) and the eastern part of the Cregganbaun shear zone (Leake et al. 1993).

Meath The ore at Navan mine comprises fine layers of sphalerite and coarser-grain crystalline galena, with varying amounts of barite, marcasite and pyrite. It is a world-class carbonate-hosted epigenetic syn-sedimentary replacement deposit. In the 1990s it was the largest zinc producer in Europe and rated fifth largest worldwide. Pre-production ore reserves of 69 million tonnes grading 10.1% zinc were reported in 1985 and the later discovery of a southwestern extension has taken the total to over 90 million tonnes. By 2004, production had been 57.7 million tonnes grading 8.3% zinc and 2.1% lead (Ashton et al. 2004, Ashton 2005). Small crystals of honey-colour sphalerite occur in cavities within massive ore and a heavily twinned, totally recrystallized, collomorphic variety has also been documented. Sphalerite is noted to contain appreciable calcium and to replace zinc carbonates during diagenesis. Four generations of sphalerite have been identified in the U-lens (southwestern extension) with the first three colloform generations becoming lighter in colour, from dark, almost opaque, brown, through shades of orange brown to pale yellowish brown. The fourth generation produced blocky, pale-yellow to yellowish-brown translucent crystals (Schultz 1971, Boyce et al. 1983, Kucha & Wieczorek 1984, Andrew & Ashton 1985, Ashton et al. 1986, Fontboté & Amstutz 1986, Dickinson & Pattrick 1987, Ryback & Moreton 1993, Anderson et al. 1998, Blakeman et al. 2002, Peace et al. 2003, Wilkinson 2003). The Tatestown–Scallanstown deposit (3.6 million tonnes at 6.9% Zn + Pb), near Navan is regarded as a satellite to the major Navan orebody. Sphalerite, galena and marcasite–pyrite dominate the mineral assemblage (Andrew & Poustie 1986). Stratigraphically equivalent beds to those at Navan host a sub-economic barite–galena–sphalerite vein deposit at Drumlerry, near Oldcastle (Brand & Emo 1986).

Monaghan Sphalerite occurs with chalcopyrite and galena in carbonate veinlets crosscutting arsenopyrite–stibnite-bearing veins once worked for antimony at Clontibret mine, Tullybuck. Sphalerite–galena-bearing quartz veins also occur in the local area (Morris 1984, Morris et al. 1986, Ryback & Moreton 1993, Steed & Morris 1997). Sphalerite is reported from various other mines in the area: Annaglogh (Cu–Pb–Zn); Coolartragh (Pb); Croaghan (Ba–Pb–Zn); Hope mine, Cornalough–Carrickagarvan (Ag–Ba–Pb–Zn); Lemgare (Ba–Pb–Zn); Tamlat (Ag–Pb–Zn); Tassan (Ag–Pb–Zn); Tonagh (Pb–Zn) (Cole 1922, Morris 1984).

Sligo Abbeytown lead–zinc mine, near Ballysadare, worked a dolomitized limestone-replacement deposit in which dark reddish-brown sphalerite (containing chalcopyrite inclusions) crystallized early and was more abundant than galena. Later sphalerite is honey colour and occurs in calcite veinlets (Cole 1922, Fernandez 1957, O'Brien 1959, Schultz 1971, Halliday & Mitchell 1983, Hitzman 1986, Ryback & Moreton 1992).

Tipperary The Silvermines district is notable for its large stratiform sedimentary-exhalative lead–zinc-barite deposit (estimated in 1978 to contain 4.6 million tonnes of ore with 6.5% zinc, 2.5% lead and 30 ppm silver), but considerable early mining activity was focused on mineralized fault breccias and fracture fillings close to a major fault system. Sphalerite is described from the following: Ballygown South mine – early iron-rich sphalerite containing exsolved pyrrhotite, and later light-yellow sphalerite in alternate colloform bands with marcasite–pyrite (Russell 1907b, Rhoden 1959); Ballynoe (Magcobar) barite open pit – minor sphalerite (fibrous reddish-brown) and galena (Schultz 1971, Mullane & Kinnaird 1998); Gortnadyne mine – dark-brown (iron-rich) and light-yellow sphalerite (Russell 1907b, Rhoden 1959); Mogul mine – stratiform ore zones consisting of variable abundances of galena, pyrite and sphalerite. One remarkable 3 m-long cavity found in 1978 produced

Sphalerite crystal (variety honey-blende), 2.5 mm across, from Mogul Mine, Silvermines, Co. Tipperary, Ireland. (© D. I. Green)

many fabulous, sharp, lustrous, gem-like crystals of honey-colour sphalerite up to 1.4 cm across, associated with exceptional crystals of bournonite and galena (Schultz 1971, Zakrzewski 1989, Ryback & Moreton 1991, Wolfe 1996a, Moreton 1999, Green & Moreton 2001, Green et al. 2005a); Shallee East and West mines – dark-brown (iron-rich) and light-yellow sphalerite (Rhoden 1959); and more generally by other authors – O'Brien (1959), Morrissey et al. (1971), Taylor & Andrew (1978, 1979), Samson & Russell (1983), Taylor (1984), Russell (1990), Banks et al. (2002) and Reed & Wallace (2004). Another carbonate-hosted deposit (22 million tonnes of ore with 11.5% zinc, 1.9% lead and 26 grams per tonne silver) occurs at Lisheen, between Thurles and Rathdowney, where sphalerite is found with galena, marcasite, pyrite and other (minor) sulphides in three stratabound sulphidic lenses. Sulphides replace black matrix breccia and, where cavities have developed, these are infilled with colloform sulphide masses intergrown with calcite or dolomite (Hitzman 1992, Hitzman et al. 2002, Fusciardi et al. 2004, Wilkinson et al. 2005).

At Gortdrum mine, near Tipperary town, epigenetic mineralization related to the intrusion of two high-level basic intrusions has produced a dominantly chalcopyrite–Hg-rich tennantite orebody in which sphalerite occurs as an accessory species (Morrissey et al. 1971, Steed 1983).

Tyrone Gold-bearing quartz–carbonate veins at Curraghinalt, near Gortin, Sperrin Mountains, contain pyrite and subsidiary arsenopyrite, bornite, chalcopyrite, native copper, galena, sphalerite and tetrahedrite–tennantite (Clifford 1986, Earls et al. 1989, Clifford et al. 1990). Further investigations of vein mineralization at Curraghinalt, and at Cavanacaw, near Omagh, confirm these observations and identify at least two generations of sphalerite (some intergrown with chalcopyrite) and a suite of Ag–Au–Bi tellurides (Cliff & Wolfenden 1992, Parnell et al. 2000). Sphalerite also occurs as inclusions in alluvial gold from Curraghinalt Burn (Chapman et al. 2000c). Minor sphalerite and galena are reported from quartz veins at Cashel (Leyshon & Cazalet 1976).

Waterford Sphalerite and galena, both rare at this locality, occur in copper-rich quartz veins cutting Ordovician rocks at Knockmahon, 22 km SW of Waterford town (Holdsworth 1834, Moreton 1991).

Westmeath At Ballinalack, 15 km NW of Mullingar, a carbonate-hosted lead–zinc deposit has formed as discrete pods within a series of reef knolls. It is spatially related to the Ballinalack fault and has a style of mineralization similar to that at Tynagh in Co. Galway. Sphalerite occurs in dark-brown collomorphic bands, followed by a fine-grain very pale (almost honey-blende) variety. Pyrite and subordinate galena are associated species in a gangue dominated by calcite and containing barite and dolomite (Schultz 1971, Patterson 1970, Jones & Brand 1986). Shallow-water carbonates host another lead–zinc deposit at Moyvoughly, near Moate, 12 km N of Clara. Sphalerite occurs in stratiform bands and in crosscutting veinlets associated with barite, galena and pyrite. Massive sulphides formed by the replacement of limestone and consist of sphalerite and pyrite with minor galena making up the high-grade sections of the orebody (Patterson 1970, Poustie & Koucha 1986, Kucha 1988).

Wexford The early miners found it difficult to separate fine-grain sphalerite from intergrown galena in quartz veins at Barrystown mine, Bannow Estuary, and Caim, near Enniscorthy (O'Brien 1959, Halliday & Mitchell 1983).

Wicklow The orebody at Avoca mine is a volcanogenic exhalative sulphide deposit with considerable chalcopyrite and pyrite hosted in Ordovician volcanic rocks associated with black chloritic schists and also in stringer ore. Sphalerite and galena occur in sub-economic amounts in both ore types at the margins and top of the mineralized areas (Cole 1922, O'Brien 1959, Schultz 1971, Wheatley 1971, Badham 1978, Downes & Platt 1978, Sheppard 1980, McArdle 1989, 1993, Ryback & Moreton 1993). A fine-grain purple ore, composed of a mixture of galena and sphalerite (+ chalcopyrite) at the northeastern end of the Avoca district, was named kilmacooite (after the local townland) before its composite nature was revealed (Cole 1922, Milner & McArdle 1992). A related Cu–Pb–Zn–(Bi–Au) deposit in the Goldmine River valley, 10 km along strike from Avoca, occurs in metamorphosed BIF rocks and contains minor dark sphalerite with chalcopyrite disease subsidiary to (in order of abundance) magnetite, chalcopyrite, pyrite and hematite (Ixer et al. 1990). Other occurrences along the southeastern margin of the Leinster batholith are at: Ballingen, between Tinahely and Aughrim – sphalerite with inclusions of chalcopyrite, pyrrhotite or stannite in cassiterite–scheelite-bearing quartz veins, and in altered microtonalite (Steiger & Bowden 1982, 1986, McArdle & Kennan 1988, Gallagher 1989); and Luganure mines, Glendalough – sphalerite in a quartz gangue (O'Brien 1959, Halliday & Mitchell 1983, Ryback & Moreton 1993).

Offshore territories

Isle of Man Galena–sphalerite-bearing veins cutting metamorphosed slates and grits of the Cambro-Ordovician Manx Group were exploited at the Foxdale and Laxey mines, where a combined total of 500 000 tonnes of lead–zinc concentrates were produced between 1845 and 1938. Minor chalcopyrite and malachite were associated species. Classic specimens of sphalerite with crystallized quartz were recovered from the Laxey mines (Rudler 1905, Dewey & Eastwood 1925, Russell 1949b, Dunham et al. 1978, Alderton et al. 1995). At Snaefell mine, galena and sphalerite occur in a calcite–dolomite–quartz gangue, with minor chalcopyrite, pyrite and pyrrhotite (Lamplugh 1903).

Guernsey Iron-poor sphalerite is reported on a specimen of silver-rich galena from a calcite vein at Bouet, St Peter Port (Mourant 1939). Minor iron-rich sphalerite and molybdenite occur in chalcopyrite–pyrrhotite-bearing veinlets cutting diorite at Baubigny, near St Sampson (Ixer 1992).

Jersey Abundant sphalerite and lesser galena in carbonate veins cutting Brioverian sediments at Le Pulec were mined in about 1871. The sphalerite is iron poor and it contains inclusions of chalcopyrite and rod-shape tetrahedrite. Arsenopyrite and pyrite are associated species (Mourant & Warren 1934, Mourant 1961, Ixer & Stanley 1980, Stanley & Ixer 1982, Ixer 1992). Granite of the Northwest granite complex hosts many mineral veins: Le Doüet de la Mer, St Ouën – small quantities of arsenopyrite, galena, pyrite and sphalerite in a small carbonate vein (Robinson 1934, Mourant 1961, Ixer 1980, 1992); Gigoulande quarry, St Mary – minor sphalerite in a chalcopyrite–pyrite–quartz vein cutting granite (Ixer 1980, 1992); L'Etacq quarry – sphalerite in pegmatites and quartz–calcite veins (Mourant 1961, Ixer 1980, 1992); Les Handois and Mont Mado – rare sphalerite (Mourant 1961).

Sphalerite has also been found at: Bouley Bay, Trinity, northeast coast of Jersey – traces of sphalerite and galena in copper-enriched flow-banded rhyolite (Ixer 1980); German underground hospital, St Peter's Valley, St Lawrence – minor sphalerite (with chalcopyrite disease) and galena in a chalcopyrite–pyrite–quartz vein (Ixer 1980) and iron-poor twinned zoned sphalerite forming aggregates to 700 μm in a calcite–galena–pyrrhotite vein (Robinson 1947, Ixer 1990b); La Corbière, southwest Jersey – chalcopyrite, galena and anhedral iron-poor sphalerite in pyrite–quartz veins (Ixer 1980); L'Ouaisné Bay, southwest Jersey – veins containing calcite, chalcopyrite, fluorite, galena, hematite, magnetite and sphalerite (Mourant 1961, Ixer 1980); near Tesson Mill – minor sphalerite in veins (Mourant 1961).

sphene

A synonym of titanite.

spiessglanzocker

A synonym of cervantite and of stibiconite.

SPINEL GROUP

A large group of cubic oxides with the general formula AB_2O_4, where A = Co, Cu, Fe^{2+}, Ge, Mg, Mn^{2+}, Ni, Ti, Zn; and B = Al, Cr^{3+}, Fe^{2+}, Fe^{3+}, Mg, Mn, Ti, V^{3+}. Chromite, gahnite, hercynite, jacobsite, magnesiochromite, magnetite, qandilite, spinel and ulvöspinel are members of the group found in Britain or Ireland. Another member of the group, coulsonite, requires confirmation.

spinel

SPINEL GROUP

cubic

$MgAl_2O_4$

A very common magnesium–aluminium oxide that forms three series: with magnesiochromite, gahnite and hercynite. Spinel is found as a high-temperature accessory species in igneous rocks (e.g. basalts and peridotites), in regionally metamorphosed aluminium-rich schists and in contact-metamorphosed limestones or hornfels. Unless quantitative data are available, extensive solid-solution variation can make identification of this species difficult. This is compounded when an author does not make it clear whether it is spinel the species or spinel the group that is being referred to.

In Cornwall, olive-colour anhedral spinel occurs in recrystallized anhydrous peridotite at Countybridge quarry, Goonhilly Downs, Lizard Peninsula (Green 1964). Data presented by Davies (1983) show high chromium and iron contents in spinel from peridotite at Countybridge quarry, Gwenter, at Kernawas on Goonhilly Downs, and at Coverack. In some cases the magnesium content is relatively low and the species name cannot be spinel. The green spinel (pleonaste) found in banded greenstone hornfels in the metamorphic aureole of the Land's End granite at Kenidjack, NE of Cape Cornwall, St Just (Tilley & Flett 1930, Tilley 1935), is also misnamed, as data in Chinner & Fox (1974) show the species to be hercynite. Spinel is noted from other locations around the Land's End granite (Mitropoulos 1984) and as an uncommon accessory species in the Dartmoor granite, Devon (Brammall & Harwood 1925), but confirmatory data are lacking.

Lavas erupted from the volcanic vent on Calton Hill, Taddington, Derbyshire, contain spinel-lherzolite xenoliths (Hamad 1963, Donaldson 1978, Macdonald et al. 1984). Although the spinel is described as picotite (a chromium-rich variety of spinel), it is dominated by magnesium and hence is spinel *sensu stricto*.

Spinel is not confirmed in Wales; however, chrome spinel has been found in peridotite within the diatreme exposed at Great House and Glen Court, near Usk, Gwent (Bevins 1994). Spinel is also noted in a hornblende picrite unit of the Rhiw layered intrusion, Lleyn Peninsula, Gwynedd (Hawkins 1970), and in an altered dolerite from the Carn Menyn, Mynydd Preseli area of Dyfed (Thorpe et al. 1991).

Very little data exist to confirm spinel in Scotland and in many cases it is a chrome spinel that is being described. It is possible that some chrome spinel can be spinel *sensu stricto* (if $Fe/(Mg+Fe^{2+})>0.5$ and $Cr/(Cr+Al)<0.5$), so the occurrences are listed here, but, without supporting data, only a tentative identification can be made. If chrome spinel is reported to have dominant Fe^{2+} and Cr, it should more correctly be described as chromite and consequently is not listed here. The remaining Scottish descriptions of spinel include:

FIFE Lherzolite inclusions containing chrome spinel occur in Carboniferous–Permian vents at Coalyard Hill and Ruddons Point, near Elie (Chapman 1976). DUMFRIES & GALLOWAY Spinel occurs in hornfels xenoliths within marginal rocks of the Cairnsmore of Carsphairn granite complex (Deer 1937b). GRAMPIAN Chrome spinel in cumulate assemblages within the Arthrath mafic intrusion (Fletcher et al. 1997); accessory spinel in an iron-rich cordierite (sekaninaite) hornfels at Gallowhill Brae, Arnage (Read 1929); spinel in silica-poor hornfels within the aureole of the Morven–Cabrach basic intrusion, near Glenbuchat (Allen 1970); spinel in migmatitic gneiss associated with biotite, cordierite, garnet, plagioclase (oligoclase) and quartz at Ardlethen and Arnage, Ythan Valley (Read 1952). HIGHLAND Spinel is confirmed in siliceous dolomite from Allt Guibhsachain in the northeast corner of the Ballachulish granite aureole (Ferry 1996); confirmed spinel forms octahedra and clumps in the marbles of Glenmore of Glenelg (Clough & Pollard 1899, Livingstone 2002); confirmed spinel occurs in pyroxenite xenoliths in a Permo-Carboniferous diatreme associated with an alkaline dolerite dyke on the island of Eilean Shona, NW of Ardnamurchan (Upton et al. 2003); bluish-green spinel occurs in calcium–magnesium-rich silicate hornfels close to the contact with diorite gneiss in Coire nam Muc, 4 km WNW of Corran ferry, Ardgour district (Drever 1939, 1940); brown or greenish-brown spinel octahedra are enclosed in calcite, gehlenite or monticellite within a gabbro/limestone contact zone at Camas Mòr, Isle of Muck (Tilley 1947); spinel is associated with clinohumite, diopside, phlogopite, quartz and troilite in impure marble from Sgiath Bheinn, S of Còsag, Glenelg, Lochaber (Davidson 1995a); spinel is associated with grossular, vesuvianite and xanthophyllite at the contact of limestone with the Beinn an Dubhaich granite, Broadford area, Isle of Skye (Tilley 1949, 1951, Holness 1992); both spinel and magnesiochromite are described as chrome spinel in picritic basalts from the northeastern side of An Cruachan and from Sgùrr Beag, Isle of Skye, although the data do not approach end-member compositions (Bell & Williamson 1994); chrome spinel (abbreviated to chromite) and a Cr-poor spinel are found in ultramafic lithologies in the Cuillin Complex, Isle of Skye (Bell & Claydon 1992, Pirrie et al. 2000a); chrome spinel inclusions occur in olivine within a layered picritic dyke at Suisnish Point, Isle of Skye (Parsons 1976); chrome spinel data from the mafic–ultramafic complex on the Isle of Rum, collected by Henderson & Suddaby (1971), indicate that spinel, chromite and magnesiochromite are all present. STRATHCLYDE Confirmed spinel occurs in pyroxenite xenoliths in Permo-Carboniferous alkali basalts at Auchengarth and Waterside, near Muirkirk (Upton et al. 2003); deep-brown chrome spinel (picotite) mantles bottle-green spinel (ceylonite) in wehrlite belonging to the Girvan–Ballantrae ophiolite complex to the NE of Knockormal (Bloxam & Allen 1959); spinel, sometimes replacing staurolite, occurs in Precambrian metasediments caught up in the intrusion of lamprophyre dykes on the shores of Loch Lomond (Dempster & Bluck 1991); chrome spinel occurring as inclusions in olivine within Tertiary basalts on the Isle of Mull appears to be rich in Cr or Fe^{3+} (Kerr 1998). WESTERN ISLES Confirmed spinel (green and with a high iron content) occurs in zones of alteration within the Western gabbro, St Kilda (Hirta) (Harding et al. 1984); confirmed spinel forms euhedral crystals, up to 5 mm across, in Archaean metamorphosed limestones from the Rodel area of South Harris (Davidson 1943); confirmed iron-rich spinel is an accessory constituent of sapphirine-bearing garnet–kyanite/sillimanite–orthopyroxene granulites to the S and SW of Roineabhal, South Harris (Baba 2003); spinel reported to co-exist with ilmenite in the Shiant Isles Main Sill is actually titaniferous magnetite (Gibb & Henderson 1996, Henderson et al. 2000). SHETLAND Chrome spinel occurs in a metamorphosed ultrabasic sheet (now mainly magnesite and talc) at Cunningsburgh, Mainland (Smith et al. 1992). Chrome spinel from the Shetland ophiolite complex has a wide range of compositions, with chromite and magnesiochromite being most common. However, spinel is found in some dunite pods within harzburgite (Gass et al. 1982, Prichard & Lord 1988).

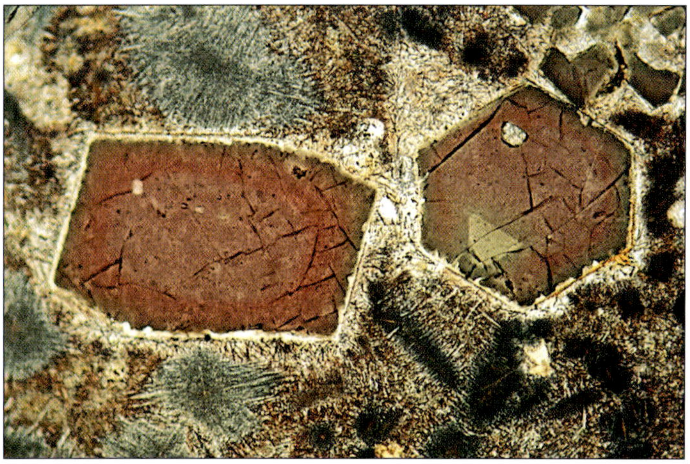

Euhedral crystals of spinel in a cordierite-rich xenolith from the composite sill at Nun's Pass, Carsaig, Isle of Mull, Strathclyde, Scotland; width 1.3 mm, viewed in PPL. (© K. Gardner)

In Ireland, spinel is described from the chalk/dolerite contact zone at Scawt Hill, Larne, Co. Antrim. Spinel is a prominent mineral found in hybrid rocks formed by the assimilation of chalk by the doleritic magma and is associated with calcite, gehlenite (melilite), larnite, merwinite, perovskite and spurrite. In thin section the spinel appears as deep-brown or greenish-brown octahedra. Its high Fe^{3+} content places it as an intermediate composition between spinel and magnetite (Tilley 1929b, Tilley & Harwood 1931). Such compositions must be characteristic of chalk/dolerite contact zones, as very similar compositions are reported from the contaminated dolerite plug at Carneal, Co. Antrim. Here, spinel forms opaque or brown octahedra up to 0.25 mm across, which are referred to as ferrian chlorospinel (Sabine et al. 1968, Sabine & Young 1975). At Brockley on Rathlin Island, Co. Antrim, dark-green well formed spinel occurs with cordierite and mullite in porcellanite representing a completely recrystallized bauxitic lithomarge metamorphosed by a dolerite plug (Wilson & Robbie 1966). Elsewhere in Ireland, spinel is found in: desilicated corundum–magnetite–orthopyroxene–spinel xenoliths in gabbro at Cashel Hill, Connemara,

Co. Galway (Ahmed-Said & Leake 1990); contact-metamorphosed calc-silicate rocks around a syenitic intrusion at Barnavave, Carlingford, Co. Louth, where it is variably associated with apatite, calcite, custerite, diopside, grossular, melilite, monticellite, phlogopite, spurrite, tilleyite and wollastonite (Osborne 1932a,b); spinel peridotite xenoliths within a dolerite dyke at Inver, Co. Donegal (Gallagher & Elsdon 1990, Shaw & Edgar 1997).

spinelle ruby

A red chromium-rich gem variety of spinel.

spionkopite

hexagonal $Cu_{39}S_{28}$

A rare copper sulphide, found S of the Askrigg Block section of the Northern Pennine orefield in minor Cu–(As–Sb) mineralization at Scaleber Bridge, 3 km SSE of Settle, Yorkshire. The mineralization is in dolomitized limestones and consists of azurite, calcite, chalcopyrite, malachite and quartz, with minor amounts of enargite and tetrahedrite-group species. Samples of calcite host sulphide inclusions and contain flecks of spionkopite (1–10 μm wide) as an alteration product of chalcopyrite (Ixer & Stanley 1998). Spionkopite has also been found in void-infilling carbonate-associated copper mineralization at Judkins quarry, Nuneaton, Warwickshire. It occurs in trace amounts as lamellar grains partially or totally replacing geerite or forming optically continuous multiple grains in bornite (Ixer et al. 2005a).

At Cliff, Unst, Shetland, Scotland, sulphide-bearing dunite pods in harzburgite contain arsenic-rich chromite with minor heazlewoodite, millerite, pentlandite and trace amounts of native copper, gold, silver and spionkopite. The locality is also noted for extremely high palladium and platinum concentrations dominated by sperrylite and by platinum and palladium tellurides (Prichard et al. 1994).

Epigenetic copper–dolomite association ore at the Great Orme mines, Llandudno, Gwynedd, Wales, is found adjacent to N–S faults in limestone that is now dolomitized. The primary ore species chalcopyrite, marcasite and pyrite are partially altered to banded limonite and malachite accompanied by thin rims of covellite, digenite, djurleite, spionkopite, and trace amounts of photosensitive silver minerals. Additionally, supergene calcite is interbanded with botryoidal limonite accompanied with native copper, malachite, spionkopite, tenorite and hydrocarbons (Ixer & Davies 1996, Ixer & Stanley 1996, Ixer 2001).

Spionkopite may be more common than the entries above suggest, as so-called "blaubleibender covellite" has now been characterized as a mixture of spionkopite and yarrowite (Ryback & Tandy 1992, Bevins 1994). Examples include: Clevedon, Avon (Starkey 1986c, Ixer et al. 1993); Alderley Edge ridge, Cheshire (Braithwaite 1994); Seathwaite Tarn, near Coniston, Cumbria (Stanley & Criddle 1979); Parys Mountain, Anglesey, Gwynedd, Wales (Southwood & Bevins 1995); Matta Taing, West Mainland, Shetland, Scotland (Hall & Donovan 1978); Ballycoog, Goldmine River valley, near Avoca, Ireland (Ixer et al. 1990); Bouley Bay, Trinity, Jersey (Ixer 1980, 1992); German underground hospital, St Peter's Valley, St Lawrence, Jersey (Ixer 1990); Sark's Hope mine, Little Sark (Ixer & Stanley 1983).

spodumene

PYROXENE GROUP

monoclinic $LiAlSi_2O_6$

A relatively uncommon lithium aluminosilicate that occurs in lithium-rich granitic pegmatites and aplites. Spodumene is noted together with a remarkable number of other species from the Meldon aplite quarry, near Okehampton, Devon (Kingsbury 1964). Specimens containing spodumene reputedly from Meldon and now in the Kingsbury collection at the NHM were described in more detail by Drysdale (1985, 1991), but, given the doubt about the provenance of other Kingsbury specimens (Ryback et al. 1998, 2001), further confirmation is desirable. Later work on the Meldon aplite does not confirm spodumene (von Knorring & Condliffe 1984).

In Ireland, spodumene occurs in a belt of lithium pegmatites that extends along the eastern flank of the Tullow Lowlands unit of the Leinster batholith. The surface expression of these pegmatites is almost exclusively that of boulder accumulations, with very little outcrop. Often the pegmatites are in close proximity to marginal granites displaying crush structures and greisen. At Seskinnamadra, W of Tomduff Hill, Co. Carlow, hundreds of pegmatite boulders containing spodumene have been found. Spodumene crystals are commonly fibrous, pinkish and often bent. Some of the spodumene (kunzite) is pink and glassy (Steiger & von Knorring 1975). An attractive purplish-pink variety of spodumene, as well as the more normal white variety, occurs in a predominantly albitic pegmatite containing lesser amounts of manganese-rich apatite, cassiterite, columbite–tantalite, microcline, muscovite, quartz and spessartine in the vicinity of Rathnagerragh Castle, near Myshall, Co. Carlow (Steiger & von Knorring 1975, Scoon 1978). At Aclare House, near Myshall, boulders of albite–spodumene pegmatite contain spodumene laths up to 35 cm long associated with albite, microcline, muscovite and quartz, and accessory amounts of arsenopyrite, manganese-rich apatite, bertrandite, beryl, cassiterite, chalcopyrite, columbite–tantalite, fluorite, lithiophilite, pyrite, spessartine, tourmaline and zircon. In a finer-grain saccharoidal albite unit, white, grey or greyish-green spodumene occurs with porphyroblastic quartz and yellowish-green muscovite (Steiger & von Knorring 1975, Scoon 1978, Kennan et al. 1986, McArdle et al. 1986b). At Moylisha, near Shillelagh, Co. Wicklow, the lithium pegmatites contain spodumene laths up to 10 cm long, irregularly arranged in a matrix of feldspar (albite and microcline), muscovite, quartz with accessory cassiterite, columbite, garnet (almandine–spessartine), hiddenite, microlite and iron oxides. Some samples consist of 70% spodumene. Kunzite, the lilac-pink gem variety of spodumene, is also noted from here (Steiger & von Knorring 1975, Scoon 1978, Luecke 1981). At Stranakelly, near Shillelagh, Co. Wicklow, lithium pegmatites contain spodumene laths up to 30 cm long, but in associated greisen it is pseudomorphed by purple and pink lepidolite. Cassiterite, manganotantalite, microlite and spessartine are notable accessory species in the greisen (Scoon 1978, Luecke 1981, O'Connor et al. 1991). Early reports of spodumene in granite at Killiney, Co. Dublin (Thomson 1836, Greg & Lettsom 1858), are not confirmed in later works, although the record is repeated in Gaines et al. (1997).

spurrite

monoclinic $Ca_5(SiO_4)_2(CO_3)$

A rare calcium silicate carbonate, dimorphous with paraspurrite and found in the chalk/dolerite contact zone at Scawt Hill, Larne, Co. Antrim, Ireland. Spurrite is the most prominent mineral found in hybrid rocks formed from the assimilation of chalk by the doleritic magma and it is associated with calcite, larnite (especially), gehlenite (or melilite), merwinite, perovskite and spinel. It usually occurs as a finely crystalline mass of white colour, but elongate crystals up to 12 mm long are also found. Under the microscope, spurrite is colourless and it often shows distinctive polysynthetic twinning (Tilley 1929a,b). In a thin calcitic crust lining a cavity of a spurrite

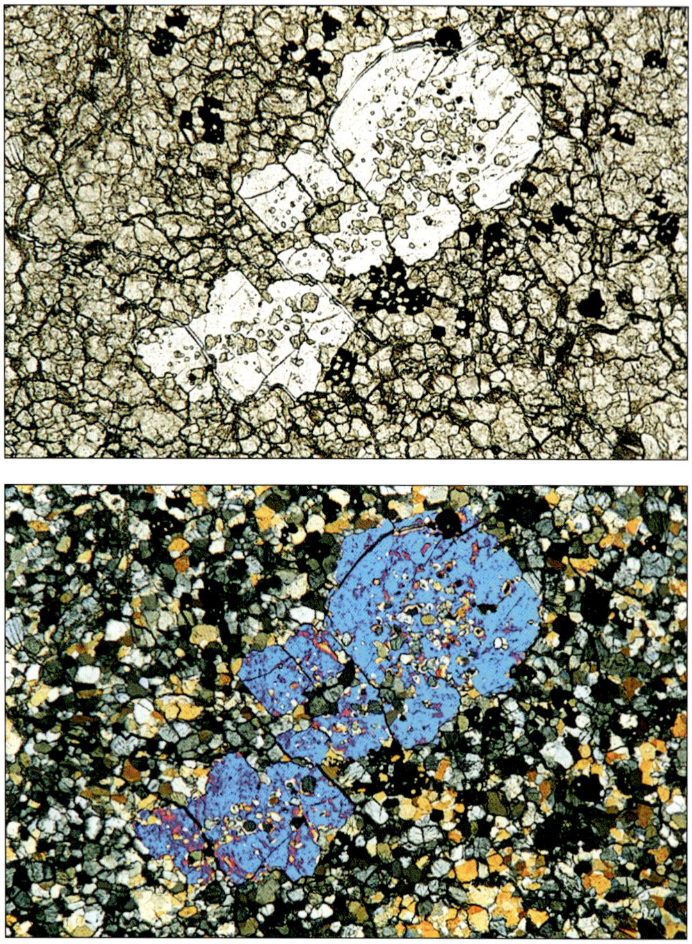

Large porphyroblast of clear spurrite enclosing grains of larnite in a matrix of granular larnite; Scawt Hill, near Larne, Co. Antrim, Ireland; width 2.25 mm; viewed in PPL (upper) and XPL (lower). (© K. Gardner)

contact rock, spurrite is replaced by afwillite (Tilley 1930a), and, in spurrite marbles adjacent to the hybrid rocks, spurrite is, partially to totally, replaced by fibrous scawtite (Tilley 1938b). The unusual character of the Scawt Hill assemblages is probably attributable to the high temperature prevailing during metamorphism (Tilley & Harwood 1931). In another chalk/dolerite contact zone associated with a Tertiary volcanic plug at Carneal, Larne, Co. Antrim, the chalk is converted by intense contact metamorphism to assemblages containing larnite, magnetite, merwinite, scawtite, spinel, spurrite, titanite and wollastonite, whereas the dolerite is desilicated to melilite- and wollastonite-bearing pyroxene-rich alkaline hybrid rocks. Spurrite is most commonly found in rocks containing larnite and spinel or magnetite, poikiloblastically cementing larnite prisms in both metasomatized chalk and dolerite (Henmi et al. 1973, Sabine & Young 1975). Thermally metamorphosed limestones to the south of Barnavave, Carlingford, Co. Louth, contain assemblages in which common spurrite is associated with calcite, grossular, hillebrandite, melilite, tilleyite, vesuvianite and wollastonite; the spurrite forms subhedral or rounded grains and irregular plates. There is some evidence to suggest that spurrite is partially replaced by tilleyite (Nockolds & Vincent 1947). At Killala Bay, near Inishcrone, Co. Sligo, thermal metamorphism of limestone by basalt–dolerite dykes has produced a suite of spurrite-rich rocks in which bredigite, calcite, cuspidine, gehlenite, kilchoanite, larnite, rankinite, tilleyite and wollastonite are associated species (Nawaz 1974).

In Scotland, extreme thermal metamorphism of Liassic limestones by a gabbroic intrusion near Camphouse, Ardnamurchan, Highland Region, has produced rock types in which there is extensive development of bredigite, calcite, gehlenite, grossular, kilchoanite, larnite, merwinite, monticellite, rankinite, rustumite, spurrite and wollastonite (Agrell 1951, 1965). At Camas Mòr, Isle of Muck, Highland Region, another gabbro/limestone contact zone contains spurrite-rich assemblages in metasomatized limestones. Spurrite forms irregular spongy crystals enclosing gehlenite or as isolated grains in calcite. Merwinite and monticellite are closely associated species and in one spurrite–gehlenite assemblage there is late replacement of spurrite by tilleyite (Tilley 1947). Spurrite is also reported from a banded cherty nodule in a high-temperature calcic skarn near the contact of the Cuillin gabbro complex at Camasunary Bay, Strathaird Peninsula, Isle of Skye, Highland Region. Associated major species include garnet (andradite–grossular), melilite, tilleyite, vesuvianite and wollastonite (Beard & Drake 2007).

STANNITE GROUP

Tetragonal sulphides and selenides with the general formula A_3BX_4, where A = Ag, Cd, Cu, Fe, Hg, Zn; B = As, Ge, In, Sb, Sn; and X = S, Se. Famatinite, kuramite, luzonite and stannite are members of the group found in Britain or Ireland.

stannite

STANNITE GROUP
tetragonal Cu_2FeSnS_4
TL Wheal Rock (West Kitty), St Agnes, Cornwall, England

A relatively uncommon copper–iron–tin sulphide, dimorphous with ferrokësterite and found in tin-bearing hydrothermal-vein deposits. Stannite is in fact only one member of a complex group of Cu–Fe–Sn–Zn sulphides and many of the reported occurrences of stannite from Cornwall and Devon do not actually consist of this mineral. Ferrokësterite, kësterite, kuramite, mawsonite and stannoidite are all related species found in this region. The problem of identification is exacerbated by abundant intricate intergrowths of the various species and distinguishing stannite from ferrokësterite or kësterite in hand specimen is often impossible (Alderton 1991). The term "stannite" has also been used to describe both a mixture of cassiterite and quartz (Greg & Lettsom 1858) and a mixture of cassiterite and chalcopyrite (Collins 1892). Apparently it is only at the type locality that pure stannite has been found (although it is possible that zinc was never analyzed); elsewhere, zinc is a significant additional element (Collins 1892, Alderton 1991).

In southwest England, the best examples of stannite are found in greisen-bordered vein swarms located in small granite stocks. At Cligga Head, Perranzabuloe, Cornwall, zinc-rich stannite occurs with arsenopyrite, cassiterite, löllingite, sphalerite and wolframite, and minor native bismuth, chalcopyrite and pyrite. Ferrokësterite, kësterite and stannoidite have also been found at this site – the best for collecting stannite-related minerals (Russell & Vincent 1952, Polkinghorne 1954, Jackson et al. 1977, Moore & Howie 1984, Hosking et al. 1987, Kissin & Owens 1989, Alderton 1991). Stannite is a minor phase at St Michael's Mount, Marazion, Cornwall, with stannoidite and kësterite being the more common Cu–Fe–Sn–Zn sulphides; otherwise the ore assemblage is very similar to that at Cligga Head (Davison 1920b, Hosking 1954c, Moore & Howie 1984, Alderton 1991, Dominy et al. 1995). Stannite is also noted in Devon (without comment) from Hemerdon, Plympton, and from Great Consols mine, Tavistock Hamlets (Alderton 1991).

Stannite has been confirmed in disseminated Cu–Fe–Sn–Zn sulphide mineralization within the Hensbarrow stock of lithium-mica topaz granite (part of the composite St Austell granite complex). Here it occurs in pegmatitic vugs within the roof zone of the granite and is associated with mawsonite and stannoidite (Manning 1983, Alderton 1991). At Gunheath china-clay pit (including Bunny mine), many greisen-derived specimens of so-called stannite have been shown to be either kuramite or stannoidite, although minor stannite is also present (Russell & Vincent 1952, Hawkes et al. 1987, Hosking et al. 1987, Alderton 1991, Williamson et al. 1997, 2002).

Stannite is reported from other Cornish locations by Russell & Vincent (1952), Hosking et al. (1987), Alderton (1991) and other workers, but the identifications deserve further analytical confirmation: Barrier mine, Gwennap (Collins 1892); Botallack mine, St Just (Greg & Lettsom 1858, Bancroft & Weller 1993); Cameron quarry, St Agnes (Hosking & Camm 1985); Carn Brea mines, near Redruth, where it was so abundant that it was sold as an ore of copper (Greg & Lettsom 1858, Rudler 1905, Robson 1945a); Darlington, Marazion; Dolcoath mine, Camborne; East Pool and Agar mine, Illogan, near Redruth (Brown 1934); Fowey Consols, Tywardreath, St Blazey (Lewis 1997); Geevor mine, near Pendeen, St Just; Hingston Down Consols mine, 2 km W of Gunnislake, Calstock; Holmbush mine, Callington; Kilifreth mine, Kenwyn (Robson 1949b); Mount Wellington mine, Twelveheads, 7 km W of Truro (Kettaneh & Badham 1978); Mulberry mine, Lanivet, St Austell (Dominy et al. 1995); Newlyn quarry, near Penzance; Penberthy Croft mine, St Hilary (Camm & Merry 1991b); Poldice mine, St Day; South Crofty mine, Camborne (Dominy et al. 1994, LeBoutillier 1996); Redmoor–Kit Hill (Newall & Newall 1989); Stenna Gwyn mine, St Stephen-in-Brannel, St Austell (Greg & Lettsom 1858, Rudler 1905, Weiss 1994); Trevoole mine, near Camborne; West Wheal Providence, St Erth; Wheal Agar, Illogan, near Redruth – one confirmed analysis (Springer 1968); Wheal Busy, St Day; Wheal Jane, Kea, near Truro (Dunham et al. 1978, Rust 1982d); Wheal Kine (Polberro mine), St Agnes; Wheal Primrose (West Kitty), St Agnes (Greg & Lettsom 1858); Wheal Scorrier (Scorrier Consols mine), St Agnes (Greg & Lettsom 1858); and Wherry mine, Penzance (Alderton 1991).

Stannite is not known from Wales or Scotland. Rare stannite–kësterite, described from a silicified sulphide-rich greisen at Gairnshiel Bridge, Glen Gairn, Grampian, Scotland, is closer in composition to kësterite than to stannite (Webb et al. 1992).

In Ireland, stannite has been reported as a trace constituent of greisen-type mineralization associated with the Main Leinster granite at Ballingen, between Tinahely and Aughrim, Co. Wicklow. Arsenopyrite, pyrrhotite and scheelite are the major sulphide species present (Steiger & Bowden 1982, 1986, McArdle & Kennan 1988, Gallagher 1989). Stannite may also have been found in the Cronebane mines, Avoca, Co. Wicklow (Greg & Lettsom 1858).

In the Channel Islands, stannite occurs in a granite-hosted polymetallic vein assemblage dominated by chalcopyrite, molybdenite, pyrite and sphalerite at L'Etacq (southeast quarry), northwest Jersey (Ixer 1980, 1992).

stannite (of Garby and others)

In addition to its use as a species name, "stannite" has been applied to mixtures of cassiterite with chalcopyrite or feldspar or quartz (TL Wheal Primrose, Cornwall, England).

stannoidite

orthorhombic $Cu_8(Fe,Zn)_3Sn_2S_{12}$

An uncommon copper–iron–zinc–tin sulphide found in the Hensbarrow stock at Gunheath china-clay pit, Carthew, St Austell, Cornwall, where it is disseminated within non-megacrystic lithium-mica granite. Stannoidite with a dark metallic lustre, together with two other tin-bearing sulphides (mawsonite and stannite), occurs as spots up to 2 cm across mixed with silicates. Stannoidite forms irregular composite grains with stannite and also occurs as exsolution lamellae in stannite. Additionally, cassiterite occurs in vugs in association with the stannite–stannoidite assemblage. A second paragenesis comprises bornite–mawsonite–stannoidite. Albite, lepidolite, topaz and zinnwaldite are common associates, and digenite, ilmenorutile and tennantite are reported rarely. The disseminated tin-sulphide mineralization is thought to be related to large-scale post-magmatic processes that involved meteoric water as well as fluid–rock interaction with the granite and with country rocks (Manning 1983, Weiss 1994). Veins containing iron oxides and phosphates crosscut the kaolinized granite at Gunheath china-clay pit and contain stannoidite, stannite and cassiterite (var. varlamoffite) often associated with regions of complex intergrowths of barbosalite, chalcosiderite–turquoise, dufrénite, fluorapatite, leucophosphite and quartz. If stannoidite borders an open cavity in these veins, it is overgrown with tufts of brochantite (Watson 1987, Elton 1998).

Silvery-grey kësterite mantled by flesh-colour stannoidite; the steel-grey bladed mineral is ferberite. St Michael's Mount, Marazion, near Penzance, Cornwall; width 2.8 mm; viewed in reflected light. (© A. G. Tindle)

Stannoidite is considered to be an abundant tin sulphide in the small granitic stock of St Michael's Mount, Marazion, Cornwall, where it is found in mineralized veins in close association with kësterite, either replacing it or exsolving from it (Moore & Howie 1984). A lack of chemical and X-ray data resulted in stannoidite being misidentified as a pink variety of stannite or mawsonite by Hosking (1968). A comprehensive review of tin sulphides in southwest England is provided by Alderton (1991).

At Gairnshiel Bridge, Glen Gairn, Grampian, Scotland, stannoidite is reported from a quartz–wolframite (hübnerite) vein and silicified wallrock assemblage cutting a zinnwaldite-bearing lithium-enriched granite (Livingstone 2002).

staurolite

monoclinic (pseudo-orthorhombic) $Fe_2^{2+}Al_9Si_4O_{23}(OH)$

A common iron aluminosilicate hydroxide found in pelitic schists, gneisses and argillaceous rocks regionally metamorphosed to amphibolite grade. Staurolite is frequently used with the Al_2O_5 polymorphs (andalusite, kyanite and sillimanite) to estimate the pressure and temperature of metamorphism. It also occurs as a detrital mineral and as such is reported from: Cretaceous and Eocene sands at Holden's Wood, Brentwood, Essex; Hayes Common and Plumstead, Kent (now Greater London); Kennel Wood, Shirley, Surrey (now Greater London); and Northwood, Middlesex (now Greater London). It is associated with a range of detrital species: andalusite, apatite, epidote, garnet, hornblende, ilmenite, kyanite, magnetite, rutile, sillimanite, spinel, tourmaline and zircon (Davies 1915).

Staurolite forms small subhedral prisms (<0.25 mm) in anthophyllite–cordierite-bearing granulites of the Old Lizard Head Series, N of the main Lizard gabbro at Pistil Ogo, near Polpeor, Lizard Peninsula, Cornwall (Tilley 1937b).

In Wales, staurolite has been found only in glacial deposits exposed around Ty'n-y-caeau, near Menai, Anglesey, Gwynedd (Greenly 1919), and at Ludchurch, Dyfed (Griffiths & Stuart 1940).

Scotland

Staurolite is a fairly common constituent of alluvial panned concentrates collected on St Kilda (Hirta), Western Isles, where it is accompanied by many of the species listed above and anatase, andalusite, clear blue corundum, chevkinite, diopside, fergusonite, goyazite, titanite and xenotime (Harding et al. 1984). Mackie (1928) found a few grains of staurolite at Inchbae, Highland Region, and at Pitsligo and Rothes Burn, both Grampian, during an examination of the heavy-mineral contents of Caledonian granites, and he concluded that his samples had been contaminated with sedimentary material.

Grampian Reports of staurolite include: rich-brown or amber staurolite in quartz veins exposed at the upper fork of Baharm Burn, near Auchlunkart House, 8 km W of Keith (Heddle 1901); small brown staurolite is sporadically found in andalusite–cordierite schist around Methlick, Ythan Valley (Read 1952); Lower Dalradian staurolite schist along the western margin of the Morven–Cabrach basic intrusion, near Strathdon (Allan 1970); Dalradian chloritoid–plagioclase–staurolite schist near Limpet Burn, N of Stonehaven (Baltatzis & Wood 1977); Dalradian pelitic schists contain ragged staurolite with sigmoidal inclusion trails at Portsoy, and the species is also noted from other locations along the lineament to Tomintoul (Beddoe-Stephens 1990); the Dalradian Queen's Hill Formation contains a horizon of unusual composition, where ragged staurolite porphyroblasts are found in semi-pelitic gneiss, E of Aultonrea, S of Ballater (Goodman et al. 1990); Dalradian garnet–staurolite–mica schist containing sodium-rich biotite from 400 m W of Craig of Finnoch (Dempster & Jackson 1996).

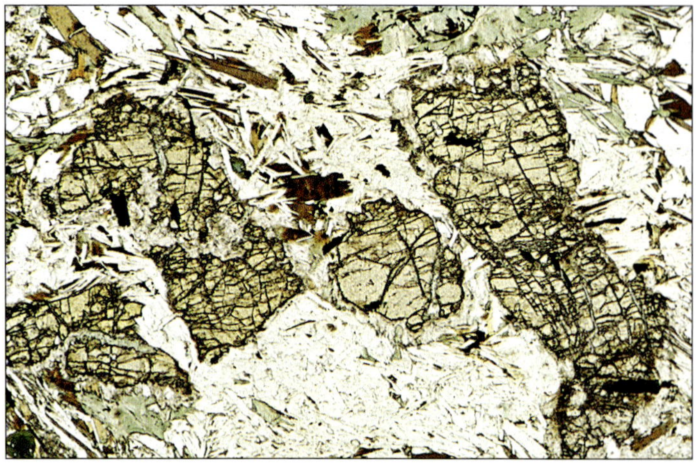

Fragmented yellow porphyroblasts of staurolite in a muscovite (colourless), biotite (brown) and chlorite (green) schist. Portsoy, Grampian, Scotland; width 4.4 mm; viewed in PPL (upper) and XPL (lower). (© K. Gardner)

Highland A part of the Lewisian complex, termed the Cnoc an t-Sidhean supracrustals, near Stoer, contain brown gneisses in which quartz-free layers (<300 mm thick) host cm-size staurolite, corundum and kyanite porphyroblasts in a white-mica matrix (Cartwright et al. 1985). A restricted facies of Moine Series regionally metamorphosed pelitic hornfels containing staurolite and kyanite is noted from W of Loch a' Chaoruinn, 5 km SSW of Carn Chuinneag (Harker 1954a). Staurolite is a minor constituent (occurring only as inclusions) in kyanite schists of the Glen Urquhart serpentinite-bearing metamorphic complex, W of Loch Ness, near Drumnadrochit (Rock et al. 1986b). Heddle (1901) describes small orange-yellow crystals of staurolite in a biotite–edenite rock from W of the Free Church of Milton in Glen Urquhart. Staurolite is an important constituent of a garnet–sillimanite gneiss of the Strathy Complex, found at Strathy lighthouse, near Melvich (Burns et al. 2004).

Strathclyde Precambrian metasediments caught up in the intrusion of lamprophyre dykes exposed on the shores of Loch Lomond, contain occasional crystals of staurolite together with chlorite, corundum, magnetite and sillimanite in essentially biotite–feldspar–garnet rocks. Staurolite is bright rust-yellow and it forms small tabular randomly orientated grains, unlike regional metamorphic staurolite found in Dalradian metasediments (Dempster & Bluck 1991).

Tayside Reports of staurolite include: coarsely crystalline kyanite–staurolite schist on the Crags of Cannes, at the head of Glen Isla, containing abundant staurolite crystals up to 3 cm long (Snelling 1957); Dalradian staurolite–biotite–garnet–muscovite–plagioclase–quartz schist at Elf Hillock, Inchdowrie, and staurolite–

kyanite–muscovite–sillimanite schist at Corrie Carlowie, both Glen Clova (Chinner 1961, 1965); garnet–kyanite–mica–staurolite schist between Weem and Grandtully, near Pitlochry (Sivaprakash 1981); a zinc-rich variety of staurolite occurs in Dalradian kyanite–staurolite gneiss within the southern contact aureole of the Glen Doll diorite (Schumacher 1985).

Shetland Rich-brown staurolite crystals, with a high lustre and up to 1 cm long, are reported from three localities in gneiss near the southern end of Loch of Cliff, Unst: above Wood Wick beach, on the hill of Snuegar and on the hill between Wood Wick and Loch of Cliff (Heddle 1878a). Later work confirms staurolite in a wide belt of garnet–kyanite gneiss belonging to the Valla Field Group outcropping over much of western Unst (Read 1934b). Staurolite has also been found at three other locations: in argillaceous schists on the summit of Saxa Vord, Unst, associated with chloritoid, chlorite, kyanite, muscovite and quartz (Read 1932); in mica schist at Bixter Voe, Mainland (Heddle 1901); and in magnetite gneiss at Vidlin Ness, Mainland (Garson & May 1976).

Ireland

In Ireland, staurolite, exhibiting sector twinning and compositional zonation, occurs in the aureole of Ardara granite, along the shore of Gweebara Bay, Co. Donegal. Crystals are often small (<0.4 mm in length), anhedral and fringed with biotite, but reach 1 mm in length 700–900 m from the contact. Their poikiloblastic texture suggests rapid growth during contact metamorphism (Akaad 1956, Smellie 1974a,b). SE of Portnoo harbour, near Naran (and within the Ardara aureole), zinc-rich staurolite breaks down to form zinc-rich hercynite, cordierite and sillimanite (Atkin 1978). Staurolite and garnet are present in regionally metamorphosed Dalradian pelitic rocks within the aureole of the Easky adamellite, N of Tobercurry, Ox Mountains, Co. Mayo and Co. Sligo. Both species are corroded because of resorption that occurred during emplacement of the adamellite (Yardley & Long 1981). Other Irish occurrences of staurolite include: brownish-black crystals in a hard talcose rock at Glenmalur, Co. Wicklow (Greg & Lettsom 1858); poikiloblastic staurolite with abundant quartz inclusions in schists adjacent to the Aclare spodumene pegmatite, near Myshall, Co. Carlow (Whitworth 1992); and as a detrital component in the Ordovician Upper Derrylea Formation to the S of Clew Bay, Co. Mayo (Mange et al. 2003).

steatite

A massive form of talc. Steatite is also confused with saponite (var. soapstone), a member of the smectite group.

steel ore

A synonym of galena and of siderite.

stellerite

ZEOLITE GROUP

orthorhombic $Ca_4(Al_8Si_{28})O_{72}\cdot 28H_2O$

A relatively common hydrated calcium aluminosilicate found as colourless, transparent or translucent, platy crystals (1–10 mm across) in granite, rhyolite, andesite and basalt at many localities in Ireland. These include: Dunseverick, near the Giant's Causeway, Co. Antrim; the Belfast escarpment, Giant's Causeway; Tardree Mountain, Co. Antrim; Mourne Mountains, Co. Down (Ryback et al. 1988); and, less precisely, on a museum specimen labelled "Giant's Causeway, Co. Antrim" (Nawaz 1988). Stellerite has also been found along the South Antrim Escarpment to the SW of Belfast in basalt at Castle Robin Primary School quarry and in a basaltic andesite dyke at Slievenagravery quarry, SW of Hannahstown (Adair 1987).

Olivine basalts of Old Red Sandstone (Devonian) age outcropping beneath the lighthouse at Todhead Point, 9 km S of Stonehaven, Grampian, Scotland, contain amygdales up to 1 cm across, consisting of brick-red prismatic stellerite and mordenite. Laumontite is also noted (Dyer et al. 1993).

On the Isle of Skye, Highland Region, stellerite occurs as white wheatsheaf crystals up to 3 cm long associated with calcite and laumontite in amygdales within Tertiary basalts at the classic zeolite locality – Moonen Bay, Duirinish Peninsula (Green & Todd 2001). About 17 km away, in a roadcut at Edinbane, similar amygdales in basalt contain calcite, erionite, heulandite, stellerite and stilbite (Savage 1995). Superb cabinet specimens of stellerite and stilbite were once collected from this locality (Green & Todd 2001).

stellite

TL Kilsyth, near the Forth of Clyde canal, Strathclyde, Scotland

Possibly pectolite or wollastonite.

stephanite

orthorhombic Ag_5SbS_4

Lustrous black stephanite on crystallized ankerite, siderite and quartz; Wheal Newton, Harrowbarrow, Cornwall; crystal is 37 mm along the diagonal and 18 mm thick. (© The Natural History Museum, London)

A rare silver sulphantimonite found on a specimen of indistinctly crystallized acanthite (argentite) and native silver from Wheal Ludcott, St Ive, near Liskeard, Cornwall, where it forms short black prisms (Davies 1866). However, a far more attractive crystal of stephanite, perched on a matrix of siderite and quartz, came from Wheal Newton, Harrowbarrow, SW of Gunnislake. (Lewis 1882, Rudler 1905). A third specimen of stephanite, intergrown with pyrargyrite, came from Wheal Boys, Trewetha, St Endellion, Cornwall (Prior 1890). Despite many references to stephanite in later works, it is these three specimens to which the authors appear to be referring (Collins 1892, Robson 1949b, Russell 1949b, Hosking 1951, Rust 1982a, 1983a). The silver mineralization at Wheal Ludcott was dominated by acanthite (argentite) and hosted by a galena-bearing quartz–siderite lode up to 1 m wide. Fluorite, proustite, pyrargyrite, pyrite and sphalerite are associates (Dines 1956, Rust 1983a).

Stephanite is also described by Ixer & Stanley (1987) from samples in the Kingsbury collection at the NHM. Originally described as having been collected from Tynebottom mine, NW of Garrigill, near Alston, Cumbria, this provenance is now considered dubious (Ryback et al. 1998, 2001).

At Tynagh mine, near Killimor, Co. Galway, Ireland, epigenetic Cu–Pb–Zn–(Ag) mineralization contains a minor sulphosalt assemblage that includes bournonite, enargite, meneghinite, polybasite, pyrargyrite and stephanite. They occur with galena and tennantite–tetrahedrite (Schultz 1966a).

Stephanite and a wide range of other sulphide species are described from Sark's Hope mine, Little Sark (Mourant & Warren 1934), but the species is not reported by later workers (e.g. Ixer & Stanley 1983). Further confirmation is desirable.

sternbergite?

orthorhombic $AgFe_2S_3$

A rare silver–iron sulphide described by Ixer & Stanley (1987) and Ryback & Tandy (1992) from samples in the Kingsbury collection at the NHM. Originally described as having been collected from Tynebottom mine, NW of Garrigill, near Alston, Cumbria, this provenance is now considered dubious (Ryback et al. 1998, 2001). Unless further discoveries are made, this species cannot be considered native to Britain or Ireland.

stevensite

SMECTITE GROUP

monoclinic $(Ca,Na)_xMg_{3-y}Si_4O_{10}(OH)_2$

A questionable calcium–sodium–magnesium silicate hydroxide found as an alteration product of pectolite and forming radiating groups up to about 2 cm in diameter associated with calcite, prehnite and talc (var. kerolite) in cavities and veins in gabbro at Dean quarry, near St Keverne on the Lizard Peninsula, Cornwall (Elton et al. 1997). Golley & Williams (1995) note stevensite from Trevassack, also on the Lizard Peninsula. Stevensite forming amygdale-like structures after pectolite is reported from the Whin Sill at Barrasford quarry, near Gunnerton, Northumberland. Stevensite occurs in two forms: the first is fibrous, has an earthy lustre, variable colours (cream, greenish, brownish or black) and completely pseudomorphs pectolite; the second is amorphous, has a dull lustre and is usually bluish white. The amygdale-like structures can be dominated by either stevensite or pectolite, and may contain calcite and datolite, all capped with chlorite (Randall 1959).

Stevensite is noted from Gimlet quarry, Pwllheli, Gwynedd, Wales (T. F. Cotterell pers. comm. 2003).

Two specimens of pectolite (var. walkerite) in the NMS and originally collected from Corstorphine Hill, Edinburgh, were found to be coated with a brown gel-like substance that, after extensive analysis, proved to be stevensite. Subsequent examination of other pectolite samples in the museum confirmed stevensite on samples from Lendalfoot and the Isle of Mull, both Strathclyde (Davidson 1989).

stibiconise

A synonym of stibiconite.

STIBICONITE GROUP

Cubic oxides with the general formula $A_{1-2}B_2O_6(O,OH,F)$, where A = Ag, Bi, Ca, Cu, Fe^{2+}, Fe^{3+}, K, Mn^{2+}, Na, Pb, Sb^{3+}; and B = Fe^{3+}, Sb^{5+}, Ti. Bindheimite and stibiconite are members of the group found in Britain or Ireland. Another member of the group, roméite, requires confirmation.

stibiconite

STIBICONITE GROUP

cubic $Sb^{3+}Sb^{5+}_2O_6(OH)$

A rare antimony oxide hydroxide that occurs as an oxidation product of other antimony-bearing minerals, especially stibnite, which it may replace entirely. Russell (1949b) reports stibiconite from five localities in Cornwall: Hatt mine trial, Botusfleming, near Saltash – several small fragments of stibnite associated with stibiconite; Cutlinwith farm, Landrake with St Erney – bladed masses of stibnite altered to stibiconite with traces of kermesite and valentinite; Dizzard (Duchy & Stuckey United) mine, Poundstock – a single specimen of bladed stibnite associated with minor stibiconite, dickite and quartz; Trenute mine, near Lezant – stibnite with traces of stibiconite associated with appreciable pyrite, and ankerite, arsenopyrite, dolomite, quartz and sphalerite; Wheal Boys, Trewetha, St Endellion – 11 specimens in the Woodward collection at the University of Cambridge collected from mines in the parish of St Endellion include bindheimite, jamesonite, stibiconite and stibnite; Wheal Prosper (Lanhadron), St Ewe, near Mevagissey – fine bladed needles of stibnite with much canary-yellow and white stibiconite together with pyrite and sphalerite. Stibiconite is also noted from five other Cornish locations: Bounds Cliff, E of Port Isaac (Clayton et al. 1990); Gunver Head, near Padstow (Schlomann & Weiss 1998); Trebullett mine, near Lezant; Wheal Leigh, Pillaton (Golley & Williams 1995); and Trevinnick, near St Endellion (Greg & Lettsom 1858). Stibiconite forms yellow encrustations that occupy small cavities in the Meldon aplite, near Okehampton, Devon (von Knorring & Condliffe 1984).

An antimony-rich quartz vein in the aureole of the Skiddaw granite, at Coombe Height, near the head of Wet Swine Gill, Caldbeck Fells, Cumbria, contains yellow and grey earthy crusts of bindheimite and stibiconite. The vein contains an early As–Fe–Sb primary assemblage (mainly stibnite and berthierite) and a later antimony–lead assemblage (zinkenite with lesser fülöppite and semseyite). Other minor phases in the vein include native antimony, claudetite, jamesonite, senarmontite, sphalerite and native sulphur (Fortey et al. 1984, Leppington & Green 1998, Neall & Green 2001b). A quartz vein at Robin Hood mine, near Bassenthwaite, contains stibiconite with dickite, sphalerite, stibnite and valentinite (Russell 1949b, Young 1987a, Green & Briscoe 2002). The occurrence of stibiconite from Wanthwaite mine, St John's in the Vale, near Threlkeld (Young 1987a), requires further confirmation as it is based on specimens in the Kingsbury collection at the NHM. Serious doubts about the provenance of other Kingsbury specimens said to be from this locality have been expressed by Ryback et al. (1998, 2001).

At Bwlch mine, near Deganwy, Gwynedd, Wales, a quartz and chalcedonic chert-filled lode contains stibnite and semseyite, together with small quantities of anglesite, ankerite, bindheimite, galena, pyrite, sphalerite and thin yellow resinous crusts of stibiconite (Russell 1944, 1949b). This occurrence requires confirmation.

In Scotland, at Louisa mine, Glendinning, Glenshanna Burn, 13 km NNW of Langholm, Dumfries & Galloway, a mineralized quartz vein dominated by stibnite intimately mixed with galena and semseyite also contains stibiconite and valentinite (Russell 1949b). At Knipe (Fountainhead) mine antimony mine, Hare Hill, SE of New Cumnock, beautiful long narrow blades of stibnite associated with kermesite and soft earthy yellow-to-white stibiconite were recovered from the dumps (Russell 1949b, Macpherson & Livingstone 1982). The antimony mineralization is derived from a 30–45 cm-wide quartz–stibnite vein that outcrops near the centre of the small Knipe–Hare Hill granite mass (Williams 1965, 1973).

Stibiconite is reported from only one location in Ireland – at Lisnageenly and Mucklin, Silvermines, Co. Tipperary – where it occurs as a rusty yellow alteration product of stibnite (Russell 1949b).

stibioluzonite

An antimony-rich variety of luzonite; also a synonym of famatinite.

stibiopalladinite

hexagonal Pd_5Sb_2

A rare palladium antimonide found in chromite-rich dunite at Cliff, Baltasound–Haroldswick area, Unst, Shetland, Scotland. Stibiopalladinite (generally <10 μm across) and sperrylite are the two main PGE species found here in association with chalcopyrite, cobaltite, heazlewoodite, millerite, nickeline, orcelite, pentlandite and pyrrhotite (Gunn et al. 1985, Gunn & Styles 2002). From the same locality, Prichard & Tarkian (1988) describe palladium antimonides of irregular shape that, on the basis of available data, were considered to be stibiopalladinite, mertieite-I or mertieite-II. However, Prichard et al. (1986, 1994) would appear to confirm them to be stibiopalladinite. In the Loch Borralan alkaline pyroxenite–syenite complex, Assynt, Highland Region, possible stibiopalladinite occurs rarely with more common isomertieite, sperrylite and bismuth–palladium tellurides (often with chalcopyrite, galena and hessite) in late brittle fractures and along grain boundaries (Styles et al. 2004). Farther south, in the Thornhill Basin, 20 km N of Dumfries, Dumfries & Galloway, a single alluvial gold grain is reported to contain a stibiopalladinite rim (Leake & Cameron 1996, Leake et al. 1997c, Gunn & Styles 2002).

stiblite

A synonym of stibiconite.

stibnite

orthorhombic Sb_2S_3

A common antimony sulphide that forms a series with bismuthinite and is dimorphous with metastibnite. Stibnite is found in hydrothermal veins formed over a wide range of temperatures, and is the most important ore of antimony.

The largest group of antimony deposits in Britain is in north Cornwall, particularly in the Port Isaac area, where the mineralization is stratabound and confined to Middle Devonian to Lower Carboniferous volcano-sedimentary basinal successions. Minor shear structures and quartz veining developed during late Variscan deformation, at which time many antimony–lead–sulphosalt deposits formed. Stibnite is one of the less common antimony species found in the Bounds Cliff vein system, 2 km E of Port Isaac, St Endellion, where it is found along intergrain fractures and vein intersections with galena and other antimony-bearing species – boulangerite, bournonite, jamesonite, plagionite and tetrahedrite (Clayton et al. 1990). At Poltreworgy mine, near Port Isaac, jamesonite was found at the top of the lode, but at depth the ore changed to coarse-blade and fine-grain stibnite (Dewey 1920, Russell 1949b, Hosking 1951, Dines 1956). In the same area, stibnite is also reported from: Tresungers mine, S of Trewetha (Russell 1949b); Trevinnick, SE of Port Isaac – in a jamesonite–stibnite lode, where the stibnite is said to be gold rich (Dewey 1920, 1925); Wheal Boys (Trewetha mine), Trewetha, St Endellion – beautifully crystallized stibnite (Greg & Lettsom 1858, Dewey 1920, Russell 1949b, Hosking 1951, Dines 1956, Penhallurick 1997); Treore mine, Port Gaverne, St Teath (Dewey 1920, Hosking 1951, Dines 1956, Camm 1995).

Stibnite occurrences are reported from around Landrake in east Cornwall: Cutlinwith farm, W of Landrake, St Erney (Russell 1949b); Hatt mine trial, Botusfleming, NW of Saltash (Russell 1949b); Tredinnick mine, W of Landrake – stibnite was the main ore species and quite good specimens could be found on the mine dumps (Russell 1949b); Wheal Leigh, Pillaton, 4 km NW of Landrake – many fine specimens were found on the mine dumps (Russell 1949b).

Other Cornish occurrences of stibnite include: Dizzard (Duchy & Stuckey United) mine, Poundstock – a single specimen was found associated with stibiconite (Russell 1949b); Fowey Consols mine, Tywardreath, St Blazey – stibnite (antimonite) found in a cross-course vein (Dewey 1920, Lewis 1997); Pengenna mine, SW of St Teath (Hosking 1951); Trebullett Hill mine, near Lezant – stibnite forming bladed masses intergrown with sphalerite, and needles associated with dolomite or ankerite (Dewey 1920, Russell 1949b); Trenute mine, near Lezant, 6 km S of Launceston (Dewey 1920, Russell 1949b, Hosking 1951, Dines 1956); Wheal Herland, Gwinear (Ball et al. 1982); Wheal Prosper (Lanhadron mine), St Ewe, near Mevagissey – fine bladed needles of stibnite up to 7 mm long with stibiconite (Russell 1949b).

Stibnite (antimonite) was said to be found sparingly as fibrous masses and acicular crystals associated with silver-rich galena in the Combe Martin mines, north Devon (Hall 1868b, Dunham et al. 1978, Rust 1982a).

Unconfirmed reports of stibnite from Gregory mine, Ashover, and Rutland mine, Matlock Bath, both in Derbyshire, possibly refer to a steel-ore variety of galena (Ford & Sarjeant 1964, Ford et al. 1993).

One of the most interesting discoveries in recent years has been that of an antimony-rich quartz vein within the aureole of the Skiddaw granite, near the head of Wet Swine Gill, Coombe Height, Caldbeck Fells, Cumbria. Here, stibnite occurs with berthierite and a suite of supergene species that includes bindheimite, claudetite, senarmontite, stibiconite, native sulphur and possibly kermesite (Fortey et al. 1984, Leppington & Green 1998, Neall & Green 2001b, Green & Briscoe 2002). Elsewhere in Cumbria, stibnite is reported from: Robin Hood mine, Bassenthwaite (Greg & Lettsom 1858, Postlethwaite 1913, Russell 1949b, Davidson & Thomson 1951, Young 1987a, Green & Briscoe 2002); Troutbeck railway station (Ward 1876, Dewey 1920, Russell 1949b); St Sunday Crag mines, Helvellyn area (Davidson & Thomson 1951, Firman 1978); and Wanthwaite mine, St John's in the Vale, near Threlkeld (Postlethwaite 1913). Reports of stibnite from Carrock mine, Grainsgill, Caldbeck Fells (Kingsbury & Hartley 1956b, Hartley 1984), Embleton quarry, near Cockermouth, and Wanthwaite mine, St John's in the Vale, near Threlkeld (Young 1987a), require further confirmation, as they appear to be based on specimens in the Kingsbury collection at the NHM. The work of Ryback et al. (1998, 2001) indicates that the provenance of many Kingsbury samples is in doubt. Certainly, stibnite was not reported by Ball et al. (1985a) in their thorough investigation of the ore mineralization at Carrock mine. However, there are vague earlier descriptions of stibnite from Carrock Fell (Greg & Lettsom 1858, Postlethwaite 1913), so the Kingsbury occurrence may be valid.

Plumose aggregate of stibnite prisms to 7 cm in quartz, with creamy stibiconite. Hare Hill, New Cumnock, Dumfries & Galloway, Scotland. (© The Trustees of the National Museums of Scotland)

Prior to 1864, antimony was mined from irregular patches and quartz veinlets in an Ordovician nodular ignimbrite at Bwlch farm, near Deganwy, Gwynedd, Wales. The ore consisted of coarsely bladed masses, up to 30 mm across, of stibnite associated with minor anglesite, ankerite, bindheimite, galena, jamesonite, marcasite, plagionite, pyrite, rosiaite, semseyite, sphalerite, stibiconite, zinkenite and possibly robinsonite (Russell 1944, 1949b, Bevins et al. 1988, Ryback & Francis 2001). Two other Welsh stibnite occurrences are documented by Readwin (1888) from the Dolgellau goldbelt, Gwynedd, and by Bevins (1994) on a museum specimen collected from the shore between Rhôs-mynach mine and Dulas, Anglesey, Gwynedd (no other specimens have been found from this location).

In Scotland, stibnite is the primary antimony mineral found in quartz–(calcite) veins within brecciated Silurian slates at Louisa (Glendinning) mine, Glendinning, Glenshanna Burn, 13 km NNW of Langholm, Dumfries & Galloway. Ore microscopy has revealed a three-stage hydrothermal paragenesis, comprising:

- early crystallization of arsenopyrite and pyrite
- overgrowth of galena, sphalerite and stibnite
- late (minor) formation of barite, chalcopyrite, galena and sphalerite.

Native gold is associated with the arsenopyrite. Other species recorded from the mine include: ankerite, bournonite, cinnabar, jamesonite, semseyite, stibiconite, tennantite, tetrahedrite, valentinite and rarely cervantite (Greg & Lettsom 1858, Heddle 1901, Herbert Smith 1919, Dewey 1920, Russell 1949b, Williams 1973, Gallagher et al. 1983, Ball et al. 1985b, Duller et al. 1997, Livingstone 2002). Excellent stibnite specimens have also been recovered from Knipe (Fountainhead) mine, Hare Hill, 5 km SE of New Cumnock, Dumfries & Galloway. Plumose aggregates of stibnite prisms up to 7 cm long, altering to creamy stibiconite, occur in quartz veinstone in a zoned As–Cu–Pb–Sb–Zn assemblage hosted by a late-Caledonian granodiorite intrusion. Native gold is associated with the sulphides (arsenopyrite in particular) at this location (Greg & Lettsom 1858, Heddle 1901, Dewey 1920, Russell 1949b, Williams 1973, Samson & Banks 1988, Naden & Caulfield 1989, Boast et al. 1990, Livingstone 2002). A very unusual occurrence of bladed and fine-grain stibnite associated with fluorite is noted from Maisley Limestone quarry, Keith, Grampian (Greg & Lettsom 1858, Heddle 1901, Russell 1949b, Livingstone 2002). Smith et al. (1978a) report microscopic stibnite in borehole material from an area of stratiform chalcopyrite-rich mineralization around the Abhainn Srathain mines, S of Meall Mòr, South Knapdale, Strathclyde.

Undoubtedly, the Clontibret mines (at Tullybuck and Lisglassan), near Milltown, Co. Monaghan, are the most important stibnite localities in Ireland, although much of this interest is because traces of gold have been found in the antimony ore. Mineralization is fault controlled and consists of dominant gold-rich arsenopyrite, pyrite and stibnite within quartz–carbonate veins in greywacke host rocks. Excellent crystals of stibnite up to 20 cm long are reported from Lisglassan (Cole 1922, Russell 1949b, McCannell 1957, O'Brien 1959, Halliday & Mitchell 1983, Morris 1984, Morris et al. 1986, McArdle 1989, Ryback & Moreton 1993, Steed & Morris 1997). About 800 m N of Clontibret, at Bryanlitter, a similar gold-rich lode contains stibnite with disseminated, but locally abundant, arsenopyrite and pyrite. Native antimony forms inclusions in the stibnite (Steed & Morris 1997). Elsewhere in Ireland, stibnite is reported from: Carrahin mine, Co. Clare (Halliday & Mitchell 1983); Castleshane mine, near Monaghan, Co. Monaghan (Russell 1949b); Hope mine, Cornalough–Carrickagarvan, Co. Monaghan (Morris 1984); and Connary mine, near Avoca, Co. Wicklow (Cole 1922, Russell 1949b). Further confirmation is desirable for all of these occurrences.

Coarsely crystalline stibnite, with minute crystals of pyrite in grey quartz, is noted from Niarbyl mine, Dalby, Isle of Man (Greg & Lettsom 1858, Russell 1949b). At Traie Vrish, SE of the Niarbyl, a trial adit into the cliff was made in 1893–1894 to look for an extension of the antimony deposit, but this was unsuccessful. Boulders of stibnite can still be found on the shore at Traie Vrish (Lamplugh 1903, Starkey 1998a).

In the Channel Isles, minor amounts of stibnite occur as discrete 20 μm-long lanceolate crystals, often forming sheaf-like aggregates in galena–sphalerite-bearing carbonate veins cutting Precambrian Brioverian sediments close to the granite quarries at Le Pulec, northwest coast of Jersey. Stibnite overgrows arsenopyrite, bournonite, chalcopyrite and galena (Ixer & Stanley 1980, Ixer 1992). Stibnite is briefly reported from L'Etacq quarry, Jersey (Lacroix 1913, Rothwell & Mason 1992), and Sark's Hope Pb–(Ag) mine, Little Sark (Mourant & Warren 1934).

stichtite?

trigonal — HYDROTALCITE GROUP — $Mg_6Cr_2(CO_3)(OH)_{16}\cdot 4H_2O$

A rare hydrated magnesium–chromium carbonate hydroxide, dimorphous with barbertonite and reported from an antigorite–serpentinite mass close to talc schists near Hoo Field, Cunningsburgh, Mainland, Shetland, Scotland. Fibrous stichtite is reported to form conspicuous rose-pink patches in aggregates to 0.5 cm associated with antigorite, chromite, dolomite and iron oxide. Stichtite replaces original chromite grains (Read & Dixon 1932). Macpherson & Livingstone (1982) suggest that the mineral may be barbertonite instead. This requires further confirmation.

stilbite (series)

monoclinic — ZEOLITE GROUP — $(Ca_{0.5},K,Na)_9(Al_9Si_{27})O_{72}\cdot 28H_2O$

A common hydrated calcium–potassium–sodium aluminosilicate whose name has been given series status in the zeolite classification of Coombs et al. (1997). There are now two stilbite species – stilbite-Ca and stilbite-Na – but these names do not appear in older publications and, until chemical data become available, their species status cannot be established. Stilbite is a low-temperature hydrothermal mineral found in amygdales and cavities in basalts, andesites and various metamorphic rocks.

Small sheaf-like aggregates and distinct colourless glassy crystals of stilbite up to 6 mm long are associated with apatite, muscovite (gilbertite), orthoclase, quartz and tourmaline in cavities within coarse pegmatitic granite at Colcerrow quarry, Luxulyan, Cornwall (Russell 1910b). Stilbite with a similar paragenesis is noted from the Carnmenellis granite at Trolvis quarry, near Longdowns (Hosking 1954a,b), and from the nearby Chywoon quarry, where, within a granitic pegmatite, stilbite forms spherulitic masses to 8 mm and single well terminated crystals up to 3 mm long (Moulding & Hooper 2005). At Dean quarry, 2 km SE of St Keverne, Lizard Peninsula, stilbite forms minute brilliant colourless transparent crystals, usually < 2 mm in length. It grows on calcite, natrolite or prehnite in white veins cutting gabbros of variable texture (Seager 1971, Wolters & Wolters 1986). The mineralization is post-metamorphic and it extends to Porthkerris Point on the Lizard, where small scattered bladed crystals of stilbite have been found in hydrothermal veins cutting amphibolites and granulites (Seager 1978b). Robson (1949b) and Golley & Williams (1995) note stilbite at Parc Bean Cove, Mullion, Lizard Peninsula. Stilbite was described from prehnite veins cutting greenstone between Wheal Cock and the Crowns Rock, Botallack, St Just (Carne 1822, Greg & Lettsom 1858), but Russell (1910b) was unable to confirm the occurrence.

In the northern Pennines, stilbite forms colourless lustrous tabular prisms up to 1.5 mm long, lining open fissures in chloritized doleritic pegmatite at High Force quarry, Upper Teesdale, Co. Durham. It is accompanied by analcime and chabazite (Young et al. 1991a).

Stilbite occurs most commonly in Scotland, particularly in the Tertiary basalts on the Isle of Skye, Highland Region, where vitreous, white, creamy or pale-pink crystals have been found. Radiating bundles of plates, single crystals and sheaves (up to 1.25 cm across) are reported, the latter commonly in veins and large cavities over 0.6 m in diameter, especially in the lavas of the Beinn Edra and Ramascaig groups. Stilbite is associated with chabazite, mesolite and thomsonite (King 1976). Localities include: Allt Ribhein, Fiskavaig Bay (Mackenzie 1957); An Cròcan, Loch Brittle (King 1976); Beinn na Gréine, near Portree (King 1976); Biod a' Mhurain, southern Duirinish (Wolfe 1997); Camas na h-Uamda, Duirinish (Wolfe 2000); Edinbane (Savage 1995, Green & Todd 2001); Flodigarry, Quiraing (Passaglia et al. 1978, Green & Todd 2001); northern lip of Loch Eynort (Green & Todd 2001); Moonen Bay, N of Dunvegan (Green & Todd 1996); Oisgill Bay, N of Moonen Bay (Green & Todd 2001); Sgùrr nam Boc, between Loch Brittle and Loch Eynort (Livingstone 1990b, Pearce & Pearce 1991, Green & Wood 1996, Green 1998a,b, Green & Todd 2001); Sgùrr nam Fiadh (Wolfe 1997); the Storr, N of Portree (Anderson 1851, Greg & Lettsom 1858, Gottardi & Galli 1985); Talisker Point (Greg & Lettsom 1858, Green & Wood 1996). As well as describing many locations on the Isle of Skye, Heddle (1901) also reports stilbite from the isles of Canna and Sanday, and from W of Uamh Fhraing on the Isle of Eigg. Elsewhere in Highland Region, stilbite occurs with harmotome and montmorillonite in miarolitic cavities within syenite on Sgòrr Chaonasaid, near the northeast contact of the Ben Loyal igneous complex, 7 km S of Tongue (von Knorring & Dearnley 1960b, Gallagher et al. 1971).

Returning to Tertiary basalt occurrences, but in Strathclyde, stilbite is described from Maol nan Damh on the southwestern shoulder of Ben More, Isle of Mull (Currie 1898, McLintock 1910b), and is said to have sporadic occurrence in the mesolite zone of the Mull lavas (Walker 1971). At 'S Airde Beinn, 5 km SW of Tobermory, Isle of Mull, stilbite–heulandite-bearing amygdaloidal basalt has been subjected to thermal metamorphism during the intrusion of a volcanic plug, and larnite, melilite, rankinite and wollastonite have crystallized (Cann 1965a). Temporary exposures of basalt uncovered during road widening to the S of Tobermory in 1998 revealed vesicles containing well formed pinkish stilbite associated with agate, analcime, calcite and heulandite (Green & Todd 2001). Heddle (1901) reports stilbite from the Treshnish Isles and the Isles of Arran, Kerrera and Staffa.

Hematite-stained stilbite, 12 mm long, from Touch Hills, between Glasgow and Stirling, Central Region, Scotland. (© D. I. Green)

The Carboniferous Clyde Plateau basalts around Glasgow contain many examples of stilbite: red pointed crystals embedded in calcite at Bowling quarry, 2 km WNW of Old Kilpatrick (Tschernich 1992); exceptional specimens of bright-red pointed crystals associated with heulandite and quartz at Cochno, NW of Bearsden, and Lang Craigs, Kilpatrick Hills, near Dumbarton (Tschernich 1992); colourless to reddish-brown crystals up to 0.5 mm across at Hartfield Moss, 9 km WSW of Barrhead (Meikle 1990); golden-brown to colourless crystals up to 2 cm long associated with calcite, chabazite and mesolite in large pockets at Kilmalcolm, 10 km SE of Greenock (Tschernich 1992); red stilbite from Loch Cray and Loch Humphrey (Heddle 1880a); colourless transparent well formed doubly terminated crystals up to 1.5 mm across at Reilly quarry, 3 km NNE of Houston (Meikle 2003). Excellent red crystals associated with analcime, barite, calcite, heulandite, natrolite and quartz have also been found along the shore of Earlsburn (Earlston) reservoir, Touch Hills, Central Region (Green & Todd 2001).

Isolated occurrences of stilbite are reported from other regions of Scotland, including: in a basaltic dyke on the shore of Caolas an Scarp, Hushinish, North Harris, Western Isles, forming surface coatings with chabazite (Dyer et al. 1996); as large scarlet crystals near Kinneff church, off the Bervie to Stonehaven road, Grampian (Heddle 1901); large red sheaves in olivine basalt at Todhead Point, 9 km S of Stonehaven, Grampian (Heddle 1901), have now been shown to be stellerite (Dyer et al. 1993); and at Glen Farg, 16 km S of Perth, Tayside Region (Greg & Lettsom 1858).

In Ireland, stilbite is frequently noted from the Tertiary basalts of Co. Antrim. It forms colourless-to-white pointed sheaf-like aggregates and single crystals up to 3 cm long associated with analcime, calcite, chabazite, heulandite and natrolite in red weathered tops and bottoms of various lava flows and was found at 9% of the zeolite localities studied by Walker (1960a). Stilbite is abundant in quarries along an escarpment SW of Belfast, from Collin Glen to Collin Mountain, and at Aughrim Hill and White Mountain (Walker 1960a); it is particularly abundant at Groganstown quarry, SW of Collin Mountain (Adair 1987). A vesicular basalt at Kane's quarry, Boghill, Glengormley, near Belfast, contains platy colourless golden or olive-green sheaf-like bundles of stilbite up to 15 mm long. Nearby cavities contain chabazite, cowlesite, erionite (or offrétite), fluorapophyllite, gyrolite, heulandite, okenite and thomsonite (Tschernich 1992). Stilbite occurs sparsely in lavas on the edge of, or near, the

Bow-ties and a radiating fan of stilbite. Sgùrr nam Boc, Isle of Skye, Highland Region, Scotland; 45 cm tall. (© The Trustees of the National Museums of Scotland)

Garron Plateau (Walker 1951, 1960a). Stilbite is one of the zeolites described from Island Magee, near Larne (Tomkeieff 1934), but this is disputed by Tschernich (1992). Along the Giant's Causeway, on the north coast of Co. Antrim, stilbite forms colourless to pale-green flat-top rectangular crystals and radiating hemispheres, both commonly associated with celadonite, chabazite, heulandite, phillipsite, quartz and rarely paulingite (Tschernich 1992). A pyroxenite found near the margins of the Scawt Hill dolerite intrusion, near Larne, contains vesicles infilled with major amounts of thomsonite and minor amounts of analcime, calcite, natrolite and stilbite (Tilley & Harwood 1931). Greg & Lettsom (1858) provide the following additional locations for stilbite: four on the north coast of Co. Antrim – Ballintoy, Benmore Head, Conical Hill at Ballycastle, and Portrush; and Bruce's Castle, Rathlin Island.

Tertiary lavas extend into Co. Londonderry, and at Knockcloghrim, NW of Castledawson, stilbite crystals up to 5 mm long are found with calcite, chabazite and thomsonite along the western edge of the basalt outcrop (Tschernich 1992).

In Co. Down, slender colourless prisms of stilbite up to 10 mm long, and black platy aggregates up to 2 cm across, are found with natrolite on feldspar, mica and smoky quartz in granitic pegmatites associated with the Mourne Mountains granites at Lindsay's Leap quarry, Slieve Donard Mountain, near Newcastle (Nockolds & Richey 1939, Tschernich 1992, Green & Moreton 2001).

stilbite-Ca

ZEOLITE GROUP

monoclinic $NaCa_4(Al_9Si_{27})O_{72}\cdot 28H_2O$

An hydrated sodium–calcium aluminosilicate found in amygdaloidal basalts on the Isle of Skye, Highland Region, Scotland. EMP data indicate that stilbite-Ca occurs on Beinn na Gréine to the SW of Portree and An Cròcan along the north shore of Loch Brittle (King 1976). Other occurrences are confirmed on the Isle of Skye from Loch Ainort (D. I. Green sample and XRD, Tindle unpub. EMP data) and Sgùrr nam Boc (NMS sample and XRD, Tindle unpub. EMP data).

At Lindsay's Leap quarry in the Mourne Mountains, Co. Down, Ireland, drusy cavities in granite contain colourless-to-brown mm-size flat-topped transparent prismatic-to-tabular crystals of stilbite-Ca forming radiating groups that are clearly later than most of the other cavity species (Green et al. 2005a).

stilpnomelane

monoclinic or triclinic $(K,Ca,Na)(Fe,Mg,Al)_8(Si,Al)_{12}(O,OH)_{36}\cdot nH_2O$

A relatively common, but locally abundant, hydrated potassium–iron–magnesium aluminosilicate hydroxide typically found in metamorphosed iron-rich igneous rocks. Stilpnomelane is described from a single specimen collected at Penberthy Croft mine, St Hilary, Cornwall, where it forms bright-green translucent prismatic subhedral crystals up to 0.2 mm long (Betterton 1997). Betterton et al. (2000) suggest that this occurrence could represent sample contamination or an analytical error, but note that stilpnomelane does occur commonly in iron-rich greenschist-facies metamorphic rocks at this location. A second Cornish location is Mullion Island on the Lizard Peninsula (Golley & Williams 1995). Characteristic specimens of stilpnomelane forming veins with quartz and green chlorite were collected on the dumps of the Paddy End section of Coniston mine, Eskdale, Cumbria (Russell 1925). In the same general area, the Bonser vein (3 km NW of Coniston), once the most productive source of copper in the Lake District, contains a mineral assemblage in which stilpnomelane occurs with arsenopyrite, native bismuth, bismuthinite, calcite, chalcopyrite, chlorite, cosalite, dolomite, joséite, laitakarite, magnetite, marcasite, pyrite, pyrrhotite, quartz and sphalerite. Stilpnomelane forms fine acicular crystals up to 200 µm long and matted aggregates up to 1 mm long. It overgrows chlorite and also occurs in interstitial areas in magnetite (Stanley & Vaughan 1982b).

Across the whole of Wales, Ordovician metabasic rocks subjected to prehnite–pumpellyite or greenschist-facies metamorphism contain sporadic stilpnomelane. However, it is most abundant in more fractionated rocks, such as meta-andesites and meta-dacites, and in pegmatitic patches within metabasic intrusions, such as in the St David's Head and Treseisyllt intrusions in Dyfed and from the summit of Cadair Idris, Snowdonia, Gwynedd (Roach 1969, Bevins & Roach 1982, Bevins & Rowbotham 1983). At St David's Head, stilpnomelane forms abundant radiating aggregates of brown needle-like crystals in granophyric gabbro (Bevins 1994). Stilpnomelane is present only in the ferrodolerite horizon (the most fractionated part) of the Tal y Fan intrusion, a subconcordant sheet of altered olivine dolerite 7 km SW of Conwy, Gwynedd (Merriman et al. 1986, Bevins & Merriman 1988). In the Cregenen granophyre, near Arthog, Gwynedd, stilpnomelane forms sharply terminated laths in bow-tie aggregates or radiating sheafs (Bloxam & Price 1961).

At Pen-yr-allt iron mine, near Llanfrothen, Penrhyndeudraeth, Gwynedd, Wales, dense-brown brittle stilpnomelane occurs in Ordovician pisolitic ironstones adjacent to an igneous mass. Because of its colour, it was originally mistaken for biotite (Hallimond 1924, Pulfrey 1933). Similar ironstone occurrences of stilpnomelane are reported from three other localities in Gwynedd: Cross Foxes, Dolgellau (Bevins 1994); Betws Garmon, mid-way between Snowdon and Caernarfon, Snowdonia (Pulfrey 1933); and from 3 km SW of Drws-y-nant station, at Tyllau Mwn mine, near Dolgellau (Matthews & Scoon 1964).

In Scotland, Dalradian metabasic rocks (epidiorites) subjected to low-grade metamorphism at Cnoc-na-Faire, near Carsaig, N of Tayvallich, Strathclyde, contain muddy-brown biotite, and stilpnomelane that in thin section is pleochroic from gold to golden brown (Wiseman 1934). Stilpnomelane has since been shown to occur extensively in greenschist-facies metabasic rocks in Tayvallich, Knapdale, and at Port Ellen on the Isle of Islay. At Crinan, north Tayvallich, abundant stilpnomelane occurs in plates, fibres and sheaves intergrown with or enclosed in amphibole porphyroblasts and patches of chlorite. Both ferro- and ferri-stilpnomelane varieties are recognized (Graham 1976). Stilpnomelane is also present as a rare brown microphenocryst phase in the Tertiary riebeckite trachyte intrusion exposed at the southern tip and the summit (Mullach Mòr) of Holy Island, Lamlash Bay, Isle of Arran, Strathclyde (Tindle unpub. EMP data). Around Loch Katrine in the Aberfoyle district, Central Region, stilpnomelane has a sporadic distribution in regionally metamorphosed Dalradian greywackes (Mather 1970).

Stilpnomelane and minnesotaite are considered to be late diagenetic species in the Lower Carboniferous Tynagh Iron Formation, near Killimor, Co. Galway, Ireland. Stilpnomelane forms platy or needle-shape crystals up to 150 µm long or unorientated interlocking aggregates. It is difficult to distinguish from chlorite, but in thin section its strong pleochroism from pale yellow to deep olive brown or medium to dark green make it distinctive (Schultz 1966a).

stilpnosiderite

A synonym of goethite.

Delicate pink pyramidal crystals of stokesite, up to 7.5 mm long, associated with platy muscovite. Kessel Downs (Halvasso) quarry, near Longdowns, Cornwall. (© D. I. Green)

stokesite

orthorhombic $CaSnSi_3O_9 \cdot 2H_2O$

TL Roscommon Cliff, St Just, Cornwall, England

Named after Sir George Gabriel Stokes FRS (1819–1903), Irish mathematician, physicist, Lucasian Professor at Cambridge University and president of the Royal Society; born at Skreen, Co. Sligo. Stokesite is a very rare hydrated calcium–tin silicate initially discovered on a single specimen in the Joseph Carne (1782–1858) collection in the Sedgwick Museum, University of Cambridge. The specimen comprises crystallized axinite containing a shallow cavity in which there is a single colourless transparent crystal of stokesite about 10 mm long; the crystal resembles the selenite variety of gypsum (Hutchinson 1900). No further specimens were found until 1975, when a small crystal about 4 mm long was found in a discordant tin-rich calc-silicate vein dominated by grossular and axinite at Wheal Cock Zawn, Roscommon Cliff, St Just. Remarkably, in the same year, stokesite was also found at Kessel Downs (Halvasso) quarry, near Longdowns, Cornwall. Several crystals up to 7 mm long were found in two boulders of granite containing drusy pegmatitic patches, that had fallen from the northern wall of the quarry. Initially thought to be bertrandite, analysis showed them to be stokesite. The steep pyramidal crystals are clear, colourless to pale pink, and occur as either individuals or radiating groups on albite and orthoclase associated with apatite, muscovite and tourmaline (Couper & Barstow 1977, Couper & Clark 1977). See also Gay & Rickson (1960), Alderton & Jackson (1978), Hubbard (1989), Wolloxall (1989), Bancroft & Weller (1993).

stolzite

tetragonal $PbWO_4$

A lead tungstate, dimorphous with raspite and visually similar to wulfenite. It is found as platy crystals coated with duftite on spoilheaps produced during the mining of galena at Brandy Gill, Caldbeck Fells, Cumbria (Davidson & Thomson 1951, Young 1987a). The stolzite crystals reach 10 mm in length in poorly defined platy aggregates; and, less commonly, stolzite forms smaller but well defined sharp tabular crystals and simple prisms. Cooper & Stanley (1990) report that, in addition to dumps on the upper western level, it has also been found *in situ* in the old level. Associated species are bayldonite, "calcio-duftite", cerussite, malachite and mimetite. A vein outcrop on the eastern side of Brandy Gill contains small honey-yellow crystals that were once thought to be wulfenite, but on analysis all but one were shown to be stolzite (Rothwell & Mason 1982). Stolzite from Brandy Gill has also been mentioned by other authors (Claringbull 1951, Kingsbury & Hartley 1955, Hartley 1984, Green 1986, Neall & Green 2001a). Analysis of small orange-yellow pseudo-cubic crystals up to 0.4 mm across on mimetite–pyromorphite from lead veins in Poddy Gill, Caldbeck Fells, show that they are stolzite, although other visually indistinguishable crystals in this assemblage are wulfenite (Cooper & Stanley 1990). Further examination of material from Poddy Gill shows that yellow molybdenum-rich stolzite overgrows wulfenite (D. I. Green specimen, Tindle unpub. EMP data). At Carrock mine, Caldbeck Fells, stolzite has been found on spoilheaps on the north side of Grainsgill. It forms very thin cm-size incomplete plates in a gossany quartz matrix

Pseudo-cubic orange crystals of stolzite to 0.4 mm, with pale-green pyromorphite; Poddy Gill, Caldbeck Fells, Cumbria. (© M. P. Cooper)

and has a very indistinct appearance (Young 1987a, Cooper & Stanley 1990, Green & Briscoe 2002). Early reports of stolzite from Force Crag mine, Braithwaite, near Keswick, Cumbria (Greg & Lettsom 1858, Postlethwaite 1913), are considered doubtful by later workers (Davidson & Thomson 1951, Young et al. 1986), but the species still appears in modern listings (Tyler 2005).

In Cornwall, Cheesewring quarry, SW of Henwood, Linkinhorne, and Hingston Down Consols mine, 2 km W of Gunnislake, Calstock, are two locations where stolzite has recently been identified. At the former locality it occurs in cavities within wolframite, as inconspicuous buff tabular crystals up to 0.4 mm long associated with ferritungstite, mpororoite and scorodite. At the latter locality it occurs in drusy cavities in granite associated with orthoclase and schorl, where it forms elongated pale-brown bipyramidal crystals up to 0.5 mm long. Supergene oxidation of primary lead and tungsten minerals in a near-surface environment is the most likely mechanism of formation (Green et al. 1996a).

In Scotland, molybdenum-rich stolzite is confirmed from lead–zinc veins at Whyte's Cleuch, Wanlockhead, Dumfries & Galloway (Livingstone 1992), and, at Gairnshiel Bridge, Glen Gairn, Grampian, it is noted to replace wolframite in a hübnerite-bearing quartz vein cutting zinnwaldite granite (Webb et al. 1992).

STOTTITE GROUP

Tetragonal hydroxides with the general formula $AB(OH)_6$, where $A = Fe^{2+}$, Fe^{3+}, Mn^{2+}, Na; and $B = Ge^{4+}$, Sn^{4+}, Sb^{5+}. Jeanbandyite is the only species of the group found in Britain or Ireland.

strashimirite

monoclinic $Cu^{2+}_8(AsO_4)_4(OH)_4 \cdot 5H_2O$

A very rare hydrated copper arsenate hydroxide found as a secondary mineral in the oxidized zone of arsenic–copper-bearing mineral deposits. At Wheal Gorland, St Day, 2 km E of Redruth, Cornwall, strashimirite is said to be pale blue and compact (Ryback & Tandy 1992). The world's best liroconite crystal was collected from this mine and it sits on a base of pale-green strashimirite (Bancroft & Weller 1993); the specimen is now on show in the Royal Cornwall Museum, Truro. At Ting Tang mine, Carharrack, the dumps of Morcom's shaft contain pale-blue crystalline spherules of strashimirite in a supergene assemblage that also contains liroconite and parnauitite (Bruce & Aubrey-Jones 1998). At Wheal Unity, St Day, Gwennap, a third Cornish occurrence of strashimirite is noted by Golley & Williams (1995).

In Cumbria, strashimirite has been found as pale-blue to bluish-green crusts composed of minute lath-like crystals from the 60-fathom level dumps of Sandbed mine, Caldbeck Fells (Green et al. 2006a).

Strashimirite forms pale lime-green to blue-green crusts and sprays of acicular crystals in cavities within a native copper–cuprite deposit near the Precambrian/Triassic unconformity at New Cliffe Hill quarry, Stanton under Bardon, Leicestershire (Hubbard et al. 2005).

Pale bluish-green fluffy spherules of strashimirite are documented by Sarp et al. (1987) and Ryback & Tandy (1992) from Tynagh mine, near Killimor, Co. Galway, Ireland.

strengite

VARISCITE GROUP

orthorhombic $Fe^{3+}PO_4 \cdot 2H_2O$

A rare hydrated iron phosphate, dimorphous with phosphosiderite, that forms a series with variscite and is found as minute glassy spherules on wolframite from an exposure of the Harding vein at Carrock mine, Grainsgill, Caldbeck Fells, Cumbria (Cooper & Stanley 1990). Strengite from nearby Burdell Gill (Kingsbury 1963, Hartley 1984, Young 1987a, Cooper & Stanley 1990, Ryback & Tandy 1992) and from Wheal Phoenix, Linkinhorne, Cornwall (Kingsbury 1963, Ryback & Tandy 1992, Golley & Williams 1995), is less certain, as the specimens were supposedly collected by A. W. G. Kingsbury. Ryback et al. (1998, 2001) suggest that the provenance of many Kingsbury samples is in doubt, and further confirmation is desirable.

Strengite is also noted from Gravel Hill mine, Perranzabuloe, Cornwall, and as clusters of pale-lilac needles with cacoxenite from Ballycormick, near Shanagolden, Co. Limerick, Ireland (Ryback & Tandy 1992). It has also been identified using XRD and energy-dispersive EMP techniques as small euhedral transparent purple crystals, in veins at Hingston Down quarry, 2 km W of Gunnislake, Calstock, Cornwall (D. I. Green pers. comm. 2004).

stromeyerite

orthorhombic AgCuS

A rare silver–copper sulphide found in the Glen of Aherlow Cu–(Ag) orebody, near Galbally, Co. Limerick, Ireland. A suite of sulphides (bornite, chalcocite and chalcopyrite, together with minor galena and sphalerite) occurs as disseminations within Carboniferous Limestone, as discrete blebs within calcite veins and as cemented material in breccias. Intensity of mineralization is related to the degree of deformation of the host limestone. Stromeyerite has been found in two samples, where it forms grains up to 25 μm long in close association with chalcocite and native silver. The deposit has some epigenetic characteristics and probably formed because of submarine fumarolic activity, with mineralization being derived from volcanogenic solutions (Cameron & Romer 1970, Schultz 1971).

At Gortdrum mine, near Tipperary town, Co. Tipperary, epigenetic mineralization related to the intrusion of two high-level basic intrusions has produced a dominantly chalcopyrite–Hg-rich tennantite orebody in which stromeyerite has been found as an accessory mineral, along with cinnabar, cobaltite, gortdrumite, sphalerite, wittichenite and native amalgam (var. kongsbergite). Stromeyerite is present mainly along dislocations in chalcocite (Morrissey et al. 1971, Steed 1983, 1986, Duane 1986). Banded chalcopyrite–pyrite and chalcopyrite–galena–sphalerite ores forming a limestone-hosted polymetallic deposit at Muckross mine, Killarney, Co. Kerry, contain minor cobaltite, molybdenite and stromeyerite. On nearby Ross Island, stromeyerite occurs in thin chalcopyrite–tennantite veinlets associated with minor cobaltite and trace amounts of arsenopyrite, bornite, galena, marcasite, molybdenite, pyrite and sphalerite (Kinnaird et al. 2002).

Stromeyerite is a rare constituent of bornite–chalcocite–tennantite ore from the carbonate-hosted stratabound lead–zinc deposit at Lisheen mine, Moyne, Co. Tipperary. It is one of the last species to form and occurs as blebs (<30 μm across) and veinlets cutting bornite, and as exsolution lamellae in chalcocite (S. Moreton sample, Tindle unpub. EMP data).

Stromeyerite and a wide range of other sulphide species are described from Sark's Hope mine, Little Sark (Mourant & Warren 1934), but the species is not reported by later workers (e.g. Ixer & Stanley 1983). Confirmation is desirable for this occurrence and for stromeyerite on specimens in the Kingsbury collection at the NHM, said to have been found at Pengenna mine, St Kew, Cornwall (Ryback et al. 1998).

stromnite

TL Stromness, Orkney Isles, Scotland

A mixture of barite and strontianite.

strontianite

ARAGONITE GROUP

orthorhombic $SrCO_3$

TL Strontian, Highland Region, Scotland

An uncommon strontium carbonate usually found in low-temperature hydrothermal veins cutting limestone where it is associated with barite, calcite or celestine. Less commonly it is found in hydrothermal metallic veins.

In the Northern Pennine orefield, lustrous, pale-green to colourless, radiating, fibrous or prismatic strontianite has been found close to the Whin Sill along a witherite vein system at Settlingstones and Stonecroft mines, near Hexham, Northumberland (Trestrail 1931, Ineson 1972, Young 1985b, 1987c). Reports of strontianite from Greenlaws mine, S of Daddry Shield, Weardale, Co. Durham (Sherlock & Hollingsworth 1938), are discredited by Young (1985b), who suggests that calcite was misidentified. Rare white to pale-green divergent sprays of acicular strontianite up to 5 mm long in cavities in barite are noted from the south end of West High Cross vein, part of Brownley Hill mine, Nenthead, near Alston, Cumbria. The vein consists mainly of barite and calcite, together with some ankerite (Wood 1991, Green et al. 2000). Fine crystallized specimens of strontianite have been recovered from Merryfield mine, near Pateley Bridge, Nidderdale, Yorkshire (Greg & Lettsom 1858, Melmore 1943, Dunham & Wilson 1985, Young 1987c, Wood 1993). Spectacular well crystallized strontianite has more recently been recovered from lead veins on the Sun Hush level of the Gunnerside Gill mines, Swaledale, Yorkshire. White-to-colourless, sharply terminated, steep-sided pyramids of strontianite form dense crystal sprays to 4 cm across in groups up to 15 cm across. Other occurrences in the Swaledale area are discussed by Young (1987c) and other authors, and include: from dumps at the entrance of the Brandy Bottle incline in Flincher Gill, related to workings on the Friarfold vein (Ineson & Young 1994); on the Bunting, Priscilla and Sir Francis levels in Gunnerside Gill (Wood 1993); Land End mine, W of Keld; Old Rake, Hard Level Gill; Victoria level and Hard level, Old Gang mine (Sherlock & Hollingworth 1938, Dunham & Wilson 1985, Ineson & Young 1994).

Isolated occurrences of strontianite are reported from a few other locations in England: in celestine deposits to the E of Leechpool farm, near Yate, Avon (Faithfull 1988); Old Potts Gill mine, Caldbeck Fells, Cumbria (a claim made by A. W. G. Kingsbury and doubted by Green & Briscoe 2002); spherulitic calciostrontianite in veins and nodules in celestine-bearing limestone at Durlston Head, Swanage, Dorset

(Salter & West 1965); colourless crystals associated with celestine at Asquith Boulevard, Leicester, Leicestershire (Cooper 1991); Ecton, Staffordshire (Watson 1860) – not found since and perhaps a misidentification of celestine (Ford et al. 1993).

Strontianite is a rare species in Wales, with specimens in the NMW from Benallt mine, Rhiw, and Nant mine, Llanfaelrhys, near Rhiw, Gwynedd, and also from Pennant mine, near St Asaph, Clwyd (Bevins 1994). Globular aggregates of strontianite to 0.3 mm have been recovered from cavities in fallen blocks of limestone on Penarth beach, South Glamorgan (Wolfe 1996a). A radiating mass of white acicular strontianite crystals up to 1 mm long from Penarth Head is noted by Bevins (1994).

Although the type locality for strontianite is given as Strontian (a small hamlet near Loch Sunart), the actual location is unquestionably one of the lead mines located about 8 km to the N (e.g. Bellsgrove, Corrantee, Fee Donald, Middleshop and Whitesmith mines), where it occurs in barite-rich veins adjacent to altered basaltic dykes (Wilson & Flett 1921). Strontianite was the first new species to be described from Scotland; its discovery and the element strontium derived from it are documented by Livingstone (2002). Examples now in the NMS include an exceptional apple-green specimen with 12 cm-long radiating glassy prisms from Whitesmith mine, and an 8 cm-radius brown specimen from Fee Donald mine. Fibrous veinlets and acicular crystals of strontianite coloured white, light green, pink, yellow, grey, brownish or colourless are noted by Winrow (1986b). Reworking of the area between Bellsgrove and Whitesmith mines for barite in the 1980s produced rare cm-size green or brown acicular crystals of strontianite, but not comparable in quality to the fine brewsterite, calcite and harmotome recovered at the same time (Green & Todd 2001, Green 2005a). Fist-size balls of radiating strontianite have been found at Whitesmith mine (Wolfe 2006).

Strontianite has been found on the dumps of Muirshiel barite mine, 10 km NW of Lochwinnoch, NE of Largs, Strathclyde. It is often associated with cockscomb barite and has developed various habits, including grey-to-white spheroidal clusters up to

Sprays of strontianite from Sir Francis level, Swaledale, Yorkshire; the specimen is 8 × 5.5 cm. (© M. P. Cooper)

Green and brown colour zoned strontianite from Whitesmith mine, near Strontian, Highland Region, Scotland. Spray at the top of the image is 15 mm in length. (© D. I. Green)

10 mm across, steep pseudo-hexagonal pyramids and curved crystals. It also occurs in white to pale-green fibrous veins (Stephenson & Coats 1983, Todd & Laurence 1989). At Trearne quarry, near Beith, Strathclyde, cavities within brachiopod shells contain white tufts of strontianite associated with calcite, and white spheroids associated with fluorite (Todd 1989, Green & Todd 2001). Other occurrences of strontianite in Scotland include: in syenite at Beinn Bhreac, 2 km E of Tongue, Highland Region (Heddle 1883, 1901); in veins and interstices of breccia at Stromness lead mine and along the beach at Point of Ness, Mainland, Orkney (Greg & Lettsom 1858, Heddle 1901, Wilson & Flett 1921, Mykura 1976); in the Cnoc nan Cuilean syenite, near Ben Loyal, Highland Region (Gallagher et al. 1971); in syenite at Glen Dessary, W of Loch Arkaig, Highland Region (Fowler 1992). Strontianite was also found at Leadhills, Strathclyde, but unquestionably the specimens originally came from Strontian (Heddle 1901, Brown 1927).

In Ireland, strontianite has been found at the Giant's Causeway, Co. Antrim (Greg & Lettsom 1858), and at Goldenbridge and near Finglas Bridge, Co. Dublin (Ryback & Moreton 1993).

strunzite?

triclinic (pseudo-hexagonal) $Mn^{2+}Fe^{3+}_2(PO_4)_2(OH)_2 \cdot 6H_2O$

A rare hydrated manganese–iron phosphate hydroxide reported in cavities within goethite from Gravel Hill mine, Perranzabuloe, Cornwall. Strunzite forms straw-yellow sprays of acicular crystals up to 2 mm in length (Ryback & Tandy 1992). This mineral has also been reported as ferrostrunzite (Weiss 1989). EMP analysis (Tindle unpub. data) shows that the mineral cannot be strunzite (as Fe/Mn = 15–20), but, as the technique cannot measure Fe^{2+} and Fe^{3+} contents, it is not possible to name the species – it could be ferristrunzite or ferrostrunzite.

Radiating fan of strontianite crystals, 17 cm across; Strontian, Highland Region, Scotland. (© M. P. Cooper)

strüverite

Burke (2006), reporting on an IMA-CNMNC-approved mass discreditation of species, states that strüverite is formally discredited, and compositions are now to be referred to as tantalum-bearing rutile. See rutile for examples from Wales and Ireland.

struvite

orthorhombic $(NH_4)MgPO_4 \cdot 6H_2O$

A rare hydrated ammonium–magnesium phosphate found in pre-AD 1060 soil horizons from archaeological excavations in High Street and Fishamble Street, Dublin, Ireland. Struvite forms 6–20 mm fragmented tabular amber-yellow to smoky-brown translucent-to-opaque crystals (Ryback et al. 1988).

stützite

hexagonal $Ag_{5-x}Te_3$

A very rare silver telluride found at Lochy Trib (Eas Daimh), W of Tyndrum, Strathclyde, Scotland. Stützite forms inclusions in an alluvial gold (electrum) grain, associated with other gold grains containing acanthite, gersdorffite and hessite inclusions. An origin related to igneous activity in the area is proposed (Chapman et al. 2000c).

sulfoborite

A synonym of sulphoborite.

sulfur

A synonym of sulphur.

sulphato-carbonate of baryta

TL Brownley Hill mine, Nenthead, near Alston, Cumbria, England

A mixture – an overgrowth of witherite on barite.

sulphoborite

orthorhombic $Mg_3B_2(SO_4)(OH)_8(OH,F)_2$

A rare magnesium borosulphate fluorhydroxide found in the Upper Halite–Anhydrite and Upper Anhydrite subzones of the Permian Lower Evaporites in the Fordon F.3 borehole, Humberside (13 km S of Scarborough, Yorkshire). The Permian evaporite sequence underlies much of North Yorkshire and extends considerably beneath the North Sea. At Fordon the sequence is 430 m thick. Sulphoborite forms colourless diamond-shape euhedral porphyroblasts up to 8 mm long. It is vitreous, has good cleavage, and contains corroded relics of anhydrite and strings of carbonate granules. Undoubtedly, sulphoborite replaces anhydrite mainly from reaction with brines percolating downwards during evaporite formation (Stewart 1963, 1965).

sulphur

orthorhombic S

A relatively uncommon native element, dimorphous with rosickýite and often formed as a result of low oxidation potential and highly acidic chemical reactions in sulphide-bearing mineral deposits.

On the beach near the Pier hotel, Clevedon, Avon, pebbles of barite contain a wide range of other species, including native sulphur (Starkey 1984). A large infilled mineralized cave within Carboniferous Limestone at Stancombe quarry, near Flax Bourton, Bristol, contains native sulphur and greenockite as part of a secondary assemblage related to barite–calcite–galena veins (Zachariades 1993). At Stancombe quarry, near Flax Bourton, Bristol, native sulphur occurs with anglesite, cerussite, calamine and rare greenockite in fractured galena pods within barite–calcite veins cutting a Carboniferous Limestone cave system. Prismatic crystals of native sulphur are very rare, but, when found, occur as distorted bipyramids and twinned crystals (Alabaster 1983).

Very pale-yellow rounded crystals of sulphur have been found in partly oxidized galena at Engine Vein mine, Alderley Edge ridge, Cheshire (Braithwaite 1994).

At Penberthy Croft mine, St Hilary, Cornwall, native sulphur forms subhedral translucent yellowish crystals up to 3 mm across with cerussite, covellite, galena and quartz (Betterton et al. 2000), and in partially altered galena it occurs with laurionite and phosgenite (Wallace 1998). Sulphur has been found at Poldice mine, near St Day, and an impure variety with greyish-yellow colouration was found in cavities within pyrite at Nangiles mine, near Twelveheads, Kea (Garby 1848, Greg & Lettsom 1858). A cliff exposure at Gunver Head, in the area of Padstow Consols mine, N of Trevone, produced crystals of native sulphur associated with anglesite, cerussite, cumengéite and phosgenite in an oxidized galena–quartz veinlet (Braithwaite & Merry 1996).

A quartz–antimony vein within the aureole of the Skiddaw granite, near the head of Wet Swine Gill, Coombe Height, Caldbeck Fells, Cumbria, contains a suite of supergene species that includes bindheimite, senarmontite, stibiconite, native sulphur and possibly kermesite (Fortey et al. 1984, Leppington & Green 1998, Neall & Green 2001b). Native sulphur is also noted as rare yellow crusts in oxidized galena at Brae Fell mine, Dale Beck Valley (Bridges et al. 2006), as pale-yellow globular crystals up to 0.25 mm across on galena from Silver Gill (Green et al. 2005b), and as transparent pale-yellow globular crystals to 0.5 mm at both Driggith and Sandbed mines (Green et al. 2006a). Young (1987a) reports native sulphur on altered galena at Whiteoak Beck, High Nook, Loweswater. Native sulphur in the Northern Pennine orefield usually occurs with corroded galena or other sulphides, as follows:

CUMBRIA Brownley Hill mine, Nenthead, near Alston (Clark & Rust 1993, Bridges & Young 1998, Green et al. 2000); Dryburn Washpool vein, Tynebottom mine, NW of Garrigill, near Alston (Bridges & Young 1998); Dufton mine, near Appleby (Greg & Lettsom 1858, Bridges & Young 1998); Smittergill Head mine, Melmerby Fell, Alston Moor (Bridges & Young 1998). DURHAM Closehouse mine, Lunedale; Drypry vein, N of Wolsingham; Harehope Gill mine, White Kirkley, Weardale (Bridges & Young 1998). NORTHUMBERLAND Burntshieldhaugh vein, Devil's Water, NW of Blanchard; Shildon mine, Blanchland (Bridges & Young 1998).

In Derbyshire, weathering of sulphides has produced yellow coatings of native sulphur at several locations: with well developed crystals of anglesite and wulfenite in a small flat of galena in an abandoned fluorite working above Ball Eye quarry in the Via Gellia, near Cromford (Bridges 1983); Bonsall Leys (Ford et al. 1993); classic bipyramidal habit and with well developed crystal faces to 1 mm on galena at Crich quarry (Wilson et al. 1995); with anglesite as an alteration product of galena at Dene quarry, Cromford, and Netherwater mine, Hazlebadge (Braithwaite & Ryback 1963b, Braithwaite 1983); associated with pyrite at Millclose mine, Wensley, near Matlock (Sarjeant 1957a); minute yellowish-green crystals with cerussite and anglesite on altered galena at Milltown quarry, near Ashover (Rust 1991c); associated with sphalerite at Moury mine, near Tideswell (Mostaghel 1985); on galena at Wirksworth and at Bole Hill, Cromford (Greg & Lettsom 1858).

Native sulphur has been described from the Aislaby (Eskdale No. 2) borehole, Eskdale, near Whitby, Yorkshire, where it is a minor component in Permian evaporites. It forms aggregates up to 1.5 mm across and encloses corroded anhydrite relics (Stewart 1949). Another borehole (F.1) at Fordon, Humberside, also contains minor native sulphur (Stewart 1963). At Billingham anhydrite mine (under the ICI factory site), Cleveland, native sulphur, together with epsomite, gypsum, halite, lepidocrocite, pyrite, sylvite, talc and thenardite, have formed as efflorescences on the walls and floor of the abandoned workings (Raymond 1959). In a study of the Lower Magnesian Limestone of northeast Yorkshire and Cleveland, Raymond (1962) confirms that native sulphur is commonly found in small amounts at Whitby and occurs as late growths in vugs associated with anhydrite, calcite or dolomite.

In Wales, native sulphur has been described from galena-bearing veins at Nant-y-blaidd mine, near Llanrhaeadr-ym-Mochnant, Clwyd (Wedd et al. 1929), and in quarries at Dan-y-graig and Ochrwyth, near Risca, Gwent (Alabaster 1983). Native sulphur was once found in the gossan capping to the Parys Mountain copper deposit on Anglesey, Gwynedd, but this has long since been removed (Greenly 1919). Visually identified native sulphur occurs with covellite in cavities within galena at Frongoch mine (between Devil's Bridge and Pont-rhyd-y-groes) and Penrhiw mine, Ystumtuen, both in the Central Wales orefield, Dyfed (Mason & Green 1995, Green et al. 1996b).

Native sulphur was formed when a heap of shaly mine waste spontaneously ignited at Emily coal pit, Arniston, Lothian, Scotland. It formed powdery sublimates in the cooler parts of the heap, together with mascagnite, tschermigite and an unnamed aluminium sulphate (Shand 1910). At Craignish, Isle of Mull, Strathclyde, pale-yellow native sulphur was found in a 13 cm-wide cavity within a porphyritic Tertiary dyke (Livingstone & Baird 1991). Sulphur also forms a coating on altered galena at Tyndrum mine, Central Region (Braithwaite & Ryback 1963b).

In Ireland, native sulphur, and gypsum, limonite and morenosite, occur in cavities within ore material at St Patrick's copper mine, Avoca, Co. Wicklow (King & Evans 1964). Earlier reports are from Brook Lodge and near Glan, both Co. Galway, and Castlecara, Co. Mayo (Greg & Lettsom 1858).

sulphuret of bismuth
A synonym of bismuthinite.

sulphuret of molybdenum
A synonym of molybdenite.

sunstone
A synonym of orthoclase, feldspar group.

surinamite
monoclinic $Mg_3BeAl_4Si_3O_{16}$

A rare magnesium–beryllium aluminosilicate reported from a Lewisian sapphirine-bearing potassium-feldspar–kyanite–orthopyroxene–quartz gneiss, 0.5 km S of Roineabhal summit, South Harris, Western Isles, Scotland. It forms pale-blue tabular crystals up to 1 mm long, surrounded by cordierite, or tiny grains enclosed in kyanite. Inclusions of a silica-rich sapphirine enclosed in surinamite are presumed to be khmaralite (Baba 1999, 2003, 2004, Baba et al. 2000). Surinamite is also reported in sapphirine-bearing gneiss close to Loch a' Sgùrr, near Finsbay (Fionnsabhagh), South Harris (Green & Todd 2001).

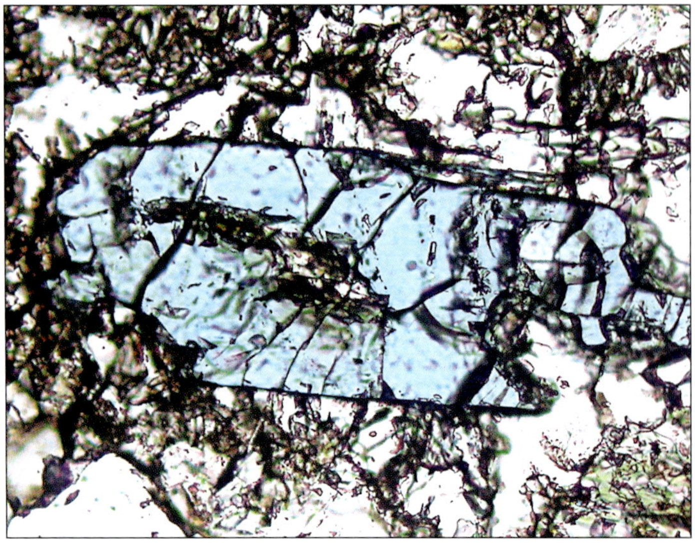

Pale-blue surinamite mantled by colourless cordierite in a pyroxene granulite from the South Harris igneous complex, Western Isles, Scotland; width 0.5 mm; viewed in PPL. (© S. Baba)

susannite
trigonal $Pb_4(SO_4)(CO_3)_2(OH)_2$

TL Susanna mine, Leadhills, Strathclyde, Scotland

A rare lead sulphate carbonate hydroxide, trimorphous with leadhillite and macphersonite, and found in the oxidized zone of hydrothermal lead deposits. At Susanna mine, it is said to form translucent, transparent or opaque, grey, pale-green, yellow or dark-brown crystals in association with caledonite, cerussite, lanarkite, leadhillite and pyromorphite (Greg & Lettsom 1858). As early distinction between susannite and leadhillite proved difficult, it is possible that some of these species may have been misidentified. A specimen collected from the Susanna vein and now at the NHM contains pale-green crystals of susannite up to 2 mm long associated with cerussite and leadhillite (Livingstone 2002). At nearby Leadhills Dod, susannite occurs with chrysocolla and malachite (Heddle 1901), and a specimen in the NMS contains equant greenish-brown susannite associated with caledonite, cerussite, leadhillite, macphersonite, mattheddleite and pyromorphite (Livingstone & Sarp 1984, Livingstone et al. 1987). Apple-green crystals of susannite up to 1 cm long are also noted from the Glengonnar mine, Leadhills (Gillanders 1981).

Susannite has been found at a few locations in the Caldbeck Fells, Cumbria. At Red Gill mine, distinctive aquamarine susannite forms small acute rhombohedral crystals with caledonite, cerussite, leadhillite and mattheddleite (Cooper & Stanley 1990). Chenite, elyite, linarite, macphersonite, queitite and scotlandite are other supergene species reported from Red Gill (Green & Briscoe 2002). A specimen from

Susannite crystal (2.1 mm tall), associated with mattheddleite; Red Gill, Caldbeck Fells, Cumbria. (© D. I. Green)

Roughton Gill, now in the Ludlam collection at the NHM, contains pale-blue equant-to-tabular susannite to 10 mm embedded in acicular caledonite (Cooper & Stanley 1990). In higher Roughton Gill on the flank of Balliway Rigg, near Thief Gills, small crystals and crusts of susannite with anglesite, cerussite and leadhillite have been found (Green & Briscoe 2002). At Short Grain lead mine, cavities in barite once occupied by galena contain pale aquamarine crystals of susannite with caledonite, cerussite, leadhillite and mattheddleite (Cooper et al. 1988, Cooper & Stanley 1990, Green & Briscoe 2002). Pale-green tabular-to-elongated crystals of susannite associated with leadhillite have been identified on an ore pile at Silver Gill (Green et al. 2005b). A sixth Caldbeck Fells occurrence of susannite from Brae Fell mine (Cooper & Stanley 1990) is less certain, as the specimen is from the Kingsbury collection at the NHM; Bridges et al. (2006) suggest that it is a 19th-century Leadhills–Wanlockhead specimen. The provenance of many specimens in this collection has recently been brought into question (Ryback et al. 1998, 2001). Susannite is also noted from Kinniside lead mine, Cleator Moor, Cumbria (Green & Briscoe 2002).

Susannite crystal, 4 mm across, embedded in a velvety mass of blue caledonite; Roughton Gill mine, Caldbeck Fells, Cumbria. (© The Natural History Museum, London)

The Central Wales orefield in Dyfed is another area where susannite is reported from many locations. It is found as tabular-to-blocky colourless crystals up to 0.8 mm long associated with caledonite, in brecciated galena–sphalerite-bearing quartz veinstone from Bwlchglas mine, near Tal-y-bont (Rust 1996c). Groups of rod-like to prismatic crystals of susannite up to 0.3 mm long occur in association with bechererite, caledonite, hemimorphite, hydrozincite, hydrocerussite, leadhillite and schulenbergite at Esgair Hir Pb–(Cu) mine, Ceulanymaesmawr (Rust & Mason 1988, Rust et al. 2003). Susannite occurs as yellowish pseudo-hexagonal tabular crystals up to 1 mm long and colourless pseudo-hexagonal crystals up to 0.6 mm across at Llechweddhelyg mine, near Clawddmelyn, between Bont-goch and Cwm-symlog. Their morphology is so similar to leadhillite that the two species cannot be confirmed visually (Rust 1990c). Frongoch mine, between Devil's Bridge and Pont-rhyd-y-groes is perhaps the finest Welsh locality for susannite. It is relatively common as a late-stage species in the post-mining basic sulphate assemblage, where it forms pseudo-hexagonal crystals up to 2 mm across, coloured white with green, blue or yellow tints, or occasionally toffee brown. Some crystals show step-like growth patterns and overgrowths of several different forms (Green et al. 1996b). At Hendrefelen mine, Ysbyty Ystwyth, susannite has been found on the mine dump along with caledonite and mimetite (Clark & Rust 1993). At Eaglebrook (Nantycagl) mine, Ceulanymaesmawr, susannite is associated with bechererite and hemimorphite, and at Dylife mine, N of Hafren Forest, Central Wales orefield, Powys, susannite occurs with bechererite, caledonite, hydrocerussite, langite, lautenthalite, namuwite and schulenbergite (Rust et al. 2003).

svanbergite?

trigonal $SrAl_3[(P,S)O_4]_2(OH,H_2O)_6$

A very rare hydrated strontium–aluminium phosphate sulphate hydroxide reported from Wheal Coates, St Agnes, Cornwall (Kingsbury 1964b, Embrey 1978). Given the doubts about the provenance of many Kingsbury specimens (Ryback et al. 1998, 2001), further confirmation is required. Unless further discoveries are made, this species cannot be considered native to Britain or Ireland.

sweetite

tetragonal $Zn(OH)_2$

TL Milltown quarry, near Ashover, Derbyshire, England

Named after Jessie May Sweet (1901–1979), senior scientific officer and curator of the mineral collection at the British Museum (Natural History); birth registered at Wisbech, Cambridgeshire. Sweetite is a very rare zinc hydroxide, trimorphous with ashoverite and wülfingite, and is found at the type locality in cavities within an oxidized vein. It forms colourless or whitish bipyramids up to 1 mm long, scattered over the surface of colourless fluorite cubes. Associated species include ashoverite, calcite, cerussite, hemimorphite, hydrozincite and smithsonite. Hydrocerussite, weddellite and ettringite may be seen on the same specimens, but not in direct

Crystal of sweetite, 1.5 mm across; Milltown quarry, Ashover, Derbyshire. (© M. P. Cooper)

association. Anglesite, cerussite, galena, hydrocerussite and litharge are found in other cavities in the oxidized vein (Clark et al. 1984, 1988, Rust 1991c).

swinestone

A variety of calcite that emits a foetid odour when broken.

switzerite?

monoclinic $(Mn^{2+},Fe^{2+})_3(PO_4)_2\cdot 7H_2O$

A rare hydrated manganese–iron phosphate reported from manganese-rich quartz veins cutting Ordovician greywackes and mudstones that have been thermally metamorphosed by the Skiddaw granite at Burdell Gill, Caldbeck Fells, Cumbria. Within the veins, there are cavities and clefts that are partially filled with manganese oxides (cryptomelane, pyrolusite and romanèchite) and a wide range of supergene species that includes beudantite, carminite, lepidocrocite, mimetite, parasymplesite, pharmacosiderite, plumbogummite, rhodochrosite, scorodite, switzerite and vivianite. Switzerite is reported as divergent sprays of colourless to pale-brown obliquely terminated blades or as thin films of pale fibrous material on hairline-fracture surfaces in gossanous quartz, associated with barium-pharmacosiderite (Neall & Leppington 2003). According to Gaines et al. (1997) switzerite crystals must be completely hydrated, as when exposed to air they may disintegrate explosively to metaswitzerite.

syelite

A synonym of bastite (chrysotile ± lizardite or talc or anthophyllite).

sylvanite

monoclinic $(Au,Ag)_2Te_4$

A very rare gold–silver telluride found in a brecciated gold-bearing quartz–pyrite vein (Halliday's vein), Tyndrum, Central Region, Scotland, where it occurs as ≤20 μm-diameter inclusions in pyrite with cervelleite (described as phase C: see cervelleite), petzite and an unidentified Ag–Te–S phase. These gold and silver minerals are associated with native gold (electrum with high silver content), chalcopyrite, galena and sphalerite. The gold–silver mineralization is thought to be attributable to hydrothermal activity related to copper–molybdenum mineralization associated with Late Caledonian granites (Pattrick et al. 1988, Curtis et al. 1993).

In Wales, sylvanite has been found in a late-stage hydrothermal vein at Gwynfynydd mine, near Dolgellau, Gwynedd. It occurs along with acanthite replacing chalcopyrite and pyrrhotite (Pryor 1988).

Young (1987a) briefly documents sylvanite as a constituent of an arsenopyrite-rich vein at Carrock mine, Grainsgill, Cumbria.

sylvinite

A mixture of halite and sylvite.

sylvite

cubic KCl

A locally abundant potassium chloride found in the Middle Evaporites of the Yorkshire potash field from Eskdale No. 2 borehole at Aislaby, Eskdale, near Whitby, Yorkshire. Sylvite occurs in all samples from the halite–sylvite zone, where it forms irregular grains up to 38 mm in length. Where free from a multitude of hematite inclusions, it is translucent milky white; otherwise it is usually red. It is visually similar to halite in hand specimen, although its astringent taste is distinctive. Sylvite is also a widespread minor component of the Upper Evaporites from this borehole, where it is associated with halite and rinneite (Stewart 1951a,b). In the Eskdale No. 3 borehole at Sleights, a yellow variety of sylvite, associated with halite and limonite, is thought to have arisen from the decomposition of rinneite, and, in the Eskdale No. 4 borehole at Sneaton, opalescent crystals of sylvite occur in a bed of halite and also form a 10 cm-thick vein of pure sylvite (Armstrong et al. 1951). Abundant sylvite is reported from the Middle Evaporites in four other boreholes in the Yorkshire potash field: F.1 Robin Hood's Bay, F.2 Staintondale, F.3 Little Beck and F.4 Hawsker. In all cases the sylvite appears to have formed during secondary redistribution of the salts, replacing carnallite, halite and rinneite at various times (Stewart 1956), although primary crystallization of sylvite is proposed for a single layer of halite–sylvite rock (Stewart 1965). Summaries of the Yorkshire evaporite deposits are given by Dunham (1960) and Stewart (1965).

At Boulby potash mine, Loftus, Cleveland (but still part of the Yorkshire potash field), sylvite is a major constituent of the Boulby Potash, sometimes forming pure white beds several metres thick, but more commonly stained red by hematite and intimately mixed with halite. The mine works sylvite for fertilizer production. Pure

Interpenetrant cubes of sylvite from Boulby mine, Cleveland; width 1 cm; SEM mosaic image. (© E. Condliffe, University of Leeds)

white sylvite often forms the core of borate nodules found in the mine (Bottrell et al. 1996, Green & Freier 1996). At Billingham anhydrite mine (under the ICI factory site), Cleveland, sylvite, together with epsomite, gypsum, halite, lepidocrocite, pyrite, native sulphur, talc and thenardite, have formed as efflorescences on the walls and floor of the abandoned workings (Raymond 1959).

In Scotland, sylvite is reported within fluid inclusions found in rocks from the Caledonian Lagalochan intrusive complex, near Loch Melfort, Strathclyde, Scotland (Livingstone 2002). Halite, sylvite and an opaque phase (probably magnetite) are also noted as daughter crystals in fluid inclusions within clinopyroxene from a skarn at the margin of the Beinn an Dubhaich granite, near Broadford, Isle of Skye (Smith et al. 2004).

symesite

triclinic $Pb_{10}(SO_4)O_7C_4\cdot H_2O$

TL Merehead (Torr Works) quarry, Cranmore, Somerset, England

Named after Robert Symes (b. 1937), keeper of minerals at the NHM, in recognition of his many contributions to the mineralogy of the ore deposits of southwest England; born in Harrow, Middlesex. Symesite is an hydrated lead–sulphate oxychloride related to the nadorite family of lead-sheet minerals. It is found in the oxidized zone of a Carboniferous Fe–(Cu–Pb) deposit, where it occurs as pink crystal blebs up to 2 mm long and as pink crystalline aggregates up to 10 mm in diameter. Associated species are blixite, cerussite, chloroxiphite, coronadite, hematite, hydrocerussite, mereheadite, paralaurionite, parkinsonite and pyrolusite (Welch et al. 2000). The lead oxychlorides are thought to have formed by interaction of primary galena with low-temperature, metalliferous brines along fractures cutting the original Fe–Mn–(Cu–Pb) deposit (Symes & Embrey 1977, Humphreys et al. 1980), although it is possible that they precipitated directly from a cooling brine (Welch et al. 2000). A detailed discussion of the manganese mineralization at Merehead quarry is provided by Turner (2007) – see mereheadite.

symplesite

triclinic $Fe^{2+}_3(AsO_4)_2\cdot 8H_2O$

A rare hydrated iron arsenate, dimorphous with parasymplesite and first described from Wanthwaite mine, St John's in the Vale, near Threlkeld, Cumbria, where it occurs with beudantite, dundasite, gold, hydrozincite, löllingite, pyrite, sphalerite and stibnite (Young 1987a). The discovery of symplesite was based on samples in the Kingsbury collection at the NHM, a collection known to contain incorrectly provenanced material (Ryback et al. 1998, 2001). Probable symplesite, forming pale-greenish flat radiating aggregates on quartz, is reported from the dumps below Nether Row Brow (Lower level), Caldbeck (Cooper & Stanley 1990, Ryback & Tandy 1992). Young et al. (1992b) also note symplesite from a temporary exposure of the vein worked at Nether Row Brow, in the track immediately S of the collapsed level portal. Here, symplesite occurs as crusts of greenish-white to pale-brown spherules up to about 0.5 mm across, composed of very delicate radiating fibrous crystals. At Sandbeds Gill mine, near Bassenthwaite, symplesite forming tufts of sage-green acicular crystals up to 0.2 mm long is noted from a chalcopyrite–galena mineralized quartz vein (Thomson 1997b, Green et al. 2003a). Finally, two other occurrences of symplesite in Cumbria are reported by Green et al. (2003a): Carrock mine, Caldbeck Fells, where poorly developed thin prismatic crystals (up to 0.5 mm long and in patches up to 1.5 mm across) of sage-green symplesite encrust partially oxidized arsenopyrite; and Tynebottom mine, NW of Garrigill, near Alston, where a dark-green aggregate of somewhat rounded bladed crystals (up to 1 mm long) on a cellular limonitic quartz matrix has been identified as an intergrowth of symplesite and parasymplesite. Symplesite appears to have formed by recent (post-mining) supergene processes at most of the above localities.

synchysite-(Ce)

monoclinic (pseudo-hexagonal) $Ca(Ce,La)(CO_3)_2F$

A rare calcium–cerium-dominant carbonate fluoride found in accessory amounts in quartz veins throughout the Alston Block part of the Northern Pennine orefield, where it is by far the most common REE-bearing mineral. It has been identified at five locations near Rookhope, Weardale, Co. Durham – Cambokeels, Groverake, Redburn, Stotfield Burn and Whiteheaps mines – and around Alston in Cumbria – Great Sulphur vein, Alston Moor; Rotherhope mine, Leadgate; and Tynebottom mine, NW of Garrigill. In an early chalcopyrite–pyrite–quartz association, it forms acicular to tabular crystals, 20–100 μm long and 2–10 μm wide, overgrowing monazite or xenotime and enclosing small bismuthinite inclusions. In a later chalcopyrite–fluorite–galena–pyrrhotite–quartz–sphalerite–(Co–Fe–Ni sulpharsenide) assemblage, it occurs as radiating crystals growing out from silicified wallrocks into fluorite or as 100–200 μm long and 20 μm-wide discrete crystals totally enclosed within fluorite. These assemblages probably arose during or some time after intrusion of the buried Weardale granite (Brown et al. 1987a, Ixer et al. 1996, Ixer 2003).

Wirth & Green (2002) report very rare red translucent-to-transparent hexagonal crystals of synchysite-(Ce) from miarolitic cavities in granite from the Shap granite quarry, Cumbria.

In Wales, synchysite-(Ce) occurs (also very rarely) at Tanygrisiau, Blaenau Ffestiniog, Gwynedd, where it forms white tabular crystals up to 0.5 mm long in alpine-type quartz veins produced as a result of low-grade metamorphism (Green & Middleton 1996, Green 2005a). Minute disc-shape crystals of synchysite-(Ce) are reported from Dolyhir quarry, Old Radnor, Powys. They are found in very small fissures associated with anatase, quartz and, quite commonly, ewaldite (T. F. Cotterell and D. I. Green pers. comms 2006). Synchysite-(Ce) recorded from Manod quarry, near Blaenau Ffestiniog, requires verification (Bevins 1994).

synchysite-(Y)

monoclinic (pseudo-hexagonal) $Ca(Y,Ce)(CO_3)_2F$

A rare calcium-yttrium-dominant carbonate fluoride found in carbonatites, granitic pegmatites, and low-temperature alpine-type veins.

Synchesite-(Y) is reported from an altered Ordovician rhyolite found in Beddgelert Forest, Snowdonia, Gwynedd, Wales, where it occurs in accessory amounts as radiating sheaves intergrown with quartz and potassium feldspar (Howells et al. 1991, Bevins 1994).

Cluster of intergrown pseudo-hexagonal crystals of synchysite-(Ce) to 2.5 mm, with anatase (black) on quartz; Cwmorthin quarry, NW of Tanygrisiau, near Blaenau Ffestiniog, Gwynedd, Wales. (© D. I. Green)

szaibelyite/szaibélyite

monoclinic $MgBO_2(OH)$

A very rare magnesium borate hydroxide that forms a series with sussexite and is found in a contact-skarn where the Tertiary Beinn an Dubhaich granite has intruded dolomite at Kilbride and at No. 1 Prospect, Kilchrist, both near Broadford, Isle of Skye, Highland Region, Scotland. At Kilchrist, szaibelyite occurs in a carbonate–chondrodite–ludwigite–magnetite skarn. Szaibelyite occurs as matted fibres or plates around intergrown masses of magnetite and ludwigite: it formed during metasomatic replacement of dolomite when boron-rich fluids were expelled from the granite. Datolite, fluoborite, harkerite and ludwigite are associated boron-rich species, and clinohumite, cuspidine and diopside are other more common species found in the skarns (Tilley 1948b, 1951).

szomolnokite

KIESERITE GROUP

monoclinic $Fe^{2+}SO_4 \cdot H_2O$

A rare hydrated iron sulphate found as white powdery efflorescences at Crumlin colliery, Ebbw Vale, Gwent, Wales. It usually forms by precipitation from acid solutions such as mine waters and is typically associated with pyrite (Bevins 1994).

T

tacharanite

monoclinic $Ca_{12}Al_2Si_{18}O_{51}\cdot 18H_2O$

TL Portree, Isle of Skye, Highland Region, Scotland

An uncommon hydrated calcium aluminosilicate found in vesicles in an olivine dolerite 1 km north of Portree. Tacharanite is just one of many species filling vesicles in the area, others being calcite, gyrolite, mesolite, saponite, thomsonite, tobermorite and xonotlite. Often, two or more species are found in the same amygdale, an example being one with a tacharanite–tobermorite core and a gyrolite–mesolite rim. Tacharanite is a dense white mineral that, on exposure to air, was said to break down to tobermorite and gyrolite (Sweet 1961). On the Isle of Mull, Strathclyde, and across the Sound of Mull on the Morvern Peninsula, Highland Region, tacharanite occurs with gyrolite, okenite, pectolite, reyerite, tobermorite and xonotlite, and with a wide range of zeolite species in amygdales within Tertiary olivine basalts (Walker 1971). At Binhill quarry, Cairnie, near Huntly, Grampian, fibrous tacharanite and scawtite occur sporadically in fracture-filled veins of calcite, prehnite or xonotlite cutting gabbro (Livingstone 1974).

Pale pinkish-fawn tacharanite is one of many hydrated and secondary minerals found in angular amygdales up to 1 cm across in hybrid pyroxenite and coarse olivine basalt from Carneal, Larne, Co. Antrim, Ireland. The rocks are in a chalk/dolerite contact zone associated with a Tertiary volcanic plug. Other associated species include allophane, calcite, gyrolite, hibschite (syn. hydrogrossular), perovskite, portlandite, thomsonite, tobermorite and xonotlite (Henmi et al. 1973, Sabine & Young 1975).

A crystallographic investigation of tacharanite, using material from many of the above locations, is provided by Cliff et al. (1975), who concluded that, although it shares important characteristics with tobermorite, it does not break down to form tobermorite and gyrolite.

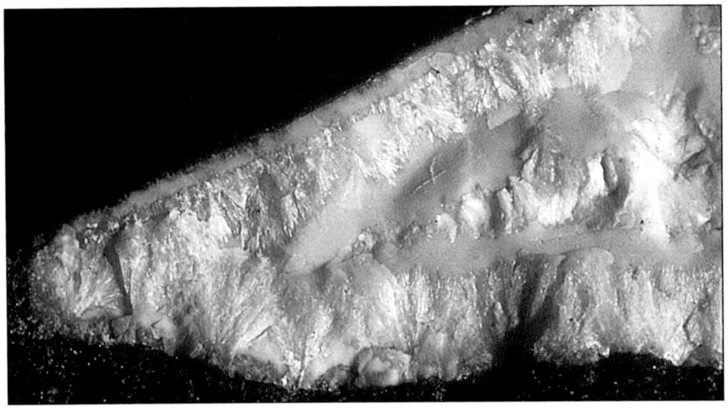

Infilled amygdale of tacharanite in Tertiary basalt from Staffin, Isle of Skye, Highland Region, Scotland; width 25 mm. (© The Trustees of the National Museums of Scotland)

tachyhydrite?

trigonal $CaMg_2Cl_6\cdot 12H_2O$

A rare hydrated calcium–magnesium chloride thought to be present in pale-yellow efflorescences that formed on the walls of the Billingham anhydrite mine (under the ICI factory site), Cleveland. Tachyhydrite is one of the secondary minerals that have grown in recent years by evaporation of water from the Upper Magnesian Limestone. Other species confirmed in brightly coloured efflorescences are epsomite, gypsum, halite, lepidocrocite, pyrite, sylvite, thenardite and talc; ilsemannite and tamarugite are also probably present (Raymond 1959).

talc

monoclinic or triclinic $Mg_3Si_4O_{10}(OH)_2$

A common magnesium silicate hydroxide that forms during hydrothermal alteration of olivine, pyroxene and amphibole in ultrabasic rocks. Talc is also found in greenschist-facies schists derived from magnesium-rich rocks or, less commonly, in thermally metamorphosed dolomitic limestone. There is some doubt about the correct identification of talc in early literature, particularly when obsolete names are used, as some authorities consider steatite to be a variety of talc (Clark 1993), whereas others suggest that it can be clinochlore, pyrophyllite or saponite (de Fourestier 1999, Bayliss 2000).

On the Lizard Peninsula, Cornwall, talc has been found at four locations: with chrysotile and tremolite in veins cutting serpentinite, in a quarry to the N of Church Cove, Landewednack (Hall & Jackson 1975); at Pol Cornick, near Mullion, where it occurs in calcite–minnesotaite veinlets cutting serpentinized peridotite (Clark 1970a); with chlorite, chrysotile, magnesite, magnetite and tremolite all replacing serpentinized peridotite at the Rill, Kynance Cove (Clark 1967); and at Pentreath beach and the Devil's Frying-pan (Golley & Williams 1995). Additionally, microcrystalline anhedral masses of talc (var. kerolite) associated with calcite, pectolite and prehnite have been found filling cavities and veins in gabbro at Dean quarry, near St Keverne on the Lizard Peninsula (Elton et al. 1997). Power et al. (1997) report that both massive talc and pale green-to-white vein talc are present throughout the Lizard peridotite. Well crystallized talc was once found in the aureole of the St Austell granite at Tremore quarry, Lanivet, SW of Bodmin, in a chalcopyrite-bearing vein that also contained chlorite, fluorite and sphalerite (Barrow & Thomas 1908). Talc is also present with chlorite and dolomite in the alteration assemblage of other basic and ultrabasic intrusions in Cornwall, at: Duporth (duporthite is a mixture of talc and chlorite); New quarry, Polyphant; and Greystone, Lezant, near Launceston (Power & Scott 1995).

Talc is a widespread minor component of the Permian evaporite sequence discovered in the Aislaby (Eskdale No. 2) borehole, Eskdale, near Whitby, Yorkshire. The mineral occurs in specimens from the halite–anhydrite zone, the polyhalite zone and the lower anhydrite zone. It occurs as soft flexible thin plates up to 5 mm in diameter and is invariably found along the cleavages and intergrain boundaries of halite grains (Stewart 1949, 1951b). Later borings proved talc in evaporites from No. 3 borehole at Sleights, in Eskdale (Armstrong et al. 1951), from the F.1 borehole at Fordon, Humberside (Stewart 1963), and from the M3 borehole at Wilton, near Guisborough, Cleveland (Raymond 1962). Stewart (1965) suggests that talc formed during geothermal metamorphism. Talc is also noted as an efflorescence on the walls of passageways in the abandoned anhydrite mine under the ICI factory site at Billingham-on-Tees, Cleveland (Raymond 1959).

Many early descriptions of talc or talc-like minerals are probably misidentifications, for example Brandy Gill in the Caldbeck Fells, the Egremont area and the Shap granite, all in Cumbria (Young 1987a). However, a pale-green cross-fibred veinlet in grey mudstone from Holehouse Gill, Ulpha, Cumbria, does appear to contain talc. At Lady's Rake mine, along Harwood Beck, Co. Durham, talc occurs in unusual magnetite-rich specimens containing calcite, galena, nickeline and sphalerite, with minor gersdorffite, pyrrhotite and ullmannite (Green & Briscoe 2002).

Talc has been reported from a few locations in Gwynedd, Wales. It forms thin veins cutting the Penmaenmawr intrusion, near Conwy, Lleyn Peninsula (Sargent 1925), and, at Dolfrwynog mine, near Llanfachreth, it occurs with chlorite in altered basic rocks along the course of chalcopyrite–gold–pyrite lodes (Dewey & Eastwood 1925). Because of the problems of identification outlined above, the early descriptions of talc from various locations on Anglesey (Greenly 1919) and the Dolgellau goldbelt (Readwin 1888) deserve further investigation. In the Four Mile Bridge area of Holy Island, Anglesey, talc and chlorite occur in dense white rocks formed during the alteration of serpentinite along faults in the vicinity of gabbros (Maltman 1977). Elsewhere in Wales, talc occurs in calcite–quartz veins cutting altered basaltic lavas at Llanelwedd quarry, near Builth Wells, Powys (Metcalfe 1990), and replacing olivine in gabbros from the St David's Head intrusion, Carn Llidi, Dyfed (Roach 1969).

Books of talc to 6 cm on serpentinite; Queyhouse quarry, Quoys, Unst, Shetland, Scotland. (© The Trustees of the National Museums of Scotland)

Perhaps the most important economic occurrences of talc in Britain or Ireland are on Shetland. At Queyhouse quarry, Quoys, W of Baltasound, Unst, 250 000 tonnes of talc were produced between 1930 and 1975. It occurs on the western margin of the Unst serpentinite in a talc–magnesite rock formed by metamorphism associated with thrusting (Dunham et al. 1978). Substantial talc–magnesite deposits also occur in areas of metamorphic rocks at Cunningburgh, 13 km SSW of Lerwick, and Fethaland in the N, both on Mainland, and near Hesta Ness on Fetlar (Berridge et al. 1969, Bain et al. 1971). Earlier descriptions of talc from Unst are provided by Heddle (1901), who notes magnificent apple-green foliated masses at North Cross Geo, Haroldswick, and both colourless and apple-green foliated masses at the Taing of Norwick. Read (1934a) describes talc-rich lenses up to 4 m long within gneiss at Burgar Stack on the west coast of Burra Firth, northwest Unst. Other more recent descriptions of talc in Scotland include the following.

FIFE Chapel quarry, near Kirkcaldy (Phemister & MacGregor 1942, Guppy 1956). GRAMPIAN Damhead quarry, 2 km SSW of Portsoy (Harry & Patterson 1949). HIGHLAND SW of Achmelvich, An Fharaid Mhòr, near Lochinver (Droop 1994); Ardintoul, N of Glenelg (Berridge 1969, Dunham et al. 1978); Glenmore of Glenelg (O'Hara 1960); Stoer, near Lochinver (Berridge 1969). On the Isle of Skye: Ard Ghunel (Matthews 1967); aureole of the Beinn an Dubhaich granite, Broadford area (Tilley 1948b,c, 1951, Hoersch 1981, Holness 1992); Camas Bar-avaig (Matthews 1967). STRATHCLYDE Boyleston quarry, SW of Barrhead (Meikle 1989a); McInroy's Point, SW of Gourock, Firth of Clyde (Herriot 1971). TAYSIDE Corrycharmaig mine, Glen Lochay, NW of Killin (Guppy 1956, Dunham et al. 1978, Hawson & Hall 1987, Power & Pirrie 2000). WESTERN ISLES Loch Meurach, South Harris (Robinson & Dossett 2002).

A good white talc interbedded in schists has been worked at Crohy Head, Lough Gartan, Co. Donegal, Ireland, and a pale-green talc is noted in metamorphic rocks on the islands of Inishbofin and Inishshark in Co. Galway (O'Brien 1959). Another discovered talc deposit at Westport, Co. Mayo, is briefly mentioned by Ryback & Moreton (1993).

talcite

TL Wicklow, Ireland

A synonym of damourite; a variety of muscovite, mica group. Talcite was also used as an early name for talc.

tallingite

TL Botallack mine, St Just, Cornwall, England

Named after Richard Talling (1820–1883), a mineral dealer who found many new species, including botallackite, churchite, devilline, langite, ludlamite and woodwardite. Talling was born in Lostwithiel, Cornwall, and had a "Fancy Repository" on the corner of Fore Street and Queen Street, Lostwithiel. Tallingite is a synonym of connellite.

talmessite

FAIRFIELDITE GROUP

triclinic $Ca_2Mg(AsO_4)_2 \cdot 2H_2O$

An uncommon hydrated calcium–magnesium arsenate that forms a series with gaitite and is found within the metamorphic aureole of the Dartmoor granite at Ivy Tor mine, Belstone, Meldon area, near Okehampton, Devon. A piece of veinstone from the mine dump, when broken open, revealed pink dolomite gangue and primary löllingite containing a cavity lined with white radiating spherules of talmessite, often mixed with rauenthalite or phaunouxite. The talmessite spherules are either silky and up to 0.5 mm across or they form more irregular waxy aggregates up to 1 mm across. Other minerals described from the mine include arsenopyrite, axinite, native bismuth, bismuthinite, chalcopyrite, garnet, marcasite, quartz, scheelite and unspecified cobalt-rich species (Braithwaite et al. 1996).

tamaite

monoclinic $(Ca,K,Ba,Na)_xMn_6(Si,Al)_{10}O_{24}(OH)_4 \cdot nH_2O$

A rare hydrated calcium–potassium–barium–sodium–manganese aluminosilicate hydroxide found in lenses of manganese ore within Ordovician mudstones at Benallt mine, Rhiw, Lleyn Peninsula, Gwynedd, Wales. Tamaite occurs as small micaceous patches in thin veins cutting the manganese ore. It is compositionally zoned, with parts of crystals having eggletonite, ganophyllite or tamaite compositions (T. F. Cotterell pers. comm. 2006). Previously, tamaite had been reported from Benallt mine, but under the name "Ca-rich ganophyllite" (Kato 1980, Matsubara et al. 2000).

tamarite

TL Wheal Tamar, Bere Ferrers, Devon, England

A synonym of chalcophyllite.

tamarugite

monoclinic $NaAl(SO_4)_2 \cdot 6H_2O$

An uncommon hydrated sodium–aluminium sulphate found at Hurlet, Paisley, Strathclyde, Scotland, where it forms white fluffy encrustations in association with alunogen and potassium alum (Macpherson & Livingstone 1982).

In England, tamarugite has been found over a restricted area at Compton Chine on the southwest coast of the Isle of Wight. It forms dominantly yellow, red and occasionally white globular encrustations covering the vertical surface of disturbed beds of bluish-black pyritiferous argillaceous siltstones of the Lower Greensand Group of Lower Cretaceous age. Oxidation of pyrite in the siltstone, coupled with the action of sea spray, allowed the tamarugite to form (King 1998a). Earlier reports of possible tamarugite from the now abandoned Billingham mine (under the ICI factory site), Cleveland (Raymond 1959, Stewart 1965), remain unconfirmed.

tangeite?

orthorhombic $CaCu(VO_4)(OH)$

Tangeite was once described as a synonym of calciovolborthite. Burke (2006), reporting on an IMA-CNMNC-approved mass discreditation of species, states that calciovolborthite is formally discredited and that compositions are either tangeite or vésigniéite. British reports of calciovolborthite now require further confirmation; some could be tangeite.

tantalite

See ferrotantalite and manganotantalite.

tapiolite

Now divided into ferrotapiolite and manganotapiolite. See ferrotapiolite.

tarnowitzite

A mixture of aragonite and cerussite, both aragonite group.

tavistockite

TL Tavistock, Devon, England

A synonym of apatite (or wavellite), but the aluminium and H_2O in the original analysis were probably derived from kaolinite, with which the apatite was intimately mixed. Formally discredited by Embrey & Fejer (1969).

telargpalite

cubic $(Pd,Ag)_3Te$

A rare palladium–silver telluride, identified in a layered mafic–ultramafic intrusion on the Isle of Rum, Highland Region, Scotland. Within certain chromitite layers of the Eastern Layered Series, telargpalite occurs in association with species identified as braggite, cooperite, hongshiite, irarsite, isoferroplatinum, keithconnite, kotulskite, laurite, moncheite, palladoarsenide, platarsite, sperrylite, telluropalladinite and the native species palladium and platinum. Most of these grains are 0.2–2 µm in diameter, although for some species grains > 15 µm are common. The majority are in contact with sulphide minerals such as bornite, chalcopyrite or pentlandite. The chromitite layers were formed during magmatic fractionation and crystal settling of an open-system magma chamber, modified by magma mixing and liquid immiscibility of sulphide and silicate melts (Power et al. 2000a).

tellurbismuth

A synonym of tellurobismuthite.

tellurium

trigonal Te

Tellurium is a rare species found in a hydrothermal vein deposit at Calliachar Burn, 4 km from Aberfeldy, Tayside Region, Scotland. It occurs in trace amounts in quartz–sulphide ± gold veins cutting Upper Dalradian gritty quartzites, quartz–mica schists and amphibolites. It crystallizes along galena cleavage planes during the partial alteration of hessite, in association with benleonardite, gold (electrum) and melonite (Ixer et al. 1997).

Tellurium has also been tentatively identified in the West Sgaorishal ultramafic plug, Isle of Rum, Highland Region, Scotland, where discrete PGMs, especially kotulskite and sperrylite, occur within disseminated sulphide (chalcopyrite–pentlandite) mineralization. Less commonly, rounded blebs of froodite, gold (electrum), sperrylite, tellurium, and unidentified Bi–Pt–Te and Pd–Sb–Te phases are also found. Grainsize is up to 10 µm, but more often about 2 µm (Power et al. 2003).

tellurobismuthite

TETRADYMITE GROUP

trigonal Bi_2Te_3

A rare bismuth telluride that forms a series with tellurantimony and is found at mines in the Dolgellau goldbelt, Gwynedd, Wales: Caegwian; Cefn Coch; Clogau, Bontddu; Ffridd-Goch, Llanfachreth; and Vigra. At Clogau mine, tellurobismuthite forms brilliant silver flakes up to 1.5 mm across in a calcite–quartz matrix with associated galena, other telluride species and occasionally native gold (Gilbey 1968, Bevins 1994). Two telluride assemblages were recognized by Naden (1988) in a detailed mineralogical investigation of the Clogau mine; first, a tellurobismuthite-dominated assemblage associated with minor galena, native gold and tellurides of silver and lead, and, secondly, a sulphide-dominated telluride assemblage (native bismuth, bismuthinite, hessite, pilsenite and tetradymite) with minor galena and rare native gold. Mason et al. (2002) illustrate characteristic symplectite-like intergrowths of galena, tellurobismuthite and tetradymite from Clogau mine, and complex galena–tellurobismuthite intergrowths with the very rare species aleksite are illustrated by Bevins & Stanley (1990). Kingsbury (1965c) and Embrey (1978) suggest that early reports of tetradymite from the Clogau and Vigra mines were misidentifications of tellurobismuthite, but tetradymite is confirmed in later works (Cook et al. 2007).

In Scotland, tellurobismuthite is reported as a minor component of copper–nickel mineralization near the base of a diorite intrusion at Talnotry, near Newton Stewart, Dumfries & Galloway. It is associated with argentopentlandite, chalcopyrite, gersdorffite, goethite, gold, marcasite, molybdenite, nickeline, pentlandite, pyrite, pyrrhotite, sphalerite and violarite (Stanley et al. 1987). Along the Strathclyde/Dumfries & Galloway border, in the Leadhills area, a single alluvial gold grain is reported to contain tellurobismuthite inclusions (Leake et al. 1998).

Tellurobismuthite is also noted from Gwavas quarry, 1 km NW of Helston, Cornwall (Siddiqui 1964).

telluropalladinite

monoclinic Pd_9Te_4

A rare palladium telluride, identified in a layered mafic–ultramafic intrusion on the Isle of Rum, Highland Region, Scotland. Within certain chromitite layers of the Eastern Layered Series, telluropalladinite is found in association with species positively or tentatively identified as braggite, cooperite, hongshiite, irarsite, isoferroplatinum, keithconnite, kotulskite, laurite, moncheite, palladoarsenide, platarsite, potarite, sperrylite, telargpalite and the native species palladium and platinum. Most of these grains are 0.2–2 µm in diameter, although for some species grains > 15 µm are also present. The majority are in contact with sulphide minerals such as bornite, chalcopyrite or pentlandite. The chromitite layers were formed during magmatic fractionation and crystal settling of an open-system magma chamber, modified by magma mixing and liquid immiscibility of sulphide and silicate melts (Butcher et al. 1999, Power et al. 2000a). Telluropalladinite has also been identified in chalcopyrite-rich dykes associated with the West Sgaorishal ultramafic plug on Rum. It occurs with froodite, gold (var. electrum), keithconnite, kotulskite, paolovite, native silver, sperrylite and Ag–Bi–Pd–Te phases, usually as rounded anhedral blebs enclosed within chalcopyrite or pentlandite. Grainsize varies between > 1 µm and < 40 µm, although the majority of grains are < 5 µm (Power et al. 2003).

temagamite

orthorhombic Pd_3HgTe_3

A rare palladium–mercury telluride. A single inclusion was found in one of 120 placer gold grains analyzed from Tod Burn, near the Whiteadder reservoir, Lammermuir Hills, Lothian, Scotland. Tiny 5 µm-diameter inclusion observed in a gold-rich alloy (98.5% Au, 1.2% Pd and 0.3% Ag), together with various unidentified polymetallic selenides, tellurides and selenotellurides of copper, mercury and palladium. Associated species forming inclusions in other gold grains from this site include bohdanowiczite, clausthalite, naumannite, oosterboschite, tiemannite and umangite, together with various selenide, telluride and selenotellurides containing bismuth (R. J. Chapman pers. comm. 2007).

tennantite

TETRAHEDRITE GROUP

cubic $Cu_6Cu_4(Fe,Zn)_2(As,Sb)_4S_{13}$

TL Cornwall, England

Named after Smithson Tennant FRS (1761–1815), a chemist who discovered iridium and osmium, and was professor of chemistry at the University of Cambridge; born in Selby, Yorkshire. Tennantite is a relatively uncommon copper–iron–zinc arsenide antimonide (As dominant) that forms a series with tetrahedrite (antimony dominant) and is found in hydrothermal veins in association with other sulphide species.

Early reports of tennantite are from Wheal Jewel and Wheal Unity, near St Day, Cornwall, where it was found in "great perfection" (Garby 1848, Greg & Lettsom 1858). It was also found in copper-rich lodes in the metamorphic aureole of the Bodmin Moor granite, at South Caradon mine, St Cleer, near Liskeard, where it occurred with azurite, bornite, chalcocite, chalcopyrite, cuprite, fluorite, malachite and tenorite (Shambrook 1982). During a study of disseminated tin-sulphide mineralization in the Hensbarrow stock of non-megacrystic lithium-mica granite (part of the St Austell granite complex), Manning (1983) identified tennantite along with bornite, cassiterite, digenite, mawsonite, stannite and stannoidite. Tennantite is also mentioned by Weiss (1994) in a review of stockwork mineralization, greisen zones and quartz veins at Gunheath china-clay pit, which now includes part of the Hensbarrow stock. Other Cornish locations for tennantite include: Dolcoath and North Roskear mines, Camborne (Greg & Lettsom 1858); Levant mine, near Pendeen, St Just (Wolloxall 1987); Carn Brea, Cook's Kitchen and Tincroft mines, Illogan; Tresavean mine, Gwennap; Trevascus and Relistian mines, Gwinear; Wheal Jane, Kea; Geevor mine, near Pendeen, St Just; and Hingston Down quarry, Calstock (all Golley & Williams 1995). Tennantite has been tentatively identified as a minor constituent of the sulphide ore at Wheal Gorland, St Day, 2 km E of Redruth (Rust 1981b).

Close to the beach, near the Pier hotel, Clevedon, Avon, vein-style mineralization cutting Upper Triassic dolomitic limestone breccias comprises banded barite with minor tennantite, chalcopyrite, enargite, galena (most abundant) and sphalerite, together with their alteration products. Under the microscope, tennantite is pale green and it displays orange-to-red internal reflections. It can be intergrown with sphalerite or it forms euhedral crystals overgrowing chalcopyrite (Starkey 1986c, Ixer et al. 1993).

In the Cheshire Basin at Bickerton, S of Warrington, tennantite is reported from barite-dominated redbed-type mineralization closely associated with marginal basin faults. The paragenetic sequence is: early pyrite (bravoite and nickeloan pyrite) occasionally with Cu–Fe–Ni sulpharsenide rims, followed by bornite and tennantite (the most abundant sulphides) in which chalcopyrite is found as lamellae. Covellite is a secondary replacement species (Naylor et al. 1989).

An silver–zinc-rich variety of tennantite is reported from the dumps of Lockridge (Goldstreet) mine, SW of Bere Alston, Devon. It occurs with galena, tetrahedrite and intergrown sphalerite–wurtzite (Grguric & Nickel 2006).

An early discovery of tennantite–tetrahedrite (fahlerz) from Fell End Clouds (also called Clouds mine), Ravenstonedale, 10 km SSW of Kirkby Stephen, Cumbria (Dakyns et al. 1891), has more recently been confirmed as tennantite (Small 1982). It was found in grey copper ore in a calcite–fluorite stockwork within limestone at High Skennerskeugh level. Starkey (1987) indicates that the main ore minerals at nearby Stennerskeugh Clouds includes galena and tennantite with minor chalcopyrite in a barite gangue, and notes duftite as a rare supergene species at the site. About 10 km away, at Higher Longrigg mine, Hartley Birkett, Kirkby Stephen, grey massive zinc-rich tennantite is associated with adamite (cuproadamite) and it occurs in chalcopyrite–galena-rich barite veins cutting limestone (Braithwaite 1988). Other Cumbrian occurrences of tennantite include: Birk Fell mine, Tilberthwaite (Stanley 1979); Dale Head mine, Newlands Valley (Stanley & Vaughan 1980); Holehouse Gill mine, Ulpha (Young 1987a); and Paddy End mine, Coniston, Eskdale (Stanley 1979, Young & Johnson 1985). Tennantite and tennantite–tetrahedrite are also reported from Driggith and Old Potts Gill mines, Caldbeck Fells, on the evidence of A. W. G. Kingsbury (Hartley 1984, Cooper & Stanley 1990). However, as many samples in the Kingsbury collection at the NHM are incorrectly provenanced (Ryback et al. 1998, 2001), doubts remain about tennantite from these locations. Green et al. (2006a) were unable to locate the Kingsbury specimens from Driggith mine for re-evaluation and they suggest that the description should be treated with caution.

Isolated reports of tennantite from elsewhere in England include: Alderley Edge ridge, Cheshire (Ixer & Budd 1998); Bicknoller quarry, near Watchet, Quantock Hills, Somerset (Criddle & Stanley 1993); and Scaleber Bridge, 3 km SSE of Settle, Yorkshire (Ixer & Stanley 1998). Tennantite–tetrahedrite is also reported from Judkins quarry, Nuneaton, Warwickshire (Ince et al. 1991).

Wales

Gwynedd Minor tennantite occurs within the chalcopyrite and to a lesser extent the galena–sphalerite portions of the Parys Mountain copper deposit, Anglesey. It forms 20 µm rims between chalcopyrite and galena, or chalcopyrite and sphalerite, and is also found as isolated anhedral 400 µm-diameter grains in sphalerite (Wheatley 1971b, Sivaprakash 1977, Pointon & Ixer 1980, Ridgway 1980, Southwood 1982, Bevins & Southwood 1995). At Great Orme mine, Llandudno, tennantite occurs as a trace constituent of a polymetallic ore assemblage comprising chalcopyrite, galena, marcasite, pyrite and siegenite, with minor azurite, cobaltite/gersdorffite, malachite and sphalerite (Ixer & Stanley 1996, Ixer 2001). Significant porphyry mineralization containing tennantite was recently exposed in a roadcut in the Coed-y-Brenin Forest. Bornite, covellite, malachite and tyrolite are associated species, the latter two originating by weathering (Armstrong et al. 2003).

Powys Minor barite–chalcopyrite–galena–tennantite vein mineralization has been reported from Dolyhir quarry, Old Radnor (Wolfe 2000, 2004), and more generally from the Old Radnor inlier of Silurian age (Parnell & Eakin 1989).

Scotland

Central An alluvial gold grain from the Ochil Hills contains inclusions of possible silver-rich tennantite (Leake et al. 1993).

Dumfries & Galloway Brecciated country rocks overlying granodiorite sheets associated with a porphyry dyke swarm contain tennantite in copper–molybdenum-rich mineralization at Black Stockarton Moor (Fortey 1980, Lowry et al. 1997); tennantite is a minor species associated with stibnite–sphalerite veining at Louisa mine, Glendinning, Glenshanna Burn, 13 km NNW of Langholm (Gallagher et al. 1983, Duller et al. 1997); spots of relict tennantite, native bismuth and uraninite occur with a suite of rare secondary alteration products in mineralized veins at Needle's Eye, SW of Caulkerbush (Braithwaite & Knight 1990).

Highland Irregular inclusions of a zinc-rich variety of tennantite is relatively common as shaped (~50 µm) in galena and sphalerite within calcite–fluorite–quartz veins associated with a quartz monzonite intrusion at Ratagain, Loch Duich, near Glenelg (Alderton 1986, 1988b).

Strathclyde Microscopic tennantite forms inclusions in galena and sphalerite associated with gold mineralization in the Lagalochan subvolcanic complex, near Kilmelford (Harris et al. 1988); closely associated arsenopyrite and tennantite have been found in gold-mineralized quartz–carbonate veins associated with the Fore Burn diorite–granodiorite complex, Tairlow plantation, near Straiton (Allen et al. 1982, Charley et al. 1989).

Tayside Individual occurrences of tennantite and arsenopyrite are reported from the Foss and Ben Eagach barite deposits, near Aberfeldy (Moles 1985); arsenic-rich tennantite is found as inclusions in pyrite within granitic lenses intruded into the Tomnadashan diorite, Loch Tay (Pattrick 1984); gold-bearing quartz–sulphide veins in Urlar Burn, Aberfeldy, contain rare tennantite rims between chalcopyrite and galena (Ixer et al. 1997).

Shetland Tennantite is reported on west Mainland from Matta Taing, Walls and Sandness, where it is associated with bornite, chalcopyrite, covellite, galena, pyrite and sphalerite in diagenetically formed sulphide nodules (< 20 µm) in the Devonian Upper Melby Fish Bed (Hall & Donovan 1978).

Ireland

Clare Tennantite occurs in a disseminated copper–silver sulphide deposit (mainly arsenopyrite, chalcopyrite, pyrite and galena) within argillaceous limestones and shales, also in calcite–sulphide veins at Ballyvergin (Schultz 1971, Andrew 1986).

Cork Massive milky quartz veins, often tens of metres wide, contain chalcopyrite ore with minor tennantite–tetrahedrite and bornite at the Berehaven (Bearhaven) mines, Allihies (Moreton & Aspen 1993), including Mountain mine (Schultz 1971); thin quartz veins cutting redbeds host bornite–chalcopyrite–galena mineralization, with minor chalcocite, pyrite and tennantite, at Cappaghglass (Kinnaird et al. 2002); the Tullacondra copper–silver deposit, NNW of Mallow, is hosted by Carboniferous limestones, shales and sandstones, and contains considerable amounts of very finely disseminated tennantite in a silver-rich zone (Wilbur & Carter 1986).

Galway Limestone-hosted Cu–Pb–Zn–(Ag) mineralization at Tynagh mine, near Killimor, includes tennantite as one of the primary ore species (Schultz 1966b, 1971, Boast 1971, Morrissey et al. 1971, Russell 1975, Boast et al. 1981a, Clifford et al. 1986).

Kerry At Ardtully, tennantite occurs with bornite–chalcopyrite–idaite and traces of cobaltite in epigenetic veins hosted by dolomitic limestones (Kinnaird et al. 2002); a limestone-hosted stratabound deposit at Crow Island within Lough Leane comprises bornite–chalcopyrite–tennantite and galena–sphalerite mineralization – rare tennantite is intergrown with other sulphides or forms inclusions in chalcopyrite, pyrite or sphalerite (Kinnaird et al. 2002, Ixer 2004); tennantite is also noted from the nearby limestone-hosted copper-polymetallic deposits at Ross Island and Muckross mine, Killarney National Park (Moreton et al. 1999, Kinnaird et al. 2002, Ixer & Pattrick 2003, Nex et al. 2003); at Leaghillaun, Bearhaven (Berehaven) Peninsula, tennantite occurs as silvery masses intergrown with chalcopyrite in a quartz vein hosted by Devonian sandstones (Moreton & Green 2006b).

Limerick Minor amounts of arsenopyrite, chalcopyrite and tennantite occur with major galena, pyrite and sphalerite in a complex limestone-hosted deposit at Courtbrown, on the south shore of the River Shannon, near Ballysteen (Grennan 1986, Reed & Wallace 2001).

Meath The carbonate-hosted lead–zinc deposit at Navan is dominated by galena, marcasite, pyrite and sphalerite mineralization, but both chalcopyrite and tennantite are also found (Wilkinson 2003).

Tyrone Gold-bearing quartz veins hosted by Dalradian metasediments contain pyrite with subsidiary arsenopyrite, bornite, chalcopyrite, native copper, galena, sphalerite, tennantite (near end member) and tetrahedrite at Curraghinalt, near Gortin, Sperrin Mountains (Clifford 1986, Earls et al. 1989, McCaffrey & Johnston 1996, Parnell et al. 2000).

Tipperary Tennantite (partly mercury-bearing) occurs in a Cu–(Ag–Hg) deposit within fractured redbeds at Gortdrum mine, near Tipperary town (Morrissey et al. 1971, Schultz 1971, Steed 1986, Duane et al. 1986, Green & Moreton 2001, Braithwaite & Wilson 2003) – Halliday & Mitchell (1983) note a bornite–chalcocite–tennantite assemblage in altered shale here; the carbonate-hosted Lisheen Pb–Zn–(Ag)

Mercurian tennantite on ferroan dolomite from Gortdrum mine, Co. Tipperary, Ireland; height of crystal 7 mm. (© D. I. Green)

deposit, between Thurles and Rathdowney, consists of three stratabound sulphidic lenses dominated by galena, marcasite, pyrite and sphalerite, in which tennantite occurs with arsenopyrite, bornite, boulangerite, bournonite, chalcopyrite, gersdorffite, gratonite, jordanite, nickeline, pararammelsburgite and seligmannite (Strashimirov et al. 1999, Hitzman et al. 2002, Ixer 2002, Fusciardi et al. 2004, Wilkinson et al. 2005); a stratabound, mainly marcasite–pyrite, but also galena–sphalerite-rich orebody at Silvermines, is variously described as containing tennantite–tetrahedrite (Morrissey et al. 1971, Schultz 1971), tennantite (Taylor & Andrew 1978, Moreton 1999) or argentotennantite (Zakrzewski 1989), although only the last author presents data to confirm the identification.

Offshore territories

In the Channel Isles, tennantite with variable composition and containing up to 20 wt% silver or up to 4.6 wt% lead has been found in a Pb–(Ag)-enriched brecciated vein cutting foliated hornblende granite at Sark's Hope mine, Little Sark (Bishop et al. 1977, Ixer & Stanley 1983, Ixer 1992).

tenorite

monoclinic $Cu^{2+}O$

A relatively uncommon copper oxide found as a supergene oxidation product of copper-rich orebodies. Tenorite occurs at Penberthy Croft mine, St Hilary, Cornwall, as black earthy crystalline aggregates infilling small cavities (up to 1.2 mm across) in cuprite. Associated species are aurichalcite, brochantite and connellite (Betterton 1996e), or dolomite and malachite (Rust 1982b). It occurs in oxidized vein material at Wheal Gorland, St Day, 2 km E of Redruth, associated with cuprite (Rust 1981b). At Wheal Pollard, N of St Cleer, in the Caradon and Phoenix mining district, a 12 cm-wide lode containing tenorite with chalcopyrite, fluorite and pyrite was found cutting tourmalinized granite and an elvan dyke (Shambrook 1982). Dewey (1925) comments that the lodes of many mines in the Caradon group contained azurite (chessylite), native copper, cuprite, malachite and tenorite near the surface, whereas, at lower levels, bornite, chalcocite and chalcopyrite appeared. Other locations in Cornwall where tenorite is noted include: Wheal Buller, near Redruth; Wheal Trefusis, Redruth (both Greg & Lettsom 1858); Botallack mine, St Just; Dolcoath mine, Camborne; East Crinnis mine, St Austell; Fowey Consols mine, Tywardreath, St Blazey; Great Wheal Fortune, Breage; Great Wheal Vor, Breage; Old Gunnislake mine, Calstock; Relistian mine, Gwinear; Wheal Basset, near Redruth; Wheal Jewel, St Day, Gwennap; Wheal Providence, near Carbis Bay, St Ives; Wheal Speed, St Just (all Collins 1892); Marke Valley mine, Caradon Hill, 3 km NE of St Cleer; South Caradon mine, St Cleer, near Liskeard (both Shambrook 1982); West Caradon mine, St Cleer, Liskeard (Lawrence 1990); Phoenix United mines, Linkinhorne, and Ting Tang mine, Gwennap (Golley & Williams 1995).

In Devon, tenorite occurs with cuprite in association with unusual sheets of native copper and concretions enriched in uranium–vanadium within Permian redbeds at Littleham Cove, Budleigh Salterton (Milodowski et al. 2000, 2002).

In Leicestershire, tenorite (syn. melaconite) has been found in supergene copper mineralization at Bardon Hill quarry, where it is associated with cuprite; and Clouds Hill quarry, near Breedon on the Hill, associated with native copper, goethite and malachite (King 1973). Tenorite also occurs as 5–150 μm-long radiating sheaves within cuprite in void-infilling copper mineralization near the Precambrian/Triassic unconformity at New Cliffe Hill quarry, Stanton under Bardon (Ixer et al. 2005a).

Tenorite overgrowths are reported on mineralized specimens of cuprite found in Triassic Mercia Mudstone Group sandstones to the W of Moorfield farm, N of Redditch, Hereford & Worcester (Old et al. 1991).

Tenorite is noted, but not described, from the mines around Ecton Hill, Staffordshire (Robey & Porter 1972), and from Watershaw Rake, Longstone, Derbyshire (Ford & Sarjeant 1964a). At Moons Hill quarry, NE of Shepton Mallet, Mendip Hills, Somerset, tenorite occurs with chrysocolla, covellite, cuprite and malachite, coating joint planes or in thin veins and patches adjacent to joints (Kingsbury 1941). It is also reported from Potts Gill barite mine, Caldbeck Fells, Cumbria, where it was recovered from a copper-rich vein together with brochantite, chrysocolla and malachite (Davidson & Thomson 1951). Greg & Lettsom (1858) list tenorite from Hay Gill and Roughton Gill, both in the Caldbeck Fells, and Firman (1978) notes it in copper-rich quartz veins at Levers Water, NW of Coniston.

In Wales, tenorite (syn. melaconite) occurred with anglesite, barite, native copper and tetrahedrite in a gossan (now mined away) associated with the Great Lode at Parys mine, Parys Mountain, Anglesey, Gwynedd (Greenly 1919, Dewey & Eastwood 1925). Supergene alteration of copper–dolomite ore at the Great Orme mines, Llandudno, Gwynedd, has produced early calcite crystals in association with native copper, covellite, cuprite, digenite, spionkopite and tenorite, all overgrowing limonite pseudomorphs after chalcopyrite; but, as redox conditions changed, later generations of calcite are associated with the growth of limonite, botryoidal malachite and manganese oxides (Ixer & Davies 1996, Ixer & Stanley 1996, Ixer 2001).

Black earthy coatings of tenorite are produced through the supergene alteration of chalcopyrite and cuprite in a few mines in the Central Wales orefield of Dyfed and Powys. These include: Darren mine, Pen-bont Rhydybeddau, Dyfed; Esgair Fraith mine, near Tal-y-bont, Dyfed; and Snowbrook (Nantyreira) mine, Powys (Jones & Moreton 1977). Black bituminous masses of tenorite, up to 5 mm across, coat chalcocite and chalcopyrite and at Esgair Hir mine, Ceulanymaesmawr, Dyfed (Rust & Mason 1988).

Early reports of tenorite (syn. melaconite) from Leadhills, Strathclyde, Scotland, by Greg & Lettsom (1858) and Heddle (1901) are unsubstantiated and, according to Macpherson & Livingstone (1982), require confirmation. Tenorite is further noted from Cu–Pb–(Ag)-rich quartz–carbonate veins in the South Bradda mines, N of Port Erin Bay, Isle of Man. Atacamite, cerussite, chalcopyrite, native copper, cuprite, galena, malachite and pyrite also occur in the lode (Dewey & Eastwood 1925). Blocks of quartz–sulphide veinstone collected on the beach below the mine add arsenopyrite, pharmacosiderite and scorodite (and possibly bismuthinite and sideronatrite) to the species found in the veins (Starkey 1998b). Tenorite is noted from Laxey mine on the Isle of Man (Lamplugh 1903).

tephorite

Mis-spelling of tephroite (Dines 1956).

tephroite

OLIVINE GROUP

orthorhombic $Mn^{2+}_2SiO_4$

A generally uncommon manganese silicate that forms a series with fayalite and is an important constituent of the manganese ores of Treburland mine, 2 km SE of Fivelanes, Altarnun, Cornwall, where it is found associated with bustamite, manganese-rich hedenbergite, rhodonite and spessartine. The ore lies within, or in contact with,

Granular tephroite in an interstitial matrix of calcite from Treburland manganese mine, Altarnun, Cornwall; width 2.75 mm; viewed in PPL (upper) and XPL (lower). (© K. Gardner)

a highly calcareous calc-silicate rock, originally an impure limestone that has been thermally metamorphosed by the adjacent Bodmin Moor granite pluton. The manganese ore was probably deposited as rhodochrosite prior to metamorphism. Tephroite forms very dark-brown fine-grain crystalline masses (or rarely cleavage crystals up to 3 cm long) often bordering rhodonite and mixed with pinkish-buff bustamite (Russell 1946d, Tilley 1946). Light-brownish tephroite, associated with rhodonite and pyrrhotite, and from a similar paragenesis to Treburland, is also reported from the aureole of the Dartmoor granite, at Meldon (British Railways quarry complex), near Okehampton, Devon (Howie 1965, Dearman 1966a).

In Wales, tephroite is described from the Benallt manganese mine, at Rhiw on the Lleyn Peninsula, Gwynedd, where it occurs intimately associated with alleghanyite. The ore consists of a complex mixture of manganese silicates (alleghanyite, pennantite, spessartine, tephroite) and other subsidiary manganese and iron species (ganophyllite, jacobsite, pyrochroite, pyrophanite). Tephroite forms olive-green composite narrow blade-like crystals up to 2 cm long (Campbell Smith et al. 1944, Dunham et al. 1978, White & Hyde 1982).

terra ponderosa aerata

A synonym of witherite.

tesselite

A variety of apophyllite with a tesselated appearance in polarized light.

TETRADYMITE GROUP

Trigonal, selenides and tellurides with the general formula A_2X_3, where A = Bi, Sb; and X = S, Se, Te. Tellurobismuthite and tetradymite are members of the group found in Britain or Ireland.

tetradymite

TETRADYMITE GROUP

trigonal Bi_2Te_2S

A rare bismuth sulphotelluride, said to be found within quartz veins at Brandy Gill mine, Caldbeck Fells, Cumbria, where it occurs in foliated masses with a brilliant metallic lustre on fresh surfaces. Tetradymite is reported with native bismuth, bismuthinite (sulphuret of bismuth) and molybdenite (Greg & Lettsom 1858, Postlethwaite 1913). Unfortunately, these records are thought to be misidentifications of ingodite or joséite (Cooper & Stanley 1990). There is also some doubt as to whether tetradymite or "grünlingite" occurs in quartz–wolframite veins associated with greisen at Carrock mine, Grainsgill, Caldbeck Fells (Hitchen 1934, Davidson & Thomson 1951, Eastwood et al. 1968, Young 1987a), although Clarke (1974) presents an EMP analysis that could be of tetradymite. Tetradymite requires confirmation in the Caldbeck Fells.

At Clogau mine in the Dolgellau goldbelt, Gwynedd, Wales, complex shear-zone-hosted mesothermal gold-lode mineralization has a four-stage paragenetic sequence, one of which is an Ag–Au–Bi–Sb–Te bonanza-type assemblage, where tetradymite is associated with native gold. Symplectite-like intergrowths of galena, tellurobismuthite and tetradymite are characteristic, and aleksite, native bismuth, bismuthinite, boulangerite, bournonite, hedleyite, nagyágite and pilsenite are also found (Mason et al. 2002). Tetradymite is further described from Clogau mine by Gilbey (1968), Naden (1988) and Bevins (1994), and much earlier by Smyth (1862), Forbes (1867), Readwin (1888), Andrew (1910) and Dewey & Eastwood (1925). Kingsbury (1965c) and Embrey (1978) suggest that the early reports of tetradymite were misidentifications of tellurobismuthite, but Cook et al. (2007) confirm its presence in association with aleksite, galena, tellurobismuthite and an unnamed phase (corresponding to $PbBi_4Te_4S_3$).

In Scotland, quartz veins cutting monzogranite at Loch Shin, 2 km W of Lairg, Highland Region, contain a bismuth sulphosalt assemblage in which tetradymite occurs with aikinite, berryite, gladite, hammarite, lindströmite, krupkaite and pekoite. The monzogranite also contains low-grade disseminated fluorite–molybdenite–pyrite mineralization (Lowry et al. 1994). Farther south, at Borland Glen in the Ochil Hills, Central Region, four alluvial gold grains have been found containing 10–12 µm-diameter tetradymite and pyrite inclusions. Other gold grains from here contain inclusions of galenobismutite, pilsenite, tennantite and wittichenite (Leake et al. 1993, Chapman et al. 2000c).

TETRAHEDRITE GROUP

Cubic sulphides, selenides and tellurides with the general formula $A_{12}B_4X_{13}$, where A = Ag, Cu, Fe, Hg, Zn; B = As, Sb, Te; and X = S, Se, Te. Argentotennantite, freibergite, tennantite and tetrahedrite are members of the group found in Britain or Ireland.

tetrahedrite

TETRAHEDRITE GROUP

cubic $Cu_6Cu_4(Fe,Zn)_2(Sb,As)_4S_{13}$

A relatively common copper–iron–zinc antimonide arsenide (antimony dominant) that forms two series, with tetrahedrite (As dominant) and with freibergite, and is found in hydrothermal veins or contact-metamorphic deposits.

England

Cheshire Tetrahedrite with high silver and zinc content forms small (5–15 µm across) green-grey anhedral crystals intergrown with sphalerite, all enclosed in galena, at Engine Vein mine and Stormy Point, Alderley Edge ridge. The mineralization is in faulted Triassic sandstones (Ixer & Vaughan 1982, Braithwaite 1994).
Cornwall Fine examples of tetrahedrite associated with bournonite were once found at Herodsfoot mine, near Liskeard (Collins 1904b, Rudler 1905), and large tetrahedral crystals up to 35 mm on edge are reported from Crinnis mine, St Austell (Greg & Lettsom 1858). In the Port Isaac–St Endellion mining district, tetrahedrite is one of the species found in polymetallic stratiform and vein deposits within Middle Devonian to Lower Carboniferous volcano-sedimentary basinal successions. The deposits include: the Bounds Cliff vein system, E of Port Isaac, where tetrahedrite is partially replaced by bournonite and possibly meneghinite; the Pentire vein system, near New Polzeath, St Minver, where unorientated rounded or irregular blebs of silver-bearing minerals occur as intergrowths in galena and include tetrahedrite commonly associated with polybasite or antimonpearceite; and the Port Quin vein system (at Port Quin and Reedy Cliff), where native gold has been found in tetrahedrite alteration zones. Tetrahedrite is also noted from Poltrewgorgy mine, Treore mine and Tregilden quarry, near Pendoggett, all in this same district (Russell 1944, 1949b, Clayton et al. 1990, Camm 1995). Tetrahedrite also occurs with bournonite, chalcopyrite, galena, native gold, pyrite and sphalerite in shear zones within brecciated and carbonatized (metasomatized) basaltic volcanics near Port Quin (Clayton 1993). Dark-grey lustrous metallic tetrahedrite crystals up to 15 mm on edge, and associated with calcite and dolomite, have been collected from Greystone quarry, Lezant, near Launceston (Elton et al. 2004). Other reports of tetrahedrite include: Crowns lode at Botallack mine, St Just (Hosking 1951); as a detrital species in the Fowey Estuary (Pirrie et al. 2002); flattened prisms up to 9 mm long at Gunver Head, near Padstow (Schlomann & Weiss 1998); brilliant crystals at Levant mine, St Just (Greg & Lettsom 1858, Wolloxall 1987); associated with bournonite on one specimen collected between Swanpool and Maenporth beaches (Dean 1983); well formed crystals up to 2 mm long in a primary Cu–Pb–Sb sulphide-rich vein at Newporth beach, 2 km SW of Falmouth (Dean 1982, Dean et al. 1983); in jamesonite–bindheimite-bearing quartz veins at Tolpetherwin quarry, Two Bridges, South Petherwin,

Tetrahedrite, 10 mm across, coated with a thin crust of chalcopyrite; Herodsfoot mine, Liskeard, Cornwall. (© M. P. Cooper)

and in Cu–Pb–Sb ore from West Tamar mine on West Town farm, Botus Fleming (Russell 1949b); silver-rich tetrahedrite and bournonite in copper ore from Wheal Anna Maria, Mawnan (Camm 1995); in dump material from the galena-bearing quartz lode at Wheal Golden, Perranzabuloe (Merry & Braithwaite 1995); massive at Wheal Prosper, near Falmouth (Greg & Lettsom 1858). Rust (1982a) considers polytelite to be a variety of tetrahedrite in a description of Silver Vein mine, Lostwithiel, but this could equally well be freibergite. A report of tetrahedrite from Pengenna (Trewethen) mine, St Kew (Kingsbury 1965), remains unproven, as Ryback et al. (1998) suggest that the provenance of many Kingsbury samples requires further confirmation.

Cumbria At Driggith mine, Caldbeck Fells, silver-rich tetrahedrite forms as anhedral grains (<200 μm across), both on quartz and as inclusions in galena (Stanley & Vaughan 1981). Other Caldbeck occurrences include: inclusions in sphalerite from Carrock mine, Grainsgill (Thimmaiah 1956), and small masses in granular quartz on the dumps of Nether Row Brow (Cooper & Stanley 1990). Tennantite–tetrahedrite is reported from the Driggith–Sandbeds outcrop workings and from higher Roughton Gill (Hartley 1984), but this requires confirmation, as the report could be based on Kingsbury specimens with dubious provenance (see Ryback et al. 1998). Tetrahedrite does occur at Sandbed mine as inclusion-free masses, up to 20 μm across, surrounded by chalcopyrite and larger rounded masses to several mm (Green et al. 2006a). There is also a Kingsbury connection with tetrahedrite from Embleton quarry, near Cockermouth (Young 1987a). Analytical confirmation was desirable for the tetrahedrite described from Clouds and Hartley mines, near Kirkby Stephen (Daykyns 1891, Davidson & Thomson 1951). Since then, Small (1982) has confirmed both tennantite from Clouds, and tetrahedrite from chalcopyrite-bearing veins at the nearby Cumpston Hill (Hanging Lund Scar), Mallerstang, 11 km S of Kirkby Stephen. At Eagle Crag mine, Patterdale, tetrahedrite occurs as tetrahedral crystals up to 15 mm across, which are usually weathered. In rare instances, sharp lustrous crystals have been found associated with chalcopyrite and galena in quartz veinstone (Stanley 1979, Stanley & Vaughan 1981, Green & Briscoe 2002). Conspicuous macroscopic, well crystallized tetrahedrite is associated with chalcopyrite, galena and sphalerite in the Ruthwaite Lodge vein, Grisedale (Young 1986). Other reports are from: Greenside mine, near Glenridding, Patterdale (Firman 1978); Goldscope mine, Newlands Valley, near Keswick (Young 1987a); Paddy End mine, Coniston, Eskdale (Russell 1925); Piers Gill, 1 km N of Scafell summit, Wasdale (Aldridge 2005); and Tilberthwaite Gill (Stanley & Vaughan 1981).

Derbyshire Tetrahedrite associated with pyrite (bravoite) has been found in chalcopyrite ore at Millclose mine, Wensley, near Matlock (Bannister 1940), and is also noted from Oxclose mine, Snitterton, W of Matlock (Mostaghel 1984).

Devon A complex lode system that runs parallel to the margin of the nearby Dartmoor granite in the Teign Valley, near Hennock, contains chalcopyrite, galena, pyrite, sphalerite and tetrahedrite in a barite–quartz gangue (Beer & Ball 1977, Ball et al. 1985b). Fault-bounded antimony–gold mineralization at Loddiswell mine, 3 km N of the village of Loddiswell, contains silver-rich tetrahedrite with native gold inclusions (Stanley et al. 1990b, Camm 1995). Silver-rich tetrahedrite is reported from the dumps of Lockridge (Goldstreet) mine, SW of Bere Alston. It occurs with galena, tennantite and intergrown sphalerite–wurtzite (Grguric & Nickel 2006). Fahlerz is a name used to describe tetrahedrite or tennantite from: Bere Alston mines; Britannia and Molland mines, near North Molton; Combe Martin mines; and Tavistock (all Hall 1868b).

Durham Tetrahedrite is of only minor importance in the Northern Pennine orefield. It has been found associated with pyrrhotite and sphalerite in a pyrite-dominated ore at Closehouse barite mine, SW of Middleton-in-Teesdale, Lunedale, and with arsenopyrite, gersdorffite and ullmannite in a more complex vein and metasomatic chalcopyrite–galena–marcasite–pyrite–sphalerite ore found in the Teesdale Valley (Vaughan & Ixer 1980). Microveins and small pegmatites within the Weardale granite, sampled in a borehole at Rookhope, contain minor amounts of tetrahedrite and other sulphides (Ixer 1986, Brown et al. 1987a).

Somerset Tetrahedrite is a rare sulphide found with chalcocite, chalcopyrite and galena at Higher Pitts mine, Priddy, Mendip Hills, a locality better known for the oxychlorides chloroxiphite, diaboléite and mendipite (Kingsbury 1941, Burr 1996).

Warwickshire Limited amounts of tetrahedrite–tennantite form <2 mm lustrous black crystals encrusting chalcopyrite at Judkins quarry, Nuneaton, where copper mineralization occurs close to the Precambrian/Triassic unconformity (Ince et al. 1991). Ixer et al. (2005a) describe void-infilling mineralization from this location that includes tetrahedrite as a late-forming minor constituent overgrowing chalcopyrite into voids or cementing chalcopyrite and altered copper-poor bornite.

Wales

Dyfed Tetrahedrite is uncommon in Wales, and most reports are from the Central Wales orefield (in the Tal-y-bont–Goginan area), where it is usually part of the A1-c: early complex polymetallic assemblage of Mason (1997). Chalcopyrite and galena are the major components of the assemblage, and minor constituents are bournonite, cobaltite, millerite, pentlandite, pyrite, siegenite, tučekite and ullmannite. Where visible in hand specimen, tetrahedrite forms dark-grey granular masses up to 4 cm across, occupying cavities in quartz-cemented breccia. In thin section it is frequently intergrown with bournonite, and both occur as inclusions in galena. In many cases, tetrahedrite is silver rich and it approaches freibergite in composition. Localities include: Alltycrib mine, Tal-y-bont; Bwlchrhennaid mine, near Goginan; Cwmerfin mine; Darren mine, Pen-bont Rhydybeddau; Eaglebrook (Nantycagl) mine, Ceulanymaesmawr; Erglodd mine, N of Tal-y-bont; Esgair Fraith mine, near Tal-y-bont; Goginan mine; Nantyrarian mine; Pengraigddu mine; and South Darren mine (Rust & Mason 1988, Bevins 1994, Mason 1997, 1998). Elsewhere in Dyfed, it is uncertain if the fahlerz (syn. fahl-ore) found at Cystanog mine, near Carmarthen (Carruthers et al. 1915), is tennantite or tetrahedrite.

Gwynedd Tetrahedrite (forming a solid-solution series with tennantite) is confirmed from the Parys Mountain copper deposit, Anglesey, where it has a restricted occurrence within the chalcopyrite and to a lesser extent the galena–sphalerite portions of the deposit. It forms isolated grains up to 400 μm across, or inclusions and veinlets in sphalerite, or 20 μm-thick rims between chalcopyrite and galena, or chalcopyrite and sphalerite. EMP data suggest that it is less common than tennantite (Greenly 1919, Wheatley 1971b, Sivaprakash 1977, Pointon & Ixer 1980, Ridgway 1980). In the Dolgellau area, tetrahedrite is of minor importance, but is noted from: Clogau mine, Bontddu (Naden 1988); Foel Ispri mine, as inclusions in galena (Gilbey 1968); and Gwynfynydd mine, as inclusions in sphalerite (Gilbey 1968, Rice & Sharp 1976, Allen & Jackson 1985, Leake et al. 1993). Additionally, the polytelite from Tyddyn Gwladys mine in the Dolgellau goldbelt, described by Forbes (1868), is probably silver-rich tetrahedrite, particularly as Naden (1988) found tetrahedrite on material from this location in the Kingsbury collection at the NHM.

Scotland

Borders Alluvial gold grains from the vicinity of Glengaber Burn between Moffat and Peebles contain inclusions of tetrahedrite and many other species. In the tributary of Craigierigg Burn, the inclusion assemblage includes the species arsenopyrite, chalcopyrite, galena, gersdorffite, pyrite, sphalerite, tetrahedrite, an unnamed nickel sulphosalt and palladium–silver tellurides (Leake et al. 1993, Chapman et al. 2000a,c).

Central Lead–zinc vein and breccia mineralization associated with the Tyndrum–Glen Fyne fault contains a very unusual variety of tetrahedrite, which is both silver and cadmium rich. It occurs as inclusions in galena, particularly from the Hard vein of Tyndrum main mine, where inclusions reach 100 μm across (Pattrick 1978, 1985, Ridgway 1980). A few km down stream from this mineralization, in the Crom Allt river near Tyndrum, Pattrick & Hall (1984) found a mineralized boulder that also contained tetrahedrite–tennantite inclusions in galena. Tetrahedrite is noted in the Ochil Hills from Airthrey and the Blairlogie mines, although there is some doubt about the identification (Greg & Lettsom 1858, Heddle 1901, Dickie & Forster 1974).

Dumfries & Galloway Minor amounts of tetrahedrite and a suite of other As–Cu–Fe–Pb–Sb sulphides are reported from sphalerite–stibnite-bearing quartz veins at Glendinning (Louisa) mine, Glenshanna Burn, 13 km NNW of Langholm, and from boreholes in the surrounding area (Gallagher et al. 1983, Duller et al. 1997). Tetrahedrite is also noted in a vein gold prospect at Moorbrock Hill, near Carsphairn (Naden & Caulfield 1989). An analysis labelled tetrahedrite from the Black Stockarton Moor subvolcanic complex, to the NE of Kirkcudbright (Lowry et al. 1997), is actually tennantite. Minor tetrahedrite occurs as 5–7 μm inclusions in arsenopyrite within an Ag–As–Cu–Pb–Sb–Zn assemblage hosted by a late-Caledonian diorite–granodiorite intrusion at Knipe (Fountainhead) mine, Hare Hill, SE of New Cumnock (Naden & Caulfield 1989, Boast et al. 1990).

Highland Along a stream section in Gleann a' Chaolais, altered granite belonging to the Ballachulish igneous complex contains minor tetrahedrite in a chalcopyrite–molybdenite–pyrite stockwork (Fortey 1980).

Lothian Tetrahedrite occurs with galena, sphalerite and minor chalcopyrite in a vein gold prospect at Stobshiel, S of Haddington (Naden & Caulfield 1989).

Strathclyde The Fore Burn dioritic complex, Tairlow plantation, near Straiton, contains arsenopyrite–(chalcopyrite–pyrite)-bearing quartz–carbonate veins that host both native gold and native silver. Within this assemblage, tetrahedrite typically occurs within massive chalcopyrite, which it replaces along fractures and cleavages (Allen et al. 1982, Charley et al. 1989). A Caledonian diatreme–porphyry breccia complex at Lagalochan, near Kilmelford, also contains tetrahedrite together with precious metals in a zoned shear-related arsenopyrite–(chalcopyrite)–galena–hessite–pyrite–sphalerite suite that forms veins and disseminations in gouge and rockflour matrix. Freibergite and an unnamed bismuth–lead telluride are associated species (Harris et al. 1988).

Tayside Mineralization at Tomnadashan, Loch Tay, is associated with granitic lenses intruded into a small Caledonian diorite body and it comprises chalcopyrite, pyrite, tennantite and tetrahedrite as disseminations and irregular veinlets. In some veins, masses of tetrahedrite reach 5 cm across, although much smaller blebs and inclusions in pyrite are perhaps more common (Heddle 1901, Dunham et al. 1978, Fortey 1980, Ridgway 1980, Pattrick 1984, Livingstone 2002). At Calliachar Burn, 4 km from Aberfeldy, quartz–sulphide ±gold veins cutting Upper Dalradian gritty quartzites, quartz–mica schists and amphibolites contain zoned tetrahedrite crystals included along the grain boundaries of galena or forming small intergrowths with hessite ± benleonardite ± electrum. Tetrahedrite also occurs in the nearby Urlar Burn quartz–sulphide veins. In both localities, tetrahedrite is subsidiary to chalcopyrite, galena ± pyrite ± sphalerite (Ixer et al. 1997).

Shetland Heddle (1901) reports tetrahedrite in chalcopyrite–malachite-bearing veins in Old Red Sandstone from Sandlodge mine on Mainland.

Ireland

Cavan Tetrahedrite and galena are subordinate to sphalerite in a small vein cutting pyritic carbonaceous shales between Muff and Tonyhull townlands (O'Donnell 1980, Morris 1984).

Clare A disseminated sulphide deposit in argillaceous limestones and shales at Ballyvergin, Ballyvergin Townland, contains minor tennantite–tetrahedrite in a predominantly chalcopyrite–pyrite ore (Schultz 1971).

Cork Devonian redbeds host significant mineralization in the SW of Ireland. These include: chalcopyrite-bearing quartz veins, containing minor pyrite and tetrahedrite, that form a major copper deposit once worked at the Berehaven (Bearhaven) mines (including Mountain mine), Allihies (O'Brien 1959, Schultz 1971, Halliday & Mitchell 1983, Reilly 1986, Moreton & Aspen 1993, Kinnaird et al. 2002); a quartz vein at Ballycummisk, near Rossbrin Cove, dominated by molybdenite and tetrahedrite, the latter replaced by later bornite and chalcopyrite (Ni Wen et al. 1999, Kinnaird et al. 2002); minor arsenopyrite–bournonite–chalcopyrite–galena–meneghinite–tetrahedrite-bearing veins at Dhurhode (Ni Wen et al. 1991, 1996, 1999, Kinnaird et al. 2002); a chalcopyrite–tetrahedrite vein at Gortavallig (Cole 1922); arsenopyrite, bornite, chalcopyrite, galena and tetrahedrite disseminated and in minor veins at Kilkilleen (Ni Wen et al. 1999); and minor occurrences of tetrahedrite also occur at Cappagh, Crookhaven and Rooska West (Ni Wen et al. 1999). In some of these publications, tetrahedrite is used as an abbreviation for the tennantite–tetrahedrite solid-solution series, thus making positive identification of tetrahedrite uncertain.

Galway Chalcopyrite, galena, sphalerite and tennantite–tetrahedrite were the economic sulphides sought at Tynagh mine, near Killimor (Morrissey et al. 1971).

Kerry Tetrahedrite is reported with bornite, calcite and dolomite in limestone at Ardtully (Cloontoo) mine, near Kenmare (Cole 1922, Halliday & Mitchell 1983). At Ross Island mine, Killarney, tetrahedrite occurs with chalcopyrite and bornite in pockets, irregular patches and strings within limestone (Russell 1927b).

Kilkenny Stratabound replacement lead–zinc mineralization predominates at Galmoy, near Kilkenny, but associated with it is fracture-fill copper–silver mineralization in which chalcopyrite and tennantite–tetrahedrite occur (Doyle et al. 1992).

Limerick Fragments of quartz gangue containing bornite and tetrahedrite are reported from Pallaskentry mine (Cole 1922).

Longford A cadmium-rich sphalerite deposit in Carboniferous sandstones, conglomerates and sandy limestones at Keel, Ardagh, contains silver-rich tetrahedrite as minute inclusions in sphalerite (Patterson 1970, Morrissey et al. 1971, Slowey 1986).

Mayo Gold-bearing quartz veins in the Cregganbaun shear zone at Cushinyen contain aggregates of tetrahedrite together with bournonite, chalcopyrite, galena, pyrite and sphalerite in mm-size flecks in the main lode (Thompson et al. 1992). Tetrahedrite forms inclusions in alluvial gold grains from Croagh Patrick and the Lugbeg River (Chapman et al. 2000c).

Monaghan Tetrahedrite forms inclusions in pyrite within the arsenopyrite–gold–quartz–stibnite vein deposit at Clontibret mine (Lisglassan–Tullybuck deposit), Tullybuck (Morris 1984, Morris et al. 1986).

Tipperary The Cu–(Ag–Hg) deposit at Gortdrum mine, near Tipperary town, contains tennantite–tetrahedrite (in part, mercury bearing) and other sulphides in dolomitized intrusive breccias (Schultz 1971, Duane et al. 1986). Faulted dolomitized limestones and sandstone breccias host predominantly lead–zinc mineralization in the Silvermines district and include: abundant silver-rich tetrahedrite (in places), barite and chalcopyrite at Gortnadyne (Gorteenadiha) mine, near Silvermines, and minor silver-rich tetrahedrite with barite and galena at Shallee mine (Russell 1907b). Rhoden (1959) reports a multitude of inclusions of boulangerite, bournonite and tetrahedrite in galena from Shallee mine. At Mogul mine, fine specimens of gem-like dolomite, galena, pyrite and sphalerite, with rare tetrahedrite and hemimorphite, were recovered from a large cavity found in 1978 (Ryback & Moreton 1991). The deposits are also discussed more generally by Morrissey et al. (1971), Schultz (1971), Taylor & Andrew (1978), Zakrzewski (1989) and Moreton (1999).

Tyrone Members of the tennantite–tetrahedrite solid-solution series are widely distributed within gold- and sulphide-bearing quartz veins in metamorphosed Dalradian rocks at Curraghinalt, near Gortin in the Sperrin Mountains (Earls et al. 1989, McCaffrey & Johnston 1996, Parnell et al. 2000). At nearby Cavanacaw, near Omagh, silver-rich tetrahedrite is the only sulpharsenide found in similar quartz veins (Parnell et al. 2000). Traces of tetrahedrite were also found in pyrite-dominated gold–quartz veins at the Lack deposit, WSW of Omagh (Cliff & Wolfenden 1992). Chapman et al. (2000c) report tetrahedrite in alluvial gold from Attagh Burn in the Sperrin Mountains.

Waterford A fine-grain cluster of sulphides in veinstone from Tankardstown mine, Bunmahon has been found to contain small (<20 μm) grains of silver-rich tetrahedrite associated with chalcopyrite and a cobalt arsenide mineral (Moreton 2005a).

Westmeath Carbonate-hosted lead–zinc mineralization at Ballinalack, 15 km NW of Mullingar, contains minor millerite associated with pyrite (bravoite) in an ore dominated by barite, galena, marcasite–pyrite and sphalerite, all in a calcite gangue (Jones & Brand 1986).

Wicklow Tetrahedrite is a rare constituent of a low-grade sulphide replacement deposit (dominated by chalcopyrite, galena, pyrite and sphalerite) in schistose tuffs, lavas and acid intrusives at St Patrick's mine, Avoca (King & Evans 1964). Russell (1949b) reports a small patch of tetrahedrite on a specimen from Cronebane (Magpie) mine, Ovoca.

Offshore territories

Isle of Man Massive tetrahedrite very rich in silver (and referred to as polytelite) was once recovered from Foxdale lead mine (Forbes 1867, Russell 1949b).

Jersey Tetrahedrite forms inclusions in chalcopyrite and pyrite within a carbonate vein cutting granite at Le Doüet de la Mer, St Ouën (Ixer 1980, 1992). At Le Pulec on the northwest coast of Jersey, more substantial mineralization is hosted by the Jersey Shale Formation, where three galena- and sphalerite-bearing carbonate veins were discovered. Here tetrahedrite is silver rich and it forms intergrowths with chalcopyrite, thin veinlets within sphalerite and large (<500 μm) subhedral grains around sphalerite (Ixer & Stanley 1980, Stanley & Ixer 1982, Ixer 1992). Tetrahedrite is an accessory mineral in net-veined aplogranite from La Moye (Ixer 1988), and small molybdenite-bearing calcite–quartz veins at L'Etacq quarry also contain minor tetrahedrite (Ixer 1980, 1992).

Sark The Pb–(Ag) lode worked at Sark's Hope mine, Little Sark, contains members of the tennantite–tetrahedrite series that approach freibergite compositions because of their high silver content (Ixer & Stanley 1983, Ixer 1992).

tetranatrolite

tetragonal $(Na,Ca)_{16}(Al_{19}Si_{21})O_{80}\cdot 16H_2O$

An hydrated sodium–calcium aluminosilicate found at Hills Port, S of the Gobbins, Island Magee, Co. Antrim, Ireland. Throughout this part of Co. Antrim, chabazite, gonnardite, heulandite, natrolite, paranatrolite, tetranatrolite, thomsonite and other zeolite species infill vesicles in Tertiary basalts (Nawaz 1988a). Tetranatrolite is fibrous and it forms white-to-pink spherules up to 3 cm across, intimately associated with thomsonite (Ryback et al. 1988). Tetranatrolite was considered to be a product of dehydration of paranatrolite and was given doubtful status. It has structural characteristics similar to gonnardite (Coombs et al. 1997). Tetranatrolite was later discredited and the mineral confirmed to be gonnardite (Gottiardi & Galli 1999).

thalite

A synonym of saponite, smectite group.

thaumasite

ETTRINGITE GROUP

hexagonal $Ca_6Si_2(CO_3)_2(SO_4)_2(OH)_{12}\cdot 24(H_2O)$

Thaumasite is a rare complex hydrated calcium silicate carbonate sulphate hydroxide. At High Sedling mine, Cowshill, Weardale, Co. Durham, it has been found as crusts of soft white spherules, up to 0.25 mm across, on weathered mudstone (Young et al. 1998). At Shap Blue quarry, Cumbria, thaumasite occurs as blocky hexagonal crystals, up to 0.7 mm across, encrusting prehnite crystals. It probably formed during a late phase of hydrothermal mineralization that also produced apophyllite, laumontite and pectolite (Young et al. 1998). Young et al. (1985c) report thaumasite intermixed with thenardite from a gypsum quarry in the Edale Shales near Kirkby Thore, Cumbria. A specimen of thaumasite from Embleton quarry, near Cockermouth, Cumbria, supposedly found by A. W. G. Kingsbury and now at the NHM, is documented by Embrey (footnote in Knill 1960). Ryback et al. (1998) have shown that some samples in the Kingsbury collection have been incorrectly provenanced, and the status of the thaumasite sample is yet to be confirmed.

Thaumasite has been collected on the Glasdir mine dumps, near Dolgellau, Gwynedd, where it forms white powdery acicular crystals in a mineralized breccia or post-mining cemented aggregate (Bevins 1994).

Thaumasite is also recorded from Shetland at Colla Firth, north Mainland, where it is accompanied by gyrolite, pectolite and xonotlite in serpentinite veins (Butler et al. 1960).

In Ireland, patches of soft white acicular crystals of thaumasite up to 3 mm long have been found in a borehole cutting a Tertiary dolerite sill near Flush Hall, Ballyalton, Newtownwards, Co. Down (Knill 1960).

theisite

orthorhombic $Cu_5Zn_5As_2O_8(OH)_{14}$

A copper–zinc arsenate hydroxide found at Tynagh mine, near Killimor, Co. Galway, Ireland. Theisite is found in an unconsolidated partially oxidized cap enriched in other supergene minerals above a primary sulphide orebody consisting predominantly of galena, sphalerite and pyrite. Altogether 104 species have been described from this locality, including some new to Ireland (bayldonite, beudantite, chalcophyllite, hedyphane, olivenite, parnauite and strashimirite) and some new to both Britain and Ireland (clinotyrolite, orthoserpierite and theisite) (Ryback & Moreton 1992).

thenardite

orthorhombic Na_2SO_4

An uncommon sodium sulphate forming efflorescences on the walls of the Billingham anhydrite mine (under the ICI factory site), Cleveland. Thenardite is one of the secondary minerals that have grown in recent years by evaporation of water from the Upper Magnesian Limestone. Other species confirmed are epsomite, gypsum, halite, lepidocrocite, pyrite, sylvite and talc; ilsemannite, tachyhydrite and tamarugite are also probably present (Raymond 1959). Post-mining efflorescences on mudstone from the workings of the Mountfield mine, Battle, Sussex, contain colourless-to-white loosely aggregated thenardite associated with gypsum seams (Young et al. 1985). Specimens of mirabilite from a gypsum quarry in the Eden Shales near Kirkby Thore, Cumbria, have altered during storage to white powdery thenardite (Young et al. 1985c). Perhaps the most remarkable occurrence of thenardite is from a gravel pit at Wingmoor farm, Bishop's Cleeve, Gloucestershire, where Lower Jurassic mudstones had been exposed. The mudstones were covered with a white powdery coating of thenardite (King 1997).

theophrastite

BRUCITE GROUP

trigonal $Ni(OH)_2$

TL Hagdale quarry, Unst, Shetland, Scotland (also a TL in Greece)

A rare nickel hydroxide found encrusting chromitite at the type locality. Theophrastite intimately mixed with zaratite was initially described as a bluish-green unnamed nickel hydroxide by Macpherson & Livingstone (1974) and formally described later by Livingstone & Bish (1982). At the same time, the species was found and described by Greek authors, hence its name being derived from that of Theophrastus (the Greek philosopher). Theophrastite forms aggregates of thin platy crystals and may be either bright emerald green or pale bluish green, with the overall appearance of a gel-like coat. Specimens also contain chlorite and goethite, and lesser amounts of heazlewoodite, honessite, pentlandite and reevesite. Breakdown of the rare primary nickel sulphides in the chromite deposit allowed theophrastite to form as an alteration product (Livingstone & Bish 1982).

thinolite

A pseudomorph of calcite after an unknown mineral, possibly ikaite.

thiospinel

A general term for compounds AB_2S_4 with the spinel structure.

thomsonite (series)

ZEOLITE GROUP

orthorhombic $Ca_2Na(Al_5Si_5)O_{20}{\cdot}6H_2O$

TL Old Kilpatrick, near Dumbarton, Strathclyde, Scotland

Named after Thomas Thomson (1773–1852), professor of chemistry at the University of Glasgow and author of *Outlines of mineralogy and chemistry* (1836), who was editor of the journal in which the first description was published; born in Crieff, Perthshire (part of Tayside Region), Scotland. Thomsonite is a common hydrated calcium–sodium aluminosilicate, whose name has been given series status in the zeolite classification of Coombs et al. (1997). It is often found in low-silica basalts and less commonly in contact-metamorphic zones. Thomsonite-Ca is reported from Britain and Ireland, but most descriptions lack the species name.

Thomsonite was said to be plentiful in the neighbourhood of Kilpatrick and Dumbarton, and, although terminated crystals of the species are rare, some were found near Kilpatrick (Greg & Lettsom 1858). At Bowling quarry, 2 km WNW of Old Kilpatrick, thomsonite is associated with veins of prehnite and forms radiating groups of acicular white crystals, which are sometimes barred with red (Heddle 1901). At Hartfield Moss, 7 km NE of Beith, SW of Glasgow, Strathclyde, thomsonite is a common species that forms chalky, white or colourless, fibrous or compact, radial aggregates and it is also noted with prehnite and natrolite as colourless well formed crystals up to 30 mm long. Analcime, harmotome, heulandite, laumontite and stilbite are other zeolite species found at this locality (Meikle 1990). Thomsonite is relatively abundant in altered basaltic lavas at Loanhead quarry, near Beith, Strathclyde, where it occurs in amygdales up to 70×140 mm. Radial groupings of crystals or individual well formed, clear terminated crystals up to 30 mm long occur in association with analcime, calcite, harmotome and natrolite (Meikle 1989b). Adjacent to a basaltic dyke cutting the lavas at Loanhead, thomsonite is found in vesicles closely associated with analcime and a low-temperature type of garnet (Livingstone 1989a). Superb cabinet specimens of white stellate thomsonite have since been found at this locality (Green & Todd 2001). Museum-quality specimens of thomsonite have also been recovered from Boyleston quarry, SW of Barrhead (Hall et al. 1989). In all of the above examples, thomsonite is hosted in Carboniferous basalts belonging to the Clyde Plateau Lavas.

Spray of prismatic terminated thomsonite crystals to 65 mm; Loanhead quarry, near Beith, Strathclyde, Scotland. (© D. I. Green)

Thomsonite is also found in basaltic lavas and dykes of Tertiary age on the mainland and on offshore islands along the west coast of Scotland. Allison (1936) notes its occurrence with analcime and natrolite in dolerite and crinanite (alkali dolerite) dykes at Craignish Point, end of the Craignish Peninsula, SW of Kilmelford, Strathclyde. One particular porphyritic dyke was found to contain a single large cavity, 10–13 cm across, containing goethite, jarosite and native sulphur, whereas other smaller vesicles contain analcime and thomsonite (Livingstone & Baird 1991). The dykes are related to the intrusion of the now-eroded remnants of a large central volcano centred over the Isle of Mull. On Mull itself, Walker (1971) discovered a zonal sequence of zeolite species in amygdaloidal basaltic lavas to the W and NW of the central intrusive complex, and thomsonite is reported from the mesolite, laumontite and prehnite zones. The species also occurs on the adjacent mainland in the Morvern Peninsula, Highland Region. On An Gearna, southeastern slopes of Ben More, Isle of Mull, vesicular basalts contain rare thomsonite forming sheaf-like growths intimately associated with albite, prehnite and scolecite. Thomsonite is also occasionally found in basalts on Beinn Fhada, to the N of Ben More, associated with albite, epidote and prehnite (McLintock 1910b).

The Tertiary basalts of the Isle of Skye, are well known for zeolite species, and thomsonite appears to be the most common, being found in 94% of localities visited by King (1976). Many species including thomsonite are reported from Moonen Bay, on the west coast of the Duirinish Peninsula, where it forms spherical aggregates of lathlike crystals, some reaching 1 cm in diameter, and commonly associated with apophyllite, or more rarely with chabazite, mesolite or stilbite (Green & Todd 1996). A detailed investigation of the minerals from a weathered vein cutting basalt, 2 km N of Uig village, proved that uigite is an intimate mixture of gyrolite and the faröelite variety of thomsonite (Sweet 1959).

Dyer et al. (1996) found thomsonite in basaltic dykes and sills on the Isle of Arran, Strathclyde, at Bennan Head, Breadlebane, Kingscross Point and also Whiting Bay. Additionally, these authors report thomsonite in basaltic dykes on the Isle of Islay (Kildalton Cross and on the foreshore near the Bunnahabhain distillery) and the Isle of Colonsay (Uragaig, Kiloran Bay), both Strathclyde. Thomsonite is the main zeolite species in the amygdaloidal top of a crinanite sill on the Shiant Isles, Western Isles. It formed along with analcime and phillipsite during post-magmatic deuteric alteration of the primary felsic assemblages (Gibb & Henderson 1996).

In Highland Region, other parageneses for thomsonite are reported. It occurs with analcime, apophyllite and prehnite in contact-metamorphosed Moine Series banded limestones in the Ardgour district, on the north side of Loch Linnhe, a few km SW of Fort William and Ben Nevis (Drever 1939). At Camphouse, Ardnamurchan, an augite diorite contaminated with Liassic limestone contains thomsonite as a secondary species. It is accompanied by analcime, prehnite, xonotlite and hydrogarnets (Agrell 1951). Thomsonite pseudomorphs feldspathoid in an alkali syenite termed "ledmorite", found in the Ledbeg River by the Elphin Road, Assynt. The locality is close to Loch Borralan (Tilley 1958). Other locations in Scotland include the following.

CENTRAL Mugdock tunnel, 2 km S of Strathblane (Heddle 1901). HIGHLAND (MOSTLY ON THE ISLE OF SKYE) Allt Coir' a' Ghobhainn, near Drynoch (Livingstone 1988); Allt Ribhein, Fiskavaig Bay (Mackenzie 1957); Dunvegan (Heddle 1880b); Echoing rock, near Fir Bhreugach, 2 km N of the Quiraing (Sweet 1960); Leac na' Fionn, Quiraing, N of Portree; Rubha na h-Airde Glaise, 3 km NE of Portree; the Storr, N of Portree (all Heddle 1901); N of Portree on the Staffin road (Sweet 1961); Quiraing, N of Portree (Pearce 1996a). The faröelite variety of thomsonite is reported from Uamh Fhraing, Isle of Eigg (Heddle 1901). STRATHCLYDE Bishopton (Glasgow–Greenock railway) tunnel and Bowling quarry, Kilpatrick Hills, near Glasgow (Heddle 1901, Hey 1932a, Mourant 1933, Tschernich 1992); Boyleston quarry, SW of Barrhead (Heddle 1887a); Cairn na Burgh More, Cairn na Burgh Beg and Lunga, three of the Treshnish Isles; Gryfe waterworks tunnel, 4 km S of Greenock; Loch Thom, S of Greenock; the Slacks, Kilpatrick Hills, N of Old Kilpatrick (all Heddle 1901); Kilmalcolm, 10 km SE of Greenock (Heddle 1901, Gottardi & Galli 1985); Lochwinnoch (Tschernich 1992); Port Glasgow (Greg & Lettsom 1858, Heddle 1901, Gottardi & Galli 1985); Reilly quarry and landfill site, Houston (Wolfe 1997).

In Ireland, thomsonite is a common species found in the Tertiary amygdaloidal basalts of Co. Antrim. At the Gobbins, Island Magee, near Larne, fibrous thomsonite occurs in the same vesicles as the more common chabazite and is possibly mixed with gmelinite and levyne (Tomkeieff 1934). In a review of the zeolite species found in Antrim lavas, Walker (1959a) describes white or pink thomsonite of the faröelite variety forming radiating aggregates of closely packed acicular crystals. It occurs in 53% of the 670 localities examined by Walker (1960a), who also notes a pale-brown variety (with an oolitic appearance) composed of compact spheres (1–2 mm diameter) with a radiating structure, found near Randalstown and to the N of Templepatrick. Ashcroft (1916) reports thomsonite from ENE of Cregganboy farm, near Kinbane, 5 km NW of Ballycastle, and Greg & Lettsom (1858) describe it with analcime from the Ball, northwest end of Rathlin Island, off the coast of Antrim. In the basalt quarries around Ballymena, thomsonite occurs with chabazite, gyrolite and mesolite, together with fine cabinet specimens of apophyllite (Green & Moreton 2001).

Thomsonite is a common species in two chalk/dolerite reaction zones in Co. Antrim. At Scawt Hill it is found in melilite-rich hybridized dolerite, where it occurs in vesicles associated with calcite and scawtite (Tilley 1930b). In olivine dolerite close to the chalk contact, analcime, calcite, natrolite, stilbite and thomsonite are all found infilling vesicles. The first two species are also found as alteration products of interstitial plagioclase feldspar (Tilley & Harwood 1931). At Carneal, Larne, Co. Antrim, thomsonite is a very common alteration product in pyroxenites, pyroxene-rich dolerite, gabbro and a melilite–titanaugite rock, which are hybridized after having assimilated chalk. It replaces basic plagioclase feldspar and infills vesicles (Sabine & Young 1975). Thomsonite has been found in the contact aureole of dolerite dykes at Blind Rock, SW of Mount Charles, Co. Limerick (Nawaz 1979), and at Killala Bay, Inishcrone, Co. Sligo (Nawaz 1974). Other reports are from vesicles in the basalts of Co. Antrim: Carnmoney Hill, Ballyduff and Collinward, Belfast (Hey 1932a); Parkgate quarry, Templepatrick; Dundrod; Ladyhill quarry, Tardree, near Antrim town; Ballynulto quarry, near Broughshane (all Tschernich 1992); Agnews Hill, near Portrush; Black Cave, near Larne; and Island Magee (all Hintze 1897).

In Co. Londonderry, thomsonite forms colourless blocky prisms, grey-bladed cavity linings and white scattered blades within vesicular basalt at Moneymore, brownish-white hemispheres at Downhill and Magilligan, and blades covered in calcite at Legannon in the Sperrin Mountains (Tschernich 1992). Fibrous thomsonite and natrolite are the main amygdale species in basalt at Sounding Hill quarry, NNW of Magherafelt (Old 1972).

Altered gabbro found as glacial erratics on Thurstaston beach, Wirral, Merseyside, contains white translucent sprays of well terminated acicular thomsonite (Rothwell & Dickinson 1989).

In Wales, thomsonite, associated with analcime, infills vesicles in a Tertiary olivine-dolerite dyke in central Snowdonia, Gwynedd (Williams 1924). Bevins (1994) notes thomsonite in monchiquitic basanite dyke rocks from Great House and Glen Court, near Usk, Gwent.

Colourless blades of thomsonite are recorded from Fort Joli, Guernsey (Mourant 1939).

thomsonite-Ca

ZEOLITE GROUP

orthorhombic $Ca_2Na(Al_5Si_5)O_{20}\cdot 6H_2O$

Thomsonite-Ca is the new name for the most common species of the thomsonite series, and many (if not most) occurrences listed above are probably of this species. This is confirmed for examples from Tertiary basalts on the Isle of Skye, Highland Region, Scotland, at the Storr, Quiraing and Uig, and Carboniferous basalts in Strathclyde at Bishopton, Bowling, Dumbarton, Kilpatrick and Lochwinnoch, and in Central Region at Mugdock (Heddle 1901). EMP data reported by King (1976) indicate that thomsonite-Ca occurs in Tertiary basalts at other locations on the Isle of Skye: Allt Mòr, W of Loch Brittle; Beinn na Greine, near Portree; Ben Horneval, 5 km N of Roskhill; Braon a' Mheallain, 3 km NE of Bracadale; and Flashader, Loch Greshornish, 3 km NE of Edinbane. A large crinanite–(alkali dolerite) sill exposed at Kingscross Point and Clauchlands Point, both Lamlash Bay, Isle of Arran, Strathclyde, contains accessory amounts of analcime, mesolite and thomsonite-Ca infilling interstices between earlier-formed pyroxene and plagioclase feldspar crystals (Tindle unpub. EMP data). Thomsonite-Ca has also been found on low ground below Creag Brimishgan on the shoreline of Loch Na Keal, Isle of Mull, Strathclyde (S. A. Rust sample, Tindle unpub. EMP data).

Thomsonite-Ca is also found in the vesicles of hydrothermally altered Tertiary basaltic lavas at the White Rocks, near Portrush, Co. Antrim, Ireland (Robert 2001). Elsewhere in Co. Antrim, thomsonite-Ca occurs in: amygdaloidal basalts from Ladyhill quarry, Tardree, NE of Antrim town; Maxwell's quarry, Scotstown Road area, 5 km from Ballymena, off the main road to Larne; Magheramorne quarry, near Larne; Port-na-Boe, Giant's Causeway (H. J. Foy samples, D. I. Green XRD, Tindle unpub. EMP data). Adair (1987) lists other Tertiary basalt areas (and specific locations): South Antrim Escarpment to the W and SW of Belfast (Aughrim Plantation quarry; Belshaws quarry; Castle Robin Primary School quarry; Hightown quarry; Sale Corner quarry; Slievenagravery quarry; White Mountain quarry; Wolf Hill quarry); Galboly River section, near Garron Point; Giant's Causeway area (Dunseverick Slip; Portnabrock; Portnaweelan; Portninish); and Island Magee area (Black Cave; Brown's Bay). Thomsonite-Ca also occurs at Moneymore quarry, near Magherafelt, Lough Neagh, Co. Londonderry, where it has a blocky or oolitic (scoulerite) habit (Adair 1987).

Erratic blocks of altered gabbro from a gravel pit at Burscough, SE of Southport, Lancashire, contain cavities infilled with radiating thomsonite-Ca (D. I. Green sample, Tindle unpub. EMP data) similar to that found on Thurstaston beach, Wirral, Merseyside (Rothwell & Dickinson 1989).

thomsonite-Na

ZEOLITE GROUP

A few EMP analyses of thomsonite reported from the Isle of Skye, Highland Region, Scotland, by King (1976) from Tertiary basalts at Flashader, Loch Greshornish, 3 km NE of Edinbane and from along the coast N of Talisker Bay, have more sodium than calcium and probably should be called thomsonite-Na. However, thomsonite-Na has not been formally described or approved by the IMA-CNMNC.

thomsonite (of Squires)

TL Little Caton, Norwich, Norfolk, England

An uncertain mineral; a silicate and carbonate of calcium and magnesium.

thorianite

cubic ThO_2

An uncommon thorium oxide that forms a series with uraninite and has been found as an accessory phase in the Cnoc nan Cuilean intrusion (part of the Ben Loyal syenite complex), Lettermore, Highland Region, Scotland. Thorianite is associated with allanite, apatite and thorite in the more mafic syenites (Gallagher et al. 1971). At Talnotry, near Newton Stewart, Dumfries & Galloway, relatively abundant apatite and zircon crystals (<100 µm long) containing thorianite and uraninite inclusions (<10 µm in diameter) occur in a small copper–nickel mineralized diorite intrusion. Barite, monazite, thorianite and uraninite also form discrete grains, especially in the more leucocratic fractions (Power et al. 2004).

In the Presteigne area of Powys, Wales, thorianite is associated with thorite and it occurs as inclusions (2–5 µm across) within spherical bitumen nodules in the Folly Sandstone of Llandovery age. Sandstone-hosted thorium-enriched bitumen mineralization is unusual, although in this instance is large enough to contain an estimate of over 500 tonnes of thorium along a 2 km-long outcrop (Parnell & Eakin 1989).

thorite

tetragonal $(Th,U)SiO_4$

An uncommon thorium–uranium silicate, dimorphous with huttonite and found in felsic igneous rocks, in particular granite pegmatites.

Thorite occurs in the North Tincroft lode in New Cook's Kitchen mine, Camborne, Cornwall, where it forms tiny inclusions (<5 µm) in monazite within arsenopyrite-rich ore (LeBoutillier et al. 2000). At Porth Nanven, near Cape Cornwall, thorite is associated with monazite-(Ce) in a siderophyllite granite from the Land's End granite pluton (Müller et al. 2006). In Cumbria (near the border with Yorkshire), accessory amounts of apatite, thorite, xenotime and zircon are reported from the buried Wensleydale granite sampled in the Raydale borehole, 7 km SSE of Hawes (Brown et al. 1987a). Thorite (Basham et al. 1982) and a mixed brownish-yellow thorite–thorogummite phase (Webb & Brown 1984a) are noted from the Shap granite, Cumbria.

In Wales, thorite occurs in Ordovician ashflow tuff from Ramsey Island, Dyfed, and as overgrowths on xenotime in Ordovician mudstones from the Aberangell area, Gwynedd (Bevins 1994). The rhyolites of Snowdonia contain accessory amounts of thorite as a secondary species. Other accessory minerals in the rhyolites include aeschynite, allanite, apatite, lanthanite, larsenite–fayalite, monazite, parisite, synchysite, titanite, xenotime and zircon, and possibly gadolinite (Howells et al. 1991). An exceptional occurrence of thorite is described from sub-mm-scale bitumen nodules in the Llandovery Folly Sandstone at Nash Wood, near Presteigne, Powys. The source of the thorium is unknown (Parnell & Eakin 1989).

Thorite (var. orangite) occurs in a pegmatitic vein associated with galena, lepidocrocite (syn. rubinglimmer), magnetite, microcline (var. amazonite), titanite (syn. sphene) and topaz (later shown to be phenakite) at Beinn Laoghal, Sgòrr Chaonasaid, on the north slope of the Ben Loyal syenite complex, near Lettermore, Highland Region, Scotland. Thorite is brownish black to clove brown, and vitreous when fresh, but is resinous when weathered (orangite). It forms square prismatic crystals with pyramidal terminations and, less commonly, octahedral crystals. The giant erratic boulder at Beinn Bhreac, 2 km E of Tongue, thought to be derived from Beinn Laoghal (7 km to the S), also contains thorite (Heddle 1883, 1901, Macpherson & Livingstone 1982). The large Chaipaval (Ceapabhal) granitic pegmatite, near Northton (Taobh Tuath), South Harris, Western Isles, contains an accessory and minor mineral assemblage that includes sparse thorite with allanite, beryl, columbite, gahnite, kasolite, monazite, pyrite, spessartine, topaz, tourmaline and uraninite, and possibly betafite. Yellow thorite is also found with thorogummite as small inclusions (<0.6 mm) in biotite within the nearby Sletteval pegmatite (von Knorring & Dearnley 1959b, 1960a). Thorite is noted from the Tertiary Conachair granite on St Kilda (Hirta), Western Isles, but the accompanying chemical formula (ThO_2) suggests that the authors (Harding et al. 1982) mistook it for thorianite.

Thorite has been reported as an accessory mineral at many locations during investigations into the radioelement potential of Caledonian granites across Scotland: Mount Battock, Bennachie and Cairngorm granites, all Grampian (Webb & Brown 1984b); Meall Odhar, South Cruachan, Central Starav and Outer Starav granites, all part of the Etive granite complex, Strathclyde; and Cairngorm granite, Grampian (all Barritt 1983); and Helmsdale granite and Cnoc nan Cuilean syenite, Ben Loyal, both Highland Region (Gallagher et al. 1971). The last authors also note allanite, euxenite and thorite in pegmatites exposed near Kinlochbervie, Highland Region. In Dumfries & Galloway, thorite occurs in carbonate veins cutting diorite to the W of Loch Dee in the Loch Doon granite complex (Tindle 1982), and the species has also been found in granite from the Cairnsmore of Carsphairn intrusion (Tindle et al. 1988). At Needle's Eye, SW of Caulkerbush, Dumfries & Galloway, granodiorite close to carbonate-hosted uraninite-bearing veins contains rare 1–10 µm-long crystals of thorite (uranothorite) associated with other accessory species: apatite, monazite, xenotime and zircon. Some thorite is altered to a rhabdophane- or thorogummite-like mineral (Basham et al. 1989).

In Ireland, monazite, thorite, uraninite and zircon are accessory phases in the Costelloe Murvey leucogranite, an evolved unit of the Galway granite batholith, south Connemara, Co. Galway. In thin section the thorite forms euhedral crystals with a square cross section and a size range of 0.05–0.15 mm (Feely & Madden 1988, Feely et al. 1989).

thorogummite

tetragonal $ThU[(SiO_4),(OH)_4]$

An uncommon thorium silicate hydroxide found in the Lewisian granitic pegmatite at Sletteval, South Harris, Western Isles, Scotland. Thorogummite is glassy red or earthy brown and it forms many small inclusions (3–6 mm across) in biotite. Yellow thorite is closely associated with the thorogummite. Allanite, apatite, magnetite, spessartine and zircon are also found in minor amounts in the pegmatite (von Knorring & Dearnley 1960a).

thucholite, thucolite

A mixture of uraninite with carbonaceous matter. At Laxey lead–zinc mine, Isle of Man, thucholite is an anthracite-like coal-black material of friable character and it contains uraninite and sulphide (mainly chalcopyrite) inclusions in a hydrocarbon complex (Davidson & Bowie 1951).

thuringite

An iron-rich variety of chamosite, chlorite group.

tiemannite

SPHALERITE GROUP

cubic HgSe

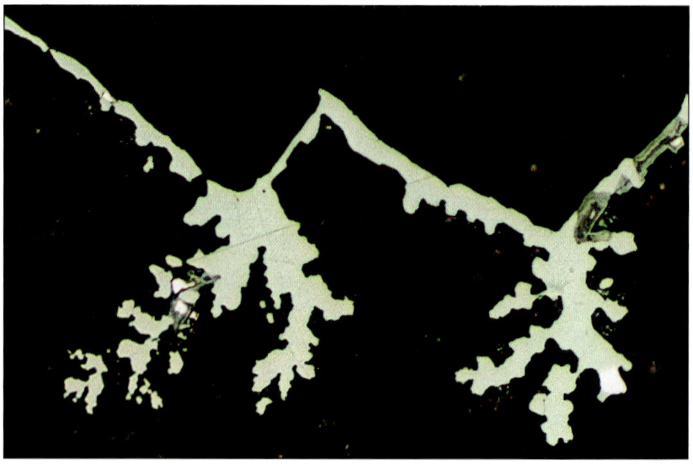

Skeletal tiemannite in calcite. Hope's Nose, Torquay, Devon; width 300 µm; viewed in reflected light. (© C. J. Stanley, The Natural History Museum, London)

A rare mercury selenide found in a gold-bearing carbonate vein cutting Middle Devonian limestones at Hope's Nose, Torquay, Devon. The selenide assemblage includes clausthalite, eucairite, fischesserite, klockmannite, naumannite, penroseite, trüstedtite, tyrrellite and umangite in a gangue of dominantly calcite (Stanley et al. 1990a, Camm 1995). Tiemannite, clausthalite and unidentified selenide species are also found as inclusions in alluvial gold grains from the South Hams district of Devon. The selenides were deposited during a late phase of gold growth from a distinct type of solution that was saline, oxidizing and at relatively low temperature (Leake et al. 1991, 1993).

Of 1794 alluvial gold grains examined by Leake et al. (1998) from the Leadhills area, along the Strathclyde/Dumfries & Galloway border, Scotland, one grain is reported to contain tiemannite inclusions. Parnell (1988) describes tiemannite and an unnamed bismuth–silver selenide occurring as sub-rounded 1 μm-diameter inclusions in pyrobitumen from a spoilheap at the Silver Glen mine, Alva, Ochil Hills, Central Region. Tiemannite is the commonest inclusion species in alluvial gold from the Mauchline Basin, NW of Cumnock, Strathclyde. It is associated with copper selenides, selenides of silver, bismuth, cadmium and lead, and palladium telluride (Leake et al. 1997b).

tiger eye
Quartz pseudomorphous after crocidolite. The name refers to the mineral's yellowish-brown colour and chatoyant lustre.

tile ore
A massive variety of cuprite.

tilleyite
monoclinic $Ca_5Si_2O_7(CO_3)_2$

Named after Cecil Edgar Tilley FRS (1894–1973), professor of mineralogy and petrology in the Department of Mineralogy and Petrology at the University of Cambridge; born in Adelaide, South Australia. Tilleyite is a rare calcium silicate carbonate found in thermally metamorphosed limestones to the south of Barnavave, Carlingford, Co. Louth, Ireland. It is associated with calcite, grossular, hillebrandite, melilite, spurrite, vesuvianite and wollastonite, and it forms subhedral or rounded crystals, or irregular plates, in calcite-rich rocks (Nockolds & Vincent 1947).

In limestone that has been thermally metamorphosed by two basalt-dolerite dykes at Killala Bay, near Inishcrone, Co. Sligo, Ireland, tilleyite is replaced by killalaite in spurrite-bearing rocks (Nawaz 1974). Similarly, at Carlingford, Co. Louth, interaction of gabbro with limestone has produced assemblages that include cuspidine, eudialyte, kilchoanite, monticellite, neptunite, spurrite and tilleyite (Ryback & Moreton 1993).

At Camas Mòr, Isle of Muck, Highland Region, Scotland, another gabbro/limestone contact zone containing tilleyite is reported by Tilley himself (Tilley 1947). In a thick band of dark-grey resinous-lustre gehlenite rock on the western contact of the gabbro, sponge-like crystals (up to 1.5 mm across) of tilleyite occur in aggregates intergrown with gehlenite. Agrell (1965) reports tilleyite in Scotland, this time from Kilchoan, Ardnamurchan, Highland Region. As with the other occurrences, tilleyite forms in a thermally metamorphosed limestone, but at Kilchoan it is found in late-stage veins and, with calcite or wollastonite, it appears to replace spurrite and kilchoanite. Tilleyite is also reported from a banded cherty nodule in a high-temperature calcic skarn near the contact of the Cuillin gabbro complex at Camasunary Bay, Strathaird Peninsula, Isle of Skye, Highland Region. Associated major species include garnet (andradite–grossular), melilite, spurrite, vesuvianite and wollastonite (Beard & Drake 2007).Tilleyite produces bright orange-red cathodoluminescence (Long & Agrell 1965). A vague report of tilleyite from the isles of Rum and Skye is given by Anthony et al. (1995).

tin?
tetragonal Sn

Tin is a very common term used in many mineralogical and mining texts. In every case, the use of the word "tin" refers to the element and not the species. The major ore mineral found in tin mines is cassiterite. Native tin does not occur in Britain or Ireland.

tin pyrites
A synonym of stannite.

tin-sphene
A synonym of malayaite, titanite group.

tinstone
A synonym of cassiterite.

tin-white cobalt
A synonym of smaltite, an arsenic-deficient variety of skutterudite.

titanaugite
A titanium-rich variety of augite, pyroxene group.

titanic iron
A synonym of ilmenite or pseudorutile.

titanite
monoclinic $CaTiSiO_5$

A very common calcium–titanium silicate found as an accessory mineral in intermediate and felsic plutonic rocks, pegmatites and alpine-type veins. It also develops in thermally and low-grade regionally metamorphosed rocks, and in some skarn deposits. Titanite often forms euhedral wedge-shape or diamond-shape cross sections and is referred to in early works as "sphene".

England
Cornwall Skarn assemblages are developed in the aureole around the Land's End granite at Gryll's Bunny and the Crowns, Botallack, St Just. They contain tin-bearing species or varieties of cassiterite, epidote, malayaite, pargasite and titanite, and have been referred to as tin floors because of their sub-horizontal sheet-like form (Jackson 1974, van Marcke de Lummen 1985, 1986b). Small white rosettes of titanite are also recorded with wickmanite in cavities in a large axinite pod in a nearby skarn deposit on the south side of Wheal Cock Zawn (Hubbard 1989). Elsewhere in the Land's End granite aureole, titanite occurs in thermally metamorphosed hornblende–plagioclase-bearing dolerites and calc-silicate hornfels, in which axinite, clinozoisite, diopside, epidote, grossular, plagioclase (labradorite) and zoisite are also found (Mitropoulos 1984).
Cumbria Titanite is always present and often abundant in porphyritic granite from the Shap granite quarry. It is also found as well formed dark-brown crystals up to 0.75 mm long associated with orthoclase in miarolitic cavities in the granite. Calcite, bavenite, epidote, milarite and scheelite are other cavity species (Harker & Marr 1891, Rastall & Wilcockson 1915, Russell 1925, Webb & Brown 1984a, Wirth 1995, Wirth et al. 2001, Wirth & Green 2002). Deuteric alteration has produced carbonate, chlorite, epidote, sericite and titanite in an augite–phyric microdiorite from Scawgill Bridge, near Whinlatter Pass (Fortey & Nancarrow 1990).
Devon Small yellowish crystals of titanite were collected with anatase from the Virtuous Lady mine, Buckland Monachorum (Greg & Lettsom 1858, Hall 1868b), but none could be found later (Russell 1913a). However, titanite (syn. grothite) is a common accessory mineral in biotite–muscovite granite collected in the vicinity of Birch Tor and Vitifer mines, Dartmoor (Alderton 1988), and is described as very abundant in tourmaline-rich nodules within granite at Hamel Down on Dartmoor (Brammall & Harwood 1925). Albite–chlorite–epidote schists at Start Point contain accessory amounts of titanite associated with epidote and chlorite (Tilley 1923).
Leicestershire Separation of accessory minerals from granodioritic rocks in the Mountsorrel district showed titanite to occur at Brazil Wood, Cocklow quarry, Kinchley, Main Mountsorrel quarry, Rubbish Tip quarry and Swithland reservoir, in all of which it forms irregular almost colourless to deep-brown crystals up to 0.5 mm across (Taylor 1934).
Merseyside Attractive wedge-shape lemon-yellow euhedral crystals of titanite have been found in glacial erratics on Thurstaston beach, Wirral. Exceptional crystals reach 8 mm in length (Rothwell & Dickinson 1989).

Wales
Titanite is a widely developed low-grade metamorphic mineral in rocks of basaltic composition, usually forming anhedral granules up to 0.1 mm across replacing primary iron–titanium oxides. Bevins & Rowbotham (1983) report 12 such occurrences: Clwyd – Hendre quarry, Pandy, near Glyn Ceiriog; Dyfed – Garn Fawr, near Strumble Head; Garn Turn Rocks, E of Wolf's Castle; St David's Head intrusion; Y Garn, near Pontiago; Gwynedd – Aberdaron, Lleyn Peninsula; Cadair Idris; Cwm Eigiau, 6 km W of Trefriw; Dolgellau; Hafodty, 3 km ESE of Penmaenmawr; Llanfachraeth, Anglesey; and Powys – Builth Wells. Titanite is also an extensively developed accessory mineral in the outer marginal zones of the Tal y Fan metabasite intrusion, E of Bangor, Gwynedd, where it replaces ilmenite and forms intergrowths with actinolite and chlorite as a replacement of clinopyroxene (Bevins & Merriman 1988).

The rhyolitic lavas of Snowdonia contain titanite as a primary accessory mineral, with pink and colourless crystals noted from Capel Curig (Williams 1922) and tin-rich euhedral sector-zoned crystals found at Beddgelert Forest (Howells et al. 1991, Bevins 1994). The margins of the Crogenen granophyre, near Dolgellau, also contain abundant titanite (Cox & Wells 1920), but the granular form suggests a secondary origin. Titanite is also recorded from alpine-type veins at Fron-oleu, Prenteg, Tremadog, where pinkish-brown crystals up to 6 mm long have been found (Greg & Lettsom 1858, Starkey & Robinson 1992).

Scotland

Central Titanite occurs in copper–zinc-mineralized chloritic schists at Creag Bhocan, SSW of Tyndrum (Scott et al. 1988).

Dumfries & Galloway Titanite is a characteristic accessory mineral in the Criffell granite complex. It forms: reddish-brown crystals at Aird's Point, near New Abbey, and at the Beeswing and Kippford granite quarries; pale-amber and reddish-brown crystals up to 4.5 mm long at Craignair granite quarry; and pale yellowish-brown crystals up to 3 mm long at Loch Kinder granite quarry (Williams 1965). McMullen & Todd (1990) add Kinharvie Burn, 6 km W of New Abbey, as a further Criffell titanite location, and Basham et al. (1989) record it at Needle's Eye, SW of Caulkerbush. Titanite is also documented from diorite, granodiorite and granite units of the Loch Doon granite complex, where it forms pale-brown, interstitial or euhedral diamond-shape crystals up to 1 mm long (Tindle 1982).

Grampian In a study of the geothermal potential of the Eastern Highland granites, Webb & Brown (1984b) noted an allanite–titanite association in granites with <74 wt% SiO_2. These include the Cairngorm northeast granite (see also Barritt 1983) and microgranite, the central part of the Mount Battock granite and the eastern part of the Ballater granite. Apatite, thorite, uraninite and zircon are associated accessory minerals. Titanite is so abundant in calc-silicate rocks in the Dalradian Water of Tanar Limestone Formation, between the Ballater and Lochnagar granite complexes, that it constitutes a major mineral (Goodman et al. 1990). At Glen Lui, 2 km SSE of Derry Lodge, Cairngorms, titanite is a very common accessory mineral in a series of granulites, including an unusual aegirine granulite. It shows considerable variation in colour (colourless, grey or reddish), mode of occurrence (blebs, lozenges and aggregates) and, in some associated pegmatitic veinlets, two generations are noted (McLachan 1951).

Highland Titanite is found as an accessory species in aegirine syenite exposed between Ledmore and Ledbeg, and in silicocarbonatite to the S of Ledmore. Both are related to the nearby Loch Borralan alkali syenite complex, Assynt (Teall 1900, Shand 1910b, Young et al. 1994b). The Canisp porphyry is another aegirine-bearing rock in Assynt, where beautiful euhedral titanite has been reported (Sabine 1953). The Ben Loyal syenite complex contains reddish-brown anhedral granular titanite at Cnoc nan Cuilean (King 1942) and the species is also noted at Lettermore quarry on the shore of Loch Loyal, and with allanite, magnetite and microcline (amazonite) in pegmatitic veins in the cliffs of Sgòrr Chaonasaid (von Knorring & Dearnley 1960b). The Glen Dessary syenite, W of Loch Arkaig, contains titanite as an important uranium-bearing phase, together with accessory apatite, allanite, monazite, zircon and zirconolite (Fowler 1981, 1992, Fowler & Williams 1986).

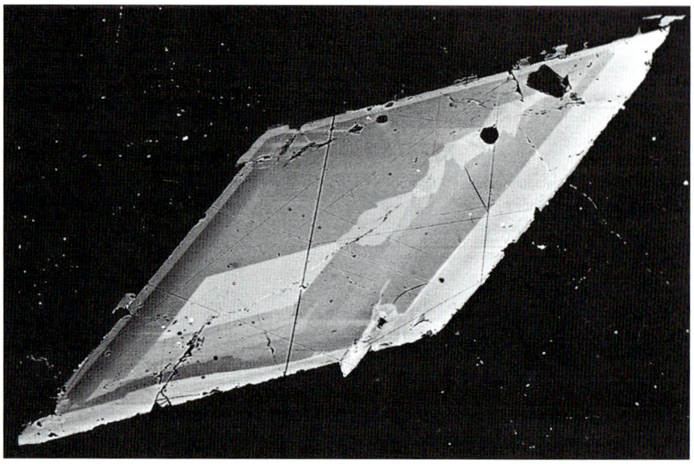

Complex zonation in a basal section of titanite from the Strontian granite. Highland Region, Scotland; width 1.0 mm; BSE image. (© B. A. Paterson)

Euhedral titanite in granodiorite from the Morvern–Strontian granite complex, Highland Region, Scotland; width 0.9 mm; viewed in PPL (upper) and XPL (lower). (© K. Gardner)

Titanite is also a widely distributed accessory mineral in Caledonian diorite–granite complexes, examples being the quartz diorites of the Ben Nevis complex (Haslam 1968) and, most spectacularly, various units of the Strontian complex and diorite from the Ratagain intrusion, where complex oscillatory and sector zoning has been observed (Paterson et al. 1989, Paterson & Stephens 1992). Uranium–lead age dating of a brown titanite from the Strontian complex has shown that the intrusion crystallized 423–412 million years ago; the same method produced an age of 427 million years for an augite–biotite diorite from the Rubha Mòr intrusion, Kentallen, near Fort William (Rogers & Dunning 1991).

Uranium–lead age dating of titanite has also been applied to regional metamorphism studies, with an age of 737 million years established from a Moine Series (Morar Group) titanite-rich calc-silicate pod, close to Loch Eilt and the Sound of Arisaig. Euhedral titanite is thought to have formed during prograde regional metamorphism by the partial or complete alteration of detrital iron–titanium oxides and rutile (Tanner & Evans 2003). A further example, forming dark-brown subhedral grains, is from a tuff on the Isle of Muck that has been dated at 62 million years old (Chambers et al. 2005).

Other metamorphic examples of titanite include: Moine Series schists composed essentially of biotite and titanite at many locations in Ardnamurchan and in the Lower Psammitic Group of Morar and South Knoydart to the E of Mallaig (Butler 1962); an unusual pyroxene–titanite-rich xenolith (possibly a skarn) in diorite of the Glenelg–Ratagain igneous complex, found on the banks of the Glenmore River, NNE of Suardalan (Nicholls 1951a); a titanite-rich band associated with an amphibolite sill at Fionn Allt, 6 km WNW of the head of Loch Shin (Read 1931b); titanite and apatite are particularly abundant in Moinian quartzose amphibolites in the White Bridge area of Glen Garry to the SW of Fort Augustus (Highton 1994); the Scottish Mineral Collection at the NMS holds a clove-brown lozenge-shape crystal of titanite, 1.0×2.7 cm, in a biotite–garnet–quartz matrix from close to the mica prospect at Strathpeffer (Livingstone 2002). A 50 mm-long pale-yellow crystal of titanite and dark-brown pea-size crystals with perfect form (from Shinness on the eastern shore of Loch Shin) were considered to be the finest in Britain (Heddle 1882c).

Strathclyde Titanite is a common accessory mineral in the Caledonian Etive granite complex to the E and NE of Ben Starav and it occurs most abundantly in the North Cruachan adamellite and the Outer Starav biotite granite (Barritt 1983). It is also reported in an appinitic diorite from Garabal Hill and in appinite from the Arrochar intrusion (Rogers & Dunning 1991). At Coire Buidhe, Glen Creran, titanite forms yellow lozenge-shape euhedral crystals, up to 6 mm long, in granite and throughout a skarn zone (Ingram et al. 1993). A highly fractured imperfect crystal of titanite in quartz associated with epidote, garnet and molybdenite from this locality, now in the collection of the NMS, is considerably larger at 40 mm long (Livingstone 2002).

Vesicles in Clyde Plateau basalt from Hartfield Moss, 9 km WSW of Barrhead, contain colourless-to-pale greyish-green titanite up to 0.5 mm across, associated with feldspar and probable offrétite (Meikle 1990). Gem-like pale yellowish-green titanite is also found in vesicular basalt at Loanhead quarry, Beith (Robinson & King 1993). Amphibolite related to the Girvan–Ballantrae ophiolite complex at Knockormal contains abundant deep resin-brown titanite associated with albite, apatite, crossite and epidote (Bloxam & Allen 1959). On the northern tip of the Isle of Arran, authigenic titanite has formed as a secondary pore-filling cement in Permian sandstones from the Cock of Arran area and is probably derived from hydrothermal fluids associated with the intrusion of the Arran granite (Hole et al. 1992, van Panhuys-Sigler & Trewin 1990).

Tayside Titanite occurs with allanite, apatite, epidote and zircon as accessory species in the Middle Dalradian Ben Lawers Schist and metabasites in the Aberfeldy area (Coats et al. 1984, Moles 1985). Pinkish-brown intergrown crystals of titanite up to 8 mm long have been found covering an area of 50 mm across in a quartz vein cutting limestone at Shierglas quarry, SE of Blair Atholl (R. E. Starkey pers. comm. 2007).

Western Isles On St Kilda (Hirta), titanite has been found in a gabbroic pegmatite at Glen Bay, where it is an accessory species with allanite, apatite, epidote, zircon and zirkelite. The nearby Glen Bay granite contains a lanthanum-enriched titanite associated with allanite, apatite, chevkinite and zircon, and a granitic sheet at Na h-Eagan contains small radiating sheaves of titanite associated with chlorite (Harding et al. 1982, 1984). A meta-acid pegmatite exposed in a roadside quarry at North Ford, North Uist, produced clove-brown resinous-to-vitreous euhedral titanite crystals up to 3 cm long, which are now at the NMS (Livingstone 2002). A meta-quartz microdiorite dyke associated with Lewisian ortho- and paragneisses at Garry-a-siar (Gearraidh a Siar), Benbecula, contains up to 6% modal titanite (Fettes et al. 1992).

In addition to the locations above, Heddle (1901) documents many titanite localities throughout Scotland, divided into the categories: diorite, granite, granite veins, syenitic granite and syenite, gneiss, and primitive limestone.

Ireland

Antrim Chalk/dolerite contact zones near Larne at Carneal (Sabine & Young 1975) and Scawt Hill (Tilley & Harwood 1931) contain titanite as an accessory component.

Cork A magnesium-rich variety of titanite associated with diopside and pargasite occurs in an amphibole–pyroxenite nodule collected from a megacryst-bearing pyroclastic pipe at Black Ball Head, near Cahermore (Pracht & Kinnaird 1995).

Donegal Crystals of titanite are major constituents of contact-skarn rocks between granite and limestone at Annagarry, 4 km SSW of Bunbeg, and euhedral crystals up to 1.5 mm long are reported from diopside-bearing pegmatite dykes to the S of Bunbeg quay (Haughton 1862, Whitten 1951, Ryback & Moreton 1992). Titanite also occurs in hornblende–plagioclase skarn mantling limestone inclusions in granite at Ballynacarrick (Gindy 1953) and is an important accessory species in garnet–vesuvianite–wollastonite veins found in the limestone quarries of Lough Anure, Thorr area (Pitcher 1950). Titanite is a relatively abundant accessory mineral in the granite on the southern shore of Aran Island (Ryback & Moreton 1992).

Galway Euhedral magmatic crystals of titanite and secondary titanite associated with epidote, garnet and prehnite occur in granite making up the bulk of the Galway batholith (Leake 1998).

Louth Accessory amounts of pale-brown titanite occur in hybridized limestone and aegirine-augite-bearing syenite at Barnavave, Carlingford (Osborne 1932b, Nockolds 1937, 1944, 1950).

Wicklow The Aughrim portion of the Leinster batholith contains colourless-to-pink pleochroic titanite (Smithson 1932). At Glendalough, a mixed mica–quartz diorite rock is considered to contain xenocrysts of clinozoisite, epidote, hornblende and titanite, the latter forming crystals with varying habit – moulded on ilmenite or clinozoisite or as needle-shape grains (Thomson 1908).

Offshore territories

Alderney Pale-brown titanite forms wedge-shape crystals associated with hornblende in the Bibette Head granite (Nockolds 1932).

Jersey Brown crystals of titanite up to 2 cm across are reported from dioritic pegmatites at Ronez and Samarès, and in granite at La Moye. At Côtil Point, small white crystals of titanite are associated with an actinolite–epidote–garnet assemblage in andesite (Mourant 1961).

titanoferrite

A synonym of ilmenite.

titanomagnetite

A titanium-rich variety of magnetite.

tobermorite

orthorhombic $Ca_5Si_6O_{16}(OH)_2{\cdot}nH_2O$

TL Tobermory, Isle of Mull, Strathclyde, Scotland

An uncommon hydrated calcium silicate hydroxide found totally filling vesicles in Tertiary basaltic lavas in a cliff exposure immediately to the north of the pier at Tobermory. Tobermorite is translucent, very pale pinkish-white, and frequently surrounded by a thin zone of pale-blue massive mesolite. It is also found in cliffs near the lighthouse N of Tobermory and along the shore towards Bloody Bay (Heddle 1880b). Claringbull & Hey (1952) showed that the mineral lining vesicles at Tobermory was, in fact, thomsonite. Livingstone (2002) notes that Heddle also collected tobermorite from Gribun, Loch na Keal, Isle of Mull. In a review of zeolite distribution patterns on Mull and the adjacent Morvern Peninsula, Walker (1971) found 37 unspecified localities (out of a total of 520) containing tobermorite. Speculating on the apparent rarity of tobermorite, he suggested that this may be attributable to its formation deep in a basalt pile, with few accumulations of basalt being sufficiently thick, or eroded deeply enough, for the tobermorite-bearing rocks to be exposed. At Ardtornish Bay on the Morvern Peninsula, Currie (1905) found reddish-white massive tobermorite (apparently on mesolite) in amygdaloidal basalt.

The Heddle collection in the NMS contains specimens of tobermorite from the Isle of Skye, Highland Region, including: Sgùrr nam Boc, Loch Einort; Geoda Tuill, Loch Einort; and Dunan Thearna Sgùrr, Loch Brittle (Livingstone 2002). Near the Allt Coir a' Ghobainn River close to Drynoch, a reyerite–tobermorite assemblage has been found in an amygdaloidal olivine basalt. Tobermorite forms highly fibrous radiating sheaves or rosettes, the latter up to 8 mm in diameter. Calcite, chlorite, pectolite and thomsonite are minor associated species (Livingstone 1988, 1989b). Just to the N of Portree, fine-grain to distinctly fibrous tobermorite occurs with calcite, gyrolite, mesolite, saponite, tacharanite, thomsonite and xonotlite in fine-grain, often composite, amygdales within olivine dolerite. For instance, a core of tobermorite may have a rim of mesolite, gyrolite or thomsonite. Another specimen contains a tobermorite core surrounded by a zone of tacharanite enclosing radiating spherules of gyrolite, in turn surrounded by mesolite (Sweet 1961).

Elsewhere in Scotland, tobermorite is reported from a single vesicle a few cm across within a basaltic volcanic plug of probable Carboniferous age on Castle Hill, 3 km WSW of Kilbirnie, Strathclyde. Tobermorite occurs as a white-to-pink mass with small acicular crystals discernible and is associated with pectolite and prehnite (Webb 1971).

Tobermorite occurs in a somewhat different environment in Ireland, where it is found in cavities in altered dolerite within the chalk/dolerite contact zone at Ballycraigy, Larne, Co. Antrim. It is associated with scawtite in larnite-bearing rocks and it occurs as minute radiating aggregates of fibres that readily alter to a fine white powder. The instability of tobermorite from Ballycraigy required McConnell (1954) to store it in the laboratory over silica gel (a desiccant) to prevent it hydrating. Further

Tobermorite in amygdaloidal basalt from Portree, Isle of Skye, Highland, Scotland; width 33 mm. (© The Trustees of the National Museums of Scotland)

examination of the Ballycraigy tobermorite by Taylor (1959) showed that it can transform to xonotlite under hydrothermal conditions. Contact rocks associated with the Carneal volcanic plug, Co. Antrim, contain white tobermorite in small (5 mm in diameter) amygdales in basalt surrounded by thomsonite. It is also found with calcite in scattered amygdales within a melilite–titanaugite rock and in a pyroxene-rich dolerite, again associated with thomsonite. In a complex scawtite–xonotlite–(hibschite) rock, tobermorite forms veins with xonotlite and small brown patches associated with hibschite. It is also associated with allophane and tacharanite in a hybrid pyroxenite. In thin section it can be colourless, pale fawn or turbid brown (Sabine & Young 1975).

tochilinite

monoclinic or triclinic $6(FeS)\cdot 5[Mg(OH)_2]$

A rare iron–sulphur– magnesium hydroxide found in sulphide-rich serpentinized ultramafic bodies and noted from an unspecified location on the Lizard Peninsula, Cornwall (Anthony et al. 1990). This is a reference to the unnamed fibrous iron–magnesium sulphide found in veinlets cutting peridotite at Pol Cornick, near Mullion. Tochilinite is associated with calcite, minnesotaite, talc and valleriite (Clarke 1970a).

todorokite

monoclinic $(Na,Ca,K,Ba,Sr)_{1-x}(Mn,Mg,Al)_6O_{12}\cdot 3\text{–}4H_2O$

An uncommon hydrated sodium–manganese oxide found as a hydrothermal alteration or weathering product of other manganese-rich species. Todorokite occurs as aggregates of lustrous silvery-brown scales in cavities within highly oxidized dolomite at Gerrans Bay, Cornwall (Wolfe 2006). Boreholes in fracture-mineralized Permo-Triassic St Bees Sandstone from the Sellafield area, Cumbria, contains fine-grain todorokite forming sooty coatings lining the pore walls of the sandstone, or as dendritic films and coatings on fracture surfaces. Todorokite and other manganese oxides (goethite, hollandite and romanèchite), with which it is admixed, are derived mainly from the dissolution and oxidative decomposition of iron–manganese-rich calcite host-rock cements and vein fillings (Milodowski et al. 1998).

In Wales, todorokite is the main supergene alteration product of Cambrian bedded manganese ore at the Haffoty mines, near Barmouth, and at Afon Cwm-mynach in the Harlech Dome area of Gwynedd, where it occurs with pyrite, rhodochrosite, spessartine, plus minor magnetite, hematite, chlorite and pyrite, and rare rhodonite (Glasby 1974, Allen & Jackson 1985, Bennett 1987).

Todorokite is reported from a ferromanganese sediment at Noblehouse, near West Linton, Borders, Scotland. The sediment is principally composed of hematite and quartz, and also contains rare goethite, montmorillonite and illite. Birnessite and minor todorokite are present in the more manganese-rich samples. An exhalative origin is proposed, similar to that found in ophiolite complexes, suggesting that the sequence (sediments and underlying basalts) formed at an active spreading centre (Nicholson 1989a,b, 1992). Spherical manganese nodules, up to 3 cm in diameter, from the upper 10–15 cm of the recent sediments of Loch Fyne, Strathclyde, contain todorokite as the principal mineral phase and are said to be similar to shallow-water todorokite-rich nodules found elsewhere in the world (Calvert & Price 1970).

Coating of tiny todorokite crystals on 5 mm-long aragonite crystals found in boulders on the beach near Port Quin, Cornwall. (© D. I. Green)

Wafer-thin todorokite from Gerrans Bay, near Porthscatho, Cornwall; width 490 μm; SEM image. (© D. I. Green)

topaz

orthorhombic $Al_2SiO_4(F,OH)_2$

A relatively uncommon aluminosilicate fluorhydroxide found in granite (in veins and cavities), granite pegmatites, rhyolite and in greisen deposits formed from high-temperature volatile-rich hydrothermal fluids. Topaz is also found as a detrital species. In many cases, topaz is a secondary mineral resulting from the alteration of feldspar, especially in greisen assemblages.

England

Cornwall In Cornwall (and Devon), some of the intrusions that form part of the Hercynian granite batholith contain minor topaz, particularly the fluorine- and lithium-enriched plutons and pegmatites: *Land's End granite:* Topaz granite occurs in rubble at Botallack Head, St Just (Manning 1986); accessory andalusite, topaz and tourmaline have been found in fine-grain granite near Knill's monument, Carbis Bay, St Ives (Robson 1945b). *Tregonning–Godolphin granite:* Subhedral to euhedral topaz is a minor component of amblygonite-bearing leucogranite in the roof zone of the Tregonning granite at Carn Clodgy, near Rinsey (Stone & George 1978); small, yellowish, doubly terminated, prismatic crystals of topaz up to 3 mm long and associated with cassiterite were once found at Great Wheal Fortune, Breage, near Porthleven (Russell 1924, Dominy et al. 1995); topaz is a common accessory mineral in the granite at Lesceave Por, near Trewavas Head, Breage (Russell 1924); subhedral to euhedral topaz is a minor component of amblygonite-bearing leucogranite associated with pegmatite at Megiliggar Rocks, near Breage (Hosking 1954b, Stone 1969, 1984, Stone & George 1978); topaz in the Tregonning granite is discussed by Manning (1986) and Stone (1992). *Carnmenellis granite:* Pegmatite developed in cavities within the granite contains topaz crystals in aggregates up to 6 mm across at Trolvis quarry, near Longdowns. Apatite, bertrandite, fluorite and tourmaline are associated species (Hosking 1954a); topaz is noted as a rare species in fine-grain biotite–muscovite granite at Crowan Beacon, Crowan (Russell 1924); topaz is recorded from the granite quarries of Constantine and Mabe, near Falmouth (Garby 1848). *St Austell granite:* Small opaque yellowish-to-white short prismatic crystals of topaz up to 4 mm long associated with cassiterite in narrow veins cutting kaolinized granite at Beam mine (now part of Goonbarrow china-clay pit), St Austell (Russell 1924). Later discoveries at Goonbarrow include bluish-white topaz associated with wolframite (Weiss 1994), topaz with cassiterite (Grigg 1986), and thin topaz-bearing veins also containing apatite, arsenopyrite, cassiterite, fluorite, löllingite, pyrite, quartz, tourmaline and wolframite (Bray & Spooner 1983); partially kaolinized granite and greisen are exceptionally rich in topaz at Bunny (St Austell Hills) mine, Bugle, near St Austell (Garby 1848, Russell 1924); opaque white short prismatic crystals of topaz associated with cassiterite and tourmaline and muscovite (gilbertite) at Conce Moor, Luxulyan, near Lostwithiel (Russell 1924); minor topaz in metasomatically altered granite (luxullianite) at Luxulyan quarry, near Lostwithiel (Lister 1978); topaz is associated with apatite, cassiterite, ferrocolumbite, ilmenite, ilmenorutile, monazite-(Ce) and zircon in the waste material remaining after kaolinite extraction at Goonvean and Rostowrack china-clay pits, W of Nanpean, St Austell (Scott et al. 1998); the Hensbarrow stock is a topaz-bearing granite affected by greisenization and is exposed in Gunheath china-clay pit, Carthew, St Austell (Hodkinson & Clark 1977, Hawkes et al. 1987, Weiss 1994, Williamson et al. 1997, 2002); topaz occurs in kaolinized granite at Trethosa china-clay pit, Carthew (Manning & Exley 1984, Psyrillos et al. 1998); at Wheal Remfry china-clay pit, 2 km from Indian Queens, near St Austell, a quartz–topaz–tourmaline vein produced transparent elongated topaz crystals up to 6 mm long and, in one small area, topaz-lined pods were found associated with a bismuth-rich assemblage consisting of bismutite, smrkovecite and waylandite (Grigg 1986, Hacker 1997, Rust 1998, S. A. Rust specimen, Tindle unpub. EMP data). Topaz is also an accessory component of hydrothermal breccia at this locality (Wilson & Long 1983); a single specimen of yellowish prismatic topaz, 5 mm long, from Wheal Maudlin, Lanlivery, is at the NHM (Russell 1924); general descriptions of topaz granites within the St Austell granite complex are provided by Richardson (1923), Manning & Hill (1990) and Manning et al. (1996). *Bodmin Moor granite:* Topaz occurs with tourmaline in a quartz–muscovite (sericite) assemblage adjacent to a wolframite lode at Hawkswood mine, North Hill, Bodmin Moor, 12 km N of Liskeard (Alderton et al. 1980). Ghosh (1927) reports accessory amounts of topaz forming flat, well formed, rectangular tablets, partially or almost completely enclosing tourmaline in fine-grain granite between Hill Tor and Newel Tor. *Minor intrusions:* White coarsely crystalline and well defined yellowish prismatic crystals of topaz, occasionally doubly terminated and up to 1 cm long have been found with muscovite (gilbertite), tourmaline and wolframite in disconnected masses within porphyritic granite at Belowda Beacon (mine), near Roche, N of St Austell (Russell 1924, Chaudhry & Howie 1970, Manning & Exley 1984); coarsely crystalline greenish-white masses of topaz occur with lithium mica, quartz, scorodite and wolframite (possibly a greisen) at Castle-an-Dinas wolframite mine, St Columb Major – colourless-to-white crystals of topaz up to 6 mm long occur with quartz in sediments forming the walls of the wolframite lode and the species also occurs with fluorite in the lode itself (Davison 1920a, Russell 1924, Hey et al. 1938, Polkinghorne 1954); microscopic accessory topaz occurs in greisen bordering cassiterite–wolframite veins and it forms crystals up to 6 mm long on joint faces in altered granite at Cligga Head, Perranzabuloe – colourless, white and yellow-stained topaz crystals up to 7 mm in length were also found in a solid but cavity-rich topaz–tourmaline vein (Scrivenor 1903b, Russell 1924, Chick 1987); topaz occurs in certain small cassiterite–wolframite-bearing quartz veins with greisen walls cutting granite at St Michael's Mount, Marazion. Colourless doubly terminated prismatic crystals of topaz, up to 6 mm long, associated with deep-blue apatite and lithium mica from this locality, are in the Russell collection at the NHM (Greg & Lettsom 1858, Davison 1920b, Russell 1924, Hosking 1951, 1954c, Moore 1977, Alderton & Moore 1981, Dominy et al. 1995). Other topaz occurrences associated with minor intrusions include: Mulberry mine, Lanivet; St Mewan Beacon, near St Austell; Trevaunance mine and Wheal Kine, both St Agnes (all Russell 1924); Redmoor–Kit Hill (Newall 1994, Newall & Newall 1989); Seal Hole mine, St Agnes (Collins 1892); and Wheal Kitty, St Agnes (Kingsbury 1954).

Cumbria Topaz is a constituent of As–Bi–F–Mo-rich greisen associated with the Eskdale granite at Water Crag, near Devoke Water (Simpson 1934, Young 1985d, Young et al. 1986). Related occurrences of fluorite-bearing greisen contain small patches of pale-cream saccharoidal topaz at Bull How, SW of Scale Bridge, and up to 30% topaz at Dawsonground Crags, W of Scale Bridge. Anhedral topaz crystals, up to 1cm across, are noted in greisen from Birker Beck, Whincop (Young 1985d).

Devon The Meldon aplite dyke is an iron–lithium-rich granitic intrusion, containing accessory topaz, that is emplaced into the metamorphic aureole of the Dartmoor granite. It also contains significant amounts of topaz associated with albite, lepidolite, orthoclase, petalite and quartz in crosscutting pegmatitic veinlets (Teall 1888b, Worth 1920, Chaudhry & Howie 1970, 1973a, Chaudhry 1971, von Knorring & Condliffe 1984). Topaz is briefly mentioned as an uncommon species in Dartmoor granite, such as that found on Hamel Down (Brammall & Harwood 1925), Lee Moor china-clay pit, near Wotter, southwest Dartmoor (Knox & Jackson 1990), and Whitstone quarry, Bovey Tracey (Hodge 1989). On Lundy island, topaz forms irregular grains in granite and aplite veins with pegmatitic margins. Drusy cavities in the pegmatite contain well shaped pale-blue or colourless topaz together with albite, apatite, beryl, fluorite, smoky quartz and tourmaline (McLintock & Hall 1912, Dollar 1941, Edmonds et al. 1979, Thorpe et al. 1990).

Leicestershire Very scarce small irregular fragments of topaz were recovered from heavy-mineral assemblages extracted from the Mountsorrel granodiorite and associated rocks from Cocklow quarry and the Main Mountsorrel quarry, Mountsorrel area (Taylor 1934).

Topaz in a miarolitic cavity in granite from Lundy, Bristol Channel, off the north coast of Devon; the crystal is 6 mm tall. (© D. I. Green)

Wales

Dyfed Topaz is one of the detrital minerals found in boulder clay at Ludchurch (Griffiths & Stuart 1940).

Gwynedd Topaz has been found in a biotite–garnet rock in the aureole of the Bwlch y Cywion intrusion, Snowdonia, and also (from the same intrusion) in concentrates derived from heavy minerals (Williams 1930b).

Scotland

Grampian Miarolitic cavities in granite exposed in cliffs on the eastern side of Loch Avon, Cairngorm Mountains, contain topaz, and in the nearby River Avon, fist-size water-worn topaz crystals were once found, of a colour described by Heddle (1901) as fine blue. Some of the largest Scottish topaz crystals have been found on Beinn a' Bhuird, Cairngorm Mountains; they are pale blue and weigh up to 0.5 kg (Heddle 1901, Rudler 1905). Other examples are documented by Dunlop (2005). Cabinet specimens of topaz and beryl with feldspar and smoky quartz have been collected from pegmatites on Ben Macdui (Green & Todd 2001) and one panned stream-sediment concentrate from Luibeg Burn, Derry Cairngorm, was found to contain topaz (Beer & Bennett 1988, Haynes 1993). The above occurrences are all associated with the Cairngorm granite complex.

At Gairnshiel Bridge in Glen Gairn, W of Ballater, topaz forms anhedral pale-blue lustrous crystals up to 1.5 cm across in a cassiterite–wolframite-bearing quartz vein assemblage and marginal greisen within zinnwaldite-bearing granite (Tindle & Webb 1989, Webb et al. 1992, Green & Todd 2001, Livingstone 2002).

Highland Pegmatitic veins in syenite at Sgòrr Chaonasaid, Ben Loyal, contain topaz, associated with albite, magnetite, microcline (amazonite) and tourmaline (Heddle 1901).

Strathclyde Small colourless crystals of gem-quality topaz occur in granite boulders (from the Isle of Arran) at Machrihanish Bay, near Campbeltown, Kintyre Peninsula (Heddle 1901, Berridge 1969, Livingstone 2002); equant crystals up to 2 mm across line drusy cavities in microgranite at Brodick Bay and Glenshant Hill on Arran (Macgregor et al. 1972, Livingstone 2002). Pale-blue to yellow topaz within drusy cavities in the Northern granite on Arran is noted by Emeleus & Gyopari (1992).

Western Isles Topaz possibly occurs in the pegmatite at Chaipaval (Ceapabhal), 3 km NW of Northton (Taobh Tuath), South Harris (Heddle 1901, von Knorring & Dearnley 1959b).

Topaz in a miarolitic cavity within granite; Diamond Rocks, Mourne Mountains, Co. Down, Ireland; height of crystal 2.2 mm. (© D. I. Green)

Ireland

Down Miarolitic cavities in granite between Slieve Bingian and Slieve Corragh, Mourne Mountains, have produced beautiful specimens of beryl, chrysoberyl, fluorite and topaz, and a greenish-yellow tourmaline described as peridot (Haughton 1856); at Diamond Rocks in the Mourne Mountains, miarolitic cavities in granite contain colourless transparent topaz up to 15 mm long associated with beryl, feldspar, mica and smoky quartz. Some of the topaz and beryl from Diamond Rocks is particularly fine (Wolfe 1996b, Green & Moreton 2001); greisen veins near the margins of the Mourne granite to the SW of Newcastle contain bluish-green mica, quartz and topaz (Nockolds & Richey 1939); at Lindsay's Leap quarry in the Mourne Mountains, drusy cavities in granite contain uncommon colourless transparent blocky euhedral crystals of topaz up to 4 mm long (Green et al. 2005a); and beautifully crystallized topaz from the Mourne granites is associated with well formed zinnwaldite and occurs in ten different forms (Greg & Lettsom 1858, Sollas 1890).
Wicklow Alluvial topaz is noted from the NE of Croghan Kinshela (Forbes 1869).

Offshore territories

Topaz occurs in the mineralized aplogranite at La Moye, Jersey (Ixer 1988).

topazolite

A variety of andradite, garnet group with the colour and transparency of topaz.

torbernite

AUTUNITE GROUP
tetragonal $Cu^{2+}(UO_2)_2(PO_4)_2{\cdot}8\text{–}12H_2O$

A relatively uncommon hydrated copper uranyl phosphate found in the oxidized zone of some uranium-rich copper deposits. Early work did not fully discriminate between autunite and torbernite, and the term "uranite" was applied to both species (Greg & Lettsom 1858). Visually, torbernite and metatorbernite are often difficult to distinguish from one another and identification is further complicated by solid solution between torbernite and zeunerite end members.

Torbernite forms beautiful emerald-green square thin tabular crystals and was said to occur in the shallow parts of almost every tin and copper mine in Cornwall (James 1947). At South Terras mine, St Stephen-in-Brannel, green plates of torbernite and yellow autunite were found in the Uranium lode, where they coated pitchblende and vein quartz (Collins 1912, Robertson & Dines 1929, Dines 1956). Deep-green square plates of torbernite up to 1.5 mm across are found with fluorite and limonite in fractures within topaz granite at Gunheath china-clay pit, Carthew, St Austell (Weiss 1994). Pegmatitic veins at this locality have been found containing rare torbernite, together with apatite, arsenopyrite, columbite, opal, topaz, tourmaline, turquoise and zinnwaldite (Hodkinson & Clark 1977). A pegmatitic vug in the granite at Gunheath contained torbernite as a late-formed species, associated with florencite-(Ce), fluorapatite, fluorite and goyazite (Elton & Grigg 1996). In the Hensbarrow section of the pit, torbernite growing on crusts of morinite was found in drusy quartz veinlets cutting kaolinized granite boulders (Elton et al. 1996). Fine specimens containing blocky transparent emerald-green torbernite crystals up to 2 mm across have been collected from the nearby Blackpool china-clay pit (Pearce 1996b, Green 2005a). Visually indistinguishable torbernite and zeunerite have partially dehydrated to produce translucent to opaque metatorbernite and metazeunerite on specimens from Ting Tang mine (Morcom's shaft), Carharrack, Gwennap, Cornwall (Bruce & Aubrey-Jones 1998).

Other Cornish occurrences of torbernite include: Carnmeal mine, E of Godolphin; Goodaver mine, St Cleer; Kit Hill United, Stoke Climsland; Wheal Betsy, Tremore, Lanivet (all Ball et al. 1982); Cheesewring quarry, SW of Henwood (Bowman 1911); Chywoon quarry, near Longdowns (Moulding & Hooper 2005); Cligga Head, Perranzabuloe (Robson 1949b); Egloshellen (Gwindra) mine, St Stephen (Dines 1956, Ball et al. 1982); Fowey Consols mine, St Blazey, Tywardreath (Lewis 1997); Great Wheal Fortune, near Porthleven (Wolfe 1996b); Gunnislake and Old Gunnislake mines, Calstock (Garby 1848 [uranite], Collins 1892, Ball et al. 1982); Loe Warren Zawn and North Zawn, Botallack, St Just (Elton & Hooper 1995a,c, 1996); Ponsanooth, near Penryn, and Withiel mine, Withiel (James 1947); St Agnes mine; South Wheal Frances, Illogan; Wheal Providence, near Carbis Bay, St Ives; Wheal Jewell – several mines in Cornwall (and Devon) have this name; Wheal Trenwith, St Ives (all Collins 1892); St Ives Consols (Robson 1945b); Stenna Gwyn mine, St Stephen-in-Brannel, St Austell (Collins 1892, Ball et al. 1982, Embrey & Symes 1987, Weiss 1994); Tincroft mine, Illogan (Collins 1892, Embrey & Symes 1987); Tolcarne mine, Camborne (Greg & Lettsom 1858 [uranite], Collins 1892, Ball et al. 1982, Embrey & Symes 1987); Wheal Alice, Grampound Road (Harrison & Taylor 1966, Ball et al. 1982); Wheal Basset and South Wheal Basset, near Redruth, Illogan (Garby 1848 [uranite], Collins 1892, Wolfe 1996a); Wheal Buller, near Redruth (Garby 1848 [uranite], Collins 1892, Embrey & Symes 1987); Wheal Edward, St Just (Garby 1848 [uranite], Collins 1892, Wolloxall 1988, Bancroft & Weller 1993); Wheal Gorland, St Day, 2 km E of Redruth (Collins 1892, Kingsbury 1954, Rust 1981b); Wheal James, Withiel (Greg & Lettsom 1858 [uranite], Collins 1892, Ball

Deep-green square plates of torbernite to 3 mm, altered around the edges to pale-green metatorbernite; Carn Brea, Illogan, Cornwall. (© M. P. Cooper)

et al. 1982); Wheal Owles, Penwith, St Just (Wolloxall 1988); Wheal Remfry china-clay pit, St Austell (Grigg 1986); Wheal Unity, St Day, Gwennap (Greg & Lettsom 1858 [uranite], Collins 1892). Robson (1954) states that torbernite, well known from many localities, is actually metatorbernite.

In Devon, mineralized veins containing torbernite have been found cutting Dartmoor granite at Bedford United mines, near Tavistock (Greg & Lettsom 1858, Hall 1868b), and Merrivale quarry (Ball et al. 1982).

Torbernite forms minute tabular crystals in cavities within native copper–cuprite veinstone near the Precambrian/Triassic unconformity at New Cliffe Hill quarry, Stanton under Bardon, Leicestershire (Hubbard et al. 2005).

In Scotland, subhedral yellowish-green plates (approximately 1 mm across) of torbernite and metatorbernite have been found in a single boulder from Balmullo quarry, SW of Luthrie, Fife (Moreton 2006). Torbernite is also described from fractures in Cu–Mo–U–hydrocarbon mineralized granite in the valley of Ord Burn, N of Helmsdale, Highland Region (Tweedie 1979). In nearby Ousdale Burn, uranium-rich arkose boulders contain metatorbernite (Gallagher et al. 1971). Further confirmation is desirable for the last two occurrences.

In Ireland, pyrite and rare torbernite (altering to metatorbernite) have been found on the joint surfaces of a beryl-bearing greisen lens in the Rosses granite (part of the Donegal granite batholith), Sheskinarone, 11 km WSW of Bunbeg, Co. Donegal (Burke et al. 1964, O'Connor 1986, Ryback & Moreton 1992). Uranium-rich quartz breccias, veins and greisen around Borris, in the Tullow pluton (southern end of the Leinster granite batholith), Co. Carlow, contain autunite, fluorite and torbernite (O'Connor 1986, Ryback & Moreton 1993).

tosudite

monoclinic $Na_{0.5}(Al,Mg)_6(Si,Al)_8O_{18}(OH)_{12}\cdot 5H_2O$

An uncommon hydrated sodium–aluminium–magnesium silicate hydroxide clay mineral composed of a regular 1:1 interstratification of dioctahedral chlorite with smectite. Tosudite is one of a few ordered interstratified clay minerals, others being brinrobertsite (pyrophyllite–smectite), corrensite (chlorite–vermiculite or chlorite–smectite) and rectorite (illite–smectite). These minerals are generally the metastable intermediate products of a sequence of reactions involving end-member clay minerals (Dong et al. 2002). Tosudite occurs in the clay fraction of planar to cross-stratified micaceous fine- to medium-grain Old Red Sandstone exposed in low cliffs on the west bank of the River Severn to the N of Lydney Harbour, near the Forest of Dean, Gloucestershire. It is associated with other clay minerals, including dioctahedral chlorite, illite and kaolinite, and probably has a diagenetic origin (Garvie 1992).

totaigite

TL Totaig, Highland Region, Scotland

A mixture of olivine and serpentine.

TOURMALINE GROUP

Trigonal borosilicates with the general formula $WX_3Y_6(BO_3)_3Si_6O_{18}(O,OH,F)_4$, where W = Ca, K, Na; X = Al, Fe^{2+}, Fe^{3+}, Li, Mg, Mn^{2+}; and Y = Al, Cr^{3+}, Fe^{3+}, V^{3+}. Dravite, elbaite, feruvite, foitite, magnesiofoitite, rossmanite, schorl and uvite are members of the group found in Britain or Ireland. Additionally, Henry & Hawthorne (1999) have proposed a new classification of the tourmaline group, suggesting that the general formula for the tourmaline group is better represented by $XY_3Z_6[T_6O_{18}][BO_3]_3V_3W$, where X=Ca, Na, □; Y=Li, Mg, Fe^{2+}, Mn^{2+}, Al, Cr^{3+}, V^{3+}, Fe^{3+}, (Ti^{4+}); Z=Mg, Al, Fe^{3+}, Cr^{3+}, V^{3+}; T=Si, Al, (B); B=B,(□); V=OH, O; and W=OH, F, O. The classification has found wide usage, despite its not having been submitted to the IMA-CNMNC. All tourmaline data from Britain and Ireland have had structural formula and names calculated using this revised classification (Tindle et al. 2002) and this adds fluor-elbaite* and fluor-schorl* to the list of tourmaline species. In line with the concept of "named" minerals devised for the amphibole group (Burke & Leake 2005), these two minerals are considered to be "named tourmalines" until they are formally described. There are, of course, many occurrences of tourmaline where the specimens have not been analyzed, or for which original data are not available. Of these, perhaps the most notable are those once found in pegmatitic cavities within the Dartmoor granite at Woolley farm, near Bovey Tracey, Devon. Crystals are doubly terminated and reach 9 cm across and 8 cm in length (Embrey & Symes 1987). They are probably schorl.

towanite

TL Wheal Towan, near Porthtowan, Cornwall, England

A synonym of chalcopyrite.

travertine

A fine-grain variety of calcite formed as a deposit from springs or rivers.

trembathite

trigonal $(Mg,Fe^{2+})_3B_7O_{13}Cl$

A rare magnesium–iron borate chloride, dimorphous with boracite and described from Boulby potash mine, Loftus, Cleveland, where it occurs in nodules within anhydritic shale overlying halite–sylvite rock of the third Zechstein evaporite cycle. Milne et al. (1977) refer to it as iron-boracite. Isolated euhedral brown subtranslucent blocky crystals of the mineral, exceptionally up to 2 cm long, in a pink sylvinite matrix are perhaps the world's best and are referred to as ericaite by Green & Freier (1994, 1996). Later work has shown that ericaite is unstable at room temperatures, and the mineral is now considered to represent a solid-solution series between congolite and trembathite (Burns & Carpenter 1996, Schindler & Hawthorne 1998, Green 2005b).

tremolite

AMPHIBOLE GROUP

monoclinic $\square Ca_2Mg_5Si_8O_{22}(OH)_2$

A relatively common calcium–magnesium silicate hydroxide that forms a series with actinolite and ferro-actinolite, and is found in contact-metamorphosed siliceous sediments and greenschist-facies metamorphic rocks derived from ultramafic or magnesium-rich carbonate rocks. Where chemical data are available (a few locations in Scotland), identification is based on the 1997 classification of the amphibole group (Leake et al. 1997a, Mogessie et al. 2004); otherwise, optical determinations are described. Because early reports describe tremolite as a synonym for a fibrous or glassy form of hornblende (Greg & Lettsom 1858), some of these descriptions may be erroneous by modern standards.

England

Cornwall Serpentinization of peridotite at the Rill, Kynance Cove, Lizard Peninsula, has produced a chlorite–chrysotile–magnesite–magnetite talc tremolite assemblage (Clark 1967). In a related occurrence in a quarry to the N of Church Cove, Landewednack, veins in serpentinite contain good specimens of fine-grain chrysotile, talc and tremolite (Hall & Jackson 1975).

Cumbria Metamorphosed calcareous breccia of the Upper Coniston Limestone at Wasdale Head, Shap (Harker & Marr 1891), contains abundant colourless tremolite and augite. Fibrous tremolite is an alteration product of metamorphosed diabase at Carrock mine, Grainsgill, Caldbeck Fells (Davidson & Thomson 1951).

Devon Metasomatism of calcareous shales around the margins of the Meldon aplite, near Okehampton, has produced fine-grain flinty skarns containing axinite, diopside, quartz and tremolite (von Knorring & Condliffe 1984).

Wales

Gwynedd Tremolite occurs in carbonate-rich rocks formed by serpentinization and carbonatization of a suite of gabbro and peridotite intrusions on Anglesey and the adjacent Holy Island. These include tremolite schists and marbles from: Bodrwyn, near Cerrigceinwen; Cefn-coch mill; near Fadog; Gareg-lwyd House; Mynachdy; N of Bronddel; near Pwllpillo; Rhyd-y-Bont Creek; and Tyddyn-dai and Tyfry, both near Pentraeth (Greenly 1919). Later work indicates that the most intense development of tremolite was where altered gabbros had been metasomatized along fault planes (Maltman 1977).

Scotland

Grampian Tremolite and actinolite are confirmed in mafic bands (thought to be a calc-silicate hornfels that developed from an exhalative sulphidic dolomitic chert) within a Dalradian Fe–Pb–Zn mineralized cherty quartzite in cliffs SE of Loch Kander, near Braemar. Associated species include armenite, clinozoisite, diopside and hyalophane (Fortey et al. 1993). Amphibolite-facies regional metamorphism of metabasalts and ultrabasic rocks from the Dalradian Blackwater Formation, near Cabrach, has produced actinolite, magnesiohornblende and tremolite as alteration products of primary clinopyroxene (Macdonald et al. 2005). Pale greenish-grey fibrous asbestiform tremolite is confirmed at Portsoy, where it occurs in narrow veins cutting gabbro–serpentinite rocks (Heddle 1878c).

Highland Tremolite is confirmed in metamorphosed Lewisian ultramafic rocks to the SW of An Fharaid Mhòr, Achmelvich, near Lochinver. It occurs with dolomite, quartz and talc (Droop 1994). Magnificent tremolite is confirmed from limestone quarries at Shinness on the eastern shore of Loch Shin (Heddle 1878c). Silvery-white to pale-green specimens, now in the NMS, are up to 28 cm long, with individual prisms up to 1 cm wide (Livingstone 2002). These two authors also confirm tremolite

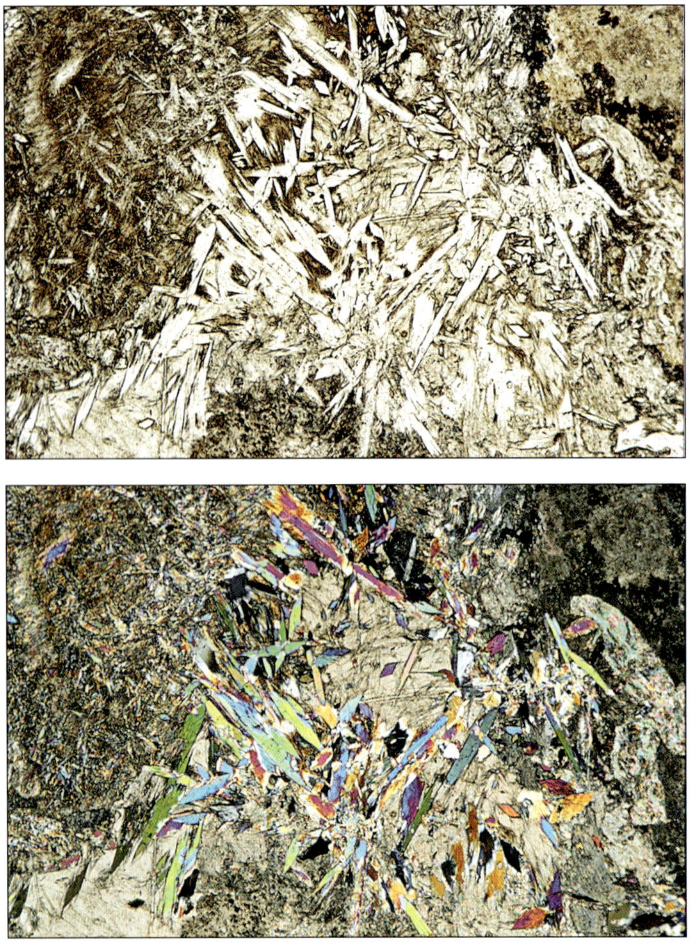

Tremolite forming a dense aggregate of small prisms and fine needles in a marble from the Durness Limestone, near Kilchrist, Isle of Skye, Highland Region, Scotland; width 3.4 mm; viewed in PPL (upper) and XPL (lower). (© K. Gardner)

in impure metamorphosed limestones from Milltown in Glen Urquhart, with both yellow and bluish-black varieties being reported. Tremolite and actinolite form rhythmic overgrowths over hornblende in appinites associated with pipes of explosion breccia at Ardsheal Hill, Back Settlement, Gleann Chàrnan and Kiloran Bay, all near Kentallen. The zonation represents growth under alternating physical conditions during the episodic release of trapped volatiles (Bowes et al. 1964). Metasomatized chert nodules within dolomite in the contact aureole of the Beinn an Dubhaich granite aureole contain tremolite (confirmed), diopside, forsterite and talc. The minerals appear successively in reaction rims around the nodules and have been used to trace metamorphic reactions in calc-silicate assemblages (Harker 1904, Hoersch 1981, Holness 1992). Similar reactions involving tremolite have also been recognized in metamorphosed siliceous dolomites in the aureole of the Ballachulish granite (Ferry 1996). Tremolite is further noted in a chalcopyrite-rich (plagioclase–clinopyroxene) dyke associated with the West Sgaorishal ultramafic plug on the Isle of Rum (Power et al. 2003). Tremolite data presented by Sills (1982) from Scourian ultramafic granulites at Drumbeg, Assynt, are actually actinolite. A roadside exposure at Auchtertyre, near Eilean Donan Castle, produced dark-green elongate crystals of tremolite up to 12 cm long, in a talcose matrix associated with sheared and distorted bright metallic crystals of pyrite (R. E. Starkey sample, Tindle unpub. EMP data).

Strathclyde Tremolite is confirmed in marble from Iona, off the coast of the Isle of Mull and (unconfirmed) as large independent and reticulated crystals in limestone from the Isle of Tiree (Heddle 1901). On Creinch, an island in Loch Lomond, close to the Highland Boundary fault, tremolite is an alteration product of a tough green volcaniclastic rock (Henderson et al. 1983).

Tayside At Corrycharmaig, Glen Lochay, NW of Killin, serpentinization of an ultrabasic intrusion within a regional terrain of Dalradian schists has produced a rock with over 90% antigorite, together with minor amounts of chlorite, dolomite, magnesite, tremolite and opaque oxides (Harrison 1985, Hawson & Hall 1987). Fine specimens of tremolite, forming plumose and radiating crystals, were once found in a quarry at Marble Lodge, Glen Tilt, E of Blair Atholl (Heddle 1901). Well crystallized edenite is described from boulders in a stream bed SW of Trinafour, near Kinloch Rannoch (Green & Todd 2001), but later analysis confirmed only tremolite (S. Moreton sample, Tindle unpub. EMP data).

Western Isles Tremolite and tschermakite are reported to occur in sheared parts of the Western gabbro on St Kilda (Hirta) (Harding et al. 1984) and in granitized and metamorphosed peridotite exposed between Rubha Sgeir nan Sgarbh and Loch Meurach in Langavat Valley, South Harris (Livingstone 1976a) – although the tremolite data from South Harris now classify as parvo-cannilloite and parvo-magnesiohornblende.

Shetland Very pale watery greyish-green tremolite is confirmed from a diallage rock at Balta on Unst, where it is described as the purest and most flexible amianthus in Scotland. It is a soft asbestiform type of tremolite and it forms fibres up to 10 cm long (Heddle 1878c). Tremolite is also confirmed from a hornblende gneiss on the banks of Nidister, Hillswick (Heddle 1878c), and from a sheared sulphide-bearing pyroxenite from E of Sandwick Bay (Maynard 1993), both on Unst.

Ireland

Donegal Fibrous slightly greenish tremolite is one of the main skarn minerals (others are diopside, grossular, vesuvianite and wollastonite) found in the limestone quarries of Lough Anure, Thorr area (Pitcher 1950).

Galway Lozenge-shape grains of tremolite have been identified in diopside-rich veins cutting Dalradian metasediments of the Connemara Limestone Formation in Connemara (Yardley et al. 1991). At Derryclare Lough, Connemara "marble" occasionally contains radiating clusters of compact fibrous tremolite (Ryback & Moreton 1992).

Wicklow Tremolite is associated with albite, prehnite, titanite and well formed clinozoisite in veins cutting amphibolite at the top of Camaderry Mountain, between Glendasan and Glendalough (Johnston 1949, Ryback & Moreton 1993). Tremolite overgrowths on hornblende in a hornblende peridotite are also described from this area (Thomson 1908).

tridymite

monoclinic (pseudo-hexagonal) SiO_2

A rare silicon oxide, polymorphous with coesite, cristobalite, quartz and stishovite. High-temperature β-tridymite is stable between 870°C and 1470°C, but can exist metastably below that temperature; α-tridymite is triclinic and is stable to 117°C. Tridymite forms in contact-metamorphosed sandstones and can be deposited from hot gases in vesicles or can occur as phenocrysts in felsic volcanic rocks.

Tridymite is confirmed from a roadside cutting in a dolerite sill between Mealt and Lealt, Trotternish Peninsula, Isle of Skye, Highland Region, Scotland. It is a component of opal-CT that occurs as a botryoidal coating on calcite (Livingstone 1993). Previous reports in the Scottish literature of what was thought to be tridymite are considered to be unsubstantiated, as the mineral in question appears to have inverted to quartz over time and is now represented by paramorphs (Macpherson & Livingstone 1982); however, these occurrences are still worthy of description. Harker (1950) drew attention to the effects of high-grade metamorphism on Torridonian sandstones around the margins of eucrite and peridotite intrusions on the Isle of Rum, Highland Region, and he reported total replacement of original quartz by tridymite during this process. A partially melted Torridonian arkose close to a gabbroic plug at Allt Bealach Mhic Nèall on Rum has developed a fine-grain spherulitic texture composed of plagioclase feldspar, relict quartz, and quartz paramorphs after tridymite (Holness 2002). Sedimentary xenoliths within gabbro at Rubha Carrach on the northwest coast of Ardnamurchan, Highland Region, contain cryptocrystalline turbid feldspar and minute lath-shape crystals of quartz that probably pseudomorph tridymite (Wells 1951). Lower Carboniferous sedimentary rocks have been intruded by and caught up as xenoliths in Tertiary dolerite dykes at McInroy's Point, on the shore of the Firth of Clyde, 3 km SW of Gourock, Strathclyde. They are thermally metamorphosed and are now represented by an assemblage of glass, quartz paramorphs after tridymite, cordierite, pyroxene and biotite (±garnet). The so-called tridymite forms many acicular crystals (0.02–0.06 mm long, by 0.002 mm broad) embedded in turbid matter (Herriot 1971).

Tridymite is also reported from two igneous parageneses in Scotland. The Coire Uaigneich granophyre, on the southeastern shoulder of Blaven towards An Stac, Isle of Skye, Highland Region, contains so-called tridymite, both as a phenocryst phase and as a component of the groundmass. Phenocrysts average 0.5 mm in length and have a platy habit (Wager et al. 1953). At Craignure, on the Isle of Mull, Strathclyde, plates of tridymite are reported in a granophyric rock (termed a craignurite) from the central acidic component of a composite layered intrusion. The rock also contains skeletal crystals of plagioclase feldspar (andesine–oligoclase) and acicular augite (Skelhorn 1962).

Thin hexagonal plates of tridymite were discovered in cavities within rhyolite at Tardree Mountain, Co. Antrim, Ireland. Apatite, biotite, epidote, andesine feldspar (clinoclase), magnetite, quartz and sanidine are associated species (von Lasaulx 1879). Tridymite has since been confirmed from here by Macpherson (1983) and Nawaz (1983a). The olivine-dolerite plug at Tieveragh, Co. Antrim, has intruded and thermally metamorphosed ashy Old Red Sandstone to produce fused rocks (vitric buchite) close to the contact. A mineral assemblage that includes cordierite, enstatite, feldspar, hematite, magnetite, osumilite-(Mg) and "tridymite" has resulted (Chinner & Dixon 1973). Another olivine-dolerite plug on the hill of Tievebulliagh, 5 km WSW of Cushendall, Co. Antrim, intrudes Tertiary basalts and contains a large block of metamorphosed lithomarge (porcellanite) in which so-called tridymite occurs in association with magnetite and as isolated crystals in cordierite. Other minerals in the porcellanite include corundum, cristobalite, plagioclase feldspar, mullite, quartz, rutile and sapphirine, (Agrell & Langley 1958).

Paramorphs of quartz after tridymite are described from a banded rhyolite sampled at Les Platons, northeast coast of Jersey, Channel Isles (Mourant 1961).

triphylite

orthorhombic $LiFe^{2+}PO_4$

A rare lithium–iron phosphate that forms a series with lithiophilite and is found in complex granitic pegmatites in the layered roof zone of the Tregonning granite at Megiliggar Rocks, near Breage, Cornwall. Green triphylite is intergrown with triplite and alters to alluaudite (Stone & George 1983).

triplite

monoclinic $(Mn^{2+},Fe^{2+})_2(PO_4)(F,OH)$

Triplite is a rare manganese–iron phosphate fluorhydroxide that forms a series with zwieselite and is found associated with löllingite at the top of the main granite sheet in the layered roof zone of the Tregonning granite at Megiliggar Rocks, near Breage, Cornwall (George et al. 1981, Stone & George 1983).

In Scotland, brown glassy irregular grains of triplite are described from the lepidolite–tourmaline-bearing pegmatite at Glenbuchat, Grampian (Macpherson & Livingstone 1982).

triploidite

monoclinic $(Mn^{2+})_2(PO_4)(OH)$

A rare manganese phosphate hydroxide that forms a series with wolfeite and is found at Wheal Owles, Penwith, St Just, Cornwall. Triploidite was discovered on an old specimen of neotocite at the NHM. It forms small (< 1 mm), yellow, orange-pink or orange-red, globular aggregates with a radial structure and was deposited in a cavity on the botryoidal surface of neotocite (Embrey 1978, Clark & Couper 1979).

tristramite

RHABDOPHANE GROUP

hexagonal $(Ca,U^{4+},Fe^{3+})(PO_4,SO_4)\cdot 2H_2O$

TL Wheal Trewavas, Trewavas Head, Breage, Cornwall, England

Named after Tristram, a medieval figure associated with the Arthurian legend and a possible resident of the discovery area. Tristramite is a rare hydrated calcium–uranium–iron phosphate sulphate that occurs as a late-stage replacement or matrix to brecciated uraninite (var. pitchblende) and is associated with sulphides in hydrothermal quartz veins related to Hercynian granites in Cornwall. In addition to the type locality, tristramite is also reported from Cornwall at: Wheal Alice, Grampound Road; Wheal Buller, near Redruth; and Wheal Providence, near Carbis Bay, St Ives. Tristramite forms fine-grain aggregates of acicular–to–fibrous crystals with an overall greenish body colour and is commonly associated with chalcopyrite, galena, marcasite and sphalerite (Atkin et al. 1983). Other localities in Cornwall include Trequeen and also Wheal Damsel, Gwennap (Anthony et al. 2000).

trögerite

AUTUNITE GROUP

monoclinic or tetragonal $H_3O(UO_2)(AsO_4)\cdot 3H_2O$

A rare hydrated hydronium uranyl arsenate found in a supergene assemblage of Cu–Pb–U minerals associated with a uraninite-bearing vein near the base of the seacliff at Loe Warren Zawn, W of Botallack, St Just, Cornwall. Trögerite forms flattened radiating groups of bright yellowish-green transparent acicular crystals at the northwestern end of the exposure; associated species include boltwoodite, cuprite and rutherfordine. Formation by the action of surface waters and sea spray on uraninite, chalcocite and chalcopyrite is proposed (Elton & Hooper 1995a). Trögerite with erythrite is noted from South Terras mine, St Stephen-in-Brannel, Cornwall (Robson 1954).

A uranium-rich bitumen from Dalbeattie, Dumfries & Galloway, Scotland, contains fracture fills and inclusions of uraninite (pitchblende) occasionally coated in trögerite (Eakin & Gize 1992). The actual location may be from along the coast at Marbruie, near Dalbeattie, where trögerite is found in a radioactive vein consisting of native bismuth, hydrocarbons, quartz, uraninite (pitchblende) and minor chalcopyrite and pyrite. Associated secondary minerals include atacamite, bismoclite, boltwoodite and connellite (Darnley 1962a). Additionally, Basham et al. (1989) describe a greenish-yellow coating of fine platelets in a uraninite (pitchblende)-bearing carbonate vein at Needle's Eye, SW of Caulkerbush, as a uranium–arsenic mineral, possibly trögerite. Hydrocarbon samples from Alva, Ochil Hills, Central Region, contain trögerite, uraninite and zeunerite (Parnell 1988c).

troilite

hexagonal FeS

An uncommon iron sulphide found with pyrrhotite in marble from Sgiath Bheinn, S of Còsag, Glenelg, Lochaber, Highland Region, Scotland. Troilite forms tiny bronze-colour flecks and grains up to 2 mm across. Associated species are clinohumite, diopside, phlogopite, quartz and spinel (Davidson 1995a). Sulphide droplets in the unit 11/12 chromite band, Eastern Layered Series, Isle of Rum, Highland Region, contain chalcopyrite and pentlandite, with minor bornite, chalcocite, native copper, cubanite, digenite, gold (electrum) and pyrrhotite (Dunham & Wilkinson 1985). These authors state that the pyrrhotite is best described as troilite.

trona

monoclinic $Na_3(CO_3)(HCO_3)\cdot 2H_2O$

An uncommon hydrated sodium carbonate bicarbonate found as a post-mining efflorescence in the mine under the ICI factory site at Billingham-on-Tees, Cleveland. The efflorescence has resulted from upward seepage of water from the Upper Magnesian Limestone. Other minerals confirmed in efflorescences at the mine include epsomite, gypsum, halite, lepidocrocite, pyrite, sulphur, sylvite, talc and thenardite; apthitalite, arsenopyrite, ilsemannite, molybdenite, tachyhydrite and tamarugite may also be present (Raymond 1959, Stewart 1965).

trüstedtite

LINNAEITE GROUP

cubic Ni_3Se_3

A very rare nickel selenide, dimorphous with wilkmanite and found in a gold-bearing carbonate vein cutting Middle Devonian limestones at Hope's Nose, Torquay, Devon. The selenide assemblage includes clausthalite, eucairite, fischesserite, klockmannite, naumannite, penroseite, tiemannite, tyrrellite and umangite in a gangue dominated by calcite. Trüstedtite is the earliest selenide to form and it occurs as irregular lath-like relicts (5–50 μm long) in tyrrellite (Stanley et al. 1990a).

tschermakite

AMPHIBOLE GROUP

monoclinic $\square Ca_2(Mg_3AlFe^{3+})(Si_6Al_2)O_{22}(OH)_2$

A relatively common calcium–magnesium–iron aluminosilicate hydroxide that forms a series with ferrotschermakite and is found in medium- to high-grade metamorphic rocks such as amphibolites, in eclogites and in ultramafic igneous rocks. Where possible, identification is based on the 1997 classification of the amphibole group (Leake et al. 1997a, Mogessie et al. 2004).

England

Cornwall A tin-enriched variety of tschermakite occurs in a calc-silicate body (representing metasomatically altered metabasalt) in the aureole of the Land's End granite at Chycornish Carn, St Just (Alderton & Jackson 1978). A finely banded plagioclase lherzolite from the Lizard ophiolite complex at Countybridge quarry, Goonhilly Downs, Lizard Peninsula, contains both pargasite and tschermakite (Davies 1983).

Wales

Gwynedd On the Lleyn Peninsula, tschermakite is reported from gabbros of the Rhiw intrusion (Bevins 1994) and ultrabasic gneisses exposed at Parwyd (Horák 1993). In both cases the data are unpublished.

Scotland

Central So-called hornblende schists from Aberfoyle Forest, near Balmaha, Loch Lomond (near the Highland Boundary fault), contain acicular green edenite, magnesiohornblende and lesser amounts of both pargasite and tschermakite (Henderson et al. 1983).

Highland Tschermakite occurs in a Devonian appinitic rock (possibly a feeder for the basic Lorne lavas) from the Beinn Ghlas intrusion in the northern part of the Lorne Plateau, to the E of Oban (Groome & Hall 1974). Data from a Lewisian epidote–garnet–hornblende gneiss from Cruard, Isle of Skye, suggest that alumino-tschermakite is present (Matthews & Cheeney 1968), but re-analysis of the same material using a different method (Leake 1968) produced a tschermakite composition. An amphibolitized eclogite from Beinn a' Chapuill, Glenelg, contains bluish-green tschermakite replacing omphacite (Alderman 1936, O'Hara 1960). A Lewisian hornblende gneiss from Glaschoille House, Knoydart, contains pale yellowish-green to medium bluish-green pleochroic tschermakite (Lambert 1963).
Strathclyde A garnet–hornblende clinopyroxenite, previously described as eclogite, exposed between Knockormal and Carleton Mains farms, contains tschermakite and is interpreted as a fragment of lower crust or upper mantle. It is clearly part of the Girvan–Ballantrae ophiolite complex (Smellie & Stone 1984). A hornblende-bearing garnet metapyroxenite containing pargasite and tschermakite, exposed near Knocklaugh, is from the aureole of this complex (Treloar et al. 1980).
Tayside An investigation of amphibole compositions from Dalradian Argyll and Southern Highland Group metabasites around and to the NE of Aberfeldy, and centred on the classic area of Barrovian metamorphic zones that developed during the Grampian orogeny, shows that tschermakite is restricted to the sillimanite and andalusite zones, with aluminotschermakite found in the garnet and kyanite zones (Zenk & Schulz 2004). Both green and brown amphiboles from hornblende schist xenoliths, and from large crystals in an appinite from the Glen Tilt igneous complex, E of Blair Atholl, have tschermakite compositions (Deer 1938, 1953).
Western Isles Tschermakite and tremolite are reported to occur in sheared parts of the Western gabbro on St Kilda (Hirta) (Harding et al. 1984), but compositional data are unpublished.
Shetland A fibrous-looking green (black on a weathered surface) amphibole from a diorite on the northeast side of the entrance of Trista Voe, Fetlar, is parvo-tschermakite (Heddle 1878c).

Ireland
Galway The Doughruagh intrusion in Connemara consists of quartz-rich gneisses in which zoned amphibole crystals have tschermakite cores and cummingtonite rims, all formed as alteration products of primary igneous species (Angus & Kanaris-Sotiriou 1982).

tschermigite
cubic $(NH_4)Al(SO_4)_2{\cdot}12H_2O$

A rare hydrated ammonium–aluminium sulphate found as colourless crystalline patches within massive opaque white epsomite on a NMW specimen from the Two-Feet-Nine seam, Deep Navigation pit, Treharris, Mid Glamorgan, Wales (T. F. Cotterell pers. com. 2008). Tschermigite requires further confirmation from Emily coal pit, Arniston, Lothian, Scotland (Shand 1910a, Macpherson 1983).

TSUMCORITE GROUP
Monoclinic and triclinic arsenates, phosphates, sulphates and vanadates, with the general formula $(M1)(M2)_2(XO_4)_2(OH,H_2O)_2$, where $M1$ = Pb, Ca, Na; $M2$ = Cu, Zn, Fe^{3+}, Co, Mn^{3+}; and X = As, P, V, S. Gartrellite and mawbyite are members of the group found in Britain or Ireland.

tsumebite
BRACKEBUSCHITE GROUP
monoclinic $Pb_2Cu(PO_4)(SO_4)(OH)$

A rare lead–copper phosphate sulphate hydroxide found in association with brochantite on the 90-fathom-level dumps at Roughton Gill, Caldbeck Fells, Cumbria. Tsumebite forms apple-green spheroidal masses with arsentsumebite rims (Wolfe 1996a, Green & Briscoe 2002, Tindle et al. 2006). Yellowish-green granular crusts on quartz with intermediate arsentsumebite–tsumebite compositions are reported from Carrock Fell, Cumbria (Embrey 1978, Young 1987a).

tsumoite
trigonal BiTe

A very rare bismuth telluride found at Urlar Burn, 5 km from Aberfeldy, Tayside Region, Scotland, in quartz–sulphide ±gold veins cutting Upper Dalradian gritty quartzites, quartz–mica schists and amphibolites. Tsumoite is one of several telluride species forming inclusions in bismuth-bearing galena, others being altaite, benleonardite, coloradoite, gold (electrum), melonite and volynskite (Ixer et al. 1997). Along the Strathclyde/Dumfries & Galloway border, in the Leadhills area, two alluvial gold grains are reported to contain tsumoite inclusions (Leake et al. 1998).

Apple-green spherules of tsumebite (with arsentsumebite exteriors) to 1 mm; emerald-green prismatic brochantite is an associated species; Roughton Gill mine, Caldbeck Fells, Cumbria. (© D. I. Green)

tučekite
HAUCHECORNITE GROUP
tetragonal $Ni_9Sb_2S_8$

A very rare nickel sulphantimonide found in hydrothermal veins in a cluster of mines within the Central Wales orefield, Dyfed: Darren mine, Pen-bont Rhydybeddau; Eaglebrook (Nantycagl) mine, Ceulanymaesmawr; Erglodd mine, N of Tal-y-bont; Esgair Hir mine, Ceulanymaesmawr; Esgair Fraith mine, near Tal-y-bont; Henfwlch mine, between Cerrig yr Hafan and Nant-y-moch reservoir, E of Tal-y-bont; and Hyddgen mine, near Nant-y-moch reservoir. Tučekite occurs as blocky elongated tetragonal crystals (up to 0.5 mm long) included in galena. Associated species are chalcopyrite, millerite, siegenite and ullmannite (also electrum at Eaglebrook mine). When observed through an ore microscope, tučekite is one of the most colourful ore minerals ever found (Mason 1997, 1998).

tuesite
TL Tweed, Borders, Scotland

Named after *Tuesa*, the Latin name for the River Tweed. Tuesite is a synonym of lithomarge.

tulameenite
tetragonal Pt_2FeCu

A very rare platinum–iron–copper alloy that forms a series with ferronickelplatinum and is found in the Leadhills area, close to the Strathclyde/Dumfries & Galloway border, Scotland. A single alluvial gold grain is reported to contain tulameenite inclusions (Leake et al. 1998).

tungstate of iron
A synonym of ferberite.

tungstate of lime
A synonym of scheelite.

tungstite
orthorhombic $WO_3{\cdot}H_2O$

An uncommon hydrated tungsten oxide found in tungsten–molybdenum-rich quartz veinstone within greisen associated with the Skiddaw granite at the Carrock mine site, Grainsgill, Caldbeck Fells, Cumbria. Supergene oxidation has produced secondary yellow powdery tungstite and molybdite as replacement products of wolframite and molybdenite, respectively (Finlayson 1910, Hitchen 1934, Davidson & Thomas 1948, Hartley 1984, Young 1987a, Cooper & Stanley 1990). This locality is referred to as Brandy Gill by Greg & Lettsom (1858), who describe the mineral as wolframine. A yellow earthy crust of tungstite is also noted on wolframite from six Cornish localities: Cligga Head, Perranzabuloe; Drakewalls mine, Callington; East Pool mine, Illogan, near Redruth; South Crofty mine, Camborne; Poldice mine, E of St Day, Gwennap; and Wheal Friendship, SE of St Day, Gwennap (Greg & Lettsom 1858, Rudler 1905, Robson 1949b, 1954).

Reflected-light images of tučekite showing its anisotropy in cross-polarized light (analyzer rotated at different angles). Associated species are galena (grey) and chalcopyrite (yellow). Eaglebrook (Nantycagl) mine, Ceulanymaesmawr, Dyfed, Wales; width 0.6 mm. (© J. S. Mason)

Near Gairnshiel Bridge, Glen Gairn, Grampian, Scotland, bright-yellow tungstite infills a rectangular cavity, 3×2 mm, in a quartz–hübnerite vein cutting greisenized zinnwaldite granite (Livingstone 1993).

turgite

Incorrect spelling of turite; a synonym of hydrohematite. Noted as isolated cubic crystals in clay slate from the Isle of Kerrera and to the E of Oban, Strathclyde (Heddle 1882a).

TURQUOISE GROUP

Triclinic phosphates, with the general formula $AB_6(PO_4)_xPO_3(OH)_{2-x}(OH)_8\cdot4H_2O$, where A = Ca, Cu^{2+}, Fe^{2+}, Zn; and B = Al, Fe^{3+}, Cr^{3+}. Chalcosiderite, faustite, planerite and turquoise are members of the group found in Britain or Ireland. Aheylite, requires confirmation and coeruleolactite is now discredited.

turquoise

TURQUOISE GROUP

triclinic $Cu^{2+}Al_6(PO_4)_4(OH)_8\cdot4H_2O$

A relatively common hydrated copper–aluminium phosphate hydroxide that forms a series with chalcosiderite and is closely related to the now-discredited rashleighite (which is distinctly green and has an iron-rich intermediate composition between turquoise and chalcosiderite). A review of turquoise–chalcosiderite and related species is given by Braithwaite (1981). Henwoodite (also a discredited name) is identical in composition to turquoise and was first described from Wheal Phoenix, Linkinhorne, Cornwall, where it formed sky-blue globular masses on limonite. Associated species include andrewsite (discredited), chalcosiderite and dufrénite (Collins 1876a, 1892, 1911a). Blue turquoise and rashleighite occur at Bunny mine, St Austell, as layers up to 4 cm thick in cassiterite and wolframite lodes cutting kaolinized granite (Russell & Vincent 1952). Larger and better specimens of turquoise were found at Gunheath china-clay pit near by, almost certainly in an extension of one of the Bunny mine lodes. Beautiful blue turquoise occurred in thick impersistent bands up to 7 cm thick, sometimes covering prismatic crystals of quartz (Russell 1952). Blocks of turquoise weighing up to 1 kg were recovered from a "turquoise lode" in the northern part of Gunheath between 1976 and 1983 and beautiful sparkling blue pseudo-rhombohedral crystals of turquoise up to 0.05 mm across were recovered from a quartz-lined vug in 1984 (Grigg 1986, Weiss 1994). At Hensbarrow china-clay pit (now part of Gunheath pit), small clear discrete crystals of turquoise, 0.1–0.25 mm long, have been found lining cavities in kaolinized granite in association with fluorite, muscovite, schorl and wavellite (Russell 1952, Braithwaite 1981, King 2002). Massive, pale-blue to deep sky-blue turquoise is found in quartz veins at this locality and also forms pale-blue crusts on joint surfaces (Hawkes et al. 1987). At Wheal Remfry china-clay pit, 2 km from Indian Queens, near St Austell, pale-blue crystals of turquoise, up to 0.08 mm across, growing on tourmaline, were found in a single small cavity within kaolinized granite (Braithwaite 1981). Rare blue turquoise is also noted from the Castle-an-Dinas wolfram mine, St Columb Major, Cornwall (Hey et al. 1938, Russell 1952).

The turquoise described from Driggith mine, Caldbeck Fells, Cumbria (Hartley 1984), is based on a specimen of dubious provenance in the Kingsbury collection at the NHM. The specimen is an exact match with material from Castle-an-Dinas mine (Green et al. 2006a). The turquoise reported from Goonvean china-clay pit, St Austell (Kingsbury 1959b), may also be dubious.

Well formed sharp crystals of turquoise up to 0.3 mm across; Gunheath china-clay pit, Carthew, near St Austell, Cornwall. (© M. P. Cooper)

In Ireland, turquoise occurs as translucent bluish-green coatings, 1 mm thick, in patches up to 2×10 cm, coating joint surfaces in black and dark-grey shales at the base of coastal cliffs in a small bay NW of Doon East, 1 km N of Ballybunnion, Co. Kerry. The shales contain pyritiferous and carbonaceous bands, and in the local area are known to contain arsenopyrite, chalcanthite, chalcopyrite, crandallite, melanterite, pyrite and wavellite (Ryback & Moreton 1993, Moreton et al. 1995). Turquoise also occurs in a brecciated slaty gossan at Ballycormick, near Shanagolden, Co. Limerick, where it coats vuggy white quartz and is associated with cacoxenite, crandallite and wavellite (Ryback & Moreton 1991, Moreton et al. 1995). Upper Carboniferous phosphate-rich black shales at Grouse Lodge quarry, 5 km S of Shanagolden, Co. Limerick, contain turquoise (bluish-green coatings on joints and fracture filling brown nodular masses) together with gibbsite, kaolinite, koninckite, planerite and wavellite (Moreton & Green 2005).

tush

A Derbyshire name for red or pink barite.

tyreeite

TL Isle of Tyree, Strathclyde, Scotland

A mixture of talc, mica and hematite.

tyrolite

monoclinic $Ca_2Cu_9(AsO_4)_4(CO_3)(OH)_8 \cdot 11H_2O$

An uncommon hydrated calcium–copper arsenate carbonate hydroxide, originally thought to have orthorhombic symmetry, but now shown to be monoclinic with two polytypes, -1*M* and -2*M* (Krivovichev et al. 2006). Tyrolite-1*M* is thought to be identical to clinotyrolite.

Tyrolite is found in copper–lead-rich mineralized faults and porous conglomeratic units of Triassic New Red Sandstone at Wood mine and West mine, Alderley Edge ridge, Cheshire. It is a secondary mineral that forms fans of bladed crystals partially pseudomorphed by a copper silicate presumed to be chrysocolla (Braithwaite 1994). Carlon (1979) also notes its occurrence in the nearby Engine Vein mine.

Bluish-green tyrolite and chrysocolla occur in fractures in diorite close to a native copper–cuprite deposit near the Precambrian/Triassic unconformity at New Cliffe Hill quarry, Stanton under Bardon, Leicestershire (Hubbard et al. 2005). Tight fractures in Precambrian volcaniclastic rocks at Bardon Hill quarry, Coalville, are also known to contain tyrolite. It forms deep bluish-green cleavage masses associated with azurite, chrysocolla, cuprite, malachite and zálesíite (Ince 2007).

A report of tyrolite from Penberthy Croft mine, St Hilary, Cornwall (Golley & Williams 1995), was not substantiated by Betterton et al. (2000). Tyrolite described from Potts Gill mine, Caldbeck Fells, Cumbria (Hartley 1984, Young 1987a), is based on specimens in the Kingsbury collection at the NHM and may have dubious provenance (see Ryback et al. 1998, 2001).

Bevins (1994) notes tyrolite as radiating lath-like blue crystals up to 2 mm long, on joints associated with chrysocolla and iron oxides from Gwaithyrafon mine, Cwmsymlog, SE of Tal-y-bont, Central Wales orefield, Dyfed. Exceedingly rare bright-blue flattened circular hemispherical and botryoidal masses of a tyrolite-like mineral are described from a trial level, 2 km NW of Bontddu, near Dolgellau, Gwynedd (Saich & Rust 1987). In 1995 a barite–chalcopyrite–galena–tennantite vein was temporarily exposed at Dolyhir quarry, Old Radnor, near Presteigne, Powys. Secondary species found at the site include anglesite, azurite, cerussite, native copper, cuprite, harmotome, malachite, olivenite and tyrolite (Wolfe 2000).

Uranium-bearing veins in the Needle's Eye area, SW of Caulkerbush, Dumfries & Galloway, Scotland, host a range of supergene species that includes bright-green sheaves of tyrolite up to 1 mm long, together with cornubite, kahlerite, lavendulan, meta-autunite, metakahlerite, metazeunerite, mixite and zeunerite (Braithwaite & Knight 1990, Green & Todd 2001). At Alva mine, Silver Glen, Ochil Hills, Central Region, tyrolite occurs with annabergite and erythrite as a coating on clinosafflorite, which in turn coats native silver (Green & Todd 2001). Loose blocks of country rock have been found in Silver Glen encrusted with bluish-green radiating pearly blades of tyrolite (Livingstone 1993).

In Ireland, tyrolite is one of 104 species reported from the large lead–zinc mine at Tynagh, near Killimor, Co. Galway (Ryback & Moreton 1992, Green & Moreton 2001). Radiating groups of bladed tyrolite with tetrahedrite are also described from the old dumps of Hollyford mine, 9 km NNE of Tipperary town (Ryback et al. 1988). However, the best tyrolite in Britain or Ireland comes from the unusual Cu–(Ag–Hg) deposit at Gortdrum mine, near Tipperary town, where lustrous radiating blades up to 7 mm long have been found (Ryback & Moreton 1991, Braithwaite & Wilson 2003). At Leaghillaun, Bearhaven (Berehaven) Peninsula, Co. Kerry, a chalcopyrite–tennantite-bearing quartz vein hosted by Devonian sandstones has produced an oxidized suite of supergene minerals that includes tufts and sheaves of green tyrolite up to 5 mm long, associated with azurite, chrysocolla, clinoclase, cornwallite, malachite and zálesíite (Moreton & Green 2006b).

Upper Radiating sprays of green tyrolite, up to 2 mm in diameter, from Gwaithyrafon mine, Cwmsymlog, Dyfed, Wales. (© National Museum of Wales, Cardiff; photographer M. P. Cooper)

Lower Spherules of green tyrolite, 5 mm across, associated with blue azurite; Dolyhir quarry, Old Radnor, near Presteigne, Powys, Wales. (© D. I. Green)

tyrrellite

LINNAEITE GROUP

cubic $(Cu,Co,Ni)_3Se_4$

A very rare copper–cobalt–nickel selenide found in a gold-bearing carbonate vein cutting Middle Devonian limestones at Hope's Nose, Torquay, Devon. The selenide assemblage includes clausthalite, eucairite, fischesserite, klockmannite, naumannite, penroseite, tiemannite, trüstedtite and umangite, all in a gangue dominated by calcite. Tyrrellite forms anhedral inclusions (50–100 μm) in clausthalite and penroseite, and also euhedral inclusions in clausthalite, klockmannite and umangite (Stanley et al. 1990a).

tyuyamunite

orthorhombic $Ca(UO_2)_2V_2O_8 \cdot 5\text{–}8H_2O$

A rare hydrated calcium uranyl vanadate found in disused mineworkings of the Cargodna section of Wheal Owles, St Just, Cornwall. Tyuyamunite occurs as earthy crusts and small canary-yellow microcrystalline aggregates. The aggregates form sheaves and fan-like groups of acute petaline crystals, with individual crystals up to 5–10 μm long. Associated species include nováčekite, uranophane and zeunerite. Tyuyamunite is a supergene species found in cavities within quartz, which crystallized below a localized patch of aragonite in the roof of the stope (Elton & Hooper 1994).

U

ugrandite

An old group name for andradite–grossular–uvarovite garnets.

uigite

TL Uig, Isle of Skye, Highland Region, Scotland

A mixture of gyrolite and thomsonite, both zeolite group (see Heddle 1882b).

ullmannite

COBALTITE GROUP

cubic NiSbS

A relatively uncommon nickel sulphantimonite that forms a series with willyamite and is usually found in hydrothermal veins.

Ullmannite (<5 μm-diameter grains) has been found in quartz–fluorite veinstone on the dumps of North Devon United (East Wheal Friendship) mine, Peter Tavy, Devon, within the metamorphic aureole of the Dartmoor granite. It is intimately associated with cassiterite, gersdorffite, native bismuth, bismite, parkerite, sphalerite, stannite and an unnamed copper–lead selenide sulphide (Rumsey & Savage 2005).

Ullmannite is reported from localities in the Alston Block of the northern Pennines. At Brownley Hill lead mine, Nenthead, near Alston, Cumbria, a 1 m-long cavity on a minor vein called Tatters String was found to contain ullmannite as cubic metallic-grey crystals on galena, associated with bottinoite, cerussite and sulphur (Clark & Rust 1993, Green et al. 2000). Microscopic amounts of ullmannite and gersdorffite associated with nickeline, pyrite and pyrrhotite have been recovered from magnetite-rich ore on the dumps of Lady's Rake mine, along Harwood Beck, Teesdale, Co. Durham. Mineralizing fluids from the Whin Sill circulating through the Teesside fault system is thought to have produced this assemblage, and a similar origin is proposed for nickel mineralization at Settlingstones lead mine, near Hexham, Northumberland, where ullmannite also occurs (Young et al. 1985b). Cubic crystals of ullmannite were found in nickeline (niccolite) ore associated with sphalerite and witherite during the mining of galena at Settlingstones mine (Russell 1927, Trestrail 1931). Ixer (1986) notes that ullmannite forms rims around pyrite (bravoite) and skutterudite at this mine. At New Brancepeth colliery, 7 km W of Durham, alstonite and a small amount of tin-white ullmannite were found in a barite–witherite fissure vein cutting Coal Measures strata. The ullmannite forms rough cubes (to 2 cm) in groupings reaching 6 cm across. Small octahedra of ullmannite up to 3 mm across, and parallel intergrowths of galena and ullmannite forming individual crystals, are also noted (Spencer 1910, Smythe 1947). A more detailed examination showed that minor amounts of nickeline and pyrite (bravoite) are associated with the ullmannite, which is both zoned and selectively replaced by galena (Ixer 1986). Other occurrences of ullmannite include: Tynebottom mine, NW of Garrigill, near Alston, Cumbria; Nenthead, near Alston; and Teesdale Valley, Co. Durham (Vaughan & Ixer 1980).

Ullmannite is reported from a single locality in the English Lake District, at Barrow mine, Braithwaite, SW of Keswick, Cumbria, where it forms inclusions in galena in association with native antimony, boulangerite and bournonite (Stanley & Vaughan 1981).

Examination of dump material derived from the Cu–Pb–Zn mines of the Central Wales orefield in Dyfed (and Powys) has shown ullmannite to be widely distributed. At Esgair Hir mine, Ceulanymaesmawr, ullmannite forms euhedral inclusions (typically 20–40 μm across) in galena and occasionally is found as discrete crystals, up to 0.7 mm across, along the boundaries of chalcopyrite (Rust & Mason 1988). Quartz veinstone at Hendrefelen mine, Ysbyty Ystwyth, contains a primary assemblage of calcite, chalcopyrite, galena and ullmannite, the last occasionally forming euhedral crystals within cavities. Crystals of silvery grey ullmannite to 3 mm are sometimes coated by aggregates of bottinoite; on one specimen, bottinoite has grown on bindheimite-replaced ullmannite (Rust 1990a, Clark & Rust 1993). Bottinoite is an alteration product of ullmannite at Mynyddgorddu mine, W of Bont-goch, Ceulanymaesmawr (Clark & Rust 1993). Mason (1997, 1998) recognizes ullmannite in two distinct paragenetic assemblages termed the A1 early complex and A2 late

Intergrown crystals of ullmannite on quartz, associated with golden chalcopyrite; Hendrefelen mine, Ysbyty Ystwyth, Dyfed, Wales; width 2.5 mm. (© A. G. Tindle)

simple assemblages, both associated with phases of brecciation. It is the A2 assemblage that is of most interest to specimen mineralogists, as rich specimens, in which ullmannite forms interpenetrant cubes to 10 mm on edge, are found in this assemblage. A2 localities include: Bryntail mine, W of Llanidloes; Brynyrafr mine, near Nant-y-moch reservoir; Cefngwyn mine, near Aberystwyth; Dolclettwr mine, Tre'r-ddol; Elgar mine, 3 km S of Tal-y-bont; Esgair Hir mine, Ceulanymaesmawr; Frongoch mine, between Devil's Bridge and Pont-rhyd-y-groes; Goginan mine (also A1 assemblage present); Grogwynion mine, Pont-rhyd-y-groes; Gwaithgoch mine, 2 km W of Pont-rhyd-y-groes; Mynyddgorddu mine, W of Bont-goch, Ceulanymaesmawr; Nantiago mine, near Hafren Forest; Nantymwyn mine, NE of Llandovery; Pandy mine, Tre'r-ddol; Ynystudur mine, Tre'r-ddol. In A1 assemblages, ullmannite is found only as microscopic inclusions in galena.

In Scotland, ullmannite is reported from a boulder of mineralized breccia recovered from the Crom Allt river, near Tyndrum, Central Region. The assemblage includes rare, small (<20 μm) euhedral crystals of ullmannite, together with chalcopyrite, siderite, sphalerite, and galena with exsolved tennantite–tetrahedrite lamellae (Pattrick & Hall 1984). Alluvial gold grains containing inclusions of ullmannite are noted from Craigierigg Burn, Borders and Shortcleugh Water, Leadhills, Strathclyde (Leake et al. 1998, Chapman et al. 2000c).

Ullmannite is a trace constituent of an As–Co–Cu–Ni assemblage found near major faults within the carbonate-hosted Lisheen Pb–Zn–(Ag) deposit, between Thurles and Rathdowney, Co. Tipperary, Ireland. Associated species are bournonite, cobaltite–gersdorffite, nickeline, pararammelsburgite, pyrite (bravoite) and tennantite, with traces of arsenopyrite, barite, chalcopyrite and hematite (Ixer 2002, Fusciardi et al. 2004).

ulvöspinel?

SPINEL GROUP

cubic $TiFe^{2+}_2O_4$

A titanium–iron oxide noted in accessory amounts in the Cambir dolerite, on St Kilda (Hirta), Western Isles, Scotland. The oxide grains in the dolerite are mostly magnetite, with variable contents of either ilmenite lamellae or ulvöspinel (Harding et al. 1984). Identification was optical and requires confirmation (Livingstone 1993).

umangite

tetragonal Cu_3Se_2

A very rare copper selenide found in a gold-bearing carbonate vein cutting Middle Devonian limestones at Hope's Nose, Torquay, Devon. The selenide assemblage

Reddish umangite containing euhedral inclusions of white tyrrellite; the flame-like textured mineral with radial fractures (top right) is klockmannite. Hope's Nose, Torquay, Devon; width 800 µm; viewed in reflected light. (© C. J. Stanley, The Natural History Museum, London)

includes clausthalite, eucairite, fischesserite, klockmannite, naumannite, penroseite, tiemannite, trüstedtite and tyrrellite in a gangue of dominantly calcite. Umangite forms anhedral grains up to 200 µm across included in clausthalite. It also covers larger areas in calcite, where it contains inclusions of tyrrellite and is replaced by klockmannite (Stanley et al. 1990a).

uraconite

A uranium sulphate, possibly uranopilite or zippeite, or a mixture of both.

uralite

Actinolite pseudomorphous after pyroxene.

uraninite

cubic UO_2

A relatively common uranium oxide found in accessory amounts in granite and syenite pegmatites. Uraninite is also found in tin-bearing veins, in As–Ag–Bi–Co–Ni and other sulphide veins, and in some sandstones. Pitchblende is a botryoidal or massive variety of uraninite.

In Cornwall, uraninite is a ubiquitous magmatic accessory mineral and the major carrier of uranium in unaltered granite. Biotite granite and crosscutting microgranite from the Carnmenellis pluton contain euhedral uraninite crystals, 10–500 µm across, as inclusions in biotite, muscovite or tourmaline (Jefferies 1985c). Hydrothermal alteration of such crystals liberated uranium to provide a source for later pitchblende vein mineralization. At South Terras mine, St Stephen-in-Brannel (the only mine in Britain worked primarily for uranium) pitchblende occurred in cross-course veins within killas near the southern margin of the St Austell granite complex. At surface the first ores were green and yellow oxidation products (autunite, torbernite or zippeite), but at depth unaltered black resinous pitchblende ore was recovered. The pitchblende was associated with abundant chalcopyrite and pyrite, together with lesser amounts of arsenopyrite, bornite, chalcocite, covellite, galena, marcasite, and traces of bismuth, chromium, cobalt and nickel (Robertson & Dines 1929, Gregory 1945, James 1947, Dines 1956, Pockley 1964, Ball et al. 1982). Darnley et al. (1965) report a crystallization sequence of pyrite–quartz, pitchblende and siderite in borehole material. Pitchblende also occurs in some Cornish copper and tin lodes, such as at Geevor mine, near Pendeen, St Just, at the western end of the Land's End granite, where it is found with arsenopyrite, bornite, chalcopyrite, chalcocite and covellite (Dines 1956, Ball et al. 1982). Wheal Trenwith, St Ives, another once-important uranium producer on the northeastern part of the Land's End granite, contained pitchblende in association with cassiterite and chalcocite (James 1947). Ball et al. (1982) reports bismuth, galena, molybdenite and sphalerite as later associates.

Many other Cornish uraninite (mostly pitchblende) occurrences are noted by James (1947), Ball et al. (1982) and other workers: uraninite at Bosworgey borehole, near Fraddam (Ball & Basham 1979, 1984); Botallack (Loe Warren Zawn), St Just (Elton & Hooper 1995a,c); Carn Brea mines, near Redruth; Carnmarth, Gwennap; Carnmeal mine, E of Godolphin; thorium-poor uraninite in biotite granite from the Carnmenellis pluton, SE of Camborne (Jefferies 1984, 1985a); Crow Hill mine, St Stephen; Dolcoath mine, Camborne (Hosking 1951); East Pool mine, near Redruth, Illogan; East Wheal Lovell, Wendron; Egloshellen (Gwindra) mine, St Stephen (all Darnley et al. 1960); uraninite at Goonvean and Rostowrack china-clay pits, W of Nanpean, St Austell (Scott et al. 1998); Great Dowglas mine, Sticker, near St Austell; Gunheath china-clay pit, Carthew, St Austell (Hawkes et al. 1987); Illogan mine, near Redruth; Lamorna quarry, St Just (Ball & Basham 1979, Charoy 1986); Luxulyan quarry, near Lostwithiel (Ball et al. 1982, Ball & Basham 1979); Pednandrea mine, Redruth; Pendarves mine, Camborne; Roskrow United mine, Ponsanooth, and Wheal Alice, Grampound Road (Harrison & Taylor 1966); Rosemanowas quarry, Carnmenellis (Ball & Basham 1979); Roskrow United mine, Ponsantooth (Collins 1892); uraninite in two-mica granite covering 70% of the St Austell granite complex (Manning et al. 1996); St Agnes mine; St Aubyn and Grylls, Breage; St Austell Consols mine, St Austell (Collins 1892); St Ives Consols, St Ives; uraninite at South Crofty mine, Camborne (Lucas & Zaghloul 1960, Ball & Basham 1979); South Tresavean mine, Falmouth (Rust 1982a, Hosking 1951); South Wheal Basset, near Redruth, Illogan; Tincroft mine, near Redruth (Greg & Lettsom 1858); Ting Tang mine, Carharrack (Greg & Lettsom 1858, Collins 1892); Tolcarne mine, Camborne; Tregune mine, Altarnun; Tregurtha Downs, Goldsithney; Tresavean mine, Gwennap, near Redruth; West Jewel mine, St Day; West Wheal Peevor, Redruth; Wheal Basset, near Redruth, Illogan (Collins 1892); Wheal Betsy, Tremore, Lanivet; Wheal Bray, Altarnun (Darnley et al. 1965, Harrison & Taylor 1966); Wheal Buller, near Redruth (Collins 1892); Wheal Carharrack, St Day; Wheal Edward, St Just (Collins 1892, Elton et al. 1994b); Wheal Gorland, St Day, 2 km E of Redruth; Wheal Herland, Gwinear; Wheal James, Withiel; Wheal Owles, Penwith, St Just (Collins 1892, Pockley 1964, Elton et al. 1994b); Wheal Providence, near Carbis Bay, St Ives (Greg & Lettsom 1858, Collins 1892); Wheal Speed, St Ives; Wheal Trewavas, Trewavas Head, Breage (Atkin et al. 1983); Wheal Unity, St Day, Gwennap; Wherry mine, Penzance; Withiel mine, Withiel (Collins 1892).

Sadly, the results of age determinations of pitchblende from South Terras and Wheal Owles (Pockley 1964, Darnley et al. 1965) are to be treated with caution, as the samples were provided by A. W. G. Kingsbury and their provenance is suspect (Ryback et al. 2001).

Uraninite is briefly mentioned as a rare accessory inclusion species in biotite or secondary tourmaline from the Dartmoor granite. It is accompanied by apatite, monazite, xenotime and zircon (Knox & Jackson 1990, Ward et al. 1992). A very similar assemblage is also found in the Tertiary Lundy granite, off the north coast of Devon, although the mica is lithium-rich siderophyllite and the uraninite has an appreciable thorium content (Thorpe et al. 1990, 1995). Pitchblende and uranium secondary species are associated with copper mineralization at Kingswood mine, near Buckfastleigh (Darnley et al. 1965, Ball et al. 1982, Durrance & Laming 1982), and at Littleham Cove, near Budleigh Salterton, coffinite, pitchblende and cobalt–nickel arsenides occur in uranium–vanadium-rich nodules within the Permian Littleham Mudstone Formation (Milodowski et al. 2002).

Elsewhere in England, uraninite occurs in various parageneses: microscopic inclusions in hydrocarbons within hydrothermal lead–zinc veins cutting Carboniferous Limestone at Windy Knoll, Castleton, Derbyshire; Breedon Cloud, near Shepshed, and Staunton Harold, Ticknall, Ashby-de-la-Zouch, both Leicestershire (Parnell 1988c); 0.5–2 µm-diameter inclusions in xenotime-(Y), associated with bismuthinite and hematite at Groverake lead mine, near Rookhope, Weardale, Co. Durham (Ixer 2003); accessory uraninite with apatite, monazite and zircon in the Weardale granite sampled in the Rookhope borehole, Weardale, Co. Durham (Brown et al. 1987a); accessory uraninite in Skiddaw granite and associated greisen at Carrock, Grainsgill, Caldbeck Fells, Cumbria (Dawson & Harrison 1966, Buckley 1984, Ball et al. 1985a); accessory thorium-rich uraninite in granite from the Shap granite quarry, Cumbria (Basham et al. 1982, Webb & Brown 1984a, Wirth & Green 2002).

Uraninite forms 1–10 µm inclusions in solid hydrocarbons within low-temperature hydrothermal mineralized veins in Carboniferous Limestone in the Ty Gwyn copper mine, Great Orme's Head, Llandudno, Gwynedd, Wales, and from the lead–zinc orefields at Halkyn and Minera, both in Clwyd (Eakin 1989, Parnell 1988c). Specific localities are named by Eakin (1989) as Graig quarry, Halkyn, Hendre quarry, Llaramon quarry, Olwyn Goch shaft, Pant quarry and Pant-y-pwll quarry. Eakin & Gize (1992) report uraninite with a probable brannerite rim in bitumens from Llandulas Limestone quarry, near Colwyn Bay, Clwyd.

Reports of uraninite in Scotland are from widely dispersed locations and parageneses, as follow:

CENTRAL Uraninite-bearing veins postdate the main lead–zinc mineralization at Tyndrum Main mine. Acanthite (argentite), barite, calcite, chalcopyrite, galena, kasolite, liebigite, safflorite, sphalerite, tetrahedrite and uranospinite are associated species (Harrison 1956, Darnley et al. 1960, Darnley 1962b, Gallagher et al.

Colloform uraninite with trace amounts of rounded chalcopyrite (top left); Tyndrum, Central Region, Scotland; width 600 µm; viewed in reflected light. (© R. A. Ixer)

1971, Ineson & Mitchell 1974b, Ridgway 1980, Pattrick 1985). At Alva mine, Silver Glen, Ochil Hills, uraninite has been found with tiemannite and an unnamed bismuth–silver selenide, all as < 1 mm-diameter inclusions in pyrobitumen (Parnell 1988a). DUMFRIES & GALLOWAY Uraninite and chalcopyrite coat joint surfaces in hornfels adjacent to hematite–quartz veins close to the Criffell granodiorite contact at Beeswing (Gallagher et al. 1971); irregular inclusions of pitchblende up to 50 µm in diameter occur in hydrocarbon from Dalbeattie (Eakin & Gize 1992); uraninite is said to be abundant in a suite of highly radioactive veins in the Needle's Eye area, SW of Caulkerbush, in the southern aureole of the Criffell granodiorite. It is associated with a wide range of As–Bi–Co–Cu–Fe–Pb species and many uranium-rich secondary species (Williams 1965, Miller & Taylor 1966, Roberts et al. 1988, Basham et al. 1989, Braithwaite & Knight 1990, Green & Todd 2001); uraninite and traces of chalcopyrite in a clay-filled joint-infilling in hornfels at Potterland, N of Auchencairn, Criffell (Gallagher et al. 1971); uraninite and thorianite form discrete grains and inclusions (< 10 µm across) in zircon within mineralized diorite at Palnure Burn, Talnotry, 5 km NE of Newton Stewart (Power et al. 2004). GRAMPIAN Uraninite occurs with a wide range of other accessory species (andalusite, apatite, beryl, garnet, monazite, thorite, topaz, xenotime, zircon, and iron–titanium oxides) in the Cairngorm granite (Webb & Brown 1984b, Harrison 1988). HIGHLAND Uraninite, brannerite and xenotime occur in reduction spots within Devonian Old Red Sandstone at Dingwall (Parnell & Eakin 1987, Hoffman 1993); uraninite, secondary uranium minerals and barite occur in Lower Old Red Sandstone arkose in Ousdale Burn, Helmsdale (Bowie & Gallagher 1970, Gallagher et al. 1971, Dunham et al. 1978). STRATHCLYDE Dalradian-hosted lead–zinc mineralization at Inverneil mine, S of Adrishaig, Knapdale contains hydrocarbon–pitchblende complexes associated with chalcopyrite and löllingite (Miller & Taylor 1966, Gallagher et al. 1971, Parnell 1988c). WESTERN ISLES Lewisian granitic pegmatites on South Harris: uncommon uraninite, sometimes forming inclusions in ferrocolumbite within the Chaipaval (Ceapabhal) pegmatite, 3 km NW of Northton (Taobh Tuath). Kasolite, monazite, thorite, zircon and a pyrochlore mineral (probably betafite) are associated radioelement-bearing species; black or dark-green uraninite forms inclusions in feldspar, magnetite or monazite in the Loch a' Sgùrr pegmatite, near Finsbay (Fionnsabagh). Allanite, monazite, thorite, uranophane and zircon are associated radioelement-bearing species (von Knorring & Dearnley 1959b, 1960a).

In Co. Galway, Ireland, monazite, thorite, uraninite and zircon are reported as accessory species in the Costelloe Murvey leucogranite, Connemara (Feely & Madden 1988, Feely et al. 1989). Uraninite is also present in accessory amounts in the Main Donegal granite pluton, typically in NE–SW trending zones within biotite pegmatites. Near Fintown and extending between Croaghleheen and Lough Muck, Co. Donegal, euhedral uraninite cubes (0.2–2 mm across) and microbotryoidal pitchblende veinlets are found in pegmatite and associated schists respectively (O'Connor et al. 1984, Legg et al. 1985, O'Connor 1986, Ryback & Moreton 1992). At Sheshkinarone, in the Rosses granite (a part of the Donegal granite complex), rare uraninite, metatorbernite and torbernite have been found in beryl-bearing greisen (Burke et al. 1964, O'Connor 1986). The rare occurrence of uraninite (pitchblende) mineralization, with minor brannerite and possibly coffinite associated with base-metal ore mineralization at Gortdrum mine, near Tipperary town, has allowed direct dating of the deposit to be made and provides evidence for a major tectonothermal event that occurred during the Carboniferous as a result of the later closing stages of the proto-Atlantic. The uraninite exhibits a variety of textures indicative of colloidal deposition, including concentric banding and radial structures (Duane et al. 1986, Duane 1988). Pitchblende has been identified on microfracture surfaces in chlorite schist close to the Tullow pluton (a part of the Leinster granite batholith) at Curraghlane, Co. Carlow, and thin quartz–pitchblende veins are noted at Cullentragh, on the eastern contact of the Blackstairs pluton (O'Connor 1986).

Hydrocarbons discovered in the chalcopyrite zone at Laxey lead–zinc mine, E of Snaefell, Isle of Man, have been shown to contain anhedral uraninite grains (up to 21 µm across) that are partially replaced by pitchblende. A hydrothermal origin is proposed (Davidson & Bowie 1951, Parnell 1988c, Eakin 1989, Eakin & Gize 1992).

uranite

A general term for autunite, meta-autunite or torbernite.

uranophane

monoclinic $Ca(UO_2)_2[SiO_3(OH)]_2 \cdot 5H_2O$

A relatively common hydrated calcium–uranium silicate hydroxide, dimorphous with uranophane-beta and found in the Loch a' Sgùrr pegmatite, NE of Sletteval, near Finsbay (Fionnsabhagh), South Harris, Western Isles, Scotland. Uranophane occurs as a yellow powdery film on fracture surfaces and is associated with an accessory mineral assemblage of allanite, epidote, magnetite, monazite, spessartine, thorite, uraninite and zircon (von Knorring & Dearnley 1959b, 1960a).

On mainland Scotland, the seacliffs between Balcary Point and Caulkerbush, S of Dalbeattie, Dumfries & Galloway, host uranium-rich veins containing pale-yellow earthy films and yellow botryoidal crusts composed of acicular crystals of uranophane. Details of all the veins are not reported, although along this 20 km stretch of coastline, the following species (many with a supergene origin) have all been found: annabergite, atacamite, bismoclite, native bismuth, bismutite, boltwoodite, bornite, brochantite, calcite, chalcopyrite, connellite, cornubite, covellite, dolomite, erythrite, eulytite, gypsum, hematite, kahlerite, langite, lavendulan, löllingite, meta-autunite, metakahlerite, metazeunerite, mixite, nickeline, nováčekite, pharmacosiderite, pyrite, quartz (smoky variety), schoepite, sphalerite, tennantite, tyrolite, uraninite (var. pitchblende), vandendreisscheite, walpurgite, wulfenite and zeunerite (Miller & Taylor 1966, Braithwaite & Knight 1990, Green & Todd 2001). One vein, near Needle's Eye, SW of Caulkerbush, contains abundant delicate yellow needles of uranophane associated with uraninite, pitchy bitumen, shiny orange spheres of vandendreisscheite, and orange-yellow schoepite (Braithwaite & Knight 1990). Uranophane and uraninite are found as inclusions in hydrocarbon samples from Alva, Ochil Hills, Central Region (Parnell 1988c).

At West Wheal Owles, St Just, Cornwall, mineralized fissure veins contain a supergene assemblage of uranium species in which uranophane forms mats and spherules of divergent crystals associated with compreignacite and cuprosklodowskite. Torbernite, tyuyamunite, uraninite and uranopilite are also found at this site. At nearby Wheal Edward, uranophane is associated with brochantite, calcite, compreignacite, connellite and cuprosklodowskite on a dark mineralized hornfels that also contains chalcopyrite, potassium feldspar, pitchblende and siderite (Elton & Hooper 1994, Elton et al. 1994b).

Uranophane and rutherfordine are said to have been found in schist below the Old Red Sandstone unconformity at Boherkyle, near Graiguenamanagh, Co. Kilkenny, Ireland (Ryback & Moreton 1993), but this requires confirmation.

uranopilite

monoclinic $(UO_2)_6(SO_4)(OH)_{10} \cdot 12H_2O$

A rare hydrated uranium sulphate hydroxide found with compreignacite, cuprosklodowskite, torbernite and uraninite at Wheal Owles, Penwith, St Just, Cornwall. Uranopilite is a secondary mineral and forms orange encrustations on altered uraninite-bearing veinstone (Frondel 1952, Ryback & Tandy 1992, Elton et al. 1994b). Further occurrences of uranopilite are noted by Golley & Williams (1995) from Wheal Edward (another mine in the Wheal Owles sett) and from the nearby Geevor mine (near Pendeen).

uranospathite

tetragonal $(Al,\square)(UO_2)_2F(PO_4)_2 \cdot 20(H_2O,F)$

TL Redruth, Cornwall, England

A rare hydrated aluminium uranyl fluorphosphate found at the type locality where the Carnmenellis granite is in contact with Devonian sandstones. Mineralization consists of three main species – chalcopyrite, chalcocite and cassiterite – with accessory amounts of arsenopyrite, bassetite, pyrite, uranospathite, vochtenite and wolframite. Uranospathite forms thin yellow rectangular plates that, if twinned, form cross-shape crystals. Non-reversible dehydration is a problem when storing the species and it needs to be kept at low temperature or in a humid atmosphere, or it will decompose to a phase identical to sabugalite (Hallimond 1915, Frondel 1954, Walenta 1978). Gaines et al. (1997) and Anthony et al. (2000) give the locality as Basset mines, Redruth. An examination of microfractures in kaolinized granite from Wheal Martin china-clay pit, St Austell, Cornwall, showed they are lined with fluorite, schorl and a mineral thought to be uranospathite (Tindle unpub. EMP data).

uranospinite

tetragonal — AUTUNITE GROUP — $Ca(UO_2)_2(AsO_4)_2{\cdot}10H_2O$

Clusters of pale-green platy uranospinite crystals to 0.5 mm across, from Wheal Edward, St Just, Cornwall. (© D. I. Green)

A rare hydrated calcium uranyl arsenate found in the aureole of the Land's End granite at Wheal Edward, St Just, Cornwall. Uranospinite is intimately mixed with sodium-uranospinite and it forms pale-green crystals. The two species commonly overgrow zeunerite (D. I. Green pers. comm. 2007).

Uranospinite also occurs as a component of a late uranium-rich vein hosted by Dalradian metasediments at Tyndrum Main mine, Central Region, Scotland. It is found in vuggy cavities, where it is associated with pink calcite, uraninite and sulphides. Chalcopyrite, galena, marcasite, pyrite and safflorite are part of the primary assemblage and they are found with the secondary uranium minerals kasolite, liebigite, uranospinite and zeunerite, and secondary erythrite. The secondary species probably formed during post-depositional oxidation of the deposit (Darnley 1962b, Pattrick 1985).

uranthorite

A uranium-rich variety of thorite.

urquhartite

TL Milton, Glen Urquhart, Highland Region, Scotland

A name for a wholly undefined mineral in the gneiss at Milton (Hey 1961).

uvarovite?

cubic — GARNET GROUP — $Ca_3Cr_2(SiO_4)_3$

An uncommon calcium–chromium silicate that forms a series with grossular and is described from thermally metamorphosed banded limestones of the Moine Series, Ardgour district, on the north side of Loch Linnhe and a few km SW of Fort William, Highland Region, Scotland. Green uvarovite, mantled by an opaque chromium-rich phase and colourless garnet, is reported from from a single location in banded rocks whose assemblage may also contain biotite, calcite, chlorite, orthoclase, plagioclase, prehnite, pyroxene, pyrrhotite, scapolite, titanite, tremolite or quartz (Drever 1939). Unfortunately, there does not appear to be any confirmatory evidence to support this identification, and further evidence is therefore desirable, particularly as a green garnet originally described as uvarovite from Hagdale quarry, Unst, Shetland, was shown when analyzed to be a chromium-rich grossular (Macpherson & Livingstone 1982).

uvite

trigonal — TOURMALINE GROUP — $CaMg_3Al_6(BO_3)_3Si_6O_{18}[(OH)_3O]$

A rare calcium–magnesium–aluminium borosilicate hydroxide, found in quartz–tourmaline vein samples collected along the shoreline at Carrick Du and Clodgy Point, near St Ives, Cornwall. The samples were not *in situ*, but were probably derived from the nearby Carrick Du mine. At both localities, zoned dravite, showing pale-blue to pale-brown pleochroism in thin section, is the major tourmaline species present and this is sometimes rimmed with uvite (Tindle unpub. EMP data). At Gryll's Bunny, near Botallack, St Just, Cornwall, metasomatic alteration of amphibolite and actinolitic greenstone has produced coarsely granular (≤ 5 mm) tourmaline-rich rocks as part of a hydrothermal event induced by the emplacement of the Land's End granite. In thin section, blue to pale-brown tourmaline (black in hand-specimen) is strongly zoned and mainly schorl or dravite in composition. Uvite and feruvite occur less commonly, either as discrete crystals or as rims to dravite–schorl crystals. In some cases, all four species are found within a single crystal (Tindle unpub. EMP data). The tourmaline-rich explosion breccia at Wheal Remfry china-clay pit, near Indian Queens, St Austell, contains small subhedral crystals of uvite with fluorapatite and magnetite inclusions and thin dravite–schorl rims (S. A. Rust sample, Tindle unpub. EMP data).

Uvite forms sporadic clots within altered gabbro, 300 m SW of the summit of Hanter Hill, Powys, Wales. The bluish-black clots are up to 15 mm across and consist of a single crystal or group of crystals in near-parallel orientation. Individual crystals exhibit well defined colour zones, becoming darker outwards, and on the basis of a single chemical analysis is reported to be of dravite composition with an unusually high calcium content (Holgate 1977). In view of the zonation and the fact that the analysis is based on a bulk mineral-separate (and hence prone to contamination), the presence of this species in Wales requires further investigation.

uytenbogaardtite

tetragonal — Ag_3AuS_2

A rare silver–gold sulphide found in a 60 cm-thick polyphase quartz vein cemented by ankerite, within the Scafell dacite at Piers Gill, 1 km N of Scafell summit, Wasdale, Cumbria. Uytenbogaardtite has been found as a single 3 μm-diameter inclusion in sphalerite. It occurs with acanthite (argentite), native gold (electrum), pyrargyrite, tetrahedrite and an unknown selenium–silver mineral in an Ag–As–Au–Cu–Pb–Zn assemblage similar in many respects to that found in a porphyry-copper system (Aldridge 2005).

V

vaesite

cubic PYRITE GROUP NiS_2

A rare nickel sulphide that forms a series with cattierite and is found in Ni–Pb–Zn ore with a barite and minor calcite gangue in sandstones and shales close to the Whin Sill at Settlingstones mine, near Hexham, Northumberland. Small quantities of vaesite, violarite and millerite form successive rims around nickeline that has partially altered to rammelsbergite. Other associated species are galena, gersdorffite, marcasite, pyrite (bravoite), skutterudite, sphalerite and ullmannite (Ixer 1986, 1990a, Brown et al. 1987a). Vaesite was identified by optical methods (it has a distinctive violet-brown colour in reflected light), but confirmatory analytical data are desirable.

Vaesite occurs with chalcopyrite, jordanite and tennantite, all as inclusions in galena, within the carbonate-hosted Lisheen Pb–Zn–(Ag) deposit, between Thurles and Rathdowney, Co. Tipperary, Ireland (Strashimirov et al. 1999, Fusciardi et al. 2004).

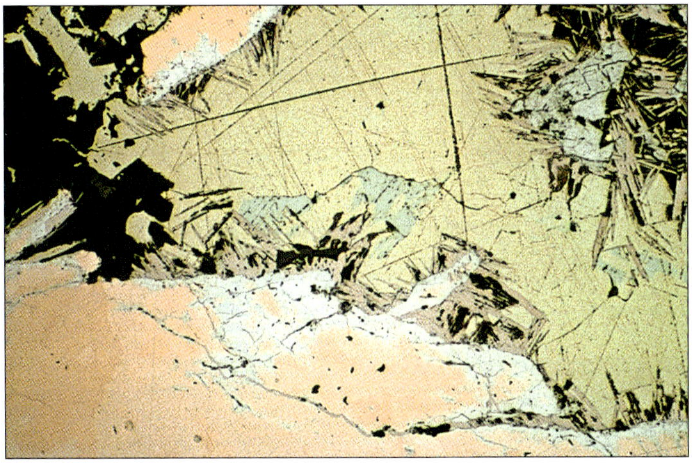

Acicular crystals of violet-brown vaesite (centre and top right), associated with nickeline (pink, bottom left) and altered along grain edges to very pale-green rammelsbergite. A central vein of yellow pyrite intergrown with marcasite (pale green) completes the assemblage. Settlingstones mine, near Hexham, Northumberland; width 500 μm; viewed in reflected light. (© R. A. Ixer)

valentinite

tetragonal Sb_2O_3

A rare antimony oxide, dimorphous with senarmontite and found by Russell (1949b) at four localities:

- In an antimony-rich lode at Bodannon mine, St Endellion, Cornwall, with jamesonite (apparently the main ore species), anglesite, bindheimite, nadorite, senarmontite and a single specimen of native silver.
- Traces of valentinite and kermesite were found with stibiconite on bladed masses of stibnite from a 30 cm-wide quartz–stibnite vein cutting Upper Devonian slates and grits at Cutlinwith farm, Landrake with St Erney, Cornwall.
- Rare valentinite was found with bournonite or tetrahedrite in a quartz–ankerite–dolomite–siderite lode at Wheal Leigh, Pillaton, Cornwall, where many good specimens of stibnite were once recovered.
- At Robin Hood mine, Hause, Bassenthwaite, Cumbria, valentinite was found as minute radiating aggregates of silky white crystals on a single specimen of stibnite from a quartz vein cutting an altered dolerite dyke.

Well crystallized valentinite occurs in cavities within an antimony-rich quartz vein from the aureole of the Skiddaw granite, at Coombe Height, near the head of Wet Swine Gill, Caldbeck Fells, Cumbria. The vein contains an early As–Fe–Sb primary assemblage (mainly stibnite and berthierite) and a later Pb–Sb assemblage

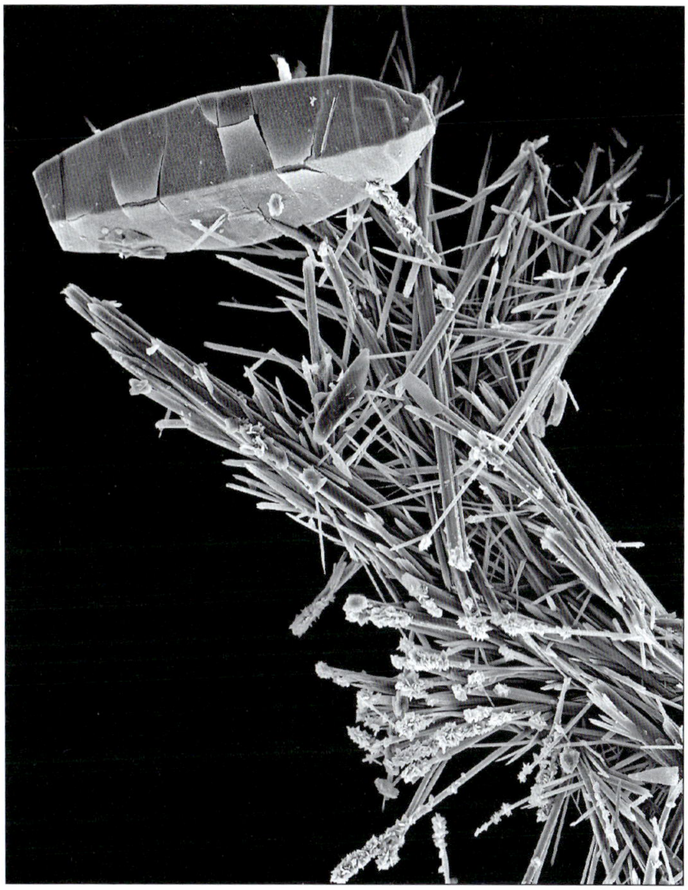

A tower of stibnite needles supporting a single blocky crystal of valentinite from Wet Swine Gill, Caldbeck Fells, Cumbria; width 200 μm; SEM image. (© D. I. Green)

(zinkenite with lesser fülöppite and semseyite). Other minor phases in the vein include native antimony, claudetite, jamesonite, senarmontite, sphalerite and native sulphur (Fortey et al. 1984, Leppington & Green 1998, Neall & Green 2001b). The valentinite has an unusual red colouration thought to be caused by inclusions or coatings of metastibnite (Green & Briscoe 2002). The valentinite reported by Young (1987a) from Wanthwaite mine, St John's in the Vale, near Threlkeld, Cumbria, is based on material from the Kingsbury collection at the NHM. The occurrence warrants further investigation, as some specimens in this collection appear to have incorrect provenance (Ryback et al. 1998, 2001). The valentinite described with bindheimite, nadorite and senarmontite in cavities within massive jamesonite at Trevinnick mine, St Kew, Cornwall (Kingsbury 1957c), also requires confirmation. Additionally, valentinite is noted as minute radiating aggregates with dickite and stibnite from Robin Hood mine, Bassenthwaite, Cumbria (Young 1987a); without comment from two locations in Cornwall – Port Quin beach, St Endellion, and Treore mine, St Teath (Golley & Williams 1995); and as boat-shape translucent crystals to 4 mm on edge at Pengenna mine, St Kew, Cornwall (Wolfe 2006).

In Scotland, at Louisa mine, Glendinning, Glenshanna Burn, Westerkirk, 13 km NNW of Langholm, Dumfries & Galloway, a mineralized quartz vein, dominated by stibnite intimately mixed with galena, was found to contain drusy cavities lined with crystals of ankerite, calcite, cervantite, pyrite, semseyite, sphalerite and valentinite. The valentinite forms groups of minute acicular crystals (Heddle 1901, Herbert Smith 1919, Williams 1973). At Knipe (Fountainhead) antimony mine, SE of New

Cumnock, Dumfries & Galloway, valentinite forms dirty-white crusts as a supergene alteration product of stibnite. Cervantite and kermesite are other oxidation products replacing stibnite. The antimony mineralization is derived from a 30–45 cm-wide quartz–stibnite vein that outcrops near the centre of the small Knipe–Hare Hill granite mass (Williams 1965, 1973).

At Traie Vrish, SE of the Niarbyl, near Dalby, Isle of Man, sizeable water-worn boulders of pyrite–stibnite-bearing quartz veinstone can be found close to a cliffside adit that was driven in 1893–94. Valentinite occurs as an alteration product of stibnite and is restricted in occurrence to cavities and fracture surfaces. It forms distinctive colourless-to-white acicular crystals and spherical crystal aggregates up to 2 mm in diameter. It is associated with patches of an unidentified red mineral that elsewhere has been inconclusively called kermesite (Starkey 1998a).

valleriite

hexagonal $4(Fe,Cu)S{\cdot}3(Mg,Al)(OH)_2$

A rare iron–copper sulphide magnesium–aluminium hydroxide found at Pol Cornick, SW of Mullion, Lizard Peninsula, Cornwall, where it occurs in veinlets of calcite–minnesotaite–talc cutting serpentinized peridotite of the Lizard ultramafic body (Clark 1970a).

In Scotland, valleriite is found in an allivalite (a pyroxene-free gabbro) from Huntly, Grampian, where it occurs as lamellae associated with small patches of sulphides composed of chalcopyrite, pentlandite and pyrrhotite (Babkine & Conquéré 1968). A mineral with many of the optical properties of valleriite, and associated with genkinite and a platinum–rhodium-rich antimonide, is described from Harold's Grave, Baltasound–Haroldswick area, Unst, Shetland, Scotland. However, it may be a platinum carbonate (Prichard et al. 1994).

vanadinite

APATITE GROUP

hexagonal $Pb_5(VO_4)_3Cl$

A rare lead vanadate chloride found in the oxidized zone of lead-bearing deposits, the vanadium being leached from wallrock silicates. Vanadinite has been found in the lead lode of Greystone quarry, Lezant, near Launceston, Cornwall, where it has grown on manganese oxides as drusy crusts of orange-red curved hexagonal crystals, usually less than 1 mm across (Elton et al. 2004). It has also been reported from Penberthy Croft mine, St Hilary, Cornwall (Camm & Merry 1991b), but does not actually occur there (Elton et al. 2004).

Vanadinite prisms to 5.5 mm in length. Wanlockhead, Dumfries & Galloway, Scotland. (© M. P. Cooper)

Lustrous vanadinite from the Leadhills–Wanlockhead orefield, Scotland; width 4 cm. (© D. I. Green)

Vanadinite is reported from small opencast workings in Carboniferous Limestone at Wesley Theological College (Wesley mine), Westbury-on-Trym, Avon. Lemon-yellow, opaque crystals have been found encrusting the walls of a cavity in a large calcite geode, and, elsewhere in the deposit, vanadinite replaces parkinsonite (Alabaster 1989b). Longwood quarry, Long Ashton, Avon, is a second location in the Bristol district, where small (1 mm) lustrous orange-brown transparent crystals of vanadinite have been found associated with cerussite, in a deposit of manganese oxides within Carboniferous Limestone (Alabaster 1989b). Analysis of vanadium-rich mimetite from Merehead (Torr Works) quarry, Cranmore, Mendip Hills, Somerset (working Carboniferous Limestone) has revealed minute spots (approx. 20 µm diameter) with V > As, thus identifying the mineral as vanadinite (Rumsey 2007).

At Judkins quarry, NW of Nuneaton, Warwickshire, colourless to pale-brown vanadinite was found as small, dull to highly lustrous, hexagonal prisms, up to 0.5 mm long, in a zone of mineralization just below the Precambrian/Cambrian unconformity (the exposure has now been quarried away). The vanadinite is associated with mottramite and calcite (Ince et al. 1990). At Newhurst quarry, Shepshed, Leicestershire, vanadinite forms colourless to pale-brown transparent hexagonal crystals associated with mottramite (Ince 2005). A few isolated specimens of vanadinite, forming corroded bipyramidal crystals up to 6 mm long, have also been found at New Cliffe Hill quarry, Stanton under Bardon, Leicestershire (Ince 2001, Hubbard et al. 2005).

Reports of vanadinite from the mines on Alderley Edge ridge, Cheshire, are considered doubtful in the absence of any evidence of identification (Braithwaite 1994). Serious doubts have also been expressed about the provenance of vanadinite specimens from Brandy Gill, Carrock mine, Ingray Gill and Old Potts Gill mine, all in the Caldbeck Fells, Cumbria (Ryback et al. 2001, Green & Briscoe 2002). The specimens have been described by Kingsbury & Hartley (1956a), Hartley (1984), Young (1987a) and Cooper & Stanley (1990). Virtually all these descriptions are based on dubious Kingsbury samples. The only independently confirmed vanadinite from these localities is that found as creamy radiating fibrous crystals on quartz from the upper levels of Brandy Gill (Cooper & Stanley 1990). Ryback et al. (2001) state that the Kingsbury specimens from Brandy Gill specimens are consistent with intentional falsification of the locality and they suggest that the specimens actually came from Mammoth mine, Tiger, Pinal County, Arizona, USA. However, vanadinite is confirmed from Arm o' Grain, Caldbeck Fells, where it occurs in a quartz vein as white hexagonal plates and elongated hexagonal epimorphs up to about 1 mm long, sometimes overgrown by mottramite (Green et al. 2006b).

Discrete euhedral-to-anhedral pale lemon-yellow crystals of vanadinite, associated with barite, calcite, dolomite, iron-rich calcite, manganocalcite and pyrobelonite in a vuggy braunite–hausmannite ore have been found at the disused manganese mine at Tŷ Coch mine, Porthcawl, Mid Glamorgan, Wales. Vanadinite also forms yellow encrustations in brecciated manganese ore (Criddle & Symes 1977).

Exquisite specimens of orange-brown vanadinite have been collected from High Pirn mine on the Belton Grain vein, Wanlockhead, Dumfries & Galloway, Scotland. Complex hexagonal prismatic crystals, richly coating the surfaces of some specimens, reach 8 mm in length. Cream, yellowish-brown and orange-brown resinous globules of vanadinite, associated with lead-rich apatite and pyromorphite, are also found (Heddle 1901, Gillanders 1981, Livingstone 1994a,b, 2002). Mention of vanadinite from "Hegh-pirn of the Susannah mine at Wanlock Head, Dumfriesshire" by Greg & Lettsom (1858) is somewhat confusing, as Susanna mine is in neighbouring Leadhills, Strathclyde.

Extremely rare pale reddish-brown vanadinite, forming euhedral hexagonal prisms up to 0.4 mm long and associated with greenockite and mottramite, is found as coatings on prehnite and calcite from Loanhead quarry, near Beith, Strathclyde. Feathery aggregates of hexagonal crystals, cream to pale straw in colour, on purplish-red tuff also occur at this quarry (Meikle 1989).

Vanadinite has been found with bismutite, wolframite and wulfenite in panned concentrates from stream sediments within a mineralized area of the Etive granite complex to the E and NE of Ben Starav, Strathclyde. However, none of these species has been found in bedrock samples (Haslam et al. 1985).

vandenbrandeite?

triclinic $Cu^{2+}(UO_2)(OH)_4$

A rare copper–uranium oxide hydroxide reported from Wheal Edward, St Just, Cornwall (Golley & Williams 1995). Requires further confirmation.

vandendriesscheite

orthorhombic $Pb_{1.5}(UO_2)_{10}O_6(OH)_{11}\cdot 11H_2O$

A rare hydrated lead uranyl oxide hydroxide found half-way up a cliff, in a highly uranium-rich fissure vein close to Needle's Eye, SW of Caulkerbush, Dumfries & Galloway, Scotland. This and other similar veins are in the aureole of the Criffell granodiorite. Vandendriesscheite forms aggregates of orange spheres less than 0.1 mm across and is associated with schoepite and uranophane. Other supergene minerals from this vicinity, produced during the alteration of primary As–Bi–Co–Cu–Fe–Pb–U minerals, include eulytite, kahlerite, lavendulan, metakahlerite and walpurgite (Macpherson & Livingstone 1982, Braithwaite & Knight 1990).

At Loe Warren Zawn, on the coast to the west of Botallack, St Just, Cornwall, an assemblage of supergene Cu–Pb–U mineralization contains vandendriesscheite, boltwoodite and rutherfordine closely associated with uraninite. Vandendriesscheite forms thin dark-yellow crusty patches on a matrix rich in uraninite and chalcocite (Elton & Hooper 1995a).

A vug (3 mm across) containing pale-brown needles of vandendreisscheite from near Needle's Eye, SW of Caulkerbush, Dumfries & Galloway, Scotland. (© D. I. Green)

vanoxite?

crystal system? $V_4^{4+}V_2^{5+}O_{13}\cdot 8H_2O$?

A poorly characterized hydrated vanadium oxide, possibly present in black vanadium-rich nodules within Mercia Mudstone (Keuper Marl) at Glen Parva and Thurmaston (King 1973).

VARISCITE GROUP

Orthorhombic arsenates and phosphates with the general formula $AXO_4\cdot 2H_2O$, where A = Al, Fe^{3+}, Cr^{3+}, In^{3+}; and X = As, P. Mansfieldite, scorodite, strengite and variscite are members of the group found in Britain or Ireland.

variscite

VARISCITE GROUP

orthorhombic $AlPO_4\cdot 2H_2O$

A rare hydrated aluminium phosphate, dimorphous with metavariscite, that forms a series with strengite and is found on wavellite at Hensbarrow china-clay pit, St Austell, Cornwall, where it forms aggregates of sharp pale-green transparent crystals up to 3 mm in length. The specimen was not found *in situ* and may have been bulldozed from the adjacent Gunheath china-clay pit (Barstow 1982). At Gunheath, colourless variscite forms spherulitic aggregates of intergrown crystals (to 2 mm in diameter) with metavariscite cores. Variscite and metavariscite typically occur on a matrix of granite pegmatite or coarse quartz–tourmaline vein material. Cassiterite occurs in the matrix of several specimens. Wavellite is a common and intimate associate, and turquoise occurs on variscite on one specimen. Variscite with an unusual equant habit was found on one specimen as crusts of single or intergrown crystals on the surface of drusy quartz, which formed a thin veinlet in a small granite boulder. Here the crystals were pale green, dull green, translucent to opaque, and up to 1 mm in length (Elton 1996). Vugs in tourmaline-bearing quartz greisen at Gunheath contain off-white globular aggregates of variscite 0.5–1 mm in diameter and, additionally, white, apple-green and pale-blue varieties have also been noted (Weiss 1994).

In the Lake District, turquoise to pale-green botryoidal variscite occurs with arsenopyrite, quartz, scorodite and wavellite from the Long Comb (Cobalt) mine, Scar Crag, Causey Pike, Cumbria (Young 1987a).

Variscite spherules to 1.1 mm from High Down quarry, Heddon, near South Molton, Devon. (© D. I. Green)

At Pwlldu Head, near Bishopston, Gower Peninsula, West Glamorgan, Wales, a single specimen of sea-green wavellite within Upper Carboniferous limestone, shales and siliceous sandstones was found to host small (<1 mm) colourless to orange-yellow rosettes of variscite (Plant & Jones 2001). Along the beach here, Dean & Cotterell (2003) found variscite in close association with cacoxenite, crandallite and wavellite on chert thought to have originated from the limestone quarry at Pwlldu Head.

Confirmation is required for variscite reported from: Treore mine, Port Gaverne, near Port Isaac, St Teath, Cornwall; High Down quarry and Filleigh & Pitts Cleave quarry, Tavistock, Devon (all claimed by A. W. G. Kingsbury – Ryback et al. 1998); and Laharran quarry, Minane Bridge, Co. Cork, Ireland (Ryback & Moreton 1991). Variscite has since been confirmed from High Down quarry (D. I. Green pers. comm. 2006).

variolite

A rock name, also used to describe a variety of orthoclase, feldspar group.

varlamoffite

amorphous $(Sn,Fe)(O,OH)_2$

Varlamoffite is a tin–iron oxyhydroxide that has previously been considered to be a variety of cassiterite. After heat treatment (at about 1200°C) varlamoffite has an XRD pattern identical to that of cassiterite (Taylor et al. 1970). It has been given questionable status by the IMA.

At Cligga mine, Perranzabuloe, Cornwall, varlamoffite forms a soft, creamy-yellow, powdery coating (up to 25 mm thick) on specimens of greisen and is associated with mica, quartz and tourmaline. It has also been found on quartz-wolframite veinstone at this location. At Bunny mine, St Austell, Cornwall, varlamoffite is ivory yellow to primrose yellow and is intimately associated with rashleighite and turquoise in layers up to 75 mm thick on cassiterite- and wolframite-bearing veinstone. Cream-buff to brownish-yellow varlamoffite with a coarse colloidal texture is reported from the dumps of Kit Hill Consols mine, Stoke Climsland, Cornwall (Russell & Vincent 1952).

Creamy-yellow earthy varlamoffite associated with quartz; Cligga mine, Perranzabuloe, Cornwall; height of specimen is 55 mm. (© A. G. Tindle)

varvicite

TL Warwickshire, England

A synonym of pyrolusite, rutile group.

vaterite

hexagonal $CaCO_3$

TL Ballycraigy, Larne, Co. Antrim, Ireland

A rare calcium carbonate, trimorphous with aragonite and calcite, and identified in a carbonated calcium silicate hydrogel complex that pseudomorphs larnite at the type locality. Vaterite is very finely fibrous or platy and is unstable in the presence of water at room temperature and pressure, dissolving readily with concomitant precipitation of calcite (McConnell 1960).

vauquelinite

monoclinic $Pb_2Cu^{2+}(CrO_4)(PO_4)(OH)$

A very rare lead–copper chromate phosphate hydroxide found as a late-stage secondary mineral at Greystone quarry, Lezant, Cornwall, where Cu–Pb–Zn vein mineralization cuts a complex metasedimentary sequence. Vauquelinite was found on specimens in the Maurice Grigg collection at Manchester Museum, as sulphur-yellow aggregates of lath-like crystals on malachite. Analysis showed it to be arsenic-rich vauquelinite (Elton et al. 2004).

Vauquelinite was described from Wanlockhead, Dumfries & Galloway, Scotland, by Davies (1877) and also listed later by Heddle (1901) and Brown (1927). However, Davies's original material was investigated by Bannister (1933) and shown to be mottramite.

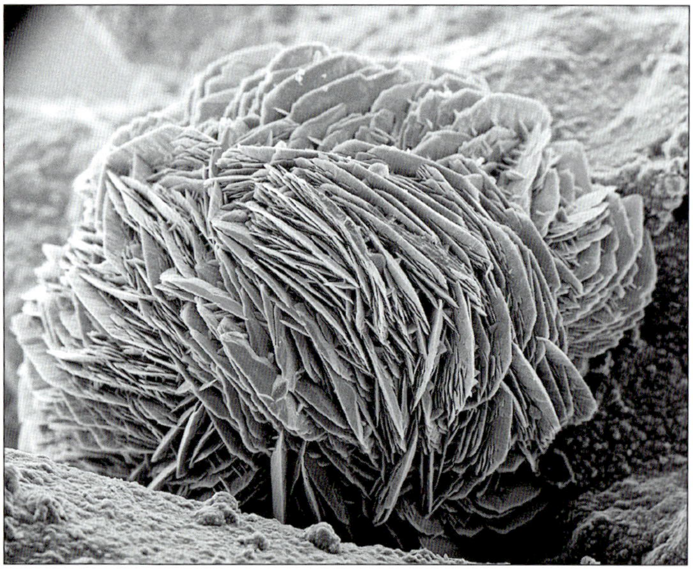

Rosette of vauquelinite on vein quartz from Greystone quarry, Lezant, near Launceston, Cornwall; width 400 μm; SEM image. (© J. J. Hooper)

veatchite-p

monoclinic $Sr_2[B_5O_8(OH)]_2B(OH)_3{\cdot}H_2O$

A rare hydrated strontium borate hydroxide, trimorphous with veatchite and veatchite-A (Clark 1993), and found in the Permian Lower Evaporites of the Yorkshire potash field from the Eskdale No. 2 borehole at Aislaby, Eskdale, near Whitby, Yorkshire. Veatchite-p forms pearly or vitreous, colourless transparent crystals of acicular and platy habit. The needles are up to 3 mm long and 0.5 mm thick. The plates reach 2×0.5 mm and several crystals are doubly terminated. Euhedral inclusions of anhydrite have been found in one crystal. The main species associated with veatchite-p are anhydrite and halite, with minor magnesite, pyrite and quartz. The species was originally described as veatchite (Stewart et al. 1954), but was later shown to be veatchite-p (Beevers & Stewart 1960). Additionally, veatchite-p is noted from Boulby potash mine, Loftus, Cleveland (Anthony et al. 2003). Mandarino & Back (2004) consider there to be only two polytypes – veatchite-A and p-veatchite).

velvet ore

A synonym of cyanotrichite.

verbeekite

monoclinic $PdSe_2$

A rare palladium selenide found in a gold-bearing carbonate vein in Middle Devonian limestones at Hope's Nose, Torquay, Devon. Verbeekite occurs in a single grain 25 µm in diameter, enclosed in native gold, associated with possible oosterboschite, cerussite and an unnamed Pd_2HgSe_3 phase (Paar et al. 1998). It was initially described as unknown species "uk2" until verbeekite was formally described from Shaba Province in the Democratic Republic of Congo (Roberts et al. 2002).

verdelite

A green variety of elbaite, tourmaline group.

vermiculite

monoclinic $(Mg,Fe^{2+},Al)_3(Si,Al)_4O_{10}(OH)_2{\cdot}4H_2O$

A group of silicates with the general formula given above. An examination of soil derived from a biotite-rich quartz gabbro in the Strathdon area, Grampian, Scotland, found that gibbsite and kaolinite, together with aluminous chlorite–vermiculite, are the natural weathering products of biotite (Wilson 1966). This is the Glenbuchat locality listed in Livingstone (2002), who also notes vermiculite in an altered ultrabasic pod at Rodel, South Harris, Western Isles. Macpherson (1983) reports an unsubstantiated English occurrence of vermiculite at Walney Island, near Barrow, Lancashire. An occurrence of vermiculite in serpentinites within Moine granulites, northeast Ox Mountains, Co. Sligo, Ireland (Lemon 1966), deserves further investigation.

vernadite

hexagonal $(Mn^{4+},Fe^{3+},Ca,Na)(O,OH)_2{\cdot}nH_2O$

An uncommon hydrated manganese–iron–calcium–sodium hydroxide found in a manganese-enriched fraction of clay material from an iron–manganese pan in poorly drained brown forest soil from the Blairmore Estate, near Huntly, Grampian, Scotland (Russell et al. 1984a).

verrucite

TL Sandy Braes, Co. Antrim, Ireland

A variety of mesolite, zeolite group – reddish-brown, compact and warty.

vésigniéite

monoclinic $BaCu^{2+}_3(VO_4)_2(OH)_2$

A rare barium–copper vanadate hydroxide found as a secondary mineral in copper-rich quartz–carbonate veins at Newhurst quarry, near Shepshed, Leicestershire. The veins cut granophyric diorite and hornfelsed volcanic tuffs, and contain a primary assemblage of bornite, chalcocite and chalcopyrite. In the upper portions of the veins, supergene processes have produced a suite of copper minerals and goethite. Polysynthetic twins of yellowish-green vésigniéite are found as tiny rosettes up to 2.4 mm across in cavities within a bornite–chalcocite vein (King & Wilson 1976, Ince 2005). Similar greenish crystal aggregates have been identified in Leicestershire from Bardon Hill quarry, Coalville, and New Cliffe Hill quarry, Stanton under Bardon (Hubbard & Green 2001). The best specimens from New Cliffe Hill are among the finest examples of the species in Britain (Hubbard et al. 2005). Elsewhere in England, vésigniéite has been noted from Merehead (Torr Works) quarry, Cranmore, Somerset, but further investigation showed it to be chrome-yellow paint, presumably from blast marking (R. Turner pers. comm. 2007).

Unusual hexagonal crystals (up to 3 mm across) of vésigniéite, perhaps epimorphous after an earlier generation of vésigniéite; New Cliffe Hill quarry, Stanton under Bardon, Leicestershire. (© D. I. Green)

Interlocking plates of bright-green vésigniéite on pale-blue chrysocolla. New Cliffe Hill quarry, Stanton under Bardon, Leicestershire; width 5 mm. (© D. I. Green)

Vésigniéite is also found at the small abandoned Castletown mine, 4 km SSE of Lochgilphead, Strathclyde, Scotland. Quartz–sulphide veinstone in a copper-rich portion of the mine contains a supergene assemblage of botallackite, clinoatacamite, connellite, dickite, hematite and malachite, together with uncommon anilite and yellow-green spherules of vésigniéite. The vésigniéite occurs in narrow fractures, forming radial aggregates that are mostly <0.3 mm across (Green & Todd 2001, Hubbard & Green 2001).

Burke (2006), reporting on an approved mass discreditation of species by the IMA-CNMNC, states that calciovolborthite is formally discredited, and compositions are either tangeite or vésigniéite. British reports of calciovolborthite now require further confirmation.

vesuvianite

tetragonal $Ca_{19}(Al,Mg,Fe)_{13}Si_{18}O_{68}(O,OH,F)_{10}$

A relatively common calcium–aluminium–magnesium–iron silicate fluorhydroxide formed in skarn rocks during contact or regional metamorphism of limestone or in garnetized gabbros and serpentinites.

At Tyland quarry, Camelford, Cornwall, contact metamorphism surrounding the Bodmin Moor granite has produced colourless to pale yellowish-green or yellowish-brown prisms of vesuvianite up to 5 mm in length in calc-silicate rocks also containing chlorite, epidote, garnet and quartz (Barrow & Thomas 1908). Still in the aureole of the Bodmin Moor granite, but to the S in the Caradon area, a considerable amount of vesuvianite was found in a small quarry at Trenuth Mill. Near by, at South Yolland farm, blocks of vesuvianite forming poorly developed crystals up to 6 mm long occur scattered around the margins of fields (Barrow & Thomas 1909). At Treburland manganese mine, 2 km SE of Fivelanes, Altarnun, Bodmin Moor, Cornwall, rare vesuvianite (idocrase) forms long columnar brownish-yellow prisms up to 6.5 cm long, embedded in a dark-grey calc-silicate rock, in which grossular, pyrrhotite, rhodonite and wollastonite are associated species (Russell 1946d). More recently, purple vesuvianite and white-to-colourless grossular have been found at Kennack Sands, Lizard Peninsula, Cornwall, where they have been produced by calcium metasomatism of gabbroic dykes (Hooper 1997, Robinson 1997b). Botallack, St Just, is another Cornish location for vesuvianite (Golley & Williams 1995).

Contact-metamorphic and metasomatic effects in Lower Carboniferous rocks around the Dartmoor granite atMeldon railway quarry, near Okehampton, Devon, have produced datolite–grossular–vesuvianite veins crosscutting calc-silicate skarns (El Sharkawi & Dearman 1967), and associated limestones have been converted to diopside–grossular–vesuvianite hornfels (El Sharkawi & Dearman 1966). Vesuvianite is also recorded from the dumps of Red-a-ven mine, Meldon, in skarn material

A prismatic vesuvianite crystal from Delnabo quarry, Glen Gairn, Grampian, Scotland; 2 mm long. (© D. I. Green)

containing many species, including helvine (Kingsbury 1961). Ryback et al. (1998) cast doubt on the provenance of many samples collected by Kingsbury and suggest that they were deliberately mislabelled.

Although generally uncommon, light-brown crystals of vesuvianite (idocrase) are sometimes found in the metamorphic aureole of the Shap granite at Wasdale Head, Shap, Cumbria. Locally (in metamorphosed Lower Coniston Limestone) the species occurs in abundance, where it is associated with lime garnet (probably grossular) and augite (Harker & Marr 1891).

Grass-green porphyroblasts of vesuvianite in gneiss are reported from Porth Helygyn, Anglesey, Gwynedd, Wales (Greenly 1919), but recent work has failed to confirm this (Bevins 1994).

In Scotland, vesuvianite is common at a few Grampian localities: Crathie quarry, Creag Mhòr quarries and Leac Ghorm, all near Balmoral Castle; lustrous rich brown and rarely apple-green crystals, sometimes up to 14 cm long, at Delnabo quarry, Glen Gairn; Mammie Hill, E of Gairnshiel Lodge, Glen Gairn; brown and yellowish-green at Monaltrie, near the Pass of Ballater (all Heddle 1901); pale-brown long prismatic crystals associated with grossular at Dorback Lodge, Braes of Abernethy, W of Tomintoul (Walker & Davidson 1935); with diopside, grossular and wollastonite in calc-silicate hornfels at Pollagach Burn, E of Pannanich Hill, E of Ballater (Crane 1987). In all cases the vesuvianite occurs in a contact-metamorphic environment adjacent to granitic intrusions of Caledonian age. Vesuvianite–wollastonite-bearing marble of Dalradian age at Kinnaird Head, Fraserburgh, formed slightly differently, being related to the intrusion of garnetiferous aplites and pegmatites (Kneller 1987, Johnson et al. 2000). Euhedral crystals of vesuvianite (up to 10 mm in length) have been found along the foreshore at Fraserburgh (Green & Todd 2001).

Boron–fluorine metasomatism of dolerite dykes occurred during the intrusion of the Tertiary Beinn an Dubhaich granite, near Broadford, Isle of Skye, Highland Region, and produced diopside–vesuvianite-rich veinlets in garnetized dolerite skarn veinlets. Similar veinlets at Kilchrist Prospect No. 1 extend into the granite and contain diopside–hedenbergite, fluorite, andradite–grossular and vesuvianite, and accessory allanite (orthite) or epidote (Tilley 1951, Smith et al. 2002). A marginal alkali pyroxene-bearing granite is also reported from here, adjacent to metasomatized dolomitic limestone. At the actual contact, a thin skarn zone containing clinopyroxene, grossular, spinel, vesuvianite and xanthophyllite has developed. Wollastonite is further noted in a similar skarn assemblage at the western end of the intrusion to the ESE of Camas Malag (Tilley 1949). Vesuvianite is also reported from a high-temperature calcic skarn near the contact of the Cuillin gabbro complex at Camasunary Bay, Strathaird Peninsula, Isle of Skye. Associated major species include calcite, garnet (andradite–grossular), melilite and wollastonite (Beard & Drake 2007). Symplectite-bearing nodules in the Ardgour marble at Coire Dubh, near Corran ferry, Highland Region, are noted to contain vesuvianite and 23 other minerals, including anorthite, augite, clinozoisite, meionite, prehnite and zoisite (Drever 1936). Rare vesuvianite, associated with grossular, is more generally described from regionally metamorphosed Moine Series impure limestones in the Ardgour district, to the SW of Fort William (Drever 1939).

At Coire Buidhe, Glen Creran, Strathclyde, vesuvianite forms distinctive brown flat and striated crystals up to 15 mm long in cavities and quartz masses within skarn deposits. These are related to the intrusion of the Cruachan granite into metamorphosed impure limestones of Dalradian age (Ingram et al. 1993). The complex assemblage also contains andradite, epidote, magnetite, powellite, scheelite and titanite, some in well formed crystals (Green & Todd 2001).

Contact metamorphism of calc-silicate rocks around granitic intrusions has produced vesuvianite at some Irish locations. At Barnavave, Carlingford, Co. Louth, it is found in a suite of rocks variably containing apatite, calcite, custerite, diopside, grossular, melilite, monticellite, phlogopite, spinel (pleonaste), spurrite, tilleyite and wollastonite (Osborne 1932a,b, Nockolds 1944, 1947). A remarkable series of grossular–vesuvianite–wollastonite veins occur in limestone quarries at Lough Anure, in the Thorr area of Co. Donegal. These skarn minerals exhibit a zonal arrangement, and a central quartz–feldspar dykelet is present in some cases. Diopside, titanite (sphene) and tremolite are associated species (Pitcher 1950). In the 1980s, one such quarry (now a private garden) produced sprays of lustrous vesuvianite crystals up to 5 cm long (Green & Moreton 2001). The Dalradian calc-silicate hornfels to the E of Loughin Island, Bunbeg, 7 km NNW of Lough Anure, are hosts to grossular–vesuvianite skarns that form boudinaged lenses up to 50 mm thick. The metasomatism is related to a series of pegmatite dykes associated with the nearby Donegal granite batholith (Greg & Lettsom 1858, Whitten 1951). At Carrickavrickan, near Curran Hill, Dungloe, Co. Donegal, vesuvianite forms subhedral crystals up to 8 cm long associated with nearly euhedral grossular crystals (Green & Moreton 2001). Zoned grossular–vesuvianite–wollastonite skarn assemblages are further described from limestone inclusions in granite at Ballynacarrick, Co. Donegal (Gindy 1953). Calc-silicate rocks in the northern aureole of the Ardara granite complex, near Maas, Co. Donegal, contain vesuvianite together with calcite, diopside, epidote, grossular, potash feldspar, quartz and zoisite in bands, up to 3 m thick, interbedded with sillimanite hornfels (Akaad 1956). In Connemara, Co. Galway, vesuvianite is found in grossular–wollastonite skarn at Fountain Hill (Ahmed-Said & Leake 1996) and in similar skarns on the coast W of Grallagh, where attractive crystals of diopside, grossular and vesuvianite associated with wollastonite are noted (Ryback & Moreton 1992).

veszelyite

monoclinic $(Cu^{2+},Zn)_3(PO_4)(OH)_3 \cdot 2H_2O$

A rare hydrated copper–zinc phosphate hydroxide found in cavities in quartz on the dump of Waygate shaft on Straitstep vein, Wanlockhead, Dumfries & Galloway, Scotland. Veszelyite is a supergene species that forms euhedral, lustrous, transparent-to-translucent dark-blue crystals up to 5 mm across (but mostly ≤ 1 mm). Pale-blue

Bipyramidal veszelyite crystal, 2 mm across, on hemimorphite; Straitstep vein, Wanlockhead, Dumfries & Galloway, Scotland. (© D. I. Green)

Bipyramidal veszelyite crystals to 0.5 mm on hemimorphite; Straitstep vein, Wanlockhead, Dumfries & Galloway, Scotland. (© M. P. Cooper)

pseudomorphs of chrysocolla after veszelyite up to 9 mm are also reported. Associated secondary minerals are aurichalcite, brochantite, hemimorphite and malachite. The locality is now grassed over (Green 1990, Green & Todd 2001, Green 2005a).

Veszelyite has also been identified on two specimens at the NHM that were collected in Cornwall over a hundred years ago. One is thought to come from Wheal Gorland, St Day, 2 km E of Redruth (Rumsey 2005).

villarsite

Weathered forsterite, olivine group.

violarite

LINNAEITE GROUP

cubic $Fe^{2+}Ni^{3+}_2S_4$

An uncommon iron–nickel sulphide found in a talc–carbonate unit forming part of the basal thrust of the Shetland ophiolite complex near Loch of Cliff, Baltasound–Haroldswick area, Unst, Shetland, Scotland. Violarite forms part of a low-temperature alteration assemblage with pyrrhotite, nickeline and secondary magnesite (Maynard et al. 1997).

At Talnotry, Newton Stewart, Dumfries & Galloway, violarite is one of the last ore minerals to have crystallized in a small body of copper–nickel mineralization developed at the base of an altered diorite sheet. Stanley et al. (1987) have constructed the following paragenetic sequence of ore minerals for the deposit: gersdorffite, nickeline, tellurobismuthite, molybdenite, pyrrhotite, pentlandite, chalcopyrite, gold, argentopentlandite, sphalerite, pyrite, marcasite, violarite and goethite.

In Wales, violarite is reported from the Esgair Hir mine, Ceulanymaesmawr, Central Wales orefield, Dyfed, where it occurs in chalcopyrite–galena–quartz veins along with bournonite, cobalt pentlandite, gudmundite, linnaeite (later shown to be siegenite), millerite, tetrahedrite and ullmannite (Rust & Mason 1988).

Occurrences of violarite in England are restricted to the Alston Block, where it is a component of late barite–fluorite–galena–sphalerite mineralization at Nenthead, Cumbria, and barite–galena–sphalerite–witherite mineralization at Settlingstones mine, near Hexham, Northumberland. A third occurrence (possibly dubious because it is based on samples that Kingsbury claimed to have collected) is from Tynebottom mine, NW of Garrigill, near Alston, Cumbria (Ixer 1986, Ixer & Stanley 1987).

vishnevite

CANCRINITE GROUP

hexagonal $Na_8(AlSiO_4)_6O_{24}(SO_4)\cdot 2H_2O$

An hydrated sodium aluminosilicate sulphate, carbonate and chloride, found in a road quarry at Allt a' Mhuilinn, Loch Borralan, Assynt, Highland Region, Scotland. The mineral was originally described as sulphatic cancrinite (Stewart 1941), but renamed vishnevite by Livingstone (1993). Light-blue to light greyish-blue vishnevite was found in pegmatites and veins in borolanite (a type of alkali syenite) and it forms large cleavable masses associated with andradite, orthoclase and a dark mica, plus accessory allanite, calcite, titanite and zircon. Ventura et al. (2007) note Loch Borralan vishnevite is enriched in CO_3.

vitreous copper

A synonym of chalcocite.

VIVIANITE GROUP

Monoclinic arsenates and phosphates with the general formula $A^{2+}_3(XO_4)_2\cdot 8H_2O$, where A^{2+} = Co, Fe, Mg, Mn, Ni, Zn; and X = As, P. Annabergite, erythrite, köttigite, parasymplesite and vivianite are members of the group found in Britain or Ireland.

vivianite

VIVIANITE GROUP

tetragonal $Fe^{2+}_3(PO_4)_2\cdot 8H_2O$

TL probably Wheal Kine, St Agnes, Cornwall, England

Named after John Henry Vivian FRS (1785–1855), a pupil of Abraham Gotlobb Werner (the influential German geologist who founded Neptunism), and later an industrialist and member of parliament for Truro; born in Truro, Cornwall. It was once thought to have been named after J. G. Vivian, an English mineralogist (Embrey 1994). Vivianite is an hydrated iron phosphate found in the gossans of metallic ore deposits and in clays, especially when associated with bones or other organic matter.

At the probable type locality, vivianite forms crystals (the finest ever found in Britain or Ireland) up to 50 mm long on pyrrhotite, in association with siderite. Fine specimens are also reported from Wheal Falmouth, Kea, and from Parknoweth, near St Just (Greg & Lettsom 1858). At Wheal Jane, Kea, vivianite occurs as broad inky-black bladed crystals up to 2 cm long, aggregated over a pyrite-rich matrix, and as rare rod-like crystals up to 1.5 cm long associated with pyrite and siderite in quartz crystal cavities (Rust 1982d). Cronstedtite and ludlamite were also found with vivianite at Wheal Jane in about 1886 (Collins 1911a). At Megiliggar Rocks, near Porthleven, the roof of an aplite–pegmatite complex at the margin of the Tregonning–Godolphin granite contains a suite of rare phosphate species that includes vivianite, thought to be a late alteration (weathering) product of triplite. Other phosphates found are alluaudite, amblygonite, fluorapatite, graftonite, lithiophilite, ludlamite, natromontebrasite and triphylite (George et al. 1981, Stone & George 1983). Vivianite is listed from a few other Cornish locations: Botallack and Wheal Edward, St Just; Wheal Gorland, St Day, 2 km E of Redruth; Wheal Kine (Wheal Kind), NW of St Agnes (Anthony et al. 2000); Wheal Maudlin, 3 km NW of Lostwithiel (Collins 1892, Robson 1949b). It is also reported from Devon: Wheal Betsy, near Tavistock and Wheal Friendship (Devon Friendship), Mary Tavy (Hall 1868b, Collins 1892).

Vivianite–parasymplesite is reported from Burdell Gill, Caldbeck Fells, Cumbria (Hartley 1984), but the location is doubted by Neall & Leppington (2003). The specimen, now in the Kingsbury collection at the NHM, may be part of the provenance problem outlined by Ryback et al. (1998, 2001).

Vivianite can also form soft pulverescent masses, sometimes found in peat mosses, with bog iron ore. It is often pale when first dug up, but acquires its Berlin-blue colour after exposure to air (Rudler 1905). Greg & Lettsom (1858) report vivianite in alluvial sediments at: Bury St Edmunds, Suffolk; Clifton, near Bristol; Toxteth Park, Liverpool; and the Isle of Dogs, Kent. Smith (1994) reports more sedimentary occurrences in Yorkshire; in York at Lendal Bridge and Coppergate, and in Humberside at Howden and Huggate Wold, Pocklington. Vivianite was also found in a trial boring in postglacial clay at Swalwell, Tyne & Wear (Bridges & Young 1998). At Cassington, 5 km NW of Oxford, temporary exposures of Oxford Clay yielded vivianite in clays associated with grey calcareous nodules, where it formed bright-blue films and blue spherical aggregates up to 4 mm in diameter (Atkins 1995). Extremely fine vivianite is also present in concretions within London Clay at Southleigh landfill site, near Emsworth, Hampshire, where it is locally abundant, forming needle-shape crystals to 5 mm long, often in radiating masses (Bone 1999).

Vivianite is also commonly found on fossil bones, examples being: encrustations

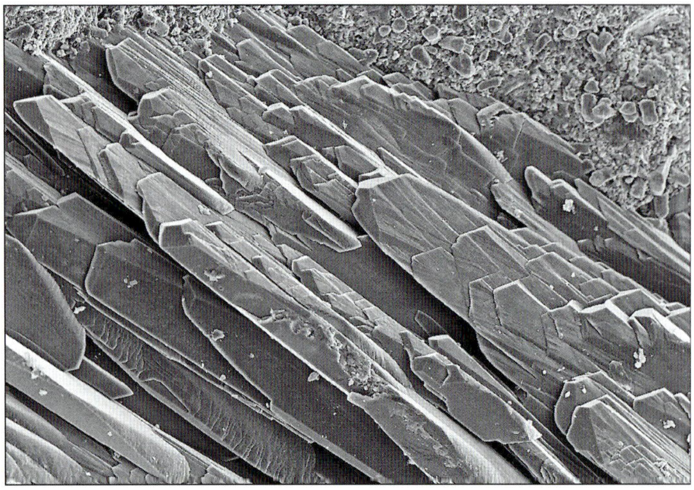

Above Deep-blue needle-shape crystals of vivianite lining fractures in a concretion within London Clay from Southleigh landfill site, near Emsworth, Hampshire; width 500 µm; SEM image. (© E. Condliffe, University of Leeds)

Left Dark-blue elongate crystals of vivianite from Southleigh landfill site, near Emsworth, Hampshire; width 8 mm. (© R. E. Starkey)

on bones and horns at Ballagh, Isle of Man, and fine crystals in the interior of antlers of an Irish elk of unknown provenance (Greg & Lettsom 1858); a coating on the denticles of weathered fossil fish at Steeplehouse quarry, near Wirksworth, Derbyshire (Ford et al. 1993); and fine-grain radiating zoned managanese-rich crystals forming dark spots in a feldspathic sandstone hone excavated from Rose Lane, Canterbury, Kent (R. A. Ixer pers. comm. 2007). Vivianite is further reported on wood, metal, bone and other artefacts of Roman origin at Vindolanda, Northumberland (Birley & Young 1993).

In Wales, vivianite occurs on a fragment of iron found in a peat bog at Valley, Anglesey, Gwynedd, and in boulder clay from Threapwood, Clwyd (Bevins 1994).

In Scotland, vivianite has been found in two metamorphic segregation pods occurring in striped kyanite–sillimanite-bearing Moine gneiss at the entrance to Gleann Còsaidh, Loch Quoich, Highland Region, where it developed as a secondary mineral on a suite of phosphates, including graftonite, jahnsite, johnsomervilleite, mitridatite, phosphosiderite and rockbridgeite (Livingstone 1980). In Lothian, it is disseminated in carbonaceous clays and bones at the Head of Haugh Burn, Cauldhame, Linlithgow, and at the Royal Park, Edinburgh. Tabular vitreous crystals of vivianite occur in bones, up to 1.5 cm long at Loch Lee and up to 5 mm long at Kilmaurs, both Strathclyde. It forms a blue coating on fossil fish at Banniskirk quarry, Highland Region (Livingstone 2002).

Vivianite is confirmed from Ballybetagh bog, Co. Dublin, Ireland (Moss 1876), and from a borehole at Ballyline Bridge, Co. Kilkenny, where it occurs as granules, coatings on organic particles, and vein fillings in shrinkage cracks in a dark-grey Pleistocene clay (Coxon & Flegg 1985). In Jersey, it has been found in recent clays, silts and sands at St Hélier, St Ouën's Bay and near Tesson Mill (Mourant 1961).

vochtenite

monoclinic $(Fe^{2+},Mg)Fe^{3+}(UO_2)(PO_4)_4(OH)\cdot 12–13H_2O$

TL Wheal Basset, Illogan, near Redruth, Cornwall, England

A very rare hydrated iron–magnesium uranyl phosphate hydroxide found at the type locality, at the contact of the Carnmenellis granite with Devonian sandstones. Mineralization consists of three main species – chalcopyrite, chalcocite and cassiterite – with accessory amounts of arsenopyrite, bassetite, pyrite and wolframite. Brown vochtenite with a bronzy lustre occurs as crystal aggregates with a pseudo-quadratic outline, consisting of small (0.5–1.0 mm) subparallel crystals (Zwaan et al. 1989).

volborthite

monoclinic $Cu_3^{2+}V_2^{5+}O_7(OH)_2\cdot 2H_2O$

A rare hydrated copper vanadate hydroxide found encrusting the surface of spherical vanadium-rich nodules and lenses of gypsum within the Triassic Keuper Marl at the Butterley and Blaby brick pit, Glen Parva, Leicestershire. Volborthite forms microscopic yellowish-green crusts rimming pale bluish-green sponge-like reniform masses of calciovolborthite. Etching of primary gypsum by meteoric water produced calcium sulphate in solution; this, in combination with vanadium leached from the nodules, enabled the two vanadates to precipitate (King & Dixon 1971). Crusts of oily green to yellowish-green volborthite have also been found coating fractures in diorite at New Cliffe Hill quarry, Stanton under Bardon, Leicestershire (Hubbard et al. 2005). Volborthite forms dark-green rosettes and yellow-green balls up to 0.4 mm long associated with metazeunerite on the dumps of Cliff lode, Wheal Edward, St Just, Cornwall (N. Hubbard pers. comm. 2006).

volkonskoite?

SMECTITE GROUP

monoclinic $Ca_{0.3}(Cr^{3+},Mg,Fe^{3+})_2(Si,Al)_4O_{10}(OH)_2\cdot 4H_2O$

An hydrated calcium–chromium–magnesium–iron aluminosilicate hydroxide. A specimen labelled volkonskoite from Murlough Bay, Co. Antrim, Ireland, is in the British Micromount Society reference collection. Volkonskoite was visually identified as pale-green crusts and veins mixed with dark-green nontronite. However, its composition could not be confirmed on a duplicate sample from the same location (H. J. Foy sample, Tindle unpub. EMP data).

voltaite

cubic $K_2Fe_5^{2+}Fe_3^{3+}Al(SO_4)_{12}\cdot 18H_2O$

An hydrated potassium–iron–aluminium sulphate found in a railway cutting, W of Stanely, near Paisley, Strathclyde, Scotland. Voltaite forms small clusters of minute black crystals associated with coquimbite, halotrichite, pyrite and römerite. In thin section, voltaite is pale green (Macpherson & Livingstone 1982).

voltzite

Wurtzite mixed with an organometallic zinc compound.

volynskite

orthorhombic $AgBiTe_2$

A rare silver–bismuth telluride found at Urlar Burn, 5 km from Aberfeldy, Tayside Region, Scotland, in quartz–sulphide ±gold veins cutting Upper Dalradian gritty quartzites, quartz–mica schists and amphibolites. Galena at Urlar Burn contains many telluride inclusions, and here volynskite forms intergrowths with hessite in association with altaite, benleonardite, coloradoite and gold (electrum) (Ixer et al. 1997). A single alluvial gold grain containing volynskite inclusions was found by Leake et al. (1998) in the Leadhills area, along the Strathclyde/Dumfries & Galloway border. Its origin is described as being of mesothermal shear-zone type.

vreckite

TL Beinn Bhreac (Ben Bhreck), E of Tongue, Highland Region, Scotland

A synonym of bhreckite (Heddle 1879c). A compound with uncertain composition, perhaps calcium-rich clinochlore.

W

wad

The name has been used for graphite, but most commonly is used for hydrated manganese oxides such as cryptomelane, manganite, pyrolusite and romanechèite.

wairakite

ZEOLITE GROUP

monoclinic or tetragonal $Ca(Al_2Si_4)O_{12}\cdot 2H_2O$

An uncommon hydrated calcium aluminosilicate that possibly forms a series with analcime. Wairakite has been described in amygdales within Carboniferous Clyde Plateau basalts that have undergone zeolite-facies burial metamorphism at two localities in the Midland Valley of Scotland (both Strathclyde): at Loanhead quarry, near Beith, wairakite is mantled by analcime, and, at Milngavie, 11 km N of Glasgow, wairakite occurs associated with laumontite (Evans 1987, Livingstone 1993).

wakefieldite-(Ce)

tetragonal $(Ce^{3+},Pb^{2+},Pb^{4+})VO_4$

A rare cerium–lead vanadate found as detrital grains in Devonian redbeds at Gamrie Bay, Gardenstown, 11 km E of Banff, Grampian, Scotland. Wakefieldite-(Ce) occurs with other detrital species; apatite, garnet, monazite, xenotime, zircon and an unnamed vanadium oxide, in medium- to coarse-grain cross-bedded sandstones. Reduction spots in the sediments also contain the vanadium mica, roscoelite (van Panhuys-Sigler et al. 1996).

walkerite

TL Corstorphine Hill, Edinburgh, Lothian, Scotland

Named after Revd John Walker (1731–1803), professor of natural history at the University of Edinburgh; born in Canongate, Edinburgh. Walkerite is a variety of pectolite or a mixture of pectolite and stevensite. See Heddle (1800b).

walpurgite

triclinic $(BiO)_4(UO_2)(AsO_4)_2\cdot 2H_2O$

A very rare hydrated bismuth uranyl arsenate found in a native bismuth-rich vein close to the top of the cliff at Needle's Eye, SW of Caulkerbush, Dumfries & Galloway, Scotland. This and other similar veins are in the aureole of the Criffell granodiorite. Alteration rims surrounding relict native bismuth consist of elongated bright-yellow transparent prismatic crystals of walpurgite (< 1 mm) associated with bright-green radiating needles of mixite and small brown spherules of bismutite. Other supergene minerals from this vicinity, produced during the alteration of primary As–Bi–Co–Cu–Fe–Pb–U minerals, include eulytite, kahlerite, lavendulan, metakahlerite and schoepite (Miller & Taylor 1966, Macpherson & Livingstone 1982, Braithwaite & Knight 1990, Green & Todd 2001).

wapplerite

A possible mixture of rösslerite and pharmacolite.

wardite?

tetragonal $NaAl_3(PO_4)_2(OH)_4\cdot 2H_2O$

An uncommon hydrated sodium–aluminium phosphate hydroxide similar in composition to natromontebrasite. A study of the composition and structure of natromontebrasite resulted in its discreditation and it was concluded that it is a mixture of OH-rich amblygonite with lacroixite and subordinate wardite (Fransolet et al. 2007). Wardite is not confirmed in Britain or Ireland, but natromontebrasite has been described from complex granitic pegmatites in the layered roof zone of the Tregonning granite at Megiliggar Rocks, near Breage, Cornwall (Stone & George 1983).

waringtonite

TL Cornwall, England

Named after Sir Warrington Wilkinson Smyth FRS (1817–1890), mining engineer and geologist, who was professor of mining and mineralogy at the Royal School of Mines, London and president of both the Geological Society of London and the Royal Geological Society of Cornwall. Among his published works were *A year with the Turks* (1854) and *A treatise on coal and coalmining* (1867). He was born in San Giovanni in the kingdom of Naples, Italy.

washingtonite

A probable mixture of hematite and ilmenite.

water spar

A local (Yorkshire) name for an intergrowth of witherite and strontianite.

wavellite

orthorhombic $Al_3(PO_4)_2(OH,F)_3\cdot 5H_2O$

TL High Down quarry, Heddon, 2 km WSW of Filleigh, near South Molton, Devon, England

Named after William Wavell FRS (b. 1750 or 1751, d. 1829), an English physician of Horwood Parish, Devon, who in some sources is said to have discovered the mineral, although an earlier discovery by John Hill is also claimed (Green et al. 2007). Old reference works give the locality as Barnstaple, the nearest town of any size (e.g. Greg & Lettsom 1858). Wavellite is an uncommon but locally abundant hydrated aluminium phosphate fluorhydroxide found as a secondary mineral in low-grade aluminous metamorphic rocks or rarely in hydrothermal veins. At the type locality, wavellite occurs in crevices in black slate, where it forms small hemispherical masses composed of radiating fibres of acicular crystals, of white, grey, brown or bluish colour (Rudler 1905). Embrey & Symes (1987) describe a specimen from the type locality, now in the collection of the NHM, as forming large botryoidal spheres up to 5 cm in diameter (average nearer 1 cm), coloured pale green on one side and blackened by carbonaceous inclusions on the other. Variscite occurs at the centre of some spheres. A thorough review of wavellite in Britain and Ireland is provided by Green et al. (2007).

In Cornwall, wavellite has been found in tin–tungsten bearing quartz–tourmaline veins in Hensbarrow china-clay pit (now part of Gunheath pit), 1 km W of Stenalees, St Austell. It forms creamy-white fibrous crystallized crusts lining cavities, associated with local infillings of cassiterite (varlamoffite) and turquoise (Braithwaite

Cross sections through wavellite spherules. High Down quarry, Heddon, near South Molton, Devon; width 30 mm. (© Royal Cornwall Museum, Royal Institution of Cornwall; photographer D. I. Green)

Radiating aggregate of colourless wavellite, 30 mm across, intimately intergrown with pale-blue turquoise; Gunheath china-clay pit, St Austell, Cornwall. (© D. I. Green)

1981, Barstow 1982, Hawkes et al. 1987). Colourless to snow-white flat radiating crystalline aggregates of wavellite up to 10 mm long are noted from Gunheath pit (Weiss 1994). Elton (1996) found sprays of acicular wavellite crystals up to 3 mm across intimately associated with variscite at this locality. Golley & Williams (1995) add Wheal Remfry china-clay pit, 2 km from Indian Queens, near St Austell, as a further location for the species. Wavellite (associated with the rare species fluellite) is documented from the Goonvean china-clay pit (Kingsbury 1954, 1959b, 1964b); however, the work of Ryback et al. (1998, 2001) indicates that the provenance of many Kingsbury samples is dubious, and further investigation is desirable. Grigg (1986) mentions wavellite from this locality, but it is not known if this is an independent discovery or a note on the Kingsbury description. Wavellite is confirmed from Stenna Gwyn mine, St Stephen-in-Brannel, St Austell, although the species had earlier been described as hydrargillite and tavistockite (Greg & Lettsom 1858, Embrey & Fejer 1969, Weiss 1994). Kingsbury (1954, 1959b, 1964b) also claimed to have found wavellite in a block of greisen within an old ruined building at this mine. Four other locations in Cornwall where wavellite has been found are: the Castle-an-Dinas wolfram mine, St Columb Major (Russell 1948); Kit Hill, Stoke Climsland and Old Beam mine, near Roche (Hall 1868a, Collins 1892); Hingston Down quarry, 2 km W of Gunnislake (Wolfe 1996a,b).

Wavellite forms white to pale-green silky radiating crystals with scorodite and variscite at Scar Crag cobalt mine, Causey Pike, 4 km SW of Braithwaite, Cumbria (Young 1987a, Green & Briscoe 2002).

Greg & Lettsom (1858) report wavellite from the Shiant Isles, 8 km E of the Isle of Lewis, Western Isles, Scotland. The discovery is also detailed by Walker (1930), but the species is not confirmed (here or anywhere else in Scotland) by later workers (Macpherson & Livingstone 1982, Livingstone 2002).

In Wales, wavellite appears to have been found first at Cil Ifor Hill (Cilifor Top), near Llanrhidian, Gower Peninsula, West Glamorgan (Logan 1837). Sea-green acicular crystals of wavellite were then also found on the clifftop at Pwlldu Head, Gower (Strahan 1907, North 1916). Strahan (1907) describes discs of radiating crystals of wavellite in joints within bedded clays, limestones and cherts between Clyne Common and Bishopston, on Gower; during road widening here, Trueman (1930) noted the species as botryoidal masses filling the spaces between chert clasts in fault zones. More recently, typical radiating aggregates of acicular white (or rarely lime-green) crystals of wavellite (to 25 mm) have been found in scree deposits on Cil Ifor Hill and beach deposits related to Pwlldu quarry. No wavellite-bearing rocks are now exposed *in situ* (Wolfe 1997, Plant & Jones 2001). At Pwlldu quarry, wavellite is associated with cacoxenite, crandallite and variscite (Dean & Cotterell 2003, Green 2005a). Dixon (1921) reports wavellite in Carboniferous chert at Gumfreston and Lydstep Lodge, Dyfed.

Spherules of wavellite found in fractures within a phosphatic sediment at Pwlldu beach, the Gower, West Glamorgan, Wales; width 40 mm. (© D. I. Green)

Botryoidal wavellite found along joint planes in shale at Laharran quarry, Minane Bridge, Co. Cork, Ireland; the specimen is 45 mm tall. (© D. I. Green)

Fine wavellite has been found at Laharran quarry, Minane Bridge, Co. Cork, Ireland, where attractive mint-green hemispheres up to 20 mm across and beautiful radiating aggregates of lustrous terminated prisms occur in the joints of Carboniferous shale (Ryback & Moreton 1991, Wolfe 1997, 2006, Green & Moreton 2001, Green et al. 2007). Wavellite is also reported from Doon East, 1 km N of Ballybunnion, Co. Kerry, where it is locally abundant in some of the Carboniferous shale bands. Here, pale-green rosettes of wavellite up to 3 cm across occur on fracture surfaces or joints, associated with crandallite and turquoise (Ryback & Moreton 1993, Moreton et al. 1995, Green et al. 2007). Fine specimens of botryoidal wavellite collected from this area are in the Russell collection at the NHM (Francis & Ryback 1987). At Ballycormick, near Shanagolden, Co. Limerick, silky white wavellite occurs with cacoxenite in a brecciated slaty gossan. Crandallite and turquoise are also found at this locality (Ryback & Moreton 1991). Light-green or attractive translucent bluish-green mammillated crystalline crusts of wavellite, forming rosettes up to 8 mm in diameter, occur on joint surfaces in Carboniferous shale at Grouse Lodge quarry, near Shanagolden. Gibbsite, kaolinite, koninckite, planerite and turquoise are also found at this location (Moreton & Green 2005, Wolfe 2006, Green et al. 2007). Other descriptions of wavellite in Ireland include: Doolin phosphate mine, the Burren, Co. Clare (Green et al. 2007); Carrigaline, near Minane Bridge, and Traeton (actually Tracton), both Co. Cork (Hall 1868a); Clonmel (beautiful dark apple-green botryoidal masses) and Springfield Abbey, both Co. Tipperary (Greg & Lettsom 1858, Rudler 1905); and Booley Bay, Co. Wexford (Du Noyer 1865).

A reference to wavellite from "Northumberland; on sandstone near Newcastle" (Greg & Lettsom 1858) is almost certainly an error and probably should be Newcastle, Co. Limerick, Ireland, from where Hall (1868a) reports two locations (Cahermoyle and Lisgordan) for wavellite in Coal Measures rocks.

waylandite

trigonal $BiAl_3(PO_4)_2(OH,H_2O)_6$

Named after Edward James Wayland (1888–1966), a London-born geologist and first director of the Uganda Geological Survey. Waylandite is a rare hydrated bismuth–aluminium phosphate hydroxide found on museum specimens originally from Restormel (Royal) iron mine, N of Lostwithiel, Cornwall. It forms colourless or brownish crystals less than 0.5 mm long on a matrix of brown bismutite, or forms bluish crystalline bands cutting the bismutite. Most grains show twinning, cyclic in some cases (Clark et al. 1986).

In the Hensbarrow part of Gunheath china-clay pit, Carthew, St Austell, Cornwall, granitic pegmatites at the margin of the biotite granite and near to crosscutting mineralized quartz and quartz–tourmaline veins contain a suite of phosphate species that includes apatite, crandallite, goyazite, mitridatite, rockbridgeite and waylandite. The pegmatites are lithium rich and contain accessory amounts of columbite–tantalite, fluorite, ilmenorutile, topaz and uraninite (Hawkes et al. 1987). Additionally, on a Russell Society field excursion to this site, Jeremy Hooper uncovered rare cyrilovite and leucophosphite, together with waylandite, wavellite and chalcosiderite in iron-stained veins (Griffiths 1997). At the western end of the Hensbarrow granite in the Wheal Remfry china-clay pit, 2 km from Indian Queens, near St Austell, a topaz vein produced bismuth mineralization (mainly bismutite–waylandite) mineralization towards the bottom of a topaz-rich pod. The waylandite occurs as drusy crusts of colourless to pale greyish-blue crystals up to 0.3 mm across and as a pale-brown boxwork in cavities up to 10 mm. Rare individual white to clove-brown pseudo-cubic crystals up to 0.5 mm are also noted (Rust 1998). Tiny greenish-blue crystals in cavities within goethite from Phoenix United mines (Stowe's section), Linkinhorne, are reported to have a composition close to that of waylandite (Ryback & Tandy 1992). Anthony et al. (2000) list waylandite from Levant mine and Wheal Owles, St Just.

Waylandite from Wheal Phoenix, Linkinhorne, Cornwall; width 170 µm; SEM image. (© D. I. Green)

websterite

TL Newhaven, Sussex, England

Named after Thomas Webster (1773–1844), professor of geology at University College London; born in Kirkwall, Orkney Isles, Scotland. Websterite is a synonym of aluminite.

weddellite

tetragonal $Ca(C_2O_4)\cdot 2H_2O$

Transparent pale-brown weddellite, 0.4 mm across, attached to a doubly terminated quartz crystal; Milltown quarry, Ashover, Derbyshire. (© M. P. Cooper)

A very rare calcium oxalate hydrate found in the zinc-hydroxide zone and in boulders from the main vein at Milltown quarry, near Ashover, Derbyshire. Weddellite forms colourless-to-brownish low bipyramids modified by a prism or as tan prismatic crystals up to 1.5 mm long. It is associated with ettringite and calcite, less commonly with ashoverite, cerussite, litharge, sweetite, or hydrocerussite on fluorite. Its origin is probably attributable to interaction between oxalic acid derived from decomposing plants and the underlying limestone (Rust 1983b, 1991c, Ford et al. 1993).

wehrlite

A mixture of pilsenite and hessite.

weilerite

trigonal $BaAl_3(SO_4)(AsO_4)(OH)_6$

A poorly described mineral that is not recognized as a species by Mandarino & Back (2004). Its composition was redefined by Scott (1987) and a further revision is likely when the IMA subcommittee report on the alunite group minerals is published.

weilite?

triclinic $Ca(AsO_3OH)$

A rare calcium arsenate hydroxide found coating brickwork and plasterwork on an arsenic calciner at Roseworthy, near Camborne, Cornwall. Weilite occurs with gypsum and the arsenic-bearing species arsenolite, haidingerite and pharmacolite, and together they form large undulating sheets several metres wide and up to 5 cm thick. Confirmation is by XRD (M. R. Power pers. comm. 2004). In this case, weilite is not a mineral and, more correctly, it should be regarded as an anthropogenic substance or synthetic equivalent (Nickel 1995).

Weilite is also reported from Wanthwaite mine, St John's in the Vale, near

Threlkeld, Cumbria (Young 1987a, Ryback & Tandy 1992). However, these descriptions are based on samples in the Kingsbury collection at the NHM and their provenance is in doubt. Indeed, some Kingsbury samples labelled as if they were from this location are now thought to have originated in Lavrion, Greece (Ryback et al. 1998, 2001, Green & Briscoe 2002).

weissigite

A variety of orthoclase, or possibly microcline, feldspar group.

wellsite

monoclinic $(Ba,Ca,K_2)(Al_2Si_6)O_{16}\cdot 6H_2O$

An uncommon hydrated barium–calcium–potassium aluminosilicate found at Benanouran Head, Giant's Causeway, Co. Antrim, Ireland. Wellsite forms colourless, whitish or orange-tinted twinned crystals (<5 mm) resembling phillipsite, in cavities in tholeiitic basalt (Adair 1987, Ryback et al. 1988). In the reclassification of the zeolite group (Coombs et al. 1997), wellsite was considered to be a mixture of barium-rich phillipsite-Ca and calcium-rich harmotome, and was therefore discredited.

wernerite

An intermediate composition between marialite and meionite, both scapolite group.

wheel ore

A synonym of bournonite.

whewellite

monoclinic $CaC_2O_4\cdot H_2O$

Named after William Whewell FRS (1794–1866), English natural scientist, philosopher, and professor of mineralogy at the University of Cambridge; born in Lancaster. Whewellite is a rare calcium oxalate hydrate found at the lichen/rock interface on serpentinite colonized by *Lecanora atra* at Mill of Johnston, 6 km WSW of Insch, Grampian, Scotland. It is a species subordinate to glushinskite and it forms a white powder beneath the lichen, associated with traces of serpentine and quartz (Wilson et al. 1980). Whewellite also forms a creamy-white layer at the lichen/rock interface on various rock types on the west coast of the Isle of Mull, Strathclyde, Scotland (Livingstone 1993).

A description of whewellite from Mealbank quarry, Ingleton, Yorkshire, by Ryback & Tandy (1992) and based on a specimen in the Kingsbury collection at the NHM, is now thought to have doubtful provenance. Ryback et al. (2001) suggest that the specimen may have come from Zwickau, Saxony, Germany.

white arsenic

A synonym of arsenolite.

white copper ore

An arsenic-rich variety of marcasite (kyrosite).

WHITEITE GROUP

Monoclinic phosphates with the general formula $AB^{2+}CX_2(PO_4)_4(OH)_2\cdot 8H_2O$, where A = Ca, Mn^{2+}; B^{2+} = Mg, Mn, Fe, Zn; C = Mg, Fe^{2+}; and X = Al, Fe^{3+}. Jahnsite-(CaMnFe) is the only member of the group found in Britain or Ireland.

white lead

A synonym of hydrocerussite.

white mica

Usually refers to muscovite, but has also been applied to celadonite, margarite and phlogopite.

white mundic

A synonym of arsenopyrite.

white pyrites

A synonym of marcasite.

white vitriol

A synonym of goslarite.

whitmoreite?

ARTHURITE GROUP

monoclinic $Fe^{2+}Fe^{3+}_2(PO_4)_2(OH)_2\cdot 4H_2O$

A rare hydrated iron phosphate hydroxide described from Gravel Hill mine, Perranzabuloe, Cornwall (Golley & Williams 1995). It was identified by XRD, but no further details are given. Field-trip notes compiled by members of the South-West branch of the Russell Society list the occurrence as probable.

wicklowite

TL Co. Wicklow, Ireland

A doubtful mineral, possibly a vanadate of lead; also called vichlovite.

wickmanite

SCHOENFLIESITE GROUP

cubic $Mn^{2+}Sn^{4+}(OH)_6$

A rare manganese–tin hydroxide, dimorphous with tetrawickmanite and found in a discordant calc-silicate body within the metamorphic aureole of the Land's End granite, at Wheal Cock Zawn, W of Botallack, St Just, Cornwall. Wickmanite occurs as orange-yellow octahedral crystals (up to 1 mm on edge) within small cavities in a large axinite pod. Associated species include apatite, chalcopyrite (partially altered to botallackite), grossular, orthoclase (var. adularia), pargasite and titanite (Hubbard 1988, Anon. 1989b).

widenmannite

orthorhombic $Pb_2(UO_2)(CO_3)_3$

A rare lead uranyl carbonate found in a localized patch of Cu–Pb–U mineralization near the base of the seacliff at Loe Warren Zawn, W of Botallack, St Just, Cornwall. Widenmannite has formed where surface water draining from rubble at the base of the cliff has trickled over veinlets carrying chalcocite, chalcopyrite and pyrite, and has precipitated an assemblage of supergene carbonate and phosphate minerals on exposed surfaces and joint planes near a patch of uraninite mineralization. Widenmannite crystals are transparent, colourless to pale greenish yellow, and lath-like. The laths form mats encrusting hornfels or are arranged in stellate groups; they are rare as spheroidal aggregates. Dewindtite is associated with widenmannite in some specimens and together they appear to be the last formed of the uranium-bearing species at this site. Other associated species include cerussite, kasolite, malachite and intermediate members of the torbernite–zeunerite series (Elton & Hooper 1995c). A second occurrence of widenmannite is reported from North Zawn, 800 m to the SW of Loe Warren Zawn, where it is found on calcite–quartz–sulphide boulders near a mineralized lode. Here it is associated with sodium-zippeite in or near areas of altered galena. Again it is a very late-formed mineral. It is suggested that both widenmannite and sodium-zippeite formed by the action of the sea spray and rainwater on the mineralized host rock, as uranium secondary minerals are not found on hornfels elsewhere (Elton & Hooper 1996).

Lath-like crystals of widenmannite from Loe Warren Zawn, St Just, Cornwall; width 50 μm; SEM image. (© N. J. Elton)

willemite?

trigonal Zn_2SiO_4

A zinc silicate reported from Wanlockhead, Dumfries & Galloway, Scotland by Glen & Young (1876). A very doubtful occurrence (Macpherson 1983).

williamsite

A variety of antigorite, kaolinite–serpentine group, described by Heddle (1878a, 1901) as an apple-green lamellar serpentine.

winchite

AMPHIBOLE GROUP
monoclinic $\square(CaNa)Mg_4(Al,Fe^{3+})Si_8O_{22}(OH)_2$

Named after Howard James Winch (1877–1964), a geologist who discovered the mineral in India; born in Woodford, Essex. Winchite is a relatively common calcium–sodium–magnesium–iron–aluminosilicate hydroxide that forms a series with ferrowinchite and is found in blueschist-facies mica schists.

Green cores to blue amphiboles from the Mona Complex at the Marquis of Anglesey's monument, near Llanfairpwllgwyngyll, Anglesey, Gwynedd, Wales, were described as "hornblende" by Greenly (1919). However, an investigation of similar material from there, and from temporary roadcuts made during construction of the Llanfairpwllgwyngyll bypass, concluded that the green cores consist of barroisite, ferrobarroisite and winchite, whereas the rims are pale-blue glaucophane (Gibbons & Gyopari 1986). The current amphibole classification (Leake et al. 1997a, Mogessie et al. 2004) confirms winchite on the basis of these data. Actinolite, alumino-ferro-winchite, aluminowinchite, ferroglaucophane, glaucophane and katophorite are also found in these rocks (Gyopari 1984).

In Scotland, winchite occurs in metasomatized (fenitized) quartzites at Four Burns and Loyne, near Loch Borralan, Assynt, Highland Region. The nearby Borralan syenite complex was the source of the strongly alkaline fluids required to produce fenitization of the surrounding rocks. The petrography of these rocks shows green actinolite developed in the early stages of metasomatism and this was later replaced by blue varieties that include magnesioriebeckite, riebeckite and winchite (Woolley et al. 1972). These authors describe winchite as crossite (magnesioriebeckite type), but this term was discredited in the 1997 reclassification of the amphibole group.

Winchite occurs in amphibolites and schists associated with the Girvan–Ballantrae ophiolite complex to the N of Knockormal, Strathclyde – with barroisite, alumino-winchite, edenite, ferro-edenite, glaucophane and pargasite (Bloxam & Allen 1959, A. Livingstone unpub. EMP data).

withamite

TL Glencoe, Highland Region, Scotland

Named after Henry Thomas Maire Witham (1779–1844), a botanist and palaeontologist who in 1824 discovered the mineral in Glencoe, Scotland (his name was Silvertop prior to his marriage to the heiress Eliza Witham of Headlam, Durham). Withamite has been described as a red-to-yellow variety of epidote and a manganese-deficient piemontite.

witherite

ARAGONITE GROUP
orthorhombic $BaCO_3$
TL Alston Moor, Cumbria, England

Named after William Withering FRS (1741–1799), an English physician, mineralogist and natural historian who first noted the mineral; born in Wellington, Shropshire. Alston Moor is generally accepted as the type locality, although Anglezark Moor mine, east side of the Liverpool waterworks, near Chorley, Lancashire, is also a contender (Turner 1963). Witherite is a relatively common barium carbonate found in low-temperature hydrothermal veins and is typically associated with barite, calcite and galena.

With over a million tonnes of witherite mined (almost the whole world's supply), the Alston Block section of the Northern Pennine orefield was by far the most important region for the mineral. From 1873 until 1969 the world's leading producer was Settlingstones mine, 5 km N of Haydon Bridge, near Hexham, Northumberland. Ore was extracted from a 1 km-long oreshoot close to the Whin Sill. The deposit consists chiefly of greyish crystalline masses of witherite, together with yellow and brownish-white transparent varieties. When crystallized, witherite forms pseudo-hexagonal bipyramids with rough faces and horizontal striations. Barite, bournonite, calcite, galena, harmotome, nickeline, pyrite, sphalerite, strontianite and ullmannite are associated species (Rudler 1905, Russell 1927b, Trestrail 1931, Ineson 1972, 1976, Young & Bridges 1984, Dunham 1990). Beautiful pure and translucent crystals, coloured white or with a pale yellowish-green tint, were found at the nearby Fallowfield mine, where they occurred in pockets within another galena–witherite vein (Dunham 1959a, 1990, Briscoe 1986b, Cook 2007). Other occurrences of witherite in the northern Pennines are often associated with barytocalcite (or rarely alstonite) and are documented by Young (1985c) and others.

Cumbria Nine locations around Alston: Ayle Burn mine (Goodchild 1884, Davidson & Thomson 1951); with fine barytocalcite at Blagill mine, Nenthead (Goodchild 1884, Davidson & Thomson 1951, Bridges & Green 2006b); pseudo-hexagonal crystals to 3 cm, with flat pyramidal terminations on old-time specimens and as sharp lustrous pseudo-hexagonal bipyramids, exceptionally to 1 cm in a brecciated sulphide-rich vein at: Brownley Hill mine, Nenthead (Green et al. 2000); Clarghyll mine, 3 km N of Alston; Middlecleugh Second Sun vein, SE of Nenthead; Mounthooly Plantation, 2 km N of Alston; squat, slightly convex, hexagonal prisms of opaque chalky-white impure witherite at Nentsberry Haggs mine (Davidson & Thomson 1951, Bancroft 1973, Young & Bridges 1984, Dunham 1990, Moore 1996, Green & Briscoe 2002, Bridges & Green 2006b); abundant, coarsely crystalline, white to pale-cream masses up to 75 mm across from Scaleburn and Boulder End Cross veins in Rampgill mine, Nenthead (Bunting 1994, Green & Briscoe 2002, Bridges & Green 2006b); and Taylor's Groove mine, 2 km N of Alston. Four locations near Appleby: Hilton mine, Scordale (Davidson & Thomson 1951, Ixer 1986); Murton mines, Scordale; Silverband mine, Milburn (Davidson & Thomson 1951); and Stakebeck mine, Dufton Fell. Also Loppysike mine, east slope of Great Dun Fell, and in the Lake District at Greenside mine, near Glenridding, Patterdale (Davidson & Thomson 1951, Young 1987a). The occurrence of witherite at Potts Gill barite mine, Caldbeck Fells (Kingsbury & Hartley 1956a, Hartley 1984) requires further confirmation.

DURHAM Coldberry Sun vein, Grasshill Common; Dubbysike mine, near Harwood Beck; Flushiemere mine, Newbiggin Common; Foster's Hush and Lunehead mine, Lunedale, both along the Brough to Middleton road (Young et al. 1985a, Green & Briscoe 2002); Greenhurth mine, Upper Teesdale. In the Durham coalfield, witherite is noted from: Brandon colliery; Craghead colliery, SE of Stanley; New Brancepeth colliery, SW of Brandon – forming nodular masses with a radiating structure and scaly surface, and as white or grey, almost opaque, lenticular crystals to 4 cm in diameter. Alstonite and ullmannite are associated species (Spencer 1910, Smythe 1922, Dunham 1990); three major ore shoots of almost pure witherite at South Moor colliery, Stanley (Dunham 1959a, 1990, Dunham et al. 1978); Tanfield Moor colliery, White-le-Head (Dunham 1990); and Ushaw Moor colliery, W of Durham (Dunham 1990). NORTHUMBERLAND In addition to Fallowfield and Settlingstones mines, four locations N of Hexham: Greyside mine, near Settlingstones; Langley Barony mine, 2 km NW of Haydon Bridge; Stonecroft mine, near Settlingstones (Young & Bridges 1984); Waterhouse mine, 3 km E of Haydon Bridge. Eight locations in West Allendale: Blackpool Bridge level; Greenleycleugh South vein; Hearty Cleugh mine; Hesleywell Hush; Keirsleywell mine, near Mohope Burn; Longcleugh mine, Whitewalls Burn; Mohopehead mine, Mohope Burn; and Scraithole mine, near Carrshield (Green & Briscoe 2002). Three other locations: Farneycleugh mine, 6 km N of Nenthead; Wellhope level and Wellhopehead mine, both N of Nenthead (Young & Bridges 1984).

Witherite is also found in the Askrigg Block section of the Northern Pennine orefield, particularly in the North Swaledale mineral belt, near Gunnerside, Yorkshire, where galena was extracted from low-temperature veins hosted by Carboniferous chert, limestone and sandstone. At Hard level, radiating fibrous crystals of witherite (often deeply weathered) occur on the mine dumps with small amounts of aurichalcite, barite, fluorite, galena and strontianite (Young 1987b, Ineson & Young 1994). An intergrowth (known locally as water spar) of witherite and strontianite was found in barite–calcite–galena veins at the nearby Victoria level (Ineson & Young 1994), and elsewhere similar assemblages are also noted along the Friarfold–Surrender and Old Rake veins (Young et al. 1989). The extensive dumps near Surrender mine have produced witherite associated with barite, calcite, galena, sphalerite and traces of chalcopyrite (Dunham & Wilson 1985). The Brandy Bottle inclined adits were driven to access the Friarfold vein, and excellent specimens of yellowish-white compact or radiating crystalline witherite, together with aurichalcite, barite, cinnabar, fluorite, hemimorphite and strontianite, can now be collected from the many associated spoilheaps (Young 1987b,c, Ineson & Young 1994). At Bunting level, Gunnerside Gill, fracture surfaces of radiating witherite are covered with stellate aggregates of aurichalcite (Young 1987b). Brown mammillated witherite associated with strontianite is reported from the Sir Francis level; and, along the Lownathwaite North vein, well crystallized witherite occurs in vugs within massive witherite (Wood 1993). Witherite and strontianite are present locally within barite–calcite–fluorite gangue in the Old Gang mines and, in a section known as Lucky String, the rare aluminofluoride species, gearksutite, has also been found (Green 2003). Elsewhere in Yorkshire, witherite was encountered at Lolly mine and Providence Prosperous, both in Nidderdale, and Virgin (Virgin Moss) and Whitaside mines, both on East Bolton Moor, Wensleydale. Beautiful well developed pseudo-hexagonal crystals of witherite were once collected from Danby level, W of Langthwaite (Dunham 1959b, Dunham & Wilson 1985, Young et al. 1992b).

An occurrence of witherite in galena-bearing barite–calcite veins is reported from

Encrustation of small crystals and bipyramidal ribbed crystals of witherite to 45 mm; Fallowfield mine, Hexham, Northumberland. (© M. P. Cooper)

Rorrington (West Snailbeach) mine (Engine shaft), S of Meadowtown, Shropshire (Starkey et al. 1994). Greg & Lettsom (1858) had earlier noted witherite at Snailbeach mine, but Richards (1992) could not confirm this. Doubtful occurrences of witherite are noted from Hoptonwood quarry, Derbyshire (Ford et al. 1993) and Engine Vein mine, Alderley Edge ridge, Cheshire (Braithwaite 1994).

In Wales, abundant witherite within galena–sphalerite-bearing barite veins is reported from two localities in the Powys section of the Central Wales orefield. At Pen-y-Clun mine, NW of Llanidloes, large masses of witherite contain cavities in which well developed crystals of colourless-to-white translucent-to-transparent witherite occur (Morgan & Starkey 1991), and at Gorn mine, E of Llanidloes witherite is intergrown with barite (Carruthers et al. 1915, Jones & Moreton 1977, Bevins 1994). Other occurrences of witherite in Powys include: Aberdaunant, Bryntail and Van mines, all near Llanidloes (Carruthers et al. 1915, Jones & Moreton 1977, Hughes 1991, Mason 1997); Craig Rhiwarth, Craig-y-Mwyn, Cwm Orog (fine specimens) and Llangynog mines, all near Llangynog (Carruthers et al. 1915, Dunham et al. 1978, Bevins 1994); acicular-prismatic crystals up to 10 mm long in brecciated Precambrian argillite and associated with alstonite, ewaldite and harmotome at Dolyhir quarry, Old Radnor, near Presteigne (Wolfe 2004, Green et al. 2005d).

Witherite occurs in pockets associated with barite in galena–sphalerite-bearing veins at Pennant mine, 5 km E of St Asaph, Clwyd (Carruthers et al. 1915, Archer 1959), and in Mid Glamorgan it occurs in cavities within Triassic Dolomite Conglomerate at Cefn Parc mine, Llantrisant, associated with alstonite, calcite, barite, galena, pyrite and sphalerite (Alabaster 1990, Bevins 1994).

The first report of witherite in Scotland was from the western branch of Glencrieff mine, Wanlockhead, Dumfries & Galloway, where grey or white globular specimens, up to 18 cm across, were found in a 4.5 m-long cavity within a barite–calcite–galena–quartz vein. A large assortment of beautiful pseudomorphous specimens of barite after witherite were also found. Witherite is also noted as massive crystalline aggregates from Wembley shaft, N of Leadhills, Strathclyde (Brown 1919, 1927, Wilson & Flett 1921, Temple 1955). Witherite has been found as platy elongated tabular crystals to 4 mm and pseudo-hexagonal tabular crystals with barytocalcite on the dumps of Hilderston silver mine, Bathgate Hills, Lothian (Meikle 1994, Green & Todd 2001), and as small inclusions within pyrite in the Foss barite deposit, near Aberfeldy, Tayside Region (Moles 1985, Livingstone 2002).

Greg & Lettsom (1858) note witherite from Foxdale lead mine on the Isle of Man.

wittichenite

orthorhombic Cu_3BiS_3

An uncommon copper sulphobismuthite found in hematized and brecciated quartz–chlorite veins (now dump material) from the Upper level of the abandoned copper mine at Seathwaite Tarn, near Coniston, Cumbria. Bornite, digenite and djurleite are the main sulphide species in an assemblage that contains the following minor components: arsenopyrite, chalcopyrite, cobaltite, covellite and pyrite, and blaubleibender covellite (now recognized as intermixed spionkopite and yarrowite). Wittichenite is associated with microcrystalline hematite and can occur along grain boundaries and fractures in digenite and djurleite. Graphic intergrowths of wittichenite (up to 0.2 mm across) with digenite and bornite are also noted (Criddle & Stanley 1979, Stanley & Criddle 1979).

In Cornwall, there is scant mention of wittichenite, although it is noted from Geevor mine, near Pendeen, St Just (Golley & Williams 1995), and Levant mine near by (Wolloxall 1987). Additionally, "cupreous bismuth" occurring at Botallack and Levant mines, St Just, and at Wheal Buller, near Redruth, is thought to be wittichenite (Collins 1892) and stated to be wittichenite by Robson (1949b).

At Borland Glen, Ochil Hills, Central Region, Scotland, two small inclusions, close in composition to wittichenite, are reported from a single gold grain found in a panned concentrate of alluvium (Leake et al. 1993, Chapman et al. 2000c). Later work in the area confirmed wittichenite inclusions in alluvial gold at Borland Glen, and from Balquandy, Glenclach and Lamb Burns (Chapman et al. 2005).

In Ireland, late Hercynian mineralized veins cutting folded Devonian and

Pale wittichenite replacing digenite (blue) and bornite (orange). The irregular needle-shape crystals are hematite. Seathwaite Tarn copper mine, Cumbria; width 280 µm; viewed in reflected light. (© C. J. Stanley, The Natural History Museum, London)

Carboniferous rocks of the Munster Basin at Dhurode, West Carbery mining district, Co. Cork have also been shown to contain wittichenite. The coarse-grain copper ore is complex and in addition contains native bismuth, bornite, bournonite, chalcocite, chalcopyrite, cubanite, digenite, djurleite, "enargite-group minerals", galena, meneghinite and tetrahedrite, with rare molybdenite, pyrite and sphalerite, in a chlorite–quartz–siderite gangue (Ni Wen et al. 1991, 1999). Kinnaird et al. (2002) describe thick laterally extensive quartz veins at Dhurode, containing a late sulphide assemblage that includes native bismuth, bornite, djurleite, chalcocite, molybdenite and rare wittichenite. Elsewhere in the Munster Basin, at Letter and Mount Gabriel, both near Ballydehob, wittichenite is found in different, although related, parageneses: as a minor constituent of stratiform disseminated mineralization and partially replaced by bismuthinite and chalcopyrite during a later supergene alteration event (Ni Wen et al. 1999).

Wittichenite is also reported from Gortdrum mine, near Tipperary town, Co. Tipperary, where it occurs as an accessory constituent with native amalgam, cobaltite, gortdrumite, sphalerite and stromeyerite. The dominantly chalcopyrite and mercury-rich tennantite mineralization is hosted by calcite and iron-rich dolomite veinlets within shattered massive bioclastic limestones, and is thought to have formed because of the high geothermal gradient associated with the Carboniferous volcanic activity in the area. Bornite, chalcocite, cinnabar and pyrite are minor components of the ore (Steed 1983, 1986).

wodginite

monoclinic $MnSnTa_2O_8$

A rare manganese–tin–tantalum oxide found in a panned concentrate of alluvium collected in the grounds of Aclare House, near Myshall, Co. Carlow, Ireland. Wodginite, forming angular grains up to 200 µm long and ferrotapiolite are minor components of the concentrate that also contains common cassiterite, ferrocolumbite, manganocolumbite, manganotantalite and less-common ferrotantalite and monazite. Such an assemblage is typical of lithium-enriched granite pegmatites and the source of the grains is most likely to be the nearby Aclare spodumene pegmatite (R. A. Ixer sample, Tindle unpub. EMP data).

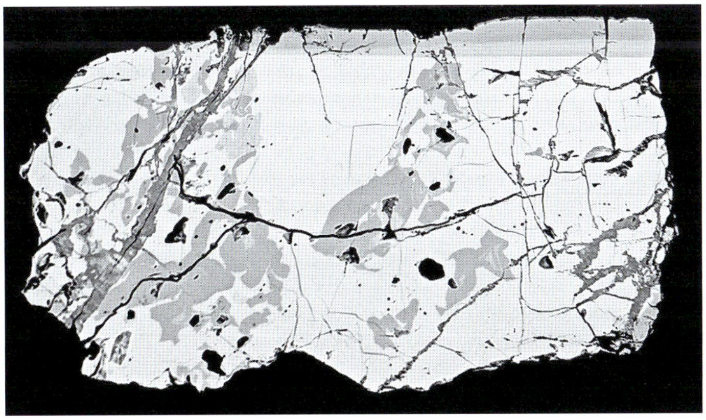

Wodginite with a discontinuous zoned rim and patchy replacement (darker grey) associated with the later fracturing. Ultimately derived from nearby spodumene pegmatites, the grain was one of many collected in a panned concentrate from the grounds of Aclare House, near Myshall, Co. Carlow, Ireland. Width 400 µm; BSE image. (© A. G. Tindle)

wolfram

A synonym of wolframite.

wolframine

A synonym of tungstite and also wolframite.

wolframite

An iron–manganese tungstate that has lost its species status in favour of ferberite and hübnerite (the iron- and manganese-rich end members). Characteristically it forms in granite pegmatites, greisen and high-temperature quartz veins associated with granite intrusions. Individual crystals are often strongly zoned with compositions crossing the species boundary. Wolframite is the main tungsten ore mineral, hence there are many references to it in the literature. Unfortunately, because of a lack of chemical data, most occurrences cannot be given species names.

England

Cornwall Wolframite is associated with all the granite intrusions of southwest England. The most important mines producing wolfram, a common term for wolframite-bearing ore, are detailed in Dines (1956) and Brooks (2001). *Land's End granite:* Balleswidden mine, E of St Just (Polkinghorne 1954, Dominy et al. 1995); Botallack sett (inc. Wheal Cock, Wheal Owles, Wheal Edward), St Just (Bancroft & Weller 1993); Wheal Providence, near Carbis Bay, St Ives (Robson 1954). *St Michael's Mount granite:* A small isolated granite cupola near Marazion (Rudler 1905, Davison 1920b, Russell 1924, Hosking 1951, 1954c, Hosking & Polkinghorne 1954, Polkinghorne 1954, Hall & Jackson 1975, Jackson & Rankin 1976, Moore & Howie 1978, Moore & Moore 1979, Moore & Swart 1979, Swart & Moore 1982, Alderton 1991, Campbell & Panter 1990, Dominy et al. 1995). *Tregonning–Godolphin granite:* Godolphin mine (Hosking & Polkinghorne 1954); Great Wheal Fortune, near Porthleven (Dominy et al. 1995); Wheal Tremayne, Gwinear (Hosking & Polkinghorne 1954). *Carn Brea–Carn Marth–Carnmenellis granites:* Balmynheer mine, Wendron (Stephens 1940b, Dominy et al. 1996a); Bosworgey borehole, near Fraddam (Ball & Basham 1984); Carn Brea mines, near Redruth – brilliant crystals (Garby 1848, Greg & Lettsom 1858, Rudler 1905, Hosking 1951, Hosking & Polkinghorne 1954); Carn Marth, Gwennap (Polkinghorne 1954); Chasewater, near Redruth (Moore & Howie 1978); Copper Bottom mine dump, near Camborne (Goode & Taylor 1980); Dolcoath mine, Camborne (Hosking 1951); East Pool and Agar mine, Illogan, near Redruth (Rudler 1905, Dewey 1925, Hosking 1951, Hosking & Polkinghorne 1954, Ball et al. 1982, Beer & Ball 1987); Illogan mine, near Redruth (Ball et al. 1982); Killifreth mine, Kenwyn, near Redruth (Davison 1920a, Dewey 1925, Bawden 1929, Hosking & Polkinghorne 1954); Mount Wellington mine, Twelveheads (Kettaneh & Badham 1978); New Roskear mine, Camborne (Robson 1954); Park-an-Chy mine, Redruth (Brooks 2001); Pednandrea mine, Redruth (Golley & Williams 1995); Poldice mine, St Day (Dewey 1925); St Day United mines, Gwennap (Russell 1911a); South Crofty mine, Camborne (Dewey 1925, Hosking 1949, 1951, Polkinghorne 1954, Hosking & Trounson 1959, Clark 1970c, Taylor & Hosking 1970, Moore & Howie 1978, Ball et al. 1982, Scrivener et al. 1986, Beer & Ball 1987, Wolters & Wolters 1987, Dominy et al. 1994, LeBoutillier 1996, Wolfe 1998); Tincroft mine, Illogan (Rudler 1905, Dewey 1925, Hosking & Polkinghorne 1954, Polkinghorne 1954); Ting Tang mine, Carharrack (Bruce & Aubrey-Jones 1998); Wheal Basset, Illogan (Hosking & Polkinghorne 1954, Zwaan et al. 1989); Wheal Gorland, St Day, 2 km E of Redruth (Collins 1911a, Hosking & Polkinghorne 1954, Rust 1981b, Fisher 1988, Brooks 2001); Wheal Jane, Kea, near Truro (Rayment et al. 1971, Dunham et al. 1978, Rust 1982d, Holl & Bromley 1988); Wheal Peevor, Redruth (Hosking & Polkinghorne 1954, Brooks 2001); Wheal Pendarves, SW of Camborne (Scrivener et al. 1986). *St Agnes granite:* Cligga Head, Perranzabuloe (Rudler 1905, Russell 1924, Hosking 1951, Russell & Vincent 1952, Hosking & Polkinghorne 1954, Hosking & Trounson 1959, Hall 1971, Hall & Jackson 1975, Jackson et al. 1977, Moore & Howie 1978, Moore & Swart 1979, Swart & Moore 1982, Manning 1986, Beer & Ball 1987, Chick 1987, Campbell & Panter 1990, Brooks 2001). *St Austell granite:* Bunny (St Austell Hills) mine, Bugle (Collins 1911a, Russell 1948, Dines 1956, Weiss 1994, Brooks 2001); Castle-an Dinas, St Columb Major – Cornwall's premier wolfram mine (Davison 1920a, Dewey 1925, Russell 1924, 1948, Hey et al. 1938, Hosking & Polkinghorne

1954, Polkinghorne 1954, Hosking & Trounson 1959, Ball et al. 1985b, Manning 1986, Beer & Ball 1987, Shrimpton 1993, Brooks 2001); Goonbarrow china-clay pit, near Bugle (Bray & Spooner 1983); Great Beam, near Roche (Dewey 1925); Gunheath china-clay pit (inc. Hensbarrow section), Carthew (Barstow 1982, Manning 1983, Ince 1986, Hawkes et al. 1987, Alderton 1991, Weiss 1994, Elton & Grigg 1996, Williamson et al. 1997, 2002); Mulberry mine and openworks, Lanivet (Le Neve Foster 1878a, Russell 1924, Dewey 1925, Hosking 1951, Starkey 1986a, Dominy et al. 1995); Stenna Gwyn mine, St Stephen-in-Brannel (Greg & Lettsom 1858, Ball et al. 1982, Weiss 1994); Wheal Maudlin, Lostwithiel (Garby 1848, Kingsbury 1961). *Bodmin Moor granite:* Buttern Hill alluvial deposits and mine, northern edge of Bodmin Moor (Dewey 1925, Hosking & Polkinghorne 1954, Polkinghorne 1954, Shrimpton 1993); Cannaframe and Gazeland mines, Bodmin Moor (Brooks 2001); Cheesewring quarry, SW of Henwood, Linkinhorne (Bowman 1911); Halvana mine, Bodmin Moor (Brooks 2001); Hawkswood mine, North Hill (Hosking & Polkinghorne 1954, Polkinghorne 1954, Alderton et al. 1980, Brooks 2001); Herodsfoot mine, SW of Liskeard (Polkinghorne 1954, Beer & Ball 1987); Minions Cotts, south-east Bodmin Moor (Hosking & Polkinghorne 1954); Penhale quarry, St Cleer and Trewint Marsh, Altarnun – alluvial deposits (Hosking & Polkinghorne 1954, Polkinghorne 1954); Trebartha Lemarne mine, near North Hill, and Treburland mine, 1.5 km SE of Fivelanes, Altarnun (Hosking & Polkinghorne 1954, Brooks 2001); Wheal Vincent, near Trewint, Altarnun (Hosking & Polkinghorne 1954, Polkinghorne 1954, Dunham et al. 1978, Brooks 2001). *Kit Hill–Gunnislake granites:* Drakewalls mine, Calstock – fibrous and lamellar crystals (Greg & Lettsom 1858); Deer Park mine, N of Kit Hill, near Luckett (Polkinghorne 1954); Gunnislake Clitters mine, Calstock and Downgate mine, 1.5 km N of Callington, near Kit Hill (Polkinghorne 1954, Brooks 2001); Dimson mine, Gunnislake (Dewey 1925); Hawkmoor mine, Gunnislake (Brooks 2001); Hingston Down Consols mine, 2 km W of Gunnislake, Calstock (Green et al. 1996a, Brooks 2001); Holmbush mine, Kelly Bray, Callington (Hosking & Polkinghorne 1954, Newall 1994); Kelly Bray and Redmoor mines, Kelly Bray (Newall & Newall 1989, Newall 1994); Kit Hill Consols, Stoke Climsland (Dewey 1925, Hosking & Polkinghorne 1954, Polkinghorne 1954, Wolfe 1998, Brooks 2001); Old Gunnislake mine, Calstock (Ball et al. 1982, Turner & Bannon 1992); Silver Valley mine, S of Kit Hill (Hosking & Polkinghorne 1954, Polkinghorne 1954).

Elsewhere, wolframite is noted in intertidal sediments of the Camel, Fowey and Gannel estuaries (Pirrie et al. 2000b, 2002).

Cumbria Wolframite occurs only in the Caldbeck Fells. The most important location is Carrock mine, Grainsgill, where wolframite–(scheelite) was mined from quartz veins associated with a greisenized part of the Skiddaw granite (Finlayson 1910, Hitchen 1934, Dunham & Hollingworth 1947, Davidson & Thomson 1951, Eastwood 1959, Eastwood et al. 1968, Shaw 1970, Shepherd et al. 1976, Dunham et al. 1978, Firman 1978, Fortey 1980, Beddoe-Stephens & Fortey 1981, Young 1987a, Tyler 2003). Early reports of wolframite from Brandy Gill and Carrock Fells by Greg & Lettsom (1858), Bristow (1861), Hall (1868a) and Kendall (1884) probably relate to the site of Carrock mine. Wolframite is also noted from Burdell Gill (Appleton & Wadge 1976, Neall & Leppington 2003) and Dry Swine Gill (Shaw 1970).

Devon *Dartmoor granite:* Bedford Consols mine, near Tavistock (Dewey 1925, Hosking & Polkinghorne 1954, Polkinghorne 1954, Dunham et al. 1978); Hemerdon Bal, Plympton (Terrell 1920, Polkinghorne 1954, Dunham et al. 1978, Keats 1981, Durrance & Laming 1982, Ball et al. 1985b, Beer & Ball 1987, Brooks 2001); Wheal Friendship (Devon Friendship), Mary Tavy (Braithwaite 1982b); Wheal Russell, Tavistock (Hosking & Polkinghorne 1954).

Elsewhere, wolframite is noted in veins cutting the Tertiary granite on Lundy island, off the north Devon coast (Bromley & Holl 1986).

Wales

Gwynedd Small grains of cassiterite and a tungsten mineral (possibly wolframite) were found as inclusions in magnetite from quartz veins cutting the Ordovician Snowdon Volcanic Group at Shadow Gully, Cwm Tregalan, S of Snowdon (Seenan 1988, Colman & Appleby 1991).

Scotland

Grampian An unusual niobium-rich wolframite with a mainly hübnerite composition has been found in a quartz vein cutting zinnwaldite granite within the Glen Gairn granite complex, at Gairnshiel Bridge (Tindle & Webb 1989, Webb et al. 1992, Green & Todd 2001). One blade of wolframite has been found in a quartz vein associated with the Middleton granite, along the northern edge of Knockinglews Plantation, near Middleton, between Chapel of Garioch and Inverurie. Beryl, molybdenite, pyrite and scheelite are found in other quartz veins near by (Colman et al. 1989).

Strathclyde Panned concentrates from stream sediments within an area of molybdenite mineralization in the Central Starav granite (part of the Etive granite complex) to the E and NE of Ben Starav, showed the presence of bismutite, vanadinite, wolframite and wulfenite, but none were found *in situ* (Haslam et al. 1985).

Ireland

Galway Small amounts of wolframite are noted from iron-sulphide-dominated stratabound sulphide lenses and disseminations in the Dalradian Lakes Marble Formation in Connemara. Molybdenite and scheelite are associated species (McArdle et al. 1986a). The nearby Galway and Oughterard granites are potential sources of metals required to produce these species.

Waterford Wolframite is confirmed by XRD from Grallagh Upper (Ryback 1992).

Wicklow Grains of wolframite have been found in gold-rich sands from the north-eastern side of Croghan Kinshela. Blue corundum (sapphire), cassiterite and native platinum are also reported (Forbes 1869, Ryback & Moreton 1993).

wollastonite

monoclinic or triclinic $CaSiO_3$

Named after William Hyde Wollaston FRS (1766–1828), a physician, chemist and mineralogist who discovered palladium and rhodium; born at East Dereham, Norfolk. Wollastonite is a common calcium silicate found in thermally metamorphosed siliceous carbonates, including skarn deposits, along the margins of igneous intrusions. At Treburland mine, 2 km SE of Fivelanes, Altarnun, Cornwall, thin layers of wollastonite, forming radiating aggregates of white silky fibres, occur in a calc-silicate assemblage that formed at the margin of the Bodmin Moor granite. It is associated with axinite, calcite, chlorite, diopside, garnet, pyrite, pyrrhotite, vesuvianite (idocrase) and later-formed zeolite-group minerals (Russell 1946d, Kingsbury 1961). Wollastonite is briefly mentioned from four other Cornish locations: Castle-an-Dinas mine, St Columb Major (Davison 1920a); Clicker Tor quarry, Menheniot (Robson 1949b); North Roskear mine, Camborne (Collins 1911a); and Wheal Mary (Wheal Ding section), Lanivet, near Bodmin (Kingsbury 1961).

In the northwestern part of the metamorphic aureole around the Dartmoor granite, such as at Red-a-ven mine, Meldon, near Okehampton, Devon, siliceous limestone has been transformed into a coarse-grain white fibrous wollastonite-bearing hornfels. Datolite, löllingite and malayaite are associated species (El Sharkawi & Dearman 1966, Dearman 1966b). Good specimens of wollastonite have also been found more recently at Meldon quarry (Robinson 1997a).

In Wales, wollastonite has been confirmed in impure Carboniferous Limestone adjacent to the Plas Newydd dyke, Anglesey, Gwynedd. It forms radiating sprays of needle-like crystals, up to 0.5 mm long, associated with calcite and vesuvianite (T. F. Cotterell pers. comm. 2007).

Wollastonite occurs at Kilchoan, Ardnamurchan, Highland Region, Scotland, in nodular or concretionary limestones that have been thermally metamorphosed by dolerite and gabbroic intrusions. It forms porphyroblasts about 1 cm in length and matrix crystals up to 0.1 mm long in a tough grossular–kilchoanite–vesuvianite-bearing rock interbanded with calcite–spurrite marble. Other associated species include åkermanite, cuspidine, larnite, merwinite, monticellite and rustumite (Agrell & Gay 1961, Agrell 1965). An iron-rich variety of wollastonite occurs with schorlomite and titanaugite from a pegmatite near Camphouse, Ardnamurchan (Howie & Woolley 1968). This is probably the contaminated pyroxenite produced by interaction of a gabbroic intrusion and Jurassic limestones, noted earlier by Agrell (1951). Wollastonite is also reported from a high-temperature calcic skarn near the contact of the Cuillin gabbro complex at Camasunary Bay, Strathaird Peninsula, Isle of Skye, Highland Region. Associated major species include calcite, garnet (andradite–grossular), melilite and vesuvianite (Beard & Drake 2007). Wollastonite is described from four other locations in this region: widespread in limestones of the Moine Series associated with diorite–gneiss, Ardgour district (Drever 1936, 1939); in limestone contact zones of the Camas Mòr gabbroic intrusion, Isle of Muck (Tilley 1947); in veins cutting anthophyllite–biotite rocks near the Free Church of Milton, Glen Urquhart (Heddle 1893b); and in primary skarns within the aureole of the Beinn an Dubhaich granite, Broadford area, Isle of Skye (Tilley 1948b, 1949, 1951). Iron wollastonite, reported from Camas Malag, Broadford area (Tilley 1948a), was later renamed ferrobustamite (Rapoport & Burnham 1973).

At 'S Airde Beinn, 5 km SW of Tobermory, Isle of Mull, Strathclyde, basalts in the aureole of a volcanic plug have been thermally metamorphosed. One type of amygdale found to the N of the plug contains radiating aggregates of colourless plates of gyrolite with smaller amounts of analcime, natrolite and thomsonite, but metamorphism has replaced the gyrolite with reyerite, and this in turn has been replaced by wollastonite (Cann 1965a).

Fibrous wollastonite from a calc-silicate hornfels at the margin of the Ballater granite, Pollagach Burn, E of Pannanich Hill, Deeside, Grampian, Scotland; width 3 cm. (© D. I. Green)

In the Grampian Region of Scotland, wollastonite is most often reported from the aureoles of Caledonian granite intrusions. At Pollagach Burn, E of Pannanich Hill, E of Ballater, Deeside, wollastonite occurs as felted mats and radiating aggregates intergrown with large crystals of grossular and vesuvianite in calc-silicate hornfels adjacent to the Ballater granite (Hutchinson 1933, Crane 1987). In metamorphosed limestones around the Dorback granite and particularly 0.4 km from Fae, Braes of Abernethy, 9 km W of Tomintoul, wollastonite is a major constituent of calcite–grossular–vesuvianite-bearing calc-silicate rocks and marbles (Walker & Davidson 1935). Contamination of a pegmatite magma with limestone is proposed to explain the white tabular wollastonite occurring with orthoclase in fissure veins at Shallowplough quarry, near Ellon (Read 1925). Other locations include: Corntulloch in Deeside, Crathie limestone quarry near Balmoral, and Delnabo quarry in Glen Gairn (Heddle 1901); and Kinnaird Head, Fraserburgh (Kneller 1987, Johnson et al. 2000).

Perhaps the most unusual wollastonite reported is that found as inclusions in alluvial gold from Glengarry, Tayside Region, Scotland (Chapman et al. 2000c).

Wollastonite occurs with xonotlite as a thin mantle around flint nodules in larnite–spurrite rocks in the contact zone between thermally metamorphosed chalk and the dolerite plug at Scawt Hill, Larne, Co. Antrim, Ireland (Tilley 1929b, Tilley & Harwood 1931, Tilley & Alderman 1934). It is also found in contaminated dolerite at this locality, but here it has atypical optical properties that led Tilley (1937d) to suggest it was iron wollastonite, a composition now formally described as the species ferrobustamite. Thermal metamorphism of calcareous aureole rocks by an unusual syenite intrusion in the area around Barnavave, Carlingford, Co. Louth, has produced a suite of rocks in which wollastonite occurs with diopside, grossular and vesuvianite at the junction of impure limestone and calcareous silt or in other assemblages where wollastonite is associated with augite, calcite, melilite or spurrite. Wollastonite presents a striking appearance when it forms large bladed crystals up to 75 mm long associated with well formed grossular. In certain bands it produces beautiful radiating fibrous structures. Late veins carrying a little wollastonite show replacement by pectolite (Osborne 1932b, Nockolds 1937, 1944, 1947, 1950).

Elsewhere in Ireland, wollastonite occurs in: xenoliths of flint-bearing chalk within dolerite at Carneal, Larne, Co. Antrim (Sabine et al. 1968, Sabine & Young 1975); rock type unspecified, but reported with garnet and diopside from Aphort, Aran Island, Co. Donegal (Russell 1946a); grossular–vesuvianite–wollastonite assemblages in metamorphosed xenoliths of limestone within granite at Ballynacarrick, Co. Donegal (Gindy 1953); remarkable diopside–grossular–tremolite–vesuvianite–wollastonite skarns in the limestone quarries of Lough Anure, Thorr area, Co. Donegal (Pitcher 1950); calcite, diopside, wollastonite skarn interbedded with garnet–vesuvianite-bearing rocks to the S of Bunbeg quay, E of Loughlin Island, Co. Donegal (Whitten 1951); rock type unspecified, but occurring in white fibrous masses at Dunmore Head, Mourne Mountains, Co. Down (Greg & Lettsom 1858); diopside, calcite, andradite–grossular, vesuvianite, wollastonite skarns from the aureole of the Omey granite on Fountain Hill, west Connemara, Co. Galway (Ahmed-Said & Leake 1996); attractive diopside–grossular–vesuvianite–wollastonite skarns to the W of Grallagh, Co. Galway (Ryback & Moreton 1992); and thermally metamorphosed limestone adjacent to two dolerite dykes at Killala Bay, near Inishcrone, Co. Sligo (Nawaz 1974).

wollastonite (of Thomson and the Edinburgh mineralogists)

A term once used to describe pectolite (TL Kilsyth, Forth & Clyde canal, Strathclyde, Scotland) or prehnite (TL Castle Hill, Edinburgh), but now a species name.

wölsendorfite

orthorhombic $Pb_7(UO_2)_{14}O_{19}(OH)_4{\cdot}12H_2O$

A very rare complex hydrated lead uranyl hydroxide found in a localized patch of Cu–Pb–U mineralization near the base of the seacliff at Loe Warren Zawn, W of Botallack, St Just, Cornwall. Wölsendorfite has formed where surface water draining from rubble at the base of the cliff has trickled over veinlets carrying chalcocite, chalcopyrite and pyrite, and has precipitated an assemblage of supergene carbonate and phosphate minerals on exposed surfaces and joint planes near a patch of uraninite mineralization. Wölsendorfite forms thin orange fibrous crusts composed of radiating prismatic or lath-like crystals 0.01 mm long and 0.001 mm wide. One specimen is associated with boltwoodite; others at the southeastern end of the outcrop are associated with dewindtite, kasolite, pyrite and zeunerite (Elton & Hooper 1995a).

wood copper

A massive variety of olivenite.

wood opal

A variety of opal formed during the fossilization of wood.

woodruffite

monoclinic $Zn_2Mn^{4+}_5O_{12}{\cdot}4H_2O$

An uncommon hydrated zinc–manganese oxide found in patchy exposures of iron and manganese oxides in the Lecht Valley, near Tomintoul, Grampian, Scotland, especially at the site of the old ironstone mine. The deposit is mainly composed of goethite and cryptomelane, with occasional patches of hematite. Black fine-grain woodruffite occurs in minor amounts within veinlets of manganese oxides, along with chalcophanite, lithiophorite, pyrolusite and ramsdellite. The Lecht mineralization is derived from weathering of the local Dalradian sequence, and ponding of the resulting Fe–Mn–Zn solutions produced a seepage-bog type deposit in which the permeable breccia acted as a channel-way for these supergene solutions. Diagnostic lithogeochemical anomalies in the area also indicate the existence of a concealed and related Pb–Zn–(Ag) exhalative deposit with a potential for hidden gold mineralization (Nicholson 1987a, 1989a,b, 1992, Nicholson & Anderton 1989).

Other occurrences of woodruffite are reported from Ben Eagach, near Aberfeldy, Tayside Region, where it occurs as a weathered oxide crust on a manganese-rich Dalradian carbonate horizon in association with groutite (Nicholson 1989b, 1992), and as black crusts on calcite on a single specimen from Corrantee mine, near Strontian, Highland Region (D. I. Green pers. comm. 2006).

wood tin

A variety of cassiterite showing repeated light- and dark-brown bands, somewhat similar to that found in many varieties of agate. See cassiterite.

woodwardite

trigonal $(Cu,Al)_9(SO_4)_2(OH)_{18}{\cdot}nH_2O$

TL Cornwall, England

Named after Samuel Pickworth Woodward (1821–1865), English naturalist, palaeontologist, professor of botany and natural history (including geology) at the Royal Agricultural College, Cirencester, and author of *A manual of the mollusca* (three parts: 1851, 1853, 1856); born in Norwich, Norfolk. Woodwardite is a rare hydrated copper–aluminium sulphate hydroxide found in stalactitic forms of minute botryoidal aggregates from an undisclosed type locality somewhere in Cornwall. Woodwardite is a soft greenish turquoise-blue translucent mineral with a dull lustre (Church 1866a). More recently, woodwardite has been found on weathered blocks of partially cemented slate-rich veinstone on the dumps of Penberthy Croft mine, St Hilary, where it forms botryoidal crusts associated with birnessite, brochantite, chalcopyrite, connellite, devilline, dolomite, goethite, langite and wroewolfeite. Woodwardite is typically a post-mining product (Betterton 1989, Betterton et al. 2000).

Botryoidal crust of pale-blue woodwardite with minor birnessite; Penberthy Croft mine, St Hilary, Cornwall; width 5 mm. (© D. I. Green)

A specimen of woodwardite from Botallack, St Just, is noted in the catalogue of the NHM (Bancroft & Weller 1993), and Golley & Williams (1995) also note it from South Caradon mine, St Cleer, and Wheal Cock, St Just. Anthony et al. (2003) add Wheal Hazard, near Botallack, St Just, as a further location.

In Wales, an XRD pattern similar to that of woodwardite is reported from pale-blue microbotryoidal coatings found at Eaglebrook (Nantycagl) mine, Ceulanymaesmawr, Central Wales orefield, Dyfed (Braithwaite 1982b, Jones 1983). Material described as woodwardite from Simdde Dyllhuan, Drwys-y-coed, Nantlle, Gwynedd, and from Drwys-y-coed mine has been described as the copper analogue of glaucocerinite (Nickel 1976, Raade et al. 1985, Bevins 1994), but later work confirms it to be the newly described species, hydrowoodwardite (Witzke 1999).

wooldridgeite

orthorhombic $Na_2CaCu_2(P_2O_7)_2\cdot10(H_2O)$

TL Judkins quarry, Nuneaton, Warwickshire, England

Named after James Wooldridge (1923–1995), a fervent amateur mineralogist, micromounter and gemologist from Fernhill Heath, Worcestershire, who discovered the mineral in 1989; born in Birmingham. Wooldridgeite is a rare transparent bluish-green hydrated sodium–calcium–copper pyrophosphate that occurs as both intergrown crystalline aggregates and as isolated well formed crystals 50–200 μm long. It is associated with calcite scalenohedra, botryoidal clots of chalcopyrite, corroded bornite pseudomorphs after chalcocite, and the desert-rose variety of barite. At Judkins quarry it formed along fractures close to the Precambrian/Cambrian unconformity during a low-temperature hydrothermal event (Hawthorne et al. 1999a). Wooldridgeite from Judkins quarry had previously been described as an unknown by Ince et al. (1991), and another unknown forming blue blades on chrysocolla from Roughton Gill mine, Caldbeck Fells, Cumbria (Cooper & Stanley 1990) appears to be the same species.

wroewolfeite

monoclinic $Cu^{2+}_4(SO_4)(OH)_6\cdot2H_2O$

An uncommon hydrated copper sulphate hydroxide found as a secondary mineral in the oxidized zone of copper deposits. Wroewolfeite has been found on the dumps in the eastern section of Penberthy Croft mine, St Hilary, Cornwall, where it forms small light-blue scales of poorly formed crystals usually less than 2 mm long. It has formed on joints and in cavities in weathered brecciated slates and dolomite vein material as a result of alteration of chalcopyrite (Barstow 1983). Associated species include birnessite, brochantite, connellite, goethite and woodwardite (Betterton 1989). On rare occasions, wroewolfeite forms translucent greenish-blue prismatic crystals up to 1 mm long, with narrow chisel-shape terminations (Betterton et al. 2000). Wroewolfeite is noted without description from East Gunnislake mine, Calstock, Cornwall (Anthony et al. 2003). At Wheal Friendship (Devon Friendship), Mary Tavy, Devon, wroewolfeite occurs on the mine dumps as beautiful greenish-blue to sky-blue sharp transparent crystals (1–2 mm long) scattered on open joints in slaty vein material. A pale-blue variety of wroewolfeite reaching 5 mm in length is noted from a single specimen, where it forms radiating aggregates of flattened tabular crystals implanted on a joint in quartz (Braithwaite 1982b, Barstow 1983).

Intergrown rhombic bipyramidal wooldridgeite; Judkins quarry, Nuneaton, Warwickshire; width 300 μm; SEM image. (© M. Rothwell)

At Ladywell mine, Shelve, Shropshire, oxidation of chalcopyrite–galena–sphalerite ore has produced a wide range of supergene species that includes wroewolfeite as well as anglesite, brochantite, cerussite, devilline, goethite, jarosite, langite, linarite and pyromorphite. Wroewolfeite forms greenish-blue bladed-to-tabular crystals up to 0.5 mm in length associated with langite and linarite (Dunn & Rouse 1975, Braithwaite 1982b). Braithwaite (1982b) also notes wroewolfeite from the Browngill vein of Tynebottom mine, NW of Garrigill, near Alston, Cumbria, again as a supergene alteration product of primary chalcopyrite–galena–sphalerite ore. Rust (1987a) suggests that wroewolfeite and other secondary sulphate species (anglesite, barite, brochantite, devilline, gypsum, linarite and serpierite) at Tynebottom mine formed as circulating groundwater passing through limestone became saturated with salts of various metals that were then precipitated on the floor of the mine passageways. A post-mining origin is also suggested for the wroewolfeite found at Red Gill and Silver Gill mines (Cooper & Stanley 1990, Green et al. 2005b) and Driggith and Sandbed mines (Green et al. 2006a), all Caldbeck Fells, Cumbria.

In Wales, wroewolfeite forming clear sharp greenish-blue blades with rectangular terminations postdates brochantite and linarite at Eaglebrook (Nantycagl) mine, Ceulanymaesmawr, Central Wales orefield, Dyfed (Braithwaite 1982b, Jones 1983). Elsewhere in the same region, wroewolfeite is found at Esgair Hir mine (Rust & Mason 1988), Frongoch mine, between Devil's Bridge and Pont-rhyd-y-groes (Green et al. 1996b), and, as a rare species, at Penrhiw mine, Ystumtuen (Mason & Green 1995). At Drws-y-Coed mine, near Nantlle, Snowdonia, the ore minerals chalcopyrite, galena, pyrite, pyrrhotite and sphalerite were worked from four lodes cutting Upper Cambrian and Ordovician sediments. Wroewolfeite occurs in the supergene assemblage as greenish-blue blades 0.1–1 mm long on calcite, associated with amorphous copper sulphates (Dunn & Rouse 1975, Braithwaite 1982b). Bevins (1994) notes aggregates of blue wroewolfeite with malachite on shale from Glasdir mine, Llanfachreth, near Dolgellau, Gwynedd.

West Blackcraig mine, near Newton Stewart, Dumfries & Galloway, Scotland, worked lead and zinc ore from a vein containing patches of chalcopyrite, galena and sphalerite. Specimens from the dumps carry these primary sulphides together with their characteristic alteration products, aurichalcite, brochantite, cerussite, hemimorphite, hydrozincite, linarite and malachite. Wroewolfeite forming tiny dark-blue

WULFENITE

Wroewolfeite crystals to 1 mm; Sheefry mine, Co. Mayo, Ireland. (© M. P. Cooper)

grains or bladed crystal aggregates was found on two specimens associated with langite, posnjakite and serpierite (Livingstone et al. 1976). Anthropogenic wroewolfeite is also common as transparent-to-translucent blue crystals on slag material from the Meadowfoot smelter, Wanlockhead, Dumfries & Galloway (Green 1987a).

The long-abandoned Pb–(Ag) Sheefry mine, 11 km S of Croagh Patrick, Co. Mayo, appears to be the only location in Ireland where wroewolfeite has been found. It occurs as clusters of well formed crystals up to 1 mm across associated with anglesite and linarite. Sheefry mine is also noted for fine millerite (part of the primary ore) and post-mining supergene species – annabergite, erythrite and gypsum (Moreton & Jackson 1991, Ryback & Moreton 1992, Green & Moreton 2001).

wulfenite

tetragonal $PbMoO_4$

A relatively common lead–molybdenum oxide found as a secondary mineral in the oxidized zone of hydrothermal lead deposits. For many years wulfenite was considered to be a rare species in Britain and Ireland, but over 70 localities are now known, mainly as a result of the increased use of stereo microscopes by amateur collectors.

At Greystone quarry, Lezant, near Launceston, Cornwall, fine small specimens of wulfenite occur as orange-red, yellow and buff bipyramidal crystals and yellowish tablets up to 1 mm across, associated with cerussite, pyromorphite and occasionally dundasite (Cooper & Green 1993, Grigg 1996, Wolfe 2000, Elton et al. 2004). Transparent orange crystals, up to 0.7 mm long with a sharp bipyramidal habit and associated with pyromorphite, have been found in a galena-bearing vein at Wheal Golden, Perranzabuloe (Merry & Braithwaite 1995, Wolfe 2006). These authors also note bright-orange equant bipyramidal wulfenite crystals up to 0.3 mm across and tabular crystals up to 1 mm long in cavities within galena–quartz veinstone on the dumps of Wheal Mary Ann, Menheniot. Blocky orange crystals of wulfenite crystals up to 0.4 mm long, with steep pyramidal faces and pinacoidal terminations associated with corkite and pyromorphite, are reported from Wheal Ludcott, St Ive, near Liskeard (Green & Merry 1997), and three other occurrences in Cornwall are recorded without description by Golley & Williams (1995): Garras mine, St Erme; Redmoor mine, Callington; and Polzeath Consols, St Minver. A specimen of intermediate stolzite–wulfenite composition supposedly collected by A. W. G. Kingsbury and said to come from Holmbush mine, near Callington (Rothwell & Mason 1992), requires further confirmation.

A major review of wulfenite undertaken by Rothwell & Mason (1992) drew on the work of others and summarized many new localities in England and elsewhere, as follows.

England

Avon Pale lemon-yellow to orange-yellow tabular crystals, up to 1 mm long, forming patchy encrustations on the walls of cavities in manganese ore, and also as very well formed crystals in chalcopyrite–galena veins at Coombe Farm quarry, Westbury-on-Trym, Bristol (Alabaster 1978, 1989b, Green 1986); smooth-sided uncorroded thin tabular, square rectangular or octagonal plates, up to 150 µm across, in drusy manganese-rich ore at Wesley mine, in the grounds of Wesley Theological College, Westbury-on-Trym (Alabaster 1989b).

Cheshire On Alderley Edge ridge: thin yellow plates in quartz cavities, greenish-grey pseudo-cubic crystals on malachite and minute acute bipyramidal crystals on pyromorphite at Engine Vein mine; tiny orange-yellow tabular to rectangular prismatic crystals at West mine (both Carlon 1979 and Braithwaite 1994); and rich encrustations of minute tabular crystals at Wood mine (Braithwaite1994).

Cumbria *Caldbeck Fells:* Small yellow plates associated with pyromorphite at Arm o' Grain (Cooper & Stanley 1990, Green et al. 2006b); small (<1.5 mm) thin "honey-yellow" plates, often near-circular in outline, associated with pyromorphite–mimetite on iron-stained quartz at Brandy Gill mine (Russell 1936c, Davidson & Thomson 1951, Kingsbury & Hartley 1955, Hartley 1984, Green 1986, Cooper & Stanley 1990); minute orange steeply bipyramidal crystals in a vein exposure above High level, Mexico mine, and tabular yellow plates and bipyramids up to 2 mm long on the dumps of Low level, Todd Gill (Cooper & Stanley 1990); tabular pseudo-cubic, or bipyramidal orange-yellow crystals on pyromorphite at Poddy Gill, although many analyzed specimens have proved to be stolzite (Davidson & Thomson 1951, Braithwaite et al. 1963a, Starkey 1981, Green 1986, Cooper & Stanley 1990); cherry-red plates on quartz on the 90-fathom level dump of Roughton Gill mine, but possibly dropped there by careless collectors after visiting Silver Gill, where wulfenite occurs as bipyramids and plates up to 1 mm across associated with cerussite, malachite or pyromorphite (Cooper & Stanley 1990, Green et al. 2005b). Descriptions of wulfenite from Carrock mine, Driggith mine and Ingray Gill (Fortey 1980, Hartley 1984, Young 1987a, Cooper & Stanley 1990) are mostly based on specimens in the Kingsbury collection at the NHM and their provenance requires further confirmation. Green et al. (2006a) suggest Kingsbury's claimed occurrence of wulfenite at Driggith mine must be treated with great caution as the provenance is doubtful. *Elsewhere in the Lake District and West Cumbria:* Resin-colour tabular crystals to 1 mm at Burtness Wood, Buttermere (Clark 1963); one tabular orange crystal, pale greenish-yellow plates up to 2 mm across associated with hemimorphite and mm-size white to butterscotch-yellow blocky to thick-tabular crystals scattered on mimetite or hemimorphite at Eagle Crag mine, Patterdale (Green & Briscoe 2002); smoky-brown to yellow, thin tabular crystals on fluorite associated with barite, chalcopyrite and galena at Hartsop Hall mine, Hartsop, Patterdale (Russell 1944, Green 1986); sharp bright-orange blocky crystals on cerussite and small yellow acute pyramidal and tabular crystals at Myers Head mine, Hartsop (Green & Briscoe 2002); small pale-brown tabular and bar-shape crystals up to 1 mm long with pyromorphite at Coombe Beck, Buttermere; blocky orange crystals on pyromorphite and square brown plates up to 1 mm across at Kinniside mine, Cleator Moor (Green & Briscoe 2002); pale-brown tabular crystals up to 1 mm long in crusts of pyromorphite on galena from the Ruthwaite Lodge vein, Grisedale, SW of Glenridding, Ullswater – anglesite, cerussite and harmotome are associated species (Young 1986, Todhunter 2007); and orange, yellow and white blocky crystals in miarolitic cavities at Shap granite quarry (Wirth & Green 2002).

Derbyshire Yellow or orange-tinted bipyramids to 0.3 mm and larger waxy-brown crystals with complex zonal growth patterns at Ball Eye mine, Bonsall (Green 1986); colourless to pale orange-yellow truncated pyramids associated with galena in fluorite workings at Ball Eye quarry, Bonsall (Bridges 1983, Ford et al. 1993); Bonsall Leys, near the Pike Lane opencast (Ford et al. 1993); white, orange-yellow and reddish-orange crystals up to 2 mm long, some tabular, others steeply bipyramidal, associated with barite, cerussite, fluorite and rare native sulphur at Crich quarry, Crich (Wilson et al. 1995); colourless to intense-yellow tabular and bipyramidal crystals often associated with manganese-rich wad, also cubic crystals at Isolation mine, Snelston, near Ashbourne (Green 1986, Ford et al. 1993, Jackman 1996); twinned and striated tetragonal prisms coating fluorite near Whirlpool Rising in Speedwell Cavern, Castleton (Quirk 1985); orange-yellow bipyramidal crystals up to 2 mm long associated with barite and fluorite at Ladywash mine, Eyam (Green 1986); orange-red crystals up to 2.5 mm long associated with fluorite at a location near Sacheverel farm, Winster; minute yellow crystals on fluorite at Speedwell mine, Castleton; wulfenite requires confirmation at Odin mine, Castleton (all Ford et al. 1993).

Leicestershire Single highly lustrous pale-orange crystals up to 1.3 mm across on cerussite from a phreatic cave in dolomitized limestone at Breedon quarry (Cloud Hill quarry), Breedon on the Hill (King 1991); associated with barite on one specimen from Earl Ferrers lead mine, Staunton Harold; waxy golden-yellow prismatic or bipyramidal crystals up to 0.7 mm long, associated with cerussite on oxidized galena at Tickow Lane mine, Shepshed (Green 1986); thin pale-yellow to orange square plates, pale-brown to brown tetragonal bipyramidal crystals, and small yellow-orange tabular crystals, the latter in Triassic sediments close to the Precambrian/Triassic unconformity at Newhurst quarry, Shepshed (Ince 2005).

Nottinghamshire Bulwell Pottery Co. quarry and as colourless lustrous plates associated with malachite at the north end of the Annesley railway tunnel, Kirkby in Ashfield (Deans 1961, Green 1986); with cerussite, galena and hydrocarbons in borehole cores at Springfield and associated with galena and hydrocarbons at Limekiln Wood, Newstead, near Mansfield (Deans 1961).

Shropshire Tiny orange bipyramids at Shelve mine, Shelve.

Somerset Deep-red to bright-orange platy crystals up to 1.5 mm across associated with cerussite and hydrocerussite in pockets of manganese-rich wad within a calcite vein at Batts Combe quarry, near Cheddar, Mendip Hills (Eastwood 1983, Green 1986); unusually large (reaching 3 cm across) and forming rough pale "wine-yellow" crystals on the outside of nodules of aragonite at Higher Pitts farm, Priddy, Mendip Hills (Spencer & Mountain 1923, Kingsbury 1941, Green 1986, Alabaster 1989b, Burr 1996); pale-brown platy crystals forming groupings up to 2.5 cm long and commonly associated with cerussite, paralaurionite and parkinsonite in manganese-rich pods at Merehead (Torr Works) quarry, Cranmore (Alabaster 1975, 1989b, Symes & Embrey 1977, Green 1986, Symes et al. 1994, Turner 2007).

Staffordshire Rare pale-yellow elongated bipyramidal crystals up to 0.75 mm long in oxidized barite–galena matrix at Waterbank mine, Ecton Hill (Rust 1985, Green 1986, Rust 1995c); tabular and bipyramidal crystals up to 0.5 mm long at the Lum, Ecton Hill (Cooper 1987a).

Wales

Dyfed Orange bipyramids up to 0.5 mm long at Alltycrib mine, Tal-y-bont; orange to orange-red crystals up to 2 mm long with a variety of habits and associated with pyromorphite at Bwlchglas mine, near Tal-y-bont (Braithwaite 1982a, Green 1986, Mason 2004); yellow-orange tabular crystals on hemimorphite in cavities within brecciated mudstone at Bwlchrhennaid mine, near Goginan; rare bright-orange bipyramidal crystals in quartz-veined mudstone at Cwmdarren mine, Trefeirig; thin yellow plates up to 0.75 mm long on cerussite at Darren mine, Pen-bont Rhydybeddau; single brown tabular crystal associated with pyromorphite at Eaglebrook (Nantycagl) mine, Ceulanymaesmawr; orange-yellow tablets and elongated bipyramids to 1 mm on pyromorphite at East Glogfach mine, Ysbyty Yswyth (Ryback & Saville 1967, Green 1986); tabular transparent-to-orange crystals up to 1.5 mm long on cerussite and pyromorphite at Elgar mine, 3 km S of Tal-y-bont (Mason 2004); rare brownish bipyramids up to 0.5 mm long on pyromorphite at Esgair Fraith mine, near Tal-y-bont; rare orange to orange-brown bipyramids to 1 mm on pyromorphite at Esgair Hir mine, Ceulanymaesmawr (Rust & Mason 1988, Mason 2004); exceptionally rare elongated crystals associated with pyromorphite at Frongoch mine, between Devil's Bridge and Pont-rhyd-y-groes (Green et al. 1996b); yellow crystals on pyromorphite at Glogfawr mine, Devil's Bridge (Jones & Moreton 1977); single orange platy tablet, 0.5 mm across, on cerussite and pyromorphite at Henfwlch mine, between Cerrig yr Hafan and Nant-y-moch reservoir, E of Tal-y-bont; very rare bright-yellow tabular crystals in vuggy quartz associated with goethite and pyromorphite at Llechweddhelyg mine, near Clawddmelyn, between Bont-goch and Cwmsymlog (Jones 1987, Mason 2004); brown-to-orange tabular crystals to 1 mm on pyromorphite at Mynyddgorddu mine, W of Bont-goch, Ceulanymaesmawr (Clark & Rust 1993); zoned yellow-to-green tabular crystals to 2 mm and orange

Tabular wulfenite, 5 mm on edge, from Llechweddhelyg mine, Bont-goch, near Clawddmelyn, Dyfed, Wales. (© National Museum of Wales, Cardiff; photographer M. P. Cooper)

Orange bipyramidal wulfenite associated with pale-green pyromorphite from Bwlchglas mine, near Tal-y-bont, Dyfed, Wales; width 0.3 mm. (© National Museum of Wales, Cardiff; photographer M. P. Cooper)

bipyramids on pyromorphite at Pandy mine, Tre'r-ddol (Mason 2004); single minute brown bipyramidal crystal at Penybanc mine, Tal-y-bont; and minute brownish bipyramidal crystals on pyromorphite on the mine dump and pale-yellow bipyramidal crystals up to 0.3 mm underground at Ystrad Einion mine, near Eglwysfach (Mason & Rust 1997). All these localities are in the Central Wales orefield, which extends into Powys.

Mid Glamorgan Bevelled rectangular plates in compact manganese ore with barite and calcite at Tŷ Coch farm, near Porthcawl (Braithwaite & Lamb 1986, Green 1986).

Powys Caramel-colour bipyramidal crystals up to 0.5 mm long associated with goethite and pyromorphite at Aberdaunant mine, near Llanidloes; minute yellow crystals on pyromorphite at Dyfngwm mine, Machynlleth (Jones & Moreton 1977); yellow microcrystals at Dylife mine, N of Hafren Forest (Jones & Moreton 1977, Moreton 1982); rare lemon-yellow tabular crystals up to 1 mm partially embedded in malachite at Geufron mine, near Llanidloes; and tabular and bipyramidal crystals up to 1.5 mm long associated with pyromorphite at South Nantycar mine, near Rhayader. Apart from the Central Wales orefield localities above, wulfenite is also found at Cwm Orog mine, Llangynog, where it forms orange-yellow plates up to 1 mm long on barite (Cooper 1987a), and at Dolyhir quarry, Old Radnor, near Presteigne, where it forms tiny elongated pale-yellow crystals (T. F. Cotterell pers. comm. 2007).

Scotland

Central Minute pyramidal wulfenite crystals have been found on a single block of oxidized galena from the ancient lead workings at Tyndrum (Wolfe 2006).

Dumfries & Galloway In panned concentrates from Bail Hill, 5 km NW of Sanquhar (Haynes 1993); greyish-brown crystals up to to 0.5 mm long around Needle's Eye, SW of Caulkerbush (Braithwaite & Knight 1990, Green & Todd 2001); tabular crystals on hemimorphite at Clatteringshaws reservoir, New Galloway; colourless to pale-yellow tabular crystals at Drumruck lead mine, near Gatehouse of Fleet (Livingstone 1992); Kings Laggan mine, N of the Cairnsmore of Fleet pluton (Haynes 1993); small translucent chrome-yellow tablets at Lauchentyre mine, W of Gatehouse of Fleet (Heddle 1901, Livingstone 1992, Haynes 1993); dark-orange thick tabular plates on gossany quartz at Pibble mine, 5 km NE of Creetown (Williams 1973, Green 1986); and Margaret's vein and Straitstep vein (Waygate shaft), Wanlockhead (Livingstone 1992, Wolfe 1996b).

Grampian Well formed orange bipyramids up to 1 mm long on quartz and thin yellow tabular crystals at Abergairn mine, near Ballater (Livingstone 1992, Green & Todd 2001); colourless-to-buff bipyramids and purplish-grey crystals associated with Ag–Bi–Mo–Sn–W mineralized zinnwaldite-bearing greisen at Gairnshiel Bridge, Glen Gairn (Livingstone 1992, 2002, Webb et al. 1992, Green & Todd 2001); and in cavities in granite at the Pass of Ballater, N of Ballater (Livingstone 1992, Green & Todd 2001). The Grampian occurrences are unusual in being granite related; most other occurrences are associated with lead-vein mineralization.

Highland In panned concentrates from the southeastern margin of the Grudie granite, 5 km SW of Lairg (Gallagher et al. 1971, Haynes 1993); small orange-red crystals, usually acutely pyramidal, associated with pyromorphite at Struy mines, Loch na Mèine, midway between Strath Glass and Strath Farrar, 3 km SW of Struy, near Inverness (Russell 1946b, Gallagher et al. 1971, Green 1986, Livingstone 1992, Haynes 1993, Green & Todd 2001).

Strathclyde Five localities in the Leadhills orefield: orange bipyramidal crystals on pyromorphite at East Stayvoyage vein; orange truncated bipyramidal crystals at Hopeful vein; orange-red tabular crystals with pyromorphite and cerussite at Horner's vein; minute orange or orange-yellow bipyramids on yellow pyromorphite at Raik vein, E of Broad Law (Green 1986, Livingstone 1992); and bright-orange bipyramids or tablets on pyromorphite at Susanna vein (Rust 1983c, Livingstone 1992). Wulfenite has also been found in panned concentrates from stream sediments within a mineralized area of the Etive granite complex, to the E and NE of Ben Starav (Haslam et al. 1985).

Ireland

Galway Tynagh mine, near Killimor (Green & Moreton 2001).

Monaghan Minute rounded bipyramids associated with barite, cerussite and galena at Lemgare mine, between Monaghan and Castleblaney (Ryback & Moreton 1993).

Wicklow Orange blocky to tabular crystals to 2 mm at Glendalough (Green & Moreton 2001).

Offshore territories

Jersey Well shaped bipyramidal "honey-yellow" crystals in a molybdenite-bearing quartz vein at L'Etacq (Lacroix 1913, Mourant & Warren 1934, Mourant 1961).

wülfingite

orthorhombic $Zn(OH)_2$

A rare zinc hydroxide, trimorphous with ashoverite and sweetite, and found in an oxidized vein exposure at Milltown quarry, Ashover, Derbyshire. Wülfingite has been found on only one specimen, where it forms star-like clusters of small (< 1 mm) colourless bladed crystals associated with ashoverite, smithsonite and calcite on iron-stained fluorite (Clark et al. 1988, Rust 1991c).

WURTZITE GROUP

Hexagonal or trigonal sulphides and selenides with the general formula *AX*, where *A* = Cd, Fe, Mn, Zn; and *X* = S, Se. Greenockite and wurtzite are members of the group found in Britain or Ireland.

wurtzite

WURTZITE GROUP

hexagonal (Zn,Fe)S

A rare zinc–iron sulphide, trimorphous with matraite and sphalerite, and found at Wheal Unity, St Day, Gwennap, Cornwall, as needles collectively forming botryoidal overgrowths. Less precisely, wurtzite is also recorded from Liskeard, Cornwall, as a thin coating of radial needles with additional dark minerals intergrown, and at Tavistock, Devon, as large radial fibres on pyrite stalactites and as crystalline aggregates (Nölting 1890). The dumps of Lockridge (Goldstreet) mine, SW of Bere Alston, Devon, contain specimens of intergrown sphalerite–wurtzite with a characteristic

Hexagonal plates of wurtzite from Dolyhir quarry, Old Radnor, Powys, Wales; width 4 mm. (© D. I. Green)

"ice fern" texture, associated with galena, tennantite and tetrahedrite (Grguric & Nickel 2006). Golley & Williams (1995) list additional locations in Cornwall: Gunheath china-clay pit, Carthew, St Austell; and Wheal Ellen, St Agnes; Wheal Jane, Kea, near Truro.

Wurtzite is also noted without comment from Gregory mine, Ashover, Derbyshire (Smith 1982), and with some doubt from three other Derbyshire locations: Fall Hill (Gate) quarry, Ashover; Millclose mine, Wensley, near Matlock; and Royal mine, Matlock (Ford et al. 1993). Rare wurtzite is said to form brownish-black pyramidal crystals in a fluorite vein at Fall Hill (Gate) quarry (Spence 1982). At Callow Hill quarry, Pontesbury, Shropshire, superb dark-brown twinned crystals of wurtzite (up to 0.5 mm across) and sphalerite have been found lining cavities in massive wurtzite from a lead–zinc-rich vein associated with a dolerite dyke that cuts the Ordovician Mytton Flags (N. Hubbard and R. E. Starkey pers. comms 2007).

Platy crystals in thin ewaldite-bearing fractures within the Yat Wood Formation (Precambrian basement) from Dolyhir quarry, Old Radnor, Powys, Wales, are confirmed to be sphalerite paramorphs after wurtzite, with a little residual wurtzite. These fractures also contain anatase, calcite, harmotome and quartz (T. F. Cotterell and D. I. Green pers. comms 2006).

In Scotland, wurtzite forms medium- to dark-brown lath-like crystals (up to 0.4 mm long) within septarian nodules in fishbeds in Lower Carboniferous shales at Wardie, Edinburgh (Macpherson & Livingstone 1982).

X

xanthiosite?

monoclinic $Ni_3(AsO_4)_2$

A nickel arsenate first reported from South Terras mine, St Stephen-in-Brannel, Cornwall (Davis et al. 1965). However, the occurrence is based on samples provided by A. W. G. Kingsbury, and their provenance is now considered to be very doubtful. Ryback et al. (2001) suggest that the specimens came from Johanngeorgenstadt, Saxony, Germany. Xanthiosite can no longer be considered a British or Irish species.

xanthochroite

A synonym of greenockite.

xanthoconite

monoclinic Ag_3AsS_3

A rare silver sulpharsenide, dimorphous with proustite and found in a very high-grade pod of polymetallic silver mineralization cutting pyrite mineralization in the B-zone of the Silvermines orebody, Co. Tipperary, Ireland. The silver mineralization includes acanthite, argyrodite, miargyrite, proustite, smithite and xanthoconite, together with barite, galena and sphalerite (Gasparrini 1978, Taylor & Andrew 1979). Taylor (1984) also reports geocronite–jordanite, gersdorffite and pyrargyrite from the same zone. A comprehensive review of mineralization in the Silvermines district is provided by Moreton (1999).

Xanthoconite, forming euhedral red crystalline aggregrates up to 1.5 mm across, has been found in a tennantite- and realgar-bearing calcite vein, at Dolyhir Quarry, Old Radnor, near Presteigne, Powys, Wales (D. I. Green XRD, pers. comm. 2006).

Red crystalline aggregrate of xanthoconite, *c.* 2 mm across, in calcite. Dolyhir quarry, Old Radnor, Powys, Wales. (© D. I. Green)

xantholite

TL Milltown, Glen Urquhart near Loch Ness, Highland Region, Scotland

A misidentification – actually impure kaolinite (see Heddle 1879a). Xantholite is also a synonym of polyadelphite (a variety of andradite, garnet group).

xanthophyllite

A synonym of clintonite, mica group.

xanthosiderite

A synonym of goethite.

xanthoxenite

triclinic $Ca_4Fe^{3+}_2(PO_4)_4(OH)_2\cdot3H_2O$

A rare hydrated calcium–iron phosphate hydroxide listed without detail from Hendre quarry, Pandy, near Glyn Ceiriog, Clwyd, Wales (Bernard & Hyršl 2004).

xenotime-(Y)

tetragonal YPO_4

A rare yttrium phosphate that forms a series with chernovite-(Y) and is found as an accessory constituent in granitic rocks and pegmatites, in rhyolites, and also as a detrital mineral in placer deposits. Many authors simply describe the species as xenotime. There is no evidence to suggest that the much less common species xenotime-(Yb) occurs in Britain or Ireland. Because of xenotime's resistance to erosion, it often occurs as a detrital phase in alluvial deposits, whose source is in many cases granite related. In England, it is recorded at two locations in Cornwall – the Camel and Gannel estuaries (Pirrie et al. 2000b) and the Fowey estuary (Pirrie et al. 2002); and at seven locations in Yorkshire – Earl's Crag, Silsden; Grimbald's Crag, Knaresborough; Keighley Moor; Otley Chevin; Plumpton Rocks, Knaresborough; Rumbles Moor, above Ilkley; and Steeton Moor (Gilligan 1919). In Scotland it is described from: Luibeg Burn, Derry Cairngorm, Grampian (Haynes 1974, 1993, Beer & Bennett 1988), and St Kilda (Hirta), Western Isles (Harding et al. 1984).

As part of an investigation of the radioactive accessory mineral assemblage in biotite granite from the Carnmenellis pluton in Cornwall, Jefferies (1985a) examined 1200 individual grains and found that only 1% were xenotime (other species identified were monazite 48%, zircon 48% and uraninite 3%). Textural evidence suggests that they all crystallized at an early magmatic stage. The rarity of xenotime-(Y) is again emphasized by Ward et al. (1992), who used an EMP to point-count and produce a modal analysis of biotite granite from the compositionally zoned Dartmoor granite, Devon. Values of 0.0001–0.0008% were obtained for xenotime-(Y) and confirmed it to be the rarest known species in the intrusion. This study also noted that xenotime-(Y) usually forms minute euhedral inclusions in biotite, associated with monazite, zircon and, less commonly, brockite, epidote and uraninite. The accessory-mineral assemblage of the Tertiary peraluminous leucogranite on Lundy island, off the north coast of Devon, is remarkably similar to the Hercynian granites in Cornwall and Devon. Xenotime-(Y) forms subhedral tabular crystals up to 80 μm in length, with some internal zoning and optical properties very similar to zircon. When a crystallization sequence can be established, the sequence of crystallization is zircon, xenotime-(Y), monazite (Thorpe et al. 1990). A modal abundance for xenotime-(Y) in the Lundy granite was calculated to be 0.00873% (Thorpe et al. 1995).

Bismuthinite-bearing quartz veins from the Alston Block of the Northern Pennine orefield, close to or above the Rookhope and Tynehead cupolas of the buried Weardale granite, contain bastnäsite, synchysite, monazite and xenotime-(Y) as rare accessory species. The last two species form discrete 20–40 μm equant crystals that overgrow bismuthinite, whereas texturally later synchysite and bastnäsite form acicular to tabular inclusions in fluorite or quartz and locally overgrow bismuthinite (Ixer et al. 1996). Specific localities where xenotime-(Y) has been found include Rotherhope mine, Leadgate, near Alston, Cumbria, and three localities in Co. Durham: Groverake mine, 4 km SE of Allenheads; Redburn mine, 1 km W of Rookhope; and Whiteheaps mine, 4 km N of Rookhope. At these locations, xenotime-(Y) forms 5–10 μm-diameter equant crystals, sometimes in aggregates to 100 μm, all within a quartz matrix. It also forms mixed aggregates with monazite to 200 μm across and in turn it is overgrown by chalcopyrite, pyrite and synchysite (Ixer 2003). Xenotime is said to be common (although this is surely a relative term) in the buried Wensleydale granite sampled in the Raydale borehole, Wensleydale, Yorkshire (Brown et al. 1987a). It is also noted in a borehole that has sampled the Skiddaw granite adjacent to the River Caldew, Caldbeck Fells, Cumbria (Webb & Brown 1984a).

In Wales, xenotime overgrows zircon in altered rhyolites from Cerrig Cochion and it forms small dispersed grains in altered rhyolite at Castell Yr Arddu, both Snowdonia, Gwynedd (Howells et al. 1991, Bevins 1994). It also occurs in rhyolitic tuffs on Ramsey Island, Dyfed (Bevins 1994). Well crystallized transparent golden

Xenotime-(Y) from Cwmorthin quarry, Blaenau Ffestiniog, Gwynedd, Wales; width 390 µm; SEM image. (© D. I. Green)

tetragonal prisms of xenotime-(Y), up to 0.5 mm long and closely associated with anatase and monazite-(Ce), have been recovered from Hendre quarry, Pandy, near Glyn Ceiriog, Clwyd (Starkey et al. 1991).

Xenotime-(Y) has been reported as an accessory mineral in Caledonian granites throughout Scotland: Criffell granodiorite and associated uraninite-bearing carbonate vein mineralization at Needle's Eye, SW of Caulkerbush, Dumfries & Galloway (Basham et al. 1989); Ballater granite, Grampian (Webb & Brown 1984b); Cairngorm granite, Grampian (Barritt 1983, Webb & Brown 1984b, Harrison 1988); and Helmsdale granite, Highland Region (Gallagher et al. 1971). Xenotime-(Y) and monazite-(Ce) also occur in an aplite dyke on Craigmawhannal, crosscutting the Loch Doon granite complex, Strathclyde (Tindle & Pearce 1981, Tindle 1982), and in the Glen Gairn granite complex at Gairnshiel, near Ballater, Grampian, xenotime occurs as an early crystallizing phase in zinnwaldite and sulphide-enriched greisen (Webb et al. 1992). The Tertiary aegirine–riebeckite granite of Rockall island, 440 km W of northern Skye, NW of Scotland, also contains xenotime, although this has been found only in crushed rock powder (Sabine 1960).

A more unusual occurrence of xenotime is in uranium and REE-enriched bitumens within Devonian Old Red Sandstones in Highland Region outcropping near Craig Well, Dingwall, and between Lothbeg and Helmsdale, where it occurs with brannerite and uraninite in reduction spots. The enrichments occurred post-depositionally, by absorption from groundwaters (Hofmann 1983, Parnell 1985).

xonaltite

A synonym of xonotlite.

xonotlite

monoclinic or triclinic $Ca_6Si_6O_{17}(OH)_2$

An uncommon calcium silicate hydroxide found where granite has intruded dolomite at Strath Suardal, near Broadford, Isle of Skye, Highland Region, Scotland. Veinlets of xonotlite, diopside, pectolite and wollastonite cut chert nodules in metasomatized dolomite (Tilley 1951). To the north of Portree, also on Skye, pinkish-white xonotlite has been described from the core of a vesicle within dolerite. The rim of this vesicle consisted of gyrolite and mesolite. Other vesicles in dolerite typically contain calcite, gyrolite, mesolite, saponite, tacharanite, thomsonite or tobermorite, and Sweet (1961) suggests that tacharanite may be the parent mineral from which xonotlite formed. Heddle (1893b) notes xonotlite from Geo da Tuill, Loch Eynort, Isle of Skye. On the Isle of Mull, Strathclyde, and across the Sound of Mull on the Morvern Peninsula, Highland Region, xonotlite occurs with gyrolite, okenite, pectolite, reyerite, tacharanite, tobermorite and a wide range of zeolite species in amygdales within Tertiary olivine basalts (Walker 1971). Xonotlite had previously been described by Heddle (1882a) from Kilfinnichen on the shore of Loch Scridain, and

Delicate fibres of xonotlite in a 1 cm-wide vein within gabbro from Binhill quarry, Cairnie, near Huntly, Grampian, Scotland. (© The Trustees of the National Museums of Scotland)

Loch na Keal on Mull. At Camphouse, Ardnamurchan, Highland Region, contamination of an augite diorite by limestone has allowed the formation of xonotlite as a secondary mineral, along with analcime, prehnite, thomsonite and hydrogarnet (Agrell 1951). Pink xonotlite is associated with tacharanite (and scawtite) at Binhill quarry, Cairnie, Huntly, Grampian, where gabbro is cut by fracture-filled veins containing calcite, xonotlite or prehnite. Other hydrated calcium silicates (e.g. gyrolite and laumontite) are also present in pegmatite veins in the same quarry (Livingstone 1974). At Colla Firth, north Mainland, Shetland, xonotlite is reported from veins in serpentinites, accompanied by gyrolite, pectolite and thaumasite (Butler et al. 1960). Heddle (1880b) briefly describes xonotlite in Tertiary basalts from the Isle of Mull, Strathclyde.

In the chalk/dolerite contact zone at Scawt Hill, Larne, Co. Antrim, Ireland, xonotlite forms aggregates of fibrous radiating crystals enclosing blades of wollastonite up to 3 mm long and associated with flint nodules in the chalk. Other species in this part of the contact zone include afwillite, calcite, ettringite, gyrolite, larnite, merwinite, perovskite, spinel and spurrite (Tilley & Harwood 1931, Tilley & Alderman 1934). At Carneal, Larne, Co. Antrim, in another chalk/dolerite contact zone associated with a Tertiary volcanic plug, xonotlite forms the groundmass of complex calc-silicate assemblages, or veins them in other rocks. In a quartz–wollastonite–xonotlite rock, representing metasomatized flint, xonotlite forms radiating fibrous veinlets with some spherulitic crystallization around wollastonite. Other hydrated and secondary minerals associated with xonotlite are allophane, calcite, gyrolite, hibschite (hydrogrossular), perovskite, portlandite, tacharanite, thomsonite and tobermorite (Henmi et al. 1973, Sabine & Young 1975). Taylor (1959) showed experimentally that tobermorite from Ballycraigy, Larne, Co. Antrim, converts to xonotlite at 380 °C and 250 atmospheres.

Brown-to-green fibrous xonotlite associated with colourless acicular-to-tabular wollastonite; Scawt Hill, near Larne, Co. Antrim, Ireland; width 3.4 mm; viewed in PPL. (© K. Gardner)

Y

Yarrowite illustrating extreme reflection pleochroism from pale blue (vertical cleavage) to deep-blue (horizontal cleavage). Cannington Park, Bridgwater, Somerset; width 220 µm; viewed in oil-immersion reflected light. (© C. J. Stanley, The Natural History Museum, London)

yarrowite

hexagonal Cu_9S_8

A rarely described copper sulphide found as grains up to 100 µm across in the chalcocite–covellite–tetrahedrite assemblage at Cannington Park, near Bridgwater, Somerset. Here, yarrowite replaces bornite and chalcocite, and is probably widespread as an alteration product of the copper sulphides. At Rigg Head quarry, Borrowdale, Cumbria, yarrowite occurs with digenite, chalcocite and covellite as alteration rims around chalcopyrite, galena and sphalerite (Green et al. 2006).

Yarrowite has been found in void-infilling carbonate-associated copper mineralization at Judkins quarry, Nuneaton, Warwickshire. It occurs in trace amounts as 1–2 µm-wide rims around crazed copper-poor bornite and idaite, or 5–20 µm-diameter rosettes of radiating crystals replacing bornite and to a lesser extent chalcopyrite (Ixer et al. 2005a).

Fine-grain yarrowite occurs as a post-mining alteration product of Cu–Pb–Zn ore from Crow Island, near Muckross, Killarney, Co. Kerry, Ireland. It forms 5–20 µm-wide alteration rims around chalcopyrite, and 10–60 µm rims around galena (Ixer 2004).

Yarrowite was previously referred to as "blaubleibender covellite", but this has now been characterized as a mixture of yarrowite and spionkopite (Ryback & Tandy 1992, Criddle & Stanley 1993; C. J. Stanley pers. comm. 2003). Examples of "blaubleibender covellite" include: Clevedon, Avon (Starkey 1986c, Ixer et al. 1993); Alderley Edge ridge, Cheshire (Ixer & Vaughan 1982, Braithwaite 1994); Seathwaite Tarn, near Coniston, Cumbria (Stanley & Criddle 1979); Loddiswell mine, 3 km N of Loddiswell, Devon (Stanley et al. 1990b); Parys Mountain, Anglesey, Gwynedd, Wales (Southwood & Bevins 1995); Matta Taing, west coast of Mainland, Shetland, Scotland (Hall & Donovan 1978); Ballycoog, Goldmine River valley, near Avoca, Co. Wicklow, Ireland (Ixer et al. 1990); Bouley Bay, Trinity, Jersey (Ixer 1980, 1992); German underground hospital, St Peter's Valley, St Lawrence, Jersey (Ixer 1990); Sark's Hope mine, Little Sark, Channel Isles (Ixer & Stanley 1983).

yellow arsenic

A synonym of orpiment.

yellow copper or yellow ore

A synonym of chalcopyrite.

yellow ochre

A synonym of limonite.

ytterite

A synonym of both gadolinite and tengerite.

yttrotantalite

A true species name, but once used to describe a yellow variety of fergusonite.

yugawaralite?

ZEOLITE GROUP

monoclinic or triclinic $Ca_2(Si_6Al_2)O_{16}{\cdot}4H_2O$

A rare hydrated calcium aluminosilicate, tentatively identified in a chalcopyrite-rich (clinopyroxene–plagioclase) dyke associated with the West Sgaorishal ultramafic plug, on the Isle of Rum, Highland Region, Scotland. The dyke is heavily bleached and altered along its centre, and it contains rare vugs infilled with zeolites. XRD analysis suggests that yugawaralite and thomsonite are the two species present in the vugs, together with quartz. The dyke is also notable for its PGM assemblage with froodite, gold (electrum), keithconnite, kotulskite, paolovite, native silver, sperrylite, telluropalladinite and Ag–Bi–Pd–Te phases all identified (Power et al. 2003).

Z

Hair-like masses of zálesíite up to 0.1 mm long. New Cliffe Hill quarry, Stanton under Bardon, Leicestershire; SEM image. (© D. I. Green)

zálesíite

MIXITE GROUP

hexagonal $CaCu_6(AsO_4)_2(AsO_3OH)(OH)_6{\cdot}3H_2O$

A rare hydrated calcium–copper arsenate hydroxide that forms as a result of weathering of chalcopyrite and cobalt arsenides (Sejkora et al. 1999).

Radiating sprays of highly elongated to acicular pale-green to apple-green hexagonal prisms to 0.6 mm from Driggith and Sandbed mines, Caldbeck Fells, Cumbria, have been unequivocally identified as zálesíite. Malachite is a common associate (D. I. Green pers. comm. 2006). Zálesíite has been identified on specimens from carbonate-rich veins in diorite surrounding a deposit of native copper and cuprite at New Cliffe Hill quarry, Stanton under Bardon, Leicestershire. It occurs as sprays and mats of minute hair-like pale-blue to blue-green crystals with volborthite and malachite on dolomite, and as minute green hair-like crystals with chrysocolla (Hubbard et al. 2005). Greenish-yellow crystals of zálesíite are also noted from the nearby Bardon Hill quarry (Ince 2007).

Zálesíite is listed without detail from Penberthy Croft and Croft Gothal mines, St Hilary, Cornwall (Bernard & Hyršl 2004). However, if this discovery is the same as the "Ca-agardite" noted by Golley & Williams (1995), then it must be viewed with suspicion, as it has since been shown that there is no calcium present in these samples (J. Betterton pers. comm. 2005).

At Leaghillaun, Bearhaven (Berehaven) Peninsula, Co. Kerry, Ireland, a chalcopyrite–tennantite-bearing quartz vein hosted by Devonian sandstones has produced an oxidized suite of supergene minerals that includes rare zálesíite forming pale-green acicular crystals to 0.5 mm. Associated species include azurite, chrysocolla, clinoclase, cornwallite, malachite and tyrolite (Moreton & Green 2006).

zaratite

cubic $Ni_3(CO_3)(OH)_4{\cdot}4H_2O$

An uncommon hydrated nickel carbonate hydroxide found in fine specimens at Hagdale quarry, Unst, Shetland, Scotland, where it replaces pentlandite in chromite-rich serpentinite. The zaratite is emerald green, vitreous and translucent (Heddle 1878a, 1901). Livingstone & Bish (1982) recognized that the zaratite was intimately mixed with another phase, which they named theophrastite. From a suite of samples collected in the Cliff area, NW of Baltasound, Unst, Gunn et al. (1985) found veins of zaratite cutting chromitite. In the same sample, sperrylite, laurite and unnamed Pd, Sb, Te, Ir, Rh, Pt, Ru, As and S phases were also found.

Yellowish-green zaratite forming a thin film on millerite and associated with chalcopyrite, galena, retgersite and siderite is reported from a clay–ironstone septarian nodule collected at Newdigate colliery, Bedworth, SW of Nuneaton, Warwickshire (Walker 1988). Zaratite is also thought to be present as an emerald-green coating on millerite in minor calcite veins at the Waterbank end of Clayton adit, part of the Ecton Hill mines, Ecton, Staffordshire (Starkey 1983).

zavaritskite?

monoclinic BiOF

A rare bismuth oxyfluoride reported with ferrimolybdite, rooseveltite and scorodite as supergene alteration products of arsenopyrite, native bismuth, bismuthinite, fluorite and molybdenite mineralization in topaz-rich greisen from the margin of the Eskdale granite at Water Crag, near Devoke Water, Cumbria. Both rooseveltite and zavaritskite require further confirmation (Young 1985d,e).

ZEOLITE GROUP

A very large group of silicates having framework structures containing open cavities in the form of channels and cages. The crystallography of the group varies greatly between species and there are many crystal systems represented. The group was subject to a major classification overhaul by Coombs et al. (1997) and consequently there are many references to the group where new species names are not quoted. Those species and series identified in Britain or Ireland include: analcime, brewsterite (series), chabazite (series), clinoptilolite (series), cowlesite, edingtonite, epistilbite, erionite (series), faujasite (series), ferrierite (series), garronite, gismondine, gmelinite (series), gobbinsite, gonnardite, harmotome, heulandite (series), laumontite, levyne (series), mesolite, mordenite, natrolite, offrétite, paulingite (series), phillipsite (series), pollucite, scolecite, stellerite, stilbite (series), thomsonite and wairakite. Another zeolite species, leucite, may once have formed, but now appears to be totally altered. Yugawaralite is tentatively identified. Paranatrolite and tetranatrolite would be members of this group if they were better characterized. A comprehensive listing of zeolite localities is given by Wilson et al. (2003).

zeunerite

AUTUNITE GROUP

tetragonal $Cu^{2+}(UO_2)_2(AsO_4)_2{\cdot}12H_2O$

A rare hydrated copper uranyl arsenate found in the oxidized zone of arsenic-bearing hydrothermal uranium deposits.

Zeunerite occurs in the Cargodna section of Wheal Owles, Penwith, St Just, Cornwall. Zeunerite is intimately associated with tyuyamunite (Elton & Hooper 1994). Near by, on the coast at Loe Warren Zawn, W of Botallack, green rectangular crusts and spheroidal aggregates of zeunerite form part of a supergene assemblage associated with a Cu–Pb–U mineralized vein within the aureole of the Land's End granite. There is a complete spread of compositions between end-member zeunerite through a range of intermediate compositions to essentially end-member torbernite (Elton & Hooper 1995a). Zeunerite is also described from the dumps of Morcom's shaft, Ting Tang mine, Carharrack, Cornwall, where it is associated with sulphide-rich (arsenopyrite, chalcopyrite, pyrite) veins cutting hornfels near the margins of the Carn Marth granite (Bruce & Aubrey-Jones 1998). Zeunerite is further noted from eight other Cornish locations: Cligga Head mine, Perranzabuloe (R. E. Starkey pers. comm. 2004); Great Wheal Fortune, near Porthleven (Wolfe 1996b); Roughtor mine, St Cleer (Golley & Williams 1995); South Terras mine, St Stephen-in-Brannel (Robson 1954); Wheal Basset, Illogan, near Redruth (Wolfe 1996a); Wheal Edward, St Just (Golley & Williams 1995); Wheal Gorland, Gwennap, near Redruth (James 1947, Robson 1949b, 1954, Kingsbury & Hartley 1960, Rust 1981b); and Wheal Maid, Gwennap (Golley & Williams 1995).

In Scotland, zeunerite occurs as an alteration product of uraninite (pitchblende) and in highly uranium-rich fissure veins close to Needle's Eye, SW of Caulkerbush, Dumfries & Galloway. The veins are in the aureole of the Criffell granodiorite. Zeunerite (and metazeunerite) also forms green platy crystals up to 0.2 mm across, associated with lavendulan on joint and cavity surfaces. Other supergene minerals from this vicinity, produced during the alteration of primary As–Bi–Co–Cu–Fe–Pb–U minerals, include cornubite, eulytite, kahlerite, malachite, metakahlerite, mixite, schoepite, tyrolite, uranophane, vandendriesscheite, walpurgite and wulfenite (Miller & Taylor 1966, Basham et al. 1989, Braithwaite & Knight 1990, Green & Todd

Tabular zeunerite associated with pale-blue connellite; Cligga Head, Perranzabuloe, Cornwall; width 4mm. (© A. G. Tindle)

2001). Zeunerite is one of the secondary species (others being erythrite, kasolite, liebigite and uranospinite) that occurs with pink calcite, uraninite and sulphides in vuggy cavities, as a component of late uranium-rich mineralization hosted by Dalradian metasediments at Tyndrum Main mine, Central Region, Scotland. The secondary species probably formed during post-depositional oxidation of the deposit (Darnley 1962b, Pattrick 1985).

zeuxite

TL Wheal Unity, St Day, Gwennap, Cornwall

A green or brownish-green variety of tourmaline reported from Carn Marth, near Redruth; Great Work mine, SW of Godolphin Cross and Wheal Unity (all in Cornwall), and from near Okehampton, Devon (Collins 1876c).

zinc?

hexagonal Zn

A very rare native metal species described with a suite of PGMs as inclusions in chromite grains in chromitite from Quoys, Unst, Shetland, Scotland. The possibility cannot be discounted that the mineral is a zinc oxide, as oxygen was never analyzed (Gass et al. 1982, H. M. Prichard pers. comm. 2004). The occurrence is referenced by Gunn et al. (1985). Further confirmation is desirable.

zinc blende

A synonym of sphalerite.

zinckenite

A synonym of zinkenite.

zinc spar

A synonym of smithsonite, calcite group.

zincolivenite

orthorhombic $CuZn(AsO_4)(OH)$

A copper–lead arsenate hydroxide whose composition has been noted in Britain and Ireland for many years, and which had previously been called cuproadamite (a copper-rich variety of adamite). Cuproadamite was discredited as part of an IMA-CNMNC-approved mass discreditation of species (Burke 2006). It was suggested that the mineral should be referred to as copper-bearing adamite. However, this has now changed, with the IMA-approved formal description of zincolivenite (Chukanov et al. 2007). This work sets a compositional range of $Cu_{0.5}Zn_{1.5}$ to $Cu_{1.5}Zn_{0.5}$ for the species and lists the type as locality the Kamariza mine dumps, Lavrion, Atikki Prefecture, Greece.

In Britain and Ireland, copper-bearing adamite has been found at: Penberthy Croft, St Hilary, Cornwall (D. Moulding specimen, D. I. Green XRD, Tindle unpub. EMP data); Driggith and Sandbed mines, Caldbeck Fells, Cumbria (Green et al. 2006a); Dolyhir quarry, Old Radnor, near Presteigne, Powys, Wales (D. I. Green XRD, Tindle unpub. EMP data); Tynagh mine, near Killimor, Co. Galway, Ireland (Ryback 1988a, Ryback & Moreton 1992, D. I. Green XRD, Tindle unpub. EMP data). It is likely that this is all zincolivenite. See also adamite and cuproadamite.

zinkenite

hexagonal $Pb_9Sb_{22}S_{42}$

An uncommon lead sulphantimonite found in an antimony-rich quartz vein within the aureole of the Skiddaw granite, near the head of Wet Swine Gill, Coombe Height, Caldbeck Fells, Cumbria. The vein contains an early As–Fe–Sb primary assemblage (comprising arsenopyrite, berthierite and stibnite) and a later antimony–lead assemblage (including fülöppite, semseyite and zinkenite). Zinkenite forms patches up to 15mm wide, which may enclose small bundles of stibnite. Other minor phases in the vein include native antimony, jamesonite, senarmontite, sphalerite and sulphur (Fortey et al. 1984, Leppington & Green 1998).

In Wales, zinkenite has been reported from an epithermal deposit in nodular ignimbrite at Bwlch mine, near Deganwy, S of Llandudno, Gwynedd. Zinkenite and other antimony-bearing minerals (jamesonite, plagionite, semseyite, stibnite, and possible boulangerite and robinsonite) typically occur in irregular patches, and in thin veins and stringers up to 2cm in width, associated with quartz (Bevins et al. 1988).

At Glendinning mine, Glenshanna Burn, Dumfries & Galloway, Scotland, zinkenite forms irridescent finely felted crystal aggregates associated with stibnite in antimony-rich veins. Galena, semseyite and sphalerite are also recorded (Macpherson & Livingstone 1982).

Reports of zinkenite from Carrock mine, Grainsgill, Caldbeck Fells, Cumbria (Kingsbury & Hartley 1956b, Hartley 1984), require further confirmation. Unfortunately, this appears to be the only British zinkenite locality listed by Anthony et al. (1990) and Bernard & Hyršl (2004).

zinnkies

TL Wheal Rock, St Agnes, Cornwall, England

A synonym of stannite.

zinnwaldite (series)

MICA GROUP

monoclinic $K(Al,Fe,Li)_3(Si,Al)_4O_{10}(OH)F$

A compositional range of dark trioctahedral micas containing lithium on, or close to, the siderophyllite–polylithionite join. Zinnwaldite does not define an end-member composition and is therefore not a true species. However, it has been elevated to series status in the IMA reclassification of the mica group (Rieder et al. 1998) and, as such, is included in *Fleischer's glossary of mineral species* (Mandarino & Back 2004). Zinnwaldite is usually found in lithium-rich granites, granite pegmatites, tin-bearing greisen deposits or high-temperature quartz veins.

In Cornwall, zinnwaldite is found in the Hensbarrow topaz-bearing non-megacrystic lithium-mica granite, between Carthew and Stenalees. This is the youngest and one of the most chemically evolved plutons in the St Austell granite complex. The present outcrop is considered to be close to the roof of the intrusion, because of the presence of many pegmatites and vugs, and from the miarolitic cavities in the granite. A section of the pluton has been worked for its kaolinite content and now forms part of Gunheath china-clay pit. Pegmatites cutting the granite contain transparent bronze-brown to copper-brown scaly lamellar aggregates of zinnwaldite to 2cm across, growing at right angles to the contact. The Hensbarrow stock and adjacent granites have been affected by greisenization, commonly at the margins of tourmalinized veins, and here zinnwaldite is recognized by its pale light-brown to brownish silver-grey colour (Hodkinson & Clark 1977, Barstow 1982, Stone et al. 1988, Henderson et al. 1989, Alderton 1991, Weiss 1994). Together with albite, cassiterite (varlamoffite), columbite, ilmenorutile, muscovite, qitianlingite, stannite and zircon, zinnwaldite also forms inclusions in topaz within greisen at Gunheath (Williamson et al. 1997). Other reports of zinnwaldite associated with the St Austell granite complex include: a general discussion in Manning et al. (1996); Bodelva china-clay pit, St Blazey Gate, E of St Austell (Mueller et al. 1999); Goonbarrow china-clay pit, near Bugle (Bray & Spooner 1983); Goonvean china-clay pit, SE of Treviscoe; Kernick china-clay pit, near Treviscoe; Prosper china-clay pit, near Nanpean; Littlejohn china-clay pit (all Henderson et al. 1989); Rostowrack china-clay pit (Stone et al. 1988); Trelavour Downs (Cundy et al. 1960, Hogg & Meads 1970), but said to be lithian siderophyllite by Stone et al. (1988); and Trethosa china-clay pit, Carthew (Psyrillos et al. 1998). A small granite boss at Castle-an-Dinas, St Columb

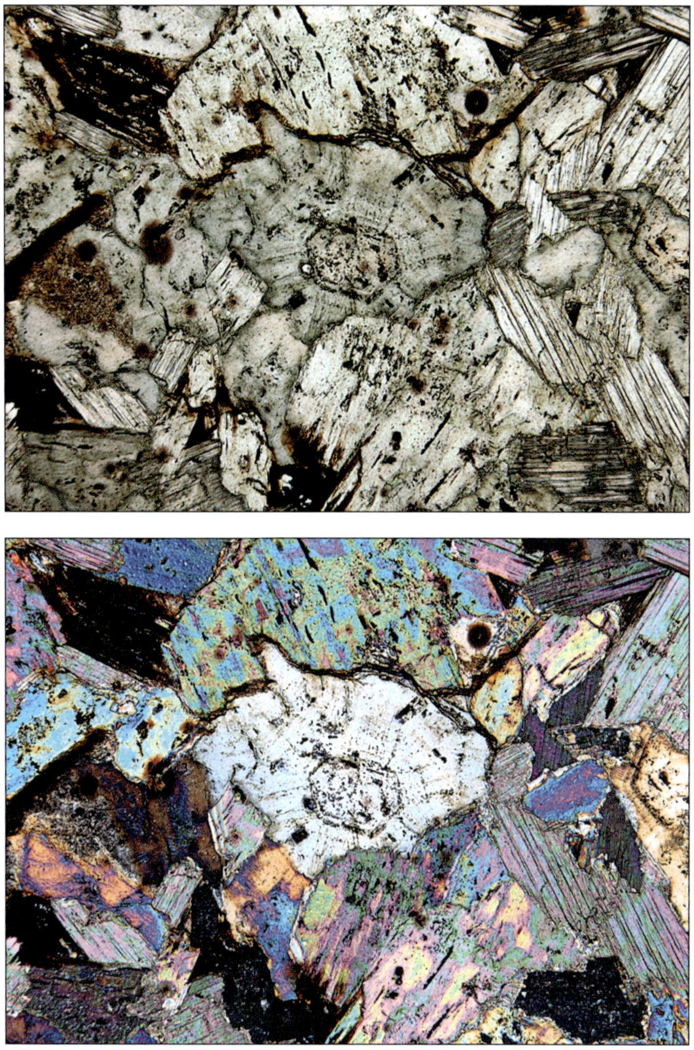

Zinnwaldite-rich vein from an area of tin–tungsten mineralization; Gairnshiel, Glen Gairn, Grampian, Scotland; width 2.8 mm; viewed in PPL (upper) and XPL (lower). (© A. G. Tindle)

Major (representing a northern extension of the St Austell granite complex), contains zinnwaldite associated with wolframite vein mineralization (Hey et al. 1938, Manning & Hill 1990). Another small granite boss outcrops at Cligga Head, on the north coast of Cornwall; it hosts a tin-bearing greisen containing a lithium-rich mica (protolithionite) with a composition intermediate between siderophyllite and zinnwaldite (Hall 1971).

The Tregonning granite, between Breage and Rinsey, is another Cornish zinnwaldite–topaz–(albite)-bearing intrusion. Zinnwaldite forms ragged grains interfingering potassium feldspar in granite: at the eastern end of Praa Sands; in a quarry NNW of Lower Trevurvas, Rinsey West Cliff; and at Trewavas Head, all near Breage. In the roof complex exposed at Megiliggar Rocks, there is a transition between zinnwaldite and lepidolite compositions. Intermediate siderophyllite–zinnwaldite compositions are noted at hornfels contacts (Stone 1984, 1992, Stone et al. 1988). Zinnwaldite is also noted in an albite–zinnwaldite granite dyke related to the Land's End granite pluton at Porthmeor Cove, W of St Ives, Cornwall (Müller et al. 2006).

In Scotland, zinnwaldite (pale green and showing zonation in thin section) occurs in a small intrusion of lithium–fluorine-enriched granite within the Glen Gairn granite complex, at Gairnshiel Bridge, 10 km NW of Ballater, Grampian. In parts the granite is pervasively greisenized and is host to polymetallic Ag–Bi–Mo–Sn–W (hübnerite-dominated) vein mineralization (Hall & Walsh 1972, Tindle & Webb 1989, Webb et al. 1992). Zinnwaldite is associated with beryl, microcline, muscovite and quartz (cairngorm) in granite from the Pass of Ballater (Heddle 1901) and is also found in granite to the N of Sgòrr Buide and on the southern slopes of Morven, both along the northwestern margin of the Ballater intrusion (Tindle unpub. EMP data).

Well formed and zoned crystals of zinnwaldite, varying in colour from dark green to pale violet, are described from topaz-bearing miarolitic granite at an undisclosed locality in the Mourne Mountains, Co. Down, Ireland (Sollas 1890, Gaines et al. 1997). Analysis of granite from Diamond Rocks in the Mourne Mountains confirms the earlier descriptions (S. A. Rust sample, Tindle unpub. EMP data). At Lindsay's Leap quarry, also in the Mourne Mountains, drusy cavities in granite contain dark-green masses of a muscovite or zinnwaldite-type mica associated with chlorite (Green et al. 2005a).

At Mont Mado, Jersey, Channel Isles, a lithium-bearing mica, occurring in geodes as prisms up to 3 cm high, is probably zinnwaldite (Mourant 1961).

ZIPPEITE GROUP

Orthorhombic uranyl sulphates with the general formula $A(UO_2)_2(SO_4)(OH)_3 \cdot H_2O$, or $B_4(UO_2)_6(SO_4)_3(OH)_{10} \cdot 4H_2O$, or $C_2(UO_2)_6(SO_4)_3(OH)_{10} \cdot 16H_2O$, where A = K, B = Na, and C = Co, Mg, Ni, Zn. Sodium-zippeite and zippeite are members of the group found in Britain or Ireland.

zippeite

monoclinic — ZIPPEITE GROUP — $K(UO_2)_2(SO_4)(OH)_3 \cdot H_2O$

A rare hydrated potassium uranyl sulphate hydroxide reported only from Cornwall, where it occurs as delicate acicular yellow crystals that typically develop as encrustations on mine walls through supergene alteration of uranium ore minerals such as uraninite (pitchblende). Zippeite is found in the Radium lode of South Terras mine, St Stephen-in-Brannel, with other uranium-rich species: autunite, torbernite, uraninite (pitchblende) and associated with arsenopyrite, chalcopyrite, galena and pyrite (Robertson & Dines 1929). At Wheal Owles, near Botallack, St Just, zippeite occurs as very pale-yellow botryoidal crusts (<2 mm) on the joint surfaces of greenstone country rock (Wolloxall 1988). Other occurrences include: Botallack mine, St Just; Clijah mine, Redruth; Dolcoath mine, Camborne; Gunnislake mine, Calstock, near Gunnislake; Pednandrea mine, Redruth; St Michael's Mount, Marazion (all James 1947); Callington mine, near Liskeard and Restormel iron mine, N of Lostwithiel (Collins 1892, Greg & Lettsom 1858); Carharrack mine, St Day (Robson 1949b); Levant mine, St Just (Roberts et al. 1990); New Roskear mine, Camborne (Robson 1949b); Old Gunnislake mine, Calstock, near Gunnislake (Ball et al. 1982); South Crofty mine, Camborne (LeBoutillier 1996); Wheal Buller, near Redruth (Greg & Lettsom 1858, Collins 1892, James 1947, Robson 1949b); Wheal Edward, near Botallack, St Just (Rudler 1905).

zircon

tetragonal — $ZrSiO_4$

A very common zirconium silicate found in accessory amounts in a wide range of igneous and metamorphic rocks. Zircon also occurs as a detrital mineral in sedimentary rocks and alluvial deposits.

Heavy-mineral assemblages, collected from the Camel, Fowey and Gannel estuaries in Cornwall, contain zircon in cassiterite-enriched intertidal sediments contaminated by detritus from the many nearby tin and copper mines (Pirrie et al. 2000b, 2002). The Millstone Grit in Yorkshire (particularly micaceous and arenaceous shales, fine-grain grits, sandstones and flags) contains minor detrital zircon forming water-clear euhedral (but rounded) and zoned crystals. Anatase, barite, brookite, garnet, ilmenite, magnetite, monazite, pyrite, rutile, topaz, tourmaline and xenotime are associated heavy minerals in these rocks. Localities include: Bowland Knotts, near Slaidburn; Bramley Fall quarry, Newlay; Clayton Wood and Hawkesworth Wood quarries, Horsforth; Earl's Crag and Netherwood Plantation quarry, Silsden; East Keswick, Bardsey; Fountains Fell; Grassington Moor; Great Almescliffe Crags, near Weeton, Wharfedale; Great Whernside, Kettlewell; Grimbald's Crag and Plumpton Rocks, Knaresborough; Ingleborough; Keighley Moor; Old Cote Moor, Arncliffe; Penyghent; Rumbles Moor, above Ilkley; Snosty quarry, Bardsey; Steeton Moor; Tannery borehole, Meanwood, Leeds; the Strid, Bolton Woods, Wharfedale; and Whitehall quarry, Butcher Hill, Horsforth (all Gilligan 1919). Detrital zircon is also found in Cretaceous and Eocene sands at: Faringdon, Berkshire; Holden's Wood, Brentwood, Essex; Hayes Common and Plumstead, Kent (now Greater London); Kennel Wood, Shirley, Surrey (now Greater London); Northwood, Middlesex (now Greater London); Shotover Hill, Oxfordshire (all Davies 1915); and in Northumbrian beach sands at Alnmouth, at Budle Bay near Bamburgh, and at Hauxley Haven, near Amble (Gallagher 1974). Abundant detrital zircon is found in Carboniferous sandstones E of the River Taff, in Mid Glamorgan (Heard 1922), and in sandy boulder clay at Ludchurch, Dyfed (Griffiths & Stuart 1940). Zircon is a constituent of alluvial panned concentrates from St Kilda (Hirta), Western Isles, Scotland, where it is accompanied by anatase, andalusite, clear-blue corundum, chevkinite, diopside, fergusonite, garnet, goyazite, kyanite, sillimanite, spinel, titanite, tourmaline and xenotime (Harding et al. 1984). A simpler detrital assemblage containing pyrope and zircon is noted in beach sand from Elie Ness, Fife (Livingstone 2002). Boswell (1927) describes a purple variety of detrital zircon in sedimentary rocks from many widely dispersed locations over Britain and Ireland.

Other (mainly igneous and metamorphic) occurrences of accessory zircon include the following.

England

Cornwall (and Devon) Zircon is a common accessory species within the Hercynian granite batholith: Bosworgey, near Fraddam – zircon occurs in a concealed northern extension of the Tregonning–Godolphin granite (Ball & Basham 1984); Carn Brea granite and associated veins, near Redruth – zircon is associated with apatite, fluorite and pyrite (Robson 1945a); Carnmenellis granite, to the SE of Camborne – accessory mineral assemblage includes zircon together with apatite, monazite-(Ce), xenotime-(Y) and thorium-poor uraninite (Jefferies 1984, 1985a, Charoy 1986): Dartmoor granite, Devon – zircon forms minute euhedral inclusions mostly within biotite and is one of 24 accessory species reported from this intrusion (Brammall & Harwood 1925, Ward et al. 1992); Goonvean and Rostowrack china-clay pits, W of Nanpean, St Austell – zircon is associated with apatite, cassiterite, ferrocolumbite, ilmenite, ilmenorutile, monazite-(Ce) and topaz in the waste material remaining after kaolinite extraction (Scott et al. 1998); Gunheath china-clay pit (Hensbarrow section), Carthew, St Austell – zircon forms inclusions in topaz within greisen (Williamson et al. 1997); along the western margin of the Land's End granite – zircon is reported with fluorapatite, monazite-(Ce), monazite-(Nd) and thorite (Müller et al. 2006); St Michael's Mount, Marazion – zircon occurs with accessory apatite, cassiterite, fluorite, magnetite and topaz in porphyritic granite (Hosking 1954c). Additionally, euhedral, short pale-brown prismatic grains of zircon are recorded as an accessory constituent of plagiogranite belonging to the Lizard ophiolite, as found at Porthkerris (Clark et al. 1998). The Tertiary granite on Lundy island, off the north Devon coast also contains accessory zircon (Dollar 1941).

Derbyshire Within xenoliths in the Great Rocks Dale dyke, near Buxton (Cope 1933).

Leicestershire Zircon is abundant and extremely varied in gabbro, diorite and granodiorite from the Mountsorrel district (Taylor 1934).

Wales

Dyfed Pristine crystals in rhyolite from the Fishguard Volcanic Group (Bevins 1982); in Ordovician silicic volcanic rocks at St David's Head (Bevins et al. 1991); in porphyritic rhyolite on Ramsey Island (Kokelaar et al. 1985).

Gwynedd In many of the silicic volcanic rocks found in Snowdonia and including rhyolite, where zircon occurs as a primary species with allanite, apatite and titanite (Bromley 1969, Howells et al. 1991); in Lower Palaeozoic microtonalitic and granitoid rocks across the Lleyn Peninsula (Croudace 1982).

Scotland

Zircon is ubiquitous in the Caledonian granites of Scotland, being found in heavy-mineral concentrates collected from many intrusions (Mackie 1928). Investigations into the radioelement potential of some of these granites show that they contain accessory amounts of uranium-rich zircon associated with monazite, xenotime or, less commonly, uraninite: Mount Battock, Bennachie and Cairngorm granites, all Grampian (Webb & Brown 1984b); Meall Odhar, south Cruachan, Central Starav and Outer Starav granites, Etive, Strathclyde, and all units of the Cairngorm granite complex, Grampian (Barritt 1983). A fission-track study of the Loch Doon granite complex, part in Strathclyde and part in Dumfries & Galloway, graphically illustrates the variable uranium contents of allanite, apatite, monazite, thorite, titanite and zircon (Tindle 1982). A fission-track technique was also used to examine small (<100 µm) zircon crystals associated with apatite, baddeleyite and zirconolite in ultrabasic cumulate rocks from the Rum layered intrusion (Williams 1978) and zircon associated with allanite, apatite and titanite from the Glen Dessary syenite, W of Loch Arkaig (Fowler 1981, 1992), both Highland Region. BSE imaging is another revealing technique that has been applied to the study of zircon from the Cluanie intrusion, Glen Shiel, Highland Region. It enabled Paterson et al. (1989) to show that these zircons have complex oscillatory zonation patterns and multiple phases of growth and overgrowth. BSE imaging combined with uranium–lead isotopic dating of euhedral zircon (~200–300 µm in length) from the Glen Sanda granite, Strontian, Highland Region, confirms a complex crystallization history and indicates that refractory zircon cores were incorporated into the magma, perhaps during partial melting. These were later overgrown by melt-precipitated zircon (Paterson et al. 1992). Another study of this nature is centred on the Ballachulish and Glen Coe igneous complexes, Highland Region, where uranium–lead dating of zircon has been used to establish precise ages of both complexes. The study also examined the metamorphic behaviour of uranium-bearing accessory baddeleyite, monazite and zircon in the aureole of the Ballachulish complex (Fraser et al. 2004).

The REE content of accessory zircon in certain Tertiary epigranites (Beinn Dearg Mhòr, Glamaig and Marsco) on the Isle of Skye is markedly different from detrital zircon found in nearby Torridonian arkose and the Coire Uaigneich granophyre (which have elevated yttrium, REE and thorium contents). Because of this, it is suggested that the granophyre may have formed during partial melting of Torridonian sediments (Exley 1980). Miarolitic cavities at several localities in the granite of the Red Cuillin on Skye contain well crystallized zircon and other accessory species – fluorite, gadolinite, kainosite-(Y) and scheelite (Wolfe 2000, Green & Todd 2001). Similar miarolitic cavities in the Tertiary granite of Northern Arran contain zircon with allanite-(Ce), fergusonite-(Y), gadolinite-(Y) and smoky quartz (Hyslop et al. 1999, Green & Todd 2001).

Zircons from Lewisian-age tonalitic granulite and trondhjemitic gneiss exposed in the Highland Region of Scotland have been shown (using cathodoluminescence imaging) to contain relict cores from a Late Archaean protolith and complex oscillatory zoned rims formed during magmatic growth and metamorphic recrystallization. A uranium–lead isotopic technique, using an ion microprobe capable of dating micron-size regions of individual zircon crystals, has allowed a complex chronology to be established to explain the origin of these rocks. Samples are from Badcall Bay, Gruinard Bay and Lochinver (Love et al. 2004). Other descriptions of accessory zircon from Scotland include the following.

Dumfries & Galloway Zircon occurs: with allanite-(Ce), apatite, ilmenite, monazite-(Ce), titanite and xenotime-(Y) in the Loch Doon granite complex, part of which

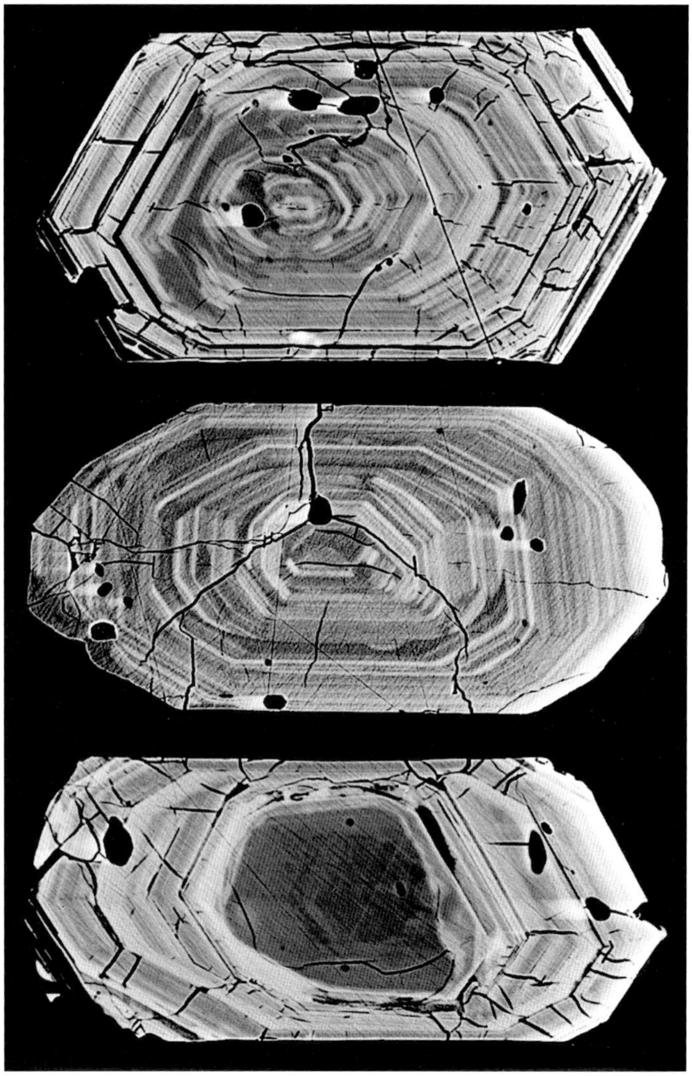

Oscillatory zoned zircon from the Strontian granite, Highland Region, Scotland; the lowest grain has a relict core. Width 900 µm; BSE image. (© B. A. Paterson)

is in Strathclyde (Tindle 1982); with allanite-(Ce), apatite, titanite, thorite and thorium-rich uraninite in the zoned diorite–granodiorite–granite complex at Cairnsmore of Carsphairn (Tindle et al. 1988).

Fife Zircon occurs as megacrysts up to 5 mm long in Carboniferous volcanics within the volcanic vent at Elie Ness (Macintyre et al. 1981).

Grampian Zircon occurs with anatase, cassiterite, chrysoberyl, columbite, fergusonite, ilmenite, ilmenorutile, monazite, topaz and xenotime in stream-sediment panned concentrates collected from Luibeg Burn, near Ben Macdui, and derived from the Cairngorm granite complex (Beer & Bennett 1988). Oscillatory zoned zircon in Dalradian metasedimentary rocks from 2 km NE of Stonehaven have been used to examine the timing of metamorphic fluid infiltration and showed that some crystals were of Archean age (Breeding et al. 2004).

Highland Zircon occurs: with allanite (orthite), apatite, ilmenite, pyrite and titanite in the Cnoc nan Cuilean intrusion, Ben Loyal syenite complex (King 1942, Gallagher et al. 1971); as metamorphically grown clear or pale-hyacynth structureless rounded crystals in Lewisian granulites from the Upper Badcall Bay and Kylestrome area and in ultramafic gneisses at First Inlet, Scourie (Pidgeon & Bowes 1972, Kinny & Friend 1994); as corroded grains with anhedral zoning in monzodiorite, elongate unzoned crystals in quartz diorite, stout crystals with strong euhedral zoning in granite belonging to the Ballachulish igneous complex (Weiss & Troll 1989); as trace amounts of zircon, baddeleyite, geikielite, qandilite and rutile in metamorphosed siliceous dolomites within the aureole of the Ballachulish granite complex (Ferry 1996); as an accessory species with apatite, magnetite and titanite in the Ben Nevis granite complex (Haslam 1968); unusual, pale-pink sector-zoned zircon up to 700 µm in length, from the granodioritic Tollie Gneiss of Lewisian age exposed between Gairloch and Poolewe (Hoffman & Long 1984); as clear colourless crystals in pyroxene–mica diorite from the Ratagain complex, Loch Duich, near Glenelg, and slightly cracked needles in the Strontian granodiorite (Rogers & Dunning 1991); as subhedral prismatic grains intergrown with pargasite and plagioclase in eclogite from the inlier of presumed Lewisian age at Letterfairn, Glenelg (Brewer et al. 2003); as perfectly transparent ruby-red crystals in granular quartz schist from E of An Gorm-loch, SE of Ben Hope (Heddle 1901); in relative abundance in quartzofeldspathic pegmatite veins within gabbros of the Outer Layered Series of the Cuillin intrusion, Isle of Skye (Hamilton et al. 1998).

Oxygen-isotope analysis of zircons from the Tertiary Igneous Province, including 14 granite bodies on the Isle of Skye, shows that wallrock melting by adjacent mafic plutons is an important process in the formation of these granites (Monani & Valley 2001).

Strathclyde Clear colourless euhedral prisms and pyramids of zircon averaging 0.3 mm in length occur within the Ailsa Craig microgranite, Firth of Clyde (Harrison et al. 1987). Apatite and zircon are common accessory species in the Etive granite complex to the E and NE of Ben Starav (Haslam et al. 1985). Clear needles and euhedral prisms of zircon occur in coarse appinitic diorite from the Garabal Hill–Glen Fyne granite complex, W of the head of Loch Lomond (Rogers & Dunning 1991).

Euhedral zircon crystals, up to 15 µm, in a Cuillin epigranite, Eas a' Bhradain, W of Loch Ainort, Isle of Skye, Highland Region, Scotland; SEM image. (© D. I. Green)

Western Isles In the Lewisian granitic pegmatites at Chaipaval (Ceapabhal), 3 km NW of Northton (Taobh Tuath) and Loch a' Sgùrr, near Finsbay (Fionnsabhagh), both South Harris (von Knorring & Dearnley 1959b, 1960a); associated with accessory allanite, apatite, chevkinite, epidote, titanite and zirconolite in a Tertiary acidic pegmatite cutting gabbro at Glen Bay, St Kilda (Hirta) (Harding et al. 1982, 1984); and as delicate pink transparent crystals in granite from Brann a' Bharra, N of Balallan, Isle of Lewis (Heddle 1901).

Ireland

In Ireland, zircon has often been reported from Caledonian granite complexes. The Costelloe Murvey leucogranite, Connemara, Co. Galway, is one example where accessory amounts of tiny yttrium-rich zircon crystals (< 15 µm) are found with monazite, thorite, uraninite and unidentified Nb–Th–Ti–U phases (Feely & Madden 1988, Feely et al. 1989). Other intrusions within the Galway granite complex contain accessory allanite, apatite, titanite and zircon (Coats & Wilson 1971). Extraordinarily elongate zircon is reported from the Leinster granite, where singly or doubly terminated zoned crystals reach lengths of 1500 µm, but are only 12–50 µm in diameter (Gupta 1973). Zircon was found in every sample examined from the northern portion of the Leinster granite in Co. Wicklow, and heavy-mineral concentrates show it to be relatively abundant at Aghavannagh (where it develops imperfect zonation) and Glenmacnass (Smithson 1932). A lithium-enriched cassiterite pegmatite at Aclare, near Myshall, Co. Carlow, contains minor zircon together with manganese-rich apatite, beryl, columbite–tantalite and spodumene (Scoon 1978). Zircon is also described from biotite–cordierite hornfels in the aureole of the Cashel gabbro, N and NW of Cashel Hill, Connemara. Rounded to sub-rounded brown crystals up to 90 µm in length are thought to be detrital, whereas pink crystals (≥ 60 µm) are more euhedral, have well preserved terminations, and are considered to have a metamorphic origin (Ahmed-Said & Leake 1990).

Offshore territories

On Jersey, macroscopic crystals of zircon have been found in granite pegmatites at L'Etacq, La Moye, Les Handois, Mont Mado and Mont Orgueil (Mourant 1961). Zircon forms zoned prismatic crystals in the granite at L'Ouaisné Bay, southwest Jersey, and is described as abundant in granite from the southeast quarry at L'Etacq (Ixer 1980). Minor zircon is also found in the contaminated granite of Bibette Head, Alderney (Nockolds 1932).

zirconolite

orthorhombic, monoclinic, trigonal or metamict $CaZrTi_2O_7$

A rare calcium–zirconium–titanium oxide reported only from Scotland.

In the layered ultrabasic intrusion on the Isle of Rum, Highland Region, it forms small (< 60 µm) interstitial grains in various cumulate rocks. In thin section, zirconolite is dark reddish brown and is found associated (or intergrown) with apatite, baddeleyite, chrome spinel, ilmenite and zircon (Williams 1978).

In the Glen Bay gabbro on St Kilda (Hirta), Western Isles, zirconolite (zirkelite) occurs within an elongate pegmatite pod consisting of allanite, apatite, augite, biotite, epidote, ferro-edenite, magnetite, ilmenite, quartz, orthoclase, titanite and zircon. The pegmatite probably formed from the last residues of a basaltic liquid that concentrated niobium, yttrium, zirconium and REEs as it crystallized. Zirconolite forms anhedral or acicular crystals up to 1 mm long (Harding et al. 1982, 1984).

The Glen Dessary syenite, W of Loch Arkaig, Highland Region, provides a further occurrence of zirconolite. As the syenitic magma differentiated during crystallization, it developed an accessory mineral assemblage of allanite, apatite, titanite, zircon and zirconolite. Zirconolite forms small (< 10 µm) euhedral inclusions in alkali-feldspar phenocrysts and is deep orange-brown in thin section (Fowler & Williams 1986, Fowler 1992).

A fourth occurrence of zirconolite is in the Cnoc Rhaonastil dolerite intrusion on Islay, Strathclyde, which represents a single body of alkali olivine basalt magma that differentiated *in situ*. Late-stage alkali differentiates of this magma produced syenites with a minor phase assemblage of aenigmatite, zirconium-bearing aegirine, calcium catapleiite, fluorapatite, monazite and zirconolite (Preston et al. 1998a, 2000a, 2000b).

zirkelite

The cubic mineral $(Ti,Ca,Zr)O_{2-x}$, as classified by Bayliss et al. (1989). Prior to this classification, zirkelite was synonymous with zirconolite.

zoisite

EPIDOTE GROUP

orthorhombic $Ca_2Al_3Si_3O_{12}(OH)$

A common calcium aluminosilicate hydroxide, dimorphous with clinozoisite and typically found in low- to medium-grade regionally metamorphosed rocks. It is also found in eclogite and blueschist-facies metamorphic rocks.

At Dinas Head, adjoining Trevose Head, 8 km W of Padstow, Cornwall, accessory amounts of dravite and zoisite occur in quartz–albite rocks (termed adinoles), that were formed when an intrusion of dolerite metasomatically altered slate (Agrell 1939, 1941). Zoisite (and epidote) also occur at Tolpetherwin quarry, Two Bridges, South Petherwin, Cornwall, where greenstone (dolerite) is cut by bindheimite–jamesonite-bearing quartz veins (Russell 1949b).

Reviewing the Mountsorrel granodiorite and associated igneous rocks of Leicestershire, Taylor (1934) notes zoisite as a secondary accessory species in gabbro at Swithland reservoir; and four locations in granodiorite at Kinchley, Cocklow quarry, South of Cocklow Wood and in the Main Mountsorrel quarry. It is always less abundant than epidote.

Thin tabular imbricated crystals of zoisite occur with axinite in quartz veins cutting the Penmaenmawr intrusion, near Conwy, Wales (Sargent 1925). On the northern shores of the Menai Straits, near Menai Bridge, Anglesey, and at a locality N of Hendrefor, Gwynedd, lawsonite-bearing blueschists are noted to contain accessory amounts of apatite, epidote, hematite, magnetite, white mica, titanite (sphene) and zoisite (Gibbons & Mann 1983). Zoisite is further described from the gabbroic margins of dolerite intrusions at Bwlch y Ddeufaen, Conwy Valley, Snowdonia, Gwynedd, where it occurs in vesicles with calcite, chlorite, clinozoisite and prehnite (Jenkins & Ball 1964).

In Scotland, zoisite occurs in considerable amounts (modally 16%) within small pegmatitic feldspar-rich veins at Loch Vallarip (Balla-riop), Rodelpark area (Pairc Roghadail), South Harris, Western Isles. Allanite, epidote, muscovite, quartz and scapolite are associated species (von Knorring & Dearnley 1960a). In Highland Region, zoisite forms: yellowish-white crystals in garnetiferous schist on the western slopes of Meallan Rairigidh, Fannich; lustrous crystals to 50 mm long on the east side of Carn na Criche, 6 km N of Loch Fannich; lustrous brown crystals N of Raven Rock, 3 km NW of Strathpeffer (Heddle 1901) and brown and white varieties at Loch Garve, near Dingwall (Hamilton Bell 1884). Zoisite has also been found in limestone quarries in Highland Region: whitish zoisite embedded in calcite, associated with actinolite, quartz and rare scapolite at Glen Urquhart, near Gartally; green crystals associated with grossular (cinnamonstone) at Achnagonalin, Grantown; pale-brown lustrous crystals in quartz associated with biotite, chlorite and diopside (sahlite) at Laggan, Dulnain Bridge (all Heddle 1901). Zoisite has also been described from calc-silicate rocks around Loch Eilt, near the Sound of Arisaig, Highland Region, as part of a study of regional metamorphism and deformation of Precambrian Moine rocks (Tanner & Evans 2003).

Interlocking blades of zoisite from Easter Laggan, Dulnain Bridge, near Grantown-on-Spey, on the Grampian/Highland boundary, Scotland; width of specimen 12 cm. (© The Trustees of the National Museums of Scotland)

A small Moinian amphibolitized metabasite intrusion, 4 km S of Tomatin, Grampian, has inclusions of zoisite in garnet. The rock also contains diopside–plagioclase symplectites, suggesting that it may once have been an eclogite (Baker 1986). Metamorphosed limestones found as xenoliths in the Dorback granite, Braes of Abernethy, 9 km W of Tomintoul, contain a wide range of metamorphic minerals, including clinozoisite, diopside, epidote, grossular, brown mica, vesuvianite (idocrase), wollastonite and zoisite (Walker & Davidson 1935).

Zoisite is intimately associated with garnet in altered gabbros adjacent to serpentinites belonging to the Girvan–Ballantrae ophiolite complex, on the southeast slopes of Byne Hill, Strathclyde. Epidote, diopside, pectolite and prehnite are thought to have formed at the same time as zoisite during a late hydrothermal event. Diopside–zoisite veins also cut the serpentinite (Bloxam 1954). Heddle (1901) notes zoisite at two other locations in Strathclyde: Sgòrr na h-Ulaidh, Glen Creran, 7 km S of Glencoe, and in the banks of the stream in Coire Muide, 1 km SW of Clachaig, S of Glencoe.

Zoisite is a minor species that occurs with vesuvianite (idocrase) in calc-silicate rocks associated with feldspathic dykes cutting skarn rocks adjacent to an alkali granite at Lough Anure, Thorr area, Co. Donegal, Ireland (Pitcher 1950). Another calc-silicate rock, in the northern aureole of the Ardara granite complex, near Maas, Co. Donegal, contains zoisite together with calcite, diopside, epidote, grossular, potash feldspar, quartz and vesuvianite in bands, up to 3 m thick, interbedded with sillimanite hornfels (Akaad 1956). Allagour quarry and Ballinascorney House reservoir, both Tallaght Hills, Co. Dublin, are locations where zoisite is reported from chert-like adinoles (metasomatically altered pelitic metasediments). The mineral assemblage is representative of the chlorite-schist facies and in addition to zoisite it contains albitic plagioclase, chlorite, leucoxene dust, quartz and sericite (Angus & Kanaris-Sotiriou 1995). A sulphide-rich zoisite–amphibolite is noted from Glendalough, Co. Wicklow (Thomson 1908), but it appears to be clinozoisite rather than zoisite that is being described. The zoisite said to occur at Hollyhill, near Strabane, Co. Tyrone (Greg & Lettsom 1858), is also thought to be clinozoisite.

Accessory zoisite with a minutely granular habit, but occasionally in larger grains is found in the contaminated granite of Bibette Head, Alderney, Channel Isles (Nockolds 1932).

zunyite?

cubic $Al_{13}Si_5O_{20}(OH,F)_{18}Cl$

An hydrated aluminosilicate fluorhydroxide chloride. Supposedly found with alunite, dickite, gypsum and thaumasite at Embleton quarry, near Cockermouth, Cumbria, by Kingsbury in 1956. The assemblage is reported in a quartz–carbonate vein cutting a quartz-dolerite sill (Embrey, footnote in Knill 1960). This is a very dubious British occurrence and requires re-evaluation (Ryback et al. 1998).

zwieselite

monoclinic $(Fe^{2+},Mn^{2+})_2(PO_4)F$

A rare iron–manganese fluorphosphate that forms a series with triplite and is found as dark-brown masses in pegmatite at Megiliggar Rocks, Tremearne, near Breage, Cornwall (Ryback & Tandy 1992).

APPENDIX 1
TABLES

Table A1 British and Irish species named after British or Irish citizens, and those with British or Irish affiliation.

Species	Name	Type locality
aikinite	Arthur Aikin (1773–1854)	Russia
allanite-(Ce)	Thomas Allan (1777–1833)	Greenland
apjohnite	James Apjohn (1796–1886)	Mozambique
arthurite	Jointly Sir Arthur Edward Ian Montagu Russell (1878–1964) and Arthur William Gerald Kingsbury (1906–1968)	Hingston Downs Consols, Calstock, Cornwall, England
babingtonite	William Babington (1756–1833)	Norway
baddeleyite	Joseph Baddeley	Sri Lanka
bannisterite	Frederick Allen Bannister (1901–1970)	Benallt mine, Lleyn Peninsula, Gwynedd, Wales
barstowite	Richard William Barstow (1947–1982)	Bounds Cliff, St Endellion, Cornwall, England
bayldonite	John Bayldon	Penberthy Croft, St Hilary, Cornwall, England
bobkingite	Robert King (b. 1923)	New Cliffe Hill quarry, Stanton under Bardon, Leicestershire, England
braggite	Sir William Henry Bragg (1862–1942) and his son William Lawrence Bragg (1890–1971)	South Africa
brammallite (series)	Alfred Brammall (1879–1954)	Llandybie, Dyfed, Wales
brewsterite-Sr[1]	Sir David Brewster (1781–1868)	Strontian, Highland Region, Scotland
brianyoungite	Brian Young (b. 1947)	Brownley Hill mine, Nenthead, nr Alston, Cumbria, England
brinrobertsite	Brinley Roberts (b. 1932)	750 m NE of Penrhyn Castle, nr Bangor, Gwynedd, Wales
brookite[1]	Henry James Brooke (1771–1857)	Twll maen grisal, Fron-oleu, Prenteg, nr Tremadog, Gwynedd, Wales
brucite	Archibald Bruce (1777–1818)	USA
chenevixite	Richard Chenevix (1774–1830)	Wheal Gorland, St Day, 2 km E of Redruth, Cornwall, England
childrenite	John George Children (1777–1852)	Near Tavistock, Devon, England
chrisstanleyite	Christopher John Stanley (b. 1954)	Hope's Nose, Torquay, Devon, England
churchite-(Y)	Arthur Herbert Church (1834–1915)	Tretoil mine, Lanivet, 8 km NW of Lostwithiel, Cornwall, England
connellite	Arthur Connell (1794–1863)	Wheal Providence, St Day, Gwennap, Cornwall, England
dickite	Allan Brugh Dick (1833–1926)	Trwynbychan, nr Amlwch, Anglesey, Gwynedd, Wales
edingtonite	Thomas Edington (1783–1859)	quarry 1 km NE of Old Kilpatrick, nr Glasgow, Strathclyde, Scotland
fergusonite-(Y)	Robert Ferguson (1767–1840)	Greenland
forsterite	Adolarius Jacob Forster (1739–1806)	Italy
geikielite	Sir Archibald Geikie (1835–1924)	Sri Lanka
greenockite	Charles Murray Cathcart (1783–1859) – Lord Greenock	Bishopton railway tunnel, 7 km WNW of Renfrew, Strathclyde, Scotland
harkerite	Alfred Harker (1859–1939)	Camas Malag, nr Broadford, Isle of Skye, Highland Region, Scotland
heulandite-Ca[1]	John Henry Heuland (1778–1856)	Unknown TL; Heuland was of German origin, but spent much of his life in Britain
hollandite	Sir Thomas Henry Holland (1868–1947)	India
hollingworthite	Sydney Ewart Hollingworth (1899–1966)	South Africa
humite[1]	Sir Abraham Hume (1749–1838)	Italy
hydrowoodwardite	Samuel Pickworth Woodward (1821–1865)	Germany (but identical to incompletely described material from Simdde Dyllhuan, Drwys-y-Coed, nr Nantlle, Gwynedd, Wales)
jamesonite	Robert Jameson (1774–1854)	St Endellion, Cornwall, England
johnsomervilleite	John M. Somerville (1908–1978)	Gleann Còsaidh, Loch Quoich, Knoydart, Highland Region, Scotland
lawsonite	Andrew Cowper Lawson (1861–1952)	USA
ludlamite	Henry Ludlam (1824–1880)	Wheal Jane, Kea, nr Truro, Cornwall, England
macphersonite	Harry Gordon Macpherson (1925-2001)	Leadhills Dod, Leadhills, Strathclyde, Scotland
mattheddleite	Matthew Forster Heddle (1828–1897)	Leadhills Dod, Leadhills, Strathclyde, Scotland
mawbyite	Sir Maurice Alan Edgar Mawby (1904–1977)	Australia
mawsonite	Sir Douglas Mawson (1882–1958)	Australia
millerite	William Hallowes Miller (1801–1880)	Germany
parkinsonite	Frank Reginald Derek Parkinson (1928–1993)	Merehead quarry (Torr Works), Cranmore, Mendip Hills, Somerset, England
pennantite	Thomas Pennant (1726–1798)	Benallt mine, Lleyn Peninsula, Gwynedd, Wales
pentlandite	Joseph Barclay Pentland (1797–1873)	Craignure mine, nr Inveraray, Strathclyde, Scotland
phillipsite-Ca[1]	William Phillips (1775–1828)	USA
roscoelite	Henry Enfield Roscoe (1833–1915)	USA
russellite	Sir Arthur Edward Ian Montagu Russell (1878–1964)	Castle-an-Dinas mine, St Columb Major, Cornwall, England
rutherfordine	Ernest Rutherford, 1st Baron Rutherford of Nelson (1871–1937)	Tanzania
smithite	George Frederick Herbert Smith (1872–1953)	Switzerland
smithsonite	James Smithson (1765–1829)	Derbyshire and the Mendip Hills, Somerset, England
stokesite	Sir George Gabriel Stokes (1819–1903)	Roscommon Cliff, St Just, Cornwall, England
sweetite	Jessie May Sweet (1901–1979)	Milltown quarry, nr Ashover, Derbyshire, England
symesite	Robert Symes (b. 1937)	Merehead quarry (Torr Works), Cranmore, Mendip Hills, Somerset, England
tennantite (also argentotennantite)	Smithson Tennant (1761–1815)	Cornwall, England
thomsonite-Ca[1]	Thomas Thomson (1773–1852)	Old Kilpatrick, nr Dumbarton, Strathclyde, Scotland
tilleyite	Cecil Edgar Tilley (1894–1973)	USA
tristramite	Sir Tristram of Arthurian legend	Wheal Trewavas, Trewavas Head, Breage, Cornwall, England

Table A1 continued

vivianite	John Henry Vivian (1785–1855)	Wheal Kine, St Agnes, Cornwall, England?
wavellite	William Wavell (1750–1829)	High Down quarry, Heddon, 2 km WSW of Filleigh, nr South Molton, Devon, England
waylandite	Edward James Wayland (1888–1966)	Uganda
whewellite	William Whewell (1794–1866)	USA
winchite[1]	Howard James Winch (1877–1964)	India
witherite	William Withering (1741–1799)	Alston Moor, Cumbria, England
wollastonite	William Hyde Wollaston (1766–1828)	Romania
woodwardite	Samuel Pickworth Woodward (1821–1865)	Cornwall, England
wooldridgeite	James Wooldridge (1923–1995)	Judkins quarry, Nuneaton, Warwickshire, England
Kingsbury claim: spencerite	Leonard James Spencer (1870–1959)	Canada

1. These people have also given their names to other species (or named species) found in Britain or Ireland: alumino-ferrowinchite, aluminowinchite*, clinohumite, ferriwinchite, ferri-ferrowinchite*, ferrowinchite, heulandite-Ba, heulandite-K, heulandite-Sr, manganhumite, pseudobrookite, phillipsite-K, phillipsite-Na. Additional non-British or Irish species include: heulandite-Na, parvowinchite and thomsonite-Sr. * Indicates a named species; see amphibole group on p. 33 for further details. Species in this footnote do not have British or Irish type localities.

Table A2 British and Irish species named after British or Irish locations.

Species	Type locality
alstonite	Alston Moor, Cumbria, England (TL Brownley Hill mine, Alston; also Fallowfield, Hexham, Northumberland)
anglesite	Anglesey, Gwynedd, Wales (TL Parys mine)
ashoverite	Ashover, Derbyshire, England
bassetite	Basset mine, Redruth, Cornwall, England
birnessite	Birness, Aberdeenshire, Grampian Region, Scotland
botallackite	Botallack mine, St Just, Cornwall, England
caledonite	Caledonia - historical name of Scotland (TL Leadhills, Strathclyde, Scotland)
corkite	County Cork, Ireland (TL Glandore iron mine)
cornubite	From Cornubia, the medieval Latin name for Cornwall (TL Wheal Carpenter, nr Hayle, Gwinear, Cornwall, England)
cornwallite	County of Cornwall, England
cymrite	Cymru, the Welsh name for Wales (TL Benallt mine, Lleyn Peninsula, Gwynedd, Wales)
epsomite	Epsom, Surrey, England
garronite	Garron Plateau, Co. Antrim, Ireland
gobbinsite	Gobbins, Island Magee, Co. Antrim, Ireland (TL Hills Port)
gortdrumite	Gortdrum mine, nr Tipperary, Co. Tipperary, Ireland
hydroscarbroite	Scarborough, Yorkshire, England (TL South Bay)
kilchoanite	Kilchoan, Ardnamurchan, Highland Region, Scotland
killalaite	Killala Bay, nr Inishcrone, Co. Sligo, Ireland
lanarkite	Lanarkshire (part of Strathclyde), Scotland (TL Susanna mine, Leadhills, Strathclyde)
larnite	Larne, Co. Antrim, Ireland (TL Scawt Hill)
leadhillite	Leadhills, Strathclyde, Scotland
liskeardite	Liskeard, Cornwall, England
lizardite	The Lizard, Cornwall, England (TL Kennack Cove)
macaulayite	Macaulay Institute for Soil Research, Craigiebuckler, Aberdeen, Grampian Region, Scotland founded by Thomas Bassett Macaulay (1860-1942) (TL Bennachie car park, 7 km WNW of Inverurue, Grampian Region, Scotland)
matlockite	Matlock, Derbyshire, England (TL Cromford)
mendipite	Mendip Hills, Avon, England (TL Churchill)
mereheadite	Merehead quarry, Cranmore, Mendip Hills, Somerset, England
mottramite	Mottram St Andrew, Cheshire, England
mullite	Isle of Mull, Strathclyde, Scotland (TL Seabank Villa)
namuwite	National Museum of Wales, Cardiff, where it was discovered on a specimen originally labelled "hydrozincite (coloured by malachite) . . ." (TL Aberllyn lead mine, 2 km N of Betws-y-coed, Gwydyr Forest, Gwynedd, Wales)
redgillite	Silver Gill, Caldbeck Fells, Cumbria, England (nr Red Gill where it was first found)
scarbroite	Scarborough, Yorkshire, England (TL South Bay)
scawtite	Scawt Hill, Larne, Co. Antrim, Ireland
scotlandite	Scotland (TL Susanna vein, Leadhills, Strathclyde, Scotland)
strontianite	Strontian, Highland Region, Scotland
susannite	Susanna mine, Leadhills, Strathclyde, Scotland
tobermorite	Tobermory, Isle of Mull, Highland Region, Scotland

Two other species have names derived from British locations: clinotobermorite and paralstonite. Their type localities are in Japan and the USA respectively. Paralstonite is reported from Dolyhir quarry, Old Radnor, near Presteigne, Powys, Wales.

Table A3 Species with British or Irish type localities, but without citizen or locality names.

Species	Type locality
banalsite	Benallt mine, Lleyn Peninsula, Gwynedd, Wales
barytocalcite	Blagill mine, Alston Moor, Cumbria, England
bazirite	Rockall island, Western Isles, Scotland
bredigite	Scawt Hill, Larne, Co. Antrim, Ireland
bournonite	Wheal Boys, Trewetha, St Endellion, Cornwall, England
chalcosiderite	Wheal Phoenix mine, Linkinhorne, nr Liskeard, Cornwall, England
chenite	Susanna mine, Leadhills, Strathclyde, Scotland
chloroxiphite	Higher Pitts Farm, Priddy, Mendip Hills, Somerset, England
clinoclase	Wheal Gorland, St Day, 2 km E of Redruth, Cornwall, England
dellaite	Kilchoan, Ardnamurchan, Highland Region, Scotland
devilline	Cornwall, England
diaboléite	Higher Pitts Farm, Priddy, Mendip Hills, Somerset, England
ferrobustamite	Camas Malag, nr Broadford, Isle of Skye, Highland Region, Scotland
ferrokësterite	Cligga mine, Perranzabuloe, Cornwall, England
fluellite	Stenna Gwyn mine, N of St Stephen-in-Brannel, nr St Austell, Cornwall, England
gmelinite (series)	Little Deer Park, Glenarm, Co. Antrim, Ireland - co-type with Vicenza, Italy
gyrolite	The Storr, 10 km N of Portree, Isle of Skye, Highland Region, Scotland
hydrobasaluminite	Lodge Pit, Irchester, nr Wellingborough, Northamptonshire, England
hydrocalumite	Scawt Hill, Larne, Co. Antrim, Ireland
langite	St Blazey and St Just, Cornwall, England
lanthanite-(Ce)	Britannia mine, Snowdonia, Gwynedd, Wales
liroconite	Wheal Gorland, St Day, 2 km E of Redruth, Cornwall, England
olivenite	Carharrack mine, Gwennap, Cornwall, England
osumilite-(Mg)	Tieveragh, Co. Antrim, Ireland
pharmacosiderite	Carharrack mine, Gwennap and Tincroft mine, Pool, Illogan, both Cornwall, England
plattnerite	Leadhills, Strathclyde, Scotland (probably)
portlandite	Scawt Hill, Larne, Co. Antrim, Ireland
rankinite	Scawt Hill, Larne, Co. Antrim, Ireland
rhabdophane-(Nd)	Fowey Consols mine, Tywardreath, St Blazey, Cornwall, England?
rustumite	Kilchoan, Ardnamurchan, Highland Region, Scotland
saponite	Lizard Point, Cornwall, England
stannite	Wheal Rock (West Kitty), St Agnes, Cornwall, England
tacharanite	Portree, Isle of Skye, Highland Region, Scotland
uranospathite	Redruth, Cornwall, England
vaterite	Ballycraigy, Larne, Co. Antrim, Ireland
vochtenite	Wheal Basset, Illogan, nr Redruth, Cornwall, England
Kingsbury claims:	
aerugite	TL said to be South Terras mine, St Stephen-in-Brannel, Cornwall, England; but now considered to be from Saxony, Germany (Ryback et al. 2001)
xanthiosite	

Table A4 British and Irish varieties and discredited species named after British or Irish citizens. Further details of these minerals can be found in the species section of *MBI*.

Mineral	Name
andrewsite	Thomas Andrews (1813–1885)
beekite	Henry Beeke, Dean of Bristol (1751–1837)
breadalbanite	John Campbell (1796–1862) second Marquis of Breadalbane
collieite	John Norman Collie (1859–1942)
comptonite	Spencer Joshua Alwyne Compton (1790–1851)
cotterite	Miss E. Cotter
davidsonite	James Davidson (1792–1841)
diackethyst	Revd Diack
doranite	Patrick Doran (1779–1880)
dudgeonite	Patrick Dudgeon (1817–1895)
enysite	John Samuel Enys (1797–1872)
evansite	Brooke Evans (1797–1862)
gibsonite	Alexander Gibson (1800–1867)
gilbertite	Davies Gilbert (1767–1839)
gregorite	Revd William Gregor (1761–1817)
grovesite	Arthur William Groves (1903–1956)
haughtonite	Samuel Haughton (1821–1897)
henwoodite	William Jory Henwood (1805–1875)
herschelite	Sir John Frederick William Herschel (1792–1871)
hibbertite	Samuel Hibbert-Ware (1782–1848)
hullite	Edward Hull (1829–1917)
johnstonite	Finlay Weir Johnston (1796–1855)
kirwanite	Richard Kirwan (1733–1812)
lehuntite	Captain Lehunt
lettsomite	William Garrow Lettsom (1805–1887)
livesite	Sir George Thomas Livesey (1834–1908)
loganite	Sir William Edmond Logan (1798–1875)
lyollite	Sir Charles Lyell (1797–1875)
murchisonite	Sir Roderick Impey Murchison (1792–1871)
rashleighite	Philip Rashleigh (1729–1811)
scoulerite	John Scouler (1804–1871)
tallingite	Richard Talling (1820–1883)
walkerite	Revd John Walker (1731–1803)
waringtonite	Sir Warington Wilkinson Smyth (1817–1890)
websterite	Thomas Webster (1773–1844)
withamite	Henry Thomas Maire Witham (1779–1844)
Organic	
bernalite	John Desmond Bernal (1901–1971)
cloustonite	Revd Charles Clouston (1800–1884)
hatchettite	Charles Hatchett (1765–1847)
Artificial	
heddlite	Matthew Forster Heddle (1828–1897)

Table A5 British or Irish varieties and discredited species with a British or Irish type locality, but not a citizen or locality name. Further details of these minerals can be found in the species section of *MBI*.

Mineral	Type locality
alumyte	Glenarm and Ballintoy, Co. Antrim, Ireland
astrapia	Isle of Sheppey, Kent, England
basaluminite	Irchester, Northamptonshire, England
bicalcareo-carbonate of barytes	Brownley Hill, Nenthead, nr Alston, Cumbria, England
calciopalygorskite	Strontian, Highland Region, Scotland
campylite	Dry Gill, Caldbeck Fells, Cumbria, England
chalilite	Donegore Mountains, Sandy Brae, Co. Antrim, Ireland
chlorophaeite	Isle of Rum, Highland Region, Scotland
Cl-tyretskite	Boulby mine, Cleveland, England
cotton-stone	Isle of Skye, Highland Region, Scotland
crucilite	Clonmell, Co. Waterford, Ireland
daphnite	Penzance, Cornwall, England
dolianite	Knock Station, Grampian Region, Scotland
eosite	Leadhills, Strathclyde, Scotland
erythrite (of Thomson)	Kilpatrick Hills, nr Glasgow, Strathclyde, Scotland
ferrite	Between Gleniffar and Boyleston, Strathclyde, Scotland
ferropericlase	Broadford area, Isle of Skye, Highland Region, Scotland
galactite	Glen Farg, Tayside Region, Scotland
grünlingite	Carrock Fell, Cumbria, England (probably)
harringtonite	Portrush and the Skerries, Co. Antrim, Ireland
hydro-glockerite	Parys Mountain, Anglesey, Gwynedd, Wales
hydromuscovite	Ogofau, nr Pumsaint, Dyfed, Wales
isopyre	St Just, Cornwall, England
isostannite	Cligga Head, Perranzabuloe, Cornwall, England
kilmacooite	Magpie and Connary mines, Avoca, Co. Wicklow, Ireland
koodilite	Co. Down, Ireland
mesolitine	Newtown Crumlin, Clough, Co. Antrim, Ireland
ortho-antigorite	Unst, Shetland, Scotland
parastilbite	Kilkeel, Co. Down, Ireland
plinthite	Co. Antrim, Ireland
plumbo-aragonite	Leadhills, Strathclyde, Scotland
plumbocalcite	Wanlockhead, Dumfries & Galloway, Scotland
plumbonacrite*	Wanlockhead, Dumfries & Galloway, Scotland
prasolite	Kilpatrick Hills, nr Glasgow, Strathclyde, Scotland
reichite	Alston Moor, Cumbria, England
rhodalite	Co. Antrim, Ireland
rock silk	Tayport, Fife Region, Scotland
sappare	Botriphnie, Grampian Region, Scotland
silicite	Co. Antrim, Ireland
sphaerodialogite	Gwynedd, Wales
stannite (of Garby and others)	Wheal Primrose, Cornwall, England
stellite	Kilsyth, nr the Forth of Clyde canal, Strathclyde, Scotland
sulphato-carbonate of baryta	Brownley Hill mine, Alston, Cumbria, England
talcite	Wicklow, Ireland
varvicite	Warwickshire, England
verrucite	Sandy Braes, Co. Antrim, Ireland
wollastonite (of the Edinburgh mineralogists)	Castle Hill, Edinburgh, Lothian Region, Scotland
wollastonite (of Thomson)	Kilsyth, Forth & Clyde canal, Strathclyde Region, Scotland
xantholite	Milltown, nr Loch Ness, Highland Region, Scotland
zeuxite	United mines, Cornwall, England
zinnkies	Wheal Rock, St Agnes, Cornwall, England

* Plumbonacrite is a confirmed species, but its occurrence at the type locality requires confirmation.

Table A6 British and Irish varieties and discredited species named after British locations. Further details of these minerals can be found in the species section of *MBI*.

Mineral	Type locality
abriachanite	Abriachan, Grampian Region, Scotland
agnesite	Wheal Coates, St Agnes, Cornwall, England
antrimolite	County Antrim, Ireland
balvraidite	Balvraid, Glen Beg, Highland Region, Scotland
bhreckite	"Ben Bhreck", Beinn Bhreac, Highland Region, Scotland
bowlingite	Bowling quarry, Dumbartonshire, Strathclyde, Scotland
breadalbanite	Breadalbane, nr Killin, Central Region, Scotland
bromlite, brownlite	Brownley (Bromley) Hill, Nenthead, nr Alston, Cumbria, England
bruiachite	Loch Bruiach, Highland Region, Scotland
cairngorm	Cairngorm Mountains, Grampian Region, Scotland
cardenite	Carden Wood, Grampian Region, Scotland
cathkinite	Cathkin quarry, nr Glasgow, Strathclyde, Scotland
churchillite	Churchill, Mendip Hills, Somerset, England
cliftonite	Clifton, Avon, England
cluthalite	Kilpatrick Hills, Dumbarton, Strathclyde, Scotland (Clutha is an old name for the Clyde Valley, Scotland)
condurrite	Condurrow mine, Camborne, Cornwall, England
craigtonite	Craigton, Hill of Fare, Grampian Region, Scotland
cromfordite	Cromford, Matlock, Derbyshire, England
derbyshire spar	Derbyshire, England
devonite	Devon, England
duporthite	Duporth, St Austell, Cornwall, England
ellonite	Ellon, Grampian Region, Scotland
erinite	Named after Ireland (Erin) (TL Giant's Causeway, Co. Antrim, Ireland)
fargite	Glen Farg, Tayside Region, Scotland
francolite	Wheal Franco, Walkhampton, nr Tavistock, Devon, England
glottalite	Glotta – a name for the Clyde, Scotland
gorlandite	Wheal Gorland, St Day, 2 km E of Redruth, Cornwall, England
haytorite	Hay Tor, Devon, England
holmbushite	Holmbush mine, Callington, Cornwall, England
hovite	Hove, Brighton, Sussex, England
inverarite	Inverary, Strathclyde, Scotland
jarrowite	Jarrow, River Tyne, Durham, Co. Durham, England
kilbrickenite	Monanoe (Kilbricken) mine, Quin, Co. Clare, Ireland
killinite	Killiney Bay, Co. Dublin, Ireland
ledikite	Ledikin, Grampian Region, Scotland
leedsite	Leeds, Yorkshire, England
manaccanite	Manaccan, Cornwall, England (TL Tregonwell Mill)
matlockite (of Chapman)	Matlock, Derbyshire, England (TL Cromford level)
mornite	Mourne Mountains, Co. Down, Ireland
morvernite	Strontian, Highland Region, Scotland
penwithite	Penwith, St Just, Cornwall, England (TL Wheal Owles)
polyphant stone	Polyphant, SW of Launceston, Cornwall, England
ratholite	Ratho quarries, Edinburgh, Lothian Region, Scotland
redruthite	Redruth, Cornwall, England
restormelite	Restormel mine, nr Lostwithiel, Cornwall, England
rosstrevorite	Rosstrevor, Co. Down, Ireland
rubislite	Rubislaw, Aberdeen, Grampian Region, Scotland
scyelite	Loch Scye, Caithness, Highland Region, Scotland
skiagite	Glen Skiag, Highland Region, Scotland
stromnite	Stromness, Orkney Isles, Scotland
tamarite	Wheal Tamar, Cornwall, England
tavistockite	Tavistock, Devon, England
totaigite	Totaig, Highland Region, Scotland
towanite	Wheal Towan, nr St Agnes, Cornwall, England
tuesite	From the Latin *tuesa* – Tweed (TL in Borders Region, Scotland)
tyreeite	Isle of Tyree, Strathclyde, Scotland
uigite	Uig, Isle of Skye, Highland Region, Scotland
urquhartite	Glen Urquhart, Highland Region, Scotland (TL Milton)
wicklowite	County Wicklow, Ireland
Artificial	
conistonite	Coniston, Cumbria, England

Table A7 Species named after British or Irish citizens (or who have connections to these countries), but not found (or confirmed) in Britain or Ireland.

Species	Name
agrellite	Stuart Olof Agrell (1913–1996), professor of mineralogy and petrology, University of Cambridge; born in Ruislip, Middlesex
allanpringite	Allan Pring (b. 1956), British–Australian crystal chemist and professor at the South Australian Museum, Adelaide, Australia; born in Hawarden, Clwyd, Wales
cobaltarthurite	Sir Arthur Edward Ian Montague Russell (1878–1964), English mineral collector of Swallowfield Park, Reading, Berkshire, and Mr Arthur William Gerald Kingsbury (1906–1968), research assistant to the reader in mineralogy at the University of Oxford; born in East Meone, nr Petersfield, Hampshire
ashcroftine-(Y)	Frederick Noel Ashcroft (originally Friedrich Noel Ashcroft Fleischmann; 1878–1949), mineralogist and benefactor to the Natural History Museum, London; born in Wavertree, Liverpool
bakerite	Sir Richard C. Baker (1858-1937), of Nutfield, Surrey, mining director of the Pacific Coast Borax Company in California, and president of the Tonopah and Tidewater Railroad, who discovered the mineral. The town of Baker in San Bernardo County, California, USA, is also named in his honour.
barrerite	Richard Maling Barrer FRS (1910–1996), founding father of zeolite chemistry and professor of physical chemistry at Imperial College, London; all his professional life was spent in Britain, but he was born in Wellington, New Zealand
bernalite (also the name of an olefinic bitumen)	John Desmond Bernal FRS (1901–1971), lecturer in structural crystallography and assistant director of research at the University of Cambridge, and later professor of physics at Birkbeck College, University of London; born in Brookswatson, Nenagh, Co. Tipperary, Ireland
bleasdaleite	Revd John Ignatius Bleasdale FRS (1822–1884), Catholic clergyman, chancellor of the archdiocese of Melbourne, Australia, and a founder member of both the Royal Society of Victoria and the Melbourne Microscopical Society; born in Lancashire
bowieite	Stanley Hay Umphray Bowie FRS (b. 1917), geologist at the British Geological Survey and Atomic Energy Authority; professor of applied geology, Strathclyde University, Scotland, and Imperial College, London; born in Parkhall Bixter, Shetland Isles, Scotland
bracewellite	Smith Bracewell (1899–1979), director of the British Guiana (now Guyana) Geological Survey, who gave the first description of the assemblage containing it
brindleyite	George William Brindley (1905–1983), English–American mineralogist and professor of mineral sciences, Pennsylvania State University, USA; born in Stoke on Trent, Staffordshire
calderite	James Calder of Nairn, shipping agent, banker, sheriff, amateur astronomer, geologist and an early writer on the geology of India
carmichaelite	Ian Stuart Edward Carmichael FRS (b. 1930), professor of geology, University of California, Berkeley, and joint author of *Igneous petrology* (1974); born in Hampstead, London
chadwickite	Sir James Chadwick FRS (1891–1974), nuclear physicist and Lyon Jones Professor of Physics at the University of Liverpool; born in Bollington, nr Macclesfield, Cheshire
chapmanite	Edward John Chapman (1821–1904), professor of mineralogy, UCL, and professor of mineralogy and geology, University College, Toronto, author of *Practical mineralogy* (1843); born in London
claringbullite	Sir Gordon Frank Claringbull (1911–1990), keeper of mineralogy and director of the Natural History Museum, London; born in Whetstone, London
coulsonite (also magnesiocoulsonite)	Arthur Lennox Coulson (b. 1898), geologist, Indian Geological Survey
crawfordite	Adair Crawford FRS (1748–1795), physician and chemist who practised medicine in Edinburgh from 1777 and discovered strontium salts in 1790; born in Crumlin, Co. Antrim, Ireland
crichtonite	Sir Alexander Crichton FRS (1763–1856), physician to the tsar of Russia and collector of minerals and fossils; born in Newington, Edinburgh, Lothian Region, Scotland
criddleite	Alan John Criddle (1944–2002), ore mineralogist at the Natural History Museum, London; born in Cardiff, South Glamorgan, Wales
crookesite	Sir William Crookes FRS (1832–1919), physicist and chemist who discovered thallium in 1861; born in London

Table A7 continued

davyne (also quadridavyne)	Sir Humphry Davy FRS, baronet (1820–1899), chemist who isolated and named several chemical elements and invented the miners lamp; born in Penzance, Cornwall
dawsonite	Sir John William Dawson FRS (1820–1899), principal of McGill University, Montreal, Canada, who studied natural history at the University of Edinburgh and became a fellow of the Royal Society; born of a Scottish mother in Pictou, Nova Scotia, Canada
deerite	William Alexander Deer FRS (b. 1910), professor of mineralogy and petrology, University of Cambridge, joint author of the series of books titled *Rock-forming minerals* (1962–present); born in Rusholme, Manchester
dittmarite	William (Wilhelm) Dittmar FRS (1833–1892), professor of chemistry at Anderson's College, University of Glasgow, Strathclyde, Scotland; long-time British resident, born in Umstadt, nr Darmstadt, Germany
drysdallite	Alan Roy Drysdall (b. 1933), director of the Geological Survey of Zambia; born in Southampton
earlandite	Arthur Earland (1866–1958), civil servant and oceanographer; born in Lewisham, Kent
embreyite	Peter Godwin Embrey (b. 1929), keeper of minerals at the Natural History Museum, London; born in Birmingham
emeleusite	Charles Henry Emeleus FRS (b. 1930), reader in the Department of Earth Sciences, University of Durham; born in Belfast
fergusonite-(Ce) [also fergusonite-beta-(Ce), fergusonite-beta-(Nd), fergusonite-beta-(Y)]	Robert Ferguson FRS (1767–1840), mineral collector, politician (lord lieutenant of the County of Fife and member of parliament), and fellow of the Royal Societies of London and Edinburgh, who resided in Raith House, nr Kirkcaldy, Fife
fermorite	Sir Lewis Leigh Fermor FRS (1880–1954), director of the Geological Survey of India and author of *Memoir on the manganese deposits of India* (1909); born in Peckham, London
ferriwhittakerite	Eric James William Whittaker (b. 1921), pioneer in amphibole crystal chemistry, University of Oxford
frankhawthorneite	Frank Christopher Hawthorne (b. 1946), professor of crystallography and mineralogy at the University of Manitoba, Canada; born in Bristol
gatehouseite	Bryan Michael Kenneth Gatehouse (b. 1932), English crystal chemist at Monash University, Melbourne; born in Poole, Dorset
gerstleyite	James Mack Gerstley (b. 1907), English–American industrial executive, Pacific Borax Company, California; born in Hampstead, London
grayite	Anton Gray (1896–1978), chief geologist and vice-president (exploration) at the Kennecott Copper Corporation, then director of Anglo-Transvaal Consolidated Investment Company Ltd, also a senior mining advisor to the United Kingdom Atomic Energy Authority
gregoryite	John Walter Gregory FRS (1864–1932), geologist, geographer, explorer, and professor of geology at the University of Melbourne, Australia, and the University of Glasgow; an early pioneer of the volcanism and structure of the East African rift system; born in Bow, London
hallimondite	Arthur Francis Hallimond (1890–1968), mineralogist, specialist in secondary uranium minerals and author of *Manual of the polarising microscope* (1948); birth registered in Guisborough, Cleveland
hannayite	James Ballantine Hannay (1855–1931), lecturer in chemistry at Owen's College, University of Manchester; best known for a discredited synthesis of diamond; born in Glasgow
hatchite	Frederick Henry Hatch (1864–1932), geologist and mining engineer, lecturer at Christ's College, Cambridge, president of the Institute of Mining and Metallurgy, and author of *The textbook of petrology – igneous rocks* (1891) and *The geology of South Africa* (1905); born in Bayswater, London
haxonite	Howard James Axon (1924–1992), expert in the metallurgy of meteorites and reader at the Manchester Materials Science Centre, University of Manchester; born in Retford, Nottinghamshire
henryite	Norman Fordyce McKerron Henry (1909–1983), mineralogist at St John's College and Department of Mineralogy and Petrology, University of Cambridge; born in Aberdeen
herbertsmithite (also smithite)	George Frederick Herbert Smith (1872–1953), crystallographer and keeper of minerals at the Natural History Museum, London, who discovered paratacamite, a species related to herbertsmithite, author of *Gem-stones and their distinctive characters* (1912); born in Edgbaston, Birmingham
heulandite-Na	John Henry Heuland (1778-1856), mineral dealer and collector, based in London, who was also secretary of the Geological Society of London; born "Johannes Heinrich" in Bayreuth, Germany.
heyite	Max Hutchison Hey (1904–1984), chemist, mineralogist and principal scientific officer at the Natural History Museum, London; compiler of the *Index of minerals*; born in Leyland, Lancashire
hopeite (also parahopeite)	Thomas Charles Hope FRS (1766–1844), professor of chemistry and medicine, University of Edinburgh, who discovered the element strontium; born in Edinburgh
howieite	Robert Andrew Howie (b. 1923), professor of mineralogy, King's College, University of London, joint author of the series of books titled *Rock-forming minerals* (1962–present); born in Emberton, nr Olney, Buckinghamshire
hügelite	Baron Freidrich Maria Aloys François Charles von Hügel (1852–1925), philosopher, theologian and biblical critic; born in Florence, Italy, of a Scottish mother
hutchinsonite	Arthur Hutchinson FRS (1866–1937), professor of mineralogy, University of Cambridge, and designer of the *Hutchinson stereographic projector*; birth registered in Kensington, London
hydroxylhumite	Sir Abraham Hume FRS (1749–1838), English connoisseur and collector of works of art, precious stones and minerals; born in Berkeley Square, London
joesmithite	Joseph Victor Smith (1928–2007), English–American mineralogist, petrologist, and Louis Block Professor Emeritus in the Department of the Geophysical Sciences at the University of Chicago, Illinois, USA; born in Fritchley, nr Crich, Derbyshire
kentbrooksite (also ferrokentbrooksite and carbokentbrooksite)	Charles Kent Brooks (b. 1939), reader at the Geological Institute, University of Copenhagen, Denmark; born in Westmorland (part of Cumbria)
kingstonite	Gordon Andrew Kingston (b. 1939), senior lecturer in the Department of Geology, University of Wales, Cardiff (kingstonite was named after him for his contribution to PGE mineralogy); born in Kenton, Middlesex
knorringite	Oleg von Knorring (1915–1994), reader in mineral chemistry, University of Leeds; born in Novgorod, Russia, but a long-time resident of Leeds
leakeite (also fluor-ferroleakeite and potassicleakeite)	Bernard Elgey Leake (b. 1932), amphibole mineralogist and professor of geology, University of Glasgow; born in Grimsby, Lincolnshire
liveingite	George Downing Liveing FRS (1827–1924), professor of chemistry, University of Cambridge; born at Nayland, Suffolk
livingstonite	David Livingstone (1813–1873), Scottish explorer of Africa and medical missionary; born in Blantyre, Lanarkshire (Strathclyde), Scotland
lonsdaleite	Dame Kathleen (Yardley) Lonsdale FRS (1903–1971), pacificist, crystallographer and professor of chemistry at the University of London; born in Newbridge, Ireland
mackayite	John William Mackay (1831–1902), Irish–American mine operator on the Comstock lode, Nevada, USA – a famous gold–silver deposit; born in Dublin, Ireland (see Manter 1950)
manganbabingtonite	William Babington FRS (1757–1833), physicist, mineralogist and founder member of the Geological Society of London; born in Port Glenone, nr Coleraine, Co. Antrim, Ireland
marrite	John Edward Marr FRS (1857–1933), professor of geology at the University of Cambridge and president of the Geological Society of London; born in Poulton, Lancashire
metavivianite	John Henry Vivian (1785–1855), a pupil of Abraham Gotlobb Werner (the influential German geologist who founded Neptunism) and member of parliament for Truro; born in Truro, Cornwall
meurigite	Professor Sir John Meurig Thomas KB FRS (b. 1932), eminent British crystal chemist at the University of Cambridge; born in Llanelli, Dyfed, Wales
miersite	Sir Henry Alexander Miers FRS (1858–1942), Wayneflete Professor of Mineralogy, University of Oxford; principal of the University of London; vice-chancellor of the University of Manchester and professor of crystallography; born in Rio de Janeiro, Brazil

Table A7 continued

mountainite	Edgar Donald Mountain (1901–1985), mineralogist, geologist and professor of geology, Rhodes University, Grahamstown, South Africa; born in Camberwell, London
muirite	John Muir (1838–1914), geologist and explorer – the father of American Conservation and founder of Yosemite National Park; born in Dunbar, East Lothian, Scotland
nagelschmidtite	Gunter Nagelschmidt (1906–1980), German-born chemist, mineralogist, pioneer in the application of X-ray diffraction and electron microscopy to dust problems in industrial health and chairman of the British Medical Research Council's Industrial Pulmonary Diseases Committee
naldrettite	Anthony James Naldrett (b. 1933), professor of geology, University of Toronto, past president of the Mineralogical Society of Canada and the International Mineralogical Association, a specialist in magmatic sulphide ore deposits and author of *Magmatic sulfide deposits: geology, geochemistry and exploration* (2004); born in Hampstead, London
oldhamite	Thomas Oldham FRS (1816–1878), Irish geologist and director of the Indian Geological Survey (1850–1876); born in Dublin, Ireland
painite	Arthur Charles Davy Pain (1901–1971), English gemologist who collected the first two crystals from Ohn Gaing village, Mogok District, Burma
parkerite	Robert Lüling Parker (1893–1973), professor of mineralogy, Swiss Federal Institute of Technology, Zürich, Switzerland; born in Marylebone, London
pattersonite	Arthur Lindo Patterson (1902–1966), pioneer X-ray crystallographer who held British, American and Canadian citizenship; born at Nelson, New Zealand, of British parents.
petewiliamsite	Peter Allan Williams (b. 1950), reader in chemistry at the University of Wales and later professor of minerals and materials at the School of Natural Sciences, University of Western Sydney, Australia; born in Sydney, but whose ancestors are of Cornish descent, with many arriving in Australia during the early gold rushes. Some family members were responsible for the opening up of Penberthy Croft, Wheal Fortune and Wheal Vor, all in Cornwall.
playfairite	John Playfair FRS (1748–1819), geologist and professor of mathematics, University of Edinburgh; born in Benvie, nr Dundee
ransomite	Frederick Leslie Ransome (1868–1935), professor of economic geology at the University of Arizona, then California Institute of Technology (Caltech) and consultant on mine litigation, irrigation problems and dam-site location; born in Greenwich, England
rhodesite	Cecil John Rhodes (1853–1902), British colonial statesman and financier, founder of the DeBeers Mining Company; born in Bishop Stortford, Hertfordshire
rosemaryite (also ferrorosemaryite)	Rosemary Wyllie (née Blair) (b. 1932), lecturer in architectural history and interior design at the Harrington Institute of Interior Design in Chicago, and the managing editor of the University of Chicago's *Journal of Geology*; born in Hull, Yorkshire (Humberside)
roweite	George Rowe (1868–1947), English–American mineral collector of Franklin, New Jersey, USA; born in Camborne, Cornwall
scandiobabingtonite	William Babington FRS (1757–1833), physicist, mineralogist and a founder member of the Geological Society of London; born in Port Glenone, nr Coleraine, Co. Antrim, Ireland
selwynite	Alfred Richard Cecil Selwyn FRS (1824–1902), geologist and the founding director of the Geological Survey of Victoria, Australia; born in Kilmington, Somerset
shandite	Samuel James Shand (1882–1957), professor of petrology, Columbia University, New York; born in Edinburgh
sharpite	Major Robert Richard Sharp (1881–1956), English engineer, prospector and soldier, who discovered the Shikolobwe uranium deposit, Zaire in 1915
sheldrickite	George Michael Sheldrick FRS (b. 1942), professor of structural chemistry, University of Göttingen, Germany; born in Huddersfield, Yorkshire
simpsonite	Edward Sydney Simpson (1875–1939), mineralogist and assayer with the Geological Survey of Western Australia; later government mineralogist and analyst; also author of *A key to mineral groups, species and varieties* (1932); born in Woollahra, Sydney, of English and Irish parents
soddyite	Frederick Soddy FRS (1877–1956), radiochemist and professor of chemistry at the universities of Aberdeen and Oxford; born in Eastbourne, Sussex
sorbyite	Henry Clifton Sorby FRS (1826–1908), chemist, geologist, microscopist and metallographer; regarded as the founder of metallography; born in Woodbourne, Sheffield, Yorkshire
spencerite	Leonard James Spencer FRS (1870–1959), keeper of minerals at the Natural History Museum, London, and editor of *Mineralogical Magazine*; born in Worcester, England
stanleyite	Sir Henry Morton Stanley [formerly John Rowlands] (1841–1904), Welsh–American journalist who, on assignment from the *New York Herald*, found David Livingstone in Africa in 1871; born in Denbigh, Clwyd, Wales
taaffeite (also magnesiotaaffeite)	Count Edward Charles Richard Taaffe (1898–1967), gemologist of Dublin; born in Bohemia (Czechoslovakia) – father was the 12th Viscount of Corran, Baron of Ballymote, Co. Sligo, Ireland
teallite	Sir Jethro Justinian Harris Teall FRS (1849–1924), director of the Geological Survey of Great Britain and Ireland, author of *British petrography* (1888); born in Northleach, Gloucestershire
teschemacherite	Edward Frederick Teschemacher (1761–1863), chemist who first described the mineral from guano deposits in South Africa; born in London
thomasclarkite-(Y)	Thomas Henry Clark FRS (1893–1996), professor at McGill University, Montreal, Canada; geologist who mapped the Quebec Applachian Mountains and co-authored with C. W. Stearn *The geological evolution of North America* (1960); born in London
thomsonite-Sr	Thomas Thomson FRS (1773–1852), professor of chemistry at the University of Glasgow, who was editor of the journal in which the first description was published; born in Crieff, Perthshire (part of Tayside Region), Scotland
trechmannite	Charles Otto Trechmann (1851–1917), English crystallographer and analytical chemist of Castle Eden, Co. Durham; author of over 80 books on archaeology and geology; born in Hartlepool, Co. Durham
trevorite	Major Tudor Gruffydd Trevor (1865–1958), Welsh–South African geologist and mining inspector for the area around Pretoria, Transvaal, South Africa
truscottite	Samuel John Truscott (1870–1950), professor of mining, Royal School of Mines, London; born in Fowey, Cornwall
tyrellite	Joseph Burr Tyrrell (1858–1957), Canadian explorer, surveyor, mining geologist and president of the Northern Canadian Mining Corporation; born in Weston, Ontario, Canada, of an Irish father
vaughanite	David John Vaughan (b. 1946), professor of mineralogy and director of the Williamson Research Centre for Molecular Environmental Science at the University of Manchester; born in Newport, Gwent, Wales
vincentite	Ewart Albert Vincent (b. 1919), professor of geology, University of Oxford; born in Aylesbury, Buckinghamshire
wadeite	Arthur Wade (1878–1951), petroleum geologist who discovered the mineral; born in Halifax, Yorkshire
weloganite	Sir William Edmond Logan FRS (1798–1875), geologist and first director of the Geological Survey of Canada; born in Montreal, Canada, of Scottish parents
wyllieite (also ferrowyllieite)	Peter John Wyllie FRS (b. 1930), professor of petrology and geochemistry at the University of Chicago and later professor of geology at the California Institute of Technology, author of *The dynamic Earth* (1971) and *The way the Earth works* (1976); born in London
zincowoodwardite	Samuel Pickworth Woodward (1821–1865), English naturalist and geologist; born in Norwich, Norfolk
zussmanite	Jack Zussman (b. 1924), professor of mineralogy, University of Manchester, joint author of the book series *Rock-forming minerals* (1962–present); born in London

Nationality could not be established for those who gave names to the following species: afwillite, harrisonite, jamesite, mathewrogersite, o'danielite, osbornite, parsonite, stottite and tarbuttite.

Table A8 Varieties and discredited species named after British or Irish citizens (or who have connections to these countries), but not found in Britain or Ireland.

Species	Name
barklyite	Sir Henry Barkly FRS (1843–1890), colonial governor of British Guiana, then Jamaica, and then Victoria, Australia; born in Highbury, Middlesex
bowmanite	Herbert Lister Bowman (1874–1942), Waynflete Professor of Mineralogy and Crystallography, University of Oxford; born in Kensington, London
bucklandite	William Buckland FRS (1784–1856), professor of geology at Corpus Christi College, Oxford, and dean of Westminster, author of *Geology and mineralogy considered with reference to natural theology* (1836); born in Axminster, Devon
chalmersite	George Chalmers FRS (1856–1928), superintendent of the Morro Velho gold mine, Minas Gerais, Brazil; born at Flushing Mylor, Cornwall
darwinite	Charles Robert Darwin FRS (1809–1882), geologist, naturalist and author of *The origin of species* (1859); born in Shrewsbury, Shropshire (see Herbert 2005, Rudwick 2005)
daviesite	Thomas Davies (1837–1892), mineralogist and assistant in the Department of Mineralogy, Natural History Museum, London; born in the parish of St Pancras, London
dunhamite	Sir Kingsley Dunham FRS (1910–2001), professor of geology at the University of Durham and director of the British Geological Survey; born in Sturminster Newton, Dorset
ferrosmithsonite, mangansmithsonite	James Smithson FRS (1765–1829), English chemist and mineralogist, son of the duke of Northumberland and founder of the Smithsonian Institution in Washington DC
forbesite	David Forbes FRS (1828–1876), geological chemist, mining engineer and fellow of the Royal Society; born in Douglas, on the Isle of Man
gieseckite	Sir Charles Lewis Giesecke (formerly Johann Georg Metzler) (1761–1833), professor of mineralogy at Trinity College Dublin, Ireland; born in Augsburg, Bavaria
greenovite	George Bellas Greenough FRS (1778–1855), London-born geologist, mapmaker, one of the founder members of the Geological Society of London and author of *A critical examination of the first principles of Geology* (1819)
hislopite	Stephen Hislop (1817–1863), naturalist and Free Church of Scotland missionary in Nagpur, central India; born in Duns, Berwickshire (part of Borders), Scotland
hunterite	Revd Robert Hunter (1823–1897), geologist, lexicographer and Free Church of Scotland missionary; born in Newburgh, Fife Region, Scotland
johnstonotite	Robert Mackenzie Johnston (1845–1918), government statistician and registrar general of Tasmania; born in one of the fishing villages in the Black Isle, Highland, Scotland
juddite	John Wesley Judd FRS (1840–1916), English mineralogist and petrologist, professor of geology, Royal School of Mines, Imperial College, London; born in Portsmouth, Hampshire
kaneite	Sir Robert John Kane FRS (1809–1890), chemist, president of Queen's College, Cork, and author of *Elements of chemistry* (1841) and *The industrial resources of Ireland* (1844); born in Dublin
kennedyite	William Quarrier Kennedy FRS (1903–1979), professor of geology at the University of Leeds; born in Bridge of Weir, Renfrewshire (part of Strathclyde), Scotland
lewisite	William James Lewis FRS (1847–1926), professor of mineralogy at the University of Cambridge; born in Llanwyddelan, Montgomeryshire (part of Powys), Wales
maclureite	William (originally James) Maclure (1763–1840), the father of American geology – he made the first geological map of the USA; born in Ayr, Strathclyde, Scotland
maitlandite	Andrew Gibb Maitland (1864–1951), government geologist of the Geological Survey of Western Australia and author of *Summary of the geology of Western Australia* (1919); born in Birkby, nr Huddersfield, Yorkshire
maskelynite	Mervyn Herbert Nevil Story Maskelyne FRS (1823–1911), professor of mineralogy, University of Oxford and keeper of minerals, Natural History Museum, London; born at Basset Down House, nr Wroughton, Wiltshire
maufite	Herbert Brantwood Maufe (Muff) (1879–1946), director of the Geological Survey of Southern Rhodesia; born in Ilkley, Yorkshire
nuttallite	Thomas Nuttall (1786–1859), English–American naturalist and author of books ranging from botany to ornithology; born in Settle, Yorkshire
owenite	David Dale Owen (1807–1860), chemist, medical doctor, and state geologist of Arkansas, Indiana and Kentucky, USA (a master of reconnaissance geology); born at Braxfield House, nr New Lanark, Strathclyde, Scotland
oxychildrenite	John George Children FRS (1777–1852), a pioneer mineral chemist, astronomer, zoologist, keeper of natural history at the British Museum (1822–1840) and one-time secretary of the Royal Society (Gunther 1978); born in Tonbridge, Kent.
oxyhowieite	Robert Andrew Howie (b. 1923), professor of mineralogy, King's College, University of London
parabayldonite	John Bayldon – physician and friend of A. H. Church, who first described bayldonite from its type locality in Cornwall
parajamesonite	Robert Jameson FRS (1774–1854), Scottish mineralogist and professor of natural history at the University of Edinburgh; born in Leith, nr Edinburgh
paravivianite	John Henry Vivian (1785–1855), a pupil of Abraham Gotlobb Werner (the influential German geologist who founded Neptunism) and member of parliament for Truro; born in Truro, Cornwall
percylite	John Percy FRS (1817–1889), metallurgist who performed the first analysis of the material; born in Nottingham
piddingtonite	Henry Piddington (1797–1858), meteorologist, curator of the Museum of Economic Geology in Calcutta, and author of *The horn-book for the law of storms for the Indian and China Seas* (1844); born in Lewes, Sussex
priorite	George Thurland Prior FRS (1862–1936), mineral chemist and keeper of minerals at the Natural History Museum, London; born in Oxford
pseudophillipsite	William Phillips FRS (1775–1828), London-born mineralogist and a founder member of the Geological Society of London
pseudowavellite	William Wavell FRS (b. 1750 or 1751, d. 1829), English physician of Horwood Parish, Devon
sollyite	Richard Harrison Solly (1851–1925), London-born mineralogist and crystallographer of the Natural History Museum, London and the University of Cambridge, author of *An elementary introduction to mineralogy* (1894)
steadite	John Edward Stead FRS (1851–1923), metallographer and early ore microscopist who worked on Cleveland iron ores and became president of the Cleveland Institution of Engineers; he also held honorary doctorates at the universities of Leeds, Manchester and Sheffield
stewartite	James Stewart (1855–1944), engine driver at the De Beers Company, Kimberley, South Africa; born in Dalkeith, Midlothian (part of Lothian Region), Scotland
turnerite	Charles Hampden Turner (b. *c.* 1780–1856), mineral collector of Rook's Nest, Godstone, Surrey
wiltshireite	Revd Thomas Wiltshire (1826–1902), London-born professor of geology and mineralogy at King's College, London
youngite	Professor John Young (1835–1902), of the University of Glasgow; born in Edinburgh
Artificial	
carnegieite	Andrew Carnegie (1835–1919), founder of the Carnegie Institute of Washington; born in Dunfermline, Fife, Scotland.

Nationality could not be established for those who gave names to the following minerals: edwardite, hortonite and sloanite.

Table A9 Boundary changes.

Country	*MBI* regions (admin. regions post-1974)	Old county names (pre-1974)	Unitary authorities (from April 1997)
England	Avon	S part of Gloucestershire, N part of Somerset	Bath & North East Somerset, Bristol, South Gloucestershire, North Somerset
	Cleveland	N part of North Riding of Yorkshire, SE part of Durham	Hartlepool, Middlesborough, Redcar & Darlington, Stockton-on-Tees
	Cumbria	Cumberland, Westmorland, part of Lancashire around Barrow in Furness	Cumbria
	Durham	Durham (Tyne & Wear was part of Durham)	Darlington, Durham
	Humberside	East Riding of Yorkshire	East Riding of Yorkshire, Kingston upon Hull, North Lincolnshire, North East Lincolnshire
	Leicestershire	Leicestershire, Rutland	Leicestershire, Rutland
	Sussex (*MBI* combines East & West Sussex)	Sussex	East & West Sussex
	Tyne & Wear	NE part of Durham	Gateshead, Newcastle upon Tyne, Sunderland, North & South Tyneside
	Yorkshire (*MBI* combines North, West and SouthYorkshire)	Yorkshire (includes North and West Riding)	North Yorkshire and York; Bradford, Calderdale, Kirklees, Leeds & Wakefield; Barnsley, Doncaster, Rotherham, Sheffield
Wales	Clwyd	Denbighshire, Flintshire, NE part of Merionethshire	Conwy (excluding SW part), Denbighshire, Flintshire, Wrexham
	Dyfed	Cardiganshire, Carmarthenshire, Pembrokeshire	Carmarthenshire, Ceredigion, Pembrokeshire
	Mid Glamorgan	part of Glamorgan	Bridgend, Merthyr Tydfil, Rhondda Cynon Taff
	South Glamorgan	part of Glamorgan	Cardiff, The Vale of Glamorgan
	West Glamorgan	part of Glamorgan	Neath Port Talbot, Swansea
	Gwent	Monmouthshire	Blaenau Gwent, Caerphilly, Monmouthshire, Newport, Torfaen
	Gwynedd (includes Anglesey)	Anglesey, Caernarvonshire, Merionethshire	Isle of Anglesey, Gwynedd, SW part of Clwyd
	Powys	Brecknockshire, Montgomeryshire, Radnorshire	Powys
Scotland	Borders	Berwickshire, Peebles-shire, Roxburghshire, Selkirkshire, part of Edinburgh	Scottish Borders
	Central	Clackmannanshire, Stirling, SW part of Perthshire	Clackmannanshire, Falkirk, Stirling
	Dumfries & Galloway	Dumfries-shire, Kirkudbrightshire, Wigtownshire	Dumfries & Galloway
	Fife	Fife	Fife
	Grampian	Aberdeenshire, Banffshire, Elginshire, Kincardineshire	Aberdeenshire, City of Aberdeen, Moray
	Highland (includes Inner Hebrides i.e. Isles of: Canna, Eigg, Muck, Raasay, Rona, Scalpay, Skye, Soay)	Caithness, Inverness-shire, Nairn, Ross & Cromarty, N part of Argyllshire (including Ardgour, Ardnamurchan, Morvern and Sunart)	Highland (includes all Inner Hebrides)
	Lothian	Edinburgh, Haddingtonshire, Linlithgowshire	City of Edinburgh, East Lothian, Midlothian, West Lothian
	Strathclyde (includes the Inner Hebridean Isles of Coll, Colonsay, Gigha, Islay, Jura, Kerrara, Lismore, Luing, Mull, Oronsay, Scarba, Treshnish and additionally Ailsa Craig and the Isles of Arran and Bute)	Ayrshire, Argyllshire (but excluding N part - Ardgour, Ardnamurchan, Morvern and Sunart), Arran & Bute, Dumbartonshire, Lanarkshire, Renfrewshire (also includes all Inner Hebridean Isles in Strathclyde)	Argyll & Bute (includes the Inner Hebridean Isles of Coll, Colonsay, Gigha, Islay, Jura, Kerrara, Lismore, Luing, Mull, Oronsay, Scarba, Treshnish and additionally the Isle and Bute), East Ayrshire, North Ayrshire (includes the Isle of Arran), South Ayrshire (includes Ailsa Craig), East & West Dumbartonshire, City of Glasgow, Inverclyde, North & South Lanarkshire, Renfrewshire, East Renfrewshire
	Tayside Region	Forfar, Kinross-shire, Perthshire	Angus, City of Dundee, Perth & Kinross
	Western Isles	Outer Hebrides (includes the Isles of Lewis, Barra, Benbecula, Harris, North and South Uist, Flannan Isles, St Kilda)	Western Isles or Comhairle nan Eilean Siar
	Orkney Isles (includes Hoy, Mainland, North & South Ronaldsay, Sanday, Stromsay, Westray)	No change	No change
	Shetland Isles (includes Bressay, Fetlar, Foula, Mainland, Unst, Whalsay, Yell)	No change	No change
Northern Ireland	Antrim	Antrim	Antrim, Ballymena, Ballymoney, Belfast, Carrickfergus, Larne, N part of Lisburn, Moyle, Newtownabbey
	Armagh	Armagh	Armagh, W part of Craigavon, W part of Newry & Down
	Down	Down	Ards, Banbridge, Castlereagh, E part of Craigavon, Down, North Down, S part of Lisburn, E part of Newry & Mourne
	Fermanagh	Fermanagh	No change
	Londonderry	Londonderry	Coleraine, Derry, Limavady, Magherafelt
	Tyrone	Tyrone	Cookstown, Dungannon, Omagh, Strabane
Republic of Ireland	Dublin	Dublin	City of Dublin, South Dublin, Fingal, Rathdown
	Tipperary	Tipperary	North & South Tipperary

APPENDIX 2
MUSEUM COLLECTIONS

Britain and Ireland have many fine museum collections, but, despite there being a thriving mineralogical community eager to view specimens, some museums have recently closed, and others have changed from institutions that display specimens to become mere store rooms. However, it is still possible to view specimens not on public display by making special arrangements with the museums. Details of the finest British and Irish collections are now given. It should also be noted that many of these museums have excellent foreign or gemstone collections (or both).

England

Bath: Royal Literary and Scientific Institution

The Bath Royal Literary and Scientific Institution was founded in 1824 for the advancement of science, literature and art in Bath. The collection includes reference specimens of the more common minerals and many rarer ones; its particular strength is material from southwest England. The number of specimens exceeds 4000 and the collection was mainly assembled in the 19th century, with some 20th-century additions, including a collection from Merehead (Torr Works) quarry, Somerset, donated by Chris Alabaster (University of Bristol) in the early 1980s. As part of a virtual-museum project, mineral specimens from the collection can be viewed online at www.brlsi.org/virtmu/minerals/virtmineralogy.cfm.

Birmingham: Lapworth Museum, University of Birmingham

Founded in 1880 and centred on the William McClean collection, acquired in 1915. The museum is named after Charles Lapworth, the first professor of geology at Mason College, the forerunner of the University of Birmingham, and it holds 12 000 mineral specimens, including agates collected by Charles Lapworth, the Cornish mineral collection of William Murdock (1754–1839), the J. G. Goodchild (1844–1906) collection of Cumbrian minerals, and the Robert Jasper More (d. 1836–1903) collection of Shropshire minerals. Specimens are from many locations worldwide, with particularly fine examples from the British mining areas of Cornwall, Cumbria, Shropshire and Leadhills–Wanlockhead in Scotland. Free access is provided to the general public, and the collection can also be investigated online (mimsy.bham.ac.uk).

Bristol: City Museum and Art Gallery

The Bristol City Museum and Art Gallery was founded in 1823, initially from the collection of the Bristol Institution for the Advancement of Science, Literature and the Arts. Most of the mineral collection on display was destroyed by bombing in 1940 (although other specimens in storage were not). Bristol's collection of 15 000 mineral specimens is worldwide in scope, but with particular emphasis on minerals from the Mendip Hills, southwest England and the local Bristol area (especially celestine and related evaporites).

Cambridge: Sedgwick Museum, University of Cambridge

The Sedgwick Museum was founded *c.* 1818, with the mineral collection established in 1823 when the university bought the personal collection of the first professor of mineralogy, E. D. Clarke. Four collections make up about half the specimens: those of Sir Abraham Hume (acquired 1841), Henry Brooke (1857), Thomas Wiltshire (1897) and Joseph Carne (1899). The Carne collection includes many fine Cornish pieces collected in the first half of the 19th century. There are over 40 000 specimens in the mineral collection, almost all of which are in store, although there is a small mineral gallery in the Sedgwick Museum. British specimens on show include: calcite, aragonite, fluorite and mimetite from Cumbria; calcite, fluorite and pyromorphite from Cornwall; and calcite and galena from Derbyshire. The Sedgwick Museum also holds: the geological collection of John Woodward (1665–1728), which includes ores from the Mendip Hills and mid-Wales; the Harker petrological collection of 150 000 rock specimens, among them unusual assemblages from contact-metamorphosed limestones in Scotland and Ireland; and the rocks collected by Charles Darwin during his voyage on *HMS Beagle*. The staff of the museum will respond to queries about the collections, but the documentation is such that it is not always possible to obtain answers quickly. Access to specimens in storage can be given by arrangement – at least two weeks' notice is required.

Cornwall: Camborne School of Mines, University of Exeter, Penryn

The Camborne Museum was founded in 1891 as a teaching resource for Cornish miners studying at the School of Mines. Donations came from local mine owners, and the Robert Hunt collection was incorporated in the 1950s. The museum holds 15 000 mineral specimens, many of which formed a very impressive public mineral display when housed at Pool, Redruth. In 2004 the Camborne School of Mines transferred to its new site, where a subset of the original display is now on view. Although the museum is not generally open to the public, group visits may be arranged by contacting the Camborne School of Mines via cornwall@exeter.ac.uk. A virtual museum containing images of minerals, along with a considerable amount of other information on the Cornubian orefield and the geology of Cornwall, can be toured at www.ex.ac.uk/geomincentre.

Cornwall: Cornwall Geological Museum, Penzance

The Cornwall Geological Museum (Royal Geological Society of Cornwall) was founded in 1814 with a mineral collection mostly amassed in the 19th century. Donations were made by Joseph Henry Collins, Humphrey Davy, Davies Gilbert, Sir Charles Lemon, Clement Le Neve Foster and Warington Wilkinson Smyth. The collection consists primarily of southwest England and other British localities. The museum had a large, interesting and attractive mineral gallery, and held 6500 mineral specimens, but it is now closed and the collections have been put in store at the British Geological Survey headquarters, Keyworth, Nottinghamshire.

Cornwall: Geevor Tin Mine Heritage Centre, Pendeen, near St Just

Opened in August 1993. It is the largest preserved mining site in Britain and is centred on one of Cornwall's last working tin mines. It worked under various names from the early 1770s until mining was halted in 1990, when there was a slump in the price of tin. A small but attractive mineral collection of local species is on display in one of the mine buildings.

Cornwall: The Royal Cornwall Museum, Truro

The Royal Cornwall Museum was founded in 1818 essentially to house a collection of Cornwall and Devon minerals belonging to the Royal Institution of Cornwall. Some 2000 specimens were catalogued by Joseph Henry Collins between 1878 and 1881, but since then the collection has grown to 13 000 specimens. The core of the collection is that of Philip Rashleigh (1729–1811) – certainly one of the finest 18th-century mineral collections in existence. In terms of its historical value, quality and the rarity of its specimens, it is of international importance. Other acquisitions include the collection of James Wickett (1841–1921), a selection from the Williams of Caerhays collection acquired in 1893, and many donated pieces from William Sargent and Richard Barstow (1947–1982). Key specimens from the Nick Talbot collection were added in 2002. The collection is displayed in the specimen-rich Rashleigh gallery, in traditional wooden cabinets donated by early benefactors of the institution.

Cumbria: Tullie House Museum, Carlisle

The Tullie House Museum was founded in the 1870s with the donation of the collection of Professor Robert Harkness. Other donations followed, including the J. W. Branston collection of Cumbrian minerals. The local-minerals collection is of considerable quality, and contains some 2000 specimens. This reflects the fact that Cumbria was one of the richest metal-mining areas in Britain. In particular, the Caldbeck Fells near Carlisle are renowned worldwide for their mineral wealth and have produced some of the finest mineral specimens ever found. The museum collection holds a good range of specimens, some of superb quality, from the Caldbeck Fells, Northern Pennine and West Cumbrian orefields. Many of the minerals were collected during the 20th century from mines that are now closed. Additions to the collection continue to be made as gifts from local amateur and professional mineralogists. Important recent acquisitions include the J. D. Ingham collection. About 60 specimens of Cumbrian minerals are displayed on cast glass blocks designed and arranged by a glass artist. They are grouped by mining area – Northern Pennines, Caldbeck Fells, etc. In addition, the museum houses a collection of about 1000 examples of foreign minerals and an important geological collection of approximately 10 000 specimens.

Cumbria: Kendal Museum

Kendal Museum is the new home of the John Hamer collection of over 2000 mineral specimens. Northern England minerals are its main strength, although beautiful rubies from Burma, and tourmaline from Brazil, are added attractions. The collection was rescued in 2003 by Carol Davies (the curator of natural history at the museum), aided by members of the Westmorland Geological Society. Once all the samples had been catalogued, a new well crafted display was created, opening to the general public in October 2005. It forms part of the Lake District Natural History gallery. It is a traditional museum display of the sort that is sadly missing in some larger institutions with finer collections. Specimens not on display can be examined by appointment with the curator.

Derbyshire: Buxton Museum and Art Gallery, Derbyshire

The Buxton museum has a collection in excess of 1000 mineral specimens, which includes representative examples of the Peak District's lead-mining heritage. Fine crystallized specimens of calcite, fluorite and galena are from the famous mining fields of Castleton, Wirksworth, Matlock, Eyam and Hucklow Edge. Northern Pennine material is also well represented, including splendid specimens of fluorite, galena, quartz and sphalerite from the Alston and Weardale mining fields. Noteworthy too are barite, dolomite, hematite and quartz combinations from the Cumbrian iron mines. There are examples from the 19th-century Cornish copper and tin mines, including famous former mining areas such as Botallack, Caradon Hill and Gwennap. Certain international locations are also represented: native copper from Lake Superior, and malachite from Australia and Siberia. Some of the fine non-British specimens carry the labels of the well known dealer F. H. Butler, of Brompton Road, London, who was in business until 1927. Alongside this collection, the museum houses extensive collections representing the geological history of the Peak District. The museum also holds the best public collection of worked Ashford Black Marble artefacts in the country. Using locally sourced limestone of the highest quality, containing inclusions of bitumen, 19th-century Derbyshire craftsmen developed techniques of inlaying, decorating and polishing, to create a range of ornaments, from table tops to jewellery. With an eye to the contemporary high Victorian taste, these ornaments were decorated with local rocks and minerals, either in geometric patterns or in flowers and butterflies, and reached the wider public audience through being displayed at events such as the Great Exhibition of 1851. Some specimens of the mineral collection are on permanent display, including Blue John (worked and unworked), calcite and stalactitic barite. The reserve collection is currently undergoing a review, with support from the Horrocks bequest and the Manchester Geological Association. It is anticipated that more specimens will be placed on permanent display as a reminder of 2000 years of lead mining and its impact on the community. Visitors wishing to access the reserve collection should contact the museums manager for an appointment.

Derbyshire: Devonshire Collection, Chatworth House, near Bakewell,

A collection assembled nearly 200 years ago by Georgiana, Duchess of Devonshire (1757–1806), with additions by "Hart", the 6th Duke, made *c.* 1817–1827. The collection has been reassembled and restored by members of the Russell Society (Cooper 2005) and it includes superb old-time Derbyshire and Cornish specimens. Part of the collection is on public display in Chatsworth House.

Derbyshire: Derby Museum

Derby Museum was founded 1836 and is host to a 2500-sample mineral collection. The collection is worldwide in scope and it covers a large spectrum of mineral species, including famous Derbyshire specialities such as Blue John (fluorite) and matlockite. Sadly, much of the original geology collection was lost in a flood in the 1930s.

Devon: Torquay Museum

Torquay Museum was founded in 1844 by the Torquay Natural History Society and is now owned and run by the Torquay Museum Society. The museum holds a collection of more than 3000 minerals from all over the world. Devon and Cornwall are well represented with about 800 specimens that include examples of dendritic gold from Hope's Nose in Torquay. Important individual collections include those of R. N. Worth and A. H. Ogilvie. Display of the collection is limited to just a few specimens on permanent view in the Time Ark Natural History Gallery, but the collection is documented using Modes database software and is available to view on request.

Durham: North of England Lead Mining Museum, Killhope, Weardale

Founded in 1984, the museum has a permanent display of North Pennine minerals, mainly from Lindsay Greenbank's collection. Of related interest is the national collection of spar boxes (Forbes 2004). These are mineral cabinets made by lead miners in the late 19th century, and some are on permanent display. The museum has a small reserve collection not on display.

Lancashire: Bolton Museum and Archive Service

Bolton began acquiring natural-science and ethnography specimens in about 1852 (following the Libraries Act), but it was not until 1884 that the first museum, the Chadwick, opened. In 1947 the present museum opened in the town centre; the Chadwick was demolished in 1956. The mineral collection holds about 3000 specimens (200 of which are on display) forming a general reference collection. The great majority of these are northern-counties minerals. In total the museum holds 20 000 geological specimens, most of which are palaeontological.

Leicester: City Museums (New Walk Museum and Art Gallery)

The New Walk Museum and Art Gallery was originally founded in 1839 and it incorporates the geological, mineralogical and palaeontological collections of the Leicester Literary and Philosophical Society. Its earlier names were the Leicester Town Museum (1849), City of Leicester Museum and Art Gallery (1919) and Leicestershire Museums, Art Galleries and Records Service (1974). The museum holds 10 000 mineral specimens, with the vast majority in storage.

Leicester: Department of Geology, University of Leicester

The museum in the Department of Geology, University of Leicester was founded in 1951 to service the teaching, assessment and research activities. The collection contains material gathered by staff and students, as well as material obtained by purchase, gift and exchange. The collections are worldwide in scope, with particular local and British strengths. Among more historical material, the collection of Edwin Beer (1879–1986) is notable. In addition to extensive systematic collections, there are significant economic suites, with well located specimens, plus thin and polished sections from a very wide range of UK and foreign mines and mineral deposits, many no longer accessible. The museum holds about 25 000 mineral specimens, most of which are well documented and catalogued on computer. The museum is open to the public by prior appointment.

Liverpool: World Museum (National Museums Liverpool)

The World Museum was founded in 1851, although most of the collection was destroyed in the bombing of May 1941. Important surviving historical collections are the Derby collection of cut agates, and specimens belonging to T. S. Traill and G. H. Morton. Minerals from the north of England are a speciality, with specimens acquired from the Greenbank and Sutcliffe collections. The museum holds 14 500 mineral specimens. A small display of minerals is open to the public, but the remainder of the collection is available to view only by prior appointment.

London: The Natural History Museum (NHM)

Founded in 1753, the NHM now holds one of the world's most comprehensive mineral collections. The museum originated from the extensive collections of Sir Hans Sloane (1660–1753), which formed the basis of the British Museum. The British Museum (Natural History) became a separate entity in 1881 when the natural-science collections were separated from the ethnographical and other collections, and were moved to a purpose-built building in South Kensington. The museum became completely autonomous in 1963. In 1985 it merged with the Geological Museum of the Institute of Geological Sciences (now the British Geological Survey). Its name was formally changed from British Museum (Natural History) to Natural History Museum in 1992.

The collection of minerals numbers about 173 000, and rock specimens double that number. The collection is particularly strong in specimens from the early days of mineral collecting and, although there is an emphasis on British material, it has a worldwide coverage. Since its inception in 1753 around the small collection of "Chrystals, Spars and Pretious Stones" of Hans Sloane, the mineral collection has grown by the addition of many other collections and individual specimens, although very little was added between 1753 and 1798. In 1799 the collection grew with the purchase of about 7000 specimens from Charles Hatchett (1765–1847), which included items from the Russian nobleman Count Apollos Moussin-Poushkin (1760–1805), and with the bequest of Revd Clayton Mordaunt Cracherode (1730–1799), a former museum trustee who donated 868 fine specimens. The museum appointed its first curator of minerals, Charles Konig, at this time. In 1810, the collection of Charles Francis Greville (1749–1809), son of the Earl of Warwick, was purchased at the enormous cost of £13 727 (equivalent to the price of a house at today's values). This collection included the excellent collection of Ignatz von Born (after whom bornite is named). In 1816, the collection of 14 000 specimens belonging to Baron F. C. von Beroldingen was purchased. Other important collections added were: 2000 Vesuvian specimens from Teodoro Monticelli (1759–1846); a fine suite of silver ores from the Harz Mountains mining district, presented by George IV; a collection of *c.* 9000 specimens started by Thomas Allen and his son Robert, and added to by Robert Philips Greg (acquired 1860); and *c.* 3250 specimens (acquired 1865) assembled by Count Nikolai Koksharov (1818–1892). Although there were many acquisitions after this, there were few large collections added until the 1960s. Two exceptions were *c.* 17 000 specimens transferred from the Museum of the Geological Society in 1911 (although many of these were rocks), and the fine Swiss collection of Frederick Noel Ashcroft, consisting of about 7000 specimens, given over a period of some 20 years from about 1918. In 1964, upon the death of Sir Arthur Russell, his extensive collection of about 12 000 British mineral specimens was bequeathed. This is recognized as one of the finest collections of British minerals ever assembled and it includes many specimens from earlier collections (notably from Philip Rashleigh, George Croker Fox, Warington Wilkinson Smyth, etc.). In 1968, the museum purchased about 1100 fine specimens of worldwide origin from Charles Locke Key and Frederick Ludlow Smith. The most recent major acquisition was the result of the merger of the Geological Museum with the BM(NH) in 1985, when the entire mineral and gemstone collections of some 35 000 specimens were transferred (along with some rocks, building stones, meteorites and ores). The Geological Museum also handed over the historically important collection of Henry Ludlam (1824–1880). This, and the collections of Ashcroft and Russell, are kept as satellite collections, separate from the main mineral collection. The main collection is housed in the original 1881 gallery and in the original wooden cases. About 8000 examples are on permanent display, covering about 60% of known mineral species from a wide diversity of world locations. In all, the collection has examples of about 75% of the known 4000+ mineral species, and new examples are acquired as they become available.

The mineral collection today is seen as an international resource, able to supply material to *bone fide* researchers, as well as being a national treasure of many of the finest specimens ever to have been obtained, especially from British mines. A catalogue of type-mineral specimens held in the museum has been published (Clark et al. 1995) and is shortly to be updated. The NHM also holds the dubious distinction of being custodian of the Arthur Kingsbury collection. A description of the current NHM mineral galleries is provided by White (2006).

In November 2007 the museum opened a new gallery called "The Vault". The display is well designed, well lit and well labelled, and contains an excellent selection of impressive specimens, both natural crystals and cut gemstones. Most notable are a beryl (var. morganite) from Afghanistan, wire gold from Venezuela, rhodochrosite from Sweet Home mine in Colorado, South African sperrylite, a suite of 296 coloured diamonds, and the "Devonshire Emerald" (a 6 cm tall, 1384 carat crystal from Colombia and now belonging to the Chatsworth collection). British specimens include Hope's Nose native gold, and a siderite "box" epimorph from the Virtuous Lady mine in Devon (Bancroft 1973).

Manchester: Manchester Museum, University of Manchester

The Manchester Museum was founded in 1821 by the Manchester Society for the Promotion of Natural History. Mineral specimens were part of a general natural-history collection that was amalgamated with the Manchester Geological Society collection in about 1850. There were problems maintaining the collection in the following decade and it was with some relief that it passed to Owen's College, which in 1867 became the Victoria University of Manchester. For more than a century the mineral collection was the responsibility of the keepers of geology and, as they all had a preference for palaeontology, it was rather neglected. A post of keeper of mineralogy was established in 1992 and the collection has expanded considerably since then. It currently holds 22 000 mineral specimens, including excellent material from the mines of northern England. About 300 specimens, including most of the best British material, are displayed in the rocks and minerals gallery (opened in 2003).

Middlesborough: Dorman Museum

The Dorman Museum was founded in 1904. It is well endowed with historical mineral specimens from the Northern Pennine orefield (particularly fluorite) and it includes site-reference specimens from Co. Durham, Cumbria, Northumberland, and to a lesser extent Yorkshire. Notable collections include those of Richard Barstow, Lindsay Greenbank and James Ianson (19th-century European specimens). The collections of Cleveland County Museum Service and Darlington Tubwell Row Museum were transferred to the Dorman in the 1990s. The museum now holds about 5500 mineral specimens. There is a small permanent display of world minerals and regularly changing displays that incorporate mineral specimens, such as Earth in Space (March 2008), which explores themes of planet formation, geological time, evolution and extinction, and incorporates a mineral mine.

Newcastle-upon-Tyne: Hancock Museum

The Hancock Museum was founded in the early 19th century with donations from William Hutton (1797–1860), Thomas Sopwith, Christopher Cookson and the Charlton Brothers of Bellingham, Northumberland. Later donations are by Brian Young and Trevor Bridges. The museum's strength is minerals from the Northern Pennine orefield and it is host to a 7000-sample mineral collection. From April 2006 the museum was closed for 2–3 years for refurbishment.

Norwich: Norwich Castle Museum

Norwich Castle Museum was opened on the current site in 1894 after being transferred from premises of the Norfolk and Norwich Literary Institution. Almost half of the mineral collection was collected between 1825 and 1894, during the life of the former Norfolk and Norwich Museum. The collection is typical of a county museum and it contains over 3000 specimens (including 370 different species and many fine individual specimens). Around one third of the collection has site data, often down to an individual mine; for another third, only general locality data is known, such as the county or country in which they were collected. Many specimens are derived from mining areas of the UK and others are worldwide, notably Australia, Canada and continental Europe. Donors contributing significant numbers of British specimens included: Robert Fitch (Cornwall and Cumbria); John Taylor, a 19th-century mining engineer (Cornwall and Wales); Philip C. Cambridge (East Anglia and the Midlands); H. F. Euren (Cornwall); Revd G. R. Leathes; Revd E. E. Montford; Revd C. Paroissien; T. Southwell; and Edward Stanley (Bishop of Norwich). A small sample of the collection is on display at the museum. The remainder of the collection is in offsite storage pending the establishment of a permanent home.

Nottinghamshire: British Geological Survey, Keyworth

Six years after the establishment of the Geological Survey in 1835, it opened its first museum, the Museum of Economic Geology, in Craig's Court, Whitehall, London. Pressure of space necessitated a move first to Jermyn Street and then to South Kensington. In 1985, most of the collections were moved to the survey's new headquarters at Keyworth, but the mineral and gemstone collections were donated to the British Museum (Natural History), so that they could remain on display in London. The Survey soon appreciated the loss and began a new mineral collection for research and display purposes. This collection now includes the Seager, Small, Thorman and Royal Geological Society of Cornwall collections, among others, and numbers over 26 500 specimens. It is predominantly a research collection and academic visitors are welcome, as are organized parties. The Geological Survey office in Edinburgh also maintains a significant mineral collection.

Oxford: University Museum of Natural History

The Oxford University Museum of Natural History was founded in 1860 and was established to bring together under one roof the natural-history collections of the Ashmolean Museum, Clarendon Museum and some of the colleges, along with those of all of the university's scientific teaching and research collections. The museum holds in excess of 30 000 mineral specimens, as well as gemstones, rocks, meteorites and mineralogical instruments. Important holdings of British minerals are in the collections of Richard Simmons (acquired 1848), Sir Henry Enfield Roscoe (1906), George Allen (1908), Sir Ford North (1909), Hugo Müller (1915), and Morrison Thomas (1984). Added to these are specimens collected by past and present museum staff, and generous donations from people actively collecting British minerals. About 680 minerals are on display, of which approximately 180 are British. A public-access catalogue of the mineral collection can be searched from the museum's website, which also gives information about opening hours, special events and details of how to visit the reserve collections (www.oum.ox.ac.uk).

Plymouth City Museum and Art Gallery

Opened on the current site in 1910, the Plymouth collection was built up during the 20th century and it mainly consists of southwest England specimens. Based on collections transferred from Devonport Museum in 1924, the dominant parts include: those of Sir John St Aubyn (1758–1839); a British and worldwide collection arranged by Count de Bournon in 1815, with specimens from Richard Green, Lord Bute and W. Babington; Col. Sir William Serjeant (1857–1930), with a substantial cassiterite collection from nearly 160 Cornish and Devon mines; and Richard Barstow (1947–1982), a Cornish mineral dealer, with over 2000 specimens, including important rarities and a large library of cassiterite from southwest England. The collection holds about 10 000 specimens from 500 species. Displays are limited, forming part of a general natural-history gallery; the remainder is held in store.

Sheffield: City Museum

The Sheffield City Museum was founded in 1875, based on the collections of the Sheffield Literary and Philosophical Society, W. M. Bragge, Earl Fitzwilliam, E. F. Newton, J. W. Puttrell and Colonel J. W. Rimmington (1832–1909). Minerals from Derbyshire and southwest England are well represented. Displays of Derbyshire minerals are the most notable mineral exhibits. The museum holds 4000 mineral specimens.

Somerset: Wells and Mendip Museum, Wells

Wells and Mendip Museum was founded in 1893. Its mineral collections reflect the locality's importance to the development of geological science, especially the contributions of pioneer geologists working in the Bath and Wells area, such as William Smith, William Buckland and Henry De la Beche. The Ensor collection, purchased in 1894, includes many fine mineral specimens. It was augmented in the early 20th century by fossils and minerals donated by local enthusiasts, especially members of the Wells Natural History and Archaeological Society. These are complemented by a specimen collection assembled by Thomas Willcox, the last manager of Priddy lead mines. Throughout its history, the museum has had links with local extractive industries and has acquired many artefacts from them. The museum's research collections extend to over 30 000 items and include distinctive local specimens such as barite from Frome, celestine from Westbury-sub-Mendip, and mendipite, a rare lead oxychloride unique to the area around the village of Green Ore. Displays give particular emphasis to lead mining activity in the area since Roman times.

Sunderland: Museum and Art Gallery

Founded in 1846 and host to a 3700-sample mineral collection that includes the E. Backhouse (d. 1808–1879) and C. T. Trechmann (1885-1965) collections. The minerals are worldwide in scope and cover a large spectrum of species, including famous Northern Pennine orefield specialities such as alstonite, barite, barytocalcite, fluorite, galena and witherite. Permian evaporites and Zambian copper ores are also well represented. Only a small selection of samples are on display, but the others can be viewed by appointment.

Surrey: Haslemere Educational Museum

The Haslemere Educational Museum was founded in 1888 by Sir Jonathan Hutchinson (1828–1913), an eminent Quaker surgeon. The original collection was formed of items gathered from his travels and from auction houses around the country. The collections have since developed to number approximately 400 000 items that relate to the broad themes of natural history, geology and human history. The mineral collection is mainly from British localities and it holds over 5000 specimens from over 200 species. The collection includes common and widespread mineral species, and samples of common metal ores, as well as a wider range of more uncommon minerals. Well crystallized mineral specimens of note from the Britain include calcite from the north of England and fluorite from various localities in northern England and Cornwall. Rare minerals include chalcophyllite, olivenite and pharmacosiderite. There are 50 mineral specimens on public display showing different varieties of quartz, along with sulphides, native elements (copper, gold and sulphur), oxides, carbonates, halides, sulphates, phosphates, molybdates, tungstates, arsenates and silicates. The main mineral collectors represented at the museum are: Sir Archibald Geikie (1835–1924), director-general of the Geological Survey, president of the Geological Society, and president of the Royal Society; John Edward Lee (1808–1887); John Clarke Hawkshaw (1841–1921); and Charles Percy Richards (1851–1941) – with minerals mainly from the china-clay pits of Cornwall. The majority of the minerals are held in storage, but a public geology gallery displays about 1% of the collection. The reserve collection can be viewed by appointment.

York: Yorkshire Museum

Founded in 1822 by the Yorkshire Philosophical Society, the Yorkshire Museum contains a mineral collection derived from worldwide and classic localities, including many specimens from 19th-century collectors. Of particular note is the collection of northern-counties minerals from famous sites in the northern Pennines. The museum holds 8000 mineral specimens; these are not on permanent display, but regularly changing displays see a turnover of minerals on show to the public.

Wales

Cardiff: Amgueddfa Cymru – National Museum Wales (NMW)

The NMW was founded in 1912, originally based on specimens in the old Cardiff Museum, but added to with the acquisition in 1927 of the G. J. Williams collection of Welsh minerals and including specimens from the collection of David Pennant, son of the famous naturalist Thomas Pennant. More recently the collection has been enhanced by acquisition of the R. J. King mineral collection (over 10 000 registered, including material from the Henry Harwood collection), which comprises specimens from worldwide locations, with particularly fine fluorite from northern England and a comprehensive cassiterite holding. The Welsh mineral holding was enhanced by specimens collected during the MINESCAN mine and mineral-site survey project. The mineral collection comprises approximately 33 000 specimens, with over 500 on display, and all held on electronic inventory. An excellent image-rich website complements the collection: www.museumwales.ac.uk/en/mineralogy_of_wales/.

Scotland

Aberdeen: University of Aberdeen

A collection of showcase specimens, gemstones and systematic minerals is assembled in the Department of Geology and Petroleum Geology. It includes the collections of Revd J. M. Gordon (1849–1922) and T. C. Phemister (1902–1982). The systematic collection occupies 11 oak cabinets and is estimated to contain over 10 000 specimens.

Dumfries & Galloway: Dumfries Museum

Dumfries Museum was founded in 1835 as an astronomical observatory and museum. It hosts a small but interesting assemblage of minerals based on the collections of Patrick Dudgeon (1817–1895) and Thomas Boyle Grierson (1818–1889). The collection includes excellent specimens from the Leadhills–Wanlockhead mining district, including well crystallized aragonite on a chalcopyrite–quartz matrix, a large specimen of barite on sphalerite and quartz, a blocky galena group and a plate of crystalline sphalerite.

Dumfries & Galloway: Wanlockhead Museum of Scottish Lead Mining

The Wanlockhead Museum Trust was founded in 1974 and now includes a miners' library, miners' cottages, an underground mine and a visitor centre with a well organized display of minerals from the Leadhills–Wanlockhead orefield. Over a hundred mineral specimens are on display, all housed in modern well lit cabinets. Special displays include many examples of local alluvial gold, examples of primary and secondary species, and, of particular interest, specimens of galena with unusual crystal form.

Edinburgh: National Museums of Scotland

The National Museums of Scotland came into being in 1985 with the merger of the Museum of the Society of Antiquities (founded 1780) and the Industrial Museum of Scotland (founded 1854 and later to become the Royal Museum of Scotland). In 1997 the new Museum of Scotland was opened and, at the same time, the Royal Museum of Scotland shortened its name to the Royal Museum. The Royal Museum was once the home of a fine display of Scottish minerals, but sadly, apart from a few exceptions, the collection is now in store and will remain there for the foreseeable future. The museum holds 60 000 mineral specimens in total, including world-class specimens from the Leadhills–Wanlockhead and Strontian orefields. The museum is in a transitional period, during which time the Royal Museum masterplan aims to develop new displays devoted to the natural world for a "modern audience"; it is hoped that these displays will open in 2011. Until 2011 the mineral collection has moved off site, but access is possible by appointment. White (2006) discusses the pending problems for the mineral collection, emphasizing the importance of having specimens on display.

Edinburgh: Cockburn Museum, Grant Institute, University of Edinburgh

The Cockburn Museum was founded in 1873 as "a museum for the teaching of geology" by Professor Archibald Geikie, the holder of the first chair of geology at the University of Edinburgh. Major donations of special and rare specimens (particularly minerals) over the years have included extremely valuable reference material for research investigations, as well as some beautiful specimens for display. These include: the Hall collection of the first experimental apparatus for simulating melting and other processes in rocks, and minerals from Sir James Hall (1761–1832) of Dunglass, Lothian – a friend of James Hutton; Thomas Brown of Waterhaughs and Lanfine, Ayrshire (part of Strathclyde); James Currie and James Davidson of Summerville, near Dumfries, Dumfries & Galloway; as well as the Lyell rock, mineral and fossil collection donated by Sir Charles Lyell. This extensive collection of geological specimens and historical objects reflects Edinburgh's prominent position in geological sciences since the time of James Hutton (1726–1797). The collection holds 110 000 mineral specimens, although only a few are on display; the remainder can be viewed by prior appointment.

Glasgow: Kelvingrove Art Gallery and Museum, Glasgow Museums

Kelvingrove Art Gallery and Museum was originally founded as the City Industrial Museum in 1870, which in 1901 merged with the McLellan Galleries to form a major part of the Glasgow International Exhibition. Mineral collections built up rapidly, particularly in 1896 with the purchase of the collection of David Corse Glen (1824–1892).

Glasgow: Hunterian Museum, University of Glasgow

The Hunterian Museum was opened in 1807, based on the very fine, and mostly intact, 18th-century collection of William Hunter (1718–1783), who made his fortune in London as a teacher of anatomy and a physician to royalty. Other major collections include that of Thomas Brown (1774–1853), Frederick Eck (1806–1884), Alexander Thoms (1837–1925), James "Paraffin" Young (1811–1883) and Frank Rutley (1842–1904, author of *Elements of mineralogy*), all of which are accompanied by original archival materials. The collections are worldwide in their scope, but particular strengths are greenockite and zeolites from the Scottish Carboniferous, Leadhills–Wanlockhead minerals, and some very fine Scottish sapphires. The museum holds 30 000 mineral specimens. As of 2007, there are over 200 mineral specimens on display. Collections not on display, and archival materials, are accessible by appointment at the museum's Thurso Street facility.

St Andrews: University of St Andrews

The mineral collection is now in the Department of Geography and Geosciences, but originated in the museum of the Literary and Philosophical Society of St Andrews. The museum was founded as a joint venture with the university in 1838 by Sir David Brewster. Included is a teaching collection (on display), a systematic reference collection of over 1200 specimens, and the collection of C. F. Davidson (1911–1967), a specialist in radioactive minerals and ore genesis.

Ireland

Belfast: Ulster Museum

The Ulster Museum was founded in 1821 based on the collections of the Belfast Natural History and Philosophical Society. It contains a fine collection of Irish zeolites (>600 specimens) and incorporates the mineral collections of J. Grainger (1890), J. Hanna (1926) and the Geological Survey of Northern Ireland, including original Geological Survey of Ireland material from the 19th century. There are 360 specimens on display. The museum holds 12 000 mineral specimens.

Cork: Geological Museum

The first president of University College Cork was Sir Robert Kane, appointed in 1849; he was also concurrently director of the Museum of Economic Geology in Dublin. A priority of the new university was to establish a geological museum. A large collection of minerals was gathered within a few years, but some of these were lost in the late 1950s when the museum was closed to make space for administration. A new geological museum was set up in the Department of Geology in 1981. It has, at present, over 4000 mineral specimens and was named the Cork Geological Museum in 1990. Some specimens are on loan to a variety of small museums in Co. Cork. The collection houses the type specimen of cotterite, a rare variety of quartz; it was found near Mallow, in Co. Cork, by a Miss Cotter, and named after her by Professor Harkness, who was professor of geology at the time; she donated the specimen to the museum in 1876. The museum houses several donated mineral collections, such as the Lady Windle and the Mary O'Donoghue collections. It is mainly a teaching museum but is open to the public by appointment.

Dublin: National Museum of Ireland – Natural History

The National Museum of Ireland was founded in 1792 when the Irish Parliament provided the Royal Dublin Society with funds to purchase the European mineral collection of Nathanael Gottfried Leske (1751–1786), then being offered for sale in Germany. Additions include the collections of Sir Charles Lewis Giesecke (1761–1833), Baron Henry Joy (d. 1861), the Royal Dublin Society, Museum of Irish Industry, Museum of Science and Art, and purchases from British, German and North American mineral dealers in the 19th century. The museum has been open to the public since 1857. Current holdings are approximately 10 000 mineral specimens (2000 are Irish). Collections are mostly in storage and can be visited by appointment.

Dublin: Geological Museum, Trinity College

The Geological Museum at Trinity College, Dublin, was founded in 1777 as the Dublin University Museum. The collections are now housed in a smaller gallery and store situated in the Museum Building on the Trinity College campus. By 1807 the university museum housed a collection of 1089 minerals, as well as some petrological specimens. In 1823 the college purchased approximately 600 mineral specimens from the Honourable George Knox (1865–1827, member of parliament for the University of Dublin). Many of these specimens are difficult to recognize in the present collection, but Knox's catalogues indicate that they came from the Americas, particularly the British colonies of the Caribbean, and from Brazil, Scandinavia, Siberia, Greenland and Saxony. Other specimens were collected from Britain and from parts of Ireland, including the Giant's Causeway, the Mourne Mountains, County Down (mainly beryl) and Dublin. Other donations and bequests were made by Sir Charles Lewis Giesecke (1761–1833) and John Joly, professor of geology (1897–1933). The latter collection contains few Irish specimens, but includes zeolite minerals from the Giant's Causeway, and beryl crystals from the granite of the Dublin Mountains. The museum is best regarded as a research and teaching museum, and its displays are viewed by university students, school pupils and the general public. Material is utilized for undergraduate teaching programmes and for academic research. Today the collections number 7000 mineralogical specimens, of which only 5% are on display. A comprehensive account of the history of the Geological Museum and its holdings is given in Wyse Jackson (1992).

Galway: James Mitchell Museum, National University of Ireland

The James Mitchell Museum was founded in 1852. The core of the museum is the type, figured and backup material for William King's Monograph of the Permian Fossils of England, published by the Palaeontographical Society in 1850. Restoration of the main gallery of the museum in its original mid-19th-century style was completed in 1992. As a whole, the museum contains some 15 000 rock, mineral and fossil specimens from all over the world, and includes 1750 minerals. Mineral displays include Irish mineral deposits, minerals from Connemara and Co. Galway, the Eleanor Miles collection of gemstones and worldwide specimens, the Dave McDougall collection of mainly Irish ore minerals, and more general (and rare these days for a museum) systematic displays. In total, over 750 mineral and rocks specimens are on general display. The museum is a true treasure – a place where the display of specimens is of prime importance and where the fabric of the building is as interesting as the contents of the cabinets.

GLOSSARY

accessory minerals Minerals present in a rock in such small quantities that they are of negligible importance when considering the overall composition of the rock.
adinole An albite–quartz rock formed when slate is metasomatically altered during the intrusion of a basic magma.
adit A horizontal tunnel into a hillside to draw the water from mineworkings and to provide access and ventilation.
alkaline rock Igneous rock rich in sodium and potassium, and low in calcium.
alluvial deposit Detrital material that has been eroded, transported and deposited by streams.
amorphous A mineral with no definite crystal structure, i.e. no regular arrangement of atoms.
amygdale A gas cavity in a volcanic rock infilled with secondary minerals such as calcite, quartz or zeolite group species.
augen Large lenticular eye-shape mineral grains or aggregates found in metamorphic rocks such as gneiss (*augen* is German for eyes).
aureole The zone around an igneous intrusion in which the heat of the intrusion has caused thermal metamorphism. The size of the aureole depends on the nature of both the intrusion and the country rock.
authigenic minerals Minerals that are formed in place rather than having been transported and deposited there; a term usually applied to some biologically or chemically formed sedimentary rocks.

basic rock Silica-poor igneous rock containing feldspars, which are generally more calcic than sodic.
batholith A large deep-seated subterranean intrusive mass of igneous rock, typically granite, outcropping over at least 100 km^2 and extending to unknown depth.
black tin Unrefined tin ore ready for smelting, and synonymous with cassiterite.
blister copper An unusual form of chalcopyrite in which layers of fibrous crystals develop botryoidal masses.
bole A primitive type of smelting hearth, usually sited on hilltops.
botryoidal A smooth rounded form, similar to a bunch of grapes.
boxwork A skeletal framework of intersecting plates that form in a pre-existing cavity, often after earlier minerals have been removed by groundwater-related processes.
breccia A cemented mass of rocks composed of angular rock fragments.
buchite A glassy hornfels produced by partial melting of a sedimentary rock close to the contact of an igneous intrusion, or by partial or total melting of a xenolith.

calcine To heat (roast) a substance to a high temperature but below the melting or fusing point, causing loss of moisture, reduction or oxidation, and the decomposition of carbonates and other compounds. Arsenic was once produced from arsenopyrite-rich ore using calciners consisting of furnaces connected to long flues, on whose walls white powdery arsenic soot accumulated.
calcrete Soil cemented by calcium carbonate.
Caledonian An orogeny that occurred during the Cambrian and continued into the Ordovician period.
capillary A mineral habit consisting of flexible and thread-like crystals.
carbona A type of granite-hosted cassiterite-bearing replacement body found in the Cornubian orefield. Carbonas are characterized by the pervasive hydrothermal alteration of the granite to chloritic, sericitic and tourmalinitic assemblages.
carbonatite An igneous rock containing more than 50% of carbonate minerals (mostly ankerite, calcite or dolomite).
caunter A lode that diverges at a considerable angle from the direction of other lodes.
chalcopyrite disease Microscopic blebs or orientated rods of chalcopyrite found in sphalerite produce a spotted texture not unlike a skin disease (e.g. chicken pox). Exsolution and diffusion processes have been proposed to explain its formation.
chatoyancy An optical reflectance effect, seen in certain gemstones such as tiger eye, that arises from a fibrous structure (or is attributable to fibrous inclusions) in a stone.
colloform A more-or-less spherical form similar to botryoidal.
concretion Concretions form in sedimentary rocks, usually early in the burial history of the sediment. They are often ovoid or spherical in shape and may have a nucleus containing minerals such as ankerite, barite, dolomite, gypsum, marcasite, pyrite or siderite.
cortex An external layer, like bark or rind.
crinanite An alkaline variety of dolerite, containing accessory amounts of analcime interstitially between earlier-formed plagioclase feldspar and pyroxene crystals.
cromlech A Welsh term for a prehistoric monument consisting of monoliths encircling a mound.
cross course A fault that crosses the vein being followed and which usually displaces it.
cross cut A tunnel driven at right angles to the direction of a vein.
cumulate rock An igneous rock forming the lower ultramafic layers of an intrusion, where the early-formed crystals settled during slow cooling at the bottom of the magma chamber.
cupola A relatively small isolated body of plutonic rock (often granite) that joins to a large body at depth. Granite cupolas sometimes contain tin or tungsten.

Dalradian A group of metamorphosed sediments laid down between 800 and 600 million years ago (Ma). These rocks are Precambrian and Cambrian in age.
detrital Grains of rocks, or more usually minerals, derived by erosion or weathering from pre-existing rocks.
deuteric alteration Produced by the reaction of primary magmatic minerals with watery solutions derived from the same magmatic source.
diatomaceous earth A naturally occurring soft chalk-like sedimentary rock that easily crumbles into a fine white to off-white powder. Its main constituents are diatoms – microscopic unicellular marine algae with siliceous cell walls.
diatreme A pipe-like volcanic conduit formed at great depth and filled with debris produced during the explosive release of volatile-rich gases.
dimorphous A mineral with similar chemical composition but different crystal form.
dressing floor A place where ore minerals are separated from gangue.
druse A crust of small crystals lining a cavity or mineral vein.
dyke A sheet-like intrusion of igneous rock, normally of intermediate grainsize, that cuts discordantly through the surrounding rock.

eclogite A rare metamorphic rock resembling basalt in chemical composition, formed by regional metamorphism under conditions of very high temperature and pressure.
efflorescence A normally whitish powdery coating of submicroscopic minerals forming an encrustation on rock surfaces.
element A substance consisting wholly of atoms of the same atomic number, listed in the periodic table; there are over a hundred known elements, of which the first 92, up to and including uranium, occur naturally.
elvan A Cornish mining term for a dyke rock with a granitic composition.
endo-, exo- Inside or within, and outside or external. In a geological context, used to describe types of skarn deposits found at the margins of some igneous bodies – (endoskarn – within the body; exoskarn – external to the body).
epigenetic Formed at or near the Earth's surface.
epimorph A mineral or crystal that encloses another (syn. perimorph).
epithermal Mineral veins and ore deposits formed at shallow depth from ascending hot solutions (temperature range 50–200°C) mainly derived from magmatic sources.
euhedral A term applied to grains displaying fully developed crystal form.
exsolution The separation of a homogeneous solid solution into distinct crystal phases. High-temperature alkali feldspar (e.g. orthoclase) often develops a perthitic texture when sodium-rich feldspar (plagioclase series) segregates out during slow cooling.

fathom A measure equivalent to 6 feet or nearly 2 metres.
felsic rock Pale igneous rock, with a particularly high feldspar $\pm$ quartz content.
fenite A syenite produced by alkali metasomatism in the contact zone around a carbonatite intrusion.
filiform Exhibiting many hair-like or threadlike filaments.
flat See **replacement flat deposit**.
fractionation The segregation of minerals in a body of cooling magma as a result of their crystallization at different temperatures, successive crops of crystals sinking into the magma and accumulating at the base of the intrusion.

framboidal Spheroidal aggregates of crystals resembling a raspberry.

gangue The unwanted part of an ore deposit, both the minerals that are left unexcavated and those that are separated from the ore-mineral concentrate after excavation.
geo An inlet in a coastal cliff face created by wave erosion. Used in the Orkney and Shetland Isles, NE Scotland.
glomeroporphyritic Phenocrysts of the same species clustered together in porphyritic igneous rocks.
gossan An iron-rich deposit overlying a sulphide ore deposit, often caused by oxidation of pyrite and chalcopyrite as downward-percolating waters remove copper and sulphur. "Gossan" derives from the Cornish for "blood" – referring to the rusty colour of the deposit.
granodiorite Igneous rock similar to granite but containing more mafic minerals and more feldspars.
greisen Granite isochemically altered by pneumatolysis.

hemimorphic A crystal with polar symmetry, having two ends with different crystal form.
holotype A single specimen from which all the data for the original description were obtained.
honestone A rock suitable for sharpening cutting implements.
hopper crystal A fast-grown crystal with fully developed edges, but with interior spaces that are hollowed out in steps, giving the impression material that had been removed.
hush A man-made cut down a hillside using a stream of water to reveal bedrock and any veins. The stream would be dammed back and then released.
hydrocarbons Organic compounds consisting primarily of hydrogen and carbon, as occur in petroleum, natural gas and coal; a few hydrocarbons are minerals.
hypothermal Hydrothermal ore deposits formed in a high-temperature environment (300–500°C).

ignimbrite A type of pyroclastic rock with a streaky appearance, formed by the welding together of hot tuff material deposited during an explosive volcanic eruption.
inclusion A small crystal totally enclosed within a larger host crystal. Fluid inclusions represent trapped liquids.
intercumulus crystals In certain magmas (e.g. those producing peridotite) the first minerals to crystallize are particularly dense and they settle out to accumulate at the bottom of the magma chamber as cumulate crystals. The surrounding magma eventually cools and crystallizes around them to form intercumulus crystals.
intersertal Angular interstices filled with glass, in rocks with interlocking granular textures; usually found in basalts and dolerites.
interstitial In a petrographic context, the space between early-formed closely spaced crystals. Interstitial spaces can be filled with late-forming (or interstitial) minerals.
iron pan A type of hardpan rich in iron oxide – a secondary accumulation of hard calcareous material forming layers near the surface of stony soils in arid and semi-arid regions.
isotope A form of an element having the same atomic number as other forms of the same element, but a different atomic weight, because of a different number of neutrons in the nucleus.

jig A box-like device used to separate ore minerals from gangue by shaking them up and down on a sieve in water.

killas A Cornish term for low-grade metamorphic rocks: phyllites, slaty shales and mudstones with subordinate bands of grit and conglomerate that were deformed and metamorphosed during the Variscan orogeny.

lamprophyre A group of (dyke-like) igneous rocks characterized by a porphyritic texture and containing phenocrysts of dark minerals such as biotite–phlogopite or pyroxene.
level Horizontal tunnels usually set at vertical intervals, e.g. 10 fathoms, 20 fathoms.
lineament An extensive linear topographical feature that may reflect the trend of some underlying structure, often detectable only from aerial photography.
lithomarge An aluminium-rich clay-like rock with a smooth texture.
lode A regular principal vein.

mafic rock A rock predominantly composed of ferromagnesian rock-forming silicates (e.g. augite).
mammillated Rounded and mutually interfering masses, similar, but more flattened than a botryoidal habit.
mesostasis The last-formed interstitial material of an igneous rock; often glassy or finely crystalline.
mesothermal Hydrothermal deposits formed at temperatures of 200– 300°C.
metamict Uranium- or thorium-rich minerals can become metamict when the original crystal structure is disrupted by radiation damage. The original external morphology is essentially unchanged.
metasomatism A type of metamorphism where existing minerals are transformed totally or partially into new minerals by the replacement of their chemical constituents. It often occurs when granite-derived hydrothermal fluids interact with surrounding host rocks.
miarolitic cavity A small irregular cavity found in granitic rocks and sometimes containing well formed secondary minerals.
mineral An element or chemical compound that is normally crystalline and which has been formed as a result of geological processes.
minette A lamprophyric rock consisting of biotite or phlogopite phenocrysts in a feldspar–mica–pyroxene matrix.
Mississippi Valley type (MVT) MVT ore deposits are epigenetic stratabound carbonate-hosted sulphide bodies composed predominantly of sphalerite and galena. These deposits account for 35% of the world's lead and zinc resources. The deposits occur mainly in dolostone as open-space fillings, collapse breccias or as replacement of the carbonate host rock. Less commonly, sulphide and gangue minerals occupy primary carbonate porosity. The deposits are epigenetic, having been emplaced after lithification of the host rock.
modal analysis An estimate of rock composition in terms of the proportions of minerals present.
monchiquite An alkaline variety of lamprophyre consisting of phenocrysts of amphibole, mica and pyroxene, in a glassy groundmass containing analcime or nepheline.
myrmekite An intergrowth of plagioclase feldspar and vermicular quartz.

opencast or opencut A place where ore is removed from a shallow deposit at the surface.
ophitic A texture found in igneous rocks in which randomly orientated plagioclase feldspar crystals are partially or wholly surrounded by a single large pyroxene crystal.
ore An aggregate of minerals and rocks from which it is commercially worthwhile to extract minerals (usually metals). An ore has two parts: the gangue (the unwanted rock and minerals) and the desired ore mineral itself.
ore-shoot The part of the vein containing workable ore.
oscillatory zoning A pattern of growth found in some minerals (e.g. columbite group, plagioclase feldspar, pyroxene group) when rapidly changing magmatic or hydrothermal conditions occur prior to consolidation. Individual zones have differing compositions, which oscillate in regular (e.g. A,B,C,D,A,B,C,D) or irregular ways.
oxidized zone The upper part of an orebody that has been modified by a solution of oxygen-bearing surface waters, soil acids and carbon dioxide. It extends from the surface to the water table.

paragenesis An association of minerals, related in space and time.
paramorph A type of pseudomorph in which one mineral is replaced by another with the same chemical composition (e.g. rutile after brookite).
peralkaline rocks Igneous rocks having $Al_2O_3 < (Na_2O+K_2O)$ in molecular proportions. Such rocks typically contain aegirine or riebeckite.
peraluminous rocks Igneous rocks having $Al_2O_3 > (CaO+Na_2O+K_2O)$ in molecular proportions. Such rocks typically contain almandine–spessartine garnet, corundum, muscovite, topaz or tourmaline.
perlitic texture Found in glassy and devitrified igneous rocks and consisting of cracks, of curved to spherical shape, caused by contraction from cooling.
perthite A parallel to subparallel intergrowth of potassium and sodium feldspar. Usually potassium feldspar is the host mineral from which the sodium feldspar has exsolved.
phantom crystal A crystal within a crystal. The outlines of the early-formed crystal are visible though the transparent outer crystal.
phenocryst A large crystal in a finer-grain groundmass within porphyritic igneous rocks. Phenocrysts are crystals of the first minerals to form as magma cools and begins to solidify.
pig of lead A mass of lead formed when the molten metal is poured into a trough with branches coming off it. Each branch ends in a depression that, when filled, becomes a pig of solid metal. When viewed from above, the shapes look like pigs feeding from a trough.
plane-polarized light Light in which the waves vibrate in only one plane (ordinary light vibrates in all directions perpendicular to the direction of propagation of the ray).
pleochroism A term used in thin-section descriptions of minerals when minerals exhibit a variation in colour as they are rotated in plane-polarized light on a microscope stage. Pleochroism occurs when certain wavelengths of light are absorbed from incident white light.
pluton A large-scale mass of igneous rock that has consolidated at great depth beneath the Earth's surface (smaller than a batholith).
pneumatolysis Alteration of a rock by the introduction of new elements, especially volatiles.
poikilitic A texture produced in igneous rocks when small variably orientated early-formed crystals are enclosed in a larger crystal.
poikiloblastic A metamorphic texture consisting of small, usually rounded, crystals that are enclosed within much larger crystals.

point counting A way of estimating the modal proportions of minerals in a thin section of rock. It can be performed mechanically with a device called a point counter and a motorized stage on a microscope, or by taking a photograph of the thin section and calculating the proportions using computer software.
polymorphs Minerals with the same chemical composition, but different crystal structure, e.g. marcasite and pyrite, or calcite and aragonite.
polytype A mineral with the same chemical formula and the same structural sub-units as another, but with a different stacking sequence of the sub-units.
porcellanite A hard dense rock similar in appearance to unglazed porcelain. It can be an impure variety of chert containing clay and calcareous matter, or a tough contact metamorphosed hornfels formed from a lateritic soil horizon.
porphyroblast A compound name derived from porphyritic and -blastic. The name is applied to large well formed crystals in a metamorphic rock formed during the process of metamorphism.
prograde metamorphism A process of increasing temperature and pressure, which causes new minerals to replace older ones.
provenance The source location of a rock or mineral.
pseudomorph A mineral replaced by another mineral by a process of chemical alteration without any change in the external form.
pyribole A collective term to describe members of the pyroxene and amphibole groups.
pyroclastic rock A rock made up of particles thrown into the air by volcanic eruption.

rake A mineral vein disposed more or less vertically (larger than a scrin).
redbeds An assemblage of sedimentary rocks characterized by a red colour resulting from formation in a highly oxidizing environment, the redness being attributable to the presence of iron in a ferric, rather than ferrous, state.
reniform A bean or leaf shape resembling that of a kidney.
replacement flat deposit Horizontal ore bodies produced when limestone is extensively replaced adjacent to mineralized veins.
residual orebody The part of a deposit left behind after surface processes have acted on a primary sulphide deposit; equivalent to the oxidized zone of a deposit.
rise A shaft driven upwards.

saddle dolomite Crystals of dolomite with pronounced curved rhombohedral faces.
saussuritization Replacement of plagioclase in basalts and gabbros, by a fine-grain aggregate of albite, calcite, epidote, muscovite (sericite), zoisite and a range of zeolite species.
scalenohedron A pyramidal form under the rhombohedral system.
scrin A thin vertical mineral vein usually branching out of a rake.
secondary mineralization Any type of mineralization where a primary deposit is altered and which results in the formation of a different mineral assemblage. Commonly, such alteration is confined to a zone of near-surface weathering.
sedimentary exhalative deposit A lenticular (stratiform) deposit formed at the interface between volcanic and sedimentary units when metals precipitate out of hydrothermal solutions.
selvedge Usually a fragment of host rock incorporated into the margins of an igneous intrusion, but also used to describe the early-crystallized remnants in a composite vein.
septarian nodules Ovoid-shape concretions containing angular cavities or cracks (septaria) often infilled with barite, calcite or siderite.
sett An area of ground leased for mining.
scrin A small vertically orientated mineral vein often branching out of a rake, sometimes in groups (or swarms).
showing A minor occurrence of ore minerals, too small for economic exploitation.
sinter Opaline siliceous material deposited by hot springs or geysers.
skarn A calcium-rich silicate rock derived from the thermal metamorphism and alteration of impure limestone and dolomite adjacent to an igneous intrusion.
slag Waste material produced in smelting.
smalt A deep-blue paint and ceramic pigment produced by pulverizing a glass made of silica, potash and cobalt oxide.
smelting The process whereby metal is extracted from its ore.
sough An adit or level specifically driven to drain a mine.
sövite Coarse-grain calcite–carbonatite.
sphenoidal Although having tetragonal symmetry, characteristically sphenoidal crystals of chalcopyrite resemble tetrahedra.
stamps Machines using heavy weights to crush ore.
stock A discordant body of intrusive igneous rock – like a pluton.
stockwork A mineral deposit in the form of a network of veinlets in host rock.
stope A working where the vein is cut away between levels.
stratabound deposit A mineral deposit whose form is lenticular or bedded.
strike The direction in which a horizontal line is drawn on a plane.The strike is important in determining the direction in which to measure the dip of rock strata. Strike is always at right angles to the direction of dip.
stringer A very narrow vein.
stromatolite Laminated calcareous sedimentary formations produced by lime-secreting cyanobacteria.
sublimate A substance that can pass directly from a solid state to a vapour without forming a liquid.
sump A small shaft, usually sunk from one level to another for access and to improve ventilation. Also, the bottom of an engine shaft.
supergene Secondary processes (especially oxidation caused by weathering) resulting from the action of descending water on primary minerals.
symplectite An intergrowth of two minerals, in which one is contained as worm-like inclusions in the other.

tailings Finely ground waste material from mineral processing.
tteschenite A dark alkaline igneous rock (dolerite) often found in sill-like bodies. Essential constituents are augite and plagioclase feldspar. Analcime, magnetite (or ilmenite) and Na-rich amphibole (pargasite) are also present in minor amounts. Teschenite is a synonym of crinanite.
thunder eggs Agate-filled nodules that form in some types of volcanic rocks. As molten lava cools, steam gases become trapped and form bubbles. Silica minerals often crystallize within these bubbles or grow crystal fibres that radiate outwards from the centre (spherules). Sometimes gas pressure can force the partially solidified spherulite apart and form a central hollow. Silica gels or clays filling the cavity can produce a typical star-shape pattern.
tin floor A thinly layered tourmaline-bearing rock containing the tin-ore mineral cassiterite.
toadstone A general term applied to olivine dolerite or volcanic rocks such as ash and tuff. The rock may be altered to a green clay or be fragmental.
tonstein Kaolinitic bands containing characteristic fossils from short-lived but widespread marine invasions.
touchstone A hard mineral such as jasper, formerly used to test the quality of gold or silver by comparing the streak left on the stone by one of these metals with that of a standard alloy.
transmitted light Light that is passed through a transparent or translucent medium – in mineralogy, usually through a thin section of rock using a polarizing microscope.
tuff A rock composed of consolidated volcanic ash. Varieties include crystal tuffs, lithic tuffs, lapilli tuffs and vitric tuffs.

ultramafic rock An igneous rock containing less than 45% SiO_2. Ultramafic rocks are mostly plutonic and contain virtually no quartz or feldspar, their main constituents being ferromagnesian silicates, oxides and sulphides.
umber An impure, earthy form of iron or manganese oxides, sometimes used as a brown or red pigment.
unconformity A substantial break or gap in the geological record in which a rock layer is not overlain by another in the same stratigraphical succession.
uralitization A relatively common alteration process in basic igneous rocks, where pyroxene is replaced by fibrous amphibole.

vein A mineral formation that is of far greater extent in two dimensions than in the third. Sheet-like fissure veins occur where fissures formed in the rock become filled with minerals. Ladder veins form in series of fractures in, for example, dykes. Saddle veins are the shape of a lens: concave below and convex above. Veins containing economically important ores are termed lodes. Many veins have a hydrothermal origin.
veinstone The valueless non-metalliferous mineral or rock material that surrounds the ore minerals in a vein; it is often barite, calcite or quartz.
vesicle A small cavity in a volcanic rock resulting from the presence of a bubble of gas in the magma trapped during solidification.
volatile component A component of magma whose vapour pressure is high enough to allow it to be concentrated in a gaseous or vaporous phase.
vug A natural cavity in a vein often lined with beautiful crystals.

white tin Refined metallic tin – a term historically associated with tin mining in Cornwall.

xenolith An inclusion of a foreign rock embedded in an igneous body; either a block of country rock or a block of the igneous rock itself that solidified earlier than its host and has different in composition.

zawn A Cornish term for a steep-sided coastal inlet, where softer rock has been preferentially eroded by the sea. Some zawns are fault controlled.

BIBLIOGRAPHY

Abdul-Samad, F. A., J. H. Thomas, P. A. Williams, R. F. Symes 1982. Chemistry of formation of lanarkite, Pb_2OSO_4. *Mineralogical Magazine* **46**, 499–501.

Adair, B. G. I. 1987. *Zeolites and zeolitisation of the Tertiary basalts of northeastern Ireland.* PhD thesis, Queen's University, Belfast.

Adam, G-J. 1869. *Tableaux minéralogique.* Paris.

Adam, W. 1846. *Gem of the Peak* (4th edn), Mozley, Derby. Ashbourne, Derbyshire: Moorland Publishing [1973: 5th edn reprinted with biographical notes by T. D. Ford].

Adye, E. H. 1907. *Modern lithology illustrated and defined: for the use of university, technical and civil-service students.* Edinburgh: W. & A. K. Johnson.

Agrell, S. O. 1939. The adinoles of Dinas Head, Cornwall. *Mineralogical Magazine* **25**, 305–337.

——— 1941. Dravite-bearing rocks from Dinas Head, Cornwall. *Mineralogical Magazine* **26**, 81–93.

——— 1951. A gabbro–limestone contact near Camphouse, Ardnamurchan. *American Mineralogist* **36**, 1080.

——— 1965. Polythermal metamorphism of limestones at Kilchoan, Ardnamurchan. *Mineralogical Magazine* **34**, 1–15.

Agrell, S. O. & P. Gay 1961. Kilchoanite, a polymorph of rankinite. *Nature* **189**, 743.

Agrell, S. O. & J. M. Langley 1958. The dolerite plug at Tievebulliagh, near Cushendall, Co. Antrim. *Royal Irish Academy, Proceedings* **59B**, 93–127.

Aherne, S., N. A. Reynolds, D. J. Burke 1992. Gold mineralization in the Silurian and Ordovician of south Mayo. See Bowden et al. (1992: 39–49).

Aherne, S., N. A. Reynolds, D. J. Burke, K. N. O'Keeffe, K. M. Verbruggen 1989. Gold mineralization in Silurian rocks south of Clew Bay, County Mayo, Ireland. *Institution of Mining and Metallurgy, Transactions B* **98**, 60–62.

Ahmed, A. A. & B. E. Leake 1978. The Inishdawros meta-peridotite, Callow, Ballyconneely, Connemara, western Ireland. *Mineralogical Magazine* **42**, 69–74.

——— 1990. S-type granite formation in the Dalradian rocks of Connemara, western Ireland. *Mineralogical Magazine* **54**, 1–22.

——— 1996. The conditions of metamorphism of a grossular–wollastonite–vesuvianite skarn from the Omey granite, Connemara, western Ireland, with special reference to the chemistry of vesuvianite. *Mineralogical Magazine* **60**, 541–50.

Ahmed-Said, Y. & P. W. G. Tanner 2000. P–T conditions during emplacement, and D2 regional metamorphism, of the Ben Vuirich granite, Perthshire, Scotland. *Mineralogical Magazine* **64**, 737–53.

Ainsworth, W. 1834. *An account of the caves of Ballybunnion, County of Kerry, with some mineralogical details.* Dublin: William Curry.

Akaad, M. K. 1956. The northern aureole of the Ardara pluton of County Donegal. *Geological Magazine* **93**, 377–92.

Akizuki, M., Y. Kudoh, T. Kuribastashi 1966. Crystal structures of the {011}, {610} and {010} growth sectors in brewsterite. *American Mineralogist* **51**, 1501–506.

Al-Agha, M. R., S. D. Burley, C. D. Curtis, J. Esson 1995. Complex cementation textures and authigenic mineral assemblages in Recent concretions from the Lincolnshire Wash (east coast, UK) driven by Fe (0) to Fe (II) oxidation. *Geological Society of London, Journal* **152**, 157–71.

Alabaster, C. 1975. Some copper, lead and manganese minerals from Merehead quarry, East Mendip. *Bristol Naturalists' Society, Proceedings* **34**, 76–104.

——— 1976. Post Inferior Oolite mineralization at Whatley quarry, east Mendip. *Bristol Naturalists' Society, Proceedings* **35**, 73–84.

——— 1977. An occurrence of brucite at Merehead quarry, Cranmore, Somerset. *Mineralogical Magazine* **41**, 406–408.

——— 1978. A new wulfenite locality, near Bristol. *Mineralogical Magazine* **42**, 298–300.

——— 1980. Analcime from the Lower Lias on the west Somerset coast. *Mineralogical Magazine* **43**, 761–4.

——— 1983. Native sulphur from Stancombe quarry, Flax Bourton and Dan-y-graig quarry, Risca (South Wales). *Bristol Naturalists' Society, Proceedings* **43**, 13–27.

——— 1985a. Melanotekite from the Bristol district. *Bristol Naturalists' Society, Proceedings* **45**, 11–16.

——— 1985b. Plattnerite from the Bristol district. *Bristol Naturalists' Society, Proceedings* **45**, 17–22.

——— 1989a. Alumohydrocalcite from the Bristol district. *Russell Society, Journal* **2**(2), 5–10.

——— 1989b. The Wesley mine: a further occurrence of manganese-oxide-hosted lead oxychloride minerals in the Bristol district. *Russell Society, Journal* **2**(2), 29–47.

——— 1990a. Alstonite and barytocalcite from Llantrisant, South Wales, and barytocalcite from Holwell, Mendip Hills, England. *Russell Society, Journal* **3**(1), 1–14.

——— 1990b. New occurrences of secondary aluminium minerals from the Bristol district. *Russell Society, Journal* **3**(2), 49–59.

Alberti, A., G. Cruciani, I. Dauru 1995. Order/disorder in natrolite-group minerals. *European Journal of Mineralogy* **7**, 501–508.

Alderman, A. R. 1935. Almandine from Botallack, Cornwall. *Mineralogical Magazine* **24**, 42–8.

——— 1936. Eclogites from the neighbourhood of Glenelg, Inverness-shire. *Geological Society of London, Quarterly Journal* **92**, 488–530.

Alderton, D. H. M. 1978. Fluid inclusion data for lead–zinc ores from southwest England. *Institution of Mining and Metallurgy, Transactions B* **87**, 132–5.

——— 1986. Hessite and electrum from the Ratagain intrusion, northwest Scotland. *Mineralogical Magazine* **50**, 179.

——— 1988a. Calc-silicate minerals from the Dartmoor granite. *Mineralogical Magazine* **52**, 527–9.

——— 1988b. Ag–Au–Te mineralization in the Ratagain complex, northwest Scotland. *Institution of Mining and Metallurgy, Transactions B* **97**, 171–80.

——— 1991. Stannite revisited. *Russell Society, Journal* **4**(1), 17–23.

Alderton, D. H. M. & R. S. Harmon 1991. Fluid inclusion and stable-isotope evidence for the origin of mineralizing fluids in southwest England. *Mineralogical Magazine* **55**, 605–611.

Alderton, D. H. M. & N. J. Jackson 1978. Discordant calc-silicate bodies from the St Just aureole, Cornwall. *Mineralogical Magazine* **42**, 427–34.

Alderton, D. H. M. & F. Moore 1981. New determinations of tin and tungsten in granites from southwest England. *Mineralogical Magazine* **44**, 354–6.

Alderton, D. H. M., D. Mitchell, R. Giddens 1995. Gratonite from the Isle of Man. *Mineralogical Magazine* **59**, 573–6.

Alderton, D. H. M., J. A. Pearce, P. J. Potts 1980. Rare earth element mobility during granite alteration: evidence from southwest England. *Earth and Planetary Science Letters* **49**, 149–65.

Aldridge, J. 2005. *Characterization of the mineralization occurring at Wasdale Head, Lake District.* MSc thesis, Camborne School of Mines, University of Exeter.

Alexander, R. W. S., J. B. Dawson, E. M. Patterson, R. L. Hervig 1986. The megacryst and inclusion assemblage from the Black Rock vent, Ayrshire. *Scottish Journal of Geology* **22**, 203–212.

Al-Hermezi, H. M. 1985. Qandilite, a new spinel end-member, Mg_2TiO_4, from the Qala–Dizeh region, northeast Iraq. *Mineralogical Magazine* **49**, 739–44.

Al-Salah, S., R. Fuge, W. J. Rea 1977. The geochemistry of some biotites from the Dartmoor granite. *Ussher Society, Proceedings* **4**, 37–48.

Allan, W. C. 1970. The Morven–Cabrach basic intrusion. *Scottish Journal of Geology* **6**, 53–72.

Allen, P. 2002. Field meeting report: Hemerdon Bal, Plympton, Devon. *Russell Society, Newsletter* **41**, 31.

Allen, P. M. & G. D. Easterbrook 1978. Mineralized breccia pipe and other intrusion breccias in the Harlech Dome, North Wales. *Institution of Mining and Metallurgy, Transactions B* **87**, 157–61.

Allen, P. M. & A. A. Jackson 1985. *Geology of the country around Harlech (sheet 135 and part of sheet 149).* Memoir, British Geological Survey, Keyworth.

Allen, P. M., D. C. Cooper, M. E. Parker, G. D. Easterbrook, H. W. Haslam 1982. *Mineral exploration in the area of the Fore Burn igneous complex, southwestern Scotland.* Report 55, Mineral Reconnaissance Program, Institute of Geological Sciences, London.

Allison, A. 1936. The Tertiary dykes of the Craignish area, Argyll. *Geological Magazine* **73**, 73–87.

Allport, S. 1871. On the microscopical structure and composition of a phonolite from the Wolf Rock. *Geological Magazine* **8**, 247–50.

——— 1874. Note on the phonolite from the Wolf Rock. *Geological Magazine* **11**, 462–3.

Alwan, A. K. & P. A. Williams 1979. Nickeloan hydrozincite: a new variety. *Mineralogical*

Magazine **43**, 397–8.

Amiri-Garroissi, K. 1977. Origin of montmorillonite in the early Jurassic shales of NW Scotland. *Geological Magazine*, **114**, 281–90.

Anderson, I. K., J. H. Ashton, A. J. Boyce, A. E. Fallick, M. J. Russell 1998. Ore depositional processes in the Navan Zn–Pb deposit, Ireland. *Economic Geology* **93**, 535–63.

Anderson, F. W. & K. C. Dunham 1966. *The geology of northern Skye.* Memoir, Institute of Geological Sciences, Edinburgh.

Anderson, J. G. C. & S. Elder 1935. Apophyllite from Traprain Law, East Lothian. *Mineralogical Magazine* **24**, 90–91.

Anderson, T. 1851. Description and analysis of gyrolite, a new mineral species. *Philosophical Magazine (4th series)* **1**, 111–13.

Anderson, W. & J. A. Smythe 1942. An occurrence of millerite in the Durham Coal Measures. *Geological Magazine* **79**, 220–24.

Anderton, J. G. C. & A. Lamont 1935. The geology of the Glasgow district from its quarries. *Quarry Managers' Journal* **18**, 216.

Andrew, A. R. 1910. The geology of the Dolgelley gold belt, North Wales [parts I–III]. *Geological Magazine* **47**, 159–71, 201–211, 261–71.

Andrew, C. J. 1986. The geological setting and style of mineralization at Ballyvergin, County Clare. See Andrew et al. (1986: 475–80).

Andrew, C. J. & J. H. Ashton 1985. Regional setting, geology and metal distribution patterns of Navan orebody, Ireland. *Institution of Mining and Metallurgy, Transactions B* **94**, 66–93.

Andrew, C. J. & A. Poustie 1986. Syndiagenetic or epigenetic mineralization – the evidence from the Tatestown zinc–lead prospect, Co. Meath. See Andrew et al. (1986: 281–96).

Andrew, C. J., R. W. A. Crowe, S. Finlay, W. M. Pennell, J. F. Pyne (eds) 1986. *Geology and genesis of mineral deposits in Ireland.* Dublin: Irish Association for Economic Geology.

Angus, N. S. & R. Kanaris-Sotiriou 1982. Autometasomatic gneisses of the Currywongaun–Doughruagh syntectonic intrusion, Connemara, Ireland. *Mineralogical Magazine* **46**, 411–20.

——— 1995. Adinoles revisited: hydrothermal Na(Ca) metasomatism of pelite screens adjacent to tholeiitic dykes in the Dublin terrane, Ireland. *Mineralogical Magazine* **59**, 367–82.

Angus, N. S. & R. Middleton 1985. Compositional variation in högbomites from north Connemara, Ireland. *Mineralogical Magazine* **49**, 649–54.

Annels, A. E. & D. E. Roberts 1989. Turbidite-hosted gold mineralization at the Dolaucothi gold mines, Dyfed, Wales, United Kingdom. *Economic Geology* **84**, 1293–314.

Anon. 1974. *Scourie.* Molybdenite, Project AE3, Mineral Exploration & Investments Grant Act 1972, Institute of Geological Sciences, Edinburgh.

——— 1989a. Mineral news. *UK Journal of Mines and Minerals* **6**, 2.

——— 1989b. Mineral news. *UK Journal of Mines and Minerals* **7**, 8–12.

——— 1993. [Front-cover illustration]. *Mineralogical Record* **24**(4).

Anthony, J. W., R. A. Bideaux, K. W. Bladh, M. C. Nichols 1990. *Handbook of mineralogy, vol. I: elements, sulphides, sulphosalts.* Tucson: Mineral Data Publishing.

——— 1995. *Handbook of mineralogy, vol. II: silica, silicates.* Tucson: Mineral Data Publishing.

——— 1997. *Handbook of mineralogy, vol. III: halides, hydroxides, oxides.* Tucson: Mineral Data Publishing.

——— 2000. *Handbook of mineralogy, vol. IV: arsenates, phosphates, vanadates.* Tucson: Mineral Data Publishing.

——— 2003. *Handbook of mineralogy, vol. V: borates, carbonates, sulphates.* Tucson: Mineral Data Publishing.

Anwar, Y. M. 1955. A clinopyroxene from the granophyre of Meall Dearg, Skye. *Geological Magazine* **92**, 367–73.

Appleby, A. & P. J. Williams 1988. Epigenetic tourmalinite and arsenopyrite in the aureole of the Leinster batholith, southeast Ireland. *Mineralium Deposita*, **23**, 247–55.

Appleton, J. D. & A. J. Wadge 1976. *Investigation of tungsten and other mineralization associated with the Skiddaw granite near Carrock mine, Cumbria.* Report 7, Mineral Reconnaissance Program, Institute of Geological Sciences, London.

Archer, A. A. 1959. The distribution of non-ferrous ores in the Lower Palaeozoic rocks of North Wales. See Higham (1959: 259–76).

Archer, A. A. & R. W. Elliot 1965. The occurrence of olivine dolerite dykes near Llanrwst, North Wales. *Geological Survey of Great Britain, Bulletin* **23**, 145–53.

Armbruster, T. 2002. Revised nomenclature of högbomite, nigerite and taaffeite minerals. *European Journal of Mineralogy* **14**, 389–95.

Armbruster, T., P. Bonazzi, M. Akasaka, V. Bermanec, C. Chopin, R. Gieré, S. Heuss-Assbichler, A. Liebscher, S. Menchetti, Y. Pan, M. Pasero 2006. Recommended nomenclature of epidote-group minerals. *European Journal of Mineralogy* **18**, 551–67.

Armstrong, G., K. C. Dunham, C. O. Harvey, P. A. Sabine, W. F. Waters 1951. The paragenesis of sylvine, carnallite, polyhalite and kieserite in Eskdale borings Nos 3, 4 and 6, northeast Yorkshire. *Mineralogical Magazine* **29**, 667–89.

Armstrong, R., R. J. Herrington, M. A. Savage 2003. Tennantite and tyrolite from the Coed-y-Brenin forest, North Wales. *Russell Society, Journal* **8**(1), 18.

Arthurs, J. W. & G. Earls 2004. Minerals. See Mitchell (2004: 255–72).

Arthurton, R. S. & J. E. Hemingway 1972. The St Bees evaporites – a carbonate–evaporite formation of Upper Permian age in west Cumberland, England. *Yorkshire Geological Society, Proceedings* **38**, 43–8.

Arthurton, R. S. & A. J. Wadge 1981. *Geology of the country around Penrith (sheet 24).* Memoir, Institute of Geological Sciences, London.

Arthurton, R. S., I. C. Burgess, D. W. Holliday 1978. Permian and Triassic. See Moseley (1978: 189–206).

Artioli, G. & H. Foy 1994. Gobbinsite from Magheramorne quarry, Northern Ireland. *Mineralogical Magazine* **58**, 615–20.

Artioli, G. & E. Galli 1999. Gonnardite: re-examination of holotype material and discreditation of tetranatrolite. *American Mineralogist* **84**, 1445–50.

Ashby, D. F. 1946. Pyroxenes from the Lower Carboniferous basalts of the Old Pallas area, Co. Limerick. *Mineralogical Magazine* **27**, 195–6.

Ashcroft, F. N. 1916. The natrolite occurrence near Kinblane (White Head), County Antrim. *Mineralogical Magazine* **17**, 305–308.

Ashton, J. H. 2005. The Navan carbonate-hosted Zn–Pb deposit, Ireland. *Ore Geology Reviews* **27**, 270.

Ashton, J. H., D. T. Downing, S. Finlay 1986. The geology of the Navan Zn–Pb orebody. See Andrew et al. (1986: 243–58).

Ashton, J. H., M. P. Holdstock, J. F. Geraghty, W. G. O'Keeffe, N. Martinez, W. Peace, M. E. Philcox 2004. The Navan orebody – discovery and geology of the South West extension. See Kelly et al. (2004: 405–436).

Ashton, J. H., A. Black, J. Geraghty, M. Holdstock, E. Hyland 1992. The geological setting and metal distribution patterns of Zn–Pb–Fe mineralization in the Navan Boulder Conglomerate. See Bowden et al. (1992: 171–211).

Ashworth, J. R. 1976. Petrogenesis of migmatites in the Huntly–Portsoy area, northeast Scotland. *Mineralogical Magazine* **40**, 661–82.

Ashworth, J. R. & G. A. Chinner 1978. Coexisting garnet and cordierite in migmatites from the Scottish Caledonides. *Contributions to Mineralogy and Petrology* **65**, 379–94.

Atencio, D., J. M. V. Coutinho, S. Graeser, P. A. Matioli, L. A. D. Menzies Filho 2004. Lindbergite, a new Mn oxalate dihydrate from Boca Rica mine, Galiléia, Minas Gerais, Brazil, and other occurrences. *American Mineralogist* **89**, 1087–1091.

Atherton, M. P. 1968. The variation in garnet, biotite and chlorite composition in medium-grade pelitic rocks from the Dalradian, Scotland, with particular reference to the zonation in garnet. *Contributions to Mineralogy and Petrology* **18**, 347–71.

——— 1980. The occurrence and implications of chloritoid in a contact aureole andalusite schist from Ardara, County Donegal. *Royal Dublin Society, Journal of Earth Sciences* **3**, 101–109.

Atherton, M. P. & M. S. Brotherton 1974. Metamorphic index minerals in the eastern Dalradian. *Scottish Journal of Geology* **17**, 185–221.

Atherton, M. P., M. H. Naggar, W. S. Pitcher 1975. Kyanite in some thermal aureoles. *American Journal of Science* **275**, 432–43.

Atkin, B. P. 1978. Hercynite as a breakdown product of staurolite from within the aureole of the Ardara pluton, Co. Donegal, Eire. *Mineralogical Magazine* **42**, 237–9.

Atkin, D., I. R. Basham, J. F. W. Bowles 1983. Tristramite, a new calcium uranium phosphate of the rhabdophane group. *Mineralogical Magazine* **47**, 393–6.

Atkins, F. B. 1995. Sulphide, sulphate, phosphate and carbonate mineral assemblages in the Oxford Clay, Cassington–Yarnton area, Oxfordshire, England. *Russell Society, Journal* **6**(1), 41–5.

Aveline, W. T. & T. McK. Hughes 1888. *The geology of the country around Kendal, Sedbergh, Bowness and Tebay (sheet 98NE)* (2nd edn revised and enlarged by A. Strahan). Memoir, Geological Survey of Great Britain and Ireland, London.

Avison, M. 1984. The Loch Guitane volcanic complex, County Kerry – a preliminary survey. *Irish Journal of Earth Sciences* **6**, 127–36.

Baba, S. 1999. Sapphirine-bearing orthopyroxene–kyanite/sillimanite granulites from South Harris, NW Scotland: evidence for Proterozoic UHT metamorphism in the Lewisian. *Contributions to Mineralogy and Petrology* **136**, 33–47.

——— 2003. Two stages of sapphirine formation during prograde and retrograde metamorphism in the Palaeoproterozoic Lewisian Complex, South Harris, NW Scotland. *Journal of Petrology* **44**, 329–54.

——— 2004. Palaeoproterozoic UHT metamorphism in the Lewisian Complex and North Atlantic region. *Journal of Mineralogical and Petrological Sciences* **99**, 202–212.

Baba, S., E. S. Grew, C. K. Shearer, J. W. Sheraton 2000. Surinamite: a high-temperature metamorphic beryllosilicate from Lewisian sapphirine-bearing kyanite–orthopyroxene–quartz–potassium feldspar gneiss at South Harris, NW Scotland. *American Mineralogist* **85**, 1474–84.

Babkine, J. & F. Conquéré 1968. Nouvelles données sur la composition de la mackinawite dans les roches basiques et ultrabasiques. *Académie des Sciences, Paris, Comptes Rendus Serie D* **267**, 267–70.

Bachmann, H-G., W. Birch, E. R. Coogan, R. B. Cook, M. P. Cooper, K. M. Dunnell and 12 other authors 2003. *Gold, the noble mineral.* East Hampton, Connecticut: Lapis International.

Back, M. E. & J. A. Mandarino 2008. *Fleischer's glossary of mineral species 2008.* Tucson: Mineralogical Record.

Badham, J. P. N. 1978. Slumped sulphide deposits at Avoca, Ireland, and their significance (contributed remarks by J. W. Platt). *Institution of Mining and Metallurgy, Transactions B* **87**, 21–6, 184–8.

Bain, D. C. 1970. Plumbogummite-group minerals from Mull and Morvern. *Mineralogical Magazine* **37**, 934–8.

Bain, D. C., P. F. S. Ritchie, D. R. Clark, D. M. L. Duthie 1980. Geochemistry and mineralogy of weathered basalt from Morvern, Scotland. *Mineralogical Magazine* **43**, 865–72.

Bain, J. A., D. A. Briggs, F. May 1971. Geology and mineralogical appraisal of an extensive talc–magnesite deposit in the Shetlands. *Institution of Mining and Metallurgy, Transactions B* **80**, 77–84.

Baker, A. J. 1986. Ecolgitic amphibolites from the Grampian Moines. *Mineralogical Magazine* **50**, 217–21.

Ball, D. F. 1964. Gibbsite in altered granitic rock in North Wales. *Nature* **204**, 673–4.

——— 1969. Basaluminite from Cambrian rocks near Harlech. *Mineralogical Magazine* **37**, 291–3.

Ball, T. K. & I. R. Basham 1979. Radioactive accessory minerals in granites from southwest England. *Ussher Society, Proceedings* **4**, 437–48.

——— 1984. Petrogenesis of the Bosworgey granitic cusp in the southwest England tin province and its implications for ore mineral genesis. *Mineralium Deposita* **19**, 70–77.

Ball, T. K. & D. J. Bland 1985. The Cae Coch massive sulphide deposit, Trefriw, North Wales. *Geological Society of London, Journal* **142**, 889–98.

Ball, T. K., I. R. Basham, U. McL. Michie 1982. Uraniferous vein occurrences of southwest England – paragenesis and genesis. In *Vein types and similar U deposits in rocks younger than the Proterozoic*, 113–58. International Atomic Energy Authority, Vienna.

Ball, T. K., N. J. Fortey, T. J. Shepherd 1985a. Mineralization at the Carrock Fell tungsten mine, northern England: paragenetic, fluid-inclusion and geochemical study. *Mineralium Deposita* **20**, 57–65.

Ball, T. K., R. A. Nicholson, D. Peachey 1985b. Gas geochemistry as an aid to detection of buried mineral deposits. *Institution of Mining and Metallurgy, Transactions B* **94**, 181–8.

Ball, V. 1887. On the existing records as to the discovery of a diamond in Ireland in the year 1816. *Royal Geological Society of Ireland, Journal* **7**, 163–6.

Ball Clay Heritage Society 2003. *The ball clays of Devon and Dorset.* St Austell: Cornish Hillside Publications.

Baltatzis, E. 1979. Distribution of Fe and Mg between garnet and biotite in Scottish Barrovian metamorphic zones. *Mineralogical Magazine* **43**, 155–7.

Baltatzis, E. & C. Katagas 1981. Margarite pseudomorphs after kyanite in Glen Esk, Scotland. *American Mineralogist* **66**, 213–16.

Baltatzis, E. & B. J. Wood 1977. The occurrence of paragonite in chloritoid schists from Stonehaven, Scotland. *Mineralogical Magazine* **41**, 211–16.

Balzer, R., S. Bronko, M. P. Cooper, R. Duthaler, S. Dyl, M. Gray and 10 other authors 2003. *Calcite: the mineral with the most forms.* East Hampton, Connecticut: Lapis International.

Bancroft, P. & S. Weller 1993. Cornwall's famous mines. *Mineralogical Record* **24**(4), 259–83.

Banks, D. A., A. J. Boyce, I. M. Samson 2002. Constraints on the origins of fluids forming Irish Zn–Pb–Ba deposits: evidence from the composition of fluid inclusions. *Economic Geology* **97**, 471–80.

Bannister, F. A. 1932. The distinction of pyrite from marcasite. *Mineralogical Magazine* **23**, 179–87.

——— 1933. The identity of mottramite and psittacinite with cupriferous descloizite. *Mineralogical Magazine* **23**, 376–86.

——— 1934. The crystal structure and optical properties of matlockite (with a chemical analysis by M. H. Hey). *Mineralogical Magazine* **23**, 587–97.

——— 1936. Ettringite from Scawt Hill, Co. Antrim. *Mineralogical Magazine* **24**, 324–9.

——— 1940. Bravoite from Mill Close mine, Derbyshire. *Mineralogical Magazine* **25**, 609–614.

——— 1943. Brammallite (sodium illite) a new mineral from Llandebie, South Wales. *Mineralogical Magazine* **26**, 304–307.

Bannister, F. A. & S. E. Hollingworth 1948. Two new British minerals. *Nature* **162**, 565.

Bannister, F. A., M. H. Hey, W. Campbell Smith 1955. Grovesite, the manganese-rich analogue of berthierine. *Mineralogical Magazine* **30**, 645–7.

Bannister, F. A., M. H. Hey, G. F. Claringbull 1950. Connellite, buttgenbachite and tallingite. *Mineralogical Magazine* **29**, 280–86.

Baqri, R. & F. Hodson 1979. Apatite in South Wales anthracite. *Geological Magazine* **116**, 483–6.

Barber, J. P. & B. W. D. Yardley 1985. Conditions of high-grade metamorphism in the Dalradian of Connemara, Ireland. *Geological Society of London, Journal* **142**, 87–96.

Baring-Gould, S. 1888. A gold-washing apparatus. *Devonshire Association, Transactions* **20**, 376–7.

Barnes, R. P. & F. R. Andrews 1981. Pumpellyite–actinolite grade regional metamorphism in south cornwall. *Ussher Society, Proceedings* **5**, 139–46.

Barnett, G. W. T. 1959. Lead in Islay. See Higham (1959: 65–76).

Barnicoat, A. C. & M. J. O'Hara 1979. High-temperature pyroxenes from an ironstone at Scourie, Sutherland. *Mineralogical Magazine* **43**, 371–5.

Barnicoat, A. C., I. Cartwright, M. J. O'Hara 1987. Kyanite in the mainland Lewisian Complex. *Scottish Journal of Geology* **23**, 209–213.

Baron, M. & J. Parnell 2005. Fluid evolution in base-metal sulphide mineral deposits in the metamorphic basement rocks of southwest Scotland and Northern Ireland. *Geological Journal* **40**, 3–21.

Barressi, A. A., P. Orlandi, M. Pasero 2007. History of ardennite and the new mineral ardennite-(V). *European Journal of Mineralogy* **19**, 581–7.

Barritt, S. D. 1983. *The controls of radioelement distribution in the Etive and Cairngorm granites: implications for heat production.* PhD thesis, Open University, Milton Keynes.

Barrow, G. 1898. On the occurrence of chloritoid in Kincardineshire. *Geological Society of London, Quarterly Journal* **54**, 149–56.

Barrow, G. & H. H. Thomas 1908. On the occurrence of metamorphic minerals in calcareous rocks in the Bodmin and Camelford areas, Cornwall. *Mineralogical Magazine* **15**, 113–23.

——— 1909. Some additional localities for idocrase in Cornwall. *Mineralogical Magazine* **15**, 238–40.

Barstow, R. W. 1982. Variscite from Hensbarrow china-clay works, St Austell. *Mineralogical Magazine* **46**, 512.

——— 1983. Wroewolfeite in southwest England. *Mineralogical Magazine* **47**, 240–41.

Barstow, R. W. & M. Cooper 1982. Churchite from Wheal Pendarves, Camborne, Cornwall. *Mineralogical Magazine* **46**, 402–403.

Bartlett, N. 1980. The copper mines of west county Cork. *Gems* **12**, 10.

Barton, M. E. 1963. The petrology of a teschenite sill at Landywood, Staffordshire. *Geological Magazine* **100**, 533–50.

Basham, I. R., T. K. Ball, B. Beddoe-Stephens, U. M. Mitchie 1982. Uranium-bearing accessory minerals and granite fertility, I:I studies of granites from the British Isles. *Symposium on uranium exploration methods*, 398–413.Vienna: International Atomic Energy Authority.

Basham, I. R., A. E. Milodowski, E. K Hyslop, J. M. Pearce 1989. *The location of uranium in source rocks and sites of secondary deposition at the Needle's Eye natural analogue site, Dumfries & Galloway.* Report DOE/RW/89/367, Department of the Environment, London.

Basta, E. Z. 1957. Accurate determination of the cell dimensions of magnetite. *Mineralogical Magazine* **31**, 431–54.

Batchelor, R. A., D. C. Armstrong, M. McDonald 1992. Composition of fluids in quartz: discrimination of magma pulses in a Caledonian granitoid. *Mineralogical Magazine* **56**, 335–42.

Bateson, J. H. & C. C. Johnson 1984. An interpretation of regional geochemical drainage basin data from the Dent–Ingleton area of the Askrigg block. *Yorkshire Geological Society, Proceedings* **45**, 91–7.

Battey, M. H. & A. A. Moss 1962. Powellite from Traprain Law, Haddingtonshire, Scotland. *Mineralogical Magazine* **33**, 158–61.

Bau, M., R. L. Romer, V. Lüders, P. Dulski 2003. Tracing element sources of hydrothermal mineral deposits: REE and Y distribution and Sr–Nd–Pb isotopes in fluorite from MVT deposits in the Pennine orefield, England. *Mineralium Deposita* **38**, 992–1008.

Bawden, E. H. 1929. Killifreth mine, Cornwall. *Mining Magazine* **12**, 279–86.

Bayliss, P. 1975. Nomenclature of the trioctahedral chlorites. *Canadian Mineralogist* **13**, 178–80.

——— 2000. *Glossary of obsolete mineral names.* Tucson: Mineralogical Record.

Bayliss, P., F. Mazzi, R. Munno, T. J. White 1989. Mineral nomenclature: zirconolite. *Mineralogical Magazine* **53**, 565–9.

Bazley, A. 1997. Diamond in a patchwork of colours. *Gem and Jewellery News* **6**, 30.

Beard, A. D. & S. M. Drake 2007. A melilite-bearing high-temperature calcic skarn, Camasunary Bay, Isle of Skye. *Scottish Journal of Geology* **43**, 57–67.

Beddoe-Stephens, B. 1990. Pressures and temperatures of Dalradian metamorphism and the andalusite–kyanite transformation in the northeast Grampians. *Scottish Journal of Geology* **26**, 3–14.

Beddoe-Stephens, B. & N. J. Fortey 1981. Columbite from the Carrock Fell tungsten deposit. *Mineralogical Magazine* **44**, 217–23.

Beer, K. E. & T. K. Ball 1977. Barytes mineralization in the Teign Valley, Devon. *Institution of Mining and Metallurgy, Transactions B* **86**, 91–5.

——— 1987. Tungsten mineralization and magmatism in southwest England. *Chronique de la Recherche Minière* **487**, 53–62.

Beer, K. E., T. K. Ball, D. C. Cooper, A. D. Evans, R. C. Jones, K. E. Rollin, J. M. C. Tombs

1992. *Mineral investigations in the Teign Valley, Devon, part 2: base metals.* Report 123, Mineral Reconnaissance Program, British Geological Survey, Keyworth.

Beer, K. E. & M. J. Bennett 1988. *Geochemistry of sediments from the Lui drainage, Braemar, Grampian.* Report 96, Mineral Reconnaissance Program, British Geological Survey, Keyworth.

Beer, K. E., A. J. Burley, J. M. C. Tomb 1975. The concealed granite roof in southwest Cornwall. *Institution of Mining and Metallurgy, Transactions B* **84**, 24–5.

Beer, K. E., G. S. Kimbell, M. J. Bennett 1989. *Skarn-type copper mineralization in the vicinity of Belstone Consols mine, Okehampton, Devon.* Report 101, Mineral Reconnaissance Program, British Geological Survey, Keyworth.

Beevers, C. A. & F. H. Stewart 1960. *p*-Veatchite from Yorkshire. *Mineralogical Magazine* **32**, 500.

Bell, B. R. & R. V. Claydon 1992. The cumulus and post-cumulus evolution of the chrome spinels in ultrabasic layered intrusions: evidence from the Cuillin igneous complex, Isle of Skye, Scotland. *Contributions to Mineralogy and Petrology* **112**, 242–53.

Bell, B. R. & I. T. Williamson 1994. Picritic basalts from the Palaeocene lava field of west-central Skye, Scotland: evidence for parental magma compositions. *Mineralogical Magazine* **58**, 347–56.

Bell, B. R., R. V. Claydon, G. Rogers 1994. The petrology and geochemistry of cone sheets from the Cuillin igneous complex, Isle of Skye: evidence for combined assimilation and fractional crystallization during lithospheric extension. *Journal of Petrology* **35**, 1055–1094.

Belson, E., D. I. Green, A. G. Tindle 2005. The first British occurrence of arsenogoyazite at Wheal Cock, St Just, Cornwall. *UK Journal of Mines and Minerals* **26**, 27–8.

Bemrose, H. H. A. 1904. Quartzite dykes in Mountain Limestone near Snelston (Derbyshire). *Geological Society of London, Quarterly Journal* **60**, 364–71.

Benham, A. J., F. M. McEvoy, K. E. Rollin 2004. Potential for stratiform massive sulphide mineralization in southwest England. *Applied Earth Science* **113**, 227–46.

Benjamin, R. E. K. 1968. An axinite–epidote–tourmaline vein-cutting amphibolite, western Connemara, Eire. *Mineralogical Magazine* **36**, 747–50.

Bennett, J. (ed.) 1995. *Minera lead mines and quarries.* Wrexham: Wrexham Maelor Borough Council.

Bennett, J. & R. Vernon 2002. *Metal mines of Llanengan.* Warrington: Gwydyr Mines Publications.

Bennett, M. A. 1987a. *The Cambrian manganese deposits of North Wales.* PhD thesis, University of Leeds.

——— 1987b. Genesis and diagenesis of the Cambrian manganese deposits, Harlech, North Wales. *Geological Journal* **22**, 7–18.

Bernard, J. H. & J. Hyršl 2004. *Minerals and their localities.* Prague: Granit.

Berridge, N. G. 1969. *A summary of the mineral resources of the "crofter counties" of Scotland, comprising Argyllshire, Caithness, Inverness-sshire, Orkney and Shetland, Ross and Cromarty, and Sutherland.* Report 69.5, Institute of Geological Sciences, London.

Berry, L. G. 1940. Studies of mineral sulpho-salts, II: jamesonite from Cornwall and Bolivia. *Mineralogical Magazine* **25**, 597–608.

Bertrand, M. E. 1878. Sur le leadhillite de Matlock. *Académie des Sciences Paris, Comptes Rendus* **86**, 348.

Berzelius, J. J. 1819. *Nouveau système de minéralogie.* Paris: Méquignon-Marvis [originally published in Swedish, Stockholm, 1814].

Betterton, J. 1989. Birnessite from Penberthy Croft mine, Cornwall – a second English occurrence. *Russell Society, Journal* **2**(2), 48.

——— 1996a. Monazite-(La) and anatase from Penberthy Croft mine, St Hilary, Cornwall. *UK Journal of Mines and Minerals* **16**, 13.

——— 1996b. Schulenbergite, a second Cornish occurrence. *UK Journal of Mines and Minerals* **16**, 13.

——— 1996c. Mattheddleite, the first Cornish occurrence. *UK Journal of Mines and Minerals* **16**, 15.

——— 1996d. Monazite-(Ce) from Croft Gothal mine, Conwall. *UK Journal of Mines and Minerals* **17**, 16–17.

——— 1996e. Tenorite from Penberthy Croft mine, Cornwall. *UK Journal of Mines and Minerals* **17**, 17.

——— 1997. Stilpnomelane from Penberthy Croft mine, a second Cornish occurrence. *UK Journal of Mines and Minerals* **18**, 15.

——— 1998. Bismuthinite from Penberthy Croft mine, Cornwall. *UK Journal of Mines and Minerals* **19**, 11.

Betterton, J., D. J. Green, C. Jewson, J. Spratt, P. Tandy 1998. The composition and structure of jeanbandyite and natanite. *Mineralogical Magazine* **62**, 707–712.

Betterton, J., D. I. Green, J. Mason 2000. Famous mineral localities: Penberthy Croft mine, St Hilary, Cornwall, England. *UK Journal of Mines and Minerals* **20**, 7–37.

Beudant, F. S. 1832. *Traité élémentaire de minéralogie,* vol. 2. Paris: Verdière.

Bevins, R. E. 1978. Pumpellyite-bearing basic igneous rocks from the Lower Ordovician of North Pembrokeshire, Wales. *Mineralogical Magazine* **42**, 81–3.

——— 1979. *The geology of the Strumble Head–Fishguard region, Dyfed, Wales.* PhD thesis, University of Keele.

——— 1982. Petrology and geochemistry of the Fishguard volcanic complex, Wales. *Geological Journal* **17**, 1–21.

——— 1985. Pumpellyite-dominated metadomain alteration at Builth Wells, Wales – evidence for a fossil submarine hydrothermal system? *Mineralogical Magazine* **49**, 451–6.

——— 1994. *A mineralogy of Wales.* Geological Series 16, National Museum of Wales, Cardiff.

Bevins, R. E. & J. M. Horák 1985a. A first occurrence of laumontite in Wales. *Russell Society, Journal* **1**(3), 78–9.

——— 1985b. Siegenite in clay–ironstone nodules from the South Wales coalfield. *Russell Society, Journal* **1**(3), 83–5.

Bevins, R. E. & J. S. Mason 1997. *Welsh metallophyte and metallogenic evaluation project: Dyfed and Powys.* Science Report 156, Countryside Council for Wales, Cardiff.

Bevins, R. E. & R. J. Merriman 1988. Compositional controls on coexisting prehnite-actinolite and prehnite-pumpellyite facies assemblages, in the Tal y Fan metabasite intrusion, North Wales: implications for Caledonian metamorphic field gradients. *Journal of Metamorphic Geology* **6**, 17–39.

Bevins, R. E. & R. A. Roach 1982. Ordovician igneous activity in southwest Dyfed. In *Geological excursions in Dyfed, southwest Wales*, M. G. Bassett (ed.), 65–80. National Museum of Wales, Cardiff.

Bevins, R. E. & G. Rowbotham 1983. Low-grade metamorphism within the Welsh sector of the paratectonic Caledonides. *Geological Journal* **18**, 141–67.

Bevins, R. E. & C. J. Stanley 1990. Aleksite, a lead bismuth sulfotelluride: a second world occurrence from the Dolgellau gold belt, Gwynedd, Wales. *Russell Society, Journal* **3**(2), 67–9.

Bevins, R. E., D. H. M. Alderton, J. M. Horák 1988. Lead–antimony mineralization at Bwlch mine, Deganwy, Wales. *Mineralogical Magazine* **52**, 391–4.

Bevins, R. E., G. J. Lees, R. A. Roach 1991. Ordovician bimodal volcanism in southwest Wales: geochemical evidence for petrogenesis of the silicic rocks. *Geological Society of London, Journal* **148**, 719–29.

Bevins, R. E., G. J. Lees, R. A. Roach, G. Rowbotham, P. A. Floyd 1994. Petrogenesis of the St David's Head layered intrusion, Wales: a complex history of multiple magma injection and *in situ* crystallization. *Royal Society of Edinburgh, Transactions: Earth Sciences* **85**, 91–121.

Bevins, R. E., R. Metcalfe, D. Robinson 1993. Temporal variations in water–rock interaction as a control on pumpellyite formation. *Terra Abstracts* **5**, 413.

Bevins, R. E., P. O'Brien, S. Turgoose, P. A. Williams 1982a. A new find of nickeloan hydrozincite in the Parc mine, North Wales. *Russell Society, Journal* **1**(1), 19–21.

Bevins, R. E., G. Rowbotham, F. S. Stephens, S. Turgoose, P. A. Williams 1985. Lanthanite-(Ce), (Ce, La, Nd)$_2$(CO$_3$)$_3$.8H$_2$O, a new mineral from Wales, UK. *American Mineralogist* **70**, 411–13.

Bevins, R. E., R. F. Symes, J. M. Horak, V. A. D. Holyer 1987. Hetaerolite from Eastern Cliff, Kennack Sands, Cornwall: the first British occurrence. *Mineralogical Magazine* **51**, 172–3.

Bevins, R. E., S. Turgoose, P. A. Williams 1982b. Namuwite, (Zn,Cu)$_4$SO$_4$(OH)$_6$.H$_2$O, a new mineral from Wales. *Mineralogical Magazine* **46**, 51–4.

Bick, D. E. 1982. *The old copper mines of Snowdonia.* Newent, Gloucestershire: The Pound House.

——— 1986a. The mines and minerals of Newent, Gloucestershire. *Russell Society, Journal* **1**(4), 114–18.

——— 1986b. *Frongoch lead & zinc mine* (2nd edn). British Mining 30, Northern Mine Research Society, Sheffield.

——— 1988. *The old metal mines of mid-Wales, part 3: Cardiganshire – north of Goginan* (2nd edn). Newent, Gloucestershire: The Pound House.

——— 1990. *The old metal mines of mid-Wales, parts 4 & 5: west Montgomeryshire, Aberdovey, Dinas Mawddwy and Llangynog* (2nd edn). Newent, Gloucestershire: The Pound House.

——— 1991. *The old metal mines of mid-Wales, part 6: a miscellany.* Newent, Gloucestershire: The Pound House.

——— 1992. *The old metal mines of mid-Wales, parts 1 & 2: Cardiganshire – south of Goginan* (revised edn). Newent, Gloucestershire: The Pound House.

Bill, H., J. Sierro, R. Lacroix 1967. Origin of colouration in some fluorites. *American Mineralogist* **52**, 1003–1008.

Billington, G. 1995. Field meeting report: Vaynor quarry, Cefn Coed, Merthyr Tydfil, Mid Glamorgan. *Russell Society, Newsletter* **26**, 12.

——— 1997a. Field meeting report: Vaynor quarry, Merthyr, Glamorgan. *Russell Society, Newsletter* **30**, 15.

——— 1997b. Field meeting report: South Wales coast, Ogmore–Penarth, South Glamorgan. *Russell Society, Newsletter* **30**, 18.

——— 1997c. Field meeting report: Welsh Weekend. *Russell Society, Newsletter* **31**, 16.

Birley, R. & B. Young 1993. Vivianite of post-Roman origin from Vindolandia, Northumberland. *Natural History Society of Northumberland, Transactions* **56**, 65–7.

Bish, D. L. & A. Livingstone 1981. The crystal chemistry and paragenesis of honessite and hydrohonessite: the sulphate analogues of reevesite. *Mineralogical Magazine* **44**, 339–43.

Bishop, A. C. & W. J. French 1982. Nature and origin of meladiorite layers in northern Guernsey, Channel Islands. *Mineralogical Magazine* **46**, 301–321.

Bishop, A. C. & C. H. Key 1983. Nature and origin of layering in the diorites of southeast Jersey, Channel Isles. *Geological Society of London, Journal* **140**, 921–37.

Bishop, A. C., A. G. Couper, A. E. Mourant 1982. Re-examination of the alleged occurrence of wollastonite and epistilbite in Jersey. *Mineralogical Magazine* **46**, 504–506.

Bishop, A. C., A. J. Criddle, A. M. Clark 1977. Plumbian tennantite from Sark, Channel Islands. *Mineralogical Magazine* **41**, 59–63.

Black, H. S. 1976. Silica amygdale minerals from Craigahulliar. *Irish Naturalists' Journal* **18**, 278.

Blackburn, W. H., W. H. Dennen, P. I. Russell 1997. *Encyclopedia of mineral names.* Ottawa: Mineralogical Society of Canada.

Blake, J. F. 1888. The occurrence of a glaucophane-bearing rock in Anglesey. *Geological Magazine* **25**, 125–7.

Blakeman, R. J., J. H. Ashton, A. J. Boyce, A. E. Fallick, M. J. Russell 2002. Timing of interplay between hydrothermal and surface fluids in the Navan Zn + Pb orebody, Ireland: evidence from metal distribution trends, mineral textures, and $\delta^{34}S$ analyses. *Economic Geology* **97**, 73–91.

Bloxam, T. W. 1954. Rodingite from the Girvan–Ballantrae complex, Ayrshire. *Mineralogical Magazine* **30**, 525–8.

——— 1958. Pumpellyite from South Ayrshire. *Mineralogical Magazine* **31**, 811–13.

——— 1964. Hydrogrossular from the Girvan–Ballantrae complex, Ayrshire. *Mineralogical Magazine* **33**, 814–15.

Bloxam, T. W. & J. B. Allen 1959. Glaucophane schist, eclogite, and associated rocks from Knockormal in the Girvan–Ballantrae complex, south Ayrshire. *Royal Society of Edinburgh, Transactions: Earth Sciences* **64**, 1–27.

Bloxam, T. W. & M. H. J. Dirk 1988. The petrology and geochemistry of the St David's granophyre and the Cwm Bach rhyolite, Pembrokeshire, Dyfed. *Mineralogical Magazine* **52**, 563–75.

Bloxam, T. W. & N. B. Price 1961. Stilpnomelane in North Wales. *Nature* **190**, 525–6.

Blyth, F. G. H. 1942. Intrusive rocks of the Shelve area, south Shropshire. *Geological Society of London, Quarterly Journal* **97**, 169–204.

——— 1948. Pyroxene from the Squilver dolerite, south Shropshire. *Mineralogical Magazine* **28**, 380–83.

——— 1949. A "luxullianite" from east Dartmoor. *Geological Magazine* **86**, 175–9.

Boast, A. M. 1979. *A textural and isotopic study of Irish base-metal mineralization of Lower Carboniferous age, with specific reference to the Tynagh deposit.* PhD thesis, Imperial College, University of London.

Boast, A. M., M. L. Coleman, C. Halls 1981a. Textural and stable isotopic evidence for the genesis of the Tynagh base metal deposit, Ireland. *Economic Geology* **76**, 27–55.

Boast, A. M., M. Harris, D. Steffe 1990. Intrusive-hosted gold mineralization at Hare Hill, Southern Uplands, Scotland. *Institution of Mining and Metallurgy, Transactions B* **99**, 106–112.

Boast, A. M., I. G. Swainbank, M. L. Coleman, C. Halls 1981b. Lead isotope variation in the Tynagh, Silvermines and Navan base-metal deposits, Ireland. *Institution of Mining and Metallurgy, Transactions B* **90**, 115–19.

Bøggild O. B. 1922. Re-examination of some zeolites. *Royal Danish Academy of Sciences and Letters* **4**, 1–42.

Boland, M. B., J. A. Clifford, A. H. Meldrum, A. Poustie 1992. Residual base metal and barite mineralization at Silvermines, Co. Tipperary, Ireland. See Bowden et al. (1992: 247–60).

Bolton, W. A. 1996. Malachite from Dyserth quarry, Clwyd, North Wales. *Russell Society, Newsletter* **28**, 24–5.

Bone, D. A. 1999. The London Clay Formation (Early Eocene) at Southleigh landfill site, near Emsworth, Hampshire (with a note on Lidsey landfill site, near Bognor Regis, West Sussex). *Tertiary Research* **19**, 91–9.

Bonney, T. G. 1877. On the serpentine and associated rocks of the Lizard district. *Geological Society of London, Quarterly Journal* **33**, 884–924.

——— 1878. On some mica traps from Kendal and Sedbergh districts. *Geological Society of London, Quarterly Journal* **35**, 165–80.

——— T. G. 1888. On a peculiar variety of hornblende from Mynydd Mawr, Caernarvonshire. *Mineralogical Magazine* **8**, 103–107.

Bor, L. 1950. Pisanite from Parys Mountain, Anglesey. *Mineralogical Magazine* **29**, 63–7.

Boswell, P. G. H. 1927. On the distribution of purple zircon in British sedimentary rocks. *Mineralogical Magazine* **21**, 310–17.

——— 1928. The rarer detrital minerals of British sedimentary rocks. *Geological Society of Glasgow, Transactions* **18**, 129–48.

Bottrell, S. H. 1991. Sulphur isotope evidence for the origin of cave evaporites in Ogof y Daren Cilau, South Wales. *Mineralogical Magazine* **55**, 209–210.

Bottrell, S. H. & M. D. B. Morton 1992. A reinterpretation of the genesis of the Cae Coch pyrite deposit, north Wales. *Geological Society of London, Journal* **149**, 581–4.

Bottrell, S. H., M. A. Leosson, R. J. Newton 1996. Origin of brine inflows at Boulby potash mine, Cleveland, England. *Institution of Mining and Metallurgy, Transactions B* **105**, 159–64.

Bouch, J. E., J. Naden, T. J. Shepherd, J. A. McKervey, B. Young, A. J. Benham, H. J. Sloane 2006. Direct evidence of fluid mixing in the formation of stratabound Pb–Zn–Ba–F mineralization in the Alston Block, North Pennine orefield (England). *Mineralium Deposita* **41**, 821–35.

Boulton, W. S. 1911. On a monchiquite intrusion in the Old Red Sandstone of Monmouthshire. *Geological Society of London, Quarterly Journal* **67**, 460–76.

Bowden, A. A., G. Earls, P. G. O'Connor, J. F. Pyne (eds) 1992. *The Irish minerals industry 1980–1990.* Dublin: Irish Association of Economic Geology.

Bowell, R. J. 1992. Supergene copper mineral assemblages at Botallack, St Just, Cornwall. *Russell Society, Journal* **4**(2), 45–53.

Bowen, N. L., J. W. Greig, E. G. Zies 1924. Mullite, a silicate of alumina. *Journal of Washington Academy of Sciences* **14**, 183–91.

Bowes, D. R. & A. M. Hopgood 1969. Composition of a dolerite dyke, Mingulay, Outer Hebrides. *Mineralogical Magazine* **37**, 427.

Bowes, D. R., E. D. Kinloch, A. E. Wright 1964. Rhythmic amphibole overgrowths in appinites associated with explosion breccias in Argyll. *Mineralogical Magazine* **33**, 963–73.

Bowie, S. H. U., D. Ostle, M. J. Gallagher 1970. Uranium reconnaissance in northern Scotland. *Institution of Mining and Metallurgy, Transactions B* **79**, 180–82.

Bowler, C. M. L. & G. A. Kingston 1971. Mineralization in the Triassic rocks of the Llantrisant area, Glamorgan. In *1970 Gregynog Mineral Exploitation Colloquium Report*, 30–34. Cardiff: University College.

Bowles, J. F. W. & D. J. Morgan 1984. The composition of rhabdophane. *Mineralogical Magazine* **48**, 146–8.

Bowman, H. L. 1900. On monazite and associated minerals from Tintagel, Cornwall. *Mineralogical Magazine* **12**, 358–62.

——— 1903. Note on the refractive indices of pyromorphite, mimetite and vanadinite. *Mineralogical Magazine* **13**, 324–9.

——— 1911. On the occurrence of bertrandite at the Cheesewring quarry, near Liskeard, Cornwall. *Mineralogical Magazine* **16**, 47–50.

——— 1914. Note on calcite from the chalk at Corfe Castle, Dorset. *Mineralogical Magazine* **17**, 135–7.

Boyce, A. J., R. Anderton, M. J. Russell 1983. Rapid subsidence and early Carboniferous base-metal mineralization in Ireland. *Institution of Mining and Metallurgy, Transactions B* **92**, 55–65.

Boyce, A. J., A. E. Fallick, T. J. Fletcher, M. J. Russell, J. H. Ashton 1994. Detailed sulphur isotope studies of Lower Palaeozoic-hosted pyrite below the giant Navan Zn + Pb mine, Ireland: evidence of mass transport of crustal S to a sediment-hosted deposit. *Mineralogical Magazine* **58A**, 109–110.

Bradshaw, P. M. D. 1967. Distribution of selected elements in feldspar, biotite and muscovite in relation to mineralization. *Institution of Mining and Metallurgy, Transactions B* **76**, 137–48.

Bradshaw, R. & F. C. Phillips 1967. X-ray studies on natural fabrics, I: growth fabrics in hematite kidney ore and in fibrous calcite. *Mineralogical Magazine* **36**, 70–77.

Braithwaite, R. S. W. 1981. Turquoise crystals from Britain and a review of related species. *Mineralogical Record* **12**, 349–53.

——— 1982a. Pyromorphite, wulfenite and other minerals from the Bwlch-Glas mine, central Wales. *Mineralogical Record* **13**(3), 151–3.

——— 1982b. Wroewolfeite in Britain. *Mineralogical Record* **13**(3), 167–70, 174.

——— 1983. Minerals of the Derbyshire orefield. *Mineralogical Record* **14**(1), 15–24.

——— 1987. Ceruleite: infrared spectroscopy and a new locality in Cornwall. *Mineralogical Magazine* **51**, 738–41.

——— 1988. Cuproadamite and tennantite from Higher Longrigg mine, Hartley Birkett, Cumbria. *Mineralogical Magazine* **52**, 552–3.

——— 1994. Mineralogy of the Alderley Edge–Mottram St Andrew area, Cheshire, England. *Russell Society, Journal* **5**(2), 91–102.

Braithwaite, R. S. W. & B. V. Cooper 1982. Childrenite in southwest England. *Mineralogical Magazine* **46**, 119–26.

Braithwaite, R. S. W. & H. Corke 1980. Kidwellite from Cornwall. *Mineralogical Magazine* **43**, 952–3.

Braithwaite, R. S. W. & T. B. Greenland 1966. Rare mineral discovered at Matlock Bath, Derbyshire. *The Times* (17 June 1966).

Braithwaite, R. S. W. & J. R. Knight 1968. Serpierite from Ecton, Staffordshire. *Mineralogical Magazine* **36**, 882.

——— 1990. Rare minerals, including several new to Britain, in supergene alteration of U–Cu–As–Bi–Co mineralization near Dalbeattie, south Scotland. *Mineralogical Magazine* **54**, 129–31.

Braithwaite, R. S. W. & R. P. H. Lamb 1986. Wulfenite from Tŷ Coch, Glamorgan (Powys), South Wales. *Mineralogical Magazine* **50**, 180–81.

——— 2007. Childrenite from the English Lake District. *Russell Society, Journal* **10**, 50–51.

Braithwaite, R. S. W. & M. G. H. Merry 1996. Diaboléite from Gunver Head, Cornwall, England. *Russell Society, Journal* **6**(2), 99.

Braithwaite, R. S. W. & R. Pritchard 1983a. Infrared spectroscopic analysis of the olivinite–adamite series, and of phosphate substitution in olivenite. *Mineralogical Magazine* **47**, 51–7.

——— 1983b. Nakauriite from Unst, Shetland. *Mineralogical Magazine* **47**, 84–5.

Braithwaite, R. S. W. & G. Ryback 1963a. Rosasite, aurichalcite and associated minerals from the Heights of Abraham, Matlock Bath. *Mineralogical Magazine* **33**, 441–9.

——— 1963b. Sulphur in altered galena from Dene quarry, Cromford and Netherwater mine, Hazlebadge, Derbyshire, from Engine Vein, Alderley Edge, Cheshire and Tyndrum mine, Tyndrum, Perthshire [exhibit]. Notice 121, Mineralogical Society, London.

——— 1988. Philipsburgite from the Caldbeck Fells and kipushite from Montana, and their infrared spectra. *Mineralogical Magazine* **52**, 529–33.

——— 1994. Reichenbachite from Cornwall and Portugal. *Mineralogical Magazine* **58**, 449–54.

——— 1995. Beyerite, $(Ca,Pb)(BiO)_2(CO_3)_2$, a mineral new to the British Isles, from Cumbria, England. *Russell Society, Journal* **6**(1), 51–2.

Braithwaite, R. S. W. & J. I. Wilson 2001a. Millerite and associated minerals from Brixham, Devon, England. *Russell Society, Journal* **7**(2), 87–8.

——— 2001b. Botallackite, including good crystals from Ireland. *Russell Society, Journal* **7**(2), 96.

——— 2003. Well formed crystals of erythrite, with mercurian silver, lavendulan and other minerals from Gortdrum mine, Tipperary, Ireland. *Russell Society, Journal* **8**(1), 35–6.

——— Braithwaite, R. S. W. & J. I. Wilson 2007. An unusual fluorescent fluorite from Derbyshire. *Russell Society, Journal* **10**, 52–4.

Braithwaite, R. S. W., M. P. Cooper, A. D. Hart 1989a. Queitite, a mineral new to Britain, from the Caldbeck Fells, Cumbria. *Mineralogical Magazine* **53**, 508–509.

Braithwaite, R. S. W., A. Dyer, R. P. H. Lamb 2001. Gismondine-Ba, a zeolite from the weathering of slags. *Russell Society, Journal* **7**(2), 83–5.

Braithwaite, R. S. W., W. T. Flowers, R. N. Haszeldine, M. Russell 1973. The cause of the colour of Blue John and other purple fluorites. *Mineralogical Magazine* **39**, 401–411.

Braithwaite, R. S. W., T. B. Greenland, G. Ryback 1963a. Wulfenite from Poddy Gill, Caldbeck Fells. *Mineralogical Magazine* **33**, 720.

——— 1963b. A mercury mineral from the British Isles. *Nature* **200**, 1004.

——— 1964a. Celestine and aurichalcite from Clayton adit, Ecton Hill, Wetton, Staffordshire [exhibit: general meeting of the Mineralogical Society, University of Manchester, 25 September 1963]. *Mineralogical Magazine* **33**, lxxxvi.

Braithwaite, R. S. W., M. G. H. Merry, B. Jackson 1996. Talmessite, with rauenthalite or phaunouxite from Ivy Tor mine, Devon, England. *Russell Society, Journal* **6**(2), 100–101.

Braithwaite, R. S. W., W. H. Paar, J. E. Chisholm 1989. Phurcalite from Dartmoor, southwest England, and its identity with "nisaite" from Portugal. *Mineralogical Magazine* **53**, 583–9.

Braithwaite, R. S. W., G. Ryback, T. B. Greenland 1964b. Stalactitic hydrozincite from Ecton mines, Wetton, Staffordshire [exhibit: general meeting of the Mineralogical Society, University of Manchester, 25 September 1963]. *Mineralogical Magazine* **33**, lxxxvi.

Brammall, A. 1926. Gold and silver in the Dartmoor granite. *Mineralogical Magazine* **21**, 14–20.

Brammall, A. & S. Bracewell 1936. Variability of garnets in granites. *Mineralogical Magazine* **24**, 254–6.

Brammall, A. & H. F. Harwood 1923. The occurrence of rutile, brookite, and anatase on Dartmoor. *Mineralogical Magazine* **20**, 20–26.

——— 1924. The occurrence of a gold-bearing pegmatite on Dartmoor. *Mineralogical Magazine* **20**, 201–211.

——— 1925. Tourmalinization in the Dartmoor granite. *Mineralogical Magazine* **20**, 319–30.

——— 1927. The temperature of formation for tourmaline, rutile, brookite and anatase in the Dartmoor granite. *Mineralogical Magazine* **21**, 205–220.

——— 1932. The Dartmoor granites: their genetic relationships. *Geological Society of London, Quarterly Journal* **88**, 171–237.

Brammall, A. & B. Rama Rao 1936. The variable composition of cordierite in the Dartmoor granites. *Mineralogical Magazine* **24**, 257–9.

Brammall, A., J. G. C. Leech, F. A. Bannister 1937. The paragenesis of cookeite and hydromuscovite associated with gold at Ogofau, Carmarthenshire. *Mineralogical Magazine* **24**, 507–521.

Brammall, A., J. G. C. Leech, W. J. Dunstall 1940. Montmorillonite in fuller's earth, Nutfield, Surrey. *Geological Magazine* **77**, 102–112.

Bramall, V. 1921. The mining, manufacture and uses of barytes in the neighbourhood of Appleby, Westmorland. *Institute of Mining Engineers, Transactions* **61**, 42–5.

Brand, S. F. & G. T. Emo 1986. A note on Zn–Pb–Ba mineralization near Oldcastle, County Meath. See Andrew et al. (1986: 297–304).

Bray, C. J. & E. T. C. Spooner 1983. Sheeted vein Sn–W mineralization and greisenization associated with economic kaolinization, Goonbarrow china-clay pit, St Austell, Cornwall, England: geologic relationships and geochronology. *Economic Geology* **78**, 1064–1089.

Brearley, A. J. 1986. An electron optical study of muscovite breakdown in pelitic xenoliths during pyrometamorphism. *Mineralogical Magazine* **50**, 385–97.

——— 1987. A natural example of the disequilibrium breakdown of biotite at high temperature: TEM observations and comparison with experimental kinetic data. *Mineralogical Magazine* **51**, 93–106.

——— 1988. Chloritoid from low-grade pelitic rocks in North Wales. *Mineralogical Magazine* **52**, 394–6.

Brearley, A. J. & P. E. Champness 1986. Magnetite exsolution in almandine garnet. *Mineralogical Magazine* **50**, 621–33.

Breeding, C. M., J. J. Ague, M. Grove, A. L. Rupke 2004. Isotopic and chemical alteration of zircon by metamorphic fluids: U–Pb age depth-profiling of zircon crystals from Barrow's garnet zone, northeast Scotland. *American Mineralogist* **89**, 1067–1077.

Brewer, T. S., C. D. Storey, R. R. Parrish, S. Temperley, B. F. Windley 2003. Grenvillian age decompression of eclogites in the Glenelg–Attadale inlier, northwest Scotland. *Geological Society of London, Journal* **160**, 565–74.

Bridges, T. F. 1982. An occurrence of nickel minerals in the Hilton mine, Scordale, Cumbria. *Russell Society, Journal* **1**(1), 33–9.

——— 1983. An occurrence of wulfenite, anglesite and native sulphur in the opencast above Ball Eye quarry, Cromford, Derbyshire. *Russell Society, Journal* **1**(2), 48–51.

——— 1987. Serpierite and devilline from the Northern Pennine orefield. *Yorkshire Geological Society, Proceedings* **46**, 169.

——— 1990. Cinnabar from Machen quarry, Machen, South Wales. *Russell Society, Journal* **3**(2), 85.

——— 1997a. Field meeting report: Coldstone quarry, Pateley Bridge. *Russell Society, Newsletter* **31**, 13.

——— 1997b. Field meeting report: Middleton Common, Teesdale, Co. Durham. *Russell Society, Newsletter* **31**, 15.

——— 2002. Field meeting report: Coldstone quarry, Greenhow, North Yorkshire. *Russell Society, Newsletter* **41**, 32.

——— 2003. Cotunnite from the beach at Clevedon, Avon, England. *Russell Society, Journal* **8**(1), 36–7.

Bridges, T. F. & D. I. Green 2005a. Gearksutite, a second British occurrence, from the Ball Eye opencast, Cromford, Derbyshire. *Russell Society, Journal* **8**(2), 75–7.

——— 2005b. An update of the supergene mineralogy of Hilton mine, Scordale, Cumbria. *Russell Society, Journal* **8**(2), 108–110.

——— 2006a. The first occurrence of köttigite from Hilton mine, Scordale, Cumbria. *Russell Society, Journal* **9**, 3.

——— 2006b. Baryte replacement by barium carbonate minerals. *Russell Society, Journal* **9**, 73–82.

Bridges, T. F. & M. E. Smith 1983. Phosgenite and matlockite in Derbyshire. *Russell Society, Journal* **1**(2), 7–15.

Bridges, T. F. & H. Wilkinson 2003. Epimorphs of quartz after fluorite from the Rampgill, Coalcleugh and Barneycraig mine system, Nenthead, Cumbria, England. *Russell Society, Journal* **8**(1), 38–9.

——— 2005. Epimorphs of quartz after fluorite from the Rampgill, Coalcleugh and Barneycraig mine system, Nenthead, Cumbria, England – an update. *Russell Society, Journal* **8**(2), 112.

Bridges, T. F. & B. Young 1998. Supergene minerals of the Northern Pennine orefield – a review. *Russell Society, Journal* **7**(1), 3–14.

Bridges, T. F., D. I. Green, M. S. Rumsey 2006. A review of the mineralogy of Brae Fell mine, Caldbeck Fells, Cumbria. *Russell Society, Journal* **9**, 39–44.

Brindley, G. W. 1951. The crystal structure of some chamosite minerals. *Mineralogical Magazine* **29**, 502–525.

——— 1982. Chemical compositions of berthierines – a review. *Clay and Clay Minerals* **30**, 153–5.

Brindley, G. W. & K. Robinson 1946. The structure of kaolinite. *Mineralogical Magazine* **27**, 242–53.

——— 1947. Note on the occurrence of anatase in some fireclay deposits. *Mineralogical Magazine* **27**, 244–7.

Brindley, G. W. & R. F. Youell 1953. Ferrous chamosite and ferric chamosite. *Mineralogical*

Magazine **30**, 57–70.
Brindley, J. C. 1957. The aureole rocks of the Leinster granite in south Dublin, Ireland. *Royal Irish Academy, Proceedings* **57B**, 1–18.
Briscoe, P. 1986a. Red Gill mine. *UK Journal of Mines and Minerals* **1**, 2–4.
——— 1986b. Fallowfield mine. *UK Journal of Mines and Minerals* **1**, 12–13.
——— 1986c. Heights and the Green Hole. *UK Journal of Mines and Minerals* **1**, 37.
——— 2006. Unusual banded barite from Barras End mine, Swaledale, North Yorkshire. *UK Journal of Mines and Minerals* **27**, 11–14.
Briscoe, P. J. & D. I. Green 1989. The copper trials of Great Sled Dale, North Yorkshire. *UK Journal of Mines and Minerals* **7**, 45–6.
Bristow, C. M. & C. S. Exley 1994. Historical and geological aspects of the china-clay industry of southwest England. *Royal Geological Society of Cornwall, Transactions* **21**, 247–314.
Bristow, H. W. 1861. *A glossary of mineralogy.* London: Longman, Green, Longman and Roberts.
Bromley, A. V. 1964. Allanite in the Tan-y-Grisiau microgranite, Merionethshire, North Wales. *American Mineralogist* **49**, 1747–52.
——— 1969. Acid plutonic igneous activity in the Ordovician of North Wales. *The Precambrian and Lower Palaeozoic rocks of Wales*, A. Wood (ed.), 387–408. Cardiff: University of Wales Press.
Bromley, A. V. & J. Holl 1986. Tin mineralization in southwest England. In *Mineral processing at a crossroads*, B. A. Wills, R. W. Barley (eds), 195–262. Dordrecht: Martinus Nijhoff.
Brooke, H. J. A. 1823. On some undescribed minerals. *Quarterly Journal of Literature, Science and Art, London* **16**, 274–5.
——— 1824. On baryto-calcite. *Annals of Philosophy* **8**, 114–16.
Brooke, J. 1994a. *The tin streams of Wendron.* Truro: Twelveheads Press.
Brooke, M. F. 1984. Copiapite – a new location. *Gems and Mineral Realm* **16**(2), 29–31.
——— 1986. Devilline and other minerals from Wheal Friendship, Mary Tavy, Devon, UK. *Russell Society, Journal* **1**(4), 123–4.
——— 1994b. Fluorite at Hampstead Farm quarry, Chipping Sodbury, Avon: a new British occurrence. *Russell Society, Journal* **5**(2), 103–107.
Brooks, C. K., P. Henderson, J. G. Ronsbo 1981. Rare-earth partition between allanite and glass in the obsidian of Sandy Braes, Northern Ireland. *Mineralogical Magazine* **44**, 157–60.
Brooks, T. 2001. *Castle-an-Dinas 1916–1957.* St Austell: Cornish Hillside Publications.
——— 2004. *Devon's last metal mine – Great Rock shiny ore mine.* St Austell: Cornish Hillside Publications.
Broughton, D. G. 1967. Tin workings in the eastern district of the parish of Chagford, Devon. *Geologists' Association, Proceedings* **78**, 447–62.
Browell, E. J. J. 1861. Description and analysis of an undescribed mineral from Jarrow Slake. *Tyneside Naturalists' Field Club, Transactions* **5**, 103–104.
Brown, A. G. 1967. *Base metal mineralization at Ballyvergin and Maghera, Co. Clare, Eire.* MSc thesis, University of Leicester.
Brown, G., J. A. Catt, A. H. Weir 1969. Zeolites of the clinoptilolite–heulandite type in sediments of southeast England. *Mineralogical Magazine* **37**, 480–88.
Brown, G. C., R. A. Ixer, J. A. Plant, P. C. Webb 1987a. Geochemistry of granites beneath the north Pennines and their role in orefield mineralization. *Institution of Mining and Metallurgy, Transactions B* **96**, 65–76.
Brown, G. M. 1954. A suggested igneous origin for the banded granular hornfelses within the hypersthene gabbro of Ardnamurchan, Argyllshire. *Mineralogical Magazine* **30**, 529–33.
Brown, J. C. 1934. Lagerstättliche und erzmikroskopische Untersuchung der Zinnerzgänge der East Pool mine bei Redruth in Cornwall. *Neues Jahrbuch für Mineralogie, Abhandlungen* **68**, 298–336.
Brown, M. J., P. M. Allen, D. C. Cooper, D. G. Cameron, S. F. Pease, B. Beddoe-Stephens, A. D. Evans 1987b. *Volcanogenic mineralization in the Treffgarne area, southwest Dyfed, Wales.* Report 86, Mineral Reconnaissance Program, British Geological Survey, Keyworth.
Brown, P. E. 1956. The Mourne Mountains granites – a further study. *Geological Magazine* **93**, 72–84.
Brown, R. 1919. The mines and minerals of Leadhills. *Dumfriesshire and Galloway Natural History and Antiquarian Society, Transactions* **6**, 124–37.
——— 1927. More about the mines and minerals of Wanlockhead and Leadhills. *Dumfriesshire and Galloway Natural History and Antiquarian Society, Transactions* **13**, 58–79.
Bruce, I. & D. Aubrey-Jones 1998. Famous mineral localities: Ting Tang mine, Gwennap, Cornwall. *UK Journal of Mines and Minerals* **19**, 18–23.
Bryce, J. 1833. List of the simple minerals of the counties of Down, Antrim and Derry. *Philosophical Magazine (3rd series)* **3**, 83–7.
Buckley, A. 1984. Anomalous uranium in superficial deposits overlying the Wasdale Head granite, Cumbria. *Yorkshire Geological Society, Proceedings* **45**, 431–42.
——— 2005. *The story of mining in Cornwall: a world of payable ground.* Fowey: Cornwall Editions.
Buckley, H. A., J. C. Bevan, K. M. Brown, L. R. Johnson, V. C. Farmer 1978. Glauconite and celadonite: two separate mineral species. *Mineralogical Magazine* **42**, 373–82.
Budd, P., D. Gale, R. A. Ixer, R. G. Thomas 1994. Tin sources for prehistoric bronze production in Ireland. *Antiquity* **68**, 518–24.
Buist, D. S., A. M. M. Gadalla, J. White 1966. Delafossite and the system Cu–Fe–O. *Mineralogical Magazine* **35**, 731–41.
Bunting, R. 1994. Witherite from Scaleburn vein, Nenthead, Cumbria, England. *Russell Society, Journal* **5**(2), 118–19.
Burke, E. A. J. 2006. A mass discreditation of GQN minerals. *Canadian Mineralogist* **44**, 1557–60.
——— 2008. Tidying up mineral names: an IMA-CNMNC scheme for suffixes, hyphens and diacritical marks. *Mineralogical Record* **49**(2), 131–5.
Burke, E. A. J. & B. E. Leake 2005. "Named Amphiboles": a new category of amphiboles recognized by the International Mineralogical Association (IMA) and a defined sequence for the use of prefixes in amphibole names. *American Mineralogist* **90**, 516–17.
Burke, K. C., M. A. Cunningham, M. J. Gallagher, J. R. Hawkes 1964. Beryl in the Rosses granite, northwest Ireland. *Economic Geology* **59**, 1539–50.
Burns, I. M., M. B. Fowler, R. A. Strachan, P. B. Greenwood 2004. Geochemistry, petrogenesis and structural setting of the meta-igneous Strathy complex: a unique basement block within the Scottish Caledonides? *Geological Magazine* **141**, 209–223.
Burns, P. C. & M. A. Carpenter 1996. Phase transitions in the series boracite–trembathite–congolite. *Canadian Mineralogist* **34**, 881–92.
Burr, P. S. 1992. Notes on the history of phosgenite and matlockite from Matlock, England. *Mineralogical Record* **23**(5), 377–86.
——— 1996. Famous mineral localities: the Higher Pitts mine, Mendip Hills, Somerset, England. *Mineralogical Record* **27**(4), 245–59.
Busrewil, M. T., R. J. Pankhurst, W. J. Wadsworth 1975. The origin of the Kennethmont granite–diorite series, Insch, Aberdeenshire. *Mineralogical Magazine* **40**, 363–76.
Butcher, A. R., D. Pirrie, H. M. Prichard, P. Fisher 1999. Platinum-group mineralization in the Rum layered intrusion, Scottish Hebrides, UK. *Geological Society of London, Journal* **156**, 213–16.
Butcher, N. J. D. & J. D. Hedges 1987. Exploration and extraction of structurally and lithostratigraphically controlled fluorite deposits in the Castleton–Bradwell areas of the Southern Pennine orefield, England. *Institution of Mining and Metallurgy, Transactions B* **96**, 149–55.
Butler, A. J., E. A. Jobbins, P. J. Adams, F. W. Dunning, B. R. Young 1960. *Summary of progress of the Geological Survey of Great Britain and the Museum of Practical Geology for the year 1959.* London: HMSO.
Butler, B. C. M. 1961. Metamorphism and metasomatism of rocks of the Moine Series by a dolerite plug in Glenmore, Ardnamurchan. *Mineralogical Magazine* **32**, 866–97.
——— 1962. Biotite- and sphene-rich rocks in the Moine Series of Ardnamurchan. *Geological Magazine* **99**, 173–82.
Butler, F. H. 1887. Note on francolite. *Mineralogical Magazine* **7**, 164.
——— 1907. On the occurrence of silver ore in the Perran mine, Perranuthnoe, Cornwall. *Mineralogical Magazine* **14**, 385–8.
Butterfield, J. A. 1939. Brookite in the Millstone Grit of Yorkshire. *Geological Magazine* **76**, 220–28.

Caillère, S. & S. Hénin 1949. Occurrence of sepiolite in the Lizard serpentines. *Nature* **163**, 962.
Calas, G. 1972. On the blue colour of naturally banded fluorites. *Mineralogical Magazine* **38**, 977–9.
Callender, R. M. & J. Macaulay 1984. *The ancient metal mines of the Isle of Islay, Argyll.* British Mining 24, Northern Mine Research Society, Sheffield.
Calvert, J. 1853. *The gold rocks of Great Britain and Ireland.* London: Chapman & Hall.
Calvert, S. E. & N. B. Price 1970. Composition of manganese nodules and manganese carbonates from Loch Fyne, Scotland. *Contributions to Mineralogy and Petrology* **29**, 215–33.
Cámara, F., R. Oberti, C. Chopin, O. Medenbach 2006. The arrojadite enigma, I: a new formula and a new model for the arrojadite structure. *American Mineralogist* **91**, 1249–59.
Cameron, D. E. & D. M. Romer 1970. Denison copper–silver deposit at Aherlow, County Limerick, Ireland. *Institution of Mining and Metallurgy, Transactions B* **79**, 171–3.
Cameron, I. B. & R. I. Lawson 1979. An occurrence of paratacamite in southwest Scotland. *Mineralogical Magazine* **43**, 547.
Cameron, J. 1951. The geology of Hemerdon wolfram mine, Devon. *Institution of Mining and Metallurgy, Transactions B* **61**, 1–14.
Camm, G. S. 1995. *Gold in the Counties of Cornwall and Devon.* St Austell: Cornish Hillside Publications.
Camm, G. S. & S. C. Dominy 1997. Metasediment-hosted tin mineralization in the Indian Queens area, mid-Cornwall. *Ussher Society, Proceedings* **9**, 211–14.
Camm, G. S. & K. F. G. Hosking 1985. Stanniferous placer development on an evolving landsurface with special reference to places near St Austell, Cornwall. *Geological Society of*

London, Journal **142**, 803–813.

Camm, G. S. & M. Merry 1991a. Neufund der besten europaischen Bayldonitstufen von der Typlokalitat in Cornwall. *Lapis* **16**(9), 37–41.

——— 1991b. Bayldonite and its associates from Penberthy Croft, Cornwall. *UK Journal of Mines and Minerals* **9**, 6–15.

Campbell, A. & K. S. Panter 1990. Comparison of fluid inclusions in coexisting (cogenetic?) wolframite, cassiterite, and quartz from St Michael's Mount and Cligga Head, Cornwall, England. *Geochimica et Cosmochimica Acta* **54**, 673–81.

Campbell, R. & J. W. Lunn 1925. Chlorophaeite in the dolerites (tholeiites) of Dalmahoy and Kaimes Hills, Edinburgh. *Mineralogical Magazine* **20**, 435–40.

Campbell Smith, W. 1913. The mineral collection of Thomas Pennant (1726–1798). *Mineralogical Magazine* **16**, 331–42.

——— 1945. Banalsite crystals from Wales. *Mineralogical Magazine* **27**, 63–4.

——— 1947. Minerals [exhibit: general meeting of the Mineralogical Society, Burlington House, London, 24 January 1946]. *Mineralogical Magazine* **27**, xxxvii.

Campbell Smith, W. & F. A. Bannister 1948. Ganophyllite from the Benallt mine, Rhiw, Caernarvonshire. *Mineralogical Magazine* **28**, 343–52.

Campbell Smith, W. & G. F. Claringbull 1947. Pyrophanite from the Benallt mine, Rhiw, Carnarvonshire. *Mineralogical Magazine* **28**, 108–110.

Campbell Smith, W., F. A. Bannister, M. H. Hey 1944. Banalsite, a new barium feldspar from Wales. *Mineralogical Magazine* **27**, 33–46.

——— 1946. Pennantite, a new manganese-rich chlorite from Benallt mine, Rhiw, Caernarvonshire. *Mineralogical Magazine* **27**, 217–20.

Cann, J. R. 1965a. The metamorphism of amygdales at 'S Airde Beinn, northern Mull. *Mineralogical Magazine* **34**, 92–106.

——— 1965b. Gyrolite and reyerite from 'S Airde Beinn, northern Mull. *Mineralogical Magazine* **35**, 1–4.

Cann, J. R. & Banks, D. A. 2001. Constraints on the genesis of the mineralization of the Alston block, Northern Pennine orefield, northern England. *Yorkshire Geological Society, Proceedings* **53**, 187–96.

Canning, J. C., P. J. Henney, M. A. Morrison, J. W. Gaskarth 1996. Geochemistry of late Caledonian minettes from northern Britain: implications for the Caledonian subcontinental lithospheric mantle. *Mineralogical Magazine* **60**, 221–36.

Cantrill, T. C., R. L. Sherlock, H. Dewey 1919. *Iron ores – sundry unbedded ores.* Special Reports on the Mineral Resources of Great Britain IX, Memoir, Geological Survey of Great Britain, London.

Cattell, A. & P. Williams 1988. Anorthosite xenoliths in Loch Maree Group metabasalts. *Scottish Journal of Geology* **24**, 201–213.

Carlon, C. J. 1979. *The Alderley Edge mines.* Altrincham: John Sherratt.

Carmichael, I. S. E. 1960a. The pyroxenes and olivines from Tertiary acid glasses. *Journal of Petrology* **1**, 309–336.

——— 1960b. The feldspar phenocrysts of some Tertiary acid glasses. *Mineralogical Magazine* **32**, 587–608.

Carne, J. 1822. On the mineral productions and the geology of the parish of St Just. *Royal Geological Society of Cornwall, Transactions* **2**, 290–358.

Carruth, N. 1981. Pharmacosiderite – colour changes. *Mineral Realm* **1**(2), 4.

Carruthers, R. G. & A. Strahan 1923. *Lead and zinc ores of Durham, Yorkshire and Derbyshire, with notes on the Isle of Man.* Special Reports on the Mineral Resources of Great Britain XXVI, Memoir, Geological Survey of Great Britain, London.

Carruthers, R. G., T. Eastwood, G. V. Wilson, R. W. Pocock, D. A. Wray 1915. *Barytes and witherite.* Special Reports on the Mineral Resources of Great Britain II, Memoir, Geological Survey of Great Britain, London.

Carruthers, R. G., R. W. Pocock, D. A. Wray 1916. *Fluorspar.* Special Reports on the Mineral Resources of Great Britain IV, Memoir, Geological Survey of Great Britain, London.

Cartwright, I., W. R. Fitches, M. J. O'Hara, A. C. Barnicoat, S. O'Hara 1985. Archaean supracrustal rocks from the Lewisian near Stoer, Sutherland. *Scottish Journal of Geology* **21**, 187–96.

Castro, A. & W. E. Stephens 1992. Amphibole-rich polycrystalline clots in calc-alkaline granitic rocks and their enclaves. *Canadian Mineralogist* **30**, 1093–1112.

Cattermole, P. J. 1976. The crystallization and differentiation of a layered intrusion of hydrated alkali-olivine basalt parentage at Rhiw, North Wales. *Geological Journal* **11**, 45–70.

Caunt, S. 1984. Geochemical aspects of the Threlkeld microgranite, Cumbria. *Leeds Geological Association, Transactions* **10**, 88–100.

Chambers, B. (ed.) 1992. The plain man's guide to the rocks and minerals of Killhope. In *Men, mines and minerals of the north Pennines,* B. Chambers (ed.), 1–6. Killhope, Durham: Friends of Killhope.

Chambers, L. M., M. S. Pringle, R. R. Parrish 2005. Rapid formation of the Small Isles Tertiary centre constrained by precise $^{40}Ar/^{39}Ar$ and U–Pb ages. *Lithos* **79**, 367–84.

Chapman, N. A. 1976. Inclusions and megacrysts from undersaturated tuffs and basanites, east Fife, Scotland. *Journal of Petrology* **17**, 472–98.

Chapman, N. A. & R. Powell 1976. Origin of anorthoclase megacrysts in alkali basalts. *Contributions to Mineralogy and Petrology* **58**, 29–35.

Chapman, R. J. & R. C. Leake 2005. Evidence for the historical exploitation of possible bedrock gold in the upper Mennock Water, Wanlockhead, Dumfries and Galloway. *Scottish Journal of Geology* **41**, 135–40.

Chapman, R. J., R. C. Leake, J. D. Floyd 2000a. Regional variation in gold mineralization in the vicinity of the Glengaber Burn, Scottish Borders. *Scottish Journal of Geology* **36**, 165–76.

Chapman, R. J., R. C. Leake, N. R. Moles 2000b. The use of microchemical analysis of alluvial gold grains in mineral exploration: experiences in Britain and Ireland. *Journal of Geochemical Exploration* **71**, 241–68.

Chapman, R. J., R. C. Leake, N. R. Moles, G. Earls, C. Cooper, K. Harrington, R. Berzins, J. D. Floyd 2000c. The application of microchemical analysis of alluvial gold grains to the understanding of complex local and regional gold mineralization: a case study in the Irish and Scottish Caledonides. *Economic Geology* **95**, 1753–73.

Chapman, R. J., M. H. Shaw, R. C. Leake, B. Jackson 2005. Gold mineralization in the central Ochil Hills, Perthshire, UK. *Applied Earth Science* **114**, 53–64.

Charley, M. J., R. E. Hazleton, S. J. Tear 1989. Precious-metal mineralization associated with Fore Burn igneous complex, Ayrshire, southwest Scotland. *Institution of Mining and Metallurgy, Transactions B* **98**, 48–9.

Charoy, B. 1979. Greisenisation, minéralisation et fluides associés à Cligga Head, Cornwall (sud-ouest de l'Angleterre). *Bulletin de Minéralogie* **102**, 633–41.

——— 1986. The genesis of the Cornubian batholith (southwest England): the example of the Carnmenellis pluton. *Journal of Petrology* **27**, 571–604.

Chaudry, M. N. 1971. Feldspars from the Meldon aplite, Devonshire, England. *Mineralogical Magazine* **38**, 179–85.

Chaudhry, M. N. & R. A. Howie 1970. Topaz from the Meldon aplite, Devonshire. *Mineralogical Magazine* **37**, 717–20.

——— 1973a. Lithium aluminium micas from the Meldon aplite, Devonshire, England. *Mineralogical Magazine* **39**, 289–96.

——— 1973b. Muscovite (gilbertite) from the Meldon aplite. *Ussher Society, Proceedings* **2**, 480–81.

——— 1976. Lithium tourmalines from the Meldon aplite, Devonshire, England. *Mineralogical Magazine* **40**, 747–51.

Chaudhry, M. N. & A. Mahmood 1979. Types of distribution of the minerals of the Meldon aplite, Devonshire. *Mineralogical Magazine* **43**, 307–309.

Chew, D. M., J. S. Daly, L. M. Page, M. J. Kennedy 2003. Grampian orogenesis and the development of blueschist-facies metamorphism in western Ireland. *Geological Society of London, Journal* **160**, 911–24.

Chew, K. J. 1978. Crystalline manganese oxides in till, from Bridge of Don, Aberdeen. *Scottish Journal of Geology* **14**, 329–34.

Chick, L. 1987. Present-day collecting at Cligga Head, Cornwall. *UK Journal of Mines and Minerals* **3**, 33–4.

Chinner, G. A. 1961. The origin of sillimanite in Glen Clova, Angus. *Journal of Petrology* **2**, 313–23.

——— 1965. The kyanite isograd in Glen Clova, Angus, Scotland. *Mineralogical Magazine* **34**, 132–43.

——— 1967. Chloritoid and the isochemical nature of Barrow's zones. *Journal of Petrology* **8**, 268–82.

——— 1974. Dalradian margarite: a preliminary note. *Geological Magazine* **111**, 75–8.

Chinner, G. A. & P. D. Dixon 1973. Irish osumilite. *Mineralogical Magazine* **39**, 189–92.

Chinner, G. A. & J. S. Fox 1974. The origin of cordierite–anthophyllite rocks in the Land's End aureole. *Geological Magazine* **111**, 397–408.

Chisholm, J. E., G. C. Jones, O. W. Purvis 1987. Hydrated copper oxalate, moolooite, in lichens. *Mineralogical Magazine* **51**, 715–18.

Chopin, C. R. Oberti, F. Cámara 2006. The arrojadite enigma, II: compositional spaces, new members, and nomenclature of the group. *American Mineralogist* **91**, 1260–70.

Chukanov, N. V., S. Möckel, R. K. Raszvetaeva, A. E. Zadov 2003. Greifensteinit aus Sachsen, Österreich, Cornwall, USA, Brasilien und Kasachstan. *Lapis* **28**(4), 39.

Chukanov, N. V., D. Yu. Pushcharovsky, N. V. Zubkova, I. V. Pekov, M. Pasero, S. Merlino, S. Möckel, M. Kh. Rabadanov, D. I. Belakovskiy 2007. Zincolivenite $CuZn(AsO_4)(OH)$: a new adamite-group mineral with ordered distribution of Cu and Zn. *Doklady Earth Sciences* **415A**(6), 841–5.

Church, A. H. 1865a. Notes on a Cornish mineral of the atacamite group. *Chemical Society, Journal (series 2)* **3**, 212–15.

——— 1865b. Chemical researches on some new and rare Cornish minerals. *Chemical Society, Journal (series 2)* **3**, 259–68.

——— 1866a. Chemical researches on new and rare Cornish minerals. *Chemical Society, Journal (series 2)* **19**, 131–5.

——— 1866b. Chloropal in Cornwall. *The Chemical News* **14**, 71.

——— 1870. Chemical researches on new and rare Cornish minerals. *Chemical Society, Journal (series 2)* **23**, 3–5.

Clapham, R. C. & J. Dalglish 1864. On minerals and salts found in coal pits. *Report of the British Association for the Advancement of Science for 1863* **37**(2), 37–9.

——— 1884. On minerals and salts found in coal pits. *North of England Institute of Mining Engineering, Transactions* **13**, 219–26.

Claringbull, G. F. 1951. New occurrences of duftite. **29**, 609–614.

——— 1952. Nacrite from Groby, Leicestershire. *Mineralogical Magazine, Transactions* **29**, 973.

Claringbull, G. F. & M. H. Hey 1952. A re-examination of tobermorite. *Mineralogical Magazine* **29**, 960–62.

——— 1953. A re-examination of churchite. *Mineralogical Magazine* **30**, 211–17.

Claringbull, G. F., M. H. Hey, R. J. Davies 1959. Cornubite, a new mineral isomorphous with cornwallite. *Mineralogical Magazine* **32**, 1–5.

Clark, A. H. 1967. Mackinawite from the Lizard, Cornwall. *Mineralogical Magazine* **36**, 614–16.

——— 1970a. A probable second occurrence of Jambor's "fibrous iron sulfide". *American Mineralogist* **55**, 283–4.

——— 1970b. Early beryllium-bearing veins, South Crofty mine, Cornwall. *Institution of Mining and Metallurgy, Transactions B* **79**, 173–5.

——— 1970c. Manganese–iron ratios in wolframite, South Crofty mine, Cornwall: a discussion. *Economic Geology* **65**, 889–92.

Clark, A. H., D. J. Scott, H. A. Sandeman, A. V. Bromley, E. Farrar 1998. Siegenian generation of the Lizard ophiolite: U–Pb zircon age data for plagiogranite, Porthkerris, Cornwall. *Geological Society of London, Journal* **155**, 595–8.

Clark, A. M. 1993. *Hey's mineral index (mineral species, varieties and synonyms)*. London: Chapman & Hall, on behalf of the Natural History Museum.

Clark, A. M. & A. G. Couper 1979. End-member triploidite from Cornwall. *Mineralogical Magazine* **43**, 179–80.

Clark, A. M. & A. J. Criddle 1982. Palladium minerals from Hope's Nose, Torquay, Devon. *Mineralogical Magazine* **46**, 371–7.

Clark, A. M. & E. E. Fejer 1976. Zoned genthelvite from the Cairngorm Mountains, Scotland. *Mineralogical Magazine* **40**, 637–9.

Clark, A. M. & S. A. Rust 1993. Bottinoite, a mineral new to Britain. *Mineralogical Magazine* **57**, 543–4.

Clark, A. M., A. G. Couper, P. G. Embrey, E. E. Fejer 1986. Waylandite: new data, from an occurrence in Cornwall, with a note on "agnesite". *Mineralogical Magazine* **50**, 731–3.

Clark, A. M., A. J. Easton, M. Mount 1978. A study of the neotocite group. *Mineralogical Magazine* **42**, 279–80.

Clark, A. M., E. E. Fejer, A. G. Couper, G. C. Jones 1984. Sweetite, a new mineral from Derbyshire. *Mineralogical Magazine* **48**, 267–9.

Clark, A. M., E. E. Fejer, A. G. Couper, O. von Knorring, R. W. Turner, R. W. Barstow 1983. Iron-rich roscherite from Gunnislake, Cornwall. *Mineralogical Magazine* **47**, 81–3.

Clark, A. M., E. E. Fejer, G. Cressey, P. C. Tandy 1988. Ashoverite, a new mineral, and other polymorphs of $Zn(OH)_2$ from Milltown, Ashover, Derbyshire. *Mineralogical Magazine* **52**, 699–702.

Clark, A. M., D. A. Smith, R. F. Symes, P. C. Tandy, A. D. Hart 1995. *Catalogue of type mineral specimens in the Natural History Museum, London*. London: Natural History Museum.

Clark, C. M. 2007. Tourmaline: structural formula calculations. *Canadian Mineralogist* **45**, 229–37.

Clark, L. 1963. *The geology and petrology of the Ennerdale granophyre, its metamorphic aureole and associated mineralization*. PhD thesis, University of Leeds.

Clarke, G. M. 1974. *The mineralogy and geochemistry of tellurium with special reference to the bismuth sulphotellurides*. MSc thesis, University College, Cardiff.

Clarke, P. D. & W. J. Wadsworth 1970. The Insch layered intrusion. *Scottish Journal of Geology* **6**, 7–25.

Claxton, C. W. 1963. An occurrence of regionally metamorphosed Precambrian schists in southwest Pembrokeshire. *Geological Magazine* **100**, 219–23.

Clayton, R. E. 1993. Basalt carbonatisation and its implication for antimony mineralization in north Cornwall. *Ussher Society, Proceedings* **8**, 105–111.

Clayton, R. E., R. C. Scrivener, C. J. Stanley 1990. Mineralogical and preliminary fluid inclusion studies of lead–antimony mineralization in north Cornwall. *Ussher Society, Proceedings* **7**, 258–62.

Clayton, T. 1980. Hydrobasaluminite and basaluminite from Chickerell, Dorset. *Mineralogical Magazine* **43**, 931–7.

Cliff, D. C. & M. Wolfenden 1992. The Lack gold deposit, Northern Ireland. See Bowden et al. (1992: 65–75).

Cliff, G., J. A. Gard, G. W. Lorimer, H. F. W. Taylor 1975. Tacharanite. *Mineralogical Magazine* **40**, 113–26.

Clifford, J. A. 1986. A note on gold mineralization in County Tyrone. See Andrew et al. (1986: 45–7).

Clifford, J. A. & G. Earls 1982. *Annual report on prospecting licences 78NI 1c–1h for year ending 23rd August 1982*. Exploration Department, Ulster Base Metals Ltd, Dublin.

Clifford, J. A., A. H. Meldrum, R. T. G. Parker, G. Earls 1990. 1980–1990: a decade of gold exploration in Northern Ireland and Scotland. *Institution of Mining and Metallurgy, Transactions B* **99**, 133–8.

Clifford, J. A., P. Ryan, H. Kucha 1986. A review of the geological setting of the Tynagh orebody, Co. Galway. See Andrew et al. (1986: 419–40).

Clifford, T. N. 1959. A preliminary note on chromian mica in an ankerite–quartz–pyrite rock from southern Ross-shire, Scotland. *Mineralogical Magazine* **32**, 178–80.

Clifford, T. N., D. C. Rex, R. Green, A. P. le Roex, H. S. Pienaar, D. Bühmann 1999. Chromian illite–ankerite–quartz parageneses from the Kintail district of southern Ross-shire, Scotland. *Mineralogical Magazine* **63**, 37–52.

Clipstone, D. & P. Roycroft 1992. Detrital magmatic muscovite from the Lower Carboniferous of southwest Ireland: buried granites uncovered? *Geological Society of London, Journal* **149**, 163–6.

Clough, C. T. 1907. *The geological structure of the Northwest Highlands of Scotland*. Memoir, Geological Survey of Great Britain, Edinburgh.

Clough, C. T. & W. Pollard 1899. On spinel and forsterite from the Glenelg limestone (Inverness-shire). *Geological Society of London, Quarterly Journal* **55**, 372–80.

Coats, J. S. & J. R. Wilson 1971. The eastern end of the Galway granite. *Mineralogical Magazine* **38**, 138–51.

Coats, J. S., N. J. Fortey, M. J. Gallagher, A. Grout 1984. Stratiform barium enrichment in the Dalradian of Scotland. *Economic Geology* **79**, 1585–95.

Coats, J. S., C. G. Smith, N. J. Fortey, M. J. Gallagher, F. May, W. J. McCourt 1980. Stratabound barium-zinc mineralization in Dalradian schist, Aberfeldy, Scotland. *Institution of Mining and Metallurgy, Transactions B* **89**, 110–22.

Coats, J. S., C. G. Smith, M. J. Gallagher, F. May, W. J. McCourt, M. E. Parker, N. J. Fortey 1981. *Stratabound barium–zinc mineralization in Dalradian schist near Aberfeldy, Scotland*. Report 40, Mineral Reconnaissance Program, Institute of Geological Sciences, London.

Cockbain, A. G. 1968. Lead apatite solid-solution series. *Mineralogical Magazine* **36**, 1171–3.

Cole, G. A. J. 1891. On occurrences of riebeckite in Britain. *Mineralogical Magazine* **9**, 222–6.

——— 1922. *Memoir of localities of minerals of economic importance and metalliferous mines in Ireland*. Dublin: Mining Heritage Society of Ireland [1998: facsimile].

Collie, J. N. 1879. On the celestine and baryto-celestine of Clifton. *Mineralogical Magazine* **2**, 220–22.

Collins, J. H. 1871. *A handbook to the mineralogy of Cornwall and Devon*. London: Longmans.

——— 1876a. Note on new minerals from West Phoenix mine, and from St Agnes. *Mineralogical Magazine* **1**, 11–15.

——— 1876b. Note on the occurrence of scorodite, pharmacosiderite and olivenite in greenstone at Terras mine, St Stephens. *Mineralogical Magazine* **1**, 16–17.

——— 1876c. Note on the occurrence of achroite at Rock Hill, in the parish of St Austell, Cornwall, and on the black tourmaline of the same locality. *Mineralogical Magazine* **1**, 55–8.

——— 1877a. Remarks on gramenite from Smallacombe, and on the chloropal group of minerals. *Mineralogical Magazine* **1**, 67–71.

——— 1877b. Note on certain black quartz crystals from Boscaswell Downs, Cornwall. *Mineralogical Magazine* **1**, 115–16.

——— 1877c. Note on duporthite, a new asbestiform mineral. *Mineralogical Magazine* **1**, 226–7.

——— 1878a. Preliminary note on penwithite, a new Cornish mineral. *Mineralogical Magazine* **2**, 91.

——— 1878b. Notes on Cornish mineral localities. *Mineralogical Magazine* **2**, 92–4.

——— 1879a. Additional notes on penwithite. *Mineralogical Magazine* **3**, 89–90.

——— 1879b. Note on christophite from St Agnes. *Mineralogical Magazine* **3**, 91–2.

——— 1880. On some Cornish tin stones and tin capels [parts I & II]. *Mineralogical Magazine* **4**, 1–20, 103–116.

——— 1881. Catalogue of the minerals in the museum of the Royal Institution of Cornwall. *Royal Institution of Cornwall, Journal* **2**, 32.

——— 1887. On the nature and origin of clays: the composition of kaolinite. *Mineralogical Magazine* **7**, 205–214.

——— 1892. *A handbook to the mineralogy of Cornwall and Devon* (2nd edn). Truro: D. Bradford Barton [1969: facsimile].

——— 1904a. The precious metals in the west of England. *Royal Institution of Cornwall, Journal* **16**, 103–119.

——— 1904b. Notes on the principal lead-bearing lodes of the West of England. *Royal Geological Society of Cornwall, Transactions* **12**, 683–718.

——— 1911a. A list of minerals found in Cornwall and Devon, with notes supplementary to the author's "Handbook". *Royal Institution of Cornwall, Journal* **18**, 425–61.

——— 1911b. Additional notes on wood-tin. *Mineralogical Magazine* **16**, 30–34.

——— 1912. *Observations on the West of England mining region, being an account of the mineral deposits and economic geology of the region.* Transaction 14, Royal Geological Society of Cornwall, Penzance.

Collins, R. S. 1942. Cummingtonite and gedrite from Sutherland. *Mineralogical Magazine* **26**, 254–9.

——— 1972. *Barium minerals.* Mineral Dossier 2, Mineral Resources Consultative Committee, Institute of Geological Sciences, London.

Colman, T. B. & A-K. Appleby 1991. Volcanogenic quartz–magnetite–hematite veins, Snowdon, North Wales. *Mineralogical Magazine* **55**, 257–62.

Colman, T. B., K. E. Beer, D. G. Cameron, G. S. Kimbell 1989. *Molybdenum mineralization near Chapel of Gairioch, Inverurie, Aberdeenshire.* Report 100, Mineral Reconnaissance Program, British Geological Survey, Keyworth.

Colvine, R. J. L. 1968. Pyrope from Elie, Fife. *Scottish Journal of Geology* **4**, 283–6.

Coombs, D. S. and 17 other members of the IMA zeolite subcommittee 1997. Recommended nomenclature for zeolite minerals: report of the subcommittee on zeolites of the International Mineralogical Association, Commission on New Minerals and Mineral Names. *Canadian Mineralogist* **35**, 1571–606.

Connybeare, W. D. & W. Phillips 1822. *Outlines of the geology of England and Wales.* London: W. Phillips.

Cook, D. R. 1976. *The geology of the Cairnsmore of Fleet granite and its environs, southwest Scotland.* PhD thesis, University of St Andrews.

Cook, N. J., C. I. Ciobanu, C. J. Stanley, W. H. Paar, K. Sundblad 2007. Compositional data for Bi–Pb tellurosulfides. *Canadian Mineralogist* **45**, 417–35.

Cook, R. B. 2007. Witherite, Fallowfield, Northumberland, England. *Rocks and Minerals* **82**, 222–6.

Coomaraswamy, A. K. 1903. Observations on the Tiree Marble, with notes on others from Iona. *Geological Society of London, Quarterly Journal* **59**, 91–104.

Cooper, D. C., I. R. Basham, T. K. Smith 1983. On the occurrence of an unusual form of monazite in panned stream sediments in Wales. *Geological Journal* **18**, 121–7.

Cooper, D. C., D. G. Cameron, B. Young, J. D. Cornwell, D. J. Bland 1991. *Mineral exploration in the Cockermouth area, Cumbria, part 1: regional surveys.* Technical Report WF/91/4, British Geological Survey, Keyworth.

Cooper, D. C., M. K. Lee, N. J. Fortey, A. H. Cooper, C. C. Rundle, B. C. Webb, P. M. Allen 1988. The Crummock Water aureole: a zone of metasomatism and source of ore metals in the English Lake District. *Geological Society of London, Journal* **145**, 523–40.

Cooper, M. P. 1987a. New finds. *British Micromount Society Newsletter* **20**, 6–9.

——— 1987b. Cassiterite pseudomorphs after orthoclase from Wheal Coates, St Agnes, Cornwall. *UK Journal of Mines and Minerals* **3**, 30–32.

——— 1991a. Micromounts. *UK Journal of Mines and Minerals* **10**, 3.

——— 1991b. The British Micromount Society's reference collection of British minerals. *UK Journal of Mines and Minerals* **10**, 36–7.

——— 1995. Crystallized enargite from the Gregory mine, Ashover, Derbyshire. *UK Journal of Mines and Minerals* **15**, 28–9.

——— 2001. Keeping it in the family: the Humphreys, Forsters and Heulands. *Matrix* (spring), 3–31.

——— 2004. Henry Ludlam. In *Oxford dictionary of national biography.* Oxford: Oxford University Press.

——— 2005. The Devonshire mineral collection of Chatsworth House: an 18th century survivor and its restoration. *Mineralogical Record* **36**(3), 239–72.

——— 2006. *Robbing the sparry garniture: 200 years of British mineral dealers, 1750–1950.* Tucson: Mineralogical Record.

Cooper, M. P. & D. I. Green 1993. Mineral news. *UK Journal of Mines and Minerals* **12**, 26–32.

Cooper, M. P. & C. J. Stanley 1990. *Minerals of the English Lake District: Caldbeck Fells.* London: British Museum (Natural History).

——— 1991a. Die Mineralien der Caldbeck Fells, Cumberland, England. *Lapis* **22**(3), 13–34.

——— 1991b. Pyromorphite group minerals from the Caldbeck Fells, Cumbria, England. *Mineralogical Record* **22**(2), 105–121.

Cooper, M. P., D. I. Green, R. S. W. Braithwaite 1988. The occurrence of mattheddleite in the Caldbeck Fells, Cumbria: a preliminary note. *UK Journal of Mines and Minerals* **5**, 21.

Cooper, M. R. & R. H. Hunter 1995. Precision serial lapping, imaging and three-dimensional reconstruction of minus-cement and post-cementation intergranular pore systems in the Penrith Sandstone of northwestern England. *Mineralogical Magazine* **59**, 213–20.

Cope, F. W. 1933. A tholeiite dyke near Buxton, Derbyshire. *Geological Magazine* **70**, 414–22.

Cope, J. C. W. & R. E. Bevins 1993. The stratigraphy and setting of the Precambrian rocks of the Llangynog inlier, Dyfed, South Wales. *Geological Magazine* **130**, 101–111.

Cornwell, J. D., J. P. Busby, T. B. Colman, G. E. Norton, N. J. P. Smith 2001. The Snelston Ridge: evidence for a concealed ridge of mineralized Dinantian rocks to the south of the Derbyshire dome. *Yorkshire Geological Society, Proceedings* **53**, 257–64.

Corrie, K. D. & D. I. Green 2003. The barite and supposed celestine from the Triassic mudstone of Sidmouth, Devon. *UK Journal of Mines and Minerals* **23**, 27–35.

Cosgrove, M. E. & F. Hodson 1963. Jarosite from Meal Bank quarry, Ingleton, Yorkshire. *Yorkshire Geological Society, Proceedings* **34**, 81–90.

Cotterell, T. F. 2006a. Hydronium jarosite from Bryncastell mine, Dolgellau, Gwynedd, Wales. *Russell Society, Journal* **9**, 65–7.

——— 2006b. The first British occurrence of cesàrolite at Eaglebrook mine, Ceulanymaesmawr, Ceredigion, Wales. *UK Journal of Mines and Minerals* **27**, 36–8.

——— 2006c. Caryopilite and pyroxmangite from Nant mine, Nant-Y-Gadwen, Llanfaelrhys, Pen Llyn, Gwynedd, Wales. *UK Journal of Mines and Minerals* **27**, 51–3.

——— 2007. Cesàrolite from Frongoch mine, Devil's Bridge, Ceredigion, Wales. *Russell Society, Journal* **10**, 55–56.

Cotterell, T. F. & A. Dean 2007. The first occurrence of paralstonite at Dolyhir quarry, Old Radnor, Powys, Wales. *UK Journal of Mines and Minerals* **28**, 31–5.

Cotterell, T. F. & J. S. Mason 2003. Chalcophanite from Drosgol mine, Ponterwyd, Ceredigion: the first Welsh occurrence. *UK Journal of Mines and Minerals* **23**, 39–40.

Cotterell, T. F. & P. K. Todhunter 2007. Corkite and hinsdalite from Frongoch mine, Devil's Bridge, Ceredigion, Wales, including evidence to suggest that orpheite is a variety of hinsdalite. *Russell Society, Journal* **10**, 57–64.

Cotterell, T. F., J. S. Mason, D. I. Green 2003. Brianyoungite from Elgar mine and Frongoch mine, Ceredigion, Wales. *Russell Society, Journal* **8**(1), 33–4.

Cotter-Howells, J. & R. Giddens 1990. Pyromorphite: a secondary lead mineral. *Peak District Mines Historical Society, Bulletin* **11**, 21–2.

Couper, A. G. & R. W. Barstow 1977. Rediscovery of stokesite crystals in Cornwall, England. *Mineralogical Record* **8**, 294–7.

Couper, A. G. & A. M. Clark 1977. Stokesite crystals from two localities in Cornwall. *Mineralogical Magazine* **41**, 411–14.

Cowdry, S. 1995. Stalactitic galena from Lamb Bottom, Mendip Hills, Somerset, England. *Russell Society, Journal* **6**(1), 37–40.

——— 1996. Stalactitic galena in the Mendip Hills, Somerset, England: another 17th century reference. *Russell Society, Journal* **6**(2), 101.

——— 2002a. Field meeting report: Dolyhir quarry, Old Radnor. *Russell Society, Newsletter* **41**, 31.

——— 2002b. Field meeting report: Wick quarry. *Russell Society, Newsletter* **41**, 38.

Cox, A. H. & A. K. Wells 1920. The Lower Palaeozoic rocks of the Arthog–Dolgelley district (Merionethshire). *Geological Society of London, Quarterly Journal* **76**, 254–324.

Coxon, P. & A. Flegg 1985. A Middle Pleistocene interglacial deposit from Ballyline, Co. Kilkenny. *Royal Irish Academy, Proceedings* **85B**, 107–120.

Crampton, C. B. 1959. Analysis of heavy minerals in certain drift soils of Yorkshire. *Yorkshire Geological Society, Proceedings* **32**, 69–82.

Crane, A. 1987. The geology of Deeside. See Trewin et al. (1987: 227–42).

Cressey, G. 1987. Skarn formation between metachalk and agglomerate in the Central ring complex, Isle of Arran, Scotland. *Mineralogical Magazine* **51**, 231–46.

Criddle, A. J. & C. J. Stanley 1979. New data on wittichenite. *Mineralogical Magazine* **43**, 109–113.

——— (eds) 1993. *The quantitative data file for ore minerals* (3rd edn). London: Chapman & Hall.

Criddle, A. J. & R. F. Symes 1977. Mineralization at Tŷ Coch, Glamorgan (Mid Glamorgan), Wales: the second occurrence of pyrobelonite. *Mineralogical Magazine* **41**, 85–90.

Critchley, H. 1995. Field meeting report: Hendre quarry, Pandy, Glyn Ceiriog, Clwyd. *Russell Society, Newsletter* **26**, 16.

Critchley, M. F. 1979. A geological outline of the Ecton copper mines, Staffordshire. *Peak District Mines Historical Society, Bulletin* **7**(4), 177–91.

Crook, T. 1912. On the frequent occurrence of ankerite in coal. *Mineralogical Magazine* **16**, 219–23.

Croudace, I. W. 1982. The geochemistry and petrogenesis of the Lower Paleozoic granitoids of the Lleyn Peninsula, North Wales. *Geochimica et Cosmochimica Acta* **46**, 609–622.

Crowe, R. W. A. 1986. The stratigraphic and structural setting of Zn–Ba–Pb mineralization at Newtown Cashel, Co. Longford. See Andrew et al. (1986: 331–40).

Crowley, S. F., S. H. Bottrell, M. D. B. McCarthy, J. Ward, B. Young 1997. δ^{34}S of Lower Carboniferous anhydrite, Cumbria and its implications for barite mineralization in the northern Pennines. *Geological Society of London, Journal* **154**, 597–600.

Crummy, J., A. J. Hall, R. S. Haszeldine, I. K. Anderson 1997. Potential for epithermal gold mineralization in east and central Sutherland, Scotland: indications from River Brora headwaters. *Institution of Mining and Metallurgy, Transactions B* **106**, 9–14.

Cumming, G. A. 1935. The structural and volcanic geology of the Elie–St Monance district, Fife. *Edinburgh Geological Society, Transactions* **13**, 340–65.

Cummins, W. A. 1969. Excursion to the Nuneaton area, Warwickshire. *Mercian Geologist* **3**, 89–92.

Cundy, E. K., W. Windle, I. H. Warren 1960. The occurrence of zinnwaldite in Cornwall. *Clay Minerals Bulletin* **4**, 151–6.

Cunningham, M. A. 1950. *Sheefry mine*. Mine Records, Geological Survey of Ireland, Dublin.

Currie, J. 1898. The minerals of the Tertiary eruptive rocks of Ben More, Mull. *Edinburgh Geological Society, Transactions* **7**, 223–9.

——— 1905. Note on some new localities for gyrolite and tobermorite. *Mineralogical Magazine* **14**, 93–5.

Curtis, C. D. 1976. "Unmixed" Ca^{2+}/Mg^{2+} saponite at Calton Hill, Derbyshire. *Clay Mineralogy* **11**, 85–9.

Curtis, C. D., M. J. Pearson, V. A. Somogyi 1975. Mineralogy, chemistry, and origin of a concretionary siderite sheet (clay–ironstone band) in the Westphalian of Yorkshire. *Mineralogical Magazine* **40**, 385–93.

Curtis, M. T. 1981. The mineralization of Southfields quarry, Chipping Sodbury, Avon. *Bristol Naturalists' Society, Proceedings* **40**, 36–44.

Curtis, S. F., R. A. D. Pattrick, G. R. T. Jenkin, A. E. Fallick, A. J. Boyce, J. E. Treagus 1993. Fluid inclusion and stable isotope study of fault-related mineralization in Tyndrum area, Scotland. *Institution of Mining and Metallurgy, Transactions B* **102**, 39–47.

Dagger, G. W. 1977. Controls of copper mineralization at Coniston, English Lake District. *Geological Magazine* **114**, 195–202.

Dakyns, J. R., R. J. Tiddeman, J. G. Goodchild 1897. *The geology of the country between Appleby, Ullswater and Haweswater*. Memoir, Geological Survey of Great Britain and Ireland, London.

Darnley, A. G. 1962a. *Pitchblende from Sandyhills, Kircudbrightshire*. Report 20, Age Determination Unit, Atomic Energy Division, Geological Survey of Great Britain, London.

——— 1962b. *A new analysis of pitchblende from Tyndrum, Perthshire*. Report 21, Age Determination Unit, Atomic Energy Division, Geological Survey of Great Britain, London.

Darnley, A. G., T. H. English, O. Sprake, E. R. Preece, D. Avery, 1965. Ages of uraninite and coffinite from southwest England. *Mineralogical Magazine* **34**, 159–76.

Darnley, A. G., G. H. Smith, T. R. D. Chandler, D. F. Dance, E. R. Preece 1960. Ages of uraninite from two British localities. *Mineralogical Magazine* **32**, 654–8.

Davey, J. C. 1948. The geology of South Tolgus shaft. *Royal Geological Society of Cornwall, Transactions* **17**, 361–7.

Davidson, C. F. 1932. The geology of Moncreiffe Hill, Perthshire. *Geological Magazine* **69**, 452–64.

——— 1943. The Archaean rocks of the Rodil district, South Harris, Outer Hebrides. *Royal Society of Edinburgh, Transactions: Earth Sciences* **61**, 71–112.

Davidson, C. F. & S. H. U. Bowie 1951. On thucholite and related hydrocarbon–uraninite complexes. *Geological Survey of Great Britain, Bulletin* **3**, 1–19.

Davidson, P. J. 1989. Stevensite, a smectite-group mineral from Corstorphine Hill, Edinburgh. *Scottish Journal of Geology* **25**, 63–7.

——— 1995a. Troilite from Glenelg, Highland Region, Scotland: the first British Isles occurrence. *Russell Society, Journal* **6**(1), 52–3.

——— 1995b. Rutheniridosmine from the Ochil Hills, Scotland: the first British Isles occurrence. *Russell Society, Journal* **6**(1), 54–5.

Davidson, W. F. & N. Thomson 1951. Some notes on the minerals of Cumberland and Westmorland. *The North Western Naturalist* **23**, 136–54.

Davies, E. C. & T. W. Bloxam 1990. Petrology and geochemistry of late Precambrian volcanic rocks of the St David's area, Pembrokeshire, South Wales (UK). *Geologie en Minjbouw* **69**, 407–416.

Davies, G. M. 1915. Detrital andalusite in Cretaceous and Eocene sands. *Mineralogical Magazine* **17**, 218–20.

Davies, G. R. 1983. *The isotopic evolution of the British Lithosphere*. PhD thesis, Open University, Milton Keynes.

Davies, T. 1866. Crystallized stephanite and argentite from Cornwall. *Geological Magazine* **3**, 432.

——— 1867a. Researches in British mineralogy: senarmontite from Cornwall. *Geological Magazine* **4**, 192.

——— 1867b. Researches in British mineralogy. *Geological Magazine* **4**, 575–6.

——— 1877. Notes on vauquelinite from Scotland and cantonite from Cornwall. *Mineralogical Magazine* **1**, 112–14.

Davis, A. G. 1967. The mineralogy and phase equilibirum of Keuper Marl. *Quarterly Journal of Engineering Geology* **1**, 25–38.

Davis, R. J. & M. H. Hey 1964. Arthurite, a new copper–iron arsenate from Cornwall. *Mineralogical Magazine* **33**, 937–41.

——— 1969. The cell contents of arthurite redetermined. *Mineralogical Magazine* **37**, 520–21.

Davis, R. J., M. H. Hey, A. W. G. Kingsbury 1965. Xanthiosite and aerugite. *Mineralogical Magazine* **35**, 72–83.

Davison, E. H. 1920a. On the geology of Castle-an-Dinas wolfram mine. *Geological Magazine* **57**, 347–51.

——— 1920b. On the geology of St Michael's Mount. *Royal Geological Society of Cornwall, Transactions* **15**, 313–21.

——— 1921. Note on the pegmatite from Carn Marth. *Royal Geological Society of Cornwall, Transactions* **15**, 333–6.

——— 1922. *Section on the examination of veinstones*. Report, Tin and Tungsten Research Board, Department of Scientific and Industrial Research, London.

Dawson, J. & R. K. Harrison 1966. Uraninite in the Grainsgill greisen, Cumberland. *Geological Survey of Great Britain, Bulletin* **25**, 91.

Dawson, J., J. D. Floyd, P. R. Phillip 1979. *A mineral reconnaissance in the Abington–Biggar–Moffat area, south-central Scotland*. Report 28, Mineral Reconnaissance Program, Institute of Geological Sciences, London.

Day, E. L. 1979. Visit to the Redburn fluorspar mine in Weardale. *Cumberland Geological Society, Proceedings* **4**, 45–6.

Day, F. H. 1928. Some notes on the minerals of Caldbeck Fells. *Carlisle Natural History Society, Transactions* **4**, 66–79.

Dean, A. C. 1982. First British occurrence of cumengéite. *Russell Society, Journal* **1**(1), 40–41.

——— 1983. A new locality for phosgenite. *Russell Society, Journal* **1**(2), 19–20.

Dean, A. C. & T. F. Cotterell 2003. Cacoxenite and crandallite from Pwlldu beach, Bishopton, Gower, Swansea, South Wales: the first Welsh occurrence. *Russell Society, Journal* **8**(1), 30–32.

Dean, A. C., R. F. Symes, J. H. Thomas, P. A. Williams 1983. Cumengéite from Cornwall. *Mineralogical Magazine* **47**, 235–6.

Deans, T. 1934. The spherulitic ironstones of West Yorkshire. *Geological Magazine* **71**, 49–65.

——— 1961. A galena–wulfenite–asphaltite horizon in the Magnesian Limestone of Nottinghamshire. *Mineralogical Magazine* **32**, 705–715.

Deans, T. & H. G. C. Vincent 1938. Francolite from sedimentary ironstones of the Coal Measures. *Mineralogical Magazine* **25**, 135–9.

Dearman, W. R. 1965. Bismuth and molybdenite from Meldon quarry, near Okehampton, Devon. *Ussher Society, Proceedings* **1**, 196–8.

——— 1966a. Rhodonite from Meldon, Okehampton, Devonshire. *Mineralogical Magazine* **35**, 991–4.

——— 1966b. Datolite from Meldon, Okehampton, Devonshire. *Mineralogical Magazine* **35**, 1017–1018.

Dearman, W. R. & G. F. Claringbull 1960. Bavenite from the Meldon aplite quarries, Okehampton, Devonshire. *Mineralogical Magazine* **32**, 577–8.

Dearman, W. R. & J. M. Jones 1967. Millerite from Boldon colliery, County Durham. *Natural History Society of Northumberland, Transactions* **16**, 193–6.

Dearnley, R. 1961. Humite and chondrodite in a Lewisian crystalline limestone from South Harris, Outer Hebrides. *Mineralogical Magazine* **32**, 910–11.

——— 1962. Diopside–orthoclase–hornblende rocks from the Lewisian paragneisses of South Harris, Outer Hebrides. *Geological Magazine* **99**, 27–9.

——— 1963. The Lewisian Complex of South Harris; with some observations on the metamorphosed basic intrusions of the Outer Hebrides, Scotland. *Geological Society of London, Quarterly Journal* **119**, 243–312.

Deer, W. A. 1937a. Note on a pegmatitic hornblende from the Carsphairn complex. *Geological Magazine* **74**, 359–61.

——— 1937b. The marginal rocks of the Cairnsmore of Carsphairn complex. *Geological Magazine* **74**, 361–76.

——— 1937c. The composition and paragenesis of the biotites of the Carsphairn igneous complex. *Mineralogical Magazine* **24**, 495–502.

——— 1938. The composition and paragenesis of the hornblendes of the Glen Tilt complex, Perthshire. *Mineralogical Magazine* **25**, 56–74.

——— 1953. The diorites and associated rocks of the Glen Tilt complex. *Geological Magazine* **90**, 27–35.

Deer, W. A., R. A. Howie, J. Zussman 1997. *Rock-forming minerals: double chain silicates*, vol. 2B (2nd edn). London: Longman.

——— 2001. *Rock-forming minerals: framework silicates – feldspars*, vol. 4A (2nd edn). London: Geological Society.

——— 2004. *Rock-forming minerals: framework silicates – silica minerals, feldspathoids and the zeolites*, vol. 4B (2nd edn). London: Geological Society.

de Fourestier, J. 1999. *Glossary of mineral synonyms*. Ottawa: Mineralogical Society of Canada.

de Lapparent, J. 1934. Boehmite and diaspore in the bauxitic clays of Ayrshire. *Summary of progress of the Geological Survey of Great Britain and the Museum of Practical Geology*, part II, 1–7. London: Geological Survey of Great Britain.

Dempster, T. J. 1992. Zoning and recrystallization of phengitic micas: implications for metamorphic equilibration. *Contributions to Mineralogy and Petrology* **109**, 526–37.

Dempster, T. J. & B. J. Bluck 1991. Xenoliths in the lamprophyre dykes of Lomondside: constraints on the nature of the crust beneath the southern Dalradian. *Scottish Journal of Geology* **27**, 157–65.

Dempster, T. J. & R. A. Jackson 1996. Na-biotite in Dalradian pelitic schists, Angus. *Scottish Journal of Geology* **32**, 173–7.

Dempster, T. J., P. W. G. Tanner, P. Ainsworth 1994. Chemical zoning of white micas: a record of fluid infiltration in the Oughterard granite, western Ireland. *American Mineralogist* **79**, 536–44.

De Nul, R. 2005. The first Cornish occurrence of elyite at Greystone quarry, Lezant. *UK Journal of Mines and Minerals* **25**, 35–6.

Derham, J. M. & M. Feely 1988. A K-feldspar breccia from the Mo–Cu stockwork deposit in the Galway granite, west of Ireland. *Geological Society of London, Journal* **145**, 661–7.

Des Cloizeaux, A. L. 1880. *La Société Française de Minéralogie et de Cristallographie, Bulletin* **3**, 170–71.

Dewey, H. 1921. *Lead, silver–lead and zinc ores of Cornwall, Devon and Somerset.* Special Reports of the Mineral Resources of Great Britain XXI, Memoir, Geological Survey of Great Britain, London.

——— 1923. *Copper ores of Cornwall and Devon.* Special Reports on the Mineral Resources of Great Britain XXVII, Memoir, Geological Survey of Great Britain, London.

——— 1925. The mineral zones of Cornwall. *Geologists' Association, Proceedings* **36**, 107–135.

Dewey, H. & C. E. N. Bromehead 1915. *Tungsten and maganese ores.* Special Reports on the Mineral Resources of Great Britain I, Memoir, Geological Survey of Great Britain, London.

Dewey, H. & H. C. Dines 1923. *Tungsten and maganese ores* (3rd edn). Special Reports on the Mineral Resources of Great Britain I, Memoir, Geological Survey of Great Britain, London.

Dewey, H. & T. Eastwood 1925. *Copper ores of the Midlands, the Lake District and the Isle of Man.* Special Reports on the Mineral Resources of Great Britain XXX, Memoir, Geological Survey of Great Britain, London.

Dewey, H., J. S. Flett, G. V. Wilson 1920. *Arsenic and antimony ores.* Special Reports on the Mineral Resources of Great Britain XV, Memoir, Geological Survey of Great Britain, London.

Dewey, J. F. 1967. The structural and metamorphic history of the Lower Palaeozoic rocks of central Murrisk, County Mayo, Eire. *Geological Society of London, Quarterly Journal* **123**, 125–55.

Diack, A. L. 1979. Quartz gemstones of Scotland. *Lapidary Journal* **32**, 2618–23.

Dick, A. 1888. On kaolinite. *Mineralogical Magazine* **8**, 15–27.

——— 1908. Supplementary note on the mineral kaolinite. *Mineralogical Magazine* **15**, 124–7.

Dickie, D. M. & C. W. Forster 1974. *Mines and minerals of the Ochils.* Alloa: Clackmannanshire Field Studies Society.

Dickinson, C. & R. A. D. Pattrick 1987. A TEM investigation of optical variations in sphalerite. *Mineralogical Magazine* **51**, 127.

Dickson, J. A. D. 1980. Artificial colouration of fluorite by electron bombardment. *Mineralogical Magazine* **43**, 820–22.

Dickson, J. A. D. & C. Barber 1977. Chemical variation in a partially dolomitized Visean limestone bed, Isle of Man. *Mineralogical Magazine* **41**, 145–53.

Dines, H. G. 1956. *The metalliferous mining region of southwest England.* Economic memoir (reprinted with corrections, 1988), Geological Survey of Great Britain, London.

——— 1958. The West Shropshire mining region. *Geological Survey of Great Britain, Bulletin* **14**, 1–43.

Dines, H. D. & M. A. Edmunds 1929. *The geology of the country around Aldershot and Guildford.* Memoir, Geological Survey of Great Britain, London.

Dines, H. G., S. C. A. Holmes, J. A. Robbie 1954. *The geology of the country around Chatham.* Memoir, Geological Survey of Great Britain, London.

Dixon, E. E. L. 1921. *The geology of the South Wales coalfield, part XIII: the country around Pembroke and Tenby.* Memoir, Geological Survey of Great Britain, London.

Dixon, E. E. L., J. Maden, F. M. Trotter, S. E. Hollingworth, L. H. Tonks 1926. *The geology of the Carlisle, Longtown and Silloth district.* Memoir, Geological Survey of Great Britain, London.

D'Lemos, R. S. 1992. Magma mingling and melt modification between granitic pipes and host diorite, Guernsey, Channel Islands. *Geological Society of London, Journal* **149**, 709–720.

Dollar, A. T. J. 1932. Age problems of the Lundy granite. *Geological Magazine* **69**, 265–8.

——— 1941. The Lundy complex: its petrology and tectonics. *Geological Society of London, Quarterly Journal* **97**, 39–77.

Dominy, S. C. & G. C. Camm 1996. The nature and exploitation of narrow tin-bearing veins: a case study from South Crofty mine, Cornwall, UK. In *Memoirs 1996*, H. M. Martell (series ed.), 150–70. British Mining 57, Northern Mine Research Society, Sheffield.

——— 1997. Controls of ore localization in tin-bearing veins: a review. *Ussher Society, Proceedings* **9**, 241–9.

——— 1998. Geology and hydrothermal development of Bostraze–Balleswidden kaolin deposit, Cornwall, United Kingdom. *Institution of Mining and Metallurgy, Transactions B* **107**, 148–57.

Dominy, S. C. & R. F. G. Phelps 1996. Gwynfynydd's gold veins. *Mining Magazine* (December), 325–33.

Dominy, S. C., G. S. Camm, M. A. Bussell, R. C. Scrivener, C. Halls 1995. A review of tin stockwork mineralization in the southwest England orefield. *Ussher Society, Proceedings* **8**, 368–73.

Dominy, S. C., G. S. Camm, M. A. Bussell 1996a. Characteristics and development of carbona-style replacement tin mineralization in west Cornwall. *Ussher Society, Proceedings* **9**, 97–104.

Dominy, S. C., R. F. G. Phelps, A. Bussell, C. L. Guard 1996b. Geology and exploitation of complex gold-bearing veins in the Gwynfynydd mine, Dolgellau, North Wales, UK. In *Memoirs 1996*, H. M. Martell (series ed.), 70–91. British Mining 57, Northern Mine Research Society, Sheffield.

Dominy, S. C., R. F. G. Phelps, G. S. Camm 1996c. Geological controls on gold-grade distribution in Chidlaw link zone, Gwynfynydd mine, Dolgellau, North Wales. *Institution of Mining and Metallurgy, Transactions B* **105**, 151–8.

Dominy, S. C., R. C. Scrivener, N. Le Boutellier, M. A. Bussell, C. Halls 1994. Crosscourses in South Crofty mine, Cornwall: further studies of paragenesis and structure. *Ussher Society, Proceedings* **8**, 237–41.

Donaldson, C. H. 1978. Petrology of the uppermost upper mantle deduced from spinel lherzolite and harzburgite nodules at Calton Hill, Derbyshire. *Contributions to Mineralogy and Petrology* **65**, 363–77.

Dong, H., D. R. Peacor, R. J. Merriman, S. J. Kemp 2002. Brinrobertsite: a new RI interstratified pyrophyllite/smectite-like clay mineral: characterization and geological origin. *Mineralogical Magazine* **66**, 605–617.

Dossett, I. & D. I. Green 1998. Atacamite, botallackite and cumengéite from Abersoch, Gwynedd, Wales. *Russell Society, Journal* **7**(1), 37–8.

Dossett, I., D. I. Green, T. F. Cotterell 2007. Powellite from Benallt mine, Lleyn Peninsula, Gwynedd: a first Welsh occurrence. *Russell Society, Journal* **10**, 54–5.

Douch, H. I. 1964. *East Wheal Rose: the history of Cornwall's greatest lead mine.* Truro: Bradford Barton.

Doughty, P. S., A. M. Clark, G. C. Jones 1982. A platinum nugget found in Northern Ireland. *Irish Naturalists' Journal* **20**, 490–91.

Downes, K. M. J. & J. W. Platt 1978. The Avoca–Ballard mineralized belt, County Wicklow. *Royal Dublin Society, Journal of Earth Sciences* **1**, 121–33.

Doyle, E., A. A. Bowden, G. V. Jones, G. A. Stanley 1992. The geology of the Galmoy zinc–lead deposits, Co. Kilkenny. See Bowden et al. (1992: 211–26).

Draper, L. & P. Draper 1990. *The Raasay iron mine – where enemies became friends.* Dingwall, Ross-shire: L. & P. Draper.

Drever, H. I. 1936. Symplectite-bearing nodules in the Ardgour marble, Argyllshire. *Geological Magazine* **73**, 448–68.

——— 1939. A petrological study of the limestones in the Moine Series of Ardgour, Argyllshire. *Geological Magazine* **76**, 501–518.

——— 1940. The geology of Ardgour, Argyllshire. *Royal Society of Edinburgh, Transactions: Earth Sciences* **60**, 141–70.

Drever, H. I. & R. Johnston 1965. New petrographical data on the Shiant Isles picrite. *Mineralogical Magazine* **34**, 194–203.

Droop, G. T. R. 1994. Triple-chain pyriboles in Lewisian ultramafic rocks. *Mineralogical Magazine* **58**, 1–20.

Droop, G. T. R. & M. Moazzen 2007. Contact metamorphism and partial melting of Dalradian pelites and semipelites in the southern sector of the Etive aureole. *Scottish Journal of Geology* **43**, 155–79.

Drugman, J. 1915. On childrenite from Crinnis mine, Cornwall. *Mineralogical Magazine* **17**, 193–201.

——— 1927. On beta-quartz twins from some Cornish localities. *Mineralogical Magazine* **21**, 366–82.

Drysdale, D. J. 1985. Petalite and spodumene in the Meldon aplite, Devon. *Mineralogical Magazine* **49**, 758–9.

——— 1991. Lithium aluminium silicate minerals and pollucite from Meldon, Devon and San Piero in Campo, Elba. *Papers, Department of Geology, University of Queensland* **12**, 286–93.

Duane, M. J. 1988. Genesis, mineralogy and geochemistry of uranium in the Gortdrum stratiform copper deposit, Ireland. *Mineralium Deposita* **23**, 50–57.

Duane, M. J., H. J. Welke, H. L. Allsopp 1986. U–Pb age for some base-metal sulfide deposits in Ireland: genetic implications for Mississippi Valley-type mineralization. *Geology* **14**, 477–80.

Dudgeon, P. 1876. Historical notes on the occurrence of gold in the south of Scotland. *Min-*

eralogical Magazine **1**, 21–8.
——— 1895. Occurrence of mispickel in the Stewartry of Kirkudbright. *Mineralogical Magazine* **11**, 15.
Duffin, W. J. & J. Goodyear 1960. A thermal and X-ray investigation of scarbroite. *Mineralogical Magazine* **32**, 353–62.
Dufrénoy, A. 1856. *Traité de minéralogie* (1st edn), vol. 2 (p. 549). Paris: Carilian-Goeury et Vor Dalmont.
Duller, P. R., M. J. Gallagher, A. J. Hall, M. J. Russell 1997. Glendinning deposit – an example of turbidite-hosted arsenic–antimony–gold mineralization in the Southern Uplands, Scotland. *Institution of Mining and Metallurgy, Transactions B* **106**, 119–34.
Dunham, A. C. & W. J. Wadsworth 1978. Cryptic variation in the Rhum layered intrusion. *Mineralogical Magazine* **42**, 347–56.
Dunham, A. C. & F. C. F. Wilkinson 1985. Sulphide droplets and the unit 11/12 chromite band: a mineralogical study. *Geological Magazine* **122**, 539–48.
Dunham, K. C. 1934. The genesis of the North Pennine ore deposits. *Geological Society of London, Quarterly Journal* **90**, 689–720.
——— 1937. The paragenesis and colour of fluorite in the English Pennines. *American Mineralogist* **22**, 468–78.
——— 1944. The production of galena and associated minerals in the northern Pennines; with comparative statistics for Great Britain. *Institution of Mining and Metallurgy, Transactions B* **53**, 181–252.
——— 1947. Excursion to Durham and the northern Pennines. *Mineralogical Magazine* **28**, 45–51.
——— 1948. *Geology of the Northern Pennine orefield, vol. 1: Tyne to Stainmore*. Economic Memoir, Geological Survey of Great Britain, London.
——— 1952. *Fluorspar* (4th edn). Special Reports on the Mineral Resources of Great Britain IV, Memoir, Geological Survey of Great Britain, London.
——— 1959a. Non-ferrous mining potentialities of the northern Pennines. See Higham (1959: 115–47).
——— 1959b. Epigenetic mineralization in Yorkshire. *Yorkshire Geological Society, Proceedings* **32**, 1–30.
——— 1960. Syngenetic and diagenetic mineralization in Yorkshire. *Yorkshire Geological Society, Proceedings* **32**, 229–84.
——— 1967. Mineralization in relation to the pre-Carboniferous basement rocks, northern England. *Yorkshire Geological Society, Proceedings* **36**, 195–201.
——— 1984. Classification, age and origin of the Cumbrian hematite. *Cumberland Geological Society, Proceedings* **4**, 259–66.
——— 1990. *Geology of the Northern Pennine orefield, vol. 1: Tyne to Stainmore* (2nd edn). Memoir, British Geological Survey, Keyworth.
Dunham, K. C. & H. G. Dines 1945. *Barium minerals in England and Wales*. Wartime pamphlet 46, Geological Survey of Great Britain, London.
Dunham, K. C. & S. E. Hollingworth 1947. Excursion to Penrith and the North West. *Mineralogical Magazine* **28**, 248–54.
Dunham, K. C. & C. J. Stubblefield 1944. The stratigraphy, structure and mineralization of the Greenhow mining area, Yorkshire. *Geological Society of London, Quarterly Journal* **100**, 209–268.
Dunham, K. C. & A. A. Wilson 1985. *Geology of the Northern Pennine orefield, vol. 2: Stainmore to Craven*. Memoir, British Geological Survey, Keyworth.
Dunham, K. C., K. E. Beer, R. A. Ellis, M. J. Gallagher, M. J. C. Nutt, B. C. Webb 1978. United Kingdom. In *Mineral deposits of Europe, vol. 1: northwest Europe*, S. H. U. Bowie, A. Kvalheim, H. W. Haslam (eds), 263–317. London: Institution of Mining and Metallurgy and Mineralogical Society.
Dunham, K. C., G. F. Claringbull, F. A. Bannister 1948. Dickite and collophane in the Magnesian Limestone. *Mineralogical Magazine* **28**, 338–42.
Dunham, K. C., B. Young, G. A. L. Johnson, T. B. Colman, R. Fossett 2001. Rich silver-bearing lead ores in the northern Pennines? *Yorkshire Geological Society, Proceedings* **53**, 207–212.
Dunlop, B. S. N. 2005. *Cairngorm stones*. Grantown: Museum & Heritage Trust.
Dunn, P. J. 1990. Andrewsite and laubmannite formally discredited. *American Mineralogist* **75**, 1197–9.
Dunn, P. J. & J. A. Mandarino 1988. Formal definitions of type mineral specimens. *Mineralogical Record* **19**, 227–8.
Dunn, P. J. & R. C. Rouse 1975. Wroewolfeite, a new copper sulphate hydroxide hydrate. *Mineralogical Magazine* **40**, 1–5.
Dunn, P. J. & W. E. Wilson 1978. Nomenclature revisions in the apophyllite group: hydroxylapophyllite, apophyllite, fluorapophyllite. *Mineralogical Record* **9**(3), 95–8.
Dunn, P. J., D. R. Peacor, J. E. Nelen, R. A. Ramik 1983. Ganophyllite from Franklin, New Jersey; Pajsberg, Sweden; and Wales: new chemical data. *Mineralogical Magazine* **47**, 563–6.
Du Noyer, G. V. 1865. *Explanation to accompany sheets 167, 168, 178 and 179 of the maps, and sheet 13 of the longitudinal sections of the Geological Survey of Ireland counties of Waterford, Wexford, Kilkenny and Tipperary*. Dublin: Alexander Thoms.
Durrance, E. M. & M. C. George 1976. Metatyuyamunite from the uraniferous–vanadiniferous nodules in the Permian marls and sandstones of Budleigh Salterton. *Ussher Society, Proceedings* **3**, 435–40.
Durrance, E. M. & D. J. C. Laming (eds) 1982. *The geology of Devon*. Exeter: University of Exeter Press.
Dutton, A., P. J. Fasham, D. A. Jenkins, A. E. Caseldine, S. Hamilton-Dyer 1994. Prehistoric copper mining on the Great Orme, Llandudno, Gwynedd. *Prehistoric Society, Proceedings* **60**, 245–86.
Dwerryhouse, A. R. 1909. Intrusive rocks in the neighbourhood of Eskdale. *Geological Society of London, Quarterly Journal* **65**, 55–80.
Dyer, A. & O. Wilson 1988. Zeolites. *UK Journal of Mines and Minerals* **4**, 17–21.
Dyer, A., A. S. Abdel Gawad, M. Mikhail, H. Enamy, M. Afshang 1991. The natural zeolite, laumontite, as a potential material for the treatment of aqueous nuclear wastes. *Journal of Radioanalytical and Nuclear Chemistry* **154**, 265–76.
Dyer, A., H. Enamy, O. M. Wilson, H. Foy 1987. Doranite – a magnesium rich analcime. *Zeolites* **7**, 283–6.
Dyer, A., D. I. Green, A. G. Tindle, C. D. Williams 2006. Clinoptilolite and heulandite from Wheal Hazard, St Just, Cornwall. *Russell Society, Journal* **9**, 68–72.
Dyer, A., R. Waughman, C. Williams 2003. Calcium analcime from the Totherick, Ayrshire, Scotland: an unusual crystal form with trapezoidal faces. *Russell Society, Journal* **8**(1), 26–9.
Dyer, A., O. M. Wilson, H. Enamy, C. D. Williams 1993. Stellerite from Todhead Point, Grampian Region, Scotland. *Mineralogical Magazine* **57**, 540–42.
Dyer, A., O. M. Wilson, C. D. Williams, J. Smethurst 1996. Zeolite occurrences in basaltic dykes on the Western Isles of Scotland. *Russell Society, Journal* **6**(2), 89–92.
Dyer, A., O. M. Wilson, B. Young 2001. Natrolite from Calton Hill, Derbyshire, England. *Russell Society, Journal* **7**(2), 89.

Eagar, R. M. C. & F. M. Broadhurst 1959. Alderley Edge, Cheshire. In *The area around Manchester*, 18–22. Guide 7, Geologists' Association, London.
Eakin, P. A. 1989. Isotopic and petrographic studies of uraniferous hydrocarbons from around the Irish Sea Basin. *Geological Society of London, Journal* **146**, 663–73.
Eakin, P. A. & A. P. Gize 1992. Reflected-light microscopy of uraniferous bitumens. *Mineralogical Magazine* **56**, 85–99.
Earl, B. 1994. *Cornish mining*. St Austell: Cornish Hillside Publications.
——— 1996. *The Cornish arsenic industry*. Camborne: Penhellick Publications.
Earls, G., J. A. Clifford, A. H. Meldrum 1989. Curraginalt gold deposit, County Tyrone, Northern Ireland. *Institution of Mining and Metallurgy, Transactions B* **98**, 50–51.
Earls, G., R. T. G. Parker, J. A. Clifford, A. H. Meldrum 1992. The geology of the Cononish gold–silver deposit, Grampian Highlands of Scotland. See Bowden et al. (1992: 89–103).
Earp, J. R. 1958. Mineral veins of the Minera–Maeshafn district of North Wales. *Geological Survey of Great Britain, Bulletin* **14**, 44–69.
Eastwood, N. J. 1983. A new locality for wulfenite on the Mendips. *Russell Society, Journal* **1**(2), 25–6.
Eastwood, T. 1921. *Lead and zinc ores of the Lake District*. Special Reports on the Mineral Resources of Great Britain XXII, Memoir, Geological Survey of Great Britain, London.
——— 1959. The Lake District mining field. See Higham (1959: 149–74).
Eastwood, T., E. E. L. Dixon, S. E. Hollingworth, B. Smith 1931. *The geology of the Whitehaven and Workington district (sheet 28)*. Memoir, Geological Survey of Great Britain, London.
Eastwood, T., S. E. Hollingworth, W. C. C. Rose, F. M. Trotter 1968. *Geology of the country around Cockermouth and Caldbeck (sheet 23)*. Memoir, Institute of Geological Sciences, London.
Edgar, A. D., A. Mottana, N. D. Macrae 1969. The chemistry and cell parameters of omphacites and related pyroxenes. *Mineralogical Magazine* **37**, 61–74.
Edge, A. B. 1920. Siliceous sinter from Lustleigh, Devon. *Mineralogical Magazine* **19**, 10–13.
Edmond, B. A. 1983. *Report on stratabound lead–zinc–silver mineralization on Blocks A & B, Islay House Estate, Islay, Argyllshire, Scotland*. Project AE243, Mineral Exploration & Investment Grant Act 1972, Institute of Geological Sciences, Edinburgh.
Edmonds, E. A., A. Whittaker, B. J. Williams 1985. *The geology of the country around Ilfracombe and Barnstaple (sheets 277 and 293)*. Memoir, British Geological Survey, Keyworth.
Edmonds, E. A., B. J. Williams, R. T. Taylor 1979. *Geology of Bideford and Lundy island (sheet 292, New Series, with sheets 275, 276, 291 and part of sheet 308)*. Memoir, Institute of Geological Sciences, London.
Edwards, R. P. 1976. Aspects of trace metal and ore distribution in Cornwall. *Institution of Mining and Metallurgy, Transactions B* **85**, 83–90.
Eeckhout, S. G., C. Corteel, E. Van Coster, E. De Grave, P. De Paepe 2004. Crystal-

chemical characterization of tourmalines from the English Lake District: electron-microprobe analyses and Mössbauer spectroscopy. *American Mineralogist* **89**, 1743–51.

Elliot, R. B. 1956. The Eskdalemuir tholeiite and its contribution to an understanding of tholeiite genesis. *Mineralogical Magazine* **31**, 245–54.

Ellis, R. A. 1977. Disseminated copper in a Caledonian calc-alkaline intrusion, Argyllshire, Scotland. *Institution of Mining and Metallurgy, Transactions B* **86**, 52–4.

Ellis, R. A., G. R. Marsden, N. J. Fortey 1978. *Disseminated sulphide mineralization at Garbh Achadh, Argyllshire, Scotland.* Report 23, Mineral Reconnaissance Program, Institute of Geological Sciences, London.

Elsden, J. V. 1904. On the age of the Llyn–Padarn dykes. *Geological Society of London, Quarterly Journal* **60**, 372–88.

Elsdon, R. 1975. Manganoan ilmenite from the Leinster granite, Ireland. *Mineralogical Magazine* **40**, 419–21.

El Sharkawi, M. A. H. & W. R. Dearman 1966. Tin-bearing skarns from the northwest border of the Dartmoor granite, Devonshire, England. *Economic Geology* **61**, 362–9.

——— 1967. Metasomatism of cherts on the northwest margin of Dartmoor, Devonshire. *Mineralogical Magazine* **36**, 260–73.

Elton, N. J. 1996. Variscite and metavariscite from Gunheath china-clay pit, St Austell, Cornwall. *Mineralogical Magazine* **60**, 671–2.

——— 1997. Hydroxyl-chlorapatite from the St Austell granite, Cornwall, England. *Mineralogical Magazine* **61**, 719–20.

——— 1998. Barbosalite, frondelite and other phosphates from Gunheath china-clay pit, St Austell, Cornwall, England. *Russell Society, Journal* **7**(1), 15–18.

Elton, N. J. & M. J. E. Grigg 1996. Calcian goyazite and florencite-(Ce) from Gunheath china-clay pit, St Austell, Cornwall. *Russell Society, Journal* **6**(2), 81–3.

Elton, N. J. & J. J. Hooper 1992. Andersonite and schröckingerite from Geevor mine, Cornwall: two species new to Britain. *Mineralogical Magazine* **56**, 124–5.

——— 1993. Sodium-zippeite from Geevor mine, St Just, Cornwall. *Mineralogical Magazine* **57**, 352–4.

——— 1994. Tyuyamunite from St Just, Cornwall. *Russell Society, Journal* **5**(2), 114–15.

——— 1995a. Supergene U–Pb–Cu mineralization at Loe Warren, St Just, Cornwall, England. *Russell Society, Journal* **6**(1), 17–26.

——— 1995b. A second occurrence of mahlmoodite, from Cornwall, England. *Mineralogical Magazine* **59**, 166–8.

——— 1995c. Widenmannite from Cornwall, England: the second world occurrence. *Mineralogical Magazine* **59**, 745–9.

——— 1996. A further occurrence of widenmannite in Cornwall, England. *Russell Society, Journal* **6**(2), 98.

Elton, N., D. I. Green, J. Hooper, S. Weiss 2004. Famous mineral localities: Greystone quarry, Lezant, Cornwall. *UK Journal of Mines and Minerals* **24**, 9–28.

Elton, N. J., J. J. Hooper, L. Coggon 1992. Cuprosklodowskite from Geevor mine, Cornwall. *Russell Society, Journal* **4**(2), 66–7.

——— 1996. Morinite from Gunheath china-clay pit, St Austell, Cornwall. *Mineralogical Magazine* **60**, 517–18.

Elton, N. J., J. J. Hooper, V. A. D. Holyer 1997. An occurrence of stevensite and kerolite in the Devonian Crousa gabbro at Dean quarry, the Lizard, Cornwall, England. *Clay Mineralogy* **32**, 241–52.

——— 1998. Al-free gyrolite from the Lizard, Cornwall, England. *Mineralogical Magazine* **62**, 271–2.

Elton, N. J., J. J. Hooper, A. E. Jeal 1994a. Nováčekite and metanováčekite from Cornwall. *Mineralogical Magazine* **58**, 513–14.

Elton, N. J., J. J. Hooper, G. Ryback 1994b. Compreignacite: a second occurrence, from St Just, Cornwall. *Mineralogical Magazine* **58**, 339–41.

Embrey, P. G. 1953. Note on the occurrence of beryl and löllingite at the New Consols mine, Stoke Climsland, Cornwall. *Mineralogical Magazine* **30**, 259–62.

——— 1973. Arthur William Gerald Kingsbury (1906–1968). *Mineralogical Magazine* **39**, 1–3.

——— 1978. Fourth supplementary list of British minerals. *Mineralogical Magazine* **42**, 169–78.

——— 1994. John Henry Vivian (1785–1855): the probable eponym of vivianite. *Russell Society, Journal* **5**(2), 77–81.

Embrey, P. G. & E. E. Fejer 1969. Tavistockite and bialite discredited. *Mineralogical Magazine* **37**, 123–7.

Embrey, P. G. & R. F. Symes 1987. *Minerals of Cornwall and Devon.* London: British Museum (Natural History).

Emeleus, C. H. & B. R. Bell 2005. *The Palaeogene volcanic districts of Scotland* [British Regional Geology series]. Keyworth: British Geological Survey.

Emeleus, C. H. & M. C. Gyopari 1992. *British Tertiary Volcanic Province.* London: Chapman & Hall.

Emeleus, C. H., A. C. Dunham, R. N. Thompson 1971. Iron-rich pigeonites from acid rocks in the Tertiary igneous province of Scotland. *American Mineralogist* **56**, 940–51.

Emo, G. T. 1986. Some considerations regarding the styles of mineralization at Harberton Bridge, Co. Kildare. See Andrew et al. (1986: 461–70).

England, R. W. 1992. The genesis, ascent, and emplacement of the northern Arran granite, Scotland: implications for granite diapirism. *Geological Society of America, Bulletin* **104**, 606–614.

English, H. 1851. *The mining almanack for 1849, being a yearly compendium of information on general science, tabular, and with other statistical details relating to the mining interests of Great Britain.* London: Simpkin, Marshall, on behalf of the Mining Journal Office.

Ercit, T. S. 2002. The mess that is "allanite". *Canadian Mineralogist* **40**, 1411–19.

Ertl, A., U. Kolitsch, S. Prowatke, M. D. Dyar, D. J. Henry 2006. The F-analogue of schorl from Grasstein, Trentino – South Tyrol, Italy: crystal structure and chemistry. *European Journal of Minerapogy* **18**, 583–8.

Essene, E. J., C. E. Henderson, A. Livingstone 2006. The missing sulphur in mattheddleite, sulphur analysis of sulphates, and paragenetic relations at Leadhills, Scotland. *Mineralogical Magazine* **70**, 265–80.

Evans, A. M. 1977. Copper–molybdenum mineralization in the Ballachulish granite, Argyllshire, Scotland. *Institution of Mining and Metallurgy, Transactions B* **86**, 152–3.

Evans, A. M. & A. El-Nikhely 1982. Some palaeomagnetic dates from the west Cumbrian hematite deposits, England. *Institution of Mining and Metallurgy, Transactions B* **91**, 41–3.

Evans, A. M. & N. D. M. Evans 1977. Some preliminary palaeomagnetic studies of mineralization in the Mendip orefield. *Institution of Mining and Metallurgy, Transactions B* **86**, 149–51.

Evans, A. M. & R. J. King 1962. Palygorskite in Leicestershire. *Nature* **194**, 860.

Evans, C. D. R. 1968. *Geological succession and structure of the area east of Bethesda.* PhD thesis, University of Wales, Aberystwyth.

Evans, L. J. 1987. *Low-grade regional metamorphism of Palaeozoic rocks in the Midland Valley of Scotland.* PhD thesis, University of St Andrews.

Ewart, A. 1962. Hydrothermal alteration in the Carrock Fell area, Cumberland, England. *Geological Magazine* **99**, 1–8.

Exley, C. S. & K. M. Edmondson 1993. Aspects of alkali feldspars in Cornubian granites. *Ussher Society, Proceedings* **8**, 117–23.

Exley, C. S. & M. Stone 1964. The granitic rocks of South West England. In *Present views of some aspects of the geology of Cornwall and Devon*, K. F. G. Hosking, G. L. Shrimpton (eds), 132–84. Penzance: Royal Geological Society of Cornwall.

Exley, R. A. 1980. Microprobe studies of REE-rich accessory minerals: implications for Skye granite petrogenesis and REE mobility in hydrothermal systems. *Earth and Planetary Science Letters* **48**, 97–110.

Eyles, V. A. 1952. *The composition and origin of the Antrim laterites and bauxites.* Memoir, Geological Survey of Northern Ireland, Belfast.

Fairbairn, R. A. 1993. *The mines of Alston Moor.* British Mining 47, Northern Mine Research Society, Sheffield.

——— 1996. *Weardale mines.* British Mining 56, Northern Mine Research Society, Sheffield.

——— 2000. *Allendale, Tynedale and Derwent lead mines.* British Mining 65, Northern Mine Research Society, Sheffield.

——— (ed.) 2003. *Fluorspar in the north Pennines.* Killhope, Durham: Friends of Killhope.

Fairchild, I. J. 1977. Phengite spherules from the Dalradian Bonahaven Formation, Islay, Scotland: glauconitized microfossils. *Geological Magazine* **114**, 355–64.

Faithfull, J. W. 1988. Strontianite from near Yate (Avon). *Russell Society, Journal* **2**(1), 1–3.

——— 2007. A Scottish "diamond" specimen rediscovered and re-identified: the first Scottish occurrence of piemontite. *Scottish Journal of Geology* **43**, 33–40.

Faithfull, J. W. & N. Hubbard 1988. Coffinite from Gypsy Lane brickpit, Leicester. *Russell Society, Journal* **2**(1), 25–8.

——— 2001. Otavite from Bishopton, Renfrewshire, Scotland. *Russell Society, Journal* **7**(2), 94–5.

Faithfull, J. W. & F. Ince 1992. An unusual occurrence of mottramite from Peldar Tor, Leicestershire. *Russell Society, Journal* **4**(2), 59–62.

Farrow, C. M., A. Herriot, B. E. Leake 1982. A Carboniferous arfvedsonite–aegirine trachyte from West Kilbride, Scotland. *Mineralogical Magazine* **46**, 399–401.

Fawcett, J. J. 1965. Alteration products of olivine and pyroxene in basalt lavas from the Isle of Mull. *Mineralogical Magazine* **35**, 55–71.

Fearnsides, W. G. 1910. Excursion to North Wales. *Geologists' Association, Proceedings* **21**, 368–90.

Feely, M. & J. S. Madden 1988. Trace element variation in the leucogranites within the main Galway granite, Connemara, Ireland. *Mineralogical Magazine* **52**, 139–46.

Feely, M., E. McCabe, C. T. Williams 1989. Th and REE bearing accessory minerals in a high heat production (HHP) leucogranite within the Galway granite, western Ireland. *Institution of Mining and Metallurgy, Transactions B* **98**, 27–32.

Ferguson, C. C. & S. I. Al-Ameen 1985. Muscovite breakdown and corundum growth at anomalously low fH_2O: a study of contact metamorphism and convective fluid movement around the Omey granite, Connemara, Ireland. *Mineralogical Magazine* **49**, 505–514.

Ferguson, C. C. & P. K. Harvey 1979. Thermally overprinted Dalradian rocks near Cleggan, Connemara, western Ireland. *Geologists' Association, Proceedings* **90**, 43–50.

Ferguson, L. K., A. E. Fallick, I. Allison 1998. Tourmaline in a sub-Cambrian palaeosol on the Proterozoic Lewisian rocks of northwest Scotland. *Geological Society of London, Journal* **155**, 725–31.

Fermor, L. L. 1951. On a discovery of copper ore in the Torridonian rocks of Sutherland. *Geological Magazine* **88**, 215–18.

Fernandez, L. L. 1957. *The geology and mineralization of Abbeytown mine, Co. Sligo, Eire.* BSc thesis, Royal School of Mines, London.

Ferry, J. M. 1996. Three novel isograds in metamorphosed siliceous dolomites from the Ballachulish aureole, Scotland. *American Mineralogist* **81**, 485–94.

Fettes, D. J., J. R. Mendum, D. I. Smith, J. V. Watson 1992. *Geology of the Outer Hebrides: sheets (solid edition) Lewis and Harris, Uist and Barra (Scotland).* Memoir, British Geological Survey, Keyworth.

Field, R. 1989. Colour in Scottish agates. *UK Journal of Mines and Minerals* **7**, 13–17.

——— 2005. Ardownie/Ethibeaton quarry und Ayrshire. *Mineralien Welt* **16**(3), 66–72.

Finlayson, A. M. 1910. The ore-bearing pegmatites of Carrock Fell and the genetic significance of tungsten ores. *Geological Magazine* **47**, 19–28.

Firman, R. J. 1953a. *Metamorphism and metasomatism around the Shap and Eskdale granites.* PhD thesis, University of Durham.

——— 1953b. On the occurrence of nacrite at Shap, Westmorland. *Mineralogical Magazine* **30**, 199–200.

——— 1957. Fissure metasomatism in volcanic rocks adjacent to the Shap granite, Westmorland. *Geological Society of London, Quarterly Journal* **113**, 205–222.

——— 1972. The Shap area: Excursion to the northern Lake District. *Geologists' Association, Proceedings* **83**, 449–50.

——— 1977. Lake District copper [in Correspondence]. *Geological Magazine* **114**, 483.

——— 1978. Epigenetic mineralization. See Moseley (1978: 226–41).

Firsoff, V. A. 1971. *Gemstones of the British Isles.* Edinburgh: Oliver & Boyd.

Firth, J. N. M. 1971. *The mineralogy of the South Wales coalfield.* PhD thesis, University of Bristol.

Fisher, A. 1988. Wheal Gorland – a personal recollection of visits made during 1976 and 1977. *UK Journal of Mines and Minerals* **4**, 32–4.

Fisher, J. 2004. Fluorite from the Northern Pennine orefield, England. *Rocks and Minerals* **79**, 378–97.

——— 2006. Classic nineteenth century fluorite specimens from Weardale, Durham: a mineralogical mystery. *UK Journal of Mines and Minerals* **27**, 25–8.

Fisher, J. & L. Greenbank 2003. The Rogerley mine, Weardale, County Durham, England. *UK Journal of Mines and Minerals* **23**, 9–20.

Fisk, J., D. Smith, M. Vincent 1982. Trevinnick mine, Cornwall. *Mineral Realm* **1**(5), 10–23.

Fitton, J. G. 1972. The genetic significance of almandine–pyrope phenocrysts in the calc-alkaline Borrowdale Volcanic Group, Northern England. *Contributions to Mineralogy and Petrology* **36**, 231–48.

Fleet, M. E. 2003. *Rock-forming minerals: sheet silicates – micas.* (vol. 3A, 2nd edn). London: Geological Society.

Fleischermann, F. N. A. 1910. On the occurrence of gyrolite in County Antrim. *Mineralogical Magazine* **15**, 288–93.

Fleming, P. 2000. Lakeland's mining heritage – copper. In *Lakeland's mining heritage – the last 500 years*, A. Cameron (ed.), 89–114. Alston: Cumbria Amenity Trust Mining History Society.

Fletcher, C. J. N. 1988. Tidal erosion, solution cavities and exhalative mineralization associated with the Jurassic unconformity at Ogmore, South Glamorgan. *Geologists' Association, Proceedings* **99**, 1–14.

Fletcher, C. J. N. & B. Young 1988. Rosasite from Bute quarry, Mid Glamorgan. The first reported occurrence in Wales. *Russell Society, Journal* **2**(1), 19–23.

Fletcher, C. J. N., I. G. Swainbank, T. B. Colman 1993. Metallogenic evolution in Wales: constraints from lead isotope modelling. *Geological Society of London, Journal* **150**, 77–82.

Fletcher, L. 1921. The meteoritic stone seen to fall near Crumlin, Co. Antrim, on September 13, 1902. *Mineralogical Magazine* **19**, 149–62.

Fletcher, T. A. & C. M. Rice 1989. Geology, mineralization (Ni–Cu) and precious metal geochemistry of Caledonian mafic and ultramafic intrusions near Huntly, northeast Scotland. *Institution of Mining and Metallurgy, Transactions B* **98**, 185–200.

Fletcher, T. A., A. J. Boyce, A. E. Fallick, C. M. Rice, R. L. F. Kay 1997. Geology and stable isotope study of Arthrath mafic intrusion and Ni–Cu mineralization, northeast Scotland. *Institution of Mining and Metallurgy, Transactions B* **106**, 169–78.

Flett, J. S. 1900. The Trap dykes of the Orkneys. *Royal Society of Edinburgh, Transactions: Earth Sciences* **39**, 865–905.

Flight, W. 1883. Two new aluminous mineral species, evigtokite and liskeardite. *Mineralogical Magazine* **5**, 266–8.

Flindall, R. & A. Hayes 1976. *The caverns and mines of Matlock Bath, 1: the Nestus mines; Rutland and Masson caverns.* Buxton, Derbyshire: Moorland Publishing.

Flinn, D. & A. Pentecost 1995. Travertine-cemented screes on the serpentinite seacliffs of Unst and Fetlar, Shetland. *Mineralogical Magazine* **59**, 259–65.

Flörke, O. W., U. Flörke, U. Giese 1984. Moganite, a new microcrystalline silica mineral. *Neues Jahrbuch für Mineralogie, Abhandlungen* **149**, 325–6.

Floyd, P. A. & G. Rowbotham 1982. Chemistry of primary and secondary minerals in titaniferous brown amphibole-bearing greenstones from north Cornwall. *Ussher Society, Proceedings* **5**, 296–303.

Fontboté, L. & G. C. Amstutz 1986. Ore textures and structures resulting from diagenetic crystallization processes in ore deposits, and their use in exploration: a short review and an example from the Navan mine, Ireland. See Andrew et al. (1986: 685–91).

Foord, E. E., M. R. Oakman, C. H. Maxwell 1985. Durangite from the Black Range, New Mexico, and new data on durangite from Durango and Cornwall. *Canadian Mineralogist* **23**, 241–6.

Forbes, D. 1867. Researches in British mineralogy. *Philosophical Magazine (4th series)* **34**, 329–54.

——— 1868. Researches in British mineralogy, II: polytelite. *Philosophical Magazine (4th series)* **35**, 171–84.

——— 1869. Researches in British mineralogy, III: native gold. *Philosophical Magazine* (4th series) **37**, 321–32.

Forbes, I. 2004. *Secret worlds: spar boxes of the northern Pennines.* Killhope, Durham: Lead Mining Museum.

Ford, T. D. 1955. Blue John fluorspar. *Yorkshire Geological Society, Proceedings* **30**, 35–60.

——— 1963. The occurrence of halloysite in Derbyshire. *Clay Minerals Bulletin* **5**, 302–307.

——— 1967. Some mineral deposits of the Carboniferous Limestone of Derbyshire. In *Geological excursions in the Sheffield region and the Peak District National Park*, E. Neves, C. Downie (eds), 53–75. Sheffield: University of Sheffield.

——— 1969. The Blue John fluorspar deposits of Treak Cliff, Derbyshire, in relation to the boulder bed. *Yorkshire Geological Society, Proceedings* **37**, 153–7.

——— 1979. Blue John: Derbyshire's unique gem. *Gems* **11**, 13–23.

——— 2000. *Derbyshire Blue John.* Ashbourne, Derbyshire: Landmark Publishing.

Ford, T. D. & P. R. Ineson 1971. The fluorspar mining potential of the Derbyshire ore field. *Institution of Mining and Metallurgy, Transactions B* **80**, 186–210.

Ford, T. D. & R. J. King 1965. Epigenetic bedded galena–barite deposits in the Golconda mine, Brassington, Derbyshire. *Economic Geology* **60**, 1686–702.

Ford, T. D. & J. H. Rieuwerts (eds) 2000. *Lead mining in the Peak District.* Ashbourne: Landmark Publishing.

Ford, T. D. & W. A. S. Sarjeant 1964a. The Peak District mineral index. *Peak District Mines Historical Society, Bulletin* **2**, 122–50.

——— 1964b. The "stalactitic" barytes of Derbyshire. *Yorkshire Geological Society, Proceedings* **34**, 371–86.

Ford, T. D., W. A. S. Sarjeant, M. E. Smith 1993. The minerals of the Peak District of Derbyshire. *UK Journal of Mines and Minerals* **13**, 16–55.

Fortey, N. J. 1978. *Mineral parageneses of the Harding Vein tungsten deposit, Carrock Fell mine, Cumbria.* Report 228, Mineralogy Unit, Institute of Geological Sciences, London.

——— 1980. Hydrothermal mineralization associated with late Caledonian intrusions in northern Britain: preliminary comments. *Institution of Mining and Metallurgy, Transactions B* **89**, 173–6.

Fortey, N. J. & B. Beddoe-Stephens 1982. Barium silicates in stratabound Ba–Zn mineralization in the Scottish Dalradian. *Mineralogical Magazine* **46**, 63–72.

Fortey, N. J. & U. McL. Michie 1978. Aegirine of possible authigenic origin in Middle Devonian sediments in Caithness, Scotland. *Mineralogical Magazine* **42**, 439–42.

Fortey, N. J. & P. H. A. Nancarrow 1990. Chromian minerals in microdiorite from the English Lake District. *Russell Society, Journal* **3**(1), 15–22.

Fortey, N. J. & C. G. Smith 1986. Stratabound mineralization in Dalradian rocks near Tyndrum, Perthshire. *Scottish Journal of Geology* **22**, 377–93.

Fortey, N. J., J. S. Coats, M. J. Gallagher, C. G. Smith, P. G. Greenwood 1993. New stratabound barite and base metals in Middle Dalradian rocks near Braemar, northeast Scotland. *Institution of Mining and Metallurgy, Transactions B* **102**, 55–64.

Fortey, N. J., A. H. Cooper, P. J. Henney, T. Colman, P. H. A. Nancarrow 1994. Appinitic intrusions in the English Lake District. *Mineralogy and Petrology* **51**, 355–75.

Fortey, N. J., J. D. Ingham, B. R. H. Skilton, B. Young, T. J. Shepherd 1984. Antimony mineralization at Wet Swine Gill, Caldbeck Fells, Cumbria. *Yorkshire Geological Society, Proceedings* **45**, 59–65.

Fortey, N. J., P. H. A. Nancarrow, M. J. Gallagher 1991. Armenite from the Middle Dalradian of Scotland. *Mineralogical Magazine* **55**, 135–8.

Förtsch, E. B. 1967. Plumbogummite from Roughton Gill, Cumberland. *Mineralogical Magazine* **36**, 530–37.

Fowler, A. 1943. On fluorite and other minerals in Lower Permian rocks of south Durham. *Geological Magazine* **80**, 41–51.

——— 1956. Minerals in the Permian and Trias of northeast England. *Geologists' Association, Proceedings* **67**, 251–65.

——— 1959. The non-ferrous minerals of Northern Ireland. See Higham (1959: 27–34).

Fowler, M. B. 1981. Uranium content, distribution, and migration in the Glendessary syenite, Inverness-shire. *Mineralogical Magazine* **44**, 443–8.

——— 1992. Elemental and O–Sr–Nd isotope geochemistry of the Glen Dessary syenite, northwest Scotland. *Geological Society of London, Journal* **149**, 209–220.

Fowler, M. B. & C. T. Williams 1986. Zirconolite from the Glen Dessary syenite: a comparison with other Scottish zirconolites. *Mineralogical Magazine* **50**, 326–8.

Foy, H. J. 1992. Paulingite from Northern Ireland. *UK Journal of Mines and Minerals* **11**, 24.

Foy, P. 1997. Field meeting report: Tynebottom mine, Garrigill. *Russell Society, Newsletter* **31**, 10.

Francis, G. H. 1955. Gedrite from Glen Urquhart, Inverness-shire. *Mineralogical Magazine* **30**, 709–716.

Francis, J. G. & G. Ryback 1987. Chalcomenite from Ballybunnion, Co. Kerry, Eire. *Mineralogical Magazine* **51**, 751.

Francis, P. 1972. The occurrence of native copper in the Tertiary basalts of Co. Antrim. *Irish Naturalists' Journal* **17**, 274–6.

Fransolet, A-M., F. Fontan, Ph. de Parseval 2007. Natromontebrasite: a discredited mineral species. *Canadian MIneralogist* **45**, 391–6.

Fraser, G. L., D. R. M. Pattison, L. M. Heaman 2004. Age of the Ballachulish and Glencoe igneous complexes (Scottish Highlands), and paragenesis of zircon, monazite and baddeleyite in the Ballachulish aureole. *Geological Society of London, Journal* **161**, 447–62.

Freeman, S. R., R. W. H. Butler, R. A. Cliff, D. C. Rex 1998. Direct dating of mylonite evolution: a multi-disciplinary geochronological study from the Moine thrust zone, northwest Scotland. *Geological Society of London, Journal* **155**, 745–58.

French, W. J. 1976. The origin of leucodiorites associated with the appinitic intrusions of County Donegal. *Yorkshire Geological Society, Proceedings* **41**, 107–126.

French, W. J., M. D. Hassan, J. E. Westcott 1977. A celadonite–vermiculite series from the volcanic rocks of the Ochils, Stirlingshire. *Mineralogical Magazine* **41**, 481–5.

Frenzel, A. 1875. Mineralogisches, 1: Descloizit und Vanadinit. *Neues Jahrbuch für Geologie und Palaontologie* **50**, 673–86.

Frick, L. R., D. D. Lambert, S. J. Reeves, L. M. Heaman 1994. Platinum-group element and osmium isotope evidence for the origin of Scourie dykes. *Mineralogical Magazine* **58A**, 290–91.

Friend, J. N. & J. P. Allchin 1940. Colloidal gold as a colouring principle in minerals. *Mineralogical Magazine* **25**, 584–96.

Frondel, C. 1952. Studies of uranium minerals (X): uranopilite. *American Mineralogist* **37**, 950–59.

——— 1954. Bassetite and uranospathite. *Mineralogical Magazine* **30**, 343–53.

——— 1972. Jacob Forster (1739–1806) and his connections with forsterite and palladium. *Mineralogical Magazine* **38**, 545–50.

Furness, R. R. 1965. The petrography and provenance of the Coniston Grits east of the Lune Valley, Westmorland. *Geological Magazine* **102**, 252–60.

Fusciardi, I. P., J. F. Güven, D. R. A. Stewart, V. Carboni, J. J. Walsh 2004. The geology and genesis of the Lisheen Zn–Pb deposit, Co. Tipperary, Ireland. See Kelly et al. (2004: 455–82).

Gages, A. 1860. On the formation of orpiment in a mass of sulphate of barytes, found interstratified in the Carboniferous Limestone near Silvermines, County of Tipperary. *Geological Society of Dublin, Journal* **8**, 243–4.

Gaines, R. V., H. C. W. Skinner, E. E. Foord, B. Mason, A. Rosenzweig 1997. *Dana's new mineralogy: the system of mineralogy of James Dwight Dana and Edward Salisbury Dana* (8th edn). New York: John Wiley.

Gallagher, M. J. 1963. Lamprophyre dykes from Argyll. *Mineralogical Magazine* **33**, 415–30.

——— 1964. Rock alteration in some mineralized basic dykes in Britain. *Institution of Mining and Metallurgy, Transactions B* **73**, 825–40.

——— 1970. Galena–fluorite mineralization near Lairg, Sutherland. *Institution of Mining and Metallurgy, Transactions B* **79**, 182–4.

——— 1974. Rutile and zircon in Northumbrian beach sand. *Institution of Mining and Metallurgy, Transactions B* **83**, 97–8.

——— 1984. Barite deposits and potential of Scotland. *Institution of Mining and Metallurgy, Transactions B* **93**, 130–32.

Gallagher, M. J. & R. T. Smith 1975. *Molybdenite mineralization in Precambrian rocks near Lairg, Scotland.* Report 3, Mineral Reconnaissance Program, Institute of Geological Sciences, London.

Gallagher, M. J., U. McL. Michie, R. T. Smith, L. Haynes 1971. New evidence of uranium and other mineralization in Scotland. *Institution of Mining and Metallurgy, Transactions B* **80**, 150–73.

Gallagher, M. J., C. G. Smith, J. S. Coats, P. G. Greenwood, B. J. Chackfield, N. J. Fortey, P. H. A. Nancarrow 1989a. *Stratabound barium and base-metal mineralization in middle Dalradian metasediments near Braemar, Scotland.* Report 104, Mineral Reconnaissance Program, British Geological Survey, Keyworth.

Gallagher, M. J., R. T. Smith, J. D. Peacock, L. Haynes 1974. Molybdenite mineralization in Precambrian rocks near Lairg, Scotland. *Institution of Mining and Metallurgy, Transactions B* **83**, 81–7.

Gallagher, M. J., P. Stone, A. E. S. Kemp, M. G. Hills, R. C. Jones, R. T. Smith, D. Peachey, B. P. Vickers, M. E. Parker, K. E. Rollin, B. R. H. Skilton 1983. *Stratabound arsenic and vein antimony mineralization in Silurian greywackes at Glendinning, south Scotland.* Report 59, Mineral Reconnaissance Program, Institute of Geological Sciences, London.

Gallagher, M. J., P. Stone, P. R. Duller 1989b. Gold-bearing arsenic–antimony concentrations in Silurian greywackes, south Scotland. *Institution of Mining and Metallurgy, Transactions B* **98**, 58–60.

Gallagher, S. & R. Elsdon 1990. Spinel lherzolite and other xenoliths from a dolerite dyke in southwest Donegal. *Geological Magazine* **127**, 177–80.

Gallagher, V. 1988. Coupled substitutions in schorl–dravite tourmaline: new evidence from southeast Ireland. *Mineralogical Magazine* **52**, 637–50.

——— 1989. Geological and isotopic studies of microtonalite-hosted Sn–W mineralization in southeast Ireland. *Mineralium Deposita* **24**, 19–28.

Gallagher, V. & P. S. Kennan 1992. Tourmaline on the margin of the Leinster granite, southeast Ireland. *Irish Journal of Earth Sciences* **11**, 131–50.

Gallagher, V., A. J. Boyce, A. E. Fallick, P. J. Möhr 1992. An isotopic study of the Harberton Bridge Fe–Zn–Pb deposit, County Kildare, and its implications for the metallogenesis in Ireland. See Bowden et al. (1992: 261–72).

Galli, E., R. Rinaldi, C. Modena 1981. Crystal chemistry of levynes. *Zeolites* **1**, 157–60.

Gallois, R. W. 2003. The distribution of halite (rock salt) in the Mercia Mudstone Group (mid- to late Triassic) in southwest England. *Geoscience in Southwest England* **10**, 383–9.

Galwey, A. K., K. A. Jones, R. Reed, D. Dollimore 1979. The blue colouration in banded fluorite (Blue John) from Castleton, Derbyshire, England. *Mineralogical Magazine* **43**, 243–50.

Gamble, J. A. 1982. The mineral chemistry of dolerites and gabbros from the central complex of Slieve Gullion, northeast Ireland. *Mineralogical Magazine* **46**, 103–110.

Garby, J. 1846. Notice of the occurrence of gold in a cross course in Cornwall. *Royal Geological Society of Cornwall, Transactions* **6**, 265–6.

——— 1848. A catalogue of minerals found in Cornwall, with their localities. *Royal Geological Society of Cornwall, Transactions* **7**, 72–92.

Garfield, L. 2002. Field meeting report: Dolyhir quarry, near Kington. *Russell Society, Newsletter* **41**, 38.

Garnett, C. S. 1923a. On a peculiar chlorite-rock at Ible, Derbyshire. *Mineralogical Magazine* **20**, 60–64.

——— 1923b. The "toadstone-clays" of Derbyshire. *Mineralogical Magazine* **20**, 151–7.

Garson, M. S. & F. May 1976. Copper mineralization at Vidlin, Shetland. *Institution of Mining and Metallurgy, Transactions B* **85**, 153–7.

Garson, M. S., J. S. Coats, N. M. S. Rock, T. Deans 1984. Fenites, breccia dykes, albitites, and carbonatitic veins near the Great Glen fault, Inverness, Scotland. *Geological Society of London, Journal* **141**, 711–32.

Garvie, L. A. J. 1992. Diagenetic tosudite from the lowermost St Maughan's Group, Lydney Harbour, Forest of Dean, UK. *Clay Mineralogy* **27**, 507–513.

Gasparrini, C. 1978. *Identification and description of the silver-bearing minerals in ten sulphide samples from Ireland.* Unpublished report prepared for Noranda Mines Ltd by Mimmet Scientific Ltd, Toronto.

Gass, I. G., C. R. Neary, H. M. Prichard, I. D. Bartholemew 1982. *The chromite of the Shetland ophiolite: a re-appraisal in the light of new theory and techniques.* Report for the Commission of European Communities, Department of Earth Sciences, Open University, Milton Keynes.

Gay, P. 1965. An X-ray powder method for the estimation of (K, Ba) feldspars. *Mineralogical Magazine* **34**, 204–213.

Gay, P. & K. O. Rickson 1960. X-ray data on stokesite. *Mineralogical Magazine* **32**, 433–5.

Gay, P. & N. N. Roy 1968. The mineralogy of the potassium–barium feldspar series, III: subsolidus relationships. *Mineralogical Magazine* **36**, 914–32.

Gayer, R. A. & A. J. Criddle 1970. Mineralogy and genesis of the Llanharry iron ore deposits, Glamorgan. *Proceedings of the 9th Commonwealth Mining and Metallurgy Congress, London, 1969* (vol. 2), M. J. Jones (ed.), 605–626.

Gehring, A. U. 1990. Diagenesis of ferriferous phases in the Northampton Ironstone in the Cowthick quarry near Corby (England). *Geological Magazine* **127**, 169–76.

George, M. C., M. Stone, E. E. Fejer, R. F. Symes 1981. Triplite from the Megiliggar Rocks, Cornwall. *Mineralogical Magazine* **44**, 236–8.

Gerrard, S. 2000. *The early British tin industry.* Stroud, Gloucestershire: Tempus Publishing.

Ghosh, P. K. 1927. Petrology of the Bodmin Moor granite (eastern part), Cornwall. *Mineralogical Magazine* **21**, 285–309.

——— 1934. The Carnmenellis granite: its petrology, metamorphism and tectonics. *Geological Society of London, Quarterly Journal* **90**, 240–76.

Gibb, F. G. F. 1972. A differentiated ultrabasic sheet on Sgùrr Dearg, Isle of Skye. *Mineralogical Magazine* **38**, 811–18.

Gibb, F. G. F. & C. M. B. Henderson 1978a. Possible higher pressure relics within titaniferous augites in a basic sill. *Geological Magazine* **115**, 55–62.

——— 1978b. The petrology of the Dippin Sill, Isle of Arran. *Scottish Journal of Geology* **14**, 1–27.

——— 1996. The Shiant Isles Main Sill: structure and mineral fractionation trends. *Mineralogical Magazine* **60**, 67–97.

——— 2006. Chemistry of the Shiant Isles Main Sill, northwest Scotland, and wider implications for the petrogenesis of mafic sills. *Journal of Petrology* **47**, 191–230.

Gibbons, W. 1981. Glaucophanic amphibole in a Monian shear zone on the mainland of North Wales. *Geological Society of London, Journal* **138**, 139–43.

Gibbons, W. & M. Gyopari 1986. A greenschist protolith for blueschist on Anglesey, UK. In *Blueschists and eclogites,* B. W. Evans & E. H. Brown (eds), 217–28. Memoir 164, Geological Society of America, Boulder, Colorado.

Gibbons, W. & A. Mann 1983. Pre-Mesozoic lawsonite in Anglesey, North Wales: preservation of ancient blueschists. *Geology* **11**, 3–6.

Gibson, C. 1979. *Primary dispersion patterns for selected elements in the sulphide ores, host rocks and iron formation at Tynagh, Co. Galway, Ireland.* PhD thesis, University of Cardiff.

Gilbert, J. 1878. Notice of a vugh in St Ives Consols mine. *Royal Geological Society of Cornwall, Transactions* **9**, 158–60.

Gilbey, J. W. G. 1968. *The mineralogy, paragenesis and structure of the ores of the Dolgellau gold belt, Merionethshire, and associated wallrock alteration.* PhD thesis, University of London.

Gill, M. C. 1987. *The Yorkshire and Lancashire lead mines: a study of lead mining in the South Craven and Rossendale districts.* British Mining 33, Northern Mine Research Society, Sheffield.

——— 1993. *The Grassington mines.* British Mining 46, Northern Mine Research Society, Sheffield.

——— 1994. *The Wharfdale mines.* British Mining 49, Northern Mine Research Society, Sheffield.

——— 1998. *The Greenhow mines.* British Mining 60, Northern Mine Research Society, Sheffield.

——— 2004. *Swaledale: its mines and smelt mills.* Ashbourne: Landmark Publishing.

Gill, W. D., F. I. Khalaf, M. S. Massoud 1977. Clay minerals as an index of the degree of metamorphism of the carbonate and terrigenous rocks in the South Wales coalfield. *Sedimentology* **24**, 675–91.

Gillanders, R. J. 1981. The Leadhills–Wanlockhead district, Scotland. *Mineralogical Record* **12**(4), 235–50.

Gillen, C. 1987a. Dalradian of the Portsoy district. See Trewin et al. (1987: 71–8).

——— 1987b. Huntly, Elgin and Lossiemouth. See Trewin et al. (1987: 148–59).

Gilligan, A. 1919. The petrography of the Millstone Grit of Yorkshire. *Geological Society of London, Quarterly Journal* **75**, 251–94.

Gindy, A. R. 1953. Progressive replacement of limestone inclusions in granite at Ballynacarrick, Co. Donegal. *Geological Magazine* **90**, 152–8.

Glasby, G. P. 1974. A geochemical study of the manganese ore deposits of the Harlech Dome, North Wales. *Journal of Earth Sciences* **8**, 445–50.

Gleeson, S. A., J. J. Wilkinson, H. F. Shaw, R. J. Herrington 2000. Post-magmatic hydrothermal circulation and the origin of base metal mineralization, Cornwall, UK. *Geological Society of London, Journal* **157**, 589–600.

Glen, D. C. & J. Young 1876. List of minerals and rock specimens found in the central, southern and western districts of Scotland. In *Catalogue of west Scottish fossils,* J. Armstrong, J. Young, D. Robertson (eds), 156–64. British Association for the Advancement of Science, Glasgow.

Goldring, D. C. & D. A. Greenwood 1977. Fluorite mineralization associated with metasomatic hematite deposits at Beckermet mine, Cumbria. *Institution of Mining and Metallurgy, Transactions B* **86**, 157.

——— 1990. Fluorite mineralization at Beckermet iron ore mine, Cumbria, north England. *Institution of Mining and Metallurgy, Transactions B* **99**, 113–19.

Golley, P. & R. Williams 1995. *Cornish mineral reference manual.* Truro: Endsleigh.

Goodchild, J. G. 1875. Wulfenite at Caldbeck Fell. *Geological Magazine* **12**, 565–6.

——— 1882. Contribution towards a list of the minerals occurring in Cumberland and Westmorland. *Cumberland Association for the Advancement of Literature and Science, Transactions* **7**, 101–203.

——— 1884. Contribution towards a list of the minerals occurring in Cumberland and Westmorland, part II. *Cumberland Association for the Advancement of Literature and Science, Transactions* **8**, 189–204.

——— 1885. Contribution towards a list of the minerals occurring in Cumberland and Westmorland, part III. *Cumberland Association for the Advancement of Literature and Science, Transactions* **9**, 175–99.

——— 1901. *Supplement to Heddle's "The mineralogy of Scotland",* J. G. Goodchild (ed.), 193–212. Edinburgh: David Douglas.

Goode, A. J. J. & R. T. Taylor 1980. *Intrusive and pneumatolytic breccias in southwest England.* Report 80/2, Institute of Geological Sciences, London.

Goodman, B. A., M. L. Berrow, J. D. Russell 1988. Transformation of poorly crystalline oxides during boiling with NaOH to concentrate iron oxides from soils. *Journal of Soil Science* **39**, 87–98.

Goodman, S. & M. A. Lappin 1996. The thermal aureole of the Lochnagar complex: mineral reactions and implications from thermal modelling. *Scottish Journal of Geology* **32**, 159–72.

Goodman, S., A. G. Leslie, W. A. Ashcroft, A. Crane 1990. *The geology of the central part of sheet 65E (Ballater): contribution to the memoir.* Technical Report WA/90/59, British Geological Survey, Keyworth.

Goodyear, J. & W. J. Duffin 1961. An X-ray examination of an exceptionally well crystallized kaolinite. *Mineralogical Magazine* **32**, 902–907.

Gordon, W. T. 1922. Native gold at Torquay, Devonshire. *Nature* **109**, 583.

Gottardi, G. & E. Galli 1985. *Natural zeolites.* Berlin: Springer.

Gough, D. 1963. *Geology of the Greenside lead mine, Cumberland.* PhD thesis, University of London.

——— 1965. Structural analysis of ore shoots at Greenside lead mine, Cumberland, England. *Economic Geology* **60**, 1459–72.

Gough, J. W. 1930. *The mines of Mendip.* Oxford: Oxford University Press.

Goulding, J. 1995. Field meeting report: Vaynor quarry, Cefn Coed, Merthyr Tydfil, Mid Glamorgan. *Russell Society, Newsletter* **26**, 15.

——— 1997a. Field meeting report: Whatley quarry, near Frome, Somerset. *Russell Society, Newsletter* **31**, 13.

——— 1997b. Field meeting report: Derbyshire weekend. *Russell Society, Newsletter* **31**, 15.

Goulding, J. & M. T. Price 1995. Well crystallized fraipontite from Machen quarry, Mid Glamorgan: a first Welsh locality. *Russell Society, Journal* **6**(1), 50.

Graham, C. M. 1976. A note on some Dalradian stilpnomelanes and their oxidation. *Mineralogical Magazine* **40**, 467–72.

Gramaccioli, C. 1955. La presenza di un vanadato di uranio – identificabile con la carnotite – nella miniera inglese di South Terras (Cornovaglia). *Natura (Milan)* **46**, 30–32.

Grantham, D. R. 1928. The petrology of the Shap granite. *Geologists' Association, Proceedings* **39**, 299–331.

Green, A. H. & A. Strahan 1887. *The geology of the Carboniferous Limestone, Yoredale rocks and Millstone Grit of North Yorkshire (parts of sheets 88SE, 81NE, 81SE, 72NE, 82NW, 82SW and 71NW)* (2nd edn). Memoir, Geological Survey of Great Britain and Ireland, London.

Green, C. P. & M. J. Eden 1971. Gibbsite in the weathered Dartmoor granite. *Geoderma* **6**, 315–17.

Green, D. H. 1964. The petrogenesis of the high-temperature peridotite intrusion in the Lizard area, Cornwall. *Journal of Petrology* **5**, 134–8.

Green, D. I. 1986. UK wulfenite. *UK Journal of Mines and Minerals* **1**, 28–9.

——— 1987a. The minerals of the Meadowfoot smelter. *UK Journal of Mines and Minerals* **2**, 3–9.

——— 1987b. The copper mines of Middleton Tyas, North Yorkshire. *UK Journal of Mines and Minerals* **3**, 28–9.

——— 1988. The minerals of Rubha a' Mhill, Islay, Scotland. *UK Journal of Mines and Minerals* **4**, 29–31.

——— 1989. Scotlandite from Higher Roughton Gill, Caldbeck Fells, Cumbria. *Mineralogical Magazine* **53**, 653.

——— 1990. Veszelyite – a mineral new to Britain, from Wanlockhead, Scotland. *UK Journal of Mines and Minerals* **8**, 6–7.

——— 1991. Lyons show 1990. *UK Journal of Mines and Minerals* **9**, 48–9.

——— 1993. British shows: the Rock Exchange, Bakewell, October 1992. *UK Journal of Mines and Minerals* **12**, 27–8.

——— 1995a. Editorial – personal labels. *UK Journal of Mines and Minerals* **15**, 2–3.

——— 1995b. Carbonate cyanotrichite from Wedding Cave mine, Bwlchgwyn, Clwyd. *UK Journal of Mines and Minerals* **15**, 19.

——— 1998a. Editorial – the zeolite group. *UK Journal of Mines and Minerals* **19**, 2–4.

——— 1998b. The Sussex show. *UK Journal of Mines and Minerals* **19**, 7–8.

——— 1998c. Phosgenite: a first Irish occurrence from Killiney, Co. Dublin. *UK Journal of Mines and Minerals* **19**, 12.

——— 2003. Gearksutite from the Old Gang mines, Swaledale, North Yorkshire, England. *Russell Society, Journal* **8**(1), 41–3.

——— 2004. Anthropogenic humboldtine from Cornwall, England. *Geological Curator* **8**, 33–6.

——— 2005a. Digital combination photography: a technique for producing improved images of microscopic minerals. *UK Journal of Mines and Minerals* **25**, 5–18.

——— 2005b. The iron and magnesium borate minerals from Boulby mine, Redcar and Cleveland, England. *UK Journal of Mines and Minerals* **25**, 31–2.

Green, D. I. & V. Betz 2006. Digital combination photography of micro-minerals using bellows lenses. *UK Journal of Mines and Minerals* **27**, 15–24.

Green, D. I. & P. J. Briscoe 2002. Twenty years in minerals: the classic areas of northern England. *UK Journal of Mines and Minerals* **22**, 3–42.

Green, D. I. & M. D. Freier 1994. Borate nodules from the English Zechstein at Boulby mine, Cleveland. *UK Journal of Mines and Minerals* **14**, 4–8.

——— 1996. The Boulby mine, Cleveland, England. *Mineralogical Record* **27**(3), 163–70.

Green, D. I. & N. Hubbard 1996. Chenite from Driggith mine, Cumbria. *UK Journal of Mines and Minerals* **16**, 15.

Green, D. I. & D. McCallum 1993. Millerite from Brownley Hill mine, Nenthead, Cumbria. *UK Journal of Mines and Minerals* **12**, 22–4.

——— 2005. Kainosite-(Y) from the Strontian mines, Highland region, Scotland. *UK Journal of Mines and Minerals* **26**, 23–6.

Green, D. I. & M. G. H. Merry 1997. Wulfenite from Wheal Ludcott, Cornwall. *UK Journal of Mines and Minerals* **18**, 15.

Green, D. I. & D. Middleton 1996. Alpine-type vein minerals from Tanygrisiau, Gwynedd. *UK Journal of Mines and Minerals* **16**, 30–33.

Green, D. I. & S. Moreton 2001. Twenty years in minerals: Ireland. *UK Journal of Mines and Minerals* **21**, 29–36.

Green, D. I. & J. R. Nudds 1998. Harmotome from Brownley Hill mine, Nenthead, Cumbria, England. *Russell Society, Journal* **7**(1), 31–3.

Green, D. I. & J. C. Spence 1988. The minerals of Vale Road quarry, Mansfield Woodhouse, Nottinghamshire. *UK Journal of Mines and Minerals* **5**, 28–30.

Green, D. I. & J. G. Todd 1996. Zeolites and related minerals from Moonen Bay, Isle of Skye, Scotland. *UK Journal of Mines and Minerals* **16**, 21–7.

——— 2001. Twenty years in minerals: Scotland. *UK Journal of Mines and Minerals* **21**, 9–27.

Green, D. I. & M. Wood 1996. Epistilbite from the Isle of Skye, Scotland. *Russell Society, Journal* **6**(2), 85–7.

Green, D. I. & B. Young 2006. Hydromagnesite and dypingite from the Northern Pennine orefield, Northern England. *Yorkshire Geological Society, Proceedings* **56**, 151–4.

Green, D. I., R. Bell, S. Moreton 2005a. Drusy cavity minerals including the first Irish danalite from Lindsay's Leap, Mourne Mountains, Co. Down. *UK Journal of Mines and Minerals* **25**, 25–30.

Green, D. I., J. Betterton, R. Bell, T. Neall 2001. Ramsbeckite with schulenbergite from Cornwall and Cumbria, England. *Russell Society, Journal* **7**(2), 86–7.

Green, D. I., T. F. Bridges, M. P. Cooper, N. Thomson 2005b. A review of the supergene mineralization at Silver Gill, Caldbeck Fells, Cumbria. *Russell Society, Journal* **8**(2), 85–97.

Green, D. I., T. F. Bridges, T. Neall 2005c. The occurrence and formation of galena in supergene environments. *Russell Society, Journal* **8**(2), 66–70.

Green, D. I., T. F. Bridges, M. S. Rumsey, A. G. Tindle 2006a. A review of the mineralization at Driggith and Sandbed mines, Caldbeck Fells, Cumbria. *Russell Society, Journal* **9**, 4–38.

Green, D. I., T. F. Cotterell, I. Jones, D. Cox, R. Cleevely 2007. Wavellite: its discovery and occurrences in the British Isles. *UK Journal of Mines and Minerals* **28**, 11–30.

Green, D. I., N. Hubbard, T. F. Cotterell 2005d. The first British occurrence of ewaldite, at Dolyhir quarry, Old Radnor, Powys. *UK Journal of Mines and Minerals* **25**, 23–4.

Green, D. I., N. Hubbard, C. Jewson 1996a. Stolzite from two Cornish localities. *UK Journal of Mines and Minerals* **16**, 14.

Green, D. I., C. M. Leppington, T. Neall 2003b. Gartrellite, a mineral new to Britain from Low Pike, Caldbeck Fells, Cumbria, England. *Russell Society, Journal* **8**(1), 43–4.

Green, D. I., D. McCallum, M. Wood 1997. Supergene Cu, Pb, Zn and Ag minerals from Force Crag mine, Coledale, Cumbria. *UK Journal of Mines and Minerals* **18**, 10–14.

——— 2000. The Brownley Hill mine, Alston Moor district, Cumbria, England. *Mineralogical Record* **31**(3), 231–50.

Green, D. I., T. Neall, T. F. Cotterell, C. M. Leppington 2003a. Symplesite and parasymplesite from Cumbria and Cornwall, England. *Russell Society, Journal* **8**(1), 16–17.

Green, D. I., T. Neall, A. G. Tindle 2005e. The composition of fülöppite from Wet Swine Gill, Caldbeck Fells, Cumbria. *Russell Society, Journal* **8**(2), 101–102.

Green, D. I., S. A. Rust, J. S. Mason 1996b. Classic mineral localities, Frongoch mine, Dyfed. *UK Journal of Mines and Minerals* **17**, 29–38.

Green, D. I., A. G. Tindle, T. F. Bridges, T. Neall 2006b. A review of the mineralization at Arm o' Grain, Caldbeck Fells, Cumbria. *Russell Society, Journal* **9**, 44–53.

Green, D. I., A. G. Tindle, C. Jewson 2005f. The first British occurrence of philipsbornite at Wheal Unity, Gwennap, Cornwall. *UK Journal of Mines and Minerals* **25**, 37–8.

Green, D. I., A. G. Tindle, S. Moreton 2005g. Brewsterite-Ba and harmotome from the Wicklow lead mines, Co. Wicklow, Ireland. *Irish Journal of Earth Sciences* **23**, 101–106.

Green, D. I., B. Young, R. A. Ixer 2006c. Beaverite from Cumbria and Yorkshire. *Yorkshire Geological Society, Proceedings* **56**, 155–8.

Green, G. W. 1958. The central Mendip lead–zinc orefield. *Geological Survey of Great Britain, Bulletin* **14**, 70–90.

Green, J. F. N. 1913. *The older Palaeozoic succession of the Duddon Estuary.* London: Hayman, Christy & Lilly.

——— 1915. The garnets and streaky rocks of the English Lake District. *Mineralogical Magazine* **17**, 207–217.

——— 1917. The age of the chief intrusions of the Lake District. *Geologists' Association, Proceedings* **28**, 1–30.

Green, K. 1984. Mineral collecting in Britain. *Lapidary Journal* **38**, 418–25.

Greenly, E. 1902. The origin and associations of the jaspers of southeastern Anglesey. *Geological Society of London, Quarterly Journal* **58**, 425–40.

——— 1919. *The geology of Anglesey* [2 vols]. Memoir, Geological Survey of Great Britain, London.

Greenwood, D. A. 1977. Fluorspar mining in the northern Pennines. *Institution of Mining and Metallurgy, Transactions B* **86**, 181–90.

Greenwood, H. W. 1919. The Trias of the Macclesfield district, with notes on its relation to the adjacent Carboniferous rocks and to the Trias of the Midlands. *Liverpool Geological Society, Proceedings* **12**, 325–38.

Greg, R. P. 1860. On several new British minerals. *Philosophical Magazine* (4th series) **19**, 13–14.

Greg, R. P. & W. G. Lettsom 1858. *Manual of the mineralogy of Great Britain and Ireland.* London: John van Voorst (reprinted with additions, 1977, Lapidary Publications, Kent).

Gregory, J. W. & D. Leitch 1928. The nickel–cobalt ores of Talnotry, Kirkudbrightshire. *Institution of Mining and Metallurgy, Transactions B* **37**, 178–95.

Gregory, M. 1945. Production of radium in Cornwall. *Royal Geological Society of Cornwall, Transactions* **17**, 306–312.

Greig, D. C. 1988. *Geology of the Eyemouth district (sheet 34).* Memoir, British Geological Survey, London.

Grennan, E. F. 1986. Geology and genesis of the Courtbrown Pb–Zn–Ag deposit. See Andrew et al. (1986: 449–56).

——— 1992. The Glangevlin gypsum deposit, Co. Cavan. See Bowden et al. (1992: 317–26).

Grew, E. S., P. M. Adams, R. Kirby, M. Wiedenbeck 1999. Harkerite and associated minerals in marble and skarn from Crestmore quarry, Riverside County, California and Cascade Slide, Adirondack Mountains, New York. *Canadian Mineralogist* **37**, 227–96.

Grguric, B. A. 1998. Sector phosphorescent gypsum (selenite) crystals from Mepal Bridge, Cambridgeshire, England. *Russell Society, Journal* **7**(1), 19–22.

——— 2005. Mineral specimen mortality. *Mineralogical Record* **36**(4), 383.

Grguric, B. A. & E. H. Nickel 2006. Acicular wurtzite and sphalerite from the Lockridge mine, Bere Alston, Devon. *UK Journal of Mines and Minerals* **27**, 39–40.

Grice, J. D. 2005. The structure of spurrite, tilleyite and scawtite, and relationships to other silicate–carbonate minerals. *Canadian Mineralogist* **43**, 1489–500.

Griffith, A. E. & H. E. Wilson 1982. *Geology of the country around Carrickfergus and Bangor (sheet 29).* Memoir, Geological Survey of Northern Ireland, Belfast.

Griffiths, A. 1997. Field meeting report: Gunheath china-clay pit, St Austell, Cornwall. *Russell Society, Newsletter* **31**, 13–14.

Griffiths, J. C. 1939. The mineralogy of the glacial deposits of the region between the rivers Neath and Towy, South Wales. *Geologists' Association, Proceedings* **50**, 433–62.

Griffiths, J. C. & A. Stuart 1940. An occurrence of detrital diaspore in South Wales. *Geological Magazine* **77**, 74–6.

Grigg, M. 1986. Minerals of the china-clay pits. *UK Journal of Mines and Minerals* **1**, 45–7.

——— 1996. Wulfenite from Greystones quarry, Cornwall. *UK Journal of Mines and Minerals* **16**, 13.

Groome, D. R. & A. Hall 1974. The geochemistry of the Devonian lavas of the northern Larne Plateau, Scotland. *Mineralogical Magazine* **39**, 621–40.

Groves, A. W. 1928. The identification of dumortierite as grains: dumortierite in Cornish granite. *Mineralogical Magazine* **21**, 489–92.

——— 1952. *Wartime investigations into the haematite and manganese ore resources of Great Britain and Northern Ireland.* Item 20.703, Permanent Records of Research & Development, Ministry of Supply, London.

Gunn, A. G. & M. H. Shaw 1992. *Platinum-group elements in the Huntly intrusion, Aberdeenshire, northeast Scotland.* Report 124, Mineral Reconnaissance Program, British Geological Survey, Keyworth.

Gunn, A. G. & M. T. Styles 2002. Platinum-group element occurrences in Britain: magmatic, hydrothermal and supergene. *Institution of Mining and Metallurgy, Transactions B* **111**, 2–14.

Gunn, A. G., R. C. Leake, M. T. Styles, J. H. Bateson 1985. *Platinum-group element mineralization in the Unst ophiolite, Shetland.* Report 73, Mineral Reconnaissance Program, British Geological Survey, Keyworth.

Gunn, A. G., M. T. Styles, D. Stephenson, M. H. Shaw, K. E. Rollin 1990. *Platinum-group elements in ultramafic rocks of the Upper Deveron Valley, near Huntly, Aberdeenshire.* Report 115, Mineral Reconnaissance Program, British Geological Survey, Keyworth.

Gunther, A. E. 1978. John George Children FRS (1777–1852) of the British Museum: mineralogist and reluctant keeper of zoology. *British Museum (Natural History), Bulletin* **6**(4), 75–108.

Guppy, E. M. 1944. Boracite from a boring at Aislaby, Yorkshire. *Mineralogical Magazine* **27**, 51–3.

——— 1956. *Chemical analyses of igneous rocks, metamorphic rocks and minerals 1931–1954* (pp. 59–73). Memoir, Geological Survey of Great Britain, London.

Gupta, L. N. 1973. Unusual zircons from the Leinster pluton. *Mineralogical Magazine* **39**, 253–5.

Gyopari, M. 1984. *A study of blueschist mineral chemistry and a new look at the Penmynydd/Gwna boundary in southeast Anglesey.* MPhil thesis, University of Wales.

Haacke, D. F. & P. A. Williams 1981. Stability of plumbonacrite. *Journal of Inorganic and Nuclear Chemistry* **43**, 406.

Hacker, C. M. 1997. Rutile from Wheal Remfry china-clay pit, Indian Queens, St Enoder, Cornwall. *UK Journal of Mines and Minerals* **18**, 16.

Hadfield, G. S. & H. C. M. Whiteside 1936. The Borrowdale Volcanic Series of High Pike and the adjoining Low Rigg microgranite. *Geologists' Association, Proceedings* **47**, 42–64.

Haggerty, R. 1995. The mineralization of the Llanrwst orefield, North Wales. *UK Journal of Mines and Minerals* **15**, 5–10.

Haggerty, R., R. A. Cliff, S. H. Bottrell 1996a. Pb-isotope evidence for the timing of episodic mineralization in the Llanrwst and Llanfair–Talhaiarn orefields, North Wales. *Mineralium Deposita* **31**, 93–7.

Haggerty, R., B. M. Rohl, P. D. Budd, N. H. Gale 1996b. Pb-isotope evidence on the origin of the West Shropshire orefield, England. *Geological Magazine* **133**, 611–17.

Hall, A. 1965a. The occurrence of prehnite in appinitic rocks from Donegal, Ireland. *Mineralogical Magazine* **35**, 234–6.

——— 1965b. The origin of accessory garnet in the Donegal granite. *Mineralogical Magazine* **35**, 628–33.

——— 1966. The feldspars of the Rosses granite complex, Donegal, Ireland. *Mineralogical Magazine* **35**, 975–82.

——— 1971. Greisenization in the granite of Cligga Head, Cornwall. *Geologists' Association, Proceedings* **82**, 209–230.

——— 1979. Rodingites in the Lizard complex. *Ussher Society, Proceedings* **4**, 269–73.

——— 1982. The Pendennis peralkaline minette. *Mineralogical Magazine* **45**, 257–66.

——— 1993a. The influence of secondary alteration on the ammonium content of granites, exemplified by the Rosses Complex of Donegal. *Mineralogical Magazine* **57**, 591–8.

Hall, A. & N. J. Jackson 1975. Summer field meeting in west Cornwall. *Geologists' Association, Proceedings* **86**, 95–102.

Hall, A. & J. N. Walsh 1971. The beryls of the Rosses district, Donegal. *Mineralogical Magazine* **38**, 328–34.

——— 1972. Zinnwaldite granite from Glen Gairn, Aberdeenshire. *Scottish Journal of Geology* **8**, 265–7.

Hall, A. J. 1993b. Stratiform mineralization in the Dalradian of Scotland. See Pattrick & Polya (1993: 38–101).

Hall, A. J. & R. N. Donovan 1978. Origin of complex sulphide nodules related to diagenesis of lacustrine sediments of Middle Devonian age from the Shetland Islands. *Scottish Journal of Geology* **14**, 289–99.

Hall, A. J., D. Banks, A. E. Fallick, P. J. Hamilton 1989. An hydrothermal origin for copper-impregnated prehnite and analcime from Boylestone quarry, Barrhead, Scotland. *Geological Society of London, Journal* **146**, 701–713.

Hall, A. J., A. J. Boyce, A. E. Fallick 1988. A sulphur isotope study of iron sulphides in the Late Precambrian Dalradian Easdale Slate Formation, Argyll, Scotland. *Mineralogical Magazine* **52**, 483–90.

Hall, A. J., P. McConville, A. J. Boyce, A. E. Fallick 1994. Sulphides with high δ^{34}S from the Late Precambrian Bonahaven dolomite, Argyll, Scotland. *Mineralogical Magazine* **58**, 486–90.

Hall, G. W. 1993c. *Metal mines of southern Wales* (2nd edn). Privately published.

Hall, S. 1930. The geology of the Godolphin granite: a study of the coastal geology between Perranuthnoe and Looe Pool. *Geologists' Association, Proceedings* **41**, 117–47.

Hall, T. M. 1868a. *The mineralogist's directory; or, a guide to the principal mineral localities of Great Britain and Ireland.* London: Edward Stanford.

——— 1868b. On the mineral localities of Devonshire. *Devon Association for the Advancement of Science, Transactions* **2**, 332–46.

——— 1877. Remarks on the distribution of minerals in north Devon. *Mineralogical Magazine* **1**, 212–14.

Halliday, A. N. & J. G. Mitchell 1976. Structural, K–Ar and ^{40}Ar–^{39}Ar age studies of adularia K-feldspars from the Lizard Complex, England. *Earth and Planetary Science Letters* **29**, 227–37.

——— 1983. K–Ar ages of clay concentrates from Irish orebodies and their bearing on the timing of mineralization. *Royal Society of Edinburgh, Transactions: Earth Sciences* **74**, 1–14.

Hallimond, A. F. 1914. Optically uniaxial augite from Mull. *Mineralogical Magazine* **17**, 97–9.

——— 1915. On bassetite and uranospathite, new species hitherto classed as autunite. *Mineralogical Magazine* **17**, 221–36.

——— 1920. Meta-torbernite I, its physical properties and relation to torbernite. *Mineralogical Magazine* **19**, 43–7.

——— 1922. On glauconite from the Greensand near Lewes, Sussex: the constitution of glauconite. *Mineralogical Magazine* **19**, 330–33.

——— 1924. On stilpnomelane in North Wales. *Mineralogical Magazine* **20**, 193–7.

——— 1939. Note on a specimen of white beryl from Beam mine, St Austell, Cornwall. *Mineralogical Magazine* **25**, 351–2.

——— 1947. Pyroxenes, amphibole, and mica from the Tiree Marble. *Mineralogical Magazine* **28**, 230–43.

Hallsworth, C. R., A. Livingstone, A. C. Morton 1992. Detrital goldmanite from the Palaeocene of the North Sea. *Mineralogical Magazine* **56**, 117–20.

Halsall, T. J. 1980. Occurrence of copper mineralization on Isle of Arran, Scotland. *Institution of Mining and Metallurgy, Transactions B* **89**, 156–7.

Hamad, S. El D. 1963. The chemistry and mineralogy of the olivine nodules of Calton Hill, Derbyshire. *Mineralogical Magazine* **33**, 483–97.

Hambly, W. 1897. The conglomerates of Cornwall and the banket of South Africa. Report 65 (pp. 97–103), Royal Cornwall Polytechnic Society, Falmouth.

Hamilton, M. A., D. G. Pearson, R. N. Thompson, S. P. Kelley, C. H. Emeleus 1998. Rapid eruption of Skye lavas inferred from precise U–Pb and Ar–Ar dating of the Rum and Cuillin plutonic complexes. *Nature* **394**, 260–63.

Hamilton Bell, W. 1884. Note on new locality for zoisite. *Mineralogical Magazine* **6**, 109–110.

Hamilton Jenkin, A. K. 1964. *Mines and miners of Cornwall, vol. 9: Padstow, St Columb and Bodmin.* Truro: Bradford Barton.

——— 1967. *Mines and miners of Cornwall, vol. 13: the Lizard–Falmouth–Mevagissey.* Truro: Bradford Barton.

——— 1969. *Mines and miners of Cornwall, vol. 15: Calstock, Callington and Launceston.* Truro: Bradford Barton.

——— 1981. *Mines of Devon, north and east of Dartmoor.* Exeter: Devon Library Services.

Hamilton, J. R. & J. F. Lawrence 1970. *Men and mining on the Quantocks.* Bracknell: Town & Country Press.

Hannay, J. B. 1877. On bowlingite, a new Scottish mineral. *Mineralogical Magazine* **1**, 154–7.

Harding, R. P., R. J. Merriman, P. H. A. Nancarrow 1982. A note on the occurrence of chevkinite, allanite, and zirkelite on St Kilda, Scotland. *Mineralogical Magazine* **46**, 445–8.

——— 1983. Zr-rich pyroxenes and glauconitic minerals in the Tertiary alkali granite of Ailsa Craig. *Scottish Journal of Geology* **19**, 219–27.

——— 1984. *St Kilda: an illustrated account of the geology.* Report 16.7, British Geological Survey, Keyworth.

Harker, A. 1888a. Notes on the geology of Mynydd Mawr and the Nantlle Valley. *Geological Magazine* **25**, 221–6.

——— 1888b. Additional note on the blue hornblende of Mynydd Mawr. *Geological Magazine* **25**, 455–6.

——— 1889. *The Bala Volcanic Series of Caernarvonshire.* Cambridge: Cambridge University Press.

——— 1895. Carrock Fell: a study in the variation of igneous rock masses, part II: the Carrock Fell granophyre; part III: the Grainsgill greisen. *Geological Society of London, Quarterly Journal* **51**, 125–48.

——— 1904. *Tertiary igneous rocks of Skye* (pp. 144–51). Memoir, Geological Survey of Great Britain and Ireland, Edinburgh.

——— 1950. *Metamorphism: a study of the transformations of rock masses* (3rd edn). London: Methuen.

——— 1954a. *Harker's "Petrology for students"* (8th edn revised by C. E. Tilley, S. R. Nockolds, M. Black). Cambridge: Cambridge University Press.

Harker, A. & J. E. Marr 1891. The Shap granite and the associated igneous and metamorphic rocks. *Geological Society of London, Quarterly Journal* **47**, 266–328.

Harker, R. I. 1954b. Further data on the petrology of the pelitic hornfelses of the Carn Chuinneag–Inchbae region, Ross-shire, with special reference to the status of almandine. *Geological Magazine* **91**, 445–62.

——— 1962. The older ortho-gneisses of Carn Chuinneag and Inchbea. *Journal of Petrology* **3**, 214–37.

Harker, R. I. & T. Leahy 2001. The significance of a hastingsite granite in the Carn Chuinneag–Inchbea Complex. *Scottish Journal of Geology* **37**, 53–7.

Harker, R. S. 1973. Fluorite with cassiterite. *Mineralogical Record* **4**, 94–5.

Harkness, R. 1866. On the metamorphic and fossiliferous rocks of the county of Galway. *Geological Society of London, Quarterly Journal* **22**, 506–513.

——— 1878. On cotterite, a new variety of quartz. *Mineralogical Magazine* **2**, 82–4.

Harley, S. 1999. *In the Bewick vein: the story of a Northumbrian lead mine.* Haydon Bridge, Northumberland: Honeybrook Press.

Harris, M., E. A. Kay, M. A. Widnall, E. M. Jones, G. B. Steele 1988. Geology and mineralization of the Lagalochan intrusive complex, western Argyll, Scotland. *Institution of Mining and Metallurgy, Transactions B* **97**, 15–21.

Harrison, D. J. 1985. Mineralogical and chemical appraisal of Corrycharmaig serpentinite intrusion, Glen Lochay, Perthshire. *Institution of Mining and Metallurgy, Transactions B* **94**, 147–51.

Harrison, R. K. 1956. *Report on the preliminary survey of radioactivity in the Tyndrum mining area.* Report 186, Atomic Energy Division, Geological Survey of Great Britain, London.

——— 1973. *Mineralogy Unit annual report for 1973* (pp. 123–4), Geochemical Division, Institute of Geological Sciences, London.

——— 1975. Jasperoid and chalcedonic concretions in anhydrite, Sandwith mine, St Bees, Cumbria. *Geological Survey of Great Britain, Bulletin* **52**, 55–9.

Harrison, R. K. & K. Taylor 1966. Radian anglesite from Wheal Speed, Cornwall. *Geological Survey of Great Britain, Bulletin* **25**, 41–8.

Harrison, R. K., N. Aitkenhead, B. R. Young, P. F. Dagger 1975a. Goethite from Hindlow, Derbyshire. *Geological Survey of Great Britain, Bulletin* **52**, 51–4.

Harrison, R. K., R. A. Old, M. T. Styles, B. R. Young 1983. *Coffinite in nodules from the Mercia Mudstone group (Triassic) of the IGS Knowle borehole, West Midlands.* Report 83.10, Institute of Geological Sciences, London, 13–16.

Harrison, R. K., N. J. Snelling, R. J. Merriman, G. E. Morgan, A. J. J. Goode 1977. The Wolf Rock, Cornwall: new chemical, isotopic age and palaeomagnetic data. *Geological Magazine* **114**, 249–64.

Harrison, R. K., P. Stone, I. B. Cameron, R. W. Elliot, R. R. Harding 1987. *Geology, petrology and geochemistry of Ailsa Craig, Ayrshire.* Report 16.9, British Geological Survey, Keyworth.

Harrison, R. K., A. E. Tresham, B. R. Young, R. I. Lawson 1975b. Concretionary concentrations of the rarer elements in Permo-Triassic redbeds of southwest England. *Geological Survey of Great Britain, Bulletin* **52**, 1–26.

Harrison, S. & J. Fuller 1987. Gold from Hope's Nose, Torquay, Devon, England. *Mineralogical Record* **18**(1), 85–8.

Harrison, T. N. 1988. Magmatic garnets in the Cairngorm granite, Scotland. *Mineralogical Magazine* **52**, 659–67.

——— 1990. Chemical variation in micas from the Cairngorm pluton, Scotland. *Mineralogical Magazine* **54**, 355–66.

Harrison, W. J. 1894. New localities for the minerals brookite, natrolite and barytes. *Geological Magazine* **31**, 567.

——— 1897. An occurrence of prehnite in Wales. *Mineralogical Magazine* **11**, 198.

Harry, W. T. 1951. On a cupriferous Lewisian paragneiss. *Mineralogical Magazine* **29**, 542–4.

Harry, W. T. & E. M. Patterson 1949. Unusual dolomite from Portsoy, Banff. *Mineralogical Magazine* **28**, 714–17.

Hart, C. E. 1944. Gold in Dean Forest. *Bristol and Gloucestershire Archaeological Society, Journal* **65**, 98–104.

Harte, B. 1987. Glen Esk Dalradian, Barrovian metamorphic zones. See Trewin et al. (1987: 193–210).

Hartley, E. G. J. 1900. On the constitution of the natural arsenates and phosphates, part III: plumbogummite and hitchcockite. *Mineralogical Magazine* **12**, 223–33.

Hartley, J. 1959. Coronadite from Cumberland. *Mineralogical Magazine* **32**, 343–4.

——— 1984. A list of minerals associated with the ore deposits of the Caldbeck Fells, Cumbria. *Leeds Geological Association, Transactions* **10**, 22–39.

Hartley, J. J. 1932. The volcanic and other igneous rocks of Great and Little Langdale. *Geologists' Association, Proceedings* **43**, 32–69.

Hartley, W. N. 1882. The analysis of rhabdophane, a new British mineral. *Chemical Society, Journal* **41**, 210–18.

Harvey, P. K. & C. C. Ferguson 1973. Spherically arranged inclusions in post-tectonic garnet porphyroblasts. *Mineralogical Magazine* **39**, 85–8.

Harwood, G. M. & F. W. Smith 1986. Mineralization in Upper Permian carbonates at outcrop in eastern England. In *The English Zechstein and related topics*, G. M. Harwood & D. B. Smith (eds), 103–111. Special publication 22, Geological Society, London.

Haslam, H. W. 1968. The crystallization of intermediate and acid magmas at Ben Nevis, Scotland. *Journal of Petrology* **9**, 84–104.

——— 1970. Appinite xenoliths and associated rocks from the Ben Nevis igneous complex. *Geological Magazine* **107**, 341–56.

——— 1971. Andalusite in the Mullach nan Coirean granite, Inverness-shire. *Geological Magazine* **108**, 97–102.

——— 1975. *Geochemical survey of stream waters and stream sediments from the Cheviot area.* Report 75.6, Institute of Geological Sciences, London.

——— 1986. Pyroxenes and coexisting minerals in the Cheviot granite. *Mineralogical Magazine* **50**, 671–4.

Haslam, H. W. & G. S. Kimbell 1981. *Disseminated copper–molybdenum mineralization near Ballachulish, Highland Region.* Report 43, Mineral Reconnaissance Program, Institute of Geological Sciences, London.

Haslam, H. W., D. G. Cameron, M. F. Miller 1985. *Disseminated molybdenum mineralization in the Etive plutonic complex in the western Highlands of Scotland.* Report 76, Mineral Reconnaissance Program, British Geological Survey, Keyworth.

Haslett, S. K. 1992. Petrology of a monchiquite from the Welsh Borderlands. *Mercian Geologist* **13**, 43–6.

Hatch, F. H., A. K. Wells, M. K. Wells 1972. *Petrology of the igneous rocks* (13th edn). London: George Allen & Unwin.

Haughton, S. 1856. Experimental researches on the granites of Ireland. *Geological Society of London, Quarterly Journal* **12**, 171–202.

——— 1862. Experimental researches on the granites of Ireland, III: on the granites of Donegal. *Geological Society of London, Quarterly Journal* **18**, 403–420.

Hawkes, J. R. & H. E. Wilson 1975. The Portrush sill, County Antrim, Northern Ireland. *Geological Survey of Great Britain, Bulletin* **51**, 1–19.

Hawkes, J. R., P. M. Harris, J. Dangerfield, G. E. Strong, A. E. Davis, P. H. A. Nancarrow, A. D. Francis, C. V. Smale 1987. *The lithium potential of the St Austell granite.* Report 19.4, British Geological Survey, Keyworth.

Hawkes, J. R., R. J. Merriman, R. R. Harding, D. P. F. Darbyshire 1975. Rockall Island: new geological, petrological, chemical and Rb–Sr age data. In *Expeditions to Rockall 1971–72*, R. K. Harrison (ed.), 15–51. Report 75.1, Institute of Geological Sciences, London.

Hawkes, L. & J. A. Smythe 1935. Ankerites of the Northumberland coalfield. *Mineralogical Magazine* **24**, 65–75.

Hawkins, T. R. W. 1970. Hornblende gabbros and picrites at Rhiw, Caernarvonshire. *Geological Journal* **7**, 1–24.

Hawson, C. A. & A. J. Hall 1987. Middle Dalradian Corrycharmaig serpentinite, Perthshire, Scotland: an ultramafic intrusion. *Institution of Mining and Metallurgy, Transactions B* **96**, 173–7.

Hawthorne, F. C. & D. J. Henry 1999. Classification of the minerals of the tourmaline group. *European Journal of Mineralogy* **11**, 201–215.

Hawthorne, F. C. & R. Oberti 2006. On the classification of amphiboles. *Canadian Mineralogist* **44**, 1–21.

Hawthorne, F. C., M. A. Cooper, D. I. Green, R. E. Starkey, A. C. Roberts, J. D. Grice 1999a. Wooldridgeite, $Na_2CaCu_2(P_2O_7)_2(H_2O)_{10}$: a new mineral from Judkins quarry, Warwickshire, England. *Mineralogical Magazine* **63**, 13–16.

Hawthorne, F. C., M. A. Cooper, J. D. Grice, A. C. Roberts, N. Hubbard 2002. Description and crystal structure of bobkingite, $Cu_5^{2+}Cl_2(OH)_8(H_2O)_2$, a new mineral from the New Cliffe Hill quarry, Stanton under Bardon, Leicestershire, UK. *Mineralogical Magazine* **66**, 301–311.

Hawthorne, F. C., J. B. Selway, A. Kato, S. Matsubara, M. Shimizu, J. D. Grice, J. Vajdak 1999b. Magnesiofoitite, $\square(Mg_2Al)Al_6(Si_6O_{18})(BO_3)_3(OH)_4$, a new alkali-deficient tourmaline. *Canadian Mineralogist* **37**, 1439–43.

Haynes, L. 1974. *The identification of heavy minerals derived from tin-bearing rocks in the Cairngorms.* Unpublished report 142, Mineralogy Unit, Institute of Geological Sciences, London.

——— 1993. W–Sn–Mo–Bi–Ag mineralization associated with zinnwaldite-bearing granite from Glen Gairn, Scotland – discussion of Webb et al. 1992 *Trans IMM* **101**, 59. *Institution of Mining and Metallurgy, Transactions B* **102**, 211–14.

Heaman, L. M. & J. Tarney 1989. U–Pb baddeleyite ages for the Scourie dyke swarm, Scotland: evidence for two distinct intrusion events. *Nature* **340**, 705–708.

Heaney, P. J. & J. E. Post 1992. The widespread distribution of a novel silica polymorph in microcrystalline quartz varieties. *Science* **255**, 441–3.

Heard, A. 1922. The petrology of the Pennant Series, east of the River Taff. *Geological Magazine* **59**, 83–94.

Heddle, M. F. 1857. On mesolite and faröelite (mesole). *Philosophical Magazine (4th series)* **13**, 50–55.

——— 1877. Chapters on the mineralogy of Scotland. Chapter second – the felspars.

Royal Society of Edinburgh, Transactions: Earth Sciences **28**, 197–271.

——— 1878a. Orkney and Shetland: Island of Unst. *Mineralogical Magazine* **2**, 12–35.

——— 1878b. The geognosy and mineralogy of Scotland: Islands of Uya, Haaf Grunay, Fetlar, and Yell. *Mineralogical Magazine* **2**, 106–133.

——— 1878c. Chapters on the mineralogy of Scotland. Chapter fourth – augite, hornblende, and serpentinous change. *Royal Society of Edinburgh, Transactions: Earth Sciences* **28**, 453–555.

——— 1879a. Pilolite, an unrecognised species. *Mineralogical Magazine* **2**, 206–219.

——— 1879b. The geognosy and mineralogy of Scotland: Mainland [Shetland], part II, Hillswick Ness. *Mineralogical Magazine* **3**, 18–56.

——— 1879c. Preliminary notice of substances which may prove to be new minerals. *Mineralogical Magazine* **3**, 57–68.

——— 1879d. On haughtonite: a new mica. *Mineralogical Magazine* **3**, 72–84.

——— 1879e. Chapters on the mineralogy of Scotland. Chapter fifth – the micas; with description of haughtonite, a new mineral species. *Royal Society of Edinburgh, Transactions: Earth Sciences* **29**, 1–46.

——— 1879f. Chapters on the mineralogy of Scotland. Chapter sixth – chloritic minerals. *Royal Society of Edinburgh, Transactions: Earth Sciences* **29**, 55–118.

——— 1880a. On a new face of stilbite, from two localities. *Mineralogical Magazine* **4**, 44.

——— 1880b. Preliminary notes of substances which may prove to be new minerals. *Mineralogical Magazine* **4**, 117–23.

——— 1881. The geognosy and mineralogy of Scotland: Sutherland [part I]. *Mineralogical Magazine* **4**, 197–254.

——— 1882a. Minerals new to Britain. *Mineralogical Magazine* **5**, 1–25.

——— 1882b. On some ill determined minerals. *Mineralogical Magazine* **5**, 26–31.

——— 1882c. The geognosy and mineralogy of Scotland: Sutherland part I. *Mineralogical Magazine* **5**, 71–106.

——— 1883. The geognosy and mineralogy of Scotland: Sutherland part IV. *Mineralogical Magazine* **5**, 133–89.

——— 1887a. On the occurrence of greenockite at a new locality. *Mineralogical Magazine* **7**, 133–7.

——— 1887b. On a form of calcite from Helim, Sutherland. *Mineralogical Magazine* **7**, 138.

——— 1889a. On dudgeonite, hydroplumbite, plumbonacrite and plattnerite. *Mineralogical Magazine* **8**, 200–203.

——— 1889b. On the identity of bruiachite and fluor. *Mineralogical Magazine* **8**, 274–7.

——— 1891. On the occurrence of sapphire in Scotland. *Mineralogical Magazine* **9**, 389–90.

——— 1893a. On new localities for zeolites. *Geological Society of Glasgow, Transactions* **9**, 72–9.

——— 1893b. On pectolite and okenite from new localities, the former with new appearances. *Geological Society of Glasgow, Transactions* **9**(2), 241–55.

——— 1901. *The mineralogy of Scotland*, 2 vols. Edinburgh: David Douglas.

Hemingway, J. E. & G. P. Riddler 1982. Basin inversion in North Yorkshire. *Institution of Mining and Metallurgy, Transactions B* **91**, 175–86.

Henderson, C. M. B. & F. G. F. Gibb 1977. Formation of analcime in the Dippin Sill, Isle of Arran. *Mineralogical Magazine* **41**, 534–7.

Henderson, C. M. B., F. G. F. Gibb, K. A. Foland 2000. Mineral fractionation and pre- and post-emplacement processes in the uppermost part of the Shiant Isles Main Sill. *Mineralogical Magazine* **64**, 779–90.

Henderson, C. M. B., J. S. Martin, R. A. Mason 1989. Compositional relations in Li-micas from southwest England and France: an ion- and electron-microprobe study. *Mineralogical Magazine* **53**, 427–49.

Henderson, P. & P. Suddaby 1971. The nature and origin of the chrome spinel of the Rhum layered intrusion. *Contributions to Mineralogy and Petrology* **33**, 21–31.

Henderson, W. G., N. J. Fortey, C. E. Johnson, A. Grout 1983. *Mineral reconnaissance at the Highland Boundary with special reference to the Loch Lomond and Aberfoyle areas.* Report 61, Mineral Reconnaissance Program, Institute of Geological Sciences, London.

Hendry, J. P., M. J. Pearson, N. H. Trewin, A. E. Fallick 2006. Jurassic septarian concretions from NW Scotland record interdependent bacterial, physical and chemical processes of marine mudrock diagenesis. *Sedimentology* **53**, 537–65.

Henley, S. 1971. Hedenbergite and sphalerite from the Perran Iron lode, Cornwall. *Ussher Society, Proceedings* **2**, 329–34.

Henmi, C., I. Kusachi, K. Henmi, P. A. Sabine, B. R. Young 1973. A new mineral bicchulite, the natural analogue of gehlenite hydrate, from Fuka, Okayama prefecture, Japan, and Carneal, County Antrim, Northern Ireland. *Mineralogical Journal* **7**, 243–51.

Henry, N. F. M. 1935. Some data on the iron-rich hypersthenes. *Mineralogical Magazine* **24**, 221–6.

Henslow, J. S. 1822. Geological description of Anglesea. *Cambridge Philosophical Society, Transactions* **1**, 259–452.

Henson, F. A. 1956. On the occurrence of micaceous hematite in the Hennock Lustleigh area, eastern Dartmoor. *Geologists' Association, Proceedings* **67**, 87–102.

Henwood, W. J. 1855. *Four lectures on geology and mining.* Leeds: *Mining Journal.*

——— 1871. Notice of the Sark's Hope mine, Sark. *Royal Geological Society of Cornwall, Transactions* **8**, 530–39.

——— 1874. On the detrital tin ore of Cornwall. *Royal Institution of Cornwall, Journal* **4**, 191–254.

Herbert, S. 2005. *Charles Darwin, geologist.* Ithaca: Cornell University Press.

Herbert Smith, G. F. 1919. Semseyite from Dumfriesshire. *Mineralogical Magazine* **18**, 354–9.

Herbert Smith, G. F., F. N. Ashcroft, G. T. Prior 1916. Chabazite and associated minerals from County Antrim. *Mineralogical Magazine* **17**, 274–304.

Herd, R. K., B. F. Windley, D. Ackermand 1984. Grandidierite from a pelitic xenolith in the Haddo House complex, northeast Scotland. *Mineralogical Magazine* **48**, 401–406.

Herriot, A. 1968. A riebeckite-bearing trachyte from north Ayrshire. *Geological Society of Glasgow, Proceedings* **109**, 33–4.

——— 1971. A xenolithic dyke at McInroy's Point, Gourock, Renfrewshire. *Scottish Journal of Geology* **7**, 153–61.

Heslop, M. K. 1912. A preliminary note on the uniaxial augites of the north of England igneous dykes. *University of Durham Philosophical Society, Proceedings* **4**, 172–4.

Hey, M. H. 1932a. Studies on the zeolites, part II: thomsonite (including faröelite) and gonnardite. *Mineralogical Magazine* **23**, 51–125.

——— 1932b. Studies on the zeolites, part III: natrolite and metanatrolite. *Mineralogical Magazine* **23**, 243–9.

——— 1933. Studies on the zeolites, part V: mesolite. *Mineralogical Magazine* **23**, 421–47.

——— 1934. Studies on the zeolites, part VI: edingtonite. *Mineralogical Magazine* **23**, 483–94.

——— 1936. Studies on the zeolites, part IX: scolecite and metascolecite. *Mineralogical Magazine* **24**, 227–53.

——— 1954. A new review of the chlorites. *Mineralogical Magazine* **30**, 277–92.

——— 1960. Glottalite is chabazite. *Mineralogical Magazine* **32**, 421–2.

——— 1961. Twenty-second list of new mineral names. *Mineralogical Magazine* **32**, 941–91.

——— 1973. Mineral analysis and analysts. *Mineralogical Magazine* **39**, 4–24.

Hey, M. H., F. A. Bannister, A. Russell 1938. Russellite, a new British mineral. *Mineralogical Magazine* **25**, 41–55.

Hibbert, A., W. H. Johnson, J. J. Adams 1940. *Survey of iron ore resources of west Cumberland.* Whitehaven: Cumberland Development Council.

Higham, F. (ed.) 1959. *The future of non-ferrous mining in Great Britain and Ireland.* London: Institution of Mining and Metallurgy.

Highton, A. J. 1994. A re-evaluation of "metasedimentary xenoliths" in the West Highlands granitic gneiss of Inverness-shire. *Scottish Journal of Geology* **30**, 39–49.

Hill, E. & T. G. Bonney 1890. On the northwest region of Charnwood Forest, with other notes. *Geological Society of London, Quarterly Journal* **42**, 101–108.

Hill, I. G., R. H. Worden, I. G. Meighan 2000. Geochemical evolution of a palaeolaterite: the Interbasaltic Formation, Northern Ireland. *Chemical Geology* **166**, 65–84.

Hill, J. A. & K. C. Dunham 1968. The barytes deposits at Closehouse, Lunedale, Yorkshire. *Yorkshire Geological Society, Proceedings* **36**, 351–72.

Hintze, C. 1897. *Handbuch der Mineralogie,* zweiter Band: *Silicate und Titanate.* Leipzig: Verlag Von Veit.

——— 1904. *Handbuch der Mineralogie*, erster Band, Abteilung 1: *Elemente und Sulfide* (1229–33). Berlin: Verlag von Veit.

Hirst, D. M. & F. W. Smith 1974. Controls on barite mineralization in the Lower Magnesian Limestone of the Ferryhill area, County Durham. *Institution of Mining and Metallurgy, Transactions B* **83**, 49–55.

Hitchin, C. S. 1934. The Skiddaw granite and its residual products. *Geological Society of London, Quarterly Journal* **90**, 158–200.

Hitzman, M. W. 1986. Geology of the Abbeytown mine, Co. Sligo, Ireland. See Andrew et al. (1986: 341–54).

Hitzman, M. W. & D. W. Beaty 2004. The Irish Zn–Pb–(Ba–Ag) orefield. See Kelly et al. (2004: 499–532).

Hitzman, M. W. & D. Large 1986. A review and classification of the Irish carbonate-hosted base metal deposits. See Andrew et al. (1986: 217–38).

Hitzman, M. W., P. O'Connor, E. Shearley, C. Schaffalitzky, D. W. Beaty, J. R. Allan, T. Thompson 1992. Discovery and geology of the Lisheen Zn–Pb–Ag prospect, Rathdowney Trend, Ireland. See Bowden et al. (1992: 227–46).

Hitzman, M. W., P. B. Redmond, D. W. Beaty 2002. The carbonate-hosted Lisheen Zn–Pb–Ag deposit, County Tipperary, Ireland. *Economic Geology* **97**, 1627–55.

Hobson, D. M. 1972. Boulangerite from Port Gaverne, Cornwall. *Mineralogical Magazine* **38**, 767–8.

Hobson, G. V. 1959. Barytes in Scotland with special reference to Gasswater and Muirshiel mines. See Higham (1959: 85–97).

Hodge, W. G. 1989. An investigation into my garden wall. *UK Journal of Mines and Minerals* **6**, 27.

Hodkinson, I. P. & A. M. Clark 1977. Columbite from St Austell, Cornwall. *Mineralogical Magazine* **41**, 131–2.

Hoersch, A. L. 1981. Progressive metamorphism of the chert-bearing Durness Limestone in the Beinn an Dubhaich aureole, Isle of Skye, Scotland: a re-examination. *American Mineralogist* **66**, 491–506.

Hofmann, B. A. 1993. Organic matter associated with mineralized reduction spots in red beds. In *Bitumens in ore deposits*, J. Parnell, H. Kutcha, P. Landais (eds), 362–78. Berlin: Springer.

Hoffman, J. F. & J. V. P. Long 1984. Unusual sector zoning in Lewisian zircons. *Mineralogical Magazine* **48**, 513–17.

Hogarth, D. D. 1977. Classification of the pyrochlore group. *American Mineralogist* **62**, 403–410.

Hogg, C. S. & R. E. Meads 1970. The Mössbauer spectra of several micas and related minerals. *Mineralogical Magazine* **37**, 606–614.

Hogg, C. S., P. J. Malden, R. E. Meads 1975. Identification of iron-containing impurities in natural kaolinites using the Mössbauer effect. *Mineralogical Magazine* **40**, 89–96.

Holdsworth, J. H. 1834. On the geology of the district of the Knockmahon mines, in the County of Waterford. *Geological Society of Dublin, Journal* **1**, 85–99.

Hole, M. J. & M. A. Morrison 1992. The differentiated dolerite boss, Cnoc Rhaonastil, Islay: a natural experiment in the low-pressure differentiation of an alkali olivine-basalt magma. *Scottish Journal of Geology* **28**, 55–69.

Hole, M. J., N. H. Trewin, J. Still 1992. Mobility of the high field strength, rare-earth elements and yttrium during late diagenesis. *Geological Society of London, Journal* **149**, 689–92.

Holgate, N. 1951. On crossite from Anglesey. *Mineralogical Magazine* **29**, 792–8.

——— 1973. Dichroic pigment layers in Blue John fluorite. *Mineralogical Magazine* **39**, 363–6.

——— 1977. Tourmaline from amphibolitized gabbro at Hanter Hill, Radnorshire. *Mineralogical Magazine* **41**, 124–7.

Holl, C. J. & A. V. Bromley 1988. Application of mineralogy to the benefication of a tin-sulphide ore. *Minerals Engineering* **1**, 19–30.

Holland, E. G. 1981. *Coniston copper mines: a field guide to the mines in the copper orefield at Coniston in the English Lake District.* Milnthorpe, Cumbria: Cicerone Press.

Holliday, D. W. 1973. Early diagenesis in nodular anhydrite rocks. *Institution of Mining and Metallurgy, Transactions B* **82**, 81–4.

Hollingworth, S. E. & F. A. Bannister 1950. Basaluminite and hydrobasaluminite, two new minerals from Northamptonshire. *Mineralogical Magazine* **29**, 1–17.

Hollis, J. A., S. L. Harley, R. W. White, G. L. Clarke 2006. Preservation of evidence for prograde metamorphism in ultrahigh-temperature, high-pressure kyanite-bearing granulites, South Harris, Scotland. *Journal of Metamorphic Geology* **24**, 263–79.

Holmes, A. H. & A. A. Smales 1955. Monazite from Bodmin Moor, Cornwall. *Royal Society of Edinburgh, Transactions: Earth Sciences* **63**, 115–29.

Holness, M. B. 1992. Metamorphism and fluid infiltration of the calc-silicate aureole of the Beinn an Dubhaich granite, Skye. *Journal of Petrology* **33**, 1261–93.

——— 2002. Spherulitic textures formed during crystallization of partially melted arkose, Rum, Scotland. *Geological Magazine* **139**, 651–63.

Holness, M. B. & A. E. Fallick 1994. Palaeohydrology of the Beinn an Dubhaich aureole, Skye, and inferred mechanisms of fluid infiltration in carbonates. *Mineralogical Magazine* **58a**, 428–9.

Holub, F. V., H. Klápová, B. J. Bluck, D. R. Bowes 1984. Petrology and geochemistry of post-obduction dykes of the Ballantrae complex, southwest Scotland. *Royal Society of Edinburgh, Transactions: Earth Sciences* **75**, 211–23.

Holyer, V. A. D. 1996. The minerals of Countybridge quarry, Goonhilly Downs, the Lizard, Cornwall. *UK Journal of Mines and Minerals* **17**, 25–8.

——— 1997. Field meeting report: Porthoustock quarry, St Keverne, Cornwall. *Russell Society, Newsletter* **31**, 15–16.

Hooper, J. J. 1997. Field meeting report: Eastern Cliff, Kennack Sands, the Lizard, Cornwall. *Russell Society, Newsletter* **31**, 16.

Horák, J. M. 1993. *The Late Precambrian Coedana and Sarn complexes, northwest Wales – a geochemical and petrological study.* PhD thesis, University of Wales, Cardiff.

Horák, J. M. & W. Gibbons 1986. Reclassification of blueschist amphiboles from Anglesey, North Wales. *Mineralogical Magazine* **50**, 533–5.

Horne, J. & J. J. H. Teall 1892. On boralanite – an igneous rock intrusive in the Cambrian limestone of Assynt, Sutherland, and the Torridon sandstone of Ross-shire. *Royal Society of Edinburgh, Transactions: Earth Sciences* **37**, 163–78.

Hosking, K. F. G. 1949. Fissure systems and mineralisation in Cornwall. *Royal Geological Society of Cornwall, Transactions* **18**, 9–49.

——— 1951. Primary ore deposition in Cornwall. *Royal Geological Society of Cornwall, Transactions* **18**, 309–356.

——— 1954a. The pegmatites of Trolvis quarry, Carnmenellis, Cornwall. *Geological Magazine* **91**, 273–84.

——— 1954b. Cornish pegmatites and bodies with pegmatite affinities. *Royal Geological Society of Cornwall, Transactions* **18**, 411–55.

——— 1954c. The vein system of St Michael's Mount, Cornwall. *Royal Geological Society of Cornwall, Transactions* **18**, 493–509.

——— 1970. The nature of the primary tin ores of the southwest of England. *International Tin Council, London* **3**, 1155–243.

Hosking, K. F. G. & G. S. Camm 1981. The nature and genesis of supergene galena and certain other associated species from Wheal Hope, Perranzabuloe, Cornwall. *Camborne School of Mines, Journal* **81**, 33–42.

——— 1985. The occurrence of cassiterite and other species of economic interest in the greisenized granite porphyry of Cameron quarry, St Agnes, Cornwall. In *High heat production (HHP) granites, hydrothermal circulation and ore genesis*, 517–33 [proceedings of a conference held at St Austell, September 1985].

Hosking, K. F. G. & J. P. R. Polkinghorne 1954. The significance of the variable iron content of the "wolframites" of the West of England. *Royal Geological Society of Cornwall, Transactions* **18**, 526–234.

Hosking, K. F. G. & J. H. Trounson 1959. The mineral potential of Cornwall. See Higham (1959: 355–69).

Hosking, K. F. G., G. J. Stanley, G. S. Camm 1987. The occurrence, nature and genesis of wood tin in southwest England. *Royal Geological Society of Cornwall, Transactions* **21**, 153–212.

House, A. 2002. Field meeting report: Whatley quarry, near Frome, Somerset. *Russell Society, Newsletter* **41**, 37–8.

Houston, R. S. 1912. Notes on the mineralogy of Renfrewshire. *Paisley Naturalists' Society, Transactions* **1**, 1–88.

Houston, W. J. 1964. Weardale lead. *Mine and Quarry Engineering* **30**, 98–107.

Howard, F. T. 1895. The geology of Barry Island. *Cardiff Naturalists' Society, Transactions* **50**, 532–5.

——— 1899. Note on the lead deposits of the eastern fringe of the South Wales coalfield. *Cardiff Naturalists' Society, Transactions* **30**, 46–7.

Howard, H. 1996. Romanesque and gothic wall paintings in Winchester cathedral: an unusual use and alteration of vivianite. *Russell Society, Journal* **6**(2), 93–6.

Howell, J. 1878. The geology of Brighton, part II. *Geologists' Association, Proceedings* **5**, 80–90.

Howells, M. F., A. J. Reedman, S. D. G. Campbell 1991. *Ordovician (Caradoc) marginal basin volcanism in Snowdonia (northwest Wales).* Keyworth: British Geological Survey.

Howie, R. A. 1963. Kaersutite from the Lugar sill, Ayrshire, and from alnöite breccia, Alnö Island, Sweden. *Mineralogical Magazine* **33**, 718–20.

——— 1964. Some orthopyroxenes from Scottish metamorphic rocks. *Mineralogical Magazine* **33**, 903–911.

——— 1965. Bustamite, rhodonite, spessartine and tephroite from Meldon, Okehampton, Devon. *Mineralogical Magazine* **34**, 249–55.

——— 1975. Tourmaline. *Ussher Society, Proceedings* **3**, 209–213.

Howie, R. A. & J. N. Walsh 1981. Riebeckitic arfvedsonite and aenigmatite from Ailsa Craig microgranite. *Scottish Journal of Geology* **17**, 123–8.

Howie, R. A. & A. R. Woolley 1968. The role of titanium and the effect of TiO_2 on the cell size, refractive index, and specific gravity in the andradite–melanite–schorlomite series. *Mineralogical Magazine* **36**, 775–90.

Howie, R. A., E. Pegram, J. N. Walsh 1982. The content of rare earth elements in English fluorites: a preliminary study. *Russell Society, Journal* **1**(1), 22–5.

Hubbard, N. 1989. Wickmanite from Whealcock Zawn, Botallack, Cornwall. *Mineralogical Magazine* **53**, 388.

——— 1991. Phosgenite, the first Welsh occurrence. *Russell Society, Journal* **4**(1), 35.

Hubbard, N. & H. Enamy 1993. Edingtonite from Squilver quarry, Shropshire, England. *Mineralogical Magazine* **57**, 349–51.

Hubbard, N. & D. I. Green 2001. Vésigniéite from Castletown mine, Lochgilphead, Argyll and Bute, a first Scottish occurrence. *Russell Society, Journal* **7**(2), 93.

——— 2005. Cotunnite from Penrhyn Du mine, Gwynedd, Wales. *Russell Society, Journal* **8**(2), 106–107.

Hubbard, N., S. Burchmore, D. I. Green 2005. A remarkable supergene mineral suite from New Cliffe Hill quarry, Stanton under Bardon, Leicestershire. *UK Journal of Mines and Minerals* **26**, 31–42.

Hubbard, N., D. I. Green, T. F. Cotterell 2007. Arsentsumebite from Dolyhir quarry, Old Radnor, Powys. *UK Journal of Mines and Minerals* **28**, 9–10.

Huggett, J. M. & J. Cuadros 2005. Low-temperature illitization of smectite in the late Eocene and early Oligocene of the Isle of Wight (Hampshire Basin), UK. *American Mineralogist* **90**,

1192–202.

Hughes, S. J. S. 1981. *The Cwmystwyth mines.* British Mining 17, Northern Mine Research Society, Sheffield.

——— 1988. *The Goginan mines.* British Mining 35, Northern Mine Research Society, Sheffield.

——— 1989a. *The Darren mines.* British Mining 40, Northern Mine Research Society, Sheffield.

——— 1989b. The history of Ystrad Einion mine. *UK Journal of Mines and Minerals* **6**, 12–14.

——— 1989c. Bwlchglas mine, Dyfed, Wales. *UK Journal of Mines and Minerals* **7**, 20–28.

——— 1990. Llywernog mine and museum, Dyfed, Wales. *UK Journal of Mines and Minerals* **8**, 30–37.

——— 1991. The Van mine, Llanidloes, Powys, Wales. *UK Journal of Mines and Minerals* **9**, 16–23.

Hughes, T. M. M. 1889. On the manner of occurrence of beekite and its bearing upon the origin of siliceous beds of Palaeozoic age. *Mineralogical Magazine* **8**, 265–71.

Hughes, W. J. 1959. The non-ferrous mining possibilities of Central Wales [orefield]. See Higham (1959: 277–94).

Humphreys, D. A., J. H. Thomas, P. A. Williams, R. F. Symes 1980. The chemical stability of mendipite, diaboléite, chloroxiphite, cumengéite, and their relationships to other secondary lead(II) minerals. *Mineralogical Magazine* **43**, 901–904.

Hussak, E. & G. T. Prior 1895. Lewisite and zirkelite, two new Brazilian minerals. *Mineralogical Magazine* **11**, 80–88.

Hutchings, W. M. 1889. On the occurrence of ottrelite in the phyllites of north Cornwall. *Geological Magazine* **26**, 214–20.

Hutchinson, A. 1900. On stokesite, a new mineral containing tin from Cornwall. *Mineralogical Magazine* **12**, 274–81.

——— 1903. The chemical composition and optical characters of chalybite from Cornwall. *Mineralogical Magazine* **13**, 209–216.

Hutchinson, A. & W. Campbell Smith 1912. On sericite from North Wales and on penninite and labradorite from Ireland. *Mineralogical Magazine* **16**, 264–71.

Hutchinson, A. G. 1933. The metamorphism of the Deeside limestone, Aberdeenshire. *Royal Society of Edinburgh, Transactions: Earth Sciences* **57**, 557–92.

Hutchinson, D. 2001. *The origin and distribution of platinum-group mineralization in the oceanic upper mantle: a comparison of samples from the Lizard and Troodos ophiolite complexes, and dredge samples from the Tonga Trench.* PhD thesis, University of Cardiff.

Hutton, C. O. 1938. On the nature of withamite from Glencoe, Scotland. *Mineralogical Magazine* **25**, 119–24.

Hyslop, E. K., R. J. Gillanders, P. G. Hill, R. D. Fakes 1999. Rare-earth-bearing minerals fergusonite and gadolinite from the Arran granite. *Scottish Journal of Geology* **35**, 65–9.

Ikin, N. P. & R. S. Harmon 1983. Mineralogy and petrology of the Highland Border Suite serpentinites. *Mineralogical Magazine* **47**, 301–310.

Ince, F. 1986. The first British occurrence of cyrilovite and a second British locality for chalcosiderite. *Russell Society, Journal* **1**(4), 127–8.

——— 1991. The mineralogy of Wapping mine, Matlock Bath, Derbyshire. *Russell Society, Journal* **4**(1), 25–34.

——— 2001. Vanadinite in Leicestershire. *Russell Society, Newsletter* **39**, 16–17.

——— 2005. The mineralogy of Newhurst quarry, Shepshed, Leicestershire. *Russell Society, Journal* **8**(2), 47–65.

——— 2007. The mineralogy of Bardon Hill quarry, Coalville, Leicestershire. *Russell Society, Journal* **10**, 27–39.

Ince, F., R. E. Starkey, J. Wooldridge 1991. The geology and mineralogy of Judkins quarry, Warwickshire. *UK Journal of Mines and Minerals* **10**, 8–13.

Ineson, P. R. 1972. Alteration of the Whin Sill adjacent to baryte–witherite mineralization, Settlingstones mine, Northumberland. *Institution of Mining and Metallurgy, Transactions B* **81**, 67–72.

——— 1976. Ores of the northern Pennines. In *Handbook of stratabound and stratiform ore deposits*, K. H. Wolf (ed.), 197–230. Amsterdam: Elsevier.

Ineson, P. R. & F. A. M. Al-Kufaishi 1970. The mineralogy and paragenetic sequence of Long Rake vein at Raper mine, Derbyshire. *Mercian Geologist* **3**, 337–51.

Ineson, P. R. & J. G. Mitchell 1974a. K–Ar isotopic age determinations from some Lake District mineral localities. *Geological Magazine* **111**, 521–37.

——— 1974b. K–Ar age determinations from some Scottish mineral localities. *Institution of Mining and Metallurgy, Transactions B* **83**, 13–18.

Ineson, P. R. & B. Young 1994. The North Swaledale mineral belt around Gunnerside. See Scrutton (1994: 51–7).

Ineson, P. R., R. T. Richardson, G. H. Wood 1972. A baryte–galena vein in the Magnesian Limestone at Whitwell, Derbyshire. *Yorkshire Geological Society, Proceedings* **39**, 139–50.

Ineson, P. R., S. G. Walters, R. M. Simon 1983. The petrography and geochemistry of the Waterswallows sill, Buxton, Derbyshire. *Yorkshire Geological Society, Proceedings* **44**, 341–54.

Ingram, S. I. 1994. Agates from Ardownie quarry, Monifieth, Tayside, Scotland. *UK Journal of Mines and Minerals* **14**, 17–20.

Ingram, S. M., D. G. Anderson, J. G. Todd 1993. Molybdenite from Coire Buidhe, Glen Creran, Argyll. *UK Journal of Mines and Minerals* **12**, 4–8.

——— 1996. Powellite and scheelite from Coire Buidhe, Argyll. *UK Journal of Mines and Minerals* **16**, 14.

Ingram, S., J. G. Todd, D. Anderson 1992. Pyrite from Goat quarry, an exceptional Scottish occurrence. *UK Journal of Mines and Minerals* **11**, 8–10.

Irving, A. 1891. Note on the occurrence of melanterite in the Upper Eocene strata of the Thames Basin. *Mineralogical Magazine* **9**, 392–3.

Irving, J. 1930. Four "felstone" intrusions in central Berwickshire. *Geological Magazine* **67**, 529–41.

Irwin, H. 1981. On calcic dolomite–ankerite from the Kimmeridge Clay. *Mineralogical Magazine* **44**, 105–107.

Ixer, R. A. 1974a. Distribution and varieties of bravoite in Derbyshire. *Peak District Mines Historical Society, Bulletin* **5**, 333–4.

——— 1974b. The mineralogy and paragenesis of a fluorspar flat at Masson Hill, Matlock, Derbyshire. *Mineralogical Magazine* **39**, 811–15.

——— 1978. The distribution of bravoite and nickeliferous marcasite in central Britain. *Mineralogical Magazine* **42**, 149–50.

——— 1980. The ore minerals of Jersey. *Société Jersiase Annual Bulletin* **22**, 443–51.

——— 1986. The ore mineralogy and parageneses of the lead–zinc–fluorite–baryte orefields of the English Pennine and Mendip Hills. In *Mineral parageneses*, J. R. Craig, R. D. Hagni, W. Kiesl, I. M. Lange, N. V. Petrovskaya, T. N. Shadlun, G. Udubasa, S. S. Augustithus (eds), 179–210. Athens: Theophrastus Publications.

——— 1988. Copper mineralization from La Moye. *Société Jersiase Annual Bulletin* **24**, 505–506.

——— 1990a. *Atlas of opaque and ore minerals in their associations.* Milton Keynes: Open University Press.

——— 1990b. Mineralization within the German underground hospital. *Société Jersiase Annual Bulletin* **25**, 357–9.

——— 1992. Mining, mineralization and tax evasion in the Channel Islands. *Open University Geological Society, Journal* **13**(2), 16–23.

——— 2001. An assessment of copper mineralization from the Great Orme mine, Llandudno, North Wales, as ore in the Bronze age. *Yorkshire Geological Society, Proceedings* **53**, 213–19.

——— 2002. *Report on ore samples from Lisheen mine.* Unpublished report, Lisheen mine County Tipperary, Ireland.

——— 2003. The distribution of rare earth elements in North Pennine fluorspar and fluorite. *UK Journal of Mines and Minerals* **23**, 21–6.

——— 2004. The petrography of the zinc–lead–copper ores at Crow Island, Killarney, Ireland. *UK Journal of Mines and Minerals* **24**, 29–34.

Ixer, R. A. & P. Budd 1998. The mineralogy of Bronze Age copper ores from the British Isles: implications for the composition of early metalwork. *Oxford Journal of Archaeology* **17**, 15–41.

Ixer, R. A. & J. Davies 1996. Mineralization at the Great Orme copper mines, Llandudno, North Wales. *UK Journal of Mines and Minerals* **17**, 7–14.

Ixer, R. A. & R. A. D. Pattrick 2003. Copper–arsenic ores and Bronze Age mining and metallurgy, with special reference to the British Isles. In *Mining and metal production through the ages*, P. Craddock & J. Lang (eds), 9–21. London: British Museum Press.

Ixer, R. A. & C. J. Stanley 1980. Mineralization at Le Pulec, Jersey, Channel Islands. *Mineralogical Magazine* **43**, 1025–1029.

——— 1983. Silver mineralization at Sark's Hope mine, Sark, Channel Islands. *Mineralogical Magazine* **47**, 539–45.

——— 1987. A silver–nickel–cobalt mineral association at Tynebottom mine, Garrigill, near Alston, Cumbria. *Yorkshire Geological Society, Proceedings* **46**, 133–9.

——— 1996. Siegenite-bearing assemblages found at the Great Orme mine, Llandudno, North Wales. *Mineralogical Magazine* **60**, 978–82.

——— 1998. Enargite group minerals from Scaleber Bridge, North Yorkshire, England. *Russell Society, Journal* **7**(1), 41–2.

Ixer, R. A. & R. Townley 1979. The sulphide mineralogy and paragenesis of the South Pennine orefield. *Mercian Geologist* **7**, 51–63.

Ixer, R. A. & D. J. Vaughan 1982. The primary ore mineralogy of the Alderley Edge deposit, Cheshire. *Mineralogical Magazine* **46**, 485–92.

——— 1993. Lead–zinc–fluorite–baryte deposits of the Pennines, North Wales and the Mendips. See Pattrick & Polya (1993: 355–418).

Ixer, R. A., P. McArdle, C. J. Stanley 1990. Primary gold mineralization within metamorphosed iron ores, southeast Ireland. *Geological Survey of Ireland Bulletin* **4**, 221–6.

Ixer, R. A., R. A. D. Pattrick, C. J. Stanley 1997. Geology, mineralogy and genesis of gold mineralization at Calliachar–Urlar Burn, Scotland. *Institution of Mining and Metallurgy, Transactions B* **106**, 99–108.

Ixer, R. A., R. A. D. Pattrick, R. E. Starkey 1993. Lead–zinc–copper–arsenic–baryte mineralization from Clevedon, near Bristol, England. *Russell Society, Journal* **5**(1), 23–30.

Ixer, R. A., C. J. Stanley, D. J. Vaughan 1979a. Cobalt-, nickel-, and iron-bearing sulpharsenides from the North of England. *Mineralogical Magazine* **43**, 389–95.

Ixer, R. A., A. G. Tindle, R. Chapman 2005a. Algodonite and domeykite from New Cliffe Hill quarry, Stanton under Bardon, Leicestershire. *UK Journal of Mines and Minerals* **26**, 43–51.

Ixer, R. A., A. G. Tindle, D. Liss, A. D. Chambers 2005b. On the occurrence of ilvaite in the pegmatitic facies of the Whin Sill, Teesdale, North Pennines, England. *UK Journal of Mines and Minerals* **25**, 39–46.

Ixer, R. A., P. Turner, B. Waugh 1979b. Authigenic iron and titanium oxides in Triassic redbeds (St Bees Sandstone), Cumbria, northern England. *Geological Journal* **14**, 179–92.

Ixer, R. A., B. Young, C. J. Stanley 1996. Bismuth-bearing assemblages from the Northern Pennine orefield. *Mineralogical Magazine* **60**, 317–24.

Jackman, P. S. 1996. Mineralogy of Snelston mine, near Ashbourne, Derbyshire, England. *Russell Society, Journal* **6**(2), 75–80.

Jackson, B. 1982. An occurrence of gem quality elbaite from Glenbuchat, Aberdeenshire, Scotland. *Journal of Gemmology* **18**, 121–5.

——— 1984. Sapphire from Loch Roag, Isle of Lewis, Scotland. *Journal of Gemmology* **19**, 336–42.

——— 1990. Queitite, a first Scottish occurrence. *Scottish Journal of Geology* **26**, 57–8.

Jackson, N. J. 1974. Gryll's Bunny, a "tin floor" at Botallack. *Ussher Society, Proceedings* **3**, 186–8.

——— 1978. The Halvosso pegmatites. *Ussher Society, Proceedings* **4**, 190–91.

Jackson, N. J. & A. H. Rankin 1976. Fluid inclusion studies at St Michael's Mount. *Ussher Society, Proceedings* **4**, 430–34.

Jackson, N. J., A. N. Halliday, S. M. F. Sheppard, J. G. Mitchell 1982. Hydrothermal activity in the St Just mining district, Cornwall, England. In *Metallization associated with acid magmatism*, A. M. Evans (ed.), 137–79. Chichester: John Wiley.

Jackson, N. J., J. McM. Moore, A. H. Rankin 1977. Fluid inclusions and mineralisation at Cligga Head. *Geological Society of London, Journal* **134**, 343–9.

Jambor, J. L., J. E. Dutrizac, A. C. Roberts, J. D. Grice, J. T. Szymanski 1996. Clinoatacamite, a new polymorph of $Cu_2(OH)_3Cl$, and its relationship to paratacamite and "anarakite". *Canadian Mineralogist* **34**, 61–72.

James, C. C. 1947. Uranium ores in Cornish mines. *Royal Geological Society of Cornwall, Transactions* **17**, 256–68.

Jameson, R. 1840. Notice of greenockite, a new mineral species of the Order Blende. *Edinburgh New Philosophical Journal* **28**, 390–92.

Jassim, R. Z., R. A. D. Pattrick, M. J. Russell 1983. On the origin of the silver + copper + cobalt + baryte mineralization of Ochil Hills, Scotland: a sulphur isotope study. *Institution of Mining and Metallurgy, Transactions B* **92**, 213–16.

Jeans, C. V. 1978. The origin of the Triassic clay assemblages of Europe with special reference to the Keuper Marl and Rhaetic of parts of England. *Royal Society of London, Philosophical Transactions* **289**, 549–639.

Jefferies, N. L. 1984. The radioactive accessory mineral assemblage of the Carnmenellis granite, Cornwall. *Ussher Society, Proceedings* **6**, 35–40.

——— 1985a. The distribution of the rare earth elements within the Carnmenellis pluton, Cornwall. *Mineralogical Magazine* **49**, 495–504.

——— 1985b. The origin of sillimanite-bearing pelitic xenoliths within the Carnmenellis pluton, Cornwall. *Ussher Society, Proceedings* **6**, 229–36.

——— 1985c. Uraninite within the Carnmenellis pluton, Cornwall. In *High heat production (HHP) granites, hydrothermal circulation and ore genesis* (C. Halls chairman), 163–8. London: Institution of Mining and Metallurgy, London.

Jeffrey, C. A. 1997. Replacement mineralization styles and breccia dome formation at Dirtlow Rake fluorite–barite deposit, Castleton, England. *Institution of Mining and Metallurgy, Transactions B* **106**, 15–23.

——— 2001. Supergene processes in galena-replacement mineralization from the Longstone Edge vein system, Derbyshire. *Yorkshire Geological Society, Proceedings* **53**, 197–206.

Jenkins, D. A. & D. F. Ball 1964. Pumpellyite in Snowdonian soils and rocks. *Mineralogical Magazine* **33**, 1093–1096.

Jenkins, D. A. & D. B. Johnson 1993. Abandoned metal mines: a unique mineralogical and microbiological resource. *Russell Society, Journal* **5**(1), 40–47.

Jenkins, D. A., D. B. Johnson, C. Freeman 2000. Mynydd Parys Cu–Pb–Zn mines: mineralogy, microbiology and acid mine drainage. *Environmental mineralogy: microbial interactions, anthropogenic influences, contaminated land and waste management*, J. D. Cotter-Howells, L. S. Campbell, E. Valsami-Jones, M. Batchelder (eds), 161–79. Mineralogical Society Series 9, the Society, London.

Jenkinson, H. I. 1875. *Practical guide to the English Lake District* (4th edn). London: Edward Stanford. [A chapter on mineralogy (pp. lxxi–xciii) was contributed by Bryce M. Wright Jr]

Jewson, C. 1997. Field meeting report: Hingston Down quarry, Cornwall. *Russell Society, Newsletter* **31**, 11–12.

Jiang, W-T. & D. R. Peacor 1993. Formation and modification of metastable intermediate sodium potassium mica, paragonite, and muscovite in hydrothermally altered metabasites from northern Wales. *American Mineralogist* **78**, 782–93.

Jiang, W-T., D. R. Peacor, R. J. Merriman, B. Roberts 1990. Transmission and analytical electron microscopic study of mixed-layer illite/smectite formed as an apparent replacement product of diagenetic illite. *Clay and Clay Minerals* **38**, 449–68.

Jobbins, E. A., F. G. Dimes, R. A. Binns, M. H. Hey, S. J. B. Reed 1966. The Barwell meteorite. *Mineralogical Magazine* **35**, 881–902.

Joel, E. C., J. J. Taylor, R. A. Ixer, M. Goodway 1996. Lead isotope analysis and the Great Orme mine. In *Archaeological sciences 1995*, A. Sinclair, E. Slater, J. Gowlett (eds), 123–31. Oxford: Oxbow.

Johnsen, O. & J. D. Grice 1999. The crystal chemistry of the eudialyte group. *Canadian Mineralogist* **37**, 865–91.

Johnsen, O., G. Ferraris, R. A. Gault, J. D. Grice, A. R. Kampf, I. V. Pekov 2003. The nomenclature of eudialyte-group minerals. *Canadian Mineralogist* **41**, 785–94.

Johnson, D. B., W. I. Kelso, D. A. Jenkins 1979. Bacterial streamer growth in a disused pyrite mine. *Environmental Pollution* **18**, 107–117.

Johnson, T. E., N. F. C. Hudson, G. T. R. Droop 2000. Wollastonite-bearing assemblages from the Dalradian at Fraserburgh, northeast Scotland, and their bearing on the emplacement of garnetiferous granitoid sheets. *Mineralogical Magazine* **64**, 1165–76.

Johnston, J. F. W. 1835. On the dimorphism of barytocalcite. *Philosophical Magazine (3rd series)* **6**, 1–4.

Johnston, R. W. 1949. Clinozoisite from Camaderry Mountain, Co. Wicklow. *Mineralogical Magazine* **28**, 505–515.

——— 1953. The olivines of the Garbh Eilean sill, Shiant Isles. *Geological Magazine* **90**, 161–71.

Joly, J. 1886. On the occurrence of harmotome at Glendalough, Co. Wicklow. *Royal Dublin Society, Scientific Proceedings* **5**, 165–8.

Jones, A. D. 1981. Millerite. *Mineral Realm* **1**(2), 22–6.

——— 1983. Nant-y-Cagl. *Mineral Realm* **3**(2/3), 42–76.

——— 1987. The minerals of Llechweddhelg. *UK Journal of Mines and Minerals* **3**, 25–7.

Jones, E. M., C. M. Rice, J. R. Tweedie 1983. Lower Proterozoic stratiform Cu–Zn–Au mineralization near Gairloch, Wester Ross, northwest Scotland. *Institution of Mining and Metallurgy, Transactions B* **92**, 160–61.

——— 1987. Lower Proterozoic stratiform sulphide deposits in Loch Maree Group, Gairloch, northwest Scotland. *Institution of Mining and Metallurgy, Transactions B* **96**, 128–48.

Jones, G. L. & J. Fitzsimons 1992. A pure dolomite deposit northeast of Cahir, Co. Tipperary. See Bowden et al. (1992: 327–34).

Jones, G. V. & S. F. Brand 1986. The setting, styles of mineralization and mode of origin of the Ballinalack Zn–Pb deposit. See Andrew et al. (1986: 355–76).

Jones, I. 2006. Old classics and new finds: fluorite from the United Kingdom. In *Fluorite: the collector's choice*, J. Fisher, M. Jarnot, G. Neumeier, A. Pasto, G. Staebler, T. Wilson (eds), 28–37. East Hampton, Connecticut: Lithographie.

Jones, J. 1982. Mineralogy of Bage mine. *Peak District Mines Historical Society, Bulletin* **8**(4), 260–61.

Jones, J. A. & N. J. M. Moreton 1977. *The mines and minerals of mid-Wales.* [Published privately by the authors.]

Jones, O. T. 1924. Lead and zinc ores in the slatey rocks of Britain. *Institute of Mining Engineers, Transactions* **66**, 219–42.

Kanaris-Sotiriou, R. & N. S. Angus 1979. Metasomatic reaction between acid pegmatite and orthopyroxenite at Currywongaun, Connemara, Ireland. *Mineralogical Magazine* **43**, 473–8.

Kassi, A. M. & J. A. Weir 1993. Depositional and geotectonic history of the Gala area, eastern Southern Uplands, Scotland. *Royal Society of Edinburgh, Transactions: Earth Sciences* **84**, 161–73.

Kato, T. 1980. The crystal structure of ganophyllite: monoclinic subcell. *Mineralogical Journal* **10**, 1–13.

Keats, W. 1981. *Alteration and replacement of wolframite in the Hemerdon tungsten deposit, Devon.* PhD thesis, Camborne School of Mines.

Keeling, P. S. 1956. Sepiolite at a locality in the Keuper Marl of the Midlands. *Mineralogical Magazine* **31**, 328–38.

Kelley, S. P., G. Turner, A. W. Butterfield, T. J. Shepherd 1986. The source and significance of argon isotopes in fluid inclusions from areas of mineralization. *Earth and Planetary*

Science Letters **79**, 303–318.

Kelly, J. G., C. J. Andrew, J. H. Ashton, M. B. Boland, G. Earls, L. Fusciardi, G. Stanley (eds) 2004. *Europe's major base metal deposits.* Dublin: Irish Association for Economic Geology.

Kellock, E. 1969. Alkaline basic igneous rocks in the Orkneys. *Scottish Journal of Geology* **5**, 140–53.

Kemp, S. J., C. A. Rochelle, R. J. Merriman 2005. Back-reacted saponite in Jurassic mudstones and limestones intruded by a Tertiary sill, Isle of Skye. *Clay Minerals* **40**, 263–82.

Kempe, D. R. C. 1968. Kaersutite from the minverite of Cornwall. *Mineralogical Magazine* **36**, 873–5.

Kendall, A. 1988. Aragonite in Ogof Daren Cilau. *Cave Science* **15**, 83–4.

Kendall, J. D. 1884. Mineral veins of the Lake District. *Manchester Geological Society, Transactions* **17**, 292–341.

——— 1893. *The iron ores of Great Britain and Ireland.* London: Crosby Lockwood.

Kennan, P. S. 1997. Tourmaline in the aureole of the Leinster granite, southeast Ireland: towards a model for tourmalinization around granites? *Irish Journal of Earth Sciences* **16**, 33–44.

Kennan, P. S., P. McArdle, F. M. Williams, E. Doyle 1986. A review of metal deposits associated with the Leinster granite, southeast Ireland and a model for their genesis. See Andrew et al. (1986: 201–210).

Kennedy, W. Q. 1936. An occurrence of greenalite chert in the Ordovician rocks of the Southern Uplands of Scotland. *Mineralogical Magazine* **24**, 433–6.

Kennedy, W. Q. & T. R. M. Lawrie 1943. *Commercial mica in Scotland, part 2: preliminary description of some occurrences north of the Great Glen.* Wartime pamphlet 34, Geological Survey of Great Britain, Edinburgh.

Kerr, A. C. 1998. Mineral chemistry of the Mull–Morvern Tertiary succession, western Scotland. *Mineralogical Magazine* **62**, 295–312.

Kerrick, D. M. 1987. Fibrolite in contact aureoles of Donegal, Ireland. *American Mineralogist* **72**, 240–54.

Kettaneh, Y. A. & J. P. N. Badham 1978. Mineralization and paragenesis at the Mount Wellington mine, Cornwall. *Economic Geology* **73**, 486–95.

Khan, I. H. 1972. *Geochemistry of the aureole of the Land's End granite.* PhD thesis, University of Birmingham.

Kinahan, G. H. 1889. Economic geology of Ireland. *Royal Dublin Society, Scientific Proceedings* **5**, 200–317.

Kincaid, J. 1857. A section across the Coal Beds of Leitrim. *Geological Society of Dublin, Journal* **7**, 301–302.

Kinch, E. 1888. On dufrénite from Cornwall. *Mineralogical Magazine* **8**, 112–15.

Kinch, E. & F. H. Butler 1886. On a new variety of mineral from Cornwall; with a note on the crystallographical characters by H. A. Miers. *Mineralogical Magazine* **7**, 65–70.

King, B. C. 1942. The Cnoc nan Cuilean area of the Ben Loyal igneous complex. *Geological Society of London, Quarterly Journal* **98**, 147–85.

King, P. M. 1976. *The secondary minerals of the Tertiary lavas of northern and central Skye – zeolite zonation patterns, their origin and formation.* PhD thesis, University of Aberdeen.

King, R. J. 1959. The mineralization of the Mountsorrel granodiorite. *Leicestershire Literary and Philosophical Society, Transactions* **53**, 18–30.

——— 1966. Epi-syngenetic mineralization in the English Midlands. *Mercian Geologist* **1**, 291–301.

——— 1967. The minerals of Leicestershire, section 1 – the elements. *Leicestershire Literary and Philosophical Society, Transactions* **61**, 55–64.

——— 1973. *The mineralogy of Leicestershire.* PhD thesis, University of Leicester.

——— 1982a. A new occurrence of scarbroite in Britain. *Russell Society, Journal* **1**(1), 9–18.

——— 1982b. The Boltsburn mine, Weardale, County Durham, England. *Mineralogical Record* **13**, 5–18.

——— 1983. The occurrence of galena in Leicestershire. *Russell Society, Journal* **1**(2), 27–47.

——— 1991. Lead–molybdenum mineralization in an ancient cave in north Leicestershire. *Russell Society, Journal* **4**(1), 9–12.

——— 1993. The occurrence of sphalerite in Leicestershire, England. *Russell Society, Journal* **5**(1), 15–21.

——— 1997. A remarkable occurrence of thenardite (Na_2SO_4) in Gloucestershire. *Russell Society, Newsletter* **30**, 20.

——— 1998a. Tamarugite on the Isle of Wight, UK. *Mineralogical Magazine* **62**, 371–2.

——— 1998b. The occurrence of "copper ore" chalcopyrite ($CuFeS_2$) in Leicestershire. *Leicestershire Literary and Philosophical Society, Transactions* **92**, 30–33.

——— 1999. Minerals explained, 26: the garnets. *Geology Today* **15** (Jan/Feb), 34–7.

——— 2002. Minerals explained, 36: turquoise. *Geology Today* **18** (May/June), 110–14.

King, R. J. & A. M. Evans 1964. An occurrence of morenosite in Ireland. *Mineralogical Magazine* **33**, 1110–13.

King, R. J. & J. A. Dixon 1971. A new occurrence of vanadium minerals in Leicestershire. *Mineralogical Magazine* **38**, 488–92.

King, R. J. & T. D. Ford 1969. Mineral localities at the base of the Trias in Leicestershire and Derbyshire. *Mercian Geologist* **3**, 85–8.

King, R. J. & B. A. Ludlam 1969. The quest for a lost lead mine in Leicestershire. *Peak District Mines Historical Society, Bulletin* **4**, 3–28.

King, R. J. & R. N. Wilson 1976. An occurrence of vésigniéite in Leicestershire. *Mineralogical Magazine* **40**, 533–5.

Kingsbury, A. W. G. 1935. Notes of some recent finds of minerals on the Mendip Hills. *Somersetshire Archaeological and Natural History Society, Proceedings* **80**, 51–6.

——— 1941. Mineral localities on the Mendip Hills, Somerset. *Mineralogical Magazine* **26**, 67–80.

——— 1954. New occurrences of rare copper and other minerals in Devon and Cornwall. *Royal Geological Society of Cornwall, Transactions* **18**, 386–406.

——— 1956. The rediscovery of churchite in Cornwall. *Mineralogical Magazine* **31**, 282.

——— 1957a. Rockbridgeite from Cornwall and Devon. *Mineralogical Magazine* **31**, 429–30.

——— 1957b. A new occurrence of nadorite in Cornwall. *Mineralogical Magazine* **31**, 499.

——— 1957c. New occurrences of phosgenite. *Mineralogical Magazine* **31**, 500–501.

——— 1958. Two beryllium minerals new to Britain: euclase and herderite. *Mineralogical Magazine* **31**, 815–17.

——— 1959a. Two minerals new to Britain: hübnerite and adamite [exhibit: general meeting of the Mineralogical Society, at Burlington House, London, 23 January 1958]. *Mineralogical Magazine* **32**, xliv.

——— 1959b. A new occurrence of fluellite in Cornwall, and its paragenesis. *Royal Geological Society of Cornwall, Transactions* **19**, 42–51.

——— 1961. Beryllium minerals in Cornwall and Devon: helvine, genthelvite and danalite. *Mineralogical Magazine* **32**, 921–40.

——— 1963. Strengite [exhibit: general meeting of the Mineralogical Society, Burlington House, London, 7 November 1963]. *Mineralogical Magazine* **33**, lxiv.

——— 1964a. Bismutoferrite [exhibit: general meeting of the Mineralogical Society, Burlington House, London, 7 November 1963]. *Mineralogical Magazine* **33**, xc.

——— 1964b. Some minerals of special interest in southwest England. In *Present views of some aspects of the geology of Cornwall and Devon*, K. F. G. Hosking & G. L. Shrimpton (eds), 243–66. Penzance: Royal Geological Society of Cornwall.

——— 1965a. Jarosite specimens [exhibit: general meeting of the Mineralogical Society, Burlington House, London, 12 March 1964]. *Mineralogical Magazine* **35**, li.

——— 1965b. Beryllium minerals new to Britain [exhibit: general meeting of the Mineralogical Society, Burlington House, London, 5 November 1964]. *Mineralogical Magazine* **35**, lvii.

——— 1965c. Tellurbismuth and meneghinite, two minerals new to Britain. *Mineralogical Magazine* **35**, 424–6.

——— 1966. Sir Arthur Russell (memorial). *Mineralogical Magazine* **35**, 673–7.

——— 1969. Mercury minerals from northwest Yorkshire [exhibit: general meeting of the Mineralogical Society, Burlington House, London, 14 March 1968]. *Mineralogical Magazine* **37**, xlii.

Kingsbury, A. W. G. & J. Hartley 1955. On the occurrence of the rare copper molybdate, lindgrenite at Brandy Gill, Carrock Fell, Cumberland. *Mineralogical Magazine* **30**, 723–6.

——— 1956a. New occurrences of vanadium minerals (mottramite, descloizite and vanadinite) in the Caldbeck area of Cumberland. *Mineralogical Magazine* **31**, 289–95.

——— 1956b. Cosalite and other lead sulpho-salts at Grainsgill, Carrock Fell, Caldbeck, Cumberland. *Mineralogical Magazine* **31**, 296–300.

——— 1956c. Atacamite from Cumberland and Cornwall. *Mineralogical Magazine* **31**, 349–50.

——— 1957a. Childrenite from the Lake District, Cumberland. *Mineralogical Magazine* **31**, 498.

——— 1957b. New occurrences of arseniosiderite. *Mineralogical Magazine* **31**, 499–500.

——— 1957c. New occurrences of rosasite in Britain. *Mineralogical Magazine* **31**, 501–502.

——— 1957d. Carpholite from Cumberland and Cornwall. *Mineralogical Magazine* **31**, 502.

——— 1957e. Serpierite from the Lake District. *Mineralogical Magazine* **31**, 604–605.

——— 1957f. Beaverite from the Lake District. *Mineralogical Magazine* **31**, 700–702.

——— 1958. Jarosite and natrojarosite from the Lake District. *Mineralogical Magazine* **31**, 813–15.

——— 1960. Carminite and beudantite from the northern part of the Lake District and from Cornwall. *Mineralogical Magazine* **32**, 423–32.

Kingsmill, P. 1982. Selenite from Shelford quarry, Kent. *Mineral Realm* **1**(5), 24–6.

Kinnaird, J. A., R. A. Ixer, B. Barreiro, P. A. Nex 2002. Contrasting sources for lead in Cu-polymetallic and Zn–Pb mineralization in Ireland: constraints from lead isotopes. *Mineralium Deposita* **37**, 495–511.

Kinny, P. & C. Friend 1994. A new insight into Lewisian chronology. *Mineralogical Magazine* **58A**, 481–2.

Kissin, S. A. & D. R. Owens 1989. The relatives of stannite in the light of new data. *Canadian*

Mineralogist **27**, 673–88.

Kitchen, D. E. 1984. Pyrometamorphism and the contamination of basaltic magma at Tieveragh, Co. Antrim. *Geological Society of London, Journal* **141**, 733–45.

——— 1985. The partial melting of basalt and its enclosed mineral-filled cavities at Scawt Hill, Co. Antrim. *Mineralogical Magazine* **49**, 655–62.

Kneller, B. C. 1987. Dalradian of Fraserburgh. See Trewin et al. (1987: 99–105).

Kneller, B. C. & C. Gillen 1987. Aberdeen city and environs. See Trewin et al. (1987: 55–64).

Knight, J. R. & R. W. Barstow 1970. Posnjakite from Cornwall. *Mineralogical Magazine* **37**, 740.

Knill, D. C. (including footnote by P. G. Embrey) 1960. Thaumasite from Co. Down, Northern Ireland. *Mineralogical Magazine* **32**, 416–18.

——— 1969. The Permian igneous rocks of Devon. *Geological Survey of Great Britain, Bulletin* **29**, 115–38.

Knox, D. A. & N. J. Jackson 1990. Composite granite intrusions of southwest Dartmoor, Devon. *Ussher Society, Proceedings* **7**, 246–51.

Kokelaar, B. P. 1986. Petrology and geochemistry of the Rhobell volcanic complex: amphibole-dominated fractionation at an early Ordovician volcano in North Wales. *Journal of Petrology* **27**, 887–914.

Kokelaar, B. P., R. E. Bevins, R. A. Roach 1985. Submarine silicic volcanism and associated sedimentary and tectonic processes, Ramsey Island, southwest Wales. *Geological Society of London, Journal* **142**, 591–613.

Krivovichev, S. V. & P. C. Burns 2000. Crystal chemistry of basic lead carbonates, II: crystal structure of synthetic "plumbonacrite". *Mineralogical Magazine* **64**, 1069–1075.

——— 2006. The crystal structure of $Pb_8O_5(OH)_2Cl_4$, a synthetic analogue of blixite? *Canadian Mineralogist* **44**, 515–22.

Krivovichev, S. V., D. Yu. Chernyshov, N. Döbelin, T. Armbruster, V. Kahlenberg, R. Kaindl, G. Ferraris, R. Tessadri, G. Kaltenhauser 2006. Crystal chemistry and polytypism of tyrolite. *American Mineralogist* **91**, 1378–84.

Kucha, H. 1988. Zn–Pb sulphides as rim cements, filling cements and replacements of carbonate sediments, Moyvoughly, Ireland – a product of two convective cells. *Institution of Mining and Metallurgy, Transactions B* **97**, 64–76.

Kucha, H. & A. Wieczorek 1984. Sulfide–carbonate relationships in the Navan (Tara) Zn–Pb deposit, Ireland. *Mineralium Deposita* **19**, 208–216.

Lacey, E. 1862. On specimens of galena from Axe Edge near Buxton. *Manchester Geological Society, Transactions* **3**, 135.

Lacroix, A. 1913. *Minéralogie de la France et de ses colonies* [2nd supplement to vol. 5]. Paris: Béranger.

Laffoley, N. 1998. Minerals from Sark's Hope mine. *Transactions La Société Guernsey* **24**, 273–83.

Lambert, R. St J. 1963. Mineral reactions under polymetamorphic conditions at Lewisian–Moine contacts. In *The British Caledonides*, M. R. W. Johnson & F. H. Stewart (eds), 177–88. Edinburgh: Oliver & Boyd.

Lamplugh, G. W. 1903. *Economic geology of the Isle of Man, with special reference to the metalliferous mines*. Memoir, Geological Survey of Great Britain and Ireland, London.

Laouar, R., A. J. Boyce, A. E. Fallick, B. E. Leake 1990. A sulphur isotope study on selected Caledonian granites of Britain and Ireland. *Geological Journal* **25**, 359–69.

Larter, R. C., A. J. Boyce, M. J. Russell 1981. Hydrothermal pyrite chimneys from the Ballynoe barite deposit, Silvermines, Co. Tipperary, Ireland. *Mineralium Deposita* **16**, 309–318.

Law, A. D. 1982. Studies of the orthoamphiboles, III: hydroxyl spectra of gedrites. *Mineralogical Magazine* **45**, 63–71.

Lawrence, D. J. D., B. C. Webb, B. Young, D. E. White 1986. *The geology of the late Ordovician and Silurian rocks (Windemere Group) in the area around Kentmere and Crook*. Report 18.5, British Geological Survey, Keyworth.

Lawrence, J. A. 1990. An occurrence of djurleite at West Caradon mine, Cornwall. *Russell Society, Journal* **3**(2), 81–4.

Lawton, A. 1994. Beaverite from Alderley Edge, Cheshire. *UK Journal of Mines and Minerals* **14**, 9.

Leake, B. E. 1960. Compilation of chemical analyses and physical constants of natural cordierites. *American Mineralogist* **45**, 282–98.

——— 1967. Zoned garnet from the Galway granite and its aplites. *Earth and Planetary Science Letters* **3**, 311–16.

——— 1968. Amphibole from the Moine nappe in Skye. *Scottish Journal of Geology* **4**, 287–9.

——— 1972. Garnetiferous striped amphibolites from Connemara, western Ireland. *Mineralogical Magazine* **38**, 649–65.

——— 1974. The crystallization history and mechanism of emplacement of the western part of the Galway granite, Connemara, western Ireland. *Mineralogical Magazine* **39**, 498–513.

——— 1978. Nomenclature of amphiboles. *Mineralogical Magazine* **42**, 533–63.

——— 1998. Widespread secondary Ca garnet and other Ca silicates in the Galway granite and its satellite plutons caused by fluid movements, western Ireland. *Mineralogical Magazine* **62**, 381–6.

Leake, B. E. & Ahmed Y. Said 1994. Hornblende barometry of the Galway batholith, Ireland: an empirical test. *Mineralogy and Petrology* **51**, 243–50.

Leake, B. E., and 21 other authors 1997a. Nomenclature of amphiboles: report of the subcommittee on amphiboles of the International Mineralogical Association, Commission on New Minerals and Mineral Names. *Mineralogical Magazine* **61**, 295–321.

Leake, B. E., and 10 other authors 2004. Nomenclature of amphiboles: additions and revisions to the International Mineralogical Association's amphibole nomenclature. *European Journal of Mineralogy* **16**, 191–6.

Leake, R. C. & M. J. Brown 1979. Porphyry-style copper mineralization at Black Stockarton Moor, southwest Scotland. *Institution of Mining and Metallurgy, Transactions B* **88**, 177–81.

Leake, R. C. & D. G. Cameron 1996. *Exploration for gold in the Thornhill Basin, southern Scotland*. Data release 19, Mineral Reconnaissance Program, British Geological Survey, Keyworth.

Leake, R. C., H. A. Auld, P. Stone, C. E. Johnson 1981. *Gold mineralization at the southern margin of the Loch Doon granitoid complex, southwest Scotland*. Report 46, Mineral Reconnaissance Program, Institute of Geological Sciences, London.

Leake, R. C., D. J. Bland, C. Cooper 1993. Source characterization of alluvial gold from mineral inclusions and internal compositional variation. *Institution of Mining and Metallurgy, Transactions B* **102**, 65–82.

Leake, R. C., D. J. Bland, M. T. Styles, D. G. Cameron 1991. Internal structure of Au–Pd–Pt grains from south Devon, in relation to low-temperature transport and deposition. *Institution of Mining and Metallurgy, Transactions B* **100**, 159–78.

Leake, R. C., D. G. Cameron, D. J. Bland, M. T. Styles, N. J. Fortey 1997b. *The potential for gold mineralization in the British Permian and Triassic redbeds and their contacts with underlying rocks*. Report 144, Mineral Reconnaissance Program, British Geological Survey, Keyworth.

Leake, R. C., D. G. Cameron, D. J. Bland, M. T. Styles, K. E. Rollin 1989. Exploration for gold in south Devon, England. *Institution of Mining and Metallurgy, Transactions B* **98**, 76–7.

Leake, R. C., D. G. Cameron, R. C. Scrivener, D. J. Bland 1994. *Exploration for gold in the Crediton trough, Devon*. Report 134., Mineral Reconnaissance Program, British Geological Survey, Keyworth.

Leake, R. C., R. J. Chapman, D. J. Bland, E. Condliffe, M. T. Styles 1997c. Microchemical characterization of alluvial gold from Scotland. *Institution of Mining and Metallurgy, Transactions B* **106**, 85–98.

Leake, R. C., R. J. Chapman, D. J. Bland, P. Stone, D. G. Cameron, M. T. Styles 1998. The origin of alluvial gold in the Leadhills area of Scotland: evidence from interpretation of internal chemical characteristics. *Journal of Geochemical Exploration* **63**, 7–36.

Leat, P. T. & R. S. Thorpe 1986. Ordovician volcanism in the Welsh Borderland. *Geological Magazine* **123**, 629–40.

Le Bas, M. J. 1965. The contamination of a gabbro by Carboniferous Limestone at Carlingford, Co. Louth. *Mineralogical Magazine* **34**, 292–302.

LeBoutillier, N. G. 1996. *South Crofty mine geology and mineralization*. Camborne: Penhellick Publications.

LeBoutillier, N. G., R. K. Shail, C. Jewson 2000. The cassiterite–anatase–ilmenite–monazite–sulphide–chlorite assemblage of North Tincroft lode. *Ussher Society, Proceedings* **10**, 50–57.

Lecornu, L. 1879. Memoirs sur le Calcaire Carbonifère et les filons du plomb du Derbyshire. *Annales des Mines, Series 7* **15**, 5–53.

Lee, G. W. 1920. The Mesozoic rocks of Applecross, Raasay and northeast Skye. In *The iron ores of Scotland*, M. Macgregor, G. W. Lee, G. V. Wilson (eds), 198. Special Reports on the Mineral Resources of Great Britain XI, Memoir, Geological Survey of Great Britain, Edinburgh.

Lee, M. R., K. A. Waldron, I. Parsons 1995. Exsolution and alteration microtextures in alkali feldspar phenocrysts from the Shap granite. *Mineralogical Magazine* **59**, 63–78.

Lees, G. M. & A. H. Taitt 1946. The geological results of the search for oilfields in Great Britain. *Geological Society of London, Quarterly Journal* **102**, 255–317.

Legg, I. C., J. F. Pyne, C. Nolan, P. McArdle, A. M. Flegg, P. J. O'Connor 1985. *Mineral localities in the Dalradian and associated igneous rocks of County Donegal, Republic of Ireland, and of Northern Ireland*. Report Series 85.3 (Mineral Resources), Geological Survey of Ireland, Dublin.

Lemon, G. G. 1966. Serpentinites in the northeast Ox Mountains, Eire. *Geological Magazine* **103**, 124–37.

Le Neve Foster, C. 1876. Notes on new minerals and mineral localities in Cornwall and Devon. *Mineralogical Magazine* **1**, 8–10.

——— 1877. On some new mineral localities in Cornwall and Devon. *Mineralogical Maga-*

zine **1**, 73–6.

——— 1878a. On some tin stockworks in Cornwall. *Geological Society of London, Quarterly Journal* **34**, 654–9.

——— 1878b. Remarks on the lode at Wheal Mary Ann, Menheniot. *Royal Geological Society of Cornwall, Transactions* **9**, 152–7.

——— 1878c. Remarks upon the tin deposits of East Wheal Lovell. *Royal Geological Society of Cornwall, Transactions* **9**, 167–76.

——— 1882. On the occurrence of cobalt ores in Flintshire. *Royal Geological Society of Cornwall, Transactions* **10**, 107–112.

Leppington, C. M. & D. I. Green 1998. Antimonian claudetite from Wet Swine Gill, Caldbeck Fells, Cumbria, England. *Russell Society, Journal* **7**(1), 36–7.

Leslie, A. G. 1987. Southern contact of the Insch mass, and the Bennachie granite. See Trewin et al. (1987: 179–84).

——— 1988. A chloritoid-bearing paragenesis in the Macduff Slates of Central Buchan. *Scottish Journal of Geology* **24**, 223–32.

Lettsom, W. G. 1878. On rhabdophane, a new mineral. *Philosophical Magazine (5th series)* **13**, 527–8.

Leverett, P. & P. A. Williams 2007. Buttgenbachite from Cornwall: a single-crystal X-ray study. *Russell Society, Journal* **10**, 47–50.

Lewis, A. 1990. Underground exploration of the Great Orme Copper Mines. In *Early Mining in the British Isles*, P. Crew, S. Crew (eds), 5–20. Plas Tan y Blwch.

Lewis, H. P. 1931. A sandstone with fluorspar cement and other sandstones from West Cumberland. *Geological Magazine* **68**, 543–57.

Lewis, J. S. 1997. *A richly yielding piece of ground – the story of Fowey Consols mine near St Blazey*. St Austell: Cornish Hillside Publications.

Lewis, W. J. 1882. On a crystal of stephanite from Wheal Newton. *Cambridge Philosophical Society, Proceedings* **4**(4), 240–45.

——— 1967. *Lead mining in Wales*. Cardiff: University of Wales Press.

Lewis, W. J. & A. L. Hall 1900. On some remarkable composite crystals of copper pyrites from Cornwall. *Mineralogical Magazine* **12**, 324–32.

Leyshon, P. R. & P. C. D. Cazalet 1976. Base-metal exploration programme in Lower Palaeozoic volcanic rocks, Co. Tyrone, Northern Ireland. *Institution of Mining and Metallurgy, Transactions B* **85**, 91–9.

Libbrecht, K. 2006. *Ken Libbrecht's field guide to snowflakes*. St Paul, Minnesota: Voyageur Press.

Liferovich, R. P., A. J. Locock, R. H. Mitchell, A. K. Shpachenko 2006a. The crystal structure of stronalsite and a redetermination of the structure of banalsite. *Canadian Mineralogist* **44**, 533–46.

Liferovich, R. P., R. H. Mitchell, D. R. Zozulya, A. K. Shpachenko 2006b. Paragenesis and composition of banalsite, stronalsite and their solid solution in nepheline syenite and ultramafic alkaline rocks. *Canadian Mineralogist* **44**, 929–42.

Lindsey, C. R. 1905. Note on the occurrence of brookite in the Cleveland Ironstone. *Mineralogical Magazine* **14**, 96–8.

Linnaeus, C. 1735. *Systema naturae . . .* Stockholm. [1802: translation by W. Turton, London].

Lister, C. J. 1978. Luxullianite *in situ* within the St Austell granite, Cornwall. *Mineralogical Magazine* **42**, 295–7.

——— 1979. Quartz-cored tourmalines from Cape Cornwall and other localities. *Ussher Society, Proceedings* **4**, 402–418.

Livingstone, A. 1967. A garnet peridotite and a garnet–amphibole pyroxenite from South Harris, Outer Hebrides, and their bearing on the South Harris eclogite facies status. *Mineralogical Magazine* **36**, 380–88.

——— 1974. An occurrence of tacharanite and scawtite in the Huntly gabbro, Aberdeenshire. *Mineralogical Magazine* **39**, 820–21.

——— 1976a. A metamorphosed, layered alpine-type peridotite in the Langavat Valley, South Harris, Outer Hebrides. *Mineralogical Magazine* **40**, 493–9.

——— 1976b. Julgoldite, new data and occurrences: a second recording. *Mineralogical Magazine* **40**, 761–3.

——— 1976c. The paragenesis of spinel– and garnet–amphibole lherzolite in the Rodel area, South Harris. *Scottish Journal of Geology* **12**, 293–300.

——— 1980. Johnsomervilleite, a new transition-metal phosphate mineral from the Loch Quoich area, Scotland. *Mineralogical Magazine* **43**, 833–6.

——— 1988. Reyerite, tobermorite, calcian analcime and bytownite from amygdales in a Skye basalt. *Mineralogical Magazine* **52**, 711–13.

——— 1989a. Low-temperature, hydrothermal garnet associated with zeolites, from basalt lavas near Beith, Ayrshire. *Mineralogical Magazine* **53**, 125–9.

——— 1989b. A calcian analcime–bytownite intergrowth in basalt, from Skye, Scotland, and calcian analcime relationships. *Mineralogical Magazine* **53**, 382–5.

——— 1990a. Matthew Forster Heddle (1828–1897), famous Scottish mineralogist. *Russell Society, Journal* **3**(2), 61–5.

——— 1990b. Copper–aluminium analogues of hydrohonessite and honessite, and woodwardite relationships. *Mineralogical Magazine* **54**, 649–53.

——— 1991. The zinc analogue of ktenasite from Smallcleugh and Brownley Hill mines, Nenthead, Cumbria. *Russell Society, Journal* **4**(1), 13–15.

——— 1992. The composition of stolzite from Wanlockhead and wulfenite from various Scottish localities. *Russell Society, Journal* **4**(2), 55–7.

——— 1993. Glossary of Scottish mineral species – an update. *Scottish Journal of Geology* **29**, 87–101.

——— 1994a. Analyses of calcian phosphatian vanadinite, and apatite high in lead from Wanlockhead, Scotland. *Russell Society, Journal* **5**(2), 124–6.

——— 1994b. An apatite high in lead from Wanlockhead, Strathclyde Region, Scotland. *Mineralogical Magazine* **58**, 159–63.

——— 2002. *Minerals of Scotland: past and present*. Edinburgh: National Museum of Scotland.

Livingstone, A. & W. J. Baird 1991. An occurrence of native sulphur near Craignish, Argyllshire. *Scottish Journal of Geology* **27**, 74.

Livingstone, A. & D. L. Bish 1982. On the new mineral theophrastite, a nickel hydroxide, from Unst, Shetland, Scotland. *Mineralogical Magazine* **46**, 1–5.

Livingstone, A. & P. E. Champness 1993. Brianyoungite, a new mineral related to hydrozincite, from the north of England orefield. *Mineralogical Magazine* **57**, 665–70.

Livingstone, A. & N. Cogger 1966. A new British locality for beudantite: Sandford Hill, Somerset. *Mineralogical Magazine* **35**, 1013–1016.

Livingstone, A. & H. G. Macpherson 1983. Fifth supplementary list of British minerals (Scottish). *Mineralogical Magazine* **47**, 99–105.

Livingstone, A. & H. Sarp 1984. Macphersonite, a new mineral from Leadhills, Scotland, and Saint-Prix, France – a polymorph of leadhillite and susannite. *Mineralogical Magazine* **48**, 277–82.

Livingstone, A., T. F. Bridges, R. E. Bevins 1990. Schulenbergite and namuwite from Smallcleugh mine, Nenthead, Cumbria. *Russell Society, Journal* **3**(1), 23–4.

Livingstone, A., M. T. Frost, P. Suddaby 1973. Conjugate cummingtonite and hornblende at Sgeir nan Sgarbh, South Harris, Outer Hebrides. *Mineralogical Magazine* **39**, 464–6.

Livingstone, A., H. G. Macpherson, B. Jackson 1976. Wroewolfeite and other langite-group minerals from Blackcraig, Kirkcudbrightshire. *Mineralogical Magazine* **40**, 893–4.

Livingstone, A., G. Ryback, E. E. Fejer, C. J. Stanley 1987. Mattheddleite, a new mineral of the apatite group from Leadhills, Strathclyde Region. *Scottish Journal of Geology* **23**, 1–8.

Llewellyn, P. G. & R. Stabbins 1970. The Hathern Anhydrite Series, Lower Carboniferous, Leicestershire, England. *Institution of Mining and Metallurgy, Transactions B* **79**, 1–15.

Lloyd, D. & R. W. Barstow 1983. A new occurrence of aurichalcite and hemimorphite from Cornwall. *Russell Society, Journal* **1**(2), 23–4.

Logan, W. E. 1837. [In a report in the proceedings of the 2nd annual meeting, p. 23]. Swansea Philosophical and Literary Institution, Swansea.

Long, J. V. P. & S. O. Agrell 1965. The cathodoluminescence of minerals in thin section. *Mineralogical Magazine* **34**, 318–26.

Long, J. V. P. & J. D. C. McConnell 1959. A mineralogical application of X-ray absorption microspectroscopy: the hydration of larnite. *Mineralogical Magazine* **32**, 117–27.

Lord, R. A. & H. M. Prichard 1997. Exploration and origin of stratigraphically controlled platinum-group element mineralization in crustal-sequence ultramafics, Shetland ophiolite complex. *Institution of Mining and Metallurgy, Transactions B* **106**, 179–93.

Lord, R. A., H. M. Prichard, C. R. Neary 1994. Magmatic platinum-group element concentrations and hydrothermal upgrading in Shetland ophiolite complex. *Institution of Mining and Metallurgy, Transactions B* **103**, 87–106.

Loughlin, S. C. 1999. Shap Fell Crags. In *Caledonian igneous rocks of Great Britain*, D. Stephenson, R. E. Bevins, W. J. Wadsworth (eds), 211–15. London: Chapman & Hall, on behalf of the Joint Nature Conservation Committee and the British Geological Survey.

Love, L. G., M. L. Coleman, C. D. Curtis 1983. Diagenetic pyrite formation and sulphur isotope fractionation associated with a Westphalian marine incursion, northern England. *Royal Society of Edinburgh, Transactions: Earth Sciences* **74**, 165–82.

Love, G. J., P. D. Kinny, C. R. L. Friend 2004. Timing of magmatism and metamorphism in the Gruinard Bay area of the Lewisian Gneiss Complex: comparisons with the Assynt terrane and implications for terrane accretion. *Contributions to Mineralogy and Petrology* **146**, 620–36.

Loveland, P. J. & V. C. Bendelow 1984. Celadonite-aluminous-glauconite, an example from the Lake District, UK. *Mineralogical Magazine* **48**, 113–17.

Lowe, H. J. 1900. Natrolite from the Coverack district. *Royal Geological Society of Cornwall, Transactions* **12**, 336–7.

Lowry, D. 1991. *The genesis of Late Caledonian granitoid-related mineralization in northern Britain*. PhD thesis, University of St Andrews.

——— 1993. First occurrences of matildite ($AgBiS_2$) associated with Caledonian intrusives in Scotland. *Mineralogical Magazine* **57**, 751–4.

Lowry, D., A. J. Boyce, A. E. Fallick, W. E. Stephens 1997. Sources of sulphur, metals and fluids in granitoid-related mineralization of the Southern Uplands, Scotland. *Institution of Mining and Metallurgy, Transactions B* **106**, 157–68.

Lowry, D., W. E. Stephens, D. A. Herd, C. J. Stanley 1994. Bismuth sulphosalts within quartz veining hosted by the Loch Shin monzogranite, Scotland. *Mineralogical Magazine* **58**, 39–47.

Lowther, J. M., A. B. Balding, F. M. McEvoy, S. Dunphy, P. MacEoin, A. A. Bowden, P. McDermott 2004. The Galmoy Zn–Pb orebodies: structure and metal distribution – clues to the genesis of the deposits. See Kelly et al. (2004: 437–53).

Lucas, M. & Z. M. Zaghloul 1960. Uraninite concentrate from South Crofty mine, Camborne, Cornwall. *Royal Geological Society of Cornwall, Transactions* **19**, 113–15.

Luecke, W. 1981. Lithium pegmatites in the Leinster granite (southeast Ireland). *Chemical Geology* **34**, 195–233.

Luff, K. 1996. The Sussex mineral show, 1995. *UK Journal of Mines and Minerals* **16**, 7.

Lyle, P. & J. Preston 1993. Geochemistry and volcanology of the Tertiary basalts of the Giant's Causeway area, Northern Ireland. *Geological Society of London, Journal* **150**, 109–120.

Lynas, B. D. T. 1973. The Cambrian and Ordovician rocks of the Migneint area, North Wales. *Geological Society of London, Journal* **129**, 481–503.

Macadam, W. I. 1885. On the chemical composition of graphite from Knock, Aberdeenshire. *Mineralogical Magazine* **6**, 172.

——— 1886a. On the chemical composition of the mineral bruiachite, found by Mr Thomas D. Wallace, of Inverness, at Loch Bhruithaich, Inverness-shire. *Mineralogical Magazine* **7**, 42.

——— 1886b. Further note on agalmatolite in Gairloch Parish, Ross-shire. *Mineralogical Magazine* **7**, 74.

MacAlister, D. A. 1909. Note on the association of cassiterite and specular iron in the lodes of Dartmoor. *Geological Magazine* **46**, 402–409.

MacDonald, D. J., F. C. Hawthorne, J. D. Grice 1993. Foitite, $\square[Fe^{2+}_2(Al, Fe^{3+})]Al_6(Si_6O_{18})(BO_3)_3(OH)_4$, a new alkali-deficient tourmaline: description and crystal structure. *American Mineralogist* **78**, 1229–303.

Macdonald, R. & H. E. Belkin 2002. Compositional variation in minerals of the chevkinite group. *Mineralogical Magazine* **66**, 1075–1098.

Macdonald, R., D. J. Fettes, D. Stephenson, C. M. Graham 2005. Basic and ultrabasic volcanic rocks from the Argyll Group (Dalradian) of northeast Scotland. *Scottish Journal of Geology* **41**, 159–74.

Macdonald, R., K. N. Gass, R. S. Thorpe, I. G. Gass 1984. Geochemistry and petrogenesis of the Derbyshire Carboniferous basalts. *Geological Society of London, Journal* **141**, 147–59.

Macdonald, R., D. Millward, B. Beddoe-Stephens, J. Laybourn-Parry 1988a. The role of tholeiitic magmatism in the English Lake District: evidence from dykes in Eskdale. *Mineralogical Magazine* **52**, 459–72.

Macdonald, R., N. M. S. Rock, C. C. Rundle, O. J. Russell 1986. Relationships between late Caledonian lamprophyric, syenitic, and granitic magmas in a differentiated dyke, southern Scotland. *Mineralogical Magazine* **50**, 547–57.

Macdonald, R., L. Wilson, R. S. Thorpe, A. Martin 1988b. Emplacement of the Cleveland dyke: evidence from geochemistry, mineralogy, and physical modelling. *Journal of Petrology* **29**, 559–83.

MacGregor, A. G. 1944. *Barytes in central Scotland.* Wartime pamphlet 38, Geological Survey of Great Britain, Edinburgh.

——— 1960. Garleton Hills and Traprain Law. In *Edinburgh geology – an excursion guide*, G. H. D. Grant & E. K. Walton (eds), 61–9. Edinburgh: Oliver & Boyd.

MacGregor, A. G. & F. R. Ennos 1922. The Traprain Law phonolite. *Geological Magazine* **59**, 514–23.

MacGregor, A. G., A. Herriot, B. C. King 1972. *Excursion guide to the geology of Arran.* Glasgow: Geological Society of Glasgow.

MacGregor, M. & J. Phemister 1972. *Geological excursion guide to the Assynt district of Sutherland* (3rd edn). Edinburgh: Edinburgh Geological Society.

Macintyre, R. M., R. A. Cliff, N. A. Chapman 1981. Geochronological evidence for phased volcanic activity in Fife and Caithness necks, Scotland. *Royal Society of Edinburgh, Transactions: Earth Sciences* **72**, 1–7.

Mackay, R. A. 1959. The Leadhills Wanlockhead mining district. See Higham (1959: 49–64).

Mackenzie, K. J. D. & J. M. Green 1971. The cause of colouration in Derbyshire Blue John banded fluorite and other blue banded fluorites. *Mineralogical Magazine* **38**, 459–70.

Mackenzie, R. C. 1957. Saponite from Allt Ribhein, Fiskavaig Bay, Skye. *Mineralogical Magazine* **31**, 672–80.

Mackenzie, R. C. & A. A. Milne 1953. The effect of grinding on micas, I: muscovite. *Mineralogical Magazine* **30**, 178–85.

Mackenzie, R. C., G. F. Walker, R. Hart 1949. Illite occurring in decomposed granite at Ballater, Aberdeenshire. *Mineralogical Magazine* **28**, 704–713.

Mackie, W. 1925. Dumortierite in British rocks. *Edinburgh Geological Society, Transactions* **11**, 352.

——— 1928. The heavier accessory minerals in the granites of Scotland. *Edinburgh Geological Society, Transactions* **12**, 22–40.

MacLeod, G. 1992. Zoned manganiferous garnets of magmatic origin from the Southern Uplands of Scotland. *Mineralogical Magazine* **56**, 115–16.

Macpherson, H. G. 1983. References for, and updating of, L. J. Spencer's first and second supplementary lists of British minerals. *Mineralogical Magazine* **47**, 243–57.

——— 1989. *Agates.* London: British Museum (Natural History).

Macpherson, H. G. & A. Livingstone 1974. Nickel hydroxides from Unst, Shetland. *Mineralogical Magazine* **39**, 718–20.

——— 1982. Glossary of Scottish mineral species 1981. *Scottish Journal of Geology* **18**, 1–47.

Magraw, D. 1978. New boreholes into Permian beds off Northumberland and Durham. *Yorkshire Geological Society, Proceedings* **42**, 157–83.

Mallard, M. E. 1890. Sur la lussatite, nouvelle variété minérale cristallisée de silice. *La Société Français Minéralogie, Bulletin* **13**, 63–6.

Mallet, J. W. 1859. On brewsterite. *Philosophical Magazine (4th series)* **18**, 218–20.

Maltman, A. J. 1977. Serpentinites and related rocks of Anglesey. *Geological Journal* **12**, 113–27.

Mandarino, J. A. 1999. *Fleischer's glossary of mineral species 1999.* Tucson: Mineralogical Record.

Mandarino, J. A. & M. E. Back 2004. *Fleischer's glossary of mineral species 2004.* Tucson: Mineralogical Record.

Mange, M. A., J. F. Dewey, D. T. Wright 2003. Heavy minerals solve structural and stratigraphic problems in Ordovician strata of the western Irish Caledonides. *Geological Magazine* **140**, 25–30.

Manning, D. A. C. 1983. Disseminated tin sulphides in the St Austell granite. *Ussher Society, Proceedings* **5**, 411–16.

——— 1986. Contrasting styles of Sn–W mineralization in peninsular Thailand and southwest England. *Mineralium Deposita* **21**, 44–52.

——— 1991. Chemical variation in tourmalines from southwest England. *Ussher Society, Proceedings* **7**, 327–32.

Manning, D. A. C. & C. S. Exley 1984. The origins of late-stage rocks in the St Austell granite – a re-interpretation. *Geological Society of London, Journal* **141**, 581–91.

Manning, D. A. C. & P. I. Hill 1990. The petrogenetic and metallogenic significance of topaz granite from the southwest England orefield. In *Ore-bearing granite systems: petrogenesis and mineralizing processes*, H. J. Stein & J. L. Hannah, 51–69. Special Paper 246, Geological Society of America, Boulder, Colorado.

Manning, D. A. C., P. I. Hill, J. H. Howe 1996. Primary lithological variation in the kaolinized St Austell granite, Cornwall, England. *Geological Society of London, Journal* **153**, 827–38.

Manning, W. 1959. The Parys and Mona mines in Anglesey. See Higham (1959: 313–27).

Manter, E. H. 1950. *Rocket of the Comstock.* Caldwell, Idaho: Caxton Printers.

Masheder, R. & A. H. Rankin 1988. Fluid inclusion studies on the Ecton Hill copper deposits, north Staffordshire. *Mineralogical Magazine* **52**, 473–82.

Maskelyne, N. S. 1875. On andrewsite and chalkosiderite. *Chemical Society, Journal (series 2)* **13**, 586–91.

——— 1878. A new mineral. *Nature* **18**, 426.

Mason, B. 1968. Kaersutite from San Carlos, Arizona, with comments on the paragenesis of this mineral. *Mineralogical Magazine* **36**, 997–1002.

Mason, J. S. 1997. Regional polyphase and polymetallic vein mineralization in the Caledonides of the Central Wales orefield. *Institution of Mining and Metallurgy, Transactions B* **106**, 135–43.

——— 1998. Tučekite, a mineral new to Britain, and other rare ore minerals from the Central Wales orefield. *UK Journal of Mines and Minerals* **19**, 30–36.

——— 2004. The development of supergene lead mineralization in Central Wales. *UK Journal of Mines and Minerals* **24**, 35–46.

Mason, J. S. & D. I. Green 1995. Supergene minerals including exceptional ramsbeckite from Penrhiw mine, Ystumtuen, Dyfed. *UK Journal of Mines and Minerals* **15**, 21–7.

——— 1996. Supergene copper mineralization *in situ* at Lodge Park copper trial, Dyfed. *UK Journal of Mines and Minerals* **17**, 18–23.

Mason, J. S. & S. J. S. Hughes 1990. Geology of the Darren district. In *The Darren mines*, S. J. S. Hughes (ed.), 131–41. British Mining 40, Northern Mine Research Society, Sheffield.

Mason, J. S. & S. A. Rust 1997. The mineralogy of Ystrad Einion mine, Dyfed, Wales. *UK Journal of Mines and Minerals* **18**, 33–6.

Mason, J. S. & K. W. Williamson 2001. Filiform pyrite from the Prince Edward mine, Dolgellau gold-belt, Gwynedd. *UK Journal of Mines and Minerals* **21**, 37–8.

Mason, J. S., R. E. Bevins, D. H. M. Alderton 2002. Ore mineralogy of the mesothermal gold lodes of the Dolgellau gold belt, North Wales. *Institution of Mining and Metallurgy, Transactions B* **111**, 203–214.

Mason, J. S., W. R. Fitches, R. E. Bevins 1999. Evidence of a pre-tectonic origin for the auriferous vein-type mineralization of the Dolgellau gold belt, North Wales. *Institution of Mining and Metallurgy, Transactions B* **108**, 45–52.

Mathavan, V. & D. R. Bowes 2005. Multiple growth history of porphyroblasts in Barrovian metamorphism of Dalradian albite schists near Loch Lomond, southwest Scottish Highlands. *Scottish Journal of Geology* **41**, 175–88.

Mather, J. D. 1970. The biotite isograd and the lower greenschist facies in the Dalradian rocks of Scotland. *Journal of Petrology* **11**, 253–75.

Matley, C. A. 1928. The Pre-Cambrian complex and associated rocks of southwestern Lleyn (Caernarvonshire). *Geological Society of London, Quarterly Journal* **84**, 440–504.

Matley, C. A. & T. S. Wilson 1946. The Harlech Dome, north of the Barmouth Estuary. *Geological Society of London, Quarterly Journal* **102**, 1–40.

Matsubara, S., R. Miyawaki, T. Tiba, H. Imai 2000. Tamaite, the Ca-analogue of ganophyllite, from the Shiromaru mine, Okutama, Tokyo, Japan. *Journal of Mineralogical and Petrological Sciences* **95**, 79–83.

Matthews, D. W. 1967. Zoned ultrabasic bodies in the Lewisian of the Moine nappe in Skye. *Scottish Journal of Geology* **3**, 17–33.

Matthews, D. W. & R. F. Cheeney 1968. The metamorphic evolution of the Moine nappe in Skye. *Scottish Journal of Geology* **4**, 20–30.

Matthews, D. W. & J. H. Scoon 1964. Notes on a new occurrence of stilpnomelane from North Wales. *Mineralogical Magazine* **33**, 1032–1037.

Mawe, J. 1802. *Mineralogy of Derbyshire.* London: William Phillips.

Max, M. D. & V. Talbot 1986. Molybdenum concentrations in the western end of the Galway granite and their structural setting. See Andrew et al. (1986: 177–85).

Max, M. D., P. J. Treloar, J. A. Winchester, M. J. Oppenheim 1983. Cr-mica from the Precambrian Erris complex, northwest Mayo, Ireland. *Mineralogical Magazine* **47**, 359–64.

Mayall, M. J. 1979. The clay mineralogy of the Rhaetic transgression in Somerset – environmental and stratigraphical implications. *Ussher Society, Proceedings* **4**, 303–311.

Maynard, J. 1993. *The genesis of sulphides in the Shetland ophiolite.* PhD thesis, Open University, Milton Keynes.

Maynard, J., H. M. Prichard, R. A. Ixer, R. A. Lord, I. P. Wright, C. T. Pillinger, P. McConville, A. J. Boyce, A. E. Fallick 1997. Sulphur isotope study of Ni–Fe–Cu mineralization in the Shetland ophiolite. *Institution of Mining and Metallurgy, Transactions B* **106**, 215–26.

Mazdab, F. K. 2003. The diversity and occurrence of potassium-dominant amphiboles. *Canadian Mineralogist* **41**, 1329–44.

McArdle. P. 1981. The country rocks flanking the Leinster granite between Aughrim and Ballymurphy. *Geological Survey of Ireland, Bulletin* **3**, 85–95.

——— 1984. *The geological setting of mineralization along the southeast margin of the Tullow Lowlands Unit of the Leinster granite.* PhD thesis, National University of Ireland.

——— 1989. Geological setting of gold mineralization in the Republic of Ireland. *Institution of Mining and Metallurgy, Transactions B* **98**, 7–12.

——— 1993. Evolution and preservation of volcanogenic sulphides at Avoca, southeast Ireland. *Institution of Mining and Metallurgy, Transactions B* **102**, 149–63.

McArdle, P. & P. S. Kennan 1987. The distribution, genesis and potential of tungsten, lithium and associated metal deposits in the southeast margin of the Leinster granite. *Geological Survey of Ireland, Bulletin* **4**, 27–40.

——— 1988. Controls on the occurrence and distribution of tungsten and lithium deposits on the southeast margin of the Leinster granite, Ireland. In *Mineral deposits within the European Community*, J. Boissonnas & P. Omenetto (eds), 199–209 (Special Publication 6, Society for Geology Applied to Mineral Deposits). Berlin: Springer.

——— 1989. Geological setting of tantalum-bearing mineralization in the Caledonides of southeast Ireland. In *Lanthanides, tantalum and niobium – mineralogy, geochemistry, characteristics of primary ore deposits, prospecting, processing and applications*, P. Möller, P. Černý, F. Saupe (eds), 257–60. Berlin: Springer.

——— 1992. Deformational and stratigraphic influences on mineralization in southeast Ireland. *Mineralium Deposita* **27**, 213–18.

McArdle, P. & W. P. Warren 1987. Iron formation as a bedrock source of gold in southeast Ireland and its implication for exploration. *Institution of Mining and Metallurgy, Transactions B* **96**, 195–200.

McArdle, P., M. Fitzell, M. G. Oosterom, P. J. O'Connor, P. S. Kennan 1989. Tourmalinite as a potential host rock for gold in the Caledonides of southeast Ireland. *Mineralium Deposita* **24**, 154–9.

McArdle, P., J. H. Morris, P. R. R. Gardiner 1987. *Gold in Ireland: the potential for mineralization.* Report Series 87.1 (Mineral Resources), Geological Survey of Ireland, Dublin.

McArdle, P., N. Reynolds, C. Schaffalitzky, A. M. Bell 1986a. Controls on mineralization in the Dalradian of Ireland. See Andrew et al. (1986: 31–43).

McArdle, P., F. M. Williams, E. M. Doyle, P. S. Kennan, G. Moore-Lewy, V. Gallagher, P. J. O'Connor 1986b. *The metallogenic relationships of mineralization along the southeast margin of the Leinster batholith.* Report Series 86.2 (Mineral Resources), Geological Survey of Ireland, Dublin.

McCaffrey, K. J. W. & J. D. Johnston 1996. Fractal analysis of a mineralized vein deposit: Curraghinalt gold deposit, Country Tyrone. *Mineralium Deposita* **31**, 52–8.

McCallien, W. J. 1932. The Kainozoic igneous rocks of Kintyre. *Geological Magazine* **69**, 49–61.

McConnell, J. D. C. 1954. The hydrated calcium silicates riversideite, tobermorite and plombièrite. *Mineralogical Magazine* **30**, 293–305.

——— 1955. The hydration of larnite (β-Ca_2SiO_4) and bredigite (α_1-Ca_2SiO_4) and the properties of the resulting gelatinous mineral plombierite. *Mineralogical Magazine* **30**, 672–80.

——— 1957. Gold in Ireland. Press release, 29 January 1957, Can-Erin Mines Ltd, Toronto.

——— 1960. Vaterite from Ballycraigy, Larne, Northern Ireland. *Mineralogical Magazine* **32**, 535–44.

McCready, A. J. & J. Parnell 1998. A Phanerozoic analogue for Witwatersrand-type uranium mineralization: uranium–titanium–bitumen nodules in Devonian conglomerite/sandstone, Orkney, Scotland. *Institution of Mining and Metallurgy, Transactions B* **107**, 89–97.

McElduff, B. & J. D. Johnston 1989. Bedrock occurrence of platinum group minerals (PGMs) in Ireland. In *Irish Association of Economic Geology Annual Review*, 111.

McFadzean, A. 1988. The mines of Lindal Moor, Cumbria. *UK Journal of Mines and Minerals* **5**, 15–17.

——— 1989. The development of mining methods on Lindal Moor, Cumbria. *UK Journal of Mines and Minerals* **6**, 24–6.

McIntosh, R. P. 2001. Mica quarrying and processing in Scotland during the Second World War. *Russell Society, Journal* **7**(2), 71–4.

McKerrow, W. S. & F. B. Atkins 1989. *Isle of Arran* (2nd edn). Guide 32, Geologists' Association, London.

McKervey, J. A., A. G. Gunn, M. T. Styles 2007. Platinum-group elements in Ordovician magmatic Ni–Cu sulfide prospects in northeast Scotland. *Canadian Mineralogist* **45**, 335–53.

McKie, D. 1963. The högbomite polytypes. *Mineralogical Magazine* **33**, 563–80.

McLachlan, G. R. 1951. The aegirine-granulites of Glen Lui, Braemar, Aberdeenshire. *Mineralogical Magazine* **29**, 476–95.

McLintock, W. F. P. 1910a. On datolite from the Lizard district, Cornwall. *Mineralogical Magazine* **15**, 407–413.

——— 1910b. On zeolites and associated minerals from the Tertiary lavas around Ben More, Mull. *Royal Society of Edinburgh, Transactions: Earth Sciences* **51**, 1–33.

——— 1923. On the occurrence of petalite and pneumatolytic apatite in the Meldon aplite, Okehampton, Devonshire. *Mineralogical Magazine* **20**, 140–50.

McLintock, W. F. P. & T. C. F. Hall 1912. On topaz and beryl from the granite of Lundy island. *Mineralogical Magazine* **16**, 294–301.

McMullen, M. J. & J. G. Todd 1990. Mineralization of the Kinharvie Burn, southwest Scotland. *UK Journal of Mines and Minerals* **8**, 43–5.

McNamara, M. J. 1963. Paragonite from the Dalradian. *Nature* **197**, 1193.

McPherson, G. & T. Lamb 1921. Platinum-bearing rocks in the Lizard district. *Geological Magazine* **58**, 512–14.

Meighan, I. G. & R. A. Ixer 1980. A further note on the molybdenite mineralization at Croreagh quarry, County Down. *Irish Naturalists' Journal* **20**, 78.

Meighan, I. G., J. L. Gill, J. C. Neeson 1979. An occurrence of molybdenite in the Newry igneous complex, County Down. *Irish Naturalists' Journal* **19**, 426–7.

Meikle, T. K. 1989a. The secondary mineralogy of the Clyde plateau lavas, Scotland, part 1: Boyleston quarry. *Russell Society, Journal* **2**(2), 11–14.

——— 1989b. The secondary mineralogy of the Clyde plateau lavas, Scotland, part 2: Loanhead quarry. *Russell Society, Journal* **2**(2), 15–21.

——— 1990. The secondary mineralogy of the Clyde plateau lavas, Scotland, part 3: Hartfield Moss. *Russell Society, Journal* **3**(2), 43–7.

——— 1992a. Growth of acanthite on native silver from the Clyde plateau lavas, Scotland. *Russell Society, Journal* **4**(2), 67–9.

——— 1992b. Greenockite from Bishopton, the type locality – a review. *UK Journal of Mines and Minerals* **11**, 4–6.

——— 1994. Native silver from Hilderston mine, West Lothian, Scotland. *Russell Society, Journal* **5**(2), 83–90.

——— 2003. The secondary mineralogy of the Clyde Plateau lavas, Scotland, part 4: Reilly quarry. *Russell Society, Journal* **8**(1), 19–22.

Meikle, T. K. & J. G. Todd 1995. Silica-rich edingtonite and associated minerals from Loanhead quarry, Beith, Strathclyde – a new Scottish locality. *Russell Society, Journal* **6**(1), 27–30.

Mello, J. M. 1875. *Handbook to the geology of Derbyshire.* London: Bemrose.

Melmore, S. 1920. The metamorphism of the Carrock Fell gabbro: with a note on the origin of the sulphide veins of the Caldbeck Fells. *Geological Magazine* **57**, 266–8.

——— 1943. Letters in possession of the Yorkshire Philosophical Society. *North Western Naturalist* **18**, 148.

Menuge, J. F., M. Feely, C. O'Reilly 1997. Origin and granite alteration effects of hydrothermal fluid: isotopic evidence from fluorite veins. *Mineralium Deposita* **32**, 34–43.

Mercy, E. L. P. & M. J. O'Hara 1965. Websterite from Glenelg, Inverness-shire. *Scottish Journal of Geology* **1**, 282–4.

——— 1968. Nepheline normative eclogite from Loch Duich, Ross-shire. *Scottish Journal of Geology* **4**, 1–9.

Merriman, R. J. & B. Roberts 1985. A survey of white mica crystallinity and polytypes in pelitic rocks of Snowdonia and Llyn, North Wales. *Mineralogical Magazine* **49**, 305–319.

Merriman, R. J., R. E. Bevins, T. K. Ball 1986. Petrological and geochemical variations within the Tal y Fan intrusion: a study of element mobility during low-grade metamorphism with implications for petrotectonic modelling. *Journal of Petrology* **27**, 1409–436.

Merry, M. G. H. & R. S. W. Braithwaite 1995. Wulfenite from Cornwall, England. *Russell Society, Journal* **6**(1), 47–8.

Merry, M. G. H. & S. Weiss 2005. Axinit – neufund in Cornwall, Sudwest-England. *Lapis* **30**(4), 32–4.

Metcalfe, R. 1990. *Fluid/rock interaction and metadomain formation during low-grade metamorphism in the Welsh marginal basin.* PhD thesis, University of Bristol.

Metcalfe, R., R. E. Bevins, D. Robinson 1994. Theoretical constraints on calcium silicate domain formation and fluid flow during alteration of basic igneous rocks. *Mineralogical Magazine* **58a**, 601–602.

Meyer, H. O. A. 1965. Revision of the stratigraphy of the Permian evaporites and associated strata in northwestern England. *Yorkshire Geological Society, Proceedings* **35**, 71–89.

Michel, C. 1994. A regular pattern of chert within Lower Carboniferous limestones at the Benbulben Range, northwest Ireland. *Mineralogical Magazine* **58a**, 611–12.

Michie, U. McL. 1972. Further evidence of uranium mineralization in Orkney. *Institution of Mining and Metallurgy, Transactions B* **81**, 53–4.

Midgley, H. G. 1951. A serpentine mineral from Kennack Cove, Lizard, Cornwall. *Mineralogical Magazine* **29**, 526–30.

——— 1958. A further occurrence of phosgenite. *Mineralogical Magazine* **31**, 883.

Miers, H. A. 1884. The crystallography of bournonite. *Mineralogical Magazine* **6**, 59–79.

——— 1889a. Calcites from the neighbourhood of Egremont, Cumberland. *Mineralogical Magazine* **8**, 149–53.

——— 1889b. Mineralogical notes. Polybasite; aikinite; quartz; cuprite; the locality of turnerite. *Mineralogical Magazine* **8**, 204–209.

——— 1894. Spangolite. *Mineralogical Magazine* **10**, 273–7.

——— 1897. On some British pseudomorphs. *Mineralogical Magazine* **11**, 263–85.

——— 1929. *Mineralogy: an introduction to the scientific study of minerals* (2nd edn). London: Macmillan.

Miers, H. A. & G. T. Prior 1891. Danalite from Cornwall. *Mineralogical Magazine* **10**, 10–14.

Mighall, T. M., J. P. Grattan, S. Timberlake, J. A. Lees, S. Forsyth 2002. An atmospheric pollution history for lead–zinc mining from the Ystwyth Valley, Dyfed, mid-Wales, UK as recorded by an upland blanket peat. *Geochemistry: Exploration, Environment and Analysis* **2**, 175–84.

Mikkelsen, A., A. B. Andersen, S. B. Engelsen, H. C. B. Hansen, O. Larsen, L. H. Skibsted 1999. Presence and dehydration of ikaite, calcium carbonate hexahydrate, in frozen shrimp shell. *Journal of Agriculture and Food Chemistry* **47**, 911–17.

Miller, J. M. & K. Taylor 1966. Uranium mineralization near Dalbeattie, Kirkudbrightshire. *Geological Survey of Great Britain, Bulletin* **25**, 1–17.

Millward, D., B. Beddoe-Stephens, B. Young 1999. Pre-Acadian copper mineralization in the English Lake District. *Geological Magazine* **136**, 159–76.

Milne, J. K., M. J. Saunders, P. J. E. Wood 1977. Iron-boracite from the English Zechstein. *Mineralogical Magazine* **41**, 404–406.

Milner, A. L. & P. McArdle 1992. Gold mineralization in Ordovician volcanic rocks at Kilmacoo, Co. Wicklow: its exploration and geological controls. See Bowden et al. (1992: 51–64).

Milodowski, A. E., M. R. Gillespie, J. Naden, N. J. Fortey, T. J. Shepherd, J. M. Pearce, R. Metcalfe 1998. The petrology and paragenesis of fracture mineralization in the Sellafield area, west Cumbria. *Yorkshire Geological Society, Proceedings* **52**, 215–41.

Milodowski, A. E., G. E. Strong, K. S. Wilson, S. Holloway, A. H. Bath 1987. *Diagenetic influences on the aquifer properties of the Permo-Triassic sandstones in the East Yorkshire and Lincolnshire Basin.* Investigation of the Geothermal Potential of the UK, British Geological Survey, Keyworth.

Milodowski, A. E., M. T. Styles, V. L. Hards 2000. *A natural analogue for copper waste canisters: the copper–uranium mineralised concretions in the Permian mudrocks of south Devon, United Kingdom.* Technical Report TR-00-11, Swedish Nuclear Fuel and Waste Management Co., Stockholm.

Milodowski, A. E., M. T. Styles, M. S. A. Horstwood, S. J. Kemp 2002. *Alteration of uraniferous and native copper concretions in the Permian mudrocks of south Devon, United Kingdom.* Technical Report TR-02-09, Swedish Nuclear Fuel and Waste Management Co., Stockholm.

Mitchell, J. G. & P. R. Ineson 1975. Potassium–argon ages for the graphite deposits and related rocks of Seathwaite, Cumbria. *Yorkshire Geological Society, Proceedings* **40**, 413–18.

Mitchell, W. I. 2004. Permian. See Mitchell (2004: 125–32).

——— (ed.) 2004. *The geology of Northern Ireland.* Belfast: Geological Survey of Northern Ireland.

Mitropoulos, P. 1984. Rare-earth element distribution in the metabasic rocks of the Land's End granite aureole, southwest England. *Mineralogical Magazine* **48**, 495–505.

Moazzen, M. & G. T. R. Droop 2005. Application of mineral thermometers and barometers to granitoid igneous rocks: the Etive complex, W. Scotland. *Mineralogy and Petrology* **83**, 27–53.

Moazzen, M., G. T. R. Droop, B. Harte 2001. Abrupt transition in H_2O activity in the melt-present zone of a thermal aureole: evidence from H_2O contents of cordierites. *Geology* **29**, 311–14.

Mogessie, A., K. Ettinger, B. E. Leake 2004. AMPH-IMA04: a revised HyperCard program to determine the name of an amphibole from chemical analyses according to the 2004 International Mineralogical Association scheme. *Mineralogical Magazine* **68**, 825–30.

Moles, N. R. 1983. Sphalerite composition in relation to deposition and metamorphism of the Foss stratiform Ba–Zn–Pb deposit, Aberfeldy, Scotland. *Mineralogical Magazine* **47**, 487–500.

——— 1985. *Geology, geochemistry and petrology of the Foss stratiform barite–base metal deposit and adjacent Dalradian metasediments, near Aberfeldy, Scotland.* PhD thesis, University of Edinburgh.

——— 2007. Edingtonite, harmotome and other vein minerals at the Conlig–Whitespots lead mines, Co. Down, Northern Ireland. *Russell Society, Journal* **10**, 40–46.

Moles, N. R. & R. Nawaz 1996. Harmotome associated with Tertiary dyke intrusion in mineralized breccias at Newtownards, Northern Ireland. *Irish Journal of Earth Sciences* **15**, 145–53.

Moles, N. R., D. Smyth, C. E. Maher, E. H. Beattie, M. Kelly 2004. Dispersion of cerussite-rich tailings and plant uptake of heavy metals at historical lead mines near Newtownards, Northern Ireland. *Applied Earth Science* **113**, 21–30.

Monani, S. & J. W. Valley 2001. Oxygen isotope ratios of zircon: magma genesis of low $\delta^{18}O$ granites from the British Tertiary Igneous Province, western Scotland. *Earth and Planetary Science Letters* **184**, 377–92.

Money, M., B. Randall, B. Turner 1995. The geology of the North Tyne and Saughtree. See Scrutton (1995: 114–26).

Monk, P. K. 1982. Baryte and other minerals from the Isle of Sheppey, Kent. *Russell Society, Journal* **1**(1), 5–7.

Moody, K. M. 1953. The space-group and cell-dimensions of rankinite. *Mineralogical Magazine* **30**, 79.

Moorbath, S. 1962. Lead isotope abundance studies on mineral occurrences in the British Isles and their geological significance. *Royal Society of London, Philosophical Transactions* **254A**, 295–360.

Moore, B. 1988. Lead and zinc mining in the Ewden Valley, South Yorkshire. *UK Journal of Mines and Minerals* **4**, 25–7.

Moore, D. J. 1979. An unusual calcite from Muirshiel, Renfrewshire. *Mineralogical Magazine* **43**, 446–8.

Moore, F. 1977. The occurrence of topaz-rich greisens in St Michael's Mount, Cornwall. *Ussher Society, Proceedings* **4**, 49–56.

——— 1980. *The geology and mineralization of St Michael's Mount and Cligga Head granites, Cornwall.* PhD thesis, University of London.

Moore, F. & R. A. Howie 1978. On the application of the hübnerite:ferberite ratio as a geothermometer. *Mineralium Deposita* **13**, 391–7.

——— 1979. Geochemistry of some Cornubian cassiterites. *Mineralium Deposita* **14**, 103–107.

——— 1984. Tin-bearing sulphides from St Michael's Mount and Cligga Head, Cornwall. *Mineralogical Magazine* **48**, 389–96.

Moore, F. & D. J. Moore 1979. Fluid inclusion study of mineralization at St Michael's Mount, Cornwall. *Institution of Mining and Metallurgy, Transactions B* **88**, 57–60.

Moore, F. & P. K. Swart 1979. The uranium content of ore minerals from St Michael's Mount and Cligga Head, Cornwall. *Ussher Society, Proceedings* **4**, 432–6.

Moore, P. & J. Pearce 1997. The Sussex Mineral and Lapidary Society, 25 years on. *UK Journal of Mines and Minerals* **18**, 8–9.

Moore, T. 1996. What's new in minerals: Tucson show 1996. *Mineralogical Record* **27**(3), 213–23.

More, A. P., D. J. Vaughan, J. R. Ashworth 1991. Banded sphalerite from the North Pennine orefield. *Mineralogical Magazine* **55**, 409–416.

Moreton, N. J. M. 1982. Metal mines of the British Isles, 1: Dylife (Powys). *Russell Society, Journal* **1**(1), 8.

Moreton, S. 1991. Secondary copper minerals from the Knockmahon area, County Waterford, Republic of Ireland. *Russell Society, Journal* **4**(1), 36–7.

——— 1993. The gold rushes of Ireland. *UK Journal of Mines and Minerals* **12**, 42–4.

——— 1996. The Alva silver mine, Silver Glen, Alva, Scotland. *Mineralogical Record* **27**(6), 405–414.

——— 1998. Penrith show. *UK Journal of Mines and Minerals* **19**, 8.

——— 1999. The Silvermines district, County Tipperary, Ireland. *Mineralogical Record* **30**(2), 99–106.

——— 2005a. The mineralogy of the Bunmahon copper mines, Co. Waterford. *Mining Heritage Trust of Ireland, Journal* **5**, 11–14.

——— 2005b. Mud-filled fluorite vugs in Greenlaws mine, Weardale, Co. Durham. *Russell Society, Journal* **8**(2), 71–5.

——— 2006. A mineralogical tour of the kingdom of Fife. *UK Journal of Mines and Minerals* **27**, 29–35.

——— 2007. *Bonanzas and Jacobites: the story of the Silver Glen.* Edinburgh: National Museum of Scotland Enterprises.

Moreton, S. & P. Aspen 1992. Connellite from Colvend, Kirkudbrightshire, and brochantite from Auchencat Burn, Dumfriesshire. *Russell Society, Journal* **4**(2), 63–4.

——— 1993. Botallackite from Dooneen mine, Allihies, County Cork, Republic of Ireland. *Russell Society, Journal* **5**(1), 31–2.

Moreton, S. & D. I. Green 2005. Phosphate minerals including the first Irish koninckite and planerite from Grouse Lodge quarry, Shanagolden, Co. Limerick. *Russell Society, Journal* **8**(2), 78–80.

——— 2006a. A zeolite-bearing skarn at Barnahallia, Connemara, Co. Galway, Ireland. *Russell Society, Journal* **9**, 90–92.

——— 2006b. Copper minerals including the first Irish clinoclase and zálesíite from Leaghillaun, Co. Kerry. *UK Journal of Mines and Minerals* **27**, 41–4.

Moreton, S. & B. Jackson 1991. Annabergite and wroewolfeite: two minerals new to Ireland. *Irish Journal of Earth Sciences* **11**, 177–80.

Moreton, S., P. Aspen, D. I. Green, S. M. Ingram 1998. The silver and cobalt mineralization near Alva, Central Region, Scotland. *Russell Society, Journal* **7**(1), 23–30.

Moreton, S., P. Davidson, D. I. Green, G. Ryback 1999. Unusual supergene minerals from Muckross mine, Co. Kerry. *Irish Journal of Earth Sciences* **17**, 19–24.

Moreton, S., D. I. Green, A. G. Tindle 2006. Manganese oxide minerals from veins in the Leinster granite. *Irish Journal of Earth Sciences* **24**, 29–36.

Moreton, S., G. Ryback, P. Aspen 1995. Basaluminite, hydronium jarosite, metasideronatrite and sideronatrite – four sulphate minerals new to Ireland – from Ballybunnion, County Kerry. *Irish Journal of Earth Sciences* **14**, 1–5.

Morgan, D. & R. Starkey 1991. Harmotome from the Pen-y-Clun mine, Llanidloes, Dyfed, Wales. *UK Journal of Mines and Minerals* **10**, 4–7.

Morgan, D. J., S. St J. Warne, S. B. Warrington, P. H. A. Nancarrow 1986. Thermal decomposition reactions of caledonite and their products. *Mineralogical Magazine* **50**, 521–6.

Morgan, W. C. 1967. Genthelvite and bertrandite from the Cairngorm Mountains, Scotland. *Mineralogical Magazine* **36**, 60–63.

Morimoto, N. 1988. Nomenclature of pyroxenes. *Mineralogical Magazine* **52**, 535–50.

Morse, G. (ed.) 2002. *Minerals of the Mendips* (2nd edn). Occasional Paper 18, British Micromount Society.

Morris, J. H. 1984. *The metallic mineral deposits of the Lower Palaeozoic Longford Down inlier, in the Republic of Ireland.* Report Series 84.1 (Mineral Resources), Geological Survey of Ireland, Dublin.

Morris, J. H., G. M. Steed, D. G. Wilbur 1986. The Lisglassan–Tullybuck deposit, County Monaghan: Sb–As–Au vein mineralization in Lower Palaeozoic greywackes. See Andrew et al. (1986: 103–120).

Morrison, J. 1918. The Shap minor intrusions. *Geological Society of London, Quarterly Journal* **74**, 116–44.

Morrissey, C. J. 1970. *The mineralogy, structure and origin of the lead–zinc–copper residual orebody at Tynagh, Co. Galway, Ireland.* PhD thesis, University of London.

——— 1972. A quasi-framboidal form of syn-sedimentary pyrite. *Institution of Mining and Metallurgy, Transactions B* **81**, 55–6.

Morrissey, C. J., G. R. Davis, G. M. Steed 1971. Mineralization in the Lower Carboniferous of central Ireland. *Institution of Mining and Metallurgy, Transactions B* **80**, 174–85.

Mortimore, R. N., C. J. Wood, R. W. Gallois 2001. *British Upper Cretaceous stratigraphy.* Geological Conservation Review Series 23, Joint Nature Conservation Committee, Peterborough.

Moseley, F. 1960. The succession and structure of the Borrowdale Volcanic rocks southeast of Ullswater. *Geological Society of London, Quarterly Journal* **116**, 55–84.

——— (ed.) 1978. *The geology of the Lake District.* Occasional Publication 3, Yorkshire Geological Society.

Moss, A. A., E. E. Fejer, P. G. Embrey 1969. On the X-ray identification of amblygonite and montebrasite. *Mineralogical Magazine* **37**, 414–22.

Moss, R. J. 1876. Report on the exploration of Ballybetagh Bog. *Royal Irish Academy, Proceedings* **2**, 547–52.

Mostaghel, M. A. 1984. *Trace elements in sulphide minerals and their genesis in the South Pennine orefield.* PhD thesis, University of Leicester.

——— 1985. Sphalerite mineralogy and trace element geochemistry in the South Pennine orefield. *Yorkshire Geological Society, Proceedings* **45**, 207–218.

Moulding, D. & J. Hooper 2005. Pegmatite minerals from Chywoon quarry, Cornwall. *UK Journal of Mines and Minerals* **26**, 5–22.

Moulding, D. & A. G. Tindle 2005. Columbite from Chywoon quarry in the Carnmenellis granite, Cornwall. *Russell Society, Journal* **8**(2), 103–106.

Mountain, E. D. 1924. Calcite crystals from Holywell, North Wales. *Mineralogical Magazine* **20**, 212–16.

Mourant, A. E. 1933. The dehydration of thomsonite. *Mineralogical Magazine* **23**, 371–5.

——— 1939. *The Lukis geological collection: catalogue of the Lukis and Island Museum (Guernsey)* (pp. 21–32). St Peter Port: States Ancient Monuments Committee.

——— 1961. The minerals of Jersey. *La Société Jersiaise, Bulletin* **18**, 69–90.

——— 1985. The identification of two fibrous sulphosalts from l'Etacq, Jersey. *Mineralogical Magazine* **49**, 137–8.

Mourant, A. E. & J. P. Warren 1934. Minerals and mining in the Channel Islands. *La Société Guernesiaise, Reports and Transactions* **12**, 73–88.

Moxon, T. J. 1991. On the origin of agate with particular reference to fortification agate found in the Midland Valley, Scotland. *Chemie der Erde* **51**, 251–60.

——— 1996. *Agate: microstructure and possible origin.* Doncaster: Terra Publications.

Müller, A., R. Seltmann, C. Halls, W. Siebel, P. Dulski, T. Jeffries, J. Spratt, A. Kronz 2006. The magmatic evolution of the Land's End pluton, Cornwall and associated pre-enrichment of metals. *Ore Geology Reviews* **28**, 329–67.

Müller, A., B. J. Williamson, M. Smith 2005. Origin of quartz cores in tourmaline from Roche Rock, Cornwall. *Mineralogical Magazine* **69**, 381–401.

Mueller, G. 1951. *A genetical and geochemical survey of the Derbyshire mineral deposits.* PhD thesis, University of London.

——— 1954a. The theory of genesis of oil through hydrothermal alteration of coal type substances within certain Lower Carboniferous strata of the British Isles. In *19th International Geological Congress Algiers 1952, comptes rendus*, fascicule 12, 279–328.

——— 1954b. The distribution of coloured varieties of fluorites within the thermal zones of the Derbyshire mineral deposits. *19th International Geological Congress Algiers 1952, comptes rendus*, fascicule 15, 523–39.

Mueller, S., P. W. Scott, M. J. Evans 1999. Kaolinization, mineralization and structures in biotite granite at Bodelva, St Austell, Cornwall. *Geoscience in Southwest England* **9**, 310–17.

Muir, I. D. 1953. Quartzite xenoliths from the Ballachulish granodiorite. *Geological Magazine* **90**, 409–428.

Muir, R. O. & J. M. Ridgway 1975. Sulphide mineralization of the continental Devonian sediments of Orkney (Scotland). *Mineralium Deposita* **10**, 205–215.

Mullane, M. M. & J. A. Kinnaird 1998. Synsedimentary mineralization at Ballynoe barite deposit, near Silvermines, Co. Tipperary, Ireland. *Institution of Mining and Metallurgy, Transactions B* **107**, 48–61.

Mundy, D. & R. Arthurton 1994. The Craven fault zone – Malham to Settle. See Scrutton (1994: 30–41).

Mykura, W. 1972. Igneous intrusions and mineralisation in Fair Isle, Shetland Islands. *Geological Survey of Great Britain, Bulletin* **41**, 33–51.

——— 1976. *British regional geology: Orkney and Shetland.* Edinburgh: HMSO (Institute of Geological Sciences).

Mykura, W. & J. Phemister 1976. *The geology of western Shetland (sheet 127 and parts of sheets 125, 126 and 128).* Memoir, Institute of Geological Sciences, Edinburgh.

Mykura, W. & B. R. Young 1969. *Sodic scapolite (dipyre) in the Shetland Islands.* Report 69.4, Institute of Geological Sciences, London.

Nadeau, P. H., W. J. Wilson, W. J. McHardy, J. M. Tait 1984. Interparticle diffraction: a new concept for interstratified clays. *Clay Mineralogy* **19**, 757–69.

Naden, J. 1988. *Gold mineralization in the Caledonides of the British Isles with reference to the Dolgellau gold belt and the Southern Uplands of Scotland.* PhD thesis, University of Aston in Birmingham.

Naden, J. & J. B. D. Caulfield 1989. Fluid inclusion and isotopic studies of gold mineralization in the southern Uplands of Scotland. *Institution of Mining and Metallurgy, Transactions B* **98**, 46–8.

Nagelschmidt, G. & D. Hicks 1943. The mica of certain Coal Measures shales of South Wales. *Mineralogical Magazine* **26**, 297–303.

Nancarrow, P. H. A. 1985. *Vanadiniferous nodules from the Littleham Marl near Budleigh Salterton, Devon.* Mineralogy and Petrology Report 85.12, British Geological Survey, Keyworth.

Nataraj, T. S. 1967. *Glaucophanic metamorphism in Anglesey.* PhD thesis, University of Leeds.

Nawaz, R. 1974. Killalaite, a new mineral from Co. Sligo, Ireland. *Mineralogical Magazine* **39**, 544–8.

——— 1977. A second occurrence of killalaite. *Mineralogical Magazine* **41**, 546–8.

——— 1978a. An occurrence of okenite in the Tertiary basalts of Northern Ireland. *Irish Naturalists' Journal* **19**, 249.

——— 1978b. Two minerals, hydromagnesite and offrétite, new to Ireland. *Irish Naturalists' Journal* **19**, 249.

——— 1979. Petrogenesis of the contact rocks, Blind Rock dyke, County Donegal. *Royal Irish Academy, Proceedings* **79B**, 177–89.

——— 1982. A chemical classification scheme for the gismondine group zeolites. *Irish Naturalists' Journal* **20**, 480–83.

——— 1983a. Space group of a low tridymite, Tardree Mountain, Northern Ireland. *Mineralogical Magazine* **47**, 567.

——— 1983b. New data on gobbinsite and garronite. *Mineralogical Magazine* **47**, 567–8.

——— 1984. New data on cowlesite from Northern Ireland. *Mineralogical Magazine* **48**, 565–6.

——— 1988a. Gonnardite and disordered natrolite-group minerals: their distinction and relations with mesolite, natrolite and thomsonite. *Mineralogical Magazine* **52**, 207–219.

——— 1988b. A note on occurrence and optical orientation of brewsterite. *Mineralogical Magazine* **52**, 416–17.

Nawaz, R. & J. F. Malone 1982. Gobbinsite, a new zeolite mineral from Co. Antrim, N. Ireland. *Mineralogical Magazine* **46**, 365–9.

Nawaz, R. & G. Ryback 1989. Re-examination of kirwanite: a ferri-ferro-hornblende from Co. Down, Northern Ireland. *Mineralogical Magazine* **53**, 253–5.

Naylor, H., P. Turner, D. J. Vaughan, A. J. Boyce, A. E. Fallick 1989. Genetic studies of red-bed mineralization in the Triassic of the Cheshire Basin, northwest England. *Geological Society of London, Journal* **146**, 685–99.

Naylor, P. 1981. *Discovering lost mines.* Princes Risborough: Shire Publications.

Neall, T. 2004. Apatite crystals from the Coal Measures of Northern England. *UK Journal of Mines and Minerals* **24**, 47.

——— 2005. Cassedanneite from Lady Anne Hopetoun shaft, Leadhills, Scotland. *UK Journal of Mines and Minerals* **26**, 29–30.

Neall, T. & D. I. Green 2001a. Raspite from Carrock mine, Caldbeck Fells, Cumbria, England: a first British occurrence. *Russell Society, Journal* **7**(2), 91.

——— 2001b. Parasymplesite, a mineral new to Britain, from Wet Swine Gill, Caldbeck Fells, Cumbria, England. *Russell Society, Journal* **7**(2), 92–3.

——— 2005. Kegilite, a first British occurrence, from Wanlockhead, Dumfries & Galloway Region, Scotland. *Russell Society, Journal* **8**(2), 110–11.

Neall, T. & C. M. Leppington 1994. Cinnabar from the Caldbeck Fells, Cumbria, England. *Russell Society, Journal* **5**(2), 123–4.

——— 2003. New investigation of the mineralization in Burdell Gill, Caldbeck Fells, Cumbria. *Russell Society, Journal* **8**(1), 1–9.

Neall, T., D. I. Green, T. F. Bridges 2006a. Scotlandite in the British Isles: its occurrence, morphology and paragenesis. *Russell Society, Journal* **9**, 83–9.

Neall, T., A. G. Tindle, D. I. Green 2006b. The first British occurrence of arsendescloizite at Sandbed mine, Caldbeck Fells, Cumbria. *UK Journal of Mines and Minerals* **27**, 45–7.

Neall, T., B. Young, E. K. Hyslop, R. D. Fakes 1993. Namibite from Buckbarrow Beck, Cumbria: the first British occurrence. *Mineralogical Magazine* **57**, 754–5.

Neary, C. R. & H. M. Prichard 1985. Molybdenum mineralization on Unst, Shetland Isles. *Scottish Journal of Geology* **21**, 197–204.

Neumann, H. 1950. Pseudomorphs of pyrrhotine after pyrite in the Ballachulish slates. *Mineralogical Magazine* **29**, 234–8.

Newall, P. S. 1994. An integrated geochemical approach to investigate the concealed mineralization at the Redmoor Sn/W sheeted-vein deposit, east Cornwall, England. *Journal of Southeast Asian Earth Sciences* **10**, 109–130.

Newall, P. S. & G. C. Newall 1989. Use of lithogeochemistry as an exploration tool at Redmoor sheeted-vein complex, east Cornwall, southwest England. *Institution of Mining and Metallurgy, Transactions B* **98**, 162–74.

Newman, D. 1971. Recent developments in mineral exploration in the Northwest Highlands and Islands of Scotland. *Institution of Mining and Metallurgy, Transactions B* **80**, 276–88.

Newton, E. F. 1937. Petrography of some English fuller's earths, and the rocks associated with them. *Geologists' Association, Proceedings* **48**, 175–97.

Nex, P. A., J. A. Kinnaird, R. A. Ixer 2003. Localized ductile thrusting north of the Variscan Front, Ross Island, southwest Ireland. *Geological Journal* **38**, 15–29.

Nicholls, G. D. 1951a. An unusual pyroxene-rich xenolith in the diorite of the Glenelg–Ratagain igneous complex. *Geological Magazine* **88**, 284–95.

——— 1951b. The Glenelg–Ratagain igneous complex. *Geological Society of London, Quarterly Journal* **106**, 309–344.

——— 1958. Autometasomatism in the Lower Spilites of the Builth Volcanic Series. *Geological Society of London, Quarterly Journal* **114**, 137–62.

Nicholls, R. A. 1979. *Geochemistry and clay mineralogy of some Silurian sedimentary rocks, Dyfed, S. W. Wales.* PhD thesis, University of Bristol.

Nicholson, C. & R. C. Scrivener 2000. Minerals in Devon – past, present and future. *Devon Association for the Advancement of Science, Report and Transactions* **132**, 61–74.

Nicholson, K. 1983. *Manganese mineralization in Scotland.* PhD thesis, University of Strathclyde.

——— 1986. Mineralogy and geochemistry of manganese and iron veins, Arndilly, Banffshire. *Scottish Journal of Geology* **22**, 213–24.

——— 1987a. Ironstone–gossan discrimination: pitfalls of a simple geochemical approach – a case study from northeast Scotland. *Journal of Geochemical Exploration* **27**, 238–57.

——— 1987b. Rhodochrosite from Islay, Argyllshire and Dalroy, Inverness-shire, Scotland. *Mineralogical Magazine* **51**, 677–80.

——— 1988a. An ancient manganese-iron deposit of freshwater origin, Islay, Argyllshire. *Scottish Journal of Geology* **24**, 175–87.

——— 1988b. Manganese-cemented Quaternary gravels: birnessite, lithiophorite and quenselite from an example in Wigtownshire. *Scottish Journal of Geology* **24**, 194–200.

——— 1988c. Birnessite from Gourock, Renfrewshire, Scotland. *Mineralogical Magazine* **52**, 416.

——— 1989a. Manganese oxides from the British Isles. *Mineralogical Magazine* **53**, 637–8.

——— 1989b. Manganese minerals in Scotland. *Scottish Journal of Geology* **25**, 125–42.

——— 1989c. Early Devonian geothermal systems in northeast Scotland: exploration targets for epithermal gold. *Geology* **17**, 568–71.

——— 1990. Stratiform manganese mineralisation near Inverness, Scotland: a Devonian sublacustrine hot-spring deposit? *Mineralium Deposita* **25**, 126–31.

——— 1992. Genetic types of manganese oxide deposits in Scotland: indicators of paleo-ocean-spreading rate and a Devonian geochemical mobility boundary. *Economic Geology* **87**, 1301–309.

Nicholson, K. & R. Anderton 1989. The Dalradian rocks of the Lecht, northeast Scotland: stratigraphy, faulting, geochemistry and mineralization. *Royal Society of Edinburgh, Transactions: Earth Sciences* **80**, 143–57.

Nicholson, K. & D. Banks 1987. Magnetite, pyrrhotine and pentlandite from the Leadhills–Wanlockhead mining district, Scotland. *Mineralogical Magazine* **51**, 175–6.

Nicholson, K. & G. P. Durant 1986. Fluorite–baryte–calcite–dolomite–iron–manganese mineralization at Craigmuscat quarry, Gourock: a historical review. *Scottish Journal of Geology* **22**, 299–302.

Nickel, E. H. 1976. New data on woodwardite. *Mineralogical Magazine* **40**, 644–7.

——— 1995. The definition of a mineral. *Canadian Mineralogist* **33**, 689–90.

Nickel, E. H. & J. A. Mandarino 1987. Procedures involving the IMA Commission on New Minerals and Mineral Names, and guidelines on mineral nomenclature. *Canadian Mineralogist* **25**, 353–77.

Nickel, E. H. & M. C. Nichols 1991. *Mineral reference manual.* New York: Van Nostrand Reinhold.

Nickless, E. F. P., S. J. Booth, P. N. Mosley 1975. Celestite deposits of the Bristol area. *Institution of Mining and Metallurgy, Transactions B* **84**, 62–4.

——— 1976. *The celestite resources of the area northeast of Bristol, with notes on occurrences north and south of the Mendip Hills and in the Vale of Glamorgan.* Mineral Assessment Report 25, Institute of Geological Sciences, London.

Ni Wen, J. R. Ashworth, R. A. Ixer 1991. Evidence for the mechanism of the reaction producing a bournonite–galena symplectite from meneghinite. *Mineralogical Magazine* **55**, 153–8.

Ni Wen, A. J. Boyce, A. E. Fallick, J. R. Ashworth, R. A. Ixer 1996. The genesis of quartz veins by metamorphic remobilization from stratiform redbed deposits, southwest County Cork, Ireland. *Mineralogy and Petrology* **57**, 73–89.

Ni Wen, R. A. Ixer, J. A. Kinnaird, J. R. Ashworth, P. A. Nex 1999. Ore petrology and metamorphic mobilization of copper ores in redbeds, southwest County Cork, Ireland. *Institution of Mining and Metallurgy, Transactions B* **108**, 53–63.

Noall, C. 1999. *Botallack.* Truro: Dyllansow Truran.

Nockolds, S. R. 1931. The Dhoon (Isle of Man) granite: a study in contamination. *Mineralogical Magazine* **22**, 494–509.

——— 1932. The contaminated granite of Bibette Head, Alderney. *Geological Magazine* **69**, 433–51.

——— 1937. Contributions to the petrology of Barnavave, Carlingford, IFS, 2: an occurrence of quartz-bearing syenite and its xenoliths. *Geological Magazine* **74**, 125–32.

——— 1938a. Contributions to the petrology of Barnavave, Carlingford, IFS, 3: on some hybrids from the E and SE slopes of Barnavave Mountain. *Geological Magazine* **75**, 469–79.

——— 1938b. On the occurrence of acmite in the riebeckite-microgranite of Mynydd Mawr, Caernarvonshire. *Mineralogical Magazine* **25**, 35–7.

——— 1941. The Garabal Hill–Glen Fyne igneous complex. *Geological Society of London, Quarterly Journal* **96**, 451–511.

——— 1944. Contributions to the petrology of Barnavave, Carlingford, IFS, 4: some limestone xenoliths enclosed in the junction hybrids. *Geological Magazine* **81**, 88–94.

——— 1950. On the occurrence of neptunite and eudialyte in quartz-bearing syenites from Barnavave, Carlingford, Ireland. *Mineralogical Magazine* **29**, 27–33.

Nockolds, S. R. & R. L. Mitchell 1946. The geochemistry of some Caledonian plutonic rocks: a study in the relationship between the major and trace elements of igneous rocks and their minerals. *Royal Society of Edinburgh, Transactions: Earth Sciences* **61**, 533–75.

Nockolds, S. R. & J. E. Richey 1939. Replacement veins in the Mourne Mountains granites, N. Ireland. *American Journal of Science* **237**, 27–47.

Nockolds, S. R. & H. G. C. Vincent 1947. On tilleyite and its associated minerals from Carlingford, Ireland. *Mineralogical Magazine* **28**, 151–8.

Nölting, J. 1890. Über das Verhältniss der sogenannten Schalenblende zur regulären Blende und zum hexagonalen Wurtzit [abstract]. *Zeitschift für Krystallographie und Mineralogie* **17**, 220–22.

North, F. J. 1916. The minerals of Glamorgan. *Cardiff Naturalists' Society, Transactions* **49**, 18–51.

North, F. J. & W. E. Howarth 1928. On the occurrence of millerite and associated minerals in the Coal Measures of South Wales. *South Wales Institute of Engineers, Proceedings* **44**, 325–48.

Notholt, A. J. G. & D. E. Highley 1971. *Fluorspar*. Mineral Dossier 1, Mineral Resources Consultative Committee, London.

Notholt, A. J. G., D. E. Highley, R. R. Harding 1985. Investigation of phosphate (apatite) potential of Loch Borralan igneous complex, Northwest Highlands, Scotland. *Institution of Mining and Metallurgy, Transactions B* **94**, 58–65.

Nutt, M. J. C. 1979. The Haweswater complex. In *The Caledonides of the British Isles reviewed*, A. L. Harris, C. H. Holland, B. E. Leake (eds), 727–33. Edinburgh: Scottish Academic Press, on behalf of the Geological Society of London.

Nutter, G. 1953. *A spectrographic and geochemical study of the trace elements in the Archaean pegmatites of northwest Sutherland*. PhD thesis, University of Leeds.

Oakley, K. P. 1934. Phosphatic calculi in Silurian polyzoa. *Royal Society, Proceedings* **116**, 296–314.

O'Brien, M. V. 1959. The future of non-ferrous mining in Ireland. See Higham (1959: 5–34).

O'Connor, P. J. 1986. Uranium mineralization in the Irish Caledonides. See Andrew et al. (1986: 157–75).

O'Connor, P. J. & A. Poustie 1986. Geological setting of, and alteration associated with, the Charlestown mineral deposit. See Andrew et al. (1986: 89–101).

O'Connor, P. J., V. Gallagher, P. S. Kennan 1991. Genesis of lithium pegmatites from the Leinster granite margin, southeast Ireland: geochemical constraints. *Geological Journal* **26**, 295–305.

O'Connor, P. J., H. Högelsberger, M. Feely, D. C. Rex 1993. Fluid inclusion studies: rare-earth element chemistry and age of hydrothermal fluorite mineralization in western Ireland – a link with continental rifting? *Institution of Mining and Metallurgy, Transactions B* **102**, 141–8.

O'Connor, P. J., C. B. Long, I. R. Basham, I. G. Swainbank, B. Beddoe-Stephens 1984. Age and geological setting of uranium mineralization associated with the main Donegal granite. *Institution of Mining and Metallurgy, Transactions B* **93**, 190–94.

O'Donnell, M. 1980. *The geology of the Kill area, Co. Cavan*. BSc thesis, National University of Ireland.

——— 1994. *The colour dictionary of gemstones and minerals*. London: Orbis Publishing.

Offer, D. J. 1995. *Investigation of the Loch Roag sapphire deposit*. MSc thesis, University of Leicester.

O'Hara, M. J. 1960. A garnet–hornblende–pyroxene rock from Glenelg, Inverness-shire. *Geological Magazine* **97**, 145–56.

——— 1961. Petrology of the Scourie dyke, Sutherland. *Mineralogical Magazine* **32**, 848–65.

Old, R. A. 1972. Amygdale minerals at Sounding Hill quarry, Magherafelt, Co. Londonderry. *Irish Naturalists' Journal* **17**, 177–80.

Old, R. A., R. J. O. Hamblin, K. Ambrose, G. Warrington 1991. *Geology of the country around Redditch (sheet 183)*. Memoir, British Geological Survey, Keyworth.

Oliver, G. J. H. 1978. Prehnite–pumpellyite facies metamorphism in County Cavan, Ireland. *Nature* **274**, 242–3.

Oliver, G. J. H. & J. K. Leggett 1980. Metamorphism in an accretionary prism: prehnite–punpellyite facies metamorphism of the Southern Uplands of Scotland. *Royal Society of Edinburgh, Transactions: Earth Sciences* **71**, 235–46.

Oliver, R. L. 1956. The origin of garnets in the Borrowdale Volcanic Series and associated rocks, English Lake District. *Geological Magazine* **93**, 121–35.

——— 1961. The Borrowdale Volcanic and associated rocks of the Scafell area, English Lake District. *Geological Society of London, Quarterly Journal* **117**, 377–417.

Oppenheim, M. J., P. M. Brück, R. Elsdon, A. Weaver, W. P. Warren 1977. Eskolaite, Cr_2O_3, from County Wicklow, Ireland. *Mineralogical Magazine* **41**, 402–403.

O'Raghallaigh, C., M. Feely, P. McArdle, C. MacDermot, M. Geoghegan, R. Keary 1997. *Mineral localities in the Galway Bay area*. Report Series 97.1 (mineral resources), Geological Survey of Ireland, Dublin.

O'Reilly, J. P. 1891. On the occurrence of idocrase in the County of Monaghan. *Royal Irish Academy, Proceedings* **3**(1), 446–50.

——— 1897. On the micas of the Three Rock Mountain, Co. Dublin. *Mineralogical Magazine* **11**, 199–210.

Orme, G. R. & W. W. M. Brown 1963. Diagenetic fabrics in the Avonian limestones of Derbyshire and North Wales. *Yorkshire Geological Society, Proceedings* **34**, 51–66.

Orme, G. R. & T. D. Ford 1970. Polyphase mineralization in chert from the Ashford black marble mine, Derbyshire. *Yorkshire Geological Society, Proceedings* **38**, 163–73.

Osborne, G. D. 1932a. On the occurrence of custerite and monticellite in metamorphosed limestone from the Carlingford district, Co. Louth, Ireland. *Geological Magazine* **69**, 61–2.

——— 1932b. The metamorphosed limestones and associated contaminated rocks of Carlingford district, Co. Louth. *Geological Magazine* **69**, 209–233.

Otty, E. 1982. The Central Wales mining field. *Mineral Realm* **1**(6), 26–30.

——— 1986. The Llanrwst mines and Cae Coch. *UK Journal of Mines and Minerals* **1**, 5–6.

Paar, W. H., R. S. W. Braithwaite, T. T. Chen, P. Keller 1984. A new mineral, scotlandite ($PbSO_3$) from Leadhills, Scotland: the first naturally occurring sulphite. *Mineralogical Magazine* **48**, 283–8.

Paar, W. H., K. Mereiter, R. S. W. Braithwaite, P. Keller 1986. Chenite, $Pb_4Cu(SO_4)_2(OH)_6$, a new mineral, from Leadhills, Scotland. *Mineralogical Magazine* **50**, 129–35.

Paar, W. H., A. C. Roberts, A. J. Criddle, D. Topa 1998. A new mineral, chrisstanleyite, $Ag_2Pd_3Se_4$, from Hope's Nose, Torquay, Devon, England. *Mineralogical Magazine* **62**, 257–64.

Pabian, R., B. Jackson, P. Tandy, J. Cromartie 2006. *Agates: treasures of the Earth*. London: Natural History Museum.

Palacz, Z. A. & S. R. Tait 1985. Isotopic and geochemical investigation of Unit 10 from the Eastern Layered Series of the Rhum intrusion, northwest Scotland. *Geological Magazine* **122**, 485–90.

Parker, R. T. G., J. A. Clifford, A. H. Meldrum 1989. The Cononish gold–silver deposit, Perthshire, Scotland *Institution of Mining and Metallurgy, Transactions B* **98**, 51–2.

Parnell, J. 1981. Genesis of the graphite deposit at Seathwaite, Borrowdale, Cumbria. *Geological Magazine* **119**, 511–12.

——— 1983. Stromatolite-hosted mineralization in the Oil Shale Group, Scotland. *Institution of Mining and Metallurgy, Transactions B* **92**, 98–9.

——— 1985. Uranium/rare earth-enriched hydrocarbons in Devonian sandstones, northern Scotland. *Neues Jahrbuch für Mineralogie,* Monatshefte 3, 132–44.

——— 1988a. Mercury and silver–bismuth selenides at Alva, Scotland. *Mineralogical Magazine* **52**, 719–20.

——— 1988b. The mineralogy of redbed uranium–vanadium mineralization in the Permo-Triassic of Belfast. *Irish Journal of Earth Sciences* **9**, 119–24.

——— 1988c. Mineralogy of uraniferous hydrocarbons in Carboniferous-hosted mineral deposits, Great Britain. *Uranium* **4**, 197–218.

Parnell, J. & P. Eakin 1987. The replacement of sandstones by uraniferous hydrocarbons: significance for petroleum migration. *Mineralogical Magazine* **51**, 505–515.

——— 1989. Thorium-bitumen mineralization in Silurian sandstones, Welsh Borderlands. *Mineralogical Magazine* **53**, 111–16.

Parnell, J. & T. Janaway 1990. Sulphide-mineralized algal breccias in a Devonian evaporitic lake system, Orkney, Scotland. *Ore Geology Reviews* **5**, 445–60.

Parnell, J., and 10 other authors 2000. Regional fluid flow and gold mineralization in the Dalradian of the Sperrin Mountains, Northern Ireland. *Economic Geology* **95**, 1389–416.

Parslow, G. R. 1976. The Suisnish layered dyke. *Mineralogical Magazine* **40**, 683–93.

Parsons, C. E. 1897. The deposit at Mill Close lead mines, Darley Dale. *Institute of Mining Engineers, Transactions* **12**, 115–21, 570–72.

Parsons, I. & R. Boyd 1971. Distribution of potassium feldspar polymorphs in intrusive sequences. *Mineralogical Magazine* **38**, 295–311.

Parsons, L. M. 1922. Dolomitization in the Carboniferous Limestone of the Midlands. *Geological Magazine* **59**, 51–63, 104–116.

Passaglia, E. & G. Gottardi 1973. Crystal chemistry and nomenclature of pumpellyites and julgoldites. *Canadian Mineralogist* **12**, 219–23.

Passaglia, E., G. Artioli, A. Gualtieri 1998. Crystal chemistry of the zeolites erionite and offrétite. *American Mineralogist* **83**, 577–89.

Passaglia, E., E. Galli, L. Leoni, G. Rossi 1978a. The crystal chemistry of stilbites and stellerites. *Bulletin de Minéralogie* **101**, 368–75.

Passaglia, E., A. F. Gualtieri, E. Marchi 2001. The crystal chemistry of paulingite. *European Journal of Mineralogy* **13**, 113–19.

Passaglia, E., D. Pongiluppi, G. Vezzalini 1978b. The crystal chemistry of gmelinites. *Neues Jahrbuch für Mineralogie,* Monatshefte 7, 310–24.

Patel, A. R. & J. Koshy 1969. Growth spirals on baryte. *Mineralogical Magazine* **37**, 296–8.

Paterson, B. A., W. E. Stephens, D. A. Herd 1989. Zoning in granitoid accessory minerals as revealed by backscattered electron imagery. *Mineralogical Magazine* **53**, 55–61.

Paterson, B. A., W. E. Stephens, G. Rogers, I. S. Williams, R. W. Hinton, D. A. Herd 1992. The nature of zircon inheritance in two granite plutons. *Royal Society of Edinburgh, Transactions: Earth Sciences* **83**, 459–71.

Patrick, D. J. 1980. *Mineral investigations at Carrock Fell, Cumbria, part 1: geophysical survey.* Report 33, Mineral Reconnaissance Program, Institute of Geological Sciences, London.

Patterson, E. M. 1951. An occurrence of quartz-trachyte among the Tertiary basalt lavas of northeast Ireland. *Royal Irish Academy, Proceedings* **53B**, 265–87.

Patterson, J. M. 1970. *Geology and mineralization of the Keel area, County Longford, Republic of Ireland.* PhD thesis, University of London.

Pattison, D. R. M. 1987. Variations in Mg/(Mg+Fe), F, and (Fe, Mg)Si = 2Al in pelitic minerals in the Ballachulish thermal aureole, Scotland. *American Mineralogist* **72**, 255–72.

Pattison, S. R. 1854. On auriferous quartz-rock in north Cornwall. *Geological Society of London, Quarterly Journal* **10**, 246–9.

——— 1855. A day in the North Devon mineral district. *Royal Geological Society of Cornwall, Transactions* **7**, 223–7.

Pattrick, R. A. D. 1978. Microprobe analyses of cadmium-rich tetrahedrites from Tyndrum, Perthshire, Scotland. *Mineralogical Magazine* **42**, 286–8.

——— 1984. Sulphide mineralogy of the Tomnadashan copper deposit and the Corrie Buie lead veins, south Loch Tayside, Scotland. *Mineralogical Magazine* **48**, 85–91.

——— 1985. Pb–Zn and minor U mineralization at Tyndrum, Scotland. *Mineralogical Magazine* **49**, 671–81.

Pattrick, R. A. D. & R. J. Bowell 1991. The genesis of the West Shropshire orefield: evidence from fluid inclusions, sphalerite chemistry, and sulphur isotopic ratios. *Geological Journal* **26**, 101–115.

Pattrick, R. A. D. & A. J. Hall 1984. Anomalous sulphide boulder mineralization at Tyndrum, Perthshire. *Scottish Journal of Geology* **20**, 407–410.

Pattrick, R. A. D. & D. A. Polya (eds) 1993. *Mineralization in the British Isles.* London: Chapman & Hall.

Pattrick, R. A. D., A. Boyce, R. M. MacIntyre 1988. Gold–silver vein mineralization at Tyndrum, Scotland. *Mineralogy and Petrology* **38**, 61–76.

Pattrick, R. A. D., M. L. Coleman, M. J. Russell 1983. Sulphur isotopic investigation of vein lead–zinc mineralization at Tyndrum, Scotland. *Mineralium Deposita* **18**, 477–85.

Peace, W. M., M. W. Wallace, M. P. Holdstovk, J. H. Ashton 2003. Ore textures within the U lens of the Navan Zn–Pb deposit, Ireland. *Mineralium Deposita* **38**, 568–84.

Peach, B. N., W. Gunn, C. T. Clough, L. W. Hinxman, C. B. Crampton, E. M. Anderson, J. S. Flett 1912. *The geology of Ben Wyvis, Carn Chuinneag, Inchbae and the surrounding country.* Memoir, Geological Survey of Great Britain, Edinburgh.

Peach, B. N., J. Horne, W. Gunn, C. T. Clough, L. W. Hinxman, J. J. H. Teall 1907. *The geological structure of the Northwest Highlands of Scotland.* Memoir, Geological Survey of Great Britain, Edinburgh.

Peacock, M. A. 1941. Josëite, grünlingite, oruetite. *University of Toronto Studies, Geological Series* **46**, 83–105.

Pearce, A. J. 1994a. *Metalliferous mines of Shropshire, vol. 1: gazetteer.* Newport: Shropshire Caving and Mining Club.

Pearce, J. 1994b. Mineral news. *UK Journal of Mines and Minerals* **14**, 34–5.

——— 1996a. Sussex goes North. *UK Journal of Mines and Minerals* **16**, 8.

——— 1996b. Guest editorial – the British Micromount Society (1981–1996). *UK Journal of Mines and Minerals* **17**, 2–3.

Pearce, J. & P. Pearce 1991. The rediscovery of Sgùrr nam Boc, Isle of Skye, Scotland. *UK Journal of Mines and Minerals* **10**, 43–6.

Pearce, R. 1878. Note on pitchblende in Cornwall. *Royal Geological Society of Cornwall, Transactions* **9**, 103–104.

Pearson, K. & C. A. Jeffrey 1997. Low-temperature mineralization of the sub-Triassic unconformity surface and alteration of the underlying intrusions of southern Leicestershire, England. *Exploration and Mining Geology* **6**, 139–52.

Pearson, M. J. 1974. Sideritic concretions from the Westphalian of Yorkshire: a chemical investigation of the carbonate phase. *Mineralogical Magazine* **39**, 696–9.

Penhallurick, R. D. 1997. The mineral collection of the Royal Institution of Cornwall. *UK Journal of Mines and Minerals* **18**, 17–32.

Perceval, S. G. 1880. On the occurrence of celestine in New Red Marl, near Sidmouth. *Mineralogical Magazine* **3**, 255.

Percy, J. 1850. Chemical examination of lettsomite (velvet ore). *Philosophical Magazine* **36**, 100–103.

Perutz, M. 1937. "Iron rhodonite" (from slag) and pyroxmangite and their relationship to rhodonite. *Mineralogical Magazine* **24**, 573–6.

Phemister, J. 1940. Note on an occurrence of bertrandite and beryl at the South Crofty mine, Cornwall. *Mineralogical Magazine* **25**, 573–8.

Phemister, J. & A. G. MacGregor 1942. Note on datolite and other minerals in a contact-altered limestone at Chapel quarry, near Kirkcaldy, Fife. *Mineralogical Magazine* **26**, 275–82.

Phemister, J., C. O. Harvey, P. A. Sabine 1950. The riebeckite-bearing dykes of Shetland. *Mineralogical Magazine* **29**, 359–73.

Phemister, R. C., W. E. Fraser, D. H. Williamson 1960. Dalradian metamorphism and structure – Stonehaven to Aberdeen. *International Geological Congress XXI session (Copenhagen), part xiii*, B. Asklund & C. Poulsen, 352–61 [general ed. T. Sorgenfrei]. Copenhagen: Det Berlingske Bogtrykeri.

Philips, W. 1816. On the oxyd of uranium, the production of Cornwall, together with a description and series of its crystalline forms. *Geological Society of London, Transactions* **3**, 112–20.

Phillips, F. C. 1930. Some mineralogical and chemical changes induced by progressive metamorphism in the Green Bed group of the Dalradian. *Mineralogical Magazine* **22**, 239–56.

——— 1932. Crystals of brookite tabular parallel to the basal plane. *Mineralogical Magazine* **23**, 126–9.

Phillips, W., H. J. Brooke, W. H. Miller 1852. *An elementary introduction to mineralogy.* London: Longman, Brown, Green.

Phillips, W. J. 1968. The crystallization of the teschenite from the Lugar Sill, Ayrshire. *Geological Magazine* **105**, 23–34.

Pidgeon, R. T. & D. R. Bowes 1972. Zircon U–Pb ages of granulites from the Central Region of the Lewisian, northwestern Scotland. *Geological Magazine* **109**, 247–58.

Pirrie, D., M. R. Power, J. C. Ø. Andersen, A. R. Butcher 2000a. Platinum-group mineralization in the Tertiary Igneous Province: new data from Mull and Skye, Scottish Inner Hebrides, UK. *Geological Magazine* **137**, 651–8.

Pirrie, D., M. R. Power, A. Payne, G. S. Camm, P. D. Wheeler 2000b. Impact of mining on sedimentation: the Camel and Gannel estuaries, Cornwall. *Ussher Society, Proceedings* **10**, 21–8.

Pirrie, D., M. R. Power, P. D. Wheeler, A. S. Ball 2000c. A new occurrence of diagenetic simonkolleite from the Gannel Estuary, Cornwall. *Ussher Society, Proceedings* **10**, 18–20.

Pirrie, D., M. R. Power, P. D. Wheeler, A. Cundy, C. Bridges, G. Davey 2002. Geochemical signature of historic mining: Fowey Estuary, Cornwall, UK. *Journal of Geochemical Exploration* **76**, 31–43.

Pisani, F. 1864. Sur quelques nouveaux minéraux de Cornouailles. *Académie des Sciences, Paris, Comptes Rendus* **59**, 912–13.

——— 1866. Sur une nouvelle espèce du Cornouailles, la chenevixite. *Académie des Sciences, Paris, Comptes Rendus* **62**, 690–92.

Pitcher, W. S. 1950. Calc-silicate veins in the limestone of Lough Anure, Co. Donegal. *Mineralogical Magazine* **29**, 126–41.

Plant, J. A., N. Breward, M. D. Forrest, R. T. Smith 1989. The gold pathfinder elements As, Sb and Bi – their distribution and significance in the southwest Highlands of Scotland. *Institution of Mining and Metallurgy, Transactions B* **98**, 91–101.

Plant, J. A., D. G. Jones, H. W. Haslam (eds) 1999. *The Cheshire Basin.* Memoir, British Geological Survey, Keyworth.

Plant, S. P. 1997. Field meeting report: International Colliery and Park Slip West. *Russell Society, Newsletter* **31**, 9.

——— 2003. Secondary minerals from the Cwmavon Valley copper smelting slags, Glamorgan, South Wales. *Russell Society, Journal* **8**(1), 9–15.

Plant, S. P. & D. Evans 2005. Carbonate-fluorapatite from the Westphalian Coal Measures of South Wales, UK. *Russell Society, Journal* **8**(2), 98–101.

Plant, S. P. & I. E. Jones 1995. Minerals of Machen quarry, Mid Glamorgan, Wales. *Russell Society, Journal* **6**(1), 31–6.

——— 2001. Wavellite and variscite on Gower, Swansea, South Wales. *Russell Society, Journal* **7**(2), 79–81.

Platten, I. M. & S. C. Dominy 1999. Re-evaluation of quartz vein history in the Dolgellau goldbelt, North Wales, United Kingdom. *Geological Journal* **34**, 369–91.

Pluth, J. J., I. M. Steele, A. R. Kampf, D. I. Green 2006. Redgillite, $Cu_6(OH)_{10}(SO_4)\cdot H_2O$, a new mineral from the Caldbeck Fells, Cumbria, England, UK: description and crystal structure. *Mineralogical Magazine* **69**, 973–80.

Pockley, R. P. C. 1964. Four new uranium–lead ages from Cornwall. *Mineralogical Magazine* **33**, 1081–1092.

Pointon, C. M. 1980. Some environmental features of volcanogenic mineralization at Avoca,

Ireland and Parys Mountain, Anglesey, Wales. *Norges Geologiske Undersøkelse* **360**, 259–68.

Pointon, C. R. & R. A. Ixer 1980. Parys Mountain mineral deposit, Anglesey, Wales: geology and ore mineralogy. *Institution of Mining and Metallurgy, Transactions B* **89**, 143–55.

Pointer, C. M., J. R. Ashworth, P. R. Simpson 1989. Genesis of coffinite and the U–Ti association on Lower Old Red Sandstone sediments. *Mineralium Deposita* **24**, 117–23.

Poitrasson, F., S. Chenery, D. J. Bland 1996. Contrasted monazite hydrothermal alteration mechanisms and their geochemical implications. *Earth and Planetary Science Letters* **145**, 79–96.

Polkinghorne, J. P. R. 1951. Bridford baryte mine. *Royal Geological Society of Cornwall, Transactions* **18**, 240–54.

——— 1954. Tungsten in Cornwall. *Royal Geological Society of Cornwall, Transactions* **18**, 510–25.

Pollard, A. M., R. G. Thomas, P. A. Williams 1990. Connellite: stability relationships with other secondary copper minerals. *Mineralogical Magazine* **54**, 425–30.

Pollard, A. M., R. G. Thomas, P. A. Williams, R. E. Bevins, S. Turgoose 1989. Carbonatian connellite, a new variety, from the Britannia mine, North Wales, and from the Botallack mine, Cornwall. *Russell Society, Journal* **2**(2), 23–7.

Pollard, A. M., R. G. Thomas, P. A. Williams, J. Just, P. J. Bridge 1991. The synthesis and composition of georgeite and its reactions to form other secondary copper (II) carbonates. *Mineralogical Magazine* **55**, 163–6.

Poole, E. G. & B. Kelk 1971. *Calcium montmorillonite (fuller's earth) in the Lower Greensand of the Baulking area, Berkshire.* Report 71.4, Institute of Geological Sciences, London.

Poole, E. G. & A. J. Whiteman 1966. *The geology of the country around Nantwich and Whitchurch.* London: Institute of Geological Sciences.

Porteous, W. G. 1973. Metamorphic index minerals in the eastern Dalradian. *Scottish Journal of Geology* **9**, 29–43.

——— 1973. A breccia pipe in the Dalradian Series, east Kincardineshire. *Scottish Journal of Geology* **9**, 233–7.

Porter, L. 2004. *Ecton copper mines under the dukes of Devonshire – 1760–1790.* Ashbourne, Derbyshire: Landmark Publishing.

Postlethwaite, J. 1913. *Mines and mining in the Lake District* (3rd edn). Whitehaven: W. H. Moss [1975: reprinted by Michael Moon, Whitehaven).

Poustie, A. 1982. The discovery of a new mineral deposit within an Ordovician volcanic series near Charlestown, County Mayo. In *Mineral exploration in Ireland: progress and developments 1971–1981*, A. G. Brown & J. Pyne (eds), 97–107. Dublin: Irish Association of Economic Geology.

Poustie, A. & H. Kucha 1986. The geological setting, style and petrology of zinc–lead mineralization in the Moyvoughly area, Co. Westmeath. See Andrew et al. (1986: 305–318).

Power, G. M. 1968. Chemical variation in tourmalines from southwest England. *Mineralogical Magazine* **36**, 1078–1089.

Power, M. R. & D. Pirrie 2000. Platinum-group element mineralization within ultramafic rocks at Corrycharmaig, Perthshire: implications for the origin of the complex. *Scottish Journal of Geology* **36**, 143–50.

——— 2001. Platinum-group minerals in the Talnotry massive sulphide body, southwest Scotland – abstract. *Institution of Mining and Metallurgy, Transactions B* **110**, 235.

——— 2004. Platinum-group minerals within the Ballantrae complex, southwest Scotland. *Scottish Journal of Geology* **40**, 1–5.

Power, M. R. & P. W. Scott 1995. Talc–carbonate alteration of some basic and ultrabasic intrusions in Cornwall. *Ussher Society, Proceedings* **8**, 392–7.

Power, M. R., D. Pirrie, J. C. Ø. Andersen 2001. Sulphide-hosted platinum-group element mineralization, Rum, Inner Hebrides, United Kingdom [abstract]. *Institution of Mining and Metallurgy, Transactions B* **110**, 235.

——— 2003. Diversity of platinum-group element mineralization styles in the North Atlantic igneous province: new evidence from Rum, UK. *Geological Magazine* **140**, 499–512.

Power, M. R., D. Pirrie, J. C. Ø. Andersen, A. R. Butcher 2000a. Stratigraphical distribution of platinum-group minerals in the Eastern Layered Series, Rum, Scotland. *Mineralium Deposita* **35**, 762–75.

Power, M. R., D. Pirrie, J. C. Ø. Andersen, P. D. Wheeler 2000b. Testing the validity of chrome spinel chemistry as a provenance and petrogenetic indicator. *Geology* **28**, 1027–1030.

Power, M. R., D. Pirrie, J. Jedwab, C. J. Stanley 2004. Platinum-group element mineralization in an As-rich magmatic sulphide system, Talnotry, southwest Scotland. *Mineralogical Magazine* **68**, 395–411.

Power, M. R., R. K. Shail, A. C. Alexander, P. W. Scott 1997. Alteration and vein mineralization within the Lizard complex, south Cornwall: constraints on the timing of serpentinization. *Ussher Society, Proceedings* **9**, 188–94.

Pracht, M. & J. A. Kinnaird 1995. Mineral chemistry of megacrysts and ultramafic nodules from an undersaturated pipe at Black Ball Head, County Cork. *Irish Journal of Earth Sciences* **14**, 47–58.

Preston, J. 1967. A Tertiary feeder dyke in County Fermanagh, Northern Ireland. *Royal Dublin Society, Scientific Proceedings* **3A**, 1–16.

——— 1977. An unusual occurrence of quartz and amorphous silica at Carmean, Moneymore, Co. Antrim. *Geological Magazine* **114**, 389–92.

Preston, R. J., M. J. Hole, J. Bouch, J. Still 1998a. The occurrence of zirconian aegirine and calcic catapleiite ($CaZrSi_3O_9{\cdot}2H_2O$) within a nepheline syenite, British Tertiary igneous province. *Scottish Journal of Geology* **34**, 173–80.

Preston, R. J., M. J. Hole, J. Still 2000a. Exceptional REE-enrichment in apatite during the low-pressure fractional crystallization of alkali olivine basalt: an example from the British Tertiary Igneous Province. *Royal Society of Edinburgh, Transactions: Earth Sciences* **90**, 273–85.

——— 2000b. The occurrence of Zr-bearing amphiboles and their relationships with the pyroxenes and biotites in the teschenite and nepheline syenites of a differentiated dolerite boss, Islay, northwest Scotland. *Mineralogical Magazine* **64**, 459–68.

Preston, R. J., M. J. Hole, J. Still, H. Patton 1998b. The mineral chemistry and petrology of Tertiary pitchstones from Scotland. *Royal Society of Edinburgh, Transactions: Earth Sciences* **89**, 95–111.

Preston, R. J., I. Meighan, D. Simpson, M. J. Hole 2002. Mineral chemical provenance of Neolithic pitchstone artefacts from Ballygalley, County Antrim, Northern Ireland. *Geoarchaeology* **17**, 217–38.

Prichard, H. M. 1985. The Shetland ophiolite. In *The Caledonide orogen: Scandinavia and related areas*, D. G. Gee & B. A. Sturt (eds), 1173–84. Chichester: John Wiley.

Prichard, H. M. & R. A. Lord 1988. The Shetland ophiolite complex: evidence for a suprasubduction zone origin and implications for platinum-group element mineralization. In *Mineral deposits within the European Community*, J. Boissonnas & P. Omenetto (eds), 289–301. Special Publication 6, Society for Geology Applied to Mineral Deposits. Berlin: Springer.

Prichard, H. M. & M. Tarkian 1988. Platinum and palladium minerals from two PGE-rich localities in the Shetland ophiolite complex. *Canadian Mineralogist* **26**, 979–90.

Prichard, H. M., R. A. Ixer, R. A. Lord, J. Maynard, N. Williams 1994. Assemblages of platinum-group minerals and sulphides in silicate lithologies and chromite-rich rocks within the Shetland ophiolite. *Canadian Mineralogist* **32**, 271–94.

Prichard, H. M., C. R. Neary, P. J. Potts 1986. Mineral exploration in the Ballantrae complex, southwest Scotland. In *Metallogeny of basic and ultrabasic rocks*, M. J. Gallagher, R. A. Ixer, C. R. Neary, H. M. Prichard (eds), 395–414. London: Institution of Mining and Metallurgy.

Prichard, H. M., P. J. Potts, C. R. Neary 1981. Platinum-group element minerals in the Unst chromite, Shetland Isles. *Institution of Mining and Metallurgy, Transactions B* **90**, 186–8.

Prichard, H. M., P. J. Potts, C. R. Neary, R. A. Lord, G. R. Ward 1988. *Development of techniques for the determination of the platinum-group elements in ultramafic rock complexes of potential economic significance.* Unpublished report commissioned by the European Community Raw Materials Programme, contract no. MSM 089 UK (H).

Primmer, T. J. 1985. A transition from diagenesis to greenschist facies within a major Variscan fold/thrust complex in southwest England. *Mineralogical Magazine* **49**, 365–74.

Pring, W. D., J. Birch, M. Dawe, M. Taylor, M. Deliens, K. Walenta 1995. Kintoreite, $PbFe_3(PO_4)_2(OH,H_2O)_6$, a new mineral of the jarosite–alunite family, and lusungite discredited. *Mineralogical Magazine* **59**, 143–9.

Prior, G. T. 1890. On zinc sulphide replacing stibnite and orpiment: analysis of stephanite and polybasite. *Mineralogical Magazine* **9**, 9–15.

——— 1902. The identity of kilbrickenite with geochronite: and analyses of miersite, marshite, and copper-pyrites. *Mineralogical Magazine* **13**, 186–90.

——— 1906. Dundasite from North Wales. *Mineralogical Magazine* **14**, 167–9.

——— 1910. On an analcite-basalt from Rathjordan, Co. Limerick. *Mineralogical Magazine* **15**, 315–17.

Prout, S. J. & J. C. Ø. Andersen 2001. Platinum-group element occurrences within the Ben Buie intrusion, Isle of Mull, northwest Scotland. *Institution of Mining and Metallurgy, Transactions B* **110**, 235–6.

Prowse, B. A. 1988. On the morphology of barite crystals from Silverwood colliery, South Yorkshire. *UK Journal of Mines and Minerals* **4**, 3–4.

Psyrillos, A., D. A. C. Manning, S. D. Burley 1998. Geochemical constraints on kaolinization in the St Austell granite, Cornwall, England. *Geological Society of London, Journal* **155**, 829–40.

Pulfrey, W. 1933. The iron-ore oolites and pisolites of North Wales. *Geological Society of London, Quarterly Journal* **89**, 401–430.

Pryor, M. J. 1988. Geological and fluid inclusion studies at Ogofau and Gwynfynydd gold mines, Wales. *South African Journal of Geology* **91**, 450–64.

Pye, K. 1981. Marshrock formed by iron sulphide and siderite cementation in saltmarsh sediments. *Nature* **294**, 650–52.

——— 1984. SEM analysis of siderite cements in intertidal marsh sediments, Norfolk, England. *Marine Geology* **56**, 1–12.

——— 1988. An occurrence of akaganéite (β-FeOOH·Cl) in recent oxidized carbonate concretions, Norfolk, England. *Mineralogical Magazine* **52**, 125–6.

Quirk, D. G. 1985. The mineralogy and paragenesis of Speedwell mine, Castleton, Derbyshire. *Mercian Geologist* **10**, 39–60.
——— 1993. Origin of the Peak District orefield. *UK Journal of Mines and Minerals* **13**, 4–15.

Raade, G., C. J. Elliot, V. K. Din 1985. New data on glaucocerinite. *Mineralogical Magazine* **49**, 583–90.
Raistrick, A. 1936a. The copper deposits of Middleton Tyas, N Yorks. *Naturalist* **61**, 111–15.
——— 1936b. Lead smelting in the north Pennines during the 17th and 18th centuries. *University of Durham Philosophical Society, Proceedings* **9**, 164–70.
——— 1973. *Lead mining in the Mid-Pennines.* Truro: Bradford Barton.
——— 1975. *The lead industry of Swaledale and Wensleydale* [2 vols]. Buxton, Derbyshire: Moorland Publishing.
——— 1983. *Mines and miners on Malham Moor.* Littleborough, Lancashire: George Kelsall Publishing.
Ramsbottom, W. H. C., G. H. Rhys, E. G. Smith 1962. Boreholes in the Carboniferous rocks of the Ashover district, Derbyshire. *Geological Survey of Great Britain, Bulletin* **19**, 75–168.
Randall, B. A. O. 1953. A large mineralized cavity in a tholeiite dike in Northumberland. *Mineralogical Magazine* **30**, 246–53.
——— 1959. Stevensite from the Whin Sill in the region of the North Tyne. *Mineralogical Magazine* **32**, 218–25.
Randall, B. A. O. & J. M. Jones 1966. Sideronatrite from mineralized cavities in the Rising Sun colliery, Backworth, Northumberland. *Mineralogical Magazine* **35**, 983–90.
Rankin, A. H. 1977. Macro fluid inclusions in fluorite from the Northern Pennine orefield. *Institution of Mining and Metallurgy, Transactions B* **86**, 158.
Rankin, A. H. & A. J. Criddle 1985. Mineralizing fluids and metastable low-temperature inclusion brines at Llanharry iron deposit, South Wales. *Institution of Mining and Metallurgy, Transactions B* **94**, 126–32.
Rankin, A. H. & M. J. Graham 1988. Na, K and Li contents of mineralizing fluids in the Northern Pennine orefield, England, and their genetic significance. *Institution of Mining and Metallurgy, Transactions B* **97**, 99–107.
Rankin, A. H., M. F. Miller, J. S. Carter 1987. The release of trace elements and volatiles from crinoidal limestone during thermal decrepitation. *Mineralogical Magazine* **51**, 517–25.
Rapaport, P. A. & C. W. Burnham 1973. Ferrobustamite: the crystal structures of two Ca, Fe bustamite-type pyroxenoids. *Zeitschift für Krystallographie* **138**, 419–38.
Rashleigh, P. 1797. *Specimens of British minerals selected from the cabinet of Phillip Rashleigh* (part 1). London: Nicol & White.
Rastall, R. H. 1906. The Buttermere and Ennerdale granophye. *Geological Society of London, Quarterly Journal* **62**, 253–74.
——— 1910. The Skiddaw granite and its metamorphism. *Geological Society of London, Quarterly Journal* **66**, 116–41.
——— 1941. Xenoliths at Threlkeld. *Yorkshire Geological Society, Proceedings* **24**, 223–32.
——— 1942. The ore deposits of the Skiddaw district. *Yorkshire Geological Society, Proceedings* **24**, 329–43.
Rastall, R. H. & J. E. Hemingway 1949. The Yorkshire Dogger, IV: Rosedale and Farndale. *Geological Magazine* **86**, 265–78.
Rastall, R. H. & W. H. Wilcockson 1915. The accessory minerals of the granitic rocks of the English Lake District. *Geological Society of London, Quarterly Journal* **71**, 592–622.
Rawson, J. R., D. A. Carswell, D. Smallwood 2001. Garnet-bearing olivine-websterite within the Eastern Glenelg Lewisian of the Glenelg inlier, Northwest Highlands. *Scottish Journal of Geology* **37**, 27–34.
Raybould, J. G. 1974. Ore textures, paragenesis and zoning in the lead–zinc veins of mid-Wales. *Institution of Mining and Metallurgy, Transactions B* **83**, 112–19.
Rayment, B. D., G. R. Davis, J. D. Wilson 1971. Controls to mineralization at Wheal Jane, Cornwall. *Institution of Mining and Metallurgy, Transactions B* **80**, 224–37.
Raymond, L. R. 1959. Recent secondary minerals in the Billingham anhydrite mine. *Mineralogical Magazine* **32**, 172–5.
——— 1962. The petrology of the Lower Magnesian Limestone of northeast Yorkshire and southeast Durham. *Geological Society of London, Quarterly Journal* **118**, 39–64.
Read, D., D. C. Cooper, J. M. McArthur 1987. The composition and distribution of nodular monazite in the Lower Palaeozoic rocks of Great Britain. *Mineralogical Magazine* **51**, 271–80.
Read, H. H. 1925. A diopside-bearing pegmatite near Ellon in Aberdeenshire. *Edinburgh Geological Society, Transactions* **11**, 353–6.
——— 1929. Occurrence of iron-rich cordierite in Aberdeenshire. *Geological Magazine* **66**, 547–52.
——— 1931a. On corundum-spinel xenoliths in the gabbro of Haddo House, Aberdeenshire. *Geological Magazine* **69**, 446–53.
——— 1931b. *The geology of central Sutherland.* Memoir, Geological Survey of Great Britain, Edinburgh.
——— 1932. On quartz–kyanite rocks in Unst, Shetland Islands, and their bearing on metamorphic differentiation. *Mineralogical Magazine* **23**, 317–28.
——— 1934a. On zoned associations of antigorite, talc, actinolite, chlorite, and biotite in Unst, Shetland Islands. *Mineralogical Magazine* **23**, 519–40.
——— 1934b. The metamorphic geology of Unst in the Shetland Islands. *Geological Society of London, Quarterly Journal* **90**, 637–88.
——— 1952. Metamorphism and migmatisation in the Ythan Valley, Aberdeenshire. *Edinburgh Geological Society, Transactions* **15**, 265–79.
Read, H. H. & B. E. Dixon 1932. On stichtite from Cunningsburgh, Shetland Islands. *Mineralogical Magazine* **23**, 309–316.
Read, H. H. & I. S. Double 1935. On the occurrence of chondrodite in the Glenelg limestone of Inverness-shire. *Mineralogical Magazine* **24**, 84–9.
Readwin, T. A. 1864. On the recent discovery of gold near Bala Lake, Merionethshire. *Report of the British Association for the Advancement of Science for 1862*, 86–7.
——— 1888. *Gold in Wales.* 1–12. London.
Reed, C. P. & M. W. Wallace 2001. Diagenetic evidence for an epigenetic origin of the Courtbrown Zn–Pb deposit, Ireland. *Mineralium Deposita* **36**, 428–41.
——— 2004. Zn–Pb mineralization in the Silvermines district, Ireland: a product of burial diagenesis. *Mineralium Deposita* **39**, 87–102.
Reedman, A. J., T. B. Colman, S. D. G. Campbell, M. F. Howells 1985. Volcanogenic mineralization related to the Snowdon Volcanic Group (Ordovician), Gwynedd, North Wales. *Geological Society of London, Journal* **142**, 875–88.
Reid, C. & J. B. Scrivenor 1906. *The geology of the country near Newquay (sheet 346).* Memoir, Geological Survey of Great Britain, London.
Reid, C., G. Barrow, H. Dewey 1910. *The geology of the country around Padstow and Camelford (sheets 335 and 336).* Memoir, Geological Survey of Great Britain, London.
Reilly, T. A. 1986. A review of vein mineralization in southwest County Cork, Ireland. See Andrew et al. (1986: 513–44).
Reynolds, D. L. 1936. Demonstrations in petrogenesis from Kiloran Bay, Colonsay, I: the transfusion of quartzite. *Mineralogical Magazine* **24**, 367–407.
Rhoden, H. N. 1959. Mineralogy of the Silvermines district, County Tipperary, Eire. *Mineralogical Magazine* **32**, 128–39.
Rhodes, F. H. T. 1954. Insects from a travertine deposit in Upper Teesdale, County Durham. *University of Durham Philosophical Society, Proceedings* **11**, 131–5.
Rice, A. H. N. 1993. Textural and twin sector zoning and displacement of graphite in chiastolite and pyralspite and grandite garnets in the Variscides of southwest England. *Ussher Society, Proceedings* **8**, 124–31.
Rice, C. M. 1973. Chemical weathering on the Carnmenellis granite. *Mineralogical Magazine* **39**, 429–47.
Rice, C. M. & B. Davies 1979. Copper mineralization associated with an appinite pipe in Argyll, Scotland. *Institution of Mining and Metallurgy, Transactions B* **88**, 154–60.
Rice, C. M. & J. M. Williams 1969. A Mössbauer study of biotite weathering. *Mineralogical Magazine* **37**, 210–15.
Rice, C. M., and 13 other authors 1995. A Devonian auriferous hot spring system, Rhynie, Scotland. *Geological Society of London, Journal* **152**, 229–50.
Rice, R. & G. J. Sharp 1976. Copper mineralization in the forest of Coed-y-Brenin, North Wales. *Institution of Mining and Metallurgy, Transactions B* **85**, 1–13.
Richards, L. 1992. Snailbeach lead mine – reclamation and conservation. *Earth Science Conservation* **31**, 8–12.
Richardson, S. W. 1968. The petrology of the metamorphosed syenite in Glen Dessary, Inverness-shire. *Geological Society of London, Quarterly Journal* **124**, 9–51.
Richardson, W. A. 1920. The fibrous gypsum of Nottinghamshire. *Mineralogical Magazine* **19**, 77–95.
——— 1921. The micropetrography of the rock-gypsum of Nottinghamshire. *Mineralogical Magazine* **19**, 196–207.
——— 1923. A micrometric study of the St Austell granite. *Geological Society of London, Quarterly Journal* **79**, 547–76.
Richey, J. E. 1927. The structural relations of the Mourne granites (Northern Ireland). *Geological Society of London, Quarterly Journal* **83**, 653–88.
Richey, J. E., H. H. Thomas, E. B. Bailey, J. B. Simpson, V. A. Eyles, G. W. Lee 1930. *The geology of Ardnamurchan, northwest Mull and Coll (sheet 51 and part of sheet 52).* Memoir, Geological Survey of Great Britain, London.
Ridgway, J. M. 1980. Silver. *Mineral Resources Consultative Committee* 25, 1–70. London: HMSO.
Řídkošil, T., J. Sejkora, V. Šrein 1996. Smrkovecite, monoclinic $Bi_2O(OH)(PO_4)$, a new mineral of the atelestite group. *Neues Jahrbuch für Mineralogie*, Monatshefte 3, 97–102.
Ridley, I. 1971. The petrology of some volcanic rocks from the British Tertiary Province: the

islands of Rhum, Eigg, Canna and Muck. *Contributions to Mineralogy and Petrology* **32**, 251–66.

Rieder, M., P. Povondra, J. Fryda 1995. Coexisting biotite and muscovite: an example from a Moinian mica schist at Glenfinnan, Scottish Highlands. *Mineralogy and Petrology* **53**, 63–74.

Rieder, M., and 14 other members of the IMA mica subcommittee 1998. Nomenclature of the micas. *Canadian Mineralogist* **36**, 905–912.

Rieuwerts, J. H. 1960. The mines of Calver, Coombs Dale and Longstone Edge. *Peak District Mines Historical Society, Bulletin* **1**(3), 3–6.

——— 1963. Gold in Lathkill Dale. *Peak District Mines Historical Society, Bulletin* **5**, 48–9.

Roach, R. A. 1969. The composite nature of the St David's Head and Carn Llidi intrusions of north Pembrokeshire. In *The Precambrian and Lower Palaeozoic rocks of Wales*, A. Wood (ed.), 409–433. Cardiff: University of Wales Press.

Robert, C. 2001. Hydrothermal alteration processes of Tertiary lavas of Northern Ireland. *Mineralogical Magazine* **65**, 543–54.

Roberts, A. C., W. H. Paar, M. A. Cooper, D. Topa, A. J. Criddle, J. Jedwab 2002. Verbeekite, monoclinic $PdSe_2$, a new mineral from the Musonoi Cu–Co–Mn–U mine, near Kolwezi, Shaba Province, Democratic Republic of Congo. *Mineralogical Magazine* **66**, 173–9.

Roberts, B. 1979. *The geology of Snowdonia and Llyn: an outline and field guide.* Bristol: Adam Hilger.

——— 1981. Low-grade and very low-grade regional metabasic Ordovician rocks of Llyn and Snowdonia, Gwynedd, North Wales. *Geological Magazine* **119**, 189–200.

Roberts, B. & R. J. Merriman 1990. Cambrian and Ordovician metabentonites and their relevance to the origins of associated mudrocks in the northern sector of the Lower Palaeozoic Welsh marginal basin. *Geological Magazine* **127**, 31–43.

Roberts, B., J. A. Evans, R. J. Merriman, M. Smith 1989. Discussion on low grade metamorphism of the Welsh Basin Lower Palaeozoic succession: an example of diastathermal metamorphism? *Geological Society of London, Journal* **146**, 885–90.

Roberts, D. 1986. *Geochemistry of recent freshwater manganese tunnel deposits.* PhD thesis, University of Wales.

Roberts, D. E. 1983. Metasomatism and the formation of greisen in Grainsgill, Cumbria, England. *Geological Journal* **18**, 43–52.

Roberts, G. E. 1860. *The rocks of Worcestershire.* London: Joseph Masters.

Roberts, P. D., T. K. Ball, P. J. Hooker, A. E. Milodowski 1988. A uranium geochemical study at the natural analogue site of Needle's Eye, SW Scotland. *Materials Research Society, Proceedings* **127**, 933–40.

Roberts, W. L., T. J. Campbell, G. R. Rapp 1990. *Encyclopedia of minerals* (2nd edn). London: Chapman & Hall.

Robertson, T. 1945. *Potash felspar.* Wartime pamphlet 44, Geological Survey of Great Britain, London.

Robertson, T. & H. G. Dines 1929. The South Terras radium deposit, Cornwall. *Mining Magazine* **12**, 147–53.

Robey, J. A. & L. Porter 1972. *The copper and lead mines of Ecton Hill, Staffordshire.* Buxton, Derbyshire: Moorland Publishing, on behalf of the Peak District Mines Historical Society.

Robinson, A. J. 1933. Annual geological report. *Société Jersiaise Annual Bulletin* **12**, 100–102.

——— 1934. Annual geological report. *Société Jersiaise Annual Bulletin* **12**, 237–40.

——— 1947. Annual geological report. *Société Jersiaise Annual Bulletin* **14**, 279–80.

Robinson, B. W. & P. R. Ineson 1979. Sulphur, oxygen and carbon isotope investigations of lead–zinc–barite–fluorite–calcite mineralization, Derbyshire, England. *Institution of Mining and Metallurgy, Transactions B* **88**, 107–117.

Robinson, D., R. A. Nicholls, L. J. Thomas 1980. Clay mineral evidence for low-grade Caledonian and Variscan metamorphism in southwestern Dyfed, South Wales. *Mineralogical Magazine* **43**, 857–63.

Robinson, G. W. & V. T. King 1993. What's new in minerals? *Mineralogical Record* **24**(5), 381–95.

Robinson, G. W., V. T. King, E. Asselborn, F. Cureton, R. Tschernich, R. Sielecki 1992. What's new in minerals? *Mineralogical Record* **23**(5), 423–38.

Robinson, J. 1997a. Field meeting report: Meldon CAMASS quarry, near Okehampton, Devon. *Russell Society, Newsletter* **30**, 16.

——— 1997b. Field meeting report: Kennack Sands, the Lizard, Cornwall. *Russell Society, Newsletter* **30**, 16–17.

Robinson, J. & I. Dossett 2002. Field meeting report: North West trip to the Isles of Harris & Skye. *Russell Society, Newsletter* **41**, 40–43.

Robson, J. 1945a. Geology of Carn Brea. *Royal Geological Society of Cornwall, Transactions* **17**, 208–221.

——— 1945b. The geology of the St Ives district. *Royal Geological Society of Cornwall, Transactions* **17**, 272–83.

——— 1949a. The geology of the Land's End peninsula. *Royal Geological Society of Cornwall, Transactions* **17**, 427–54.

——— 1949b. Cornish mineral index. *Royal Geological Society of Cornwall, Transactions* **17**, 455–75.

——— 1954. Cornish mineral index amendments. *Royal Geological Society of Cornwall, Transactions* **18**, 406–410.

Rock, N. M. S. 1977. A new occurrence of fenite in the Loch Borrolan alkaline complex, Assynt. *Mineralogical Magazine* **41**, 529 and M7.

——— 1984. *New types of hornblendic rocks and prehnite-veining in the Moines west of the Great Glen, Inverness-shire.* Report 83.3, Institute of Geological Sciences, London.

Rock, N. M. S., C. Cooper, J. W. Gaskarth 1986a. Late Caledonian subvolcanic vents and associated dykes in the Kirkudbright area, Galloway, southwest Scotland. *Yorkshire Geological Society, Proceedings* **46**, 29–37.

Rock, N. M. S., R. Macdonald, S. E. Drewery, R. J. Pankhurst, M. Brook 1986b. Pelites of the Glen Urquhart serpentinite metamorphic complex, west of Loch Ness (anomalous local limestone–pelite successions within the Moine outcrop: III). *Scottish Journal of Geology* **22**, 179–202.

Rodgers, K. A. & G. Cressey 2001. The occurrence, detection and significance of moganite (SiO_2) among some silica sinters. *Mineralogical Magazine* **65**, 157–67.

Rodgers, P. R. 1975. *Yorkshire minerals.* Clapham, Lancaster: Dalesman Publishing.

Rogers, C. 1968. *A collector's guide to minerals, rocks and gemstones in Cornwall and Devon.* Truro: Bradford Barton.

Rogers, G. & G. R. Dunning 1991. Geochronology of appinitic and related granitic magmatism in the W Highlands of Scotland: constraints on the timing of transcurrent fault movement. *Geological Society of London, Journal* **148**, 17–27.

Rogers, P. J. 1974. Bravoite inclusions in fluorites from the Derbyshire and Askrigg orefields. *Peak District Mines Historical Society, Bulletin* **5**, 333.

——— 1977. Fluid inclusion studies in fluorite from the Derbyshire orefield. *Institution of Mining and Metallurgy, Transactions B* **86**, 128–32.

——— 1978. Fluid inclusion studies on fluorite from the Askrigg Block. *Institution of Mining and Metallurgy, Transactions B* **87**, 125–31.

Rollinson, H. R. 1979. Ilmenite–magnetite geothermometry in trondhjemites from the Scourian complex of northwest Scotland. *Mineralogical Magazine* **43**, 165–70.

——— 1980. Mineral reactions in a granulite facies calc-silicate rock from Scourie. *Scottish Journal of Geology* **16**, 153–64.

Roscoe, H. E. 1876. On two new vanadium minerals. *Royal Society (London), Proceedings* **25**, 109–112.

Rose, W. C. C. & K. C. Dunham 1977. *Geology and hematite deposits of south Cumbria: economic memoir for 1:50 000 geological sheet 58, and southern part of sheet 48.* Memoir, Institute of Geological Sciences, London.

Ross, C. S. & P. F. Kerr 1930. Dickite, a kaolin mineral. *American Mineralogist* **15**, 34–9.

Rothstein, A. T. V. 1972. Spinels from the Dawros peridotite, Connemara, Ireland. *Mineralogical Magazine* **38**, 957–60.

Rothwell, M. & J. Dickinson 1989. Minerals in glacial erratics from Thurstaston beach, Wirral, Merseyside. *UK Journal of Mines and Minerals* **6**, 38–43.

Rothwell, M. & J. Mason 1992. Wulfenite in the British Isles. *UK Journal of Mines and Minerals* **11**, 30–41.

Rottenbury, F. J. 1974. *Geology, mineralogy and mining history of the metalliferous mining areas of Exmoor.* PhD thesis, University of Leeds.

Rowbotham, G. & G. J. Lees 1990. The Wolf's Cave gabbro – alkaline olivine-basaltic magmatism in Jersey, Channel Islands. *Ussher Society, Proceedings* **7**, 251–7.

Roycroft, P. D. 1989. Zoned muscovite from the Leinster granite, SE Ireland. *Mineralogical Magazine* **53**, 633–5.

Roycroft, P. D. & P. S. Kennan 2000. Muscovite megacrysts in the Leinster granite, southeast Ireland: the record of a very complex growth history. *Irish Journal of Earth Sciences* **18**, 101–112.

Roydhouse, P. & D. I. Green, 2003. Copper minerals from Flamborough Head, East Yorkshire. *UK Journal of Mines and Minerals* **23**, 36.

Rudashevsky, N. S., A. M. McDonald, L. J. Cabri, T. F. D. Nielsen, C. J. Stanley, Yu. L. Kretzler, V. N. Rudashevsky 2004. Skaergaardite, PdCu, a new platinum-group intermetallic mineral from the Skaergaard intrusion, Greenland. *Mineralogical Magazine* **68**, 615–32.

Rudler, F. W. 1905. *A handbook to a collection of the minerals of the British Islands, mostly selected from the Ludlam collection in the Museum of Practical Geology.* London: HMSO.

Rudwick, M. J. S. 2005. *Lyell and Darwin, geologists: studies in the Earth sciences in the Age of Reform.* Aldershot: Ashgate.

Rumsey, M. S. 2005. Mineralogical note: veszelyite from Cornwall. *Russell Society, Newsletter* **46**, 4.

——— 2007. The first British occurrence of preisingerite and rooseveltite from Carrock mine, Caldbeck Fells, Cumbria. *UK Journal of Mines and Minerals* **28**, 51–3.

Rumsey, M. S. & M. Savage 2005. The first British occurrence of parkerite at North Devon

United mine, Peter Tavy, Devon. *UK Journal of Mines and Minerals* **25**, 19–22.

Rumsey, M. S., T. F. Bridges, J. Spratt 2007. The first British occurrence of petitjeanite at Carrock mine, Caldbeck Fells, Cumbria. *UK Journal of Mines and Minerals* **28**, 47–50.

Rushton, D. R. A. 1972. Arsenates of copper from Shetland. *Mineralogical Magazine* **38**, 626–7.

Russell, A. 1905. Minerals [exhibit: general meeting of the Mineralogical Society, Burlington House, London, 15 November 1904]. *Mineralogical Magazine* **14**, xxvii.

——— 1907a. On the occurrence of linarite and caledonite in County Wicklow. *Mineralogical Magazine* **14**, 348–9.

——— 1907b. Note on the mines and minerals of the Silvermines district, Co. Tipperary. *Mineralogical Magazine* **13**, 350–53.

——— 1910a. On the occurrence of the rare mineral carminite in Cornwall. *Mineralogical Magazine* **15**, 285–7.

——— 1910b. Notes on the occurrence of zeolites in Cornwall and Devon. *Mineralogical Magazine* **15**, 377–84.

——— 1911a. On the occurrence of phenacite in Cornwall. *Mineralogical Magazine* **16**, 55–62.

——— 1911b. An occurrence of the barium–feldspar celsian in North Wales. *Nature* **16**, 180.

——— 1912a. Prehnite from the Lizard district, Cornwall. *Mineralogical Magazine* **16**, 217–18.

——— 1912b. Notes on the occurrence of dundasite in Derbyshire and Co. Galway. *Mineralogical Magazine* **16**, 272–3.

——— 1913a. An account of the minerals found in the Virtuous Lady mine, near Tavistock. *Mineralogical Magazine* **17**, 1–14.

——— 1913b. Notes on the occurrence of bertrandite at some new localities in Cornwall. *Mineralogical Magazine* **17**, 15–21.

——— 1920a. On the occurrence of phenacite and scheelite at Wheal Cock, St Just, Cornwall. *Mineralogical Magazine* **19**, 19–22.

——— 1920b. On the occurrence of cotunnite, anglesite, leadhillite and galena on fused lead from the wreck of the fireship *Firebrand* in Falmouth harbour, Cornwall. *Mineralogical Magazine* **19**, 64–8.

——— 1922. On a discovery of pitchblende at Kingswood mine, Buckfastleigh, Devon. *Geological Magazine* **59**, 183.

——— 1924. Topaz from Cornwall, with an account of its localities. *Mineralogical Magazine* **20**, 221–36.

——— 1925. A notice of the occurrence of native arsenic in Cornwall; of bismuthinite at Shap, Westmorland; and of smaltite and niccolite at Coniston, Lancashire. *Mineralogical Magazine* **20**, 299–304.

——— 1926. Report on the southern excursion to Devon and Cornwall of the Mineralogical Society – jubilee celebration. *Mineralogical Magazine* **21**, 124–8.

——— 1927a. On the occurrence of the rare mineral nadorite in Cornwall, and of beraunite in Co. Cork, Ireland. *Mineralogical Magazine* **21**, 272–4.

——— 1927b. Notice of an occurrence of niccolite and ullmannite at the Settlingstones mine, Fourstones, Northumberland; and of serpierite at Ross Island mine, Killarney, Co. Kerry, Ireland. *Mineralogical Magazine* **21**, 383–7.

——— 1929. On the occurrence of native gold at Hope's Nose, Torquay, Devonshire. *Mineralogical Magazine* **22**, 159–62.

——— 1936a. Notes on the occurrence of fluorite in Aberdeenshire and Banffshire. *Mineralogical Magazine* **24**, 307–317.

——— 1936b. Barytes crystals from the Manvers Main colliery, Wath-upon-Dearne, near Rotherham, Yorkshire. *Mineralogical Magazine* **24**, 318–20.

——— 1936c. Notes on the occurrence of wulfenite at Brandy Gill, Cumberland, and of leadhillite at Drumruck mine, Kirkcudbrightshire. *Mineralogical Magazine* **24**, 321–3.

——— 1944. Notes on some minerals either new or rare to Britain. *Mineralogical Magazine* **27**, 1–10.

——— 1946a. Minerals [exhibit: general meeting of the Mineralogical Society, Burlington House, London, 24 January 1946]. *Mineralogical Magazine* **27**, xxxvii–xxxviii.

——— 1946b. An account of the Struy lead mines, Inverness-shire, and of wulfenite, harmotome, and other minerals which occur there. *Mineralogical Magazine* **27**, 147–54.

——— 1946c. Baryte and fluorite from Loch Bruicheach, Beuly, Inverness-shire. *Mineralogical Magazine* **27**, 155–6.

——— 1946d. On rhodonite and tephroite from Treburland manganese mine, Altarnun, Cornwall; and on rhodonite from other localities in Cornwall and Devonshire. *Mineralogical Magazine* **27**, 221–35.

——— 1948. On rashleighite, a new mineral from Cornwall, intermediate between turquoise and chalcosiderite. *Mineralogical Magazine* **28**, 353–8.

——— 1949a. The Wherry mine, Penzance: its history and its mineral productions. *Mineralogical Magazine* **28**, 517–33.

——— 1949b. An account of the antimony mines of Great Britain and Ireland with lists of the minerals found therein [unpublished transcript, NHM]. *Mineralogical Magazine* **29**, lix.

——— 1952. On the occurrence of turquoise in Cornwall. *Mineralogical Magazine* **29**, 909–912.

Russell, A. & G. F. Claringbull 1956. Ceruleite from Wheal Gorland, Gwennap, Cornwall [paper presented at general meeting of the Mineralogical Society, Burlington House, London, 9 June 1955]. *Mineralogical Magazine* **31**, xlv.

Russell, A. & A. Hutchinson 1922. On laurionite and paralaurionite from Cornwall (Report and Proceedings of Min. Soc. meeting 10th Jan. 1922). *Geological Magazine* **59**, 182–3.

——— 1927. On laurionite and associated minerals from Cornwall. *Mineralogical Magazine* **21**, 221–8.

Russell, A. & E. A. Vincent 1952. On the occurrence of varlamoffite (partially hydrated stannic oxide) in Cornwall. *Mineralogical Magazine* **29**, 817–26.

Russell, G. E. 1990. *Silvermines: the mine and the company.* Limerick: McKerns Printing Works.

Russell, J. D., A. Birnie, A. R. Fraser 1984a. High-gradient magnetic separation (HGMS) in soil clay-mineral studies. *Clay Mineralogy* **19**, 771–8.

Russell, J. D., A. R. Fraser, A. Livingstone 1984b. The infrared absorption spectra of the three polymorphs of $PbSO_4(CO_3)_2(OH)_2$ (leadhillite, susannite, and macphersonite). *Mineralogical Magazine* **48**, 295–7.

Russell, M. J. 1975. Lithogeochemical environment of the Tynagh base-metal deposit, Ireland, and its bearing on ore deposition. *Institution of Mining and Metallurgy, Transactions B* **84**, 128–33.

Rust, S. A. 1981a. Wheal Gorland, part 1: the mine. *Mineral Realm* **1**(2), 10–18.

——— 1981b. Wheal Gorland, part 2: the minerals. *Mineral Realm* **1**(3), 6–19.

——— 1982a. Silver minerals in Cornwall and Devon. *Mineral Realm* **1**(4), 20–28.

——— 1982b. Connellite. *Mineral Realm* **1**(5), 31–4.

——— 1982c. Cu_2S. *Mineral Realm* **1**(6), 12–13.

——— 1982d. Wheal Jane. *Mineral Realm* **2**(1), 9–20.

——— 1982e. Mount Wellington mine. *Mineral Realm* **2**(2), 9–10.

——— 1982f. Chalcopyrite from Devon and Cornwall. *Mineral Realm* **2**(2), 14–15.

——— 1982g. Wheal Jane. *Mineral Realm* **2**(2), 27.

——— 1983a. Wheal Wrey and Ludcott United. *Mineral Realm* **2**(6), 10–24.

——— 1983b. Weddellite – a new British mineral species from Derbyshire. *Mineral Realm* **3**(1), 22.

——— 1983c. Notes on crystals of wulfenite from Leadhills, Scotland. *Mineral Realm* **3**(2/3), 11–12.

——— 1983d. Notes on micro-specimens found at the Waterbank mine, Ecton, Staffordshire. *Mineral Realm* **3**(2/3), 29–32.

——— 1985. New finds. *British Micromount Newsletter* **13**, 6.

——— 1987a. Post-mining minerals of Tynebottom mine, Cumbria. *UK Journal of Mines and Minerals* **2**, 21–2.

——— 1987b. Secondary minerals from Wetgrooves mine, Aysgarth, Yorkshire. *UK Journal of Mines and Minerals* **3**, 14.

——— 1990a. Ullmannite from Hendrefelen mine, Ysbyty Ystwyth, Dyfed, Wales. *UK Journal of Mines and Minerals* **8**, 47.

——— 1990b. Mattheddleite from the Darren mine, Dyfed, Wales. *UK Journal of Mines and Minerals* **8**, 47–8.

——— 1990c. Susannite and sulphatian schmiederite from Llechwedd Helyg mine, Tir-y-Mynach, Dyfed, Wales. *UK Journal of Mines and Minerals* **8**, 48.

——— 1990d. Filiform pyrite from Castell mine, Ponterwyd, Dyfed, Wales. *UK Journal of Mines and Minerals* **8**, 48–9.

——— 1991a. Siegenite from Erglodd mine, Talybont, Dyfed, Wales. *UK Journal of Mines and Minerals* **9**, 4.

——— 1991b. Ktenasite, from Smallcleugh mine, Nenthead, Cumbria, the first British occurrence. *UK Journal of Mines and Minerals* **9**, 5.

——— 1991c. The minerals of Milltown quarry, Ashover, Derbyshire. *UK Journal of Mines and Minerals* **10**, 14–19.

——— 1992. Ramsbeckite, the first three British occurrences. *UK Journal of Mines and Minerals* **11**, 24.

——— 1994. Chenite from Llechwedd Helyg mine, Tir-y-Mynach, Dyfed, Wales. *UK Journal of Mines and Minerals* **14**, 9.

——— 1995a. Bismutite from Penberthy Croft mine, St Hilary, Cornwall. *UK Journal of Mines and Minerals* **15**, 18.

——— 1995b. Laurionite from Nantycagl mine, Ceulanymaesmawr, Dyfed, Wales. *UK Journal of Mines and Minerals* **15**, 19.

——— 1995c. Supergene minerals from Waterbank mine, Ecton, Staffordshire. *UK Journal of Mines and Minerals* **15**, 30–35.

——— 1996a. Millerite from Penberthy Croft mine, Cornwall. *UK Journal of Mines and Minerals* **16**, 15.

——— 1996b. Bismoclite, a first Cornish occurrence at Croft Gothal mine. *UK Journal of Mines and Minerals* **17**, 17.

——— 1996c. Susannite and caledonite *in situ* in Central Wales. *UK Journal of Mines and Minerals* **17**, 18.

——— 1996d. Hildagoite and arsenolite, two new occurrences at Penberthy Croft mine, Cornwall. *UK Journal of Mines and Minerals* **17**, 18.

——— 1997. Chenite from Esgair Hir mine, Talybont, Dyfed. *UK Journal of Mines and Minerals* **18**, 16.

——— 1998. Bismuth minerals from Wheal Remfy pit, Cornwall. *UK Journal of Mines and Minerals* **19**, 11.

Rust, S. A. & D. I. Green 2005. Olsacherite, the first British occurrence, from Waterbank mine, Ecton, Staffordshire. *UK Journal of Mines and Minerals* **25**, 33–4.

Rust, S. A. & J. S. Mason 1988. The minerals of Esgair Hir mine, Bwlch-y-Esgair Hir, Ceulanymaesmawr, Dyfed, Wales. *UK Journal of Mines and Minerals* **5**, 35–43.

——— 1993. Minerals from the Frongoch mine, mid-Wales. *British Micromount Newsletter* **39**, 18–19.

——— 1994. An unusual occurrence of arsenate minerals at Gwaith-yr-afon mine, Dyfed, Wales. *Russell Society, Journal* **5**(2), 109–113.

Rust, S. A. & D. Rust 1987. Micro minerals from Dyfngwm mine. *UK Journal of Mines and Minerals* **2**, 28–32.

Rust, S. A., R. Bell, D. I. Green 2003. The occurrence of bechererite in the British Isles. *UK Journal of Mines and Minerals* **23**, 37–8.

Rust, S. A., D. Moulding, D. I. Green 2007. Bechererite from Leadhills, south Lanarkshire and the Charlotte United mines, Perranuthnoe, Cornwall. *UK Journal of Mines and Minerals* **28**, 7–8.

Ryback, G. 1988a. Mimetite from Tynagh mine, Co. Galway. *Irish Naturalists' Journal* **22**, 491–2.

——— 1988b. Alumohydrocalcite from Scarborough, North Yorkshire, and Weston Favell, Northamptonshire. *Russell Society, Journal* **2**(1), 9–12.

——— 1992. *Glossary of Irish minerals*. Unpublished manuscript.

Ryback, G. & J. G. Francis 2001. Rosiaite from Bwlch mine, Deganwy, Conwy, Wales. *Russell Society, Journal* **7**(2), 88.

——— 1991. Microminerals from Ireland, part 1: the southwest (Munster). *UK Journal of Mines and Minerals* **10**, 22–7.

——— 1992. Microminerals from Ireland, part 2: the northwest (Connacht and Donegal). *UK Journal of Mines and Minerals* **11**, 42–7.

Ryback, G. & S. Moreton 1993. Microminerals from Ireland, part 3: the East (Leinster, Cavan and Monaghan), with a post-script on western Ireland. *UK Journal of Mines and Minerals* **12**, 36–41.

Ryback, G. & G. Savile 1967. Wulfenite from Ysbyty Ystwyth, Cardiganshire. *Mineralogical Magazine* **36**, 457–9.

Ryback, G. & P. C. Tandy 1992. Eighth supplementary list of British Isles minerals (English). *Mineralogical Magazine* **56**, 261–75.

Ryback, G., A. M. Clark, C. J. Stanley 1998. Re-examination of the A. W. G. Kingsbury collection of British minerals at the Natural History Museum, London. *Geological Curator* **6**, 317–22.

Ryback, G., A. D. Hart, C. J. Stanley 2001. A. W. G. Kingsbury's specimens of British minerals, part 1: some examples of falsified localities. *Russell Society, Journal* **7**(2), 51–69.

Ryback, G., R. Nawaz, E. Farley 1988. Seventh supplementary list of British Isles minerals (Irish). *Mineralogical Magazine* **52**, 267–74.

Ryde, C. 1987. Coldstones quarry. *UK Journal of Mines and Minerals* **2**, 19.

——— 1988. Mineral news. *UK Journal of Mines and Minerals* **5**, 9–11.

Sabine, P. A. 1950. The optical properties and composition of the acmitic pyroxenes. *Mineralogical Magazine* **29**, 113–25.

——— 1952. The ledmorite dike of Achmelvich, near Lochinver, Sutherland. *Mineralogical Magazine* **29**, 827–32.

——— 1953. The petrography and geological significance of the post-Cambrian minor intrusions of Assynt and the adjoining districts of northwest Scotland. *Geological Society of London, Quarterly Journal* **109**, 137–72.

——— 1960. The geology of Rockall, north Atlantic. *Geological Survey of Great Britain, Bulletin* **16**, 156–78.

——— 1970. *Perlitic obsidian at Sandy Braes, Co. Antrim: its devitrification and volumetric relationships*. Report 70/7, Institute of Geological Sciences, London.

Sabine, P. A. & B. R. Young 1971. *Mineralogy Unit annual report for 1971* (pp. 102–103). Geochemical Division, Institute of Geological Sciences, London.

——— 1975. Metamorphic processes at high temperature and low pressure: the petrogenesis of the metasomatized and assimilated rocks of Carneal, Co. Antrim. *Royal Society of London, Philosophical Transactions* **280A**, 225–69.

Sabine, P. A., S. T. Rye, R. R. Skelhorn 1972. Calcium-montmorillonite clay from Mull, Argyllshire. *Institution of Mining and Metallurgy, Transactions B* **81**, 49–50.

Sabine, P. A., G. A. Sergeant, B. R. Young 1968. Ferrian chlorospinel from Carneal, Co. Antrim. *Mineralogical Magazine* **36**, 948–54.

Sabine, P. A., M. T. Styles, B. R. Young 1982. *Gehlenite, an exomorphic mineral from Carneal, Co. Antrim, Northern Ireland*. Report 82/1 (pp. 61–3), Institute of Geological Sciences, London.

——— 1985. The nature and paragenesis of natural bredigite and associated minerals from Carneal and Scawt Hill, Co. Antrim. *Mineralogical Magazine* **49**, 663–70.

Sagar-Fenton, M. & S. B. Smith 2005. *Serpentine*. Truro, Cornwall: Truran Books.

Saich, D. A. & S. A. Rust 1987. Micro minerals from a trial level in Wales. *UK Journal of Mines and Minerals* **3**, 3–4.

Salmon, S. 1991. Amphiboles from the igneous complex at Sorel Point, Jersey, Channel Islands – reflectors of acid–basic magma interaction. *Ussher Society, Proceedings* **7**, 333–7.

Salter, D. L. & I. M. West 1965. Calciostrontianite in the basal Purbeck beds of Durlston Head. *Mineralogical Magazine* **35**, 146–50.

Samson, I. M. & D. A. Banks 1988. Epithermal base-metal vein mineralization in the Southern Uplands of Scotland: nature and origin of the fluids. *Mineralium Deposita* **23**, 1–8.

Samson, I. M. & M. J. Russell 1983. Fluid inclusion data from Silvermines base-metal–baryte deposits, Ireland. *Institution of Mining and Metallurgy, Transactions B* **92**, 66–71.

Sandell, E. B., M. H. Hey, D. McConnell 1939. The composition of francolite. *Mineralogical Magazine* **25**, 395–401.

Sandeman, H. A., Y. Chen, A. H. Clark, E. Farrar 1995. Constraints on the P–T conditions and age of emplacement of the Lizard ophiolite, Cornwall: amphibole–plagioclase thermobarometry and $^{40}Ar/^{39}Ar$ geochronology of basal amphibolites. *Canadian Journal of Earth Sciences* **32**, 261–72.

Sanders, I. S. & D. H. Doff 1991. A blue sodic beryl from southeast Ireland. *Mineralogical Magazine* **55**, 167–72.

Sargent, H. C. 1915. The Penmaenmawr intrusion. *Geological Magazine* **52**, 15–27.

——— 1916. Axinite veins in the Penmaenmawr porphyrite. *Geological Magazine* **53**, 5–7.

——— 1918. On a spilitic facies of Lower Carboniferous lava-flows in Derbyshire. *Geological Society of London, Quarterly Journal* **73**, 11.

——— 1924. Notes on the petrology of Penmaenmawr Mountain (part 1). *Liverpool Geological Society, Proceedings* **14**, 82–9.

——— 1925. Notes on the petrology of Penmaenmawr Mountain (part II). *Liverpool Geological Society, Proceedings* **14**, 123–42.

Sarjeant, W. A. S. 1956a. The mineralogy of Ecton Hill. *Sheffield University Geological Society, Journal* **2**(3), 87–92.

——— 1956b. Some Derbyshire mineral localities. *Sheffield University Geological Society, Journal* **2**(4), 114–21.

——— 1957. Further Derbyshire mineral localities. *Sheffield University Geological Society, Journal* **2**(5), 172–9.

——— 1967. Fibrous chlorites in the volcanic rocks of Derbyshire. *Mercian Geologist* **2**, 85–97.

Sarjeant, W. A. S. & T. D. Ford 1993. The oakstone and other pseudostalactitic baryte of Derbyshire. *UK Journal of Mines and Minerals* **12**, 10–14.

Sarp, H., P. Perroud, A. de Haller 1987. Seconde occurrence de l'orthoserpiérite dans la mine de Tynagh (Irlande). *Archives des Sciences et Compte Rendu des Séances de la Société* [. . . des Naturalistes Genevois] **40**, 47–50.

Savage K. D. 1995. Erionite from Edinbane, Isle of Skye, Scotland. *UK Journal of Mines and Minerals* **15**, 18.

——— 1997a. Field meeting report: Cwt-y-bugail slate quarry/Manod quarry, Blaenau Ffestiniog, Gwynedd. *Russell Society, Newsletter* **30**, 15.

——— 1997b. Field meeting report: Whitesmith mine, Strontian. *Russell Society, Newsletter* **31**, 10.

Sawkins, F. J. 1966. Ore genesis in the North Pennine orefield in the light of fluid inclusion studies. *Economic Geology* **61**, 385–401.

Scarffe, A. A. 2004. *The Great Laxey mine*. Douglas: Manx Heritage Foundation.

Scheding, C. 1991. Field meeting report: AGM weekend, Truro: Wheal Alfred. *Russell Society, Newsletter* **22**, 7.

Schindler, M. & F. C. Hawthorne 1998. The crystal structure of trembathite, $(Mg_{1.55}Fe_{1.43}Mn_{0.02})B_7O_{13}Cl$, a mineral of the boracite group: an example of the insertion of a cluster into a three-dimensional net. *Canadian Mineralogist* **36**, 1195–201.

Schlomann, C. & S. Weiss 1998. Diaboleit, Cumengeit und Dundasit vom Gunver Head, bei Padstow, Cornwall. *Lapis* **23**(3), 28–33.

Schneider, H. W. 1884. On the haematite iron mines of Lower Furness. *Cumberland Association for the Advancement of Literature and Science, Transactions* **10**, 99–108.

Schnellmann, G. A. & J. D. Willson 1947. Lead–zinc mineralization in north Derbyshire. *Institution of Mining and Metallurgy, Transactions B* **56**, 549–62.

Schultz, R. W. 1966a. Lower Carboniferous cherty ironstones at Tynagh, Ireland. *Economic Geology* **61**, 311–42.

——— 1966b. The Northgate base-metal deposit at Tynagh, County Galway, Ireland. *Economic Geology* **61**, 1443–59.

——— 1968. *The geology of the primary sulphide–barite deposits at Tynagh, Ireland.* PhD thesis, Trinity College, Dublin.

——— 1971. Mineral exploration practice in Ireland. *Institution of Mining and Metallurgy, Transactions B* **80**, 238–58.

Schumacher, R. 1985. Zincian staurolite in Glen Doll, Scotland. *Mineralogical Magazine* **49**, 561–71.

Scoon, R. N. 1978. *Lithium pegmatites in the Leinster granites of southeast Ireland.* MSc thesis, University College, Cardiff.

Scotsman, I. C. 1987. Clay diagenesis in the Kimmeridge Clay Formation, onshore UK, and its relation to organic maturation. *Mineralogical Magazine* **51**, 535–51.

Scott, A. 1914a. Augite from Bail Hill, Dumfriesshire. *Mineralogical Magazine* **17**, 100–110.

——— 1914b. Barkevikite from Lugar, Ayrshire. *Mineralogical Magazine* **17**, 138–42.

Scott, B. 1967. Barytes mineralization at Gasswater mine, Ayrshire. *Institution of Mining and Metallurgy, Transactions B* **76**, 40–51.

Scott, K. M. 1987. Solid solution in, and classification of, gossan-derived members of the alunite–jarosite family, northwest Queensland, Australia. *American Mineralogist* **72**, 178–87.

Scott, P. W. & M. R. Power 1995. Duporthite: a mixture of talc and chlorite. *Ussher Society, Proceedings* **8**, 445.

Scott, P. W., N. Kimyongur, R. Middleton 1987. The mineralogy and petrology of the "Coprolite Bed" phosphorite deposit (Lower Cretaceous) at Speeton, North Yorkshire. *Yorkshire Geological Society, Proceedings* **46**, 123–31.

Scott, P. W., R. D. Pascoe, F. W. Hart 1998. Columbite–tantalite, rutile and other accessory minerals from the St Austell topaz granite, Cornwall. *Ussher Society, Proceedings* **9**, 165–70.

Scott, R. A., D. A. Polya, R. A. D. Pattrick 1988. Proximal Cu+Zn exhalites in the Argyll Group Dalradian, Creag Bhocan, Perthshire. *Scottish Journal of Geology* **24**, 97–112.

Scott, R. H. 1871. Notes on the minerals of Strontian, Argyllshire. *Geological Society of London, Quarterly Journal* **27**, 372–6.

Scrivener, R. C. & M. J. Bennett 1980. Ore genesis and controls of mineralization of the Upper Palaeozoic rocks of north Devon. *Ussher Society, Proceedings* **5**, 54–8.

Scrivener, R. C. & R. A. Edwards 1990. Field excursion to the New Red Sandstone of the eastern Crediton Trough, 3rd January 1990. *Ussher Society, Proceedings* **7**, 304–305.

Scrivener, R. C. & J. H. Walbeoffe-Wilson 1975. Alluvial tin at Colston, Buckfastleigh. *Ussher Society, Proceedings* **3**, 237–42.

Scrivener, R. C., B. V. Cooper, M. C. George 1987. Mineralogy and paragenesis of the Haytor iron-ore deposit. *Ussher Society, Proceedings* **7**, 558.

Scrivener, R. C., B. V. Cooper, M. C. George, T. J. Shepherd 1982. Gold-bearing carbonate veins in the Middle Devonian Limestone of Hope's Nose, Torquay [abstract]. *Ussher Society, Proceedings* **5**, 393.

Scrivener, R. C., T. J. Shepherd, N. Garrioch 1986. Ore genesis at Wheal Pendarves and South Crofty mine, Cornwall – a preliminary fluid inclusion study. *Ussher Society, Proceedings* **6**, 412–16.

Scrivenor, J. B. 1903a. 1. Anatase in the Trias of the Midlands of England. 2. A peculiar occurrence of magnetite in the Upper Bunter Sands. *Mineralogical Magazine* **13**, 348–52.

——— 1903b. The granite and greisen of Cligga Head (western Cornwall). *Geological Society of London, Quarterly Journal* **59**, 142–59.

Scrutton, C. T. (ed.) 1994. *Yorkshire rocks and landscape: a field guide.* Maryport, Cumbria: Ellenbank Press, on behalf of the Yorkshire Geological Society.

——— 1995 (ed.). *Northumbrian rocks and landscape: a field guide.* Maryport, Cumbria: Ellenbank Press, on behalf of the Yorkshire Geological Society.

Seager, A. F. 1953. The surface structure of crystals. *Mineralogical Magazine* **30**, 1–25.

——— 1969. Zeolites and other minerals from Dean quarry, the Lizard, Cornwall. *Mineralogical Magazine* **37**, 147–8.

——— 1971. Mineralisation and paragenesis at Dean quarry, the Lizard, Cornwall. *Royal Geological Society of Cornwall, Transactions* **20**, 97–113.

——— 1978a. Zonal dissolution in analcime and pseudomorphs of adularia after analcime from the Lizard, Cornwall. *Mineralogical Magazine* **42**, 245–9.

——— 1978b. Paragenesis of hydrothermal mineralization in amphibolites and granulites around Porthkerris Point, the Lizard, Cornwall. *Mineralogical Magazine* **42**, 509–511.

Seager, A. F. & W. F. Davidson 1952. Changes in habit during the growth of baryte crystals from the north of England. *Mineralogical Magazine* **29**, 885–94.

Seager, A. F., F. J. Fitch, J. A. Miller 1978. Dating of adularia and the relationship of hydrothermal events in the Lizard complex, Cornwall. *Geological Magazine* **115**, 211–14.

Searl, A. 1989. Saddle dolomite: a new view of its nature and origin. *Mineralogical Magazine* **53**, 547–55.

——— 1990. Complex sector zonation in ankerite: geochemical controls on crystal morphology and inter-sector element partitioning. *Mineralogical Magazine* **54**, 501–507.

Sedgwick, A. 1835. On the geological relations and internal structure of the Magnesian Limestone, etc. *Geological Society of London, Transactions (Second Series)* **3**, 37–124.

Seenan, J. A. 1988. *The geochemistry of quartz–magnetite veins in Cwm Tregalan, Snowdonia, North Wales.* MSc thesis, University Leicester.

Sejkora, J., T. Řídkošil, V. Šrein 1999. Zálesíite, a new mineral of the mixite group, from Zálesí, Rychlebské hory Mountains, Czech Republic. *Neues Jahrbuch für Mineralogie, Abhandlungen* **175**, 105–124.

Selwood, E. B., E. M. Durrance, C. M. Bristow 1998. *The geology of Cornwall and the Scilly Isles.* Exeter: University of Exeter Press.

Semmons, W. 1885. Note on a recent discovery of "connellite". *Mineralogical Magazine* **6**, 160–63.

Seymour, H. J. 1900. On the occurrence of a blue amphibole in a hornblende kersantite from Co. Down. *Geological Magazine* **37**, 257–60.

——— 1902. A (preliminary) list of the minerals occurring in Ireland. *Geological Magazine* **39**, 500–501.

Seymour-Smith, S. 1977. A visit to the Glen Sannox baryte mine, Isle of Arran. *Russell Society, Newsletter* **4**, 23–6.

Shackleton, E. H. 1966. *Lakeland geology.* Clapham, Lancaster: Dalesman Publishing.

Shaikh, A. M. 1990. A new crystal growth form of vaterite, $CaCO_3$. *Journal of Crystal Growth* **23**, 263–5.

Shambrook, H. R. 1982. *The Caradon and Phoenix mining area.* British Mining 20, Northern Mine Research Society, Sheffield.

Shand, S. J. 1910a. On a group of minerals formed by the combustion of pyritous shales in Midlothian. *Mineralogical Magazine* **15**, 403–406.

——— 1910b. On boralanite and its associates in Assynt. *Edinburgh Geological Society, Transactions* **9**, 376–416.

Sharp, F. 1997. Cave mineralization at Hampstead Farm quarry. *Russell Society, Newsletter* **30**, 21–2.

Sharpe, A. 1990. *Coastal slate quarries – Tintagel to Trebarwith.* Truro: Cornwall Archaeological Trust.

Shaw, C. J. & A. D. Edgar 1997. Post-entrainment mineral-melt reactions in spinel peridotite xenoliths from Inver, Donegal, Ireland. *Geological Magazine* **134**, 771–9.

Shaw M. H., A. G. Gunn, K. E. Rollin, M. Styles 1994. *Platinum-group element mineralization in the Loch Ailsh alkaline igneous complex, northwest Scotland.* Report 131, Mineral Reconnaissance Program, British Geological Survey, Keyworth.

Shaw, W. T. 1959. Comments on the paper "The Lake District mining field" by Eastwood, T. See Higham (1959: 219–23).

——— 1970. *Mining in the Lake counties.* Clapham, Lancaster: Dalesman Publishing.

Shearman, D. J. & A. J. Smith 1985. Ikaite, the parent mineral of jarrowite-type pseudomorphs. *Geologists' Association, Proceedings* **96**, 305–314.

Shelley, D. 1966. The significance of granophyric and myrmekitic textures in the Lundy granites. *Mineralogical Magazine* **35**, 678–92.

Shepherd, T. J. 1973. *Geochemical evidence for basement control of the West Cumberland haematite mineralisation.* PhD thesis, University of Durham.

Shepherd, T. J. & D. C. Goldring 1993. Cumbrian hematite deposits, northwest England. See Pattrick & Polya (1993: 419–45).

Shepherd, T. J. & R. C. Scrivener 1987. Role of basinal brines in the genesis of polymetallic vein deposits, Kit Hill–Gunnislake area, SW England. *Ussher Society, Proceedings* **7**, 491–7.

Shepherd, T. J., R. D. Beckinsale, C. C. Rundle, J. Durham 1976. Genesis of Carrock Fell tungsten deposits, Cumbria: fluid inclusion and isotopic study. *Institution of Mining and Metallurgy, Transactions B* **85**, 63–73.

Shepherd, T. J., J. E. Bouch, A. G. Gunn, J. A. McKervey, J. Naden, R. C. Scrivener, M. T. Styles, D. E. Large 2005. Permo-Triassic unconformity-related Au–Pd mineralization, south Devon, UK: new insights and the European perspective. *Mineralium Deposita* **40**, 24–44.

Sheppard, W. A. 1980. The ores and host-rock geology of the Avoca mines, Co. Wicklow, Ireland. *Norges Geologiske Undersogelse* **360**, 269–83.

Sherlock, R. L. 1918. The geology and genesis of the Trefriw pyrites deposit. *Geological Society of London, Quarterly Journal* **74**, 106–115.

Sherlock, R. L. & S. E. Hollingworth 1938. *Gypsum and anhydrite, celestite and strontianite* (3rd edn). Special Reports on the Mineral Resources of Great Britain III, Memoir, Geological Survey of Great Britain, London.

Shrimpton, R. D. S. 1993. The story of a mineral collection (part 1). *Russell Society, Journal* **5**(1), 5–9.

——— 1994. The story of a mineral collection (part 2). *Russell Society, Journal* **5**(2), 71–5.

Siddiqui, S. F. A. 1964. *The geology and mineralization of the area between Newlyn and Mousehole, Cornwall.* PhD thesis, University of London.

Sills, J. D. 1982. The retrogression of ultramafic granulites from the Scourian of northwest

Scotland. *Mineralogical Magazine* **46**, 55–61.

Simpson, B. 1934. The petrology of the Eskdale (Cumberland) granite. *Geologists' Association, Proceedings* **45**, 17–34.

Simpson, I. M. & F. M. Broadhurst 1969. A boulder bed at Treak Cliff, north Derbyshire. *Yorkshire Geological Society, Proceedings* **37**, 141–51.

Singh, B., M. J. Wilson, W. J. McHardy, A. R. Fraser, G. Merrington 1999. Mineralogy and chemistry of ochre sediments from an acid mine drainage near a disused mine in Cornwall, UK. *Clay Minerals* **34**, 301–317.

Sivaprakash, C. 1977. *Geochemistry of some sulphides and sulphosalts from Parys Mountain, Anglesey.* MPhil thesis, University of Aston in Birmingham.

——— 1981. Zoned garnets in some Scottish Dalradian pelites. *Mineralogical Magazine* **44**, 301–307.

Skelhorn, R. R. 1962. Quartz after tridymite in an acid intrusion from Mull. *Mineralogical Magazine* **33**, 138–44.

Slater, D. & D. E. Highley 1976. The iron-ore deposits in the United Kingdom and Northern Ireland. In *The iron-ore deposits of Europe and adjacent areas*, A. Zitzmann (ed.), 393–409. Hannover: Bundesanstalt für Geowissenschaften und Rohstoffe.

Slowey, E. P. 1986. The zinc–lead and barite deposits at Keel, County Longford. See Andrew et al. (1986: 319–30).

Small, A. T. 1977. *Mineralization of the Stainmore depression and northern part of the Askrigg Block.* PhD thesis, University of Durham.

——— 1982. New data on tetrahedrite, tennantite, chalcopyrite and pyromorphite from the Cumbrian and North Yorkshire Pennines. *Yorkshire Geological Society, Proceedings* **44**, 153–8.

Smart, T. B. 1962. The aureole of the Barnesmore granite. *Irish Naturalists' Journal* **14**, 55–9.

Smellie, J. A. T. 1974a. Compositional variation within staurolite crystals from the Ardara aureole, Co. Donegal, Ireland. *Mineralogical Magazine* **39**, 672–83.

——— 1974b. Formation of atoll garnets from the aureole of the Ardara pluton, Co. Donegal, Ireland. *Mineralogical Magazine* **39**, 878–88.

Smellie, J. L. & P. Stone 1984. "Eclogite" in the Ballantrae complex: a garnet–clinopyroxenite segregation in mantle harzburgite? *Scottish Journal of Geology* **20**, 315–27.

Smith, B. 1924. *Iron ores: haematites of west Cumberland, Lancashire and the Lake District.* Special Reports on the Mineral Resources of Great Britain VIII, Memoir, Geological Survey of Great Britain, London.

Smith, C. G. 1977. Investigation of stratiform sulphide mineralization in parts of the Dalradian of central Perthshire, Scotland. *Institution of Mining and Metallurgy, Transactions B* **86**, 50–51.

Smith, C. G. & G. R. Marsden 1977. *Report on geophysical and geological surveys at Blackmount, Argyllshire.* Report 16, Mineral Reconnaissance Program, Institute of Geological Sciences, London.

Smith, C. G. & E. R. Phillips 2002. Cummingtonite in the Dalradian of northeast Scotland. *Mineralogical Magazine* **66**, 337–52.

Smith, C. G., M. J. Gallagher, J. S. Coats, M. E. Parker 1984. Detection and general characteristics of stratabound mineralization in the Dalradian of Scotland. *Institution of Mining and Metallurgy, Transactions B* **93**, 125–33.

Smith, C. G., M. J. Gallagher, I. C. Legg, E. Grennan, N. G. Ward 1992. Prospects for small-scale mineral development in Scotland and Ireland. See Bowden et al. (1992: 287–300).

Smith, C. G., W. G. McCourt, N. J. Fortey, C. E. Johnson, M. E. Parker, J. S. Coats, U. McL. Michie 1978a. *Investigation of stratiform sulphide mineralization at Meall Mòr, South Knapdale, Argyll.* Report 15, Mineral Reconnaissance Program, Institute of Geological Sciences, London.

Smith, C. G., R. T. Smith, R. C. Leake, M. T. Styles, I. C. Legg 1996. *Mineral exploration in the Hilltown and Slieve Croob areas, County Down.* Technical report GSNI/96/2, Geological Survey of Northern Ireland, Belfast.

Smith, D. G. W. 1965. The chemistry and mineralogy of some emery-like rocks from Sithean Sluaigh, Strachur, Argyllshire. *American Mineralogist* **50**, 1982–2002.

Smith, E. G., G. H. Rhys, R. A. Eden 1967. *Geology of the country around Chesterfield, Matlock and Mansfield (sheet 112).* Memoir, Institute of Geological Sciences, London.

Smith, F. W. 1973a. Supergene native copper in the Northern Pennine orefield. *Mineralogical Magazine* **39**, 244.

——— 1973b. Fluid inclusion studies on fluorite from the North Wales orefield. *Institution of Mining and Metallurgy, Transactions B* **82**, 174–6.

Smith, F. W. & R. G. Hardy 1981. Clay minerals in the veins of the North Pennine orefield. *Clay Mineralogy* **16**, 309–312.

Smith, G. 1998. Prospects for an expanding gemstone industry. *Earthwise* **12**, 31.

Smith, H. G. 1936a. The South Hill lamprophyre, Jersey. *Geological Magazine* **73**, 87–91.

——— 1936b. New lamprophyres and monchiquites from Jersey. *Geological Society of London, Quarterly Journal* **92**, 365–81.

Smith, J. 1895. Charred coal with graphite, and discovery of diamonds in the graphite at Craigman, New Cumnock. *Geological Society of Glasgow, Transactions* **10**, 257–62.

Smith, M. E. 1958. The lead industry of the Ewden Valley, Yorkshire. *Sorby Record* **1**, 21–3.

——— 1961. Informal excursion to the Ewden Valley, Yorkshire. *Peak District Mines Historical Society, Bulletin* **1**(2), 12.

——— 1962. The Odin mine, Castleton. *Peak District Mines Historical Society, Bulletin* **1**(6), 18–23.

——— 1982. The mineralogy of Fall Hill quarry, Ashover, Derbyshire. *Russell Society, Journal* **1**(1), 26–32.

——— 1994. Vivianite from South Humberside and Yorkshire, England. *Russell Society, Journal* **5**(2), 119–20.

Smith, M. H. 1967. An occurrence of gibbsite in east Yorkshire. *Yorkshire Geological Society, Proceedings* **36**, 57–9.

Smith, M. L. & C. Frondel 1968. The related layered minerals ganophyllite, bannisterite and stilpnomelane. *Mineralogical Magazine* **36**, 893–913.

Smith, M. P., P. Henderson, T. Jeffries 2002. The formation and alteration of allanite in skarn from the Beinn an Dubhaich granite aureole, Skye. *European Journal of Mineralogy* **14**, 471–86.

Smith, M. P., P. Henderson, T. E. R. Jeffries, J. Long, C. T. Williams 2004. The rare-earth elements and uranium in garnets from the Beinn an Dubhaich aureole, Skye, Scotland, UK: constraints on processes in a dynamic hydrothermal system. *Journal of Petrology* **45**, 457–84.

Smith, P. 2002. Bazhenovite. British Micromount Society website: www.micromounters.org.uk.

Smith, R. A. 1986. *Permo-Triassic and Dinantian rocks of the Belfast harbour borehole.* Report 18.6, British Geological Survey, Keyworth.

Smith, R. T., M. J. Gallagher, N. J. Fortey 1978b. Controls of lead–zinc mineralization in basal Carboniferous rocks near Langholm, Scotland. *Institution of Mining and Metallurgy, Transactions B* **87**, 140–43.

Smith, W. D. F. 1988. *The Great Orme copper mines.* Llandudno: Creuddyn Publications.

Smith, W. W. 1959. Pseudomorphs after olivine in Markle basalt. *Mineralogical Magazine* **32**, 324–31.

Smith Aitken, W. W. 1965. An occurrence of phlogopite and its transformation to vermiculite by weathering. *Mineralogical Magazine* **35**, 151–64.

Smithson, F. 1932. The petrography of the northern portion of the Leinster granite. *Geological Magazine* **69**, 465–75.

Smithson, F. & G. Brown 1957. Dickite from sandstones in northern England and North Wales. *Mineralogical Magazine* **31**, 381–91.

Smythe, J. A. 1921. Minerals of the North Country: fluorspar. *Vasculum* **8**, 19–24.

——— 1922. Minerals of the North Country: barium minerals. *Vasculum* **8**, 90–93, 111–16.

——— 1924. Minerals of the North Country: sulphides. *Vasculum* **11**, 7–14.

——— 1927. Minerals of the North Country: miscellaneous. *Vasculum* **14**, 12–17.

——— 1933. Nickel-bearing goslarite, epsomite and melanterite from County Durham. *Vasculum* **19**, 12–17.

——— 1947. Chemical analysis of ullmannite from New Brancepeth, Co. Durham. *Mineralogical Magazine* **28**, 173–4.

Smythe, J. A. & K. C. Dunham 1947. Ankerites and chalybites from the Northern Pennine orefield and the northeast coalfield. *Mineralogical Magazine* **28**, 54–74.

Smythe, W. W. 1862. Gold mining at Clogau, North Wales. *Mining & Smelting Magazine* **91**, 177–88.

——— 1879–1884. List of minerals found in the Isle of Man. *Isle of Man Natural History and Antiquarian Society, Transactions* **I**, 143–7.

——— 1880. On the occurrence of feather ore (plumosite) in Foxdale mine. *Royal Geological Society of Cornwall, Transactions* **10**, 82–9.

Smythe, W. W., T. Reeks, F. W. Rudler 1864. *A catalogue of the mineral collections in the Museum of Practical Geology.* London: HMSO.

Snelling, N. J. 1957. A contribution to the mineralogy of chloritoid. *Mineralogical Magazine* **31**, 469–75.

Sobott, R. J. G. 1979. Hornblende twinning on $(\bar{1}01)$ in an amphibolite from Barra, Outer Hebrides, Scotland. *Mineralogical Magazine* **43**, 177–8.

Sollas, W. J. 1890. On the occurrence of zinnwaldite in the granite of the Mourne Mountains. *Royal Irish Academy, Proceedings* **1**, 379.

Solly, R. H. 1885. Fine crystals of pale-lilac calcite from Tankerville mine, near Shelve, Shropshire. *Mineralogical Magazine* **6**, 120–23.

——— 1886. Notes on minerals from Cornwall and Devon. *Mineralogical Magazine* **6**, 202–211.

——— 1889. Pseudomorphs of haematite after iron pyrites. *Mineralogical Magazine* **8**, 183–5.

——— 1891. Cassiterite, "sparable tin" from Cornwall. *Mineralogical Magazine* **9**, 199–208.

Solomon, M., T. A. Rafter, K. C. Dunham 1971. Sulphur and oxygen isotope studies in the northern Pennines in relation to ore genesis. *Institution of Mining and Metallurgy, Transac-*

tions B **80**, 259–75.

Sørensen, H. 1955. On sapphirine from west Greenland. *Meddelelser om Grønland* **137**, 1–32.

Southwood, M. J. 1982. *The geological setting of the sulphide deposits of Morfa Du, Parys Mountain, Anglesey*. PhD thesis, University College Cardiff.

Southwood, M. J. & R. Bevins 1995. Parys Mountain, the type locality for anglesite. *UK Journal of Mines and Minerals* **15**, 11–17.

Sowerby, J. 1811. *British mineralogy*, vol. IV. London: Richard Taylor.

——— 1817. *British mineralogy*, vol. V. London: Richard Taylor.

Sparrow, C. 1989. Preliminary report of the mineralogy of Gravel Hill mine, Perranzabuloe. *Newsletter 3, Russell Society South-West Branch.*

Spears, D. A. 1979. Geochemical aspects of the Santonian Chalk of Ramsgate, England, and the origin of the chert and clay minerals. *Mineralogical Magazine* **43**, 159–64.

Spence, J. 1982. Fallgate quarry, Ashover, Derbyshire. *Mineral Realm* **1**(3), 4–8.

Spencer, L. J. 1897. The "satin spar" of Alston in Cumberland; and the determination of massive and fibrous calcites and aragonites. *Mineralogical Magazine* **11**, 184–7.

——— 1898. *Supplementary list of British minerals* [reproduced in the facsimile copy of the *Manual of mineralogy of Great Britain and Ireland* by R. P. Greg & W. G. Lettsom 1858]. Broadstairs, Kent: Lapidary Publications [1977 reprint].

——— 1910. On the occurrence of alstonite and ullmannite (a species new to Britain) in a barytes–witherite vein at New Brancepeth colliery near Durham. *Mineralogical Magazine* **15**, 302–311.

——— 1942. Barium-feldspars (celsian and paracelsian) from Wales. *Mineralogical Magazine* **26**, 231–45.

——— 1958. Third supplementary list of British minerals. *Mineralogical Magazine* **31**, 787–806.

Spencer, L. J. & E. D. Mountain 1923. New lead–copper minerals from the Mendip Hills (Somerset). *Mineralogical Magazine* **20**, 67–92.

Spiller, J. 1876. On the occurrence of native calcium chloride at Guy's Cliffe, Warwickshire. *Chemical Society, Journal* **1**, 154–7.

Springer, G. 1968. Electronprobe analyses of stannite and related tin minerals. *Mineralogical Magazine* **36**, 1045–1051.

Stanley, C. J. 1979. *Mineralogical studies of copper, lead, zinc and cobalt mineralization in the English Lake District*. PhD thesis, University of Aston in Birmingham.

Stanley, C. J. & A. J. Criddle 1979. Mineralization at Seathwaite Tarn near Coniston, English Lake District: The first occurrence of wittichenite in Great Britain. *Mineralogical Magazine* **43**, 103–107.

Stanley, C. J. & R. A. Ixer 1982. Mineralization at Le Pulec, Jersey, Channel Islands: no. 1 lode. *Mineralogical Magazine* **46**, 134–6.

Stanley, C. J. & D. J. Vaughan 1980. Interpretive studies of copper mineralization to the south of Keswick, England. *Institution of Mining and Metallurgy, Transactions B* **89**, 25–30.

——— 1981. Native antimony and bournonite intergrowths in galena from the English Lake District. *Mineralogical Magazine* **44**, 257–60.

——— 1982a. Copper, lead, zinc and cobalt mineralization in the English Lake District: classification, conditions of formation and genesis. *Geological Society of London, Journal* **139**, 569–79.

——— 1982b. Mineralization in the Bonser vein, Coniston, English Lake District: mineral assemblages, paragenesis, and formation conditions. *Mineralogical Magazine* **46**, 343–50.

Stanley, C. J., A. J. Criddle, D. Lloyd 1990a. Precious and base metal selenide mineralization at Hope's Nose, Torquay, Devon. *Mineralogical Magazine* **54**, 485–93.

Stanley, C. J., C. Halls, G. S. Camm, J. James 1990b. Gold–antimony mineralization at Loddiswell, Devon, UK. *Terra Nova* **2**, 224–31.

Stanley, C. J., G. C. Jones, A. D. Hart, P. Keller, D. Lloyd 1991. Barstowite, $3PbCl_2 \cdot PbCO_3 \cdot H_2O$, a new mineral from Bounds Cliff, St Endellion, Cornwall. *Mineralogical Magazine* **55**, 121–5.

Stanley, C. J., R. F. Symes, G. C. Jones 1987. Nickel–copper mineralization at Talnotry, Newton Stewart, Scotland. *Mineralogy and Petrology* **37**, 293–313.

Stanton, W. I. 1991. The habitat and origin of lead ore in Grebe Swallet mine, Charterhouse-on-Mendip, Somerset. *Bristol Speleological Society, Proceedings* **19**(1), 43–65.

Starkey, R. E. 1981. Wulfenite from Poddy Gill. *Mineral Realm* **1**(3), 30–31.

——— 1982. Geology and mineralization at Burgam mine, Salop. *Mineral Realm* **1**(4), 13–18.

——— 1983. On the occurrence of millerite at Ecton Hill, Staffordshire. *Russell Society, Journal* **1**(2), 17–18.

——— 1984. Phosgenite from Clevedon, Avon. *Bristol Naturalists' Society, Proceedings* **44**, 13–14.

——— 1986a. Cyanotrichite from Cornwall. *Russell Society, Journal* **1**(4), 119–20.

——— 1986b. An occurrence of carbonate-cyanotrichite from Devon. *Russell Society, Journal* **1**(4), 121–2.

——— 1986c. A new British locality for beudantite: Clevedon, Avon. *Russell Society, Journal* **1**(4), 125–6.

——— 1987a. [Diaboléite]. *British Micromount Society Newsletter* (November 1987).

——— 1987b. Duftite from the Northern Pennine orefield. *Yorkshire Geological Society, Proceedings* **46**, 289.

——— 1987c. Knoydart, Scotland. *UK Journal of Mines and Minerals* **2**, 17–18.

——— 1988a. Phosgenite from Lossiemouth, Grampian Region: confirmation of the first Scottish occurrence. *Scottish Journal of Geology* **24**, 15–19.

——— 1988b. Minerals of the Isle of Skye. *UK Journal of Mines and Minerals* **5**, 18–20.

——— 1988c. British topographical mineralogy – where are we going? *UK Journal of Mines and Minerals* **5**, 22–3.

——— 1989. British mineral and gem show – London 1989. *Russell Society, Journal* **2**(2), 51–2.

——— 1990. Bertrandite from the Pass of Ballater, Grampian Region, Scotland. *Russell Society, Journal* **3**(1), 31.

——— 1995. Correction, Powellite from Coire Buidhe, Argyll. *UK Journal of Mines and Minerals* **15**, 18.

——— 1998a. Valentinite from the Isle of Man. *Russell Society, Journal* **7**(1), 35–6.

——— 1998b. Scorodite and pharmacosiderite from the Isle of Man. *UK Journal of Mines and Minerals* **19**, 12.

Starkey, R. E. & N. Hubbard 1998. Anilite, with botallackite and clinoatacamite, from Castletown mine, Lochgilpead, Scotland. *Russell Society, Journal* **7**(1), 38–9.

Starkey, R. E. & G. W. Robinson 1992. Prenteg, Tremadog, Gwynedd, Wales. *Mineralogical Record* **23**(5), 391–9.

Starkey, R. E., N. Hubbard, M. P. Bayley 1991. Mineralization at Hendre quarry, Glyn Ceiriog, Clwyd, Wales. *UK Journal of Mines and Minerals* **10**, 48–51.

Starkey, R. E., N. Hubbard, B. Young 1994. Barytocalcite and witherite from Rorrington mine, Chirbury, Salop, England. *Russell Society, Journal* **5**(2), 115–18.

Steed, G. M. 1983. Gortdrumite, a new sulphide mineral containing copper and mercury, from Ireland. *Mineralogical Magazine* **47**, 35–6.

——— 1986. The geology and genesis of the Gortdrum Cu–Ag–Hg orebody. See Andrew et al. (1986: 481–500).

Steed, G. M. & J. H. Morris 1997. Isotopic evidence for the origins of a Caledonian gold–arsenopyrite–pyrite deposit at Clontibret, Ireland. *Institution of Mining and Metallurgy, Transactions B* **106**, 109–117.

Steed, G. M., A. E. Annels, P. L. Shrestha, P. S. Tater 1976. Geochemical and biogeochemical prospecting in the area of the Ogfau gold mines, Dyfed, Wales. *Institution of Mining and Metallurgy, Transactions B* **85**, 109–117.

Steele, I. M., J. J. Pluth, A. Livingstone 1998. Crystal structure of macphersonite $Pb_4SO_4(CO_3)_2(OH)_2$: comparison with leadhillite. *Mineralogical Magazine* **62**, 451–9.

Steele, I. M., J. J. Pluth, C. J. Stanley 1999. Crystal structure of barstowite $3PbCl_2 \cdot PbCO_3 \cdot H_2O$. *Mineralogical Magazine* **63**, 901–907.

Steiger, R. & A. Bowden 1982. Tungsten mineralization in southeast Leinster, Ireland. In *Mineral exploration in Ireland: progress and developments 1971–1981*, A. G. Brown & J. Pyne (eds), 108–114. Dublin: Irish Association of Economic Geology.

——— 1986. Tungsten mineralization in southeast Leinster, Ireland. See Andrew et al. (1986: 211–16).

Steiger, R. & O. von Knorring 1975. A lithium pegmatite belt in Ireland. *Journal of Earth Sciences* (Leeds) **8**, 433–43.

Stephen, I. 1954. An occurrence of palygorskite in the Shetland Isles. *Mineralogical Magazine* **30**, 471–80.

Stephens, F. J. 1899. Recent discoveries of gold in west Cornwall. *Royal Geological Society of Cornwall, Transactions* **12**, 241–57.

——— 1940a. The south Gwennap mining district, with a portion of Baldhu and Kea. *Mining Magazine* **62**, 9–19.

——— 1940b. The Wendron mining district with notes on the flat lode and Old Wheal Vor. *Mining Magazine* **63**, 233–40.

Stephenson, D. 1983. *Hilderston mine, West Lothian: mining history and the nature of the vein mineralization as deduced from old records.* Report NL83.4, Institute of Geological Sciences, London.

Stephenson, D. & J. S. Coats 1983. *Baryte and copper mineralization in the Renfrewshire Hills, central Scotland.* Report 67, Mineral Reconnaissance Program, Institute of Geological Sciences, London.

Stewart, D. R. A. 1999. *The occurrence, characterization and genesis of melanterite at Lisheen mine, Co. Tipperary, Ireland.* MSc thesis, University of Leicester.

Stewart, F. H. 1941. On sulphatic cancrinite and analcime (eudnophite) from Loch Borolan, Assynt. *Mineralogical Magazine* **26**, 1–8.

——— 1942. Chemical data on a silica-poor argillaceous hornfels and its constituent minerals. *Mineralogical Magazine* **26**, 260–66.

——— 1946. The gabbroic complex of Belhelvie in Aberdeenshire. *Geological Society of London, Quarterly Journal* **102**, 465–98.

——— 1949. The petrology of the evaporites of the Eskdale no. 2 boring, east Yorkshire, part I: the lower evaporite bed. *Mineralogical Magazine* **28**, 621–75.

——— 1950. Note on garnet crystals from Cairnie, Aberdeenshire. *Mineralogical Magazine* **29**, 252–3.

——— 1951a. The petrology of the evaporites of the Eskdale no. 2 boring, east Yorkshire, part II: the middle evaporite bed. *Mineralogical Magazine* **29**, 445–75.

——— 1951b. The petrology of the evaporites of the Eskdale no. 2 boring, east Yorkshire, part III: the upper evaporite bed. *Mineralogical Magazine* **29**, 557–72.

——— 1954. Permian evaporites and associated rocks in Texas and New Mexico compared with those of northern England. *Yorkshire Geological Society, Proceedings* **29**, 185–235.

——— 1956. Replacements involving early carnallite in the potassium-bearing evaporites of Yorkshire. *Mineralogical Magazine* **31**, 127–35.

——— 1963. The Permian lower evaporites of Fordon in Yorkshire. *Yorkshire Geological Society, Proceedings* **34**, 1–44.

——— 1965. The mineralogy of the British Permian evaporites. *Mineralogical Magazine* **34**, 460–70.

Stewart, F. H., R. A. Chalmers, R. Phillips 1954. Veatchite from the Permian evaporites of Yorkshire. *Mineralogical Magazine* **30**, 389–92.

Stewart, G. E. 1947. Billingham mine. *Institution of Mining and Metallurgy, Transactions B* **56**, 113–23.

Stewart, M. 1932. Notes on the geology of North Rona. *Geological Magazine* **69**, 179–85.

Stimac, J. A., A. H. Clark, Y. Chen, S. Garcia 1995. Enclaves and their bearing on the origin of the Cornubian batholith, southwest England. *Mineralogical Magazine* **59**, 273–96.

Stoddard, W. W. 1876. On the occurrence of celestine in the Keuper marls, and its influence on the composition of plants. *Mineralogical Magazine* **1**, 4–7.

Stokes, A. H. 1879. The economic geology of Derbyshire. *Chesterfield and Derbyshire Institute of Mining, Civil and Mechanical Engineers, Transactions* **6**, 60–156.

Stone, M. 1969. Nature and origin of banding in the granitic sheets, Tremearne, Porthleven, Cornwall. *Geological Magazine* **106**, 142–58.

——— 1984. Textural evolution of lithium mica granites in the Cornubian batholith. *Geologists' Association, Proceedings* **95**, 29–41.

——— 1988. The significance of almandine garnets in the Lundy and Dartmoor granites. *Mineralogical Magazine* **52**, 651–8.

——— 1990. The Lundy granite: a geochemical and petrogenetic comparison with Hercynian and Tertiary granites. *Mineralogical Magazine* **54**, 431–46.

——— 1992. The Tregonning granite: petrogenesis of Li-mica granites in the Cornubian batholith. *Mineralogical Magazine* **56**, 141–55.

——— 1995. The main Dartmoor granites: petrogenesis and comparisons with the Carnmenellis and Isles of Scilly granites. *Ussher Society, Proceedings* **8**, 379–84.

——— 2000. Petrogenetic implications from biotite compositional variations in the Cornubian granitic batholith. *Mineralogical Magazine* **64**, 729–35.

Stone, M. & M. C. George 1978. Amblygonite in leucogranites of the Tregonning–Godolphin granite, Cornwall. *Mineralogical Magazine* **42**, 151–2.

——— 1983. Some phosphate minerals at the Megiliggar Rocks, Cornwall. *Ussher Society, Proceedings* **5**, 428–31.

Stone, P. & R. C. Leake 1984. Disseminated and epigenetic Pb–Zn mineralization in Ordovician mudstone, Galloway. *Scottish Journal of Geology* **20**, 181–90.

Stone, M., C. S. Exley, M. C. George 1988. Compositions of trioctahedral micas in the Cornubian batholith. *Mineralogical Magazine* **52**, 175–92.

Stone, P., A. G. Gunn, J. S. Coats, R. M. Carruthers 1986. Mineral exploration in the Ballantrae complex, southwest Scotland. In *Metallogeny of basic and ultrabasic rocks*, M. J. Gallagher, R. A. Ixer, C. R. Neary, H. M. Prichard (eds), 265–78. London: Institution of Mining and Metallurgy.

Storrie, J. 1885. Notes on the occurrence of grains of native gold in the New Red Marl strata during the boring of an artesian well at the Phoenix brewery, Working Street, Cardiff. *Cardiff Naturalists' Society, Transactions* **26**, 107–109.

Strahan, A. 1885. *The geology of the coasts adjoining Rhyl, Abergele and Colwyn (sheet 79 NW)*. Memoir, Geological Survey of Great Britain and Ireland, London.

——— 1890. *The geology of the neighbourhoods of Flint, Mold and Ruthin (sheet 79SE)*. Memoir, Geological Survey of Great Britain and Ireland, London.

——— 1899. *The geology of the South Wales coalfield, part 1: the country around Newport, Monmouthshire (sheet 249)*. Memoir, Geological Survey of Great Britain and Ireland, London.

——— 1907. *The geology of the South Wales coalfield, part VIII: the country around Swansea (sheet 247)*. Memoir, Geological Survey of Great Britain, London.

Strahan, A. & O. T. Jones 1907. *The geology of the South Wales coalfield, part 8: the country around Swansea (sheet 247)*. Memoir, Geological Survey of Great Britain, London.

Strahan, A., T. C. Cantrill, E. E. L. Dixon, H. H. Thomas 1909. *The geology of the South Wales coalfield, part X: the country around Carmarthen (sheet 229)*. Memoir, Geological Survey of Great Britain, London.

Strashimirov, S., S. Dobrev, I. Sivilova 1999. *Petrological and fluid inclusion studies of core samples from the Lisheen mine, Ireland*. Company report, University of Mining and Geology "St Ivan Rilski", Sofia, Bulgaria.

Strens, R. G. J. 1962. *The geology of the Borrowdale–Honister district (Cumberland) with special reference to mineralization*. PhD thesis, University of Nottingham.

——— 1963. Pyromorphite as a possible primary phase. *Mineralogical Magazine* **33**, 722–3.

——— 1964. Epidotes of the Borrowdale Volcanic rocks of central Borrowdale. *Mineralogical Magazine* **33**, 868–86.

——— 1965. The graphitic deposit of Seathwaite in Borrowdale Cumberland. *Geological Magazine* **102**, 393–406.

Strogen, P. & R. Elsdon 1974. Recently formed gypsum from Co. Wicklow, Ireland. *Mineralogical Magazine* **39**, 818–20.

Strong, G. E. & A. E. Milodowski 1987. Aspects of the diagenesis of the Sherwood Sandstones of the Wessex Basin and their influence on reservoir characteristics. In *Diagenesis of sedimentary sequences*, J. D. Marshall (ed.), 325–37. Oxford: Blackwell Scientific.

Stumpfl, E. F. 1961. Contribution to the study of ore minerals in some igneous rocks from Assynt. *Mineralogical Magazine* **32**, 767–77.

Styles, M. T., A. G. Gunn, K. E. Rollin 2004. A preliminary study of PGE in the late Caledonian Loch Borralan and Loch Ailsh alkaline pyroxenite–syenite complexes, northwest Scotland. *Mineralium Deposita* **39**, 240–55.

Swart, P. K. & F. Moore 1982. The occurrence of uranium in association with cassiterite, wolframite and sulphide mineralization in southwest England. *Mineralogical Magazine* **46**, 211–15.

Sweet, J. M. 1930. Notes on British barytes. *Mineralogical Magazine* **22**, 257–70.

——— 1959. A re-examination of uigite. *Mineralogical Magazine* **32**, 340–42.

——— 1960a. British gold from the Bouglise collection. *Mineralogical Magazine* **32**, 420–21.

——— 1960b. What is plinthite? *Mineralogical Magazine* **32**, 455–8.

——— 1961. Tacharanite and other hydrated calcium silicates from Portree, Isle of Skye. *Mineralogical Magazine* **32**, 745–53.

Sweeting, G. S. 1946. An outline of the geology of Ashover. *Geologists' Association, Proceedings* **57**, 117–36.

Sylvester-Bradley, P. C. & R. J. King 1963. Evidence for abiogenic hydrocarbons. *Nature* **198**, 728–31.

Symes, R. F. 1977. Datolite and apophyllite from the Mendips. *Mineralogical Magazine* **41**, 410–11.

Symes, R. F. & P. G. Embrey 1977. Mendipite and other rare oxychloride minerals from the Mendip Hills, Somerset, England. *Mineralogical Record* **8**, 298–303.

Symes, R. F., G. Cressey, A. J. Criddle, C. J. Stanley, J. G. Francis, G. C. Jones 1994. Parkinsonite, $(Pb,Mo,\square)_8O_8Cl_2$, a new mineral from Merehead quarry, Somerset. *Mineralogical Magazine* **58**, 59–68.

Symes, R. F., M. Wirth, B. Young 1990. Schultenite from the Caldbeck Fells, Cumbria: the first British occurrence. *Mineralogical Magazine* **54**, 659.

Symonds, W. S. 1872. *Records of the rocks; or, notes on the geology, natural history, and antiquities of North and South Wales, Devon & Cornwall*. London: John Murray.

Talbot, V. & M. D. Max 1984. Application of various geophysical exploration techniques to Cu and Mo mineralization in Galway granite, Ireland. *Institution of Mining and Metallurgy, Transactions B* **94**, 109–113.

Tanner, P. W. G. & J. A. Evans 2003. Late Precambrian U–Pb titanite age for peak regional metamorphism and deformation (Knoydartian orogeny) in the western Moine, Scotland. *Geological Society of London, Journal* **160**, 555–64.

Tarkian, M. & H. M. Prichard 1987. Irarsite–hollingworthite solid-solution series and other associated Ru-, Os-, Ir-, and Rh-bearing PGM's from the Shetland ophiolite complex. *Mineralium Deposita* **22**, 178–84.

Tarney, J. 1969. Epitaxic relations between coexisting pyroxenes. *Mineralogical Magazine* **37**, 115–22.

Taylor, A. 1884. On the occurrence of prehnite and other minerals in the rocks of Samson's Ribs and Salisbury Crags. *Mineralogical Magazine* **6**, 104–105.

Taylor, B. 1996. Alpine-type minerals from Bunmahon, Co. Waterford, Ireland. *UK Journal of Mines and Minerals* **17**, 17.

Taylor, H. F. W. 1959. The transformation of tobermorite into xonotlite. *Mineralogical Magazine* **32**, 110–16.

——— 1977. The crystal structure of killalaite. *Mineralogical Magazine* **41**, 363–9.

Taylor, J. H. 1934. The Mountsorrel granodiorite and associated igneous rocks. *Geological Magazine* **71**, 1–16.

——— 1950. Baryte-bearing nodules from the Middle Lias of the English East Midlands. *Mineralogical Magazine* **29**, 18–26.

Taylor, K. & R. K. Harrison 1958. Coffinite in Cornwall. *Nature* **181**, 363.

Taylor, R. G. & K. F. G. Hosking 1970. Manganese–iron ratios in wolframite, South Crofty mine, Cornwall. *Economic Geology* **65**, 47–53.

Taylor, S. 1984. Structural and paleotopographic controls of lead–zinc mineralization in the Silvermines orebodies, Republic of Ireland. *Economic Geology* **79**, 529–48.

Taylor, S. & C. J. Andrew 1978. Silvermines orebodies, County Tipperary, Ireland. *Institution of Mining and Metallurgy, Transactions B* **87**, 111–24.

——— 1979. Silvermines orebodies, County Tipperary, Ireland. *Institution of Mining and Metallurgy, Transactions B* **88**, 125.

Teall, J. J. H. 1887. On granite containing andalusite from the Cheesewring, Cornwall. *Mineralogical Magazine* **7**, 161–3.

——— 1888a. Notes on some minerals from the Lizard. *Mineralogical Magazine* **8**, 116–20.

——— 1888b. *British petrography: with special reference to igneous rocks.* London: Durlau.

——— 1891a. On an eclogite from Loch Duich. *Mineralogical Magazine* **9**, 217–18.

——— 1891b. On a micro-granite containing riebeckite from Ailsa Craig. *Mineralogical Magazine* **9**, 219–21.

——— 1900. Nepheline syenite and its associates in the northwest of Scotland. *Geological Magazine* **37**, 385–92.

——— 1907. Pre-Torridonian intrusive rocks associated with the Lewisian Gneiss. In *The geological structure of the Northwest Highlands of Scotland*, A. Geikie (ed.), 84. Memoir, Geological Survey of Great Britain, Edinburgh.

Teilhard de Chardin, P. & F. Pelletier 1911. Notes minéralogiques et géologiques sur l'île de Jersey. *La Société Jersiaise, Bulletin* **7**, 99–111.

Temple, A. K. 1954a. *The paragenetic mineralogy of the Leadhills and Wanlockhead lead–zinc deposits.* PhD thesis, University of Leeds.

——— 1954b. Rammelsbergite from the Southern Uplands of Scotland. *Mineralogical Magazine* **30**, 541–3.

——— 1955. The Leadhills–Wanlockhead lead and zinc deposits. *Royal Society of Edinburgh, Transactions: Earth Sciences* **63**, 85–113.

Tennant, S. C. & G. M. Steed 1997. Role of lithogeochemistry in reassessment of the geological setting of Parys Mountain polymetallic sulphide deposit, Anglesey, Wales. *Institution of Mining and Metallurgy, Transactions B* **106**, 144–56.

Tepley, F. J. & J. P. Davidson 2003. Mineral-scale Sr-isotope constraints on magma evolution and chamber dynamics in the Rum layered intrusion, Scotland. *Contributions to Mineralogy and Petrology* **145**, 628–41.

Terrell, E. 1920. The Hemerdon wolfram–tin mine. *Mining Magazine* **22**, 75–87.

Thimmaiah, T. 1956. *Mineralization of the Caldbeck Fells area, Cumberland.* PhD thesis, University of London.

Thirlwall, M. F. 1981. Peralkaline rhyolites from the Ordovician Tweeddale lavas, Peebleshire, Scotland. *Geological Journal* **16**, 41–4.

Thomas, H. H. 1905. On an epidote from Inverness-shire. *Mineralogical Magazine* **14**, 109–114.

——— 1909a. Detrital andalusite in Tertiary and post-Tertiary sands. *Mineralogical Magazine* **15**, 241–4.

——— 1909b. Orthite in North Wales. *Nature* **81**, 487.

——— 1922. On certain xenolithic Tertiary minor intrusions in the Island of Mull (Argyllshire). *Geological Society of London, Quarterly Journal* **78**, 229–60.

Thomas, T. M. 1968. A new occurrence of celestine, near Llantrisant, Mid Glamorgan. *Geological Magazine* **105**, 185–6.

Thompson, D. B. 1970. Alderley Edge. In *The area around Manchester*, R. M. C. Eagar, F. M. Broadhurst, J. W. Jackson (eds), 39–51. Guide 7, Geologists' Association, London.

Thompson, R. N. 1969. Tertiary granites and associated rocks of the Marsco area, Isle of Skye. *Geological Society of London, Quarterly Journal* **124**, 349–85.

——— 1974. Primary basalts and magma genesis, I: Skye, northwest Scotland. *Contributions to Mineralogy and Petrology* **45**, 317–41.

——— 1976. Alkali amphiboles in the Eocene high-level granites of Skye, Scotland. *Mineralogical Magazine* **40**, 891–3.

——— 1981. Thermal aspects of the origin of Hebridean Tertiary acid magmas, I: an experimental study of partial fusion of Lewisian gneisses and Torridonian sediments. *Mineralogical Magazine* **44**, 161–70.

Thompson, R. N. & M. B. Fowler 1986. Subduction-related shoshonitic and ultrapotassic magmatism: a study of Siluro-Ordovician syenites from the Scottish Caledonides. *Contributions to Mineralogy and Petrology* **94**, 507–522.

Thompson, S. J., C. H. Shine, C. Cooper, C. Halls, R. Zhao 1992. Shear-hosted gold mineralization in Co. Mayo, Ireland. See Bowden et al. (1992: 21–37).

Thomson, J. A. 1908. The hornblendic rocks of Glendalough and Greystones (County Wicklow). *Geological Society of London, Quarterly Journal* **64**, 475–95.

Thomson, J. S. 1883. Note on crystals of calamine from Wanlockhead, Dumfries-shire. *Mineralogical Magazine* **5**, 332.

——— 1887. Note on the presence of lead in calcites from the Leadhills district. *Mineralogical Magazine* **7**, 143–4.

——— 1892. Analysis of aragonite from Shetland. *Mineralogical Magazine* **10**, 22–3.

Thomson, N. 1997a. Schulenbergite from Nentsberry Haggs mine, Cumbria. *UK Journal of Mines and Minerals* **18**, 15.

——— 1997b. Symplesite from Sandbeds Gill mine, Cumbria. *UK Journal of Mines and Minerals* **18**, 16.

Thomson, T. 1836. *Outlines of mineralogy, geology and mineral analysis*, vols I and II. London.

Thorpe, R. S. & A. G. Tindle 1992. The petrology and petrogenesis of a Tertiary bimodal dolerite–peralkaline/subalkaline trachyte/rhyolite dyke association from Lundy (Bristol Channel, UK). *Geological Journal* **27**, 101–117.

Thorpe, R. S., M. E. Cosgrove, P. van Calsteren 1986. Rare earth element, Sr- and Nd-isotope evidence for petrogenesis of Permian basaltic and K-rich volcanic rocks from southwest England. *Mineralogical Magazine* **50**, 481–90.

Thorpe, R. S., A. G. Tindle, A. G. Gledhill 1990. The petrology, geochemistry and petrogenesis of the Tertiary Lundy granite (Bristol Channel, UK). *Journal of Petrology* **31**, 1379–406.

Thorpe, R. S., A. G. Tindle, O. Williams-Thorpe 1995. Radioelement distribution in the Tertiary Lundy granite (Bristol Channel, UK). *Geological Magazine* **132**, 413–25.

Thorpe, R. S., O. Williams-Thorpe, D. G. Jenkins, J. S. Watson 1991. The geological sources and transport of the bluestones of Stonehenge, Wiltshire, UK. *Prehistoric Society, Proceedings* **57**, 103–157.

Threadgould, R. & B. N. Young 1994. The Assynt carbonatite. *Earth Heritage* **1**, 25–6.

Thurgood, I. D. 1982. Childrenite. *Mineral Realm* **1**(4), 11–12.

Thurlow, C. 2001. *China clay from Cornwall and Devon* (3rd edn). St Austell: Cornish Hillside Publications.

Tidmarsh, W. G. 1932. The Permian lavas of Devon. *Geological Society of London, Quarterly Journal* **88**, 712–75.

Tilley, C. E. 1923. The petrology of the metamorphosed rocks of the Start area (south Devon). *Geological Society of London, Quarterly Journal* **79**, 172–204.

——— 1924. Contact metamorphism in the Comrie area of the Perthshire Highlands. *Geological Society of London, Quarterly Journal* **80**, 22–71.

——— 1925. Petrographical notes on some chloritoid rocks, II: chloritoid phyllites of the Tintagel area, North Cornwall. *Geological Magazine* **62**, 314–18.

——— 1929a. On melilite as a product of interaction of limestone and basaltic liquid. *Geological Magazine* **66**, 347–53.

——— 1929b. On larnite (calcium orthosilicate, a new mineral) and its associated minerals from the limestone contact zone of Scawt Hill, Co. Antrim. *Mineralogical Magazine* **22**, 77–86.

——— 1930a. An association of afwillite with spurrite. *Geological Magazine* **67**, 168–9.

——— 1930b. Scawtite, a new mineral from Scawt Hill, Co. Antrim. *Mineralogical Magazine* **22**, 222–4.

——— 1933. Portlandite, a new mineral from Scawt Hill, Co. Antrim. *Mineralogical Magazine* **23**, 419–20.

——— 1934. Hydrocalumite ($4CaO{\cdot}Al_2O_3{\cdot}12H_2O$), a new mineral from Scawt Hill, Co. Antrim. *Mineralogical Magazine* **23**, 607–615.

——— 1935a. The role of kyanite in the "hornfels zone" of the Carn Chuinneag granite (Ross-shire). *Mineralogical Magazine* **24**, 92–7.

——— 1935b. Metasomatism associated with the greenstone hornfelses of Kenidjack and Botallack, Cornwall. *Mineralogical Magazine* **24**, 181–202.

——— 1936. Eulysites and related rock types from Loch Duich, Ross-shire. *Mineralogical Magazine* **24**, 337–42.

——— 1937a. Pyroxmangite from Inverness-shire, Scotland. *American Mineralogist* **22**, 720–27.

——— 1937b. Anthophyllite–cordierite–granulites of the Lizard. *Geological Magazine* **74**, 300–309.

——— 1937c. The paragenesis of kyanite–amphibolites. *Mineralogical Magazine* **24**, 555–67.

——— 1937d. Wollastonite solid solutions from Scawt Hill, Co. Antrim. *Mineralogical Magazine* **24**, 569–72.

——— 1938a. The status of hornblende in low-grade metamorphic zones of greenschist. *Geological Magazine* **75**, 497–503.

——— 1938b. On scawtite pseudomorphs after spurrite at Scawt Hill, Co. Antrim. *Mineralogical Magazine* **25**, 38–40.

——— 1942. Tricalcium disilicate (rankinite), a new mineral from Scawt Hill, Co. Antrim. *Mineralogical Magazine* **26**, 190–96.

——— 1943. Polyhalite from a depth of 4,500 ft in a boring at Aislaby, Sleights, near Whitby. *Mineralogical Magazine* **26**, lvii.

——— 1946. Bustamite from Treburland manganese mine, Cornwall, and its paragenesis. *Mineralogical Magazine* **27**, 236–41.

——— 1947. The gabbro–limestone contact zone of Camas Mòr, Muck, Inverness-shire.

La Commission Géologique de Finlande, Bulletin **140**, 97–105.

——— 1948a. On iron-wollastonite in contact skarns: an example from Skye. *American Mineralogist* **33**, 736–8.

——— 1948b. Dolomite contact skarns of the Broadford area, Skye: a preliminary note. *Geological Magazine* **85**, 213–16.

——— 1948c. Earlier stages in the metamorphism of siliceous dolomites. *Mineralogical Magazine* **28**, 272–6.

——— 1949. An alkali facies of granite at granite–dolomite contacts in Skye. *Geological Magazine* **86**, 81–93.

——— 1951. The zoned contact skarns of the Broadford area, Skye: a study of boron–fluorine metasomatism in dolomites. *Mineralogical Magazine* **29**, 621–66.

——— 1958. Some new chemical data on assemblages of the Assynt alkali suite. *Edinburgh Geological Society, Transactions* **17**, 156–64.

Tilley, C. E. & A. R. Alderman 1934. Progressive metasomatism in the flint nodules of the Scawt Hill contact zone. *Mineralogical Magazine* **23**, 513–18.

Tilley, C. E. & J. S. Flett 1930. Hornfelses from Kenidjack, Cornwall. *Summary of Progress, Geological Survey of Great Britain for 1929*, part II, 24–41. London: Geological Survey of Great Britain.

Tilley, C. E. & H. F. Harwood 1931. The dolerite–chalk contact of Scawt Hill, Co. Antrim. The production of basic alkali rocks by the assimilation of limestone by basaltic magma. *Mineralogical Magazine* **22**, 439–68.

Tilley, C. E. & H. C. G. Vincent 1947. Cuspidine from dolomite contact skarns, Broadford, Skye. *Mineralogical Magazine* **28**, 90–95.

——— 1948. The occurrence of an orthorhombic high-temperature form of Ca_2SiO_4 (bredigite) in the Scawt Hill contact-zone and as a constituent of slags. *Mineralogical Magazine* **28**, 255–71.

Timberlake, S. 1988. Excavations at Parys Mountain and Nantyreira. *Archaeology in Wales* **28**, 11–17.

——— 1993. Cwmystwyth, Wales: 3500 years of mining history – some problems of conservation and recording. *Russell Society, Journal* **5**(1), 49–53.

Timberlake, S. & S. Mills 2003. The use of a portable XRF within an early nineteenth century cobalt mine on Alderley Edge, Cheshire. *UK Journal of Mines and Minerals* **23**, 41–6.

Tindle, A. G. 1982. *Petrogenesis of the Loch Doon granitic intrusion, Southern Uplands of Scotland.* PhD thesis, Open University, Milton Keynes.

Tindle, A. G. & J. A. Pearce 1981. Petrogenetic modelling of *in situ* fractional crystallization in the zoned Loch Doon pluton, Scotland. *Contributions to Mineralogy and Petrology* **78**, 196–207.

——— 1983. Assimilation and partial melting of the continental crust: evidence from the mineralogy and geochemistry of autoliths and xenoliths *Lithos* **83**, 185–202.

Tindle, A. G. & P. C. Webb 1989. Niobian wolframite from Glen Gairn, eastern Highlands of Scotland: a microprobe investigation. *Geochimica et Cosmochimica Acta* **53**, 1921–35.

——— 1990. Estimation of lithium contents in trioctahedral micas using microprobe data: application to micas from granitic rocks. *European Journal of Mineralogy* **2**, 595–610.

——— 1994. Probe-Amph – a spreadsheet program to classify microprobe-derived amphibole analyses. *Computers and Geosciences* **20**, 1201–228.

Tindle, A. G., F. W. Breaks, J. B. Selway 2002. Tourmaline in petalite-subtype granitic pegmatites: evidence of fractionation and contamination from the Pakeagama Lake and Separation Lake areas of northwestern Ontario, Canada. *Canadian Mineralogist* **40**, 753–88.

Tindle, A. G., T. F. Bridges, D. I. Green 2006. The composition of tsumebite from Roughton Gill, Caldbeck Fells, Cumbria. *UK Journal of Mines and Minerals* **27**, 48–50.

Tindle, A. G., D. W. McGarvie, P. C. Webb 1988. The relative roles of hybridization and crystal fractionation in the evolution of the Cairnsmore of Carsphairn intrusion, Southern Uplands of Scotland. *Geological Society of London, Journal* **145**, 11–21.

Tischendorf, G., H-J. Förster, B. Gottesmann 2001. Minor- and trace-element composition of trioctahedral micas: a review. *Mineralogical Magazine* **65**, 249–34.

Tischendorf, G., H-J. Förster, B. Gottsmann, M. Rieder 2007. True and brittle micas: composition and solid solution series. *Mineralogical Magazine* **71**, 285–320.

Todd, J. G. 1989. Minerals of Trearne quarry, Beith, Ayrshire, Scotland. *UK Journal of Mines and Minerals* **6**, 18–20.

Todd, J. G. & D. W. A. Laurence 1989. Muirshiel mine, central Scotland. *UK Journal of Mines and Minerals* **7**, 40–43.

Todd, J. G. & M. J. McMullen 1991. Vein minerals of the Mannoch Hill, Scotland. *UK Journal of Mines and Minerals* **9**, 35–8.

Todhunter, P. 2002. Field meeting report: Parys Mountain, Anglesey, Wales. *Russell Society, Newsletter* **41**, 32.

——— 2007. Harmotome from Ruthwaite Lodge mine, Cumbria. *UK Journal of Mines and Minerals* **28**, 45–6.

Tomkeieff, S. I. 1926. On some chloritic minerals associated with the basaltic Carboniferous rocks of Derbyshire. *Mineralogical Magazine* **21**, 73–82.

——— 1929. A contribution to the mineralogy of the Whin Sill. *Mineralogical Magazine* **22**, 100–119.

——— 1928. The volcanic complex of Calton Hill (Derbyshire). *Geological Society of London, Quarterly Journal* **84**, 703–718.

——— 1934. Differentiation in basalt lavas, Island Magee, Co. Antrim. *Geological Magazine* **71**, 501–512.

——— 1940. The dolerite plugs of Tieveragh and Tievebulliagh near Cushendall, Co. Antrim, with a note on buchite. *Geological Magazine* **77**, 54–64.

——— 1941. Metasomatism in the basalt of Haddenrig quarry near Kelso and the veining of the rocks exposed there. *Mineralogical Magazine* **26**, 45–59.

——— 1942. The Tertiary lavas of Rum. *Geological Magazine* **79**, 1–13.

Traill, J. G. 1939. The geology and development of Mill Close mine, Derbyshire. *Economic Geology* **24**, 851–89.

Treagus, J. E., R. A. D. Pattrick, S. F. Curtis 1999. Movement and mineralization in the Tyndrum fault zone, Scotland and its regional significance. *Geological Society of London, Journal* **156**, 591–604.

Trechmann, C. O. 1886. Barytes from Addiewell, West Calder, in Midlothian. *Mineralogical Magazine* **7**, 49–56.

——— 1901. Note on a British occurrence of mirabilite. *Mineralogical Magazine* **13**, 73–4.

Treloar, P. J. 1981. Garnet–biotite–cordierite thermometry and barometry in the Cashel thermal aureole, Connemara, Ireland. *Mineralogical Magazine* **44**, 183–9.

Treloar, P. J. & M. D. Max 1984. The hornfels and depths of emplacement of the Carnsore granite, southwest Ireland. *Geologists' Association, Proceedings* **95**, 185–90.

Treloar, P. J., B. J. Bluck, D. R. Bowes, A. Dudek 1980. Hornblende–garnet metapyroxenite beneath serpentinite in the Ballantrae complex of southwest Scotland and its bearing on the depth provenance of obducted oceanic lithosphere. *Royal Society of Edinburgh, Transactions: Earth Sciences* **71**, 201–212.

Tresise, G. 1991. The mineral collections of Liverpool museum. *UK Journal of Mines and Minerals* **10**, 33–5.

——— 1931. The witherite deposit of the Settlingstones mine, Northumberland. *Institution of Mining and Metallurgy, Transactions B* **40**, 56–72.

Trewin, N. H., B. C. Kneller, C. Gillen (eds) 1987. *An excursion guide to the geology of the Aberdeen area.* Edinburgh: Scottish Academic Press.

Trotter, F. M. 1944. The age of the ore deposits of the Lake District and of the Alston Block. *Geological Magazine* **81**, 223–9.

——— 1945. The origin of the west Cumbrian haematites. *Geological Magazine* **82**, 67–80.

Trotter, F. M. & S. E. Hollingworth 1932. *The geology of the Brampton district.* Memoir, Geological Survey of Great Britain, London.

Trude, K. J. & J. J. Wilkinson 2001. A mineralogical and fluid inclusion study of the Harberton Bridge Fe–Zn–Pb deposit, County Kildare, Ireland. *Geological Society of London, Journal* **158**, 37–46.

Trueman, A. E. 1930. Wavellite in the cherts of Bishopton. *Swansea Scientific and Field Naturalists' Society, Proceedings* **1**, 98–9.

Trythall, R. J. B. 1988. *The Mid-Ordovician oolitic ironstones of North Wales.* DPhil thesis, Luton College of Higher Education.

Tschernich, R. W. 1992. *Zeolites of the world.* Phoenix: Geoscience Press.

Tucker, M. E. 1977. The marginal Triassic deposits of South Wales: continental facies and palaeogeography. *Geological Journal* **12**, 169–88.

Tuffnell, J. 1997. Field meeting report: Penlee quarry, Newlyn, Cornwall. *Russell Society, Newsletter* **31**, 17.

Turner, D. C. & A. L. Harris 1971. A note on ground mica resources in Scotland. *Institution of Mining and Metallurgy, Transactions B* **80**, 119–21.

Turner, J. S. 1963. The type locality of witherite. *Mineralogical Magazine* **33**, 431–2.

Turner, G. & M. P. Bannon 1992. Argon isotope geochemistry of inclusion fluids from granite-associated mineral veins in southwest and northeast England. *Geochimica et Cosmochimica Acta* **56**, 227–43.

Turner, N. 1990. The mineral collection of the Natural History Museum, Wollaton Hall, Nottingham. *UK Journal of Mines and Minerals* **8**, 15–17.

Turner, P. & R. A. Ixer 1977. Diagenetic development of unstable and stable magnetization in the St Bees Sandstone (Triassic) of northern England. *Earth and Planetary Science Letters* **34**, 113–24.

Turner, R. 2007. A mechanism for the formation of the mineralized Mn deposits at Merehead quarry, Cranmore, Somerset, England. *Mineralogical Magazine* **70**, 629–53.

Tweedie, J. R. 1979. Origin of uranium and other metal enrichments in the Helmsdale granite, eastern Sutherland, Scotland. *Institution of Mining and Metallurgy, Transactions B* **88**, 145–53.

Tyler, I. 2003. *Carrock and the mines of Skiddaw and Blencathra.* Keswick, Cumbria: Blue Rock Publications.

——— 2005. *Force Crag: the history of a Lakeland mine.* Keswick, Cumbria: Blue Rock

Publications.

Tyler, I. M. & J. R. Ashworth 1981. Garnet zoning and re-equilibration in the Strontian area, Scotland. *Mineralogical Magazine* **44**, 293–300.

Tyler, P. A. 2004. The Parys Mountain volcanogenic sulphide deposit, North Wales: a new interpretation. See Kelly et al. (2004: 127–54).

Tyler, R. C. & B. C. King 1967. The pyroxenes of the alkaline igneous complexes of eastern Uganda. *Mineralogical Magazine* **36**, 5–21.

Tyrrell, G. W. 1928a. *The geology of Arran*. Memoir, Geological Survey of Great Britain, Edinburgh.

——— 1928b. On some dolerite sills containing analcite-syenite in Central Ayrshire. *Geological Society of London, Quarterly Journal* **84**, 540–67.

Tyson, L. O. 1989. *A history of the manor and lead mines of Marrick, Swaledale*. British Mining 38, Northern Mine Research Society, Sheffield.

——— 1995. *The Arkengarthdale mines*. British Mining 53, Northern Mine Research Society, Sheffield.

Tyson, L. O., I. M. Spensley, R. F. White 1995. *The Grinton mines*. British Mining 51, Northern Mine Research Society, Sheffield.

Upton, B. J. G., P. Aspen, N. A. Chapman 1983. The upper mantle and deep crust beneath the British Isles: evidence from inclusions in volcanic rocks. *Geological Society of London, Journal* **140**, 105–121.

Upton, B. G. J., P. Aspen, R. W. Hinton 2003. Garnet-pyroxenite xenoliths and pyropic megacrysts in Scottish alkali basalts. *Scottish Journal of Geology* **39**, 169–84.

Upton, B. G. J., P. Aspen, D. C. Rex, F. Melcher, P. Kinny 1998. Lower crustal and possible shallow mantle samples from beneath the Hebrides: evidence from a xenolithic dyke at Gribun, western Mull. *Geological Society of London, Journal* **155**, 813–28.

Upton, B. G. J., A. C. Skovgaard, J. McClurg, L. Kirstein, M. Cheadle, C. H. Emeleus, W. J. Wadsworth, A. E. Fallick 2002. Picritic magmas and the Rum ultramafic complex, Scotland. *Geological Magazine* **139**, 437–52.

Upton, B. G. J., D. Stephenson, R. M. Ellam 2007. Silica-undersaturated trachytic rocks of central Scotland. *Scottish Journal of Geology* **43**, 143–53.

Uttley, S. 1986. Halite from South Yorkshire. *UK Journal of Mines and Minerals* **1**, 42.

——— 1987a. Barite from Silverwood colliery. *UK Journal of Mines and Minerals* **3**, 21.

——— 1987b. Smallcleugh melanterite. *UK Journal of Mines and Minerals* **2**, 27.

——— 1988. Mineralization at the Goldthorpe/Hickleton complex, South Yorkshire. *UK Journal of Mines and Minerals* **4**, 35–7.

——— 1989. Pike Lane opencut, Winster Moor, Derbyshire. *UK Journal of Mines and Minerals* **7**, 4–6.

van de Kamp, P. C. 1969. The Silurian volcanic rocks of the Mendip Hills, Somerset; and the Tortworth area, Gloucestershire, England. *Geological Magazine* **106**, 542–53.

van Marcke de Lummen, G. 1985. Mineralogy and geochemistry of skarn deposits in the Land's End aureole, Cornwall. *Ussher Society, Proceedings* **6**, 211–17.

——— 1986a. Fluor-bearing hydro-andradite from an altered basalt in the Land's End area, southwest England. *Bulletin de Minéralogie* **109**, 613–16.

——— 1986b. Tin-bearing epidote from skarn in the Land's End aureole, Cornwall, England. *Canadian Mineralogist* **24**, 411–15.

van Panhuys-Sigler, M. & N. H. Trewin 1990. Authigenic sphene cement in Permian sandstones from Arran. *Scottish Journal of Geology* **26**, 139–44.

van Panhuys-Sigler, M., N. H. Trewin, J. Still 1996. Roscoelite associated with reduction spots in Devonian redbeds, Gamrie Bay, Banffshire. *Scottish Journal of Geology* **32**, 127–32.

Varvill, W. W. 1959. The future of lead–zinc and fluorspar mining in Derbyshire. See Higham (1959: 175–203).

Vaughan, D. J. 1969. Zonal variation in bravoite. *American Mineralogist* **54**, 1075–1083.

Vaughan, D. J. & R. A. Ixer 1980. Studies of sulphide mineralogy of North Pennine ores and its contribution to genetic models. *Institution of Mining and Metallurgy, Transactions B* **89**, 99–110.

Velde, D. 1971a. Les lamprophyres à feldspath alcalin et biotite: minettes et roches voisines. *Contributions to Mineralogy and Petrology* **30**, 216–39.

——— 1971b. A note on an analcite-bearing lamproite from Devonshire. *Geological Magazine* **108**, 201–204.

Ventura, G. D., F. Bellatreccia, G. C. Parodi, F. Cámara, M. Piccinini 2007. Single-crystal FTIR and X-ray study of vishnevite, ideally $[Na_6(SO_4)][Na_2(H_2O)_2](Si_6Al_6O_{24})$. *American Mineralogist* **92**, 713-21.

Versey, H. C. 1939. The petrography of the Permian rocks in the southern part of the Vale of Eden. *Geological Society of London, Quarterly Journal* **95**, 275–98.

Vezzalini, G., G. Artioli, S. Quartieri, H. Foy 1992. The crystal chemistry of cowlesite. *Mineralogical Magazine* **56**, 575–9.

Vipan, P. G. L. 1959. Lead and zinc mining in southwest England. See Higham (1959: 337–51).

Virgo, D. & M. Ross 1973. Pyroxenes from Mull andesites. In *Annual report of the director of the Geophysical Laboratory, 1972–1973*, 535–40. Carnegie Institution, Washington DC.

Visscher, H. 1971. *The Permian and Triassic of the Kingscourt outlier, Ireland*. Special Report 1, Geological Society of Ireland, Dublin.

Vivian, S. 1885. The hematite deposits of the southern outcrop of the Carboniferous Limestone of South Wales. *South Wales Institute of Engineers, Transactions* **14**, 164–75.

Vivian, W. 1859. On arborescent native copper in the Great Ormeshead, North Wales. *Geological Society of London, Quarterly Journal* **15**, 109–110.

von Knorring, O. 1951. A note on the occurrence of columbite in the Meldon aplite, Devonshire. *Mineralogical Magazine* **29**, 799–801.

von Knorring, O. & E. Condliffe 1984. On the occurrence of niobium–tantalum and other rare-element minerals in the Meldon aplite, Devonshire. *Mineralogical Magazine* **48**, 443–8.

von Knorring, O. & R. Dearnley 1959a. Molybdenite associated with Laxfordian gneisses at Loch Stack, Sutherlandshire. *Mineralogical Magazine* **32**, 344–5.

——— 1959b. Niobium–zirconium–thorium–uranium and rare-earth minerals from the pegmatites of South Harris, Outer Hebrides. *Nature* **183**, 255–6.

——— 1960a. The Lewisian pegmatites of South Harris, Outer Hebrides. *Mineralogical Magazine* **32**, 366–78.

——— 1960b. A note on a nordmarkite and an associated rare-earth mineral from the Ben Loyal syenite complex, Sutherlandshire. *Mineralogical Magazine* **32**, 389–91.

von Lasaulx, A. 1879. On the tridymite–quartz trachyte of Tardree Mountain and on the olivine gabbro of the Carlingford Mountains. *Royal Geological Society of Ireland, Journal* **15**, 30–38.

Wadsworth, W. J. 1986. Silicate mineralogy in the later fractionation stages of the Insch intrusion, northeast Scotland. *Mineralogical Magazine* **50**, 583–95.

——— 1988. Silicate mineralogy of the Middle Zone cumulates and associated gabbroic rocks from the Insch intrusion, northeast Scotland. *Mineralogical Magazine* **52**, 309–322.

——— 1991. Silicate mineralogy of the Belhelvie cumulates, northeast Scotland. *Mineralogical Magazine* **55**, 113–19.

——— 1994. The peridotite plugs of northern Rum. *Scottish Journal of Geology* **30**, 167–74.

Wager, L. R. 1932. The geology of the Roundstone district, County Galway. *Royal Irish Academy, Proceedings* **41B**, 46–72.

Wager, L. R. & E. A. Vincent 1962. Ferrodiorite from the Isle of Skye. *Mineralogical Magazine* **33**, 26–36.

Wager, L. R., D. S. Weedon, E. A. Vincent 1953. A granophyre from Coire Uaigneich, Isle of Skye, containing quartz paramorphs after tridymite. *Mineralogical Magazine* **30**, 263–76.

Wagner, C. & D. Velde 1985. Mineralogy of two peralkaline, arfvedsonite-bearing minettes: a new occurrence of Zn-rich chromite. *Bulletin de Minéralogie* **109**, 173–87.

Walenta, K. 1978. Uranospathite and arsenuranospathite. *Mineralogical Magazine* **42**, 117–28.

Walker, D. R. G. 1988. Retgersite, $NiSO_4{\cdot}H_2O$: a new British occurrence from Warwickshire. *Russell Society, Journal* **2**(1), 29–30.

Walker, F. 1930. The geology of the Shiant Isles (Hebrides). *Geological Society of London, Quarterly Journal* **86**, 355–98.

——— 1932. An albitite from Ve Skerries, Shetland Isles. *Mineralogical Magazine* **23**, 239–42.

——— 1935. The late Palaeozoic quartz-dolerites and tholeiites of Scotland. *Mineralogical Magazine* **24**, 141–59.

Walker, F. & C. F. Davidson 1935. Marginal and contact phenomena of the Dorback granite. *Geological Magazine* **72**, 49–63.

Walker, F. & E. M. Patterson 1959. A differentiated boss of alkali dolerite from Cnoc Rhaonastil, Islay. *Mineralogical Magazine* **32**, 140–52.

Walker, F., E. A. Vincent, R. L. Mitchell 1952. The chemistry and mineralogy of the Kinkell tholeiite, Stirlingshire. *Mineralogical Magazine* **29**, 895–908.

Walker, F. D. L., M. R. Lee, I. Parsons 1995. Micropores and micropermeable texture in alkali feldspars: geochemical and geophysical implication. *Mineralogical Magazine* **59**, 505–534.

Walker, G. P. L. 1951. The amygdale minerals in the Tertiary lavas of Ireland, I: the distribution of chabazite habits and zeolites in the Garron plateau area, County Antrim. *Mineralogical Magazine* **29**, 773–91.

——— 1959a. The amygdale minerals in the Tertiary lavas of Ireland, II: the distribution of gmelinite. *Mineralogical Magazine* **32**, 202–217.

——— 1959b. Some observations on the Antrim basalts and associated dolerite intrusions. *Geologists' Association, Proceedings* **70**, 179–205.

——— 1960a. The amygdale minerals in the Tertiary lavas of Ireland, III: regional distribution. *Mineralogical Magazine* **32**, 503–527.

——— 1960b. The amygdale minerals in the Tertiary lavas of Ireland, IV: the crystal habit of calcite. *Mineralogical Magazine* **32**, 609–618.

——— 1962a. Garronite, a new zeolite, from Ireland and Iceland. *Mineralogical Magazine* **33**, 173–86.

——— 1962b. Low-potash gismondine from Ireland and Iceland. *Mineralogical Magazine* **33**, 187–201.

——— 1971. The distribution of amygdale minerals in Mull and Morvern (western Scotland). In *Studies in Earth sciences* [W. D. West commemorative vol.], T. V. V. Murty & S. S. Roa (eds), 181–94.

Walker, G. P. L. & J. V. Ross 1954. A xenolithic monchiquite dyke near Glenfinnan, Inverness-shire. *Geological Magazine* **91**, 463–72.

Wallace, P. 1998. Laurionite, a second Cornish occurrence from Penberthy Croft mine. *UK Journal of Mines and Minerals* **19**, 11.

Wallace, T. 1884. On kyanite localities in the North. *Mineralogical Magazine* **6**, 106–107.

——— 1885. Notes on Northern minerals. *Mineralogical Magazine* **6**, 168–70.

Wallace, W. 1865. On the growth of flos ferri or coralloidal arragonite. *Geological Society of London, Quarterly Journal* **21**, 413–21.

Wallerius, J. G. 1747. *Mineralogia, eller mineral-riket.* Stockholm: L. Salvius.

Walsh, J. N. 1975. Clinopyroxenes and biotites from the Centre III igneous complex, Ardnamurchan, Argyllshire. *Mineralogical Magazine* **40**, 335–45.

Walshaw, R. D., J. F. Menuge, S. Tyrrell 2006. Metal sources of the Navan carbonate-hosted base metal deposit, Ireland: Nd and Sr isotope evidence for deep hydrothermal convection. *Mineralium Deposita* **41**, 803–819.

Ward, C. D., J. M. McArthur, J. N. Walsh 1992. Rare earth element behaviour during evolution and alteration of the Dartmoor granite, southwest England. *Journal of Petrology* **33**, 785–815.

Ward, G. R. 1983a. Cerussite from White Rake, Tideswell, Derbyshire. *Mineral Realm* **3**(1), 30.

——— 1983b. Bertrandite from Hingston Down quarry, Calstock, Cornwall. *Ussher Society, Proceedings* **5**, 485–6.

Ward, J. C. 1876. *The geology of the northern part of the English Lake District (sheet 101SE).* Memoir, Geological Survey of Great Britain and Ireland, London.

——— 1877. Notes on the occurrence of chlorite among the Lower Silurian volcanic rocks of the English Lake District. *Mineralogical Magazine* **1**, 119–22.

Ward, P. 1982. Gold in North Wales. *Mineral Realm* **2**(1), 30–34.

Warren, P. T., D. Price, M. J. C. Nutt, E. G. Smith 1984. *Geology of the country around Rhyl and Denbigh (sheets 95 and 107 and parts of 94 and 206).* Memoir, British Geological Survey, Keyworth.

Warrington, G. 1965. The metalliferous mining district of Alderley Edge, Cheshire. *Mercian Geologist* **1**, 111–29.

Waters, R. A. & D. J. D. Lawrence 1987. *Geology of the South Wales coalfield, part III: the country around Cardiff (sheet 263)*, 3rd edn. Memoir, British Geological Survey, Keyworth.

Waterston, C. D. 1953. An occurrence of harmotome in northwest Ross-shire. *Mineralogical Magazine* **30**, 136–8.

Watling, R. J. 1974. Identification and analysis of lattice-held mercury in sphalerite from Keel prospect, County Longford, Ireland. *Institution of Mining and Metallurgy, Transactions B* **83**, 88–94.

Watson, J. V. 1948. Late sillimanite in the migmatites of Kildonan, Sutherland. *Geological Magazine* **85**, 149–62.

Watson, J. J. W. 1860. Notes on the metalliferous saddles or ore-bearing beds in the contorted strata of the Lower Carboniferous beds of certain parts of north Derbyshire and north Staffordshire. *The Geologist* **3**, 357–69.

Watson, R. 1987. What's new in minerals. *UK Journal of Mines and Minerals* **3**, 37.

——— 1988. Mineral news. *UK Journal of Mines and Minerals* **4**, 38–9.

Watson, S. 1900. Recent mineral deposits and their relation to vein formation. *Weardale Nature and Field Club, Transactions* **1**, 57–61.

Watt, G. R., P. Wright, S. Galloway, C. McLean 1997. Cathodoluminescence and trace element zoning in quartz phenocrysts and xenocrysts. *Geochimica et Cosmochimica Acta* **61**, 4337–48.

Waymouth, C., P. Thornely, W. H. Taylor 1938. An X-ray examination of mordenite (ptilolite). *Mineralogical Magazine* **25**, 212–16.

Webb, A. B. St J. 1971. Tobermorite from Castle Hill near Kilbirnie, Ayrshire. *Mineralogical Magazine* **38**, 253.

Webb, P. C. & G. C. Brown 1984a. *Lake District granites: heat production and related geochemistry.* Investigation of the Geothermal Potential of the UK, British Geological Survey, Keyworth.

——— 1984b. *The eastern Highlands granites: heat production and related geochemistry.* Investigation of the Geothermal Potential of the UK, British Geological Survey, Keyworth.

Webb, P. C., A. G. Tindle, R. A. Ixer 1992. W–Sn–Mo–Bi–Ag mineralization associated with zinnwaldite-bearing granite from Glen Gairn, Scotland. *Institution of Mining and Metallurgy, Transactions B* **101**, 59–72.

Wedd, C. B., B. Smith, W. B. R. King, D. A. Wray 1929. *The country around Oswestry (sheet 137).* Memoir, Geological Survey of Great Britain, London.

Weis, P. L., I. Friedman, J. P. Gleason 1981. The origins of epigenetic graphite: evidence from isotopes. *Geochimica et Cosmochimica Acta* **45**, 2325–32.

Weiss, S. 1989. Aktuelle Neufunde aus Cornwall. *Lapis* **14**, 40–41.

——— 1994. The minerals of the china-clay pits in Cornwall. *UK Journal of Mines and Minerals* **14**, 21–33.

——— 2000. Blauer skorodit von Hemerdon: "specimen mining" im sudwesten England. *Lapis* **25**(10), 34–40.

Weiss, S. & M. Merry 2007. Arsenflorenceit von Hemerdon Ball, Devon, England. *Lapis* **32**(2), 33.

Weiss, S. & G. Troll 1989. The Ballachulish igneous complex, Scotland: petrography, mineral chemistry and order of crystallization in the monzodiorite–quartz diorite suite and in the granite. *Journal of Petrology* **30**, 1069–1115.

Weiss, S., J. Hooper, N. Elton, D. I. Green 2001. Seltene Mineralien aus dem Greystone quarry, Cornwall, Grossbritannien. *Lapis* **26**(4), 19–33.

Welch, M. D., M. A. Cooper, F. C. Hawthorne, A. J. Criddle 2000. Symesite, $Pb_{10}(SO_4)O_7Cl_4(H_2O)$, a new PbO-related sheet mineral: description and crystal structure. *American Mineralogist* **85**, 1526–33.

Welch, M. D., A. J. Criddle, R. F. Symes 1998. Mereheadite, $Pb_2O(OH)Cl$: a new litharge-related oxychloride from Merehead quarry, Cranmore, Somerset. *Mineralogical Magazine* **62**, 387–93.

Wells, M. K. 1951. Sedimentary inclusions in the hypersthene-gabbro, Ardnamurchan, Argyllshire. *Mineralogical Magazine* **29**, 715–36.

Wells, M. K., A. C. S. Smith, J. F. W. Bowles 1974. Carbonate enrichment at the margins of a basic dyke, Ardnamurchan. *Mineralogical Magazine* **39**, 514–24.

Werner, A. G. 1774. *Von den aüsserlichen Kennzeichen der Fossilien.* Leipzig, Germany: S. L. Crusius (translated by A. Carozzi as *Of the outward characteristics of the fossils.* Urbana, Illinois: University of Illinois Press (1962).

——— 1817. *Letztes Mineral-System.* Freiberg: Graz und Gerlach.

West, I. M., A. Brandon, M. Smith 1968. A tidal flat evaporite facies in the Visean of Ireland. *Journal of Sedimentary Petrology* **38**, 1078–1093.

Wheatley, C. J. V. 1971a. Aspect of metallogenesis within the Southern Caledonides of Great Britain and Ireland. *Institution of Mining and Metallurgy, Transactions B* **80**, 211–13.

——— 1971b. *Economic geology of the Avoca mineralization belt, southeast Ireland and Parys Mountain, Anglesey.* PhD thesis, University of London.

Wheeler, J., L. S. Mangan, D. J. Prior 2004. Disequilibrium in the Ross of Mull contact metamorphic aureole, Scotland: a consequence of polymetamorphism. *Journal of Petrology* **45**, 835–53.

White, J. S. 2006. Mineral museums update. *Mineralogical Record* **37**(4), 323–9.

White, T. J. & B. G. Hyde 1982. Electron microscope study of the humite minerals, II: Mn-rich specimens. *Physics and Chemistry of Minerals* **8**, 167–74.

Whitehead, D. 1971. Base metal sulphides at Tynagh. In *The genesis of base metal deposits in Ireland*, D. Skevington (ed.). Irish Geological Society, Galway Symposium Proceedings. Dublin.

Whitehead, P. J. P. 1973. Some further notes on Jacob Forster (1739–1806), mineral collector and dealer. *Mineralogical Magazine* **39**, 361–3.

Whittaker, E. J. W. & J. Zussman 1956. The characterization of serpentine minerals by X-ray diffraction. *Mineralogical Magazine* **31**, 107–126.

Whitten, E. H. T. 1951. Cataclastic pegmatites and calc-silicates skarns near Bunbeg, Co. Donegal. *Mineralogical Magazine* **29**, 737–56.

——— 1954. Two arfvedsonite rhyolite intrusions from Cloghaneely, Co. Donegal. *Mineralogical Magazine* **30**, 393–9.

Whitworth, M. P. 1992. Petrogenetic implications of garnets associated with lithium pegmatites from southeast Ireland. *Mineralogical Magazine* **56**, 75–83.

Whitworth, M. P. & M. Feely 1994. The compositional range of magmatic Mn-garnets in the Galway granite, Connemara, Ireland. *Mineralogical Magazine* **58**, 163–68.

Whitworth, M. P. & A. H. Rankin 1989. Evolution of fluid phases associated with lithium pegmatites from southeast Ireland. *Mineralogical Magazine* **53**, 271–84.

Wilbur, D. G. & J. S. Carter 1986. Cu–Ag mineralization at Tullacondra, Mallow, Co. Cork. See Andrew et al. (1986: 501–508).

Wilcockson, W. H. 1932. Igneous rocks and mineralization of the High Peak of Derbyshire. *Geologists' Association, Proceedings* **43**, 184–91.

Wilkinson, J. J. 2003. On diagenesis, dolomitization and mineralization in the Irish Zn–Pb orefield. *Mineralium Deposita* **38**, 968–83.

Wilkinson, J. J., S. L. Eyre, A. J. Boyce 2005. Ore-forming processes in Irish-type carbonate-hosted Zn–Pb deposits: evidence from mineralogy, chemistry, and isotopic composition of sulphides at the Lisheen mine. *Economic Geology* **100**, 63–86.

Willan, R. C. R. 1980. Stratabound sulphide mineralization in the Dalradian Supergroup of the Grampian Highlands, Scotland. *Norges Geologiske Undersogelse* **300**, 241–58.

Willan, R. C. R. & M. L. Coleman 1983. Sulphur isotope study of the Aberfeldy barite, zinc, lead deposit and minor sulphide mineralization in the Dalradian metamorphic terrain, Scotland. *Economic Geology* **78**, 1619–56.

Willan, R. C. R. & A. J. Hall 1980. Sphalerite geobarometry and trace-element studies on stratiform sulphide from McPhun's Cairn, Loch Fyne, Argyll, Scotland. *Institution of Mining and Metallurgy, Transactions B* **89**, 31–40.

Williams, C. J. 1995. *Great Orme mines*. British Mining 52, Northern Mine Research Society, Sheffield.

Williams, C. T. 1978. Uranium-enriched minerals in mesostasis areas of the Rhum layered pluton. *Contributions to Mineralogy and Petrology* **66**, 29–39.

Williams, D. 1924. On two olivine-dolerite dykes in Snowdonia. *Liverpool Geological Society, Proceedings* **14**, 38–47.

Williams, F. M., W. A. Sheppard, P. McArdle 1986. Avoca mine, County Wicklow: a review of geological and isotopic studies. See Andrew et al. (1986: 71–82).

Williams, H. 1922. The igneous rocks of the Capel Curig district (North Wales). *Liverpool Geological Society, Proceedings* **13**, 166–201.

——— 1927. The geology of Snowdon (North Wales). *Geological Society of London, Quarterly Journal* **83**, 346–431.

——— 1930a. The Snowdon district: report of the Easter field meeting of the Geologists' Association. *Geologists' Association, Proceedings* **41**, 190–205.

——— 1930b. The geology between Nant Peris and Nant Ffrancon (Snowdonia). *Geological Society of London, Quarterly Journal* **86**, 191–232.

Williams, J. 1965. Further notes on mineralogy in Dumfries and Galloway. *Dumfriesshire and Galloway Natural History and Antiquarian Society, Transactions* **42**, 14–23.

——— 1973. The antimony, bismuth, molybdenum and tungsten minerals of southwest Scotland. *Dumfriesshire and Galloway Natural History and Antiquarian Society, Transactions* **50**, 1–4.

Williams, M. E. 2002. Lateral variation within a hiatal-condensed bed and the controls on the formation and occurrence of berthierine ooids: a case study from the Middle Jurassic of northeast England, UK. *Geological Journal* **37**, 279–96.

Williams, P. J. 1986. Petrology and origin of iron-rich silicate–magnetite–quartz rocks from Flowerdale near Gairloch, Wester Ross. *Scottish Journal of Geology* **22**, 1–12.

——— 1987. Metasomatic phenomena adjacent to a granite pegmatite, Garry-a-siar, Benbecula, Outer Hebrides. *Mineralogical Magazine* **51**, 735–8.

Williams, R. A. 1991. *The Berehaven copper mines*. British Mining 42, Northern Mine Research Society, Sheffield.

Williams, W. 1802. *Observations of the Snowdon mountains*. Oxford: S. Collingwood.

Williamson, B. J., J. Spratt, J. T. Adams, A. G. Tindle, C. J. Stanley 2000. Geochemical constraints from zoned hydrothermal tourmalines on fluid evolution and Sn mineralization: an example from fault breccias at Roche, southwest England. *Journal of Petrology* **41**, 1439–53.

Williamson, B. J., C. J. Stanley, J. J. Wilkinson 1997. Implications from inclusions in topaz for greisenization and mineralization in the Hensbarrow topaz granite, Cornwall, England. *Contributions to Mineralogy and Petrology* **127**, 119–28.

Williamson, B. J., J. J. Wilkinson, P. F. Luckham, C. J. Stanley 2002. Formation of coagulated colloidal silica in high-temperature mineralizing fluids. *Mineralogical Magazine* **66**, 547–53.

Williamson, D. H. 1953. Petrology of chloritoid and staurolite rocks north of Stonehaven, Kincardineshire. *Geological Magazine* **90**, 353–61.

Willies, L., K. Gregory, H. Parker 1989. *Millclose – the mine that drowned*. Cromford, Derbyshire: Scarthin Books.

Wilmot, R. D. & B. Young 1985. Aluminite and other aluminium minerals from Newhaven, Sussex: the first occurrence of nordstrandite in Great Britain. *Geologists' Association, Proceedings* **96**, 47–52.

Wilson, A. A., G. A. Sergeant, B. R. Young, R. K. Harrison 1966. The Rowhurst tonstein, north Staffordshire, and the occurrence of crandallite. *Yorkshire Geological Society, Proceedings* **35**, 421–7.

Wilson, A. C. 1975. An occurrence of sapphire in the Land's End granite, Cornwall. *Geological Survey of Great Britain, Bulletin* **52**, 61–3.

Wilson, G. C. & J. V. P. Long 1983. The distribution of lithium in some Cornish minerals: ion microprobe measurements. *Mineralogical Magazine* **47**, 191–9.

Wilson, G. V. & J. S. Flett 1921. *The lead, zinc, copper and nickel ores of Scotland*. Special Reports on the Mineral Resources of Great Britain XVII, Memoir, Geological Survey of Great Britain, Edinburgh.

Wilson, G. V., T. Eastwood, R. W. Pocock, D. A. Wray, T. Robertson 1922. *Barytes and witherite* (3rd edn). Special Reports on the Mineral Resources of Great Britain II, Memoir, Geological Survey of Great Britain, London.

Wilson, H. E. 1971. Review of recent mineral development in Northern Ireland. *Institution of Mining and Metallurgy, Transactions B* **80**, 139–41.

Wilson, H. E. & J. A. Robbie 1966. *Geology of the country around Ballycastle*. Memoir, Geological Survey of Northern Ireland, Belfast.

Wilson, J. R. & J. S. Coats 1972. Alkali feldspars from part of the Galway granite, Ireland. *Mineralogical Magazine* **38**, 801–810.

Wilson, J. I., M. J. Colledge, R. S. W. Braithwaite 1995. Wulfenite: a new find in Derbyshire, England. *Russell Society, Journal* **6**(1), 48–9.

Wilson, M. J. 1966. The weathering of biotite in some Aberdeenshire soils. *Mineralogical Magazine* **35**, 1080–1093.

——— 1974. Hydrothermal formation and alteration of laumontite in hornblende schist near Huntly, Aberdeenshire. *Mineralogical Magazine* **39**, 608–609.

Wilson, M. J. & D. Jones 1984. The occurrence and significance of manganese oxalate in *Pertusaria corallina* (Lichenes). *Pedobiologia* **26**, 373–9.

Wilson, M. J. & J. M. Tait 1977. Halloysite in some soils from northeast Scotland. *Clay Mineralogy* **12**, 59–66.

Wilson, M. J., M. L. Berrow, W. J. McHardy 1970. Lithiophorite from the Lecht mines, Tomintoul, Banffshire. *Mineralogical Magazine* **37**, 618–23.

Wilson, M. J., P. D. Cradwick, M. L. Berrow, W. J. McHardy, J. D. Russell 1976. Nickeloan pyroaurite from Leslie, Aberdeenshire. *Mineralogical Magazine* **40**, 447–51.

Wilson, M. J., D. Jones, J. D. Russell 1980. Glushinskite, a naturally occurring magnesium oxalate. *Mineralogical Magazine* **43**, 837–40.

Wilson, M. J., J. D. Russell, J. M. Tait, D. R. Clark, A. R. Fraser 1984. Macaulayite, a new mineral from northeast Scotland. *Mineralogical Magazine* **48**, 127–9.

Wilson, M. J., J. M. Tait, D. R. Clark, A. R. Fraser, I. Stephen 1981. A swelling hematite/layer silicate complex in weathered granite. *Clay Mineralogy* **16**, 201–277.

Wilson, O. M., A. Dyer, H. Enamy 2003. *Zeolites of the British Isles*. Occasional paper 15, British Micromount Society, London.

Winrow, A. 1986a. Hilton mine. *UK Journal of Mines and Minerals* **1**, 31–6.

——— 1986b. The mines of Strontian. *UK Journal of Mines and Minerals* **1**, 40–43.

Wirth, M. M. 1989. Native silver from Red Gill mine, Caldbeck Fells, Cumbria. *Russell Society, Journal* **2**(2), 49.

——— 1995. Bavenite from Shap granite quarry, Cumbria, England. *Russell Society, Journal* **6**(1), 51.

——— 2003. Minerals from Orrock quarry, Fife Region, Scotland. *Russell Society, Journal* **8**(1), 23–5.

Wirth, M. M. & D. I. Green 2002. The Shap granite quarry, Cumbria. *UK Journal of Mines and Minerals* **22**, 43–52.

Wirth, M. M., D. I. Green, G. Ryback 2001. Milarite from Shap granite quarry, Cumbria, England, with a discussion of milarite occurrences in Cornwall and Devon. *Russell Society, Journal* **7**(2), 90–91.

Wiseman, J. D. H. 1934. The central and southwest Highland epidiorites: a study in progressive metamorphism. *Geological Society of London, Quarterly Journal* **90**, 354–417.

Witzke, T. 1999. Hydrowoodwardite, a new mineral of the hydrotalcite group from Königswalde near Annaberg, Saxony, Germany and other localities. *Neues Jahrbuch für Mineralogie*, Monatshefte 2, 75–86.

Wolfe, M. 1996a. Mineral news. *UK Journal of Mines and Minerals* **16**, 4–12.

——— 1996b. Mineral news. *UK Journal of Mines and Minerals* **17**, 4–6.

——— 1997. Mineral news. *UK Journal of Mines and Minerals* **18**, 4–6.

——— 1998. Mineral news. *UK Journal of Mines and Minerals* **19**, 5–8.

——— 2000. Mineral news. *UK Journal of Mines and Minerals* **20**, 4–6.

——— 2004. Mineral news. *UK Journal of Mines and Minerals* **24**, 4–8.

——— 2006. Mineral news. *UK Journal of Mines and Minerals* **27**, 6–10.

Wolloxall, T. 1981. Some micro-mount minerals from the Red Gill mine. *Mineral Realm* **1**(2), 5–9.

——— 1987. Levant mine. *UK Journal of Mines and Minerals* **2**, 24–7.

——— 1988. Wheal Owles. *UK Journal of Mines and Minerals* **5**, 7–8.

——— 1989. Famous mineral localities: Botallack, Cornwall. *UK Journal of Mines and Minerals* **6**, 4–11.

Wolters, J. & A. Wolters 1986. Dean quarry. *UK Journal of Mines and Minerals* **1**, 20–27.

——— 1987. South Crofty mine – Cornwall. *UK Journal of Mines and Minerals* **3**, 18–20.

Wood, M. 1991. Strontianite from Brownley Hill mine, Nenthead, Cumbria. *Russell Society, Journal* **4**(1), 35–6.

——— 1993. Strontianite from Gunnerside Gill, Swaledale, North Yorkshire. *UK Journal of Mines and Minerals* **12**, 16–20.

Woodham, C., S. Finlay, R. Holman 1989. Gold exploration in Dalradian of Northern Ireland. *Institution of Mining and Metallurgy, Transactions B* **98**, 63–5.

Woodland, A. W. 1935. Spessartite in the Cambrian manganese ore of Merionethshire. *Geological Magazine* **72**, 384.

——— 1938a. Petrological studies in the Harlech Grit Series of Merionethshire, II: the petrography and petrology of some of the grits. *Geological Magazine* **75**, 440–54.

——— 1938b. Some petrological studies in the Harlech Grit Series of Merionethshire, III: the development of pyrite in the grits and mudstones. *Geological Magazine* **75**, 529–39.

——— 1939. The petrography and petrology of the Lower Cambrian manganese ore of west Merionethshire. *Geological Society of London, Quarterly Journal* **95**, 1–35.

Woods, P. J. E. 1973. Potash exploration in Yorkshire: Boulby mine pilot borehole. *Institution of Mining and Metallurgy, Transactions B* **82**, 99–106.

——— 1979. The geology of Boulby mine. *Economic Geology* **74**, 409–418.

Woodward, A. S. 1884. Note on the occurrence of evansite in east Cheshire. *Mineralogical Magazine* **5**, 333–4.

Woodward, A. S. & J. M. Mello 1881. The minerals of the Midlands. *Naturalist* **4**, 183–6.

Wooley, A. R. 1973. The pseudoleucite borolanites and associated rocks of the southeastern tract of the Borralan complex, Scotland. *British Museum (Natural History), Bulletin: Mineralogy Series* **2**, 287–333.

Woolley, A. R., R. F. Symes, C. J. Elliot 1972. Metasomatized (fenitized) quartzites from the Borralan complex, Scotland. *Mineralogical Magazine* **38**, 819–36.

Worden, R. H., P. E. Champness, G. T. R. Droop 1987. Transmission electron microscopy of the pyrometamorphic breakdown of phengite and chlorite. *Mineralogical Magazine* **51**, 107–121.

Worley, N. E. 1976. Lithostratigraphical control of mineralization in the Blende vein, Magpie mine, Sheldon near Bakewell, Derbyshire. *Yorkshire Geological Society, Proceedings* **41**, 95–106.

——— 2005. The occurrence of halite in the Permian A Bed Evaporite, Kirkby Thore, Cumbria. *Yorkshire Geological Society, Proceedings* **55**, 199–203.

Worley, N. E., T. Worthington, L. Riley 1978. The geology and exploration of Hubberdale mine, Taddington. *Peak District Mines Historical Society, Bulletin* **7**, 31–40.

Workman, G. M. 1960. Report on excursions to Matlock Bath. *Peak District Mines Historical Society, Bulletin* **1**(2), 14–19.

Worku, T. & A. Parker 1992. Occurrence of bassanite in Lower Lias rocks of the Lyme Regis area, England. *Mineralogical Magazine* **56**, 258–9.

Worth, R. H. 1920. The geology of the Meldon valleys near Okehampton, on the northern verge of Dartmoor. *Geological Society of London, Quarterly Journal* **75**, 77–114.

Wright, C. A. & J. R. Wilson 1970. Jarosite from the Eocene of the Hampshire Basin. *Mineralogical Magazine* **37**, 941.

Wurzburger, P. 1872. The hematite iron deposits of Furness. *Iron and Steel Institute, Journal* **1**, 135–42.

Wyley, J. F. 1962. Metatyuyamunite from the Permian, Budleigh Salterton, Devon [exhibit: general meeting of the Mineralogical Society, Burlington House, London, 26 January 1961]. *Mineralogical Magazine* **33**, xlvi.

Wyse Jackson, P. N. 1992. The geological collections of Trinity College, Dublin 1780–1990. *The Geological Curator* **5**, 263–74.

Yardley, B. W. D. & C. B. Long 1981. Contact metamorphism and fluid movement around the Easky adamellite, Ox Mountains, Ireland. *Mineralogical Magazine* **44**, 125–31.

Yardley, B. W. D., E. Condliffe, G. E. Lloyd, S. H. M. Harris 1996. Polyphase garnets from western Ireland: two-phase intergrowths in the grossular–almandine series. *European Journal of Mineralogy* **8**, 383–92.

Yardley, B. W. D., C. A. Rochelle, A. C. Barnicoat, G. E. Lloyd 1991. Oscillatory zoning in metamorphic minerals: an indicator of infiltration metasomatism. *Mineralogical Magazine* **55**, 357–65.

Young, B. 1984a. Siderite from the west Cumbrian hematite deposits. *Cumberland Geological Society, Proceedings* **4**, 267–8.

——— 1984b. Geology and history of Nab Gill mine, Eskdale, Cumbria. *Cumberland Geological Society, Proceedings* **4**, 269–75.

——— 1985a. Pyromorphite in the North Pennines. *Russell Society, Journal* **1**(3), 81–2.

——— 1985b. Strontianite from the Northern Pennine orefield. *Mineralogical Magazine* **49**, 762.

——— 1985c. The distribution of barytocalcite and alstonite in the Northern Pennine orefield. *Yorkshire Geological Society, Proceedings* **45**, 199–206.

——— 1985d. *Greisens and related rocks associated with the Eskdale granite, Cumbria*. Open File Report PDA2 85.2, British Geological Survey, Newcastle upon Tyne.

——— 1985e. *Mineralization associated with the Eskdale intrusion, Cumbria*. Open File Report PDA2 85.3, British Geological Survey, Newcastle upon Tyne.

——— 1986. Wulfenite from Ruthwaite Lodge, Grisdale, a new Lake District occurrence. *Cumberland Geological Society, Proceedings* **4**, 403–404.

——— 1987a. *Glossary of the minerals of the Lake District and adjoining areas*. Newcastle upon Tyne: British Geological Survey.

——— 1987b. Uncommon Pennine minerals, part 1: aurichalcite in the Yorkshire and Cumbrian Pennines. *Leeds Geological Association, Transactions* **11**(2), 1–7.

——— 1987c. Uncommon Pennine minerals, part 2: strontianite from the Yorkshire Pennines. *Leeds Geological Association, Transactions* **11**(3), 1–8.

——— 1988. Birnessite from Treburland mine, Altarnum, Cornwall. *Mineralogical Magazine* **52**, 415–16.

——— 1993. Some new occurrences of barytocalcite in the Northern Pennine orefield. *Natural History Society of Northumberland, Transactions* **56**, 61–4.

——— 1995. The Northern Pennine orefield: Weardale and Nenthead. See Scrutton (1995: 154–61).

——— 2001. Copper mineralization in the Permian Magnesian Limestone at Raisby, Coxhoe, County Durham, England. *Russell Society, Journal* **7**(2), 75–8.

——— 2005. A large glacial erratic boulder of gypsum from the Durham coast. *Natural History Society of Northumberland, Transactions* **64**, 207–210.

Young, B. & T. F. Bridges 1984. Harmotome from Northumberland. *Natural History Society of Northumberland, Transactions* **52**, 24–6.

Young, B. & A. H. Cooper 1988. The geology and mineralization of Force Crag mine, Cumbria. *Cumberland Geological Society, Proceedings* **5**(1), 5–11.

Young, B. & J. G. Francis 1989. Rosasite from West Pasture mine, Weardale, Co. Durham. *Russell Society, Journal* **2**(2), 50.

Young, B. & E. W. Johnson 1985. Langite and posnjakite from the Lake District. *Russell Society, Journal* **1**(3), 80.

Young, B. & P. H. A. Nancarrow 1988a. Millerite from the Cumbrian coalfield. *Russell Society, Journal* **2**(1), 5–7.

——— 1988b. Rozenite and other sulphate minerals from the Cumbrian coalfield. *Mineralogical Magazine* **52**, 551–2.

——— 1990. Hausmannite from west Cumbria. *Russell Society, Journal* **3**(1), 29–30.

Young, B., M. P. Boland, E. K. Hyslop 1992a. Metavoltine from Cockermouth, Cumbria. *Mineralogical Magazine* **56**, 431–2.

Young, B., T. E. Bridges, E. K. Hyslop 1994a. Leadhillite from the Northern Pennine orefield, England. *Russell Society, Journal* **5**(2), 121–3.

Young, B., T. F. Bridges, P. R. Ineson 1987. Supergene cadmium mineralization in the Northern Pennine orefield. *Yorkshire Geological Society, Proceedings* **46**, 275–8.

Young, B., M. P. Cooper, P. H. A. Nancarrow, D. I. Green 1990a. Supergene minerals from Low Pike, Caldbeck Fells, Cumbria. *Russell Society, Journal* **3**(1), 25–8.

Young, B., A. Dyer, N. Hubbard, R. E. Starkey 1991a. Apophyllite and other zeolite-type minerals from the Whin Sill of the northern Pennines. *Mineralogical Magazine* **55**, 203–208.

Young, B., R. J. Firman, R. E. Starkey 1988. An occurrence of apophyllite at Shap, Cumbria. *Mineralogical Magazine* **52**, 133.

Young, B., N. J. Fortey, P. H. A. Nancarrow 1986. An occurrence of tungsten mineralization in the Eskdale Intrusion, west Cumbria. *Yorkshire Geological Society, Proceedings* **46**, 15–21.

Young, B., N. J. Fortey, P. H. A. Nancarrow 1991b. Russellite from Buckbarrow Beck, Cumbria, England. *Russell Society, Journal* **4**(1), 3–7.

Young, B., L. Greenbank, P. H. A. Nancarrow 1990b. Alstonite *in situ* at Brownley Hill mine, Nenthead, Cumbria. *Mineralogical Magazine* **54**, 515–16.

Young, B., E. K. Hyslop, J. Baty, D. I. Green 2008. Supergene mineralization in colliery spoil at Hawthorn Hive, Co. Durham: the first record of apjohnite in Great Britain. *Natural History Society of Northumberland, Transactions* (in press).

Young, B., E. K. Hyslop, T. F. Bridges, J. Cooper 2005. New records of supergene minerals from the Northern Pennine orefield. *Natural History Society of Northumberland, Transactions* **64**, 211–14.

Young, B., E. K. Hyslop, R. A. Fairbairn 1996. Coronadite from the Northern Pennine orefield, England. *Russell Society, Journal* **6**(2), 97.

Young, B., E. K. Hyslop, T. F. Bridges, N. Hubbard 1998. Thaumasite from High Sedling mine, Weardale, County Durham, and from Shap, Cumbria, England. *Russell Society, Journal* **7**(1), 40.

Young, B., E. K. Hyslop, D. Millward 1992b. Barium-pharmacosiderite and symplesite from Caldbeck Fells, Cumbria. *Russell Society, Journal* **4**(2), 64–6.

Young, B., P. R. Ineson, T. F. Bridges, M. E. Smith 1989. Cinnabar from the northern Pennines, England. *Mineralogical Magazine* **53**, 388–9.

Young, B., A. Livingstone, N. Thomson 1992c. Fraiponite from Wensleydale, North Yorkshire. *Yorkshire Geological Society, Proceedings* **49**, 125–7.

Young, B., T. Pettigrew, T. F. Bridges 1985a. Rosasite and aurichalcite from the Northern Pennine orefield. *Natural History Society of Northumberland, Transactions* **54**, 31.

Young, B., E. R. Phillips, B. Smith 2005. Fluorite-bearing marble from Barrasford quarry, Northumberland. *Natural History Society of Northumberland, Transactions* **64**, 199–206.

Young, B., G. Ryback, R. S. W. Braithwaite, J. G. Francis 1997. Prosopite, doyleite and otavite from Coldstones quarry, Pateley Bridge, North Yorkshire. *Mineralogical Magazine* **61**, 895–7.

Young, B., M. T. Styles, N. G. Berridge 1985b. Niccolite–magnetite mineralization from Upper Teesdale, north Pennines. *Mineralogical Magazine* **49**, 555–9.

Young, B. N., I. Parsons, R. Threadgould 1994b. Carbonatite near the Borralan intrusion, Assynt. *Geological Society of London, Journal* **151**, 945–54.

Young, B. R., R. K. Harrison, G. A. Sergeant, I. P. Stevenson 1968. An unusual glauconite associated with hydrocarbon in reef limestones, near Castleton, Derbyshire. *Yorkshire Geological Society, Proceedings* **36**, 417–34.

Young, B. R., R. J. Hawkes, R. J. Merriman, M. T. Styles 1978. Bazirite, $BaZrSi_3O_9$, a new mineral from Rockall Island, Inverness-shire, Scotland. *Mineralogical Magazine* **42**, 35–40.

Young, B. R., B. Young, A. T. Raine 1985c. Thenardite (Na_2SO_4) a mineral new to Britain from Sussex and Cumbria. *Russell Society, Journal* **1**(3), 92–3.

Young, I. M. 1984b. Mixing of supernatant and interstitial fluids in the Rhum layered intrusion. *Mineralogical Magazine* **48**, 345–50.

Young, J. 1990. *Alabaster*. Matlock: Derbyshire Museum Service.

Younger, P. L. 2003. Frazer's Grove: the life and death of the last great Weardale fluorspar mine. In *Fluorspar in the North Pennines*, R. A. Fairnbairn (ed.), 47–72. Killhope, Durham: Friends of Killhope.

Zachariades, N. 1993. Cinnabar from Stancombe quarry, near Bristol, England. *Russell Society, Journal* **5**(1), 33.

Zakrzewski, M. A. 1989. Members of the freibergite–argentotennantite series and associated minerals from Silvermines, County Tipperary, Ireland. *Mineralogical Magazine* **53**, 293–8.

Zavyalov, E. N. & V. Beigizov 1981a. The new bismuth mineral ingodite, Bi_2TeS [in Russian]. *Zapiski Vsesoyuznogo Mineralogicheskogo Obshchestva* **110**, 594–600. [Translated in Mineralogical Abstracts 82M/3350.]

——— 1981b. Once more on the problem of grünlingite [in Russian]. *Zapiski Vsesoyuznogo Mineralogicheskogo Obshchestva* **110**, 633–5. [Translated in Mineralogical Abstracts 82M/3316.]

Zeh, A. & I. L. Millar 2001. Metamorphic evolution of garnet–epidote–biotite gneiss from the Moine Supergroup, Scotland, and geotectonic implications. *Journal of Petrology* **42**, 529–54.

Zenk, M. & B. Schulz 2004. Zoned Ca-amphiboles and related P–T evolution in metabasites from the classical Barrovian metamorphic zones in Scotland. *Mineralogical Magazine* **68**, 769–86.

Zhou, J. 1987. Geology of a copper-bearing intrusive suite near Kilmelford, Argyllshire, Scotland. *Institution of Mining and Metallurgy, Transactions B* **96**, 179–86.

Ziemba, T. G. P. 1975. *Mineral veins in Bannerdale Crags*. British Mining 1 (pp. 41–8), Northern Mine Research Society, Sheffield.

Zumpe, H. H. 1971. Microstructures in Cenomanian glauconite from the Isle of Wight, England. *Mineralogical Magazine* **38**, 215–24.

Zwaan, P. C., C. E. S. Arps, E. de Grave 1989. Vochtenite, $(Fe^{2+},Mg)Fe^{3+}[UO_2/PO_4]_4(OH)\cdot 12–13H_2O$, a new uranyl phosphate mineral from Wheal Basset, Redruth, Cornwall, England. *Mineralogical Magazine* **53**, 473–8.